ORGANIC ELECTROCHEMISTRY

Manuel M. Baizer (1914–1988)

ORGANIC ELECTROCHEMISTRY

An Introduction and a Guide

Third Edition, Revised and Expanded

Edited by

HENNING LUND
Aarhus University
Aarhus, Denmark

MANUEL M. BAIZER
University of California
Los Angeles, California
and
University of California
Santa Barbara, California

MARCEL DEKKER, INC. New York • Basel • Hong Kong

Library of Congress Cataloging-in-Publication Data

Organic electrochemistry : an introduction and guide / edited by
Henning Lund, Manuel M. Baizer. -- 3rd ed., rev. and expanded.
p. cm.
Includes bibliographical references and indexes.
1. Organic electrochemistry. I. Lund, Henning,
II. Baizer, Manuel M.
QD273.073 1990
547.1'37--dc20 90-3824
ISBN 0-8247-8154-6 CIP

This book is printed on acid-free paper.

MARCEL DEKKER, INC.
270 Madison Avenue, New York, New York 10016

Current printing (last digit):
10 9 8 7 6 5 4 3 2 1

PRINTED IN THE UNITED STATES OF AMERICA

Preface to the Third Edition

The field of organic electrochemistry has been growing steadily since the publication of the second edition of this book, with respect to both knowledge of the subject and the number of scientists and engineers working in the field. A couple of years ago it became apparent that an updated and expanded edition of Organic Electrochemistry would be desirable at an earlier date than originally planned.

The aim of the third edition is the same as it was for the two preceding editions, to be a guide for readers during their first steps into the field, as well as later, during the exploration of the many possibilities offered by organic electrochemistry.

The organization of this volume has been changed somewhat, but is essentially as in the two preceding editions. A few chapters have been added "Reduction of Azomethine Compounds" (Chap. 11), "Natural Products and Pharmaceuticals" (Chap. 19), "Biomass" (Chap. 20), "Photoelectrochemistry" (Chap. 34), and "Paired Electrosynthesis" (Chap. 35). A few chapters have been deleted and the relevant material incorporated into other chapters. Some chapters were completely rewritten, either because the former author had other commitments and could not update it or the development of the field required a new approach. Thus, for example, knowledge of the mechanism of electrode reactions has become more and more detailed, which made a rewriting of some of the chapters on the fundamental aspects desirable.

Manuel M. Baizer was primus motor for the preparation of Organic Electrochemistry. In spite of serious heart attacks during his last couple of years he worked hard, probably too hard, on his chapters for the third edition. At his death (July 1988) electrochemistry lost one of its pioneers and I and many others lost a highly regarded friend.

Henning Lund

Preface to the Second Edition

A decade has elapsed since the manuscripts were assembled and melded into what became the first edition of this book. This period has seen enormous advances in all aspects of organic electrochemical science and technology. Newer techniques have led to new understanding of electrode reactions. Hundreds of new syntheses have been devised. New electrode materials and forms and new cell designs have made electrochemical processing more attractive than ever before. Many new academic and industrial scientists have entered or are examining the field, stimulated at least in part (kind reviewers have said) by the widespread availability and acceptance of Organic Electrochemistry: An Introduction and a Guide. Concerns about rising costs of petroleum feedstocks and the needs for abating environmental pollution and conserving energy were only a vague malaise a decade ago but are now foreseen to be major problems until at least the end of this century.

We are therefore persuaded that the time has come for issuance of a second edition of this book. All chapters (renumbered) have been updated and revised; some have been completely rewritten; new ones have been added and others deleted. The aim remains the same: to conduct the reader through the sometimes dark labyrinths of organic electrochemical science and technology to a clear understanding of the capabilities and limitations of this methodology.

In the microcosm comprised of a group of co-authors and colleagues, ten years is a long time in the course of human events. Two of the original authors (G. Parravano and P. Iversen) unhappily have died. Some (e.g., M. Rifi) have been diverted from this field; others (e.g., L. Eberson) have taken on additional commitments which limited their availability for this project. We have been fortunate to be able to engage well-qualified replacements who, like the others, are actively engaged in research in organic electrochemistry. Most happily for the senior editor, Henning Lund, who contributed so much to the contents of the first edition, has consented to be co-editor of this one.

In view of the inevitable lapse of time between the submittal of a manuscript and its appearance in book form we have done our utmost to provide access to the most recent leading references. The literature has been monitored through 1982.

Finally, as before, all translations from French and German and all renditions in American English of manuscripts by authors for whom the latter is not the native language were done by the senior editor.

Manuel M. Baizer
Henning Lund

Preface to the First Edition

The organic chemist who wishes to employ a pressure catalytic hydrogenation in a synthetic sequence learns how to operate the necessary equipment and does not hesitate to carry out the experiment. While doing so, he is aware that formidable fundamental problems of mass transfer, adsorption, surface chemistry, kinetics, etc., are involved and that, while these are research areas in their own right, they are beyond his immediate objective. He derives his guidance probably from a book on organic chemistry. Unfortunately, however, this chemist's confrontation, if any, with electrochemistry has been first through thermodynamics or polarography or the intricacies of double-layer theory. In other words, he may well have been introduced to the discipline by way of problems which are the special areas of concern to the analytical and physical chemists for whom the notation Ox $\underset{}{\overset{ne}{\rightleftharpoons}}$ Red may serve as well for ferric-ferrous as for nitrobenzene-nitrobenzene radical ion. He may well have obtained the highest academic degree without having performed a single synthetic or kinetic organic electrochemical experiment.

This book is directed primarily to the organic chemist. Its aim is to present to him the capabilities and limitations of organic electrochemistry: Its principles, methods, problems, utility in synthesis and in elucidation of mechanisms of organic reactions, and fertility as a research area. Fundamental broad problems relating to physical phenomena before and during electron transfer are presented only in essence since excellent treatments dealing extensively and in depth with the component segments are available. The major emphasis here is upon the elucidation of the chemical events which accompany electron-transfer steps. It is recognized that this leads to only a partial description and rationalization of the phenomena; in most cases no more is possible at this stage of the science.

Organic electrochemistry has expanded in many directions in the last twenty-five years. Advances in theory, in the design of equipment, and in the employment of new techniques have brought to the subject a degree of sophistication formerly lacking and have rendered obsolete the previous monographs (now out-of-print): C. J. Brockman's Electro-organic Chemistry (1926), S. Glasstone and A. Hickling's Electrolytic Oxidation and Reduction (1935), F. Fichter's Organische Elektrochemie (1942), and M. J. Allen's Organic Electrode Processes (1958). Professor S. Swann, Jr.'s review (1956) is deliberately largely pragmatic.

The chapters chosen for inclusion in this book illustrate but do not exhaust the total scope of the subject. Specific anodic and cathodic reaction types are illustrated but are not usually, because of limitations of space, followed by comprehensive tables of data. The literature citations, on the other hand, are reasonably complete (at least to September 1970).

Each author represented here is actively engaged in research in one or another area of organic electrochemistry and has contributed to the common goal of replacing "art" by science. The original members of our group (Cauquis, Dietz, Eberson, and Lund) have seen all segments of the book while it was in various stages of preparation and have offered many valuable comments. I am particularly grateful to Dr. Eberson for his detailed constructive criticism.

It has become routine, at least for reviewers, to point out that in a multi-authored book there is overlap of material, nonuniformity of style, repetition, etc. This book was not hastily assembled, and there was adequate time to achieve uniformity if that had been desired. But the material was deliberately organized so that a given segment might appear in two or more contexts and, further, where unanimity of opinion does not exist, more than one current viewpoint might be expressed.

We all owe a tremendous debt to Miss Pat Merlenbach who, though only part time on this project, handled magnificently not only a superabundance of manuscripts but a flood of polyglot correspondence. The Monsanto Company, the Director of Central Research (Dr. B. L. Williams, Jr.), and the Section Director (Mr. H. W. Mohrman) were most kind in providing facilities, time, encouragement, and tolerance in the course of this undertaking.

Translations from French and German were made by the Editor.

Manuel M. Baizer

Contents

PREFACE TO THE THIRD EDITION iii

PREFACE TO THE SECOND EDITION v

PREFACE TO THE FIRST EDITION vii

CONTRIBUTORS xiii

ABBREVIATIONS AND SYMBOLS xv

1. Introductory Survey 1
   Manuel M. Baizer and Henning Lund

I. PRINCIPLES AND METHODS

2. Basic Concepts 11
   Christian Amatore

3. Methods for the Elucidation of Organic Electrochemical Reactions 121
   Ole Hammerich and Vernon D. Parker

4. Relations Between Micro- and Macrophenomena 207
   Christian Amatore

5. Comparison Between Electrochemical and Analogous Chemical Reactions 233
   Hans J. Schäfer

II. PRACTICAL PROBLEMS

6. Practical Problems in Electrolysis 253
   Henning Lund

III. SYNTHETIC AND MECHANISTIC ASPECTS OF CATHODIC REACTIONS OF ORGANIC COMPOUNDS CLASSIFIED BY ELECTROPHORES

7. Cathodic Reactions of Hydrocarbons 331
   Herbert Kiesele and Jürgen Heinze

8. Halogenated Organic Compounds 361
   Dennis G. Peters

9. Cathodic Reduction of Nitro and Related Compounds 401
   Henning Lund

10. Carbonyl Compounds 433
    Manuel M. Baizer

11. Reduction of Azomethine Compounds 465
Henning Lund

12. Carboxylic Acids and Derivatives 483
Lennart Eberson, James H. P. Utley, and John H. Wagenknecht

IV. SYNTHETIC AND MECHANISTIC ASPECTS OF ANODIC REACTIONS OF ORGANIC COMPOUNDS CLASSIFIED BY ELECTROPHORES

13. Anodic Oxidation of Hydrocarbons 505
Lennart Eberson, James H. P. Utley, and Ole Hammerich

14. Carboxylic Acids 535
Sigeru Torii and Hideo Tanaka

15. Nitrogen-Containing Compounds 581
Eberhard Steckhan

16. Anodic Oxidation of Oxygen-Containing Compounds 615
Ole Hammerich and Bo Svensmark

17. Anodic Oxidation of Sulfur-Containing Compounds 659
Bo Svensmark and Ole Hammerich

V. ELECTROCHEMISTRY OF CERTAIN COMPREHENSIVE CLASSES OF COMPOUNDS

18. Electrolysis of Heterocyclic Compounds 701
Henning Lund

19. Natural Products and Pharmaceuticals 765
Yoshio Ban, Tameo Iwasaki, and Hiroshi Ohmizu

20. Biomass 787
Helena L. Chum

21. Organoelemental and Coordination Compounds 809
Lorenz Walder

VI. ELECTRODE REACTIONS CLASSIFIED BY REACTION TYPE

22. Electrolytic Reductive Coupling 879
Manuel M. Baizer

23. Electrolytic Oxidative Coupling 949
Hans J. Schäfer

24. Cleavages and Deprotection 1029
Henning Lund

25. Anodic Substitution 1069
Lennart Eberson, James H. P. Utley, and Ole Hammerich

26. Anodic Fluorination 1103
W. V. Childs, Leif Christensen, F. W. Klink, and C. F. Kolpin

VII. STEREOCHEMISTRY

27. Stereochemistry of Organic Electrode Processes 1131
Tsutomu Nonaka

VIII. INDIRECT REACTIONS

28. Amalgam and Related Reductions 1199
Henning Lund

29. Electrogenerated Reagents 1217
Jacques Simonet

30. Electrogenerated Acids and Bases 1265
Manuel M. Baizer

IX. PRESENT AND FUTURE APPLICATIONS

31. Industrial Electroorganic Chemistry 1285
D. E. Danly and C. J. H. King

32. Electrochemical Polymerization 1337
B. Lionel Funt

33. Chemically Modified Electrodes and Conducting Polymers 1363
A. F. Diaz

34. Photoelectrochemistry 1397
Marye Anne Fox

35. Paired Electrosynthesis 1421
Manuel M. Baizer

36. Challenges and Opportunities 1431
Manuel M. Baizer

AUTHOR INDEX 1439

SUBJECT INDEX 1527

Contributors

CHRISTIAN AMATORE Ecole Normale Supérieure, Laboratoire de Chimie, Paris, France

YOSHIO BAN Hokkaido University, Sapporo, Japan

MANUEL M. BAIZER[†] University of California, Los Angeles, California, and University of California, Santa Barbara, California

W. V. CHILDS 3M Company, St. Paul, Minnesota

LEIF CHRISTENSEN 3M Company, St. Paul, Minnesota

HELENA L. CHUM Solar Energy Research Institute, Golden, Colorado

D. E. DANLY Monsanto Chemical Company, Pensacola, Florida

A. F. DIAZ IBM Research Division, San Jose, California

LENNART EBERSON Division of Organic Chemistry, University of Lund, Lund, Sweden

MARYE ANNE FOX The University of Texas, Austin, Texas

B. LIONEL FUNT Department of Chemistry, Simon Fraser University, Burnaby, Vancouver, British Columbia, Canada

OLE HAMMERICH H. C. Ørsted Institute, University of Copenhagen, Copenhagen, Denmark

TAMEO IWASAKI Tanabe Seiyaku Company, Ltd., Osaka, Japan

JÜRGEN HEINZE Institut für Physicalische Chemie, Universität Freiburg, Freiburg, Federal Republic of Germany

HERBERT KIESELE Universität Lübeck, Lübeck, Federal Republic of Germany

C. J. H. KING Monsanto Chemical Company, Pensacola, Florida

F. W. KLINK 3M Company, St. Paul, Minnesota

C. F. KOLPIN 3M Company, St. Paul, Minnesota

HENNING LUND Aarhus University, Aarhus, Denmark

TSUTOMU NONAKA Tokyo Institute of Technology, Yokohama, Japan

HIROSHI OHMIZU Tanabe Seiyaku Company, Ltd., Osaka, Japan

VERNON D. PARKER Utah State University, Logan, Utah

DENNIS G. PETERS Indiana University, Bloomington, Indiana

[†]Deceased

HANS J. SCHÄFER Organisch-Chemisches Institut, Wilhelms Universität, Münster, Federal Republic of Germany

JACQUES SIMONET Laboratory of Electrochemistry, University of Rennes, Rennes, France

EBERHARD STECKHAN Institut für Organische Chemie und Biochemie der Universität Bonn, Bonn, Federal Republic of Germany

BO SVENSMARK H. C. Ørsted Institute, University of Copenhagen, Copenhagen, Denmark

HIDEO TANAKA Faculty of Engineering, Okayama University, Okayama, Japan

SIGERU TORII Faculty of Engineering, Okayama University, Okayama, Japan

JAMES H. P. UTLEY Queen Mary and Westfield College, University of London, London, England

JOHN H. WAGENKNECHT Monsanto Chemical Company, St. Louis, Missouri

LORENZ WALDER Institute of Organic Chemistry, University of Bern, Bern, Switzerland

Abbreviations and Symbols

ELECTRODES

DME	Dropping mercury electrode
HMDE	Hanging mercury drop electrode
NHE	Normal hydrogen electrode
OTE	Optically transparent electrode
RDE	Rotating disc electrode
RRDE	Rotating ring-disc electrode
SHE	Standard hydrogen electrode

TECHNIQUES

CV	Cyclic voltammetry
DPSC	Double potential step chronoamperometry
LSV	Linear sweep voltammetry
dc pol, ac pol	dc, ac polarography

SOLVENTS

AN	Acrylonitrile
AcOH	Acetic acid
Ac_2O	Acetic anhydride
DME	Dimethoxyethane
DMF	Dimethylformamide
DMSO	Dimethylsulfoxide
EtOH	Ethanol
HMPA	Hexamethylphosphoramide
MeCN	Acetonitrile
MeOH	Methanol
NMP	N-Methylpyrrolidone
PC	Propylene carbonate
Py	Pyridine
TFA	Trifluoroacetic acid
THF	Tetrahydrofuran

CHEMICALS, LIGANDS, RADICALS

Ac	Acetyl
acac	Acetylacetone
aib	Tripeptide of α-aminoisobutyric acid
Ant	Anthracene
Ar	Aryl
bipy	2,2'-Bipyridine
Boc	t-Butoxycarbonyl
Bu	Butyl
Bz	Benzyl
Cbz	Benzyloxycarbonyl
COT	Cyclooctatetraene
Cp	Cyclopentadiene
Cyclam	1,4,8,11-tetraazacyclotetradecane
Dim	Diimine
dppe	1,2-bis(diphenylphosphino)ethane, $(Ph_2PCH_2)_2$
dppp	1,2-bis(diphenylphosphino)propane, $(Ph_2PCH_2)_2CH_2$
E^+	Electrophile
Et	Ethyl
EWG	Electron withdrawing group
Fc	Ferrocenyl
H_2P	Porphyrin
H_2OP	Octaethylporphyrin
Me	Methyl
MP	Metalloporphyrin
MTPP	Tetraphenylmetalloporphyrine
Naph	Naphthalene
Nu	Nucleophile
Pa	Polyacetylene
Pc	Phthalocyanine
Ph	Phenyl
phen	1,10-Phenanthroline
Pr	Propyl
R	Alkyl
Sal	Salicylaldehyde
Salen	N,N'-ethylene bis (salycylideneamine)
TBA	Tetrabutylammonium
TEA	Tetraethylammonium
THP	Tetrahydropyran
TMA	Tetramethylammonium
TPA	Tetrapropylammonium

STANDARD SUBSCRIPTS

an	anodic
cat	cathodic
d	diffusion
eq	equilibrium
f	forward
b	backward
l	limiting
p	peak
O,R	pertaining to the oxidized, reduced species

ELECTRIC, KINETIC, AND THERMODYNAMIC SYMBOLS

A	Electrode surface area
A_{ff}	Affinity (in eV)
a_S	Equivalent radius of species S
C	Concentration (see also (S))
C^b	Concentration in the bulk solution
C^o	Initial concentration of the starting material
C_S	Concentration of species S (see also (S), etc.)
$C_{x=...}$	Concentration at x=...
D	Diffusion coefficient; average diffusion coefficient
D_S	Diffusion coefficient of species S
e_{met1}	Electron in metal phase number 1, 2, etc.
E	Electrode potential vs. a given reference electrode
ΔE_f^{act}	Activation energy of a reaction between encounter pairs
E^{in} (or E^{fin})	Potential energy of the electron for any state of the initial (or final) system
E_j	Junction potential
E^o (or E_j^o)	Standard reduction potential (relative to the half reaction 1, 2, etc.)
E_E^o	Standard reduction potential of an electrode E
$E_{O/R}^o$	Standard reduction potential relative to the half reaction: $O + ne \rightleftharpoons R$
$E^{o'}$ (or $E_1^{o'}$)	Formal potential (relative to the half reaction 1, 2, etc.)
E_p	Peak potential in CV
$E_{p/2}$	Half-peak potential in CV
E_w	Any characteristic potential pertaining to a given electrochemical wave

$E_{1/2}$	Half-wave potential
$E^{\ddagger}$	Potential energy of the electron in the initial or final systems at the transition state
f_S	Activity coefficient of species S
F	Faraday
ΔG_r^f (or ΔG_p^f)	Free energy of formation of the reactant (for product) encounter pair
ΔG^o	Standard free energy change in a specified reaction
$\Delta G^{o\prime}$	Standard free energy change between encounter pairs
ΔG_{red}^o (or G_{ox}^o)	Standard free energy change for the considered half-reaction in the direction of reduction (or oxidation)
$\Delta G_{S,solv1}^o$	Standard free energy of solvation for species S in solvent 1, 2, etc.
$\Delta G_{S,rel}^o$	Standard free energy of relaxation of an excited species S in gas phase
ΔG^S	Free energy change for species S between the end of the diffuse layer and the electron transfer site
$\Delta G_{exp}^{\ddagger}$	Experimental (apparent) activation free energy
$\Delta G_f^{\ddagger}$ (or $\Delta G_b^{\ddagger}$)	Overall free activation energy for the forward (or backward) reaction
$\Delta G^{\ddagger o}$	Intrinsic activation barrier
h	Planck constant
i	Current
i^{lim}	Limiting current in steady state methods
I_p	Ionization potential (in eV)
J (or J(x))	Overall flux of particles through an elementary surface perpendicular to the x axis, at x
J_{conv}	Flux of particles through an elementary surface perpendicular to the x axis, arising from convection
J_{dif}	Flux of particles through an elementary surface perpendicular to the x axis, arising from diffusion
J_{migr}	Flux of particles through an elementary surface perpendicular to the x axis, arising from convection
$\underline{k}$	Boltzmann constant
k	Rate constant
k^{ap}	Apparent rate constant
k_f^{act} (or k_b^{act})	Rate constant of the forward (or backward) reaction between encounter pairs, for an adiabatic electron transfer
k_f^{dif} (or k_b^{dif})	Overall diffusion limited forward (or backward) rate constant
k_{elec}	Apparent first order rate constant equivalent to an electrolysis

k_f (or k_b)	Measured forward (or backward) rate constant
k^o	Intrinsic standard heterogeneous rate constant
k_t^o	True standard heterogeneous rate constant
k_{dif}^r (or k_{dif}^p)	Diffusion rate constant for the formation of the reactant (or product) pair
k_Z	Rate constant for the reaction involving the transfer of a Z moiety
k_1	Rate constant number 1, 2, etc.
$k_f^\ddagger$ (or $k_b^\ddagger$)	Apparent forward (or backward) activation rate constant
K	Equilibrium constant
K_a	Acidity dissociation equilibrium constant
K_a'	Formal acidity dissociation equilibrium constant
K_c(or K_p)	General kinetic term of consumption (or production) for the considered species
K_j	Dimensionless kinetic term for consumption and production for the considered species
L	Mean distance covered by a particle during a specified time period
l	Any specified length Free motion length for a "random walking" particle
M (or m_S)	Molar mass (of species S)
n (or n_1)	Number of electrons exchanged (for the half-reaction 1, 2, etc.)
n_{ap}	Apparent number of electrons exchanged
n_{op}	Refractive index of the solvent
N_A	Avogadro's number
N	Number of nuclear degrees of freedom
N_S (or N)	Number of molecules (of species S)
P_1	Competitive parameter number 1, 2, etc.
P_e	Probability of electron tunnelling between the initial and final systems
q_i	ith nuclear coordinate (i = 1, N)
q_i^{in} (or q_i^{fin})	Equilibrium value of the ith nuclear coordinate (i = 1, N) in the initial (or final system)
q_i	Value of the ith nuclear coordinate (i = 1, N) in the initial and final system, at the transition state
Q	Charge consumed (coulombs)
Q_t	Charge consumed after a period t of electrolysis
r	Distance between reactant, or product, centers in the encounter pairs
r_o	Radius of a disc electrode; smaller dimension of an electrode

R	Gas constant
R (or R_Ω)	Resistance of a solution
R_i (or $R_{\Omega i}$)	Electrical potential drop arising from ohmic phenomena
s	Dimensionless concentration of species S
S_{gas}	Species S in gas phase
S_o	Species S at the end of the diffuse layer
S_{solv}	Species S solvated
(S)	Activity of species S
$[S]$	Concentration of species S
$[S]_{bulk}$	Concentration of species S in the bulk solution
$(S)_{x=...}$	Activity of species S at x=...
$[S]_{x=...}$	Concentration of species S at x=...
$[S]_{yield}$	Yield of species S
S_Φ	Species S at the electron transfer site
ΔS	Entropy change
ΔS_{pair}	Entropy of formation of a pair electrode/molecule
ΔS_{rot}	Entropy of formation of a molecule pair (rigid rotator)
S^*	Excited species S in gas phase Isotopically labelled species S
t	Current time; time period
T	Absolute temperature
$v(x)$	Velocity of a fluid along the x axis
V	Volume of the bulk solution
V_i	Stabilizing energy at the transition state
W_r(or W_p)	Energy of formation of a reactant (or product) encounter pair
$W_r(x)$ (or $W_p(x)$)	Energy of formation of a loose reactant (or product) encounter pair, x being the distance between centers of the molecules
$W_\delta^{non\text{-}eq}$	Energy required to produce a non-equilibrium polarization of the solvent, corresponding to a virtual charge δ transferred
$W_\delta^{\varepsilon op}$	Energy required to produce an optical equlibrium polarization of the solvent, corresponding to a charge δ transferred
$W_\delta^{\varepsilon s}$	Energy required to produce a static equilibrium polarization of the solvent, corresponding to a charge δ transferred
X	Coordinate along the x axis Normalized reaction coordinate Distance between the centers of reacting molecules Distance from the electrode surface

X_o	Double layer thickness
X_Φ	Distance of the electron transfer site, from the electrode surface
$X^\ddagger$	Normalized reaction coordinate value at the transition state
Y	Dimensionless space variable
z (or z_S)	Charge of species S
Z	Collision frequency (homogeneous)
Z_{an}	Anodic faradaic impedance
Z_{cat}	Cathodic faradaic impedance
Z_{el}	Collision frequency (heterogeneous)

GREEK SYMBOLS

α	Transfer coefficient (heterogeneous)
β	Brönsted coefficient (homogeneous)
β_{exp}	Measured Brönsted coefficient
δ	Virtual fractional charge transferred; Diffusion layer thickness
δ_{conv}	Stagnant layer thickness
δ^{in} (or δ^{fin})	Virtual fractional charge transferred in the initial (or final) system
$\bar{\Delta}^2$	Average square displacement of a particle during a specified time period
ε_{op}	Optical dielectric constant
ε_S	Static dielectric constant
Θ	Characteristic time of an electrochemical method
Θ	Duration of an experiment
κ	Transmission coefficient
κ	Proportionality constant between velocity of a particle and the force applied to it
κ	Dimensionless equilibrium constant
λ	Dimensionless homogeneous rate constant
λ (or λ^{xy})	Reorganization energy (relative to the X/Y homogeneous couple)
λ_i (or λ_i^{xy})	Inner-shell reorganization energy (relative to the X/Y homogeneous couple)
λ_o (or λ_o^{xy})	Outer-shell reorganization energy (relative to the X/Y homogeneous couple)
λ_i^S	Inner-shell reorganization energy relative to species S
λ^{el}	Reorganization energy relative to an heterogeneous electron transfer

λ_i^{el} (or λ_o^{el})	Inner-(or outer-)shell reorganization energy relative to a heterogeneous electron transfer
Λ	Dimensionless heterogeneous rate constant
μ	Chemical potential
μ^o	Standard chemical potential
ν	Sweep rate in CV or LSV (Vs^{-1})
ξ	Dimensionless potential
τ	Dimensionless time Duration of a free motion for a "random walking" particle
Φ	Potential difference between the anode and the cathode
Φ_A	Potential difference between the auxiliary electrode and the solution
Φ_{an}	Potential difference between the anode and the solution
Φ_{cat}	Potential difference between the cathode and the solution
Φ_{CE}	Potential difference between the counter electrode and the solution
Φ_{Ref}	Potential difference between the reference electrode and the solution
Φ_S	Potential difference between the electron transfer site and the solution
Φ_W	Potential difference between the working electrode and the solution
Φ_Ω	Potential drop due to ohmic phenomena
Ψ	Dimensionless current
ω	Rotation frequency of RDE
Ω_i	Reduced force constant relative to the ith nuclear coordinate
Ω_i^{in} (or Ω_i^{fin})	Force constant relative to the ith nuclear coordinate in the initial (or final) system
Ω_o	Reduced force constant relative to the oscillator equivalent to the outer-shell medium

ORGANIC ELECTROCHEMISTRY

1

Introductory Survey

MANUEL M. BAIZER† / University of California, Los Angeles, California and University of California, Santa Barbara, California

HENNING LUND / Aarhus University, Aarhus, Denmark

I. General Considerations 1

II. Experimental Results 2
- A. Transformation of a functional group 2
- B. Substitution reactions 3
- C. Addition reactions 4
- D. Coupling reactions 5
- E. Cleavage 6
- F. Indirect electrolysis 6
- G. Polymerization 7

III. Concluding Remarks 7

References 7

I. GENERAL CONSIDERATIONS

This book is concerned almost exclusively with the liquid-phase electrochemistry of organic substances capable of undergoing electron-exchange reactions with electrodes. This capability, constantly being extended, may be defined in terms of the presence of "electrophores" within the molecule. (The term "electrophore" seems to have occurred simultaneously to a number of people and is gaining widespread use; it is meant to designate a function that renders a molecule electroreducible or electro-oxidizable).

A discussion of the experimental problems is presented in Chap. 6. In simplest terms, all the experimental procedures require in common a container that may or may not have ion-permeable membranes to separate it into several chambers, a source of (usually) direct current, a means for voltage control, an ammeter, at least one positive (anode) and one negative (cathode) electrode, and a liquid dielectric medium within which is usually dissolved a preferably inert ("supporting") electrolyte together with the usually poorly conducting substrate. Passage of current through an appropriate assemblage inaugurates a complex sequence of events.

Organic electrochemistry has relevance to analysis and synthesis of organic compounds, elucidation or confirmation of structure, investigations into the nature of catalytic activity, the study of transient intermediates, the generation of chemiluminescence, the nature of electron-transfer processes, structure-eletroactivity relationships, initiation of polymerization, synthesis of catalysts and catalyst components, degradation, biological redox systems,

†Deceased, July 1988.

and so on. It is germane to work on kinetics, reaction mechanisms, salt effects, solvation, effect of electric field on chemical reactions, support of molecular orbital theory, and other areas of inquiry. It is therefore fortunate that in this as in other developing branches of science and technology [1], investigators of a variety of talents, backgrounds, and personal and professional motivations have chosen to dedicate their efforts to a furtherance of knowledge in a variety of the components of organic electrochemistry. This pluralism has made possible simultaneous advances on many fronts. Furthermore, since this is an interdisciplinary subject most participants have brought to it expertise in more than one field. The result is that organic chemists who are active or becoming active in this area have at their disposal a spectrum of challenging problems ranging from the theoretical to the applied, one or more of which may well coincide with their special interests.

A simple example illustrates for one particular system the diverse aspects of it that are amenable to study by electrochemical means. The electrolytic reduction of nitrobenzene under specified conditions yields p-aminophenol, azoxybenzene, azobenzene, hydrazobenzene, benzidine, or aniline [2]. The detailed mechanism of the electron-transfer steps for this or for most other organic compounds is under intensive investigation. Polarography (Chap. 3) permits one to determine the cathode voltage and the number of electrons involved under given conditions of pH, for example. (Polarography of a series of substituted nitrobenzenes has been part of a larger project [3] concerning the relationship between structure and reducibility.) Cyclic voltammetry (Chap. 3) indicates that the first step of the reduction is the transfer of one electron to yield an ion radical and permits the estimation of its half-life. Electrolysis within the cavity of an electron spin resonance apparatus (Chap. 3) is an elegant means for preparing the intermediate and recording its spectrum. (The nitrobenzene anion radical may be used to initiate the polymerization [4] of acrylonitrile.) Controlled potential electrolysis (Chap. 6) of nitrobenzene makes possible the preparation of one or the other of the several products in high yield. Continuous (Chap. 31) as well as batch cells may be used [5]. The yield and the time for the completion of the reaction are profoundly influenced by rotating the cathode [6].

In order to orient the experimental organic chemist who is hesitantly first becoming acquainted with this "novel" area, it is proposed to use an unorthodox approach in this introductory survey and first to illustrate the extensive scope of organic electrochemistry by the following experimental results chosen from the areas of synthesis and polymer chemistry.

II. EXPERIMENTAL RESULTS

Some examples of synthetic applications of electrolysis are presented in this section.

A. Transformation of a Functional Group

Cathodic Preparation of N-tert-*Butylhydroxylamine Hydrochloride* [7, 8]

t-Nitrobutane (60–70 g) was reduced at a mercury cathode in 400 ml of a mixture of 1:1 concentrated hydrochloric acid-ethanol in a conventional H cell [9]. The temperature was kept at 30–40°C by having the cell in a water bath. A current of 2–3 A from an ordinary selenium rectifier (some voltage ripples are easily tolerated) was found satisfactory for about 90% of the reaction. Near the end of the reaction the current was reduced stepwise to minimize hydrogen evolution or a potentiostat was employed to keep the cathode potential at −0.90 V versus Ag/AgCl (4 N HCl). Workup by evaporation of the catholyte gave a crude hydrochloride, which on recrystallization from ethanol-diethyl ether (1:5) gave 75–85% N-tert-butylhydroxylammonium chloride, melting point 182–183°C.

Anodic Preparation of S-*Benzyl α-Toluenethiosulfonate [10]*

Dibenzyldisulfide (6.1 g) was dissolved in 300 ml acetic acid and 20 ml concentrated hydrochloric acid and electrolyzed at 35°C using a platinum anode (50 cm^2) and a similar cathode at 0.04 A cm^{-2}. After 9600 C the product was isolated in 92% yield by evaporation of the solvent.

B. Substitution Reactions

Anodic Preparation of 2,4-Dimethoxybenzonitrile [11]

1,3-Dimethoxybenzene (10^{-2} mol) was electrolyzed in a mixture of 100 ml dichloromethane, 100 ml of 1 M sodium cyanide, and 10 mmol tetrabutylammonium hydrogen sulfate dissolved in 15 ml water and neutralized with sodium hydroxide. The anode was a platinum foil (55 cm^2) and the cathode of stainless steel; a high-speed stirring device was placed in the middle of the vessel between the electrodes. The very efficient stirring kept the elctrolyte well emulsified during the run. The potential drop across the cell varied between 12 and 20 V. When starting up the electrolysis the stirrer had to be turned on for a few seconds and then off and left for half a minute and then turned on again in order to get the voltage over the cell to rise from the initial value of 4–5 V. This is probably due to a wetting phenomena.

After 2 F mol^{-1} had passed, the stirrer was switched off; the two phases separated rapidly and the organic layer was evaporated to dryness. The residue was taken up in 15 ml water and extracted with 3 × 20 ml ether. The combined ether layers were dried over magnesium sulfate and reduced in volume to about 15 ml. Finally, products were analyzed by gas-liquid chromatography (GLC) using a 2 m × 3 mm 5% neopentylglycol succinate column. During workup a solid, insoluble in ether, precipitated. After recrystallization from ethanol, the substance was shown to be 4,6-dimethoxyisophthalonitrile. The main product, 2,4-dimethoxybenzonitrile, was isolated and recrystallized from ligroin (boiling point 80–110°C), melting point 90–91°C.

*Cathodic Preparation of 1-*tert*-Butylpyrene [12]*

Pyrene (2.02 g, 0.01 mol) was reduced at −1.5 V (versus Ag/AgI, 0.1 M I^-) in N,N-dimethylformamide (DMF) containing tetrabutylammonium iodide (0.1 M) and tert-butyl chloride (20 ml). A conventional H cell with sintered glass diaphragm, a mercury cathode (35 cm^2), and a carbon anode were used; the potential was kept constant by means of a potentiostat. After the passage of 2–2.5 F mol^{-1} the reduction was finished, and chloranil

(4.92 g, 0.02 mol) was added to the catholyte. The stirred solution was allowed to stand for 2 h and then poured into water and acidified. The mixture was extracted five times with 200 ml portions of toluene. After washing and drying the toluene was evaporated. The dark red-brown oily product was cleared of tarry materials and chloranil hydroquinone on a short column (alumina) and subsequently separated by column chromatography on silica mixed with 10% caffeine. Petrol ether served as eluent (boiling point below 50°C). 1-tert-Butylpyrene (52%), melting point 100–100.5°C, 4-tert-butyl-4,5-dihydropyrene (14%), and small amounts of di-tert-butylated products were isolated.

Cathodic Preparation of 3-Trifluoromethylbenzoic Acid [13]

In an undivided cell with a stainless steel gauze cathode and a magnesium rod anode, 30 ml anhydrous DMF, 300 mg tetrabutylammonium bromide, and 3.0 ml 3-chloro-trifluoromethylbenzene were introduced. The cell was immersed in cold water to dissipate the heat evolved by the electrolysis and to keep the temperature near 5°C. After bubbling CO_2 through the solution for a few minutes to expel oxygen, a low overpressure (0.1 bar) of CO_2 was applied. A constant current (0.4 A) was used and the solution stirred during electrolysis. The reaction was followed by GLC. After electrolysis most of the solvent and the residual starting material were evaporated; the residue was acidified with 6 N aqueous hydrochloric acid (10 ml) and extracted with diethyl ether (3 × 10 ml). The organic extract was treated with 2 N aqueous sodium hydroxide (2 × 10 ml) to separate acidic and neutral products. The aqueous layer was again acidified and extracted with diethyl ether (3 × 30 ml), which was evaporated; the yield of trifluoromethylbenzoic acid was 85%.

C. Addition Reactions

Anodic Preparation of 3,3,6,6-Tetramethoxy-1,4-cyclohexadiene [14]

A 600 ml tall form beaker was equipped with a thermometer, a magnetic stirring bar, and two electrodes. A 45 mesh cylindrical platinum anode was used. Surrounding the anode was a cylindrical nickel cathode. The electrodes were held in place (distance between anode and cathode, 0.75 cm) and suspended in the beaker by means of a clamp formed from Delrin rods. The electrodes were connected to an adjustable dc power supply.

A solution consisting of 27.6 g (0.2 mol) p-dimethoxybenzene, 4.0 g potassium hydroxide, and 400 ml methanol was placed in the apparatus. The beaker and contents were cooled with a 0°C bath. The solution was electrolyzed with magnetic stirring for 6 h at a current intensity maintained at 2.0 A. The temperature of the solution varied between 8 and 14°C. During this time small amounts of methanol were added occasionally to compensate for evaporation.

After electrolysis the solution was reduced to a volume of 100 ml in a rotary evaporator and then extracted 10 times with 1000 ml portions of hexane. The hexane fractions were dried over anhydrous magnesium sulfate and the hexane evaporated on a rotary evaporator to yield white crystals, which were recrystallized from 75 ml pentane to afford 27.8–28.2 g (70–71%) of pure 3,3,6,6-tetramethoxy-1,4-cyclohexadiene, melting point 40–43°C.

Cathodic Preparation of Ethyl 3-Phenyl-4-oxopentanoate [15,16]

Ethyl cinnamate (2 ml) was reduced under nitrogen in an H cell at a mercury cathode in 150 ml of N,N-dimethylformamide containing 10 ml acetic anhydride and 10 g Al_2O_3 at −1.85 V (SCE); after dilution of the catholyte with water, neutralization, and extraction with ether, the extracted products were separated on a column of silica using acetone-petrol ether (1:9) as eluent. Isolated were ethyl 3-phenyl-4-oxopentanoate (75%) and small amounts of other products.

D. Coupling Reactions

Anodic Preparation of Dimethyl-1,14-tetradecanedioate [17]

The electrolysis cell was a multinecked round-bottomed flask. One platinum anode (4 × 5 cm) spaced between two platinum cathodes (5.5 × 2.5 cm) was used. A solution in 500 ml of refluxing methanol of 282 g (1.5 mol) monomethyl suberate and approximately 0.1 mol sodium methoxide was electrolyzed at about 1.6 A. The polarity of the electrodes was reversed periodically to remove coatings from the electrode. When the pH of the solution became greater than 7, the electrolysis was discontinued, and the solution was made acid with acetic acid and heated to remove easily volatile material in vacuo. The higher-boiling material was dissolved in a mixture of ether and benzene and the solution filtered, washed with aqueous sodium bicarbonate, and dried. After removal of the solvents, the crude product was recrystallized from methanol, 225.7 g (84%), melting point 43–45°C.

Cathodic Preparation of Diethyl Adipate from Ethyl Acrylate [18]

A divided cell (Chap. 6) was used. The membrane was an Alundum cup. The cathode was mercury and the anode platinum. The catholyte was a solution of 76.0 g ethyl acrylate containing a trace of polymerization inhibitor, 75.8 g of a 76.5% solution of methyltriethylammonium p-toluenesulfonate in 88.5 g dimethylformamide (DMF). The anolyte contained 40 ml of a 38% solution of the quaternary salt. The electrolysis was carried out at 3 A for 2 h at a temperature of 35–40°C. As product formed, additional DMF was added to the catholyte to maintain homogeneity. The pH of the catholyte was maintained at 7–9 by occasional addition of drops of acetic acid. The catholyte was then separated from Hg, diluted with water, and extracted thoroughly with methylene chloride. The washed and dried extracts were distilled first to remove solvents and excess starting material and then to recover 23.6 g product, boiling point 142–148°C (30 mm). The yield was approximately 100% based on the amount of current that had been passed.

Cathodic Preparation of 1,3-Dimethylbicyclobutane and Reductive Cyclization [19]

An H cell (Chap. 6) with sintered glass divider was used. The catholyte contained a solution of 50 g 1,3-dibromo-1,3-dimethylcyclobutane in 230 ml dimethylformamide-0.1 M lithium lithium bromide. The anolyte was 150 ml of the solvent-salt solution. Electrolysis at 0.5 A for 16 h yielded 9–15 g (55–94%) of the product, boiling point 54.5°C (760 mm). The yield was poorer when a higher concentration of starting material was used.

E. Cleavage

Anodic Deprotection of Alcohols [20]

A simple electrochemical cell was constructed from a beaker and a glass tube of about 75 ml capacity that was sealed at one end with a sintered glass frit of medium porosity. The frit was covered with a gel of 0.1 M lithium perchlorate in DMF and methylcellulose. This isolates the cathode and anode while maintaining electrical conductivity between the two. The tube was suspended in the beaker and the inside of the tube was utilized as the anodic chamber. A platinum wire cathode and a platinum mesh anode were used. The power source was a Wenking 70HVI potentiostat.

A solution of 60–85% aqueous acetonitrile was made 0.1 M in lithium perchlorate. This solution was placed in the electrochemical cell. To the anode chamber was added 4 mmol of the anisyl ether of the alcohol, and the solution was electrolyzed at +1.65 V (SCE) until the initial current of approximately 200 mA had dropped to less than 3 mA. The anolyte was concentrated in vacuo with minimal heating until the acetonitrile was removed. (The solution may be neutralized before concentration to minimize any reaction under the acidic conditions of the concentrated solution, but for most compounds no undesirable reactions were observed when this neutralization was omitted. Caution: Under no circumstances should the solution be evaporated to dryness.) The resulting aqueous suspension was saturated with sodium chloride and extracted with ether. The combined ether extract was washed with saturated sodium bisulfite solution, dried, and concentrated to yield products that were essentially pure by nuclear magnetic resonance (NMR) analysis. The residue was then appropriately purified.

Cathodic Cleavage of Tritylone Esters [21]

Tritylone decyl ether (2 g 4.68 mmol) in 50 ml of 0.1 M in LiBr methanol was electrolyzed in a divided cell (working electrode: mercury pool, counter-electrode: platinum sheet) at 20°C and controlled potential (−1.4 V versus Ag/AgCl) until 1900 A was consumed. For workup, methanol was evaporated and the residue taken up in ether. Bulb-to-bulb distillation afforded 0.6 g (3.8 mmol, 81%) of decanol; liquid chromatography yielded 1.06 g (3.9 mmol, 83%) of tritylone alcohol.

The subject of cleavage reactions is treated in Chap. 24.

F. Indirect Electrolysis

Electrochemical Oxidation of Alcohols Using Iodonium Ion as Electron Carrier [22]

Into a cell equipped with platinum or cabon electrodes was put an aqueous solution of 2.49 g (0.015 mol) of potassium iodide in 15 ml water and 0.06 mol of alcohol. When the alcohol is relatively insoluble in water or is solid at room temperature, t-butanol and/or hexane are necessary as cosolvents. The cell was cooled with a water jacket, and a constant current was passed through the two-phase solution with vigorous stirring. After 4–15 F/mol of electricity was passed, the organic layer was separated and the aqueous layer was extracted with three portions of ether. The combined organic and ethereal layers were washed with aqueous sodium thiosulfate. The products were isolated by distillation, yield 60–84%.

The subject of indirect electrolysis is treated in Chap. 29.

G. Polymerization

"Living" anionic polymers [23] were cathodically generated in an H cell (Chap. 6). Materials were rigorously purified and handled by vacuum line technique. The solvent was tetrahydrofuran, the electrolytes alkali metal tetraphenylborons or alkali metal tetra-alkylaluminums, and the monomers styrene, α-methylstyrene, or other vinyl compounds with electron-withdrawing groups. The concentration of "living ends," measured spectrophotometrically, was directly proportional to the amount of current passed. Further, the polarity of the cell could be reversed and a calculated fraction of the living ends destroyed. The surviving dianions could initiate new polymerization with the same or another monomer so that block copolymers of given composition could be made. In addition, by using an appropriate feedback device, it was possible to maintain a predetermined concentration of living ends in the polymerization of monomers (e.g., methyl methacrylate), which tend to terminate by intramolecular reaction.

The subject of electroinitiated polymerization is treated extensively in Chap. 32.

III. CONCLUDING REMARKS

In the "classic period" of electrochemistry around the turn of the century, results were obtained empirically as they were in other procedures for any other organic synthesis. Such a procedure is still a possibility, but there now exist many kinds of techniques that may help the investigator to choose the optimum conditions for an electrosynthesis.

A number of electroanalytical methods (e.g., polarography, cyclic voltammetry, derivative linear sweep voltammetry, and voltammetry at ring-disk electrodes; see Chap. 3) can be used to elucidate the mechanism of the electrochemical reaction and to determine some of the rate constants of the individual steps. Such knowledge makes it possible to select the reaction parameters on rational grounds.

In recent years the number of investigations within organic electrochemistry has been increasing; some of the developing areas are mentioned in Chap. 36. There are many reasons for this expansion: the price of electricity is foreseen to increase less rapidly than that of traditional chemicals; the electron is a nonpolluting reagent; and electrochemical reactions are easy to control automatically. For an organic chemist the study of electrochemistry also gives the "spin-off" benefit that many of the electroanalytical results (e.g., redox potentials and electron transfer reactions) may be useful for investigations of "ordinary" chemical reactions.

REFERENCES

1. J. O'M. Bockris, J. Electroanal. Chem., 9, 408 (1965).
2. S. Swann, Jr., in Technique of Organic Chemistry (A. Weissberger, ed.), 2nd ed., Vol. II, Wiley-Interscience, New York, 1956, pp. 385-523; F. Fichter, Organische Elektrochemie, Steinkopf, Dresden, 1942.
3. P. Zuman, Substituent Effects in Organic Polarography, Plenum, New York, 1967.
4. G. Mengoli, G. Farina, and E. Vianello, Eur. Polym. J., 5, 61 (1969).

5. F. Goodridge, Chem. Process Eng., 49, 93 (1968).
6. K. S. Udupa, G. S. Subramanian, and H. V. K. Udupa, J. Electrochem. Soc., 108, 373 (1961).
7. P. E. Iversen and T. B. Christensen, Acta Chem. Scand., B31, 733 (1977).
8. P. E. Iversen, J. Chem. Educ., 51, 489 (1974).
9. P. E. Iversen, J. Chem. Educ., 48, 136 (1971).
10. F. Fichter and P. Sjøstedt, Ber. Dtsch. Chem. Ges., 43, 3422 (1910).
11. L. Eberson and B. Helgée, Acta Chem. Scand., B29, 451 (1975).
12. P. E. Hansen, A. Berg, and H. Lund, Acta Chem. Scand., B30, 267 (1976).
13. M. Heintz, O. Sock, C. Saboureau, J. Perichon, and M. Troupe, Tetrahedron, 44, 1631 (1988).
14. P. Margaretha and P. Tissot, Org. Synth., 57, 92 (1977).
15. H. Lund and C. Degrand, Tetrahedron Lett., 3593 (1977).
16. T. Shono, I. Nishiguchi, and H. Ohmizu, J. Amer. Chem. Soc., 99, 7396 (1977).
17. J. D. Anderson, M. M. Baizer, and J. P. Petrovich, J. Org. Chem., 31, 3890 (1966).
18. M. M. Baizer, Tetrahedron Lett., 973 (1963).
19. M. R. Rifi, J. Amer. Chem. Soc., 89, 4442 (1967).
20. S. M. Weinreb, G. A. Epling, R. Comi, and M. Reitano, J. Org. Chem., 40, 1356 (1976).
21. C. van der Stouwe and H. J. Schäfer, Tetrahedron Lett., 2643 (1979).
22. T. Shono, Y. Matsumura, J. Hayashi, and M. Mizoguchi, Tetrahedron Lett., 165 (1979).
23. B. L. Funt, U.S. Pat. 3,448,020 (1969); Chem. Abstr., 71, 39630 (1969).

I

PRINCIPLES AND METHODS

2

Basic Concepts

CHRISTIAN AMATORE / Ecole Normale Supérieure, Laboratoire de Chimie, Paris, France

A te convien tenere altro viaggio . . .
Chè questa bestia, per la qual tu gride,
Non lascia altrui pasar per la sua via,
Ma tanto lo impedisce, che l'uccide. . . .
Dante, Divina Commedia

I. Introduction 11

II. Fundamental Aspects of Electron Transfer Reactions 13
- A. Oxidation-reduction reaction versus electron transfer reactions in organic chemistry and electrochemistry 13
- B. Standard potentials: What they mean and what they do not mean 18
- C. Mechanism and theory of outer sphere electron transfer reactions 31

III. Fundamental Aspects of Electrode Phenomena 53
- A. Monitoring a half-reaction: The electrochemical cell 53
- B. General overview of an electrode reaction 59
- C. Kinetics of heterogeneous electron transfers 60
- D. Adsorption phenomena 68
- E. Coupled chemical reactions 69

IV. Mass Transfer in Electrochemistry 80
- A. Fundamental aspects of mass transfer processes 80
- B. Steady-state electrochemical methods: Half-wave potential $E_{1/2}$ 95
- C. Transient electrochemical methods 104

Symbols and Their Definitions 109

References and Notes 114

I. INTRODUCTION

In accordance with the previously stated purposes of this book, our objective in this chapter is to familiarize organic chemists with the fundamental electrochemical concepts that support several aspects of organic electrochemistry elaborated in this book, as well as many important features of electron transfer reactions. In fact, although the term "electrochemistry" evokes for most readers the idea of chemistry at electrodes, in our opinion electron transfer chemistry constitutes a better definition when organic or organometallic electrochemistry is concerned. This is particularly obvious when one thinks of an electrode as a macrosized molecule whose ability to provide or accept electrons may be precisely adjusted by fine-tuning of the electrode potential. Similarly, electrochemical reactions involving specific interactions with the electrode material are no more than the analogs of inner sphere, ion-pairing, or complexation reactions. Thus, in many respects, electrochemical reactions do not differ basically from their homogeneous counterparts. This is even more true insofar as it concerns the chemical reactivity of electrogenerated intermediates, since they evolve in conditions that a priori do not differ from those considered in homogeneous chemistry. Indeed, the extraordinarily large electrical fields (of magnitude comparable to those at the origin of storm lightning) drop to negligible values within a few angstroms of the electrode surface, in contradiction to generally held ideas. Thus as soon as molecules have moved over a few molecular diameters from the electrode surface, they probe no special electrical effects associated with their electrochemical origin. This is an important point, although not well recognized or publicized, since it permits an easy transposition of electrochemical results to homogeneous situations, or vice versa.

Yet there are specific particularities of chemistry at electrodes. They both arise from the fact that electrodes supply and accept electrons to and from molecules dissolved in a solution. This has two important consequences dealing with mass transfer (i.e., transfer of the reagents from the homogeneous volumic space and to the heterogeneous two-dimensinal surface) and with current transport across the solution.

Transport of the current through the solution requires that rather conductive media be used, at least when high currents are considered. This is an obvious disadvantage of electrochemical methods, although in our opinion it is fully compensated by the extraordinary advantage of a precise adjustment of the driving force via the electrode potential.

Mass transfer from a volumic homogeneous region, the bulk solution, to a two-dimensional surface results in a spatial structuring of solutions in the close vicinity of the electrode surface. Although such effects are not specific to electrochemistry,* they may be thought, at first, as additional difficulties to cope with in electrochemistry. However, these spatial structurings may be used with considerable advantage to control the chemical route followed by a given intermediate and force it into a pathway that would never be followed under homogeneous conditions. Indeed this is used in most electrochemical reactions, especially in indirect reductions or oxidations, in which the proper choice of the electron transfer mediator or of the electrogenerated reagent precursor is crucial to the success of a particular reaction. In fact, the conscious (or even subconscious) use of spatial structuring of solutions is not a privilege of electrochemists: nature uses the same phenomenon in most of its reactions, as in the proton transfer "pump" crucial to ATP synthesis or in the photosynthetic chain.

From our experience the largest difficulty (or activation barrier) that "homogeneous chemists" encounter in trying to deal with the electrochemical literature is directly related to electrochemical jargon rather than to fundamental concepts. Although electrochemistry was born as a synthetic method (by Kolbe, Haber, and Fichter, among others), most of electrochemical textbooks use an analytic approach to these fundamental concepts. As a result, mathematical formulations and jargon invade most of the presentations, which often results in an effect similar to the caveat at the entrance of Dante's Inferno:

---

*They intervene as soon as heterogeneous reactants or phase boundaries are involved in the reaction medium.

Per me si va nella città dolente,
Per me si va nell' eterno dolore
Per me si va tra la perduta gente . . .
Lasciate ogni speranza, voi ch' entrate.

Dante Alighieri, Divina Commedia

Our point is not to criticize these necessary and worthwhile approaches that give to electrochemistry its ability to interpret and rationalize on a quantitative basis a large number of experimental facts that extend far beyond strictly electrochemical domains. Most of our published work (as well as the latter section of this chapter) advocates our belief in the usefulness of these physicomathematical approaches, yet we regret that they may be perceived by nonanalytical chemists as important obstacles in approaching electrochemistry. For this reason, in this chapter we take advantage of the fact that most of these precise approaches are extremely well exposed in popular electrochemical textbooks [1] to try to present most of the necessary electrochemical basis in words and concepts using as much as possible references (or antireferences) to analogous concepts of homogeneous chemistry. Thus we hope that this chapter may constitute both a whole and a guide to further and more specialized readings in electrochemistry. Yet there are three exceptions, each dealing with kinetic aspects, in which we had to break our resolution. The first two concern electron transfer theories, and the third relates to electrochemical kinetics in the presence of follow-up chemical reactions. Although these three points, and their pertinent derivations, are in our opinion very important for a fine understanding of electrochemical processes and capabilities, we suggest that the reader not interested in the mathematical body may skip the equations but nevertheless follow the corresponding text.

This chapter is divided into three parts. In the first, basic definitions and their consequences for homogeneous chemistry are presented. The second deals with the fundamental aspects of electrode phenomena, whereas the third discusses the problem of mass transfer at electrodes and its consequences for electrochemical kinetics. The particular problems and concepts associated with preparative-scale electrolysis are presented in a special chapter (Chapter 4).

II. FUNDAMENTAL ASPECTS OF ELECTRON TRANSFER REACTIONS

A. Oxidation-Reduction Reactions Versus Electron Transfer Reactions in Organic Chemistry and Electrochemistry

Oxidation-Reduction Versus Electron Transfer Reactions

In homogeneous chemistry, pure electron transfer reactions are seldom encountered. Indeed, with the exception of a few examples, electron transfers are often associated with atom or group transfers. This usually results in a confused notion of the nature of oxidation-reduction reactions. For example, the reaction of a ketone with sodium in alcohol to afford the corresponding alcohol, via the sequence in Eq. (1) [2],

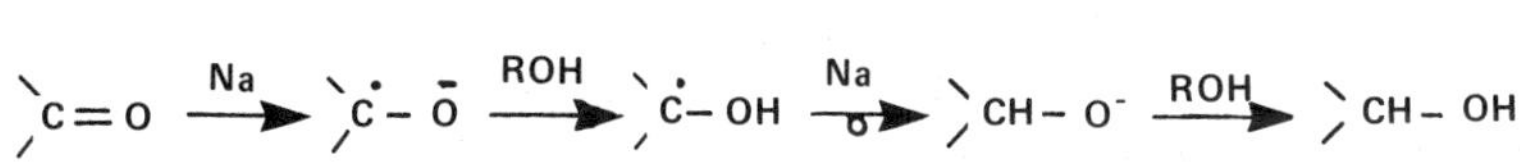

(1)

is termed a reduction. The same term is used for the Meerwein-Ponndorf-Verley reaction, which is supposed to involve a concerted cyclic transition state [Eq. (2)] [3,4]:

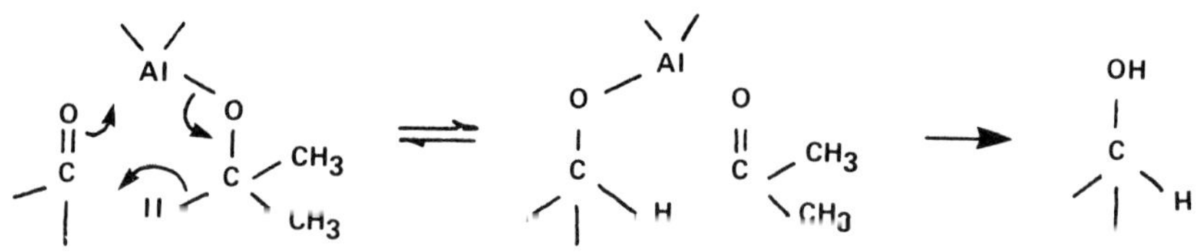

(2)

Similarly, reactions of metallic hydrides of aluminum or boron, in which hydride is transferred to the carbon [Eq. (3) [6] or (4) [7,8]], are also considered reductions [8]:

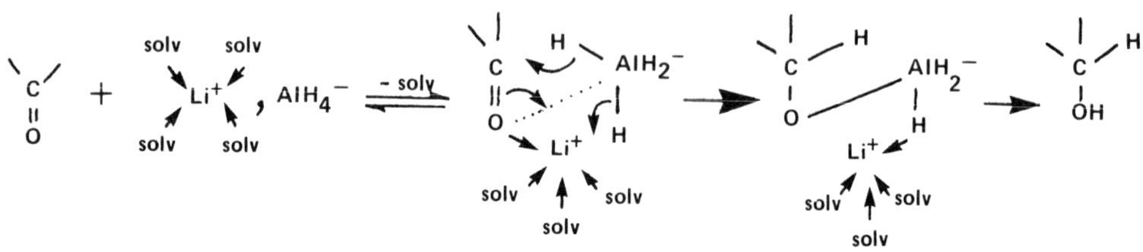

(3)

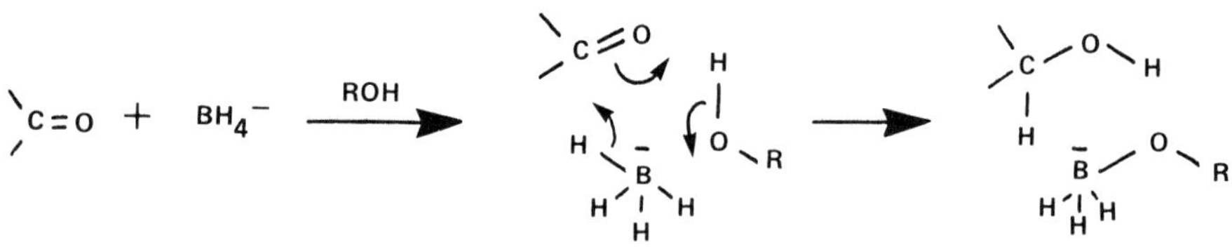

(4)

Although the preceding four reactions all obey the same stoichiometry for the carbonyl-alcohol transformation and thus involve the same variation in the oxidation state of the carbon atom, they are obviously different. Sodium reduction, in Eq. (1), involves single electron and proton transfers in a succession of separate steps; conversely, in reactions (2) through (4), groups or atoms are transferred.

The same confused notions also exist for oxidation reactions, the problem being even more subtle. For example, permanganate is presented in most introductory textbooks as a typical oxidant corresponding to the half-reaction in Eq. (5):

$$MnO_4^- + 5e + 8H^+ \rightleftharpoons Mn^{2+} + 4H_2O \tag{5}$$

This, associated with the classic one-electron oxidation of ferrous to ferric salts by permanganate, for example, may lead to the implicit notion that permanganate actually accepts electrons in a similar way as the carbonyl group in Eq. (1). That this is not always the case in practice is clearly evidenced by the mechanism of alcohol or aldehyde oxidation by permanganate to ketones or carboxylic acids. Indeed, it is generally accepted that the mechanism of acidic oxidation of aldehydes by permanganate involves no electron transfer steps but rather atom transfers as in the sequence of Eqs. (6) through (8) (where B^- is one of the bases present in the reaction medium) to afford the carboxylic acid and an unstable manganese (V) moiety, which rapidly evolves to MnO_2 or Mn^{2+} according to the pH of the solution [9].

(6)

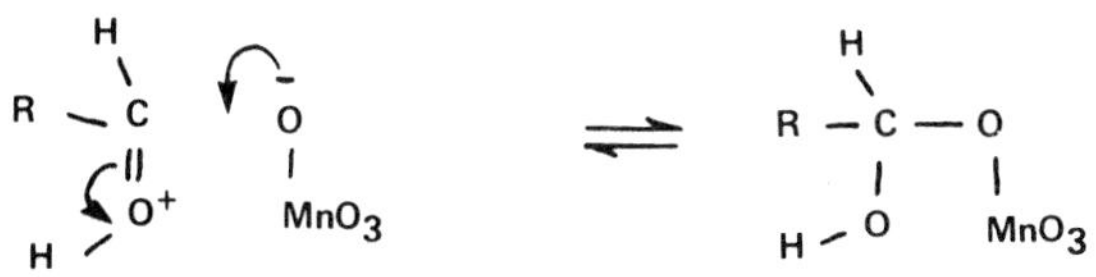
(7)

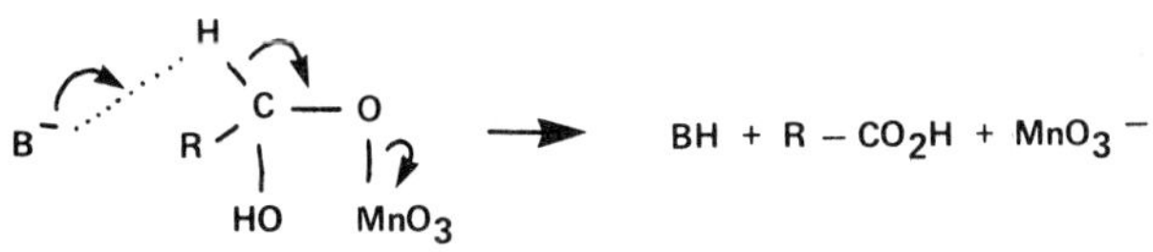
(8)

Similarly, a reaction proceeding through a sequence of one-electron transfers and chemical steps, such as pinacol formation from a ketone [Eqs. (9) and (10)],

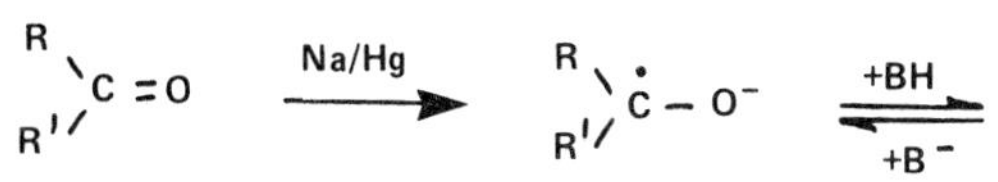
(9)

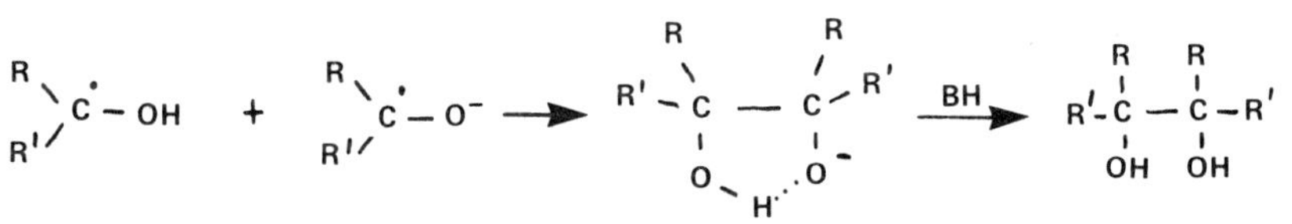
(10)

can be reversed via a sequence of steps involving atom transfers as in the classic glycol oxidation to ketones by lead tetra-acetate in Eqs. (11) through (13) [10]:

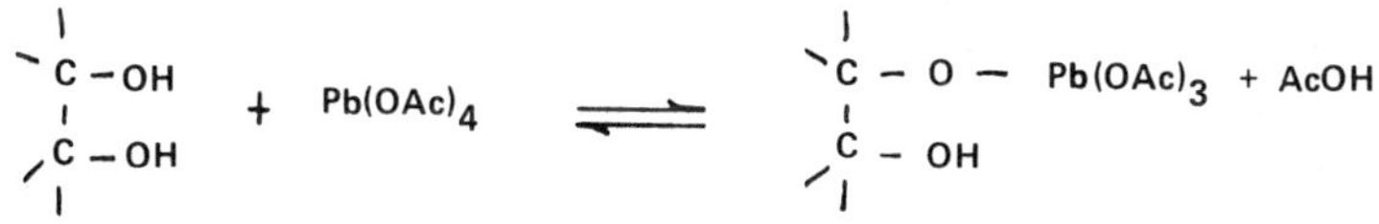
(11)

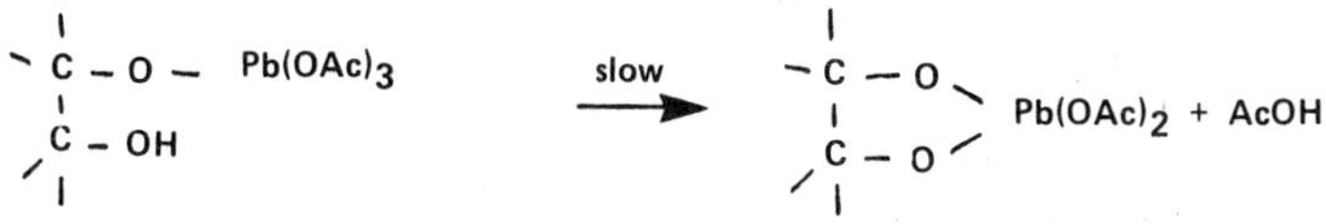
(12)

(13)

Outer Sphere and Inner Sphere Electron Transfers

The preceding examples, which were purposely restricted to well-known reactions of carbonyl and related functions, illustrate the large ambiguities associated with oxidation-reduction notions in organic chemistry. As suggested earlier, these ambiguities certainly stem from the fact that more attention is given to the overall transformation of one of the reactants (the substrate) than to the other(s). Indeed, in the ketone to alcohol reductions described, more interest is devoted to the ketone than to the coreactant [Na, $Al(OR)_3$, $Li^+AlH_4^-$, or BH_4^-]. If the complete stoichiometries are considered, it appears more clearly that all four reactions [Eqs. (1) through (4)] are totally different.

On the other hand, when one thinks in terms of electrochemical reductions or oxidations, special attention is devoted to the coreactant, that is, to the electrode that provides or accepts electrons. Thus in order to discuss or compare electrochemical reactions with their organic analogs, it is of the utmost importance to use more precise terms than the so inaccurate reduction or oxidation notions. A similar problem has been addressed in the inorganic and organometallic fields. Indeed, it was early recognized that oxidation-reduction reactions at metal centers must be classified according to two types: outer sphere or inner sphere reactions. A typical example of this dichotomy is given in Eqs. (14) and (15), which relate to chromium$^{(II)}$ oxidations by cobalt$^{(III)}$ complexes.

Outer sphere (electron transfer) reaction [11]:

$$Cr^{2+} + Co^{(III)}(NH_3)_6^{3+} \longrightarrow Cr^{3+} + Co^{(II)}(NH_3)_6^{2+}, \ldots \tag{14}$$

Inner sphere (ligand transfer) reaction [11,12]:

$$Cr^{2+} + Co^{(III)}(NH_3)_5Cl^{2+} \longrightarrow Cr^{(III)}Cl^{2+} + Co^{(II)}(NH_3)_5^{2+}, \ldots \tag{15}$$

Besides their relatively close stoichiometries, that the fundamental nature of the reactions differs is evidenced by the acceleration by a factor of approximately 6×10^9 in the rate constant when a chlorine ligand is involved in the transition state of the process.

Outer sphere electron transfers correspond to situations in which an electron is transferred without the necessity of bond formation or bond cleavage between or within the reactants. The electron is then transferred when the partners are sufficiently close to allow orbitals of suitable geometries to overlap. The energetic stabilization required is of the order of 1 kcal/mol^{-1}, which is considerably smaller than that corresponding to any bond formation. The ketone reduction by sodium metal in Eq. (1a) is considered to belong to this class of electron transfer. Similarly, owing to the usual poor ability of electrodes in establishing bonds with the electroactive molecules, most of the electrochemical electron transfers pertain to this group. Interestingly, theories have been developed for outer sphere electron transfers that allow reasonable predictions of their rate constants and activation energies (see Sec. II.C and III.C.).

Note that this definition does not imply that the products obtained upon electron transfer must be stable, but only that the electron transfer activation process does not imply any important molecular rearrangement. Indeed, one or both of the products may chemically evolve through fast follow-up reactions as in the reaction sequences (1) or (9) and (10).

Inner sphere reduction or oxidation is normally restricted to bond formation between the reactants in the transition state. For example the chromium(II) oxidation in Eq. (15) has been shown to involve a chloride bridge between the chromium and cobalt centers as depicted in Eq. (16) [12]:

$$[Cr^{2+}\cdots Cl\cdots Co^{(III)}(NH_3)_5]^{\#} \qquad (16)$$

The organic chemical reduction-oxidations presented in the reaction sequences (2), (3), (4), (7), or (11) through (13) belong to this class. Obviously, because an electrode is considered able only to supply or accept electrons, this class of reduction-oxidation reactions should not be observed in electrochemistry. Yet it must be recalled that this is not a very clear-cut problem since the existence of such group or atom transfer has been invoked to explain the role of chemical surfaces, or of additives, as in the Kolbe reaction [13].

In our opinion a third class of reduction-oxidation reactions exists, which is not clearly encompassed by one of the two just mentioned. It corresponds to the cases in which there really is an electron transfer (i.e., not a group or atom transfer) but the latter is concerted with bond breaking or formation within one of the reactants. We propose that the phrase "inner sphere electron transfer" be used to refer to such situations. A characteristic example of this class is given by the reduction of alkyl halides, represented in Eq. (17):

$$RR'R''C{-}X + e \longrightarrow RR'R''C^{\bullet} + X^{-} \qquad (17)$$

Indeed, it is now considered that carbon—halide bond cleavage is concerted with electron uptake (see Ref. 14 and Chap. 8), that is, that there is no intermediacy of an anion radical, such as $[RR'R''C{-}X]^{\bullet-}$, during the electron transfer [15]. These cases must be contrasted with those, such as the aryl halide reduction in Eq. (18),

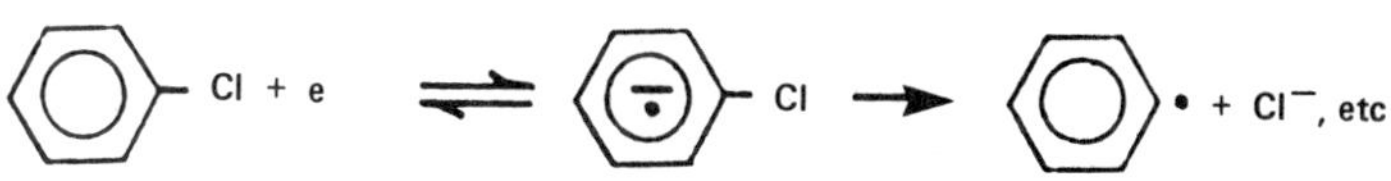

(18)

in which the electron is taken up (or lost) without drastic molecular rearrangement [14,16], to afford an anion radical (or a cation radical in oxidations), although the latter may be extremely short-lived. Formally, the carbonyl reduction in Eq. (1a) should fall in this class of reaction, since it is usually considered that the π-carbonyl bond is broken [2b] upon electron transfer to yield a ketyl anion radical [Eq. (20)].

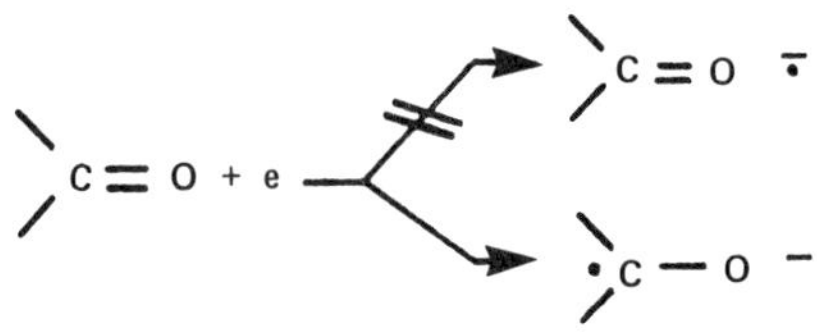

(19)

(20)

Yet the nature of the bond cleaved, especially when the latter is involved in extended delocalizations as in aromatic ketones, and the fact that the skeleton of the molecule remains intact, may explain why these reductions are generally considered as belonging to the outer sphere class.

From Table 1, which summarizes this discussion, it is seen that outer sphere and inner sphere electron transfers have their exact analogs in both fields of organic chemistry and electrochemistry. The third class, that is, inner sphere group transfers, deals with the peculiar properties of special reactants in organic chemistry or with those of special electrode materials in electrochemistry.

At this point it should be emphasized that there is a large body of outer sphere or inner sphere electron transfer reactions identified and used in electrochemistry, whereas this class of reactions is less developed in organic chemistry. Conversely, there is an extensive variety of chemicals designed for specific group or atom transfer in reduction-oxidation reactions in organic chemistry, but this is an area that is not very developed in electrochemistry. This certainly originates from the obvious fact that the reductive or oxidative strength of an electrode can be varied over a wide range and adjusted with considerable precision because of the easy control of the electrode potential. This advantage is not matched in homogeneous organic chemistry owing to the discrete number of electron transfer reagents, particularly for reductions. On the contrary, surface modifications and reactivities, especially under the conditions of the large electrical fields encountered in electrochemistry (Sec. III.A, The electrode/solution interfacial region), are not easy to design and control within the present state of the art, whereas owing to the extensive development of organic [8], inorganic, and organometallic [17] chemistry, a large variety of specific reagents is made available. However, the apparent disadvantage of electrochemistry in the latter area exists only when the electrode is considered the main reactant effecting the sought reaction. Indeed, specific chemicals (catalytic or stoichiometric reactants) that effect the desired transformation on the substrate of interest may easily be electrochemically generated as discussed in Part VIII of this book [18].

B. Standard Potentials: What They Mean and What They Do Not Mean

Standard Potentials

When considering a reaction between a possible electron donor and a possible electron acceptor, it is important to decide a priori whether it will take place. This is usually answered by comparison of the standard (reduction) potentials E^0 pertinent to each reactant. Indeed, when one considers the possible reaction in Eq. (21),

$$A + D \rightleftharpoons A^{\overline{\cdot}} + D^{\dot{+}} \qquad (21)$$

Table 1. Chemical and Electrochemical Reductions or Oxidations in Organic Chemistry

Designation	Bond breaking or formation in the transition state	Follow-up chemical reaction	Examples: Chemical	Examples: Electrochemical
Outer sphere electron transfer	No	No	Anthracene $\xrightarrow{Na}$ Anthracene$^{\overset{\cdot}{-}}$	Anthracene + e $\rightleftharpoons$ Anthracene$^{\overset{\cdot}{-}}$
			$ArCH_3 + Fe^{(III)} \rightleftharpoons ArCH_3^{\overset{+}{\cdot}} + Fe^{(II)}$	$ArCH_3 - e \rightleftharpoons ArCH_3^{\overset{+}{\cdot}}$
		Yes	$ArCH_3^{\overset{+}{\cdot}}$ + base $\longrightarrow ArCH_2^{\cdot}, \cdots$	$ArCH_3^{\overset{+}{\cdot}}$ + base $\longrightarrow ArCH_2^{\cdot}, \cdots$
Inner sphere electron transfer	Within reactants	Frequent	Wurtz reaction	$RX + e \longrightarrow R^{\cdot} + X^-$
Group or atom transfer	Between reactants		Meerwein-Ponndorf reaction	Invoked to interpret the effect of additives or electrode surface states (as in the Kolbe reaction)

the corresponding equilibrium constant K is related to the difference between the standard (reduction) potentials E^o_{A/A^-} and $E^o_{D^+/D}$ of the $A/A^{\cdot -}$ and $D^{\cdot +}/D$ couples, respectively. It follows from the expression of K,

$$K = e^{F(E^o_{A/A^-} - E^o_{D^+/D})/RT}$$

that at equilibrium Eq. (21) should be displaced toward the right-hand side when $E^o_{A/A^-} \gg E^o_{D^+/D}$ and to the left-hand side when the converse is true (however, see Sec. II.D, Relationships between ΔE^o and feasibility of an electron transfer reaction). When the difference between the values of the two standard reduction potentials is small, the reactant(s) and product(s) equilibrium concentrations obviously depend on the potential difference but also greatly on the initial composition of the solution.

According to its definition, the standard (reduction) potential of the $A/A^{\cdot -}$ couple is the standard electromotive force of a cell in which an $A/A^{\cdot -}$ electrode (where the activities of A and $A^{\cdot -}$ are made unity) is opposed to a NHE (normal hydrogen electrode) whose potential is assigned to zero by convention.

$$A + e \rightleftharpoons A^{\cdot -} \qquad (E^o) \tag{22}$$

Thus a standard reduction potential E^o is ascribed to the "half-reaction" in Eq. (22).* When the activities of A and $A^{\cdot -}$ are not unity, the potential of the $A/A^{\cdot -}$ electrode is given by the Nernst equation,

$$E = E^o + \frac{RT}{F} \ln \frac{(A)}{(A^{\cdot -})} \tag{23a}$$

as a function of these activities [20]. Note that in this discussion a "simple" or elementary electron transfer is considered. In actual practice the definition of E^o may be extended to more complex situations (Sec. II.B, Formal reduction potentials) in which the half-reaction corresponds to a series of equilibria involving at least one electron transfer step. If the pertinent half-reaction is written

$$\alpha_1 O_1 + \alpha_2 O_2 + \cdots + ne \rightleftharpoons \beta_1 R_1 + \beta_2 R_2 + \cdots$$

the potential of the corresponding electrode is given by the Nernst equation,

$$E = E^o + \frac{RT}{nF} \ln \frac{(O_1)^{\alpha 1}(O_2)^{\alpha 2}\cdots}{(R_1)^{\beta 1}(R_2)^{\beta 2}\cdots} \tag{23b}$$

---

*Because of thermodynamic and electrochemical conventions, standard potentials are defined in the direction of reduction, independently of the respective chemical stabilities of the molecules involved. Thus for the oxidation of toluene to its cation radical, E^o refers to the reduction of the highly unstable cation radical into the highly stable toluene. To overcome such a priori chemical nonsense, E^o is frequently designated as the standard oxidation potential of toluene for example. However, such a term should not be accepted according to canonical rules because it formally implies that the cell now operates in a driven mode, that is, is connected to an external

In practice there are several limitations to such measurement. Obviously it implies that both members of the half-reaction are sufficiently stable for a cell to be realized. This is a serious difficulty in organic chemistry owing to the usual great reactivities of the species formed upon electron transfers. For the most frequent cases it is then impossible to rely on reversible thermodynamic transformations to determine experimental values of standard reduction potentials. However, these important figures, or at least very precisely approximated values, can be obtained from current intensity potential curves or transient electrochemical methods as is discussed in a later section.

Reference Electrodes and Liquid Junction Potentials

Potentials are not measurable on an absolute scale, yet relative scales are easily constructed from measurements of potential differences. This is a fortiori true for electrode potentials. Thus, as explained earlier, electrode potentials are given, by convention, as their potential difference versus the NHE. Yet there are several difficulties associated with the use of the NHE. The most obvious is that, although it may be approximated via extrapolations, a real NHE cannot be constructed according to the stipulations included in its definition. The other limitations directly follow from the fact that approximated NHE are almost impossible to use in real experiments. Thus many other reference electrodes have been proposed and constructed [21]. Each of these electrodes is used by electrochemists according to historical or sentimental reasons or because a given reference electrode is more adequate for a given experimental situation owing to geometric factors or possible pollution of the solution under study.

To be considered a suitable reference electrode, an electrode must have a known and reproducible potential versus the NHE. This potential must be nearly an invariant of the current flowing through the electrode, which implies that the electrode reaction is extremely fast and the electrode reactants are extremely concentrated (see Secs. III.C, Reversible and irreversible electron transfers, and IV.B, Mechanisms involving a follow-up reaction). It should also have a small temperature coefficient and should be easily constructed in a reproducible way. A large variety of electrodes meeting these requirements to different degrees have been devised; Table 2 presents a few of the most frequently used.

Inspection of Table 2 shows that the solution composition of the electrode half-cell is critical. These solutions may differ considerably (nature of the solvent, nature and concentrations of the electrolytes or ionic members of the half-reaction, and so on) from those used in organic or organometallic electrochemistry. The two solutions, that of the experimental system under study and that in the inner reference electrode compartment, then tend to equilibrate across their interface (the liquid junction) to reach a final state in which both solutions are identical. In practice, to avoid pollution of both solutions, a third solution is usually interposed, which acts as a buffer. This "bridge" then results in the creation of two liquid junctions. At each of these junctions the physical process of equilibration is controlled

---

power supply [19]. Thus in this chapter we prefer to use the denomination standard reduction potentials, rather than the usual term "standard potential," as a reminder of the E^0 definition, although such an expression is basically a pleonasm.

Table 2. Potentials of Some Reference Electrodes

Reference electrode		E^o_{Ref}, V versus NHE[a]	
Half-cell[b]	Name	25°C	20°C
Hg/HgO, NaOH (1 N)		0.14[c]	
Ag/AgCl, KCl (s)		0.20[c]	
Ag/AgCl, KCl (1 M)		0.22[c]	0.22[c]
Hg/Hg_2Cl_2, NaCl (s)	SSCE	0.236[d]	
Hg/Hg_2Cl_2, KCl (s)	SCE	0.241[d]	0.245[d]
Hg/Hg_2Cl_2, KCl (1 M)	NCE	0.280[d]	0.281[d]
Hg/Hg_2Cl_2, KCl (0.1 M)		0.334[d]	
Hg/Hg_2SO_4, K_2SO_4 (s)		0.64[c]	
Hg/HgO, NaOH (0.1 M)		0.93[d]	

[a]For any O/R couple, E^o (versus NHE) = E^o (versus Ref) + E^o_{Ref}.
[b]In aqueous solutions; (s) indicates saturation.
[c]From Ref. 21c.
[d]From Refs. 21a and 21b.
Source: From Refs. 21.

by the diffusion of the various components, that is, depends on their mobilities, which may be extremely different [22]. In the process of equilibration there is then a trend to break the electroneutrality on both sides of the junction. Indeed, the center of positive charges tends to separate from the center of negative charges because of intrinsic differences in the mobilities of the cations and anions. This results in the creation of an electrostatic potential difference E_j across the junction, which opposes the charge separation. The system reaches then a steady-state "equilibrium" so that the electrostatic energy exactly matches that resulting from differences in chemical composition across the liquid junction. For identical solvents on both sides of the junction, models have been developed to estimate the resulting junction potentials E_j, which may reach several tens of millivolts in practice [23]. Yet in most organic or organometallic experiments, different solvents are used on both sides of the liquid junctions. Other phenomena (e.g., solvent mixing, different abilities in wetting the physical interface, usually a sintered glass frit or a glass crack, and interfacial and capillary tensions [24]) are then involved. This makes the corresponding analysis intractable or totally irrelevant to real conditions, especially when one considers the difficult reproducibility of most of these phenomena from one experimental setup to another.

In practice it is advisable to use reference electrodes involving, when possible, a solvent identical to that used in the electrochemical cell. Yet this is not always possible, since other problems may arise from solubilities or temperature dependence, for example. In any case, when potential determinations are crucial, a way to overcome these difficulties is to calibrate the potential scale by measuring the potential of a reference compound in the same solution, that is, introducing it in the investigated solution at the end of the experiment. Under such conditions, and provided that the solution composition remains invariant in the course of the experiment, a "floating reference electrode" may also be used. A silver wire of large surface area or a mercury pool, for example, constitutes such a floating electrode. Indeed, it keeps a constant, but unknown potential difference vis-à-vis the solution. Yet the resulting potential scales obtained from run to run need to be reconciled, via the calibration procedure explained earlier, since the floating reference potential is unknown and prone to vary as a function of the exact composition of the solution.

Standard Reduction Potentials Versus Ionization Potentials or Affinities

When one of the members of a given half-reaction is too unstable for measurement of the pertinent standard reduction potential, one frequently relies on the values of ionization potentials or affinities. However, one should be extremely cautious in the use of such data obtained from gas phase when applying them to reactions in solution [25].

Let us consider a simple half-reaction $O + e \rightleftharpoons R$, with a standard reduction potential E°. It must be realized that E° corresponds to (1) stable solvation states for O and R and (2) stable nuclear configurations for O and R. On the other hand, ionization potentials Ip or affinities A_{ff} correspond to nonsolvated O and R and radiative transitions (Sec. II.C), that is, to unstable nuclear configurations R* or O*, respectively, for R (affinities) or O (ionization potentials). The ensuing thermodynamic relationships between the three numbers E°, Ip, and A_{ff} are then derived from thermodynamic considerations based on the cycles in scheme 1.†

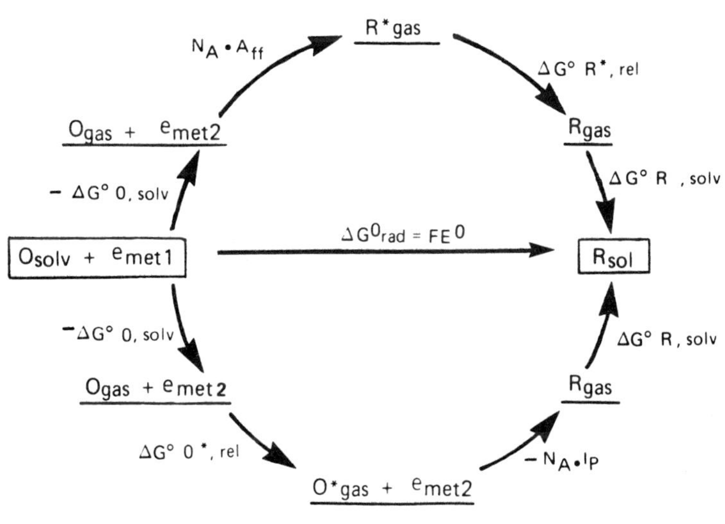

Scheme 1

†Note that no explicit attention to NHE or metal phases [26] has been given in the cycles, since these factors introduce a constant term Cst in Eq. (24) or (25), independent of the O/R couple.

Thus it follows that

$$E^\circ = -A_{ff} + \frac{(\Delta G^\circ_{0,solv} - \Delta G^\circ_{R,solv}) - \Delta G^\circ_{R^*,rel}}{F} + Cst \qquad (24)$$

and

$$E^\circ = I_P + \frac{(\Delta G^\circ_{0,solv} - \Delta G^\circ_{R,solv}) + \Delta G^\circ_{0^*,rel}}{F} + Cst \qquad (25)$$

where E°, A_{ff}, and I_P are expressed in volts and electron-volts, whereas ΔG° is in joules and relative to molar quantities. From Eqs. (24) and (25), it is seen that E° and I_P (or E° and A_{ff}) should correlate linearly, with a slope of unity for an extended series of O/R couples, only when the terms in the fraction are reasonably constant within the series. In practice this condition is almost impossible to fulfill accurately but is approximated for large delocalized molecules belonging to an identical class [27,28]. Indeed for such molecules the gain or loss of an electron introduces only small disturbances in the molecules. Similarly, there are nearly no specific solvation effects, the charge being largely delocalized. Then, provided the equivalent solvation radii are large or remain close within the series, the solvation differences are negligible [27]. Figure 1 shows that when these conditions are fulfilled, good correlations with slopes close to the unity are observed. Thus determination of E° for some of the compounds of the series allows good estimates of E° to be obtained from I_P or A_{ff} measurements. Yet it must be emphasized at this point that there should be virtually no special difficulties in the direct measurements of the unknown E° values for the O/R couples meeting all the requirements discussed here. Indeed, these conditions restrict the validity of the method to chemically reversible outer sphere electron transfers.

From this statement it is seen that the eventual practical interest of the method deals with series of redox couples that involve a too chemically

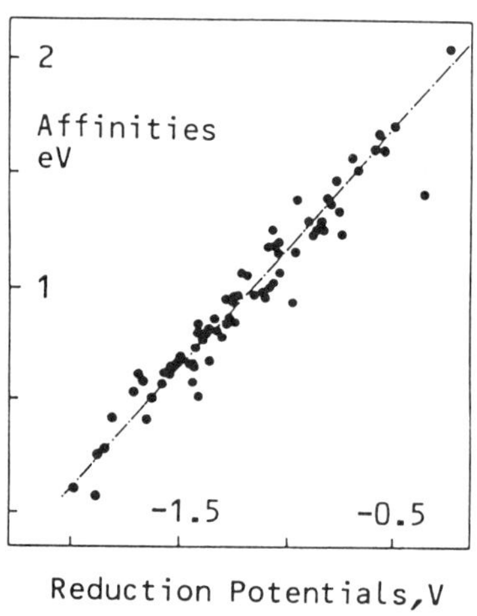

Figure 1. Correlation between affinities and reduction potentials for an extended series of polycyclic aromatic hydrocarbons. (Data from Ref. 28.)

unstable member for experimental E° to be determined easily. An example of such a case is given by the series of alkylbenzenes [29] presented in Fig. 2. Indeed, owing to the rather small size of the π system compared with the large aromatics in Fig. 1, as well as to the considerable variations in size with the number and nature of the alkyl substituent, it seems clear that the difference in solvation energies in Eq. (25) must vary within the series.

$$(\Delta G^{o}_{0,solv} - \Delta G^{o}_{R,solv}) + \Delta G^{o}_{0^*,rel} \approx -0.3FE^{o}$$

Interestingly, a good linear correlation is nevertheless observed in Fig. 2, yet with a slope of 0.7, instead of unity as in Fig. 1. From Eq. (25) it follows that the term on the left-hand side also correlates with E°, a reasonable conclusion owing to the homogeneity in the series of substituents. Yet the correlation in Fig. 2 clearly shows the danger of the blind use of such correlations, that is, of the use of Ip or A_{ff} values instead of E°. Indeed, a difference larger than 0.5 V is introduced when extrapolating E° from Ip values and assuming a unity slope as in Fig. 1.

Comparison or Extrapolation Between Solvents

Owing to the extremely large variety of solvents used in organic or organometallic chemistry, it is unlikely that all the pertinent electrochemical data can be found for a particular system and a particular solvent [30]. Thus the problem of the transposition of data obtained for one solvent to another frequently arises. As discussed for Ip or A_{ff}, this should not be done without extreme precautions. Indeed, by substraction of the two equations, akin to Eq. (25), pertinent to each solvent, one obtains readily for a given

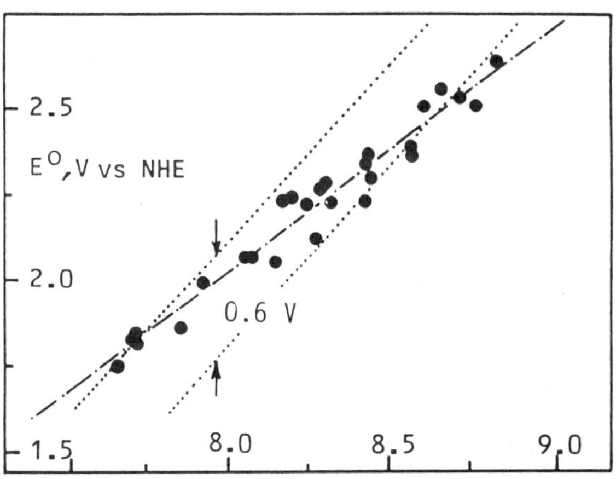

Figure 2. Correlation between ionization potentials and standard reduction potentials of the corresponding cation radicals for a series of alkylbenzenes. Two dotted lines, with slopes unity, are positioned at each end of the series to emphasize the deviations vis-a-vis the data in Fig. 1. (Data from Ref. 29.)

O/R redox system

$$E^{\circ}_{solv1} - E^{\circ}_{solv2} = \frac{(\Delta G^{\circ}_{O,solv1} - \Delta G^{\circ}_{O,solv2}) - (\Delta G^{\circ}_{R,solv1} - \Delta G^{\circ}_{R,solv2})}{F} + C'$$

where C' is a constant term independent of the O/R couple. However, the numerator, which may include specific solvation terms for each solvent, may vary considerably from one redox couple to the other. Thus on an absolute basis, extrapolations between solvents are meaningless; however, they may be used with great caution. Evidently the same is a fortiori true for the utilization of HOMO or LUMO energies [31] or Hammett-Taft and related extrathermodynamic relationships [32].

Formal Reduction Potentials

It is usually difficult to determine or impose activities, since activation coefficients are nearly always unknown for the redox systems investigated in organic or organometallic chemistry. Thus formal reduction potentials $E^{\circ\prime}$ are measured, when feasible, rather than E°. $E^{\circ\prime}$ corresponds to the electrode potential versus NHE when the concentration ratio C_O/C_R is made unity, rather than the ratio of the activities, as in the definition of E° [33]. Thus from Eq. (23), it follows that

$$E = E^{\circ} + \frac{RT}{F}\ln\frac{(O)}{(R)} = E^{\circ} + \frac{RT}{F}\ln\frac{f_O C_O}{f_R C_R}$$

that is,

$$E^{\circ\prime} = E^{\circ} + \frac{RT}{F}\ln\frac{f_O}{f_R} \qquad (26)$$

where f_J is the activity coefficient of species J. The latter being affected by ionic strength and concentrations, for example, it is predicted that $E^{\circ\prime}$ values should vary from medium to medium. This is important to remember when $E^{\circ\prime}$ values, determined under the conditions of low concentration and large ionic strength usually met in electrochemistry, need to be utilized under the conditions of low ionic strength and high concentration frequently used in organic chemistry [34].

Formal potential may also include some constant or selected components of the medium that participate in the overall redox reaction. Thus pH, and concentrations of complexing reagents, for example, are often included in $E^{\circ\prime}$ [35] when the O and R molecules are not interconnected by a simple electron transfer. To explicate this point, let us discuss a typical example related to aldehyde oxidation. The stoichiometry of the reaction in Eq. (27) involves an overall transfer of two electrons and one oxygen atom.

$$R{-}CHO + 3H_2O \rightleftharpoons R{-}CO_2H + 2e + 2H_3O^+ \qquad (E^{\circ}_1) \qquad (27)$$

Yet the acid may exist under two equivalent forms, the acid per se or its conjugated base. Thus, when the pH is such that the carboxylate anion is the stable form, Eq. (27) is better rewritten as in Eq. (28).

$$\text{R-CHO} + 4\text{H}_2\text{O} \rightleftharpoons \text{R-CO}_2^- + 2e + 3\text{H}_3\text{O}^+ \qquad (E_2^o) \tag{28}$$

From a preparative point of view, both reactions are nearly identical, since the acid and its conjugated base are two forms of the same chemical entity. From a thermodynamic point of view, however, they differ because of the involvement of the acid dissociation in Eq. (29),

$$\text{R-CO}_2\text{H} + \text{H}_2\text{O} \rightleftharpoons \text{R-CO}_2^- + \text{H}_3\text{O}^+ \qquad (K_a) \tag{29}$$

which thus corresponds to the definition of a second standard reduction potential E_2^o in Eq. (28). Yet E_1^o and E_2^o are related because of the necessity of a unique potential for the solution. Indeed from Eq. (27), the potential is obtained as in Eq. (30), whereas it is expressed as in Eq. (31) from Eq. (28).

$$E = E_1^o + \frac{RT}{2F} \ln \frac{(\text{RCO}_2\text{H})(\text{H}_3\text{O}^+)^2}{(\text{RCHO})} \tag{30}$$

$$E = E_2^o + \frac{RT}{2F} \ln \frac{(\text{RCO}_2^-)(\text{H}_3\text{O}^+)^3}{(\text{RCHO})} \tag{31}$$

Identification of both equations readily gives the relationship between E_1^o and E_2^o in Eq. (32), which shows that provided that K_a is known, knowledge of either E_1^o or E_2^o, is sufficient.

$$E_2^o = E_1^o - \frac{RT}{2F} \ln K_a \tag{32}$$

On the other hand, owing to the involvement of proton activities in Eq. (30) or (31), it is expected that the acid-aldehyde reduction potential is extremely dependent on the pH. Thus it is more convenient to separate the pH-dependent terms from those corresponding to the degree of conversion. Similarly, it is more realistic to consider the sum of the acid and its conjugated base activities to represent the oxidation products. Thus, taking into account Eq. (29), Eq. (30) is rewritten as in Eq. (33),

$$E = E_1^o + \frac{RT}{F} \ln \frac{h^3}{h + K_a} + \frac{RT}{F} \ln \frac{(\text{acid})}{(\text{aldehyde})} \tag{33}$$

where the term "acid" represents the sum of its two forms and $h = (H_3O^+)$. Similarly, concentrations may be introduced for the acid and aldehyde, rather than activities:

$$E = E^{o\prime} + \frac{RT}{F} \ln \frac{[\text{acid}]}{[\text{aldehyde}]} \tag{34}$$

This allows finally the rewriting of Eq. (33) under the form in Eq. (34), which directly gives the conversion ratio as a function of the potential and of the pH-dependent formal potential $E^{o\prime}$ in Eq. (35), where $K_a' = K_a(f_{RCO2H}/f_{RCO2^-})$, and $E_1^{o\prime} = E_1^o + (RT/2F)\ln(f_{RCO2H}/f_{RCO2^-})$:

$$E^{o\prime} = E_1^{o\prime} + \frac{RT}{2F} \ln\left[\frac{h^3}{h + K_a'}\right] \tag{35}$$

It is seen from Eq. (34) that although it is not established on definitions as solid as those for E^o, $E^{o\prime}$ nevertheless is of great practical interest since it is the reference parameter that controls the actual conversion ratio in a given situation. Indeed, at any pH or ionic strength, $E > E^{o\prime}$ requires that the acid is the major component in the system, whereas it is the aldehyde for $E < E^{o\prime}$. At $E = E^{o\prime}$, 50% of the aldehyde is oxidized. To decide which potential is needed at a given pH, or vice versa, it is then important to know the variations of $E^{o\prime}$ with pH. These variations are conveniently represented in the form of a "potential-pH" diagram and are derived from Eq. (35). Indeed, from the latter it is easily seen that $E^{o\prime}$ varies linearly with the pH as soon as the latter differs slightly from pK_a'; that is (at room temperature),

$$pH < pK_a' \qquad E^{o\prime} \approx E_1^{o\prime} - 0.06\ pH$$

$$pH = pK_a' \qquad E^{o\prime} \approx E_1^{o\prime} - 0.01 - 0.06\ pK_a'$$

$$pH > pK_a' \qquad E^{o\prime} \approx E_1^{o\prime} + 0.03\ pK_a' - 0.09\ pH$$

Such a diagram is represented in Fig. 3. The "acid" predominance zone corresponds to the upper part of the diagram, that is, to pH and potential conditions so that the point representing the system is above the $E^{o\prime}$ versus pH curve. Conversely, the aldehyde predominates when the system is located under this boundary. To illustrate the practical interest of these

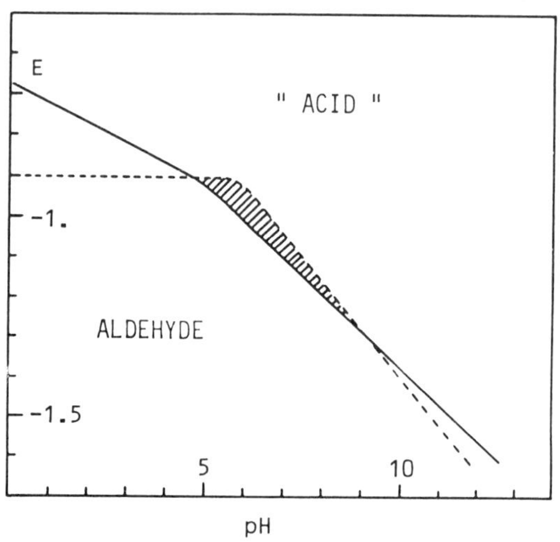

Figure 3. Potential-pH diagram for the aldehyde-acid transformation (solid line). The dashed line gives that for Ag/Ag^+ (0.1 M) when the pH is imposed by NH_3. The hatched zone corresponds to the only region where the silver mirror test can be performed.

diagrams, let us refer to the classic test of the silver mirror, which corresponds to aldehydes oxidation by the silver nitrate-ammonia reagent.

The standard reduction potential for Ag^+/Ag is 0.80 V versus NHE. Comparison with the diagram in Fig. 3 shows that the silver ion is not able to oxidize aldehydes except under basic conditions. Yet under such conditions silver oxides are formed and precipitate.

$$Ag^+ + 2NH_3 \rightleftharpoons Ag(NH_3)^+ + NH_3 \rightleftharpoons Ag(NH_3)_2^+ \tag{36}$$

Thus a particular base, ammonia, which gives stable complexes with Ag^+ in Eqs. (36), is used to bring about the basic pH. However, because of the formation of complexes in Eqs. (36), the free Ag^+ concentration decreases, which tends to lower the oxidative strength of the solution ($E = E^o_{Ag^+} + 0.06 \ln (Ag^+)$). This implies that the ammonia concentration must be adjusted precisely: sufficient for the pH to be in the required range, but not too high in order to retain a sufficient oxidizing power of the silver ion solution. In practice the best concentration is that corresponding to the titration point of the silver ion, which explains the great caution used in the preparation of the reagent $Ag(NH_3)_2^+$, NO_3^-.

Relationships Between Thermodynamic Driving Force $\Delta E°$ and Feasibility of an Electron Transfer Reaction

According to thermodynamics, the feasibility of an electron transfer reaction depends on the free energy change ΔG associated with it.

$$\alpha_1 O_1 + \beta_2 R_2 \rightleftharpoons \beta_1 R_1 + \alpha_2 O_2 \tag{37}$$

For the electron transfer in Eq. (37), which involves the two half-reactions in Eq. (38),

$$\alpha_1 O_1 + ne \rightleftharpoons \beta_1 R_1 \qquad (E_1^o) \tag{38a}$$

$$\alpha_2 O_2 + ne \rightleftharpoons \beta_2 R_2 \qquad (E_2^o) \tag{38b}$$

the free energy change is given from basic thermodynamics by

$$\Delta G = \Delta G^o + RT \ln \left[\frac{(R_1)^{\beta 1}(O_2)^{\alpha 2}}{(O_1)^{\alpha 1}(R_2)^{\beta 2}} \right]_{in.}$$

where $\Delta G° = nF(E_2^o - E_1^o)$ and the subscript in. means that the activities to be used in the bracketed term are those corresponding to the initial state of the system. A reaction is observed provided ΔG is negative or positive. The equilibrium point is then such that $\Delta G = 0$; therefore, at equilibrium the activities are such as given in Eq. (39), where the subscript eq. indicates equilibrium:

$$\left[\frac{(R_1)^{\beta 1}(O_2)^{\alpha 2}}{(O_1)^{\alpha 1}(R_2)^{\beta 2}} \right]_{eq.} = e^{(nF/RT)(E_1^o - E_2^o)} \tag{39}$$

From Eq. (39) it is deduced that for the forward electron transfer in Eq. (37) to be realized to some extent, it is necessary that the term on the left-hand side of Eq. (39) is considerably larger than unity and thus that $E_1^o > E_2^o$. Conversely, $E_1^o < E_2^o$ should lead to the conclusion that the reaction in Eq. (37) is not possible. This conclusion may not be true, however depending on the respective stabilities of the various species considered in Eq. (37).

When all reactants and products are stable, these conclusions are valid. However, when one or both products are not stable, Eq. (37) is continuously displaced to the right, even when thermodynamics predicts that the equilibrium should lie to the left-hand side. This is a situation akin to that usually considered in equilibrium displacements by physical removal (e.g., distillation, precipitation, and complexation) of one of the products. Thus one may have $E_1^o < E_2^o$ and nevertheless observe a reaction between O_1 and R_2 in Eq. (37). Characteristic examples of such situations are given elsewhere in this book, particularly in Chap. 29, since this phenomenon is the basis of redox catalysis. In such a situation the validity of Eq. (39) is not altered, provided the equilibrium remains established. Yet since (R_1) or (O_2) approaches zero, because of the follow-up chemical reaction(s), (O_1) and (R_2) may be extremely small and the bracketed term on the left-hand side of Eq. (39) remains considerably smaller than unity, fulfilling the condition imposed by $E_1^o < E_2^o$.

$$\alpha_1 O_1 + \beta_2 R_2 \rightleftharpoons \beta_1 R_1 + \alpha_2 O_2 \longrightarrow \cdots \qquad (40)$$

It is thus seen that a direct relationship exists between the thermodynamic driving force $\Delta E°$ and the feasability of an electron transfer reaction only for those special cases in which all the products and reactants are stable within the time scale considered. Yet it must be pointed out that this is seldom encountered in usual practice, owing to the generally high reactivities of the species formed upon electron transfers.

This discussion may lead to the extreme conclusion that when the products of an electron transfer reaction are unstable, as in Eq. (40), thermodynamic figures, such as $\Delta E°$, are irrelevant. From a simple point of view this is true. Yet in practice kinetic notions are implicitly involved in the discussion of the feasability of any chemical reaction. Indeed, a possible reaction whose completion requires infinite time is not considered feasible. Amazingly, it is because of kinetics that thermodynamic figures recover an interest for situations like that featured in Eq. (40). To state this point exactly, let us present and discuss the case of the oxidation of methylbenzenes by tris-(1,10-phenantroline)-iron(III) complexes (FeL_3^{3+}) in the presence of puridine bases [36].

From their standard reduction potentials, in the range of 1.35 V versus NHE [37], FeL_3^{3+} complexes should not be able to oxidize methylbenzenes, the standard potentials of which exceed 1.75 V versus NHE [29]. The reaction, however, proceeds easily in the presence of a pyridine base, according to the following stoichiometry:

$$ArCH_3 + 2FeL_3^{3+} + 2py \longrightarrow ArCH_2\text{-}py^+ + pyH^+ + 2FeL_3^{2+}$$

and was shown [36,38,39] to involve an electron transfer between the iron and the arene centers, as outlined in scheme 2. Two limiting kinetic situations

$$ArCH_3 + Fe^{III} \rightleftharpoons ArCH_3^{+\bullet} + Fe^{II} \quad (k_f, k_b) \tag{41}$$

$$ArCH_3^{+\bullet} + py \longrightarrow ArCH_2^{\bullet} + pyH^{+} \quad (k_H) \tag{42}$$

$$ArCH_2^{\bullet} + Fe^{III} \longrightarrow ArCH_2^{+} + Fe^{II} \quad (fast) \tag{43}$$

$$ArCH_2^{+} + py \longrightarrow ArCH_2\text{-}py^{+} \quad (fast) \tag{44}$$

Scheme 2.

are observed according to the respective rates of the proton transfer [Eq. (42)] or of the backward electron transfer [Eq. (41)]. When the latter is larger, the electron transfer in Eq. (41) acts as a rapid equilibrium, the proton transfer being the rate-determining step (RDS). Thus an apparent rate constant is determined, as in Eq. (45):

$$k^{ap} = k_H \frac{k_f}{k_b} \tag{45}$$

Conversely, when the proton transfer rate is considerably faster than the backward electron transfer, the RDS is the forward electron transfer and the observed rate constant is then k_f, as given in Eq. (46):

$$k^{ap} = k_f \tag{46}$$

It is thus seen that, in the first situation, although the thermodynamic interdiction has been overruled, thermodynamic figures control the apparent rate constant observed. Indeed, Eq. (45) may be rewritten as in Eq. (47):

$$k^{ap} = k_H e^{(F/RT)(E^o_{Fe} - E^o_{ArCH_3})} \tag{47}$$

Thus owing to the large magnitude of k_H (10^3–10^6 M^{-1} s^{-1} [36b–c,38,40]), the rate constant in Eq. (47) remains appreciable even for largely endergonic electron transfers (for 1,2,4,5-tetramethylbenzene, $E^o_{Fe} - E^o_{ArCH3} < -0.6$ V). Yet if the electron transfer is too endergonic, k^{ap} becomes too small for the reaction to proceed at a significant rate.

The involvement of thermodynamic figures when the forward electron transfer is the RDS is more subtle and arises because of linear or quadratic [41] relationships between activation free energies and thermodynamic driving force, which applies for outer sphere electron transfers like that in Eq. (41) (Sec. II.C, QFER).

C. Mechanism and Theory of Outer Sphere Electron Transfer Reactions

Owing to the aim and scope of the present chapter, it is impossible to discuss here all the subtleties and refinements of electron transfer theories. Yet our purpose is to give a general overview of the different concepts on which these theories are elaborated. As such this text may constitute per se a presentation of the basic features essential to a thoughtful experimental use of outer sphere electron transfer mechanisms. On the other hand, we hope it may constitute an introduction to more specialized readings [42].

In the following, we tried to rely, for the sake of simplicity, on simple physicochemical pictures that often were not involved in the original works [41,43,44]. As a consequence this text short-circuits some important but specialized theoretical aspects of the original theories.

Diffusion Limitation in Electron Transfer Reactions

A very common characteristic of electron transfer reactions is that they frequently involve activation energies lower than other chemical reactions that relate to group or atom transfers. As a consequence the activated processes may be extremely fast compared with the physical process of bringing the two reactants close together (or separating the products) so that they can react. In such cases, the overall rate constant determined does not reflect activation parameters but rather transport properties. Note that this phenomenon is implicitly recognized in homogeneous chemistry. Indeed, the "diffusion limit" rate constant, of the order of 10^9–10^{10} M^{-1} s^{-1}, corresponds to reactions that are limited by encounters of the reactants, whereas the limit related to product separation is intuitively included within part of the notion invoked in "cage separation."

Thus, in a general situation, an electron transfer between a donor D and an acceptor A must be considered in terms of, at least, the three successive elementary steps outlined in scheme 3. This representation may be even more segmented when, for example, different ion pairs are involved, such as [$A^{\overline{\cdot}}$, solvent, $D^{\dot{+}}$] or [$A^{\overline{\cdot}}$, $(solvent)_n$, $D^{\dot{+}}$]. Yet it is sufficient here to consider the simplest situation represented in scheme 3 [45]. The formation or dissociation of cages in scheme 3, each a physical process, is normally handled on physicochemical grounds. Yet following a notation by Debye their effects may be represented under the form of pseudo-rate constants: k^r_{dif} (reactant pair) and k^p_{dif} (product pair). Using this notation it follows [46] that the overall forward k_f and backward rate constants k_b are

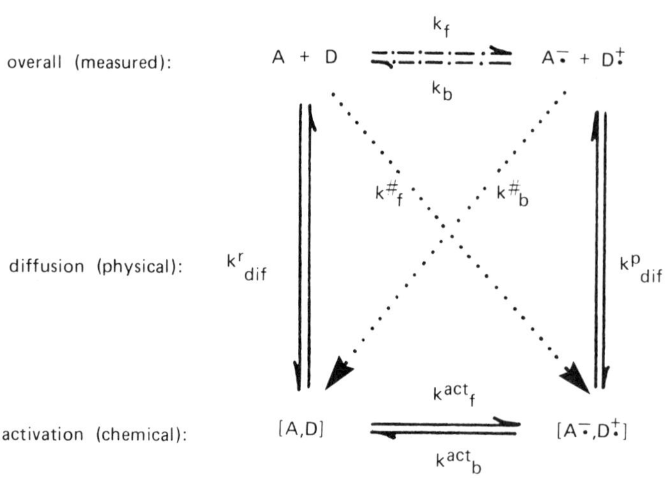

Scheme 3

given by Eqs. (48) and (49) as a function of the activation rate constants k_f^{act} and k_b^{act} relative to the pairs.

$$\frac{1}{k_f} = \frac{e^{\Delta G_r^f/RT}}{k_f^{act}} + \frac{1}{k_{dif}^r} + \frac{e^{\Delta G^o/RT}}{k_{dif}^p} \tag{48}$$

$$\frac{1}{k_b} = \frac{e^{\Delta G_p^f/RT}}{k_b^{act}} + \frac{1}{k_{dif}^p} + \frac{e^{-\Delta G^o/RT}}{k_{dif}^r} \tag{49}$$

In Eqs. (48) and (49), $\Delta G^o = F(E^o_{D+/D} - E^o_{A/A^-})$ and ΔG_r^f or ΔG_p^f represents the free energy necessary to form the pairs in scheme 3 for the reactants and products, respectively. The diffusion rate constants are given by Eqs. (50) and (51) [45], where r is the distance between the pair centers and D_r or D_p the sum of the diffusion coefficients of the reactants or products, respectively.

$$k_{dif}^r = \frac{4\pi N_A D_r}{\int_r^\infty x^{-2} e^{W_r(x)/RT}\, dx} \tag{50}$$

$$k_{dif}^p = \frac{4\pi N_A D_p}{\int_r^\infty x^{-2} e^{W_p(x)/RT}\, dx} \tag{51}$$

Both pseudo-rate constants may then be evaluated from a suitable description of the work terms $W_r(x)$ and $W_p(x)$ necessary to bring the reactant centers or those of the products, respectively, from infinity to a distance x. When the work terms are negligible, straightforward integration of Eqs. (50) and (51) yields

$$k_{dif}^r = 4\pi N_A D_r r \qquad \text{and} \qquad k_{dif}^p = 4\pi N_A D_p r$$

which corresponds to the "diffusion limit rate constants" of the order of 10^9–10^{10} M^{-1} s^{-1} first proposed by Smoluchowsky [45b]. When the work terms are not negligible compared to RT (due to, e.g., specific interactions, electrostatic interactions between charged ions, or specific ion pairs), the diffusion limit rate constant may differ appreciably from these values, being larger when the work terms are negative (attractive pairs) or considerably smaller when the work terms are positive (repulsive pairs).

Because of the symmetry in Eqs. (48) and (49) we restrict the following discussion to the forward process. It is also noted that k_f and k_b fulfill the thermodynamic condition in Eq. (52):

$$k_f = k_b\, e^{-\Delta G^o/RT} \tag{52}$$

$$k_f^{act} = k_b^{act}\, e^{-\Delta G^{o\prime}/RT} \tag{53}$$

Indeed, k_b^{act} and k_f^{act}, being true chemical rate constants, must obey Eq. (53), where $\Delta G^{o\prime}$ is the free energy change in the reaction between the pairs. When there is no other entropic contribution in the formation of the pairs besides those corresponding to the formation of rigid pairs, one has:

$$\Delta G_r^f - \Delta G_p^f \approx W_r - W_p \quad \text{and then} \quad \Delta G^{o\prime} = \Delta G^o - W_r + W_p$$

From Eq. (48) it is seen that the observed rate constant is always smaller than or equal to k_f^{dif}, defined by considering only the transport limitations in k_f expression, that is, by Eq. (54a).*

$$\frac{1}{k_f^{dif}} = \frac{1}{k_{dif}^r} + \frac{e^{\Delta G^o/RT}}{k_{dif}^p} \tag{54a}$$

When $\Delta G^o \ll RT\ln(k_{dif}^p/k_{dif}^r)$, the first term in Eq. (54a) predominates, which means that in the corresponding range one obtains $k_f^{dif} \approx k_{dif}^r$. This corresponds to the traditional diffusion limit, that is, to the case of extremely exergonic reactions. Conversely, for highly endergonic reactions, Eq. (54a) yields $k_f^{dif} \approx k_{dif}^p\, e^{-\Delta G^o/RT}$. In the latter case it might be surprising that a diffusion control may be observed, yet this apparent paradox corresponds simply to the fulfillment of the thermodynamic conditions in Eqs. (52) and (54b), the highly exergonic backward reaction being diffusion controlled.

Thus under any circumstances the overall observed rate constant k_f cannot exceed the limit of k_f^{dif} given in Eq. (54a). This limit is represented in the form of a log k_f^{dif} versus ΔG^o plot in Fig. 4. It must be noticed that since no special considerations have been given to the intrinsic nature of the overall reaction up to this point, these results are valid for any kind of bimolecular chemical reaction. Yet for reactions involving group or atom transfers, $W_r(r)$ and $W_p(r)$ may reach high values (precursor complexes) compared with electron transfer reactions [47]. As a result, the values of k_{dif}^r or k_{dif}^p may be considerably smaller than the usual figures observed in the $10^9 - 10^{10}$ M^{-1} s^{-1} range for electron transfers.

Activation control of the reaction is observed when $k_f \ll k_f^{dif}$, the latter being given in Eq. (54a). Indeed, formation of the precursor or successor complexes is then considerably faster than the activation process, which therefore limits the overall reaction. Thus it ensues from Eqs. (48) and (49) that $k_f \approx k_f^{\#}$, and $k_b \approx k_b^{\#}$,

$$k_f^{\#} = k_f^{act}\, e^{-\Delta G_r^f/RT} \quad \text{and} \quad k_b^{\#} = k_b^{act}\, e^{-\Delta G_p^f/RT} \tag{55}$$

---

*Note that, similarly, a transport limit

$$k_b^{dif} = k_f^{dif}\, e^{\Delta G^o/RT} \tag{54b}$$

is defined for the backward electron transfer. Based on scheme 3 and Eqs. (48) and (49), k_b^{dif} is given as in Eq. (54b).

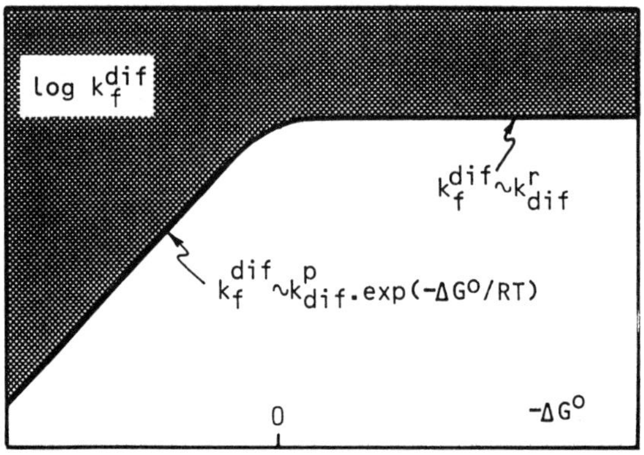

Figure 4. Variations in k_f^{dif}, as given in Eq. (54), with the driving force. The shaded area corresponds to the zone where an activation control is not achievable, the overall kinetics being under diffusion control.

where the rate constants $k_f^{\#}$ and $k_b^{\#}$, defined in Eq. (55), correspond to the usual definition of activation rate constants. Indeed they refer to the activated process, the initial states being the dissociated A, D or $A^{\overline{\cdot}}$, $D^{\dot{+}}$ pairs, respectively (compare scheme 3). Owing to this formulation, Eqs. (48) and (49) are usually rewritten under their equivalent forms in Eqs. (56),

$$\frac{1}{k_f} = \frac{1}{k_f^{\#}} + \frac{1}{k_f^{dif}} \qquad \frac{1}{k_b} = \frac{1}{k_b^{\#}} + \frac{1}{k_b^{dif}} \tag{56a}$$

$$\frac{k_f}{k_b} = e^{\Delta G^o/RT} \tag{56b}$$

where $k_f^{\#}$, $k_b^{\#}$, k_f^{dif}, and k_b^{dif} are defined in Eqs. (55) and 54a,b). Yet when using such elegant but compact formulations, it is important to recall that k_f^{dif} and k_b^{dif} are not the usual diffusion limit rate constants defined in Eqs. (50) and (51) but may be considerably smaller for endergonic reactions [compare Fig. 4 and Eqs. (54)].

In the following we focus the presentation on the activation contributions, that is, on $k_f^{\#}$ or k_f^{act}, to the overall rate constant k_f. First we discuss the nature of an electron transfer reaction. Later, models for the evaluation of the activation energy $\Delta G_f^{\#}$ are presented.

Basic Features of Electron Transfer in Solution

The fundamental role of the surrounding medium in an electron transfer reaction was early recognized by Libby [48]. Indeed, because of the large number of molecular collisions in solution, molecules are thermalized, in contrast to what is observed in gas phase (EI, Ip, and so on). Thus radiative electron transfers (that is, analogous to photochemical excitation; see Sec. II.C, Physical meaning of reorganization energy and of energy diagrams) do not take place in condensed phases. In other words, an electron cannot be transferred between two systems in which its energy is different, as shown schematically in Fig. 5. In the following a "system" designates a couple of molecules exchanging an electron, that is, A and D, or $A^{\cdot -}$ and $D^{\cdot +}$, together with the surrounding solvent. Thus the "initial system" corresponds to the reactants and the surrounding medium, in any out-of-equilibrium nuclear configuration, which may greatly differ from its equilibrium state, denoted [A, D] in scheme 3. Similarly, the "final system" is associated with the products. Because of the intrinsic difference in mass between the nuclei and the transferred electron, the Franck-Condon principle applies. In other words, nuclear motions (10^{-12} s) are frozen vis-à-vis those of the electron (10^{-15} s). Thus for any nuclear configuration of the initial (or final) system, a potential energy, termed the energy of the initial (or final) system, is associated with the electron, which probes the field of the nuclei (and other electrons). Because the electron is bound to be located in either D or A, the initial and final systems are potential energy minima as a function of the electron coordinate, as shown schematically in Fig. 5. In the physical process of being transferred the electron must then tunnel through a potential barier corresponding to the virtual states in which the electron would be between A and D. The height of this barrier is inversely related to the degree of overlapping of the A and D orbitals effecting the electron transfer. It greatly influences Pe, the probability that the electron tunnels during the time interval in which the initial and final systems have identical energies (compare Fig. 5b). This time interval, a function of nuclear motions, is nearly independent of the acceptor and the donor, so that Pe depends mainly on the height of the barrier. As soon as the energy stabilization V_i arising from overlapping orbitals is larger than approximately 1 kcal/mol, Pe is close to unity and the electron transfer is termed adiabatic. In the converse situation, the electron transfer is called nonadiabatic and Pe is considerably smaller than unity (Sec. II.C, Physical meaning of the reorganization energy and of energy diagrams).

From this presentation it is seen that the overall proability P of electron transfer is approximated by $P = Pn \cdot Pe$, where Pn is the probability that the nuclei achieve a configuration so that the electron energy is identical for the initial and final systems (Fig. 5b). In usual kinetic terms, this means that the rate constant k_f^{act} is given by $k_f^{act} = \kappa(k_f^{act})_{ad}$, where κ is the "transmission coefficient" ($\kappa = 1$ for an adiabatic electron transfer; $\kappa \ll 1$ for nonadiabatic transfers) and $(k_f^{act})_{ad}$ is the rate constant observed for an adiabatic electron transfer. The latter then depends only on nuclear motions that affect the potential energy of the electron. In usual chemical terms, $(k_f^{act})_{ad}$ is directly related to the height of the activation barrier, that is, to the energetic separation between the state where the electron may tunnel and that corresponding to the initial system at equilibrium (compare Fig. 5b), denoted [A, D] in scheme 3.

$$k_f^{act} = \kappa \frac{kT}{h} e^{-\Delta E_f^{act}/RT} \tag{57}$$

From absolute rate theory [49], k_f^{act} is then given by Eq. (57), where ΔE_f^{act} is the activation energy barrier expressed for molar quantities. From Eq. (55) it follows that

$$k_f^{\#} = \kappa \frac{kT}{h} e^{-\Delta G_r^f/RT} e^{-\Delta E_f^{act}/RT}$$

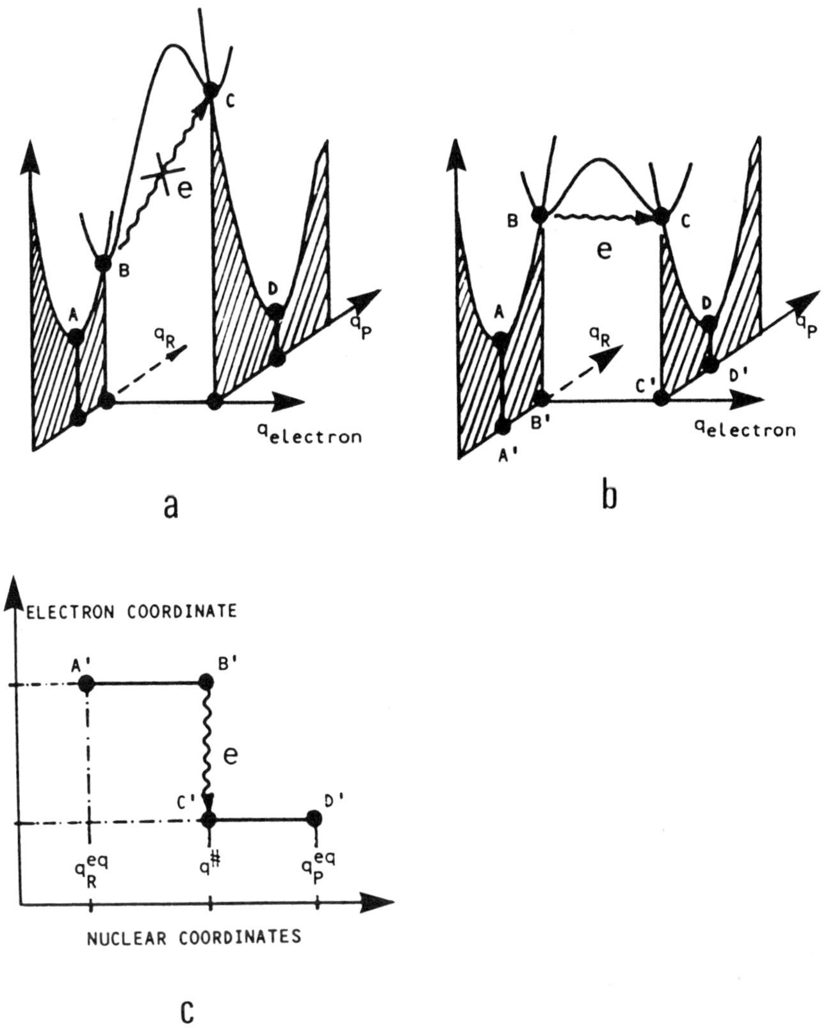

Figure 5. Schematic variation in the electron potential energy as a function of the nuclear coordinates (q_R and q_P for the reactant and product systems, respectively) and of the electron coordinate. (a) Forbidden electron transfer; (b) allowed electron transfer; (c) projection of the "system trajectory" on the electron coordinate-nuclear coordinates plane. q_R^{eq} and q_P^{eq} are the values of the nuclear coordinate at the equilibrium for the reactant or product systems, respectively, and $q^{\#}$ that at the transition state. Solid lines, variations of the nuclear coordinates; ($\rightsquigarrow$) electron tunneling.

Note that the same presentation applies to $k_b^{\#}$ and thus

$$k_b^{\#} = \kappa \frac{kT}{h} e^{-\Delta G_p^f/RT} e^{-\Delta E_b^{act}/RT}$$

It is usually more convenient to introduce explicitly the internal energy (W_r and W_p respectively) and the entropic components of the free energies ΔG_r^f and ΔG_p^f. When the formation of pairs involves no specific interaction, their entropy of formation is identical to that of a rigid rotator (ΔS_{rot}), which yields

$$\Delta G_r^f = W_r - T\,\Delta S_{rot} \quad \text{and} \quad \Delta G_p^f = W_p - T\,\Delta S_{rot}$$

and thus

$$k_f^{\#} = \kappa\left(\frac{kT}{h} e^{\Delta S_{rot}/R} e^{-W_r/RT} e^{-\Delta E_f^{act}/RT}\right)$$

and

$$k_b^{\#} = \kappa\left(\frac{kT}{h} e^{\Delta S_{rot}/R} e^{-W_p/RT} e^{-\Delta E_b^{act}/RT}\right)$$

From classic statistical thermodynamics [50] the product

$$(kT/h)e^{\Delta S_{rot}/RT}$$

is shown to be equal to the collision frequency Z [51], which is usually of the order 10^{11} M^{-1} s^{-1}:*

$$Z = 10^{-3} N_A \left[\frac{8\pi RT(m_A + m_D)}{m_A m_D}\right]^{1/2} r^2$$

One then finally obtains [43,44]

$$k_f^{\#} = \kappa\, Z\, e^{-W_r/RT} e^{-\Delta E_f^{act}/RT} \tag{58}$$

and

$$k_b^{\#} = \kappa\, Z\, e^{-W_p/RT} e^{-\Delta E_b^{act}/RT} \tag{59}$$

with $k_f^{\#} = k_b^{\#} e^{\Delta G^o/RT}$ because $\Delta G^{o\prime} = \Delta G^o - W_r + W_p \approx \Delta E_f^{act} - \Delta E_b^{act}$.

An absolute expression for κ is nearly impossible to derive in a general case owing to its close dependence on the orbital system of a specific A and

---

*In this expression m_A and m_D are the molar masses of A and D, and r, the distance between the molecule centers in the precursor or successor complexes, is expressed in angstroms, Z being in M^{-1} s^{-1}.

D couple [52]. Thus general theories consider only adiabatic electron transfers, that is, those corresponding to $\kappa = 1$. In practice, owing to the very small (less than 1 kcal/mol) interaction required, this is not a limitation for most electron transfers in organic chemistry.

The theoretical predictions of $k_f^{\#}$ then amount to evaluating the magnitude of ΔE_f^{act} in Eq. (58). Various models have been proposed for this determination, and most of them rely on harmonic approximations for the variations of energy of the initial or final systems with nuclear deformation. The basis of the harmonic assumption is that, for outer sphere electron transfers, (1) only small perturbations in bond length or bond angle are involved and (2) a large number of nuclear coordinates participate in the activation process. Owing to these two considerations it is understood that although the accumulated energies may be important (from several kilocalories per mole to a few tens of kilocalories per mole), each of the nuclear coordinates may remain close to its equilibrium value, where harmonic descriptions of the corresponding energy variations are accurate. Note in this context that since for inner sphere electron transfers (Sec. II.A, Outer sphere and inner sphere electron transfers, and Table 1) bonds are broken or created in the activation process, the harmonic assumption fails. Indeed, in such electron transfers the activation energy is concentrated within one given bond, which neeeds more detailed description of the potential energy reaction coordinate surfaces [54]. Similarly, the geometric requirements for κ to be unity are certainly more important than for outer sphere electron transfers, owing to the extreme localization of the orbitals concerned in the electron exchange. Thus in the following we limit our presentation to the harmonic model approximation.

Let us denote by q_i the ith nuclear coordinate ($i = 1-N$), q_i^{in} and q_i^{fin}, its equilibrium values in the initial and final systems, respectively, and Ω_i, the corresponding reduced force constant (i.e., $\Omega_i = 2\Omega_i^{in}\Omega_i^{fin}/(\Omega_i^{in} + \Omega_i^{fin})$, where Ω_i^{in} and Ω_i^{fin} are the force constants in the initial and final systems, respectively [43]). With these notations the energy of the initial and final systems are given in Eqs. (60) and (61):

$$E^{in} = E_{eq}^{in} + \frac{1}{2}\sum \Omega_i(q_i - q_i^{in})^2 \qquad (60)$$

$$E^{fin} = E_{eq}^{fin} + \frac{1}{2}\sum \Omega_i(q_i - q_i^{fin})^2 \qquad (61)$$

where E_{eq}^{in} and E_{eq}^{fin} are the respective equilibrium values, such as

$$E_{eq}^{fin} - E_{eq}^{in} = \Delta G^{o\prime} = \Delta G^{o} - W_r + W_p \qquad (62)$$

The energy $E^{\#}$ of the transition state must be such as $E^{\#} = E^{in} = E^{fin}$. Then by substraction of Eqs. (60) and (61) and denoting by $q_i^{\#}$ the value of q_i at the transition state, one obtains

$$0 = \Delta G^{o\prime} - \frac{1}{2}\sum \Omega_i[(q_i^{\#} - q_i^{in})^2 - (q_i^{\#} - q_i^{fin})^2] \qquad (63)$$

Equation (63) describes a hypersurface of N − 1 dimensions, which correspond to all the possible transition states (compare Fig. 6 when N = 2). Yet the transition state through which the system reacts is that of lower energy. This is obtained by derivation of Eq. (63) versus $q_i^{\#}$. From the resulting equation and Eq. (63), one shows readily by the Lagrange factor method that ΔE_f^{act} is given in Eq. (64):

$$\Delta E_f^{act} = E^{\#} - E_{eq}^{in} = \frac{\lambda}{4}\left(1 + \frac{\Delta G^{o\prime}}{\lambda}\right)^2 \tag{64}$$

where

$$\lambda = \frac{1}{2}\sum \Omega_i (q_i^{fin} - q_i^{in})^2 \tag{65}$$

is the <u>reorganization energy of the system</u>. Because λ is a sum, one may separate the terms relative to bond length or bond angle variations in the A and D molecules (i = 1, M) from those relative to the surrounding medium (i = M + 1, N). Thus λ is usually written as the sum of two contributions:

$$\lambda = \lambda_i + \lambda_o \tag{66}$$

with

$$\lambda_i = \frac{1}{2}\sum \Omega_i (q_i^{fin} - q_i^{in})^2 \quad \text{for } i = 1, M \tag{67}$$

being the <u>inner shell reorganization energy</u> and

$$\lambda_o = \frac{1}{2}\sum \Omega_i (q_i^{fin} - q_i^{in})^2 \quad \text{for } i = M + 1, N \tag{68}$$

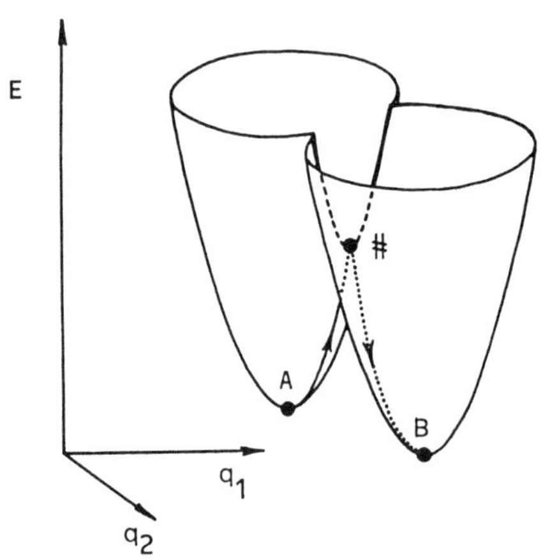

Figure 6. Schematic representation of the reaction pathway (···), for two nuclear coordinates (N = 2). The dashed line represents the location of all possible transition states.

the outer shell reorganization energy. Note, however, that these terms must not be confused with the inner sphere or outer sphere nature of an electron transfer (Sec. II.A, Outer sphere and inner sphere electron trans fers, and Table 1).

λ_i may be evaluated from x-ray and infrared (IR) data or from theoretical calculations. However, for organic outer sphere electron transfers, this contribution is usually much smaller than λ_o. In our opinion one of the greatest merits of the Marcus [43] and Levich-Dogonadze [44] theories is that they allow rather correct predictions of λ_o through simple equations. Thus for most outer sphere electron transfers, reasonably accurate values of the rate constants can be predicted.

Role of the Solvent: The Outer Shell Reorganization Energy

Owing to the usually small energetic interactions between a solvent molecule and the solutes, random fluctuations of the medium surrounding the reactants or the products constantly take place.* During these fluctuations, the solvent nuclei have out-of-equilibrium positions, which then induces nonequilibrium values for the electrostatic field probed by the electron to be transferred. Yet the electrons of the solvent instantly adjust to the local field. This then results in equilibrium values for the electronic components of the electrostatic field probed by the electron exchanged. This dichotomy between the nuclei (vibrational-orientational-translational) and the electronic polarizations arises because of large differences in motion frequencies (10^{-12} s for atoms and 10^{-15} s for electrons).

To determine the resulting energetic variations, the solvent is treated as a dielectric continuum surrounding the reactants or the products. The latter are represented by two metallized spheres of internal charges Z_A and Z_D for the reactants and $Z_A - 1$ and $Z_D + 1$ for the products. The sphere radii a_A and a_D are supposed invariant upon electron transfer† because inner shell geometric variations are usually small, although they may result in large λ_i owing to the large magnitudes of the associated force constant [compare Eq. (67)]. Figure 7 gives a schematic representation of such solvent fluctuations.

Within this model the energy associated with a nonequilibrium polarization state is determined by considering [57]: (1) for the solvent nuclei the polarization corresponds to an equilibrium state in which a virtual fractional charge δe has been transferred, whereas (2) the solvent electronic polarization still corresponds to the real charges. In practice, for the initial system, for example, this is equivalent to determining the variation in the overall electrostatic energy, that is, that associated with the static dielectric constant ε_S of the solvent when the reactant charges vary from (Z_A,

*Note that the tightly bound solvent molecules are then considered as belonging to the inner shell subsystem and their contributions to the activation energy are then included in λ_i.

†Note that more sophisticated models have been proposed to take into account the nonspherical nature of reactant and products, the eventual distribution of charges on different centers within the molecule [55], the eventual coupling between the inner shell and solvent fluctuation modes, and other factors [56]. Yet the simplest model described here is sufficient for our purposes, since it includes most of the essential features of electron transfer reactions.

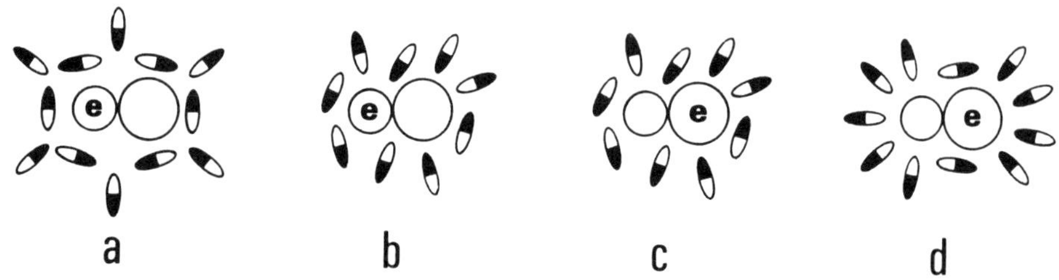

Figure 7. Schematic representation of the solvent fluctuations inducing the electron transfer. (a, d) Equilibrium states for the reactants and products, respectively. (b, c) Out-of-equilibrium states for the reactants and products at the transition state.

Z_D) to ($Z_A - \delta$, $Z_D + \delta$), and to correcting it from the electronic contribution, which is associated with ε_{op}, the optical dielectric constant of the solvent ($\varepsilon_{op} = n_{op}^2$; n_{op} being the refractive index of the solvent). From the corresponding energetic cycle represented in scheme 4, it is easily deduced that the free energy required to produce an appropriate nonequilibrium polarization is $W_\delta^{noneq} = W_\delta^{\varepsilon s} - W_\delta^{\varepsilon op}$ and can be evaluated from simple electrostatics, as in the Born solvation model [27]. Thus both $W_\delta^{\varepsilon s}$ and $W_\delta^{\varepsilon op}$ are obtained, using ε_s or ε_{op}, respectively, by the sum of the three electrostatic contributions: (1) charge of the A sphere from Z_A to $Z_A - \delta$, (2) charge of the D sphere from Z_D to $Z_D + \delta$, and (3) variation of the electrostatic interaction between the two metallized spheres A and D at distance r. Thus one obtains

$$W_\delta^{\varepsilon s} \text{ or } W_\delta^{\varepsilon op} = -\left[\frac{N_A(\delta e)^2}{8\pi\varepsilon}\right]\left(\frac{1}{a_A} + \frac{1}{a_D} - \frac{2}{r}\right)$$

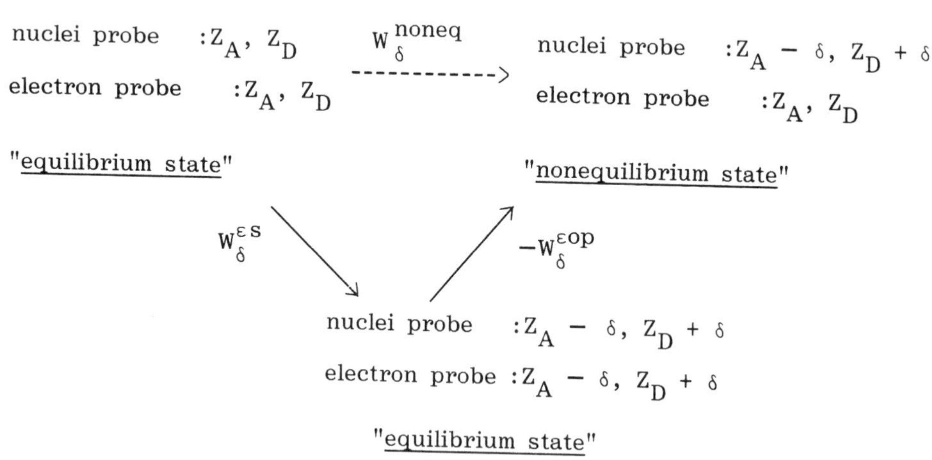

Scheme 4

where $\varepsilon = \varepsilon_s$ or ε_{op} and e is the charge of the electron. It is thus seen that for a given δ the outer shell component of the enery of the out-of-equilibrium state of the initial system is given by Eq. (69):

$$(E^{in})_o = (E^{in}_{eq})_o - \left(\frac{N_A e^2}{8\pi}\right)\left(\frac{1}{\varepsilon_s} - \frac{1}{\varepsilon_{op}}\right)\left(\frac{1}{a_A} + \frac{1}{a_D} - \frac{2}{r}\right)\delta^2 \qquad (69)$$

A similar expression would be obtained along the same lines for the final system, yet with δ replaced by $1 - \delta$. Comparison of Eqs. (69) and (60) shows that the outer shell (solvent) contributions are equivalent to a single harmonic oscillator of force constant Ω_o, given in Eq. (70),

$$\Omega_o = -\left[\frac{N_A e^2}{4\pi}\left(\frac{1}{\varepsilon_s} - \frac{1}{\varepsilon_{op}}\right)\right]\left(\frac{1}{a_A} + \frac{1}{a_D} - \frac{2}{r}\right) \qquad (70)$$

where the virtual charge transferred δ represents the elongation. It then results from Eq. (68) that

$$\lambda_o = \frac{1}{2}\Omega_o(\delta^{fin} - \delta^{in})^2 = \frac{1}{2}\Omega_o \qquad (71)$$

since $\delta^{in} = 0$ and $\delta^{fin} = 1$ for one electron transferred. The outer shell reorganization energy λ_o is thus finally obtained under the formulation [57] in Eq. (72):

$$\lambda_o = \left[\frac{N_A e^2}{8\pi}\left(\frac{1}{\varepsilon_s} - \frac{1}{\varepsilon_{op}}\right)\right]\left(\frac{1}{a_A} + \frac{1}{a_D} - \frac{2}{r}\right) \qquad (72)$$

The latter equation, involving easily acessible data, allows predictions of λ_o for most situations of interest. Note that because of the involvement of a squared term in Eq. (71), the reorganization energy for a simultaneous transfer of n electrons is n^2 times larger than that for a single electron transfer since $\delta^{fin} - \delta^{in} = n$ for n electrons transferred.

Physical Meaning of the Reorganization Energy and of Energy Diagrams

From the preceding discussion it is understood that although the real phenomenon of electron transfer affects a large number of nuclear coordinates, the overall effect is akin to what would be observed for a single generalized coordinate q. This is particularly true for the external medium contribution as established earlier. For simplification of the following discussion, let us assume that only one coordinate q is involved, which varies between q^{in} at the initial system equilibrium to q^{fin} at the final system equilibrium. Let us then introduce the normalized coordinate $x = (q - q^{in})/(q^{fin} - q^{in})$. It is seen from Eq. (60) or Eqs. (61) and (65) that E^{in} and E^{fin} are given as a function of x by

$$E^{in} = E^{in}_{eq} + \lambda x^2 \qquad (73)$$

$$E^{fin} = E^{fin}_{eq} + \lambda(1 - x)^2 \qquad (74)$$

with x = 0 corresponding to the initial system equilibrium point whereas x = 1 corresponds to the final system equilibrium point. Figure 8a represents the resulting variations in the two energies with x. It is seen from this figure that, if the electron transfer were to occur radiatively from the initial system equilibrium point, as in the gas phase (e.g., in I_p determinations), then the final system curve would be reached at the point x = 0, that is, from Eq. (74), at $E = E_{eq}^{fin} + \lambda$. The system would then relax to its equilibrium, that is, x = 1 and $E = E_{eq}^{fin}$, thus radiating an energy λ during its reorganization.

A second common interpretation of the reorganization energy is related to the activation process rather than to a comparison between the gas and the condensed phases. The activation barrier for the forward reaction is given by equating E^{in} and E^{fin} in Eqs. (73) and (74). Thus it follows that $x^{\#}$, the normalized coordinate at the transition state, is given by resolution of

$$E^{\#} = E_{eq}^{in} + \lambda(x^{\#})^2 = E_{eq}^{fin} + \lambda(1 - x^{\#})^2$$

which easily yields Eq. (75) when taking into account that $\Delta G^{o'} = E_{eq}^{fin} - E_{eq}^{in}$.

$$x^{\#} = \frac{1}{2}\left(\frac{\Delta G^{o'}}{\lambda} + 1\right) \tag{75}$$

$$\Delta E_f^{act} = E^{\#} - E_{eq}^{in} = \lambda(x^{\#})^2 = \frac{\lambda}{4}\left(1 + \frac{\Delta G^{o'}}{\lambda}\right)^2 \tag{76}$$

Introduction of this value into Eq. (73) gives the activation energy in Eq. (76) for the forward reaction.* Thus it is seen from the latter equation that $\lambda/4$ is equal to the "intrinsic activation barrier" $\Delta G_o^{\#}$, that is, to the

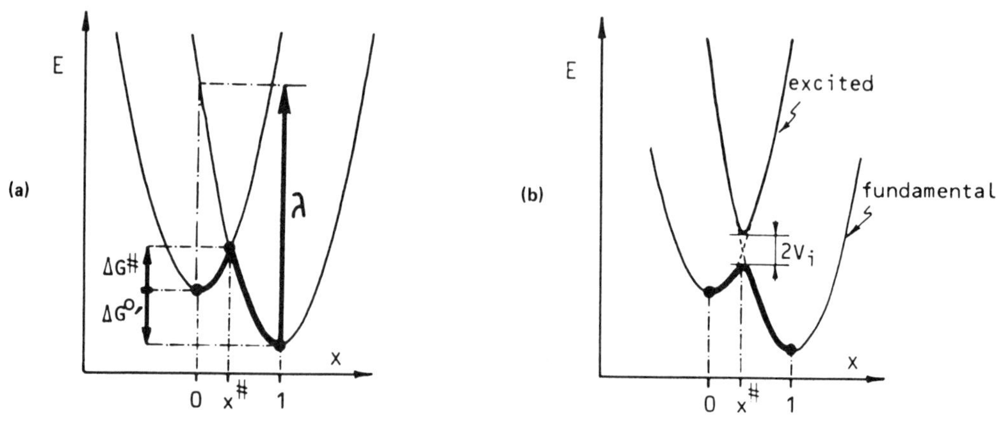

Figure 8. Potential energy-normalized reaction coordinate diagrams, without (a) or with (b) interaction between the initial and final subsystems (see text).

*Note that Eq. (76) is identical to Eq. (64).

activation energy that would be obtained for $\Delta G^{o\prime} = 0$. Thus if a series of electron transfers with virtually identical λ is considered, determination of the activation barrier for $\Delta G^{o\prime} = 0$ gives direct access to the experimental value of λ. Proponents of this second interpretation then prefer to reformulate Eqs. (64) or (76) under the equivalent form

$$\Delta E_f^{act} = \Delta G_o^{\#} \left(1 + \frac{\Delta G^{o\prime}}{4\Delta G_o^{\#}}\right)^2 \tag{77}$$

To conclude, we discuss more deeply the energetic diagrams like that presented in Fig. 8a. As noted previously, the A and D orbitals must interact in order that the electron may be transferred. This simply means that an electronic coupling is observed between the initial system and the final system. This results in mixing of the two systems, which amounts to separating the potential energy curves in Fig. 8a into a lower (or fundamental) state and an upper (or excited) state, as represented in Fig. 8b. Because the potential energies of the electron tend to be identical for the initial and final systems when the transition state is approached, the degree of mixing in this region is higher than near the bottom of each potential well. Thus the maximal effect is reached at the transition state, where a splitting of magnitude $2V_i$ is observed [58].

In practice, as soon as $2V_i > RT$, that is, $V_i > 0.25$ kcal/mol at room temperature, when the system crosses the transition point it remains on the lower curve and then reaches the final system equilibrium. This corresponds to a probability of 1 for electron transfer each time the system crosses the barrier (provided its momentum is in the right direction); that is, $\kappa = 1$.

When $2V_i \simeq RT$, when reaching the transition point the system may "jump" onto the upper curve. If so, when it goes backward it "falls" onto the lower curve, with a large probability (because of the direction of its momentum) of advancing toward the initial potential well. In such a case the electron has not been transferred and then $\kappa < 1$.

When $2V_i \ll RT$, this phenomenon occurs almost every time the system crosses the transition point, which means that $\kappa \approx 0$ and no electron transfer occurs.

From these considerations it is understood that Eqs. (64), (76), and (77), derived without considering electronic coupling, give activation energies overestimated by a term V_i and should normally be corrected. However, as noted for κ, V_i is predictable with difficulty from simple considerations and is usually not included in the activation energy.

Energies and Free Energies in Electron Transfer Theories

In the preceding description of electron transfer theories, energies were used rather than free energies. For a reaction in liquid phase ΔE is nearly identical with ΔH, so that the results obtained earlier are valid as soon as entropic variations are negligible,* and thus $\Delta E \approx \Delta H \approx \Delta G$. In practice the distinction between ΔG and ΔE is forgotten. Since outer sphere electron

---

*Note that for the transition state, the entropy corresponding to the formation of pairs is included in the models through the free energies of formation of the precursor or successor complexes (Sec. II.C, Diffusion limitation in electron transfer reactions and Basic features of electron transfers in solutions).

transfers involve small bond lengths or angle variations, this is acceptable for inner shell components of activation free energies. Concerning outer shell components, the approximation is considered not drastic and is equivalent [59] to those made for solvation models (Born) or according to the Debye-Hückel theory [1a,c]. Thus, within these approximations and that previously discussed dealing with V_i, it is considered that the activation rate constant of outer sphere electron transfers is given by [from Eq. (58)]

$$k_f^{\#} = \kappa Z e^{-\Delta G_f^{\#}/RT} \tag{78}$$

with

$$\Delta G_f^{\#} = W_r + \frac{\lambda}{4}\left(1 + \frac{\Delta G^{o\prime}}{\lambda}\right)^2 \tag{79}$$

where $\Delta G^{o\prime}$ is the free energy change between the precursor and successor complexes:

$$\Delta G^{o\prime} = F(E^o_{D^+/D} - E^o_{A/A^-}) - W_r + W_p \tag{80}$$

and λ is given in Eqs. (66), (67), and (72).

Quadratic Free Energy Relationships (QFER)

When one considers a series of related electron transfers, that is, involving acceptors (and donors) of reasonably close chemical electronic structures and radii, the reorganization energy λ should remain constant for the series. Thus it is seen from Eq. (79) that the activation energy varies as a quadratic function of the driving force, in contrast to the usual observation of linear free energy relationships in organic chemistry. As a result, the Brönsted coefficient β varies linearly with the driving force, with a value of 0.5 at $\Delta G^{o\prime} = 0$. Indeed, from Eq. (78) and a derivation of Eq. (79), one obtains

$$\beta = -RT\left(\frac{\partial \ln k_f^{\#}}{\partial \Delta G^o}\right) = \frac{1}{2} + \frac{\Delta G^{o\prime}}{2\lambda} \tag{81}$$

When $|\Delta G^{o\prime}| < \lambda$, the predictions of such an equation are in agreement with the deductions of the Hammond postulate. Indeed for exergonic reactions, β tends to zero when $\Delta G^{o\prime} \longrightarrow -\lambda$, which corresponds to an activationless reaction. Conversely, for endergonic reactions, β tends to unity when $\Delta G^{o\prime} \longrightarrow \lambda$, which features a barrierless reaction.

However, when $|\Delta G^{o\prime}| > \lambda$, β reaches negative values in the exergonic domain and is greater than unity in the endergonic case, according to Eq. (81). To recall these facts the region such as $|\Delta G^{o\prime}| > \lambda$ is usually called the abnormal or inverted region, the second term refering to the unusual decrease in rate constant with larger exergonicity of the reaction. In actual practice there is a large debate about the experimental existence of this inverted region. Moreover, more elaborate models have been proposed involving the possible participation of excited vibronic states, which show that for most organic electron transfers one should observe a normal behavior, that is [60],

$$\Delta G^{\#} \approx W_r, \text{ or } \beta = 0 \text{ for } \quad \Delta G^{o\prime} < -\lambda \tag{82}$$

$$\Delta G^{\#} \approx W_r + \Delta G^{o\prime}, \text{ or } \beta = 1 \text{ for } \quad \Delta G^{o\prime} > \lambda \tag{83}$$

although $\Delta G^{\#}$, given in Eq. (79), remains valid for $|\Delta G^{o\prime}| < \lambda$.

Similarly other models, such as the Marcus-Levine-Agmon [61] model, have been developed that predict the nonexistence of the inverted region. The latter model leads to the $\Delta G^{\#}$ variations with $\Delta G^{o\prime}$ and λ in Eq. (84):

$$\Delta G^{\#} = \Delta G^{o\prime} + \frac{\lambda}{4 \ln 2} \ln (1 + e^{-\Delta G^{o\prime} \cdot 4\ln 2/\lambda}) \tag{84}$$

Particularly interesting is that Eq. (84) respects the limits in Eqs. (82) and (83). For intermediate values of the driving force, the calculated activation barrier from Eq. (84) is very close to that obtained in the classic harmonic model developed in this section [Eq. (79)]. Thus from an experimental point of view, distinction between the two models is rather difficult or impossible [36a], although totally different mathematical expressions for $\Delta G^{\#}$ variations with $\Delta G^{o\prime}$ and λ are obtained.

To conclude, we discuss the experimental use of Eq. (81). Indeed in experimental practice one has access to overall rate constants k_f or k_b in scheme 3 and Eqs. (56). Thus the experimental Brönsted coefficient β_{exp} is determined as in Eq. (85):

$$\beta_{exp} = -RT \left(\frac{\partial \ln k_f}{\partial \Delta G^{o}} \right) \tag{85}$$

The latter becomes identical to that in Eq. (81) only when $k_f \approx k_f^{\#}$, that is, when $k_f^{\#} << k_f^{dif}$ in Eq. (54a). However, when diffusion of the reactants or products becomes the rate-determining factor, $k_f \approx k_f^{dif}$, as discussed earlier. Owing to the discreteness of homogeneous experimental data, the shift from an activation control to a diffusion control of the reaction is usually difficult to appreciate, particularly in the endergonic region. The perversity of the problem is even larger when one realizes that application of Eq. (85) to a diffusion-controlled reaction yields $\beta_{exp} = 0$ for exergonic situations and $\beta_{exp} = 1$ for endergonic, owing to the two components in the overall pseudo-rate constant k_f^{dif} in Eq. (54a). The difficulty is greater as the reorganization energy λ decreases, as evidenced by Fig. 9. This caveat [36] is conveniently represented in Fig. 10, which summarizes the various conclusions reached here in the form of a plot of $\Delta G^{o\prime}$ versus λ. Indeed it is seen from this diagram that according to the location of a given system, the physical meaning of experimental activation energies obtained by strict application of Eq. (86) [deduced from Eq. (78), assuming $\kappa = 1$] may greatly differ [36a].

$$\Delta G^{\#}_{exp} = -RT \ln \frac{k_f}{Z} \tag{86}$$

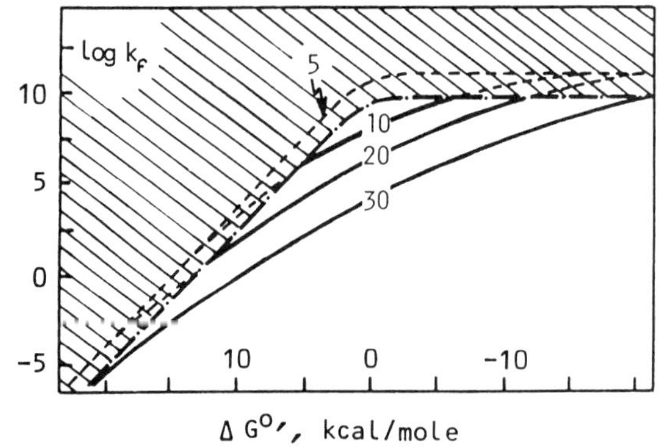

Figure 9. Variations in the observed rate constant k_f as a function of the driving force for various values of the reorganization energy (numbers on the curves in kcal/mol). The solid curves correspond to an activation control. The hatched zone corresponds to a diffusion control (compare Fig. 4).

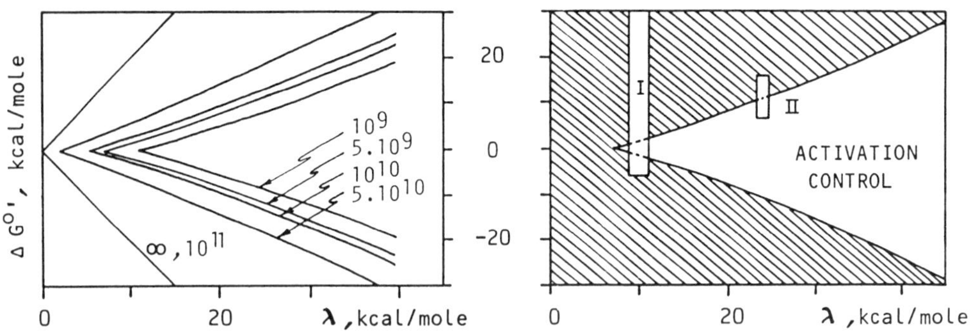

Figure 10. Relationship between the driving force and the reorganization energy for the reaction to be under activation control or diffusion control (hatched zone). The boundaries between the two zones are given for various values of $k_r^{dif} = k_p^{dif}$. The boxes correspond to the location of the experimental systems in Fig. 11(I) or 12(II).

Cross-Relationships: Evaluation of Rate Constants from Isotopic Rate Constants

An interesting consequence of the additive formulation in Eqs. (66) and (67) of the reorganization energy λ^{AD} for an A/D electron transfer is that it may be broken in three components: $\lambda_o{}^{AD}$, the outer shell reorganization, and $\lambda_i{}^A$ (and $\lambda_i{}^D$), which involve the inner shell contributions related specifically to bond length and bond angle variations in the acceptor (and the donor) center:

$$\lambda^{AD} = \lambda_o{}^{AD} + \lambda_i{}^A + \lambda_i{}^D \tag{87}$$

On the other hand, for the isotopic electron transfer, that is,

$$A^{*} + A^{\overline{\cdot}} \rightleftharpoons A^{*\overline{\cdot}} + A \qquad (\text{or} \quad D^{*} + D^{\overset{+}{\cdot}} \rightleftharpoons D^{*\overset{+}{\cdot}} + D)$$

the inner shell component for the reorganization energy becomes

$$\lambda_i^{AA} = 2\lambda_i^{A}$$

since two A centers are involved; similarly, $\lambda_i^{DD} = 2\lambda_i^{D}$ for the isotopic electron transfer relative to the donor. Thus it follows that Eq. (87) may be rewritten under the form in Eq. (88):

$$\lambda^{AD} = \lambda_o^{AD} + \frac{1}{2}(\lambda_i^{AA} + \lambda_i^{DD}) \tag{88}$$

Formally λ_o^{AD} cannot be split into two (A and D) components because of the presence of $r \approx a_A + a_D$ in Eq. (72), which acts as a coupling term. Yet a tedious but elementary algebraic transformation allows Eq. (72) to be rewritten as

$$\lambda_o^{AD} = \frac{1}{2}(\lambda_o^{AA} + \lambda_o^{DD})\left[1 + \frac{(a_A - a_D)^2}{(a_A + a_D)^2}\right] \tag{89}$$

In Eq. (89), λ_o^{AA} and λ_o^{DD} are the outer shell reorganization energies for the respective isotopic electron transfers, defined similarly as λ_o^{AD} in Eq. (72). Thus as soon as the equivalent radii for the acceptor and the donor do not differ by a factor larger than 2, one gets within about 10% precision:

$$\lambda_o^{AD} = \frac{1}{2}(\lambda_o^{AA} + \lambda_o^{DD})$$

Introduction of this latter result into Eq. (88) allows one to formulate λ^{AD} as the sum of two components relative to the individual isotopic reactions in Eq. (90):

$$\lambda^{AD} = \frac{1}{2}[(\lambda_o^{AA} + \lambda_i^{AA}) + (\lambda_o^{DD} + \lambda_i^{DD})] = \frac{1}{2}(\lambda^{AA} + \lambda^{DD}) \tag{90}$$

The operative interest of such a cross-relationship is extreme, since when associated with Eqs. (78) through (80) it allows an a priori estimation of the rate constant of any outer sphere electron transfer reaction provided the corresponding standard reduction potentials and isotopic rate constants are known. Yet a caveat in this approach is the necesity that $\kappa = 1$ for all reactions, that is, that there is always sufficient overlap between the orbitals [62].

Experimental Illustrations

The purpose of this section is not to present a compendium or a discussion of the numerous experimental illustrations of the validity of electron transfer theories, but rather to select two typical examples that illustrate the main features of these theories in organic chemistry [63].

Validity of QFER over a Wide Domain of Free Energies. The main difficulty in discussing the dependence of rate constants with driving force over a wide domain stems from the fact that, to be significant, the series must involve electron transfers with nearly constant reorganization energies. However, it is easily understood that the two requirements (ΔG° over several tens of kilocalories per mole but λ invariant) are difficult to satisfy simultaneously.

Such a series, which extends over nearly 70 kcal/mol in ΔG°, has been presented by Rehm and Weller [64]. The series corresponds to electron transfers between an excited acceptor and a donor acting as a quencher or between an excited donor and an acceptor quencher. Thus rate constants ranging from 10^6 to 2×10^{10} M^{-1} s^{-1}, have been determined for more than 60 different donor-acceptor systems in acetonitrile. Examination of the corresponding results, presented in Fig. 11, clearly shows that three domains may be defined: (1) $\Delta G^\circ < -10$ kcal/mol, where k_f is independent of the driving force, that is, where $\beta_{exp} = 0$; (2) -10 kcal/mol $< \Delta G^\circ <$ 10 kcal/mol, where β_{exp} varies from zero to unity; (3) $\Delta G^\circ > 10$ kcal/mol, where $\beta = 1$. These data, corrected for diffusion components, correspond to an activation control of the electron transfer reactions. They qualitatively agree with the predictions of Eqs. (78) and (79) for $|\Delta G^\circ| < 10$ kcal/mol but show that Eq. (79) is not valid outside this range, Eqs. (82) and (83) being more adequate. As such they have been considered a convincing experimental argument for the non-existence of the inverted region [65]. Yet, owing to the large scatter of the data, a precise determination of the validity of Eq. (79) in the intermediate ($|\Delta G^\circ| < 10$ kcal/mol) free energy range is impossible.

Validity of the QFER for Low End Ergonicities or Exergonicities. In order to test the experimental validity of the quadratic relationship between $\Delta G^{\#}$ and ΔG°, a rather small free energy domain needs to be investigated. As a consequence even small variations in λ result in too large a scatter of the data for significant quantitative conclusions to be drawn (compare the data in Fig. 11 for $|\Delta G^\circ| < 10$ kcal/mol). Thus a series of very closely related redox couples must be investigated.* A series meeting these requirements is given by the endergonic oxidation in acetonitrile of methylbenzenes by tris-(5X-phenantroline)-iron(III), X = H, Cl, NO_2, already discussed [36a,c]. Indeed, for such a series the inner shell contributions are minimal owing to the close identity of the iron (II/III) complexes [66] and the small distortions of the arene cation radical vis-à-vis the neutral (for $C_6H_6^{+\cdot}$ the inner shell reorganization energy from a nuclear configuration corresponding to benzene involves a Jahn-Teller distortion [67] and has been evaluated to approximately 3 kcal/mol by Salem [67c]. Thus the outer shell reorganization energy predicted acording to Eq. (72) is (λ_o in kcal/mol; a_{Ar}, a_{Fe}, and $r = a_{Ar} + a_{Fe}$ in Å)

$$\lambda_o = 1.67 \times 10^2 \left(\frac{1}{\varepsilon_{op}} - \frac{1}{\varepsilon_s}\right)\left(\frac{1}{a_{Ar}} + \frac{1}{a_{Fe}} - \frac{2}{r}\right)$$

---

*Note that the corresponding condition is more easily satisfied for electron transfers at electrodes, owing to the easy variation of the driving force, by adjustment of the electrode potential without modification of the redox couple.

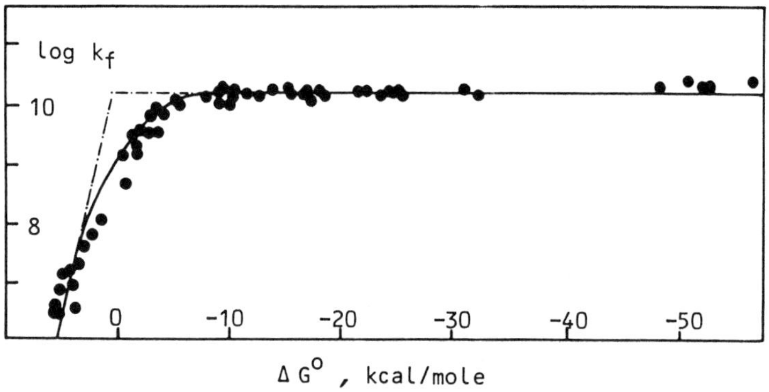

Figure 11. Linear energy relationship over an extended domain of driving force. (Experimental data from Ref. 64.)

that is, $\lambda_o \approx 21$ kcal/mol for $a_{Ar} \approx 3.5$ Å, $a_{Fe} \approx 7$ Å [68], and $\varepsilon_s = 37.5$ and $\varepsilon_{op} = n_{op}^2 = (1.34)^2$ for acetonitrile. The resulting predicted value of the global reorganization energy is then 24 kcal/mol for each couple in the series.

Figure 12a indicates how the measured rate constant k_f for the endergonic electron transfer oxidation of methylbenzene by $(phen)_3$-iron(III) complexes varies with the driving force. It is seen from these data that, provided that $\Delta G^{o\prime} < 10$ kcal/mol, a very good fit is obtained with the rate constant predicted from Eqs. (78) through (80) on the basis of an activation control of the overall reaction when using $\kappa = 1$, $Z = 10^{11}$ M^{-1} s^{-1}, and $\lambda = 24$ kcal/mol, determined earlier. However, for larger positive values of $\Delta G^{o\prime}$, the diffusion processes contribute significantly to the experimental rate constant. Yet the activation rate constant $k_f^{\#}$ may be extracted from the experimental k_f data through the use of Eq. (56a). The real activation free energy $\Delta G_f^{\#}$ is then derived through Eq. (78) and plotted as a function of $\Delta G^{o\prime} = \Delta G^o + W_p$ in Fig. 12b. Quadratic regression analysis of these data allowed their fit by a second-order polynomial expression in Eq. (91) [36a,c] (represented as the dashed line in Fig. 12b):

$$\Delta G_f^{\#} = 6.7 + 0.50\,\Delta G^{o\prime} + 8.7 \times 10^{-3}(\Delta G^{o\prime})^2 \qquad (91)$$

which is in good agreement with the predicted variations in Eq. (79) for $\lambda = 27$ kcal/mol. Indeed, development of Eq. (79) with $\lambda = 27$ kcal/mol yields

$$\Delta G_f^{\#} = 6.7 + 0.50\,\Delta G^{o\prime} + 9.3 \times 10^{-3}(\Delta G^{o\prime})^2$$

Another way to appreciate the correctness of Eq. (79) in the prediction of the experimental barrier consists in the evaluation of λ for each individual couple as presented in Table 3. Indeed, from Eq. (79), λ is obtained through the solution of the second-order equation

$$\lambda^2 + 2(\Delta G^{o\prime} - 2\Delta G_f^{\#})\lambda + (\Delta G^{o\prime})^2 = 0 \qquad (92)$$

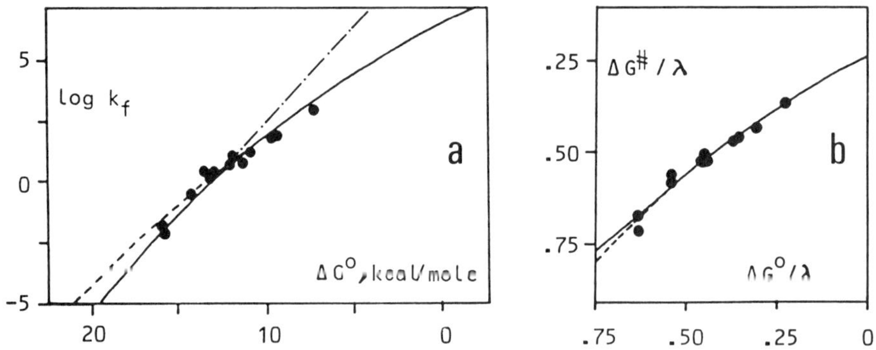

Figure 12. Quadratic free energy relationship in the low endergonic region for the oxidation of methylbenzenes by $Fe(phen)_3^{3+}$. (a) Experimental variation of the observed rate constant compared with the theoretical variations (solid curve): (---) activation or (- •--•-) diffusion controls. (b) Experimental relationship between the activation free energy and the driving force compared with the theoretical variations for λ = 27 kcal/mol in the solid curve: (---) polynomial regression in Eq. (92) [36a,c]. (Data from Ref. 36.)

Table 3. Experimental Evaluation of the Reorganization Energy λ for the Oxidation of Methylbenzenes by Tris-(5X-phenantroline)-Iron(III) Complexes

Methylbenzene	X	$\Delta G^{o\prime a}$	$\Delta G_f^{\#a}$	λ^a	$\lambda_{average}^{a,b}$
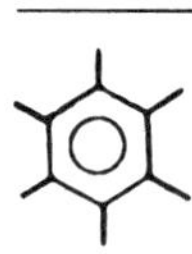	H	11.2	13.4	26.4	
	Cl	8.86	12.6	30.1	28.2 (1.8)
	NO_2	6.55	10.7	28.2	
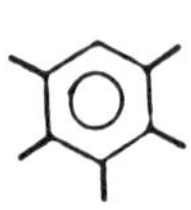	H	14.2	15.6	26.4	
	Cl	11.9	13.9	26.4	26.3 (0.4)
	NO_2	9.55	12.4	27.1	
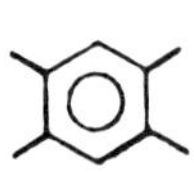	H	16.0	16.8	24.9	
	Cl	13.7	13.4	—[c]	25.1 (0.2)
	NO_2	11.4	13.3	25.2	
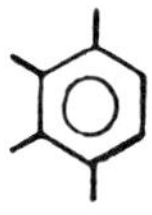	H	15.8	18.0	32.8	
	Cl	13.5	14.0	19.8	27.3 (6.7)
	NO_2	11.2	14.0	29.3	

[a]In kcal/mol.

[b]Average value for a given arene, standard deviation indicated in parentheses. For all couples λ_{av} = 27.0 (3.3) kcal/mol.

[c]Not determinable since $\Delta G_f^{\#} > \Delta G^{o\prime}$.

Source: Data from Ref. 36a.

$$\Delta G_f^{\#} = RT \ln \frac{Z}{k_f^{\#}} \tag{93}$$

where $\Delta G^{o'}$ and the experimental activation free enegy $\Delta G_f^{\#}$ from Eq. (93) are known. Thus a value of 27.0 kcal/mol (standard deviation 3.3 kcal/mol, 11 data) is obtained for the average reorganization energy, which is in very close agreement with that predicted (24 kcal/mol).

III. FUNDAMENTAL ASPECTS OF ELECTRODE PHENOMENA

A. Monitoring a Half-Reaction: The Electrochemical Cell

Monitoring a Half-Reaction

In homogeneous chemistry an electron transfer reaction involves two reactants, an acceptor A and a donor D. Thus, formally, an electron transfer reaction consists of the superimposition of two elementary chemical acts: the reduction of the acceptor and the oxidation of the donor. Yet as discussed in the first part of this chapter, since more attention is given to one of these elemental reactions, the overall reaction is termed a reduction when the acceptor is the compound of interest or an oxidation when chemical transformation of the donor is favored.

In electrochemistry the same phenomenon (essentially related to charge conservation) occurs, yet the reduction of the acceptor A occurs at one electrode (the cathode in electrolytic cells) and the oxidation of the donor D at the other (anode). Thus the kinetics* of the overall cell reaction depends on both half-reactions, in a similar way as the kinetics of an homogeneous electron transfer depends on the acceptor and the donor. However, because the two half-reactions occur at different locations, it is technologically possible to break the coupling between the two electrodes. This is done via two different approaches, based on the same principle, that take advantage of the potential distribution in an electrochemical cell, shown in Fig. 13a. From this schematic representation it is seen that the potential difference Φ between the two electrodes is the sum of three components: two of them, Φ_{cat} and Φ_{an}, are related to each electrode, and the third, Ri, arises because of the ohmic drop due to the current flow through the resistive solution. The potential variation observed at each electrode is localized within an extremely thin layer (<100 Å) of solution adjacent to the solution (Sec. III.A, The electrode/solution interfacial region), which means that the overall cell behaves like an electrical circuit composed in part of two capacitors and a resistance as sketched in Fig. 13b. However, such a simple circuit is not sufficient since it does not allow constant current to flow through the cell because of the capacitors. In practice, two additional impedances, Z_{cat} and Z_{an},† must be added in parallel with each capacitor, which correspond to the electrochemical reactions taking place at the electrodes. The latter are necessarily in parallel with the capacitors, as shown in Fig. 13c, since they use the potential difference between the electrode and the local solution as their driving force and act as derivations allowing the current to flow. Thus it is easily understood that when a given poten-

*That is, the current flowing through the cell, since it corresponds to the number of electrons consumed at each electrode per unit of time.

†Z_{cat} and Z_{an} are called faradaic impedances because of their origin.

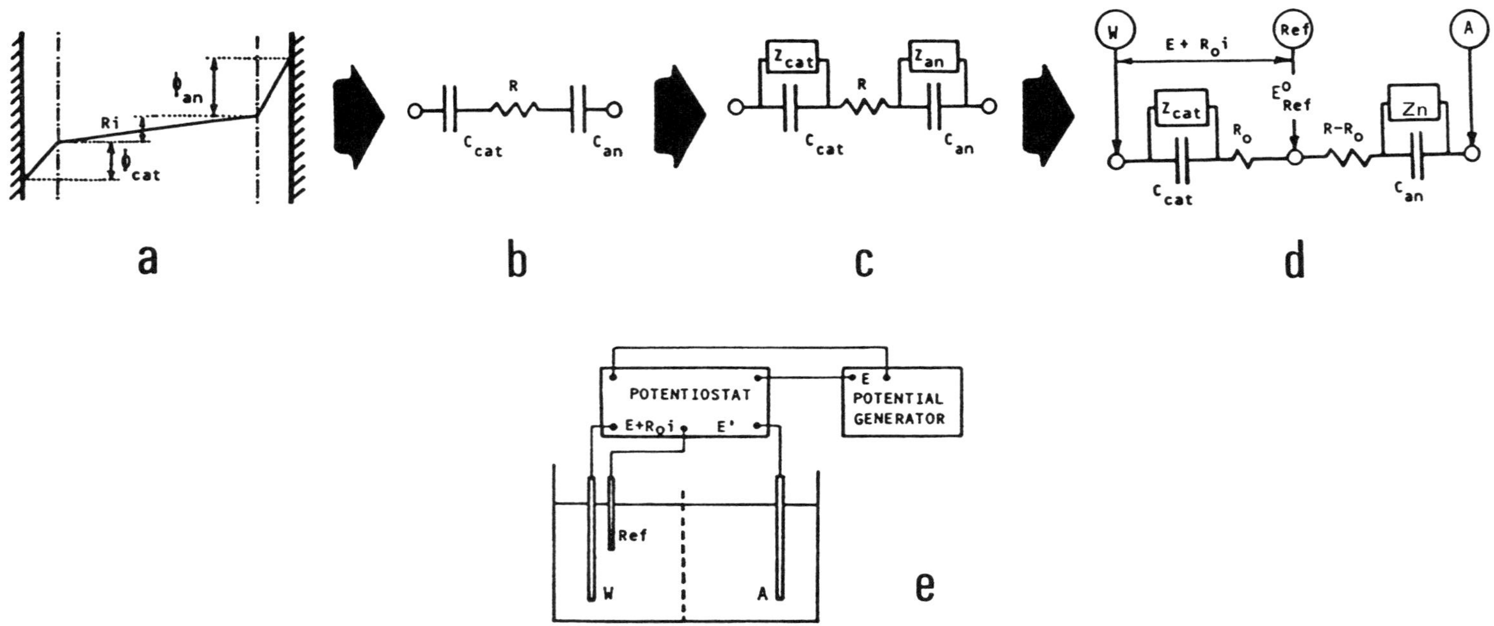

Figure 13. Representations of the potential variations in the electrochemical cell and associated electronic schemes (a–d). ($\Phi = \Phi_{cat} + Ri + \Phi_{an}$; see text). W, working electrode; Ref, reference electrode; A, auxiliary electrode. (e) Schematic description of the electrochemical cell with a potentiostat.

tial difference is imposed between the two electrodes, the three individual potential drops Φ_{cat}, Φ_{an}, and Ri must adjust in order that the currents flowing through each impedance and through the solution are equal; this results in coupling between the processes occurring at each electrode and in the solution.

Thus to individualize one of the electrodes it is necessary to control, that is, to measure or to impose, its individual potential difference versus the solution (Φ_{cat} or Φ_{an}). This is done via the introduction of a third electrode, the reference electrode Ref, placed in the solution near the electrode of interest (hereafter named the working electrode W, the other electrode being called the auxiliary electrode A). Because of its characteristics (Sec. III.C, Reversible and irreversible electron transfers), the reference electrode imposes a potential difference $\Phi_{Ref} \approx E^{o}_{Ref}$ vis-à-vis the point of the solution where it is located. Φ_{Ref} may include various components related to the redox half-reaction or junction potentials, for example, but is independent of the current flowing through the reference electrode. Such a three-electrode system, sketched in Fig. 13d, formally allows the physical individualization of the working electrode. Indeed, one may, impose an arbitrary potential difference $\Phi_W - \Phi_A$, for example, between the working and auxiliary electrode, measure the resulting $\Phi_W - \Phi_{Ref}$ potential difference, and compare it to the desired value E to be imposed. If $\Phi_W - \Phi_{Ref}$ is too small, $\Phi_W - \Phi_A$ is increased; if $\Phi_W - \Phi_{Ref}$ is too large, $\Phi_W - \Phi_A$ is decreased. Thus $\Phi_W - \Phi_{Ref}$ is progressively adjusted to the selected value E (note then that $\Phi_{Ref} - \Phi_A$ is also imposed as a result of the procedure, but its exact value is a priori of no interest). Then the current i flowing through the cell may be measured and related to the potential E. Repetition of the whole procedure allows the current-potential characteristic, that is, the (i, E) relationship, of the half-reaction taking place at the working electrode to be determined. Yet, although possible, if the whole procedure were performed manually, it would take several hours to record precisely an (i, E) characteristic. Thus pertinent electronic apparatus, called potentiostats [69], have been designed to perform the whole procedure within microseconds or less [70]. Besides their convenience, potentiostats allow the recording of transient electrochemical kinetics, which would not have been possible manually.

The second approach, although used first in electrochemistry, derives from that corresponding to three electrodes and potentiostat. It consists of using a reference electrode as the auxillary electrode. This is indeed possible, provided the size of the reference electrode is sufficiently large for the current density flowing through it to be infinitely small, although the current may be large. The potential difference between the reference-auxiliary electrode (called a counter electrode in a two-electrode device) and the solution in its near proximity is thus constant (Sec. III.C, Reversible and irreversible electron transfers). The potential difference Φ_W associated with the working electrode is then given by

$$\Phi_W = \Phi - \Phi_{CE} - Ri \tag{94}$$

which shows that, provided that $\Phi >> Ri$, Φ_W is measured (or imposed) within a constant term Φ_{CE} when the overall potential difference Φ is measured (or imposed). This second approach has been discarded vis-à-vis the three-electrode potentiostat approach because in organic electrochemistry the resistance of solutions is usually large so that the ohmic drop term Ri is not negligible under the usual conditions.* Thus the relationship between Φ_W and Φ depends on the current flowing through the cell. Note that a similar problem is also involved for the three-electrode approach, since the working and reference electrodes are separated physically by part of the solution, which introduces an ohmic drop component in the potential difference between the working electrode and the reference electrode. Yet this effect is usually smaller because the resistance R_0 involved is smaller and is eventually corrected via additional electronic circuitry in the potentiostat [69].†

The Electrochemical Cell

From this presentation it is seen that electrochemical data that pertain to a single half-reaction taking place at the working electrode can be experimentally obtained. In our opinion this is one of the greatest advantages of electrochemistry, although not really identified and recognized in other fields of chemistry. Indeed, whatever the homogeneous chemical technique used, the chemistry associated with a donor-acceptor couple necessarily depends on both members.

Another important advantage, less apparent in this presentation, is that because of the involvement of an external electrical driving force electrochemistry allows overall endergonic reactions to be performed at a considerable rate. Indeed, let us consider the a priori endergonic electron transfer reaction in Eq. (95):

$$A + D \rightleftharpoons A^{\bar{\cdot}} + D^{\dot{+}} \tag{95}$$

Thus under electrochemical conditions, A can be reduced and D oxidized provided that sufficient cathodic and anodic potentials are applied. Yet upon preparative-scale electrolysis, $A^{\bar{\cdot}}$ and $D^{\dot{+}}$ recombine to A and D, except if they react chemically:

$$\text{Cathode:} \quad A + e \rightleftharpoons A^{\bar{\cdot}} \longrightarrow \cdots$$

$$\text{Anode:} \quad D - e \rightleftharpoons D^{\dot{+}} \longrightarrow \cdots$$

Thus in the absence of chemical reaction or when undesired reactions between the follow-up product(s) may arise, it is necessary to avoid the unwanted reaction. Taking advantage of the different physical locations of the two electrodes, this is done by separating the solution into two parts (catholyte and anolyte, respectively) using a glas frit or a membrane so that the electrical connection (via ionic transport; see Sec. IV.A, Relationships between the electrochemical current and mass transfer) is not interrupted [71].

---

*However, it regains interest with the development of ultramicroelectrodes. Indeed, because of their micrometer (or smaller) sizes, the currents are of the order of nanoamperes. Thus even with R of several kilo-ohms, the ohmic drop term is totally negligible.

†In the latter case, the current is measured and a R_0i term is reinjected via a feedback loop so that the potentiostat actually imposes the potential difference $E + R_0i$ between the working electrode and the reference electrode; this results in "canceling" the ohmic drop (Fig. 13e).

Resistive Effects and Ohmic Drop: The Supporting Electrolyte-Solvent System

In the preceding part of this section we focused our attention on the electrodes, where the real electrochemical events occur. Yet as already explained, the current exchanged at the electrodes must flow through the cell, that is, through the solution between the working and counter or auxiliary electrodes. For this reason the solution must be conducting; otherwise no current flows, even when the electrode processes are extremely easy. Indeed, if the resistance of the solution is R and i the current flowing through the cell, Ohm's law indicates that the potential "lost" for conducting purposes is Ri.* In practice, ohmic drop can be overcome by increasing the potential difference between the working and auxiliary electrodes by the amount required. Yet this is not an acceptable solution because of the electrical power required and the thermal energy dissipation. Indeed, Ri is associated with an electrical power Ri^2 that is dissipated thermally in the solution and needs to be supplied by the potentiostat or the external power supply connected to the cell. To roughly quantify these effects, let us consider a laboratory electrolysis cell in which i = 1 A is required. Let us assume that R = 100 Ω, a common figure for most organic solutions. Thus Ri = 100 V and ri^2 = 0.1 kW. For an average thermal capacitance of 1 cal/K per ml and a cell of 100 ml, for example, this amounts to a temperature increase in the cell of approximately 0.25°C/s when it does not exchange energy with the external atmosphere. Thus in an actual experiment the cell must be cooled so that its temperature is maintained within reasonable limits. In other words, this means that an additional 0.1 kW power must be lost to cool the cell.

At the laboratory scale these energetic losses are easily controlled; thus the main problem relates to the potentiostat, or potential supply, power limitation, since most of the usual equipment affords between 70 and 100 W of usable electrical power. At the industrial scale, since the current intensities are considerably larger, serious problems arise from lost energy and temperature control of the cell.

To conclude this introductory section concerning ohmic drop effects, let us consider the case of transient electrochemical techniques. In this case the currents are of the order of few milliamperes or less, and thus all these effects should cancel. Unfortunately this is not the case because of the size of the working electrodes. Indeed these are often of millimetric dimensions, which amounts to a considerable narrowing of the field lines near the tip of the working electrode.† As a result the resistance increases, and most of it is concentrated near the very tip of the working electrode, even when a Luggin-Haber capillary probe is used for the reference [72]. Resistances of several kilo-ohms are frequently involved, which introduces important difficulties for an exact control of the potential of the working electrode [compare Eq. (94)]. Obviously, when one does not care about such control, this should not be perceived as a limitation, and rather resistive solutions may be used. Note also in this respect that when the working electrode is a disk of radius r_0, the resistance R varies as $1/r_0$ pro-

---

*Ri is usually said to be the "ohmic drop" to recall that this potential is not used for purely electrochemical purposes.

†The auxiliary electrode is usually of considerably larger dimensions (1 to several cm^2). Thus the equivalent conductor should be thought of as a cone rather than as a cylinder as for preparative cells.

vided the auxiliary electrode is of considerably larger dimensions. On the other hand, the current flowing through the electrode is proportional to its surface area (Sec. III.B) provided the electrode is not under diffusional steady state. Thus the ohmic drop is proportional to r_0 and decreases when the latter decreases. This explains one of the advantages of ultramicroelectrodes, at which ohmic drop effects naturally cancel since r_0 is of the order of a few micrometers [73].

This discussion and presentation of ohmic drop problems should lead to the conclusion that, with maybe an exception for ultramicroelectrodes, the resistivity of electrochemical solutions must be decreased as much as possible.

In solutions, the current is necessarily carried by dissociated ionic species. Theories of this phenomenon are well developed and presented in most introductory electrochemistry textbooks. Yet it is sufficient for our purpose here to recall that the resistance is then inversely proportional to the concentration of the dissociated electrolyte (Sec. IV.A, Relationships between the electrochemical curent and mass transfer) for the usual case of a 1:1 electrolyte (e.g., $NBu_4^+BF_4^-$ and $NEt_4^+BF_4^-$). Thus a normal means to decrease ohmic drop is to add an inert* electrolyte in large concentration. Yet as mentioned earlier, only the dissociated ions participate in the current transport since electrically neutral ion pairs play no role. Thus it is important that most of the supporting electrolyte be dissociated. Such a condition, together with the large concentrations required (approximately 0.1 M), implies that a rather dissociating solvent is used. This is the reason that, in organic electrochemistry, the preference is for solvents with rather large dielectric constants ($\varepsilon_S > 10$) and supporting electrolytes with rather large ions (e.g., $NBu_4^+BF_4^-$). Considering these conditions and that the solvent should be able to dissolve organic molecules explains why most electrochemical experiments involve a narrow class of solvents, such as acetonitrile ($\varepsilon_S = 37.5$), dimethylformamide (40), dimethylsulfoxide (46.7), and propylene carbonate (36.7) [71].

The Electrode/Solution Interfacial Region

To simplify the following discussion, we restrict it to the case of a cathode. Obviously, however, the results and phenomena directly apply to anodes, cations being replaced by anions, and vice versa.

Experimentally it was early recognized that an electrode/solution interphase behaves as a capacitor† [74]. This is because the negatively charged surface of the cathode generates a very strong electrical field that tends to attract positive ions from the solution, as sketched in Fig. 14a. The positive layer thus formed exerts, on the solution side, an electrical field of opposite direction that attracts negative ions from the solution. However, since this latter field is partly compensated by that resulting from the negative charges in the electrode, the charge density of the anionic layer is smaller than that of the cationic one. Repetition of this phenomenon results

---

*"Inert" means that the electrolyte is not involved in the electrochemical transformation occurring at the electrodes. The electrolyte is usually also called the supporting electrolyte after its function of "supporting" the current flow through the cell.

†Note that this property is used in electronics in the so-called electrochemical capacitors, which are no more than electrochemical cells without faradaic reactions.

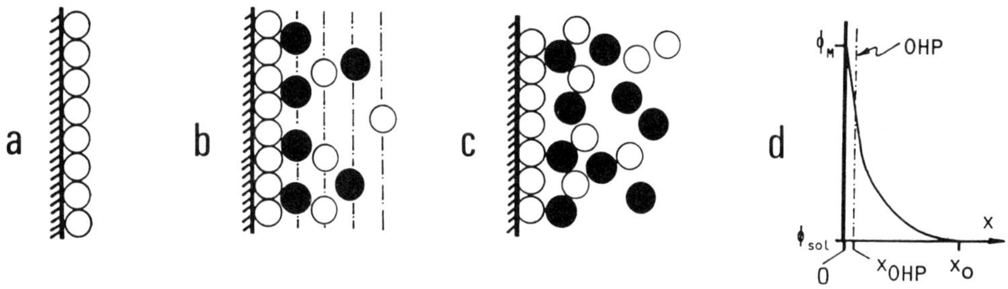

Figure 14. Schematic representation of the electrode/solution interfacial region. (a) Helmholtz model; (b) structured layer model; (c) thermally disorganized layers, and (d) resulting potential variations with the distance of the electrode; Φ_M, electrode potential; Φ_{sol}, solution potential; OHP, outer Helmholtz plane (few Å); X_o, extremity of the diffuse layer (few tens of Å); $X < X_{OHP}$, compact layer; $X_{OHP} < X < X_o$, diffuse layer.

in alternating layers of positive and negative charges, with a charge density in each layer, however, decreasing with the electrode distance, as pictured in Fig. 14b. Because of the decrease in electrical field with distance, the successive layers are less and less bound. Then, owing to thermal agitation, these layers cannot be as rigid as might be deduced from the preceding presentation and interpenetrate each other except for the first, which is tightly bound to the electrode [75]. This phenomenon is indicated by the corresponding names compact layer for the first layer and diffuse layer for the remaining part; both of them constitute the double layer and act as two capacitors in series. The capacitance (usually 20–50 $\mu F/cm^2$) and dimensions (a few angstroms for the compact layer and a few tens of angstroms for the diffuse layer) may be evaluated via the Gouy-Chapman-Stern theory [76], which is developed in most electrochemical textbooks. The main conclusions of this model, which are in general agreement with experimental observations, are the following: (1) because of the compact layer, a large potential drop arises within a few angsroms of the electrode surface (compare Fig. 14d); thus most of the potential difference between the electrode and the solution is lost at the end of the compact layer (i.e., at the outer Helmholtz plane, OHP); (2) the potential difference between the OHP and the solution results in an electrical field considerably lower than that between the electrode and the OHP, which nearly exponentially decays with distance; (3) beyond several tens of angstroms from the electrode the potential is identical to that in the bulk of the solution, except for eventual ohmic drop effects, as discussed earlier (compare Fig. 13a); (4) as a result of the third conclusion, beyond a few tens of angstroms the electrode exerts no electrostatic attraction or repulsion on any ionic species, which explains why an anion may be reduced or a cation oxidized at electrodes.

B. General Overview of an Electrode Reaction

When thinking of any electrochemical reaction, it is important to remember that the very act of electron transfer takes place at a surface, whereas the material to be reduced or oxidized is dispersed in a volumetric phase. Thus mass transfer from the bulk of the solution (or to the bulk solution, for the products) plays a central role in electrochemical processes. As such the physical processes of mass transfer (diffusion, migration, and convection; see Sec. IV.A, The physical processes of mass transfer) have received considerable attention in the electrochemical literature and are discussed separately in a later section of this chapter. For our purpose here, it is sufficient to know that mass transfer results in the creation of concentration profiles as a function of the distance from the electrode, as illustrated in Fig. 15. For most of the usual electrochemical situations these profiles extend over a few to a few tens of micrometers from the electrode surface. Beyond this region, called the diffusion layer, where mass transfer processes are effective, the solution is homogeneous (except for convective effects as in "homogeneous chemistry").

Thus a simple electrode reaction involves successively the following three steps: (1) mass transfer of the reactant from the bulk solution to the electrode; (2) electron transfer at the electrode surface; and (3) mass transfer of the product of the reaction from the electrode surface to the bulk solution. These three elementary steps may be perturbed by earlier or later chemical steps and by adsorption or desorption phenomena at the electrode surface. Each of these processes may become the rate-determining step or may interfere significantly in the overall kinetics. At first glance the situation then seems much more complicated than in homogeneous chemistry. Yet this is only an illusion because most of these steps have their exact counterparts in homogeneous chemistry. Mass transfer corresponds to the diffusion control of a chemical reaction as described. Earlier or later chemical steps are involved in most homogeneous electron transfer reactions. Adsorption and desorption are analogous to the formation of stable precursor or successor complexes. Electron transfer to or from an adsorbed species is very reminiscent of inner sphere reductions or oxidations. These analogies are not fortuitous; indeed, an electrode may be viewed as a gigantic molecule whose reducing or oxidizing power is adjustable at will.

C. Kinetics of Heterogeneous Electron Transfers

Mechanism and Theory of Outer Sphere Electron Transfers at Electrodes

In the following we consider an outer sphere electron transfer reaction involving the exchange of n electrons ($n > 0$ for a reduction; $n < 0$ for an

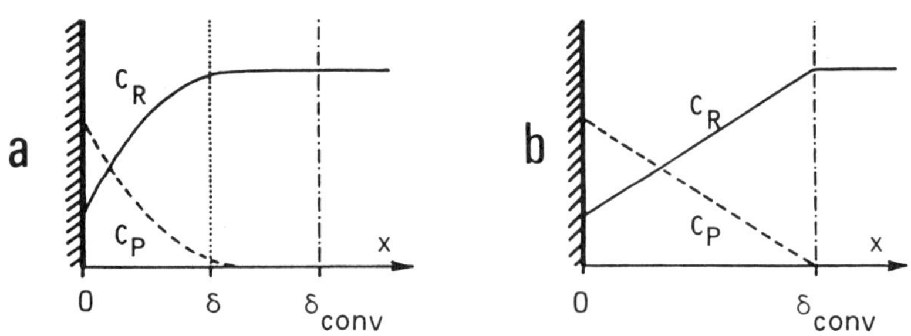

Figure 15. Concentration profiles for the reactant and the product as a function of the electrode distance X in (a) transient or (b) steady-state electrochemical techniques: δ and δ_{conv}, diffusion layer and stagnant layer thicknesses.

oxidation) in Eq. (96), taking place at an electrode whose potential is E.*

$$R + ne \rightleftharpoons P \qquad (E^{\circ}) \tag{96}$$

The electrode is supposed to have no particular interaction with R or P. Its surface area is hereafter designated by A.

Because in electrochemical reactions electron transfers occur near the electrode/solution interface, that is, at a two-dimensional surface, it is more suitable to refer to the number of moles converted per unit of time and unit of electrode surface area, dN_R/dt and dN_P/dt, rather than using concentrations as in homogeneous kinetics. Since there is no stable interaction with the electrode, $dN_R/dt = -dN_P/dt$. On the other hand, each mole of the reactant R electrolyzed corresponds to a charge nF transferred that affords the charge consumed per unit of time, that is, the current:

$$i = -nFA \frac{dN_R}{dt} = nFA \frac{dN_P}{dt} \tag{97}$$

Since the definition of E° in Eq. (96) refers to standard conditions, that is, conditions in which no external electrical field applies on the reactant or the product, it is more convenient to consider the act of electron transfer as a succession of three individual steps as outlined in scheme 5, which is reminiscent of that in scheme 3, established for the homogeneous analogous situation (Sec. II.C, Diffusion limitation in electron transfer reactions). In scheme 5, R_0 or P_0 relates to R or P in the closest plane to the electrode where no electrical potential applies, that is, at the "end" of the diffuse layer,† denoted X_0 in Fig. 14d. R_Φ and P_Φ relate to R and P at the site of electron transfer X_Φ, usually considered close to or slightly within the OHP, where an electrical potential Φ_S (i.e., the electrical potential at the electron transfer site) applies.

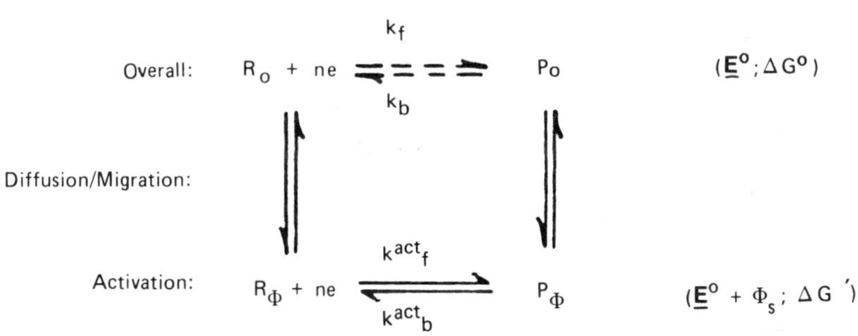

Scheme 5

*From the discussion in Sec. III.A, Monitoring a half-reaction, $E = \Phi_W - \Phi_{Ref}$. Yet when E° is given vis-à-vis the same reference, the contribution of the reference cancels in the difference E − E°. Thus in the following E is considered the electrode potential vis-à-vis that of the solution.

†Note that this limit must not be confused with that of the diffusion layer/bulk solution interface, owing to the extreme differences in size: some tens of angstroms for the double layer and some tens of micrometers for the diffusion layer.

Let $(R)_0$ and $(P)_0$ be the activities of R and P at X_0 and $(R)_\Phi$ and $(P)_\Phi$ those at $X = X_\Phi$ (the electron activity is not considered because it is implicity included in the potentials). For all the electron transfer reactions investigated up to now, the rates of the diffusion-migration steps in scheme 5 do not limit the overall process. Although this may be of importance for the prediction of the maximal rate of electron transfer at an electrode, we do not consider the possible kinetic limitation by either of these physical processes (compare the diffusion control of an homogeneous electron transfer discussed earlier). Thus, within this restriction, the existence of the diffusion-migration processes involves only thermodynamic contributions. Thus the overall rate constants in scheme 5, defined in Eq. (98),

$$\frac{i}{nFA} = \frac{-dN_R}{dt} = \frac{dN_P}{dt} = k_f(R)_0 - k_b(P)_0 \tag{98}$$

are related to the true activation constants via the relationships in Eqs. (99) and (100) where ΔG_Φ^R and ΔG_Φ^P, are the respective free energy changes for R and P between the end of the diffuse layer and the electron transfer site.

$$k_f = k_f^{act} e^{-\Delta G_\Phi^R/RT} \tag{99}$$

$$k_b = k_b^{act} e^{-\Delta G_\Phi^P/RT} \tag{100}$$

Note that because of their definitions in Eq. (98), the heterogeneous rate constants are expressed in units of length per units of time (usually in cm s^{-1} in the electrochemical literature). As for the homogeneous case, it is more convenient to separate the terms arising from electrical potential from those, ΔS_{pair}, of entropic nature, due to the formation of a pair electrode/molecule. Thus $\Delta G_\Phi^R = Z_R F \Phi_s - T\Delta S_{pair}$ and $\Delta G_\Phi^P = Z_P F \Phi_s - T\Delta S_{pair}$, where Z_R and Z_P are the respective ionic charges of R and P (note that $Z_P = Z_R - n$ because of the charge balance). On the other hand, k_f^{act} and k_b^{act} are obtained from the absolute rate constant theory as

$$k_f^{act} = \frac{kT}{h} e^{-\Delta G_f^{act}/RT}$$

$$k_b^{act} = \frac{kT}{h} e^{-\Delta G_b^{act}/RT}$$

This allows one finally to write the experimentally observed rate constants as in Eqs. (101) and (102):

$$k_f = Z_{el} e^{-Z_R F \Phi/RT} e^{-\Delta G_f^{act}/RT} \tag{101}$$

$$k_b = Z_{el} e^{-Z_P F \Phi/RT} e^{-\Delta G_b^{act}/RT} \tag{102}$$

where $Z_{el} = (kT/h)e^{\Delta S_{pair}/R}$ is shown from classic statistic thermodynamics to be related to the collision frequency of a molecule on a wall [50], that is,

$$Z_{el} = \left(\frac{RT}{2\pi m}\right)^{1/2} \tag{103}$$

in which m is the molar mass of R or P. For most organic molecules, Z_{el} is of the order of approximately 2×10^3 cm s^{-1}. Note, however, that no transmission coefficient [compare the κ terms in the homogeneous analog rate constants in Eqs. (58) and (59)] is included in Eqs. (101) and (102). Indeed, the orbital variety at an electrode surface is sufficiently wide for convenient overlap to occur with most organic molecules prone to outer sphere electron transfer, so that the adiabatic requirement is generally fulfilled. Moreover, Hale [77] estimated that electron transfers at an electrode are adabatic as soon as the electron transfer site is within 15 Å of the electrode surface, for most organic molecules.

Thus, determination of the rate constants in Eq. (101) or (102) requires only that of ΔG_f^{act}, since $\Delta G_b^{act} = \Delta G_f^{act} - \Delta G^{o\prime}$, $\Delta G^{o\prime}$ being the free energy of the electron transfer between R_Φ and P_Φ (compare scheme 5):

$$\Delta G^{o\prime} = \Delta G^{o} - Z_R F\Phi_s + Z_P F\Phi_s = nF(E - E^{o}) - nF\Phi_s$$

that is,

$$\Delta G^{o\prime} = nF(E - E^{o} - \Phi_s) \tag{104}$$

The forward activation free energy is determined along the same lines as presented for homogeneous electron transfer since both problems are strictly equivalent [43,44,78,79]. Thus, ΔG_f^{act} is given as a function of the local driving force [$\Delta G^{o\prime}$ in Eq. (104)] and of the associated reorganization energy λ^{el} by the expression in Eq. (105), which derives from Eq. (64).

$$\Delta G_f^{act} = \frac{\lambda^{el}}{4}\left(1 + \frac{\Delta G^{o\prime}}{\lambda^{el}}\right)^2 \tag{105}$$

Again λ^{el} involves two contributions, of inner shell λ_i^{el} and outer shell λ_o^{el} nature, but these are contributions that relate only to the R/P couple since the electrode is not affected by the electron transfer. λ_i^{el} may be evaluated from bond length and bond angle variations, as is its homogeneous analog in Eq. (67).

For the evaluation of λ_o^{el}, the same model as in the homogeneous case is used, yet the electrode is replaced by the electrical image of reactant for the initial system (or that of the product for the final system) through the electrode, which acts as a electrostatic mirror. However, since these images are imaginary, only the real contributions are considered for the final evaluation of λ_o^{el}, given in Eq. (106).

$$\lambda_o^{el} = N_A \frac{e^2}{8\pi}\left(\frac{1}{\varepsilon_{op}} - \frac{1}{\varepsilon_s}\right)\left(\frac{1}{a_R} - \frac{1}{r_\Phi}\right) n^2 \tag{106}$$

In Eq. (106), a_R is the equivalent radius of R (or P) and r_Φ is the distance between the centers of the R (or P) molecule and its electrical image when R (or P) is at the electron transfer site. In practice, Marcus approximates r_Φ by $2a_R$, assuming that the molecule is in contact with the electrode surface when at the electron transfer site. Thus it follows that λ_o^{el} is given in Eq. (107) [43a,b]:

$$(\lambda_o^{el})_{Marcus} = N_A \frac{e^2}{8\pi}\left(\frac{1}{\varepsilon_{op}} - \frac{1}{\varepsilon_s}\right)\frac{1}{2a_R} n^2 \tag{107}$$

Hush [78c] disagrees with the formulation in Eq. (106), considering that the solvent polarization does not affect the potential in the double layer, the compact layer acting as an electrostatic Faraday cage. Thus the incidental solvent fluctuations energetically affect only R or P, without consideration of their images [note that this amounts to considering r_Φ as infinite in Eq. (106)]. Thus Hush proposes the expression in Eq. (108) for the outer shell reorganization energy:

$$(\lambda_o^{el})_{Hush} = N_A \frac{e^2}{8\pi}\left(\frac{1}{\varepsilon_{op}} - \frac{1}{\varepsilon_s}\right)\frac{1}{a_R} n^2 \tag{108}$$

Interestingly, the Marcus heterogeneous reorganization energy $(\lambda_o^{el})_{Marcus}$ in Eq. (107) is exactly half that, λ_o^{RR}, of the homogeneous isotopic electron transfer:

$$R^* + P \rightleftharpoons P^* + R$$

whereas that proposed by Hush is identical to λ_o^{RR}. Based on this analogy, an elegant series of experiments by Kojima and Bard [80] led to the conclusion that Hush's reorganization energy is more adequate [81].

Introduction of $\lambda^{el} = \lambda_o^{el} + \lambda_i^{el}$ in Eq. (105), when taking into account Eqs. (104), (101), (102), and (98), allows the expression of the current to be predicted as in Eq. (109), where k(E) is the apparent rate constant for an electrode potential E. k(E) is given in Eq. (110) under a formulation modeled on that of the empirical Volmer and Butler law [82], where the transfer coefficient α is given in Eq. (111).

$$i = nFAk(E)[(R)_0 - (P)_0 e^{nF[E-E^\circ]/RT}] \tag{109}$$

$$k(E) = k_t^0 e^{(\alpha n - Z_R)F\Phi_s/RT} e^{-\alpha nF(E-E^\circ)/RT} \tag{110}$$

$$\alpha = \frac{1}{2} + \frac{F(E - E^\circ - \Phi_s)}{4\lambda^{el}} \tag{111}$$

The true standard heterogeneous rate constant k_t^0 is expressed in Eq. (112) as a function of λ^{el}. Yet note that usually the intrinsic standard heterogeneous rate constant k^0 in Eq. (113) is defined and used rather than k_t^0.

$$k_t^0 = Z_{el} e^{-\lambda^{el}/4RT} \tag{112}$$

$$k^0 = k_t^0 e^{(\alpha n - Z_R) F \Phi_s / RT} \tag{113}$$

Provided the exponential term in Eq. (113), called the Frumkin correction, is invariant with E, that is, when Φ_s does not depend on the electrode potential, both are constant and characterize the R/P electron transfer independently of the electrode potential. Otherwise k_t^0 is preferred. In the following we nevertheless use k^0 for simplicity in the formulations.

The Transfer Coefficient α

The formulation in Eq. (110) shows that the apparent heterogeneous rate constant k(E) depends on the potential. This dependence has been observed experimentally since the earliest times of electrode kinetics and is at the origin of the so-called Tafel plots [83]. Indeed, when the potential E is sufficiently different from E°, for example $nF(E - E^\circ) \ll 0$, the term relative to $(P)_0$ in Eq. (109) vanishes and the current is given by Eq. (114).

$$i = nFAk(E)(R)_0 = nFAk^0 e^{-\alpha nF(E-E^\circ)/RT}(R)_0 \tag{114}$$

When $(R)_0$ is maintained constant, then a plot of the logarithm of the current versus the potential yields a straight line of slope $\alpha nF/RT$, provided that α is constant. Thus such observations led Butler and Volmer, two pioneers in this area, to propose an empirical law [82], named after them, in the form of Eq. (109), with α an adjustable constant parameter. α is called the transfer coefficient after the following observation: when the driving force, that is, $-nF(E - E^\circ)$ in Eq. (114), is increased, only a fraction α of the additional energy is used to increase the apparent rate constant. Thus, α is the fraction of the driving force transferred for kinetic purposes and plays a role identical to that of the Brönsted coefficient in organic chemistry (Sec. II.C, QFER).

From Eq. (111) it is seen that, provided that λ^{el} is large enough, α remains constant with the potential and is close to 0.5. This is generally in agreement with observations dealing with reactions with low values of k_t^0 [compare the λ^{el} term in Eq. (112)]. However, for organic molecules with small inner shell contributions and rather large radii, λ^{el} ($\approx \lambda_0^{el}$) is not expected to be extremely large, and therefore variations of α with the potential should be observed. This is indeed the case in practice [84], as illustrated by the curvature of the logarithmic plot of k(E) with the potential in Fig. 16 for the reduction of 2-methyl-2-nitropropane, in DMF at the HMDE [84].

$$(CH_3)_3C{-}NO_2 + e \rightleftharpoons (CH_3)_3C{-}NO_2^{\cdot-} \tag{115}$$

Moreover, the plot of the corresponding $\alpha(E)$ variations in Fig. 16b shows that α depends linearly on the potential with a slope (0.249 V^{-1}), in close agreement with that predicted (0.210 V^{-1}) from the Marcus reorganization energy (29 kcal/mol) determined from Eq. (107).

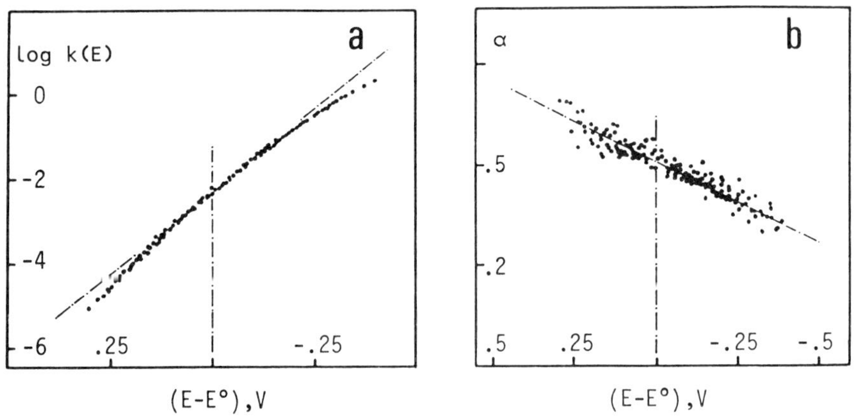

Figure 16. (a) Variations in the apparent heterogeneous rate constant k(E) in Eq. (110) with the electrode potential for the reduction of $(CH_3)_3C{-}NO_2$ in DMF at the hanging mercury drop. A dashed line with slope 0.5 is positioned at $E = E^o$ to emphasize the curvature. (b) Resulting variations in the transfer coefficient α in Eq. (111). (Experimental data from Ref. 84b.)

Reversible and Irreversible Electron Transfers: Their Role in the Meaning of Oxidation or Reduction Potentials

From the current expression in Eq. (109), it is seen that when k^0A is extremely large, the bracketed term on the far right must tend to zero in order that the current remain finite. Thus in these conditions,*

$$(R)_{x=0} \approx (P)_{x=0} e^{nF(E-E^o)/RT}$$

which may be written under the form of the classic Nernst law in Eq. (116). Thus, provided the heterogeneous rate constant is large (> 0.1 cm s^{-1}), the electron transfer is controlled entirely by thermodynamics and the relative activities of the oxidant and the reductant at the electrode surface obey the Nernst law in Eq. (116). The electron transfer is thus termed <u>reversible</u> or <u>nernstian</u>.

$$E = E^o + \frac{RT}{nF} \ln \frac{(R^0)_{x=0}}{(P)_{x=0}} \tag{116}$$

In other words, when $(R^0)_{x=0}$ and $(P)_{x=0}$ are kept constant, because of a large excess of material at the electrode surface, for example, the electrode potential is given in Eq. (116) independently of the current flow. Such an electrode is then suitable for use as a reference electrode (large k^0, large surface area A, and excess of material).

Conversely, when k^0 is extremely small, the current flow is negligible in the potential range around E°. Thus, in order that a measurable current

*Hereafter, $(R)_0$ and $(P)_0$ are considered the activities at the "electrode surface" and denoted $(R)_{x=0}$ and $(P)_{x=0}$ to recall this definition, x being the distance from the electrode.

may flow, the value of E − E° must be significantly increased [compare k(E) in Eq. (110)]. For example, in order that a cathodic current ($n > 0$) be observed, E must be considerably more negative than E°. Thus, in Eq. (109), $e^{nF(E-E^{o})/RT} \longrightarrow 0$ in the potential range where the wave is observed, and Eq. (109) can be rewritten as in Eq. (117):

$$i = nFAk^{0}e^{-\alpha nF(E-E^{o})/RT}(R)_{x=0} \tag{117}$$

Similarly, the observation of an anodic current requires that E is considerably positive to E°. Then in the potential range where an anodic current is observed, $e^{nF(E-E^{o})/RT} \longrightarrow \infty$, and Eq. (109) becomes

$$i = -nFAk^{0}e^{(1-\alpha)nF(E-E^{o})/RT}(P)_{x=0} \tag{118}$$

In these situations the current flow is entirely controlled by the kinetics of the heterogeneous electron transfer, which is then said to be <u>slow or irreversible</u>. The current-potential characteristic of the redox couple under study then reflects thermodynamic (E°) and kinetic (k^0) properties. Indeed, the driving force E − E° is used to overcome not only thermodynamic but also kinetic limitations.

In this presentation we discussed the reversible or irreversible nature of the electron transfer on an absolute basis. Yet in practice the notion of "measurable" current flow is obviously related to a given scale. In electrochemistry the usual scale referred to is based on the mass transfer rate. Thus for a given mass transfer rate, a given current density is expected to flow through the electrode. This limit is defined by the current density that would be observed if the electron transfer under investigation is nernstian. Within this context it is easily understood that any electron transfer is nernstian at very low mass transfer rates. Similarly, it should become slow or irreversible for large mass transfer rates, because then the current scale is made very high. In practice the largest mass transfer rates currently achievable with standard equipment are such that if $k^0 > 0.1$ cm s^{-1} the electron transfer may be considered "intrinsically" nernstian.

A second point not considered in the discussion is related to the possible involvement of chemical reactions. For the sake of simplicity, let us restrict the presentation to the most common stiuation in which the product P formed upon electron transfer is consumed by a fast chemical step, as outlined in scheme 6.

It is easily seen from this simple representation that as soon as the rate of the chemical removal of P is larger that that of the back electron transfer, the rate-determining step of the overall process is the forward electron transfer. Thus, independently of the intrinsic value of k^0, the Butler-Volmer law in Eq. (109) simplifies to that in Eq. (117), because $(P)_{x=0}$ is made negligible. As a result, the R/P electron transfer presents all the kinetic characteristics of a slow electron transfer [85]. However,

$$R + ne \underset{k_b}{\overset{k_f}{\rightleftharpoons}} P \longrightarrow \cdots$$

Scheme 6

since this behavior is not related to the intrinsic value of k^0, the current-potential curve may be observed in the close vicinity of E° or even positive to E° for a reduction (negative to E° for an oxidation) [81].

D. Adsorption Phenomena

Many solutes or follow-up products of electrochemical reactions may have a large tendency to adsorb on electrode surfaces [86]. This may be related to long-range electrostatic interactions, such as those discussed in the presentation of the double layer; there is then no specificity of the ions attached, particularly as concerns their chemical properties. In such cases the phenomenon is said to be nonspecific and is akin to nonspecific salt effects or ion pairing in organic chemistry. On the other hand, because of particular chemical affinities between the electrode surface and a substrate, specific adsorption occurs, which involves generally strong interactions and may result in the creation of organic layers at the electrode surface. Analogies of the latter are given by complexation reactions or specific salt effects (e.g., Li^+ with an enolate ion or Bu^-Li^+ solutions) in organic chemistry or chemisorption and physisorption in heterogeneous catalysis.

Both classes are distinct because of the nature of the electrode-adsorbed species interactions. Yet from a phenomenological point of view, a better classification seems to be that based on the possible electroactivity or conductive properties of the adsorbed material.

When adsorbed, electroinactive species usually give rise to partial or total blocking of the electrode surface, which may then not be able to effect electron transfers to other electroactive molecules. Thus depending on the nature of the electrode coverage, various phenomena may be observed, ranging from an apparent decrease in the electrode surface (large and separated islands of adsorbed species) to an apparent decrease in the electron transfer rate but with the appearance of unchanged electrode surface area (micrometric islands at micrometric distances from each other) [87]. Related also to this class is the "natural" blocking of electrodes by oxide layers or functional groups at carbon electrodes [88].

Conducting (via electrons or via permeation of electroactive species) adsorbed species usually introduce small disturbances in electrochemical behavior compared with the former class. Yet obviously the diffuse layer is extremely affected vis-à-vis the bare electrode. Thus the rate constants k^0 may be modified to a high degree* because of large changes in the Frumkin correction. Such effects may easily explain, at least on a qualitative basis, the well-known dependency of k^0 on the size of the supporting electrolyte cation (reductions) or anion (oxidations), as well as on ionic additives [89].

The third class involves electroactive adsorbed species, yet it also encompasses the cases in which, although originally inactive, the species is activated or modified in the adsorption process to give rise to the formation of electroactive adsorbed molecules. Obviously all these effects discussed for the two previous classes are observed for this class, but additional effects are observed for the adsorbed species itself. Indeed, adsorption

---

*Note that for $n = 1$, an increase by 0.2 V in Φ_S results in the multiplication of k^0 by approximately 50, when $\alpha = 0.5$ and by approximately 350 when $\alpha = 0.75$, as determined from Eq. (113).

modifies (1) the concentration of the species at the electrode surface vis-à-vis that predicted from its bulk concentration, and (2) the energetics of the electron transfer reaction. The latter point arises because of additional thermodynamic contributions to be considered in the R_0/R_Φ or P_0/P_Φ equilibria in scheme 5 and also from changes in the reorganization energies. The former point leads to a variety of behaviors, since it is obviously a function of the kinetics of adsorption-desorption phenomena. A third additional effect of electroactive adsorbed species arises from their possible ability to act as electron transfer mediators for other species that are reduced or oxidized with more difficulty. This special class (the term "electrode modification" is then more adequate), which is at present a very active area of research, is discussed in another chapter of this book (see Chap. 33).

E. Coupled Chemical Reactions

Electrochemical Versus Homogeneous Chemical Reactivities

In the preceding sections of this presentation, much attention was devoted to the essential act of electron transfer. Yet although being, for fundamental or analytic purposes, the subject of a large body of the electrochemical literature, pure electron transfer reactions generally are of lesser interest in terms of preparative electrochemistry. In the opinion of this author, electron transfer at an electrode must be considered as a particular class of activation of molecules to enhance their chemical reactivity. As such, electrode kinetics must be understood (as for other methods of chemical activation) in order to control and eventually direct the overall process to the selected target (compare Chap. 4).

From the preceding discussion, it should appear that electrochemistry affords a very facile and precise way to generate highly energetic intermediates via control of the electrode potential. Note in this respect that one electron exchanged over a potential difference of 1 V amounts to injecting 1 eV in a molecule, that is, approximately 23 kcal/mol. Owing to the possibility of performing electrochemistry over potential differences of several volts, it is easily seen that the method may involve energies comparable to those of most chemical bonds and largely beyond most activation energies. Thus highly energetic intermediates may be generated under mild and precisely controlled conditions. Another useful aspect of electrochemical generation of chemical intermediates is related to the current flow through the cell. Indeed, the current is a direct measure of the production rate of the intermediate. It ensures that the production rate, an important parameter in product selectivity, is easily controlled and adjusted.

At this point, two questions must be addressed: Are these intermediates identical to those obtained under usual chemical conditions? Do they react under conditions matching those of homogeneous experiments?

The answer to the first question is obviously positive. Indeed, since the substrates are generally neutral, the primary electron transfer affords paramagnetic intermediates (anion or cation radicals) that may undergo selected cascades of reactions to afford most of the well-identified intermediates of organic chemistry: radicals, carbocations, carbanions, carbenes, and so on. Neutral radicals may also be generated by reduction or oxidation of stable cations or anions, respectively. It is thus seen that all the basic intermediates of organic chemistry may be electrochemically generated, provided an electroactive precursor exists. Yet this is not always a serious limitation, since the intermediates sought may be eventually formed indirectly, that is, through an electroactive proreagent, as discussed elsewhere in this book (Chaps. 29 and 30).

The second question is more difficult to answer. Indeed, since the electrical perturbation extends only over few angstroms from the electrode surface, the medium in which the electrogenerated intermediates react is identical to the bulk of the solution, which favors a positive answer. On the other hand, they react under essentially nonisotropic conditions because of the existence of concentration profiles, that is, of concentration changes with distance of the electrode (see Fig. 15). Yet this is a situation identical to that encountered when heterogeneous reagents or polyphasic conditions are used in usual "homogeneous" chemistry. A more serious difference is related to the nature of the media often used under electrochemical conditions. As explained in the section devoted to ohmic drop effects, electrochemical solutions must generally be rather conducting, a requirement met by using solvents with rather high dielectric constants (see Chap. 6). Thus although solvents, such as liquid ammonia, or media with smaller dielectric constants, such as methylene chloride, THF, DME, or even arenes and alkanes, can be used [90,91], the preference of most organic electrochemists is for convenience or by tradition, to such solvents as acetonitrile and dimethylformamide. The latter are prone to intervene in the reaction paths of the electrogenerated intermediates via their nucleophilic or acidobasic properties, for example, as well as via hydrogen atom transfers to radicals [92]. The same is true for the so-called inert electrolytes, frequently quaternary ammonium salts, known for their acidic (Hofmann reaction) abilities, as well as for being relatively good hydrogen atom donors. Thus the usual electrochemical media may induce reaction paths that are quite different from those observed with identical intermediates under conventional organic conditions. We want to emphasize, however, that this is not to be viewed as a limitation of the method. Indeed, these induced reaction paths may be used with great profit, as in organic chemistry when methanol is used as the solvent to incorporate methoxy groups. Several examples of this strategy are given in various places in this book. On the other hand, as mentioned earlier, a large variety of solvent-electrolyte systems may be used in electrochemistry [71,90], although the general body of reported data pertains to a limited class of media.

Chemical Reversibility and Irreversibility: Their Role in the Meaning of Oxidation or Reduction Potentials

When considering a simple electron transfer reaction at an electrode, as in Eq. (119), one can conceive easily that there are, a priori, four limiting situations in which chemical reactivity may affect the electrochemical measurements: R may be consumed or produced and P may be consumed or produced.

$$R + ne \rightleftharpoons P \qquad (E^{o}) \tag{119}$$

In practice all four possibilities may be combined, which leads to a large variety of electrochemical mechanisms. Thus our purpose in this section cannot be quantitative if we wish to keep some generality. We want, however, to describe qualitatively the role and the effect of chemical reactions on the overall electrode kinetics. Thus for the sake of simplicity, we discuss the case of a simple EC (electrochemical-chemical steps; see Sec. III.E, Classification of coupled chemical reactions) sequence, in Eqs. (120) and (121) as a typical example:

$$R + ne \rightleftharpoons P \qquad (E^{\circ}, k^{0}, \alpha) \tag{120}$$

$$P \longrightarrow \cdots \qquad (k) \tag{121}$$

The chemical step in Eq. (121) is homogeneous and thus intervenes while P, which is formed at the electrode, is transported to the bulk of the solution via the mass transfer mechanisms discussed earlier. From this simple consideration, it is easily understood that an important factor is the relative magnitude of the rate constant of the chemical reaction vis-à-vis that of mass transfer, as outlined in scheme 7.

Obviously, when the mass transfer rate is much larger than that of the chemical steps, the transfer of P from the electrode surface to the bulk solution is not affected by the presence of kinetics. Thus the electrochemical behavior of the EC sequence in Eqs. (120) and (121) is identical to that of a simple electron transfer mechanism. The R/P electron transfer is then said to be chemically reversible* but may be nernstian or slow, according to the magnitude of k^0 in Eq. (120).

In the converse situation, three limiting cases may be observed. In the first case the electron transfer is intrinsically slow. The RDS of the overall process is then the forward electron transfer, and the redox system is said to be slow and chemically irreversible. The electrochemical wave is then observed at potentials sufficiently different from E° for $n(E - E^{\circ}) \ll 0$ (Sec. III.C, Reversible and irreversible electron transfers). The two other situations are encountered when k^0 is large enough for the electron transfer to be nernstian in the absence of the follow-up chemical step.

When, although fast vis-à-vis the mass transfer rate, the chemical reaction is slow compared with the backward electron transfer, the reduction or oxidation in Eq. (120) remains at equilibrium. The electrode potential is then given by the Nernst law in Eq. (122):

$$E = E^{\circ} + \frac{RT}{nF} \ln \frac{(R)_{x=0}}{(P)_{x=0}} \tag{122}$$

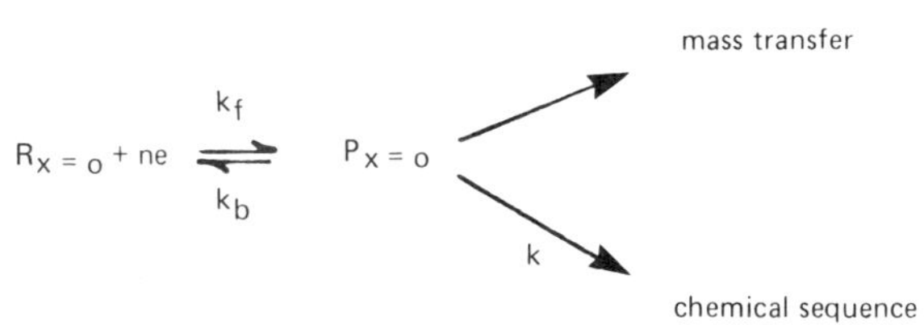

Scheme 7

*To avoid possible confusion with the "reversible" or "irreversible" nature of the electron transfer step per se, we suggest that reversibility or irreversibility be reserved to the chemical part and the electron transfer be designated by "nernstian" or "slow," respectively. In the following we use this terminology.

Yet because of the chemical reaction, the activity of P at the electrode surface is considerably decreased. From Eq. (122) it is seen that the electrode potential is then positive (for a reduction; negative for an oxidation) to that observed for the same current density, but in the absence of the follow-up reaction. As a result the current-potential characteristic is observed in a potential range positive to E° for a reduction and negative to E° for an oxidation. The system is then said to be nernstian and chemically irreversible.

In the last situation the rate of back electron transfer, although intrinsically fast, is slower than that of the chemical step. Thus as in the first situation, the RDS is the forward electron transfer, and the redox system is said to be slow and chemically irreversible. Yet since k^0 is large, the electrochemical wave is observed before E° [81,85].

From this description it is seen that, depending on the exact degree of competition between the four rates in scheme 7, the electrochemical wave is observed in potential ranges that may be positive or negative to E°, as summarized in Table 4. Without a quantitative treatment of the pertinent electrochemical data, it is almost impossible to decide the position of E° vis-à-vis the potential location E_W, where the wave is observed. This is an important caveat to remember when using potential data, such as peak potentials or half-wave potentials, instead of E° values. The experimental difficulty is even more severe in practice. Indeed, when one considers a series of related chemicals, there are great chances that E°, k^0, and k vary uniformly with respect to each other because they relate intimately to the orbital energy and characteristics of the acquired (LUMO) or lost (HOMO) electron [16,29]. Thus, the potential E_W at which the electrochemical wave is observed may correlate with E°, since it depends on the three figures. Such a case is presented in Fig. 17 for the series of akylbenzenes [29] already mentioned in this chapter, in which the peak potentials of the chemically irreversible voltammograms (at 0.1 V s^{-1}) correlate with E° with a slope close to unity.

Table 4. Nature and Location[a] of the Electrochemical Wave Observed for an EC Sequence[b] as a Function of the Mass Transfer Rate

k versus mass transfer rate	k^0 versus mass transfer rate: Small	k^0 versus mass transfer rate: Large
Small	Slow, reversible,[c] $n(E_W - E^\circ) < 0$	Nernstian, reversible,[c] $E_W \approx E^\circ$
Large	Slow, irreversible,[c] $n(E_W - E^\circ) < 0$	Nernstian or slow irreversible,[c] $n(E_W - E^\circ) > 0$

[a] E_W, any characteristic potential of the wave: $E_{1/2}$, E_p, $E_{p/2}$, $\cdots$; E°, standard reduction potential of the electron transfer step, involving an exchange of n electrons (n > 0 for a reduction; n < 0 for an oxidation).
[b] k^0, intrinsic standard heterogeneous rate constant; k, rate constant of the homogeneous follow-up chemical reaction.
[c] Chemically reversible or irreversible (see footnote page 71).

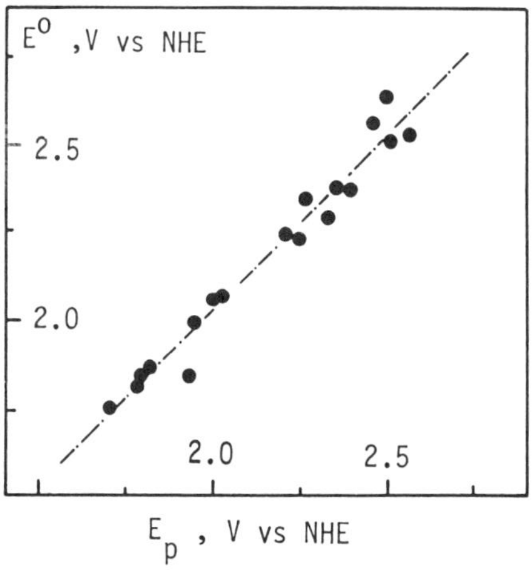

Figure 17. Correlation between E^0 and anodic peak potentials E_p of the chemically irreversible cyclic voltammograms at $v = 0.1\ Vs^{-1}$ for an extended series of alkylbenzenes. (Data from Ref. 29.)

To conclude this section we discuss the special case of follow-up equilibria, that is, when the chemical step in Eq. (121) is reversible [93].

$$R + ne \rightleftharpoons P \rightleftharpoons Z \qquad (E^o,\ k(E);\ k_f,\ k_b)$$

Obviously, when $K = k_f/k_b \ll 1$, the equilibrium lies toward P and does not affect the R/P electron transfer. In the converse situation, and when k_f and k_b are larger than the mass transfer rate, a rapid equilibrium displaced toward Z establishes, and $(P)_{x=0} \approx (Z)_{x=0}/K$. Introduction of this relation in the current rate law [Eq. (109)] then yields Eq. (123), which is identical in its formulation to a Butler-Volmer law but for an hypothetical electron transfer $R + ne \rightleftharpoons Z$, whose formal potential is $E^{o\prime} = E^o + (RT/nF) \ln K$, as evidenced by the formulation in Eq. (124).

$$i = nFAk(E)[(R)_{x=0} - (Z)_{x=0}e^{nF(E-E^o)/RT}/K] \tag{123}$$

$$i = nFAk(E)[(R)_{x=0} - (Z)_{x=0}e^{nF(E-E^{o\prime})/RT}] \tag{124}$$

This constitutes the basis of the operational significance of formal potentials introduced in Sec. II.B, Formal reduction potentials. Yet the formal analogy between Eqs. (109) and (124) should not mask the fact that k(E), that is, α, k_t^0, or k^0 in Eq. (124), remains related to the real electron transfer $R + ne \rightleftharpoons P$, that is, to E^o, not to $E^{o\prime}$ [compare Eqs. (110 through (113)].

Classification of Coupled Chemical Reactions

From this discussion, it is seen that the electrochemical behavior of a given electron transfer reaction may be extremely perturbed by the intervention of coupled chemical steps. The resulting modifications obviously depend on the exact nature of the chemical sequence associated with the electron transfer step. It is thus convenient to classify the different possible reaction schemes that may arise. Yet as in organic chemistry, for the basic mechanisms (e.g., S_N1, S_N2, E_1, and E_2) such a classification is viewed as a general frame to which real mechanisms are referred. Indeed, the latter may often be more subtle (compare, e.g., S_{Ni} or E_{1cb} vis-à-vis the mechanisms mentioned earlier) or involve cooperative or competitive combinations of the elementary schemes presented later. A second similarity with the organic basic schemes is that there is no unity in the denominations of the various sequences of electron transfers and chemical steps. Indeed, each school or laboratory has proposed its own nomenclature, which has received a different audience and may have evolved with time. In the following we use as much as possible the most frequent designations, and when not possible we refer to the nomenclature used in Savéant's group, owing to their large contributions to this area [94].

The following classification is based upon the fact that electrochemical behavior depends on the location of the electron transfer step in the overall sequence. Also, electron transfer steps are considered to involve the exchange of a single electron, that is, n = ±1 (n = 1 for a reduction and n = −1 for an oxidation), a normal situation in organic electrochemistry. Indeed, although overall stoichiometries involving the net exchange of many electrons are frequently observed, the energetic and activation requirements for a direct polyelectron transfer are too high for any to exist concurrently with a rapid succession of single electron transfer steps.

Within this frame, the most frequently encountered situations are the following.*

Preceding Chemical Reaction: CE Mechanism

$$Z \rightleftharpoons A \quad \left(k_f,\ k_b,\ K = \frac{k_f}{k_b}\right) \qquad \text{(C)}$$

$$A + ne \rightleftharpoons B \ (E^{\circ}) \qquad \text{(E)}$$

The electroactive species A is generated by a reaction, generally an equilibrium displaced toward Z, that precedes the electron transfer step. Z is the reactant introduced in the cell or the predominant form of the reactant in the reaction medium. These reaction schemes were introduced to rationalize the various electrochemical phenomena observed during the reduction of certain aldehydes in aqueous solutions. Indeed, in water, formaldehyde, π-electron-poor heterocyclic aldehydes, and a few aldehydes with strongly electron-attracting groups exist as their nonreducible hydrated form in rapid equilibrium with the reducible carbonyl:

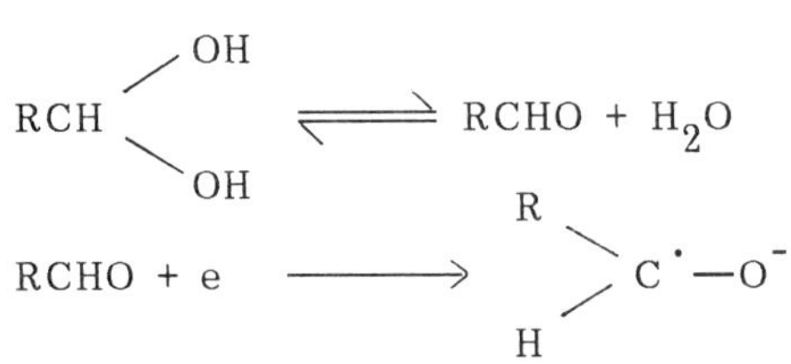

*E is a heterogeneous electron transfer; C, a chemical step; also, A, B, and so on, designate the electroactive molecules or their reaction products, whereas Y, Z, and so on, relate to coreactants or nonelectroactive species.

When the rate of the Z $\rightleftharpoons$ A interconversion is slow vis-à-vis the rate of mass transfer,* it acts as the RDS for the overall sequence. The currents observed are then abnormally low for the mass transfer rate and reactant concentration considered and are said to be kinetic currents.

Conversely, when both rates are fast, the preceding chemical reaction acts as a rapid pre-equilibrium, and (A) = K(Z). The electrochemical data are then formally identical to that of the virtual Z + ne $\rightleftharpoons$ B electron transfer, with a formal potential

$$E^{o\prime} = E^{o} - \frac{RT}{nF} \ln \frac{K}{K + 1}$$

Redox Catalysis or Mediated Electron Transfer: EC' Mechanism

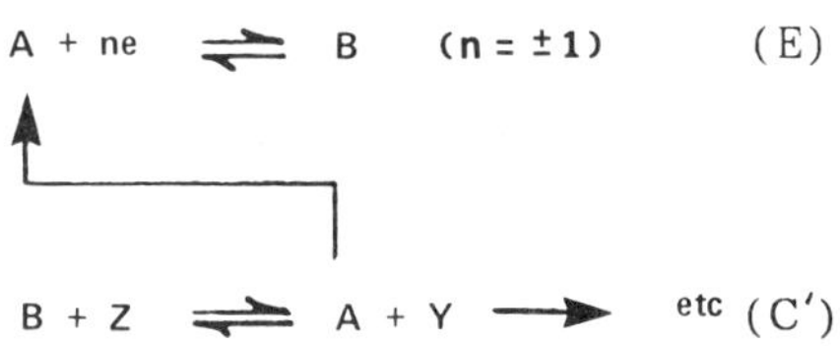

In this sequence, the A/B couple acts as an electron transfer mediator, which allows the reduction or the oxidation of the electroinactive (at the A/B wave potential) Z species. The electroactive species A is regenerated via the homogeneous electron transfer step C', which results in a considerable enhancement of the current observed for the A/B wave as soon as the C' step is rapid.

Various subclassifications exist according to the exact nature of the chemical step, which may eventually be a succession of elementary steps with formation of intermediate products. As explained earlier for the displacement of endergonic electron transfer steps, the C' step occurs because it is continuously pulled to the right by the further chemical reaction of the Y species. Note that Chap. 29 is devoted to this class of mechanisms.

Consecutive Reaction: EC Mechanism

$$A + ne \rightleftharpoons B \quad (n = \pm 1) \qquad (E)$$

$$B \longrightarrow C \qquad (C')$$

This sequence is one of the most ubiquitous in organic or organometallic chemistry and has already been discussed in preceding parts of this section. The chemical step may involve reactions (or a succession of chemical reactions) of high-order molecularities, not only first-order reactions as suggested by the notation. Classic examples are given by activated olefin electrohydrodimerization mechanisms:

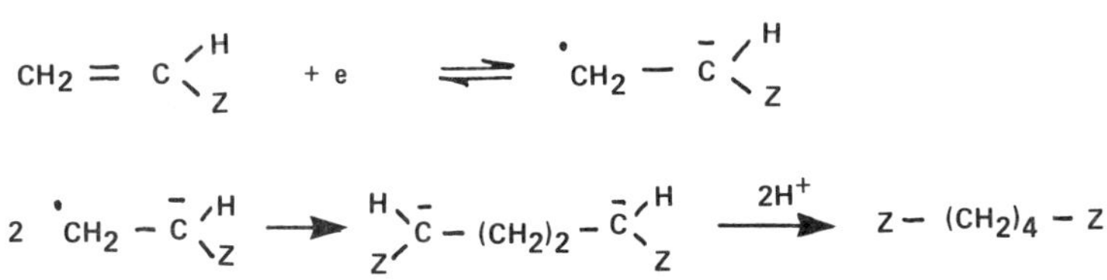

*In the following, "rapid," "fast," and "slow" always refer to the rate of mass transfer.

Multielectron Reactions: ECE and Related Mechanisms. As explained earlier, a direct exchange of many electrons at the same redox center appears to be extremely difficult because of energetic and activation requirements. From the formulation in Eq. (106), it is indeed noted that the exchange of 2e involves a reorganization energy λ_0^{el} four times larger than that for a single-electron exchange at the same center. For this reason and because of the free energy relationship in Eq. (105), a two-electron step without formation of an intermediate should correspond to a slower process than the corresponding EE sequence in Eq. (125):

$$A + 2e \rightleftharpoons A^{\overline{\cdot}} + e \rightleftharpoons A^{2-} \tag{125}$$

On the other hand, in the second step of this sequence, the electron is transferred to a molecule with an additional negative charge compared with A. Thus at least a coulombic repulsion term must be overcome, which implies that the second electron transfer is necessarily performed at a more negative potential than the first (by at least 0.250 V for an equivalent radius of 2 Å for the redox centers, and for example, $\varepsilon \approx 37.5$ for acetonitrile as the solvent). As a result, the overall EE sequence should not be observed at the potential of the A reduction but should occur in two separate waves.

However, there are two ways in which this conclusion may be invalid, and both need the relay of a chemical reaction.

The first possibility is that the reorganization energy associated with the first electron transfer is considerably larger than that of the second. Such a difference is obviously not related to the outer shell components, which are necessarily close, as evidenced by Eq. (106). Thus the origin for the differences must be sought in the inner shell reorganization energy, which implies that a profound molecular reorganization must be involved during the first electron transfer. Examples of this situation are given by the numerous cases in which two redox centers are electronically coupled in the same molecule, as in $O_2N{-}(CH_2)_n{-}NO_2$. Similarly, for inner sphere electron transfers, the molecular moiety bearing the negative charge may be lost, leaving a neutral and easily reducible species. This is so in the reduction of alkyl halides to alkanes:

$$RX + e \longrightarrow R^{\cdot} + X \qquad (E_1)$$

$$R^{\cdot} + e \longrightarrow R^{-} \qquad (E_2)$$

$$R^{-} + H^{+} \longrightarrow RH \qquad (C)$$

in which the first electron transfer is induced by the carbon-halogen bond cleavage and affords an easily reducible neutral radical [95].

However, for most organic electroactive molecules, such large reorganization energies cannot be invoked, owing to the extended delocalization of the orbitals participating in electron transfers. Yet the interposition of a chemical step between the two electron transfers results in the same effect, as soon as the chemical reaction gives rise to a species more easily reducible (or oxidizable, in oxidations) than the starting reactant.

$$A + ne \rightleftharpoons B \qquad (E^{\circ}_1) \qquad (E)$$

$$B \longrightarrow C \qquad (C)$$

$$C + ne \rightleftharpoons D \quad (E^\circ_2;\ n(E^\circ_2 - E^\circ_1) > 0) \qquad (E)$$

The intermediate formed is then reduced (or oxidized) at the electrode or in the solution owing to the large exergonicity of the resulting reaction:

$$B + C \longrightarrow A + D \quad (\Delta G^\circ = -nF(E^\circ_2 - E^\circ_1) \ll 0) \qquad (DISP)$$

In the former case (second electron transfer at the electrode), the overall sequence is termed ECE; it is designated a "DISP" (rather than an ECDisp) mechanism in the second case because the homogeneous electron transfer step is close to being a disproportionation reaction. Indeed, the latter involves an electron transfer between two chemically related molecules with the same oxidation numbers. In practice the distinction between the two mechanisms is nearly impossible on experimental grounds [96] because their rate-determining step, the $B \longrightarrow C$ reaction, is identical.

A typical example of this sequence is given by the reduction of aromatic halides:

$$ArX + e \rightleftharpoons ArX^{\bar{\cdot}}$$

$$ArX^{\bar{\cdot}} \longrightarrow Ar^{\cdot} + X^-$$

$$Ar^{\cdot} + e \longrightarrow Ar^- \quad \text{and/or} \quad Ar^{\cdot} + ArX^{\bar{\cdot}} \longrightarrow Ar^- + ArX$$

$$Ar^- + H^+ \longrightarrow ArH$$

As in the EC electrochemical sequence, a large variety of mechanisms belong to the general ECE-DISP frame and differ according to the exact molecularity of the interposed chemical step, which may also be a succession of elementary steps. Similarly, cascades of ECECE . . . sequences are possible, as in the classic reduction of nitro derivatives to hydroxylamines, which involves four electrons and four protons [97]:

$$Ar{-}NO_2 + 4e + 4ROH \longrightarrow Ar{-}NH{-}OH + 4RO^- + H_2O \qquad (126)$$

Zero-Electron Reactions: Electron Transfer Catalysis of Chemical Reactions. In the preceding ECE sequence, both electron transfers are of the same kind, that is, both reductions or oxidations, and thus their combination gives an increase in the absolute value of the number of electrons exchanged. Yet when the second electron transfer is of a nature opposite to the first, the overall number of electrons consumed is zero. Thus the sequence amounts to an electron transfer catalysis of a nonredox chemical reaction [98], as outlined in the following sequence:

$$A + ne \rightleftharpoons B \qquad (E)$$

$$B \longrightarrow C \qquad (C)$$

$$C - ne \longrightarrow D \quad \text{and/or} \quad C + A \longrightarrow D + B \qquad (E)$$

Naturally such mechanisms are observed only when the overall A ⟶ D reaction is exergonic but is considerably slower than that occurring at the level of the reduced or oxidized states B and C. Although rather recent in the electrochemical literature, these processes have been shown to be general and to encompass a large variety of chemical reactions. Among the most documented is the $S_{RN}1$ reaction, which amounts to an aromatic nucleophilic substitution. Although initially postulated by Bunnett [99a], the confirmation and the subtleties of its mechanism have been established on electrochemical grounds [100]:

$$ArX + e \longrightarrow ArX^{\bar{\cdot}}$$

$$ArX^{\bar{\cdot}} \longrightarrow Ar^{\cdot} + X^-$$

$$Ar^{\cdot} + Nu^- \longrightarrow ArNu^{\bar{\cdot}}$$

$$ArNu^{\bar{\cdot}} - e \longrightarrow ArNu \quad \text{and/or} \quad ArNu^{\bar{\cdot}} + ArX \longrightarrow ArNu + ArX^{\bar{\cdot}}$$

Overall:

$$ArX + Nu^- \longrightarrow ArNu + X^-$$

Electrochemical Reaction Mechanisms and the Principle of Microscopic Reversibility

As understood from the large variety of possible reaction sequences outlined in the preceding section, a large number of reaction paths may be invoked a priori to explain the same overall electrochemical transformation. Yet owing to the great capabilities of electrochemical techniques for the identification of reaction mechanisms, the problem of their distinction frequently arises. For example, let us consider the very common two-electron-two-proton electrochemical sequence in Eq. (127):

$$A + 2e + 2H^+ \rightleftharpoons AH_2 \qquad (127)$$

The latter is involved, for example, in the reduction of aromatics to their dihydro derivatives, of quinones to hydroquinones, carbonyls to alcohols. A priori, there are six plausible different possibilities of transferring two electrons and two protons, depending on the exact order of the steps. These six paths involve the participation of seven different intermediates: $A^{\bar{\cdot}}$, A^{2-}, $AH^{\bar{\cdot}}$, $AH^{\cdot}$, AH^+, AH_2^{2+}, and $AH_2^{+\cdot}$. It is thus seen that the discussion of the possible mechanism(s) may rapidly become an overwhelming task, particularly when more steps are involved, as in the 4e + $4H^+$ sequence in Eq. (126). Thus in order to discuss the exact route followed, a convenient representation of the different possible schemes is highly desirable. Such a representation of all possible paths is given by "square-schemes" diagrams [101], such as that shown in scheme 8 for the A/AH_2 reduction. The horizontal transformations correspond to electron exchanges, whereas the vertical transformations involve a proton transfer.

Based on chemical grounds, several routes may be ruled out in scheme 8. For example, starting from an aromatic hydrocarbon A, routes δ, ε, and φ, may be excluded, except in a superacidic medium. This is also certainly

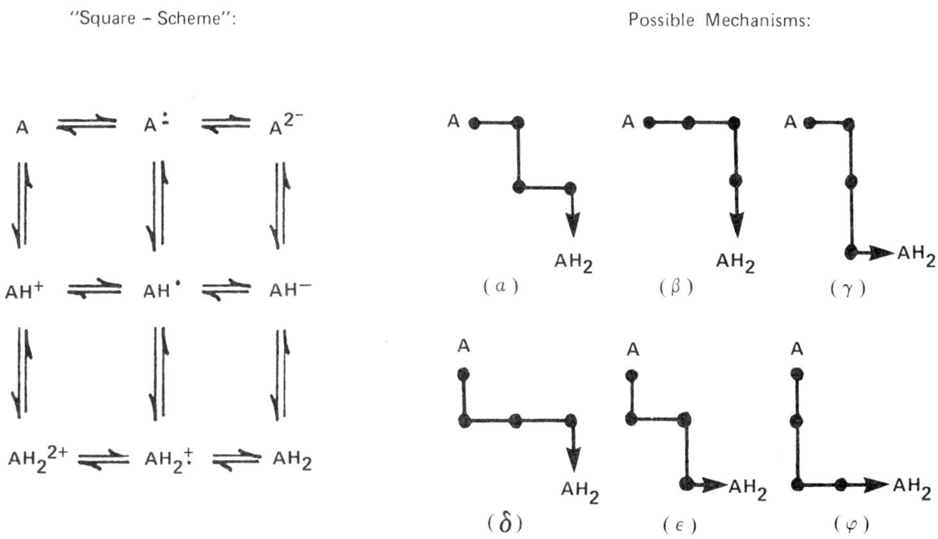

Scheme 8

true for the γ route, which involves a difficult protonation of $AH^{\cdot}$ to $AH_2^{+\cdot}$. Thus only two routes, α and β, appear chemically reasonable under the usual electrochemical conditions. In practice, both may be followed, path α prevailing under weakly acidic media and low cathodic potentials, and path β predominating in less acidic conditions and at more negative values of the potential. Note, however, that additional complications, not considered in scheme 8, involve the possible role of homogeneous electron exchanges. Thus $AH^{\cdot}$ may be reduced at the electrode but also by the strong reductant $A^{\cdot-}$.

The reverse transformation may also be considered, as in the oxidation of aromatic dihydro derivatives to their parent hydrocarbon. Then one starts from AH_2, and routes α, β, and δ must be eliminated, except in very basic media. This is also certainly true for route γ, since anion radicals $A^{\cdot-}$ usually have a marked basic character. Thus, under the usual electrochemical conditions, only routes ε and φ must be retained. Path ε predominates at less positive potentials and weakly basic conditions, whereas path φ should be observed at more positive potentials.

Consider now a given $Ar \rightleftharpoons ArH_2$ reaction under rather neutral conditions. It is inferred from the preceding discussion that route α is followed for the reduction, whereas the oxidation goes through path ε. At first glance, such a result seems to have to be rejected since it contradicts the principle of microreversibility, which claims hat both reactions should follow the same mechanism.

However, both transformations do not occur at the same electrode potentials: the reduction $Ar \longrightarrow ArH_2$ needs rather negative values, whereas the oxidation $ArH_2 \longrightarrow Ar$ is performed at rather positive values. Thus the two reactions occur under driving forces of opposite direction, with an energetic difference of approximately 100–130 kcal/mol. Thus, invoking the principle of microreversibility under such conditions is no more reasonable than invoking it when comparing the Birch reduction to the oxidation of aromatic dihydro derivatives by a strong oxidant.

III. MASS TRANSFER IN ELECTROCHEMISTRY

A. Fundamental Aspects of Mass Transfer Processes

Introduction

As mentioned in the preceding sections, mass transfer plays a crucial role in electrochemistry. The ubiquity of this phenomenon evidently arises because the electrons are exchanged at a two-dimensional surface boundary but the reactants and products are dispersed in a three-dimensional solution. Thus in the absence of mass transfer, only a small layer (of a few molecular radii thicknesses) would exchange electrons with each electrode, as is the case, for example, when the electroactive material is adsorbed at the electrode (Sec. III.D). As such, mass transfer is essential in controlling the success and rate of any preparative electrochemical reaction.

Another important feature of mass transfer processes is related to the very physical nature of the phenomenon. As such it is easily quantifiable and predictable. Thus the rate of mass transfer to and from an electrode may be determined a priori for a given electrochemical system. As a result this rate may be used as natural built-in clock by which the rate of other electrochemical processes may be measured. Such a property was apparent in our earlier discussions related to electrode kinetics (electron transfer and coupled chemical reactions). Basically is proceeds from the same idea as that frequently used in organic chemistry for relative rate constant determinations, when opposing a chemical reaction of known (or taken as the reference in a series of experiments) rate constant against a chemical reaction whose rate constant (or relative rate constant) is to be determined. Many such examples exist in the organic literature, among which are the famous "radical-clocks":

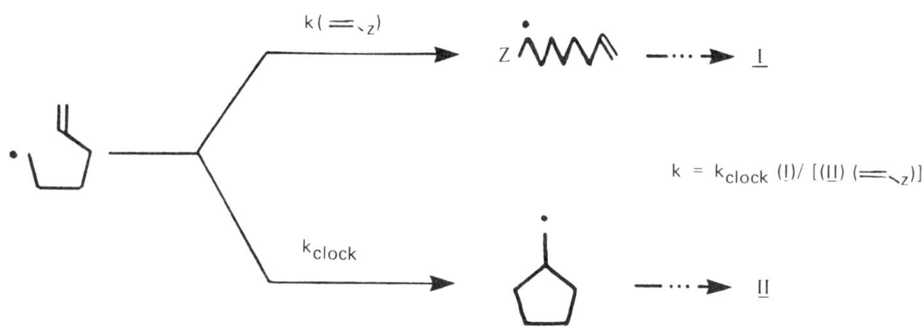

However, because of the intrinsic specifity of such chemical clocks, they cannot have a general use outside a given class of chemistry. Thus a better analogy with the electrochemical approach involving mass transfer rates would be that in which a reaction rate constant is matched to the diffusion limit rate constant that represents the physical process of bringing two reactants to distances such that they can interact chemically (Sec. II.C, Diffusion limitation in electron transfer reactions). Obviously such methods are not developed in homogeneous chemistry, principally because most chemical reactions, with maybe a few exceptions, among which are electron transfer or radical reactions, proceed with rates considerably slower than that of the diffusion limit. Another limitation to the development of such an approach is also related to the fact that diffusion limit is a rate of mass transfer (analogous to that involved in electrochemical experiments) but relates to mass transfer over atomic distances and is thus not prone to be adjusted at will.

In this respect, electrochemical mass transfer is particularly convenient. As mentioned earlier, its physical nature and its necessary involvement in any electrochemical experiment confer on it one of the characteristics of the internal clock just described. On the other hand, that it is active in the vicinity of a macroscopic object, the electrode, allows its precise control. In practice, mass transfer rates at electrodes are continuously adjustable from equivalent times of a few seconds to a few tenths of microseconds [102]. When this is coupled to the use of a diffusion limit in the nanosecond or less time scale, it explains why electrochemical transient techniques allow kinetic information to be obtained and rate constants determined for reactions whose lifetimes range from seconds to tens of picoseconds [103].

The Physical Processes of Mass Transfer: Fick's First Law

There are three physical processes involved in the transport of molecules in condensed media: convection, migration, and diffusion.

Convection may be the most intuitive and is present in all areas of chemistry; it corresponds to forced displacements of small volumic parts of the solution. The molecule is then carried within the fluid elements in which it is dissolved. The flux of particles J_{conv} through an elementary surface, that is, the number of moles of particles crossing a surface of area unity per unit of time, is then simply expressed as in Eq. (128),

$$J_{conv} = Cv(x) \tag{128}$$

where v(x) is the velocity of the fluid normal to the surface and C the concentration of the particle in the fluid element. It then follows that the mass transfer contribution is independent of the particle but is only a function of the local hydrodynamics (natural or thermal convection, forced convection, eddies, and so on). A large amount of work has already been devoted to quantifying the hydrodynamics of a solution [104]. Yet for our purposes here it is sufficient to recall two conclusions that apply in most electrochemical situations. (1) Because of shocks and interpenetration of the fluid elements in motion, concentrations are nearly identical at any point of a solution far from any wall. (2) Near any wall or solid surface, solid-liquid frictions and bouncing phenomena result in the creation of an immobile solution layer in which no convective effects occur.* Obviously the thickness δ_{conv} of this layer depends on the fluid velocity in the vicinity of the wall. Yet the resulting variation is small: thermal agitation and average vibrations result in $\delta_{conv} \approx 30-50$ μm, whereas magnetic stirring at maximum agitation results in $\delta_{conv} \approx 10-20$ μm.

---

*Note that the separation of the solution into two parts near a wall, that is, the immobile layer and the agitated homogeneous solution, is really simplistic since there is a continuous variation in the fluid velocity from the wall to the bulk [105]. Yet this dichotomous approximation is sufficient for most purposes and allows great simplifications in physicomathematical treatments of the transport problem.

From an electrochemical point of view it is easily inferred that the solution in a cell near an electrode is separable into two parts: a stagnant layer adjacent to the electrode in which no convective motions occur, and the remainder of the solution, which is homogeneous (bulk solution). Yet this is not a particularity of electrochemical methods since the same phenomena occur at any solid/liquid interface, as when metal particles (reductions by Zn or Na, for example) or any heterogeneous reagent is used in organic homogeneous chemistry, as well as in phase transfer catalysis or related methods.

The other two contributions to mass transfer arise from forces on the molecule considered. When the molecule or particle is charged and an external electrical field is applied, electrostatic forces are effective and a positive molecule "descends" the electrical potential gradient, that is, goes in the direction of lower potentials, whereas a negative ion "climbs" the potential gradient. Because of frequent shocks and bouncing against solvent molecules, the molecule reaches a limiting velocity proportional to the potential gradient times its charge z, the proportionality coefficient γ depending on the molecules shape and size, as well as on the viscosity of the medium. This phenomenon, designated migration, then corresponds to a flux J_{migr} given in Eq. (129),

$$J_{migr} = -\gamma \ zF \frac{\partial \Phi}{\partial x} C \tag{129}$$

where x is the coordinate normal to the elementary surface crossed, $\partial\Phi/\partial x$ the electrical potential gradient along the x axis, and C the local molecule concentration. Φ can have an external origin as in electrophoresis or can be the result of local changes in the distribution of other charges. In the presence of a sufficient excess of supporting electrolyte, the latter contribution is shown to be extremely small [106].

The third contribution is called diffusion, but a better name, although not used, would be "spreading." Indeed, it corresponds to the natural tendency of nonrigid objects to spread so that the object's constituents are leveled. This is well known in any field of chemistry and particularly disturbing in chromatography, since it is responsible for the enlarging of peaks with retention times. The physical origin of this phenomenon is easily understood when considering the chemical potential μ of a solute (for simplicity, concentrations and activities are assumed equal) given in Eq. (130), where μ^0 is the standard value. Thus when C is not uniformly distributed along the x axis, one obtains the chemical potential gradient in Eq. (131).

$$\mu = \mu^0 + RT \ln C \tag{130}$$

$$\frac{\partial \mu}{\partial x} = RT \frac{\partial \ln C}{\partial x} = \frac{RT}{C} \frac{\partial C}{\partial x} \tag{131}$$

This potential gradient corresponds to a force that acts on the molecule, as in an electrical potential gradient. Again this gives the molecule a limiting velocity oriented in the direction opposite to the gradient, that is, toward the lesser values of μ, and proportional to the force $-\partial\mu/\partial x$. Because of the same collisions and bouncing effects as in the migration case, the proportionality coefficient is γ, identical to that in Eq. (129). Thus the diffusion flux J_{dif} is given in Eq. (132a):

$$J_{dif} = -\gamma \frac{\partial \mu}{\partial x} C = -\gamma RT \frac{\partial C}{\partial x} \tag{132a}$$

In practice one prefers to reformulate Eqs. (129) and (132a) by introducing the diffusion coefficient $D = \gamma RT$ of the molecule in the medium considered; that is,

$$J_{dif} = -D \frac{\partial C}{\partial x} \tag{132b}$$

which is commonly designated as Fick's first law.

In the most general situation all three mass transfer modes cooperate algebraically; thus the number of moles crossing a surface of unit area per unit of time is given in Eq. (133), which constitutes an extension of the Fick's first law.

$$J(x) = J_{conv} + J_{migr} + J_{dif} = Cv(x) - CD\frac{zF}{RT}\frac{\partial \Phi}{\partial x} - D\frac{\partial C}{\partial x} \tag{133}$$

Relationships Between the Electrochemical Current and Mass Transfers

On the basis of the dichotomous representation of the solution near the electrode surface, Eq. (133) simplifies in the two domains. Within the stagnant layer and in the presence of an excess of supporting electrolyte, the two first terms are negligible and one obtains Eq. (134). Conversely, when $x > \delta_{conv}$, the solution is macroscopically homogeneous; the diffusional contribution then vanishes, and the flux is given in Eq. (135).

$$0 < x < \delta_{conv}: \quad J(x) = -D\frac{\partial C}{\partial x} \tag{134}$$

$$x < \delta_{conv}: \quad J(x) = C\left(v(x) - D\frac{zF}{RT}\frac{\partial \Phi}{\partial x}\right) \tag{135}$$

Equation (135) allows one to discuss more specifically the problem of ohmic drop. Indeed, the current density at any point in the bulk solution is given by addition of the fluxes of all ionic species times their charge:

$$\frac{i}{FA} = -\sum Z_j J_j(x) = -v(x)\sum Z_j C_j + \frac{F}{RT}\frac{\partial \Phi}{\partial x}\sum Z_j^2 C_j D_j$$

Because of electroneutrality, the first summation term on the right-hand side of this equation vanishes, which yields Eq. (136):

$$d\Phi = i\frac{RT}{\sum (Z_j^2 F^2 C_j D_j)} \frac{dx}{A} \tag{136}$$

The latter gives the potential drop $d\Phi$ associated with the current i through a space element of cross section A and thickness dx. By comparison with Ohm's law and definition of the resistivity it is seen that the resistivity ρ of the solution is given by Eq. (137). The ensuing resistance R_Ω for a solution of cross section A and length ℓ is obtained in Eq. (138).

$$\rho = \frac{1}{\sum[(Z_jF)^2C_j(D_j/RT]} \qquad (137)$$

$$R_\Omega = \frac{\rho \ell}{A} \qquad (138)$$

Such an expression justifies a posteriori the previous discussion of the ohmic drop effect in electrochemical cells (Sec. III.A, Resistive effects and ohmic drop). Indeed, for a given resistance of the solution in Eq. (138), the ohmic drop across the cell is expressed as in Eq. (139):

$$\Phi_\Omega = R_\Omega i \qquad (139)$$

Thus the larger the concentration of dissociated ions, the smaller the ohmic drop. This in turn depends on the nature and affinity of the solvent/supporting electrolyte system, as discussed earlier and in Chap. 6.

Let us now discuss more precisely the relationship between current and mass transfer at the electrode surface. From Eq. (134) it is seen that the flux of electroactive species is given, in the absence of migration, by Eq. (140), at the electrode surface (x = 0).

$$J_{x=0} = -D\left(\frac{\partial C}{\partial x}\right)_{x=0} \qquad (140)$$

$$i = -nFAJ_{x=0} = nFAD\left(\frac{\partial C}{\partial x}\right)_{x=0} \qquad (141)$$

When there is no accumulation of material at the electrode surface, this flux corresponds to an exchange of n electrons per molecule, that is, to a current given in Eq. (141).* Thus for the electrochemical reaction in Eq. (142), the reactant flux is associated with an identical, but opposite sign, product flux; that is,

$$R + ne \rightleftharpoons P \qquad (142)$$

$$i = nFAD_R\left(\frac{\partial[R]}{\partial x}\right)_{x=0} = -nFAD_P\left(\frac{\partial[P]}{\partial x}\right)_{x=0}$$

Both fluxes must also correspond to the rate of electron exchange at the electrode surface. Since this latter has already been shown to be given by the Butler-Volmer rate law in Eq. (109), the current must fulfil simultaneously the set of the three Eqs. (143) through (145).

$$i = nFAk^0e^{-\alpha nF[E-E^\circ]/RT}([R]_{x=0} - [P]_{x=0}e^{nF(E-E^\circ)/RT}) \qquad (143)$$

$$i = nFAD_R\left(\frac{\partial[R]}{\partial x}\right)_{x=0} \qquad (144)$$

*Note that a minus sign is used so that cathodic currents (n > 0) are positive and anodic currents (n < 0) are negative according to electrochemical conventions.

$$i = -nFAD_P \left(\frac{\partial [P]}{\partial x} \right)_{x=0} \tag{145}$$

Comparison of Eqs. (143) and (144) for example, allows us to discuss more quantitatively the relationship between mass transfer rate and reversibility or irreversibility of an electron transfer step (see Table 4). Indeed, by equalizing both equations, one readily obtains Eq. (146) for E = E°.

$$1 - \frac{[P]_{x=0}}{[R]_{x=0}} = \frac{D_R(\partial [R]/\partial x)_{x=0}}{k^0 [R]_{x=0}} \tag{146}$$

From this simple equation it is seen that when k^0 is much larger than the flux at the electrode, the right-hand side tends to zero, and $[P]_{x=0} \approx [R]_{x=0}$, which corresponds to the fulfilment of a Nernst equation at the electrode surface for E = E°. Conversely, when k^0 is much smaller than the demanded flux, one obtains $[P]_{x=0} > [R]_{x=0}$ or $[P]_{x=0} < [R]_{x=0}$, according to the flux sign. Thus to obtain $[P]_{x=0} \approx [R]_{x=0}$, one must impose an electrode potential E, such as

$$(1 - e^{nF[E-E^\circ]/RT})e^{-\alpha nF(E-E^\circ)/RT} = \frac{D_R(\partial [R]/\partial x)_{x=0}}{k^0 [R]_{x=0}} \tag{147}$$

that is, such that $n(E - E^\circ) \ll 0$, which is highly cathodic with respect to E° for a reduction and highly anodic versus E° for an oxidation.

The Microscopic Origin of the Diffusion Coefficient: Mass Transfer Rates and Diffusion Layer Thickness

Before concluding this section devoted to Fick's first law, we discuss the microscopic origin of the ubiquituous diffusion coefficient D [compare Eqs. (133) through (137), (140) through (142), and (144) through (147)]. In the preceding presentation, D was introduced as $D = \gamma RT$, where γ is the proportionality constant between the velocity of the molecule and the forces acting on it, based on semimacroscopic considerations.

In the absence of any applied force, a molecule is nevertheless moving because of thermal agitation. Yet owing to frequent collisions with the surrounding molecules, free motions occur only over atomic distances. The overall molecular movement then resembles a random succession of linear segments of approximately identical length ℓ, and duration τ. Over a time period $t = m\tau$, the average square displacement $\bar{\Delta}^2$ is determined via statistical equations to be $\bar{\Delta}^2 = m\ell^2 = (\ell^2/\tau)t$ [107]. Let us now suppose that a force is applied to the particle arising, for example, from a concentration gradient along the x axis. Consider a plane perpendicular to the x axis at x = 0, as shown in Fig. 18a. During a time of t seconds, a "random-walking" molecule covers a mean distance $L = (\bar{\Delta}^2)^{1/2}$. Thus those molecules from the negative axis region that may cross the plane at x = 0 during the time duration t originate from a volume element, adjacent to the plane, of thickness L. Let $C(-L/2)$ be the average concentration of molecules within this element. Similarly, those that may also cross the plane but originate from the positive axis pertain to an identical volume element of thickness L in which the average concentration is $C(+L/2)$. Since movements in a posi-

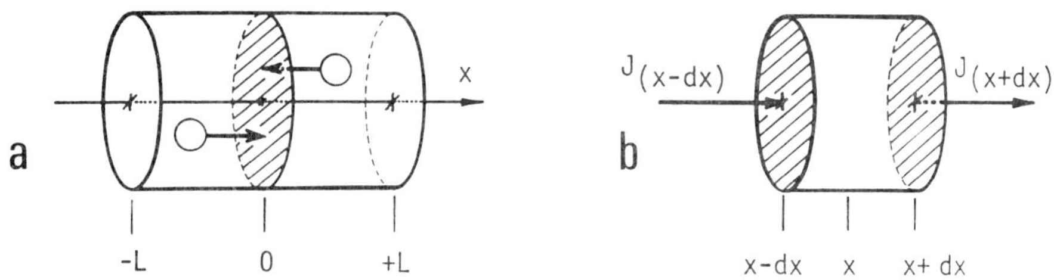

Figure 18. Symbolic representations of (a) the mass fluxes through an unit area surface, and (b) the "entering and outgoing" mass fluxes in an elementary volume of solution of unit area cross section (see text).

tive or negative direction are equally probable, only half the molecules contained in each of these elements of solution cross the plane. Thus a unit surface area of the plane at $x = 0$ is crossed by $\frac{1}{2}LC(-L/2)$ molecules moving toward positive values of x and $\frac{1}{2}LC(L/2)$ molecules moving toward negative values. The flux of molecules, that is, the balanced number of molecules that have crossed the unit surface area of the plane at $x = 0$ per unit of time and are lost (algebraically) from the layer on the negative axis side, is then

$$J_{x=0} = \frac{\frac{1}{2}LC(-L/2) - \frac{1}{2}LC(L/2)}{t}$$

Noting that $[C(L/2) - C(-L/2)]/L$ is the concentration gradient at $x = 0$, that is $(dC/dx)_{x=0}$, this equation is rewritten as

$$J_{x=0} = -\frac{L^2}{2t}\left(\frac{dC}{dx}\right)_{x=0}$$

which, owing to the definition of L, yields finally

$$J_{x=0} = -\frac{\ell^2}{2\tau}\left(\frac{dC}{dx}\right)_{x=0}$$

and thus

$$D = \frac{\ell^2}{2\tau} \qquad (148)$$

Comparison of this equation with Eq. (132b) or (134) allows the microscopic formulation of the diffusion coefficient D in Eq. (148), where ℓ is the average free motion length and τ the average time duration between two collisions. Besides its intrinsic value, such a result allows one to estimate how far a molecule has moved over an average period of time t. Indeed, from the expressions of L and D, one readily obtains Eq. (149):

$$L = (2Dt)^{1/2} \qquad (149)$$

This equation provides a very convenient means to estimate, for example, the average distance covered by a molecule during a time interval t. In other words, a molecule generated at time zero at an electrode reaches a distance of approximately $\delta = (2Dt)^{1/2}$ after a time t has elapsed, provided its movements are controlled only by diffusion. For this reason, δ is usually called the diffusion layer thickness. This figure has an extreme importance in discussing the nature of mass transfer at an electrode. Indeed, it is easily understood that when $\delta < \delta_{conv}$, the concentrations of the species generated or consumed at an electrode have their bulk values before reaching a solution region in which convection is involved. This corresponds to the shortest times of electrochemical experiments, since taking into account Eq. (149), the preceding inequality yields $t < t_{max} = \delta^2_{conv}/2D$. All the mass transfer processes at the electrode are then governed by diffusion (and possibly, at low excesses of supporting electrolyte, by migration). Conversely, when $t > t_{max}$, only an extremely small part of the molecular trajectories is controlled by diffusion, convection being the main means of mass transfer. As is elaborated in the following, this simple notion is extremely important in discriminating transient electrochemical techniques ($t \leq t_{max}$) from steady-state techniques ($t > t_{max}$), that is, diffusion-controlled from convection-controlled mass transfer.

A second important property of Eq. (149) is that it provides an estimate of the rate, in terms of a characteristic time Θ, associated with mass transfer. Indeed, this is the time Θ needed for a molecule to reach the electrode, that is, to cover the space interval in which the molecular concentration differs from that in the bulk. In transient methods this time is identical to that elapsed since the beginning of the experiment, provided that it is lower than $t_{max} = \delta^2_{conv}/2D$. For steady-state methods, the length to be covered is δ_{conv}, and thus from Eq. (149) it follows that $\Theta = \delta^2_{conv}/2D$. The rate of mass transfer can be defined as $1/\Theta$, since it is obviously equivalent to a first-order process (see Chap. 4 for a demonstration of this point). Yet in light of the previous discussion, it is preferable to think in terms of a characteristic time Θ associated with a given electrochemical method rather than in terms of mass transfer rate, although this intuitive latter notion was extremely worthwhile up to this point.*

Electrochemical Homogeneous Kinetics: Fick's Second Law

Fick's second law is based upon mass conservation at any point in time and space. To formulate this point more precisely, let us consider that the spatial distribution of the concentration C of the species of interet depends on a single direction x. † Consider now a cylindrical space element of length 2 dx and cross-sectional area A, centered at x, and let us designate C(x, t) the average concentration at time t within this element, as shown in Fig. 18b.

---

*Note also that all the preceding discussions immediately transpose when comparing the half-life times $t_{1/2}$ associated with chemical reactions to Θ. In our opinion this is more satisfactory since it avoids the necessity of defining a rate constant associated with mass transfer processes, which are essentially of a physical, not a chemical, nature.

†Note that, as for Fick's first law, all the demonstrations easily adapt to a three-dimensional space. Yet since for most electrochemical methods the linear space approximation, that is, depending only on the normal distance from the electrode, is adequate, we restrict ourselves to this simpler case in this presentation.

During a time interval dt, an algebraic flux J[x − dx, t], enters the volume element through its boundary at x − dx.* Similarly, an algebraic flux J[x + dx, t] leaves the element through the boundary at x + dx. During the same time interval, $\underline{K_p}$ dt moles of the species have been produced, and $\underline{K_c}$ dt have been consumed via chemical reactions within the volume element. Thus the number of moles of the species varies by dN, in Eq. (150), during the time duration dt.

$$dN = J(x - dx, t)A\,dt - J(x + dx, t)A\,dt + \underline{K_p}\,dt - \underline{K_c}\,dt \tag{150}$$

Dividing the dN expression by the volume of element, 2A dx, gives the concentration variation; that is

$$\frac{\partial C}{\partial t} = -\frac{J[x - dx, t] - J[x + dx, t]}{2\,dx} + \frac{\underline{K_p}}{2A\,dx} - \frac{\underline{K_c}}{2A\,dx}$$

The first term on the right-hand side is the derivative of the flux vis-à-vis x, whereas $\underline{K_p}/2A$ dx and $\underline{K_c}/2A$ dx relate to the concentration variations from chemical origin, denoted $(\delta C/\delta t)_{chem}$.

$$\frac{\partial C}{\partial t} = -\frac{\partial J}{\partial x} + \left(\frac{\partial C}{\partial t}\right)_{chem} \tag{151}$$

Then Eq. (151) simply relates the overall time dependence of the concentration to the algebraically additive mass transfer and chemical components. In this respect it is important to point out that the chemical term is identical to that obtained under homogeneous conditions for the same chemical sequence. Thus Eq. (151) constitutes a generalization of the usual kinetic rate laws to conditions in which concentrations are not homogeneous. Although developed here in the context of electrochemical techniques, it is valid in any kind of chemical situation in which concentrations are not uniform, as near any heterogeneous reactant or phase-transfer boundaries.

Under electrochemical conditions, that is, near an electrode surface, two limiting formulations are obtained for Eq. (151). For distances from the electrode larger than δ_{conv}, the solution is macroscopically homogeneous. Thus $\partial J/\partial x = 0$, and Eq. (151) simplifies to the usual kinetic rate law in Eq. (152).

$$x > \delta_{conv}: \qquad \frac{\partial C}{\partial t} = \left(\frac{\partial C}{\partial t}\right)_{chem} \tag{152}$$

$$0 < x < \delta_{conv}: \qquad \frac{\partial C}{\partial t} = D\frac{\partial^2 C}{\partial x^2} + \left(\frac{\partial C}{\partial t}\right)_{chem} \tag{153}$$

Conversely, within the stagnant layer and in the presence of a sufficient excess of supporting electrolyte, the flux is given in Eq. (134). Introduction of this latter figure in Eq. (151) yields finally Eq. (153), which is also commonly designated as Fick's second law for diffusion-reaction.

---

*Note that the terms "enter" and "leave" are used because fluxes are oriented with respect to the x axis. Thus J > 0 through a surface corresponds to a flux oriented toward positive x values, and the converse is also true.

In the derivation of Eq. (152), we neglected the quantity of the species produced or consumed at the electrode. In practice this is true for microscale electrochemical experiments but is necessarily inexact for macroscale or exaustive electrolysis. To derive the general equation, let us consider the total equation for the bulk solution. The chemical contribution is given in Eq. (152). On the other hand, a flux of the species at the convection layer limit δ_{conv} must be considered. This flux corresponds to an algebraic number dN_{elec} of moles of the species produced, given in Eq. (154),

$$dN_{elec} = (J(x = \delta_{conv}^{-},t))\ A\ dt \tag{154}$$

where A is the electrode surface area.* Dividing both members of Eq. (154) by the volume V of the bulk solution allows the reformulation of Eq. (154) as

$$\left(\frac{\partial C}{\partial t}\right)_{elec} = -\frac{DA}{V}\left(\frac{\partial C}{\partial x}\right)_{x=\delta_{conv}^{-}} \tag{155}$$

Thus the overall concentration rate law in the solution is finally expressed as in Eq. (156) in a general case.

$$x > \delta_{conv}: \quad \frac{\partial C}{\partial t} = -\frac{DA}{V}\left(\frac{\partial C}{\partial t}\right)_{x=\delta_{conv}^{-}} + \left(\frac{\partial C}{\partial t}\right)_{chem} \tag{156}$$

The absolute value of the first term on the right-hand side of Eq. (156) is at maximum equal to $(DA/V)C^0/\delta_{conv}$, where C^0 is the bulk concentration of the electroactive reactant (Sec. IV.B). The electrode comsumption or production is then equivalent to pseudo-first-order kinetics whose apparent rate constant is at maximum equal to

$$k_{elec} = \frac{DA}{V\ \delta_{conv}} \tag{157}$$

To evaluate k_{elec}, let us consider the following average figures: $D = 10^{-5}$ $cm^2\ s^{-1}$, $\delta_{conv} = 10\ \mu m = 10^{-3}$ cm, and cell volume = 100 cm^3. Then $k_{elec} = 10^{-4}$ A, when expressed s^{-1} and A is expressed in cm^2. For microscale electrochemical experiments, A is of the order of few square millimeters, that is, a few hundredths of a square centimeter; thus $k_{elec} \approx$ $10^{-6}\ s^{-1}$. This means that several hours of continuous electrolysis are needed to significantly perturb (>1%) the bulk solution. This is the reason this term is neglected in practice, Eq. (152) being used instead of the more rigorous Eq. (156). Obviously, when preparative-scale electrolysis is considered, one wants to increase k_{elec} in Eq. (157), that is, use large A/V ratios and small δ_{conv} by increasing the electrode surface area and convection within the cell.

Owing to the organization of this book, preparative-scale electrolysis is discussed separately in Chap. 4. Thus in the following we assume that Eq. (152) represents the dependence on time of the bulk solution concentration. In other words, the electrochemical behavior, in the absence of migrational

---

*The minus superscript in δ_{conv}^{-} means that the flux is that within the diffusion layer, that is, the limit of J(x), when x tends to δ_{conv}, from lower values.

contributions, is given by a set of equations analogous to that in Eq. (158), with boundary conditions at $x = 0$, depending on the electrode potential and kinetics [compare Eqs. (143) through (145)] and at $x = \delta_{conv}$, $C = C^b$, where the bulk concentration C^b is given as a function of time by Eq. (159)*:

$$0 < x < \delta_{conv}: \qquad \frac{\partial C}{\partial t} = D\,\frac{\partial^2 C}{\partial x^2} + \left(\frac{\partial C}{\partial t}\right)_{chem} \tag{158}$$

$$x > \delta_{conv}: \qquad C = C^b \quad \text{such as,} \quad \frac{dC^b}{dt} = \left(\frac{dC^b}{dt}\right)_{chem} \tag{159}$$

Dimensionless Formulation of Electrochemical Equations

The dimensionless analysis of equations is an analytic mathematical technique of frequent use in engineering. Indeed, this technique allows one to "concentrate" all the a priori independent parameters resulting in an identical effect into a single dimensionless parameter. To illustrate the principle and operational interest of the method, let us consider a general second-order homogeneous reaction, such as that in Eq. (160), the reactant concentration being C^0 at the beginning ($t = 0$) of the experiment:

$$2R \longrightarrow P \qquad (k) \tag{160}$$

Strict application of homogeneous chemical kinetics readily yields the rate law in Eq. (161), which governs the reactant concentration, as a function of time. Integration of this rate law gives the classic expression in Eq. (162) for an experiment of duration θ. Yet Eq. (162) may be rewritten under the equivalent form in Eq. (163).

$$\frac{d[R]}{dt} = -\,2k[R]^2 \tag{161}$$

$$\frac{1}{[R]} = \frac{1}{C^0} + 2k\theta \tag{162}$$

$$\frac{[R]}{C^0} = \frac{1}{1 + 2kC^0\theta} \tag{163}$$

In many respects, the formulation in Eq. (163) is more interesting than that in Eq. (162). Indeed, its left-hand side gives the instantaneous fraction of the reactant not converted to the product at any time θ. Chemically this is usually more significant than the actual concentration. Its right-hand side shows that this fraction depends on a single factor (or parameter) $2kC^0\theta$, which includes the effect of the a priori three independent factors: an intrinsic factor k and two experimental parameters C^0 and θ. Just at a

---

*Note that when a figure C depends on several independent variables x and t, its variations vis-à-vis one of these variables are given by its partial derivative $\partial C/\partial x$ with respect to the variable. When C depends on a single variable, the usual derivation is used: dC/dx. The intrinsic difference between the two kinds of derivations is indicated by the symbolic use of ∂ or d in the derivative notations.

glance it shows that doubling the concentration, for example, results in the same effect as keeping the same concentration and doubling the duration of the experiment, or increasing the temperature, so that k is doubled.

In this example the dimensionless reactant fraction $r = [R]/C^0$ and dimensionless rate parameter $\lambda = 2kC^0\theta$ were introduced by reformulating the integrated rate law in Eq. (162). However, this could have been done directly on the rate law in Eq. (161) by introducing a dimensionless time $\tau = t/\theta$ (i.e., the elapsed fraction of the overall experiment duration θ) and the reactant fraction r defined earlier. Then Eq. (161) reformulates as in Eq. (164a) and is associated with the initial condition in Eq. (164b). Integration of Eq. (164a) owing to the initial condition, yields readily Eq. (165), which is identical to Eq. (163).

$$\tau > 0: \quad \frac{dr}{r^2} = -\lambda\, d\tau \tag{164a}$$

$$\tau = 0: \quad r = 1 \tag{164b}$$

$$\tau = 1: \quad r = \frac{1}{1 + \lambda} \tag{165}$$

The interest of this second approach, that is, dimensionless analysis of the rate law itself, becomes obvious for cases in which the integration of the rate law is not achievable analytically and must be performed numerically. Indeed, the set of Eqs. (164) shows that the variable of interest, $r = [R]/C^0$, depends only on $\lambda = 2kC^0\theta$. Thus numerical integration must be performed only for a selected number of λ values to obtain a single curve giving r as a function of λ, that is, as a function of all factors k, C^0, and θ affecting the conversion yield. If one says that 10 different values of λ, for example, are needed to determine the curve with the required accuracy, the process needs 10 calculations. Yet without dimensionless analysis the same information with the same accuracy would be contained in at least 100 curves, that is, [R] as a function of θ, for 10 selected values of C^0 and for each 10 values of k, which would need 1000 calculations. Besides the intrinsic interest of having all the information contained in a single curve rather than in a hundred, the method is essentially time saving. For example, consider that a numerical integration requires 5 min; the whole dimensionless curve needs about 1 h calculation time, whereas the normal or direct method requires at least 100 h!

Application to Electrochemical Systems. As explained in the previous parts of this section, two figures tightly control electrochemical kinetics: a spatial one δ, the thickness of the diffusion layer, and a temporal one θ, the time elapsed since the beginning of the electrochemical perturbation. Thus in most situations these figures constitute adequate and suitable references for the definitions of the dimensionless space $y = x/\delta$ and time $\tau = t/\theta$. Similarly, in most electrochemical kinetic experiments, the bulk concentration C^0 of the starting material remains constant (see, however, Chap. 4 for preparative electrochemistry). It follows that $s = [S]/C^0$ constitutes a meaningful dimensionless concentration for any species S whose real concentration is [S]. Another obvious dimensionless parameter concerns the potential figures. Indeed, in nearly all electrochemical circumstances, the electrode potential per se is not important but influences the electrochemical behavior through its difference from E° times nF/RT, [Compare, e.g., the

Nernst equation of the Volmer-Butler rate law in Eq. (143).] Thus definition of the dimensionless potential as $\xi = (nF/RT)(E^\circ - E)$ appears suitable. y, τ, s, and ξ constitute the master dimensionless parameters from which the definition of all others ensues, as is shown in the following.

Let us consider first the electrode boundary conditions in Eq. (144) or (145). Introduction of $r = [R]/C^0$ and $y = x/\delta$ in Eq. (144), for example, directly affords Eq. (166):

$$i = \frac{nFADC^0}{\delta}\left(\frac{\partial r}{\partial y}\right)_0 \tag{166}$$

where the subscript 0 is used to indicate "at y = 0," which shows that a convenient dimensionless definition for the current is $\psi = i/(nFADC^0/\delta)$, which yields Eq. (167):

$$\psi = \left(\frac{\partial r}{\partial y}\right)_0 \tag{167}$$

$$\psi = \frac{k^0\delta}{D}e^{\alpha\xi}(r_0 - p_0e^{-\xi}) \tag{168}$$

$$\psi = \Lambda e^{\alpha\xi}(r_0 - p_0e^{-\xi}) \tag{169}$$

Introduction of r, $p = [P]/C^0$, $\xi = nF/RT(E^\circ - E)$, and ψ in the Volmer-Butler rate law [Eq. (143)] readily yields Eq. (168). The latter shows that a convenient dimensionless rate of electron transfer $\Lambda = k^0\delta/D$, since it compares the intrinsic value of the rate constant to that of the mass transfer process. Thus Eq. (168) reformulates as Eq. (169). Let us now examine the time- and space-dependent partial derivative equation of the kind demonstrated by Eq. (158), which describes variations in the concentration profiles in the stagnant layer adjacent to the electrode. For any species S, introducing τ, y, and s leads to reformulation of Eq. (158) as in Eq. (170):

$$0 < y < \frac{\delta_{conv}}{\delta}: \quad \frac{C^0}{\Theta}\frac{\partial s}{\partial t} = \frac{DC^0}{\delta^2}\frac{\partial^2 s}{\partial y^2} + \left(\frac{\partial [S]}{\partial \tau}\right)_{chem} \tag{170}$$

To proceed further it is necessary to define the nature of the chemical production or consumption terms in $(\partial[S]/\partial\tau)_{chem}$. For that, let us consider the general term $\underline{K_j}$ in Eq. (171),

$$\left(\frac{\partial [S]}{\partial \tau}\right)_{chem} = \sum \underline{K_j} \quad \text{with} \quad \underline{K_j} = \pm k_j[S]^{\alpha j}\,\pi[Z_1]^{\beta 1} \tag{171}$$

$$\underline{K_j} = \pm\lambda_j[s^{\alpha j}\,\pi z_1^{\beta 1}]\frac{C^0 D}{\delta^2} = \tilde{\lambda}_j\frac{C^0 D}{\delta^2} \tag{172}$$

where the + or − sign corresponds to an S production or consumption, respectively. Thus $\underline{K_j}$ reformulates as in Eq. (172), where the dimensionless rate constant λ_j is defined* as $\lambda_j = k_j(\delta^2/D)(C^0)^{\alpha j-1+\Sigma\beta 1}$. Introduction of these expressions in Eq. (170) yields, after multiplication of both members by δ^2/DC^0,

$$0 < y < \frac{\delta_{conv}}{\delta}: \qquad \frac{\delta^2}{D\Theta}\frac{\partial s}{\partial \tau} = \frac{\partial^2 s}{\partial y^2} + \sum \tilde{\lambda}_j \tag{173}$$

The formulation in Eq. (173) shows that the dimensionless parameter $\delta^2/D\Theta$ plays a crucial role in the time dependence of the S concentration profile. This parameter compares the diffusion layer thickness δ to $(D\Theta)^{1/2}$. On the other hand, we have discussed the dependence of δ on the duration of the experiment. Thus from Eq. (149), $\delta \approx (D\Theta)^{1/2}$. provided that $\delta \ll \delta_{conv}$, that is, that $\Theta \ll \delta_{conv}{}^2/D$. Then the diffusion layer extends within a small fraction of the stagnant layer and Eq. (173) reformulates as in Eq. (174),

$$0 < y < \frac{\delta_{conv}}{(D\Theta)^{1/2}} \approx \infty: \qquad \frac{\partial s}{\partial \tau} = \frac{\partial^2 s}{\partial y^2} + \sum \tilde{\lambda}_j \tag{174}$$

which shows that the concentration profile of S depends on the space as well as on the time. For this reason the corresponding electrochemical methods are said to be transient. Note also that since $\Theta = \delta^2/D$ under these conditions, the dimensionless rate constant is then $\lambda_j = k_j\Theta(C^0)^{\alpha j-1+\Sigma\beta 1}$ or $\lambda_j = k_j^{ap}\Theta(C^0)^{\alpha j-1}$.*

Conversely, when $\Theta \gg \delta^2_{conv}/D$, the diffusion layer would extend well beyond the stagnant layer, which is physically impossible. Thus one has $\delta = \delta_{cnv}$, and $\delta^2/D\Theta \approx 0$. On the other hand, $\partial s/\partial\tau$ cannot assume infinite values, except maybe at temporal discontinuities related to the exact electrochemical perturbation considered. Thus, with the exception of these discontinuities, the left-hand side member of Eq. (173) tends to zero. Thus, Eq. (173) reformulates as in Eq. (175),

$$0 < y < 1: \qquad 0 = \frac{\partial^2 s}{\partial y^2} + \sum \tilde{\lambda}_j \tag{175}$$

which shows that a steady-state concentration profile is observed. For this reason the corresponding electrochemical methods are said to be steady-state methods. Note that the expression of the dimensionless rate constants reformulates as

$$\lambda_j = k_j \frac{\delta^2_{conv}}{D}(C^0)^{\alpha j-1+\Sigma\beta 1} \qquad \text{or*} \qquad \lambda_j = k_j^{ap}\frac{\delta^2_{conv}}{D}(C^0)^{\alpha j-1}$$

---

*Note that an alternative formulation, when the $[Z_1]$ concentrations in Eq. (171) are kept constant, consists of including their contribution into the rate constant, which amounts to defining a pseudo-rate constant $k_j^{ap} = k_j\pi[Z_1]^{\beta 1}$. Thus Eq. (171) is rewritten as

$$\underline{K_j} = \pm k_j^{ap}[S]^{\alpha j}$$

which results in the definition of $\lambda_j = k_j^{ap}(\delta^2/D)(C^0)^{\alpha j-1}$. Obviously, any intermediate attitude is possible as a function of the particular experiment dealt with.

It is thus seen that dimensionless analysis of electrochemical equations allows a very straightforward classification of the electrochemical methods into two limiting classes. Transient methods (including linear sweep voltammetry, cyclic voltammetry, chronoamperometries, and chronocoulometries; see Sec. IV.C and Chap. 3) correspond to the shortest perturbation times (usually less than 10 s) and the widest convection layers (usually = 20–50 μm, as imposed by natural convection and vibrations). On the other hand, steady-state methods (e.g., polarography and rotating disk electrodes; see Sec. IV.B and Chap. 3) need δ_{conv} to be as small as possible (usually = 5–10 μm, as imposed by forced convection) and the experiment to be of long enough duration.

The preceding presentation encompassed the widest class of electrochemical methods. As such the reference figures used in the definition of the dimensionless parameters used were not specifically related to a particular method. Yet when a specific electrochemical method is considered, other equivalent definitions may be used for greater convenience. For example, in cyclic voltammetry, Θ is frequently defined as the time elapsed, $\Theta = \Delta E/v$, between two waves separated by a potential difference ΔE at a scan rate v (in V s^{-1}). Yet in linear sweep voltammetry, Θ is usually defined as $\Theta = RT/nFv$, that is, approximately the time needed to describe the rising branch of a voltammogram (see Chap. 3). Table 5 presents the most widely used dimensionless formulations.

Table 5. Dimensionless Variables for Electrochemical Kinetics

Variable		Variable	
Real	Dimensionless	Real	Dimensionless
Time t[a]	$\tau = \frac{t}{\theta}$	Rate constants k:	
Distance x[b]	$y = \frac{x}{\delta}$	First order	$\lambda = \frac{k\delta^2}{D}$
		Second order	$\lambda = k\frac{C^0\delta^2}{D}$
Concentration or activity [S]	$s = \frac{[S]}{C^0}$	ρth order	$\lambda = k(C^0)^{(\rho-1)}\frac{\delta^2}{D}$
Potential E[c]	$\xi = \frac{nF}{RT}(E^{o} - E)$	Heterogeneous k^0	$\Lambda = \frac{k^0\delta}{D}$
Current i[b]	$\psi = i\frac{\delta}{nFADC^0}$	Equilibrium constant[d] K	$\kappa = \frac{K}{(C^0)^m}$

[a]Cyclic voltammetry: $\theta = RT/nFv$; v, scan rate. Staircase methods: θ = duration of the forward pulse.
[b]Steady-state methods: $\delta = \delta_{conv}$. Transient methods, $\delta = (D\theta)^{1/2}$.
[c]E^{o}, standard reduction potential; n, number of electrons exchanged; $n > 0$ for reductions, $n < 0$ for oxidations.
[d]The dimensions of k are considered $mol^m \cdot l^{-m}$.

B. Steady-State Electrochemical Methods: Half-Wave Potential $E_{1/2}$

Pure Electron Transfer Mechanisms

In the following we consider a simple electron transfer mechanism in order to discuss quantitatively the variations in the potential location of the steady-state voltammogram of the system according to the kinetics of the heterogeneous electron transfer. In the derivation of the kinetics we consider that the solution contains only the reactant at concentration C^0 before the electrochemical experiment. Let E^o, k^0, and α be the standard reduction potential, the standard heterogeneous rate constant, and the transfer coefficient of the electron transfer in Eq. (176).

$$R + ne \rightleftharpoons P \qquad (E^o, k^0, \alpha) \tag{176}$$

For most organic electroactive molecules, the variations in shape and solvation upon electron transfer are minimal, owing to the usually large delocalization. It follows that the diffusion coefficients relative to R and P are generally very close to their average value D, which is then used to introduce a simplification in the following derivations.

From the general equation (175), the concentration profiles of R and P are solutions of the following differential equations in dimensionless variables (see Table 5):

$$\frac{d^2r}{dy^2} = 0 \qquad 0 < y < 1 \tag{177}$$

$$\frac{d^2p}{dy^2} = 0 \tag{178}$$

These equations are associated with the boundary conditions

$$\psi = \left(\frac{dr}{dy}\right)_0 = -\left(\frac{dp}{dy}\right)_0 \qquad \text{electrode } (y = 0) \tag{179}$$

$$\psi = \Lambda e^{\alpha\xi}(r_0 - p_0 e^{-\xi}) \tag{180}$$

$$r = 1 \qquad \text{convection layer boundary } (y = 1) \tag{181}$$

$$p = 0 \tag{182}$$

Integration of Eqs. (177) and (178), when taking Eq. (179) into account, readily yields

$$r = r_0 + \psi y \qquad \text{and} \qquad p = p_0 - \psi y$$

Introduction of conditions (181) and (182) into these linear concentration profiles gives the relationships between r_0, p_0, and ψ in Eq. (183). Incorporation of these latter figures into Eq. (180) and reorganization of the latter as a function of the dimensionless current ψ yields the equation of the dimensionless voltammogram in Eq. (184).

$$r_0 = 1 - \psi \qquad \text{and} \qquad p_0 = \psi \tag{183}$$

$$\psi = \frac{1}{1 + e^{-\xi} + (1/\Lambda)e^{-\alpha\xi}} \tag{184}$$

The latter shows that the current reaches two limiting values for $\xi \longrightarrow -\infty$ [i.e., for $n(E - E^\circ) \ll 0$] and $\xi \longrightarrow +\infty$ [i.e., for $n(E - E^\circ) \gg 0$]. Indeed, from Eq. (184) it ensues easily that $\psi \longrightarrow 0$ for $\xi \longrightarrow -\infty$ and $\psi \longrightarrow 1$ for $\xi \longrightarrow +\infty$, which shows that no current is observed when the potential is less cathodic than E° (for a reduction; $n > 0$) or anodic than E° (for an oxidation; $n < 0$). Conversely, at sufficient cathodic ($n > 0$) or anodic ($n < 0$) values of the electrode potential, the current tends to a plateau value ($\psi \longrightarrow 1$) given in Eq. (185):

$$i^{lim} = \frac{nFADC^0}{\delta_{conv}} \tag{185}$$

Of interest is the proportionality of i^{lim} to the bulk concentration of the reactant. Indeed, Eq. (185) demonstrates that the electrochemical current can be used to monitor the bulk concentration of any chemical species, provided the latter is electroactive. This is the basis of a large variety of chemical or biochemical sensors, as well as that of electrochemical detectors in high-performance liquid chromotography (HPLC). From Eq. (183) it is seen that the observance of the current plateau corresponds to the fact that the reactant concentration has been completely depleted ($r_0 = 0$) at the electrode, so that increasing ξ results in no further change in the concentration profiles and therefore of the current. Similarly, the limit at $\psi = 0$ corresponds to $p_0 = 0$; thus decreasing the dimensionless potential results in no further change. The observation of the two limits thus corresponds to the fact that the concentrations of R and P at the electrode are limited by 0 and C^0. This is independent of the electrode kinetics, which explains that the limiting current i^{lim} does not depend on the kinetics of the electron exchange. However, the location of the voltammogram depends on the rate of electron transfer.

When $\Lambda \gg 1$, that is, when $k^0 \gg D/\delta$,* the second term in the denominator in Eq. (184) vanishes and the voltammogram equation simplifies to Eq. (186),

$$\frac{i}{i^{lim}} = \psi = \frac{1}{1 + e^{-\xi}} \tag{186}$$

which is independent of the electron transfer rate. Moreover, introduction of r_0 and p_0, as given by Eq. (183), into Eq. (186) shows that

$$r_0 = p_0 e^{-\xi} \tag{187}$$

that is, in real variables,

$$E = E^\circ + \frac{RT}{nF} \ln \frac{[R]_{x=0}}{[P]_{x=0}}$$

---

*In practice, $D \approx 5 \times 10^{-6}$ cm^2 s^{-1} and $\delta \approx 5$ μm $= 5 \times 10^{-4}$ cm are average values; thus the inequality is satisfied for $k^0 \gg 0.01$ cm s^{-1}.

which demonstrates that the electrode potential obeys the Nernst equation. Thus the electron transfer is sufficiently fast to be controlled thermodynamically. It is then said to be fast or nernstian. At $E = E^o$, $[R]_{x=0} = [P]_{x=0}$, that is, from Eq. (183), $i/i^{lim} = 0.5$. Thus E^o may be determined as the potential for which $i/i^{lim} = 0.5$. This potential, the half-wave potential $E_{1/2}$, thus bears an important thermodynamic significance since it allows the determination of E^o. Indeed, most E^o values determined for organic redox couples have been measured through this or related techniques. Yet it must be emphasized that $E_{1/2}$ involves contributions other than E^o. In a real situation, the activity and diffusion coeficients of R and P may differ, which yields [108]

$$E_{1/2} = E^o + \frac{RT}{nF} \ln\left[\frac{f_R}{f_P}\left(\frac{D_P}{D_R}\right)^{1/2}\right]$$

However, most of the time, the corrective term to E^o is unfortunately unknown and beyond the error associated with potential measurements on an absolute (NHE) scale, mainly because of junction potential and reference electrode potential drifts. Thus in actual experiments $E_{1/2}$ and E^o (or E'^o) are generally considered identical [94].

Let us now consider the converse situation, that in which $\Lambda \ll 1$ (i.e., $k^0 \ll D/\delta$). Then Eq. (184) simplifies to Eq. (188):

$$\psi = \frac{i}{i^{lim}} = \frac{1}{1 + (1/\Lambda)e^{-\alpha\xi}} \tag{188}$$

The half-wave potential, that $E_{1/2}$ at which $i/i^{lim} = 0.5$, is then such that

$$E_{1/2} = E^o + \frac{RT}{\alpha nF} \ln \Lambda = E^o + \frac{RT}{\alpha nF} \ln \frac{k^0\delta}{D} \tag{189}$$

Since $k^0\delta/D \ll 1$, the logarithmic term in Eq. (189) is negative and $n(E_{1/2} - E^o) \ll 0$. The wave is thus located in a potential range well cathodic of E^o for a reduction process ($n > 0$) or anodic of E^o for an oxidation process. Then $E_{1/2}$ represents both thermodynamic (E^o) and kinetic contributions ($\ln k^0\delta/D$) and must not be used as a substitute for E^o.

Figure 19a presents the variations in the steady-state voltammograms for the R/P redox system as a function of $\Lambda = k^0\delta/D$ for α constant and equal to 0.5. The variations in $E_{1/2}$ with the same parameter $k^0\delta/D$ are presented in Fig. 19b for selected constant values of α. Therefore, in an actual experiment, variations in $E_{1/2}$ with δ (note that k^0 and D are intrinsic figures for a given couple R/P and a given medium) are indicative of kinetic complications and should warn against the use of $E_{1/2}$ as an estimate of E^o.

Mechanism Involving a Follow-up Reaction: EC Mechanisms

Let us first consider the case in which the initial electron transfer in Eq. (190) is sufficiently rapid to be nernstian (i.e., $\Lambda = k^0\delta/D \gg 1$ but is followed by an irreversible, pseudo-first-order, chemical reaction (that is, k may include the concentration of other reagent, when constant).

$$R + ne \rightleftharpoons P \qquad (E^o, k^0, \alpha) \tag{190}$$

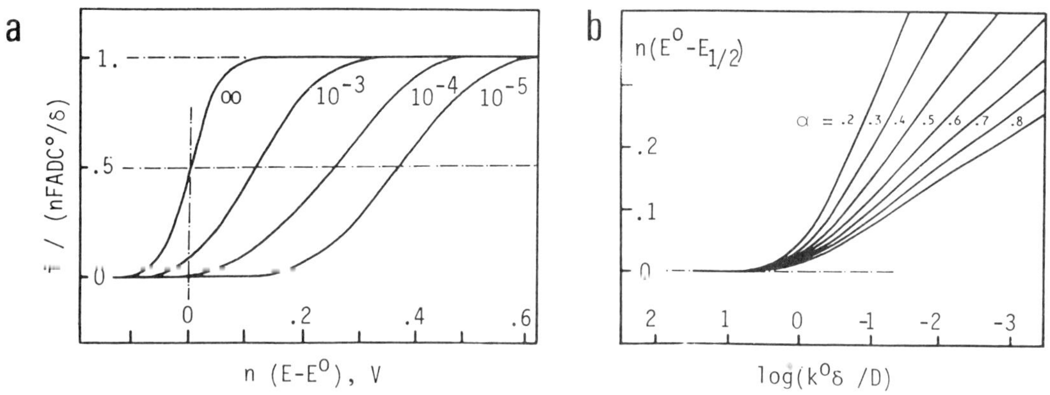

Figure 19. (a) Steady-state voltammograms of a simple redox system as a function of its intrinsic heterogeneous rate constant (number on the curves: k^0, in cm s^{-1}) for $\delta^2/D = 0.1$ s. (b) Variations in the half-wave potential $E_{1/2}$ as a function of the dimensionless rate constant $\Lambda = k^0\delta/D$ for different values of the transfer coefficient α.

$$P \longrightarrow \cdots \qquad (k) \tag{191}$$

The dimensionless equations governing the R and P concentration profiles are obtained from Table 5, as in Eqs. (192) and (193).

$$\frac{d^2r}{dy^2} = 0 \tag{192}$$

$$\frac{d^2p}{dy^2} = \lambda p \tag{193}$$

$$\left.\right\} \quad 0 < y < 1$$

For a nernstian heterogeneous electron transfer, the boundary conditions at the electrode formulate as in Eqs. (194) and (195):

$$\psi = \left(\frac{dr}{dy}\right)_0 = -\left(\frac{dp}{dy}\right)_0 \tag{194}$$

$$r_0 = p_0 e^{-\xi} \tag{195}$$

$$\left.\right\} \quad y = 0$$

[Compare Eq. (187)]. For a bulk solution containing only R, the boundary condition at the extremity of the convection layer is then

$$y = 1: \qquad r = 1, \qquad p = 0 \tag{196}$$

Integration of Eq. (192), taking into account Eqs. (194) and (196), yields Eq. (197):

$$r = r_0 + \psi y, \qquad \text{with} \qquad r_0 = 1 - \psi \tag{197}$$

From this simple equation, it is seen that, as for the preceding situations, the voltammogram has two limiting values. When $r_0 \approx 1$, $\psi \approx 0$, which corresponds to large negative values of ξ according to Eq. (195). Conversely, $\psi \approx 1$ when $r_0 \approx 0$, which is observed for large positive values of ξ as seen from Eq. (195). Thus the overall qualitative shape of the steady-state voltammogram remains identical to that observed in the absence of a chemical follow-up reaction. Yet they differ considerably on a quantitative basis, as shown by the following analysis.

Integration of Eq. (193) yields he dimensionless concentration profile of P in Eq. (198),

$$p = Ae^{(\lambda^{1/2}y)} + Be^{(-\lambda^{1/2}y)} \tag{198}$$

where A and B are two constants. These constants are determined according to the boundary conditions (194) and (196). Indeed, Eq. (194) yields by derivation (and setting y = 0) of Eq. (198):

$$\psi = -\lambda^{1/2}A + \lambda^{1/2}B \tag{199}$$

$$0 = Ae^{(\lambda^{1/2})} + Be^{(-\lambda^{1/2})} \tag{200}$$

whereas Eq. (196) affords Eq. (200) by setting y = 1 in Eq. (198). The system of two linear equations (199) and (200) with two unknowns (A and B) allows A and B to be determined as a function of ψ and λ. Thus,

$$A = \frac{\psi/\lambda^{1/2}}{1 + e^{(2\lambda^{1/2})}} \quad \text{and} \quad B = -Ae^{(2\lambda^{1/2})}$$

$$p_0 = A + B = \frac{\psi \tanh \lambda^{1/2}}{\lambda^{1/2}} \tag{201}$$

Determination of $p_0 = A + B$ from Eq. (198) is thus readily obtained as in Eq. (201). Incorporation of this latter value, as well as $r_0 = 1 - \psi$ from Eq. (197), into the Nernst equation (195), gives access to the equation of the steady-state voltammogram in Eq. (202).

$$\frac{i}{i^{\lim}} = \psi = \frac{1}{1 + e^{(-\xi^*)}} \tag{202}$$

$$\xi^* = \xi + \frac{1}{2} \ln \lambda - \ln \tanh \lambda^{1/2} \tag{203}$$

The latter is identical with that obtained for a nernstian system [compare Eq. (186)] but one in which the potential axis has been translated by a constant term (for a constant value of $\lambda = k\delta^2/D$), as evidenced by Eq. (203). Its half-wave potential then corresponds to $\xi^* = 0$ instead of $\xi = 0$. From Eq. (203) and the definition of ξ in Table 5, one obtains the expression in Eq. (204). Since one always has $\lambda^{1/2} > \tanh \lambda^{1/2}$, the corrective term to E° in Eq. (204) always has the sign of n. Thus the wave is observed in a potential range E_W such that $n(E_W - E^{\circ}) > 0$, that is, anodic to E° for a reduction ($n > 0$) and cathodic to E° for an oxidation ($n < 0$).

$$E_{1/2} = E^{\circ} + \frac{RT}{2nF} \ln \lambda - \frac{RT}{nF} \ln \tanh \lambda^{1/2} \tag{204}$$

When $\lambda \longrightarrow 0$ i.e., $k \ll D/\delta^2$, one obtains $E_{1/2} \approx E^o$, since then $\tanh \lambda^{1/2} \approx \lambda^{1/2}$, resulting in a negligible corrective term in Eq. (204). This means that the chemical reaction in Eq. (191) is too slow vis-à-vis the mass transfer rate to significantly affect the voltammogram. The system is then said to be nernstian and chemically reversible. Conversely, when $k \gg D/\delta^2$, that is, when the chemical reaction is faster than the mass transfer at the electrode, $\lambda \longrightarrow \infty$. Then, because $\tanh \lambda^{1/2} \longrightarrow 1$ for $\lambda^{1/2} \longrightarrow \infty$, Eq. (204) simplifies to Eq. (205):

$$E_{1/2} = E^0 + \frac{RT}{2nF} \ln \frac{k\delta^2}{D} \tag{205}$$

From the latter, it is seen that for n = 1 and room temperature, for example, $E_{1/2}$ shifts anodically by approximately 30 mV per unit of log $k\delta^2/D$. As for the slow electron transfer case, $E_{1/2}$ is therefore not an acceptable substitute for E^o since it incorporates kinetic contributions that then arise from the follow-up chemical reaction.

The variations in $E_{1/2}$ with $\lambda = k\delta^2/D$ result from the fact that when λ increases, the P concentration at the electrode surface decreases, as is apparent in Eq. (201)* and Fig. 20. Thus at any driving force imposed by the value of ξ, one obtains from Eq. (195) for large λ values:

$$r_0 = \frac{\psi}{\lambda^{1/2}} e^{(-\xi)} = \frac{\psi}{(k\delta^2/D)^{1/2}} e^{(-\xi)} \tag{206}$$

which shows that the reactant concentration is considerably less than that, $r_0 = \psi\, e^{(-\xi)}$, obtained at the same potential in the absence of a fast follow-

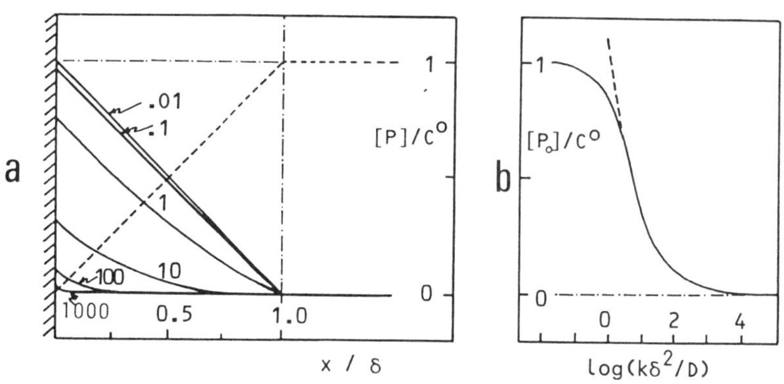

Figure 20. Steady-state electrochemical method. (a) Concentration profiles of the product obtained upon electron transfer in the EC sequence in Eqs. (190) to (191) as a function of the dimensionless chemical rate constant $k\delta^2/D$ (numbers on the solid curves). The reactant concentration is shown for comparison as the dashed line. (b) Variations in the product electrode concentration as a function of $k\delta^2/D$. The dashed curve corresponds to the approximation in Eq. (206).

---

*Indeed from this equation, when $\lambda > 7$, one obtains $p_0 \approx \psi/\lambda^{1/2}$ with an accuracy better than 1%.

up chemical step. Thus the heterogeneous electron transfer is continuously displaced by P removal in a way similar to that already observed for the corresponding homogeneous situation in Eqs. (207) and (208). Yet in the latter case, as elaborated in a previous section [Sec. II.B, Eqs. (45) and (47)], the follow-up chemical step affects the overall kinetics by a factor involving k with a unity exponent, whereas an exponent of one-half is involved in the heterogeneous equivalent case:

$$M + ne \rightleftharpoons N$$

$$N + R \rightleftharpoons M + P \qquad (K) \tag{207}$$

$$P \longrightarrow \cdots \tag{208}$$

[Compare the half-factor in Eq. (205) or the half-exponent in Eq. (206)]. This effect, which arises from the heterogeneous nature of the electrochemical process (i.e., a surface reaction vis-à-vis a volume reaction in homogeneous phases), is the basis of the efficiency of redox catalysis or mediated electron transfer (see Sec. III.E, Classification of coupled chemical reactions and also Chap. 29 mainly devoted to this topic). Thus for a given redox system, as in the sequence in Eqs. (190) and (191), the use of a redox mediator M in Eq. (207) allows the reduction of R to be performed at potentials less cathodic than $E_{1/2}$ in Eq. (205) (or the R oxidation at potentials less anodic than $E_{1/2}$) for the same electrochemical setup (i.e., an identical mass transfer rate).

Chemical Kinetics from Half-Wave Potentials: Determination of Rate Constants and Reaction Orders

The preceding two examples were presented to illustrate the simplicity of electrochemical analysis under steady-state conditions. Indeed, the kinetic derivations are very similar to those encountered in the homogeneous chemical situation, except for the replacement of the usual derivative dC/dt by a second-order derivative vis-à-vis the space variable d^2C/dy^2 as apparent in Eq. (193), for example. The existence of a second-order differentiation introduces unusual dependencies on reaction orders when compared with those observed in homogeneous kinetics. Thus if one considers the rate constant k in the chemical reaction (191) to be a pseudo-first-order rate constant, that is, $k = k_0[Z]$, where Z is a coreagent, the concentration of which is maintained constant, a dependency on $[Z]^{1/2}$ is observed [because of the involvement of k with a power of one-half in Eq. (198)] for electrochemical experiments rather than the dependency on [Z] that would be observed under homogeneous experiments. Let us consider a general case in which the product formed upon electron transfer reacts via a chemical reaction of the ρth order in P, and involving other coreagents A, B, . . . , in excess, as featured in the reaction sequence (209) and (210):

$$R + ne \rightleftharpoons P \qquad (E^{\circ}) \tag{209}$$

$$\rho P + \alpha A + \beta B + \cdots \longrightarrow \cdots \qquad (k_0) \tag{210}$$

One obtains the pseudo-rate constant $k = k_0[A]^{\alpha}[B]^{\beta}$ . . . . Thus dimensionles analysis affords the differential equation in Eq. (211), which describes the concentration profile of the P species when λ is given as in Eq. (212).

$$\frac{d^2p}{dy^2} = \rho\lambda p^{\rho} \tag{211}$$

$$\lambda = k(C^0)^{(\rho-1)} \frac{\delta^2}{D} \qquad k = k_0[A]^{\alpha}[B]^{\beta}\cdots \tag{212}$$

When λ is extremely large and the usual situation in which the bulk solution contains no P is considered, integration of Eq. (211) is performed as follows. Multiplication of both members by 2(dp/dy) dy yields Eq. (213):

$$2\frac{dp}{dy}\frac{d^2p}{dy^2}dy = 2\rho\lambda p^{\rho}\,dp \tag{213}$$

The left-hand term in Eq. (213) is a development of the derivation of $(dp/dy)^2$ vis-à-vis y. The right-hand side is the derivation of $[2\rho\lambda/(\rho + 1)]p^{(\rho+1)}$ versus p. Thus integration of Eq. (213) yields

$$\left(\frac{dp}{dy}\right)^2 = \frac{2\rho\lambda}{\rho + 1}p^{(\rho+1)} + \text{cst} \tag{214}$$

The integration constant in Eq. (214) is determined by the condition $p_0 \approx 0$ when $y \approx 1$. When λ is large this is also equivalent to saying that $dp/dy \approx 0$ when $y \approx 1$, since P exists only in a thin layer (called the kinetic layer) adjacent to the electrode (compare, e.g., Fig. 20a). The integration constant in Eq. (214) is then zero, and one obtains at the electrode surface, for y = 0,

$$\psi^2 = \frac{2\rho\lambda}{\rho + 1}p_0^{(\rho+1)} \qquad \text{since} \qquad \psi = -\left(\frac{dp}{dy}\right)_0 \tag{215}$$

On the other hand, since R is not involved in chemical reactions, one obtains as in Eq. (183), $r_0 = 1 - \psi$. Thus for the nernstian electron transfer in Eq. (209), one obtains Eq. (216) when taking Eq. (215) into account.

$$r_0 = p_0 e^{(-\xi)}$$

$$1 - \psi = \psi^{2/(\rho+1)}\left(\frac{\rho + 1}{2\lambda\rho}\right)^{1/(\rho+1)} e^{(-\xi)} \tag{216}$$

λ being a constant term, it may be included in the exponential factor, which yields the final equation of the steady-state voltammogram in Eq. (217). From this equation it is seen that the current depends only on the apparent dimensionless potential ξ^* given in Eq. (218).

$$\psi + \psi^{2/(\rho+1)}e^{(-\xi^*)} = 1 \tag{217}$$

$$\xi^* = \xi + \frac{1}{\rho + 1}\ln\frac{2\lambda\rho}{\rho + 1}$$

$$= \frac{nF}{RT}(E^{\circ} - E) + \frac{1}{\rho + 1}\ln\frac{2\lambda\rho}{\rho + 1} \tag{218}$$

From the equation $\psi = 1 - r_0$, the limiting plateau current corresponds to $\psi = 1$. Thus the half-wave potential $E_{1/2}$ is obtained by replacing ψ by 1/2 in Eq. (217), which yields

$$E_{1/2} = E^{o} + \frac{RT}{nF}\frac{1-\rho}{1+\rho}\ln 2 + \frac{RT/nF}{\rho+1}\ln\frac{2\rho}{\rho+1} + \left[\frac{RT}{nF}(\rho+1)\right]\ln\left\{k_0\,[A]^{\alpha}[B]^{\beta}\cdots(C^0)^{(\rho-1)}\frac{\delta^2}{D}\right\} \tag{219}$$

Because of the expression of the last term, it is seen that at room temperature, for example, where $RT/F \ln 10 \approx 60$ mV, $E_{1/2}$ varies by $[60\alpha(\rho+1)/n]$, $[60\beta(\rho+1)/n]$, and $[60(\rho-1)(\rho+1)/n]$ mV per unit of log [A], log [B], and log C^0, respectively. These variations then allows an easy determination of the various reaction orders of any follow-up chemical reactions, provided the initial electron transfer is fast enough and the number n of electrons exchanged is known.

Another approach consist of varying a reactant concentration and modifying the mass transfer rate δ in order that $E_{1/2}$ remain constant. The term in braces in Eq. (219) must then be kept constant; this is achieved when δ is equal to $\delta_{1/2}$, as given in Eq. (220):

$$\delta_{1/2} = \text{constant}[A]^{\alpha/2}[B]^{\beta/2}\cdots(C^0)^{(\rho-1)/2}\left(\frac{k_0}{D}\right)^{1/2} \tag{220}$$

Expression (220) shows that the slope of log $\delta_{1/2}$ versus log [A] is $\alpha/2$. Similarly, $\beta/2$ and $(\rho - 1)/2$ are the slopes observed for the variations with log [B] or log C^0, which is a convenient way to determine the reaction orders. Note that although developed here for $E_{1/2}$, this procedure may be used with any chosen potential E_ε such that i/i^{lim} has a fixed value ε. Similarly, variations in k_0/D with the temperature (i.e., activation energy determinations) can be obtained through this procedure, provided care is taken of the RT/F factors, which need to be corrected [compare Eq. (219)] (see, e.g., Chap. 3 for extensions of this method).

All these methods stem from the fact that for a given reaction sequence, such as that in Eqs. (209) and (210), which involve a single rate-determining step, all the kinetics and thus the shape and location of the voltammogram depends only on the dimensionless rate constant parameter λ in Eq. (212). As a result, any modification of the experimental conditions that keep λ constant does not modify the dimensionless voltammogram.* Thus quantitative information on the chemical mechanism [Eq. (210) may be a succession of chemical steps] is obtained without mathematical derivation, but only from dimensionless analysis (compare Chap. 3).

However, determination of the intrinsic rate constant k_0 or of the apparent ρ^{th} order rate constant k requires that the precise mathematical dependence of $E_{1/2}$, for example, on the dimensionless rate constant be determined as in Eq. (219). Indeed, the principle of all methods for determining k_0 or k consists of increasing the mass transfer rate (i.e., here decreasing δ) so that $k\delta^2/D \approx 0$ and thus $E_{1/2} \approx E^{o}$.

These methods constitute the frame on which any particular method can be elaborated. Yet in practice, the experimental difficulty is that with

---

*Yet the variations associated with the temperature may result in additional effects because of the temperature involvement in Eq. (219).

standard apparatus, δ^2/D cannot be varied over an extremely wide range. For example, with the rotating disk electrode (RDE), which is the most convenient steady-state method (with the exception of ultramicroelectrodes [109]), δ depends on the rotation frequency w of the electrode (see Chap. 3). Yet to maintain correct hydrodynamic conditions w cannot be varied, with standard apparatus, outside the range of 10 rotations per second (rps) to 1000 rps, which limits access to fast kinetics.

Transient electrochemical techniques [110] allow this range of investigation to be widely extended (from 1 s to approximately 10 ns). Indeed, the same method may be used in a time domain extending over approximately eight orders of magnitude. Besides this very important point, transient electrochemical techniques provide current-intensity and current-potential patterns that are easily recognizable, for example in cyclic voltammetry. Thus with minimal eye training, a large amount of qualitative or nearly quantitative date may be obtained just by inspection of a transient voltammogram. This is in many respects an important advantage for diagnosis of kinetics, analogous with that of IR, UV, or NMR spectroscopies for structural information. In our opinion this is one of the main reasons that transient electrochemical methods have progressively supplanted steady-state methods. Indeed, as illustrated by the preceding examples, steady-state voltammograms have generally sigmoidal shapes that are difficult to relate to a particular mechanism without quantitative analysis.

C. Transient Electrochemical Methods

Introduction: Time Hysteresis in Current Reversal Techniques

As explained earlier, in transient electrochemical methods an electrical perturbation (potential, current, charge, and so on) is imposed at the working electrode during a time period Θ (usually less than 10 s) short enough for the diffusion layer $\delta \approx (2D\Theta)^{1/2}$ to be smaller than the convection layer δ_{conv} imposed by natural convection. Thus the electrochemical response of the system investigated depends on the exact perturbation as well as on the elapsed time. This duality is apparent when one considers a double-pulse potentiostatic perturbation applied to the electrode as in the double-step chronoamperometric method.

$$R + ne \rightleftharpoons P \qquad (E^0) \qquad (n > 0) \tag{221}$$

Let us consider, for example, the simple nernstian reduction reaction in Eq. (221) and a solution containing initially only the reactant R. Before any electrochemical perturbation the electrode rest potential E_1 is made largely positive to E^o. At time zero the potential is stepped to a value E_2, sufficiently negative to E^o, so that the concentration of R is close to zero at the electrode surface. After a time Θ, the electrode potential is stepped back to E_1, so that the concentration of P at the electrode surface becomes zero. When this potentiostatic perturbation, represented in Fig. 21a, is applied in a steady-state method, the R and P concentration profiles are linear and depend only on the electrode potential but not on time, as shown in Fig. 20a (for $k \approx 0$). Yet when the same perturbation is applied in transient methods, the concentration profiles are curved and time dependent, as evidenced in Fig. 21b. Thus it is seen from this figure that a step duration at E_1, much longer than the step duration Θ at E_2, is needed for the initial concentration profiles to be restored. This hysteresis corresponds to to the propagation of the diffusion perturbation within the solution, which then keeps a "memory" of the past perturbation. This information is stored via the structuring of the concentrations in the space near the electrode as a function of the elapsed time.

The current flowing through the electrode is proportional to the gradient of the R (or the P) concentration profile at the electrode surface. Thus the progressive smoothing of the concentration profiles with the time elapsed after each discontinuity of the perturbation results in a constant decrease of the current with time, as represented in Fig. 21c.

Let us now assume that the product P formed upon the electron transfer in Eq. (221) may react chemically with a half-life $t_{1/2}$. When $\theta \ll t_{1/2}$, almost no P has time to disappear during the time duration of the experiment. The P concentration profile then remains identical to that shown in Fig. 21b, and the current observed for $t > \theta$, which corresponds to P reoxidation, is identical to that in Fig. 21c. Conversely, when $\theta \gg t_{1/2}$, nearly all P molecules produced at the electrode are consumed and the P concentration profile is flat and close to zero, except in the close vicinity of the electrode. As a result almost no oxidation current is observed for $t > \theta$, that is, during the period in which the electrode potential is stepped positive to E^o (see Fig. 21a). In the intermediate range, that is, when θ and $t_{1/2}$ are of similar orders of magnitude (generally $0.1t_{1/2} < \theta < 10t_{1/2}$), intermediate values of the oxidation current are observed as in Fig. 22. Determination of these variations allows the precise determination of the chemical follow-up sequence in which P is involved, as well as of the pertinent rate constants (see Chap. 3).

However, this large sensitivity for mechanistic analysis is earned at the expense of two important factors. First, owing to the large potential variations, the fine dependence of the system on the electrode potential around E^o is not seen. Second, and maybe more important, the current versus time variations have no important visual characteristics. Indeed, as seen

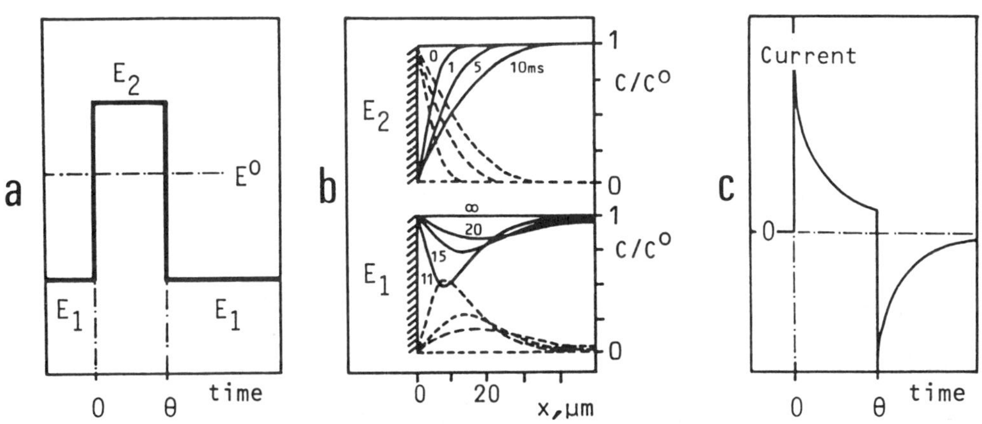

Figure 21. Transient potentiostatic electrochemical perturbation of a simple electron transfer reaction: (a) imposed potential; (b) resulting concentration profiles for the reactant (solid curves) or the product (dashed curves) for a step duration θ = 10 ms at various times from the beginning of the pulse; (c) resulting variations in the current.

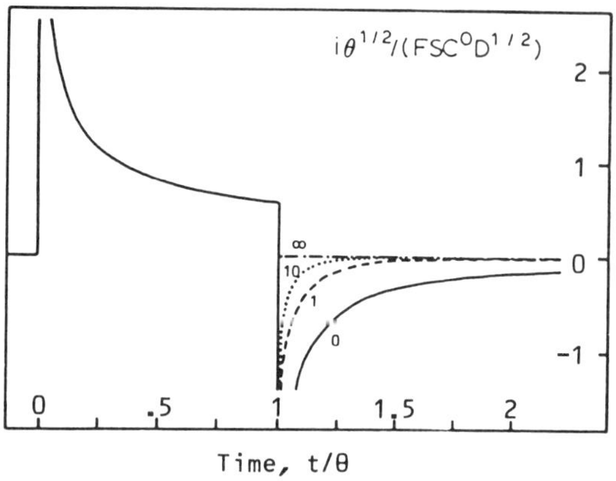

Figure 22. Variations in the current trace in a potentiostatic experiment for an EC sequence for different values of the dimensionless chemical rate constant $k\theta$ (numbers on the curves).

from Fig. 22, for example, all the curves are very similar and only their juxtaposition on the same figure allows their shapes to be compared.

For these two reasons cyclic voltammetry has progressively supplanted potentiostatic or galvanostatic methods in electrochemical kinetic investigations. A more complete description of this method is given in Chap. 3, yet for our purposes here we present the method briefly. It basically consists of applying a linear variation of the electrode potential E with time from a potential E_i, where the reactant R is not electroactive, to a potential E_f sufficient for the reactant concentration to be zero at the electrode surface. At E_f the potential is linearly scanned back to E_i, usually with an identical slope with respect to the time. The latter, v, is designated as the scan rate (in volts per second). Figure 23a presents such a potential-time dependence, together with the corresponding i – E curve* (Fig. 23b) obtained for the reduction process in Eq. (221). Eye inspection of the cyclic voltammogram in Fig. 23b shows the presence of two current peaks, in contrast to the monotonous variations of the current in Fig. 21c. These peaks result from the existence of two opposite effects that successively control the magnitude of the current. For example, for the cathodic scan in Fig. 23, the gradual decrease in R concentration at the electrode surface, when the potential is made more and more cathodic, tends to increase the current by making the R concentration profile steeper. Yet the propagation of the perturbation into the solution, as shown in Fig. 21b, tends to flatten the concentration profile, which results in a progressive decay of the current. In practice it is easily understood that the large variations in the surface concentrations for potentials in the vicinity of E° overwhelm the diffusion effect, the current then tending to rise. Yet when $[R]_{x=0}$ is close to zero, the potential variations affect the current magnitude less than the

*Note that rather than presenting the current variations versus a time axis, a current versus potential plot is preferred. Thus in Fig. 23b, the upper trace corresponds to the forward cathodic scan, whereas the lower trace corresponds to the backward anodic scan.

diffusion propagation in the solution, which explains why the current progressively decays. For the backward scan the same phenomenology applies to the product concentration $[P]_{x=0}$, which then results in the appearance of an inverted peak on the lower trace in Fig. 23b. When P reacts chemically, the magnitude of the reverse peak gradually decreases (Fig. 23c) because there is less and less P present in the solution. Again this allows determination of mechanisms and their pertinent rate constants, as explained for the double-step chronoamperometric method. From this brief presentation it is seen that an important aspect of cyclic voltammetry is that the shapes of cyclic voltammograms are extremely indicative of the chemical processes occurring at the electrode or in the solution. As such it is an extremely useful tool for kinetic diagnosis. Yet the much more complicated shapes, when compared with those in Fig. 22, for example, makes quantitative information on the current more difficult to obtain.

Transient Electrochemical Methods and Chemical Kinetics [110]

From the preceding presentation it is seen that the current variations with time or potential-time are intimately related to the concentration profiles of the species engaged in heterogeneous electron transfers at the electrode surface. Thus any kinetic or thermodynamic perturbation of these concentration profiles results in a variation of the current/time or current/potential-time transient characteristics as outlined in Fig. 22 or 23c. As was also apparent in the preceding discussion, as well as in several other places in this chapter, the degree to which these perturbations affect the concentration profiles is a function of their relative effect vis-à-vis the mass transfer rate. Dimensionless analysis again proves to be a very convenient way

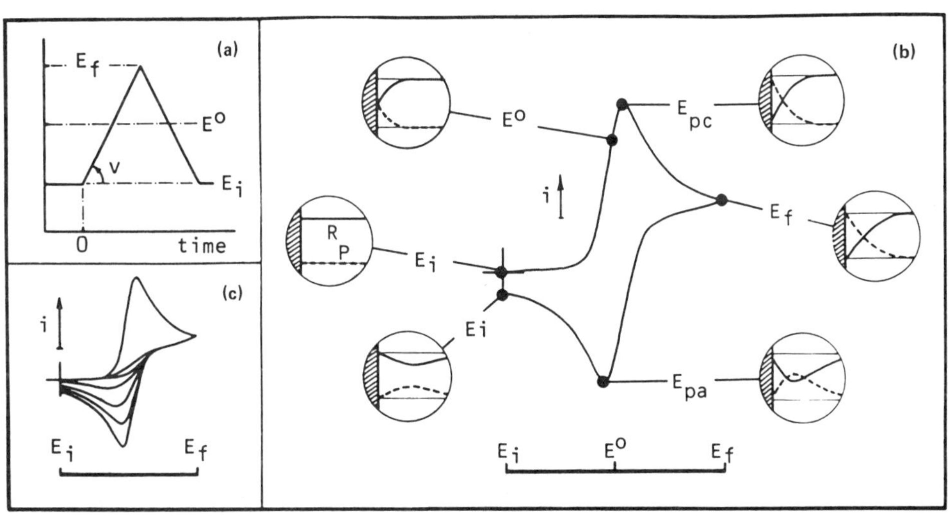

Figure 23. Cyclic voltammetry. (a) Imposed potential versus time variations. (b) Resulting transient current-potential curve for a simple electron transfer. The concentration profiles of the reactant R and product P are indicated at various characteristic potentials of the voltammogram. E_{pc} and E_{pa}, cathodic and anodic peak potentials. (c) Schematic change of the cyclic voltammogram as a function of the chemical stability of the product.

to appreciate this degree of interference through the dimensionless rate constants λ, equilibrium constants κ, or heterogeneous rate constants Λ in Table 5. Thus any experimental variation resulting in an overall constancy of these dimensionless parameters does not change the dimensionless ψ versus τ or ψ versus ξ curve, that is, the dimensionless voltammogram. This is a situation identical to that developed more extensively for the steady-state methods and is the basis of most electrochemical methods for the determination of reaction orders and rate constants.

Yet when applied to current reversal techniques, such as double-step chronoamperometry or cyclic voltammetry, these methods require that an appreciable current be observed during the backward perturbation, that is, for $t > \theta$, in potentiostatic methods or after the potential scan inversion in cyclic voltammetry. This requires that the characteristic time θ of the method is adjusted to match the half-life $t_{1/2}$ of the electrogenerated intermediate. Today, owing to the recent development of ultramicroelectrodes, θ can be routinely varied from a few seconds to a few nanoseconds [102]. Yet with basic standard electrochemical equipment, θ is usually restricted from the second to the low millisecond range. Thus for experimental situations involving faster chemical reactions, current reversal techniques are of little use. Yet, as is the case for steady-state methods, much kinetic information allowing mechanistic discriminations and reaction order determinations may be gathered from characteristic potential changes. Obviously, staircase methods, such as chronoamperometries, which are blind to these variations, are not convenient, but such methods as cyclic voltammetry or its descendants (Chap. 3) are extremely precise and adequate through peak potential or half-peak potential analysis.

Indeed, in cyclic voltammetry, peak potentials E_p play a role identical to that of half-wave potentials $E_{1/2}$ in steady-state methods. As for the latter methods, peak potentials vary linearly with the logarithm of dimensionless kinetic parameters λ or Λ in Table 5, provided these latter have values sufficiently large when compared to unity [94]. These linear variations, which may be used for determination of reaction orders, stem from the same mathematical reasons as explained in the case of $E_{1/2}$. Yet the physical reason is quite different as evidenced by the case of the simple EC sequence in Eqs. (222) and (223):

$$R + ne \rightleftharpoons P \qquad (E^o) \tag{222}$$

$$P \longrightarrow \cdots \qquad (k) \tag{223}$$

From Eq. (174), the partial derivative equation describing the concentration profile of P is written in dimensionless terms (compare Table 5):

$$0 < y: \qquad \frac{\partial p}{\partial \tau} = \frac{\partial^2 p}{\partial y^2} - \lambda p \tag{224}$$

When $\lambda = k\theta \gg 1$, the kinetic term in Eq. (224) tends to be extremely large. On the other hand, $\partial p/\partial \tau$ cannot be infinite, for obvious physical reasons (except at possible discontinuities in the potential variations). Thus the diffusion term $\partial^2 p/\partial y^2$ must compensate* for the kinetic term in order

*Note that this may be established on firm mathematical grounds, but it is beyond the scope of this presentation [94].

that $\partial p/\partial\tau$ remain finite. In other words, a quasi-steady state is reached by mutual compensation of kinetics and diffusion. As a result $\partial p/\partial\tau \ll \partial^2 p/\partial y^2$ and λp. Equation (224) thus becomes equivalent to Eq. (225), at least as concerns the derivation of the concentration profile of P.

$$0 < y: \qquad 0 \approx \frac{\partial^2 p}{\partial y^2} - \lambda p \tag{225}$$

Comparison of Eq. (225) and Eq. (193) or (211) shows that the problem is identical to that presented for steady-state methods. Thus, the same mathematical derivations show that any characteristic potential figure (E_p or half-peak potential $E_{p/2}$, and so on) varies linearly with the logarithm of $\lambda = k\Theta$. Indeed, one obtains by simple transposition in Eq. (219),

$$E_p = E^o + \left[\frac{RT}{nF(\rho + 1)}\right] \ln\,[k_0[A]^{\alpha}[B]^{\beta}\cdots(C^0)^{(\rho-1)}\Theta] + \text{cst} \tag{226}$$

for the peak potential variations relative to the EC sequence in Eqs. (227) and (228).

$$R + ne \rightleftharpoons P \qquad (E^o) \tag{227}$$

$$\rho P + \alpha A + \beta B + \cdots \longrightarrow \cdots \qquad (k_0;\ k = k_0[A]^{\alpha}[B]^{\beta}\cdots) \tag{228}$$

In Eq. (226), Θ is usually defined as $\Theta = RT/Fvn$, v being the scan rate, yet any dimensionless parameter, such as $\Theta = \Delta E/v$, where ΔE is an adequate difference of potential, is equally convenient. Then, even when k_0 cannot be determined, its temperature variations, that is, $\Delta H^{\#}$, as well as the different reaction orders ρ, α, β, . . . , may be determined through the same experimental procedures already discussed for the analogous case of $E_{1/2}$ in steady-state methods.

Extension of these approaches to much more sophisticated kinetic situations, which involve, for example, two or three routes for a key intermediate, have been developed (Ref. 94 and references therein). Relative rate constants (or relations between rate constants) may then be obtained that allow, by reference to a known value, the absolute determination of all the series. For example, rate constants for the cleavage of aromatic halide anion radicals, corresponding to a half-life from about a second to a few picoseconds, have been determined using these procedures [103].

ACKNOWLEDGMENTS

Dr. Jean-Noel Verpeaux (CNRS, ENS) is cordially thanked for his invaluable discussions and contributions during the preparation of this chapter. I am also deeply indebted to the late Prof. Manuel Baizer for having kindly revised this chapter for proper English usage.

SYMBOLS AND THEIR DEFINITIONS

A	Electrode surface area
A_{ff}	Affinity (eV)
a_S	Equivalent radius of species S

C	Concentration (see also [S], ...)
C^b	Concentration in the bulk solution
C^0	Initial concentration of the starting material
C_S	Concentration of species S (see also [S],...)
$C_{x=\cdots}$	Concentration at x = ···
D	Diffusion coefficient; average diffusion coefficient
D_S	Diffusion coefficient of species S
e_{metj}	Electron in metal phase number j
E	Electrode potential versus a given reference electrode
ΔE_f^{act}	Activation energy of a reaction between encounter pairs
E^{in} (or E^{fin})	Potential energy of the electron for any state of the initial (or final) system
E_{eq}^{in} (or E_{eq}^{fin})	Potential energy of the electron at the equilibrium state of the initial (or final) system
E_j	Junction potential
E^o (or E_1^o)	Standard reduction potential (relative to the half-reaction 1)
E_E^o	Standard reduction potential of an electrode E
$E_{O/R}^o$	Standard reduction potential relative to the half-reaction $O + ne \rightleftharpoons R$
$E^{o\prime}$ (or $E_1^{o\prime}$)	Formal potential (relative to the half-reaction 1)
E_p (or E_{pa}, E_{pc})	Peak potential in CV (E_{pa}, E_{pc}: anodic or cathodic peak potential)
$E_{p/2}$	Half-peak potential in CV
E_w	Any characteristic potential pertaining to a given electrochemical wave
$E_{1/2}$	Half-wave potential
$E^{\#}$	Potential energy of the electron in the initial or final system at the transition state
f_S	Activity coefficient of species S
F	Faraday
ΔG_r^f (or ΔG_p^f)	Free energy of formation of the reactant (or product) encounter pair
ΔG^o	Standard free energy change in a specified reaction
$\Delta G^{o\prime}$	Standard free energy change between encounter pairs
ΔG_{red}^o (or ΔG_{ox}^o)	Standard free energy change for the considered half-reaction in the direction of reduction (or oxidation)
$\Delta G_{S,solvl}^o$	Standard free energy of solvation for species S in solvent 1
$\Delta G_{S*,rel}^o$	Standard free energy of relaxation of an excited species S in gas phase
ΔG_{Φ}^S	Free energy change for species S between the end of the diffuse layer and the electron transfer site
$\Delta G_{exp}^{\#}$	Experimental (apparent) activation free energy
$\Delta G_f^{\#}$ (or $\Delta G_b^{\#}$)	Overall free activation energy for the forward (or backward) reaction
$\Delta G_o^{\#}$	Intrinsic activation barrier
h	Planck constant
i	Current
i^{lim}	Limiting current in steady-state methods (current plateau)

I_p	Ionization potential (eV)
J (or $J(x)$)	Overall flux of particles through an elementary surface (i.e., of unity area) perpendicular to the x axis at x
J_{conv}	Flux of particles through an elementary surface perpendicular to the x axis, arising from convection
J_{dif}	Flux of particles through an elementary surface perpendicular to the x axis arising from diffusion
J_{migr}	Flux of particles through an elementary surface perpendicular to the x axis arising from migration
k	Boltzmann constant
k	Chemical rate constant
k^{ap}	Apparent rate constant (i.e., including constant concentrations of coreactants)
k_f^{act} (or k_b^{act})	Rate constant of the forward (or backward) reaction between encounter pairs
$(k_f^{act})_{ad}$	Rate constant of the forward (or backward) reaction between encounter pairs for an adiabatic electron transfer
k_f^{dif} (or k_b^{dif})	Overall diffusion limit rate constant forward (or backward) rate constant
k_{elec}	Apparent first-order rate constant equivalent to an electrolysis
k_f (or k_b)	Measured forward (or backward) rate constant
k^0	Intrinsic standard heterogeneous rate constant
$k_t{}^0$	True standard heterogeneous rate constant
k_{dif}^r (or k_{dif}^p)	Diffusion rate constant for the formation of the reactant (or product) pair formation
k_Z	Rate constant for the reaction involving the transfer of a Z moiety
k_1	Rate constant number 1
$k_f^{\#}$ (or $k_b^{\#}$)	Apparent forward (or backward) activation rate constant
K	Equilibrium constant
K_a	Acidity dissociation equilibrium rate constant
K_a'	Formal acidity dissociation equilibrium rate constant
$\underline{K_c}$ (or $\underline{K_p}$)	General kinetic term of consumption (or production) for the considered species
L	Mean distance covered by a particle during a specified time period
ℓ	Any specified length; free motion length for a "random-walking" particle
m (or m_S)	Molar mass (of species S)
n (or n_1)	Number of electron exchanged (for the half-reaction 1)
n_{ap}	Apparent number of electron exchanged
n_{op}	Refractive index of the solvent
N_A	Avogadro's number
N	Number of nuclear degrees of freedom
N_S (or N)	Number of molecules (of species S)
pH	$pH = - \ln [H_3O^+]$
Pe	Probability of electron tunneling between the initial and final systems

q_i	i^{th} nuclear coordinate (i = 1, M, inner shell; i = M + 1, N, outer shell)
q_i^{in} (or q_i^{fin})	Equilibrium value of the ith nuclear coordinate (i =1, N) in the initial (or final system)
$q_i^{\#}$	Value of the ith nuclear coordinate (i = 1, N) at the transition state for the initial or final system
r	Distance between reactant or product centers in the encounter pairs
r_0	Radius of a disk electrode; smaller dimension of an electrode
R	Constant of perfect gases
R (or R_Ω)	Electrical resistance of a solution
Ri (or $R_\Omega i$)	Electrical potential drop arising from ohmic phenomena
RDS	Rate determining step
s	Dimensionless concentration of species S, i.e., s = $[S]/C^0$
S_{gas}	Species S in gas phase
S_0	Species S when located at the end of the diffuse layer
S_{solv}	Species S solvated
(S)	Activity of species S
[S]	Concentration of species S
$[S]_{bulk}$	Concentration of species S in the bulk solution
$(S)_{x=\cdots}$	Activity of species S at x = $\cdots$
$[S]_{x=\cdots}$	Concentration of species S at x = $\cdots$
S_Φ	Species S when at a location where an electrical potential Φ applies
S^*	Excited species S in gas phase; isotopically labeled species S
ΔS	Entropy change for a specified reaction
ΔS_{pair}	Entropy of formation of an electrode-molecule pair
ΔS_{rot}	Entropy of formation of a molecular pair (rigid rotator)
t	Time or time period
T	Absolute temperature
v(x)	Velocity of a fluid along the x axis
V	Volume of the bulk solution
V_i	Stabilization energy at the transition state
w	Rotation rate (rad s^{-1}) of a RDE
W_r (or W_p)	Energy formation of a reactant (or product) encounter pair
$W_r(x)$ (or $W_p(x)$)	Energy of formation of a loose reactant (or product) encounter pair, x being the distance between the centers of the molecules
W_δ^{noneq}	Energy required to produce a nonequilibrium polarization of the solvent, corresponding to a virtual charge δ transferred
$W_\delta^{\varepsilon op}$ (or $W_\delta^{\varepsilon s}$)	Energy required to produce an optical (or a static) equilibrium polarization of the solvent, corresponding to a virtual charge δ transferred
X (or x)	Coordinate along the x axis; normalized reaction coordinate; distance between the centers of reacting molecules; distance from the electrode surface
X_0	Double layer thickness

X_Φ (or x_Φ)	Distance of the electron transfer site from the electrode surface
$X^\#$ (or $x^\#$)	Normalized reaction coordinate value at the transition state
y	Dimensionless space variable
Z (or Z_S)	Charge of species S
Z (or Z_{el})	Homogeneous (or heterogeneous, i.e., at an electrode) collision frequency
Z_{an} (or Z_{cat})	Anodic (or cathodic) faradaic impedance

Greek Symbols

α	Transfer coefficient (heterogeneous)
β	Brönsted coefficient (homogeneous)
β_{exp}	Experimentally measured Brönsted coefficient
γ	Proportionality constant between velocity of a particle and the force applied to it
δ	Virtual fractional charge transferred; diffusion layer thickness
δ_{conv}	Stagnant layer thickness
δ^{in} (or δ^{fin})	Virtual fractional charge transferred in the initial (or final) system
$\bar{\Delta}^2$	Average square displacement of a particle during a specified time period
ε_{op} (or ε_s)	Optical (or static) dielectric constant
Φ	Potential difference between the anode and the cathode
Φ_A	Potential difference between the auxiliary electrode and the solution
Φ_{an} (or Φ_{cat})	Potential difference between the anode (or the cathode) and the solution
Φ_{CE}	Potential difference between the counter-electrode and the solution
Φ_{Ref}	Potential difference between the reference electrode and the solution
Φ_W	Potential difference between the working electrode and the solution
Φ_S	Potential difference between the electron transfer site and the solution
Φ_Ω	Potential drop due to ohmic phenomena
Θ	Characteristic time of an electrochemical method; duration of an experiment
κ	Transmission coefficient; dimensionless equilibrium constant
λ	Dimensionless homogeneous chemical rate constant
λ (or λ^{XY})	Reorganization energy (relative to the homogeneous X/Y couple)
λ_i (or λ_i^{XY})	Inner shell reorganization energy (relative to the homogeneous X/Y couple)
λ_o (or λ_o^{XY})	Outer shell reorganization energy (relative to the homogeneous X/Y couple)
λ_i^S	Inner shell reorganization energy relative to species S
λ^{el} (λ_i^{el} or λ_o^{el})	Reorganization energy (inner or outer shell) relative to an heterogeneous electron transfer

Λ	Dimensionless heterogeneous rate constant
$\tilde{\lambda}_j$	Dimensionless kinetic term for consumption and production for the considered species
μ	Chemical potential
μ^0	Standard chemical potential
Ω_i	Reduced force constant relative to the ith nuclear coordinate
Ω_i^{in} (or Ω_i^{fin})	Force constant relative to the ith nuclear coordinate in the initial (or final) system
Ω_o	Reduced force constant relative to the oscillator equivalent to the outer shell
ψ	Dimensionless current
τ	Dimensionless time; duration of free motion for a "random-walking" particle
ξ	Dimensionless potential

REFERENCES AND NOTES

1. Most of the concepts developed in this chapter are presented in detailed analytical form in: (a) A. J. Bard and R. L. Faulkner, Electrochemical Methods, J. Wiley and Sons, New York, 1980; (b) Laboratory Techniques in Electroanalytical Chemistry (P. T. Kissinger and W. R. Heineman, eds.), Dekker, New York, 1984; (c) J. O'M. Bockris and A. K. N. Reddy, Modern Electrochemistry, Plenum, New York, 1977, Vols. 1 and 2.
2. (a) J. W. Huffman, Acc. Chem. Res., 16, 399 (1983); (b) V. Rautenstrauch and M. Geoffroy, J. Amer. Chem. Soc., 98, 5035 (1976) and 99, 6280 (1977).
3. See, e.g., (a) V. J. Shiner, Jr., and D. Whittaker, J. Amer. Chem. Soc. 85, 2337 (1963); (b) E. W. Warnhoff, P. Reynolds-Warnhoff, and M. Y. H. Wong, J. Amer. Chem. Soc., 102, 5956 (1980).
4. See however, E. C. Ashby, A. B. Goel, and J. N. Argyropoulos, Tetrahedron Lett., 23, 2273 (1982) for the possible occurrence of paramagnetic intermediates (SET mechanisms [5]) in the reduction with metal alkoxides.
5. For an extensive review on single electron transfer (SET) mechanisms, see M. Chanon and M. L. Tobe, Angew. Chem. 21, 1 (1982).
6. E. C. Ashby and J. R. Boone, J. Amer. Chem. Soc., 98, 5524 (1976).
7. D. C. Wigfield and F. W. Gowland, J. Org. Chem., 42, 1108 (1977).
8. J. March, Advanced Organic Chemistry, J. Wiley and Sons, New York, 1985, 3rd ed., pp. 811–814.
9. See, e.g., F. Freeman, D. K. Lin, and G. R. Moore, J. Org. Chem., 47, 56 (1982).
10. (a) R. Criegee, L. Kraft, and B. Rank, Justus Liebigs Ann. Chem., 507, 159 (1933); (b) W. A. Waters, Mechanisms of Oxidation of Organic Compounds, Wiley, New York, 1964, pp. 97–106; (c) see, however, J. K. Kochi, Organometallic Mechanisms and Catalysis, Academic Press, New York, 1978, p. 108, for the possible involvement of one-electron transfer steps, with, e.g., V^{v}/V^{iv}.
11. Ref. 10c, pp. 11–13.
12. H. Taube and H. Myers, J. Amer. Chem. Soc., 76, 2103 (1954).

13. See Chap. 23 in this book, and, e.g., J. H. P. Utley, in Technique of Electroorganic Synthesis(N. L. Weinberg, ed.,), Wiley Interscience, New York, 1974, Part 1, Chap. 6.
14. See, e.g., C. P. Andrieux, J. M. Saveant, and K. B. Su, J. Phys. Chem., 90, 3815 (1986) and references therein as well as Chap. 8 in this book.
15. (a) J. M. Saveant, J. Amer. Chem. Soc., 109, 6788 (1987); (b) W. E. Wentworth, R. George, and H. Keith, J. Phys. Chem., 51, 1791 (1969); (c) M. C. R. Symons, Pure Appl. Chem., 53, 223 (1981).
16. C. P. Andrieux, J. M. Saveant, and D. Zann, Nouv. J. Chim., 8, 107 (1984).
17. See, e.g., J. K. Kochi, Ref. 10c.
18. See Chap. 28, 29, 30, and 33 in this book.
19. See Ref. 1c, pp. 1124–1129.
20. See Ref. 1a, pp. 44–62, for a thermodynamic derivation of Nernst equation.
21. (a) See Ref. 1a, pp. 52–53; (b) Reference Electrodes (D. J. G. Ives and G. J. Janz, eds.), Academic Press, New York, 1961; (c) G. Charlot, J. Badoz-Lambling, and B. Tremillon, Les Reactions Electrochimiques, Masson, Paris, 1959, pp. 147–149 and 347–348; (d) see also H. Lund, Chap. 6 in this book, for a brief discussion on reference electrodes.
22. (a) Reference 1a, pp. 62–72; (b) Ref. 1c, Vol. 1, Chap. 4; (c) J. J. Lingane, Electroanalytical Chemistry, Wiley Interscience, New York, 1958, 2nd ed., Chap. 3; (d) K. J. Vetter, Electrochemical Kinetics, Academic Press, New York, 1967.
23. As, e.g., the Henderson equation, established in Ref. 1a, p. 71.
24. See, e.g., Ref. 1c, pp. 688–717, and references therein.
25. M. E. Peover, in Electroanalytical Chemistry (A. J. Bard, ed.), Vol. 2, Dekker, New York, 1967.
26. See Ref. 1c, pp. 660–679.
27. For example, consider the evaluation of solvation energies via the Born model: (a) Ref. 1c, pp. 49–60; (b) W. M. Latimer, K. S. Pitzer, and C. M. Slansky, J. Chem. Phys., 7, 108 (1939); (c) R. M. Noyes, J. Amer. Chem. Soc., 84, 513 (1962); (d) J. F. Coetzee and J. J. Campion, ibid., 89, 2513 (1967); (e) M. Tanaka, Inorg. Chem., 15, 2325 (1976).
28. M. J. S. Dewar, J. A. Hashmall, and N. Trinajstic, J. Amer. Chem. Soc., 92, 5555 (1970).
29. J. O. Howell, J. M. Goncalves, C. Amatore, L. Klasinc, R. M. Wightman, and J. K. Kochi, J. Amer. Chem. Soc., 106, 3968 (1984).
30. For an extensive series of E^0 values, see L. Meites and P. Zuman, CRC Handbook Series in Organic Electrochemistry, CRC Press, Cleveland, 1975, Vols. I and II: and L. Meites, P. Zuman, and E. B. Rupp, ibid., Vols. III–V.
31. (a) A. Steitwieser, Jr., Molecular Orbital Theory for Organic Chemists, Wiley, New York, 1961; (b) P. Zuman, Substituent Effects in Organic Polarography, Plenum, New York, 1967; (c) R. M. Hedges and F. A. Matsen, J. Chem. Phys., 28, 950 (1958); (d) G. J. Hoijtink, Rec. Trav. Chim. Pays-Bas, 74, 1525 (1955); 77, 55 (1958); (e) R. M. Schmidt and E. Heilbronner, Helv. Chim. Acta, 37, 1453 (1954); (f) V. D. Parker, J. Amer. Chem. Soc., 98, 98 (1976).

32. (a) P. Zuman, in Progress in Physical Organic Chemistry (A. Streitwieser, Jr., and R. W. Taft, eds.), Vol. 5, Wiley Interscience, New York, 1967; (b) Ref. 31b, p. 72.
33. Ref. 1a, pp. 51–52.
34. For the dependence of activity coefficients on experimental conditions, see, e.g., (a) Ref. 1c, pp. 180–267; (b) M. C. Justice and J. C. Justice, J. Sol. Chem., 5, 543 (1976).
35. See, e.g., W. C. Mansfield, Oxidation-Reduction Potentials of Organic Systems, Williams & Wilkins, Baltimore, 1960, pp. 118–145.
36. (a) C. J. Schlesener, C. Amatore, and J. K. Kochi, J. Amer. Chem. Soc., 106, 3567 (1984); (b) ibid, 106, 7472 (1984); (c) J. Phys. Chem., 90, 3747 (1986).
37. (a) C. L. Wong and J. K. Kochi, J. Amer. Chem. Soc., 101, 5593 (1980); (b) Ref. 36a.
38. Compare Chap. 13 in this book and (a) V. D. Parker, J. Electroanal. Chem., 21, App. 1–3 (1969); (b) A. Bewick, J. M. Mellor, and B. S. Pons, Electrochim. Acta, 23, 77 (1978).
39. (a) Ref. 38b; (b) K. L. Rollick and J. K. Kochi, J. Amer. Chem. Soc., 104, 1319 (1982).
40. L. Eberson, L. Jönsson, and L. G. Wistrand, Acta Chem. Scand., B32, 520 (1978).
41. R. A. Marcus, J. Chem. Phys., 24, 4966 (1956). For quadratic relations arising from different origins, see also, e.g., (a) D. E. Magnoli and J. R. Murdoch, J. Amer. Chem. Soc., 103, 7465 (1981); (b) J. R. Murdoch and D. E. Magnoli, ibid., 104, 3792 (1982); (c) J. R. Murdoch, ibid., 105, 2159 (1983); 105, 2667 (1983); (d) J. G. Kurz, Chem. Phys. Lett., 57, 243 (1978); (e) E. J. Grunwald, J. Amer. Chem. Soc., 107, 125 (1985).
42. For a specialized and extensive review, see, e.g., N. Sutin, in Progress in Inorganic Chemistry (S. J. Lippard, ed.), Wiley Interscience, New York, 1983, Vol. 30, pp. 441–498.
43. (a) R. A. Marcus, Annu. Rev. Phys. Chem., 15, 155 (1964); (b) R. A. Marcus, J. Chem. Phys., 43, 679 (1965); (c) E. Waisman, G. Worry, and R. A. Marcus, J. Electroanal. Chem., 82, 9 (1977); (d) R. A. Marcus, Faraday Discuss. Chem. Soc., 74, 7(1982); (e) R. A. Marcus and N. Sutin, Biochim. Biophys. Acta, 811, 265 (1985).
44. (a) V. G. Levich, in Advances in Electrochemistry and Electrochemical Engineering (P. Delahay, ed.), Wiley Interscience, New York, 1966, Chap. 4; (b) M. A. Vorotynsev, R. R. Dogonadze, and A. M. Kuznetsov, Dokl. Akad. Nauk SSSR, 195, 1135 (1970); (c) E. D. German, V. G. Dvali, R. R. Dogonadze, and A. M. Kuznetsov, Electrokhimiya, 12, 639 (1976); (d) R. R. Dogonadze, and A. M. Kuznetsov, Progr. Surf. Sci., 6, 1 (1975); (e) R. R. Dogonadze, A. M. Kuznetsov, and V. G. Levich, Electrochim. Acta, 13, 1025 (1968); (f) R. R. Dogonadze, in Reactions of Molecules at Electrodes (N. S. Hush, ed.), Wiley Interscience, New York, 1971, Chap. 3.
45. (a) P. Debye, Trans. Electrochem. Soc., 82, 265 (1942); (b) Z. Smoluchowsky, Physik. Chem., 92, 129 (1917).
46. R. A. Marcus, Discuss. Faraday Soc., 129, 21 (1960).
47. Compare, e.g., Ref 36c.
48. W. F. Libby, J. Phys. Chem., 56, 863 (1952).
49. P. W. Atkins, Physical Chemistry, Oxford University Press, Oxford, 1978, pp. 907–911.

50. Ref. 49: (a) pp. 911–912; (b) pp. 800–804.
51. See, e.g., (a) G. M. Brown and N. Sutin, J. Amer. Chem. Soc., 101, 883 (1979); (b) A. M. North, The Collision Theory of Chemical Reactions in Liquids, Wiley, New York, 1964.
52. See, e.g., Refs. 44a and 44f, or N. Sutin and B. S. Brunschwig, A. C. S. Symp. Series, 198, 105 (1982) for evaluation of κ, as a function of V_i, on the basis of a Landau-Zener [53] treatment of the electron tunneling probability Pe.
53. (a) L. Landau, Phys. Z. Sowjetunion, 2, 46 (1932); (b) C. Zener, Proc. Roy. Soc., Ser. A, 137, 696 (1932) and 140, 660 (1933).
54. See, e.g., Ref. 15a for a simple approach to the problem when $\kappa = 1$.
55. For more elaborate models, see, e.g., (a) Ref. 44b; (b) R. D. Cannon, Chem. Phys. Lett., 49, 299 (1977); (c) R. D. Cannon, Electron Transfer Reactions, Butterworths, London, 1980; (d) M. E. Peover and J. S. Powell, J. Electroanal. Chem., 20, 427 (1969); (e) M. Falsig, H. Lund, L. Nadjo, and J. M. Saveant, Nouv. J. Chim., 4, 445 (1980); (f) as well as a critical review in E. D. German and A. M. Kuznetsov, Electrochim. Acta, 26, 1595 (1981).
56. See Ref. 42, pp. 486–487.
57. (a) R. A. Marcus in Ref. 41; (b) R. A. Marcus, J. Chem. Phys., 26, 867 (1957); 26, 872 (1957); (c) N. S. Hush, Trans. Faraday Soc., 57, 557 (1961).
58. See, e.g., Ref. 42, pp. 455–459, for a simple presentation of the problem on quantum mechanical grounds.
59. See, e.g., Ref. 1c, pp. 59–60 and pp. 201–202.
60. (a) S. Efrima and M. Bixon, Chem. Phys. Lett., 25, 34 (1974); (b) J. Chem. Phys., 64, 3639 (1976); (c) Chem. Phys., 13, 447 (1976).
61. (a) R. A. Marcus, J. Phys. Chem., 72, 891 (1968); (b) N. Agmon and R. D. Levine, Chem. Phys. Lett., 52, 197 (1977); (c) R. D. Levine, J. Phys. Chem., 83, 159 (1979).
62. See, e.g., Ref. 42, pp. 479–481, for a more complete discussion.
63. See, e.g., L. Eberson, Advan. Phys. Org. Chem., 18, 79 (1982).
64. D. Rehm and A. Weller, Isr. J. Chem., 8, 259 (1970).
65. See, e.g., T. J. Meyer, in Progress in Inorganic Chemistry (S. J. Lippard, ed.), Wiley Interscience, New York, 1983, Vol. 30, pp. 420–424, for a detailed discussion and pertinent references on the "inverted region."
66. B. S. Brunschwig, C. Creutz, D. H. Macartney, T. K. Sham, and N. Sutin, Faraday Discuss. Chem. Soc., 74, 113 (1982).
67. (a) T. Nakajima, A. Toyota, and M. Kataoka, J. Amer. Chem. Soc., 104, 5610 (1982); (b) M. Iwasaki, K. Toriyama, and K. Nunome, J. Chem. Soc., Chem. Comm., 320 (1983); (c) L. Salem, The Molecular Orbital Theory in Conjugated Systems, Benjamin, New York, 1966, pp. 467–485.
68. J. E. Dickens, F. Basolo, and H. M. Neumann, J. Amer. Chem. Soc., 79, 1289 (1957).
69. See, e.g., Ref. 1a, pp. 563–567, or Ref. 1b, pp. 163–192, for descriptions of related electronics.
70. J. O. Howell, W. G. Kuhr, R. E. Ensman, and R. M. Wightman, J. Electroanal. Chem., 209, 77 (1986).
71. See H. Lund, Chap. 6, this book.
72. Ref. 1a, p. 24.

73. (a) C. Amatore, in Ultramicroelectrodes (M. Fleischmann, S. Pons, D. Rolison, and P. P. Schmidt, eds.), Datatech Systems, Morganton, 1987; (b) R. M. Wightman and D. O. Wipf, in Electroanalytical Chemistry (A. J. Bard, ed.), Dekker, New York, 1989, Vol. 15. (c) see, however, S. Bruckenstein, Anal. Chem., 59, 2098 (1987), for limitation to this decrease under steady-state conditions.
74. H. L. F. von Helmholtz, Ann. Phys. (Leipzig), 89, 211 (1853) and 7, 337 (1879).
75. (a) G. Gouy, J. Phys. Radium, 9, 457 (1910); (b) D. L. Chapman, Phil. Mag., 25, 475 (1913).
76. See, e.g., Ref. 1a, pp. 500–511, as well as references quoted therein, for a detailled presentation of the various models.
77. J. M. Hale, in Reactions of Molecules at Electrodes (N. S. Hush, ed.), Wiley Interscience, New York, 1971, Chap. 4, p. 229.
78. (a) N. S. Hush, J. Chem. Phys., 28, 962 (1958); (b) Ref. 57c; (c) Electrochim. Acta, 13, 1005 (1968).
79. (a) P. P. Schmidt, in Electrochemistry, A Specialist Periodical Report, H. R. Thirsk, Senior Reporter, The Chemical Society, London, 1977, Vol. 5; (b) 1978, Vol. 6.
80. H. Kojima and A. J. Bard, J. Amer. Chem. Soc., 97, 6317 (1975).
81. Yet the problem is not so clear-cut. See, e.g., C. P. Andrieux, C. Blocman, J. M. Dumas-Bouchiat, and J. M. Saveant, J. Amer. Chem. Soc., 101, 3431 (1979).
82. (a) J. A. V. Butler, Trans. Faraday Soc., 19, 729 (1924); (b) T. Erdey-Gruz and M. Volmer, Z. Phys. Chem. (Frankfurt am Main), 150A, 203 (1930).
83. (a) J. Tafel, Z. Phys. Chem. (Frankfurt am Main), 50A, 641 (1905); (b) Ref. 1a, pp. 106–108.
84. (a) J. M. Saveant and D. Tessier, J. Electroanal. Chem., 65, 57 (1975); (b) J. Phys. Chem., 81, 2192 (1977); (c) ibid., 82, 1723 (1978); (d) Faraday Discuss. Chem. Soc., 74, 57 (1982); (e) D. Garreau, J. M. Saveant, and D. Tessier, J. Phys. Chem., 83, 3003 (1979).
85. (a) L. Nadjo and J. M. Saveant, J. Electroanal. Chem., 48, 113 (1973); (b) R. J. Klingler and J. K. Kochi, J. Amer. Chem. Soc., 104, 4186 (1982).
86. See Ref. 1a, pp. 511–519, and references therein, for a detailled presentation of adsorption thermodynamics and kinetics.
87. (a) C. Amatore, J. M. Saveant, and D. Tessier, J. Electroanal. Chem., 146, 37 (1983); (b) ibid., 147, 39 (1983), and references therein.
88. See, e.g., M. R. Deakin, K. J. Stutts, and R. M. Wightman, J. Electroanal Chem., 182, 113 (1985).
89. See, e.g., E. Ahlberg and V. D. Parker, Acta Chem. Scand., B37, 723 (1983).
90. C. K. Mann, in Electroanalytical Chemistry (A. J. Bard, ed.), Dekker, New York, 1969, Vol. 3, pp. 57–134.
91. (a) L. Geng, A. G. Ewing, J. C. Jernigan, and R. W. Murray, Anal. Chem., 58, 852 (1986); (b) R. Lines and V. D. Parker, Acta Chem. Scand., Ser. B, B31, 369 (1977); (c) J. O. Howell and R. M. Wightman, Anal. Chem., 56, 524 (1984); (d) J. Phys. Chem., 88, 3915 (1984).
92. See, e.g., (a) F. M'Halla, J. Pinson, and J. M. Saveant, J. Amer. Chem. Soc., 102, 4120 (1980); (b) J. Electroanal. Chem., 89, 347

(1978); (c) C. Amatore, F. M'Halla, and J. M. Saveant, ibid, 123, 219 (1981).
93. (a) J. M. Saveant and E. Vianello, Electrochim. Acta, 8, 905 (1963); (b) J. M. Saveant, C. P. Andrieux, and L. Nadjo, J. Electroanal. Chem., 41, 137 (1973); (c) N. Fatouros, M. Chemla, C. Amatore, and J. M. Saveant, ibid., 172, 67 (1984); (d) C. Amatore, D. Garreau, M. Hammi, J. Pinson, and J. M. Saveant, ibid., 184, 1 (1985).
94. C. P. Andrieux and J. M. Saveant, in Investigations of Rates and Mechanisms of Reactions (C. F. Bernasconi, ed.), J. Wiley and Sons, New York, 1986, Vol. 6, 41E, Part 2, Chap. 7.
95. (a) C. P. Andrieux, J. M. Saveant, and K. B. Su, J. Phys. Chem., 90, 3815 (1986); (b) Ref. 15a, and references therein.
96. (a) C. Amatore and J. M. Saveant, J. Electroanal. Chem., 86, 227 (1978); (b) ibid., 102, 21 (1979).
97. W. Smith and A. J. Bard, J. Amer. Chem. Soc., 97, 5203 (1975).
98. For a review, see, e.g., J. M. Saveant, Acc. Chem. Res., 13, 323 (1980), as well as Chap. 29 in this book.
99. (a) J. F. Bunnett, Acc. Chem. Res., 11, 413 (1978); (b) R. A. Rossi and R. H. Rossi, Aromatic Nucleophilic Substitutions by the $S_{RN}1$ Mechanism, A. C. S. Monographs, 178, ACS, Washington, D.C., 1983.
100. See, e.g., N. Alam, C. Amatore, C. Combellas, J. Pinson, J. M. Saveant, A. Thiebault, and J. N. Verpeaux, J. Org. Chem., 53, 1496 (1988) and references therein.
101. See, e.g., E. Laviron, J. Electroanal. Chem., 148, 1 (1983); ibid., 169, 29 (1984); ibid., 208, 357 (1986), for recent work on square schemes.
102. C. Amatore, A. Jutand, and F. Pflüger, J. Electroanal Chem., 218, 361 (1987).
103. C. Amatore, M. A. Oturan, J. Pinson, J. M. Saveant, and A. Thiebault, J. Amer. Chem. Soc., 107, 3451 (1985).
104. See, e.g., (a) J. S. Newman, Electrochemical Systems, Prentice-Hall Englewood Cliffs, N.J., 1973, pp. 305–339; (b) V. G. Levich, Physicochemical Hydrodynamics, Prentice-Hall, Englewood Cliffs, N.J., 1962, and references therein.
105. See, e.g., Ref. 104b, p. 60.
106. C. Amatore, M. R. Deakin, and R. M. Wightman, J. Electroanal. Chem., 220, 49 (1987).
107. See, e.g., Ref. 1a, pp. 128–130.
108. Ref. 1c, p. 1069.
109. See, e.g., (a) M. Fleischmann, F. Lasserre, J. Robinson, and D. Swan, J. Electroanal. Chem., 177, 97 (1984); (b) M. Fleischmann, F. Lasserre, and J. Robinson, ibid., 177, 115 (1984).
110. For a recent and concise review, see, e.g., D. C. Johnson, M. D. Ryan, and G. S. Wilson, Anal. Chem., 58, 33R (1986); as well as Chap. 3 in this book.

3

Methods for the Elucidation of Organic Electrochemical Reactions

OLE HAMMERICH / H. C. Ørsted Institute, University of Copenhagen, Copenhagen, Denmark

VERNON D. PARKER / Utah State University, Logan, Utah

I. Introduction 122

II. Linear Sweep Voltammetry and Cyclic Voltammetry 123
- A. Introduction 123
- B. Experimental considerations 123
- C. Charge transfer reactions 130
- D. Coupled chemical reactions 133
- E. More detailed linear sweep voltammetry analysis 135
- F. Applications of linear sweep voltammetry and cyclic voltammetry 136

III. Potential-Step and Current-Step Methods 145
- A. Chronoamperometry and double-potential-step chronoamperometry 145
- B. Chronocoulometry and double-potential-step chronocoulometry 150
- C. Chronopotentiometry and current-reversal chronopotentiometry 152

IV. Polarography 157
- A. Fundamental aspects 157
- B. Applications 161

V. Methods Based on Forced Convection 164
- A. General considerations 164
- B. The rotating disk electrode 165
- C. The rotating ring-disk electrode 171
- D. Further aspects of the application of rotating electrodes 173
- E. Application of rotating disk electrodes in homogeneous kinetics 174

VI. Coulometry 176

VII. Methods Based on the Combination of Electrochemical and Optical Techniques 179
- A. Techniques and cells for spectroelectrochemistry 179
- B. Modes of operation and applications 180

VIII. Application of ESR Spectroscopy 190

References 195

I. INTRODUCTION

The utility of electrochemical methods for determining reaction pathways of reactive intermediates is much broader than implied by the title of this chapter. The application of electroanalytic methods to mechanism studies 20 years ago was primarily in the study of the course of electrochemical reactions and was often carried out in conjunction with preparative electrochemistry. Today, this type of study is less often encountered. Instead, electrochemical methods are more likely to be used to gain information about the energetics and mechanisms of homogeneous chemical reactions. The application of electrochemical techniques in organic and physical organic chemistry has recently been reviewed [1,2]. One reason for the increased importance of electrochemical techniques in these fields is the general recognition of the importance of electron transfer reactions in organic chemistry [3].

In terms of mechanism analysis, electrochemical methods are somewhat unique in that the electrode serves not only as the means of generating the intermediate, but also the means to monitor the reactions following charge transfer. Furthermore, measurement of the reversible potential for the oxidation or reduction of a substrate provides a direct measure of the free energy of formation of the resulting intermediate, either anion or cation radical, if the substrate is a neutral organic molecule.

Our presentation of eletrochemical methods to elucidate the mechanisms of organic electrochemical reactions is no more than an introduction meant to serve as a guide for the beginner or as a brief overview for nonelectrochemists who need to know the methods available, should the need for more detailed studies arise. Recent references, including textbooks [4–7], to more detailed treatments are provided for all the techniques discussed. In addition, the reader of this chapter should also consult Chap. 2. The discussions of the various electrochemical methods are supplemented with pertinent examples.

Most serious electrochemical studies carried out today make use of on-line computers to control experiments and process data. Discussion of the use of computers in electrochemical analysis is, however, beyond the scope of this chapter. The interested reader is referred to the textbook references just given. The commercial software available for statistical treatment of data includes data-smoothing routines and fast Fourier transform (FFT) analysis as well as numerous other options. An excellent book [8] on this subject, accompanied by programs on disks, is highly recommended. Recent examples of the application of computer evaluation of experimental results include deviation pattern recognition [9] and recording of current-time profiles of the form $\Delta \ln i / \Delta \ln t$ versus t [10] for mechanistic classification as well as nonlinear regression techniques [11–13].

The electrochemical literature abounds with symbols and abbreviations. We have attempted to define all symbols and abbreviations in the text where they first appear. A few comments about the abbreviations used to designate electrode mechanisms are necessary. The notation commonly used is that due to Testa and Reinmuth [14], where e indicates heterogeneous charge transfer and c is a chemical reaction. It is helpful to distinguish between slow and fast steps by using lowercase letters for fast and a capital letter for the rate-determining step [15]. Since electron transfer reactions can be either heterogeneous or homogeneous, those taking place in solution are given the subscript h [16]. Using this notation, the familiar ECE-type mechanism (1) through (3),

$$A \pm e^- \rightleftharpoons B \qquad e \tag{1}$$

$$B \xrightarrow{k} C \qquad C \tag{2}$$

$$C + B \rightleftharpoons D + A \qquad e_h \tag{3}$$

is given the notation eCe_h, as indicated. This mechanism is often referred to as DISP1 as well.

II. LINEAR SWEEP VOLTAMMETRY AND CYCLIC VOLTAMMETRY

A. Introduction

Voltammetry in unstirred solution in which the predominant mode of mass transport is limited to diffusion is one of the most useful and easily applied methods for the study of organic electrode processes. The most significant theoretical contribution to this type of voltammetry was published in 1964 by Nicholson and Shain [17]. The decade following this seminal work was a period of intense activity in the calculation of theoretical electrode responses for linear sweep voltammetry (LSV) and cyclic voltammetry (CV). Most notable of the large volume of papers published are those of Nicholson and co-workers [18–21] and of the Savéant group [15,22–28]. The development of digital simulation by Feldberg in the late 1960s [29] had a very strong impact in that the mathematical complexity of the calculations was greatly reduced and the approach was readily adopted for simple reaction schemes without an extensive theoretical background. The incorporation of more complicated reaction schemes has been discussed recently [30]. The almost total concentration on calculations of the electrode response during the 1965–1980 period left little time for the development of the experimental aspects of the field. Consequently, both LSV and CV were in 1980 essentially qualitative methods. Efforts initiated by Parker and coworkers in the early 1980s [31–41] have been directed toward the development of LSV and CV as quantitative tools for the study of electrode kinetics. The initial efforts in this development have been reviewed [42].

B. Experimental Considerations

The fundamental variable of LSV and CV is the voltage sweep rate ν [1,2, 4–7,43]. The width of the kinetic window, both for heterogeneous charge transfer and for homogeneous follow-up reactions, is limited by the practical range of ν. The latter is determined by a number of factors. Four of the most important are cell design, potentiostat quality, precision requirements, and the uncertainty in the response due to uncompensated resistance R_u (see below). Cell design is most critical in the extremes of the operable range of ν. At low ν, care must be taken to shield the working electrode to prevent convection; at high ν the relative positions of the three electrodes must be optimized to reduce resistance problems. The quality of the potentiostat, usually reflected by the rise time, becomes critical at high ν. The other two factors are actually not separable. It is necessary to have control over the R_u problem to make precise measurements. These considerations are not directly applicable to the use of recently developed ultramicroelectrodes. Consequently, ultramicroelectrodes are treated in a separate section in this chapter.

Typical Experimental Setup

Figure 1 illustrates an arrangement of apparatus for obtaining cyclic or linear sweep voltammograms. The cell shown is a simple and convenient design consisting of a 30 ml tube fitted with a B19/26 ground glass joint. The electrode holder is made of Teflon and has holes for the three electrodes and a nitrogen inlet tube. The fixed arrangement of the electrodes is desirable to avoid changes in R_u when an electrode is taken out of the electrolyte and then replaced. Without further refinement, this cell is suitable for measurements from about 0.1 to 300 V s^{-1} using electrodes of diameters ranging from about 0.1 to 1 mm. It should be kept in mind that to use a single-compartment cell the current passed during the measurements must cause a negligible change in the solution composition. This is only true when small electrodes are used. Reference electrodes are frequently commercial aqueous calomel electrodes. However, the use of this type of electrode can cause unnecessary complications, such as the need to separate the reference and working compartments of the cell. It is much more convenient to use reference electrodes constructed in the manner described by Moe [44],

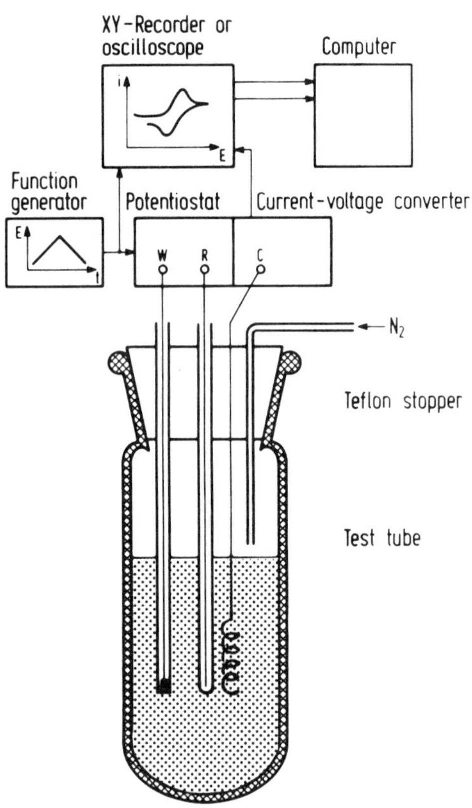

Figure 1. Experimental setup for linear sweep and cyclic voltammetry. W, working electrode; R, reference electrode; C, counterelectrode.

which can be made so that the leakage through the cracked soft glass is negligible during the time of the measurement. A discussion and leading references to the original literature concerning the instrumental aspects of voltammetry can be found in several of the monographs cited in the introduction (e.g., Refs. 5 and 7). The essential features are the potentiostat, a triangular wave function generator, and a recording device. The recording device is most conveniently a digital oscilloscope or transient recorder interfaced to a computer. Units consisting only of commercially available components have been used in our laboratories for the past 10 years [45,46].

It is often necessary to have solvent-supporting electrolyte solutions that are essentially free of reactive impurities to observe a voltammetric response for reactive intermediates. Although an elaborate vacuum line apparatus has been described for this purpose [47-49], it has been shown in general to be less effective, and certainly less convenient, than conducting the voltammetric experiments over neutral alumina [50]. The alumina technique was first described by Hammerich and Parker [51] as a means of measuring the reversible potentials for the formation of aromatic dications. The method was later extended to reversible potential measurements of anion radical reduction [52]. The method is simply to add active neutral alumina to the voltammetric cell, for example of the type shown in Fig. 1, and to mix vigorously for a few minutes by mechanical stirring or by using the nitrogen bubbler before making measurements. The method is very effective in removing the acidic, basic, electrophilic, and nucleophilic impurities normally present in trace amounts even in the most carefully purified solvent-electrolyte systems. However, one should be aware that the substrate may also be adsorbed at alumina, and therefore this approach can in general only be used in qualitative work.

Measurement Precision

The precision in electrode peak potentials E_p measured by X-Y recording is generally considered to be of the order of ±5 mV.* A significant improvement in precision can be achieved simply by measuring half-peak potentials $E_{p/2}$; using X-Y recording with a resolution of 20 mV/cm on the potential axis resulted in $E_{p/2}$ values with a precision of ±1 mV, which can be improved still further using digital recording [37]. The importance in precision for LSV measurements (see later) can be appreciated by considering the error in $dE_p/d \log \nu$ obtained from two point slopes as a function of the sweep rate range [40]. If the error in E_p is ±0.2 mV and ν is varied by only a factor of 2, the error in the LSV slope is only ±1.3 mV per decade. If it is possible to vary ν by as much as a factor of 100, the error in the LSV slope becomes the same as the error in E_p. Thus, even under these extreme circumstances a precision of ±5 mV in E_p could give rise to LSV slopes that can not distinguish processes with theoretical values of 19.7 and 29.6 mV per decade. Using X-Y recording the available sweep rate range is no more than a factor of 50, from about 10 to 500 mV s^{-1} [40]. In any case, because of the interference of heterogeneous charge transfer kinetics, it is not advisable to use ν greater than about 1 V s^{-1} unless the heterogeneous charge transfer rate constant is known to be large.

---

*This is the precision usually claimed, but in practice it is difficult to measure peaks to much better than ±10 mV.

Uncompensated Solution Resistance

A great deal has been written about the problems associated with making electrode measurements in resistive media [5,53–58]. The discussion here is limited to LSV and CV and emphasizes the practical aspects of handling the problem rather than the theoretical approaches.

The R_u problem in CV and LSV was first studied in detail in 1965 by Nicholson [54] and by deVries and van Dalen [55]. Savéant and coworkers [56], building on earlier work by Roffia and Lavachielli [57], considered the problem in detail with relation to the double-layer charging current and the potentiostat response.

The electrode potential during LSV for a negative sweep is described by

$$E(\text{measured}) = E(\text{initial}) - \nu t + iR_u \qquad (4)$$

where i is the instantaneous current. This equation suggests that the real electrode potential E(real) can be determined only under conditions in which the last term vanishes. Thus, the problem is to correct E(measured) for the contribution of iR_u. This is normally accomplished by a positive feedback circuit incorporated in the potentiostat, which adds a fraction of the current follower output to the voltage provided by the function generator. If the feedback resistance R_f is exactly equal to R_u, the iR_u term in Eq. (4) is compensated for and E(measured) is equal to E(real). The problem then is the selection of the value R_f. Although this can be accomplished by direct measurement of R_u and by more complicated procedures [5,53], a simpler procedure is desirable for the level of sophistication of most electroorganic studies. Such a simple and convenient method has been described by Evans and co-workers [58]. The feedback circuit is adjusted until the output of the current follower goes into oscillation and is then backed off slightly. This method is quite effective as long as ν is not greater than about 100 V s^{-1} with electrodes of diameters no greater than 1 mm when precise electrode potentials are not necessary.

Quantitative Methods

In order for CV and LSV to be suitable for quantitative studies, two problems must be overcome. These problems are the measurement precision and the uncertainty due to iR_u. As discussed in a later section, the use of ultramicroelectrodes greatly diminishes the iR_u problem. However, conventional LSV relationships are based upon linear diffusion to planar electrodes, a condition not fulfilled with ultramicroelectrodes at low sweep rates. Linear diffusion is approximated at planar disk electrodes with diameters of the order of 0.6 mm or greater at voltage sweep rates of about 100 mV s^{-1} or greater.

The precision problem has effectively been eliminated by recent work. The use of the first derivative of the current-potential curve allows a precision of the order of ±0.1 mV to be realized during peak-potential measurements [33-35]. Derivative cyclic voltammetry (DCV) was investigated during the 1960s by Perone and coworkers [59,60], but because of errors in the analog differentiators the method was not recommended for the measurement of electrode potentials. It has been found that a commercially available selective amplifier (Princeton Applied Research, Model 189 Selective Amplifier) can serve as an analog differentiator [61]. Development of the use of analog differentiation has been the subject of two papers by Ahlberg and Parker entitled "Quantitative Cyclic Voltammetry" [33,36]. As the title implies, this mode of voltammetry gives quantitative results for both heterogeneous [33] and homogeneous [36] kinetic studies. The method was further developed for LSV with derivative linear sweep voltammetry (DLSV) [34,35,40]. A typical cyclic voltammogram and the corresponding first derivative obtained by analog differentiation are illustrated in Fig. 2. The feature giving the high precision is that the peak potential is measured at the point where the steeply descending first derivative crosses zero. Recent work by Nielsen, Laursen, and Hammerich [62] has shown that digital filtering of the noise using FFT methods produces an LSV signal suitable for mathematical differentiation, thereby eliminating the necessity for the analog differentiator during DLSV and DCV measurements.

A procedure has been developed to determine the correct R_f to eliminate the effect of the iR_u term in Eq. (4) [32.33]. The procedure, which involves obtaining an array of data using substrate concentration and R_f settings as the primary experimental variables, was tested for consistency by determining the heterogeneous rate constant for the reduction of benzonitrile in N,N-dimethylformamide (DMF) over a wide range of ν, 10–1000 V s^{-1}, with standard deviations of less than ±10%. For most purposes we find it only necessary to use the procedure to determine the R_f setting for a particular potentiostat-electrode configuration one time and that this setting can then be applied in other determinations. The proper settings that we have found usually fall within 85–95% of the oscillation values described in the previous section.

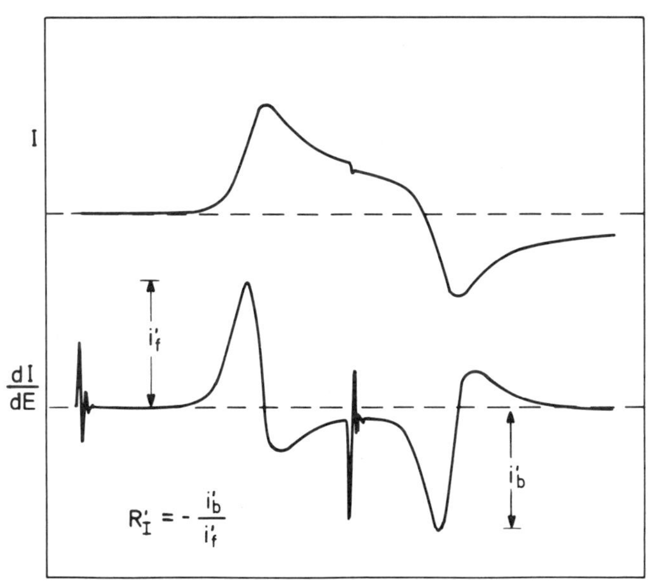

Figure 2. Cyclic voltammetry waveforms; top, current-potential (time); bottom, differentiated current-potential (time).

Ultramicroelectrodes

Although the use of ultramicroelectrodes is not restricted to any specific measurement technique, special considerations warrant treatment in a separate section. Owing to space limitations, we consider applications of ultramicroelectrodes only in the context of cyclic voltammetry. The reader is referred to a recent monograph dealing with the subject for an indication of the scope as well as a comprehensive list of references to the ultramicroelectrode literature [63].

Voltammetry at ultramicroelectrodes is a new and at present very active area of research. The first application of CV at ultramicroelectrodes was reported by Lines and Parker in 1977 [64]. The objective of the initial study was to investigate the behavior of aromatic anion radicals and dianions in poorly conducting media, such as benzene and chlorobenzene. Significantly, it was observed that the response at 25 μm electrodes did not correspond to that expected for linear diffusion, and this was attributed to an edge effect [64]. The enhanced current due to radial diffusion at the edges of ultramicroelectrodes was studied by Heinze [65] using digital simulation. The perspective in using very high voltage sweep rates during CV at ultramicroelectrodes was demonstrated by Ahlberg and Parker [66], who showed that with background subtraction the quasi-reversible reduction of anthracene to the anion radical could be observed in DMF at 10,000 V s^{-1}. Howell and Wightman used the same system with 7 μm diameter electrodes and obtained greatly improved cyclic voltammograms with minimal iR distortion at sweep rates as high as 20,000 V s^{-1} [67]. Cyclic voltammograms for the reduction of anthracene in acetonitrile (MeCN) at sweep rates up to 100,000 V s^{-1} [67] are illustrated in Fig. 3. The million V s^{-1} mark for CV at a 5 μm gold disk during reduction of anthracene in MeCN has recently been surpassed [68]. A significant feature of this work is that it was shown that iR compensation is necessary for quantitative work even when ultramicroelectrodes are used.

The definition of ultramicroelectrodes is somewhat arbitrary and with reference to the common disk geometry is currently meant to describe electrodes with diameters of about 10 μm or less. The electrochemical response at ultramicroelectrodes can be classified into two categories depending upon the time regime, that is, the voltage sweep rate during cyclic voltammetry. At low sweep rates, the faradaic current can greatly exceed that expected for linear diffusion, giving rise to enhanced faradaic-double-layer current ratios. This phenomenon, due to edge effects [63], results in increased sensitivity and allows detection at lower substrate concentrations and also provides the basis for steady-state methods involving ultramicroelectrodes. The second category of electrode response at ultramicroelectrodes occurs at high voltage sweep rates. Under these conditions, linear diffusion is operative and the ultramicroelectrode response does not differ from that of conventional electrodes. The advantage of ultramicroelectrodes in the high sweep rate regime is that iR drops are of much less importance and can become negligible as the electrode diameter is diminished.

Although applications of ultramicroelectrodes can be expected to play an important role in electrode mechanism studies in the future, for example in solvents without deliberately added supporting electrolyte, this is a very new area of research and only a few studies have been carried out. We limit our coverage of this topic to brief discussions of two applications that give an indication of the potential of cyclic voltammetry at ultramicroeletrodes in the high sweep rate regime.

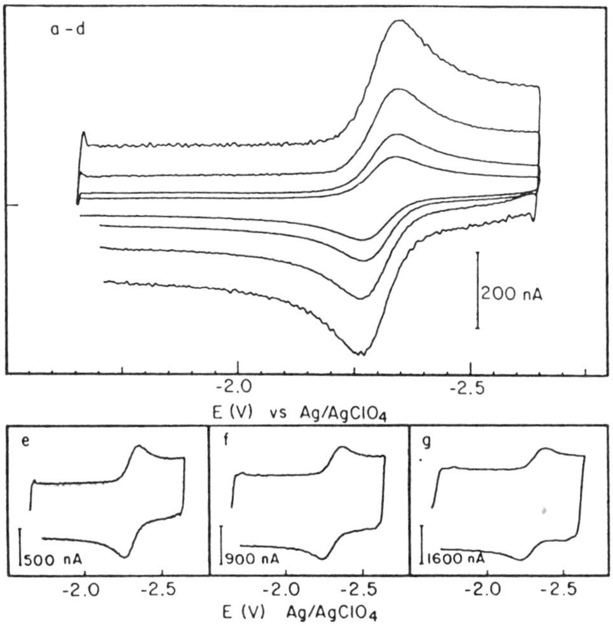

Figure 3. Cyclic voltammograms for the reduction of anthracene (2.22 mM) in MeCN with 0.6 M tetraethylammonium perchlorate at a gold microvoltammetric electrode: (a) 1000; (b) 2000; (c) 5000; (d) 10,000; (e) 20,000; (f) 50,000, and (g) 100,000 V s^{-1}. (From Ref. 67.)

Montenegro and Pletcher [69] confirmed the potential value of high-sweep-rate cyclic voltammetry to determine rate constants for rapid heterogeneous charge transfer (see later). They studied a number of oxidations and reductions at 10 μm Pt electrodes in MeCN using a two-electrode cell and simple instrumentation consisting only of a sweep generator and an operational amplifier to measure the current. The precision of heterogeneous rate constant determinations using Nicholson's [18] theoretical data is illustrated for the oxidation of ferrocene by the data in Table 1. Peak potential separations at 500, 1000, and 3000 V s^{-1} over a range of substrate concentrations resulted in rate constants equal to 1.10, 1.13, and 1.13 cm s^{-1}. The heterogeneous rate constant for the reduction of anthracene was observed to be 2.6 cm s^{-1}, slightly lower than the value reported by Howell and Wightman [67]. The discrepancy is probably due to the fact that different conditions, supporting electrolyte and electrode material, were used in the two studies rather than to experimental error. It appears that CV using ultramicroelectrodes is a superior method for the determination of heterogeneous rate constants of rapid charge transfer processes.

Fitch and Evans [70] chose a rapid first-order reaction to demonstrate the utility of rapid-scan CV at ultramicroelectrodes for the study of kinetics of coupled chemical reactions. Bianthrones exist as thermochromic isomers differing in relative conformations of the two anthronelike ring systems. The stable isomer at ambient temperature is the yellow A form, which is equilibrated with the green B form as the temperature is increased. Reduc-

Table 1. Cyclic Voltammetry Data for the Oxidation of Ferrocene in $MeCN/Bu_4NBF_4$ (0.5 M) at a 10 μm Pt Disk Electrode

	Peak potential separation at ν ($V\ s^{-1}$)		
[Ferrocene] (mM)	500	1000	3000
2	80	90	115
5	90	100	115
10	85	90	110
20	85	90	110
Average value (mV)	85	93	113
k_s ($cm\ s^{-1}$)	1.10	1.13	1.13

Source: From Ref. 69.

tion of A produces $A^{\cdot -}$, which rapidly undergoes a conformational change to the B-like anion radical. The reversible substrate-anion radical redox potentials for the two forms differ by about 230 mV for the isomers of 1,1'-dimethylbianthrone in DMF at 361 K. The rate constant for the conversion of B to A was determined as 500 s^{-1} under these conditions by comparing experimental cyclic voltammograms obtained at 200, 500, 1000, and 5000 V s^{-1} using 5 μm Pt disk electrodes to theoretical voltammograms obtained by digital simulation. Both three-electrode (potentiostat) and two-electrode configurations were used, and no difference could be detected in the response with a substrate concentration of 3.4 mM at a voltage sweep rate of 2000 V s^{-1}. The onset of distortion due to iR drop was observed at about 8000 V s^{-1} at a 5 μm Pt disk electrode during the reduction of 5.36 mM 9,10-anthraquinone in DMF containing 0.6 M tetraethylammonium perchlorate.

C. Charge Transfer Reactions

LSV and especially CV are convenient methods of studying charge transfer reactions. In this section we examine the relationship between current and potential for cases in which the product of the electrode reaction is stable in solution; that is, there are no chemical reactions coupled to charge transfer at the electrode. The reactions are conveniently classified as nernstian, quasi-reversible, or irreversible depending on the magnitude of the rate constants for reaction (5) in relationship to ν (see also Chap. 2).

$$A \pm e^- \underset{k_{hb}}{\overset{k_{hf}}{\rightleftharpoons}} B \tag{5}$$

A nernstian charge transfer is one in which the current is controlled only by the potential and diffusion, and the surface concentrations of A and B at all potentials are governed by the Nernst equation. A quasi-reversible charge transfer is characterized by mixed control of diffusion and the rate of electron transfer. These two classifications depend on ν; that is, a process that appears nernstian at low ν may become quasi-reversible at higher values of ν. The irreversible charge transfer is one in which B is not electroactive in the potential range where the measurements are made. As before, the borderline between quasi-reversible and irreversible charge transfer changes with changing ν for a particular reaction.

Nernstian Charge Transfer

In contrast to the sigmoidal current-potential curves observed in polarography (Sec. IV) or rotating electrode voltammetry (Sec. V), the distinguishing feature of the LSV wave is the presence of a peak. The differences in form are a consequence of the mode of mass transfer. Convective mass transport provides a constant source of solution with the concentration of A equal to the bulk value during polarography and rotating electrode voltammetry, and the current is dependent only on the potential at a constant rate of mass transport. On the other hand, when diffusion is the only mode of mass transport, the concentration gradient of A in the vicinity of the electrode is dependent not only on the potential but also on time, and the magnitude of the current is proportional to $\nu^{1/2}$. At potentials lower than the polarographic $E_{1/2}$, the current is very nearly that predicted from the Nernst equation, whereas at higher potentials it is dictated by diffusion. In the intermediate region of mixed control, a peak is observed at $E_{1/2} - 1.109RT/nF$. Thus for a linear sweep the current is a function of both potential and sweep rate. The peak on the reverse scan of a cyclic voltammogram is of the same shape as the forward or LSV wave.

A cyclic voltammogram for a nernstian charge transfer is illustrated in Fig. 4. If the system behaves ideally, the response satisfies the following criteria:

1. The peak potential separation ($\Delta E_p = E_p^{ox} - E_p^{red}$) is equal to $2 \times 1.109RT/nF$ or 57 mV for n = 1 at 298 K and independent of ν.
2. The peak current ratio ($-i_p^{ox}/i_p^{red}$) is unity and independent of ν.
3. The peak current, most conveniently measured on the forward sweep, increases linearly with $\nu^{1/2}$.
4. The half-peak width ($E_{p/2} - E_p$, where $E_{p/2}$ is the potential at which $i = i_p/2$) is equal to $2 \times 1.109RT/nF$.

The LSV wave conforms to the last two criteria.

A difficulty arises in determining peak-current ratios in that it is not obvious how the baseline for the reverse sweep should be determined. The most reliable procedure appears to be graphic methods described by Nicholson [71], which should be used with great care, however [72].

Quasi-Reversible Charge Transfer

The list of criteria for nernstian voltammograms does not hold for quasi-reversible systems. The peak potential separations and the half-peak widths are dependent on ν, and the peak currents do not vary linearly with $\nu^{1/2}$. The peak-current ratio is unity only under special circumstances. The reason for the breakdown in the relationships is that the current in this case is also controlled by the kinetic parameters for reaction (5).

The kinetic diffusion equations cannot be solved analytically, and it was necessary to resort to numerical methods to obtain mathematical descriptions of the voltammograms. Calculations for the LSV wave were reported by

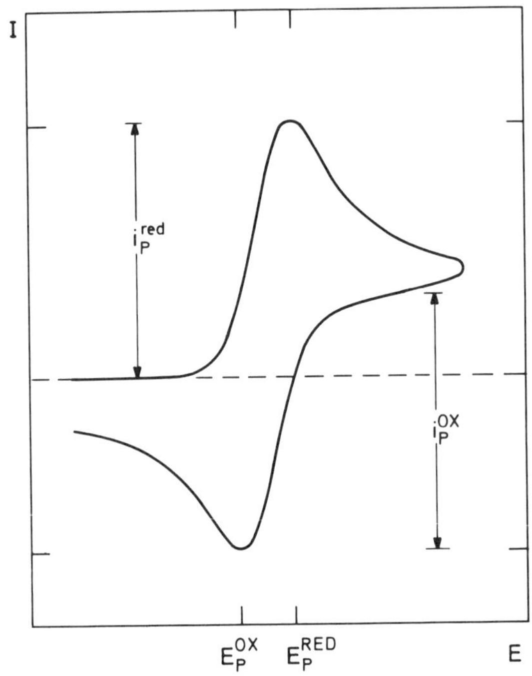

Figure 4. Cyclic voltammogram of a nernstian system.

Matsuda and Ayabe [73] and later by Nadjo and Savéant [22]. Calculations by Nicholson and Shain [18] established CV as a method for studying the kinetics of heterogeneous charge transfer. The most significant experimental observation is the peak potential, and this is dependent on the heterogeneous charge transfer rate constant k_s, which is equal to k_{hf} and k_{hb} at the reversible potential, as well as the transfer coefficient α. It is customary to describe the response in terms of normalized rate constants, which are defined in Eqs. (6) and (7) for LSV and CV, respectively.

$$\Lambda = k_s \left(\frac{D_A nF}{RT} \right)^{-1/2} \nu^{-1/2} \tag{6}$$

$$\Psi = \gamma^{\alpha} k_s \left(\frac{\pi D_A nF}{RT} \right)^{-1/2} \nu^{-1/2} \tag{7}$$

In Eqs. (6) and (7), D_A is the diffusion coefficient of A and γ in Eq. (7) equals $(D_A/D_B)^{1/2}$. In order to use LSV or CV for heterogeneous kinetic studies, working curves are usually constructed in which E_p (LSV) or ΔE_p (CV) is plotted as a function of Λ or Ψ, respectively.

Irreversible Charge Transfer

If k_{hf}, Eq. (5), becomes progressively smaller, the shape of the quasi-reversible wave continues to change, for a given value of ν, until a limiting form is reached. This limiting form, the irreversible LSV wave, is is characterized by a half-peak width of 96.2 mV at 298 K for a one-electron transfer, and E_p changes by RT/nF for each decade change in ν. When the limiting form is reached, k_{hb} is negligible at potentials at which the forward process takes place.

D. Coupled Chemical Reactions

Some of the most useful applications of LSV and CV are concerned with the study of mechanisms of the reactions of electrogenerated intermediates. A very large share of the literature that has been cited deals with this aspect of voltammetry [1,2,15,17,19–30,36–38,40,42]. In terms of the use of LSV and CV for the study of reactions coupled to charge transfer, it is convenient to classify the processes with regard to the nature of the response. A purely kinetic wave is one in which the charge transfer is nernstian and a stationary state is established by mutual compensation of the chemical reaction of the intermediate and the diffusion processes [22]. Another classification results when the rate of the chemical reaction is such that the shape of the LSV wave is affected by ν, but the charge transfer is still nernstian. Under these conditions the intermediate can be observed during CV if ν is sufficiently high. The third classification takes into account processes with homogeneous reaction rates in the range encompassed by the first two, but in this case the charge transfer is quasi-reversible.

Purely Kinetic Waves

In this case the position, but not the shape, of the LSV wave is dependent on ν. LSV is a powerful tool for the study of these processes. Very thorough theoretical analyses of the LSV response for various reaction mechanisms have been carried out [17,19–28], and a series of papers [16,40,74,75] has been devoted to assimilating the theoretical results in a form useful to the experimentalist.

During the study of homogeneous chemical reactions following the charge transfer reaction (5), three observables relate the changes in E_p to the variables ν, the bulk concentration of substrate C_A^*, and the bulk concentration of an additional reactant C_X^*. In order to determine the theoretical response for any mechanism, it is necessary to derive the rate law (8) for that mechanism:

$$\text{Rate} = k_{obs} C_A{}^a C_B{}^b C_X{}^x C_I{}^i \tag{8}$$

In Eq. (8) the superscripts are the reaction orders and I is a substance formed during the reaction that participates further in the process. Once Eq. (8) has been derived, the LSV observables can be predicted directly from Eqs. (9) through (11).

$$\frac{dE_p}{d \log \nu} = \frac{1}{b + 1} (\ln 10) \frac{RT}{nF} \tag{9}$$

$$\frac{dE_p}{d \log C_A^*} = \frac{a + b + x + i - 1}{b + 1} (\ln 10) \frac{RT}{nF} \tag{10}$$

$$\frac{dE_p}{d\ \log C_X^*} = \frac{x}{b+1} (\ln 10) \frac{RT}{nF} \tag{11}$$

The kinetic shift of the LSV wave recorded under purely kinetic conditions is dependent on the magnitude of the rate constant k_{obs} relative to ν. For instance, for the commonly encountered eCe_h scheme, (12) through (15),

$$A + e^- \rightleftharpoons A^{\cdot -} \tag{12}$$

$$A^{\cdot -} + HB \xrightarrow{k_{13}} AH^{\cdot} + B^- \tag{13}$$

$$AH^{\cdot} + A^{\cdot -} \longrightarrow AH^- + A \tag{14}$$

$$AH^- + HB \longrightarrow AH_2 + B^- \tag{15}$$

here given for a protonation reaction, the associated rate law under pseudo-first-order conditions is given by

$$\frac{d[A^{\cdot -}]}{dt} = -2k_{13}C_{HB}^*[A^{\cdot -}] = -2k_{obs}[A^{\cdot -}] \tag{16}$$

The value of k_{obs} for this mechanism is related to the kinetic shift through [22]

$$E_p - E^o = -\frac{RT}{nF}\left(0.783 - \frac{1}{2} \ln \frac{k_{obs}RT}{2\nu nF}\right) \tag{17}$$

Analogous equations have been derived for other mechanisms as well.

Mixed Sweep Rate and Kinetic Control

The characteristic features of processes in this category are (1) that the LSV relationship (9) does not hold, that is, E_p does not vary linearly with log ν, and (2) the peak current ratio during CV is less than 1.0 but is measurable.

The usual procedure for studying coupled chemical reactions by CV is to compare experimental peak current ratios to theoretical working curves calculated for the suspected reaction mechanism (rate law). Because of the difficulty in evaluating the baseline for the reverse scan, CV is a highly qualitative kinetic method. The development of DCV as a quantitative method for the study of homogeneous chemical reactions coupled to heterogeneous charge transfer has changed this situation.

Equations for obtaining rate constants from values of $\nu_{1/2}$, defined as the voltage sweep rate at which R_I' is equal to 0.5 (see Fig. 2), of the form in Eq. (18), have been reported [76]. In Eq. (18) the superscript on the substrate concentration is zero for a first-order and −1 for a second-order reaction of the electrogenerated intermediate:

$$k_{obs} = (\text{constant})\ \nu_{1/2} C_A^{\ a} \tag{18}$$

Mixed Heterogeneous and Homogeneous Kinetic Control

The limits of heterogeneous and homogeneous rate constants in relationship to ν have been established by theoretical calculations [22]. The relationships for nernstian systems developed in the preceding two sections are not valid for processes in this category. The most viable approach appears to be to compare theoretical voltammograms calculated for the appropriate heterogeneous and homogeneous rate parameters directly with experimental voltammograms. Normalized potential sweep voltammetry [38] represents an accurate and convenient method for making such comparisons. This technique is briefly discussed in the following section.

E. More Detailed Linear Sweep Voltammetry Analysis

A disadvantage of LSV and CV analyses of voltammetric data is that only peak potentials and currents are generally used; the remainder of the available data are not used. Three techniques that use the full curves are discussed in this section. Listed in decreasing order of mathematical sophistication, but not necessarily in order of usability, these are (1) semi-integral electroanalysis [79–80] and the equivalent convolution potential sweep voltammetry (CPSV) [56,81–84], (2) normalized potential sweep voltammetry (NPSV) [38,39], and (3) linear current potential analysis (LCP) [37].

Semi-integral Electroanalysis and Convolution Potential Sweep Voltammetry

The mathematics involved in these transformations are beyond the scope of this chapter, and only a qualitative description is given here. The essential feature of the methods is that semi-integration [79–80] or computing convolution integrals [56,81–84] of the voltammetric data, usually obtained in digital form, results in the transformation of the LSV wave to what has been termed a neopolarogram [78]. Logarithmic analyses can be carried out on the neopolarogram in much the same way as with classic polarography (see Sec. IV). The main emphasis of the Oldham group was on analytic applications, whereas the Savéant group has applied the method to both heterogeneous and homogeneous kinetic studies. Some of the logarithmic slopes for various electrode processes are summarized in Table 2 along with the

Table 2. Comparison of Linear Current Potential Analysis with Those of LSV and CPSV for Mechanism Assignment

Mechanism	LCP[a] (mV)	LSV[b] (mV)	CPSV[c] (mV)
eC	68.9	29.6	59.2
eCe_h	68.9	29.6	59.2
eC(dim)	53.2	19.7	39.4
eeC	35.1	14.8	29.6

[a]Linear current potential analysis, $dE/dI_N = dE/d(i/i_p)$.
[b]Linear sweep voltammetry analysis, $dE_p/d \log \nu$.
[c]Convolution potential sweep voltammetry.

corresponding LSV slope and that from LCP analysis (see below). In addition, CPSV is a convenient method for heterogeneous kinetics as long as k_s does not exceed about 0.1 cm s^{-1}.

Normalized Potential Sweep Voltammetry

This method consists of normalizing the current at all points along the LSV wave by dividing by the peak current. The potential scale is then redefined relative to that where $I_N = 0.5$ ($I_N = i/i_p$). A three-dimensional analysis in which theoretical electrode-potential data are taken as the X axis, experimental electrode-potential data as the Y axis, and I_N as the Z axis results in a direct comparison of experiment with theory when the curves are projected onto the X-Y plane. If the theoretical data properly describe the process under study, a slope of 1.000, an intercept of 0.00, and a correlation coefficient of unity result from the analysis. Thus the method involves the use of data along the entire voltammetric wave and can be used for both homogeneous [38] and heterogeneous [39] kinetic studies. An advantage of the method over CPSV is that the only requirement to use the method is that an LSV peak is observed. Thus when the neopolarogram does not reach a limiting plateau, as is often the case when there are interfering processes, CPSV cannot be applied, but NPSV can. In addition, the method is more easily applied than CPSV and comparable information is obtained by the two methods.

Linear Current-Potential Analysis

The objective of both CPSV and NPSV is to make use of more of the information hidden in the LSV wave. It is difficult to carry out either of the analyses without digital data and a computer. Thus the methods are not generally available for studying electrode processes to chemists who do not have the necessary data-retrieval systems. LCP can fill this gap. Analysis of theoretical current-potential data for nernstian and purely kinetic waves revealed that a nearly linear region exists where I_N (see earlier discussion) ranges from 0.50 to 0.75 [37]. An LCP analysis involves taking the slope dE/dI_N over data measured at intervals of $0.05I_N$ over this range. Slightly different fractions were originally reported, but this has a negligible effect on the slopes. Slopes have been reported for nernstian as well as purely kinetic waves, and some of the results are included in Table 2. An accurate LCP analysis can be made on voltammograms recorded on X-Y recorders provided that a suitable expansion is used on the potential axis. The method gives the same type of information as CPSV and NPSV and is certainly the most convenient to apply. It has the disadvantage that the potential region over which the data are used is restricted. A useful application of the method is to verify that LSV waves have the purely kinetic shape required for the LSV analysis.

F. Applications of Linear Sweep Voltammetry and Cyclic Voltammetry

The applications of LSV and CV to the study of charge transfer reactions and coupled chemical reactions are so numerous that a review of the subject is clearly beyond the scope of this chapter. The examples described here were selected to demonstrate the utility of these techniques in the study of heterogeneous and homogeneous kinetics. The first application described does not fit into this category but is included in order to demonstrate the qualitative use of CV in the study of organic electrode processes.

Use of Cyclic Voltammetry in the Study of Potential-Dependent Processes

The unsymmetrical diarylethane (I) was observed to give rise to the products II and III on anodic oxidation in either MeCN or $CH_2Cl_2-CF_3COOH$ (TFA) (4:1), whereas compound IV gave only V under similar conditions [85].

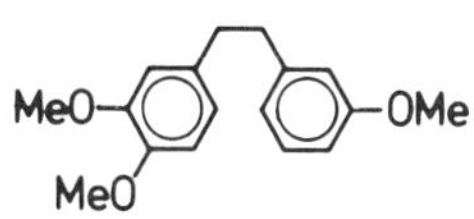

(I)

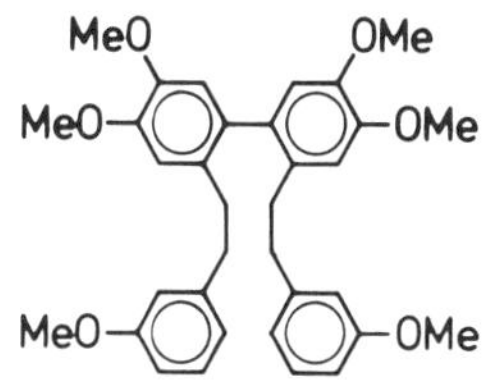

(II)

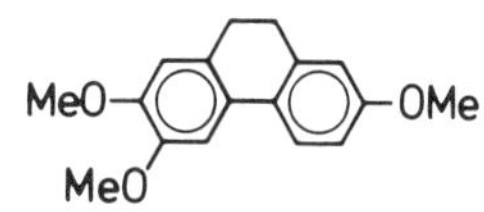

(III)

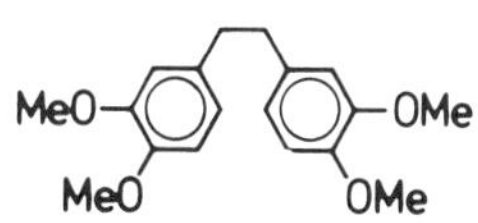

(IV)

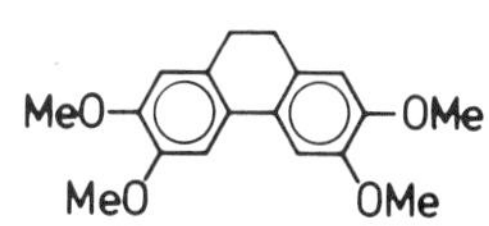

(V)

The cyclic voltammogram for the oxidation of IV in MeCN is illustrated in Fig. 5. The oxidation peak O_1 corresponds to the transfer of $3e^-$ accompanied by the formation of V in the form of the cation radical $V^{+\cdot}$. The reversible redox couple R_2-O_2 is due to $V^{+\cdot} + e^- \rightleftharpoons V$. The voltammogram for the oxidation of I under the same conditions is considerably more complex and is illustrated in Fig. 6. In this case a small oxidation peak O_2 appears close to the first, and after extending the sweep out beyond the second oxidation stage O_3, no fewer than three reduction peaks are observed. That the relative proportion of products giving rise to R_4 and R_5, respectively, is potential dependent was shown by potential-step sweep experiments in which the step potential was varied according to the potentials marked as 1, 2, 3, 4, and 5. The product giving rise to R_4 was observed to predominate at the lowest step potentials, whereas R_5 was the major reduction process after the potential had been stepped out to the potential of O_3. The CV evidence was central in establishing that II arises from coupling of $I^{+\cdot}$, whereas III is the product of intramolecular cyclization of the dication diradical of I.

Distinguishing Between Radical-Radical and Radical-Substrate Coupling Reactions by Linear Sweep Voltammetry

Upon formation of a radical ion B during a nernstian charge transfer reaction (5), a number of processes leading to dimeric products are feasible. A question of mechanistic importance is whether the reactions involve radical-radical dimerization, Eq. (19), or radical-substrate coupling, Eq. (20), followed by an electron transfer reaction, Eq. (21). The first possibility gives rise to rate law (22), whereas the second can give Eq. (23) or (24) depending on which step is rate determining.

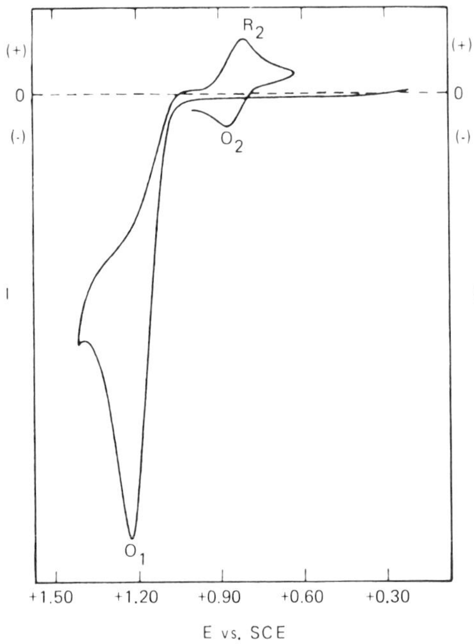

Figure 5. Cyclic voltammogram for the oxidation of 3,3',4,4'-tetramethoxybibenzyl (IV) in MeCN containing $NaClO_4$ (0.1 M). Voltage sweep rate: 86 mV s^{-1}. [From Ref. 85. Reprinted with permission from J. Amer. Chem. Soc., 95, 7132 (1973). Copyright 1973 American Chemical Society.)

$$A \pm e^- \rightleftharpoons B \tag{5}$$

$$2B \xrightarrow{k_{19}} B{-}B \tag{19}$$

$$A + B \underset{k_{-20}}{\overset{k_{20}}{\rightleftharpoons}} A{-}B \qquad \left(K_{20} = \frac{k_{20}}{k_{-20}}\right) \tag{20}$$

$$A{-}B + B \xrightarrow{k_{21}} B{-}B + A \tag{21}$$

$$\text{Rate} = k_{19}[B]^2 \tag{22}$$

$$\text{Rate} = k_{20}[A][B] \tag{23}$$

$$\text{Rate} = k_{obs}[A][B]^2 = k_{21}K_{20}[A][B]^2 \tag{24}$$

The power of LSV using the reaction order approach [74] is obvious on applying Eqs. (9) and (10) to rate laws (22) through (24). The sweep rate dependence, Eq. (9), is predicted to be 1/3 ln 10 RT/nF, 1/2 ln 10 RT/nF, and 1/3 ln 10 RT/nF for rate laws (22), (23), and (24), respectively. The

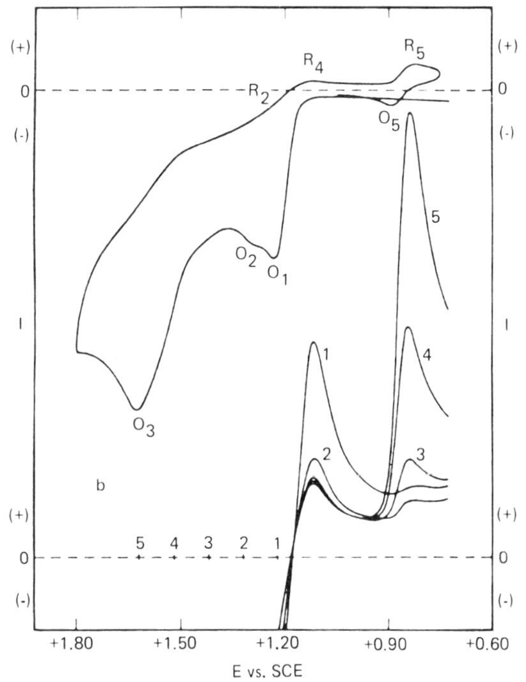

Figure 6. Cyclic and potential-step sweep voltammogram for the oxidation of 3,3',4-trimethoxybibenzyl (I) at −28°C in MeCN containing $NaClO_4$ (0.1 M). Hold potentials: 1, +1.22 V; 2, +1.32 V; 3, +1.42 V; 4, +1.52 V; 5, +1.62 V. The current amplification of the potential-step sweep voltammograms is 10 times that of the cyclic voltammogram. [From Ref. 85. Reprinted in part with permission from J. Amer. Chem. Soc., 95, 7132 (1973). Copyright 1973 American Chemical Society.]

substrate concentration dependence is predicted by Eq. (10) to be equal to (1/3)(ln 10)(RT/nF), (1/2)(ln 10)(RT/nF), and (2/3)(ln 10)(RT/nF) for the same three rate laws. (At 298 K, the value of (ln 10)(RT/nF) is equal to 59.2 mV for n = 1.)

A number of studies of this reaction type may be cited in which LSV employing quantitative measurement techniques [40] allowed conclusive mechanism assignments to be made. The anodic coupling of 4,4'-dimethoxystilbene (DMS) in MeCN was found to give rise to a number of products depending on the reaction conditions [86–88]. Rotating ring-disk [88] (see Sec. V.C) and spectroelectrochemical studies [87] (see Sec. VII) led to the proposal that the predominant reaction pathway involved dimerization of the cation radicals in the key step

$$2DMS^{\overset{+}{\cdot}} \longrightarrow \text{dimer dication} \qquad (25)$$

However, more recent LSV studies have conclusively shown that the primary step involves radical-substrate coupling. The substrate concentration dependence during LSV is the diagnostic feature capable of differentiating the two mechanisms. At 298 K, $dE_p/d \log C_A^*$ is expected to equal 19.7 mV per decade for rate law (22) and 39.4 mV per decade for rate law (24). The observed slopes were 37.4 ± 1.1 and 36.5 ± 0,3 mV per decade in the absence and the presence of methanol, respectively. LCP analysis resulted in dE/dI_N equal to 53.6 ± 0.3 mV, verifying that the process results in a purely kinetic wave for a second-order process. These data can leave little doubt that under the conditions of the voltammetric studies the primary process is the coupling between $DMS^{\cdot+}$ and substrate.

In view of the large number of mechanistic studies that have been carried out on electrohydrodimerization of activated olefins (see Chap. 22), the LSV study of this reaction is perhaps even more illustrative. Contrary to the general belief that these reactions all involve an initial radical-radical coupling, the LSV study conclusively showed that for the electrohydrodimerization of methyl cinnamate in anhydrous MeCN, the primary reaction step is the coupling of the anion radical with substrate [89]. In this case, $dE_p/d \log C_A^*$ was observed to be 36.8 ± 2.3 mV. In addition, DCV showed that the reaction orders are more consistent with rate law (24) than with rate law (22).

Distinguishing Between First- and Second-Order Reactions by Linear Sweep Voltammetry

This may seem to be an almost trivial task, in that Eq. (9) indicates that reactions first order in B will give rise to $dE_p/d \log \nu$ of $(1/2)(\ln 10) \times (RT/nF)$, whereas a reaction second order in B is accompanied by a slope of $(1/3)(\ln 10)(RT/nF)$. In fact, this is a very straightforward matter provided that the measurement precision is high enough to allow the slopes to be obtained over the very small interval of ν normally available to ensure nernstian behavior. The latter requires quantitative measurement procedures [40].

A case in which LSV has provided the deciding evidence can be found in the reduction of 9-diazofluorene ($Fl{-}N_2$). This process has been proposed [90] to involve the first-order decomposition of the anion radical to the carbene anion radical, reaction (26), but this suggestion was challenged and an anion radical dimerization mechanism [91] was suggested, reaction (27).

$$(Fl{-}N_2)^{\cdot-} \longrightarrow Fl^{\cdot-} + N_2 \tag{26}$$

$$2(Fl{-}N_2)^{\cdot-} \longrightarrow (Fl{-}N_2)_2^{2-} \tag{27}$$

A subsequent LSV study [92] showed that $dE_p/d \log \nu$ was equal to 20.7 ± 1.7 mV per decade and $dE_p/d \log C_A^*$ was equal to 19.2 ± 1.4 mV per decade when the reaction was carried out in MeCN at 23°C. Rate-determining reaction (26) leads to a prediction of 29.6 and 0 mV per decade for the two slopes, whereas reaction (27) leads to the prediction of 19.7 mV per decade for both slopes. The data are clearly consistent only with the latter.

The Protonation of Anion Radicals Derived from Aromatic Hydrocarbons

Apparently contradictory experimental results have stimulated much activity in the study of the mechanism of protonation of anion radicals derived from aromatic hydrocarbons [36,37,40,46,75,93–98]. The results of reaction order analysis by double-potential-step chronoamperometry (DPSCA) (see Sec. III.A) and DCV for the protonation of anthracene anion radical by phenol in DMF

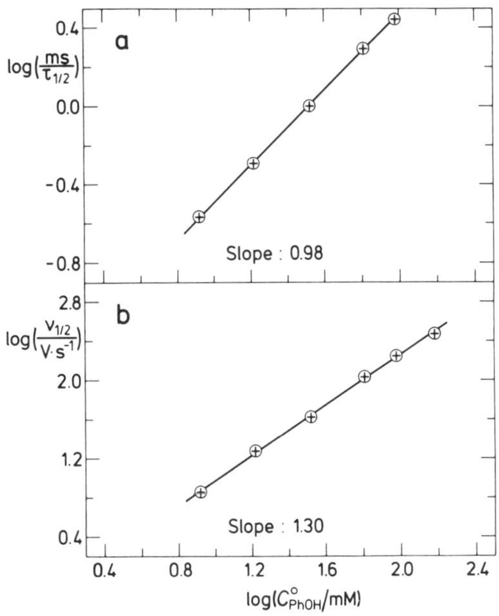

Figure 7. Reaction order analysis by (a) DPSCA and (b) DCV of the protonation of anthracene anion radical by phenol in DMF. $C_A^* = 1$ mM, T = 297 K. (From Ref. 97.)

are given in Fig. 7 [97]. The apparent reaction orders, that is, the slopes of the linear regression lines defined by the experimental points, are clearly different, and only that, ≈1.0, observed by means of DPSCA is apparently in agreement with the mechanism shown earlier. The apparent reaction order observed by means of DCV (1.2–1.3) pointed toward a rate expression higher than first order in [HB] (HB = PhOH), which was supported by LSV slopes [Eq. (11)] in the range of 32–36 mV per decade (Ref. 46 and references cited therein). (A first-order reaction is predicted to give a slope of 29.6 mV per decade and a second-order reaction a slope of 59.2 mV per decade.) It has been demonstrated [97], however, that these observations are mainly due to the differences in conversion during these very different experiments. During DPSCA the conversion is low and the reactions studied are essentially under pseudo-first-order conditions, even at moderate reagent-substrate ratios, whereas during DCV and LSV the conversion is high and the consumption of any added reagent cannot be neglected unless it is present in an excess of at least 25 (DCV). The difference in conversion is illustrated by the concentration profiles obtained by digital simulation (Fig. 8) [97]. The intuitive solution to this problem is to carry out the measurements at concentrations of HB sufficiently high to secure pseudo-first-order conditions, even for DCV and LSV. However, this approach is not feasible because of unwanted kinetic contributions from the dimer, $(HB)_2$, at high concentrations of HB. The best solution to problems of this kind seems to be to carry out the measurements under second-order conditions, that is, at similar concentrations of A and HB. Experiments carried out in DMF this way have unambiguously demonstrated [97,98] that the reaction scheme (12) through (15), is not complete and in addition, for HB = PhOH, should in-

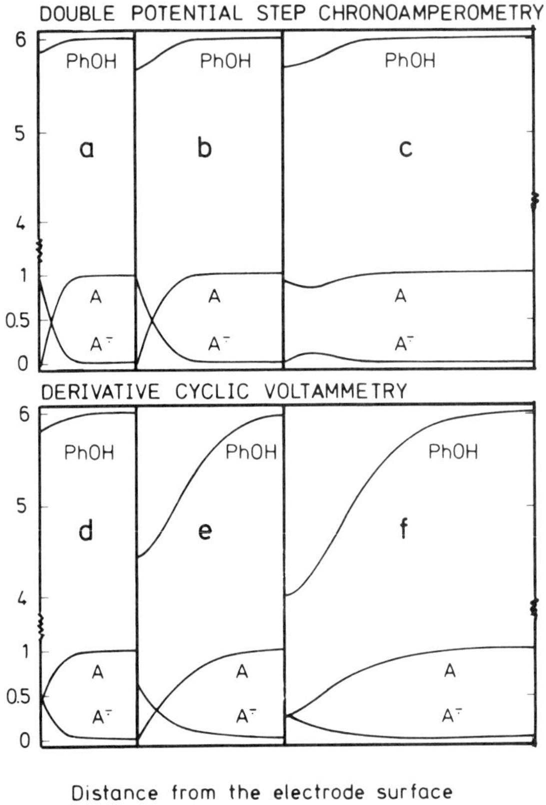

Figure 8. Simulated concentration profiles for anthracene (A), anthracene anion radical ($A^{\cdot-}$) and phenol (PhOH) at $C^*_{PhOH}/C^*_A = 6$ [mechanism (12) through (15) plus (28)]. The numbers given at the Y axis represent values of $[A]/C^*_A$, $[A^{\cdot-}]/C^*_A$, and $[PhOH]/C^*_A$, respectively. The profiles correspond to (a) $t = 0.4t_f$; (b) $t = t_f$; (c) $t = 2t_f$; (d) $t = t(i'_f)$; (e) $E = E_{switch}$; and (f) $t = t(i'_b)$. (From Ref. 97.)

clude the essentially irreversible formation of the hydrogen-bonded complex between HB and B^-:

$$HB + B^- \longrightarrow HB/B^- \tag{28}$$

The working curves for three cases, (1) mechanism (12) through (15), (2) mechanism (12) through (15) plus (28), and (3) mechanism (12) through (15) plus irreversible formation of a 2:1 hydrogen-bonded complex, are shown in Fig. 9 together with experimental points [97]. The agreement with mechanism 2 is obvious.

The value of k_{13} for the protonation of anthracene anion radical by phenol is 5×10^3 M^{-1} s^{-1} in DMF [36,93,95,98]. For a faster reaction the kinetics may be studied by measurements of the kinetic shift as discussed earlier [Eq. (17)]. However, because of the problems associated with the high concentrations of HB necessary for the validity of Eq. (17), that is, the partic-

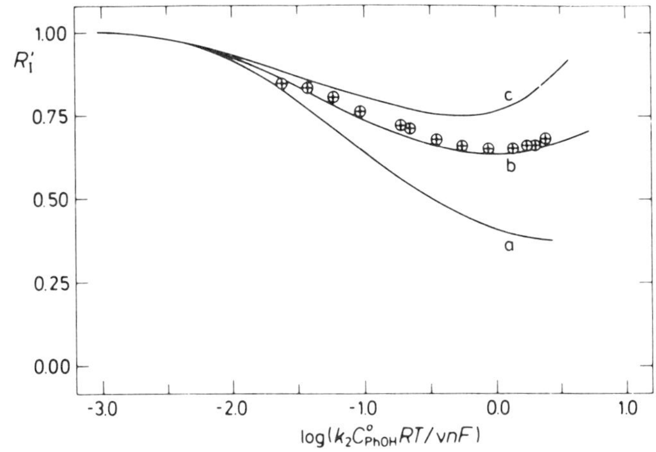

Figure 9. Derivative cyclic voltammetry working curves for (a) mechanism (12) through (15); (b) mechanism (12) through (15) plus (28); and (c) mechanism (12) through (15) plus fast and irreversible formation of a 2:1 $PhOH-PhO^-$ complex at $C^*_{PhOH}/C^*_A = 1$ together with experimental points (⊕) obtained at $C^*_{PhOH} = C^*_A = 2$ mM and T = 295 K (k_2 at the figure equals k_{13} in this chapter). (From Ref. 97.)

ipation of $(HB)_2$ in the protonation step, it was found more convenient to monitor the kinetic shift as $E_p - E_{p/4}$ under second-order conditions, where $E_{p/4}$ is the potential of the LSV wave at $i = i_p/4$ [99]. As an example of this approach, the experimental points observed for HB = PhCOOH are given in Fig. 10 together with the working curve.

Application of Convolution Potential Sweep Voltammetry

1,3-Dibenzoylpropane gives rise to the corresponding cis-diol upon cathodic cyclization in MeCN [83]. It was pointed out that a number of mechanisms are possible for this process, including cyclization of either the anion radical or the dianion [83]. To distinguish between the various possibilities by LSV, it would be necessary to conduct experiments in both buffered and unbuffered media. The latter arises because of the necessity of determining the role of the protonation steps in determining the overall kinetics of the reaction. On the other hand, it was shown that CPSV is capable of distinguishing between the various possibilities without resorting to buffering experiments [25]. The mechanism was suggested to consist of an initial disproportionation of the anion radical to the dianion, Eq. (29); the rate-determining step is either the cyclization (30) or the subsequent protonation (31).

$$2[PhCO(CH_2)_3COPh]^{\cdot -} \rightleftharpoons [PhCO(CH_2)_3COPh]^{2-} + PhCO(CH_2)_3COPh \quad (29)$$

$[PhCO(CH_2)_3COPh]^{2-} \longrightarrow$ 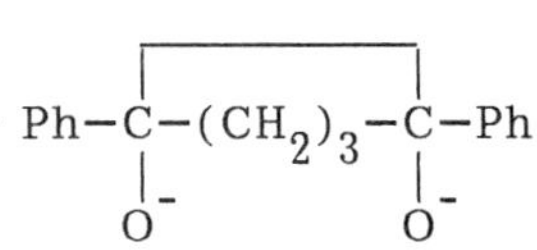 (30)

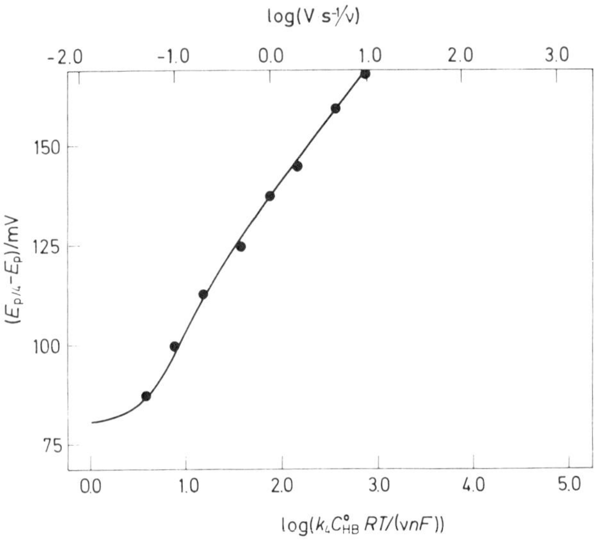

Figure 10. Theoretical data (full line) for $E_{p/4} - E_p$ at T = 298 K for mechanism (12) through (15) at $C^*_{HB}/C^*_A = 0.5$. The experimental points (•) are for A = anthracene (2 mM) and HB = benzoic acid (1 mM) obtained at the sweep rates indicated by the scale at the top of the figure (k_4 at the figure equals k_{13} in this chapter). (From Ref. 99.)

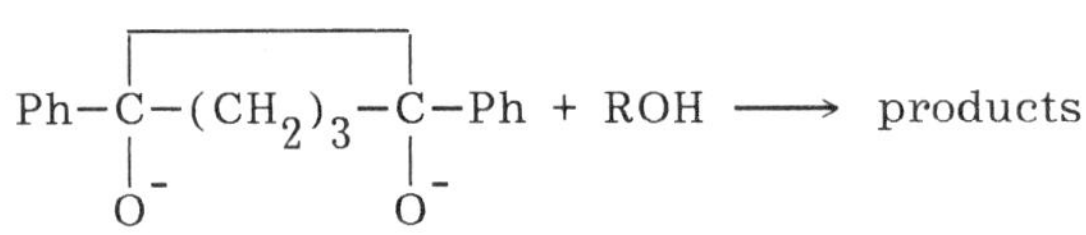

(31)

Demonstration of the Potential Dependence of the Transfer Coefficient

According to electron transfer theory [100], the transfer coefficient α is potential dependent. This problem has been investigated with the aid of a number of different experimental techniques over the past several years, and most recently both CPSV [101] and CV [102] have been used to demonstrate the phenomenon during the reduction of 2-nitro-2-methylpropane in MeCN. Both studies indicate that $d\alpha/dE$ is an approximately linear function and is equal to about 0.4 V^{-1}, depending on the reaction condition. The CV study [102] consisted of the comparison of experimental voltammograms with those calculated by digital simulation.

Summary

This brief survey of LSV and CV applications should serve to suggest the great potential of these techniques in homogeneous kinetic studies. It appears that only the surface of a large reservoir of possible applications has been sampled.

III. POTENTIAL-STEP AND CURRENT-STEP METHODS

A. Chronoamperometry and Double-Potential-Step Chronoamperometry

When the potential of a planar electrode is suddenly shifted from a value at which no current flows to a value at which the electron transfer (5) proceeds at the diffusion-controlled rate, the current-time behavior is given by Eq. (32), where A is the electrode area, C_A^* is the bulk concentration of A, D_A is the diffusion coefficient of A, and $i_c(t)$ is the double-layer charging current [4,5]

$$i(t) = \frac{nFAC_A^* D_A^{1/2}}{(\pi t)^{1/2}} + i_c(t) \tag{32}$$

The latter quantity usually decays to zero after 0.5 ms or less and may be neglected in experiments lasting 1 ms or more. According to this equation, which for $i_c(t) = 0$ is known as the Cottrell equation, the current approaches zero when the time approaches infinity. However, undisturbed linear diffusion can be maintained only over rather short time intervals unless special precautions are taken, and the measurements of current-time curves, called chronoamperometry (CA), are often complicated by additional modes of transport. Therefore, the use of properly shielded electrodes is recommended in chronoamperometric experiments exceeding approximately 1 s.

The application of CA for monitoring homogeneous chemical reactions is limited by the nature of the process associated with the heterogeneous electron transfer. As appears from Eq. (32), the only parameter that may be affected by follow-up reactions is n, the number of electrons transferred, whereas preceding chemical steps are able to influence both n and the effective concentration of electroactive substrate at the electrode surface. Thus, mechanistic schemes like the simple eC and eC(dim) cannot be characterized by this type of experiment. However, it has been demonstrated that if the potential is shifted to values insufficient for diffusion-controlled electron transfer, rate constants for eC processes may be measured [103]. Working curves have been calculated relating the apparent number of electrons exchanged n_{app} given by Eq. (33) to the dimensionless parameter log kt, where k is

$$n_{app} = \frac{it^{1/2}(\text{actual})}{it^{1/2}(\text{diff.contr})} \tag{33}$$

the first-order (or pseudo-first-order) rate constant for the chemical step (Fig. 11). However, experiments of this kind may be hard to perform with high precision because of difficulties in the accurate control of the potential in the region close to E^o.

If, during the chemical step, a product is formed that undergoes further electron transfer at the applied potential, an eCe-type process, the value of n_{app} depends on k even under conditions in which the heterogeneous electron transfer is diffusion controlled [104]. It is easily understood that n_{app} approaches unity when k decreases toward zero, whereas a value of n_{app} equal to 2 is observed for large values of k. The competition between heterogeneous electron transfer (eCe) and homogeneous electron transfer (eCe_h) in the second e step and the possibility of distinguishing between these two pathways has been thoroughly discussed by Amatore and Savéant

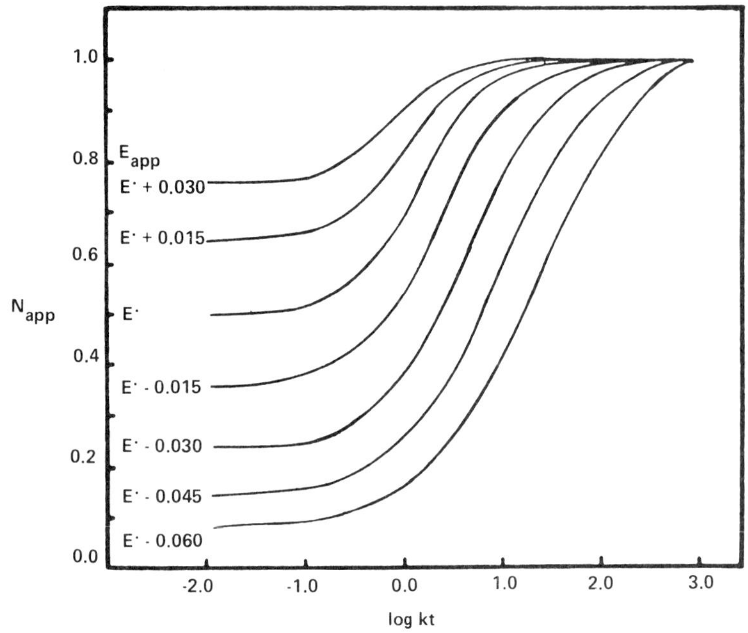

Figure 11. EC working curves for potential-dependent chronoamperometry. [From Ref. 103. Reprinted with permission from J. Phys. Chem., 76, 1666 (1972). Copyright 1972 American Chemical Society.]

[105,106]. It was concluded that the eCe_h pathway dominates the eCe pathway unless the measurement time is kept below 10^{-7} s. This is not possible in CA, and accordingly, the eCe mechanism cannot be characterized by this technique.

If the current pulse is followed by a linear potential scan, the diffusion layer can be analyzed for products [107]. By means of this technique, information about both the identity (via E_p) and the reactivity (via i_p) can be obtained. The application of the method was illustrated by the reduction of azobenzene (VI) to hydrazobenzene (VII), which under acidic conditions undergoes the benzidine rearrangement, scheme (34).

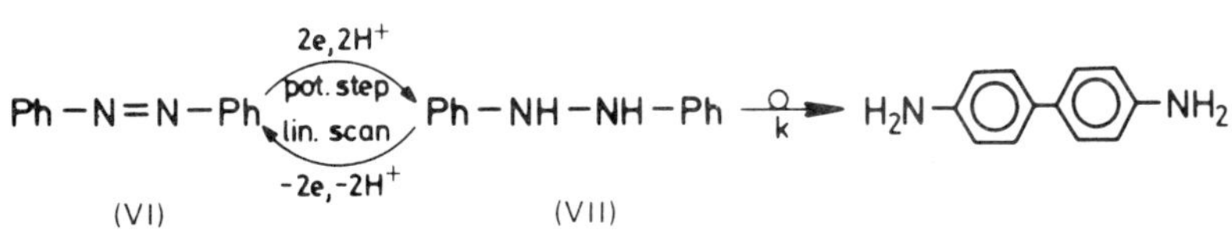

(34)

A relation between the peak height i_p and the time τ' (Fig. 12) was derived from which k could be calculated.

An extension of CA, double-potential-step chronoamperometry, involves a second potential step to a potential at which the initially formed intermediate is converted back to starting material at the diffusion-controlled rate [4,5,108]. The potential-time program and the current-time response are illustrated in Fig. 13. The current is measured twice, the first time at $t = t_f$ (i_f) and the second time usually at $t = 2t_f$ (i_b). In the absence of

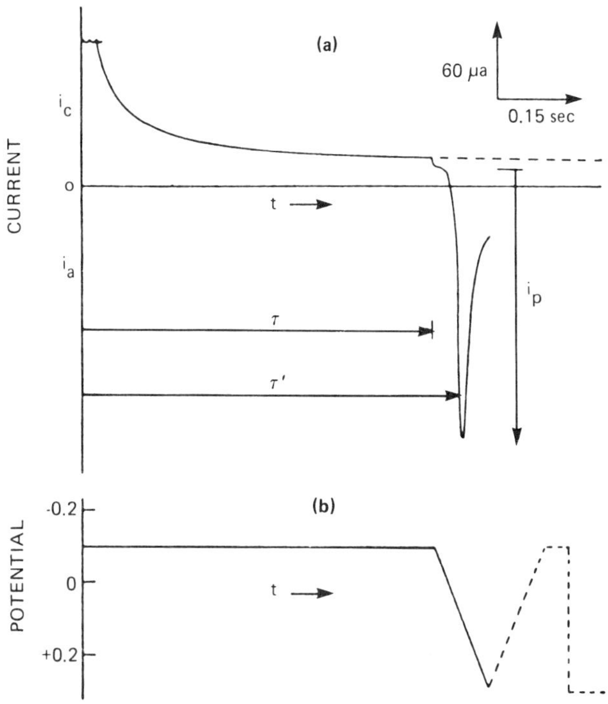

Figure 12. Potential-step linear scan voltammetry; (a) current-time response curve and (b) potential-time program. Azobenzene (2 mM) in ethanol-water (50 wt%). Scan rate: 4.15 V s^{-1}. [From Ref. 107. Reprinted with permission from J. Phys. Chem., 70, 845 (1966). Copyright 1966 American Chemical Society.]

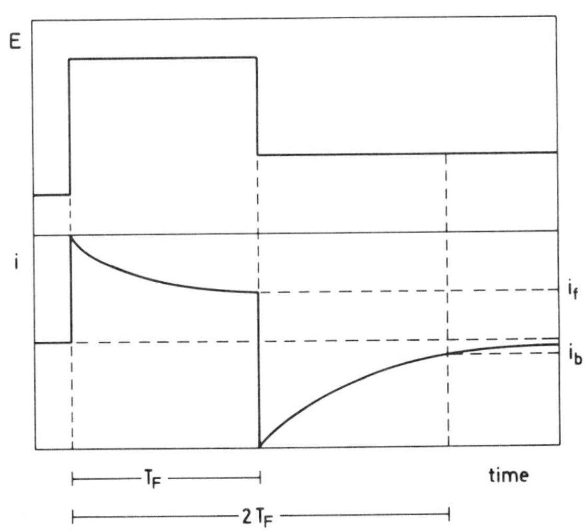

Figure 13. Double-potential-step chronoamperometry: potential-time program and current-time response.

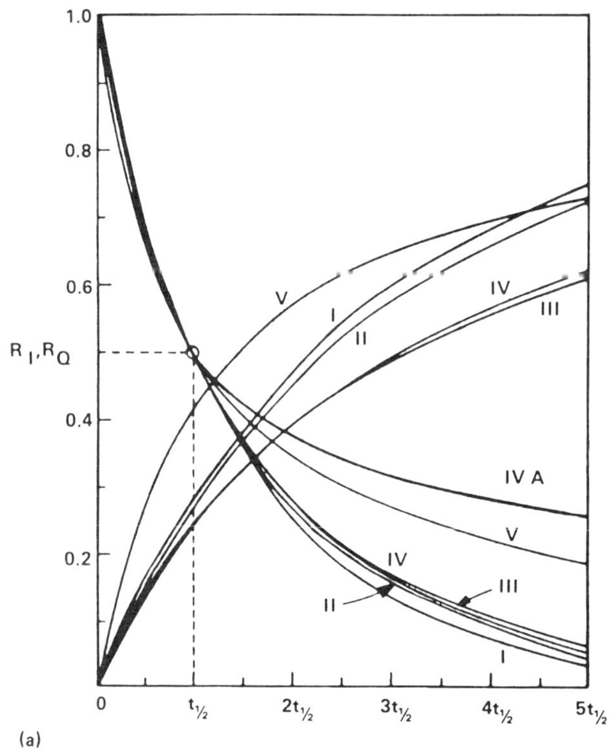

Figure 14. (a) Variation of the normalized current R_I and coulomb R_Q parameters as functions of $t_{1/2}$ for different mechanisms (I) eC; (II) eCe; (III) eeC(dim, dianion-substrate); (IV) eC(dim, anion radical-substrate)e; (IVA) variant of IV (see Ref. 109) and (V) eC(dim, anion radical-anion radical). (From Ref. 109.) (b) Double-potential-step chronoamperometry of anthracene (A) in DMF (0.1 M tetrabutylammonium tetrafluoroborate). $C_A^* =$ 1 mM. C_{PhOH}^* (mM) = 9.95 (+); 18.2 (•); 40.0 (○); 78.8 (x); 100 (□); and 200 (*). R at the figure corresponds to R_I in this chapter and $\lambda = k_{13}t_fC_{PhOH}^*$. (From Ref. 95.)

kinetic complications the ratio $-i_b/i_f$ is independent of t_f and has the value 0.2928 [4]. However, if a chemical reaction is associated with the electron transfer reaction, the ratio $-i_b/i_f$ depends on t_f. Working curves for different mechanisms may conveniently be calculated by digital simulation [29,30] and may be presented as the normalized current ratio $R_I = -i_b/(i_f \times 0.2928)$ either as a function of time measured in units of $t_{1/2}$ (the value of t_f for which R_I equals 0.5), Fig. 14a [109], or as a function of the dimensionless parameters log kt (first-order decay) or log kC_A^*t (second-order decay), Fig. 14b [93,95,97,98,110]. The value of the rate constant associated with the chemical step may be calculated by fitting experimental data to the working curve as demonstrated in Fig. 14b for the protonation of the anthracene anion radical by phenol or from the relations given in Table 3. The goodness of the fit, and thereby the mechanism assignment, may be evaluated after the data are brought into linear form by suitable normalization [111–113] or directly with the aid of a nonlinear regression procedure [11].

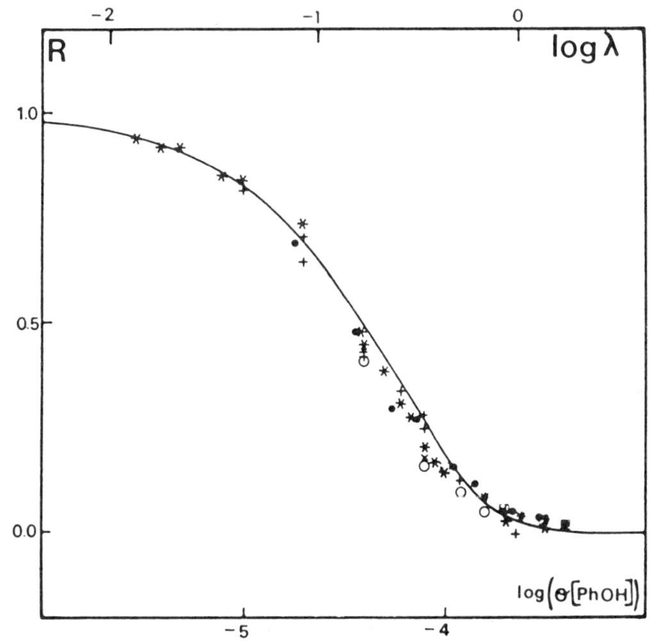

(b)

Figure 14. (Continued).

The approximate kinetic window of DPSCA at electrodes with a diameter in the 0.5–2 mm range is given by Eq. (35), where k is a first-order or pseudo-first-order rate constant:

$$0.3 < k\ (s^{-1}) < 300 \tag{35}$$

The lower limit, corresponding to $t_f \approx 1$ s, is imposed mainly by the requirement that mass transport to and from the working electrode should be re-

Table 3. Relationship Between the Rate Constant k and $t_{1/2}$ in Double-Potential-Step Chronoamperometry

Mechanism	$kt_{1/2}$
eC	0.406
eCe	0.273
eeC(dim, dianion-substrate)	$0.922/C_A^*$
eC(dim, anion radical-substrate)e	$0.690/C_A^*$
eC(dim, anion radical-anion radical)	$0.830/C_A^*$

Source: From Ref. 109.

stricted to (linear) diffusion. This limit is dependent on the size of the working electrode and increases with decreasing diameter. The upper limit, corresponding to $t_f \approx 1$ ms, depends on a number of factors, among which are the time constant of the electrochemical cell (which again depends on the size of the electrode) and the performance of the potentiostat and the data-sampling equipment. Also, this limit increases with decreasing diameter of the working electrode.

Double-potential-step chronoamperometry offers two attractive features, at least. First, since the method includes potential steps to E values at which the heterogeneous electron transfer is diffusion controlled, experiments may be performed with the iR-compensation circuit of the potentiostat switched off. Second, the perturbation of the reaction layer is very small, which means that even a small excess of reagent relative to substrate ensures pseudo-first-order conditions [97] (see Sec. II.F, The protonation of anion radicals derived from aromatic hydrocarbons). A disadvantage is that DPSCA is a "blind" technique, and current due, for example, to long-lived intermediates may erroneously be attributed to regeneration of substrate. In such cases, it may be advantageous to use triple-potential-step chronoamperometry [114] in which the current flowing during the third potential step reflects whether starting material has indeed been formed during the second step. In general, potential-step techniques should only be used together with a potential sweep technique, such as CV.

Examples illustrating the application of DPSCA include the protonation of anion radicals derived from aromatic hydrocarbons [93,95,97], the reversible dimerization of anthracene anion radicals carrying electron-withdrawing substituents in the 9-position [115,116], the decay of $Ph_2CN_2^{\bar{\cdot}}$ [117], the reaction between an electrogenerated base and an electroactive substrate [118], and the conversion of the B form to the A form for 10,10'-dimethyl-9,9'-biacridylidene [119].

B. Chronocoulometry and Double-Potential-Step Chronocoulometry

A method analogous to CA is chronocoulometry, by which the charge Q instead of the current is monitored as a function of time [4,5,120]. The method has found less widespread use than chronoamperometric methods in organic electrochemistry but offers certain advantages when, for example, an electroactive reactant is specifically adsorbed at the electrode surface [120–124]. Integration of Eq. (32) with respect to time results in Eq. (36), which illustrates that a plot of Q versus $t^{1/2}$ results in a straight line with the intercept Q_c, the charge flowing into the interfacial capacitance as a result of the potential step. This quantity contains information about the degree of substrate adsorption at the electrode surface.

$$Q(t) = \frac{2nFAC_A^*(D_A t)^{1/2}}{\pi^{1/2}} + Q_c \tag{36}$$

The application of chronocoulometry for mechanism analysis and evaluation of kinetic parameters is similar to the application of CA and is often based on visual comparison of experimental data with theoretical working curves. Alternatively, the data may be linearized before the evaluation [111,112] or evaluated by a nonlinear regression method. The anodic hydroxylation of tetrahydrocarbazoles in MeCN serves as an illustration of the

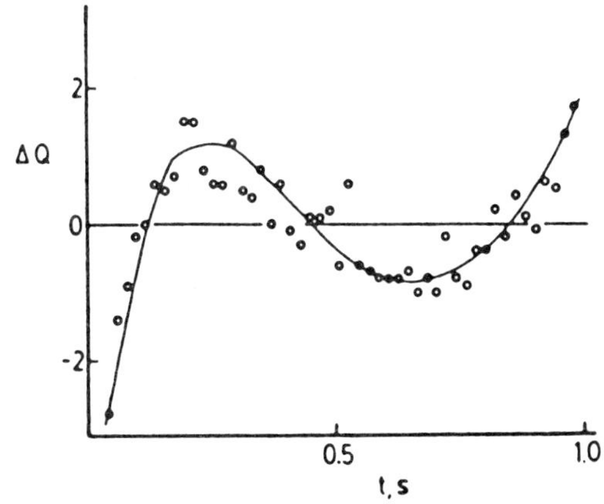

Figure 15. Deviation plot for chronocoulometric data on 0.65 mM 7-methoxy-1-carbomethoxy-1,2,3,4-tetrahydrocarbazole in MeCN (11.1 M H_2O) regressed onto the eCe_h equation. Solid line is theoretical deviation pattern for noise-free eCe data with a rate constant of 25 s^{-1}. (From Ref. 125.)

application of nonlinear regression combined with a deviation pattern recognition to discriminate between the eCe and eCe_h pathways [125]. The deviation plot for the chronocoulometric data obtained for 7-methoxy-1-carbomethoxy-1,2,3,4-tetrahydrocarbazole in MeCN containing water (11.1 M) regressed onto the eCe_h mechanism is shown in Fig. 15. The deviation plot is seen to closely follow the curve predicted for the eCe mechanism which was taken as evidence for this pathway and scheme (37):

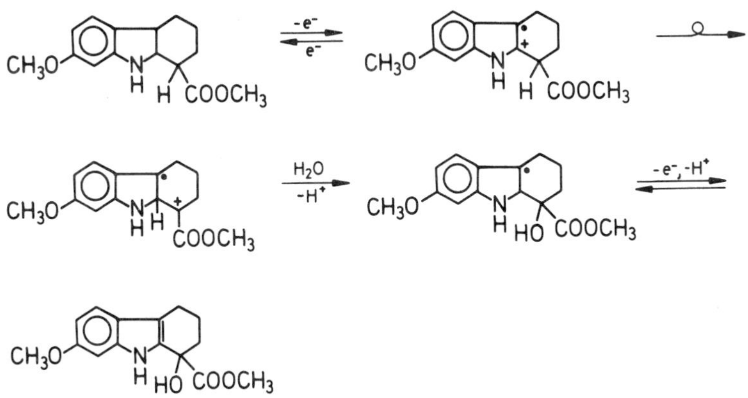

(37)

Double-potential-step chronocoulometry [4,5,120] may be used similarly to DPSCA. The working curves include the charge ratio $R_Q = Q(2t_f)/Q(t_f)$ instead of the current ratio R_I. The reductive cyclization of ethyl cinnamate illustrates the use of the technique [126,127]. Difficulties in fitting experimental data to the working curve for anion radical dimerization [126] led to the proposal that the reaction proceeds via the reversible formation of a dimeric species (VIII), which after protonation undergoes the final cyclization to IX, scheme (38). Depending on the size of the equilibrium constant K, this mechanism results in an apparent reaction order in anion

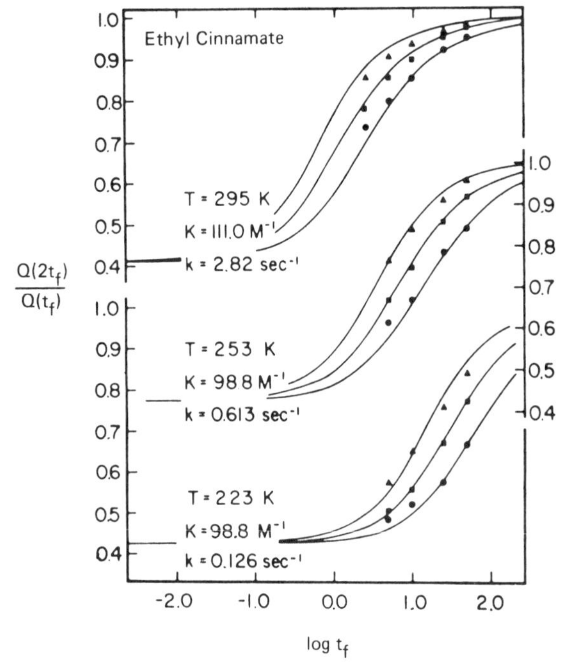

Figure 16. Double-potential-step chronocoulometry; experimental fit of three different concentrations of ethyl cinnamate to pre-equilibrium working curves. (▴) 31.6 mM; (▪) 10.0 mM; (•) 3.16 mM. (From Ref. 127. Reprinted by permission of the publisher, The Electrochemical Society, Inc.)

radical concentration between 1 and 2. Working curves and experimental points are depicted in Fig. 16.

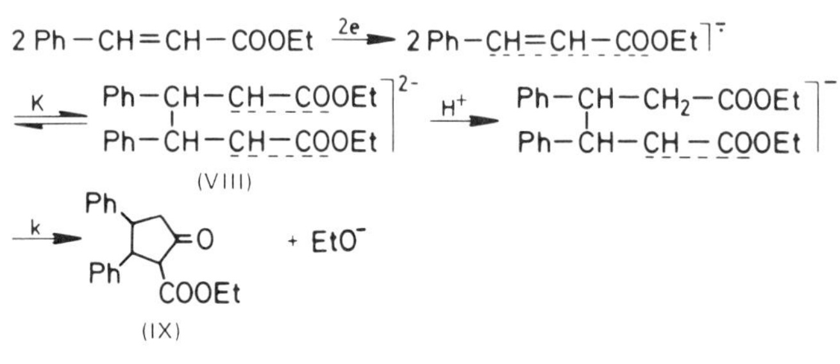

(38)

C. Chronopotentiometry and Current-Reversal Chronopotentiometry

Chronopotentiometry in its most simple form involves measurement of the potential of the working electrode as a function of time after the onset of a constant current i [4,5]. Because of the decrease in concentration of electroactive material close to the electrode, the potential changes with time, and when the concentration is sufficiently small a rapid potential shift is observed to the value of another redox system or to the value for discharge of the solvent-supporting electrolyte system (Fig. 17). The time at which

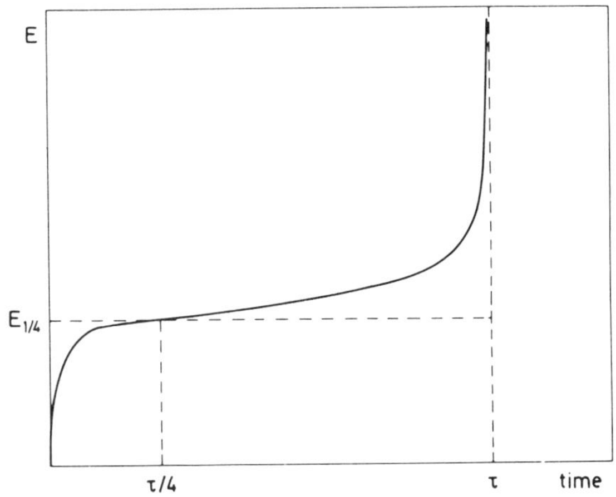

Figure 17. Chronopotentiometry; potential-time relationship after application of a constant current. $E_{1/4}$ is the potential acquired by the electrode at time $\tau/4$ and is equal to the polarographic half-wave potential $E_{1/2}$.

the rapid potential shift is observed, called the transition time τ, can, for reversible electron transfer, be calculated from the Sand equation (39):

$$\tau^{1/2} = \frac{\pi^{1/2} nFD_A^{1/2} C_A^*}{2i} \tag{39}$$

The complete chronopotentiometric wave is described by Eq. (40). If $t = \tau/4$, the third term cancels and Eq. (40) reduces to Eq. (41), which defines the quarter-wave potential $E_{\tau/4}$ and which is equal to E^o if $D_A = D_B$.

$$E = E^o + \frac{RT}{2nF} \ln \frac{D_B}{D_A} + \frac{RT}{nF} \ln \frac{\tau^{1/2} - t^{1/2}}{t^{1/2}} \tag{40}$$

$$E_{\tau/4} = E^o + \frac{RT}{2nF} \ln \frac{D_B}{D_A} \tag{41}$$

Thus, the quarter-wave potential plays the same role in chronopotentiometry [128] as the half-wave potential in polarography (see the next section). Although the theory of the technique is still being developed, for example for programmed current in conjunction with the dropping mercury electrode [129,130] and for the application of microcylinder [131] and microdisk electrodes [132], chronopotentiometry has still found very little use in mechanistic organic electrochemistry. This is primarily due to experimental difficulties in the accurate evaluation of the transition time when the potential-time curve is drawn out or when τ is small.

When applied to the analysis of kinetic data, chronopotentiometry is most often used in the current reversal mode, in which the direction of the current is changed after some time t_f (Fig. 18). When only the direction,

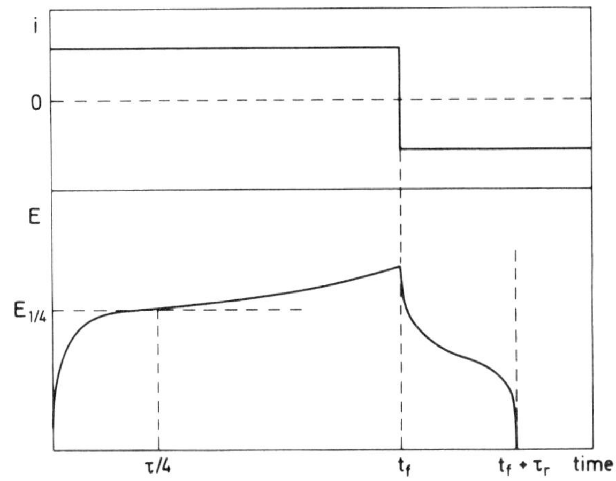

Figure 18. Current reversal chronopotentiometry; current-time program and potential-time response.

but not the magnitude of the current is changed, the reverse transition time τ_r is given by [133].

$$\tau_r = \frac{t_f}{3} \qquad (t_f \leq \tau) \tag{42}$$

Diagnostic criteria for a number of mechanisms have been summarized by Reinmuth [134]. The analysis of experimental data is most conveniently carried out with the aid of a working curve relating the parameters t_f and τ_r and the rate constant k. As an example, consider a simple first-order catalytic mechanism. The relationship between t_f, τ_r, and k for this kinetic perturbation is given by [5,135]

$$2\text{erf}\left[(k\tau_r)^{1/2}\right] = \text{erf}\left[k^{1/2}(t_f + \tau_r)^{1/2}\right] \tag{43}$$

and the associated working curve is depicted in Fig. 19. The electrochemical oxidation of p-aminophenol is believed to follow scheme (44), which may be analyzed in this way [136,137].

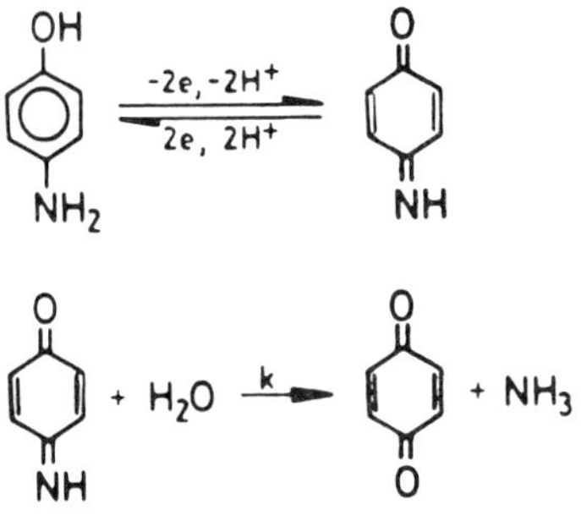

(44)

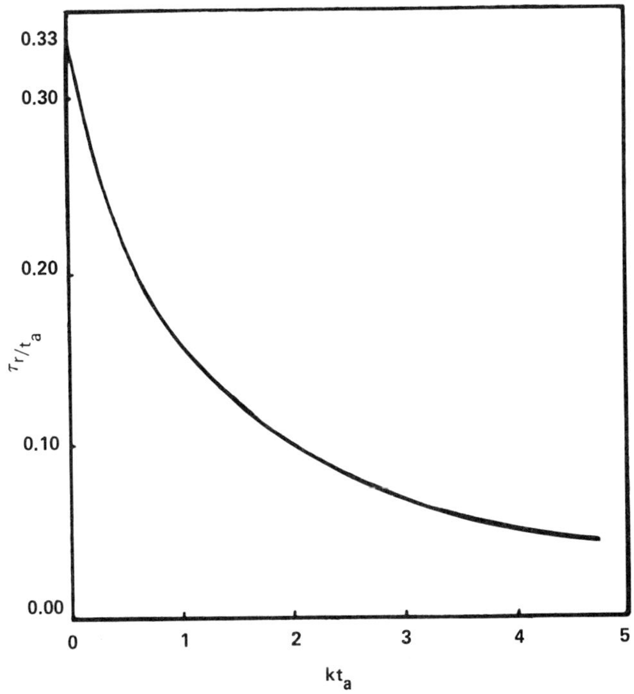

Figure 19. EC working curve for current reversal chronopotentiometry. t_a corresponds to t_f in the text. [From Ref. 137. Reprinted with permission from Anal. Chem., 32, 1512 (1960). Copyright 1960 American Chemical Society.]

Variation in t_f and calculation of the corresponding τ_r/t_f ratios provide a series of kt_f values from which k may be obtained by plotting the product kt_f against t_f.

Another example of the application of chronopotentiometry is the oxidation of tropylidene, scheme (45) [138,139], and thioxanthene [139].

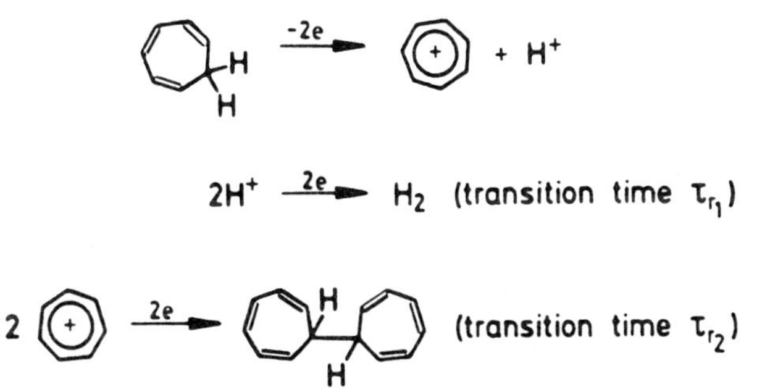

(45)

After current reversal, transition times corresponding to the two reduction processes were observed (Fig. 20). The theoretical analysis resulted in the relations given in Table 4. In Table 4 are also shown the experimental results obtained for thioxanthene, which follows a similar reaction pathway.

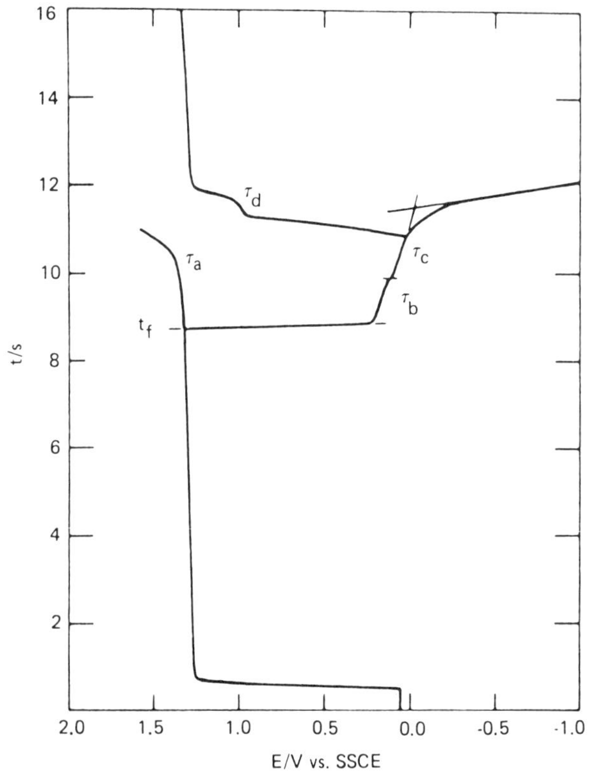

Figure 20. Chronopotentiometry with current reversal on 2 mM thioxanthene in MeCN containing TEAP (0.1 M). $-i_f = i_b = 100\ \mu A$. (From Ref. 139.)

Table 4. Transition-Time Ratios for the Mechanism Illustrated in Scheme (45)

	t_f/τ_{r1}	τ_{r2}/τ_{r1}	$t_f/(\tau_{r1} + \tau_{r2})$	Reference
Theoretical	8.0	1.67	3.0	138
Tropylidene	7.6	1.5	3.0	138
	7.90	1.62	3.03	139
Thioxanthene	8.1	1.7	3.02	139

IV. POLAROGRAPHY

A. Fundamental Aspects

"Polarography" is the term used for voltammetry with the dropping mercury electrode (DME). The electrode (Fig. 21) was developed early in this century by Heyrovsky and was a dominating tool in electroanalytical chemistry for several decades. Because of the low oxidation potential of mercury, +0.3–0.4 V versus SCE, the DME has been used almost exclusively for the study of reduction processes. Compared with solid electrodes, the DME offers a number of advantages, two of the more important being the microscopic smoothness and the continuous renewal of the electrode surface. The latter property reduces undesirable surface effects due to previous applications or electrode processes. The polarographic technique has been discussed extensively in several textbooks and reviews [4–7,140–146] to which the reader is referred for details concerning both theoretical problems and practical applications.

In Fig. 22 is shown the typical current response, a polarogram, at the DME as a result of a slow (1–10 mV s^{-1}) linear potential scan in the negative direction. The shape of the polarogram illustrates clearly the expansion of the electrode surface during the lifetime of the individual Hg drops. Two parameters are generally used to characterize the polarogram, the half-wave potential $E_{1/2}$ and the limiting current i^{lim}. In the absence of adsorption and kinetic complications, the height of the current plateau, measured as the mean current during the drop lifetime, is given by the Ilkovic equation (46):

$$\bar{i}_d = 607nm^{2/3}t^{1/6}D_A^{1/2}C_A^* \qquad (46)$$

The subscript d indicates that the surface concentration of substrate under uncomplicated conditions is dependent only on diffusion of material to the

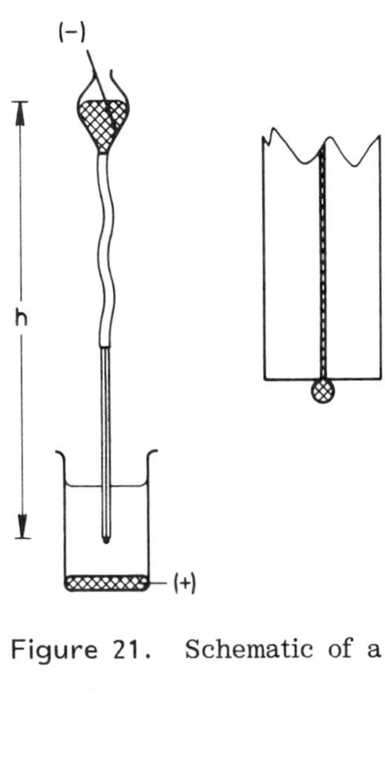

Figure 21. Schematic of a dropping mercury electrode.

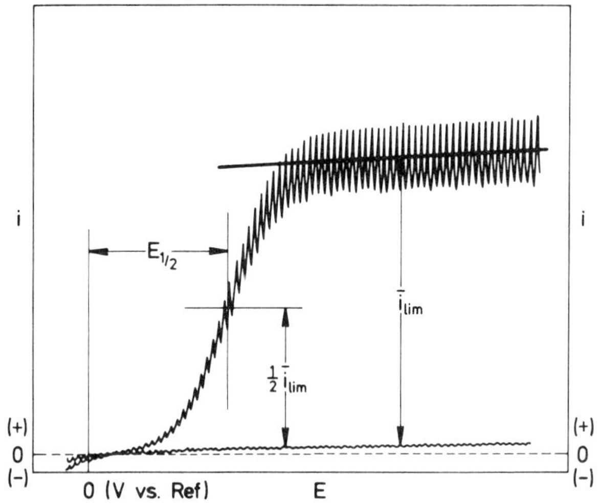

Figure 22. Polarogram for the reversible reaction $A + ne^- \rightleftharpoons B$.

electrode; m is the mass (mg) of mercury flowing out of the capillary per second, and t is the drop lifetime. Polarographic current data are frequently normalized with respect to C_A^*, $m^{2/3}$, and $t^{1/6}$ to eliminate the parameters associated with a particular experiment. The current normalized as in Eq. (47) is called the diffusion-current constant I.

$$I = \frac{\bar{i}_d}{C_A^* m^{2/3} t^{1/6}} = 607nD_A^{1/2} \tag{47}$$

If the electron transfer is reversible, the complete polarographic wave may be described by a modified Nernst equation (48) [140,142]:

$$E + E^o + \frac{RT}{nF} \ln \frac{f_A D_B^{1/2}}{f_B D_A^{1/2}} + \frac{RT}{nF} \ln \frac{\bar{i}_d - \bar{i}}{\bar{i}} \tag{48}$$

where f_A and f_B are the activity coefficients of A and B, respectively [Eq. (5)]. If $\bar{i} = 0.5$ then i_d, the third term in Eq. (48), cancels and the expression reduces to

$$E = E_{1/2} = E^o + \frac{RT}{nF} \ln \frac{f_A D_B^{1/2}}{f_B D_A^{1/2}} \tag{49}$$

which defines the half-wave potential $E_{1/2}$. When $f_A = f_B$ and $D_A = D_B$, $E_{1/2}$ equals E^o. It is furthermore seen that $E_{1/2}$ is independent of the substrate concentration. Introduction of $E_{1/2}$ in Eq. (48) results in the more convenient

$$E = E_{1/2} + \frac{\bar{i}_d - \bar{i}}{\bar{i}} = E_{1/2} + \frac{0.0591}{n} \log \frac{\bar{i}_d - \bar{i}}{\bar{i}} \tag{50}$$

The reversibility of the electron transfer reaction may be tested via this equation, which predicts that a plot of E versus log $[\bar{i}_d - \bar{i})/\bar{i}]$ results in a straight line with slope 0.0591/n V at 25°C (Fig. 23). Slopes smaller than 0.0591/n V are observed when the electrode process is irreversible, in which case E and i are related through [142,146]

$$E = \frac{RT}{\alpha nF} \ln \frac{1.39 k_s t^{1/2}}{D_A^{1/2}} + 0.9163 \frac{RT}{\alpha nF} \ln \frac{i_d - i}{i} \qquad i \geq \frac{i_d}{10} \tag{51}$$

where k_s is the heterogeneous rate constant of the electrode process at E°. It should be noted that, when Eq. (51) is used, it is the current at the end of the drop life rather than the mean current that should be measured. In analogy to the reversible case, the half-wave potential can be introduced, being the value of E when $i = 0.5 i_d$,

$$E = E_{1/2} = \frac{RF}{\alpha nF} \ln \frac{1.349 k_s t^{1/2}}{D_A^{1/2}} \tag{52}$$

and thus

$$E = E_{1/2} + 0.9163 \frac{RT}{\alpha nF} \ln \frac{i_d - i}{i} = E_{1/2} + \frac{0.0542}{\alpha n} \log \frac{i_d - i}{i} \tag{53}$$

The transfer coefficient α may be obtained from a plot of E versus log$[(i_d - i)/i]$, which should be linear with the slope of 0.0542/αn V.

Both m and t in the Ilkovic equation (46) depend on the height of the mercury reservoir h, and since m is equal to c'h and t is equal to c''/h,

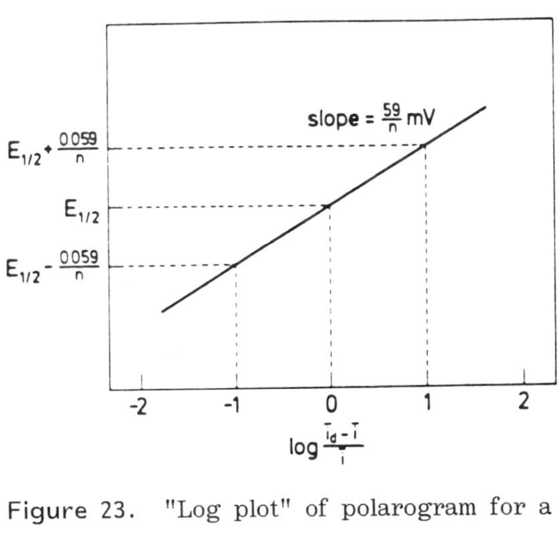

Figure 23. "Log plot" of polarogram for a reversible reaction.

where c' and c'' are constants, we have the following relationship between $\bar{i}_d$ and h:

$$\bar{i}_d = 607n(c'h)^{2/3}\left(\frac{c''}{h}\right)^{1/6} D_A^{1/2} C_A^* = (\text{const})h^{1/2} \tag{54}$$

Thus the magnitude of a purely diffusion-controlled polarographic wave is proportional to the square root of the height of the mercury reservoir. Other borderline cases that may be identified through the dependence of i^{lim} on h include electron transfer reactions limited by the rate of a preceding chemical step (kinetic waves) or by the adsorption of substrate or product (adsorption waves) (see Table 5).

A kinetic wave may arise as a result of the scheme

$$C \underset{k_b}{\overset{k_f}{\rightleftharpoons}} A \qquad \left(K = \frac{k_f}{k_b}\right)$$

$$A + ne^- \longrightarrow B$$

Provided that the equilibrium concentration of A is small and the transformation of C into A is slow, the kinetic current i_k is given by

$$\bar{i}_k = 493nD_C^{1/2}C_C^* m^{2/3} t^{2/3} \frac{k_f}{k_b^{1/2}} \tag{55}$$

The introduction of m = c'h and t = c''/h shows that i_k is independent of h, which allows kinetic currents to be distinguished conveniently from diffusion currents. If the transformation of C into A is so rapid that i_k is a significant fraction of the diffusion current for C, Eq. (56) can be used to evaluate the ratio $k_f/k_b^{1/2}$.

$$\frac{\bar{i}_k}{\bar{i}_d - \bar{i}_k} = 0.81 \frac{k_f}{k_b^{1/2}} t^{1/2} = 0.81(Kk_f t)^{1/2} \tag{56}$$

Table 5. Common Types of Polarographic Limiting Currents and Their Relationship with the Height of the Mercury Reservoir h

Type of current	Relationship with h
Diffusion current (limited by the rate of mass transfer to the electrode surface)	$i_d = (\text{const})h^{1/2}$
Kinetic current (limited by a chemical step preceding the electrode reaction)	i_k independent of h
Adsorption current (caused by adsorption of substrate or products on the DME surface)	$i_a = (\text{const})h$
Catalytic current (limited by the catalytic effect of a polrographically inactive catalyst)	i_c; no simple relationship with h

More complicated versions of the Ce scheme have been treated as well [147].

If either the substrate or the product is adsorbed on the electrode surface, an adsorption wave may be observed in addition to the diffusion-controlled wave. In product adsorption the following processes may occur:

$$A + ne^- \longrightarrow B_{ads}$$

$$A + ne^- \longrightarrow B_{soln}$$

The formation of B in adsorbed state is limited by the area of the DME. The current i_a due to this step is given by Eq. (57), where a is the area covered by one molecule of B:

$$\bar{i}_a = 13.66 m^{2/3} t^{-1/3} a^{-1} \tag{57}$$

The product $m^{2/3}t^{-1/3}$, and thus i_a, is proportional to h, which serves to identify adsorption currents.

B. Applications

Polarography has been used extensively to characterize the cathodic behavior of numerous substrates under aqueous conditions (see Refs. 140–146). Both the half-wave potential and the limiting current may be affected by the proton activity. The polarograms resulting from the reduction of anthracene (AN) in solvent systems of increasing proton activity illustrate a general trend observed during this type of experiment (Fig. 24). Under nonaqueous conditions two reversible one-electron polarographic waves are observed corresponding to the successive formation of the anion radical and

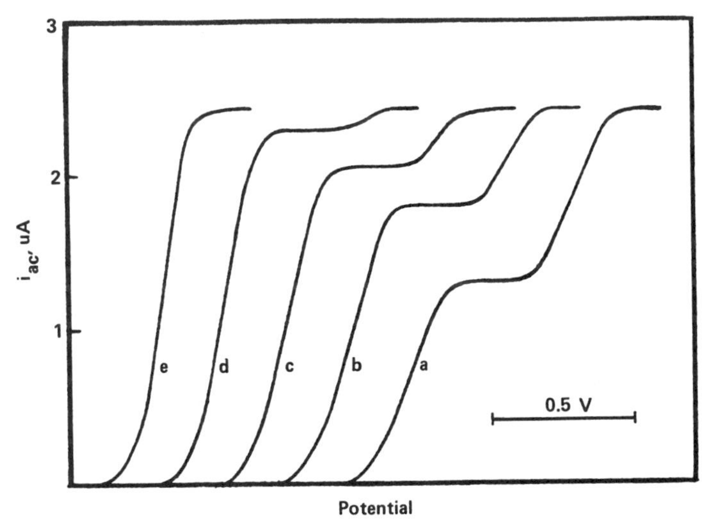

Figure 24. Effect of increasing proton activity, a through e, on the polarogram of anthracene. For the sake of clarity the various polarograms have been shifted horizontally. (From Ref. 148.)

the dianion (Fig. 24, curve a). When the water content increases, the first wave grows at the expense of the second wave, the sum of the two remaining almost constant, until finally at high proton activities the two waves have merged into a single two-electron wave (Fig. 24, curve e). This behavior can be rationalized in terms of the eCe_h scheme (58) through (61).

$$AN + e^- \longrightarrow AN^{\cdot -} \quad e \tag{58}$$

$$AN^{\cdot -} + H^+ \longrightarrow ANH^{\cdot} \quad C \tag{59}$$

$$ANH^{\cdot} + AN^{\cdot -} \longrightarrow ANH^- + AN \quad e_h \tag{60}$$

$$ANH^- + H^+ \longrightarrow ANH_2 \tag{61}$$

The increase in height of the first wave is related to the increasing importance of the chemical step (59), during which a product, $ANH^{\cdot}$, is more easy to reduce than substrate (see Ref. 148 for a thorough discussion of this important point),

In purely aqueous solution the half-wave potential for the general reversible scheme

$$A + ne^- + mH^+ \longrightarrow B$$

is given by

$$E_{1/2} = E_{1/2}(0) - 0.059 \frac{m}{n} \text{pH} \tag{62}$$

where $E_{1/2}(0)$ is the half-wave potential at pH = 0. Thus for two reaction schemes commonly met in organic electrochemistry, m = n = 1 and m = n = 2, a plot of $E_{1/2}$ versus pH should yield a straight line with the slope 59 mV/pH. Numerous examples of the variation in $E_{1/2}$ with changing pH have been published [144,149].

A complicating factor in the reduction of organic compounds under nonaqueous conditions is the formation of ion pairs between the negatively charged species resulting from the electrode process and the cation of the supporting electrolyte M^+:

$$A + e^- \rightleftharpoons A^{\cdot -} \quad (E_{1/2}) \tag{63}$$

$$A^{\cdot -} + pM^+ \rightleftharpoons (A^{\cdot -}, pM^+) \quad (K_{assoc}) \tag{64}$$

The shift in half-wave potential due to this association is given by a modification (65) of the Nernst equation [150]:

$$\Delta E_{1/2} = \frac{RT}{nF} \ln (1 + K_{assoc}[M^+]^p) \tag{65}$$

In general, the product $K_{assoc}[M^+]$ is much larger than unity and consequently a plot of $E_{1/2}$ versus log $[M^+]$ should yield a straight line from the slope of which the stoichiometric number p may be determined. Quinones and various nitro compounds have frequently been studied with this goal in mind [151–153]. Deviations form linear dependence of $E_{1/2}$ on log $[M^+]$ are observed in some cases, especially at high M^+ concentrations, which may be attributed to the formation of higher associates.

Equation (57) may be used to estimate a, the area covered by one molecule at the DME. This type of analysis was used in a study of the polarographic reduction of tropylium ion [154]. Several adsorption waves were observed depending on the concentration of substrate. From the first of these the value of a was calculated as 29 $Å^2$, in good agreement with the area of a seven-membered ring, which was estimated to cover 22 $Å^2$. Polarography has also been used to investigate whether the dimerization of an α,β-unsaturated ketone in acidic solution involves adsorbed radicals [155].

A study of the interconversion of the two geometric isomers of bianthrone, A (most stable form) and B (metastable form), in aqueous phenol [156] and nonaqueous DMF [157], scheme (66), serves as an illustration of the use of kinetic currents in polarography.

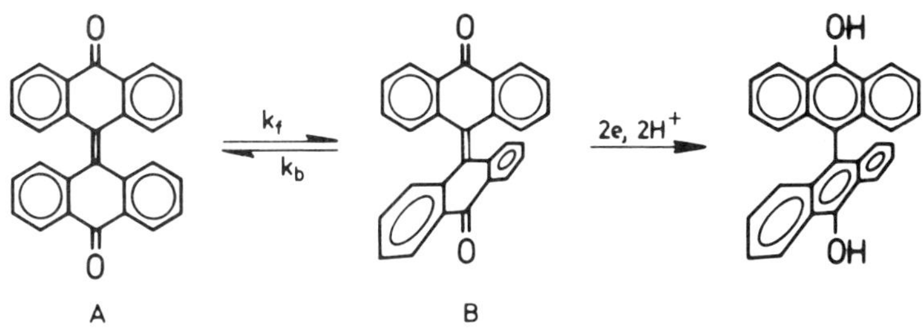

(66)

In Fig. 25 are shown the polarographic waves obtained for bianthrone in 75% aqueous phenol at three different temperatures. The values of $\bar{i}_k$ and $\bar{i}_d$ together with K values obtained by spectroscopy provided [Eq. (56)] good estimates of k_f [10^{-2} s^{-1}; see scheme (66)]. Other examples illustrating the application of polarography for the analysis of preceding chemical steps are found in Ref. [158] and in a series of papers devoted to the study of the reduction of pyridine-4-aldehyde in aqueous solution in which the aldehyde exists in equilibrium with the corresponding hydrate [147, 159].

Although polarography in the present context to some extent has lost some importance relative to new powerful techniques, the method deserves not to be forgotten. Polarography is still in many cases a competitive method, especially for the examination of, for eample, electron transfer reactions coupled with slow chemical steps, as illustrated by the study of bianthrone mentioned earlier, or electrode processes associated with adsorption phenomena.

Examples of applications of more sophisticated experimental methods based on the DME in the study of organic electrode reactions include the recording of i-t curves for single mercury drops (reduction of nitrophenols [160]), derivative polarography (reduction of azobenzene to hydrazobenzene followed by the benzidine rearrangement [13]), differential pulse polarography (hydrodimerization of 1-(4-methoxyphenyl)-1-hepten-3-one [161] and crotonaldehyde [162]), and mechanical square-wave polarography (heterogeneous and homogeneous rate constants [163]).

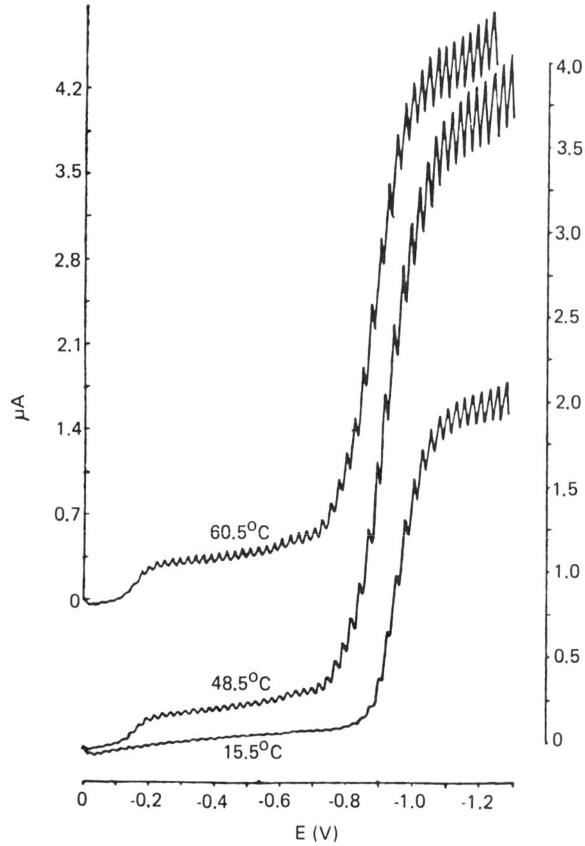

Figure 25. Polarographic curves of bianthrone (1 mM) in aqueous 75.4 wt% phenol containing NaOAc (0.5 M) at different temperatures. Potentials are versus Hg pool. (From Ref. 156.)

V. METHODS BASED ON FORCED CONVECTION

A. General Considerations

A common feature of the techniques described so far is that mass transport is caused by diffusion only; that is, it is the result of an unperturbed concentration gradient in solution. The origin of this gradient is, of course, the conversion of substrate at the electrode, possibly complicated by follow-up reactions due to the instability of the initially formed product. The time t after the electrode potential has been stepped from a value at which no reaction takes place to a value at which the electron transfer rate is diffusion controlled, the thickness of the diffusion layer δ at planar electrodes, and t are related by Eq. (67), where, as before, D_A is the diffusion coefficient of the substrate:

$$\delta = (\pi D_A t)^{1/2} \tag{67}$$

When δ becomes larger than a few tenths of a millimeter, natural convection begins to interfere and the assumption of diffusion as the only means of

Table 6. Transport-Time Characteristics of Common Types of Electrode Systems

Electrode	Well-defined transport	Calculable transport	Time characteristics
Stationary	Short times only	Short times only	Transient
Rotating wire	Yes	No	Steady state
Vibrating wire	Yes	No	Cyclic
Mercury drop	Yes	Yes	Cyclic
Rotating disk	Yes	Yes	Steady state

Source: From Ref. 164.

mass transport breaks down. For typical values of D_A ($\approx 10^{-5}$ cm^2 s^{-1}) this happens when t is of the order of 10 s, and at times larger than approximately 1 min the deviations from pure diffusion are so pronounced and unpredictable that current measurements become essentially meaningless. The factors that influence the transport in an apparently motionless solution have humorously been summarized by Albery and Hitchman in their monograph [164], and the common types of electrode systems were compared with respect to their transport and time characteristics (Table 6). By inspection of the table it becomes obvious that a rotating disk electrode (RDE) offers an important advantage over other systems in that the transport to the electrode surface at the same time is well defined and calculable. Owing to the rotation of the electrode, a strong convective force is imposed on the solution, and in comparison other forces that may influence the solution transport are small and can be neglected. Furthermore, the diffusion layer thickness for a rotating electrode is small compared with that of a stationary electrode, and taken together, these characteristics result in an experimental setup that is relatively insensitive to the nature of the surroundings. Thorough discussions of the applications of rotating disk electrodes and the associated mathematics are found in several texts [4,164–170].

B. The Rotating Disk Electrode

The electrode tip of an RDE system may consist of a cylindrical piece of conducting material (usually Pt, Au, or glassy carbon), which together with a wire or a steel rod for electrical contact is sealed into an insulating shaft (Fig. 26). When the electrode is rotated, an adjacent thin layer of the solution dragged by the electrode surface acquires the tangential velocity of the electrode. Because of centrifugal force, this is superimposed by a radical velocity, resulting in an overall flow pattern at the electrode surface, as illustrated in Fig. 27. The liquid thrown out horizontally is replaced through an upward vertical flow (Fig. 26).

The mathematical treatment on which the equations given in this section are based is valid only as long as the fluid flow is laminar. At very high rotation rates the flow becomes turbulent, and at very low rotation rates natural convection begins to play an unwanted role. The two extremes can be given in terms of the Reynolds number Re, which for the RDE is defined as

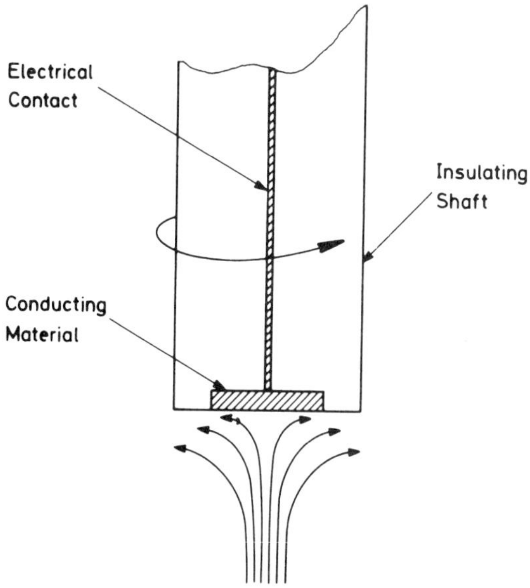

Figure 26. Rotating disk electrode. The stream lines indicate the flow below the disk.

$$Re = r^2 \omega \nu_k^{-1} \tag{68}$$

where r is the radius (cm) of the RDE, including the insulating shaft (!), ω is the angular rotation rate (rad s^{-1}), and ν_k is the kinematic viscosity coefficient ($cm^2\ s^{-1}$). Laminar flow can be maintained for Re values between 10 and 10^4, both values being approximate. For a typical commercial RDE these extremes limit the useful range of rotation rate to values roughly between 0.2 and 200 rev s^{-1}.

The tangential and radial velocities decrease with increasing distance from the electrode surface, and at some critical distance δ_0 it is sufficiently accurate to consider the laminar upward motion as the only mode of transport. The liquid layer in which it is necessary to take into account both the tan-

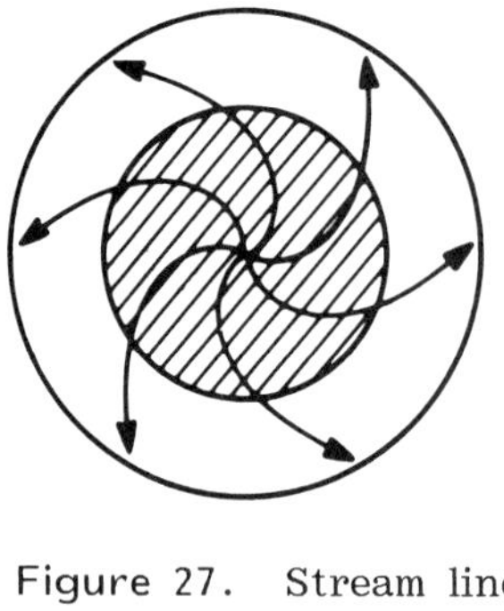

Figure 27. Stream lines at the electrode surface of a rotating disk electrode.

Table 7. Thickness of the Hydrodynamic Boundary Layer δ_o and the Diffusion Boundary Layer δ at Different Rotation Rates[a]

N (rev s^{-1})	rad s^{-1}	δ_o (cm)	δ (cm)
1	6.28	0.12	0.0064
3	18.9	0.069	0.0037
10	62.8	0.038	0.0020
30	189	0.022	0.0012
100	628	0.012	0.0006

[a]The values are based on ν_k = 0.01 cm^2 s^{-1} and D_A = 10^{-5} cm^2 s^{-1}.

gential and the radial velocity is called the hydrodynamic boundary layer, and its approximate thickness is given by

$$\delta_o = 3\nu_k^{1/2}\omega^{-1/2} \qquad (69)$$

Typical values of δ_o at various rotation rates are given in Table 7.

When an electrochemical process takes place at the electrode surface, a concentration gradient develops near the surface, resulting in diffusion as an additional mode of mass transport. The liquid layer in which the transport by diffusion is comparable to the convectional motion is called the diffusion boundary layer, and its thickness is approximately 5% of the hydrodynamic boundary layer thickness (Table 7). Within the Nernst approximation (Fig. 28), in which the concentration gradient is considered linear, the thickness δ is given by

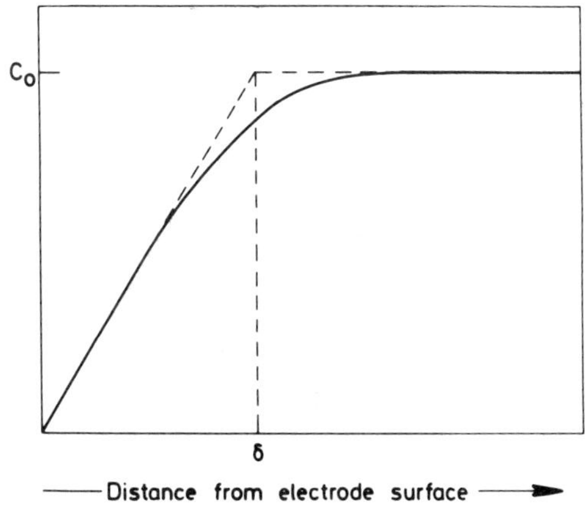

Figure 28. Concentration profile close to the electrode surface under mass transfer control. The full line represents the actual profile, and the broken line illustrates the Nernst approximation.

$$\delta = 1.61 D_A^{1/3} \nu_k^{1/6} \omega^{-1/2} \tag{70}$$

Both δ_0 and δ have the same value over the entire electrode surface, which has given rise to the description of the electrode as a uniformly accessible surface. It should be pointed out, however, that in the absence of complete mass transfer control, the current density is not uniform across the disk radius, as has been predicted [171] and experimentally verified [172, 173]. The concepts of a hydrodynamic and a diffusion boundary layer have no theoretical significance as such but serve mainly to provide a suitable model for the hydrodynamic conditions related to the rotating electrode.

At constant rotating speed the current-voltage curve for uncomplicated rapid charge transfer is similar to a polarogram (Fig. 29). The curve can be considered as consisting of three section, a, b, and c in the figure, depending on the actual value of the applied potential compared to E°. For the general case in which the process is under the mixed control of mass and charge transfer (Fig. 29b), the current is given by [174,175]

$$i = \frac{nFAC_A^* D_A}{1.61 D_A^{1/3} \nu_k^{1/6} \omega^{-1/2} + D_A/k_{hf}} = \frac{nFAC_A^* D_A}{\delta + D_A/k_{hf}} \tag{71}$$

where A is the area of the RDE (cm^2), C_A^* is the bulk concentration of substrate, and k_{hf} is the heterogeneous charge transfer rate constant at the given potential ($cm\ s^{-1}$). For a typical value of $\delta \approx 10^{-3}$ cm (see Table 7), the term D_A/k_{hf} can be neglected if $k_{hf} > 10^{-1}$ $cm\ s^{-1}$. The current is now controlled solely by the mass transfer to the electrode (Fig. 29c), and

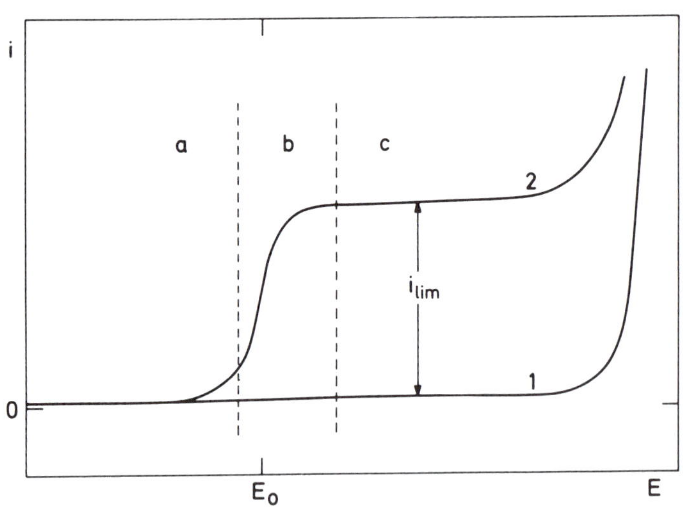

Figure 29. Current-voltage curves obtained at the rotating disk electrode: (1) without and (2) with substrate added. The indicated regions are: a, control by charge transfer; b, mixed control be charge and mass transfer; c, control by mass transfer.

$$i^{lim} = 0.62nFAC_A^*D_A^{2/3}\nu_k^{-1/6}\omega^{1/2} = \frac{nFAC_A^*D_A}{\delta} \tag{72}$$

An important consequence of this equation is the prediction that i^{lim} varies linearly with $\omega^{1/2}$, a crucial test of the quality of the RDE. On the other hand, if $k_{hf} < 10^{-4}$ cm s^{-1}, the term D_A/k_{hf} is large compared with δ, and for this situation, in which the current is controlled by the rate of the electrochemical process (Fig. 29a), Eq. (73) is valid:

$$i = nFAC_A^*k_{hf} \tag{73}$$

These relations have been discussed and experimentally tested for the Fe^{2+}-Fe^{3+} system [174]. Values of k_s, the heterogeneous charge transfer rate constant at the formal potential, up to approximately 0.1 cm s^{-1} may be measured with the aid of the RDE.

By inspection of Eqs. (71) through (73) it is seen that measurement of the rotating disk current may provide information concerning the value of n, C_A^*, D_A, ν_k, or k_{hf} if the remaining parameters are known. However, the determination of viscosity coefficients is more conveniently performed by other methods, and to the best of our knowledge the RDE has found no application in this field. Accuracy in the determination of n values depends mainly on how accurately the value of D_A for the compound in question is known. For the majority of organic compounds, diffusion coefficients in common organic solvents have not been measured, but comparison with a suitable compound of similar structure for which both n and D are known can often solve the problem (assuming the same D value for the two substrates). Since D_A appears only to the power of 2/3, the resulting n value is only moderately dependent on variations in the D_A estimate. The measurement of diffusion coefficients by means of the RDE technique has been the subject of a number of papers. This type of work has been discussed adequately in a reference already given [166] and need not be treated further here.

A major field of application of the RDE in organic electrochemistry is elucidation of the mechanistic details for reactions following or preceding heterogeneous electron transfer. One way of using the RDE in kinetic work is based on the relationship between i^{lim} and ω. In the absence of kinetic complications, a plot of i^{lim} versus $\omega^{1/2}$ should be a straight line, Eq. (72). However, if the electron transfer reaction is preceded or followed by a chemical step, the linear dependence of i^{lim} on $\omega^{1/2}$ may change to a curved relationship. For the eCe scheme the relationship between i^{lim} and ω is given by [177]

$$i^{lim} = i_1^{lim}(2 - e^{-0.834\nu_k^{1/3}k/D_A^{1/3}\omega}) \tag{74}$$

where k is the rate constant for the chemical step and i_1^{lim} is the current observed for the first electron transfer step only; that is, in the absence of the chemical reaction, or equivalently in terms of the apparent number of electrons exchanged, $n_{app} = i^{lim}/i_1^{lim}$:

$$n_{app} = 2 - e^{-0.834\nu_k^{1/3}k/D_A^{1/3}\omega} \tag{75}$$

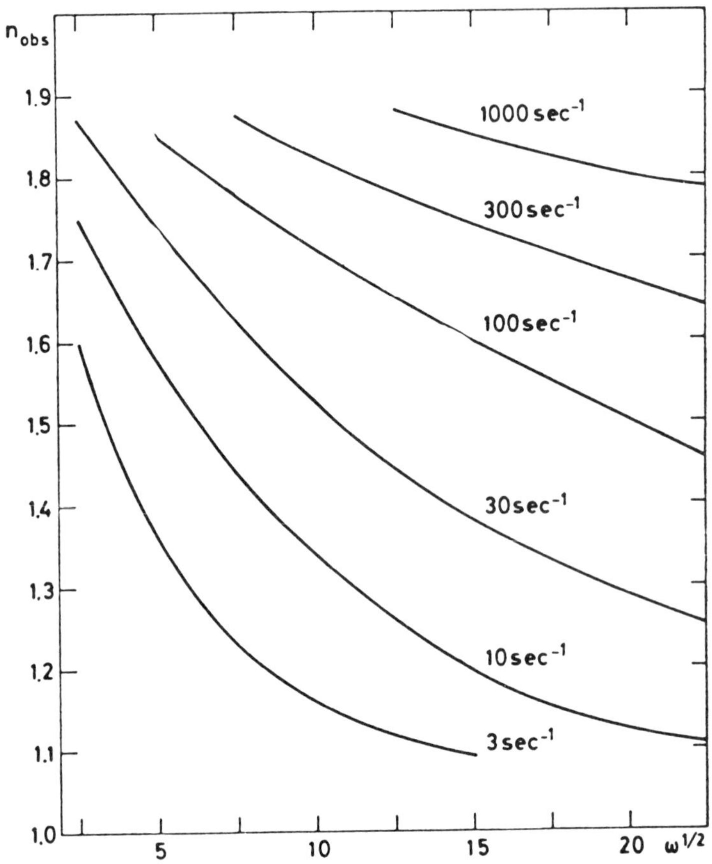

Figure 30. Rotating disk electrode working curves for the eCe_h mechanism at various values of the rate constant for the chemical step.

Other cases are treated in Ref. 166. It is easily seen that, for relatively fast chemical steps in which $k \gg \omega$, the exponential term is close to zero and the current corresponds to the exchange of two electrons. On the other hand, if $k = 0$, the current corresponds to a one-electron process. Consequently, for intermediate values of k the apparent number of electrons varies between 1 and 2. Equation (75) predicts that n_{app} trends toward smaller values with increasing ω. The exact relationship between n_{app} and ω ($\omega^{1/2}$), a working curve, can be calculated from Eq. (75) or by digital simulation [29] for any value of k provided that the remaining parameters are known. Working curves at different k values for the eCe_h case are shown in Fig. 30. The range of measurable first-order rate constants is $0.1-10^3$ s^{-1}. Typical studies in which the RDE has been used include the dimerization of the triphenylamine cation radical [178] and the cyclization of the tetraphenylethylene dication [179].

C. The Rotating Ring-Disk Electrode

At the RDE the electrode products are thrown horizontally away from the electrode surface and thus escape further detection. A solution to the need for additional information in RDE experiments was proposed by Frumkin and Nekrasov [180], who introduced the rotating ring-disk electrode (RRDE). The electrode consists of a conducting disk and a conducting ring in a concentric arrangement separated by insulating material (Fig. 31). By proper adjustment of the ring potential, initial electrode products or products from follow-up reactions may be monitored as a function of the angular rotation rate. The parameter most frequently used in RRDE experiments is the collection efficiency N, which relates the ring current i_r to the disk current i_d:

$$N = \frac{i_r}{i_d} \tag{76}$$

These observables usually have opposite signs (see below) and should be measured at potentials at which mass transport is rate determining. The value of N is less than unity, even in the absence of kinetic complications, because only a fraction of the molecules generated at the disk electrode reach the ring owing to diffusion away from the electrode. The theoretical value of N depends on the geometry of the electrode system. Of special importance is the size of the gap between the two electrodes. The value of N typically varies between 0.45 and 0.2 [181] for uncomplicated electron transfer; for example,

Disk: $A \pm e^- \longrightarrow B$

Ring: $B \pm e^- \longrightarrow A$

and is in this case independent of ω.

If species B undergoes a chemical reaction, the observed ring current, and therefore N, drops to a lower value. The fraction of the B molecules that escape detection at the ring depends on two factors, the rate of the chemical step and the time interval between the generation and detection of B, that is, on ω. Consequently, it can be expected that for a given

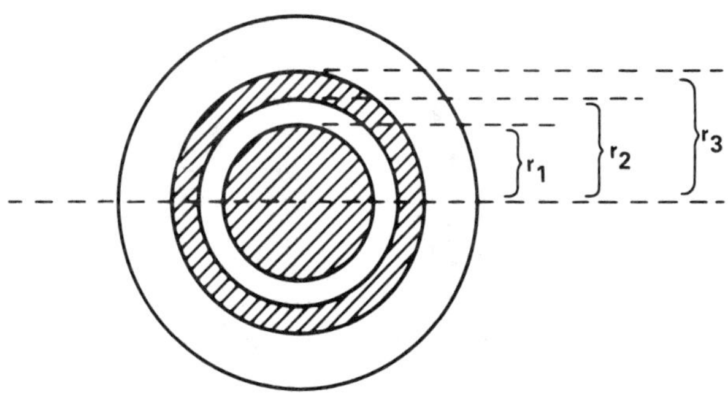

Figure 31. Schematic of a rotating ring-disk electrode.

chemical reaction the value of N increases with increasing ω and, in the limit approach, the value for uncomplicated electron transfer. Working curves for the determination of rate constants can be constructed by analogy to those already described for the simple RDE, this time relating N to ω or an expression containing this parameter. Because of the complexity of the mathematics associated with the calculation, digital simulation has been extensively used for this purpose [181,182] and a complete computer program is available [183]. This treatment has included the following cases: uncomplicated electron transfer reactions, first- and second-order homogeneous follow-up reactions to electroinactive products, eCe-type and catalytic processes, and dimerization reactions. Comparison of analytic and numerical solutions for a number of cases demonstrated satisfactory agreement between the two approaches [184]. If the mechanism is known, the rate constant can be obtained from the appropriate working curve, which for the RRDE usually relates the collection efficiency to one of the following parameters [181]:

$$XKT = (0.51)^{-2/3} k_1 \omega^{-1} \nu_k^{1/3} D_A^{-1/3} \qquad \text{(first-order kinetics)}$$

$$XKTC = (0.51)^{-2/3} k_2 \omega^{-1} \nu_k^{1/3} D_A^{-1/3} C_A^* \qquad \text{(second-order kinetics)}$$

where the k_1 and k_2 are the rate constants. The other symbols have their usual significance. However, for a specific case it may be more convenient to use a set of working curves, similar to those shown for the RDE, that relates N to $\omega^{1/2}$ (Fig. 32).

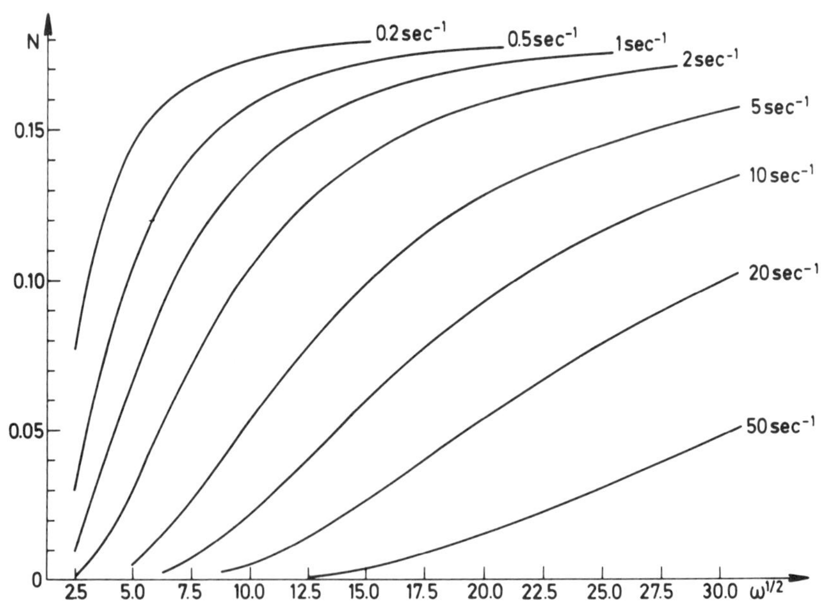

Figure 32. Rotating ring-disk electrode working curves for the eCe mechanism at various values of the rate constant k for the chemical step. The value of N corresponding to k = 0 is 0.184 for this particualr electrode.

Application of the RRDE as a diagnostic tool to explore the mechanism of an unknown reaction pathway relies, as most methods do, on the differences between the different working curves. For the RDE the working curves for different mechanisms in many cases do not deviate sufficiently to allow a distinction between two possible pathways (see, e.g., Ref. 168). The presence of the ring electrode provides extra information, and thus the RRDE is a more sensitive instrument for mechanical analysis. An example is the distinction between the anion radical-anion radical and anion radical substrate coupling mechanisms for $CS_2^{\cdot-}$ [185]. It has been pointed out, however, that even with this electrode setup the usual working curves do not allow the unequivocal discrimination between, for example, eCe-type and catalytic mechanisms [181], or the different mechanisms leading to dimerization of substrate [182]. A solution to this problem has been found that takes advantage of the possibility of varying the relative fluxes of A and B from the disk via the current density (see Ref. 182 and 186 for details and Refs. 187 and 188 for examples).

The RRDE technique can provide quantitative information for reactions having rate constants in the following approximate ranges:

First order: $10^{-1} < k_1 < 10^3\ s^{-1}$

Pseudo-first order: $10^{-1} < k_2 < 10^7\ M^{-1}\ s^{-1}$ $(k_1 = k_2C^*_{reactant})$

Second order: $10^1 < k_2 < 10^9\ M^{-1}\ s^{-1}$

D. Further Aspects of the Application of Rotating Electrodes

A further improvement of the RRDE involves splitting of the ring electrode into two semicircular segments [189]. The potential of each half of the ring electrode can be independently controlled, and thus it is possible to monitor the stream of material from the disk electrode at two different potentials. With this electrode a single intermediate can be characterized by both its oxidation and reduction at the same time, and in more complicated reaction schemes the electrode makes it possible to monitor two different intermediates in one experiment. This of course saves time, but what is more important is that the experimental error due to possible time-dependent behavior of the disk electrode can be eliminated. The split-ring-disk electrode has so far found use primarily in the study of inorganic systems [189,190], but the applications in organic electroanalytic work seem obvious.

Another design, the rotating double-ring electrode, has been proposed particularly for the study of electrogenerated chemiluminescence [191]. The electrode consists of two platinum rings embedded in Teflon and separated by three thin spacers at a very small distance, 0.05 cm. The support of fluid is provided by a cylindrical channel through the lower part of the electrode. Digital simulation was used to calculate the velocities of flow [191,192]. The agreement between experimental and simulation results was within 5–10%.

Semitransparent or semiconducting electrodes have been used extensively in studies combining electrochemistry and photochemistry-spectroscopy. Such studies are normally focused on either spectroscopic characterization of electrochemically generated intermediates (spectroelectrochemistry, see Sec. VII) or the electrochemical behavior of photochemically produced species. Whereas spectroelectrochemical studies are preferentially carried out in quiet solutions, the introduction of a semitransparent or semiconducting disk in rotating ring-disk electrodes has been very promising for the study of light-initiated electrochemical reactions. Advantage has been taken of the steady-state conditions at the RDE or RRDE, and ω is also a useful experimental variable in this case. The semitransparent electrode allows an excitation or photostimulated electron transfer reaction to take place in the diffusion layer, with subsequent detection of the photogenerated species at either the disk or the ring electrode. The technique was used successfully in a study of the details in the reduction of anthraquinones to the corresponding hydroquinones [193]. A possible intermediate in this reduction is the semiquinone radical, which can be photochemically generated by irradiation of a solution containing both the quinone and the hydroquinone. The semiquinone thus produced can be either oxidized or reduced at the electrode, and rate constants for the electrochemical reactions as well as the homogeneous recombination were measured. The theory for the semitransparent RDE was published in 1977 [194], and later studies have been centered mainly on processes of potential interest for the development of usable photogalvanic cells [195–197].

In all the applications discussed so far, the RDE and RRDE have been operated under steady-state conditions; that is, the concentration profiles and thus the current are independent of time. However, successful attempts have been made to introduce modulations of the steady state. These have been of almost all possible kinds, including current- or potential-step techniques, ac voltammetry, and ω modulation. However, it is beyond the scope of the present text to discuss these methods in detail. The general theory of transient currents at rotating ring-disk electrodes have been developed by Albery and coworkers in a series of papers [198,199]. The method involves either a current or a potential step at the disk electrode from the initial to a new value. The theory has been extended to include ac measurements [202]. Plots of i^{lim} versus $\omega^{1/2}$, Eq. (72), may be recorded automatically by suitable programming of ω [201,202]. More complicated procedures involve potentiometry at fixed $i/\omega^{1/2}$ ratios [202,203] and sinusoidal variations of $\omega^{1/2}$ superimposed on the rotation rate [204–206].

E. Application of Rotating Disk Electrodes in Homogeneous Kinetics

The decay in concentration of a slowly reacting species can conveniently be followed with the aid of the RDE [Eq. (72); $\omega^{1/2}$ = constant]. Compared with other methods of monitoring concentration variations, the RDE offers the advantage that concentrations can be followed over a range of four orders of magnitude or more in the same cell, together with the possibility of providing electrochemical information about intermediates and products. The material conversion at the electrode surface can in this context be neglected for all practical purposes. Typical examples of this type of application of the RDE are found in a series of papers concerning the reactions of aromatic and heteroaromatic cation radicals with various nucleophiles, for example the thianthrene cation radical ($TH^{+\cdot}$) with anisole (ANI) [207,208]. In general these reactions were found to be second order in cation radical concentration (Fig. 33) and inhibited by unoxidized substrate, as expected for a reaction including disproportionation of the cation radical as a key step. However, a careful investigation of the substrate dependence using the RDE demonstrated the presence of a constant term in the denominator of the rate law, the inclusion of which results in

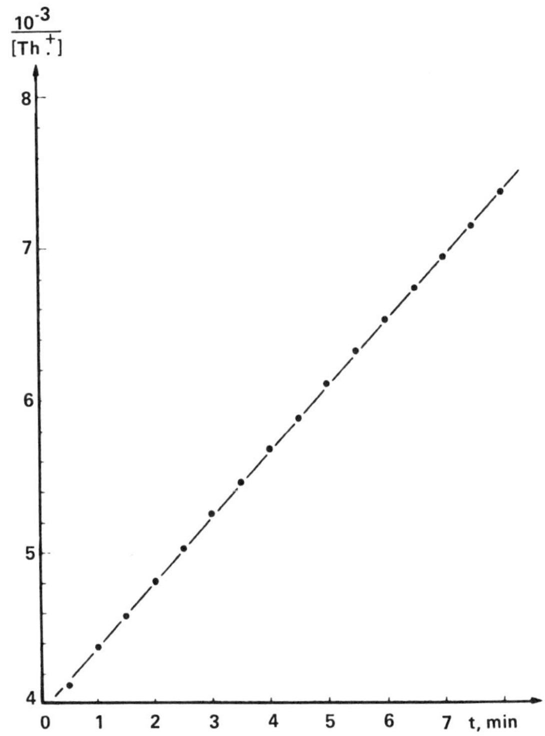

Figure 33. Second-order rate plot for the anisylation of thianthrene cation radical in CH_2Cl_2-TFA-TFAn (97:2:1). [From Ref. 207. Reprinted with permission from J. Amer. Chem. Soc., 97, 101 (1975). Copyright 1975 American Chemical Society.]

$$-\frac{d[TH^{\dot{+}}]}{dt} = \frac{2k_{80}K_{78}K_{79}[TH^{\dot{+}}]^2[ANI]}{k_{80}/k_{-79} + [TH]} \tag{77}$$

where the rate and equilibrium constants refer to the mechanism given by Eqs. (78) through (80). The reversible formation of a complex between the cation radical and the nucleophile as the first step seems to be of general importance, and a similar process has also been proposed for other reactions involving ion radicals [209].

$$TH^{\dot{+}} + AN \underset{k_{-78}}{\overset{k_{78}}{\rightleftharpoons}} (TH/AN)^{\dot{+}} \tag{78}$$

$$(TH/AN)^{\dot{+}} + TH^{\dot{+}} \underset{k_{-79}}{\overset{k_{79}}{\rightleftharpoons}} (TH/AN)^{2+} + TH \tag{79}$$

$(TH/AN)^{2+} \xrightarrow{k_{80}}$ 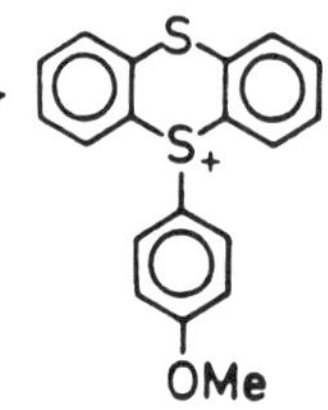 $+ H^+$ (80)

VI. COULOMETRY

An important parameter in all kinds of electrochemical work is the number of electrons n transferred during the electrode process. Values of n may be obtained from almost any electrochemical technique provided that a number of other parameters are known [Eqs. (81), (32), (72), and (46)]:

Linear sweep voltammetry: $$i_p = 2.68 \times 10^5 n^{3/2} AC_A^* D_A^{1/2} \nu^{1/2} \tag{81}$$

Chronoamperometry: $$i(t) = \frac{nFAD_A^{1/2} C_A^*}{(\pi t)^{1/2}} \quad (\text{at } i_c(t) = 0) \tag{32}$$

Rotating disk electrode : $$i^{\lim} = 0.62nFAC_A^* D_A^{2/3} \nu_k^{-1/6} \omega^{1/2} \tag{72}$$

Polarography: $$\bar{i}_d = 607nm^{2/3} t^{1/6} D_A^{1/2} C_A^* \tag{46}$$

However, it is the exception rather than the rule that all the necessary parameters are known for a particular experiment. This applies especially to the diffusion coefficient D_A, which is known only for very few substances in the solvent systems commonly used in electrochemical work. A simple solution to this kind of problem is based on the quantitative comparison of LSV and CA [210]. From Eqs. (81) and (32) it is easily seen that the ratio $R = (i_p/\nu^{1/2})/it^{1/2}$ is given by the simple expression

$$R = \frac{i_p/\nu^{1/2}}{it^{1/2}} = 4.92n^{1/2} \tag{82}$$

since all other parameters cancel. Thus, it is possible to estimate n values without the prior knowledge of D_A and A, the electrode area, provided the same experimental setup is used for both experiments. The experimental values of R for systems with known n were found to agree with the theoretical values (4.92 for n = 1 and 6.96 for n = 2) within 6%.

An alternative method for the determination of n values is based on measurement of the total amount of charge necessary for the exhaustive electrolysis of a known amount of substrate. This type of experiment, called coulometry, may be carried out either at constant potential or at constant current.

During coulometry at constant potential, the total amount of charge is obtained by integration of the current-time curve (Fig. 34) either manually or graphically or with an electronic integrator.

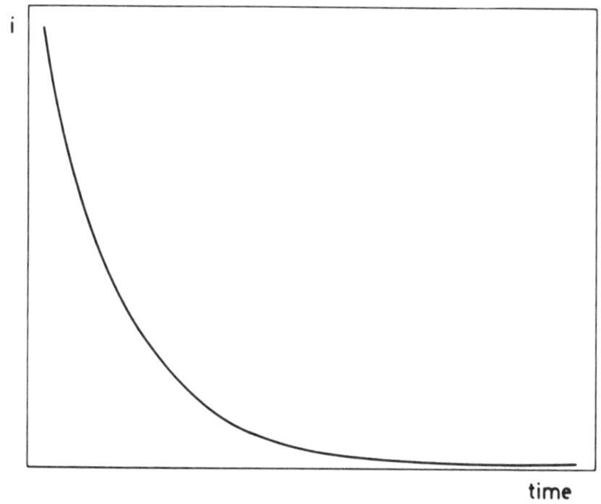

Figure 34. Current i as a function of time t in electrolysis (coulometry) at constant potential. $i = i(t = 0)e^{-D_A At/\delta V}$, where V is the volume of the solution (See Ref. 211.)

The advantage of potential control during the coulometric experiment has to be paid for by a rather long electrolysis time. The end point, $C_A^* = 0$ and $i = 0$, is in principle not reached until infinite time has elapsed. In practice, however, the electrolysis is stopped when the current has decayed to a few percent of the initial value. The error introduced in this way can be neglected for all practical purposes.

Coulometry at constant current is often considered less attractive than constant-potential coulometry. Control of the current rather than the potential has, however, a number of obvious advantages. First, the charge consumed during the reaction is directly proportional to the electrolysis time, and second, simpler electronic equipment may be used. It has been demonstrated [212] that when the current density is low the potential of the working electrode stays almost constant until 90% of the substrate is consumed. The total conversion of 0.1 mmol substrate at 25 mA requires an electrolysis time of 6.44 min at n = 1. The change in concentration during electrolysis is conveniently followed by LSV or CV. Cyclic voltammograms of a solution containing 0.1 mmol 3,4-diphenyl-1,2-dithiolylium ion (X), scheme (83) [213], after constant-current (25 mA) reduction for 0, 1, 2, 3, 4, 5, and 6 min, respectively, are shown in Fig. 35. Extrapolation of the current to $i_p = 0$ gives the time for complete conversion, which in the present case was very close to the predicted 6.44 min [n = 1, scheme (83)].

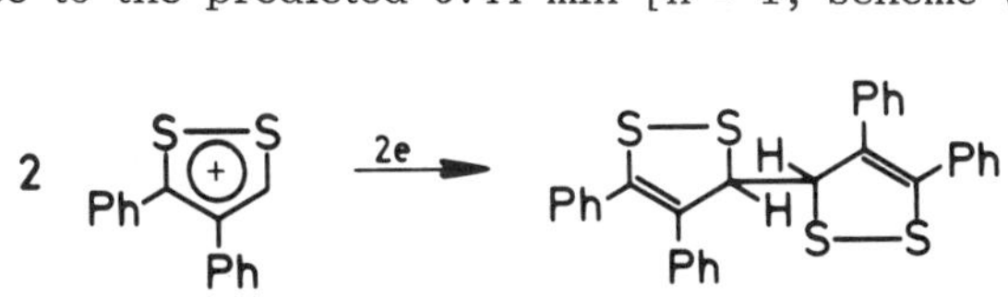

(X)

(83)

Under conditions in which the primary electrode product undergoes a slow chemical reaction, that is, $t_{1/2}$ is of the order of seconds, the value of n determined by a relatively fast technique like LSV may differ from that

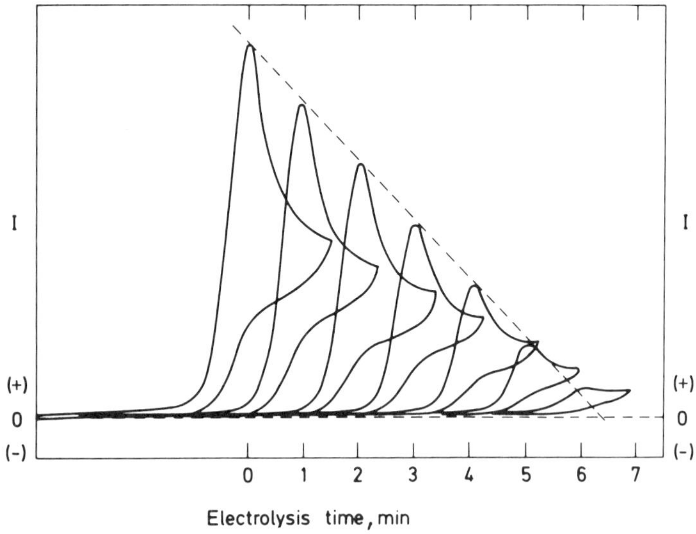

Figure 35. Constant current coulometry (25 mA) of 3,4-diphenyl-1,2-dithiolylium ion (X) (0.1 mmol). Cyclic voltammograms, horizontally displaced, are shown after reduction for 0, 1, 2, 3, 4, 5, and 6 min, respectively.

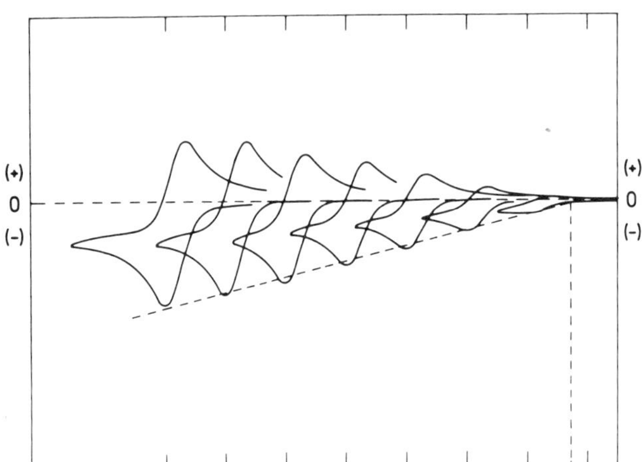

Figure 36. Constant-current coulometry (50 mA) of 2,3,5,6-tetraphenyl-1,4-dithiin (XI) (0.1 mmol). Cyclic voltammograms, horizontally displaced, are shown after oxidation for 0, 1, 2, 3, 4, 5, and 6 min, respectively.

obtained by a slow experiment like coulometry. This type of behavior was observed in the anodic oxidation of 2,3,5,6-tetraphenyl-1,4-dithiin (XI) in MeCN [214].

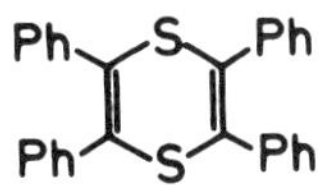

(XI)

During CV the reversible oxidation of XI to the cation radical is observed. The voltammetric curve fulfills all the criteria for a one-electron oxidation to stable product (see Sec. II). However, when constant-current coulometry was carried out as described earlier, 6.44 min was required to oxidize completely 0.1 mmol XI at 50 mA to a product electroinactive in the potential region of interest, indicating an overall two-electron process (Fig. 36). Thus, apparently contradictory results may be obtained owing to the difference in time scale between the two types of experiments (see also Chap. 4).

VII. METHODS BASED ON THE COMBINATION OF ELECTROCHEMICAL AND OPTICAL TECHNIQUES

Spectroelectrochemistry takes advantage of the possibility of characterizing and monitoring spectroscopically the intermediates and products formed as the result of the heterogeneous electron transfer reaction [4,7,215]. The major experimental problems concern both the preparation of suitable electrodes and cells and electronic detection of the absorptions, which are often extremely weak. Within the fields of organic electrochemistry and mechanism analysis, interest was earlier centered mainly on absorptions in the ultraviolet (UV)-visible region, and several reviews have been published dealing with both the theoretical and practical aspects of the technique [216–220]. However, in the last decade, attention has also been focused on the application of infrared (IR) [221–223] and resonance Raman spectroscopy [224–227]. In addition, it is possible to study the electrode surface and the interfacial region. However, this type of work is beyond the scope of the present text, and the interested reader is referred to references given in a series of reviews and recent publications [221–223,228–235].

A. Techniques and Cells for Spectroelectrochemistry

Spectroelectrochemistry can be performed in a number of ways depending on the type of electrode employed [4,7,215,221,222]. Transmission spectroscopy, by which the analyzing beam is passed through an optically transparent electrode (OTE), is illustrated in Fig. 37a. The OTE may consist of a transparent material, usually glass or quartz, covered with a thin metal film, most often Pt or Au. Transparent conducting materials, such as oxides of Sn and In, have been used as well. A variant of the OTE is the minigrid electrode, which is made from a wire mesh, usually between 50 and 1000 wires cm^{-1}. The transparency of such electrodes is due to the holes in the metal grid. Although convenient to work with, the OTE technique originally suffered from the necessity of passing the light beam through the entire cell; that is, only a fraction of the solution in the light beam was affected by the electrochemical process. This problem has now been overcome by the application of various designs of thin-layer cells [236,237].

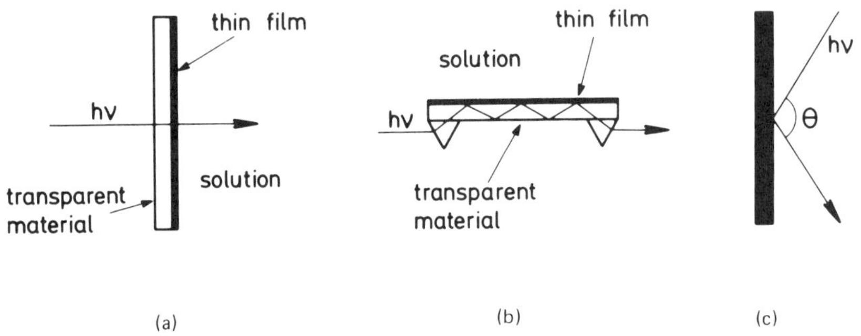

Figure 37. Electrode systems for spectroelectrochemistry: (a) optically transparent electrode, (b) electrode for internal reflection spectroscopy; and (c) electrode for specular reflectance measurements.

Another solution to this problem, internal reflection spectroscopy (IRS), takes advantage of the total reflection observed when a light beam is directed to the back of the OTE at an angle greater than the critical angle. Since the reflection takes place at the electrode/solution interface, the light beam is attenuated only by molecules close to the electrode surface. The penetration depth is about 1000 Å in a typical experiment. Because of this very small effective cell length, the electrode is most frequently used in a multiple reflection mode (Fig. 37b).

Conventional electrodes for voltammetry, such as Pt electrodes, can be used in spectroelectrochemical work. The light beam is reflected at the electrode surface and passes the solution of interest twice, which causes an improvement in sensitivity (Fig. 37c). The sensitivity of this method, called specular reflectance spectroscopy (SRS) [4,215,238], can be further increased by increasing the angle Θ to almost 180° (glancing incidence reflectance spectroscopy) [239].

In the limit, the light beam traverses the solution parallel to the electrode without reflection [240,241]. When a highly collimated beam is used, as is generated by a laser, this method may provide spectroscopic information about the concentration-distance profiles in voltammetric experiments [240]. In an interesting variant of this approach, called diffractive spectroelectrochemistry, a laser is placed relative to the cell in such a way that part of the beam strikes the electrode and part of it travels parallel to the electrode surface [242]. The light passing close to the surface, which therefore contains information about this particular region of the solution, is diffracted away from the main beam and may be analyzed by a strategically positioned photomultiplier.

Much effort has been put into the design of suitable cells. Some of the setups are depicted in Fig. 38, and many other solutions to special needs have been reported.

B. Modes of Operation and Applications

Recording of Spectral Data

Spectra of stable species are conveniently obtained by use of a sandwich-type thin-layer cell equipped with OTE. An illustrative example is the re-

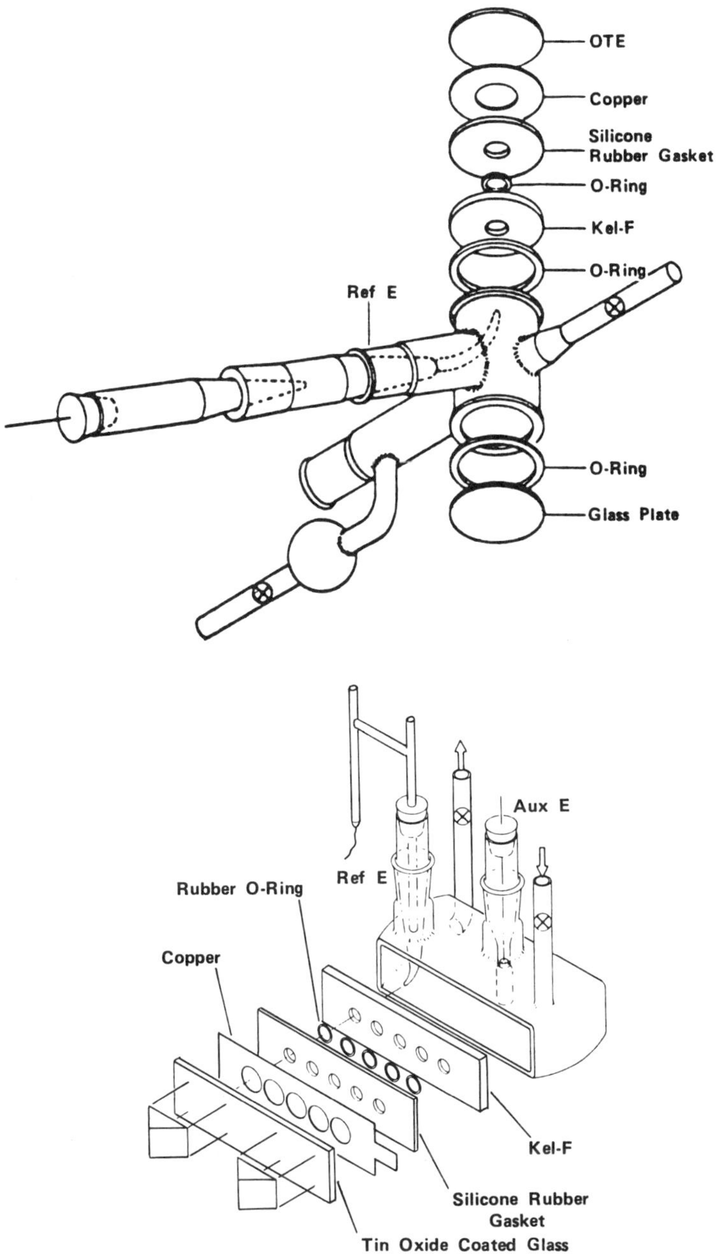

Figure 38. Examples of spectroelectrochemical cells for use in transmission spectroscopy (top) and internal reflectance spectroscopy (bottom). (From Ref. 217.)

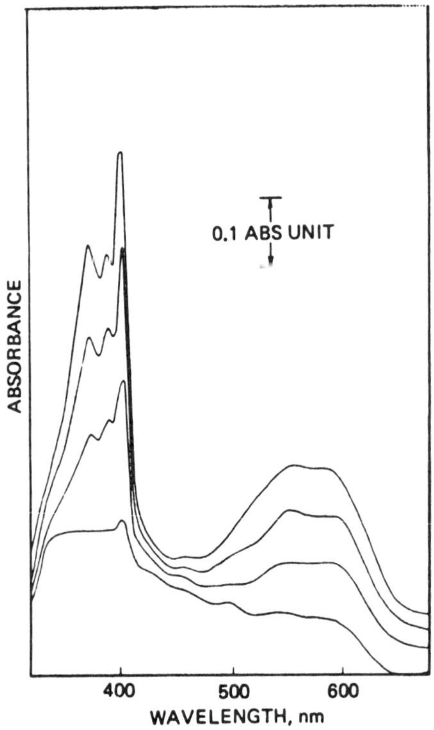

Figure 39. Spectra of 1,1'-dibenzyl-4,4'-bypyridinium cation radical taken after equal incremental reductions of the corresponding dication at an SnO_2 OTE. (From Ref. 244.)

duction of 1,1'-dibenzyl-4,4'-bipyridinium dication to the cation radical (Fig. 39) [243]. If the current efficiency is 100%, the concentration of the electrogenerated species can easily be determined from the amount of charge passed through the solution, which enables calculation of the extinction coefficient for the compound in question. The same experimental setup is useful for measurements of spectral characteristics of species for which the heterogeneous electron transfer is so slow that the direct conversion at the electrode is either inconvenient or impossible. In such cases advantage can be taken of the presence of an electron carrier, which is capable of exchanging electrons rapidly with the electrode as well as the substrate. This approach was taken in a study of the formal potentials and number of electrons associated with the reduction of cytochrome c and cytochrome c oxidase [244]. The spectral changes observed after consecutive coulometric generation of as little as 5 nanoequivalents of the electron carrier, which in this particular case was 1,1'-dimethyl-4,4'-bipyridinium cation radical, are illustrated in Fig. 40.

So far, few studies have been published concerning the spectroscopic properties of ion radicals in the IR region and even less experience has been gained in the application of IR spectroscopy for kinetic studies. Examples of IR spectra of anion radicals include those of tetracyanoethylene [245] and benzophenone (see Ref. 222 and references cited therein).

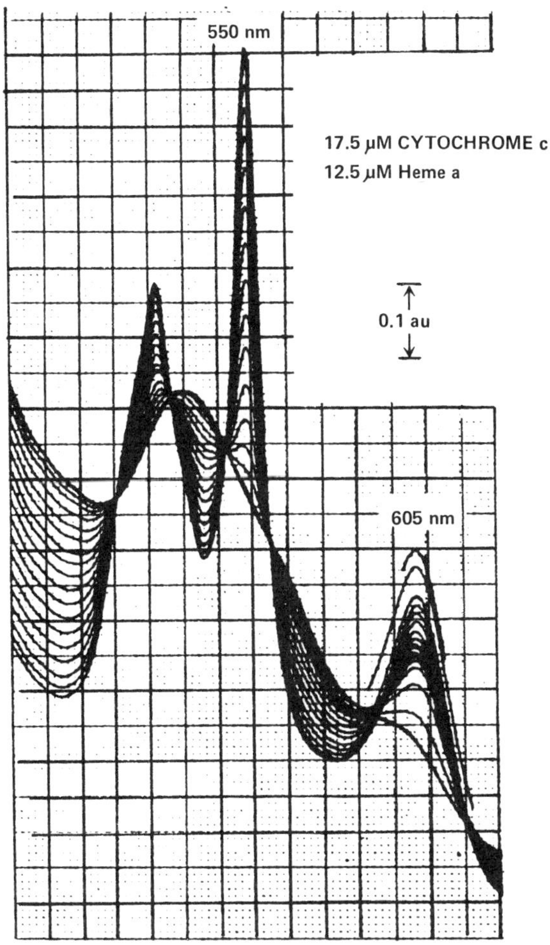

Figure 40. Spectrocoulometric titration of cytochrome c (17.5 μM) and cytochrome c oxidase by reduction with 1.1-dimethyl-4,4'-bipyridylium cation radical. Each spectrum is recorded after generation of a 5×10^{-9} equivalent of the cation radical. (From Ref. 244.)

When the heterogeneous electron transfer is followed by a chemical step, the foregoing procedures are inadequate for the recording of spectral data unless the rate of the chemical reaction is very low. Rapidly reacting species can be spectroscopically characterized by the application of either of two techniques, rapid scan spectroscopy (RSS) and modulated specular reflectance spectroscopy (MSRS), both of which involve a much more elaborate experimental setup than the spectrocoulometric procedures just described.

In the first of these techniques, RSS [246], the rapid scan may be obtained through the use of a vibrating mirror that allows a spectral region of approximately 400 nm to be covered in 1 ms. Coupled with a signal-averaging technique and computer control of the spectrometer and the data acquisition system, the relative sensitivity can be extended to 10^{-4} s for a 5 ms scan [247]. Examples of the application of RSS for the determination of spectral data of rapidly reacting intermediates are found in studies of the electrochemical oxidations of 4-methoxyphenyl-substituted olefins [247], tetraphenylethylene (XII) [248], and 3,3'-dimethylnaphthidine [249] and the reduction of 2,2'-diquinoxalyl [250].

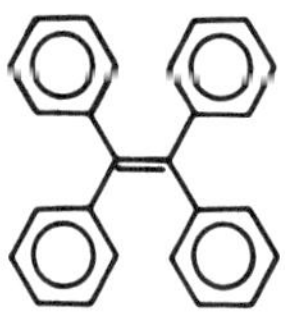

(XII)

$$\xrightarrow{-e} 12^{+\cdot}$$

$$2 \quad 12^{+\cdot} \rightleftharpoons 12^{2+} + 12$$

$$12^{2+} \xrightarrow{1.47 \times 10^{3}\ s^{-1}}$$

$$+ 2H^{+}$$

Kinetic and mechanistic (Fig. 41a) information can be gained from the analysis of the time-dependent spectra obtained when a repetitive scan technique is used [249–251]. The visible spectra taken at 0.9 s intervals during the reduction of porphyrin diacid, $P(0)H_4^{2+}$, to the phlorin monocation, $P(-II)H_5^{+}$ [251], are illustrated in Fig. 41b. From the spectra and data obtained by means of CV, the rate constant for the deprotonation step, Eq. (87), could be determined and approximate spectra for the two intermediates were estimated.

$$P(0)H_4^{2+} + e^- \longrightarrow P(-I)H_4^{+\cdot} \tag{84}$$

$$2H^+ + 2P(-I)H_4^{+\cdot} \rightleftharpoons P(0)H_4^{2+} + P(-II)H_6^{2+} \tag{85}$$

$$P(-I)H_4^{+\cdot} + e^- + 2H^+ \longrightarrow P(-II)H_6^{2+} \tag{86}$$

$$P(-II)H_6^{2+} \xrightarrow{k\,=\,0.25\ s^{-1}} P(-II)H_5^{+} + H^+ \tag{87}$$

A different approach to the same problem involves modulation of the electrode potential by a periodic function, for example, a sine wave or a pulse, MSRS [252–257]. Absorbance variations other than those that are synchronized with the modulation frequency are discriminated with the aid of a lock-in amplifier, leaving the system almost insensitive to the absorbance of products built up during the experiment. In this way it has been possible to obtain the visible spectrum of perylene anion radical in the presence of acetic acid (Fig. 42) and thianthrene cation radical in the presence of water [253]. Other pertinent examples include the oxidation of N,N-dimethyl-p-phenylenediamine [254] and 2,6-di-tert-butylanilines [256] to the corresponding cation radicals.

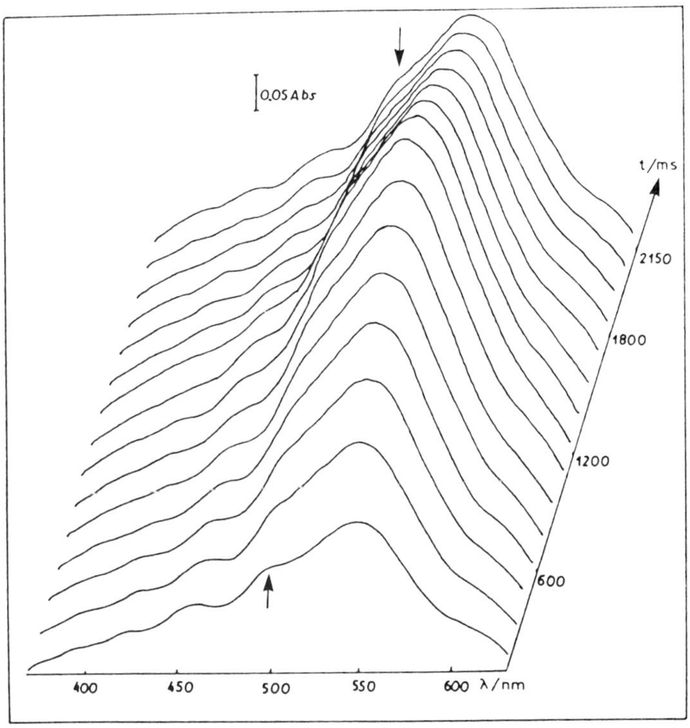

(a)

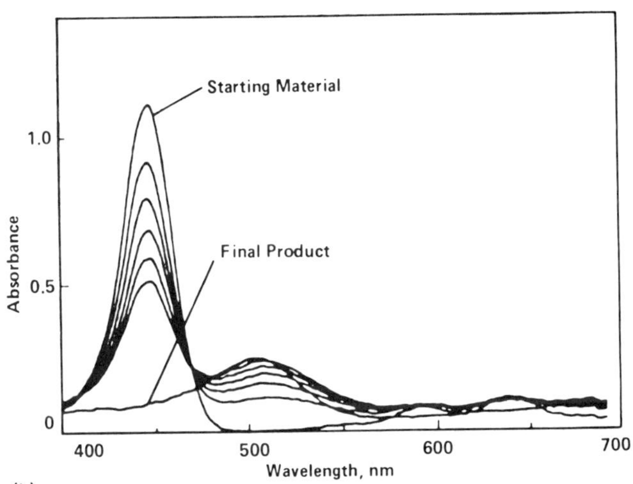

(b)

Figure 41. (a) The spectra of 3,3'-dimethylnaphthidine oxidation products recorded during chronoamperometric measurements at an SnO_2 OTE. (From Ref. 249.) (b) Time-dependent spectra for porphyrin diacid reduction in 1 M HCl. [From Ref. 251. Reprinted with permission from Anal. Chem., 51, 1139 (1979). Copyright 1979 American Chemical Society.]

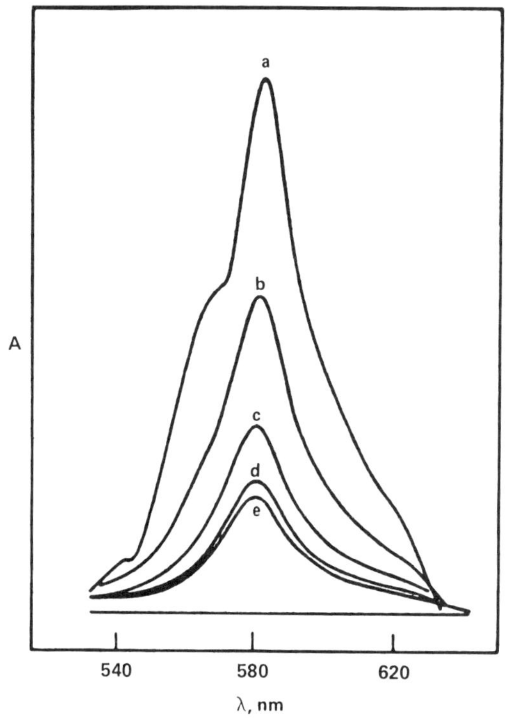

Figure 42. Visible absorption spectra of the perylene anion radical in DMF containing acetic acid (70 mM) measured by modulated specular reflectance at: a, 3 Hz; b, 10 Hz; c, 30 Hz; d, 60 Hz; and e, 80 Hz. (From Ref. 252.)

Kinetic Measurements

When spectroelectrochemistry is used as a tool in reaction kinetics it is important to know accurately the rate of generation of reactive intermediates, that is, the accurate potential of the working electrode. This requirement becomes a particular problem when an OTE is the preferred electrode because of the ohmic drop in the electrode itself. The most profitable way of operation is therefore to use a potential-step procedure by which the potential is stepped to a value sufficiently negative or positive for the heterogeneous electron transfer reaction to proceed at the diffusion-controlled rate. In transmission spectroscopy the absorbance $A_B(t)$ of the electrogenerated product B [Eq. (5)], in the absence of chemical follow-up reactions, is given by [258,259]

$$A_B(t) = \frac{2}{\sqrt{\pi}} \varepsilon_B C_A^* \sqrt{D_A t} \tag{88}$$

where ε_B is the extinction coefficient of B. If B is unstable and disappears during the experiment, the absorbance measured is smaller than that obtained in the absence of the chemical step. Expressions for $A_B(t)$ for a number of mechanistic perturbations have been calculated [259]. The results are most conveniently presented in terms of the normalized absorbance $A_B{}^N = A_B(t)/(2\varepsilon_B C_A^* \sqrt{D_A t}/\sqrt{\pi})$, which obviously is unity when B is stable, as a function of a dimensionless kinetic parameter, such as log kt or log ktC_A^*. The working curve for the eCe_h mechanism, which has been considered for the pyridination of 9,10-diphenylanthracene cation radical ($DPA^{\dot{+}}$), is illustrated in Fig. 43a.

Performance of the experiment, usually termed chronoabsorptometry, includes the recording of the absorbance-time curves for $DPA^{\dot{+}}$ in the absence

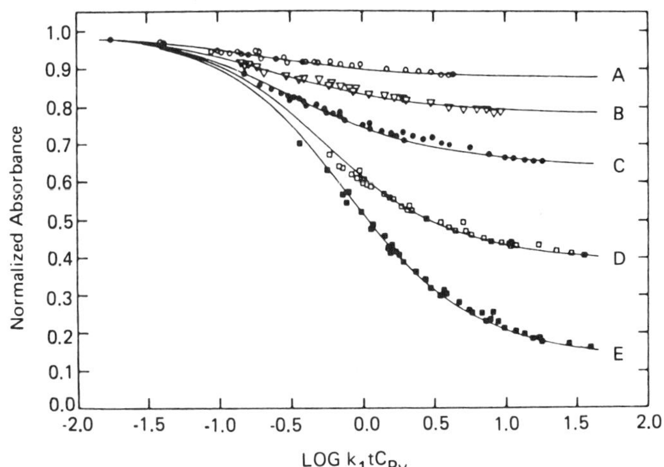

(a)

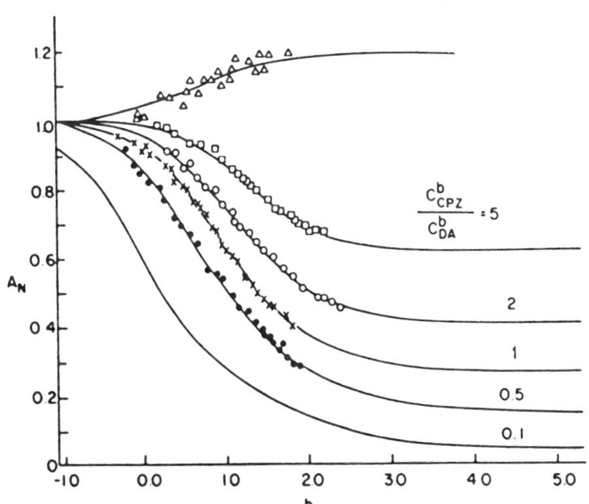

(b)

Figure 43. (a) eCe_h working curves for chronoabsorptometry. Experimental points concern the reaction between 9,10-diphenylanthracene cation radical and pyridine at various concentration ratios. (From Ref. 260.) (b) Normalized absorbance A_N versus log kC_{CPZ}^*t for the reaction between chlorpromazine (CPZ) cation radical and dopamine (DA). (From Ref. 264.)

and in the presence of pyridine from which data for the normalized absorbance-time dependence are obtained. The rate constant can be estimated by fitting these data to the working curve. In the present case [260], the rate constant for the nucleophilic attack of pyridine on $DPA^{+\cdot}$ was calculated to be $1.86 \times 10^4\ M^{-1}\ s^{-1}$.

By analogy to the extension of CA to DPSCA (see Sec. III), chronoabsorptometry has also been applied in a double-potential-step version [261, 262], and working curves for 22 (!) different mechanisms have been calculated [110].

IRS and SRS may be used by analogy to chronoabsorptometry for the determination of kinetic and mechanistic details (Fig. 43b) [263–265]. In the absence of chemical follow-up reactions, the absorbance-time dependence, again assuming diffusion-controlled heterogeneous electron transfer, for IRS is given by Eq. (89), which at times greater than approximately 1 ms reduces to Eq. (90),

$$A_B(t) = \varepsilon_B b_{eff} C_A^* \sqrt{\frac{D_A}{D_B}}\ (1 - e^{a^2 t} \operatorname{erfc}(a\sqrt{t}) \tag{89}$$

$$A_B = \varepsilon_B b_{eff} C_A^* \sqrt{\frac{D_A}{D_B}} \tag{90}$$

where $a = \sqrt{D_B}/\delta$, b_{eff} is an efficiency factor, and δ is the diffusion layer thickness (see Sec. V). The simplification (90) arises from the fact that the penetration depth of the light beam is only of the order of 1000 Å, and shortly after the potential has been switched on, the concentration profile close to the electrode is essentially in a steady-state condition, rendering the absorbance independent of time.

In potential-step chronoabsorptometry the absorbance of the electrogenerated species is followed during the electrochemical perturbation. Additional information may be gained by open-circuit relaxation (OCR), in which the absorbance decay is followed as a function of time after the working electrode has been diconnected from the potentiostat. When the electrode is connected, there is a continuous, although decreasing, generation of the reactive species, and both homogeneous and heterogeneous processes can take place during the reaction to products. However, after the electrode has been disconnected and the generation of the reactive species stops, only homogeneous electron transfer can occur. OCR spectroscopy has been discussed in detail in a number of papers [261,266–268]. In Fig. 44 is illustrated the absorbance-time curve obtained in a typical experiment, the generation of perylene anion radical in the absence (a) and in the presence (b) of phenol [269]. The stability of the perylene anion radical in the absence of the proton donor is demonstrated by the time independence of the absorbance after the electrode has been disconnected at t = 400 ms. The pulse length may be varied between approximately 100 ms and several seconds. Since the average concentration of the reactive intermediate is dependent on this parameter, pulse-length variation is important in OCR experiments. Disconnection of the electrode is usually performed with an electronic switch. The lower limit for applicable pulse lengths is not set by the electronics, but by the time necessary to discharge the double layer. However, by application of charge injection [270], together with a special diode switching technique [269], the problem arising from the double-layer discharging can be

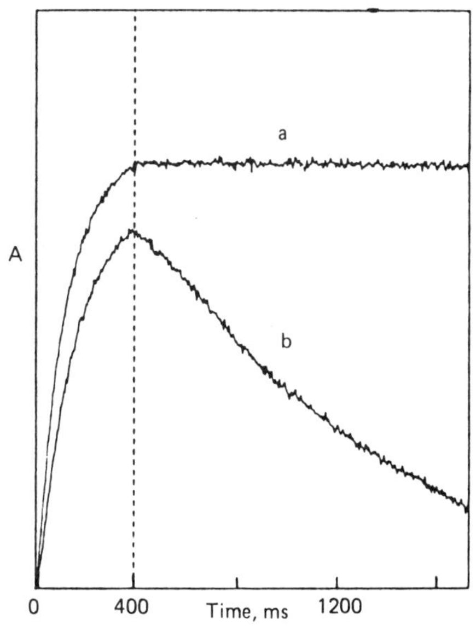

Figure 44. Absorbance-time curves for the spectroelectrochemical generation of perylene anion radical at 570 nm in DMF: (a) without and (b) with addition of phenol. The time to open circuit (dashed line) is 400 ms. (From Ref. 239.)

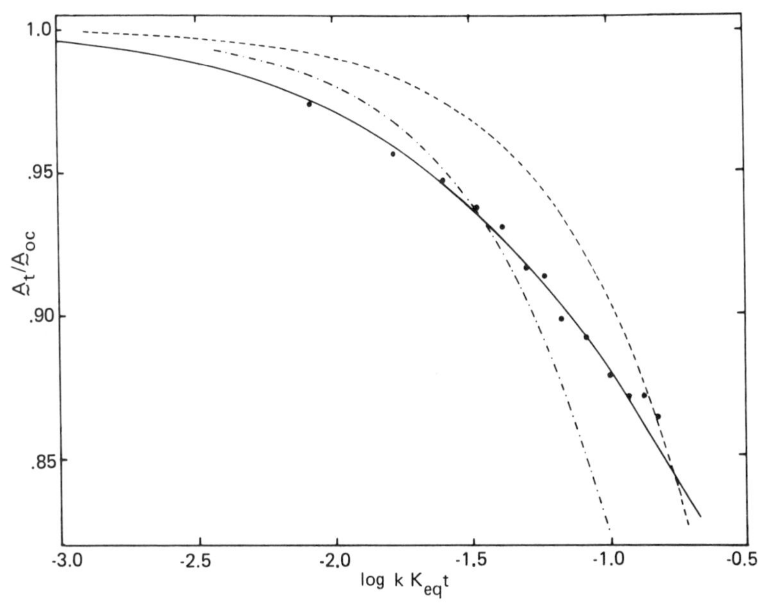

Figure 45. Working curve for open-circuit relaxation for various mechanisms: eCe (dashed line); disproportionation (solid line), and eCe_h (dash-point line). Experimental points are for the oxidative cyclization of tetraphenylethylene. (From Ref. 248.)

partly overcome and measurements can be performed in the microsecond range.

Working curves have been calculated for a number of reaction mechanisms. The experimental data are normalized by the absorption measured at the time of electrode disconnection, and the mechanism analysis and calculation of rate data then follow essentially the procedure already described for chronoabsorptometry. An illustrative example of the application of OCR is the study mentioned earlier, the intramolecular cyclization of tetraphenylethylene cation radical [248]. Working curves for three mechanisms, (1) the eCe mechanism, (2) the disproportionation mechanism, and (3) the eCe_h mechanism, together with experimental points, are shown in Fig. 45. Although it may be difficult or impossible to distinguish between mechanisms 1 and 3, the disproportionation is clearly different from the other two and allowed the conclusion that this mechanism is the most likely for the reaction.

The OCR technique has also played an important role in studies of the anodic oxidation of 4,4'-dimethoxystilbene [87] and the reaction between thianthrene cation radical with various nucleophiles [271]. The results of the latter study confirmed the conclusions arrived at by homogeneous kinetics [207].

VIII. APPLICATION OF ESR SPECTOSCOPY

Since the initially formed intermediates in organic electrode reactions are often radicals, the combination of electrochemical equipment with an electron spin resonance (ESR) spectrometer is a desirable possibility. The major practical problem encountered in designing an adequate experimental setup arises from the physical restrictions imposed on the electrochemical cell by the shape and size of the resonance cavity. Two different approaches have been taken to meet the requirements. One involves the formation of the radical species outside the magnetic field in a streaming solution that carries the elctrode products into the ESR cavity. By the other technique the radical species are formed by electrolysis in a small cell placed directly in the cavity. Both techniques have for many years been used extensively in qualitative and semiquantitative work [272,273], and the design and construction of cells have now reached such a level of sophistication that accurate quantitative experiments over a broad range of reaction rates can be carried out [274–281].

When a flow system is used in quantitative work, the transport conditions must be such that the distribution of radicals in the cell can be exactly calculated. This is most easily obtained when a simple and highly symmetrical cell is employed. A satisfactory solution to the problem was found by Albery and co-workers [274], who reported the successful use of a cylindircal cell in which a ring electrode was placed in such a way that the electrode surface was flush with the inner wall of the cell (Fig. 46). The ESR cavity was placed immediately below the electrode. Provided that the flow is sufficiently slow that laminar conditions prevail, the associated mathematics may be solved and equations relating the intensity of the ESR signal to the flow rate and a kinetic parameter containing the rate constant were presented. It was proposed that rate constants for radical decay, disproportionation, or dimerization, up to 10^8 M^{-1} s^{-1} can be measured with the aid of this cell.

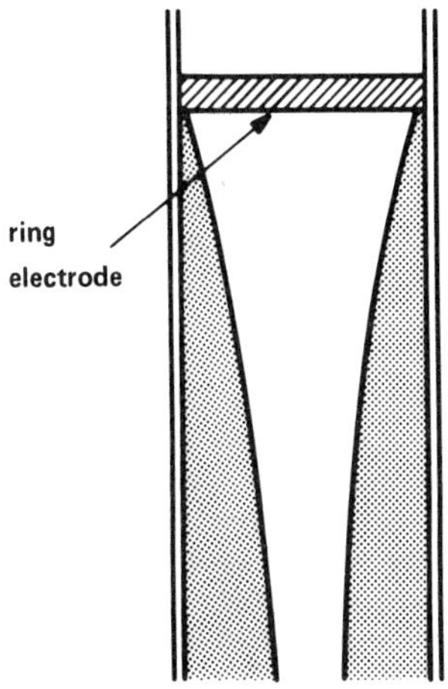

Figure 46. Tube electrode for ESR spectroscopy.

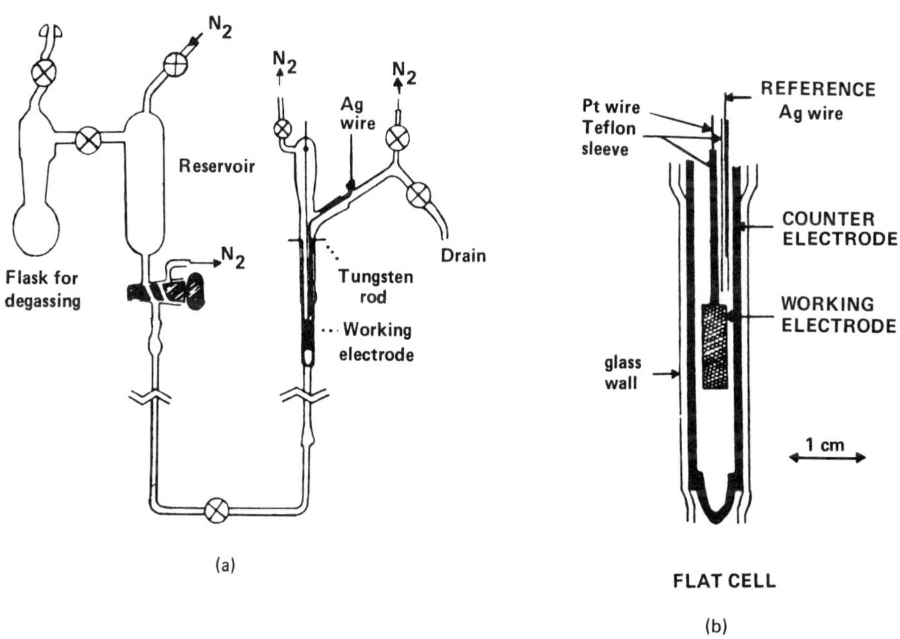

Figure 47. (a) Electrochemical cell and reservoir assembly and (b) detail of flat cell portion of call. [From Ref. 275. Reprinted with permission from J. Phys. Chem., 75, 3281 (1971). Copyright 1971 American Chemical Society.]

A cell (Fig. 47) that allowed the precise control of potential and current was designed by Goldberg and Bard, who also demonstrated the advantage of combining ESR spectroscopy with electrochemical techniques, such as CA, CV, and chronopotentiometry [275]. The latter approach was taken in a study of the reduction of cyclo-octatetraene (COT) in which it was demonstrated that the COT anion radical is stable in the presence of tetrabutylammonium ion, which had been a matter of dispute in previous work (Fig. 48) [283].

Another design for the electrochemical formation of radicals directly in the cavity involved the flow-through channel cell shown in Fig. 49 [279]. The cell was constructed of synthetic silica to avoid interference with the microwave field. The flow through this cell was demonstrated to be laminar up to Reynolds numbers less than 10^2, which allowed the derivation of quantitative relations between the ESR signal, the flow rate, and the current.

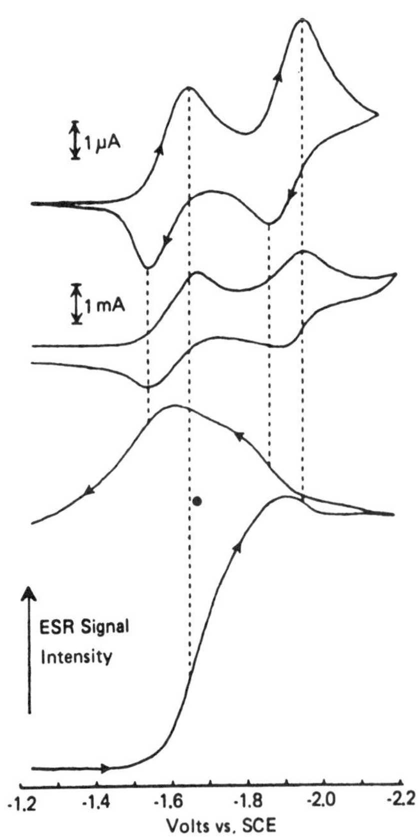

Figure 48. Cyclic voltammogram of COT in HMPA on hanging mercury drop electrode (uppermost curve). Cyclic voltammogram of COT on 15 cm^2 amalgamated gold electrode in ESR cell (central curve). ESR-potential curve recorded simultaneously with the central curve (lowermost curve). [From Ref. 283. Reprinted with permission from J. Amer. Chem. Soc., 97, 218 (1975). Copyright 1975 American Chemical Society.]

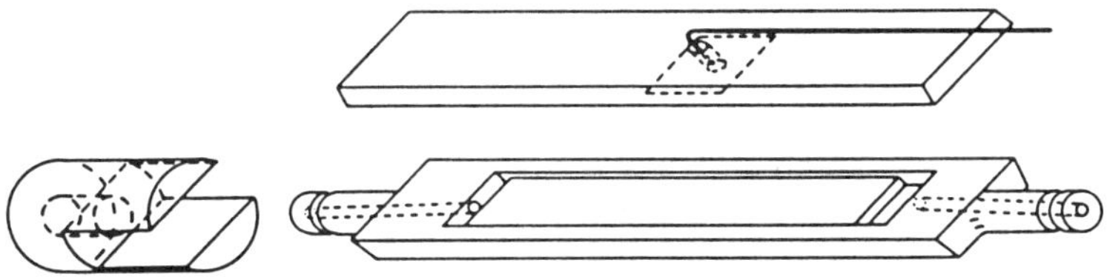

Figure 49. View of cell (unassembled), showing cover plate with electrode and lead-out wire, channel plate, and one of two PTFE mounting spacers. (From Red. 279.)

Recently, an electrochemical ESR cell of a rather unconventional design has been constructed [281]. In this cell the electrolyte is pumped through a porous frit on top of which is placed a layer of mercury. The electrolyte forms bubbles in the mercury, the concave surface of which acts as the working electrode. From the mercury electrode the solution is pumped directly into a thin-layer ESR cell. The electrochemical properties of this cell resemble those of the dropping mercury electrode.

In earlier work the most important application of combined electrochemistry and ESR spectroscopy was the characterization and identification of intermediates and products of electrode reactions. For instance, the ESR technique is particularly useful for measurement of the degree of protonation under conditions in which the radical ions take part in acid-base equilibria [284,285]. Such information may be obtained only with difficulty by other methods, but the coupling pattern of the ESR spectrum may often give the answer directly. An illustrative example is found in the anodic oxidation of 2,4,6-tri-tert-butylaniline (XIII), which, as expected, gives the cation radical as the initial electrode product, scheme (91) [284].

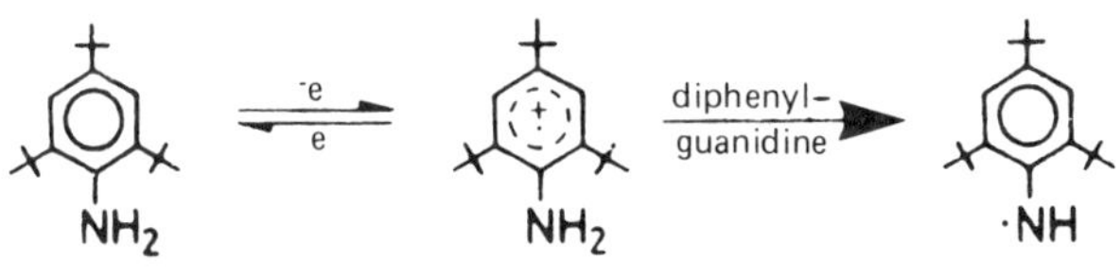

(XIII) (91)

In an aprotic solvent like MeCN or CH_3NO_2, the cation radical is stable and the ESR spectrum observed is in accordance with the reversible one-electron transfer indicated by CV. However, when the electrolysis is carried out in the presence of diphenylguanidine as a base, the ESR spectrum changes drastically and can be attributed to the presence of the neutral free radical formed by deprotonation of the cation radical.

The reductive cleavage of the carbon-halogen bond has been one of the most extensively studied reactions in organic electrochemistry. For the reduction of 4-iodonitrobenzene [286] and 4-chloroacetophenone [287] in the presence of CN^-, ESR evidence has been presented for the ultimate formation of the corresponding 4-cyanoarene anion radicals. These reactions were proposed to proceed via the formation of aryl radicals trapped by CN^-, scheme (92). Additional evidence for this reaction pathway was gained through a spin-trap experiment [288–290] (the electrochemical behavior of a number of nitroso spin-traps has been reported; see Ref. 290), by which the postulated aryl radical was trapped by α-phenyl-N-tert-butylnitrone, resulting in the formation of the corresponding nitroxide (93). The proposed mechanism thus resembles the $S_{RN}1$ scheme [291].

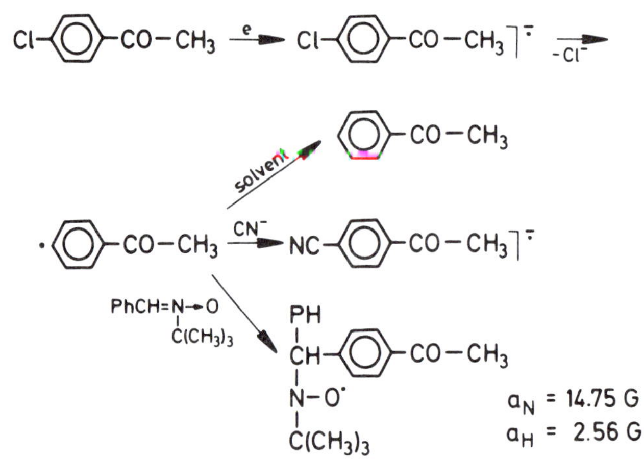

(92)

(93)

Further examples that illustrate the application of ESR spectroscopy in electrochemistry can be found in papers concerning the reduction of phosphonium and ammonium salts [292], the electron transfer reactions of a series of Weitz-type systems [293], and the oxidation of Ph_3CCOOH [294] and Ph_3COH [295], resulting in the intermediate formation of Ph_3C^+, which could be reduced to the ESR-detectable $Ph_3C^{\cdot}$ radical.

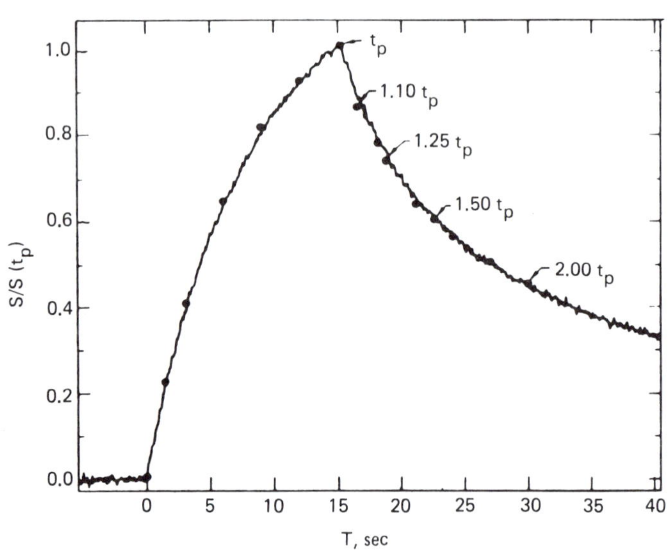

Figure 50. ESR signal-time dependence for 7.5 mM diethylfumarate. Current pulse time t_p = 15 s, i = 0.21 mA. [From Ref. 277. Reprinted with permission from J. Phys. Chem., 78, 295 (1974). Copyright 1974 American Chemical Society.]

Only a few kinetic studies in which ESR spectroscopy was used to monitor radical decay have been published. For example, rate constants for the reaction between $SO_2^{\overline{\cdot}}$ and organic halides have been determined [296] and decay curves have been used to gain kinetic information about the reactions of anion radicals of aliphatic nitrocompounds [280,297].

Signal-time behavior of the ESR response following a current pulse has been calculated by Goldberg and Bard [276] for a number of mechanisms, including first-order decomposition, radical ion dimerization, and radical ion-substrate coupling, and working curves from which rate constants can be calculated were presented. Application of this approach, which is very similar to that taken, for example, in transmission spectroelectrochemistry, was demonstrated for the reductive dimerization of a series of activated olefins (Fig. 50), a reaction that has been studied by means of a number of different electrochemical techniques. It was proposed that the major reaction pathway involved the direct coupling of two anion radicals, and as an example, the rate constant for dimethylfumarate was found to be equal to 160 ± 26 $M^{-1}\ s^{-1}$, which should be compared with the value obtained by other techniques: RRDE, 110 ± 20 $M^{-1}\ s^{-1}$; double-potential-step chronocoulometry, 117 ± 6 $M^{-1}\ s^{-1}$; and CV, 160 ± 40 $M^{-1}\ s^{-1}$. Thus, it can be concluded that ESR spectroscopy combined with electrochemical techniques provides results in satisfactory agreement with those obtained by other methods.

REFERENCES

1. V. D. Parker, Adv. Phys. Org. Chem., 19, 131 (1983).
2. V. D. Parker, in Topics in Organic Electrochemistry (A. J. Fry and W. E. Britton, eds.), Plenum Press, New York, 1986, Chap. 2.
3. L. Eberson, Adv. Phys. Org. Chem., 18, 79 (1982); L. Eberson, Electron Transfer Reactions in Organic Chemistry, Springer-Verlag, Berlin, 1987.
4. A. J. Bard and L. R. Faulkner, Electrochemical Methods, Wiley, New York, 1980.
5. D. D. MacDonald, Transient Techniques in Electrochemistry, Plenum, New York, 1977.
6. The Southampton Electrochemistry Group, Instrumental Methods in Electrochemistry, Ellis Horwood, Chichester, 1985.
7. Comprehensive Treatise of Electrochemistry (R. E. White, J. O'M. Bockris, B. E. Conway, and E. Yeager, eds.), Vol. 8, Plenum, New York, 1984.
8. W. H. Press, B. P. Flannery, S. A. Teukolsky, and W. T. Vetterling, Numerical Recipies, Cambridge University Press, Cambridge, 1986.
9. J. F. Rusling, Anal. Chem., 55, 1713, 1719 (1983).
10. A. Therdteppitak and J. T. Maloy, Anal. Chem., 56, 2592 (1984).
11. F. E. Woodard, R. D. Goodin, P. J. Kinlen, and J. H. Wagenknecht, Anal. Chem., 56, 1226 (1984).
12. T. F. Connors, J. F. Rusling, and A. Owlia, Anal. Chem., 57, 170 (1985).
13. M. H. Kim, Anal. Chem., 59, 2136 (1987).
14. A. C. Testa and W. H. Reinmuth, Anal. Chem., 33, 1320 (1961).
15. C. P. Andrieux, L. Nadjo, and J. M. Savéant, J. Electroanal. Chem., 26, 147 (1970).

16. V. D. Parker, Acta Chem. Scand., B34, 359 (1980).
17. R. S. Nicholson and I. Shain, Anal. Chem., 36, 706 (1964).
18. R. S. Nicholson and I. Shain, Anal. Chem., 37, 1351 (1965).
19. R. S. Nicholson and I. Shain, Anal. Chem., 37, 178, 190 (1965).
20. R. S. Nicholson, J. M. Wilson, and M. L. Olmstead, Anal. Chem., 38, 542 (1966).
21. M. L. Olmstead, R. G. Hamilton, and R. S. Nicholson, J. Electroanal. Chem., 41, 260 (1969).
22. L. Nadjo and J. M. Savéant, Anal. Chem., 44, 327 (1973); Anal. Chem., 48, 113 (1973).
23. C. P. Andrieux, L. Nadjo, and J. M. Savéant, Anal. Chem., 42, 223 (1970).
24. C. P. Andrieux and J. M. Savéant, Anal. Chem., 53, 165 (1974).
25. F. Ammar, C. P. Andrieux, and J. M. Savéant, Anal. Chem., 53, 407 (1974).
26. M. Mastragostino, L. Nadjo, and J. M. Savéant, Electrochim. Acta, 13, 721 (1968).
27. J. M. Savéant, Electrochim. Acta, 12, 753 (1967).
28. J. M. Savéant and E. Vianello, C.R. Acad. Sci., 256, 2597 (1963).
29. S. W. Feldberg, in Electroanalytical Chemistry (A. J. Bard, ed.), Vol. 3, Dekker, New York, 1969, p. 199.
30. M. F. Nielsen, K. Almdal, O. Hammerich, and V. D. Parker, Acta Chem. Scand., A41, 423 (1987), and references cited therein.
31. E. Ahlberg and V. D. Parker, J. Electroanal. Chem., 106, 419 (1980).
32. E. Ahlberg and V. D. Parker, J. Electroanal. Chem.,107, 197 (1980).
33. E. Ahlberg and V. D. Parker, J. Electroanal. Chem., 121, 57 (1981).
34. R. Eliason and V. D. Parker, J. Electroanal. Chem., 165, 21 (1984).
35. R. Eliason and V. D. Parker, J. Electroanal. Chem., 170, 347 (1984).
36. E. Ahlberg and V. D. Parker, J. Electroanal. Chem., 121, 73 (1981).
37. B. Aalstad and V. D. Parker, J. Electroanal. Chem., 112, 163 (1980).
38. B. Aalstad and V. D. Parker, J. Electroanal. Chem., 122, 183 (1981).
39. B. Aalstad, E. Ahlberg, and V. D. Parker, J. Electroanal. Chem., 122, 195 (1981).
40. V. D. Parker, Acta Chem. Scand., B35, 373 (1981).
41. E. Ahlberg and V. D. Parker, Acta Chem. Scand., B34, 71 (1980).
42. V. D. Parker, in Electroanalytical Chemistry (A. J. Bard, ed.), Vol. 14, Dekker, New York, 1986, p. 1.
43. S. P. Perone, Anal. Chem., 38, 1158 (1966).
44. N. S. Moe, Anal. Chem., 46, 968 (1974).
45. E. Ahlberg and V. D. Parker, Acta Chem. Scand., B34, 91 (1980).
46. M. F. Nielsen, O. Hammerich, and V. D. Parker, Acta Chem. Scand., B40, 101 (1986).
47. J. L. Sadler and A. J. Bard, J. Amer. Chem. Soc., 90, 1979 (1968).
48. J. L. Mills, R. F. Nelson, S. G. Shore, and L. B. Anderson, Anal. Chem., 43, 157 (1971).
49. S. Cheng and M. D. Hawley, J. Org. Chem., 50, 3388 (1985).
50. R. Lines, B. S. Jensen, and V. D. Parker, Acta Chem. Scand., B32, 510 (1978).
51. O. Hammerich and V. D. Parker, Electrochim. Acta, 18, 537 (1973).
52. B. S. Jensen and V. D. Parker, J. Chem. Soc. Chem. Commun., 367 (1974); J. Amer. Chem. Soc., 97, 5211 (1975).
53. D. Britz, J. Electroanal. Chem., 88, 309 (1978).

54. R. S. Nicholson, Anal. Chem., 37, 667 (1965).
55. W. T. de Vries and E. van Dalen, J. Electroanal. Chem., 10, 183 (1965).
56. J. C. Imbeaux and J.-M. Savéant, J. Electroanal. Chem., 28, 325 (1970); L. Nadjo, J.-M. Savéant, and D. Tessier, J. Electroanal. Chem., 52, 403 (1974); J.-M. Savéant and D. Tessier, J. Electroanal. Chem., 77, 225 (1977); D. Garreau and J.-M. Savéant, J. Electroanal. Chem., 86, 63 (1978).
57. S. Roffia and M. Lavachielli, J. Electroanal. Chem., 22, 117 (1969).
58. P. E. Whitson, H. W. Vanden Born, and D. H. Evans, Anal. Chem., 45, 1298 (1973).
59. S. P. Perone and T. R. Mueller, Anal. Chem., 37, 3 (1965); C. V. Evins and S. P. Perone, Anal. Chem., 39, 309 (1967).
60. S. P. Perone, in Electrochemistry: Calculations, Simulation and Instrumentation (J. S. Mattson, H. B. Mark, Jr., and H. C. MacDonald, Jr., eds.), Dekker, New York, 1972, Chap. 13.
61. E. Ahlberg, B. Svensmark, and V. D. Parker, Acta Chem. Scand., B34, 53 (1980).
62. M. F. Nielsen, S. Aa. Laursen, and O. Hammerich, submitted for publication.
63. M. Fleischmann, S. Pons, D. Rolison, and P. P. Schmidt, Ultramicroelectrodes, Datatech Systems, New York, 1987.
64. R. Lines and V. D. Parker, Acta Chem. Scand., B31, 369 (1977).
65. J. Heinze, Ber. Bunsenges. Phys. Chem., 85, 1096 (1981).
66. E. Ahlberg and V. D. Parker, Acta Chem. Scand., B33, 696 (1979).
67. J. O. Howell and R. M. Wightman, Anal. Chem., 56, 524 (1984).
68. C. P. Andrieux, D. Garreau, P. Hapiot, and J. M. Saveant, J. Electroanal. Chem., 243, 321 (1988); 248, 447 (1988).
69. M. I. Montenegro and D. Pletcher, J. Electroanal. Chem., 200, 371 (1986).
70. A. Fitch and D. H. Evans, J. Electroanal. Chem., 202, 83 (1986).
71. R. S. Nicholson, Anal. Chem., 38, 1406 (1966).
72. A. J. Bellamy, J. Electroanal. Chem., 151, 263 (1983).
73. H. Matsuda and Y. Ayabe, Z. Elektrochem., 59, 494 (1955).
74. V. D. Parker, Acta Chem. Scand., B35, 259 (1981).
75. E. Ahlberg and V. D. Parker, Acta Chem. Scand., B35, 117 (1981).
76. V. D. Parker, Acta Chem. Scand., B35, 233 (1981).
77. K. B. Oldham, Anal. Chem., 41, 1904 (1969); Anal. Chem., 44, 196 (1972).
78. K. B. Oldham and J. Spanier, J. Electroanal. Chem., 26, 331 (1970).
79. M. Greeness and K. B. Oldham, Anal. Chem., 44, 1121 (1972).
80. M. Goto and K. B. Oldham, Anal. Chem., 45, 204 (1973).
81. J. C. Imbeaux and J.-M. Savéant, J. Electroanal. Chem., 44, 169 (1973).
82. L. Nadjo, J.-M. Savéant, and D. Tessier, J. Electroanal. Chem., 44, 143 (1973).
83. C. P. Andrieux, J.-M. Savéant, and D. Tessier, J. Electroanal., 63, 429 (1975).
84. J.-M. Savéant and D. Tessier, J. Electroanal. Chem., 65, 57 (1975).
85. A. Ronlán, O. Hammerich, and V. D. Parker, J. Amer. Chem. Soc., 95, 7132 (1973).

86. B. Aalstad, A. Ronlán, and V. D. Parker, Acta Chem. Scand., B35, 247 (1981).
87. E. Steckhan, J. Amer. Chem. Soc., 100, 3536 (1978).
88. B. Burgbacher and H. Schäfer, J. Amer. Chem. Soc., 101, 7590 (1979).
89. V. D. Parker, Acta Chem. Scand., B35, 149 (1981).
90. R. N. McDonald, J. R. January, K. J. Borhani, and M. D. Hawley, J. Amer. Chem. Soc., 99, 1268 (1977).
91. D. Bethell, P. J. Galsworthy, K. L. Handoo, and V. D. Parker, J. Chem. Soc. Chem. Commun., 534 (1980).
92. V. D. Parker and D. Bethell, Acta Chem. Scand., B34, 617 (1980).
93. C. Amatore and J.-M. Savéant, J. Electroanal. Chem., 107, 353 (1980).
94. V. D. Parker, Acta Chem. Scand., B35, 349 (1981).
95. C. Amatore, M. Gareil, and J.-M. Savéant, J. Electroanal. Chem., 147, 1 (1983).
96. C. Amatore, M. Gareil, and J.-M. Savéant, J. Electroanal. Chem., 176, 377 (1984).
97. M. F. Nielsen, O. Hammerich, and V. D. Parker, Acta Chem. Scand., B41, 50 (1987).
98. M. F. Nielsen, O. Hammerich, and V. D. Parker, Acta Chem. Scand., B41, 64 (1987).
99. M. F. Nielsen and O. Hammerich, Acta Chem. Scand., B41, 668 (1987).
100. W. J. Albery, Electrode Kinetics, Clarendon Press, Oxford, 1975.
101. J.-M. Savéant and D. Tessier, J. Phys. Chem., 81, 2192 (1977).
102. D. A. Corrigan and D. H. Evans, J. Electroanal. Chem., 106, 287 (1980).
103. L. S. Marcoux and T. J. P. O'Brien, J. Phys. Chem., 76, 1666 (1972).
104. L. S. Marcoux, J. Phys. Chem., 76, 324 (1972).
105. C. Amatore and J.-M. Savéant, J. Electroanal. Chem., 102, 21 (1979).
106. C. Amatore and J.-M. Savéant, J. Electroanal. Chem., 86, 227 (1978).
107. W. M. Schwarz and I. Shain, J. Phys. Chem., 70, 845 (1966).
108. W. M. Schwarz and I. Shain, J. Phys. Chem., 69, 30 (1965).
109. W. V. Childs, J. T. Maloy, C. P. Keszthelyi, and A. J. Bard, J. Electrochem. Soc., 118, 874 (1971).
110. M. K. Hanafey, R. L. Scott, T. H. Ridgway, and C. N. Reilley, Anal. Chem., 50, 116 (1978).
111. K. B. Oldham, J. Electroanal. Chem., 145, 9 (1983).
112. J. M. R. Mellado, J. Electroanal. Chem., 209, 57 (1986).
113. V. D. Parker, Acta Chem. Scand., B37, 597 (1983); A38, 671 (1984).
114. D. A. Van Galen, M. P. Young, and M. D. Hawley, J. Electroanal. Chem., 175, 53 (1984).
115. C. Amatore, D. Garreau, M. Hammi, J. Pinson, and J.-M. Savéant, J. Electroanal. Chem., 184, 1 (1985), and references cited therein.
116. R. Eliason, O. Hammerich, and V. D. Parker, Acta Chem. Scand., B42, 7 (1988), and references cited therein.
117. D. Bethell and V. D. Parker, J. Chem. Soc., Perkin Trans. II, 841 (1982).
118. A. J. Bellamy, J. Electroanal. Chem., 158, 69 (1983).

119. E. Ahlberg, O. Hammerich, and V. D. Parker, J. Amer. Chem. Soc., 103, 844 (1981).
120. F. C. Anson and R. A. Osteryoung, J. Chem. Educ., 60, 293 (1983).
121. J. H. Christie, G. Lauer, and R. A. Osteryoung, J. Electroanal. Chem., 7, 60 (1964).
122. F. C. Anson, Anal. Chem., 38, 54 (1966).
123. J. Puy, M. Pla, F. Mas, and F. Sanz, J. Electroanal. Chem., 241, 89 (1988).
124. M. S. De Vicente, M. R. Moncelli, and R. Guidelli, J. Electroanal. Chem., 248, 55 (1988).
125. J. F. Rusling, M. Y. Brooks, B. J. Scheer, T. T.-T. Chou, S. S. Shukla, and M. Rossi, Anal. Chem., 58, 1942 (1986).
126. R. D. Grypa and J. T. Maloy, J. Electrochem. Soc., 122, 509 (1979).
127. B. M. Bezilla, Jr., and J. T. Maloy, J. Electrochem. Soc., 126, 579 (1979).
128. E. Solon and A. J. Bard, J. Amer. Chem. Soc., 86, 1926 (1964).
129. J. Galvez, J. Electroanal. Chem., 132, 15 (1982); J. Galvez and A. Molina, J. Electroanal. Chem., 146, 221 (1983); J. Galvez and R. Saura, J. Electroanal. Chem., 146, 233 (1983); J. Galvez, T. Fuente, A. Molina, and R. Saura, J. Electroanal. Chem., 146, 243 (1983); J. Galvez, J. Zapata, and A. Molina, J. Electroanal. Chem., 219, 1 (1987); J. Galvez and J. Zapata, J. Electroanal. Chem., 219, 13 (1987); J. Galvez, F. Robles, J. Zapata and S.-M. Park, J. Electroanal. Chem., 241, 33 (1988).
130. J. Galvez, A. Molina, T. Perez, and M. H. Cordoba, Anal. Chem., 56, 887 (1984); J. Galvez, M. L. Alcaraz, T. Perez, and M. H. Cordoba, Anal. Chem., 57, 2116 (1985).
131. K. Aoki, K. Honda, K. Tokuda, and H. Matsuda, J. Electroanal. Chem., 195, 51 (1985).
132. K. Aoki, K. Akimoto, K. Tokuda, H. Matsuda, and J. Osteryoung, J. Electroanal. Chem., 182, 281 (1985).
133. T. Berzins and P. Delahay, J. Amer. Chem. Soc., 75, 4205 (1953).
134. W. H. Reinmuth, Anal. Chem., 32, 1514 (1960).
135. C. Furlani and G. Morpurgo, J. Electroanal. Chem., 1, 351 (1959).
136. M. D. Hawley and R. N. Adams, J. Electroanal. Chem., 10, 376 (1965).
137. A. C. Testa and W. H. Reinmuth, Anal. Chem., 32, 1512 (1960).
138. D. H. Geske, J. Amer. Chem. Soc., 81, 4145 (1959).
139. P. T. Kissinger, P. T. Holt, and C. N. Reilley, J. Electroanal. Chem., 33, 1 (1971).
140. J. Heyrovsky and J. Kuta, Principles of Polarography, Academic Press, New York, 1966.
141. J. Kuta, in Comprehensive Treatise of Electrochemistry, (R. E. White, J. O'M. Bockris, B. E. Conway, and E. Yeager, eds.), Vol. 8, Plenum, New York, 1984, Chap. 4.
142. L. Meites, Polarographic Techniques, 2nd ed., Interscience, New York, 1965.
143. I. M. Kolthoff and J. J. Lingane, Polarography, 2nd ed., Interscience, New York, 1952.
144. C. L. Perrin, in Progress in Physical Organic Chemistry, (S. G. Cohen, A. Streitwieser, Jr., and R. W. Taft, eds.), Vol. 3, Interscience, New York, 1965, p. 165.

145. R. Guidelli, in Electroanalytical Chemistry (A. J. Bard, ed.), Vol. 5, Dekker, New York, 1971, p. 149.
146. Z. Galus, Fundamentals of Electrochemical Analysis, Ellis Hardwood, New York, 1976.
147. L. Camacho, M. Blázquez, M. Jiménez, and M. Domínguez, J. Electroanal. Chem., 172, 173 (1984); M. Blázquez, L. Camacho, M. Jiménez, and M. Domínguez, J. Electroanal. Chem., 189, 195 (1985).
148. G. J. Hoytink, in Advances in Electrochemistry and Electrochemical Engineering (H. Gerischer and C. W. Tobias, eds.), Vol. 7, Wiley-Interscience, New York, 1970, p. 221.
149. A. M. Heras, E. Munoz, J. L. Avila, L. Camacho, and J. L. Cruz, J. Electroanal. Chem., 243, 293 (1988).
150. M. E. Peover and J. D. Davies, J. Electroanal. Chem., 6, 46 (1963).
151. M. Lipsztajn, T. M. Krygowski, and Z. Galus, J. Electroanal. Chem., 49, 17 (1974).
152. A. Lasia, J. Electroanal. Chem., 102, 117 (1979).
153. T. Fujinaga, K. Izutsu, and T. Nomura, J. Electroanal. Chem., 29, 203 (1971).
154. P. Zuman and J. Chodkowski, Coll. Czech. Chem. Commun., 27, 759 (1962).
155. M. A. Almirón, L. Camacho, E. Munoz, J. L. Avila, and J. V. Sinisterra, J. Electroanal. Chem., 241, 297 (1988).
156. Z. R. Grabowski and M. S. Balasiewicz, Trans. Faraday Soc., 64, 3346 (1968).
157. O. Hammerich and V. D. Parker, Acta Chem. Scand., B35, 395 (1981).
158. R. Brdicka, V. Hanus, and J. Koutecky, in Progress in Polarography (P. Zuman and I. M. Kolthoff, eds.), Vol. 1, Interscience, New York, 1962, Chap. VII.
159. J. M. Rodríguez-Mellado, M. Blázquez, and J. J. Ruiz, J. Electroanal. Chem., 195, 363 (1985); M. Bláquez, M. Jiménez, L. Camacho, and M. Domínguez, J. Electroanal. Chem., 243, 309 (1988).
160. L. Camacho, A. Heras, J. L. Avila, and F. García-Blanco, J. Electroanal. Chem., 172, 167 (1984).
161. M. De Juan, A. M. Heras, L. Camacho, J. L. Avila, and C. Jiménez, J. Electroanal. Chem., 191, 303 (1985).
162. D. González-Arjona, M. Rueda, and J. J. Ruiz, J. Electroanal. Chem., 239, 239 (1988).
163. F. Qiang-sheng and L. Sin-rhu, Anal. Chem., 53, 1006 (1981); L. Sin-Rhu and F. Qiang-sheng, Anal. Chem., 54, 1362 (1982).
164. W. J. Albery and M. L. Hitchman, Ring-Disc Electrodes, Clarendon, Oxford, 1971.
165. A. C. Riddiford, in Advances in Electrochemistry and Electrochemical Engineering (P. Delahay, ed.), Vol. 4, Wiley-Interscience, New York, 1966, p. 47.
166. F. Opekar and P. Beran, J. Electroanal. Chem., 69, 1 (1976).
167. P. K. Adanuvor, R. E. White, and S. E. Lorimer, J. Electrochem. Soc., 134, 625 (1987).
168. R. C. Compton, R. G. Harland, P. R. Unwin, and A. M. Waller, J. Chem. Soc. Faraday Trans. 1, 83, 1261 (1987).
169. R. N. Adams, Electrochemistry at Solid Electrodes, Dekker, New York, 1969.
170. S. Bruckenstein and B. Miller, Acc. Chem. Res., 10, 54 (1977).

171. J. Newman, J. Electrochem. Soc., 113, 501 (1966).
172. W. Marathe and J. Newman, J. Electrochem. Soc., 116, 1704 (1969).
173. S. Bruckenstein and B. Miller, J. Electrochem. Soc., 117, 1044 (1970).
174. Z. Galus and R. N. Adams, J. Phys. Chem., 67, 866 (1963).
175. V. G. Levich, Physicochemical Hydrodynamics, Prentice-Hall, Englewood Cliffs, N.J., 1962.
176. K. W. Pratt, Anal. Chem., 56, 1967 (1984).
177. P. A. Malachesky, L. S. Marcoux, and R. N. Adams, J. Phys. Chem., 70, 4068 (1966).
178. L. S. Marcoux, R. N. Adams, and S. W. Feldberg, J. Phys. Chem., 73, 2611 (1969).
179. U. Svanholm, A. Ronlán, and V. D. Parker, J. Amer. Chem. Soc., 96, 5108 (1974).
180. A. N. Frumkin and L. N. Nekrasov, Dokl. Akad. Nauk SSSR, 126, 115 (1959).
181. K. B. Prater and A. J. Bard, J. Electrochem. Soc., 117, 207, 335, 1517 (1970).
182. V. J. Puglisi and A. J. Bard, J. Electrochem. Soc., 119, 833 (1972).
183. K. B. Prater, in Electrochemistry: Calculations, Simulation and Instrumentation (J. S. Mattson, H. B. Mark, Jr., and H. C. MacDonald, Jr., eds.), Dekker, New York, 1972, p. 219.
184. W. J. Albery and J. S. Drury, Trans. Faraday Soc., 68, 456 (1972).
185. P. Jeroschewski and F. Pragst, J. Electroanal. Chem., 149, 131 (1983).
186. V. J. Puglisi and A. J. Bard, J. Electrochem. Soc., 119, 829 (1972); 120, 748 (1973).
187. G. Neubert and K. B. Prater, J. Electrochem. Soc., 121, 745 (1974).
188. N. R. Armstrong, N. E. Vandenborgh, and R. K. Quinn, J. Electrochem. Soc., 122, 615 (1975).
189. B. Miller and R. E. Visco, J. Electrochem. Soc., 115, 251 (1968).
190. B. Miller, J. Electrochem. Soc., 116, 1675 (1969).
191. S. Clarenbach, E. W. Grabner, and E. Brauer, Ber. Bunsenges. Phys. Chem., 77, 908 (1973).
192. S. Clarenbach and E. W. Grabner, Ber. Bunsenges. Phys. Chem., 80, 115 (1976).
193. W. J. Albery, M. D. Archer, N. J. Field, and A. D. Turner, Disucss. Faraday Soc., 56, 28 (1974).
194. W. J. Albery, M. D. Archer, and R. G. Edgell, J. Electroanal. Chem., 82, 199 (1977).
195. W. J. Albery, W. R. Bowen, F. S. Fisher, and A. D. Turner, J. Electroanal. Chem., 107, 1, 11 (1980).
196. W. J. Albery, P. N. Bartlett, W. R. Bowen, F. S. Fisher, and A. W. Foulds, J. Electroanal. Chem., 107, 23 (1980).
197. W. J. Albery, W. R. Bowen, F. S. Fisher, A. W. Foulds, K. J. Hall, A. R. Hillman, R. G. Edgell, and A. F. Orchard, J. Electroanal. Chem., 107, 37 (1980).
198. W. J. Albery, Trans. Faraday Soc., 63, 1771 (1967); 67, 153 (1971).
199. W. J. Albery, J. S. Drury, and M. L. Hitchman, Trans. Faraday Soc., 67, 161, 166, 2162 (1971).
200. W. J. Albery, R. G. Compton, and A. R. Hillman, J. Chem. Soc., Faraday Trans. 1, 74, 1007 (1978), and references cited therein.

201. S. C. Creason and R. F. Nelson, J. Electroanal. Chem., 27, 189 (1970).
202. B. Miller and S. Bruckenstein, J. Electrochem. Soc., 117, 1032 (1970).
203. B. Miller, M. I. Bellavance, and S. Bruckenstein, J. Electrochem. Soc., 118, 1082 (1971).
204. B. Miller and S. Bruckenstein, Anal. Chem., 46, 2026 (1974).
205. K. Tokuda and S. Bruckenstein, J. Electrochem. Soc., 126, 431 (1979).
206. Y. Kanzaki and S. Bruckenstein, J. Electrochem. Soc., 126, 437 (1979).
207. U. Svanholm, O. Hammerich, and V. D. Parker, J. Amer. Chem. Soc., 97, 101 (1975).
208. U. Svanholm and V. D. Parker, J. Amer. Chem. Soc., 98, 997, 2942 (1976).
209. R. M. Wightman, J. R. Cockrell, R. W. Murray, J. F. Bunnett, and S. B. Jones, J. Amer. Chem. Soc., 98, 2562 (1976).
210. P. A. Malachesky, Anal. Chem., 41, 1493 (1969).
211. J. J. Lingane, J. Amer. Chem. Soc., 67, 1916 (1945).
212. V. D. Parker, Acta Chem. Scand., 24, 2768 (1970).
213. C. T. Pedersen and V. D. Parker, Tetrahedron Lett., 767 (1972).
214. O. Hammerich, unpublished results.
215. W. R. Heineman, J. Chem. Educ., 60, 305 (1983).
216. T. Kuwana and W. R. Heineman, Acc. Chem. Res., 9, 241 (1976).
217. T. Kuwana and N. Winograd, in Electroanalytical Chemistry (A. J. Bard, ed.), Vol. 7, Dekker, New York, 1974, p. 1.
218. R. H. Muller, ed., Advances in Electrochemistry and Electrochemical Engineering, Vol. 9, Wiley-Interscience, New York, 1973.
219. W. R. Heineman, Anal. Chem., 50, 390A (1978).
220. W. R. Heineman, F. M. Hawkridge, and H. N. Blount, in Electroanalytical Chemistry (A. J. Bard, ed.), Vol. 13, Dekker, New York, 1984, p. 1.
221. J. K. Foley and S. Pons, Anal. Chem., 57, 945A (1985).
222. K. Ashley and S. Pons, Chem. Rev., 88, 673 (1988).
223. A. Bewick, in Trends in Interfacial Electrochemistry (A. F. Silva, ed.), Reidel, Dordrecht, 1986, p. 331.
224. R. P. Van Duyne, in Chemical and Biological Applications of Lasers (L. B. Moore, ed.), Vol. 4, Academic, New York, 1979, Chap. 4.
225. R. P. Cooney, M. R. Mahoney, and A. J. McQuillan, in Advances in Infrared and Raman Spectroscopy (R. J. Clark and R. E. Hester, eds.), Vol. 8, Heyden, London, 1982, Chap. 4.
226. M. Fleischmann and I. R. Hill, in Comprehensive Treaties of Electrochemistry (R. E. White, J. O'M. Bockris, B. E. Conway, and E. Yeager, eds.), Vol. 8, Plenum, New York, 1984, Chap. 6.
227. R. T. Packard and R. L. McCreery, Anal. Chem., 59, 2631 (1987).
228. J. D. E. McIntyre and W. F. Peck, Jr., Discuss. Faraday Soc., 56, 122 (1976); D. M. Kolb, Discuss. Faraday Soc., 56, 138 (1976); B. E. Conway, H. Angerstein-Kozlowska, F. C. Ho, J. Klinger, B. MacDougall, and S. Gottesfeld, Discuss. Faraday Soc., 56, 210 (1976).
229. W. R. Heineman and P. T. Kissinger, Anal. Chem., 50, 166R (1978); 52, 138R (1980).

230. M. Fleischman, P. R. Graves, and J. Robinson, J. Electroanal. Chem., 182, 73 (1985).
231. Optical studies of adsorbed layers at interface, Symp. Faraday Soc., No. 4 (1970).
232. H. Baltruschat, N. Starud, and J. Heitbaum, J. Electroanal. Chem., 239, 361 (1988); D. J. Blackwood and S. Pons, J. Electroanal. Chem., 247, 277 (1988).
233. M.-C. Pham, F. Adami, and J.-E. Dubois, J. Electrochem. Soc., 134, 2166 (1987).
234. A. R. Hillman and E. F. Mallen, J. Electroanal. Chem., 243, 403 (1988); P. A. Christensen, A. Hamnett, and A. R. Hillman, J. Electroanal. Chem., 242, 47 (1988); S. N. Hoier, D. S. Ginley, and S.-M. Park, J. Electrochem. Soc., 135, 91 (1988).
235. P. A. Christensen, A. Hamnett, and P. R. Trevellick, J. Electroanal. Chem., 242, 23 (1988).
236. R. K. Rhodes and K. M. Kadish, Anal. Chem., 53, 1539 (1981).
237. C. W. Anderson and M. R. Cushman, Anal. Chem., 54, 2122 (1982).
238. A. S. Hinman and B. J. Pavelich, Can. J. Chem., 65, 919 (1987).
239. R. L. McCreery, R. Pruiksma, and R. Fagan, Anal. Chem., 51, 749 (1079); J. P. Skully and R. L. McCreery, Anal. Chem., 52, 1885 (1980).
240. R. Pruiksma and R. L. McCreery, Anal. Chem., 53, 202 (1981); C.-C. Jan and R. L. McCreery, Anal. CHem., 58, 2771 (1986).
241. J. D. Brewster and J. L. Anderson, Anal. Chem., 54, 2560 (1982).
242. P. Rossi, C. W. McCurdy, and R. L. McCreery, J. Amer. Chem. Soc., 103, 2524 (1981); C.-C. Jan, B. K. Lavine, and R. L. McCreery, Anal. Chem., 57, 752 (1985).
243. E. Steckhan and T. Kuwana, Ber. Bunsenges. Phys. Chem., 78, 253 (1974).
244. W. R. Heineman, T. Kuwana, and C. R. Hartzell, Biochem. Biophys. Res. Commun., 49, 1 (1972); 50, 892 (1973).
245. S. B. Khoo, J. F. Foley, C. Korzeniewski, S. Pons, and C. Marcott, J. Electroanal. Chem., 233, 223 (1987).
246. J. W. Strojek, G. A. Gruver, and T. Kuwana, Anal. CHem., 41, 481 (1969).
247. E. Steckhan and D. A. Yates, Ber. Bunsenges. Phys. Chem., 81, 369 (1977).
248. E. Steckhan, Electrochim. Acta, 22, 395 (1977).
249. M. Lapkowski and J. W. Strojek, J. Electroanal. Chem., 182, 315 (1985).
250. I. Baranowska and M. Lapkowski, J. Electroanal. Chem., 193, 205 (1985).
251. D. L. Langhus and G. S. Wilson, Anal. Chem., 51, 1134, 1139 (1979).
252. E. Ahlberg, B. Svensmark, D. P. Parker, and V. D. Parker, Acta Chem. Scand., B32, 363 (1978).
253. A. W. B. Aylmer-Kelly, A. Bewick, P. R. Cantrill, and A. M. Tuxford, Discuss. Faraday Soc., 56, 96 (1974).
254. F. Möllers and R. Memming, Ber. Bunsenges. Phys. Chem., 77, 879 (1973).
255. S. Pons and S. B. Khoo, J. Amer. Chem. Soc., 104, 3845 (1982).
256. B. Speiser, A. Rieker, and S. Pons, J. Electroanal. Chem., 147, 205 (1983).

257. K. Ashley, F. Sarfarazi, S. J. Buckland, J. K. Foley, Q. Mei, B. Halton, P. J. Stang, and S. Pons, Can. J. Chem., 65, 2062 (1987).
258. J. W. Strojek, T. Kuwana, ans S. W. Feldberg, J. Amer. Chem. Soc., 90, 1353 (1968).
259. C. Li and G. S. Wilson, Anal. Chem., 45, 2370 (1973).
260. H. N. Blount, J. Electroanal. Chem., 42, 271 (1973).
261. H. N. Blount, N. Winograd, and T. Kuwana, J. Electroanal. Chem., 74, 3231 (1970).
262. T. H. Ridgway, C. N. Reilley, and R. P. van Duyne, J. Electroanal. Chem., 67, 1 (1976).
263. R. S. Robinson, C. W. McCurdy, and R. L. McCreery, Anal. Chem., 54, 2356 (1982).
264. J. S. Mayausky and R. L. McCreery, Anal. Chem., 55, 308 (1983).
265. R. S. Robinson and R. L. McCreery, J. Electroanal. Chem., 182, 61 (1985).
266. R. F. Broman, W. R. Heineman, and T. Kuwana, Discuss. Faraday Soc., 56, 16 (1974).
267. J. F. Evans and H. N. Blount, J. Electroanal. Chem., 102, 289 (1979).
268. H. N. Blount and T. Kuwana, J. Electroanal. Chem., 27, 464 (1970).
269. E. Ahlberg, J. Halvorsen, and V. D. Parker, Acta Chem. Scand., B33, 781 (1979).
270. J. E. Davis and N. Winograd, Anal. Chem., 44, 2152 (1972).
271. A. Dusserre and M. Genies, Bull. Soc. Chim. Fr., I-282 (1979).
272. B. Kastening, in Advances in Analytical Chemistry and Instrumentation (H. W. Nürnberg, ed.), Vol. 10, Wiley, New York, 1974, p. 421; B. Kastening, in Comprehensive Treaties of Electrochemistry (R. E. White, J. O'M. Bockris, B. E. Conway, and E. Yeager, eds.), Vol. 8, Plenum, New York, 1984, Chap. 7.
273. T. M. McKinney, in Electroanalytical Chemistry (A. J. Bard, ed.), Vol. 10, Dekker, New York, 1977, p. 97; I. B. Goldberg and T. M. McKinney, in Laboratory Techniques in Electroanalytical Chemistry (P. T. Kissinger and W. R. Heineman, eds.), Dekker, New York, 1984, p. 675.
274. W. J. Albery, B. A. Coles, and A. M. Couper, J. Electroanal. Chem., 65, 901 (1975); W. J. Albery, A. T. Chadwick, B. A. Coles, and N. A. Hampson, J. Electroanal. Chem., 75, 229 (1977); W. J. Albery, R. G. Compton, A. T. Chadwick, B. A. Coles, and J. A. Lenkait, J. Chem. Soc., Faraday Trans. 1, 76, 1391 (1980).
275. I. B. Goldberg and A. J. Bard, J. Phys. Chem., 75, 3281 (1971).
276. I. B. Goldberg and A. J. Bard, J. Phys. Chem., 78, 290 (1974).
277. I. B. Goldberg, D. Boyd, R. Hirasawa, and A. J. Bard, J. Phys. Chem., 78, 295 (1974).
278. R. D. Allendorfer, G. A. Martinchek, and S. Bruckenstein, Anal. Chem., 47, 890 (1975).
279. B. A. Coles and R. G. Compton, J. Electroanal. Chem., 144, 87 (1983).
280. R. G. Compton, D. J. Page, and G. R. Sealy, J. Electroanal. Chem., 161, 129 (1984); J. Electroanal. Chem., 163, 65 (1984).
281. R. N. Bagchi, A. M. Bond, F. Scholz, and R. Stösser, J. Electroanal. Chem., 245, 105 (1988).
282. R. Hirasawa, T. Mukaibo, H. Hasegawa, N. Odan, and T. Maruyama, J. Phys. Chem., 72, 2541 (1968).

283. R. D. Allendorfer, J. Amer. Chem. Soc., 97, 218 (1975), and references cited therein.
284. G. Cauquis, G. Fauvelot, and J. Rigaudy, Bull Soc. Chim. Fr., 4928 (1968); G. Cauquis, Bull. Soc. Chim. Fr., 1618 (1968).
285. W. J. Albery, R. G. Compton, and I. S. Kerr, J. Chem. Soc., Perkin Trans. II, 825 (1981).
286. D. E. Bartak, W. C. Danen, and M. D. Hawley, J. Org. Chem., 35, 1206 (1970).
287. G. J. Gores, C. E. Koeppe, and D. E. Bartak, J. Org. Chem., 44, 380 (1979).
288. A. J. Bard, J. C. Gilberg, and R. D. Goodin, J. Amer. Chem. Soc., 96, 620 (1974).
289. E. E. Bancroft, H. N. Blount, and E. G. Janzen, J. Amer. Chem. Soc., 101, 3692 (1979).
290. G. Gronchi, P. Courbis, P. Tordo, G. Mousset, and J. Simonet, J. Phys. Chem., 87, 1343 (1983); P. Martigny, G. Mabon, J. Simonet, and G. Mousset, J. Electroanal. Chem., 121, 349 (1981).
291. J. F. Bunnett, Acc. Chem. Res., 11, 413 (1978).
292. C. K. White and R. D. Rieke, J. Org. Chem., 43, 4638 (1978).
293. S. Hünig and H. Berneth, Top. Curr. Chem., 92, 1 (1980).
294. R. G. Compton, B. A. Coles, and M. J. Day, J. Electroanal. Chem., 200, 205 (1986).
295. R. G. Compton, B. A. Coles, and M. J. Day, J. Chem. Res. (S), 260 (1986).
296. B. Kastening, B. Gastisa-Michelcic, and J. Divisek, Discuss. Faraday Soc., 56, 341 (1974).
297. J. K. Dohrmann, F. Gallusser, and H. Wittchen, Discuss. Faraday Soc., 56, 330 (1974).

4

Relations Between Micro- and Macrophenomena

CHRISTIAN AMATORE / Ecole Normale Supérieure, Laboratoire de Chimie, Paris, France

> Chemical reaction, like politics, is the art of the possible; electrochemical reaction, like military tactics, is the art of overriding the impossible by applying electrical force.
>
> R. B. MacMullin

I. Introduction 207

II. Basic Principles 208
- A. Homogeneous chemical equivalent of electrolysis rates 208
- B. Coulometry: Apparent number of electrons consumed 212
- C. Separation of waves and selectivity in electrolysis 216

III. Competitive Reaction Schemes: Optimization of Preparative-Scale Electrolysis 217
- A. Slow reaction schemes 217
- B. An elementary method for fast competitive kinetics 220

IV. Conclusion 229

Symbols and Their Definitions 230

References 231

I. INTRODUCTION

In Chap. 2, our presentation of electrochemical concepts was focused mainly on the specific aspects related to the electrode and to the solution in its close vicinity. Indeed as soon as kinetics is considered, the bulk solution may often be regarded as a concentration buffer whose only role is to maintain constant or known reactant concentrations at the external diffusion layer boundary. Obviously this is not the case when preparative electrochemistry is considered. Indeed the goal is then to change the solution composition. Yet this composition change is effected through the relay of the diffusion layer, which "transmits" the modifications achieved at the electrode to the bulk solution. Since chemical reactions may take place within the diffusion layer, as discussed in Chap. 2, the way the bulk solution is affected depends largely on the microphenomena occurring in the thin solution layer (5-20 μm, usually) adjacent to the electrode. This results in an extensive coupling between micro- and macrophenomena at electrodes, which is one of the reasons that kinetics, the study of microelectrolysis phenomena, is so developed in electrochemistry.

Yet these important relations should not conceal the large differences between micro- and macroelectrolysis. These differences, which were discussed briefly in Chap. 2, have their main origin in the necessity of passing a current through the solution between the working and auxiliary (or counter) electrodes. Obviously the ensuing problems are quite different when one considers currents in the microampere (or even nanoampere for ultramicroelectrodes) range or in the ampere range as in macroscale laboratory electrolysis.* Other changes are also introduced that originate from the differences in dilution. Under microelectrolytic conditions very dilute (usually 0.1-10 mM) solutions are considered. Thus activities and diffusion coefficients, for example, are close to their values at infinite dilution. This is not the case in macroelectrolysis since rather concentrated solutions are used. This may introduce large changes in mass transfer rates as well as in chemical reactivity, and also in such specific factors such as migration [2]. These changes in concentration, as well as the larger current densities and electrolysis times, may also result in an increased importance of adsorption, coatings, or pollution phenomena at the electrodes. This is particularly true in aqueous media in which adsorption or chemisorption may play a decisive role in the distribution and selectivity of the products as a function of the electrode material.

As understood from this discussion it is generally unrealistic to believe that microelectrolytic results transpose exactly to macroelectrolytic conditions.† However, provided one keeps these difficulties in mind, microelectrolysis results and approaches constitute extremely useful guides in tackling macroscale electrolysis. Amazingly, data obtained at ultramicroelectrodes (i.e., of micrometric or smaller dimensions) should transpose with the least difficulty since they may be obtained with concentration or current density (a few amperes per square centimeter) conditions matching those found in preparative-scale electrolysis. Moreover, they allow meaningful results to be obtained in the presence of strong migrational contributions [3,4].

II. BASIC PRINCIPLES

A. Homogeneous Chemical Equivalent of Electrolysis Rates

As explained in Chap. 2, the bulk concentrations are affected by the phenomena occurring within the diffusion layer (which is identified, in this chapter, with the stagnant layer imposed by hydrodynamics) through the general relationship in Eq. (1):

$$\frac{dC}{dt} = -\frac{DA}{V}\left(\frac{\partial C}{\partial x}\right)_{x=\delta} + \left(\frac{dC}{dt}\right)_{chem} \tag{1}$$

This equation derives from the conservation of mass fluxes at the diffusion layer/bulk solution interface [compare Eqs. (155) and (156) in Chap. 2 and

*Note that in the following we do not discuss the ultraspecific aspects of industrial-scale electrolysis, which generally cannot be dissociated from a particular process [1].

†But this is not particular to electrochemistry, since most of the above difficulties (with the exception of those related to the current) are encountered when transferring any physical organic result to preparative conditions.

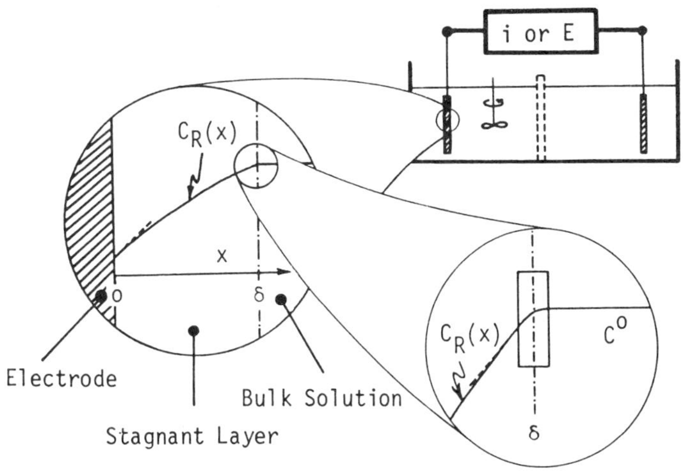

Figure 1. Schematic representation of the electrochemical cell. First insert: concentration profile of the reactant in the stagnant layer in the vicinity of the electrode. Second insert: mass transfer "black box" at the boundary between the stagnant layer and the bulk solution ($x = \delta$).

Fig. 1.) In Eq. (1) dC/dt is the overall rate of the concentration variations considered, whereas $(dC/dt)_{chem}$ is that arising from possible consumption or production of the species within the bulk solution. The electrochemical rate of production or consumption is given by the first term on the right-hand side of Eq. (1). It depends on the concentration gradient of the species at the end of the diffusion layer, that is, at $x = \delta$, as well as on the ratio of the electrode surface area to the solution volume, A/V. The larger the absolute value of this rate, the smaller the duration of the electrolysis, which is the reason that generally large electrode surface areas and high stirring rates (i.e., small stagnant layer thickness) are used for electrolysis.

Note that in Eq. (1), a single space coordinate is considered, x, the distance normal to the electrode. This necessarily implies that all electrode locations behave identically. In practice this may be approximated by using symmetrical electrode geometries, with the auxiliary electrode (or the separator in divided cells) parallel and of nearly identical size to that of the working electrode (see Chap. 6). This results in a "concentration field" as uniform as possible at the electrode surface(s). Note that this field is conveniently visualized when thinking in terms of the electrostatic field that would establish if the electrodes were those of a large capacitor, since both fields obey analogous laws [5]. However, in practice, ideal electrochemical geometries are only with difficulty compatible with other experimental constraints, so that the potential and current distribution at electrode surfaces may not be uniform. This is also true for the thickness of the stagnant layer since a uniform fluid velocity is difficult to achieve at the electrode surface. For these reasons, the first term on the right-hand side of Eq. (1) is merely considered as resulting from an averaging at the electrode surface; this is also true for the "electrode potential" E or the "electrode current density" i/A. Nevertheless, to maintain some generality and simplicity in the following presentation, we assume that the concentration profiles depend only on the distance x from the electrode surface.

The problem is then to relate the concentration gradient $(\partial C/\partial x)_{x=\delta}$ at the stagnant layer boundary to that $(\partial C/\partial x)_{x=0}$ at the electrode surface (compare Fig. 1). This is done via the integration of the differential equations, such as Eq. (175) in Chap. 2, which control the concentration profiles within the stagnant layer. In the most simple situation, the species whose bulk concentration is given in Eq. (1) undergoes no chemical reaction within the stagnant layer. Then from Eq. (175) (Chap. 2), its concentration flux is constant within the layer, and thus $(\partial C/\partial x)_{x=\delta} = (\partial C/\partial x)_{x=0} = (C - C_{x=0})/\delta$, where $C_{x=0}$ is the concentration of the species at the electrode surface. Thus Eq. (1) is transformed into Eq. (2),

$$\frac{dC}{dt} = -\frac{DA}{V\delta}(C - C_{x=0}) + \left(\frac{dC}{dt}\right)_{chem} \tag{2}$$

where the constant factor $DA/V\delta$ has the dimensions of a first-order rate constant (s^{-1}). Most of the time one is interested in exhaustive electrolysis and thus imposes an electrode potential so that $C_{x=0} = 0$, which in most cases simplifies Eq. (2) to Eq. (3) for the reactant.

$$\frac{dC}{dt} = -\frac{DA}{V\delta} C + \left(\frac{dC}{dt}\right)_{chem} \tag{3}$$

Interestingly, under such circumstances the electrochemical consumption of the species is equivalent to a pseudo-first-order chemical reaction taking place into the bulk solution, with a rate constant k_{elec} in Eq. (4) [6].

$$k_{elec} = \frac{DA}{V\delta} \tag{4}$$

Note that it may be surprising that a chemical term is considered in Eq. (3), whereas we have supposed that no chemical reaction term was involved within the stagnant layer. As explained in Chap. 2, this stems from the fact that the effect of any chemical reaction within the diffusion layer depends on the relative magnitude of $k\delta^2/D$ versus unity. For usual laboratory conditions $\delta \approx 10^{-3}$ cm and $D \approx 5 \times 10^{-6}$ cm^2 s^{-1} and then δ^2/D is of the order of 0.2 s. Thus, provided k is less than approximately 1 s^{-1} it has no tangible effect on the concentration profiles of the species, whereas it has a definite effect in the bulk solution due to the long reaction times (usually longer or comparable to half an hour). In practice it is important to decide when the simplification in Eq. (2) or (3) applies to a given experimental situation. The discussion just presented affords a simple answer to the problem. Indeed, consider the electron transfer reaction in Eq. (5), possibly followed by a chemical step in Eq. (6).

$$R + ne \rightleftharpoons P \qquad (E^\circ) \tag{5}$$

$$P \longrightarrow P' \qquad (k) \tag{6}$$

The simplification in Eqs. (2) and (3) is equivalent to saying that k is much smaller than the mass transfer rate D/δ^2, which is of the order of 5 s^{-1}. From Chap. 2, this implies that the R/P redox couple gives a chemically reversible cyclic voltammogram at a scan rate corresponding to the same mass transfer rate. From a dimensionless analysis and Table 5 in Chap. 2, this corresponds to a scan rate v such as $Fv/RT \approx D/\delta^2$, that is, v of the order 0.1 V s^{-1}. Thus, observation for identical or close experimental conditions of a chemically reversible cyclic voltammogram at $v \approx 0.1\ V\ s^{-1}$ for the redox couple of interest is sufficient to prove the validity of the assumptions leading to the homogeneous equivalent rate laws in Eqs. (7) and (8),

$$\frac{d[R]}{dt} = -k_{elec}[R] \tag{7}$$

$$\frac{d[P]}{dt} = +k_{elec}[R] + \left(\frac{d[P]}{dt}\right)_{chem} \tag{8}$$

with k_{elec} given in Eq. (4). In the converse situation, that is, when a totally irreversible cyclic voltammogram is observed, Eq. (7) still applies for R, which is not chemically affected in the sequence considered, but Eq. (8) does not reflect the variations in the bulk concentration of the product P [7]. Indeed, the observation of a chemically irreversible cyclic voltammogram shows that P is too short-lived to have any possibility of escaping the stagnant layer. However, because of mass conservation within the stagnant layer, the outgoing flux of P', the product of chemical reaction of P in Eq. (6), is identical to the incoming flux of R. Equation (8) is then replaced by $[P] \approx 0$, and Eq. (9) results, which is tantamount to the steady-state approximation in homogeneous chemistry:

$$\frac{d[P']}{dt} = +k_{elec}[R] \tag{9}$$

From this analysis and presentation it is seen that when single chemical pathways are followed, that is, when no competitive route interferes in the reaction of intermediate(s) within the diffusion layer, the electrolysis affects the different bulk concentrations in a way similar to a first-order chemical reaction, as outlined in scheme 1. This analogy affords an extremely simple way to handle the various problems that may arise during electrolysis. Indeed they then become exactly identical to those that would be observed for homogeneous chemistry, that is, as if the R/P redox reaction were performed using a homogeneous redox reagent.

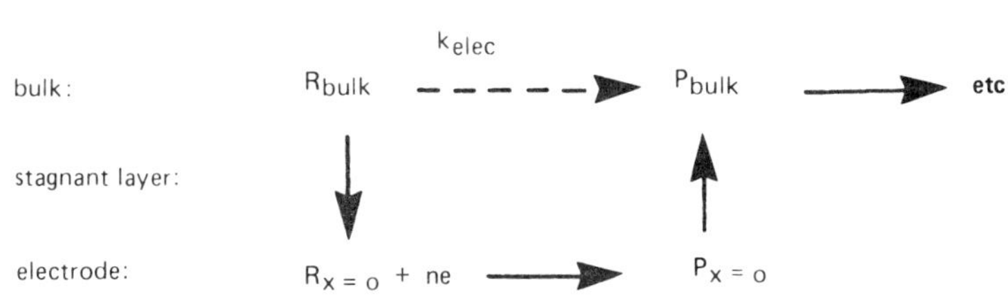

Scheme 1.

B. Coulometry: Apparent Number of Electrons Consumed

In the preceding section we have presented the analogy that exists between electrolysis and homogeneous chemistry in terms of an equivalent rate constant $k_{elec} = DA/V\delta$. Yet it may be surprising that this rate constant does not include a term related to the current. However, the current corresponds to a number of moles consumed per unit of time, that is, to an instantaneous rate, and therefore not to a rate constant. From the description outlined in scheme 1 and Eq. (141) in Chap. 2, the current exchanged at the electrode at any time is given in Eq. (10):

$$i = nFAD\left(\frac{\partial[R]}{\partial x}\right)_{x=0} = -nFAD\left(\frac{\partial[P]}{\partial x}\right)_{x=0} \tag{10}$$

When R undergoes no significant chemical reaction within the diffusion layer, its concentration profile is linear, as established previously, and Eq. (10) is rewritten as in Eq. (11):

$$i = \frac{nFAD}{\delta}([R]_{bulk} - [R]_{x=0}) \tag{11}$$

Again, when one considers the most frequent situation in which the electrode potential is such as $[R]_{x=0} = 0$, Eq. (12) is finally obtained:

$$i = \frac{nFAD}{\delta}[R]_{bulk} \tag{12}$$

Comparison to the definition of k_{elec} [Eq. (4)] affords the relationship in Eq. (13) between the instantaneous current and the equivalent rate of electrolysis [6].

$$i = nFVk_{elec}[R]_{bulk} \tag{13}$$

Note that Eqs. (12) and (13) are independent of the eventual chemical fate of R or P in the bulk solution but suppose only that R undergoes no chemical reaction within the stagnant layer.

The overall quantity of electricity Q_t consumed after a duration t of electrolysis is given by the integration versus time of Eq. (13), which affords Eq. (14):

$$Q_t = \int_0^t i\,dt = nFVk_{elec}\int_0^t [R]_{bulk}\,dt \tag{14}$$

In Eq. (14), $[R]_{bulk}$ is obtained as a function of time in Eq. (15), which follows directly from Eq. (3).

$$\frac{d[R]_{bulk}}{dt} = -k_{elec}[R]_{bulk} + \left(\frac{d[R]_{bulk}}{dt}\right)_{chem} \tag{15}$$

When the last term in Eq. (15) is negligible, R is consumed only at the electrode. Thus the charge consumption is easily obtained by considering that each mole of R converted consumes nF coulombs, that is, Q_t is given in Eq. (16), where C^0 is the initial bulk concentration of R.

$$Q_t = nFV(C^0 - [R]_{bulk}) \tag{16}$$

On the other hand, from Eq. (15), $[R]_{bulk}$ decays exponentially with time as in Eq. (17):

$$[R]_{bulk} = C^0 e^{-k_{elec}t} \tag{17}$$

At the end of electrolysis, that is for $t \gg 1/k_{elec}$, one obtains the overall charge consumption in Eq. (18). In practice Eq. (18) allows the experimental determination of n, the number of electrons consumed per molecule of R, as in Eq. (19):

$$Q = nFVC^0 \tag{18}$$

$$n = \frac{Q}{C^0FV} \tag{19}$$

But when experimentally possible, one does not proceed from Eq. (19), because when the R concentration becomes small, that is, near the end of electrolysis, residual currents arising from other electroactive species may contribute significantly to the current and then alter the charge measured. It is then preferable to rely on the plot of Q versus $[R]_{bulk}$, according to Eq. (16). Thus a linear plot should be obtained as in Fig. 2a, in which extrapolation at $[R]_{bulk} = 0$ allows a more accurate value of Q to be determined. Note that the ensuing value of n should be identical to that determined from transient or steady-state methods.

However, it happens frequently that the two figures do not agree, which is a proof that the chemical term in Eq. (15) is not negligible in preparative-

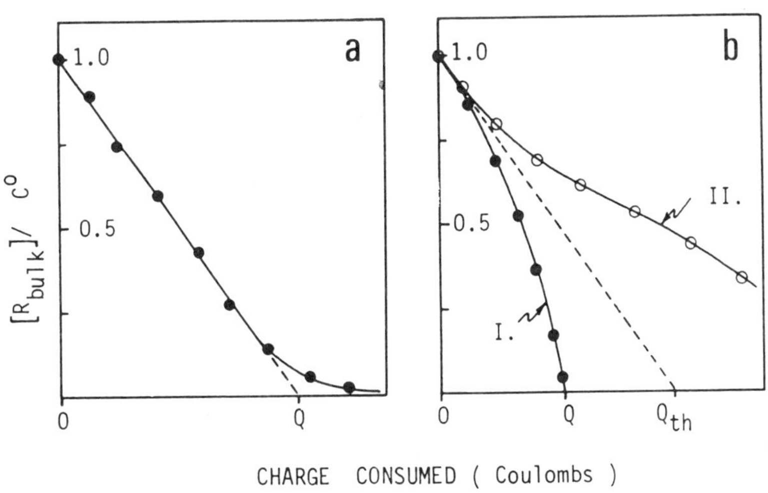

Figure 2. Variations in the bulk concentration of the reactant as a function of the charge consumed: (a) in the absence or (b) in the presence of chemical reaction(s) consuming (I) or regenerating (II) the electrolyzed species (initial concentration of the reactant C^0).

scale electrolysis. Indeed, when the latter is positive, that is, corresponds to a production of R in the bulk solution, as by a regenerating sequence as in Eq. (20),

$$\begin{aligned} &R + ne \longrightarrow P \quad \text{stagnant layer} \\ &P + Z \longrightarrow R + \cdots \quad \text{bulk solution} \end{aligned} \tag{20}$$

the instant R concentration is larger than predicted by Eq. (17). From Eq. (14) it is seen that Q is then larger than determined via Eq. (18), which results in a larger number of electrons consumed n_{ap} during the electrolysis, when compared with the net electron exchange n at the electrode identified in microelectrolysis.

$$n_{ap} = \frac{Q}{FVC^0} \tag{21}$$

Conversely, when R is consumed in the bulk solution, Q and therefore n_{ap} are smaller than predicted in Eqs. (18) and (19).

A typical example of such behavior is given by the reduction of pyrylium cations. Under steady state or moderate scan cyclic volammetry, their reduction involves the net consumption of one electron per molecule, which corresponds to the fast radical dimerization in Eq. (23).

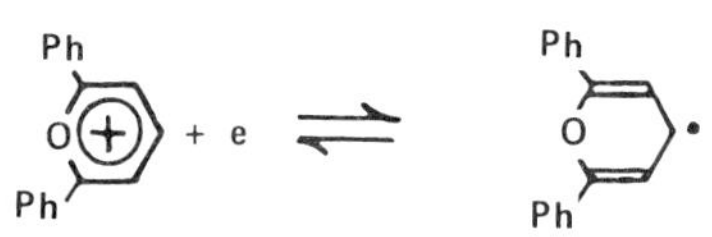

(22)

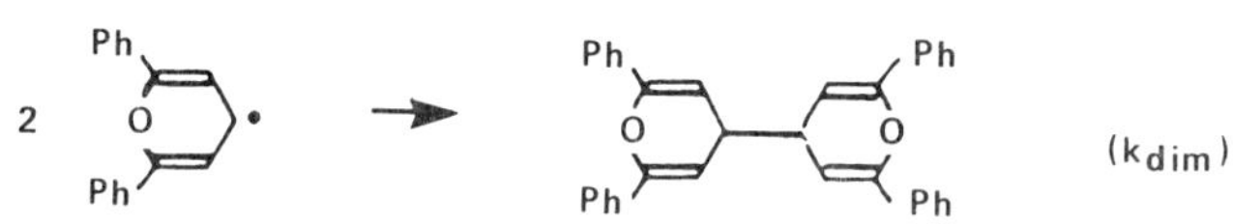

(23)

The corresponding rate constant k_{dim}, close to the diffusion limit, has been determined in acetonitrile using ultrafast cyclic volammetry in the low million volts per second [8a]. However, bulk electrolysis and coulometries indicate an overall consumption of 2/3 electrons. This was shown to correspond to the involvement of the slow hydride transfer sequence outlined in reactions (24) and (25) [9].

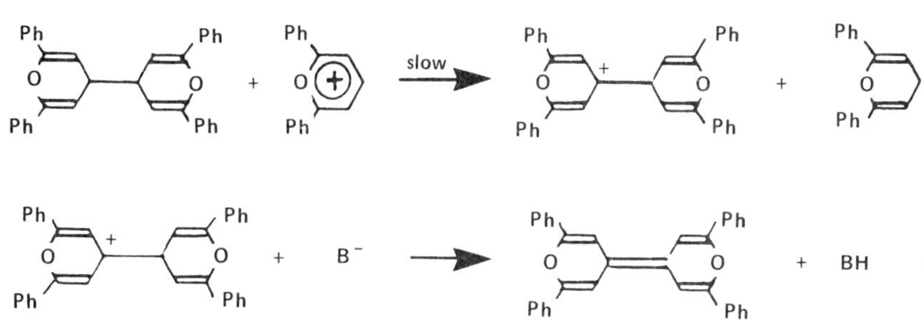

(24)

(25)

Reaction (24) is too slow to be detected in cyclic voltammetry, yet it takes place during the electrolysis to afford finally a 4H-pyran and a bipyranylidene as summarized in the balanced Eq. (26).

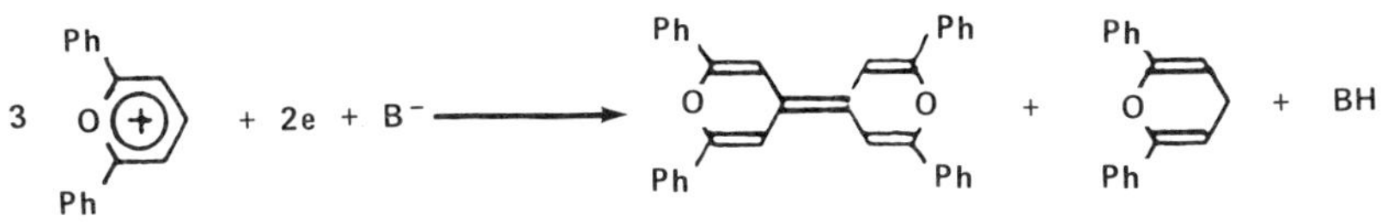
(26)

Thus three pyrylium molecules are consumed for only two electrons exchanged, resulting in $n_{ap} = 2/3$.

In this discussion we have considered the involvement of a slow follow-up reaction of the reactant as a cause for the difference between n and n_{ap}. Yet this is not the only situation possible. In most cases such disagreements are due to the involvement of an ECE kind of sequence as that featured in Eqs. (27) through (29),

$$R + ne \longrightarrow P \qquad (E^\circ) \tag{27}$$

$$P \longrightarrow R' \qquad \left(\text{slow versus } \frac{D}{\delta^2}\right) \tag{28}$$

$$R' + n'e \longrightarrow P' \qquad (E^{\circ\prime}) \tag{29}$$

as well as to their homogeneous analogs in which the heterogeneous electron transfer is replaced by a homogeneous electron transfer (disproportionation sequences). When the interposed chemical step or the disproportionation is slow enough to make no contribution within the diffusion layer, the apparent number of electrons observed in microelectrolysis is n, that is, reflects the redox process in Eq. (27). However, the chemical step may be sufficiently rapid when compared with the electrolysis duration to convert quantitatively the primary product P into R'. Thus, provided that $n(E^{\circ\prime} - E^\circ) > 0$, R' is coelectrolyzed and the apparent number of electrons exchanged is $n_{ap} = n + n'$. The absolute value of n_{ap} may then be larger than that of n when both have the same sign, or smaller when of opposite signs. Similarly, slow disproportionation reactions pulled by a fast protonation step, for example, may affect the n_{ap} value, as outlined in the following sequence:

$$R + ne \longrightarrow P$$

$$2P \rightleftharpoons R + P'$$

$$P' + Z \longrightarrow \cdots$$

These examples illustrate some of the difficulties that may be encountered when comparing the numbers of electrons exchanged in micro- or macroscale electrolysis. In a few instances these chemical complications may be detected by important curvatures of the Q versus $[R]_{bulk}$ plots as sketched in Fig. 2b. Such curvatures are observed when the degree of involvement of the follow-up chemical steps increases during electrolysis. Indeed, because of concentration effects the effective rate of the chemical step may be slow at the beginning of the electrolysis , and thus a slope corresponding to $n_{ap} = n$ is observed. However, when the pertinent concentrations of intermediates increase, the rate of their chemical reaction may become sufficient so that the intermediates reach steady-state behavior, which results in a tendency to reach a slope corresponding to $n_{ap} = n + n'$ while electrolysis progresses. Yet this supposes that the degree of participation changes during the electrolysis, which obviously is not a general fact. Thus the absence of curvature in the Q versus $[R]_{bulk}$ plots cannot be taken as evidence of no chemical complication, that is, of the fact that $n = n_{ap}$. This is important to remember when using coulometric results, that is, n_{ap} values, in the rationalization of electrochemical mechanisms.

C. Separation of Waves and Selectivity in Electrolysis

It is often alleged that the fine-tuning of electrode potential allows high selectivity in electrolysis. This is certainly true for a wide variety of electrochemical situations but may turn out totally inexact in others, as is shown in the following discussion. Indeed, let us consider from steady-state or transient electrochemical data two successive reduction waves* at E_1 and E_2 for a compound involving two reduction sites A and R: A–R. Let us assume, in addition, that electrolysis of related molecules, but without the R center, is known to occur at potentials lying in the same range where the first wave is observed. Thus an "evident" rationalization of the observed waves consists in ascribing them to the successive reductions in Eqs. (30) and (31):

$$\text{A–R} + e \longrightarrow {}^{\bar{\cdot}}\text{A–R} \longrightarrow\cdots\longrightarrow \text{B–R} \quad (k_1) \quad \text{first wave} \qquad (30)$$

$$\text{B–R} + e \longrightarrow \text{B–R}^{\bar{\cdot}} \longrightarrow\cdots\longrightarrow \text{B–P} \quad (k_2) \quad \text{second wave} \qquad (31)$$

When the lifetime of the anion radical formed upon the electron transfer in step (30) is extremely short, the mechanism in Eqs. (30) and (31) is certainly valid and the electrolysis product obtained at $E = E_1$ is B–R. Yet when this is not the case a combination of A–P and B–R products may be obtained although the electrolysis is performed at the first wave [11,12]. Indeed, an alternative route is now offered to the ${}^{\bar{\cdot}}$A–R anion radical formed in Eq. (30), which involves a possible intra- or intermolecular electron transfer to the R site to afford the A–R${}^{\bar{\cdot}}$ anion radical.

$${}^{\bar{\cdot}}\text{A–R} \rightleftharpoons \text{A–R}^{\bar{\cdot}} \longrightarrow\cdots\longrightarrow \text{A–P} \qquad (32)$$

If the latter is not highly unstable, a back electron transfer occurs to regenerate the ${}^{\bar{\cdot}}$A–R anion radical. In the converse situation A–R${}^{\bar{\cdot}}$ may evolve directly to the A–P product, which corresponds to the overall reduction of the R moiety although the first electron transfer has occurred at the A group. The observed phenomenon is thus tantamount to indirect electrolysis [13], where the A group plays the role of an intramolecular electron mediator for the reduction of the R group. Although not explicitly recognized as such, this is the case for the reduction of aromatic halides, for example, since the initial electron is transferred to the aromatic ring and then internally transferred to the carbon-halogen antibonding orbital, resulting finally in the overall reduction of the carbon-halogen bond.

---

*Transposition to oxidations is obvious, so that the discussion hereafter is developed only for a cathodic process.

In such situations the exact nature of the B–R or A–P product obtained depends on the relative chemical stabilities of the $A^{\cdot-}$ and $R^{\cdot-}$ moieties [rate constants k_1 and k_2 in Eqs. (30) and (31)], as well as on the difference between the standard reduction potentials of the two groups, as explained later. Indeed, in a general case, one must consider that the mechanism occurring at the first reduction wave is not as in Eq. (30) but may involve the sequence in Eqs. (33) through (35).

$$\text{A–R} + e \longrightarrow {}^{\cdot-}\text{A–R} \tag{33}$$

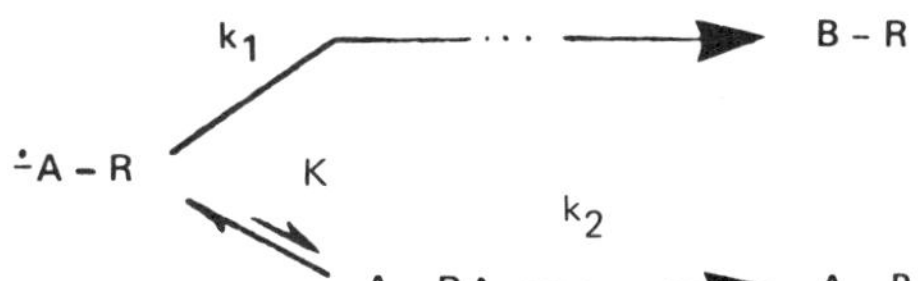

(34)

(35)

When the homogeneous electron transfer is sufficiently fast to be equilibrated (with an equilibrium constant K). the formation rate constant of the A–P product is k_2K when related to ${}^{\cdot-}$A–R.

$$\frac{[\text{B–R}]_{\text{yield}}}{[\text{A–P}]_{\text{yield}}} = \frac{k_1}{k_2 K} \tag{36}$$

$$\frac{[\text{B–R}]_{\text{yield}}}{[\text{A–P}]_{\text{yield}}} = \frac{k_1(k_2 + k_b)}{k_2 k_f} \tag{37}$$

Thus the relative yields of the B–R and A–P products are given in Eq. (36), and it is noteworthy, are independent of the electrode potential. In a more general case, the forward k_f and backward k_b rate constants of the homogeneous electron transfer must be considered, and the relative yields are given in Eq. (37)*, and are again independent of the electrode potential, provided it is kept on the first wave.

III. COMPETITIVE REACTION SCHEMES: OPTIMIZATION OF PREPARATIVE-SCALE ELECTROLYSIS

A. Slow Reaction Schemes

In electrochemistry, as in other chemical methods, situations in which a single product is obtained are seldom encountered. In most experimental cases the target product is often accompanied by unwanted side products

*Note that in Eq. (37), k_f and k_b may include a concentration term when the homogeneous electron transfer is bimolecular as in Eq. (38):

$$^{\cdot-}\text{A–R} + \text{A–R} \rightleftharpoons \text{A–R} + \text{A–R}^{\cdot-} \tag{38}$$

formed along competitive pathways. Under such conditions optimization of the target product yield is obviously desirable. Frequently this optimization is sought on the basis of chemical intuition and of "blind tests" of supposedly critical parameters. Yet it is difficult to be certain (except when a 100% yield is obtained!) that the best yield obtained is the maximum. In the following we want to show that semiquantitative predictions can be made on the basis of very simple considerations. The goal of these approaches is not to replace the necessary tests, but to provide useful guidelines for the selection of the parameters that affect the yield of the target product.

From the preceding presentation it is easily inferred that different behaviors are obtained when the "key species," that is, that at which the different chemical routes are branched, exists only within the stagnant layer or exists in the bulk solution, as schematized in Fig. 3. In the latter case, the competition problem is analogous to the homogeneous chemical situation, owing to the equivalencies established previously. For example, let us consider a reaction sequence like that in Fig. 3 (case a or b):

$$A + e \rightleftharpoons B \rightarrow C \begin{smallmatrix}\nearrow \\ \rightarrow \\ \searrow\end{smallmatrix}$$

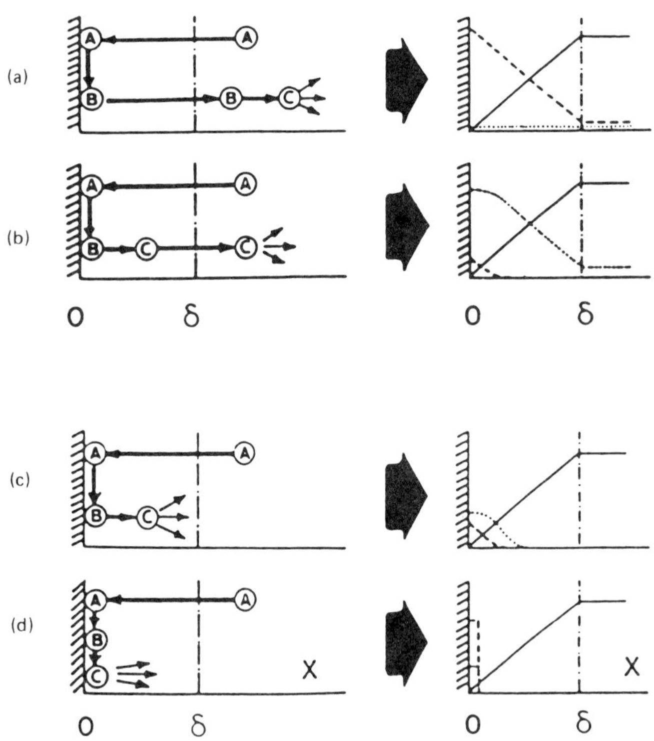

Figure 3. Schematic representation of the localization of the "branching point" for an ECC competitive sequence and resulting concentration profiles for (—) A, (---) B, and (·····) C. (a, b) Slow production of the intermediate "key species"; (c) fast production of the key species; (d) reaction in the adsorbed phase.

$$A + e \longrightarrow B \qquad (E^{o}) \tag{39}$$

$$B \longrightarrow C \qquad (k_1) \tag{40}$$

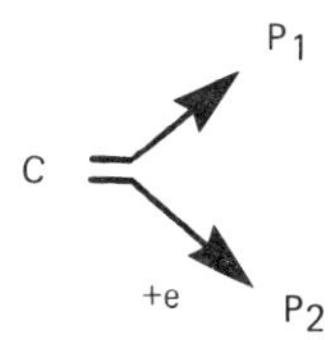

$$(k_2) \tag{41}$$

$$(E^{o\prime} > E^{o}) \tag{42}$$

For a potentiostatic electrolysis on the plateau of the A/B reduction wave, the reaction sequence in Eqs. (39) through (42) is equivalent to the homogeneous mechanism in Eqs. (43) through (46), where $k_{elec} = DA/V\delta$:

$$A \longrightarrow B \qquad (k_{elec}) \tag{43}$$

$$B \longrightarrow C \qquad (k_1) \tag{44}$$

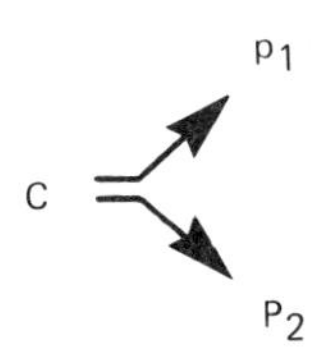

$$(k_2) \tag{45}$$

$$(k_{elec}) \tag{46}$$

thus the yield of species P_1 vis-à-vis P_2 is simply expressed as in Eq. (47).

$$\frac{[P_1]_{yield}}{[P_1]_{yield} + [P_2]_{yield}} = \frac{k_2}{k_2 + k_{elec}} \tag{47}$$

From Eq. (47) it is easily seen that maximizing the yield of P_1 amounts to decreasing the ratio k_{elec}/k_2, that is, of $DA/V\delta k_2$. Thus it obviously follows that one must decrease the electrode surface area and the stirring rate (i.e., increase δ), increase the bulk solution volume, and when possible, increase k_2. The opposite changes are needed when the target compound is P_2. Similarly, when B is the compound sought, one wants $k_{elec} \gg k_1$, that is, a large surface area for the electrode, small bulk solution volume, and high stirring rates (small δ).

Thus, owing to the equivalent homogeneous formulation, there are no specific particularities associated with optimization in electrochemistry, as soon as the branching point takes place in the bulk solution. As explained earlier, this is generally easy to decide on the basis of cyclic voltammograms of the system of interest.

Conversely, when the key species exists only in a thin reaction layer adjacent to the electrode (case c in Fig. 3), the degree of competition between the various routes is greatly influenced by the local concentration profiles of the different species involved. In such cases general theories have been established that allow a complete and quantitative description of the optimization procedure. However in the following we want to present a very simple approach to the problem.*

---

*Note that case d in Fig. 3, which relates to surface reactions between adsorbed or chemisorbed species, is not considered here because of the too specific nature of the electrode-species interactions involved.

B. An Elementary Method for Fast Competitive Kinetics

Presentation of the Method

All the results presented in the following discussion derive from the simple consideration* that when a species is involved in a branched kinetic scheme, the major pathway is the fastest; that is, the main product is that formed according to the pathway corresponding to the highest rate of consumption of the key species.

On the other hand, when the key species is formed at a given rate, the faster its consumption, the smaller its concentration. Taking into account these two obvious statements leads to the following conclusion: When two or more pathways compete for the consumption of a given species, the major pathway is that leading to the smallest concentration of this species, when each pathway is considered alone under identical experimental conditions.

When two competing mechanisms are considered, for example, a competition parameter may thus be defined as the ratio of the average concentrations of the key species obtained when each of these mechanisms is considered alone. Let us denote by C_1 and C_2 the corresponding concentrations for mechanisms 1 and 2 in scheme 2, when they do not interfere. The competition parameter $p = C_1/C_2$ is then such that $p \ll 1$ favors the occurrence

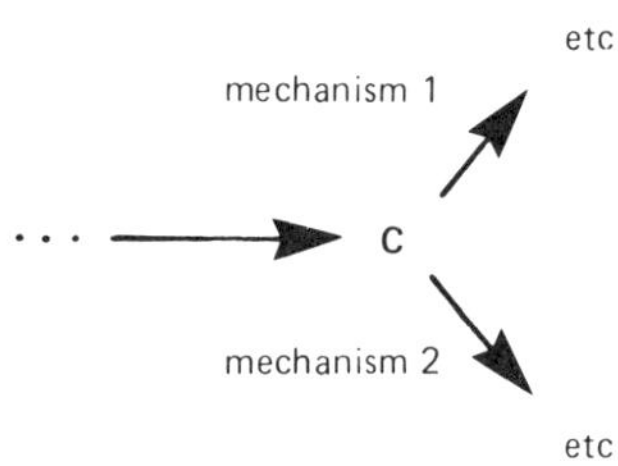

Scheme 2.

of mechanism 1 as the main pathway since this condition corresponds to $C_1 \ll C_2$, whereas mechanism 2 is observed when $p \gg 1$. In the general case the dependence of both C_1 and C_2 on the experimental conditions (concentrations of the reactants and of the coreactants, stirring rate, and so on) and intrinsic factors (rate constants and diffusion coefficients, for example) may be determined, thus leading to the expression of p as a function of these factors. The effects of the latter on the overall product distribution are then readily given in the direction and magnitude included in the p expression.

Table 1 gives the average concentrations to be used for the various intermediates involved in the most basic schemes of electrochemical mechanisms† which may allow the simple derivation of the competition parameters p for most of the possible experimental situations.

---

*Yet they are based and supported by a rigorous physicomathematical analysis. See, for example, Refs. 7a and 14-20.

†Note that each of the reactions presented in Table 1 may consist of true elemental reactions or of a kinetically equivalent series of elemental steps.

Table 1. "Average Concentrations" of the Different Intermediates Involved in the Main Electrochemical Mechanisms[a,b]

Mechanism[b]	Scheme	Average concentration[c]
EC	$A + ne \rightleftharpoons B$	$[B] = C^0/(k\theta)^{1/2}$
	$B \xrightarrow{k} C$	$[C] = C^0$
ECE	$A + ne \rightleftharpoons B$	$[B] = C^0/(k\theta)^{1/2}$
	$B \xrightarrow{k} C$	$[C] = C^0/(k\theta)^{1/2}$
	$C + ne \rightarrow D$	$[D] = C^0$
DISP	$A + ne \rightleftharpoons B$	$[B] = C^0/(2k\theta)^{1/2}$
	$B \xrightarrow{k} C$	$[C] = k/k_d$
	$B + C \xrightarrow{k_d} A + D$	$[D] = C^0$
ECC	$A + ne \rightleftharpoons B$	$[B] = C^0/(k_1\theta)^{1/2}$
	$B \xrightarrow{k_1} C$	$[C] = (C^0/k_2)(k_1/\theta)^{1/2}$
	$C \xrightarrow{k_2} D$	$[D] = C^0$
Dim 1	$A + ne \rightleftharpoons B$	$[B] = C^{02/3}(3/4k\theta)^{1/3}$
	$2B \xrightarrow{k} C$	$[C] = C^0/2$
Dim 2 (ECE)	$A + ne \rightleftharpoons B$	$[B] = C^{02/3}/(k\theta)^{1/3}$
	$B + A \xrightarrow{k} C$	$[C] = C^{02/3}/(k\theta)^{1/3}$
	$C + ne \longrightarrow D$	$[D] = C^0/2$
Dim 2 (DISP)	$A + ne \rightleftharpoons B$	$[B] = C^{02/3}/(2k\theta)/^{1/3}$
	$B + A \xrightarrow{k} C$	$[C] = (kC^0)^{2/3}/(k_d\theta^{1/3})$
	$B + C \xrightarrow{k_d} A + D$	$[D] = C^0/2$
Dim 3 (DISP)	$A + ne \rightleftharpoons B$	$[B] = (3C^0/4kK\theta)^{1/3}$
	$B + A \rightleftharpoons C$ [K][d]	$[C] = C^0K[B]$
	$B + C \xrightarrow{k_d} A + D$	$[D] = C^0/2$
ECDim	$A + ne \rightleftharpoons B$	$[B] = C^0/(k_1\theta)^{1/2}$
	$B \xrightarrow{k_1} C$	$[C] = (C^0/2k_2)^{1/2}(k_1/\theta)^{1/4}$
	$2C \xrightarrow{k_2} D$	$[D] = C^0/2$

[a]The concentrations are given for an electrolysis on the plateau of the A/B wave. C^0, bulk concentration of A; $\theta = \delta^2/D$.
[b]Usual names, or taken from the nomenclature developed in Savéant's group.
[c]These concentrations are valid provided each resulting figure is lower than approximately $C^0/10$.
[d]Rapid equilibrium; K, equilibrium constant.

In the preceding discussion we have considered for simplicity that only two pathways were possible for the key species. Thus a single parameter p is defined. However, when more routes are branched for the same species, the situation is a priori more difficult to handle. For three competitive routes, for example, Table 1 affords three concentrations C_1, C_2, and C_3, and thus three parameters $p_1 = C_1/C_2$, $p_2 = C_1/C_3$, and $p_3 = C_2/C_3$ are defined, which are all affected by the experimental and intrinsic parameters. Thus it is a priori difficult to appreciate the effect resulting of the variation of one of the experimental parameters on the overall competition. Yet it is first noticed that the consideration of only two parameters suffices to describe the competition since they are all given by ratios or products of the two others, as $p_3 = p_2/p_1$. Thus a convenient presentation of the competition problem consists in the construction of a diagram in which log p_1 is the ordinate axis and log p_2 is the abscissa axis, as in Fig. 4. In such a diagram, "kinetic zones" corresponding to the predominance of one of the three possible mechanisms may be defined in a similar way as "thermodynamic stability zones" corresponding to the predominance of a species are defined, for example as a function of the solution potential and pH (Chap. 2). Thus mechanism 1, which corresponds to $p_1 \ll 1$ and $p_2 \ll 1$, is located in the lower left corner of Fig. 4, that is, $\log p_1 < 0$ and $\log p_2 < 0$. Mechansim 2 is observed for $p_1 \gg 1$ and $p_3 = p_2/p_1 \ll 1$, and its kinetic zone is then such that $\log p_1 > 0$ and $\log p_1 > \log p_2$. Similarly, the kinetic zone corresponding to the predominance of mechanism 3 is such that $p_3 \gg 1$ and $p_2 \gg 1$, that is, $\log p_2 > 0$ and $\log p_2 > \log p_1$. Then three boundary lines delimiting the zone of predominance of each mechanism are constructed in Fig. 4. Near one of these boundaries the two delimited mechanisms compete without interference from the third. The general competition, that is, that involving the three mechanisms, is observed near the intersection of the three boundary lines in the central region of Fig. 4. When such a diagram has been constructed the effect of each experimental parameter is easily seen by construction of a "compass card," such as that shown in the

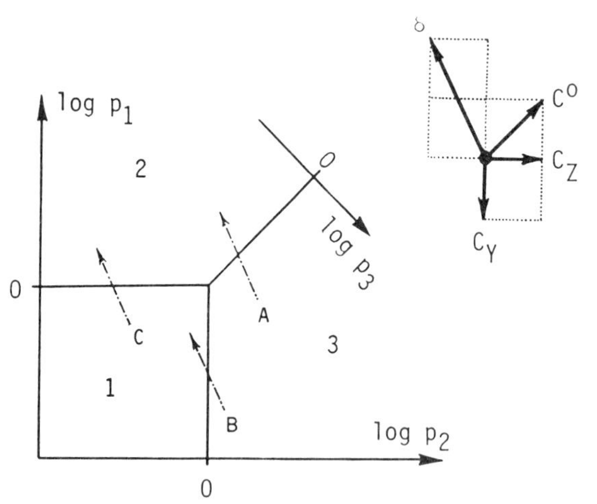

Figure 4. Kinetic zone diagram and corresponding "compass card" for a competition between three hypothetical mechanisms, as discussed in the text.

upper right corner of Fig. 4. For example, let us assume that $p_1 \propto C^0\delta^2/C_y$ and $p_2 \propto C^0C_z/\delta$. Thus a multiplication of δ by 10, for example, results in an increase of 2 in log p_1 and a decrease by 1 in log p_2. A vector corresponding to the "δ effect" can then be constructed in the diagram, which has a projection -1 in the abscissa and +2 in the ordinate. Similarly other vectors can be drawn for C^0, C_y, and C_z to constitute the compass card shown in the upper right corner of Fig. 4. Such a compass allows easy visualization of the effect of any variation in the experimental parameters. For example, it is seen in Fig. 4 that when the system is at point A, an increase in δ results in a trend to shift from mechanism 3 to mechanism 2, whereas the same variation results in a tendency to pass from mechanism 3 to 1 when starting from location B or from mechanism 1 to 2 when starting from point C. Such different effects induced by the same experimental variation would have been puzzling and difficult to rationalize, whereas they are easily understood from the kinetic zone diagram and its associated compass card. Similarly, a compass card showing the effect of each instrinsic parameter may be constructed.

In the following discussion we want to show how this simple method may be used in practical situations, either to rationalize a series of results or to decide a priori the worthiness of an experimental strategy. To maintain some homogeneity in the presentation, the two aspects are discussed on the basis of examples pertaining to the electrochemistry of aryl halides.

Predictive Value of the Method

This aspect is developed by addressing the following question: Is there any chance that biaryls may be formed, via aryl radical dimerization, during the reductive electrolysis of aryl halides?

The reduction of aryl halides is known [21] to afford a frangible anion radical $ArX^{\cdot-}$, which yields a σ-aryl radical by cleaving off the halide ion in Eq. (49).

$$ArX + e \longrightarrow ArX^{\cdot-} \qquad (E^\circ) \qquad (48)$$

$$ArX^{\cdot-} \longrightarrow Ar^{\cdot} + X^- \qquad (k_1) \qquad (49)$$

Although these σ-aryl radicals are prone to undergo a facile reduction at the electrode [Eq. (50)] or in solution as in Eq. (51), one may think of the possibility of impeding these normal pathways to favor radical dimerization [Eqs. (48), (49), and (53)].

$$Ar^{\cdot} + e \longrightarrow Ar^- \qquad (E^{\circ\prime} > E^\circ) \qquad (50)$$

$$Ar^{\cdot} + ArX^{\cdot-} \longrightarrow Ar^- + ArX \qquad (k_{dif}) \qquad (51)$$

$$Ar^- + BH \longrightarrow ArH + B^- \qquad (52)$$

$$2\,Ar^{\cdot} \longrightarrow Ar{-}Ar \qquad (k_{dim}) \qquad (53)$$

Indeed, the dimerization of these σ radicals is expected to be extremely fast and possibly close to diffusion control. Then owing to its bimolecular nature, a concentration effect should favor this step vis-à-vis the reduction at the electrode. On the other hand, increasing the rate of the anion radical

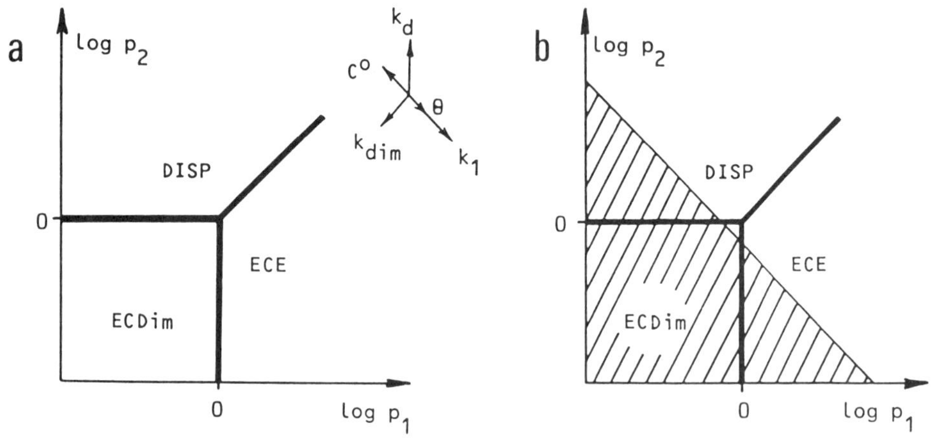

Figure 5. Kinetic zone diagram for an ECDim/ECE/DISP competitive sequence [see text, Eqs. (48) through (53), and Table 2 for definitions of p_1, p_2, and p_3; note that for the example discussed in the text $k_d = k_{dif}$]. The hatched zone in b corresponds to the region without experimental validity.

cleavage in Eq. (49) should decrease its concentration with a concomitant increase in the radical concentration, both effects being in favor of the duplicating step.*

The average concentrations for the key species Ar· to be used in the following analysis are obtained from Table 1, where the role of Ar· is symbolized by C in the three competing sequences: ECE [Eqs. (48) through (50) and (52)], DISP (Eqs. (48), (49), (51), and (52)], and ECDim [Eqs. (48), (49), and (53)]. This allows the three competition parameters to be obtained as in Table 2: p_1 (ECDim/ECE), p_2(ECDim/DISP), and $p_3 = (p_1/p_2)$ (DISP/ECE) from the ratios of the respective average concentrations. Following this procedure a kinetic zone diagram and a compass card are then constructed in Fig. 5a. Such a diagram shows that these considerations on the effect of C^0 or of the cleavage rate constant k_1 of $ArX^{\bar{\cdot}}$ are valid only when the point representing the system is in the ECE zone (C^0 effect) or in the DISP zone (k_1 effect). When starting from an ECDim zone, however, an increase in C^0 results in a tendency to shift toward the DISP zone and an increase in k_1 favors reduction via an ECE mechanism. The concentration effect is difficult to predict on an intuitive basis, yet that of k_1 could have been intuitively guessed. Indeed, when k_1 increases, Ar· is produced closer and closer to the electrode surface and then has less and less possibility to dimerize before reaching the electrode to be reduced. As for C^0, the effect of the mass transfer rate (i.e., of $\theta = \delta^2/D$ in the compass card) was almost impossible to predict on an intuitive basis.

To illustrate a further useful aspect of the method, let us discuss the experimental validity of the ECDim zone in Fig. 5a. Indeed, the representation in Fig. 5a supposes that any values are possible for p_1 and p_2. Yet

*Note that Ar· is prone to undergo other facile reaction paths (see later). The concurrent routes in Eqs. (50) through (52) then constitute the minimal competitive sequence to be considered.

Table 2. Competition Parmeters for an ECDim/ECE/DISP Competing Sequence

Average concentration of $Ar^{\cdot}$[a]		Competition parameter		Major pathway[d]	
Mechanism	$[Ar^{\cdot}]$[b,c]	Sequence	Competition parameter	$p_i \ll 1$	$p_i \gg 1$
ECDim	$(C^0/2k_{dim})^{1/2}(k_1/\theta)^{1/4}$	ECDim/ECE	$p_1 = (Ar^{\cdot})_{ECDim}/(Ar^{\cdot})_{ECE}$	ECDim	ECE
			$p_1 = k_1^{3/4}\theta^{1/4}/(2k_{dim}C^0)^{1/2}$		
ECE	$C^0/(k_1\theta)^{1/2}$	ECDim/DISP	$p_2 = (Ar^{\cdot})_{ECDim}/(Ar^{\cdot})_{DISP}$	ECDim	DISP
			$p_2 = k_d C^{0 1/2}/k_1^{3/4}(2k_{dim})^{1/2}\theta^{1/4}$		
DISP	k_1/k_d	DISP/ECE	$p_3 = (Ar^{\cdot})_{DISP}/(Ar^{\cdot})_{ECE}$	DISP	ECE
			$p_3 = p_1/p_2$		

[a]Taken from Table 1.
[b]k_1, rate constant of the aromatic halide anion radical cleavage; k_{dim}, dimerization rate constant of the σ-aryl radicals; k_d, homogeneous electron transfer rate constant (usually $k_d = k_{dif}$, the diffusion-controlled rate constant).
[c]C^0, ArX bulk concentration; $\theta = \delta^2/D$.
[d]Both pathways are observed when the value of p_i is close to unity (i.e., when ~ $0.1 \le p_i \le 10$).

in an actual case this is not true since the product $p_1p_2 = k_{dif}/2k_{dim}$, is necessarily larger than 1/2 because k_{dim} cannot exceed the diffusion limit k_{dif}. This implies that the hatched zone in Fig. 5b, which corresponds to $p_1p_2 < 0.5$, has no experimental validity. Thus, no system can be located in this zone regardless of the values of the experimental (C^0, θ, and δ) or intrinsic parameters (D, k_{dif}, k_{dim}, or k_1); this shows that almost all the ECDim zone has no experimental significance. Although this method does not allow the maximum yield of biaryl to be determined, it leads to the conclusion that it is necessarily poor and requires extremely precise conditions to be achieved. A complete and accurate analysis [17] of the problem establishes that this maximum yield is of the order of 10-20% at best, which is in complete agreement with the maximum experimental yields in the range of 7%.

However, this does not mean that biaryls cannot be obtained via aryl halide reduction but only that this strategy is not experimentally valid as soon as $Ar^{\cdot}$ is more easily reducible than the parent halide. Yet other approaches to the synthetic problem may be found. An obvious one consists in suppressing the facile reductions of $Ar^{\cdot}$ by using an electrophore as the leaving group X-, which will be the site of the initial electron uptake [Eq. (48)]. $Ar^{\cdot}$ may not then be reducible at the reduction potential of the parent pseudohalide, as in reaction (54), which is shown in benzylic series [22].

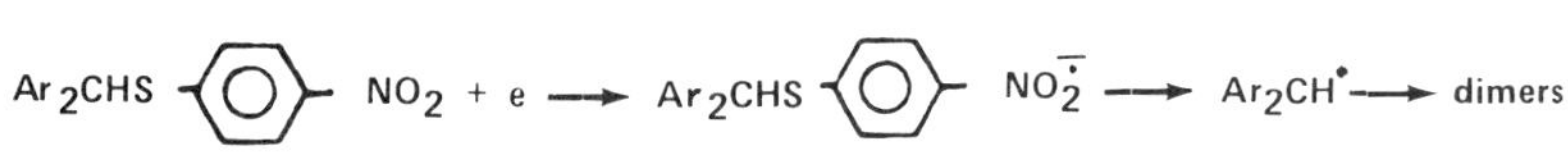

(54)

Another approach consists, for example, in using an ArX or a coreactant Ar'X, prone to react with $Ar^{\cdot}$ via the Ullman kind of reaction in Eq. (55) [23,24]:

$$Ar^{\cdot} + ArX \longrightarrow [Ar{-}ArX]^{\cdot} \longrightarrow \cdots \qquad (55)$$

Naturally, each of these possible approaches, as well as many others that may be conceived, may be tested "on paper" following this method to decide its possible validity and the best experimental conditions to favor it. For example, application of this "on paper" strategy shows that only moderate yields are obtained for the process in Eq. (56), when directly reducing the aryl halide at the electrode, whereas interesting yields (i.e., from 80 to 70%) are possible by using a redox mediator for performing the initial ArX reduction [24c]:

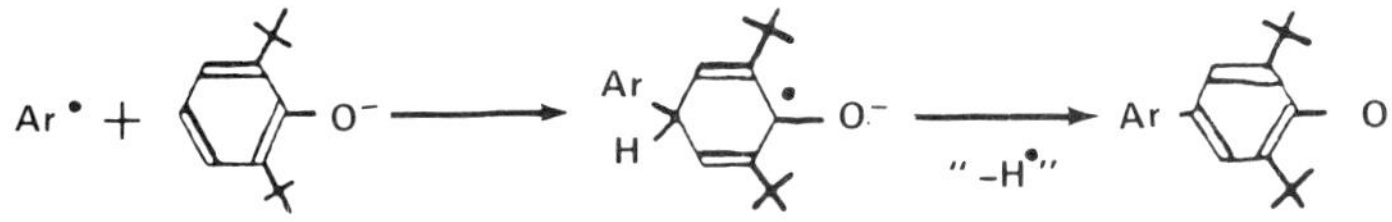

(56)

Rationalization of a Series of Results

Figure 6 presents the variations in the yield in monodeuterated arene obtained when an aromatic halide is reduced in acetonitrile in the presence of 10% D_2O [25].

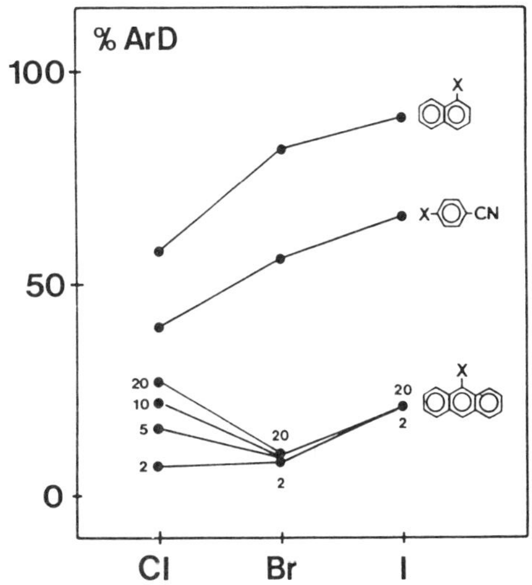

Figure 6. Experimental variations in the yield of monodeuterated arene ArD as obtained by reduction of the corresponding halide in acetonitrile in the presence of 10% D_2O and 0.1 M $LiClO_4$. The number on the curves are the initial concentrations (mM) of the halide. (Data from Ref. 25.)

$$ArX + 2e \xrightarrow{ACN,\ 10\%\ D_2O} yArD + (1 - y)ArH \qquad (57)$$

The overall stoichiometry of the reaction shown in Eq. (57) corresponds to the formation of ArD and ArH via two competitive pathways. ArD is obtained along the sequence presented in Eqs. (48) through (51), the resulting σ-aromatic anion being deuterated by D_2O, as in Eq. (58):

$$Ar^- + D_2O \longrightarrow ArD + DO^- \qquad (58)$$

ArH is formed via an H atom transfer from the ACN solvent to the aryl radical formed upon the mechanism in Eqs. (48) and (49), as outlined in the following sequence [25]*:

$$Ar^{\cdot} + CH_3CN \longrightarrow ArH + {}^{\cdot}CH_2CN \qquad (k_H) \qquad (59)$$

$${}^{\cdot}CH_2CN(+e \text{ or } ArX^{\dot{-}}) \longrightarrow {}^{-}CH_2CN \qquad (60)$$

$${}^{-}CH_2CN + D_2O \rightleftharpoons CH_2DCN + DO^- \qquad (61)$$

The data in Fig. 6 show that for the 1-halonaphthalene or 4-halobenzonitrile series a continuous variation is observed; that is, the yield in ArD increases in the series Cl << Br < I, in agreement (at least for the monotonic variation) with usual chemical expectations. But for the 9-anthracene

*The scrambling in Eq. (61) has been shown to play a minimal role during the electrolysis on the basis of experimental isotopic studies [25].

series a nonuniform variation is observed as a function of the halide. Moreover the ArD yields obtained for the 9-chloroanthracene are extremely dependent on the concentration or the stirring rate, whereas the effects of these factors are minimal for the bromo derivative and totally negligible for the iodo derivative or for the 1-halonaphtalene or 4-halobenzonitrile series. Such a priori puzzling experimental results may, however, be easily rationalized through the preceding method.

From Table 1 the two parameters corresponding to the competition between ECE [reactions (48), (49), (50), and (58)] or DISP [reactions (48), (49), (51), and (58)] sequences and the ECC sequence [reactions (48), (49), and (59) through (61)] are obtained as follows*:

$$p_1 = \frac{k_1}{k_H} \qquad \text{ECC/ECE} \qquad (62)$$

$$p_2 = \frac{C^0}{\theta^{1/2}} \frac{k_{dif}}{k_H k_1^{1/2}} \qquad \text{ECC/DISP} \qquad (63)$$

$$p_3 = \frac{p_1}{p_2} \qquad \text{DISP/ECE} \qquad (64)$$

This allows the construction of the kinetic zone diagram in Fig. 7a according to the previously explained procedure. The location of the experimental systems considered in Fig. 6 indicates that the benzonitrile or naphthalene derivatives undergo an ECE/ECC competition without interference from the DISP route. Thus, as indicated by the compass card in Fig. 7a or by the formulation of p_1 in Eq. (62), C^0 or $\theta = \delta^2/D$ has no effect on the overall ArD yield, in agreement with the experimental observations. Moreover, for a given aromatic moiety, k_H is constant and does not depend on the halide whereas k_1 increases in the order Cl $\ll$ Br $<$ I. Thus $p_1 = k_1/k_H$ increases monotonically in the series, which explains the uniform variations in Fig. 6. The 9-iodo anthracene also undergoes an ECC versus ECE competition, whereas an ECC/DISP competition is observed for the 9-chloro derivative. From the compass card in Fig. 7a it is then seen that C^0 or θ considerably affects the ArD yield for the 9-chloroanthracene whereas there is no effect of these parameters on the iodo derivative. For the bromide, an intermediate situation is observed but more shifted toward predominance of an ECC than for the iodide or chloride cases, which explains the minimum observed in Fig. 6 as well as the small variations in the ArD yield with concentration or stirring rate changes. It is thus seen that the simple approach presented here is sufficient to rationalize qualitatively all the a priori puzzling observations in Fig. 6. But a correct and thorough analysis of the problem on a physicomathematical basis [15] may be achieved and leads to a more quantitative description of these facts. The result of such an analysis is shown in Fig. 7b, under the form of predicted variations

---

*Note that for the H atom transfer pathway, steps (60) and (61), which occur after the branching point [Eq. (59)], play no role in the ArD yield. Thus the sequence of reactions (48), (49), and (59) through (61) behaves like an ECC sequence in regard to the competition between the H$^\cdot$ versus (+e, +D_2O) routes.

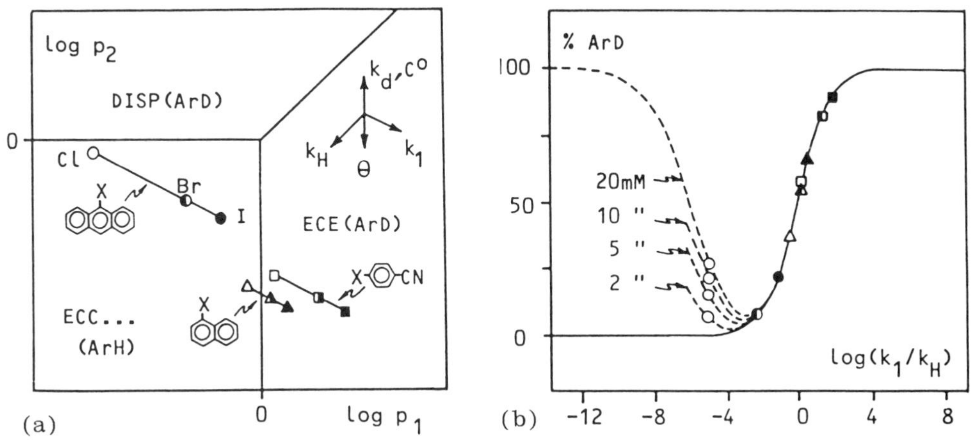

Figure 7. (a) Kinetic zone diagram for the ECC/ECE/DISP competitive sequence in the ArD versus ArH formation under the experimental conditions of Fig. 6 [see text and Eqs. (48) through (51) and (58) through (61); note that for the example discussed $k_d = k_{dif}$]. (b) Theoretical variations in the yield of ArD as a function of the concentration of ArX (numbers on the curves) and of the rate constants ratio k_1/k_H. (Experimental data from Ref. 25.)

in the ArD yield, for the systems in Fig. 6. Although more quantitative information is then obtained, it is seen that all the trends observed in Fig. 7b have been qualitatively predicted via the above approach, which requires no sophisticated derivations and can be handled "on the back of an envelope."

IV. CONCLUSION

This presentation was designed to discuss the relationships between micro- and macroscale electrolysis. Besides the intrinsic factors arising from scaling-up problems (larger concentrations, larger current densities, and longer reaction times), we have tried to show how the results of microscale electrolysis could be used with large profit, either in devising adequate experimental conditions or in the rationalization of product distributions. However, it is important to decide in such cases if the branching point leading to the different products takes place in the bulk solution or occurs in the very close vicinity of the electrode. Indeed in the first case the problem must be examined in terms of its homogeneous equivalent, the electrolytic reactions being replaced by first-order reactions. Conversely, in the second case an electrochemical analysis taking into account the concentration profiles in the vicinity of the electrode must be performed to solve quantitatively the distribution problem. However, in such cases, and when only semiquantitative predictions are needed, a very simple approach based upon competition parameters and kinetic zone diagrams proves to be considerably useful.*

---

*Note that this presentation considers only one branching point, but when several are present the same analysis must be applied at each branching point.

ACKNOWLEDGMENTS

Drs. J. -N. Verpeaux (ENS and CNRS) and A. Jutand (ENS and CNRS) are thanked for helpful discussions during the elaboration of this text, as well as for their critical reviews of its earliest versions.

SYMBOLS AND THEIR DEFINITIONS

A	Electrode surface area
C	Concentration (see also [S] and others)
C^b	Concentration in the bulk solution
C^0	Initial concentration of the starting material
C_S	Concentration of species S (see also [S] and others)
$C_{x=\cdots}$	Concentration at x = ⋯
D	Diffusion coefficient; average diffusion coefficient
D_S	Diffusion coefficient of species S
E	Electrode potential versus a given reference electrode
E^o (or E_1^o)	Standard reduction potential (relative to the half-reaction 1)
$E^o_{O/R}$	Standard reduction potential relative to the half-reaction: $O + ne \rightleftharpoons R$
F	Faraday
i	Current
k	Chemical rate constant
k_{elec}	Apparent first-order rate constant equivalent to an electrolysis
k_f(or k_b)	Measured forward (or backward) rate constant
k_Z	Rate constant for the reaction involving the transfer of a Z moiety
k_1	Rate constant number 1
K	Equilibrium constant
n (or n_1)	Number of electron exchanged (for the half-reaction 1)
n_{ap}	Apparent number of electron exchanged
p_1	Competition parameter number 1
Q	Charge consumed (coulombs)
Q_t	Charge consumed after a period t of an electrolysis
(S)	Activity of species S
[S]	Concentration of species S
$[S]_{bulk}$	Concentration of species S in the bulk solution
$(S)_{x=\cdots}$	Activity of species S at x = ⋯
$[S]_{x=\cdots}$	Concentration of species S at x = ⋯
$[S]_{yield}$	Yield in species S
V	Volume of the bulk solution
x	Coordinate along the x axis; distance from the electrode surface

Greek Symbols

δ	Diffusion layer thickness
θ	Characteristic time of an electrochemical method; duration of an experiment

REFERENCES

1. See, e.g., R. Jansson (a) A.I.C.H.E. Symp. Ser., 79 (229), 92 (1983); 79 (229), 119 (1983); (b) in Sel. Contrib. Symp. Recent Adv. Electrochem. Cell Des. (R. E. White, ed.), Plenum, New York, 1984, pp. 175-195.
2. J. S. Newman, Electrochemical Systems, Prentice Hall, Englewood Cliffs, N.J., 1973, pp. 239-253.
3. See descriptions of ultramicroelectrodes characteristics in, e.g., (a) R. M. Wightman and D. O. Wipf, in Electroanalytical Chemistry (A. J. Bard, ed.), Vol. 15, Dekker, New York, 1989; (b) Ultramicroelectrodes (M. Fleischmann, S. Pons, D. Rolison, and P. P. Schmidt, eds.), Datatech Systems, Morganton, N.C., 1987; (c) R. M. Wightman, Science, 240, 215 (1988).
4. (a) C. Amatore, M. R. Deakin, and R. M. Wightman, J. Electroanal. Chem., 220, 49 (1987); (b) M. I. Montenegro and D. Pletcher, J. Electroanal. Chem., 248, 229 (1988).
5. See, e.g., (a) Ref. 2, pp. 340-352; (b) A. J. Bard and R. L. Faulkner, Electrochemical Methods, Wiley and Sons, New York, 1980, pp. 375-377; (c) G. L. Booman and W. B. Holbrook, Anal. Chem., 35, 1793 (1963); 37, 795 (1965).
6. (a) A. J. Bard and K. S. V. Santhanam, in Electroanalytical Chemistry (A. J. Bard, ed.), Vol. 4, Dekker, New York, 1970, p. 220;(b) J. J. Lingane, J. Amer. Chem. Soc., 67, 1916 (1945).
7. (a) C. Amatore and J. M. Savéant, J. Electroanal. Chem., 123, 189 (1981); (b) H. Wendt, Angew. Chem., 21, 256 (1982); (c) H. Wendt and V. Plzak, J. Electroanal. Chem., 154, 13 (1983); (d) V. Plzak and H. Wendt, J. Electroanal. Chem., 154, 29 (1983); (e) V. Plzak and H. Wendt, J. Electroanal. Chem., 180, 185 (1984); (f) J. M. Savéant, J. Electroanal. Chem., 236, 31 (1987).
8. (a) C. Amatore, A. Jutand, and F. Pflüger, J. Electroanal. Chem., 218, 361 (1987); (b) C. Fabre, R. Fugnitto, and H. Strzelecka, C. R. Acad. Sci., Paris, Ser. C, 282, 175 (1976); (c) F. Pragst, R. Ziebig, U. Seydewitz, and G. Driesel, Electrochim. Acta, 25, 341 (1980).
9. C. Amatore, A. Jutand, F. Pflüger, C. Jallabert, H. Strzelecka and M. Veber, Tetrahedron Lett., 30, 1383 (1989).
10. See, e.g., (a) V. D. Parker, K. Nyberg, and L. Eberson, J. Electroanal. Chem., 22, 150 (1969); (b) V. D. Parker and L. Eberson, J. Chem. Soc. Chem. Commun., 340 (1969); (c) A. J. Bard and J. Phelps, J. Electroanal. Chem., 68, 313 (1976); for examples in aromatic oxidations, or (d) M. Szwarc, Acc. Chem. Res., 5, 169 (1972), for reductions.
11. See, e.g., J. Simonet and H. Lund, Acta Chem. Scand., B31, 909 (1977), for a typical experimental example.
12. For a discussion, see D. Pletcher, J. Electroanal. Chem., 179, 263 (1984).
13. See, e.g., (a) H. Lund and J. Simonet, J. Electroanal. Chem., 65, 205 (1975), for an early synthetic example.
14. C. Amatore and J. M. Savéant, J. Electroanal. Chem., 123, 203 (1981).
15. C. Amatore, F. M'Halla, and J. M. Savéant, J. Electroanal. Chem., 123, 219 (1981).
16. C. Amatore, J. Pinson, J. M. Savéant, and A. Thiébault, J. Electroanal. Chem., 123, 231 (1981).

17. C. Amatore and J. M. Savéant, J. Electroanal. Chem., 125, 1 (1981).
18. C. Amatore and J. M. Savéant, J. Electroanal. Chem., 125, 23 (1981).
19. C. Amatore and J. M. Savéant, J. Electroanal. Chem., 126, 1 (1981).
20. C. Amatore and J. M. Savéant, J. Amer. Chem. Soc., 103, 5021 (1981).
21. See, e.g., C. P. Andrieux, J. M. Savéant, and D. Zann, Nouv. J. Chim., 8, 107 (1984).
22. G. Farnia, M. G. Severin, G. Capobianco, and E. Vianello, J. Chem. Soc. Perkin Trans. II, 1 (1978).
23. J. Grimshaw, R. Hamilton, and J. Trocha-Grimshaw, J. Chem. Soc. Perkin Trans., I, 229 (1982).
24. (a) C. Amatore, C. Combellas, J. Pinson, J. M. Savéant, and A. Thiébault, J. Chem. Soc. Chem. Commun., 7 (1988); (b) N. Alam, C. Amatore, C. Combellas, J. Pinson, J. M. Savéant, A. Thiébault, and J. N. Verpeaux, J. Org. Chem., 53, 1496 (1988); (c) N. Alam, C. Amatore, C. Combellas, A. Thiébault, and J. N. Verpeaux, Tetrahedron Lett., 28, 6171 (1987).
25. F. M'Halla, J. Pinson, and J. M. Savéant, J. Amer. Chem. Soc., 102, 4120 (1980).

5

Comparison Between Electrochemical and Analogous Chemical Reactions

HANS J. SCHÄFER / Organisch-Chemisches Institut, Wilhems Universität, Münster, Federal Republic of Germany

I. Introduction 233
II. Comparison of Electrochemical and Analogous Chemical Reactions with Regard to Their Mechanism 234
  A. Some general considerations: The steps of the electrochemical and chemical mechanisms 234
  B. More detailed comparisons of electrochemical reactions and analogous chemical reactions 236
III. Comparison of Electrochemical and Analogous Chemical Reactions in Synthesis 240
  A. Introduction 240
  B. CC bond-forming reactions 240
  C. Functional group interconversions 241
  D. Some general comments 244
  References 245

I. INTRODUCTION

Electrochemistry is the combination of a heterogeneous electron transfer at an electrode with a chemical reaction. By this electron transfer the reactive intermediates of dissociative processes (cations, anions, and radicals) [1] and radical ions are formed. The differences in chemical reactions are due to the way in which these reactive intermediates are generated. In electrochemistry the electrode transfers the electron; in chemical reactions this occurs from a chemical reductant or oxidant (see Chap. 2) [2]. Furthermore, carbocations, radicals, and carbanions are chemically generated by dissociation [3], homolysis [4], or deprotonation [5]. The reaction conditions, the solvents, and the reactants encountered by the electrochemically or chemically formed intermediates can differ substantially and thus can lead to diverse reaction pathways and products. These points are considered in Sec. II by comparing the mechanisms.

Electrochemistry is also an esteemed method in organic synthesis (see Parts III-VIII) [6]. Often, when other synthetic methods fail, electrochemistry is chosen as a last possibility, which leads to success but also to disappointment due to the high expectancy. To achieve a more realistic picture the advantages and disadvantages of chemical and electrochemical conversions are compared in Sec. III on synthesis.

Additionally there are features of electrochemistry that cannot be compared because they are unique to this field and have no counterpart in conventional chemistry. They result from the combination of electron transfer with a chemical conversion. This way reactive intermediates can be generated, which otherwise are not or are in a more limited extent accessible, such as radical ions. Mechanisms can be studied in otherwise not accessible depth from time-resolved electrical signals of starting materials, products, or intermediates (see Chap. 3) [7]. Electrical energy can be reversibly stored in batteries [8] or gained in fuel cells [9]. Electrochemical sensors can transform small concentrations of substances specifically and depending on the quantity into electrical signals [10].

II. COMPARISON OF ELECTROCHEMICAL AND ANALOGOUS CHEMICAL REACTIONS WITH REGARD TO THEIR MECHANISM

A. Some General Considerations: The Steps of the Electrochemical and Chemical Mechanisms

If one looks at an electrochemical and a chemical reaction, three general steps can be distinguished: (1) The approach of the substrate to the chemical oxidant or reductant by forming covalent or ionic bonds, in electrochemistry the diffusion to and sometimes adsorption at the electrode; (2) generation of the reactive intermediate by electron transfer or thermal or photochemical dissociation; and (3) the follow-up reaction of the intermediate to the product.

Bond Formation Between Substrate and Chemical Reductant or Oxidant, Diffusion, and Adsorption

In chemical oxidation or reduction the redox redox reagent and the substrate often form a covalent or ionic bond, for example an ester in chromic acid oxidation [11] or selenium dioxide oxidation of ketones and aldehydes [12]. In electrochemical processes the substrate must diffuse from the bulk of the solution to the electrode and compete there with other components of the electrolyte by competitive adsorption for a position at the electrode surface [13]. The next step is then generation of the reactive intermediate by electron transfer or a dissociative process.

Reactive Intermediates by Electron Transfer and Dissociative Processes

Chemical electron transfers are mostly those of the inner sphere type, whereas electrochemical transfers belong to those of the outer sphere type (see Chap. 2) [2a]. This means that chemical oxidations or reductions have lower activation energies than analogous electrochemical conversions, thereby generating less reactive and thus more selective intermediates. Their transition states are more sensitive to stereochemical and electronic influences, which also leads to more selective conversions. At some oxide electrodes, such as the $Ni(OH)_2$ electrode [14], the oxidation occurs as inner sphere electron transfer by hydrogen abstraction. Also at doped titanium anodes this seems to be partially the case [15]. It cannot be definitely excluded that also in some oxidations at platinum or graphite anodes higher valency oxides at the surface act as inner sphere electron transfer agents. "Electrochemical" inner sphere electron transfers are increasingly developed by attaching polymer-bound electrocatalysts as mediators to the electrode surface [16a].

By electron transfer, radical ions, anions, cations, and radicals can be generated as reactive intermediates:

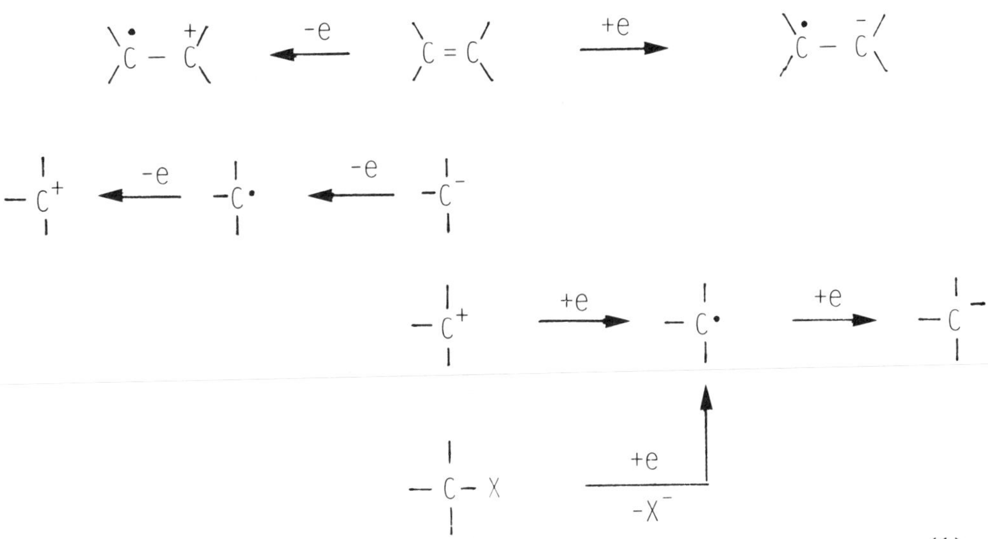

(1)

Radical ions, up to now, seem to be mostly products of electron transfer [16b]. Anions can also be generated by deprotonation [5], cations by S_N1 or A_E processes [3] and radicals by homolysis [4]. When radicals are generated by thermal or photochemical homolysis they are formed pairwise, so that cage combinations predominate, but in electron transfer the radicals are formed singly, whereby radical transfer reactions are favored.

The Follow-up Reaction and Product Formation

Distinctions in the follow-up reactions of electrochemical and chemical conversions are due to different concentrations of the reactive intermediates, different reaction media and additives, and the possibility of further oxidation or reduction. In electrochemical reactions the intermediates generated at the electrode surface diffuse into the bulk solution. Owing to their high reactivity, they react in a fairly thin reaction layer that adheres to the electrode and thus their concentration is higher than in homogeneous reactions [17], where they are uniformly spread over the medium. This can influence the product distribution. Radical additions initiated by electron transfer often lead in homogeneous solution to polymers [18], but in electrochemical reactions the termination of radical polymerization, that is, the formation of additive monomers and dimers, preponderates [19]. In electrochemical reactions the electrode surface can be depleted of the electroactive substrate at high current densities. This way reactions between the reactive intermediate and the educt can be diminished [20].

The adsorption of supporting electrolyte or other additives can create a special microenvironment in the reaction layer near the electrode that can favor certain pathways. Thus the use of tetraalkylammonium ions as supporting electrolyte depletes the electrode surface of water and thus favors the hydrodimerization of acrylonitrile [21]. The addition of strychninium salt and its adsorption promote an asymmetric induction in the reduction of the prochiral 2-acetylpyridine to 1-(2-pyridyl)ethanol [22].

The proximity of the electrode to the reaction layer can lead to follow-up oxidations and reductions of intermediate products. These are much less pronounced in homogeneous solution.

Each electron transfer generates a proton or Lewis acid at the anode and a hydroxyl ion or Lewis base at the cathode. This creates strongly acidic or basic media in the reaction layers adhering to the anode or cathode, whereby acid- [23] or base-catalyzed reactions [24] can be induced (see Chap. 30).

These more general considerations are subsequently illustrated by some studies in which data for electrochemical reactions and analogous chemical reactions are available.

B. More Detailed Comparisons of Electrochemical Reactions and Analogous Chemical Reactions

Oxidative Nucleophilic Aromatic Substitution

Eberson has compared the isomer ratios for the acetoxylation of aromatic substrates at the anode with those from different oxidants (Table 1) [25]. The anodic reaction proceeds by an initial electron transfer, thus forming a radical cation, which is then attacked by acetate ion. The homogeneous reactions, which come closest to the results of the anodic oxidation, are the $Co(III)W_{12}O_{40}^{5-}$ and the Ag(II) oxidation. These and also the Ce(IV) oxidant, which still exhibits some similarities in the isomer distribution, have been characterized as electron transfer oxidants. Whereas the Pd(II)-catalyzed acetoxylation has been shown to proceed via organopalladium species [26]. The isomer distributions are drastically different from those of ET-mediated mechansims, predominantly meta orientation being observed.

Table 1. Isomer Ratios for Oxidative Acetoxylation of Aromatic Hydrocarbons

	Anisole		Naphthalene	Nuclear/side chain	
Oxidant	o/p	m/p	1-/2-	Mesitylene	p-Xylene
Pt anode[a]	2.3	0.12	24	4.4	0.9
Ag(II)[b]	2.2	0.032	19	2.3	—
Co(III)W[c]	1.3	0.1	99	0.41	<0.01
Ce(IV)[d]	0.82	<0.01	24	0.7	
Pd(II)[e]	0.5	49	1.3	<0.01	0.66

[a]In HOAc-0.5 M NaOAc, 20°C.
[b]In HOAc-0.5 M NaOAc at 40°C.
[c]Co(III) $W_{12}O_{40}^{5-}$ in HOAc-H_2O (4:1), 0.1 M KOAc at 102°C.
[d]Ce(IV) ammonium nitrate or acetate in HOAc at 20-70°C.
[e]In HOAc at 115°C.

The Oxidative Decarboxylation of Carboxylic Acids

Kolbe electrolysis also allows some comparisons with analogous homogeneous reactions with regard to dimerization, substitution, or addition reactions of the generated radicals.

Photolytic or thermal decarboxylation of diacylperoxides is a source of alkyl radicals similar to Kolbe electrolysis. The anodic oxidation of propionate has been compared with the thermal decomposition of dipropionyl peroxide [27]. Examination of the yields shows that reaction between radicals is favored in the electrolytic process, whereas in peroxide decomposition CH abstraction from the solvent or the substrate occurs to a higher extent. This illustrates the effect of the higher radical concentration at the electrode.

Photolysis at low temperatures decreases the extent of the CH abstraction due to its higher activation energy. Additionally, the lower mobility of the radicals in the solid increases the cage reaction of the pairwise generated radicals. This way unsymmetrical diacylperoxides form unsymmetrical dimers with less than 1% of the symmetrical product [28]. Peroxides from carboxylic acids, being optically active at the α-carbon, couple photochemically in the solid with only a minor loss of optical activity [29].

In Kolbe electrolysis, coupling of the radicals generated from different carboxylates occurs statistically [30], and the chirality at the α-carbon is totally lost [31,32].

The easy follow-up oxidation of intermediates at the electrode is illustrated by the oxidation of trimethylacetic acid. The decarboxylation of carboxylate by peroxidisulfate/Ag(I) is the best known way to simulate the Kolbe anodic oxidation in a homogeneous solution [33]. The oxidation of trimethylacetic acid with $S_2O_8^{2-}$/Ag(I) in the presence of protonated chinolinium affords in a S_RAr substitution a high yield of t-butylated product [Eq. (2)] [34].

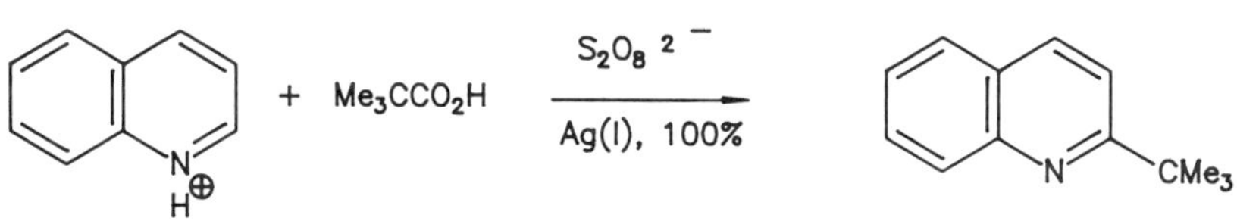

(2)

$$\mathrm{Me_3CCO_2H} \xrightarrow[-\mathrm{CO_2}]{-e} \mathrm{Me_3C\cdot} \xrightarrow{-e} \mathrm{Me_3C^+} \xrightarrow{\mathrm{CH_3CN}} \mathrm{Me_3C{-}NHCOCH_3} \qquad (3)$$

The anodic oxidation of trimethylacetic acid in acetonitrile, however, affords t-butylacetamide owing to further oxidation of the intermediate t-butyl radical to the t-butyl cation and its Ritter reaction with MeCN [Eq. (3)] [35].

The reaction of styrene with carbomethoxymethyl radicals generated in homogeneous solution and by Kolbe electrolysis has been compared (Table 2) [36]:

Table 2. Addition of the $MeO_2CCH_2\cdot$ Radical Generated by Kolbe Electrolysis and Homogeneous Chemical Reactions to Styrene

Method	Relative yields (%)				Total yield (%)
[Eq. (4)]	III	IV	V	VI	of III–VI
a	16.3	15.1	5.7	62.9	43
b	Trace	Trace	—	100	1.5
c	10	5	—	85	2.3
d	13.0	14.8	1.8	70.4	38.4

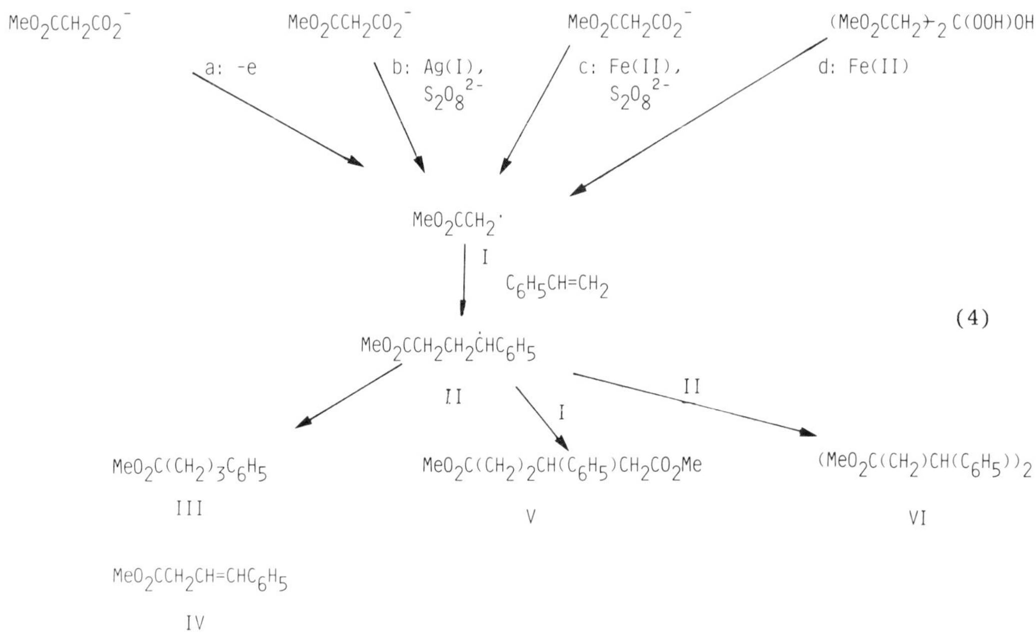

(4)

The methoxycarbonylmethyl radical I is obtained by Kolbe electrolysis from methylmalonate [Eq. (4a)], and in homogeneous solution by peroxidisulfate oxidation mediated by Ag(I) or Fe(II) [Eq. (4b, c)] and reductive fragmentation of the hydroperoxide of dimethyl acetone dicarboxylate [Eq. (4d)]. I forms with styrene the adduct II, which reacts by disproportionation to III and IV, by coupling with I to V and by dimerization to VI. The yields with $S_2O_8^{2-}$ are low due to the low reaction temperature that was applied for comparison with the other two reactions. The yields and product ratios for the electrolysis and the reductive fragmentation correspond surprisingly well, and no extensive polymerization was found for the homogeneous reaction.

Other Examples

Further comparisons of electrochemical and homogeneous conversions can be found for the oxidative coupling of phenols and phenol ethers [37] or in indirect electrolyses (see Chap. 29) [38] in which the electrochemically activated mediators react in homogeneous solution with the substrate. Very similar results are often obtained in heterogeneous and anodic oxidations, as with nickel peroxide [39] and with the nickel hydroxide electrode [14].

The Birch reduction of aromatic hydrocarbons or of double bonds with alkali metals in liquid ammonia or amines [40] resembles in yield and selectivity the cathodic reduction with lithium halide as supporting electrolyte, for example [41].

In reductive acylation and dimerization, the cathode is often superior to dissolving metal or radical anion reductions. So α,β-unsaturated ketones or esters can be acylated in high yield to 1,4-dicarbonyl compounds at the mercury cathode [42], but the corresponding reaction with sodium in tetrahydrofuran (THF) fails [43]. Reductive acylation of double bonds is possible in high yield, however, using vitamin B_{12} as mediator [44]. Here cobalt-alkyl complexes play an important role as intermediates.

In the reductive dimerization of methyl cinnamate to a cyclopentanone [Eq. (5)], similar yields are found at the cathode [45] and with metals (sodium, THF, and TBAI, −78°C) [43].

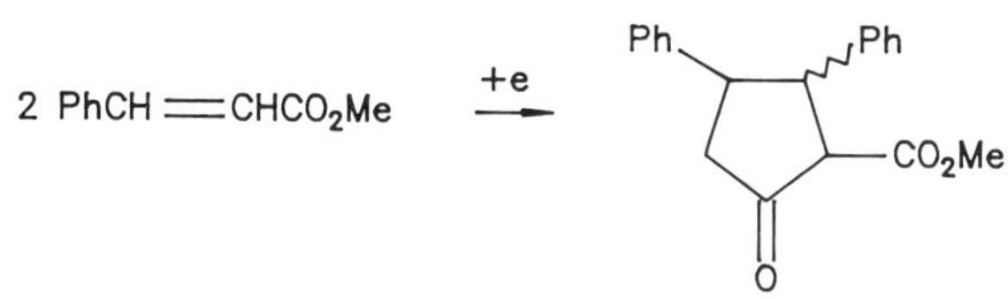

(5)

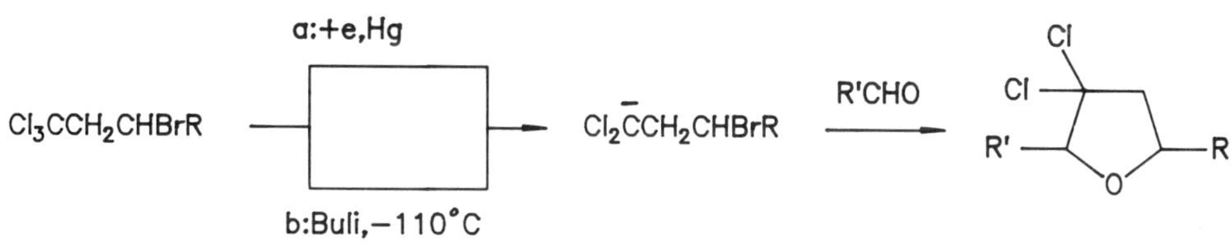

(6)

Because of the potential selective conversion at the electrode, halides can be reduced at the cathode to carbanions in the presence of carbonyl compounds, which are reduced at more cathodic potentials. This way labile carbanions can be obtained and reacted under conditions in which the same species generated by a metalorganic route would decompose. For example, trichlorobromoalkane can be cathodically converted in the presence of aldehydes to a dichloromethyl anion at 0°C [route A, Eq. (6)] and be trapped to form a dichlorotetrahydrofuran, but for the metallorganic route [route B, Eq. (6)] a reaction temperature of −110°C is necessary [46].

III. COMPARISON OF ELECTROCHEMICAL AND ANALOGOUS CHEMICAL REACTIONS IN SYNTHESIS

A. Introduction

For the synthesis of organic compounds two types of reaction are used: CC bond-forming reactions for the construction of large structures from small units and functional group interconversion for introduction of the functionality into the target molecule and activation of the building blocks [47].

B. CC Bond-Forming Reactions

The CC bond-forming reactions can be classified as polar reactions, in which an activated donor reacts selectively with an acceptor or vice versa; radical reactions, which are increasingly applied in coupling or addition because they do not interfere with polar groups in the molecule and thus save steps as no protections are needed [48]; pericyclic reactions, especially cycloadditons, for the construction of ring compounds with high stereoselectivity; and transition metal-catalyzed reactions.

The first two reaction types involve reactive intermediates, namely radical ions, carbocations, carbanions, and radicals, that can be easily prepared in wide variety at the electrode. Therefore, the electrode is especially suited for these two types of reaction.

Pericyclic electrochemical reactions are less developed. They involve chain reactions with a radical cation as chain transferring step or the generation of reactive dienophiles (see Chap. 23, Sec. V). Transition metal complexes are increasingly applied in electrochemistry as electrocatalysts for reductive carboxylation [49,50], acylation or alkylation [44], or activation [51], for example.

Reactive acceptors are anodically available as radical cations in a wide variety by the oxidative "Umpolung" of donors. This way two donors can be coupled in one step, if one of them is converted to an acceptor by ET. Chemically two or more steps are required for this operation. Thus electrochemistry can save steps in synthesis. The selectivity in mixed anodic couplings, however, needs to be further improved. Of more general synthetic value seem to be the generation of latent immonium ions for electrophilic addition [52] and the oxidative coupling of olefins [53]. Chemically there is, on the other hand, a highly developed experience and a large arsenal of reagents to execute selective S_E and A_E reactions with a wide choice of substrates. High selectivities have been achieved there, for example with silicium as terminator [54].

Reactive donors can be generated cathodically by reductive cleavage of halides to carbanions or by reduction of double bonds to radical anions. By reductive Umpolung, two acceptors can be dimerized in one step, such as acrylonitrile to adipodinitrile, which is chemically not possible. Two different acceptors can be selectively coupled either by controlled potential electrolysis (CPE) [42,55] or with cobalt mediators [44].

On the other hand, highly developed chemical techniques are available to generate and to react activated donors of different functionalities and carbon structure by hydrogen-, halide- or metal-metal interchange [5].

Furthermore, in these cases low temperatures and inert solvents, necessary to handle reactive and labile carbanions and to achieve high selectivities, can be applied.

Radicals can be generated at the anode by oxidation of carbanions, alkyl borates, alkoxides, and carboxylates (see Chap. 23, Sec. V) and at the cathode by reduction of protonated carbonyl compounds or onium salts (see Chap. 10). Thereby a wide choice of different radical structures can be mildly and simply obtained from readily available precursors. These radicals are especially suited for coupling reactions because these are favored by the high radical concentrations at the electrode. Compared with these possibilities chemical methods for radical coupling are less general and limited to less easily accessible starting materials, such as azo and peroxy compounds [4].

The electrogenerated radicals can also be trapped by olefins to yield additive dimers and additive monomers. The product distribution can in part be controlled by the current density and olefin concentration [Eq. (7), paths a and b] and by the electrode potential or the substituent of the olefin [Eq. (7), path c) (see Chap. 23, Sec. V).

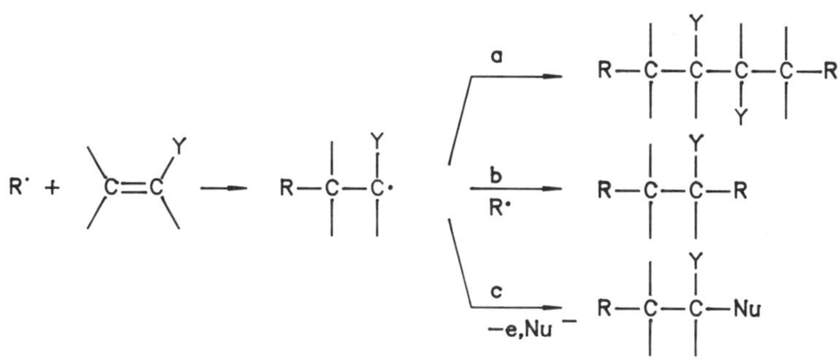

(7)

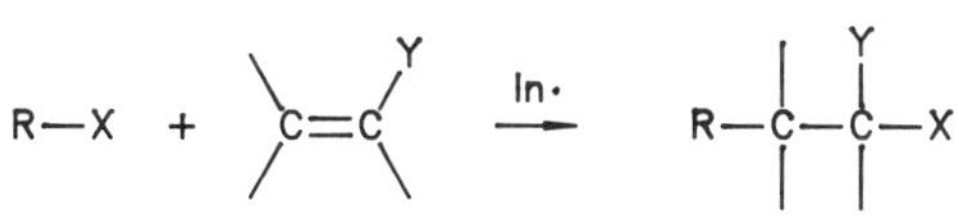

(8)

Chemically these types of adducts are not or are much less easily accessible, and the products are those of a radical chain addition [Eq. (8)] [4]; one CC and one C–H or C–heteroatom bond is formed, but electrochemically two carbon–carbon bonds are obtained.

C. Functional Group Interconversions

Electrochemical functional group interconversions can be classified into four groups:

1. Anodic substitution and its cathodic counterpart cathodic cleavage
2. Anodic dehydrogenation and cathodic hydrogenation
3. Anodic addition and cathodic elimination
4. Anodic cleavage, where the cathodic reversal would be a CC bond-forming reaction, such as hydrodimerization

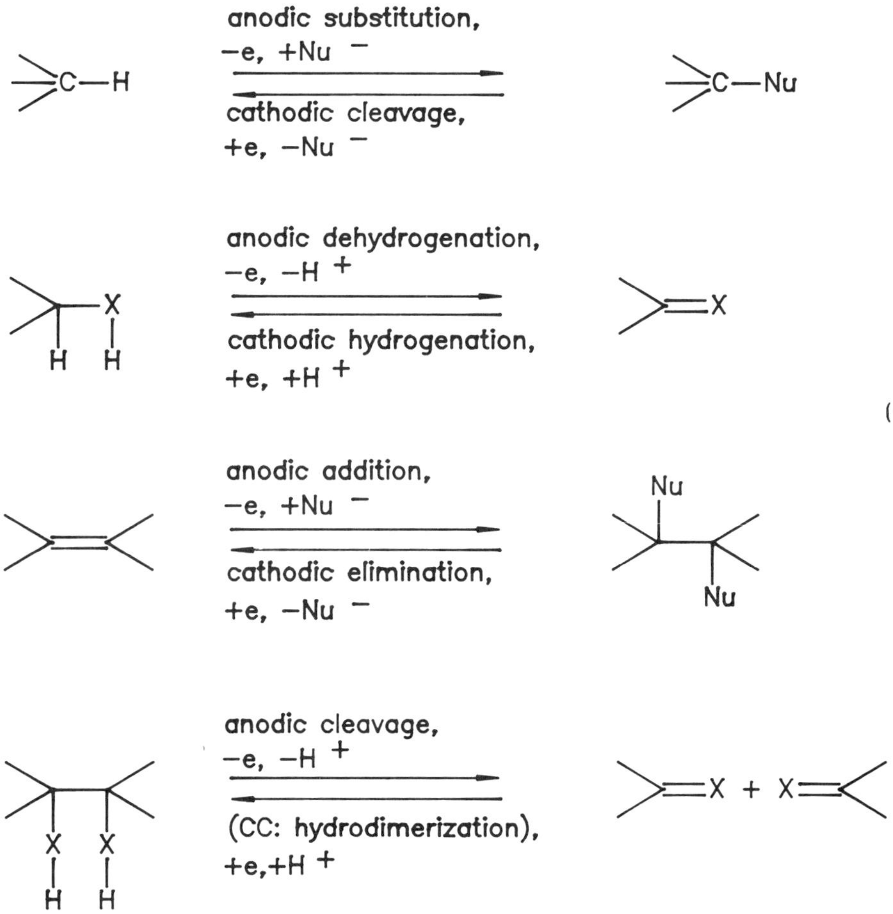

(9)

For the anodic substitution of unactivated CH bonds, some fairly selective reactions for tertiary CH bonds in hydrocarbons and γ−CH bonds in esters or ketones are available [56,57]. However, in some cases a better control of follow-up oxidations remains to be developed. Chemically a number of selective reactions are available, such as ozone on silica gel for tertiary CH bonds [58], the Barton or Hoffmann-Loeffler-Freytag reaction for γ−CH bonds [59], $(i\text{-prop})_2NCl/H^+$ [60], and template-directed oxidations for remote CH bonds [61].

For activated CH bonds a wealth of chemical methods is available, such as singlet oxygen, N-bromosuccinimide, $Br_2/h\nu$, $Pb(OAc)_4$, or SeO_2. Anodically fair yields are obtained for CH bonds activated by vinyl, phenyl, or alkoxy, although sometimes problems of regioselectivity occur. The α oxidation of N-acyl-N-alkylamines leading to synthetically very interesting precursors of immonium cations [52] has no chemical counterpart.

For the nucleophilic substitution of aromatic CH bonds, electrochemistry has a clear advantage over chemical conversions. Although with the former nucleophiles like OR, OCOR, Br, and CN can be introduced simply in one step [62], chemically multistep procedures are necessary to achieve the same goal.

Carboxylic acids can be regarded as CH bonds with an inserted CO_2 group. They can be converted by non-Kolbe electrolysis into products in which the CO_2H group is replaced by a nucleophile or a double bond (see Chap. 14) [63]. This conversion can be combined with specific rearrangements or fragmentations [64]. The comparable chemical conversions—Hundsdieker or Kochi reaction [65] or Hofmann degradation [66]—have much more limited applications.

The cathodic cleavage of C–X to C–H bonds can be applied to a variety of X, such as Hal, NR_3^+, PR_3^+, OAc, OCH_3, OTos, $OPO(OEt)_2$, or epoxides. Generally good yields, an often high potential selectivity, and sometimes good stereoselectivities are obtained (see Chaps. 8 and 24). Chemical reactions for similar conversions that are well worked out, especially with regard to yield and selectivity, are the reduction of tosylates with $LiAlH_4$ and of halides with Bu_3SnH [67] or radical deoxygenation [68].

For the dehydrogenation of HC–XH bonds, for example of alcohols to ketones, aldehydes to carboxylic acids, and amines to nitriles, there is a wealth of anodic reactions, such as the nickel hydroxide electrode [14], indirect electrolysis with I^-, NO_3^-, or thioanisole [69], RuO_2/Cl^- [70], and chemical oxidations [Cr(VI), MnO_2, MnO_4^-, DMSO/Ac_2O, Ag_2O/Celite, and O_2/Pt] [71] for specific applications available. The advantages of the electrochemical conversion is a lower price, a more easy scale-up, and fewer pollution problems.

The hydrogenation of C=X and C=C double bonds can be achieved cathodically and with a number of chemical reagents, such as catalytic hydrogenation, metal hydrides, and dissolving metals. Here the best method must be selected for each individual case. The advantages of the electrochemical method are the potential selective conversion at the cathode, the possibility of reducing carbon–heteroatom double bonds in the presence of C=C bonds, or the <u>trans</u> hydrogenation of C=C bonds in contrast to the <u>cis</u> hydrogenation with catalytic hydrogen.

Anodic addition affords products, some of which are of industrial interest, such as propylene oxide [72], diols, or 1,4-dimethoxyfuran [73]. To improve the selectivity, sometimes mediators like Tl^{3+}, Pd^{2+}, Ce^{4+}, or Ru^{3+} are used (see Chap. 29) [38]. Chemical reagents that serve a similar purpose include MnO_4^-, OsO_4/morpholinoxide, Hal_2, and peracids [74].

Reductive elimination can remove a variety of vicinal nucleophiles to form double bonds:

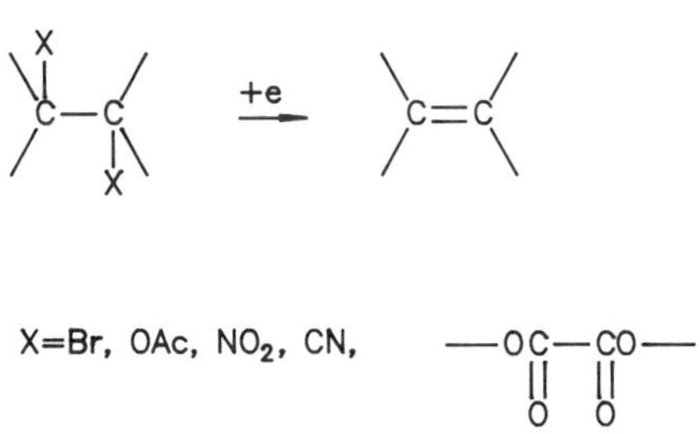

(10)

At the cathode a chemoselective elimination depending on the degree of alkylation of the 1,2-dihalide is possible [75]. Chemically these reductions are more limited in scope, they can be conducted with I^- in DMF, for example, with Zn, Mg, or Cr^{2+} [76]. Although the cleavage of 1,2-diols can be simply achieved at the nickel hydroxide electrode [14], chemically the more expensive reagents $Pb(OAc)_4$ or IO_4^- must be used.

The conversion of heteroatoms to different oxidation states, for example $NH_2 \rightarrow NO_2$, $-S- \rightarrow -SO-$, $-SO_2-$, or $NO_2 \rightarrow NHOH$, NH_2 is possible at the electrode. Of advantage, especially in the reduction of nitro functions, is the potential selective conversion to certain oxidation states (see Chaps. 9, 16, and 17).

By controlled potential electrolysis protecting groups can be selectively removed. This makes electrochemical deprotection (see Chap. 24) [77,78] a valuable supplement to chemical deprotection [79]. Additionally, less interference with other functional groups is encountered than in chemical deprotection.

D. Some General Comments

The supposed need for sophisticated equipment is often the reason that electrochemistry is less frequently applied than it could be. However, more than 90% of electrolyses—especially the first exploratory experiments—can be conducted in simple and readily available equipment. A beaker with a stirring bar serves as a reaction vessel. Quite often a diaphragm is unnecessary as most anodic oxidations can be conducted in an undivided cell. Graphite plates, graphite felt, and a platinum foil fastened to an inert support are used as anodes and a steel or lead plate or a mercury pool as cathode. The silver/silver chloride reference electrode is prepared by anodically chlorinating a silver wire. The cell voltage is supplied from an ac source, a voltage regulator, and a rectifier, and the potential between the working- and the reference-electrode is read from a high resistance voltmeter.

The little knowledge necessary to conduct the electrolysis is easily obtained. The oxidation or reduction potential of the substrate is determined with the equipment just mentioned. The electrolyte is then chosen in such a way that its decomposition potential exceeds that of the oxidation or reduction potential of the substrate. The electrode material is selected so that it does not oxidatively or reductively dissolve at the applied potential. This procedure can be applied—with only minor changes—to a wide variety of substrates and conversions. It is to be compared with chemical reactions in which if the modern selective reagents are frequently applicable to only a limited number of substrates and often must be prepared in multistep syntheses, and, if acquired from a manufacturer, the price of these chemical reagents in most cases exceeds 10- to 1000-fold that of the electrical current.

Electrochemical preparations are often easier to conduct than chemical conversions. Solubility problems, which often occur with inorganic redox reagents in organic solvents, are not encountered. On the other hand, the inertness of solvents and the lower attainable temperatures in chemical reactions cannot be achieved to this extent in electrolysis. Polar electrolytes are necessary, and the temperatures for practical reasons cannot be lowered much below −40°C in preparative-scale electrolyses.

Workup of electrolyses is often easier as no products of chemical oxidants or reductants must be separated. Additionally in electrolysis a similar workup can be used for a wide number of applications. Furthermore, the scale-up of electrochemical reactions is often easier to achieve than of chemical conversions.

The selectivity of electrochemical reactions with regard to a potential selective conversion is superior to that of chemical reactions. However, the chemoselectivity, the regioselectivity, and especially the stereoselectivity are often inferior to those of chemical reactions. The reasons are that the latter involve inner sphere electron transfers, which by nature are more selective than outer sphere electron transfer at the electrode, that lower temperatures are attainable, and that a much more profound knowledge of the control of these reactions has been accumulated. With regard to the economy of synthesis, the electrode can be superior to chemical reactions. First, the redox Umpolung saves steps because it allows the one-step coupling of two donors or two acceptors. In comparable chemical conversions two or more steps are necessary for that purpose. Second, at the electrode a synthetic building unit can be used in several types of reactivity, such as electrophile, nucleophile, or radical, by changing the reactivity via electron transfer. In chemical preparations the reagents are usually designed for only one reactivity.

A big advantage of electrochemical reactions compared with chemical is the effective contribution to pollution control. Because of the direct electron transfer, it avoids the problem of separation and waste treatment of the often poisonous end products of the chemical oxidants or reductants. Furthermore, by electrodialysis, organic acids or bases can be regenerated from their salts without the use of sulfuric acid or sodium hydroxide, for example, leading to the production of sodium salts or sulfates as waste. At the same time inorganic acids and bases, necessary for chemical production, are provided by this process. An application has been demonstrated in the preparation of methoxyacetic acid by oxidation of methoxyethanol at the nickel hydroxide electrode [80]. Finally, unwanted side products can be converted into the wanted product. The economy of the process is thereby increased, and the problem of waste separation and treatment is decreased. This is accomplished in the manufacture of chloroacetic acid by chlorination of acetic acid, for example. There the waste product dichloroacetic acid, formed by overchlorination, is cathodically converted to chloroacetic acid [81].

REFERENCES

1. R. W. Alder, R. Baker, and J. M. Brown, Mechanism in Organic Chemistry, Wiley-Interscience, New York, 1971.
2. (a) L. Eberson, Electron Transfer Reactions in Organic Chemistry, Springer-Verlag, New York, 1987; (b) Methoden der Organischen Chemie (Houben-Weyl), Vol. IV, 1a-d, G. Thieme-Verlag, Stuttgart, 1987; (c) Oxidation in Organic Chemistry, Parts A-D (K. B. Wiberg and W. S. Trahanovsky, eds.), Academic Press, New York, 1965-1982; (d) Organic Reaction Mechanisms, Chap. Oxidation and Reduction, Wiley, New York, 1965-1986.
3. G. A. Olah and P. v. R. Schleyer, eds., Carbonium Ions, Vols. I-V, Wiley-Interscience, New York, 1968-1976; P. Vogel, Carbocation Chemistry, Elsevier, New York, 1985.
4. (a) J. K. Kochi, ed., Free Radicals, Vols. 1 and 2, Wiley-Interscience,

New York, 1973; (b) B. Giese, Radicals in Organic Synthesis: Formation of Carbon-Carbon Bonds, Pergamon Press, New York, 1986.

5. (a) Methoden der Organischen Chemie (Houben-Weyl), Vol. XIII/1, 2a, G. Thieme-Verlag, Stuttgart, 1970, 1973; (b) Organic Reaction Mechanisms, Wiley, New York, 1970-1986, Carbanions and Electrophilic Aliphatic Substitution; (c) J. C. Stowell, Carbanions in Organic Synthesis, Wiley, New York, 1979; (d) R. B. Bates and C. A. Ogle, Carbanion Chemistry, Springer, New York, 1983.
6. (a) S. Torii, Electroorganic Syntheses, Part I, Oxidations, Verlag Chemie, Kodansha, 1985; (b) T. Shono, Electroorganic Chemistry as a New Tool in Organic Synthesis, Springer, New York, 1984; (c) H. J. Schäfer, Kontakte (Merck), 1987, (2)17, (3)37.
7. (a) V. D. Parker, Advan. Phys. Org. Chem., 19, 131 (1983); (b) O. Hammerich and V. D. Parker, Advan. Phys. Org. Chem., 20, 55 (1984); (c) J. Heinze, Angew. Chem., 96, 823 (1984); Angew. Chem. Int. Ed. Engl., 23, 931 (1984); (d) D. Lexa and J. M. Saveant, Acc. Chem. Res. 1983, (16)235.
8. D. Pletcher, Industrial Electrochemistry, Chap. 10, Chapman and Hall, London, 1982; F. Beck and K. J. Euler, Elektrochemische Energiespeicher, VDE-Verlag, Berlin, 1984.
9. C. H. Hamann and W. Vielstich, Elektrochemie II, Verlag Chemie-Physik, Weinheim, 1981, p. 359.
10. A. P. F. Turner, I. Karube, and G. S. Wilson, eds., Biosensors: Fundamentals and Applications, Oxford University Press, Oxford, 1987.
11. K. B. Wiberg and H. J. Schäfer, J. Amer. Chem. Soc., 91, 927, 933 (1969).
12. K. B. Sharpless and K. M. Gordon, J. Amer. Chem. Soc., 98, 300 (1976).
13. H. Fischer, J. Electroanal. Chem., 62, 163 (1975).
14. H. J. Schäfer, Top. Curr. Chem., 142, 102 (1987).
15. F. Beck and H. Schulz, Electrochim. Acta, 29, 1569 (1984).
16. (a) R. W. Murray, Electroanalytical Chemistry (A. J. Bard, ed.), Vol. 13, Dekker, New York, 1984, p. 191; (b) For another mode of formation see: L. Eberson and L. Jönsson, J. Chem. Soc. Chem. Commun., 1981, 333.
17. (a) H. J. Schäfer, Habilitation Thesis, Göttingen, 1970; (b) H. Wendt, Angew. Chem., 94, 275 (1982); Angew. Chem. Int. Ed. Engl., 21, 256 (1982).
18. H. Logemann, in Methoden der Organischen Chemie (Houben-Weyl), Vol. E20, G. Thieme Verlag, Berlin, 1987, p. 54.
19. H. Schäfer and A. Al Azrak, Chem. Ber., 105, 2398 (1972).
20. V. Plzak, H. Schneider, and H. Wendt, Ber. Bunsenges. Phys. Chem., 78, 1373 (1974).
21. M. M. Baizer and J. D. Anderson, J. Electrochem. Soc., 111, 223 (1964).
22. J. Kopilov, E. Kariv, and L. L. Miller, J. Amer. Chem. Soc., 99, 3450 (1977); For earlier examples of asymmetric induction in electrochemistry see: R. N. Gourley, J. Grimshaw and P. G. Millar, J. Chem. Soc. C, 1970, 2318; L. Horner and D. Degner, Electrochim. Acta, 13, 611 (1974).
23. (a) S. Torii, Electroorganic Syntheses, Part I, Oxidations, Verlag Chemie, Kodansha, 1985, p. 315; (b) K. Uneyama, Top. Curr. Chem., 142, 167 (1987).

24. T. Shono, Electroorganic Chemistry as a New Tool in Organic Synthesis, Springer, New York, 1984, p. 154; (b) J. H. P. Utley, Top. Curr. Chem., 142, 131 (1987).
25. L. Eberson, Electron Transfer Reactions in Organic Chemistry, Springer-Verlag, New York, 1987, p. 80.
26. L. Eberson and L. Jönsson, Acta Chem. Scand., B30, 361 (1976); Liebigs Ann. Chem., 1977, 233.
27. S. Goldschmidt, W. Leicher, and H. Haas, Liebigs Ann.Chem., 577, 153 (1952).
28. M. Feldhues and H. J. Schäfer, Tetrahedron, 41, 4195, 4213 (1985).
29. R. Lomölder and H. J. Schäfer, Angew. Chem., 99, 1282 (1987); Angew. Chem. Int. Ed. Engl., 26, 1253 (1987).
30. J. H. P. Utley, in Technique of Electroorganic Synthesis, Part I (N. L. Weinberg, ed.), Wiley, New York, 1974, p. 820.
31. L. Eberson, K. Nyberg, and R. Servin, Acta Chem. Scand., B30, 906 (1976); L. Eberson and G. Ryde-Petterson, Acta Chem. Scand., 27, 1159 (1973).
32. J. H. P. Utley and G. B. Yates, J. Chem. Soc. Perkin II, 1978, 395.
33. L. Eberson, S. Gränse, and B. Olofson, Acta Chem. Scand., 22, 2462 (1968); L. Eberson, Electrochim. Acta, 12, 1473 (1967).
34. F. Minisci, A. Citterio, and C. Giordano, Acc. Chem. Res., 16, 27 (1982).
35. L. Eberson and K. Nyberg, Acta Chem. Scand., 18, 1567 (1964).
36. H. J. Schäfer and R. Pistorius, unpublished results.
37. M. Sainsbury, Tetrahedron, 36, 3327 (1980).
38. E. Steckhan, Top. Curr. Chem., 142, 1 (1987).
39. M. V. George and U. S. Balachandran, Chem. Rev., 75, 491 (1975); E. Ross, Kontakte (Merck), 1979, 19.
40. A. J. Birch and G. S. Rao, Advan. Org. Chem., 8, 1 (1972).
41. H. W. Sternberg and D. M. Mohilner, J. Amer. Chem. Soc., 91, 4191 (1969); R. A. Benkeser, E. M. Kaiser, and R. F. Lambert, J. Amer. Chem. Soc., 86, 5272 (1964); E. Kariv-Miller, K. E. Swenson, and D. Zemach, J. Org. Chem., 48, 4210 (1983); E. Kariv-Miller, K. E. Swenson, G. K. Lehmann, and R. Andruzzi, J. Org. Chem., 50, 556 (1985).
42. (a) H. Lund and C. Degrand, Tetrahedron Lett., 1977, 3593; (b) T. Shono, I. Nishiguchi, and H. Ohmizu, J. Amer. Chem. Soc., 99, 7396 (1977).
43. A. Grosse and H. J. Schäfer, unpublished results.
44. R. Scheffold, S. Albrecht, R. Orlinski, H. R. Ruf, P. Stamouli, O. Tinembart, L. Waldner, and C. Weymuth, Pure Appl. Chem., 59, 363 (1987).
45. J. P. Petrovich, M. M. Baizer, and M. R. Ort, J. Electrochem. Soc., 116, 749 (1969); C. Z. Smith and J. H. P. Utley, J. Chem. Soc. Chem. Commun., 1981, 492.
46. H. Claus and H. J. Schäfer, Tetrahedron Lett., 26, 4899 (1985); H. Claus, Dissertation Thesis, Münster, 1986.
47. J. Fuhrhop and G. Penzlin, Organic Synthesis, Verlag Chemie, Weinheim, 1983; S. Warren, Organic Synthesis, The Disconnection Approach, Wiley, New York, 1982.
48. B. Giese, Radicals in Organic Synthesis: Formation of Carbon-Carbon Bonds, Pergamon Press, New York, 1986, p. 22.
49. G. Filardo, S.Gambino, G. S. Silvestri, A. Gennara, and E. Vianello, J. Electroanal. Chem., 177, 303 (1984).

50. J. F. Fauvarque, A. Jutand, and M. Francois, Nouv. J. Chim., 10, 119 (1986).
51. (a) C. Degrand, M. Bikrani, and B. Gantheron, J. Organometallic Chem., 299, 111 (1986), and earlier papers; (b) Y. L. Chiu, A. E. G. Santana, and J. H. P. Utley, Tetrahedron Lett., 28, 1349 (1987).
52. S. D. Ross, M. Finkelstein and R. O. C. Petersen, J. Amer. Chem. Soc., 88, 4657 (1966); K. Nyberg and R. Servin, Acta Chem. Scand., B30, 640 (1976); T. Shono, Y. Matsumura, K. Tsubata, and K. Uchida, J. Org. Chem., 51, 2590 (1986); T. Shono, Y. Matsumura, K. Uchida, and H. Kobayashi, J. Org. Chem., 50, 3243 (1985).
53. D. Koch, H. Schäfer, and E. Steckhan, Chem. Ber., 107, 3657 (1974); R. Engels, H. J. Schäfer, and E. Steckhan, Liebigs Ann. Chem., 1977, 204; H. Baltes, E. Steckhan, and H. J. Schäfer, Chem. Ber., 111, 1294 (1978).
54. K. Krohn, Nachr. Chem. Tech. Lab., 35, 358 (1987).
55. M. M. Baizer, Tetrahedron Lett., 1963, 973; J. Org. Chem., 29, 1670 (1964).
56. L. L. Miller and V. Ramachandran, J. Org. Chem., 39, 369 (1974).
57. A. Hembrock and H. J. Schäfer, Angew. Chem., 97, 1048 (1985); Angew. Chem. Int. Ed. Engl., 24, 1055 (1985).
58. E. Keinan and Y. Mazur, Synthesis, 1976, 523.
59. L. Stella, Angew. Chem., 95, 368 (1983); Angew. Chem. Int. Ed. Engl., 22, 337 (1983).
60. N. C. Deno, W. E. Billups, R. Fishbein, C. Pierson, R. Whalen, and J. C. Wyckhoff, J. Amer. Chem. Soc., 93, 438 (1971); H. J. Schäfer and E. Cramer, Fat Sci. Technol., 90, 351 (1988).
61. R. Breslow, Acc. Chem. Res., 13, 170 (1980).
62. L. Eberson and K. Nyberg, Acc. Chem. Res., 6, 106 (1973).
63. S. Torii, Electroorganic Syntheses, Part I, Oxidations, Verlag Chemie, Kodansha, 1985, p. 57.
64. R. Michaelis, U. Müller, and H. J. Schäfer, Angew. Chem., 99, 1049 (1987); Angew. Chem. Int. Ed. Engl., 26, 1026 (1987); J. H. P. Utley, in Technique of Electroorganic Synthesis, Part I (N. L. Weinberg, ed.), Wiley, New York, 1974, p. 867.
65. J. K. Kochi, J. Amer. Chem. Soc., 87, 2500 (1965); J. Org. Chem., 30, 3265 (1965).
66. J. March, Advanced Organic Chemistry, Wiley, New York, 1985, p. 863.
67. W. P. Neumann, Synthesis, 1977, 665.
68. W. Hartwig, Tetrahedron, 39, 2609 (1983).
69. T. Shono, Y. Matsumura, J. Hayashi, and M. Mizoguchi, Tetrahedron Lett., 1979, 165; 21, 1867 (1980).
70. S. Torii, T. Inokuchi, and T. Yukawa, Chem. Lett., 1984, 1063.
71. W. C. Carruthers, Some Modern Methods of Organic Synthesis, Cambridge University Press, London, 1978, p. 338.
72. K. H. Simmrock, Hydrocarbon Proc., 57, 105 (1978), Chem. Abstr., 90, 63490 (1979); Chem.-Ing.-Tech., 48, 1085 (1976).
73. D. Degner, in Technique of Electroorganic Synthesis (N. L. Weinberg and B. V. Tilak, eds.), Wiley, New York, 1982, p. 251.
74. W. C. Carruthers, Some Modern Methods of Organic Synthesis, Cambridge University Press, 1978, p. 355.
75. U. Husstedt and H. J. Schäfer, Synthesis, 1979, 964, 966.
76. E. Baciochi, in The Chemistry of Functional Groups, Suppl. D. (S. Patai and Z. Rapoport, eds.), Wiley, New York, 1983, p. 161.

77. T. Shono, Electroorganic Chemistry as a New Tool in Organic Synthesis, Springer, New York, 1984, p. 150.
78. V. G. Mairanovsky, Angew. Chem., 88, 283 (1976); Angew. Chem. Int. Ed. Engl., 15, 281 (1976); M. I. Montenegro, Electrochim. Acta, 31, 607 (1986).
79. T. W. Greene, Protective Groups in Organic Synthesis, Wiley, New York, 1981.
80. M. Steiniger and H. Voss, BASF, Ludwigshafen, Deutsches Pat. 1987, Nr. DE 3620013 A1.
81. S. Dapperheld, Angewandte Elektrochemie, Jahrestagung 1987, Dechema Monographie.

II

PRACTICAL PROBLEMS

6
Practical Problems in Electrolysis

HENNING LUND / Aarhus University, Aarhus, Denmark

I. Introduction 254

II. Cells for Electrolysis 256
- A. General considerations 256
- B. Cells with two-dimensional electrodes 260
- C. Cells with three-dimensional electrodes 268
- D. Special-purpose cells 270

III. Diaphragm Materials 270
- A. General considerations 270
- B. Porous materials 271
- C. Ion-exchange membranes 271

IV. Electrode Materials 271
- A. General considerations 271
- B. Cathode materials 273
- C. Anode materials 275
- D. Semiconductors 277
- E. Optically transparent electrodes 277
- F. Electrodes with chemically modified surfaces 277

V. Reference Electrodes 278
- A. General considerations 278
- B. Electrodes of the first kind 279
- C. Electrodes of the second kind 280
- D. Redox electrodes 282

VI. Solvents for Electrolysis 283
- A. General considerations 283
- B. Protic solvents 289
- C. Aprotic solvents 294
- D. Salts 302
- E. Supercritical fluids 303

VII. Electrolytes 304
- A. Anions 304
- B. Cations 306
- C. Buffers 307

VIII. Voltage Sources and Coulometers 309
- A. Voltage sources 309
- B. Coulometers 309

IX. Practical Hints 309
A. Circuit 310
B. Counterelectrode reactions 310
C. Removal of oxygen 311
D. Determination of n values 311
E. Workup 311
F. Removal of impurities 312
G. Detection of intermediates 312
References 312

I. INTRODUCTION

An attractive feature of the electrolytic method is that it has the possibility of controlling the activity of the reagent, the electron, over a wide range. The success or failure of an electrolysis depends to a great degree on the proper selection of experimental conditions. An intelligent choice of these requires an understanding of the principles of controlling the reaction and of the properties of the different components, such as electrolytic cells, electrodes, solvents, and electrolytes, which are part of the electrolytic equipment.

In classic electrolytic reactions the current density, measured in amperes per square decimeter, was the quantity that was controlled, possibly because it was the easiest factor to measure and keep constant. For a long time nearly all electrolytic reactions were performed with control of the current density, although Haber, as early as 1898, in his famous papers on the stepwise reduction of nitro compounds, realized that the potential of the working electrode was the proper quantity to control.

The differences in the two ways of controlling the electrolytic reaction are illustrated in Fig. 1. In this figure curve I depicts schematically the connection between the current through the cell and the potential of the working electrode. The initial solution contains two reducible compounds or one compound with two groups reducible at different potentials. When the potential at the cathode is between 0 and E_A, no electron transfer across the electrical double layer can take place and thus no current runs through the cell. If the cathode potential is made more negative, the electron transfer becomes possible; that is, the reduction of the most easily reducible compound or group starts. Between E_A and E_B the current rises in dependence on the potential, but when the value E_B has been reached, all the molecules that arrive at the electrode and that can undergo the first reduction are reduced as soon as they reach the electrode. In the potential interval E_B to E_C the current is limited by the transport of the reducible compound to the cathode. This current is called the limiting current i_d, and under fixed conditions it is proportional to the concentration of the electroactive compound.

A further change of the electrode potential in the negative direction results in the occurrence of the second electrode reaction and the current rises; a similar S-shaped curve results from this reduction. At more negative potentials a third reaction or a reduction of the medium takes place.

If a suitable current i_o [$i_o < i_d(I)$] is sent through the cell, the cathode potential assumes the value $E_o(I)$, and when $i_o < i_d(I)$ this is well below the potential (E_C) at which the second electrode reaction starts; a selective reduction thus occurs at the beginning of the electrolysis. During

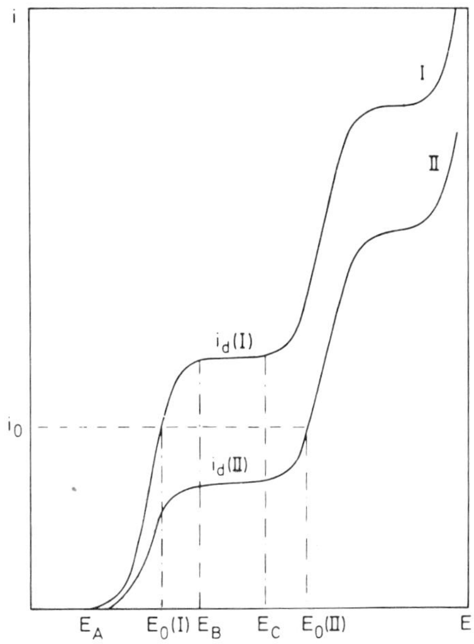

Figure 1. Schematic representation of the connection between the current and the potential of the working electrode in a solution containing a compound with two groups reducible at different potentials. Curve I, before electrolysis; curve II, after the passage of some current; I_0 is an applied current, E_0(I) and E_0(II) the potentials corresponding to i_0; i_d is the limiting current, and E_A, E_B, and E_C are applied potentials.

the electrolysis the concentration of the reducible compound, and thus its limiting current, diminishes and after a while (curve II) the limiting current becomes smaller than the applied current [$i_0 > i_d$(II)]. The cathode potential has then, by necessity, reached the value E_0(II) and at this potential the second electrode reaction also takes place; the electrolysis is no longer selective.

If the second electrode process does not interfere with the first (e.g., it may be due to the background), the method may still be acceptable in the laboratory (e.g., anodic methoxylation or an electrolytic reduction with simultaneous hydrogen evolution). For industrial processes, however, a high current yield (or current efficiency) is generally desirable and waste of current on the background is less tolerable. The current yield is the theoretical amount of electricity divided by the amount actually employed for the production of a particular substance (usually expressed as a percentage).

On the other hand, when the electrode potential is the controlled factor and is kept at a suitable value, such as E_B, the second electrode process cannot take place, and the reduction remains selective to the end. The current through the cell is never higher than the limiting current corresponding to the first electrode reaction. This means that the current decreases during the reduction and becomes very small toward the end of the reaction, as the limiting current is proportional to the concentration of the electroactive material.

Another way of keeping the potential constant is to maintain a certain concentration of the reducible compound, e.g., corresponding to the limiting current i_d (I) in Fig. 1. By continuously adding substrate and removing product and by keeping the current at a constant value smaller than i_d (I), for example, i_0, the potential stays at E_0 (I) and the reaction remains selective. This way is preferred in large-scale electrolysis.

In some reactions the required potential lies outside the decomposition potential of the medium and a certain current density is necessary to force the potential to acquire that value; a competition between transfer of electrons to the substrate and to the electrolytic medium takes place. The outcome of this competition, the current yield, depends on such factors as electrode material, current density, specific adsorption, and concentrations of the reacting species.

If the current yield is of minor importance and no further oxidation or reduction than desired occurs (except reactions of the supporting electrolyte), constant current electrolysis is simpler to use than constant potential electrolysis.

The information found in Chaps. 2–4 and the considerations presented here can be translated into practical solutions of the problems of electrolysis in many ways, some of which are discussed in the next sections. Although the emphasis in this chapter is on laboratory-scale electrolyses and in Chap. 31 on industrial-scale work, it is clear that many of the factors that enter into consideration of cell design and choice of electrode, for example, are common to both.

II. CELLS FOR ELECTROLYSIS

A. General Considerations

Most of the industrial cells discussed in Chap. 31 have been developed from laboratory prototypes mentioned in this chapter. Polarographic cells and other cells for electroanalytical purposes discussed in different monographs [1–4] have been omitted; however, some electron spin and some electroanalytical cells can be found in Chap. 3. The problems connected with the scale-up of organic electrode processes and electrochemical engineering have been treated [5–8] and are discussed in Chap. 31. Fuel cells have also been omitted here; a review has been given by Eisenberg [9].

A number of problems must be considered when designing a cell for electrolysis, such as potential distribution at the working electrode, position of the reference electrode, ohmic resistance in the circuit, mass and heat transfer, the need for a diaphragm, and the necessity of working in a closed system.

Potential Distribution

It is desirable that the potential gradient be the same all over the electrode surface so that a uniform current density distribution is obtained, provided the substrate (or depolarizer) concentration is equal at different parts of the electrode. (Depolarizer was the name given by Heyrovsky [10] for the substance that receives or gives away electrons at the electrode surface.) This can be achieved in two ways by employing concentric, cylindrical electrodes or planar, parallel electrodes of equal size. Generally only the electrode areas facing each other carry the current, so in the former arrangement only one side of the electrode is effective, whereas in the latter arrangement the working electrode may be placed between two auxiliary electrodes; this is of interest when the working electrode is of an expensive material whereas the auxiliary electrodes may be of cheaper material, as in the Kolbe reaction [11].

Mercury electrodes present special problems. It is more difficult to make electrodes of the required symmetry, and the inequalities in the distance to the anode introduced by stirring the mercury surface result in differences in the potential gradient along the electrode surface and uneven depolarization (see later). Amalgamated electrodes may be used in many cases instead of mercury.

Sometimes, especially in laboratory cells that carry only a small current. the required symmetry of the cell is overrruled by other considerations, such as convenience of operation, as in the H cells (see p. 261).

The use of small alternating voltage on a dc current, which has been claimed [12] to enhance the reaction rate and selectivity in some cases, does not require special considerations in the construction of the cell.

Position of Reference Electrode

Theoretical considerations and actual measurements of the working electrode potential under working conditions show that a uniform potential distribution is possible only with concentric electrodes of the same area [13,14]. Thus, none of the intrinsic dissymmetrical H cells with mercury electrodes, which have been so widely used, are in fact able to function with a truly constant potential at all parts of the surface. In reduction experiments [14,15] measurements with a probe electrode have shown that the parts of the electrode near the diaphragm are at more negative potentials than the control potential governed by a reference electrode placed at the mercury surface remote from the diaphragm; as expected the differences are found to be greater with increasing current. Another factor connected with stirred pool electrodes is that the distance between the tip of the reference electrode and the mercury surface varies with nonuniform stirring. Thus, the measured potential difference contains a variable component due to the ohmic voltage (iR) drop in the solution. There is thus a limitation to how exactly the potentiostat should desirably follow small changes of the control potential. In actual practice a controlled potential electrolysis (CPE) is mostly carried out in a more or less narrow potential range around some average value [16]. As expressed by Harrar and Shain [13]: "In the general case when the potential distribution is known only approximately the best approach is to place the reference electrode as near the working electrode surface as possible on the line of minimum distance between the counter and working electrodes." A controlled potential electrolysis is thus an electrolysis in which the potential at the working electrode varies within certain limits. In the author's experience a selective electrolysis requires a separation of 0.15–0.2 V in polarographic half-wave potential between two electrode reactions.

Ohmic Resistance

A low ohmic resistance of the cell is desired from several points of view; this may be achieved by having a small distance between anode and cathode, a large electrode surface, a compact cell, and a low-resistance diaphragm. A distance between anode and cathode as small as 0.1–0.2 mm has been reported [17], although usually larger distances are employed. A large electrode area may involve the use of many electrodes; a further development in this direction is the "fluidized bed" electrode (Chap. 31).

For laboratory experiments electroosmosis may be a problem, especially when working in nonaqueous solvents of high resistance. The resulting pumping through the diaphragm, most often from anode to cathode compartment, depends above all on the cell voltage [18,19] and thus on the resistance. In practice it is not easy to predict when and how much pumping will take place.

Diaphragm

Since the electrode reaction or its products at the auxiliary electrode may interfere with the desired product or the electrode reaction of the working electrode, it is very often necessary, especially in reductions, to separate the cell into anode and cathode compartments by some kind of diaphragm that allows the current to pass but suppresses mixing of anolyte and catholyte. Cells of this kind are called divided cells.

The design of the cell is simpler and the ohmic resistance is lower in an undivided cell, and diaphragms should be used only when unavoidable. In some cases the mixing of the catholyte and anolyte may be lowered sufficiently simply by pressing a porous material, such as porous plastics, against the counterelectrode and thereby dampening convection mass transfer [20].

Mass Transfer

For small cells, stirring, either mechanical or magnetic, may be sufficient to obtain a reasonable mass transfer. Pumping the substrate between the electrodes is a procedure suitable for large-scale electrolysis, as it allows a favorable position of the electrodes and the use of high current densities. The use of ultrasound may increase the rate of electrolysis; it has been used to promote the electrochemical reduction of some rather insoluble powders [21] and in some indirect reductions [22].

Special forms of mass transfer operate at the wicking electrode [23] and with foaming electrolytes [24]. In the former, that some substrates "creep" on the electrode surface is exploited to transport the substrate to the working electrode. The latter may be useful when electrolyzing gaseous substrates. If a gas evolution process occurs at the working electrode, this gives very efficient stirring. Stirring by gas bubbles may also be achieved with porous electrodes and a gaseous depolarizer in excess.

It has been found [25–28] that applying a magnetic field during an electrolysis enhances the limiting current; this is generally interpreted as a consequence of the magnetohydrodynamic effect. The enhancement factor of the limiting current in an unstirred solution may be 2-3 for a magnetic field of 8000 × G. It has been reported that during the anodic oxidation of phenylacetic acid the yield of benzaldehyde increased about 40% in the presence of a magnetic field]28].

Heat Transfer

For removal of the heat generated during electrolysis the cell may be surrounded by a cooling bath or provided with a water jacket and/or internal cooling spiral; this also permits electrolysis at constant temperature. In larger cells with metallic electrodes the electrodes can be made hollow to allow for a direct water cooling of the electrode itself. However, it is often more advisable to pump the electrolyte through an external heat-exchanger system, thus making a flow cell.

Closed System

It is simplest to use an open cell, but often it is necessary to apply a closed system—either because oxygen or water must be excluded or high pressure is necessary, or because the reaction involves poisonous reactants.

Often oxygen can be removed to a sufficient degree by bubbling an inert gas through the solution, but when working in systems in which even traces of oxygen must be avoided, all manipulations of substances must occur on a vacuum line and the cell constructed to meet this requirement.

Space-Time Yield

The space-time yield is defined as the amount of desired product divided by time and cell volume, and it is important for the economy of a process (Chap. 31). For laboratory cells space-time yields as such are of minor importance, but if a cell is designed for a process in which commercialization is a possibility, one might just as well have the conception in mind from the beginning, and it is advisable early in such an investigation to use experimental conditions akin to those to be used industrially.

Every practical cell construction represents some sort of a compromise of these demands, and nearly every research group in the field of organic electrosynthesis seems to favor its own approach to the problem of designing the ideal cell. Moreover, it must be remembered that the passage of the current through the cell is always associated with two simultaneous electrode processes: and oxidation at the anode and a reduction at the cathode. From an economic point of view, it is of course desirable that both processes be engaged in the preparation of useful products. This is, however, difficult to realize with purely organic systems (Chap. 35), but in principle an organic reduction could easily be combined with an evolution of chlorine or oxygen at the anode and an organic oxidation with production of hydrogen or sodium amalgam at the cathode. In laboratory work one is mostly satisfied by concentrating on only one process at the working electrode and arranging it in such a manner that the product(s) from the counter (auxiliary) electrode does not interfere with the reaction under investigation or do any harm to the surroundings or to the electrode itself; a hydrogen anode has, for example, been suggested as auxiliary electrode in nonaqueous media [29].

The actual design depends on the type and number of electrodes and operational principle and scale, and on the actual process. Once the relevant parameters of the electrode reactions are known, it is often possible to use a simpler cell with only two electrodes and a minimum of additional facilities for that special reaction. For exploratory work, however, it is more desirable to have a very flexible construction with many additional facilities so that a broad range of experiments can be carried out in the same cell.

It may often be advisable (for cathodic processes especially) to have in the cell the possibility of including an indicator electrode and a pH control system [30].

In the following some representative examples have been selected for more detailed discussion. References are given to constructions of similar types. Supplementary hints may be found in books by Allen [31], Tomilov et al. [32], Weinberg [33], Beck [34], and Pletcher [5]. A comparison has been made between the performance of some types of cells in a selected reaction (oxidation of organic anions) [35].

The presentation of the cells has been organized according to the following criteria: two- or three-dimensional electrodes, batch or continuous operation, and absence or presence of a diaphragm.

B. Cells with Two-Dimensional Electrodes

In "classic" preparative electrochemistry and in most research work, two-dimensional electrodes have been used, partly because it is simpler to control the different parameters at such electrodes and partly because time-space yield is not a deciding factor for laboratory cells.

Batch Cells

Undivided cells A simple piece of equipment for the performance of a simple electrolysis may consist of a beaker to contain a conducting solution of the substrate and two working electrodes connected to some dc source. The vessel is usually made of some rigid indifferent material, such as glass, but the inner wall of the container can also be made of a conducting material to serve as one of the electrodes. An undivided cell may be used when the products from the electrode reactions do not react with the other electrode or the other products. The classic Kolbe electrosynthesis is an example of a reaction that goes well in such an extremely simple "beaker cell."

Solid electrodes may be used in the form of wires, rods, sheets, bars, plates, or gauze, or a liquid, such as mercury, may be placed at the bottom of the container; electrode materials are treated in Sec. IV. The distance between the electrodes should be kept as small as possible to minimize the ohmic resistance of the cell and thus the heat generated during the passage of the current. To get a larger working area several electrodes [36] may be connected in parallel in the same manner as in a conventional storage battery, thus comprising a "tank cell." In cases (e.g., Kolbe electrolyses) in which a film may be formed on the electrode surface during the electrolysis, both electrodes may preferably be made of platinum and the surface cleaned by occasionally changing the direction of the current [37] or using pulse techniques [38,39].

In more elaborate electrolytic experiments it is often desirable to use a closed cell [40] with accessories for working under an inert atmosphere, or in vacuum [41], at elevated pressure, with reflux, or to allow the identification of gaseous products; a stirrer and a thermometer are also conveniently added. For the performance of controlled potential electrolyses with external control of the working electrode potential by a potentiostat, it is necessary to include a reference electrode, which should then be placed as near the working electrode surface [42] as possible. However, reference electrodes are not so much used in undivided cells, and it must be stressed that CPE

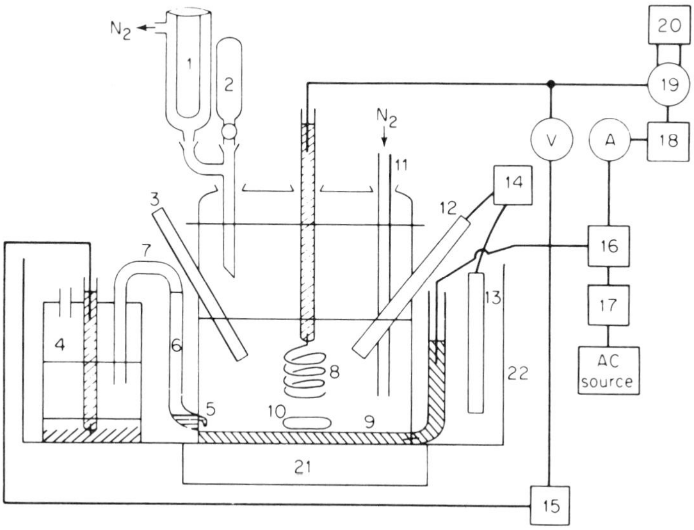

Figure 2. Apparatus for electrolysis. 1, Dry-ice acetone cooler; 2, dropping funnel; 3, thermometer; 4, saturated calomel electrode; 5, liquid junction; 6, saturated KCl aqueous solution; 7, saturated KCl-agar bridge; 8, platinum anode; 9, mercury cathode; 10, stirrer; 11, N_2 gas inlet; 12, glass electrode; 13, temperature-calibrating electrode; 14, pH meter; 15, vacuum voltmeter; 16, constant voltage supply; 17, stabilizer; 18, mercury coulometer; 19, standard resistor (1 Ω); 20, recorder; 21, magnetic stirrer; 22, thermostat. (Reprinted from Ref. 43 by courtesy of the Chemical Society of Japan.)

may be realized without the use of a three-electrode system and a potentiostat by having a sufficiently high concentration of the electroactive substance during electrolysis at constant current with partial conversion or by adding another redox system to transfer the electrons to or from the compound under investigation (see Chap. 29). Various kinds of reference electrodes are treated in a separate section.

In Fig. 2 is shown an undivided cell for controlled potential reduction under nitrogen; it was designated for the hydrodimerization of acrylonitrile [43].

Electrolytic fluorination of organic compounds is usually done in undivided cells with several alternating cathodes and anodes [44–46]; the corrosive nature of the HF medium presents special technological problems. Some of these are discussed in Chap. 26.

Divided cells

H cells In principle an H cell consists of two containers placed beside each other and connected through a diaphragm; the actual design depends on the types of electrodes and diaphragms used. The prototype for cells with mercury electrodes is the well-known Lingane cell [47] developed for use with a three-electrode system and later modified by Pasternak [48]. Figure 3 shows a modification that is easily constructed from common glassware by a

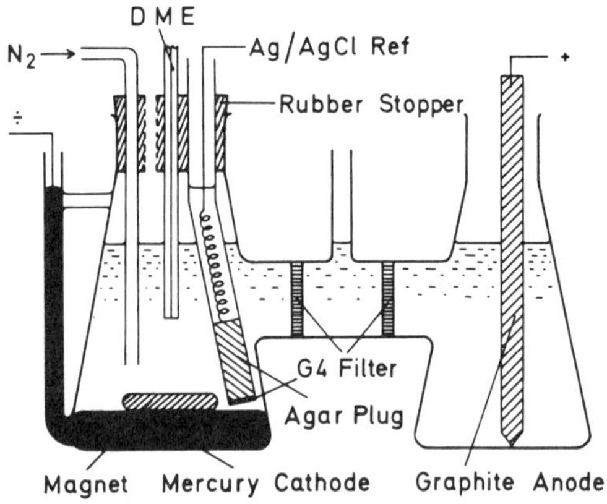

Figure 3. Semimacroscale electrolytic cell. (From Ref. 50.)

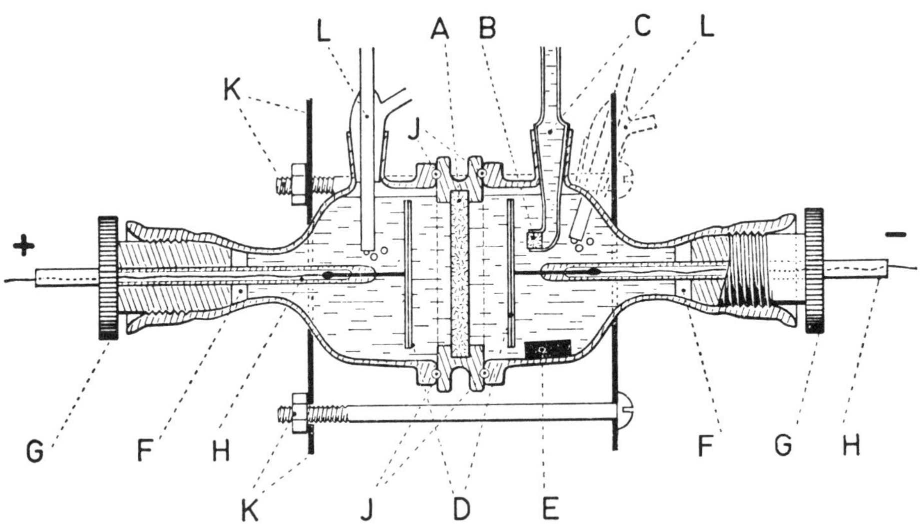

Figure 4. Electrolysis cell. A, B, fritted glass disks; C, electrolytic junction between solution and reference electrode; D, platinum grid; E, magnetic stirrer (in Teflon); F, Teflon gasket; G, Torion stopper; H, glass tube; J, ground glass surfaces; K, locking screw; L, bubbler. (From Ref. 56.)

skilled glassworker and has been used for several years in the author's laboratory [49,50]. It may also be used with a solid working electrode; for nonaqueous work the working electrode compartment may be closed and provided with standard joints. A dropping mercury electrode (DME) is included and allows the concentration of the electroactive substance to be followed during the electrolysis. The electrode surface is renewed by magnetic stirring at the same time as the solution is stirred. An agar plug may be placed in the middle compartment of the cell [36,49,50] and in the reference electrode, but this can be done only in aqueous media, and it is often more convenient to use the same solution throughout the cell. A cell of similar type [51] has been reported with two middle parts and a propeller stirrer as in the original constructions [47,48], in which a stopcock for separation of the mercury was also included. A good description of an all-glass disconnectable H cell construction has been given by Peltier and co-workers [52]. Straightforward H cells without special features have been designed by Rifi [53] and Metcalf and Whitnack [54]. A detachable H cell for membrane diaphragms has been designed by Allen [31]. In closed cells a connection between anode and cathode chamber above the level of the solution may be desirable to equalize the pressure in the system [55].

H cells with solid plane parallel electrodes of the same area have been used rather often for work in anhydrous media. They give a uniform current distribution. An example of one such cell with a reference electrode is shown in Fig. 4 [56]. It has a small distance between the electrodes, which is suitable for work in media of low electrical conductivity. A similar construction has been used by Sugino et al. [57]. A special detachable cell with thermostated hollow brass electrodes and vertical magnetic stirring has been described by Schulz and Strobel [58].

Cylindrical cells. From the preceding discussion it should be clear that it is difficult to realize a preparative cell with a strictly uniform current and potential distribution, especially with a mercury electrode with which the condition of parallel electrode surfaces of same area is not easily fulfilled. A good approach has been given by Moinet and Peltier [16], and a schematic representation of the construction is shown in Fig. 5. Special attention has been paid to the problem of renewing the mercury electrode surface with minimal variation in the distance to the reference electrode. This has been done by stirring at moderate speed with a magnet mounted on a Teflon cross turning on a short axis centered at the bottom of the vessel. Good thermal stability is secured by a cooling coil through the anode compartment connected in series with the outer bath. The cells have been proved to work satisfactorily with currents up to 3 A in a well-conducting aqueous medium.

A versatile cell [59] of the same type has been described; it could also be constructed from more simple components [60]. A cell for preparative electrolyses on a 0.1–1 mol scale with anolyte flow has been described by Iversen and Lund [61]. A simple cylindrical cell has been used by Horner and Mentrup [62] for reduction of quaternary salts; in this case a reference electrode is not necessary and the anolyte was simply tap water continuously removed by a water pump. Slightly differing constructions are given in the experimental chapters of Refs. 31–34.

When both the working and counterelectrodes are solid, concentric cylindrical electrodes separated by a cylindrical diaphragm can be made. This has been done in several published constructions. A Russian version (Fig. 6) [63–65] has the shaft of a propeller stirrer protruding through a bearing

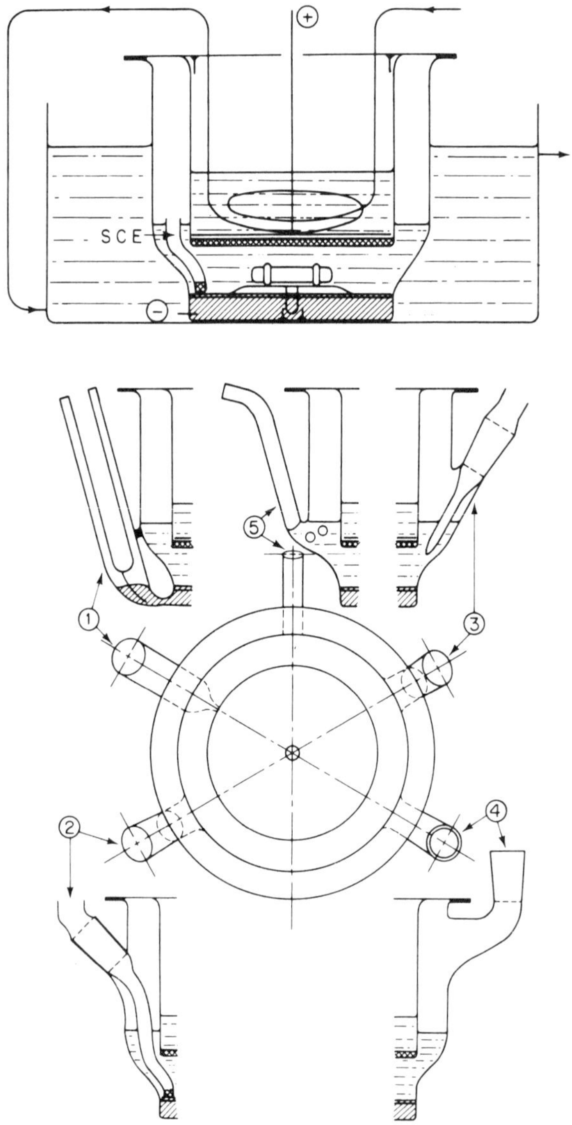

Figure 5. 1, Cathode connection; 2, SCE reference electrode; 3, thermometer; 4, available, for example, for reflux condenser; 5, gas inlet. (From Ref. 16.)

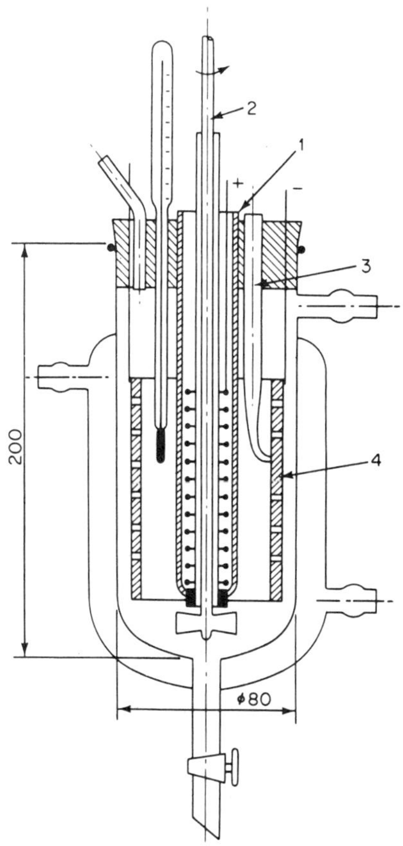

Figure 6. Electrolyzer with diaphragm: 1, diaphragm; 2, mixer; 3, electrolytic bridge; 4, cathode. (Reprinted from Ref. 65 by courtesy of the Academy of Sciences of the USSR.)

in the inner compartment; the cell in Fig. 6 includes a reference electrode and a stopcock for easy withdrawal of the electrolyzed solution.

A conventionally designed cell using the same principles has been reported by McMillan [66]; similar constructions (also without a reference electrode) have been used by Izgaryshev and Petrova [67] and by Condit [68]. An all-glass cell designed especially for oxidations in organic media has been described by Barry and Cauquis [69].

In order to increase mass transfer the solution is generally stirred; alternatively, the working electrode may be rotated [70–75] or vibrated [17]. Rotation works well for laboratory cells but is less suited for industrial cells.

Flow Cells

In principle a batch cell can easily be changed to a cell with flow operation simply by adding an external loop with a pump to recirculate the electrolyte.

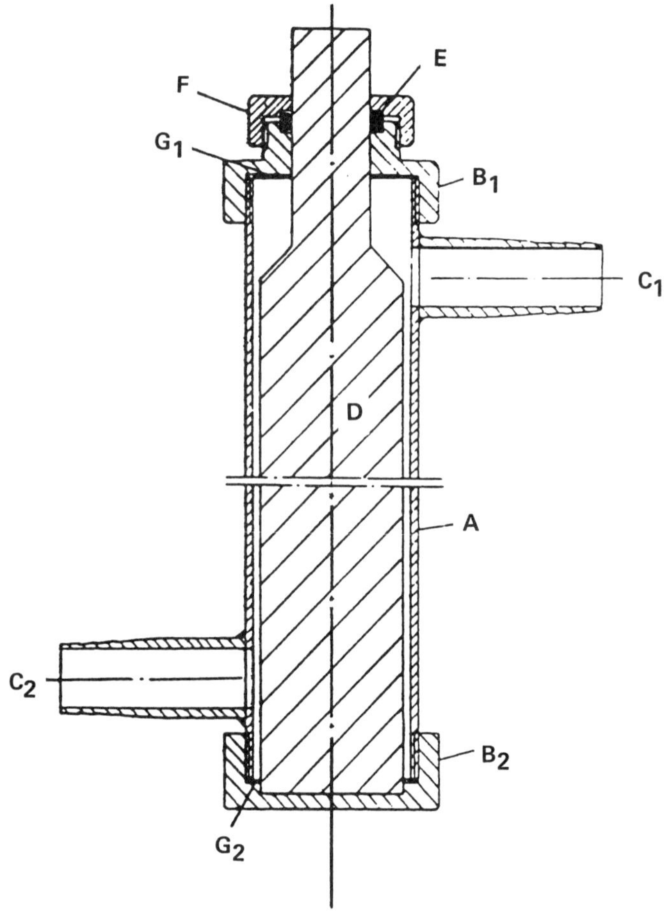

Figure 7. Cross section of concentric capillary gap cell. A, cathode; B_1 and B_2, anode support; C_1, outlet; C_2, inlet; D, anode; E, silicon rubber seal. (From Ref. 76a.)

Such a loop also enables the cell to work with a larger reservoir of electrolyte and substrate. An external heat exchanger can also conveniently be included. However, this does not change the basic concept of the cell; only the batch size may be increased.

A cell for continuous flow operation must be designed with a high electrode surface-electrolyte volume ratio, provided with a feeding system, and last but not least, connected with suitable auxiliary equipment for continuous removal of the product(s) of electrolysis and reestablishment of the electrolyte composition. The continuous workup procedure during electrolysis is somewhat inconvenient in the laboratory, and consequently small continuous flow cells have mostly been operated with recycling to a reservoir before scaling up. Large cells and their industrial applications are discussed in Chap. 31.

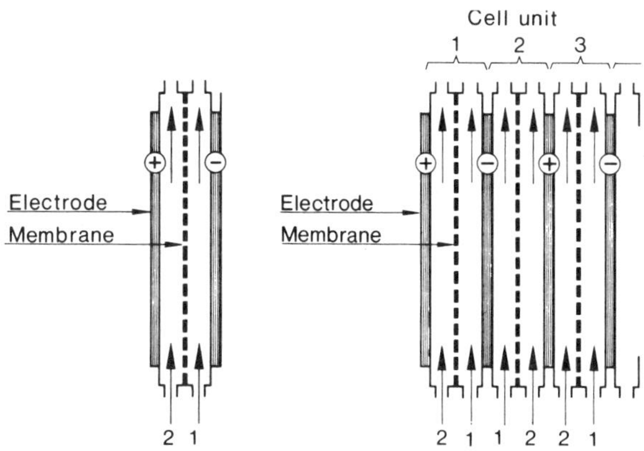

Figure 8. Filter press (plate and frame) cell: 1, catholyte; 2, anolyte.

Undivided flow cells A simple cell for large-scale laboratory electrolysis of organic compounds is shown in Fig. 7, and this concentric capillary gap cell has been used for many anodic reactions [76].

The cathode (A) is a 2 mm thick stainless steel tube 300–500 mm long with inner diameter 44 mm. The anode is a cylindrical graphite rod 40 or 43 mm in diameter; the distance between cathode and anode is thus 0.5 or 2 mm. The solution is pumped in at C_2 and leaves the cell at C_1.

Divided flow cells Filter press (plate and frame) (Fig. 8) cells may be used with an without diaphragm; the type of cell is treated in Chap. 31. Filter press cells for laboratory-scale work are now available commercially. It might be pointed out that when commercializing a process is a possibility, it might save time to switch early to a cell type akin to those usually employed in large-scale operations.

The various configurations for a reference electrode in such a parallel plate cell have been analyzed; a convenient way is to place the reference electrode in the feed line and compensate for ohmic potential drop [77].

Cells with Porous Electrodes

Porous electrodes have certain advantages and disadvantages. An advantage is that a single flow through the electrode may be sufficient to reduce the substrate completely; a disadvantage is that a control of the effective potential is difficult. Cells with porous electrodes may be used with or without a diaphragm [78–81]; in Fig. 9 is shown a cell that may be used with one, two, or three electrical circuits; it has been used for the preparation of nitroso compounds by reduction of nitro compounds to phenylhydroxylamines and oxidation of the latter. To balance the different requirements for the cathodic and anodic reactions one or two extra electrical circuits were employed.

SPE Cells

Solid polymer electrolytes (SPE) may be used in organic electrosynthesis [17,82–85]. The electrode material, usually platinum or gold, may be deposited on one or both sides of a membrane [82], such as Nafion, or a metal

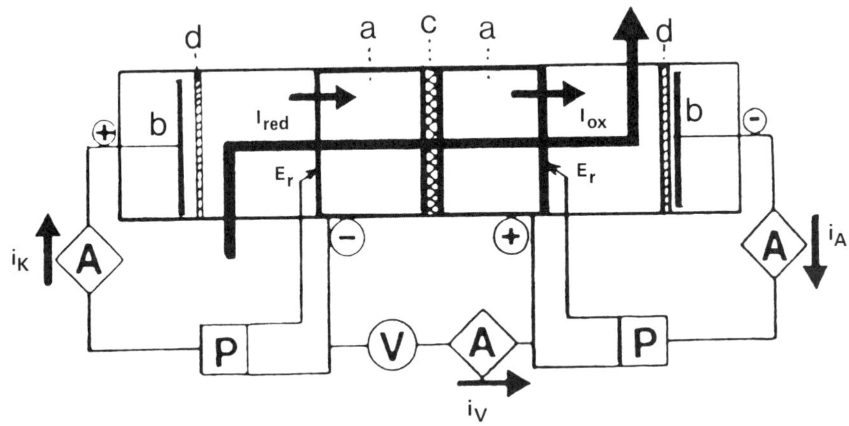

Figure 9. Schematic diagram of the cell and current distribution with one ($i_K = i_A = 0$), two ($i_K \neq 0$ and $i_A = 0$), or three ($i_K \neq 0$ and $i_A \neq 0$) electrical circuits. A, ammeter; P, potentiostat; V, power supply; a, porous working electrode; b, auxiliary counterelectrode; c, porous insulator; d, fritted glass separator; E_R, reference electrode, →, electrode flow circuit. $i_{red} = i_K + i_V$; $i_{ox} = i_A + i_V$. (Reprinted from Ref. 81 by permission of Chapman and Hall.)

grid is pressed against the membrane [83]. The deposited metal may be modified by deposition of other metals to increase the hydrogen overvoltage, for example [84].

An advantage of the use of SPE is that no supporting electrolyte is required, so reduction can be made in other solvents than are usually employed, and one of the electrodes may even be in contact with gas phase rather than with a solvent [85].

C. Cells with Three-Dimensional Electrodes

Cells with three-dimensional electrodes have bipolar electrodes; such electrodes are characterized by the feature that one part of the electrode is anode and another part cathode. This can be realized in different ways, such as the pile capillary gap cell (Beck/Guthke cell, Chap. 31) [17,86], the "Swiss roll cell" [87], and packed- and fluidized-bed cells (Chap. 31) [88–93]. These cells are developed to meet economic demands, such as high space-time yield and simplicity in construction; they are discussed in Chap. 31.

In this chapter are mentioned a few basic problems with regard to bipolar electrodes, with the packed-bed electrode taken as an example (Fig. 10).

A packed-bed reactor consists of a number of electrolytically conducting particles, which are separated from each other so they are not in direct contact. One end of the bed is connected to the anode and the other to the cathode.

It is essential for the operation of such a reactor that the conductivity of the electrolyte solution is not too high. The potential drop in the so-

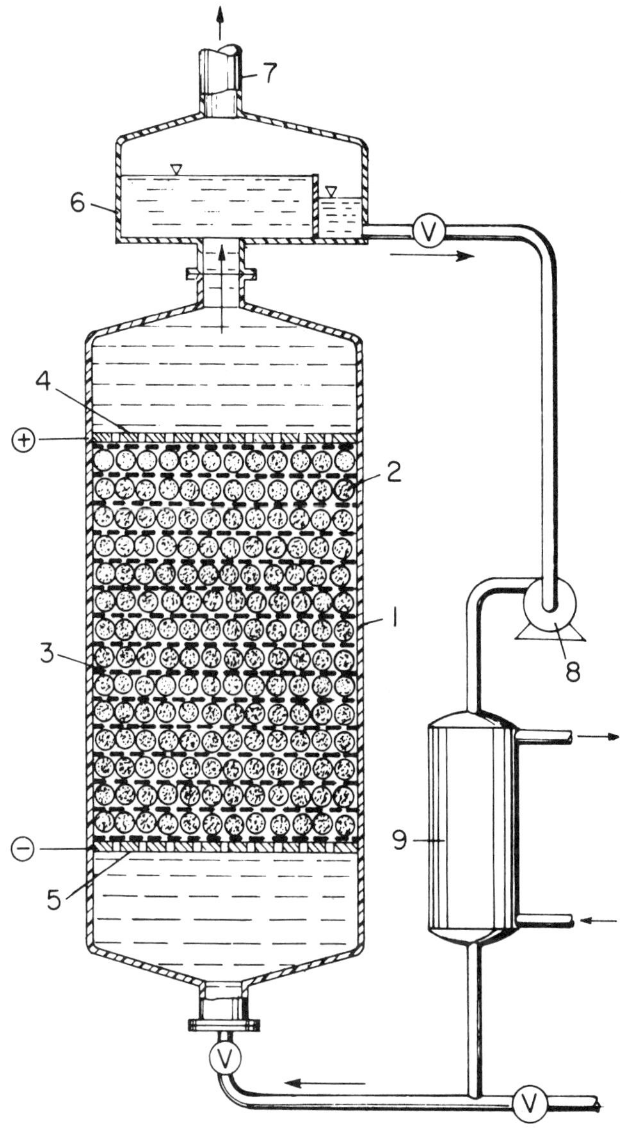

Figure 10. Schematic elevation of a bipolar packed-bed reactor. 1, shell; 2, bipolar particles; 3, separators; 4, anodic end plate; 5, cathodic end plate; 6, head tank; 7, vent; 8, pump; 9, heat exchanger. (From Ref. 88.)

lution through a distance equivalent to the diameter of a particle must be greater than the sum of the anodic and cathodic electrode potentials plus the iR drop within the conducting particle; otherwise the current will flow in the solution and no electrolysis will occur at the particles. In order to obtain such conditions a certain critical current density must be exceeded, and the ratio between faradaic current and creeping current will grow with current density.

D. Special-Purpose Cells

A number of cells described in the literature are not aimed primarily at preparative electrolysis, but as means of generating intermediates to be investigated by some physical-chemical method, generally some sort of spectroscopy; such cells are not discussed here.

For certain purposes electrolysis of thin layers is advantageous; a cell with a well-defined thin layer and a uniform potential distribution has been described [94]. Cells for electrochemistry at very low temperatures have been constructed [95].

Optically transparent electrodes (OTE) have been used for many kinds of spectroelectrochemical investigations; several aspects of this type of research have been covered in reviews [96].

Cells for electroanalysis in supercritical and near-supercritical liquids have been described [97,98].

III. DIAPHRAGM MATERIALS

A. General Considerations

As mentioned earlier, it is often desirable or even necessary to separate the anode chamber from the cathode compartment by some kind of diaphragm. The ideal material should be an electric insulator chemically inert to and preventing the interdiffusion of the anolyte and catholyte components while presenting only a little resistance to the electric current. For practical reasons the material should also possess reasonable mechanical strength and good dimensional stability. No real material can of course fulfill all of these demands simultaneously, and it is generally necessary by trial and error [99] to find a suitable compromise with which the deviations from the ideal can be tolerated for the given process. Fortunately there is a rather wide selection of available refractories, membranes, and woven materials, especially for laboratory cells for which the engineering demands are less strict than for production cells.

The diaphragm materials can be divided into two categories according to the mechanism by which the electric current is carried through the medium: porous materials and ion-exchange membranes [or nonpermselective and permselective materials (Chap. 31)]. With porous materials a more or less tight barrier is offered to the diffusion of both ions and uncharged species, and the electrical resistivity generally increases drastically with decreasing pore size. Thus there is no change in the transport number of the ions as in the case of ion-exchange membranes, which are able to discriminate between ions of opposite charge to a very high degree according to the type of membrane material. Sometimes a controlled seepage through the diaphragm may even be desirable; this problem has been discussed by MacMullin [99].

B. Porous Materials

From this discussion of cell construction it is obvious that a great variety of porous materials has been used in laboratory cells. Most common are sintered glass (especially suitable for all-glass cells) and ceramic refractories, such as Alundum, unglazed porcelain, and pottery. These materials are available in many different sizes, shapes, and porosities, which is convenient for laboratory purposes. A serious drawback is that for greater pieces of material the wall thickness must be considerably increased to obtain a satisfactory mechanical strength; this, however, results in a higher voltage loss in the diaphragm with consequent unproductive heat generation and limits the cell size in practice. The ceramic materials are also less suitable for diaphragms for prolonged operation in alkaline media, as the ceramic matrix that holds the grains together is attacked.

Materials of more or less broad utility, reported in the literature, include the following: filter paper [100], asbestos [70,71,73,101,102], cellophane [103], nylon cloth [74], and porous plastic [89,104]; other potential candidates are listed in Chap. 31. The porosity of the medium may be decreased by partially filling the voids with another material, such as agar-agar, silica gel [16], or magnesium hydroxide.

C. Ion-Exchange Membranes

In most cases the cation-exchange type (H^+ form) of membranes [105,106] has been used in electrolytic cells, but membranes enabling the anions to pass are also available. It must be remembered that if the membrane is not of completely homogeneous polymeric structure (it may be finely dispersed in some binding matrix), channels usually exist through the matrix, making it less permselective. When the product of electrolysis or the starting material is an ion with a charge opposite to that of the working electrode (e.g., reduction of trichloroacetic acid in ammonium buffer [107]), it is especially advantageous to use an ion-exchange membrane as diaphragm to avoid loss of depolarizer or product due to electrical migration out of the working electrode component. In some cases [17,81–85] the ion-exchange membrane serves both as a diaphragm and as an electrolyte.

The polyelectrolyte membranes are often the material of choice for production cells, and hence a more detailed discussion of their properties is naturally postponed to Chap. 31. Several examples of cells employing ion-exchange membranes have been given in the section on cell constructions. Ion-exchange membranes are also used for electrodialysis.

IV. ELECTRODE MATERIALS

A. General Considerations

The electrode material is an important factor in directing the course of an electrode reaction, but the influence of the electrode material is one of the least understood points in organic electrolysis. It is known that hydrogen or oxygen overvoltage, catalytic effect, adsorption of substrate to the electrode, the nature of impurities and added components, the physical state of—and even the history of—the electrode may influence a branching of an electrode process. Efforts have been made to clarify the situation, and although some knowledge has been gained, research in this area is very much needed. Besides the factors already mentioned, the resistance of the electrode toward corrosion must be considered when the electrode material is chosen.

The hydrogen overvoltage of an electrode material is an important factor in reduction in protic solvents, as it determines how negative the potential may be before the reduction of the medium starts to compete seriously with the electrode reaction of the substrate. Metals can be divided into low-, medium-, and high-overvoltage materials. In the first group is platinized platinum; in the second one smooth platinum, nickel, and copper; and in the third group tin, lead, cadmium, and mercury. The values of hydrogen overvoltage of different materials depend on the current density at which they are measured.

Hydrogen overvoltage is caused by a slow step during the electrolytic evolution of hydrogen; the reaction involves the following main steps:

$$H^+_{solv} + e^- \rightleftharpoons H_{ads} \tag{1}$$

$$H_{ads} + H^+_{solv} + e^- \rightleftharpoons H_{2_{ads}} \tag{2}$$

$$H_{2_{ads}} \rightleftharpoons H_{2_{gas}} \tag{3}$$

The first reaction [108] (1) is believed to be the slow step on mercury and other high-overvoltage metals. This means that no appreciable concentration of adsorbed hydrogen atoms is present on mercury and implies that at mercury cathodes the reduction occurs by a transfer of electrons directly to the substrate, not via adsorbed hydrogen atoms. This implication has been tested [109] and confirmed in the reduction of acrylonitrile in the presence of hydrogen ions.

At electrodes of low hydrogen overvoltage, reaction (2) is considered [110–112] the slow step, which implies that the electrode surface is covered to a certain extent with adsorbed atomic hydrogen. This is consistent with the finding that reduction at such electrodes in many cases occurs by the electrocatalytic reaction [113,114], which in many respects is similar to the well-known catalytic hydrogenation reaction.

Generally, electrodes with high hydrogen overvoltage give the most highly reduced product, but there is no firmly established connection, as the catalytic effect of the electrode material may be the factor that determines the product. Phenol, for example, is not reducible at a lead or mercury cathode, whereas it is reduced to cyclohexanol at a platinized platinum electrode in 2 N sulfuric acid [115,116].

Oxygen overvoltage plays a similar role in anodic reactions as hydrogen overvoltage in reductions, but the choice of anode materials with different oxygen overvoltages is very restricted as the overriding consideration is the corrosion resistance of the material. Smooth platinum, gold, lead dioxide, and glassy carbon are among the resistant anode materials with high oxygen overvoltage.

The catalytic effect of the electrode material on the branching of an electrode reaction is most often apparent at cathodes with low overvoltage but may also be observed at materials with high hydrogen overvoltage. In the reduction of acetone in 6 N H_2SO_4 at a platinized platinum electrode two independent paths are accessible: One leads to propane and the other to isopropyl alcohol. The rate of formation of these two products depends on the voltage and on the history of the electrode.

The catalytic effect of high-overvoltage electrode materials may be exemplified by the reduction of ketones in hydrochloric acid at mercury and cadmium or amalgamated zinc electrodes; at mercury, alcohols, pinacols, organic mercury compounds, or sometimes hydrocarbons may be formed (Chap. 10), whereas at cadmium, the reaction is to some degree analogous to the Clemmensen reduction and yields methylene compounds [117].

Some examples of the effect of electrode material on the branching of reactions are the reduction of benzyltriethylammonium nitrate in dimethylformamide (DMF) to bibenzyl or toluene at aluminum or platinum electrodes [118] and the reduction of aliphatic aldehydes [119] or alkyl arylketones [120] at different materials.

Chemical modification of the electrodes (Chap. 33) may change the catalytic effect of electrodes.

Adsorption of the substrate at the electrode plays an only partly understood role, predictably only with difficulty, in organic electrochemistry. An example of the influence of adsorption is found in the Kolbe reaction in an aqueous acetate solution at smooth platinum and gold electrodes. Although both materials have high oxygen overvoltage, the Kolbe reaction takes place only at platinum and at a potential higher than that of the oxygen evolution. Addition of indifferent electrolytes that compete with the acetate ions in the adsorption at the electrode lowers the yield of the Kolbe product. In methanol, in which adsorption is less important than in water, the Kolbe reaction proceeds both at platinum and gold anodes.

The surface may influence the catalytic activity of the electrode and the adsorption of the substrate. Sometimes in the reduction of certain nitro compounds at a tin electrode, for example, the surface changes during the electrolysis as if some of the metal has been dissolved and reduced again [121]. Such a phenomenon is also observed during the reduction of *m*-nitrobenzenesulfonic acid in a fluidized bed electrode [122]. The role of added stannous salts (Chap. 29) to the catholyte in certain reductions has a bearing on this problem.

Porous electrodes may be used to achieve a high space-time yield, as they possess a large internal surface. A consequence of using such electrodes is, however, that interior mass transfer and ohmic resistance effects may lead to a nonuniform potential distribution, which affects the selectivity of the reaction. The cell design and the adjustable parameters must thus be optimized for each reaction [78–81,123].

Impurities in the electrode material may modify its properties both in negative and positive directions; the latter may be exploited by using alloys or additions. The desired modification may be an enhanced mechanical strength, a higher catalytic activity of the electrode material, or a greater resistance toward corrosion. In some cases, however, the amount of side products is increased by the presence of impurities.

B. Cathode Materials

The choice of cathode materials is less restricted by corrosion considerations than the choice of anodes; only a few metals react so readily with the medium that their use is impractical. Some of the most popular electrode materials are mercury, lead, tin, titanium, copper, iron, aluminum, platinum, nickel, and carbon.

Mercury

Mercury has a very high hydrogen overvoltage. It is obtained commercially sufficiently pure for electrochemical purposes. Stirring of a mercury electrode provides a clean, reproducible surface, which is one of the major advantages of mercury. Another is that in electrolysis at a mercury cathode results obtained from polarography can be used for selecting the optimal conditions. On the other hand, the liquid state of mercury places certain restrictions on the design of cells employing mercury cathodes. In some cases amalgamated copper or lead electrodes may be used, the amalgamation being done either by treating the solid electrode with a solution of mercuric chloride or simply by rubbing mercury on its surface.

A mercury cathode mostly functions only as an electron donor; numerous examples of such simple reductions have been reported. In some cases, however, the electrode is attacked with the formation of organic mercury compounds. When using mercury, suitable precautions must be taken to avoid mercury poisoning and environmental pollution.

As anode, mercury has a very limited use as it is attacked at rather low anodic potentials.

Lead

Lead has been much used as a cathode material ever since the beginning of organic electrochemistry; it has a high hydrogen overvoltage and is easy to work mechanically. The purity of the material is important in some, but not in all cases, and different procedures involving successively anodic and cathodic treatments have been recommended [124] in order to obtain a pure, reproducible surface.

In many reductions a lead cathode yields the same kind of product as a mercury electrode, but in some cases it is either inferior or superior to mercury. The latter is found in the reduction in sulfuric acid of aromatic acids to the corresponding alcohols. In such cases specific adsorption of the substrate at the electrode may partly exclude the discharge of hydrogen ions and thus promote transfer of electrons to the substrate, but the nature of this (assumed) adsorption is only partly understood.

The mechanical strength of lead is low, which may be a problem in large cells. Lead plated on titanium may be used.

Tin and Copper

Tin and copper electrodes have been used mostly in reductions of nitro compounds. It has been noticed that the electrode surface of tin may change structure during an electrolysis. This has been interpreted [121] as showing that the electrode material reacts "chemically" with the substrate and the stannous (or stannic) ions are discharged subsequently. The negative potential of the cathode would counteract the dissolution of the tin metal, but no investigations using modern electroanalytical techniques have been made to settle the question.

Cadmium

The hydrogen overvoltage of cadmium is rather large and cadmium may be used, plated on steel, as an alternative to lead. Cadmium has been used as cathode material in the hydrodimerization of acrylonitrile (Chap. 31).

Silver

Silver electrodes have been employed in form of an expanded silver sheet on a plastic plate in the dehalogenation of polyhalogenated heterocyclic compounds; examples are given in Chap. 18.

Aluminum and Iron

Aluminum and iron (steel) have not been used widely in laboratory cells but are preferred in technical installations when possible (Chap. 31).

Carbon

Although carbon has mostly been used as anode material it has certain applications as cathode material. The surface of graphite (and to a lesser degree glassy carbon) contains different oxidized groups, such as carboxyl, hydroxyl, carbonyl, and quinones. In some cases such groups catalyze a specific reaction [125]; they are also essential for a chemical modification of a carbon electrode (Chap. 33).

Composite Electrodes

Such electrodes have mostly been used as anodes, but cathodic redox catalysis is also possible. In Ti/TiO_2 composite cathodes, prepared by the ceramic method, $Ti(OH)_4$ is reduced to $Ti(OH)_3$, which "chemically" reduces the substrate, such as aromatic nitro compounds [126], in an "inner sphere" electron transfer reaction.

Electrodes for Electrolytic Hydrogenation

Nickel (Raney nickel) [114,127–129], Devarda copper [130], and platinum metals have a low hydrogen overvoltage and may be used for electrocatalytic hydrogenation (Chap. 29). Regeneration of poisoned Raney nickel electrodes has been described [131].

Sacrificial Cathodes

Phosphorous [23,132], sulfur [133–135], selenium [136], and tellurium [137] electrodes may be used for incorporation of these elements in organic compounds. Sulfur and selenium are used in a mixture with graphite, whereas white phosphorous may be used melted or in contact with carbon or mercury; tellurium has sufficient conductivity to be used directly or applied on a platinum electrode.

A somewhat similar approach has been used by mixing a finely dispersed, slightly soluble organic solid into a paste with graphite powder; an aqueous electrolyte was employed. Thus, p-dinitrobenzene has been reduced to p-nitrophenylhydroxylamine by this method [138].

C. Anode Materials

The choice of anode material is very limited because most metals are attacked in the anodic reaction. In laboratory experiments platinum, gold, and carbon are most widely used; lead containing a small percentage of silver and lead dioxide can be employed in dilute sulfuric acid and iron and nickel in alkaline solution.

Platinum

Platinum is used both as microelectrode and in macroscale electrolysis; for larger electrodes platinum plated on titanium may be useful. Platinum is in aqueous solvents probably covered with an oxide at potentials more positive than about 0.75 V (SCE) [139,140], so the electrode material may actually be a platinum oxide.

In some anodic reactions, especially in aprotic media, the products obtained form a layer of tarry material on the electrode surface that (by covering it) insulates it electrically from the solution. The removal of this layer and the attainment and maintenance of a clean, reproducible surface is often a major problem in the use of noble metal anodes. Heating the platinum in a flame, treatment with nitric acid, or mechanical removal of the insulating layer have all been recommended. Sometimes it is possible to avoid the fouling of the electrode by using pulse electrolysis.

Carbon

Graphite electrodes are very useful as anodes; electrodes of natural graphite are, however, not as resistant toward corrosion as platinum electrodes and thus have a limited lifetime. Graphite is available in many forms, one resembling cloth, which is convenient for laboratory purposes, another one being glassy carbon. Glassy carbon (vitreous carbon) can be used in many cases in which platinum previously was used. The properties of glassy carbon depend above all on the temperature at which it is made [141,142]. Glassy carbon is difficult to work mechanically.

Oxidation of organic substrates at a graphite anode often results in other products than oxidation at a platinum electrode; the tendency to further oxidation of a primarily produced radical to a carbocation seems higher at carbon than at platinum. Products obtained at graphite, glassy carbon, or other forms of carbon may differ considerably [141–143].

In voltammetry it is important to secure the reproducibility of solid electrodes. In the carbon paste electrode the problem of contamination of the surface is solved by producing a fresh surface for each run.

Lead and Lead Dioxide

Pure or alloyed lead may be employed as anode in sulfuric acid; addition of 1% silver, 0.3% tin, and a little cobalt raises its resistance toward corrosion. Other metals may improve the yield of a given electrode process. Thus, addition of antimony and cadmium to a lead anode [144] is advantageous in the oxidation of o-toluenesulfonamide to o-benzoylsulfonimide (saccharin). The same effect may be obtained, however, by using an uncoated, unalloyed lead anode if Sb_2O_3 is added to the anolyte [145]. It seems possible that the dissolution of some antimony from the alloyed lead anode takes place and produces the same effect as the Sb_2O_3 in the anolyte.

In many cases the surface of the anode is oxidized to lead oxide, which thus is the real anode material. Titanium anodes coated with lead oxide have been found to have a long lifetime [146,147]. In some cases the lead dioxide, a strong oxidant in acid solution, reacts chemically with the substrate [148] and is continuously regenerated whereas it acts as an inert electrode in other oxidations. A review of the basic electrochemistry of PbO_2, mainly in relation to the lead battery, has been published [149].

Nickel

A nickel anode is in alkaline solution protected against corrosion by a layer of nickel oxides. Ni^{+3}-oxide (NiOOH) is capable of oxidizing a number of functional groups; primary alcohols may be oxidized to carboxylic acids [150–153], which is of interest for the technical production of an intermediate for vitamin C production [154]. NiOOH chemically oxidizes the substrate and is regenerated electrochemically; a large anode surface, which is realized in the Swiss-roll cell (Chap. 31), is thus advantageous.

Dimensionally Stable Anodes (DSA) and Composite Anodes

These materials [155–157], consisting of precious metal (e.g., ruthenium) oxides coated on baser metals (e.g., titanium), have already had a major impact on the chloralkali industry as replacements for graphite. There is considerable, as yet unpublished work on their utilization in organic electrochemistry. The prospects are good that a number of organic electrosyntheses that were not considered industrially viable because they required platinum will now become candidates for adoption.

Titanium/chromium oxide ($Ti/Cr_2O_3 + TiO_2$) composite electrodes, fabricated by a ceramic method [158], may be used for oxidation of organic compounds. The chromium is oxidized to CrO_3, which chemically oxidizes the substrate. The lifetime of the electrode may be improved by doping with Sb_2O_3.

Sacrificial Anodes

Anodes of aluminum [159], magnesium [160], and zinc [161] have been used as sacrificial anodes mostly in nonaqueous solvents. One advantage of using such anodes is that they can be employed in an undivided cell; the oxidation products, the metal ions, are (with the exception of Zn^{2+}) not easily reduced at the cathode. Another advantage, and often a deciding one in coupling reactions, is that the metal ions may coordinate with the anion formed at the cathode and thus "protect" it against unwanted side reactions.

D. Semiconductors

Semiconductor electrodes have recently mostly been studied in connection with photoelectrochemical cells. This aspect is dicussed in Chap. 34.

E. Optically Transparent Electrodes

Optically transparent electrodes (OTE) have been discussed thoroughly in reviews, to which the reader is referred [96].

F. Electrodes with Chemically Modified Surfaces

Recently considerable interest has been shown for electrodes on which the surface has been modified chemically (CME, chemically modified electrodes). The emphasis in the investigations has mainly been on studying the preparation, properties, and mechanism of the electron transfer, but gradually the emphasis in the field is shifting to applications. The preparation and some application of such electrodes are discussed in Chap. 33; the use of CME for induction of chirality is treated in Chap. 27.

Electrodes may also be modified by deposition of metal adatoms at potentials several hundred millivolts positive to the reversible potential for metal deposition. A submonolayer of adatoms may lower overpotentials for electron transfer processes or improve the selectivity of an electrocatalytic surface. Underpotential deposition and electrocatalysis have recently been discussed in a review [162].

V. REFERENCE ELECTRODES

A. General Considerations

A reference electrode [163,164] is a half-cell that defines a potential to which all other measurements are referred. The primary standard electrode is the standard hyrogen electrode (SHE), but as this electrode is inconvenient for practical work, other reference electrodes are used. Such reference electrodes must have a potential that changes very little and is known to within 1 mV or so. In some cases of controlled potential electrolysis it is sufficient to know the potential of the working electrode during the electrolysis within 10–20 mV, because the potential variations between different points of the electrode are of this magnitude (see earlier), and less precise electrodes may be termed comparison electrodes. In principle any electrode at the surface of which an electrochemical reaction with a large exchange current can take place may be used as a reference electrode.

Several problems are encountered when potentials in different solvents are sought compared to the potential scale in water. A variety of approaches [165–174] have been followed to attack this problem; usually the approach has been to introduce some kind of nonthermodynamic assumption, such as the supposition that certain large, monovalent ions (Rb^+, Cs^+) [165,171] or redox systems [166–168] of the charge type n/n + 1 (preferably 0/+1, such as ferrocene/ferricinium [166,167]), have a nearly equal free energy of solvation in the two solvents so that the free energy of a transfer of the reference ion is small. Recently [174] the redox couples ferrocinium/ferrocene and bis(biphenyl)chromium(I)/bis(biphenyl)chromium(O) (BCr^+/BCr) have been recommended as reference redox systems for measurements in nonaqueous solvents; however, an investigation concluded [175] that the electrochemistry of ferrocene in MeCN at microelectrodes was far from ideal, as some film formation may occur.

Another attack on these problems has been based on the midpoint between the reversible oxidation and reduction potentials of aromatic hydrocarbons [176]. Although such problems are important for certain aspects of electrochemistry, they are of minor interest for practical electrolysis, as long as the liquid junction potential between the reference electrode and the test solution can be kept constant and reproducible within acceptable limits. If it is suspected that the liquid junction potential may interfere with the measurements, internal reference systems may be added.

The ideal reference electrode is indefinitely stable, unpolarizable, easy to handle, temperature independent, has a low resistance, and does not cause contamination of the test solution.

A stable potential of a reference electrode is very important. Instability may be caused by several factors. One of the components may be unstable in the solvent. Thus, mercurous chloride is unstable in acetonitrile and a calomel electrode is unreliable in this solvent. Evaporation of solvent from a standard solution, the concentration of which determines the potential, changes the potential. A liquid junction potential may change; this is observed when an aqueous calomel electrode, connected to the test solution through a potassium chloride-agar bridge, is used for measurements of potentials in acetonitrile for periods longer than about 5 min (see later).

An unpolarizable electrode is an electrode the potential of which does not change significantly when a small current passes through it. Most of the potential-measuring devices used draw a small current from the reference electrode, depending on the resistance of the circuit. Polarization is diminished by employing a reference electrode with an electrode reaction with a high exchange current, with a large area, and with a sufficiently high concentration of the potential-determining species, so its concentration is not changed appreciably by the passage of the current.

In some reference electrodes the concentration of the potential-determining species is kept constant by maintaining a saturated solution of a suitable compound. The solubility of a compound is generally temperature dependent, and the potential dependence on temperature of such electrodes is thus higher than for other kinds of reference electrodes.

The contact between the reference half-cell and the test solution requires some cnsideration. The ohmic resistance of the electric contact should preferably be kept low and contamination of the test solution by the reference electrode and vice versa must be avoided. Usually some kind of salt bridge with or without a diaphragm is employed. The liquid junction potential is lowered if a salt, such as KCl, with nearly equal mobilities of the cation and anion is used. The diffusion between the two half-cells is discouraged by introducing some kind of diaphragm, such as a salt-saturated agar gel, a fritted glass plate, asbestos fibers, a fritted glass membrane, cracked soft glass [177], or a Teflon roll junction [178]. A considerable ohmic resistance may be introduced simultaneously, which puts certain requirements on the internal resistance of the potential-measuring device. In certain cases it has been necessary [179] to employ an electrometer, but generally a pH meter-type instrument can be used.

The reference electrodes have been divided according to the electrode reaction responsible for maintaining the constant potential: In electrodes of the first kind, the potential of the metal electrode is determined by the concentration of the metal ions; in electrodes of the second kind, the potential is determined by the concentration of an anion that forms an insoluble salt with the metal cation; and in redox electrodes, the potential of an indifferent electrode is determined by the relative concentrations of the two components of a redox system.

B. Electrodes of the First Kind

Metal electrodes of this kind require that neither the metal nor the metal cation reacts chemically with the solvent; reactive metals may, however, be used as their amalgams. Electrodes of the first kind require a standardized solution in the half-cell, which makes such electrodes less convenient to employ than most electrodes of the second kind.

The silver/silver ion electrode is a widely used reference electrode [180] in nonaqueous solvents, such as acetonitrile [181]. The electrode consists of a silver wire immersed in a solution of $AgClO_4$ or $AgNO_3$. It is connected to the test solution through a salt bridge containing $NaClO_4$ or a tetraalkylammonium perchlorate and equipped with some kind of device to diminish diffusion.

The Ag/Ag^+ electrode is reversible [182] in many solvents. The main advantage of this and similar reference electrodes is that it allows working in a system containing a single solvent. Silver ions react with certain organic solvents, such as DMF, and other metal electrodes must then be used; however, a silver cryptate electrode [Ag/0.005 M Ag Cryp(2.2)ClO_4] is stable in DMF [183].

Silver perchlorate is hygroscopic and has a tendency to explode. Handling $AgClO_4$ may be avoided by its electrolytic preparation in the reference electrode [184]: current is passed through a silver anode in an anolyte containing a soluble perchlorate until the desired silver ion concentration is reached. Silver nitrate may be used instead of silver perchlorate, although it is not as fully dissociated in acetonitrile [185]. The influence of the ion pairing on the potential has been discussed [186].

A zinc amalgam electrode has been employed as reference electrode in dimethyl sulfoxide (DMSO) [187] and ammonia [188]. The saturated zinc amalgam was prepared [189] by electrolytic deposition of zinc into a layer of pure mercury from an aqueous $ZnSO_4$ solution. The solution in equilibrium with the Zn(Hg) consisted of a saturated solution of $Zn(ClO_4)_2$, 4DMSO or $ZnCl_2$, $6NH_3$.

The use of a sodium amalgam electrode [Na(Hg)/$NaClO_4$(s)] in DMF has ben reported [187] and a similar lithium amalgam electrode has been employed in DMSO [189]. The potential of the cell, Li(Hg)/Li^+Cl^- (DMSO), has been measured for LiCl concentrations from 0.01 to 1.0 M and the Nernst relation was verified within 1 mV; the Li(Hg) electrode obeys the Tafel equation with the transfer coefficient $\alpha = 0.5$ over a current density range from 10^{-7} to 2×10^{-3} A/cm^2 [189].

A special reference electrode is the electron electrode, which can operate in solvents capable of solvating electrons, such as ammonia. A sheet of platinum suspended in a solution of sodium metal in ammonia may be used as a reference electrode [190]. The concentration of the reference solution can be varied over relatively wide limits without a change in the potential of the electrode. A sodium concentration of 0.001 M was foumd suitable [190].

C. Electrodes of the Second Kind

An electrode of the second kind may be used if the metal cation forms an insoluble salt with an anion, but from the point of establishing a satisfactory reference electrode of this kind the most interesting constant is not the solubility constant, but the equilibrium constant for the reaction

$$MeX_{(s)} + X^- \rightleftharpoons MeX_2^-$$

Calomel Electrode

The saturated calomel electrode (SCE) is probably the most widely used reference electrode. It consists of a mercury electrode in contact with a solution saturated with respect to both Hg_2Cl_2 and KCl. The half-reaction is

$$Hg_2Cl_2 + 2e^- \rightleftharpoons 2Hg + 2Cl^-$$

The activity of the potential-determining mercurous ion is maintained constant through the solubility product Hg_2Cl_2 by the constant activity of Cl^- in the saturated KCl solution, which is in equilibrium with solid KCl. The passage of current thus just changes the quantities of the solid salts, Hg_2Cl_2 and KCl, but not the composition of the soluton. The solubilities of Hg_2Cl_2 and KCl change with temperature, and the potential of the aqueous SCE versus NHE is $E = +0.242 - 7.6 \times 10^{-4} \times (t^o - 25)$.

The "normal calomel electrode" (NCE), which employs a 1 N KCl solution instead of a saturated solution, has a smaller temperature coefficient [$E = +0.280 - 2.4 \times 10^{-4} \times (t^o - 25)$ versus NHE], but because it requires a standardized solution it is less convenient than the SCE. The potential change of SCE with temperature is less than 1 mV/degree and is of no importance for controlled potential electrolysis.

Many designs of calomel electrodes exist, and several types are commercially available. Some of these are constructed for voltammetric work, whereas others are used in connection with glass electrodes. The latter type generally has a high resistance, which must be considered when the choice of voltmeter or potentiostat is made.

The use of SCE presents only a few problems in aqueous solution. For measurements in nonaqueous solvents it can be employed in two fashions: the aqueous SCE can be connected to the nonaqueous solvent through some device that hinders contamination of the solvent, or an electrode analogous to the aqueous SCE may be constructed in the nonaqueous solvent.

In many nonaqueous solvents a calomel elctrode is inapplicable. In acetonitrile, mercurous chloride, although insoluble, undergoes a disproportionation reaction in the presence of excess chloride ion according to

$$Hg_2Cl_2 + Cl^- \longrightarrow Hg^o + HgCl_3^-$$

It is often possible to employ an aqueous SCE that is connected through a conventional aqueous KCl-agar bridge to a suitable salt solution in the nonaqueous solvent, which in turn dips into the solution under investigation. In such cases gradual deposition of KCl and/or dehydrated agar at the tip of the salt bridge may cause the liquid junction potential to change and thus the potential of the electrode to drift. The drift is often negligible when the measurement is completed within about 5 min [191].

If chloride ions must be avoided, a <u>mercury/mercurous sulfate electrode</u> [$Hg/Hg_2SO_4(s)$, $K_2SO_4(s)$; E = 0.621 V versus NHE] may be employed. In alkaline solution a <u>mercury/mercuric oxide electrode</u> (E = 0.098 versus NHE) may be useful.

The saturated <u>silver/silver chloride electrode</u> [Ag/AgCl(s), KCl(s); E = 0.197 V versus NHE, E = −0.045 V versus SCE] is especially convenient as it can be constructed by placing a silver wire in a solution of saturated potassium chloride, separated from the test solution by an agar plug and a glass filter (e.g., Fig. 3). An analogous electrode may be employed in certain nonaqueous solvents, e.g., nitromethane [184], acetic acid [192], and DMF [193], but it does not work well in others, such as ammonia or acetonitrile [194], in which complexes are formed that cause the potential to drift.

The <u>copper fluoride electrode</u> has been found to be a convenient reference electrode in hydrogen fluoride [195].

The <u>thallium amalgam/thallium chloride electrode</u> [Tl(Hg)/TlCl(s), LiCl] is a stable and reversible reference electrode in such solvents as DMSO and

DMF [193,196]. Similarly, the lead amalgam/lead chloride electrode [193, 197] may be employed, but this electrode takes a longer time to reach a stable potential than the Tl(Hg)/TlCl electrode.

In dimethylformamide, a cadmium amalgam electrode [193,198] gives a reproducible potential that is stable and well poised. The cadmium amalgam was prepared [198] by heating cadmium with pure mercury for several minutes and adding, after cooling, sufficient mercury to make a slurry, which was filtered through a pinhole in filter paper. The half-cell was : Cd(Hg)/ $CdCl_2$(s), $CdCl_2-H_2O$(s), NaCl(s), DMF. The variation of cell potential with water content has been examined.

D. Redox Electrodes

In a redox electrode both components of the redox system are in solution, and their relative concentrations determine the potential of an inert electrode immersed in the system. Such electrodes have been used as reference electrodes only in special cases, because the potential is dependent on two concentrations, the ratio of which must be kept constant. The primary standard reference electrode, the hydrogen electrode, has been used as practical reference electrode in certain cases, as in HF [199,200].

Fc^+/Fc and BCr^+/BCr

Ferricinum ion/ferrocene (Fc^+/Fc) and bis(biphenyl)chromium(I)/bis(biphenyl)chromium(O) (BCr^+/BCr) have been proposed [174] as reference redox systems for nonaqueous electrochemistry. Both materials have been found suitable for practical referencing, for example, in THF [201], covering complementary potential ranges.

Chloranil Electrode

The tetrachloroquinone (chloranil)/tetrachlorohydroquinone system has been used as reference electrode in alcohols and acetic acid [202], often in the presence of an excess mineral acid. The half-cell consisted of a platinum electrode in a solution saturated with respect to both chloranil and the hydroquinone. The electrode is well poised, and at constant temperature and constant hydrogen ion activity the potential is stable.

Triiodide/Iodide Electrode

The triiodide/iodide electrode ($Pt/I_3^-/I^-$) has been recommended as reference electrode in MeCN and propylene carbonate [203]. It was found to have a higher exchange current than the Ag^+/Ag electrode; the exchange current was highest at Pt, lower at glassy carbon, and lowest at stainless steel. A Teflon salt bridge [178] was recommended.

Poly(vinylferrocene) Electrode (PVF)

A platinum electrode coated with a film of poly(vinylferrocene), which was electrodeposited in its oxidized form and later partially reduced, was found to be a stable reference electrode in MeCN (but not in DMF) [204]. Polymer films containing other redox systems may also be used.

Table 1 summarizes the suitability of the various reference electrodes in different solvents.

VI. SOLVENTS FOR ELECTROLYSIS

A. General Considerations

Electrolysis can occur only in a medium that conducts the electric current, and the nature of the solvent is important for the course of electrolytic reactions. Such factors as proton activity, usable potential range, dielectric constant, ability to dissolve electrolytes and substrates, ion pair formation, accessible temperature range, vapor pressure, viscosity, toxicity, and price must be taken into consideration when the choice of solvent is made.

Proton Activity

The availability of protons is one of the most important properties of a medium by which it can influence an electrolytic reaction, especially a reduction. Anodic reactions are generally less susceptible to the proton activity, but both the stability of cation radicals and the reactivity of some nucleophiles in anodic substitution reactions depend on proton activity.

At high proton activity a protonation of the substrate may occur prior to the charge-transfer step. A protonated molecule is reduced at a less negative potential than an unprotonated one. At medium proton activity the unprotonated molecule is reduced to a radical anion; this is rapidly protonated to a radical which generally is more easily reduced than the unreduced substrate. In aprotic media the radical anion formed in the charge transfer step may be so long-lived that it can be regarded as the product. The radical anion is normally more difficut to reduce than the unreduced substrate.

A high proton activity generally makes an oxidation occur at more positive potentials; it also diminishes the tendency to lose protons from cation radicals, which thus may have reasonable stability under strongly acidic conditions.

In aqueous solution a change in pH (1) may alter the sequence of reducibility of different "electrophores"; (2) the degree of reaction of an "electrophore" may be different in acid and alkaline solutions; (3) a branching of a reaction may be influenced by the availability of protons; and (4) the stereochemistry of the product may be pH dependent. As examples can be given: (1) the reduction of isonicotinic amide in which the carboxamide group is reduced in acid solution whereas the pyridine ring suffers reduction at high pH (Chap. 18); (2) the reduction of aromatic nitro compounds, which at low pH at a suitable potential yields an amine and, in alkaline solution, a phenylhydroxylamine (and its condensation products; Chap. 9); (3) the reduction of acrylonitrile may yield the two-electron reduction product, propionitrile, or the one-electron reaction product, adiponitrile, depending on the availability of protons (Chap. 22); (4) the stereochemistry of the product mixture of pinacols formed on electrolysis of aromatic ketones is pH dependent (Chaps. 10 and 27).

When considering an aprotic solvent, it must be recalled that it is difficult to obtain a strictly anhydrous medium. Especially in voltammetric experiments in which the concentration of the substrate is low, the residual water nearly always present in a "dry" solution may influence the results obtained. It must also be considered that the anodic limiting reaction of some anions, such as perchlorates [234,236], results in the formation of water. Protons may also be supplied by abstraction by an electrogenerated base from the nonaqueous solvent or from tetraalkylammonium ions, in the latter case by a Hofmann elimination. A convenient way to remove the

Table 1. Reference Electrodes in Different Solvents[a,b]

	SCE (aqueous)	Hg/Hg_2X_2	Ag/AgCl	Ag/Ag^+	$Li(Hg)/Li^+$	$Na(Hg)/Na^+$	$Tl(Hg)/Tl^+$
CH_3CN	+[205,206][c]	-	(-)[194]	+[165,206]			
DMF	+	(+)	(+)[193]			+[187]	+[193]
NH_3	-	(+)[205]	-	-			
$(H_2NCH_2)_2$	+	(+)[227]					
C_5H_5N		+		+[209,210]			
$((CH_3)_2N)_3PO$			-	+			
DMSO	+	-	(+)[193,211]		+[190]		+[193,196]
CH_3OH	+	+	+	+			
CH_3COOH		+[212]	+[192]				
$(CH_3CO)_2O$		+[214]		+[215]			
H_2SO_4		+[216]					
FSO_3H		+[219]		+[217][e]			
PC	+			+[220]			(+)[193]
CH_3NO_2			+[184]	+[184]			
THF				+[222,223]			
$(CH_3OCH_2)_2$				+[224,225]			
CH_2Cl_2	+						
THAB	+[226]						
HF		+[232,233]					

[a]Symbols: +, stable reference electrode; (+), usable comparison electrode; (-), dubious comparison electrode; -, unsuitable electrode.
[b]DMF = dimethylformamide; DMSO = dimethyl sulfoxide; PC = propylene carbonate, THF = tetrahydrofuran; THAB = tetrahexylammonium benzoate.
[c]Reference numbers.
[d]Poly(vinylferrocene) electrode.
[e]Au/Au^{3+}

major part of acidic impurities from "aprotic" solvents is to add active anhydrous alumina to the solution [237] or let the solution pass through a column of anhydrous alumina.

A determination of water in a "dry" solution is thus often of interest. The Karl Fischer method is unrealiable below about 20 ppm (1 mM). A gas chromatographic method has been employed [238] for determination of water in certain cyclic esters, but injection of electrolyte solutions causes accumulation of salts in the injection chamber and sometimes erratic results due to thermal decomposition of electrolytes. Analytical use of the 1.9 μm water band in the near-infrared [239] may be practical for certain solvents, such as propylene carbonate, but not for many of the commonly used solvents. A method has been described [240] in which the water reacts with lead tetraacetate in benzene; the lead dioxide thus formed is determined spectrophotometrically at about 400 nm. As little as 2.5 ppm in a 2 ml sample can be measured. The method was found to be applicable to a wide range of solvents and to certain solutions commonly employed in electrochemistry.

The two one-electron reduction waves of 2,5-diphenyl-1,3,4-oxadiazole have been used as a criterion for estimating the degree of drying of aprotic solvents [241]. When $C_{H_2O} < 3-5 \times 10^{-3}$ M the two waves had equal height and $E_{1/2}$ -1.96 and -2.33 V (SCE), respectively, whereas in insufficiently

$Zn(Hg)/Zn^{2+}$	$Cd(Hg)/Cd^{2+}$	$Pb(Hg)/Pb^{2+}$	Cu/CuF	$H_2(Pt)/H^+$	Chloranil	I_3^-/I^-	Electrons (solvated)	PVF[d]
				-[228]		+[203]		+[204]
	+[193,198]	(+)[193]						-[204]
+[207]		+ [197]					+[193]	
+[208]								
				-[229]				
+[187]				-[230]				
				+[213]	+[202]			
				+[218]				
	(-)[193]	(-)[193]		+[221]				
				+[231]				
			+[195]	+[199,200]				

purified solvents, the height of the two one-electron waves were not equal and $\Delta E_{1/2}$ decreased.

An extremely sensitive method using tritiated water as tracer for the determination of solvent water content has been developed [242] in which drying efficiency is determined by the addition of a specified amount of tritium-labeled water to a rigorously dried solvent and subsequent determination of the decrease in activity of the solvent after treatment with the drying agent.

Besides the ability to donate protons, the properties of the medium (solvent, supporting electrolyte) as hydrogen atom donor is of importance. Hence the strength of the available bonds to hydrogen must be considered.

Usable Potential Range

The useful potential range that can be obtained in a given system depends on the electrode material, the supporting electrolyte, the temperature [243], and the solvent; only the solvent is discussed here. Approximate values for the anodic and cathodic limits of some commonly used solvents are given in Table 2. The numbers can be taken only as a crude indication of the usable potential range, as the latter depends on the criteria used.

Many solvents are available that are so difficult to reduce that it is the discharge of the supporting electrolyte that sets the cathodic limit for the

Table 2. Dielectric Constants and Accessible Potential Range for Some Solvents [a,b]

	Dielectric constant	Working electrode	Reference electrode	Supporting electrolyte	Anodic limit (V)	Working electrode	Reference electrode	Supporting electrolyte	Cathodic limit (V)
Water	80	Pt	SCE	$HClO_4$	1.5	Hg	SCE	TBAP	-2.7
Methanol	33					Hg	Hg pool	TEAB	-2.2 [423][c]
H_2SO_4 (96-99%)	>84					Hg	Hg pool	None	-0.7 [216]
CH_3COOH	6.2	Pt	SCE	NaOAc	2.0 [291]	Hg	SCE	TEAP	-1.7 [424]
CH_3CN	37.5	Pt	Ag/Ag^+	$LiClO_4$	2.4 [327]	Pt	Ag/Ag^+	$LiClO_4$	-3.5 [327]
DMF	36.7	Pt	Hg pool	$LiClO_4$	1.5 [336]	Hg	Hg pool	TEAP	-3.5 [336]
NMP	32	Pt	Hg pool	$LiClO_4$	1.4 [336]	Hg	Hg pool	TEAP	-3.3 [336]
HMPA	30	Pt	Ag/Ag^+	$LiClO_4$	0.8 [327]	Hg	Ag/Ag^+	$LiClO_4$	-3.6 [343]
NH_3	23					Hg, Pt	Hg pool	TBAI	-2.3 [205]
$(H_2NCH_2)_2$	12.5	C	SCE	TEAP	0.1 [292]	Hg	SCE	TEAP	-2.65 [292]
Pyridine	13	Graphite	Ag/Ag^+	$LiClO_4$	1.4 [349]	Hg	Hg pool	$LiClO_4$	-1.7 [349]
DMSO	46.7	Pt	SCE	$NaClO_4$	0.7 [347]	Hg	SCE	TEAP	-2.8 [347]
Sulfolane	44	Pt	SCE	$NaClO_4$	2.3 [357]	Pt	SCE	$NaClO_4$	-4 [357]
PC	69	Pt	SCE	TEAP	1.7 [370]	Hg	SCE	TEAP	-2.5 [370]
CH_3NO_2	36.7	Pt	SCE	$LiClO_4$	2.7 [184]	Hg	SCE	$LiClO_4$	-1.2 [184]
THF	7.4	Pt	Ag/Ag^+	$LiClO_4$	1.8 [223]	Pt	Ag/Ag^+	$LiClO_4$	-3.6 [223]
$(CH_3OCH_2)_2$		Hg	SCE	TBAP	0.65 [224]	Hg	SCE	TBAP	-2.95 [224]
CH_2Cl_2	8.9	Pt	SCE	TBAP	1.8 [262]	Pt	SCE	TBAP	-1.7 [262]

[a]The limit for the accessible potential for a solvent depends on many factors, such as supporting electrolyte, electrode material, and magnitude of permissible current density, and the numbers given are cited as illustrative values only.
[b]DMF = dimethylformamide; NMP = N-methylpyrrolidone; HMPA = hexamethylphosphotriamide; DMSO = dimethyl sulfoxide; sulfolane = tetramethylene sulfone; PC = propylene carbonate; THF = tetrahydrofuran; TBAP, TBAI = tetrabutylammonium perchlorate or iodide, respectively; TEAP, TEAB = tetraethylammonium perchlorate or bromide, respectively.
[c]Reference numbers.

the potential; on the other hand, only few polar solvents are resistant to anodic oxidation. The reason is that most polar groups have loosely bound nonbonding electrons on oxygen or nitrogen that are easily lost at the anode. The solvents useful for anodic reactions at high potentials contain electron-attracting groups, such as nitrile, carboxy, sulfone, or nitro groups. For more easily oxidized substrates sulfoxides or dialkylamides may be used, whereas water, alcohols, and aliphatic amines have a very limited useful potential range for anodic reactions.

Dielectric Constant

The dielectric constant is of interest primarily because it influences the ohmic resistance of the medium. Solvents with high dielectric constant are preferable because salts dissociate better in such media. The dielectric constants for some solvents commonly used for electrolysis are listed in Table 2. These solvents can be divided roughly into three main groups: those with high dielectric constant ($D > 60$), those with medium dielectric constant ($20 < D < 50$), and those with low dielectric constant ($D < 13$). In the first group are solvents like water, formamide, N-methylamides, and propylene carbonate; the second one comprises compounds like acetonitrile, dimethylformamide, dimethyl sulfoxide, methanol, nitromethane, and ammonia; whereas in the last group belong solvents like acetic acid, ethylenediamine, methylamine, tetrahydrofuran, dioxane, and methylene chloride. The solvents in the last group require a higher concentration of supporting electrolyte in order to acquire a reasonable conductivity.

For acetonitrile, propylene carbonate, 4-butyrolactone, DMSO, DMF, and similar amides, the maximum conductance for fully ionized salts occurs at approximately 1 M concentration and is about $10^{-2}\ \Omega^{-1}\ cm^{-1}$, comparable to an aqueous salt solution of 0.5–1 M.

The rate constant for the heterogenous electron transfer to aromatic compounds is usually somewhat higher in MeCN than in DMF, which indicates that not only the dielectric but also the dynamic properties of the solvents influence the rate of radical anion formation [244].

Dissolving Power

The choice of solvent is often a result of a series of compromises. Very few solvents have good dissolving power for both organic substrates and inorganic salts. The good electrochemical properties of water are mared by the lack of its ability to dissolve many organic systems. This ability may be raised by using water in a mixture with an organic solvent like ethanol, acetonitrile, dimethylformamide, or dioxane, or by using a "hydrotropic" supporting electrolyte [245] like a tetraalkylammonium p-toluenesulfonate. Tetraalkylammonium salts are soluble in most polar solvents and even in less polar liquids, such as chloroform and methylene chloride. Some polar solvents, such as acetonitrile and DMF, combine good dissolving power for both organic compounds and a variety of salts.

Besides the dissolving power of a solvent toward substrate and supporting electrolyte, the ability to solvate intermediate cations and anions is also important. The chemical properties of charged species are dependent on whether there are formed tight ion pairs, solvent-separated ion pairs, or symmetrically solvated ions. Redox potentials of ions are thus dependent on solvent (se, e.g., Ref. 246). A knowledge of the donor and acceptor values [247] of a solvent is thus helpful in predicting its properties as medium. Also, the autoprotolytic constant of a solvent [248] is of interest in this respect.

A combination of the good conductivity of water and the dissolving power of an organic solvent may be exploited by using an emulsion [249–257]. A simple emulsion is useful only if the substrate has a reasonable solubility in water [257], and only a limited advantage, compared with a suspension of the substrate, is obtained. Utilization of a mediator and a phase-transfer catalyst may enhance the reaction rate considerably. In some cases the organic phase covers the electrode with a thin layer, and the electron transfer to the substrate occurs in the organic layer. Under such circumstances the aqueous electrolyte serves as a conducting medium and may furthermore be essential for protolytic or other follow-up reactions. Another aspect of the use of a two-phase system is that a water-insoluble product from a water-soluble substrate may continuously be extracted and thus avoid further reactions.

When cells with solid polymer electrolytes (SPE) are used, no electrolytes need to be dissolved in the medium, and solvents, which usually cannot be employed in electrolysis, may be used [78–84].

Temperature Range and Other Factors

It is desirable that the solvent be a liquid in a convenient temperature interval and that the vapor pressure be not too high at the working temperature to avoid excessive loss of solvent. This is not an important point since the electrolysis is often carried out in a closed system anyway. The vapor pressure may also sometimes be lowered considerably by working with a high salt concentration. Too low a vapor pressure of the solvent is inconvenient, as it may be difficult to remove the solvent during the workup.

The viscosity of the medium influences not only the mass transfer, but also the rate constant of the heterogeneous electron transfer [258]. A low viscosity is preferable both from the point of view of diffusion and from considerations of pumping in flow cells. For some kinds of electroanalytical work, however, a high viscosity is preferable. Diffusion coefficients in some solvents useful for electrolysis have been published [259].

The toxicity and odor of the solvent warrant consideration. Proper design of the apparatus and careful handling of the solvent make these properties less objectionable.

The chirality of a solvent, S,S(+)2,3-dimethoxy-1,4-bis(dimethylamino)-butane (DDB), has been used to induce chirality in reduction products [260].

Purification

Few, if any, organic solvents are obtained commercially pure enough for electrochemical work. However, when selecting a method for purification of a chosen solvent, one should realize that it is seldom necessary to lower the concentration of all impurities. An aprotic solvent used for a reduction should especially be purified for electrophiles, whereas nucleophiles (bases) should be removed from a solvent for oxidations.

Solvents obtained from different sources may have different impurities, and even in solvents from the same source the kind and concentration of impurities may change from batch to batch.

Water is always an impurity in an organic aprotic solvent; the influence of a given water concentration depends on the stability of the complexes between water and the solvent; water is, for example, less firmly bound to MeCN than to DMF or DMSO.

When one wants to remove a given impurity, there is always a danger that the reagent or method introduces other impurities or decomposition products; some solvents, such as DMF, decompose somewhat on distillation at ambient pressure.

The storage of a solvent is important, and ideally one should use the solvent immediately after purification. Some solvents are decomposed by light, and most solvents absorb some water if given the chance.

Methods for purification of some aprotic solvents and tests for impurity recommended by IUPAC have been published [261].

B. Protic Solvents

It is practical to divide solvents into two groups, protic and aprotic solvents. Protic solvents are those that have protons bonded to heteroatoms and include acids, neutral solvents, and some bases. A review of solvents useful for electrochemistry has appeared [262]. Electrochemical reactions in nonaqueous systems [263] and the chemistry of nonaqueous solvents [264] have been treated in monographs.

Acid Solvents

The electrochemistry of a protonated organic compound differs in many cases from that of the neutral molecule, and the strength of the acid necessary for protonation of the substrate must be considered. Sulfuric acid and acetic acid, often containing some water, have generally been used. When high acidities are desired, fluorosulfonic acid may be used.

Sulfuric acid Concentrated or slightly diluted sulfuric acid is in many respects an excellent medium for electrolysis. It dissolves many organic compounds, and it is sufficiently dissociated to make the addition of foreign electrolytes unnecessary.

Sulfuric acid can protonate even very weak bases, such as aromatic ethers or some aromatic hydrocarbons; it may promote the formation of carbocations either by dehydration of hydroxyl compounds or by addition of a proton to a double bond. These properties are still more strongly developed in the "superacids."

Sulfuric acid has certain disadvantages: It cannot be removed by distillation during the workup. It is, therefore, often most practical to dilute the medium with water and extract the product. If neutralization is necessary, concentrated ammonia is preferable. Another disadvantage is that sulfuric acid may oxidize or sulfonate the substrate.

Most of the electrolyses in sulfuric acid have been performed without control of the potential. In polarographic work the mercury pool has been used for anodic and cathodic reactions: As anode, platinum or lead dioxide have been employed; as cathode, lead, mercury, or platinum.

A 1:1 mixture of sulfuric acid and ethanol was a very popular medium for electrolytic reductions around the turn of the century. The mixture dissolves many organic compounds, but the leveling effect of the alcohol lowers its protonating ability considerably.

Fluorosulfonic acid This solvent is a very strong acid and may be used when stabilization of cationic species is required [217,265–270]. It has above all been used as medium for the oxidation of alkanes. It is rather resistant toward anodic oxidation, the oxidation product being peroxydisulfuryldifluoride [269].

Fluorosulfonic acid is not a conducting medium, and it is necessary to add a supporting electrolyte, such as $NaSO_3F$ or KSO_3F, or a ionizable organic compound, such as acetic acid, which acts as a strong base and forms CH_3CO^+. Antimony pentafluoride acts as an acid in FSO_3H.

As reference electrode the hydrogen electrode [218], the $Au(s)/Au^{+3}$ electrode [217], and $HgSO_4/Hg$ in HSO_3F [219] have been used, but an internal reference system, such as $perylene^+/perylene^{2+}$, has been recommended [268] to avoid unknown liquid junction potentials.

Perfluorosulfonic acids may be used similarly [271]. They may be purified by distillation and treatment with H_2O_2 [272].

Hydrogen fluoride Hydrogen fluoride has been used for electrochemical fluorination of organic compounds [273–276] and is a promising solvent for electrochemical studies. It has a high dielectric constant (80 at 0°C), is less viscous than water, is transparent for UV light to 165 nm, and is difficult to oxidize. It dissolves many metal fluorides, giving highly conducting solutions; it also dissolves many organic substances. The main disadvantages are its poisonous effect, its ability to attack glass, and its relatively low boiling point, 19.5°C. The low boiling point gives it a tendency to form bubbles, which must be taken into account in the construction of cells. This and other practical problems have been discussed [277].

Hydrogen fluoride can be handled in apparatus of suitable metals (copper, nickel, magnesium, or aluminum, which all form a protective fluoride coating, or platinum) or plastic materials [especially polypropylene, Teflon, and polyvinylidene fluoride (Viton)]; polychlorotrifluoroethylene (Kel-F) can be made into transparent windows. A capillary for a dropping mercury electrode may be made from Teflon [278]. Hydrogen fluoride is obtained commercially in steel cylinders in a purity of 99.5%. The impurities may be removed by distillation [279] or electrolysis [280]. During the electrolytic removal of water the explosive F_2O is formed, which must be taken into consideration [276].

The useful potential range in the cathodic direction is rather limited owing to the evolution of hydrogen, whereas the anodic limit is about 2.6 V (SCE). Sodium or potassium fluoride as supporting electrolyte may be used; the addition of these salts lowers the protonating power of the solvent. The solubility of NaF is about 30 g NaF per 100 g HF; other fluorides, such as LiF, KF, CsF, NH_4F, AgF, and TlF, are very soluble in liquid HF. The acidity of HF may be enhanced by the addition of BF_3 or SbF_5.

The addition of amines to hydrogen fluoride makes the fluoride ion a better nucleophile and reduces the acidity of HF; $Et_3N/3HF$ and pyridine/HF are commercially available. $Et_3N/3HF$ may be used as such or mixed with MeCN as a medium for electrochemical fluorination (Chap. 26) [281].

Several types of reference electrodes may be used. The hydrogen electrode has been shown to behave reversibly in HF [199,200] and has been employed as reference electrode both in electroanalytical and preparative work [200], and mercury fluoride [232,282] and copper fluoride [195,277] have also been used as such. A comparison of these reference electrodes pointed to the hydrogen electrode (H_2/Pd) as the most convenient [283].

Trifluoroacetic acid Trifluoroacetic acid has mainly been used as a medium for oxidations [284]. It has a dielectric constant of 8.4 [285,286] and a boiling point at 72.5°C; although it has a low conductivity, it is possible to conduct electrolysis in it without using a supporting electrolyte [287]. Its main advantage is that many cation radicals show considerable stability in this medium. Sometimes it is used in mixture with its anhydride and methylene chloride [288].

Acetic acid Acetic acid is a good solvent for many organic compounds and some inorganic salts. It has a rather low dielectric constant, which makes it necessary to have a considerable electrolyte concentration ($\geq$ 0.5 M) to obtain a reasonable conductance.

Acetic acid may be considered as solvent for acetoxylations [289–291] and for reactions that require an acidic medium other than sulfuric acid. The acidity may be enhanced by adding perchloric acid to acetic acid containing a suitable amount of acetic anhydride, which reacts with the introduced water.

Usable supporting electrolytes include NaOAc, NH_4OAc, LiCl, HCl, H_2SO_4, $HClO_4$, $NaClO_4$, Bu_4NClO_4, and Bu_4NBF_4. The choice of supporting electrolyte may influence the product distribution. Thus, anodic acetoxylation gives different results in NaOAc/HOAc and in Bu_4NBF_4/HOAc. Several types of reference electrodes work well in acetic acid. The chloranil electrode [202] behaves reversibly in this medium and so does the analog of SCE [292] [$Hg/Hg_2Cl_2(s)$, LiCl(s), HOAc]; the mercury pool [293] and the Ag/AgCl electrode [192] can also be used.

The cathodic limit is about −1.7 V (SCE), and the limiting reaction is evolution of hydrogen. The anodic range ends at about 2.0 V (SCE) when NaOAc is electrolyte [291]. The limitng reaction is probably the discharge of acetate ions.

Acetic acid may be purified by partial freezing [291] or by distilling it from CrO_3[292,294] in the presence of acetic anhydride. CrO_3 oxidizes impurities and acts as an acid catalyst for the reaction between water and acetic anhydride. An excess of acetic anhydride is necessary to force the equilibrium $2AcOH \rightleftharpoons Ac_2O + H_2O$ sufficiently to the left [295]. Modifications of the method have been described [212,293].

Other acids Methanesulfonic acid [296] and formic acid have also been used as solvent for electrolysis; superacids offer possibilities for unusual electrode reactions.

Neutral Solvents

This group comprises water, mono- and polyvalent alcohols, and monoethers of polyvalent alcohols. Often mixtures of water and alcohols are used. Such mixed solvents retain to some degree the favorable characteristics of water but raise its ability to dissolve organic substrates.

Water Water is the solvent of choice for electrolysis, unless some of its properties are undesired or inadequate. An aprotic solvent is sometimes preferable to water since a reaction mechanism of an electrode reaction in an aprotic, organic solvent is often simpler to elucidate than in water. Adsorption phenomena are often a complicating factor in aqueous solution, but much less so in many organic solvents. The usable potential range of water in the anodic direction is quite limited, and the most important factor, the solubility of many organic compounds in water, is quite unsatisfactory. The use of "hydrotropic" salts (Sec. VII.A) may enhance the solubility of organic compounds in an aqueous medium. Another way of combining the good conductivity of aqueous solutions with the better dissolving power for organic substrates is the use of emulsions [249–257], preferentially with a mediator and a phase transfer reagent.

Methanol The electrochemical behavior of methanol is rather similar to that of water. Methanol is well suited for anodic reactions of the Kolbe type and for methoxylations; for reductions it may be used when the solubility of the substrate in water is too low. The dielectric constant of methanol is fairly high ($\varepsilon = 33$); its liquid range (−98 to 64°C) is convenient, and it is easy to remove during the workup.

Quite a few electrolytes are soluble in methanol, such as NH_4Cl, LiCl, HCl, KOH, KOMe, $NaClO_4$, and tetraalkylammonium salts. For reductions a solution of HCl in MeOH is convenient as it has a high conductance and both components are easy to get rid of during the isolation of the product.

Reference electrodes that can be used in water work in most cases also in MeOH; thus the Hg/Hg(I) and Ag/Ag(I) electrodes may be used. The useful potential range for large-scale electrolysis is about the same as for water, but methanol is less satisfactory for electroanalytical work.

Methanol can in general be used as received; it can be dried by treatment with magnesium [297].

Other alcohols Ethanol and the higher alcohols have lower dielectric constant and poorer ability to dissolve electrolytes. They have been used mostly in mixture with water or sulfuric acid.

Basic Solvents

Basic, protic solvents include ammonia and primary and secondary amines. These solvents are primarily of interest for an organic electrochemist because of their ability to solvate electrons, and solvated electrons have special reducing properties (Chap. 29). They also permit reductions in a protic medium in the presence of a very strong base, the conjugate base of the solvent.

Ammonia The use of ammonia as solvent for electrochemical reactions presents some problems due to its inconvenient liquid range (−77.7 to −33.4°C); this difficulty can be coped with in different ways. The electrolysis may be made in a suitable low-temperature thermostat [298], the cell can be equipped with an efficient reflux condenser [299], a high-pressure apparatus can be used [300] (ammonia has even been used in its supercritical phase as a medium for electrochemical investigations [301]), or strong solutions (5–10 M) of various salts, such as NH_4SCN, NH_4I, $LiNO_3$, $LiClO_4$, NaSCN, and NaI, can be used [302] as they form a medium in which the vapor pressure of ammonia at ambient temperature is below 1 atm. The easy purification and low price of ammonia compared with the alternative amines favor the choice of ammonia for large-scale preparations.

Ammonia has a medium-high dielectric constant (23.7 at −36°C) and good dissolving power toward inorganic salts. It can act as both an acid and a base; the strongest possible acid in ammonia is ammonium ion and the strongest base the amide ion. The pK values of some weak acids have been determined in NH_3 at −60°C [303,304].

The combination of low proton availability together with the low temperature usually employed makes ammonia a suitable medium for electrochemical investigations of strongly basic species, such as dianions [55,305]. The low tendency to donate hydrogen atoms, due to the strength of a N−H bond, is important for many investigations [306].

Besides the electrolytes mentioned earlier, such salts as NH_4Cl, NH_4NO_3, KNO_3, and $NaClO_4$ may be used as supporting electrolyte; the tetraalkylammonium salts are generally rather sparingly soluble in ammonia. Most electrolyses in ammonia, except in voltammetric studies, have been performed without a reference electrode; as such a $Zn(Hg)/ZnCl_2$ [188] or Pb/Pb^{2+} electrode [205,298,307] may be used. In polarography the mercury pool electrode has been applied [205].

Ammonia has been employed mostly for cathodic reactions, but some oxidations [299,302] have been carried out in this medium, although the potential range in the anodic direction is quite small. The anodically limiting raction is oxidation to nitrogen and protons [308]; the cathodically limiting reaction is the transfer of electrons to the solvent, which occurs at about −2.3 V (versus Hg pool electrode) in a saturated solution of TBAI. In the electrolytic generation of solvated electrons the potential is determined by the surface concentration of electrons and no external reference electrode is needed.

Ammonia is purified by distilling it after treatment with sodium; the distillation from sodium may have to be repeated.

The threshold limit value (TLV) is 25 ppm.

Methylamine Methylamine behaves electrochemically much like ammonia but has a somewhat more convenient liquid range (−93.45 to −6.3°C), although its odor and low boiling point require special considerations. Its dielectric constant is rather low (11.4 at −10°C), but sufficient conductivity for large-scale electrolysis can be obtained using LiCl as supporting electrolyte. Like ammonia, it is both an acid and a base.

Methylamine has mainly been used in large-scale electrolysis [309,310], and no reference electrodes have been employed so far. Reference electrodes such as the $Zn(Hg)/Zn^{2+}$ electrode, which work both in ammonia and ethylenediamine, would be assumed to be applicable to methylamine solutions.

Methylamine has been employed only for reductions, especially as solvent for electrolytic generation of solvated electrons. It would be of very limited use for anodic reactions. It may be purified by distillation from sodium. Its TLV is 10 ppm.

Ethylenediamine Ethylenediamine (EDA) is a primary amine with a more convenient liquid range (11 − 117°C) than methylamine. Its dielectric constant (ε = 12) is about the same as that of methylamine. It dissolves many inorganic salts and is a better solvent than ammonia for many organic substrates.

LiCl, $NaNO_3$, and tetraalkylammonium salts can be used as supporting electrolytes. For the electrolytic generation of solvated electrons mainly LiCl has been employed [311,312]. A reversible reference electrode in EDA is the $Zn(Hg)/Zn^{2+}$ electrode [313], but the Hg pool [227] or the aqueous calomel electrode, fitted with a suitable salt bridge, is also applicable.

EDA is oxidized rather easily anodically but is resistant to reduction; the cathodic limiting current is the discharge of cations or the donation of electrons to the solvent. EDA may be purified by repeated refluxing and distillation from sodium [314]. It is extremely hydroscopic. The TLV is 10 ppm.

Other primary or secondary, aliphatic or cyclic amines may be used with similar results.

C. Aprotic Solvents

The use of aprotic solvents is of interest in many cases in which aqueous solutions lack desired or have undesired properties. The scarcity of protons makes it generally simpler to elucidate the mechanism of an electrode reaction for several reasons. The intermediate (e.g., radical anions) are more stable in the absence of protons, and the reaction scheme becomes simpler.

The scarcity of protons in aprotic solvents also makes it possible for added reagents (e.g., electrophiles) to compete successfully with protons for the intermediates.

"Dry" aprotic solvents usually contain some water, which is especially of importance in voltammetric studies in which the concentrations of water and substrate are about equal. The influence of a certain molar concentration of water depends on the solvent, the supporting electrolyte, and the substrate.

In certain solvents, such as dimethylformamide and dimethylsulfoxide, water is a rather poor proton donor [315], and other impurities may be responsible for the protonation of the basic intermediates (radical anions, anions, and dianions). During preparative experiments the impurities may be reprotonated by water or, in case tetraalkylammonium salts (except tetramethylammonium salts) are used as supporting electrolyte, by attack on the cations (Hofmann elimination). Treatment of the medium with active alumina may lower the concentration of such protonating impurities [41,237].

When choosing a drying agent one must take into account that its affinity to water is different in vacuum and in a solvent, so the activity of a drying agent may be different in various solvents.

When deuterium oxide is added to an aprotic solvent containing tetraalkylammonium salts to promote incorporation of deuterium in the product, the base-promoted Hofmann elimination may become a problem, as the deuterium oxide becomes diluted with water. This can largely be avoided by running the reduction in the presence of a suspension of dry silica, which prevents the medium from being basic enough for the elimination to occur; the deuterium oxide adsorbed on the silica is in most cases sufficiently acidic to prevent Hofmann elimination. However, when the parasitic reaction is hydrogen abstraction from the solvent or supporting electrolyte, silica has no effect.

Controlled addition of a suitable proton donor or electrophile (reductions) or nucleophile (oxidations) is often useful in determining a reaction mechanism. The strength of a proton donor may vary from perchloric acid through acetic acid and a phenol to an alcohol; C acids, such as malonic ester, or N acids, such as urea, may also be used. As bases may be used pyridine, carboxylate ions, alkoxides, or salts of malonic ester. Sometimes it is of interest to determine whether it is the basic or the nucleophilic properties of the compound that are important. The use of two bases with approximately the same pK values but widely differing in nucleophilicity, such as pyridine and a 2,6-dialkylpyridine, might answer the question.

In the absence of proton donors the "aprotic" solvent may act as a proton donor; the conjugate base of the solvent may act as a nucleophile toward substrate or intermediates.

Electrochemical phenomena may also be simplified in organic solvents, compared with aqueous solutions, because of the less important role adsorption plays in organic solvents.

Acetonitrile

Acetonitrile (MeCN) is one of the most used polar aprotic solvents for both anodic and cathodic reactions. It is an excellent solvent for many organic substrates and quite a few organic and inorganic salts; it is miscible with water. Salt solutions show a reasonably high conductivity due to its rather high dielectric constant ($\varepsilon = 37$). MeCN is somewhat toxic, with a recommended maximum air concentration of 20 ppm.

Besides having good electrochemical properties, acetonitrile is an unusually good solvent for UV spectroscopy, being transparent until about 190 nm. It is furthermore a convenient solvent for electron spin resonance, ESR, spectroscopy, which is valuable for the investigation of radicals produced during electrolysis. For use in preparative electrochemistry its liquid range (−45 to 82°C) is sufficient for most purposes and its relatively low boiling point makes the removal of the solvent easy. On gas-liquid chromatography (GLC), tailing may occur, unless a suitable polar column is used.

Acetonitrile has a wide, usable potential range in both the anodic and cathodic directions. The limit is set by the electrode reaction of the supporting electrolyte in both directions. In the presence of sodium or lithium ions their discharge sets the limit; the sodium subsequently reacts with acetonitrile, whereas lithium does not. Quaternary ammonium ions are reduced with dealkylation. Perchlorate ions, a widely used supporting electrolyte, is discharged to perchlorate radicals, which probably rapidly form oxygen and chlorodioxide radicals [236]. Transfer of protons in acetonitrile is a rather slow process. Its autoprotolysis constant is 33.2 [248].

Useful supporting electrolytes in acetonitrile are $NaClO_4$, $LiClO_4$, tetrabutylammonium salts, as well as other tetraalkylammonium salts. $AgNO_3$ and $AgClO_4$ are very soluble in acetonitrile; AgCl is somewhat soluble, especially in the presence of an excess of chloride ions.

The Ag/Ag^+ electrode [180,181], which is reversible in acetonitrile, may be used as reference electrode. Quite often an aqueous calomel electrode with a suitable liquid junction is used [205]; a comparison of different reference electrodes has been made [206]. The Hg/Hg(I) is not stable in acetonitrile, as Hg(I) decomposes [316]. A platinum electrode coated with a film of poly(vinylferrocene) is a stable reference electrode in MeCN [203].

Commercial acetonitrile usually contains impurities, such as acrylonitrile, acetic acid, aldehydes, amines, and water. Several methods of purification [317–319] have been proposed: one is to distill it several times from P_2O_5 followed by fractionation from K_2CO_3 [182]; a pure product may be obtained, but the procedure is time consuming and wasteful due to polymerization of MeCN; polymerization may be avoided by using B_2O_3 instead of P_2O_5 [320].

CaH_2 treatment of MeCN has been recommended [321], but it does not remove traces of aromatic hydrocarbons and is a surprisingly ineffective drying agent in MeCN [320]. For removal of acrylonitrile, which boils 4°C lower than MeCN treatment with NaH [322] or azeotropic distillation with ethanol [323], have been recommended.

Removal of impurities by reaction in turn with benzoyl chloride and $KMnO_4$ followed by careful distillation gives a product suitable both for electrochemical and spectroscopic work [324]. Alternatively, acetonitrile is treated with N_2O_4, then with CaH_2, fractionated under N_2 from CaH_2, and the product further purified by letting it run through a column of activated alumina [298,325].

Water may be removed from commercial MeCN by azeotropic distillation [326,327] or by treatment with molecular sieves 3A; 3A is much more efficient than 4A for drying MeCN [320]. The procedures recommended by IUPAC for purification are described in Ref. 328.

As is the case for other relatively inert solvents, it is unlikely that an all-purpose purification procedure can be devised that will give optimum results in all possible applications. Consequently, purification should be tailored to the intended use of the solvent.

Pure acetonitrile is not particularly hygroscopic. The product obtained by these procedures can be stored for several months without deterioration; the water content is approximately 1 mM. It is advisable to use the same glass apparatus repeatedly for the purification and storing of acetonitrile.

Electrolysis in MeCN, in the absence of added proton donor may produce the anion of MeCN $(CH_2CN)^-$, which may act as nucleophile toward electrophilic centers [329].

At a Pt anode the solvent system MeCN/$TBABF_4$ breaks down to an adduct, CH_3CNBF_3, whereas the cathodic breakdown mainly gives the anion of 3-amino crotonitrile; in the presence of water the anion of acetamide is also formed and the breakdown commences at much less negative potentials [330].

Other nitriles, aliphatic and aromatic, have been investigated as solvents for electrolysis. Butyronitrile has been suggested as the solvent of choice for low-temperature electrochemistry [95], and propionitrile has in some cases beem employed, as it is less hygroscopic than MeCN [331]. Benzonitrile might be considered when hydrogen atom abstraction from the solvent is undesirable.

Dimethylformamide

Dimethylformamide (DMF) is a polar solvent with a dielectric constant (ε = 37) comparable to that of acetonitrile. It is a good solvent for most organic compounds and also for many organic and inorganic perchlorates and organic fluoborates. Owing to these properties DMF has been widely used in electrochemical work. DMF is, however, susceptible to hydrolysis, which yields the easily oxidizable dimethylamine and the reducing agent formic acid.

DMF has an acceptable liquid range (−61 to 153°C), although the boiling point is near the upper limit for its convenient removal by distillation during the workup. DMF is inferior to acetonitrile as a solvent for UV spectroscopy. Contact with the skin and vapor concentrations over 10 ppm should be avoided.

For reductions DMF has a usable potential range comparable to that of acetonitrile, but it is inferior for oxidations. In the presence of inorganic ions their discharge is the cathodic limiting reaction, whereas it is more uncertain whether it is the solvent or the cation that is reduced in solutions of tetraalkylammonium ions. The anodic limitng reaction at a Pt electrode is an oxidation of DMF, which involves the removal of an electron from the amide nitrogen. Its autoprotolysis constant is 29.4 [248].

As supporting electrolyte, tetraalkylammonium perchlorates or fluoborates are mostly used, but LiCl or $NaClO_4$ may also be employed. The reference electrode may be a modified Ag/AgCl electrode [193,198], a Cd(Hg)/Cd^{2+} electrode [193,198], Na(Hg)/Na^+ electrode [187], or silver/silver cryptate electrode [183].

For many purposes the commercial reagent-grade DMF, dried with molecular sieves, can be used; DMF decomposes on distillation at atmospheric pressure. The most convenient way to purify DMF is probably to let it pass through a column [323] of alumina (e.g., Woelm Al_2O_3 activity grade one). Other methods involve drying ($CuSO_4$ [332], which also removes amines), azeotropic distillation with benzene, or percolation through molecular sieves [333] followed by fractional distillation at reduced pressure. DMF is difficult to obtain in an anhydrous form, but for most purposes a small water content is not critical. An impurity of N-methylformamide may act as a proton donor or otherwise interfere in the follow-up reactions [334]. DMF should be stored in the dark under nitrogen. The different purification methods have been compared and discussed [335].

N-Methylpyrrolidone

DMF has certain disadvantages that can be traced to its hygroscopic properties and rather easy hydrolysis; N-methylpyrrolidone (NMP), which is an N-disubstituted amide just like DMF, is less apt to hydrolyze owing to its cyclic structure. Furthermore, the product of hydrolysis is not a reducing agent as is formic acid formed from DMF. Although only a limited number of experiments in NMP have been done [336,337], it seems probable that NMP will replace DMF in a number of cases because of its higher stability [338]. Vapor concentrations over 100 ppm should be avoided.

Like DMF, NMP is of limited value for oxidations but useful for cathodic reactions. Its dielectric constant is 32, and it is able to dissolve the same types of salts and organic compounds as DMF. NMP has a liquid range of from −24 to 205°C; the useful potential range is approximately the same as for DMF. Mercury in contact with its oxidation products has been used as reference electrode; it is more stable than the Ag/Ag^+ electrode [336,337].

Other organic amides, such as formamide, N-methylformamide, N-methylacetamide, and tetramethylurea, have been investigated as solvents for electrochemical use primarily due to their high dielectric constant; none of them, however, seem to offer distinct advantages over DMF or NMP and in some respects they are inferior.

3-Methyl-2-Oxazolidinone (3M2O)

3M2O is a colorless, nonvolatile (kp_1 87–90°C) solvent with a high dielectric constant (77.5). It may be purified by vacuum distillation over BaO. With TBAP as electrolyte the cathodic limit is −2.9 V (aqueous SCE); the anodic limit at a Pt anode is +1.3 V. The suitability of 3M2O for voltammetry has been investigated [338].

Hexamethylphosphoramide (HMPA)

HMPA is a polar solvent that is remarkable for its ability to solvate electrons. Even in a mixture with ethanol (67 mol% ethanol and 33 mol% HMPA), reductions via solvated electrons may take place [339–341]. It has been suggested that HMPA is selectively adsorbed at the electrode, thus providing a layer of low proton activity near the electrode.

The liquid range of HMPA is 7.2–235°C; its dielectric constant is 30 (20°C). HMPA solvates cations strongly, whereas anions are less solvated. Unlike carboxamides, such as DMF, it is not attacked by aqueous alkali at $t < 80°C$, and it is very resistant toward nucleophilic attack [341] but forms peroxides under the influence of light and oxygen [340].

HMPA is completely miscible with water; it can be extracted from water by chloroform, which forms a complex with HMPA [186], $K(CHCl_3/H_2O) = 5.5$ [342]. HMPA is somewhat more basic than DMF. HMPA dissolves such salts as LiCl, $LiClO_4$, $NaClO_4$, and R_4NClO_4; the Ag/Ag^+ may be used as reference electrode [343]. The solvation effect of the cation of the supporting electrolyte in HMPA has been investigated [344]. Using R_4NClO_4 the accessible potential range is from +0.14 to −3.35 V (Ag/Ag^+) at a mercury electrode [340]. HMPA can be purified by fractionation under vacuum, boiling point 88–92°C (3 mm), 68–70°C (1 mm).

Reflux over BaO, followed by fractionation under vacuum, reflux over sodium, and finally vacuum distillation has been recommended for the purification [344]. Alternatively, fractional distillation over CaH_2, followed by purification through a column of activated alumina and standing over pyrene and sodium, produced a product that after distillation on a vacuum line contained only a negligible amount of impurities [345]. The purification of HMPA has been critically discussed [346].

Just like DMF, HMPA is not suitable for anodic reactions, since the amide nitrogen loses an electron easily. For cathodic reactions the useful potential range extends, in the presence of LiCl, to the value at which electrons are given off to the solution. HMPA has not yet been used widely in electrochemical experiments, and although its special properties are useful in electrochemistry it should be used only when sufficient precautions have been taken with regard to its carcinogenic and other toxic properties [347]. Tetramethylurea (TMU) may in some cases be used as a nontoxic replacement for HMPA [348]. It may be purified by distillation from CaH_2.

Pyridine

Pyridine is a reasonably strong base and a good nucleophile. Although its dielectric constant is rather low ($\varepsilon = 12$), many salts are soluble in pyridine and the solutions have good conductivity. In contradistinction to most bases, pyridine, which is a π-electron-deficient heteroaromatic compound, is rather resistant toward oxidation.

Pyridine has a liquid range from −41 to 115°C; it is miscible with water and a good solvent for many organic compounds. Such salts as $LiNO_3$, LiCl, $LiClO_4$, NaI, KSCN, and tetraalkylammonium salts are soluble in pyridine. As reference electrode the Ag/Ag^+ [210] or the Hg pool electrode can be employed. Solute-solvent interactions have been discussed [349].

The useful potential range is from about −2.3 V [210] (versus Ag/Ag^+ electrode, DME, or Bu_4NI) to + 1.4 V versus Ag/Ag^+, graphite electrode/ $LiClO_4$ [349]. The cathodic limiting reaction might be the discharge of the cation, whereas an electron abstraction from pyridine is the anodic limiting reaction, in which probably 2-pyridylpyridinium ion is formed.

Besides being used as a solvent, pyridine has been employed as a nucleophile that is oxidizable with difficulty [181]. Reagent pyridine is satisfactory for most purposes; it may be dried with molecular sieves, type

5A. A recommended procedure for obtaining pure pyridine has been published [350]. The TLV is 5 ppm.

Dimethyl Sulfoxide (DMSO)

DMSO is an excellent solvent for many inorganic salts and organic compounds. It is difficult to reduce and fairly resistant to electrolytic oxidation. Its dielectric constant is high ($\varepsilon = 47$). It thus has many of the qualities desirable for a solvent for electrolysis, and it shows promise of being one of the most important electrochemical media [351]. The liquid range is from 18 to 189°C, which makes it somewhat inconvenient to get rid of DMSO in the workup. When used as solvent for electrolysis it must be considered that DMSO is not always inert but has a fair reactivity in certain reactions. DMSO is unfit for UV spectroscopy. Its autoprotolysis constant is 31.8 [249].

DMSO has low toxicity; it is, however, transported through the skin very fast. Thus, minutes after applying a drop of DMSO to the palm of the hand its sweetish taste can be detected in the mouth. It may carry with it dissolved toxic materials that otherwise would be unable to penetrate the skin.

Many salts are soluble in DMSO, so the choice of supporting electrolyte is less restricted than in most other nonaqueous solvents. In general, perchlorates, even $KClO_4$, nitrates, and halides, are soluble, whereas fluorides, cyanides, sulfates, and carbonates are not; thus not only $NaClO_4$, LiCl, $NaNO_3$, and tetraalkylammonium salts can be used but also such salts as NH_4PF_6 and NH_4SCN. The ability of DMSO to solvate ions is also of importance in the indirect electrolytic hydrodimerization of, for example, acrylonitrile using Na(Hg) [352].

The equilibrium constant of the self-ionization of DMSO is 5×10^{-8} (25°C); the conjugate base $(CH_3SOCH_2)^-$, which is formed when DMSO acts as a proton donor during a reduction, is a rather strong base and a fairly good nucleophile, which may attack electrophilic centers or radicals [353].

The potential range of DMSO in the cathodic direction is at a Hg electrode limited by the discharge of the cation. Potassium and sodium react with DMSO, whereas lithium does not; tetraalkylammonium ions are reduced about −2.8 V (SCE) [354]. The anodic limiting reaction at a Pt electrode is the oxidation of DMSO: probably one of the nonbonding electrons is lost, but the limiting reaction has not been investigated yet.

As reference electrode [193] for polarographic work an aqueous or methanolic calomel electrode, connected to the DMSO solution with a suitable salt bridge to avoid contamination, has been used. The most stable reference electrodes in DMSO seem to be the amalgam electrodes, such as Tl(Hg)/TlCl [196] or Li(Hg)/LiCl [189].

DMSO is somewhat hygroscopic, and the main impurity in a good commercial grade of DMSO is water; the odor is due to dimethyl sulfide. For many purposes DMSO can be used as received or after drying either by shaking with CaO–MgO or passing through a column of molecular sieves. For further purification fractional distillation or crystallization may be used. It has been recommended [355] that a fractional crystallization be done followed by two fractional distillations in vacuo, one through a column packed with activated carbon and one through a spinning band column. The pure DMSO must be handled without contact with humidity.

The different purification procedures have been discussed [356].

Sulfolane

Among sulfones, sulfolane (tetrahydrothiophen-1,1-dioxide, tetramethylene sulfone) shows promise for electrochemical use. It has a wide liquid range (28.6–285°C); on dissolving 0.1 M salts in sulfolane, its melting point is depressed below 25°C. Sulfolane has a relatively high dielectric constant (43 at 20°C), it is chemically stable toward both reducing and oxidizing agents, and it has a fair dissolving power for inorganic salts. One of its drawbacks is that it is rather hygroscopic.

The accessible potential range is large in both the anodic and cathodic directions. The limiting cathodic reaction seems to be the discharge of the supporting electrolyte, whereas the limiting anodic reaction probably is the oxidation of the solvent. With $NaClO_4$ as supporting electrolyte the anodic limit at a Pt electrode is about 2.3 V and the cathodic limit about −4 (versus Ag/Ag^+ electrode) [357].

As supporting electrolyte, $NaClO_4$, $LiClO_4$, NH_4PF_6, and quaternary ammonium compounds can be used. The Ag/Ag^+ [357,358] or Ag/AgCl [359] electrode or an aqueous SCE [360] with a suitable salt bridge can be used as reference electrode.

Sulfolane may be purified by fractional crystallization followed by fractional distillations at reduced pressure [360]; alternatively, treatment with aqueous hydrogen peroxide, followed by extraction with CH_2Cl_2 and fractional distillation, may be employed [357]. The purification procedures have been discussed [361].

Sulfolane is an alternative to acetonitrile or nitromethane as a solvent for anodic reactions; in reductions it generally offers no advantages over such solvents as acetonitrile, DMF, or DMSO, but special uses may be found.

Among other sulfones, dimethyl sulfone has been used as electrochemical medium [362,363]; its high melting point, 109°C, makes it less suitable for organic compounds.

Propylene Glycol Sulfite

A series of other sulfur-containing solvents has been investigated for electrolytic use. Of these propylene glycol sulfite [364], a derivative of 1,3,2-dioxthiol, was found most suitable. It has a dielectric constant of 33, forms solutions with good conductivity with such salts as $LiClO_4$, and does not react with lithium. It has not yet been used for organic electrolysis.

Nitromethane

Nitromethane is one of the few solvents useful for anodic reactions, among them anodic coupling of aromatic hydrocarbons. Its application for cathodic reactions is limited, but in some cases in which its unreactivity toward certain active halogen compounds is valuable it may be used. The liquid range of nitromethane is −29 to 101°C; its dielectric constant is 37, but the dissociation of salts is not as high as could be expected.

Nitromethane dissolves rather few inorganic salts; $LiClO_4$, $Mg(ClO_4)_2$, and tetrabutylammonium salts can be used as supporting electrolyte. The Ag/Ag^+ or Ag/AgCl electrode can be employed as reference electrode [184].

At a platinum electrode, potentials of about 3 V (versus Ag/AgCl) can be reached, and the limiting reaction is the discharge of the supporting electrolyte. The useful range in cathodic direction is very dependent on the water content.

Nitromethane has been purified by repeated washings with bicarbonate, sulfuric acid, and water, followed by drying and fractional distillation; the water content of the product is 5–8 mM. Alternatively, fractional distillation at reduced pressure followed by repeated fractional freezing may be employed [365]. The TLV is 100 ppm.

Nitrobenzene

Nitrobenzene has been used for electrolysis [366]; it was found that certain radicals were rather stable in this solvent. Nitrobenzene has a liquid range from 5.7 to 210.9°C; Bu_4NClO_4 may be used as supporting electrolyte. An aqueous SCE separated from the solution by a suitable bridge with porous glass has been used as reference electrode. Nitrobenzene may be purified by passing it through a column of alumina followed by a distillation in vacuo.

Propylene Carbonate (PC)

PC is of potential use in organic electrocemistry [367–369]. It has a high dielectric constant (ε = 69) and a wide liquid range (−49 to 242°C), but it is more reactive than acetonitrile, and the potential limit for anodic reactions is lower than for acetonitrile. It has so far been used mostly for electrochemical generation of cation radicals [370].

The cathodic limiting reaction of propylene carbonate containing $LiClO_4$ has been shown [371–373] to be a reduction to propene and carbonate. This reductive elimination is analogous to the reduction of vicinal halogen compounds to alkenes. The anodic reaction is an oxidation of the anion [373].

As reference electrode a Li/LiCl electrode, which in the presence of excess tetrabutylammonium chloride is an electrode of the second kind, may be used [374]. A $Ag/AgClO_4$ reference electrode with a $NaClO_4$ salt bridge has been used [375]; AgCl/Ag [376] and TlX/Tl [377] reference electrodes have also been employed. The purification of PC has been discussed [378].

Benzene and Chlorobenzene

It has been possible to use these nonpolar solvents for electroanalytical measurements by employing a minute working electrode [379], an ultra-micro-electrode, or using a benzene solution containing a crown ether (15-crown-5) and sodium tetraphenylborate [380].

Tetrahexylammonium perchlorate and tetrabutylammonium tetrafluoborate have been used as supporting electrolyte in benzene and chlorobenzene, respectively. The latter was suggested [379] to be an excellent solvent for the study of reversible oxidations and reductions of aromatic compounds. The TLV for chlorobenzene is 75 ppm.

Benzene and even aliphatic hydrocarbons [82] have been used as solvents in SPE cells.

Ethers

Ethers, such as tetrahydrofuran and 1,2-dimethoxyethane, have low dielectric constants (7.4 and 7.2, respectively), and the choice of supporting electrolyte is very limited. The ethers are difficult to reduce and are inert toward many metalorganic reagents [225] that are soluble in the ethers. Ethers are thus suitable as medium for the anodic addition of Grignard reagents to olefins [381]. Dimethoxyethane and polyethylene glycol dimethyl ethers have the ability to solvate electrons [382].

As supporting electrolyte $NaClO_4$ or $LiClO_4$ may be used in both solvents; TBAI in tetrahydrofuran and TBAP in dimethoxyethane have also been employed. As reference electrode the Ag/Ag^+ electrode functions satisfactorily [223,225]. Ferrocene and bis(biphenyl)chromium have also been used as reference electrodes in THF [383].

The ethers form peroxides. These can be removed by distillation from $LiAlH_4$. They must be stored under nitrogen or distilled from $LiAlH_4$ or Na directly into the electrolytic cell. If the water content of DME or THF is not negligible, the cathode may be passivated by a layer of a hydroxide formed during electrolysis, when an alkali metal salt is used as supporting electrolyte [384]. The TLV for THF is 200 ppm.

Methylene Chloride

Electrolysis may be carried out in methylene chloride. Its dielectric constant is low (ε = 9), and only the larger tetraalkylammonium salts are soluble in methylene chloride.

Methylene chloride has been used for both anodic and cathodic reactions. The major advantages of using methylene chloride seem to be that certain cation radicals are more stable in methylene chloride than in the solvents usually employed in electrochemistry [386]. A disadvantage is that chloride ions produced at the cathode may diffuse to the anode compartment and interfere with the anodic reactions. This can partially be avoided by the addition of small amounts of acetic acid to the catholyte whereby the cathodic reaction becomes an evolution of hydrogen rather than formation of chloride ions.

In electrosynthesis in systems of two immicible liquids and a phase transfer catalyst, methylene chloride is in many cases preferred as the organic solvent.

The addition of trifluoroacetic acid (TFA) and its anhydride (TFAn) to CH_2Cl_2 enhances the stability of cation radicals in the solution. A mixture of CH_2Cl_2–TFA–TFAn of 45:1:5 has been recommended [288]. TLV is 100 ppm.

Sulfur Dioxide

Sulfur dioxide is a solvent of low nucleophilicity and may be used when nucleophilic trapping by solvent is to be avoided. Its liquid range is somewhat inconvenient (−75 to −10°C), but its dielectric constant reasonably high (17.6 at −20°C).

As supporting electrolyte Bu_4NClO_4 or Pr_4NPF_6 may be used [387–389]; a 0.2 M solution of these salts has a conductivity of about $9 \times 10^{-3}\ \Omega^{-1}\ cm^{-1}$ at −22°C. As reference electrode Ag/0.1 M $AgNO_3$ in acetonitrile has been used. The anodic range of perchlorates or hexafluorophosphates in sulfur dioxide is large. $AlCl_3$ may also be used as supporting electrolyte [390]; cation radicals are considerably stabilized in this medium.

D. Salts

In certain electrolytic media the concentration of salts is so high that the medium could be regarded as consisting of a solvated salt. This is thus the case in the strong solutions (10 M) of many salts in ammonia in which the vapor pressure of ammonia at ambient temperature is less than 1 atm. It also applies to the strong solutions of tetraalkylammonium p-toluenesulfonates in water, used sometimes in organic electrochemistry.

A genuine liquid salt is tetrahexylammonium benzoate (THAB) [226], which is a viscous liquid capable of acting as both solvent and supporting electrolyte; it has, however, a rather high resistance. Although it is a fairly polar solvent, THAB is miscible with benzene, toluene, and carbon tetrachloride; a mixture of THAB-toluene (75:25) has a rather low specific resistance of 3.8 kΩ and a much lower viscosity than pure THAB, which is about as viscous as glycerol. Relatively little water is miscible with THAB; addition of 1% by volume of water to tetrahexylammonium benzoate results in a two-phase system. THAB is soluble in acetone.

THAB is useful for reductions; the cathodic limit at a Hg electrode is about −2.6 V versus Ag/AgCl; at a platinum electrode the anodic limit is at about 0.3 V.

THAB has been prepared by neutralizing tetrahexylammonium hydroxide, obtained from the iodide by treatment with Ag_2O, with benzoic acid and evaporation of the water. The analysis of the dried product corresponds to a hemihydrate. The material is somewhat hygroscopic but can be dried by brief heating to 90°C.

Voltammetry has also been made in $(toluene)_3$tetrabutylammmonium tetrafluoroborate; the resistance of the solution is between that of DMF and THF [391] but closer to that of DMF. On addition of proton donors a low "buffer capacity" of the solvent is observed.

Aluminum chloride in mixture with an alkylpyridinium chloride, preferentially butylpyridinium chloride, has certain possibilities as a medium for studying electrochemically generated cation radicals. The 0.8:1 $AlCl_3$-$BuPy^+Cl^-$ mixture is molten at room temperature but has generally been used at 40°C [392].

A mixture of 1-methyl-3-ethylimidazolium chloride with aluminum chloride also gives an ionic liquid that may be used as electrolytic medium [393]. The medium is not quite inert; for example, quinone reacts with an excess of chloride to 2-chlorohydroquinone in an acid-catalyzed Michael addition [394]. As reference electrode an aluminum wire was used.

At higher temperature certain quaternary ammonium salts (40–150°C) [395] or dimethylamine hydrochloride (170–200°C) [396] may be used as media for electrolysis.

E. Supercritical Fluids

Supercritical fluids (SCF) have been used mainly for selective extraction of compounds; the solubility of a compound in a given solvent is in many cases vastly different under ambient and supercritical conditions. Thus supercritical water dissolves both polar and nonpolar compounds, which may be explored in electrochemistry. When temperature and pressure approach the critical values, the internal structure of the solvent is loosened and the viscosity, the dielectric constant, and the density diminish.

Cyclic voltammetry in supercritical water-0.2 M $NaHSO_4$ [97] and ammonia-0.14 M CF_3SO_3K [97,301] of some organic compounds show that this electroanalytical technique was applicable under these conditions. The behavior of phenazine in NH_3 at −40°C and under supercritical conditions, for example, was analogous; two reversible reductions were found in both cases [97]. Dimethyl carbonate has been prepared from CO and MeOH on anodic oxidation in a supercritical mixture of CO_2 and MeOH [98].

VII. ELECTROLYTES

The passage of electric current through a nonmetallic liquid is dependent on ions. The choice of electrolyte depends on such properties as its solubility, dissociation constant, mobility, discharge potential, and protic activity. It is desirable to employ a salt with high solubility, complete dissociation, high mobility of the ions, and a numerically high discharge potential. The role of adsorption of ions at the electrode is discussed elsewhere. The electrolyte may influence the rate constant for electron transfer from an electrode to a substrate.

A. Anions

The choice of anion is of most importance in anodic reactions in which the choice may influence the product distribution (Chap. 25, Sec. II.A). Unless the oxidized anion is desired in an indirect electrolytic oxidation (Chap. 29), an anion that is oxidizable with difficulty, such as perchlorate, tetrafluoborate, hexafluorophosphate, or nitrate, is chosen. In aqueous solution the choice of anion is less critical than in nonaqueous solvents. In cathodic reactions the choice of cation has highest priority and the anion is generally selected on the basis of solubility of the salt or the ability of the anion to assist in dissolving the substrate.

Perchlorate

The perchlorate ion has a high anodic discharge potential, and only in a few solvents, such as acetonitrile or nitromethane, is the perchlorate ion oxidized in preference to the solvent. When oxidized, the primarily formed perchlorate radical decomposes into oxygen and chlorodioxide radical [236].

Caution must be exercised when handling perchlorates. Silver perchlorate is explosive; grinding it in a mortar should be avoided. For use in reference electrodes it may be prepared by anodic dissolution of silver in a perchlorate medium [184]. Evaporation of solutions of organic perchlorates is discouraged but, if necessary, must occur in vacuo and at moderate temperatures. Even sodium perchlorate, which can be dried at 110°C without undue risk, may cause explosions when heated with an organic solvent.

Sodium perchlorate is soluble in many organic solvents. It is hygroscopic but can be dried at 110°C. Drying $NaClO_4,H_2O$ often results in a sintered mass that must be broken and dried. By recrystallization of $NaClO_4$ at temperatures about 80°C the anhydrous salt can be obtained, which can be dried without difficulty. Lithium perchlorate is very hygroscopic. Tetraalkylammonium perchlorates are less hygroscopic, and some of them may be recrystallized from water; they are useful for voltammetric studies in organic solvents.

Tetrafluoborate

The discharge potential of the tetrafluoborate ion is slightly more positive [397] than that of the perchlorate ion. Tetrafluoborates do not have a tendency to explode, and they are recommended instead of perchlorates for use in preparative work when the solvent must be removed during the workup.

Tetrabutylammonium tetrafluoborate is only slightly soluble in water but is soluble in organic solvents. It is prepared by mixing solutions of $NaBF_4$ with a soluble tetrabutylammonium salt. Its solubility properties are convenient for its use in preparative electrolysis. During the workup the organic solent may be diluted with water whereby the salt precipitates and is recovered for future use. It may be recrystallized from toluene; the solution is dried by azeotropic distillation (or by molecular sieves A4); the compound crystallizes slowly on cooling [398].

Hexafluorophosphate

Hexafluorophosphates are slightly more resistant toward anodic oxidation than tetrafluoborates and perchlorates [397]; in a 0.1 M solution of these salts in acetonitrile the potentials (versus Ag/0.1 M Ag^+ electrode) at a platinum disk electrode (0.12 cm^2) at 1 mA were for perchlorate, 2.48 V; for tetrafluoborate, 2.91 V; and for hexafluorophosphate, 3.02 V [397].

Trifluoromethanesulfonate

The trifluoromethanesulfonates have certain avantages over perchlorates (nonexplosive) and tetrafluoborates (even less reactive toward oxidizing agents) [399]. The tetraalkylammonium trifluoromethanesulfonates are just as soluble or more soluble in the commonly used organic solvents than the corresponding perchlorates or fluoborates. They can be prepared either by metathesis or by alkylation of a tertiary amine by an ester of trifluoromethanesulfonic acid.

Nitrate

Nitrate ion is not as resistant to oxidation as perchlorate or tetrafluoborate; in nitromethane [400] solution it is oxidized anodically to NO_2^+ and oxygen, the NO_2^+ reacts either with the water always present in "dry" organic solvents to form nitric acid or with nitrate ion to form N_2O_5. Nitrates, to a higher degree than perchlorates [401], form ion pairs in some solvents. Tetrabutylammonium nitrate can be prepared conveniently by mixing an aqueous solution of $NaNO_3$ with the commercially available tetrabutylammonium hydrogen sulfate. The salt is then extracted into methylene chloride, giving an almost quantitative yield of Bu_4NNO_3.

Aromatic Sulfonates

The use of aromatic sulfonates stems from their ability to assist the solution of organic substrates in an aqueous medium. This "hydrotropic" effect [244,402,403] is accentuated if a quarternary ammonium sulfonate, rather than an alkali metal sulfonate, is used.

The aromatic sulfonates have been employed mostly for reductions, but they would be expected to be reasonably stable toward oxidation. To obtain a reasonable hydrotropic effect rather high concentrations are used; in the original hydrodimerization reaction of acrylonitrile the medium may be regarded as either a solution of tetraalkylammonium p-toluenesulfonate in water or the salt containing some water.

Carboxylate Ion

In the Kolbe reaction and in anodic carboxylation reactions, salts of carboxylic acids function as both substrate and electrolyte. Usually the presence of other anions dimishes the yield of the Kolbe reaction, whereas the presence of anions that are oxidizable with difficulty, like bicarbonate or perchlorate ions, favors the related Hofer-Moest reaction.

For reductions in aprotic media tetraalkylammonium salts of derivatives of oxalic and formic acid may be employed; in many cases the reaction may then be performed in an undivided cell, as the reaction at the counterelectrode (oxidation of the anion to carbon dioxide) generally does not interfere with the cathodic reaction [404].

Tetramethylaluminate and Tetraphenylborate

The sodium salts of these anions are soluble in tetrahydrofuran and have been found suitable as supporting electrolyte in this solvent [405,406].

B. Cations

The discharge potential of a cation determines whether it is useful, so in practice only alkali and alkali earth metal ions together with ammonium and tetraalkylammonium ions are used. The nondischarge of a cation is a sine qua non, but in addition a choice of cation must take into consideration ion pair formation, solvation, and adsorption of particular species (e.g., see Refs. 407–409).

Lithium Ions

Many lithium salts, such as lithium perchlorate and the halides, are soluble in nonaqueous solvents. The reduction potential of Li^+ depends on the electrode and the solvent. At a mercury cathode amalgam formation takes place, whereas formation of lithium metal occurs at platinum in aprotic media. Lithium metal is less reactive than sodium, and in some solvents sodium attacks the solvent whereas lithium is unreactive. A small water content in an aprotic solvent may react with lithium (or Li^+ may react with hydroxyl ions formed at the cathode) to form lithium hydroxide, which may cover the electrode with an insoluble, insulating layer.

Sodium Ions

In aqueous buffers sodium ion is a common cation, but in nonaqueous solvents practically only $NaClO_4$ is useful. Even $NaBF_4$ is not soluble in most aprotic solvents.

Tetraalkylammonium Ions

The most commonly used quaternary ammonium salts are tetrabutylammonium perchlorate (TBAP), the tetrafluoborate (TBAT), the halides (TBACl, TBAB, and TBAI), and the corresponding tetraethylammonium salts, such as the perchlorate (TEAP), but also the tetramethyl- or tetrapropylammonium salts have been employed; the former cannot undergo a base-promoted Hofmann elimination. However, evidence has been found for the formation of trimethylammonium methylide [410]. In nonpolar solvents it may be necessary to employ tetrahexyl- or tetraoctylammonium salts. The tetraalkylammonium ions are soluble in many nonaqueous media, and they may be extracted from an aqueous solution by means of chloroform or methylene chloride [411, 412], and tetraalkylammonium salts may thus be prepared by ion extraction [412].

The discharge potential becomes slightly more negative with increasing size of the ion, but in practice there is not much gained by going beyond tetrabutylammonium ions [413]. The reduction route of these ions is discussed elsewhere (Chap. 24).

In aqueous solution the tetraalkylammonium ions are adsorbed at the electrode, where they form a layer with a low proton activity. This property is used in the commercial production of adiponitrile from acrylonitrile (Chap. 31). The adsorption of the ions depends on the potential of the electrode, the electrode material, the composition of the medium, and the size of the cation.

The rate constant for electron transfer from an electrode to a substrate depends on the size of the cation; for instance, the rate constants for electron transfer at a mercury electrode in MeCN to a substrate were always smaller, when tetra-n-heptylammonium perchlorate was the electrolyte than when tetraethylammonium perchlorate was it [414].

Quaternary ammonium salts are readily accessible; some are commercially available, and their preparation presents no problems [411,412]. Reducible impurities may be removed by electrolysis at a sufficiently negative potential.

Sulfonium Salts

Tris(dimethylamino)sulfonium tetrafluoborate ($TASBF_4$) has been investigated as a supporting electrolyte in MeCN and CH_2Cl_2. Compared with $TBABF_4$ the conductance of $TASBF_4$ in CH_2Cl_2 is slightly higher, suggesting a higher degree of dissociation of the latter [415].

Cryptates

Crown ethers and other cryptates form with several metal cations stable complexes, which generally are soluble in nonaqueous media, even in rather nonpolar solvents [358]. In some cases such complexes may be considered supporting electrolytes, especially when some of the special properties of a given cryptate are desirable also in other contexts. By complexing Ag^+ with a cryptate, the Ag/Ag $Cryp^+$ electrode becomes stable in DMF [183].

Polyelectrolytes

Polyelectrolytes having ion-exchange properties may be coated on one or both electrodes [17,82–85,416], which makes it possible to use virtually nonconducting liquids. This was discussed in Sec. II.B.

C. Buffers

Control of an electrolytic reaction often requires that the proton activity remain within acceptable limits during the electrolysis. For small-scale electrolytic preparations (less than about 10 g/liter of substrate), a sufficiently high initial concentration of buffer , acid, or base is adequate in aqueous solution; for large-scale electrolysis a controlled addition of protons during a reduction must be provided. This addition may be controlled by a pH-stat or coupled to the current integrator. In aprotic media a proton donor, electrophile, or nucleophile may play a similar role as buffers in aqueous media.

Acids

Hydrochloric acid is a convenient acid for laboratory-scale electrolysis; it is easy to remove during the workup by evaporation. The oxygen-containing acids, such as sulfuric or perchloric acid, must be neutralized before basic products can be extracted. Sulfuric acid is most conveniently neutralized by concentrated ammonia rather than by sodium hydroxide, whereby the precipitation of sodium sulfate is avoided.

Even in acetic acid perchloric acid is a strong acid. Anhydrous solutions are made by adding acetic anhydride equivalent to the water present in the acetic acid and introduced together with the perchloric acid. An excess of acetic anhydride should be avoided as this solution is a very strong acylating agent and it forms yellow condensation products on standing [417].

In aprotic media the addition of proton donors may be important. If the proton donor is too strong, protons may be reduced in preference to the substrate and a too weak proton donor may not furnish sufficient protons. Often a phenol is an acceptable compromise.

Extreme acidic conditions may be obtained in superacids, such as SbF_5 or BF_3 in HF or SbF_5 in FSO_3H.

Bases

Reductions in aqueous alkaline solution do not present many problems; the choice of cation was discussed earlier. In nonaqueous solvents bases may be generated electrolytically (Chap. 30) [418–421] or formed by reaction of the solvent with an alkali metal. The conjugate base of the solvent is the strongest base obtainable in any medium.

During oxidation in a nonaqueous solvent the removal of a proton or an attack of a nucleophile on the oxidized substrate may be a chemical step in an ECE reaction. To explore which reaction is operating one can, in parallel voltammetric experiments, use two compounds that are equal in base strength but differ widely in nucleophilicity: such a pair is pyridine and 2,6-dimethylpyridine.

Buffer Systems

The use of buffers in electrolysis to maintain a desired pH warrants some consideration. During a reduction protons are consumed at the cathode surface, and unless they are replenished rapidly, the pH in the reaction layer is considerably higher than in the bulk of the solution. A high concentration of a buffer system, which exchanges protons fast, is required to maintain a desired pH.

In aqueous solution phosphate, borate, and carboxylate buffers are applicable, but in mixed solvents the solubility of phosphate and borate buffers is low; in such mixtures carboxylate buffers may be used in slightly acid, and amine buffers in slightly alkaline medium. To avoid excessive buffer concentrations it is advisable to maintain pH in the bulk of the solution constant by means of a pH-stat.

Buffering in an aprotic medium is sometimes a problem. Proton donors, such as a suitable phenol, malonic ester, or amine salt, together with the corresponding base, may be an acceptable solution. Guanidine perchlorate [422] has been proposed as an efficient proton donor (and supporting electrolyte) in aprotic solvents, as it brings the protons into the Helmholtz double layer.

VIII. VOLTAGE SOURCES AND COULOMETERS

A. Voltage Sources

To perform a simple electrolysis under constant conditions only a dc source, which may be a battery or a rectifier, is needed. Small ripples on the voltage usually do no harm in preparative work.

For experiments under controlled potential conditions the electrode voltage may be controlled manually following the reading of a dc voltmeter with high input resistance or automatically by a potentiostat. This is a device that is able to keep the potential of the working electrode within a small range around a fixed value (versus the reference electrode) independent of the cell current. The apparatus thus reduces the cell current as the electroactive substance is consumed and also follows possible changes of the cell resistance.

Since Hickling's first description [425] many variants have been proposed [426–432] and excellent commercial models are now available from several firms. All the devices work on a feedback principle, which allows the preset voltage to be maintained within tolerances and time-spaces dependent on the particular model. Good reviews of the principles have been given [433–437]. Further diagrams and references can be found in the books mentioned earlier [1,4,31,32].

B. Coulometers

Very often it is of great interest to determine the n value of an electrode reaction or the current efficiency of an electrosynthesis. Both involve a measurement of the electricity consumption, that is, an integration of the electric current over the time of electrolysis. In constant current experiments this is of course an easy task. For controlled potential electrolysis an integrating device is included in the circuit.

Coulometers (or integrators) usually belong to one of three types—(1) electrochemical, (2) electromechanical, and (3) electronic—according to their working principle. Electrochemical coulometers are impractical to use; a description of such coulometers is found in Lingane's book [4], where also some electromechanical coulometers are discussed. The principles and scope of controlled-potential coulometry have been discussed by Harrar [435], who also discusses the instrumentation. Electronic integrators can usually be obtained from the companies that supply potentiostats.

IX. PRACTICAL HINTS

Besides the major features in electrolysis discussed here, there are smaller problems to be solved. It is, of course, easiest to consult an experienced electrochemist on such mattes, but such a person may not always turn up when wanted, so the following items are briefly mentioned: circuit, counterelectrode reaction, removal of oxygen before electrolysis, removal of impurities, detection of intermediates, integration of current, and workup.

A. Circuit

For exploratory work a cheap, manually controlled circuit, as shown in Fig. 11, may be assembled from components available in every organic laboratory. Although many organic electrosyntheses may be performed without control of the potential by means of a reference electrode circuit, it is advisable in research work always to include a potential-measuring device. A potentiostat considerably facilitates the control of the electrolysis. Many different potentiostats are now commercially available. When choosing a potentiostat, such properties as output voltage, power, response time, and input resistance in the reference circuit must be considered. Electrolysis in nonaqueous solvents generally requires a higher voltage across the cell than electrolysis in water. Potentiostats for ordinary preparative electrolysis do not need as fast a response as those intended for pulse electrolysis. The input resistance of the reference circuit must be high compared with that of the reference electrode.

The circuit should also include a coulometer and an ammeter.

B. Counterelectrode Reactions

In divided cells the reaction at the counterelectrode is seldom a problem; in aqueous solvents oxygen evolution at the anode is convenient; if chloride is desired as the anion, it is advisable to add ethanol to the anodic compartment as it reacts with the evolved chlorine. In oxidations hydrogen evolution is a convenient counterelectrode reaction; it may also be used in nonaqueous solvents [29]. In methylene chloride it has been noted that chloride ion formed by reduction of methylene chloride at the counterelectrode

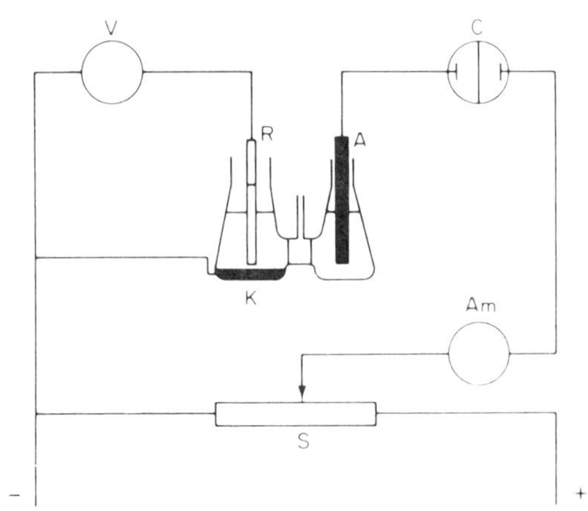

Figure 11. Circuit for constant potential reduction: K, cathode; A, Anode; R, reference electrode; V, potentiometer or pH meter; C, coulometer, Am, ammeter; S, voltage adjuster. (From Ref. 49.)

could diffuse to the anode compartment and participate in the follow-up reactions [386]; this can be avoided by adding a little acetic acid to the cathode compartment.

In undivided cells the counterelectrode reaction must be chosen with care, as the products from it obviously must not interfere with the proper electrode raction. This becomes a problem especially in nonaqueous solvents; in such media hydrogen evolution is generally an acceptable counterelectrode reaction for oxidations; for reductions the oxidation of oxalates and formates to carbon dioxide or azide to nitrogen [404] may be a solution to the problem.

In some cases sacrificial anodes (Mg, Al) may be useful; they allow working in undivided cells in nonaqueous solvents [159,160].

The ideal solution is to have two useful electrode reactions, paired electrosynthesis; this is discussed in Chap. 35.

C. Removal of Oxygen

Before electrolysis most of the dissolved oxygen may be removed by bubbling an inert gas through the solution. It is advisable, especially for small-scale work in nonaqueous solvents, to saturate the gas with the solvent before it is introduced into the cell. The inert gas, nitrogen or argon, may be obtained commercially sufficiently pure, but if not, nitrogen may be purified by passing it through a tube containing a (commercially available) copper catalyst. Previously, nitrogen has been purified by bubbling it through solutions of Cr(II) or V(II) salts or a Fieser solution [3].

Removal of traces of oxygen from a solution by bubbling an inert gas through it is virtually impossible; if such traces must be avoided, all manipulations must take place on a vacuum line and the cells constructed accordingly.

D. Determination of n Values

A direct determination of n values ($n = F/mol$ in a reaction) can be made by microcoulometry with a dropping mercury electrode either with inclusion of a coulometer [438] or by comparison of polarographic diffusion currents with a known reaction [439]. A critical review of the methods has been given by Weaver and Whitnack [440]. However, n values derived from electrolyses with macroelectrodes should be considered with some care because of the possibility of further reactions during the time of the experiment. A reliable procedure for compounds that can be dissolved fully is to follow the concentration of the electroactive substance polarographically, plot corresponding values of wave height and coulometer readings, and calculate the n value from the slope of the possibly straight line obtained during the first part of the electrolysis. If a straight line is not obtained, this indicates that secondary reactions take place.

E. Workup

Often, from the point of view of electrolysis, the choice of electrolyte or solvent is free within certain limits and the workup may than be facilitated considerably by a suitable selection of experimental conditions. For example, the reduction of a nitroalkane [61] to a hydroxylamine requires an aqueous acid medium; sulfuric acid is just as good as hydrochloride acid for the electrolysis, but whereas hydrochloric acid may be removed in vacuo at a low temperature, leaving an alkylhydroxylamine hydrochloride, the sulfuric acid must be neutralized and the free, less stable base extracted.

Salts may be removed by electrodialysis using cation and anion exchange membranes [441] or bipolar membranes [442].

F. Removal of Impurities

To obtain a meaningful value of electricity consumption it is necessary that electricity not be consumed by impurities. A convenient way to remove impurities is to preelectrolyze the catholyte at a potential somewhat more negative than that to be employed during the electrolysis of the substrate. When the background current has reached an acceptable value, the substrate is added and the electrolysis performed at the required potential.

Impurities of nucleophiles or electrophiles may be removed from aprotic solutions by passing the solution through a column of active, neutral alumina; the alumina may be activated at 400–450°C in vacuum and used immediately after.

G. Detection of Intermediates

During the electrolysis it is advisable to obtain some indication of what is going on in the cell. An indicator microelectrode in the electrolytic cell is valuable in order to follow the progress of the electrolysis and to detect intermediates. An intermediate more easily reducible than the substrate, for example, may be detected as a small prewave, and its half-wave potential or other voltammetric characteristics used for its identification. Polarographic or cyclic voltammetric curves measured during the electrolysis also present a means of monitoring the process. The combination of electrochemical and optical techniques to detect intermediates was discussed in Chap. 3.

REFERENCES

1. I. M. Kolthoff and J. J. Lingane, Polarography, 2nd ed., Wiley-Interscience, New York, 1952.
2. J. Heyrovsky and J. Kuta, Grundlagen der Polarographie, Academic Verlag, Berlin, 1965.
3. (a) I. M. Kolthoff and Y. Okinaka, in Progress in Polarography (P. Zuman and I. M. Kolthoff, eds.), Vol. II, Wiley-Interscience, New York, 1962, pp. 357–381; (b) Z. P. Zagorski, in Progress in Polarography (P. Zuman and I. M. Kolthoff, eds.), Vol. II, Wiley-Interscience, New York, 1962, pp. 549–568.
4. J. J. Lingane, Electroanalytical Chemistry, Wiley-Interscience, 2nd ed., New York, 1958.
5. (a) D. Pletcher, Industrial Electrochemistry, Chapman and Hall, London, 1982, p. 52.
6. N. L. Weinberg and B. V. Tilak, Technique of Electroorganic Synthesis, Part, III, Wiley-Interscience, New York, 1982.
7. U. Landau, E. Yeager, and D. Kortan, eds., Electrochemistry in Industry, Plenum, New York, 1982.

8. C. Wagner, in Advances in Electrochemistry and Electrochemical Engineering (C. W. Tobias, ed.), Vol. 2, Wiley-Interscience, New York, 1962, pp. 1–14.
9. M. Eisenberg, in Advances in Electrochemistry and Electrochemical Engineering (C. W. Tobias, ed.), Vol. 2, Wiley-Interscience, New York, 1962, pp. 235–291.
10. J. Heyrovsky, in Polarographie, Springer-Verlag, Vienna, 1941, p. 9.
11. T. Timmell, Sv. Kem. Tidsskr., 58, 67 (1946).
12. P. S. Fedkiw and W. Scott, Jr., J. Electrochem. Soc., 131, 1304 (1984); P. S. Fedkiw and J. C. Chao, AlChE J., 31, 1578 (1985); N. R. K. Vilambi and D. T. Chin, J. Electrochem. Soc., 134, 3074 (1987).
13. J. E. Harrar and I. Shain, Anal. Chem., 38, 1148 (1966).
14. D. Peltier and C. Moinet, Bull. Soc. Chim. Fr., 2657 (1968).
15. M. M. Baizer, J. P. Petrovich, and D. A. Tyssee, J. Electrochem. Soc., 117, 173 (1970).
16. C. Moinet and D. Peltier, Bull. Soc. Chim. Fr., 690 (1969).
17. F. Beck and H. Guthke, Chem.-Ing.-Tech., 41, 943 (1969).
18. (a) B. V. Derjaquin, S. S. Dukhin, and M. M. Kopetelova, J. Colloid Interface Sci., 38, 984 (1972); (b) I. T. Gorbachuk, V. P. Dushchenko, and B. G. Ivanitskii, Teplofiz. Teplotekh., 89 (1973); Chem. Abstr., 81, 17012 (1974).
19. P. I. Lee, J. Electrochem. Soc., 122, 151C (1975).
20. D. Matic, P. M. Robertson, and N. Ibl, Electrochim. Acta, 25, 487 (1980); P. M. Robertson, R. Oberlin, and N. Ibl, Electrochim. Acta 26, 941 (1981).
21. B. Gautheron, G. Tainturier, and C. Degrand, J. Amer. Chem. Soc., 107, 5979 (1985).
22. T. F. Connors and J. F. Rusling, Chemospher, 13, 415 (1984); Chem. Abstr., 101, 49665 (1984); J. Electroanal. Chem., 190, 213 (1985).
23. G. T. Miller, U.S. Pat. 3,361,656 (1968); Chem. Abstr., 68, 83743 (1968), presented for the Detroit Section of ECS, Nov. 21, 1968; N. M. Winslow, Trans. Electrochem. Soc., 88, 81 (1945).
24. L. Nanis and F. R. McLarnon, J. Electrochem. Soc., 117, 1527 (1970).
25. E. Z. Gak, Elektrokhimiya, 3, 89, 263 (1967).
26. T. Z. Fahidy, Electrochim. Acta, 18, 607 (1973).
27. C. Iwakura, T. Edamoto, and H. Tamura, Denki Kagaku, 52, 654 (1984); Chem. Abstr., 102, 52955 (1985).
28. T. Watanabe, Y. Tanimoto, R. Nakagaki, H. Hiromatsu, and S. Nagakura, Bull. Chem. Soc. Jpn., 60, 4166 (1987).
29. C. P. Andrieux, J. M. Dumas-Bouchiai, and J. M. Savéant, J. Electroanal. Chem., 83, 355 (1977).
30. B. Lamm and B. Samuelsson, Acta Chem. Scand., 24, 561 (1970).
31. M. J. Allen, Organic Electrode Processes, Chapman and Hall, London, 1958.
32. A. P. Tomilov, S. G. Mairanovsky, M. Ya. Fioshin, and V. A. Smirnov, Elektrohimiya Organischeskikh Soedinyenii, Khimiya, Leningrad, 1968.
33. F. Goodridge and C. J. H. King, in Technique of Electroorganic Synthesis (N. L. Weinberg, ed.), Vol. V, Part I, Wiley, New York, 1974, pp. 12–147.

34. F. Beck, Elektroorganische Chemie, Verlag Chemie, Weinheim, 1974.
35. R. E. W. Jansson and N. R. Tomov, Electrochim. Acta, 25, 497 (1980).
36. (a) W. J. Koehl, Jr., J. Amer. Chem. Soc., 86, 4686 (1964); (b) G. Dryhurst and P. J. Elving, J. Electrochem. Soc., 115, 1014 (1968).
37. R. P. Linstead, J. C. Lunt, and B. C. L. Weedon, J. Chem. Soc. 3331 (1950).
38. T. Dickinson and W. F. K. Wynne-Jones, Trans. Faraday Soc., 58, 400 (1962).
39. A. Hickling and R. Wilkins, Discuss. Faraday Soc., 45, 261 (1968).
40. S. D. Ross, M. Finkelstein, and R. C. Petersen, J. Amer. Chem. Soc., 89, 4088 (1967).
41. R. Lines, B. S. Jensen, and V. D. Parker, Acta Chem. Scand., B32, 510 (1978).
42. L. Eberson and K. Nyberg, J. Amer. Chem. Soc., 88, 1686 (1966).
43. T. Asahara, M. Seno, and T. Arai, Bull. Chem. Soc. Jpn., 42, 1316 (1969).
44. A. F. Clifford, H. K. El-Shamy, H. J. Emeléus, and R. N. Hazeldine, J. Chem. Soc., 2372 (1953).
45. T. Gramstad and R. N. Hazeldine, J. Chem. Soc., 173 (1956).
46. M. Schmeisser and P. Sartori, Chem.-Ing.-Tech., 36, 9 (1964).
47. J. J. Lingane, C. Gardner Swain, and M. Fields, J. Amer. Chem. Soc., 65, 1348 (1943).
48. R. Pasternak, Helv. Chim. Acta, 31, 753 (1948).
49. H. Lund, Oesterreich. Chem.-Z., 68, 43 (1967).
50. P. E. Iversen, J. Chem. Educ., 48, 136 (1971).
51. S. Musha, T. Wasa, and T. Naito, Bull. Chem. Soc. Jpn., 39, 1902 (1966).
52. D. Peltier, M. Le Guyader, and J. Tacussel, Bull. Soc. Chim. Fr., 2609 (1963).
53. M. R. Rifi, J. Amer. Chem. Soc., 89, 4442 (1967).
54. H. F. Metcalf and G. C. Whitnack, U.S. Pat. 2,835,631 (1958); Chem. Abstr., 52, 14394 (1958).
55. W. H. Smith and A. J. Bard, J. Amer. Chem. Soc., 97, 5203 (1975).
56. J. E. Dubois, A. Monvernay, and P. C. Lacaze, Electrochim. Acta, 15, 315 (1970).
57. K. Sugino, T. Sekine, and N. Sato, Electrochim. Technol., 1, 112 (1963); Chem. Abstr., 58, 10972 (1963).
58. R. C. Schulz and W. Strobel, Monatsh. Chem., 99, 1742 (1968).
59. A. Cisak and P. J. Elving, J. Electrochem. Soc., 110, 160 (1963).
60. A. Cisak, Rocz. Chem., 41, 809 (1967).
61. P. E. Iversen and H. Lund, Acta Chem. Scand., 19, 2303 (1965).
62. L. Horner and A. Mentrup, Liebigs Ann. Chem., 646, 49 (1961).
63. L. V. Kaabak, A. P. Tomilov, S. L. Varshavskii, and M. I. Kabachnik, Zh. Org. Khim., 3, 3 (1967).
64. A. P. Tomilov and B. L. Klyouyev, Elektrokhimiya, 2, 1405 (1966).
65. A. P. Tomilov and V. A. Klimov, Elektrokhimiya, 3, 232 (1967).
66. G. W. McMillan, U.S. Patent 2,485,982 (1949); Chem. Abstr., 44, 1836 (1950).
67. N. A. Izgaryshev and A. A. Petrova, Zh. Fiz. Khim., 24, 745 (1950).
68. P. C. Condit, Ind. Eng. Chem., 48, 1252 (1956).

69. C. Barry and G. Cauquis, Bull. Soc. Chim. Fr., 1032 (1966).
70. R. Kanakam, M. S. V. Pathy, and H. V. K. Udupa, Electrochim. Acta, 12, 329 (1967).
71. T. D. Balakrishnan, K. S. Udupa, G. S. Subramanian, and H. V. K. Udupa, Chem.-Ing.-Tech., 41, 776 (1969).
72. K. Natarajan, K. S. Udupa, G. S. Subramanian, and H. V. K. Udupa, Electrochem. Technol., 2, 151 (1964).
73. K. S. Udupa, G. S. Subramanian, and H. V. K. Udupa, J. Electrochem. Soc., 108, 373 (1961).
74. R. Ramaswamy, M. S. Venkatachalapathy, and H. V. K. Udupa, J. Electrochem. Soc., 110, 294 (1963).
75. P. N. Anantaraman, G. S. Subramanian, and H. V. K. Udupa, Ind. J. Technol., 4, 271 (1966).
76. (a) L. Eberson, K. Nyberg, and H. Sternerup, Chem. Scr., 3, 12 (1973); (b) L. Cedheim, L. Eberson, B. Helgée, K. Nyberg, R. Servin, and H. Sternerup, Acta Chem. Scand., B29, 617 (1975); (c) K. Nyberg and R. Servin, Acta Chem. Scand., B30, 640 (1976); (d) L. Eberson, J. Hlavaty, L. Jönsson, K. Nyberg, R. Servin, H. Sternerup, and L.-G. Wistrand, Acta Chem. Scand., B33, 113 (1979).
77. U. Landau, N. Weinberg, and E. Gileadi, J. Electrochem. Soc., 135, 396 (1988).
78. H. J. Landaez-Machado, A. Darchen, and C. Moinet, Electrochim. Acta, 25, 1321, 1519 (1980).
79. C. Jacob and C. Moinet, Bull. Soc. Chim. Fr., 291 (1983).
80. C. Lamoureux, C. Moinet, and A. Tallec, Electrochim. Acta, 31, 1 (1986).
81. C. Lamoureux, C. Moinet, and A. Tallec, J. Appl. Electrochem., 16, 819 (1986); C. Lamoureux and C. Moinet, Bull. Soc. Chim. Fr., 59 (1988).
82. Z. Ogumi, K. Nishio, and S. Yoshizawa, Electrochim. Acta, 26, 1779 (1981); Z. Ogumi, H. Yamashita, and K. Nishio, Electrochim. Acta, 28, 1687 (1983).
83. J. Sarrazin and A. Tallec, J. Electroanal. Chem., 137, 183 (1982); E. Raoult, J. Sarrazin, and A. Tallec, J. Appl. Electrochem., 14, 639 (1984); 15, 85 (1985); Bull. Soc. Chim. Fr., 1200 (1985); J. Membr. Soc., 30, 23 (1987).
84. Z. Ogumi, M. Inaba, S.-I. Ohashi, M. Uchida, and Z.-I. Takehara, Electrochim. Acta, 33, 365 (1988).
85. V. A. Grinberg, V. N. Zhuravleva, Yu. B. Vasil'ev, and V. E. Kazarinov, Elektrokhimiya, 19, 1447 (1983).
86. F. Beck, J. Appl. Electrochem., 2, 59 (1972).
87. P. M. Robertson, F. Schwager, and N. Ibl., J. Electroanal. Chem., 65, 883 (1975).
88. P. Gallone, Electrochim. Acta, 22, 913 (1977).
89. (a) J. R. Backhurst, J. M. Coulson, F. Goodridge, R. E. Plimley, and M. Fleischmann, J. Electrochem. Soc., 116, 1600 (1969); (b) M. Fleischmann and J. W. Oldfield, J. Electroanal. Chem., 29, 211, 231 (1971); (c) S. Germain and F. Goodridge, Electrochim. Acta, 21, 545 (1976).
90. P. LeGoff, F. Vergnes, F. Coeuret, and J. Bordet, Ind. Eng. Chem., 61, 8 (1969).

91. J. N. Hiddleston and A. F. Douglas, Electrochim. Acta, 15, 431 (1970).
92. G. Kreysa and E. Heitz, Chem.-Ing.-Tech., 48, 852 (1976).
93. A. T. S. Walker and A. A. Wragg, Electrochim. Acta, 25, 323 (1980).
94. A. P. Brown, M. Fleischmann, and D. Pletcher, J. Electroanal. Chem., 50, 65 (1974).
95. R. P. VanDuyne and C. N. Reilley, Anal. Chem., 44, 142 (1972).
96. T. Kuwana and W. R. Heineman, Acc. Chem. Res., 7, 241 (1976); T. Kuwana and N. Winograd, in Electroanalytical Chemistry (A. J. Bard, ed.), Vol. 7, Dekker, New York, 1974, p. 1; W. R. Heineman, F. M. Hawkridge, and H. N. Blount, in Electroanlytical Chemistry (A. J. Bard, ed.), Vol. 13, Dekker, New York, 1984, p. 1.
97. W. M. Flarsheim, Y.-M. Tsou, I. Trachtenberg, K. P. Johnston, and A. J. Bard, J. Phys. Chem., 90, 3857 (1986). R. M. Crooks and A. J. Bard, J. Phys. Chem., 91, 1274 (1987); J. Electroanal. Chem., 240, 253 (1988); 243, 117 (1988).
98. R. A. Dombro, Jr., G. A. Prentice, and M. A. McHugh, J. Electrochem. Soc., 135, 2219 (1988).
99. R. B. MacMullin, Electrochem. Technol., 1, 5 (1963); 2, 106 (1964).
100. T. Asahara, M. Seno, and H. Kaneko, Bull. Chem. Soc. Jpn., 41, 2985 (1968).
101. R. A. Benkeser, E. M. Kaiser, and R. F. Lambert, J. Amer. Chem. Soc., 86, 5272 (1964).
102. T. D. Balakrishnan, K. S. Udupa, G. S. Subramanian, and H. V. K. Udupa, Chem. Ind. (London), 1622 (1970).
103. J. Hranilovic, D. Koruncev, and E. Gustak, Electrochem. Technol., 6, 62 (1968).
104. J. L. Weiniger and F. F. Holub, J. Electrochem. Soc., 117, 340 (1970).
105. M. Seko, S. Ogawa, M. Yoshida, and H. Shiroki, in Electrochemische Verfahrenstechnik, Dechema Monographien, 97, 27 (1984).
106. W. G. F. Groot, in Proc. Symp. Diaphragms, Separators, and Ion-exchange Membranes, Electrochem. Soc. Meeting, Spring 1986, Boston, Proceedings Vol. 86-13, p. 1.
107. N. Urabe, T. Seyama, and W. Sakai, J. Electrochem. Soc. Jpn. (Overseas Suppl. Ed.) 28, E-69 (1960).
108. T. Erdey-Gruz and M. Volmer, Z. Phys. Chem., 150, 203 (1930).
109. L. G. Feoktistov and M. M. Gol'din, Elektrokhimiya, 4, 356 (1968).
110. J. Heyrovsky, Rec. Trav. Chim. Pays-Bas, 44, 499 (1925).
111. J. Horiuti and G. Okamoto, Sci. Pap. Inst. Phys. Chem. Res. (Tokyo) 28, 231 (1936); Chem Abstr., 30, 3332 (1936).
112. K. J. Vetter and D. Otto, Z. Electrochem., 60, 1072 (1956).
113. X. de Hemptinne and J. C. Jungers, Z. Phys. Chem., 15, 137 (1958).
114. M. Fujihira, A. Yokozawa, H. Konoshita, and T. Osa, Chem. Lett., 1089 (1982).
115. F. Fichter and R. Stocker, Ber. Dtsch. Chem. Ges., 47, 2003 (1914).
116. W. D. Bancroft and A. B. George, Trans. Amer. Electrochem. Soc., 57, 399 (1930).
117. C. Schall and W. Kirst, Z. Elektrochem., 29, 537 (1923).
118. S. D. Ross, M. Finkelstein, and R. C. Petersen, J. Amer. Chem. Soc., 92, 6003 (1970).

119. V. G. Khomyakov, A. P. Tomilov, and B. G. Soldatov, Electrokhimiya, 5, 850 (1969).
120. I. A. Avrutskaya, M. Ya. Fioshin, S. D. Kabanova, and L. E. Gerasimova, Elektrokhimiya, 5, 951 (1969).
121. A. Chilesotti, Z. Elektrochem., 7, 768 (1900).
122. F. S. Holland, CITCE Meet., Prague, 1970, Extended Abstract, p. 414.
123. (a) R. Alkire and B. Gracon, J. Electrochem. Soc., 122, 1594 (1975); (b) R. Alkire and P. K. Ng, J. Electrochem. Soc., 124, 1220 (1977); (c) J. A. Trainham and J. Newman, J. Electrochem. Soc., 124, 1528 (1977); (d) J. A. Trainham and J. Newman, J. Electrochem. Soc., 125, 58 (1978); G. M. Brown and F. A. Posey, J. Electrochem. Soc., 128, 306 (1981); (e) R. C. Alkire and R. M. Gould, J. Electrochem. Soc., 127, 605 (1980); (f) H. J. Landaez-Machado, A. Darchen, and C. Moinet, Electrochim. Acta, 25, 1321 (1980); (g) C. Moinet and J. Simonet, Sandbjerg Meet., 1981, Abstracts of Papers, p. 53.
124. J. Tafel, Ber. Dtsch. Chem. Ges., 33, 2209 (1900).
125. N. L. Weinberg, M. Lipstajn, D. J. Mazur, M. Reicher, H. R. Weinberg, and E. P. Weinberg, Stud. Org. Chem., 30, 441 (1987).
126. F. Beck and W. Gabriel, Angew. Chem., 97, 765 (1985).
127. T. Matsue, T. Yamada, H. Takahashi, and T. Osa, Bull. Soc. Chem. Jpn., 59, 3690 (1986); T. Osa, T. Matsue, A. Yokozawa, and T. Yamada, Chem. Lett., 1611 (1987).
128. L. Horner and C. Franz, Z. Naturforsch. B, 40B, 808 (1985).
129. T. Chiba, M. Okimoto, H. Nagai, and Y. Takata, Bull. Chem. Soc. Jpn., 56, 719 (1983).
130. G. Belot, S. Desjardins, and J. Lessard, Tetrahedron Lett., 25, 5347 (1984).
131. K. Park, P. N. Pintauro, M. M. Baizer, and K. Nobe, J. Appl. Electrochem., 16, 941 (1986).
132. H. Berge and G. Merkel, Wiss. Z. Wilhelm-Pieck-Univ. Rostock, Naturwiss. Reihe, 32, 80 (1983).
133. H. Berge, B. Strübing, S. Strasen, M. Aquila, and K. Peseke, Wiss. Z. Wilhelm-Pieck-Univ. Rostock, Naturwiss. Reihe, 27, 639 (1978).
134. M. L. Contreras, S. Rivas, and R. Rozas, J. Electroanal. Chem., 177, 299 (1984).
135. G. LeGuillanton, Q. T. Do, and J. Simonet, Tetrahedron Lett., 27, 2661 (1986).
136. P. Jeroschewski, W. Ruth, B. Strübing, and H. Berge, J. Prakt. Chem., 324, 787 (1982).
137. G. Merkel, H. Berge, and P. Jeroschewski, J. Prakt. Chem., 326, 467 (1984).
138. R. H. Dahn, R. J. Latham, and S. E. Mosley, J. Appl. Electrochem., 16, 213 (1986).
139. I. M. Kolthoff and N. Tanaka, Anal. Chem., 26, 632 (1954).
140. J. W. Ross and I. Shain, Anal. Chem., 28, 548 (1956).
141. Z. P. Bezrukova, Yu. B. Vasil'ev, L. S. Kanevskii, E. P. Kovsman, V. I. Lushnikov, N. T. Presnova, R. G. Pryadkina, G. A. Tarkhanov, and V. D. Chekanova, Elektrokhimiya, 12, 783 (1976).
142. L. Bjelica, R. Parsons, and R. M. Reeves, Proc. Electrochem. Soc., 1979 (publ. 1980), 80–3, 190–212; L. Bjelica, R. Parsons, and R. H. Reeves, Croat. Chem. Acta, 53, 211 (1980).

143. M. P. J. Brennan and R. Brettle, J. Chem. Soc., Perkin Trans. I, 257 (1973); R. E. Panzer and P. J. Elving, J. Electrochem. Soc., 119, 864 (1972).
144. Ger. Pat. 865,740 (1953); Chem. Abstr., 53, 292 (1959).
145. Br. Pat. 766,578 (1957); Chem. Abstr., 51, 10278 (1957).
146. D. W. Wabner, H. P. Fritz, D. Missol, R. Huss, and F. Hindelang, Z. Naturforsch., 31b, 39 (1976).
147. D. W. Wabner, R. Huss, F. Hindelang, H. P. Fritz, and D. Missol, Z. Naturforsch., 31b, 45 (1976).
148. F. Beck and W. Gabriel, J. Electroanal. Chem., 182, 355 (1985).
149. S. R. Ellis, N. A. Hampson, M. C. Ball, and F. Wilkinson, J. Appl. Electrochem., 16, 159 (1986).
150. G. Vertes, G. Horanyi, and G. Nagy, Tetrahedron, 28, 37 (1972).
151. M. Fleichmann, D. Pletcher, and K. Korinek, J. Chem. Soc., Perkin II, 1396 (1972); D. Pletcher, M. Amjad, and C. Smith, J. Electrochem. Soc., 124, 203 (1977).
152. P. M. Robertson, J. Electroanal. Chem., 111, 97 (1980).
153. H. Kaulen and H. J. Schäfer, Tetrahedron, 38, 3299 (1982); H. J. Schäfer, U. Feldhues, H. Kaulen, and H. Roholl, Stud. Org. Chem., 30, 161 (1987); H. Roholl and H. J. Schäfer, Synthesis, 408, (1987); H. J. Schäfer, Top. Curr. Chem., 142, 101 (1987).
154. P. Seiler and P. M. Robertson, Chimia, 36, 305 (1982); P. M. Robertson, P. Berg, H. Reimann, K. Schleich, and P. Seiler, J. Electrochem. Soc., 130, 591 (1983).
155. H. B. Beer, J. Electrochem. Soc., 127, 303C (1980); H. B. Beer, Neth. Appl. 6606, 302 (1966); Chem. Abstr., 67, 17379 (1967).
156. O. DeNora, Chem.-Ing.-Tech., 43, 182 (1971).
157. F. Beck and G. Csizi, Ger. Offen. 1, 915, 951 (Oct. 8, 1970); Chem. Abstr., 74, 37744 (1971).
158. F. Beck and H. Schultz, Electrochim. Acta., 29, 1569 (1984).
159. G. Silvestri, S. Gambino, and G. Filardo, Adv. Polym. Sci., 38, 28 (1981); Tetrahedron Lett., 27, 3429 (1986); Stud. Org. Chem., 30, 287 (1987) and references therein; S. Gambino, G. Filardo, and G. Silvestri, J. Appl. Electrochem., 12, 549 (1982).
160. O. Sock, M. Troupel, and J. Perichon, Tetrahedron Lett., 26, 1509 (1985); E. d'Incan, S. Sibille, and J. Perichon, Tetrahedron Lett. 27, 4175 (1986); S. Sibille, E. d'Incan, L. Leport, and J. Perichon, Tetrahedron Lett., 27, 3129 (1986); J. C. Folest, J. Y. Nedelec, and J. Perichon, Tetrahedron Lett., 28, 1885 (1987).
161. S. Sibille, E. d'Incan, L. Leport, M.-C. Massebien, and J. Perichon, Tetrahedron Lett., 28, 55 (1987).
162. G. Kokkinidis, J. Electroanal. Chem., 201, 217 (1986).
163. D. J. G. Ives and G. J. Janz, eds., Reference Electrodes, Academic, Press, New York, 1961.
164. J. N. Butler, in Advances in Electrochemistry and Electrochemical Engineering (P. Delahay and C. W. Tobias, eds.), Vol. 7, Wiley-Interscience, New York, 1970.
165. V. A. Pleskov, Usp. Khim., 16, 254 (1947); J. Phys. Chem. (USSR), 22, 351 (1948).
166. H.-M. Koepp, H. Wendt, and H. Strelow, Z. Electrochem., 64, 483 (1960).
167. H. Schneider and H. Strelow, J. Electroanal. Chem., 12, 530 (1966).
168. L. V. Nelson and R. T. Iwamoto, Anal. Chem., 33, 1795 (1961).
169. I. M. Kolthoff and F. G. Thomas, J. Phys. Chem., 69, 3029 (1965).

170. O. Popovych, Anal. Chem., 38, 558 (1966).
171. J. F. Coetzee and J. J. Campion, J. Amer. Chem. Soc., 89, 2513, 2517 (1967).
172. J. Brenet, J. Chim. Phys., 66, 1057 (1969).
173. N. Matsuura, K. Umemoto, M. Waki, Z. Takeuchi, and M. Omoto, Bull. Chem. Soc. Jpn., 47, 806 (1974).
174. G. Gritzner and J. Kůta, Pure Appl. Chem., 56, 461 (1984); Electrochim. Acta, 29, 869 (1984).
175. J. Daschbach, D. Blackwood, J. W. Pons, and S. Pons, J. Electroanal. Chem., 237, 269 (1987).
176. V. D. Parker, J. Amer. Chem. Soc., 96, 5656 (1974); 98, 98 (1976).
177. N. S. Moe, Anal. Chem., 46, 968 (1974).
178. J. F. Coetzee and C. W. Gardner, Anal. Chem., 54, 2625 (1982).
179. W. H. Tiedemann and D. N. Bennion, J. Electrochem. Soc., 117, 203 (1970).
180. V. A. Pleskov, Zh. Fiz. Khim., 22, 351 (1948).
181. H. Lund, Acta Chem. Scand., 11, 491, 1323 (1957).
182. P. Walden and E. J. Birr, Z. Phys. Chem., 144A, 269 (1929).
183. K. Igutsu, M. Ito, and E. Sarai, Anal. Sci., 1, 341 (1985).
184. G. Cauquis and D. Serve, Bull. Soc. Chim. Fr., 302 (1966).
185. H. Yeager and B. Kratochvil, J. Phys. Chem., 73, 1963 (1969).
186. B. Kratochvil, E. Lorah, and C. Garber, Anal. Chem., 41, 1793 (1969).
187. D. L. McMasters, R. B. Dunlap, J. R. Kuempel, L. W. Kreider, and T. R. Shearer, Anal. Chem., 39, 103 (1967).
188. J. Sedlet and T. DeVries, J. Amer. Chem. Soc., 73, 5808 (1951).
189. D. R. Cogley and J. N. Butler, J. Electrochem. Soc., 113, 1074 (1966).
190. R. N. Hammer and J. J. Lagowski, Anal. Chem., 34, 597 (1962).
191. J. F. Coetzee and G. R. Padmanabhan, J. Phys. Chem., 66, 1708 (1962).
192. H. W. Salzberg and M. Leung, J. Org. Chem., 30, 2873 (1965).
193. J. C. Synnott and J. N. Butler, Anal. Chem., 41 1890 (1969).
194. A. I. Popov and D. H. Geske, J. Amer. Chem. Soc., 79, 2074 (1957).
195. B. Burrows and R. Jasinski, J. Electrochem. Soc., 115, 348 (1968).
196. W. H. Smyrl and C. W. Tobias, J. Electrochem. Soc., 113, 754 (1966).
197. V. A. Pleskov and A. M. Monoszon, Acta Physicochim. USSR, 2, 615 (1935).
198. L. W. Marple, Anal. Chem., 39, 844 (1967).
199. G. I. Kaurova, L. M. Grubina, and Ts. A. Adzhemyan, Elektrokhimiya, 3, 1222 (1967).
200. A. G. Doughty, M. Fleischmann, and D. Pletcher, J. Electroanal. Chem., 51, 329 (1974).
201. S. I. Bailey, W.-P. Leung, and I. M. Ritchie, Electrochim. Acta, 30, 861 (1985).
202. J. B. Conant, L. F. Small, and B. S. Taylor, J. Amer. Chem. Soc., 47, 1959 (1925).
203. J. F. Coetzee and C. W. Gardner, Anal. Chem., 54, 2530 (1982).
204. P. J. Peerce and A. J. Bard, J. Electroanal. Chem., 108, 121 (1980); R. M. Kannuck, J. M. Bellama, E. A. Blubaugh, and R. A. Durst, Anal. Chem., 59, 1473 (1987).
205. H. A. Laitinen and C. J. Nyman, J. Amer. Chem. Soc., 70, 2241 (1948).

206. R. C. Larson, R. T. Iwamoto, and R. N. Adams, Anal. Chim. Acta, 25, 371 (1961).
207. J. Sedlet and T. DeVries, J. Amer. Chem. Soc., 73, 5808 (1951).
208. H. W. Sternberg, R. E. Markby, and I. Wender, J. Electrochem. Soc., 110, 425 (1958).
209. U. Bertocci, Z. Elektrochem., 61, 434 (1957).
210. A. Cisak and P. J. Elving, J. Electrochem. Soc., 110, 160 (1963).
211. M. C. Giordano, J. C. Barzan, and A. J. Arvia, Electrochim. Acta, 11, 741 (1966).
212. J. Cihalik and J. Šimek, Coll. Czech. Chem. Commun., 23, 615 (1958).
213. J. O'M. Bockris, Discuss. Faraday Soc., 1, 95 (1947).
214. W. B. Mather and F. C. Anson, Anal. Chem., 33, 1634 (1961).
215. V. Plichon, Bull. Soc. Chim. Fr., 262 (1964).
216. J. C. James, Trans. Faraday Soc., 47, 1240 (1951); I. Bergman and J. C. James, Trans. Faraday Soc., 50, 60 (1954).
217. G. Adhami and M. Herlem, J. Electroanal. Chem., 26, 363 (1970).
218. J. Bertram, J. P. Coleman, M. Fleischmann, and D. Pletcher, J. Chem. Soc., Perkin Trans. 2, 374 (1973).
219. P. N. Ross and P. C. Andricacos, J. Electroanal. Chem., 154, 205 (1983).
220. L. M. Mukherjee and R. G. Bates, J. Electroanal. Chem., 187, 73 (1985).
221. B. E. Conway, N. Marincic, D. Gilroy, and E. Rudd, J. Electrochem. Soc., 113, 1144 (1966).
222. J. Badoz-Lambling and M. Sato, Acta Chim. Hung., 32, 191 (1962).
223. J. Perichon and R. Buvet, Electrochim. Acta, 9, 567 (1964).
224. A. K. Hoffmann, W. G. Hodgson, D. L. Maricle, and W. H. Jura, J. Amer. Chem. Soc., 86, 631 (1964).
225. R. E. Dessy, W. Kitching, and T. Chivers, J. Amer. Chem. Soc., 88, 453 (1966).
226. C. G. Swain, A. Ohno, D. K. Roe, R. Brown, and T. Maugh, J. Amer. Chem. Soc., 89, 2648 (1967).
227. G. Schöber and V. Gutmann, Monatsh. Chem., 89, 401, 649 (1958); O. Sock, P. Lemoine, and M. Gross, Electrochim. Acta, 25, 1025 (1980).
228. G. J. Hills, Tetrahedron Lett. , 28, 433 (1987).
229. A. K. Gupta, J. Chem. Soc., 3473 (1952).
230. I. M. Kolthoff and T. B. Reddy, Inorg. Chem., 1, 189 (1962).
231. J. Perichon and R. Buvet, Bull. Soc. Chim. Fr., 3697 (1967).
232. N. Hackerman, E. S. Snavely, and L. D. Fiel, Electrochim. Acta, 12, 535 (1967).
233. Ya. N. Voitovich, V. Ya. Kazakov, and T. F. Starkova, Elektrokhimiya, 10, 404 (1974).
243. P. T. Cottrell and C. K. Mann, J. Electrochem. Soc., 116, 1499 (1969).
235. R. R. Rao, S. B. Milliken, S. L. Robinson, and C. K. Mann, Anal. Chem., 42, 1076 (1970).
236. G. Cauquis and D. Serve, J. Electroanal. Chem., 27, App. 3 (1970).
237. O. Hammerich and V. D. Parker, Electrochim. Acta, 18, 537 (1973); B. S. Jensen and V. D. Parker, J. Amer. Chem. Soc., 97, 5211 (1975).
238. R. J. Jasinski and S. Kirkland, Anal. Chem., 39, 1663 (1967).
239. R. J. Jasinski and S. Carroll, Anal. Chem., 40, 1908 (1968).

240. C. D. Thompson, F. D. Bogar, and R. T. Foley, Anal. Chem., 42, 1474 (1970).
241. V. Sh. Tsveniashvili, N. S. Khavtasi, R. L. Kharashvili, and M. V. Malashkhiya, Electrokhimiya, 22, 705 (1986).
242. D. R. Burfield, Anal. Chem., 48, 2285 (1976).
243. Y. Mugnier and E. Laviron, J. Electroanal. Chem., 110, 375 (1980).
244. Yu. Kargin, L. M. Vorontsova, L. Z. Manapova, and S. V. Kuzovenko, Elektrokhimiya, 23, 415 (1987).
245. (a) R. H. McKee and C. J. Brockmann, Trans. Electrochem. Soc., 62, 203 (1932); (b) R. H. McKee and B. G. Gerastopolou, Trans. Electrochem. Soc., 68, 329 (1935).
246. A. J. Parker, Electrochim. Acta, 21, 671 (1976).
247. V. Gutmann, Electrochim Acta, 21, 661 (1976). V. Gutmann, The Donor-Acceptor Approach to Molecular Interactions, Plenum, New York, 1978.
248. L. N. Bykova and S. I. Petrov, Zh. Anal. Khim., 27, 1076 (1972).
249. L. Eberson and B. Helgée, Chim. Scr., 5, 49 (1974).
250. L. Eberson and B. Helgée, Acta Chem. Scand., B29, 451 (1975); B31, 813 (1977); B32, 159 (1978).
251. S. R. Ellis, D. Pletcher, P. Gough, and S. R. Korn, J. Appl. Electrochem., 12, 687 (1982); 12, 694 (1982).
252. J. F. Rusling, C.-N. Shi, D. K. Gosser, and S. S. Shukla, J. Electroanal. Chem., 240, 201 (1988); J. F. Rusling, C.-N. Shi, and S. L. Suib, J. Electroanal. Chem., 245, 331 (1988).
253. Z. Ibrisagic, D. Pletcher, W. N. Brooks, and K. P. Healy, J. Appl. Electrochem., 15, 719 (1985).
254. Z. Ibrisagic, V. Misovic, I. Tabakovic, and I. Santic, J. Appl. Electrochem., 16, 907 (1986).
255. E. Laurent, G. Rauniyar, and M. Thomalla, J. Appl. Electrochem., 14, 741 (1984); 15, 121 (1985).
256. M. Fleischmann, C. L. K. Tennakoon, H. A. Bampfield, and P. Williams, J. Appl. Electrochem., 13, 593 (1983).
257. H. Feess and H. Wendt, Ber. Bunsenges. Phys. Chem., 85, 914 (1981).
258. X. Zhang and A. J. Bard, J. Amer. Chem. Soc., 107, 3719 (1985).
259. J. M. Sullivan, D. C. Hanson, and R. Killer, J. Electrochem. Soc., 117, 779 (1970).
260. D. Seebach and H. A. Oei, Angew. Chem., 87, 629 (1975).
261. J. F. Coetzee, ed., Recommended Methods for Purification of Solvents and Tests for Impurities, Pergamon, Oxford, 1982.
262. C. K. Mann, in Electroanalytical Chemistry (A. J. Bard, ed.), Vol. 3, Dekker, New York, 1969, p. 123.
263. C. K. Mann and K. K. Barnes, Electrochemical Reactions in Nonaqueous Systems, Dekker, New York, 1970.
264. J. J. Lagowski, ed., The Chemistry of Nonaqueous Solvents, Vols. I and II, Academic, New York, 1966, 1967.
265. J. Bertram, M. Fleischmann, and D. Pletcher, Tetrahedron Lett., 349 (1971).
266. J. Bertram, J. P. Coleman, M. Fleischmann, and D. Pletcher, J. Chem. Soc., Perkin Trans., II, 374 (1973).
267. F. Bobilliart, A. Thiebault, and M. Herlem, C. R. Acad. Sci., Ser. C., 278, 1485 (1974).

268. C. Pitti, F. Bobilliart, A. Thiebault, and M. Herlem, Anal. Lett., 8, 241 (1975).
269. J. P. Coleman and D. Pletcher, Tetrahedron Lett., 147 (1974).
270. A. P. Rudenko, M. Ya. Sarubin, S. F. Awerjanow, and F. Pragst, J. Prakt. Chem., 327, 191 (1985).
271. A. Germain, P. Otega, and A. Commeyrus, Nouv. J. Chim., 3, 415 (1979); B. Carre, P. L. Fabre, and J. Devynck, Bull. Soc. Chim. Fr., 255 (1987).
272. K. A. Striebel, P. C. Andricacos, E. J. Cairns, P. N. Ross, and F. R. McLarnon, J. Electrochem. Soc., 132, 2381 (1985).
273. J. H. Simons, J. Electrochem. Soc., 95, 47 (1949).
274. M. Schmeisser and P. Sartori, Chem.-Ing.-Tech., 36, 9 (1964).
275. S. Nagase, Fluorine Chem. Rev., 1, 77 (1967).
276. M. Kilpatrick and J. G. Jones, in The Chemistry of Nonaqueous Solvents (J. J. Lagowski, ed.), Vol. II, Academic, New York, 1967, p. 43 ff.
277. Yu. Ya. Vinnikov, S. P. Shavkunov, and K. N. Bil'dinov, Elektrokhimiya, 12, 1113 (1976).
278. H. P. Raaen, Anal. Chem., 34, 1714 (1962).
279. M. E. Runner, G. Baloy, and M. Kilpatrick, J. Amer. Chem. Soc., 85, 1038 (1963).
280. H. H. Rogers, S. Evans, and J. H. Johnson, J. Electrochem. Soc., 111, 701 (1964).
281. E. Laurent, B. Marquet, R. Tardivel, and A. Thiebault, Bull. Soc. Chim. Fr., 955 (1986); Shell, UK Pat. Appl. GB 2191480 (1986), Chem. Abstr. 108, 212559 (1988).
282. J. W. Sargent, A. F. Clifford, and W. R. Lemmon, Anal. Chem., 25, 1727 (1953).
283. J. S. Clarke and A. T. Kuhn, J. Electroanal. Chem., 85, 299 (1977).
284. G. Petit and J. Bessière, J. Electroanal. Chem., 25, 317 (1970); 31, 393 (1971); 34, 489 (1972); H. P. Fritz and T. Würminghausen, J. Electroanal. Chem., 54, 181 (1974); J. Chem. Soc., Perkin Trans., I, 610 (1976); Z. Naturforsch., 32b, 245 (1977). Y. H. So and L. L. Miller, Synthesis, 468 (1976).
285. W. Dannhauser and R. C. Cole, J. Amer. Chem. Soc., 74, 6105 (1952).
286. F. E. Harris and C. T. O'Konski, J. Amer. Chem. Soc., 76, 4317 (1954).
287. O. Hammerich, N. S. Moe, and V. D. Parker, J. Chem. Soc., Chem. Commun., 156 (1972).
288. O. Hammerich and V. D. Parker, Electrochim. Acta, 18, 537 (1973).
289. S. D. Ross, M. Finkelstein, and R. C. Petersen, J. Amer. Chem. Soc., 86, 4139 (1964).
290. M. Leung, J. Herz, and H. W. Salzberg, J. Org. Chem., 30, 310 (1965).
291. L. Eberson and K. Nyberg, J. Amer. Chem. Soc., 88, 1686 (1966).
292. J. Čihalik and J. Šimek, Coll. Czech. Chem. Commun., 23, 615 (1958).
293. I. Bergman and J. C. James, Trans. Faraday Soc., 48, 956 (1952).
294. K. J. P. Orton and A. E. Bradfield, J. Chem. Soc., 983 (1927).
295. J. A. Knopp, W. S. Linnell, and W. C. Child, J. Phys. Chem., 66, 1513 (1962).
296. S. Wawzonek, R. Berkey, and D. Thomson, J. Electrochem. Soc., 103, 513 (1956).
297. H. Lund and J. Bjerrum, Ber. Dtsch. Chem. Ges., 64B, 210 (1931).

298. H. A. Laitinen and C. E. Shoemaker, J. Amer. Chem. Soc., 72, 663 (1950).
299. A. F. Clifford and M. Giménez-Huguet, J. Amer. Chem. Soc., 82, 1024 (1960).
300. W. B. Schaab, R. F. Conley, and F. C. Schmidt, Anal. Chem., 33, 498 (1961).
301. G. Silvestri, S. Gambino, G. Filardo, C. Cuccia, and E. Guarino, Angew. Chem., 93, 131 (1981).
302. (a) M. H. Miles and P. M. Kellett, J. Electrochem. Soc., 115, 1225 (1968); (b) J. Electrochem. Soc., 117, 60 (1970).
303. M. Herlem, Bull. Soc. Chim. Fr., 1687 (1967).
304. M. Herlem and A. Thiebault, Bull. Soc. Chim. Fr., 383 (1970); 719 (1971).
305. A. Demortier and A. J. Bard, J. Amer. Chem. Soc., 95, 3495 (1973).
306. J. M. Savéant and A. Thiebault, J. Electroanal. Chem., 89, 335 (1978).
307. W. H. Tiedemann and D. N. Bennion, J. Electrochem. Soc., 117, 203 (1970).
308. H. Schmidt and H. Meinert, Z. Anorg. Allg. Chem., 295, 156 (1958).
309. R. A. Benkeser and E. M. Kaiser, J. Amer. Chem. Soc., 85, 2858 (1963).
310. R. A. Benkeser and S. J. Mels, J. Org. Chem., 34, 3970 (1969).
311. H. W. Sternberg, R. E. Markby, and I. Wender, J. Electrochem. Soc., 110, 425 (1963).
312. H. W. Sternberg, R. E. Markby, I. Wender, and D. M. Mohilner, J. Electrochem. Soc., 113, 1060 (1966).
313. W. B. Schaab, R. E. Bayer, J. R. Siefker, J.-Y. Kim, P. W. Brewster, and F. C. Smith, Rec. Chem. Prog., 22, 197 (1961).
314. L. M. Mukherjee and S. Bruckenstein, Pure Appl. Chem., 13, 421 (1966); M. Asthana and L. M. Mukherjee, Recommended Methods for Purification of Solvents and Tests for Impurities, Pergamon, Oxford, (1982).
315. J. R. Jezorek and H. B. Mark, J. Phys. Chem., 74, 1627 (1970).
316. E. O. Sherman and D. C. Olson, Anal. Chem., 40, 1174 (1968).
317. J. F. Coetzee, Pure Appl. Chem., 13, 429 (1966).
318. L. Carlsen, H. Egsgaard, and J. R. Andersen, Anal. Chem., 51, 1593 (1979).
319. H. Kiesele, Anal. Chem., 52, 2230 (1980).
320. (a) D. R. Burfield, K.-H. Lee, and R. H. Smithers, J. Org. Chem., 42, 3060 (1977); (b) D. R. Burfield and R. H. Smithers, J. Org. Chem., 43, 3966 (1978); (c) D. R. Burfield, G. H. Gan, and R. H. Smithers, J. Appl. Chem. Biotechnol., 28, 23 (1978); (d) D. R. Burfield, R. H. Smithers, and A. S. C. Tan, J. Amer. Chem. Soc., 46, 629 (1981).
321. J. F. Coetzee, G. P. Cunningham, D. K. McGuire, and G. R. Padmanabhan, Anal. Chem., 34, 1139 (1962).
322. G. A. Forcier and J. W. Olver, Anal. Chem., 37, 1447 (1965).
323. N. S. Moe, Acta Chem. Scand., 21, 1389 (1967).
324. J. F. O'Donnell, J. T. Ayres, and C. K. Mann, Anal. Chem., 37, 1161 (1965).
325. H. Kiesele, Anal. Chem., 53, 1952 (1981).
326. J. W. Teter and W. J. Merwin, U.S. Pat. 2,453,472 (1948); Chem. Abstr., 43, 2630 (1949).

327. J. P. Billon, J. Electroanal. Chem., 1, 486 (1960).
328. J. F. Coetzee, ed., in Recommended Methods for Purification of Solvents and Tests for Impurities, Pergamon, Oxford, 1982, p. 10.
329. E. M. Abbot, A. J. Bellamy, and J. Kerr, Chem. Ind. (London), 828 (1974), A. J. Bellamy, J. B. Kerr, C. J. McGregor, and I. S. MacKirdy, J. Chem. Soc., Perkin Trans. II, 161 (1982).
330. J. K. Foley, C. Korzeniewski and S. Pons, Can. J. Chem., 66, 201 (1988).
331. J. Heinze, Angew. Chem., 93, 186 (1981).
332. R. E. Visco, Electrochem. Soc. Meet., Dallas, Tex., May 1967.
333. S. B. Brummer, J. Chem. Phys., 42, 1636 (1965).
334. G. M. McNamee, B. C. Willett, D. M. LaPerriere, and D. G. Peters, J. Amer. Chem. Soc., 99, 1831 (1977).
335. J. Juillard, in Recommended Methods for Purification of Solvents and Tests for Impurifies (J. F. Coetzee, ed.), Pergamon, Oxford, 1981, p. 32.
336. M. Bréant, M. Bazoin, C. Buisson, M. Dupin, and J.-M. Rebattu, Bull. Soc. Chim. Fr., 5065 (1968).
337. M. Bréant and C. Buisson, J. Electroanal. Chem., 24, 145 (1970); M. Bréant, Bull. Soc. Chim. Fr., 725 (1971).
338. M. J. Kelly and W. R. Heineman, J. Electroanal. Chem., 248, 441 (1988).
339. H. W. Sternberg, R. E. Markby, I. Wender, and D. M. Mohilner, J. Amer. Chem. Soc., 89, 186 (1967); 91, 4191 (1969).
340. J.-Y. Gal and T. Yvernault, Bull. Soc. Chim. Fr., 2270 (1971); 839 (1972).
341. J.-E. Dubois and G. Dodin, Tetrahedron Lett., 2325 (1969).
342. H. Normant, Angew. Chem., 79, 1029 (1967).
343. J.-E. Dubois, P.-C. Lacaze, and A. M. de Ficquelmont, C. R. Acad. Sci., Ser. C, 262, 181, 249 (1966).
344. S. Sakura, J. Electroanal. Chem., 80, 315, 325 (1977); K. Izutsu, S. Sakura, and T. Fujinaga, Bull. Chem. Soc. Jpn., 45, 445 (1972).
345. K. Itaya, M. Kawai, and S. Toshima, J. Amer. Chem. Soc., 100, 5996 (1978).
346. T. Fujinaga, K. Izutzu, and S. Sakura, in Recommended Methods for Purification of Solvents and Tests for Impurities (J. F. Coetzee, ed.), Pergamon, Oxford, 1982, p. 38.
347. J. A. Zapp, Science, 190, 422 (1975).
348. J. F. Fauvarque, A. Jutand, and M. Francois, J. Appl. Electrochem., 18, 109 (1988).
349. P. J. Elving, J. Electroanal. Chem., 29, 55 (1971); W. R. Turner and P. J. Elving, Anal. Chem., 37, 467 (1965).
350. M. Asthana and L. M. Mukherjee, in Recommended Methods for Purification of Solvents and Tests for Impurities (J. F. Coetzee, ed.), Pergamon, Oxford, 1982, p. 44.
351. J. N. Butler, J. Electroanal. Chem., 14, 89 (1967).
352. Y. Arad, M. Levy, H. Rosen, and D. Vofsi, J. Polym. Sci. A-1, 7, 2159 (1969).
353. F. M'Halla, J. Pinson, and J. M. Savéant, J. Electroanal. Chem., 89, 347 (1978).
354. I. M. Kolthoff and T. B. Reddy, J. Electrochem. Soc., 108, 980 (1961); (b) Inorg. Chem., 1, 189 (1962).

355. R. Philippe and J.-C. Merlin, Bull. Soc. Chim. Fr., 4713 (1968).
356. C. G. Karakatsanis and T. B. Reddy, in Recommended Methods for Purification of Solvents and Tests for Impurities (J. F. Coetzee, ed.), Pergamon, Oxford, 1982, p. 25.
357. J. Desbarres, P. Picket, and R. L. Benoit, Electrochim. Acta, 13, 1899 (1968).
358. J. F. Coetzee, J. M. Simon, and R. J. Bertozzi, Anal. Chem., 41, 766 (1969).
359. J. B. Headridge, D. Pletcher, and M. Callingham, J. Chem. Soc. A, 684 (1967).
360. J. Martinmaa, Suom. Kem. B, 42, 33 (1969).
361. J. F. Coetzee, Pure Appl. Chem., 49, 211, 217 (1977); in Recommended Methods for Purification of Solvents and Tests for Impurities (J. F. Coetzee, ed.), Pergamon, Oxford, 1982, p. 16.
362. B. Bry and B. Tremillon, J. Electroanal. Chem., 30, 457 (1971); 46, 71 (1973).
363. M. Machtinger, M. J. Vuaille, and B. Tremillon, J. Electroanal. Chem., 83, 273 (1977).
364. R. E. Johnson, NASA Tech. Memo., NASA TMX-1283 (1966); NASA TMX-1380 (1966); Chem. Abstr., 66, 108836 (1967).
365. A. K. R. Unni, L. Ellias, and H. I. Schiff, J. Phys. Chem., 67, 1216 (1963).
366. L. S. Marcoux, J. M. Fritsch, and R. N. Adams, J. Amer. Chem. Soc., 89, 5768 (1967).
367. H. J. McComsey, Jr., and M. S. Spritzer, Anal. Lett., 3, 427 (1970).
368. J. Courtot-Coupez and M. L'Her, Bull. Soc. Chim. Fr., 1631 (1970).
369. D. R. Cogley, J. C. Synnott, J. N. Butler, and E. Grunwald, Electrochem. Soc. Meet., Washington, D.C., May 1971, Extended Abstract 122.
370. R. F. Nelson and R. N. Adams, J. Electroanal. Chem., 13, 184 (1967).
371. A. N. Dey and B. P. Sullivan, J. Electrochem. Soc., 117, 222 (1970).
372. G. Eichinger, J. Electroanal. Chem., 74, 183 (1976).
373. G. Eggert and J. Heitbaum, Electrochim. Acta, 31, 1443 (1986).
374. A. Caiola, G. Guy, and J. C. Sohm, Electrochim. Acta, 15, 555 (1970).
375. L. M. Mukherjee and R. G. Bates, J. Electroanal. Chem., 187, 73 (1985).
376. D. P. Boden and L. M. Mukherjee, Electrochim. Acta, 18, 781 (1973).
377. F. G. K. Baucke and C. W. Tobias, J. Electrochem. Soc., 116, 34 (1969).
378. T. Fujinaga and K. Izutsu, in Recommended Methods for Purification of Solvents and Tests for Impurities (J. F. Coetzee, ed.), Pergamon, Oxford, 1982, p. 19.
379. R. Lines and V. D. Parker, Acta Chem. Scand., B31, 369 (1977).
380. S. Nakabayashi, A. Fujishima, and K. Honda, J. Electroanal. Chem., 111, 391 (1980).
381. (a) H. J. Schäfer, Chem.-Ing.-Tech., 42, 164 (1970); (b) H. J. Schäfer and H. Küntzel, Tetrahedron Lett., 3333 (1970).
382. A. Misono, T. Osa, T. Yakamichi, and T. Kodoma, J. Electrochem. Soc., 115, 266 (1968).
383. S. I. Bailey, W. P. Leung, and I. M. Ritchie, Electrochim. Acta, 30, 861 (1985).

384. A. Caillet and G. Demange-Guerin, J. Electroanal. Chem., 40, 69 (1972).
385. J. Phelps, K. S. V. Santhanam, and A. J. Bard, J. Amer. Chem. Soc., 89, 1752 (1967).
386. K. Nyberg, Acta Chem. Scand., 24, 1609 (1970).
387. L. L. Miller and E. A. Mayeda, J. Amer. Chem. Soc., 92, 5818 (1970).
388. P. Castellonese and P. C. Lacaze, C. R. Acad. Sci., Ser. C, 274, 2050 (1972).
389. L. A. Tinker and A. J. Bard, J. Amer. Chem. Soc., 101, 2316 (1979).
390. P. C. Lacaze, J.-E. Dubois, and M. Delmar, J. Electroanal. Chem., 102, 135 (1979).
391. A. Fry and J. Touster, J. Org. Chem., 51, 3905 (1986).
392. J. Robinson and R. A. Osteryoung, J. Amer. Chem. Soc., 101, 323 (1979); J. Robinson and R. A. Osteryoung, J. Amer. Chem. Soc., 102, 4415 (1980); R. J. Gale and R. A. Osteryoung, J. Electrochem. Soc., 127, 2167 (1980).
393. M. Lipsztajn and R. A. Osteryoung, J. Electrochem. Soc., 130, 1968 (1983); Electrochim. Acta, 29, 1349 (1984); L. Janiszewska and R. A. Osteryoung, J. Electrochem. Soc., 134, 2787 (1987); 135, 116 (1988); T. A. Zawodzinski, Jr., R. T. Carlin, and R. A. Osteryoung, Anal. Chem., 59, 2639 (1987); J. F. Oudard, R. D. Allendoerfer, and R. A. Osteryoung, J. Electroanal. Chem., 241, 231 (1988).
394. F. A. Uribe and R. A. Osteryoung, J. Electrochem. Soc., 135, 378 (1988).
395. J. E. Gordon, J. Amer. Chem. Soc., 87, 4347 (1965).
396. A. Kisza, Z. Phys. Chem. (Leipzig), 237, 97 (1968).
397. M. Fleischmann and D. Pletcher, Tetrahedron Lett., 6255 (1968).
398. H. Lund, unpublished observation (1986).
399. K. Rousseau, G. C. Farrington, and D. Dolphin, J. Org. Chem., 37, 3968 (1972).
400. G. Cauquis and D. Serve, C. R. Acad. Sci., Ser. C, 262, 1516 (1966).
401. H. Yeager and B. Kratochvil, J. Phys. Chem., 73, 1963 (1969).
402. R. H. McKee, Ind. Eng. Chem., 38, 382 (1946).
403. M. M. Baizer, J. Electrochem. Soc., 111, 215 (1964).
404. R. Engels, W. J. M. van Tilborg, and C. J. Smit, Sandjerg Meet., 1981, Abstracts of Papers, p. 73.
405. B. L. Funt, D. Richardson, and S. N. Bhadani, Can. J. Chem., 44, 711 (1966).
406. N. Yamazaki, S. Nakahama, and S. Kambara, Polym. Lett., 3, 57 (1965).
407. T. Fujinaga, K. Izutsu, and T. Nomura, J. Electroanal. Chem., 29, 203 (1971).
408. M. M. Baizer and J. P. Petrovich, J. Electrochem. Soc., 114, 1023 (1967).
409. J. P. Petrovich and M. M. Baizer, J. Electrochem. Soc., 118, 447 (1971).
410. K. L. Vieira, M. S. Mubarak, and D. G. Peters, J. Amer. Chem. Soc., 106, 5372 (1984).
411. A. Brändström and K. Gustavii, Acta Chem. Scand., 113, 1215 (1969); A. Brändström, P. Berntsson, S. Carlsson, A. Djurhuus, K. Gustavii, U. Junggren, B. Lamm, and B. Samuelsson, Acta Chem. Scand., 23,

2202 (1969). A. Brändström, Preparative Ion Pair Extraction, Apotekersocieteten, Stockholm, 1974.
412. M. Makosza and E. Bialecka, Synth. Commun., 6, 313 (1976).
413. H. O. House, E. Feng, and N. P. Peet, J. Org. Chem., 36, 2371 (1971).
414. R. A. Peterson and D. H. Evans, J. Electroanal. Chem., 222, 129 (1987) and references therein.
415. J. Y. Becker, T. Berzins, B. E. Smart, and T. Fukunaga, J. Electroanal. Chem., 248, 363 (1988).
416. H. Guthke, K. Wintersberger, F. Beck, J. G. Floss, and W. Habermann, Fr. Pat. 1,476,162; Chem. Abstr., 67, 104624 (1967).
417. W. B. Mather and F. C. Anson, Anal. Chem., 33, 1634 (1961).
418. P. E. Iversen and H. Lund, Tetrahedron Lett., 3523 (1969).
419. P. E. Iversen, Tetrahedron Lett., 55 (1971).
420. M. M. Baizer, J. L. Chruma, and D. A. White, Tetrahedron Lett., 5209 (1973); R. C. Hallcher and M. M. Baizer, Liebigs Ann. Chem., 737 (1977).
421. J. H. P. Utley, in Topics in Current Chemistry (E. Steckhan, ed.), Springer-Verlag, New York, Vol. 142, 1987, p. 131.
422. R. Breslow and R. F. Drury, J. Amer. Chem. Soc., 96, 4702 (1974).
423. W. Rogers and S. M. Kipnes, Anal. Chem., 27, 1916 (1955).
424. G. B. Bachman and M. J. Astle, J. Amer. Chem. Soc., 64, 1303 (1942).
425. A. Hickling, Trans. Faraday Soc., 38, 27 (1942).
426. J. J. Lingane and S. L. Jones, Anal. Chem., 22, 1169 (1950).
427. J. Tacussel, Electrochim. Acta, 11, 381 (1966).
428. A. Bewick and O. R. Brown, J. Electroanal. Chem., 15, 129 (1967).
429. J. E. Harrar and E. Behrin, Anal. Chem., 39, 1230 (1967).
430. S. C. Creason and R. F. Nelson, J. Chem. Ed., 48, 775 (1971).
431. H. Matschiner, K. B. Otte, H. Unger, and A. Steinert, Chem. Tech. (Leipzig), 26, 106 (1974).
432. F. L. Hand and R. F. Nelson, Anal. Chem., 48, 1263 (1976).
433. G. L. Booman and W. B. Holbrook, Anal. Chem., 35, 1793 (1963).
434. A. J. Bard and K. S. V. Santhanam, in Electroanalytical Chemistry (A. J. Bard, ed.), Vol. 5, Dekker, New York, 1970, p. 215.
435. J. E. Harrar, in Electroanalytical Chemistry (A. J. Bard, ed.), Vol. 8, Dekker, New York, 1975, p. 1.
436. J. A. Von Frauenhofer and C. H. Banks, The Potentiostat and Its Applications, Butterworth, London, 1972.
437. W. M. Schwarz and I. Shain, Anal. Chem., 35, 1770 (1963); W. L. Underkofler and I. Shain, Anal. Chem., 35, 1778 (1963).
438. S. Bogan, L. Meites, E. Peters, and J. M. Sturtevant, J. Amer. Chem. Soc., 73, 584 (1951).
439. T. DeVries and J. L. Kroon, J. Amer. Chem. Soc., 75, 2484 (1953).
440. R. D. Weaver and G. C. Whitnack, Anal. Chim. Acta, 18, 51 (1958).
441. P. Sridhar and R. Palaniappan, J. Appl. Electrochem., 19, 293 (1989).
442. M. Steiniger, 15. Sandbjerg Meeting, Abstracts of Papers, p. 48 (1988).

III

SYNTHETIC AND MECHANISTIC ASPECTS OF CATHODIC REACTIONS OF ORGANIC COMPOUNDS CLASSIFIED BY ELECTROPHORES

7

Cathodic Reactions of Hydrocarbons

HERBERT KIESELE / Universität Lübeck, Lübeck, Federal Republic of Germany

JÜRGEN HEINZE / Institut für Physikalische Chemie, Universität Freiburg, Freiburg, Federal Republic of Germany

I. Introduction 331

II. Formation of Radical Anions, Dianions, and Polyanions 332
- A. Experimental aspects 332
- B. Thermodynamics 334
- C. Kinetics 342

III. Follow-up Reactions of the Reduced Species 343
- A. Homogeneous electron transfer 343
- B. Addition of electrophilic reagents and related reactions 345
- C. Dimerization and polymerization 348
- D. Intramolecular reactions 349

References 352

I. INTRODUCTION

The electrochemical reduction of pure hydrocarbons without functional groups is almost exclusively restricted to unsaturated compounds. The reason is that aliphatic hydrocarbons have extremely low electron affinities that render their reduction impossible, despite a gain of solvation energy within the stability limits of conventional solvent-electrolyte systems.

Thus, electrochemical data involving both thermodynamic and kinetic parameters of hydrocarbons are available for only olefinic and aromatic π-systems. The reduction of aromatics in particular attracted much interest in the late 1950s and early 1960s through the pioneering work of Hoijtink.

The correlation between the reduction potentials and molecular orbital energies of a series of aromatic hydrocarbons was one of the first successful applications of the Hückel molecular orbital (HMO) theory and allowed Hoijtink to develop a coherent picture of cathodic reduction [1]. The early research on this subject has been reviewed several times, emphasizing polarographic studies as well as physical chemistry [2–4].

Later, during the golden age of mechanistic chemistry, interest focused on the elucidation of reaction paths of cathodically generated species, including disproportionations [5] and chemical follow-up processes [6]. Techniques were developed and improved to measure both the kinetics of coupled chemical reactions [7] and thermodynamic parameters, such as redox potentials of higher charged aromatics [8]. The discovery of the redox properties of conducting polymers in 1979 [9] gave new impetus to the electrochemistry of unsaturated hydrocarbons, initiating extended studies on the charge storage characteristics of these species [10].

We report here the cathodic formation and follow-up reactions of anionic species of hydrocarbons. Other generation techniques, such as reduction by solvated electrons or indirect electrolysis, are discussed in detail in Chap. 29.

II. FORMATION OF RADICAL ANIONS, DIANIONS, AND POLYANIONS

A. Experimental Aspects

The first data on the electrochemical reduction of aromatic hydrocarbons were published by Wawzonek et al. [11]. They used polarography as the measurement technique and observed in most cases a series of one- and two-electron reduction waves, clearly indicating follow-up reactions resulting from the attack of the protic solvent mixture (75% dioxane-water) on the ionic intermediates. Later, electron spin resonance (ESR) spectroscopy revealed that charged species of aromatics generated by alkali metal reduction in ethers are stable and normally possess an unpaired electron [12]. Thermodynamic reduction potentials of numerous aromatics were first measured by Hoijtink and van Schooten in 96% aqueous dioxane, again using polarography [13]. These fundamental works were decisive tests of the HMO theory, showing that the polarographic half-wave potentials vary linearly with the HMO energies of the lowest unoccupied orbitals (LUMO) of the hydrocarbons [1]. Hoijtink et al. had already noticed that most aromatics can be further reduced to their respective dianions [14]. They proposed a two-step reduction scheme in which both redox potentials 1 and 2 are on average separated by about 500 mV:

$$A \underset{}{\overset{E_1^0}{\rightleftharpoons}} A^{\bar{\cdot}} \qquad (1)$$

$$A^{\bar{\cdot}} \underset{}{\overset{E_2^0}{\rightleftharpoons}} A^{2-} \qquad (2)$$

However, their careful analysis also showed that most of the dianions were not stable in the polarographic or voltammetric time scale, even less so after bulk electrolysis, and underwent follow-up reactions with water or other electrophilic impurities, details of which are discussed in Sec. III.B.

Aprotic solvents, such as acetonitrile [15,16] or dimethylformamide [17–20], considerably improved the stability of the radical anions but normally had little effect on the reactions of the more basic dianions [19–21]. The increased irreversibility of the dianion formation is probably due to the ability of dianions to abstract protons even from the solvent or, by Hofmann elimination, from the tetraalkylammonium salts that are common supporting electrolytes in aprotic solvents [2].

Progress in electrochemical instrumentation soon stimulated the application of more elaborate measurement methods than simple dc polarography, which facilitated studies of heterogeneous kinetics and detection of follow-up reactions of the electrogenerated species. Thus, conclusions originally drawn from the shape and height of polarographic curves have been amply confirmed by straightforward diagnostic criteria in ac polarography [20,22] and above all in cyclic voltammetry [19,23–25], nowadays the standard method for mechanistic studies in organic electrochemistry [26].

Hoijtink et al.]27] also developed an alternative method of generating anionic species, which was improved by Szwarc et al. [28]. The technique involves potentiometric titration of aromatic compounds with a standard solution of Na-biphenylide. The extremely negative reduction potential of biphenyl assures that most of the common aromatics can be reduced to at least their respective radical anions. The values of the thermodynamic reduction potentials are generally obtained from the potentiometric titration curve. As all experiments are usually carried out in ethereal solutions, such as tetrahydrofuran (THF) or dimethoxyethane, problems of follow-up processes are less severe. Recently, Gross and Schindewolf [29] reported the potentiometric titration of aromatics using solvated electrons in liquid ammonia.

A fundamental improvement in the facilities for studying electrode processes of reactive intermediates was the purification technique of Hammerich and Parker [8a]. They used neutral highly activated alumina suspended in the solvent electrolyte system as a scavenger of spurious impurities. Thus it was possible to generate a large number of dianions of aromatic hydrocarbons in common electrolytic solvents containing tetraalkylammonium ions. It was the first time that such dianions were stable in the time scale of slow-sweep voltammetry. As the presence of alumina in the solvent-electrolyte system may produce adsorption effects at the electrode, or in some cases chemisorption and decomposition of the electroactive species, Kiesele constructed a new electrochemical cell with an integrated alumina column [30].

Heinze modified the technique by purifying the solvent by transferring it under high vacuum from the electrochemical cell to superactive alumina in a separate vessel [26]. Such sophisticated methods made it possible to generate reversibly polyanions up to hexaanions of aromatic and olefinic hydrocarbons (Fig. 1) [26,31].

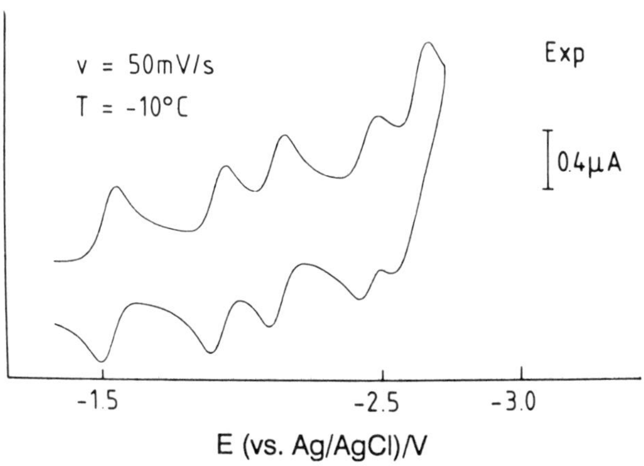

Figure 1. Cyclic voltammogram of the reduction of decacyclene indicating hexaanion formation in THF-$NaBPh_4$ under "superdry" conditions (see Ref. 26).

B. Thermodynamics

For purpose of description, the electrochemistry of hydrocarbons may usefully be classified in three categories—benzenoid, nonbenzenoid, and olefinic hydrocarbons—each of which exhibits characteristic properties upon reduction.

Benzenoid hydrocarbons have been studied in greatest detail [2,4,32]. In aprotic solvents they can be reversibly reduced to their respective anions without difficulty. Recently, even the electrochemical reduction of benzene in dimethoxyethane has been described [33]. In many cases the electrogenerated radical anions are stable enough to allow the simultaneous measurements of their ESR spectra [34–36]. The most striking feature of benzenoid hydrocarbons is the excellent correlation between their thermodynamic reduction potentials and the predictions of semiempirical π-electron theories especially of the Hückel approximation (HMO). As the thermodynamic reduction potentials are a measure of the electron affinities of the respective compounds, they can be compared with the theoretically calculated energies in the simple molecular orbital picture, in which additional electrons have been added to antibonding MOs of the π-systems. Therefore, assuming that the solvation energies for a series of aromatic hydrocarbons are constant, there should be a linear correlation between the thermodynamic reduction potentials (half-wave potentials) and the energies m_{m+1} of the lowest unoccupied MO (LUMO):

$$E_{1/2} = -bm_{m+1} + C \tag{3}$$

Independent voltammetric and polarographic measurements carried out in different solvents, such as 2-methoxyethanol, 96% dioxane, and dimethylformamide, confirm the relationship through excellent linear correlations with slopes b of approximately 2.40 [1,17,37,38]. Rather surprisingly, the differences in half-wave potentials of hydrocarbons from one solvent to another are very small. This constancy in energy values as well as slopes of correlation lines in widely varying solvents and supporting electrolytes implies that solvations energies, provided that they are not small, change in the same way from system to system.

As already observed by Hoijtink et al. [14], nearly all benzenoid hydrocarbons can be reduced to their respective dianions; only benzene and naphthalene are exceptions. In all experiments, these second reduction steps appear approximately -0.55 ± 0.10 V negatively to the first reduction, provided that the supporting electrolytes used were tetraalkylammonium salts. Therefore, these reduction potentials were also correlated with the LUMO energies of the HMO model [3]. It was suggested that the energy difference of 0.55 eV corresponds to the repulsion energy between both electrons in LUMOs of the dianions [1], despite the differences in their structures. On the other hand, quantum mechanical calculations show that the repulsion energies are much larger. Dewar has calculated values in the range of 5 eV [39]. The discrepancy between experiment and theory results from the fact that ion pairing effects and solvation influences have been neglected in the calculation. Experimental data clearly reveal that counterion effects that efficiently shield the negative excess charge have in the past been underestimated and are considerably stronger for di- and polyanions than for radical anions [40–42].

In polar solvents, such as DMF or acetonitrile, the interaction increases in the order tetrabutylammonium < K < Na < Li, and consequently, reduction potentials shift in a positive direction [6,43]. Obviously, ion pairing is greatest for the small lithium ions, in agreement with the prediction of Born's equation [44]. In the case of solvents with low dielectricity constants the pattern is different, and ion pairing becomes dominant as the radius of alkali cations increases [42,45,46]. The reasons for this behavior have not yet been studied in detail, but it has been proposed that in ethereal solvents the solvation of small cations remains stronger than that of larger ones, and therefore ion pairing of potassium should be more pronounced than that of lithium.

Ion pairing effects may considerably influence disproportionation mechanisms that involve homogeneous redox reactions of anions to their respective dianionic and neutral species [47]:

$$2A^{\overline{\cdot}} \rightleftharpoons A^{2-} + A \tag{4}$$

Disproportionation mechanisms have been proposed for protonation reactions and intramolecular rearrangements (see Secs. III.B and D) [48,49]. Their prominent feature is that follow-up processes at the level of the dianion can take place even at potentials corresponding to radical anion formation. To evaluate data for disproportionation reactions it is necessary to know the value of the disproportionation equilibrium constant:

$$K_d = \frac{[A^{2-}][A]}{[A^{\overline{\cdot}}]^2} \tag{5a}$$

$$\frac{RT}{nF} \ln K_d = E_2^0 - E_1^0 \tag{5b}$$

This can be determined from the difference in reversible potentials of the couples $A/A^{\overline{\cdot}}$ and $A^{\overline{\cdot}}/A^{2-}$.

In the case of benzenoid aromatics K_d values range between 10^{-9} and 10^{-13}, provided that tetraalkylammonium salts have been used as supporting electrolytes [8b]. In solvents of low dielectricity constant additional effects are observed, showing influences of the supporting electrolyte concentration and of the nature of the cations [6]. In the tetraalkylammonium series the strongest (contact) ion pairs are formed by Et_4N^+, and the K_d is largest for that cation [8b,50].

Dramatic changes in the disproportionation constants occur when alkali cations are used instead of tetraalkylammonium ions. Typically, the potentials of the radical anion formation is less affected than that of the dianion formation. In the presence of alkali cations ΔE^0 shifts may reach values of more than 600 mV, which corresponds to an increase in the K_d constant of more than 10 orders of magnitude [47].

The electrochemistry of nonbenzenoid hydrocarbons has attracted much interest because their structures offer unusual insights into π-electron systems, which undergo geometric changes upon reduction and obey both the Hückel 4n and 4n + 2 rules.

The most widely studied examples are cyclooctatetraene (COT, I) and its derivatives. In such conventional aprotic solvents as DMF, DMSO, or acetonitrile containing tetraalkylammonium salts, two distinct one-electron reduction waves are observed at approximately −1.64 and −1.80 versus SCE, with ΔE^0 separations varying from −130 to −240 mV [41,51–54]. In THF and NH_3

this separation reduces further [55–57], and in the presence of all alkali salts [58,59] even two-electron reduction waves with positive ΔE^0 differences were obtained, indicating large disproportionation constants. The unusually small separation of the two redox steps in comparison with the data of benzenoid aromatics was ascribed to the fact that the planar COT dianion forms a 4n + 2 π-electron system that is stabilized by its gain in Hückel resonance energy [60]. Obviously, strong ion pairing effects additionally favor the formation of the dianion [58,59]. On the other hand, it was argued that a negative shift of the first reduction step, leading to small ΔE^0 values, is caused by energy requirements accompanying the transition from the tube-shaped to the planar molecule. This is supported by electrochemical results obtained with methyl-substituted COT derivatives. Thus, all methyl-substituted derivatives due to a steric barrier are harder to reduce than I itself, and in 1,2,3,4-tetramethylcyclooctatetraene (II) reduction occurs at the extremely negative potential of −3.6 V (HMPA versus SCE) [61]. In the case of phenyl-substituted COTs, the steric demand is by and large energetically compensated through the planarization of the system and the subsequent resonance interaction between the phenyl moieties and the COT ring. Therefore, the reduction potentials are similar to that of unsubstituted I. In addition, the resonance-stabilizing effect upon the dianion is sufficient to shift the second redox potential E_2^0 positively to E_1^0, thus producing a single two-electron reduction wave [53].

The redox behavior of a number of higher annulenes than I has been studied during the past 15 years. The evaluation of their electronic properties has attracted much interest because it offers a good comparison of the fundamental differences between 4n and 4n + 2 π systems. As predicted by the simple MO theories, the transfer of an electron to a [4n]annulene should be energetically favorable because the electron is inserted into a low-lying nonbonding orbital. The further reduction to the dianion leads to a stabilized (aromatic) 4n + 2 π system, which therefore should also be easily

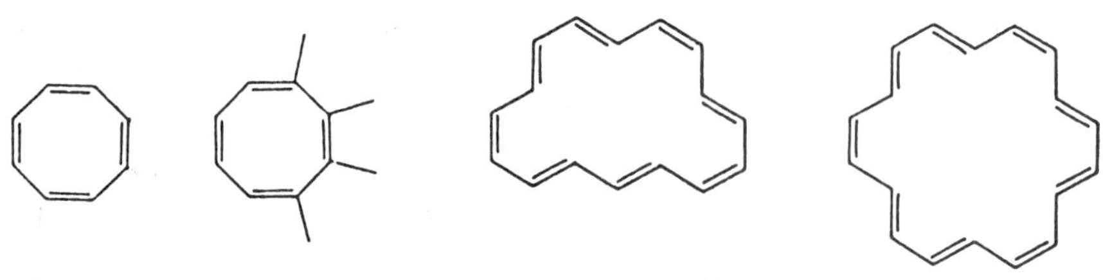

(I) (II) (III) (IV)

accessible. By contrast, during the reduction of an aromatic [4n + 2]annulene, electrons are injected into a LUMO with high energy, and in the second reduction step an unstable [4n]dianion is generated. Consequently, the reduction of [4n]annulenes should be observed at relatively positive potentials with small ΔE^0 separations for the dianion formation, and the reduction of the [4n + 2]annulenes should occur at more negative potentials with large ΔE^0 separations for the dianion formation. This is exactly what is observed in experiment (see Table 1). Although benzene [33], the classical Hückel aromatic with 4 × 1 + 2 = 6 π electrons, is reduced at −3.42 V (versus Ag/AgCl), the reduction of the [8]annulene COT occurs at −1.18 V. Similarly, the [16]annulene (III) [62,63] is more easily reduced than the corresponding [18]annulene (IV) [64], although the reduction of the larger π systems should be more favorable for electrostatic reasons.

For larger annulenes even the reduction to tetraanions is possible [42, 65,66]. This is especially favorable for neutral 4n + 2 species, which, after the injection of four electrons, re-form a stabilized 4n + 2 π system. On the other hand, the existence of four excess charges in one molecule gives rise to strong electron repulsion. It can be shielded only by ion pairing, which drives the formation of the tetraanion to a potential sufficiently positive for reduction to occur within the stability range of the solvent-electrolyte system. Thus, the electrochemical reduction of acepleiadylene (V) to its tetraanion was only achieved in THF in the presence of $LiBPh_4$ as supporting electrolyte [42]. Attempts to generate the same species in the presence of tetrabutylammonium ions were unsuccessful.

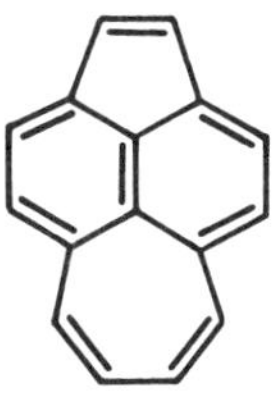

(V)

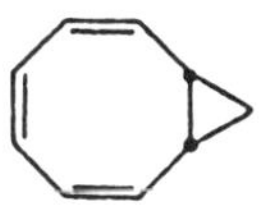

(VI)

(VII)

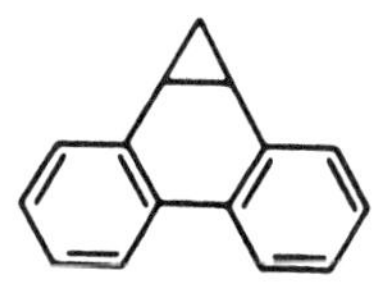

(VIII)

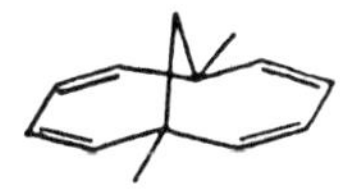

(IX)

Several studies have been published concerning the problem of homoaromaticity and its influence on reduction potentials. The term "homoaromaticity" is applied to cyclic resonance-stabilized systems in which a carbon-carbon double bond is replaced by a cyclopropane ring, or a saturated carbon atom is introduced into the conjugated chain. Paquette et al. investigated the reduction of cis-bicyclo-[6.1.0]-nona-2,4,6-triene (VI) and similar derivative [68]. They found that it is reduced (−2.55 and −2.79 V) more negatively than I but more easily than cyclooctatriene (VII) (−2.77 V), thus proving a degree of homoconjugative stabilization in the radical anion of VI. Analogously, dibenzonorcaradiene (VIII) is easier to reduce than biphenyl but harder to reduce than phenanthrene [69]. The electrochemical reduction of 1,6-dimethylbicyclo[4,4,1]undeca-2,4,7,9-tetraene (IX) was also interpreted in terms of homoconjugation in the anions. The reduction appears to proceed via an ECE scheme in which the initial reduction produces an unstable radical anion that undergoes a structural change producing another radical anion with a conjugatively stabilized π system. The second electron transfer occurs more easily than the first, thus producing a typical "two-electron" reduction [70].

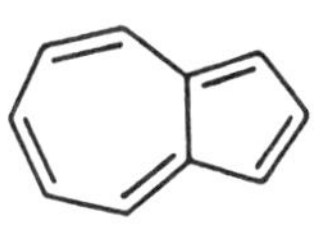

(X)

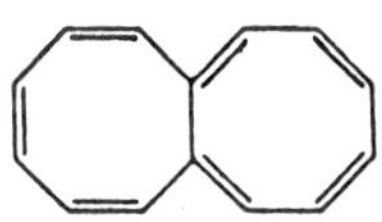

(XI)

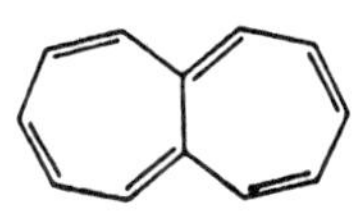

(XII)

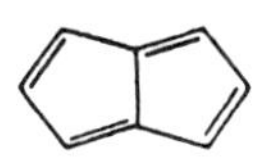

(XIII)

Of the condensed nonbenzenoid aromatics, azulene (X) has gained greatest popularity [71]. Although it is an isomer of naphthalene, its electrochemical behavior differs markedly from its benzenoid counterpart. Azulene is reversibly reduced at −1.62 V (versus Ag/AgCl), but the reduction of naphthalene takes place at −2.53 V [72]. Furthermore, the anion azulene is extremely stable against the attack of protons [73]. The bicyclic octa-

Table 1. Redox Potentials E^0 for the Reduction of Aromatic and Olefinic Hydrocarbons

Compounds	E^0_{R/R^-} (V)	$E^0_{R^-/R_2^-}$ (V)	$E^0_{R_3^-/R_4^-}$ (V)	$E^0_{R^-/R_2^-}$ (V)	Reference
Benzene[a]	-3.42	—	—	—	33, 72
Naphthalene[a]	-2.53	—	—	—	72
Anthracene[a]	-2.04	-2.64	—	—	72
Chrysene[b]	-2.27	-2.77	—	—	8b
Corenene[b]	-2.07	-2.72	—	—	8b
1,2-Benzanthracene[b]	-2.05	-2.67	—	—	8b
Phenanthrene[a]	-2.49	-3.13	—	—	72
Triphenylene[a]	-2.42	-2.97	—	—	72
Pyrene[a]	-2.29	-2.91	—	—	72
Biphenyl[c]	-2.68	-3.18	—	—	10
Bianthryl[a]	-1.92	-2.14	-2.82	-3.17	72
Decacyclene[c]	-1.74	-2.14	-2.35	-2.68 (-2.88)	26

Acepleiadylene[c]	-1.85	-2.51	-3.11	-3.14	42
COT[d]	-1.81	-1.94	—	—	53
[12]Annulene[e]	-1.35	-2.00	—	—	67
1,7-Methanol[12]annulene[f]	-1.51	-1.84	—	—	75
15,16-Dihydropyrene([14]Annulene)[g]	-2.22	-3.01	—	—	60
[16]Annulene[f]	-1.27	-1.56	—	—	62
[18]Annulene[f]	-1.60	-1.94	—	—	64
Heptalene[f]	-1.49	-2.19	—	—	75

[a]Cyclic voltammetry was performed at a Pt electrode with solutions of 10^{-3}–10^{-4} M in substrate. All potentials are expresses in V versus Ag/AgCl; solvent: dimethylamine-TBABr, temperature between −40 and −65°C.
[b]DMF-Me_4NBr, potentials versus Ag/AgCl (corrected from SCE).
[c]THF-$NaBPh_4$ or $LiBPh_4$, potentials versus Ag/AgCl.
[d]ACN-TEAP, potentials versus Ag/AgCl (corrected from SCE).
[e]THF-$TBAClO_4$, potentials versus Hg pool.
[f]DMF-$TBAClO_4$, potentials versus Ag/AgCl (corrected from SCE).
[g]DMF-$TBAClO_4$, potentials versus Ag/0.1 M $AgNO_3$.

lene (XI) has 14 π electrons. ¹H-NMR data show that it possesses a non-planar structure with polyolefinic properties and therefore resembles its monocyclic relative COT. Polarography and cyclic voltammetry reveal that it is reversibly reduced to the radical anion at $E_{1/2}$ = −1.67 V and, rather surprisingly, in a three-electron process at $E_{1/2}$ = −1.70 V to its stable tetraanion [74]. This unusual behavior can be explained only by assuming that the energy gain from delocalization in the planar 18 π system is higher than torsional and electronic repulsions. Other bicyclic systems, such as heptalene (XII) [75] and pentalene (XIII) [76], can also be reduced to their respective anions. Nevertheless, in comparison with equivalent monocyclic π systems, such as I, their reduction behavior is not yet properly understood.

(XIV) (XV) (XVI) (XVII)

A further group of nonbenzenoid aromatics is the series of odd-membered cations and anions, such as cyclopropenium (XIV) and tropylium cations (XV), as well as cyclopentadienyl (XVI) and cyclononatetracenyl anions (XVII). Regarding the arguments for the properties of Hückel-like [4n + 2] π systems, all these molecules should be energetically stabilized. Obviously, this is not fulfilled in all cases. The tropylium cation (XV) can be reduced in a one-electron step to the tropyl radical even at E_p = +0.06 V versus SCE [77]. The radical is unstable and rapidly dimerizes to bitropyl. The heptaphenyl tropylium radical is stable on the voltammetric time scale, but decays slowly producing a dimeric species involving coupling via the phenyl groups [78]. Similarly, 2,3-diphenylethyl-cyclopropenyl is irreversibly reduced in CH_3CN at E_p = −0.04 V. On the other hand, the reduction of trimethylcyclopropenyl occurs at −1.32 V, which is in better agreement with the prediction of the Hückel theory [79–81]. Otherwise, relatively positive reduction potentials seem to be typical for most carbocations [82]. The easy reducibility is probably caused by the excess positive charge.

The discovery of the metallike properties of conducting polymers has once again focused attention on the oxidation and reduction characteristics of aromatic systems. It turns out that most of these conducting materials consist of chainlike connected carbocyclic or heterocyclic aromatics [83–85].

The simplest molecules in these series are dimers, followed by oligomers of increasing chain length, and polymers. As the current-voltage curves of polymers are difficult to interpret, quantitative information on the redox properties of such systems were preferentially obtained from reduction experiments with dimers and defined oligomers. Redox data on dimers are available for biphenyl [10], bianthryl [86], biazulenyl [87], bicyclooctatetraenyl [88], and dimeric [14]annulenes [89]. All species can be reduced to at least their respective dianions. In the case of bianthryl- and bicyclooctatetraene even the formation of tetraanions has been observed. In general, the observed redox potentials of the monoanion formation differ significantly from those of the monomeric parent compound and shift to less negative values.

This is evidence of conjugative stabilization in the charged oligomeric units [90]. On the other hand, large ΔE^0 separations between the redox potentials of the respective mono- and dianions indicate strong electron repulsion between both electrophores. A typical example is biphenyl, which is reduced to the radical anion at −2.68 V and to the dianion at −3.18 V [10], whereas the reduction of the monomeric benzene occurs at −3.42 V [33]. Obviously, the biphenyl anion has gained a considerable amount of conjugative stabilization energy, but in the dianion strong electron-electron repulsion dominates. Voltammetric measurements carried out with several defined oligomers of the p-phenylene vinylene [31] and the p-phenylene [10] series, respectively, clearly demonstrate that the reduction properties of such oligomers and polymers depend on the chain length of the systems. Three effects are especially significant. First, the potentials of already existing redox states shift to less negative values when the next higher homolog is reduced. Obviously, the redox energies of different states gradually approach a common convergence limit with increasing chain length. Second, the redox states degenerate pairwise with increasing chain length, and third, in agreement with expectations, adding successive monomeric subunits in the molecular chain enlarges the number of accessible redox states. However, the energetic gap increases strongly between the lowest and highest charged states. From these results it becomes clear that in conducting polymers, such as poly-p-phenylene, a reasonable number of energetically low-lying redox states are degenerated, followed by redox states with increasingly higher energies [10]. This explains the broad, plateaulike waves that are so often characteristic of the potential range following the peaklike main wave in voltammetric experiment.

Within the hydrocarbons containing olefinic double bonds, polyacetylene (PA) is the most thoroughly studied system. The great interest results from its extremely high conductivity, up to 100,000 s/cm, which emerges on oxidative or reductive charging of the polymer [83,84,91]. As already found by MacDiarmid and colleagues [9], PA can be reversibly reduced to a polyanionic material. Its thermodynamically stable form is the trans configuration. In a recent publication [92] it was shown that cis-PA is irreversibly isomerized to the trans form upon reduction at E_p = −1.80 V versus Ag/AgCl (liquid NH_3, NaI). The voltammetric experiments prove that the isomerization only takes place via the charged cis segments of PA. The charged cis component is thermodynamically unstable and transforms through a fast chemical step to the trans form (EC mechanism), the redox potential of which is shifted in the positive direction ($E^0_{trans} > E^0_{cis}$).

Similar stability conditions are also valid for monomeric olefins. The best known example is that of the dialkyl maleate-fumarate system [93]. Again, the reduction potential of the cis form is more negative than that of the trans form (ΔE^0 = 0.28 V), and reduction of the cis-maleate induces the irreversible isomerization to the trans-fumarate. The redox potentials of cis- and trans-stilbene with values of −2.18 and −2.15 V versus SCE (DMF, TBAI), are almost equal, and therefore the electron transfer-induced isomerization is slow (see Sec. III.D) [94,95].

A further interesting point concerns unusually small differences between the reduction potentials of mono- and dianions of some molecules containing olefinic double bonds. Although in trans-stilbene the formation of the dianion occurs approximately 500 mV negatively to the radical anion formation [72,95], for tetraphenylethylene the standard potentials for the R/R^- couple

(E^0_{R/R^-}) and the R^-/R^{2-} couple ($E^0_{R^-/R^{2-}}$) are very closely spaced. The ΔE^0 separation amounts in HMPA to −138 mV, in DMF to −35 mV [96−98], and reaches in ACN even positive values of approximately 150 mV [98]. Consequently, the disproportionation constant K_d varies within five orders of magnitude, a phenomenon mainly ascribed to increasing ion pairing on going from HMPA to ACN [98]. Nevertheless, intermolecular phenomena alone are not sufficient to explain the dramatic decrease of the ΔE^0 separation in tetraphenylethylenes in comparison with stilbene. It is now generally accepted that structural changes involving twisting of the ethylenic bond and accompanying the dianion formation are the main reason for the energetic stabilization of the dianion of tetraphenylethylene [99]. Similar effects also observed with 9,9'-bifluorenylidene [100].

C. Kinetics

Conversion of an oxidized species into the reduced form and vice versa requires the reorganization of the solvent in the immediate neighborhood of the reactant, together with some structural changes within the reactant. In the case of heterogeneous charge transfer, additional double layer effects are operative, which depend, inter alia, on the type and concentration of the supporting electrolyte, but may also be influenced by the electrode material used as well as adsorption phenomena. Theoretical concepts have been developed by Hush, Marcus, and Dogonadze et al. [101−103]. Applications of the Marcus theory to problems in organic electrochemistry have been discussed by Eberson [104].

As most studies on the reduction of hydrocarbons were carried out in aprotic solvents in the presence of excess supporting electrolyte, double-layer influences on electron transfer kinetics were usually regarded as less important [2,105]. Normally, the rates of heterogeneous electron transfer to aromatic and olefinic hydrocarbons are high. It is assumed that the activation barrier is mainly caused by the solvent reorganization. Such reactions are termed outer sphere processes. Reductions of systems that require in addition a large conformational energy change in the transition state are rare [106]. The most widely discussed example from this class is the COT molecule (I), which exhibits a slow heterogeneous electron transfer for the anion formation, but the rate between the anion and the dianion is significantly faster [51,52,59]. In the literature it has been suggested that upon reduction the tube-shaped ring passes through at least a partially flattened transition state to form a planar anion radical. The experimentally determined activation energy (10−11 kcal/mol), which is similar to the activation energy of ring inversion (13.7 kcal/mol) [105], has been interpreted as supporting evidence [105]. However, independent studies of Fry et al. [41], and Parker et al. [55] have shown that the rate constant for the first reduction step increases as the cation size of the supporting electrolyte decreased from tetraheptylammonium to tetramethylammonium. This contradicts the assumption of pure structural reorganization effects during the reduction of I and points to adsorption phenomena of the electrolyte cations. It seems that the influence of structural reorganization on the activation process was overestimated in earlier work.

Tetraphenylethylene is another example of a slow charge transfer. Here, the radical anion formation is fast, whereas the second charge transfer is slow [96]. In agreement with the thermodynamic findings, the second charge transfer suggests that considerable structural reorganization occurs at the activation barrier of the second reduction step. Nor is it yet clear to what extent solvent reorganization as well influences the activation energy [96].

III. FOLLOW-UP REACTIONS OF THE REDUCED SPECIES

With one or even more electrons in antibonding orbitals, reduced hydrocarbons are highly reactive species capable of both inter- and intramolecular reactions. The preferred pathway of the follow-up reaction not only depends on the electronic and steric structure of the reduced species but also on its chemical environment, especially on the counterion and the solvent.

A. Homogeneous Electron Transfer

The most elementary follow-up reaction is the homogeneous electron transfer (ET) to another solution species, which may be identical with the donor molecule itself [47].

The equilibria of homogeneous ET reactions are governed by the standard potentials of the involved redox couples and are easily calculated with given data according to Eqs. (6) through (9):

$$A_1^{\overline{\cdot}} + A_2 \overset{K}{\rightleftharpoons} A_1 + A_2^{\overline{\cdot}} \qquad (6)$$

$$A_1 + e^- \overset{E_1^0}{\rightleftharpoons} A_1^{\overline{\cdot}} \qquad (7)$$

$$A_2 + e^- \overset{E_2^0}{\rightleftharpoons} A_2^{\overline{\cdot}} \qquad (8)$$

$$\frac{RT}{F}\ln K = E_2^0 - E_1^0 \qquad (9)$$

The kinetics are much more complex and depend on the reorganization of the molecular framework [107], the solvation shell, and the electrostatic interaction. A semiquantitative estimation of rate constants may be obtained with the well-known Marcus equation [108]. The calculated data compare quite well with experimental values. Most of the experimental hydrocarbon data have been provided by Szwarc and his school [5]. The state of the art has recently been discussed in an excellent review [104].

Homogeneous ET always has to be taken into account in dealing with electrode reactions, as was pointed out by Marcoux et al. [109]. It is of special importance when the heterogeneous ET is slow, providing an additional and more effective pathway to reduced species. The homogeneous rate constant for the second charge transfer in I similar to the heterogeneous charge transfer is considerably larger than the first. The difference greatly depends on the experimental conditions. Recently it has been shown by cyclic voltammetry in superdry THF-$TBAPF_6$ that the first ET can be catalyzed by the dianion of COT itself in a homogeneous synproportionation, giving rise to a catalytic spike. The mechanism has been confirmed by digital simulation [56].

Quite often the resulting species undergoes a fast follow-up reaction. Thus, homogeneous ET becomes a crucail step in many electrode reactions.

A well-known example is the cathodic reduction of organic halides. At mercury, heterogeneous electron transfer to most halides is slow. This kinetic barrier may be circumvented by the addition of aromatic hydrocarbons, such as phenanthrene [110], anthracene [111], and naphthalene [110]. These compounds are easily reduced at the cathode and transfer their excess electron to the organic halide. The resulting halide anion radical undergoes a fast follow-up reaction. The whole reaction sequence is

$$A + e^- \rightleftharpoons A^{\bar{\cdot}} \quad (10)$$

$$A^{\bar{\cdot}} + PhCl \rightleftharpoons A + PhCl^{\bar{\cdot}} \quad (11)$$

$$PhCl^{\bar{\cdot}} \longrightarrow Ph^{\cdot} + Cl^- \quad (12)$$

$$Ph^{\cdot} + A^{\bar{\cdot}} \rightleftharpoons Ph^- + A \quad (13)$$

$$Ph^- + H^+ \rightleftharpoons PhH \quad (14)$$

Regeneration of hydrocarbon in Eqs. (11) and (13) makes the process catalytic.

The reduction of organic halides in the presence of aromatic hydrocarbons has been the subject of detailed kinetic studies that provide rate constants for the homogeneous ET [112] and the follow-up reaction [113]. The theoretical basis for this kind of experiment (homogeneous redox catalysis) was laid by Saveant's group in a series of papers in 1978–1980 [114–116]. Homogeneous ET also plays an important role in the protonation of anion radicals [117], and it is also an essential step in the sigmatropic rearrangement in 1-arylindenes [118]. Both reactions are discussed here.

When an anion radical undergoes heterogeneous ET, formation of the neutral molecule in the ground state is strongly favored over formation of an excited state [119]. No such restriction applies to homogeneous ET, which, if sufficiently exothermic, may yield excited states of hydrocarbons. One may naively suppose that an electron is removed from the bonding molecular orbital of highest energy to give either the first excited singlet or triplet; electrochemiluminescence [120] may then occur.

The emission observed is usually the fluorescent band of the hydrocarbon, corresponding to the decay of the first excited singlet. The energy released by homogeneous electron transfer is given by the difference between the redox potentials of donor and acceptor plus a small entropy term [121] and, in many cases of electrochemiluminescence, falls short of the energy needed to populate the singlet state directly. In these "energy-deficient" cases there is sufficient energy to populate the lowest triplet state, and singlets can then be produced by diffusion-controlled triplet-triplet annihilation. Emissions observed at wavelengths other than that of the main singlet have been ascribed to excited dimers (eximers) [122,123], excited charge-transfer complexes (exiplexes) [124,125], and fluorescent products of radical ion decay. Acceptors in electrochemiluminescence may be the corresponding hydrocarbon radical cations, added alkyl halides [126] or benzoyl peroxide [127], or adventitious impurities, which need be present at only 10^{-7} M levels. The experimental technique for the method is demanding [128]. Research attention is turning from the energy balance to the detailed mechanistic steps between electron transfer and photonic emission to the question of the quantum efficiency of the process.

B. Addition of Electrophilic Reagents and Related Reactions

Protonation

Under protic conditions aromatic hydrocarbons and compounds with activated double bonds usually undergo Birch-like reactions [11,129]. The reaction sequence has been elucidated by the classic work of Hoijtink [13,14,20,130, 131], who used HMO theory to rationalize both chemical and electrochemical steps.

The anion radical produced by heterogeneous ET is monoprotonated to give a Wheland-like π radical. It readily accepts another electron because its bonding or nonbonding SOMO (single occupied molecular orbital) always has a lower energy than the antibonding LUMO of the parent hydrocarbon. The second ET preferably occurs by disproportionation [117]. In a fast follow-up reaction the resulting carbanion takes another proton to eventually yield the final dihydroproduct:

$$A + e^- \rightleftharpoons A^{\overline{\cdot}} \qquad (15)$$

$$A^{\overline{\cdot}} + H^+ \rightleftharpoons AH^{\cdot} \qquad (16)$$

$$AH^{\cdot} + e^- \rightleftharpoons AH^- \qquad (17)$$

or

$$AH^{\cdot} + A^{\overline{\cdot}} \rightleftharpoons AH^- + A \qquad (18)$$

$$AH^- + H^+ \longrightarrow AH_2 \qquad (19)$$

If the reduction potential of the resulting dihydroproduct is sufficiently positive it can undergo another reduction cycle. One of many examples is provided by the reduction of benz(a)anthracene in 75% aqueous dioxane [11,13,14,132].

Because of its general importance to organic electrochemistry, the reaction scheme just outlined has been the subject of detailed mechanistic studies [18,133–142]. As a model reaction, the protonation of anthracene anion radicals by phenol in dimethylformamide has been selected. Of five limiting kinetic variants, ECE_{rev}, ECE_{irr}, DISP 1, DISP 2, and DISP 3, the favored pathway was found to be DISP 1 [117]. It involves the protonation of the anion radical as the rate-determining step followed by homogeneous ET between the anion radical and the protonated anion radical, yielding the monohydrogenated anion, which is itself rapidly and irreversibly protonated to the final dihydrogenated product. This mechanism has been disputed because of apparent deviations from predicted reaction orders [134–138]. Recently, it turned out that the inconsistencies are largely due to the formation of homoconjugated complexes between phenol and the phenolate anion [140,141]. Thus, DISP 1 now seems to be generally accepted.

The rate-determining step of the DISP 1 mechanism, the protonation of the radical anion, largely depends on its electronic structure. As a guideline, LUMO energies of the parent hydrocarbon may be used [37,143]. The attempt to correlate rate constants with highest local charge densities failed [144]. Recently, Eberson et al. suggested the application of Dewar-Zimmermann rules [145].

Under highly protic conditions the major products of cathodic reductions of cyclic conjugated hydrocarbons are usually 1,4-dihydroderivatives [11,13, 51,146]. In 2-methoxyethanol, for example, naphthalene yields 1,4-dihydronaphthalene [146] and COT provides mainly 1,3,6-cyclooctatriene [51].

These findings are in contrast to thermodynamics favoring the formation of conjugated 1,2 isomers. Therefore, kinetic control must be assumed. The site selectivity for the first as well as the second protonation step has been predicted by HMO theory. One approach refers to local charge densities [13,147]; the other one uses localization energies as a reactivity index [148]. In any event the inductive effect of the methylene group, formed in the first protonation step, has to be taken into account if two sites provide comparable reactivity indices.

These techniques proved to be successful for the majority of the hydrocarbons, but they failed for some compounds, such as phenanthrene, terphenyl, or quaterphenyl. The failure has been interpreted by the effect of the electric field on the charge distribution [149]. Actually, these molecules have strongly anisotropic polarizabilities. In a more recent study it was demonstrated that the isomeric ratio also depends on the counterion [150].

Under nominally aprotic conditions, 1,2 protonation dominates. Reduction of naphthalene in anhydrous acetonitrile containing tetraethylammonium p-toluenesulfonate yields 1,2-dihydronaphthalene, which is subsequently reduced to tetralin [151]. Similarly, reduction of I in anhydrous DMF gives 1,3,5-cyclooctatriene almost exclusively [51]. The formation of the thermodynamically more stable products is most probably due to base-catalyzed isomerization.

An interesting situation arises from the reduction of CH-acidic hydrocarbons because these compounds can undergo self-protonation. Actually, a voltammetric investigation of 1,3-diphenyl-2-methylindene and 4,5-methylenephenanthrene in DMF-TBAP or DMSO-TBAP clearly indicated a DISP 1 mechanism, analogous to that described earlier [152]. On the other hand, reduction of fluorene [153] or 1,2,3,4-tetraphenyl-1,3-cyclopentadiene [154] results in a homolytic cleavage of the CH bond (see later).

To study the stereochemistry of protonation reactions, substituted indenes have been cathodically reduced in DMF-TBAP with added water or phenol [155,156]. In the presence of water a formal "anti" addition of protons was observed, whereas addition of phenol led to the prevalent formation of products, formally deriving from "syn" protonation. Obviously, steric effects and/or acidities of proton donor and the dihydroproduct play an important role. A detailed discussion of stereochemical aspects is found in Chap. 27.

Quite often, ion pairing causes a substantial positive shift of the reduction potential. For electrostatic reasons, the shift is especially large for the formation of higher valency ions. Therefore, with increasing interaction of the counterion, di- and polyanion formation becomes thermodynamically more favorable. Under these conditions cathodic reduction immediately produces dianions via disproportionation and heterogeneous ET. Because of their high basicity the dianions readily undergo protonation. Such a dianion mechanism was observed when tetraphenylethylene was reduced in acetonitrile-TEAP in the presence of water or alcohols. Kinetic measurements led to a mixed order rate law, rationalized by the existence of a hydrogen-bonded complex between the ion-paired dianion and the proton donor [98].

The classical Hoijtink mechanism and the dianion mechanism have been observed at electrodes with a high hydrogen overvoltage, such as mercury. If mercury is replaced by platinum with its low hydrogen overvoltage, a radical pathway seems to be favored [157a], which is closely related to catalytic hydrogenations of hydrocarbons. Spectroelectrochemical experiments provided evidence for an additional hydride mechanism [157b]:

$$2H^+ + 2e \rightleftharpoons H_{2\,ads} \tag{20}$$

$$H_{2\,ads} + e^- \rightleftharpoons H^- + H^{\cdot}_{ads} \tag{21}$$

$$A + H^- \rightleftharpoons AH^- \tag{22}$$

These results illustrate that the reaction sequence and the stereochemistry of cathodic hydrogenations is controlled by the solvent, electrolyte, proton donor, electrode material, and other factors. Thus, with increasing mechanistic knowledge, electrochemistry offers the chance to realize highly selective hydrogenations.

The addition of protons certainly is the most common "nucleophilic" reaction of reduced hydrocarbon species because of the almost ubiquitous availability of proton donors, which may be wanted or not. However, under strictly aprotic conditions it is also possible to add other electrophilic reagents, such as alkyl halides, acyl halides, CO_2, and SO_2.

Alkylation

The addition of alkyl halides to aromatic anion radicals, generated by alkali metal reduction in ethereal solvents, was known in the 1950s [158] and was reviewed by Garst in 1971 [159]. The first electrochemical analog was observed by Lund et al. [160]. These authors cathodically reduced hydrocarbons, such as naphthalene, anthracene, stilbene [110,111], and perylene [112], in the presence of alkyl halides and isolated hydrogenated and alkylated products. Similar reactions are observed when the halides are replaced by ammonium or sulfonium [161].

These alkylations can be looked upon as aliphatic nucleophilic substitutions, usually thought to proceed via S_N1, S_N2, or hybrids of these mechanisms. However, in recent years more and more evidence for a single electron transfer (SET) mechanism, represented in Eqs. (23) through (29), was obtained [162], and it was suggested that S_N2 and SET are just limiting cases of the same single electron transfer mechanism [163]. The SET pathway involves first a transfer of an electron from the nucleophile to the electrophile followed by bond formation, whereas the S_N2 reaction involves a synchronous shift of a single electron and bond formation.

Because of these fundamental aspects, the mechanism of cathodically induced alkylations has been the subject of detailed studies [112,164]. In a stereochemical investigation it was found that racemization is much more effective than inversion. This result was interpreted as evidence of competition between the SET pathway and the S_N2 mechanism, with SET the more important route [164]. The SET mechanism and the competing hydrogenation are represented in Eqs. (23) through (29):

$$A + e^- \rightleftharpoons A^{\cdot-} \tag{23}$$

$$A^{\cdot-} + BX \longrightarrow A + B^{\cdot} + X^- \tag{24}$$

$$A^{\cdot-} + B^{\cdot} \longrightarrow AB^- \tag{25}$$

$$AB^- + H^+ \longrightarrow ABH \tag{26}$$

$$A^{\cdot-} + BX \longrightarrow AH^{\cdot} + B(-H) + X^- \tag{27}$$

$$AH^{\cdot} + A^{\cdot-} \rightleftharpoons AH^- + A \tag{28}$$

$$AH^- + H^+ \rightleftharpoons AH_2 \tag{29}$$

Experimental rate constants, obtained from the reaction of radical anions with tertiary and secondary alkyl halides, correlated quite well with redox potentials as predicted by Marcus theory, whereas the reaction rates of primary halides were found to be somewhat higher [112].

From these results it was concluded that nucleophiles with high oxidation potentials ("hard" nucleophiles) require much stabilization by bond formation in the transition state, whereas nucleophiles with low oxidation potentials ("soft" nucleophiles) have a transition state closer to the pure SET transition state [112].

Acylation

Formally related reactions are observed when anthracene [165] or arylolefines [166] are reduced in the presence of carbonic acid derivatives, such as anhydrides, esters, amides, or nitriles. Under these conditions mono- or diacylated compounds are obtained. It is interesting to note that the yield of acylated products largely depends on the counterion of the reduced hydrocarbon species. It is especially high when lithium is used, which is supposed to prevent hydrodimerization by ion pair formation. In contrast to alkylation, acylation is assumed to prefer a S_N2 mechanism. However, it is not clear if the radical anion or the dianion is the reactive species. The addition of nitriles is usually followed by hydrolysis of the resulting ketimines [166].

Addition of CO_2 and SO_3

In the pioneering papers of Wawzonek et al. [15,167], it was demonstrated that CO_2 can be added to cathodically reduced hydrocarbons to yield dihydrodicarboxylates. Examples of this kind of reaction include naphthalene [167,168], anthracene, [15], 9,10-diphenylanthracene [15], phenanthrene [167], butadiene [168], stilbene [15], and diphenyl acetylene [168]. The mechanism is assumed to be analogous to that for protonation, the essential steps being nucleophilic addition by an anion radical and subsequently by an anion [169]. 1,4 Addition to naphthalene suggests that carboxylation is kinetically controlled; formation of dihydromonocarboxylates indicates competition with protonation. Mechanisms involving CO_2^- must be considered for those hydrocarbons that are reduced at more negative potentials than CO_2 [170], and for the electroinactive norbornadiene, which gives 3-nortricyclenecarboxylic acid [171]. Chemically formed CO_2^- is known to undergo typical radical addition reactions [172]. Like CO_2, SO_3 adds to reduced anthracene to give 9,10-dihydroanthracene bissulfonate [173].

C. Dimerization and Polymerization

Cathodically reduced hydrocarbons not only undergo homogeneous ET and nucleophilic attack but also coupling reactions resulting in hydrodimerization and polymerization.

Reduction of stilbene [15] or diphenylacetylene [167] in DMF yields 1,2,3,4-tetraphenylbutane, whereas phenanthrene [167] provides 9,9',10,10'-tetrahydro-9,9'-biphenanthrene. Hydrodimerization was also observed with benzalfluorene [174]. If DMF is replaced by acetonitrile, protonation completely dominates hydrodimerization [15]. In carefully dried ethers, using alkali or alkaline earth metals salts as supporting electrolyte, 1,1-diphenylethylene can be reduced cathodically to give stable solutions of 1,1,4,4-tetraphenylbutane dianions [175]. These dianions can be cleaved by flash photolysis in the presence of excess 1,1-diphenylethylene to give transient anion radicals of 1,1-diphenylethylene. Kinetic analysis of the subsequent recombination confirmed the postulated radical-radical mechanism; the rate constant was found to be $0.5 \times 10^9\ M^{-1}\ s^{-1}$ [176]. 9-Cyanoanthracene undergoes a reversible electrodimerization with almost no side reactions, making a perfect model compound [177–182]. Again, it turned out that the dimerization follows a radical-radical mechanism. It is also interesting to note that water accelerates the dimerization. This effect was rationalized by specific solvation [182].

Many alkenes, activated by electron-withdrawing groups, readily undergo hydrodimerizations. The best known example is the electrodimerization of acrylonitrile, the base of the commercial Monsanto process [183]. This type of reaction was the subject of particularly active mechanistic investigations and is treated in detail in Chap. 22. Evidence is presented there that the essential step is a coupling of two radical anions (radical-radical route). On the other hand, it cannot be excluded that nucleophilic attack of a radical anion on a neutral olefin also takes place (radical-substrate route) [184].

Sterically less demanding arylalkenes and dienes undergo not only dimerization but also polymerization. Styrene is polymerized in ethers by alkali metal reduction [185] or addition of cumyl potassium [186]. The mechanism of ET induced polymerization was extensively studied by Szwarc and his school [187]. It turned out that the first step is dimerization of the styrene anion radical, usually obtained by addition of sodium naphthalene. Under aprotic conditions the resulting dianion adds to monomers, forming polymeric "living anions." It is interesting to note that the rate of the polymerization largely depends on the counterion.

With conventional techniques and electrolytes it was not possible to obtain living anions because they are rapidly protonated by tetraalkylammonium salts and residual water. The first report of the production of living polymers by an electrolytic method must be attributed to Yamasaki et al. [188], who used tetrahydrofuran as solvent and $LiAlH_4$ or $NaAl(C_2H_5)_4$ as electrolyte for the polymerization of α-methylstyrene. A similar technique was used to polymerize styrene as well as derivatives [189–193]. The suggested mechanism agrees with the pathway described earlier. A more detailed discussion of electrochemical polymerization is given in Chap. 32.

D. Intramolecular Reactions

For many years, intramolecular reactions, such as conformational changes, bond cleavage, bond formation, and valence isomerizations, have been observed only when hydrocarbons were reduced with alkali metals in ethereal solvents. In most electrochemical experiments these reactions were dominated by the electrophilic processes described earlier. However, progress in experimental techniques [8,26,30] has made these reactions accessible to electroanalytical investigations, providing new mechanistic insight.

Conformational Changes

In recent years it has become more and more obvious that ET is frequently accompanied by conformational changes. The interconversion may precede or follow ET; the majority of these processes must be classified as CE (as yet this mechanism has not been observed in the hydrocarbon series), EC, EEC, or ECE mechanisms, whereas only a few systems follow an EE pathway. An example of and EC process is the interconversion of the anion radical of cis-stilbene, which is quite slow on the voltammetric time scale [94,95], whereas cis-azobenzene anion radicals isomerize very rapidly [194]. Tetraphenylethylene undergoes two closely separated reversible additions of one electron [97,195]. The small difference of standard potentials, equivalent to a high disproportionation constant [see Eq. (5)], has been interpreted as interconversion from an almost planar to an orthogonal conformation when going from the neutral molecular to the dianion [196]. The thermodynamics and kinetics of ET-induced interconversions of substituted tetraphenylethylenes have been studied in great detail [107]. Reduction of 1,6-dimethylbicyclo[4.4.1]undeca-2,4,7,9-tetraene appears to proceed via an ECE scheme in which the initial reduction gives an unstable radical anion that undergoes a structural change giving another anion with a conjugatively stabilized π system [70]. Conformational changes concurrent with ET (EE pathway) are observed upon reduction of I, as cited earlier. This emerging field was recently reviewed by Evans and O'Connell [107].

Bond Breaking and Bond Making Reactions

Reductive cleavage of carbon-carbon bonds was observed in the 1920s by Ziegler and Thielemann upon alkali metal reduction of diarylalkanes in ethereal solvents [197]. As was shown by Lagendijk and Szwarc [198] for 1,2-di(α-naphthyl)ethane, the primary anion radical undergoes homogeneous disproportionation supported by ion pairing. The resulting dianion decomposes by the fission of the $CH_2{-}CH_2$ bonds into the salts of α-naphthylmethyl carbanions. Similarly, 9,9'-bianthryl can be cathodically cleaved into anthracene and 9,9'-dihydroanthracene plus small amounts of reduced dimers. The dianion mechanism is quite slow, whereas the tri- and tetraanions are supposed to decay rapidly [199].

ESR spectroscopic investigations of the anion radical of fluorene indicated a first-order decay, and it was concluded that the CH bond undergoes homolytic cleavage [200]. Voltammetric studies of fluorene in DMF reached the same conclusion [153]. Electrochemical data of 1,2,3,4-tetraphenyl-1,3-cyclopentadiene, obtained by cyclic voltammetry in "superdry" THF-$NaBPh_4$ solutions at $-20°C$, are compatible with only first-order decay, as was shown by digital simulation [154]. However, with increasing temperature an acid-base mechanism described by Amatore et al. [152] becomes dominant. These examples demonstrate that even CH bonds can be cleaved by ET.

The thermal and photochemical ring opening reactions of cyclobutane are the classic examples of pericyclic processes [201]. In 1975, Bauld et al. described an ET-induced analog [202]. They observed that benzo- and phenanthrocyclobutene undergoes ring opening upon alkali metal reduction and suggested and ECE pathway. A voltammetric study of cis- and trans-1,2-dihydro-1,2-diphenylcyclobutan(1)phenanthrene in THF-$NaBPh_4$ confirmed this mechanism. However, it turned out that the rate of ring opening largely depends on the counterion. If ion pairing is prevented by the addition of 15-crown-5, the reaction rate slows dramatically. At $-50°C$ the anion radical of the Z isomer becomes stable on the voltammetric time scale, whereas the dianion exhibits a fast ring opening [49].

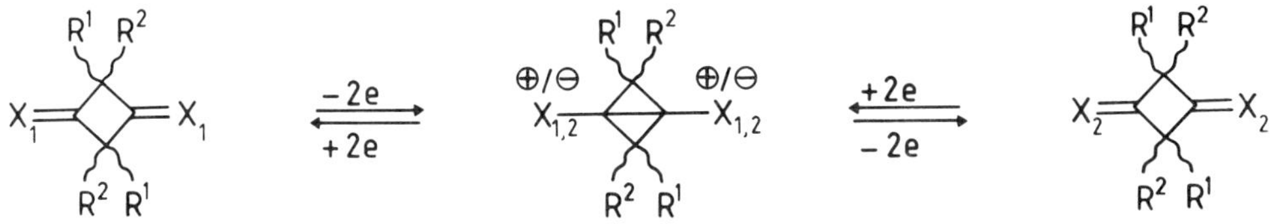

(30)

Two-electron reduction [56] or oxidation [203] of 1,3-dimethylidenecyclobutanes yields bicyclo[1.1.0]butanes (Eq. (30)). Cyclic voltammetry of 2,4-di-9H-fluoren-9-yliden-1,1,3,3-tetramethylcyclobutane in DMF at low temperatures demonstrated that bond formation proceeds via an EEC mechanism; the rate constant was found to be 20 s^{-1} [56].

ET-induced cycloadditions of polycyclic olefins and cycloreversions of cyclobutane species have been studied by ESR spectroscopy [204]. An electrochemical investigation is under way [205].

1-Arylindenes undergo sigmatropic 1,5 shifts, induced oxidatively [206], thermally [207], photochemically [208], and reductively [209]. The reductive rearrangement of 1,1,3-triphenylindene, yielding the dianion of 1,2,3-triphenyl-2H-indene, has been the subject of a voltammetric investigation in superdy THF-$NaBPh_4$ [118]. The reaction follows an EEC or EDisp C pathway (Scheme 1).

At room temperature the rearranging dianion is formed via homogeneous disproportionation, thermodynamically favored by ion pairing and kinetically supported by the fast follow-up reaction. When ion pairing is suppressed by application of low temperatures or the addition of sodium-complexing

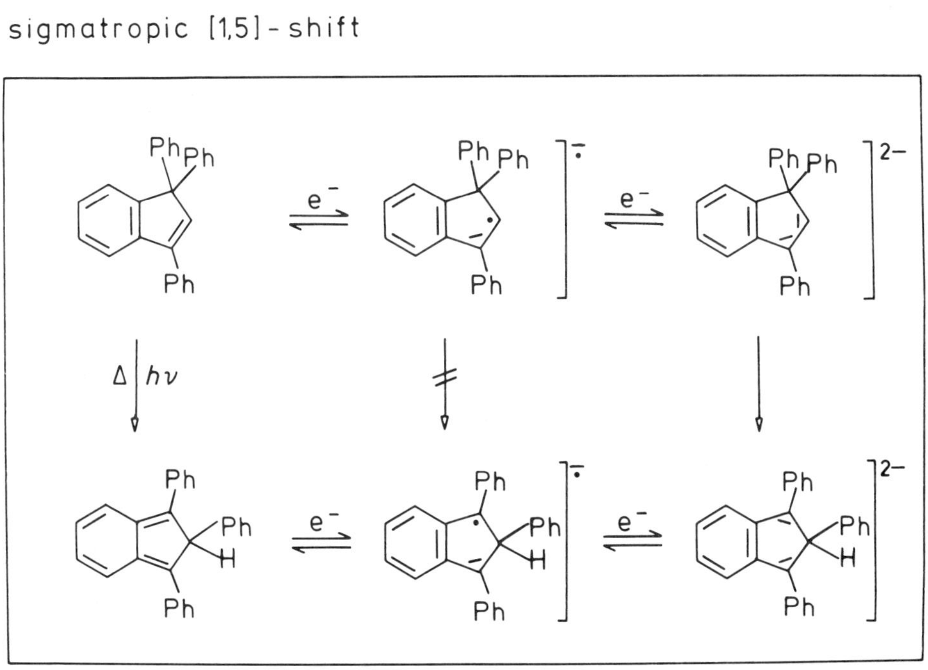

Scheme 1.

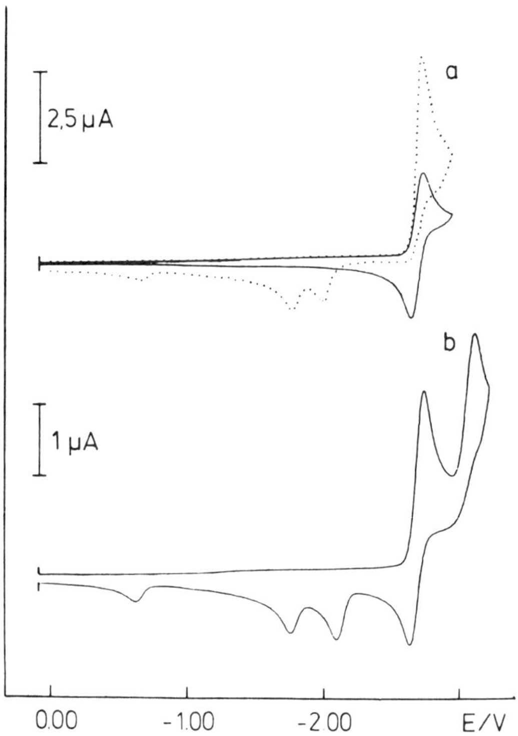

Figure 2. Cyclic voltammograms of the reduction of 1,1,3-triphenylindene in THF-$NaBPh_4$ (dotted line) and THF-$NaBPh_4$-15-crown-5 at 0°C (c = 1.5 mM, v = 100 mV s^{-1}): (a) restricted scan, (b) extended scan.

15-crown-5, the anion radical becomes persistent and the reactive dianion species must be generated via heterogeneous ET at much more negative potentials (Fig. 2). By replacing 1,1,3-triphenylindene with 1-methyl-1-phenylindene or 1,2-dihydro-1,1'-spirobiindene, the anion radical undergoes a quite unusual cleavage of the alkyl-aryl bond, whereas the dianion rearranges in the same manner as triphenylindene [210].

REFERENCES

1. G. J. Hoijtink, Rec. Trav. Chim. Pays-Bas, 74, 1525 (1955).
2. M. E. Peover, in Electroanalytical Chemistry (A. J. Bard, ed.), Vol. 2, Dekker, New York, 1967.
3. G. J. Hoijtink, in Advances in Electrochemistry and Electrochemical Engineering (P. Delahay, ed.), Vol. 7, Wiley-Interscience, New York, 1970.
4. M. E. Peover, in Reactions of Molecules at Electrodes (N. S. Hush, ed.), Wiley, New York, 1971, pp. 259ff.
5. M. Szwarc and J. Jagur-Grodzinski, in Ions and Ion Pairs in Organic Reactions (M. Szwarc, ed.), Vol. 2, Chap. 1, Wiley-Interscience, New York, 1974.

6. N. L. Holy, Chem. Rev., 74, 243 (1974).
7. D. H. Evans, Acc. Chem. Res., 10, 313 (1977).
8. (a) O. Hammerich and V. D. Parker, Electrochim. Acta, 18, 537 (1973); (b) B. S. Jensen and V. D. Parker, J. Amer. Chem. Soc., 97, 5211 (1975).
9. D. MacInnes, Jr., M. A. Druy, P. J. Nigrey, D. P. Nairns, A. G. MacDiarmid, and A. J. Heeger, J. Chem. Soc., Chem. Commun., 1981, 317.
10. J. Heinze, M. Störzbach, and J. Mortensen, Ber. Bunsenges. Phys. Chem., 91, 960 (1987).
11. (a) S. Wawzonek and H. A. Laitinen, J. Amer. Chem. Soc., 64, 1765, 2365 (1942); (b) S. Wawzonek and F. W. Fan, J. Amer. Chem. Soc., 68, 2541 (1946).
12. (a) D. Lipkin, D. E. Paul, J. Townsend, and S. I. Weissman, Science, 117, 534 (1953); (b) S. I. Weissman, J. Townsend, D. E. Paul, and G. E. Pake, J. Chem. Phys., 21, 2227 (1953).
13. G. J. Hoijtink and J. van Schooten, Rec. Trav. Chim. Pays-Bas, 71, 1089 (1952); 72, 691, 903 (1953).
14. G. J. Hoijtink, J. van Schooten, E. De Boer, W. Aalbersberg, Rec. Trav. Chim. Pays-Bas, 73, 355 (1954).
15. S. Wawzonek, R. Berkey, E. W. Blaha, and E. M. Runner, J. Electrochem. Soc., 102, 235 (1955).
16. A. C. Allison, T. A. Gough, and M. E. Peover, Nature, 197, 764 (1963).
17. A. Streitwieser and I. Schwager, J. Phys. Chem., 66, 2316 (1962).
18. R. Pointeau, Ann. Chim., 7, 669 (1962).
19. K. S. V. Santhanam, A. J. Bard, J. Amer. Chem. Soc., 88, 2669 (1964).
20. A. C. Aten, C. Buthker, and G. J. Hoijtink, Trans. Faraday Soc., 55, 324 (1959).
21. R. Dietz and B. E. Larcombe, J. Chem. Soc., B, 1369 (1970).
22. A. C. Aten and G. J. Hoijtink, Z. Phys. Chem., 21, 182 (1959).
23. T. J. Katz, W. H. Reinmuth, and D. E. Smith, J. Amer. Chem. Soc., 84, 802 (1962).
24. T. A. Gough and M. E. Peover, Polarography-1964, Macmillian, London, 1966, p. 1017.
25. R. Dietz and M. E. Peover, Trans. Faraday Soc., 62, 3535 (1966).
26. J. Heinze, Angew. Chem., 96, 823 (1984); Angew. Chem. Int. Ed. Engl., 23, 831 (1984).
27. G. J. Hoijtink, E. de Boer, P. H. van Meji, and W. P. Weijland, Rec. Trav. Chim.Pays-Bas, 74, 277 (1955); 75, 487 (1956).
28. (a) V. Slates and M. Szwarc, J. Phys. Chem., 69, 4124 (1965), (b) J. Jagur-Grodzinski, M. Field, S. L. Yang, and Szwarc, J. Phys. Chem. 69, 628 (1965).
29. W. Gross and U. Schindewolf, Ber. Bunsenges. Phys. Chem., 85, 112 (1981).
30. H. Kiesele, Anal. Chem., 53, 1952 (1981).
31. J. Heinze, J. Mortensen, K. Müllen, and R. Schenk, J. Chem. Soc., Chem. Commun., 1987, 701.
32. A. J. Fry, Synthetic Organic Electrochemistry, Chap. 7, Harper and Row, New York, 1972.
33. J. Mortensen and J. Heinze, Angew. Chem., 96, 65 (1984); Angew. Chem. Int. Ed. Engl., 23, 84 (1984).

34. D. H. Geske and A. Maki, J. Amer. Chem. Soc., 82, 2671 (1960).
35. R. E. Sioda, D. O. Cowan, and W. S. Koski, J. Amer. Chem. Soc., 89, 230 (1967).
36. R. D. Allendorfer, L. L. Miller, M. E. Larscheid, and R. Chang, J. Org. Chem., 40, 97 (1975).
37. I. Bergman, Trans. Faraday Soc., 50, 829 (1954).
38. P. H. Given, J. Chem. Soc., 1958, 2684.
39. M. J. S. Dewar and C. de Llano, J. Amer. Chem. Soc., 91, 789 (1969).
40. A. J. Fry, L. L. Chung, and V. Boekelheide, Tetrahedron Lett., 5, 445 (1974).
41. A. J. Fry, C. S. Hutchins, and L. L. Chung, J. Amer. Chem. Soc., 97, 591 (1975).
42. J. Mortensen, and J. Heinze, Tetrahedron Lett., 26, 415 (1985).
43. M. E. Peover and J. D. Davies, J. Electroanal. Chem., 6, 46 (1963).
44. J. O'M. Bockris and A. K. N. Reddy, Modern Electrochemistry, Plenum Press, New York, 1970.
45. K. H. J. Buschow, J. Dieleman, and G. J. Hoijtink, J. Chem. Phys., 42, 1993 (1965).
46. D. N. Bhattacharyya, C. L. Lee, J. Smid, and M. Szwarc, J. Phys. Chem., 69, 608 (1965).
47. M. Szwarc, Acc. Chem. Res., 5, 169 (1972).
48. A. Rainis, R. Tung, and M. Szwarc, J. Amer. Chem. SOc., 95, 659 (1973).
49. H. Kiesele, Euchem Conference on Pericyclic Reactions, Farrara, 1982; Euchem Conference on Organic and Metalorganic Electrochemistry, Dun Loaghaire, 1983.
50. L. A. Avaca and A. Bewick, J. Electroanal. Chem., 41, 405 (1973).
51. (a) R. D. Allendoerfer and P. H. Rieger, J. Amer. Chem. Soc., 87, 2336 (1965), (b) R. D. Allendoerfer, J. Amer. Chem. Soc., 97, 218 (1975).
52. B. J. Huebert and D. E. Smith, J. Electroanal. Chem., 31, 333 (1971).
53. R. D. Rieke and R. A. Copenhafer, J. Electroanal. Chem., 56, 409 (1974).
54. L. A. Paquette, G. D. Ewing, S. Traynor, and J. M. Gardlik, J. Amer. Chem. Soc., 99, 6115 (1977).
55. B. S. Jensen, A. Ronlan, and V. D. Parker, Acta Chem. Scand., B29, 394 (1975).
56. J. Heinze, M. Dietrich, K. Hinkelmann, K. Meerholz, and F. Rashwan, DECHEMA-Monographie, 112, 61 (1989).
57. L. A. Paquette, S. V. Levy, R. H. Meisinger, R. K. Russell, and M. Oku, J. Amer. Chem. Soc., 96, 5806 (1974).
58. H. Lehmkuhl, S. Kintopf, and E. Janssen, J. Organometal. Chem., 56, 41 (1973).
59. W. H. Smith and A. J. Bard, J. Electroanal. Chem., 76, 19 (1977).
60. A. J. Fry, J. Simon, T. Tashiro, T. Yamamoto, R. H. Mitchell, T. W. Dingle, R. V. Williams, and R. Mahedevan, Acta Chem. Scand., B37, 445 (1983).
61. L. A. Paquette, J. M. Photis, and G. D. Ewing, J. Amer. Chem. Soc., 97, 3538 (1975).
62. J. F. M. Oth, H. Baumann, J.-M. Gilles, and G. Schröder, J. Amer. Chem. Soc., 94, 3498 (1972).

63. D. Tanner, O. Wennerström, and E. Vogel, Tetrahedron Lett., 23, 1221 (1982).
64. J. F. M. Oth, E. P. Woo, and F. Sondheimer, J. Amer. Chem. Soc., 95, 7337 (1973).
65. K. Müllen, W. Huber, T. Meul, M. Nakagawa, and M. Iyoda, J. Amer. Chem. Soc., 104, 5403 (1982).
66. (a) K. Ankner, B. Lamm, B. Thulin, and O. Wennerström, Acta Chem. Scand., B32, 155 (1978); (b) W. Huber, K. Müllen, and O. Wennerström, Angew. Chem., 92, 636 (1980); Angew Chem. Int. Ed. Engl., 19, 624 (1980).
67. J. F. M. Oth and G. Schröder, J. Chem. Soc., (B) 1971, 904.
68. L. Anderson, M. J. Broadhurst, and L. A. Paquette, J. Amer. Chem. Soc., 95, 2198 (1973).
69. R. D. Allendoerfer, L. L. Miller, M. E. Larschied, and R. Change, J. Org. Chem. 40, 97 (1975).
70. W. Huber, K. Müllen, R. Busch, W. Grimme, and J. Heinze, Angew. Chem., 94, 294 (1982); Angew. Chem. Int. Ed. Engl., 21, 301 (1982).
71. K. Hafner, Angew. Chem., 70, 419 (1959).
72. K. Meerholz, and J. Heinze, J. Amer. Chem. Soc., 777, 2325 (1989).
73. A. J. Fry, in Topics in Organic Electrochemistry (A. J. Fry and W. E. E. Britton, eds.), Plenum Press, New York, 1986.
74. K. Müllen, J. F. M. Oth, H.-W. Engels, and E. Vogel, Angew. Chem., 91, 251 (1979), Angew. Chem. Int. Ed. Engl., 18, 229 (1979).
75. F. M. Oth, K. Müllen, K. Königshofen, J. Wassen, and E. Vogel., Helv. Chim. Acta, 57, 2387 (1974).
76. R. W. Johnson, J. Amer. Chem. Soc., 99, 1461 (1977).
77. (a) S. I. Zhdanov, Z. Phys. Chem. (Leipzig),235 (1958); (b) P. H. Plesch and A. Stasko, J. Amer. Chem. Soc. B, 1971, 2052.
78. J. M. Lead, T. Teherane, and A. J. Bard, J. Electroanal. Chem., 9, 275 (1978).
79. R. Breslow, W. Bahary, and W. Reinmuth, J. Amer. Chem. Soc., 83, 1763 (1961).
80. R. W. Johnson, T. Widlanski, and R. Breslow, Tetrahedron Lett., 1976, 4685.
81. (a) R. Breslow and W. Chu, J. Amer. Chem. SOc., 95, 411 (1973); (b) M. R. Wasielewski and R. Breslow, J. Amer. Chem. Soc., 98, 4222 (1976).
82. (a) G. Kothe, W. Sümmermann, H. Baumgärtel, and H. Zimmermenn, Tetrahedron Lett., 1969, 2185; (b) H. Volz and W. Lotsch, Tetrahedron Lett., 1969, 2275.
83. T. A. Skotheim, ed., Handbook of Conducting Polymers, Dekker, New York, 1986.
84. J. E. Frommer and R. R. Chance, in Encyclopedia of Polymer Science and Engineering (M. Grayson and J. Kroschwitz, eds.), Wiley, New York, 1986, pp. 462.
85. J. Heinze, Top. Curr. Chem. 1989.
86. (a) J. Mortensen and J. Heinze, J. Electroanal. Chem., 175, 333 (1984); (b) K. Itaya, A. J. Bard, and M. Szwarc, Z. Phys. Chem. N. F., 112, 1 (1978).
87. F. Gerson, J. Lopez, and A. Metzger, Helv. Chim. Acta, 63, 2135 (1980).
88. L. A. Paquette, G. D. Ewing, and S. Traynor, J. Amer. Chem. Soc., 98, 279 (1976).

89. A. J. Fry, T. A. Powers, K. Müllen, and W. Irmen, Tetrahedron Lett., 26, 5879 (1985).
90. A. F. Diaz, J. Crowley, J. Bargon, G. P. Gardini, and J. B. Torrance, J. Electroanal. Chem., 121, 355 (1981).
91. H. Naarmann and N. Theophilou, Synth. Met., 22, 1 (1987).
92. J. Heinze, R. Bilger, and K. Meerholz, Ber. Bunsenges. Phys. Chem., 92, 1266 (1988).
93. L.-S. R. Yeh and A. J. Bard, J. Electrochem. Soc., 124, 189 (1977).
94. R. Dietz and M. Peover, Discuss. Faraday Soc., 45, 154 (1968).
95. B. S. Jensen, R. Lines, P. Pagsberg, and V. D. Parker, Acta Chem. Scand., B31, 707 (1977).
96. M. Grezesczuk and D. E. Smith, J. Electroanal. Chem., 162, 189 (1984).
97. T. Troll and M. M. Baizer, Electrochim. Acta, 19, 951 (1974).
98. G. Farnia, F. Maran, and G. Sandona, J. Chem. Soc., Faraday Trans., 1, 82, 1885 (1986).
99. W. E. Britton, in Topics in Organic Chemistry (A. J. Fry and W. E. Britton, eds.), Plenum Press, New York, 1986, p. 227.
100. D. H. Evans and R. W. Busch, J. Amer. Chem. Soc., 104, 5057 (1982).
101. N. S. Hush, Trans. Faraday Soc., 57, 557 (1961).
102. R. A. Marcus, Can. J. Chem., 37, 155 (1959).
103. R. R. Dogonadze, A. M. Kuznetsov, and A. A. Chernenko, Russ. Chem. Rev., 34, 759 (1965).
104. L. Eberson, Adv. Phys. Org. Chem., 18, 7 (1982).
105. F. A. L. Anet, J. Amer. Chem. Soc., 84, 671 (1962).
106. J. M. Hale, in Reactions of Molecules at Electrodes (N. S. Hush, ed.), Wiley, New York, 1971, pp. 229.
107. D. H. Evans and K. O'Connell, in Electroanalytical Chemistry, Vol. 14 (A. J. Bard, ed.), Dekker, New York, 1986, p. 13.
108. R. A. Marcus, J. Phys. Chem., 67, 853 (1963).
109. L. S. Marcoux, R. N. Adams, and S. W. Feldberg, J. Phys. Chem., 73, 2611 (1969).
110. J. Simonet, M.-A. Michel, and H. Lund, Acta Chem. Scand., B29, 489 (1975).
111. E. Hobolth and H. Lund, Acta Chem. Scand., B31, 395 (1977).
112. T. Lund and H. Lund, Acta Chem. Scand., B40, 470 (1986).
113. C. P. Andrieux, C. Blocman, J. M. Dumas-Bouchiat, F. M'Halla, and J. M. Saveant, J. Amer. Chem. Soc., 101, 3806 (1979).
114. C. P. Andrieux, J. M. Dumas-Bouchiat, and J. M. Saveant, J. Electroanal. Chem., 87, 39 (1978); 87, 55 (1978); 88, 43 (1978); 113, 1 (1980).
115. C. P. Andrieux, J. M. Duman-Bouchiat, F. M'Halla, and J. M. Saveant, J. Electroanal. Chem., 113, 19 (1980).
116. C. P. Andrieux, C. Blocman, J. M. Dumas-Bouchiat, and J. M. Saveant, J. Amer. Chem. Soc., 101, 3431 (1979).
117. C. Amatore, M. Gareil, and J. M. Saveant, J. Electroanal. Chem., 147, 1 (1983).
118. H. Kiesele, Habilitationsschrift, University of Konstanz, 1981, UNESCO Workshop and Sandbjerg Meeting on Organic Electrochemistry, Sandbjerg, 1981, Abstracts of papers, pp. 23–25.
119. R. A. Marcus, J. Chem. Phys., 43, 2654 (1965); Electrochim. Acta, 13, 1217 (1968).

120. L. R. Faulkner and A. J. Bard, in Electroanalytical Chemistry, Vol. 10 (A. J. Bard, ed.), Dekker, New York, 1977, pp. 1–95.
121. G. J. Hoijtink, Discuss. Faraday Soc., 45, 14 (1968).
122. E. A. Chandros, J. W. Longworth, and R. E. Visco, J. Amer. Chem. Soc., 87, 3259 (1965).
123. H. Tachikawa and A. J. Bard, Chem. Phys. Lett., 26, 568 (1974).
124. A. Weller and K. Zachariasse, Chem. Phys. Lett., 10, 590 (1971).
125. S. M. Park and A. J. Bard, J. Amer. Chem. Soc., 97, 2978, (1975); S. M. Park, M. T. Paffett, and G. H. Daub, J. Amer. Chem. Soc., 99, 5393 (1977).
126. T. M. Siegel and H. B. Mark, J. Amer. Chem. Soc., 99, 5393 (1977).
127. D. L. Atkins and R. L. Birke, Chem Phys. Lett., 29, 428 *1874); T. D. Santa-Cruz, D. L. Atkins, and R. L. Birke, J. Amer. Chem. Soc., 98, 1677 (1976).
128. G. H. Brilmyer and A. J. Bard, J. Electrochem. Soc., 127, 104 (1980).
129. A. J. Birch and R. G. Subba, Adv. Org. Chem., 8, 1 (1972).
130. G. J. Hoijtink, Rec. Trav. Chim. Pays-Bas, 73, 895 (1954).
131. G. J. Hoijtink, Rec. Trav. Chim. Pays-Bas, 76, 885 (1957).
132. S. Valcher and A. M. Ghe, J. Electroanal. Chem., 55, 407 (1974).
133. C. Amatore and J. M. Saveant, J. Electroanal. Chem., 107, 353 (1980).
134. B. Aalstadt and V. D. Parker, J. Electroanal. Chem., 112, 163 (1981).
135. B. Aalstadt and V. D. Parker, J. Electroanal. Chem., 121, 73 (1981).
136. E. Ahlberg and V. D. Parker, Acta Chem. Scand., B35, 117 (1981).
137. V. D. Parker, Acta Chem. Scand., B35, 349 (1981).
138. V. D. Parker, Acta Chem. Scand., B35, 373 (1981).
139. C. Amatore, M. Gareil, and J. M. Saveant, J. Electroanal. Chem., 176, 377 (1984).
140. M. F. Nielsen, O. Hammerich, and V. D. Parker, Acta Chem. Scand., B40, 101 (1986).
141. M. F. Nielsen, O. Hammerich, and V. D. Parker, Acta Chem. Scand. B41, 50 (1987).
142. M. F. Nielsen, O. Hammerich, and V. D. Parker, Acta Chem. Scand., B41, 64 (1987).
143. I. Bergman, Trans. Faraday Soc., 52, 690 (1952).
144. A. J. Fry and A. Schuettenberg, J. Org. Chem., 39, 2452 (1974).
145. L. Eberson, Z. Blum, B. Helge, and K. Nyberg, Tetrahedron, 34, 731 (1978).
146. I. Bergman, Polarograpy 1964, MacMillan, London, 1966, p. 925.
147. A. Streitwieser and S. Suzuki, Tetrahedron, 16, 153 (1961).
148. E. Heilbronner and H. Bock, Das HMO-Modell und seine Anwendung, Verlag Chemie, Weinheim, 1968.
149. G. J. Hoijtink, Rec. Trav. Chim. Pays-Bas, 76, 775 (1961).
150. P. E. Hansen, O. Blaabjerg, and A. Berg, Acta Chem. Scand., B32, 720 (1078).
151. A. Misono, T. Osa, and T. Yamagishi, Bull. Chem. Soc. Jpn., 41, 2921 (1968).
152. C. Amatore, G. Capobianco, G. Farnia, G. Sandona, J. M. Saveant, M. G. Severin, and E. Vianello, J. Amer. Chem. Soc., 107, 1815 (1985).

153. (a) A. Lagu, H. B. Mark, Jr., and J. R. Jezorek, J. Org. Chem., 42, 1063 (1977). (b) K. J. Borhani, and M. D. Hawley, J. Electroanal. Chem., 101, 407 (1979).
154. H. Kiesele, D. H. Evans, 84. Meeting of the Deutsche Bunsengesellschaft, May 1985, Abstract D 43.
155. A. D. Moro, G. Farnia, F. Marcuzzi, and G. Melloni, Nouv. J. Chim., 4, 3 (1980).
156. G. Capobianco, G. Farnia, G. Sandona, F. Marcuzzi, and G. Melloni, J. Electroanal. Chem., 134, 363 (1982).
157. (a) F. Beck and H. Gerischer, Z. Elektrochem., Ber. Bunsenges. Phys. Chem., 65, 504 (1961). (b) S. Pons and S. B. Khoo, J. Amer. Soc., 104, 3845 (1982).
158. D. J. Morantz and E. Warhurst, Trans. Faraday Soc., 51, 1375 (1955).
159. J. F. Garst, Acc. Chem. Res., 4, 400 (1971).
160. H. Lund, M.-A. Michel, and J. Simonet, Acta Chem. Scand., B28, 900 (1974).
161. P. Martigny and J. Simonet, J. Electroanal. Chem., 101, 275 (1979).
162. W. Kaim, Acc. Chem. Res., 18, 160 (1985).
163. (a) A. Pross and S. S. Shaik, Acc. Chem. Res., 16, 363 (1983). (b) A. Pross, Acc. Chem. Res., 18, 212 (1985).
164. E. Hebert, J.-P. Mazaleyrat, Z. Welvart, L. Nadjo, and J.-M. Saveant, Nouv. J. Chim., 9, 75 (1985).
165. H. Lund, Acta Chem. Scand., B31, 424 (1977).
166. R. Engels and H. J. Schäfer, Angew. Chem. Int. Ed. Engl., 17, 460 (1978);R. Engels, Dissertation University Münster (1978); B. Brauer, Dissertation University Münster (1982).
167. S. Wawzonek and D. Wearring, J. Amer. Chem. Soc., 81, 2067 (1959).
168. J. W. Loveland, U.S. Pat. 3,032,489 (1962); Chem. Abstr., 57, 4470 (1962); W. C. Niekam, U.S. Pat. 3,334,045 (1967); Chem. Abstr., 68, 8809 (1968).
169. R. Dietz and M. E. Peover, Discuss. Faraday Soc., 45, 167 (1968).
170. H. Lund and J. Simonet, J. Electroanal. Chem., 65, 205 (1975).
171. D. A. Tyssee, J. H. Wagenknecht, M. M. Baizer, and J. L. Chruma, Tetrahedron Lett., 4809 (1972).
172. A. J. L. Beckwith and R. O. C. Norman, J. Chem. Soc., B, 400 (1969).
173. J. W. Loveland, U.S. Pat. 3,214,356 (1965); Chem. Abstr., 64, 671d (1966).
174. M. M. Baizer and J. D. Anderson, J. Org. Chem., 30, 1348 (1965).
175. L. D. McKeever and R. Waack, J. Organometal. Chem., 17, 142 (1969).
176. H. C. Wang, E. D. Lillie, S. Slomkowski, G. Levin, and M. Szwarc, J. Amer. Chem. Soc., 99, 4612 (1977).
177. A. Yildiz and H. Baumgärtel, Ber. Bunsenges. Phys. Chem., 81, 1177 (1977).
178. O. Hammerich and V. D. Parker, Acta Chem. Scand., B35, 381 (1981).
179. C. Amatore, J. Pinson, and J. M. Saveant, J. Electroanal. Chem., 137, 143 (1982).
180. C. Amatore, J. Pinson, and J. M. Saveant, J. Electroanal. Chem., 139, 193 (1982).
181. J. M. Saveant, Acta Chem. Scand., B37, 365 (1983).
182. C. Amatore, D. Garreau, M. Hammi, J. Pinson, and J. M. Saveant, J. Electroanal. Chem., 184, 1 (1985).

183. F. Beck, Angew. Chem. Int. Ed. Engl., 11, 760 (1972); M. M. Baizer and J. P. Petrovich, in Progress in Physical Organic Chemistry (A. Streitweiser and R. W. Taft, eds.), Vol. 7, Wiley-Interscience, New York, 1970, pp. 189–227.
184. M. Figeys and H. P. Figeys, Tetrahedron, 24, 1097 (1968).
185. W. Schlenk, J. Appenrodt, A. Michael, and A. Thal, Ber. Dtsch. Chem. Ges., 47, 473 (1914).
186. K. Ziegler and K. Bähr, Ber. Dtsch. Chem. Ges., 61, 253 (1928).
187. M. Szwarc, Ber. Bunsenges. Phys. Chem., 67, 763 (1963); Carbanions. Living Polymers and Electron Transfer Processes, Wiley, New York, 1968; in Progress in Physical Organic Chemistry (A. Streitwieser and R. W. Taft, eds.), Vol. 6, Interscience, New York, 1968, p. 323.
188. N. Yamazaki, S. Nakahama, and S. Kambara, J. Polym. Sci. Part B, 3, 57 (1965).
189. B. L. Funt and S. W. Laurent, Can. J. Chem., 42, 2728 (1964).
190. B. L. Funt and S. N. Bhadani, Can. J. Chem., 42, 2733 (1964).
191. B. L. Funt, D. Richardson, and S. N. Bhadani, Can. J. Chem., 44, 711 (1966).
192. J. D. Anderson, J. Polym. Sci., Part A-1, 6, 3185 (1968).
193. B. L. Funt, S. N. Bhadani, and D. Richardson, J. Polym. Sci. Part A-1, 4, 2871 (1966).
194. E. Laviron and Y. Mugnier, J. Electroanal. Chem., 93, 69 (1978); P. Netta and H. Levanon, J. Phys. Chem., 81, 2288 (1977).
195. A. J. Bard and J. Phelps, J. Electroanal. Chem., 25, App. 2 (1970); B. L. Funt and D. Gray, J. Electrochem. Soc., 117, 1020 (1970).
196. J. F. Garst, J. G. Pacifici, and E. R. Zabolotny, J. Amer. Chem. Soc., 88, 3872 (1966).
197. K. Ziegler and F. Thielmann, Ber. Dtsch. Chem. Ges., 56, 1740 (1923).
198. A. Lagendijk and M. Szwarc, J. Amer. Chem. Soc., 93, 5359 (1971).
199. O. Hammerich and J. M. Saveant, J. Chem. Soc., Chem. Commun., 938 (1979).
200. D. Casson and B. J. Tabner, J. Chem. Soc., (B), 887 (1969).
201. R. B. Woodward and R. Hoffmann, The Conservation of Orbital Symmetry, Verlag Chemie, Weinheim, 1970.
202. N. L. Bauld, J. Cessac, F. R. Farr, and R. Holloway, J. Amer. Chem. Soc., 98, 4561 (1976).
203. M. Horner and S. Hünig, J. Amer. Chem. Soc., 99, 6122 (1977).
204. K. Müllen and W. Huber, Helv. Chim. Acta, 61, 1310 (1978).
205. J. Heinze and K. Müllen, private communication, 1988.
206. L. L. Miller and E. A. Mayeda, J. Amer. Chem. Soc., 92, 5818 (1970).
207. L. L. Miller and R. F. Boyer, J. Amer. Chem. Soc., 93, 650 (1971).
208. L. L. McCullough, Acc. Chem. Res., 13, 270 (1980).
209. W. Schenk and E. Bergmann, Liebigs Ann. Chem., 463, 125 (1928); K. Ziegler and F. Crossman, Ber. Dtsch. Chem. Ges., 62, 1768 (1929); L. L. Miller and R. F. Boyer, J. Amer. Chem. Soc., 93, 646 (1971).
210. H. Kiesele, Angew. Chem. Int. Ed. Engl., 22, 254 (1983).

8
Halogenated Organic Compounds

DENNIS G. PETERS / Indiana University, Bloomington, Indiana

I. Introduction 362

II. Monohalogenated Alkanes 362
- A. Reduction at mercury cathodes 363
- B. Reduction at carbon cathodes 365
- C. Reduction at other electrodes 366
- D. Electrolyte effects 366

III. Polyhalogenated Alkanes 367
- A. gem-Dihaloalkanes, gem-trihaloalkanes, and carbon tetrachloride 367
- B. Vicinal dihaloalkanes and trihaloalkanes 368
- C. α,ω-Dihaloalkanes 368

IV. Halogenated Alkenes and Alkynes 369
- A. Vinyl halides 369
- B. Allyl halides 370
- C. Allenic and acetylenic halides 370

V. Benzyl Halides and Related Compounds 373

VI. Alicyclic Halides 375
- A. Monohalogenated species 375
- B. Dihalogenated species 376

VII. Halogenated Aromatic Compounds 379
- A. Monohalobenzenes 379
- B. Polyhalobenzenes 380
- C. Halogenated nitrobenzenes and cyanobenzenes 381
- D. Halogenated aryl carbonyls 382
- E. Halogenated biphenyls 384

VIII. Acyl and Phenacyl Halides 384

IX. Aliphatic α-Halocarbonyl Compounds 387

X. Halogenated Heterocyclic Compounds 388
- A. Heterocyclic compounds with one nitrogen atom 388
- B. Heterocyclic compounds with two nitrogen atoms 389
- C. Other heterocyclic species 389

XI. Indirect or Catalytic Reduction of the Carbon-Halogen Bond 390
- A. Electrogenerated radical anions as mediators 391
- B. Transition metal complexes as mediators 392
- C. Reduction of carbon-halogen bonds at chemically modified electrodes 393

References 394

I. INTRODUCTION

There continues to be immense interest in the electrochemical reduction of halogenated organic compounds, first with respect to mechanistic details about the electrolytic cleavage of carbon-halogen bonds and the chemical reactions that accompany the electron transfer process, and second with respect to the discovery and development of practical electrosyntheses based on the electrolysis of these compounds. In previous editions of this volume [1,2], considerable information about the electrochemical behavior of halogenated organic species has appeared. In addition, there have been at least two excellent and thorough reviews [3,4] containing a wealth of material on both mechanistic and synthetic aspects of the electrochemistry of halogenated organic compounds. Because the length of this chapter is necessarily limited and because the aforementioned discussions are readily available and should be consulted, an effort is made here to avoid excessive duplication and to focus more extensively on the most recent work, especially that concerning organic electrosynthesis.

II. MONOHALOGENATED ALKANES

It is now accepted that, at an inert cathode, the electrochemical behavior of any simple alkyl monohalide (RX) can usually be rationalized according to the mechanistic scheme

$$RX + e \rightleftharpoons [RX^{\cdot -}]$$
$$[RX^{\cdot -}] \longrightarrow R^{\cdot} + X^{-}$$
$$R^{\cdot} + e \longrightarrow R^{-}$$
$$2R^{\cdot} \xrightarrow{\text{combination}} R_2$$
$$2R^{\cdot} \xrightarrow{\text{disproportionation}} RH + R(-H)$$
$$R^{\cdot} + SH \longrightarrow RH + S^{\cdot}$$
$$R^{-} + RX \longrightarrow R_2 + X^{-}$$
$$R^{-} + HB \longrightarrow RH + B^{-}$$
$$RX + B^{-} \xrightarrow{E2} R(-H) + X^{-} + HB$$
$$RX + B^{-} \xrightarrow{S_N2} RB + X^{-}$$

Thus, via electrogenerated alkyl radicals or carbanions, it is possible to obtain dimers (R_2), alkanes (RH), and olefins [R(-H)]. Moreover, olefins can arise from the base-promoted dehydrohalogenation of RX, and products (RB) derived from displacement of halide by a base (B^-) can form. For certain electrode materials, such as mercury, lead, and tin, as discussed later in this chapter, the alkyl radical can interact strongly with the cathode metal, which can enhance the reduction of the alkyl monohalide and can result in the formation of organometallic compounds. In general, tertiary alkyl halides are easier to reduce than secondary alkyl halides, which are in turn easier to reduce than primary alkyl halides. Obviously, the identity of the halogen atom governs the ease of reductive cleavage of the carbon-halogen bond; iodides are easier to reduce than bromides, chlorides are so difficult to reduce that they often appear not to be electroactive, and only a single report [5] of the reduction of an alkyl fluoride (polytetrafluoroethylene) has been published. Finally, the existence of the radical anion [$RX^{\cdot-}$], formed by the addition of a single electron to a simple alkyl monohalide, has never been demonstrated; Savéant and coworkers [6] have presented an authoritative discussion of reasons such a species should not be expected for simple alkyl monohalides as well as why radical anions of aromatic halides are intermediates in the electrochemical reduction of the latter compounds.

A. Reduction at Mercury Cathodes

Primary Alkyl Monohalides

Studies of the reduction of a number of primary alkyl monoiodides at mercury electrodes have been undertaken in a variety of media. In EtOH containing TBAI as supporting electrolyte, electrolyses of iodoethane and 1-iodobutane afford virtually quantitative yields of diethylmercury and dibutylmercury, respectively [7]. In DMF containing various tetraalkylammonium perchlorates, 1-iododecane exhibits stepwise reduction first to give the decyl radical and then to give the decyl carbanion [8]. A preparative-scale electrolysis at a potential corresponding to the first stage of reduction for 1-iododecane leads to formation of didecylmercury in a yield of at least 99%. It is evident that the ease of reduction of alkyl iodides as well as the production of dialkylmercury compounds is attributable to interaction of electrogenerated alkyl radicals with the mercury cathode:

$$RI + e \longrightarrow R^{\cdot} + I^{-}$$

$$R^{\cdot} \longrightarrow R^{\cdot}(ads)$$

$$R^{\cdot}(ads) + Hg \longrightarrow RHg^{\cdot}(ads)$$

$$2RHg^{\cdot}(ads) \longrightarrow R_2Hg + Hg$$

When 1-iododecane is electrolyzed in DMF containing an excess of water and at a potential corresponding to the second stage of reduction, decane (52%) and 1-decene (24%) are the principal products, as well as small amounts of 1-decanol and several telomeric species most likely derived from reactions between decyl radicals and the solvent [9]. Removal of most of the water from the system results in a higher yield of decane (71%) and a lower quantity of 1-decene (4%) along with the same quantities of telomers; these data reveal that water is deprotonated by the decyl carbanion and that the resulting hydroxide ion causes the dehydrohalogenation of 1-iododecane to produce 1-decene. If another proton donor (diethyl malonate) is introduced into the system, the reaction between 1-iododecane and hydroxide ion is blocked. Interestingly, none of the "dimeric" species, eicosane, was found among the products. Electrolysis of 1-iododecane in the presence of a high concentration of deuterium oxide and at potentials that cause the generation of carbanions results in the formation of decane that is monodeuterated to the extent of at least 80%.

Bilewicz and Osteryoung [10] have examined the electrochemical behavior of iodoethane, 1-iodobutane, and 1-iododecane at mercury cathodes in MeCN containing several different tetraalkylammonium salts and have obtained further mechanistic insights into the processes associated with the production of dialkylmercury compounds and with the protonation of alkyl carbanions. These workers found voltammetric characteristics for the alkyl iodides that mimic those seen in DMF. On the basis of these observations, one can conclude that electrolyses of these alkyl iodides in either MeCN or DMF should give comparable product distributions; because MeCN is easier to purify than DMF, the former should be preferable for preparative-scale electrolyses.

Primary alkyl monobromides usually undergo a one-step, two-electron reduction to the corresponding alkyl carbanions, and the carbanion subsequently reacts to form alkanes and alkenes. Electrolytic reduction of 1-bromodecane in DMF containing a tetraalkylammonium salt affords a product distribution nearly the same as that obtained from the electrolysis of 1-iododecane at very negative potentials [9]. However, Fry [11] has reported that the electrochemical reduction of 1-bromohexane in DMF containing TEABr yields hexane (70%), 1-hexene (12%), dihexylmercury (2%), and 1-hexanol (7%), so there is some ambiguity about the behavior of primary alkyl monobromides. A pair of papers by Wagenknecht and co-workers [12, 13], dealing with the reduction of bromoethane and 1-bromobutane in DMF containing TEABr and saturated with carbon dioxide, reveals that dialkylmercury species are produced; from 1-bromobutane, roughly equal quantities of dibutylmercury, dibutyl oxalate, and butyl valerate arise via the following proposed scheme:

$$C_4H_9Br + e + Hg \longrightarrow C_4H_9Hg^{\cdot} + Br^-$$

$$2C_4H_9Hg^{\cdot} \longrightarrow (C_4H_9)_2Hg + Hg$$

$$2CO_2 + 2e \longrightarrow 2CO_2^{\cdot -} \longrightarrow {}^-O-\overset{\overset{\displaystyle O}{\|}}{C}-\overset{\overset{\displaystyle O}{\|}}{C}-O^-$$

$${}^-O-\overset{\overset{\displaystyle O}{\|}}{C}-\overset{\overset{\displaystyle O}{\|}}{C}-O^- + 2C_4H_9Br \longrightarrow C_4H_9O-\overset{\overset{\displaystyle O}{\|}}{C}-\overset{\overset{\displaystyle O}{\|}}{C}-OC_4H_9 + 2Br^-$$

$$CO_2^{\cdot -} + C_4H_9Br \longrightarrow [C_4H_9CO_2^{\cdot}] + Br^-$$

$$[C_4H_9CO_2^{\cdot}] + e \longrightarrow C_4H_9CO_2^-$$

$$CO_2^{\cdot -} + C_4H_9^{\cdot} \longrightarrow C_4H_9CO_2^-$$

$$C_4H_9CO_2^- + C_4H_9Br \longrightarrow C_4H_9CO_2C_4H_9 + Br^-$$

Note that this sequence of reactions features a one-electron reduction of 1-bromobutane, instead of the usual two-electron cleavage of the carbon-bromine bond.

Secondary Alkyl Monohalides

There has been only one detailed investigation [14] of the electrochemical behavior of secondary alkyl monohalides. In DMF containing tetraalkylammonium perchlorates, 2-iodooctane undergoes stepwise reduction, whereas 2-bromooctane exhibits just one stage of reduction. At a potential corresponding to the uptake of one electron by 2-iodooctane, the electrogenerated sec-octyl radical (1) disproportionates to yield octene, 1-octene, and 2-octene, (2) combines to form 7,8-dimethyltetradecane, and (3) interacts with mercury to give di-sec-octylmercury. At an appropriate potential, the sec-octyl carbanion can be formed from either 2-iodooctane or 2-bromooctane; from both compounds, one obtains the same roster of electrolysis products as before, except that the yield of octane is approximately doubled and the quantities of diorganomercury and dimeric species are much lower.

Tertiary Alkyl Monohalides

Polarographic and voltammetric studies [15–17] of the electrochemical behavior of tert-butyl iodide and tert-butyl bromide have revealed that, depending on the solvent and electrolyte, the iodide undergoes stepwise reduction and the bromide exhibits either a single two-electron reduction or a pair of one-electron transfers.

Vieira and Peters [18,19] examined the cyclic voltammetric behavior of tert-butyl bromide in DMF containing tetraalkylammonium perchlorates; because the bromide undergoes stepwise reduction under these conditions, it was possible to determine the electrolysis products derived from either the tert-butyl radical or carbanion. Essentially identical yields of isobutane (40%) and isobutylene (40%) are formed via disproportionation of the tert-butyl radical, and 2,2,3,3-tetramethylbutane (10%) arises via a radical combination process. At more negative potentials at which the tert-butyl carbanion is generated, equal yields of isobutane (47%) and isobutylene (47%) are obtained via the sequence of reactions that involves (1) protonation of the carbanion by a donor acid and (2) dehydrohalogenation of unreduced starting material by the conjugate base of the proton donor. An intriguing finding in this study was the discovery that, in the presence of TEAP (but neither TMAP nor TBAP), pivalaldehyde (14%) is formed. Apparently, the tert-butyl carbanion attacks DMF to displace dimethylamide:

$$(CH_3)_3C^- + HCON(CH_3)_2 \longrightarrow (CH_3)_3CCHO + N(CH_3)_2^-$$

This process may have synthetic usefulness and is worthy of further exploration to assess its possible generality.

B. Reduction at Carbon Cathodes

A key difference between the electrochemical behavior of alkyl monohalides at mercury and carbon cathodes is that dialkylmercury compounds cannot form at the latter. Another advantage of carbon cathodes, as Lambert and Ingall [20] demonstrated, is that it is possible to observe the cleavage of carbon-chlorine bonds in such compounds as 1-chloropentane, 1-chlorohexane, 2-chlorobutane, and tert-pentyl chloride in DMF containing TEAP; however, electrolysis products were not identified.

Wagenknecht [13] studied the electrochemical reduction of 1-bromobutane at graphite in carbon dioxide-saturated DMF that contained TEABr; in contrast to the result cited earlier for mercury electrodes, the reduction of 1-bromobutane is a two-electron process, and the resulting butyl carbanion reacts to form butane, 1-butene, octane, butyl valerate, butyl-N,N-dimethyloxamate, and dibutyl-2-methylmalonate. In a separate earlier investigation, Kaabak and coworkers [21] electrolyzed iodoethane and 1-bromobutane at graphite in DMF and reported that alkanes, olefins, dimers, and solvent-derived products are formed.

Kinetics of the electrochemical reduction of a series of monohalogenated alkanes in DMF containing $TBABF_4$ has been investigated by Savéant and coworkers [6,22]. These workers reported that 1-bromobutane, 2-bromobutane, tert-butyl bromide, and 1-iodobutane undergo one-step, two-electron reductions, whereas 2-iodobutane and tert-butyl iodide exhibit a pair of one-electron reduction waves.

Detailed studies [23] of the electrolysis products derived from the reduction of primary, secondary, and tertiary alkyl iodides and bromides have been undertaken in DMF containing tetraalkylammonium salts. When 1-iododecane or 1-bromodecane is electrolyzed, the principal products are decane and 1-decene, along with some 1-decanol; however, a yield of eicosane as high as 37% has been obtained from the alkyl iodide in the presence of TEAP. Secondary alkyl halides, such as 2-iodooctane and 2-bromooctane, yield a mixture of octane (30–35%), 1-octene (14–19%), 2-octene (9–19%), 2-octanol (1–2%), and 7,8-dimethyltetradecane (18–34%); tert-butyl bromide affords isobutane (48%), isobutylene (42%), and 2,2,3,3-tetramethylbutane (8%).

C. Reduction at Other Electrodes

Methyl, ethyl, propyl, and butyl halides can be reduced at lead and tin cathodes to produce, respectively, tetraalkyllead and tetraalkyltin compounds in excellent yield. Other metals that have been similarly employed to prepare organometallic species include bismuth, gallium, indium, and thallium. Feoktistov [2] and Hawley [3] have presented discussions, with accompanying tables and references, of these electrosyntheses. In addition, electrolytic production of tetraalkyllead species from Grignard reagents through the use of sacrificial lead anodes is discussed in Chap. 31.

Sock et al. [24] have described an efficient method for the synthesis of carboxylic acids by constant-current reduction of alkyl bromides at various electrodes (e.g., platinum, gold, stainless steel, and graphite) in the presence of carbon dioxide in such solvents as DMF or a mixture of THF and HMPA.

D. Electrolyte Effects

Decreasing the size of the tetraalkylammonium cation of the supporting electrolytes commonly used for electrolyses of alkyl halides has a profound influence on the potential at which the carbon-halogen bond is reductively cleaved. Many workers [9,10,14,17,18,25] have noted that the carbon-halogen bond becomes more difficult to reduce as the size of the tetraalkylammonium cation increases. Fry and Krieger [17] determined relative electron transfer rates in DMSO for both isobutyl and tert-butyl bromide (and for bromobenzene and iodobenzene) as a function of the identity of the tetraalkylammonium cation; the trends in electron transfer rates appear to depend on (1) shifts in the location of the outer Helmholtz plane with cation size and (2) intrinsic changes in the electron transfer process caused by different electrolytes. For any attempted electrosynthesis, it is important to recognize that sizable shifts in the potential needed to reduce a carbon-halogen bond occur when different supporting electrolytes are used.

Because alkyl carbanions are such potent bases, they are readily protonated by almost any source of hydrogen ions, including adventitious water in the solvent. However, in the absence of added proton donors, the tetraalkylammonium cations themselves serve as acids for the carbanions; deprotonation of tetraethylammonium and tetra-n-butylammonium ions leads, via the Hofmann elimination, to the corresponding trialkylamine and olefin. When tetramethylammonium cations comprise the supporting electrolyte, there is evidence [26,27] that a proton is released to yield trimethylammonium methylide; it would be interesting to explore whether this nitrogen ylide can be electrogenerated under conditions that are synthetically useful.

III. POLYHALOGENATED ALKANES

A. gem-Dihaloalkanes, gem-Trihaloalkanes, and Carbon Tetrachloride

Such compounds as 1,1-dichloro-2,2-bis(p-chlorophenyl)ethane and 1,1-dichloro-2-hydroxy-2,2-bis(p-chlorophenyl)ethane can be caused to undergo stepwise dehalogenation during a set of controlled potential electrolyses at a mercury cathode in water-EtOH media containing potassium nitrate [28]. Similarly, species possessing the trichloromethyl moiety, such as 1,1,1,5-tetrachloropentane, 1,1,1-trichloro-5-cyanopentane, and 1,1,1,5-tetrachloro-3-pentene, can be electrolytically converted in aqueous MeOH to their less chlorinated analogs [29], as exemplified by the following sequence of reactions:

$$Cl_3C(CH_2)_4Cl + 2e + H^+ \longrightarrow Cl_2CH(CH_2)_4Cl + Cl^-$$

$$Cl_2CH(CH_2)_4Cl + 2e + H^+ \longrightarrow Cl(CH_2)_5Cl + Cl^-$$

Semmelhack and Heinsohn [30] have utilized controlled potential electrolysis for the selective removal of gem-polyhaloethoxy protecting groups from carboxylic acids.

Carbon tetrachloride undergoes stepwise reduction at a mercury cathode in DMF containing TEABr [31]. Baizer and Chruma [32] demonstrated that electrolysis of carbon tetrachloride in DMF or methylene chloride containing TEAOTs and chloroform at a potential corresponding to the first stage of reduction can be employed to generate the short-lived trichloromethyl anion, which can undergo coupling reactions with electrophiles, such as acrylonitrile, ethyl acrylate, and diethyl fumarate. Karrenbrock and Schäfer [33] found that the electrogenerated trichloromethyl anion reacts with both aldehydes and ketones to yield 1,1,1-trichloroalkan-2-ols. More recently, Shono and coworkers [34–36] have developed efficient procedures for the conversion of a variety of aldehydes into 1,1,1-trichloroalkan-2-ols by using the electrogenerated trichloromethyl carbanion; chloroform, present as a proton donor, provides a continuous (catalytic) supply of the reactive species.

B. Vicinal Dihaloalkanes and Trihaloalkanes

Reduction of these compounds usually involves the uptake of a pair of electrons with elimination of two halide ions to yield an olefin. For example, Casanova and Rogers [37] demonstrated that controlled potential electrolysis of 1,2-dibromo-2-phenylpropane at a mercury cathode in DMF containing $TBABF_4$ produces 2-phenylpropene; in the same study, reduction of meso-2,3-dibromobutane was found to yield trans-2-butene only, whereas cis-2-butene is formed exclusively from dl-2,3-dibromobutane. In an investigation of the reduction of dl-1,2-dibromo-1,2-diphenylethane at mercury electrodes in DMF, Lund and Holboth [38] observed that the ratio of cis- to trans-stilbene depends on the identity and the concentration of the supporting electrolyte; these workers suggest that formation of an ion pair between smaller solvated cations and the anion generated via two-electron scission of a carbon-bromine bond favors the production of the trans isomer. A series of papers concerning the electrochemical reduction of polyhalogenated ethanes (including 1-bromo-2-chloro-1,1,2-trifluoroethane, 1,2-dibromo-1,1,2,2-tetrafluoroethane, and 1,1,2,2-tetrabromoethane) in EtOH and water-EtOH mixtures has been published by Feoktistov, Gol'din, and co-workers [39–43].

In considering the mechanism for the two-electron reduction of vicinal dihalides, Casanova and Rogers [37] as well as Fry [44] postulate the occurrence of a concerted process in which both carbon-halogen bonds are partially broken as a carbon-carbon double bond is being partially formed. On the other hand, it has been proposed on the basis of nonelectrochemical evidence [45] that a vicinal dihalide undergoes one-electron reduction to a radical anion, which loses the first halide ion to form a neutral radical; then, in a subsequent event, the neutral radical accepts another electron to give a carbanion that eliminates the second halide ion to yield an olefin.

In a study of the reduction of meso- and dl-3,4-dibromohexane and of meso- and dl-2,5-dimethyl-3,4-dibromohexane, Brown et al. [46] observed that the meso species are easier to reduce than the corresponding dl isomers. Although trans-2,5-dimethylhex-3-ene is formed nearly quantitatively from both meso- and dl-2,5-dimethyl-3,4-dibromohexane in liquid ammonia, the latter starting material is more difficult to reduce because the two bromine atoms are gauche in the transition state, whereas the transition state for reduction of the meso isomer has the two bromines in an anti configuration. For the reduction of 3,4-dibromohexane in DMF at less negative potentials, the meso isomer affords exclusively trans-hex-3-ene and the dl species gives only cis-hex-3-ene; however, at more negative potentials, the stereoselectivity is lost, and the trans-cis product ratios are 50:50 for the meso isomer and 70:30 for the dl species. Evans and co-workers [47,48] have utilized cyclic voltammetry to study the conformational equilibrium for a number of vicinal dibromides as a function of temperature.

C. α,ω-Dihaloalkanes

Virtually all synthetic and mechanistic information dealing with the electrochemical reduction of these compounds is based on the behavior of α,ω-dibromoalkanes. It would be interesting to probe the mechanism of the electrolytic reduction and to explore other electrosyntheses through studies of α,ω-diiodoalkanes as well as α,ω-dihaloalkanes possessing two different halogens.

Pioneering work by Rifi [49–51] centered on electrolytic reduction at a mercury pool of a family of α,ω-dibromoalkanes in DMF containing TBAP.

Cyclopropane was obtained exclusively from 1,3-dibromopropane, and a mixture of cyclobutane (25%) and n-butane (75%) was formed from 1,4-dibromobutane; somewhat different results were obtained by Wiberg and Epling [52]. From 1,5-dibromopentane and 1,6-dibromohexane, only straight-chain alkanes were produced. Although not strictly α,ω-dihaloalkanes, dl- and meso-2,4-dibromopentane have been investigated in DMSO containing TEABr by Fry and Britton [53,54], who concluded, as did Wiberg and Epling [52], that the reduction of these compounds at mercury occurs in a stepwise fashion and involves intramolecular cyclization of a bromocarbanion intermediate.

A study by Casanova and Rogers [55] of the reduction of compounds ranging from 1,4-dibromobutane to 1,12-dibromododecane at mercury in DMF containing tetraalkylammonium salts revealed that dialkylmercury species can be synthesized in high yield by means of controlled potential electrolysis; however, constant current electrolysis of the same starting materials produced no organomercury compounds, only a variety of acyclic and alicyclic hydrocarbons in low yield.

Other publications on the subject of α,ω-dihaloalkanes pertain to the polarography of 1,4-dibromobutane [56], the electrosynthesis of phenylcyclopropane and cyclopropanol [57], the reduction of 1,3-dibromopropane and 1,4-dibromobutane at mercury and aluminum [58], and the synthesis and mechanism of formation of polymeric alkylmercury compounds [59].

IV. HALOGENATED ALKENES AND ALKYNES

A. Vinyl Halides

Reduction of a vinyl halide involves the uptake of one electron followed by the loss of a halide ion to yield a vinyl radical

$$R_2C{=}CXR + e \longrightarrow R_2C{=}\dot{C}R + X^-$$

which is further reduced and protonated:

$$R_2C{=}\dot{C}R + e + H^+ \longrightarrow R_2C{=}CHR$$

In addition, the vinyl radical can undergo rearrangement, prior to accepting an electron and a proton, ultimately to afford either the cis or trans alkene as a final product. Among the species that have been investigated are aryl-substituted vinyl bromides [60], cis- and trans-3-iodo-3-hexene [61], bromomaleate and bromofumarate [62], and aryl-substituted 3-chloroacrylonitriles [63].

Vicinal dibromoalkenes and diiodoalkenes undergo reduction, with loss of two halide ions, to yield the corresponding alkynes that (depending on the presence of activating substituents) may be further reduced to olefins and saturated species; work by Elving and coworkers [64] on the electrochemistry of diethyl dibromofumarate and diethyl dibromomaleate illustrates this behavior. Unlike their dibromo and diiodo analogs, vicinal dichloroalkenes show more complicated electrochemical characteristics; in some cases the two carbon-chlorine bonds just undergo successive two-electron, one-proton reductive cleavage [65], whereas in other situations the reduction leads to an alkyne [66].

B. Allyl Halides

A thorough examination of the reduction of allyl bromide and allyl chloride has been conducted by Baizer and co-workers [32,67]. Polarography of allyl bromide in DMSO containing TEAP indicates stepwise reduction to the radical and to the carbanion, whereas allyl chloride exhibits a single two-electron reduction wave [67]. In DMF containing TEAOTs or TEABr, each of these allyl halides undergoes a two-electron cleavage of the carbon-halogen bond at a mercury cathode, and the resulting carbanion can react with (1) a proton donor to give propene, (2) unreduced starting material to yield 1,5-hexadiene, (3) carbon dioxide to produce the carboxylate (which, in turn, can attack the allyl halide), and (4) other species, such as acrylonitrile and ethyl acrylate [32].

In MeCN containing TBAP, allyl bromide and allyl iodide both interact chemically with a mercury electrode to form allylmercury halides that are electroactive; Bard and Merz [68] have demonstrated that the latter compounds undergo reduction to afford diallylmercury, which is itself electroactive. In addition, Bard and Merz reported that the electrochemical behavior of allyl bromide and allyl iodide at platinum is complicated by adsorption of halide ions liberated upon reduction of the parent compounds. These allyl halides exhibit well-defined voltammetric responses at carbon cathodes. On the basis of the effect of proton donors on peak currents, along with information gained from identification and quantitation of the electrolysis products, it was concluded that in MeCN the allyl halides are reduced by accepting two electrons to form the allyl anion and that the allyl radical is not involved as an intermediate.

Brillas and Costa [69] investigated the reduction of trans-1,4-dibromo-2-butene and trans-1,4-dichloro-2-butene at mercury in DMF containing TEAP and reported that 1,3-butadiene is formed as an intermediate product. Doupeux and Simonet [70,71] have described the electrochemical behavior of a number of polyhalogenated allyl halides in DMF.

C. Allenic and Acetylenic Halides

Morgan [72] has indicated that controlled potential reduction of 6-iodo-1-hexene at mercury in DMF containing TEAP leads mainly to the diorganomercury compound along with small amounts of methylcyclopentane, 1-hexene, and 1,5-hexadiene and a trace of 1,2-bis(cyclopentyl)ethane. In view of what is discussed later in this section, it would be interesting to reduce 6-iodo-1-hexene at a carbon electrode to avoid production of the diorganomercury species and to promote formation of more methylcyclopentane via intramolecular cyclization of the electrogenerated radical intermediate.

Several investigations [73–75] have dealt with the electrochemistry of acetylenic halides that upon reduction can cyclize intramolecularly. In DMF containing tetraalkylammonium perchlorates, electrolysis of 6-iodo-1-phenyl-1-hexyne at mercury gives a distribution of products that is potential dependent [73]. At certain potentials, the acetylenic iodide is converted to a radical

$$C_6H_5C{\equiv}C(CH_2)_4I + e \longrightarrow C_6H_5C{\equiv}C(CH_2)_4^{\cdot} + I^-$$

which appears to be adsorbed onto and to interact with mercury to give di(6-phenyl-5-hexyn-1-yl)mercury in a yield greater than 65%:

$$C_6H_5C{\equiv}C(CH_2)_4^{\cdot} + Hg \longrightarrow C_6H_5C{\equiv}C(CH_2)_4Hg^{\cdot}(ads)$$

$$2C_6H_5C{\equiv}C(CH_2)_4Hg^{\cdot}(ads) \longrightarrow [C_6H_5C{\equiv}C(CH_2)_4]_2Hg + Hg$$

Alternatively, the radical can cyclize and abstract a hydrogen atom (probably from the solvent, SH) to form benzylidenecyclopentane:

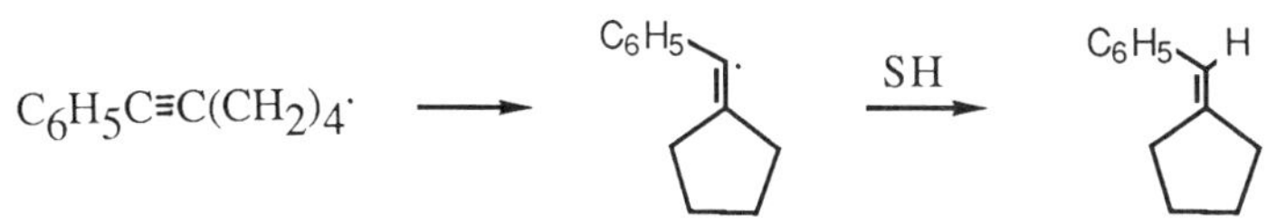

At more negative potentials, the starting material is reduced to the carbanion

$$C_6H_5C{\equiv}C(CH_2)_4I + 2e \longrightarrow C_6H_5C{\equiv}C(CH_2)_4^{-} + I^{-}$$

which is protonated by residual water or by the supporting electrolyte cation to produce 1-phenyl-1-hexyne. A modest quantity of 1-phenyl-5-hexen-1-yne is obtained via dehydrohalogenation of the starting material, and a small amount of 1-phenyl-1-hexyn-6-ol arises from attack of hydroxide ion (from water) on the parent compound.

Unlike 1-bromodecane, which forms no didecylmercury upon being reduced [9], 6-bromo-1-phenyl-1-hexyne when electrolyzed at mercury [73] gives a diorganomercury species in substantial yield. In general, the distribution of products derived from the electrolysis of 6-bromo-1-phenyl-1-hexyne is similar to that obtained from 6-iodo-1-phenyl-1-hexyne, although the bromo compound shows a greater tendency to display a carbanionic character upon being reduced. Because some commercially available DMF contains a significant concentration of *N*-methylformamide, electrogenerated carbanions can deprotonate the amide, and unreduced starting material can react with the conjugate base of the amide to form *N*-methyl-*N*-(6-phenyl-5-hexyn-1-yl)formamide:

$$C_6H_5C{\equiv}C(CH_2)_4^{-} + HC({=}O){-}N(H){-}CH_3 \longrightarrow C_6H_5C{\equiv}C(CH_2)_3CH_3 + HC({=}O){-}\underline{N}{-}CH_3$$

$$HC({=}O){-}\underline{N}{-}CH_3 + C_6H_5C{\equiv}C(CH_2)_4X \longrightarrow HC({=}O){-}N(CH_3){-}(CH_2)_4C{\equiv}CC_6H_5 + X^{-}$$

A normal pulse polarogram for the reduction of 1-iodo-5-decyne at mercury in DMF containing TMAP shows two waves of identical height, signaling stepwise formation of a radical and a carbanion intermediate, respectively,

$$C_4H_9C{\equiv}C(CH_2)_4X + e \longrightarrow C_4H_9C{\equiv}C(CH_2)_4^{\cdot} + X^{-}$$

$$C_4H_9C{\equiv}C(CH_2)_4^{\cdot} + e \longrightarrow C_4H_9C{\equiv}C(CH_2)_4^{-}$$

whereas 1-bromo-5-decyne shows a single pulse polarographic wave with a height equal to the combined heights of the two waves for the iodo species [74]. Preparative-scale electrolyses of both compounds at mercury pool cathodes afford a roster of products whose yields vary with potential. At not too negative potentials, electrogeneration of the radical intermediate results in the formation of di(5-decyn-1-yl)mercury and pentylidenecyclopentane. However, because the rate of intramolecular cyclization of the 5-decyn-1-yl radical is approximately 40 times slower than cyclization of the 6-phenyl-5-hexyn-1-yl radical, more diorganomercury species and less carbocycle are formed from 1-halo-5-decynes [74] than from 6-halo-1-phenyl-1-hexynes [73]. At more negative potentials, the carbon-halogen bond is reductively cleaved to give the carbanion intermediate, which is protonated by water to yield 5-decyne:

$$C_4H_9C{\equiv}C(CH_2)_4^- + H_2O \longrightarrow C_4H_9C{\equiv}C(CH_2)_3CH_3 + OH^-$$

Then the resulting hydroxide ion can attack unreduced starting material to form 5-decyn-1-ol

$$C_4H_9C{\equiv}C(CH_2)_4X + OH^- \longrightarrow C_4H_9C{\equiv}C(CH_2)_4OH + X^-$$

or can promote dehydrohalogenation of the starting material to form 1-decen-5-yne:

$$C_4H_9C{\equiv}C(CH_2)_4X + OH^- \longrightarrow C_4H_9C{\equiv}C(CH_2)_2CH{=}CH_2 + H_2O + X^-$$

When a proton donor (diethyl malonate) is added to the system, the last two reactions are blocked, so that the yield of 5-decyne is increased at the expense of 1-decen-5-yne and 5-decen-1-ol.

If 1-iodo-5-decyne is electrolyzed at a vitreous carbon cathode in DMF containing tetraalkylammonium perchlorates [75], the yield of pentylidenecyclopentane is increased approximately 10-fold and there are small rises in the quantities of 5-decyne and 1-decen-5-yne, in comparison to results obtained with a mercury electrode—all because the production of diorganomercury species (as described in the preceding paragraph) is impossible. Cyclic voltammetric and chronoamperometric evidence indicates that 1-iodo-5-decyne and 1-bromo-5-decyne both undergo one-step, two-electron cleavage of the carbon-halogen bond at a carbon electrode. However, because 1-bromo-5-decyne is more difficult to reduce than its iodo analog, the radical intermediate (which must form along the pathway to the carbanion) persists long enough when derived from 1-iodo-5-decyne to undergo intramolecular cyclization; this behavior explains why pentylidenecyclopentane can be produced in a yield of at least 60% from 1-iodo-5-decyne but in a yield of no better than 6% from 1-bromo-5-decyne.

It would be interesting to electrolyze 6-iodo-1-phenyl-1-hexyne at a carbon electrode in DMF; on the basis of the preceding discussion, one would expect to obtain benzylidenecyclopentane nearly quantitatively. In addition, it may be possible to effect the electroreductive intramolecular cyclizations of other acetylenic halides, such as 5-iodo-1-phenyl-1-pentyne, 1-iodo-4-nonyne, and 7-iodo-1-phenyl-1-heptyne.

V. BENZYL HALIDES AND RELATED COMPOUNDS

Investigations of the electrochemical behavior of benzyl iodide at mercury have led to the conclusion that the mechanism of reduction involves an electroinitiated chain reaction with formation of benzylmercuric iodide [76,77]:

$$C_6H_5CH_2I + e \longrightarrow [C_6H_5CH_2I^{\cdot-}] \longrightarrow C_6H_5CH_2^{\cdot} + I^-$$

$$C_6H_5CH_2^{\cdot} + Hg \longrightarrow C_6H_5CH_2Hg^{\cdot}$$

$$C_6H_5CH_2Hg^{\cdot} + C_6H_5CH_2I \longrightarrow C_6H_5CH_2Hg^+ + C_6H_5CH_2^{\cdot} + I^-$$

$$C_6H_5CH_2Hg^+ + I^- \longrightarrow C_6H_5CH_2HgI$$

Analogous processes may occur, although more slowly, for benzyl bromide and benzyl chloride [78–80].

At mercury cathodes, controlled potential electrolyses of benzyl bromide in MeCN [80] or DMF [32] and of benzyl chloride in aqueous EtOH [80] afford bibenzylmercury and toluene (and sometimes a small amount of bibenzyl) in yields that depend on the potential. Formation of these products could involve reduction of the benzylmercuric halide (or the benzylmercury cation) to give the benzylmercury radical

$$C_6H_5CH_2HgX + e \longrightarrow C_6H_5CH_2Hg^{\cdot} + X^-$$

which undergoes either disproportionation

$$2C_6H_5CH_2Hg^{\cdot} \longrightarrow (C_6H_5CH_2)_2Hg + Hg$$

or further reduction and protonation:

$$C_6H_5CH_2Hg^{\cdot} + e \longrightarrow C_6H_5CH_2^- + Hg$$

$$C_6H_5CH_2^- + H^+ \longrightarrow C_6H_5CH_3$$

Alternatively, the parent compound may be reduced directly to the benzyl radical, which can (1) accept another electron to give the benzyl anion or (2) combine with mercury to form the benzylmercury radical [81], each of which reacts as shown. Furthermore, Baizer and Chruma [32] have argued that the benzyl carbanion could arise via two-electron reduction of either benzylmercuric halide or benzyl halide itself; then attack of the carbanion on benzylmercuric halide would yield dibenzylmercury, and reaction between the carbanion and benzyl halide would lead to production of bibenzyl. Reduction of benzyl chloride at mercury in DMF containing tetraethylammonium halide and in the presence of carbon dioxide provides evidence for generation of the benzyl carbanion; benzyl phenylacetate [32] and phenylacetic acid [77] are obtained as electrolysis products. Chapter 23 offers further information about these and other reductive couplings.

It is interesting that reduction of o- and p-nitrobenzyl halides at a mercury or platinum cathode gives mainly the bibenzyl derivative, whereas nitrotoluene is a minor product [81–85]; these species are proposed to arise from the nitrobenzyl radical formed by rapid scission of a halide ion from the electrogenerated radical anion. On the other hand, electrolysis of m-nitrobenzyl halides results predominantly in the production of m-nitrotoluene [83].

Koch et al. [86] have examined the electrochemical reduction of benzyl halides at both platinum and carbon cathodes in MeCN containing TEAP. In a controlled potential electrolysis of benzyl iodide at platinum, the products are toluene, bibenzyl, and hydrocinnamonitrile. Depending on potential, toluene results from hydrogen atom abstraction by a benzyl radical or from protonation of a benzyl carbanion, whereas either coupling of benzyl radicals or attack of a benzyl carbanion on unreduced starting material is responsible for bibenzyl. Hydrocinnamonitrile arises from the reaction of the electrogenerated benzyl anion with the solvent:

$$C_6H_5CH_2^- + CH_3CN \longrightarrow C_6H_5CH_3 + {}^-CH_2CN$$

$${}^-CH_2CN + C_6H_5CH_2I \longrightarrow C_6H_5CH_2CH_2CN + I^-$$

In the presence of carbon dioxide or diethyl malonate, some of the benzyl anion is converted, respectively, into benzyl phenylacetate or the diethyl ester of benzyl malonate.

Shono et al. [87] have demonstrated that controlled potential electrolysis of benzyl chloride or (1-chloroethyl)benzene at carbon in MeCN containing TEAOTs and in the presence of various acyl chlorides gives alkyl benzyl ketones in moderate yield; mechanistically, it appears that the benzyl chloride undergoes two-electron reduction to produce a carbanion that attacks the acyl chloride:

$$C_6H_5CH_2Cl + 2e \longrightarrow C_6H_5CH_2^- + Cl^-$$

$$C_6H_5CH_2^- + C_3H_7COCl \longrightarrow C_6H_5CH_2\overset{\overset{\displaystyle O}{\|}}{C}C_3H_7 + Cl^-$$

A study of the electroreduction of 1-bromo-1-phenylethane by Casanova and coworkers [88] has shown that, at mercury cathodes in DMF containing $TEABF_4$, the product distribution is dependent upon potential. At less negative potentials, the principal products are meso-2,3-diphenylbutane (40%), dl-2,3-diphenylbutane (40%), and phenylethane (18%); it appears that an organomercury radical is involved as an intermediate and that final products arise from a one-electron process. At very negative potentials, generation of an anionic intermediate via two-electron reduction of the carbon-bromine bond leads exclusively to phenylethane. Through the use of optically active 1-bromo-1-(deuterioethyl)benzene, some intriguing conclusions were drawn about stereochemical inversion during the two-electron reduction and protonation process; potent proton donors lead to inversion of the optically active center, whereas poor (or nonexistent) proton donors favor racemization.

Research by Fry and Powers [89] has provided new insight into the mechanism of formation of bibenzyl from the electrochemical reduction of benzyl bromides. Using (1-bromo-2,2-dimethylpropyl)benzene in DMF containing lithium perchlorate, these workers have shown that electrolysis of the carbon-bromine bond gives a potential-dependent distribution of products.

At least negative potentials, the major products are meso- and dl-2,2,5,5-tetramethyl-3,4-diphenylhexane, along with both head-to-tail and tail-to-tail dimers and neopentylbenzene, all of which are derived from a radical intermediate. At potentials where the carbon-bromine bond undergoes two-electron cleavage, the carbanion intermediate is protonated to yield neopentylbenzene, whereas a small quantity of α-phenylneopentanol is formed via dehydrohalogenation.

Other substituted benzyl halides whose electrolysis products have been identified and quantitated include benzhydryl halides [90,91] and the family of α,α'-dibromoxylenes [92]. Such compounds as trifluoromethylbenzene, chlorodifluoromethylbenzene, and dichlorofluoromethylbenzene [93] can be electrolyzed at a mercury cathode in DMF containing TBAI to afford a series of species, each possessing one less halogen atom than its precursor. Similarly, trichloromethylbenzene undergoes stepwise reduction with loss of chloride in DMF [31,94], and an electrolysis at a potential corresponding to the first stage of reduction leads to the formation of 1,2-diphenyl-1,1,2,2-tetrachloroethane (63%), diphenylacetylene (25%), 1,2-dichloro-1,2-diphenylethene (7%), and dichloromethylbenzene (2%) [94]; however, reduction of trichloromethylbenzene in DMF containing $TBABF_4$ in the presence of acetic anhydride gives, besides the preceding products, 2-acetoxy-1-chloro-1-phenyl-1-propene and 1,2-diacetoxy-1-phenyl-1-propene [95]. Cathodic addition of trichloromethylbenzene to ketones gives α,β-unsaturated ketones [96]. Methyl esters of trifluoromethylbenzoic acid have been defluorinated by reduction at lead cathodes in MeOH [97,98].

Amino acids have been synthesized via the coupling reaction between Schiff's bases and the carbanions generated by the reduction of benzyl halides [99,100]. Reduction of benzyl chloride at a mercury pool cathode in DMF containing TBABr and in the presence of the Schiff's base formed from benzylamine and benzyl pyruvate leads to the production of α-methylphenylalanine in up to 86% yield, after hydrogenolysis of the electrolyzed solution over palladium-charcoal. This method has been utilized to prepare α-methyl-β-(3,4-dihydroxyphenyl)alanine, an antihypertensive agent, by reductive coupling of the aforementioned Schiff's base with 3,4-methylenedioxybenzyl chloride, followed first by the hydrogenolysis step and second by hydrolysis with hydrochloric acid.

VI. ALICYCLIC HALIDES

A. Monohalogenated Species

Considerable early research dealing with the reduction of this family of compounds centered on investigations of substituted and optically active 1-bromocyclopropanes and 1-iodocyclopropanes [101–103]. A principal concern was whether electrolytic cleavage of the carbon-halogen bond at mercury and carbon leads to both radical and carbanion intermediates; one would expect to observe racemization if a radical is formed, but retention of configuration should occur if a carbanion is generated. A discussion of the results and their possible implications regarding the mechanism of reduction of the carbon-halogen bond in these species has been presented by Mann and Barnes [104]. Tallec and coworkers [105] have investigated the effect of the identity of the supporting electrolyte cation on the reduction of derivatives of 1-bromo-1-carboxy-2,2-diphenylcyclopropane; preferential retention of configuration is observed in the presence of ammonium ion, whereas inversion occurs when tetraalkylammonium salts are employed. Other papers by the same research group include a study of the effect of adsorbed alkaloids on the stereochemistry of reduction of 1-bromo-1-carboxy-2,2-diphenylcyclopropane [106] and two reports pertaining to the reduction of optically active derivatives of 2,2-dibromocyclopropane [107,108].

Other simple monobromocycloalkanes ranging from cyclobutyl bromide to cyclohexadecyl bromide have been examined by means of polarography in DMF containing TEABr [15,109], but no information about the electrolysis products is available.

Polarographic data have been compiled for a variety of monohalogenated bicyclic compounds, including 1-bromo- and 2-bromonorbornane [17,110–112], 1-bromobicyclo[2.2.2]octane [110,111], 1,7,7-trimethyl-3-halobicyclo[2.2.1]heptan-2-ones [113], and derivatives of 1-bromo- and 2-bromo-7-oxabicyclo[2.2.1]heptanes [112]. A detailed investigation has been undertaken of the electrochemical reduction of 1-iodonorbornane and 1-bromonorbornane at mercury in DMF containing tetraalkylammonium salts [114]. Controlled potential electrolyses of the 1-halonorbornanes suggest that the dominant process is two-electron cleavage of the carbon-halogen bond, which leads to norbornane; only a small quantity of bis(1-norbornyl)mercury is obtained. At a platinum cathode, the reductions of 1-chloro-, 1-bromo-, and 1-iodonorbornane afford mainly norbornane along with a little 1,1'-binorbornyl [115].

B. Dihalogenated Species

Fry and Reed [116] have studied the electrochemical behavior of 2,2-dichloronorbornane, 2-exo-bromo-2-endo-chloronorbornane, and 2-endo-bromo-2-exo-chloronorbornane at mercury cathodes in DMF. Polarograms exhibit a single wave, and coulometry indicates that the reduction involves net uptake of two electrons per molecule of starting material. All three compounds undergo electrolytic conversion to a mixture of nortricyclene and endo-2-chloronorbornane; for example, in DMF containing TEABr and 0.35 M water, the process for 2-endo-bromo-2-exo-chloronorbornane can be represented as

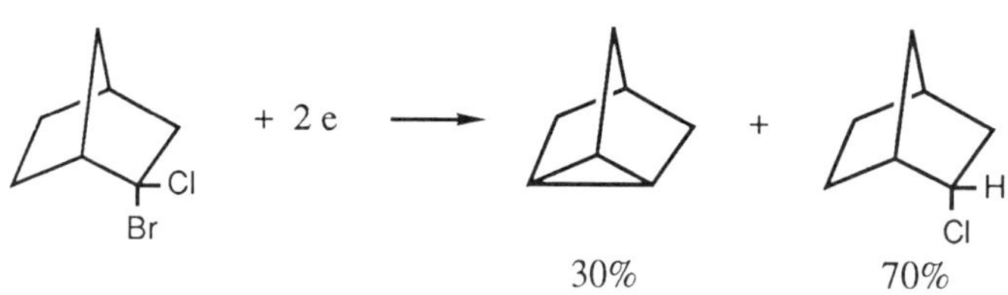

Organomercury species were neither detected as intermediates nor found as products in the electrolyses, and no exo-2-chloronorbornane was observed. Although the distribution of products depends on the identity of the supporting electrolyte and on the concentration and strength of added proton donors, the same product ratio is obtained for each starting material for a particular set of experimental conditions. Fry and Reed concluded that the reduction of each compound involves stereochemically preferred attack by the electrode upon the norbornane nucleus from the exo direction; the electrogenerated 2-chloronorbornan-2-yl carbanion can be protonated more easily from the less hindered exo direction to yield endo-2-chloronorbornane, or the carbanion can eject chloride ion to afford the carbene precursor of nortricyclene. Chapter 27 contains further discussion about the stereochemistry of this and other processes.

Two papers have been published concerning the reduction of several 7,7-dihalobicyclo[4.1.0]heptanes [117,118]; for these compounds, formation of the less stable endo-7-halobicyclo[4.1.0]heptane is preferred over the exo isomer. Semmelhack et al. [119] employed electrochemical reduction of 1,2-dimethyl-4,5,6,6-tetrachlorospiro[2.3]hexadiene at mercury in DMF containing lithium perchlorate and deuterium oxide to prepare 1,2-dimethyl-4,5,6-trichloro-6-deuteriospiro[2.3]hexadiene. At mercury cathodes in aqueous EtOH containing TMACl, it is possible to perform highly efficient reductions of polyfluorocyclohexadienes [120]:

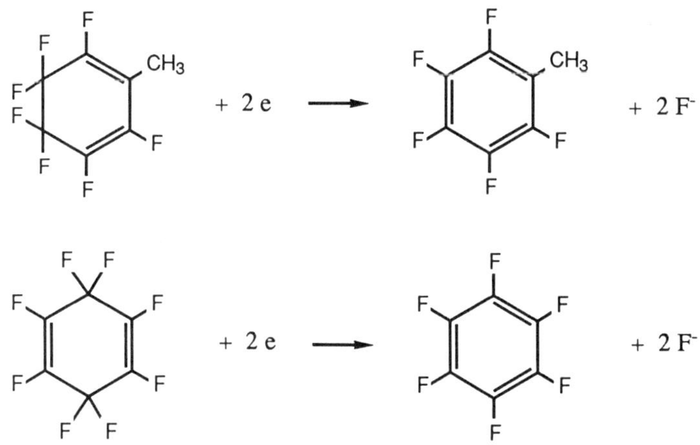

According to Rifi [49,51], electrolysis of 1-bromo-3-chlorocyclobutane at mercury in DMF containing lithium bromide yields bicyclo[1.1.0]butane (60%), cyclobutane (20%), and cyclobutene (20%):

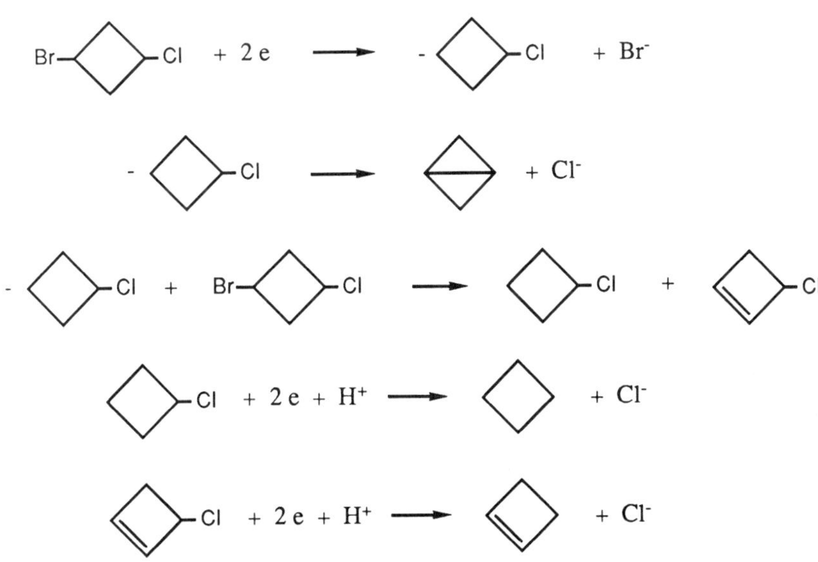

A similar pattern of electroactivity prevails for 1,3-dibromo-1,3-dimethylcyclobutane.

Either constant current [121] or controlled potential [122] electrolytic reduction at mercury of 1-bromo-4-chlorobicyclo[2.2.0]hexane at low temperature in DMF containing a tetraethylammonium salt affords $\Delta^{1,4}$-bicyclo[2.2.0]hexene, which can be trapped with cyclopentadiene:

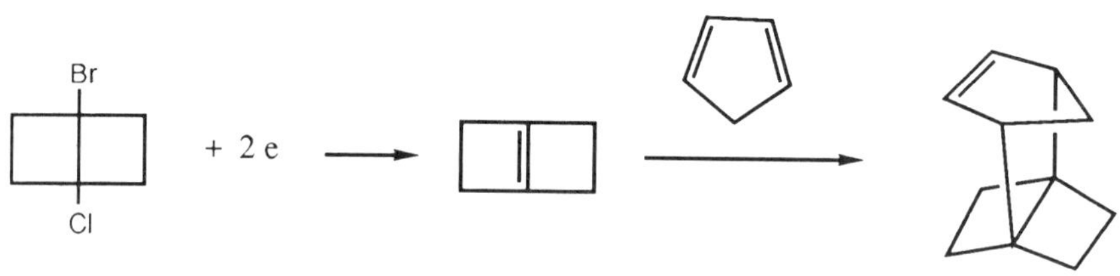

Rifi [51], as well as Wiberg et al. [123], electrolyzed 1,4-dibromobicyclo[2.2.2]octane in DMF containing a tetraalkylammonium salt; in the latter investigation, chlorine gas was introduced into the system and 1,4-dichlorobicyclo[2.2.2]octane was isolated in small yield (12%). According to Wiberg and coworkers, the appearance of the dichloro compound is a clear indication that [2.2.2]propellane is formed transiently during the electrolysis:

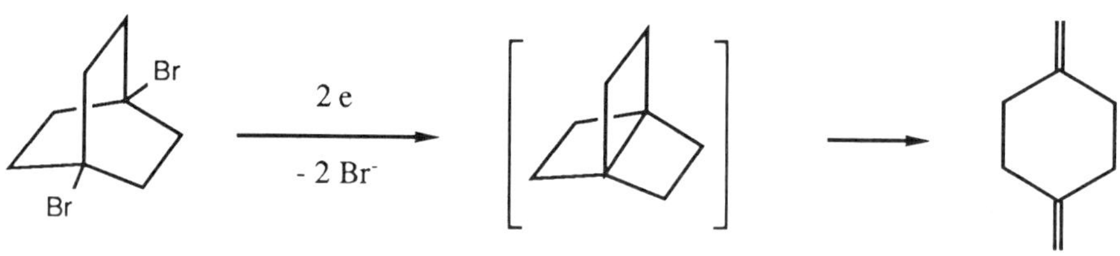

Rifi [51] has reported that electrolysis of 1,5-dibromobicyclo[3.2.1]octane leads to the stable [3.2.1]propellane; Leibzon and colleagues [124] found that the reduction of 3,7-dibromo-3,7-dinitrobicyclo[3.3.1]nonane affords 3,7-dinitroadamantane in excellent yield (75–80%).

In an investigation of the reduction of 1,4-dihalonorbornanes at platinum in DMF containing TEABr, Wiberg and his collaborators [115] observed that the electrolysis products are norbornane and 1,1'-binorbornyl, but no compelling evidence was seen for the formation of any [2.2.1]propellane. When 1,4-dihalonorbornanes are electrolyzed at mercury cathodes in DMF [125,126], major products are norbornane and bis(1-norbornyl)mercury, together with a small amount of 1,1'-binorbornyl (and, at some potentials, 1-iodonorbornane is derived from 1,4-diiodonorbornane). More importantly, three key observations suggest that [2.2.1]propellane is an intermediate in the reduction of 1,4-dihalonorbornanes: (1) norbornane is derived from 1,4-dihalonorbornanes via a process that does not involve 1-halonorbornanes, (2) bis(1-norbornyl)mercury is produced by reduction of 1,4-dihalonorbornanes but not 1-halonorbornanes, and (3) a species is generated by two-electron reduction of 1,4-dihalonorbornanes that undergoes oligomerization.

VII. HALOGENATED AROMATIC COMPOUNDS

A. Monohalobenzenes

Electrochemical reduction of monohalobenzenes is generally conceded to involve irreversible two-electron scission of the carbon-halogen bond followed by rapid protonation of the phenyl carbanion:

$$C_6H_5X + 2e + H^+ \longrightarrow C_6H_6 + X^-$$

However, an extensive series of publications by Savéant and coworkers [6, 127] has provided quantitative evidence that in nonaqueous media a more intricate mechanistic scheme for the reductive cleavage of these compounds must be considered. Accordingly, one can postulate a one-electron transfer to give a radical anion

$$C_6H_5X + e \rightleftharpoons C_6H_5X^{\dot{-}}$$

which undergoes decomposition to yield the phenyl radical:

$$C_6H_5X^{\dot{-}} \xrightarrow{k} C_6H_5^{\cdot} + X^-$$

In general, there is a linear correlation between the logarithm of the rate constant k for scission of the electrogenerated radical anion and the standard potential for the $C_6H_5X-C_6H_5X^{\dot{-}}$ couple; the more negative the standard potential, the larger is the rate constant [6]. Once formed, the phenyl radical can accept an electron

$$C_6H_5^{\cdot} + e \longrightarrow C_6H_5^-$$

or can react homogeneously with the initially generated radical anion

$$C_6H_5^{\cdot} + C_6H_5X^{\dot{-}} \longrightarrow C_6H_5^- + C_6H_5X$$

and the resulting phenyl anion can be protonated by residual water in the solvent, by the supporting electrolyte, or by the solvent:

$$C_6H_5^- + H^+ \longrightarrow C_6H_6$$

Other possible reactions include (1) abstraction of a hydrogen atom from solvent by the phenyl radical, (2) homogeneous reduction of a solvent radical by the halobenzene radical anion, and (3) heterogeneous reduction of a solvent radical at the electrode surface.

To substantiate the validity of the preceding mechanism, Savéant and his colleagues have measured the rates of heterogeneous electron transfer to the carbon-halogen bond and have investigated the kinetic parameters for the homogeneous reduction of halobenzenes by electrogenerated redox catalysts. In recent years, this approach to elucidating the kinetics of reduction of halogenated compounds has been extended to a wide variety of species, as perusal of references cited in the above-mentioned papers will reveal.

From the viewpoint of electrosyntheses based on the reduction of monohalobenzenes, three recent studies deserve attention. Matsue et al. [128]

have shown that reduction of iodobenzene and bromobenzene at a mercury cathode in carbon dioxide-saturated DMF containing TEAP leads to production of benzoic acid in up to 40% yield:

$$C_6H_5X + 2e + CO_2 + H^+ \longrightarrow C_6H_5CO_2H + X^-$$

Benzene is the major by-product. To improve the yield of benzoic acid, it would be preferable to employ a supporting electrolyte that is a poorer proton donor than TEAP, so that the electrogenerated phenyl anion intermediate can react with carbon dioxide before being protonated to form benzene. Electrolysis at a mercury cathode in diglyme containing a mixture of $TBABF_4$ and dimethylpyrrolidinium tetrafluoborate has been utilized by Kariv-Miller and Vajtner [129] to convert fluorobenzene to benzene. A low concentration (0.001–0.01 M) of the dimethylpyrrolidinium salt is essential to the success of the reductive defluorination; the dimethylpyrrolidinium cation (DMP^+) acts catalytically by being reduced to an amalgam that reacts with fluorobenzene to produce the radical anion

$$DMP^+ + e + nHg \longrightarrow DMP(Hg)_n$$

$$C_6H_5F + DMP(Hg)_n \longrightarrow C_6H_5F^{\bar{\cdot}} + DMP^+ + nHg$$

which undergoes carbon-fluorine bond breakage, followed by further reduction and protonation to afford benzene:

$$C_6H_5F^{\bar{\cdot}} \longrightarrow C_6H_5^{\cdot} + F^-$$

$$C_6H_5^{\cdot} + e \longrightarrow C_6H_5^-$$

$$C_6H_5^- + H_2O \longrightarrow C_6H_6 + OH^-$$

Savéant and coworkers [130] have demonstrated that electrolysis of a variety of substituted aryl halides in the presence of a suitable olefin and a redox catalyst results in the formation of arylated addition compounds, because an electrogenerated aryl radical is trapped by the added olefinic species. For example, the reduction of 4-chlorobenzonitrile at platinum in liquid ammonia containing 4-chlorostyrene and 4,4'-bipyridyl affords 1-(p-chlorophenyl)-2-(p-cyanophenyl)ethane in 64% yield, and electrolysis of 4-chlorobenzonitrile at mercury in DMF containing styrene and naphthonitrile gives 1-phenyl-2-(p-cyanophenyl)ethane in up to 80% yield.

B. Polyhalobenzenes

Benzene bearing two different halogens can be reduced in a stepwise fashion. For example, electrolysis of p-fluoroiodobenzene yields fluorobenzene [131], and reduction of p-bromoiodobenzene gives bromobenzene in 98% yield [132]; of course, the monohalobenzenes can be further reduced. In DMSO containing TEABr, polychlorobenzenes can be selectively dechlorinated by means of controlled potential electrolysis at a mercury cathode [133], as exemplified by the following reactions:

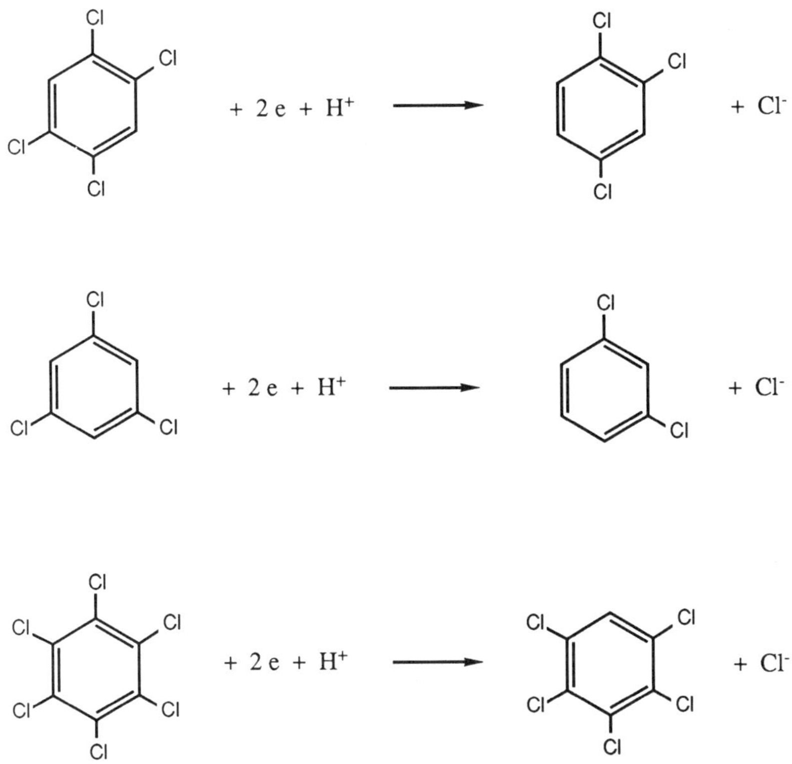

For the electrolysis of o-dibromobenzene in DMF containing tetraalkylammonium salts, benzyne has been proposed as an intermediate for reductions at both mercury [134] and platinum [135]:

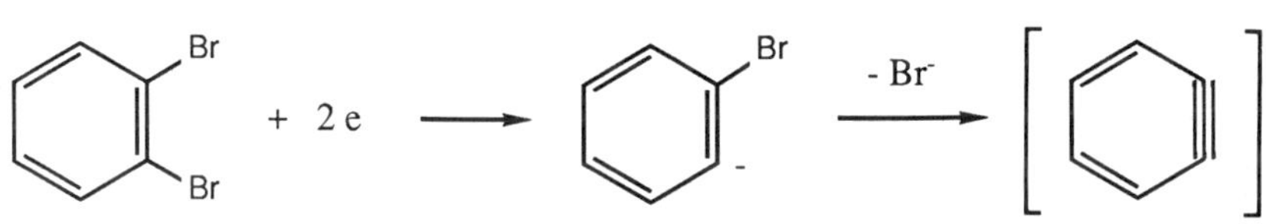

Kariv-Miller and Vajtner [129] discovered that, for electrolysis of 1,3-difluorobenzene at mercury in diglyme containing $TBABF_4$, the proportions of benzene, fluorobenzene, and unreduced starting material depend on the concentrations of water and dimethylpyrrolidinium tetrafluoborate added to the system; the yield of fluorobenzene is at least 85% for a low current density (1.25 mA/cm^2) and for 0.5% water and 0.01 M dimethylpyrrolidinium tetrafluoborate.

C. Halogenated Nitrobenzenes and Cyanobenzenes

As emphasized earlier, radical anions formed by addition of a single electron to a monohalobenzene are extremely short-lived; a halide ion is rapidly expelled and the resulting phenyl radical is quickly reduced. On the other hand, for a variety of mono- and polyhalogenated nitrobenzenes and cyanobenzenes, the electrogenerated radical anions are often sufficiently stable that their ESR spectra can be recorded. In general, the stability of these radical anions is least for iodo compounds and greatest for fluoro species, and ortho halogens are expelled much more rapidly than para halogens, which are in turn lost faster than meta halogens. Hawley [3] has reviewed much of the available information about the voltammetric characteristics of halogenated nitrobenzenes and cyanobenzenes and about the effects of experimental conditions (potential, proton donors, hydrogen atom-donating properties of solvents, and temperature) on the stability and reactions of the radical anions and their decomposition products.

D. Halogenated Aryl Carbonyls

Product distributions have been reported for controlled potential electrolyses of pentafluorobenzaldehyde, pentafluorobenzoic acid, and pentafluorobenzamide at mercury in aqueous media containing either sulfuric acid or $TEABF_4$ [136]. Reduction of the aldehyde affords both pentafluorobenzyl alcohol and 2,3,5,6-tetrafluorobenzyl alcohol. Pentafluorobenzoic acid is initially converted to a mixture of 2,3,5,6-tetrafluorobenzoic acid (73%), pentafluorobenzyl alcohol (6%), and 2,3,5,6-tetrafluorobenzyl alcohol (20%), but at a more negative potential, the last two products are obtained in a 1:2 ratio. It is noteworthy that pentafluorobenzaldehyde and 2,3,5,6-tetrafluorobenzaldehyde are easier to reduce than their benzoic acid analogs; therefore, the aldehydes cannot be isolated from electrolyses of the acids. In a sulfuric acid medium, controlled potential electrolysis of pentafluorobenzamide gives pentafluorobenzyl alcohol (58%) and 2,3,5,6-tetrafluorobenzyl alcohol (20%).

Fry et al. [132] observed that electrolysis of m-bromoacetophenone at a mercury cathode in DMF containing TEABr affords acetophenone virtually quantitatively, whereas reduction of p-bromo-γ-chlorobutyrophenone gives γ-chlorobutyrophenone in 96% yield.

According to M'Halla et al. [137], halogenated benzophenones can behave in two different ways in nonaqueous solvents, such as DMF, DMSO, and MeCN. For such compounds as m-fluoro-, p-fluoro-, and m-chlorobenzophenone, a pair of electrons is apparently involved in the reduction, and benzophenone is produced quantitatively. However, in the electrolysis of p-chloro-, m-bromo-, and p-bromobenzophenone, fewer than two electrons per molecule are transferred, and the yield of benzophenone is significantly less than 100%. In a detailed investigation of the electrochemical reduction of p-bromobenzophenone in perdeuterated DMSO, Savéant and coworkers [137] found that the resulting benzophenone is almost fully monodeuterated; this observation indicates that the dominant pathway for reduction of p-bromobenzophenone is a one-electron cleavage of the carbon-bromine bond, followed by hydrogen atom transfer from the solvent:

Subsequently, the solvent-derived radical can be reduced or can reoxidize the original radical anion of p-bromobenzophenone; in either case, the conjugate base of the solvent can attack the benzophenone radical to give a set of substituted side products that cause the final yield of benzophenone to fall well below 100% and the apparent coulometric n value to be lower than 2.

It has been observed [138,139] that reduction of N-methyl-N-(p-methoxyphenyl)bromobenzamide and its iodo analog at mercury in DMF containing TPAP results in one-electron reductive cleavage of the carbon-halogen bond to yield a radical intermediate

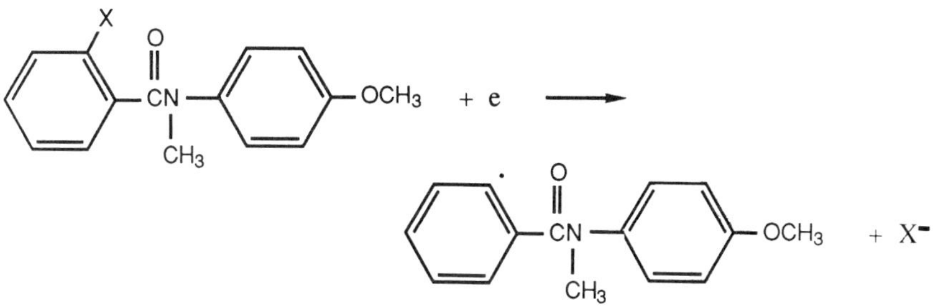

which can undergo (1) further reduction and protonation to afford the dehalogenated benzamide or (2) intramolecular cyclization or rearrangement followed by hydrogen atom abstraction:

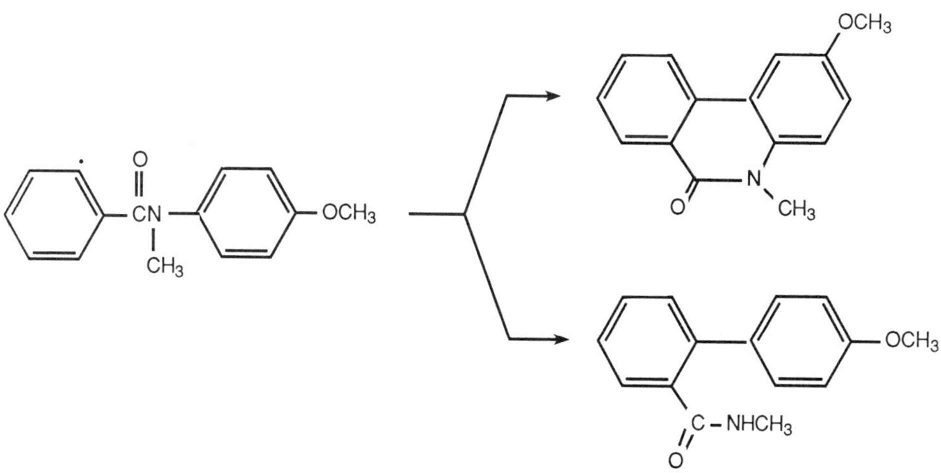

E. Halogenated Biphenyls

Farwell et al. [140,141] have investigated the electrochemical behavior of a large number of mono- and polychlorobiphenyls at mercury in DMSO containing TEABr. Although these compounds do exhibit sequential two-electron removal of chlorine, there is a tendency (because of the varied pattern of chlorine substitution) for more than just one electrolysis product to be formed at the same time; sometimes two chlorines are expelled simultaneously. In another study, Maruyama and Murakami [142] have reported that chlorinated biphenyls are reduced in DMF via a process involving one-electron cleavage of a carbon-chlorine bond to yield a chloride ion and an aryl radical, which subsequently abstracts a hydrogen atom from solvent.

Rusling and Arena [143] employed cyclic voltammetry to probe the mechanism of reduction of halogenated biphenyls (ArX) such as 4-bromo-, 4,4'-dibromo-, 3,4-dichloro-, and 2,2',5,5'-tetrachlorobiphenyl at a mercury electrode in DMF containing tetraalkylammonium iodides. Results obtained by these workers are consistent with a mechanism that involves

$$\mathrm{ArX} + e \longrightarrow \mathrm{ArX}^{\overline{\cdot}}$$

as the rate-limiting step, followed by a sequence of reactions identical with those discussed earlier in this chapter for the reduction of monohalobenzenes [6,127]:

$$\mathrm{ArX}^{\overline{\cdot}} \longrightarrow \mathrm{Ar}^{\cdot} + \mathrm{X}^{-}$$

$$\mathrm{Ar}^{\cdot} + e \longrightarrow \mathrm{Ar}^{-}$$

$$\mathrm{Ar}^{\cdot} + \mathrm{ArX}^{\overline{\cdot}} \longrightarrow \mathrm{Ar}^{-} + \mathrm{ArX}$$

$$\mathrm{Ar}^{-} + \mathrm{H}^{+} \longrightarrow \mathrm{ArH}$$

$$\mathrm{Ar}^{\cdot} + \mathrm{SH} \longrightarrow \mathrm{ArH} + \mathrm{S}^{\cdot}$$

VIII. ACYL AND PHENACYL HALIDES

Barba and co-workers have investigated the electrochemical behavior of benzoyl chloride [144], 1-naphthoyl chloride [144], 2-naphthoyl chloride [144], phthaloyl chloride [145], and cinnamoyl chloride [146]. These compounds undergo a one-electron reduction to generate the acyl radical

$$\mathrm{R\overset{\overset{\displaystyle O}{\|}}{C}Cl} + e \longrightarrow \mathrm{R\overset{\overset{\displaystyle O}{\|}}{C}{}^{\cdot}} + \mathrm{Cl}^{-}$$

which provides the basis for a number of novel electrosyntheses.

In dry acetone containing lithium perchlorate, electrolysis of benzoyl chloride [144] at a mercury pool cathode gives the acyl radical

$$\mathrm{C_6H_5\overset{\overset{\displaystyle O}{\|}}{C}Cl} + e \longrightarrow \mathrm{C_6H_5\overset{\overset{\displaystyle O}{\|}}{C}{}^{\cdot}} + \mathrm{Cl}^{-}$$

which dimerizes to yield bibenzoyl:

$$2C_6H_5\overset{O}{\overset{\|}{C}}{}^{\cdot} \longrightarrow C_6H_5\overset{O}{\overset{\|}{C}}-\overset{O}{\overset{\|}{C}}C_6H_5$$

Two-electron reduction of bibenzoyl leads to production of 1,2-diphenyl-1,2-ethenediolate

$$C_6H_5\overset{O}{\overset{\|}{C}}-\overset{O}{\overset{\|}{C}}C_6H_5 + 2e \longrightarrow C_6H_5\overset{^-O}{\overset{|}{C}}=\overset{O^-}{\overset{|}{C}}C_6H_5$$

which is acylated by two molecules of starting material to give cis- and trans-stilbenediol dibenzoate in a 1:2 ratio:

$$C_6H_5\overset{^-O}{\overset{|}{C}}=\overset{O^-}{\overset{|}{C}}C_6H_5 + 2C_6H_5\overset{O}{\overset{\|}{C}}Cl \longrightarrow C_6H_5\overset{O}{\overset{\|}{C}}-O(C_6H_5)C=C(C_6H_5)O-\overset{O}{\overset{\|}{C}}C_6H_5 + 2Cl^-$$

Similar reactions are seen for both 1-naphthoyl and 2-naphthoyl chloride [144].

Electrolysis of phthaloyl chloride [145] at mercury proceeds in much the same fashion as that of benzoyl chloride, except that the ethenediolate intermediate reacts intramolecularly to produce trans-biphthalyl:

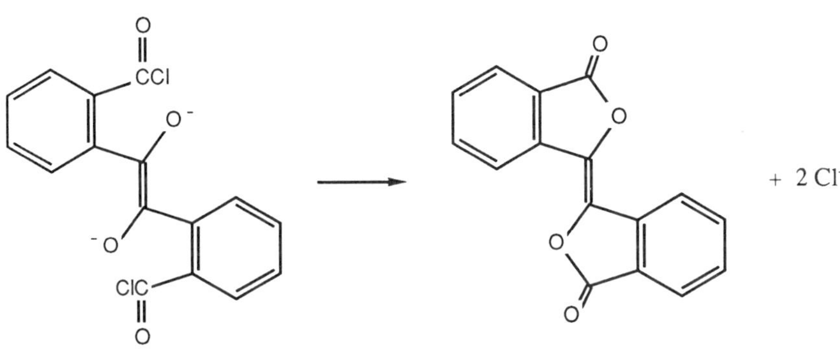

Cinnamoyl chloride [146] has been found to accept one electron at a mercury cathode in acetone containing lithium perchlorate to form a radical

$$C_6H_5CH=CH\overset{O}{\overset{\|}{C}}Cl + e \longrightarrow C_6H_5CH=CH\overset{O}{\overset{\|}{C}}{}^{\cdot} + Cl^-$$

that presumably abstracts a hydrogen atom to give an intermediate ketene, which dimerizes:

$$2\ C_6H_5CH_2CH=C=O \longrightarrow$$

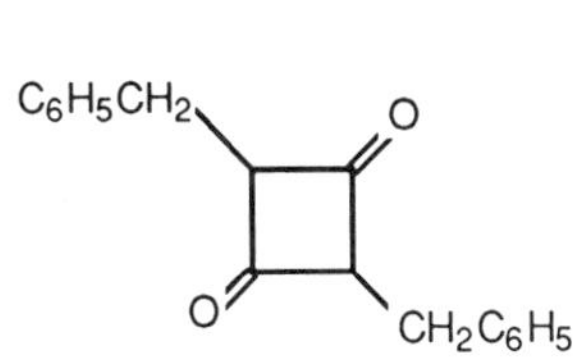

A study of the reduction of benzoyl chloride and benzoyl fluoride at carbon and platinum cathodes in MeCN containing TEAP has been carried out by Cheek and Horine [147]. Large-scale electrolysis of benzoyl chloride at carbon yields cis- and trans-stilbenediol dibenzoate, as reported earlier [144], but this product is reducible to diphenylacetylene at a more negative potential. Benzoyl fluoride is more difficult to reduce than benzoyl chloride; major electrolysis products derived from the acyl fluoride are benzyl benzoate (15%) and diphenylacetylene (15%), along with stilbenol benzoate and other (polymeric) species of high molecular weight.

In all published research pertaining to the electrochemistry of acyl halides, no evidence has been presented for two-electron reduction to the acyl anion; the latter species, if it can be generated, could yield (1) an aldehyde in the presence of a suitable proton donor, (2) a ketone in the presence of a reactive alkyl halide, (3) a diketone in the presence of an acyl halide, or (4) an α-hydroxy ketone in the presence of a ketone.

A series of papers concerning the electrochemistry of phenacyl bromides has been published by Barba and collaborators [148–152]. The controlled potential reduction of phenacyl bromide at mercury in DMF containing lithium perchlorate apparently involves two-electron cleavage of the carbon-bromine bond to form an anion

$$C_6H_5\overset{\displaystyle O}{\overset{\|}{C}}CH_2Br + 2e \longrightarrow C_6H_5\overset{\displaystyle O}{\overset{\|}{C}}CH_2^- + Br^-$$

that reacts with excess starting material to give an intermediate

$$C_6H_5\overset{\displaystyle O}{\overset{\|}{C}}CH_2^- + C_6H_5\overset{\displaystyle O}{\overset{\|}{C}}CH_2Br \longrightarrow \left[C_6H_5\overset{\displaystyle O}{\overset{\|}{C}}CH_2\underset{\displaystyle C_6H_5}{\underset{|}{C}}\overset{O}{\diagup\!\!\diagdown}CH_2 \right] + Br^-$$

which eliminates water to afford 2,4-diphenylfuran in 80% yield [148,149]:

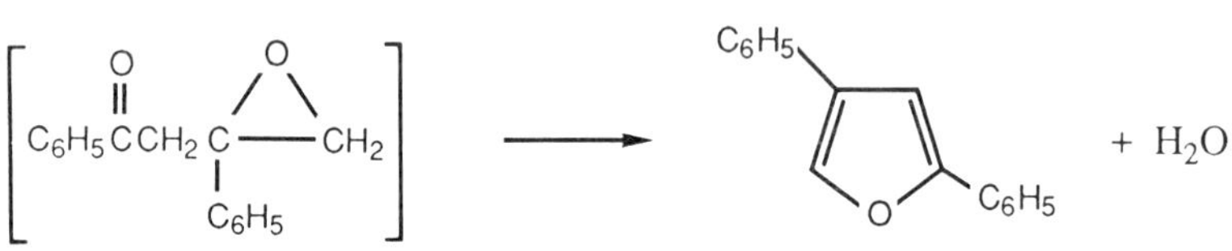

Other compounds that have been similarly synthesized are 2,4-bis(4-methoxyphenyl)furan, 2,4-bis(4-biphenylyl)furan, 2,4-bis(4-bromophenyl)furan, 2,4-bis(4-chlorophenyl)furan, and 2,4-bis(4-nitrophenyl)furan. When MeOH (a potent proton donor) is added to the system containing the unsubstituted phenacyl bromide [149], the principal products are acetophenone (53%) and 2-bromo-1,3-diphenyl-3,4-epoxy-butan-1-one (37%).

By protecting the carbonyl group of a phenacyl bromide through reaction with semicarbazide hydrochloride, by reducing the resultant semicarbazone at mercury in DMF containing lithium perchlorate to obtain a dimer, and by hydrolyzing the electrolytically formed dimer, one can synthesize a variety of 2,5-diarylfurans in excellent yield [150]. In addition, Barba and coworkers [151] demonstrated that reductive dimerization of the semicarbazone derivative of phenacyl bromide

$$2C_6H_5\overset{\overset{\displaystyle NNHCONH_2}{\|}}{C}CH_2Br + 2e \longrightarrow C_6H_5\overset{\overset{\displaystyle NNHCONH_2}{\|}}{C}CH_2CH_2\underset{\underset{\displaystyle NNHCONH_2}{\|}}{C}C_6H_5 + 2Br^-$$

can be followed by refluxing in DMF to form 3,6-diphenylpyridazine in 97% yield:

As pointed out by Barba and his colleagues [151], this method is applicable for the preparation of a variety of 3,6-diarylpyridazines.

In DMF containing lithium perchlorate, controlled potential reduction of α-bromopropiophenone at mercury gives 1,4-diphenyl-2,3-dimethylbutan-1,4-dione in 65% yield [152]. Electrolysis of the starting material in the presence of benzoyl chloride affords only 1,3-diphenyl-2-methylpropan-1,3-dione, which establishes that electrolysis of α-bromopropiophenone proceeds via two-electron scission of the carbon-bromine bond.

IX. ALIPHATIC α-HALOCARBONYL COMPOUNDS

Numerous aliphatic α-haloaldehydes have been studied polarographically in water-dioxane containing lithium chloride [153,154]. Electrochemical reduction of these compounds is influenced greatly by the equilibrium involving the hydrated and unhydrated forms of the aldehyde. In general, each α carbon-halogen bond is reductively cleaved in a two-electron, one-proton process and the final nonhalogenated aldehyde is reduced at a more negative potential [155].

Dirlam et al. [156] performed controlled potential reduction of 2-bromo-3-pentanone at mercury in aqueous DMF containing lithium chloride; a 2:1 mixture of 3-pentanone and 1-hydroxy-3-pentanone was obtained. In the same investigation, electrolysis of α,α'-dibromoacetone in the presence of benzoate afforded a mixture of products resulting from both reductive cleavage of the carbon-bromine bond and S_N2 displacement of bromide by benzoate. At a mercury cathode in an acetic acid-sodium acetate buffer, electrochemical reduction of branched dibromo ketones, such as 2,4-dibromo-2,4-dimethyl-3-pentanone, affords α-acetoxy ketones in excellent yield; less highly substituted species, such as 4,6-dibromo-5-nonanone, result in simple reductive cleavage of both carbon-bromine bonds [157].

Reduction of bis(α-bromocyclopropyl) ketone in an acetic acid-sodium acetate buffer can, depending on experimental conditions, afford a mixture of α-bromocyclopropyl cyclopropyl ketone (76%) and dicyclopropyl ketone (14%); in DMF containing acetic acid and sodium acetate, α-bromocyclopropyl cyclopropyl ketone is formed exclusively [158]. In an earlier paper, Fry and Scoggins [159] reported that the electrolysis of 2,4-dibromo-2-methylpentan-3-one gives 2,2-dimethylcyclopropanone as a transient intermediate. More recently, van Tilborg and co-workers [160] demonstrated that the low-temperature reduction of alkylated α,α'-dihaloketones to cyclopropanones (isolated as hemiacetals or hemiacylals) can be accomplished with highly alkylated starting materials; for example, electrolysis of 2,4-dibromo-2,4-dimethylpentan-3-one in the presence of methanol affords 1-methoxy-2,2,3,3-tetramethylcyclopropanol in 50–80% yield. Furthermore, electrochemical cyclization of 1,3-dihaloketones whose carbonyl moieties are protected as acetals, acylals, and aminals can produce the corresponding cyclopropanone derivatives; thus, reduction of 2,2-bis(chloromethyl)-1,3-dioxolane gives the corresponding spirocyclopropane in 75% yield.

In related work, Barba and coworkers [161] have examined the electrochemical behavior of α,α-dibromodibenzoylmethane at both platinum and mercury cathodes in DMF containing lithium perchlorate. For both of the electrode materials, the only electrolysis product obtained is α-bromodibenzoylmethane; reduction of the carbon-bromine bond appears to be a two-electron, one-proton process at platinum, whereas at mercury the reduction occurs via chemical insertion of mercury into a carbon-bromine bond, followed by one-electron cleavage of the carbon-mercury bond.

In aqueous media, α-halocarboxylic acids as well as their esters seemingly undergo two-electron, one-proton reductive cleavage [162–164]. Baizer and Chruma [32] demonstrated that electrolysis of ethyl bromoacetate at mercury in slightly wet DMF containing TEAOTs yields mainly ethyl acetate (51%) and diethyl succinate (20%), although small quantities of 1,2,3-tricarbethoxycyclopropane (7%), 1,2,3-tricarbethoxypropane (6%), and diethyl fumarate (4%) are formed. When ethyl trichloroacetate is reduced in the presence of acrylonitrile, 1-carbethoxy-1-chloro-2-cyanocyclopropane (80%) and 1-carbethoxy-1,1,3-trichloro-3-cyanopropane (20%) are produced.

X. HALOGENATED HETEROCYCLIC COMPOUNDS

Later in this volume, Chap. 18 presents a discussion of the electrochemical behavior of heterocyclic species; however, it is appropriate here to offer a short summary of the reduction of carbon-halogen bonds found in several classes of heterocyclic compounds. Many halogenated heterocyclic substances that have been examined display behavior similar to that of aryl halides; the carbon-halogen bond is reductively cleaved in a process involving two electrons and a proton to give a hydrogenated product.

A. Heterocyclic Compounds with One Nitrogen Atom

Polarographic investigations of a number of halogenated pyridines have been undertaken in aqueous media [165–167]; such compounds as 4-chloropyridine, 2-bromopyridine, and 3-iodopyridine, in either their unprotonated or protonated forms, undergo two-electron cleavage of the carbon-halogen bond, followed by uptake of a proton to give pyridine. In DMSO containing TPAP, reduction of 2-chloro-3-nitropyridine initially yields a relatively stable radical anion [168], which at a more negative potential is dechlorinated to form the 3-nitropyridine radical anion.

Chambers and coworkers [169] have examined the reduction of some polyhalopyridines in DMF containing $TEABF_4$. Controlled potential electrolysis of pentafluoropyridine at mercury affords perfluoro-4,4'-bipyridyl in 49% yield, whereas reduction of the same starting material in the presence of a proton donor (hydroquinone) gives 2,3,5,6-tetrafluoropyridine in 46% yield. On the other hand, electrolysis of pentachloropyridine in the absence of a proton donor results in the formation of 2,3,5,6-tetrachloropyridine (57%) and bis(tetrachloropyridyl)mercury (17%); apparently, the comparatively bulky ortho chlorine substituents inhibit dimerization of the intermediate radical anions. When pentachloropyridine is electrolyzed in the presence of carbon dioxide, reductive coupling of the starting material does take place, and 3-carboxyperchloro-4,4'-bipyridyl seems to be the predominant product. There has been a recent report [170] of the preparation of 2,3,5,6-tetrachloropyridine in 93% yield by controlled potential electrolysis of pentachloropyridine at a stainless steel cathode in aqueous MeOH containing zinc chloride. Kyriacou [171] has electrolyzed 2,3,5,6-tetrachloropyridine at a silver cathode in a water-THF mixture containing sodium acetate to prepare 2,3,5-trichloropyridine.

Alwair and Grimshaw [172] have reported that, in DMF containing TPAP, 6-chloroquinoline undergoes one-electron reduction to its radical anion, which immediately fragments and then accepts a hydrogen atom from the solvent to form quinoline (which is itself reducible at a more negative potential to a radical anion). Lund and coworkers [173] investigated the behavior of numerous chlorinated quinolines in DMF containing TBAI and discovered that, once a radical anion is generated, a carbon-halogen bond on the heterocyclic ring is cleaved more quickly than a carbon-chlorine bond on the fused benzene ring. Furthermore, radical anions of chloroquinolines that are not rapidly dehalogenated can react with added carbon dioxide to form carboxylated starting materials; unstable radical anions of chloroquinolines lose chloride to give neutral radicals that are further reduced before the reaction with carbon dioxide.

According to Edamura et al. [174], reduction of tetrachloro-2-picolinic acid at a silver electrode in an aqueous medium yields 3,4,6-trichloro-2-picolinic acid, which can be further reduced to 3,6-dichloro-2-picolinic acid.

B. Heterocyclic Compounds with Two Nitrogen Atoms

A survey [172] of the electrochemistry of 6-haloquinoxalines in DMF containing TPAP has revealed that 6-chloroquinoxaline can be converted into a stable radical anion, whereas the bromo and iodo analogs accept a single electron to yield radical anions that undergo fragmentation and hydrogen atom abstraction to form quinoxaline.

A voltammetric study [172] of the reductions of 2-chloro- and 2-bromophenazine in DMF suggests that these compounds are reversibly reduced to stable radical anions; however, the radical anion of 2-iodophenazine decomposes to yield iodide ion and phenazine.

Lund [175] observed that reduction of 4-chloroquinazoline in an alcoholic phosphate buffer results in two-electron cleavage of the carbon-chlorine bond with formation of quinazoline in essentially quantitative yield.

C. Other Heterocyclic Species

Laviron [165] investigated the polarography of a series of halogenated thiazoles, which were found to be reduced by uptake of two electrons and one proton to yield thiazole; Iversen [176] synthesized 1,3-thiazole in over 90% yield by reducing 2-bromo-1,3-thiazole at a mercury pool in a water-EtOH medium containing hydrochloric acid.

In water-EtOH and water-DMF mixtures, brominated and iodinated thiophenes exhibit polarographic waves corresponding to two-electron, one-proton cleavage of the carbon-halogen bond [177–179]. Justice and Hall [180] have reported that reduction of a substituted 3-chlorobenzo[b]thiophene can be used to synthesize the corresponding 3-hydrobenzo[b]thiophene.

Casadei et al. [181] have investigated the electrochemical reduction of 3-halo-β-lactams and 3,3-dihalo-β-lactams. In DMF containing TEAP, trans-3-bromo- and trans-3-chloro-1,4-diphenylazetidin-2-one undergo two-step reduction; the first wave is attributable to reductive cleavage of the carbon-halogen bond and the second wave is due to reduction of the β-lactam itself. Three stages of reduction are exhibited by 3,3-dichloro-1,4-diphenylazetidin-2-one, so that stepwise removal of the chlorines is possible. In the presence of carbon dioxide, it is noteworthy that each of the trans-3-halo-1,4-diphenylazetidin-2-ones can be nearly quantitatively converted into trans-3-carboxy-1,4-diphenylazetidin-2-one; in the absence of carbon dioxide, the major electrolysis product is the dehalogenated β-lactam, but a small amount of the ring-opened α,β-unsaturated amide is obtained.

XI. INDIRECT OR CATALYTIC REDUCTION OF THE CARBON-HALOGEN BOND

One of the characteristics of the direct electrolytic cleavage of the carbon-halogen bond is the intrinsic irreversibility of the process. In other words, the potential needed to reduce a carbon-halogen bond heterogeneously at an electrode is often much more negative than the thermodynamically reversible potential. However, it is possible to reduce carbon-halogen bonds chemically with electron transfer mediators—species that are generated at potentials more positive than those required for direct reduction of the carbon-halogen bond. Among the earliest reported applications of this concept is a paper by Lund et al. [182], who investigated the catalytic reduction of bromobenzene by the electrogenerated radical anion of chrysene. In DMF containing TBAI, chrysene (denoted here by Ar) undergoes reversible one-electron reduction at mercury to give the radical anion:

$$\mathrm{Ar} + e \rightleftharpoons \mathrm{Ar}^{\overline{\cdot}}$$

Because bromobenzene is 360 mV more difficult to reduce than chrysene, the carbon-bromine bond cannot be reduced directly at the potential used to generate the radical anion of chrysene. However, the radical anion does react chemically and rapidly with bromobenzene in solution

$$\mathrm{Ar}^{\overline{\cdot}} + C_6H_5Br \longrightarrow \mathrm{Ar} + C_6H_5Br^{\overline{\cdot}}$$

and the radical anion of bromobenzene decomposes, is subsequently reduced by a second aromatic radical anion, and is then protonated:

$$C_6H_5Br^{\overline{\cdot}} \longrightarrow C_6H_5^{\cdot} + Br^-$$

$$C_6H_5^{\cdot} + \mathrm{Ar}^{\overline{\cdot}} \longrightarrow C_6H_5^- + \mathrm{Ar}$$

$$C_6H_5^- + H^+ \longrightarrow C_6H_6$$

With this brief introductory example in mind, it is convenient to consider three categories of methods that have already been or still await to be exploited for electrosyntheses based on indirect, catalytic reduction of the carbon-halogen bond.

A. Electrogenerated Radical Anions as Mediators

From the previously mentioned work by Lund and co-workers [182] and from other publications [183–187], it is evident that the rate constant for the electron transfer reaction between an electrogenerated radical anion (mediator) and a halogenated organic compound (substrate) increases as the difference between standard potentials for reduction of the mediator and substrate decreases.

In a study of the reaction between different alkyl halides and the electrogenerated naphthalene radical anion in DMF containing TEABr, Sease and Reed [184] observed that only alkyl chlorides, such as 1-chlorohexane and 6-chloro-1-hexene, are catalytically reduced; 1-chlorohexane gives only n-hexane (80%), whereas 6-chloro-1-hexene affords methylcyclopentane (25%) and 1-hexene (60%). Neither alkyl bromides nor alkyl fluorides are reduced under the extant conditions. In an earlier paper [183], the reactions of 1-bromobutane and 1-chlorobutane with electrogenerated radical anions of both trans-stilbene and anthracene in DMSO were examined by means of cyclic voltammetry. Britton and Fry [185] utilized a potential step method to elucidate the kinetics of the electron transfer reaction between 1-chlorooctane and the phenanthrene radical anion in DMF containing TEABr.

Another aspect of the reactions between alkyl halides and electrogenerated aromatic radical anions has been probed by Simonet et al. [188]. Once a single electron has been transferred to the alkyl halide to form an alkyl radical ($R^{\cdot}$), the radical can react with another radical anion

$$R^{\cdot} + Ar^{\overline{\cdot}} \longrightarrow R^{-} + Ar$$

and then be protonated to form RH (as mentioned earlier) or the radical can bond with the radical anion to yield an alkylated aromatic hydrocarbon:

$$R^{\cdot} + Ar^{\overline{\cdot}} \longrightarrow ArR^{-} \xrightarrow{H^{+}} ArRH$$

An excellent example of the latter process is seen in the results of an electrolysis of stilbene in DMF containing TBAI and tert-butyl chloride [188]; the electrogenerated radical anion of stilbene initially converts tert-butyl chloride to the tert-butyl radical, which interacts with the stilbene radical anion to afford a 3:2 mixture of 3,3-dimethyl-1,2-diphenylbutane and 1-phenyl-2-(4'-tert-butylphenyl)ethane. In the same investigation, Simonet and coworkers [188] found that reduction of naphthalene in the presence of tert-butyl chloride results in a mixture of 1- and 2-tert-butyldihydronaphthalenes; in addition, electrogeneration of the radical anion of anthracene in a system containing methyl chloride gives both 9-methyl- and 9,10-dimethyl-9,10-dihydroanthracene.

Other studies of the reactions of alkyl halides with electrogenerated radical anions have led to publications on the reduction of diketones in the presence of methyl halides [189], the reduction of nitrobenzene in the presence of alkyl halides [190], the reductive coupling of 1,2- and 1,3-dihaloalkanes with anthracene [191], the reductive coupling of phenylacetylene with alkyl halides [192], the electron transfer between meso- and dl-1,2-dichloro-1,2-diphenylethane and electrogenerated radical anions [193], and the reductive coupling of tert-butyl bromide with azobenzene, quinoxaline, and anthracene [194]. In the last paper, a thorough discussion of the rates of electron transfer between various alkyl bromides and electrogenerated radical anions can be found.

Recently, studies of the reactions between halogenated organic species and electrogenerated radical anions have been extended to include both unsubstituted and substituted benzyl chloride [195] as well as vinyl bromides, 2-bromoindene, and 7,7-dichlorobicyclo[4.1.0]heptane [196].

B. Transition Metal Complexes as Mediators

One of the earliest reports dealing with the catalyzed reduction of alkyl halides by an electrogenerated transition metal complex is found in work by Sofranko et al. [197]. These investigators discovered that rhodium(I) coordinated to a pair of 1,2-bis(diphenylphosphino)ethane ligands, symbolized below as $Rh(diphos)_2^+$, can undergo reversible one-electron reduction at mercury to the corresponding rhodium(0) species

$$\mathrm{Rh(diphos)_2^+ + e \rightleftharpoons Rh(diphos)_2}$$

which can reduce catalytically the carbon-halogen bond of chloro-, bromo-, and iodocyclohexane to give a distribution of products dependent on the identities of the halocyclohexane as well as the solvent. For example, in benzonitrile, the products obtained from chlorocyclohexane (which cannot be reduced directly) are cyclohexane, cyclohexene, bicyclohexyl, and bis(cyclohexyl)mercury; in MeCN, iodocyclohexane gives only the first three compounds, whereas bromocyclohexane affords no bicyclohexyl.

A series of papers by Pletcher and coworkers [198–202] has focused on catalysis of the reduction of various bromoalkanes by square-planar nickel(I) complexes that are electrogenerated from the corresponding nickel(II) species; ligands coordinated to the parent nickel(II) complexes included Schiff's bases and macrocyclic tetraamines. Mechanistically, it appears that the nickel(I)-ligand species produced by reduction of the parent nickel(II)-ligand complex, Ni(II)L,

$$\mathrm{Ni(II)L + e \rightleftharpoons Ni(I)L}$$

forms an organonickel(III) intermediate with a bromoalkane:

$$\mathrm{Ni(I)L + RBr \longrightarrow R{-}Ni(III)L{-}Br}$$

Once formed, the intermediate can decompose, can accept an electron, or can undergo attack by unreduced bromoalkane [201]:

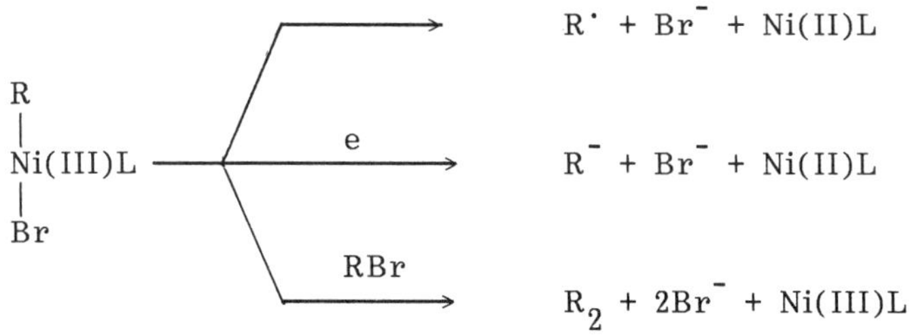

In the third of these pathways for destruction of the intermediate, the liberated nickel(III)-ligand species is subsequently reduced to the original nickel(II)-ligand complex. Gosden et al. [198] performed controlled potential reductions of two nickel(II)-ligand complexes at platinum in MeCN containing $TBABF_4$ and a bromoalkane (1-bromooctane, 2-bromooctane, or 2-bromo-2-methylheptane). Among the products are alkanes, alkenes, and dimers that one expects from both alkyl radicals and alkyl carbanions; the distribution of products depends on the identity of the nickel(II)-ligand species, and small amounts of alkylated solvent (MeCN) are seen in some cases. Pletcher and his collaborators [201] have demonstrated that unsaturated molecules, such as acrylonitrile, ethyl acrylate, and 1-buten-3-one, can insert themselves into the nickel-carbon bond of the organonickel(III) intermediate before it decomposes and that this process provides an opportunity for several attractive electroorganic syntheses.

Because acyclic and alicyclic monohaloalkanes, especially chloroalkanes, are difficult to reduce directly at an electrode, it is not surprising that earliest efforts to use transition metal complexes as mediators for electron transfer were aimed at reduction of simple haloalkanes. However, many new and elegant electrosyntheses based on the use of transition metal complexes to effect reductive cleavage of the carbon-halogen bond in more complicated molecules now appear regularly in the chemical literature. A comprehensive review by Torii [203] has summarized many recent developments in the field that involve complexes of nickel, cobalt, palladium, and chromium. A few of these approaches to electrosynthesis are as follows: (1) preparation of biphenyl in 80% yield by reduction of iodobenzene with electrogenerated bis(triphenylphosphine)nickel(0) [204]; (2) coupling of aromatic halides and alkenes with the aid of electrogenerated bis(triphenylphosphine)nickel(0) [205]; (3) reduction of alkyl halides and reductive elimination of a β-haloethyl protecting group of an acid with vitamin B_{12} derivatives as electrocatalysts [206]; (4) coupling of two identical aryl halides (or cross-coupling of two different aryl halides) into biaryls by use of electrogenerated triphenylphosphinepalladium(0) [207]; (5) coupling of benzylic and allylic halides by electrogenerated chromium(II) [208]; and (6) reductive coupling of various organic halides by electrogenerated bis(2,2'-bipyridine)nickel(0) [209].

C. Reduction of Carbon-Halogen Bonds at Chemically Modified Electrodes

An area worthy of far greater exploration is the development of chemically modified electrodes [210,211] that incorporate the various kinds of mediators described earlier for reductive cleavage of carbon-halogen bonds. In principle, there appear to be two major ways by which suitable chemically modified electrodes can be fabricated. First, electron transfer mediators can be covalently attached, via suitable bridging groups, to an electrode surface. Second, electron transfer mediators can be imbibed or physically entrapped into polymer-coated electrodes or can be a chemical component of the polymer coating itself. It is to be hoped that the use of chemically modified electrodes for practical electrosynthesis will receive much more attention.

REFERENCES

1. M. R. Rifi, in Organic Electrochemistry (M. M. Baizer, ed.), 1st ed., Dekker, New York, 1973, p. 279.
2. L. G. Feoktistov, in Organic Electrochemistry (M. M. Baizer and H. Lund, eds.), 2nd ed., Dekker, New York, 1983, p. 259.
3. M. D. Hawley, in Encyclopedia of Electrochemistry of the Elements (A. J. Bard and H. Lund, eds.), Vol. XIV, Dekker, New York, 1980.
4. J. Y. Becker, in The Chemistry of Functional Groups, Supplement D (S. Patai and Z. Rappoport, eds.), Wiley, New York, 1983, p. 203.
5. D. J. Barker, D. M. Brewis, and R. H. Dahm, Electrochim. Acta, 23, 1107 (1978).
6. C. P. Andrieux, J. M. Savéant, and K. B. Su, J. Phys. Chem., 90, 3815 (1986).
7. O. R. Brown and K. Taylor, J. Electroanal. Chem., 50, 211 (1974).
8. D. M. La Perriere, W. F. Carroll, Jr., B. C. Willett, E. C. Torp, and D. G. Peters, J. Amer. Chem. Soc., 101, 7561 (1979).
9. G. M. McNamee, B. C. Willett, D. M. La Perriere, and D. G. Peters, J. Amer. Chem. Soc., 99, 1831 (1977).
10. R. Bilewicz and J. Osteryoung, J. Electroanal. Chem., 226, 27 (1987).
11. A. J. Fry, Synthetic Organic Electrochemistry, Harper and Row, New York, 1972, p. 172.
12. D. A. Tyssee, J. H. Wagenknecht, M. M. Baizer, and J. L. Chruma, Tetrahedron Lett., 4809 (1972).
13. J. H. Wagenknecht, J. Electroanal. Chem., 52, 489 (1974).
14. M. S. Mbarak and D. G. Peters, J. Org. Chem., 47, 3397 (1982).
15. F. L. Lambert and K. Kobayashi, J. Amer. Chem. Soc., 82, 5324 (1960).
16. A. K. Hoffmann, W. G. Hodgson, D. L. Maricle, and W. H. Jura, J. Amer. Chem. Soc., 86, 631 (1964).
17. A. J. Fry and R. L. Krieger, J. Org. Chem., 41, 54 (1976).
18. K. L. Vieira and D. G. Peters, J. Electroanal. Chem., 196, 93 (1985).
19. K. L. Vieira and D. G. Peters, J. Org. Chem., 51, 1231 (1986).
20. F. L. Lambert and G. B. Ingall, Tetrahedron Lett., 3231 (1974).
21. L. V. Kaabak, A. P. Tomilov, S. L. Varshavskii, and M. I. Kabachnik, Zh. Org. Khim., 3, 3 (1967).
22. C. P. Andrieux, I. Gallardo, J. M. Savéant, and K. B. Su, J. Amer. Chem. Soc., 108, 638 (1986).
23. J. A. Cleary, M. S. Mubarak, K. L. Vieira, M. R. Anderson, and D. G. Peters, J. Electroanal. Chem., 198, 107 (1986).
24. O. Sock, M. Troupel, and J. Perichon, Tetrahedron Lett., 26, 1509 (1985).
25. K. P. Butin, N. A. Belokoneva, A. A. Zenkin, I. P. Beletskaya, and O. A. Reutov, Dokl. Akad. Nauk SSSR, 211, 878 (1973).
26. A. Merz and G. Thumm, Justus Liebigs Ann. Chem., 1526 (1978).

27. K. L. Vieira, M. S. Mubarak, and D. G. Peters, J. Amer. Chem. Soc., 106, 5372 (1984).
28. I. Rosenthal and R. J. Lacoste, J. Amer. Chem. Soc., 81, 3268 (1959).
29. M. Nagao, N. Sato, T. Akashi, and T. Yoshida, J. Amer. Chem. Soc., 88, 3447 (1966).
30. M. F. Semmelhack and G. E. Heinsohn, J. Amer. Chem. Soc., 94, 5139 (1972).
31. S. Wawzonek and R. C. Duty, J. Electrochem. Soc., 108, 1135 (1961).
32. M. M. Baizer and J. L. Chruma, J. Org. Chem., 37, 1951 (1972).
33. F. Karrenbrock and H. J. Schäfer, Tetrahedron Lett., 1521 (1978).
34. T. Shono, H. Ohmizu, S. Kawakami, S. Nakano, and N. Kise, Tetrahedron Lett., 871 (1981).
35. T. Shono, N. Kise, A. Yamazaki, and H. Ohmizu, Tetrahedron Lett., 1609 (1982).
36. T. Shono, N. Kise, and T. Suzumoto, J. Amer. Chem. Soc., 106, 259 (1984).
37. J. Casanova and H. R. Rogers, J. Org. Chem., 39, 2408 (1974).
38. H. Lund and E. Holboth, Acta Chem. Scand., B30, 895 (1976).
39. M. M. Gol'din, L. G. Feoktistov, V. R. Polishchuk, and L. S. German, Elektrokhimiya, 7, 916 (1971).
40. M. M. Gol'din, L. G. Feoktistov, A. P. Tomilov, and K. M. Smirnov, Zh. Obshch. Khim., 42, 2561 (1972).
41. L. G. Feoktistov and M. M. Gol'din, Zh. Obshch. Khim., 43, 515 (1973).
42. L. G. Feoktistov and M. M. Gol'din, Zh. Obshch. Khim., 43, 520 (1973).
43. M. M. Gol'din, L. G. Feoktistov, V. R. Polishchuk, and L. S. German, Elektrokhimiya, 9, 67 (1973).
44. A. J. Fry, Fortschr. Chem. Forsch., 34, 1 (1972).
45. J. F. Garst, J. A. Pacifici, V. D. Singleton, M. F. Ezzel, and J. I. Morris, J. Amer. Chem. Soc., 97, 5242 (1975).
46. O. R. Brown, P. H. Middleton, and T. L. Threlfall, J. Chem. Soc. Perkin Trans. II, 955 (1984).
47. A. J. Klein and D. H. Evans, J. Amer. Chem. Soc., 101, 757 (1979).
48. K. M. O'Connell and D. H. Evans, J. Amer. Chem. Soc., 105, 1473 (1983).
49. M. R. Rifi, J. Amer. Chem. Soc., 89, 4442 (1967).
50. M. R. Rifi, Tetrahedron Lett., 1043 (1969).
51. M. R. Rifi, Coll. Czech. Chem. Commun., 36, 932 (1971).
52. K. B. Wiberg and G. A. Epling, Tetrahedron Lett., 1119 (1974).
53. A. J. Fry and W. E. Britton, Tetrahedron Lett., 4363 (1971).
54. A. J. Fry and W. E. Britton, J. Org. Chem., 38, 4016 (1973).
55. J. Casanova and H. R. Rogers, J. Amer. Chem. Soc., 96, 1942 (1974).
56. J. A. Dougherty and A. J. Diefenderfer, J. Electroanal. Chem., 21, 531 (1969).
57. R. Gerdil, Helv. Chim. Acta, 53, 2100 (1970).
58. O. R. Brown and E. R. González, J. Electroanal. Chem., 43, 215 (1973).
59. L. A. Avaca, E. R. González, and N. R. Stradiotto, Electrochim. Acta, 22, 225 (1977).
60. L. L. Miller and R. Riekena, J. Org. Chem., 34, 3359 (1969).
61. A. J. Fry and M. A. Mitnick, J. Amer. Chem. Soc., 91, 6207 (1969).

62. P. J. Elving, I. Rosenthal, J. R. Hayes, and A. J. Martin, Anal. Chem., 33, 330 (1961).
63. G. LeGuillanton and A. Daver, Bull. Soc. Chim. Fr., 724 (1973).
64. I. Rosenthal, J. R. Hayes, A. J. Martin, and P. J. Elving, J. Amer. Chem. Soc., 80, 3050 (1958).
65. W. H. Jura and R. J. Gaul, J. Amer. Chem. Soc., 80, 5402 (1958).
66. J. N. Seiber, J. Org. Chem., 36, 2000 (1971).
67. J. P. Petrovich and M. M. Baizer, Electrochim. Acta, 12, 1249 (1967).
68. A. J. Bard and A. Merz, J. Amer. Chem. Soc., 101, 2959 (1979).
69. E. Brillas and J. M. Costa, J. Electroanal. Chem., 69, 435 (1976).
70. H. Doupeux and J. Simonet, Bull. Soc. Chim. Fr., 1219 (1972).
71. H. Doupeux and J. Simonet, C. R. Acad. Sci. Paris, 276, 101 (1973).
72. W. L. Morgan, Ph.D. Thesis, Wesleyan University, Middletown, Connecticut, 1972, p. 51; Diss. Abstr. Int., B32, 3224 (1971).
73. B. C. Willett, W. M. Moore, A. Salajegheh, and D. G. Peters, J. Amer. Chem. Soc., 101, 1162 (1979).
74. R. Shao, J. A. Cleary, D. M. La Perriere, and D. G. Peters, J. Org. Chem., 48, 3289 (1983).
75. R. Shao and D. G. Peters, J. Org. Chem., 52, 652 (1987).
76. N. S. Hush and K. B. Oldham, J. Electroanal. Chem., 6, 34 (1963).
77. S. Wawzonek, R. C. Duty, and J. H. Wagenknecht, J. Electrochem. Soc., 111, 74 (1964).
78. L. W. Marple, L. E. Hummelstedt, and L. B. Rogers, J. Electrochem. Soc., 107, 437 (1960).
79. L. B. Rogers and A. J. Diefenderfer, J. Electrochem. Soc., 114, 942 (1967).
80. O. R. Brown, H. R. Thirsk, and B. Thornton, Electrochim. Acta, 16, 495 (1971).
81. J. Grimshaw and J. S. Ramsay, J. Chem. Soc., B, 60 (1968).
82. G. Klopman, Helv. Chim. Acta, 44, 1908 (1961).
83. J. G. Lawless, D. E. Bartak, and M. D. Hawley, J. Amer. Chem. Soc., 91, 7121 (1969).
84. D. E. Bartak, T. M. Shields, and M. D. Hawley, J. Electroanal. Chem., 30, 289 (1971).
85. D. E. Bartak and M. D. Hawley, J. Amer. Chem. Soc., 94, 640 (1972).
86. D. A. Koch, B. J. Henne, and D. E. Bartak, J. Electrochem. Soc., 134, 3062 (1987).
87. T. Shono, I. Nishiguchi, and H. Ohmizu, Chem. Lett., 1021 (1977).
88. R. B. Yamasaki, M. Tarle, and J. Casanova, J. Org. Chem., 44, 4519 (1979).
89. A. J. Fry and T. A. Powers, J. Org. Chem., 52, 2498 (1987).
90. Y. Matsui, T. Soga, and Y. Date, Bull. Chem. Soc. Jpn., 44, 513 (1971).
91. F. M. Triebe, K. J. Borhani, and M. D. Hawley, J. Amer. Chem. Soc., 101, 4637 (1979).
92. F. H. Covitz, J. Amer. Chem. Soc., 89, 5403 (1967).
93. H. Lund and N. J. Jensen, Acta Chem. Scand., B28, 263 (1974).
94. J. P. Gisselbrecht and H. Lund, Acta Chem. Scand., B39, 773 (1985).
95. J. P. Gisselbrecht and H. Lund, Acta Chem. Scand., B39, 823 (1985).
96. M. Steiniger and H. J. Schäfer, Angew. Chem., 94, 75 (1982).
97. J. P. Coleman, H. G. Gilde, J. H. P. Utley, and B. C. L. Weedon, Chem. Commun., 738 (1970).

98. J. P. Coleman, Naser-ud-din, H. G. Gilde, J. H. P. Utley, B. C. L. Weedon, and L. Eberson, J. Chem. Soc., Perkin Trans. II, 1903 (1973).
99. T. Iwasaki and K. Harada, J. Chem. Soc., Chem. Commun., 338 (1974).
100. T. Iwasaki and K. Harada, J. Chem. Soc., Perkin Trans. I, 1730 (1977).
101. R. Annino, R. E. Erickson, J. Michalovic, and B. McKay, J. Amer. Chem. Soc., 88, 4424 (1966).
102. C. K. Mann, J. L. Webb, and H. M. Walborsky, Tetrahedron Lett., 2249 (1966).
103. J. L. Webb, C. K. Mann, and H. M. Walborsky, J. Amer. Chem. Soc., 92, 2042 (1970).
104. C. K. Mann and K. K. Barnes, Electrochemical Reactions in Nonaqueous Systems, Dekker, New York, 1970, pp. 206–211.
105. R. Hazard, S. Jaouannet, E. Raoult, and A. Tallec, Nouv. J. Chim., 6, 325 (1982).
106. R. Hazard, S. Jaouannet, and A. Tallec, Tetrahedron, 38, 93 (1982).
107. R. Hazard, S. Jaouannet, and A. Tallec, Electrochim. Acta, 28, 1095 (1983).
108. R. Hazard, S. Jaouannet, and A. Tallec, Bull. Soc. Chim. Fr., II-263 (1983).
109. J. Krupička, J. Závada, and J. Sicher, Coll. Czech. Chem. Commun., 30, 3570 (1965).
110. J. W. Sease, P. Chang, and J. L. Groth, J. Amer. Chem. Soc., 86, 3154 (1964).
111. F. L. Lambert, A. H. Albert, and J. P. Hardy, J. Amer. Chem. Soc., 86, 3155 (1964).
112. K. P. Butin, N. A. Belokoneva, V. N. Eldikov, N. S. Zefirov, I. P. Beletskaya, and O. A. Reutov, Izv. Akad. Nauk SSSR, Ser. Khim., 254 (1969).
113. N. Limosin and E. Laviron, Bull. Soc. Chim. Fr., 4189 (1969).
114. W. F. Carroll, Jr., and D. G. Peters, J. Org. Chem., 43, 4633 (1978).
115. K. B. Wiberg, W. F. Bailey, and M. E. Jason, J. Org. Chem., 41, 2711 (1976).
116. A. J. Fry and R. G. Reed, J. Amer. Chem. Soc., 94, 8475 (1972).
117. A. J. Fry and R. H. Moore, J. Org. Chem., 33, 1283 (1968).
118. R. E. Erickson, R. Annino, M. D. Scanlon, and G. Zon, J. Amer. Chem. Soc., 91, 1767 (1969).
119. M. F. Semmelhack, R. J. DeFranco, and J. Stock, Tetrahedron Lett., 1371 (1972).
120. A. M. Doyle, A. E. Pedler, and J. C. Tatlow, J. Chem. Soc. C, 2740 (1968).
121. J. Casanova and H. R. Rogers, J. Org. Chem., 39, 3803 (1974).
122. K. B. Wiberg, W. F. Bailey, and M. E. Jason, J. Org. Chem., 39, 3803 (1974).
123. K. B. Wiberg, G. A. Epling, and M. Jason, J. Amer. Chem. Soc., 96, 912 (1974).
124. V. N. Leibzon, A. S. Mendkovich, T. A. Klimova, M. M. Krayushkin, S. G. Mairanovskii, S. S. Novikov, and V. V. Sevast´yanova, Elektrokhimiya, 11, 349 (1975).
125. W. F. Carroll, Jr., and D. G. Peters, Tetrahedron Lett., 3543 (1978).

126. W. F. Carroll, Jr., and D. G. Peters, J. Amer. Chem. Soc., 102, 4127 (1980).
127. C. P. Andrieux, C. Blocman, J. M. Dumas-Bouchiat, and J. M. Savéant, J. Amer. Chem. Soc., 101, 3431 (1979).
128. T. Matsue, S. Kitahara, and T. Osa, Denki Kagaku, 50, 732 (1982).
129. E. Kariv-Miller and Z. Vajtner, J. Org. Chem., 50, 1394 (1985).
130. Z. Chami, M. Gareil, J. Pinson, J. M. Savéant, and A. Thiebault, Tetrahedron Lett., 639 (1988).
131. S. Wawzonek and J. H. Wagenknecht, J. Org. Chem., 28, 239 (1963).
132. A. J. Fry, M. A. Mitnick, and R. G. Reed, J. Org. Chem., 35, 1232 (1970).
133. S. O. Farwell, F. A. Beland, and R. D. Geer, J. Electroanal. Chem., 61, 303 (1975).
134. S. Wawzonek and J. H. Wagenknecht, J. Electrochem. Soc., 110, 420 (1963).
135. N. Egashira, I. Sakurai, and F. Hori, Denki Kagaku, 54, 282 (1986).
136. F. G. Drakesmith, J. Chem. Soc., Perkin I, 184 (1972).
137. F. M'Halla, J. Pinson, and J. M. Savéant, J. Electroanal. Chem., 89, 347 (1978).
138. J. Grimshaw and J. Trocha-Grimshaw, Tetrahedron Lett., 993 (1974).
139. J. Grimshaw and J. Trocha-Grimshaw, J. Chem. Soc., Perkin II, 215 (1975).
140. S. O. Farwell, F. A. Beland, and R. D. Geer, Anal. Chem., 47, 895 (1975).
141. S. O. Farwell, F. A. Beland, and R. D. Geer, J. Electroanal. Chem., 61, 315 (1975).
142. M. Maruyama and K. Murakami, Nippon Kagaku Kaishi, 536 (1976).
143. J. F. Rusling and J. V. Arena, J. Electroanal. Chem., 186, 225 (1985).
144. A. Guirado, F. Barba, C. Manzanera, and M. D. Velasco, J. Org. Chem., 47, 142 (1982).
145. A. Guirado, F. Barba, and J. Martín, Synthetic Commun., 13, 327 (1983).
146. A. Guirado, F. Barba, and J. Martín, Electrochim. Acta, 29, 587 (1984).
147. G. T. Cheek and P. A. Horine, J. Electrochem. Soc., 131, 1796 (1984).
148. F. Barba, M. D. Velasco, and A. Guirado, Synthesis, 625 (1981).
149. F. Barba, M. D. Velasco, and A. Guirado, Electrochim. Acta, 28, 259 (1983).
150. F. Barba, M. D. Velasco, and A. Guirado, Synthesis, 593 (1984).
151. F. Barba, M. D. Velasco, A. Guirado, and N. Moreno, Synth. Commun., 15, 939 (1985).
152. F. Barba, M. D. Velasco, A. Guirado, and I. Barba, Electrochim. Acta, 30, 1119 (1985).
153. A. Kirrman and P. Federlin, Bull. Soc. Chim. Fr., 944 (1958).
154. P. Federlin, Bull. Soc. Chim. Fr., 949 (1958).
155. P. J. Elving and C. E. Bennett, J. Electrochem. Soc., 101, 520 (1954).
156. J. P. Dirlam, L. Eberson, and J. Casanova, J. Amer. Chem. Soc., 94, 240 (1972).
157. A. J. Fry and J. J. O'Dea, J. Org. Chem., 40, 3625 (1975).
158. A. J. Fry and J. T. Andersson, J. Org. Chem., 46, 1490 (1981).

159. A. J. Fry and R. Scoggins, Tetrahedron Lett., 4079 (1972).
160. W. J. M. van Tilborg, R. Plomp, R. de Ruiter, and C. J. Smit, Rec. Trav. Chim. Pays-Bas, 99, 206 (1980).
161. F. Barba, M. D. Velasco, N. Moreno, and A. Aldaz, Electrochim. Acta, 32, 1507 (1987).
162. P. J. Elving, I. Rosenthal, and M. K. Kramer, J. Amer. Chem. Soc., 73, 1717 (1951).
163. I. Rosenthal, C. H. Albright, and P. J. Elving, J. Electrochem. Soc., 99, 227 (1952).
164. P. J. Elving and J. M. Markowitz, J. Electrochem. Soc., 101, 195 (1954).
165. E. Laviron, Bull. Soc. Chim. Fr., 2350 (1961).
166. J. Holubek and J. Volke, Coll. Czech. Chem. Commun., 27, 680 (1962).
167. R. F. Evilia and A. J. Diefenderfer, J. Electroanal. Chem., 22, 407 (1969).
168. P. T. Cottrell and P. H. Rieger, Mol. Phys., 12, 149 (1967).
169. R. D. Chambers, W. K. R. Musgrave, C. R. Sargent, and F. G. Drakesmith, Tetrahedron, 37, 591 (1980).
170. C. K. Bon, A. J. Kamp, and T. J. Sobieralski, U.S. Pat. 4,592,810 (1986); Chem. Abstr., 105, 69063 (1986).
171. D. Kyriacou, Eur. Pat. Appl. 18,069 (1980); Chem. Abstr., 94, 54933 (1981).
172. K. Alwair and J. Grimshaw, J. Chem. Soc., Perkin Trans. II, 1811 (1973).
173. P. Fuchs, U. Hess, H. H. Holst, and H. Lund, Acta Chem. Scand., B35, 185 (1981).
174. F. Y. Edamura, D. Kyriacou, and J. Love, U.S. Pat. 4,217,185 (1980); Chem. Abstr., 94, 22193 (1981).
175. H. Lund, Acta Chem. Scand., 18, 1984 (1964).
176. P. E. Iversen, Synthesis, 484 (1972).
177. S. G. Mairanovskii and A. D. Filonova, Elektrokhimiya, 1, 1044 (1965).
178. M. Person and R. Mora, Bull. Soc. Chim. Fr., 521 (1973).
179. M. Person and R. Mora, Bull. Soc. Chim. Fr., 528 (1973).
180. R. M. Justice, Jr., and D. A. Hall, U.S. Pat. 4,588,484 (1986); Chem. Abstr., 105, 69060 (1986).
181. M. A. Casadei, F. M. Moracci, and A. Inesi, J. Chem. Soc., Perkin Trans. II, 419 (1986).
182. H. Lund, M. A. Michel, and J. Simonet, Acta Chem. Scand., B28, 900 (1974).
183. S. Margel and M. Levy, J. Electroanal. Chem., 56, 259 (1974).
184. J. W. Sease and R. C. Reed, Tetrahedron Lett., 393 (1975).
185. W. E. Britton and A. J. Fry, Anal. Chem., 47, 95 (1975).
186. S. Bank and D. A. Juckett, J. Amer. Chem. Soc., 97, 567 (1975).
187. S. Bank and D. A. Juckett, J. Amer. Chem. Soc., 98, 7742 (1976).
188. J. Simonet, M. A. Michel, and H. Lund, Acta Chem. Scand., B29, 489 (1975).
189. J. Simonet and H. Lund, Bull. Soc. Chim. Fr., 2547 (1975).
190. J. H. Wagenknecht, J. Org. Chem., 42, 1836 (1977).
191. E. Hobolth and H. Lund, Acta Chem. Scand., B31, 395 (1977).
192. M. Tokuda, T. Taguchi, O. Nishio, and M. Itoh, J. Chem. Soc., Chem. Commun., 606 (1976).

193. T. Lund, S. U. Pedersen, H. Lund, K. M. Cheung, and J. H. P. Utley, Acta Chem. Scand., B41, 285 (1987).
194. T. Lund and H. Lund, Acta Chem. Scand., B40, 470 (1986).
195. T. Lund and H. Lund, Acta Chem. Scand., B41, 93 (1987).
196. N. Gatti, W. Jugelt, and H. Lund, Acta Chem. Scand., B41, 646 (1987).
197. J. A. Sofranko, R. Eisenberg, and J. A. Kampmeier, J. Amer. Chem. Soc., 101, 1042 (1979).
198. C. Gosden, K. P. Healy, and D. Pletcher, J. Chem. Soc., Dalton Trans., 972 (1978).
199. K. P. Healy and D. Pletcher, J. Organometal. Chem., 161, 109 (1978).
200. C. Gosden and D. Pletcher, J. Organometal. Chem., 186, 401 (1980).
201. J. Y. Becker, J. B. Kerr, D. Pletcher, and R. Rosas, J. Electroanal. Chem., 117, 87 (1981).
202. C. Gosden, J. B. Kerr, D. Pletcher, and R. Rosas, J. Electroanal. Chem., 117, 101 (1981).
203. S. Torii, Synthesis, 873 (1986).
204. Y. Rollin, M. Troupel, J. Perichon, and J. F. Fauvarque, J. Chem. Res., (S)322 (1981).
205. Y. Rollin, G. Meyer, M. Troupel, J. F. Fauvarque, and J. Perichon, J. Chem. Soc., Chem. Commun., 793 (1983).
206. R. Scheffold and E. Amble, Angew. Chem. Int. Ed. Engl., 19, 629 (1980).
207. S. Torii, T. Inokuchi, and T. Yukawa, J. Org. Chem., 50, 5875 (1985).
208. J. Wellmann and E. Steckhan, Synthesis, 901 (1978).
209. Y. Rollin, M. Troupel, D. G. Tuck, and J. Perichon, J. Organometal. Chem., 303, 131 (1986).
210. R. W. Murray, in Electroanalytical Chemistry (A. J. Bard, ed.), Vol. 13, Dekker, New York, 1984, p. 191.
211. M. Fujihira, in Topics in Organic Electrochemistry (A. J. Fry and W. E. Britton, eds.), Plenum Press, New York, 1986, p. 255.

9

Cathodic Reduction of Nitro and Related Compounds

HENNING LUND / Aarhus University, Aarhus, Denmark

I. Introduction 401

II. Aliphatic Nitro Compounds 402
- A. Nitroalkanes 402
- B. Nitroso compounds 409
- C. Alkylhydroxylamines 409
- D. Azoxy and azo compounds and hydrazines 410
- E. Nitric acid esters 411

III. Aromatic Nitro Compounds and Derivatives 411
- A. Mononitrobenzenes 411
- B. Dinitrobenzenes 419
- C. Nitroso compounds 421
- D. Phenylhydroxylamines 422
- E. Azoxybenzenes 422
- F. Azobenzenes 422
- G. Hydrazobenzenes 423
- H. Diazonium salts 424

IV. N-Nitramines, Nitrosamines and Azides 424
- A. N-nitramines 424
- B. N-nitrosamines 424
- C. N-nitrosamides 425
- D. Azides 425

References 426

I. INTRODUCTION

The electrolytic reduction of nitro compounds has played an important role in the development of organic electrochemistry in the "classic" period about the turn of the century [1]. It was the investigation of the stepwise reduction of nitrobenzene that led Haber [2] to the idea that the deciding factor in determining the product was the potential of the working electrode, and he was also the first to apply the principle to organic electrolysis.

In this chapter there will be discussed the reduction of aliphatic and aromatic as well as N-nitro compounds and the further reduction of the products obtained, such as nitroso, azoxy, azo compounds, and hydroxylamines. The electrochemistry of these compounds has been treated in reviews [3].

II. ALIPHATIC NITRO COMPOUNDS AND DERIVATIVES

A. Nitroalkanes

Primary and Secondary Nitroalkanes

Primary and secondary nitroalkanes are reducible in acid and neutral solutions, whereas the formation of the anion of the aci form in alkaline solution generally prevents their reduction in this medium. Electrolysis of nitroalkanes in acid solution at low temperature yields as main product the corresponding alkylhydroxylamine [4–7], which is further reducible only in a rather narrow pH region, around pH 7. The electrolytic reduction of nitroalkanes in hydrochloric acid is probably the method of choice for the preparation of alkylhydroxylamines; evaporation of the catholyte after reduction yields the alkylhydroxylamine hydrochloride in a rather pure form [6]. In Table 1 the yields of alkylhydroxylamines and amines from the reduction of some nitroalkanes are given.

A small amount of alkylamine and of a carbonyl compound is formed during the reduction. The hydroxylamine is not reduced to amine under these conditions. Reduction of N-methylhydroxylamine [8] has been attempted at −1.15 V (SCE) in hydrochloric acid at 20–30°C for 16 h. Hydrogen was evolved, but no methylamine was detected. Equation (1) has been suggested to explain the side products obtained [7,8]:

$$
\begin{array}{ccccc}
\mathrm{RR'CHNO_2} & \xrightarrow[\mathrm{2e\,+\,2H^+}]{\mathrm{H^+}} & \mathrm{[RR'CHNO]H^+} & \xrightarrow{\mathrm{2e\,+\,2H^+}} & \mathrm{RR'CH\overset{+}{N}H_2OH} \\
\mathrm{(I)} & & \mathrm{(II)} & & \mathrm{(III)} \\
 & & \downarrow & & \\
 & & \mathrm{[RR'C{=}NOH]H^+} & \xrightarrow{\mathrm{4e\,+\,4H^+}} & \mathrm{RR'CH\overset{+}{N}H_3} \\
 & & \mathrm{(IV)} & & \\
 & & \mathrm{H_2O}\ \downarrow & & \\
 & & \mathrm{RR'CO + H_3\overset{+}{N}OH} & &
\end{array}
\qquad (1)
$$

The tautomerization of the nitroso derivative (II) to the oxime (IV) competes with the further reduction, and any parameter that retards the

Table 1. Results from Electrolytic Reduction of Nitroalkanes at 20°C

R	Amount of RNO_2 (g)	Analysis of catholyte % RNHOH	% RNH_2	% RNO_2	Current yield (%)	% Isolated RNHOH
Me	28.2	89	3.6	1.2	92.6	77
Et	31.1	82	9.9	0.8	92.9	59
1-Pr	40.1	83	5.9	1.2	91.6	62
2-Pr	39.7	86	4.6	0.8	90.5	81

Source: From Ref. 6.

competing chemical reaction raises the yield of the alkylhydroxylamine (III). An increase in temperature would be expected to favor the tautomerization reaction compared to the further reduction, and reduction at a low temperature would thus be essential for the preparation of alkylhydroxylamines.

Accordingly, it has been found that α-hydroxynitroalkanes at 65–85°C are reduced to the amino alcohols [9], whereas at 0–35°C the hydroxylamino alcohol is the product [10].

The electrochemical reduction of suitable nitro compounds in aqueous methanolic hydrochloric acid has been used for the preparation of β-hydroxylaminophophoryl compounds [$HONHCH_2CHR-P(O)R^1R^2$] [11].

In aqueous alkaline solution of low ionic strength the transformation of a nitroalkane into the unreducible anion is sufficiently slow to allow cyclic voltammetry of the nitroalkane. It has been shown [12] that the reduction occurs in two steps, the first one leading to the radical ion.

In nitroalkanes substituted α to the nitro group with an unreducible electron-attracting group the α-hydrogen dissociates more easily, and the rearrangement (II) → (IV) proceeds at a higher rate; a relatively high proportion of amine must thus be expected. Thus, whereas a single carboxylic ester or sulfamido group α to the nitro group [13–15] sometimes (but not always) makes the α-hydrogen too labile for the electrolytic preparation of the hydroxylamine, the presence of two carboxylic ester groups always does. It has thus been impossible to prepare 2-hydroxylaminomalonic ester by reduction of the nitro compound. The reduction of 2-nitro-2-alkylmalonic esters, however, produces the expected 2-alkyl-2-hydroxylaminomalonic ester on reduction [16]; here the tautomerization to the oxime is excluded as no α-hydrogen atoms are available.

The form of the polarographic wave of certain nitro compounds, such as 1-(p-bromophenyl)-2-nitroethanol, depends on the buffer concentration; at low buffer concentrations the wave is steeper than at higher concentrations. This has been explained by a catalytic effect of the reduction product on the protonation of the nitro compound, which makes the nitro group more easily reducible. Some compounds of a similar structure, such as ephedrine, have a similar catalytic effect on the protonation, whereas others, such as cyclohexylamine, have not [17].

Protonation of the anion of the aci form, the nitronate anion, takes place at the highest rate on the oxygen; the aci form is unstable and either tautomerizes to the nitroalkane or is hydrolyzed to a carbonyl compound and N_2O. The reduction of an aci nitro compound, prepared from the anion, is thus generally difficult to study, but its polarographic behavior can be investigated during the reduction of α-halonitroalkanes.

9-Nitrofluorene anion, however, may be acidified to the relatively stable aci form, which in acid solution is reduced to 9-aminofluorene; in DMF CV shows the reduction to fluorenone oxime and further reduction through the imine to 9-aminofluorene [18].

Anodic oxidation of the nitroalkane anion to a radical may lead to dimerization [19], addition to unsaturated systems [20], or substitution in aromatic compounds [21]; these reactions are treated in Chap. 23.

Tertiary Nitroalkanes

These compounds differ from the primary and secondary nitroalkanes in that the intermediate nitroso derivative cannot rearrange into an oxime and this side raction is avoided. The yield of alkylhydroxylamine is thus high. A small amount of amine (<3%) is, however, usually found [8], but it is not yet clear how this is formed.

The nitroso intermediate can in this case be detected directly in the reduction mixture by its bluish green color and its polarographic wave; the nitroso compound is, as most other nitroso derivatives, more easily reducible than the corresponding nitro compound.

In nonaqueous aprotic solvents, such as dimethoxyethane [22] or acetonitrile [23,24], the reduction product from tertiary nitroalkanes is the anion radical. Cyclic voltammetric data of 2-nitro-2-methylpropane showed that the electrochemical rate constant was rather low and depended on the size of the supporting electrolyte cation; the electrochemical transfer coefficient α was found to be potential dependent [25]. The nitro-t-butyl anion radical is rather unstable (half-life of 0.66 s) and decomposes into nitrite ion and t-butyl radical. Continued electrolysis results in the formation of di-t-alkyl nitroxide radical [22,24], Eq. (2):

$$R_3CNO_2 \xrightarrow{\bar{e}} R_3CNO_2^{\overline{\cdot}} \longrightarrow R_3C^{\cdot} + NO_2^{-}$$
$$R_3C^{\cdot} + R_3CNO_2^{\overline{\cdot}} \longrightarrow (R_3C_2)_2NO_2^{-} \xrightarrow{+H^+} (R_3C)_2\dot{N}O + {}^{\cdot}OH \qquad (2)$$

In acid solution the primary step is probably also formation of a radical anion, but in this medium the latter is protonated and further reduced more rapidly than the decomposition of the radical anion; in more acidic solution a preprotonation may take place.

α, β-*Unsaturated Nitro Compounds*

The reduction of an α,β-unsaturated nitro compound [26], such as β-nitrostyrene (V), occurs in two steps. The first reduction of (V) is a four-electron reaction that yields phenylacetaldoxime (VI); the reaction has been formulated as a reduction of the nitro group followed by a rearrangement of the unsaturated hydroxylamine (VII), but might also be regarded as a 1,4-reduction of the intermediate unsaturated nitroso compound. There is no conclusive evidence for either route, but the former has been chosen here, as the protonation of the heteroatom would be expected to be faster than protonation of the carbon atom, Eq. (3):

$$\underset{(V)}{C_6H_5CH{=}CHNO_2} \xrightarrow[-H_3O]{4\bar{e} + 4H^+} \underset{(VII)}{C_6H_5CH{=}CHNHOH} \longrightarrow \underset{(VI)}{C_6H_5CH_2CH{=}NOH} \qquad (3)$$

Reduction in acid solution of nitroalkenes prepared from carbonyl compounds and nitroalkanes has been employed for the synthesis of oximes and carbonyl compounds [27,28] with a longer carbon chain.

The reduction of derivatives of α-nitroacrylic acid to amino acids follows a similar route; the amine seems for α-nitro carboxylic acids to be formed partly through the oxime and partly, at a more negative potential, through another intermediate [29].

Reduction of 1-nitro-1-alkenes in DMF in the presence of $TiCl_4$ affords nitriles in good yield; this is a method to transform aldehydes to one-carbon elongated nitriles through reaction with nitromethane [30].

α-Substituted (and α, β-Unsaturated) Nitroalkanes

β-Bromo-β-nitrostyrene (VIII) [31] gives a six-electron polarographic wave at about the same potentials as β-nitrostyrene (V), but the wave of VIII is not followed by a second wave as is the wave of V. Reduction of VIII in 0.6 N H_2SO_4 containing dioxane at E −0.4 V yielded benzyl cyanide and some benzaldehyde; if the reduction was performed in aqueous methanol, some $C_6H_5CH(OCH_3)CN$ was formed. Polarographic data for the nitrile oxide, $C_6H_5CH_2CNO$ [32,33], show that this compound is not an intermediate in the reduction, Eq. (4):

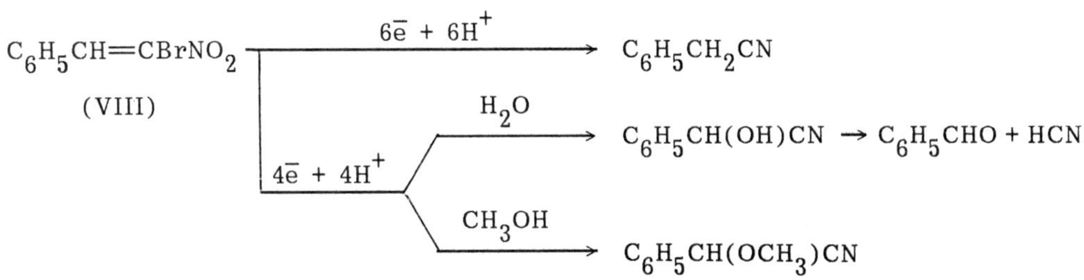

(4)

The reduction mechanism is not fully elucidated [34].

A similar reduction route is found for the reduction of β,β-dinitrostyrene.

α-Substituted Nitroalkanes

Substituents at the α-carbon atom of nitroalkanes may influence their electrode reactions considerably. The influence of electronegative groups on the rate of the nitroso to oxime transformation was mentioned. If an electronegative group is a good leaving group, it may be reductively removed; substituents of this type are halogens and nitro groups. When a nitro group is reductively lost as a nitrite ion, a further reaction between nitrous acid and a suitable substrate must always be considered a side reaction.

Polynitroalkanes

2,2-Dinitropropane is in alkaline solution reduced in a two-electron reaction [35–38] to 2-nitropropane anion; in slightly acid solution the nitronic acid is reduced to acetone oxime. 1,1-Dinitroethane behaves differently, and a wave height corresponding to n = 5–6 is found in acid solution. The following explanation has been suggested [38], Eq. (5):

$$CH_3CH(NO_2)_2 \xrightarrow{2\bar{e}} CH_3CH{=}NO_2^- + NO_2^- \xrightarrow{2H^+} CH_3CH{=}NO_2H + HNO_2 \longrightarrow$$

$$\underset{\text{(IX)}}{CH_3C(NO_2){=}NOH} \xrightarrow{4\bar{e} + 4H^+} \underset{\text{(X)}}{CH_3C(NHOH){=}NOH} \qquad (5)$$

The ethyl nitrolic acid (IX) has been shown [39] to be electrolytically reducible to N-hydroxyacetamidoxime (X), which can be oxidized anodically to ethyl nitrosolic acid. In alkaline solution ethyl nitrolic acid (IX) forms an azo derivative (XI) [40] in low yield according to Eq. (6):

$$\underset{\text{(IX)}}{CH_3C(NOH)NO_2} \xrightarrow{4\bar{e} + 4H^+} \tfrac{1}{2}\underset{\text{(XI)}}{CH_3C(NOH)N{=}NC(NOH)CH_3} \tag{6}$$

Tetranitromethane [41,42] is reduced polarographically to trinitromethane; this is, in turn reduced at pH 6 to the rather unstable dihydroxyguanidine [37], which in alkaline solution is rearranged and further reduced to hydroxyguanidine.

Pseudonitroles

In pseudonitroles, such as 2-nitro-2-nitrosopropane [38], the first reduction in acid solution is a two-electron reaction that results in the loss of the nitro group as a nitrite ion. Protonation of the primarily formed anion on oxygen forms an oxime, Eq. (7):

$$(CH_3)_2C(NO)NO_2 \xrightarrow{2\bar{e}} NO_2^- + [(CH_3)_2CNO]^- \xrightarrow{H^+} (CH_3)_2C{=}NOH \tag{7}$$

α-Halogenated Nitroalkanes

The halogen atom in α-halogenated nitroalkanes is generally removed at a less negative potential than that necessary for the reduction of the nitro group [38,43]; thus, controlled potential reduction of 2-chloro-2-nitropropane (pH 0.25; E −0.6 V, t = 10°C) gave acetone in 98% yield [38], formed by hydrolysis of the nitrolic acid.

1-Bromo-1-nitroethane (XII) shows three polarographic waves in acid solution. The half-wave potential of the first wave is pH independent, and the third wave corresponds to a reduction of acetaldoxime]38]. A reduction at a potential corresponding to the second polarographic wave yielded 60% acetaldoxime and 30% acetaldehyde, Eqs. (8) through (10):

1. wave:

$$\underset{\text{(XII)}}{CH_3CHBrNO_2} \xrightarrow{2\bar{e}} CH_3CH{=}NO_2^- + Br^- \xrightarrow[H^+]{\text{slow}} CH_3CH_2NO_2$$

$$CH_3CH{=}NO_2^- \xrightarrow[\text{fast}]{H^+} CH_3CH{=}NO_2H \tag{8}$$

2. wave:

$$\underset{\text{(XIII)}}{CH_3CH{=}NO_2H} \xrightarrow{2\bar{e} + 2H^+} CH_3CH{=}NOH + H_2O \tag{9}$$

3. wave:

$$CH_3CH{=}NOH \xrightarrow[4\bar{e} + 4H^+]{H^+} CH_3CH_2\overset{+}{N}H_3 + H_2O \tag{10}$$

The second wave thus indicates the electrolytic behavior of a nitrolic acid (XIII).

A similar reduction route is found for dihalogenated nitroalkanes such as 1,1-dichloro-1-nitroethane [38]. This compound shows two polarographic two-electron waves. The second wave does not correspond to the reduction of 1-chloro-1-nitroethane, which is the isolated product from a controlled potential reduction (E −0.2 V, pH 1) at the potential of the first wave, but rather to a reduction of the chloronitrolic acid, $CH_3CCl{=}NO_2H$. Accordingly, a reduction at the plateau of the second wave (E −1.0 V, pH 0.2, t = 5°C) produced chloroacetaldoxime, which was further characterized by its transformation into acetonitriloxide at pH 6.

In aprotic medium reduction of α-halogenated nitroalkanes yields dimers: 2-chloro-2-nitropropane thus yields 2,3-dimethyl-2,3-dinitrobutane [44] in good yield. The reaction has been suggested to be a two-electron reduction to the anion, which in a nucleophilic reaction attacks the substrate.

The reduction of trichloronitromethane may give methylamine [45] or methylhydroxylamine [46], but by working with a cooled tin cathode in 35% sulfuric acid at t < 5°C it was possible to obtain dichloroformoxime [47] as a product. By employing a suitable extraction procedure the yield of this compound could be raised [48].

A polarographic investigation of trichloronitromethane [38] shows a stepwise removal of chlorine under these conditions before the reduction of the nitro group occurs. Whether the difference in reduction path reflects the difference in the electrode material, electrode potential, or medium is not known. A formulation for the formation of dichloroformoxime could be Eq. (11):

$$CCl_3NO_2 \xrightarrow{2\bar{e}} Cl^- + CCl_2{=}NO_2^- \xrightarrow{H^+} CCl_2{=}NO_2H \xrightarrow{2\bar{e} + 2H^+} CCl_2{=}NOH \tag{11}$$

Fluorotrinitromethane is in the first step reduced to fluorodinitromethane, which in turn is reduced to fluoronitromethane [49,50]; on reduction at the first step [51], chlorotrinitromethane is mainly reduced to chlorodinitromethane, whereas the main product from bromotrinitromethane is trinitromethane.

Nitrohydrazones

The phenylhydrazone of nitroacetaldehyde (XIV) [52] gives at all pH values a single polarographic wave with a height corresponding to n = 2.9–3.5. Electrolysis of the compound in 0.05 N sulfuric acid (E −0.9 V) consumed 3.7 F/mol, and the hydrazooxime (XV) was detected as product by its anodic wave, which is caused by a two-electron oxidation to the azooxime (XVI), Eq. (12):

$$\underset{\text{(XIV)}}{CH_3C(NO_2){=}NNHC_6H_5} \xrightarrow[-H_2O]{2\bar{e} + 2H^+} \underset{\text{(XVI)}}{[CH_3C(NOH)N{=}NC_6H_5]}$$

$$2\bar{e} + 2H^+ \downarrow\uparrow -2\bar{e}-2H^+ \tag{12}$$

$$\underset{\text{(XV)}}{CH_3C(NOH)NHNHC_6H_5}$$

The corresponding hydrazooxime, obtained by reduction of the nitrohydrazone $O_2NC_6H_4NHN{=}C(NO_2)C_6H_5$ from benzaldehyde and o- or p-nitrophenylhydrazine, is further reducible [53] to the amidrazone; this is in analogy with the finding that benzamidoxime is reducible to benzamidine [54].

β-Substituted Nitro Compounds

The polarographic behavior of β-hydroxynitroalkanes is sometimes complicated in acid solution by the dehydration to the more easily reducible nitroalkene [55,56]. Nitroalkenes are also obtained from α,β-dibromonitroalkanes on reduction [31]. Reduction in DMF of acylated β-hydroxynitroalkanes produces alkenes [57]; vicinal dinitroalkanes are reduced similarly.

Reduction of α-nitro-β-hydroxybutyric acid to D,L-threonine [58] and similar reactions [29,59] have been investigated for the preparation of amino acids. The reduction to amine passes predominantly through the hydroxylamino derivative and only to a smaller degree through the oxime.

2,3-Dinitroalkenes-2 like 2,3-dinitrobutene-2 show [60] a two-electron polarographic wave at all pH values. In acid solution it is followed by another wave. A preparative reduction also indicates n = 2 and a mixture of 2-nitro-2-butene and diacetylmonoxime is the product in acid solution, whereas the former is the major product at higher pH. At pH 6 the reaction is as in Eq. (13):

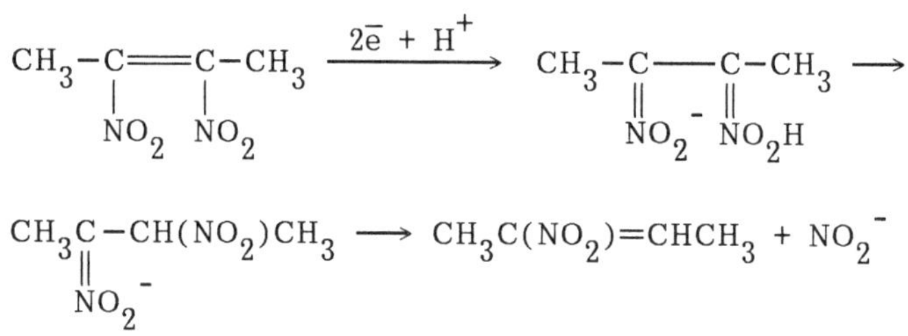

(13)

In β-carbonylnitroalkanes the nitro group is more easily reducible than the carbonyl group [61,62], whereas in 2-bromo-2-nitroacetophenone [61] the bromine is most easily reduced. A pH < 5, 2-nitroacetophenone shows a kinetic polarographic wave followed by another wave; the sum of the waves corresponds to n = 4. The kinetic wave is unusual in that its height grows with pH. The explanation is that the more easily reducible species is the enolate, which is favored by a higher pH; at low pH the nitro ketone is reduced as in Eq. (14):

$$C_6H_5COCH_2NO_2 \rightleftharpoons C_6H_5C(OH){=}CHNO_2 \xrightarrow[-H_2O]{4e^- + 4H^+} [C_6H_5COCH_2NHOH] \qquad (14)$$

The product from the electrolysis has not been identified; possibly a condensation of a primarily formed ketoalkylhydroxylamine to a pyrazine derivative takes place.

The polarographic behavior of some β-nitrooximes has been investigated [62,63] but the reduction is not fully elucidated. Reduction of trans-chloro-(2-nitrocyclohexyl)mercury yields cyclohexene in a two-electron reduction [57].

B. Nitroso Compounds

Aliphatic primary and secondary nitroso compounds are not stable but tautomerize to oximes. In nitrosocyclohexane the special steric conditions favor a dimerization rather than a tautomerization. The dimer is reduced polarographically in a six-electron reaction to N,N'-dicyclohexylhydrazine [64].

Tertiary nitroso compounds are stable and are generally reduced similarly to the corresponding nitro compound, but somewhat more easily. The reduction of mono- and dihalogenated nitrosoalkanes also goes analogously to that of the corresponding nitroalkanes; thus 2-chloro-2-nitrosopropane [38] gives acetoxime on reduction (pH 3, E −0.5 V) in 98% yield, Eq. (15):

$$(CH_3)_2CClNO \xrightarrow{2\bar{e}} [(CH_3)_2CNO^- + Cl^- \xrightarrow{H^+} (CH_3)_2C{=}NOH \qquad (15)$$

and 1,1-dichloro-1-nitrosoethane produces chloroacetaldoxime in 76% yield on electrolytic reduction.

Oximes are tautomeric with nitroso compounds. Oximes are generally reduced to amines in a reaction in which the N−O bond is cleaved before the hydrogenation of the C=N double bond occurs. In acid solution the reaction is as in Eq. (16) [54,65]:

$$RR'C{=}NOH \underset{}{\overset{H^+}{\rightleftharpoons}} [RR'C{=}NOH]H^+ \xrightarrow[-H_2O]{2\bar{e} + 2H^+} RR'C{=}NH_2 \xrightarrow{2\bar{e} + 2H^+} RR'CH\overset{+}{N}H_3 \qquad (16)$$

The reduction of oximes is discussed Chap. 11.

C. Alkylhydroxylamines

Aliphatic hydroxylamines are not easily reduced electrolytically. The polarographic wave of such compounds is not well defined and is visible only in a narrow pH region, around pH 7. In acid solution the wave is masked by hydrogen evolution, and the hydroxylamines are catalytically active in that reaction. It has been suggested [66] that a branching in the reaction takes place after the uptake of the first electron, which leads either to a further reduction of the radical or to hydrogen evolution, Eq. (17):

$$RNHOH + H^+ \rightleftharpoons R\overset{+}{N}H_2OH \xrightarrow{\bar{e}} R\dot{N}H_2OH$$

$$R\dot{N}H_2OH \begin{cases} \xrightarrow{\bar{e}} RNH_2 + OH^- \\ \xrightarrow{H^+ + \bar{e}} RNHOH + H_2 \end{cases} \qquad (17)$$

In acid solution the catalytic reaction is fast and only traces (if any) of amine are formed on reduction. Possibly by lowering the cathode potential and allowing a copious evolution of hydrogen, more amine could be obtained [67]. Alkylhydroxylamines are, however, reducible [68] to amines by electrolytically generated Ti^{3+} or Fe^{2+}.

Alkylhydroxylamines give a well-defined anodic wave in alkaline solution [6,69]. Anodic oxidation of hexafluorodimethylhydroxylamine to the stable free radical, hexafluorodimethyl nitroxide, is a convenient way of preparing this radical [70].

D. Azoxy and Azo Compounds and Hydrazines

Aliphatic azoxy compounds may be prepared from nitroalkanes by electrolytic reduction to alkylhydroxylamines as described earlier followed by anodic oxidation in alkaline solution. Azoxymethane and azoxycyclopropane have been prepared by this method [71].

Another method is the reduction of aliphatic azodioxy compounds [72], obtained by addition of nitrosyl chloride to alkenes, to azoxy compounds.

Reduction of aliphatic azoxy compounds gives the symmetrical hydrazines in good yield. This makes it possible to prepare such hydrazines from nitroalkanes in a three-step synthesis by reversing the current twice but without isolating the intermediates [71], Eq. (18):

$$RNO_2 \xrightarrow[-H_2O]{4\bar{e} + 4H^+} RNHOH \xrightarrow{-2\bar{e}-2H^+} [RNO] \xrightarrow[-H_2O]{RNHOH} \qquad (18)$$

$$RN{=}\overset{\overset{O}{\uparrow}}{N}R \xrightarrow[-H_2O]{4\bar{e} + 4H^+} RNHNHR$$

A synthesis of substituted bicyclic azo compounds utilizes the electrochemical cleavage (Chap. 24) of a bis-(2,2,2-trichloroethyl)dicarbamate in DMF followed by ferricyanide oxidation. In this way acid-, base-, and heat-sensitive azo compounds may be prepared, Eq. (19) [73]:

$$\begin{matrix} -N-COOCH_2CCl_3 \\ | \\ -N-COOCH_2CCl_3 \end{matrix} \xrightarrow[(2)\ K_3Fe(CN)_6]{(1)\ 4e^-,\ DMF} \begin{matrix} -N \\ \| \\ -N \end{matrix} + 2CO_2 + 2CH_2{=}CCl_2 \qquad (19)$$

Anodic oxidation may be used instead of the chemical oxidation; the reaction may be run as a paired synthesis.

Aliphatic azo compounds may be obtained on anodic oxidation of *N*,*N*'-dialkylhydrazines; amines or their derivatives may be oxidized to azo compounds in some cases. Anodic oxidation at a graphite electrode of *tert*-butylamine containing some aqueous sodium perchlorate thus yields 2,2'-azoisobutane together with some azoxy-*tert*-butane and minor amounts of nitro-*tert*-butane [74].

Anodic oxidation of the anions of *N*,*N*'-dialkylsulfamides in methanol at carbon or platinum anodes yields azoalkanes according to Eq. (20) [75]:

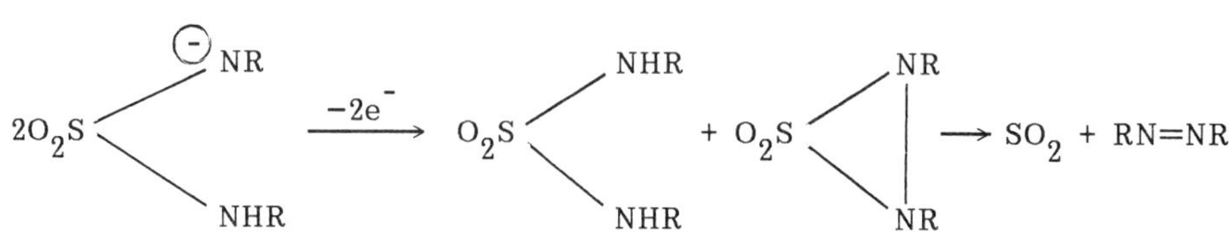

XVII (20)

A phosphorous analog of aliphatic azo compounds, bis[tris-(trimethylsilyl)methyl]diphosphene, $(Me_3Si)_3C{-}P{=}P{-}C(SiMe_3)_3$, shows a rapid reversible one-electron reduction to its anion radical at a mercury-coated platinum electrode in DME, MeCN [76], or THF [77]. The arseneous analog is reduced similarly [77] at a potential about 0.5 V less negative; the anion radical is less stable than that of the phosphorous compound.

Aliphatic diazo compounds are discussed in Chap. 11.

E. Nitric Acid Esters

Such esters are reduced polarographically in a pH-independent two-electron wave to nitrite ion and alcohol, Eq. 21 [78]:

$$R{-}O{-}NO_2 \xrightarrow{2\bar{e}} RO^- + NO_2^- \xrightarrow{H^+} ROH \qquad (21)$$

In DMF CV of benzyl nitrate gives an irreversible peak followed by a peak of one of the products, benzaldehyde, the other products being benzyl alcohol and nitrite. The primarily formed anion radical loses nitrite, and the benzyloxy radical abstracts a hydrogen atom, presumably from the substrate or another benzyloxy radical [79].

III. AROMATIC NITRO COMPOUNDS AND DERIVATIVES

Numerous investigations [1] have been made on the reduction of aromatic nitro compounds and a number of different reaction products have been obtained. This is caused more by differences in the chemical follow-up reactions rather than by the electrochemical steps.

The transfer of the first electron to a nitrobenzene in an aprotic solvent produces an anion radical. In the anion radical of an aliphatic nitro compound the charge on the nitro group is virtually unity, whereas in the anion radicals from nitrocyclopropanes [80], nitroolefins [80,81], and nitrobenzenes [82–84] a certain delocalization of the charge occurs according to ESR measurements. The degree of localization in the aromatic ring can be varied, particularly by twisting the nitro group out of the plane of the ring by introducing suitable substituents.

A. Mononitrobenzenes

Polarographically nitrobenzene is reduced in aqueous solution at low pH in two waves, and at high pH in one wave. In the presence of a surfactant or higher concentrations of organic compounds, such as alcohol, the first wave is a four-electron wave and the second one a two-electron wave [85]. In the absence of surfactants the height of the first wave corresponds to $4 < n < 5$, but the sum of the first and second wave remains constant, $n = 6$ [85]. This has been interpreted as resulting from two competitive reduction routes, one of them being a four-electron reduction to phenylhydroxylamine and further two-electron reduction at a more negative potential of this compound, and the other one going through a coupling occurring at the surface of the electrode between a protonated nitrobenzene and phenylhydroxylamine to azoxybenzene and further reduction of this compound. The polarographic reduction may be represented by scheme 1 [85,86].

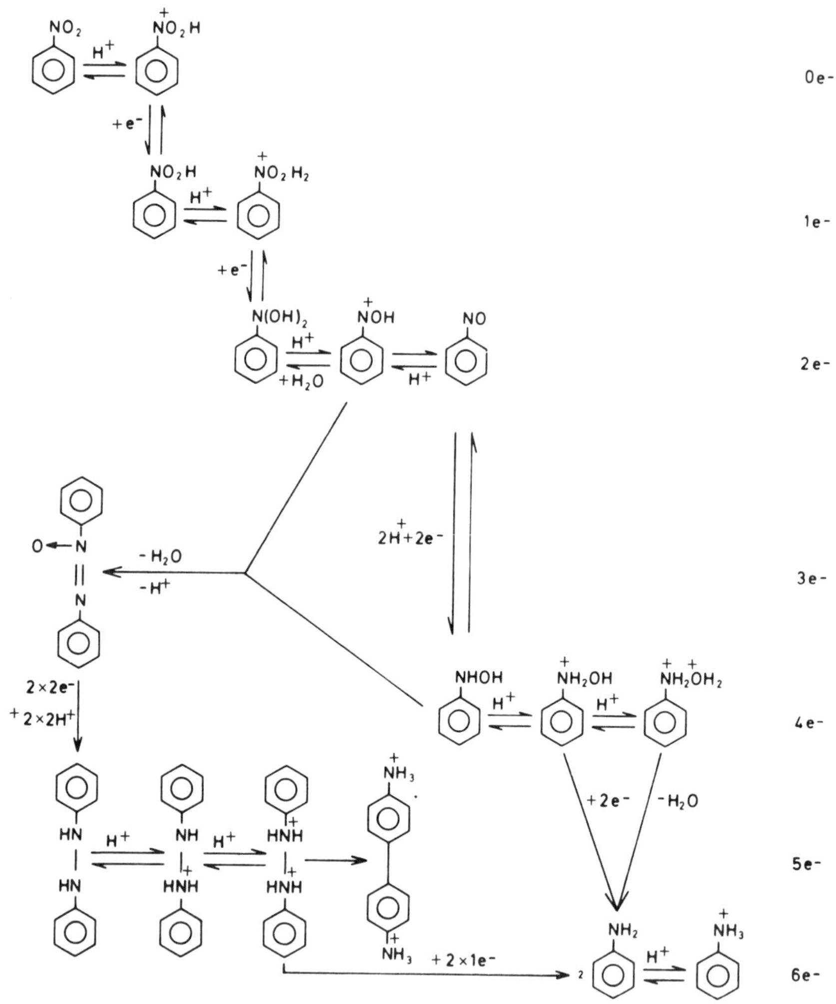

Scheme 1.

CV of nitrobenzene in neutral aqueous buffer shows at mercury [87] and gold [88] a four-electron reduction to phenylhydroxylamine, which on the anodic scan is oxidized to nitrosobenzene; on the second cathodic sweep the reduction of nitrosobenzene is observed. The response at glassy carbon, although qualitatively similar, is shifted substantially from the standard redox behavior, and at platinum no peak was observed before the discharge of protons [88].

In preparative reductions higher concentrations of organic solvents and/or substrate are present and the surface reaction, the coupling, would not be expected to play a prominent role in acid solution. A reduction at a potential of the first wave would thus result in the formation of a species, here represented by $[C_6H_5\overset{+}{N}H_2OH]$ (XVIII) at the electrode surface, and the diversity of products obtained can be explained by the reactivity of XVIII. The different reaction paths open to XVIII and the products resulting from these processes are discussed later.

In the scheme is suggested a two-electron intermediate, $ArN(OH)_2$, a dihydroxylamine; it can be experimentally detected when Ar is an electron-attracting aryl group [89,90]. Controlled potential reduction at low temperatures at pH 0–2 permits the detection of the nitroso compound, formed by an irreversible loss of water from the dihydroxylamine. The electron-attracting group (e.g., protonated pyridyl, p-nitrophenyl) helps to stabilize the hydrated form just as in the case of reduction of acid derivatives to hydrated aldehydes (Chap. 18).

In acidic solution in the presence of an arenesulfinic acid the nitrosoarene adds the sulfinic acid in preference to water [91]; the addition product is stable toward further reduction and oxidation and the arenesulfinic acid may thus be used as a protecting group during the reduction of a nitro compound and during an oxidation of a hydroxylamine to a nitroso compound, thus avoiding the formation of azoxycompounds. The protecting group may be removed by alkaline hydrolysis.

In aprotic solvents, such as DMF [92] and ammonia [93], nitrobenzene is reduced reversibly to the anion radical and at a more negative potential to the dianion; in the absence of electrophiles the dianion is stable, but if impurities are present, the second wave becomes irreversible. In some cases the anion radical of a nitroarene may dimerize; 9-nitroanthracene anion radical dimerizes thus to the dianion of 9,9'-dinitro-9,9',10,10'-tetrahydro-9,9'-biantryl [94].

Anion radicals of nitroarenes may be used as electron transfer reagents (Chap. 29); this ability is also employed in the preparation of chemically modified electrodes, as by using a poly(vinylnitroarene), for electrocatalytic reactions (Chap. 33).

Desorption

When the electrode potential is such that a further reduction is excluded and no catalyst is present to promote a dehydration reaction, the phenylhydroxylamine is obtained by desorption of the four-electron reduction product. The medium of choice should have pH 3–9 and not contain strong nucleophiles; a mixture of ammonia and ammonium chloride [95] or an acetate buffer [96] is often suitable. The electrode material is of minor importance as long as the cathode potential is controlled in some way; the pH must not be too low as rearrangements occur or too high as azoxybenzene then is formed. The temperature must be kept low.

Substitution

In strongly acid solution different substitution reactions may occur; the reduction product XVIII, protonated on nitrogen and/or oxygen, has a tendency to lose water, and this loss may be assisted by the attack of a nucleophile on the aromatic ring. Whether the loss of water occurs before the attack by the nucleophile or in a concerted reaction is not known yet. Any species that can act as a nucleophile in the strongly acidic solution may be employed for this kind of cathodic substitution reaction. The reaction may be represented by Eq. (22):

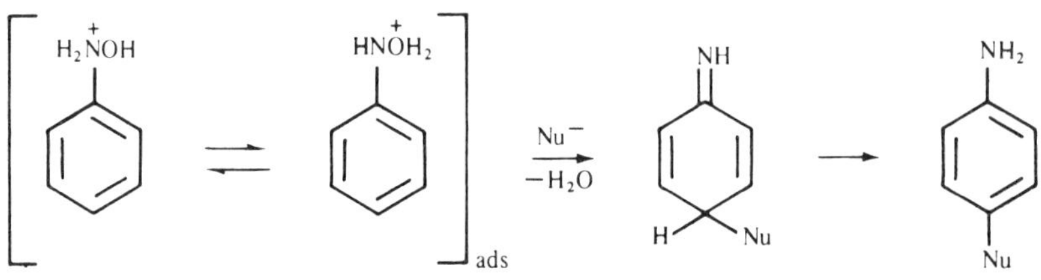

(22)

XVIII

It is well known that phenylhydroxylamine is similarly rearranged in acid solution, but the rate is somewhat slower that that occurring during the electrolytic reduction. This finding has been expressed here by representing the four-electron reduction product as adsorbed phenylhydroxylamine.

In the reduction of nitrobenzene in 4 N hydrochloric acid a mixture of p- and o-chloroaniline and p-aminophenol in the proportion 3:1:1 is found [97,98]; in sulfuric acid the hydrogen sulfate ion is too weak a nucleophile to compete with the water as nucleophile and the resulting product is p-aminophenol. Substituents capable of delocalizing the positive charge left by the loss of water from the protonated reduction product (in the position ortho to the hydroxylamine group) greatly facilitate the attack para to the hydroxylamino group. The methoxy group is excellent in this respect and o-nitroanisole is easily substituted cathodically, Eq. (23):

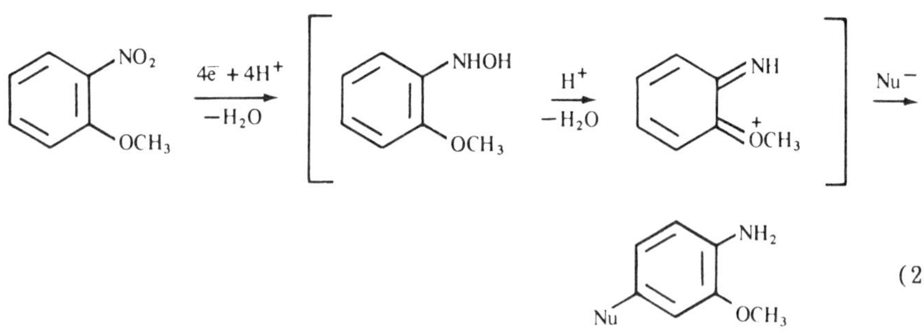

(23)

The electrolytic preparation of p-aminophenol was first studied by Gattermann [99]. A wide variety of conditions for the reaction has been described in different patents [100] and the reaction is still under study [101–103]. The reaction is usually run in a fairly strong sulfuric acid at an elevated temperature; if ethanolic sulfuric acid is used, p-phenetidine is the product [104]. The electrochemical preparation of p-aminophenol ought to be competitive with the chemical reaction.

A complication for the preparation of p-aminophenol (XIX) from phenylhydroxylamine, regardless of how the latter is prepared, is that XIX may act as a nucleophile in competition with water with the formation of 4,4'-diaminodiphenyloxide (XX) [105]; the yield of XX increases with increasing concentrations of XIX. XX is carcinogenic and must be removed carefully from pharmaceutical preparations of XIX to N-acetyl-4-aminophenol, paracetamol.

When a substituent is para to the hydroxylamino group an ipso substitution may take place; this may result in a loss of the substituent or a rearrangement. Examples are given in Eqs. (24) and (25).

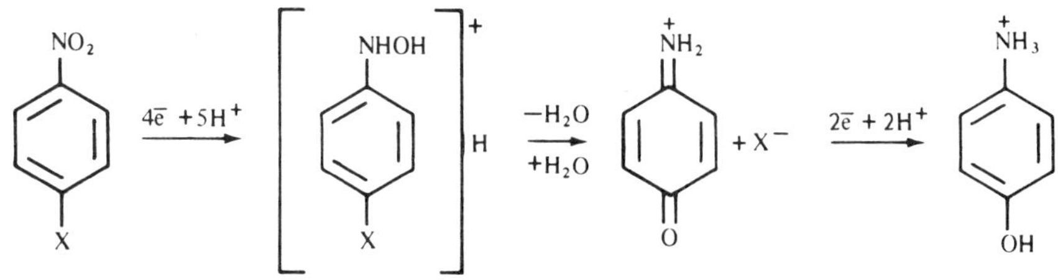

(24)

A loss of substituent, Eq. (24), is found when X = −Cl [106] or $-OCH_3$ [107,108] and has also been reported to occur during the reduction of p-nitrobenzoic acid in strong sulfuric acid [109].

p-Nitrotoluenes may form different products. In 25% sulfuric acid at 90°C, p-nitrotoluene is reduced with four electrons and the product is 2,5-dihydroxytoluene (XXI) [106]. The reaction has been formulated as Eq. (25):

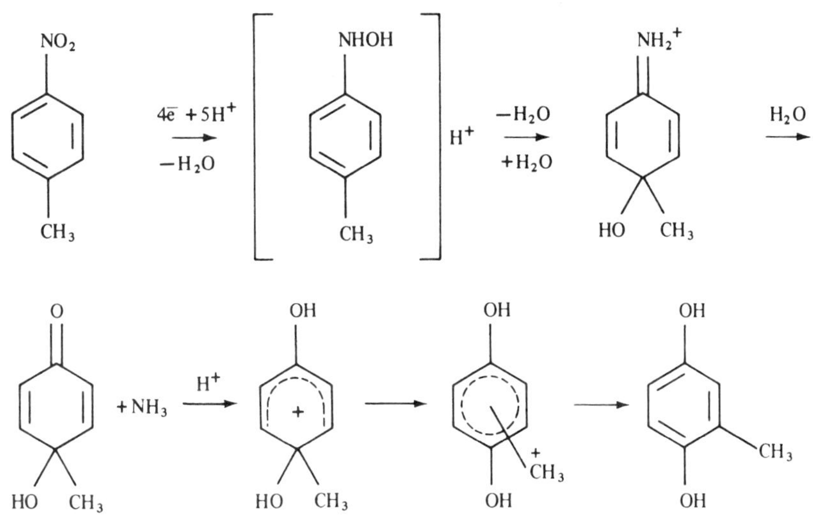

(25)

XXI

A similar reaction was found [106] in the reduction of 2-methyl-5-nitrobenzoic acid.

p-Methylthionitrobenzene is reduced to the hydroxylamine, which may react in the usual ways; besides those already discussed a rearrangement of the p-methylthiophenylhydroxylamine to p-aminophenylmethylsulfoxide [110, 111] may take place (yield 35–40%), Eq. (26):

$$CH_3SC_6H_4NHOH \xrightarrow{[H^+]} CH_3SOC_6H_4NH_2 \tag{26}$$

Side-Chain Reactions

p-Alkyl nitroarenes may on reduction lose a proton from the side chain either during the acid-catalyzed rearrangement of the arylhydroxylamine or from the parent compound, from which an electrogenerated base (EGB) may abstract one of the relatively acidic protons α to the nitroaryl.

In concentrated sulfuric acid p-nitrotoluene is reported to give 4-amino-2'-methyl-5'-nitrodiphenylmethane (XXII) [99,112], which has been verified [113]. Gattermann and Koppert [112] suggested that the primarily formed p-tolylhydroxylamine rearranged into p-aminobenzyl alcohol (XXIII), which then attacked unreduced p-nitrotoluene. It seems more likely that the intermediate is a carbocation (XXIV), which on addition of water would form XXIII, and on addition of p-nitrotoluene would give XXII, Eq. (27):

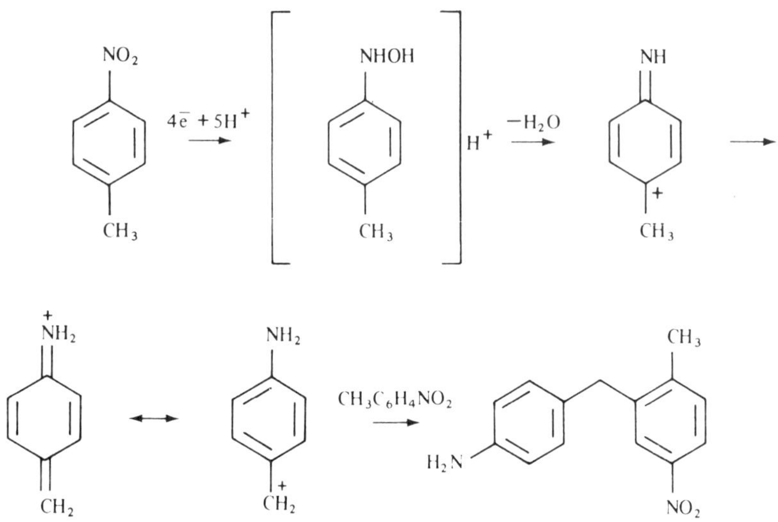

(27)

(XXIV) (XXII)

The electrophilic reagent (XXIV) may attack any aromatic compound present [113]; when this is too reactive, it may, however, be attacked by the solvent rather than by XXIV.

An analogous reaction occurs during the reduction of 4-nitroisopropylbenzene in sulfuric acid, which in a four-electron reduction forms 5-amino-3-(4-aminophenyl)-1,1,3-trimethylindan (XXV); the initially formed phenylhydroxylamine loses water to the equivalent of p-amino-α-methylstyrene, which dimerizes to the indan derivative [113].

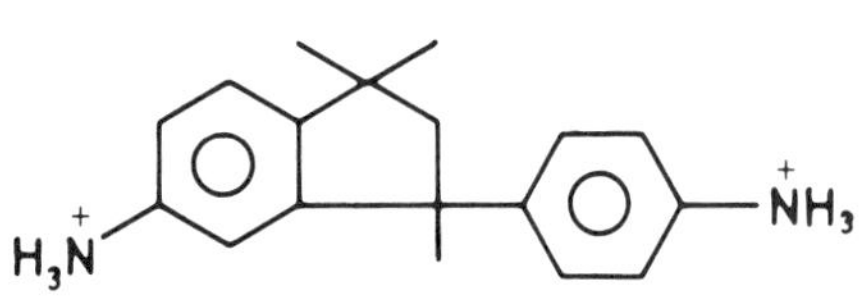

XXV

The reduction of o- and p-nitrotoluene and p-ethylnitrobenzene in DMF in the presence of dioxygen to o- and p-nitrobenzoic acid and p-nitroacetophenone plus 1-(p-nitrophenyl)ethanol, respectively, has been suggested [114] to proceed via a proton or hydrogen atom abstraction from the side chain followed by reaction of this species with superoxide ion and/or dioxygen to the isolated products.

A similar EGB-induced reaction may take place during the reduction of o-nitrotoluene in DMF in the presence of paraformaldehyde to 2-(2-nitrophenyl)ethanol and 2-(2-nitrophenyl)propane-1,3-diol [115].

Condensation Reactions

Phenylhydroxylamine may act as a nucleophile toward any electrophilic center present. As it is the unprotonated phenylhydroxylamine that acts as a nucleophile, a lowering of pH will diminish the concentration of the nucleophile; a higher acidity may, however, make the electrophilic center more reactive, so the overall effect of a change in pH is a function of the properties of both centers.

Nitrosobenzene may act as an electrophilic center, and in alkaline solution phenylhydroxylamine reacts with it with formation of azoxybenzene [116–118]. The reaction may go through an initial, homogeneous one-electron transfer followed by a radical coupling [118]. This reaction occurs generally during the reduction of simple nitrobenzenes in alkaline solution and makes azo- and hydrazobenzenes as well as benzidines easily available. Azoxybenzenes may, however, also be formed in acid solution, Eq. (28) [85,90,107]:

$$2C_6H_5NO_2 \xrightarrow[-2H_2O]{6\bar{e} + 6H^+} [C_6H_5NO + C_6H_5NHOH] \longrightarrow C_6H_5\overset{+}{N}(O^-)=NC_6H_5 + H_2O \quad (28)$$

The preparation of azoxy compounds is possible under a variety of conditions, and a number of patents [119] have been issued covering these processes. The yields are generally fair to good (40–90%); substituents ortho to the nitro group retard the formation of azoxybenzenes. A rotating cathode has been employed for this reaction [120,121].

An intramolecular azoxy coupling has been used for the preparation of heterocyclic compounds; this aspect is treated in Chap. 18.

Carbonyl groups react readily with phenylhydroxylamine, and this reaction, in which nitrones are formed, may be used to trap the intermediate phenylhydroxylamine [122]. The reaction may take place intramolecularly with the formation of a heterocyclic N-oxide or hydroxamic acid (Chap. 18). The condensation of nitrosobenzene with benzenesulfinic acid [91] was mentioned.

A special kind of condensation reaction occurs as a side reaction during the reduction of o-nitroaniline [106,123]: A phenazine derivative is formed. In the reaction (Chap. 18) a double Michael addition of o-phenylenediamine to the intermediate o-quinonediimine takes place; other nucleophiles might under suitable conditions attack the diimine.

Reduction of aromatic nitro compounds in aprotic medium in the presence of acetic anhydride [124] or alkyl halides [125] leads in high yield to O,N-diacylated or dialkylated arylhydroxylamines, respectively.

Reduction and Oxidation

In acid solution the protonated phenylhydroxylamine is further reducible at a potential somewhat more negative than that required for the reduction of the nitro compound. Sometimes addition of salts, such as Sn^{2+} or Ti^{4+} [126,

127], facilitates the reduction; this may be regarded as an indirect reduction and is treated in Chap. 29.

If an amino or hydroxyl group is situated *ortho* or *para* to the nitro group, the intermediate phenylhydroxylamine very rapidly loses water to form the easily reducible quinone mono- or diimine. This loss of water takes place fastest in acid and alkaline solution and somewhat more slowly at pH 5, which is evident from the polarographic behavior of *p*- and *o*-nitrophenol [128]. At pH 5 the loss of water is so slow that the four-electron product diffuses away from the electrode before it has lost the water. Reduction of *o*-nitrophenol in DMF containing acetic anhydride gives the diacetylated *o*-hydroxylaminophenol [124]. As mentioned, alkoxy groups *para* to the nitro group may be lost by substitution and *p*-aminophenol is the product. A similar reaction may occur in the *ortho* series [128] and in the hydroxypyridine series [129]. The reduction of *p*-nitrophenol to *p*-aminophenol has been reinvestigated [130,131].

A *para* alkyl substituent may function in the same manner as a hydroxyl group in promoting loss of water from the phenylhydroxylamine and stabilizing the quinonoid intermediate. 5-Nitroacenaphthene is thus in acid solution reduced to 5-aminoacenaphthene in a six-electron reduction, Eq. (29) [132]:

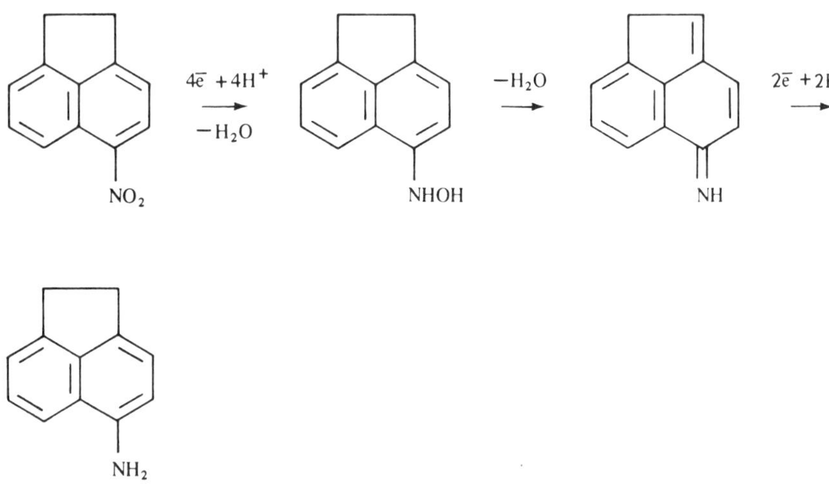

(29)

Reduction of a substituted nitrobenzene follows generally one of the reaction paths described, as the nitro group usually is the more easily reducible group. Nitrobenzyl halides [133,134], however, are reduced in the side chain rather than in the nitro group, and dinitrobibenzyl and nitrotoluene are prominent products (see Chap. 8).

The reduction of 4-nitrobenzoic acid through 4-hydroxylaminobenzoic acid (XXVI) to 4-aminobenzoic acid under industrial conditions [135] has been investigated in 3 M H_2SO_4; it was found that especially the reduction of XXVI was accelerated by raising the temperature to 60°C.

The use of the 4-nitrobenzyloxycarbonyl group as protecting group is discussed in Chap. 24.

Phenylhydroxylamines may be reoxidized anodically to the nitroso compound; this may be used in ring-closure reactions (Chap. 18) [136]. If the reduction of the nitro compound and the reoxidation of the phenylhydroxylamine is done in a cell with two closely spaced porous electrodes, a condensation to azoxybenzene may be avoided [136].

Mononitro derivatives of aromatic heterocyclic compounds and condensed aromatic systems behave as described for benzene derivatives. As in the benzene series, any prediction of the most likely reaction route for a given compound must be based upon considerations of the reactivity of the compound and its reduction products in the different types of reactions possible under a given set of parameters.

B. Dinitrobenzenes

Dinitrobenzenes are in principle reduced like the mononitrobenzenes and the same chemical follow-up reactions may occur, but generally not all of the possible reduction products are observed. In substituted dinitrobenzenes also the question arises as to which of the two nitro groups is reduced first.

o-*Dinitrobenzenes*

In dilute mineral acid solution o-dinitrobenzenes may be reduced to an o-nitrophenylhydroxylamine or further to the o-phenylenediamine [137,138]. The first step in the reduction of 1-X-2,3-dinitrobenzenes leads to 1-X-2-nitro-3-hydroxylaminobenzene (X = H, COOH, Br, CH_3, OC_2H_5), whereas 1-X-3,4-dinitrobenzenes in the first reduction form either 1-X-3-nitro-4-hydroxylaminobenzenes (X = COOH) or 1-X-3-hydroxylamino-4-nitrobenzenes (X = Cl, CH_3, OCH_3) [138], as in Eqs. (30) and (31):

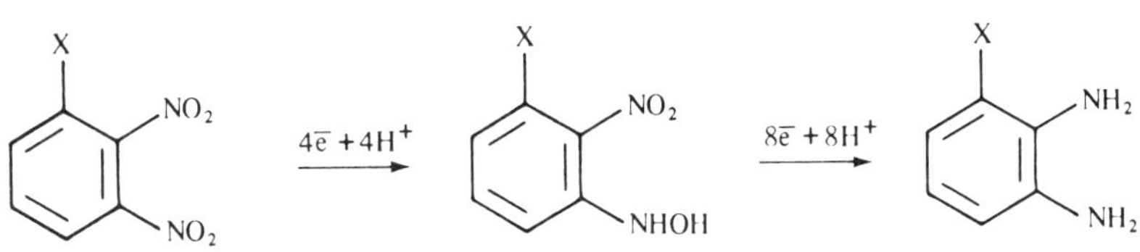

(30)

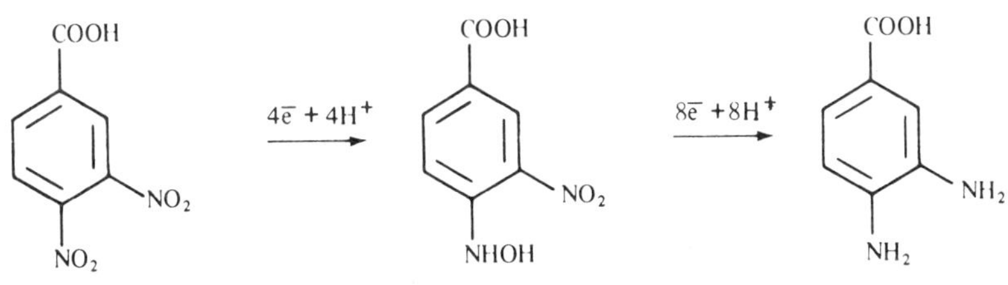

(31)

The differences in cathode potential between the first and second reduction is 0.2–0.3 V; a selective reduction thus requires good control of the cathode potential and a suitable cell design.

The hydroxylamino group may react further. A Gattermann reaction has, however, been observed only in the reduction of 2,3-dinitrotoluene. When a carboxyl group is neighbor to the hydroxylamino group, an isoxazolone may be formed; this reaction has been used as an indication of which nitro group is most easily reduced in dinitrobenzoic acids [133].

Reduction of o-dinitrobenzene in alkaline solution at the potential of the first polarographic wave (−0.7 V) also yields o-nitrophenylhydroxylamine [137]. In slightly acid solution some phenazine derivatives may be formed [123].

m-*Dinitrobenzenes*

Reduction in acid solution at room temperature generally may yield a nitrophenylhydroxylamine, a phenylene dihydroxylamine, or a phenylenediamine [108,137,139]. In alkaline solution it was found difficult [137] to effect a selective reduction of one of the nitro groups.

For unsymmetrically substituted m-dinitrobenzenes (1-X-2,4-dinitrobenzenes) the first reduction occurs either at the 2-nitro group (X = Cl, Br, I) or at the 4-nitro group (X = COOH, CH_3, C_2H_5) [108]. These differences may be caused by inductive effects and by steric inhibition of coplanarity with the benzene ring of the nitro group or of other substituents. In 2,4-dinitrochlorobenzene the inductive effect from the chlorine on the 2-nitro group probably is responsible for its easier reduction; in the first four-electron reduction 2-hydroxylamino-4-nitrochlorobenzene is formed; in the second four-electron step, 2,4-dihydroxylaminochlorobenzene; and in the last four-electron reduction, 2,4-diaminochlorobenzene.

Whereas a dinitroanisole, like 3,5-dinitroanisole, is reduced similarly, the 2,4- and 2,6-dinitroanisoles may or may not lose the alkoxy group, depending on the potential [108]. The reduction of 2,4-dinitroanisole in a mixture of ethanol and N H_2SO_4 follows this scheme, Eq. (32):

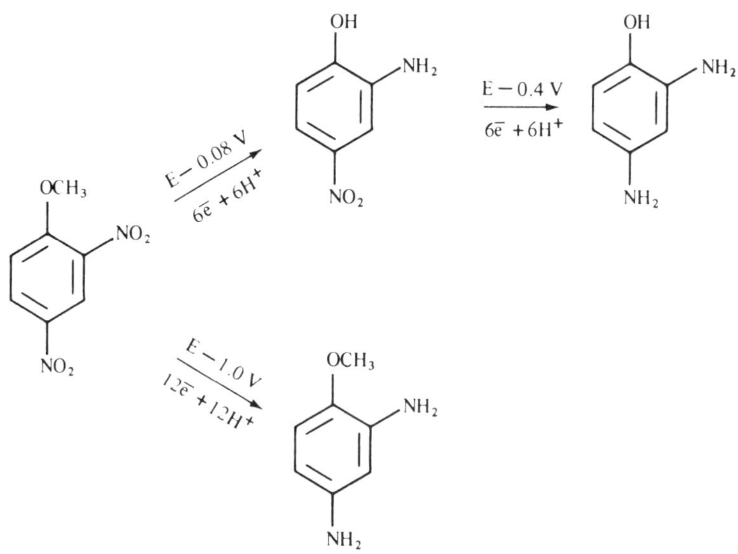

(32)

In neutral solution (ammonium acetate buffer; E −0.5 V) 2-hydroxylamino-4-nitrosoanisole is formed; at E −0.72 V the reduction consumes eight electrons.

In DMSO m-dinitrobenzene is reduced to m-nitrophenylhydroxylamine in the presence of a proton donor, whereas the radical anion, which is reduced further in a second step, is obtained in the absence of protons [140].

p-*Dinitrobenzenes*

Substituted p-dinitrobenzenes may in acid solution be reduced selectively to a nitrophenylhydroxylamine [137,141,142]; further reduction leads to the p-phenylenediamine; p,p'-dinitroazoxybenzene has been shown to be an intermediate [86]. 1-X-2,5-Dinitrobenzenes are reduced first in the 2-nitro group when X = COOH, $CONH_2$, or Br, whereas the 5-nitro group is first attacked when X = CH_3 or OCH_3 [142].

In alkaline solution the polarographic reduction shows [137,141] a two-electron wave (E −0.35 V) followed by a ten-electron wave (E −1.6 V). The two-electron reduction product $(O_2NC_6H_4NO_2)^{2-}$ is not stable and undergoes secondary chemical reactions; one of the reactions is probably a disproportionation to dinitrobenzene and nitrophenylhydroxylamine. Other intermediates, such as nitrosoaniline, have been suggested [141], but their presence has not been proved.

Dinitronaphthalenes

Several dinitronaphthalenes have been investigated polarographically [143, 144]. The first wave is usually a four-electron reduction. The 1,2- and 1,4-dinitronaphthalenes behave like o- and p-dinitrobenzenes, whereas 2,3-dinitronaphthalene does not exhibit a conjugation effect similar to that of o-dinitrobenzene. Nitro groups in α-positions are reduced in preference to those in β-positions.

The synthesis of 1,8-diaminonaphthalene by electrolysis of a suspension of 1,8-dinitronaphthalene in sulfuric acid containing a titanium salt gives a yield and purity similar to the synthesis by catalytic hydrogenation [145]. Technical and economic comparisons show that for large productions (several hundred tons per year) the catalytic hydrogenation is advantageous, whereas for smaller productions (several tons per year) the flexibility of electrolysis has some economic advantages.

Other dinitro compounds with nitro groups in different rings or one in a side chain may form heterocyclic compounds on reduction; such reactions are treated in Chap. 18.

C. Nitroso Compounds

Nitroso compounds are reduced like the nitro compounds, but at a less negative potential [146]; nitrosobenzene can thus not be isolated from a reduction of nitrobenzene. The presence of nitrosobenzene as an intermediate in the reduction of nitrobenzene can be shown by trapping it with a suitable nucleophile; for activated nitro compounds the hydrated nitroso compound may have sufficient stability to be detected [90]. The reaction with phenylhydroxylamine to form azoxybenzene has been discussed; Haber [2] used hydroxylamine and coupled the resulting diazo compound with β-naphthol. Other nucleophiles might be used to trap the nitroso compound and produce products that might have synthetic value. Some applications for the preparation of heterocyclic compounds are discussed in Chap. 18.

By reducing a nitrobenzene in acid solution in a flow cell in which it is reduced to phenylhydroxylamine during passage through a porous cathode and immediately after oxidized in a porous anode, it is possible to obtain nitrosobenzenes in good yield [147].

The presence of a nitroso intermediate in nonaqueous solution may also be deduced from results of cyclic voltammetry. Trifluoromethylnitrobenzenes [148] show two polarographic steps in DMF; the first one is due to a diffusion-controlled one-electron transfer producing a stable anion radical; the second one is a two-electron ECE process resulting in a substituted nitrosobenzene anion radical [149]. Addition of hydroquinone as a proton donor enhances the formation of the nitroso intermediate [148].

In cyclic voltammetry measurements of DMF solutions of nitroso-, azoxy-, and azobenzene using the hanging mercury drop electrode, it was found that besides the usual peaks of redox processes an additional, new system of peaks was observed. Its properties were nearly identical for all three substrates. The most probable interpretation seems to be to attribute this additional system of peaks to a reaction between adsorbed azobenzene dianion and mercury; at graphite such additional peaks were not observed, Eq. (33) [150]:

$$C_{12}H_{10}N_2{}^{2-},2M^+ + Hg - 2e^- \rightleftharpoons (C_{12}H_{10}N_2{}^{2-},Hg^{2+}) + 2M^+ \qquad (33)$$

D. Phenylhydroxylamines

Polarographic data show that below pH 1 the doubly protonated phenylhydroxylamine is the species reduced at the dropping mercury electrode (DME); between pH 5 and 7 the monoprotonated molecule is reduced. The unprotonated molecule is not reducible [151]. Phenylhydroxylamine is for the main part reduced to aniline, but according to the polarographic data a certain degree of dimerization of an intermediate radical takes place. As discussed, phenylhydroxylamines may react chemically in several ways.

E. Azoxybenzenes

Azoxybenzenes are reduced in a four-electron reduction to hydrazobenzenes and are generally more difficult to reduce than the azobenzenes; in this respect the system azoxybenzene-azobenzene is similar to that of nitrobenzene-nitrosobenzene. The analogy can be extended: Further reduction of azobenzene yields hydrazobenzene, which can be reoxidized to azobenzene and which easily rearranges in acid solution; analogously, phenylhydroxylamine can be reoxidized to nitrosobenzene and can be rearranged to p-aminophenol or similar products in acid solution.

The reduction of azoxybenzene to hydrazobenzene is of preparative interest [118,120], due to the easy conversion of hydrazobenzenes to benzidines and similar compounds. Several patents [152] describe the conversion of nitrobenzene to azoxybenzene and further to hydrazobenzene; environmental considerations are in favor of the electrochemical reduction compared with the metal powder reductions, and the method stands a good chance to be competitive in the future.

F. Azobenzenes

Azobenzenes may be prepared by oxidation of hydrazobenzenes; additionally, azobenzenes may be obtained in low to fair yield by anodic oxidation of aromatic amines at a rotating platinum screen cylinder anode in aqueous DMF [153].

The reduction of azobenzenes may follow two routes, either a reversible two-electron reduction or a four-electron reaction with cleavage of the nitrogen-nitrogen bond. The latter reaction is the predominant one for p-hydroxy- and p-aminoazobenzenes [154–156]. In this case the reaction may be formulated as in Eq. (34):

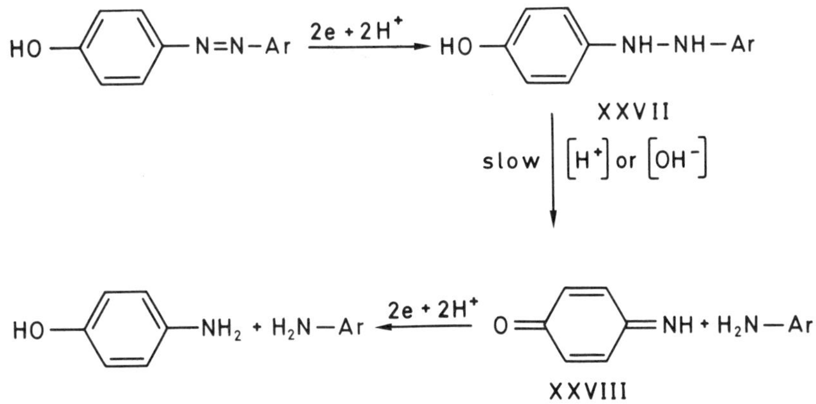

(34)

The cleavage of the nitrogen-nitrogen bond of XXVII may be catalyzed by both acids and bases [157,158]; when Ar = C_6H_5, the acid catalysis is fast enough to ensure a four-electron polarographic wave, whereas a two-electron wave is observed in alkaline solution [154]. Under preparative conditions and at elevated temperatures (50–90°C) the reduction consumes 4 F/mol [159]. Possibly XXVII is able to reduce XXVIII in solution to p-aminophenol (XXVII), thereby being oxidized to p-hydroxyazobenzene. If Ar is strongly electron withdrawing, the rate of cleavage in alkaline solution is enhanced and the height of the polarographic wave approaches that of n = 4 [157,158].

The system azobenzene-hydrazobenzene is one of the rather few organic redox couples that behave reversibly or very nearly so at the dropping mercury electrode [160]. In nonaqueous solvents, like DMF, azobenzenes are reduced in two, one-electron steps: The first one produces the radical anion, whereas the second one yields a hydrazobenzene dianion. This dianion is easily protonated; it has a tendency to decompose into arylhydrazine [161].

The stability of the anion radicals of the azo compound make them useful as electron transfer reagents, either in solution or bonded to an electrode surface. These aspects are treated in Chaps. 29 and 33, respectively. The preparation of heterocyclic compounds by reduction of azo compounds in the presence of electrophiles is discussed in Chap. 18.

The hydrazobenzene dianion is a strong base, the monoanion less so. The reduction of azobenzene and substituted azobenzenes has been used for an electrolytic generation of a base in DMF, which has been employed in some base-catalyzed reactions. This aspect is discussed in Chap. 30.

The azo group is more easily reducible than most other functional groups and is reduced close to that of a nitro group. In aqueous alkaline media 2-nitroazobenzenes are reduced to 2-arylbenzotriazole-N-oxides [162,163] (and further at more negative potentials, Chap. 18), and 2-nitrohydrazobenzene has been reported to be the first intermediate [163]. In DMF [164], 1,3-dihydroxy-2-phenylbenzotriazole is formed from 2-nitroazobenzene and 2-nitrosoazobenzene has been suggested to be an intermediate.

G. Hydrazobenzenes

Hydrazobenzene is not polarographically reducible at pH > 4, and in acid solution polarographic curves are difficult to obtain due to the fast rearrangement to benzidine. Polarographic curves of azo- and azoxybenzene show at pH < 4 a second reduction (n = 2) corresponding to the formation of aniline. From the pH dependence of the half-wave potentials and a logarithmic analysis of the curves it is proposed that only the diprotonated hydrazobenzene cations undergo reduction, whereas the monoprotonated cation exhibits a catalytic effect on the hydrogen evolution reaction [165].

H. Diazonium Salts

Aryldiazonium salts are reducible at a mercury cathode. Polarograms in acid solution show a pH-independent one-electron wave followed by a larger pH-dependent wave [166,167]. Reduction on the plateau of the first wave [168] showed n = 1 and only arylmercury salts were isolated, indicating the initial formation of a radical.

During a preparative reduction at a potential of the second wave care must be taken to avoid the reaction between the diazonium salt and the product, phenylhydrazine. By using a flow cell with a small volume-cathode area ratio and low temperatures the loss by this side reaction was diminished [169]. In the reduction of p-methoxyphenyldiazonium chloride to p-methoxyphenylhydrazine in dilute hydrochloric acid an amalgamated lead electrode was found to give the best yield [170]. Generally, the best yields of arylhydrazine are obtained from diazonium salts having electron-donating substituents in the aryl group and using low substrate concentrations [171].

Reduction of PhN_2^+ at Cu, Fe, or Ti electrodes in the presence of alkenes gave phenylated products; at a Pt electrode no addition products were formed [172].

IV. N-NITRAMINES, NITROSAMINES, AND AZIDES

A. N-Nitramines

Protonated N-nitramines are reduced in a single six-electron reaction to the unsymmetric hydrazine [173–177] together with some amine, Eq. (35) [177]:

$$RR'N{-}NO_2 \xrightarrow{6e\ +\ 6H^+} RR'NNH_2 + 2H_2O \tag{35}$$

In alkaline solution most N-nitro derivatives of primary amines form an unreducible anion $[R{-}N{-}NO_2]^-$, with nitraminopyridines as exceptions [177], whereas nitramines of secondary amines are reduced in two steps [174]. The first reduction consumes 2 F/mol and produces an N-nitrosamine. It is noteworthy that N-nitrosamines in alkaline solution are more difficult to reduce than the corresponding N-nitramines; it seems to be the only class of nitro compounds in which the nitro group is more easily reducible than the nitroso group.

Reduction of nitroguanidine in a 2 M ammonium sulfate solution at a mercury cathode at −1.0 V yields nitrosoguanidine, which precipitates; addition of nickel ions increases the yield [178].

B. N-Nitrosamines

N-Nitrosamines are reducible in both acid and alkaline solutions. Polarographic data show [179–181] that the protonated compound is reduced in a four-electron reaction, whereas the unprotonated nitrosamine consumes only 2 F/mol. Controlled potential reduction has proved [179] the following reactions, Eqs. (36) and (37):

$$\text{Acid solution:}\quad RR'NNO \xrightarrow[4\bar{e} + 4H^+]{H^+} RR'N\overset{+}{N}H_3 + H_2O \qquad (36)$$

$$\text{Alkaline solution:}\quad RR'NNO \xrightarrow{2\bar{e} + 2H^+} RR'NH + \tfrac{1}{2}N_2O + \tfrac{1}{2}H_2O \qquad (37)$$

The electrochemical reduction of N-nitrosamines in acid solution is a convenient and general synthetic route to dialkylated, unsymmetrical hydrazines [182]. Quite a few nitrosamines were reduced to hydrazines in the classic period, but later investigations [183,184] also indicate a continued interest in the reaction.

The nitrosation of secondary amines is reversible in acid solution, so N-nitrosamines may be hydrolized at low pH. Such hydrolysis is especially important for diaryl nitrosamines but also significant for aryl alkyl nitrosamines. The NO^+ liberated in the hydrolysis may react with the reduction product from the N-nitrosamine, the unsymmetrical hydrazine, with formation of amine and N_2O, Eq. (38) [185,186].

$$RR'NNH_2 + NO^+ \longrightarrow RR'NH + N_2O + H^+ \qquad (38)$$

Anodic oxidation of N-nitrosamines [187] in acetonitrile in the presence of oxygen results in N-nitramines and some β-ketonitrosamines.

C. N-Nitrosamides

N-Nitrosamides behave in acid solution like N-nitrosamines; they are reduced to acylated hydrazines. In this way alkylhydrazines may be prepared from primary amines by the following reactions, Eq. (39) [188]:

$$RNH_2 \xrightarrow[-HCl]{CH_3COCl} RNHCOCH_3 \xrightarrow[-H_2O]{HNO_2} RN(NO)COCH_3 \xrightarrow{4\bar{e} + 4H^+}$$

$$RN(NH_2)COCH_3 \xrightarrow{H_2O} RNHNH_2 \qquad (39)$$

Similarly, nitro and nitroso derivatives of urea and guanidine have been reduced to semicarbazides and aminoguanidines in acid solution [176,189–191]. In aprotic media (DMF) the primarily formed anion radical of N-nitrosourea gives rise to a self-protonation step; the deprotonated substrate cleaves easily into several products [192].

D. Azides

Azides may in protic solvents be reduced in two different ways, n = 2 in both cases. The usual reaction is a cleavage to amine and dinitrogen [193, 194]. Phenyl and benzyl azide are thus reduced according to Eq. (40):

$$R-\bar{N}-\overset{+}{N}\equiv N + 2e + 2H^+ \longrightarrow R-NH_2 + N_2 \tag{40}$$

Azide groups at activated methylene groups may be reductively lost with formation of azide ion and a methyl group [193]. Phenacylazide and 4-azidomethylpyridine are thus reduced to acetophenone and 4-methylpyridine, respectively, Eq. (41):

$$R-CH_2N_3 + 2e + H^+ \longrightarrow R-CH_3 + N_3^- \tag{41}$$

In aprotic media α-azidostyrenes have been reduced in the presence of acetic anhydride to mono- and diacetylated enamines and further reduced compounds [195]: α-Azidocinnamic esters may under similar conditions be reduced to N,N-diacetylated α-aminocinnamic ester or, by careful addition of a proton donor, to α-aminocinnamic ester, Eq. (42) [196]:

$$C_6H_5CH{=}C(N_3)-COOCH_3 \xrightarrow[-N_2]{2e + 2[H^+]} C_6H_5CH{=}C(NH_2)COOCH_3 \tag{42}$$

REFERENCES

1. F. Fichter, Organische Elektrochemie, Steinkopff, Dresden, 1942.
2. F. Haber, Z. Elektrochem., 4, 506 (1898); Z. Phys. Chem., 32, 193 (1900).
3. A. J. Bard and H. Lund, eds., Encyclopedia of Electrochemistry of the Elements, Vol. 13, Dekker, New York, 1979.
4. P. Pierron, Bull. Soc. Chim. Fr., 780 (1899).
5. M. W. Leeds and G. B. L. Smith, J. Electrochem. Soc., 98, 129 (1951).
6. P. E. Iversen and H. Lund, Acta Chem. Scand., 19, 2303 (1965).
7. M. Masui, H. Sayo, and K. Kishi, Tetrahedron, 21, 2831 (1965).
8. P. E. Iversen and H. Lund, Tetrahedron Lett., 4027 (1967).
9. G. W. McMillan, U.S. Pat. 2,485,982 (1949); Chem. Abstr., 44, 1836 (1950).
10. G. B. L. Smith and M. W. Leeds, U.S. Pat. 2,589,623 (1952); Chem. Abstr., 46, 4937 (1952).
11. H. Matschiner, H. Teichmann, H. Tanneberg, and A. Weight, Ger. (East) Pat. DD 211.116 (1984); Chem. Abstr., 102, 132265 (1985).
12. (a) V. N. Leibzon, A. S. Mendkovich, S. G. Mairanovskii, T. A. Klimova, M. M. Krayushkin, S. S. Novikov, and V. V. Sevost'yanova, Dokl. Akad. Nauk SSSR, 229, 98 (1976); Chem. Abstr., 85, 183958 (1976); (b) Elektrokhimiya, 12, 1481 (1976).
13. M. Masui and C. Yijima, J. Chem. Soc., 1101 (1963).
14. M. Masui, H. Sayo, and K. Kishi, Chem. Pharm. Bull., 12, 1397 (1964).
15. B. Loev and F. Dowalo, Tetrahedron Lett., 781 (1969).
16. P. E. Iversen, unpublished observation (1968).
17. V. N. Leibzon, V. M. Belikov, and S. G. Mairanovskii, Elektrokhimiya, 4, 290 (1968).
18. G. Inzelt, M. Juul, and H. Lund, to be published.
19. S. Wawzonek and T.-Y. Su, J. Electrochem. Soc., 120, 745 (1973).

20. H. J. Schäfer, Chem.-Ing.-Tech., 41, 179 (1969).
21. M. T. Ismail, M. F. El-Zohry, and A. A. Abdel-Wahab, J. Appl. Electrochem., 15, 469 (1985); J. Electrochem. Soc. India, 34, 97 (1985).
22. A. K. Hoffmann, W. G. Hodgson, D. L. Maricle, and W. H. Jura, J. Amer. Chem. Soc., 86, 631, 639 (1964).
23. H. Sayo, Y. Tsukitani, and M. Masui, Tetrahedron, 24, 1717 (1968).
24. H. Sayo and M. Masui, Tetrahedron, 24, 5075 (1968).
25. D. A. Corrigan and D. H. Evans, J. Electroanal. Chem., 106, 287 (1980); 233, 161 (1987).
26. (a) M. Masui and H. Sayo, Pharm. Bull., 4, 332 (1956); (b) M. Masui, H. Sayo, and Y. Nomura, Pharm. Bull., 4, 337 (1956).
27. T. Shono, H. Hamaguchi, H. Mikami, H. Nogusa, and S. Kashimura, J. Org. Chem., 48, 2103 (1983).
28. S. Torii, H. Tanaka, and T. Katoh, Chem. Lett., 607 (1983).
29. (a) E. V. Zaporozhets, I. A. Avrutskaya, M. Ya. Fioshin, K. K. Babievskii, and V. M. Belikov, Elektrokhimiya, 8, 1243, 1809 (1972); 9, 72, 270, 1127 (1973); (b) I. A. Avrutskaya, K. K. Babievskii, V. M. Belikov, E. V. Zaporozhets, I. M. Filatova, and M. Ya. Fioshin, Elektrokhimiya, 9, 1652 (1973); (c) I. M. Filatova, I. A. Avrutskaya, V. T. Novikov, Ts. B. Ran'ko, K. K. Babievskii, V. M. Belikov, and M. Ya. Fioshin, Izv. Akad. Nauk SSSR, Ser. Khim., 1360 (1975); Chem. Abstr., 83, 123200 (1975); (d) V. T. Novikov, I. A. Avrutskaya, and M. Ya. Fioshin, Elektrokhimiya, 12, 1487 (1976).
30. A. Sera, H. Tani, I. Nishiguchi, and T. Hirashima, Synthesis, 631 (1987).
31. O. Convert and J. Armand, C. R. Acad. Sci., Ser. C, 265, 1486 (1967).
32. J. Armand, Bull. Soc. Chim. Fr., 882 (1966).
33. J. Armand, P. Soychay, and F. Valentini, Bull. Soc. Chim. Fr., 4585 (1968).
34. J. Armand and O. Convert, Coll. Czech. Chem. Commun., 36, 351 (1971).
35. M. Masui and H. Sayo, J. Chem. Soc., 4773, 5325 (1961).
36. M. Masui and H. Sayo, J. Chem. Soc., 1733 (1962).
37. S. Deswarte and J. Armand, C. R. Acad. Sci., Ser. C, 258, 3865 (1964).
38. J. Armand, Bull. Soc. Chim. Fr., 543 (1966).
39. J. Armand and P. Souchay, Chim. Anal., 44, 239 (1962).
40. J. Armand, P. Bassinet, and P. Souchay, C. R. Acad. Sci., Ser. C, 270, 555 (1970).
41. S. G. Mairanovskii and V. A. Petrosyan, Elektrokhimiya, 2, 115 (1966).
42. V. A. Petrosyan, V. I. Slovetskii, S. G. Mairanovskii, and A. A. Fainzil'berg, Elektrokhimiya, 6, 1595 (1970).
43. N. Limosin and E. Laviron, Bull. Soc. Chim. Fr., 4189 (1969).
44. J. Armand, J. Pinson, and J. Simonet, Anal. Lett., 4, 219 (1971).
45. F. Fichter, Organische Elektrochemie, Steinkopff, Dresden, 1942, p. 192.
46. F. Fichter, Organische Elektrochemie, Steinkopff, Dresden, 1942, p. 151.
47. H. Brintzinger, H. W. Ziegler, and E. Schneider, Z. Elektrochem., 53, 109 (1949).

48. U.S. Pat. 2,918,418 (1959); Chem. Abstr., 54, 11774 (1960).
49. V. A. Petrosyan, S. G. Mairanovskii, L. V. Okhlobystina, V. I. Slovetskii, and A. A. Fainzil'berg, Izv. Nauk SSSR, Ser. Khim., 1690 (1971); Chem. Abstr., 75, 157610 (1971).
50. V. V. Paranov, V. A. Petrosyan, and V. I. Slovetskii, Izv. Akad. Nauk SSSR, Ser. Khim., 567 (1977); Chem. Abstr., 86, 179491 (1977).
51. V. A. Petrosyan, V. I. Slovetskii, and A. A. Fainzil'berg, Izv. Akad. Nauk SSSR. Ser. Khim., 2027 (1973); Chem. Abstr., 80, 22111 (1974); Izv. Akad. Nauk SSSR, Ser. Khim., 286 (1980); Chem. Abstr., 92, 214609 (1980).
52. J. Armand, B. Furth, J. Kossanyi, and J.-P. Morizur, Bull Soc. Chim. Fr., 2499 (1968).
53. S. Kwee and H. Lund, Acta Chem. Scand., 23, 2711 (1969).
54. H. Lund, Acta Chem. Scand., 13, 249 (1959).
55. I. Ya. Lubyanitskii, P. M. Zaitsev, and Z. V. Zaitseva, Elektrokhimiya, 1, 990 (1965).
56. S. Deswarte, Bull. Soc. Chim. Fr., 522 (1969).
57. A. Petsom and H. Lund, Acta Chem. Scand., B34, 614 (1980); unpublished observation, 1980.
58. V. T. Novikov, L. A. Ratnikova, I. A. Avrutskaya, M. Ya. Fioshin, V. M. Belikov, and K. K. Babievskii, Elektrokhimiya, 12, 1066 (1976).
59. V. T. Novikov, I. A. Avrutskaya, M. Ya. Fioshin, V. M. Belikov, and K. K. Babievskii, Elektrokhimiya, 12, 1061 (1976).
60. J. Armand and O. Convert, C. R. Acad. Sci., Ser. C, 268, 842 (1969).
61. S. Deswarte and J. Armand, C. R. Acad. Sci., Ser. C, 258, 2865 (1964).
62. S. Deswarte, Bull. Soc. Chim. Fr., 534 (1969).
63. S. Deswarte, Bull. Soc. Chim. Fr., 545 (1969).
64. R. Schindler, W. Lüttke, and L. Holleck, Chem. Ber., 90, 157 (1957).
65. P. Souchay and S. Ser, J. Chim. Phys., 49, C172 (1952).
66. M. Heyrovský and S. Vavřička, Coll. Czech. Chem. Commun., 34, 1204 (1969).
67. I. A. Avrutskaya and M. Ya. Fioshin, Elektrokhimiya, 1, 1345 (1965).
68. G. Feroci and H. Lund, Acta Chem. Scand., B30, 651 (1976).
69. P. E. Iversen and H. Lund, Anal. Chem., 41, 1322 (1969).
70. A. P. Tomilov, Yu. D. Smirnov, and A. F. Videiko, Elektrokhimiya, 2, 603 (1966).
71. P. E. Iversen, Chem. Ber., 104, 2195 (1971).
72. G. Belot and C. Degrand, Tetrahedron Lett., 153 (1976).
73. R. D. Little and G. L. Caroll, J. Org. Chem., 44, 4720 (1979); M. M. Baizer and R. D. Little, unpublished observation, 1988.
74. A. U. Blackham, S. Kwak, and J. L. Palmer, J. Electrochem. Soc., 122, 1081 (1975).
75. (a) R. Bauer and H. Wendt, Angew. Chem., 90, 214 (1978); (b) R. Bauer and H. Wendt, Angew. Chem., 90, 390 (1978); (c) Z. Ali, R. Bauer, W. Schoen, and H. Wendt, J. Appl. Electrochem., 10, 97 (1980).
76. M. Culcasi, G. Gronchi, J. Escudie, C. Couret, L. Pujol, and P. Tordo, J. Amer. Chem. Soc., 108, 3130 (1986).
77. A. J. Bard, A. H. Cowley, J. E. Kilduff, J. K. Leland, N. C. Norman, M. Pakulski, and G. A. Heath, J. Chem. Soc., Dalton Trans., 249 (1987).

78. (a) F. Kaufman, H. J. Cook, and S. M. Davis, J. Amer. Chem. Soc., 74, 4997 (1952); (b) M. S. Rusakowa, Ya. I. Tur'yan, and R. F. Ustavshchikov, Electrokhimiya, 1, 854 (1965).
79. H. Lund, unpublished observation, 1987.
80. A. I. Prokof'ev, V. M. Chibrikin, O. A. Yuzhakova, and R. G. Kostyanovskii, Izv. Akad. Nauk SSSR, Ser. Khim., 1105 (1966).
81. A. Berndt, Angew. Chem., 79, 240 (1967).
82. T. Kitagawa, T. Layloff, and R. N. Adams, Anal. Chem., 36, 925 (1964).
83. P. H. Rieger and G. K. Fraenkel, J. Chem. Phys., 39, 609 (1963).
84. M. E. Peover and J. S. Powell, J. Electroanal. Chem., 20, 427 (1969).
85. M. Heyrovský and S. Vavřička, J. Electroanal. Chem., 28, 409 (1970).
86. M. Heyrovský, S. Vavricka, and L. Holleck, Coll. Czech. Chem. Commun., 36, 971 (1971).
87. G. L. McIntire, D. M. Chiappardi, R. L. Casselberry, and H. N. Blount, J. Phys. Chem., 86, 2632 (1982).
88. I. Rubinstein, J. Electroanal. Chem., 183, 379 (1985).
89. A. Darchen and C. Moinet, J. Electroanal. Chem., 61, 373 (1975); 68, 173 (1976).
90. A. Darchen and C. Moinet, J. Chem. Soc., Chem. Commun., 487 (1976); J. Electroanal. Chem., 78, 81 (1977).
91. A. Darchen and C. Moinet, J. Chem. Soc., Chem. COmmun., 820 (1976).
92. B. S. Jensen and V. D. Parker, J. Chem. Soc., Chem. Commun., 367 (1974).
93. W. H. Smith and A. J. Bard, J. Amer. Chem. Soc., 97, 5203 (1975).
94. A. Mendkovic, L. V. Michalchenko, and V. P. Gultyai, J. Electroanal. Chem., 224, 273 (1987).
95. F. Haber and C. Schmidt, Z. Phys. Chem., 32, 272 (1900).
96. K. Brand, Ber. Dtsch. Chem. Ges., 38, 3077 (1905).
97. W. Löb, Ber. Dtsch. Chem. Ges., 29, 1896 (1896).
98. K. Elbs and F. Silbermann, Z. Elektrochem., 7, 590 (1902).
99. L. Gattermann, Ber. Dtsch. Chem. Ges., 26, 1844 (1893).
100. F. Fichter, Organische Elektrochemie, Steinkopff, Dresden, 1942, p. 156; Ger. Pat. 1,066,589 (1959); Chem. Abstr., 55, 15418 (1961); Fr. Pat. 1,416,966 (1965); Chem. Abstr., 64, 6561 (1966); Rom. RO 72382 (1980); Chem. Abstr., 96, 151408 (1982).
101. B. B. Dey, T. R. Govindachari, and S. C. Rajargopalan, J. Sci. Ind. Res. (India), 4, 559, 569, 574 (1946); Chem. Abstr., 40, 4965 (1946).
102. (a) C. L. Wilson and H. V. K. Udupa, J. Electrochem. Soc., 99, 289 (1952); (b) G. S. Krishnamurthy, H. V. K. Udupa, and B. B. Dey, J. Sci. Ind. Res. (India), 15B, 47 (1956); Chem. Abstr., 50, 15289 (1956); (c) K. Jayaraman, K. S. Udupa, and H. V. K. Udupa, Trans. SAEST, 12, 143 (1978).
103. (a) J. Marquez and D. Pletcher, J. Appl. Electrochem., 10, 567 (1980); (b) F. Goodridge and M. A. Hamilton, Electrochim. Acta, 25, 487 (1980).
104. C. Häussermann, Chem.-Ztg., Chem. App., 17, 129 (1893).
105. J. H. Wagenknecht, Am. Electrochem. Soc. Meeting, Cincinnati, May 1984, extended abstract, p. 458.
106. M. Le Guyader, Bull. Soc. Chim. Fr., 1848 (1966).

107. M. Le Guyader, Bull. Soc. Chim. Fr., 1858 (1966).
108. A. Tallec, Ann. Chim., 3, 347 (1968).
109. A. A. Clement, Amer. Chem. J., 16, 511 (1894).
110. M. Le Guyader and A. Darchen, C. R. Acad. Sci., Ser. C, 267, 1352 (1968).
111. A. Darchen, Dissertation, Rennes, 1969, p. 22.
112. L. Gattermann and K. Koppert, Ber. Dtsch. Chem. Ges., 26, 2810 (1893).
113. A. Petsom and H. Lund, Acta Chem. Scand., B34, 693 (1980).
114. H. Sagae, M. Fujihira, T. Osa, and H. Lund, Chem. Lett., 793 (1977).
115. S. Torii, H. Tanaka, and Y. Murakami, Jpn. Kokai Tokkyo Koho JP 60 200,993 (1985); Chem. Abstr., 104, 158153 (1986).
116. (a) W. Löb, Z. Electrochem., 7, 335 (1900); (b) K. Elbs and C. Wipplinger, Z. Electrochem., 7, 133, 141 (1900).
117. C. Kerns, Trans. Electrochem. Soc., 62, 183 (1932).
118. L. J. Janssen and E. Barendrecht, Electrochim. Acta, 26, 699, 1831 (1981).
119. F. Fichter, Organische Elektrochemic, Steinkopff, Dresden, 1942, p. 165.
120. K. S. Udupa, G. S. Subramanian, and H. V. K. Udupa, J. Electrochem. Soc., 108, 373 (1961); Indian Pat. 68,869 (1961); Chem. Abstr., 56, 12670 (1962).
121. T. D. Balakrishnan, K. S. Udupa, G. S. Subramanian, and H. V. K. Udupa, Chem.-Ing.-Tech., 41, 776 (1969).
122. L. Gattermann, Ber. Dtsch. Chem. Ges., 29, 3040 (1896).
123. M. Breant and J. C. Merlin, Bull. Soc. Chim. Fr., 53 (1964).
124. L. H. Klemm, P. E. Iversen, and H. Lund, Acta Chem. Scand., B28, 593 (1974); H. Lund, unpublished observation (1976).
125. J. H. Wagenknecht, J. Org. Chem., 42, 1836 (1977); J. H. Wagenknecht and G. V. Johnson, J. Electrochem. Soc., 134, 2754 (1987).
126. (a) P. N. Anantharaman and H. V. K. Udupa, J. Electrochem. Soc., (India), 27, 217 (1978); (b) M. Noel, P. N. Anantharaman, and H. V. K. Udupa, Electrochim. Acta, 25, 1083 (1980); M. Noel, P. N. Anantharaman, and H. V. K. Udupa, J. Appl. Electrochem., 12, 291 (1982).
127. H. V. K. Udupa, Stud. Org. Chem., 30, 423 (1987).
128. (a) D. Stocesova, Coll. Czech. Chem. Commun., 14, 615 (1949); (b) B. B. Dey, B. R. Pai, and R. K. Maller, J. Sci. Ind. Res. (India), 10B, 175 (1951).
129. M. Person, T. Francais-Habert, and D. Beau, C. R. Acad. Sci., Ser. C, 279, 379 (1974).
130. (a) H. V. K. Udupa and M. V. Rao, Electrochim. Acta, 12, 353 (1967); (b) P. N. Anantharaman and H. V. K. Udupa, Trans. SAEST, 15, 41 (1980).
131. A. Darchen and D. Peltier, Bull. Soc. Chim. Fr., 401 (1972).
132. S. L. Gupta and N. Kishore, Electrochim. Acta, 15, 1367 (1970).
133. J. G. Lawless, D. E. Bartak, and M. D. Hawley, J. Amer. Chem., Soc., 91, 7121 (1969); D. E. Bartak and M. D. Hawley, J. Amer. Chem. Soc., 94, 640 (1972).
134. P. Peterson, A. K. Carpenter, and R. F. Nelson, J. Electroanal. Chem., 27, 1 (1970).

135. F. Lapicque and A. Storck, J. Appl. Electrochem., 16, 825 (1986); Electrochim. Acta, 32, 709 (1987).
136. R. Hazard and A. Tallec, Bull. Soc. Chim. Fr., 679 (1975); 433 (1976); C. Lamoureux, C. Moinet, and A. Tallec, J. Appl. Electrochem., 16, 819 (1986).
137. L. Holleck and H. Schmidt, Z. Elektrochem., 59, 56 (1955).
138. A. Tallec, Ann. Chim., 3, 164 (1968).
139. A. Tallec, Ann. Chim., 4, 67 (1969).
140. J. S. Dunning and D. N. Bennion, J. Electrochem. Soc., 117, 485 (1970).
141. L. Holleck and H. Schmidt, Z. Electrochem., 59, 1039 (1955).
142. A. Tallec, Ann. Chim., 3, 155 (1968).
143. G. M. LoPresti, S.-R. Huang, and A. A. Reidlinger, J. Electrochem. Soc., 115, 1135 (1968).
144. M. Jubault and E. Raoult, C. R. Acad. Sci., Ser. C, 268, 2046 (1969).
145. J. Chaussard, R. Rouget, and M. Tassin, J. Appl. Electrochem., 16, 803 (1986).
146. L. Holleck and R. Schindler, Z. Electrochem., 60, 1138 (1956).
147. C. Lamoureux and C. Moinet, Bull. Chem. Soc. Fr., 59 (1988).
148. W. N. Greig and J. W. Rogers, J. Amer. Chem. Soc., 91, 5495 (1969).
149. W. Kemula and R. Dioda, Bull. Acad. Pol. Sci., 10, 507 (1962).
150. M. Lipsztajn, T. M. Krygowski, and Z. Galus, J. Electroanal. Chem., 81, 347 (1977).
151. M. Heyrovský, S. Vavřička, L. Holleck, and B. Kastening, J. Electroanal. Chem., 26, 399 (1970).
152. F. Fichter, Organische Elektrochemie, Steinkopff, Dresden, 1942, p. 174 ff.
153. S. Wawzonek and T. W. McIntyre, J. Electrochem. Soc., 119, 1350 (1972).
154. H. A. Laitinen and T. J. Kneip, J. Amer. Chem. Soc., 78, 736 (1956).
155. R. Hazard and A. Tallec, Bull. Soc. Chim. Fr., 2917 (1971).
156. L. Holleck, D. Jannakoudakis, and A. Wildenau, Electrochim. Acta, 12, 1523 (1967).
157. T. M. Flornece, D. A. Johnson, and G. E. Batley, J. Electroanal. Chem., 50, 113 (1984).
158. T. M. Florence, J. Electroanal. Chem., 52, 115 (1974).
159. H. Lund, PCT Int. Appl. WO 86 03194 (1986); Chem. Abstr., 105, 190654 (1986).
160. B. Nygard, Ark. Kemi, 26, 167 (1966).
161. J. L. Sadler and A. J. Bard, J. Amer. Chem. Soc., 90, 1979 (1968).
162. H. Lund and S. Kwee, Acta Chem. Scand., 22, 2879 (1968).
163. R. Hazard and A. Tallec, Bull. Soc. Chim. Fr., 2917 (1971).
164. M. Studničkova, V. N. Flerov, O. Fischer, and M. Potaček, J. Electroanal. Chem., 187, 297, 307 (1985).
165. L. Holleck, S. Vavřička, and M. Heyrovský, Electrochim. Acta, 15, 645 (1970).
166. E. R. Atkinson, C. E. Garland, and A. F. Butler, J. Amer. Chem. Soc., 75, 983 (1953).
167. R. M. Elofson, Can. J. Chem., 36, 1209 (1958).
168. J. K. Kochi, J. Amer. Chem. Soc., 77, 3208 (1955).

169. P. Rüetschi and G. Trümpler, Helv. Chim. Acta, 36, 1649 (1953).
170. T. A. Volodina, I. A. Avrutskaya, O. G. Ginzburg, E. F. Bystrikova, M. Ya. Fioshin, and N. N. Suvorov, USSR Pat. 473,710; Chem. Abstr., 83, 78852 (1975).
171. O. Orange, C. Hamet-Elfakir, and C. Caullet, J. Electrochem. Soc., 128, 1889 (1981).
172. N. I. Ganushchak, N. D. Obushak, E. P. Kovel'chuk, G. V. Trifonova, Zh. Obshch. Khim., 54, 2334 (1984).
173. H. J. Backer, Rec. Trav. Chim. Pays-Bas, 31, 142 (1912).
174. E. Laviron and P. Fournari, Bull. Soc. Chim. Fr., 518 (1966).
175. E. Laviron, P. Fournari, and M. Greusard, Bull. Soc. Chim. Fr., 1255 (1967).
176. E. Laviron, P. Fournari, and G. Refalo, Bull. Soc. Chim. Fr., 1024 (1969).
177. H. Lund and S. K. Sharma, Acta Chem. Scand., 26, 2329 (1972).
178. H. F. Metcalf and G. C. Whitnack, U.S. Pat. 2,835,631 (1958); Chem. Abstr., 52, 14394 (1958).
179. H. Lund, Acta Chem. Scand., 11, 990 (1957).
180. L. Holleck and R. Schindler, Z. Elektrochem., 62, 942 (1958).
181. F. Pulidori, G. Borghesani, C. Bighi, and R. Pedriali, J. Electroanal. Chem., 27, 385 (1970).
182. P. E. Iversen, Acta Chem. Scand., 25, 2337 (1971).
183. U.S. Pat. 2,916,426 (1959); Chem. Abstr., 54, 6370 (1960); Ger. Pat. 1,085,535 (1960); Chem. Abstr., 56, 8562 (1962); Fr. Pat. 1,186,962 (1959); Chem. Abstr., 56, 320 (1962).
184. P. E. Iversen, Chem. Ber., 105, 358 (1972); R. Pachori, S. C. Mishra, and R. A. Mishra, J. Electrochem. Soc. India, 34, 99 (1985); P. A. K. Nedungadi, A. Gupta, S. K. Mukherji, and K. Zutshi, J. Electrochem. Soc. India, 35, 203 (1986).
185. G. Jacob, C. Moinet, and A. Tallec, Electrochim. Acta, 27, 1417 (1982).
186. G. Jacob, C. Moinet, and A. Tallec, Electrochim. Acta, 28, 635 (1983).
187. M. Masui, K. Nose, A. Tanaka, E. Yamakawa, and H. Ohmori, Chem. Pharm. Bull., 29, 3758 (1981).
188. U.S. Pat. 3,267,011; Chem. Abstr., 66, 2188 (1967).
189. M. Yamashita and K. Sugino, J. Electrochem. Soc., 104, 100 (1957).
190. M. S. V. Pathy, Electrochem. Technol., 3, 94 (1965).
191. V. C. Spreter and E. Briner, Helv. Chim. Acta., 32, 215 (1949).
192. M.-T. Escot, A.-M. Martre, P. Pouillen, and P. Martinet, Bull. Chem. Soc. Fr., 548 (1986).
193. H. Lund, Oesterr. Chem. Ztg., 68, 43 (1967).
194. N. I. Malyugina, L. N. Vertyulina, N. K. Koryttsev, and A. V. Oleinik, Zh. Obschch. Khim., 44, 497 (1974).
195. D. Knittel, Monatsh. Chem., 115, 523 (1984).
196. D. Knittel, Monatsh. Chem., 115, 1335 (1984); 116, 1133 (1985); 117, 1185 (1986); D. Knittel and V. S. Rao, Monatsh. Chem., 119, 223 (1988); B. Geist and D. Knittel, Monatsh. Chem., 119, 571 (1988).

10
Carbonyl Compounds

MANUEL M. BAIZER† / University of California, Los Angeles, California, and University of California, Santa Barbara, California

I. Nonolefinic Aldehydes and Ketones 433
A. General remarks on mechanisms 433
B. Reduction products 436
II. Olefinic Carbonyl Compounds 451
A. α,β-Unsaturated carbonyl compounds 451
B. Allylic tautomers of α,β-unsaturated carbonyl compounds 456
C. Carbonyl compound with more remote unsaturation 456
References 456

I. NONOLEFINIC ALDEHYDES AND KETONES

A. General Remarks on Mechanisms

This section includes in addition to aliphatic aldehydes and ketones also aryl aldehydes, aryl alkyl, and diaryl ketones. The aryl groups, when present, act as electron-withdrawing (delocalizing) groups and facilitate the reduction (Ref. 1 has extensive tables) but rarely participate in the chemistry that follows the electron transfer step. [In Ref. 2, Shono et al. report an intramolecular reductive coupling of $Ph(CH_2)_3COCH_3$ to a hexahydronaphthalene.

The carbonyl group is an electrophore: all compounds containing it in free form are electroreducible. However, if the carbonyl group is "masked," e.g., by hydration or hemiacetal or acetal formation, it becomes nonreducible; reduction (kinetically controlled) occurs of only the quantity of free carbonyl compound in equilibrium with the "protected" form. This is strikingly illustrated in the reduction of glucose [3]. On the other hand, carbonyl derivatives (Chap. 11) in which the >C=O bond is replaced by a >C=N− bond are more easily reducible than the parent carbonyl compounds. Thioketals can be anodically cleaved to regenerate the parent ketone (see Chap. 24).

Depending on the solvent-electrolyte system and the other constituents of the medium used (protic or aprotic, presence or absence of potential proton donors,* and hardness of the cation of the supporting electrolyte) and on the rate of follow-up reactions of the intermediates, the reduction of carbonyl compounds may involve (1) two distinct one-electron stages, (2) one, two-electron stage, or (3) one, one-electron transfer. This triad of possibilities may be most easily understood if we regard (2) and (3) as "perturbations" of (1), the "idealized" case:

---

†Deceased

*Sometimes even a $-CH_2-$ group adjacent to the carbonyl may act as a proton source (Chap. 30).

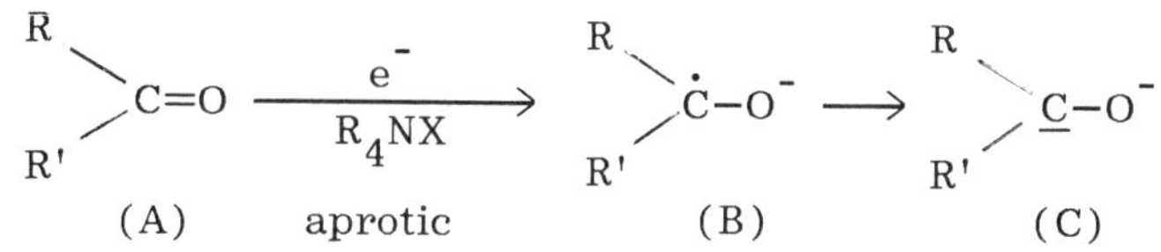

R = hydrocarbyl, R' = hydrocarbyl or H

Protonation of (B) or its generation in the presence of a "hard" cation like Li^+ leads to a transitory free radical (D), R(R')C–OM (M = H or Li), which is usually reduced more easily than (A). The result is category 2 unless the rate of dimerization of (D) is so rapid that a situation of category 3 arises.

In category (1) protonation, if it occurs with (B) and (C), follows the electron transfer step(s). There may, however, be preprotonation of the carbonyl compound, in which case a cationic species is formed that upon reduction yields a radical:

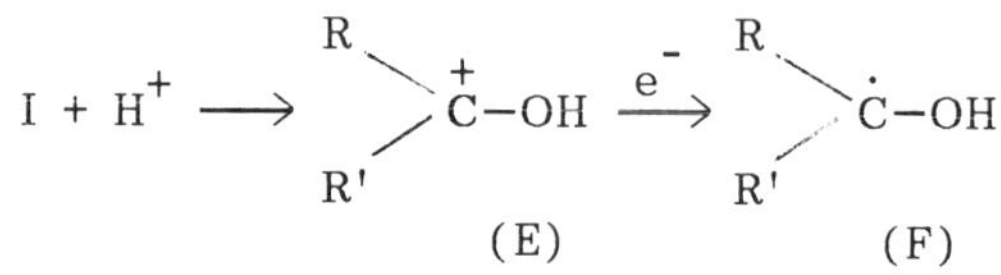

(E) is usually more easily reduced than (A). For example, the first reduction of acetone in aprotic media occurs in 80% EtOH (capable of postprotonation) approximately −2.5 V, in 1 N H_2SO_4 (capable of preprotonation) at −1.2 V (all V versus SCE).

These varieties of reduction patterns are summarized in Scheme 1. The dashed lines refer to reactions found only in special cases. Some reactions, such as $+e^- + H^+$, are represented only as an overall step without regard to the actual sequence, which is often not known. Reactions may occur at the electrode surface or in the bulk of the solution.

In some cases one should take into account keto-enol equilibria in the reduction of carbonyl compounds; one or both forms may be electroreducible [4].

In general the first electrochemical step is a reversible transfer of an electron from the electrode to the carbonyl compound (before or after protonation). A correlation of the reduction potentials with the energy of the n,π*-triplet transitions and a correlation of the electron transfer rate with the value of the electron spin density at a reactive center (C=O) group have been established [5,6]. The radicals formed from aromatic compounds are relatively stable, and many of their properties can be investigated by several methods, including electrochemical ones [7–14]. The radicals formed from aliphatic compounds are usually impossible to detect [15–19]. The following data: half-life of the acetone anion radical in aqueous 2-propanol is 72 μs, whereas that of the acetophenone anion radicals is 1.5 ms [20], can illustrate the difference between the aliphatic and aromatic compounds.

At present the data concerning intermediates and their properties are very fragmentary. Thus the rate constant of the electron transfer k is found to be equal to 0.4 ± 0.1 cm/s for benzaldehyde on mercury in 75% ethanol in the presence of 0.1 N Bu_4NOH and the formal, reversible potential of the redox system benzaldehyde/anion radical is equal to −1.592 ± 0.008 V (SCE) [21]. The rate constant of the dimerization of benzaldehyde anion radicals in sulfolane is found to be 2.4×10^3 M^{-1} s^{-1}; the correspond-

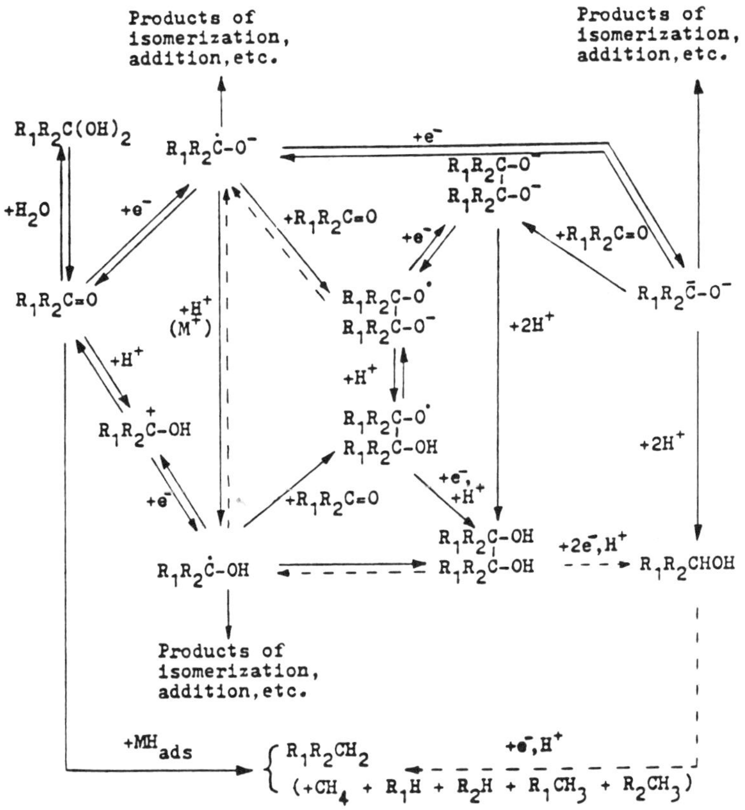

Scheme 1.

ing value for p-cyanobenzaldehyde is 45 liters/mol s [22]. The rate constant of the reaction $R^{\bar{\cdot}} + R \rightarrow {}^{-}R{-}R\cdot$ in the case of p-cyanobenzaldehyde in DMSO is found to be 1.45 mol^{-1} s^{-1} and the one for the reaction $2R^{\bar{\cdot}} \rightarrow {}^{-}R{-}R^{-}$ is 28.6 M^{-1} s^{-1} [23]. Similar data have been compiled in a review [17].

References 8 and 24–25 are important for the interpretation of various stages of the process of reduction of carbonyl compounds; they concern mainly aromatic compounds.

The formation of ion pairs and aggregates is of great importance in nonaqueous solutions [46,47]. Dianions may play an important role in the reduction of carbonyl compounds [44,48–50].

The existing information about the influence of adsorption of initial substrates, intermediates, and components of solutions on the reaction route is insufficient. Adsorption of carbonyl compounds, mainly aliphatic ketones, has been studied mainly on Hg [51–53], Bi [54], and Pt [55]. On mercury adsorption is greatest at small negative surface charge, it is increased with increasing length of the chain and on going from alkaline to acid media, and is decreased on going from aqueous to nonaqueous (alcoholic) solutions. On bismuth the free energy of adsorption is smaller than that on mercury, on the average approximately 1.8 kJ/mol. On platinum the greatest value of adsorption is observed at potential +0.2 V. In some cases it has been established that the adsorbed molecules undergo no reduction [36,40,55,56]; dimerization of radicals takes place on the electrode surface [36,57,58], and adsorbed substances induce optical activity of products [59–62]. The electrocatalytic nature of reduction processes has also been demonstrated by the dependence of the cis-trans ratio of cyclohexanol isomers on the nature of the electrode metal [63]. Adsorption of intermediates sometimes resembles chemisorption and leads to formation of organometallic compounds [34,65–67], although it is necessary to take into account a possibility of reaction of carbanions with the cations of the electrode metal [68]. Hydrocarbon formation by electroreduction of many carbonyl compounds, especially in acid media, probably goes through the stage of surface organometallic intermediates [49,65,67,69–71].

It should also be noted that whereas underivatized alcohol groups, as in $CH_3CH(OH)CH(OH)CH_3$, cannot be directly reduced electrochemically, OH on a carbon atom adjacent to a carbonyl group, such as $CH_3COCH(OH)CH_3$, can be reduced (Chap. 24). Carbonyl compounds bearing other electrophores—often more easily reducible than the carbonyl group itself—are treated elsewhere in this book in the chapters dealing with the reduction of the second electrophore [74].

B. Reduction Products

Multiplicity of Products

Reduction of nonderivatized carbonyl compounds can, as indicated in Scheme 1, depending on the structure of the substrate and the conditions used in the electrolysis (cathode material, supporting electrolyte, proticity and pH of medium, and concentration), result in the several types of products listed (for completeness) in Scheme 1. However, in this section we discuss in detail only reduction to alcohols and hydrocarbons. The formation from carbonyl compounds of organoelemental compounds (Chap. 21) and of glycols, pinacols, and other reductively coupled products as well as intramolecular couplings (cyclizations) (Chap. 22) is treated in the chapters indicated.

Guides to Selectivity

What does the accumulated experience of many years of work in this area teach us about the means of obtaining a particular type of product mentioned earlier largely or, ideally, exclusively? Here are some general principles. There are exceptions, particularly if other electrophoric groups are present in the same molecule. There are unfortunately contradictions in the literature. There are not fully elucidated phenomena. For example, both Pb and Hg are high overvoltage cathodes but reduction of acetone in acid media at the former yields the pinacol and isopropyl alcohol but at Hg no pinacol is formed, only the alcohol and some propane. With these caveats we may assert the following.

The formation of alcohols is favored by using high hydrogen overvoltage cathodes and alkaline aqueous media. (Caution: Too high alkalinity promotes aldol condensations). Aprotic media can also serve.

Reductive couplings of aldehydes to glycols and of ketones to pinacols are favored by high hydrogen overvoltage cathodes and acidic conditions, high concentration of substrate and increasing chain length in aliphatic aldehydes. Aromatic carbonyl compounds are more easily hydrodimerized than their aliphatic analogs. (Caution: Too low a pH may promote the pinacol-pinacolone rearrangement in situ.) In aprotic media the presence of Li^+ also favors the formation of hydrodimers. Many other reductive couplings are known in which the reduced carbonyl compound reacts with electrophiles. 1,3-Diketones have been converted to cyclic 1,2-diols. It must be borne in mind that H, if present on the intervening C atom, is acidic and that the parent compound may then be a proton donor toward its own derived reduced intermediate (Chap. 22).

Hydrocarbon formation is favored in acidic media at low current densities, at catalytic cathodes, and at Cd. The keto group in polyfunctional compounds, such as keto esters, can be reduced to methylene.

Reduction of carbonyl compounds at Pb, Hg, and a few other cathodes can lead to the formation of organometallics. The latter are sometimes intermediates toward the final products, hydrocarbons.

Alcohols

Reductions to alcohols of some representative aldehydes and ketones are given in Tables 1 and 2, respectively. It is evident from the wide variety of conditions that have been used why the preceding statements had to be so circumspect.

Substituted aldehydes (e.g., aldol or glyceraldehyde) form the corresponding alcohols cleanly if the substituents themselves are not reduced; if they are, they may be reduced earlier or simultaneously [130].

Again, it is noted (e.g., 4-hydroxy-3-methyl-2-pentanone [109] and 4-dimethylamino-2-butanone [131]) that nonreducible groups do not interfere with the reduction of the carbonyl. Whereas cyclic β-ketonitriles are smoothly converted to β-hydroxynitriles, compounds with alkyl substituents at the α-carbon atom undergo partial loss of cyanide [132]. Rupture of a C—N bond proceeds also in 2-amino-2-phenylacetophenones in acid but not in alkaline media [137]. In α-amino ketones rupture of the C—N bond proceeds only in the presence of conjugated aromatic keto groups, but not in alicyclic compounds (Chap. 24) [109].

Alkyl-aromatic and aromatic ketones readily form pinacols on electroreduction, but quite often they can be quantitatively reduced to alcohols instead. For the latter purpose cathodes with low hydrogen overvoltage should be employed under galvanostatic or potentiostatic conditions.

In the reduction of substituted ketones the substituent itself, depending on its reduction potential, may also be reduced so that the ultimate product obtained may be the result of intramolecular follow-up chemical reactions (see Chap. 9). Thus, o-nitrobenzophenone yields 3-phenyldihydroanthranil, a dehydration ring-closed product of the intermediate α-hydroxylaminobenzhydrol.

Reduction of compounds with adjacent carbonyl groups is complex. Thus, electroreduction at controlled potentials of methylglyoxal on mercury in buffers (pH 7) yields first hydroxyacetone, then at somewhat more negative potentials hydroxypropanol, and at still more negative potentials acetone and 1,2-propanediol [134]. Diacetyl can be reduced to acetoin, and in amine-containing buffers the latter can be reduced to 2-butanone and 2,3-butanediol [134,135a].

Under other conditions (Pd cathode, 1–10% HCl or H_2SO_4) it is reduced in about 60% yield to acetoin [137]. Polarography of biacetyl at glassy C in DMF shows two one-electron waves and dimerization by an EC mechanism at the first wave [138].

Table 1. Some Electroreductions of Saturated Aldehydes to Alcohols: $-CHO + 2e^- + 2H^+ \longrightarrow -CH_2OH$

Aldehyde	Cathode; E (V)/(SCE)	Medium; conditions	Yield (current efficiency) (%)	References
Formaldehyde	Hg	Na_2CO_3; 1 A/dm^2	(90)	72
Acetaldehyde	Pb		85.0 (91.8)	73
Acetaldehyde	Sn	Phosphate buffer, pH 9; 3–7°C	(81.5)	74
Propionaldehyde	Pb	Neutral or weakly alkaline; 20°C, 4 A/dm^2	(96.5)	74
Propionaldehyde	Cd, Zn, Hg, C, Cu, Ni, Ag, Al		(11.2, Al)	75
Butyraldehyde	Hg			76
Isobutyraldehyde	Hg			77
Valeraldehyde	Cd, Pb, Zn	Acid		78
2-Ethylhexanal	Cu	Pyridine, DMF, or ROH; H_2, 10–150 kg/cm^2, 0.25 A/dm^2	≈90	79
Glyceraldehyde	Hg	Na_2SO_4, pH 7–8; 10°C, 0.2 g/liter $FeSO_4$		80
Chloral hydrate[a]	Hg			81
α-Hydroxyisobutyraldehyde	Hg	pH 2.5	b	82
α-Hydroxyisobutyraldehyde	Hg	Phosphate buffer, pH 6	c	82
Streptomycin[d]	Hg			82
Streptomycin[e]	Zn	pH 6.8–7; 0.01–0.02 A/dm^2	≈100 (low)	83
Saccharose	Pb or amalgams	Neutral or alkaline or weakly acid	Good	76, 84

Glucose[f]	Pb, Hg			90
Glucose[g]	Graphite	Dil. H_2SO_4		85–91
Glucose[g]	Pb/Hg	Alkaline		85, 91
Glucose[h]	Raney Ni	Slightly acidic	≃100	92
Mannose[g]				90
Mannose[i]	Cd			90
Mannose[g]	Graphite	Dil. H_2SO_4		85, 91
Mannose[g]	Pb/Hg	Alkaline		85, 91
Benzaldehyde	Ni			93
Benzaldehyde	Hg; −0.2	pH 6.0–6.5; high conc, K^+, low conc. of aldehyde	≃100	94
2,4-Dimethylbenzaldehyde			70–75	95
3,5-Dimethylbenzaldehyde			70–75	95
p-Methoxybenzaldehyde			70–75	95
p-Isopropylbenzaldehyde			70–75	95
Salicylaldehyde	Hg	$NaHCO_3$ sol. + CO_2 under pressure or acetate buffer	93	96
Vanillin	Hg	$NaHCO_3$ sol. + CO_2 under pressure or acetate buffer	92	96, 135b
Piperonal	Hg	$NaHCO_3$ sol. + CO_2 under pressure or acetate buffer	92	96
α-Phenyl-α-formylacetonitrile[j]	Hg	EtOH, LiCl	50–90	97

Table 1. (Continued).

Aldehyde	Cathode E (V)/(SEC)	Medium; conditions	Yield (current efficiency) (%)	References
Furfural	Zn	pH 8.5	40–72	98
Furfural[k]	Pd/Pt	1 N H_2SO_4; 0.1 A/dm^2	19	97
Furfural[l]	Cd; −0.9 (NHE)	Phosphate buffer, pH 6	≈60 (≈50)	99

[a]Chlorine reductively removed first.
[b]Isobutanol formed at very negative potentials.
[c]Isobutyraldehyde formed at less negative potentials.
[d]Product dihydrostreptomycin and dihydrodesoxystreptomycin.
[e]Product dihydrostreptomycin.
[f]Product sorbitol and mannitol as major and minor products, resp.
[g]Forms mannitol (as does fructose).
[h]No mannitol was detected. This reduction was paired with the simultaneous indirect oxidation of glucose in an undivided flow-through cell [92].
[i]Forms sorbitol.
[j]Unsaturated nitrile is the final product, because corresponding alcohol is unstable.
[k]Tetrahydro- (61%) and dihydro- (6%) furyl alcohols.
[l]Pentadiols (13%) are also obtained.

Reduction at Hg of quaternized 2-pyridino-1,3-indandiones involves first the 1,3-indandionyl moiety [139].

Benzyl is reduced to stilbenediol; the latter changes slowly to benzoin [24,40,141]. The ratio of cis- to trans-stilbenediols is greatly dependent on potential, composition of the solution, pH, temperature, and time [142–144]. Enediol salts were isolated after the reduction of ninhydrin [146]. In 2-aroyl-6-aminocyclohexanones-1 the aromatic keto group is reduced first [145]. Acetylacetone [140] undergoes a Tafel rearrangement on reduction in acidic medium; dimerization also occurs.

Acetonylacetone [147], as does ethyl-α-butylacetoacetate [148], produces in acid media degradation products, such as acetaldehyde and acetone in the first case and 2-heptanone and 2- and 3-octanone in the second case together with normal products (2,5-hexanediol and 2,5-dimethyltetrahydrofuran and ethyl-α-butyl-β-hydroxybutyrate, respectively).

The stereochemistry of the reduction of carbonyl compounds to alcohols [149] is not considered here (see Chap. 27); it should be noted only that adsorption on an electrode of an optically active substance [126,127] or chemical bonding of the latter to the surface [150] is capable of providing conditions for asymmetric synthesis.

Polyhydroxycarbonyl compounds, biomass-derived (e.g., glucose or fructose), reduce well. In fact, the manufacture of a mixture of sorbitol and mannitol was at one time an outstanding commercial success (Chap. 31). More recent work in which the reduction of glucose was paired with the simultaneous oxidation of gluconate gave only 68% sorbitol with 28% current efficiency using the best (Zn/Hg) of a series of cathodes tried [92a]. However, the yields and current efficiencies were almost 100% at Raney Ni. An undivided flow cell system was used. There was no formation of the isomeric mannitol.

This brings up the question of "catalytic cathodes." To a semanticist, who is as welcome to read this book as anyone else, the term may seem to be a redundancy since in all but the few cases in which the cathode is purposely sacrificial (as in the formation of mercurials by the reduction of carbonyl compounds or organolead compounds by the reduction of organic halides at Pb, Chap. 8), all cathodes are catalytic in that they mediate the transfer of electrons from the power source to a component of the substrate/solvent-electrolyte system without being at the end themselves chemically changed. However, as commonly used "catalytic cathodes" denotes largely those that are composed of materials that in a proper state of subdivision are catalysts of nonelectrochemical hydrogenations [151].

While catalytic cathodes have been used occasionally in the past, they have come into greater prominence this past decade. Particularly germane to this chapter is the reduction of carbonyl compounds to alcohols [152] and to hydrocarbons (see Sec. Hydrocarbons). Chiba et al [152a] reduced a variety of ketones and of aromatic aldehydes to alcohols at Raney Ni cathodes. Aliphatic aldehydes could not be used "since they formed aldol adducts in this medium" (NaOMe + MeOH). However, glucose was reduced quantitatively to sorbitol at pH 5–7 using similar cathodes [92b]. Belot et al. [152b] have used both Raney Ni and Devarda Cu in catalytic hydrogenations; at the former cathode benzophenone was potentiostatically converted 97% to benzhydrol with 100% current efficiency. Ni^0 (generated in situ) on graphite or vitreous C has been used to reduce Ph_2CO and $PhCOCH_3$ to alcohols and hydrocarbons.

Table 2. Reductions of Nonolefinic Ketones to Alcohols: $RR'C{=}O + 2e^- + 2H^+ \longrightarrow RR'CHOH$

Ketone	Cathode; E (V) (SCE)	Medium; conditions	Yield (current efficiency) (%)	References
Acetone				See text
Acetone	Hg; −1.2	1 N H_2SO_4	(71)	100
Acetone	Ir; +0.1 (RHE)	1 N H_2SO_4	(80)	101
2-Butanone	Pb	$NaHCO_3$	High	102
2-Butanone[a]	Pb	Acid		103
2-Butanone	Zn	Alkaline		104, 105
2-Butanone	Cu	Pyridine, DMF, or AlkOH: H_2, 10–15 kg/cm^2, 0.25 A/dm^2	≈90	79
2-Pentanone	Pb[b]			106
3-Pentanone	Pb[b]			106
3-Pentanone	Zn	Alkaline	High	105
3-Hexanone	Pb[b]			106, 107
Pinacolone	Zn	Alkaline	Major	105
Pinacolone	Cd	Acid	Low	108
4-Hydroxy-3-methyl-2-pentanone			66	109
4-Dimethylamino-2-butanone			62.5	110
β-Aminoketones		Acid		111
Menthone				112, 113
Camphor[c]				114
Camphorcarboxylic acids[d]	Hg			115, 116

3-Ketocamphoronic acid[e]				116
Phenylglyoxalic acid	Hg	DMF, $EtNClO_4$		117
Cyclohexanone	Zn, Pb, Tl	Alkaline	Major	105
Cyclohexanone	Cu	Pyridine, DMF, or ROH; H_2, 10–150 kg/cm^2, 0.25 A/dm^2	Ca. 90	79
2-Methylcyclohexanone	Hg, Pb[f]			118
2-Methylcyclohexanone[g]	Cu			118
2-Methylcyclohexanone[h]	Ni			118
2-Methylcyclohexanone[i]	Hg	H_2O + MeOH, acid or alkaline	46–47	119
2,6-Dimethylcyclohexanone	Hg, Pb[j]			118
Adamantandione	Hg	DMF + 10% H_2O, Et_4NBr		119
2,6-Noradamantandione	Hg	DMF + 10% H_2O, Et_4NBr		119
Acetophenone	Raney Ni		73	120
Acetophenone	Cu	Pyridine, DMF, or ROH; H_2, 10–150 kg/cm^2, 0.25 A/dm^2	≈90	79
Acetophenone	Hg; −2.1 (MSE)	Acetate buffer + MeOH	≈60	60
2-Anilino-2-phenylacetophenone	Hg	50% MeOH, pH 4	Major	121
α-Trifluoroacetophenone	Hg; > −1.5	MeCN, Et_4NClO_4	Major	122
α-Methyldesoxybenzoin[k]	Hg	pH 8	92	123
Benzophenone	Hg	Acid		103
Benzophenone	Hg	KOAc; 0.02 A/dm^2, 80–85°C	97	103
Benzophenone	Cu	Pyridine, DMF, propanol, H_2, 10–150 kg/cm^2, 0.25 A/dm^2	≈90	79
p,p'-Dimethylaminobenzophenone		Acid		124

Table 2. (Continued)

Ketone	Cathode; E (V) (SCE)	Medium; conditions	Yield (current efficiency) (%)	References
o-Nitrobenzophenone[l]	Pb	Alkaline	86	125
2-Acetylpyridine	Hg; −0.8	Acetate buffer + EtOH	(73−80)	126
2-Acetylpyridine	Hg; −1.1	Acetate buffer + EtOH	48	127
3-Acetylpyridine	Hg; −1.35	Acetate buffer + EtOH	72	127
4-Acetylpyridine	Hg; −1.0	Acetate buffer + EtOH	40	127
3-Methyl-4-phenyl-4-pyrazol-inone		NaAcO + $NaHSO_4$	46	128
2-Ethyl-1-methyl-3-piperidone			10	111
2-Benzoylthiophene	Hg	MeCN + PhOH	83	129

[a]Pinacol formation also occurs.
[b]Electrolytically deposited.
[c]Producrs: borneol and isoborneol.
[d]Mixtures of isomers formed.
[e]Product: hydroxy acid (lactone).
[f]Trans isomer formed.
[g]Cis isomer formed.
[h]Mixed isomers, trans predominant.
[i]Isomer ratio depends on current density, among other factors.
[j]Single product formed.
[k]Erythro-1,2-diphenyl-1-propanol formed.
[l]Product is β-phenyldihydroanthranil.

There has been much work done on modification of catalytic metals for use in nonelectrochemical reactions by the incorporation of chiral compounds [153]. Some chemically modified electrodes (CMEs) have been used for asymmetric induction (Chaps. 27 and 33). An interesting relevant recent example is the reduction of methylacetoacetate using a Raney Ni cathode modified by (R,R) tartaric acid. The expected alcohol was obtained in 90% chemical yield and 28.5% optical purity [123].

Catalytic cathodes have been made by incorporating microparticles of catalytic metals in a poly(pyrrole-viologen) film on carbon felt. Among other materials 3-methoxybenzaldehyde and 4-isopropyl-2-cyclohexenone have been reduced at high current efficiencies when the metal is Pd [154]. See also Ref. 155.*

Aromatic aldehydes readily electroreduce to the corresponding alcohols [75,95,96,175,176,178], but they are more prone than aliphatic aldehydes to produce glycols, particularly on cathodes of high hydrogen overvoltage under galvanostatic conditions. In 5% sulfuric acid-alcohol on Pt, Ni, and Cu, the alcohols are the main reduction products; the yields decrease with increasing temperature. Benzaldehyde has quantitatively been converted to benzyl alcohol on Hg [94].

The carbonyl group of heterocyclic aldehydes is easily reduced to carbinol. However, resinification (e.g., furfural) in acid may be a serious side reaction. Heterocyclic aldehydes exist more in the nonreducible hydrated gem-diol form than do aliphatic and aromatic aldehydes and must therefore first be converted to the nonhydrated form. In strongly alkaline media heterocyclic aldehydes are reduced largely to glycols (Chap. 18).

Ketones are reduced to secondary alcohols (Table 2) under conditions analogous to those used for reduction of aldehydes.

The reduction of acetone has been studied particularly extensively [69, 100,103,106,114,160–165]. In acid media at Hg and Pt the current efficiencies for the formation of 2-propanol are 25–71%; pinacol, propane, and organometallics are almost always produced as by-products. In alkaline media at the same electrodes the latter two by-products are absent [46,160–166] but the yield of pinacol increases [167]. Satisfactory yields of 2-propanol on Zn and Cd in sodium acetate solution [103,168], on Zn in a mixture of sodium acetate and tetraborate [169], and on Hg in sodium chloride-sodium sulfate solutions [162] have been obtained.

Reduction of acetone to the alcohol also occurs on catalytic cathodes. On Pt in 6 N H_2SO_4, the reduction is accompanied by propane formation at low polarizations and on an electrode freshly activated by anodic polarization [170]. On Ni-Al alloys isopropyl alcohol may be obtained in 98% yield at 93% current efficiency using current densities up to 20 A/dm^2 [156].

Ordinary temperatures, high current densities, and, according to some data [157,158], organic solvents facilitate the reduction of acetone to isopropyl alcohol under galvanostatic conditions. Potentiostatic reduction of ketones may be necessary in some cases when the alcohol is reducible [159,171].

---

*Ni microparticles electrodeposited on a Nafion film covering graphite or a Ni plate are even more active than Raney Ni for the electrocatalytic reduction of 2-cyclohexen-1-one.

Table 3. Reduction of Carbonyl to Methylene: $>C{=}O + 4e^- + 4H^+ \longrightarrow -CH_2-$

Carbonyl	Cathode; E (V) (SCE)	Medium; conditions	Yield (current efficiency) (%)	References
Formaldehyde	Cu, Ag		Low	73
Acetaldehyde	Cu, Ag		Low	73
Acetaldehyde	Pt/Pt; −0.06 (NHE)	1 N $HClO_4$	High	173
Propanal	Cd			75
Propanal	Pt/Pt; −0.06 (NHE)	1 N $HClO_4$	High	173
Isobutyraldehyde	Pb		To 40	77
Valeraldehyde[a]	Cd, Pb, Zn		13	78
Enanthaldehyde[a]	Cd			75
Benzaldehyde				84, 175
2,4-Dimethylbenzaldehyde				175
3,5-Dimethylbenzaldehyde				175
o-, m-, and p-Tolualdehydes				175
Salicylaldehydes	Zn			96
p-Hydroxybenzaldehyde	Zn, Cd			96
Cumaldehyde				75
Anisaldehyde				175, 176
Protocatecualdehyde	Cd			75
Vanillin	Cd			75
Piperonal	Cd, Zn			75, 96

p-Dimethylaminobenzaldehyde				177
p-Chlorobenzaldehyde	Cu			175
Furfural	Zn		6	98
Acetone	Cd		100	65, 69, 75, 168
Acetone	Pt[b]			170
Acetone	Pt; +0.1 (NHE)	1 N H_2SO_4	(91)	101
Acetone	Zn, Zn/Hg, Al		High	65, 69
Acetone	Pb, Hg		25	65, 69, 168
Acetone	Sn, Ag		Very low	69
2-Butanone	Cd			179
2-Butanone	Zn			168
3-Pentanone	Cd		70	180
2-Pentanone	Cd-Bi			180
2-Pentanone	Cd	H_2SO_4	To 75	78
2-Pentanone	Zn	H_2SO_4	60	78
2-Pentanone	Pb	H_2SO_4	17	78
2-Pentanone	Hg	H_2SO_4	20	78
2-Pentanone	Zn, Pb, Hg	HOAc	0	181
Pinacolone	Cd		83.6	108
5-Methyl-2-hexanone	Cd	H_2SO_4	83.5	183
Acetopropanol	Cd	H_2SO_4	50	183
Diacetone alcohol	Pb	H_2SO_4		184

Table 3. (Continued).

Carbonyl	Cathode; E (V) (SCE)	Medium; conditions	Yield (current efficiency) (%)	References
Keto esters and derivatives				See text
Alicyclic ketones				See text
Acetophenone	Pt, Pd, Bi		High	185
Benzophenone			High	186–188
Benzyl-subst. phenylketones	Zn/Hg	H_2SO_4	50–90	96, 188
Benzoin				188
Deoxybenzoin				188
1-Acetylnaphthalene		Acid		189
α-Naphthyl phenyl ketone	Zn/Hg			188
2-Ethyl-1-methyl-3-piperidone[c]	Cu, Pb, Cd	H_2SO_4		183
2-Alkyl-2-hydroxy-1-methyl-5-pyrrolidinones				190
3-(N-Dimethylaminoethyl)-4-oxoquinolizidine			40	183

[a]Yield of hydrocarbon increases with temperature and current density.
[b]Propane is the only product if the electrode has been previously anodically polarized.
[c]Depending on the temperature, various ratios of 1-methyl-2-propylpyrrolidine, 1-methyl-2-ethyl-3-hydroxypiperidine, and N-methylhepthylamine are formed.

Coupled Products

All types of reductively coupled products obtained, among others, are discussed in Chap. 22.

Hydrocarbons

The carbonyl group of aldehydes and especially of ketones can often be reduced electrochemically to methylene in rather high yields [172–174]. Hydrocarbons are usually formed in acidic media—sulfuric acid. Representative examples are presented in Table 3.

The reduction of acetone to propane has been studied very extensively. In addition to the data cited it should be mentioned that when cathodes of different metals are used at the same potential the chemical yields of propane differ substantially, although the total current efficiency for the reduction products varies only by a factor of 2. For example, on mercury the chemical yield and current efficiency, respectively, for propane are 2.2 and 73%; on lead, 25.2 and 86%; on cadmium, 100 and 90%; on zinc, 97 and 86%; and on aluminum, 100 and 45%. The yield of propane increases with decreasing cathode potential. The propane yield on cadmium significantly increases with illumination [65].

Some aromatic ketones can be reduced in two stages: first to pinacols and then at more negative potentials to hydrocarbons (with rupture of C—O and central C—C bonds) [191]. The ability of some secondary alcohols (benzhydrol, 1-biphenylethanol, and others) to reduce in aprotic solutions opens the possibility of direct reduction of the corresponding ketones to hydrocarbons [159].

The keto group of keto esters can also be converted to methylene by simultaneous reduction of the ester group oxygen and elimination of the alkoxy group.

It is not quite clear whether the Tafel rearrangement—a migration of the aceto group from the α-carbon to the carboxyl group carbon—always takes place. Thus, reduction of acetoacetic ester gave butane [192]; that of ethylacetoacetic ester n-hexane instead of methylpentane [193]. Methylbenzylacetoacetic ester gave 1-(1^3-methopentyl)benzene [192–194]. The reduction product of benzylacetoacetic ester was stated to be β-benzylbutane [195], 1-n-butyl-4-methylbenzene [193], or n-amylbenzene [193–196]. On lead and cadmium in aqueous alcohol-sulfuric acid solution at elevated temperature similar transformations are observed in reduction of propyl-isopropyl, n-butyl-isobutyl, sec-butyl, and diethyl-acetoacetic esters and methyl esters of methylpropyl and methylisopropylacetoacetic acid. The yields approach 50%.

Whereas a reduction of ethyl-α-butylacetoacetate leads to the Tafel rearrangement products, this route is not suitable for preparing cycloheptane or cyclooctane derivatives [148].

Under the same conditions the reduction of the diethyl ester of α,α'-diacetylsuccinic acid provided the diethyl ester of α,α'-diethylsuccinic acid [192]. The N-methylamide of β-ketocapric acid was reduced on lead in sulfuric acid to capric acid N-methyl amide (43.5%) and N-methyl-n-octylamine (12.1%). Moreover, 1-methyl-2-n-propylpiperidone (6.8%) and N-methylconiine were also obtained [197].

The keto group of alicyclic compounds including steroids can also be reduced to methylene. For example, from menthone menthane is produced in sulfuric acid on lead and tin (50%) and on cadmium and mercury (90%) [113, 188,198,199].

The keto group of 5 α-androstanol-17-β-one-3, the methyl ester of $\Delta^{1,3,5(10)}$-estratrienol-3-one-17, and the acetate of 17α,20:20,21-bismethylenedioxy-Δ^5-pregnenol-3β-one-19 were reduced (85–97%) on lead in aqueous dioxane-sulfuric acid solution. By this method deuterium may be incorporated at C_{17}, C_{19}, and C_{20}; labeling by ordinary methods is rather difficult [200]. Chenodeoxycholic acid has been prepared by reduction of the corresponding ketone by reduction at Cd, Cd alloys, or Cd coated electrode in acid [247].

Quite often the reduction of the carbonyl group to methylene results in by-products, particularly unsaturated hydrocarbons. The yields of the latter are negligible and the reaction has no preparative significance. Reduction of isobutyraldehyde led to isobutene [77]; pinacolone to 2,3-dimethylbutene-2 [201]; and benzaldehyde [202], and benzoin [188] to stilbene. Reduction of substituted benzaldehydes also gives the respective substituted stilbenes [95,176].

The advantages of the electrochemical reduction of carbonyl to methylene must be assessed individually when alternative inexpensive "chemical" methods are available (e.g., Clemmensen, Wolf-Kishner/Huang Minlon). For example, the electrochemical Clemmensen reaction is illustrated by the reduction of acetophenone to ethylbenzene [172]. Platinized Pt was used under acidic conditions; a periodic anode pulse of 1 s every 10 s increased the yield of hydrocarbon relative to alcohol.

Stimulated perhaps by the desire to convert biomass-derived materials to fuels and fuel additives, work in this area has increased greatly in this past decade. Horányi's many studies are cited in Ref. 202. Lund and Lund, using iodide catalysis, a Hg cathode, and 2–6 N HCl as medium, have reduced 5-hydroxymethylfurfural to 2,5-hexadione; intermediates were 5-methylfurfural and 2,5-dimethylfuran [136].

Organoelemental and Coordination Compounds

These are discussed in Chap. 21. Suffice it to say that the use of these types of compounds as catalysts has increased enormously in the last few years.

Polymers

Early attempts to initiate polymerization of vinyl monomers by using cathodic products of intermediates from acetophenone were unsuccessful [203]. However, it was later shown [204] that if benzophenone is reduced in DMF with Et_4NClO_4 as supporting electrolyte acrylonitrile and methyl acrylate can be polymerized. The mechanism involved the disproportionation of benzophenone radical anion to parent compound and its dianion; the dianion transferred an electron to the monomer to initiate the process.

Electropolymerization may be of several types: (1) formation of an initiator requires the minimum amount of electricity and falls into the category that Savéant has called "electrochemistry without electricity"; (2) polymerization requiring steady input of current, as in condensation ECP; and (3) ECP of precomplexed monomer (see Chap. 32).

The reduction of carbonyl derivatives is treated in Chap. 11.

II. OLEFINIC CARBONYL COMPOUNDS

This section concerns both α,β-unsaturated carbonyl compounds and those with more remote double bonds. The reductions may involve (1) the carbonyl function alone, (2) the double bond alone, (3) both, or (4) reductive coupling (Chap. 22). It should be noted that if the γ-position has a carbon atom bearing at least one H, such a molecule may function as a carbon acid toward electrogenerated bases (EGBs, Chap. 30). If the γ-substituent also has an electron-withdrawing group, the pK_a is, as expected, lower than when it does not. This is illustrated by the dimerization (not hydrodimerization) of cyclohexene-2-one and analogously of dialkyl glutaconates; in both cases the EGBs are provided by the reduction of a small amount of the parent compound (Table 4).

When allylic compounds $CH_2{=}CHCH_2{-}EWG$ are reduced in aqueous media so that the region in the vicinity of the cathode becomes very alkaline, they isomerize to $CH_3CH{=}CHEWG$ and yield the reduction products expected from the latter. This has been demonstrated [206] when EWG is CN or COOR but not yet when the EWG is CO.

Olefinic compounds in which the unsaturation is more remote than the β,γ-position are usually reduced at the carbonyl group since isolated double bonds are not directly reducible electrochemically. However, it has been shown more recently that when the starting materials are of suitable structure intramolecular attack of the carbonyl anion radical may occur to yield cyclic compounds (Chap. 22).

A. α,β-Unsaturated Carbonyl Compounds

The complexities involved in the reduction of carbonyls without conjugated olefinic unsaturation are compounded in the case of α,β-unsaturated carbonyls because the latter are simultaneously aldehydes or ketones, activated olefins, or acetylenes. Electrolytic reduction of

$$>\underset{4}{C}=\underset{3}{\overset{|}{C}}-\underset{2}{\overset{|}{C}}=\underset{1}{O}$$

may cause: (1) saturation of the 3,4-double bond; (2) reduction of the carbonyl to alcohol; (3) both 1 and 2; (4) coupling of two molecules through C-4 or through C-2, or through C-2 of one molecule and C-4 of another; (5) formation of oligomers and polymers by sequential coupling [207]; and (6) formation of organometallic compounds.

Table 4. Some Dimerizations Promoted by Reduction of Starting Material[a]

Parent compound	E_{cath} (V)	% Reduced	Yield of dimer
Cycloxene-2-one	−1.90	0.13	65
$ROCOCH_2CH{=}CHCOOR$	−2.10	20	50

[a]Versus SCE. Hg cathode, divided cell, DMF solvent, and $PrNBF_4$ as supporting electrolyte.

The reduction mechanism for α,β-unsaturated carbonyls depend critically on acid-base reactions preceding, accompanying, and following electron transfer and importantly on the cathode material, the cation of the supporting electrolyte [29], and the concentration of substrates.

It is possible to analyze the processes of electroreduction of unsaturated carbonyls using the scheme for saturated carbonyls (see earlier), but in the present case reactions of isomerization, dimerization, and other transformation of intermediates are much more important: specifically, it should be remembered that protonation of anion radicals is possible both at carbon and oxygen atoms.

An outstanding "early" mechanistic study was reported by Pasternak [24]. More recent and more comprehensive investigations have issued from the Zuman and the Kargin groups [208] (aqueous systems) and from Simonet [209] and Wawzonek [210] and their respective co-workers (nonaqueous systems). Fortunately, polarographic studies have often been accompanied by macroelectrolyses and identification of intermediate and final products. The research of Wiemann and his students [211] have placed major emphasis on the synthetic aspects of these reductions and a comparison of electrochemical with nonelectrochemical procedures.

Electrode Mechanisms

It has been pointed out [208a] that valid generalizations in this area could be arrived at only after many types of representative compounds had been studied and many examples of each type given. However, the following statements may provide a guide through the maze of results that have accumulated in the course of many years of work:

1. The general principle [212] that, when an electroreducible group is conjugated with olefinic groups, the molecule becomes under comparable conditions more easily reducible is well illustrated for α,β-unsaturated carbonyls in Table 5.
2. The multiple C—C bond is almost certainly reduced before the carbonyl group.
3. As is the case with saturated carbonyls, the preprotonated molecule (I) is reduced anodically of its parent (II); likewise, after the first reduction the protonated neutral radical (III) is more easily reduced than the anion radical (IV):

$H[RCH{=}CHCOR']^+$	$RCH{=}CHCOR'$	$RCH_2CH_2\dot{C}R'$ with $-OH$ on C	$RCH_2CH_2\dot{C}R'$ with $-O^-$ on C
I	II	III	IV

The following species must therefore be considered in protic media

$H[RCH{=}CHCOR']\cdot$	$RCH_2CH{=}C{-}R'$ with $-O^-$ on C	RCH_2CH_2CR' with ${=}OH^+$ on C	RCH_2CH_2COR'
V	VI	VII	VIII

4. The hydrated carbonyl group (e.g., in an aldehydrol) is not electroreducible; reversion to the unhydrated form is both acid and base catalyzed.

Table 5. Comparison of $E_{1/2}$ of α,β-Unsaturated Carbonyls with Those of Corresponding Saturated Carbonyls

Compound	$-E_{1/2}$[a]	Conditions	Reference
CH_3CH_2CHO	1.8	0.1 M LiCl	207
$CH_2{=}CHCHO$	1.5	pH 7–11	213
$CH_3CH{=}CHCHO$	1.25	0.1 N NH_4Cl	214
$CH_3(CH{=}CH)_xCHO$	1.05 to 0.7	pH 7; n = 2 to 5, resp.	215
$C_6H_5CH_2CH_2CHO$	1.4	0.1 N NH_4Cl	216
$C_6H_5CH{=}CHCHO$	0.8	0.1 N NH_4Cl	216
$CH_3CH_2COCH_3$	2.25	0.025 N Me_4NI	217
$CH_2{=}CHCOCH_3$	1.42	0.1 N KCl	218
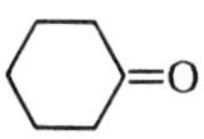	2.45	0.05 N Et_4NI, 75% dioxane	219
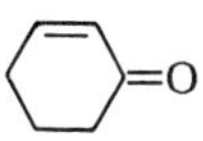	1.55	0.1 N NH_4Cl	220

[a]First wave, referred to SCE.

5. A series of four sequential one-electron reductions exists in the most ideal examples; often the first two or the latter two may be merged into single two-electron steps (see Table 6).
6. In aprotic systems the electron transfer steps (often irreversible) are intimately linked with acid-base reactions (reversible) so that CEEC and EECC reactions, for example, are involved (in which C is a reaction with a proton and E is the transfer of one electron). A monovalent cation like Li^+ (also Mg^{2+}) can sometimes play the role of a proton in neutralizing an anion radical [221]. See Zuman's analyses [208a–c].
7. Substituent effects are generally correlated with Hammett or Yukawa-Tsuno equations [222]. It should be noted that introduction of substituents into the aryl group at C-3 of 1,3-diaryl-1-propyn-3-one has a more pronounced effect than their introduction at the C-1 aryl group [64].
8. In aprotic media, since no H^+ reactions intervene between the first two $1e^-$ stages, the anion radical and the dianion intermediates are separately discernible. The acidity of R_4N^+ has been questioned [223]. Since the starting materials are subject to anion attack by anion radicals and dianions (and form oligomers and polymers), the diffusion currents found polarographically are lower than would be expected in the absence of these complications (see Chap. 3). Two, one-electron waves were shown by benzalacetone, benzalacetophenone, chalcone [208e,f], and cinnamaldehyde; crotonaldehyde gave only the first one-electron reduction [210]. In another study [209], methyl vinyl ketone showed two one-electron waves but mesityl oxide, isophorone, and ethylidene acetone only one each. Presumably in the latter cases, the second waves were as negative as the discharge potential of the supporting cation.

Table 6. Polarographic Reduction Phenomena at Various pH Ranges

pH Range[a]	Series reduced[b]	Observations	Mechanisms[b]
<4	I and VII	Shifts of $E_{1/2}$ with pH; mercurial at first wave; no dimers; the final neutral radical may dimerize, form mercurials, etc.	H^+, e^-, e^-, H^+, H^+, e^-
5–7	I and VIII	First 2–e^- wave (or two 1–e^-) followed by 2–e^- wave	H^+, e^-, e^-, H^+, e^-, H^+, e^-, H^+, e^-, H^+, or H^+, e^-, e^-, H^+, e^-, H^+, H^+, e^-
8–9	II and VIII	II to VIII, VIII to alcohol	e^-, H^+, e^-, H^+, e^-, H^+, e^-, H^+
>11 or 12	II	Two 1-electron steps	e^-, e^-, H^+_, H^+, e^-, e^-, H^+, H^+ or e^-, e^-, H^+, H^+, e^-, H^+, e^-, H^+

[a]The situation at intermediate ranges has also been analyzed.
[b]Starting with II.
Source: From Ref. 208.

9. In protic media, at Hg, the quinone (IX)-hydroquinone (X) couple is the classic example of a fast (reversible) system.

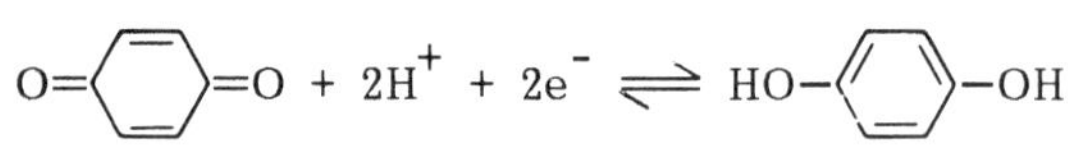

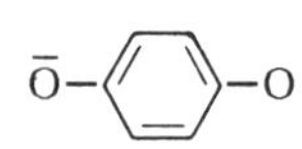

(IX) (X) (XI)

The $E_{1/2}$ values for quinone, therefore, have unqualified thermodynamic significance and correlate very well with oxidation potentials determined by other means [224], such as potentiometry.

In alkaline medium the intermediate semiquinone (XI) formed by one-electron reduction of IX has long been knwon [225]; the radical anion (XI) may undergo radical dimerizarion (→ quinhydrone) or may disproportionate to IX and X.

Polarography has been used extensively as an analytical method for quinones-hydroquinones [224]; the applications are summarized and compiled periodically in Analytical Chemistry Annual Reviews [226].

The electrochemistry of quinone imines and p-aminophenols is discussed in Chap. 18.

The electrochemical behavior of quinones in aprotic media has been summarized [227]. As might be expected from the earlier discussions, the reductions of quinones (e.g., IX) to semiquinones (e.g., XI) in a one-electron process are generally separately observable from the second reduction to a dianion. Again, even though the medium is aprotic, a "hard" acid cation of the supporting electrolyte may play the same role as H^+. Thus Eggins [228] found that IX is reduced in two $1e^-$ waves when Et_4NClO_4 in CH_3CN is used as the medium but that the reduction approached a one $2e^-$ wave when $LiClO_4$ was used as the electrolyte. This is reminiscent of the polarographic behavior of benzil in DMF [221]. A similar study with naphthoquinones has been reported [229].

A number of α- and β-substituted anthraquinones has been reduced at the dropping mercury electrode in DMF to the radical anion stage. There was good correlation between $E_{1/2}$ and the energy of the lowest vacant molecular orbital (Hückel approximation) [230].

The properties of anion radicals formed by electroreduction of trialkylcyclohexadiene-2,5-ones and methylenequinones have been studied by various methods, including ESR [231–232].

The burgeoning physical chemistry of semiquinones—and more generally of semidione radical ions—has been summarized by Russell [233].

Synthetic Aspects

Reduction to saturated carbonyl compounds and saturated alcohols This type of reduction can be carried out under a variety of conditions and often gives very good yields (e.g., 1-piperonylidene 2-heptanone → 60% 1-(3,4-methylenedioxyphenyl)-3-octane [234] and benzalacetone → 90% benzylacetone [235], but the same results can usually be achieved by catalytic means. The special advantages of the electrolytic method are more apparent in the case of reductive coupling reactions.

However, polarographic, mechanistic, and preparative studies of this type of reaction continue: gas and liquid phase reduction of acrolein to 2-propene-1-ol (→ CH_3CH_2CHO and later PrOH [236]); reduction of 2,3-diphenylindenone to the cis- and trans-indanones [237]; monomeric reduction of tetraphenylcyclopentadiene [238]; and dependence of the reduction of trans-PhCH=CHCOOMe to saturated product or hydrodimer on the cation of the supporting electrolyte and on the presence of surface-active agents [239]. A newer methodology for the selective reduction of the double bond involves the use of catalytic cathodes, such as Raney Ni or Pd/C [123,240].

Trapping of intermediates Since relatively stable radical anions and anions are formed in the course of the reduction sequence in anhydrous media, it has been possible to trap the intermediates by conducting the electrolysis in the presence of CO_2. Thus, reduction of benzalacetophenone at a Hg cathode using DMF and KI—which limited the reduction to the first wave—while CO_2 was bubbling through the solution [210] yielded two carboxylates, XII and XIII. It is proposed that XII was formed by the sequence of a $1e^-$ reduction-carbonation-$1e^-$ reduction carbonation (ECEC mechanisms) that obtained in the reductive carbonation of stilbene (Chap. 7). Acidification of XII, 2-benzoyl-3-phenylsuccinic acid, caused loss of the carboxyl α to the carbonyl and yielded α-phenyl-β-benzoylpropionic acid as the isolated product. To explain the formation of XIII it is suggested that the monocarboxylate radical resulting from the first EC sequence either before or after reduction adds to a molecule of starting material to yield an intermediate, which cyclizes to XII, 3-benzoyl-1-carboxy-4,5-diphenyl-2-hydroxycyclopentane.

Reductive alkylation of anthraquinone in LiCl/DMF at a Hg cathode using alkyl or aralkyl halides yielded 32–91% of the expected diethers [241].

Unstable 1-allenols, which isomerize to α,β-unsaturated ketones, are formed as intermediates in the reduction of acetylenic ketones [242–245] in acid media; the isomerization rates are 50 times as high as that of the corresponding enols to saturated ketones [246].

In alkaline medium indenones are directly reduced to indanones but in acid medium via enols which isomerize [245].

B. Allylic Tautomers of α,β-Unsaturated Carbonyl Compounds

The electrochemical behavior of these types of compounds has been mentioned. The most interesting feature is that the hydrodimers obtained from them are identical to those from the corresponding α,β-unsaturated compounds; allyl cyanide yielded 3,4-dimethyladiponitrile [206]. It is interesting to note that when ethyl 3-butenoate was electrolyzed to partial conversion, all the recovered ester was ethyl crotonate.

C. Carbonyl Compounds with More Remote Unsaturation

The novel work that has been published [205] involves cyclizations, that is, intramolecular reductive coupling, presented in Chap. 22.

REFERENCES

1. D. H. Evans, in Encyclopedia of Electrochemistry of the Elements, (A. J. Bard and H. Lund, eds.), vol. 12, Dekker, New York, 1978.
2. T. Shono, N. Kise, T. Suzumoto, and T. Morimoto, J. Amer. Chem. Soc., 108, 4676 (1986).
3. H. L. Chum and M. M. Baizer, The Electrochemistry of Biomass and Derived Materials, ACS Monograph 183, American Chemical Society, Washington, D.C., 1985, p. 224 ff.
4. O. W. Meshkova, W. D. Dimitrieva, and V. D. Bezuglyi, Zh. Obshch. Khim., 45, 684 (1975); G. Le Guillanton, Bull. Soc. Chim. Fr., 3458 (1973); J. P. Segretario, N. Sleszynski, and P. Zuman, J. Electroanal. Chem., 214, 259 (1986).
5. R. O. Loutfy and R. O. Loutfy, J. Phys. Chem., 76, 1650 (1972); 77, 336 (1973).
6. Y. Ueno and Y. Ono, Bull. Chem. Soc. Jpn., 48, 3585 (1975).
7. P. Valenta, Coll. Czech. Chem. Commun., 25, 855 (1960).
8. W. Kemula, Z. R. Grabowski, and M. Kalinowski, Coll. Czech. Chem. Commun., 25, 3306 (1960).
9. D. E. Austen, P. H. Given, D. S. Ingram, and M. E. Peover, Nature, 182, 1784 (1958).
10. P. Favero, G. Giacometti, P. I. Nordio, and G. Rigatti, Nuovo Cimento, 24, 21 (1962); Chem. Abstr., 57, 3002 (1962).
11. N. Steinberger and G. K. Fraenkel, J. Chem. Phys., 40, 723 (1964).
12. G. A. Russell, E. T. Strom, E. R. Talaty, and St. A. Weiner, J. Am. Amer. Chem. Soc., 88, 1998 (1966).
13. L. N. Nekrasov and A. D. Korsun, Elektrokhimiya, 4, 489, 539, 996 (1968); 5, 212 (1969).

14. R. A. Gavar, Y. A. Bender, L. C. Baumane, and J. P. Stradins, Zh. Obshch. Khim., 43, 1584 (1973).
15. W. N. Epimakhov, Elektrokhimiya, 6, 322, (1970).
16. A. P. Tomilov and Yu. D. Smirnov, in Elektrokhimicheskie Processy s Uchastiem Organicheskikh Veshchestv, Nauka, Moscow, 1970, p. 56.
17. R. M. Powers and R. A. Day, J. Org. Chem., 24, 722 (1959).
18. A. Yamura, T. Sekine, and K. Sugino, J. Electrochem. Soc. Jpn., 34, 110 (1966).
19. A. Albisson, G. Mousset, and J. Simonet, C. R. Acad. Sci. Ser. C, 272, 646 (1971).
20. M. Grätzel, A. Henglein, and K. M. Bansal, Ber. Bunsenges. Phys. Chem., 77, 6 (1973).
21. J. W. Hayes, I. Ruzic, D. E. Smith, G. L. Booman, and J. R. Delmastro, J. Electroanal. Chem., 51, 269 (1974).
22. N. R. Armstrong, N. E. Vanderborgh, and R. K. Quinn, J. Electrochem. Soc., 122, 615 (1975).
23. L.-S. Ray Yeh, J. Electroanal. Chem., 84, 159 (1977).
24. R. Pasternak, Helv. Chim. Acta, 31, 753 (1948).
25. P. J. Elving and J. T. Leone, J. Amer. Chem. Soc., 80, 1021 (1958).
26. S. G. Mairanovskii and V. N. Pavlov, Zh. Fiz. Khim., 38, 1804 (1963).
27. I. A. Korshunov, A. S. Kirillova, and Z. B. Kuznetsova, Zh. Fiz. Khim., 24, 251 (1950).
28. S. G. Mairanovskii, Izv. Akad. Nauk SSSR, Otd. Khim. Nauk, 2140 (1961).
29. M. Ashwort, Coll. Czech. Chem. Commun., 13, 229 (1948).
30. B. Kastening and L. Holleck, Z. Elektrochem., 63, 166 (1959).
31. K. Ito, Rev. Polarogr. (Kyoto), 15, 97 (1968).
32. S. G. Mairanovskii, J. Electroanal. Chem., 4, 166 (1962).
33. S. G. Mairanovskii and V. N. Pavlov, Elektrokhimiya, 1, 226 (1967).
34. P. Zuman, D. Barnes, and A. Ryvolova-Kejharova, Disc. Faraday Soc., 45, 202 (1968).
35. R. M. Powers and R. A. Day, J. Amer. Chem. Soc., 80, 806 (1958).
36. V. N. Pavlov, Y. M. Zolotovitskii, S. G. Mairanovskii, and G. A. Tedoradze, Elektrokhimiya, 1, 427 (1965).
37. J. P. Stradins and W. W. Teraud, Izv. Akad. Nauk. Latv. SSR, Ser. Khim., 169 (1964); Chem. Abstr, 61, 12953 (1964).
38. J. P. Stradins, Electrochim. Acta, 9, 711 (1964).
39. A. A. Pozdeeva and S. I. Zhdanov, Proc. III Int. Congr. Polarogr., Southampton, 1964.
40. Yu. M. Kargin and V. Z. Kondranina, Izv. Akad. Nauk SSSR, Ser. Khim., 2205 (1973).
41. Yu. M. Kargin, V. Z. Kondranina, and N. I. Semakhina, Izv. Akad. Nauk SSSR, Ser. Khim., 226 (1973).
42. Yu, M. Kargin, N. I. Semakhina, and V. Z. Kondranina, Zh. Obshch. Khim., 45, 2258 (1975).
43. F. Ammar, L. Nadjo, and J.-M. Savéant, J. Electroanal. Chem., 48, 146 (1973).
44. C. P. Andrieux, J.-M. Savéant, and D. Tessier, J. Electroanal. Chem., 63, 429 (1975).
45. Nguen Suan and L. N. Nekrasov, Elektrokhimiya, 10, 1362, 1752 (1974).
46. A. Lasia, J. Electroanal. Chem., 42, 253 (1973).
47. M. D. Ryan and D. H. Evans, J. Electroanal. Chem., 67, 333 (1976).

48. L. N. Vykhodceva and L. N. Nekrasov, Elektrokhimiya, 14, 1439 (1978).
49. A. P. Tomilov and B. L. Klyuev, in Itogi Nauki. Elektrokhimiya, 1967, VINITI, Moscow, 1969, p. 200; Chem. Abstr., 72, 38095 (1970).
50. A. Mendkovich and V. P. Gul'tyai, J. Electroanal. Chem., 169, (1984); A. Mendkovich and A. I. Rusakov, J. Electroanal. Chem., 209, 43 (1986).
51. J. Dojlido, Z. Galus, and Lj. Jeftič, J. Electroanal. Chem., 62, 433 (1975).
52. B. Behr and M. Drogowska, J. Electroanal. Chem., 82, 317 (1977).
53. O. Yu. Gusakova, B. B. Damaskin, and N. W. Fedorovich, Elektrokhimiya, 9, 1134 (1973).
54. Yu. I. Erlikh and U. V. Pal'm, Elektrokhimiya, 10, 1866 (1974).
55. A. D. Semenova, N. W. Kropotova, and G. D. Wowchenko, Elektrokhimiya, 10, 961 (1974).
56. K. Wettig, Elektrokhimiya, 3, 269 (1967).
57. V. J. Puglisi, G. L. Clapper, and D. H. Evans, Anal. Chem., 41, 279 (1969).
58. S. G. Mairanovskii, Kineticheskie i Kataliticheskie Toki v Polarografii, Nauka, Moscow, 1966, p. 208.
59. I. A. Titova, I. M. Levinson, V. G. Mairanovskii, and A. B. Ershler, Elektrokhimiya, 9, 424 (1973).
60. E. Kariv, H. A. Terni, and E. Gileadi, Electrochim. Acta, 18, 433 (1973).
61. L. Horner and D. Degner, Electrochim. Acta, 19, 611 (1974).
62. J. Grimshaw and J. S. Ramsey, J. Chem. Soc. C, 653 (1966).
63. T. Nonaka, S. Wachi, and T. Fuchigama, Chem. Lett., 47, (1977).
64. J. P. Stradins, V. S. Pisareva, and S. P. Korshunov, Zh. Org. Khim., 13, 788 (1977).
65. T. Sekine, A. Yamura, and K. Sugino, J. Electrochem. Soc., 112, 439 (1965).
66. T. Arai and T. Ogura, Bull. Chem. Soc. Jpn., 33, 1018 (1966).
67. L. Holleck and D. Marquarding, Naturwissenschaften, 49, 468 (1962).
68. (a) A. P. Tomilov and I. N. Brago, in Progress Elektrokhimii Organicheskich Soedinenii, Vol. 1, Nauka, Moscow, 1969, p. 208.
69. E. Müller, Z. Elektrochem., 33, 253 (1927).
70. R. Kalvoda and G. K. Budnikov, Coll. Czech. Chem. Commun., 28, 838 (1963).
71. G. K. Budnikov and R. Kalvoda, Zh. Obshch. Khim., 35, 1131 (1965).
72. P. G. Russell, N. Kovac, S. Srinivasan, and M. Steinberg, J. Electrochem. Soc., 124, 1329 (1977).
73. M. Y. Kagan and V. G. Kryukov, Chim. Prom. (Russ.), 10, 34 (1933); Chem. Abstr., 27, 3403 (1933).
74. V. G. Khomyakov, A. P. Tomilov, and B. G. Soldatov, Elektrokhimiya, 5, 850, 853 (1969).
75. W. Schepps, Ber. Dtsch. Chem. Ges., 46, 2564 (1913).
76. P. Condit, U.S. Patent 2,537,304 (1951), Chem. Abstr., 45, 2341 (1951).
77. S. Swann, Jr., E. Onstott, and F. Baastad, Trans. Electrochem. Soc., 102, 113 (1955).
78. S. Swann, Jr., and E. W. Field, Trans. Electrochem. Soc., 72, 327 (1937).

79. G. Filardo, M. Galluzzo, B. Giannici, and R. Ercoli, J. Chem. Soc. Dalton Trans., 1787 (1974).
80. A. P. Tomilov, A. A. Sergo, and S. L. Varshawskii, Elektrokhimiya, 1, 1126 (1965).
81. P. Federlin, Bull Soc. Chim. Fr., 949 (1958).
82. K. Tsuji, Agric. Biol. Chem., 25, 432, 915 (1961).
83. N. E. Khomutov, V. V. Skornyakov, and T. P. Fadeeva, Zh, Fiz. Khim., 38, 102 (1964).
84. N. Belen'kaya and I. Belozerskii, Zh. Obshch. Khim., 19, 1664 (1949).
85. H. J. Creighton, Trans. Electrochem. Soc., 75, 289 (1939); J. Electrochem. Soc., 99, 127 (1952).
86. M. Wolfrom, W. Binkley, C. Spencer, and B. Lew, J. Amer. Chem. Soc., 73, 3357 (1951).
87. P. Regna, J. Amer. Chem. Soc., 69, 246 (1947).
88. H. Creighton and R. Hales, U.S. Patent 2,458,895 (1949); Chem. Abstr., 43, 2104 (1949).
89. K. Sugino and S. Hayashi, J. Chem. Soc. Jpn., 65, 458 (1944).
90. E. Parker and S. Swann, Jr., Trans. Electrochem. Soc., 92, 343 (1947).
91. A. Findlay and V. H. Williams, Trans. Faraday Soc., 17, 453 (1922).
92. (a) P. N. Pintauro, D. K. Johnson, K. Park, M. M. Baizer, and K. Nobe, J. Electrochem. Soc., 132, 1850 (1985); (b) K. Park, P. N. Pintauro, M. M. Baizer, and K. Nobe, J. Appl. Electrochem., 16, 941 (1986).
93. X. Hemptinne and J. C. Jungers, Z. Phys. Chem. (Frankfurt), 15, 137 (1958).
94. S. G. Mairanovskii, T. Y. Rubinskaya, V. P. Gul'tyai, and I. V. Proskurovskaya, Izv. Akad. Nauk SSSR, Ser. Khim., 2360 (1976).
95. H. D. Law, J. Chem. Soc., 1512 (1906); 748 (1907); 1115 (1911).
96. G. Shima, Mem. Coll. Sci. Univ. Kyoto, A11, 1, 407, 419 (1928); A12, 69, 327 (1929).
97. D. V. Sokol'skii, M. S. Erzhanova, N. A. Zibrova, and B. B. Kamardinova, Elektrokhimiya, 12, 1388 (1976).
98. W. O. Reichsfeld and S. K. Kudrna, Tr. Leningr. Tekhnol. Inst. Lensoveta (Russ.), 110 (1956).
99. M. E. Manzhelei, V. I. Spektor, V. Y. Botsan, and M. N. Timotin, Zh. Obshch. Khim., 46, 1328 (1976).
100. H. Kita, S. Ishikura, and A. Katayama, Electrochim. Acta, 19. 555 (1974).
101. A. D. Semenova, N. V. Kropotova, and G. D. Vovchenko, Elektrokhimiya, 10, 1233 (1974).
102. J. Tafel, Ber. Dtsch. Chem. Ges., 45, 448 (1912).
103. K. Elbs and K. Brand, Z. Elektrochem., 7, 644 (1901); 8, 783 (1902).
104. A. P. Tomilov and M. I. Kalitina, Zh. Prikl. Khim., 38, 1574 (1965).
105. A. P. Tomilov and L. A. Ignat'eva, J. Appl. Chem. USSR, 38, 2651 (1965).
106. O. Slotterbeck, Trans. Elektrochem. Soc., 92, 377 (1947).
107. J. Goldberger and F. Tondler, Monatsh. Chem., 26, 1474 (1906).
108. J. P. Wibaut, H. Hoog, S. L. Langendijk, J. Oberhoff, and J. Smittenberg, Rec. Trav. Chim. Pays-Bas, 58, 329 (1939).
109. L. P. Kyrides, J. Amer. Chem. Soc., 55, 3431 (1933).
110. E. A. Steck and W. Boehme, J. Amer. Chem. Soc., 74, 4511 (1952).
111. N. J. Leonard, S. Swann, Jr., and H. L. Dryden, J. Amer. Chem. Soc., 74, 2871 (1952).

112. M. Matsui and S. Shimizu, Mem. Coll. Sci. Kyoto Imp. Unive., 4, 245 (1920); Centralblatt, 93, 764 (1922 IV).
113. C. Schall and W. Kirst, Z. Elektrochem., 29, 537 (1923).
114. J. Tafel and K. Schmitz, Z. Electrokhim., 8, 289 (1902).
115. J. Bredt and W. H. Perkin, J. Prakt. Chem., 89, 209 246 (1914).
116. J. Bredt, J. Prakt. Chem., 84, 786 (1911).
117. M. Ohmori and M. Takagi, Bull. Chem. Soc. Jpn., 50, 773 (1977).
118. P. Anziani, A. Anbry, and R. Cornubert, C. R. Acad. Sci., 225, 878 (1947).
119. A. S. Mendkovich, V. N. Leibzon, S. G. Mairanovskii, M. M. Krayushkin, T. A. Klimova, S. S. Novikov, and V. V. Sevost'yanova, Izv. Akad. Nauk SSSR, Ser. Khim., 1866 (1978).
120. B. Sakurai and T. Arai, Bull. Chem. Soc. Jpn., 28, 93 (1955).
121. J. Armand and L. Boulares, Bull. Soc. Chim. Fr., 711 (1975).
122. C. P. Andrieux and J.-M. Savéant, Bull. Soc. Chim. Fr., 2090 (1973).
123. T. Osa, T. Matsue, A. Yokozawa, and M. Fujihira, Denki Kagaku, 53, 104 (1985).
124. F. Escherich and M. Moest, Z. Elektrochem., 8, 840 (1902).
125. S. Swann, Jr., Trans. Electrochem. Soc., 69, 316 (1936).
126. J. Hermolin, J. Kopilov, and E. Gileadi, J. Electroanal. Chem., 71, 245 (1976); J. Kopilov, S. Shatzmiller, and E. Kariv, Electrochim. Acta, 21, 535 (1976).
127. R. N. Gourly, J. Grimshaw, and P. G. Millar, J. Chem. Soc. C, 2318 (1970).
128. F. Fichter and H. Montmollin, Helv. Chim. Acta, 5, 256 (1922).
129. P. Foulatier, J.-P. Salaün. and C. Caullet, C. R. Acd. Sci., Ser. C, 279, 679 (1974).
130. J. J. Novoa, An. Quim. Ser. A, 82(1), 89 (1986); Chem. Abstr., 106, 49474d (1987).
131. N. J. Leonard, S. Swann, Jr., and H. L. Dryden, J. Amer. Chem. Soc., 74, 2871 (1952).
132. G. Le Guillanton and M. Lamont, Nouv. J. Chim., 2, 157 (1978).
133. J. Armand and L. Boulares, Bull Chem. Soc. Fr., 711 (1975).
134. M. Federonko, J. Königstein, and K. Linek, Coll. Czech. Chem. Commun., 32, 1497, 3998 (1967).
135. (a) M. M. Baizer, T. Nonaka, K. Park, Y. Saito, and K. Nabe, J. Appl. Electrochem., 14, 197 (1984); (b) J.-J. Jow and T. C. Chou, Electrochim. Acta, 32, 311 (1987).
136. T. Lund and H. Lund, Extended Abstract No. 330, Elctrochemical Society Meeting 84-1, Cincinnati, Ohio, May 6–11, 1984; Acta Chem. Scand. , B39, 429 (1985).
137. B. Mueller, H. J. Dietz, H. Matschiner, S. Engelmann, and H. H. Herzberg, Ger. (East), Pat. DD237,683, 23 July 1986; Chem. Abstr., 106, 75019d (1987).
138. S. Thangavelu, M. Noel, and K. S. Udupa, Electrochim. Acta, 31(10), 1563 (1986).
139. J. Benders, V. Kadysh, E. Lavrinovics, and J. P. Stradins, Electrochim. Acta, 31, 1369 (1986).
140. A. D. Thomsen and H. Lund, Acta Chem. Scand., 25, 1576 (1971).
141. Y. M. Kargin and V. Z. Kondranina, Izv. Akad. Nauk SSSR, Ser. Khim., 365 (1973).
142. A. Vincenz-Chodkovska and Z. R. Grabowski, Electrochim. Acta, 9, 789 (1964).
143. E. Bauer, J. Electroanal. Chem., 14, 351 (1967).

144. H. E. Stapelfeldt and S. P. Perrone, Anal. Chem., 40, 815 (1968); H. E. Stapelfeldt and S. P. Perrone, Anal. Chem., 41, 623 (1969).
145. V. Ya. Vegnere, J. P. Stradins, and Ya. Ya. Ozol, Zh. Obshch. Khim., 43, 242 (1973).
146. L. Holleck and O. Lehmann, Monatsh. Chem., 92, 499 (1961).
147. M. Lacan, I. Tabaković, and Z. Ceković, Croat. Chem. Acta, 47, 587 (1975).
148. S. Wawzonek and D. E. Durham, J. Electrochem. Soc., 123, 500 (1976).
149. (a) D. J. Brown and L. Horner, Justus Liebigs Ann. Chem., 77, (1977); (b) I. A. Titova, I. M. Levinson, V. G. Mairanovskii, and A. B. Ershler, Sov. Electrochem., 9, 408 (1973); (c) J. Kopilov, E. Kariv, and L. L. Miller, J. Amer. Chem. Soc., 99, 3450 (1977); (d) D. Seebach, H. A. Oei, and H. Daum, Chem. Ber., 110, 2316 (1977).
150. B. F. Watkins, J. R. Behling, E. Kariv, and L. L. Miller, J. Amer. Chem. Soc., 97, 3549 (1975).
151. F. Beck, Chem.-Ing. Tech., 48, 12, 1096 (1976).
152. (a) T. Chiba, M. Okimoto, H. Nagai, and Y. Takata, Bull. Soc. Chem. Jpn., 56, 719 (1983); (b) G. Belot, S. Desjardins, and J. Lessard, Tetrahedron Lett., 25, 5347 (1984); (c) M. J. Lain and D. Pletcher, Electrochim. Acta, 32, 109 (1987).
153. U. Izumi, Advan. Catalysis, 32, 215 (1983).
154. L. Coche and J. C. Moutet, J. Am. Chem. Soc., 109, 6887 (1987).
155. T. Yamada, T. Osa, and T. Tomakazu, Chem. Lett., 1611 (1987).
156. I. V. Kirilyus, R. P. Silkina, and D. V. Sokol'skii, Kat. Deponir. Rukopisei, Estestw. i Tech. Nauki. Tekh., VINITI, Moscow, 1970, Wyp. 18, No. 631–69; Chem. Abstr., 77, 108727 (1972).
157. V. A. Smirnov, in Uspekhi Elektrokhimii Organicheskikh Soedinenii, Nauka, Moscow, 1966, p. 238; Chem. Abstr., 67, 17123 (1967).
158. A. P. Tomilov, S. G. Mairanovskii, M. Ya. Fioshin, and V. A. Smirnov, Elektrokhimiya Organicheskikh Soedinenii, Khimiya, Leningrad, 1968, p. 214.
159. H. Lund, H. Doupeux, M. A. Michel, G. Mousset, and J. Simonet, Electrochim. Acta, 19, 629 (1974).
160. C. Wilson and K. Wilson, Trans. Electrochem. Soc., 80, 151 (1941).
161. A. S. Oganesyan and L. I. Antropov, Dokl. Akad. Nauk Arm. SSR, 21, 81 (1955).
162. G. Hennig and G. Kimball, J. Phys. Chem., 12, 415 (1944).
163. J. Haggerty, Trans. Electrochem. Soc., 56, 421 (1929).
164. G. Kirchhof and A. Stepanov, Khim-Farm. Prom. (Russ.), 1, 21 (1932); Chem. Abstr., 27, 5651 (1933).
165. H. Hibber and R. R. Read, J. Amer. Chem. Soc., 46, 983 (1924).
166. N. G. Smirnova, W. A. Smirnov, and L. I. Antropov, Tr. Novocherk. Politekh. Inst. Ordzhonikidze, 79, 43 (1959); Chem. Abstr., 54, 24011 (1960).
167. H. Kita, S. Ishikura, and A. Katayama, Electrochim. Acta, 20, 441 (1975).
168. J. Tafel, Z. Elektrochem., 17, 972 (1911).
169. V. G. Khomyakov and A. P. Tomilov, Zh. Prikl. Khim., 36, 378 (1963).
170. X. de Hemptinne and K. Schunk, Trans. Faraday Soc., 65, 591 (1969).
171. M. K. Polievktov, I. G. Markova, and A. F. Oleinik, Khim. Geterotsikl. Soedin., 165 (1977); Chem. Abstr., 87, 22078 (1977).
172. D. Pletcher and M. Rasaq, Electrochim Acta, 26, 819 (1981).

173. G. Horányi and M. Novák. Acta Chim. Acad. Sci. Hung., 75, 369 (1973).
174. S. Swann, jr., Bull Cent. Electrochem. Res. Inst., 2, 6 (1955).
175. H. Kaufmann, Z. Elektrochem., 2, 365 (1896); 4, 461 (1898).
176. J. Tafel and W. Schepps, Ber. Dtsch. Chem. Ges., 44, 2148 (1911).
177. C. R. Clemo and J. M. Smith, J. Chem. Soc., 2423 (1923).
178. W. S. Rapson and R. Robinson, J. Chem. Soc., 1537 (1935).
179. S. Swann, jr., R. W. Benoliel, L. R. Lyons, and W. H. Paul, Trans. Electrochem. Soc., 79, 83 (1941).
180. H. J. Read, Trans. Electrochem. Soc., 80, 8 (1941).
181. T. J. Curphy, L. D. Trivedi, and T. Layloff, J. Org. Chem., 39, 3831 (1974).
182. J. Tafel, Ber. Dtsch. Chem. Ges., 42, 3146 (1909).
183. S. Ohki and Y. Noike, J. Pharm. Soc. Jpn., 78, 129 (1958).
184. R. R. Read and F. A. Fletcher, Trans. Electrochem. Soc., 47, 93 (1925).
185. J. Mizuguchi and S. Matsumoto, J. Pharm. Soc. Jpn., 38, 129 (1958).
186. K. Elbs and K. Brand, Z. Elektrochem., 7, 644 (1901); 8, 783 (1902).
187. E. Müller, Z. Elektrochem., 16, 236 (1910).
188. G. Shima, Mem. Coll. Sci. Kyoto Univ., A13, 315 (1930).
189. J. Grimshaw and E. J. F. Rea, J. Chem. Soc., C, 2628 (1967).
190. R. Lukes and J. Gorocholinskii, Coll. Czech. Chem. Commun., 8, 223 (1936).
191. M.-A. Michel, J. Simonet, G. Mousset, and H. Lund, Electrochim. Acta, 20, 143 (1975).
192. J. Tafel and W. Jürgens, Ber. Dtsch. Chem. Ges., 42, 2548 (1909).
193. J. Tafel, Ber. Dtsch. Chem. Ges., 45, 437 (1912).
194. H. Stenzl and F. Fichter, Helv. Chim. Acta, 20, 846 (1937).
195. J. Tafel and H. Hahl, Ber. Dtsch. Chem. Ges., 40, 3317 (1907).
196. H. Stenzl and F. Fichter, Helv. Chim. Acta, 17, 669 (1934).
197. R. Lukes and V. Dudek, Coll. Czech. Chem. Commun., 27, 467 (1962).
198. C. Schall, Z. Elektrochem., 38, 31 (1932).
199. G. H. Keats, J. Chem. Soc., 2003 (1937).
200. L. Throop and L. Tökes, J. Amer. Chem. Soc., 89, 4789 (1967).
201. J. P. Wibaut, H. Hoog, S. L. Langendijk, J. Oberhoff, and J. Smittenberg, Recl. Trav. Chim., Pays-Bas, 58, 329 (1939).
202. G. Horányi, Electrochim. Acta, 31, 1095 (1986).
203. H. Z. Friedlander, S. Swann, Jr., and C. S. Marvel, J. Electrochem. Soc., 100, 408 (1953).
204. G. Mengoli and G. Vidotto, Macromol. Chemie, 129, 73 (1969).
205. T. Shono, I. Nishiguchi, H. Ohmizu, and M. Mitani, S. Amer. Chem. Soc., 100, 545 (1978).
206. M. R. Ort and M. M. Baizer, J. Org. Chem., 31, 1646 (1966).
207. R. Bieber and G. Trumpler, Helv. Chim. Acta, 30, 2000 (1947); 31, 5 (1948).
208. (a) P. Zuman, D. Barnes, and A. Ryvolova-Kejharova, Discuss. Faraday Soc., 45, 202 (1968); (b) V. Toure, M. Levy, and P. Zuman, J. Electroanal. Chem., 56, 285 (1974); (c) P. Zuman and L. Spritzer, J. Electroanal. Chem., 69, 433 (1976); (d) V. N. Nikulin, N. M. Kargina, and Yu. M. Kargin, Zh. Obshch. Khim., 43, 2712 (1973); (e) V. N. Nikulin, N. M. Kargina, and Yu. M. Kargin, Zh. Obshch. Khim., 44, 2520 (1974); (f) V. N. Nikulin, N. M. Kargina, Yu. M. Kargin,

W. I. Sawin, and S. F. Pivovarova, Zh. Obshch. Khim., 47, 1595 (1977); (g) Yu. M. Kargin, V. Z. Latypova-Kondranina, N. M. Kargina, and S. F. Pivovarova, Zh. Obshch, Khim., 48, 2746 (1978).
209. J. Simonet, Bull Soc. Chim. Fr., 1533 (1970).
210. S. Wawzonek and A. Gundersen, J. Electrochem. Soc., 111, 324 (1964); J. C. Johnston, J. D. Faulkner, L. Mandell, and R. A. Day, Jr., J. Org. Chem., 41, 2611 (1976).
211. J. Wiemann, Bull. Soc. Chim. Fr., 2545 (1964).
212. J. Heyrovsky, A Polarographic Study of Electrokinetic Phenomena. Actual Sci. Ind. No. 90, Hermann, Paris, 1934; Polarographie, theoretische Grundlagen, praktische Ausführung und Anwendungen der Electrolyse mit der tropfenden Quecksilberelektrode. Springer-Verlag, Vienna, 1941, p. 185.
213. R. W. Moshier, Ind. Eng. Chem., Anal. Ed., 15, 107 (1943).
214. H. Adkins and F. W. Cox, J. Amer. Chem. Soc., 60, 1151 (1938).
215. (a) M. Fields and E. R. Blout, J. Amer. Chem. Soc., 70, 930 (1948); (b) E. Paspaleev and K. Batzalova, Monatsh. Chem., 101, 166 (1970).
216. A. Winkel and G. Proske, Ber. Dtsch. Chem. Ges., 69, 1917 (1936).
217. M. B. Neimann and Z. V. Markina, Zavod. Lab., 13, 1177 (1947); Chem. Abstr. 43, 4185 (1949).
218. E. I. Fulmer, J. J. Kolfeubach, and L. A. Underkofler, Ind. Eng. Chem., Anal. Ed., 16, 469 (1944).
219. M. von Stackelberg and W. Stracke, Z. Elektrochem., 53, 118 (1949).
220. H. Adkins, R. M. Elofson, A. C. Rossow, and C. C. Robinson, J. Amer. Chem. Soc., 71, 3622 (1940).
221. R. H. Philip, Jr., T. Layloff, and R. N. Adams, J. Electrochem. Soc., 111, 1189 (1964).
222. S. S. Katiyar, M. Lalithambika, and G. C. Joshi, J. Electroanal. Chem., 53, 439 (1974); S. S. Katiyar, M. Lalithambika, and D. N. Dhar, J. Electroanal. Chem., 53, 449 (1974).
223. R. Lines, B. S. Jensen, and V. D. Parker, Acta Chem. Scand., B32, 510 (1978).
224. L. I Smith, I. M. Kolthoff, S. Wawzonek, and P. M. Ruoff, J. Amer. Chem. Soc., 63, 1018 (1941); I. M. Kolthoff and J. J. Lingane, Polarography, 2nd ed., Vol. II. Wiley-Interscience, New York, 1952, p. 699 ff.
225. I. M. Kolthoff and J. J. Lingane, Polarography, 2nd ed., Vol. I, Wiley-Interscience, New York, 1952, p. 153 ff.
226. D. J. Pietrzyk, Anal. Chem. Ann. Rev., 38, 278R (1966); 40, 194R (1968); 42, 139R (1970); 44, 457R (1972); 46, 52R (1974).
227. C. K. Mann and K. K. Barnes, Electrochemical Reactions in Nonaqueous Systems, Dekker, New York 1970, pp. 190–197.
228. B. R. Eggins, J. Chem. Soc., Chem. Commun., 1267 (1969).
229. T. Fujinaga, K. Izutsu, and T. Nomura, J. Electroanal. Chem., 29, 203 (1971).
230. A. E. Brodsky, L. L. Gordienko, and L. S. Degtiarev, Electrochim. Acta, 13, 1095 (1966).
231. A. I. Prokof'jev, N. N. Bubnov, S. P. Solodovnikov, T. I. Prokof'eva, A. A. Volod'kin, and V. V. Ershov, Izv. Akad. Nauk SSSR, Ser. Khim., 2337 (1975).
232. A. A. Volod'kin, V. V. Ershov, and T. I. Prokof'eva, Izv. Akad. Nauk SSSR, Ser. Khim., 1203 (1977); L. I. Kudinova, A. A. Volod'kin, V. V. Ershov, and T. I. Prokof'eva, Izv. Akad. Nauk SSSR, Ser. Khim., 1503 (1978).

233. G. A. Russell, in Radical Ions, (E. T. Kaiser and L. Kevan, eds.), Wiley-Interscience, New York, 1969, pp. 87–150.
234. S. Ishiwata and T. Nozaki, J. Pharm. Soc. Jpn., 71, 1257 (1951).
235. G. Shima, Mem. Coll. Sci. Kyoto Imp. Univ., A12, 377 (1929).
236. E. Brillas, J. J. Novoa, M. T. Pujol, and J. Virgili, An. Quim. Ser. A, 81, 52 (1985).
237. N. Takano, N. Takeno, and M. Morita, Nippon Kagaku Kaishi, 1753 (1983); Chem. Abstr., 100, 15603n (1984).
238. N. Takano, N. Takeno, and M. Morita, Denki Kagaku, 51, 483 (1983); Chem. Abstr., 100, 41558g (1984).
239. D. A. Jaeger, D. Bolikal, and B. Nath, J. Org. Chem., 52(2), 276 (1987).
240. T. Osa, T. Matsue, A. Yokozawa, and T. Yamada, Denki Kagaku, 32, 629 (1984).
241. R. A. Misra and A. K. Yadov, Electrochim. Acta, 25, 1221 (1980); H. Lund, Denki Kagaku, 45, 2 (1977).
242. S. G. Mairanovskii, V. P. Gul'tyai, L. A. Spiridonova, I. L. Mikhelashvii, and E. N. Prilezhajeva, Elektrokhimiya, 9, 1607 (1973).
243. C. Degrand, B. Grosjean, and P.-L Compagnon, Bull. Soc. Chim. Fr., 281 (1973).
244. C. Degrand and M Lacour, Bull Soc. Chim. Fr., 817 (1976).
245. C. Degrand, P.-L. Compagnon, and M. Lacour, Electrochim. Acta, 23, 705 (1978).
246. J. Sarrazin and A. Tallec, Electrochim. Acta, 22, 1189 (1977).
247. Showa Denko, K. K., Jpn. Kokai Tokkyo Koho, JP 58 32,900 [83 32900] 25 Feb. 1983; Chem. Abstr., 99, 88476n (1983).

11
Reduction of Azomethine Compounds

HENNING LUND / University of Aarhus, Aarhus, Denmark

I. Introduction 465

II. Derivatives of Ammonia 466
  A. Imines and Schiff bases 466
  B. Immonium salts 469
  C. Azomethine derivatives of acids 469

III. Derivatives of Hydrazine 470
  A. Azines 470
  B. Hydrazones 471
  C. Acylated hydrazones 473

IV. Derivatives of Hydroxylamine 475
  A. Oximes 475
  B. Alkylated and acylated oximes 476
  C. Amide oximes 477

V. Diazoalkanes 477

References 479

I. INTRODUCTION

In this chapter the reduction of compounds with a carbon-nitrogen double bond is discussed; besides the classic carbonyl derivatives, such as Schiff bases, hydrazones, and oximes, diazoalkanes are also treated.

The reduction of azomethine compounds depends on their structure and the experimental conditions, but most of the reactions can be described by relatively few schemes. In acidic solutions compounds of the type RR'C=N−Y are reduced with saturation of the double bond if Y is carbon, but if Y is a heteroatom (oxygen or nitrogen), the reduction follows Eqs. (1) through (3) [1]:

$$\mathrm{RR'C{=}N{-}Y} + \mathrm{H}^+ \rightleftharpoons \mathrm{(RR'C{=}N{-}Y)H}^+ \tag{1}$$

$$\mathrm{(RR'C{=}N{-}Y)H}^+ + 2e^- + 2\mathrm{H}^+ \longrightarrow \mathrm{RR'C{=}\overset{+}{N}H_2} + \mathrm{HY} \tag{2}$$

$$\mathrm{RRC{=}\overset{+}{N}H_2} + 2e^- + 2\mathrm{H}^+ \longrightarrow \mathrm{RR'CH{-}\overset{+}{N}H_3} \tag{3}$$

In alkaline solution many of the azomethine compounds are not reducible, but if the azomethine group is conjugated with suitable groups, the carbon-nitrogen double bond may be reduced.

In aprotic media, such as DMF or MeCN, cleavage of the N—Y bond is the usual reaction, but in some cases the double bond is saturated. For derivatives of aldehydes a special reaction may proceed when Y is a good leaving group; a nitrile is formed as a result of an elimination of HY possibly induced by an electrogenerated base.

Formally, compounds of the type RR'C=N—Y resemble vinylic compounds RR'C=CR"Y, where Y is a leaving group. It has been suggested [2,3] on the basis of a rather low reorganization energy λ connected with the electron transfer to vinyl halides that the first electron was added to the π^*-orbital of the π-electron system rather than to the σ^*-orbital of the C—Y single bond. Further work may show whether the analogy is more than formal and that the first electron is added to the π-electron system of RR'C=N—Y, which may be only the C=N double bond, or to the N—Y single bond.

It may be pointed out that the kinship of azomethine compounds, RR'C=NY, with vinyl halides, RR'C=CR"Y, is substantiated by the fact that vinyl halides of the type RCH=CR"Y form acetylenic compounds RC≡CR" by a base-induced elimination on reduction in aprotic medium [2,3], analogously to the formation of nitriles from RCH=NY under similar conditions [4].

From the analogy with vinyl compounds two factors are expected to be of prime importance for the reduction of an azomethine compound, the electron affinity of the π-electron system and the leaving group ability of Y^-; the latter is usually inversely proportional to the acidity of HY. Protonation of RR'C=NY would facilitate the reduction of the compound both by enhancing the electron affinity of the system and by increasing the leaving group ability substantially by changing the leaving group from Y^- to the much weaker base HY.

A review on the electrochemistry of azomethine compounds gives leading references up to early 1968 [5]; the reduction of oximes [6] and hydrazones has been reviewed more recently [7,8]. In this chpater the azomethine compounds are treated in four sections according to whether they are derivatives of ammonia, hydrazine, hydroxylamine, or a diazoalkane.

II. DERIVATIVES OF AMMONIA

A. Imines and Schiff Bases

The azomethine group is generally more easily reducible than the carbonyl group from which it is derived. Imines and Schiff bases of aldehydes and ketones are reducible in a pH region that depends on their structure; imines from simple, aliphatic ketones are reducible only in alkaline solution. The compounds are easily hydrolyzed, and in most cases the equilibrium [Eq. (4)] lies far to the left.

$$>C{=}O + HNRR' \rightleftharpoons >C(OH)NRR' \overset{H^+}{\rightleftharpoons} >C{=}\overset{+}{N}RR' + H_2O \qquad (4)$$

The easy hydrolysis of the imine interferes with its polarographic or preparative reduction in aqueous solution. Polarographically inactive unconjugated ketones may, however, be determined polarographically in solutions with a large excess of ammonia [1,9,10]. Preparative reduction of such imines may also be made by reducing a solution of the ketone in the presence of a large excess of ammonia or amine, if the potential is held at the reduction potential of the imine. Using this method cyclohexane was reduced in the presence of an excess methylamine to methylcyclohexylamine [1] and 2,2,6,6-tetramethyl-4-oxopiperidine (I) reduced to 4-amino-2,2,6,6-tetramethylpiperidine (II) in a solution of ammonia and ammonium chloride, Eq. (5) [11]. When high concentrations of I are used, some of the product (II) condenses with I to a reducible azomethine compound, and varying amounts of secondary amine, $(C_9H_{18}N)_2NH$, are formed.

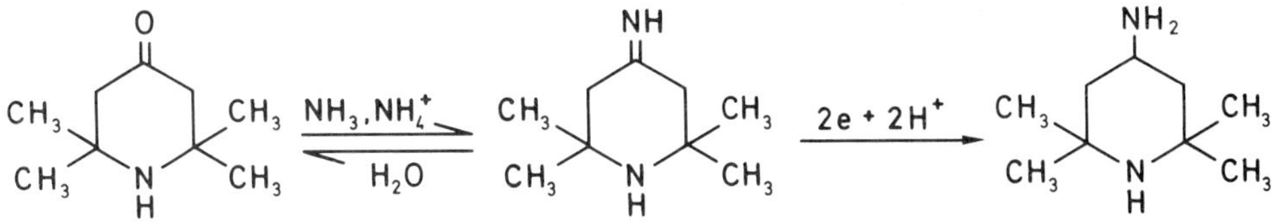

I II

(5)

The reaction is rather general for ketones with ammonia, primary and even secondary amines [10]; amino acids [12,13] have thus been prepared from α-ketoacids by reductive amination.

In a paired electrolysis amines have been N-alkylated in alcoholic solution. At the Pt(Pt) anode the alcohol was oxidized to an aldehyde that condensed with the amine to an imine; the imine was reduced at the cathode to a secondary amine. Benzylamine was thus electrolyzed to benzylethylamine in 96% yield (65% current yield) in ethanol containing 10 vol% of water [14].

Although only a few imines are stable in aqueous solution, reduction in cold 50% sulfuric acid at a lead cathode may give the amine [15,16]. In this connection it can be mentioned that 2,4-dihydroxybenzophenone oxime on reduction on the first wave in acid solution gives an imine stable enough to be isolated [17]. Schiff bases of some other benzophenone derivatives are stable in acid solution; their reduction is a straightforward saturation of the azomethine group [18,19].

Some Schiff bases of aromatic aldehydes and ketones give two, one-electron waves in aqueous acidic solution; at higher pH the waves merge to one, two-electron wave in much the same manner as carbonyl compounds [1, 20]. The product of the first one-electron reduction is a radical that may either dimerize or may be reduced further at a more negative potential. The yield of dimer depends on a number of factors, including proton availability, supporting electrolyte, cathode material, and concentration of Schiff base [21]. If the reduction is performed using a chiral supporting electrolyte, some chiral induction is observed in the secondary amine, but not in the dimer [22]; such reactions are discussed in Chap. 27, where the stereochemistry of the reduction of some Schiff bases is also discussed [23].

In MeCN most Schiff bases give a single two-electron wave; the anion radicals of imines are more basic than the corresponding ketyl anions and are able to abstract a proton from the medium [24]. Aliphatic ketimines behave similarly in DMF [23], whereas some aldehyde anils give two peaks in CV, depending on the sweep rate; the mechanism of the two-electron reduction of Schiff bases probably involves a solution electron transfer from the anion radical to the protonated anion radical [24,25]. In DMF the anion radicals from fluorenone imine and anil are stable on the time scale of CV; the dianions have half-lives of less than 1 ms and abstract a proton to give the anion of the fluorenyl amine [26].

Mixed couplings with Schiff bases have been described; reduction of benzylidene aniline (III) in the presence of acrylonitrile gave a mixture containing 1,5-diphenyl-2-pyrrolidone [27] and other products. Reductive coupling reactions are discussed in Chap. 22.

In nonaqueous solution azomethine compounds are usually stable, and several types have been reductively coupled with alkyl halides or carbon dioxide. Benzophenone anil (IV) gives thus on reduction in DMF-TBAI in the presence of methyl chloride a mixture of N- and C-methylated and N,C-dimethylated products, Eq. (6) [28]:

$$\underset{\text{IV}}{Ph_2C{=}NPh} \xrightarrow[CH_3Cl]{2e} \underset{57\%}{Ph_2CH{-}N(CH_3)Ph} + \underset{25\%}{Ph_2C(CH_3){-}NHPh} + \underset{23\%}{Ph_2C(CH_3){-}N(CH_3)Ph} \qquad (6)$$

Benzil dianil is under similar conditions reduced to a mixture of <u>cis</u> (12%) and <u>trans</u> (88%) 1,2-di-(<u>N</u>-methylanilino)-1,2-diphenylethane; interestingly enough, the <u>cis</u> compound seems to be the thermodynamically stable isomer [29]; the structure was proved by x-ray analysis [30]. In these couplings the reduced azomethine compound reacts with the alkyl halide in a nucleophilic displacement reaction.

Sterically hindered halides may be used in some cases; thus reduction of benzylidenephenethylamine in the presence of 2-bromonorbornane in DMF-TBAI gives <u>N</u>-(norborn-2-ylbenzyl)phenethylamine (33%), <u>N</u>-benzyl-<u>N</u>-norborn-2-ylphenethylamine (26%), and some <u>N</u>-benzylphenethylamine [31].

Another type of mechanism operates in the synthesis of amino acids by electroreductive coupling of alkyl halides with Schiff bases in DMF; in these reactions the alkyl halide chosen is reduced at less negative potentials than the azomethine derivative of an α-ketoester, and the coupling is proposed to be a nucleophilic addition of an alkyl anion [Eq. (7)] to the azomethine compound, Eq. (8); the benzyl group is afterward reductively removed, Eq. (9) [32,33]:

$$RX \xrightarrow{e} R^{\cdot} + X^- \xrightarrow{e} R^- + X^- \qquad (7)$$

$$PhCH_2N{=}C(R')COOR'' \xrightarrow[(2)\ H^+]{(1)\ R^-} PhCH_2NHCR(R')COOR \qquad (8)$$

$$PhCH_2NHCRR'{-}COOR'' \xrightarrow[H^+]{H_2,\ Pd(C)} PhCH_3 + H_2NCRR'{-}COOH \qquad (9)$$

Amino acids may also be prepared from azomethine compounds by reducing them in the presence of carbon dioxide [34–38]. Thus reduction of benzylidene aniline (III) in molten tetraethylammonium <u>p</u>-toluenesulfonate saturated with carbon dioxide at 140°C gave <u>N</u>-phenyl-phenylglycine (60% yield) [34], Eq. (10). In DMF a similar yield was obtained together with small amounts of a dimer and some phenylbenzylamine [38].

$$C_6H_5CH{=}NC_6H_5 \xrightarrow[CO_2]{2e\ +\ H^+} C_6H_5CH(NHC_6H_5)COO^- \tag{10}$$

III

B. Immonium Salts

Immonium salts have been used as model compounds for the study of the dimerization reaction in aprotic media, partly because no protons are involved in the reaction. It was shown that the electrodimerization process for immonium cations was purely radical [39,40].

Radicals from reduction of derivatives of benzophenone immonium salts are rather stable under aprotic conditions, whereas less conjugated radicals dimerize at the central carbon with rates depending on the conjugation [41].

From an electrochemical reduction of some cyclic immonium salts of the type N-cyclohexylidenepyrrolidinium fluoborate in aprotic media, the enamine and the saturated tertiary amine were the major products [42]. It was suggested that the reaction was a one-electron transfer followed by a rate-determining radical-radical disproportionation involving hydrogen atom transfer.

C. Azomethine Derivatives of Acids

Imidic esters are easily hydrolyzed in aqueous solution, but some are stable enough for a polarographic investigation in a cold solution. Reduction at a lead cathode in cold 1 M sulfuric acid of ethyl benzimidate gave benzylamine in 76% yield [43]. The reduction can probably be described by Eq. (11):

$$RC(OR'){=}\overset{+}{N}H_2 \xrightarrow{2e\ +\ 2H^+} RCH(OR')\overset{+}{N}H_3 \xrightarrow{-R'OH}$$

$$RCH{=}\overset{+}{N}H_2 \xrightarrow{2e\ +\ 2H^+} RCH_2\overset{+}{N}H_3 \tag{11}$$

Reduction in an aprotic medium such as MeCN would probably be advantageous.

S-Methylisothiobenzanilide is polarographically reducible in two steps; the product from the first two-electron reduction could be hydrolyzed to benzaldehyde, so the reduction route is probably analogous to Eq. (11) [5].

Amidines of aliphatic acids are generally not reducible in buffered solution. Some amidines derived from aromatic acids are reducible at very negative potentials and only in a narrow pH interval from slightly acid to alkaline solution. The reduction is similar to that of imidic esters, Eq. (11), and yields the amine [44].

Aromatic isothiocyanates have been investigated polarographically; in acid solution a four-electron wave or two poorly separated two-electron waves are found, whereas a single two-electron wave occurs in alkaline solution [45,46]. Controlled potential reduction of phenylisothiocyanate (V) in alkaline solution yielded thioformanilide (VI), Eq. (12) [45]. VI in acid solution is reduced at the second wave of V.

$$C_6H_5N{=}C{=}S \xrightarrow{2e\ +\ 2H^+} C_6H_5NHCH{=}S \tag{12}$$

V VI

III. DERIVATIVES OF HYDRAZINE

A. Azines

Benzalazine is in slightly acid solution reduced to benzylamine in a six-electron reduction, whereas it forms benzaldehyde benzylhydrazone in a two-electron reaction in alkaline solution [1]. Benzaldehyde benzylhydrazone is reduced in acid solution to benzylamine in a four-electron reaction at potentials slightly less negative than those of benzalazine, so it is not quite clear whether benzaldehyde benzylhydrazone is an intermediate in acidic solution or whether there is initially a cleavage of the central nitrogen-nitrogen bond.

It has been suggested [47] that 2-benzoylpyridine azine is reduced via the azo and hydrazine derivatives, rather than through the route suggested previously for benzalazine [1]. It would be rather exceptional if a hydrazine were reducible at the same potential as the azine.

Unsymmetrical azines of an aromatic and an aliphatic carbonyl compound, such as benzophenone cyclohexanone azine, are reduced in slightly acidic solution with cleavage of the nitrogen-nitrogen bond and saturation of the aromatic azomethine group, Eq. (13), whereas the aliphatic imine is then hydrolyzed, Eq. (14) [48]; unsymmetrical cyclic azines, such as 5-H-2,3-benzodiazepines, are reduced similarly, usually with ring contraction.

$$\mathrm{Ph_2C{=}N{-}N{=}CRR' + 4e + 5H^+ \longrightarrow Ph_2CH\overset{+}{N}H_3 + RR'C{=}NH} \tag{13}$$

$$\mathrm{RR'C{=}NH + H_2O \longrightarrow RR'C{=}O + NH_3} \tag{14}$$

Electrochemical reduction of 2,2,6,6-tetramethyl-4-oxopiperidine (I) azine or a mixture of I and hydrazine at a lead cathode in neutral medium gave a nearly quantitative yield of 4-amino-2,2,6,6-tetramethylpiperidine II [49].

Aromatic azines may be reduced in aprotic media in the presence of carbon dioxide to amino acids; thus ArCR=N—N=CRAr can be reduced (R = H) in MeCN saturated with CO_2 at −1.7 to −2.0 V (SCE) to N-arylideneamino-2-arylglycines or at −2.1 to −2.2 V to N-unsubstituted arylglycines [50], Eq. (15):

$$\mathrm{ArCH{=}N{-}N{=}CHAr} \xrightarrow[\mathrm{CO_2,\ 2e\ =\ 2H^+}]{-1.7\ \text{to}\ -2.0\ \mathrm{V}} \mathrm{ArCH(COOH)N{-}N{=}CHAr}$$

$$\mathrm{ArCH{=}N{-}N{=}CHAr} \xrightarrow[\mathrm{2CO_2,\ 6e\ +\ 6H^+}]{-2.1\ \text{to}\ -2.2\ \mathrm{V}} \mathrm{2\ ArCH(NH_2)COOH} \tag{15}$$

In DMF, benzalazine (VII) shows two, one-electron steps in CV. Preparative reduction of VII in the presence of an alkylating agent yields alkylated bis-hydrazines (VIII) or the dialkylated hydrazone (IX) [51], Eq. (16). An assignment of the stereochemistry of the bis-hydrazones, obtained by reduction of VII followed by protonation, could be achieved by cyclization with aldehydes to imidazolidines.

$$\underset{\text{VII}}{PhCH{=}N{-}N{=}CHPh} \xrightarrow[RX]{e} \underset{\text{VIII}}{Ph{-}CH{=}N{-}\underset{R}{N}{-}\underset{Ph}{CH}{-}\underset{Ph}{CH}{-}\underset{R}{N}{-}N{=}CHPh}$$

$$+ \underset{\text{IX}}{PhCH{=}N{-}\underset{R}{N}{-}\underset{R}{CHPh}} \qquad (16)$$

In some cases the anion radicals of azines in DMF are stable enough to allow their ESR spectra to be obtained. The anion radical of fluorenone azine is thus stable for months in DMF [52,53]. In the absence of a proton donor the anion radical is reduced to the dianion, which is stable on the time scale of CV. In the presence of a relatively strong proton donor, such as hexafluoro-2-propanol, fluorenone fluorenylhydrazine is obtained [53]; this is analogous to the reduction of benzalazine in aqueous alkaline solution [1].

B. Hydrazones

The electrochemistry of hydrazones, azines, and other derivatives of hydrazine is complicated by the possibility of tautomerization. The following forms must be considered [54,55]:

$$\underset{\text{X}}{RCH_2{-}CH_2{-}N{=}NR'} \rightleftharpoons \underset{\text{XI}}{RCH_2CH{=}NNHR''} \rightleftharpoons \underset{\text{XII}}{RCH{=}CHNHNHR'}$$

The hydrazone (XI) is the predominant form, and in acid and neutral solution no indication of either the azo- (X) or the ene-hydrazine form (XII) has been found. In alkaline solution, however, studies with tritium [56] have shown exchange of both the aldehyde hydrogen and of hydrogen atoms at the α-carbon, indicating the presence of both the forms (X and XII). The equilibrium constant in 0.1 M ethanolic KOH for the equilibrium ω-benzene-azotoluene ⇌ benzaldehyde phenylhydrazone is about 10^8, but for monoalkylhydrazones of aliphatic carbonyl compounds a higher proportion of the azo form is found. Acetaldehyde propylhydrazone is thus in t-butanol containing 0.02 M potassium t-butylate at 100°C in equilibrium with 3.7% of the azo form [57].

The different tautomeric forms may be reduced at different potentials, and the removal from the equilibrium of one of the forms by reduction leads to a continuous formation of the easiest reducible tautomer in an attempt to restore the equilibrium. The rate of the formation of the reducible tautomer is thus of importance for the electrode reaction, and electrochemical techniques that may be slow compared with the (usually unknown) rate of tautomerization are therefore of limited value in determining the position of the tautomeric equilibria.

The o- and p-phenylazophenols in alkaline solution and azo groups conjugated with certain heterocyclic nuclei may in some respects be treated as hydrazones; these are, however, discussed in Chap. 9 under azo compounds.

Phenylhydrazones of aromatic carbonyl compounds are generally reducible in acid solution in a four-electron reduction, which leads to the formation of

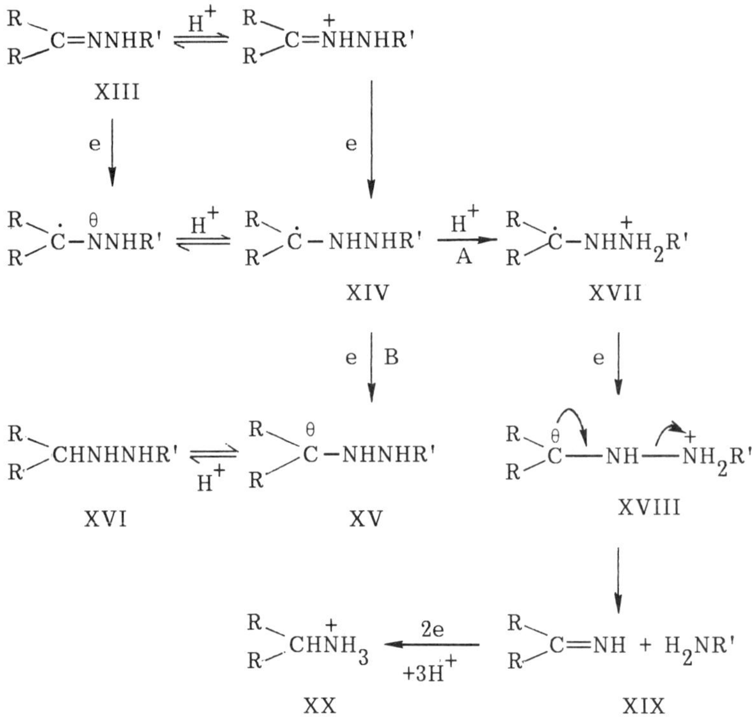

Scheme 1

aniline and another amine; in alkaline solutions only α,β-unsaturated phenylhydrazones are reducible, and the reduction leads in a two-electron reaction to a saturated phenylhydrazone [1]. Scheme 1 shows the reduction route in aqueous solution that has been suggested [1,58].

Polarographic evidence suggests that in acid solution a protonation occurs before the electron transfer; this is supported by the polarographic reduction of some quaternized hydrazones [59]. An argument in favor of this reaction sequence (scheme 1) is that the hydrazine (XVI) is not reducible under the conditions used for the reduction of the phenylhydrazone. Another argument is that the imine (XIX) is reduced at about the same potential or more easily than the phenylhydrazone.

After the uptake of a proton and an electron the radical (XIV) thus formed may accept either a proton (route A) or an electron (route B). In route A the protonated radical (XVII) accepts an electron, which results in a cleavage of the nitrogen-nitrogen bond with aniline as leaving group. The imine thus formed (XIX) is protonated and reduced further. In route B the radical (XIV) forms the carbanion (XV) on electron uptake, but as the aniline anion is a poor leaving group, no cleavage of the nitrogen-nitrogen bond occurs and the hydrazine (XVI) is formed on protonation. (In the scheme H^+ stands for any proton source.)

Phenylhydrazones of aliphatic ketones are not reducible in aqueous solution; however, trimethylhydrazonium salts of such ketones are reducible in neutral and alkaline media, Eq. (17) [60]:

$$C_6H_5CH_2C(CH_3)=N\overset{+}{N}(CH_3)_3 + 4e + 3H^+ \longrightarrow C_6H_5CH_2CH(NH_2)CH_3 + (CH_3)_3N \quad (17)$$

Phenylhydrazones of aromatic aldehydes have been investigated polarographically in DMF [61,62]; they show two waves, the first being a one-electron wave. Reduction at the limiting current of the first wave of the phenylhydrazones of benzoin and salicylaldehyde showed aniline as a product.

Fluorenone fluorenylhydrazone may be reduced in two ways in DMF, depending on the availability of protons. If the anion radical is not protonated by either an added proton donor or starting material, decomposition occurs by carbon-nitrogen bond cleavage to give fluorene and fluorenone hydrazone. In the presence of a proton donor reduction with cleavage of the nitrogen-nitrogen bond occurs; the final product is 9-aminofluorene [53].

Reduction of 1-arylidenetrimethylhydrazinium salts in DMF yields benzonitrile. The reaction is proposed to proceed via an electron uptake and cleavage of the N–N bond followed by a loss of a hydrogen atom to benzonitrile [4]. An alternative possibility is that the base produced during the reduction eliminates a proton and trimethylamine. In this connection it is of interest that the dimethylhydrazone of nitrobenzaldehyde undergoes base-induced cleavage to the nitrile at potentials of the second wave or by an electrochemically generated base, the anion of 4-nitro-N-methylimidazole [63]; other benzaldehyde hydrazones behave similarly [63].

Several substituted phenylhydrazones and other hydrazones have been investigated [65–67]; nitro groups are generally reduced before the hydrazone, but at a suitable potential the phenylhydrazone part of the molecule is reduced according to the general scheme.

C. Acylated Hydrazones

Acylated hydrazones of aromatic carbonyl compounds are reducible in acid solution following the usual scheme; anisylidene benzohydrazide was thus at pH 4.25 reduced to 4-methoxybenzylamine and benzamide, Eq. (18) [68]:

$$CH_3OC_6H_4CH=NNHCOC_6H_5 \xrightarrow[\text{pH } 4.25]{4e + 5H^+} CH_3OC_6H_4CH_2\overset{+}{N}H_3 + H_2NCOC_6H_5 \quad (18)$$

In alkaline medium benzylidene benzohydrazide is reduced in a two-electron reaction to N-benzoyl-N-benzylhydrazine, Eq. (19):

$$C_6H_5CH=NNHCOC_6H_5 \xrightarrow[\text{pH } 12]{2e + 2H^+} C_6H_5CH_2NHNHCOC_6H_5 \quad (19)$$

The reduction of nicotinoyl- and isonicotinoylhydrazones of several carbonyl compounds has been investigated. The reaction involves the expected reduction of the hydrazone and also the reduction of the amide to an alcohol [69].

The reaction schemes (18) and (19) usually seem to be followed in protic solvents; for example, benzhydrylidene hydrazides of disubstituted glycolic acids [70] and bis-benzylidene succinic hydrazide derivatives [71] have been reported to be reduced according to Eqs. (18) and (19).

In DMF, acylated hydrazones of aromatic aldehydes are reduced first to the acylated benzylhydrazine (XXI) in two, one-electron steps. At a more negative potential XXI is reduced further [62]. Calculations of the LUMO of the hydrazone fragment indicate a difference between acylhydrazones and arylhydrazones.

Tosylhydrazones behave as weak acids on reduction in DMF at a platinum electrode; the first peak of fluorenone tosylhydrazone (XXII) is due to reduction of a proton with formation of the tosylhydrazone anion (XXIII) [72], Eq. (20):

$$\underset{\text{XXII}}{\text{Fl=NNHTs}} + e \rightleftharpoons \text{Fl=NNHTs}^{\dot{-}} \longrightarrow \tfrac{1}{2}H_2 + \underset{\text{XXIII}}{\text{Fl=N}\bar{\text{N}}\text{Ts}} \qquad (20)$$

A similar formation of bases and hydrogen has also been described for both carbon and nitrogen acids [73].

Compound XXIII is reduced at a more negative potential in a reaction reversible on the time scale of CV, but the dianion radical reacts further on the coulometric time scale. The product distribution is dependent on the pK_A of the added proton donors. The following scheme is suggested, Eqs. (21) through (23):

$$\text{Fl=N}\bar{\text{N}}\text{Ts} + e \rightleftharpoons \text{Fl=NNTs}^{2\dot{-}} \xrightarrow[\text{slow}]{[H^+]} \text{Fl}^{\cdot}\text{-NH}\bar{\text{N}}\text{Ts} \qquad (21)$$

$$\text{Fl}^{\cdot}\text{-NH}\bar{\text{N}}\text{Ts} \longrightarrow \text{Fl=NH} + \text{TsN}^{\dot{-}} \qquad (22)$$

$$\text{Fl=NH} + 2e + H^+ \longrightarrow \text{FlNH}_2^- \qquad (23)$$

The nitrene anion radical $TsN^{\dot{-}}$ is proposed to be an intermediate; it may either abstract a hydrogen atom or be protonated and further reduced, depending on the proton donor.

Several polarographic investigations of Girard hydrazones, semi-, and thiosemicarbazones have been made [1,74–82], but relatively few controlled potential reductions have been reported [1,78–82]. From these it appears that in acid solution a four-electron reduction is found and that the first step is a hydrogenolysis of the nitrogen-nitrogen bond; in some cases a catalytic hydrogen evolution is found [82]. In strongly acid solution there is often a tendency found to a stepwise reduction. Thus, benzaldehyde thiosemicarbazone [78] gives at pH 0 two waves at −0.66 and −0.73 V (SCE); when alcohol is present, the separation of the waves is poor or not visible at all. In alkaline solution some of the compounds are reducible, and then a two-electron reduction is found that results in a saturation of the carbon-nitrogen double bond to the hydrazine. At intermediate pH values both reduction paths may be followed simultaneously and the relative importance of the two routes depends on many factors, such as pH, buffer concentration, temperature, and solvent. Thus, benzaldehyde and benzophenone semicarbazone [1], p-acetamidobenzaldehyde thiosemicarbazone [78], p-acetamidobenzaldehyde S-methylthiosemicarbazone [79], and cyclopentanone and benzophenone Girard hydrazone [79] all yielded the expected amine at low pH; at high pH values benzophenone Girard-D hydrazone [79] and benzaldehyde semicarbazone [81] were reduced in high yield to the acylated hydrazine; and at medium pH several of the compounds gave mixtures of the amine and the hydrazine [1,78,79].

Benzaldehyde thiobenzoylhydrazone (XXIV) is reduced in acid solution according to the general reduction scheme in which the initial step is the hydrogenolysis of the nitrogen-nitrogen bond. At pH < 2 three polarographic waves are found [59], and on the basis of controlled potential reductions the following steps were suggested:

$$\text{1. Wave:}\quad \underset{\text{XXIV}}{C_6H_5CH{=}NNHCSC_6H_5} \xrightarrow[2e\,+\,2H^+]{H^+} C_6H_5CH{=}\overset{+}{N}H_2 + H_2NCSC_6H_5 \tag{24}$$

$$\text{2. Wave:}\quad C_6H_5CH{=}\overset{+}{N}H_2 \xrightarrow{2e\,+\,2H^+} C_6H_5CH_2\overset{+}{N}H_3 \tag{25}$$

$$\text{3. Wave:}\quad C_6H_5CSNH_2 \xrightarrow[2e\,+\,2H^+]{H^+} C_6H_5CH(SH)\overset{+}{N}H_3 \tag{26}$$

followed by hydrolysis and reduction of benzaldimine and benzaldehyde.

In alkaline solution a two-electron reduction is found that results primarily in a saturation of the carbon-nitrogen double bond; the reducible species is probably the following azomethine compound, $C_6H_5CH{=}NN{=}C(S^-)C_6H_5$.

The reduction of cyclic acylated hydrazones like pyridazinones [59] and phthalazinones [83] is discussed in Chap. 18.

The anodic oxidation of hydrazones and semicarbazones to heterocyclic compounds is discussed in Chap. 18.

IV. DERIVATIVES OF HYDROXYLAMINE

A. Oximes

Among the azomethine derivatives of hydroxylamine the ordinary oximes are the most important, and many polarographic and electrolytic investigations of oximes have been made [1,84,85]. Most oximes are reducible in acid solution, only few in alkaline solution; polarographic data prove that it is the protonated oxime that is the reducible species in acid and neutral solutions. Classic reductions in sulfuric acid at lead cathodes or controlled potential reductions in acid solution at mercury cathodes yield the amine in a four-electron reduction. As the hydroxylamine corresponding to the oxime is not reducible under the conditions employed, it has been suggested [1] that the reduction proceeds as in Eq. (27):

$$RR'C{=}NOH \overset{H^+}{\rightleftharpoons} (RR'C{=}NOH)H^+ \xrightarrow{2e\,+\,2H^+} (RR'C{=}\overset{+}{N}H_2) + H_2O \xrightarrow{2e\,+\,2H^+} RR'CH\overset{+}{N}H_3 \tag{27}$$

As the site of the protonation at the species to which the transfer of electrons occurs is not known, both a protonation at N and at O may be considered.

The reduction at a lead cathode of 2,2,6,6-tetramethyl-4-hydroxyiminopiperidine as such or prepared in situ is an excellent method for the preparation of II [49].

Cyclic α,β-unsaturated oximes, such as testosteronepropionate oxime, are reduced in two, two-electron steps in acid solution. The first reduction yields the imine, which is hydrolyzed to the unsaturated ketone. Reduction at the potential of the second wave yields the unsaturated amine [1].

The postulated intermediate, the imine, can in most cases not be isolated or detected by classic polarography. The reduction of 2,4-dihydroxybenzophenone oxime occurs [87], however, in two steps, and reduction at the potential of the first wave produced the imine, which in this case was sufficiently stable to permit its isolation.

If the reaction path shown is as general as the available evidence suggests, attempts to reduce protonated oximes electrolytically to hydroxylamines are not likely to succeed. Some unprotonated oximes, such as benzaldoxime and benzophenone oxime, are reducible in not too strongly alkaline solution; the oxime anion is not reducible. An investigation of the reduction of benzaldoxime in alkaline solution [81] showed that some benzylhydroxylamine is formed under these conditions.

Monoximes of α-diketones are reduced to α-aminoketones through an intermediate enaminol [88]; dioximes of α-diketones are reduced to the diamines through diimines and endiamines.

Polarographically there is a difference between the syn- and the anti-oxime in unbuffered solutions having tetraalkylammonium ions as supporting electrolyte [89]. syn-Benzaldoxime and other syn-oximes give two waves in this medium; the first wave at −1.84 V (SCE) of syn-benzaldoxime is kinetically controlled, whereas the second one at −2.2 V (SCE) is diffusion controlled. anti-Benzaldoxime gives only one wave [at −1.84 V (SCE)] which is diffusion controlled. It has been suggested [89] that the two waves of the syn form reflect the tautomeric equilibrium $>C{=}NOH \rightleftharpoons >C{=}\overset{+}{N}HO^-$.

In some cases stereoselective electrochemical reduction of oximes has been found [90–92]; this is discussed in Chap. 27.

Aldoximes may be transformed into nitriles via nitrile oxides, obtained by indirect anodic oxidation of the aldoxime in methanol. followed by cathodic reduction of the nitrile oxide [92].

In DMF some oximes have been investigated. They are reduced in an irreversible four-electron reduction. anti-Benzaldoxime is thus in the presence of phenol reduced to benzylamine, whereas no benzylamine is obtained in carefully dried DMF. Under aprotic conditions some N,N'-dibenzylhydrazine was obtained [93]. The anion of benzaldoxime gives in CV a peak about 400 mV more negative than that of the oxime; the electrode reaction has not yet been established [60].

B. Alkylated and Acylated Oximes

Oximes may be alkylated at nitrogen or oxygen; both types are electrolytically reducible [1,94,95] in their protonated form. The nitrones are generally also reducible in alkaline solution, whereas the O-methyl ethers of oximes are not reducible in ordinary media containing alkali metal ions. In solutions containing quaternary ammonium *p*-toluenesulfonates, some *O*-alkylated oximes give a polarographic wave.

N-alkyl-substituted oximes both of aliphatic and aromatic aldehydes are reduced in a single four-electron wave in slightly acid or alkaline solution [1,94,95]. *N*-phenylbenzaldoxime is reduced in two or more steps; the first two-electron reduction results in the formation of benzalaniline, which in acid solution is reduced in two, one-electron reactions [94].

The reduction of the *O*-methyl oximes [94] is probably analogous to the reduction of the oximes in that the nitrogen-oxygen bond is cleaved before the saturation of the carbon-nitrogen double bond. The similarity in the reduction in acid solution of the oxime and its alkylated derivatives is understandable when the close resemblance between the electroactive forms is considered:

$$\mathrm{RCH{=}\overset{+}{N}H(OH)} \qquad \mathrm{RCH{=}\overset{+}{N}R'(OH)} \qquad \mathrm{RCH{=}\overset{+}{N}H(OR')}$$

Cyclic voltammetry of acylated *syn*-benzaldoxime (XXV) shows three peaks; the two first are irreversible, whereas the third is reversible and is found at the potential of benzonitrile (XXVI). Electrolysis of an acylated *syn*-benzaldoxime in DMF without addition of a proton donor yields some XXVI and other products. XXVI is probably formed by an elimination from XXV catalyzed by the base (EGB) generated during the reduction [96]; this is analogous to the reduction of vinyl halides [2,3].

$$\underset{\text{XXV}}{\mathrm{PhCH{=}NOCOR}} + \mathrm{EGB} \longrightarrow \underset{\text{XXVI}}{\mathrm{PhCN}} + \mathrm{RCOO^-} + \mathrm{EGBH^+} \tag{28}$$

N-acylated hydroxylamines may exist in several tautomeric forms; the protonated hydroxamic acid may be reducible. Protonation of a hydroxamic acid takes place mainly on oxygen to $\mathrm{R{-}C(OH){=}\overset{+}{N}HOH}$ [97]. Aromatic hydroxamic acids with electron-attracting groups, such as *p*-cyanobenzhydroxamic acid or isonicotinic hydroxamic acid, are reducible to the amides [98].

C. Amide Oximes

Aromatic amide oximes are reduced in acid solution similarly to other reducible derivatives of hydroxylamine [1,99]; the first step is a reductive loss of oxygen. The reduction of benzamide oxime thus yields benzamidine [1], Eq. (29):

$$\mathrm{PhC(NH_2){=}NOH} \overset{H^+}{\rightleftharpoons} \mathrm{PhC(NH_2){=}\overset{+}{N}HOH} \xrightarrow[2H^+]{2e} \mathrm{PhC(NH_2){=}\overset{+}{N}H_2} + \mathrm{H_2O} \tag{29}$$

V. DIAZOALKANES

A special group of azomethine derivatives is diazoalkanes, $\mathrm{RR'C{=}\overset{+}{N}{=}\overset{-}{N}{:}}$. They have been investigated in some detail electrochemically after it was shown that dissociative electron attachement to diazoalkanes was a convenient way of preparing carbene anion radicals in the gas phase.

Reduction of diazoalkanes, such as 9-diazofluorene [100] or diazophenylmethane [101,102], in DMF or MeCN gives a high yield of the corresponding azine and sometimes the ketone and other compounds as minor products, Eq. (30). The proposal that a carbene anion radical was a detectable intermediate in the reaction was later proved unlikely [103,104].

$$2RR'CN_2 \xrightarrow{e} RR'C{=}N{-}N{=}CRR' + N_2 \qquad (30)$$

The reaction is a chain reaction with a dimerization of the diazoalkane anion radical as the first chemical step; the structure of the two kinds of intermediate dimerized dianions detected by CV is not settled [105].

A carbonyl group neighboring the diazo group changes the reaction route as the carbene anion radical formed in the unimolecular decomposition of the anion radical of the diazo compound has a longer lifetime than those obtained from diazoalkanes. The reduction of diethyl diazomalonate in DMF or MeCN in the presence of a proton donor, such as hexafluoro-2-propanol, guanidinium perchlorate, or benzoic acid, has been suggested [106] to follow Eqs. (31) through (34) with formation of diethyl malonate, n = 2. (See also Chap. 3.)

$$(EtOOC)_2C{=}N_2 + e \rightleftharpoons \underset{\text{XXVII}}{(EtOOC)_2C{=}N_2^{\dot{-}}} \qquad (31)$$

$$(\text{XXVII}) \longrightarrow N_2 + (EtOOC)_2C^{\dot{-}} \qquad (32)$$

$$(EtO_2C)_2C^{\dot{-}} + HA \longrightarrow (EtO_2C)_2CH^{\cdot} + A^- \qquad (33)$$

$$(EtO_2C)_2CH^{\cdot} + (\text{XXVII}) \longrightarrow (EtO_2C)_2CH^- \xrightarrow[-A^-]{HA} (EtO_2C)_2CH_2 \qquad (34)$$

The diazocompounds, such as azibenzil [107], 2,3,4,5-tetraphenyldiazocyclopentadiene [108], diazodimedone [109], and diethyl diazomalonate [106], whose anion radicals decompose to relatively stable carbene anion radicals, are all derived from compounds with relatively low pK_A values (from 13 to 18 in DMSO). It thus seems essential for the lifetime of the carbene anion radical that the negative charge can be accommodated in the structure [110].

In aqueous solution diazoacetophenone has been investigated by polarography and coulometric reduction; between pH 6 and 8 three waves were found with n = 6, 2, and 2. The following reactions were proposed [111], Eqs. (35) through (37):

$$C_6H_5COCH{=}\overset{+}{N}{=}\overset{-}{N} + 7H^+ + 6e \longrightarrow C_6H_5COCH_2NH_2 + \overset{+}{N}H_4 \qquad (35)$$

$$C_6H_5COCH_2NH_2 + 3H^+ + 2e \longrightarrow C_6H_5COCH_3 + \overset{+}{N}H_4 \qquad (36)$$

$$C_6H_5COCH_3 + 2H^+ + 2e \longrightarrow C_6H_5CHOHCH_3 \qquad (37)$$

REFERENCES

1. H. Lund, Acta Chem. Scand., 13, 249 (1959).
2. N. Gatti, W. Jugelt, and H. Lund, Acta Chem. Scand., B41, 000 (1987).
3. N. Gatti, S. U. Pedersen, and H. Lund, Acta Chem. Scand., B42, 11 (1988).
4. V. K. Ivanova, L. N. Orlova, V. I. Savin, and Yu. P. Kitaev, Izk. Akad. Nauk SSSR, Ser. Khim., 2474 (1974).
5. H. Lund, in The Chemistry of the Carbon-Nitrogen Double Bond, (S. Patai, ed.), Wiley-Interscience, New York, 1970, pp. 505–564.
6. M. Ya. Fioshin, V. T. Novikov, and I. A. Avrutskaya, Tr.-Mosk. Khim.-Technol. Inst. D. I. Mendeleeva, 117, 33 (1981); Chem. Abstr., 97, 161872 (1982).
7. Yu. P. Kitaev and B. I. Buzykin, Hydrazones, Nauka, Moscow, 1974.
8. H. M. Fahmy, Ann. Chim. (Rome), 75, 457 (1985).
9. P. Zuman, Nature, 175, 585 (1950); P. Zuman, Coll. Czech. Chem. Commun., 15, 839 (1950).
10. T. Pienemann and H. J. Schäfer, Synthesis, 1005 (1987).
11. H. Lund, unpublished results (1986).
12. B. Ohtani, K. Nakagawa, S. Nishimoto, and T. Kagiya, Chem. Lett., 1917 (1986).
13. E. A. Jeffery and A. Meisters, Aust. J. Chem., 31, 73 (1978).
14. E. A. Jeffery, O. Johansen, and A. Meisters, Aust. J. Chem., 31, 79 (1978).
15. K. Brand, Ber. Dtsch. Chem. Ges., 42, 3461 (1909).
16. H. D. Law, J. Chem. Soc., 154 (1912).
17. H. Lund, Acta Chem. Scand., 18, 563 (1964).
18. H. Lund, Acta Chem. Scand., 14, 359 (1960).
19. H. Lund, P. Lunde, and F. Kaufmann, Acta Chem. Scand., 20, 1631 (1966).
20. L. Holleck and B. Kastening, Z. Elektrochem., 60, 127 (1956); B. Kastening, L. Holleck, and G. A. Melkonian, Z. Elektrochem., 60, 130 (1956).
21. L. Horner and D. H. Skaletz, Liebigs Ann. Chem., 1210 (1975).
22. L. Horner and D. H. Skaletz, Liebigs Ann. Chem., 1365 (1977).
23. A. J. Fry and R. G. Reed, J. Amer. Chem. Soc., 91, 6448 (1969).
24. C. P. Andrieux and J. M. Savéant, J. Electroanal. Chem., 33, 453 (1971).
25. J. E. Kuder, H. W. Bibson, and D. Wychick, J. Org. Chem., 40, 875 (1975).
26. J. H. Barnes, F. M. Triebe and M. D. Hawley, J. Electroanal. Chem., 139, 395 (1982).
27. M. M. Baizer, J. D. Anderson, J. H. Wagenknecht, M. R. Ort, and J. P. Petrovich, Electrochim. Acta, 12, 1377 (1967).
28. H. Lund and J. Simonet, C. R. Acad. Sci., Ser. C., 275, 837 (1972). H. Lund and J. Simonet, Bull. Soc. Chim. Fr., 1843 (1973).
29. J. Simonet and H. Lund, Bull. Soc. Chim. Fr., 2547 (1975).
30. G. A. P. Dalgaard and R. G. Hazell, Acta Crystallogr., B34, 3099 (1978).
31. U. Hess and S. Czapla, Z. Chem., 25, 334 (1985).
32. T. Iwasaki and K. Harada, J. Chem. Soc., Chem. Commun., 338 (1974).
33. T. Iwasaki and K. Harada, J. Chem. Soc., Perkin 1, 1730 (1977).

34. N. L. Weinberg, A. K. Hoffmann, and T. B. Reddy, Tetrahedron Lett., 2271 (1971).
35. U. Hess, Z. Chem., 20, 148 (1980).
36. U. Hess and M. Ziebig, Pharmazie, 37, 107 (1982).
37. S. Kato, Jpn. Kokai Tokkyo Koho, 80-06405, 80-11126 (1980); Chem. Abstr., 93, 7860, 26119 (1980).
38. U. Hess and R. Thiele, J. Prakt. Chem., 324, 385 (1982).
39. C. P. Andrieux and J. M. Savéant, Bull. Soc. Chim. Fr., 4671 (1968).
40. C. P. Andrieux and J. M. Savéant, J. Electroanal. Chem., 26, 223 (1970).
41. C. P. Andrieux and J. M. Savéant, J. Electroanal. Chem., 28, 446 (1970).
42. J. P. Kerr and P. E. Iversen, Acta Chem. Scand., B32, 405 (1978).
43. H. Wenker, J. Amer. Chem. Soc., 57, 772 (1935).
44. P. O. Kane, Z. Anal. Chem., 173, 50 (1960).
45. H. Lund, Coll. Czech. Chem. Commun., 25, 3313 (1960).
46. A. M. Kardos, J. Volke, and P. Kristian, Coll. Czech. Chem. Commun. 30, 931 (1965).
47. M. A. Gomez Nieto, M. D. Luque de Castro, and M. Valcarcel, Electrochim. Acta, 28, 1725 (1983).
48. R. Fuhlendorff and H. Lund, Acta Chem. Scand., B42, 52 (1988).
49. I. I. Surov, I. A. Avrutskaya, E. Yu. Kodintseva, and M. Ya. Fioshin, Elektrokhimiya, 20, 1276 (1984); M. Ya. Fioshin, I. A. Avrutskaya, I. I. Surov, and V. T. Novikov, Coll. Czech. Chem. Commun., 52, 182 (1987).
50. S. Kato, Jpn. Kokai Tokkyo Koho, 80-14841 (1980); Chem. Abstr., 93, 176065 (1980).
51. T. Troll, G. W. Ollmann, and H. Leffler, Tetrahedron Lett., 4241 (1979).
52. Yu. P. Kitaev, V. Kh. Ivanova, V. I. Savin, and L. I. Trifonova, Izv. Akad. Nauk SSSR, Ser. Khim., 2179 (1972); Yu. P. Kitaev, V. Kh. Ivanova, A. Sh. Mukhtarov, L. N. Orlova, and A. Ladygin, Izv. Akad. Nauk SSSR, Ser. Khim., 72 (1974).
53. F. M. Triebe and M. D. Hawley, J. Electroanal. Chem., 125, 421 (1981).
54. Yu. P. Kitaev and A. E. Arbuzov, Izv. Akad. Nauk SSSR, Otd. Khim. Nauk, 1037 (1957).
55. Yu. P. Kitaev and T. V. Troepol'skaya, Izv. Akad. Nauk SSSR, Otd. Khim. Nauk, 454, 465 (1963).
56. H. Simon and W. Moldenhauer, Chem. Ber., 100, 1949 (1967).
57. B. V. Ioffe and V. S. Stopskij, Tetrahedron Lett., 1333 (1968).
58. H. Lund and E. T. Jensen, Acta Chem. Scand., 24, 1867 (1970).
59. H. Lund, Disucss. Faraday Soc., 45, 193 (1968).
60. H. Lund, unpublished observation, 1987.
61. T. V. Troepol'skaya, E. N. Munin, and Yu. P. Kitaev, Izv. Akad. Nauk SSSR, Ser. Khim., 982 (1979).
62. Yu. P. Kitaev, T. V. Troepol'skaya, L. V. Ermolaeva, and E. N. Munin, Izv. Akad. Nauk SSSR, Ser. Khim., 1736 (1985).
63. Yu. M. Kargin, V. Z. Latypova, M. Yu. Kitaeva, A. A. Vafina, R. M. Zaripova, and A. V. Il'yasov, Izv. Akad. Nauk SSSR, Ser. Khim., 2410 (1984); Yu. M. Kargin, M. Yu. Kitaeva, V. Z. Latypova, R. M. Zarepova, and A. V. Il'yasov, Izv. Akad. Nauk SSSR, Ser. Khim., 607 (1988).

64. Yu. P. Kitaev and I. M. Skrebkova, Zh. Obshch. Khim., 38, 509 (1967).
65. Yu. P. Kitaev and I. M. Skrebkova, Zh. Obshch. Khim., 38, 1204 (1967).
66. Yu. P. Kitaev, I. M. Skrebkova, and L. I. Maslova, Izv. Akad., Nauk SSSR, Ser. Khim., 2194 (1970).
67. A. A. Vafina, A. V. Il'yasov, I. D. Morozova, I. M. Skrebkova, and Yu. P. Kitaev, Izv. Akad. Nauk SSSR, Ser. Khim., 1731 (1972).
68. H. Lund, Electrochim. Acta., 28, 395 (1983).
69. Yu. P. Kitaev and G. K. Budnikov, Izv. Acad. Nauk SSSR, Ser. Khim., 554, 562 (1967).
70. G. S. Posyagin, E. Yu. Posyagin, and I. S. Berdinskii, Zh. Obshch. Khim., 56, 1361 (1986).
71. H. Y. F. El-Baradie, M. M. Ghoneim, R. M. Issa, and L. M. Madkour, Bull. Electrochem., 3, 23 (1987).
72. D. A. Van Galen, J. H. Barnes, and M. D. Hawley, J. Org. Chem., 51, 2544 (1986).
73. A. V. Bukhtiarov, V. N. Golyshin, A. P. Tomilov, and O. V. Kuz'min, Zh. Obshch. Khim., 58, 857 (1988); A. V. Bukhtiarov, V. V. Mikheev, A. V. Lebedev, A. P. Tomolov, and O. V. Kuz'min, Zh. Obshch. Khim., 59, 796 (1989).
74. V. Prelog and O. Häfliger, Helv. Chim. Acta, 33, 2088 (1949).
75. M. Brezina, V. Volkova, and J. Volke, Coll. Czech. Chem. Commun., 19, 894 (1954).
76. P. Souchay and M. Graizon, Chim. Anal., 36, 85 (1954).
77. Yu. P. Kitaev, G. K. Budnikov, and A. E. Arbuzov, Izv. Akad. Nauk SSSR, Otd. Khim. Nauk, 824 (1961); Yu. P. Kitaev and G. K. Budnikov, Zh. Obshch. Khim., 33, 1396 (1963).
78. Y. Asahi, Chem. Pharm. Bull. (Tokyo), 11, 930 (1963).
79. M. Masui and H. Ohmori, Chem. Pharm. Bull. (Tokyo), 12, 877 (1964).
80. B. Fleet and P. Zuman, Coll. Czech. Chem. Commun., 32, 2066 (1967).
81. H. Lund, Tetrahedron Lett., 3651 (1968).
82. D. Fleury and M. B. Fleury, Coll. Czech. Chem. Commun., 36, 331 (1971).
83. H. Lund, Coll. Czech. Chem. Commun., 30, 4237 (1965).
84. P. Souchay and S. Ser, J. Chim. Phys., 49, C172 (1952).
85. R. M. Elofson and J. G. Atkinson, Can. J. Chem., 34, 4 (1956).
86. H. J. Gardner and W. P. Georgans, J. Chem. Soc., 4180 (1956).
87. H. Lund, Acta Chem. Scand., 18, 563 (1964).
88. J. Armand, L. Boulares, J. Pinson, and P. Souchay, Bull. Soc. Chim. Fr., 1918 (1971); J. Armand, P. Bassinet, and L. Boulares, C. R. Acad. Sci., 227, 695 (1973).
89. (a) N. Tütülkoff and St. Buduroff, C. R. Acad. Bulg. Sci., 6(3), 5 (1953); (b) N. Tütülkoff and E. Paspaleev, C. R. Acad. Bulg. Sci., 14, 159 (1961); (c) E. Paspaleev, Monatch. Chem., 97, 230 (1966).
90. A. J. Fry and J. H. Newberg, J. Amer. Chem. Soc., 89, 6374 (1987); M. Jubault, E. Raoult, J. Armand, and L. Boulares, J. Chem. Soc., Chem. Commun., 250 (1977).
91. S. Abe, T. Fuchigami, and T. Nonaka, Chem. Lett., 1033 (1983).

92. T. Shono, Y. Matsumura, K. Tsubata, T. Kamada, and K. Kishi, J. Org. Chem., 54, 2249 (1989).
93. G. A. Vagina, T. V. Troepol'skaya, and Yu. P. Kitaev, Izv. Akad. Nauk SSSR, Ser. Khim., 2488 (1983).
94. P. Zuman and O. Exner, Coll. Czech. Chem. Commun., 30, 1832 (1965).
95. T. Kubota, H. Miyanzaki, and Y. Mori, Bull. Chem. Soc. Jpn., 40, 245 (1967).
96. R. Fuhlendorff and H. Lund, unpublished observation, 1987.
97. O. Exner and B. Kakác, Coll. Czech. Chem. Commun., 28, 1656 (1963).
98. H. Lund, Talanta, 12, 1065 (1965).
99. J. Mollin and F. Kasparek, Coll. Czech. Chem. Commun., 25, 451 (1960); 26, 2438 (1961).
100. R. N. McDonald, J. R. January, K. J. Borhani, and M. D. Hawley, J. Amer. Chem. Soc., 99, 1268 (1977).
101. R. N. McDonald, K. J. Borhani, and M. D. Hawley, J. Amer. Chem. Soc., 100, 995 (1978).
102. R. N. McDonald, M. F. Triebe, J. R. January, K. J. Borhani, and M. D. Hawley, J. Amer. Chem. Soc., 102, 7867 (1980).
103. D. Bethell, P. J. Galsworthy, K. L. Handov, and V. D. Parker, J. Chem. Soc., Chem. Commun., 534 (1980).
104. V. D. Parker and D. Bethell, Acta Chem. Scand., B35, 691 (1981).
105. V. D. Parker and D. Bethell, J. Amer. Chem. Soc., 108, 895 (1986).
106. D. A. Van Galen, M. P. Young, M. D. Hawley, and R. N. McDonald, J. Amer. Chem. Soc., 107, 1465 (1985).
107. D. Bethell, L. J. McDowall, and V. D. Parker, J. Chem. Soc., Chem. Commun., 308 (1984).
108. D. Bethell and V. D. Parker, J. Chem. Res.(S) 116 (1987).
109. D. A. VanGalen and M. D. Hawley, J. Electroanal. Chem., 218, 227 (1987).
110. D. Bethell and V. D. Parker, J. Amer. Chem. Soc., 108, 7194 (1986); Acc. Chem. Res., 21, 400 (1988).
111. D. M. Coombs and L. L. Leveson, Anal. Chim. Acta, 30, 209 (1964).

12
Carboxylic Acids and Derivatives

LENNART EBERSON / Division of Organic Chemistry, University of Lund, Lund, Sweden

JAMES H. P. UTLEY / Queen Mary and Westfied College, University of London, London, England

JOHN H. WAGENKNECHT / Monsanto Chemical Company, St. Louis, Missouri

I. Introduction 483

II. Cathodic Reduction of Carboxylic Acids 483

III. Cathodic Reduction of Esters, Lactones, Anhydrides, and Acyl Halides 487

IV. Cathodic Reduction of Amides, Lactams, Imides, and Hydrazides 492

V. Cathodic Reduction of Nitriles 495

References 496

I. INTRODUCTION

This chapter is concerned with the cathodic reduction of carboxylic acids and their derivatives, that is, esters, amides, anhydrides, acyl halides, hydrazides, nitriles, and corresponding thio derivatives. Cyclic derivatives of substituted carboxylic and polycarboxylic acids, such as lactones, lactams, imides, and anhydrides, are also included. Only those transformations in which the functional group itself is involved are discussed.

Cathodic conversions of the carboxyl group and its derivatives have been long known, and tabulations and discussions of these reactions can be found in several places [1–5]. Most of the earlier work on these processes involved constant-current reduction with careful selection and control of solvent, temperature, pH, and electrode material. More recently, mechanistic studies of these processes have been made using modern electroanalytical techniques but our knowledge in this area is far from complete.

II. CATHODIC REDUCTION OF CARBOXYLIC ACIDS

The cathodic reduction of a carboxyl group can give the carboxylate anion, the aldehyde, the alcohol, or the hydrocarbon in one-, two-, four-, or six-electron reductions, respectively, Eqs. (1) through (4):

$$RCO_2H \xrightarrow{1e^-} RCO_2^- + 1/2\, H_2 \qquad (1)$$

$$RCO_2H \xrightarrow[2H^+]{2e^-} RCH(OH)_2 \longrightarrow RCHO + H_2O \qquad (2)$$

$$RCO_2H \xrightarrow[4H^+]{4e^-} RCH_2OH + H_2O \qquad (3)$$

$$RCO_2H \xrightarrow[6H^+]{6e^-} RCH_3 + 2H_2O \qquad (4)$$

Successful utilization of the reactions (2) through (4) is dependent on being able to avoid the proton reduction reaction (1).

It is generally difficult to stop the reduction at the aldehyde stage since aldehydes are usually reduced at less negative potentials than the corresponding acids. Certain "tricks" discussed later, however, often allow one to isolate the aldehyde form.

Aliphatic and aromatic acids that are not strongly activated by electron-withdrawing groups and do not contain more easily reducible groups give well-developed polarographic waves in aqueous [6] or nonaqueous [7,8] solvents, but this is usually due to reduction of protons to hydrogen. As an example, the hydrogen wave from acetic acid in acetonitrile-0.1 M tetraethylammonium perchlorate appears at −2 V (SCE) and in aqueous medium in the intermediate pH region at −1.76 V (SCE). Controlled potential electrolysis (CPE) of acetic acid in acetonitrile at −2.3 V (SCE) demonstrates that the electrode process is actually the reduction of protons. No second wave is found in the accessible potential region of these media, indicating that $RCOO^-$, the species left after reduction of the proton, is very difficult to reduce. Using this reaction, esters may be prepared under quite mild conditions, Eq. (5) [9]:

$$RCO_2H \xrightarrow[R'_4N^+]{1e^-} 1/2H_2 + RCO_2^-\, R'_4N^+ \xrightarrow{R''X} RCO_2R'' + R'_4NX \qquad (5)$$

Unless activated, therefore, it is unlikely that carboxylic acids will be easily reduced in slightly acidic or alkaline solution or in nonaqueous media. In strongly acidic medium the situation may be different since protonation of the carboxyl group may lead to the formation of a species capable of accepting electrons into the carbon-oxygen bonds.

For reduction of dicarboxylic acids in acetone solution, polarographic half-wave potentials and limiting currents are dependent on concentration in a way that suggests that at low concentration the acids are sufficiently dissociated for reduction of solvated protons to predominate, whereas at higher concentrations the more difficult reduction of undissociated acid is observed [10].

Acids containing certain activating groups can be reduced without interference from hydrogen evolution. 1-Ethyl-4-carboxypyridinium ion (I, as the bromide) is such a case, having the very strong electron-withdrawing quaternary ammonium ion as part of the aromatic ring [11]. The carboxyl group is the most easily reducible group at pH values below 7. The half-wave potential of I moves toward more negative potentials with increasing pH in such a way that the $E_{1/2}$-pH curve can be approximated by three lines: the first (pH 0.5–1.5) with a slope of 0.029 V/pH, the second (pH 2–7) with a slope of 0.068 V/pH, and the third (pH 8–13) with a slope of 0.007 V/pH. The total variation in $E_{1/2}$ is between −0.74 V in strongly acidic and -1.22 V in alkaline solution. Hydrogen evolution takes place at yet more negative potentials. From this pH dependence it was concluded that from pH 8 to 13 the reducible species contains the carboxyl group as the carboxylate ion and below 7 exists in its undissociated form. Since the pK of the acid is 1.75, the carboxyl group must be formed in a recombination reaction at the electrode surface in the pH region 2–7, as indicated by the slope of the $E_{1/2}$-pH line in this interval [12].

Controlled potential electrolysis of I in slightly acidic buffers at −1.0 V (SCE) gives the corresponding aldehyde in good yield, provided that the reaction is stopped at 2 F/mol (prolonged reaction times led to a further slow reduction of the aldehyde, suggesting a small equilibrium concentration of free aldehyde and a slow rate of conversion of the hydrated form to free aldehyde). The best yields of aldehyde are obtained in the pH region 2–5. At alkaline pH no aldehyde can be detected, possibly because the heterocyclic ring system becomes the more easily reducible group of the molecule. The corresponding N-methyl derivative has been shown to reduce to the dianion radical in strongly basic ethanolic solution [13]. Aldehyde formation is also negligible at pH < 0.

In contrast, reductions of N-methylpyridine 2- and 3-carboxylic acids in acidic solution led only to the corresponding alcohols [14], although the difference may have been that reductions of the latter were performed until all carboxylate was consumed, whereas in the earlier work electrolysis was stopped after 2 F/mol was passed.

Aldehydes are usually more easily reduced than carboxylic acids, but as in the previous case the aldehyde may exist in aqueous solution predominantly as the hydrate, Eq. (2). The hydrates are reduced at more negative potentials than carboxylic acids. Several other systems have been found to behave similarly [15,16]: pyridine-2- and 4-carboxylic acid, imidazole-2-carboxylic acid, and thiazole-2-carboxylic acid. In the last case the yield is low due to competing reduction of the ring. Yields of aldehyde are better at low temperatures due to the slowing of the dehydration reaction.

In summary, it seems [15] that the following requirements must be fulfilled for the reduction of a carboxyl group to an aldehyde function to take place: (1) the carboxyl group must be activated by an electron-withdrawing group; (2) the carboxyl group must be the most easily reducible group in the molecule (which automatically excludes the nitro group and a number of nitrogen-containing heteroaromatic groups as activating groups); and (3) the aldehyde formed must be protected against further reduction by formation of a nonreducible derivative, such as a hydrate or hemiacetal. The last criterion is probably the most important and stringent. It is important to note that aldehydes are not direct cathodic reduction products but that the hydrated form is first produced. Subsequent dehydration leads to the electroactive aldehyde.

In many cases, dehydration to the aldehyde occurs at a rate slow enough to allow the hydrated form to escape from the electrode surface. Dehydration in the bulk allows other schemes for protecting the aldehyde.

Other acids that fulfill these requirements and can be reduced cathodically to aldehydes are oxalic acid to glyoxylic acid (a reaction studied in great detail including process development [17]), and lactic acid [3]. Phthalic acid, also an activated acid, cannot be reduced to an aldehyde in in neutral or weakly acidic electrolytes at a lead cathode but instead is converted into phthalide [18].

Reduction of phthalic acid in acidic solution (e.g., 5% aqueous H_2SO_4) results in aromatic ring reduction, giving 1,2-dihydrophthalic acid [19–24], a reaction believed to have been practiced commercially [25].

Hydration is not the only way in which an aldehyde intermediate may be protected against reduction; the aldehyde may be converted in situ into a borate complex of the hydrate [26–29] or into a bisulfite addition complex [30,31]. There is compelling evidence that borate complexes, such as II in Eq. (6), in the case of salicylic acid reduction, are intermediates. The salt (II) has been isolated and characterized [29], and in the presence of proton donors it undergoes four-electron reduction; a plausible mechanism is suggested in Eq. (6):

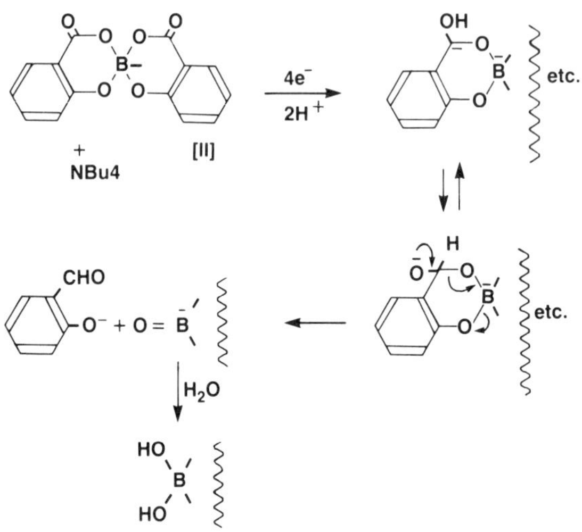

(6)

The reduction of salicylic acid to salicylaldehyde is believed to be practiced commercially in India and the USSR [25]. Two-phase systems may also be used so that the aldehyde is rapidly removed from aqueous solution by extraction into an organic phase [26–28]. It was shown that correct choice of buffer and a two-phase system allows reductions of aromatic acids to aldehydes in acceptable yield (>50%); in addition the likelihood of such success is greater the lower the pK_a value of the acid and the lower the $E_{1/2}$ value for reduction of the corresponding methyl ester [28].

Four-electron reductions of aromatic carboxylic acids occur when reactions are carried out in strongly acidic solutions at lead or mercury cathodes, giving excellent yields of the corresponding benzyl alcohols [21,32–34]. The reaction is easily performed on a large scale. For example, the adaptation of a filter press cell for continuous reduction of benzoic acid has been described [35].

A procedure for the cathodic reduction of anthranilic acid to the corresponding alcohol is described in Organic Synthesis [34]. Because the carboxylic acid group is so difficult to reduce, the method cannot normally be used for acid substituted with easily reducible functions or groups. For instance, pentafluorobenzoic acid loses fluorine by cathodic cleavage at mercury, in 20% sulfuric acid, at −1.20 V (SCE); 2,3,5,6-tetrafluorobenzoic acid is obtained in 75% yield [36]. However, at potentials more negative than −1.30 V reduction of the carboxylic acid group becomes significant and a mixture of the pentafluoro- and tetrafluorobenzyl alcohols is produced [Eq. (7)]. In very strongly acidic medium (50% $HClO_4$) only pentafluorobenzyl alcohol is produced.

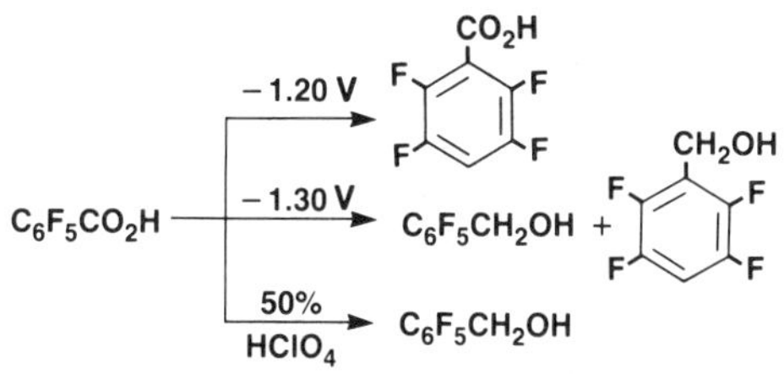

(7)

Aliphatic carboxylic acids are reducible to the alcohol in low yield in strongly acidic electrolytes. Thus, phenylacetic acid is reduced to β-phenylethanol in 25–35% yield at a lead cathode in 50% sulfuric acid-ethanol [37] and butyric acid to butanol in 80% sulfuric acid in 6.5% yield [38]. The latter conversion can also be performed in 17% yield in 7% aqueous sodium hydroxide, despite that the carboxylate form of the acid, which is reducible only with difficulty, predominates in bulk solution, suggesting an electrocatalytic reduction involving specific interaction of the substrate and electrode surface.

The six-electron reductions of carboxylic acids to methyl groups are known, such as the reduction of pyridine-2- and 4-carboxylic acids to the picolines in strongly acidic medium [39].

III. CATHODIC REDUCTION OF ESTERS, LACTONES, ANHYDRIDES, AND ACYL HALIDES

Nonactivated esters are reduced with difficulty unless strongly acidic solutions are used. Unsaturated esters, such as α,β-unsaturated and aromatic esters, are reduced relatively easily because of the activating effect of the strongly electron-withdrawing ester group for reduction of the attached unsaturated function. In these cases the products of reduction may include those of hydrodimerization and hydrogenation of the unsaturated function (e.g., aromatic ring) as well as those of reduction of the ester function. The possibilities are summarized in Eqs. (8) through (12), and subsequent discussion focuses on reduction of the ester function.

$$RCO_2R' \xrightarrow{1e^-} RCO_2R'^{\cdot -} \quad (8)$$

$$RCO_2R' \xrightarrow[H^+]{2e^-} RCO_2^- + R'H \quad (9)$$

$$RCO_2R' \xrightarrow[2H^+]{2e^-} RCH(OH)OR' \longrightarrow RCHO + HOR' \quad (10)$$

$$RCO_2R' \xrightarrow[4H^+]{4e^-} RCH_2OH + R'OH \quad (11)$$

$$RCO_2R' \xrightarrow[4H^+]{4e^-} RCH_2OR' + H_2O \quad (12)$$

The first step in the cathodic reduction of an ester in acidic medium leads to an aldehyde, presumably via the hemiacetal, Eq. (10). The aldehyde, if not stabilized in the form of an electroinactive addition compound, is then reduced in a second step to form the corresponding carbinol [40] (or sometimes also the pinacol) [41]. However, a competing pathway for ester reduction in strongly acidic medium is the formation of ethers, Eq. (12), sometimes in good yields [21,32,42,43].

Voltammetric data for ester reductions are available for several aromatic esters [44–47], and in particular cyclic voltammetry shows clearly that in the absence of proton donors reversible formation of anion radical occurs [44]. In DMF solution the peak potential for reduction of methylbenzoate is −2.29 V (versus SCE); for comparison dimethylterephthalate reduces at −1.68 V and phthalic anhydride at −1.25 V [4]. Half-wave potentials for reduction of aromatic carboxylate esters in an unbuffered solution of pH 7.2 are linearly correlated with σ values [44]; electron-withdrawing substituents in the ring or alkoxy group shift $E_{1/2}$ toward less negative potentials. Generally, esters seem to be more easily reducible than the parent carboxylic acids. Anion radicals of methyl, ethyl, and isopropyl benzoate have been detected by ESR spectroscopy upon cathodic reduction of these esters in acetonitrile-tetrapropylammonium perchlorate [45]. The anion radicals of several anhydrides, including phthalic anhydride, have similarly been studied [48].

Benzoate esters are reduced under aprotic conditions initially, as mentioned earlier, to the corresponding anion radical. The anion radicals are not stable for long periods of time, however, and generally cleave, giving benzoate and the alkyl radical, Eq. (13) [49–51].

$$ArCO_2R \xrightarrow{1e^-} ArCO_2R^{\cdot -} \longrightarrow ArCO_2^- + R\cdot \qquad (13)$$

The rate of cleavage of the anion radical to benzoate and free radical is higher as the stability of the free radical increases. Thus, the anion radical of tert-butylbenzoate cleaves according to Eq. (13) much faster than the corresponding anion radical of methylbenzoate. The free radical thus formed may undergo a variety of reactions, including substitution onto the aromatic ring of the starting ester [51].

It has also been reported that reduction of benzoate esters in dry acetonitrile leads to solvent interaction resulting in benzoylacetonitrile in good yield, Eq. (14) [52]:

$$ArCO_2R \xrightarrow[\substack{\text{Reduction}\\ CH_3CN}]{\text{Electro}-} ArCOCH_2CN \qquad (14)$$

The cathodic reduction of esters has recently been used for a variety of deprotection reactions. For example, 2-methylene-9,10-anthraquinone esters have been reduced in DMF to free the corresponding carboxylate, Eq. (15) [53]:

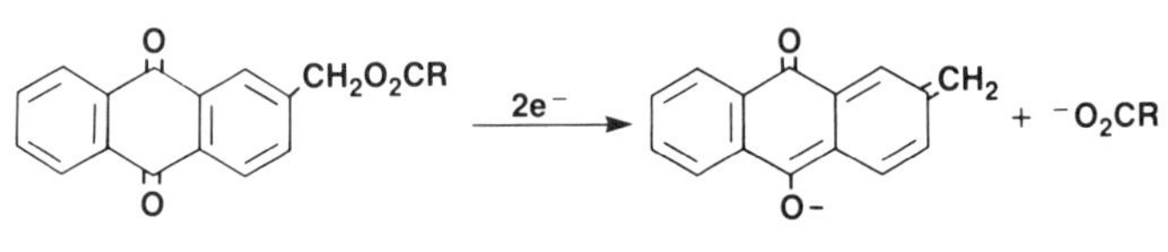

(15)

The p-nitrobenzyl ester of cephalosporin has been cleaved cathodically to yield cephalosporin [54]. The widely used benzyloxycarbonyl protecting group for amines is cathodically removed, giving free amine, CO_2, and toluene, Eq. (16) [55]:

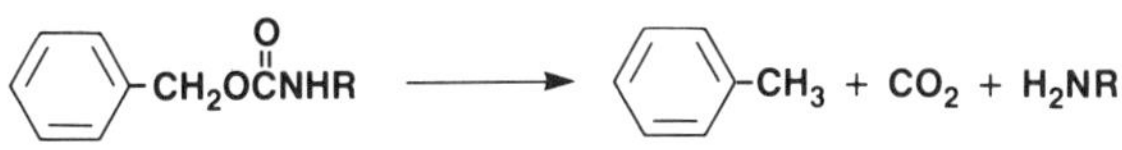

(16)

The reduction of certain esters results in formation of carboxylate and the reduced form of the alcohol precursor fragment [56–59]. For example, the cathodic cleavage at mercury of vitamin A acetate gives axerophthene in good yield, Eq. (17) [60]:

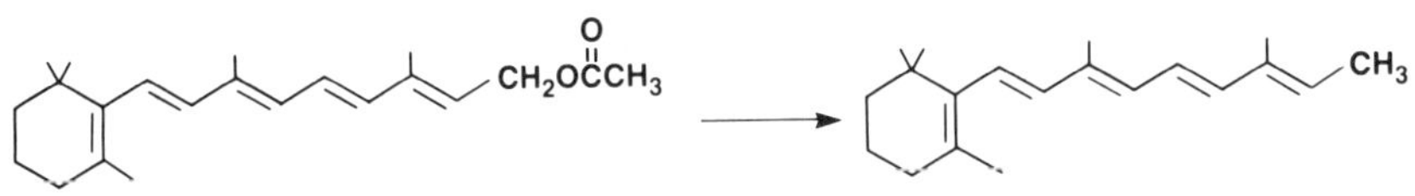

(17)

Mixed esters of oxalate containing one benzylic substituent are reduced in good yield using in situ trans-esterification to the corresponding hydrocarbon and oxalate half-ester [61]. Esters of vicinal diols (including cyclic carbonate esters) [62] and β-halo esters [63], are often reduced to the corresponding olefin [61,64].

Nonactivated esters are reduced only slowly if at all in strongly acidic medium at a lead cathode [42], whereas alkyl benzoates are reduced to a mixture of benzyl alcohol and alkyl benzyl ether [42,43], and alkyl 4-imidazolecarboxylates similarly are reduced to a mixture of the corresponding alcohol and ether [65]. Phenyl and benzyl benzoate give predominantly ethers [21,32], as do also ethyl o-chloro- and m-bromobenzoate [43].

For activated esters the problem of halting reduction at the aldehyde stage is similar to that faced in the reduction of carboxylic acids; the aldehyde must be converted in situ into an electroinactive form. In one case, that of reduction of ethyl 2-thiazolecarboxylate [16], the strongly hydrated aldehyde may be obtained in about 70% yield, Eq. (18):

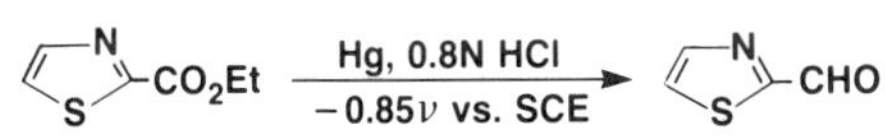

(18)

Presumably for similar reasons cathodic reduction of diethylterephthalate and diethyl furan- and thiophene-2,5-dicarboxylate [44–46] in a buffered ethanolic solution of "pH 4.2" affords the aldehyde ester under CPE conditions (in the interval between −1.50 and −1.65 V). Diethyl oxalate [66] is reduced at a mercury cathode in acidic medium to give the ethyl hemiacetal of ethyl glyoxylate (38–50% yield). In contrast, diethyl and monoethyl phthalate gave phthalide under similar conditions [19,67].

An interesting $S_{RN}1$ reaction occurs on cathodically induced rearrangement of S,S-diphenylbenzene-1,2-dicarbothiolate, forming 3,3-bis(phenylthio)phthalide, Eq. (19) [68]:

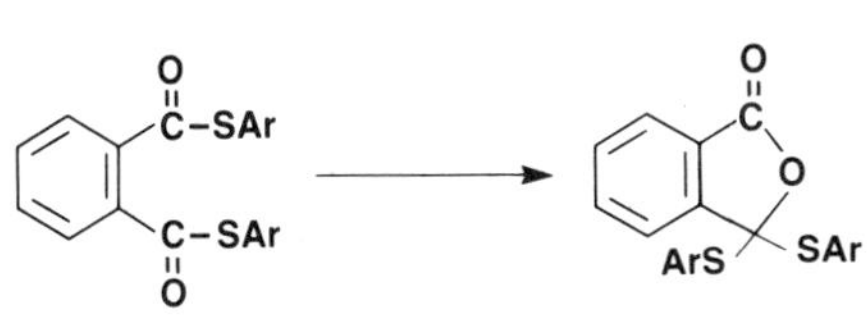

(19)

In aprotic conditions diethyl oxalate is reduced reversibly, and its relatively stable anion radical has been well characterized by ESR spectroscopy [69,70]. Not all oxalate esters give reversible reduction; the anion radicals of several (e.g., dibenzyl oxalate) undergo rapid fragmentation [61,71].

Ethenetetracarboxylate esters may be reduced to the anion radical or stable dianion states depending on the cathode potential. Both have been used extensively as electrogenerated bases [72]. Interestingly, ethanetetracarboxylate esters do not give stable anion radicals but cleave to give the corresponding malonate anions [73].

Methyl 2-naphthoate is reduced [74] in the presence of boric acid in methanolic solution at a mercury cathode to form 2-naphthalene methanol in 19% yield, a mixture of diastereomeric pinacols from further reduction of the intermediate aldehyde in 19% yield, and a mixture of methyl 1,2- and 1,4-dihydro-2-naphthoates in 42% yield. Methyl 1-naphthoate under the same conditions [75] gives methyl 1,4-dihydro-1-naphthoate (43%), a 4,4'-dimer of the 1,4-dihydro derivative (20%), di-1-naphthylglyoxal (5%), and 1-naphthalene methanol (5%). Because mercury cathodes are used with sodium salt electrolytes it is possible that electrogenerated sodium amalgam is the reducing agent.

2-Alkyl-substituted acetoacetic esters undergo an interesting rearrangement reaction (the Tafel rearrangement) upon reduction at a lead cathode in an acidic medium, Eq. (20), with complete reduction of the functional groups [76–81]. The cathodic reduction of β-diketones gives analogous rearrangements, and because these are known to involve cyclopropanediol intermediates [82], it is tempting to suggest that the intermediate (III) is involved in the Tafel rearrangement. In partial support of Eq. (20) there is evidence, although not by isolation, that 2,3-octanedione is one of the many by-products of the Tafel rearrangement of ethyl 2-butylacetoacetate [81].

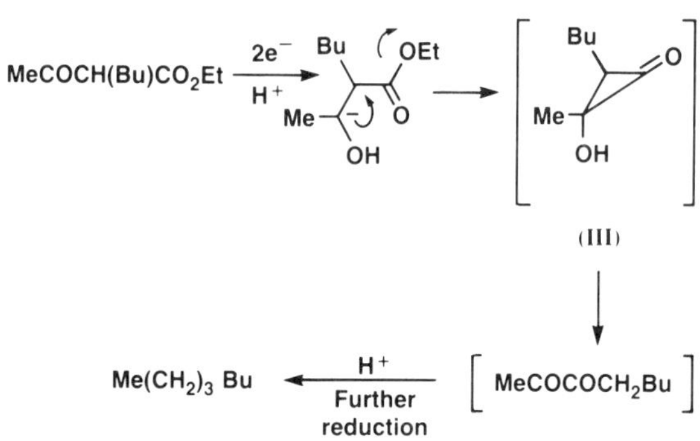

(20)

Lactones are in principle reducible to hydroxyaldehydes, and when applied to lactones of sugar acids the method works well because cyclic acetal or ketal formation stabilizes the carbonyl function and prevents overreduction [3,83–87]. In particular D-ribose may be prepared from the cathodic reduction of D-ribono-γ-lactone, Eq. (21); the optimum conditions for this reaction include careful buffering of the electrolyte [84–87].

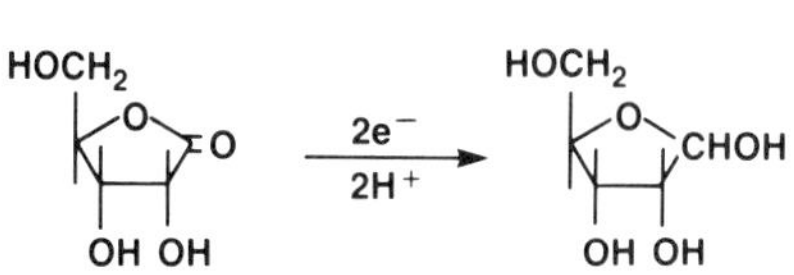

(21)

Thiol esters of aromatic acids, ArC(O)SR, are cathodically reduced with fission of the S—CO bond, affording the thiol in good yield [88], for example L-cysteine in 83% yield from S-benzoyl-L-cysteine at a mercury cathode in DMF-water-tetramethylammonium chloride. The electroreduction of aryl thioesters [containing —C(O)SR, —C(S)OR, and —C(S)SR functions] has been studied, and some interesting products have been formed [89,90], Eqs. (22) through (24). Trithiocarbonates are reduced to the tetraalkyltetrathioethylenes [91].

$$4\text{ØCSSR} \xrightarrow{6e^-} \text{ØC}\equiv\text{CØ} + 2\text{ØCS}_2^- + 4\text{RS}^- \qquad (22)$$

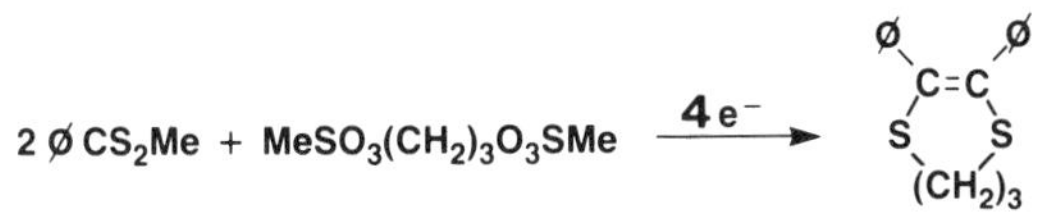

(23)

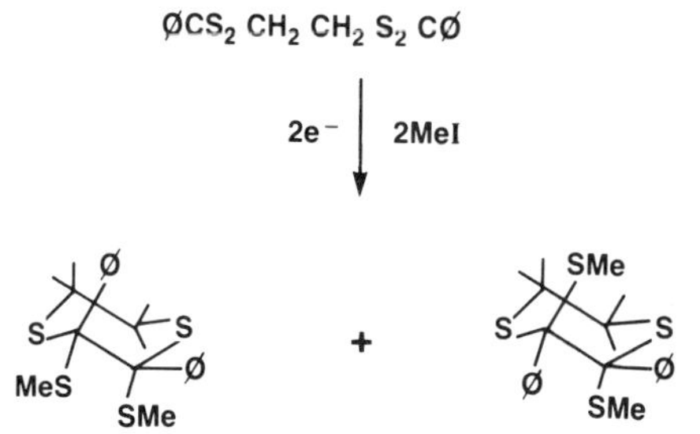

(24)

Anhydrides have received very little attention as substrates for cathodic reduction. Polarography of cyclic anhydrides [92] in acetonitrile and DMF displayed two, one-electron waves for the most easily reduced anhydrides (pyromellitic, dibromopyromellitic, and tetrachlorophthalic anhydride). For 10 mono- and dianhydrides of aromatic carboxylic acids the $E_{1/2}$ values for reduction in acetonitrile have been measured and listed [93].

Controlled potential electrolyses of phthalic and pyromellitic anhydrides at the potential of the first wave gave highly colored, oxygen-sensitive solutions, which later were shown to give well-resolved ESR spectra of the corresponding radical ions [48]. Phthalic anhydride has been reduced to phthalide on a preparative scale [94], but the reaction conditions employed (ethanol-water-ammonium carbonate) indicate that monoethyl phthalate must have been the actual substrate. An interesting reduction of 3,6-diphenylphthalic anhydride in the presence of trimethylsilyl chloride gives two types of dimeric species, Eq. (25) [95]:

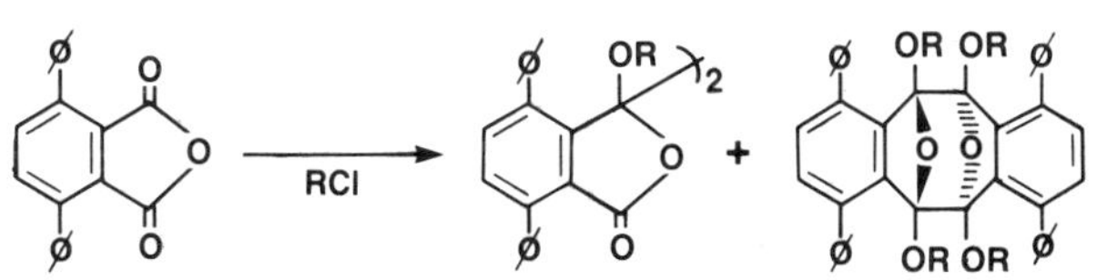

(25)

R = Me_3Si

There have been a few reports describing electroreduction of acyl halides. Benzoyl chloride when reduced (Ep = −1.82 V versus Ag/0.1 M $AgNO_3$) in aprotic media gives good yields (85%) of stilbenediol dibenzoate [96].

This seems to be a general reaction for a variety of acyl chlorides [97]. The reduction of cinnamoyl chloride leads to 2,4-dibenzyl-1,3-cyclobutadione, Eq. (26) [98]:

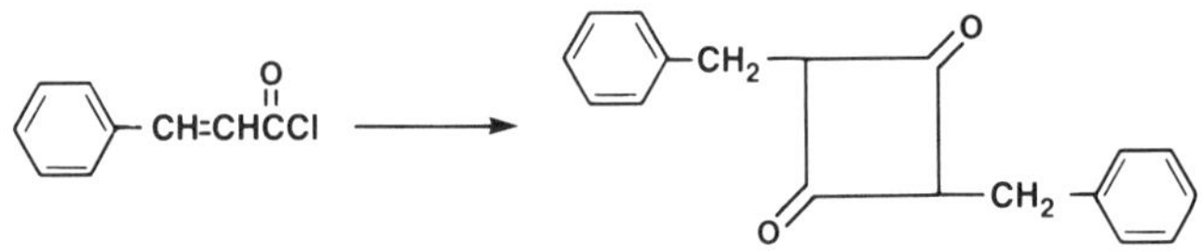

(26)

An interesting coupling of acyl halide with Michael acceptors involves an organometallic intermediate and photochemical step as well as an electroreduction. The electrochemical reduction, however, is used to reduce the intermediate cobalt oxidation state [91].

IV. CATHODIC REDUCTION OF AMIDES, LACTAMS, IMIDES, AND HYDRAZIDES

The carbonyl group of amides, lactams, and imides is reducible at lead and mercury cathodes in strongly acidic media, usually to the corresponding methylene compound. In media of lesser acidity cathodic cleavage may occur, and depending on the structure of the starting material and the detailed reaction conditions, the products outlined in Eq. (27) are accessible by cathodic reduction.

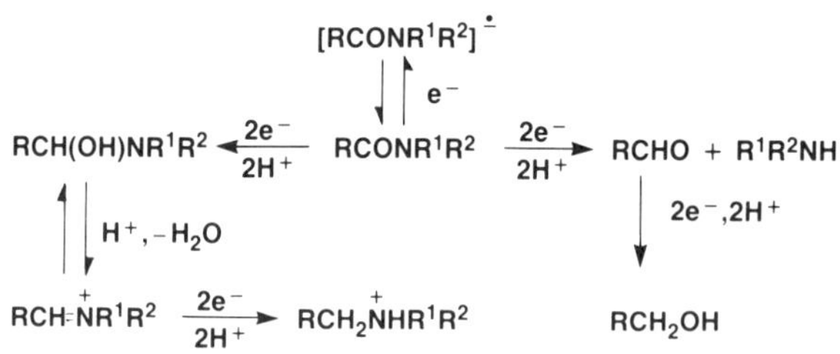

(27)

Voltammetric and ESR spectroscopic studies using aprotic conditions show that several amides give reversible one-electron reduction to relatively stable anion radicals, for example, for N,N-dimethylbenzamide [4], N-alkylphthalimides [49,100–102], and oxamides [103]. In a similar fashion the electrogeneration of anion radicals from sulfur analogs [containing the $RC(S)NR^1R^2$, $RR^1NC(S)CC(S)NR^2R^3$, and $ArC(S)CONR^1R^2$ functions] has been thoroughly investigated [104–106].

The polarography of pyridine-4-carboxamide and some of its derivatives [107], as well as carboxamides of other heterocyclic systems [15,16], has been carefully studied in aqueous medium over the whole pH range. Pyridine-4-carboxamide (IV) displays a polarographic wave with $E_{1/2}$ ranging from about −0.6 V at pH 0 to −1.3 V at pH 13, the change being related to the extent of protonation. Controlled potential electrolysis of IV in 0.8 M hydrochloric acid at −0.80 V (SCE) gave a good yield of pyridine-4-aldehyde, whereas at pH 4.5, 4-pyridine methanol was formed.

Substitution of hydrogen by phenyl in the amide group changes the product composition considerably; electrolysis of N-phenyl-4-phenyl-4-pyridine carboxamide in 1 M hydrochloric acid at −0.75 V gives a mixture of 4-pyridinemethanol and 4-anilinomethylpyridine [Eq. (28)]:

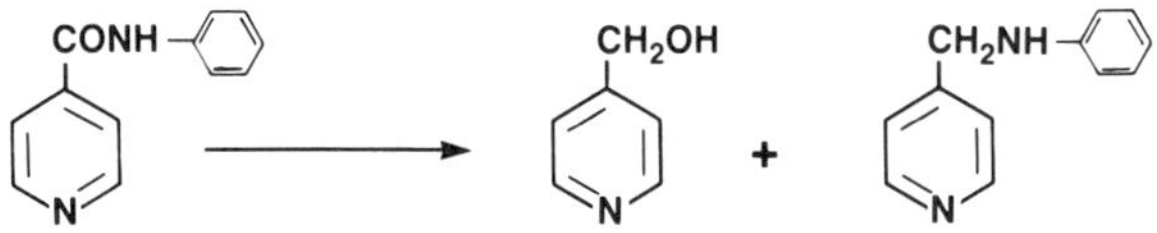

(28)

Pyridine-4-thiocarboxamide is reduced in two, two-electron steps in acidic medium; at the potential of the second wave, reduction gives 4-aminomethylpyridine. These three cases show that the preparative outcome of an amide reduction is strongly dependent on the substituents at the amide group and that the changes manifest themselves in the chemical step [Eq. (27)]. For amides of aromatic carboxylic acids the reduction potential may be gleaned from a comprehensive compilation [108].

A large number of preparative applications of amide group reductions have been compiled and discussed by other authors [1–5]. Generally, these reactions are performed at lead cathodes in aqueous sulfuric acid or ethanolic aqueous sulfuric acid and result in good yields of the corresponding aminomethyl compounds. Since the media employed are so strongly acidic it is probable that the electroactive species is the protonated amide in these cases. Simple aliphatic amides of the general formula $RCONH_2$ have not been studied; however, a number of N,N-diphenyl-substituted aliphatic amides have been successfully reduced. Benzamides give good yields of benzylamines, N,N-disubstitution favoring amine formation also in these cases. Thioamides can be used as substrates with similar results.

In alkaline solution, N-substituted benzamides undergo cathodic cleavage to form benzyl alcohol and an amine [108,109].

The reduction of thiobenzamides in anhydrous CH_3CN leads to fair yields of β-dialkylaminocinnamonitriles via a complex mechanism involving electrogenerated cyanomethyl anions [52]. However, thiobenzamides when reduced in the presence of such alkylating agents as alkyl halides or dimethyl sulfate give excellent yields of the corresponding α-aminothioethers [Eq. (29)]. Use of chlorotrimethylsilane as the alkylating agent, however, results in loss of sulfur to form a C−Si bond, Eq. (30) [110]:

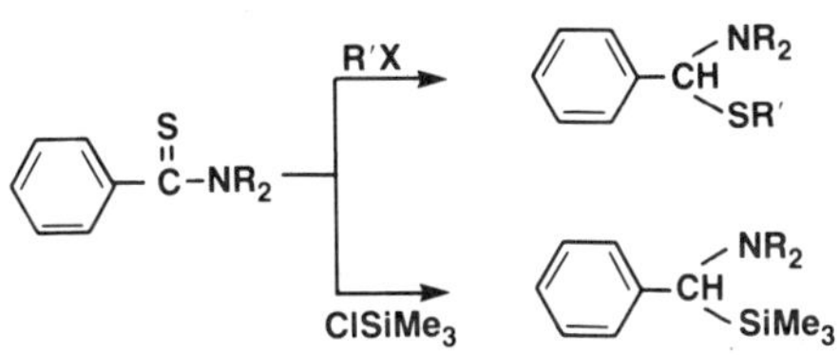

(29)

(30)

In amine solvents the reduction of amides by electrogenerated solvated electrons gives the corresponding alcohol product in the absence of a proton donor [111]. If a proton donor, such as ethanol, is added then good yields of aldehyde may be formed.

Amides containing at least one hydrogen on the nitrogen are sufficiently acidic in aprotic media to evolve H_2 on reduction at a Pt cathode, resulting in the corresponding amide anion [Eq. (31)], which has been used as an electrogenerated base [112]. This reaction in the presence of alkylating agents leads to N-alkylation of the amide [112,113]. The reaction also occurs for carbamates.

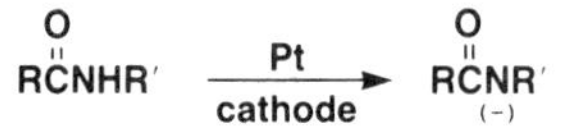

(31)

Phthalimide and substituted phthalimides are relatively easily reduced [Ep/2 in DMF ≃ −1.4 V (versus SCE)]. In alkaline or aprotic solution two, one-electron reduction steps are observed in voltammetric experiments [114–117]; anion radical formation at the first wave has been confirmed by ESR spectroscopy [48,115,116]. Considerable attention has been given to the detailed mechanism of reduction of phthalimide derivatives, Eq. (32) [114–117]:

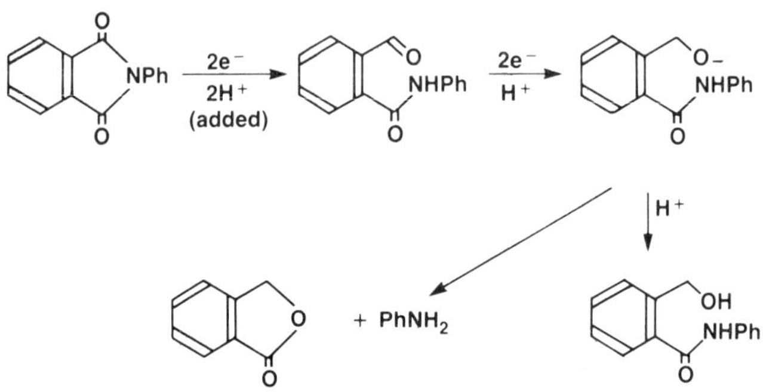

(32)

Reduction of N-methylphthalimide in the presence of dimethyl sulfate or trimethylsilyl chloride led to a dimeric product, Eq. (33) [118]:

(33)

The electroreduction of N-phenylphthalimide in the presence of trimethylsilyl chloride formed a bis-silylether, Eq. (34):

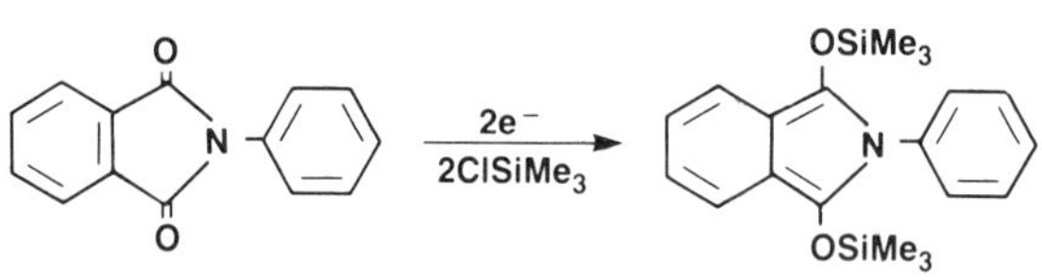

(34)

O-acetyl-2-hydroxymethylbenzanilide is formed as a result of reduction [116] of N-phenylphthalimide in DMF in the presence of acetic anhydride. In alkaline solution phthalimide is reduced to a dimer epoxide, Eq. (35) [119]:

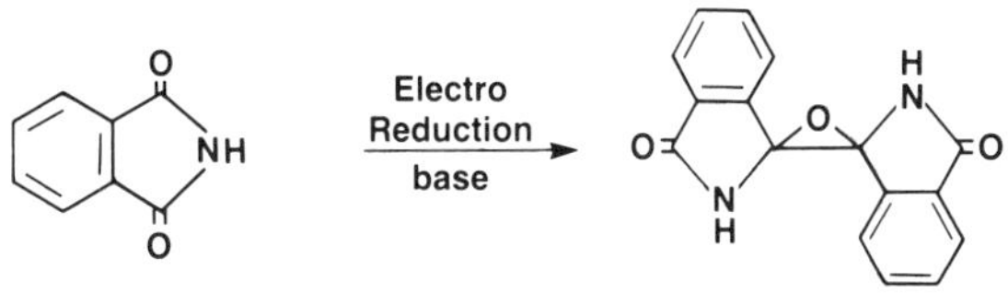

(35)

In macroscale electrolyses, lactams or fully saturated cyclic amines are obtained in good yields from imides in strongly acidic electrolytes [1–5]; for example, N-methylsuccinimide is reduced in good yield to N-methylpyrrolidone at a lead cathode in sulfuric acid solution [120].

The reduction of acyl hydrazines has been investigated very little. Polarography of 4-pyridine carboxylic acid hydrazide in acid and neutral solution shows that it is reduced in two, two-electron steps [121]. The result of controlled potential electrolysis is cleavage of the N—N bond to form 4-pyridine-carboxamide, which then undergoes reduction to the aldehyde in the second step. Other hydrazides of heterocyclic carboxylic acids behave analogously.

The N,N-diphenylhydrazone of benzoyl chloride has been reduced to form benzonitrile and diphenylamide [122].

V. CATHODIC REDUCTION OF NITRILES

Both aliphatic and aromatic nitriles are reduced in acidic solution to give amines, Eq. (36), whereas in neutral or alkaline medium cleavage of the C—CN bond takes place, Eq. (37);

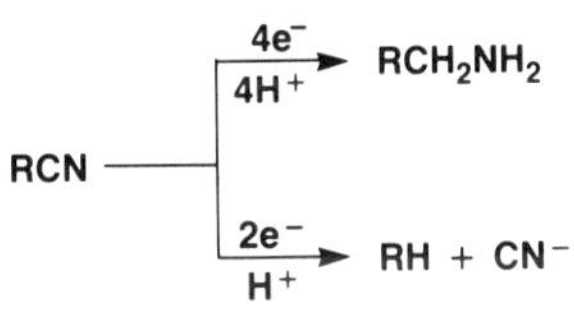

(36)

(37)

Reactions (36) and (37) have been studied polarographically [123]. In acid solution, aromatic nitriles of the type p-$RCOC_6H_4CN$ (R=H or CH_3) undergo a four-electron reduction at −0.70 V (SCE) with formation of the corresponding benzylamine. The protonated nitrile is the electroactive species. In neutral medium, nitriles of the type XC_6H_4CN (X=p-COO^-, p-SO_2NH_2, and m-SO_2NH_2) give two-electron reduction with formation of cyanide ion and C_6H_5X. This duality in mechanism has been thoroughly studied in the case of 2- and 4-cyanopyridine [124,125], and it was shown that both pH and electrode material play an important role in determining the mechanism.

The electrochemical cleavage of nitriles can also be performed in anhydrous amine media [126], making possible C—CN bond cleavage in aliphatic nitriles, such as cycloheptane carbonitrile and heptane carbonitrile. This reaction was interpreted as due to reductive cleavage by solvated electrons. The other reaction has also been accomplished with electrogenerated solvated electrons [127].

A number of aromatic nitriles including benzonitrile have been reduced electrochemically in DMF-tetrapropylammonium perchlorate to give solutions of relatively stable anion radicals whose ESR spectra have been recorded [128]. The anion radical of benzonitrile in dry DMF slowly loses cyanide to form phenyl radical, which is reduced to phenyl anion, which abstracts a proton from the electrolyte [129]. Phthalonitrile, which gives two, one-electron waves on polarography in this medium, undergoes decyanation on CPE at the potential of the second wave. There is recent evidence that the anion radicals of 1,3-dicyanobenzene and 1,3,5-tricyanobenzene may dimerize and lose cyanide to form biphenyl derivatives [130].

The preparative use of nitrile reduction has been primarily for the synthesis of amines [1,2,131–136]. This is done in acidic medium and gives fair to good yields. Electrocatalytic hydrogenation of nitriles is also sometimes efficient [137]; in these cases nickel or palladium black, deposited on the cathode, promote reduction by electrolytically generated hydrogen.

Imino ether hydrochlorides can be reduced to amines in good yields at lead cathodes in aqueous sulfuric acid [138]. Cathodic reduction of nitriles in ethanolic sulfuric acid should be a useful procedure, therefore, with the first step in situ formation of an imino ether salt.

Acetonitrile has been reduced at a tin cathode to give low yields of tetramethyl tin [139].

REFERENCES

1. F. Fichter, Organische Elektrochemie, Steinkopff, Dresden, 1942.
2. M. R. Rifi, in Technique of Electroorganic Synthesis (N. L. Weinberg, ed.), Part 2, Wiley-Interscience, New York, 1974 Chap. 8.
3. F. D. Popp and H. P. Schultz, Chem. Rev., 62, 19 (1962).
4. J. P. Coleman, in The Chemistry of Acid Derivatives, Suppl. B. (S. Patai, ed.), Wiley-Interscience, New York, 1979, pp. 782–824.
5. L. Eberson and K. Nyberg, in Encyclopedia of Electrochemistry of the Elements (A. J. Bard and H. Lund, eds.), Vol. 12, Dekker, New York, 1978, pp. 261–238.
6. I. A. Korshunov, Z. B. Kuznetsova, and M. K. Shchenikova, Zh. Fiz. Khim., 23, 1292 (1949); Chem. Abstr., 44, 2873 (1950).
7. J. F. Coetzee and I. M. Kolthoff, J. Amer. Chem. Soc., 79, 6110 (1957).
8. I. M. Kolthoff and T. B. Reddy, J. Electrochem. Soc., 108, 980 (1961).
9. T. Awata, M. M. Baizer, T. Nonaka, and T. Fuchigami, Chem. Lett., 371 (1985).
10. A. P. Kreshkov, S. L. Rogatinskaya, and G. N. Emirdzhanova, Zh. Obshch. Khim., 47, 1379 (1977).
11. H. Lund, Acta Chem. Scand., 17, 972 (1963).
12. P. Zuman, Fortschr. Chem. Forsch., 12, 1 (1969).
13. V. Volkeova, K. Klima, and J. Volke, Electrochim. Acta, 23, 1215 (1978).
14. M. T. Carter, G. N. Kamau, and J. F. Rusling, J. Electroanal. Chem., 170, 265 (1984).
15. P. E. Iversen and H. Lund, Acta Chem. Scand., 21, 279 (1967); 21, 389 (1967).
16. O. R. Brown, J. A. Harrison, and K. S. Sastry, J. Electroanal. Chem., 58, 387 (1975).
17. F. Goodridge, K. Lister, R. E. Plimley, and K. Scott, J. Appl. Electrochem., 10, 55 (1980).

18. P. R. Jones, Chem. Rev., 63, 461 (1963).
19. S. Ono, Nippon Kagku Zasshi, 75, 1195 (1954); Chem. Abstr., 51, 12704 (1957).
20. P. C. Condit, Ind. Eng. Chem., 48, 1252 (1956); Fr. Pat. 1,516,599 (1968); Chem. Abstr., 70, 92677 (1969).
21. C. Mettler, Ber. Dtsch. Chem. Ges., 39, 2933 (1906).
22. F. Somlo, Z. Electrochem., 35, 769 (1929).
23. H. Suter, F. Beck, and A. Hrubesch, U.S. Pat. 3,471,381 (1969); Chem. Abstr., 70, 92677 (1969); H. Suter, H. Nohe, F. Beck, W. Bruegel, and H. Aschenbrenner, U.S. Pat. 3,542,656 (1970); Chem. Abstr., 72, 74127 (1970).
24. S. A. Cerefice and E. K. Fields, J. Org. Chem., 39, 971 (1974).
25. T. R. Beck et al., A Survey of Organic Electrolytic Processes, Nat. Tech. Infor. Serv. ANL/OEPM-79-5, 1979.
26. C. Mettler, Ber. Dtsch. Chem. Ges., 41, 4148 (1908).
27. S. Ono and T. Yamauchi, Bull. Chem. Soc. Jpn., 25, 404 (1952).
28. J. H. Wagenknecht, J. Org. Chem., 37, 1513 (1972).
29. J. C. Hoffmann, P. M. Robertson, and N. Ibl, Tetrahedron Lett., 3433 (1972).
30. J. A. May and K. A. Kobe, J. Electrochem. Soc., 97, 183 (1950).
31. B. B. Dey and H. V. K. Udupa, Proc. 6th Meet. Int. Comm. Electrochem. Thermodyn. Kinet., 1955, p. 465; Chem. Abstr., 50, 6221 (1956).
32. C. Mettler, Ber. Dtsch. Chem. Ges., 38, 1747 (1905).
33. F. Mayer and F. A. English, Justus Liebigs Ann. Chem., 417, 69 (1929); F. Mayer, W. Schäfer, and J. Rosenbach, Arch. Pharm., 267, 581 (1929).
34. G. C. Coleman and H. L. Johnson, Org. Synth., Coll. Vol. 3, 60 (1955).
35. G. Koranyi, L. Redey, and G. Palffy, Acta Chim. Acad. Sci. Hung., 100, 305 (1979).
36. P. Carrahar and F. G. Drakesmith, J. Chem. Soc., Chem. Commun., 1562 (1968).
37. S. Ono and T. Hayashi, Bull. Chem. Soc. Jpn., 26, 232 (1953).
38. M. Masuno, T. Asahara, S. Kuroiwa, K. Shimizu, and J. Nakano, J. Chem. Soc. Jpn., Ind. Chem. Sec., 52, 151 (1949); Chem. Abstr., 45, 1884 (1951).
39. J. P. Wibaut and H. Boer, Rec. Trav. Chim. Pays-Bas, 68, 72 (1949).
40. G. Le Guillanton, Bull. Soc. Chim. Fr., 1699 (1974).
41. L. Horner and H. Honl, Justus Liebigs Ann. Chem., 2036 (1977).
42. J. Tafel and G. Friedrichs, Ber. Dtsch. Chem. Ges., 37, 3187 (1904).
43. C. Mettler, Ber. Dtsch. Chem. Ges., 37, 3692 (1904).
44. J. Nakaya, H. Kinoshita, and S. Ono, Nippon Kagaku Zasshi, 78, 935 (1957); Chem. Abstr., 53, 21276 (1959); P. Zuman, Substituent Effects in Polarography, Plenum, New York, 1967, p. 61.
45. M. Hirayama, Bull. Chem. Soc. Jpn., 40, 1822 (1967); A. V. Il'yasov, Y. M. Kargin, Y. L. Levin, I. D. Morozova, N. N. Sotnikova, Dokl. Akad. Nauk SSSR, 179, 1141 (1968); Izv. Akad. Nauk SSSR, Ser. Khim., 5, 1030 (1968).
46. J. Nakaya and S. Ono, J. Chem. Soc. Jpn., Pure Chem. Sec., 74, 907 (1953); Chem. Abstr., 48, 8082 (1954).

47. G. C. Whitnack, J. Reinhart, and E. C. Gantz, Anal. Chem., 27, 359 (1955).
48. R. E. Sioda and W. S. Koski, J. Amer. Chem. Soc., 89, 475 (1967).
49. V. P. Gul'tyai, T. Y. Rubinskaya and L. M. Korotaeva, Izv. Akad. Nauk SSSR, Ser. Khim., 1677 (1982).
50. V. P. Gul'tyai, A. S. Mendkovich, A. P. Churilina, T. Y. Rubinskaya, and L. M. Korotaeva, Izv. Akad. Nauk SSSR, Ser. Khim., 91 (1984).
51. J. H. Wagenknecht, R. D. Goodin, P. J. Kinlen, and F. E. Woodard, J. Electrochem. Soc., 131, 1559 (1984).
52. L. Kistenbrugger, P. Mischke, J. Voss, and G. Wiegand, Justus Liebigs Ann. Chem., 461 (1980).
53. R. L. Blankespoor, A. N. K. Lau, and L. L. Miller, J. Org. Chem., 49, 4441 (1984).
54. D. A. Hall, U.S. Pat. 4,379,739 (1983); Chem. Abstr., 98, 224210 (1983).
55. H. L. S. Maia, M. J. Medeiros, M. I. Montenegro, and D. Pletcher, J. Electroanal. Chem., 200, 363 (1986); M. A. Casadei and D. Pletcher, Synthesis, 1118 (1987).
56. M. Ochiai, O. Aki, A. Morimoto, T. Ikada, K. Shinozaki, and Y. Asahi, J. Chem. Soc., Perkin I, 258 (1974).
57. R. E. Erickson and C. M. Fischer, J. Org. Chem., 35, 1604 (1970).
58. J. P. Coleman, H. G. Gilde, J. H. P. Utley, and B. C. L. Weedon, J. Chem. Soc., Chem. Commun., 738 (1970).
59. V. G. Mairanovskii, N. F. Loginova, and S. Y. Mel'nik, Elektrokhimiya., 9, 1174 (1973).
60. J. G. Gourney, M. Hodler, B. Terem, and J. H. P. Utley, J. Chem. Soc., Chem. Commun., 779 (1976).
61. N. Islam, D. W. Sopher, and J. H. P. Utley, Tetrahedron, 43, 959 (1987); 2741 (1987); K. G. Ellis, N. Islam, D. W. Sopher, J. H. P. Utley, H. L. Chum, and M. Ratcliff, J. Electrochem. Soc., 134, 3058 (1987).
62. J. H. Riley, D. W. Sopher, J. H. P. Utley, and D. J. Walton, J. Chem. Res. (S), 326 (1982).
63. M. F. Semmelhack and G. E. Heinsohn, J. Amer. Chem. Soc., 94, 5139 (1972).
64. C. Adams, N. M. Kamkar, and J. H. P. Utley, J. Chem. Soc., Perkin II, 1767 (1979).
65. W. L. Mendelson, U.S. Pat. 4,055,573 (1977); Chem. Abstr., 88, 50861 (1978).
66. W. Oroshnik and P. E. Spoerri, J. Amer. Chem. Soc., 63, 3338 (1941).
67. W. M. Rodionov and V. C. Zvorukina, Bull. Soc. Chim. Fr., 5, 840 (1938).
68. K. Praefcke, C. Weichsel, M. Falsig, and H. Lund, Acta Chem. Scand., B34, 403 (1980).
69. J. Voss, Tetrahedron, 27, 3753 (1971).
70. J. Voss, K. Thimm, and L. Kistenbruegger, Tetrahedron, 33, 259 (1977).
71. J. H. P. Utley, and D. W. Sopher, J. Chem. Soc., Chem. Commun., 134 (1981); D. W. Sopher, Ph.D. thesis, London University, 1980.
72. R. C. Hallcher and M. M. Baizer, Justus Liebigs Ann. Chem., 737,

(1977); R. C. Hallcher, M.M. Baizer, and D. A. White, U.S. Pat. 4,072,583 (1978); Chem. Abstr., 88, 143499 (1978).
73. D. A. White and J. H. Wagenknecht, J. Electrochem. Soc., 128, 1470 (1981).
74. G. T. Mondodoev, N. M. Przhiyalgovskaya, and V. N. Belov, J. Org. Chem. USSR, 1, 1257 (1965).
75. G. T. Mondodoev, N. M. Przhiyalgovskaya, and V. N. Belov, J. Org. Chem. USSR, 1, 2048 (1965).
76. T. Tafel and H. Hahl, Ber. Dtsch. Chem. Ges., 40, 3312 (1907).
77. T. Tafel and W. Jurgens, Ber. Dtsch. Chem. Ges., 42, 2548 (1909).
78. T. Tafel, Ber. Dtsch. Chem. Ges., 45, 452 (1912).
79. H. Stenzel and F. Fichter, Helv. Chim. Acta, 17, 674 (1934).
80. H. Stenzel, F. Fichter, and H. Arni, Helv. Chim. Acta., 19, 392 (1936).
81. S. Wawzonek and J. E. Durham, J. Electrochem. Soc., 123, 500 (1976).
82. T. J. Curphey, C. W. Ammelotti, T. P. Layloff, R. L. McCartney, and J. H. Williams, J. Amer. Chem. Soc., 91, 2817 (1969).
83. M. Fedoronko, Advan. Carbohyd. Chem., 29, 107 (1974).
84. I. A. Avrutskaya, M. Y. Fioshin, E. V. Gromova, and V. T. Novikov, Elektrokhimiya, 8, 434 (1972); Sov. Electrochem., 8, 425 (1972).
85. M. Y. Fioshin, I. A. Avrutskaya, L. A. Muzychenko, E. V. Gromova, and V. T. Novikov, Elektrokhimiya, 8, 609 (1972); Sov. Electrochem., 8, 593 (1972).
86. S. Sugusawa and S. Matsumoto, U.S. Pat. 3,312,608 (1967); Chem. Abstr., 60, 12096 (1964).
87. Y. Takahashi, T. Sekine, and Y. Kikuchi, Jpn. Kokai, 73, 28, 410 (1973); Chem. Abstr., 79, 79132 (1973).
88. Farbwerke Hoechst A.-G., Neth. Pat. Appl., 6,602,261 (1966); Chem. Abstr., 66, 38237 (1967).
89. M. Falsig and H. Lund, Acta Chem. Scand., B34, 585 (1980).
90. J. Voss, C. Von Bulow, T. Drews, and P. Mischke, Acta Chem. Scand., B37, 519 (1983).
91. M. Falsig and H. Lund, Acta Chem. Scand., B34, 545 (1980).
92. M. Peover, Trans. Faraday Soc., 58, 2370 (1962).
93. F. E. Pykhtina, N. P. Glukhoedov, I. E. Kardash, and A. N. Pravednikov, Zh. Fiz. Khim., 52, 1217 (1978).
94. B. Sakurai, Bull. Soc. Chem. Jpn., 7, 127 (1932).
95. T. Troll, G. W. Ollmann, and H. Leffler, Angew. Chem., Int. Ed. Engl., 23, 622 (1984).
96. G. T. Cheek and P. A. Horine, J. Electrochem. Soc., 131, 1796 (1984).
97. L. Horner and K. Dickerhof, Chem. Ber., 116, 1603 (1983).
98. A. Guirado, F. Barba, and J. Martin, Electrochim. Acta, 29, 587 (1984).
99. R. Scheffold and R. Orlinski, J. Amer. Chem. Soc., 105, 7200 (1983).
100. R. Woods, Electrochim. Acta, 15, 815 (1970).
101. M. Hirayama, Bull. Chem. Soc. Jpn., 40, 1557 (1957).
102. S. F. Nelsen, J. Amer. Chem. Soc., 89, 5256 (1967).
103. J. Voss, Tetrahedron, 28, 2627 (1972).
104. J. Voss, Justus Liebigs Ann. Chem., 1231 (1974).

105. J. Voss, Justus Liebigs Ann. Chen., 1220 (1974).
106. W. Schmueser and J. Voss, J. Chem. Res. (S), 149 (1978).
107. H. Lund, Acta Chem. Scand., 17, 2325 (1963).
108. L. Horner and R. J. Singer, Justus Liebigs Ann. Chem., 723 (1969).
109. L. Horner and H. Neumann, Neth. Pat. Appl. 6,509,235 (1967); Chem. Abstr., 66, 105195 (1967).
110. J. Voss, G. Wiegand, and K. Hulsmeyer, Chem. Ber., 118, 4806 (1985).
111. R. A. Benkeser, H. Watanabe, S. J. Mels, and M. A. Sabol, J. Org. Chem., 35, 1210 (1970).
112. T. Shono, S. Kashimura, and H. Nogusa, Chem. Lett., 425, (1986); J. Org. Chem., 49, 2043 (1984).
113. R. D. Goodin and J. P. Chupp, U.S. Pat. 4,493,755 (1985); Chem. Abstr., 102, 122075 (1985).
114. A. Ryvolova-Kejharova and P. Zuman, Coll. Czech. Chem. Commun., 36, 1019 (1971).
115. G. Farnia, A. Romanin, G. Capobianco, and F. Torzo, J. Electroanal. Chem., 33, 31 (1971).
116. D. W. Leedy and D. L. Muck, J. Amer. Chem. Soc., 93, 4264 (1971).
117. J. Tirouflet, R. Robin, and M. Guyard, Bull. Soc. Chim. Fr., 571 (1956).
118. T. Troll and G. W. Ollmann, Electrochim. Acta, 29, 467 (1984).
119. J. D. Porter, S. Fletcher, and R. G. Barradas, J. Electrochem. Soc., 126, 1693 (1979).
120. A. B. Ershler, T. S. Orekhova, and I. M. Levinson, Elektrokhimiya, 15, 520 (1979).
121. H. Lund, Acta Chem. Scand., 17, 1077 (1963).
122. L. A. Taran, B. I. Buzykin, O. Y. Mironova, N. G. Gazetdinova, and Y. P. Kitaev, Izv. Akad. Nauk SSSR, Ser. Khim., 1983, 166.
123. P. Zuman and O. Manousek, Coll. Czech. Chem. Commun., 34, 1580 (1969).
124. J. Volke and A. M. Kardos, Coll. Czech. Chem. Commun., 33, 2560 (1968).
125. J. Volke and V. Skala, J. Electroanal. Chem., 36, 383 (1972).
126. P. G. Arapakos and M. K. Scott, Tetrahedron Lett., 1975 (1968).
127. A. P. Tomilov, S. E. Zabusova, and N. M. Alpatova, Elektrokhimiya, 18, 504 (1982).
128. P. H. Rieger, I. Bernal, W. H. Reinmuth, and G. F. Franekel, J. Amer. Chem. Soc., 85, 683 (1963).
129. A. M. Romaini, A. Gennaro, and E. Vianello, J. Electroanal. Chem., 88, 175 (1978).
130. M. Sertel, A. Yildiz, R. Gambert, and H. Baumgartel, Electrochim. Acta, 31, 1287 (1986).
131. K. Ishifuku, H. Sakurai, and H. Okamoto, Jpn. Pat. 180,563 (1949); Chem. Abstr., 46, 3432 (1952).
132. S. Sawa, Jpn. Pat. 5019 (1951); Chem. Abstr., 47, 2617 (1953).
133. A. P. Tomilov, S. L. Varshavskii, and Y. D. Smirnov, USSR Pat. 132, 214 (1960); Chem. Abstr., 55, 6214 (1961).
134. T. Nonaka and K. Sugino, J. Electrochem. Soc., 114, 1255 (1967); Y. D. Smirnov, L. I. Saltykova, and A. P. Tomilov, Zh. Prikl. Khim., 43, 1620 (1970).

135. Y. D. Smirnov and A. P. Tomilov, Zh. Org. Khim., 5, 864 (1969); Chem. Abstr., 71, 38299 (1969).
136. T. L. Antonova, L. N. Ivanovskaya, M. Y. Fioshin, and N. N. Savushknia, Elektrokhimiya, 23, 606 (1987); Sov. Electrochem., 23, 560 (1987).
137. V. Khrishnan, A. Muthukumarau, K. Raghupathy, and H. V. K. Udupa, Trans. SAEST, 13, 161 (1978); Chem. Abstr., 91, 46346 (1979). V. Khrishnan, K. Raghupathy, and H. V. K. Udupa, J. Electroanal. Chem. Interfacial Electrochem., 88, 433 (1978).
138. H. Wenker, J. Amer. Chem. Soc., 57, 772 (1935).
139. M. Fleischmann, G. Mengoli, and D. Pletcher, J. Electroanal. Chem., 43, 308 (1973).

IV

SYNTHETIC AND MECHANISTIC ASPECTS OF ANODIC REACTIONS OF ORGANIC COMPOUNDS CLASSIFIED BY ELECTROPHORES

13

Anodic Oxidation of Hydrocarbons

LENNART EBERSON / Division of Organic Chemistry, University of Lund, Lund, Sweden

JAMES H. P. UTLEY / Queen Mary and Westfield College, University of London, London, England

OLE HAMMERICH / H. C. Ørsted Institute, University of Copenhagen, Copenhagen, Denmark

I. Introduction 505

II. General Mechanistic Considerations 506
  A. Direct oxidation mechanism 506
  B. Indirect oxidation mechanism 507

III. Voltammetric Studies 508
  A. Simple aromatic hydrocarbons 508
  B. Alkyl aromatic hydrocarbons 511
  C. Alkanes, alkenes, and alkadienes 514

IV. Follow-up Reactions 519
  A. Accessible pathways 519
  B. Reactions of hydrocarbon cation radicals with intentionally added nucleophiles (addition, substitution, and electron transfer) 520
  C. Coupling reactions 523
  D. Indirect oxidations 528
  E. Conclusions 528

References 528

I. INTRODUCTION

The purpose of this chapter is to summarize the electrochemical behavior of hydrocarbons, with special reference to the electrode processes involved in the selective oxidation of hydrocarbons to form one or several defined organic products. Thus, the technically important problem of oxidizing hydrocarbons to carbon dioxide and water at catalytic electrodes in fuel cells is not dealt with here. This subject has been reviewed elsewhere [1].

For many years the study of hydrocarbon electrochemistry at anodes was confined to preparative studies of aromatic and alkyl aromatic hydrocarbons in water-containing electrolytes under constant-current conditions [2–4]. Usually, complex mixtures are obtained.

The use of nonaqueous solvents led to significant advances in the study of the anodic behavior of hydrocarbons. A pioneering study by Lund [5] of the voltammetry of aromatic hydrocarbons in MeCN-$NaClO_4$ at the rotating

Pt electrode was followed by a large number of electrochemical investigations in nonaqueous solvents. Concurrent preparative studies under controlled potential conditions, although not always very helpful in clarifying the nature of the initial electrochemical step, have nevertheless been of great help in finding out about the nature of the chemical reactions following electron transfer; much of this work is described in Chap. 25.

The gathering of fundamental information concerning the initial electron transfer, such as standard potentials for cation radical formation, was earlier greatly hindered by the occurrence of rapid reactions of the intermediates. Reaction with impurities in the solvent and/or rapid proton loss are common reasons for not observing reversible formation of cation radicals; the consequent irreversibility is in cyclic voltammetry (CV) associated with a shift in the peak potential, in the negative direction, from that for reversible oxidation. Considerable progress has been made in countering these difficulties. The use of acidic solutions greatly slows proton loss [6,7], the rigorous drying of solvents precludes rapid hydroxylation of cationic species [8,9], and the use of easily purified nonnucleophilic solvents, such as benzene and chlorobenzene [10] or liquid SO_2 [11,12], also permits the reversible formation of several cation radicals and dications.

The application of phase-selective second harmonic ac (SHAC) voltammetry to organic systems has proved fruitful [13,14]. This method is capable of measuring reversible potentials even when the electrogenerated intermediate undergoes moderately rapid reaction. Another approach for obtaining reversible potentials for the formation of reactive cation radicals is the application of CV with sweep rates in the kilovolt per second range [15–19], which recently has become popular owing to progress in the construction of the required microelectrodes. In several cases it has proved possible to apply specular reflectance spectroscopy (see Chap. 3) to the identification of electrogenerated cationic species and to follow their decay [20–24].

The general properties and reactions of cation radicals and dications in solution have been described in several review articles [25–28]. Apart from Chap. 25, the reader of this chapter may also find material of interest in Chaps. 23, 26, 32, and 34.

II. GENERAL MECHANISTIC CONSIDERATIONS

A. Direct Oxidation Mechanism

Following the postulate that electrons are always transferred one by one [29], the primary electrode process in the direct oxidation of a hydrocarbon is the formation of a cation radical, Eq. (1). However, because of the high chemical reactivity of cation radicals, a fast reaction normally occurs with a nucleophilic and/or basic species present in the electrolyte, resulting in the formation of either an addition product (I) between cation radical and nucleophile or an elimination product (II) by proton loss from the cation radical to a base, Eqs. (2) and (3):

$$\text{R-H} \xrightarrow{-e^-} \text{R-H}^{+\cdot} \qquad (1)$$

$$\text{R-H}^{+\cdot} \xrightarrow{\text{Nu}^-} \underset{\text{I}}{\text{R}^{\cdot}}\begin{smallmatrix}\diagup \text{Nu}\\ \diagdown \text{H}\end{smallmatrix} \tag{2}$$

$$\text{R-H}^{+\cdot} + \text{B} \longrightarrow \underset{\text{II}}{\text{R}\cdot} + \text{BH}^+ \tag{3}$$

where Nu^- is a nucleophile and B is a base. In either case the neutral radical formed may have an oxidation potential lower than that of the original substrate, making possible the transfer of a second electron, as shown in Eqs. (4) and (5):

$$\text{R}^{\cdot}\begin{smallmatrix}\diagup \text{Nu}\\ \diagdown \text{H}\end{smallmatrix} \xrightarrow{-e^-} \text{R}^{+}\begin{smallmatrix}\diagup \text{Nu}\\ \diagdown \text{H}\end{smallmatrix} \quad \text{or} \quad \text{R}^{\cdot}\begin{smallmatrix}\diagup \text{Nu}\\ \diagdown \text{H}\end{smallmatrix} + \text{R-H}^{+\cdot} \longrightarrow \text{R}^{+}\begin{smallmatrix}\diagup \text{Nu}\\ \diagdown \text{H}\end{smallmatrix} + \text{R-H} \tag{4}$$

$$\text{R}\cdot \xrightarrow{-e^-} \text{R}^+ \quad \text{or} \quad \text{R}\cdot + \text{R-H}^{+\cdot} \longrightarrow \text{R}^+ + \text{R-H} \tag{5}$$

Finally, the cations formed are converted into neutral products by a second chemical step with a nucleophile and/or base in the electrolyte. These processes correspond to the well-established ECE type of mechanisms (see Chap. 2).

Thus in many cases voltammetric studies on hydrocarbon oxidation give information only about the total number of electrons (two) transferred in the ECE-type process, not about the number of electrons (most frequently one, but sometimes two) transferred in the first electrochemical step. Preparative studies, which are often run in the presence of nucleophile or base, give little information on this point.

Alternative mechanisms are the disproportionation mechanism (6), which may be designated ECC, and the EEC mechanism (7). These two mechanisms, for which the key intermediate is the dication, are no longer thought to constitute major reaction pathways for anodic oxidation of alternant aromatic hydrocarbons (AAH) or alkyl-substituted AAH (see later) under common experimental conditions, but have been found to be operative for tetraarylethylenes, for example [30—32]. This difference in the reaction pattern of AAHs and tetraarylethylenes is due to the low reactivity of the cation radicals of the latter and at the same time favorable values of their disproportionation equilibrium constants K_d.

$$\text{R-H} \xrightarrow{-e^-} \text{R-H}^{+\cdot} \qquad 2\text{R-H}^{+\cdot} \overset{K_d}{\rightleftharpoons} \text{R-H} + \text{R-H}^{2+} \qquad \text{R-H}^{2+} \longrightarrow \text{products} \tag{6}$$

$$\text{R-H} \xrightarrow{-e^-} \text{R-H}^{+\cdot} \qquad \text{R-H}^{+\cdot} \xrightarrow{-e^-} \text{R-H}^{2+} \qquad \text{R-H}^{2+} \longrightarrow \text{products} \tag{7}$$

B. Indirect Oxidation Mechanism

Indirect oxidation mechanisms have been proposed in some cases in which voltammetric studies do not give unambiguous information about the nature of the electroactive species. Often they are assumed to involve the oxidation of the anion of the supporting electrolyte to form an inorganic radical, Eq. (8), which subsequently either attacks a weak C-H bond in the organic substrate, such as a C-H bond in the benzylic position of an alkyl aromatic hydrocarbon, or oxidizes the substrate in an electron transfer process, Eq. (9):

$$X^- \xrightarrow{-e^-} X\cdot \tag{8}$$

$$R\text{-}H + X\cdot \longrightarrow R\cdot + H\text{-}X \quad \text{or} \quad R\text{-}H + X\cdot \longrightarrow R\text{-}H^{+}\cdot + X^- \tag{9}$$

It is sometimes difficult to distinguish this type of mechanism from a direct one (see also Chap. 29).

III. VOLTAMMETRIC STUDIES

A. Simple Aromatic Hydrocarbons

Anodic voltammetry of hydrocarbons was originally performed in an electrolyte consisting of MeCN with a perchlorate salt as the supporting electrolyte at a rotating Pt electrode [5,33–35]. This electrolyte is relatively inert chemically and has a high anodic limit, allowing the determination of half-wave potentials $E_{1/2}$ up to around 2.1 V versus the Ag/Ag^+ (0.1 M) reference electrode. Generally, aromatic hydrocarbons display a well-developed wave under these conditions, although in some cases waves have been described as slightly deformed or deformed. In several cases two anodic waves separated by 0.2–0.7 V appear in the voltammograms.

The major voltammetric methods employed in the last two decades have been CV, derivative cyclic voltammetry (DCV), linear sweep voltammetry (LSV), and SHAC voltammetry. Voltammetry at a rotating disk electrode (RDE) has been performed in a few cases (see, e.g., Refs. 21 and 22 and Chaps. 2 and 3 for discussions of the fundamentals and the application of voltammetric techniques in general).

In cyclic voltammetric work the data are usually presented as anodic peak potentials E_{pa} or half-peak potentials $E_{p/2}$. The latter quantity may more easily be measured to a high degree of precision owing to the steepness of the voltammetric wave in the potential region corresponding to $E_{p/2}$. For many aromatic hydrocarbons irreversible oxidation is observed. However, fast sweep CV and SHAC voltammetry allow in some cases access to values for the reversible oxidation potentials E^{rev}. The application of voltammetric methods and the various data that can be obtained are exemplified for some representative aromatic hydrocarbons in Table 1. A more comprehensive collection of such information is available [18,19,36].

It has long been realized that excellent correlations exist between $E_{1/2}$ or E_{pa} data and Hückel molecular orbital parameters [34,35,42,43], ionization potentials (IP) [34,35,44,45], charge transfer transition energies [34], and positions of p bands in ultraviolet (UV) absorption spectra of the hydrocarbons [5,34]. The basis and limitations of such correlations have been examined critically [13,37,46]. For AAHs, the slopes of correlation lines between reversible oxidation and reduction potentials and IP and electron affinities (EA) are close to unity and have been used to suggest that in MeCN solvation energies of the hydrocarbon radical ions are constant throughout the series [37]. It was further concluded that the solvation energies of such species are equal and of the order of -2 eV. Phase-selective SHAC voltammetry has been used to evaluate the temperature depend-

Table 1. Oxidation Potentials of Aromatic Hydrocarbons by Polarography ($E_{1/2}$) or Cyclic Voltammetry (E_{pa})

Compound	Polarography, $E_{1/2}$ (V)		Cyclic voltammetry, E_{pa} or $E_{p/2}$ (V)	
	SCE[a]	Ag/Ag$^+$	SCE	Ag/Ag$^+$
Benzene			2.62[b] (ip)	
Toluene		1.98[c]; 1.93[d]		1.93[e]
p-Xylene		1.56[c]		1.64[e]
Mesitylene	1.90	1.55[c]		1.69[e]; 1.72[f] (ip)
Durene	1.62	1.29[c]		1.41[e]; 1.70[g]
Pentamethylbenzene	1.62	1.28[c]		1.33[e]; 1.63[g]
Hexamethylbenzene	1.52	1.16[c]		1.20[e]; 1.50[g]
Naphthalene	1.72	1.34[c]; 1.31[d]	1.87[a]	
Anthracene	1.20[f]	0.84[h]	1.31[i]; 1.40[j,k]	1.13[l]
9,10-Diphenylanthracene		0.86[h]	1.20[i]; 1.27[k]	
Biphenyl	1.91[f]	1.48[c]		
Fluorene	1.65	1.32[m]; 1.25[c]		
Perylene	1.0		1.09[a]	

[a] AcOH-NaOAc [38].
[b] MeCN-Bu_4NClO_4; (ip) irreversible peak [18].
[c] MeCN-$NaClO_4$ [5].
[d] MeCN-$NaClO_4$ [34].
[e] CF_3COOH-$(CF_3CO)_2O$-Bu_4NClO_4 [19].
[f] MeCN-Bu_4NBF_4; (ip) irreversible peak.
[g] MeCN-Bu_4NClO_4 [19].
[h] MeCN-$NaClO_4$ [39].
[i] CH_2Cl_2-AcOH-$(CF_3CO)_2O$-$NaClO_4$ [40].
[j] MeCN-Bu_4NBF_4-Al_2O_3 [37].
[k] CH_2Cl_2-CF_3COOH-$(CF_3CO)_2O$-Bu_4NBF_4 [6].
[l] C_6H_5Cl-Bu_4NBF_4 [10].
[m] MeCN-$(CF_3CO)_2O$-Bu_4NBF_4 [41].

ence of E^{rev} for a number of aromatic hydrocarbons [14] and hence the entropy changes associated with the formation of the cation radicals, Eq. (10):

$$\frac{dE^{rev}}{dT} = \frac{\Delta S}{F} \tag{10}$$

The values of $-\Delta S_{273.2}$ were observed to be in the range 5–20 e.u. and were attributed to reflect mainly differences in solvation. A slight increase in $-\Delta S_{273.2}$ was observed with increasing concentration of the supporting electrolyte and in going through the solvent series PC < DMF < MeCN.

Early voltammetric work was interpreted in terms of two-electron oxidation with the formation of dications. Controlled potential electrolysis of anthracene in the presence of pyridine (PYR) gave 9,10-dipyridination, which was held to confirm the intermediacy of dications [5]; in some cases, however, products were isolated that were best explained in terms of one-electron transfer [38]. It has since become clear that the ECE mechanism, with the second electron transfer taking place at the anode or in solution via the hydrocarbon cation radical, is the common mode of oxidation, with the EEC and disproportionation mechanisms relatively uncommon (see later).

A much used approach to the study of the primary electrochemical step involves the use of model compounds whose corresponding cation radicals (and in some cases dications) are relatively stable. Substituted anthracenes, particularly 9,10-diphenylanthracene (9,10-DPA), have featured largely in such work [6,11,17,40,47–52]. On slow sweep CV in MeCN, purified without critical regard to dryness, 9,10-DPA shows a one-electron reversible oxidation peak at $E_{pa} = 1.22$ V versus SCE, followed by a second one-electron irreversible peak at approximately 1.6 V. The employment of sweep rates in the kilovolts per second region [17] or of acidic [6,40], dry [40], or nonnucleophilic media [11] allows the observation of consecutive and fully reversible one-electron peaks.

The blue solution obtained by electrolysis of 9,10-DPA at 1.5 V versus SCE gives rise to a well-resolved ESR spectrum of the cation radical, identical to that recorded from a solution of the hydrocarbon in concentrated sulfuric acid. Similarly, the ESR spectra of the cation radicals of perylene and 9,10-dimethylanthracene (9,10-DMA) can be observed in solutions obtained after electrolysis of the parent hydrocarbons [48]. It is, however, appropriate to point out that in the absence of electrochemical evidence regarding the number of electrons transferred in a particular process, the mere observation of an ESR spectrum from an electrolyte during or after electrolysis is not unequivocal evidence for a one-electron transfer. The cation radical may have been formed in other, more complex processes.

It was at one time thought possible that many anodic oxidation reactions, including the pyridination of 9,10-DPA, involved the attack by nucleophiles on dications, small stationary concentrations of which were at equilibrium with the initially formed cation radicals, Eq. (6) [53]. This view was challenged [54], but real progress could not be made on resolving the matter until thermodynamically meaningful electrochemical data became available. In practice this meant the measurement of peak potentials for the reversible formation of cation radicals and dications; for a given substrate the differences in peak potentials can be used to calculate disproportionation equilibrium constants K_d, Eq. (11):

$$K_d = e^{[(E_1^0 - E_2^0)nF/(RT)]} \approx e^{[(E_{pa}^1 - E_{pa}^2)nF/(RT)]} \tag{11}$$

When such data became available (see, e.g., Refs. 6, 8, and 55) they confirmed that the estimation of K_d values from peak potentials for irreversible oxidation was grossly misleading. For instance, cyclic voltammetry of 9,10-DPA in relatively impure MeCN suggests a K_d value of approximately 4×10^{-6} [53]; data from the reversible system give a calculated K_d value of 10^{-8}. Measurements on substrates that give more stable cation radicals show that K_d values are very solvent sensitive, tending to be lower in nonnucleophilic and acidic media than in, say, nitrile solvents [8]. It is now clear that unless successive electron transfers are closely spaced, the concentrations of dication formed as a result of disproportionation are neglibible and, given the rate limit set by diffusion for attack by nucleophiles, cannot account for the observed rates of many anodic reactions.

Another approach is to use fast sweep techniques [15,17] in the case of moderately reactive cation radicals.

A third approach is to reduce the nucleophilicity of the solvent below that of MeCN (a nucleophile of about the same strength as acetic acid). The limiting factor is then the nucleophilicity of the substrate itself (see Sec. IV.C and Chaps. 23 and 32). Thus in $PhNO_2$-Pr_4NClO_4 [56] and/or CH_2Cl_2-Bu_4NClO_4 [57], 9,10-DPA, 9,10-DMA, 1,3,6,8-tetraphenylpyrene, rubrene, and tetracene undergo a one-electron oxidation to stable cation radicals. Attention has already been drawn to the use of CF_3COOH-$(CF_3CO)_2O$, alone [8,58] or mixed with CH_2Cl_2 [6,40,59–61], PhCl [10], and liquid SO_2 [11,12,62]. In the latter solvent even the reversible formation of the tetracation of 9,9'-bianthryl has been observed [12].

Finally, the stability of cationic intermediates, and hence their detection by voltammetric methods, is, as expected, enhanced at low temperatures [6,63–65]. An additional aspect of low-temperature conditions is the possibility of preparing stable salts of highly reactive cation radicals as demonstrated by the formation of a solid with the composition $(ArH)_2PF_6$ by oxidation of ArH in dichloromethane between −30 and −70°C with Bu_4NPF_6 as supporting electrolyte [66–68].

B. Alkyl Aromatic Hydrocarbons

The voltammetry of methylbenzenes has been studied extensively as examples of a reaction in which the initial electron transfer is followed by proton loss [7,18,19,20–24,69–76]. Cyclic voltammetry of these compounds at low sweep rates in MeCN typically gives two irreversible oxidation peaks, both corresponding to the transfer of two electrons. The first of these oxidation processes results in the formation of the corresponding benzyl cation, which in the presence of a suitable base is believed to be formed in an ECE-type process with the chemical step deprotonation of the initially formed cation radical, Eq. (3) (see later). The fate of the benzyl cation depends on the composition of the solvent. For instance, in CF_3COOH or CH_2Cl_2-CF_3COOH the intermediate pentamethylbenzyl cation, formed by oxidation of hexamethylbenzene, is converted to $ArCH_2OOCCF_3$. Further oxidation of this product leads to the cation radical, which is stable on the time scale of slow sweep CV [7]. In moist MeCN, preparative-scale electrolysis of alkyl aromatics at the first oxidation potential gives moderate to high yields of the corresponding N-benzylacetamides, which are themselves relatively easily oxidized. In some cases the corresponding bibenzyls were also isolated [70,71].

The mechanistic details of the anodic alkylbenzene oxidation have been under intense investigation. In the early stages of this research the major issue was the order of the electron and proton transfer steps leading to the benzyl cations and whether the second electron transfer reaction takes place at the anode or in solution. By the application of a combination of voltammetry and specular reflectance spectroscopy [20–24,72] it has been possible to monitor the electrode surface by UV-visible spectroscopy during electrolysis at controlled potential or to monitor the absorption at a predetermined λ_{max} value as a function of potential. In this way absorption spectra, which were suggested to arise from the initial cation radical (III), the benzyl radical (IV) formed by deprotonation, Eq. (12), the benzyl cation (V), and the N-benzylnitrilium ions (VI) have been recorded in MeCN, which in turn has led to methods of measuring the concentration profiles of these intermediates as a function of time and potential.

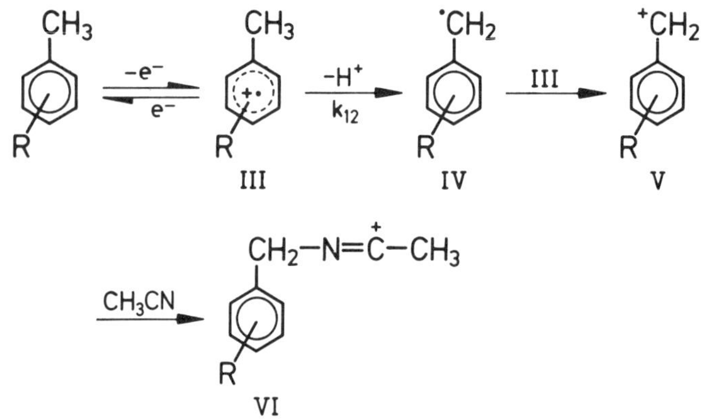

(12)

This approach has been extended to the oxidation of hexamethylbenzene at the potential of the second voltammetric wave; the product of such oxidation is side-chain bisacetamidation at the 1,3-positions, Eq. (13). The total analysis of the data from this and other electrochemical techniques was found to be in agreement with mechanism (12) and allowed the evaluation of the deprotonation rate constants k_{12} for a number of methylbenzene cation radicals.

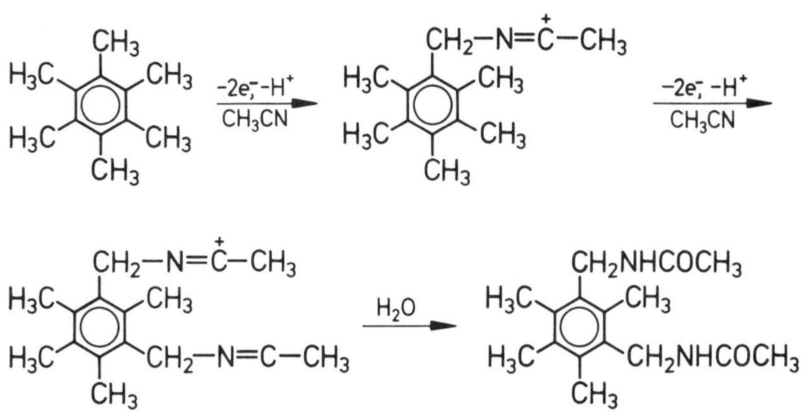

(13)

The values of k_{12} measured in this way varied from 100 s^{-1} (1,3,5-trimethylbenzene) to 870 s^{-1} (1,2,4-trimethylbenzene).

On the other hand, reoxidation of the cation radical of hexamethylbenzene in MeCN at −40°C could not be observed by CV [73], which led to the conclusion that the lifetime was less than 10^{-4} s, corresponding to a pseudo-first-order rate constant approximately one order of magnitude higher than the range already reported.

The exact nature of the base is not clear in these studies, and the differences in the rate constants may be attributed to differences in solvent quality. More insight into the nature of the proton transfer reaction has been gained through studies in MeCN of the homogeneous electron transfer from a series of methylbenzenes to Fe(III) complexes followed by proton transfer from the resulting methylbenzene cation radicals to a series of pyridine derivatives [19,74]. For hexamethylbenzene cation radical the rate constants for proton transfer varied between 4×10^2 and 9×10^5 M^{-1} s^{-1} in going from 2-fluoropyridine to 2,4,6-trimethylpyridine. The rate of deprotonation followed a general Brønsted relationship with a slope of 0.26, which suggested an overall free energy change in the exergonic region of the driving force. This together with a deuterium kinetic isotope effect that was observed to decrease with increasing driving force for proton transfer seems to indicate an early transition state in which the proton transfer has not proceeded beyond the symmetrical situation.

However, recent linear sweep voltammetry and derivative cyclic voltammetry studies have cast some doubt on the general validity of the mechanism, including simple deprotonation of the cation radicals [75]. Experimental results obtained in MeCN indicated a rate law, Eq. (14), second order in cation radical. This together with the unusual observation that the apparent rate constant decreased with increasing temperature at t > 20°C resulting in a curved Arrhenius plot and a deuterium kinetic isotope effect that decreased with decreasing temperature pointed toward a reaction scheme including the reversible dimerization of the cation radical followed by the irreversible formation of $Ar\text{-}CH_2^+$, $Ar\text{-}CH_3$, and a proton, Eqs. (15) and (16).

$$\frac{-d[ArCH_3^{+\cdot}]}{dt} = k_{14}[ArCH_3^{+\cdot}]^2 \tag{14}$$

$$2ArCH_3^{+\cdot} \rightleftharpoons (ArCH_3^{+\cdot})_2 \tag{15}$$

$$(ArCH_3^{+\cdot})_2 + B \longrightarrow ArCH_2^{\cdot} + ArCH_3 + BH^+ \tag{16}$$

The structure of this dimer is not known. The implications of these unexpected results on the conclusions drawn from previous studies, in particular the work based on spectroelectrochemical measurements, is not yet clear. However, it seems plausible that the second-order reaction is indeed the favorable pathway in the absence of suitable bases like PYR.

The simplest compound to study with respect to deprotonation of the cation radical is hexamethylbenzene because the nucleus is blocked to competing substitution reactions. For other alkyl aromatics an alternative reaction is available, that is, attack by starting material on the cation radical or benzyl cation. The balance between these pathways has been studied by the combination of LSV and specular reflectance spectroscopy for durene and mesitylene. The conclusions were that proton loss is slower from the cation radical of mesitylene than from that of durene [21,71]. Consequently, bimolecular reaction with substrate resulting in formation of the dimer (VII) is more likely for mesitylene than for durene, which will be acetamidated after proton loss and further oxidation, Eq. (17):

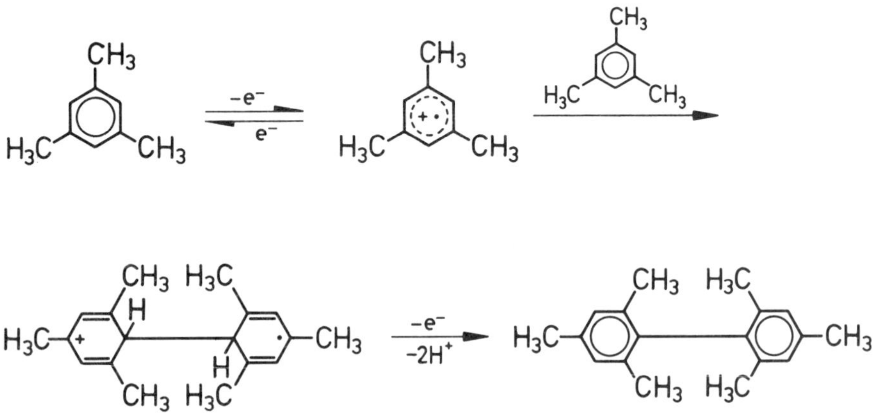

VII

(17)

The application of CV in the kilovolt per second range has allowed the determination of E^{rev} for oxidation of the methylbenzenes [19] (see Table 1). The correlation of E^0_{Ar} with IP was found to be linear with a slope of 0.71, Eq. (18),

$$E^0_{Ar} = 0.71IP - 3.67 \qquad (18)$$

considerably less than unity, indicating that the free energy change (internal and solvation) associated with the electron transfer process decreases with approximately 10 kcal/mol in going from hexamethylbenzene to toluene.

C. Alkanes, Alkenes, and Alkadienes

The anodic limit of MeCN containing a perchlorate salt is not high enough to allow a more general study of aliphatic hydrocarbons, except in a few favorable cases [34,77,78]. With the introduction of tetrafluoroborates and hexafluorophosphates [79] as supporting electrolytes the anodic limit in MeCN can be extended considerably, so that certain saturated aliphatic hydrocarbons can be studied voltammetrically. The application of ultramicroelectrodes has even made it possible to record voltammograms for short-chain alkanes in MeCN in the absence of intentionally added supporting electrolyte [80]. Tables 2 through 4 list oxidation potentials for a number of aliphatic and alicyclic hydrocarbons. The tables are organized to include a representative range of alkanes, alkenes, and strained hydrocarbons; common conditions for voltammetry are displayed, as are the relationships between oxidation potentials measured against different reference electrodes. From the diffusion currents it can be estimated that two electrons are involved in the majority of these oxidations [77,79,81].

It is apparent from Tables 1 and 2 that, apart from the use of tetrafluoroborates and hexafluorophosphates, extended anodic limits can be obtained at low temperatures [82] or by the use of such acidic solvents as

Table 2. Oxidation Potentials of Alkanes[a]

Compound	V versus Ag/Ag^+	V versus Pd/H_2
_n_-Alkanes	3.4[b]; 3.5 (E_{pa})[c]	—
	2.7[d]; 3.5–3.9[e]	
	1.14–1.46 (versus Ag/AgCl)[f]; 1.7–2.3[g]	
2,2-Dimethylbutane	3.28[b]; 3.90 (E_{pa})[c]	—
2-Methylpentane	3.01[b]	1.68[h]
Cyclopropane	3.41[e]	—
Cyclopentane	—	2.01[h]
Cyclohexane	—	1.77[h]
4-_tert_-Butylcyclohexane	3.00 (E_{pa})[c]	—
Adamantane	2.36 ($E_{p/2}$)[i]	—
	1.75 ($E_{p/2}$ versus Ag wire)[j]	
	2.38 ($E_{p/2}$)[k]	
1-_tert_-Butyladamantane	2.33 ($E_{p/2}$)[k]	—
	1.40 ($E_{p/2}$ versus Ag wire)[j]	

[a]$E_{1/2}$ values unless otherwise stated.
[b]MeCN-Et_4NBF_4 [79].
[c]MeCN-Et_4NBF_4, -45°C [82].
[d]MeCN-Bu_4NBF_4 [83].
[e]MeCN [80].
[f]CF_3SO_3H-H_2O [100].
[g]HF-BF_3/BF_4^- ($H_0 \simeq -15.6$).
[h]FSO_3H-CH_3COOH [79].
[i]MeCN-$LiClO_4$ [84].
[j]CF_3COOH-Bu_4NBF_4 [81].
[k]MeCN-Bu_4NBF_4 [85].

CF_3COOH [83,92,93], FSO_3H [79,94–98], HF [99], and CF_3SO_3H (without supporting electrolyte) [100]. In addition to the electrolyte systems included in the tables, hydrocarbon oxidation has been realized in nitromethane, nitroethane, propylene carbonate, sulfolane, and dichloromethane [101]. The observed potentials, invariably for irreversible oxidation, vary in a reasonably predictable manner. Good correlations are often found with IP

Table 3. Half-wave Potentials and Peak Potentials for Oxidation of Alkenes, Aryl Alkenes, and Alkadienes

Compound	$E_{1/2}$[a] (V versus Ag/Ag^+)	$E_{1/2}$[b] (V versus SCE)	Compound	E_{pa}[c] (V versus Ag/AgCl)
Ethylene	2.90		Styrene	1.68
1-Alkenes	2.7–2.8		Indene	1.56
2-Alkenes	2.2–2.3		_cis_-Stilbene	
Cyclohexene	2.05; 1.95[d]	2.14	_trans_-Stilbene	1.40
1,3-Butadiene	2.09		1,3-Butadiene	2.0
Norbornadiene		1.54	Norbornadiene	1.62
Norbornene		2.02	2-Phenylnorbornene	1.1[e]
1,4-Cyclohexadiene	≃1.3	1.74	Biadamantylidene	1.65[f]

[a] $MeCN-Et_4NBF_4$ [79] unless otherwise stated.
[b] $MeCN-LiClO_4$ [86].
[c] $MeOH-NaClO_4$ [87] unless otherwise stated.
[d] $MeCN-NaClO_4$ [34].
[e] $MeCN-Bu_4NClO_4$ (V versus Ag/Ag^+) [88].
[f] $CH_2Cl_2-CF_3COOH-(CF_3CO)_2O-Bu_4NBF_4$ (V versus SCE) [89]. (The potentials of a comprehensive series of related compounds are listed in Ref. 89.)

Table 4. Oxidation Potentials of Strained Hydrocarbons

Compound	$E_{1/2}$ (V versus SCE)	Compound	$E_{1/2}$ (V versus SCE)
	>2.5[a]	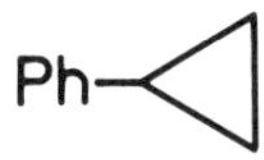	1.87[b]
	2.05[a]	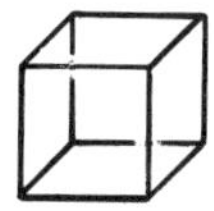	1.73[c]
	2.12[c]		0.91[c]
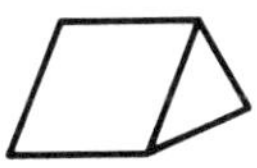	1.91[c]		

[a]MeCN-Et_4NBF_4 [90].
[b]MeCN-$LiClO_4$ [91].
[c]MeCN-$LiClO_4$ [78].

[78,81,102] and σ^+ parameters [86,87,103]. For a series of nine highly strained polycyclic alkanes, some of which are depicted in Table 4, the correlation between $E_{1/2}$ and IP is given by Eq. (19) [78]:

$$E_{1/2} = 0.78\text{IP} - 4.88 \qquad (19)$$

It is of interest to note that the slope of the correlation line in this case, 0.78, is very close to that, 0.71, observed for the series of methylbenzenes [see Eq. (18)]. Although the values of $E_{1/2}$ for the present series represent a highly irreversible process, it is most likely that the follow-up reactions for these very strained hydrocarbons, which are expected to be extremely fast, result in the maximum kinetic shift of the half-wave potential for the entire series of compounds, and accordingly, the slope, 0.78, may be close to that for the correlation between the experimentally inaccessible reversible potentials and IP. For the methylbenzenes [19] it was concluded that the free energy change was mainly associated with solvation, which in the simple Born approximation depends only on the size of the cation radical in a given solvent, here MeCN. Considering now that the methylbenzenes and the strained aliphatic hydrocarbons just referred to have approximately the same size, it seems reasonable that comparable values of the slope should be found for these two series of structurally unrelated compounds.

For alkanes (Table 2) there remains some doubt about the first step of the overall reaction; in MeCN the final products are usually the N-alkylacetamides. In neutral solution, at extreme anodic potentials, it is difficult to decide between direct, Eqs. (1), (2), and/or (3), and indirect electron transfer, Eqs. (8) and (9). For oxidation in MeCN-BF_4^- solutions, the variation in potentials is best explained in terms of the direct mechanism. However, it was found that a similar mixture of 4- and 5-acetamido derivatives was obtained from the oxidation in MeCN of methyl hexanoate in the presence of Cl^-, ClO_4^-, BF_4^-, or ClO_2 [104]. Voltammetry showed that the first three electrolytes, at least, are oxidized prior to the substrate. This suggests indirect oxidation and further reaction via homogeneous electron transfer or hydrogen abstraction, Eqs. (20) and (21) (see also Ref. 105 for a careful product study of the related oxidation of 2-octanone). A similar indirect oxidation mechanism involving hydrogen abstraction by electrogenerated nitrate radical has recently been proposed for the electrolysis of linear alkanes in t-BuOH-H_2O mixtures containing HNO_3 and saturated with O_2 [106].

$$X^- \xrightarrow{-e^-} X\cdot \ (X{=}Cl^-,\ ClO_4^-,\ BF_4^-) \quad (20)$$

$$X\cdot + CH_3(CH_2)_4COOCH_3 \xrightarrow{-HX} CH_3\dot{C}H(CH_2)_3COOCH_3$$
(+ the 4-isomer)

$$\xrightarrow{-e^-,\ +\ CH_3CN,\ H_2O} CH_3\underset{\displaystyle NHCOCH_3}{\underset{|}{C}H}(CH_2)_3COOCH_3 \quad (21)$$
(+ the 4-isomer)

In strongly acidic solution it has been proposed that electron transfer is from the protonated alkane, Eqs. (22) through (24) [97,98]:

$$RH + FSO_3H \rightleftharpoons FSO_3^- + RH_2^+ \quad (22)$$

$$RH_2^+ \xrightarrow{-e^-} R\cdot + 2H^+ \quad (23)$$

$$R\cdot \xrightarrow{-e^-} R^+ \longrightarrow \text{products} \quad (24)$$

The CV of alkanes in FSO_3H was studied as function of added AcOH (a base under these conditions) and/or $KFSO_3$; diffusion-controlled two-electron oxidation was found and the current-potential curves shifted as base was added. Similar results for cyclohexane oxidation in CF_3COOH-FSO_3H, CF_3COOH-CH_3SO_3H, and CH_2Cl_2-FSO_3H were also interpreted in terms of initial protonation of the alkane [92], The experimental basis for this conclusion has been challenged. First it was claimed that the addition of base assisted alkane oxidation, which made it unlikely that oxidation of protonated alkane was involved [94]. Second, for cyclopentene, no oxidation peak could be observed for a solution of the alkene in FSO_3H; ^1H-NMR spectroscopy showed that in such a solution the cyclopentene was substantially converted by proton addition into the secondary carbenium ion [95]. It was suggested that this cation should be oxidized at a potential similar to those claimed for the oxidation of protonated alkanes. The issue has been confused further by the use of different reference electrodes and by variations in voltammograms with temperature, which are difficult to interpret unambiguously, as well as nonelectrochemical reactions of the alkanes in these highly acidic media [99]. However, there now seems to be general agreement on a mechanism involving electron transfer from unprotonated RH involving steps (1), (3), and (5). Results of voltammetry of alkenes under less acidic conditions at Pt seems to indicate the participation of surface oxides in the oxidation process [107].

Usually, precautions to avoid molecular oxygen are not taken in voltammetric oxidation studies. However, the cation radicals produced during such experiments may react with O_2, as illustrated by the extensive work on the oxidation of biadamantylidene and a series of related compounds [89,108,109]. In the absence of O_2, reversible oxidation to the cation radical may be observed. However, if O_2 is introduced in the electrochemical cell, reaction takes place between the alkene cation radical and O_2, resulting in the formation of the corresponding dioxetane derivative in a chain reaction (25) and (26). Similar behavior has been observed for other alkanes [110].

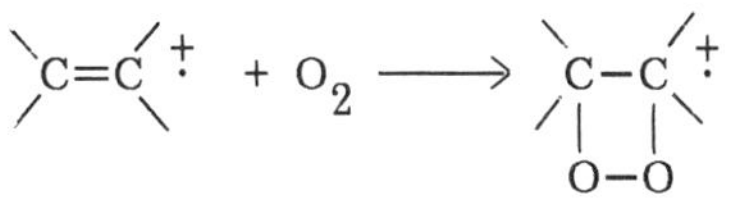

(25)

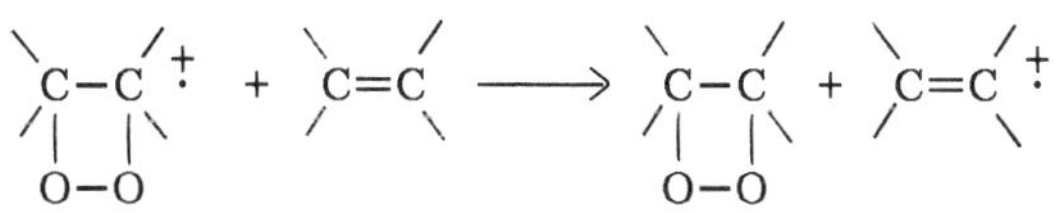

(26)

Several aryl alkenes show reversible or quasi-reversible behavior on oxidation [31,32,111]. Tetra-p-anisylethylene, for instance, may be converted into a stable dication in MeCN [31,32] or dichloromethane [32]. The results of CV in MeCN suggest that reversible two-electron transfer was being observed; controlled potential coulometry supported this. In CH_2Cl_2, however, it proved possible to observe [32] separately successive one-electron transfers and to show by coulometry that at 0.95 V (versus SCE) 0.9 F/mol was consumed, whereas at 1.20 V the value was 2.1 F/mol.

Oxidation of tetraphenylethylene (TPE) in CF_3COOH or CH_2Cl_2-CF_3COOH [30,111] gives a cation radical that is stable on the CV time scale but which undergoes second-order decomposition according to the rate law, rate = $k_{obs}[TPE^{\dot{+}}]^2/[TPE]$. The product is 9,10-diphenylphenanthrene, and the reaction seems to be a clear-cut case of disproportionation followed by intramolecular cyclization (see later).

IV. FOLLOW-UP REACTIONS

A. Accessible Pathways

A major conclusion of the discussion in Sec. III is that for a variety of model compounds the primary step is one-electron transfer to form a cation radical, and it is postulated that this is valid for all hydrocarbon oxidations. The several pathways by which the cation radical may be transformed into stable products are the ECE pathway (1) through (5) and the ECC and EEC mechanisms (6) and (7), with the dication as the kinetically active species. In some cases there is also the possibility that an indirect oxidation mechanism (8) and (9) operates, possibly followed by electrochemical or chemical steps.

It is convenient here briefly to summarize the overall reactions possible in hydrocarbon oxidations. These can be divided into substitution, Eq. (27), addition, Eq. (28), and coupling reactions (29) and (30), which cover the large majority of the reaction types.

$$R\text{-}H + Nu^- \xrightarrow[-2e^-]{} R\text{-}Nu + H^+ \qquad \text{substitution} \tag{27}$$

$$\gt C{=}C\lt \; + \; 2Nu^- \xrightarrow[-2e^-]{} Nu{-}\overset{|}{\underset{|}{C}}{-}\overset{|}{\underset{|}{C}}{-}Nu \qquad \text{addition} \tag{28}$$

$$2R{-}H \xrightarrow[-2e^-]{} R{-}R + 2H^+ \qquad \text{coupling with elimination} \tag{29}$$

$$2\gt C{=}C\lt \; + \; 2Nu^- \xrightarrow[-2e^-]{} Nu{-}\overset{|}{\underset{|}{C}}{-}\overset{|}{\underset{|}{C}}{-}\overset{|}{\underset{|}{C}}{-}\overset{|}{\underset{|}{C}}{-}Nu \qquad \text{coupling with addition} \tag{30}$$

In Eqs. (27), (28), and (30), Nu^- denotes a nucleophile, such as AcO^-, MeO^-, H_2O, or MeCN (all of these do not necessarily work in all processes, though). For alkylaromatic hydrocarbons containing at least one α-hydrogen atom, one can distinguish between nuclear and side-chain substitution, the latter reaction resulting in substitution of one α-hydrogen atom in the alkyl group.

A detailed discussion of these reaction types can be found in Chaps. 23 and 25, whereas in this chapter we are concerned mainly with the elucidation of the sequence of reactions leading to each of these reaction types.

B. Reactions of Hydrocarbon Cation Radicals with Intentionally Added Nucleophiles (Addition, Substitution, and Electron Transfer)

Convincing evidence has been put forward that the ECE-type mechanism (1), (2), and (4) followed by reaction of the carbocation with a second nucleophile is the reaction pathway in the oxidative addition of two molecules of a pyridine derivative to 9,10-DPA, Eq. (31) [112–116]. By studying the number of electrons transferred (n_{obs}) in the oxidation of 9,10-DPA in the presence of different nucleophiles at the rotating disk electrode as a function of the rotation rate ω, it has been possible to show [112] that the strength of the nucleophile strongly affects the dependence of n_{obs} on ω. The reaction scheme summarizes the underlying idea:

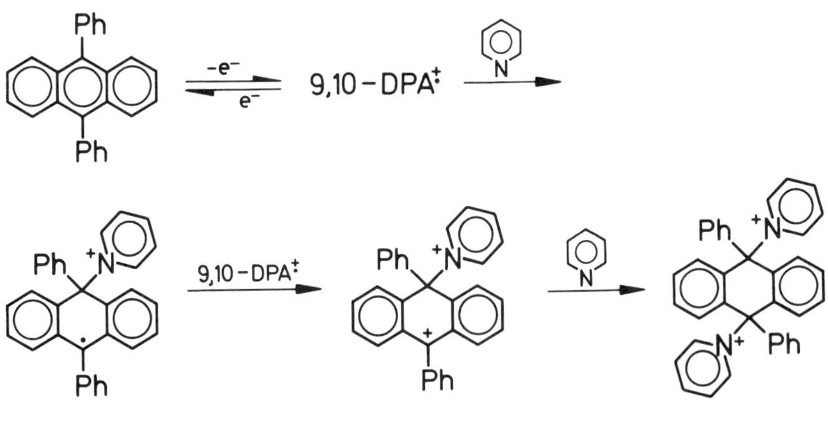

(31)

Approximate pseudo-first-order rate constants for the rate determining attack by the nucleophile on 9,10-DPA$^{+\cdot}$ could be obtained by fitting the experimental data for n_{obs} versus ω to the working curve calculated for the ECE mechanism. The results, shown in Table 5 as relative rates, dem-

Table 5. Relative Rates for the Pseudo-First-Order Reaction Between 9,10-Diphenylanthracene Cation Radical and Pyridine Derivatives

Nucleophile	Relative rate
4-Methylpyridine	330
Pyridine	100
4-Acetoxypyridine	10
4-Cyanopyridine	1

Source: From Ref. 112.

onstrate that the rates follow the order of nucleophilicity of the pyridines [this conclusion is not affected by the fact that the working curves were calculated for a mechanism at which the second electron transfer was assumed to take place at the anode instead of in solution as in Eq. (31)].

The rate law for the reaction between 9,10-DPA$^{+\cdot}$ and PYR at the millimolar concentration level is given by Eq. (32). The value of k_{32} has been determined by spectroelectrochemical methods [113–115] and by homogeneous kinetics [114,115,116] and is in the range of $2–4 \times 10^4\ M^{-1}\ s^{-1}$ in MeCN at room temperature. However, at much lower concentrations (0.01 mM), stopped-flow experiments demonstrated that the reaction was following a rate law (33) second order in 9,10-DPA$^{+\cdot}$ and first order in PYR. These data indicate a transition from rate-determining nucleophilic attack at high concentrations of 9,10-DPA$^{+\cdot}$ to rate-determining electron transfer at low concentrations [115]. The same rate law was found for attack by Cl^- [117].

$$\frac{-d[9,10\text{-DPA}^{+\cdot}]}{dt} = k_{32}[9,10\text{-DPA}^{+\cdot}][\text{PYR}] \tag{32}$$

$$\frac{-d[9,10\text{-DPA}^{+\cdot}]}{dt} = k_{33}[9,10\text{-DPA}^{+\cdot}]^2[\text{PYR}] \tag{33}$$

The rate constant for the corresponding reaction with rubrene cation radical is approximately one order of magnitude lower [18]. The solvent used in these studies, MeCN, is in itself a nucleophile, but much too weak to interfere kinetically in the presence of pyridines. For instance, the pseudo-first-order rate constant for the reaction between unsubstituted anthracene and MeCN has been estimated as $125\ s^{-1}$ [119].

Attack on cation radicals may be nucleophilic or basic. This distinction has been illustrated by a cyclic voltammetric study of 9,10-DPA and 9,10-DMA [120]. The relative reactivities of different reagents toward the cation radical of 9,10-DPA or 9,10-DMA was monitored by measuring the reduction current for the cation radical as a function of concentration or added reagent. The results of a series of such experiments involving lutidines with differing steric requirements are summarized in Table 6.

9,10-DMA$^{+\cdot}$ shows but a small difference in reactivity toward the three reagents, indicating that the attack of the reagent is on a proton of one of the methyl groups, a relatively unhindered position; that is, it acts as

Table 6. Relative Rates for Attack by Lutidines on the Cation Radicals of 9,10-Diphenylanthracene and 9,10-Dimethylanthracene

Cation radical	3,5-Lutidine	2,5-Lutidine	2,6-Lutidine
9,10-DPA$^{+\cdot}$	37	6.5	1.0
9,10-DMA$^{+\cdot}$	1.7	1.3	1.0

Source: From Ref. 120.

a base in analogy with what has been described for the oxidation of methylbenzenes in the presence of pyridine derivatives. This is supported by the isolation of a dimer (VIII) from the CPE run, Eq. (34):

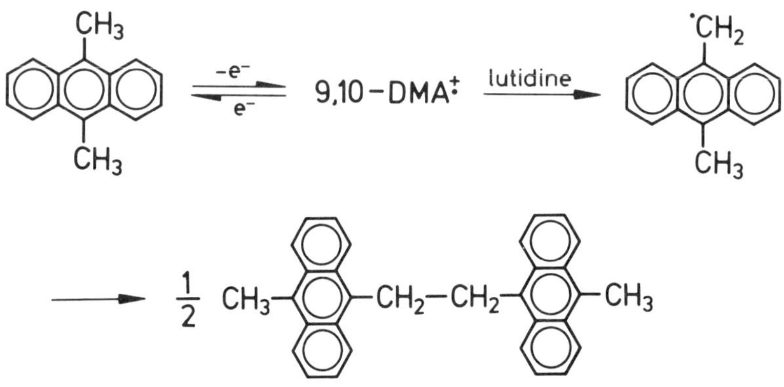

VIII

(34)

On the other hand, in the reaction with 9,10-DPA$^{+\cdot}$, only nucleophilic action is possible and hence a large difference in reactivity between the sterically unhindered 3,5-lutidine and the sterically hindered 2,6-lutidine is observed for attack on the positive 9-carbon atom.

Cation radicals are strong one-electron transfer oxidants as well, and thus one-electron oxidation of nucleophiles can sometimes take place. The rules determining the competition between electron transfer and electrophilic interaction are only vaguely known [121]. As may be expected, the difference in redox potential between cation radical and nucleophile is important [114]. For cation radicals with an E^0 value in the range of 1.3–1.5 V versus NHE, it seems that there is an upper limit of E^0 ($Nu^\cdot/Nu^-$) between 1.8 and 2.0 V versus NHE for electron transfer to take place. Nucleophiles that are oxidizable with more difficulty, such as AcO^-, react with bond formation. Cyanide ion, for which conflicting product results have been reported [114,122,123], is difficult to oxidize [124], with an estimated E^0 ($CN^\cdot/CN^-$) of 2.2 V versus NHE in MeCN, and should belong to the latter group.

At the detailed mechanistic level much less is known about the reactions taking place during coelectrolysis of nucleophiles and unsaturated hydrocarbons with π systems less extended than, for example, those for anthracene derivatives. However, in a number of cases the products isolated after anodic oxidation are structurally analogous to those already described, although the selectivity in general is low. For instance, oxidation of p-xylene in MeOH-NaOMe gave the 1,4-addition product 3,6-dimethoxy-3,6-dimethyl-1,4-cyclohexadiene (mixture of cis and trans products, approximately 20%) together with side-chain-substituted products [125]. For even less extended π systems, such as 1,3-dienes, and allenes, mixtures of addition and substitution products are formed [126,127], in cases even accompanied by dimers and trimers [128].

C. Coupling Reactions

In media of low nucleophilicity unsaturated hydrocarbons may themselves be the most nucleophilic species present and thus give rise to a variety of coupling products. Solvents that have been used to this effect include CH_2Cl_2-Bu_4NBF_4 [129] and the same solvent mixed with strong acids like CF_3COOH, CH_3SO_3H, and CF_3SO_3H [130]. A major product from such oxidation of mesitylene is bimesityl (VII) and, from durene, the heptamethyldiphenylmethane (IX); no diphenylmethane is formed from mesitylene or biphenyl from durene.

In CH_2Cl_2 solution some nuclear chlorination takes place as a result of anodic oxidation of chlorine ion displaced from the solvent; this is effectively suppressed in acidic solution. The nucleophilicity of the solvent is also important in determining the distribution of products; coupling takes place only in nonnucleophilic media, and the array of results summarized for durene in scheme (35) are best explained in terms of an ionic mechanism (only the major products are shown).

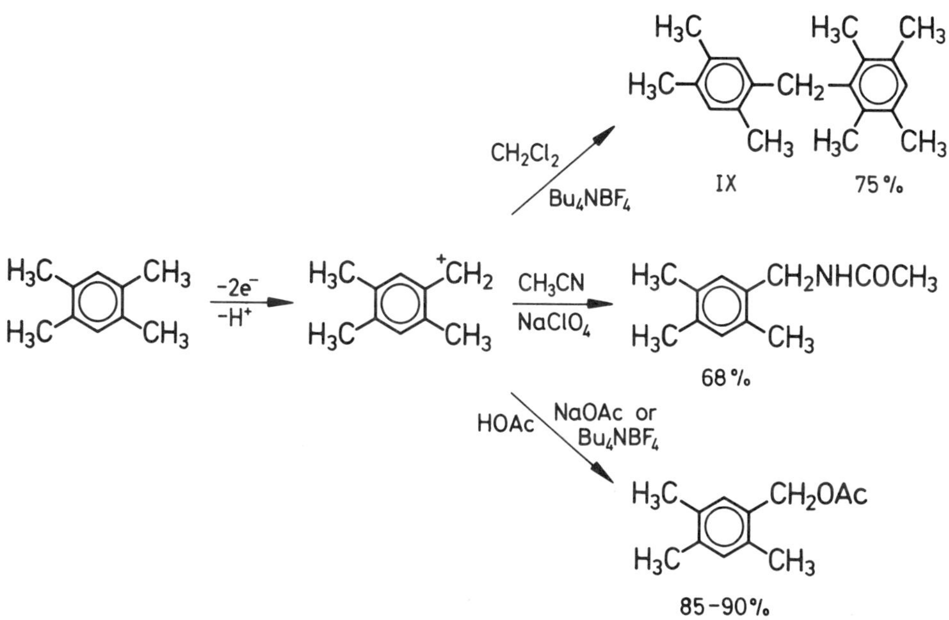

(35)

A rationalization of the results of mesitylene is given in Eq. (17). Attack of cation radical on mesitylene rather than dimerization of two cation radicals is preferred since the yield of VII increases appreciably with increasing concentration of mesitylene but is not affected by changes in the current density. However, it should be pointed out that some authors have suggested that dimer formation may take place as a coupling between two cation radicals, as in the formation of 6,6'-bibenzo[a]pyrene from benzo[a]-pyrene [131].

The anodic oxidation of anthracene in MeCN-$NaClO_4$ leads to the formation of a dimeric product, bianthrone (X), in addition to anthraquinone [132]. Bianthrone is the major product when the electrolysis is carried out at low water concentration (<50 mM) and at low anode potential [1.0 V versus Ag/Ag^+ (0.1 M)]; addition of an excess of an aliphatic alcohol gives identical results. At higher water concentrations and/or at higher anode potential (1.4 V) the amount of anthraquinone increases. Coulometric studies under conditions favoring the formation of bianthrone have given conflicting results, in that both two- [132] and three-electron [133] oxidations have been observed. The two-electron oxidation was explained by formation of 9-hydroxyanthracene (XI) as the primary electrolysis product, which then would undergo air oxidation as its tautomer, 9-anthrone (XII), during workup to form bianthrone. The three-electron oxidation process appears more likely, however, since it was found that 9-anthrone is not converted into bianthrone during workup and that bianthrone actually is the product directly isolated from the electrolysis mixture [133].

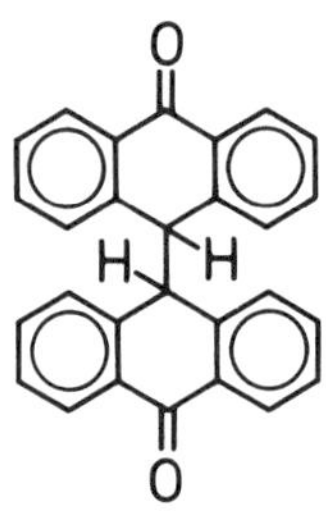

X

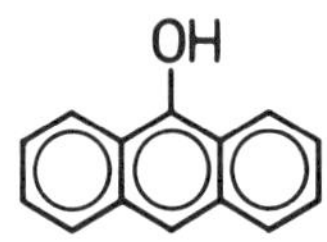

XI

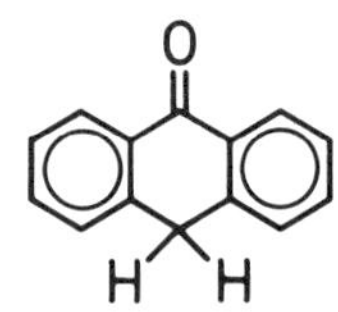

XII

The intramolecular coupling observed for tetraphenylethylene in MeCN-Et_4NClO_4 [134] and in CH_2Cl_2-CF_3COOH [30,111] has been shown to proceed via a dication formed by disproportionation of the cation radical, Eq. (36):

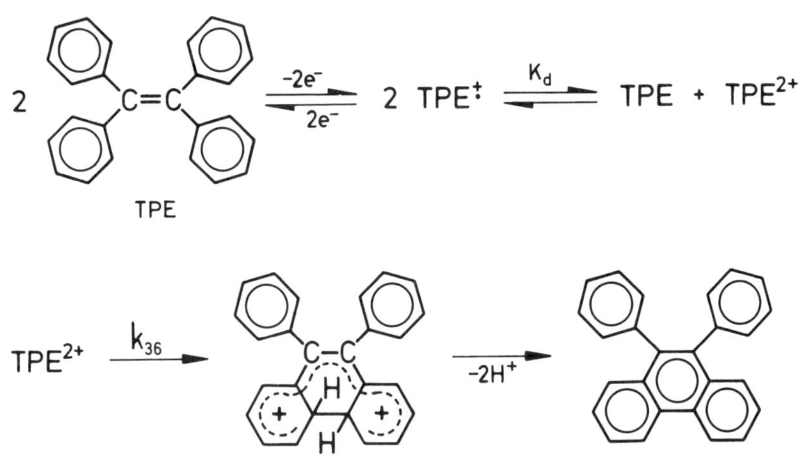

(36)

The rate constant k_{36} for the coupling reaction at 298 K was determined by derivative LSV and found to be equal to $3.2 \times 10^3\ s^{-1}$. The Arrhenius activation energy was equal to 11.4 kcal/mol, and this relatively high value was believed to reflect charge-charge repulsion in the transition state. The activation entropy at 298 K was -6 cal/K per mol, indicating solvent reorganization around the more localized charges in the transition state. Electrocatalytic oxidation of the less substituted 1,1-diphenylethylene (XIII) leads to the intermolecular coupling product 1,1,4,4-tetraphenyl-1,3-butadiene (XIV). This product has only intermediate existence under the reaction conditions and is further oxidized to the dication, which upon reaction with water undergoes a 1,2-phenyl shift, resulting finally in 1,2,4,4-tetraphenyl-3-buten-1-one (XV), Eq. (37) [135], reminiscent of the 1,2-shifts observed during anodic oxidation of 1-phenyl- and 1,4-diphenylnapthalene in acidic DMF [136].

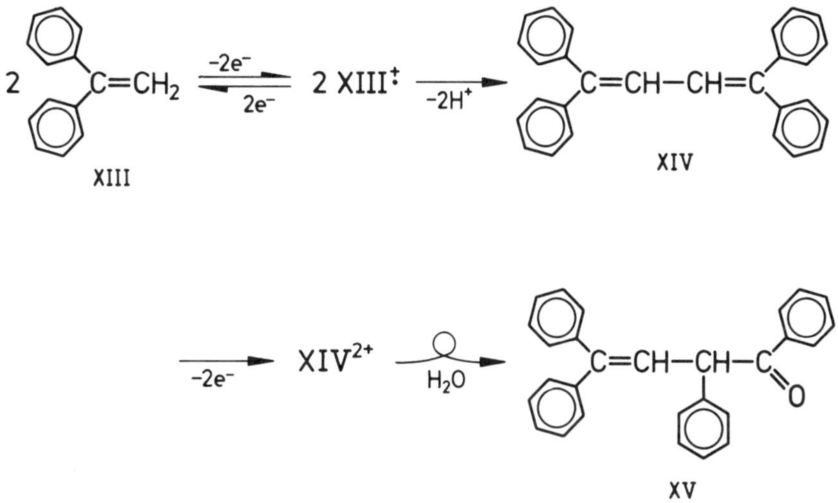

(37)

Coupling accompanied by addition is a useful reaction [87,137,138] taking place in the electrolysis of suitable substituted alkenes (aryl alkenes and butadiene) in an electrolyte consisting of MeOH-NaOMe or MeOH-NaI. The reaction is favored by the use of a carbon anode [137]. This reaction involves the rather common situation that the supporting electrolyte is discharged at lower potentials than the half-wave potential of the organic substrate, so that one would immediately suspect an indirect oxidation mechanism. The electrochemical evidence that is available, combined with considerations of the considerable dependence of product distribution on electrolyte and anode potential, is used to suggest [87] the mechanism, exemplified in Eq. (38) for oxidation of styrene (XVI) (See Chap. 23 for further details):

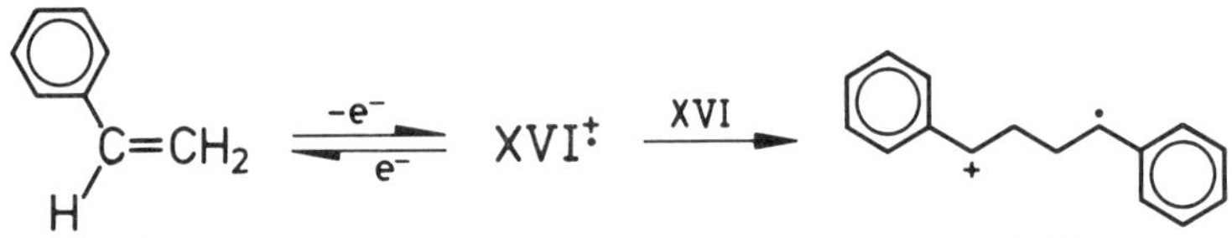

XVI XVII

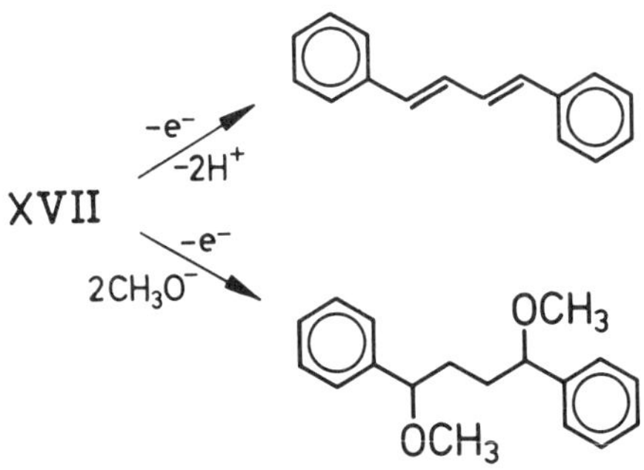

(38)

Anodically stimulated rearrangements of strained polycyclic aliphatic hydrocarbons have been reported in a few cases. For instance, tetra-tert-butyltetrahedrane (XVIII) is oxidized to the cation radical of tetra-tert-butylcyclobutadiene, Eq. (39) [139], and diaryl-bis-homocubanes (XIX) undergo 2 + 2 cycloreversion in an anodically initiated cation radical chain reaction (40) [140].

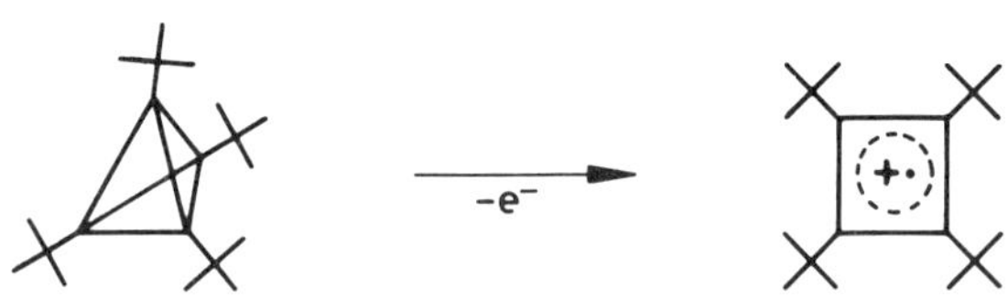

XVIII

(39)

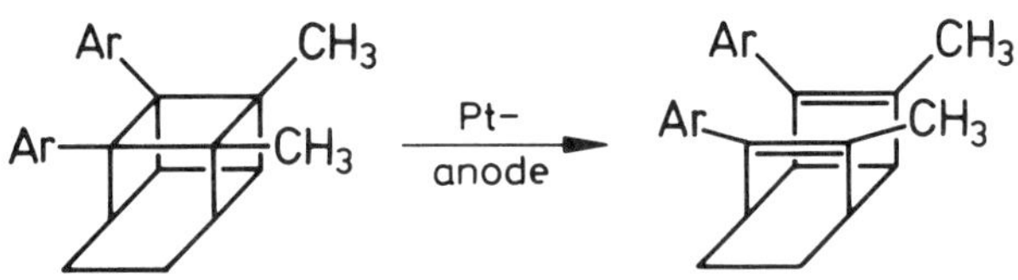

XIX

(40)

The driving force in these reactions is relief of steric strain, which also plays a central role in the anodic oxidation of cyclopropane derivatives. Of particular interest is the anodic oxidation of 1,1,2,2-tetraphenylcyclopropane, which via the 1,1,3,3-tetraphenylpropenyl cation (XX) is converted to either 1,3,3-triphenylidene (XXI) or tetraphenylallene (XXII), depending on the basicity of the solvent system, Eq. (41) [103]:

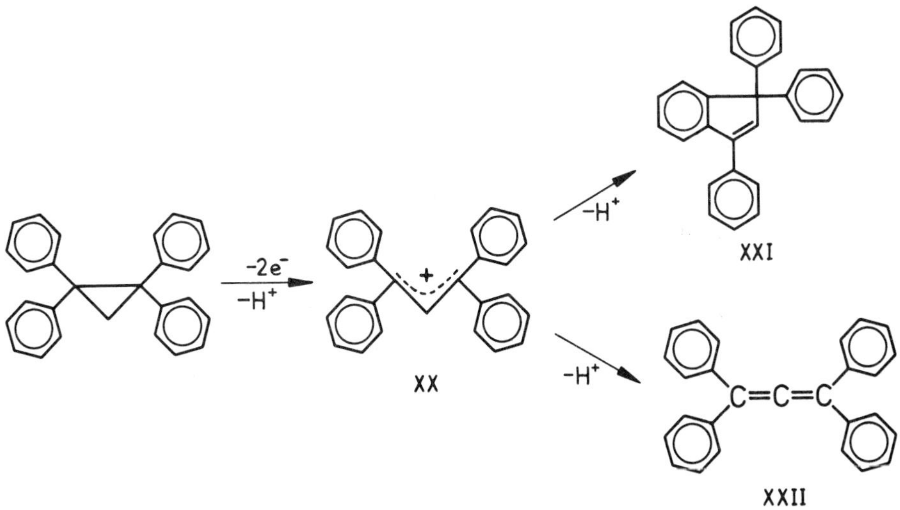

(41)

The anodically promoted cycloaddition of the indene (XXIII) [141] is mechanistically related to the 2 + 2 cycloreversion of XIX; the product (XXV) is not formally that of oxidation, and its formation has been rationalized as a 2 + 2 cycloaddition involving a cation radical (XXIV) in a chain reaction (42). Similarly, the anodic oxidation of 1,3-cyclohexadiene results in formation of the Diels-Alder product, although in very low yield. The major product is a polymer linked through the 1,4-positions [142].

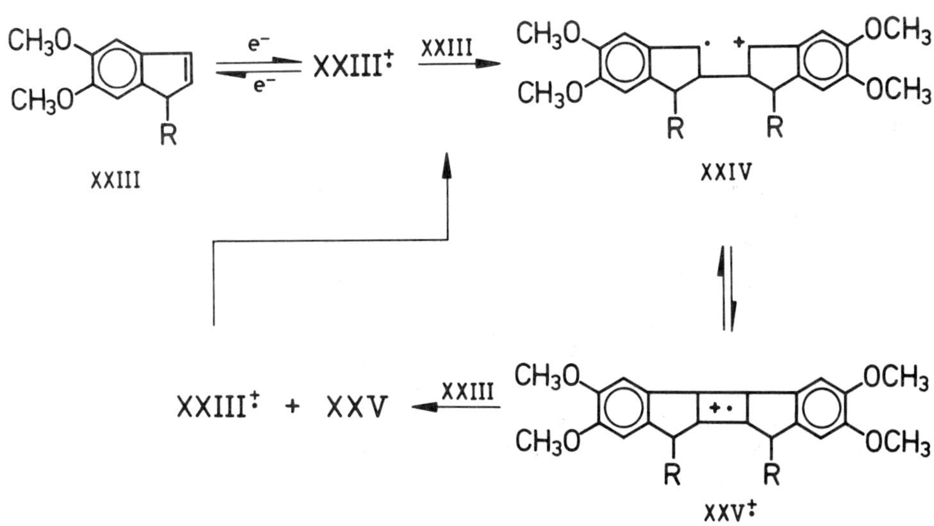

(42)

D. Indirect Oxidations

The situation in which the supporting electrolyte and/or the solvent is oxidized at a lower potential than the organic substrate is frequently encountered. Anodic methoxylation and cyanation are two typical cases for which homolytic processes involving methoxy and cyano radicals, respectively, have been invoked [143,144]. However, it can be shown that in cyanation [145–147] and at least some cases of methoxylation [148] one must work at an anode potential around or higher than the half-wave potential of the organic substrate to get any substitution product. At lower potentials, current passes through the electrolyte but no substitution takes place. In some cases this method does not give unequivocal results, for example in the side-chain acetoxylation of alkyl aromatic compounds in AcOH-Me_4NNO_3 (Ref. 149 and previous papers in this series and Ref. 150). It is also possible that direct and indirect oxidation occur simultaneously (see Chap. 25).

E. Conclusions

In most cases, hydrocarbon oxidations occur via an initial one-electron transfer to the anode to form a cation radical. The further fate of the cation radical depends on its reactivity toward the other reagents present. By suitable blocking of reactive sites and/or extensive delocalization of the positive charge the cation radical may be stable for long periods or at least during the time scale of slow sweep voltammetry. In the majority of cases, the cation radical is very reactive as an electrophile, a proton donor, or an electron acceptor and interacts with the nucleophiles or bases present; cases are also known in which it can react as a radical, that is, in dimerization processes. The factors that govern the reactivity of cation radicals are becoming increasingly better understood as a result of kinetic work using conventional and voltammetric methods and also by theoretical studies.

REFERENCES

1. E. J. Cairns, in Advances in Electrochemistry and Electrochemical Engineering (C. W. Tobias and P. Delahay, eds.), Vol. 8, Wiley-Interscience, New York, 1971, p. 337; Fuel Cells: Trends in Research and Applications (A. J. Appleby, ed.), Hemisphere Publ., New York, 1987.
2. F. Fichter, Organische Elektrochemie, Steinkopff, Dresden, 1942.
3. M. J. Allen, Organic Electrode Processes, Reinhold, New York, 1958.
4. N. L. Weinberg and H. R. Weinberg, Chem. Rev., 68, 449 (1968).
5. H. Lund, Acta Chem. Scand., 11, 1323 (1957).
6. O. Hammerich and V. D. Parker, J. Amer. Chem. Soc., 96, 4289 (1974).
7. U. Svanholm and V. D. Parker, Tetrahedron Lett., 471 (1972).
8. O. Hammerich and V. D. Parker, Electrochim. Acta, 18, 537 (1973).
9. R. Lines, B. S. Jensen, and V. D. Parker, Acta Chem. Scand., B32, 510 (1978).

10. R. Lines, and V. D. Parker, Acta Chem. Scand., B31, 369 (1977).
11. A. J. Bard and L. A. Tinker, J. Amer. Chem. Soc., 101, 2316 (1979).
12. M. Dietrich, J. Mortensen, and J. Heinze, Angew. Chem., Int. Ed. Engl., 24, 508 (1985).
13. V. D. Parker, Pure Appl. Chem., 51, 1021 (1979).
14. M. Svaan and V. D. Parker, Acta Chem. Scand., B35, 559 (1981); B38, 751, 759 (1984).
15. E. Ahlberg and V. D. Parker, Acta Chem. Scand., B33, 696 (1979).
16. J. F. Coetzee, Pure Appl. Chem., 58, 1091 (1986).
17. J. O. Howell and R. M. Wightman, J. Phys. Chem., 88, 3915 (1984).
18. J. O. Howell, J. M. Goncalves, C. Amatore, L. Klasinc, R. M. Wightman, and J. K. Kochi, J. Amer. Chem. Soc., 106, 3968 (1984).
19. C. J. Schlesener, C. Amatore, and J. K. Kochi, J. Phys. Chem., 90, 3747 (1986).
20. A. Bewick, G. J. Edwards, and J. M. Mellor, Tetrahedron Lett., 4685 (1975).
21. A. Bewick, G. J. Edwards, J. M. Mellor, and B. S. Pons, J. Chem. Soc., Perkin Trans. 2, 1952 (1977).
22. A. Bewick, G. J. Edwards, and J. M. Mellor, Liebigs Ann. Chem., 41 (1978).
23. A. Bewick, J. M. Mellor, and B. S. Pons, Electrochim. Acta, 23, 77 (1978).
24. A. Bewick, J. M. Mellor, and B. S. Pons, Electrochim. Acta, 25, 931 (1980).
25. A. J. Bard, A. Ledwith, and H. J. Shine, Advan. Phys. Org. Chem., 13, 155 (1976).
26. O. Hammerich and V. D. Parker, Advan. Phys. Org. Chem., 20, 55 (1984).
27. J. L. Courtneidge and A. G. Davies, Acc. Chem. Res., 20, 90 (1987).
28. R. M. Pagni, Tetrahedron, 40, 4161 (1984).
29. J. Weiss, Proc. R. Soc. Lond., 128 (1954); N. N. Semenov, Some Problems in Chemical Kinetics and Reactivity, Vol. 1, Princeton University Press, Princeton, N.J., 1958, p. 198.
30. B. Aalstad, A. Ronlán, and V. D. Parker, Acta Chem. Scand., B36, 199 (1982).
31. V. D. Parker, K. Nyberg, and L. Eberson, J. Electroanal. Chem., 22, 150 (1969); V. D. Parker and L. Eberson, J. Chem. Soc., Chem. Commun., 340 (1969).
32. A. J. Bard and J. Phelps, J. Electroanal. Chem., 25, App. 2 (1970); 68, 313 (1976).
33. J. W. Loveland and G. R. Dimeler, Anal. Chem., 33, 1196 (1961).
34. W. C. Neikam, G. R. Dimeler, and M. M. Desmond, J. Electrochem. Soc., 111, 1190 (1964); W. C. Neikam and M. M. Desmond, J. Amer. Chem. Soc., 86, 4811 (1964).
35. E. S. Pysh and N. C. Yang, J. Amer. Chem. Soc., 85, 2124 (1963).
36. K. Nyberg, Encyclopedia of Electrochemistry of the Elements, Vol. 11 (A. J. Bard and H. Lund, eds.), Dekker, New York, 1978, p. 43.
37. V. D. Parker, J. Amer. Chem. Soc., 98, 98 (1976).
38. L. Eberson and K. Nyberg, J. Amer. Chem. Soc., 88, 1686 (1966).

39. H. Siegerman, in Technique of Electroorganic Synthesis (N. L. Weinberg, ed.), Wiley, New York, 1970, p. 667.
40. U. Svanholm and V. D. Parker, J. Chem. Soc., Perkin Trans. 2, 1567 (1976).
41. Z. Blum, L. Cedheim, and L. Eberson, Acta Chem. Scand., B31, 662 (1977).
42. G. J. Hoijtink, Rec. Trav. Chim. Pays-Bas, 77, 555 (1958).
43. R. Zahradnik and V. Párkányi, Talanta, 12, 1289 (1965).
44. M. J. S. Dewar, J. A. Hashmall, and N. Trinajstic, J. Amer. Chem. Soc., 92, 5555 (1970).
45. L. L. Miller, G. D. Nordblom, and E. A. Mayeda, J. Org. Chem., 36, 916 (1972).
46. M. E. Peover, in Electroanalytical Chemistry (A. J. Bard, ed.), Vol. 2, Dekker, New York, 1967, p. 1.
47. R. E. Visco and E. A. Chandross, J. Amer. Chem. Soc., 86, 5350 (1964).
48. P. A. Malachesky, L. S. Marcoux, and R. N. Adams, J. Phys. Chem., 70, 2064 (1966).
49. M. E. Peover and B. S. White, J. Electroanal. Chem., 13, 93 (1967).
50. R. E. Sioda and W. S. Koski, J. Amer. Chem. Soc., 87, 5573 (1965).
51. R. E. Sioda, J. Phys. Chem., 72, 2322 (1968).
52. V. D. Parker, Acc. Chem. Res., 17, 243 (1984).
53. L. S. Marcoux, J. Amer. Chem. Soc., 93, 537 (1971).
54. V. D. Parker and L. Eberson, J. Amer. Chem. Soc., 92, 7488 (1970); V. D. Parker, J. Electroanal. Chem., 36, App. 8 (1972).
55. E. E. Bancroft, J. E. Pemberton, and H. N. Blount, J. Phys. Chem., 84, 2557 (1980).
56. L. S. Marcoux, J. M. Fritsch, and R. N. Adams, J. Amer. Chem. Soc., 89, 5766 (1967).
57. J. Phelps, K. S. V. Santhanam, and A. J. Bard, J. Amer. Chem. Soc., 89, 1752 (1967).
58. H. P. Fritz and H. Gebauer, Z. Naturforsch., 33b, 702 (1978).
59. O. Hammerich and V. D. Parker, J. Electroanal. Chem., 38, App. 9 (1972).
60. U. Svanholm and V. D. Parker, J. Amer. Chem. Soc., 98, 2942 (1976).
61. A. Ronlán and V. D. Parker, J. Chem. Soc., Chem. Commun., 33 (1974).
62. L. L. Miller and E. A. Mayeda, J. Amer. Chem. Soc., 92, 5818 (1970).
63. R. P. Van Duyne and C. N. Reilley, Anal. Chem., 44, 142, 153, 158, (1972).
64. K. Bechgaard and V. D. Parker, J. Amer. Chem. Soc., 94, 4747 (1972).
65. L. Byrd, L. L. Miller, and D. Pletcher, Tetrahedron Lett., 2419 (1972).
66. H. P. Fritz, H. Gebauer, P. Friedrich, P. Ecker, R. Artes, and U. Schubert, Z. Naturforsch., 33b, 498 (1978).
67. L. Eberson and F. Radner, Acta Chem. Scand., B40, 71 (1986).
68. C. Kröhnke, V. Enkelmann, and G. Wegner, Angew. Chem., 92, 941 (1980); V. Enkelmann, B. S. Morra, C. Kröhnke, G. Wegner, and J. Heinze, Chem. Phys., 66, 303 (1982).

69. A. E. Coleman, H. H. Richtol, and D. A. Aikens, J. Electroanal. Chem., 18, 165 (1968).
70. L. Eberson and B. Olofsson, Acta Chem. Scand., 23, 2355 (1969).
71. V. D. Parker, J. Electroanal. Chem., 21, App. 1 (1969).
72. A. Bewick, G. J. Edwards, and J. M. Mellor, Tetrahedron Lett., 4685 (1975); Electrochim. Acta, 21, 1101 (1976).
73. A. Barek, E. Ahlberg, and V. D. Parker, Acta Chem. Scand., B34, 85 (1980).
74. C. J. Schlesener, C. Amatore, and J. K. Kochi, J. Amer. Chem. Soc., 106, 3567, 7472 (1984).
75. V. D. Parker and M. Tilset, J. Amer. Chem. Soc., 108, 6371 (1986); V. D. Parker, Acta Chem. Scand., B39, 227 (1985).
76. L. Eberson, L. Jönsson, and L.-G. Wistrand, Acta Chem. Scand., B32, 520 (1978).
77. D. H. Geske, J. Amer. Chem. Soc., 81, 4145 (1959).
78. P. G. Gassman and R. Yamaguchi, Tetrahedron, 38, 1113 (1982).
79. M. Fleischmann and D. Pletcher, Tetrahedron Lett., 6255 (1968); J. Bertram, M. Fleischmann, and D. Pletcher, Tetrahedron Lett., 349 (1971).
80. J. Cassidy, S. B. Khoo, S. Pons, and M. Fleischmann, J. Phys. Chem., 89, 3933 (1985).
81. G. J. Edwards, S. R. Jones, and J. M. Mellor, J. Chem. Soc., Perkin Trans. 2, 505 (1977).
82. T. Siegel, L. L. Miller, and J. Y. Becker, J. Chem. Soc., Chem. Commun., 341 (1974).
83. D. B. Clark, M. Fleischmann, and D. Pletcher, J. Chem. Soc., Perkin Trans. 2, 1578 (1973).
84. V. R. Koch and L. L. Miller, Tetrahedron Lett., 693 (1973).
85. A. Bewick, G. J. Edwards, S. R. Jones, and J. M. Mellor, J. Chem. Soc., Perkin Trans. 1, 1831 (1977).
86. T. Shono, A. Ikeda, J. Hayashi, and S. Hakozaki, J. Amer. Chem. Soc., 97, 4261 (1975).
87. R. Engels, H. J. Schäfer, and E. Steckhan, Liebigs Ann. Chem., 204 (1977); H. Baltes, E. Steckhan, and H. J. Schäfer, Chem. Ber., 111, 1294 (1978).
88. M. A. Fox and R. Akaba, J. Amer. Chem. Soc., 105, 3460 (1983).
89. S. F. Nelsen, D. L. Kapp, R. Akaba, and D. H. Evans, J. Amer. Chem. Soc., 108, 6863 (1986).
90. T. Shono and Y. Matsumura, Bull. Chem. Soc. Jpn., 48, 2861 (1975).
91. T. Shono and Y. Matsumura, J. Org. Chem., 35, 4157 (1970).
92. H. P. Fritz and T. Würminghausen, J. Electroanal. Chem., 54, 181 (1974).
93. H. P. Fritz and T. Würminghausen, J. Chem. Soc., Perkin Trans. 2, 610 (1976).
94. F. Bobilliart, A. Thiebault, and M. Herlem, C. R. Acad. Sci., Ser. C, 287, 1485 (1974).
95. S. Pitti, M. Herlem, and J. Jordan, Tetrahedron Lett., 3221 (1976).
96. S. Pitti, M. Cerles, A. Thiebault, and M. Herlem, J. Electroanal. Chem., 126, 163 (1981).
97. J. P. Coleman and D. Pletcher, J. Electroanal. Chem., 87, 111 (1978).
98. J. Bertram, J. P. Coleman, M. Fleischmann, and D. Pletcher, J. Chem. Soc., Perkin Trans. 2, 374 (1973).

99. P.-L. Fabre, J. Devynck, and B. Tremillon, Tetrahedron, 38, 2697 (1982).
100. B. Carré, P.-L. Fabre, and J. Devynck, Bull. Soc. Chim. Fr., 255 (1987).
101. D. B. Clark, M. Fleischmann, and D. Pletcher, J. Electroanal. Chem., 42, 133 (1973).
102. L. L. Miller, V. R. Koch, T. Koenig, and M. Tuttle, J. Amer. Chem. Soc., 95, 5075 (1973).
103. D. D. M. Wayner and D. R. Arnold, Can. J. Chem., 64, 100 (1986).
104. L. L. Miller and M. Katz, J. Electroanal. Chem., 72, 329 (1976).
105. S. Hammerum and O. Hammerich, Tetrahedron Lett., 5027 (1979).
106. R. Tomat and A. Rigo, J. Appl. Electrochem., 16, 8 (1986).
107. C. E. Alciaturi and C. M. Marschoff, Electrochim. Acta, 25, 353 (1980); C. E. Alciaturi, J. Márquez, O. P. de Márquez, and C. M. Marschoff, Electrochim. Acta, 25, 357 (1980).
108. E. L. Clennan, W. Simmons, and C. W. Almgren, J. Amer. Chem. Soc., 103, 2098 (1981).
109. S. F. Nelsen, D. L. Kapp, F. Gerson, and J. Lopez, J. Amer. Chem. Soc., 108, 1027 (1986); S. F. Nelsen, Acc. Chem. Res., 20, 269 (1987).
110. M. Tsuchiya, R. Akaba, S. Aihara, H. Sakuragi, and K. Tokumaru, Chem. Lett., 1727 (1986).
111. U. Svanholm, A. Rónlan, and V. D. Parker, J. Amer. Chem. Soc., 96, 5108 (1974).
112. G. Manning, V. D. Parker, and R. N. Adams, J. Amer. Chem. Soc., 91, 4584 (1969).
113. H. N. Blount, J. Electroanal. Chem., 42, 271 (1973).
114. J. F. Evans and H. N. Blount, J. Amer. Chem. Soc., 100, 4191 (1978).
115. J. F. Evans and H. N. Blount, J. Electroanal. Chem., 102, 289 (1979); J. Phys. Chem., 83, 1970 (1979).
116. U. Svanholm and V. D. Parker, Acta Chem. Scand., B27, 1454 (1973).
117. J. F. Evans and H. N. Blount, J. Org. Chem., 41, 516 (1976).
118. P. A. Bolhuis, J. Electroanal. Chem., 110, 285 (1980).
119. O. Hammerich and V. D. Parker, J. Chem. Soc., Chem. Commun., 245 (1974).
120. V. D. Parker and L. Eberson, Tetrahedron Lett., 2839 (1969).
121. L. Eberson, Electron Transfer Reactions in Organic Chemistry, Springer-Verlag, Heidelberg, 1987, p. 94.
122. G. Cauquis and D. Serve, Bull. Soc. Chim. Fr., II-145 (1979).
123. L. Eberson and B. Larsson, Acta Chem. Scand., B41, 367 (1987).
124. R. G. Pearson, J. Amer. Chem. Soc., 108, 6109 (1986).
125. F. Barba, A. Guirado, and I. Barba, J. Org. Chem., 49, 3022 (1984).
126. T. Shono and A. Ikeda, Chem. Lett., 311 (1976).
127. J. Y. Becker, Israel J. Chem., 26, 196 (1985).
128. M. Katz, Oe. Saygin, and H. Wendt, Electrochim. Acta, 19, 193 (1974).
129. K. Nyberg, Acta Chem. Scand., 24, 1609 (1970).
130. L. Eberson, K. Nyberg, and H. Sternerup, Acta Chem. Scand., 27, 1679 (1973).
131. L. Jeftic and R. N. Adams, J. Amer. Chem. Soc., 92, 1332 (1970).

132. E. J. Majeski, J. D. Stuart, and W. E. Ohnesorge, J. Amer. Chem. Soc., 90, 633 (1968); J. Chem. Soc., Chem. Commun., 353 (1970).
133. V. D. Parker, J. Chem. Soc., Chem. Commun., 1131 (1969); L. Eberson and V. D. Parker, J. Chem. Soc., Chem. COmmun., 1290 (1970); V. D. Parker, Acta Chem. Scand., 24, 2757 (1970).
134. J. D. Stuart and J. Ohnesorge, J. Amer. Chem. Soc., 93, 4531 (1971).
135. M. Genies, J.-C. Moutet, and G. Reverdy, Electrochim. Acta, 26, 931 (1981).
136. G. A. Bhat, M. Periasamy, and M. V. Bhatt, Tetrahedron Lett., 3097 (1979).
137. H. J. Schäfer and E. Steckhan, Angew. Chem., Int. Ed. Engl., 8, 518 (1969); H. Schäfer, Chem.-Ing.-Tech., 42, 164 (1970).
138. B. Belleau and Y. K. Au-Young, Can. J. Chem., 47, 2117 (1969).
139. M. A. Fox, K. A. Cambell, S. Hünig, H. Berneth, G. Maier, K.-A. Schneider, and K.-D. Malsch, J. Org. Chem., 47, 3408 (1982).
140. Y. Takahashi, K. Sato, T. Miyashi, and T. Mukai, Chem. Lett., 1553 (1984).
141. L. Cedheim and L. Eberson, Acta Chem. Scand., B30, 527 (1976).
142. S. E. Nigenda, D. M. Schleich, S. C. Narang, and T. Keumi, J. Electrochem. Soc., 134, 2465 (1987).
143. K. Sasaki, H. Urata, K. Uneyama, and S. Nagaura, Electrochim. Acta, 12, 137 (1967).
144. K. Koyama, T. Susuki, and S. Tsutsumi, Tetrahedron, 23, 2675 (1966); T. Inoue, K. Koyama, T. Matsuoka, and S. Tsutsumi, Bull. Chem. Soc. Jpn., 40, 162 (1967).
145. V. D. Parker and B. E. Burgert, Tetrahedron Lett., 4065 (1965).
146. L. Ebeson and S. Nilsson, Discuss. Faraday Soc., 45, 242 (1968).
147. S. Andreades and E. W. Zahnow, J. Amer. Chem. Soc., 91, 4181 (1969).
148. N. L. Weinberg and T. B. Reddy, J. Amer. Chem. Soc., 90, 91 (1968); N. L. Weinberg, J. Org. Chem., 33, 4326 (1968).
149. S. D. Ross, M. Finkelstein, and R. C. Petersen, J. Org. Chem., 35, 781 (1970).
150. K. Nyberg, Acta Chem. Scand., 24, 473 (1970); 25, 3246 (1971).

14
Carboxylic Acids

SIGERU TORII AND HIDEO TANAKA / Faculty of Engineering, Okayama University, Okayama, Japan

I. Introduction 535

II. Experimental Conditions 538
- A. Constant current density electrolysis (electrode potential and current density) 538
- B. Electrode material 539
- C. Solvent, electrolyte, and additives 539
- D. Other variables (temperature, pressure, and magnetic field) 540
- E. Conclusion 541

III. Radical Reaction (One-Electron Oxidation) 542
- A. Kolbe coupling reaction 542
- B. Kolbe cross-coupling reaction 544
- C. Decarboxylative radical addition 549
- D. Miscellaneous 551

IV. Carbenium Ion Reaction (Two-Electron Oxidation) 555
- A. Acyloxylation, alkoxylation, and acetamidation 555
- B. Olefin formation 564
- C. Rearrangement 569
- D. Formation of aldehydes, ketones, acetals, and α,β-enones 571
- E. Miscellaneous 573

References 574

I. INTRODUCTION

The coupling reaction of alkyl radicals provided by electrodecarboxylation of alkanoates gives the Kolbe dimer, Eq. (1):

$$2\ \text{R-COO}^- \longrightarrow \text{R-R} + 2\ CO_2 + 2e^- \tag{1}$$

The Kolbe reaction sounds historical in electroorganic chemistry but is still alive as an important synthetic tool in today's organic synthesis.

Since the discovery of the Kolbe reaction [1], a vast number of investigations have been carried out to unveil the features of its synthetic potentiality. The first finding of the versatility by Wurts (1855) [2] is that the electrolysis of two different carboxylates gives the mixed dimer together with two possible symmetrical dimers, Eq. (2):

$$\text{R-COO}^- + \text{R'-COO}^- \longrightarrow \text{R-R'} + \text{R-R} + \text{R'-R'} + 2\ CO_2 + 2e^- \tag{2}$$

The most versatile extension of the Kolbe reaction was done by Brown and Walker (1891) [3], which opened a simple entry to the synthesis of a variety of dicarboxylic acids, Eq. (3):

$$2\ ROCO\text{-}(CH_2)_n\text{-}COO^- \longrightarrow ROCO\text{-}(CH_2)_{2n}\text{-}COOR + 2\ CO_2 + 2e^- \qquad (3)$$

Actually, this modification, the Brown-Walker reaction, is a well-established method for the synthesis of long-chain dicarboxylic acids, Eqs. (4) and (5):

$$R\text{-}COO^- + R'\text{-}OCO\text{-}(CH_2)_n\text{-}COO^- \longrightarrow R\text{-}(CH_2)_n\text{-}COOR' + 2\ CO_2 + 2e^- \qquad (4)$$

$$ROCO\text{-}(CH_2)_m\text{-}COO^- + R'OCO\text{-}(CH_2)_n\text{-}COO^- \longrightarrow ROCO\text{-}(CH_2)_{m+n}\text{-}COOR' + 2\ CO_2 + 2e^- \qquad (5)$$

Brown and Walker first proposed the generally accepted mechanism of the Kolbe reaction, which involves the initial discharge of carboxylates at the anode followed by decarboxylation and subsequent combination of the resulting radicals leading to the Kolbe dimer [3]. The radical formed may also undergo disproportionation to afford olefins and alkanes as the result of hydrogen abstraction, Eq. (6). Actually, olefins and alkanes are found as by-products.

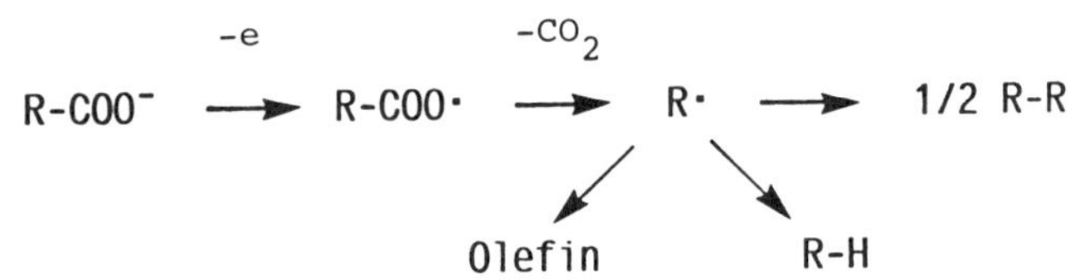

(6)

In the 1960s, a more promising expansion of the electrodecarboxylation reaction in terms of synthetic utility was recorded in which a carbenium intermediate R^+ formed at the anode plays an important role [Eq. (7)] [4]. Depending on the structural characteristics of the carboxylates and/or electrochemical variables, the cation intermediate R^+ may undergo the so-called non-Kolbe reactions, for example substitution, deprotonation, C—C bond cleavage, and rearrangement to provide alcohols, ethers, esters, amides, olefins, and others.

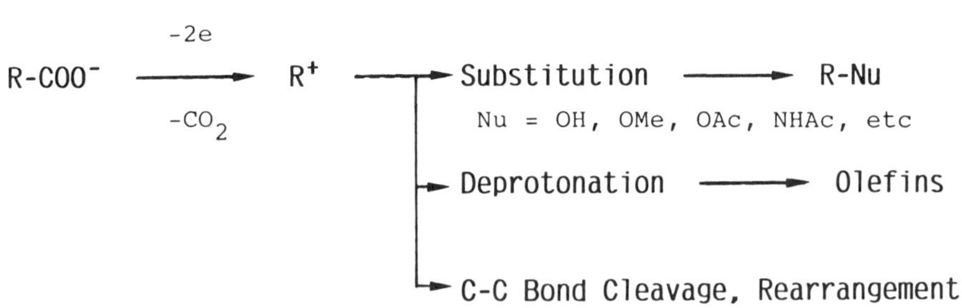

(7)

The electrochemical hydroxylation and esterification in aqueous media, the so-called Hofer-Moest reaction, initially were regarded as a side reaction, but sometimes become the major reaction in the aqueous Kolbe reaction [5]. Such puzzling phenomena encountered in the early Kolbe reaction can be rationalized on the basis of the assumption that was just described.

In the past two decades, more sophisticated electrolysis conditions suited for the Kolbe dimerization, for hydrogen abstraction, for substitution, for rearrangement, and for other radical- or cation-induced reactions have accumulated for a number of carboxylic acids and have been used for many synthetic purposes. Most recently, in the field of the two-electron transfer process, significant progress has been made in exploiting preparative use of the cation intermediates for the functionalization, such as alkoxylation, acetoxylation, acetamidation, formation of olefins and carbonyl groups, and in rearrangements. The concept of modern electrodecarboxylation of carboxylic acids tends to be far from that of the old-fashioned Kolbe reaction. Nowadays, electrodecarboxylation is available as a potential tool for building up the functional groups in complex molecules. Indeed, recent progress in this field enables us to design various synthetic equivalents to, for example, carbonyl groups, α,β-enone, and double bonds by virtue of electrodecarboxylation. Some selected examples are listed in Table 1. Today's synthetic potentialities of the electrogenerated carbenium intermediates are discussed in Sec. IV.

Table 1. Synthetic Equivalents Based on Electrodecarboxylation[a]

Target molecule	Synthetic equivalent	Reference
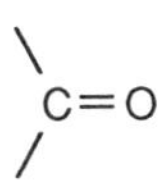	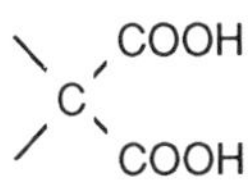	6
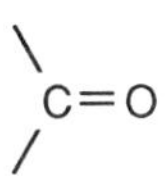	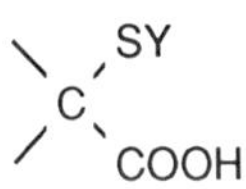	7–10
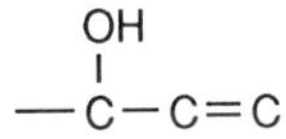	C=C–C–COOH	11, 12
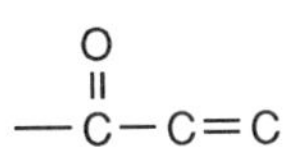	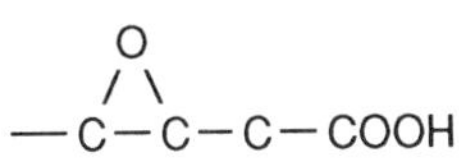	11
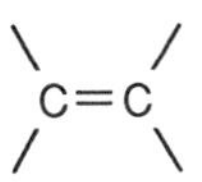	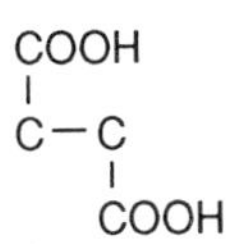	13–17

[a]Y = Ph or benzothiazolyl.

II. EXPERIMENTAL CONDITIONS

Electrodecarboxylation has been examined in a variety of electrolysis systems, and a proper combination of solvent, supporting electrolyte, electrode material, pH, concentration of salt and substrate, and other variables has been chosen according to the synthetic requirements in each case. As briefly mentioned before, the electrodecarboxylation may provide two competing pathways that give rise to either radical $R^{\cdot}$ or R^+ intermediates, leading to various products. The optimum conditions for promoting a particular reaction have been accumulated. Thus far, one can in many cases design a suitable electrolysis system, making the reaction product selective via either the radical or cation pathway.

A. Constant Current Density Electrolysis (Electrode Potential and Current Density)

The Kolbe reaction is usually performed with constant current density, and the control of the anode potential by using a potentiostat apparatus is seldom essential. A high current density, higher than 0.25 and often up to 1.0 A/cm^2, is in general recommended for the Kolbe dimerization since the rate of the radical combination should be proportional to the concentration of the radical species per unit area of the electrode. It has been reported that if the anode potential is plotted against the logarithm of the current density (Tafel plots) (Fig. 1), there is a special value of current density at which the anode potential jumps from 2.1 to 2.4 V [18]. At a current density

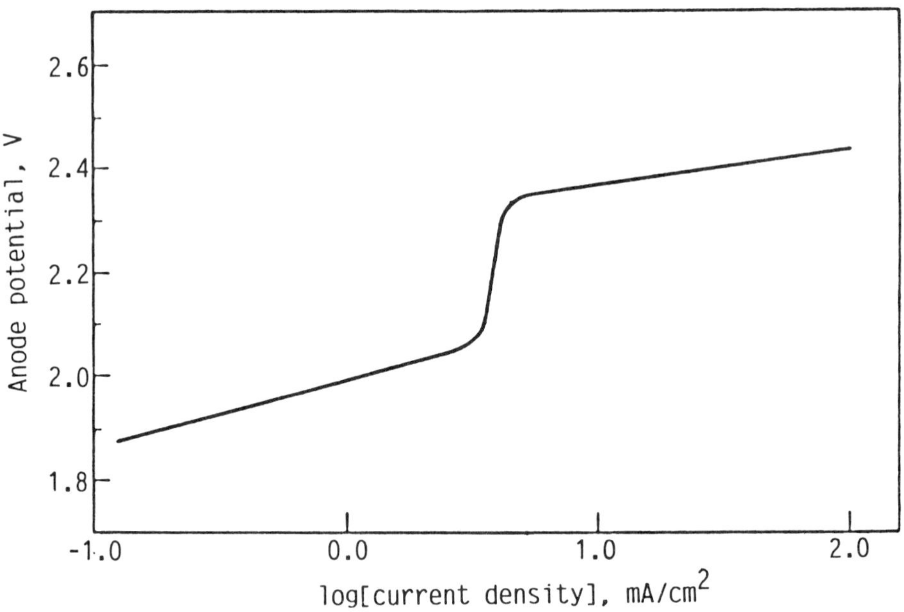

Figure 1. Anode potential (versus NHE) as a function of log [current density] in an H_2O–AcONa (0.5 M) system.

higher than that causing such a potential jump, discharge of the carboxylates proceeds almost exclusively and may completely suppress undesirable reactions that normally take place at a lower potential, such as oxygen evolution from water.

The critical potentials, at which the decarboxylation starts, have been compiled for a variety of carboxylic acids; it is normally observed in a region of 2.0–2.8 V (versus NHE) [19]. In practice, however, it is not necessary to control the anode potential since the anode potential may shift to more positive values than the critical potential even if the electrolysis is operated at a constant current density of around 1 mA/cm^2.

The radical $R^{\cdot}$ thus generated may undergo a further one-electron oxidation to give a cation R^+. Owing to the high discharge potential of carboxylates, easily oxidizable radicals are believed to couple prior to the second electron transfer. Easily oxidizable functional groups, such as amine, sulfide, and formyl, attached to the carbon atom bearing a carboxylic group, cannot survive under the electrolysis conditions. The decarboxylation of carboxylic acids may proceed at lower discharge potentials than usual in systems combining a heteroatom and a carboxylic acid. In such systems, concerted decarboxylations may proceed. The details are discussed in the following sections.

B. Electrode Material

The influence of anode materials on the electrodecarboxylation in aqueous solutions has been well documented; platinum, iridium, and vitreous carbon electrodes favor the coupling product; graphite and porous carbon anodes facilitate the two-electron process, giving cation intermediates. In contrast, in nonaqueous solvents, a variety of anode materials may be employed. The choice of electrode material is less critical, although platinum is still recommended for the one-electron process and carbon for the two-electron process. Other electrode materials, such as gold, palladium, lead dioxide [20], iron-impregnated carbon [21], and shungite [22], are useful in some cases for the Kolbe coupling.

The nature of the cathode material is not critical in Kolbe reactions. In most cases, neither starting material nor products are reducible at less negative potentials than that required for hydrogen evolution.

C. Solvent, Electrolyte, and Additives

Neutral and slightly acidic media have been recommended for the Kolbe reaction. To keep the electrolyte acidic during electrolysis, the reaction is simply performed in an aqueous solution of the acid, neutralized by an alkali metal hydroxide or trialkylamine to an extent of 2–5%. Since hydroxyl ions are formed at the cathode at the same rate as carboxylate ions are consumed at the anode, the concentration of carboxylate ion remains approximately constant during the whole run. Water is now seldom used as a solvent in Kolbe oxidations. Methanol or aqueous methanol is regarded as the best choice as a solvent [23]. DMF and wet acetonitrile can also be used in particular cases [24–26]. Special attention should be paid to assessing the solvent system for the generation and reaction of carbenium intermediates under electrolysis conditions [23]. A variety of solvent systems elaborated for the electrodecarboxylation has been compiled [27]. On the basis of a

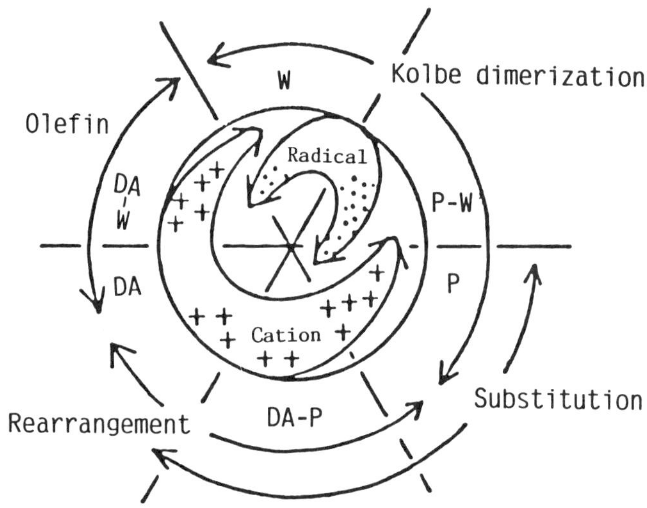

Figure 2. Correlation between reaction type, solvent, and reactive intermediate: W, water; P, protic solvent; DA, dipolar aprotic solvent.

rough statistical treatment of experimental findings, the combinations or interrelations among three kinds of solvent [water (W), protic solvent (P), and dipolar aprotic solvent (DA)], two reactive intermediates (cation and radical), and four typical reactions (Kolbe dimerization, substitution, rearrangement, and olefin formation) are schematically illustrated in Fig. 2 [27]. One can select the first choice of the solvent system that may be suited to undertaking certain reactions. For example, a dipolar aprotic solvent-water (DA-W type), such as pyridine-water, should be the first choice when aiming to obtain olefins as the exclusive electrodecarboxylation product.

Most of electrodecarboxylations have been carried out with partially neutralized carboxylic acid. Alkaline and alkaline earth metal as well as ammonium (pyridinium) carboxylates work efficiently as supporting electrolytes. Some metal ions (such as Fe^{2+} and Co^{2+}) are found to favor selectively radical reactions in electrodecarboxylation. Addition of certain salts, such as perchlorate, fluoroborate, sulfate, dihydrogen phosphate, bicarbonate, and fluoride, tends to inhibit the radical reaction and favor the formation of cation intermediates [28–31]. The remarkable effects of the salts are well explained in terms of competitive adsorption between the anions and carboxylates.

D. Other Variables (Temperature, Pressure, and Magnetic Field)

Temperature is not usually a critical process variable but may improve viscosity and mass transport. An increase in temperature generally results in an increased yield of the Kolbe dimer [32,33]; however, increasing it over 50°C causes in some cases a total change in the product obtained [34,35]. For example, the electrolysis of ω-bromoundecanoic acid in methanol at 50°C gives the Kolbe dimer (64%), but at 65°C methyl Ω-bromoundecanoate (71%) is obtained, Eq. (8):

$$\underset{64\%}{Br(CH_2)_{20}Br} \xleftarrow{50\ ^{\circ}C} Br(CH_2)_{10}COO^- \xrightarrow{60\ ^{\circ}C} \underset{71\%}{Br(CH_2)_{10}COOMe} \qquad (8)$$

Under high-pressure conditions, volatile species, such as hydrogen, carbon dioxide, and in some cases olefin alkane, are accumulated on the anode surface, which may affect the product distribution [33,36]. An aqueous Kolbe electrolysis of an alkanoic acid (in the range C_4-C_6) tends to give higher yields of the Kolbe dimers when run at elevated pressure (100 kPa) than an equivalent experiment at atmospheric pressure.

A magnetic field may effect the flow of electron and ion, resulting in perturbation of the mass transport process near the electrode [37]. For example, in the decarboxylation of phenylacetic acid, the presence of a magnetic field increases the yield of aldehyde, presumably because the supply of molecular oxygen from the bulk solution to the electrode surface is enhanced.

E. Conclusion

Experimental variables affecting the course of the electrolytic decarboxylation of carboxylic acids are summarized in Table 2. For the Kolbe dimerization, the conditions specified for a one-electron process are recommended; otherwise the reaction through carbenium ion (non-Kolbe reaction) may occur predominantly. It should be emphasized that even under the conditions most favorable for the Kolbe dimerization, the cation-derived products are usually formed to some extent or, in particular cases, as a major product, depending on the structure of the employed carboxylic acid.

Table 2. Effect of Experimental Variables on Electrodecarboxylation

Variable	One-electron process		Two-electron process
Current	High current density		Low current density
Electrode material	Pt electrode		Carbon electrode
Reaction media	Acidic	Neutral	Basic
	Water (W)		Dipolar aprotic (DA)
	Protic (P)		DA-W
	W-P		P
Additive	None		ClO_4^-, SO_4^{2-}, HCO_3^-, and others
Temperature	<50°C		Not critical

III. RADICAL REACTION (ONE-ELECTRON OXIDATION)

A. Kolbe Coupling Reaction

The Kolbe reaction can be regarded as a chemical process involving the control of a one-electron oxidation reaction. Thus, the generated radicals R$^{\cdot}$ undergo the so-called Kolbe coupling, hydrogen abstraction, and in some cases radical addition to olefins, each of which allows many useful synthetic applications. Further one-electron oxidation leading to a carbenium ion is another reaction pathway that normally competes with the reactions mentioned. Preparative use of the Kolbe coupling reported up to 1970s has been well documented in several excellent reviews [18,27,38–43]. Here some characteristic features of the early reactions as well as recent advances in this field, emphasizing the work in the 1980s, are discussed.

Some results on Kolbe couplings are listed in Table 3. The electrolysis of carboxylic acids is usually carried out in an alkaline protic solvent-(Pt electrodes) system. Most frequently, MeOH-MeONa, MeOH–H_2O-NaOH, MeOH-KOH, MeOH–Py-Et_3N, and MeOH-Py–Na-salt systems are used as electrolysis media. Aprotic solvents, such as DMSO-NaH, MeCN-NaOH, and

Table 3. Kolbe Dimerization

Carboxylic acid	Conditions	Yield (%)	Reference
$CH_3(CH_2)_nCOOH$ n = 5-15	MeOH, EtOH	60-90	42
$ROOC(CH_2)_nCOOH$ R = Me, Et; n = 2-16	MeOH, EtOH	43-80	44
$AcO(CH_2)_nCOOH$ n = 3-5	MeOH	73-83	44
$RCH(COOEt)COOH$ R = alkyl, H	MeOH	61-74	44
$FCH_2(CH_2)_nCOOH$ n = 3-9	MeOH	45-69	44
$CH_3CH_2CO(CH_2)_4COOH$	MeOH	75	45
$CH_3(CH_2)_7CH{=}CH(CH_2)_7COOH$		75	46
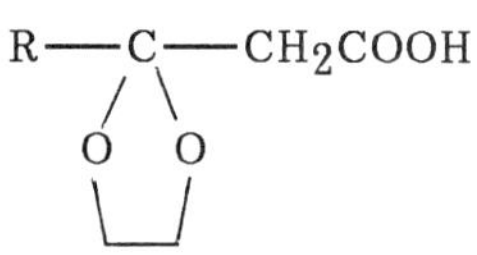 R = alkyl	MeOH–MeOK–(Pt)	63-69	47
$CH_3CO(CH_2)_2COOH$	MeOH-MeONa–(Pt)	–	48
$HOCH_2CH(CF_3)COOH$	MeOH–MeONa–(Pt)	95	49

Py–H_2O-Et_3N, are also utilized and, in particular cases, acetone–H_2O-NaOH is usable.

The proper choice of the electrolysis medium may provide significant product selectivity in each reaction. For instance, in the electrolysis of cyclopropane carboxylic acids (I), Eq. (9), a dramatic change in products is observed [50]; in a pyridine–H_2O-Et_3N-(Pt) system the Kolbe dimer (II) is mainly obtained, but in a MeOH-MeONa-(Pt) system predominantly cyclopropane (III) is formed via proton abstraction. The cyclopropane carboxylic acid (IV), however, undergoes decarboxylative coupling even in an MeOH-MeONa-(Pt) system to afford the dimer (V) as a major product along with the ester (VI), Eq. (10) [51]:

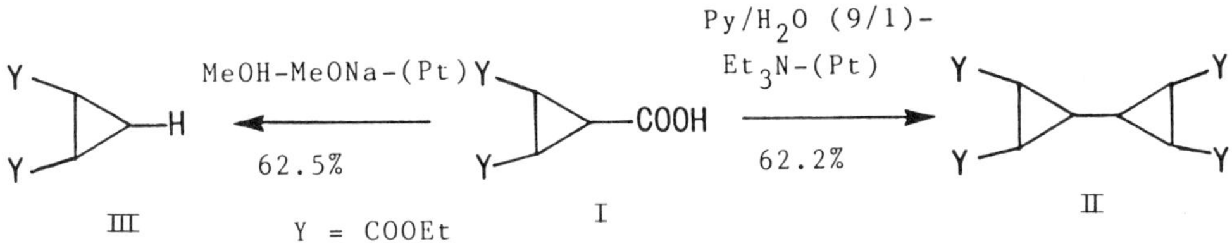

(9)

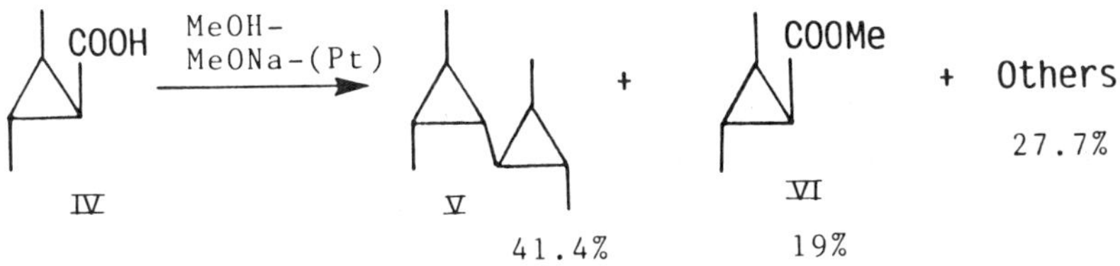

(10)

Many functional groups, such as halogen, acetal, ketone, and acetate, and in some cases alcohol, aldehyde, and amide, can survive during the Kolbe decarboxylation [44]. Double bonds isolated from the carboxylic acid group are not usually affected under the Kolbe electrolysis conditions. The β,γ-unsaturated carboxylic acids (VII) give the isomeric 1,5 dienes (IX–XI) via the allyl radicals (VIII), Eq. (11) [52]. Through this reaction, the configuration of the double bonds is almost totally (>90%) retained.

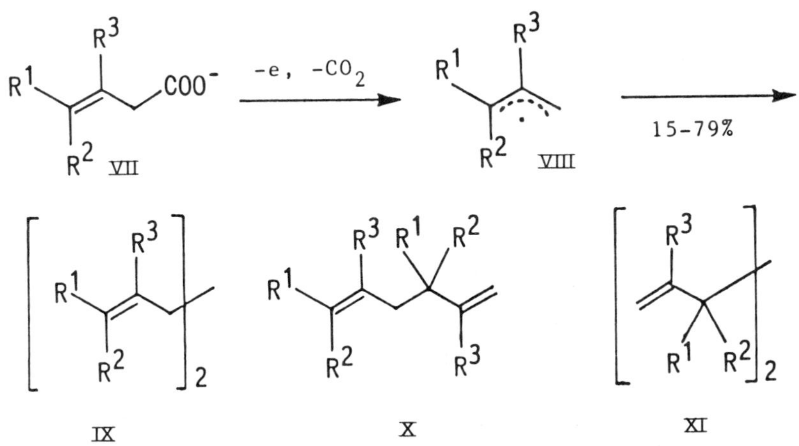

(11)

The synthetic potentiality of the Kolbe dimerization has been well documented for a variety of symmetrical target molecules. For example, the dimerization of acids (XII and XIII) is the key step of the pentacyclosqualene [53] and α-onocerin [54] syntheses. A large-scale production of sebasic acid has been realized by the Kolbe dimerization of methyl adipate half-ester [55].

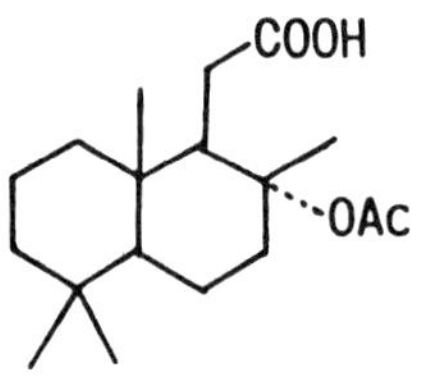

XII

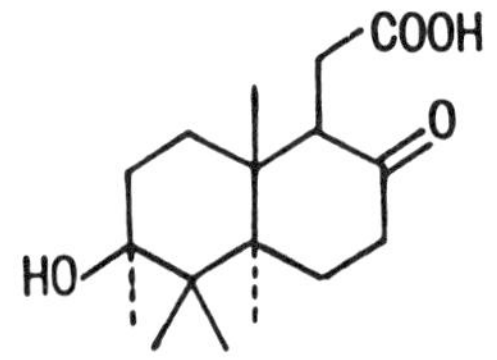

XIII

The Kolbe reaction of pyruvic acid and phenylglycoxylic acid (XIV), (R = CH_3 and Ph) affords the corresponding diketones (XV) in high yields, Eq. (12) [56,57]. Interestingly, electroluminescence can be observed during the electrolysis, which probably can be attributed to phosphorescence due to the generation of excited triplets of α-diketones in the acyl radical recombination.

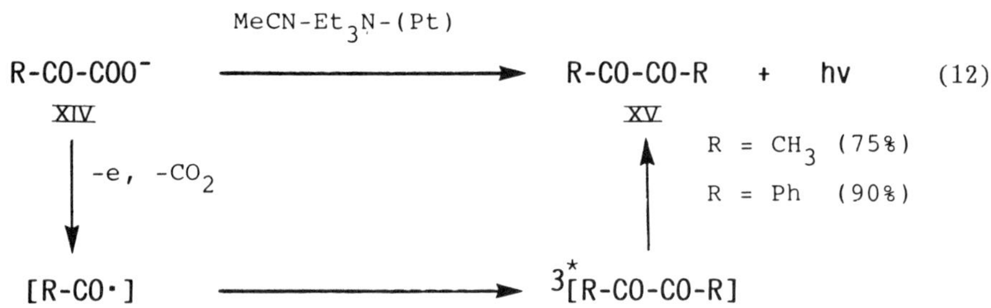

Electrodecarboxylative polymerization of dicarboxylic acids has been examined [58–61]. Under the Kolbe electrolysis conditions, dicarboxylic acids (XVI) (n = 4–8), such as adipic acid and sebasic acid, give rise to polymer along with oligomeric carboxylic acids and oligomeric hydrocarbons. The polymer produced has the character of a cross-linked and branched polyethylene having ester and methoxyl groups. The polymer formation is explained by assuming an α,ω-biradical intermediate (XVIII), Eq. (13):

$$\underset{\text{XVI}}{HOOC\text{-}(CH_2)_n\text{-}COOH} \xrightarrow[-CO_2]{-2e} \underset{\text{XVII}}{\cdot(CH_2)_n\cdot} \longrightarrow \text{Polymer} \qquad (13)$$

B. Kolbe Cross-Coupling Reaction

Electrolysis of two different carboxylic acids may give a cross-coupling product. The product is, however, always accompanied by two symmetrical dimers owing to the statistical coupling of each radical intermediates R$^\cdot$ and

R'˙, Eq. (14). Although a few attempts toward the selective cross-coupling, for example square pulse electrolysis of unsymmetrical acyl peroxide [62–64], have been made, these are still far from the goal. The disadvantageous but unavoidable formation of the two symmetrical dimers can be reduced to a single product when one of them is used in 5- to 10-fold excess [65].

$$R\text{-}COO^- + R'\text{-}COO^- \longrightarrow R\text{-}R' + R\text{-}R + R'\text{-}R' \qquad (14)$$

The Kolbe cross-coupling may permit a variety of applications in natural product synthesis. By using this method, carotenoids, fatty acids, muscone, fumulene [42], and pheromones [43] have been synthesized. Reported results of Kolbe cross-couplings are summarized in Table 4.

Under the cross-coupling conditions, the double bond of unsaturated acids generally retains its configuration. Partial Z/E isomerization of olefins is, however, observed in the electrolysis of Z-4-enoic acids [79]. The electrolysis of Z-3-methyl-4-octenoic acid (XVIII) with acetic acid gives a mixture of Z and E isomers of the cross-coupling product (XIX) (9:1–7:3) together with XX and XXI, Eq. (15).

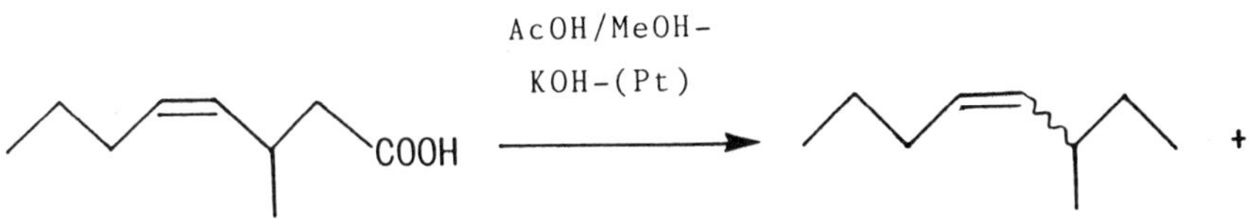

XVIII XIX (19-50%) E/Z = 9/1-7/3

XX (5-30%) XXI (2-21%) (15)

This can be well understood by assuming a Z/E isomerization via a reversible ring closure to cyclopropylcarbinyl radicals (XXII).

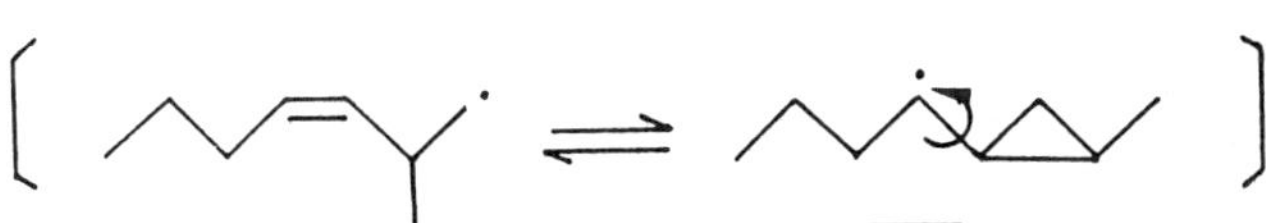

XXII

Z-n-Enoic acids (XXIII) (n > 5) fully retain the olefin configuration in the Kolbe electrolysis, but cyclic products (XXV) are obtained, to some extent, for 6- and 7-enoic acids (XXIII) (n = 4 and 5), Eq. (16).

Table 4. Kolbe Cross-Coupling

Carboxylic acids[a]	Conditions	Product (yield, %)	Reference
$HOOCCH_2\overset{CH_3}{C}HCH_2COOMe + CH_3(CH_2)_{16}COOH$	MeOH-hexane-MeOK-(Pt)	$CH_3(CH_2)_{17}\overset{CH_3}{C}HCH_2COOMe$ (25%)	66
$CbzNH\underset{COOMe}{CH}(CH_2)_2COOH + BocNH\underset{COOtBu}{CH}(CH_2)_2COOH$	MeOH–Py-MeONa-(Pt)	$CbzNH\underset{COOMe}{CH}(CH_2)_4\underset{COOtBu}{CH}NHBoc$	67
$MeOOC(CH_2)_3COOH + BocNH\underset{COOtBu}{CH}(CH_2)_2COOH$	MeOH–Py-MeONa-(Pt)	$MeOOC(CH_2)_5\underset{COOtBu}{CH}NHBoc$	67
$HO(CH_2)_{18}\overset{CH_3}{C}HCH_2COOH + MeOOC(CH_2)_3COOH$ (1:6)	MeOH-KOH-(Pt)	$HO(CH_2)_{18}\overset{CH_3}{C}H(CH_2)_4COOMe$ (31)	68
$CH_3CH_2\overset{CH_3}{C}H(CH_2)_2COOH + THPO(CH_2)_7C{\equiv}C(CH_2)_2COOH$ (1:1)	MeOH-KOH-(Pt)	$CH_3CH_2\overset{CH_3}{C}H(CH_2)_4C{\equiv}C(CH_2)_7OTHP$ (20.2)	69
$RCH_2C{\equiv}C(CH_2)_3COOH + MeOOC(CH_2)_nCOOH$ R = Me, Et, Pr	MeOH-KOH-(Pt)	$RCH_2C{\equiv}C(CH_2)_{n+3}COOMe$ (45-59)	70
$MeOOC(CH_2)_2CH{=}CH(CH_2)_2COOH + CH_3(CH_2)_7COOH$ (1:10)	MeOH-KOH-(Pt)	$MeOOC(CH_2)_2CH{=}CH(CH_2)_9CH_3$ (48)	71
$CH_3(CH_2)_7CH{=}CH(CH_2)_{10}COOH + CH_3CH_2COOH$ (1:10)	MeOH-MeONa-(Pt)	$CH_3(CH_2)_7CH{=}CH(CH_2)_{11}CH_3$ (59)	65

$CH_3(CH_2)_3CH{=}CH(CH_2)_2COOH$ + $MeOOC(CH_2)_3COOH$ (1:10)	MeOH-KOH-(Pt)	$CH_3(CH_2)_3CH{=}CH(CH_2)_5COOMe$ (55)	72
$C_2H_5CH{=}CHCH_2COOH$ + $CH_3CO(CH_2)_2COOH$ (5:1) (*trans*)	MeOH-MeOK-(Pt)	$CH_3CO(CH_2)_3CH{=}CHC_2H_5$ (33) (*trans*)	73
$CH_3(CH_2)_{17}CH(CH_3)CH_2COOH$ + $MeOOC(CH_2)_3COOH$ (1:6)	MeOH-KOH-(Pt)	$CH_3(CH_2)_{17}CH(CH_3)(CH_2)_4COOMe$ (66)	66
$CH_3(CH_2)_{17}CH(CH_3)(CH_2)_4COOH$ + $CH_3COCH(CH_3)(CH_2)_3COOH$ (1:5)	MeOH-KOH-(Pt)	$CH_3(CH_2)_{17}CH(CH_3)(CH_2)_7CH(CH_3)COCH_3$ (42.5)	66
$Br(CH_2)_{10}COOH$ + $MeOOC(CH_2)_7COOH$ (5:1)	MeOH-KOH-(Pt)	$Br(CH_2)_{17}COOMe$ (76)	68
$HO(CH_2)_{17}COOH$ + $HOOCCH_2CH(CH_3)CH_2COOMe$ (2:1)	MeOH-KOH-(Pt)	$HO(CH_2)_{18}CH(CH_3)CH_2COOMe$ (47)	68
$CH_3(CH_2)_{16}{-}COOH$ + $HOOCCH_2CH(CH_3)CH_2COOMe$	MeOH-MeOK	$CH_3(CH_2)_{17}{-}CH(CH_3)CH_2COOMe$ (25)	74
C_2H_5COOH + $C_5H_{11}COOH$	H_2O-Na_2SO4/ NaOH-(Pt)	C_4H_{10}, C_7H_{16}, $C_{10}H_{22}$ (9.2) (42)	75
C_3H_7COOH + $C_5H_{11}COOH$	H_2O-Na_2SO/ NaOH-(Pt)	C_6H_{14}, C_8H_{18}, $C_{10}H_{22}$ (total 26.8)	75

Table 4. (Continued)

Carboxylic acids[a]	Conditions	Product (yield, %)	Reference
$HOCH_2CH(CF_3)COOH + R(CH_2)_2COOH$ (1:6)	MeOH-MeONa-(Pt)	$HOCH_2CH(CF_3)(CH_2)_2R$ R = Cl(70), Me(50), OH(14)	76
$CF_3COOH + EtOCOCH_2COOH$	H_2O-KOH-(Pt)	CF_3CH_2COOEt (46)	77
$CF_3COOH + CD_3COOH$	KOAc-d3-(Pt)	CF_3CD_3 (68)	78

[a]Cbz = benzyloxycarbonyl; Boc = t-butoxycarbonyl; THP = tetrahydropyranyl.

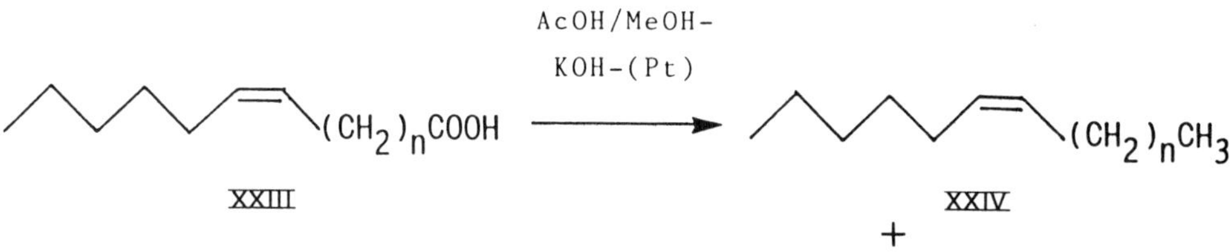

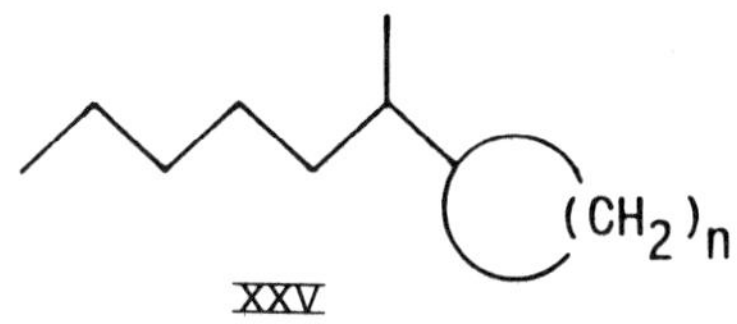

Compound XXIII	Yield, %	
n	XXIV	XXV
3	68	--
4	46	21
5	65	5
6	72	--
7	73	--

(16)

C. Decarboxylative Radical Addition

Electrolysis of carboxylates (XXVI) in the presence of olefins (XXVII) may afford the radical addition products XXIX, XXX, XXXII, and XXXIII. The plausible reaction pathways are illustrated in Eq. (17). The radical R$^{\cdot}$ generated by electrodecarboxylation of XXVI first combines with the olefin to

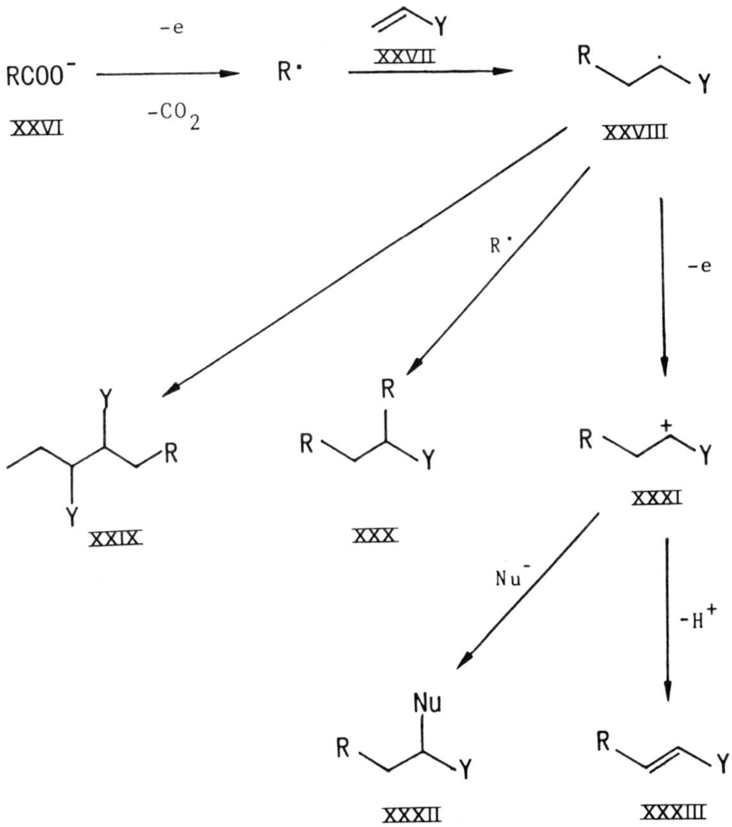

(17)

give the radical intermediates (XXVIII), which may provide the dimers (XXIX) or the radical coupling products (XXX). Further one-electron oxidation of XXVIII may provide cations (XXXI), which subsequently react with the nucleophiles Nu^- or liberate H^+ to give substituted products (XXXII) or olefins (XXXIII).

The radical addition to the olefins (XXVII) generally gives rise to a mixture of products but, in some particular cases, gives reasonable yields of the radical adducts (XXIX and XXX). Some typical examples are shown in Table 5. Both styrene (XXVII, Y = phenyl) and conjugated dienes (XXVII, Y = vinyl) afford the corresponding radical adducts in fairly good yields. Addition of the electrogenerated trifluoromethyl radical $CF_3^{\cdot}$ to olefins has been found to be a good procedure for the preparation of trifluoromethylated products. The electrolysis of partially neutralized trifluoroacetic acid (XXVI, R = CF_3) and olefins (XXVII) generally gives a mixture of mono- and bis-trifluoromethylated derivatives, types XXIX and XXX (R = CF_3), and it is quite hard in some cases to separate the products in a pure state. Recently, however, several devices have been made to overcome such difficulties. ω-Trifluoromethylalkanoic acids (XXXIV and XXXV) [83,84], β-trifluoromethylaldehyde (XXXVI) [85], α-trifluoromethylketones (XXXVII) [86], and 5-trifluoromethyluracil (XXXVIII) [87] have thus been synthesized efficiently.

The cross-coupling of the electrogenerated trifluoromethyl radical with active methylene compounds may open another access to trifluoromethylated products. Electrolysis of octyl acetoacetate (XXXIX) in an $MeCN-H_2O$-TFA-(Pt) system gives a mixture of octyl 2-trifluoromethylacetoacetate (XL) and octyl 3,3,3-trifluoropropionate (XLI), the latter of which is probably produced by deacetylation of XL with a base electrogenerated at the cathode, Eq. (18) [92]:

$$\underset{\text{XXXIX}}{CH_3CO\text{-}CH_2\text{-}COOR} + CF_3COOH \xrightarrow{-e,\ -CO_2} \underset{\text{XL}}{CH_3CO\text{-}\overset{\displaystyle CF_3}{\overset{|}{C}H}\text{-}COOR}$$

$$\xrightarrow{\text{EG Base}} \underset{\text{XLI}}{CF_3CH_2\text{-}COOR} \qquad (18)$$

R = Octyl

Intramolecular radical addition may furnish a simple and straightforward access to cyclic compounds. Mixed coupling of 6-alkenoic acids (XLII) and various carboxylic acids (XLIII) successfully gives cyclic compounds (XLIV), Eq. (19) [93,94].

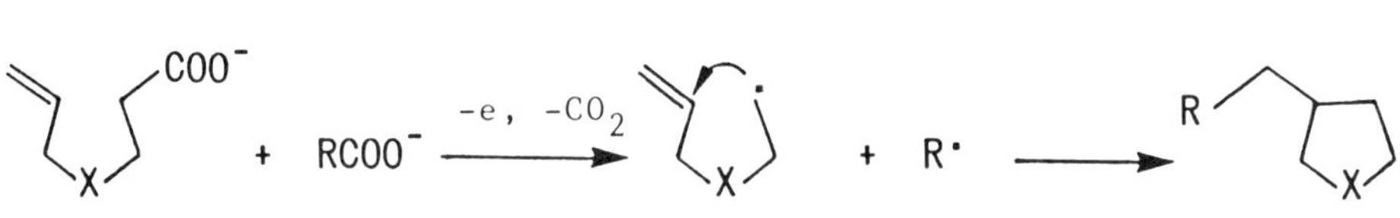

XLII XLIII XLIV

X: O, $NCOCH_3$, NCOH (19)

R: CH_3, $n\text{-}C_5H_{11}$, $(CH_2)_4COOCH_3$, etc

A stereoselective synthesis of prostagrandin precursors (XLVII has been performed by coelectrolysis of XLV and XLVI, Eq. (20) [95].

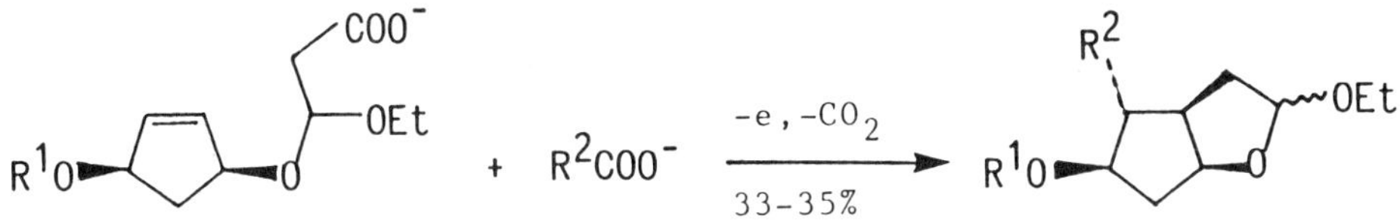

XLV   XLVI   XLVIIa : R^1=H; R^2=CH_3
XLVIIb : R^1=Ac; R^2= $(CH_2)_2COOCH_3$

(20)

D. Miscellaneous

Further applications of the radical-induced reactions in the Kolbe decarboxylation reactions involve hydrogen abstraction and coupling with in situ generated radical species from additives. For example, hydrogen abstraction by radical intermediates derived from paraconic acids (XLVIII) occurs selectively in MeOH-MeONa-Fe powder-(Pt) and MeOH-MeONa-(C) systems, Eq. 21 [20]. The effect of Fe powder is significant since butenolides (L) are obtained exclusively in an MeOH-MeONa-(Pt) system.

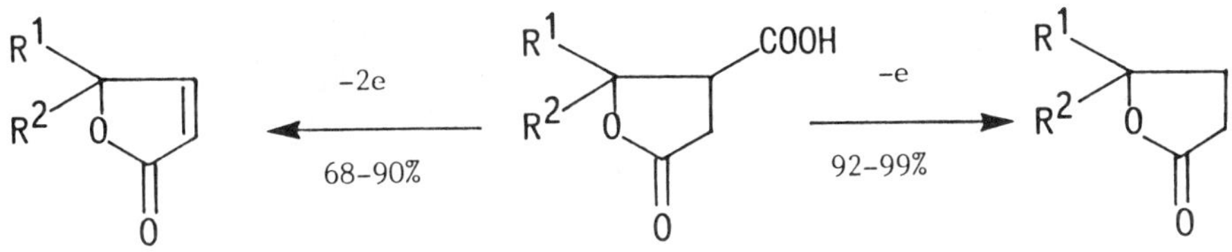

L   XLVIII   XLIX

(21)

Electrolysis of carboxylic acids in the presence of nitrate [96,97], difluoroamide [98], azide [99], and bromide ions [100] has been examined in which radical species generated from both carboxylic acids and additive anions are formally coupled to give substituted products, Eq. (22):

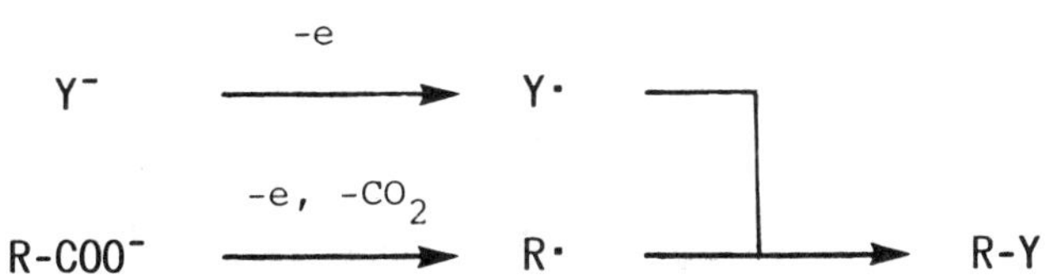

(22)

$Y^- = NO_3^-,\ F_2N^-,\ N_3^-,\ Br^-$

Table 5. Decarboxylative Olefin Addition

Carboxylic acid	Olefin	Product (yield, %)	Reference
EtOCOCOOH	$CH_2=CH-CH=CH_2$	$[EtOCOCH_2CH=CHCH_2-]_2$ (66)	80
$MeOCOCH_2COOH$	$CH_2=CHPh$	$[MeOCOCH_2CH_2CH(Ph)-]_2$ (38)	80
$MeOCO(CH_2)_4COOH$	$CH_2=CH-CH=CH_2$	$[MeOCO(CH_2)_4CH_2CH=CHCH_2-]_2$ (47) + $[=CHCH_2(CH_2)_4COOMe]_2$ (47)	81
CH_3COOH	$CH_2=C(CHO)CH_3$	$[CH_3CH_2C(CHO)(CH_3)-]_2$ (80)	82
CF_3COOH	$CH_2=CH(CH_2)_6COOH$	$CF_3(CH_2)_8COOH$[a] (16) XXXIV	83
CF_3COOH	$CH_2=CHCOOH$	$CF_3CH_2CH(COOH)_2$[b] (8) XXXV	84
CF_3COOH	$CH_2=CHCH_2OH$	$CF_3(CH_2)_2CHO$[c] (25) XXXVI	85
CF_3COOH	$CH_2=C(CH_3)OAc$	$CF_3CH_2COCH_3$ (17) XXXVII	86

CF_3COOH	$CH_2{=}CHCOOR$ R = Me, Et	$[CF_3CH_2CH(COOR)-]_2$ (24-50)		77, 87–89
CF_3COOH	$CH_2{=}CHCONH_2$	$CF_3CH_2CH(CF_3)CONH_2$ (37)	$CF_3CH{=}CHCONH_2$ (9)	89
CF_3COOH	$CH_2{=}CHCN$	$CF_3CH_2CH(CF_3)CN$ (5)	$[CF_3CH_2CH(CN)-]_2$ (30)	88
CF_3COOH	$CH_2{=}CHSO_2R$ R = Ph, Et	$CF_3CH_2CH(CF_3)SO_2R$ (24, 25)	$[CF_3CH_2CH(SO_2R)-]_2$ (22, 22)	89
CF_3COOH	EtOCOCH=CHCOOEt	EtOCOCH(CF_3)CH(CF_3)COOEt (47)	[EtOCOCH(CF_3)CH(COOEt)-]_2 (27)	87
CF_3COOH	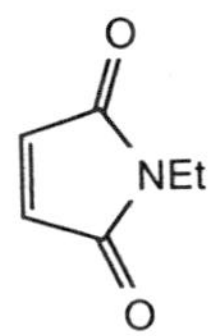	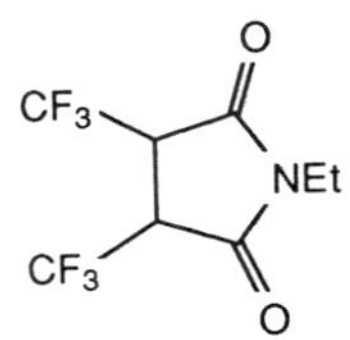 (43.8)		87
CF_3COOH	$CH{\equiv}C(CH_2)_3CH_3$	$CF_3CH{=}C(CF_3)(CH_2)_3CH_3$		88

Table 5. (Continued)

Carboxylic acid	Olefin	Product (yield %)	Reference
CF_3COOH	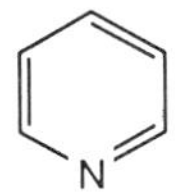	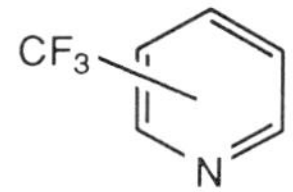 (7) α-β-γ = 27:38:35	90
CF_3COOH	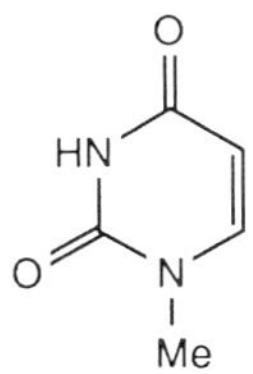	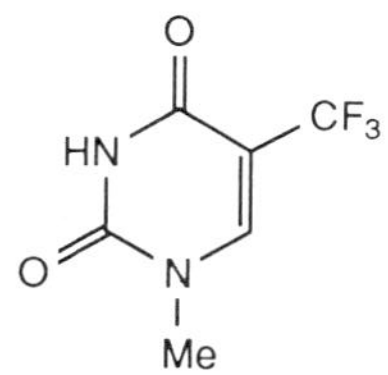 (40)	87
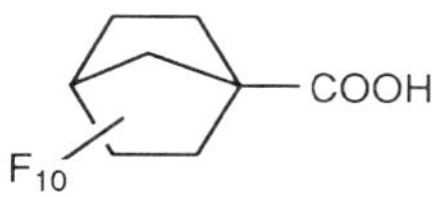	$CH_2CHCOOMe$	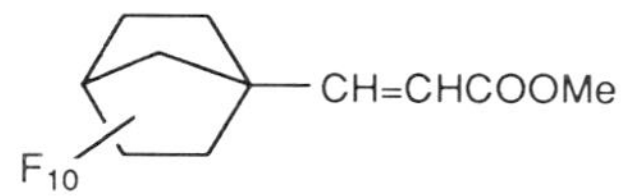 (0.4 g)	91
		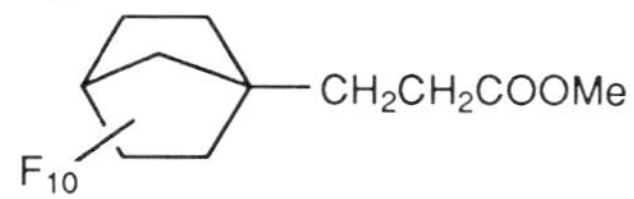 (0.3 g)	
		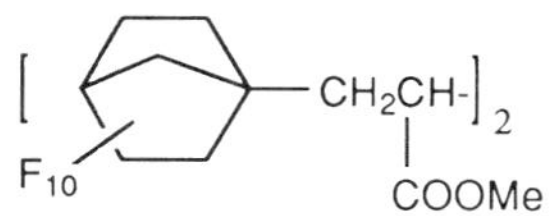 (0.8 g)	

[a]After hydrogenation (Pd/C).
[b]After hydrogenation (Pd/C) and hydrolysis (aqueous NaOH).
[c]After hydrolysis (aqueous H_2SO_4).

Substitution of an aldehyde hydrogen with methyl has been performed by electrolysis of acetic acid in the presence of aldehydes yielding methyl ketones, Eq. (23) [101,102]:

$$\text{Ph-CHO} + \text{CH}_3\text{COOH} \xrightarrow{-e,\ -\text{CO}_2} \text{Ph-CO-CH}_3 \tag{23}$$

$$\text{C}_6\text{H}_{13}\text{-CHO} + \text{CH}_3\text{COOH} \xrightarrow{-e,\ -\text{CO}_2} \text{C}_6\text{H}_{13}\text{-CO-CH}_3$$

Initiation of polymerization of olefins with radicals formed in the Kolbe electrolysis has been studied to some extent [103–107].

IV. CARBENIUM ION REACTION (TWO-ELECTRON OXIDATION)

The two-electron process in the electrodecarboxylation of carboxylates has been intensively investigated from the mechanistic and synthetic viewpoints. Significant progress of such carbenium reactions has been recognized in alkoxylation, acetoxylation, acetamidation, olefin formation (β-elimination), rearrangement, C–C bond cleavage, and so on. Typical reaction patterns of the electrogenerated cation intermediates R^+ and their synthetic potentialities are discussed in the following paragraphs.

A. Acyloxylation, Alkoxylation, and Acetamidation

Sufficiently stable electrogenerated carbenium ions can be trapped with nucleophiles or nucleophilic solvents. Acetoxylation and methoxylation proceed by electrolysis in either acetic acid or methanol. Acetamidation occurs in wet acetonitrile, in which the nucleophilic attack of the nitrile group on the carbenium ions R^+ tends to give iminium cations; subsequent hydrolysis with water in the medium completes the reaction, Eq. (24):

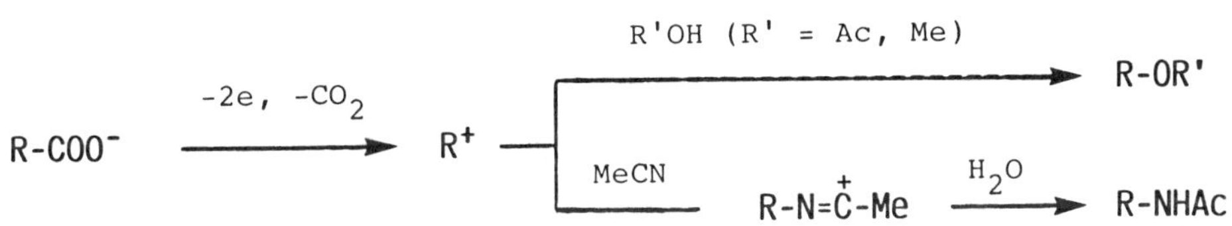

(24)

Electrogenerated carbenium ions, which are stabilized by neighboring hetero atoms, that is, oxygen nitrogen, sulfur, and others, can lead smoothly to the corresponding acetoxylated and/or methoxylated product, Eq. (25). This method can play an important role in synthetic applications. Some typical examples are shown in Table 6.

Table 6. Acyloxylation and Alkoxylation of α-Hetero-Substituted Carboxylic Acids

Carboxylic acids[a]	Conditions	Products (yield %)		Reference
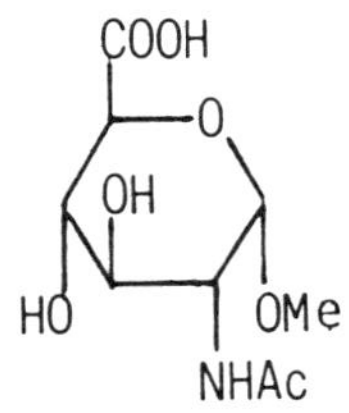	$AcOH-Et_3N-(Pt)$	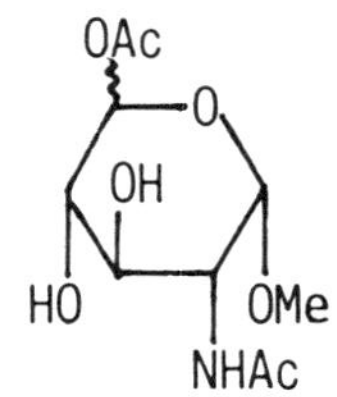	(43)	108
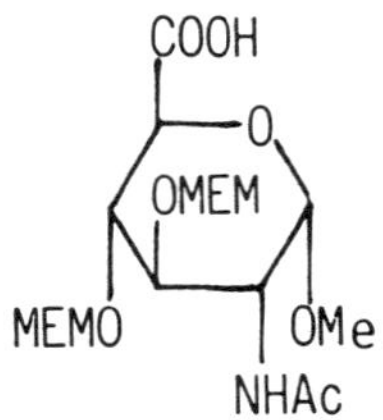	$AcOH-Et_3N-(Pt)$	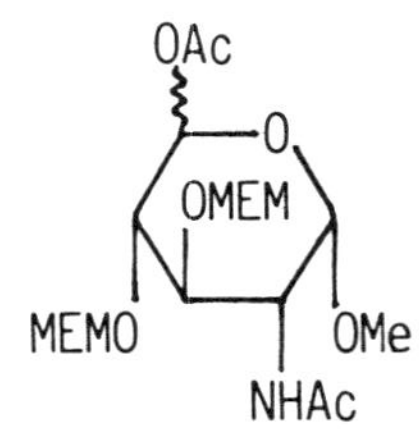	(89)	108
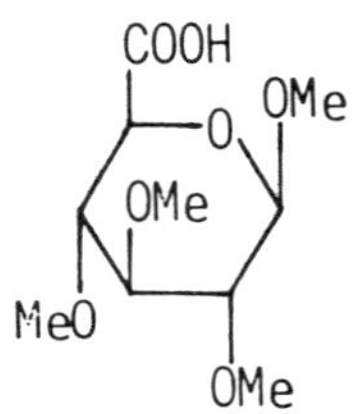	$MeOH-NaClO_4-(Pt)$	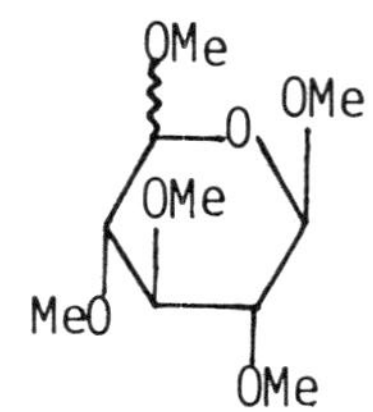	(96)	109

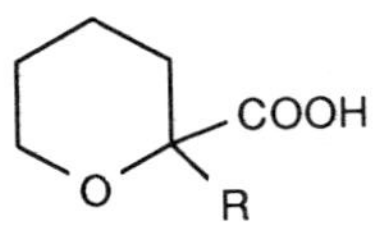 R: $CH_2CH{=}CH_2$, $CH_2C(CH_3){=}CH_2$, etc.	MeOH–K_2CO_3–Et_4NClO_4–(Pt)	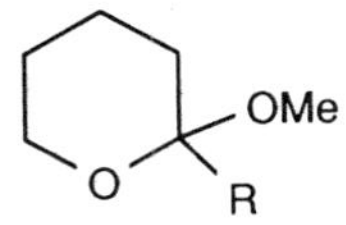	(67-90)	110
R_1–COO–C(R^2)(R^3)–COOH R^1 = Ph, CH_3, OC_2H_5 R^2, R^3 = H, CH_3, Ph, –$(CH_2)_5$–	MeOH–(C) or MeCN–Et_3N–(C)	R^1–COO–C(R^2)(R^3)–OMe R^2–CO–R^3, etc.		111, 112
EtO–C(CH_3)–COO^-	MeCN–(C)	EtO–CH(CH_3)–COO–CH(CH_3)–OEt	(81)	113
(a) 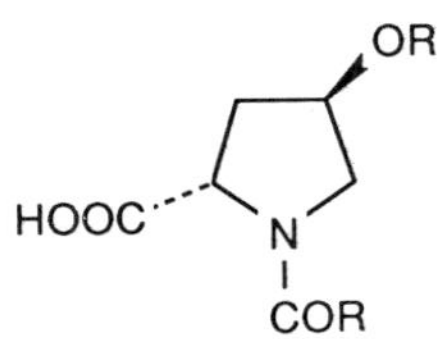 R = H, Si(–)(+) R' = Me, OMe	MeOH–Et_3N–(Pt)	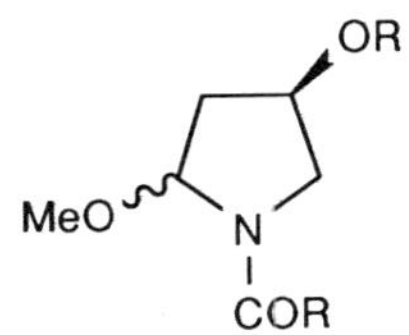	(89-97)	114, 115

Table 6. (Continued)

Carboxylic acids[a]	Conditions	Products (yield %)		Reference
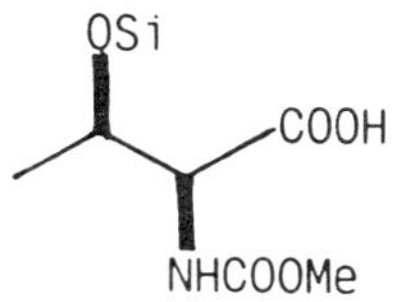	$MeOH-Et_3N-(Pt)$	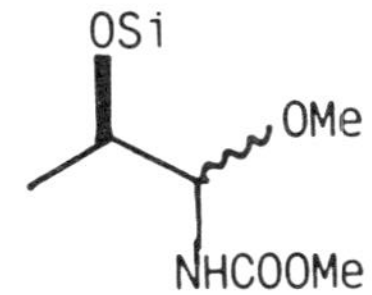	(75)	116
(b) 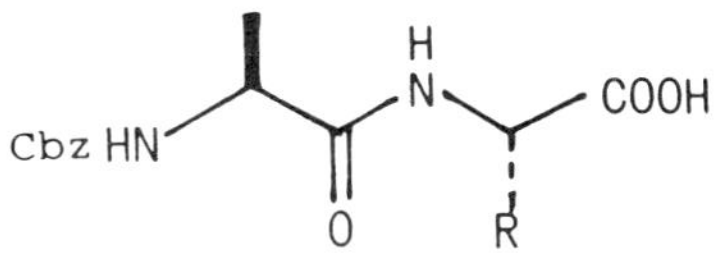 R = H, i-Pr	$MeOH-Et_3N-(Pt)$	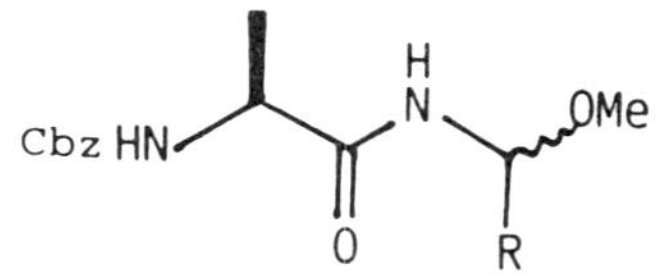	(95)	116
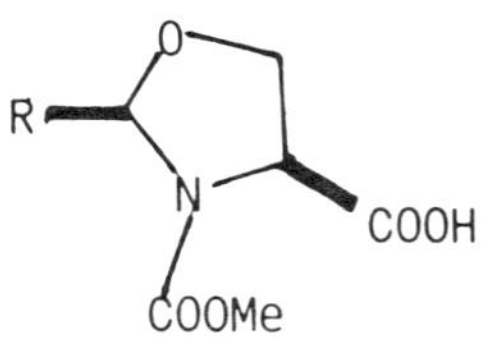 R = i-Pr, t-Bu	AcOH	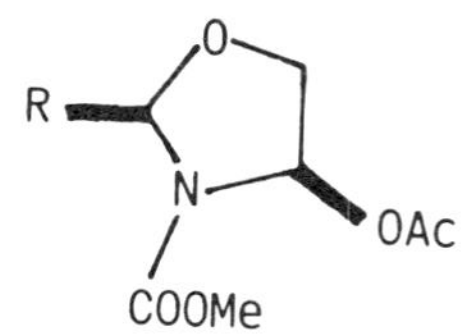	(85-87)	116
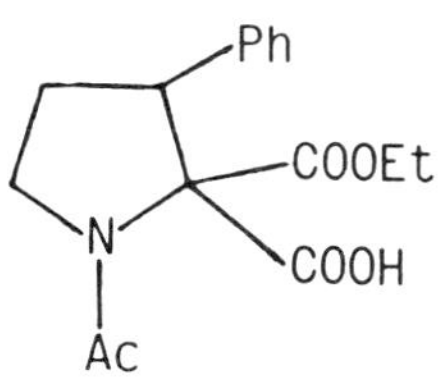	MeOH-MeONa-(C)	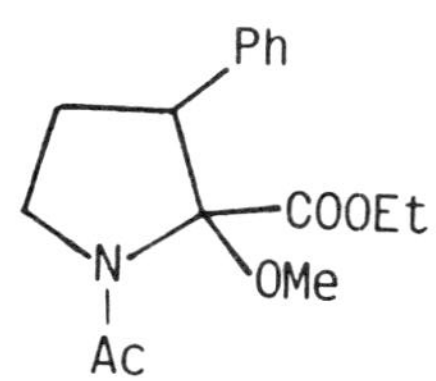	(98)	117, 118

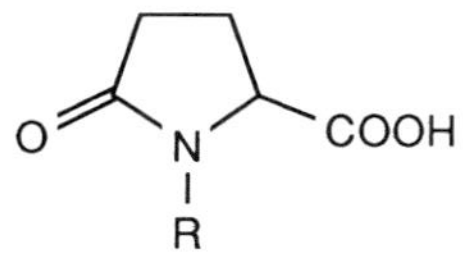 R = H, C_4H_9	MeOH–MeONa–(C)	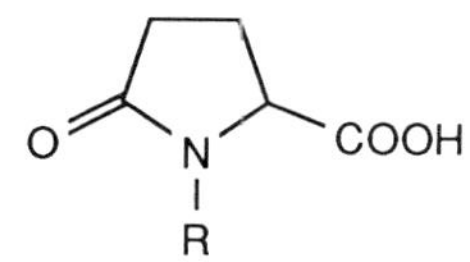	(75-98)	119
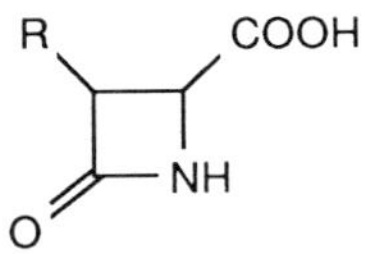 R = H, Et	MeCN–AcOH–AcONa–(Pt)	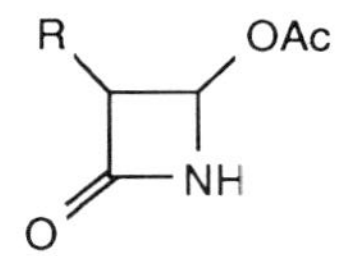	(76)	120
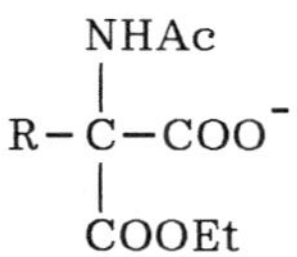 R = H, alkyl, allyl, etc.	R'OH–(C)	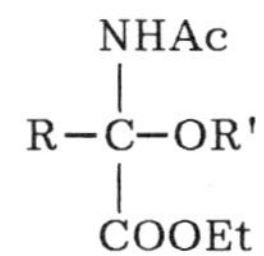 R' = Me, Et, i–Pr	(79-97)	121, 122

[a]MEM = 2-methoxyethoxymethyl; Cbz = benzyloxylcarbonyl.

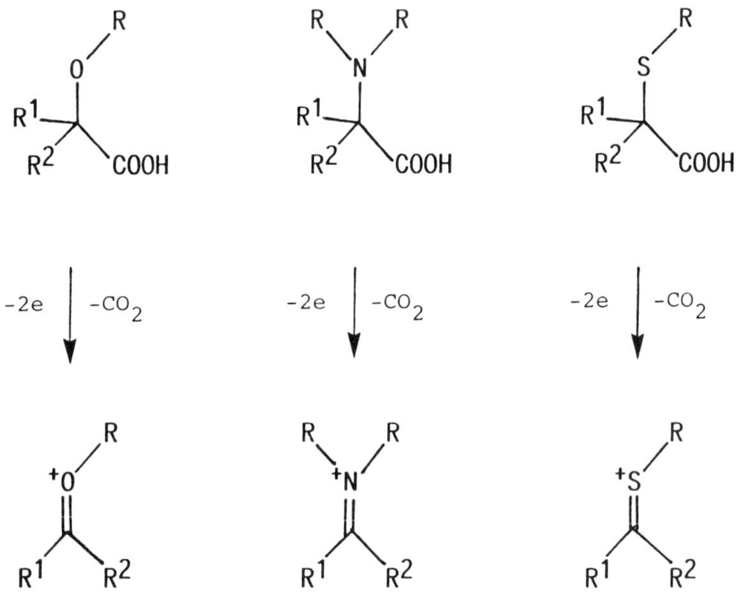

(25)

α-O-Substituted carboxylic acids can give mixed acetals; the electrolysis of uronic acid derivative (LI) in methanol thus affords the mixed acetals (LII), Eq. (26). The method has been succesfully applied to aminocyclitol synthesis as well as to the selective fission of glucuronide linkages in oligoglycosides (LIV), Eq. (27) [123].

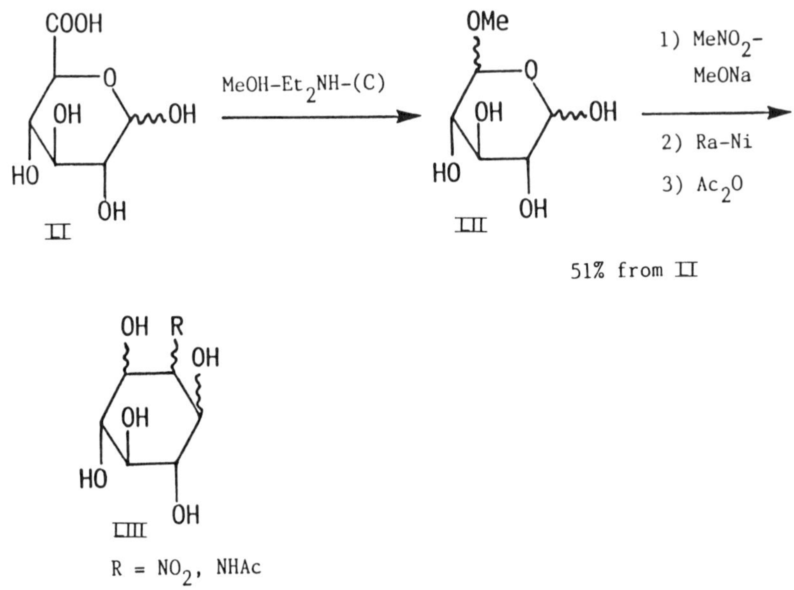

(26)

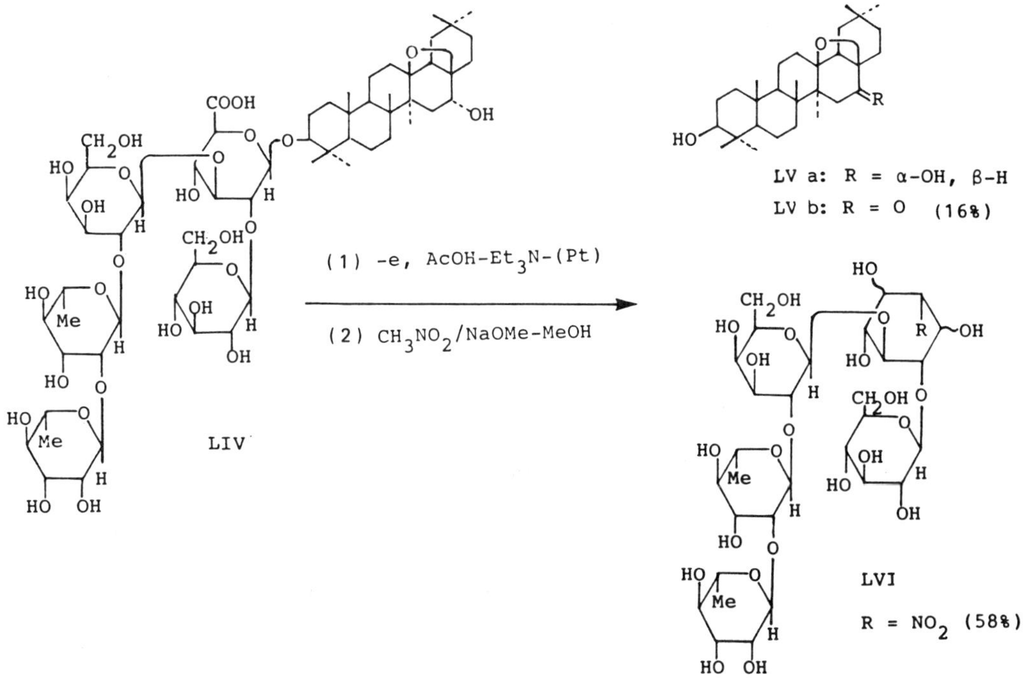

(27)

Electrolysis of the dicarboxylic acid (LVII) in a MeOH-MeONa-(Pt) system and subsequent workup affords the keto acid (LVIII), a precursor of chrysanthemic synthesis, in 86% yield via a segment of intramolecular acyloxylation and acid-promoted hydrolysis, Eq. (28) [124]:

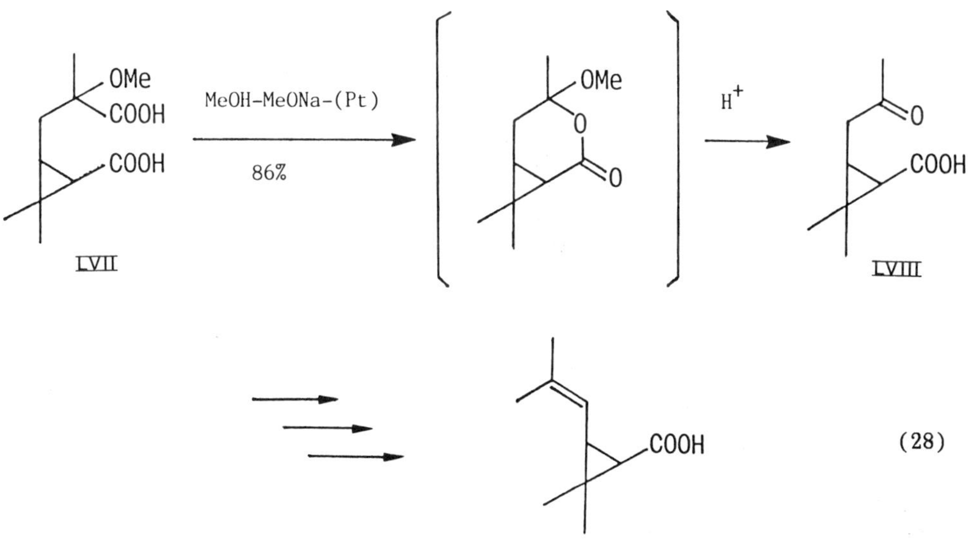

(28)

Chrysanthemic Acid

Decarboxylative acetoxylation and methoxylation of α-aminocarboxylic acids proceed smoothly, yielding synthetically useful intermediates. For example, the versatile intermediate (LX) for the synthesis of thienamycine is prepared by electrodecarboxylative acetoxylation of 4-carboxy-2-azetidinone (LIX) in an AcOH−MeCN-AcONa-(Pt) system, Eq. (29) [120]:

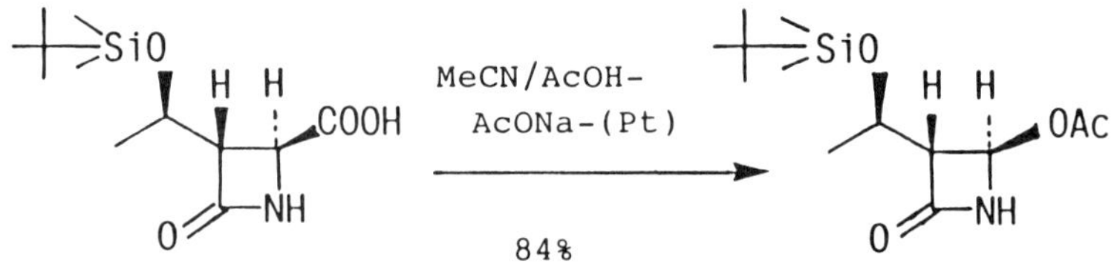

LIX LX (29)

Electrolysis of N-carbamoylaspartic acid or N-ethoxycarbamoylaspargine in MeOH-MeONa-(C) affords the corresponding methoxylated products [125]. 5-Fluorouracil derivatives (LXIII), a potent antitumor agent, can be prepared via electrolytic methoxylation of N-acylazacycloalkane-2-carboxylic acids (LXI) in MeOH-MeONa-(C) and subsequent condensation of (LXII) with 2,4-bis-(trimethylsilyl)-5-fluorouracil (TMS-5-FU), Eq. (30) [119,126]. Recently, the decarboxylative methoxylation of α-amino acids has been extended to prepare useful chiral building blocks [114−116].

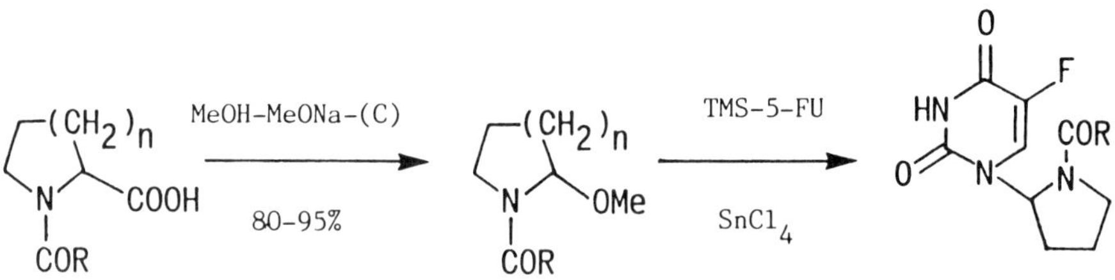

LXI LXII LXIII

$n = 1, 2$; $R = H, CH_3, CH_2NHCOOBn$ (30)

Electrolysis of α-sulfenyl carboxylic acids in protic solvents gives ketones presumably via replacement of the carboxyl group with alkoxy or acetoxyl groups and a subsequent two-electron oxidation of the sulfenyl moiety, Eq. (31). (This discussion is postponed to Sec. D.)

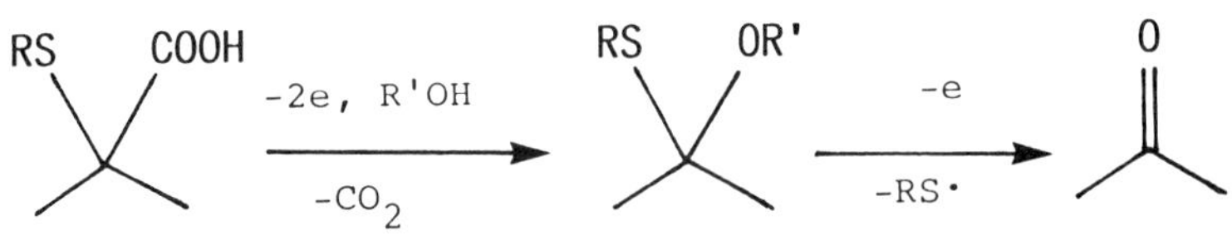

(31)

Further selected examples of the decarboxylative acetoxylation and methoxylations are listed in Table 7. The electrogenerated stable carbenium ions, such as benzylic, allylic, homoallylic, and tertiary carbocation intermediates, are likely to be trapped by suitable nucleophiles, yielding the corresponding acetates, methyl esters, and, in some cases, intramolecular acyloxylation products.

Table 7. Electroacyloxylation and Alkoxylation

Carboxylic acid	Conditions	Products (yield %)	Reference
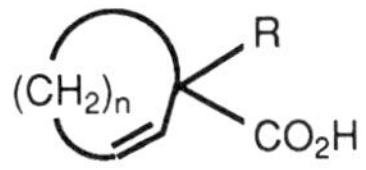 R = CH_3:phenyl, hexyl cis 3-hexenyl	AcOH/AcOEt/ tBuOH−Et_3N−(Pt)	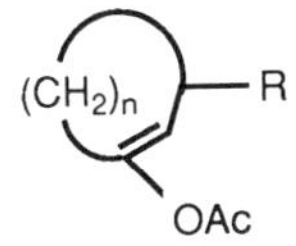	11, 12
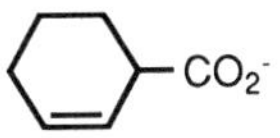	MeOH	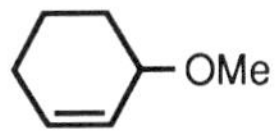 (100)	13, 17 127
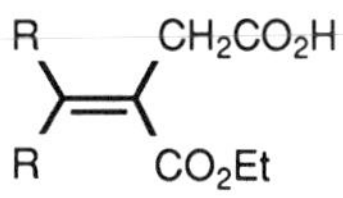 R:CH_3 Ph	MeOH−NaOMe−(C)	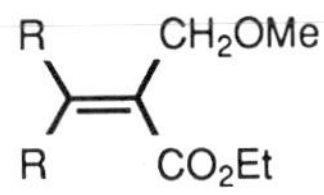 (22-31)	128
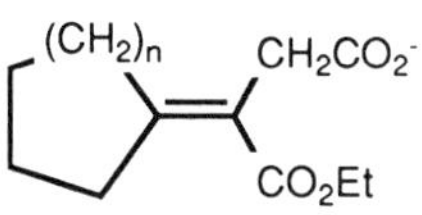 n:1,2	MeOH−NaOMe−(C)	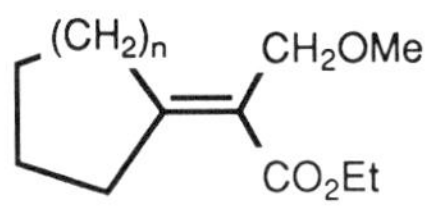 (35, 28)	128
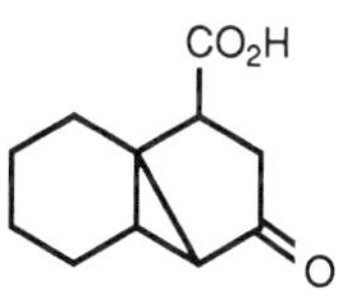	AcOH−tBuOH− Et_3N−(Pt)	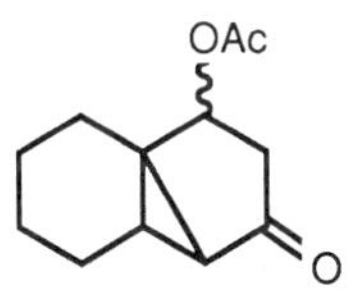 (72)	129, 130
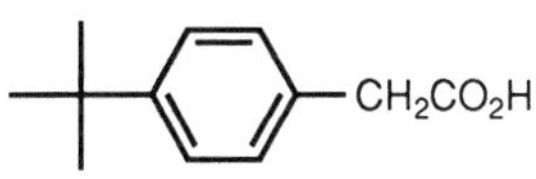		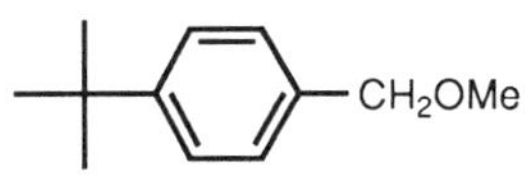	13, 127 30
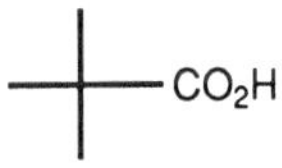	MeOH−NaOMe− CO_2−(Pt)	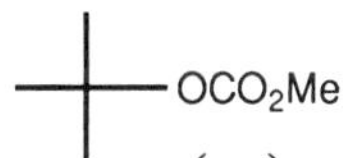 (78) (--)	131
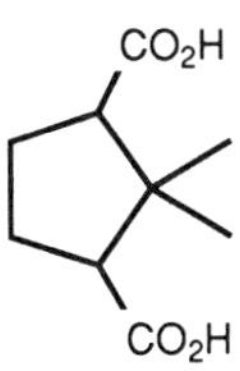	MeOH−(Pt)	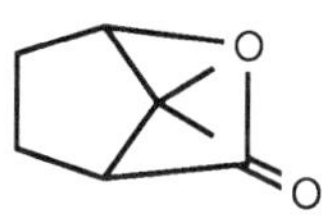 (77)	132

B. Olefin Formation

The electrodecarboxylation becomes a good procedure for the formation of carbon-carbon double bonds, which involves roughly two types of reactions: (1) decarboxylative discharge of carboxylates by releasing a vicinal leaving group, Eq. (32), and (2) initial discharge of hetero atoms at the β-position to carboxylic acid, which causes an elimination of the carboxyl group in a concerted manner, Eq. (33).

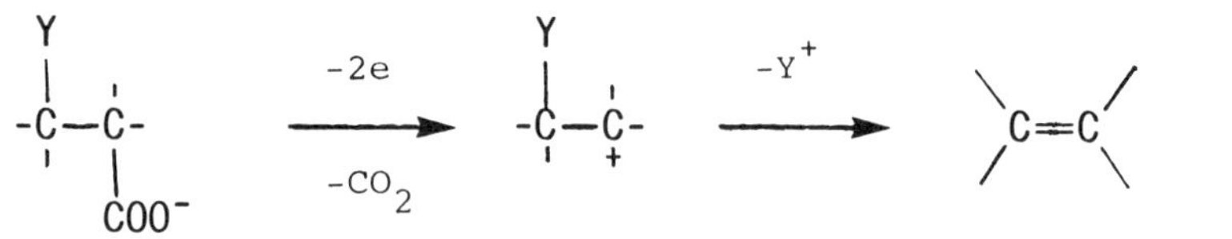

Y = H, COOH, $SiMe_3$, CH_2OH, etc (32)

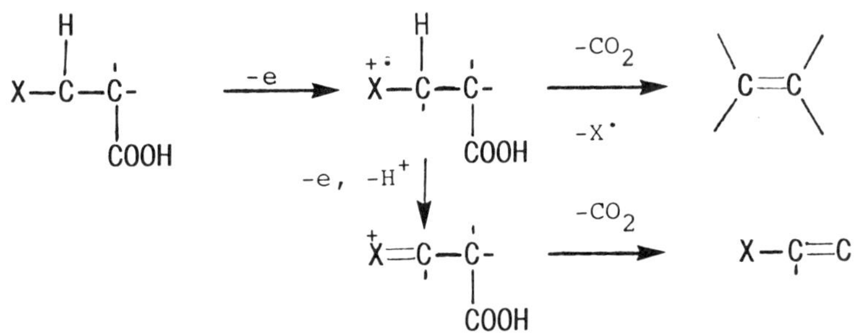

(33)

Various sets of functional groups have been devised for both types of decarboxylative carbon-carbon double-bond formation. Typical examples of a set of functional groups together with their electrolysis conditions are summarized in Table 8.

Table 8. Decarboxylative Carbon-Carbon Double-Bond Formation

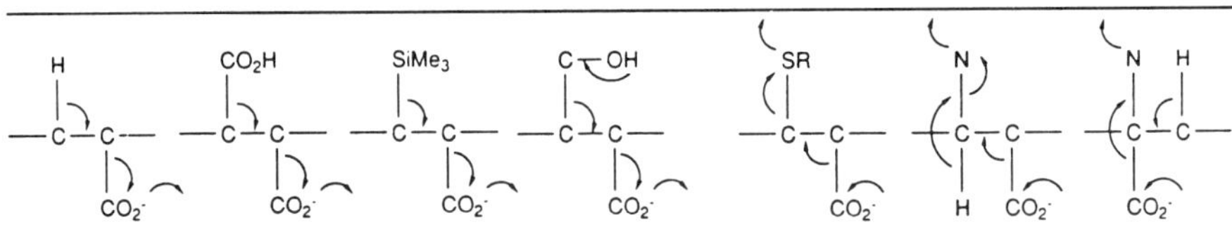

Direct discharge of carboxylates				Concerted decarboxylation		
Py/H_2O	Py/H_2O	MeCN	MeOH	H_2O	H_2O	MeCN
Et_3N	Et_3N	EtOH	Et_3N	Et_3N	KOH	Et_4NBF_4
(Pt)	(C)	KOH	(Pt)	$LiClO_4$	(C)	(Pt)
		(C)		(Pt)		
References						
21	13–17	133	134, 14	135	136	137

In such olefin formation reactions, an electrodecarboxylation-deprotonation sequence is also involved. Electrochemical aromatization of the nonconjugated dienes (LXIV and LXVI) can be successfully performed by this method, Eqs. (34) and (35) [138]. Decarboxylation of the γ-lactone α- and β-carboxylic acids (LXVIII and LXX) in a Py–H_2O-Et_3N-(Pt or C) proceeds smoothly, yielding the corresponding butenolides (LXIX and LXX), respectively, Eqs. (36) and (37) [139,140]. In a similar fashion, 2-substituted cyclopentenones (LXXIII) have been obtained starting from 2-carboxycyclopentanones (LXXII), Eq. (38) [141].

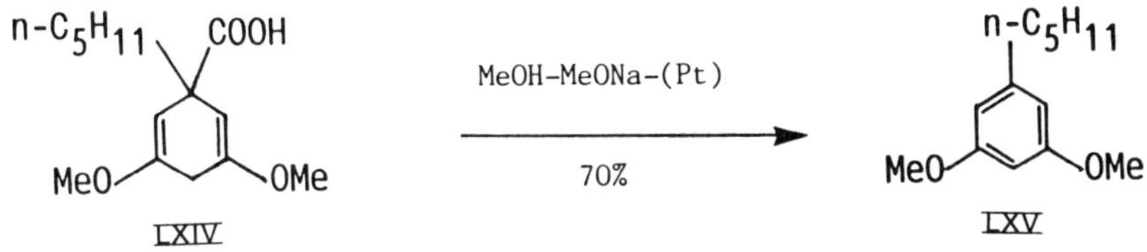

(34)

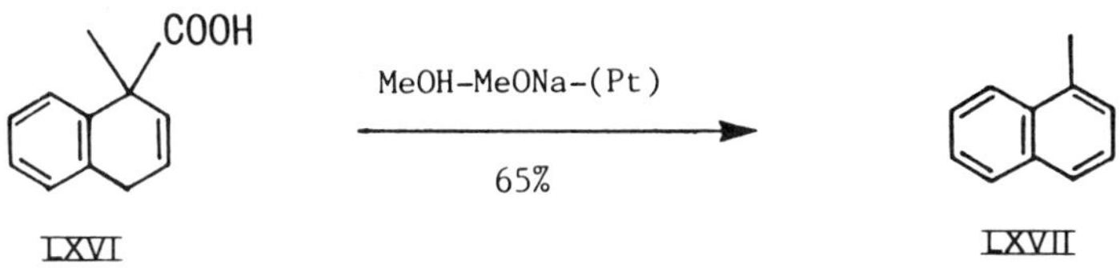

(35)

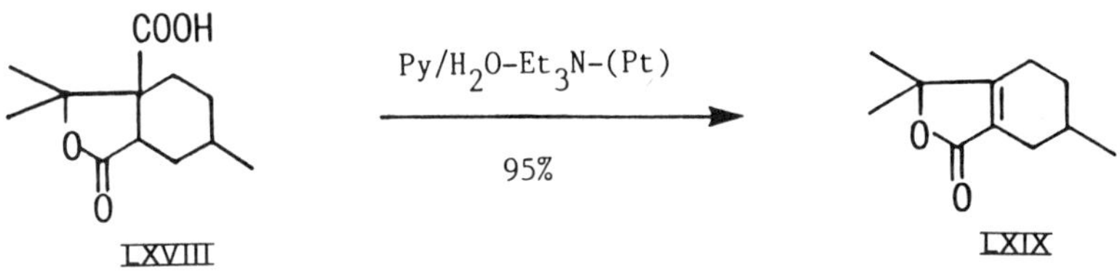

(36)

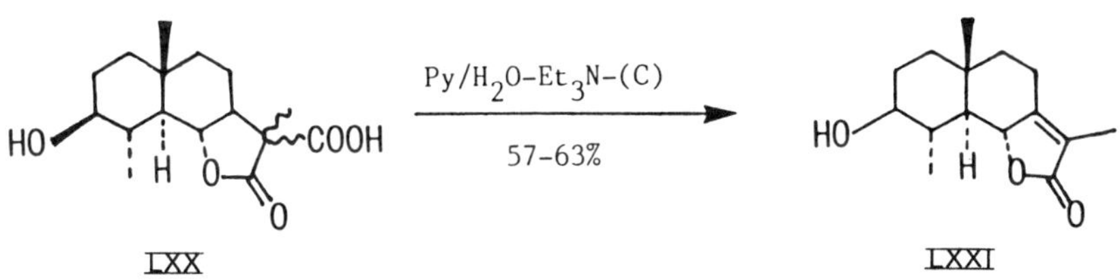

(37)

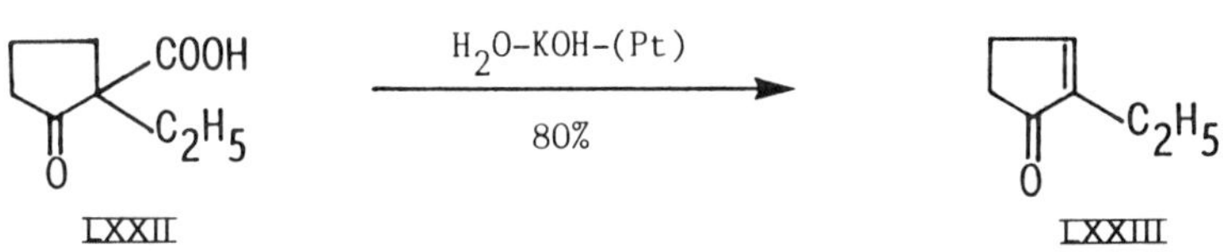

(38)

Double decarboxylation of 1,2-dicarboxylic acids may take place smoothly in an aqueous 90% pyridine-Et_3N-(Pt) electrolysis system to give olefins in fair yields [142–144]. Some representative examples are shown in Eqs. (39) [14], (40) [13,15,16], and (41) [144]:

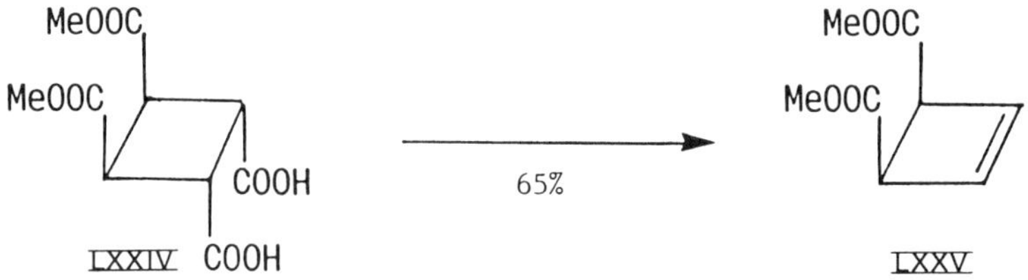

LXXIV LXXV (39)

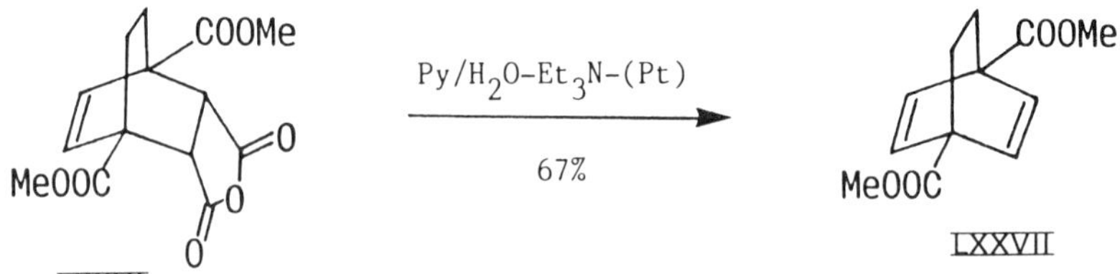

LXXVI LXXVII (40)

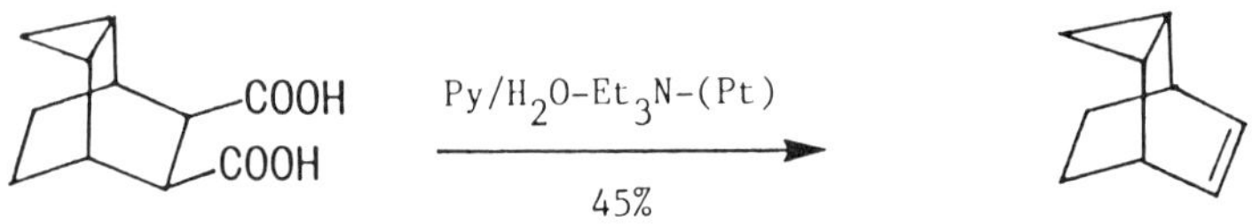

LXXVIII LXXIX (41)

A new device for olefin formation via decarboxylation-desilylation has been achieved successfully by electrolysis of β-trimethylsilyl carboxylic acids in an MeCN–EtOH-KOH-(C) system [133]. For example, norbornadiene (LXXXI) and cyclohexadiene (LXXXIII) have been obtained in good yields, Eqs. (42) and (43):

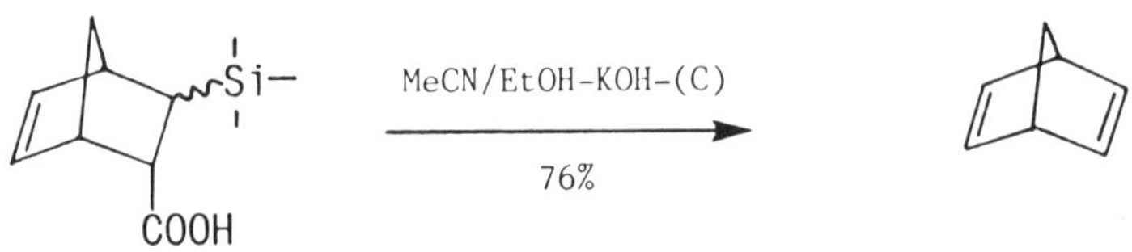

LXXX LXXXI (42)

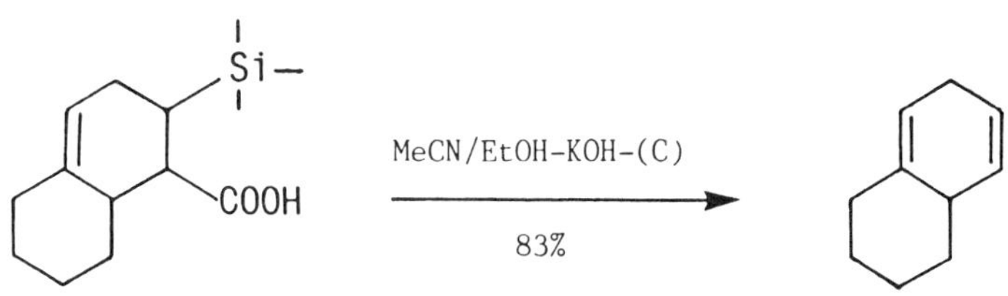

LXXXII LXXIII (43)

Electrolysis of γ-hydroxy carboxylic acid may give rise to decarboxylative β,γ-bond fission leading to olefinic ketones. An example is given in Eq. (44) [134]:

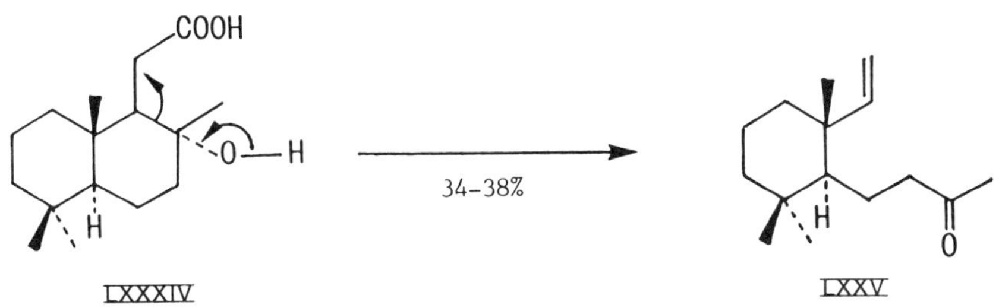

(44)

The α-exo-methylene-γ-lactone framework has been successfully constructed via two electrosynthetic pathways, that is, both by the direct and by the concerted decarboxylation processes as mentioned earlier [Eq. (45)] [135]. The electrodecarboxylation of LXXXVIII is probably initiated by a one-electron oxidation of the sulfur atom, giving first the cation radical (XC) and subsequently a concerted elimination of the thiyl radical and carbon dioxide to LXXXVII. On the other hand, the electrochemical decarboxylation of LXXXVI involves an E1-type elimination of a proton from the cation intermediate (LXXXIX) generated from direct two-electron oxidation of the carboxyl group. The latter method generally requires a higher oxidation potential than that required for the concerted method. Therefore, the concerted electrodecarboxylation method becomes more advantageous especially when the substrates or products are unstable under oxidative conditions.

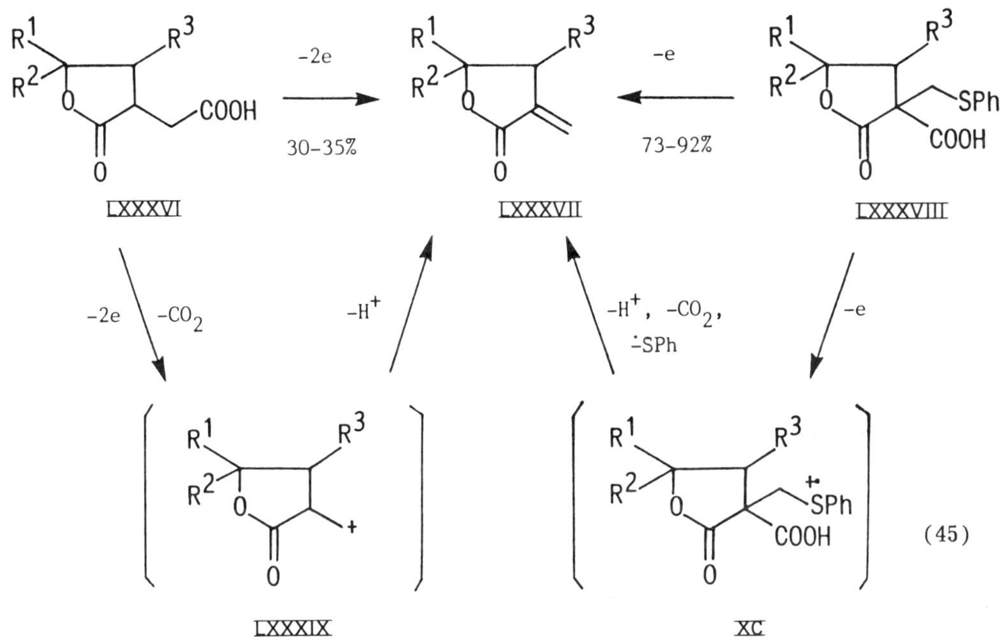

(45)

The decarboxylation started by the oxidation of a nitrogen atom promises versatile synthetic applications for making unsaturated nitrogen-containing heterocyclic frameworks. Pyrrole [118], uracil [136,145–147], thymine [136], 2-pyrazoline [137], β-carboline [148], and dihydroquinoline [149], useful intermediates for drug syntheses, have been prepared by using the electrodecarboxylation procedure. Some typical examples are given in Eqs. (46) through (51):

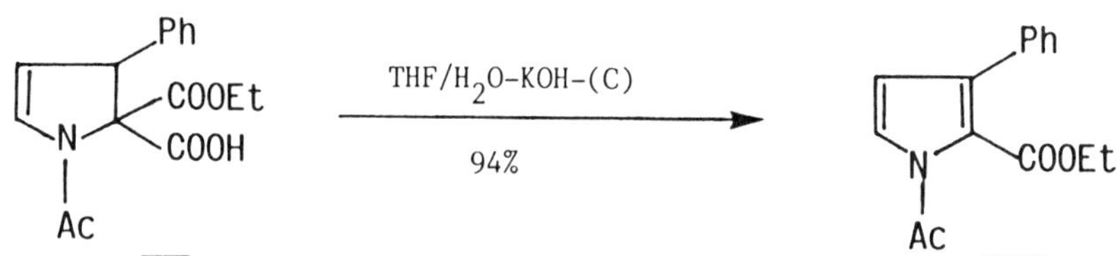

XCI XCII (46)

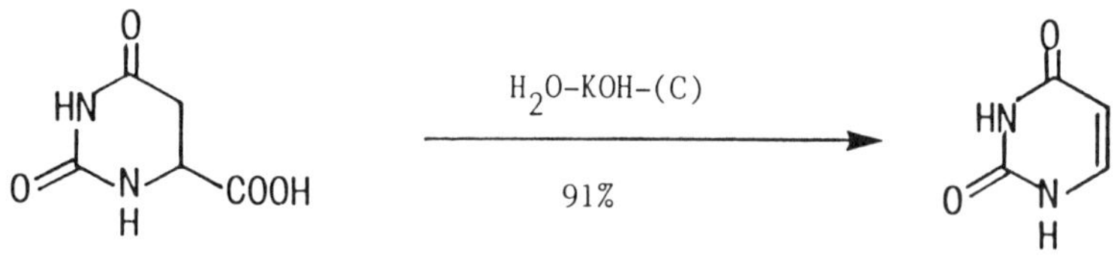

XCIII XCIV (47)

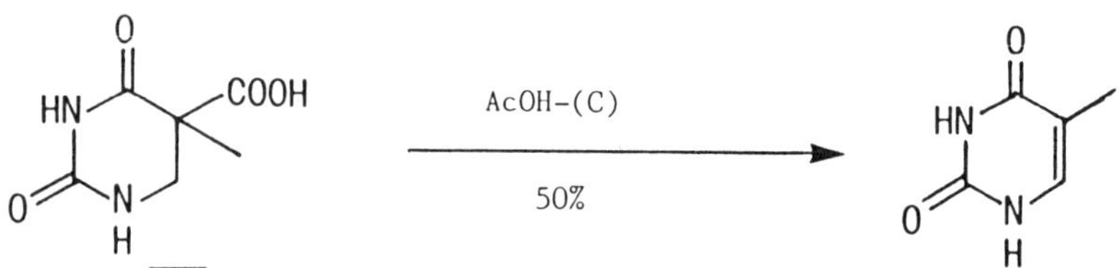

XCV XCVI (48)

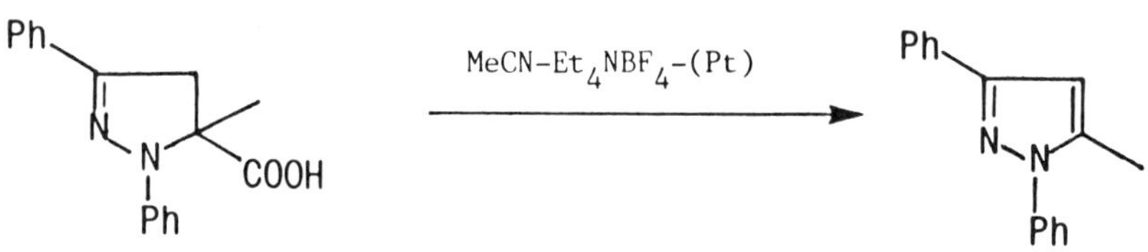

XCVII XCVIII (49)

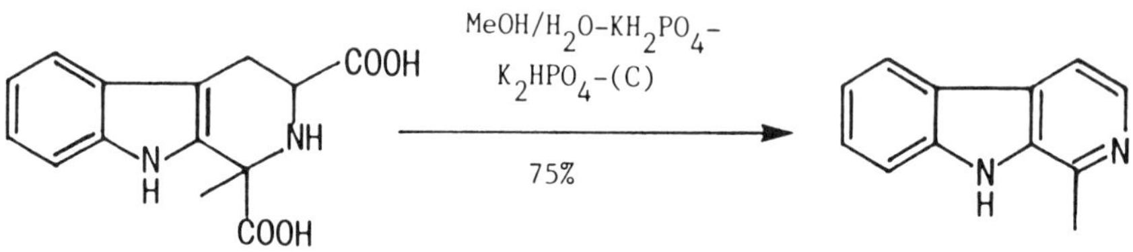

XCIX C (50)

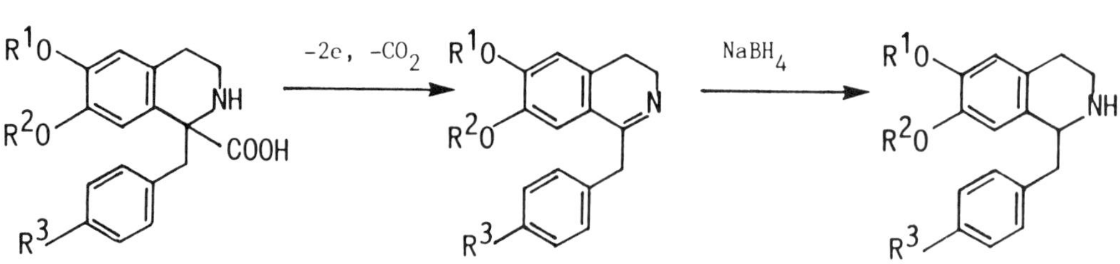

CI CII CIII

(51)

C. Rearrangement

A carbocation formed by an electrodecarboxylation process is expected to be different in nature from carbocations generated in homogeneous media, for example by solvolysis or deamination [150,151]. It is very likely that the carbocation produced in the electrodecarboxylation should be highly reactive, possessing an unsolvated vacant p-orbital, and having no counter ion in the near vicinity. In this context, the electrogenerated carbocations may behave as "hot" carbocations. Actually, certain carboxylic acids bearing not only a suitably oriented β-substituent but also a heteroatom linked to the β-carbon are liable to undergo a carbocation rearrangement. Some reactions are exemplified as follows.

The electrochemical rearrangement of paraconic acids (CIV), Eq. (52), γ-carboxy-δ-lactones (CVI), Eq. (53) [152], and β-keto carboxylic acid ethylene acetals (CVIII), Eq. (54) [153,154] has been known to proceed smoothly. γ-Keto carboxylic acids (CX) and acetals (CXII) undergo decarboxylative rearrangement to afford esters (CXI and CXIII), Eqs. (55) and (56) [155,156].

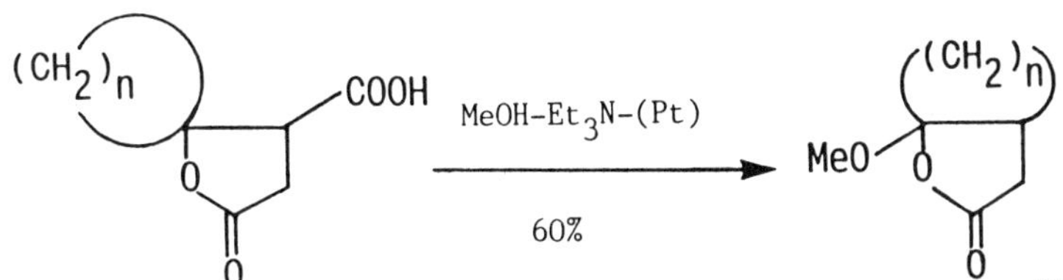

CIV CV (52)

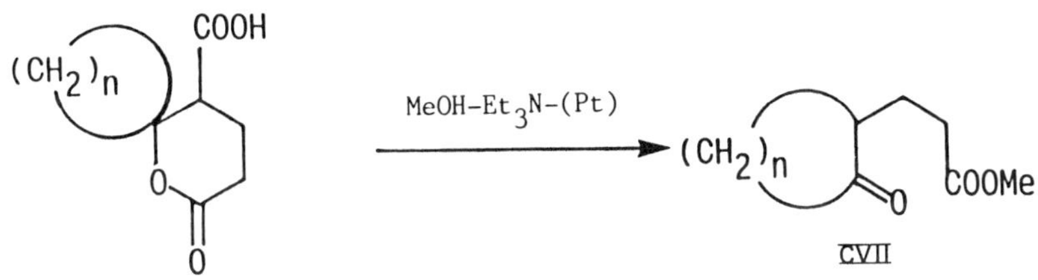

CVI CVII (53)

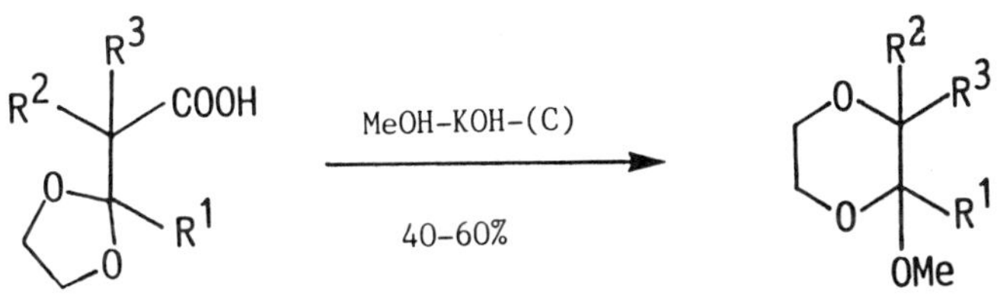

CVIII CIX (54)

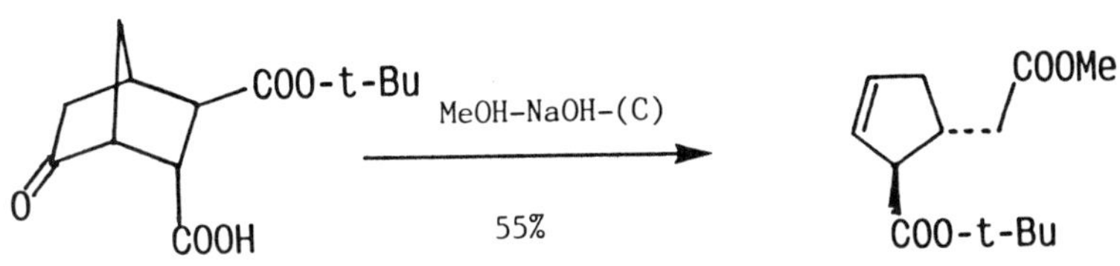

CX CXI (55)

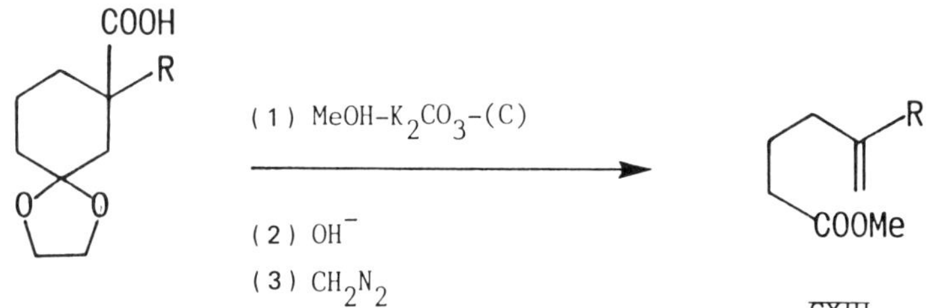

CXII CXIII

(56)

The electrodecarboxylation of β-hydroxy carboxylic acids, easily accessible by Reformatsky reaction on cyclic ketones, furnishes a promising route for the ring enlargement of cyclic ketones; for example, electrolysis of 1-hydroxy-2-cyclotetradecenylacetic acid (CXIV) in an MeOH-KOH-(C) system gives the sp^2 carbon migration product (CXV, 30%) together with the sp^3 carbon migration product (CXVI, 5%), Eq. (57) [157]:

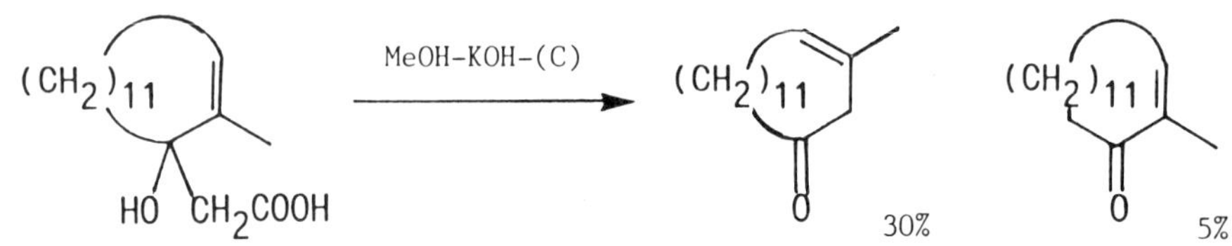

CXIV CXV CXVI

(57)

A Wagner-Meerwein rearrangement via a bridged carbenium ion has been observed in the electrodecarboxylation of bicyclo[2.2.1]heptane-2-carboxylic acids (CXVIIa, CXVIIb, and CXVIIc), leading to the formation of the same products (CXVIII and CXIX), Eq. (58) [158]. This method has been successfully applied to synthesis of useful intermediates for natural product synthesis. Some examples are given in Eqs. (59) and (60) [159–163].

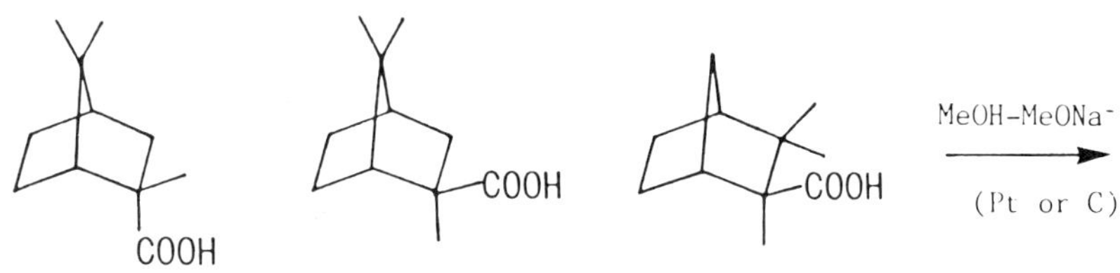

CXVII a CXVII b CXVII c

(58)

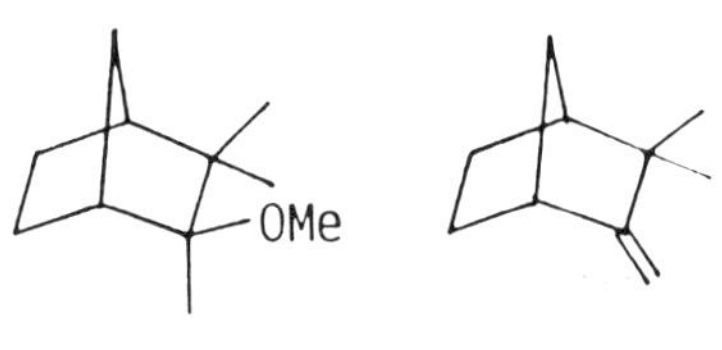

CXVIII CXIX

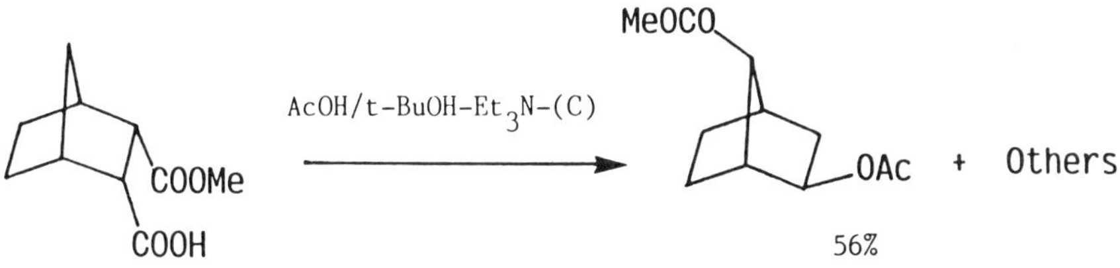

CXX CXXI (59)

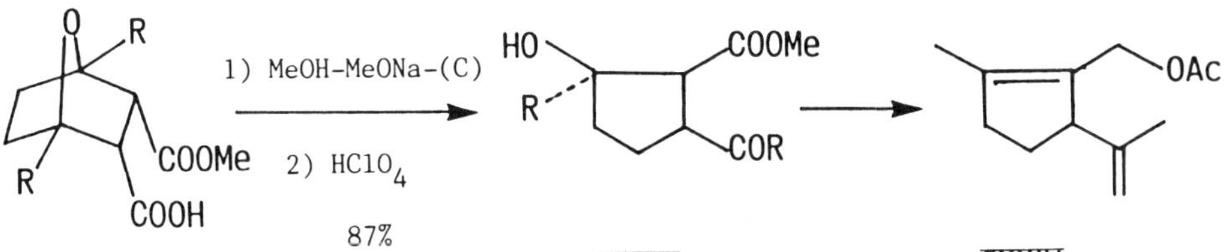

CXXII CXXIII CXXIV

(60)

D. Formation of Aldehydes, Ketones, Acetals, and α,β-Enones

Electrodecarboxylation procedures for preparing aldehydes, ketones, acetals, and enones have been intensively investigated. The following sets of functional groups are found to be electrosynthetically equivalent to carbonyl groups:

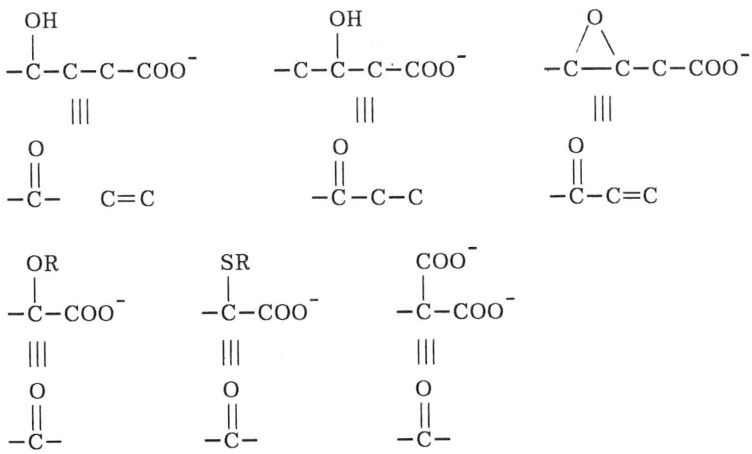

As mentioned in preceding sections, γ- and β-hydroxy carboxylic acids undergo decarboxylative rearrangement to afford ketones. An additional example is given in Eq. (61), in which the electrodecarboxylation of a γ-hydroxy carboxylic acid moiety in a bicyclo[4.4.0]decane system (CXXV) gives a 10-membered nonconjugated enone (CXXVI) [164]. On the other hand, the electrolysis of an α,β-epoxy carboxylic acid moiety arranged in the same ring system (CXXVII) provides a conjugated enone (CXXVIII) in good yield, Eq. (62) [11]. Electrodecarboxylation of α-hydroxy or α-alkoxy carboxylic acids proceeds smoothly, affording aldehydes [165] or acetals (see preceding sections).

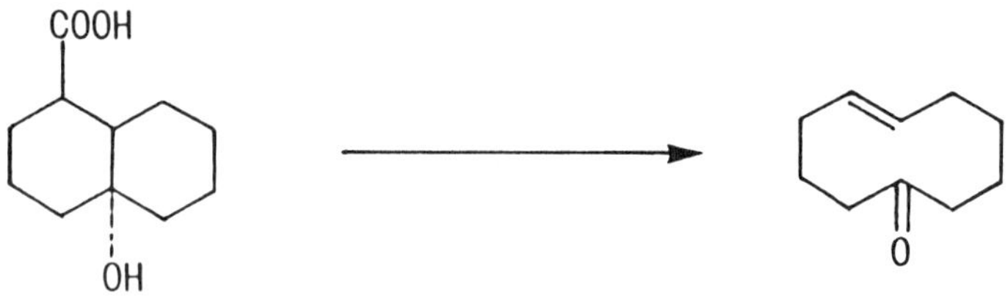

CXXV CXXVI (61)

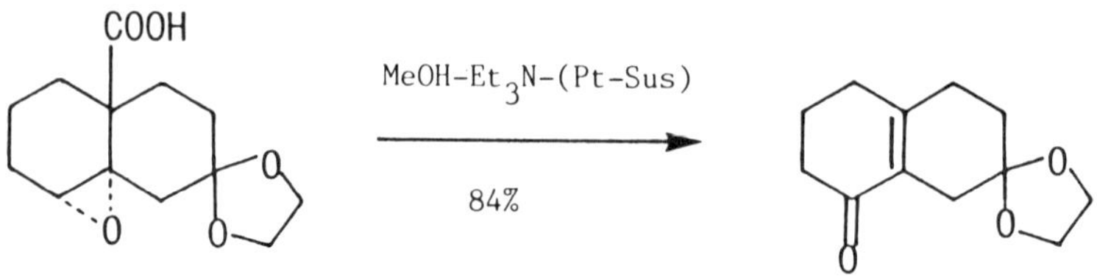

CXXVII CXXVIII (62)

α-Sulfenylcarboxylic acids can be transformed into aldehydes and ketones via electrodecarboxylation. For example, the electrolysis of α-phenylthiocarboxylic acid (CXXIX) gives aldehyde (CXXXI) in an H_2O-NaOH-(Pt) system and the corresponding acetal (CXXX) in an MeOH-$LiClO_4$-(Pt) system, Eq. (63) [9]. Cyclopentadecanone (CXXXIII) is prepared by electrolysis of α-phenylthio- and methylthio-substituted cyclopentadecanecarboxylic acids (CXXXII), Eq. (64) [8].

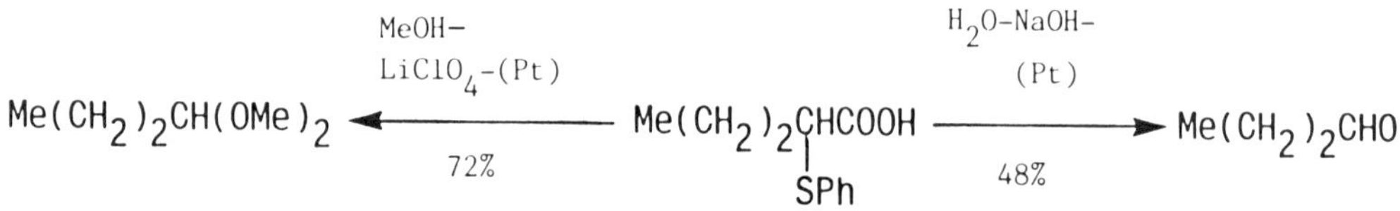

CXXX CXXIX CXXXI

(63)

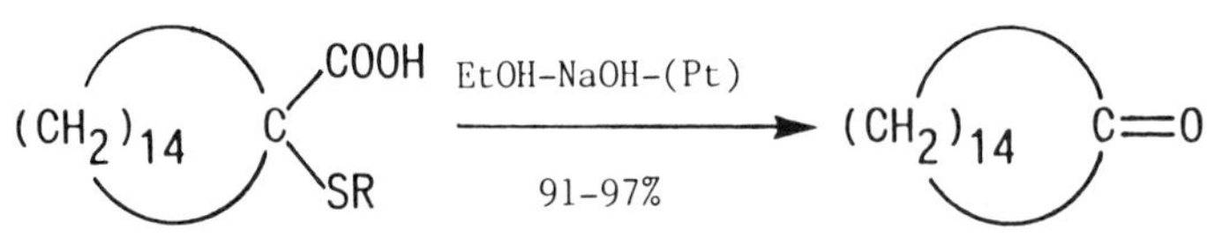

CXXXII CXXXIII (64)

The α-sulfenyl group can be replaced by a benzothiazolylthio group, as demonstrated in the synthesis of the 1,4-diketone (CXXXIV), a precursor of <u>cis</u>-jasmone, Eq. (65) [10]. The geminal dicarboxylic acid (CXXXVI) can also be transformed into the 1,4-diketone (CXXXIV) by electrochemical decarboxylation, Eq. (66) [6], suggesting that the substituted malonic acid group is synthetically equivalent to a carbonyl group.

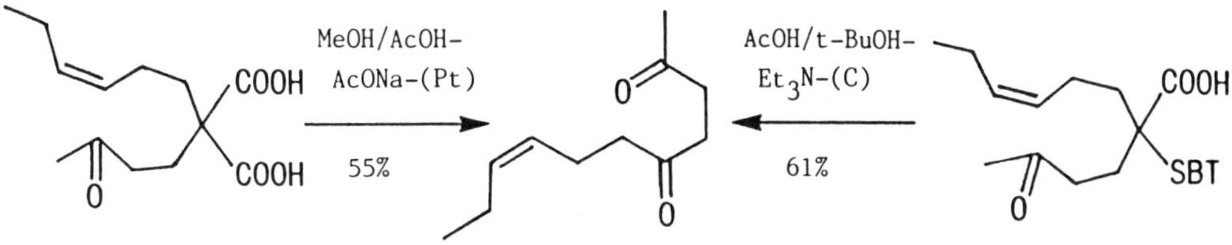

CXXXVI CXXXIV CXXXV

(66) (65)

E. Miscellaneous

Further synthetically meaningful reactions induced by electrochemical decarboxylation are briefly discussed in this section. Potent stereoselective routes to methyl <u>cis</u>- and <u>trans</u>-4,4-dimethoxy-2-butenoates (CXXXVIII and CXXXIX) and Methyl 4,4-dimethoxybutanoate (CXL) from 2-furoic acid (CXXXVII) have been achieved mostly by simple selection of an appropriate electrolyte, that is, MeOH-Et_3N-Et_4NClO_4-(Pt) for CXXXVIII; MeOH-Et_3N-Et_4NBr-(Pt) for CXXXIX, and MeOH-Et_3N-Et_4NBr-(C) for CXL, respectively, Eq. (67) [166,167]:

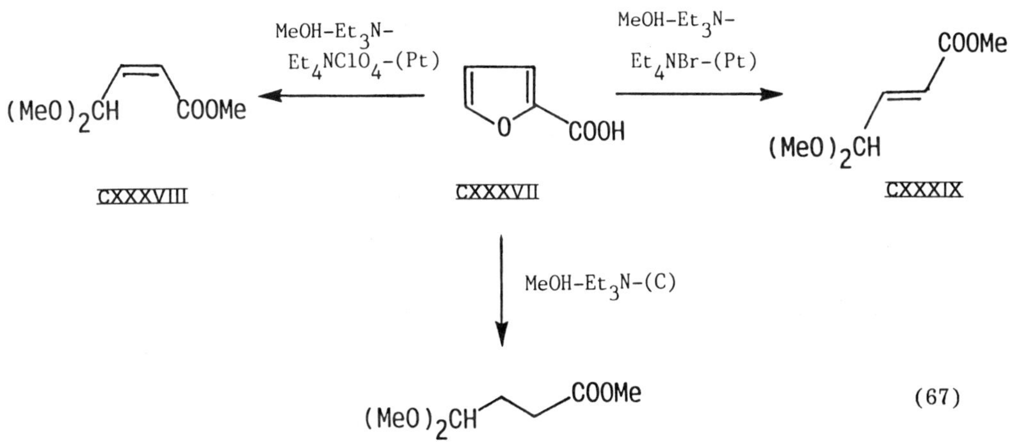

CXXXVIII CXXXVII CXXXIX

(67)

CXL

Electrodecarboxylation followed by C—C bond cleavage has been investigated for three- and four-membered ring systems. The dichlorocyclopropane carboxylic acid (CXLI) undergoes decarboxylation in either acetic acid or methanol to give the ring-opened acetate (CXLII) or methyl ether (CXLIII) in good yields, Eq. (68) [168]:

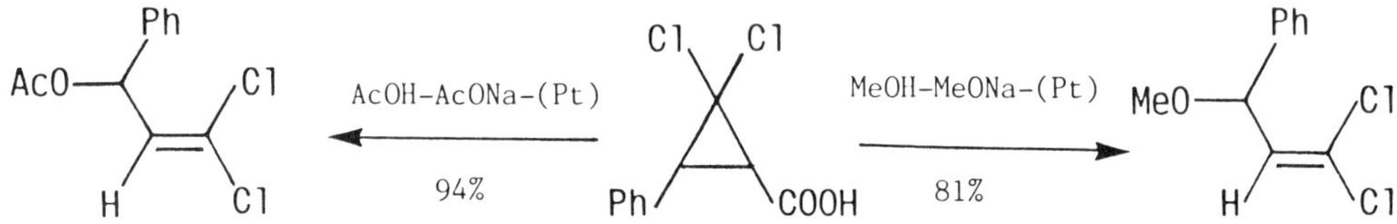

CXLII CXLI CXLIII

(68)

The double bond in alicyclic β-keto acid enol ethers (CXLIV and CXLVI) can be cleaved by electrodecarboxylation, Eqs. (69) and (70) [169]:

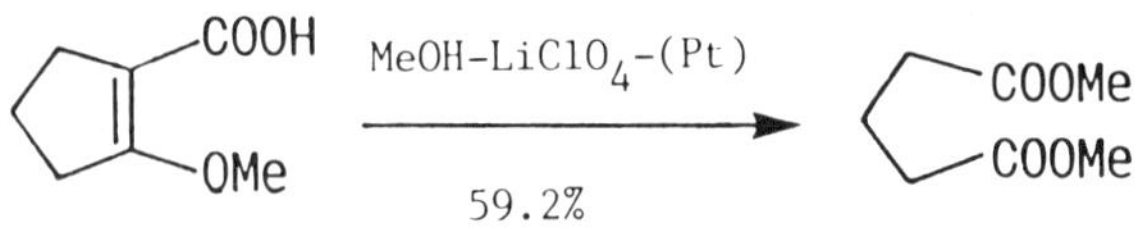

CXLIV CXLV

(69)

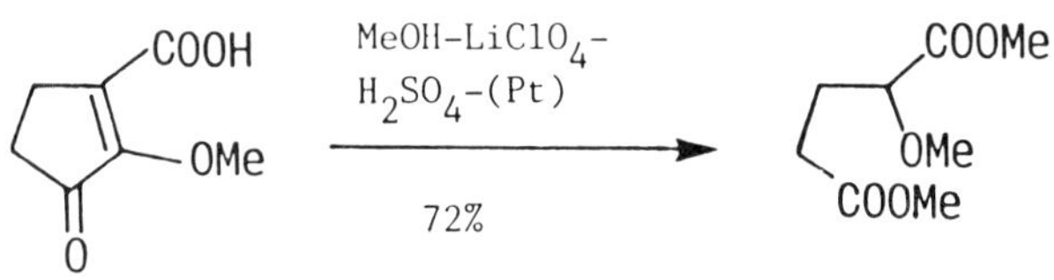

CXLVI CXLVII

(70)

REFERENCES

1. H. Kolbe, Ann. Chem. Pharm., 69, 257 (1849).
2. A. Wurts, Ann. Chim. Phys., 44, 291 (1985).
3. A. C. Brown and J. Walker, Justus Liebigs Ann. Chem., 261, 107 (1981).
4. C. Walling, Free Radical in Solution, Wiley, New York, 1957, p. 581.
5. H. Hofer and M. Moest, Ann. Chem., 323, 284 (1902).
6. J. Nokami, T. Yamamoto, M. Kawada, M. Izumi, N. Ochi, and R. Okawara, Tetrahedron Lett., 1047 (1979).
7. K. Uneyama, S. Torii, and S. Oae, Bull. Chem. Soc. Jpn., 44, 815 (1971).
8. J. Nokami, M. Matsuura, T. Sueoka, Y. Kusumoto, and M. Kawada, Chem. Lett., 1283, (1978).
9. J. Nokami, M. Kawada, R. Okawara, S. Torii, and H. Tanaka, Tetrahedron Lett., 1045 (1979).
10. S. Torii, H. Tanaka, Y. Kobayashi, J. Nokami, and M. Kawata, Bull. Chem. Soc. Jpn., 52, 1553 (1979).
11. S. Torii, T. Inokuchi, K. Mizuguchi, and M. Yamazaki, J. Org. Chem., 44, 2303 (1979).
12. S. Torii, H. Tanaka, T. Inokuchi, and K. Tomozane, Bull. Chem. Soc. Jpn., 55, 3947 (1982).

13. R. C. Kelly and I. Schletter, J. Amer. Chem. Soc., 95, 7156 (1973).
14. J. H. Leftin, D. Redpath, A. Pines, and E. Gol-Av, Israel J. Chem., 11, 75 (1973).
15. C. B. Warren and J. J. Bloomfield, J. Org. Chem., 38, 4011 (1973).
16. H. Sliwa and Y. Le Bot, Tetrahedron Lett., 4129 (1977).
17. J. H. P. Utley and G. B. Yates, J. Chem. Soc., Perkin Trans., II, 395 (1978).
18. T. Dickinson and W. F. K. Wynne-Jones, Trnas. Faraday Soc., 58, 382, 388, 400 (1962).
19. L. Eberson, Chemistry of the Carboxyl Group (S. Patai, ed.), Wiley, London, 1969, Chap. 2.
20. N. Sato, T. Sekine, and K. Sugino, J. Electrochem. Soc., 115, 242 (1968).
21. S. Torii, T. Okamoto, and H. Tanaka, J. Org. Chem., 39, 2486 (1974).
22. Y. B. Vasil'ev, A. M. Skundin, V. A. Grinberg, A. Z. Zaidenberg, V. E. Kazarinov, Y. K. Kalinin, and E. F. Dyukkiev, Elektrokhimiya, 22, 1529 (1986); UDC 541.138.2.
23. R. N. Renaud and O. E. Sullivan, Can. J. Chem., 51, 772 (1973).
24. M. Finkelstein and R. C. Petersen, J. Org. Chem., 25, 136 (1960).
25. L. Rand and A. F. Mohar, J. Org. Chem., 30, 3156 (1965).
26. D. L. Muck and E. R. Wilson, J. Electrochem. Soc., 117, 1358 (1970).
27. S. Torii, Electroorganic Syntheses, Part I, Oxidations Methods and Applications, Kodansha and VCH, Tokyo, 1985, p. 52.
28. B. C. L. Weedon, Q. Rev. (Lond.), 6, 380 (1952).
29. J. H. P. Utley, Techniques of Electroorganic Synthesis (N. L. Weinberg, ed.), Part 1, Wiley, New York, 1974, Chap. 6.
30. J. P. Coleman, J. H. P. Utley, and B. C. L. Weedon, J. Chem. Soc., Chem. Commun., 438, (1971).
31. J. P. Coleman, R. Lines, J. H. P. Utley, and B. C. L. Weedon, J. Chem. Soc., Perkin Trnas., II, 1064 (1974).
32. J.-W. Shaw, T. Nonaka, and T.-C. Chou, J. Chin. Inst. Chem. Eng., 16, 245 (1985); Chem. Abstr., 104, 12183r (1986).
33. G. I. Shul'zhenko, E. P. Kovsman, and G. N. Freidlin, Elektrokhimiya, 21, 195 (1985); UDC 621.357.1:547.
34. R. G. Woolford, Can. J. Chem., 40, 1846 (1962).
35. R. G. Woolford, W. Arbic, and A. Rosser, Can. J. Chem., 42, 1788 (1964).
36. J. E. Sanderson, P. F. Levy, L. K. Cheng, and G. W. Barnard, J. Electrochem. Soc., 130, 1844 (1983).
37. T. Watanabe, Y. Tanimoto, R. Nakagaki, M. Hiramatsu, and S. Nagakura, Bull. Chem. Soc. Jpn., 60, 4166 (1987).
38. M. M. Baizer, Organic Electrochemistry, Dekker, New York, 1973.
39. F. Beck, Elektroorganische Chemie, Verlag Chemie, Weinheim, 1974.
40. N. L. Weinberg, Technique of Electroorganic Synthesis, Parts I and II, Wiley-Interscience, New York, 1974.
41. S. D. Ross, M. Finkelstein, and E. J. Rudd, Anodic Oxidation, Academic Press, New York, 1975.
42. B. C. L. Weedon, Adv. Org. Chem., 1, 1 (1963).
43. H. J. Schäfer, Angew. Chem., Int. Ed. Engl., 20, 911 (1981).
44. L. Eberson and K. Nyberg, Encyclopedia of Electrochemistry of the Elements (A. J. Bard and H. Lund, eds.), Dekker, New York, 1978, Chap. XII-2.

45. F. Fichter and S. Lurie, Helv. Chim. Acta, 16, 885 (1933).
46. J. Peterson, Z. Elektrochem., 18, 710 (1912).
47. J. Einhorn, J.-L. Soulier, C. Bacquet, and D. Lelandais, Can. J. Chem., 61, 584 (1983).
48. U. R. Joshi and P. A. Limaye, Biovigyanam, 11, 101 (1985); Chem. Abstr., 103, 23598K (1985).
49. T. Kubota, R. Aoyagi, H. Sando, M. Kawasumi, and T. Tanaka, Chem. Lett., 1435 (1987).
50. K. Kimura, S. Horie, I. Minato, and Y. Odaira, Chem. Lett., 1209 (1973).
51. L. B. Rodewald and M. C. Lewis, Tetrahedron, 27, 5273 (1971).
52. R. F. Garwood, N. Vd Din, C. J. Scott, and B. C. L. Weedon, J. Chem. Soc., Perkin Trnas., I, 2714 (1973).
53. E. J. Corey and R. R. Sauers, J. Amer. Chem. Soc., 81, 1739 (1959).
54. G. Stork, A. Meisels, and J. E. Davies, J. Amer. Chem. Soc., 85, 3419 (1963).
55. F. Beck, Electrochim. Acta, 18, 359 (1973).
56. J. de la Fuente, E. A. Lissi, and R. Rozas, Can. J. Chem., 62, 2117 (1984).
57. H. G. Thomas, S. Kessel, and E. Muller, Chem. Ber., 119, 2173 (1986).
58. A. Gozlan and A. Zilkha, Eur. Polym. J., 23, 515 (1987).
59. A. Gozlan and A. Zilkha, Eur. Polym. J., 23, 133 (1987).
60. A. Gozlan and A. Zilkha, Eur. Polym. J., 23, 393 (1987).
61. A. Gozlan and A. Zilkha, Eur. Polym. J., 23, 143 (1987).
62. M. Feldhues and H. J. Schäfer, Tetrahedron Lett., 22, 433 (1981).
63. M. Feldhues and H. J. Schäfer, Tetrahedron, 41, 4195 (1985).
64. M. Feldhues and H. J. Schäfer, Tetrahedron, 41, 4213 (1985).
65. A. K. Yadav and P. Tissot, Helv. Chim. Acta, 67, 1698 (1984).
66. W. Seidel, H. J. Schäfer, Chem. Ber., 113, 451 (1980).
67. R. F. Nutt, R. S. Strachan, D. F. Veber, and F. W. Holly, J. Org. Chem., 45, 3078 (1980).
68. U. Jensen-Korte and H. J. Schäfer, Liebigs Ann. Chem., 1532 (1982).
69. U. Jensen and H.-J. Schäfer, Chem. Ber., 114, 292 (1981).
70. W. Seidel and H.-J. Schäfer, Chem. Ber., 113, 3898 (1980).
71. H. Klünenberg and H. J. Schäfer, Angew. Chem., Int. Ed. Engl., 17, 47 (1978).
72. W. Seidel, J. Knolle, and H. J. Schäfer, Chem. Ber., 110, 3544 (1977).
73. J. Knolle and H. J. Schäfer, Angew. Chem., Int. Ed. Engl., 14, 758 (1975).
74. R. Rossi, A. Carpita, and M. Chini, Tetrahedron, 41, 627 (1985).
75. P. F. Levy, J. E. Sanderson, and L. K. Cheng, J. Electrochem. Soc., 131, 773 (1984).
76. T. Kubota, R. Aoyagi, H. Sando, M. Kawasumi, and T. Tanaka, Chem. Lett., 1435 (1987).
77. R. N. Renaud and P. J. Champagne, Can. J. Chem., 53, 529 (1975).
78. R. N. Renaud and D. E. Sullivan, Can. J. Chem., 50, 3084 (1972).
79. M. Huhtasaari, H. J. Schäfer, and Luftmann, Acta Chem. Scand., B37, 537 (1983).

80. H. J. Schäfer, R. Pistorius, Angew. Chem., Int. Ed. Engl., 11, 841 (1972).
81. M. Y. Fioshin, L. A. Salmin, L. A. Mirkind, and A. G. Kornienko, Zh. Obshch. Khim., 10, 594 (1965).
82. M. Chkir and D. Lelandais, J. Chem. Soc., Chem. Commun., 1369 (1971).
83. N. Muller, J. Org. Chem., 49, 2826 (1984).
84. N. Muller, J. Org. Chem., 51, 263 (1986).
85. N. Muller, J. Org. Chem., 49, 4559 (1984).
86. N. Muller, J. Org. Chem., 48, 1370 (1983).
87. R. N. Renaud, P. J. Champagne, and M. Savard, Can. J. Chem., 57, 2627 (1979).
88. C. J. Brookes, P. L. Coe, D. M. Owen, A. E. Pedler, and L. C. Tatlow, J. Chem. Soc., Chem. Commun., 323 (1974); C. J. Brookes, P. L. Coe, A. E. Pedler, and J. C. Tatlow, J. Chem. Soc., Perkin Trans., I, 202 (1978).
89. K. Uneyama, K. Ueda, and H. Nanbu, Abstract of 172nd Electrochemical Society Meeting, Honolulu, Hawaii, 1987, 87(2), 1847.
90. J. H. P. Utley and R. J. Holman, Electrochim. Acta, 21, 987 (1976).
91. P. L. Coe, J. H. Sleigh, and J. C. Tatlow, J. Chem. Soc., Perkin Trans., I, 217 (1980).
92. K. Uneyama and K. Ueda, Chem. Lett., 853 (1988).
93. M. Huhtasaari, H. J. Schäfer, and L. Becking, Angew. Chem., Int. Ed. Engl., 23, 980 (1984).
94. L. Becking and H. J. Schäfer, Tetrahedron Lett., 29, 2797 (1988).
95. L. Becking and H. J. Schäfer, Tetrahedron Lett., 29, 2801 (1988).
96. F. Fichter, Organiche Elektrochemie, Steinkopff, Dresden, 1942.
97. V. Ohman, Z. Elektrochem., 42, 862 (1936).
98. G. A. Ward and C. M. Wright, J. Amer. Chem. Soc., 86, 4333 (1964).
99. C. M. Wright and G. A. Ward, Abstract of Electrochemical Society Meeting, Electroorganic Division, Vol. 2, 111 (1965).
100. G. S. Krasnikova, L. S. German, and I. L. Kvunyants, Izv. Akad. Nauk SSSR, Ser. Khim., 2127 (1972).
101. A. Takeda, S. Torii, and H. Oka, Tetrahedron Lett., 1781 (1968).
102. A. Takeda, S. Wada, S. Torii, and Y. Matsui, Bull. Chem. Soc. Jpn., 42, 1047 (1969).
103. S. Goldschmidt and E. Stöckl, Chem. Ber., 85, 630 (1952).
104. W. B. Smith and H. Gilde, J. Amer. Chem. Soc., 82, 659 (1960).
105. J. W. Breitenbach and C. Srua, Pure Appl. Chem., 4, 245 (1962).
106. W. B. Smith and D. T. Manning, J. Polym. Sci., 59, 545 (1962).
107. B. L. Funt and F. D. Williams, J. Polym. Sci., 1B, 181 (1963).
108. M. Yoshikawa, T. Kamigauchi, Y. Ikeda, and I. Kitagawa, Chem. Pharm. Bull., 29, 2582 (1981).
109. I. Kitagawa, T. Kamigauchi, H. Ohmori, and M. Yoshikawa, Chem. Pharm. Bull., 28, 3078 (1980).
110. P. G. M. Wuts and C. Sutherland, Tetrahedron Lett., 23, 3987 (1982).
111. H. G. Thomas and S. Kessel, Chem. Ber., 118, 2777 (1985).
112. H. G. Thomas, J. Gabriel, J. Fleischhauer, and G. Raabe, Chem. Ber., 116, 375 (1983).
113. H. G. Thomas and K. Katzer, Tetrahedron Lett., 887 (1974).
114. P. Renaud and D. Seebach, Synthesis, 424 (1986).

115. P. Renaud and D. Seebach, Helv. Chim. Acta, 69, 1704 (1986).
116. P. Renaud and D. Seebach, Angew. Chem., Int. Ed. Engl., 25, 843 (1986).
117. H. Horikawa, T. Iwasaki, K. Matsumoto, and M. Miyoshi, J. Org. Chem., 43, 335 (1978).
118. M. Miyoshi, K. Matsumoto, T. Iwasaki, and H. Horikawa, Jpn. Kokai, 76,149,225, 22, Dec. 1976; Chem. Abstr., 87, p. 39838a (1977).
119. T. Nishitani, H. Horikawa, T. Iwasaki, K. Matsumoto, I. Inoue, and M. Miyoshi, J. Org. Chem., 47, 1706 (1982).
120. M. Mori, K. Kagechika, K. Tohjima, and M. Shibasaki, Tetrahedron Lett., 29, 1409 (1988); M. Mori, K. Kagechika, and M. Shibasaki, private letter, 1988.
121. T. Iwasaki, H. Horikawa, K. Matsumoto, and M. Miyoshi, J. Org. Chem., 42, 2419 (1977).
122. H. Horikawa, T. Iwasaki, K. Matsumoto, and M. Miyoshi, Tetrahedron Lett., 191 (1976).
123. I. Kitagawa, M. Yoshikawa, T. Kamigauchi, K. Shirakawa, and Y. Ikeda, Chem. Pharm. Bull., 29, 2571 (1981).
124. T. Inokuchi, H. Tsujiyama, M. Kusumoto, and S. Torii, Chem. Express, 3, 623 (1988).
125. T. Iwasaki, H. Horikawa, K. Matsumoto, and M. Miyoshi, Tetrahedron Lett., 4799 (1979).
126. T. Iwasaki, H. Horikawa, K. Matsumoto, and M. Miyoshi, J. Org. Chem., 44, 1552 (1979).
127. G. E. Hawders, J. H. P. Utley, and G. B. Yates, J. Chem. Soc., Chem. Commun., 305 (1973).
128. A. K. Yadav, S. C. Mishra, and R. A. Misra, Electrochim. Acta, 31, 507 (1986).
129. S. Torii, T. Okamoto, G. Tanida, H. Hino, and Y. Kitsuya, J. Org. Chem., 41, 166 (1976).
130. S. Torii and T. Okamoto, Bull. Chem. Soc. Jpn., 49, 771 (1976).
131. R. Brettle, M. A. Khan, and J. D. Rowbottom, J. Chem. Soc., Perkin Trans., I, 2927 (1981).
132. F. Barba and A. Gwrado, unpublished results, 1977.
133. D. Hermeling and H. J. Schäfer, Angew. Chem., Int. Ed. Engl., 23, 233 (1984).
134. E. J. Corey and R. R. Sauers, J. Amer. Chem. Soc., 81, 1743 (1959).
135. S. Torii, T. Okamoto, and T. Oida, J. Org. Chem., 43, 2294 (1978).
136. K. Matsumoto, H. Horikawa, T. Iwasaki, and M. Miyoshi, Chem. Ind., 920 (1978).
137. A. Diaz, J. Org. Chem., 42, 3949 (1977).
138. J. Slobbe, J. Chem. Soc., Chem. Commun., 82 (1977).
139. T. Kawamata, S. Inayama, and K. Sata, Chem. Pharm. Bull., 28, 277 (1980).
140. S. Inayama, T. Kawamata, and N. Shimizu, Chem. Pharm. Bull., 28, 282 (1980).
141. D. Lelandais and M. Chikir, C. R. Acad. Sci. Paris, Ser. C, 281, 731 (1975).
142. P. Radick, R. Klem, S. Spurlock, J. Sims, E. E. van Tammelen, and T. Witesides, Tetrahedron Lett., 5177 (1968).
143. H. Plieninger and W. Lehnert, Chem. Ber., 100, 2427 (1967).
144. H. H. Westberg, and H. J. Dauben, Tetrahedron Lett., 5123 (1968).

145. T. Iwasaki, H. Horikawa, K. Matsumoto, and M. Miyoshi, Tetrahedron Lett., 4799 (1978).
146. R. N. Zolotar and V. B. Avilov, Zh. Anal. Khim., 32, 1895 (1977).
147. T. Iwasaki, H. Horikawa, Y. Ozaki, and K. Matsumoto, Nippon Kagaku Kaishi, 1826 (1984); Chem. Abstr., 102, 131798p (1985).
148. J. M. Bobbitt and J. P. Willis, J. Org. Chem., 45, 1978 (1980).
149. J. M. Bobitt, and T. Y. Cheng, J. Org. Chem., 41, 443 (1976).
150. E. J. Corey, N. L. Bauld, R. T. LaLonde, J. Casanoba, Jr., and E. T. Kaiser, J. Amer. Chem. Soc., 82, 2645 (1960).
151. W. J. Koehl, J. Amer. Chem. Soc., 86, 4686 (1964).
152. S. Torii and T. Inokuchi, unpublished work.
153. D. Lelandais, C. Bacquet, and J. Einhorn, J. Chem. Soc., Chem. Commun., 194 (1978).
154. D. Lelandais, C. Bacquet, and J. Einhorn, Tetrahedron, 37, 3131 (1981).
155. R. Michaelis, U. Muller, and H. J. Schäfer, Angew. Chem., Int. Ed. Engl., 26, 1026 (1987).
156. P. G. M. Wuts and M.-C. Cheng, J. Org. Chem., 51, 2844 (1986).
157. T. Shono, T. Hayashi, H. Omoto, and M. Matsumura, Tetrahedron Lett., 2667 (1977).
158. G. E. Gream and C. F. Pincombe, Aust. J. Chem., 27, 589 (1974).
159. S. Torii, H. Tanaka, and T. Mandai, J. Org. Chem., 40, 2221 (1975).
160. T. Imagawa, N. Murai, T. Akiyama, and M. Kawanisi, Tetrahedron Lett., 1691 (1979).
161. T. Akiyama, T. Fujii, H. Ishiwari, T. Imagawa, and M. Kawanisi, Tetrahedron Lett., 1978, 2165.
162. T. Imagawa, T. Sonobe, H. Ishiwari, T. Akiyama, and M. Kawanisi, J. Org. Chem., 45, 2005 (1980).
163. T. Imagawa, S. Sugita, T. Akiyama, and M. Kawanishi, Tetrahedron Lett., 22, 2569 (1981).
164. P. S. Wharton, G. A. Hiegel, and R. V. Coombs, J. Org. Chem., 28, 3217 (1963).
165. K. Doi, H. Tsuruta, and T. Yoshida, Jpn. Kokai, 77 27,739, March 02, 1977; Chem. Abstr., 87, P 102065p (1977).
166. H. Tanaka, Y. Kobayashi, and S. Torii, J. Org. Chem., 41, 3482 (1976).
167. T. Iwasaki, T. Nishitani, H. Horikawa, and I. Inoue, J. Org. Chem., 47, 3799 (1982).
168. A. Takeda, S. Wada, and Y. Murakami, Bull. Chem. Soc. Jpn., 44, 2729 (1971).
169. S. Torii, H. Tanaka, T. Fukuoka, and S. Hirai, Bull. Chem. Soc. Jpn., 46, 2534 (1973).

15
Nitrogen-Containing Compounds

EBERHARD STECKHAN / Institut für Organische Chemie und Biochemie der Universität Bonn, Bonn, Federal Republic of Germany

I. Introduction 581

II. Aliphatic Amines 582
- A. Primary amines 582
- B. Secondary and tertiary amines 583

III. Aromatic Amines 586
- A. Aniline derivatives 587
- B. Diphenylamines 590
- C. Triphenylamines 592
- D. Phenylenediamine and aminophenols 594

IV. Imines, Oximes, Hydrazones, and Semicarbazones 594

V. Enamines, Enamino Esters and Ketones, Enamides, and Enecarbamates 597

VI. N-Heterosubstituted Compounds 599

VII. Amino Alcohols and Amino Acids 600

VIII. Amides, Carbamates, Lactams, and Related Compounds 601
- A. Anodic oxidation of amides, carbamates, and lactams 602
- B. Anodic oxidation of N-acyl and N-carboalkoxy α-amino acid esters 604

References 607

I. INTRODUCTION

Nitrogen-containing compounds having lone-pair electrons are usually relatively easy to oxidize. The anodic behavior of such compounds has therefore been studied quite intensively. However, the ease of oxidation is not always paralleled by the selectivity of the reaction. For example, electrochemical oxidation of aliphatic and aromatic amines in many cases leads to a variety of products, but the anodic oxidation of amides or carbamates results in the formation of only one major product. Therefore the synthetic application of amine oxidation is found more seldom than that of amide and carbamate oxidation.

II. ALIPHATIC AMINES

The oxidation potentials of aliphatic amines depend on their structure. Primary amines (E_p^{ox} between 1.6 and 1.7 V versus NHE) are more difficult to oxidize than secondary or tertiary amines (E_p^{ox} between 1.3 and 1.0 V versus NHE). The electrochemical oxidation of simple aliphatic amines is quite complex and may lead to a variety of products. Mechanistic scheme 1, based on studies by Mann and others [1-4], gives a general survey of the possible reaction pathways. In the first step the amine (I) is oxidized to the cation radical (II), which may decompose by C–N bond cleavage to produce a relatively stable carbenium ion (IV), which may solvolyse, add to a nucleophile, or polymerize. Usually, however, II is deprotonated and further oxidized to give the iminium ion (V). V can hydrolyze to give the aldehyde (VI) and amine or ammonia (VII), or it adds to a nucleophile, giving substitution product (VIII). Under basic conditions using certain metal oxide electrodes V (R = H) may be further deprotonated and oxidized finally to yield the nitrile (IX).

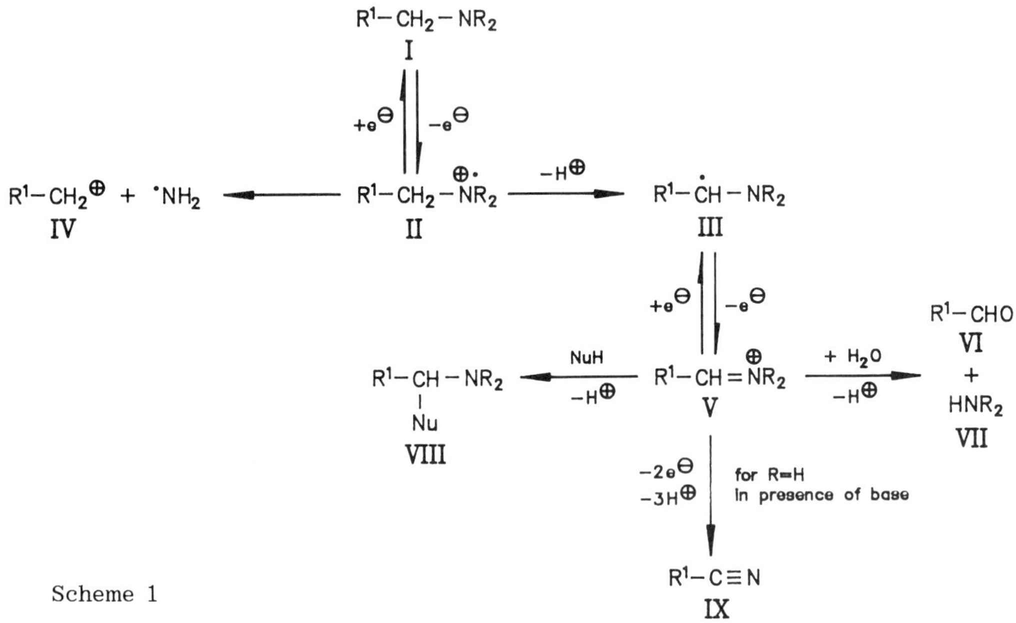

Scheme 1

Recently, however, Nelsen and Ippoliti [5] pointed out that the usually assumed extremely high acidity of the cation radical (II), at least in the case of trialkylamines, does not exist so that the cation radicals do not usually decompose by deprotonation if bases no stronger than R_3N are the only ones present. Instead, II (R = alkyl) shows exceptionally weak C(α)–H bonds. Therefore hydrogen atom abstraction from II leading to V may be the first step in its decomposition.

A. Primary Amines

Anodic oxidation of primary amines at platinum or carbon electrodes is not of significant preparative interest. The common products in acetonitrile are the aldehydes (VI), NH_4^+, nitrogen from the oxidation of ammonia, protonated starting material, and hydrocarbons [1,2]. Valuable, however, is the anodic oxidation of primary amines in basic aqueous media at silver oxide, copper oxide, or, more effectively, at nickel hydroxide electrodes to form the nitriles (IX) in good to excellent yields [6-8]. In this case the primary step in the reaction mechanism is hydrogen atom abstraction at C(α) by the intermediately formed nickel oxide hydroxide, which is formed at about 0.63 V (NHE).

$$NH_2-CH_2-(CH_2)_8-CH_2-NH_2 \xrightarrow[\substack{0.1\ N\ KOH \\ \text{in } \underline{t}\text{-butanol-water (1:1)}}]{NiOOH,\ -e^-} N{\equiv}C-(CH_2)_8-C{\equiv}N \quad (90\%)$$

α-Branched primary amines are oxidized to ketones presumably via hydrolysis of the intermediate imines. Alanine is cleaved to acetonitrile; 6-aminohexanoic acid yields the cyanocarboxylic acid in excellent yield.

Primary amines may be oxidized by a two-electron process to give hydrazines, which generally are further oxidized to azo compounds. As the oxidation potential of the primary amines is higher than that of the hydrazine, the latter very often is largely decomposed during the reaction. The amide ion, however, which may be obtained in the presence of a base, is easier to oxidize than the amine so that under these conditions azo compounds are generated in moderate yields [9].

$$n\text{-}BuNH_2 \xrightarrow[LiClO_4;\ n\text{-}BuLi]{Pt\ anode;\ THF} n\text{-}Bu-N{=}N-n\text{-}Bu \quad (33\%)$$

<u>tert</u>-Butylamine is coupled to the azo compound (41%) also in the absence of a base using an aqueous electrolyte [10], because in the absence of α-hydrogens further oxidative degradation is hindered.

B. Secondary and Tertiary Amines

In acetonitrile secondary and tertiary aliphatic amines are degraded according to scheme 1 to the aldehydes (VI), the lower amines (VII), and protonated starting material. The lower amines may be further degraded along the same pathways [1,4]. The anodic dealkylation of aliphatic tertiary amines was also studied in aqueous alkaline solution [11]. There it was found that the relative amount of dealkylation in unsymmetrical amines is predominantly governed by the number of α-protons and their acidity. A synthetically very useful reaction sequence would be the anodic oxidation of I via the iminium ion (V) to the substitution product (VIII). The addition of nucleophiles to V in the case of purely aliphatic amines is, however, possible only in a few cases. For example cyanide ion may be used as nucleophile, giving α-cyanation products in moderate yields [12].

$$(CH_3-\underset{\substack{|\\CH_3}}{CH})_2N-CH_3 \xrightarrow[NaCN]{anode;\ MeOH-H_2O\ (1:1)} (CH_3-\underset{\substack{|\\CH_3}}{CH})_2N-CH_2-CN \quad (55\%)$$

In the case of tribenzyl and trimethylamine, α-substitution products could be obtained using diethylmalonate or diethylphosphonate as nucleophiles, if 2,4,6-collidine is used as base in an acetonitrile electrolyte [13]. In this way dialkylaminomethyl phosphonates and malonates could be obtained in good yield.

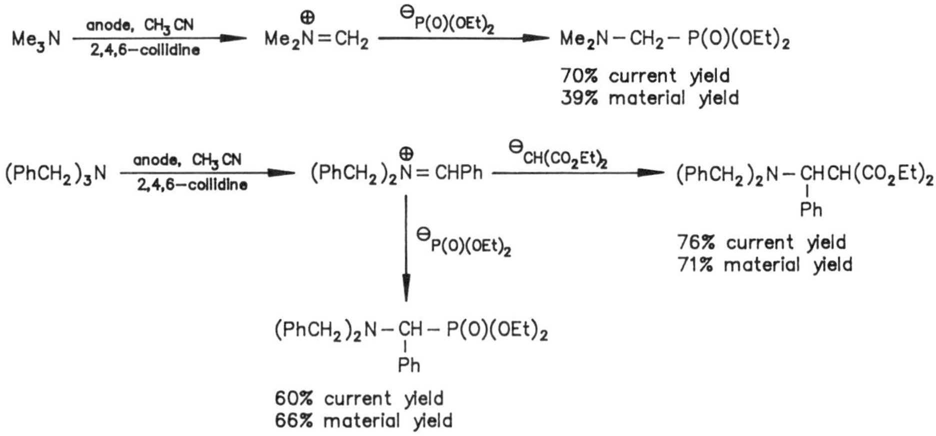

If methanol is used as nucleophile, the substitution products are usually not stable and decompose to the aldehyde and lower amine. In the case of N,N-dimethylbenzylamine (X), however, two isomeric substitution products (XI and XII) could be obtained in a ratio of 4:1 [14,15].

$$\underset{\text{X}}{\text{Ph-CH}_2\text{-N(CH}_3)_2} \xrightarrow[\text{MeOH-KOH}]{\text{Anode, } -e^-} \underset{\text{XI}}{\text{Ph-CH}_2\text{-N(CH}_3)\text{-CH}_2\text{OCH}_3} + \underset{\text{XII}}{\text{Ph-CH(OCH}_3)\text{-N(CH}_3)_2}$$

The reason for the unexpected predominant substitution at the methyl group instead of the benzyl group was a matter of discussion [16-18].

Shono et al. [19] suggested that the microsomal oxidation of tertiary amines and amides occurs via the same intermediates as the anodic oxidation in methanol solution. The dealkylation of N-methyl-N,N-dialkyl amines by both methods results in almost the same N-demethylation–N-dealkylation ratio.

Lithium N-alkyl-N-4-pentenylamide ions obtained from secondary pentenylamines may be anodically oxidized to form aminyl radicals that undergo cyclization to form cis-N-alkyl-1-methylpyrrolidines [20]. Otherwise lithium amides of secondary amines undergo anodic dimerization to form hydrazines in moderate yields [9]. Hydrazines are also generated if secondary amines are electrochemically oxidized in the presence of an alkali hydroxide [21,22].

Aziridines, which may be regarded as cyclic secondary or tertiary amines containing a β-amino substituent, undergo ring cleavage by anodic oxidation. Thus in the electrochemical oxidation at a platinum anode in a methanol solution 2-ethyl-2-phenylaziridine (XIII) forms propiophenone dimethylacetal (XVI, 50%), propiophenone (XVII, 6%), and XVIII (12%) [23]. The reaction is supposed to proceed via the intermediate azaallyl cation (XIV), leading to imine (XV). This is supported by the fact that XV can be trapped by LAH reduction to give N-methyl-N-(1-phenylpropyl)amine. Hydrolysis of XV may lead to propiophenone, and its further two electron oxidation and deprotonation forms XVIII. XVI may be generated either by proton-catalyzed acetalization of propiophenone or by oxidation of XV.

At a potential of 1.1-1.3 V versus Ag/Ag^+ (0.1 M) N-benzylaziridine (XIX) is opened to form a cyclic tetramer (XX), which may be explained by an electron transfer chain mechanism, Eqs. (1) through (5) [24]. The same reaction may also be performed by redox catalysis using tris-(4-bromophenyl)amine as mediator at a potential of about 0.75 V versus Ag/Ag^+. The possibility of catalysis by electrogenerated acid was excluded.

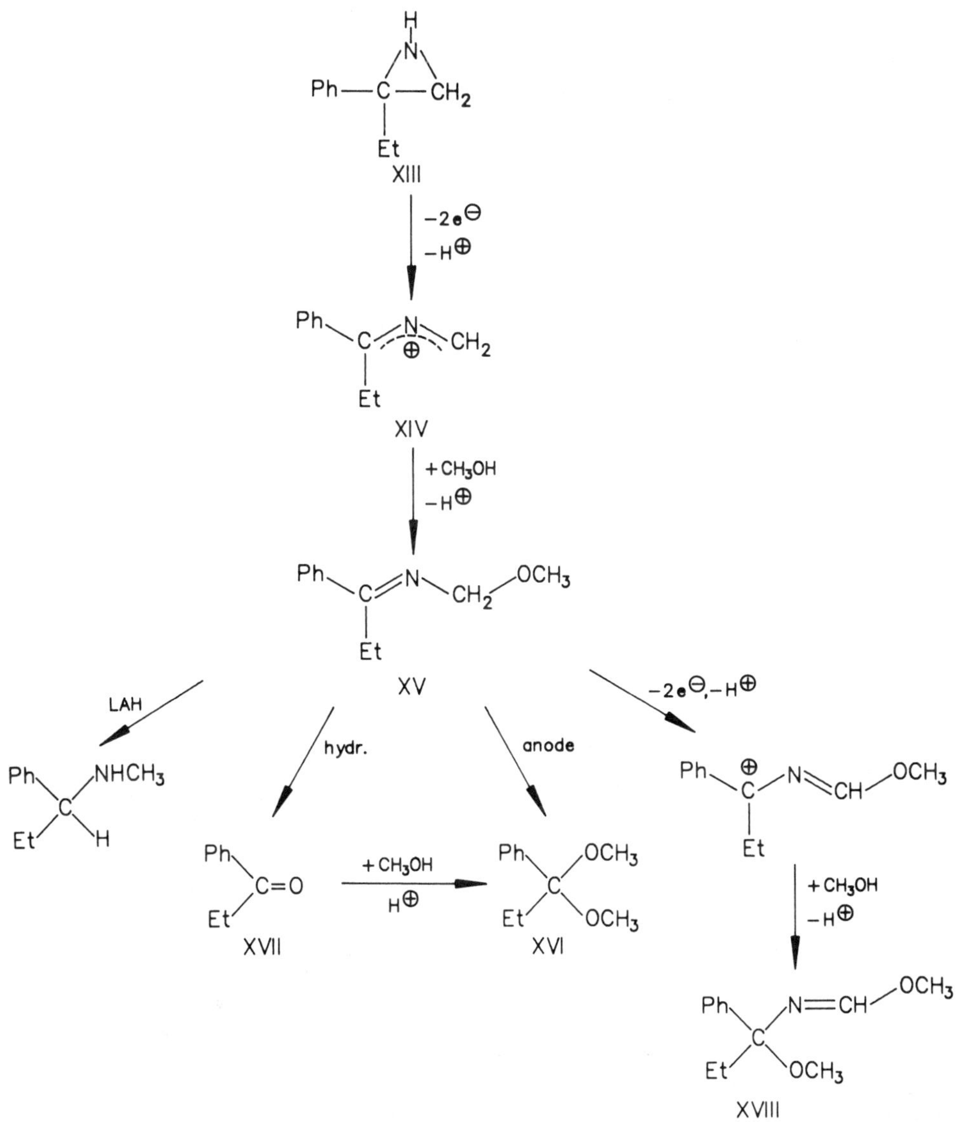

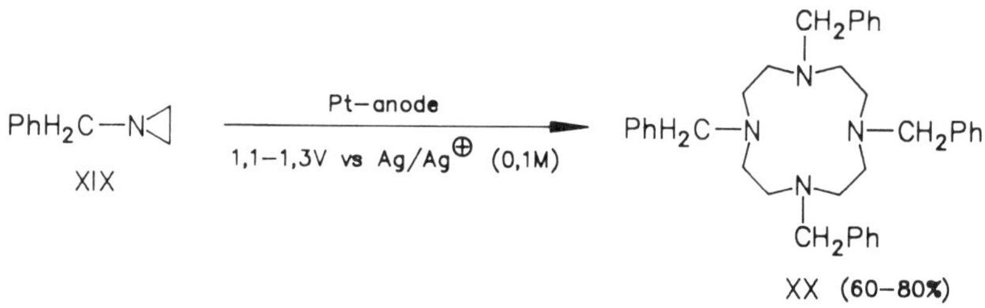

$$XIX \xrightarrow{-e^-} XIX^{+\cdot} \quad (1)$$

$$XIX^{+\cdot} \xrightarrow{\text{fast}} XIX^{+\cdot}_{open} \quad (2)$$

$$XIX^{+\cdot}_{open} \xrightarrow{+XIX} \xrightarrow{+XIX} \xrightarrow{+XIX} XX^{+\cdot}_{open} \quad (3)$$

$$XX^{+\cdot}_{open} \xrightarrow{\text{ring closure}} XX^{+\cdot} \quad (4)$$

$$XX^{+\cdot} + XIX \longrightarrow XX + XIX^{+\cdot} \quad (5)$$

Indirect electrooxidation of cyclic aziridines using NaCl or NaBr as a redox catalyst resutls in C(1)–C(2) bond cleavage with formation of the corresponding keto nitriles. This reaction is explained by the intermediate generation of an azaallenyl cation, which is hydrated to the α-hydroxy imine. Further oxidation by "Cl^+" then would lead to the open-chain keto N-chloroimine, which by HCl elimination forms the keto nitrile; its hydrolysis leads to the keto aldehyde as a side product [25].

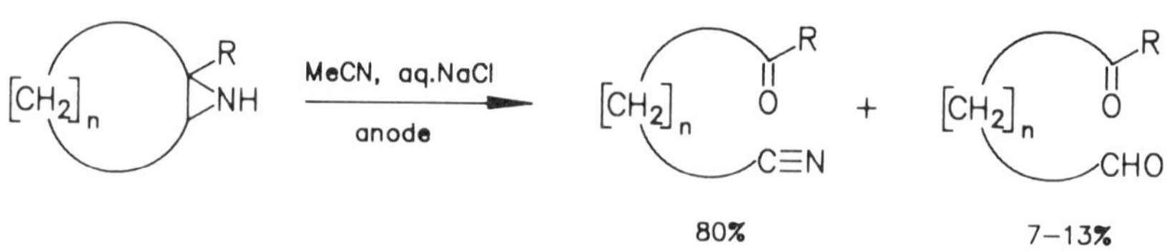

III. AROMATIC AMINES

Aromatic amines are usually very easy to oxidize. Unsubstituted, N-alkyl-, and N,N-dialkyl anilines show anodic potentials in the range of 0.4-1.0 V versus NHE. Diphenylamines cover a broader spectrum of potentials depending on their substitution pattern. Triphenylamines have oxidation potentials between 0.7 and 2.00 V versus NHE. A large number of potentials for aromatic amines has been tabulated [26]. The anodic oxidation of aromatic amines is quite complex, leading to a variety of products depending on their structure and the electrolysis conditions. Products may vary largely by using aqueous or nonaqueous, neutral, acidic, or basic conditions. In general, the anodic oxidation of tertiary aromatic amines shows a higher selectivity in product formation, thus being more valuable for synthetic purposes. The anodic oxidation of aromatic amines has been reviewed several times in detail [27-29]. Therefore mainly typical reaction patterns and more recent developments are discussed here.

A. Aniline Derivatives

Aniline and its derivatives may form three types of products (scheme 2). If no para substituent is present, tail-to-tail coupling forms benzidine derivatives (XXI); anilines and N-alkylanilines also can undergo head-to-tail coupling to give 4-aminodiphenylamines (XXII) even in the presence of certain para substituents, one of which is eliminated; head-to-head coupling forming azo compounds (XXIII) is possible with N-unsubstituted anilines.

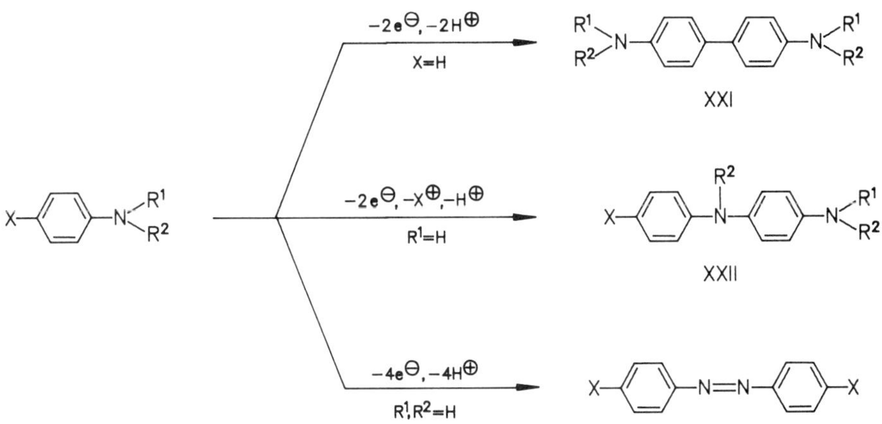

Scheme 2

Primary Aniline Derivatives and N-Alkylanilines

A generalized reaction for anilines and N-alkylanilines based on results by Hand and Nelson [30] is shown in scheme 3.

The primary intermediate formed by one-electron abstraction is the cation radical XXIV, which can undergo either dimerization or deprotonation. Dimerization and subsequent deprotonation leads to the benzidine (XXI), which is easier to oxidize than the starting material. Thus in a reversible two-electron oxidation the stable diimine species XXV is formed. Dimerization is favored at high current densities, leading to high cation radical concentrations and low concentrations of the parent molecule at the electrode surface. At low current densities, high concentrations of starting material, and in the presence of a strong enough base, deprotonation is favored to form a resonance stabilized radical. This undergoes C−N coupling with a parent molecule, which after loss of an electron and a proton leads to the 4-aminodi-phenylamine (XXII). This compound is easily oxidized to the diimine form, which may be hydrolyzed to give para-benzoquinone and the parent amine. The influence of the reaction parameters on the results of electrolyses of aniline and N-alkylanilines is given in Table 1 [30].

The formation of p-aminodiphenylamine is supposed to be the key intermediate in the formation of a dark green precipitate at the electrode surface during continued electrolysis of acidic aniline solutions. This has been characterized as an oligomer of aniline, for example as the octamer emeraldine

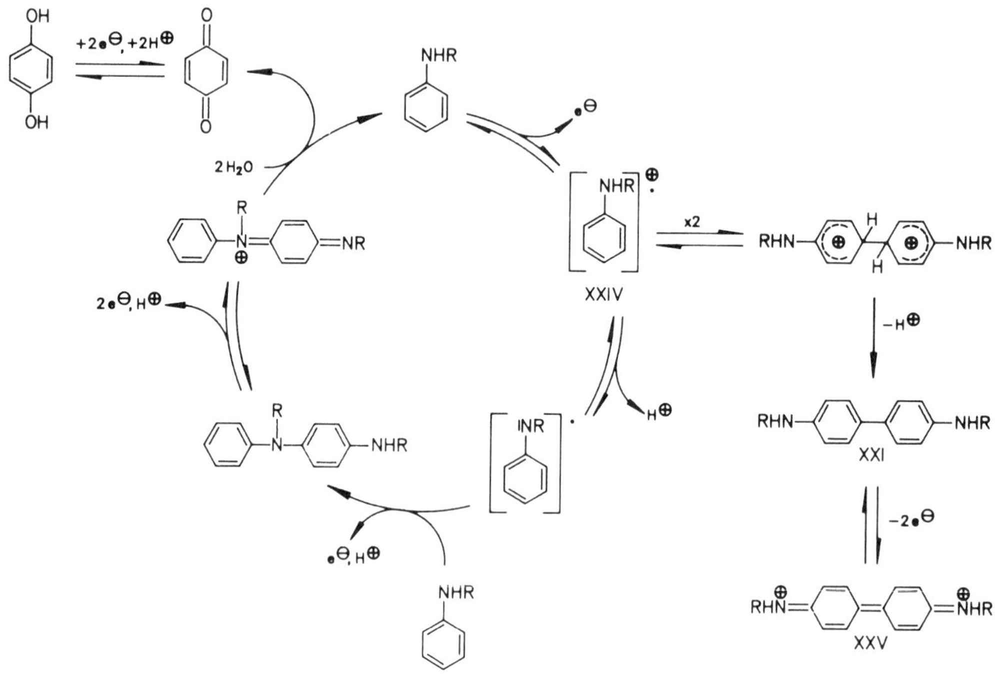

Scheme 3

Table 1. Influence of Electrolysis Conditions on Product Distribution for Anodic Oxidation of Aniline and N-Alkylanilines[a]

Compound	Electrolysis conditions	Products (%)[b]
Aniline	MeCN-TEAP, low current density	4-Aminodiphenylamine (40)
Aniline	0.05 M H_2SO_4, high or low current density	p-Benzoquinone (90-100)
Aniline	6 M H_2SO_4, high current density	p-Benzoquinone (80-90) Benzidine (10-20)
N-ethylaniline	MeCN-TEAP, low current density, low parent concentration	N,N'-diethylbenzidine (60-70)
N-ethylaniline	MeCN-TEAP, high current density	N,N'-diethylbenzidine (70-80)
N-ethylaniline	MeCN-TEAP, low current density, high parent concentration	N-ethyl-4-ethylaminodiphenyl-amine (70) N,N' diethylbenzidine (15)
N-ethylaniline	6 M H_2SO_4, high current density	p-Benzoquinone (40) N,N'-diethylbenzidine (60)
N-tert-butyl-aniline	MeCN-TEAP, high current density	N,N'-di-tert-butylbenzidine (100)

[a]Controlled potential electrolyses.
[b]Based on consumed parent compound.
Source: From Ref. 30.

formed by a cascade of head-to-tail condensations [31,32]. Hand and Nelson, however, explained it as a mixture of mainly quinhydrone with a small amount of benzidine salt [30]. Today the electropolymerization of aniline under strongly acidic conditions is one intensively studied way to form the conducting polymer poly-aniline (see Chaps. 32 and 33) [33].

Para-substituted anilines undergo head-to-head coupling in acetonitrile in the presence of pyridine [34] to give the azo compounds (XXIII). Hydrazo compounds are supposed to be the intermediates. Under acidic conditions anilines with para substituents, like methoxy, ethoxy, chloro, or carboxy groups, react via head-to-tail coupling giving the corresponding 4'-substituted p-amino diphenylamines (XXII) in the partially oxidized form (XXVI):

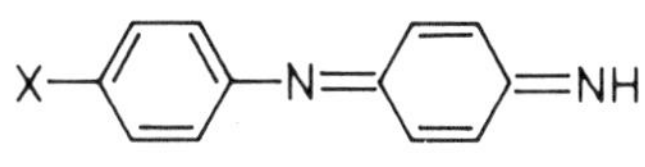

XXVI

The leaving group is supposed to be the anion X^-, thus forming the product already in an oxidized state [35]. para-Aminophenols are intensively studied. They react by a simple EC mechanism to form the quinone imine, which readily undergoes hydrolysis to p-benzoquinone [36].

N,N-*Dialkylanilines*

The most thoroughly studied compound of this class is N,N-dimethylaniline. The details of the anodic oxidation of N,N-dialkylanilines are well documented [37]. If the para position of the aromatic ring is unsubstituted, tail-to-tail coupling via the intermediate cation radical with formation of the benzidines (scheme 2, XXI) dominates in acetonitrile or under acidic conditions. Benzidine, which is easier to oxidize than the starting material, is further oxidized to the corresponding quinoidal diimino dication.

In contrast, nucleophilic substitution in α-position to nitrogen by cyanide or methoxide predominates if the electrolysis is carried out in acetonitrile in presence of cyanide [38] or methoxide, respectively, in basic methanol [39,40].

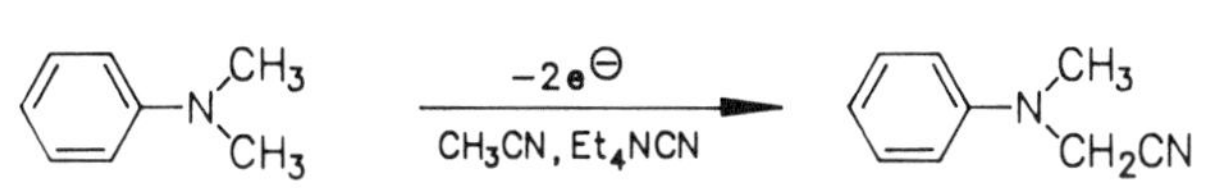

19% cy

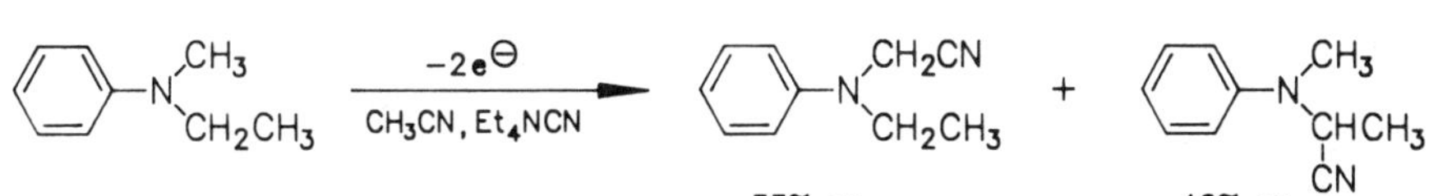

33% cy 16% cy

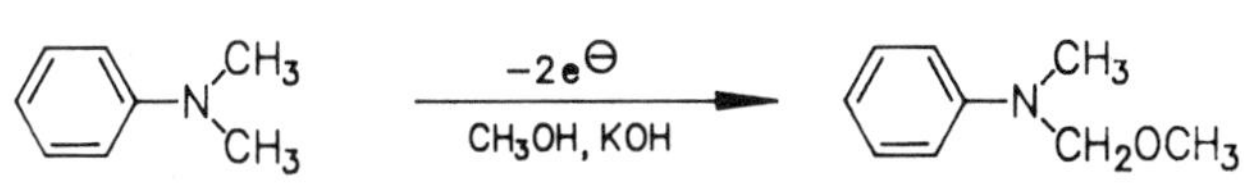

73%

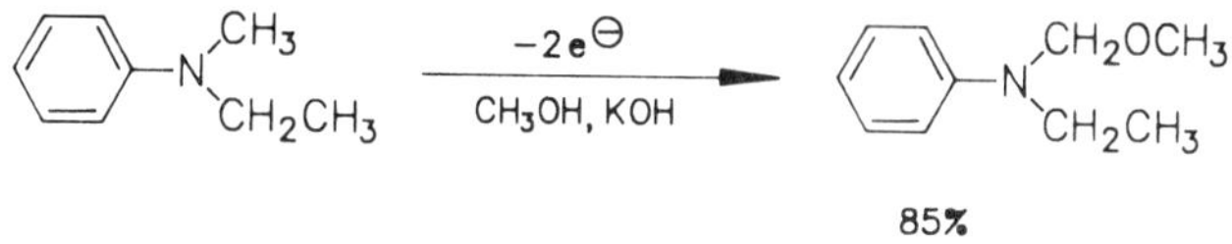

This reaction, which is supposed to occur via the iminium ion, has interesting application in organic synthesis. Thus, the α-methoxy group may be eliminated in the presence of a Lewis acid, regenerating the iminium ion, which can be trapped in situ by nucleophiles. If electron-rich alkenes, such as trimethylsilyl enol ethers, are used as C-nucleophiles in presence of $TiCl_4$, substituted tetrahydroquinolines are obtained in good to moderate yields [40].

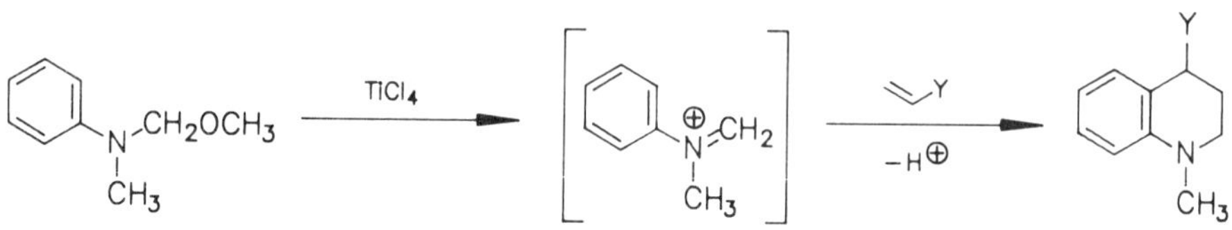

Further anodic oxidation of α-methoxy-N,N-dimethylaniline in methanol leads to the formation of α,α'-dimethoxy-N,N-dimethylaniline, which forms julolidine derivatives by reaction with two molecules of vinyl ether in presence of $TiCl_4$ [40].

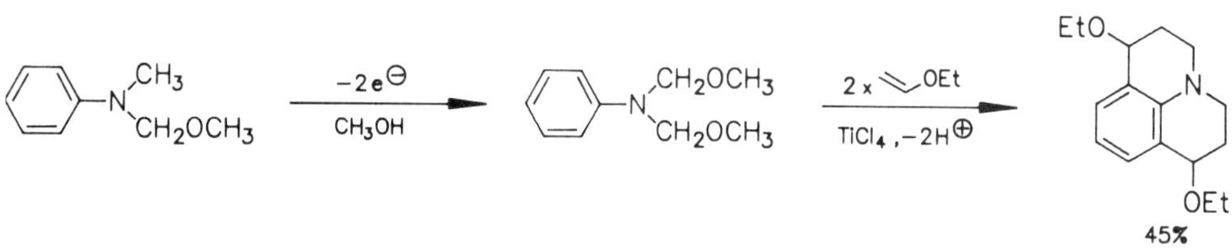

The reaction of ring-substituted N,N-dialkylanilines depends on the type and position of the substituents. With halogen substituents in the 2- or 3-position the corresponding substituted benzidines are formed as usual [27]. Bromo, iodo, and carboxy substituents in the 4-position are eliminated to give the N,N,N,N-tetraalkylbenzidine [27,41]. 4-Alkoxy substituents result in the formation of 4-dimethylaminophenol if the reaction is carried out in acetonitrile [42]. In aqueous solution, 4-hydroxy-N,N-dimethylaniline is oxidized to the quinone imine or its hydrolysis products p-benzoquinone and dimethylamine [43]. A methyl substituent in the p-position as in N,N-dimethyl-p-toluidine leads to di(p-dimethylaminophenyl)-1,2-ethane [42,44] by coupling via the p-methyl group. Similar to tribenzyl and trimethyl amine (see Sec. II.A), α-substitution can be obtained with diethylphosphonate in presence of collidine, giving (N-methyl,N-p-tolylaminomethyl)diethylphosphonate [13].

B. Diphenylamines

The major pathways for the anodic oxidation of diphenylamines are N,N or aryl,aryl coupling, N,aryl coupling, and nucleophilic substitution in the aromatic ring. Which pathway is actually followed depends on the reaction conditions and the substitution pattern of the substrate. With regard to the latter one must distinguish between p-unsubstituted p-mono- and p-disubstituted diphenylamines.

Diphenylamine has been oxidized under various conditions (scheme 4). In acetonitrile tail-to-tail coupling via the cation radical, forming N,N'-diphenylbenzidine, is favored (pathway a) [45-47]. In the presence of

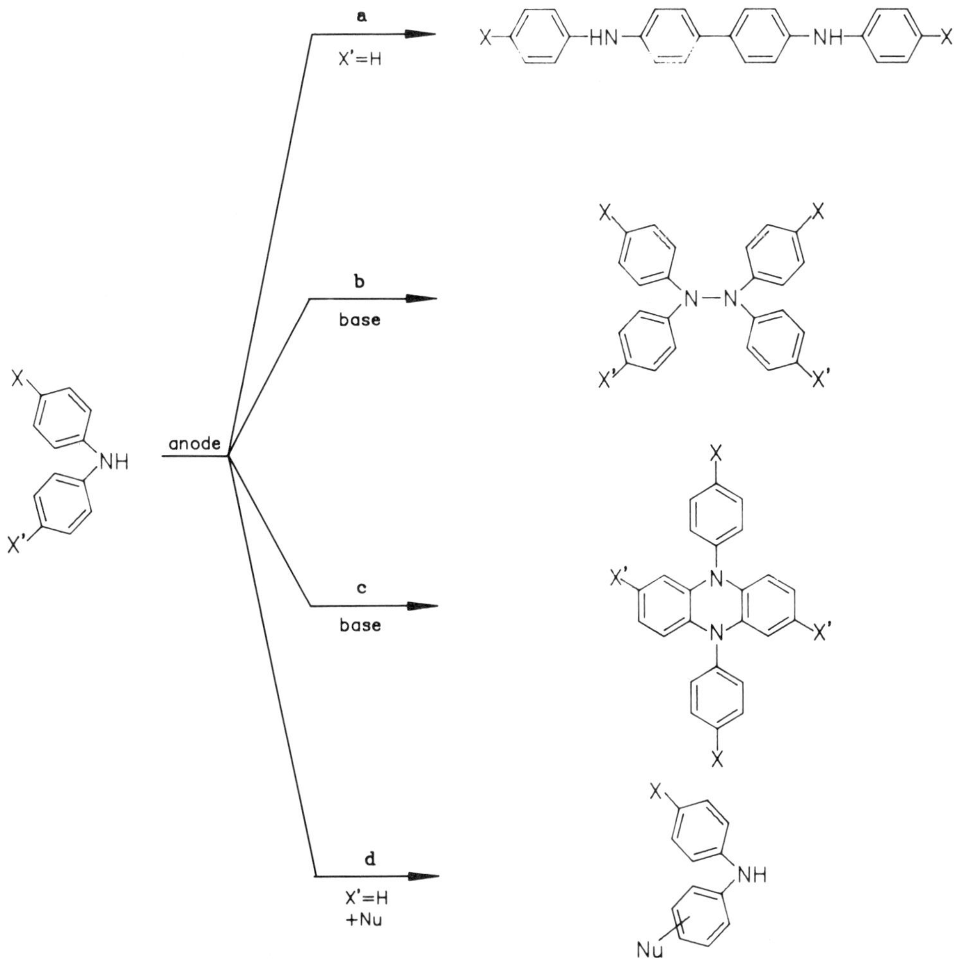

Scheme 4

cyanide or acetate acting as a base, tetraphenylhydrazine is formed in high yield via coupling of the diphenylaminyl radicals (pathway b) [47]. These radicals may be formed either by deprotonation of the initial cation radical or by deprotonation of diphenylamine to the amide ion and subsequent oxidation to the N-radical.

The reaction of p-mono- and p-disubstituted diphenylamines has carefully been studied by Serve and colleagues [46-50]. In acetonitrile media p-monosubstituted diphenylamines may form the following products: N,N'-diarylbenzidines via aryl,aryl coupling of the cation radicals (pathway a),

5,10-diaryl-5,10-dihydrophenazines and N,N,N'-triaryl-p-phenylenediamines via N,aryl coupling of a diarylnitrenium ion with the neutral substrate (pathway c), or tetraarylhydrazines via N,N coupling of the aminyl radicals (pathway b) [47]. The last two compounds are observed only in basic media. p,p'-Disubstituted diphenylamines form quite stable cation radicals during anodic oxidation in acetonitrile under neutral conditions, but in the presence of pyridine-like bases the formation of diaryldihydrophenazine compounds (pathway c) is favored. In presence of stronger bases tetraarylhydrazines are generated [48-50]. The cyclization to the phenazine has also been studied by Berkenkotter and Nelson [51], who identified the o-phenylenediamine as intermediate in this reaction.

Aromatic substitution reactions are also possible with diphenylamines (pathway d). In methanol solution containing sodium cyanide, diphenylamine is electrolyzed, giving 4-cyanodiphenylamine in 61% yield [52].

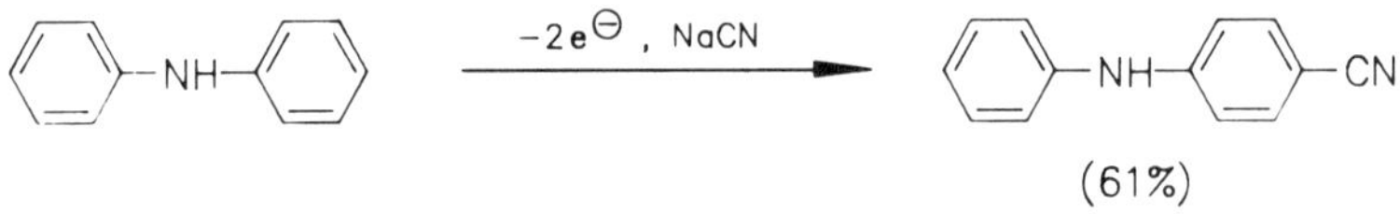

Similarly, a number of other substituted diphenylamines have been cyanated [52]. In methanol containing methoxide, diphenylamine could be methoxylated in the o- or p-position only in low yield [53]. In the presence of water p,p'-disubstituted diphenylamines containing at least one p-methoxy- or p-hydroxy substituent are oxidized to the p-benzoquinoneimine [48,49, 54]. By anodic oxidation of polysubstituted diphenylamines in acetonitrile containing TEACN, cyanated N-aryl-cyclohexadieneimines are obtained [49, 55].

Tertiary diphenylamines like N-ethyl-di-p-substituted diphenylamines can be cyclized anodically in acetonitrile to give the corresponding carbazoles. This reaction occurs via the intermediate dication of the diphenylamine [56].

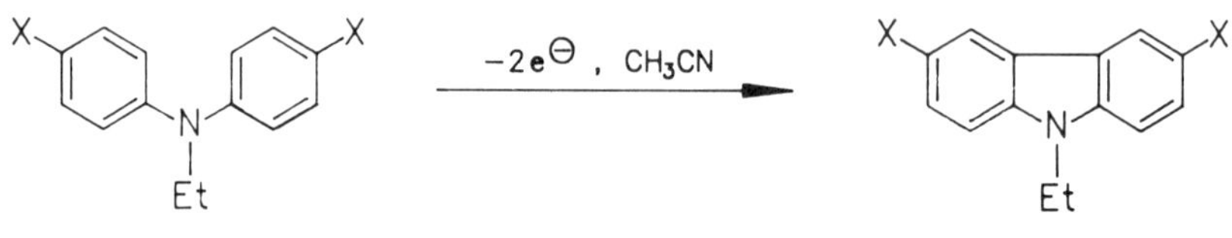

C. Triphenylamines

The electrochemical oxidation of triphenylamines has been studied in detail [57-59]. With partially p-substituted compounds tetraphenylbenzidines are formed. If all p-positions are blocked, very stable cation radicals are formed in acetonitrile and sometimes even in nucleophilic media. If the oxidation is performed to the dication stage, cyclization to the carbazoles is observed in several cases, as in the case of N-alkyl-p,p'-disubstituted diphenylamines [56]. As the cation radicals of triaryl amines are very stable, if all para positions of the aryl rings are blocked toward the attack of nucleophiles, they can be applied as organic redox catalysts for the indirect electrochemical oxidation (see Chap. 29). This was first demonstrated in the oxidation of cyanide ions by electrogenerated trianisylamine cation radicals [60]. The advantage of triarylamines as mediators is the possibility of varying their oxidation potentials over a wide range by proper selection of the ortho and para substituents. Now about 40 different tri-p-substituted triarylamines covering a potential range from about 0.8 to 2.0 V versus NHE are known [58,61-64]. Quite a number of them have been applied as redox catalysts in numerous indirect electrochemical reactions [61,62], allowing oxidations at quite low potentials and with interesting selectivities. These include deprotection of alcohols [65] and carboxylic acids [66] from their substituted benzyl ethers and esters, cleavage of carbon-sulfur [67] bonds, side-chain oxidations of alkyl aromatic compounds [68], benzyl alcohol oxidations [69], and α-methoxylations of aliphatic ethers and acetals [70]. The stability of the triarylamine cation radicals in redox catalytic reactions strongly depends on their substitution pattern and the reaction medium, and also on the overall rate of the catalytic reaction. If the catalytic cycle is slow, higher stationary concentrations of the cation radicals favor side reactions. For example, 4-bromo- and 4-iodo-substituted cation radicals can undergo 4,4' coupling under loss of the halo substituents, leading to the benzidine cation or dication radical, respectively [71]. This type of debromodimerization has been observed with tris(4-bromophenyl)ammoniumyl in acetonitrile at 0°C [72]. Similarly, tris-(2,4-dibromophenyl) amine undergoes successive slow exchange of the p-bromo against p-chloro substituents, if it is oxidized in methanol-dichloromethane in an undivided cell. This reaction may proceed as a nucleophilic ipso substitution in the para position of the cation radical [73] in which the cathodically formed Cl^- (by reduction of CH_2Cl_2) acts as the nucleophile.

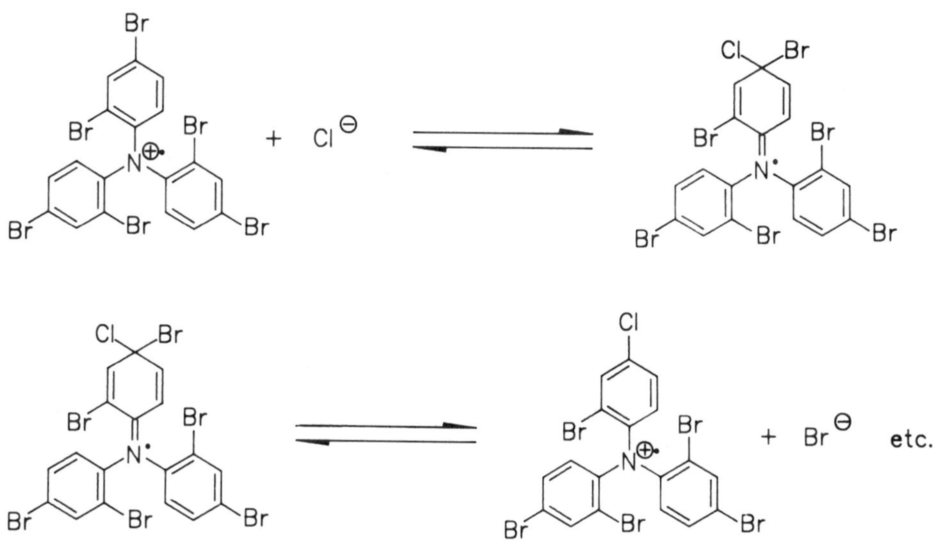

In the absence of dichloromethane the radical cations of tris-(2,4-dibromophenyl)amine can undergo several thousand turnovers without significant loss of material. This is different in the case of tris(4-bromophenyl)amine, which was found to form (2-methoxy-4-bromophenyl)-bis-(4-bromophenyl)-amine by anodic nuclear substitution, if used in pure methanol [74]. In the presence of acetate the corresponding acetoxy compound has been isolated [72]. It is interesting to note that in the case of bromo-substituted triphenylamines the electronic and steric effects of the *o*- and *p*-substituents could be discriminated by voltammetric studies. In addition, a linear relationship between the redox potentials of these amines and the long-wavelength UV/VIS absorbance maxima of the cation radicals is observed [63]. The correlation of the long-wavelength absorbance maxima of a large variety of *p*-substituted triarylamines with the Hammett σ-constants was unsuccessful and was called non-Hammett-behavior by Walter [75]. A very good linear relationship is, however, obtained if the redox potentials of the *p*-substituted triarylamines are correlated with σ/σ^+ values (for electron-donating groups σ^+ values must be used). To obtain a value for the reaction constant ρ, the redox potentials must be converted into a ratio of equilibrium constants. This has been done by using Eq. (6), which was derived from the Nernst equation:

$$\log \frac{K}{K_0} = \frac{1}{0.059}\,(E^\circ - E) = \sigma\rho \tag{6}$$

As a standard, triphenylamine with $E^\circ = 1.17$ V was used. In this way for ρ a value of -11.28 with a correlation coefficient of 0.993 was obtained [76]. The strongly negative value of ρ demonstrates the high senditivity of the cation radicals toward the electronic effects of the *p*-substituents.

Voltammetric studies of macrobicyclic host molecules bearing triphenylamine units as trifunctional spacers demonstrate the influence of steric hindrance and strain upon the redox potentials [77,78].

D. Phenylenediamine and Aminophenols

The anodic oxidation of phenylenediamines has been reviewed by Adams [79]. If unsubstituted at the nitrogens the two-electron oxidation leads to the quinone diimine. This compound undergoes either hydrolysis to the quinone imine and benzoquinone, or a 1,4-addition of a nucleophile, for example the parent phenylenediamine itself, to the quinoidal systems occurs. Further oxidation of the products may take place. In acetonitrile, the one-electron oxidation to the cation radical predominates [80]. Under these conditions *p*-phenylenediamine also leads to 1,4-coupling products [81,82]. *N*-substituted phenylenediamines form more stable cation radicals. For example, tetrakis-(*p*-bromophenyl)-*p*-phenylenediamine ($E^\circ = 0.91$ V versus NHE) and tetrakis-(2,4-dibromophenyl)-*p*-phenylenediamine ($E^\circ = 0.94$ V versus NHE) in acetonitrile even show reversible behavior for the second oxidation step to the dication [83].

Similarly to *p*-phenylenediamine, *p*-aminophenol undergoes two-electron oxidation to the quinone imine followed by hydrolysis to *p*-benzoquinone [84].

IV. IMINES, OXIMES, HYDRAZONES, AND SEMICARBAZONES

In the anodic oxidation of imines addition of a nucleophile to the carbon of the C=N double bond usually takes place. If the nucleophile is water, hydrolytic cleavage occurs. For example, controlled potential oxidation of *N*-benzylidene-*p*-anisidines in acetonitrile at the first voltammetric wave yielded the protonated anisidine. At the second wave benzaldehyde and protonated *p*-benzoquinone imine are formed by attack of residual water as nucleophile [85].

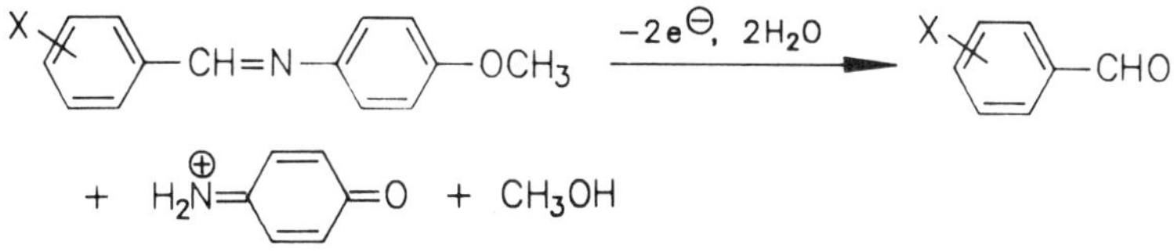

Instead, in the presence of pyridine, the oxidation at the first wave resulted in the pyridination not at the α-carbon but in the ortho position of the aromatic ring [86,87].

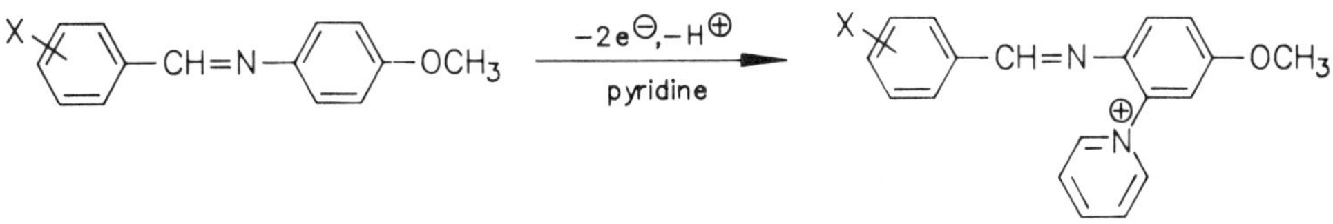

The formation of formyl-(N-methoxymethyl)phenylalkyl amines during the anodic oxidation of primary phenylakyl amines in methanol-KOH is explained by the intermediate generation of imines that are further oxidized (scheme 5). The imines are formed by reaction of the amine with anodically generated formaldehyde [88].

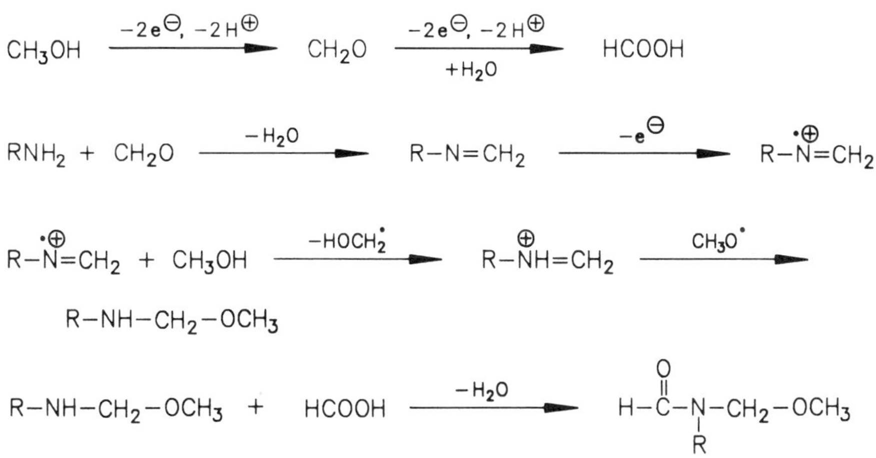

Scheme 5

As in the case of imines, during the oxidation of arylhydrazones nucleophilic attack occurs at the α-carbon (scheme 6). If the nucleophile is water and R^1 is an aryl substituent, cleavage with formation of the parent ketone,

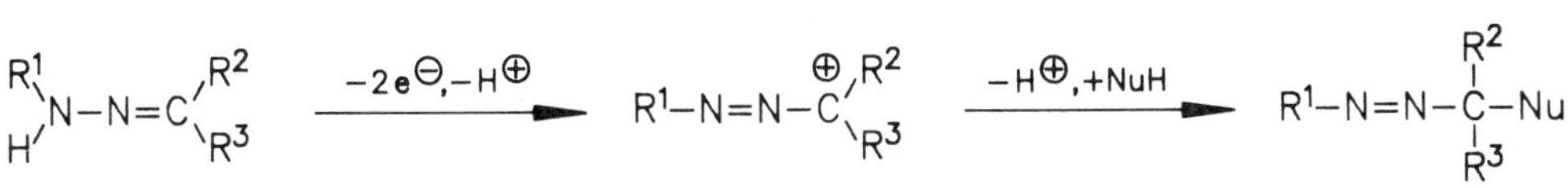

Scheme 6

nitrogen, and ArH is observed [89]. At platinum anodes and under aprotic conditions (CH_3CN-$LiClO_4$), benzophenone hydrazones afford exclusively benzophenone azines; in MeOH-NaOMe benzophenone dimethyl acetals are formed. At graphite anodes (MeOH-NaOMe) diphenyldiazomethane is formed as intermediate (scheme 7) [90].

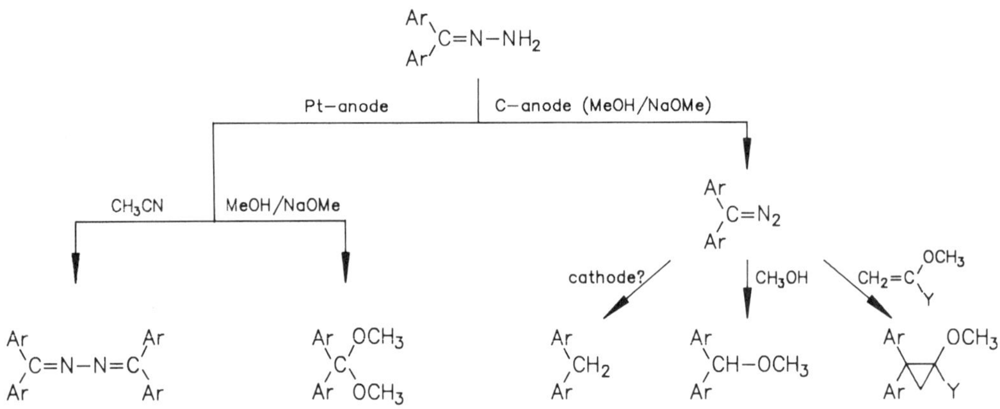

Scheme 7

Cleavage of oximes to the parent carbonyl compound may be performed under mild conditions by anodic oxidation in CH_3CN-H_2O (scheme 8) [91].

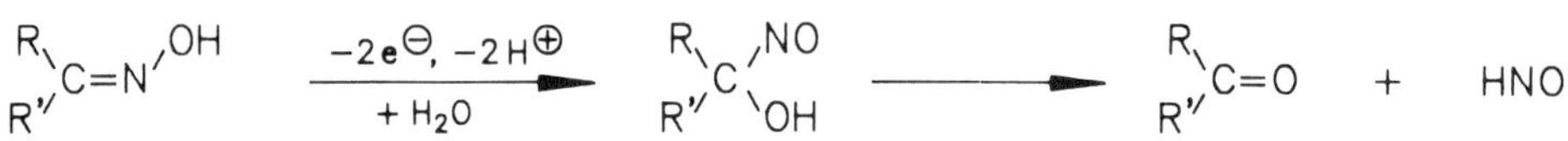

Scheme 8

Anodic oxidation of azomethine, hydrazone, oxime, and semicarbazone structures has been used to initiate intramolecular cyclization under formation of such heterocycles as triazoles [92], oxadiazoles [93], triazolinones [93], benzoxazoles [94,95], benzimidazoles [94,95], and pyrazoles (see Chap. 18) [96].

Oxidation of a N-phenylhydrazone in the presence of pyridine leads to the formation of a s-triazolo[4,3-a]pyridinium salt (scheme 9) by attack of pyridine as a nucleophile on the intermediate nitrilimine [97].

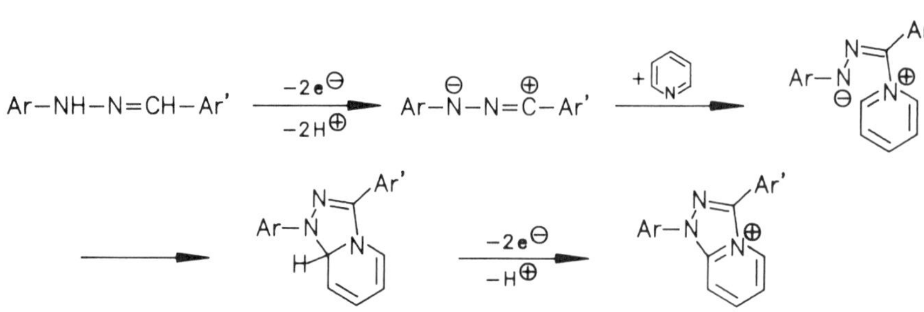

Scheme 9

V. ENAMINES, ENAMINO ESTERS AND KETONES, ENAMIDES, AND ENECARBAMATES

The anodic behavior of N-substituted alkenes can be described as the oxidation of an electron-rich double bond. Tetraamino-substituted alkenes are extremely easily oxidized. Tetrakis(dimethylamino)ethylene exhibits two reversible one-electron processes at −0.75 and −0.61 V versus SCE at a dropping mercury electrode in acetonitrile [98]. The anodic behavior of N,N-dimethylaminoalkenes has been studied intensively by cyclic voltammetry and ESR spectroscopy [99]. The anodically (E^o = 0.48 V versus SCE) generated cation radical of 1,1-bis-(N,N-dimethylamino)ethylene is shown to undergo CC coupling, forming 1,1,4,4-tetrakis-(N,N-dimethylamino) butadiene, which subsequently is further oxidized to its dication at −0.8 V [99,100]. With vicinal diamino ethylenes two reversible one-electron oxidations are usually observed [101], but geminal diamino ethylenes exhibit an irreversible behavior [99]. CC coupling can also be observed in the anodic dimerization of enamino ketones and esters leading to symmetrically substituted pyrrols like XXVII [102]:

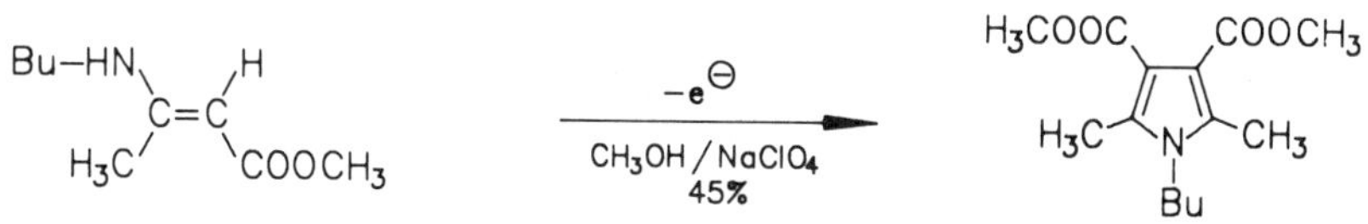

XXVII

Coupling is also observed in the case of 2-cyano-2-phenylvinylamines. It takes place between the phenyl ring in the position para to the enamino group and the C−2 of the enamine, leading to diphenylmethane derivatives [103]. Dimerization or trimerization takes place, too, with 2-phenyl- or 2-cyano-substituted N-methyl-N-phenyl-1,2-diphenylvinylamines by coupling between the p-positions of the N-phenyl and 2-phenyl substituents. The reaction strongly depends on the medium and the base [104,105]. In presence of external or internal nucleophiles, anodic oxidation of amino alkenes may lead to the products of nucleophilic substitution. Thus, electrochemical oxidation of morpholino cycloalkenes in methanol-methoxide leads to the formation of the compounds methoxylated in the allyl (XXVIII) and vinyl (XXIX) positions [106].

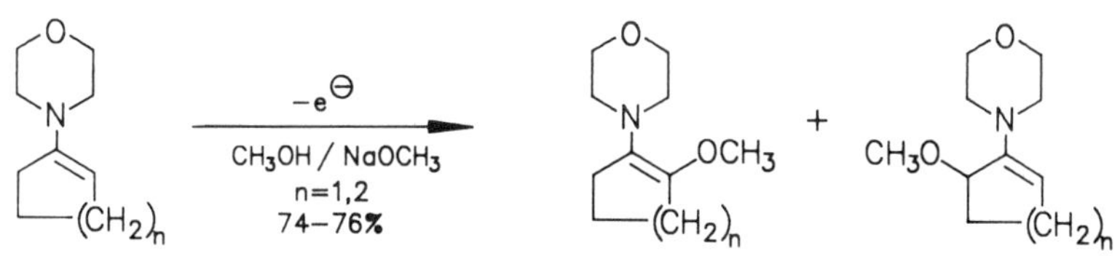

XXIX XXVIII

Similarly, carbanions from 1,3-dicarbonyl compounds can act as nucleophiles [107]. Addition to the double bond of enecarbamates is observed during anodic oxidation in methanol in the presence of an ammonium halide, leading to the product of halomethoxylation (XXX) [108].

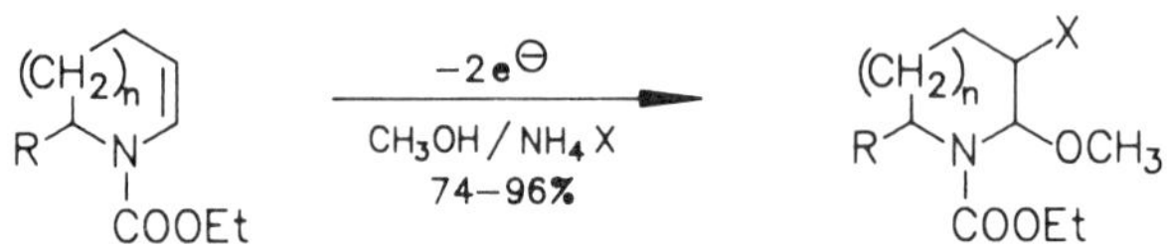

XXX

By electrochemical oxidation of enamines, prepared in situ from an aldehyde and a secondary amine and using KI as mediatory system, β-ketoamines were formed in 30-60% yield [109]. The reaction is explained by the formation of a iodohydrin as an intermediate.

$$RCH_2CHO + R'_2NH \xrightarrow[KI, H_2O-t-BuOH]{-e^-} R-CO-CH_2-NR'_2$$

Anodic oxidation of enamide (XXXI) in acetonitrile-methanol-Na_2CO_3 leads to the formation of the dimethoxylated starting material (XXXI) and the methoxy-substituted oxazoline (XXXIII) [110], which is formed by intramolecular nucleophilic attack of the benzoyl oxygen. Under similar conditions, enamide (XXXIV) undergoes intramolecular cyclization with formation of benzazepine structures XXXV and XXXVI [110].

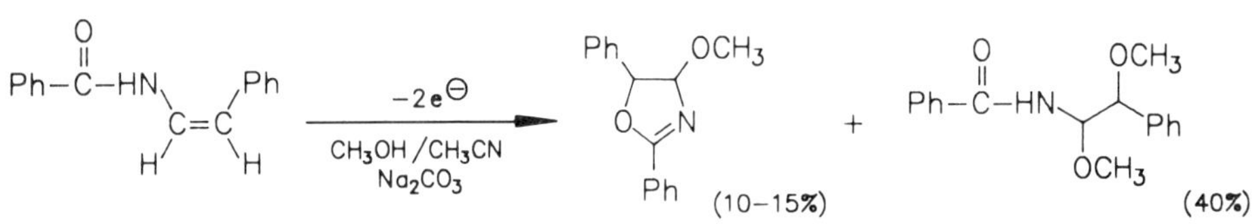

XXXI XXXIII XXXII

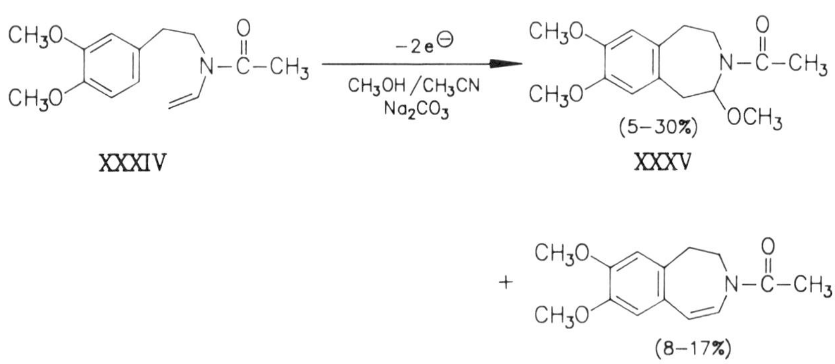

XXXIV XXXV XXXVI

Intramolecular cyclization is a general reaction in the anodic oxidation of substituted amino alkenes. Thus, the N-methyl-N-phenyl-1,2,2-triphenylvinylamine mentioned earlier leads to a new 3H-indole by anodic oxidation in the presence of 2,6-lutidine [104,105]. The corresponding enediamine undergoes an electrolytic double cyclization to form an indolooxazolidine [104]. The formation of isoquinolines (XXXVII) and benzazepiens may also be obtained by anodically initiated intramolecular cyclization of N-benzyl and β-phenethyl enaminones [111].

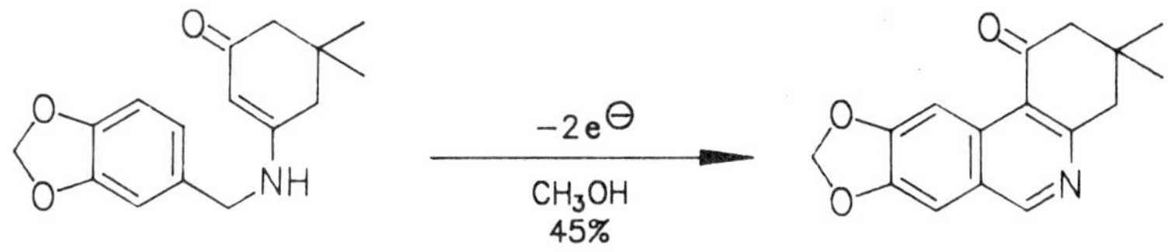

XXXVII

VI. N-HETEROSUBSTITUTED COMPOUNDS

Aromatic hydroxylamines are easily electrochemically oxidized to the nitroso compounds, which subsequently may combine with the starting material to give azoxy compounds [112]. Benzylhydroxylamines are anodically transformed to the oximes in good yields because of the facile deprotonation in benzylic position [113]. Electrolysis of methylhydroxylamine leads to the formation of the corresponding oxime and azoxymethane via the intermediate nitroso compound [114]. An interesting acylation and ethylaminocarbonylation reaction of amines and alcohols is observed if benzo- or acetohydroxamic acids or N-ethyl-N'-hydroxyurea is electrochemically oxidized in acetonitrile in the presence of the nucleophiles [115].

$$\text{R-CO-NH-OH} + \text{NuH} \xrightarrow[-2H^+]{-2e^-} \text{R-CO-Nu} + \text{HNO}$$

$$\text{EtNH-CO-NH-OH} + \text{NuH} \xrightarrow[-2H^+]{-2e^-} \text{EtNH-CO-Nu} + \text{HNO}$$

Similarly to hydroxylamines, alkoxyamines are oxidized to the nitroso compounds under loss of the alkyl group [116]. N-Alkoxycarbamates form dimers by N,N coupling, if anodically oxidized in acetonitrile in the presence of collidine [117].

The electrochemical oxidation of N-nitrosodialkylamines leads to the formation of the corresponding nitramines together with β-ketonitrosamines in which the carbonyl oxygen results from dissolved O_2 [118].

The anodic oxidation of 2-nitrophenylsulfenamides, which are easily obtained by the reaction of amines or amino acid esters with 2-nitrophenylsulfenyl chloride to form the corresponding 2-nitrophenylsulfenimines can be realized by either direct or indirect electrochemical methods. In most cases the direct electrochemical method is preferred because of the convenient reaction procedure and the simple workup [119].

$$o\text{-}NO_2\text{-Ph-S-NH-CHR}^1R^2 \xrightarrow[CH_2Cl_2\text{-TBAP-2,6-lutidine}]{-2e^-} o\text{-}NO_2\text{-Ph-S-N=CR}^1R^2 \quad 50\text{-}95\%$$

In the case of the double NPS-protected lysine methyl ester, oxidation selectively takes place at the ε-sulfenamide function under spontaneous cyclization via the α-sulfenamido group as nucleophile. Only one diastereoisomer of the N,N'-bis-(NPS)-protected methyl-6-amino-piperidine-2-yl-carboxylate seems to be formed. The transformation of α-amino alkanoates and diaryl or dialkyl disulfides to the aryl or alkyl sulfenimine derivatives can be performed electrochemically using $MgBr_2$ as mediatory system. This reaction can proceed in one step or start from the corresponding sulfenamide [120].

VII. AMINO ALCOHOLS AND AMINO ACIDS

Amino alcohols may be oxidized at either the hydroxy or the amino function. Oxidation at the hydroxy function leads to the amino acid. This reaction can be performed at lead dioxide or platinum anodes in aqueous sulfuric acids with moderate material and current yields [121]. In the case of the oxidation of 3-phenyl-2-N-acetylaminopropanol to acetylphenylalanine, an indirect electrochemical ex-cell process was proposed using anodically generated $Cr_2O_7^{2-}$ as reagent [122]. Oxidation under basic or neutral conditions usually starts at the nitrogen atom. β-Amino alcohols, similarly to 1,2-diols and 1,2-diamino compounds, undergo cleavage to aldehydes or ketones. Thus, 1-(N,N-dimethylaminomethyl)-p-methoxybenzyl alcohol is oxidized in acetonitrile to yield p-methoxybenzaldehyde, dimethylamine, and formaldehyde (scheme 10) [123]; ephedrine is decomposed in a aqueous buffer of pH 10 mainly to benzaldehyde, acetaldehyde, and N-methylamine [124]. Other β-alkanolamines behave similarly [125,126]. These reactions are explained by a fragmentation of the primarily formed amino cation radical into the aldehyde or ketone, and after further oxidation, an iminium ion is formed that subsequently hydrolyzes to aldehyde and amine. In a very similar manner (scheme 10), the o-NPS derivatives of serine and

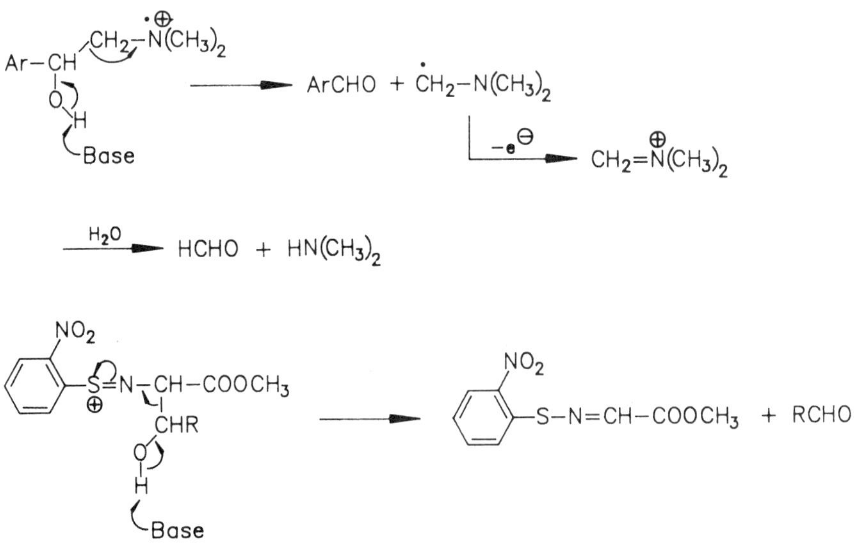

Scheme 10

threonine are fragmented under loss of formaldehyde or acetaldehyde to give the ortho-nitrophenylsulfenimine of glycine methyl carboxylate [119].

Anodic oxidation of N-benzyl-N-methylethanolamine leads to the formation of a mixture of 3-methyl-2-phenyl- and 3-benzyloxazolidine by intramolecular attack of the hydroxy group at the intermediate iminium ion, mainly in benzylic position [127]. The typical fragmentation of β-amino alcohols occurs as a side reaction in this case, giving formaldehyde and N-benzyl-N-methoxymethyl-N-methylamine.

The indirect anodic oxidation of 2-amino-1-cycloalkanols using NaCl as redox catalyst in aqueous acetonitrile leads to the formation of 1,ω-keto nitriles (50-60%) under cleavage of the C—C bond. The β-hydroxy-N-chloroamine is supposed to be the intermediate (scheme 11) [25].

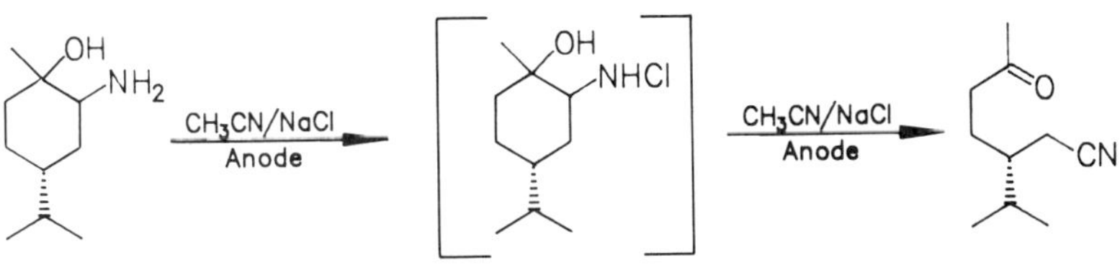

Scheme 11

The electrochemistry of amino acids has been studied in strong acid solutions. In general, the compounds are broken down to carboxylic acids, aldehydes, ammonia, and carbon dioxide. The results are reviewed by Weinberg and Weinberg [28]. The anodic oxidation mechanism has been studied in pH 10 buffer solution. Decarboxylation accompanied by C—N bond cleavage is the main reaction process [129]. The synthetically interesting Hofer-Moest decarboxylations of N-protected amino acids and α-amino malonic half-esters [128] is not treated in this chapter.

VIII. AMIDES, CARBAMATES, LACTAMS, AND RELATED COMPOUNDS

The anodic oxidation of amides, lactams, and carbamates is of considerable synthetic value because, in contrast to aliphatic amines, the products of α-oxidation are quite stable and can be used effectively, for example, in amidoalkylation reactions (see Chap. 23) [130-132]. The application of this reaction sequence has been used intensively during the past several years, but the anodic oxidation of amides and carbamates has been known for a much longer time. As early as 1954, electrooxidation of a number of amides and lactams was studied by Mizuno and Nanya in acidic aqueous media [133]. Under these conditions, dealkylation by scission of the carbon-nitrogen bond is observed, leading to carboxylic acids, amino acids, ketones, and imides. O'Donnell and Mann [134] investigated the cyclic voltammetry of a number of aliphatic amides in acetonitrile. Electrode potentials of about 2.0 V for primary amides, 1.8 V for secondary amides, and 1.2-1.5 V versus SCE for tertiary amides were found. In all cases a one-electron oxidation occurs. Finckelstein and Ross and colleagues [135-138], and later Nyberg and Eberson and colleagues [139-142] and Shono [130], showed that in the presence of alcohols, acetic acid, or formic acid the anodic oxidation of N-alkyl and N,N-dialkyl amides leads to stable α-alkoxylated and α-acyloxylated products [135-141,143].

Intramolecular anodic coupling of diaryl amides is a synthetically interesting reaction that may be applied to the synthesis of alkaloidal structures [143]. However, this aryl coupling is initiated by the oxidation of an aromatic ring (aryl ether or phenol), not by oxidation of the nitrogen. Therefore this is not discussed further in this chapter.

A. Anodic Oxidation of Amides, Carbamates, and Lactams

The mechanism for the anodic α-oxidation of amides [135,136] and carbamates [144] in the presence of such nucleophiles as water, alcohol, or carboxylic acid starts by one-electron oxidation to form the cation radical (XXXVIII) at the nitrogen. This is followed by loss of a second electron and a proton to give the iminium ion (XXXIX) to which subsequently the nucleophile is added.

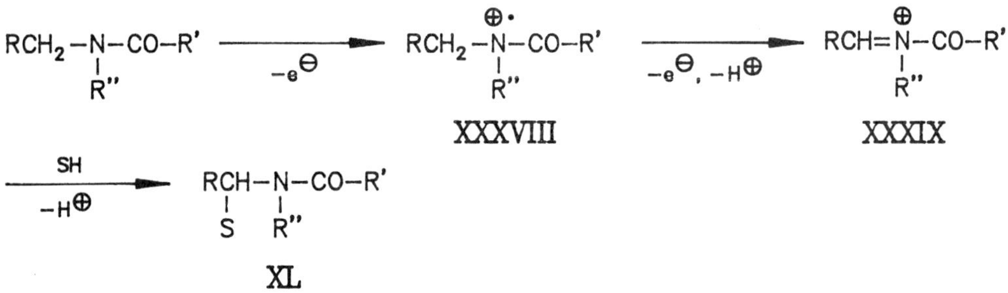

XXXVIII XXXIX XL

In the case of SH=H_2O, the product is unstable and dealkylation occurs with formation of aldehyde and amide. Formyloxylation has been explained through hydrogen atom abstraction by the formyloxy radical, formed by initial anodic oxidation of formate [135]. Similarly, the oxidation in the presence of nitrate ions has been formulated [136].

The thus formed α-alkoxylated and acyloxylated products (XL) may be used as interesting synthetic building blocks by either Lewis acid or proton-catalyzed exchange of the alkoxy group by C or heteroatomnucleophiles (path a, amidoalkylation) or via elimination under the formation of enecarbamates [144,145] and enamides [142,144,146] followed by reaction with electrophiles or nucleophiles (path b).

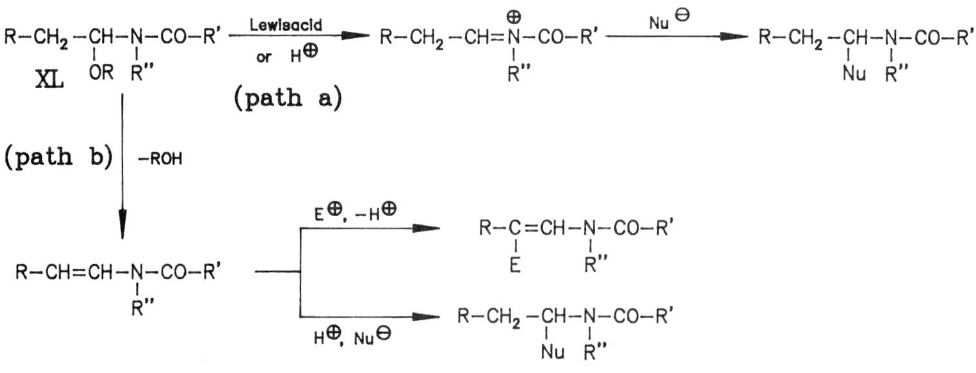

Many interesting synthetic applications have been described [130-132], especially in the field of alkaloid synthesis [147].

The anodic oxidation of amides and carbamates can usually be performed under constant-current conditions in an undivided cell at platinum anodes using R_4NBF_4 of KPF_6 as supporting electrolyte.

Polyalkoxylation of amides is more difficult to obtain than monoalkoxylation. In the case of N,N-dimethylformamide and N-methyl-N-methoxymethylformamide it can be performed in methanol-TBABF$_4$ at Pt anodes [138]. Another example is the 2,5-dimethoxylation of N-acetylpyrrolidine [148].

The methoxylation of N-ethyl carboxamides has been studied especially intensively as the N-(α-methoxyethyl)amides produced may be transformed to the technically interesting vinyl amides by methanol elimination [146]. Glassy carbon anodes together with graphite or steel cathodes and tetramethylammonium methylsulfate as conducting salt proved to be the best conditions.

Lactams could also be hydroxylated [149] or alkoxylated [150,151] in the α-position. In the case of five- and six-membered N-alkyl lactams, oxidation occurs regioselectively at the endocyclic carbon in α-position to nitrogen and at the exocyclic α-carbon with seven-membered rings [149]. This procedure has also been applied to intramolecular cyclizations (scheme 12) [152].

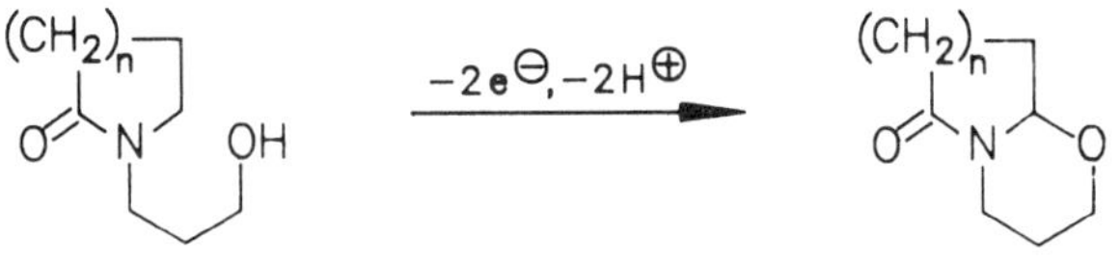

Scheme 12

The anodic oxidation of N-alkyl- and N-benzyl-β-lactams in methanol at Pt anodes in the presence of Et_4NBF_4 as electrolyte leads to the methoxylation of both the endo- and exocyclic carbon in α-position to nitrogen. The exocyclic carbon was usually more easily oxidized than the endocyclic carbon. With a tertiary exocyclic α-carbon the formation of the endooxidation product was increased. N-benzyl-β-lactams were predominantly oxidized in the exocyclic α-position. In the case of N-4-methoxybenzyl-β-lactam, exocyclic oxidation occurred with total regioselectivity [153].

Similarly to carboxamides, sulfonamides [154,155] and amidophosphates [155] may be α-methoxylated at the anode. p-Tolysulfonamides of primary amines undergo a rearrangement during anodic oxidation in methanol in the presence of KBr and KOH; this leads to α-amino aldehydes [156]. Bromide ion is involved as inorganic redox catalyst (scheme 13).

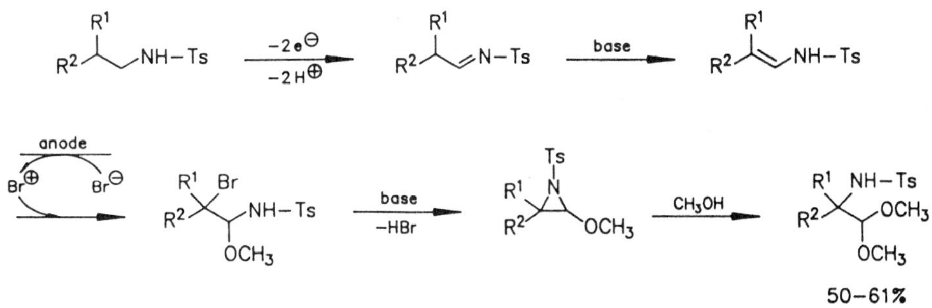

Scheme 13

Another interesting bromide ion-catalyzed reaction is the anodic oxidation (undivided cell) of secondary formamides in methanol leading to urethanes (XLIII) [157]. This reaction proceeds via the intermediate N-bromo amide (XLI), which by elimination of HBr forms the isocyanate (XLII), which is attacked by methanol.

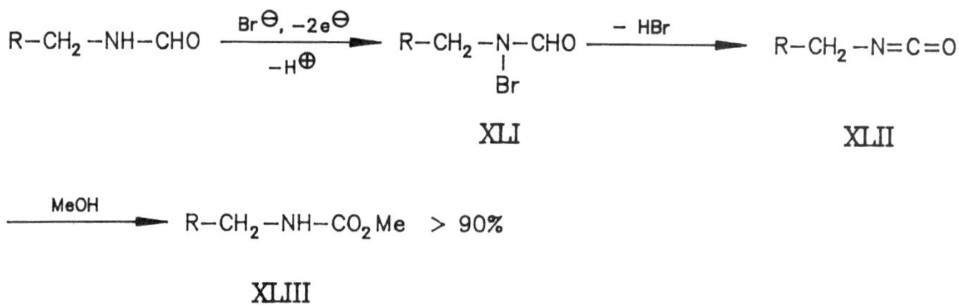

Thus, a phosgene-free technical synthesis of urethanes is made possible. Urethanes can be used as stable isocyanate equivalents.

If the direct anodic oxidation of tertiary formamides, which do not possess an α-hydrogen, is performed, the formyl function is transformed into a carboxymethyl group [158].

Regioselectivity and stereochemistry have been studied in the case of the anodic methoxylation of N-acylpiperidines and N-acylmorpholines [159]. Because of steric constraints imposed by the intermediate formation of the planar N-acyliminium ions, substitution is directed to the secondary carbon rather than to the tertiary position. The same reason accounts for the axial methoxylation in this reaction.

B. Anodic Oxidation of N-Acyl and N-Carboalkoxy α-Amino Acid Esters

Electrochemical oxidation in the α-position of protected α-amino acid esters and dipeptides is synthetically interesting as the α-carbon is transformed into an electrophilic center by the redox Umpolung principle. Direct electrochemical oxidation of N-acyl amino acid esters is not successful, presumably because of the electron-withdrawing effect of the ester function. Indirect anodic oxidation with application of NaCl as redox catalyst, however, is possible [160-162]. In methanol using an undivided cell the α-methoxylation of protected amino acids occurs according to the following scheme (scheme 14).

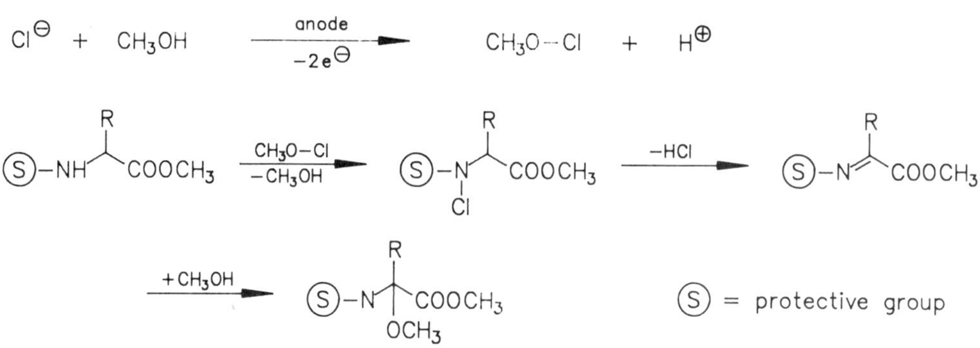

Scheme 14

In this way, N-carbomethoxylated branched α-amino acid esters and α-amino β-lactams have been methoxylated in 83-91% material yield at Pt anodes in the presence of 0.5 equivalents of NaCl after consumption of 10 F/mol. In the case of phenylglycine, 31 F/mol was necessary to obtain the α-methoxylated product in 62% yield [160]. Under these conditions, the glycine derivative was α-dimethoxylated in 95% material yield [160]. Steckhan et al. showed that in the presence of very high concentrations of chloride ions, already after the consumption of 3 F/mol, over 90% turnover of the protected amino acid esters could be obtained [162]. In the case of N-acyl glycine methylester, a 5:1 ratio of mono- and dimethoxylated product could be obtained. N-acyl phenylglycine methylester could be methoxylated in 98% yield after the consumption of 3 F/mol. Obviously, high chloride concentrations diminish the competing methanol oxidation. The reaction rate increases with increasing electron-withdrawing effect of the N-protecting group and decreases with increasing electron-donating ability of the substituents in α-position. N-acylamino malonates are faster, but N-acylamino methylphosphonates react more slowly than the corresponding N-acylamino acid esters.

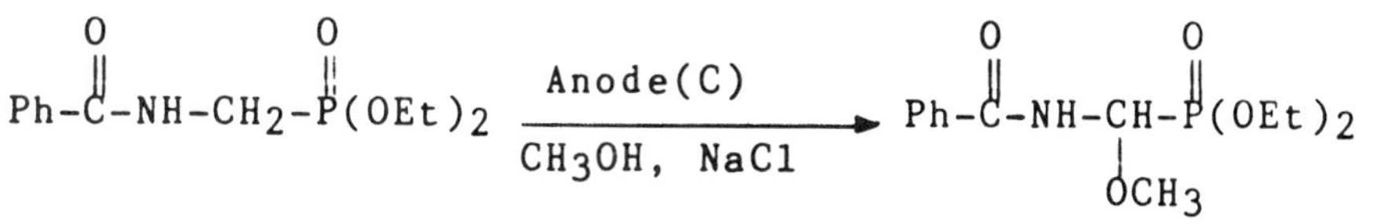

41 % (isol. yield)

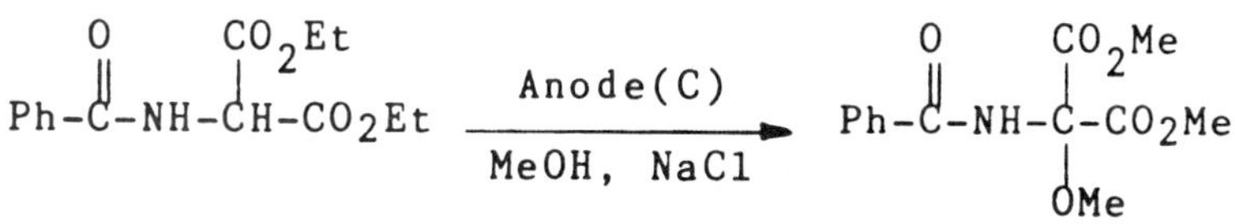

65 %

These effects can best be interpreted assuming HCl elimination from the N-chloro amino acid intermediate to be the rate-determining step [162].

Similarly, N-acyl or N-carboalkoxylated dipeptides can be anodically oxidized in the α-position to the nitrogen using NaCl as mediator [162]. In the case of dipeptides, however, two regioisomers may be obtained by oxidation at the N- or the C-terminal amino acid. The reaction sequence is shown in scheme 15.

The results show that α-methoxylation occurs selectively and in high yields at the C-terminal amino acid, if N-benzoyl protecting groups without electron-donating substituents are used. The Z-protecting group predominantly leads to oxidation at the C terminus accompanied by small amounts of products from oxidation at the N terminus and some hydantoin formation. The Boc-protecting group leads to the selective methoxylation at the C terminus if glycine is the C-terminal amino acid.

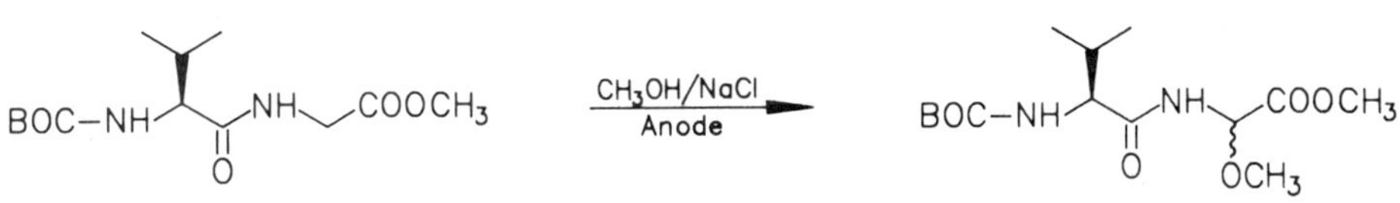

(70%)

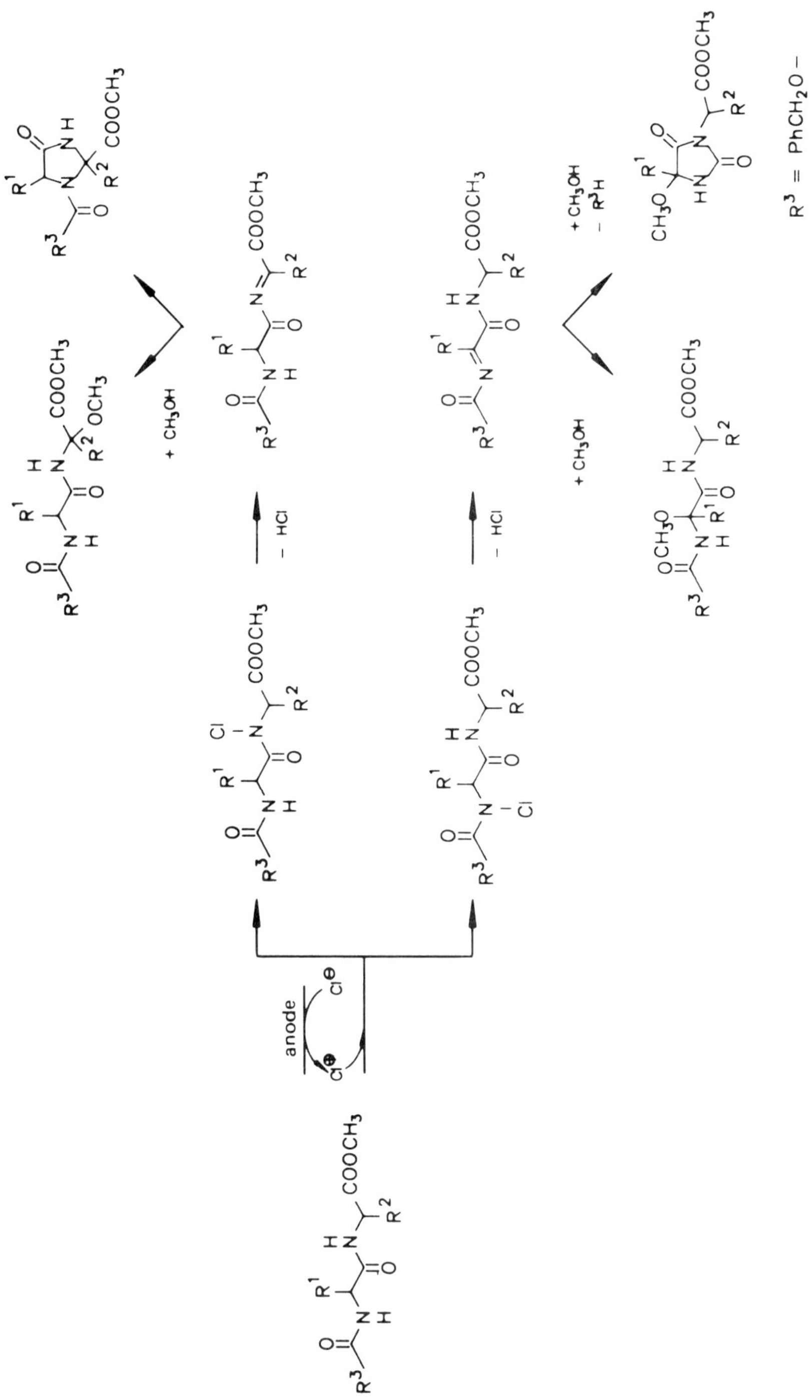

Scheme 15

In the case of branched C-terminal amino acids, methoxylation at the C terminus is accompanied by formation of imidazolidine-4-ones by α-oxidation at the C terminus followed by intramolecular attack of the nitrogen of the N-terminal amino acid. This reaction can be performed in high yields if the electrooxidation is run in acetonitrile-5% methanol [162].

In the case of basic amino acids, such as lysine or ornithine, the regioselectivity of the direct and indirect electrochemical oxidation can be demonstrated. Direct anodic oxidation of N,N'-dimethoxycarbonylated L-lysine and L-ornithine in methanol selectively resulted in the methoxylation at the ω-amino group; indirect electrochemical oxidation in the presence of NaCl selectively formed the α-methoxylated α-amino acid ester (scheme 16) [161,163,164].

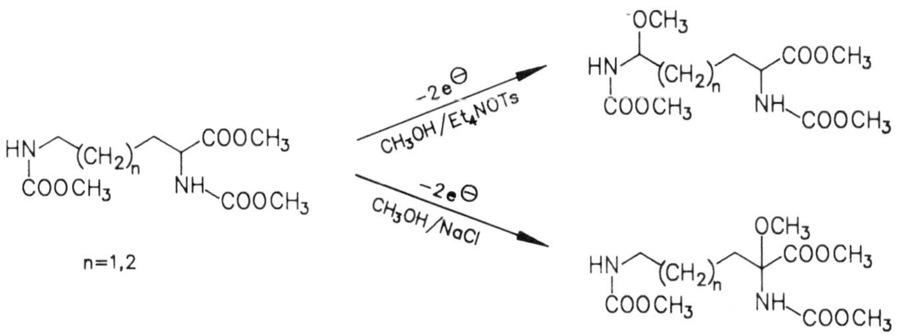

Scheme 16

The species thus formed are very useful for the synthesis of optically active alkaloids [147,163,164].

REFERENCES

1. C. K. Mann and K. A. Barnes, Electrochemical Reactions in Nonaqueous Systems, Dekker, New York, 1970, pp. 259–296.
2. K. A. Barnes and C. K. Mann, J. Org. Chem., 32, 1474 (1967).
3. P. J. Smith and C. K. Mann, J. Org. Chem., 33, 316 (1968).
4. P. J. Smith and C. K. Mann, J. Org. Chem., 34, 1821 (1969).
5. S. F. Nelsen and J. T. Ippoliti, J. Amer. Chem. Soc., 108, 4879 (1986).
6. H. J. Schäfer, Topics Curr. Chem., 142, 117 (1987).
7. M. Fleischmann, K. Korinek, and D. Pletcher, J. Chem. Soc., Perkin Trans., II, 1396 (1972); J. Electroanal. Chem., 31, 39 (1971).
8. U. Feldhues and H. J. Schäfer, Synthesis, 145 (1982).
9. R. Bauer and H. Wendt, Angew. Chem., 90, 214 (1978); Angew. Chem., Int. Ed. Engl., 17, 202 (1978).
10. A. U. Blackham, S. Kwak, and J. L. Palmer, J. Electrochem. Soc., 122, 1081 (1975).
11. M. Masui and H. Sayo, J. Chem. Soc. (B), 1593 (1971).
12. T. Chiba and Y. Takata, J. Org. Chem., 42, 2973 (1977).

13. G. Bidan and M. Genies, Tetrahedron, 37, 2297 (1981).
14. N. L. Weinberg and E. A. Brown, J. Org. Chem., 31, 4058 (1966).
15. N. L. Weinberg, J. Org. Chem., 33, 4326 (1968).
16. P. J. Smith and C. K. Mann, J. Org. Chem., 33, 316 (1968).
17. J. E. Barry, M. Finkelstein, E. A. Mayeda, and S. D. Ross, J. Org. Chem., 39, 2695 (1974).
18. J. Y. Becker, L. L. Miller, and F. R. Stermitz, J. Electroanal. Chem. 68, 181 (1976).
19. T. Shono, T. Toda, and N. Oshino, J. Amer. Chem. Soc., 104, 2639 (1982).
20. M. Tokuda, Y. Yamada, T. Takagi, and H. Suginome, Tetrahedron Lett., 26, 6085 (1985).
21. B. L. Laube, M. A. Asirvatham, and C. K. Mann, J. Org. Chem., 42, 670 (1977).
22. T. Fuchigami, T. Iwaoka, T. Nonaka, and T. Sekine, Bull. Chem. Soc. Jpn., 53, 2040 (1980).
23. P. G. Gassman, I. Nishiguchi, and H. Yamamoto, J. Amer. Chem. Soc., 97, 1600 (1975).
24. R. Kossai and J. Simonet, Tetrahedron Lett., 21, 3575 (1980).
25. S. Torii, T. Inokuchi, S. Takagishi, E. Sato, and H. Tsujiyama, Chem. Lett., 1469 (1987).
26. R. C. Reed and R. M. Wightman, Encyclopedia of Electrochemistry of the Elements, Organic Section, Vol. XV (A. J. Bard and H. Lund, eds.), Dekker, New York, 1984, pp. 3–53.
27. R. F. Nelson, Technique of Electroorganic Synthesis, Part 1, in Techniques of Chemistry (N. L. Weinberg, ed.), Vol. V, Chap. 9, Wiley-Interscience, New York, 1974.
28. N. L. Weinberg and H. R. Weinberg, Chem. Rev., 68, 475–485 (1968).
29. C. K. Mann and K. A. Barnes, Electrochemical Reactions in Nonaqueous Systems, Dekker, New York, 1970, pp. 260–269.
30. R. L. Hand and R. F. Nelson, J. Amer. Chem. Soc., 96, 850 (1974); J. Electrochem. Soc., 125, 1059 (1978).
31. D. M. Mohilner, R. N. Adams, and W. J. Argersinger, Jr., J. Amer. Chem. Soc., 84, 3618 (1962).
32. M. Breitenbach and K. H. Heckner, J. Electroanal. Chem., 29, 309 (1971).
33. J. Heinze, Topics Curr. Chem., 152 (1989), in press.
34. S. Wawzonek and T. McIntyre, J. Electrochem. Soc., 114, 1025 (1967).
35. J. Bacon and R. N. Adams, J. Amer. Chem. Soc., 90, 6596 (1968).
36. R. N. Adams, Electrochemistry at Solid Electrodes, Dekker, New York, 1969, pp. 336–344.
37. R. N. Adams, Electrochemistry at Solid Electrodes, Dekker, New York, 1969, pp. 351–353; Acc. Chem. Res., 2, 179.
38. S. Andreades and E. W. Zahnow, J. Amer. Chem. Soc., 91, 4181 (1969).
39. N. L. Weinberg and E. A. Brown, J. Org. Chem., 31, 4058 (1966); N. L. Weinberg and T. B. Reddy, J. Amer. Chem. Soc., 90, 91 (1968).
40. T. Shono, Y. Matsumura, K. Inoue, H. Ohmizu, and S. Kashimura, J. Amer. Chem. Soc., 104, 5753 (1982).
41. R. F. Nelson, Ph.D. Thesis, Kansas University, 1966.

42. R. Hand, M. Melicharek, D. I. Scoggins, R. Stotz, A. K. Carpenter, and R. F. Nelson, Coll. Czech. Chem. Commun., 36, 842 (1971).
43. M. F. Marcus and M. D. Hawley, J. Electroanal. Chem., 18, 175 (1968).
44. M. Melicharek and R. F. Nelson, J. Electroanal. Chem., 26, 201 (1970).
45. V. Dvorák, I. Nemec, and J. Zyka, Microchem. J., 12, 99 (1967).
46. G. Cauquis, J. Cognard, and D. Serve, Tetrahedron Lett., 4645, (1971).
47. D. Serve, Electrochim. Acta, 21, 1171 (1976).
48. G. Cauquis, H. Delhomme, and D. Serve, Electrochim. Acta, 21, 557 (1976).
49. D. Serve, J. Amer. Chem. Soc., 97, 432 (1975).
50. G. Cauquis, H. Delhomme, and D. Serve, Tetrahedron Lett., 4113, (1971).
51. P. Berkenkotter, and R. F. Nelson, J. Electrochem. Soc., 120, 346 (1973).
52. Y. Yoshida and T. Fueno, J. Org. Chem., 37, 4145 (1972).
53. Y. Yoshida and T. Fueno, J. Org. Chem., 41, 731 (1976).
54. D. W. Leedy and R. N. Adams, J. Amer. Chem. Soc., 92, 1646 (1970).
55. D. Serve, Bull. Soc. Chim. Fr., 1993 (1976).
56. R. Reynolds, L. L. Line, and R. F. Nelson, J. Amer. Chem. Soc., 96, 1087 (1974).
57. R. N. Adams, Electrochemistry at Solid Electrodes, Dekker, New York, 1969, pp. 353–356.
58. E. T. Seo, R. F. Nelson, J. M. Fritsch, L. S. Marcoux, and R. N. Adams, J. Amer. Chem. Soc., 88, 3498 (1966).
59. R. F. Nelson and R. N. Adams, J. Amer. Chem. Soc., 90, 3025 (1968).
60. L. Papouchado, R. N. Adams, and S. W. Feldberg, J. Electroanal. Chem., 21, 408 (1969).
61. E. Steckhan, Angew. Chem., Int. Ed. Engl., 25, 683 (1986).
62. E. Steckhan, Topics Curr. Chem., 142, 1 (1987).
63. W. Schmidt and E. Steckhan, Chem. Ber., 113, 577 (1980).
64. K. H. Grosse Brinkhaus, S. Dapperheld, T. Esch, and E. Steckhan, unpublished results.
65. W. Schmidt and E. Steckhan, Angew. Chem., Int. Ed. Engl., 17, 673 (1978); 18, 801 (1979); 18, 802 (1979); T. Esch, Diplom Thesis, Bonn, 1983; Ph.D. Thesis, Bonn, 1987.
66. S. Dapperheld and E. Steckhan, Angew. Chem., Int. Ed. Engl., 21, 780 (1982); Angew. Chem., Suppl. 1730 (1982).
67. M. Platen and E. Steckhan, Tetrahedron Lett., 21, 511 (1980); Chem. Ber., 117, 1679 (1984); Liebigs Ann. Chem., 1563 (1984); S. Dapperheld, Ph.D. Thesis, Bonn, 1984; S. Töteberg-Kaulen, Ph.D. Thesis, Bonn, 1986; N. Schultz, S. Töteberg-Kaulen, S. Dapperheld, J. Heyer, M. Platen, and K. Schumacher, Recent Advances in Electroorganic Synthesis, Proceedings of the 1st International Symposium on Electroorganic Synthesis, Kodansha, Tokyo, Elsevier, Amsterdam, 1987, pp. 127–135.
68. K. H. Grosse Brinkhaus, E. Steckhan, and D. Degner, Tetrahedron, 42, 533 (1986).

69. K. H. Grosse Brinkhaus, E. Steckhan, and W. Schmidt, Acta Chem. Scand., B37, 499 (1983).
70. K. D. Ginzel, E. Steckhan, and D. Degner, Tetrahedron, 43, 5797 (1987).
71. E. Ahlberg, B. Helgée, and V. D. Parker, Acta Chem. Scand., B34, 187 (1980).
72. L. Eberson and B. Larsson, Acta Chem. Scand., B40, 210 (1986).
73. K. H. Grosse Brinkhaus, Ph.D. Thesis, Bonn, 1984.
74. S. Dapperheld and E. Steckhan, unpublished results, 1983.
75. R. I. Walter, J. Amer. Chem. Soc., 88, 1923 (1965).
76. S. Dapperheld and E. Steckhan, unpublished results, 1983.
77. H. Schrage, J. Franke, F. Vögtle, and E. Steckhan, Angew. Chem., Int. Ed. Engl., 25 (1986).
78. H. Schrage, F. Vögtle, and E. Steckhan, J. Incl. Phen., 6, 157 (1988).
79. R. N. Adams, Electrochemistry at Solid Electrodes, Dekker, New York 1969, pp. 356–361.
80. R. C. Reed and R. M. Wightman, Encyclopedia of Electrochemistry of the Elements, Organic Section, Vol. XV (A. J. Bard and H. Lund, eds.), Dekker, New York, 1984, pp. 37–53.
81. V. Solis, T. Iwasita, and M. C. Giordani, J. Electroanal. Chem., 73, 91 (1976).
82. O. Fischer and V. Komarek, Coll. Czech. Chem. Commun., 41, 2320 (1976).
83. T. Esch, Ph.D. thesis, Bonn, 1987.
84. M. D. Hawley and R. N. Adams, J. Electroanal. Chem., 10, 376 (1965).
85. M. Masui and H. Ohmori, J. Chem. Soc., Perkin Trans., 2, 1112 (1973).
86. M. Masui and H. Ohmori, J. Chem. Soc., Perkin Trans., 2, 1882 (1972).
87. H. Ohmori, A. Matsumoto, and M. Masui, J. Chem. Soc. Perkin Trans., 2, 347 (1980).
88. U. Hess and M. Ziebig, Pharmazie, 36, 253 (1981); Z. Chem., 20, 444 (1980).
89. E.-C. Lin and M. R. Van de Mark, J. Chem. Soc., Chem. Commun., 1176 (1982).
90. T. Chiba, M. Okimoto, H. Nagai, and Y. Takata, J. Org. Chem., 48, 2968 (1983).
91. M. R. Van de Mark and A. M. Sconavacca, Org. Prep. Proc. Int., 13, 395 (1981).
92. S. Crljenak, I. Tabakovic, D. Jeremic, and I. Gaon, Acta Chem. Scand., B37, 527 (1983).
93. O. Hammerich and V. D. Parker, J. Chem. Soc., Perkin Trans., 1, 1718 (1972).
94. M. Lacan, K. Jakopcic, V. Rogic, S. Damani, O. Rogic, and I. Tabakovic, Synth. Commun., 4, 219 (1974).
95. M. Lacan, V. Rogic, I. Tabakovic, D. Galijas, and T. Solomon, Electrochim. Acta, 28, 199 (1983).
96. I. Tabakovic, M. Lacan, and S. Damani, Electrochim. Acta, 21, 621 (1976).
97. I. Tabakovic and S. Crljenak, Heterocycles, 16, 699 (1981).

98. K. Kuwata and D. H. Geske, J. Amer. Chem. Soc., 86, 2101 (1964).
99. J. M. Fritsch, H. Weingarten, and J. D. Wilson, J. Amer. Chem. Soc., 92, 4038 (1970).
100. J. M. Fritsch and H. Weingarten, J. Amer. Chem. Soc., 90, 793 (1968).
101. C. P. Andrieux and J. M. Savéant, J. Electroanal. Chem., 28, 339 (1970).
102. D. Koch and H. Schäfer, Angew. Chem., Int. Ed. Engl., 12, 245 (1973).
103. M. Masui, C. Ueda, T. Moriguchi, T. Michida, M. Kataoka, and H. Ohmori, Chem. Pharm. Bull., 32, 1392 (1984).
104. M. Cariou, R. Carlier, and J. Simonet, Bull. Chem. Soc. Chim. Fr., 781 (1986); J. Chem. Soc., Chem. Commun., 876 (1983).
105. M. Cariou and J. Simonet, J. Chem. Soc., Chem. Commun., 1146 (1984).
106. T. Shono, Y. Matsumura, H. Hamaguchi, T. Imanishi, and K. Yoshida, Bull. Chem. Soc. Jpn., 51, 2179 (1978).
107. T. Chiba, M. Okimoto, H. Nagai, and Y. Takata, J. Org. Chem., 44, 3519 (1979).
108. S. Torii, T. Inokuchi, F. Akahoshi, and M. Kubota, Synthesis, 242 (1987).
109. T. Shono, Y. Matsumura, J. Hayashi, M. Usui, S.-I. Yamani, and K. Inoue, Acta Chem. Scand., B37, 491 (1983).
110. B. Wierlemann and E. Steckhan, unpublished results, 1984.
111. W. Eilenberg and H. J. Schäfer, Tetrahedron Lett., 25, 5023 (1984).
112. S. Ozaki and M. Masui, Chem. Pharm. Bull., 26, 1364 (1978).
113. C. Degrand and C. Millet, Bull. Soc. Chim. Fr., 2488 (1975).
114. P. E. Iversen, Chem. Ber., 104, 2195 (1971).
115. M. Masui and S. Ozaki, Tetrahedron Lett., 23, 2867 (1982).
116. H. Sayo, S. Ozaki, and M. Masui, Chem. Pharm. Bull., 23, 1702 (1975); G. Cauquis, B. Chabaud, and Y. Gohee, Tetrahedron Lett., 2583 (1977).
117. S. Ozaki and M. Masui, Chem. Pharm. Bull., 27, 357 (1979).
118. M. Masui, K. Nose, S. Terauchi, E. Yamakawa, J. Jeong, C. Ueda, and H. Ohmori, Chem. Pharm. Bull., 33, 2721 (1985); M. Masui, K. Nose, A. Tanaka, E. Yamakawa, and H. Ohmori, Chem. Pharm. Bull., 29, 3758 (1981).
119. J. Heyer, S. Dapperheld, and E. Steckhan, Chem. Ber., 121, 1617 (1988); N. Schultz, S. Töteberg-Kaulen, S. Dapperheld, J. Heyer, M. Platen, K. Schumacher, and E. Steckhan, Stud. Org. Chem. (Amsterdam), 30, 127 (1987).
120. S. Torii, H. Tanaka, S.-I. Hamano, N. Tada, J. Nokami, and M. Sasaoka, Chem. Lett., 1823 (1984).
121. S. Mizuno, Y. Hirata, and K. Takei, J. Electrochem. Soc. Jpn., Overseas Ed., 29, 31 (1961); Chem. Abstr., 60, 11889 (1964); Nagoya Kogyo Kaigáju Gakuho, 14, 48 (1962); Chem. Abstr., 60, 11889 (1964).
122. H. P. Gogolin and H. Wendt, Chem.-Ing.-Tech., 48, 797 (1976).
123. L. C. Portis, J. T. Klug, and C. K. Mann, J. Org. Chem., 39, 3488 (1974).
124. M. Masui, Y. Kamada, and S. Ozaki, Chem. Pharm. Bull., 28, 1619 (1980).

125. T. Shono, H. Hamaguchi, Y. Matsumura, and K. Yoshida, Tetrahedron Lett., 3625 (1977).
126. M. Masui, Y. Kamada, E. Sasaki, and S. Ozaki, Chem. Pharm. Bull., 30, 1234 (1982).
127. N. L. Weinberg and E. A. Brown, J. Org. Chem., 31, 4054 (1966).
128. H. Horikawa, T. Iwasaki, K. Matsumoto, and M. Miyoshi, Tetrahedron Lett., 191 (1976); J. Org. Chem., 43, 335 (1978); 42, 2419 (1977); Bull. Chem. Soc. Jpn., 52, 826 (1979); P. Renaud and D. Seebach, Synthesis, 424 (1986); Angew. Chem., Int. Ed. Engl., 25, 843 (1986); Helv. Chim. Acta, 69, 1704 (1986); D. Seebach and G. Stucky, Angew. Chem., Int. Ed. Engl., 27, 1351 (1988).
129. M. Masui, Y. Kamada, T. Iizuka, and S. Ozaki, Chem. Pharm. Bull., 32, 4740 (1984).
130. T. Shono, Tetrahedron, 40, 824 (1984); T. Shono, Electroorganic Chemistry as a New Tool in Organic Synthesis, Springer-Verlag, Berlin, 1984, pp. 63–77.
131. S. Torii, Electroorganic Syntheses, Methods and Applications, Part I, Oxidations, Kodansha, Tokyo, 1985, pp. 171–180.
132. H. U. Reissig, Nachr. Chem. Tech. Lab., 34, 656 (1986).
133. S. Mizuno and S. Nanya, Bull. Nagoya Inst. Technol., 6, 275 (1954); Chem. Abstr., 49, 13803i (1955); S. Mizuno, J. Electrochem. Soc. Jpn., Overseas ed., 28, 280 (1960); 29, 27, 31, 33, 112 (1961).
134. J. F. O'Donnell and C. K. Mann, J. Electroanal. Chem., 13, 157 (1967).
135. S. D. Ross, M. Finkelstein, and R. C. Petersen, J. Amer. Chem. Soc., 86, 2745 (1964); J. Org. Chem., 31, 128 (1966); J. Amer. Chem. Soc., 88, 4657 (1966).
136. E. J. Rudd, M. Finkelstein, and S. D. Ross, J. Org. Chem., 37, 1763 (1972).
137. M. Finkelstein and S. D. Ross, Tetrahedron, 28, 4497 (1972).
138. M. Finkelstein, K. Nyberg, S. D. Ross, and R. Servin, Acta Chem. Scand., B32, 182 (1978).
139. L. Cedheim, L. Eberson, B. Helgée, K. Nyberg, R. Servin, and M. Sternerup, Acta Chem. Scand., B29, 617 (1975).
140. K. Nyberg and R. Servin, Acta Chem. Scand., B30, 640 (1976).
141. L. Eberson, J. Hlavaty, L. Jonsson, K. Nyberg, R. Servin, H. Sterenrup, and L.-G. Wistrand, Acta Chem. Scand., B33, 113 (1979).
142. K. Nyberg, Synthesis, 545 (1976).
143. M. Sainsbury and J. Wyatt, J. Chem. Soc., Perkin Trans., 1, 108 (1979); 661 (1976); E. Kotani, N. Takeuchi, and S. Tobinaga, J. Chem. Soc., Chem. Commun., 550 (1973); E. Kotani, M. Kitazawa, and S. Tobinaga, Tetrahedron, 30, 3027 (1984); H. Klünenberg and H. J. Schäfer, Tetrahedron Lett., 23, 4581 (1982); C. Schlegel and H. J. Schäfer, (1986) unpublished data; T. Shono, Topics Curr. Chem., 148, 131 (1988).
144. T. Shono, T. Hamaguchi, and H. Matsumura, J. Amer. Chem. Soc., 97, 4264 (1975); H. R. Buhmann, D. Arlt, and H. P. Müller (Bayer A. G.), Ger. Offen. 2,842,760 (1980); Chem. Abstr., 93, 94823r (1980).
145. T. Shono, Y. Matsumura, K. Tsubata, and Y. Sugihara, Tetrahedron Lett., 23, 1201 (1982); J. Amer. Chem. Soc., 104, 6697 (1982); T. Shono, Y. Matsumura, O. Onomura, and Y. Yamamoto, Tetrahedron Lett., 28, 4073 (1987); S. Torii, T. Inokuchi, and M. Kubota, J.

Org. Chem., 50, 4157 (1985); S. Torii, T. Inokuchi, F. Akahoshi, and M. Kubota, Synthesis, 242 (1987).
146. M. Mitzlaff, J. Cramer, and R. Pistorius, Chem.-Ing.-Tech., 54, 601 (1982); Ger. Offen. 2,919,756 (Hoechst A. G., 1980); Chem. Abstr., 94, 92641b (1980).
147. T. Shono, Topics Curr. Chem., 148, in press (1988).
148. M. Mitzlaff, K. Warning, and H. Jensen, Liebigs Ann. Chem., 1713 (1978).
149. M. Okita, T. Wakamatsu, and Y. Ban, J. Chem. Soc., Chem. Commun., 749 (1979); K. Irie, M. Okita, T. Wakamatsu, Y. Ban, and J. Bouveau, Nouv. J. Chim., 4, 275 (1980).
150. T. Shono, Y. Matsumura, and H. Hamaguchi, The 6th Symp. on Oxidation, The Chemical Society of Japan, 1972, abstracts, p. 43.
151. K. Warning and M. Mitzlaff, Tetrahedron Lett., 1563 (1979); M. Mitzlaff, K. Warning, and H. Rehling, Synthesis, 315 (1980).
152. M. Okita, T. Wakamatsu, M. Mori, and Y. Ban, Heterocycles, 14, 1089 (1980); M. Okita, T. Wakamatsu, and Y. Ban, Heterocycles, 20, 401 (1983).
153. M. Okita, M. Mori, T. Wakamatsu, and Y. Ban, Heterocycles, 23, 247, 317 (1985); M. Mori, Stud. Org. Chem. (Amsterdam), 30, 89 (1987).
154. S. D. Ross, M. Finkelstein, and E. J. Rudd, J. Org. Chem., 37, 2387 (1972).
155. T. Shono, Y. Matsumura, K. Tsubata, K. Uchida, T. Kanazawa, and K. Tsuda, J. Org. Chem., 49, 3711 (1984).
156. T. Shono, Y. Matsumura, S. Katoh, K. Inoue, and Y. Matsumato, Tetrahedron Lett., 27, 6083 (1986).
157. D. Degner, H. Hannebaum, and M. Steiniger, (BASF A.-G). Ger. Offen. DE 3,529,531 (1985); Chem., Abstr., 106, P146070b (1987), DECHEMA Monographie, Vol. 112, Organische Elektrochemie-Angewandte Elektrothermie, Verlag Chemie, Weinheim, 1989, pp. 375–379.
158. Z. Blum and K. Nyberg, Acta Chem. Scand., B35, 743 (1981).
159. D. D. Palasz, J. H. P. Utley, and J. D. Hardstone, J. Chem. Soc., Perkin Trans. 2, 807 (1984).
160. T. Shono, Y. Matsumura, and K. Inoue, J. Org. Chem., 48, 1388 (1983).
161. T. Shono, Y. Matsumura, and K. Inoue, J. Chem. Soc., Chem. Commun., 1169 (1983).
162. E. Steckhan, K.-D. Ginzel, J. Heyer, A. Papadopoulos, and B. Lewall, DECHEMA Monographie, Vol. 112, Organische Elektrochemie–Angewandte Electrothermie, Verlag Chemie, Weinheim, 1989, pp. 415–427; A. Papadopoulos, J. Heyer, K.-D. Ginzel, E. Steckhan, Chem. Ber., 122, (1989), in print.
163. K. Irie, K. Aoe, T. Tanaka, and S. Saito, J. Chem. Soc., Chem. Commun., 633 (1985).
164. T. Shono, Y. Matsumura, K. Tsubata, and K. Uchida, J. Org. Chem., 51, 2590 (1986).

16

Anodic Oxidation of Oxygen-Containing Compounds

OLE HAMMERICH and BO SVENSMARK / H. C. Ørsted Institute, University of Copenhagen, Copenhagen, Denmark

I. Introduction 615

II. Anodic Oxidation of Phenols, Hydroquinones, and Derivatives 616
- A. Simple phenols, hydroquinones, and hydroquinone monoethers and monoesters 616
- B. Aryl ethers and aryl esters 628

III. Anodic Oxidation of Alcohols and Derivatives 635
- A. Benzyl alcohols and related compounds 635
- B. Aliphatic alcohols and aldehydes 638
- C. Carbohydrates and related bifunctional compounds 644

IV. Anodic Oxidation of Other Aliphatic Oxygen-Containing Compounds 647
- A. Ethers without other functional groups 647
- B. Unsaturated ethers 648
- C. Alkoxyalcohols 648
- D. Peroxides 649

References 650

I. INTRODUCTION

The period between the second and third edition of this book has been a time of active research in the area of anodic oxidation of oxygen-containing compounds and in particular as regards the products and mechanisms of oxidation of phenols and aliphatic alcohols.

Contrary to the common belief during the dawn of organic electrochemistry, it is now well established that oxidation of phenolic compounds may indeed lead to valuable products. A major reason for this success is the understanding that these reactions are very sensitive to the experimental conditions and that different routes may be followed as the result of what at first glance may seem to be only minor changes of the solvent, supporting electrolyte, and electrode material.

The increasing interest in fuel cells and electrochemical detectors has advanced knowledge of the mechanistic details of the anodic oxidation of certain aliphatic alcohols and carbohydrates, in particular methanol and glucose, and the basic characteristics of these reactions are now much better understood than only a few years ago.

Because of this progress, it has been possible to give a more general presentation of some of the material in the chapter, but as in the previous editions of the book much work in the area belongs primarily in other chapters and is therefore not treated here. To get a more complete picture of the anodic oxidation of oxygen-containing compounds, the reader should also consult Chaps. 14, 18, 20, 23, 24, 25, 29, and 33.

Since this chapter is intended as a guide, not as a critical review, we have purposely omitted references to work from which, in our opinion, the conclusions are not well founded and work that does not illustrate typical behavior of the title compounds. The presentation is at places rather brief to avoid too lengthy a chapter, and only the essentials of some of the reactions are covered. We have tried to compensate for this by giving leading references instead.

II. ANODIC OXIDATION OF PHENOLS, HYDROQUINONES, AND DERIVATIVES

A. Simple Phenols, Hydroquinones, and Hydroquinone Monoethers and Monoesters

The Primary Oxidation Steps

The anodic oxidation of phenols and related aromatic hydroxy compounds (I) takes place according to one or both of the two routes presented in Eq. (1) dependent on the acid-base properties of the solvent-electrolyte system [1–7].

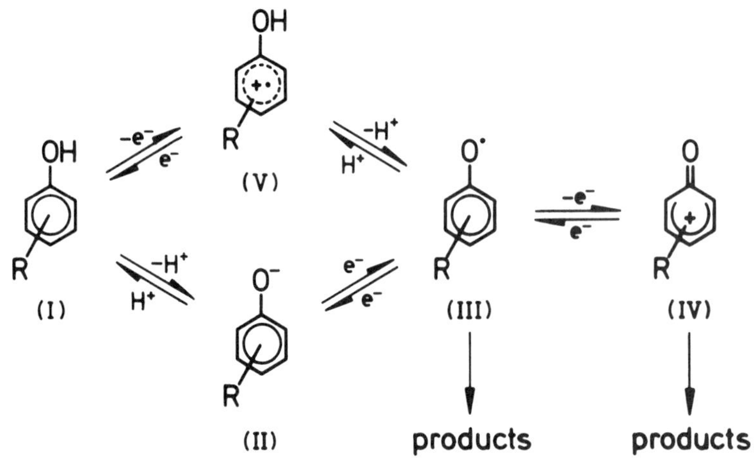

(1)

The most simple and straightforward behavior during cyclic voltammetry (CV) is observed for solutions containing only phenolate ions (II) carrying bulky substituents like tert-butyl or aryl in the 2-, 4-, and 6-positions. In these cases the oxidation proceeds as a reversible one-electron process leading to the corresponding phenoxy radicals (III) [5–9], which are stable on the time scale of a conventional voltammetric experiment [5–10]. The peak potentials E_{pa} are in the range -0.3 to -0.1 V versus SCE (see Table 1).

Table 1. Examples of Voltammetric Peak Potentials for the Reversible Oxidation of Simple Phenols, Phenolate Ions, and Phenoxy Radicals

Compound	E_{pa} (V versus SCE)	Solvent system	Reference
Phenols			
2,4,6-Trimethyl	1.65	CH_2Cl_2-$HFSO_3$	21
4-Phenyl	1.59	CH_2Cl_2-$HFSO_3$	21
4-Methoxy	1.49	CH_2Cl_2-$HFSO_3$	21
Hydroquinone	1.3	CH_2Cl_2-$HFSO_3$	22
Phenolate ions			
2,4,6-Tri-tert-butyl	-0.27	MeCN, Me_4NOH	6
4-Isopropyl-2,6-di-tert-butyl	-0.31	MeCN, Me_4NOH	6,7
4-(2-Hydroxy-2-propyl)-2,6-di-tert-butyl	-0.16	MeCN, Me_4NOH	6
4-(4-Dimethylaminophenyl)-2,6-di-tert-butyl	-0.33	MeCN, Et_4NClO_4	8
Phenoxy radicals			
2,4,6-Tri-tert-butyl	1.2[a]	MeCN, Me_4NOH	6
4-Phenyl-2,6-di-tert-butyl	1.10	MeCN, Et_4NClO_4	11
4-(4-Methoxyphenyl)-2,6-di-tert-butyl	0.79	MeCN, Et_4NClO_4	11
4-(4-Dimethylaminophenyl)-2,6-di-tert-butyl	0.26	MeCN, Et_4NClO_4	8

[a]Complete reversibility of the couple was not observed.

Extension of the voltammetric sweep into the 0.3–1.0 V region results in further oxidation of the phenoxy radicals to phenoxonium ions (IV, Table 1) [6,8,11–13].

Phenol cation radicals (V) are strongly acidic species with pK_a values in the range -2 to 0 [14–18] unless the positive charge is stabilized by the presence of electron-donating groups, like N,N-dimethylamino [8,19] or methoxy [20]. Consequently, the reversible oxidation of simple phenols to the corresponding cation radicals takes place only under conditions sufficiently acidic to suppress deprotonation of the cation radicals as has been observed in CH_2Cl_2-$HFSO_3$ at -50°C [21,22]. The values of E_{pa} observed during CV in the presence of $HFSO_3$ were between 1.3 and 1.65 V (Table

1). Under less acidic or aprotic conditions the primarily formed cation radicals deprotonate rapidly to the neutral radicals III, which owing to their lower oxidation potentials (see Table 1) may be further oxidized to the corresponding phenoxonium ions (IV). This second electron transfer may take place at the electrode and/or in solution, Eq. (2), dependent on the experimental conditions [23]:

(2)

Products Obtained from the Phenoxy Radicals

The phenoxy radicals derived from phenols carrying substituents less bulky than isopropyl in the 4-position exist in oxygen-free solution in equilibrium with the corresponding dimers Eq. (3) [5,7,9]:

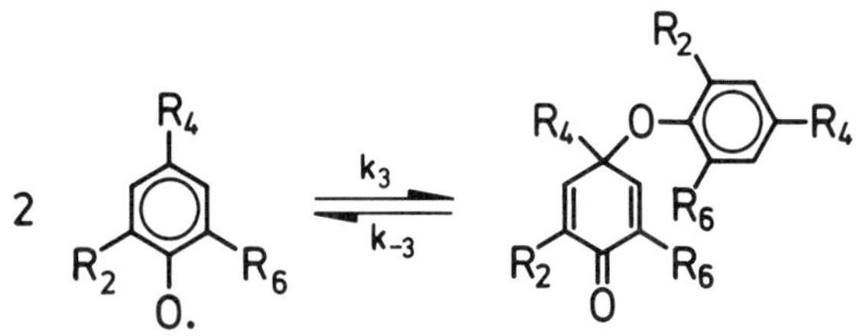

(3)

Values of k_3 and K_3 ($=k_3/k_{-3}$) have been determined by CV in MeCN at room temperature for the cases in which R_2 = R_6 = tert-Bu and R_4 = Me, Et, and Bu [9]. The results (Table 2) were found to be in good agreement with those obtained earlier by flash photolysis in CCl_4 [24].

Table 2. Rate and Equilibrium Constants for the Reversible Dimerization of 4-Substituted 2,6-Di-tert-Butylphenoxy Radicals Determined by Cyclic Voltammetry[a]

R_4	k_3 ($M^{-1} \cdot s^{-1}$)	K_3 (M^{-1})
Me	1.0×10^8	1.6×10^6
Et	3.2×10^6	2.0×10^5
Bu	2.5×10^7	1.6×10^5

[a]In MeCN at 18–21°C.
Source: From Ref. 9.

However, on a time scale more extended than that of CV, quite different products are usually obtained as a result of anodic oxidation of phenolate ions. Phenoxy radicals bearing alkyl substituents with an α-hydrogen atom, such as isopropyl, in the 4-position disproportionate readily in solution to equimolar amounts of the parent phenol and the corresponding quinone methide (VI), Eq. (4), which in turn slowly tautomerizes to the alkenylphenolate ion (VII), Eq. (5), under the alkaline conditions usually employed [7]. At the applied potential, the anion (VII) is further oxidized to the radical, which dimerizes to VIII, Eq. (6). The relative importance of the steps in Eqs. (4) through (6) is strongly dependent on both the structure of the 4-alkyl group and the composition of the solvent (see Ref. 7 for details).

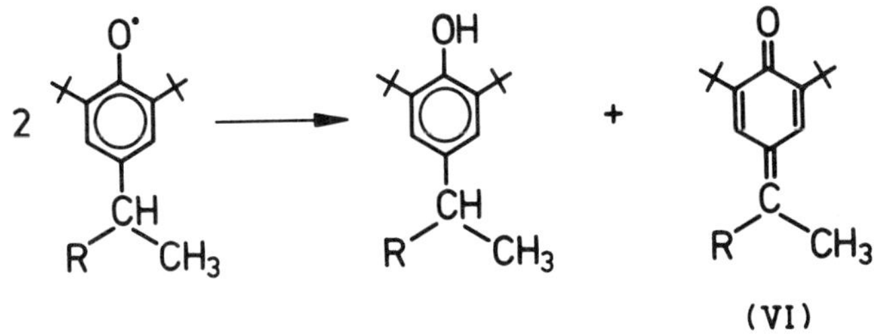

(4)

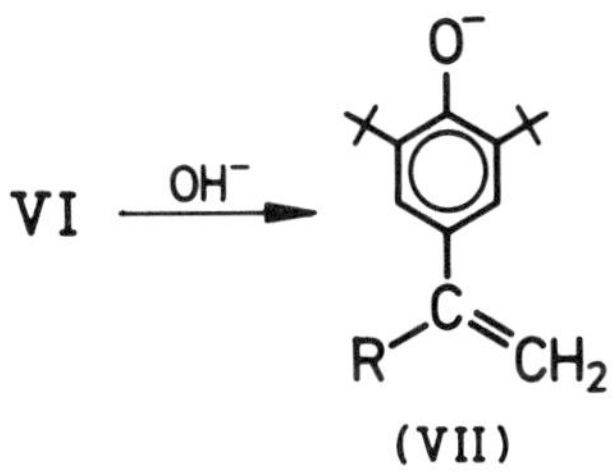

(5)

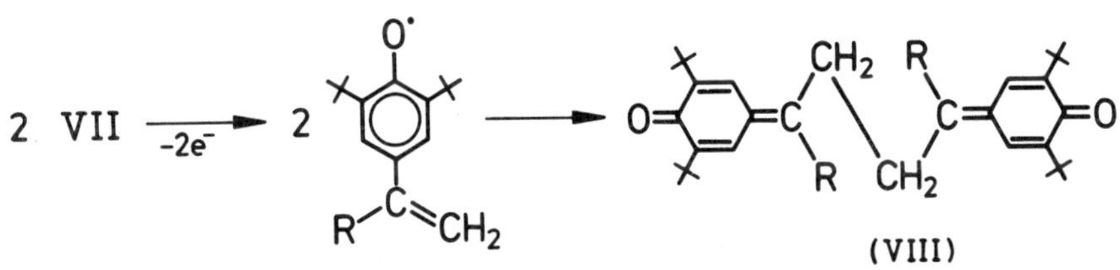

(6)

Biphenyl derivatives are the products formed when the parent phenols are even less substituted, probably through the reversible formation of a C-C dimer, which tautomerizes irreversibly to the final product [4,25]. For example, oxidation of vanillinate in MeCN was found to give the C-C dimer (IX) in 65% yield, Eq. (7) [26]:

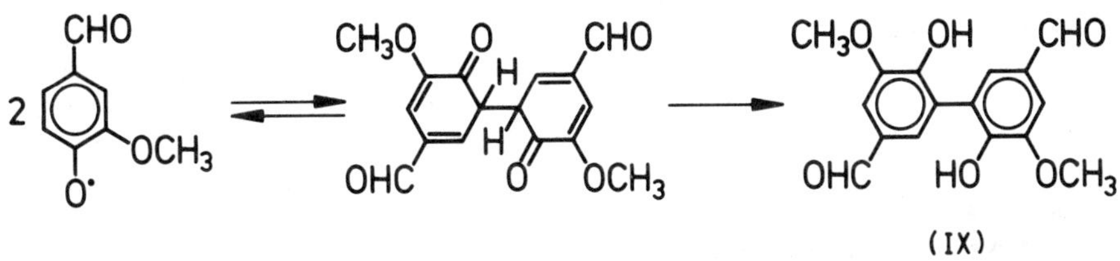

(7)

Biphenyls may also be obtained as products when the phenol itself, not the conjugate base, is the starting material [27–31] (see also references cited in Ref. 27), and it has been proposed that the dimers in these cases may also be the result of phenoxy radical coupling [28–31]. The formation of the conducting polymeric material obtained by anodic oxidation of phenol itself in MeCN [32] probably involves phenoxy radicals as well.

In general, the structures of the dimers are difficult to predict because of the interplay between electronic and steric factors and the effects of temperature, solvent, and substrate concentration [33–39].

Dimers, which are formally peroxides, are formed when the oxidation is carried out in the presence of oxygen, Eq. (8) [40]:

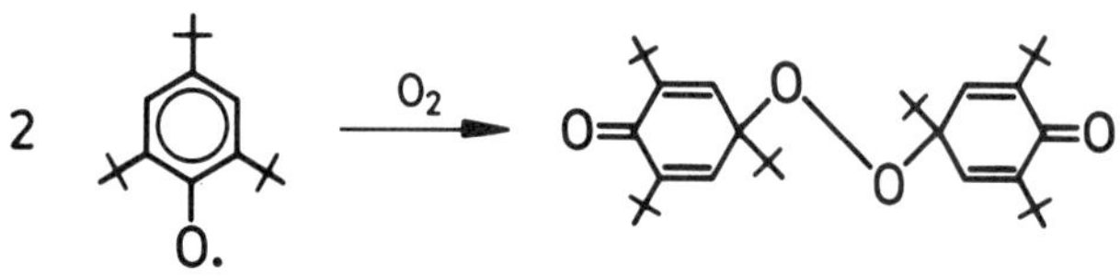

(8)

Further details on oxidative coupling reactions are discussed in Chap. 23.

Products Resulting from Deprotonation of Phenoxonium Ions

Phenoxonium ions containing a p-hydroxy group are formally protonated p-benzoquinones and as such very acidic [3,4,22,41–43]. Consequently, the dominating follow-up reaction is deprotonation to the corresponding p-benzoquinone derivatives (X), Eq. (9) (see Ref. 44, and references cited therein), which are also the products isolated after exhaustive electrolysis [27].

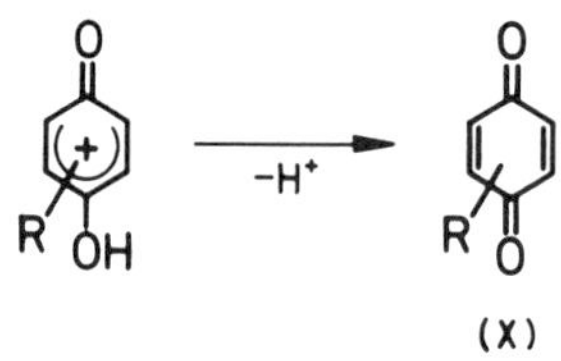

(X)

(9)

The electrochemical kinetics of the hydroquinone-quinone system has been studied under conditions in which the proton transfer reactions can be considered equilibrium processes [45].

A preparatively interesting variation of Eq. (9) is observed for the anodic oxidation of hydroquinone bearing an electron-withdrawing substituent (XI) (R = $-NO_2$, $-CHO$, or $-COCH_3$) in the 2-position [46,47]. Under aqueous conditions or in aprotic solvents containing water, the primarily formed quinone (XII) undergoes nucleophilic addition of water, resulting in the formation of a trihydroxybenzene derivative (XIII) that is further oxidized to the corresponding quinone (XIV), Eq. (10). Similar reactions have been reported to take place during the anodic oxidation of hydroquinone in MeCN in the presence of pyridine, in which case N-(2,5-dihydroxyphenyl)pyridinium ion is formed [48], and of 4-dimethylaminophenol in aqueous solution [49]. In the latter example the nucleophile is not water but dimethylamine formed in situ by hydrolysis of electrochemically generated 4-dimethylaminobenzoquinoneimine cation.

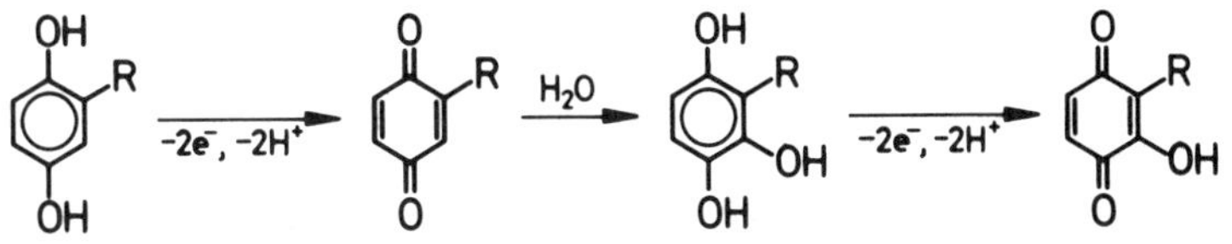

(XI) (XII) (XIII) (XIV) (10)

More exhaustive dehydrogenation may result when the starting material is an aromatic dihydroderivative. For example, vitamin K_3 (XV) has been obtained in an almost quantitative yield by anodic oxidation of 2-methyl-5, 8-dihydro-1,4-naphthalenediol in MeCN containing t-BuOH, Eq. (11) [50]:

(XV) (11)

The anodic oxidation of catechol derivatives proceeds analogously to the oxidation of the hydroquinones, resulting in formation of o-quinones [47,51–54]. However, the o-quinones are in general not the final products because of the high reactivity of these compounds, and in the absence of a suitable trapping agent the overall reaction may lead to rather complicated product mixtures including dimeric compounds and their oxidation products. An example of the successful application of an o-quinone trapping agent is the oxidation of catechol in the presence of 4-hydroxycoumarin, Eq. (12) [51]:

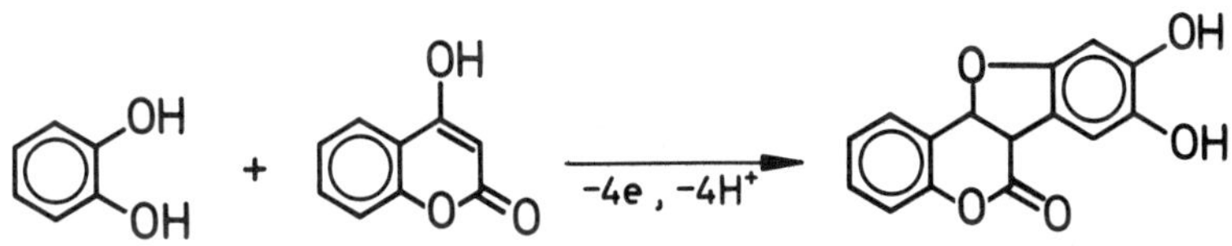

(12)

Deprotonation of alkyl-substituted phenoxonium ions to the corresponding quinone methides in MeCN has been proposed in a number of cases to account for the formation of different types of products [26,55,56]. However, it has been demonstrated [57] that the formation of the quinone methide is a very unlikely process, at least in one case [26], and it is not clear whether deprotonation of alkylphenoxonium ions to quinone methides in general takes place under the conditions of most anodic oxidation reactions.

Products Resulting from the Reaction Between Phenoxonium Ions and Nucleophilic Reagents

The predominant reaction of phenoxonium ions formed by oxidation of 2,4,6-trisubstituted phenols is nucleophilic attack by the solvent or solvent impurities (water), intentionally added nucleophiles, or the parent phenol [5,6,11,26,29–31,40,58–89]. The primary products are 2- or 4-substituted cyclohexadienones, Eq. (13),

Table 3. Yields and Experimental Conditions for Formation of Dienones and Quinones from Simple Phenols[a]

Starting material	Product	Yield (%)	Anode	Solvent system	Reference
Phenol	Dienone[b]				
Phenol	4,4-Dimethoxy	46	C	MeOH, $LiClO_4$, $NaHCO_3$	31
4-Methyl	4-Methyl-4-hydroxy	50	PLE[c]	H_2SO_4 (2 N)	30
2,4-Dimethyl	4-Methyl-4-hydroxy	44	PLE[c]	MeCN-H_2SO_4 (2 N)	30
2,6-Dimethyl	4,4-Dimethoxy	28	Pt	MeOH, $LiClO_4$, $NaHCO_3$	31
2,6-Di-tert-butyl	4,4-Dimethoxy	78	C	MeOH, $LiClO_4$, $NaHCO_3$	31
2,4,6-Trimethyl	4-Methyl-4-hydroxy	81	PLE[c]	MeCN-H_2SO_4 (2 N)	30
2,6-Di-tert-butyl-4-methyl	4-Methyl-4-methoxy	95	C	MeOH, $LiClO_4$, $NaHCO_3$	31
2,6-Di-tert-butyl-4-methyl	4-Methyl-4-hydroxy	86	Pt	MeCN-H_2O-2,6-DMP,[d] $LiClO_4$	58
2,6-Di-tert-butyl-4-methyl	4-Methyl-4-acetoxy	91	Pt	MeCN-H_2O, $LiClO_4$, NaOAc	58
2,6-Di-tert-butyl-4-methyl	4-Methyl-4-anisyl	81	Pt	MeCN-anisole, Bu_4NBF_4	64
2,6-Di-tert-butyl-4-ethyl	4-Ethyl-4-hydroxy	93	Pt	MeCN-H_2O, $NaClO_4$	59
2,4,6-Tri-tert-butyl	4-tert-Butyl-4-hydroxy	97	Pt	MeCN-H_2O-2,6-DMP,[d] $LiClO_4$	58
2,4,6-Tri-tert-butyl	4-tert-Butyl-4-methoxy	93, 95	Pt	MeCN-MeOH, ClO_4^-	58, 59
2,4,6-Tri-tert-butyl	4-tert-Butyl-4-acetoxy	94	Pt	MeCN-H_2O, NaOAc	58

2,6-Di-*tert*-butyl-4-phenyl	4-Phenyl-4-(Gly-OH)	90	Pt	CH_2Cl_2-2,6-DMP[d], Et_4NBF_4	61
2,6-Di-*tert*-butyl-4-phenyl	4-Phenyl-4-(L-Ala-OH)	94	Pt	CH_2Cl_2-2,6-DMP[d], Et_4NBF_4	61
2,6-Di-*tert*-butyl-4-phenyl	4-Phenyl-4-(L-Pro-OH)	57	Pt	CH_2Cl_2-2,6-DMP[d], Et_4NBF_4	61
2,4,6-Triphenyl	4-Phenyl-4-hydroxy	87	Pt	MeCN-H_2O, $NaClO_4$	59
2,4,6-Triphenyl	4-Phenyl-4-methoxy	80	Pt	MeCN-MeOH, $NaClO_4$	59
Phenol	Quinone				
Phenol	*p*-Benzoquinone	37	Pt	MeOH, $LiClO_4$, $NaHCO_3$	31
2-Methyl	2-Methyl	61	PLE[c]	H_2SO_4 (2 N)	30
3,5-Dimethyl	3,5-Dimethyl	84	PbO_2	H_2SO_4 (2 N)	88
2,6-Dimethyl	2,6-Dimethyl	56	PLE[c]	H_2SO_4 (2 N)	30
2,6-Dimethyl	2,2',6,6'-Tetramethyldiphenoquinone	90	Pt	MeCN, $LiClO_4$	89
2,6-Di-*tert*-butyl	2,2',6,6'-Tetra-*tert*-butyldiphenoquinone	85	Pt	MeCN, $LiClO_4$	89
2,4,6-Tri-*tert*-butyl	2,6-Di-*tert*-butyl-*p*-benzoquinone	80	Pt, C	MeCN-H_2O, $NaClO_4$	62

[a]The contents of this table are not exhaustive. Other examples may be found in the papers cited.
[b]Only the 4-substituents are given. Substituents in other positions are unchanged.
[c]Polished Lead Electrode
[d]2,6-Dimethylpyridine

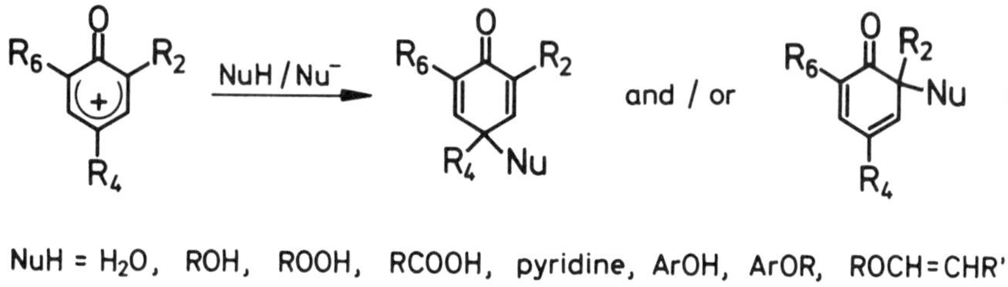

(13)

which may be isolated, often in good yields (see Table 3), if the anodic oxidation is carried out in aqueous or alcoholic solution or in aprotic solvents in the absence of strong acids. Water [30,58,59], alcohols [26,31,58–60], carboxylic acids or carboxylate ions [5,58,59], amino acids [61], pyridine [62,63], aromatic hydrocarbons carrying electron-donating substitutents, Eq. (14) [64], and alkene derivatives, Eq. (15) [65], have successfully been used as nucleophiles.

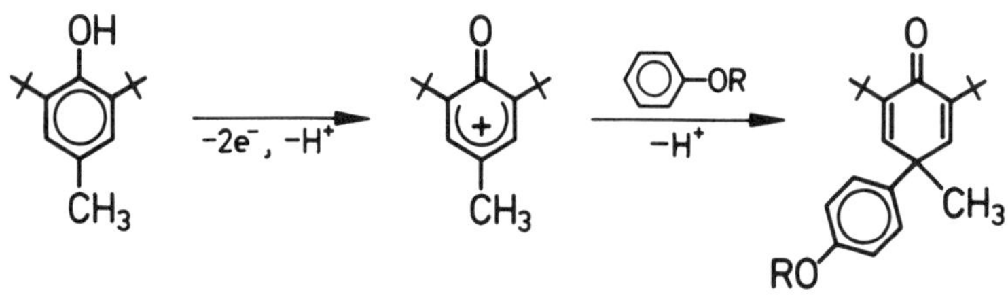

(14)

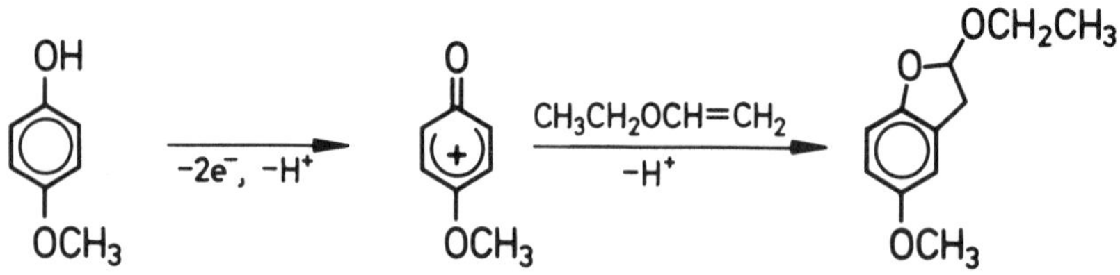

(15)

Under strongly acidic conditions dienones undergo acid-catalyzed rearrangement, resulting in the reestablishment of the aromatic nucleus as illustrated in Eq. (16) (Ref. 6 and references cited therein). If the nucleophilic reagent is water, the product of this rearrangement is, as in Eq. (16), a hydroquinone derivative (XVI), which in general is further oxidized to the corresponding quinone under the reaction conditions, Eq. (17).

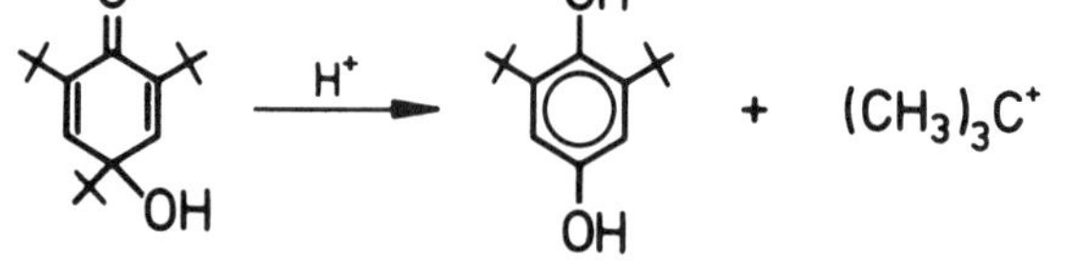

(XVI)

(16)

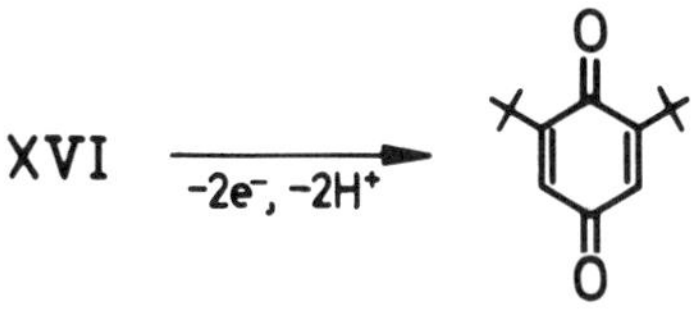

(17)

When R_4 = H the dienones are so sensitive toward acid that even the protons released during the anodic oxidation in aqueous solution are sufficient to stimulate the rearrangement to the hydroquinone. In these cases quinones may be obtained in good yields from the corresponding phenols (Table 3).

If special precautions are taken, such as moderate conversion [66] or addition of Cu(II) salts to the solvent system [67], the further oxidation of the hydroquinone can be suppressed so effectively that the direct oxidation of phenol to hydroquinone, for example, is a process of economical interest. More extensive oxidation results in mixtures of p-benzoquinone and maleic acid and, ultimately, in the formation of carbon dioxide [68].

If the nucleophilic center is part of the parent phenol as in (XVII), (XVIII), and (XIX), the anodic oxidation results in formation of new heterocyclic ring systems, Eqs. (18) and (19) [69,70], or in spirodienones (XX), which upon treatment with trifluoroacetic acid, for example, are converted to new biphenyl derivatives, Eq. (20) [71,72]. The example given in Eq. 20 is only one of many in this field of chemistry. Other examples of intramolecular coupling reactions are discussed in Chap. 23.

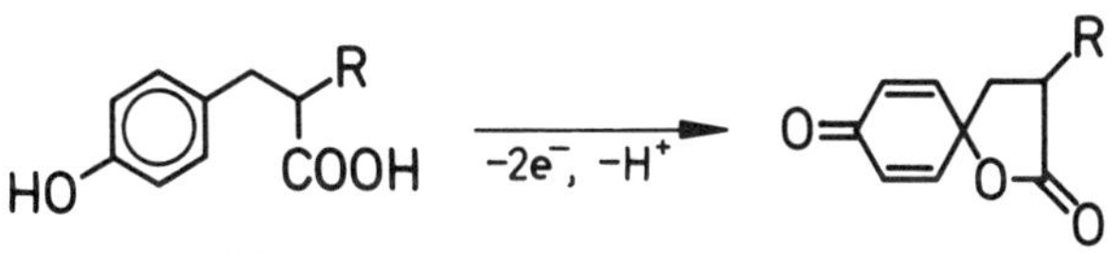

(XVII)

(18)

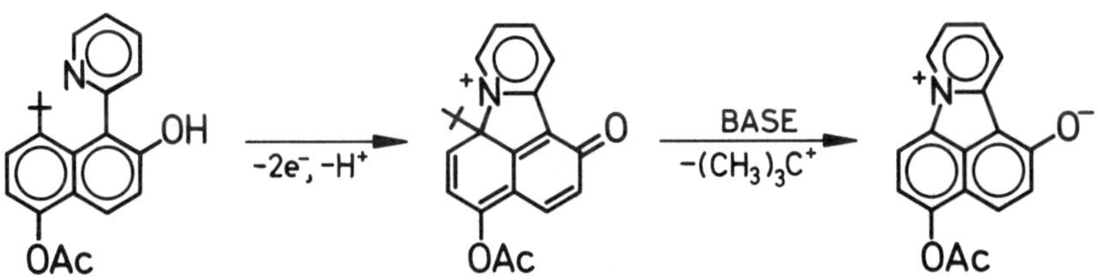

(XVIII)

(19)

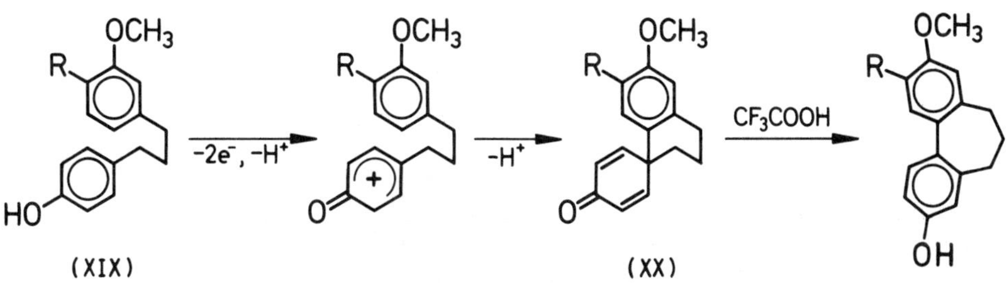

(XIX) (XX)

(20)

When MeCN is used as the solvent in the absence of intentionally added nucleophiles, the solvent itself may react with the phenoxonium ion in a Ritter-type reaction. The result is formation of benzoxazole derivatives, Eq. (21), which may be isolated in good yields [73].

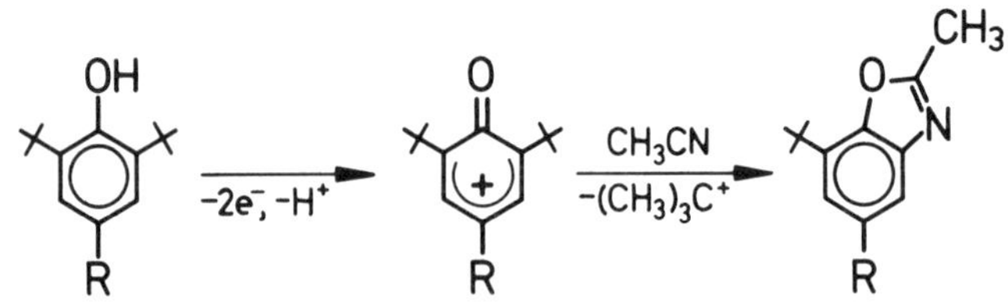

(21)

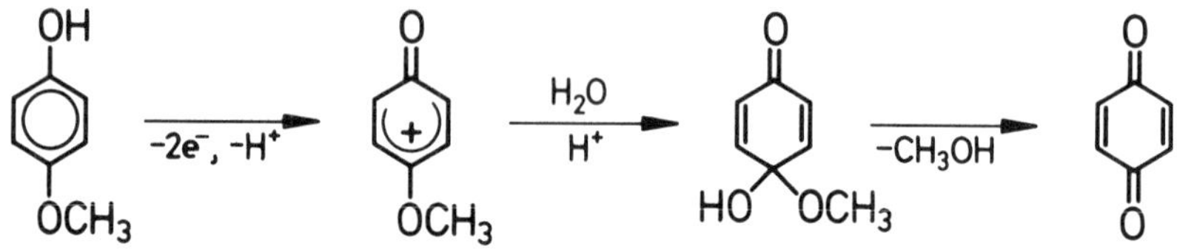

(XXI)

(22)

Anodic oxidation of phenols bearing methoxy groups in the 2- or 4-positions in the presence of water gives o- or p-benzoquinones or products derived from the quinones [28,74–76]. The intermediate dienone (XXI) is unstable under these conditions and eliminates methanol, Eq. (22), which has been detected by application of a specially constructed anode connected directly to a mass spectrometer [74].

A class of alkoxyphenols that has attracted special attention are the α-tocopherol model compounds (XXII) [77–79]. The application of CV has shown that XXII undergoes a "reversible" two-electron oxidation to the dication, Eq. (23) [77]:

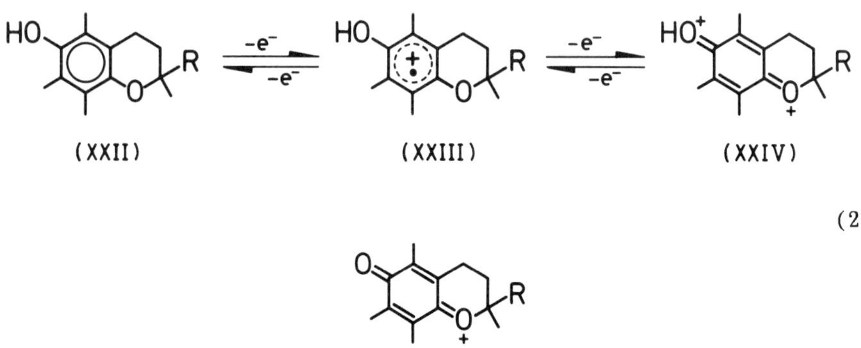

(XXII) (XXIII) (XXIV)

(XXV)

(23)

Spectral measurements in the presence of strong acid indicated that the structure of the dication was indeed XXIV [78]. Both the cation radical (XXIII) and the corresponding phenoxonium ion (XXV), which could be prepared in MeCN, were found to be remarkably stable.

Hydroquinone monoesters behave similarly to the monoethers upon anodic oxidation in aqueous solutions [80,81], ultimately giving quinones. However, when an alcohol is used as the solvent anodic transacylation may be observed, Eq. (24) [82,83]:

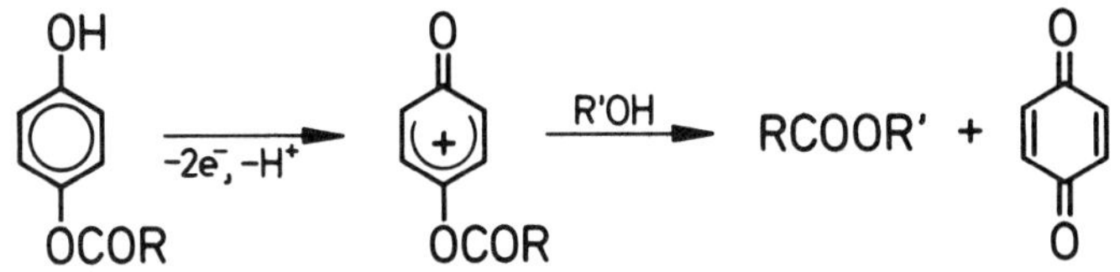

(24)

Anodic oxidation of phenols with alkenyl substituents give rise to mixtures of structurally fairly complicated products, Eqs. (25) through (27) [84–87]:

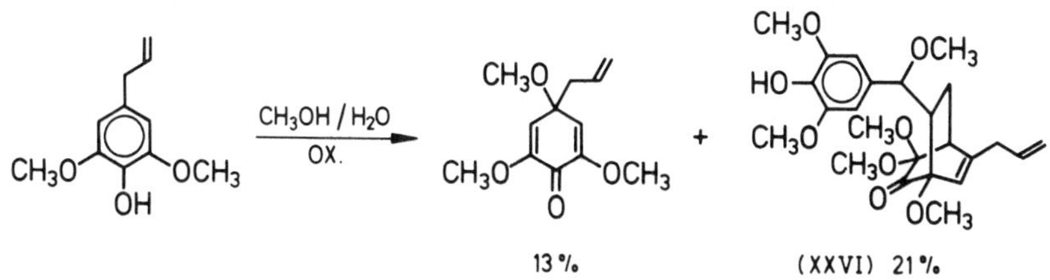

(25)

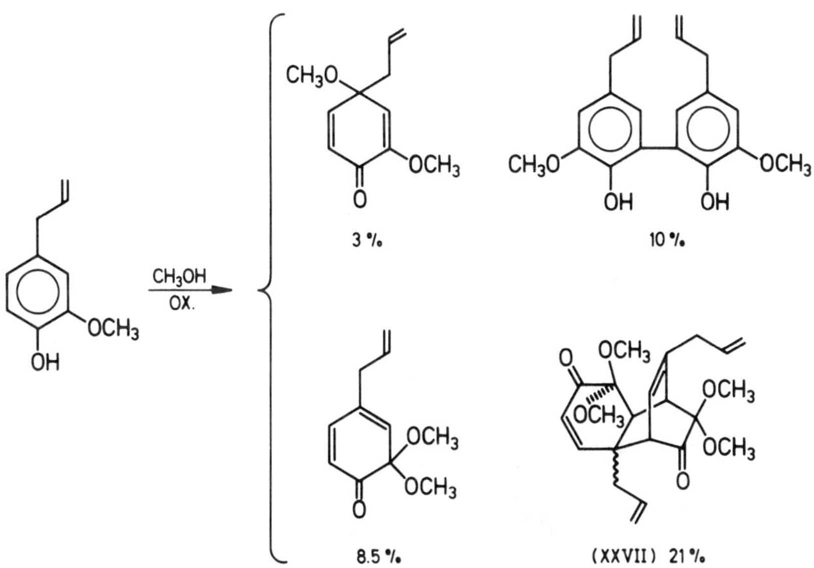

(26)

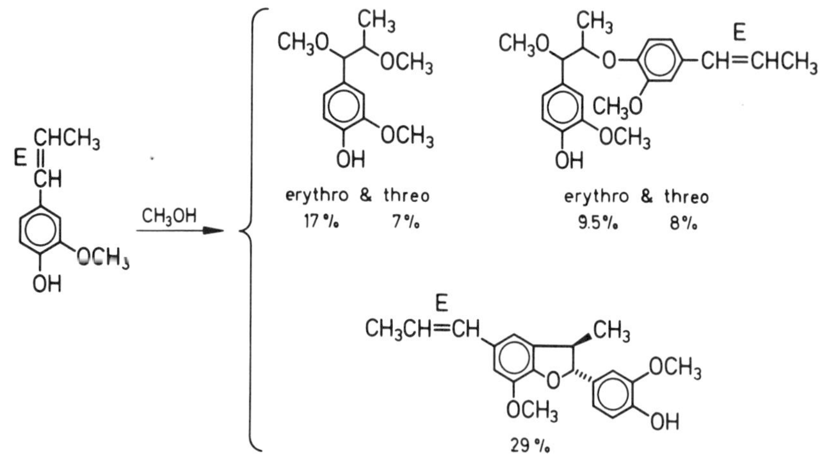

(27)

However, although the yields of even the major products are only modest, the reactions are of synthetic interest because of the multistep procedures necessary to prepare these products by conventional synthetic procedures. Products derived from both 2- and 4- substituted phenoxonium ions are observed along with compounds like XXVI and XXVII, which have their origin in Diels-Alder reactions of the o-dienones.

B. Aryl Ethers and Aryl Esters

The effect of conformation on stability and reactivity has been studied for the cation radicals and dications related to a series of methoxy-substituted biphenyls [90]. The oxidation potentials were found to increase in going from XXVIII to XXXII, Eq. (28), reflecting the increasing steric hindrance to coplanarity of the two phenyl rings.

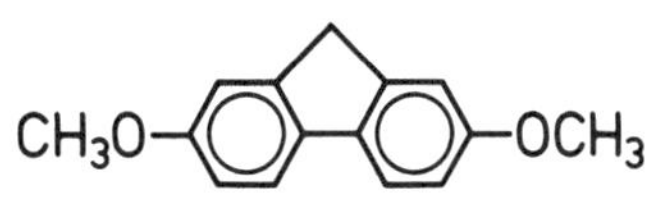

(XXVIII)

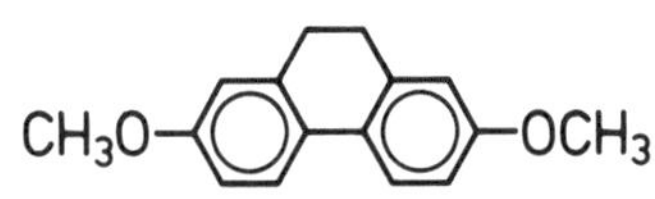

(XXIX)

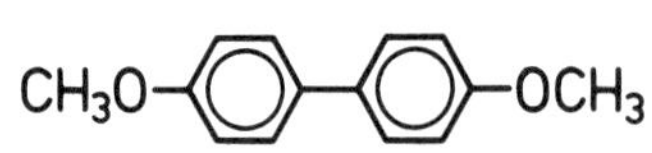

(XXX)

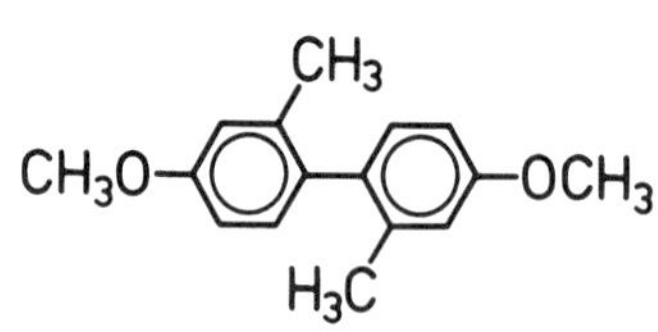

(XXXI)

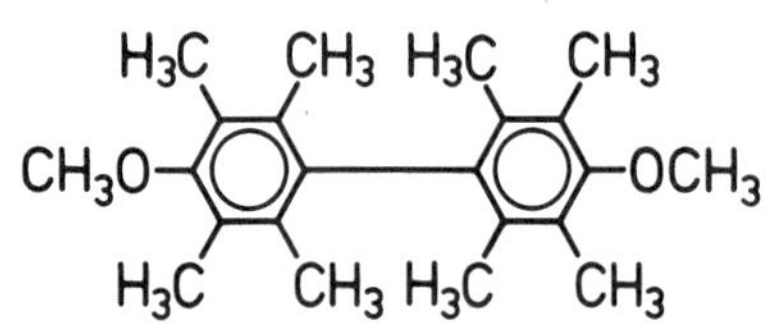

(XXXII)

(28)

In XXVIII the phenyl rings are held in a nearly coplanar conformation owing to the presence of the methylene bridge. For XXIX the preferred conformation is one in which the two rings are twisted slightly out of plane. The free rotation around the biphenyl linkage in XXX allows a large out-of-plane angle, resulting in a lower probability for the coplanar conformation. The compounds XXXI and XXXII illustrate the effect of orthosubstitution with the expected result. However, in the case of XXXII an additional conformational effect contributes to determining the oxidation potential. Because of the presence of two methyl groups ortho to the methoxy substituent, the latter is forced out of the plane of the phenyl ring, which diminishes the electron-donating effects of the methoxy groups and consequently oxidation to the cation radical is made more difficult.

The reversible formation of dications was observed for XXVIII, XXIX, and XXX and for the hexamethoxytriphenylene (XXXIII), Eq. (29), and even the trication has been observed during CV [91].

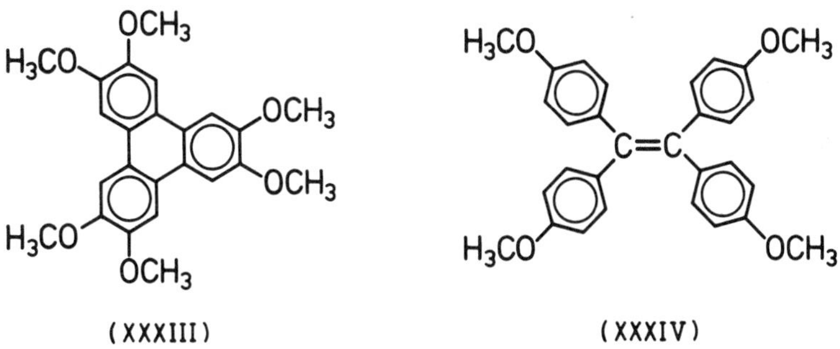

(29)

The voltammetry of tetra-p-anisylethylene (XXXIV), Eq. (29), has been carefully investigated [92,93] and the results demonstrate the effect of solvent polarity on the disproportionation equilibrium for the cation radical. The application of CV in CH_2Cl_2 showed two successive one-electron transfer couples separated by only 90–100 mV. When MeCN was gradually added to the solvent the potential difference decreased and it was shown that the difference was linearly related to the proportion of MeCN in the solvent mixture. In pure MeCN the potential difference is reduced to only 7 mV.

The cation radicals of methoxy-substituted aromatic hydrocarbons are among the most stable derived from simple organic structures. In spite of this inherent stability they are, as intuitively expected, susceptible to nucleophilic attack. The reactions are of two kinds; substitution (see Chap. 25) and addition. The prototypes of the latter reaction are the anodic oxidation of 1,4-, 1,3-, and 1,2-dimethoxybenzene in methanol containing KOH, usually 1%, resulting in formation of the bis-quinone acetals (XXXV, XXXVI, and XXXVII), Eqs. (30) through (32) [94,95]:

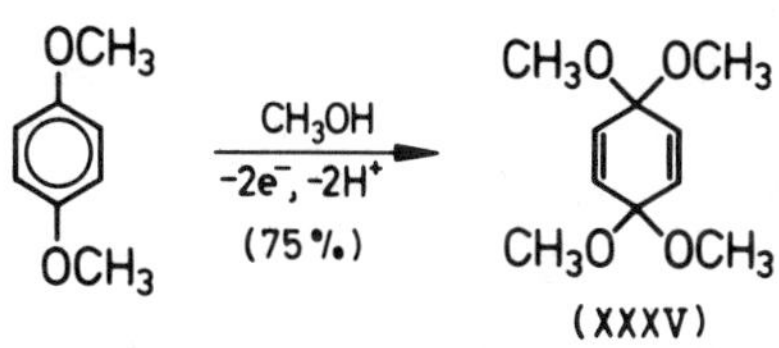

(30)

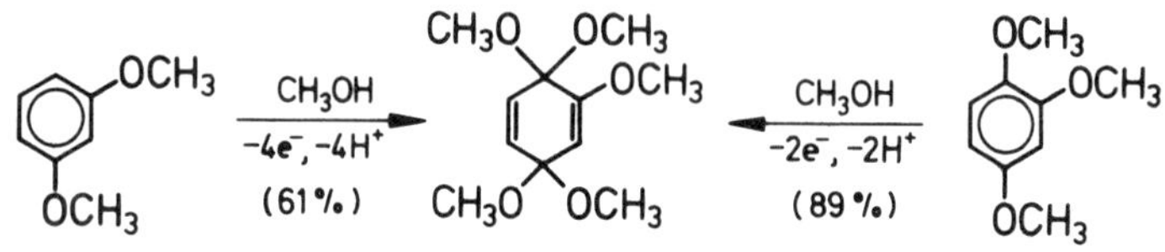

(XXXVI)

(31)

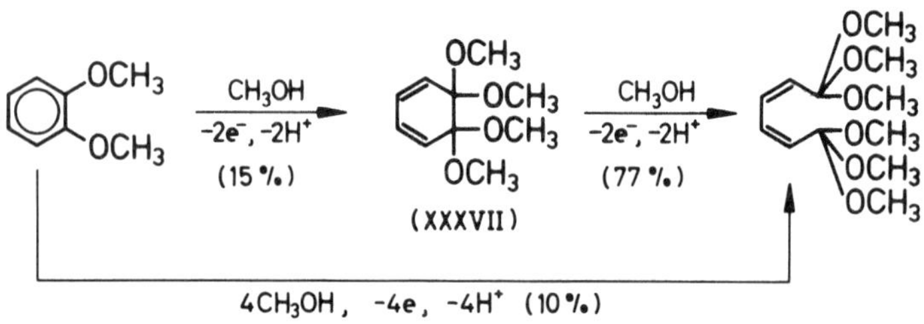

(32)

Further oxidation of XXXVII causes ring cleavage accompanied by the introduction of two additional methoxy groups. This work has stimulated a thorough investigation of the synthetic scope of the reaction [96–112] (see also references cited in Refs. 108 and 109).

The addition appears to a general process for 1,4-dialkoxybenzenes as it was demonstrated by anodic oxidation of a series of compounds (XXXVIII) in which R was equal to Et, Pr, i-Pr, Bu, i-Bu, and s-Bu [102]. The corresponding 3,6-dialkoxy-3,6-dimethoxycyclohexa-1,4-dienes (XXXIX), Eq. (33), were obtained in good yields in all cases and were shown to be approximately equimolar mixtures of the cis and trans isomers.

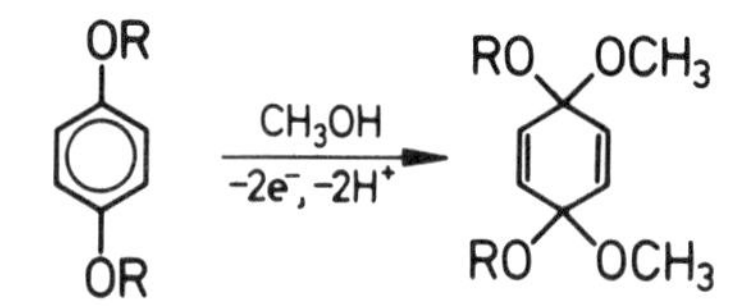

cis / trans ≈ 1/1

(XXXVIII) (XXXIX)

(33)

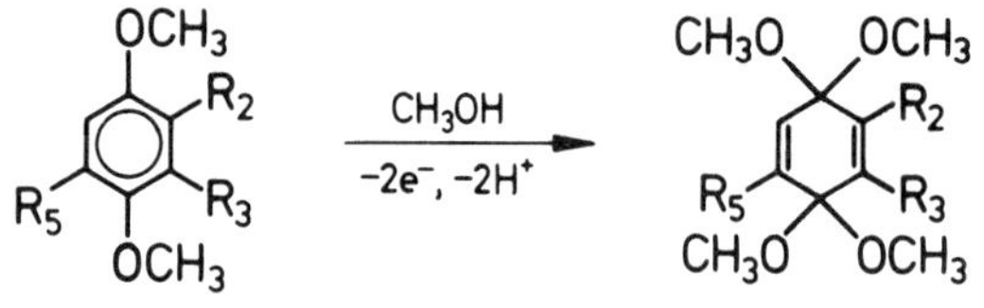

(34)

R = Et, Pr, i-Pr, Bu, i-Bu & s-Bu

The effect of substitution has been the subject of a detailed investigation [103]. Alkyl and halogen substituents do not affect the practical applications of the reaction. However, precautions should be taken when the substituents themselves are electroactive. In the case of oxidizable substituents it may be necessary to protect the substituents before the anodic oxidation, and a divided cell is necessary when the aromatic starting material carries

Table 4. Yields of Substituted Tetramethoxycyclohexadienes Formed by Anodic Oxidation of Substituted 1,4-Dimethoxybenzenes in Methanol[a]

R_2	R_3	R_5	Yield (%)	Current efficiency (%)	Cell type
Br	H	H	78	55	Undivided
Br	H	Br	58	—	Undivided
CH_3	H	H	80	74	Undivided
CH_3	CH_3	CH_3	63	66	Undivided
$CHOHCH_3$	H	H	50	32	Undivided
$CH_2CH{=}CH_2$	II	II	81	71	Undivided
$NHCOCH_3$	H	H	17	78	Divided
$CH_2CH_2COOCH_3$	H	H	61	80	Divided
$CH{=}CHCOOCH_3$	H	H	46	49	Divided
CHO	H	H	59[b]	59[b]	Divided

[a]At platinum anodes unless otherwise indicated.
[b]At vitreous carbon.
Source: From Ref. 103.

reducible substituents like $-CHO$ and $-CH{=}CHCOOCH_3$. Typical yields are given in Table 4, where the substituents refer to Eq. 34.

The addition reaction is believed to follow one or more of the three pathways shown in Eq. (35).

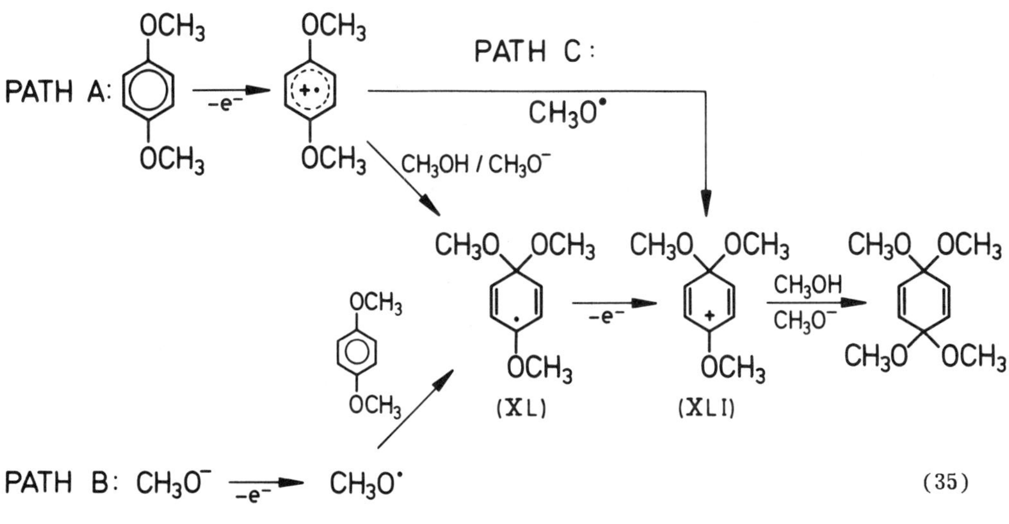

(35)

In pathway A, 1,4-dimethoxybenzene is first oxidized to the cation radical, which undergoes nucleophilic attack by methanol-methoxide. The neutral free radical (XL) formed in this way is then further oxidized to the corresponding cation (XLI), which finally reacts with methanol-methoxide. However, the radical XL may also be formed by pathway B in which the first step is one-electron oxidation of methoxide ion to the corresponding radical. The intermediate radical XL is then formed by reaction between the methoxy radical and 1,4-dimethoxybenzene. In pathway C the cation XLI is formed directly from the 1,4-dimethoxybenzene cation radical and methoxy radical.

It is not straightforward to distinguish between these three pathways. Although the reversible oxidation potential of substrates like 1,4-dimethoxybenzene are well defined and not too difficult to measure in most cases, it is a completely different matter with the methanol-methoxide mixture, the oxidation potential of which is strongly dependent on the reaction conditions. The oxidation potential of the methanol-methoxide mixture depends among other factors on the actual proton activity, which again is influenced by the water content, the electrode material, and the temperature, which influence the adsorption of the methoxy radicals formed in the electron transfer step and the supporting electrolyte concentration. Therefore, unless the aromatic starting material is much easier to oxidize than the methanol-methoxide mixture, in which case pathway A seems to be followed, one must rely on other types of experimental evidence, such as detailed product analysis as a function of systematic variation of the experimental parameters, to establish the mechanism. These studies have occasionally been made even more difficult owing to the observation of different products or product ratios under apparently identical conditions. For example, the doubly intramolecular reaction product (XLIII), Eq. (36), was reported [104] to be formed in good yield by oxidation of (XLII) in methanol containing 1% KOH, whereas a later report [105] claims the formation of the unsymmetrical product (XLIV), Eq. (37).

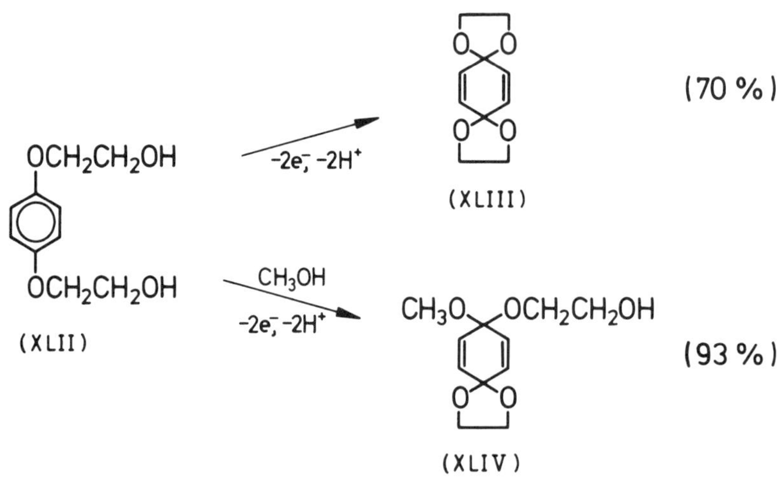

(36)

(37)

The latter product was taken as evidence for a reaction following pathway B. More detailed mechanistic discussions are found in [106,107].

Other aromatic compounds substituted similarly to 1,4-dimethoxynaphthalenes, Eq. (38) [103–107], behave analogously to the 1,4-dimethoxybenzenes, whereas naphthalenes substituted with methoxy groups in other positions often give rise to complicated product mixtures. The anodic 1,4-addition of two methoxy groups has also been reported for furan and 2-alkylfurans and provides a useful route to the synthetically important 2,5-dimethoxy-2,5-dihydrofurans, Eq. (39) [108–110].

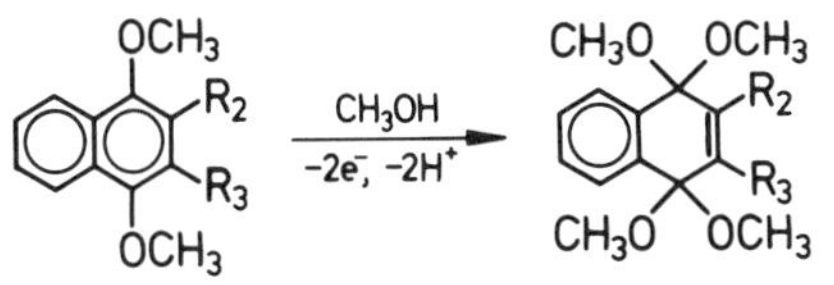

(38)

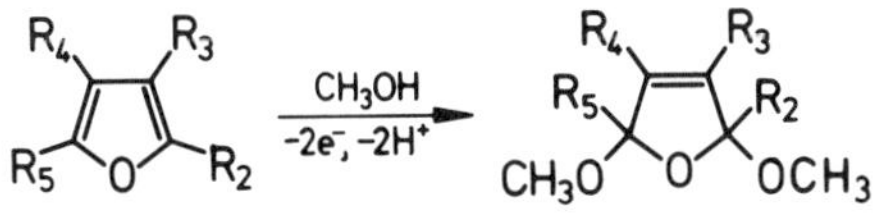

(39)

When 4-aryl- and 4-alkylalkoxybenzenes are subjected to anodic oxidation in MeOH-KOH, a different class of cyclohexadienes (XLV–XLVI) is obtained in 70–95% yield depending on the nature of the substituents, Eqs. (40) and (41) [111,112]:

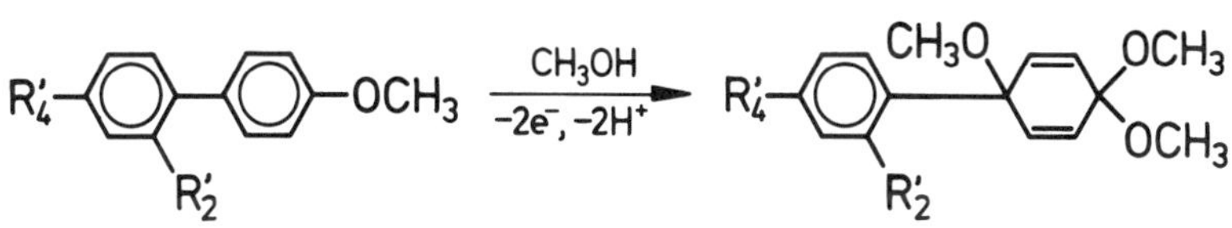

(40)

(XLV)

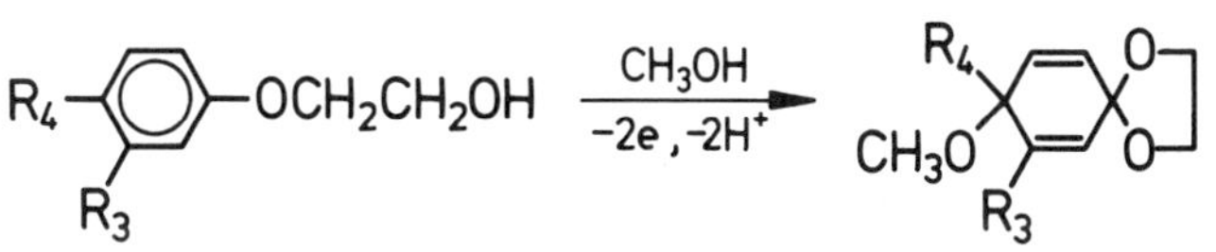

(41)

(XLVI)

However, the current yields are often low, but it has been shown [112] that it is possible to improve the efficiency by a factor of approximately 4 by application of KF as supporting electrolyte instead of KOH.

When the NaCN is used as supporting electrolyte, cyanide ion competes effectively with MeOH-MeO^- in the nucleophilic attack on the intermediate cations, resulting in mixed addition products, Eqs. (42) and (43) [113,114]:

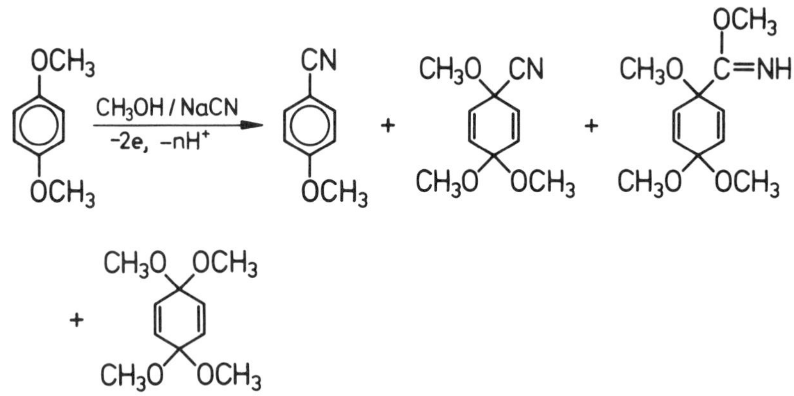

(42)

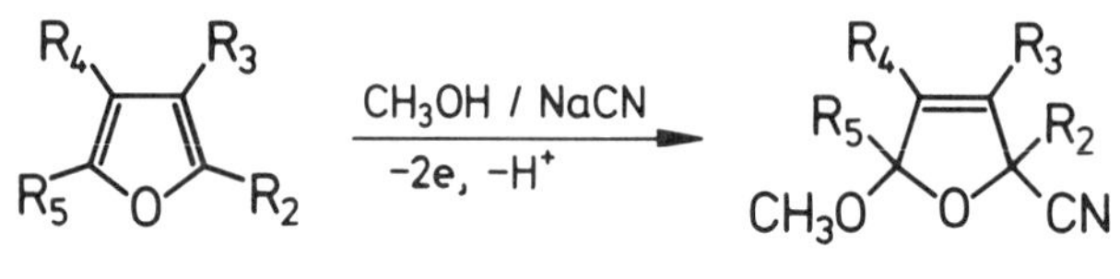

(43)

For hydroquinone silyl ethers the electrochemical oxidation takes a completely different route that involves cleavage of the O-Si bond and formation of the corresponding quinone, Eq. (44) [155]. Cleavage also competes favorably with nucleophilic addition when highly substituted 1,4-dimethoxybenzenes are oxidized anodically as was observed for 2,3,5,6-tetramethyl-1,4-dimethoxybenzene, Eq. (45) [116].

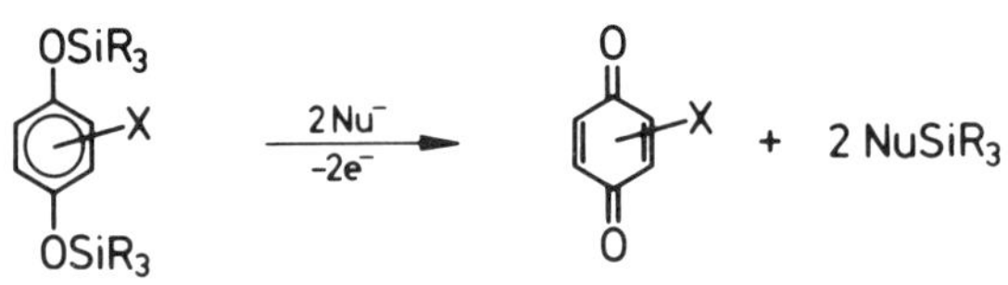

(44)

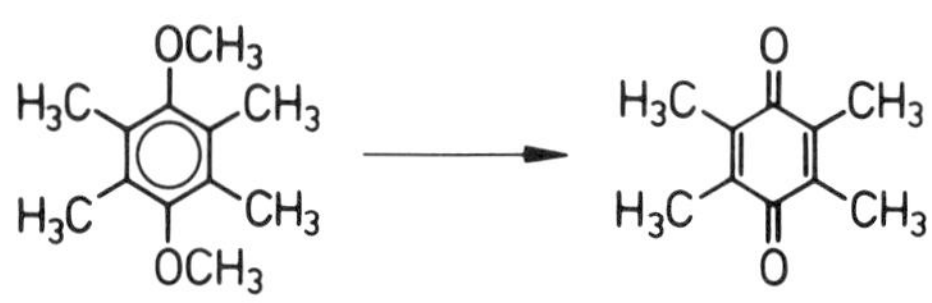

(45)

A reaction analogous to the aromatization of the vitamin K_3 precursor already described is the anodic oxidation of the 9,10-dihydrophenanthrene derivative (XLVII), which is converted almost quantitatively to the corresponding phenanthrene (XLVIII), Eq. (46) [117]:

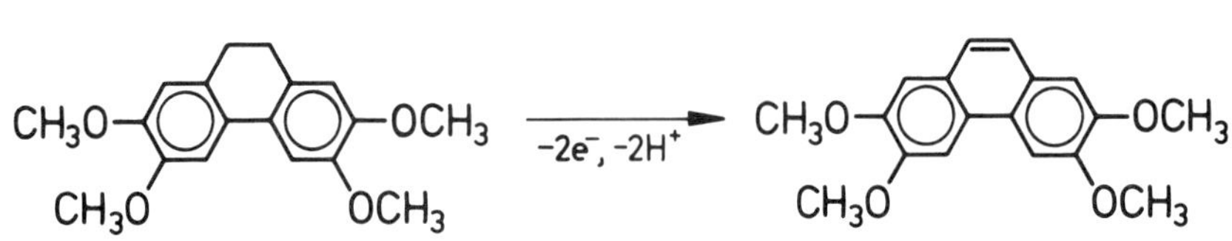

(46)

(XLVII) (XLVIII)

Other aspects of the (electrochemical) oxidation of alkyl aryl ethers are discussed in Ref. 118.

Quinones are also the final products formed during the electrochemical oxidation of hydroquinone diesters [116,119].

III. ANODIC OXIDATION OF ALCOHOLS AND DERIVATIVES

A. Benzyl Alcohols and Related Compounds

Primary benzyl alcohols can be oxidized anodically in MeCN to the corresponding aldehydes [120], a reaction that presumably proceeds according to an ECE-type mechanism, Eq. (47) [121,122]. Yields are improved when the solvent contains a base like pyridine or sodium carbonate in the absence of which polymeric material may be formed. The effect of the base is not clear at the detailed mechanistic level, but it may assist the deprotonation of the primarily formed cation radical (IL) and at the same time the presence of base prevents the acid-catalyzed dehydration of the starting material to the benzyl cation, which, if it is formed in MeCN, is trapped in a Ritter reaction, resulting in the amide (LII) as a by-product, Eq. (48).

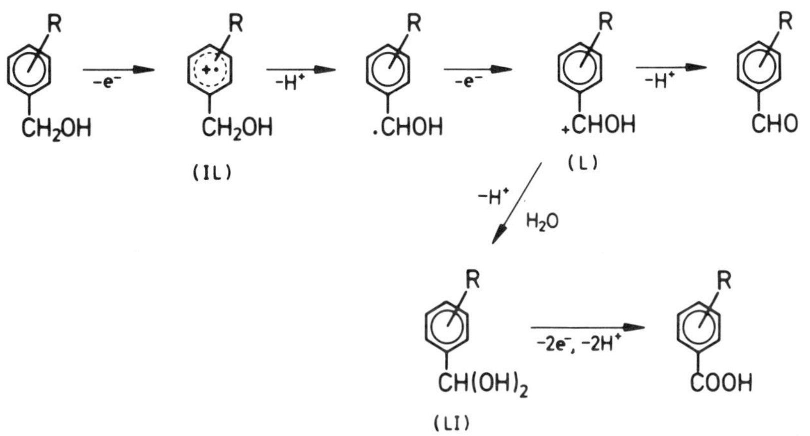

(47)

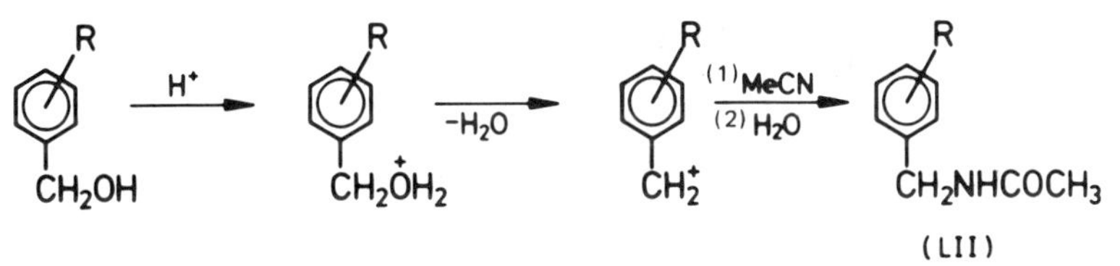

(48)

Another by-product, benzoic acid, has been suggested to arise from nucleophilic attack by water on the carbocation (L) followed by further oxidation of the diol (LI) to the benzoic acid [122,123].

The material yields of the benzaldehydes are considerably improved (>90%; see Table 5) if the oxidation process is mediated by redox catalysts like triarylamines [124], 2,2,5,5-tetramethyl-3-pyrrolin-1-oxyl derivatives [125], or osmium complexes [126] [see Eq. (49) for structures].

Table 5. Examples of Yields and Conditions for the Indirect Oxidation of Benzylalcohols to the Corresponding Benzaldehydes

Substituent	Mediator	Solvent	Base	Yield (%)	Current efficiency (%)	Reference
p-H	LIII	MeCN	Na_2CO_3	95	65	124
p-H	LIII	MeCN	Lutidine	5	<5	124
p-H	LIV	MeCN	Na_2CO_3	85	52	125
p-H	LIV	MeCN	Lutidine	49	66	125
p-H	LV	CH_2Cl_2	Na_2CO_3	—	94	126
p-CH_3	LIV	MeCN	Na_2CO_3	71	56	125
p-CH_3	LIV	MeCN	Lutidine	66	80	125
p-CH_3O	LVI	MeCN	Lutidine	96	80	124
p-CH_3O	LIV	MeCN	Na_2CO_3	85	63	125
p-CH_3O	LIV	MeCN	Lutidine	78	77	125
p-Cl	LIII	MeCN	Na_2CO_3	98	61	124
p-Br	LIII	MeCN	Na_2CO_3	90	50	124

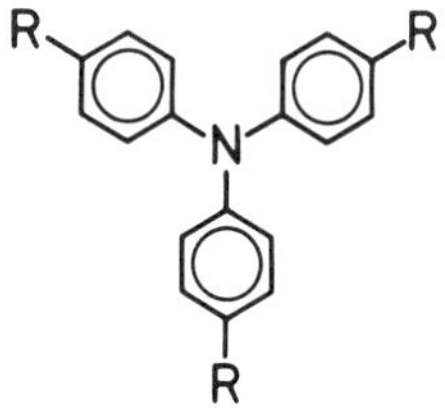

(LIII) R=Br
(LVI) R=H

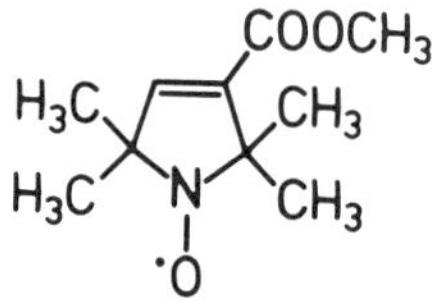

(LIV)

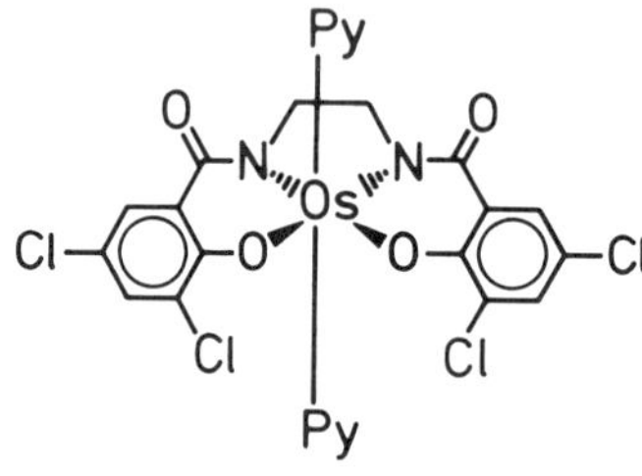

(LV)

(49)

It is of interest to note (1) that the material yields in general are higher when Na_2CO_3 instead of 2,4,6-collidine is used as the base, (2) that the current efficiency is almost 100% with the complex LV as mediator, and (3) that mediators as different as LIII through LVI and LIV give almost identical yields. In the case of LIII throught LVI the reaction was proposed to proceed as shown in Eq. (50), which is identical to that given in Eq. (47)

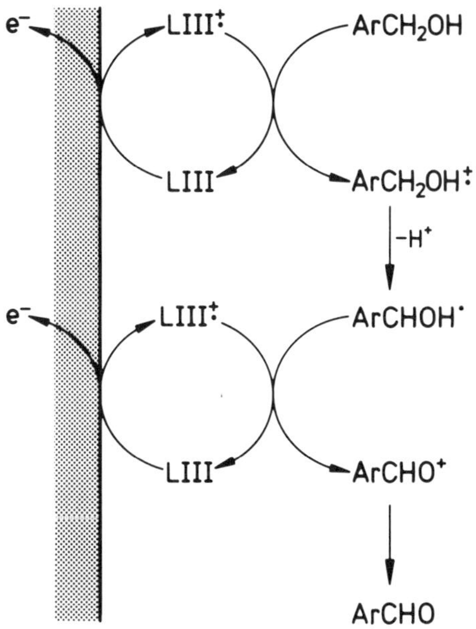

(50)

except that the mediator has now taken the place of the electrode in the first electron transfer step and the electrode and/or the primary formed cation radical in the second electron transfer step.

When the I^-/"I^+" couple is used as the mediator system (see Sect. III.B) with tert-butylalcohol as the cosolvent, benzyl benzoate is formed in modest to good yields [127].

The mechanism of anodic oxidation of benzyl alcohols bearing alkyl substituents in the α-position, that is, secondary alcohols of the general structure ArCHOHR, depends on the nature of R. When R = Me and Et the reaction in MeCN leads to the corresponding ketone as intuitively expected, Eq. (51).

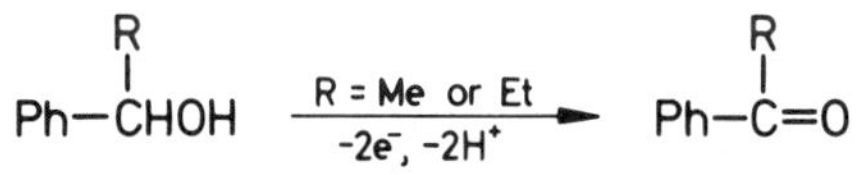

(51)

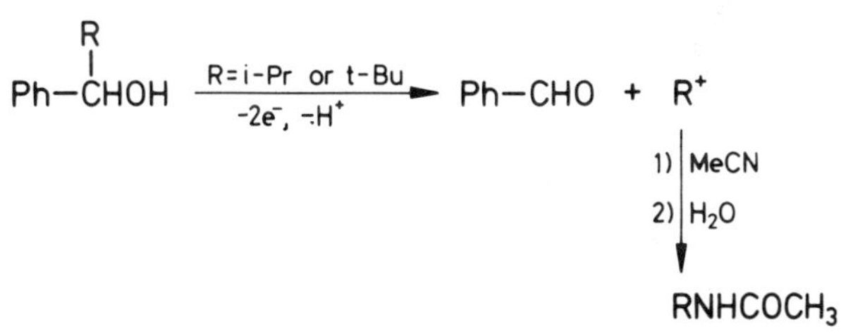

(52)

However, when R = i−Pr or t−Bu the carbon-alkyl bond is cleaved, resulting in formation of the arylaldehyde and the amide $RNHCOCH_3$, Eq. (52) [128]. Cleavage is observed independently of the structure of the substituents for tertiary alcohols, ArCOHRR' [128]. The relative leaving group ability was the expected

Me < Et < i–Pr < t–Bu

Related to these reactions is the anodic oxidation of tertiary alkylbenzylcarbinols in MeCN [129], which proceeds by cleavage of a C–C bond, leaving a benzyl radical and an aliphatic ketone, Eq. (53):

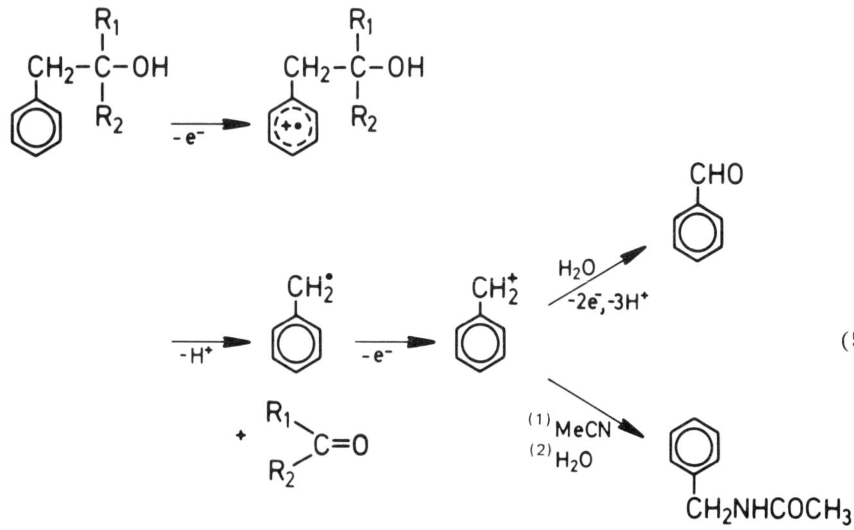

(53)

The benzyl radical is further oxidized under these conditions to the benzyl cation, which is trapped by water or MeCN as shown in Eq. (53).

Cleavage of the central C–C bond is the predominant reaction during the anodic oxidation of benzpinacol in alkaline solution [130]. The reaction, which resulted in the formation of benzophenone, was believed to involve the monoanion of the pinacol since oxidation was observed to take place only at pH values higher than 8.

Cleavage, this time of the C–O bond, is also observed when dibenzyl ethers are subjected to anodic oxidation [131,132].

The reader is referred to Chap. 24 for mechanistic details of these and other anodic cleavage reactions.

B. Aliphatic Alcohols and Aldehydes

Oxidation in Aqueous Solution

In general it is true that the mechanisms of anodic oxidation of saturated compounds are known in much less detail than the mechanisms for oxidation of aromatic compounds or compounds carrying substituents with a formal π bond. The aliphatic alcohols are not exceptions to this statement [133–135]. First, the primary intermediates formed in the anodic oxidation of alcohols or the corresponding bases are too short-lived to be detectable on the time scale of voltammetric experiments. Second, the heterogeneous electron transfer is slow and supposedly rate determining in most cases, which makes kinetic studies of the follow-up reactions impossible. Third, the oxidation process is as a rule, at least in aqueous solution, accompanied by adsorption of reactants and/or intermediates, which makes the reactions strongly dependent on both the material and the pretreatment of the anode. In particular, the surface oxide layers inevitably present (or formed) at most anode materials are intimately involved in the oxidation process. Altogether the mechanism of the oxidation process is very complex and very little definitive can be said about the first electron and proton transfer steps. However, the results of recent work directed toward the characterization of the adsorbed intermediate(s) have spread some light on the later steps in the oxidation process.

For the oxidation of the standard substrate, methanol, in acidic aqueous solution at Pt electrodes the sequence of steps given in Eqs. (54) through (58) is a widely accepted working hypothesis for the description of the overall reaction leading to carbon dioxide, Eq. (59) [136–139]. (The slashes indicate adsorbed species.)

$$CH_3OH + Pt \longrightarrow Pt/CHO + 3H^+ + 3e^- \tag{54}$$

$$H_2O + Pt \longrightarrow Pt/OH + H^+ + e^- \tag{55}$$

$$Pt/CHO + Pt/OH \longrightarrow Pt/CO + H_2O + Pt \tag{56}$$

$$H_2O + Pt \longrightarrow Pt/OH + H^+ + e^- \qquad = (55)$$

$$Pt/CO + Pt/OH \longrightarrow Pt/COOH + Pt \tag{57}$$

$$H_2O + Pt \longrightarrow Pt/OH + H^+ + e^- \qquad = (55)$$

$$Pt/COOH + Pt/OH \longrightarrow CO_2 + H_2O + 2Pt \tag{58}$$

$$CH_3OH + H_2O \longrightarrow CO_2 + 6H^+ + 6e^- \quad \text{(overall)} \tag{59}$$

Advanced studies involving combined electrochemical and spectroscopic techniques—electrochemically modulated infrared reflectance spectroscopy (EMIRS) [140], polarization-modulated Fourier transform infrared spectroscopy (PMFTIRS) [141], secondary ion mass spectroscopy (SIMS) [142], on-line mass spectroscopy [143], and electrochemical thermal desorption mass spectroscopy (ECTDMS) [144]—have demonstrated the presence of adsorbed formyl radicals, Pt/CHO, Eq. (54), and/or carbon monoxide, Pt/CO, Eq. (56), at the Pt anodes during methanol oxidation.

The oxidation process has been found to be inhibited by the presence of foreign anions in low concentration [145], the reason being blockage of the active sites at the electrode surface. The degree of inhibition was discussed within the concept of hard and soft anions and was found to be negligible in the presence of hard anions like nitrate and fluoride, whereas soft anions like iodide and bromide had a very strong inhibiting effect. The opposite effect, catalysis of the oxidation process, has been observed when foreign metal atoms, so-called ad-atoms, such as gold and ruthenium, or metal ions are present at the electrode surface [146–148]. It is believed that the cooperative effect is due to the different abilities of the metal atoms to interact with the intermediates formed during the oxidation process. This work has been extended to the application of dispersed Pt + Ru alloy clusters supported on carbon black, which show a very high catalytic efficiency, ≃ 200 mA· cm^{-2} at 0.4 V, and a limiting current of more than 1 A· cm^{-2} [149].

Electrode materials other than Pt may be used successfully for methanol oxidation. For example, it has been demonstrated that tungsten molybdenum carbide anodes with an atomic ratio Mo/(W + Mo) equal to 0.2 [150], ruthenium dioxide-coated titanium anodes [151], and iridium anodes [152] effectively catalyze the reaction.

Higher aliphatic primary alcohols are oxidized to the corresponding aldehydes or carboxylic acids in aqueous acid [27,153], but little is known about the mechanistic details. These reactions also seem to involve adsorbed intermediates, the structures of which depend on the complexity of the carbon skeleton (Ref. 154 and references cited therein). Of particular importance is the number of hydrogen atoms in the α-position [155]. Many of the experimental results point toward a mechanism involving the oxidation of the alcohol in an adsorbed state by adsorbed hydroxy radicals, Eq. (55) [156–158].

Secondary alcohols are converted almost quantitatively to the corresponding ketones when subjected to anodic oxidation in aqueous acid [27, 159,160]. We are not aware of detailed mechanistic studies.

The anodic oxidation of methanol in alkaline solution has been studied less intensively [161–164]. (See also references cited in Ref. 161). The reaction at Pt appears to follow a mechanism somewhat different from that in acidic solution, Eqs. (60) through (67) [161], the key steps being successive hydrogen abstraction reactions by adsorbed hydroxyl radicals, leading finally to carbonate and/or formate ion, Eqs. (68) and (69). Small amounts of Pb or Bi atoms at the electrode surface were found to increase the current densities; Cd and Tl were observed to have the opposite effect [161].

$$\mathrm{Pt + OH^- \rightleftharpoons Pt/OH + e^-} \tag{60}$$

$$\mathrm{Pt + CH_3OH \rightleftharpoons Pt/CH_3OH} \tag{61}$$

$$\mathrm{Pt/CH_3OH + Pt/OH \longrightarrow Pt/CH_3O + Pt + H_2O} \tag{62}$$

$$\mathrm{Pt/CH_3O + Pt/OH \longrightarrow Pt/CH_2O + Pt + H_2O} \tag{63}$$

$$\mathrm{Pt/CH_2O + Pt/OH \longrightarrow Pt/CHO + Pt + H_2O} \tag{64}$$

$$\mathrm{Pt/CHO + Pt/OH \longrightarrow Pt/CO + Pt + H_2O} \tag{65}$$

$$\mathrm{Pt/CO + Pt/OH \longrightarrow Pt/COOH + Pt} \tag{66}$$

$$\mathrm{Pt/COOH + OH^- \longrightarrow Pt/OH + HCOO^-} \tag{67a}$$

$$\mathrm{Pt/COOH + Pt/OH + 2OH^- \longrightarrow CO_3^{2-} + 2Pt + 2H_2O} \tag{67b}$$

$$\mathrm{CH_3OH + 8OH^- \longrightarrow CO_3^{2-} + 6e^- + 6H_2O} \quad \text{and/or} \tag{68}$$

$$\mathrm{CH_3OH + 5OH^- \longrightarrow HCOO^- + 4e^- + 4H_2O} \tag{69}$$

$$\mathrm{CH_2OH{-}CH_2OH + Au/OH \longrightarrow Au/CHOH{-}CH_2OH + H^+ + e^-} \tag{70}$$

The oxidation of 2-propanol, 2-deuterio-2-propanol, and octadeuterio-2-propanol on gold in 1 M KOH demonstrated that the current densities were smaller by a factor of 6.8 for the two deuterated alcohols compared with the nondeuterated substrate [165]. These results seem to indicate that the cleavage of the bond to the hydrogen atom in the 2-position is the rate-determining step. Other mechanistic details were not given. However, in the oxidation of ethylene glycol to oxalate on gold in aqueous alkaline solution [166] it was proposed that surface oxides (adsorbed hydroxyl radicals) are involved in determining the rate of hydrogen abstraction, Eq. (70).

The yields of the products obtained in aqueous solution may be improved considerably when indirect instead of direct anodic oxidation is applied (see Chap. 29). The mediators may either be present in solution or bound to the anode surface.

A number of water-soluble mediator systems have been used successfully for the oxidation of aliphatic alcohols and diols in aqueous solution, including $Cr_2O_3/Cr_2O_7^{2-}$ [167], IO_3^-/IO_4^- [168], and ruthenium complexes of the type $[Ru(trpy)(bpy)(H_2O)]^{2+}$ (trpy = 2,2',2''-terpyridine and bpy = 2,2'-bipyridine) shuttling between oxidation states II and IV [169]. The mediator may also be attached to the electrode surface, for example Pt-CO clusters, $[Pt_3(CO)_6]_n^{2-}$ (n = 3 or 5), attached to graphite disks [170].

Another type of surface-bound mediator are the high valent metal oxides of Cr(VI) [171], Ni(III) [172–177], Co(III), Ag(II), and Cu(III) [178,179] (see also references cited in Ref. 175). The nickel oxides have received particular attention (see Table 6). The reaction includes the anodic oxidation of $Ni(OH)_2$ to NiO(OH), Eq. (71), followed by rate-determining chemical oxidation of the alcohol by NiO(OH) to the corresponding carboxylate (primary alcohols) or ketone (secondary alcohols) with regeneration of $Ni(OH)_2$.

$$4\,Ni(OH)_2 + 4\,OH^- \xrightarrow[-e^-,\,-4H_2O]{} 4\,NiO(OH) \xrightarrow{RCH_2OH,\,OH^-} 4\,Ni(OH)_2 + RCOO^- \qquad (71)$$

Oxidation of Aliphatic Alcohols Under Nonaqueous Conditions

The anodic oxidation of water-free primary aliphatic alcohols follows in general a route different from that observed in aqueous solution [180–182]. The composition of the product mixture depends on the nature of the anode material, the supporting electrolyte, and the length of the aliphatic carbon chain. When the oxidation of methanol or ethanol was carried out at Pt electrodes in the neat solvent and with Bu_4NBF_4 or $NaClO_4$ as supporting electrolyte, the major products isolated were the corresponding acetals, 1,1-dialkoxyalkanes (LVIII). The formation of the acetals was believed to involve the intermediate carbocation (LVII), Eq. (72) [180]. The yields were reported in the range of 50–90%. The higher alcohols gave more complicated reaction mixtures containing 1,1-dipropoxymethane from the oxidation of propanol, for example, which indicates partial cleavage of the carbon chain. However, these additional complexities were not observed during an independent study of the oxidation of butanol [182]. When the supporting electrolyte was instead the sodium salt of the alcohol, the products were the corresponding aldehydes, Eq. (73) [180,181].

Table 6. Yields of Carboxylic Acids from the Indirect Oxidation of Aliphatic Alcohols at Nickel Hydroxide Anodes[a]

Substrate	Temperature (°C)	Yield (%)	Current density (mA/cm^2)	Reference
Ethanol	50	94–98	10	176
	Room	98	—	177
1-Propanol	50	95–99	10	176
	Room	95	—	177
1-Butanol	50	96–99	10	176
	70	92	2.4	174
	70	85	16	174
1-Hexanol	25	91	16	174
1-Octanol	70	89	16	174
1-Nonanol	70	89	16	174
1-Decanol	70	87	16	174
1-Octadecanol	70	77	16	174
3-Hexen-1-ol	25	22	16	174
4-Hepten-1-ol	25	82	16	174
4-Nonen-1-ol	70	68	16	174
2-Propyn-1-ol	5	51	16	174

[a]The reader is referred to the original literature for experimental details.

$$RCH_2OH \longrightarrow \underset{(LVII)}{R\overset{+}{C}HOH} \xrightarrow[-H^+]{RCH_2OH} \underset{\displaystyle OH}{R\underset{|}{C}HOCH_2R}$$

$$\xrightarrow[-H_2O]{RCH_2OH} \underset{(LVIII)}{RCH(OCH_2R)_2} \qquad (72)$$

$$2\,RCH_2O^- \longrightarrow RCHO + RCH_2OH + 2e^- \qquad (73)$$

Oxidation of neat butanol-containing lithium tetrafluoroborate also gave the aldehyde as the major product [182].

Esters, $RCOOCH_2R$, are formed in addition to the acetals and may even be the major products when MeCN is added as a cosolvent [182], but the composition of the product mixture in this case is also dependent on the electrode material and the supporting electrolyte. Although aliphatic alcohols are difficult to oxidize, it has been shown that the oxidation peak potential can measured by CV in MeCN [183]. The potentials were found to decrease with increasing chain length, the extremes being 2.73 V (methanol) and 2.46 V (pentanol) versus the ferrocene/ferrocenium couple.

As was observed in aqueous solution, the yields of products, esters from primary alcohols and ketones from secondary, may be improved by application of indirect oxidation, for example via X^-/"X^+" (X^- = halide ion)* [184,185], $NO_3^-/NO_3^{\cdot}$ [186] or polymeric 4-vinylpyridine (see Chap. 29) [187].

Oxidation of Aldehydes and Ketones

The conversion of aldehydes, RCHO, to carboxylic acids by electrochemical methods has received only limited attention in the last decades [168,169]. (Examples from the older literature may be found in Ref. 27). However, if the oxidation is carried out indirectly (see Chap. 29) in methanol or mixtures of water and higher alcohols, R'OH, in the presence of KI or KBr, high yields of the mixed esters, $RCOOH_3$ or RCOOR', are formed [188]. The reactions were proposed to involve the initial formation of the acetal followed by halogenation and elimination of HX, Eq. (74):

$$\mathrm{RCHO} + \mathrm{R'OH} \rightleftharpoons \underset{\textstyle \mathrm{OH}}{\mathrm{RCH}}\mathrm{-OR'} \xrightarrow[-\mathrm{H}^+]{"\mathrm{X}^+"} \underset{\textstyle \mathrm{OX}}{\mathrm{RCH}}\mathrm{-OR'} \xrightarrow{-\mathrm{HX}} \mathrm{RCOOR'} \qquad (74)$$

The anodic oxidation of ketones results in cleavage of a C—C bond if the ketone contains an α-alkyl group (see Chap. 24) or substitution (see Chap. 25) [189]. Mixtures of acetamides are obtained when long-chain ketones are subjected to anodic oxidation in MeCN [190,191]. The first step of the reaction is believed to involve intramolecular hydrogen abstraction. A careful mass spectroscopic study [192] of the products resulting from anodic oxidation of 4,4-, 5,5-, and 6,6-dideuterio-2-octanone in MeCN demonstrated (1) that the acetamides derived from the 4,4-dideuterio-2-octanone all retained the two deuterium atoms in the 4-position, (2) that only monodeuterated acetamides were derived from 5,5-dideuterio-2-octanone, Eq. (75), and (3) that 6,6-dideuterio-2-octanone gave acetamides with two deuterium atoms in different positions, Eq. (76).

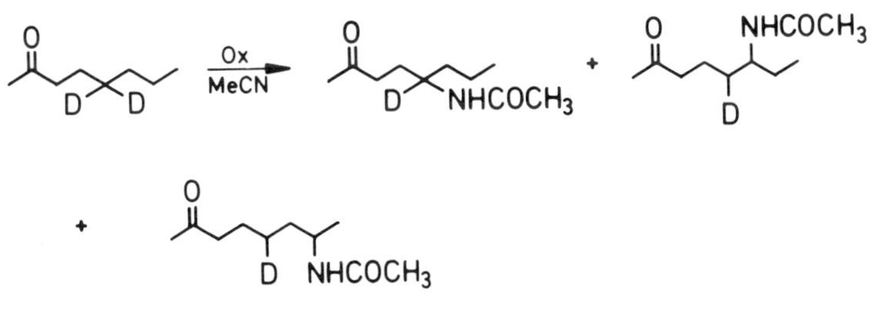

(75)

*The indirect oxidations via X^- are usually formulated as reactions involving X^+, in most cases without experimental evidence.

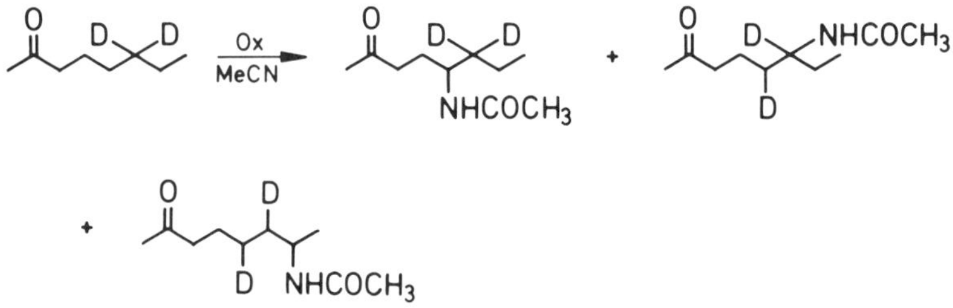

(76)

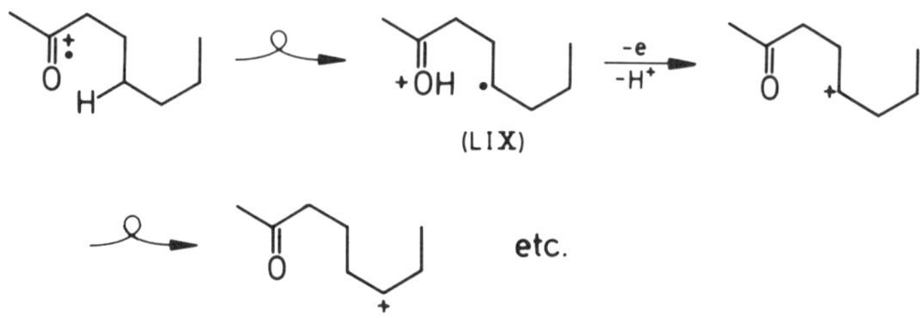

(77)

These results clearly indicated that the first chemical step in the anodic oxidation of aliphatic long-chain ketones was a site-specific hydrogen abstraction that was suggested to proceed via a six-membered transition state reminiscent of the McLafferty and Norrish type II reactions. The carbocation resulting from further oxidation of the distonic cation radical (LIX) suffers extensive hydride shifts, Eq. (77), before the nucleophilic attack by MeCN, and mechanistic conclusions based on only product analysis should accordingly be made with great care.

C. Carbohydrates and Related Bifunctional Compounds

Oxidation of Glucose and Other Aldoses

The idea of developing glucose/oxygen fuel cells to power artificial hearts or implanted cardiac pacemakers as well as sensors for the automatic monitoring of glucose concentration levels in vivo in patients suffering from diabetes mellitus (see refs. 193 and 194 for references) have stimulated much activity in the field of glucose oxidation. With these goals in mind a substantial part of the work has been barried out under attempted physiological conditions, such as modified Krebs-Ringer buffers, in cases containing intentionally added biomolecules (see later). These reaction conditions impose new complexities on the oxidation process in addition to those already discussed for the oxidation of aliphatic alcohols. For example, it has been demonstrated that the oxidation of glucose at Pt electrodes is seriously inhibited by the presence of urea, uric acid, amino acids, ascorbic acid, and chloride ions in the aqueous buffer solutions and by the products of oxidation [195–201].

Analysis of the products formed by oxidation of glucose shows that glucono-δ-lactone (LX) or gluconic acid (LXI) produced by hydrolysis of the lactone are the major products [Eq. (78); only the β-form of glucose is shown; see also Ref. 202) [194,200,201].

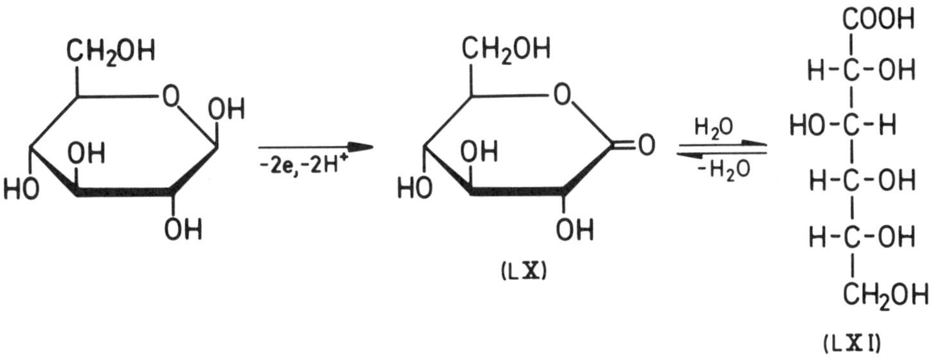

(78)

Formation of the aldonic acid is typical for oxidation of aldoses and has been observed in numerous other cases [27]. A spectacular example from the chemical industry is the anodic oxidation of diacetone-L-sorbose at Ni anodes [see Eq. (71)] to a 95% yield of the vitamin C precursor, diacetone-2-keto-L-gulonic acid [203]. Further oxidation of aldonic acid is accompanied by decarboxylation, resulting in the formation of pentoses [27]. A related example is the indirect oxidation of 1,2-O-isopropylidene-α-D-glucofuranose to 1,2-O-isopropylidene-α-D-xylo-5-hexulo-furanurono-6,3-lactone nonohydrate in aqueous calcium iodide solution [204].

Glucose gives rise to three distinct oxidation peaks at Pt working electrodes during CV, as observed by several workers [193,194,196,198, 200,201,205]. Although different views regarding the mechanistic details of the processes associated with these three peaks have been expressed, the essentials may be summarized as follows. The first of the three peaks, which is observed in the hydrogen adsorption region, is attributed to the oxidation of hydrogen formed by dehydrogenation of adsorbed glucose in the hemiacetal form [194,198,205]. The second oxidation wave corresponds to the further oxidation of the adsorbed product of dehydrogenation and probably involves adsorbed hydroxyl radicals, and the third wave is believed to reflect the regeneration of the Pt-oxides consumed during the oxidation of glucose diffusing from the bulk of the solution to the electrode surface [198]. The oxidation on gold electrodes appears to follow a similar route [198].

Recent progress in the development of glucose sensors has been based on coupling of the glucose-glucose oxidase system [206] with suitable redox catalysts, such as ferrocene monocarboxylic acid [207] and hexacyanoruthenate ion [208].

Other aspects of the oxidation of polyhydroxy compounds related to biomass are found in Chap. 20.

Oxidation of Ascorbic Acid and Related Compounds

The anodic oxidation of ascorbic acid (LXII) to dehydroascorbic acid (LXIII) proceeds as a "reversible" $2e^-$, $2H^+$ process in aqueous solution at the mercury electrode [209–211] and involves the monoanion as the electroactive species in most pH regions.

More complicated behavior, including adsorption, is observed at Pt electrodes [212,213]. Results from CV, controlled potential electrolysis, and potential step measurements in 1 M H_2SO_4 indicated the reversible formation of the dehydroascorbic acid anion radical in an adsorbed state (LXIV) as the first step, Eq. (79), followed by one-electron oxidation to adsorbed dehydroascorbic acid (LXV). Dissociation of this compound from the electrode surface is the rate-determining step, which is followed by formation of the dihydrate (LXVI), Eq. (80):

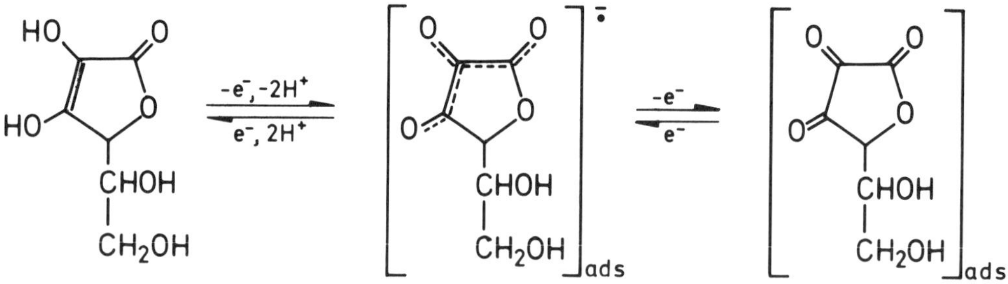

(LXII) (LXIV) (LXV)

(79)

(LXV) 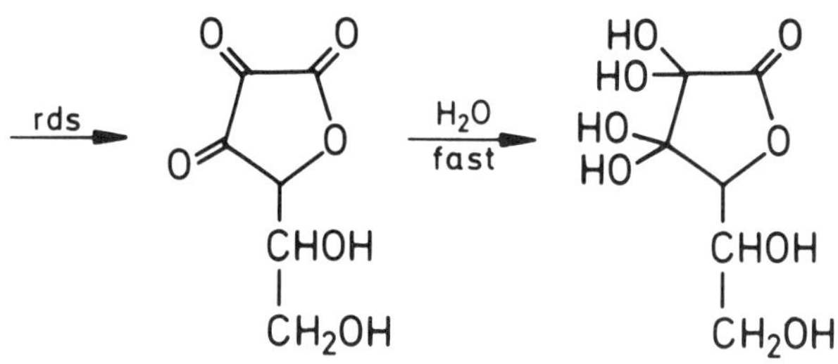

(LXIII) (LXVI)

(80)

The rate constant for hydration in aqueous solution at pH = 7.2 has been determined by CV and DPSC to be 1.35 $\times$ 10^3 s^{-1} (average of two almost identical values) [211].

Much work has been put into the search for suitable catalysts for the reaction at both Pt and C anodes. Successful attempts include oxidation at Pt in the presence of Bi^{3+}, Pb^{2+}, Hg^{2+}, Tl^+, Ag^+, and Sb^{3+} [214], and oxidation at glassy carbon with metal oxides dispersed at the surface [215] or quinhydrone adsorbed at the surface [216]. Catalytic carbon surfaces have also been obtained by electrochemical pretreatment of the anodes [217, 218]. Homogeneous redox catalysis of ascorbic acid oxidation has been observed after the addition of N,N,N',N'-tetramethyl-p-phenylenediamine to the voltammetric solution [219]. The application of this catalyst allowed the simultaneous detection of ascorbic acid and NADH, since ascorbic acid is catalyzed at the first oxidation peak of the catalyst and NADH at the second oxidation peak. Finally, reference should be made to series of papers in which C anodes carrying inorganic or polymeric organic coatings as catalysts have been used. The examples include polyacrylamide and acrylamide-vinylpyridine copolymeric gels [220], polymerized vinylferrocene [221], polypyrrole [222], and films made from prussian blue [223]. (For details about the application of modified electrodes, see Chap. 33.)

The 4,5-dihydroxy-4-cyclopentene-1,2,3-trione (LXVII), Eq. (81), is structurally related to ascorbic acid (LXII) and dehydroascorbic acid (LXIII). The dianions of LXVII (= LXVIII^{2-}) and the dicyanomethylene analogs of the type LXIX all give rise to reversible one-electron oxidation waves in DMF during CV at a Au electrode, Eq. (82) [224].

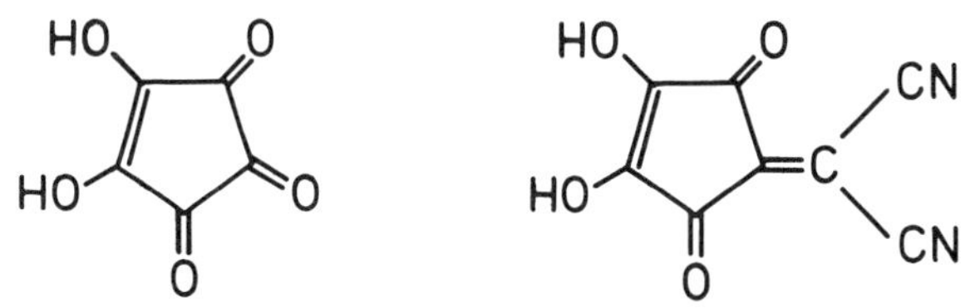

(LXVII) (LXIX) (81)

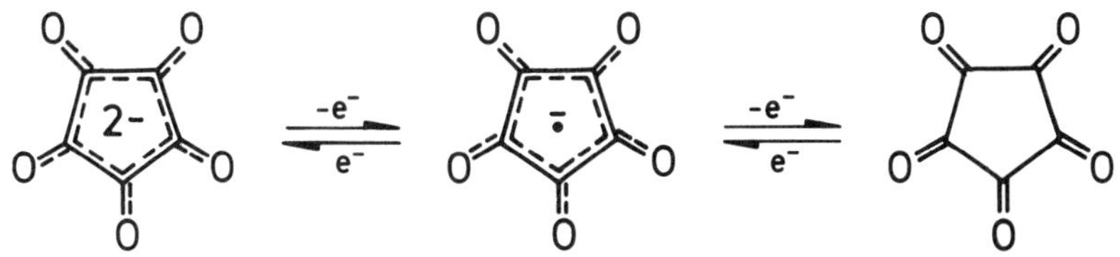

(LXVIII^{2-}) ($LXVIII^{\cdot-}$) (LXVIII) (82)

The second wave, corresponding to the formation of the neutral species, for example the aesthetic structure (LXVIII) from the anion radical, was only partially reversible if more than one oxo group was substituted by dicyanomethylene, reflecting the limited stability of these latter compounds.

IV. ANODIC OXIDATION OF OTHER ALIPHATIC OXYGEN-CONTAINING COMPOUNDS

A. Ethers Without Other Functional Groups

The predominant reaction of aliphatic ethers during anodic oxidation is α-substitution, for example hydroxylation or methoxylation, except when the ether linkage is part of a small ring system in which case the oxidation process is accompanied by bond cleavage (see Chaps. 24 and 25).

Oxidation of cyclic aliphatic ethers like tetrahydrofuran, tetrahydro-2H-pyran, and 1,4-dioxane at Pt anodes in 1 M H_2SO_4 gives the 2-hydroxy derivatives, that is, the cyclic hemiacetals, as the main products (Ref. 225 and references cited therin). Further oxidation of the hemiacetals derived from the cyclic monoethers results in the formation of the corresponding lactones, in contrast to the ring cleavage observed for the cyclic hydroxy diethers. Formation of the lactones were only observed at PtO_2 anodes.

When aliphatic ethers are subjected to anodic oxidation in MeOH containing $NaOCH_3$ as supporting electrolyte, α-methoxylated products (acetals) are obtained in modest yields (10–30%) [226]. It is not clear whether the first step is formation of the ether cation radical or the reaction proceeds via hydrogen abstraction by a radical formed during oxidation of the solvent supporting electrolyte system.

Benzophenone is formed almost quantitatively when tetraphenylethylene oxide is oxidized anodically in MeCN [122]. The reaction mechanism most likely involves cleavage of the central C—C bond at the cation radical stage followed by further oxidation of the now distonic cation radical to the dication. Reaction with water leads to a hemiacetal, which upon further reaction with water yields benzophenone as the final product.

B. Unsaturated Ethers

A reaction analogous to the anodic oxidation of anisole derivatives described in Sec. II.B is the formation of 1,2-dimethoxy-substituted aliphatic compounds by oxidation of vinyl ethers in MeOH-KOH [227–230]. Despite that the electrochemical addition of methoxy groups to aliphatic double bonds as a rule is accompanied by the formation of coupled products (see also Chap. 23) [231], the reaction is synthetically useful and may give acceptable product yields, usually around 40–80%, Eqs. (83) and (84):

$$CH_2{=}CH{-}OC_2H_5 \xrightarrow[2\,CH_3OH]{-2e^-,\,-2H^+} CH_3OCH_2{-}CH(OCH_3){-}OC_2H_5 \quad (68\%) \tag{83}$$

$$Ph{-}C(OCH_3){=}C(OCH_3){-}Ph \xrightarrow[2\,CH_3OH]{-2e^-,\,-2H^+} Ph{-}C(OCH_3)_2{-}C(OCH_3)_2{-}Ph \tag{84}$$

C. Alkoxyalcohols

Anodic oxidation of 2-methoxyethanol in MeCN gives rise to a variety of products, including formaldehyde and LXXI through LXXIV, Eq. (85), the relative amounts of which depend on both the anode material and the supporting electrolyte [232]. The overall reaction scheme is necessarily very complex, but it was suggested that the first steps of the mechanism involve the formation of the cation LXX followed by reaction with starting material to give LXXI, which could be converted to LXXII catalyzed by the protons liberated in Eq. (85):

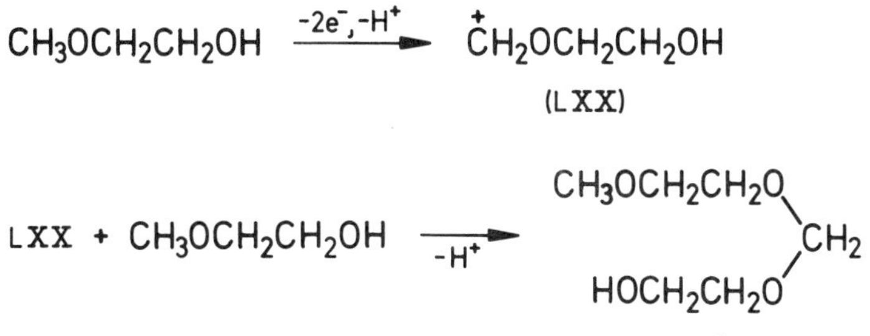

(LXX)

(LXXI) (85)

$(CH_3OCH_2CH_2O)_2CH_2$ (LXXII)

$CH_3OCH_2OCH_2OCH_2CH_2OCH_3$ (LXXIII)

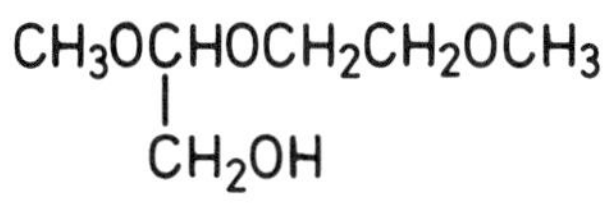

(LXXIV)

D. Peroxides

Very little is known about the anodic oxidation of aliphatic peroxides. A voltammetric study of a series of bicyclic peroxides (LXXV–LXXX), Eq. (86) [233], revealed that the values of E_{pa} in the solvent mixture CH_2Cl_2-TFA-TFAn (20:1:1) were higher than 2 V versus SCE.

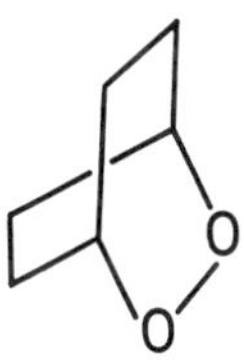

(LXXV)

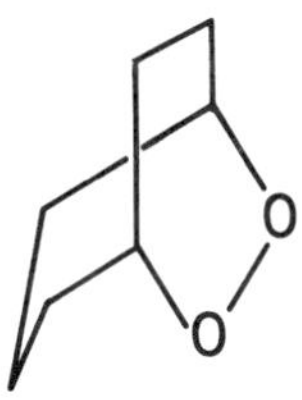

(LXXVI)

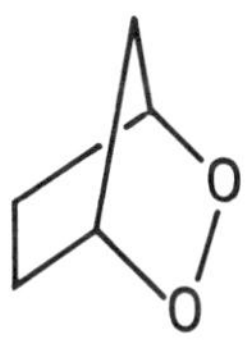

(LXXVII)

(86)

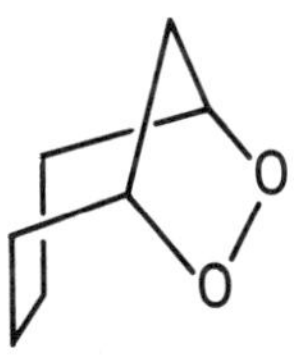

(LXXVIII)

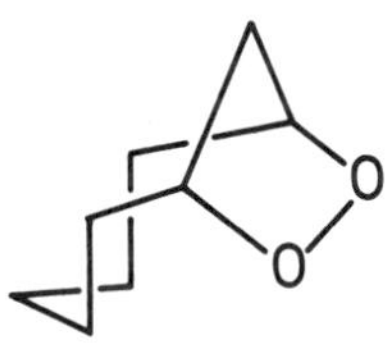

(LXXIX)

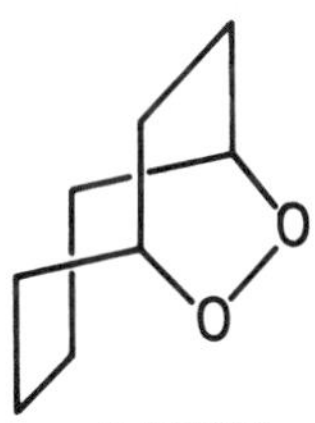

(LXXX)

The stability of the corresponding cation radicals decreased in the order given in Eq. (86), with no decomposition observed on the time scale of slow sweep CV for LXXV and no reverse current observed for LXXX. Completely irreversible behavior is also observed for the noncyclic di-tert-butylperoxide [234].

REFERENCES

1. F. P. Bub, K. Wisser, W. J. Lorenz, and W. Heimann, Ber. Bunsenges. Phys. Chem., 77, 823 (1973).
2. D. H. Evans, Acc. Chem. Res., 9, 313 (1977).
3. A. Ben Hadid, P. Rimmelin, J. Sommer, and J. Devynck, J. Chem. Soc., Perkin Trans. II, 269 (1982).
4. S. I. Bailey, I. M. Ritchie, and F. R. Hewgill, J. Chem. Soc., Perkin Trans. II, 645 (1983).
5. F. W. Steuber and K. Dimroth, Chem. Ber., 99, 258 (1966).
6. J. A. Richards, P. E. Whitson, and D. H. Evans, J. Electroanal. Chem., 63, 311 (1975).
7. J. A. Richards and D. H. Evans, J. Electroanal. Chem., 81, 171 (1977).
8. B. Speiser and A. Rieker, J. Electroanal. Chem., 102, 373 (1979).
9. D. H. Evans, P. J. Jimenez, and M. J. Kelly, J. Electroanal. Chem., 163, 145 (1984).
10. Yu. V. Vodzinskii, A. A. Vasil'eva, I. A. Korshunov, and G. A. Abakumov, Elektrokhimiya, 7, 24 (1971).
11. B. Speiser and A. Rieker, J. Chem. Res. (S), 314; (M), 3601 (1977).
12. V. D. Pokhodenko and E. P. Platonova, Elektrokhimiya, 10, 789 (1974).
13. E. P. Platonova, V. D. Pokhodenko, and E. A. Negoda, Elektrokhimiya, 12, 1644 (1976).
14. E. J. Land and G. Porter, Proc. Chem. Soc., 84 (1960).
15. E. J. Land, G. Porter, and E. Strachan, Trans. Faraday Soc., 57, 1885 (1961).
16. W. T. Dixon and D. Murphy, J. Chem. Soc., Faraday Trans. II, 72, 1221 (1976).
17. W. T. Dixon and D. Murphy, J. Chem. Soc., Faraday Trans. II, 74, 432 (1978).
18. D. M. Holton and D. Murphy, J. Chem. Soc., Faraday Trans. II, 75, 1637 (1979).
19. B. Speiser and A. Rieker, J. Electroanal. Chem., 110, 231 (1980).
20. A. Nilsson, U. Palmquist, A. Ronlán, and V. D. Parker, J. Amer. Chem. Soc., 97, 3540 (1975).
21. O. Hammerich, V. D. Parker, and A. Ronlán, Acta Chem. Scand., B30, 89 (1976).
22. O. Hammerich and V. D. Parker, Acta Chem. Scand., B36, 63 (1982).
23. C. Amatore and J. M. Savéant, J. Electroanal. Chem., 86, 227 (1978).
24. E. J. Land and G. Porter, Trans. Faraday Soc., 59, 2016 (1963).
25. L. R. Mahoney and S. A. Weiner, J. Amer. Chem. Soc., 94, 585, 1412 (1972).
26. F. J. Vermillion, Jr., and I. A. Pearl, J. Electrochem. Soc., 111, 1392 (1964).
27. N. L. Weinberg and H. R. Weinberg, Chem. Rev., 68, 449 (1968).
28. L. Papouchado, J. Bacon, and R. N. Adams, J. Electroanal. Chem., 24, App. 1 (1970).
29. L. Papouchado, R. W. Sandford, G. Petrie, and R. N. Adams, J. Electroanal. Chem., 65, 275 (1975).
30. A. Nilsson, A. Ronlán, and V. D. Parker, J. Chem. Soc., Perkin Trans. I, 2337 (1973).

31. A. Nilsson, U. Palmquist, T. Pettersson, and A. Ronlán, J. Chem. Soc., Perkin Trans. I, 696 (1978).
32. N. Oyama, T. Ohsaka, T. Hirokawa, and T. Suzuki, J. Chem. Soc. Chem. Commun., 1133 (1987).
33. D. R. Armstrong, R. J. Breckenridge, C. Cameron, D. C. Nonhebel, P. L. Pauson, and P. G. Perkins, Tetrahedron Lett., 24, 1071 (1983).
34. R. A. Anderson, D. T. Dalgleish, D. C. Nonhebel, and P. L. Pauson, J. Chem. Res. (S), 12; (M), 201 (1977).
35. D. T. Dalgleish and D. C. Nonhebel, and P. L. Pauson, J. Chem. Res. (S), 14; (M), 232 (1977).
36. R. A. Anderson, D. C. Nonhebel, and P. L. Pauson, J. Chem. Res. (S), 15; (M), 232 (1977).
37. R. A. Anderson, D. T. Dalgleish, D. C. Nonhebel, and P. L. Pauson, J. Chem. Res. (S), 16; (M), 265 (1977).
38. D. T. Dalgleish, N. P. Forrest, D. C. Nonhebel, and P. L. Pauson, J. Chem. Soc., Perkin Trans. I, 584 (1977).
39. D. R. Armstrong, C. Cameron, D. C. Nonhebel, and P. G. Perkins, J. Chem. Soc., Perkin Trans. II, 563, 569, 575, 581, 587 (1983).
40. A. B. Suttie, Tetrahedron Lett., 953 (1969).
41. J. Badoz-Lambling and G. Damange-Guerin, Anal. Lett., 2, 123 (1969).
42. W. J. Sep, J. W. Verhoeven, and T. J. de Boer, Tetrahedron, 35, 2161 (1979).
43. A. Yamagishi and M. Yamada, Bull. Chem. Soc. Jpn., 49, 371 (1976).
44. V. D. Parker, Electrochim. Acta, 18, 519 (1973).
45. E. Laviron, J. Electroanal. Chem., 164, 213 (1984).
46. L. Papouchado, G. Petrie, J. H. Sharp, and R. N. Adams, J. Amer. Chem. Soc., 90, 5620 (1968).
47. L. Papouchado, G. Petrie, and R. N. Adams, J. Electroanal. Chem., 38, 389 (1972).
48. H. Lund, Studier over Elektrodereaktioner i organisk polarografi og voltammetri, Aarhus, 1961, p. 137.
49. M. F. Marcus and M. D. Hawley, J. Electroanal. Chem., 18, 175 (1968).
50. S. Torii, H. Tanaka, and S. Nakane, Bull. Chem. Soc. Jpn., 55, 1673 (1982).
51. Z. Grujic, I. Tabakovic, and M. Trkovnik, Tetrahedron Lett., 4823 (1976).
52. M. D. Ryan, A. Yueh, and W.-Y. Chen, J. Electrochem. Soc., 127, 1489 (1980).
53. H.-Y. Cheng, E. Strope, and R. N. Adams, Anal. Chem., 51, 2243 (1979).
54. G. M. Proudfoot and I. M. Ritchie, Aust. J. Chem., 36, 885 (1983).
55. C. Iwakura, M. Tsunaga, and H. Tamura, Electrochim. Acta, 17, 1391 (1972); C. Steelink and W. E. Britton, Tetrahedron Lett., 2869 (1974).
56. H. Ohmori, C. Ueda, Y. Tokuno, H. Maeda, and M. Masui, Chem. Pharm. Bull., 33, 4007 (1985).
57. V. D. Parker and A. Ronlán, J. Electroanal. Chem., 30, 502 (1971).
58. A. Ronlán and V. D. Parker, J. Chem. Soc. (C), 3214 (1971).
59. A. Rieker, E.-L. Dreher, H. Geisel, and M. H. Khalifa, Synthesis, 851 (1978).
60. A. Nilsson, A. Ronlán, and V. D. Parker, Tetrahedron Lett., 1107 (1975).

61. M. H. Khalifa, G. Jung, and A. Rieker, Angew. Chem., 92, 739 (1980).
62. G. Popp and N. C. Reitz, J. Org. Chem., 37, 3646 (1972).
63. H. Ohmori, A. Matsumoto, and M. Masui, J. Chem. Soc., Chem. Commun., 407 (1978).
64. A. Ronlán, J. Chem. Soc., Chem. Commun., 1643 (1971).
65. Y. Shizuri, K. Nakamura, and S. Yamamura, J. Chem. Soc., Chem. Commun., 530 (1985).
66. F. H. Covitz, U.S. Pat. 3,509,031 (1970); Chem. Abstr., 73, 115824w (1970).
67. G. C. Jones and D. A. Payne, U.S. Pat. 3,994,788 (1976); Chem. Abstr., 86, 35759g (1977).
68. H. Sharifian and D. W. Kirk, J. Electrochem. Soc., 133, 921 (1986).
69. A. I. Scott, P. A. Dodson, F. McCapra, and M. B. Meyers, J. Amer. Chem. Soc., 85, 3702 (1963).
70. G. Popp, J. Org. Chem., 37, 3058 (1972).
71. U. Palmquist, A. Ronlán, and V. D. Parker, Acta Chem. Scand., B28, 267 (1974).
72. U. Palmquist, A. Nilsson, V. D. Parker, and A. Ronlán, J. Amer. Chem. Soc., 98, 2571 (1976).
73. E.-L. Rieker, Chem. Ber., 115, 288 (1982).
74. M. Petek, S. Bruckenstein, B. Feinberg, and R. N. Adams, J. Electroanal. Chem., 42, 397 (1973).
75. D. W. Leedy, J. Electroanal. Chem., 45, 467 (1973).
76. W. E. Britton and C. Steelink, Tetrahedron Lett., 2873 (1974).
77. V. D. Parker, J. Amer. Chem. Soc., 91, 5380 (1969).
78. U. Svanholm, K. Bechgaard, and V. D. Parker, J. Amer. Chem. Soc., 96, 2409 (1974).
79. M. F. Marcus and M. D. Hawley, J. Org. Chem., 35, 2185 (1970).
80. C. A. Chambers and J. Q. Chambers, J. Amer. Chem. Soc., 88, 2922 (1969).
81. E. P. Meier, J. Q. Chambers, C. A. Chambers, B. R. Eggins, and C.-S. Liao, J. Electroanal. Chem., 33, 409 (1971).
82. B. R. Eggins and R. Harwood, J. Electroanal. Chem., 83, 347 (1977).
83. R. W. Johnson, M. D. Bednarski, B. F. O'Leary, and E. R. Grover, Tetrahedron Lett., 22, 3715 (1981).
84. M. Iguchi, A. Nishiyama, Y. Terada, and S. Yamamura, Tetrahedron Lett., 4511 (1977).
85. M. Iguchi, A. Nishiyama, Y. Terada, and S. Yamamura, Chem. Lett., 451 (1978).
86. M. Iguchi, A. Nishiyama, M. Hara, Y. Terada, and S. Yamamura, Chem. Lett., 1015 (1978).
87. M. Iguchi, A. Nishiyama, H. Eto, and S. Yamamura, Chem. Lett., 939 (1981).
88. F. Fichter and M. Rinderspacher, Helv. Chim. Acta., 10, 102 (1927).
89. A. Ronlán, in Encyclopedia of Electrochemistry of the Elements (A. J. Bard and H. Lund, eds.), Vol. 11, Dekker, New York, 1978, p. 242.
90. A. Ronlán, J. Coleman, O. Hammerich, and V. D. Parker, J. Amer. Chem. Soc., 96, 845 (1974).
91. K. Bechgaard and V. D. Parker, J. Amer. Chem. Soc., 94, 4749 (1972).
92. A. J. Bard and J. Phelps, J. Electroanal. Chem., 25, App. 2 (1970).
93. U. Svanholm, B. Svensmark Jensen, and V. D. parker, J. Chem. Soc., Perkin Trans. II, 907 (1974).

94. B. Belleau and N. L. Weinberg, J. Amer. Chem. Soc., 85, 2525 (1963).
95. N. L. Weinberg and B. Belleau, Tetrahedron, 29, 279 (1973).
96. P. Margaretha and P. Tissot, Org. Synth., 57, 92 (1977).
97. E. Konz and R. Pistorius, Synthesis, 603 (1979).
98. B.-T. Groebel, E. Konz, H. Millauer, and R. Pistorius, Synthesis, 605 (1979).
99. M. G. Dolson and J. S. Swenton, J. Org. Chem., 46, 177 (1981).
100. M. J. Manning, P. W. Raynolds, and J. S. Swenton, J. Amer. Chem. Soc., 98, 5008 (1976).
101. J. S. Swenton, Acc. Chem. Res., 16, 74 (1983).
102. G. M. Elgy, W. B. Jennings, and A. E. Pedler, J. Chem. Soc., Perkin Trans. I, 1255 (1983).
103. D. R. Henton, R. L. McCreery, and J. S. Swenton, J. Org. Chem., 45, 369 (1980).
104. N. L. Weinberg and E. A. Brown, unpublished results; N. L. Weinberg, in Techniques of Chemistry (N. L. Weinberg, ed.), Vol. V, part I, Wiley, New York, 1974, pp. 259, 309.
105. P. Margaretha and P. Tissot, Helv. Chim. Acta, 58, 933 (1975).
106. A. Nilsson, U. Palmquist, T. Pettersson, and A. Ronlán, J. Chem. Soc., Perkin Trans. I, 708 (1978).
107. M. G. Dolson and J. S. Swenton, J. Amer. Chem. Soc., 103, 2361 (1981).
108. N. Elming, Adv. Org. Chem., 2, 67 (1960).
109. T. Shono, Electroorganic Chemistry as a New Tool in Organic Synthesis, Springer-Verlag, Berlin, 1984, p. 105.
110. D. Pletcher, Industrial Electrochemistry, Chapman and Hall, London, 1984, Chap. 6.
111. R. E. DeSchepper and J. S. Swenton, Tetrahedron Lett., 26, 4831 (1985).
112. M. P. Capparelli, R. E. DeSchepper, and J. S. Swenton, J. Org. Chem., 52, 4953 (1987); J. Chem. Soc., Chem. Commun., 610 (1987).
113. N. L. Weinberg, D. H. Marr, and C. N. Wu, J. Amer. Chem. Soc., 97, 1499 (1975).
114. K. Yoshida and T. Fueno, Bull. Chem. Soc. Jpn., 60, 229 (1987).
115. R. F. Stewart and L. L. Miller, J. Amer. Chem. Soc., 102, 4999 (1980).
116. V. D. Parker, J. Chem. Soc., Chem. Commun., 610 (1969).
117. A. Ronlán and V. D. Parker, J. Chem. Soc., Chem. Commun., 1567 (1970).
118. O. C. Musgrave, Chem. Rev., 69, 499 (1969).
119. E. P. Meier and J. Q. Chambers, J. Electroanal. Chem., 25, 435 (1970).
120. H. Lund, Acta Chem. Scand., 11, 491 (1957).
121. O. R. Brown, S. Chandra, and J. A. Harrison, J. Electroanal. Chem., 34, 505 (1972).
122. E. A. Mayeda, L. L. Miller, and J. F. Wolf, J. Amer. Chem. Soc., 94, 6812 (1972).
123. W. J. Bower and P. Zuman, J. Electrochem. Soc., 122, 368 (1975).
124. K.-H. Grose Brinkhaus, E. Steckhan, and W. Schmidt, Acta Chem. Scand., B37, 499 (1983).
125. A. Deronzier, D. Limosin, and J.-C. Moutet, Electrochim. Acta, 32, 1643 (1987).

126. F. C. Anson, T. J. Collins, S. L. Gipson, and T. E. Kraft, Inorg. Chem., 26, 731 (1987).
127. G. Dixit, R. Rastogi, and K. Zutshi, Electrochim. Acta, 27, 561 (1982).
128. E. A. Mayeda, J. Amer. Chem. Soc., 97, 4012 (1975).
129. M. H. Khalifa and A. Rieker, J. Chem. Res. (S), 316 (1977).
130. W. Kemula, Z. R. Grabowski, and M. Kalinowski, Coll. Czech. Chem. Commun., 25, 3306 (1960).
131. R. Lines and J. H. P. Utley, J. Chem. Soc., Perkin Trans. II, 803 (1977).
132. J. W. Boyd, P. W. Schmalzl, and L. L. Miller, J. Amer. Chem. Soc., 102, 3856 (1980).
133. G. Sundholm, in Encyclopedia of Electrochemistry of the Elements (A. J. Bard and H. Lund, eds.), Vol. 11, Dekker, New York, 1978, p. 207.
134. S. D. Ross, M. Finkelstein, and E. J. Rudd, Anodic Oxidation, Academic Press, New York, 1975.
135. K. Yoshida, Electrooxidation in Organic Chemistry, Wiley, New York, 1984.
136. V. S. Bagotzky, Y. B. Vassiliev, and O. A. Khazova, J. Electroanal. Chem., 81, 229 (1977).
137. V. S. Bagotzky, Y. B. Vassiliev, O. A. Khazova, and S. S. Sedova, Electrochim. Acta, 16, 913 (1971).
138. N. A. Hampson, M. J. Willars, and B. D. McNicol, J. Chem. Soc., Faraday Trans. I, 75, 2535 (1979).
139. B. D. McNicol, P. Attwood, and R. T. Short, J. Chem. Soc., Faraday Trans. I, 77, 2017 (1981).
140. B. Beden, C. Lamy, A. Bewick, and K. Kunimatsu, J. Electroanal. Chem., 121, 343 (1981).
141. K. Kunimatsu, J. Electroanal. Chem., 140, 205 (1982); 145, 219 (1983).
142. H. W. Buschmann, S. Wilhelm, and W. Bielstich, Electrochim. Acta, 31, 939 (1986).
143. T. Iwasita, W. Vielstich, and E. Santos, J. Electroanal. Chem., 229, 367 (1987).
144. S. Wilhelm, W. Vielstich, H. W. Buschmann, and T. Iwasita, J. Electroanal. Chem., 229, 377 (1987).
145. G. Belanger, Can. J. Chem., 50, 1891 (1972).
146. M. Watanabe and S. Motoo, J. Electroanal. Chem., 60, 259, 267, 275 (1975).
147. J. Sobkowski, K. Franaszczuk, and A. Piasecki, J. Electroanal. Chem., 196, 145 (1985).
148. M. Shibata and S. Motoo, J. Electroanal. Chem., 229, 385 (1987).
149. M. Watanabe, M. Uchida, and S. Motoo, J. Electroanal. Chem., 229, 395 (1987).
150. G. Kawamura, H. Okamoto, A. Ishikawa, and T. Kudo, J. Electrochem. Soc., 134, 1653 (1987).
151. L. D. Burke and O. J. Murphy, J. Electroanal. Chem., 101, 351 (1979).
152. A. Aramata, T. Yamazaki, K. Kunimatsu, and M. Enyo, J. Phys. Chem., 91, 2309 (1987).
153. B. Wermeckes and F. Beck, Chem. Ber., 118, 3771 (1985).
154. B. Beden, M.-C. Morin, F. Hahn, and C. Lamy, J. Electroanal. Chem., 229, 353 (1987).

155. E. Sokolova, Electrochim. Acta, 20, 323 (1975).
156. J. Willsau and J. Heitbaum, J. Electroanal. Chem., 194, 27 (1985).
157. G. Kokkinidis and D. Jannakoudakis, J. Electroanal. Chem., 153, 185 (1983).
158. A. K. Vijh, Can. J. Chem., 49, 78 (1971).
159. G. Horanyi, P. König, and I. Telcs, Acta Chim. Acad. Sci. Hung., 72, 165 (1972).
160. G. Horanyi, G. Vértes, and P. König, Naturwissenschaften, 60, 519 (1973).
161. B. Beden, F. Kadirgan, C. Lamy, and J. M. Leger, J. Electroanal. Chem., 142, 171 (1982).
162. T. Takamura and Y. Sato, Electrochim. Acta, 19, 63 (1974).
163. M.-C. Pham, F. Adami, P.-C. Lacaze, J.-P. Doucet, and J.-E. Dubois, J. Electroanal. Chem., 201, 413 (1986).
164. F. Adami, M.-C. Pham, P.-C. Lacaze, and J.-E. Dubois, J. Electroanal. Chem., 210, 295 (1986).
165. M. Beltowska-Brzezinska and W. Vielstich, Electrochim. Acta, 22, 1313 (1977).
166. W. Hauffe and J. Heitbaum, Ber. Bunsenges. Phys. Chem., 82, 487 (1978).
167. H. P. Gogolin and H. Wendt, Chem.-Ing.-Tech., 48, 797 (1976).
168. A. Yoshiyama, T. Nonaka, M. M. Baizer, and T.-C. Chou, Bull. Chem. Soc. Jpn., 58, 201 (1985).
169. M. S. Thompson, W. F. De Giovani, B. A. Moyer, and T. J. Meyer, J. Org. Chem., 49, 4972 (1984).
170. K. Machida, A. Fukuoka, M. Ichidawa, and M. Enyo, J. Chem. Soc., Chem. Commun., 1486 (1987).
171. F. Beck and H. Schultz, J. Electroanal. Chem., 229, 339 (1987).
172. G. Vértes, G. Horányi, and F. Nagy, Tetrahedron, 28, 37 (1972); Acta Chim. Acad. Sci. Hung., 68, 217 (1971).
173. H. J. Schäfer, Topics Curr. Chem., 142, 101 (1987).
174. J. Kaulen and H. J. Schäfer, Synthesis, 513 (1979).
175. P. M. Robertson, J. Electroanal. Chem., 111, 97 (1980).
176. J. Burzyk, I. Zjawiony, and A. Budniok, J. Prakt. Chem., 329, 131 (1986).
177. M. Fleischmann, K. Korinek, and D. Pletcher, J. Electroanal. Chem., 31, 39 (1971).
178. D. Pletcher, M. Fleischmann, and K. Korinek, J. Electroanal. Chem., 33, 478 (1971).
179. M. Fleischmann, K. Korinek, and D. Pletcher, J. Chem. Soc., Perkin Trans. II, 1396 (1972).
180. G. Sundholm, J. Electroanal. Chem., 31, 265 (1971).
181. G. Bélanger, J. Electrochem. Soc., 123, 818 (1976).
182. P. C. Scholl, S. E. Lentsch, and M. R. Van De Mark, Tetrahedron, 32, 303 (1976).
183. G. Sundholm, Acta Chem. Scand., 25, 3188 (1971).
184. D. A. White, J. Electrochem. Soc., 124, 1177 (1977).
185. T. Shono, Y. Matsumura, J. Hayashi, and M. Mizoguchi, Tetrahedron Lett., 165 (1979).
186. J. E. Leonard, P. C. Scholl, T. P. Steckel, S. E. Lentsch, and M. R. Van De Mark, Tetrahedron Lett., 21, 4695 (1980).
187. J. Yoshida, R. Nakai, and N. Kawabata, J. Org. Chem., 45, 5269 (1980).

188. T. Shono, Y. Matsumura, J. Hayashi, K. Inoue, F. Iwasaki, and T. Itoh, J. Org. Chem., 50, 4967 (1985).
189. J. Y. Becker, L. L. Miller, and T. M. Siegel, J. Amer. Chem. Soc., 97, 849 (1975).
190. J. Y. Becker, L. R. Byrd, and L. L. Miller, J. Amer. Chem. Soc., 96, 4718 (1974).
191. J. Y. Becker, L. R. Byrd, L. L. Miller, and Y.-H. So, J. Amer. Chem. Soc., 97, 853 (1975).
192. S. Hammerum and O. Hammerich, Tetrahedron Lett., 5027 (1979).
193. L. Marincic, J. S. Soeldner, C. K. Colton, J. Giner, and S. Morris, J. Electrochem. Soc., 126, 43 (1979).
194. S. Ernst, J. Heitbaum, and C. H. Hamann, J. Electroanal. Chem., 100, 173 (1979).
195. D. A. Gough, F. L. Anderson, J. Giner, C. K. Colton, and J. S. Soeldner, Anal. Chem., 50, 941 (1978).
196. L. Marincic, J. S. Soeldner, J. Giner, and C. K. Colton, J. Electrochem. Soc., 126, 1687 (1979).
197. J. Giner, L. Marincic, J. S. Soeldner, and C. K. Colton, J. Electrochem. Soc., 128, 2106 (1981).
198. Y. B. Vassilyev, O. A. Khazova, and N. N. Nikolaeva, J. Electroanal. Chem., 196, 105, 127 (1985).
199. E. Skou, Acta Chem. Scand., 27, 2239 (1973); Electrochim. Acta, 22, 313 (1977).
200. S. Ernst, J. Heitbaum, and C. H. Hermann, Ber. Bunsenges. Phys. Chem., 84, 50 (1980).
201. M. L. B. Rao and R. F. Drake, J. Electrochem. Soc., 116, 334 (1969).
202. S. J. Yao, A. J. Appleby, and S. K. Wolfson, Jr., Z. Phys. Chem. NF, 82, 225 (1972).
203. P. M. Robertson, P. Berg, H. Reimann, K. Schleich, and P. Seiler, J. Electrochem. Soc., 130, 591 (1983); P. Seiler and P. M. Robertson, Chimia, 36, 305 (1982).
204. P. A. Garcia, R. Velasco, and F. Barba, Chem. Script., 26, 579 (1986).
205. M. F. L. de Mele, H. A. Videla, and A. J. Arvia, Bioelectrochem. Bioenerg., 10, 239 (1983).
206. C. C. Liu, L. B. Wingard, Jr., S. K. Wolfson, Jr., S. J. Yao, A. L. Drash, and J. G. Schiller, Bioelectrochem. Bioenerg., 6, 19 (1979).
207. A. E. G. Cass, G. Davis, G. D. Francis, H. A. O. Hill, W. J. Aston, I. J. Higgins, E. V. Plotkin, L. D. L. Scott, and A. P. F. Turner, Anal. Chem., 56, 667 (1984).
208. A. L. Crumbliss, H. A. O. Hill, and D. J. Page, J. Electroanal. Chem., 206, 327 (1986).
209. M. Dominguez, A. Aldaz, and E. Sanchez-Burgos, J. Electroanal. Chem., 68, 345 (1976).
210. J. J. Ruiz, A. Aldaz, and M. Dominguez, Can. J. Chem., 55, 2799 (1977); 56, 1533 (1978).
211. S. P. Perone and W. J. Kretlow, Anal. Chem., 38, 1760 (1966).
212. M. Brezina, J. Koryta, T. Loucka, D. Marsikova, and J. Pradac, J. Electroanal. Chem., 40, 13 (1972).
213. P. Karabinas and D. Jannakoudakis, J. Electroanal. Chem., 160, 159 (1984).
214. K. Takamura and M. Sakamoto, J. Electroanal. Chem., 113, 273 (1980).

215. S. Dong and T. Kuwana, J. Electrochem. Soc., 131, 813 (1984).
216. H. Gomathi and G. P. Rao, J. Electroanal. Chem., 190, 85 (1985).
217. N. Cenas, J. Rozgaite. A. Pocius, and J. Kulys, J. Electroanal. Chem., 154, 121 (1983).
218. L. Falat and H.-Y. Cheng, J. Electroanal. Chem., 157, 393 (1983).
219. A. Kitani and L. L. Miller, J. Amer. Chem. Soc., 103, 3595 (1981).
220. J. E. Van Koppenhagen and M. Majda, J. Electroanal. Chem., 189, 379 (1985).
221. M. F. Dautartas and J. F. Evans, J. Electroanal. Chem., 109, 301 (1980).
222. R. A. Saraceno, J. G. Pack, and A. G. Ewing, J. Electroanal. Chem., 197, 265 (1986).
223. F. Li and S. Dong, Electrochim. Acta, 32, 1511 (1987).
224. L. M. Doane and A. J. Fatiadi, J. Electroanal. Chem., 135, 193 (1982).
225. B. Wermeckes, F. Beck, and H. Schulz, Tetrahedron, 43, 577 (1987); B. Wermeckes and F. Beck, Chem. Ber., 120, 1679 (1987).
226. T. Shono and Y. Matsumura, J. Amer. Chem. Soc., 91, 2803 (1969).
227. R. Couture and B. Belleau, Can. J. Chem., 50, 3424 (1972).
228. D. Koch, H. Schäfer, and E. Steckhan, Chem. Ber., 107, 3640 (1974).
229. T. Shono, Y. Matsumura, T. Imanishi, and K. Yoshida, Bull. Chem. Soc. Jpn., 51, 2179 (1978).
230. M. Cariou, J. Simonet, and J. Toupet, Tetrahedron Lett., 28, 1275 (1987).
231. M. Katz, P. Riemenschneider, and H. Wendt, Electrochim. Acta, 17, 1595 (1972).
232. S. D. Ross, J. E. Barry, M. Finkelstein, and E. J. Rudd, J. Amer. Chem. Soc., 95, 2193 (1973).
233. S. F. Nelsen, M. F. Teasley, A. J. Bloodworth, and H. J. Eggelte, J. Org. Chem., 50, 3299 (1985).
234. S. F. Nelsen, M. F. Teasley, D. L. Kapp, and R. M. Wilson, J. Org. Chem., 49, 1845 (1984).

17

Anodic Oxidation of Sulfur-Containing Compounds

BO SVENSMARK and OLE HAMMERICH / H. C. Ørsted Institute, University of Copenhagen, Copenhagen, Denmark

I. Introduction 659

II. Oxidation of Compounds Containing the Thiol or the Disulfide Group 660
- A. Oxidation of thiols and related compounds on noble metal electrodes 660
- B. Oxidation of thiols and related compounds on Hg electrodes 662
- C. Disulfides 662

III. Oxidation of Sulfides and Sulfoxides 667
- A. Dialkyl sulfides 667
- B. Thioacetals, orthothioesters, and related compounds 670
- C. Alkyl aryl sulfides and diaryl sulfides 674

IV. Oxidation of Tetrathioethylenes and Tetrathiafulvalenes 679

V. Oxidation of 1,4-Dithiins, Benzoderivatives, and Related Compounds 683
- A. The effect of structure and solvents on the oxidation potentials 684
- B. Oxidation followed by deprotonation 685
- C. Reactions of the cation radicals with nucleophiles 686

VI. Oxidation of Thiophenes 688

VII. Oxidation of Compounds Containing the Thiocarbonyl Group 689

VIII. Oxidation of Sulfonic, Sulfinic, and Sulfenic Acid Derivatives 691

References 692

I. INTRODUCTION

Organic compounds containing sulfur atoms are often treated as a single class although the chemical similarities between the different members are often very vague. Discussions of their electrochemical behavior is no exception in this respect, and several reviews dealing with the anodic oxidation of organic sulfur compounds in general [1–3], as well as the anodic behavior of selected structures [4,5], have been published in the last two decades. Reference should also be made to the work by Fichter et al. (see

Ref. 6) on the electrochemistry of organic compounds, including anodic oxidation of many sulfur compounds, work too often forgotten when the discovery of new electrochemical reactions is claimed. Material of electrochemical interest may also be found in discussions of the cation radical chemistry of sulfur compounds [7,8].

The recent interest in conducting (or superconducting) organic materials has stimulated considerable activity in the field of preparation and electrochemical characterization of easily oxidized multisulfur compounds, in particular of tetrathiafulvalenes and structurally related substances. In this and other fields of sulfur chemistry certain compound abbreviations have gained general approval, such as TTF for tetrathiafulvalene, and we occasionally use these abbreviations instead of roman numerals.

Only electrochemical behavior typical of the particular classes of compounds is included in this chapter, and we purposely omit reactions that do not reflect the presence of the sulfur atom. Compounds containing other chalcogen atoms are occasionally included in the discussion.

Because of the diversity of structures, the presentation is by necessity rather brief and only short summaries are given in most cases. Some of the electrochemical reactions of sulfur compounds belong naturally in other chapters of this book, to which reference is given. In particular, the reader is referred to the Chaps. 18, 19, and 25. Conversely, it was decided by the editors that the electrochemistry of a number of heterocyclic compounds, for example, thianthrenes, phenothiazines, thiophenes, cyclic thioacetals and bis-sulfides, tetrathiafulvalenes, and other cyclic tetrathioethylenes, should be included in the present chapter.

II. OXIDATION OF COMPOUNDS CONTAINING THE THIOL OR THE DISULFIDE GROUP

A. Oxidation of Thiols and Related Compounds on Noble Metal Electrodes

The RSH/RSSR interconversion is a classic redox couple in organic chemistry owing to the biologic importance of the cysteine-cystine system and α-lipoic acid, for example. Studies of the anodic oxidation of thiols, RSH, in aqueous solution on noble metal electrodes are complicated by adsorption phenomena and at Pt in particular by the participation of surface oxides in the oxidation process [1,9–14]. The peak potentials for oxidation observed for thiols during cyclic voltammetry vary linearly with pH by approximately 60 mV/pH at pH > pK_a(RSH) and are independent of pH at pH > pK_a(RSH). These observations, together with coulometric n values close to unity and product analyses, suggest an oxidation process involving the initial formation of the RS$^\cdot$ radical, Eq. (1), followed by dimerization to the disulfide, Eq. (2). More extensive oxidation leads to formation of the corresponding sulfonic acid, Eq. (3), probably via oxidation of adsorbed RS$^\cdot$, since RSSR is generally not electroactive in aqueous solution.

$$\text{R–SH} \xrightarrow[-e^-]{} \text{R–S}^\bullet + \text{H}^+ \qquad (1)$$

$$2\,\text{R–S}^\bullet \longrightarrow \text{R–S–S–R} \qquad (2)$$

$$\text{R–S}^\bullet \xrightarrow[-5e^-]{3H_2O} \text{R–SO}_3^- + 6\text{H}^+ \qquad (3)$$

Distinct peaks for the oxidation of RSH and RS^- to disulfides are seen in nonaqueous solution, such as DMF and MeCN [14,15].

Products containing the disulfide linkage are also formed by anodic oxidation of thio acids, Eq. (4) [16], alkylxanthates [17,18], Eq. (5), and dialkyldithiocarbamates, Eq. (6) [19,20], as well as 1,1-dicyanoethylene-2,2-dithiolates, Eq. (7) [21]. Further oxidation of the disulfide (I) leads to II, III, Eq. (8) [22], and presumably IV, which, however, was considered less stable than II and III owing to ring strain. A ring system similar to IV has been proposed to arise from oxidation of 4-(*N*,*N*-dimethylamino)dithiobenzoate via the disulfide V, (9) through (11) [23]. However, further oxidation of the structurally related tetraethylthiuram disulfide formed in Eq. (6) has been reported to yield a dication containing a five-membered ring similar to III, not the six-membered ring analogous to V^{2+} [20].

Selenophenol is oxidized to the diselenide in a fashion similar to the oxidation of thiophenol derivatives [24].

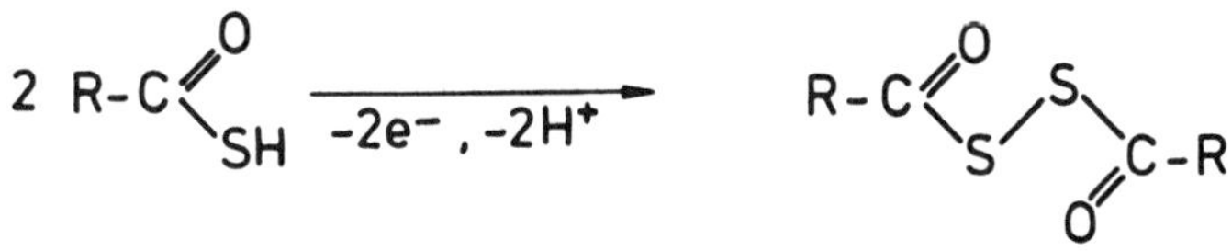

(4)

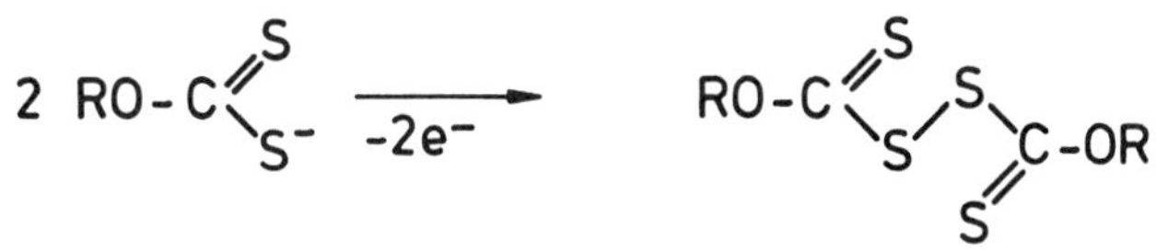

(5)

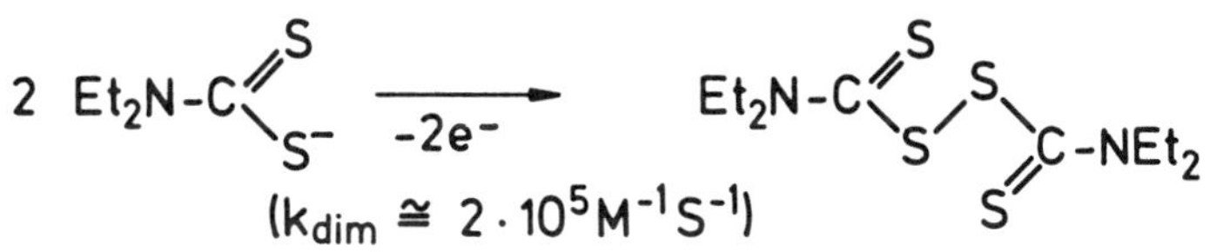

(6)

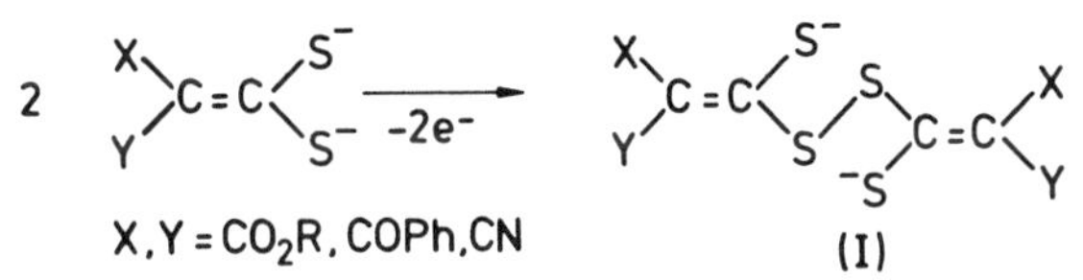

(7)

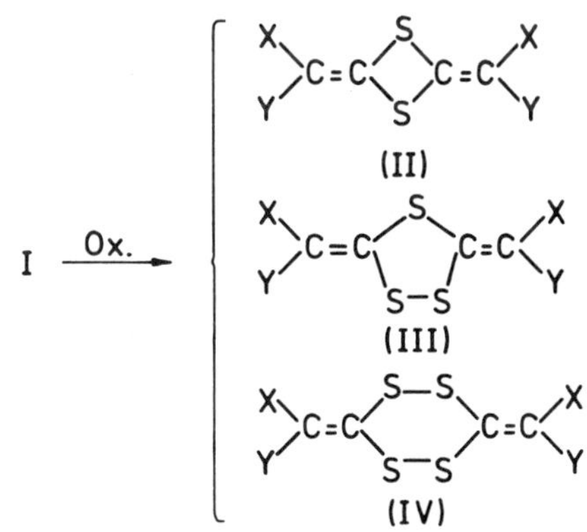

(8)

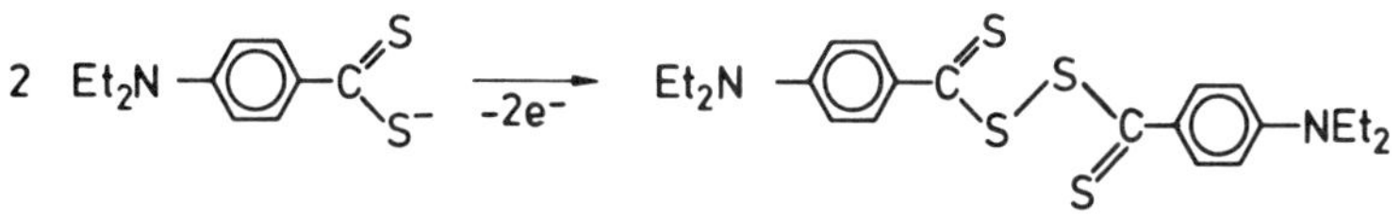 (9)

(V)

$$V \xrightarrow{-e^-} V^{+\bullet} \tag{10}$$

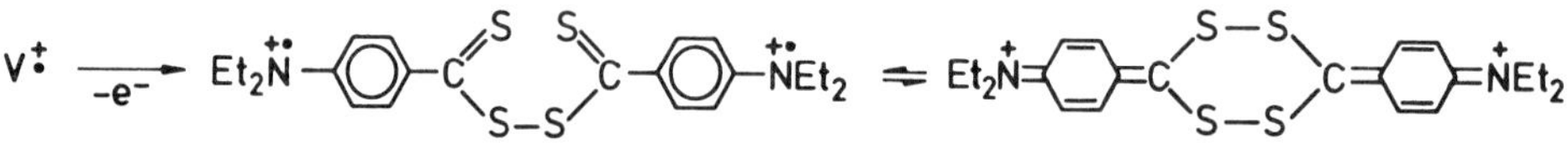

V^{2+}

(11)

B. Oxidation of Thiols and Related Compounds on Hg Electrodes

Oxidation of thiols at Hg electrodes includes the formation of alkylmercury compounds, Eqs. (12) through (14), in aqueous solution [1,9,25–27] and higher complexes under nonaqueous conditions, Eqs. (15) and (16) [28].

$$Hg + RSH \rightleftharpoons (HgSR)_{ads} + H^+ + e^- \tag{12}$$

$$(HgSR)_{ads} \rightleftharpoons HgSR \tag{13}$$

$$2\,HgSR \rightleftharpoons Hg(SR)_2 + Hg \tag{14}$$

$$Hg + 3\,PhS^- \rightleftharpoons Hg(SPh)_3^- + 2e^- \tag{15}$$

$$Hg + 2\,Hg(SPh)_3^- \rightleftharpoons 3\,Hg(SPh)_2 + 2e^- \tag{16}$$

An independent study of the oxidation of $Hg(SAr)_2$ on Pt has demonstrated the formation of ArSSAr and Hg^{2+}, presumably via the adsorbed cation radical of the mercury complex and adsorbed $ArS^{\cdot}$ radicals [29]. Preparative oxidations demonstrated the formation of the thiosulfonate, $ArSSO_2Ar$, in addition to ArSSAr. The thiosulfonate was the main product after 6 F/mol (see also Refs. 30 and 31).

Reactions analogous to those described in Eqs. (12) through (14) and (15) and (16) have been observed during the oxidation of dithiobenzoate and dithioacetate [32], alkylxanthates [33], dialkyldithiocarbamates [34,35], and various 1,1-dithiolates [36].

C. Disulfides

Aromatic compounds containing a disulfide linkage between two peri positions, such as NDS, Eq. (17), give cation radicals sufficiently stable to be observed at the time scale of slow sweep voltammetry [37,38].

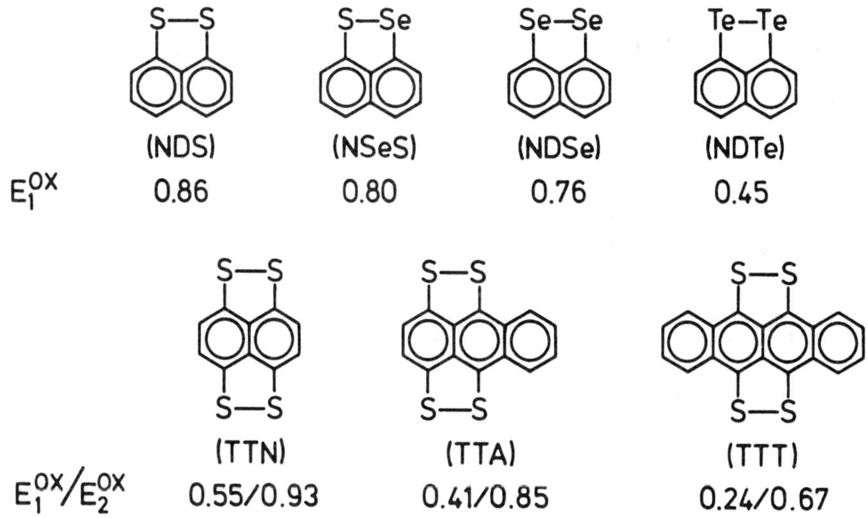

(17)

When two disulfide linkages are present in the molecule, as in TTN, even the dications are stable [39,40]. The magnitudes of the oxidation potentials have been observed to vary with structure as follows: (1) E(NDS) > E(NSeS) > E(NDSe) > E(NDTe) and (2) E(TTN) > E(TTA) > E(TTT), illustrating that the substitution of sulfur with the heavier selenium and tellurium results in compounds that are increasingly easy to oxidize and that disulfides derived from extended aromatic systems are, not unexpectedly, easier to oxidize than those derived from less extended aromatic structures.

The cation radicals of simple dialkyl and diaryl disulfides are too short-lived to be observed by voltammetric technique, although the existence of these species has been confirmed by pulse radiolysis [41] and ESR spectroscopy [42]. The products of oxidation depend markedly on the nature of R and the experimental conditions, as illustrated in Eq. (18) [29,43–47].

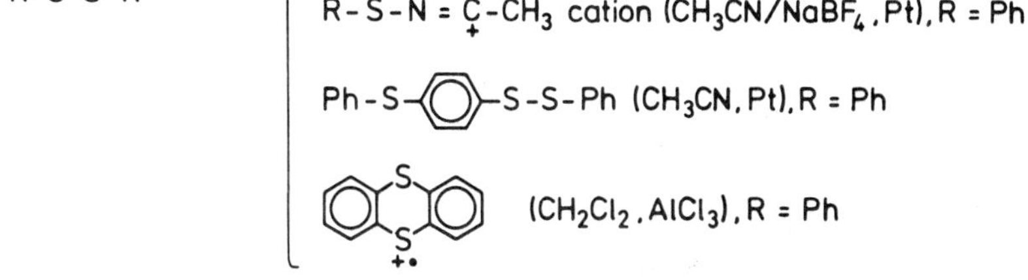

(18)

*After the electrolysis mixture had been standing for one week under N_2.

However, it is not clear whether the product claimed to be RSOSOR is actually $RSSO_2R$. It is of interest to note that oxidation of diphenyldisulfide by $AlCl_3$ in CH_2Cl_2 results in the formation of thianthrene cation radical [47]. We are not aware of the electrochemical analog of this rearrangement.

Oxidation of disulfides or diselenides in the presence of an alkene in MeCN are synthetically useful routes to 2-acetamidosulfides and selenides [46,48–51]. Typical yields are given in Table 1. The reactions were proposed to follow the mechanism shown in Eqs. (19) through (23) [46]. The reacting intermediate is believed to be RX^+ (X = S, Se), Eq. (20), which adds to the electron-rich double bond and forms a carbocation, Eq. (21), which in turn is trapped by MeCN in a Ritter-type reaction, Eq. (22). Hydrolysis finally leads to the isolable products, Eq. (23).

$$R\text{-}X\text{-}X\text{-}R \xrightarrow[-e^-]{} R\text{-}X^{+\bullet} + R\text{-}X^\bullet \qquad (19)$$

$$R\text{-}X^\bullet \xrightarrow[-e^-]{} R\text{-}X^+ \qquad (20)$$

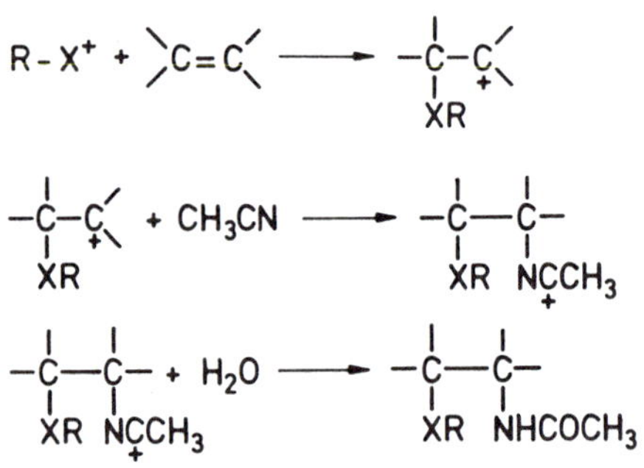

(21)

(22)

(23)

X = S or Se

Minor amounts of $PhSC_6H_4SSPh$ are formed as a by-product when RXXR = PhSSPh, presumably via electrophilic attack of PhS^+ on unoxidized starting material. With cyclic alkenes, the reaction gives almost predominantly the trans isomer of the product, and with acyclic 1-alkenes, the products containing a terminal sulfide group dominate [50].

Similarly, alkenes may be converted to 2-hydroxy, 2-methoxy, and 2-acetoxy selenides by indirect oxidation in $MeCN\text{-}H_2O$, MeOH, and HOAc, respectively, in the presence of PhSeSePh with Et_4NX (X = Cl, Br or I) as supporting electrolyte (Table 2) [52]. Further oxidation of the oxyselenide results in elimination of PhSeOH with formation of the corresponding olefin (VI), Eq. (24) [53]. This has stimulated the development of a procedure in which alkenes are oxidatively converted to allylic oxy compounds in the presence of only catalytic amounts of PhSeSePh, Eq. (24) [54].

$$PhSeSePh \xrightarrow[-2e^-]{2\,ROH} 2\,PhSeOR + 2H^+$$

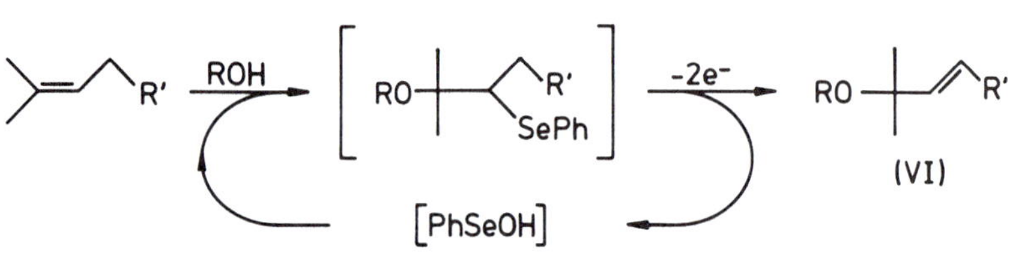

(24)

(VI)

Table 1. Examples of Yields of Acetamidosulfides and Selenides Obtained by Controlled Potential Oxidation of RXXR (X = S, Se) in the Presence of an Alkene in MeCN

RXXR	Alkene	Product	Yield (%)[a]	Reference
PhSSPh	Ethene	2-Acetamido-1-phenylthioethane	129	50
PhSSPh	1-Octene	2-Acetamido-1-phenylthiooctane	84	48, 50
PhSSPh	2-Octene	2-Acetamido-3-phenylthiooctane 3-Acetamido-2-phenylthiooctane	34 50	48
MeSSMe	Cyclohexane	_trans_-2-Acetamido-1-methylthiocyclohexane	58	48
PhSSPh	Cyclohexane	_trans_-2-Acetamido-1-phenylthiocyclohexane	50–87	48, 50
$PhCH_2SSCH_2Ph$	Cyclohexane	_trans_-2-Acetamido-1-benzylthiocyclohexane	112	50
PhSSPh	5α-Cholest-2-ene	2β-Acetamido-3α-phenylthio-5α-cholestane	58	51
PhSSPh	5β-Cholest-3-ene	3α-Acetamido-4β-phenylthio-5β-cholestane	20	51
PhSeSePh	3-Heptene	3-Acetamido-4-phenylselenoheptane 4-Acetamido-3-phenylselenoheptane	35 35	49
PhSeSePh	1-Decene	2-Acetamido-1-phenylselenodecane	99	49
PhSeSePh	Cyclopentene	_trans_-2-Acetamido-1-phenylselenocyclopentane	66	49

[a]Yields based on RXXR.

Table 2. Examples of Yields of Alkoxy, Acetoxy, and Hydroxy Selenides and the Corresponding Unsaturated Compounds Resulting from Anodic Oxidation of Alkenes in the Presence of Diphenyldiselenide

Compound	Product	R	(%)
	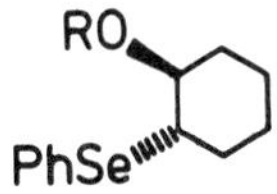	Me Ac H	96 98 96
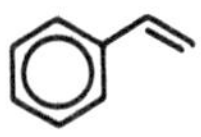	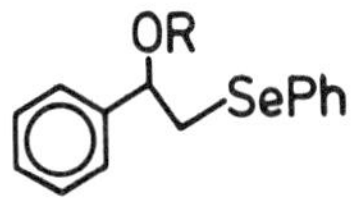	Me Ac H	91 96 92
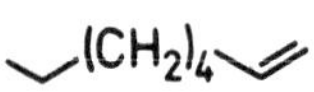	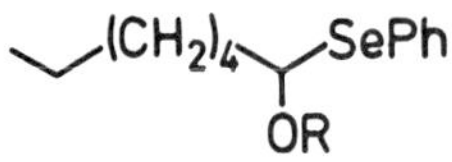	Me H	92 94
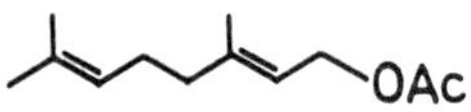	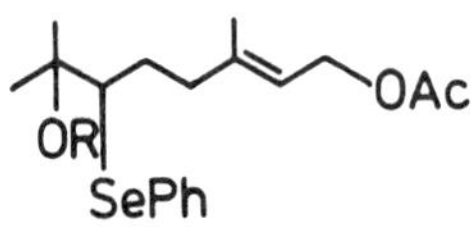	Me H	83 80
	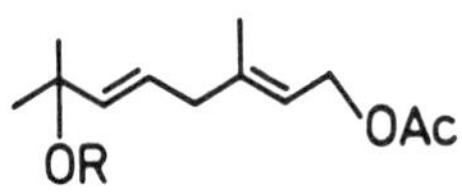	Me H	86 69
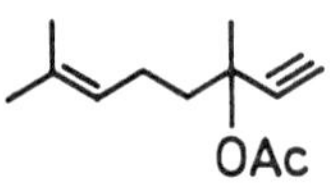	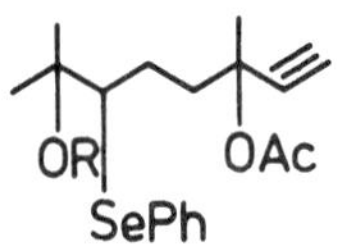	Me H	91 93
	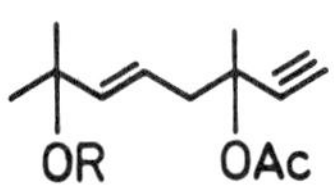	Me H	91 93
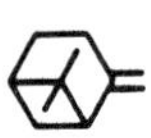	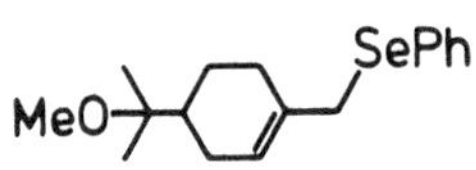		63
	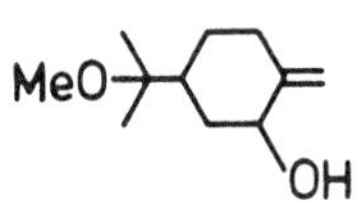		61

Source: From Refs. 52–54.

III. OXIDATION OF SULFIDES AND SULFOXIDES

A. Dialkyl Sulfides

Simple Dialkyl Sulfides

The products from oxidation of simple dialkyl sulfides in MeCN, for example, depend on the water content of the solvent. Oxidation of dimethyl sulfide in MeCN containing 1% water gave exclusively dimethyl sulfone, Eq. (25), whereas oxidation in the "dry" solvent resulted in formation of methane sulfonate ion and carbon monoxide, Eqs. (26) through (28) [55].

$$CH_3-S-CH_3 \xrightarrow[1\%\ H_2O \atop -4e^-,-4H^+]{CH_3CN} CH_3-SO_2-CH_3 \tag{25}$$

$$CH_3-S-CH_3 \xrightarrow[-e^-]{CH_3CN} CH_3-\overset{+\bullet}{S}-CH_3 \xrightarrow[-e^-,-H^+]{} CH_3-S-CH_2^+ \tag{26}$$

$$CH_3-S-CH_2^+ + CH_3-S-CH_3 \longrightarrow \underset{\text{(VII)}}{CH_3-S-CH_2-\overset{+}{S}(CH_3)_2} \tag{27}$$

$$CH_3-S-CH_2-\overset{+}{S}(CH_3)_2 \xrightarrow[\text{ox.}]{ClO_4^-} \longrightarrow CH_3SO_3^- + CO \tag{28}$$

This reaction was proposed to involve the intermediate formation of the dimethylmethylthiomethyl sulfonium ion (VII), which is slowly converted to products. The formation of dialkyl sulfones proceeds most likely through the corresponding sulfoxides [56], which may be isolated [43,45,57].

Mixtures of cis and trans isomers are obtained when substituted cyclic aliphatic sulfides are subjected to indirect anodic oxidation in the presence of, for example, Bu_4NBr, Eq. (29) [58].

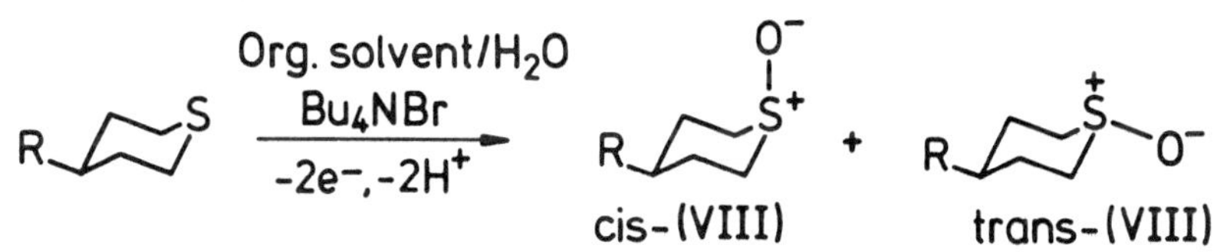

(29)

The trans isomer, trans-VIII, which is the kinetic product, dominates (>80%) in organic solvents containing water (10%), whereas the cis isomer, cis-VIII, is formed preferentially (67%) by oxidation in the presence of sulfuric acid.

Cyclic Bis-sulfides and Related Compounds

The effect of neighboring group participation in the oxidation of aliphatic sulfides has been evaluated by voltammetric studies of the di-endo- and exo,endo-norbornyl derivatives (IX and X), Eq. (30) [59].

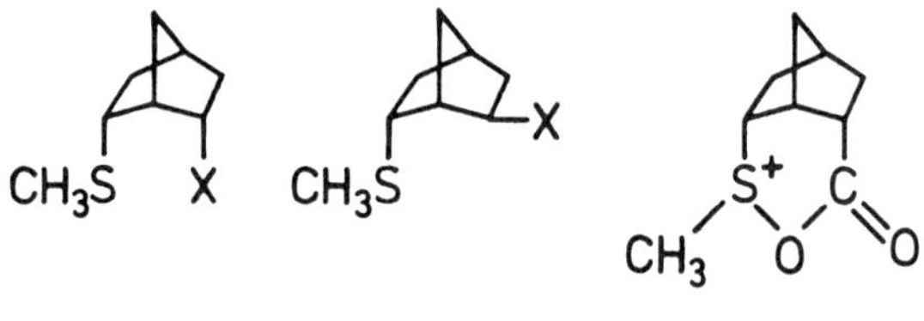

(IX) (X) (XI) (30)

The oxidation potentials of the exo isomers were found to be essentially independent of the nature of the substituent, but those of the endo isomers were strongly dependent on X. The most effective group was observed to be carboxylate, which was found to lower the oxidation potential of the sulfide by more than 600 mV compared with the corresponding exo isomer. The cyclic structure (XI) was proposed for the two-electron oxidation product [see Eq. (49); see also Ref. 60 and references cited therein for nonelectrochemical anchimerically assisted oxidations of compounds of this structure]. The protonated form (X = COOH) had only a negligible effect on the peak potential.

Neighboring group stabilization has also been observed for the formation of sulfide cation radicals during pulse radiolysis of 1-iodo-3-methylthiopropane [61], 3-methylthiopropylamine, 4-methylthiobutyric acid, and related compounds [62,63].

Similarly, the transannular interaction between the sulfur atoms in cyclic bis-sulfides of the structure XII, Eq. (31), results in stabilization of the corresponding cation radicals and dications [64].

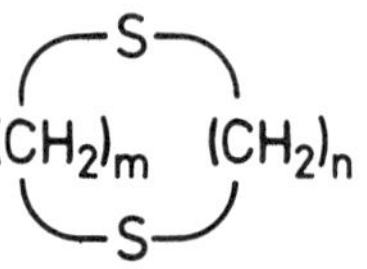

(XII)

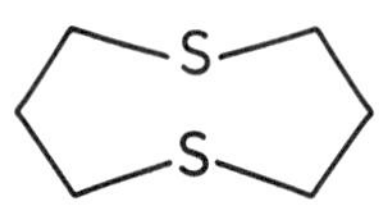

(DTCO) (= XII; m = n = 3)

$$\text{DTCO} \underset{e^-}{\overset{-e^-}{\rightleftharpoons}} \text{DTCO}^{+\bullet} \underset{e^-}{\overset{-e^-}{\rightleftharpoons}} \text{DTCO}^{2+} \qquad (31)$$

$$2\ \text{DTCO}^{+\bullet} \rightleftharpoons (\text{DTCO})_2^{2+} \quad K = 5 \cdot 10^3\ M^{-1}$$

$$(\text{DTCO})_2^{2+} \xrightarrow{2e^-} 2\ \text{DTCO}$$

Table 3. Oxidation Potentials for Aliphatic Mono- and Dithioethers in MeCN

Compound	E_p^a
1,3-Dithietane	1.13
1,4-Dithiacyclohexane	1.25
1,5-Dithiacyclooctane (DTCO)	0.34
1,5-Dithiacyclononane	0.42
1,6-Dithiacyclodecane	0.42
1,3-Dithiacyclohexane	1.14
1,4-Dithiacycloheptane	0.75
1,4-Dithiacyclooctane	0.81
7,8-Benzo-1,5-dithiacyclononane	0.48
3,4,8,9-Dibenzo-1,6-dithiacyclodecane	0.62
2,5-Dithiahexane	0.97
2,6-Dithiaheptane	0.66
2,7-Dithiaoctane	0.67
Thiolane (tetramethylene sulfide)	1.13
Thiane (pentamethylene sulfide)	1.32
4-Oxathiane	1.31
1-Thiacyclooctan-5-ol	0.72

[a]Peak potentials in V versus Ag/Ag^+ (0.1 M) in MeCN.
Source: From Ref. 64.

The results, summarized in Table 3, show that the oxidation potentials are considerably lower for the bis-sulfides than for the corresponding monosulfides for the compounds in which two five-, two six-, or a five- and six-membered fused ring may be formed as the result of the transannular interaction.

The cyclic voltammogram of DTCO, Eq. (31), exhibited the unusual behavior that the current during the reverse scan increased with decreasing sweep rate, which could be explained by assuming that the cation radical, $DTCO^{+\cdot}$, exists in equilibrium with the dimer, $(DTCO)_2^{2+}$ [65–67]. At low sweep rates the dissociation of the dimer is sufficiently fast for the observation of the reduction current for the cation radical; at high sweep rates the direct reduction of the dimer to DTCO is observed. Dimeric dications have also been observed for the cation radicals of 1,3-ditellurole (see Sec. III.B) and thianthrene and related compounds (see Sec. V.A). Intermolecular stabilization by association between the cation radicals of aliphatic sulfides and the neutral parents has been observed by pulse radiolysis [68–70].

B. Thioacetals, Orthothioesters, and Related Compounds

Thioacetals and Related Compounds

The anodic oxidation of dithioacetals (XIII), Eq. (32), in the presence of suitable nucleophiles, such as water, involves cleavage of the central carbon-sulfur bonds, resulting in the formation of the corresponding disulfide or its further oxidation products and a new carbonyl derivative (XIV), which may be converted to the parent aldehyde or ketone during workup, Eq. (33) [30,71,72].

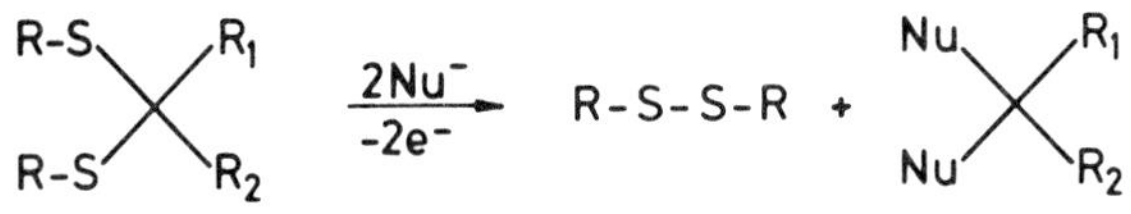

(XIII) (XIV)

(32)

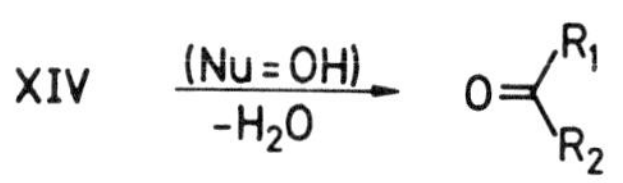

(33)

The mechanism(s) of this conversion, which has been developed into a useful protection-deprotection procedure for the carbonyl function [31,73,74], has been the subject of intensive investigations [31,75–77]. It is beyond the scope of this chapter to review all the arguments presented, and only a brief summary is given here. The oxidation of dithianes (XV) in the absence of nucleophiles was proposed [75] to follow the route depicted in Eq. (34) in which the dication (XV^{2+}), exists in equilibrium with a dimer (XVI) with a structure analogous to that of $(DTCO)_2^{2+}$, Eq. (31), formed by reaction with the neutral parent. Workup results in formation of the carbonyl compound. In contrast to this mechanism, and ECE-type route, Eq. (35), was proposed to take place in the presence of nucleophiles. The cation (XVII) could be reduced to starting material or deprotonate to the sulfoxide [77]. The 1,2-dithiolane (XVIII) formed from XVII may undergo further oxidation to the sulfoxide and the sulfone [77].

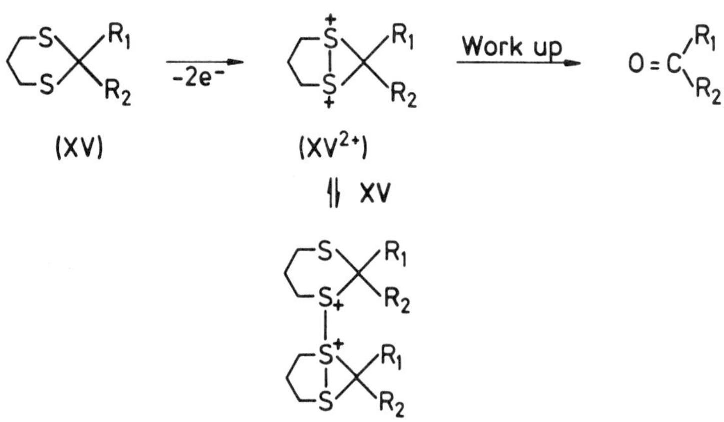

(XVI)

(34)

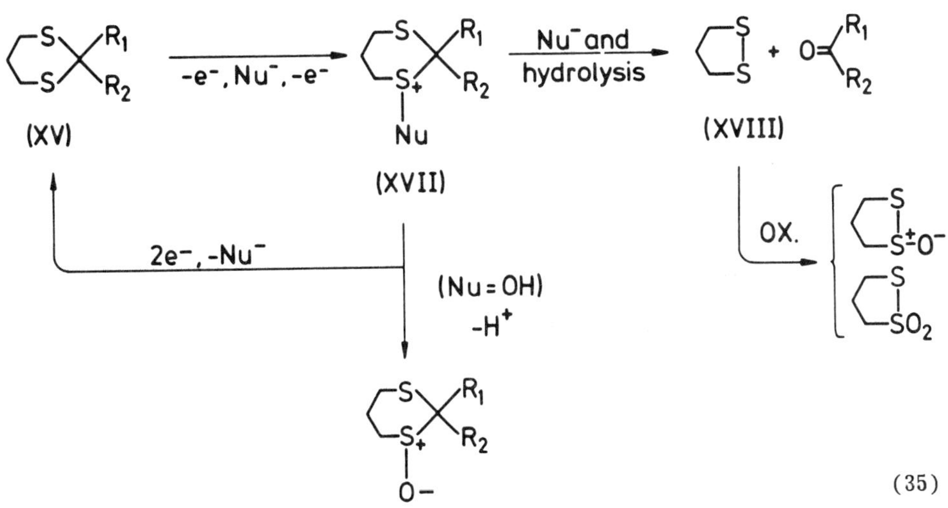

(35)

A series of oxidations of mixtures of two different substrates, Eq. (36), has been carried out to establish whether the formation of the S–S bond takes place before cleavage of the C–S bonds as implied by structures like XV^{2+} or is the result of the intermolecular coupling of RS˙ radicals [72].

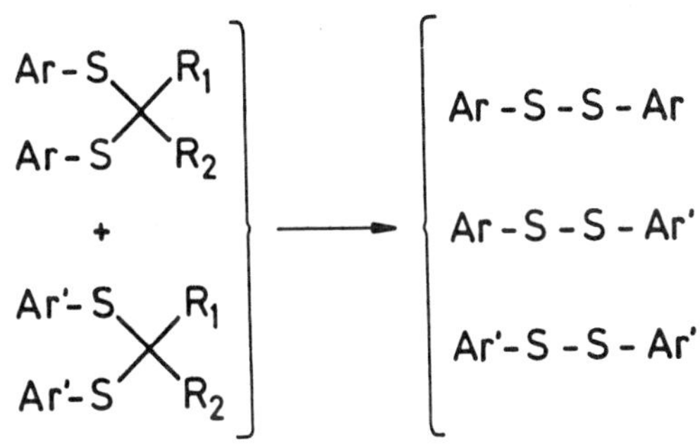

(36)

If the latter is true, it is expected that the product mixture, in addition to ArSSAr and Ar'SSAr' would contain the mixed disulfide, ArSSAr', as well, and this was found to be the case [31,75]. The results appear to be conclusive as independent experiments demonstrated that mixtures of disulfides do not interconvert under the conditions applied in the anodic oxidation [31].

When the cyclic 1,3-disulfide is an aldehyde derivative, the possibility exists that deprotonation of the initially formed cation radical may compete favorably with nucleophilic attack. This has been observed during oxidation of XIX in the presence of pyridine, Eq. (37) [78]. The resulting free radical (XX) undergoes coupling to the dimer (XXI), which under the conditions is further oxidized to the tetrathioethylene derivative (XXII). (See Sec. IV for details of the voltammetric oxidation of tetrathioethylenes.)

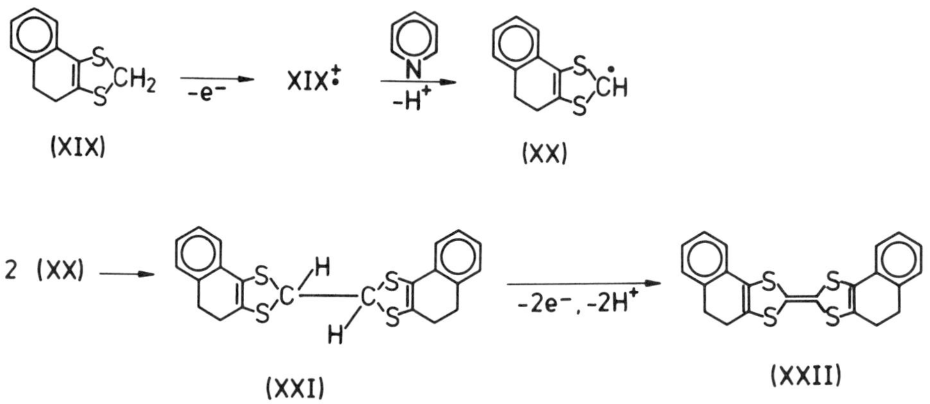

(37)

1,3-Ditellurole (XXIII) undergoes reactions analogous to those already described for the structurally related sulfur compounds, Eq. (38) [79].

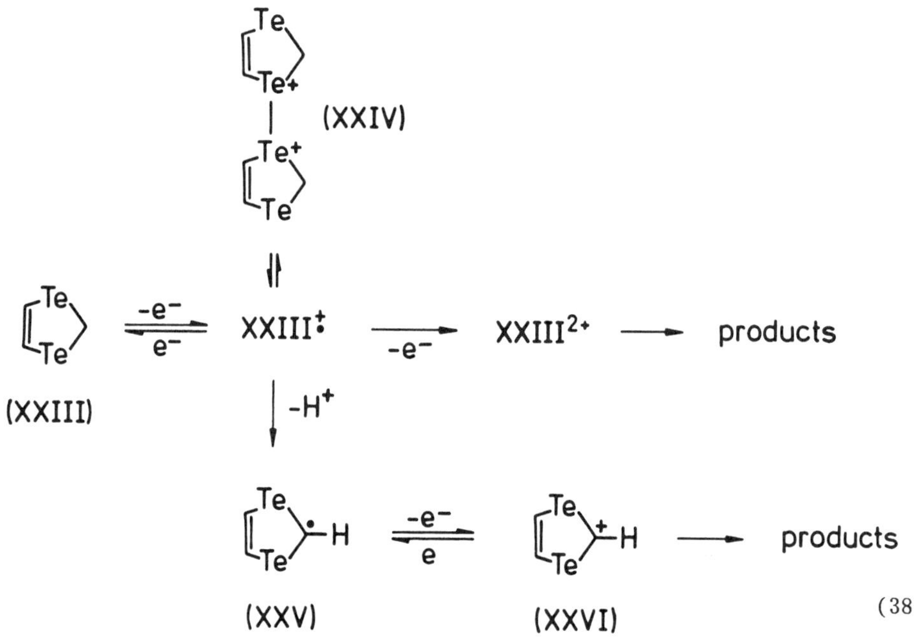

(38)

The cation radical ($XXIII^{+\cdot}$), which exists in equilibrium with the dimer (XXIV), may be futher oxidized to the corresponding dication (XXIII^{2+}), or, in a slow process, deprotonate to the neutral radical (XXV), which in turn can be oxidized to the cation (XXVI).

Orthothioesters

The mechanisms for anodic oxidation of orthothioesters are at the detailed level even more complicated than those for dithioacetals. Oxidation of the orthothiooxalate (XXVII) in MeCN was observed to result in the formation of the well-known tetrathioethylene derivatives (XXIX and XXX), Eq. (39) [80].

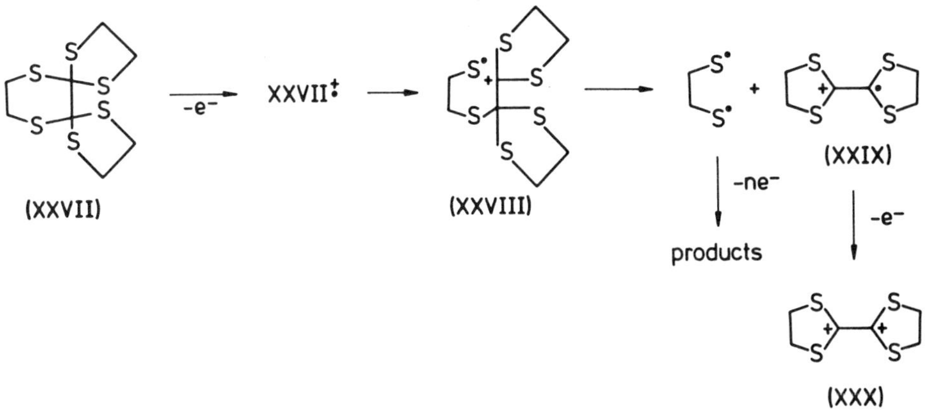

(39)

A reaction mechanism involving the successive cleavage of the two C–S bonds to the six-membered ring at the cation radical stage (XXVII$^{+\cdot}$) leading first to XXVIII has been suggested. The product (XXIX) was identified by voltammetry and ESR spectroscopy.

Even more profound rearrangements are involved in the oxidation of orthothiooxalate (XXXI), Eq. (40) [81]. In this case, the mechanism was proposed to involve the homolytic cleavage of one of the C–S bonds to the methylthio groups at the cation radical stage, resulting in the formation of a cation (XXXII), which after successive bond shifts and elimination of CH_3S^+ is converted to the tetrathioethylene derivative (XXXIII). This compound can be oxidized in two reversible one-electron transfer reactions to the dication, which in turn undergoes rearrangement to the more stable isomer (XXXIV^{2+}), Eq. (41) [81,82]. Reactions similar to those described have also been observed as the result of nonelectrochemical oxidations [83].

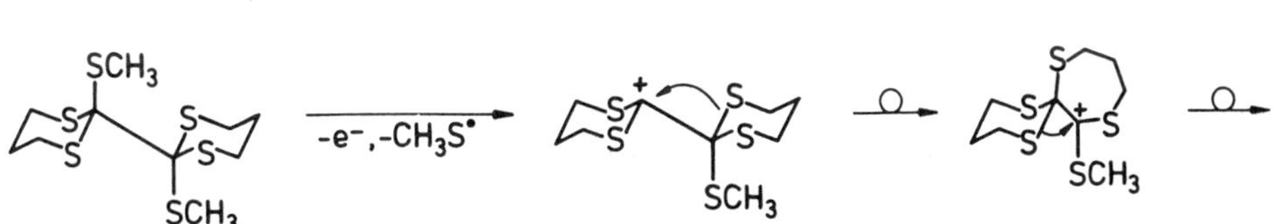

(40)

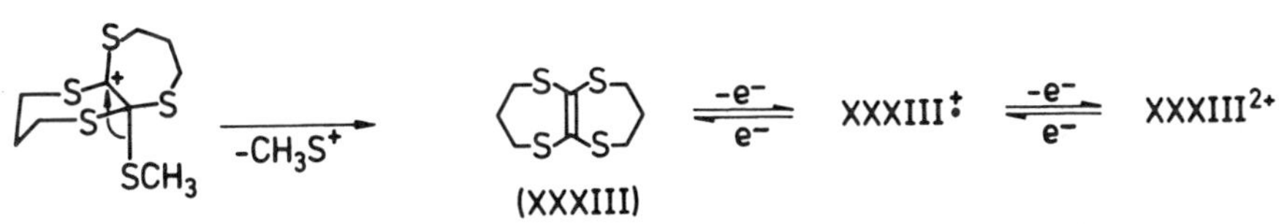

(41)

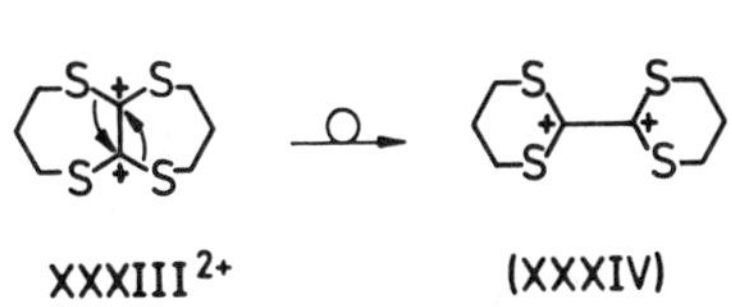

(42)

C. Alkyl Aryl Sulfides and Diaryl Sulfides

The oxidation of methylthio-substituted arenes in MeCN at the rotating Pt electrode gave rise to voltammograms that were judged to reflect the reversible formation of the corresponding cation radicals [84,85]. The half-wave potentials, some of which are summarized in Table 4, were found to correlate linearly with Hückel HOMO energies. The lifetimes of the cation radicals of 1,4-bis(methylthio)naphthalene and 1,6-bis(methylthio)pyrene were estimated from cyclic voltammetric measurements as approximately 0.2 and 0.03 s, respectively [85].

The products of anodic oxidation of aryl sulfides, Ar-S-R, are remarkably dependent on the structure of R and the water content of the solvent and supporting electrolyte system. When R is benzyl or triphenylmethyl, extensive cleavage of the S–R bond is observed, which reflects the stability of the carbocation, R^+, in these two cases, Eq. (42) [86,87]. When R is methyl or phenyl, the corresponding sulfoxides may be isolated in

Table 4. Oxidation Potentials for Methylthioarenes in MeCN

Compound	$E_{1/2}$[a]
Thioanisole	1.57
o-Bis(methylthio)benzene	1.35
m-Bis(methylthio)benzene	1.45
p-Bis(methylthio)benzene	1.19
s-Tris(methylthio)benzene	1.43
1,2,4,5-Tetrakis(methylthio)benzene	1.08
1-Methylthionaphthalene	1.32
2-Methylthionaphthalene	1.37
1,4-Bis(methylthio)naphthalene	1.07
1,5-Bis(methylthio)naphthalene	1.27
1,8-Bis(methylthio)naphthalene	1.09
2,3-Bis(methylthio)naphthalene	1.36
2,6-Bis(methylthio)naphthalene	1.10
2,7-Bis(methylthio)naphthalene	1.33
9,10-Bis(methylthio)anthracene	1.11
2,2'-Bis(methylthio)biphenyl	1.39
3,3'-Bis(methylthio)biphenyl	1.48
4,4'-Bis(methylthio)biphenyl	1.26
1,6-Bis(methylthio)pyrene	0.96

[a]In V versus SCE; from rotating Pt electrode measurements.

Source: From Refs. 84 and 85.

good yields after oxidation in solvents containing water [45,86,88–90]. Further oxidation of the sulfoxide produces the corresponding sulfone, Eq. (43) [90].

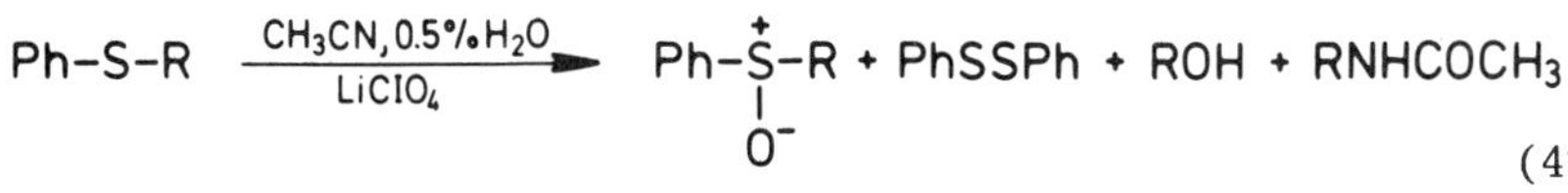

(42)

R	recovered (%)	%	%	%	%
CH_3	12	74	0	0	0
CH_2Ph	36	25	12	3	11
CPh_3	28	0	48	53	0

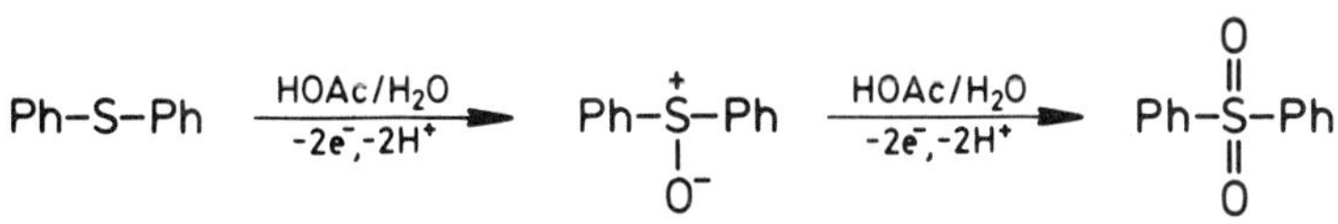

(43)

It is of interest to note that mixtures of sulfoxides with an enantiomeric excess of one of the isomers may be obtained by oxidation of Ph-S-R on electrodes the surfaces of which have been modified by covalent attachement or adsorption of chiral electroinactive molecules like amino acid or (−)-camphoric acid derivatives (see also Chap. 27) [91,92].

Oxidation of Ph-S-R (R = alkyl) in MeCN in the absence of water leads to the formation of sulfonium salts (XXXVI) [93]. The reaction was formulated as Eq. (44), but no attempts were made to determine the oxidation state, cation radical ($XXXV^{+\cdot}$), or dication (XXXV^{2+}) of the intermediates involved in formation of the new Sr–Ar bond. The formation of XXXVI (R = Ph), Eq. (44) [88,93], is in agreement with the results obtained by other workers under similar conditions [94–96]. Attempts to prepare triphenylsulfonium ion (XXXVIII, R = H), Eq. (45), by oxidation of Ph-S-Ph in the presence of benzene as the nucleophile failed [96]. However, the self-coupling product (XXXVII) could be suppressed successfully for R = OCH_3 when the concentration of anisole was kept above 0.1 M [94]. The reaction was proposed to involve disproportionation of the cation radical, Eq. (46), followed by nucleophilic attack on the dication and proton loss to yield XXXVIII, Eq. (47) (Nu = PhR) [97].

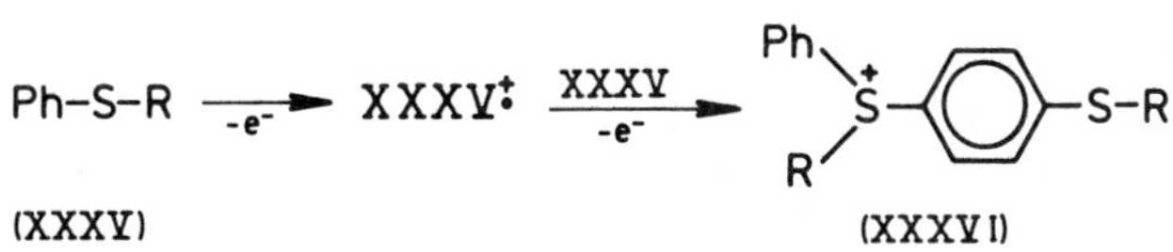

R	%
CH_3	60
$CH_2=CH-CH_2-CH_2$	67
	57
Ph	71

(44)

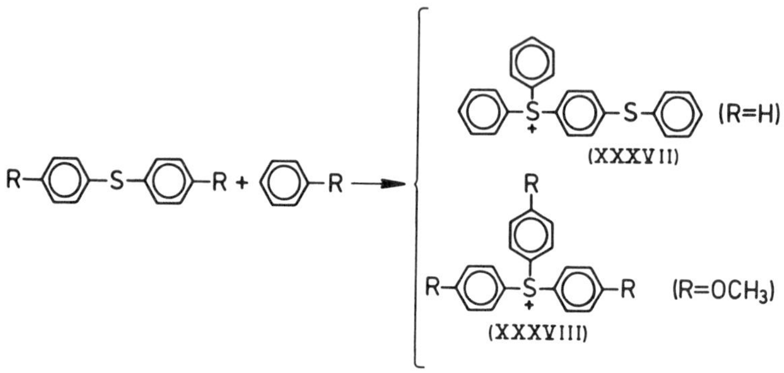

(45)

$$2\ Ar-\overset{+\bullet}{S}-Ar \rightleftharpoons Ar-\overset{2+}{S}-Ar + Ar-S-Ar \qquad (46)$$

$$Ar-\overset{2+}{S}-Ar + Nu \longrightarrow Ar-\overset{2+}{S}(Nu)-Ar \xrightarrow[-H^+]{} \text{XXXVIII} \qquad (47)$$

Anodic oxidation of diphenyl sulfoxide in MeCN produces the sulfones in 50% yield even in the absence of an oxygen source in the solvent [98]. This observation may be explained by assuming that the initially formed sulfoxide cation radical (XXXIX$^{+\bullet}$) abstracts an oxygen atom from the unoxidized starting material (XXXIX), possibly via the formation of the tetracoordinated S(IV) intermediate (XL), Eq. (48).

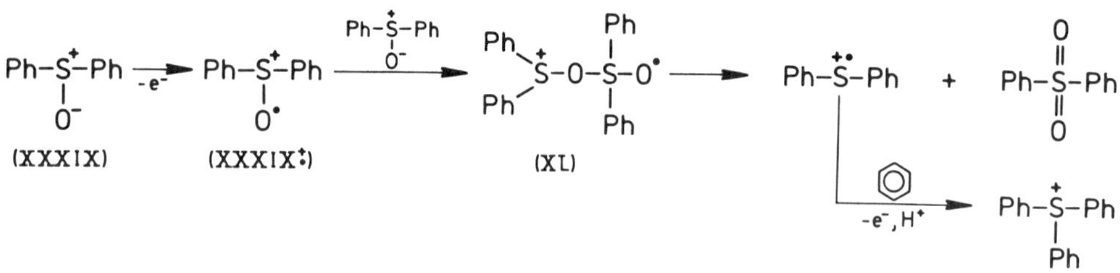

(48)

Dissociation of XL leads to formation of diphenyl sulfone and diphenyl sulfide cation radical, which in this case is produced in a solution not containing Ph_2S (see earlier) and therefore may be trapped by benzene, for example, resulting finally in triphenyl sulfonium ion.

The electrochemical behavior of tetracoordinated S(IV) compounds and of structures similar to XL has been investigated. Reduction of XLI proceeds via S–O bond cleavage to the dicarboxylate XLII, Eq. (49), which may be partially reoxidized to XLI [99]. The latter reaction indicates that anodic oxidation of suitable substrates may provide a feasible route to tetracoordinated S(IV) compounds. The oxidation potentials for XLII and similar compounds are strikingly low, 0.4–0.5 V versus Ag/Ag^+, which seems to indicate anchimerically assisted oxidation analogous to the oxidation of IX, Eq. (30).

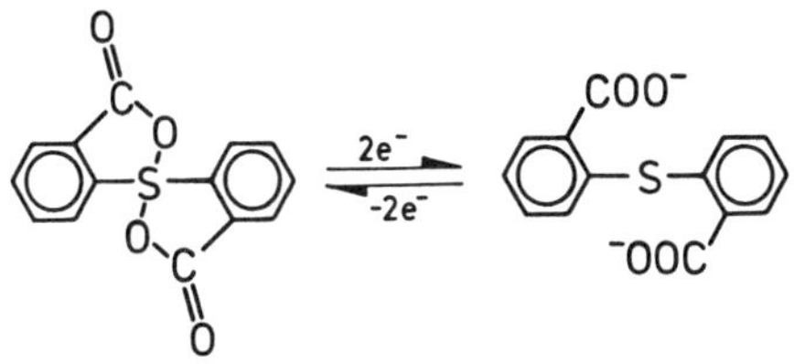

(XLI) (XLII)

(49)

The oxidation of Ar-S-R (R = alkyl) in $HOAc-Ac_2O-NaOAc$ gives mixtures of the corresponding S-oxide and the α-acetoxy derivative [100]. The latter is the Pummerer product, which normally is produced by treatment of S-oxides with Ac_2O at elevated temperatures, and therefore it was natural to attempt to carry out the anodic oxidation in the boiling solvent mixture to obtain the α-acetoxy derivative exclusively, Eq. (50). The approach worked nicely, as shown by the yields given in Table 5 [100].

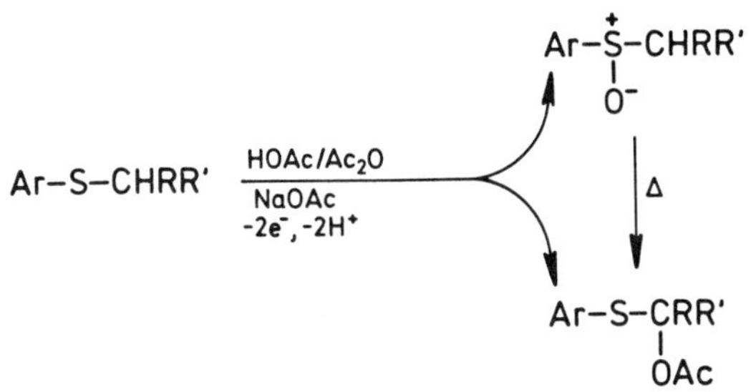

(XLIII)

(50)

Table 5. Yields of α-Acetoxylated Products (XLIII) Eq. (50), from Anodic Oxidation of Aryl Alkyl Sulfides in Boiling $HOAc-Ac_2O-NaOAc$

CHRR'	Ar	Yield (%)[a]	Current yield (%)[b]
Methyl	Phenyl	98 (70)	98
Methyl	p-Tolyl	97 (70)	91
Methyl	p-Anisyl	98 (70)	89
Methyl	p-Bromophenyl	97 (50)	93
Ethyl	Phenyl	95 (51)	63
Isopropyl	Phenyl	95 (87)	27
Benzyl	Phenyl	98 (49)[c]	87

[a]Determined by GLC. The values in parentheses are isolated yields after vacuum distillation during which some destruction was observed.
[b]Based on theoretical 2 F/mol.
[c]Crystalline material isolated by filtration.
Source: From Ref. 100.

The heat evolved during the electrolysis could be controlled by adjusting the current so that gentle reflux was maintained. Other examples of side-chain functionalization have been reported for sulfides carrying strongly electron-withdrawing substituents at the α-carbon, including methoxylation and acetoxylation of Ph-S-CH_2CF_3 [101] and methoxylation of R-S-CH_2CN [102].

When the side chain is olefinic, anodic oxidation of Ar-S-R (XLIV) provides a convenient route to 2-(alkylthio)aldehydes (XLV), Eq. (51) [103, 104]. The yields vary between 50 and 90%. The reaction mechanism, Eq. (52), has been proposed to involve the formation of the carbocation (XLVI), which rearranges to the nonclassic structure (XLVII), followed by nucleophilic attack by water giving XLV, or methanol, for example, in which case the corresponding acetal (XLVIII) is formed [104]. Electron-withdrawing substituents alter the product distribution markedly. Oxidation of the nitrile (IL) in the presence of methanol results in an addition reaction, Eq. (53), by which the mono thioacetal (L) is formed. In the absence of intentionally added nucleophiles LI is formed (34%) in a reaction that was suggested without additional evidence to involve the intermediate formation of the dimer dication (LII).

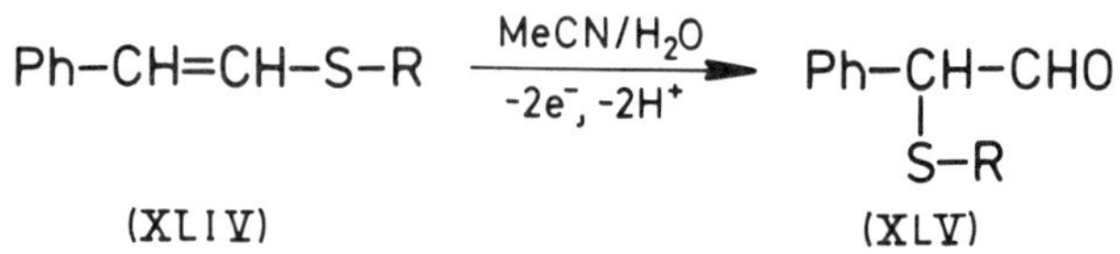

(51)

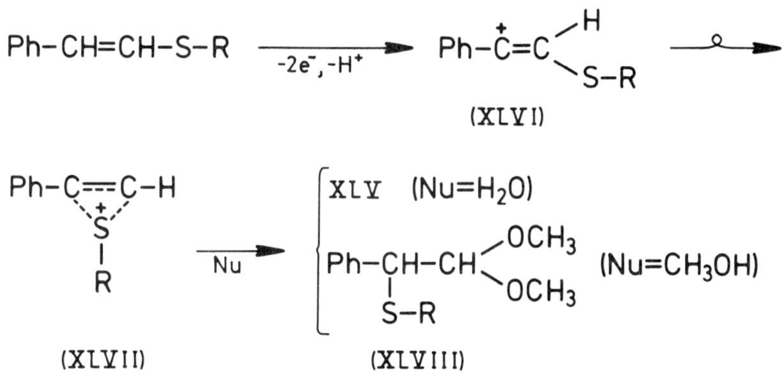

(52)

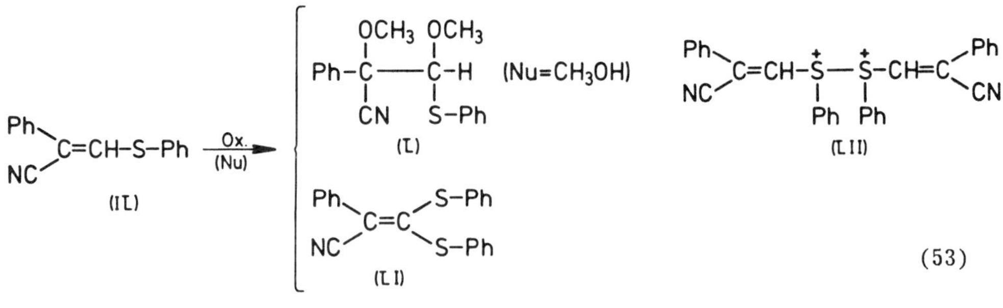

(53)

Compounds containing the Ph-S or the R-S groups have been found to undergo conversion to the corresponding methoxy or acetoxy derivatives in high yields upon anodic oxidation in methanol or acetic acid, respectively, Eq. (54) [105–108]. The fate of the sulfur function was reported in only one case, Eq. (54) (X = $COOCH_3$), in which the preparation of the methoxy derivative (LIII) was accompanied by the formation of Ph-S-S-Ph [105]. Other cleavage reactions involving Ph-S and R-S are shown in Eqs. (55) and (56) [109–111].

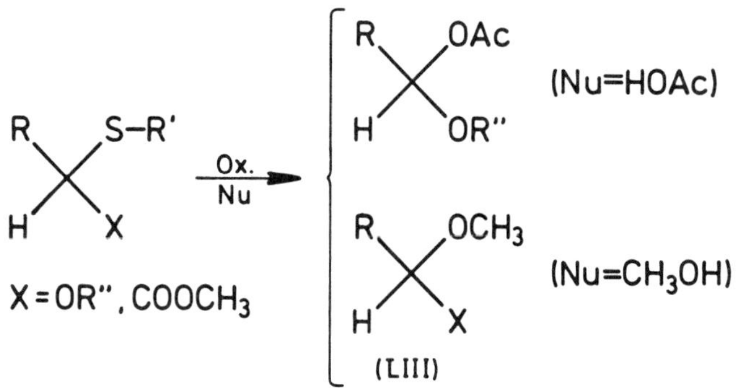

(54)

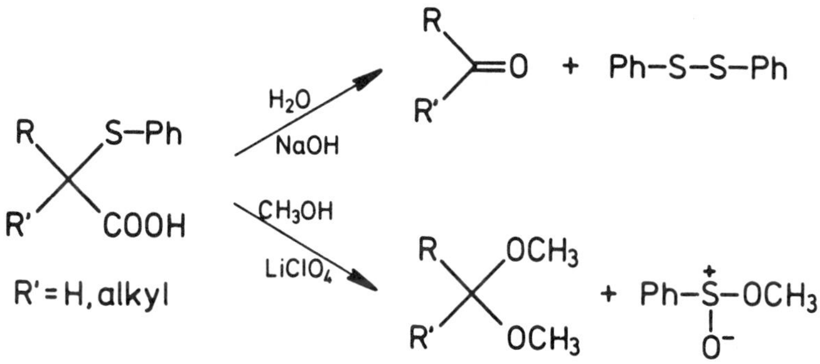

(55)

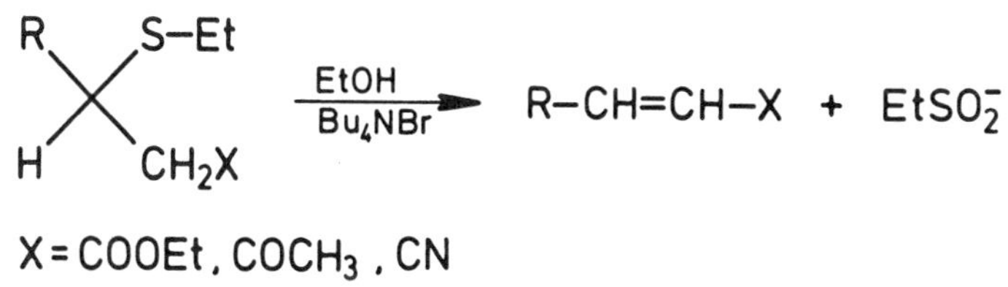

(56)

IV. OXIDATION OF TETRATHIOETHYLENES AND TETRATHIAFULVALENES

The interest in the electrochemical properties of tetrachalcogenafulvalenes (LV, LIX, LX, and LXI), Eq. (57), has increased considerably since the discovery that solid-state charge-transfer complexes involving members of this class of compounds exhibit metallike conductivity and, in cases, even superconductivity (see Refs. 112 and 113 for pertinent references).

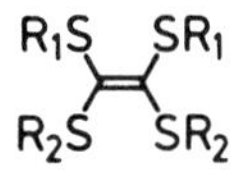
(LIV)

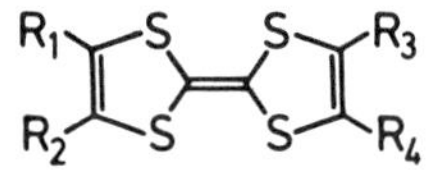
(LV)

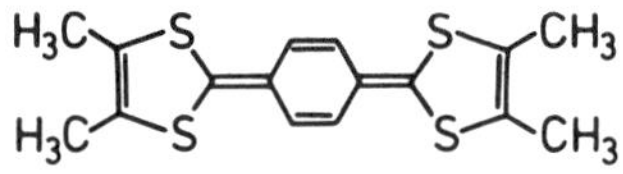
(LVI)

(57)

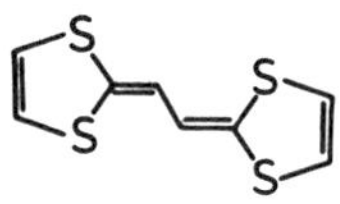
(LVII)

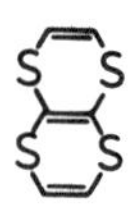
(LVIII)

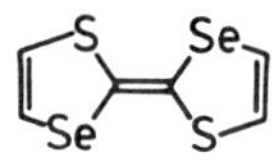
(LIX)

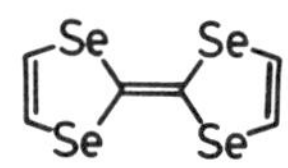
(LX)

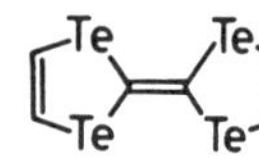
(LXI)

The voltammetric behavior of tetrachalcogenoethylenes and tetrachalcogenafulvalenes is characterized by the stability not only of the corresponding cation radicals but also of the dications; potentials for the reversible or nearly reversible successive one-electron transfers have been recorded for more than 200 compounds possessing this basic structure [5,112]. Some of these are summarized in Tables 6 and 7.

Inspection of the data in Table 6 reveals that the phenyl derivative (LIV; R = Ph) is considerably more difficult to oxidize than the corresponding alkyl derivatives [114], which most likely reflects steric hindrance of interaction between the sulfur lone-pair electrons and the ethylene π system. Another noteworthy feature of the data is the effect of the —CH=CH— bridge in tetrathiafulvalene (TTF; LIV, R_1, R_2 = CH=CH, or LV, $R_1 = R_2 = R_3 = R_4$ = H), which lowers the oxidation potential by approximately 350 mV compared with that of the saturated analog LIV (R_1, R_2 = CH_2—CH_2), illustrating the effect of the extended π network in TTF [115].

The presence of the electron-donating alkyl groups in TTF results in substrates that are more easily oxidized [117,119,120] and electron-withdrawing groups like CHO, $COCH_3$, COOH, and COOEt in substrates that are more difficult to oxidize [119,121], as intuitively expected (see Table 7). The

Table 6. Oxidation Potentials for Tetrathioethylenes in MeCN

R_1	R_2	E_1[a]	E_2[a]	Reference
C_6H_5	C_6H_5	1.2	1.4	114, 115
CH_3	CH_3	0.89	—	114
CH_3CH_2	CH_3CH_2	0.93	—	114
$(CH_3)_2CH$	$(CH_3)_2CH$	0.97	—	114
CH_2-CH_2-CH_2		0.70	0.84	115
CH_2-CH_2		0.68	1.12	115
CH=CH		0.33	0.70	115

[a]Half-wave [114] or formal [115] potentials in V versus SCE.

Table 7. Half-wave Potentials for the Oxidation of Tetrathiafulvalenes and Related Compounds to the Corresponding Cation Radicals and Dications[a]

R_1	R_2	R_3	R_4	E_1[b]	E_2[b]	Solvent	Reference
H	H	H	H	0.33	0.70	MeCN	115
H	H	H	H	0.31	0.74	MeOH-benzene	116
H	H	H	H	0.40	0.86	PhCN	117
H	H	H	H	0.47	0.81	CH_2Cl_2	118
CH_3	H	H	H	0.30[c]	0.67[c]	MeCN	119
CH_3	CH_3	H	H	0.28	0.65	MeCN	117
CH_3	H	H	CH_3	0.24	0.62	MeCN	117
CH_3	CH_3	CH_3	CH_3	0.21	0.59	MeCN	117
CH_2CH_3	CH_2CH_3	CH_2CH_3	CH_2CH_3	0.23	0.75	MeCN	120
$CH_2CH_2CH_2$		$CH_2CH_2CH_2$		0.31	0.74	MeCN	120
CH_2OH	H	H	H	0.38[c]	0.76[c]	MeCN	119
CHO	H	H	H	0.50[c]	0.86[c]	MeCN	119
$COCH_3$	H	H	H	0.44[c]	0.80[c]	MeCN	119
COOH	H	H	H	0.44[c]	0.80[c]	MeCN	119
$COOCH_2CH_3$	H	H	H	0.44[c]	0.80[c]	MeCN	119
$COOCH_2CH_3$	$COOCH_2CH_3$	H	H	0.54[c]	0.88[c]	MeCN	119

Table 7. (Continued)

R_1	R_2	R_3	R_4	E_1[b]	E_2[b]	Solvent	Reference
$COOCH_2CH_3$	$COOCH_2CH_3$	$COOCH_2CH_3$	$COOCH_2CH_3$	0.80	1.07	MeCN	121
Ph	H	H	H	0.35	0.70	MeCN	117
Ph	H	Ph	H	0.39	0.78	MeOH-benzene	116
Ph	Ph	Ph	Ph	0.41	0.77	MeOH-benzene	116
p-$CH_3OC_6H_4$	H	p-$CH_3OC_6H_4$	H	0.33	0.73	MeOH-benzene	116
p-ClC_6H_4	H	p-ClC_6H_4	H	0.43	0.79	MeOH-benzene	116
Benzo		Benzo		0.60[d]	0.94[d]	MeCN	122
LVI				-0.32	-0.11	CH_2Cl_2	123
LVII				0.18[e]	0.34[e]	MeCN	124
LVIII				0.88[d]	1.28[d]	MeCN	125
LIX				0.40	0.72	MeCN	126
LX				0.48	0.76	MeCN	126
				0.62	0.90	CH_2Cl_2	118
LXI				0.59	0.84	CH_2Cl_2	118

[a]The numbering of the substituents refers to structure LV. A more extended compilation of oxidation potentials may be found in Ref. 112.
[b]In V versus SCE.
[c]Converted from peak potentials to half-wave potentials.
[d]Converted from V versus Ag/Ag^+.
[e]Converted from V versus Ag/AgCl.

effect of Ph is almost negligible owing to the lack of coplanarity between the phenyl rings and the TTF system [116,117]. A fair correlation, $E_1(ox) = 0.10\sigma_p^+ + 0.41$, was observed for the oxidation of p-substituted diaryltetrathiafulvalenes [116]. It is of interest to note that the benzoderivative LV (R_1, R_2 = R_3, R_4 = benzo) [122] is significantly more difficult to oxidize than TTF and in that respect resembles the cyclic alkyl derivatives (Table 6). In contrast LVI is not less than 650 mV more easily oxidized than TTF [123], illustrating the gain in resonance energy associated with formation of the central aromatic ring present in the cation radical and the dication.

Comparison of the oxidation potentials for TTF (LIX, LX, and LXI) demonstrates that the tetrachalcogenafulvalenes become increasingly difficult to oxidize when S is exchanged by Se. However, going from LX to LXI, in which Se is exchanged by Te, the opposite effect is observed, which is similar to that reported for the related naphthalene series in Eq. (17). We are not aware of detailed discussions of these trends.

The potentials for formation of $TTF^{+\cdot}$ and TTF^{2+} are markedly dependent on the solvent (Table 7) but appear not to be related to solvent polarity parameters in a simple fashion (see also Sec. V).

Very little is known about the chemistry of the cation radicals and dications derived from TTF and related compounds. The lack of reactivity (or considerable stability) of these structures is also reflected by the oxidative rearrangements described earlier for cyclic orthothiocarbonates, Eqs. (39) through (41), which all result in the formation of cations possessing the tetrathioethylene structure. However, oxidation of the carboxylate ion (LV; R_1 = COO^-) has been reported to result in decarboxylation at the dication stage of the TTF. The mechanism of the reaction was found to be rather complex and to involve TTF^{2+} resulting from the decarboxylation process as an electron transfer mediator for the subsequent oxidation of a nonstoichiometric sodium salt of the starting material and the corresponding cation radical [127].

Another example of the low reactivity of TTF cation radicals is the oxidation of TTF in the presence of bromide ion, which does not lead to products resulting from nucleophilic attack but rather to formation of insoluble nonstoichiometric salts with the composition $TTFBr_x$ ($x \approx 0.8$) [128].

The acid function in LV (R_1 = COOH) has been used to immobilize the TTF structure at the surface of alkylaminesilanized RuO_2 and PtO_2 electrodes through an amide linkage [129].

V. OXIDATION OF 1,4-DITHIINS, BENZODERIVATIVES, AND RELATED COMPOUNDS

1,4-Dithiin (LXIII) and its benzoderivatives (LXIV and LXV), Eq. (58), are electron-rich compounds and accordingly form the corresponding cation radicals and dications at fairly low oxidation potentials, and in this fashion these compounds and the related structures in Eq. (58) resemble the tetrathioethylenes (Sec. IV). The cation radical of one member of the group, thianthrene (LXV, commonly abbreviated Th), has been the subject of very extensive investigations and has served as a model substrate in the early stages of cation radical chemistry (see Refs. 4 and 7 for general discussions).

(LXII) (LXIII) (LXIV) (LXV) (LXVI)

(LXVII) (LXVIII) (LXIX) (LXX)

(58)

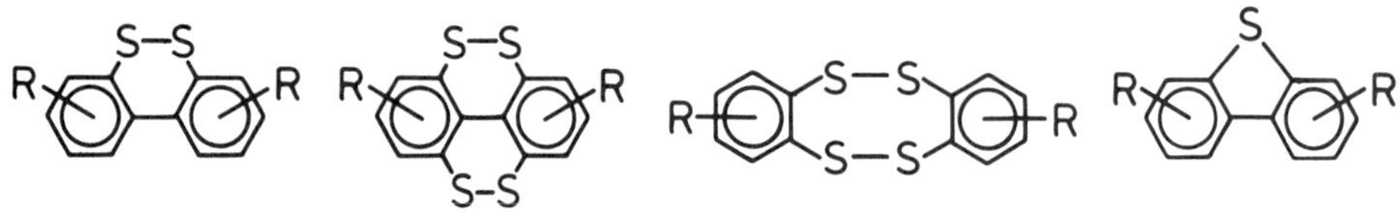

(LXXI) (LXXII) (LXXIII) (LXXIV)

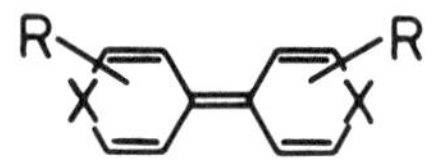

X = O, S

(LXXV)

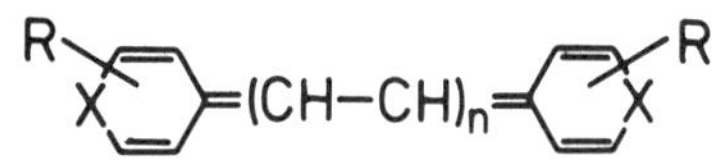

X = O, S

(LXXVI)

A. The Effect of Structure and Solvents on the Oxidation Potentials

The relation between structure and ease of oxidation follows the same trends as those reported for other organochalcogen compounds, including the naphthalene series in Eq. (17) and the tetrathioethylenes (Sec. IV); that is, oxidation becomes easier in going from O to Te as illustrated by the following three series: (1) E(LXII) > E(LXIII) [130], (2) E(LXVII) > E(LXV) > E(LXVI) [131], and (3) E(LXVII) > E(LXVIII) > E(LXIX) > E(LXX) [131], and more difficult when the CH=CH linkage becomes member of a benzene ring, for example, E(LXV) > E(LXIV) > E(LXIII) [130].

The observation that nucleophiles present in voltammetric solutions may be effectively passivated or removed when the voltammetric experiments are carried out in the presence of trifluoroacetic acid (TFA) [132], sulfuric acid [133], TFA/TFAn (TFAn = trifluoroacetic acid anhydride) [134], or activated neutral alumina [134] has made it possible to record reversible oxidation potentials for Th, for example, in a number of different solvents and from these data to calculate reliable values of the disproportionation equilibrium constant K_{disp} (Table 8). The value of K_{disp} was observed to vary by almost six orders of magnitude in passing from isobutyronitrile (largest value) to a mixture of TFA and sulfuric acid (smallest value). It is beyond the scope of this chapter to give a detailed discussion of these data, but we wish to point out that the sensitivity of K_{disp} to solvent polarity is much greater for $Th^{\cdot+}$ than for cation radicals derived from alkoxy-substituted aromatic hydrocarbons [134], for example, which most likely reflects the high local charge density at the sulfur atoms in the cations result-

Table 8. Disproportionation Equilibrium Constants for Thianthrene Cation Radical in Different Solvents

Solvent	K_{disp}[a]	Solvent	K_{disp}[a]
CH_3CN	2.3×10^{-9}	CH_2Cl_2	3.4×10^{-9}
CH_3CH_2CN	1.6×10^{-8}	PC[b]	2.4×10^{-8}
$(CH_3)_2CHCN$	7.6×10^{-8}	A[c]	6.8×10^{-11}
C_6H_5CN	4.9×10^{-9}	B[d]	2.0×10^{-12}
CH_3NO_2	1.0×10^{-9}	C[e]	6.0×10^{-14}
$C_6H_5NO_2$	2.3×10^{-9}		

[a]The disproportionation equilibrium constant calculated as $\log K_{disp} = (E_1^0 - E_2^0)/0.059$.
[b]Propylene carbonate; from Ref. 135.
[c]Solvent mixture A = CH_2Cl_2/TFA/TFAn (45/1/5).
[d]Solvent mixture B = TFA/TFAn (9/1).
[e]Solvent mixture C = TFA/H_2SO_4 (5/4); from Ref. 133.
Source: From Ref. 134 unless otherwise stated.

ing from Th. Accurate values of K_{disp} are of importance in the evaluation of the mechanistic details for the reactions between $Th^{+\cdot}$ and nucleophilic reagents and have been used to rule out the disproportionation pathway for the reaction with water, for example (see Ref. 136 for a detailed discussion).

Reversible oxidation potentials for two consecutive one-electron transfers have also been recorded for alkoxy-substituted thianthrenes and selenanthrenes [137,138], alkoxy-substituted 2,2'-(di)thiobiphenyls (LXXI, LXXII, LXXIII, and LXXIV) [138], and bis-(4H-thiopyran)s with the general structures LXXV and LXXVI [121,139].

Cation radicals of Th and the related chalcogenanthrenes exist in solution in equilibrium with dimeric species [140,141] and resemble in this respect the cation radicals of DTCO, for example [Sec. III.A and Eq. (31)].

B. Oxidation Followed by Deprotonation

Anodic oxidation of aromatic sulfur compounds containing α-hydrogen atoms frequently results in products formed by deprotonation of the positively charged intermediates. The following two examples serve to illustrate this point.

The primary reaction pathway for the oxidation of thioxanthene (LXXVII) involves the formation of the thioxanthylium ion (LXXVIII) in a ($-2e^-$, $-H^+$) process, Eq. (59) [142]. Nucleophilic attack by water results in the formation of the hydroxy derivative (LXXIX), which in turn may be further oxidized to the ketone (LXXX), the sulfoxide (LXXXI), and finally the sulfone (LXXXII) [142].

Phenothiazine (LXXXIII) is oxidized in two distinct one-electron steps in MeCN, forming the cation radical and the cation (LXXXIV), Eq. (60) [143]. Deprotonation takes place at the cation radical stage in the presence of a suitable base, resulting in a neutral free radical that dimerizes [143].

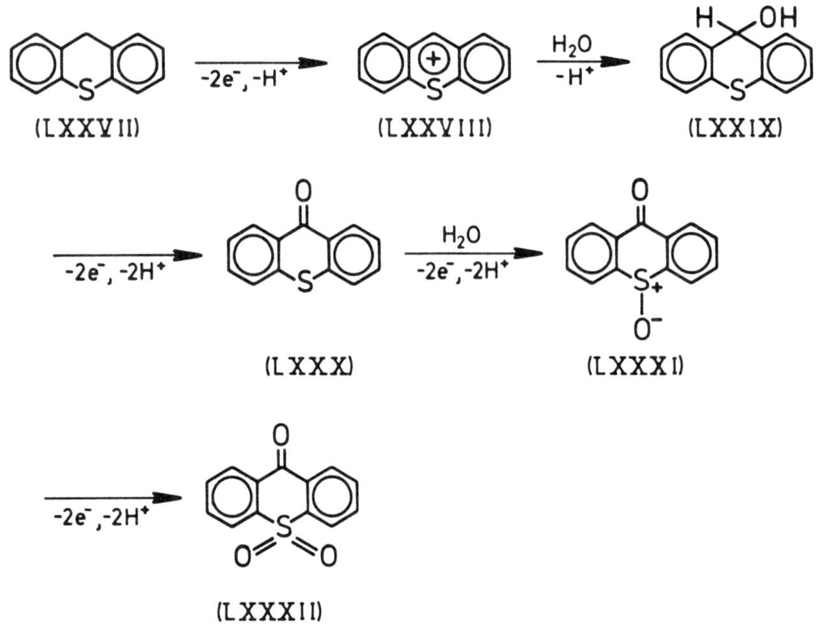

(59)

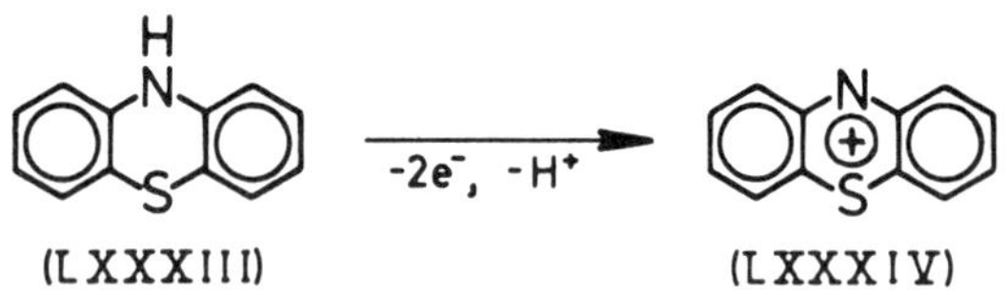

(60)

The pK_a of the phenothiazine cation radical has recently been estimated as 4–5 [144,145].

C. Reactions of the Cation Radicals with Nucleophiles

The reactions between the cation radicals of Th and some of the closely related heterocycles with a variety of nucleophiles have been studied in great detail, mainly by Shine and co-workers (see, e.g., Ref. 7 and references to follow). The products, here exemplified by reaction with $Th^{+\cdot}$ and representative nucleophiles, are with few exceptions the corresponding sulfonium salts, Eq. (61). The nucleophiles used in these studies, not all shown in Eq. (61), include superoxide ion [146], water [147–152], carboxylate ions [153,154], nitrite and nitrate ion [155,156], phosphate ion [157], ammonia [158], aliphatic amines [159,160], pyridine [155,161], organometallic compounds like dialkylmercury and dialkylzinc [162,163] and Grignard reagents [164], alkenes and alkynes [165,166], ketones [152,167,168], and various aromatic compounds including phenol and anisole [169–173]. In addition to these reactions, which all result in bond formation, electron transfer from the potential nucleophile to $Th^{+\cdot}$, for example, has been observed in a number of cases [146,174–177].

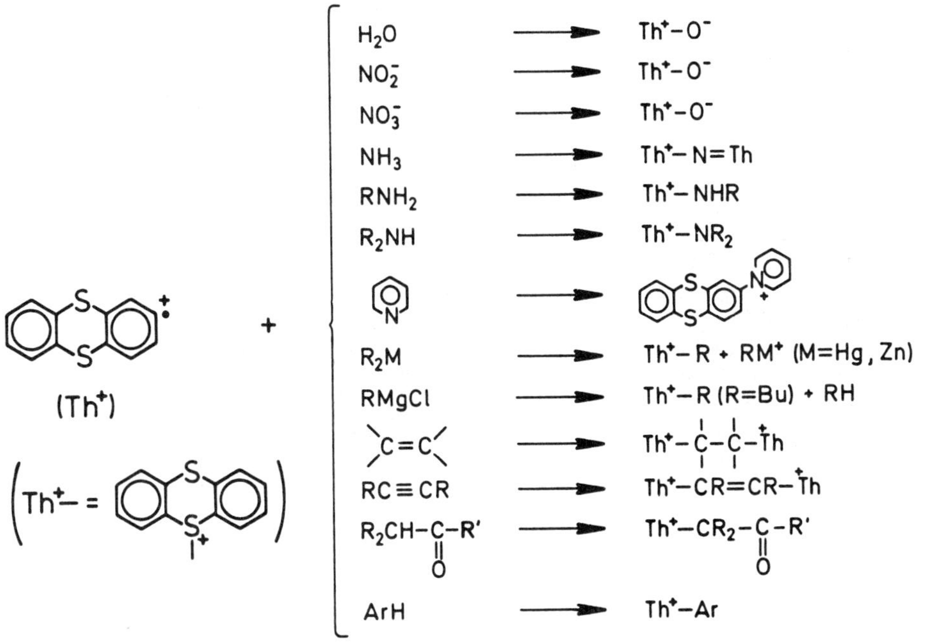

(61)

A general trend is that the reactions with nucleophiles containing oxygen result in the formation of the corresponding S-oxide and among these the reactions with water and carboxylate ions have been studied thoroughly (see Refs. 150 and 153 for leading references). A particularly nice example is the reaction between $Th^{+\cdot}$ and adenosine 5'-monophosphate–orthophosphoric acid in MeCN, which was reported to result in the formation of the high-energy phosphate, adenosine 5'-diphosphate, together with Th and the S-oxide [157]. As a result of these studies, which have occasionally led to animated discussions, there now seems to be general agreement on the basic mechanisms shown in Eqs. (62) through (65) and (66) through (68) for the reactions of $Th^{+\cdot}$ with water [150] and N-methylphenothiazine cation radical, $MPZ^{+\cdot}$, with acetate ion [153], respectively.

$$Th^{+\cdot} + H_2O \underset{}{\overset{K_{62}}{\rightleftharpoons}} Th^{\cdot}\text{-}\overset{+}{O}H_2 \qquad (62)$$

$$Th^{\cdot}\text{-}\overset{+}{O}H_2 + B \overset{K_{63}}{\rightleftharpoons} Th^{\cdot}\text{-}OH + BH^+ \qquad (63)$$

$$Th^{\cdot}\text{-}OH + Th^{+\cdot} \underset{K_{-64}}{\overset{K_{64}}{\rightleftharpoons}} Th^{+}\text{-}OH + Th \qquad (64)$$

$$Th^{+}\text{-}OH + B \xrightarrow{K_{65}} Th^{+}\text{-}O^{-} + BH^+ \qquad (65)$$

(B = base)

$$MPZ^{+\cdot} + AcO^- \underset{}{\overset{K_{66}}{\rightleftharpoons}} MPZ^{\cdot}\text{—}OAc \quad (66)$$

$$MPZ^{\cdot}\text{—}OAc + MPZ^{+\cdot} \underset{K_{-67}}{\overset{K_{67}}{\rightleftharpoons}} MPZ^{+}\text{—}OAc + MPZ \quad (67)$$

$$MPZ^{+}\text{—}OAc + AcO^- \xrightarrow{K_{68}} \text{products} \quad (68)$$

A special problem for the reactions with water is the tendency of this species to exist in aprotic solvents in equilibrium with concentration-dependent amounts of dimers, trimers, and even higher aggregates, which may result in rate expressions of high order in water, strongly dependent on the experimental conditions (see Ref. 136 for details).

Kinetic data for the reaction between $Th^{+\cdot}$ and aromatic nucleophiles, such as anisole (AnH), have been shown to be consistent with a mechanism involving initial complexation between $Th^{+\cdot}$ and AnH, Eq. (69). The complex is oxidized by uncomplexed $Th^{+\cdot}$ in a subsequent homogeneous electron transfer step to the dication complex, Eq. (70), which deprotonates under formation of the covalent sulfur-carbon bond, Eq. (71). An alternative reaction of the complex involving electron transfer, Eqs. (72) and (73), was observed to compete favorably at low concentration of AnH.

$$Th^{+\cdot} + AnH \overset{K_{69}}{\rightleftharpoons} Th^{+\cdot}/AnH \quad (69)$$

$$Th^{+\cdot}/AnH + Th^{+\cdot} \underset{K_{-70}}{\overset{K_{70}}{\rightleftharpoons}} Th^{2+}/AnH + Th \quad (70)$$

$$Th^{2+}/AnH \xrightarrow{K_{71}} Th^{+}\text{—}An + H^{+} \quad (71)$$

$$Th^{+\cdot}/AnH \underset{K_{-72}}{\overset{K_{72}}{\rightleftharpoons}} Th + AnH^{+\cdot} \quad (72)$$

$$AnH^{+\cdot} \xrightarrow{K_{73}} \text{products} \quad (73)$$

Other reactions involving aromatic nucleophiles are likely to follow a similar scheme.

VI. OXIDATION OF THIOPHENES

Little is known about the anodic oxidation of thiophenes and its derivatives, one reason being that polymerization is often the favored pathway of the primary electrode intermediates of these compounds. However, after the discovery that electrochemically formed polymers often show high electrical conductance (see, e.g., Ref. 113), the oxidation of thiophenes, for example, and related electron-rich structures has gained renewed interest. The anodic oxidation of 3-alkylthiophenes results in the formation of the 2,5-polymers, Eq. (74), which typically exhibit conductivities in the range 50–400 S/cm (see also Chap. 33) [178].

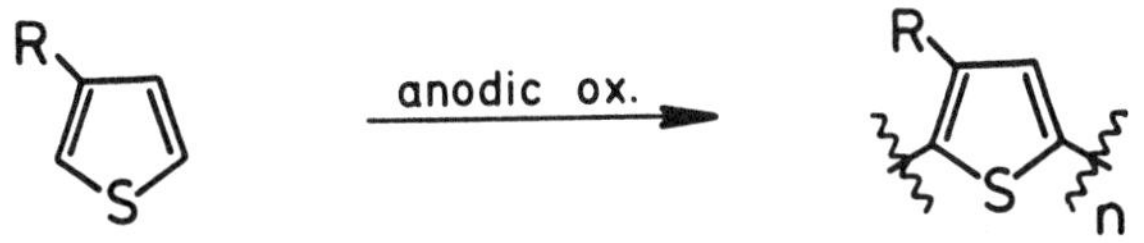

(74)

Dibenzothiophene behaves like diphenylsulfide and forms the sulfonium ion upon anodic oxidation in the absence of intentionally added nucleophiles (see Sec. III.C) [95].

Methoxylated products are formed when the oxidation of simple thiophene derivatives is carried out in methanol [179–182]. The reactions are similar to those described for 1,4-dimethoxybenzene and furan (see Chap. 16), for example, that is, 1,4-addition of two methoxy groups, with the additional feature that the sulfur atom is lost as SO_2, resulting in an open-chain acetal. In contrast, the heterocyclic ring system survives oxidation in methanol when benzothiophene is subjected to oxidation under similar conditions. Both 2,3-addition and ring-substituted products are formed [183].

VII. OXIDATION OF COMPOUNDS CONTAINING THE THIOCARBONYL GROUP

The principal reaction of cation radicals produced by the anodic oxidation of thiocarbonyl compounds appears to be the formation of dimeric dications, Eq. (75), in cases in which the positive charge can be stabilized by electron-rich heteroatoms like O, S, or N. Examples of such structures, Eq. (76), include 1,2-dithiol-3-thiones (LXXXV) [184], trithiapentalenes (not shown) [185], α-(1',2'-dithiol-3'-ylidene)acetophenones (not shown) [186], cyclic trithiocarbonates (LXXXVI) [187] (see Refs. 188 and 189), 1,3-dimethyl-2-imidazolidinethione (LXXXVII) [190], N,N,N',N'-tetraethylthiourea (LXXXVIII) [187] and 4,4'-dimethoxythiobenzophenone (LXXXIX) [187], (See also Ref. 6, pp. 95 and 119.) Similar reactions have been observed for the structurally related triphenylphosphine sulfide and O,O,O-triethyl phosphorothioate, Eq. (77) [191].

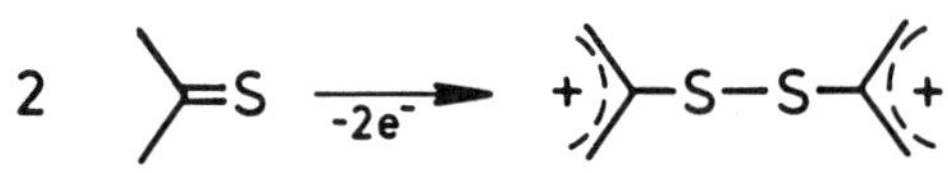

(75)

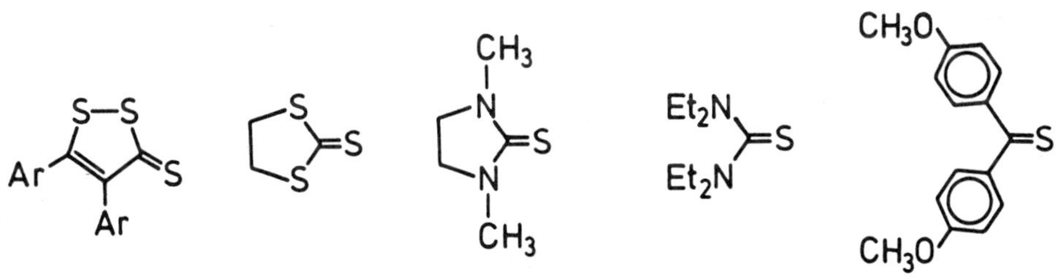

(LXXXV) (LXXXVI) (LXXXVII) (LXXXVIII) (LXXXIX)

(76)

$$X_3P{=}S \xrightarrow[-2e^-]{} X_3\overset{+}{P}{-}S{-}S{-}\overset{+}{P}X_3 \quad (77)$$

$$X = Ph, OEt$$

However, the dimeric dications are reactive species and are not the products isolated after preparative oxidations unless special precautions are taken. For example, exhaustive anodic oxidation of the cyclic trithiocarbonate (LXXXVI) in moist MeCN results in the formation of the corresponding C—O derivative [188,189]. The reaction mechanism was proposed to involve nucleophilic attack by unoxidized starting material on the primarily formed cation radicals followed by elimination of sulfur and hydrolysis, Eq. (78), in a fashion similar to that suggested to account for the formation of XCI from XC, Eq. (79) [188].

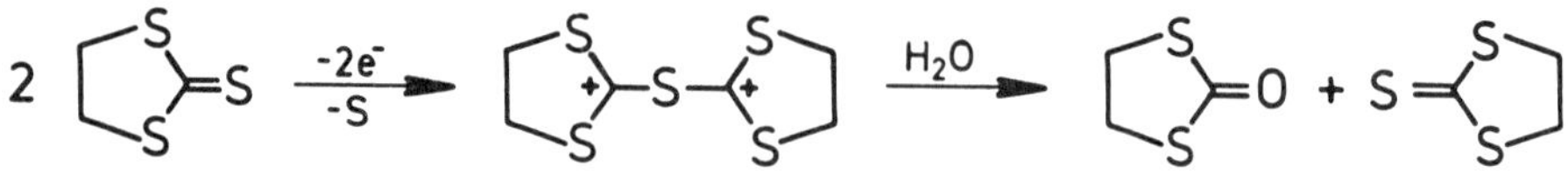

(78)

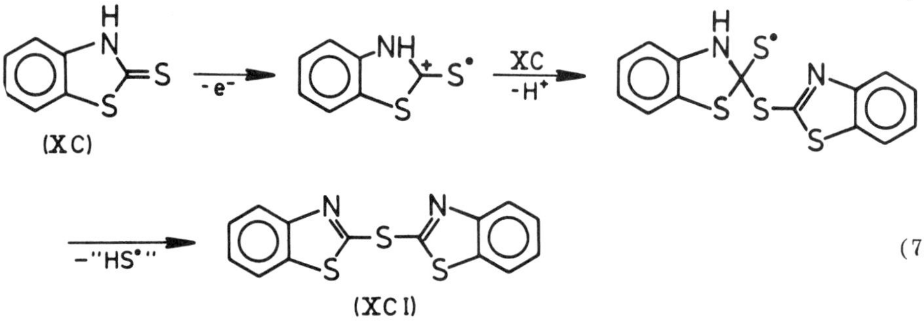

(79)

Anodic oxidation of primary alkylisothiocyanates results in the formation of dimers, XCII and XCIII, Eq. (80), which were proposed to arise via nucleophilic attack by water on the R-NCS cation radicals followed by dimerization and ring closure [192].

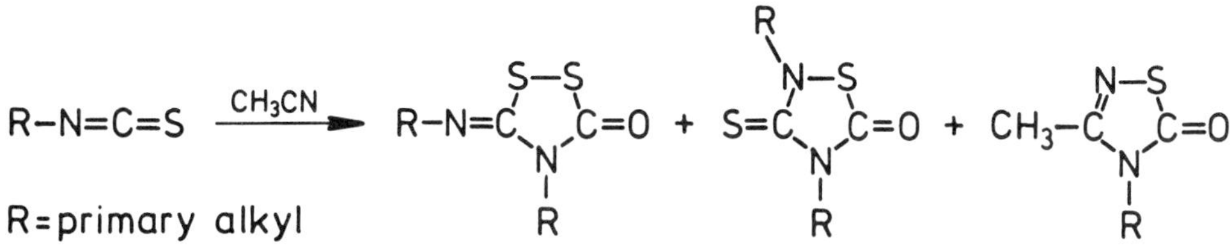

(XCII) (XCIII) (XCIV)

(80)

Water is reeliminated during the latter process and, accordingly, catalyzes the formation of the heterocyclic products. The 3-methyl-4-alkyl-1,2,4-thiadiazolidine-2-ene-5-ones (XCIV), which most likely are formed by Ritter-type reaction between MeCN and intermediate cations, were observed as by-products. When R was a tertiary alkyl group, the major reaction route involved cleavage of the R-NCS bond and formation of the corresponding acetamides, R–$NHCOCH_3$.

VIII. OXIDATION OF SULFONIC, SULFINIC, AND SULFENIC ACID DERIVATIVES

The anodic oxidation of aliphatic sulfonic acids generally leads to cleavage of the carbon-sulfur bond and production of low-molecular-weight oxidation products like CO_2, HCHO, and SO_4^{2-} (see Ref. 6, p. 99). Perfluoroalkane sulfonic acids, $R_F{-}SO_3H$, give a variety of products, among which are $R_FSO_3R_F$, R_FR_F, $(R_FSO_2)_2O$, $R_FS_2O_6R_F$, and SO_3 [193].

The anodic oxidation of aromatic sulfonic acids results mostly in ring-substituted products in which the sulfonic acid function is untouched (see Ref. 6, pp. 122–124).

Benzenesulfinate, $PhSO_2^-$, is oxidized anodically to the dimeric disulfone, $PhSO_2SO_2Ph$, Eq. (81). However, the yield is only ≃ 20% [194].

$$2\ Ph{-}SO_2^- \xrightarrow{-2e^-} Ph{-}SO_2{-}SO_2{-}Ph \qquad (81)$$

Aliphatic sulfenamides (or thiohydroxylamines), $R_2N{-}SR$, Eq. (82), are oxidized reversibly to the corresponding cation radicals in MeCN [195].

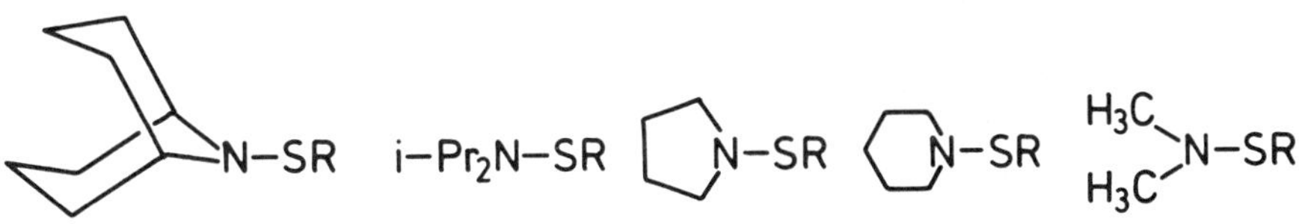

ABN

(82)

The lifetimes of the cation radicals as evaluated by the CV sweep rate necessary to achieve reversibility were observed to decrease in the order 9-ABN > $i\text{-}Pr_2N$ > $(CH_2)_5N$ ≃ $(CH_2)_4N$ > Me_2N, which seems to indicate that the follow-up reaction is deprotonation, probably with the unoxidized starting material acting as the base. The oxidation potentials were found to be in the range 0.7–1.0 V versus SCE.

Aromatic sulfenamides (XCV), give cation radicals that dimerize to biphenyl derivatives (XCVI), analogously to what is observed for aromatic amines (see Chaps. 15 and 23), Eq. (83) [196].

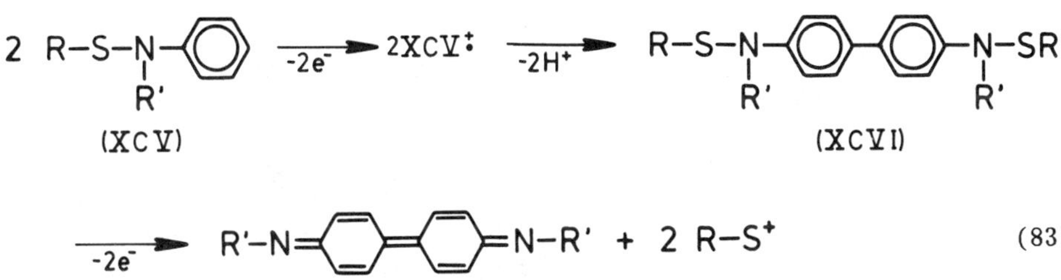

(XCV) (XCVI) (XCVII)

(83)

Further oxidation, during which the nitrogen-sulfur bond is cleaved, leads to the quinoid structure (XCVII), which in turn is oxidized to the corresponding cation radical.

REFERENCES

1. J. Q. Chambers, in Encyclopedia of Electrochemistry of the Elements (A. J. Bard and H. Lund, eds.), Dekker, New York, 1978, p. 329.
2. O. Hammerich and V. D. Parker, Sulfur Rep., 1, 317 (1981).
3. A. Swelim, A. I. A. Khodair, and F. El-Sheikh, Int. J. Sulfur Chem. B., 6, 195 (1971).
4. J. Grimshaw, in The Chemistry of the Sulfonium Group (C. J. M. Stirling and S. Patai, eds.), Wiley, New York, 1981, p. 141.
5. J. Q. Chambers, R. M. Harnden, and N. D. Canfield, in Characterization of Solutes in Nonaqueous Solvents (G. Mamantov, ed.), Plenum Press, New York, 1978, p. 115.
6. F. Fichter, Organische Elektrochemie, Steinkopff, Dresden, 1942.
7. H. J. Shine, in The Chemistry of the Sulfonium Group (C. J. M. Stirling and S. Patai, eds.), Wiley, New York, 1981, p. 523.
8. K.-D. Asmus, Acc. Chem. Res., 12, 436 (1979).
9. I. M. Kolthoff and C. Barnum, J. Amer. Chem. Soc., 62, 3061 (1940).
10. D. G. Davis and E. Bianco, J. Electroanal. Chem., 12, 254 (1966).
11. J. Koryta and J. Pradác, J. Electroanal. Chem., 17, 185 (1968).
12. N. Ossendorfová, J. Pradác, and J. Koryta, J. Electroanal. Chem., 28, 311 (1970).
13. G. Dryhurst, J. Electrochem. Soc., 117, 1113 (1970).
14. F. Magno, G. Bontempelli, and G. Pilloni, J. Electroanal. Chem., 30, 375 (1971).
15. J. K. Howie, J. J. Houts, and D. T. Sawyer, J. Amer. Chem. Soc., 99, 6323 (1977).
16. Y. Hirabayashi and T. Mazume, Bull. Chem. Soc. Jpn., 39, 1971 (1966).
17. R. Woods, J. Phys. Chem., 75, 354 (1971).
18. R. Woods, Aust. J. Chem., 25, 2329 (1972).
19. G. Cauquis and D. Lachenal, J. Electroanal. Chem., 43, 205 (1973).
20. C. Scrimager and L. J. Dehayes, Inorg. Nucl. Chem. Lett., 14, 125 (1978).
21. H.-U. Hummel and H. Procher, Z. Anorg. Allg. Chem., 537, 79 (1986).
22. R. Gompper and W. Töfl, Chem. Ber., 95, 2861 (1962).
23. G. Cauquis and A. Deronzier, J. Chem. Soc. Chem. Commun., 809 (1978).
24. F. Taboury, Ann. Chim., 15, 49 (1908).
25. W. Stricks and I. M. Kolthoff, J. Amer. Chem. Soc., 74, 4646 (1952).
26. I. M. Kolthoff, W. Stricks, and N. Tanaka, J. Amer. Chem. Soc., 77, 5211 (1955).
27. R. S. Saxena and U. S. Chaturvedi, J. Electroanal. Chem., 36, 515 (1972).
28. B. Persson and B. Nygard, J. Electroanal. Chem., 56, 373 (1974).
29. J. Simonet and P. Martinet, J. Electroanal. Chem., 88, 425 (1978).
30. J.-G. Gourcy, G. Jeminet, and J. Simonet, J. Chem. Soc., Chem. Commun., 634 (1974).
31. Q. N. Porter, J. H. P. Utley, P. D. Machion, V. L. Pardini, P. R. Schumacher, and H. Viertler, J. Chem. Soc., Perkin Trans. I, 973 (1984).
32. L. Sestili, Gazz. Chim. Ital., 102, 647 (1972).

33. A. M. Bond, A. T. Casey, and J. R. Thackeray, J. Electroanal. Chem., 48, 71 (1973).
34. W. Stricks and S. K. Chakravarti, Anal. Chem., 34, 508 (1962).
35. A. M. Bond, A. T. Casey, and J. R. Thackeray, Inorg. Chem., 12, 887 (1973).
36. A. M. Bond, A. T. Casey, and J. R. Thackeray, J. Electrochem. Soc., 120, 1502 (1973).
37. D. Dauplaise, J. Meinwald, J. C. Scott, H. Temkin, and J. Clardy, Ann. N. Y. Acad. Sci., 313, 382 (1978).
38. A. Zweig and A. K. Hoffmann, J. Org. Chem., 30, 3997 (1965).
39. T. Nogami, H. Tanaka, S. Ohnishi, Y. Tasaka, and H. Mikawa, Bull. Chem. Soc. Jpn., 57, 22 (1984).
40. F. Wudl, D. E. Schafer, and B. Miller, J. Amer. Chem. Soc., 98, 252 (1976).
41. M. Bonifacic and K.-D. Asmus, J. Chem. Soc., Perkin Trans. II, 1805 (1986).
42. H. Bock, V. Stein, and A. Semkov, Chem. Ber., 113, 3208 (1980).
43. F. Fichter and W. Wenk, Ber. Dtsch. Chem. Ges., 45, 1373 (1912).
44. G. Bontempelli, F. Magno, and G.-A. Mazzocchin, J. Electroanal. Chem., 42, 57 (1973).
45. F. Fichter and P. Sjöstedt, Ber. Dtsch. Chem. Ges., 43, 3422 (1910).
46. A. Bewick, D. E. Coe, M. Libert, and J. M. Mellor, J. Electroanal. Chem., 144, 235 (1983).
47. J. Giordan and H. Bock, Chem. Ber., 115, 2548 (1982).
48. A. Bewick, D. E. Coe, J. M. Mellor, and D. J. Walton, J. Chem. Soc., Chem. Commun., 51 (1980).
49. A. Bewick, D. E. Coe, G. B. Fuller, and J. M. Mellor, Tetrahedron Lett., 21, 3827 (1980).
50. A. Bewick, D. E. Coe, J. M. Mellor, and W. M. Owton, J. Chem. Soc., Perkin Trans. I, 1033 (1985).
51. J. M. Mellor and D. L. Bruzco de Milano, J. Chem. Soc., Perkin Trans. I, 1069 (1968).
52. S. Torii, K. Uneyama, and M. Ono, Tetrahedron Lett., 21, 2741 (1980).
53. S. Torii, K. Uneyama, and M. Ono, Tetrahedron Lett., 21, 2653 (1980).
54. S. Torii, K. Uneyama, M. Ono, and T. Bannou, J. Amer. Chem. Soc., 103, 4606 (1981).
55. P. T. Cottrell and C. K. Mann, J. Electrochem. Soc., 116, 1499 (1969).
56. M. M. Nicholson, J. Amer. Chem. Soc., 76, 2539 (1954).
57. J. Yoshida, H. Sofuku, and N. Kawabata, Bull. Chem. Soc. Jpn., 56, 1243 (1983).
58. M. Kimura, N. Kuriki, M. Inaishi, and Y. Sawaki, Tetrahedron Lett., 25, 4665 (1984).
59. R. S. Glass, J. R. Duchek, J. T. Klug, and G. S. Wilson, J. Amer. Chem. Soc., 99, 7349 (1977).
60. J. T. Doi, M. H. Goodrow, and W. K. Musker, J. Org. Chem., 51, 1026 (1986).
61. E. Anklam, H. Mohan, and K.-D. Asmus, J. Chem. Soc., Chem. Commun., 629 (1987).
62. K.-D. Asmus, M. Göbl, K.-O. Hiller, S. Mahling, and J. Mönig, J. Chem. Soc., Perkin Trans. II, 641 (1985).

63. J. Mönig, M. Göbl, and K.-D. Asmus, J. Chem. Soc., Perkin Trans. II, 647 (1985).
64. G. S. Wilson, D. D. Swanson, J. T. Klug, R. S. Glass, M. D. Ryan, and W. K. Musker, J. Amer. Chem. Soc., 101, 1040 (1979).
65. M. D. Ryan, D. D. Swanson, R. S. Glass, and G. S. Wilson, J. Phys. Chem., 85, 1069 (1981).
66. T. G. Brown, A. S. Hirschon, and W. K. Musker, J. Phys. Chem., 85, 3767 (1981).
67. W. K. Musker, Acc. Chem. Res., 13, 200 (1980).
68. M. Bonifacic and K.-D. Asmus, J. Chem. Soc., Perkin Trans. II, 758 (1980).
69. M. Göbl, M. Bonifacic, and K.-D. Asmus, J. Amer. Chem. Soc., 106, 5984 (1984).
70. M. Bonifacic and K.-D. Asmus, J. Org. Chem., 51, 1216 (1986).
71. F. Fichter and F. Braun, Ber. Dtsch. Chem. Ges., 47, 1530 (1914).
72. N. D. Canfield and J. Q. Chambers, J. Electroanal. Chem., 56, 459 (1974).
73. M. Platen and E. Steckhan, Chem. Ber., 117, 1679 (1984).
74. Q. N. Porter and J. H. P. Utley, J. Chem. Soc., Chem. Commun., 255 (1978).
75. J. Gourcy, P. Martigny, J. Simonet, and G. Jeminet, Tetrahedron, 37, 1495 (1981).
76. P. Martigny and J. Simonet, J. Electroanal. Chem., 111, 133 (1980).
77. R. S. Glass, A. Petsom, G. S. Wilson, R. Martínez, and E. Juaristi, J. Org. Chem., 51, 4337 (1986).
78. F. D. Saeva, B. P. Morgan, M. W. Fichtner, and N. F. Haley, J. Org. Chem., 49, 390 (1984).
79. M. R. Detty, N. F. Haley, R. S. Eachus, J. W. Hassett, H. R. Luss, M. G. Mason, J. M. McKelvay, and A. A. Wernberg, J. Amer. Chem. Soc., 107, 6298 (1985).
80. N. D. Canfield, J. Q. Chambers, and D. L. Coffen, J. Electroanal. Chem., 24, App. 7 (1970).
81. P. R. Moses, R. M. Harnden and J. Q. Chambers, J. Electroanal. Chem., 84, 187 (1977).
82. R. M. Harnden, P. R. Moses, and J. Q. Chambers, J. Chem. Soc., Chem. Commun., 11 (1977).
83. H. Bock, G. Brähler, U. Henkel, R. Schlecker, and D. Seebach, Chem. Ber., 113, 289 (1980).
84. A. Zweig and J. E. Lehnsen, J. Amer. Chem. Soc., 87, 2647 (1965).
85. A. Zweig, A. H. Maurer, and B. G. Roberts, J. Org. Chem., 32, 1322 (1967).
86. K. Uneyama and S. Torii, Tetrahedron Lett., 329 (1971).
87. S. Torii and K. Uneyama, Tetrahedron Lett., 4513 (1972).
88. K. Uneyama and S. Torii, J. Org. Chem., 37, 367 (1972).
89. D. S. Houghton and A. A. Humffray, Electrochim. Acta, 17, 1421 (1972).
90. A. A. Humffray and D. S. Houghton, Electrochim. Acta, 17, 1435 (1972).
91. B. E. Firth and L. L. Miller, J. Amer. Chem. Soc., 98, 8272 (1976).
92. T. Komori and T. Nonaka, J. Amer. Chem. Soc., 105, 5690 (1983).
93. S. Torii, Y. Matsuyama, K. Kawasaki, and K. Uneyama, Bull. Chem. Soc. Jpn., 46, 2912 (1973).
94. H. Wendt, Angew. Chem., Int. Ed. Engl., 21, 256 (1982).

95. G. Bontempelli, F. Magno, G.-A. Mazzocchin, and S. Zecchin, J. Electroanal. Chem., 43, 377 (1973).
96. F. Magno and G. Bontempelli, J. Electroanal. Chem., 36, 389 (1972).
97. H. Hoffelner, S. Yorgiyadi, and H. Wendt, J. Electroanal. Chem., 66, 138 (1975).
98. G. Bontempelli, F. Magno, G.-A. Mazzocchin, and R. Seeber, J. Electroanal. Chem., 55, 109 (1974).
99. C.-S. Liao, J. Q. Chambers, I. Kapovits, and J. Rabai, J. Chem. Soc., Chem. Commun., 149 (1974).
100. K. Almdal and O. Hammerich, Sulfur Lett., 2, 1 (1984).
101. T. Fuchigami, Y. Nakagawa, and T. Nonaka, Tetrahedron Lett., 27, 3869 (1986).
102. M. Kimura, K. Koie, S. Matsubara, Y. Sawaki, and H. Iwamura, J. Chem. Soc., Chem. Commun., 122 (1987).
103. A. Matsumoto, K. Suda, and C. Yijima, J. Chem. Soc., Chem. Commun., 263 (1981).
104. G. Le Guillanton and J. Simonet, Acta Chem. Scand., B37, 437 (1983).
105. S. Torii, H. Okumoto, and H. Tanaka, Chem. Lett., 617 (1980).
106. T. Mandai, H. Irei, M. Kawada, and J. Otera, Tetrahedron Lett., 25, 2371 (1984).
107. T. Mandai, H. Yasunaga, M. Kawada, and J. Otera, Chem. Lett., 715 (1984).
108. J. Yoshida and S. Isoe, Chem. Lett., 631 (1987).
109. J. Nokami, N. Matsumura, T. Sueoka, Y. Kusumoto, and M. Kawada, Chem. Lett., 1283 (1978).
110. J. Nokami, M. Kawada, R. Okawara, S. Torii, and H. Tanaka, Tetrahedron Lett., 1045 (1979).
111. M. Kimura, S. Matsubara, Y. Sawaki, and H. Iwamura, Tetrahedron Lett., 27, 4177 (1986).
112. G. Schukat, A. M. Richter, and E. Fanghänel, Sulfur Rep., 7, 155 (1987).
113. Proceedings of The International Conference of Low Dimensional Synthetic Metals, Chem. Scripta, 17, 1–230 (1981).
114. D. H. Geske and M. V. Merritt, J. Amer. Chem. Soc., 91, 6921 (1969).
115. D. L. Coffen, J. Q. Chambers, D. R. Williams, P. E. Garrett, and N. D. Canfield, J. Amer. Chem. Soc., 93, 2258 (1971).
116. G. Schukat and E. Fanghänel, J. Prakt. Chem., 327, 767 (1985).
117. M. L. Kaplan, F. Wudl, R. C. Haddon, and J. J. Hauser, Chem. Scripta, 15, 196 (1980).
118. R. D. McCullough, G. B. Kok, K. A. Lerstrup, and D. O. Cowan, J. Amer. Chem. Soc., 109, 4115 (1987).
119. D. C. Green, J. Org. Chem., 44, 1476 (1979).
120. A. Mas, J.-M. Fabre, E. Torreilles, L. Giral, and G. Brun, Tetrahedron Lett., 2579 (1977).
121. D. J. Sandman, T. J. Holmes, and D. E. Warner, J. Org. Chem., 44, 880 (1979).
122. S. Hünig, G. Kiesslich, H. Quast, and D. Scheutzow, Liebigs Ann. Chem., 310 (1973).
123. J.-M. Fabre, L. Giral, R. H. Harms, J. Heinze, and J. Mortensen, Israel J. Chem., 25, 279 (1985).
124. Z. Yoshida, T. Kawase, H. Awaji, I. Sugimoto, T. Sugimoto, and S. Yoneda, Tetrahedron Lett., 24, 3469 (1983).

125. M. Mizuno, M. P. Cava, and A. F. Garito, J. Org. Chem., 41, 1484 (1976).
126. E. M. Engler, B. A. Scott, S. Etemad, T. Penney, and V. V. Patel, J. Amer. Chem. Soc., 99, 5909 (1977).
127. K. A. Idriss and J. Q. Chambers, J. Electroanal. Chem., 109, 341 (1980).
128. J. Q. Chambers, D. C. Green, F. B. Kaufman, E. M. Engler, B. A. Scott, and R. R. Schumaker, Anal. Chem., 49, 802 (1977).
129. K. Kuo, P. R. Moses, J. R. Lenhard, D. C. Green, and R. W. Murray, Anal. Chem., 51, 745 (1979).
130. W. Schroth, R. Borsdorf, R. Herzschuh, and J. Seidler, Z. Chem., 10, 147 (1970).
131. C. Barry, G. Cauquis, and M. Maurey, Bull. Soc. Chim. Fr., 2510 (1966).
132. O. Hammerich, N. S. Moe, and V. D. Parker, J. Chem. Soc., Chem. Commun., 156 (1972).
133. O. Hammerich and V. D. Parker, J. Electroanal. Chem., 36, App. 11 (1972).
134. O. Hammerich and V. D. Parker, Electrochim. Acta, 18, 537 (1973).
135. C. Madec and J. Courtot-Coupez, J. Electroanal. Chem., 84, 169 (1977).
136. O. Hammerich and V. D. Parker, Adv. Phys. Org. Chem., 20, 55 (1984).
137. R. S. Glass, W. J. Britt, W. N. Miller, and G. S. Wilson, J. Amer. Chem. Soc., 95, 2375 (1973).
138. K.-W. Stender, G. Klar, and D. Knittel, Z. Naturforsch., 40b, 774 (1985).
139. C. H. Chen, J. J. Doney, G. A. Reynolds, and F. D. Saeva, J. Org. Chem., 48, 2757 (1983).
140. M. de Sorgo, B. Wasserman, and M. Szwarc, J. Phys. Chem., 76, 3468 (1972).
141. G. Cauquis and M. Maurey-Mey, Bull. Soc. Chim. Fr., 291 (1973).
142. P. T. Kissinger, P. T. Holt, and C. N. Reilley, J. Electroanal. Chem., 33, 1 (1971).
143. J.-P. Billon, Bull. Soc. Chim. Fr., 1784, 1923 (1960); Ann. Chim. Paris, 7, 183 (1962).
144. F. G. Bordwell and M. J. Bausch, J. Amer. Chem. Soc., 108, 2473 (1986).
145. S. A. Alkaitas, G. Beck, and M. Gratzel, J. Amer. Chem. Soc., 97, 5723 (1975).
146. W. Ando, Y. Kabe, S. Kobayashi, C. Takyu, A. Yamagishi, and H. Inaba, J. Amer. Chem. Soc., 102, 4526 (1980).
147. H. E. Imberger and A. A. Humffray, Electrochim. Acta, 18, 373 (1973).
148. H. J. Shine and Y. Murata, J. Amer. Chem. Soc., 91, 1872 (1969).
149. Y. Murata and H. J. Shine, J. Org. Chem., 34, 3368 (1969).
150. O. Hammerich and V. D. Parker, Acta Chem. Scand., B36, 421 (1982).
151. E. Vieil, M. Maurey-Mey, and G. Cauquis, Electrochim. Acta, 29, 1009 (1984).
152. Z. Uziel and J. W. Strojek, Polish J. Chem., 53, 1843 (1979).
153. O. Hammerich and V. D. Parker, Acta Chem. Scand., B37, 303 (1983).
154. J. S. Mayausky and R. L. McCreery, J. Electroanal. Chem., 145, 117 (1983).

155. H. J. Shine, J. J. Silber, R. J. Bussey, and T. Okuyama, J. Org. Chem., 37, 2691 (1972).
156. J. E. Pemberton, G. L. McIntire, H. N. Blount, and J. F. Evans, J. Phys. Chem., 83, 2696 (1979).
157. R. S. Glass, E. B. Williams, Jr., and G. S. Wilson, Biochemistry, 13, 2800 (1974).
158. H. J. Shine and J. J. Silber, J. Amer. Chem. Soc., 94, 1026 (1972).
159. H. J. Shine and K. Kim, Tetrahedron Lett., 99 (1974).
160. B. K. Bandlish, S. R. Mani, and H. J. Shine, J. Org. Chem., 42, 1538 (1977).
161. J. F. Evans and H. N. Blount, J. Org. Chem., 42, 976 (1977).
162. B. K. Bandlish, W. R. Porter, Jr., and H. J. Shine, J. Phys. Chem., 82, 1168 (1978).
163. K. Sugiyama and H. J. Shine, J. Org. Chem., 48, 143 (1983).
164. M. Soroka and H. J. Shine, Tetrahedron, 42, 6111 (1986).
165. H. J. Shine, B. K. Bandlish, S. R. Mani, and A. G. Padilla, J. Org. Chem., 44, 915 (1979).
166. K. Iwai and H. J. Shine, J. Org. Chem., 46, 271 (1981).
167. K. Kim and H. J. Shine, Tetrahedron Lett., 4413 (1974).
168. K. Kim, S. R. Mani, and H. J. Shine, J. Org. Chem., 40, 3857 (1975).
169. J. J. Silber and H. J. Shine, J. Org. Chem., 36, 2923 (1971).
170. K. Kim, V. J. Hull, and H. J. Shine, J. Org. Chem., 39, 2534 (1974).
171. U. Svanholm, O. Hammerich, and V. D. Parker, J. Amer. Chem. Soc., 97, 101 (1975).
172. U. Svanholm and V. D. Parker, J. Amer. Chem. Soc., 98, 997 (1976).
173. A. Dusserre and M. Genies, Bull. Soc. Chim. Fr., I-282 (1979).
174. S. R. Mani, Ph.D. Dissertation, Texas Tech. University, 1976. H. J. Shine, in The Chemistry of the Sulfonium Group (C. J. M. Stirling and S. Patai, eds.), Wiley, New York, 1981, p. 523.
175. U. Svanholm and V. D. Parker, J. Chem. Soc., Perkin Trans. II, 1567 (1976).
176. A. K. M. M. Hoque, A. C. Kovelesky, W.-K. Lee, and H. J. Shine, Tetrahedron Lett., 26, 5655 (1985).
177. D. H. Bae, P. S. Engel, A. K. M. M. Hoque, D. E. Keys, W.-K. Lee, R. W. Shaw, and H. J. Shine, J. Amer. Chem. Soc., 107, 2561 (1985).
178. M. Lemaire, R. Garreau, F. Garnier, and J. Roncali, New J. Chem., 11, 703 (1987).
179. M. Janda, Coll. Czech. Chem. Commun., 28, 2524 (1963).
180. M. Janda and J. Radouch, Coll. Czech. Chem. Commun., 32, 2672 (1967).
181. M. Janda and L. Paviensky, Coll. Czech. Chem. Commun., 32, 2675 (1967).
182. J. Srogl, M. Janda, I. Stibor, and Z. Salajka, Coll. Czech. Chem., Commun., 42, 1361 (1977).
183. J. Srogl, M. Janda, I. Stibor, J. Kos, V. Vyskocil, Coll. Czech. Chem. Commun., 43, 2015 (1978).
184. C. Th. Pedersen and V. D. Parker, Tetrahedron Lett., 771 (1972).
185. C. Th. Pedersen, O. Hammerich, and V. D. Parker, J. Electroanal. Chem., 38, 479 (1972).
186. C. Th. Pedersen, V. D. Parker, and O. Hammerich, Acta Chem. Scand., B30, 478 (1976).

187. R. L. Blankespoor, M. P. Doyle, D. M. Hedstrand, W. H. Tamblyn, and D. A. Van Dyke, J. Amer. Chem. Soc., 103, 7096 (1981).
188. J. Q. Chambers, P. R. Moses, R. N. Shelton, and D. L. Coffen, J. Electroanal. Chem., 38, 245 (1972).
189. P. R. Moses, J. Q. Chambers, J. O. Sutherland, and D. R. Williams, J. Electrochem. Soc., 122, 608 (1975).
190. P. Bernstein and M. N. Hull, J. Electroanal. Chem., 28, App. 1 (1970).
191. R. L. Blankespoor, M. P. Doyle, D. J. Smith, D. A. Van Dyke, and M. J. Waldyke, J. Org. Chem., 48, 1176 (1983).
192. J. Y. Becker, B. Zinger, and S. Yatziv, J. Org. Chem., 52, 2783 (1987).
193. D. Brunal, M. Oudrhiri-Hassani, H. Choukroun, and A. Germain, Nouv. J. Chim., 8, 99 (1984).
194. B. Persson, Acta Chem. Scand., B31, 88 (1977).
195. S. F. Nelsen, D. J. Steffek, G. T. Cunkle, and P. M. Gannett, J. Amer. Chem. Soc., 104, 6641 (1982).
196. H. Sayo, T. Michida, and H. Hatsumura, Chem. Pharm. Bull., 34, 558 (1986).

V

ELECTROCHEMISTRY OF CERTAIN COMPREHENSIVE CLASSES OF COMPOUNDS

18

Electrolysis of Heterocyclic Compounds

HENNING LUND / Aarhus University, Aarhus, Denmark

I. Introduction 701

II. Electrolytic Formation of Heterocyclic Systems 702
 A. Ring closure reactions 702
 B. Ring contractions 717
 C. Ring expansions 719

III. Electrolytic Reactions of Heterocyclic Systems 720
 A. Compounds with one oxygen atom 720
 B. Nuclei containing one sulfur atom 721
 C. Nuclei containing one nitrogen atom 722
 D. Compounds with one oxygen and one nitrogen atom 726
 E. Compounds with one sulfur and one nitrogen atom 727
 F. Compounds with two oxygen or two sulfur atoms 728
 G. Compounds with two nitrogen atoms 729
 H. Compounds with one oxygen and two nitrogen atoms 734
 I. Compounds with one sulfur and two nitrogen atoms 735
 J. Compounds with three nitrogen atoms 735
 K. Compounds with four nitrogen atoms 736

IV. Electrolytic Reactions of Substituted Heterocyclic Compounds 738
 A. Alkyl and alkenyl substituents 739
 B. Hydroxyl derivatives 739
 C. Aldehydes, ketones, and their derivatives 743
 D. Carboxylic acids and derivatives 744
 E. Nitrogen-containing substituents 745
 F. Sulfur-containing substituents 746
 G. Halogen derivatives 747

References 747

I. INTRODUCTION

This chapter discusses some aspects of the electrolysis of heterocyclic compounds. The emphasis is on systems in which the electrochemical properties differ from those of the corresponding carbocyclic compounds. The discussion is divided in three main sections: electrolytic formation of heterocyclic systems, electrolytic reactions of heterocyclic systems, and electrolytic reactions of substituted heterocyclic compounds.

Reviews on the electrolysis of heterocyclic compounds [1] discuss the literature in somewhat more detail than in the present chapter.

II. ELECTROLYTIC FORMATION OF HETEROCYCLIC SYSTEMS

The electrolytic reactions in which a heterocyclic system is formed are discussed in this section; the reactions are treated under the following headings:

Ring closure reactions
Ring contractions
Ring expansions

A. Ring Closure Reactions

In the electrolytic formation of heterocyclic systems the role of the current is to bring one or both of the reaction centers to a suitable oxidation state. These reactions may be classified in different ways; here they are treated according to the type of bond formed. Ring formation involving the creation of two bonds simultaneously (or nearly so) are discussed separately; when two bonds are formed in separate steps, the formation of the first bond determines the section in which it is discussed.

Ring closure reactions are intramolecular couplings; some of these are discussed in Chaps. 22 and 23. The fundamental electrode reactions are discussed in Chaps. 9 through 17.

A ring closure most often involves either the attack of an electrolytically formed nucleophile (alcohol, hydroxylamine, amine, or hydrazine) on an electrophilic center (existing or potential carbonyl group, cyano group, or nitro or nitroso group), a reaction between a nucleophile (e.g., hydroxyl or amino group or carboxylate ion) and an electrolytically generated electrophilic center (e.g., nitroso group or carbonium ion), a radical (including radical ion) coupling reaction, or a dipolar addition reaction.

The formation of a nucleophilic center is most often the result of a reduction, whereas the creation of an electrophilic center in many cases is performed by an oxidation; radicals may be formed cathodically or anodically.

Formation of Carbon-Carbon Bonds

Electrolytic formation of carbon-bonds during formation of heterocyclic compounds occurs in the reduction of ketones to pinacols, in the hydrodimerization reaction, in some radical coupling reactions, and in the oxidative coupling of activated aromatic systems.

An initial pinacol dimerization is responsible for the formation of 1,3,5,7-tetraphenyl-2,4,6-trioxatricyclo[3.3.0.1]nonane during the electrolysis of 1,3-diphenyl-1,3-propanedione in slightly acidic solution [2].

A "head-to-tail" dimerization of the initially formed radicals during the reduction o-aminoacetophenone (I) in alkaline solution leads to the formation of two stereoisomers of 3-acetyl-4-amino-7-hydroxy-7-methyl-2,6-methano-2,5,6,7-tetrahydro-1-(1H)-benzazonine (II), Eq. (1) [3]:

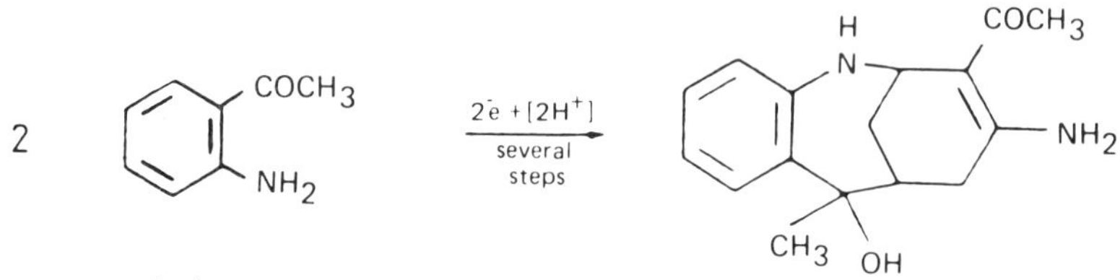

(1)

(I) (II)

Reduction of the semicarbazone of a phenacyl bromide in DMF yields the bis-semicarbazone of 1,4-diaryl-1,4-butanedione, which in DMF forms 3,6-diarylpyridazine [4] or, after treatment with aqueous acid, 2,5-diarylfuran [5]. Reduction of phenacyl bromide in DMF is reported to give 2,4-diarylfuran [6].

The hydrodimerization is also a useful method in the preparation of heterocycles [7,8]. o-Bis(β-bicarbethoxyvinylamino)benzene (III) thus yields by intramolecular reductive coupling 2,3-bis(dicarbethoxymethyl)-1,2,3,4-tetrahydroquinoxaline (IV), which on heating gives quinoxaline [7], Eq. (2). IV may be anodically oxidized to the substituted quinoxaline [9].

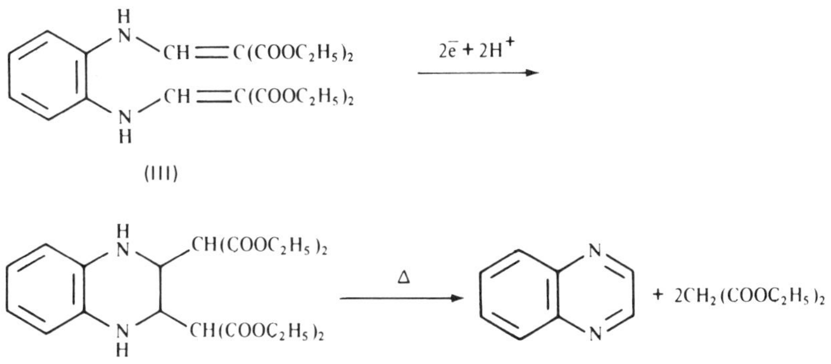

(III) (2)

(IV)

Attempts to hydrodimerize aliphatic bis-enaminones to pyrazines were less successful, as the hexahydropyrazines formed catalyze the hydrogen evolution, which then becomes the major cathodic reaction [8].

Reductive formation of a carbanion, either by reduction of a C-halogen bond or by removal of a proton by an electrogenerated base, followed by an intermolecular nucleophilic substitution, has been used in the synthesis of substituted β-lactams [10].

On reduction of aryl halides the initially formed ion radical decomposes to halide ion and aryl radical [11–14]. In appropriate substituted compounds the radical may in an intramolecular attack effect a ring closure. This is exemplified in the reductive cyclization of 5-(2-chlorophenyl)-1-phenylpyrazole to pyrazolo[1,5-f]phenanthridine [11–13] and of 1-(o-iodobenzyl)isoquinolinium methiodide (V) to aporphine in MeCN-TEABr (VI), Eq. (3) [14].

The mechanism suggested involves the loss of two hydrogen atoms so the net reaction is not a reduction; an alternative mechanism, along the lines of the electrocatalytic substitution, could also be considered.

Pyrrolidines can be obtained by a mixed Kolbe electrolysis of a β-allylaminoalkanoate with a coacid. The reaction is discussed in Chap. 14, Eq. (19).

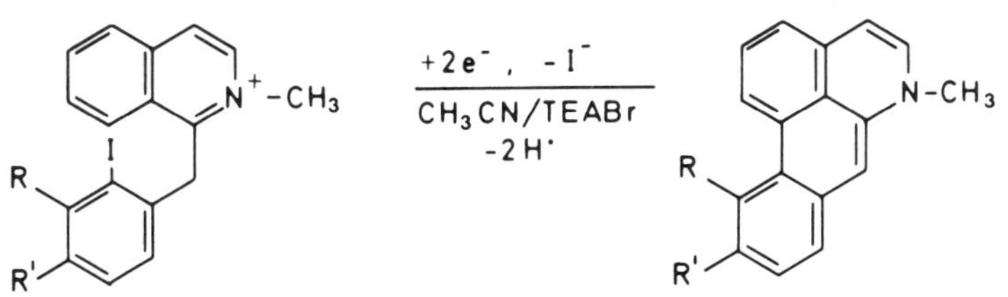

(3)

(V) (VI)

Reductive alkylation of immonium salts has been used in the synthesis of heterocyclic compounds, including different types of alkaloids, Eq. (4) [15]:

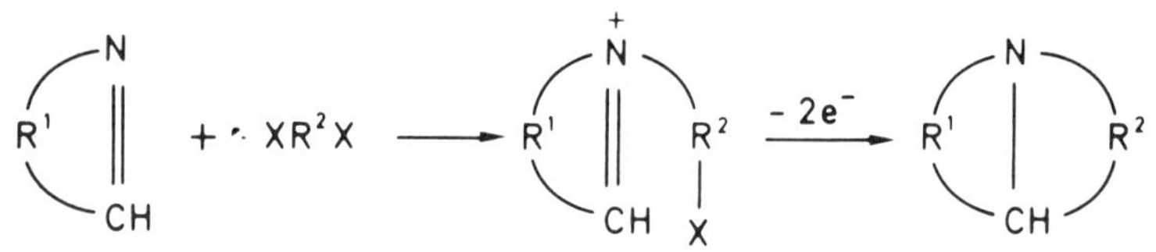

(4)

Electroreduction of 2-methyl-3,4-dihydro-6,7-dimethoxyisoquinolinium salt in the presence of 3-bromomeconine, for example, thus yields cordrastine.

The oxidative coupling of aromatic nuclei activated by alkoxy groups has been explored for the formation of natural products. Thus, laudanosine (VII), Y = NCH_3, Z = CH_2, has been oxidized [16–20] at a platinum anode in MeCN-$TMABF_4$ to O-methylflavinantine (VII), and other benzyltetrahydroisoquinolines react similarly on anodic oxidation in good to excellent yield, Eq. (5):

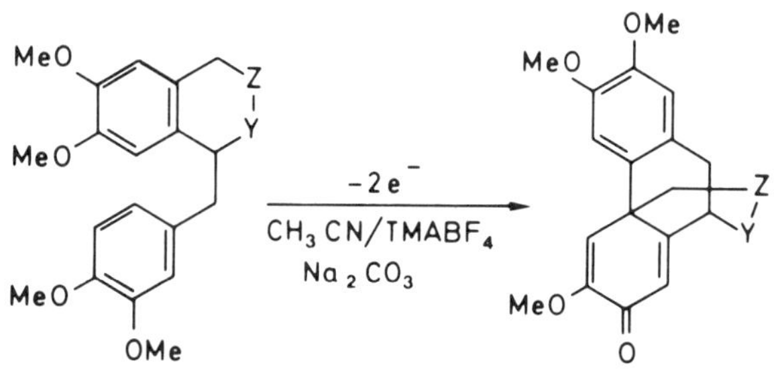

(VII) (VIII)

(5)

An anodic oxidation of 4-(3,4-dimethoxybenzyl)-6,7-dimethoxyisochroman-3-one (VII, Y = CO, Z = O) in MeCN-$NaClO_4$ containing sodium carbonate [21] resulted in a similar ring closure reaction, whereas 1-(3,4-dimethoxybenzyl)-6,7-dimethoxyisochroman-3-one (VII, Y = O, Z = CO) on a similar oxidation gave 6,12-dioxo-2,3,7-trimethoxy-6,8a,9,10-tetrahydro-9,8a-epoxyethanophenanthrene as main product rather than the expected compound with the aporphine skeleton [22]. The solvent may influence the course of such couplings [23]. Sometimes blocking of certain reactive sites is necessary to promote the desired coupling.

A number of related couplings have been reported during the synthesis of alkaloids, such as oxocrinine and cryptoplurine [24], the intramolecular coupling of diarylamines to dibenzazepine, and dibenzazocine structures [25], synthesis of galanthamine [26] and phenanthropyrrolidines [27], and the intramolecular arylation of enamines [28].

The anodic dimerization of enamine ketones or esters in MeCN-$NaClO_4$ at a graphite anode is followed by an elimination of 1 mol of amine, which results in the formation of pyrroles [29].

Anodic oxidation in DMF of the open-chain tetrapyrrole, 1',8'-dimethyl-a,c-biladiene, yielded porphyrin; one methyl group is lost from a cyclic intermediate with four conjugated pyrrole subunits by attack of a nucleophile [29a].

Formation of Carbon-Oxygen Bonds

Hydroxyl Group as Nucleophile An electrolytically formed alcohol group participating in a ring closure is found in the reduction of some phthalic acid derivatives to phthalides and of o-carboxyphenyl ketones to substituted phthalides; the latter reaction is exemplified in the reduction [30] of o-benzoyl-N,N-diethylbenzamide to 3-phenylphthalide and its C-3 dimer.

Benzyl cations, formed on anodic oxidation, may react with suitable nucleophiles. An example is the anodic oxidation in alkaline methanol of N-benzyl-N-methylethanolamine (IX), which produces three main products [31], among them 3-methyl-2-phenyloxazolidine (X) and 3-benzyloxazolidine (XII); although the primary loss of an electron probably occurs from nitrogen, the benzylic carbon becomes the positive center that attacks oxygen, Eq. (6):

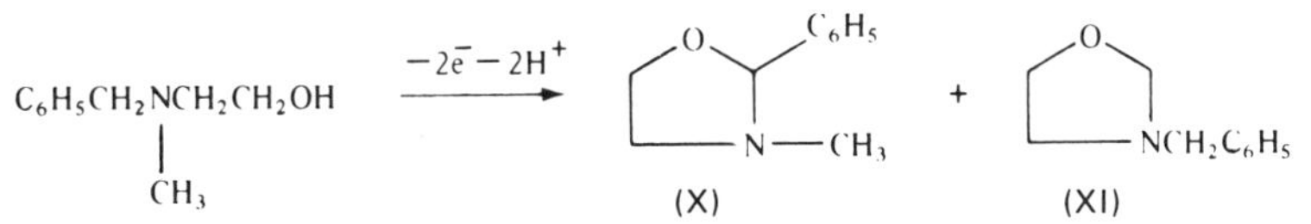

(IX) (X) (XI) (6)

Of the same type is the oxidation of (S)-2-acetamido-2-(3,4-dimethoxybenzyl)propionitrile (XII) to (4R,5S)- and (4R,5R)-4-cyano-5-(3,4-dimethoxyphenyl)-2,4-dimethyl-2-oxazoline (XIII and XIV), Eq. (7) [32[:

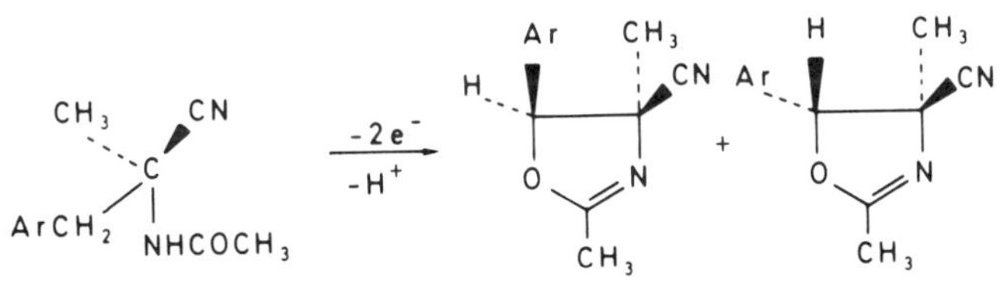

(XII) (XIII) (XIV) (7)

An amide-oxygen also acts as a nucleophile in the anodic oxidation of $RCON(Me)C_6H_4OR'$ in MeOH to N-methylbenzoxazolium perchlorate. A metastable intermediate, detected by spectroscopy, may be a quinone imine derivative [33] and the reaction analogous to the anodic substitution of hydroquinone with certain nucleophiles, such as pyridine [30]. A somehat similar reaction is found in the anodic oxidation of 4-methoxyacetic acid to a dioxaspiro compound [34].

Anodic addition of an oxygen function to a C=N or C=C double bond may be exemplified by the oxidative cyclization of aryl semicarbazones to 1,3,4-oxadiazoles in acetonitrile-acetic acid containing H_2SO_4 or, under strictly anhydrous conditions (in the presence of acetic anhydride), to triazolinones [35]. Other examples are the anodic oxidation of benzaldehyde-2-hydroxyanil to 2-phenyl-1,3-benzoxazole [36] and the indirect oxidation of 2-hydroxychalcones to flavonoids using tris(4-bromophenyl)amine in MeOH-CH_2Cl_2 as mediator [37], Eq. (8):

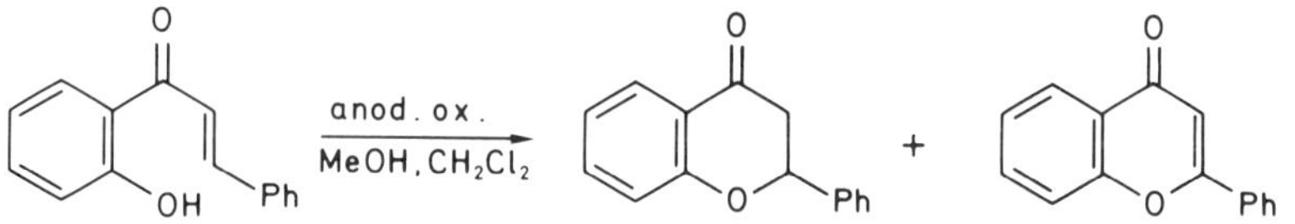

(8)

The ring closure by phenolic coupling is illustrated by the anodic coupling of catechol with 4-hydroxycoumarin to 11,12-dihydroxycoumestan (XV) in 95% yield in an undivided cell at a graphite anode [38], Eq. (9):

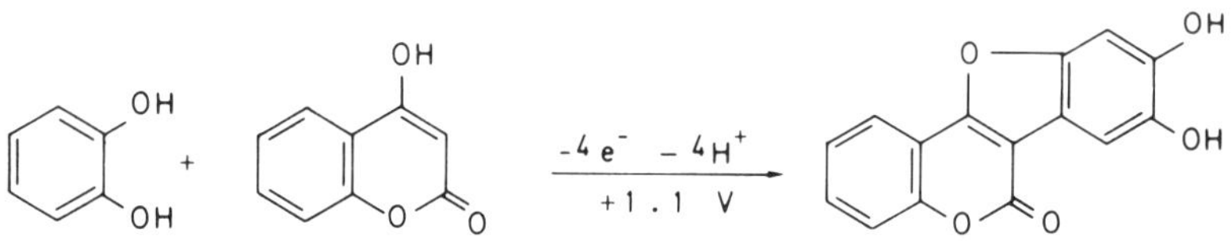

(9)

(XV)

The oxidation of some 1,5-diketones to pyrylium salts has been reported [39]; this reaction is discussed in Chap. 16.

Hydroxylamino Group as Nucleophile A ring closure of this type has been reported only in the reduction of o-nitrocarbonyl compounds to anthranils. Larger rings cannot be obtained by this method as an attack by the nitrogen atom of the intermediate hydroxylamine is the preferred mode.

o-Nitroacetophenone is reduced to methyl anthranil in sulfuric acid [40] and acetate buffer; in the latter medium the yield is quantitative. o-Nitrobenzaldehyde [40] is reduced similarly, and such a reduction may be expected in general from o-nitrocarbonyl compounds. The reduction of o-nitrobenzophenone [41] has, however, been reported to yield the dihydroanthranil. This is surprising because a reduction of an intermediate anthranil is expected to result in a hydrogenation of the nitrogen-oxygen bond rather than in saturation of the double bond, but the reduction potentials may be quite close.

In a similar reaction, α-cyano-β-nitrostyrene and α-acyl-β-nitrostilbene form isoxazoles on cathodic reduction [42].

Formation of Carbon-Nitrogen Bonds

Hydroxylamino Group as Nucleophile Aromatic nitro compounds may generally be reduced in a four-electron reduction to hydroxylamines, which in acid solution at a more negative potential can be reduced further to amines. By choosing a suitable cathode potential it is thus possible to avoid the further reduction of the hydroxylamine to the amine. When a hydroxyl or amino group is ortho or para to the nitro group, it is not feasible to isolate the hydroxylamino compound, as it is too easily dehydrated to the reducible quinone mono- or diimine. It may, however, be possible to trap the intermediate.

Reaction with carbonyl groups. A hydroxylamino group has two nucleophilic centers, the nitrogen and the oxygen atom. Whenever attack of the nitrogen atom on a carbonyl group would result in the formation of a ring having five or more members, this mode is preferred to attack by oxygen. The reaction results in a cyclic nitrone (N-oxide) or hydroxamic acid.

The reduction of o-nitrobenzalacetylacetone (XVI) in acid solution at the potential of the first polarographic wave gives the hydroxylamine, which cyclizes to the cyclic N-oxide (XVII) [43,44]; the isolated yield is 70–75%, but polarographic determination of the product in the reduced solution showed a somewhat higher yield. XVII may at a more negative potential be reduced further, but the reduction potentials of the N-oxide and the carbonyl group are rather close in acid solution, making a selective reduction of the N-oxide difficult, Eq. 10:

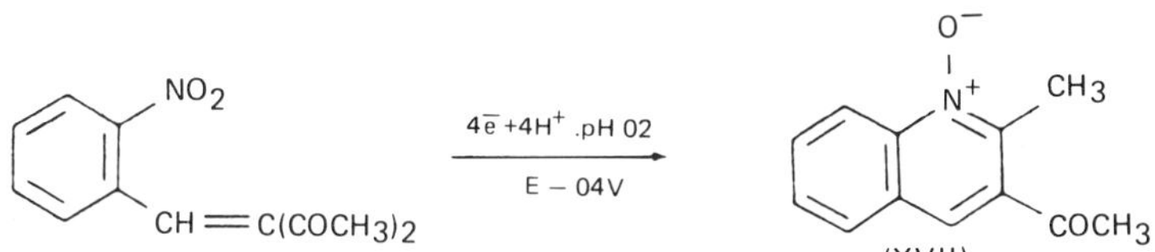

(XVI) (XVII) (10)

o-Nitrobenzalmalonic acid derivatives can be reduced in a four-electron reduction [43], and the resulting cyclic hydroxamic acid is, like many hydroxamic acids [43], further reducible in acid solution to the amine, the carbostyril.

In compounds of the type o-$NO_2C_6H_4CH{=}CYY'$ (Y,Y' = CN, COOR, COPh) the ring closure usually involves the group cis to the phenyl ring, but isomerization during the reduction has been observed [45].

Reductions at the potential of the first wave of 1-acetamino-2-nitronaphthalene, 2-acetamino-3-nitronaphthalene, and 8-acetamino-1-nitronaphthalene yield cyclic N-oxides, the two former naphthimidazole-N-oxides, and the latter 2-methylperimidine-1-oxide [46].

Several examples have been reported of similar ring closures with the formation of heterocyclic N-oxides, N-hydroxy compounds, or hydroxamic acids [47].

If the intermediate hydroxylamino group has a choice of attacking a carbonyl group with the oxygen atom to form a benzisoxazole (XIX) or reacting with a keto (XVIIIa), ester (XVIIIb), or cyano group (XVIIIc) through its nitrogen atom to a quinoline derivative (XX) in such compounds as XVIII, XVIIIa is reduced in acid solution to a quinoline-N-oxide, XVIIb predominantly to a N-hydroxyquinolone with minor amounts of XIXb, whereas XVIIIc gives mostly XIXc [48]. In ammonia buffer, however, XVIIIc yields XXc.

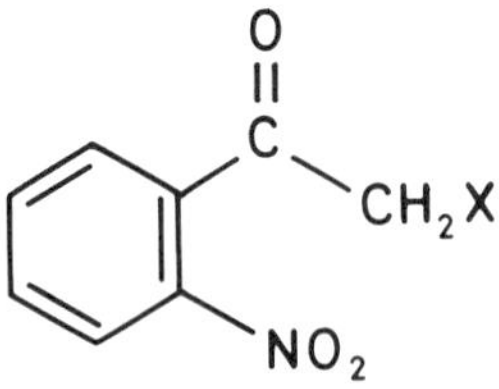

XVIII a X = $COCH_3$
b X = $COOCH_3$
c X = CN

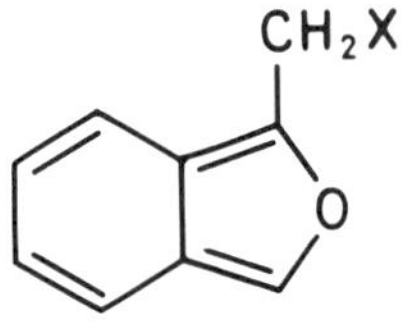

XIX a X = $COCH_3$
b X = $COOCH_3$
c X = CN

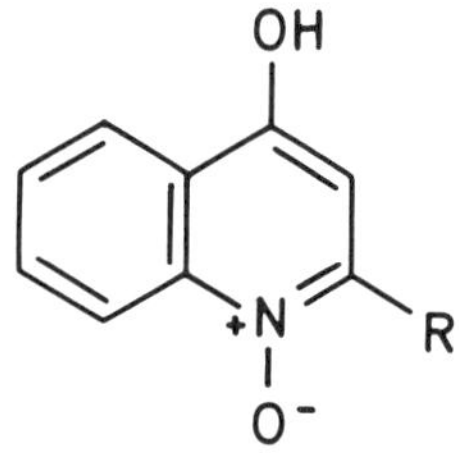

XX a R = CH_3
c R = NH_2

In the aliphatic series pyrroline-*N*-oxides [49], *N*-hydroxypyrrolidinones [49], and isoxazolones [49] may be obtained by reduction of γ-nitroketones, γ-nitroesters, and β-nitroamides, respectively.

Deprotonation by electrogenerated bases, from the probase ethyl 2-bromo-2-methylpropanoate, of β-bromoamides or β-bromohydroxamic esters has been used to form β-lactams through $N{-}C_4$ bond formation [50].

A somewhat similar ring closure in which a nitrogen anion attacks a halogenated carbon center is obtained in the anodic oxidation of dimethyl α-(ω-tosylaminoalkyl)malonates in methanol containing KI in an undivided cell. Aziridines, azetidines, and pyrrolidines may be formed in good yield [51].

Reaction with potential carbonyl groups. In principle, condensations similar to those described for carbonyl groups are possible for the corresponding azomethine compounds. An example of this kind is the reduction of 2-nitrophenylisothiocyanate (XXI) in ethanolic hydrochloric acid, which in a four-electron reduction forms 2-mercaptobenzimidazole-*N*-oxide (XXII), Eq. (11) [44]:

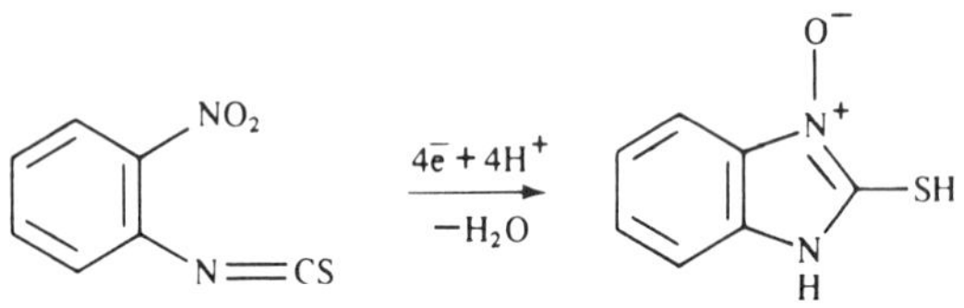

(11)

(XXI) (XXII)

2-Nitro-2'-isothiocynanatobiphenyl is in a similar manner converted to a dibenzodiazepine *N*-oxide [52].

Reaction with cyano groups. These reactions are similar to those already described; the following example, the reduction of *o*-nitrophenylthiocyanate (XXIII) [17], may illustrate this, Eq. (12):

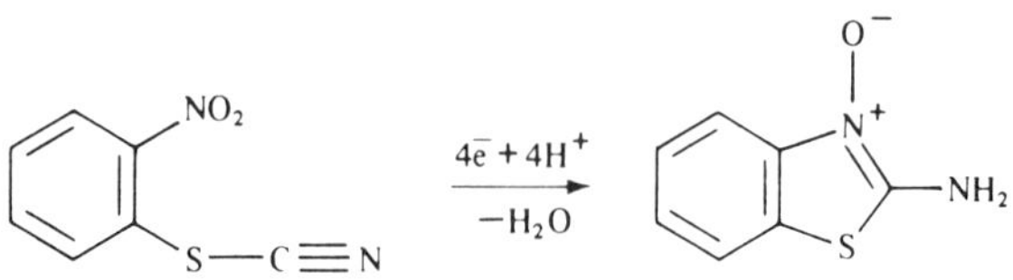

(12)

(XXIII)

In strongly acidic solution o-nitrobenzyl thiocyanate forms 2-aminobenzo-4H-3,1-thiazine-1-oxide, whereas at higher pH the reaction involves a reductive loss of SCN^- and formation of different products, including 11,12-dihydrodibenzo[c,g]diazocine and its N-oxide [53].

Amines as Nucleophiles Amines may be formed by electrolytic reduction of azomethine derivatives and nitro, nitroso, and hydroxylamino compounds. As the amine is the lowest oxidation state, control of the electrode potential is often not as critical in the preparation of amines as in the synthesis of hydroxylamines. Consequently, most of the "classic" electrolytic reductions of nitro compounds in acid solution resulted in the formation of amines.

Reaction with carbonyl groups. Cyclization of an azomethine compound occurs in the reduction of "phenolphthalein oxime" (XXIV) [54] and similar very stable anils [54], Eq. (13):

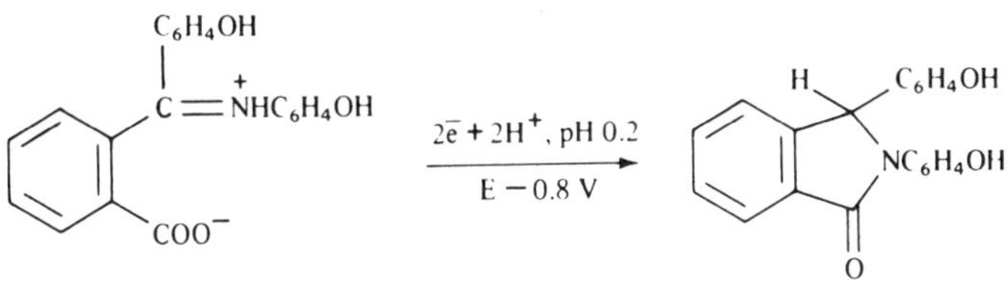

(XXIV)

(13)

The reductions of o-nitroacetanilide [55,56] to 2-methylbenzimidazole, acylated nitronaphthylamines to naphthimidazoles [46], 2-nitrophenylthiocyanate [55] to 2-aminobenzothiazole, and 2,2'-dicarboxy-6,6'-dinitrobiphenyl to 4,9-diaza-5,10-dioxo-4,5,9,10-tetrahydropyrene [57] have been reported.

Sometimes a "cathodic substitution" reaction takes place during the reduction of o-substituted nitrobenzenes; thus, the reduction of 2-nitrophenoxyacetic acid [44] in 2 N hydrochloric acid (50% ethanol) yielded 5-chloro-2H-1,4-benzoxazin-3(4H)-one rather than the expected cyclic hydroxamic acid [44]. In the presence of another nucleophile Y^- (e.g., thiocyanate) a ring substitution with this reagent occurs.

Oxidation of the methyl group in o-toluenesulfonamide at a nickel oxide electrode in aqueous carbonate leads to saccharin in 40% yield [58].

Reaction with azomethine groups. A useful synthesis of 3-phenylbenzo-1,2,4-triazine and other benzotriazines is exemplified by the reductive ring closure of o-nitrophenylazophenylnitromethane (XXV) [59]. This compound is converted in an acetate buffer by a four-electron reduction to o-nitrophenylbenzhydrazidoxime, which at a more negative potential is reduced to o-nitrophenylbenzamidrazone and further to o-aminophenylbenzamidrazone. Ring closure follows to give 1,4-dihydro-3-phenylbenzo-1,2,4-triazine (XXVI), which may be oxidized to XXVII by air or anodically. A high yield (85–96%) of this compound may be obtained in hydrochloric acid solution or acetate buffer containing DMF by reduction at the potential of the second wave, Eq. (14):

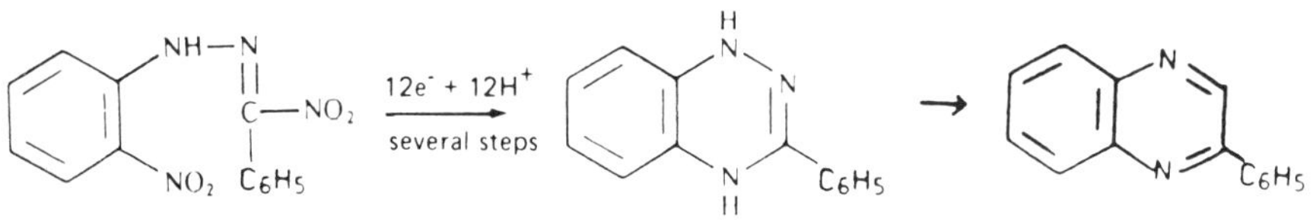

(XXV) (XXVI) (XXVII)

(14)

Other leaving groups may be used [59]; thus, thiobenz-(o-nitrophenyl)-hydrazine gives a good yield of XXVI on reduction, but benz-(o-nitrophenyl)-hydrazine does not. This is probably because the hydroxyl group is a poorer leaving group than the amino or sulfhydryl group. The ring-closed intermediate does not lose water but opens to o-hydrazinobenzanilide, which forms a benzimidazole on ring closure.

Attempts to reduce benzaldehyde-o-nitrophenylhydrazone to benzaldehyde-o-hydroxylaminophenylhydrazone, which on ring closure might form the dihydrobenzotriazine, were not successful [59] as it was not possible to avoid the further reduction to amine. A similar situation is also found in the reduction of o-nitroanilines.

It is thus essential that the central carbon be in the oxidation state of a carboxyl group and carry a good leaving group; if the central carbon atom is in a lower oxidation state, it must be oxidized before reduction of the aromatic nitro group. Benzaldehyde-o-nitrophenylhydrazone may thus be nitrosated [59] and further oxidized to o-nitrophenylazophenylnitromethane, which as described forms XXVI in two steps with consumption of altogether 12 electrons.

Many examples of oxidative ring closure by attack on an azomethine group by an amine have been reported; thus p-nitrobenzylidene-o-phenylenediamine is oxidized in MeCN-$LiClO_4$ to 2-(p-nitrophenyl)benzimidazole [36], chalcone phenylhydrazone in MeCN-C_5H_5N-$LiClO_4$ to 1,3,5-triphenylpyrazol [60], and benzylidene 2-pyridylhydrazone (XXVIII) to 3-phenyl-s-triazolo[4,3-a]pyridine (XXIX); oxazoles and imidazoles may be prepared similarly [60], Eq. (15):

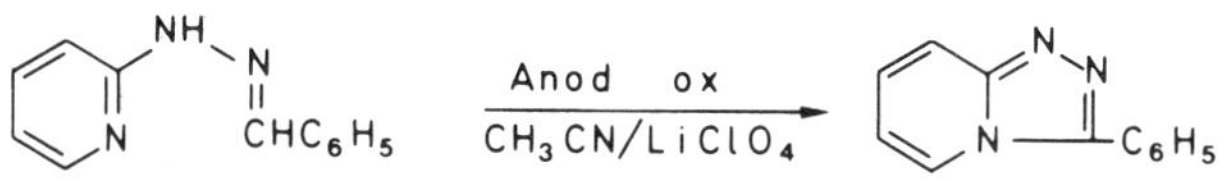

(XXVIII) (XXIX)

(15)

The reaction between an anodically produced cationic center and a tertiary amine has been explored. An example is the oxidation of 5-acetoxy-8-tert-butyl-2-hydroxy-1-(2-pyridyl)naphthalene (XXX) to the isolable perchlorate (XXXI), which slowly loses isobutylene to the zwitterion (XXXII) [61], Eq. (16):

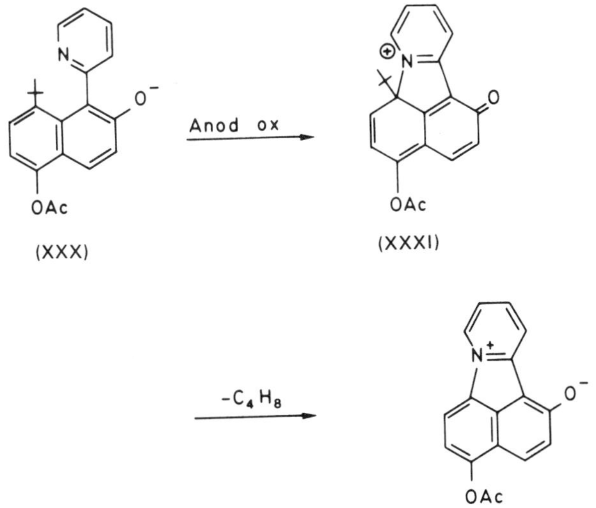

(XXX) (XXXI) (XXXII) (16)

The initial attack in the anodic oxidation of papaverine [62] probably involves a similar attack; further oxidation and dimerization leads to the isolated product, 12,12'-bis-(2,3,9,10-tetramethoxyindolo[2,1-a]isoquinolyl. An analogous reaction is the electrooxidation of a tetramethoxy-substituted 2-methyl-1-phenethyl-1,2,3,4-tetrahydroisoquinoline to a dibenzoquinolizinium derivative [63] and the oxidation of N,N'-triphenyl-o-phenylenediamine to 9,10-diphenylphenazine [64].

An oxidatively induced ring closure occurs during the oxidation of various catecholamines [65] and similar compounds [66] at the carbon paste electrode. Whereas the oxidation in molar H_2SO_4 yielded a 1,2-benzoquinone (XXXIII), sufficient free amine was present at pH 3 to undergo Michael addition to the quinone. As expected, the resulting catechol is more easily oxidizable than the starting material and is converted to the quinone, adrenochrome (XXXIV), by chemical oxidation by XXXIII, Eq. (17).

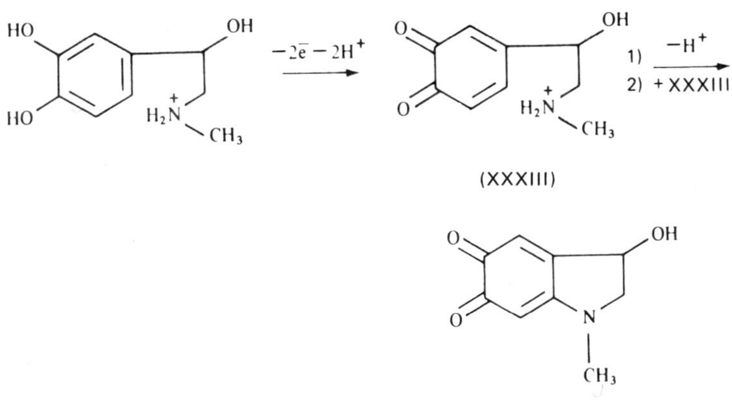

(XXXIII) (XXXIV) (17)

Anodic oxidation of phenylhydrazones generally results in a dimerization in the para position; if the para position is suitably substituted, a ring closure may take place. Thus, oxidation of para-substituted phenylhydrazones of 2-oxophenylacetonitrile yields derivatives of 1-phenyl-3-cyano-1H-indazoles [67].

Cyclic secondary amines (C_7-C_9) can be oxidized to bicyclic tertiary amines at a graphite anode in aqueous MeCN with LiBr as electrolyte. It is proposed that a positive species ("Br^+") is formed and is essential for the reaction [68].

Nitroso Groups as Electrophilic Centers Nitroso derivatives may be formed as intermediates in the reduction of nitro compounds, but as the nitroso group generally is more easily reducible than the nitro group, the lifetime of the nitroso stage is rather short. Alternatively, the nitroso compound may be prepared by anodic oxidation of a hydroxylamine at a suitable pH.

By this method N-hydroxyindoles may be prepared from derivatives of o-nitrostyrenes [69].

Activated methyl groups may react with an intermediate nitroso group in acid solution. Thus, o-nitrodimethylaniline yields on reduction in 2 N HCl a mixture of N-methylbenzimidazole and o-aminodimethylaniline [69].

Formation of Carbon-Sulfur Bonds

Carbon-sulfur bonds are formed during the reduction of carbon disulfide in aprotic medium; methylation of the product yields 4,5-bis(methylthio)-1,3-dithiole-2-thione (XXXV) [70]; if this compound is ethylated with Meerwein's reagent, followed by electrochemical reduction and thermal elimination, tetrathiomethoxy tetrathiofulvalene (XXXVI) is produced [71]. The mechanism of the formation of XXXV has been discussed [72,73], and it has been shown that carbon sulfide (CS) is not involved in the formation of XXXV [74].

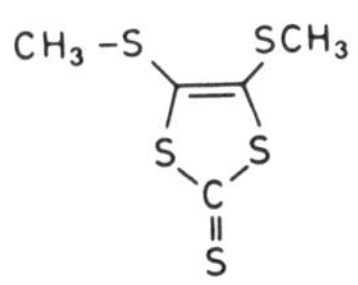

(XXXV)

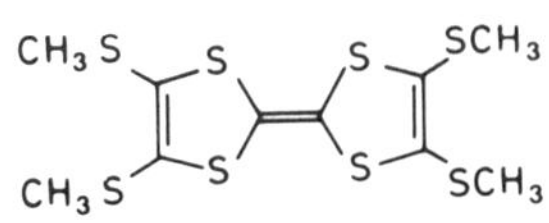

(XXXVI)

A special reaction is the reduction of certain acetylenes to thiophenes [75] at a compact sulfur/graphite electrode [76].

Thiobenzanilide, on anodic oxidation in CH_3CN-TEAP at a platinum electrode, yields 2-phenylbenzothiazole; other aromatic thiobenzamides react similarly [36]. Analogously, oxidation of N-(2-pyridyl)-thiobenzamide at controlled potential gives 2-phenyl-1,2,4-thiadiazolo[2,3-a-]pyridinium perchlorate (XXXVII) in 80% yield [36], Eq. (18):

$$2\text{—}NC_5H_4\text{—}NHCSAr \underset{+2e^- \; + H^+}{\overset{-2e^- \; - H^+}{\rightleftharpoons}}$$ 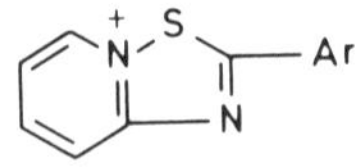 (18)

(XXXVII)

Anodic oxidation of disulfides usually gives the equivalent of RS^+, which may add to double bonds [77]; oxidation of XXXVIII in the presence of HX (X = halogen) gives XXXIX [78]; it is interesting that a five-membered, rather than a six-membered ring is formed, Eq. (19); the reaction is proposed to pass through an episulfonium ion [78].

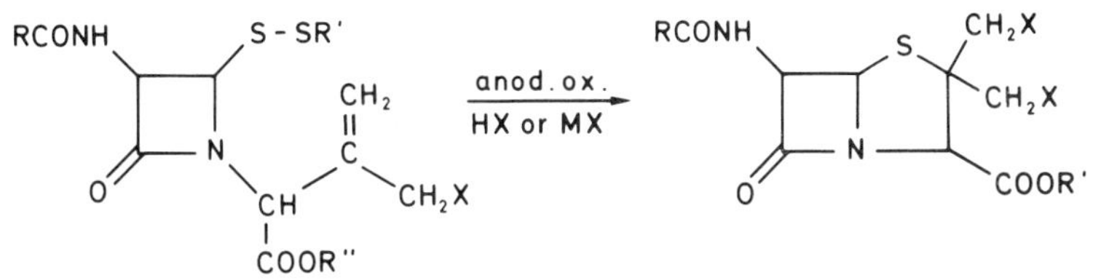

(XXXVIII) (XXXIX)

(19)

Formation of Nitrogen-Nitrogen Bonds

This type of ring closure is achieved mostly either by reaction between a hydroxylamino and a nitroso group or between a hydrazine derivative and a nitro or nitroso group.

A reductive coupling is possible whenever two aromatically bonded nitro groups are in such a position that a six- or seven-membered ring may be formed. The coupling probably occurs between a hydroxylamino group and an intermediate nitroso group with the formation of a cyclic azoxy derivative, a heterocyclic N-oxide, possibly through a radical-radical coupling (Chap. 9).

2,2'-Dinitrobiphenyl [79—82] is thus in alkaline solution at a suitable potential, reducible to benzo[c]cinnoline-N-oxide; in acid solution the reduction proceeds to 4,5-dihydrobenzo[c]cinnoline, which is very easily (e.g., by air) oxidized to benzo[c]cinnoline. A similar ring closure during the reduction of 2,2',6,6'-tetranitrobiphenyl yields N-oxides of 4,5,9,10-tetraazapyrene [83—85]. Reduction of 2,2'-dinitrodiphenylsulfide yields via 2,2'-(bishydroxylamino)diphenylsulfide in strongly acidic or in alkaline solution dibenzo[b,f]-1,4,5-thiadiazepine and its N-oxide [86]; under slightly different conditions the formation of 3-hydroxyphenothiazine has been reported [87].

1,8-Dinitronaphthalene could be expected from the coulometric data [88] to undergo ring closure on reduction, but suitable conditions have not been found yet. The diamine [89] and 8-amino-5-hydroxy-1,4-naphthoquinone-1-imine [90] have been reported as products.

When one of the nitro groups is aliphatically bonded, a coupling may take place, although the yield is often lower. 2-Nitro-1-(2'-nitrophenyl)-ethanol (XL) can in acetate buffer be reduced in two, four-electron steps. The dihydroxylamino compound (XLI) is transformed to cinnoline (XLII) by alkali and a trace of oxygen [91], Eq. (20):

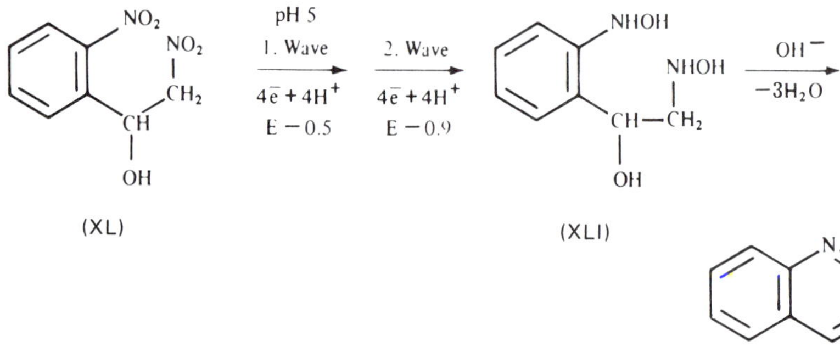

(XL) (XLI) (XLII) (20)

Indazoles may be obtained by reduction of o-nitrobenzylamines followed by reoxidation of the hydroxylamine to a nitroso group; ring closure occurs by addition of the amino group to the nitroso group followed by elimination of water [92].

Reduction of an azo compound to a hydrazine may lead to a ring closure when a nitro group is suitably situated. Thus, the reduction of 4-hydroxy-2'-nitroazobenzene (XLIII) in alkaline solution follows Eq. (21) [93–95]:

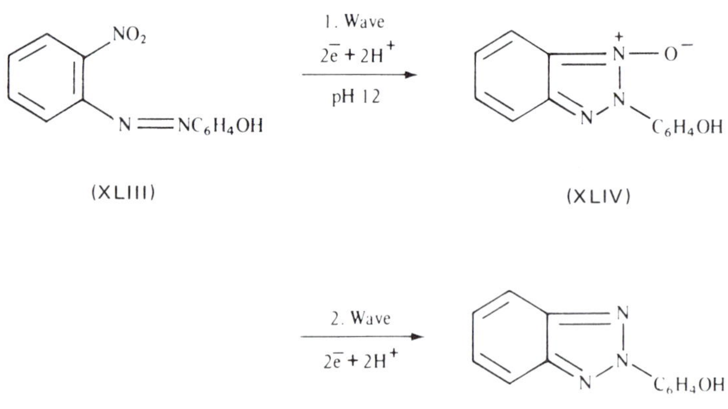

(XLIII) (XLIV) (XLV) (21)

By controlling the electrode potential it is possible to prepare the N-oxide (XLIV) in good yield [94], whereas the "classic" procedure yields the benzotriazole (XLV) [93]. In DMF 2-nitroazobenzene is reduced to 1,3-dihydroxy-2-phenylbenzotriazole [96].

Derivatives of hydrazine may also condense with a nitroso group, formed on oxidation of a hydroxylamino group; thus o-nitrobenzhydrazide forms, on reduction followed by reoxidation, benzo-1,2,3-triazinone-4 [97].

Anodic oxidation of formazanes in MeCN-TEAP at a platinum anode produces tetrazolium salts in good yield [98,99]. The ring closure probably takes place through the dication. A similar anodic coupling is the oxidation of 2-acetylpyridine phenylhydrazone to 3-methyl-1-phenyl-8-azaindazolium perchlorate [36], of benzilmonoxime phenylhydrazone to 2,4,5-triphenyl-

1,2,3-triazole-1-oxide, and of 3-hydroxyiminomethyleneamino-6-chloropyridazine to 6-chloro-s-triazole[1,5-b]pyridazine-3-oxide [60,100].

Formation of Oxygen-Nitrogen Bonds

Anodic oxidation of vicinal dioximes in MeCN-TEAP gives formazanes in fair to good yield [100—102]; in dioximes of arylglyoxal, electron-donating substituents in the aryl group direct toward the formation of 4-arylfuroxanes whereas electron-attracting substituents favor the formation of 3-arylfuroxanes.

Benzofuroxanes may be prepared by anodic oxidation of o-nitroaniline in MeCN containing iodide [103].

Intermolecular Coupling Reactions

Reductive and oxidative couplings are treated in more detail in Chaps. 22 and 23, respectively. Here are briefly mentioned some reactions that have been used in heterocyclic chemistry.

Electrochemical alkylation and acylation are well-established reactions and have been employed for the formation of heterocyclic compounds. The reduction of azobenzene [104] or azomethine compounds [105] in the presence of α,ω-dibromoalkanes leads to cyclic hydrazines, piperazines, or pyrrolidines. Similarly, reduction of nitrobenzene in the presence of 1,5-dibromopentane yields N-phenylperhydro-1,2-oxazepine [106].

Electrochemical reduction of azobenzenes in DMF in the presence of succinyl chloride gives pyridazinediones in fair yield [105]; a similar reaction with azomethine or nitroso compounds gave low or no yield of heterocyclic compounds. The reactions require dry media and unreactive solvent systems; the reaction type may be illustrated by Eq. (22):

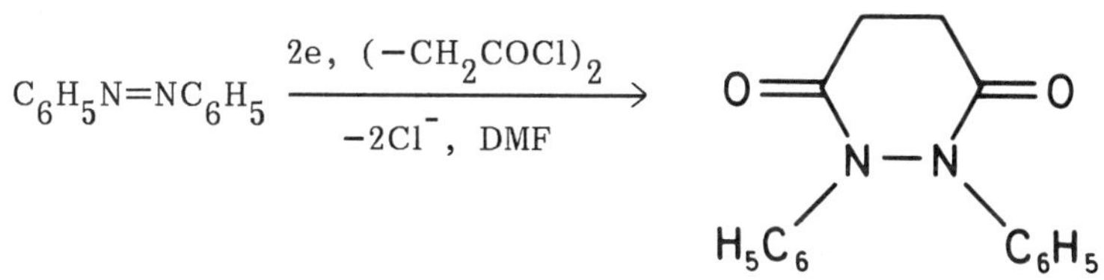

(22)

Anodic oxidation of benzylidenephenylhydrazine gives diarylnitrileimine, which may react with suitable dienophiles in a 1,3-dipolar addition reaction. Pyridine [107], some alkenes, and N-heterocycles [108] may act as dienophiles with formation of substituted 1,2,4-triazoles (XLVI) or pyrazolines, Eq. (23):

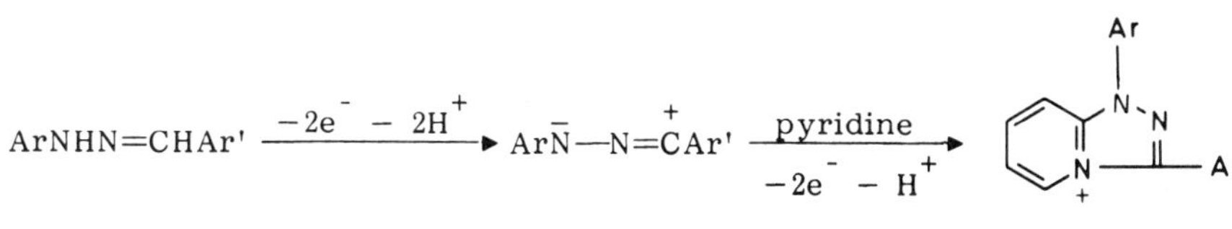

(XLVI)

(23)

Oxidation of a 1,1-disubstituted hydrazine in MeCN containing $HClO_4$ [109—111] or anhydrous HBF_4 [112] forms a diazenium cation that may react with alkenes in a cycloaddition reaction to a tetrahydrocinnoline. If both substituetns on nitrogen are aryl, an oxidative coupling between them may compete with the dipolar reaction [112].

Another type of oxidative (formal) cycloaddition is exemplified by the oxidation of 1,3-diketones in MeCN in the presence of an olefin to di- or tetrahydrofuran derivatives [113]; similar products may be obtained by reduction of 2,2-dibromo-1,3-diketones in the presence of an alkene [113].

Thus constant current oxidation of 5,5-dimethyl-1,3-cyclohexadione and styrene in MeCN (3 F/mol) yielded 2-phenyl-4-oxo-6,6-dimethyl-2,3,4,5,6,7-hexahydrobenzofuran (XLVII) in 97% yield [113], Eq. (24):

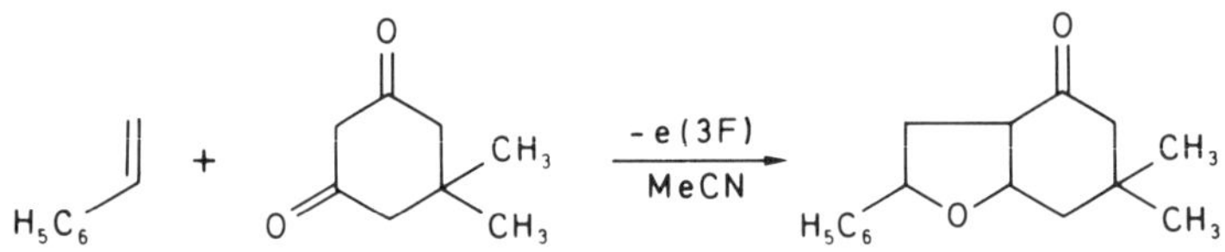

(XLVII)

(24)

Pyrrolizidines and indolizidines may be prepared by reaction between the electrogenerated anion of methyl dichloroacetate and acylated α-amino-aldehydes, obtained from amino acids. High diastereoselectivity (>99%) was observed in the reaction of some cyclic α-aminoaldehydes [114].

Acetonitrile is widely used as solvent in electrochemistry but is not always an inert solvent. In the nitrile group the nitrogen is sufficiently basic to react with carbocations and some other positively charged atoms, and the carbon may react with a nucleophilic center.

Anodic oxidation of 3,4-dimethoxyallylbenzene and iodine in MeCN containing $LiClO_4$ yielded 1,3-dimethyl-6,7-dimethoxyisoquinoline (XLVIII). The iodinated MeCN, $CH_3C^+{=}NI$, probably reacts in an aromatic, electrophilic substitution, followed by addition to the allylic double bond with ring closure, Eq. (25) [115]:

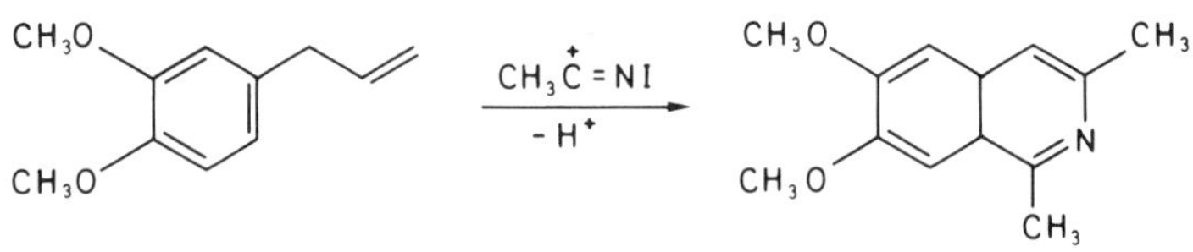

(XLVIII)

(25)

In MeCN anodic oxidation of o-toluenesulfonamides yields 3-methyl-4,5-dihydro-1,2,4-benzothiadiazepine-1,1-dioxides, probably by an initial Ritter reaction followed by ring closure [115].

Oxidation of conjugated dienes in MeCN-$NaClO_4$ in the presence of 1,3-dimethylurea gives a mixture of the possible 4,5-disubstituted-1,3-dimethylimidazolidin-2-ones in about 40% yield [116].

Many aromatic compounds yield carbonium ions on oxidation at a platinum electrode in a nonaqueous solvent like acetonitrile. 2,4,6-Tri-t-butylaniline (XLIX) yields, on anodic oxidation in acetonitrile containing pyridine, the following compounds (L and LI), Eq. (26) [117]:

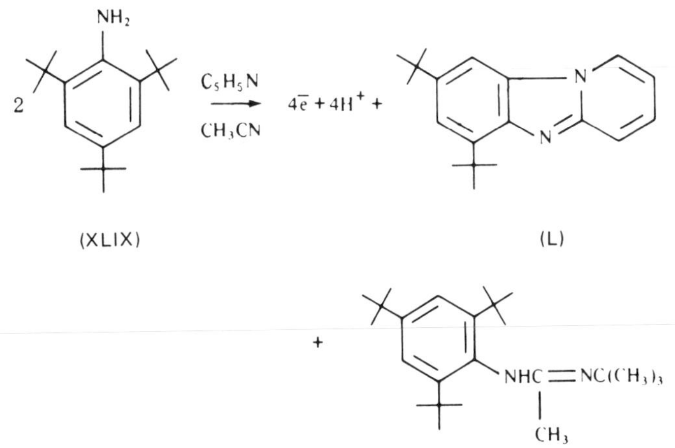

(XLIX) (L) (LI) (26)

The reaction starts with a loss of two electrons from the aromatic system of (XLIX). This is stabilized by loss of a t-butyl carbonium ion, which reacts with acetonitrile in a Ritter-type reaction, and by attack by the nucleophile pyridine. The second loss of two electrons may occur before or after the ring closure. A somewhat similar reaction is the anodic oxidation in acetonitrile of 2,6-di-t-butylphenols leading to 7-t-butylbenzoxazoles [118].

On anodic oxidation of 3,6-diisobutylpiperazine-2,5-dione in acetonitrile a compound was obtained that was suggested to be 1,6-diisopropyl-3,8-dimethyl-5H,10H-diimidazo[1,5-a:1'5'-d]pyrazine-5,10-dione, formed by 1,3-cycloaddition of the primary oxidation product to solvent [119]. Another example of inclusion of MeCN during the formation of heterocycles is the electrolysis of acetylenes in MeCN in the presence of a Co complex to pyridines [120].

Electrolysis of alkyl arenedithioates in MeCN with benzyltriethylammonium chloride as electrolyte yields 5-aryl-4-cyano-3-methylisothiazoles. It is suggested that the dithioester is chlorinated at the anode and reacts with 3-iminobutyronitrile, which is formed from MeCN by the cathodically generated base, to the isothiazole [121].

DMF may also participate in ring formation. Reduction of N-phenyl-2-bromoisobutyramide in DMF gave, besides debrominated and dehydrobrominated products, 5,5-dimethyl-2-dimethylamino-3-phenyl-4-oxazolidone [122].

B. Ring Contractions

In ring contractions occurring during electrolysis of nitrogen-containing heterocyclic compounds, a hydrogenolysis of a single bond takes place at some step, which is then followed by a ring closure. These reactions may be classified according to the single bond cleaved in the reduction. Examples of hydrogenolysis of oxygen-nitrogen, nitrogen-nitrogen, and nitrogen-carbon bonds are discussed.

Hydrogenolysis of an Oxygen-Nitrogen Bond

The initial step in the reduction of protonated oximes has been shown [123] to be a hydrogenolysis of the oxygen-nitrogen bond; in cyclic derivatives of oximes the same occurs. In 4-(4'-methoxyphenyl)-2,3-benzoxazin-1-one (LII) the reduction proceeds in two steps, and it is possible to isolate the rather stable intermediate ketimine, Eq. (27) [124]:

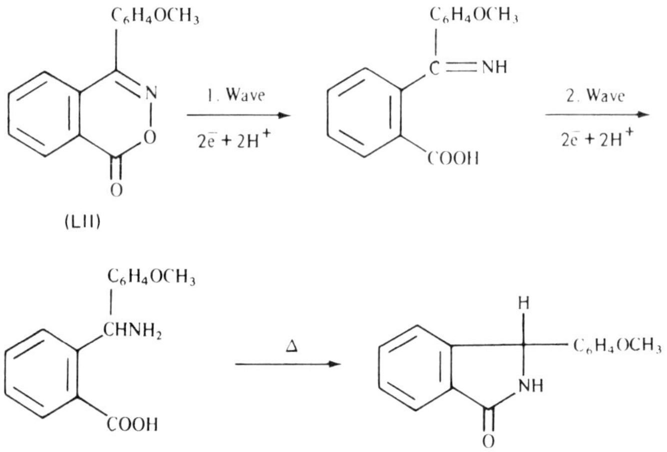

(LII)

(27)

Hydrogenolysis of a Nitrogen-Nitrogen Bond

A ring contraction involving this reaction has been found in cyclic hydrazones, hydrazides, and triazines.

The initial step in the reduction in acid solution of azines, hydrazones, and substituted hydrazones has been suggested to be a hydrogenolysis of the nitrogen-nitrogen bond [123]; the ald- and ketimine thus formed are mostly, but not always, reduced further at the potential necessary for the reductive cleavage; this depends on the nature of the substituents. This type of reduction is discussed in Chap. 11.

In acid solution cinnolines are reduced to dihydrocinnolines; a 1,4-dihydrocinnoline may be regarded as derived from a phenylhydrazone, which is reduced with cleavage of the nitrogen-nitrogen bond, Eq. (28) [79]:

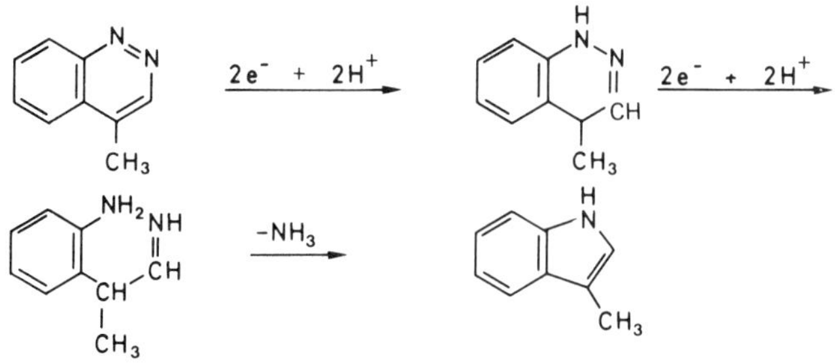

(28)

Similar ring contractions in which the nitrogen-nitrogen bond is cleaved are found in the reductive ring contractions of pyridazines to pyrroles [125,126], phthalazines to isoindoles [127], 5,6-diphenyl-1,2,4-triazinones to imidazolones [128], benzo-1,2,4-triazines to benzimidazoles [59], benzo-1,2,3-triazinone to indazolone [1a,129], and benzo-1,2,3-triazin-3-oxide to indazole [1a], benzo-2,3-diazepines to isoquinolines [130], and benzo-1-pyrano-[4,3-e]-as-triazin-3-one to benzopyranoimidazolone [131].

The reduction of phthalazinones [132] is one of the few observed examples of a reduction of a protonated compound containing >C=N—N< in which the C=N double bond has been proved to be more easily reducible than the N—N bond. Reduction at the potential of the second wave results in cleavage of the N—N bond and ring contraction. 4-Methylphthalazinone is thus reduced to 1-methylphthalimidine [132].

Hydrogenolysis of a Nitrogen-Carbon Bond

An electrolytic cleavage of such a bond in a convenient potential interval occurs only when the bond is activated, for example, by an adjacent carbonyl or azomethine group.

A ring contraction occurs during the reduction of 7-chloro-2-methylamino-5-phenyl-3H-4,5-dihydro-1,4-benzodiazepine to 6-chloro-2-methyl-4-phenyl-3,4-dihydroquinazoline [133]. It has been suggested [43] that the cleavage occurs between C-3 and N-4, although another interpretation has also been offered [133].

The reductive ring contraction of 2-phenylpyrimidines to 2-phenylpyrroles involves a cleavage of a carbon-nitrogen bond, followed by attack by a carbanion on an azomethine group [134].

Hydrogenolysis of a Carbon-Sulfur Bond

Reduction of 4-carbethoxy-2-phenyl-1,3-thiazinone-6 (LIII) in aqueous alcoholic media yields 2-carbethoxy-4-hydroxy-5-phenylpyrrole (LIV) in a 4e-process. The first step in the reduction is suggested to be the saturation of the C=N double bond; the mechanism of the reductive loss of H_2S and C—C bond formation is not yet established, Eq. (29) [135]:

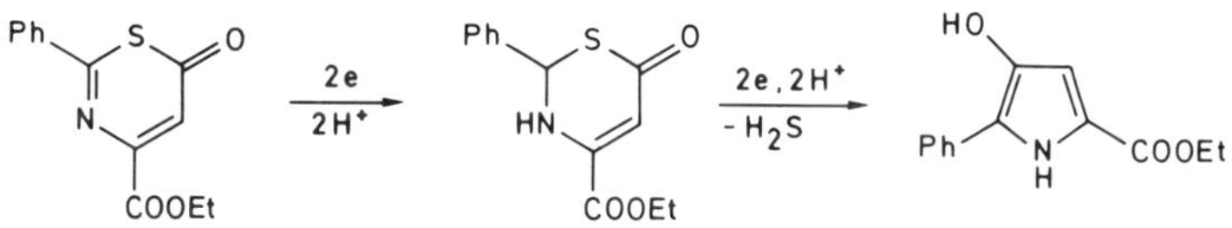

(LIII) (LIV) (29)

C. Ring Expansions

Only a few types of ring expansions involving electrolysis have been reported so far. In the preparation of ψ-phthalazinones from N-anilinophthalimides [136], the phthalimide derivative is reduced to a derivative of phthalaldehydic acid, N-anilinohydroxyphthalimidine. It has been suggested [43,

137] that this compound undergoes a ring opening to a hydrazide of phthalaldehydic acid. On ring closure the more nucleophilic nitrogen of the hydrazide attacks the aldehyde group with formation of a six-membered ring, a ψ-phthalazinone. In strongly acid solution the ψ-phthalazinone ring may be opened again and then some of the hydroxyphthalimidine forms a phenylhydrazone of phthalaldehydic acid. The phenylhydrazone would on ring closure yield 2-phenylphthalazinone; this Rowe rearrangement [137] can thus be explained by assuming that the ψ-phthalazinone is the kinetically controlled product, whereas the 2-arylphthalazinone is the thermodynamically controlled product.

A special type of ring enlargement is found in the electrolytic reduction of bicyclic α-amino ketones at a lead cathode in 30% sulfuric acid at 60°C to monocyclic products [138]. Thus, 1-ketoquinolizidine under these conditions gave 59% 5-hydroxyazacyclodecane and 4.5% 1-azabicyclo[5,3,0]decane. The initial cleavage of the carbon-nitrogen bond is analogous to that found in the reduction of derivatives of 2-aminoacetophenone [139].

In this section the anodic oxidation in nitromethane of tetraphenylpyrrole to tetraphenylpyridine [140] may be mentioned; it is discussed in Sec. III.C.

III. ELECTROLYTIC REACTIONS OF HETEROCYCLIC SYSTEMS

In this section the electrode reactions of heterocyclic compounds in which the nucleus undergoes a reduction or oxidation are discussed. Substituted compounds are included in this section only if the substituent does not become reduced or oxidized.

One advantage of using electrolytic methods in heterocyclic systems is the possibility of preparing partly reduced systems, in most cases from π-electron-deficient compounds, and it is probably the most general method for doing so.

Another useful method is the electrolytic substitution of heterocyclic compounds, which are π-electron rich or substituted with electron-donating groups.

A. Compounds with One Oxygen Atom

Furan is, as a π-electron rich compound, easily oxidized electrolytically. In acetonitrile the oxidation potential of furan is comparable to that of anisole [30]. In methanol the oxidation is an ECEC mechanism in which 2,5-dihydro-2,5-dimethoxyfuran (LV) is formed [141] in the Clauson-Kaas reaction, Eq. (30):

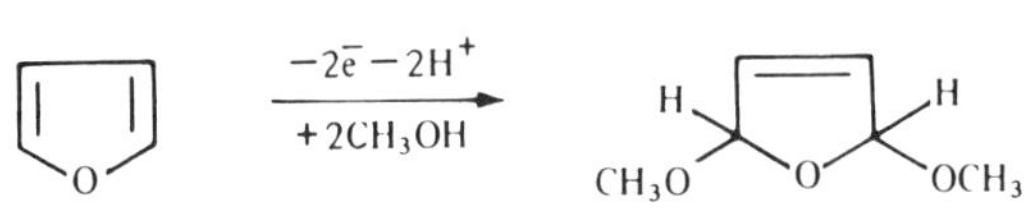

(30)

(LV)

The reaction may be performed in the presence of an excess of ammonium bromide, where the bromide ions are oxidized to bromine, which in turn oxidizes furan. The dimethoxydihydrofuran is a valuable intermediate for the preparation of various heterocyclic compounds, such as pyridines [142] and pyridazines [143–145]. The oxidation of a furan to a dimethoxydihydrofuran is thus a key step in a synthesis of pyridoxine [146].

The stereochemistry of the anodic cyanomethoxylation of 2,5-dimethylfuran [147] depends, at a platinum electrode, on the concentration of substrate and of additives; the cis-trans ratio changes from 2.89:1 to 1.46:1 on increasing the concentration of substrate from 0.01 to 0.2 M. At a carbon anode a ratio close to the equilibrium mixture 1.42:1 is obtained.

Anodic acetoxylation of furan can be done in acetic acid containing acetate [148,149] and benzoyloxylation in DMF containing benzoic acid and lithium chloride [150]. A mixture of cis- and trans-benzoyloxylated product is formed. Cyanation of 2,5-dimethylfuran gives a mixture of cis- and trans-2-cyano-5-methoxy-2,5-dimethyldihydrofuran [151–153].

Anodic oxidation of 5-alkyl-2-furoic acid in aqueous methanol yielded mainly [154] 1-butenolides and γ-keto esters, whereas oxidation of 5-acetylfuroates in methanol leads to ring-opened diesters besides the 2,5-dimethoxylated dihydrofuroates [155].

Pyrylium [156,157] and isobenzopyrylium [158] salts have been shown to be polarographically reducible in a one-electron reduction. In cyclic voltammetry in aprotic solvents, 2,4,6-trisubstituted pyrylium salts show two peaks; the shape of the peak depends on the rate of dimerization. This process occurs more rapidly at C-4 than at C-2 (C-6) [157], and the dimerization takes place spontaneously for 4-unsubstituted pyrylium salts. The equilibrium between radicals and dimer is displaced in favor of the radicals on introduction of electron-withdrawing substituents; such groups enhance the aromatic character of the radical [159]. If the reduction of pyrylium salts is made in the presence of an alkyl iodide a fair yield of the 4-alkylated 4H-pyrane is isolated with the dimer [157].

Anodic oxidation of chromone in aqueous methanol containing a halide gives 3-halo-2-methoxychromanone in a reaction that is a variant of the Clauson-Kaas reaction [160].

Electrochemical reduction of coumarin [161] affords the meso and (±) forms of a product hydrodimerized (at C-4) like many other α,β-unsaturated acid derivatives. In the presence of tertiary amines, which are not destroyed during the process but catalyze the evolution of hydrogen, a transfer of hydrogen takes place from the amine radical to the coumarin radical with formation of 3,4-dihydrocoumarin. Asymmetric amines cause asymmetric induction during the reduction of 4-methylcoumarin, Chap. 27 [162].

Voltammetric studies in DMF [163] of aryl-substituted isobenzofurans like 1,3,4,7-tetraphenylisobenzofuran show that it gives two, one-electron reduction waves and one, one-electron oxidation wave. By cyclic voltammetry (CV) it was shown that the first reduction product was stable, whereas the first oxidation product and the second reduction product were unstable.

B. Nuclei Containing One Sulfur Atom

Thiophene

Anodic oxidation in methanol of thiophene and substituted thiophenes at low temperatures (−20 to −30°C) results in methoxylation and ring opening with loss of sulfur as SO_2 [164]. From thiophene is isolated butenedialdehyde tetramethylacetal, some methoxysuccinic dialdehyde tetramethylacetal, and a small amount of methyl β-formylpropionate. In general, oxidation of thiophenes results in the formation of derivatives of α,β-unsaturated γ-dicarbonyl compounds or γ-keto esters. Cyanation of 2,5-dimethylthiophene yields mainly cis- and trans-2-cyano-5-methoxy-2,5-dimethyldihydrothiophene and 3-cyano-2,5-dimethylthiophene [165].

Thiophene is not reducible by direct electroreduction, but by indirect reduction (Chap. 29) in DMF using biphenyl radical anion as electron transfer agent it is possible to reduce thiophene to 2,5-dihydrothiophene and tetrahydrothiophene in high yield [166]; thiophenes substituted at C-2 with carboxyl, however, may be reduced by direct electrolysis to the 2,5-dihydroderivatives [167].

Benzothiophene and its alkylated derivatives afford the corresponding 2,3-dimethoxy-2,3-dihydrobenzothiophenes in the Clauson-Kaas reaction, whereas others, such as the methyl 3-methylbenzothiophene-2-carboxylates, do not [168].

The reduction of benzothiophene by "ammonium amalgams" is discussed in Chap. 28.

Benzothiophen-1,1-dioxide is polarographically reducible at about −1.3 V at pH > 2 [169]. In alkaline solution the polarographic behavior is complicated by the addition of nucleophiles to the 2,3-double bond. Controlled potential reduction in acetate buffer produced 2,3-dihydrobenzothiophen-1,1-dioxide [170].

2,4,6-Triphenylthiopyrylium salts (LVI) and some other trisubstituted thiopyrylium compounds show two reversible redox systems [159]; the anion from the second reduction of LVI has been used as electron transfer agent in indirect reduction of and coupling with alkyl halides [171].

C. Nuclei Containing One Nitrogen Atom

Aziridine Derivatives

Aziridinium salts are easily reducible, both in aqueous and nonaqueous solvents; usually, from alkylaziridinium salts, a mixture of products is obtained as a result of subsequent chemical reactions of the primary products [172]. Electroanalytical data point to a one-electron reduction followed by ring opening of the radical.

Arylaziridinium salts show a "reverse" structural effect on the polarographic half-wave potential, as electron-donating substituents make the reduction in aqueous solution easier and electron-attracting groups have the opposite effect [173]. A tetramerization of 1-benzylaziridine to a tetraazacyclododecane may be induced anodically [174].

Pyrroles and Benzopyrroles

Pyrrole is not polarographically reducible but can be reduced at a lead cathode in dilute sulfuric acid to pyrroline and further to pyrrolidine [175]. Under similar conditions 1,2-dimethylpyrroline [176] is also reduced to the pyrrolidine and indoles [177–179] to indolines or dimerized products [180]. 1-Methylindole can be reduced to 2,3-dihydro-1-methylindole in aqueous THF at very negative potentials using TBAOH as electrolyte [181]. Pyrrole may also be oxidized anodically; oxidation of pyrrole may result in the formation of polypyrrole useful for preparation of chemical modified electrodes (Chap. 33).

In acetonitrile pentaphenylpyrrole is oxidized to a rather stable cation radical [182], and tetrasubstituted pyrroles react similarly [183]; in methanolic solution 1-methylpyrrole forms 1-methyl-2,2,5,5-tetramethoxypyrroline [184].

2,3,4,5-Tetraphenylpyrrole is in nitromethane oxidized in two, one-electron steps, the first reversible. Oxidation at the potential of the first wave at low temperatures gives $n = 2$, indicating a disproportionation of the initially formed radical cation. The dication thus formed reacts with the residual water to give a hydroxylated compound that on heating rearranges to a pyrrolone. At higher temperatures the radical cation reacts with nitromethane, forming nitrite ion and a tetraphenylpyridine, reported to be 2,3,4,6-tetraphenylpyridine [140,185], but a 2,3,4,5-tetraphenylpyridine seems more likely.

In methanolic cyanide solution N-substituted pyrroles [186] are substituted in the 2-position by a nitrile group; when both 2- and 5-positions carry a methyl group, cyanation initially [187] takes place in the 2- and 5-positions, but these compounds are unstable and the final substitution occurs in the side chain [186,187]. N-substituted indoles are likewise cyanated in good yield; substitution in the 2-position dominates, although some 3-substitution takes place [186].

N-Substituted pyrrolidines (and piperidines) are, similarly to other tertiary aliphatic amines, anodically cyanated α to the nitrogen in aqueous methanolic solution containing sodium cyanide [188].

2,3-Diphenylindole forms, on anodic oxidation, a cation radical that dimerizes; the dimer is probably of the indole-indolenine type [189], dimerized at C3—C6' [190] rather than at C3—C5 [191,192]. The reaction is similar to the oxidation of 1,2,3,4-tetrahydrocarbazole [193].

N-Methyl-1,3,4,7-tetraphenylisoindole has been investigated by CV in DMF. Contrary to the isobenzofurans already discussed, it forms a stable cation radical but unstable anions [163].

Carbazoles are oxidized at controlled potential at a platinum anode in acetonitrile to cation radicals that are stable when the 3-, 6-, and 9-positions are blocked [194]. The radical cation (LVII) from carbazole dimerizes predominantly at the 3-position to 3,3-dicarbazyl, which is further oxidizable to the quinoidal dication at the potential used. In the presence of pyridine, which may cause a rapid deprotonation of the N—H proton of LVII, 9,9'-dicarbazyl is the isolated product.

Pyridine Derivatives

Pyridine is not polarographically reducible in aqueous media; in MeCN catalytic hydrogen evolution is the main reaction of protonated pyridine at the cathode [195]. Pyridine can, however, be reduced to piperidine at a lead cathode in 10% H_2SO_4 [196]; the yield is very dependent on the purity of the lead and other components. Side products are α,α-, α,γ-, and γ,γ-dipiperidyl. Pyridine may be reduced to a dimer in liquid ammonia [197]. In MeCN pyridine is reduced in a two-electron wave; dihydropyridine and cyanomethylated tetrahydropyridine are formed, the latter by attack of $(CH_2CN)^-$ on the dihydropyridine [198]. Electrochemical synthesis involving pyridine and its derivatives have recently been reviewed [1c].

The potential-limiting anodic reaction in pyridine with perchlorate as supporting electrolyte is presumably the oxidation of pyridine to an onium ion, which with excess of pyridine forms N-(2-pyridyl)pyridinium perchlorate [199].

Substituents may render the nucleus more easily reducible or oxidizable; electron-attracting substituents such as carbethoxy groups, facilitate the reduction, whereas methoxy groups, for example, make the oxidation easier. A positive charge on the nitrogen atom also facilitates the reduction of the nucleus, but the charged pyridine ring also makes the electron-attracting substituent more easily reducible. In some cases the electron reaction changes with pH; in several such instances it is found that the substituent is reduced in acid solution and the ring in alkaline medium. Polymerization may also occur, especially if C-4 is unsubstituted.

The presence of two methoxy groups on the pyridine ring makes it susceptible to anodic oxidation. In alkaline methanol the main products [184] from 2,6-dimethoxypyridine (LVIII) are 2,3,5,6-tetramethoxypyridine (LIX) and an azaquinone, probably with the structure (LX), Eq. (31).

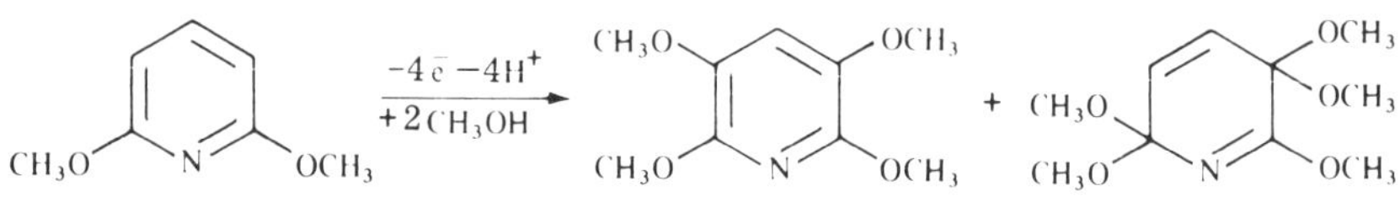

(31)

(LVIII) (LIX) (LX)

Alkyl-substituted (or arylakyl-substituted) pyridines may be oxidized to the corresponding acids in high yield in acid solution on a PbO_2 anode in a divided cell; a continuous removal of the products is an advantage [200,201].

The anodic oxidation of 4,4-disubstituted 1,4-dihydropyridines in MeCN gives a relatively stable cation radical, which catalyzes the oxidation of MeCN; some of the cation radicals lose a 4-substituent as a radical, and the nucleus is oxidized further. The oxidation of 4-monosubstituted 1,4-dihydropyridines is usually an ECE reaction, but even with such compounds a C—4 substituent may be lost; the radical abstracts a hydrogen atom from MeCN [202].

Pyridinium salts are generally polarographically reducible in neutral and alkaline solution, and dimerized dihydropyridine derivatives have been postulated as products [203—206]. The bimolecular dimerization rate constant has been found to $1 \times 10^7\ M^{-1}\ s^{-1}$ [207]. Further reduction produces the 1,4-dihydro derivative. In a few cases the structures of the dimers have been established; thus reduction of 1,2,6-trimethyl-3,5-dicarbethoxypyridinium cation yields a mixture of isomers. The less stable isomer, the 2,4'-dimer, undergoes rearrangement through two pyridil radicals into the thermally more stable 4,4'-dimer [208]. Anodic oxidation of the dimerized products, tetrahydrobipyridines, gives a mixture of alkylpyridinium salts and the dication of 1,1'-dialkyl-4,4'-bipyridine [209]. The reductive dimerization of 1-alkylpyridinium salts followed by oxidation has been used for the synthesis of the herbicide paraquat [210].

1-Alkyl-4-carbomethoxypyridinium salts are reduced in protic and aprotic media [211] to a rather stable radical; in DMF this radical is reduced in a reversible reaction to an anion that acts as a good electron-donating reagent [212]. The radicals from the 2- and 3-isomers are less stable and dimerize readily [211].

N-Methylpyridinium or N-methylquinolinium salts [213] may be oxidized to the N-methyl-2-pyridone or N-methyl-2-quinolone, respectively, in alkaline medium containing some potassium hexacyanoferrate (III). The reaction is probably an indirect electrolysis in which the hexacyanoferrate oxidizes the pseudobase and is, in turn, reoxidized at the anode.

The reduction of quaternary derivatives of nicotinic amide as model compounds for the phosphopyridine nucleotides has been subject to many polarographic [203,214–221] and electrolytic investigations. At pH 4–7 a one-electron wave is found, whereas in alkaline solution two, one-electron waves are seen. Controlled potential reduction in a phosphate buffer at the potential (−1.2 V) of the first wave of 1-methyl-3-carbamoyl or 1-benzyl-3-carbamoylpyridinium ions produces a mixture of dimers, including diastereomers of 6,6'- [215] and 4,4'- [220], in addition to unsymmetrical dimers [221].

Reduction at the potential of the second wave (−1.80 V) of NAD^+ analogs substituted at N–1 with benzyl or methyl gives primarily the 1,6-dihydro compound, which was shown by NMR spectroscopy and deuterium experiments; the 1,6-dihydro compounds are rather easily isomerized to the 1,4-dihydro compounds. 1-Phenyl-NAD^+ analogs are reduced to the 1,4-dihydro compounds [221].

Analogous results have been obtained in the electrolysis of the pyridine nucleotides [222–227]. The dimer has, however, been suggested to be 4,4'-dimers. The coenzyme activity of the reduction product varies with the experimental conditions. Some reductions yielded totally inactive products; others produced partly or fully active material. The highest activity was obtained by an indirect electrolytic reduction of TPN (Chap. 29) [227]; electrochemical reduction of methyl viologen in the presence of ferridoxin-TPN reductase caused a reduction of triphosphopyridine nucleotide (TPN) to a biochemically active product.

Electrochemical oxidation of the reduced form (NADH) of nicotinamide adenine dinucleotide and its analogs has been investigated [228]. The results are consistent with an ECE mechanism yielding the pyridinium salt. Indirect oxidation of NADH (and reduction of NAD^+) is treated in Chap. 29, and the oxidation of NADH at chemically modified electrodes is discussed in Chap. 33.

Quinoline, Isoquinoline, and Acridine

Quinoline and isoquinoline derivatives are polarographically reducible in a way similar to the pyridine compounds, but the reduction potentials required are generally less negative.

The reduction of quinoline at different electrodes in sulfuric acid to dihydro- and tetrahydroquinoline has been studied using constant current [229]. The reaction proceeds with high current efficiency at electrodes with high hydrogen overvoltage (e.g., lead), but not at platinum, which has a low hydrogen overvoltage.

At neutral pH quinoline derivatives may be dimerized in aqueous methanolic solution [230]. It is suggested that a radical-radical reaction occurs at low substrate concentration and a radical-substrate coupling at higher concentrations.

Quinolines, isoquinolines, and other heterocyclic compounds may be electrochemically carboxylated by reduction in aprotic medium in the presence of CO_2. The carboxylation is in many cases faster than the loss of chlorine from a chlorinated heterocycle, so such compounds may be, especially at low temperature, carboxylated without loss of chlorine [231].

Reduction of quinoline and isoquinoline in NH_3 [232] or DMF [233] in the presence of alkyl halides yields alkylated products.

It is of interest that a dihydroisoquinoline like 3,4-dihydropapaverine is more easily reducible than the parent compound [234]; the electrode reaction is saturation of the azomethine bond. 3,4-Dihydro-2-alkylisoquinolinium salts may be reductively benzylated at C—1 [15] in aprotic media.

Anodic oxidation of quinoline at a platinum anode in 75% sulfuric acid yields quinolinic acid (77%) [235]; other quinolines bearing substituents in the hetero ring in the 3-position may do similarly, whereas some 2- and 4-substituted derivatives form tar [236].

Acridine is polarographically reducible both in acid and alkaline solutions [237]. In strongly acid solution a single one-electron wave is found, and from a preparative reduction compounds dimerized at C—9 [30] were isolated. In less acid and in alkaline solutions two, one-electron waves are found.

Anodic oxidation of 9,9-substituted 9,10-dihydroacridines in MeCN-TBAP at a Pt electrode gives a cation radical that dimerizes to a 2,2'-biacridane system; such a dimer is further oxidizable to a quinonoid system [238].

The acridizinium and phenanthridizinium systems are reducible [239] in a manner similar to acridine.

D. Compounds with One Oxygen and One Nitrogen Atom

Oxaziridines

Oxaziridines are very easily reducible compounds, and their polarographic wave [240] begins at the dissolution wave of the mercury. In the reduction there is formed a gem-amino alcohol that at low pH loses water with the formation of a Schiff base and at higher pH loses an amine to a carbonyl compound. The Schiff base or the carbonyl compound may be further reducible.

Oxazoles and Isoxazoles

Aryl-substituted oxazoles are reducible [241] at rather negative potentials [−1.0 to −2.1 V (SCE)], and the electrode reaction consumes in most cases six electrons; 2-(1-naphthyl)-5-phenyloxazole, however, consumes two electrons.

In DMF diphenyloxazole is reduced in two one-electron reductions, the first resulting in a relatively stable anion radical, the second presumably in a dihydro derivative [242]. In MeCN 3-phenyl-substituted isoxazoles are reduced to a monoimine that is hydrolyzed to the diketone [243].

Reduction of oxazolinium salts in DMF yields the anion of the tetrahydrooxazole, which may be used as an acyl anion equivalent [244].

N-Alkylisoxazolium compounds may be reduced in neutral solution to 3-alkylamino-2-butenones [Eq. (32)] [243,245] and N-alkyl-4,5-dihydroisoxazolium salts to 4-hydroxy-2-butanones [243].

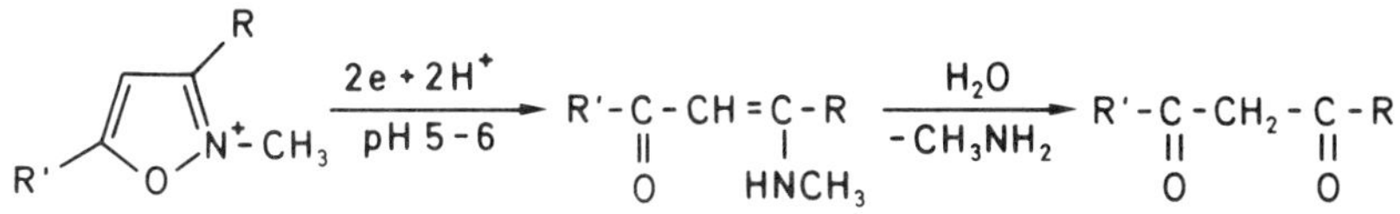
(32)

Anthranils

These compounds give in acid solution two polarographic waves, and a reduction at a potential corresponding to the foot of the first wave of 3-methylanthranil yields o-aminoacetophenone in good yield [43]. The second wave represents the reduction of this compound. In alkaline medium a single three-electron wave is found, and the reaction mixture contains several dimerized compounds [3].

Oxazines

N-Formylated 1,4-oxazines, including morpholine, can be oxidized anodically in methanolic solution to 3-methoxy derivatives, which under acidic conditions lose methanol to N-acylenamines; these are of potential synthetic interest [246]. Other acylated secondary amines, such as those from pyrrolidine, piperidine, and tetrahydroisoquinoline, behave similarly (Chap. 25) [247].

E. Compounds with One Sulfur and One Nitrogen Atom

Thiazole Derivatives

Thiazole and its simple alkyl or aryl derivatives are not polarographically reducible. Substitution in the thiazole ring with electron-attracting groups may render the nucleus reducible. Derivatives of thiazole-2-carboxylic acid are reducible in alkaline solution, and the reduction is assumed to take place in the nucleus [248].

Many thiazolium salts are polarographically reducible [249,250]; 3-methylbenzothiazolium iodide [249] is reduced in a single wave; a preparative reduction gave n between 1 and 2, and a mixture of dimer and monomer 2,3-dihydro-3-methylbenzothiazole was found. Aryl groups in the 2-position hinder dimerization [249]. With TEAP as supporting electrolyte 2-methyl-3-ethylbenzothiazolium ion gives two, one-electron waves [250].

The reduction of thiamine hydrochloride in neutral buffer is reported to involve a cleavage between the pyrimidine ring and the thiazolium ring and a ring opening of the latter; further chemical steps lead to a substituted thiourea [251].

Isothiazole Derivatives

Substituted benzenesulfonamides bearing strongly electron-attracting substituents can be reduced in slightly alkaline solution at the dropping mercury electrode to ammonia, sulfur dioxide, and a substituted benzene (Chap. 24). Saccharin is a compound of this type, and the primary electrode reaction is a cleavage of the carbon-sulfur bond; chemical reactions follow the initial cleavage.

In acid solution the reduction follows a route similar to that of phthalimide with a primary reduction of the carbonyl group [252].

Thiazines

2-Phenyl-1,3-[6H]-thiazines substituted at C−5 with an electron-withdrawing group (EWG) are reduced first at the C=N double bond and at the C=C bond at a more negative potential; if both C−4 and C−5 carry an EWG the C=C bond is reduced prior to the C=N bond [253]. If C−4 carries and EWG and C−5 an electronically "neutral" group, a cleavage of C6−S occurs with formation of a substituted benzthiamide, Eq. (33) [254]:

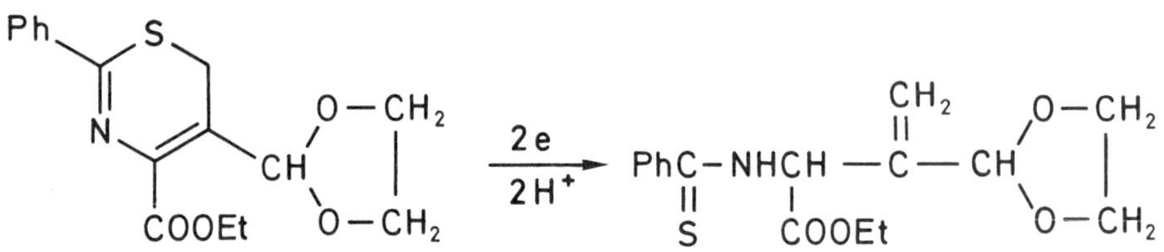

(33)

Several derivatives of phenothiazine have been found to be oxidizable at a gold anode in dilute sulfuric acid [255]. The product proved to be the sulfoxide rather than the N-oxide.

In acetonitrile phenothiazine is oxidized in two, one-electron waves; the radical cation obtained in the first oxidation loses a proton and is oxidized to the phenazothionium ion and isolated as the perchlorate [244,256] or dimerizes at nitrogen or at C−3 [257].

Reduction of 1,3-thiazinium salts in acetonitrile yield 6,6'-bi-1,3-thiazinyls; the 6,6'-bi-1,3-thiazinyls are oxidized to thiazinium salts at a platinum electrode, whereas dehydrogenation by chloranil affords 6,6'-bi-1,3-thiazinylidene [258].

F. Compounds with Two Oxygen or Two Sulfur Atoms

Ethylene trithiocarbonate is reduced in DMF to the anion radical, which loses ethylene [259]; the reaction is akin to other reductive elimination reactions. The sulfur-containing part reacts with another anion radical to a bis(trithiocarbonate) dianion.

Electrochemical reduction of 4,5-alkylthio-1,3-dithiol-2-ones in aprotic medium, followed by alkylation, gives tetraalkylthioethylenes in almost quantitative yield; the same products are formed, albeit in lower yields, on the reduction of 4,5-alkylthio-1,3-dithiol-2-thiones [260].

Voltammetric oxidation of 1,4-dioxins (LXI) and 1,4-dithiins (LXII) in acetonitrile containing perchlorate yields a cation radical in the first one-electron oxidation and a dication in the second one-electron oxidation. Both LXI and LXII are typical π-electron-rich heterocycles, whereas the dication contains six π-electrons, and the relatively high stability of the dication is connected with this "aromaticity" [261].

The monocation radical of dibenzodioxine can be prepared electrochemically and isolated as a stable perchlorate [262]; if the medium is not dry, hydroxylated and further oxidized products are obtained.

G. Compounds with Two Nitrogen Atoms

Diazirines

Diazirines are polarographically reducible in acid solution in a four-electron reduction to a gem-diamine, which is hydrolyzed to a carbonyl compound, and in alkaline medium in a two-electron reaction [263] to a diaziridine. Diaziridines are reducible only in acid solution [264] with cleavage of the N—N bond, in alkaline solution they can be oxidized anodically to diazirines [263].

The protonated diaziridine is more easily reducible than the diazirine, which has too weakly basic properties to be protonated in aqueous solution. In alkaline medium in which both are unprotonated, the diazirine is the more easily reducible. This explains why it has been difficult to obtain good yields of diaziridine by chemical and catalytic reduction of a diazirine; if such a reduction is to result in a high yield of diaziridine, it must be performed in an alkaline medium.

Di-tert-butyldiaziridinone is reduced in DMF-TEABr to N,N'-di-tert-butylurea with cleavage of the nitrogen-nitrogen bond; the presence of the electron-attracting carbonyl group makes it possible for the reductive cleavage of the unprotonated compound [265] to take place.

Pyrazoles

Pyrazolines have been investigated polarographically in a 2:2:1 mixture of DMF, benzene, and methanol with tetraethylammonium iodide as electrolyte. The polarographic data indicate a two-electron reduction of the carbon-nitrogen double bond [266]. In DMF suitably substituted pyrazoles may be reduced to Δ^4-pyrazolines [267].

The anodic oxidation of 1-arylpyrazolines has received considerable attention. In acetonitrile, without added base, the primarily formed cation radical dimerizes at C—3 to bis-(1-phenylpyrazolinyl-3) if this carbon is unsubstituted; otherwise the dimerization takes place at the p-carbon of the 1-phenyl group, if this is unsubstituted, to bis(pyrazolinyl-1)-biphenyl derivatives, which are further oxidized. In the presence of base, such as pyridine, a 1-phenylpyrazole is obtained [268—270].

Pyridazines

Pyridazines are polarographically reducible [271,272], often stepwise. The first reduction generally yields a dihydropyridazine, which can exist in different tautomeric forms, the most stable forms usually being the 1,4- and 4,5-dihydropyridazines [271]. The further reduction is complicated by the fact that the different tautomeric forms may be reduced differently and at different potentials and that the rate of transformation of a given tautomer to the most easily reducible one may or may not be fast compared to the further reduction.

The product isolated [271] from the two-electron reduction of 1-methyl-3,6-diphenylpyridazinium iodide in acid solution after extraction from neutral solution was 1,4-dihydro-1-methyl-3,6-diphenylpyridazine, but as it was found that 1,4-dihydro-3,6-diphenylpyridazine in trifluoroacetic acid was transformed to the 4,5-dihydro derivative [126], this may be the predominant tautomer in the acid solution.

The reduction in acid solution of di- and tetrahydropyridazines and pyridazinones is often best understood when their resemblance to hydrazones is considered. The reduction of such compounds has been discussed (Chap. 11), and the primary step is a hydrogenolysis of the nitrogen-nitrogen bond. The reduction of 1,4-dihydro-1-methyl-5-t-butyl-3,6-diphenylpyridazine to pyrroles has been described [125,126]; at more negative potentials pyrrolines and pyrrolidines may be obtained from pyridazines.

3,6-Diphenyl-2,3,4,5-tetrahydropyridazine (LXIII) is structurally similar to a benzylhydrazone of propiophenone, and accordingly, a four-electron polarographic wave is observed in acid solution with formation of LXIV [271]. The electrode reaction is as follows, Eq. (34):

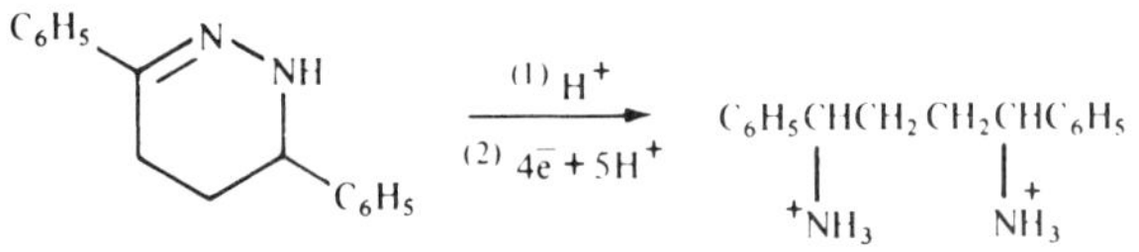

(34)

(LXIII) (LXIV)

Reduction in acetate buffer at the potential of the first wave of 3-phenyl-6-dimethylaminopyridazine yields the 4,5-dihydro derivative in high yield [271].

In a similar way other pyridazines [271], such as 3-phenyl-6-methoxypyridazine and 3-methyl-6-chloropyridazine, are reduced, but the dihydropyridazines are in these cases rather unstable and may lose methanol or hydrogen chloride, forming the corresponding 4,5-dihydropyridazinone, which being an acylated hydrazone is cleaved with ring opening and reduction or hydrolysis of the ketimine (Chap. 11).

In alkaline solution 3,6-diphenylpyridazine gives a two-electron polarographic wave that in strongly alkaline solution is followed closely by a smaller wave. A preparative reduction at the potential of the second wave yields 3,6-diphenyl-2,3,4,5-tetrahydropyridazine. It has been suggested [126] that the primary reduction product is a 1,2-dihydropyridazine that tautomerizes to the 4,5-dihydro derivative; this compound is then reduced further, analogously to the reduction of benzalazine [123]. In DMF 3,6-diphenylpyridazine may be reductively acetylated [273] or carboxylated [231].

Cinnolines

Cinnolines are generally reducible [79] in two or more steps. The first step in the reduction of aryl- or alkyl-substituted cinnolines is the formation of 1,4-dihydrocinnolines. This reduction is not reversible. Only in the case of benzo[c]cinnoline is the reduction a saturation of the nitrogen-nitrogen double bond; the reaction resembles the reduction of azobenzene to hydrazobenzene and, like this system, is nearly polarographically reversible.

1,4-Dihydrocinnolines may be regarded as cyclic phenylhydrazones, and their reduction is in accordance with the concept that only the protonated compounds are reducible, and the first two-electron step produces an amine and an imine. If the imine is reducible at the potential necessary for the hydrogenolysis of the nitrogen-nitrogen bond (Chap. 24), the imine is reduced to an amine; if not, a ring closure to an indole takes place, unless the potential is kept so negative that the further reduction of the imine can compete with the ring closure. The ring contraction to indoles has been discussed. The reduction of 1,4-dihydro-3-phenylcinnoline (LXV) proceeds by a four-electron reaction, Eq. (35):

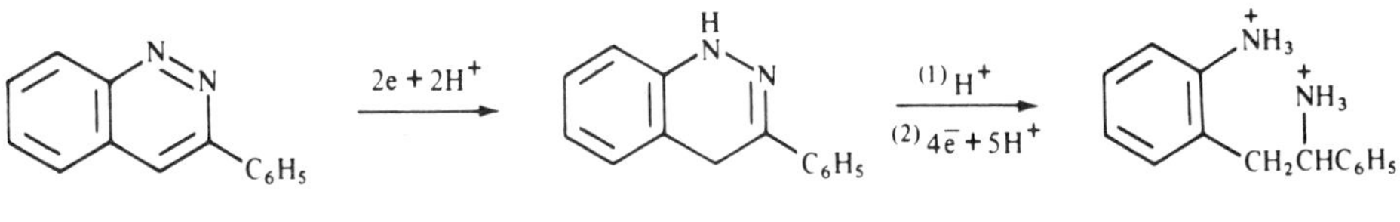

(LXV)

(35)

Derivatives of 1,2-dihydrocinnoline may be prepared by reductive acylation of cinnoline in DMF [273].

Phthalazines

These compounds are reduced in acid solution somewhat differently from cinnolines. Phthalazine is reduced polarographically in acid solution in a six-electron wave [274] presumably to o-xylene-α,α'-diamine, but 6 N hydrochloric acid [127] is required as medium if the diamine is the desired product. 1-Methylphthalazine is reduced as follows, Eq. (36):

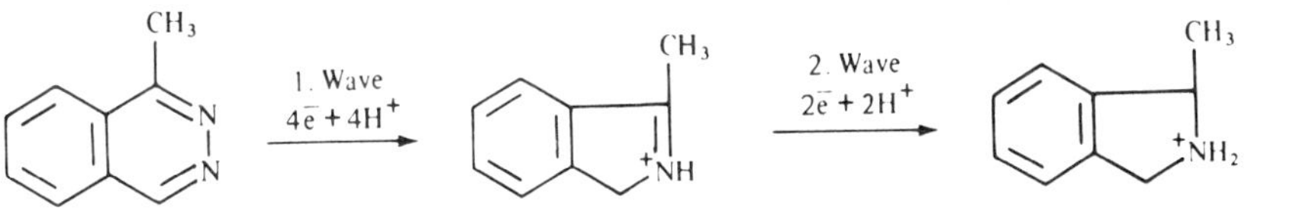

(36)

(LXVI)

By reducing at −0.65 V (SCE) it is possible to prepare 1-methylisoindole (LXVI).

4-Dimethylamino-1-phenylphthalazine can in acid solution be reduced in a four-electron reaction to 2-(1'-amino-1'-phenylmethyl)-N,N-dimethylbenzamidine [275], which at higher pH forms 1-phenyl-3-iminoisoindoline, which can be reduced to 1-phenylisoindoline (LXVII) in a four-electron reduction. From the reduction corresponding to the first wave of 4-methoxy-1-phenylphthalazine, the cyclic imidic ester 3-methoxy-1-phenylisoindole can be isolated; it may be further reduced to LXVII.

In alkaline solution phthalazine is reduced polarographically in a two-electron reduction, but in a preparative reduction it is possible to isolate mostly dimerized products, resulting from a one-electron reduction, or 1,2-dihydrophthalazine (LXVIII), depending on the conditions; LXVIII is reducible at a more negative potential to 1,2,3,4-tetrahydrophthalazine, which may be oxidized anodically to LXVIII. Thus the electrolytic method is advantageous for the preparation of LXVIII [127,276]. The dimerized products from the reduction of 2-methylphthalazinium iodide have been shown to have coupled at carbon rather than at nitrogen [127].

The reduction course of substituted phthalazines depends on the substituents; thus 1-methylphthalazine is reduced in alkaline solution to 1-methyl-3,4-dihydrophthalazine, whereas 1-phenyl-4-dimethylaminophthalazine is reduced to 1,2-dihydro-1-phenyl-4-dimethylaminophthalazine.

Pyrimidines

In acid solution pyrimidine [272–281] is reduced polarographically in two, one-electron waves; two, two-electron waves are observed in neutral medium, whereas a four-electron reduction occurs in alkaline solution. These findings have been interpreted in the following way, Eq. (37):

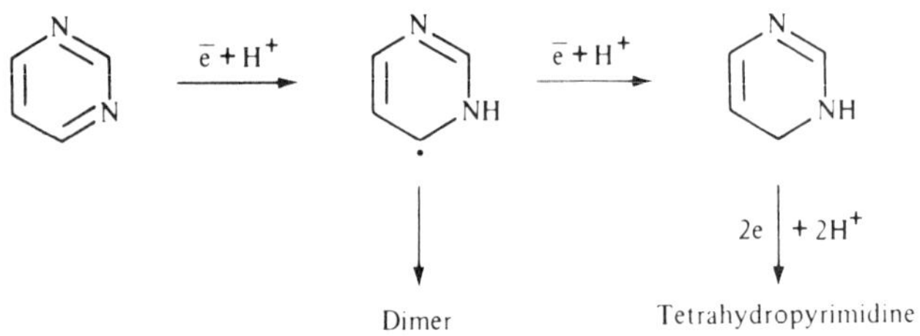

(37)

The dimer from 4-methylpyrimidine is the 6,6'-dimer [134].

The structure of the tetrahydropyrimidine has not been proved. 2-Aminopyrimidine is reduced similarly, but the dihydropyrimidine is not further reducible.

2-Phenylpyrimidines are reduced differently; they are reduced in acid solution in a four-electron reaction to 2-phenylpyrroles and ammonia [134].

Quinazolines

The polarographic behavior of quinazoline (LXIX) [282,283] is complicated by the hydration of the quinazolinium ion. The hydrated cation is not reducible, and in the acid region the wave height is partly determined by the rate of its dehydration; this has been used to find the rate of its dehydration at different pH values [282].

The reduction of LXIX resembles that of pyrimidine. In alkaline solution it is reduced stepwise to the 1,2,3,4-tetrahydroquinazoline. 4-Hydroxyquinazoline [271,284,285] is reducible in weakly alkaline and neutral solution. The reduction is analogous to the reduction of dihydroquinazoline to tetrahydroquinazoline. At higher concentrations a product dimerized at C—2 was also isolated [284–286].

Pyrazines, Quinoxalines, and Phenazines

Pyrazine and its reduction product 1,4-dihydropyrazine form a system that at a mercury electrode behaves nearly reversibly [287,288]. In acid solution two, one-electron waves are found, indicating a certain stability of the semiquinone-type one-electron reduction product. In substituted pyrazines the initially formed reduction product, the 1,4-dihydropyrazine, may tautomerize into the less easily oxidizable 1,2- or 1,6-dihydropyrazine; the rate of tautomerization is pH dependent [289].

The electrochemical reduction of 2,3-disubstituted 5,6-dihydropyrazines yields the 1,4,5,6-tetrahydropyrazines [290]. Thus, the diimine-enediamine system is analogous to the dione-enedion system. It is of interest that the 5,6-dihydropyrazines are somewhat more easily reducible than the pyrazines.

Quinoxaline (LXX) [290–296] is reduced to 1,4-dihydroquinoxaline, which may be reversibly reoxidized to LXX. 2-Methylquinoxaline is reduced to 3,4-dihydro-2-methylquinoxaline [293], whereas 2-phenylquinoxaline forms 1,2-dihydro-2-phenylquinoxaline. The 1,4-dihydroquinoxalines may be formed primarily but then tautomerize to the more stable dihydroquinoxaline. 2,3-Diphenylquinoxaline is reduced to cis-1,2,3,4-tetrahydro-2,3-diphenylquinoxaline; the 1,4-dihydro compound is transformed to the 1,2-dihydro derivative, which is further reducible [293]. 1,2-Dihydro-2,3-dimethylquinoxaline is reduced in acid solution with ring opening to an imine of 2-butanone, which is hydrolyzed to o-phenylenediamine and 2-butanone [295].

In the two-step reduction of N,N'-dialkylpyrazinium and quinoxalinium salts, the first one-electron reduction leads to a cation radical, which is the most persistent oxidation state [297].

Reduction of quinoxaline in DMF in the presence of acetic anhydride [273] produces 1,4-diacetyl-1,4-dihydroquinoxaline; reoxidation of this compound in MeCN at a Pt electrode regenerates quinoxaline with formation of acetylium ions [298]. Similarly, phenazine may be reductively alkylated to 5,10-dialkyl-5,10-dihydrophenazine and pyrazine acylated to a 1,4-dihydro derivative [299].

5,10-Dihydro-5,10-dimethylphenazine (LXXI) exhibits two successive reversible one-electron oxidation steps in acetonitrile. When LXXI is oxidized in a more nucleophilic solvent or in the presence of a nucleophile, the initially formed dication acts as a methylating agent toward the nucleophile [300] with formation of 5-methylphenazinium cation.

Naphthyridines

1,5-Naphthyridine is polarographically reducible at all pH values [301–303]. Macroscale reductions consume approximately one electron per molecule and point to a dimerization. The structure of the reduction products has not been established. The diquaternary salt of 1,5-naphthyridine is polarographically reducible [301]; 1,6-. 1,7-, 1,8-, 2,6-, and 2,7-naphthyridine behave similarly [302,303].

Diazepines and Pyridopyrimidines

Electrochemical reduction in DMF-TPAP of 6-phenyl-2,3-dihydrodiazepinium salt (LXXII) gives a pyrrolodiazepine (LXXIII) with consumption of 1 F/mol. A plausible explanation of the formation of LXXIII involves dimerization of an initially formed radical followed by intramolecular displacement of ethylenediamine [304].

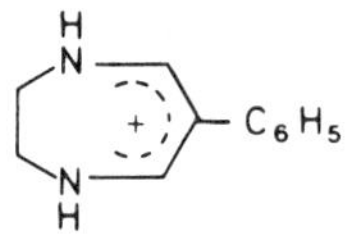

(LXXII)

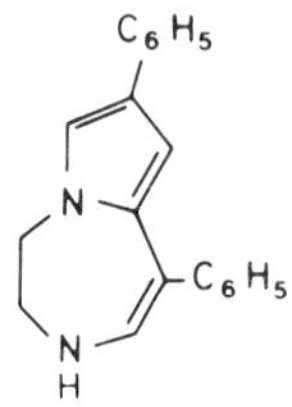

(LXXIII)

Both 1,4- [305–307] and 1,5-benzodiazepines [308] are polarographically reducible; the latter is somewhat unstable in aqueous solution, which complicates the investigations. 1,4-Benzodiazepines like 7-chloro-2-methylamino-5-phenyl-3H-1,4-diazepine-4-oxide are reducible in acid solution in three steps. The first two steps, the reduction of the N-oxide and the saturation of the benzophenone imine, are straightforward. The mechanism of the third step, a reductive ring contraction to 6-chloro-2-methyl-4-phenyl-3,4-dihydroquinazoline, is more controversial [43,133].

In acetonitrile some 1,5-benzodiazepine derivatives, prepared by condensation of o-phenylenediamine with 4,6-dimethylbicyclo[3.3.1]nona-3,6-diene-2,8-dione, have been reductively acetylated to substituted barbaralanes [309].

4-Alkyl-1-aryl-5H-2,3-benzodiazepines (LXXIV) are reduced in acidic solution in a four-electron reduction [130,310]; in dilute hydrochloric acid 1-aryl-3-alkylisoquinoline is formed [130]. In DMF, LXXIV is reduced with 2 F/mol to its 1,2-dihydro derivative.

Pyrido[1,2-a]pyrimidines are reduced in acid solution with 1 F/mol to a product dimerized at C—8 and in alkaline medium in a two electron reaction to an 8,9-dihydro derivative [311].

H. Compounds with One Oxygen and Two Nitrogen Atoms

Oxadiazoles

Aryloxadiazoles [241,312] are reducible at rather negative potentials; both four- and six-electron reactions are found, sometimes as two-step reductions. The four-electron reduction of symmetrical diaryloxadiazoles has been suggested to produce the oxadiazolidines.

Benzofurazan

This compound is polarographically reducible in the whole pH region in a six-electron reaction [313], and the product is supposedly o-phenylenediamine. At a macroelectrode some condensation reactions may be expected to occur in acid solution as side reactions just as in the case of o-dinitrobenzene reduction [314].

Sydnones

3-Phenylsydnones are polarographically reducible both in acid and in alkaline medium. In the former case six electrons are consumed per molecule, whereas at high pH, n = 4 [315]. In acid solution an α-amino acid is formed, whereas an α-hydrazino acid is the product in alkaline solution.

I. Compounds with One Sulfur and Two Nitrogen Atoms

1,3,4-Thiadiazoles

Thiadiazoles are often polarographically reducible [126,316]; not many of them have been investigated by controlled potential electrolysis. The wave height of 2,5-diphenyl-1,3,4-thiadiazole (LXXV) corresponds in acid solution to a two-electron reduction to a dihydro derivative, which is probably hydrolyzed to benzaldehyde thiobenzoylhydrazone. This compound is reducible at the reduction potential of LXXV and will thus be reduced further similarly to other acylated hydrazones (Chap. 11).

Benzo-2,1,3-thiadiazoles

These compounds and the corresponding selenodiazoles are reduced [317, 318] like the furazanes in a six-electron reduction; o-phenylenediamine and hydrogen sulfide and hydrogen selenide, respectively, are formed.

J. Compounds with Three Nitrogen Atoms

Benzotriazoles

Benzotriazoles (LXXVI) are reducible in acid solution, but only 2-substituted derivatives have been found to be reducible in alkaline solution. In acid solution the following scheme has been proposed, Eq. (38) [94]:

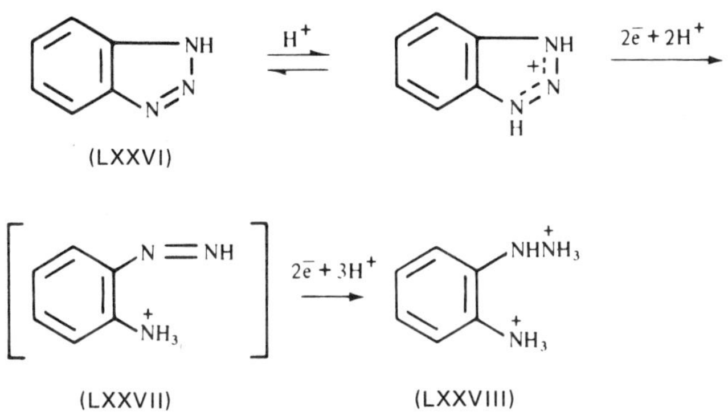

(38)

The o-aminophenyldiimide (LXXVII) cannot be isolated as it is more easily reducible than the benzotriazole, but a small concentration of it can be detected polarographically during the electrolytic reduction. o-Aminophenylhydrazine (LXXVIII) has three nucleophilic centers, and it may be a useful intermediate in the preparation of some heterocyclic compounds [94,319].

In alkaline solution 2-methylbenzotriazole is reduced to the 1,3-dihydro derivative, which may be reoxidized anodically to the starting material [94]. In acid solution the dihydrobenzotriazole forms a phenyldiimide; this compound does not cyclize in alkaline solution.

In DMF, 1-phenylbenzotriazole is reduced to a 1:1 mixture of carbazole and diphenylamine; in the presence of acetic anhydride benzotriazole is reduced in DMF to 1,2,3-triacetyldihydrobenzotriazole [320].

Triazines and Triazanaphthalenes

1,2,4-Triazines are in the first step reduced to dihydro derivatives; thus 3-(2-pyridyl)-5,6-diphenyl-1,2,4-triazine is in DMF in the presence of EtBr reduced to a mixture of diethylated products of 1,2- and 2,5-dihydro-1,2,4-triazine [321]. On reduction in MeCN, 1,3,5-triazine dimerizes to 4,4'-dihydrobitriazyl; this is hydrolyzed to tetraformamidoethane [322]. Benzo-1,2,4-triazines and dihydrobenzotriazines form a redox system that polarographically behaves nearly reversibly [59,126]. The further reduction of the dihydrobenzotriazines was discussed in Sec. IIB; 3-alkylsubstituted dihydrobenzotriazines are generally not further polarographically reducible, but the unsubstituted and 3-phenyl-substituted derivatives are.

Pyrido[2,3-b]pyrazines [323] form on reduction in aprotic medium primarily a 1,4-dihydro derivative, which may tautomerize into the 1,2- or 5,8-dihydro derivative; in the presence of acetic anhydride a 5-acetyl-5,8-dihydro derivative is isolated.

Other triazanaphthalenes are expected to be reducible, and the positions of the nitrogen atoms are the most critical factors for influencing the reduction potential.

K. Compounds with Four Nitrogen Atoms

The more nitrogen atoms inserted in an aromatic nucleus, the more easily reducible the system becomes. When a heteroaromatic system has two (or more) nitrogen atoms in a six-membered ring, the compound is reducible, at least in acid solution, and if two nitrogen atoms occupy para positions in such a ring, the redox system approaches reversibility.

Tetrazolium Salts

Tetrazolium salts are generally reduced in a four-electron reduction [324]; the first step is probably a cleavage of the nitrogen-nitrogen bond, which is followed by the reduction of a nitrogen-nitrogen double bond. Depending on the substituents the four-electron product may be further reduced. In alkaline solution formazanes may be formed.

Purine and Pteridine

The electrolytic reduction of purine, which has been reviewed [278,279,325], takes place only in acid solution and thus differs from that of pyrimidine and quinazoline. The reduction proceeds in acid solution in two, two-electron steps, the first resulting in a 3,4-dihydro derivative, whereas the tetrahydrocompound formed in the second step is hydrolyzed to formaldehyde and an imidazole derivative.

Oxidation of uric acid (LXXIX) and its N-methylated derivatives at a graphite anode produces a quinonoid product (LXXX), which adds water or another nucleophile across a C=N double bond to an unstable intermediate (LXXXI), which reacts to products. From the oxidation of uric acid can be isolated, depending on pH, allantoin, 5-hydroxyhydantoin, alloxan, and urea, Eq. (39) [326,327]:

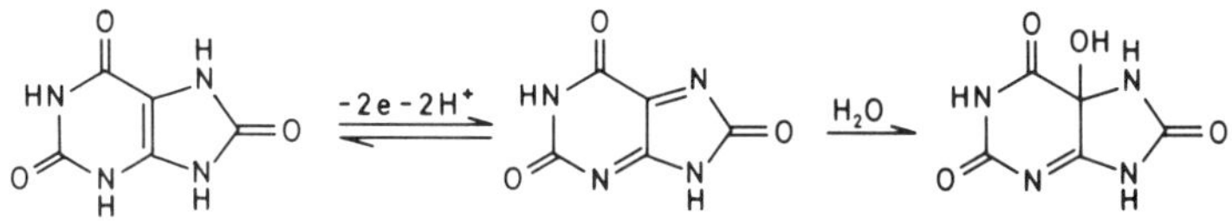

(LXXIX) (LXXX) (LXXXI)

(39)

Pteridines behave electrochemically akin to quinoxaline [328], the pyrazine ring being reduced; the 3,4-double bond may be hydrated akin to quinazoline [279,328]. In the first, reversible step of 6-methyl-2-amino-4(3H)-pteridone, (LXXXII; R = H), a 5,8-dihydro derivative (LXXXIII) is formed, which in a pH-dependent reaction tautomerizes into the more stable 7,8-dihydro compound (LXXXIV); LXXXIV is reduced in acidic and neutral solution to the 5,6,7,8-tetrahydro derivative (LXXXV), which may be oxidized to a quinonoid dihydro compound (LXXXVI); LXXXIV is formed on tautomerization of LXXXVI, Eq. (40) [329,330]:

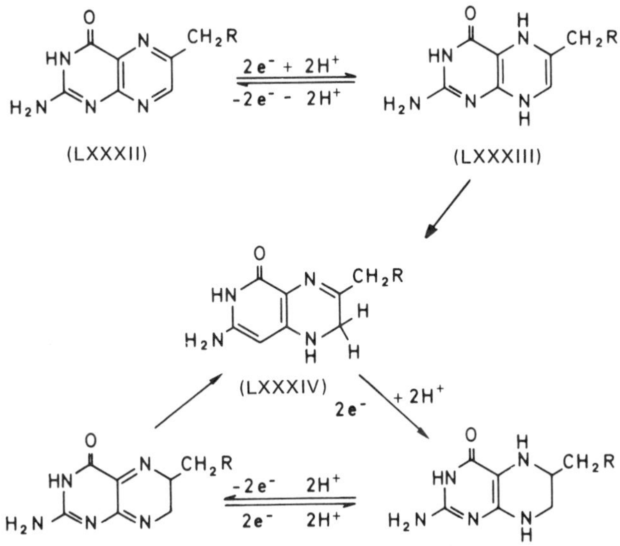

(40)

Pteridone redox couples have been used as mediators in electron transfer reactions to large biologic molecules [331].

Folic acid [LXXXII; R = $NHC_6H_4CONHCH(COOH)(CH_2)_2COOH$] is in the first step reduced as LXXXII (R = H); in acidic solution 7,8-dihydrofolic acid is reduced with cleavage of the C–9/N–10 bond to LXXXIV (R = H) [329]. In neutral medium with low buffer capacity it is possible to reduce 7,8-dihydrofolic acid to the 5,6,7,8-tetrahydrofolic acid [332].

Pyrazinopyrazines

Pyrazino[2,3-b]pyrazines are reduced in aqueous alcoholic solution to the 1,4-(5,8)-dihydro derivatives [323]; in acidic medium the two-electron wave is followed by a four-electron reduction to a hexahydro derivative. The benzo and dibenzo derivatives are similarly reduced primarily in a reversible two-electron reduction [333]; the further reduction of 5,12-dihydroquinoxalino[2,3-b]quinoxaline yields 2-(o-aminoanilinomethyl)benzimidazole [334].

Porphines

The polarographic and controlled potential reduction of porphines (LXXXVII) has been investigated, and it was shown that porphine was reduced polarographically in two, two-electron steps [335]. Reduction of meso-porphyrinedimethyl ester yielded a product that was easily reoxidized by oxygen. From the UV spectrum of the reduced solution, which was reduced in deuterium oxide, it could be shown that the reduction yielded a phlorine (LXXXVIII) rather than a chlorine (LXXXIX). Reduction of a chlorine yielded a "chlorine/phlorine."

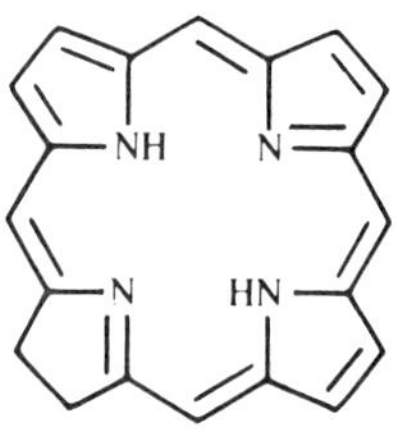

(LXXXVII)

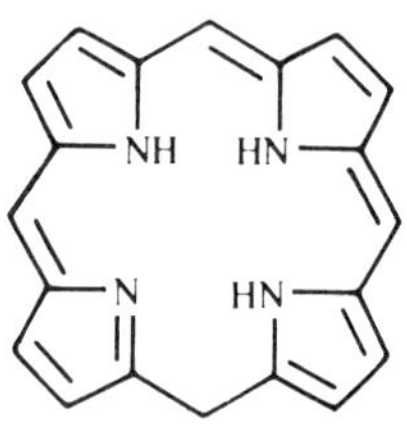

(LXXXVIII)

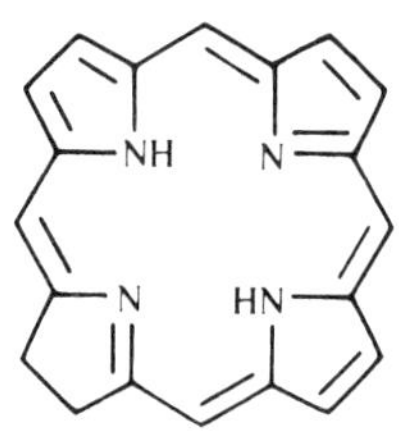

(LXXXIX)

In dimethylformamide porphines are reduced in three or four, one-electron steps, which results in the formation of mono-, di-, tri-, or tetranegative ions [336].

Anodic oxidation of porphine in alkylnitriles at a platinum electrode takes place in two, one-electron steps [337]. The loss of electrons occurs from the lone pairs on the nitrogen atoms.

Electrochemical oxidation in DMF of Zn-octaethylporphyrin in the presence of cyanide at 0.45 V (SCE) gives a quantitative yield of cyanooctaethylporphyrin; at more positive potentials di-, tri-, and tetracyanooctaethylporphyrin may be obtained [338]. Metal complexes of N-containing macrocycles are discussed in Chap. 21.

IV. ELECTROLYTIC REACTIONS OF SUBSTITUTED HETEROCYCLIC COMPOUNDS

This section treats the electrolysis of substituted heterocyclic compounds if the electrode reaction involves the substituent directly and if the electrode reaction in some way is different from what would be expected from the carbocyclic series.

A. Alkyl and Alkenyl Substituents

Alkyl derivatives of π-electron-deficient heterocyclic compounds may be oxidized anodically to carboxylic acids; alkyl and arylalkyl pyridines can be oxidized to the corresponding acids in good yields at a lead dioxide anode in aqueous sulfuric acid [200,201,339–341].

Methylpyridines and methylpyridine-N-oxides react with superoxide ions, generated electrochemically in DMF, to the carboxylic acids; the N-oxides react better than the parent compounds [342].

Hydrodimerization of activated alkenes is a well-established process. π-Electron-deficient heteroaromatic compounds activate a double bond similarly to a cyano or carbethoxy group, and in accordance with that analogy vinylpyridines can be hydrodimerized. 4-Vinylpyridine [343] forms 1,4-bis-(4-pyridyl)butane in 82% yield on electrolysis in a mildly alkaline solution containing methyltriethylammonium p-toluenesulfonate and some DMF. The mechanism is discussed in Chap. 22.

If the nucleus contains more than one nitrogen atom, the nucleus is reduced at a less negative potential than the double bond. No reports of electrolysis of such compounds have been published yet.

Ethynylpyridines are reducible polarographically [344] in a medium containing tetraethylammonium iodide, 2-ethynylpyridine at −1.72 V, and 2-methyl-5-ethynylpyridine at −2.06 V (SCE). The reduction potentials of the ethynylpyridines are very close to those of the corresponding vinyl compounds [345]. Similar results have been found for benzene derivatives [346]. The electrode reaction is thus expected to be a four-electron reduction of the ethynyl group to an ethyl group.

B. Hydroxyl Derivatives

Heterocyclic hydroxyl derivatives can be divided into two groups according to whether the hydroxyl group is exocyclic. Heterocyclic compounds having a hydroxyl substituent at the ring have the possibility of existing in different tautomeric forms [347], the hydroxy or the oxo form, and in many cases the oxo form is predominant. Nevertheless, regardless of the position of the prototropic equilibria, the compounds are treated here as "hydroxyl compounds." Phenolic compounds, for example the oxidative dimerization of 1,2,3,4-tetrahydroisoquinoline phenols [348], are not discussed.

Compounds with Hydroxyl Groups Attached to a Substituent

Hydroxyl groups are reducible when they are activated by an electron-attracting group, such as the phenacyl group [349,350]. A pyridine ring activates its substituents, but largely in the 2- and 4-positions. This is illustrated by the reduction of pyridoxol (XC), which is reducible in an ammoniacal buffer (pH 8.5–10) according to Eq. (41) [351]:

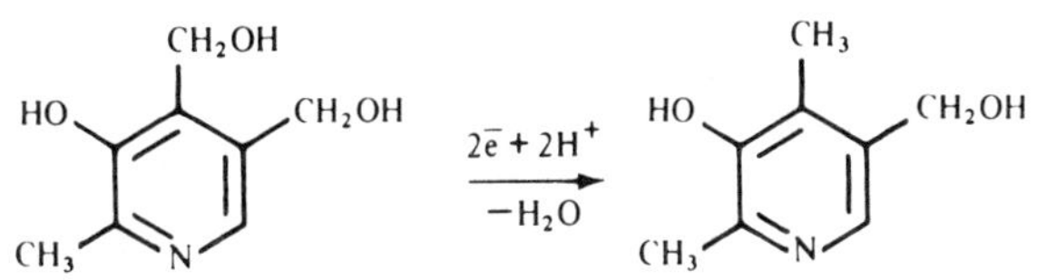

(XC)

(41)

Indirect reductive removal of hydroxyl groups in furan derivatives in hydrochloric acid containing iodide [352] is discussed in Chap. 29.

The reductive loss of the acetoxy function from the C—3 substituent in cephalosporin C derivatives [353,354] is another example of an electrochemical removal of an activated hydroxyl function, which seems rather general [355].

Compounds with Hydroxyl Groups Attached to the Ring

The reduction of such compounds generally follows one of the following routes: (1) Reduction takes place in the nucleus and may involve ring cleavage. This type is exemplified by the reduction of pyridazinones to 4,5-dihydropyridazinones followed by ring opening (Chap. 24) and is found for compounds that may be regarded as acid derivatives, such as cyclic esters, hydrazides, amides, and eneamides. (2) The carbonyl group of the oxo form is reduced. This is often followed by loss of the hydroxyl group thus formed and further reduction of the double bond. This type may be found when the oxo group is β to a heteroatom and/or α to a phenyl group and in some dihydroxy compounds. An example of this kind is the reduction of N-methylsuccinimide to N-methylpyrrolidone [356]. In maleimide derivatives the double bond is reduced prior to the carbonyl groups [357,358]. The use of pyrrolidone [359] as probase for EGB is discussed in Chap. 30.

Derivatives of Isoindole Phthalimide in acid solution is reduced polarographically [360,361] by a two-electron reduction, whereas two, one-electron waves are found in slightly alkaline medium. The two-electron reduction product [362] hydroxyphthalimidine (XCI), a derivative of phthalaldehydic acid, reacts with any nucleophile HY present. In ethanolic solution ethoxyphthalimidine is formed, and in the presence of isoindoline, 1-N-isoindolino-3-oxoisoindoline [362].

In strongly acid solution, XCI dehydrates to the easily reducible 3-oxoisoindole, which is reduced to phthalimidine; a further reduction to isoindoline may take place under conditions similar to those used for the reduction of benzamides [363–365].

Pyrimidones Pyrimidone-2 is reduced in a single, linearly pH-dependent one-electron wave; the radical formed dimerizes at position 6 to 6,6'-bis-(3,6-dihydropyrimidone-2) [366]. In nonaqueous solution the anion radical abstracts a proton from the parent compound and the radical dimerizes; the parent anion may be oxidized at the mercury electrode with formation of a mercury salt [367].

Derivatives of Cinnoline 3-Hydroxycinnoline [79] is reduced in a two-electron wave from pH 0 to 11. In strongly acid solution the main product is 1-aminooxindole, which is also obtained on reduction with zinc and sulfuric acid [368]. Reduction of an emulsion in aqueous phosphate buffer (pH 6.5) and n-butanol produced a nearly quantitative yield of 3-keto-1,2,3,4-tetrahydrocinnoline; this compound was easily reoxidized to 3-hydroxycinnoline [79]. By choosing a pH for the reduction at which the product is stable the preparation of this labile compound could be accomplished.

4-Hydroxycinnoline (XCII) [79] in acid solution is reducible in two, two-electron steps. The product from the first two-electron reduction is 4-keto-1,2,3,4-tetrahydrocinnoline (XCIII). This compound is a derivative of phenylhydrazine and can thus be oxidized anodically in alkaline solution to an azo compound (XCIV), which then tautomerizes to XCII. The carbonyl group of XCIII is activated by the phenyl ring and—in contrast to 3-keto-1,2,3,4-tetrahydrocinnoline—is thus further reducible; the reduction yields 4-hydroxy-1,2,3,4-tetrahydrocinnoline (XCV). On standing, this compound loses water with the formation of 1,4-dihydrocinnoline. If, however, the compound is oxidized in alkaline solution, the product (XCVI) loses water and forms cinnoline.

As 4-hydroxycinnoline is one of the more easily accessible derivatives of cinnoline, this reaction may be the method of choice for the preparation of cinnoline [79]. The reactions are formulated in Eq. (42):

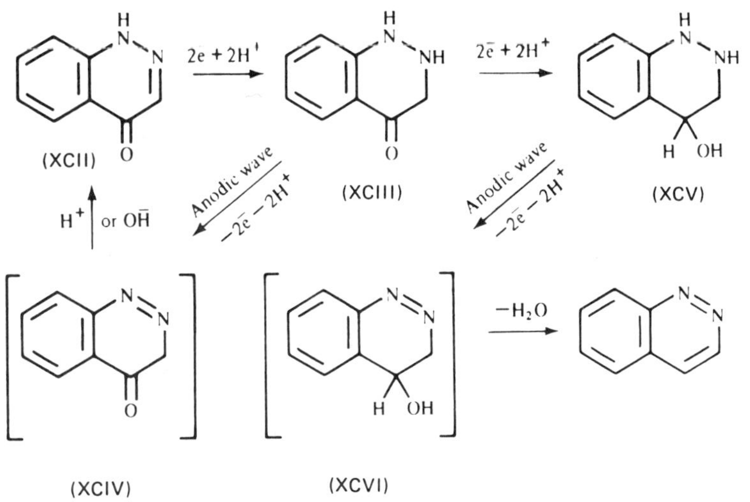

(42)

Derivatives of Phthalazine Dihydroxyphthalazines [132] like 2,3-dihydro-2,3-dimethyl-1,4-phthalazinedione are at low pH reduced to a phthalimidine in a six-electron reduction. In alkaline solution 4-hydroxy-3,4-dihydro-2,3-dimethyl-1-phthalazinone is formed in a two-electron reaction; in this medium it is not further reducible, but in acid solution it loses water to the reducible phthalazinonium compound. The product from this reaction is 3,4-dihydro-2,3-dimethyl-1-phthalazinone, which at low pH can be reduced to N-methylphthalimidine and its precursor N-methyl-2-(methylaminomethyl)benzamide.

Derivatives of Quinazoline 4-Methoxyquinazoline is in acid solution reduced to the 3,4-dihydro derivative, which eliminates methanol; the quinazoline thus formed is reduced to 3,4-dihydroquinazoline. In alkaline solution there is formed, besides quinazoline, some biquinazolyl in a reaction analogous to the benzoin condensation [369].

Derivatives of Quinoxaline Both 2-hydroxy- and 2,3-dihydroxyquinoxaline are polarographically reducible, the former by a two-electron and the latter in acid solution by a four-electron reduction [370,371]. The former is more easily reducible than the latter, which is understandable since 2-hydroxyquinoxaline (XCVIII) may be regarded as a derivative of glyoxalic acid, whereas 2,3-dihydroxyquinoxaline (XCVII) is a derivative of oxalic acid. Derivatives of aldehydes are easier to reduce than the corresponding acid derivatives, Eq. (43):

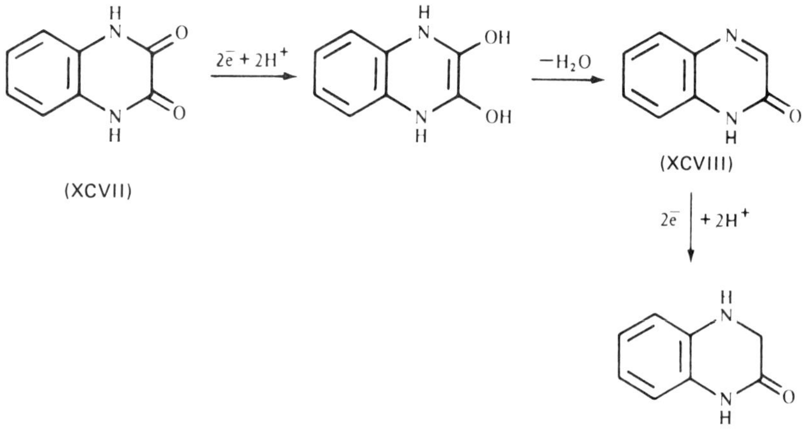

(XCVII) (XCVIII) (43)

Derivatives of Triazine The electrochemical reduction of *as*-triazin-3-ones, such as 5,6-diphenyl-*as*-triazin-3-one [372], produces a 1,4-dihydro-derivative, which tautomerizes into a 4,5-dihydro compound, a kind of cyclic semicarbazone. This compound is in acid solution (pH 1) reduced in a two-electron reaction to a mixture of the ring-contracted product 4,5-diphenylimidazol-2-one and the *trans*-1,4,5,6-tetrahydro-5,6-diphenyl-*as*-triazin-3-one; in neutral solution the *cis* derivative is obtained [372].

N-*Oxides and* N-*Hydroxy Compounds*

N-Oxides of heteroaromatic compounds are generally electrolytically reducible in acid solution [373–386]. The most studied compounds of this type are the pyridine-*N*-oxides [373,380,386]. At low pH pyridine-*N*-oxide gives a two-electron polarographic wave, but at pH > 4 the wave height diminishes, as the preprotonation step becomes too slow, Eq. (44):

$$C_5H_5\overset{+}{N}{-}O^- \underset{}{\overset{+H^+}{\rightleftharpoons}} C_5H_5\overset{+}{N}{-}OH \xrightarrow{2e^- + 2H^+} C_5H_5\overset{+}{N} + H_2O \qquad (44)$$

The assumption that it is the O-protonated *N*-oxide that is reduced is substantiated by the similarity in reduction potential between pyridine-*N*-oxide and *N*-methoxypyridinium ion in acid solution; the latter is also reducible in alkaline solution [126].

A number of different *N*-oxides have been investigated polarographically, such as *N*-oxides of quinoline [383], acridine [375], pyrazine [385], benzimidazole [381], quinoxaline [374], phenazine [377], and adenine [379], and the influence of substituents was discussed.

When the heterocyclic ring contains more than one nitrogen atom, the initial reduction of the *N*-oxide may take place in the nucleus or the reduction of the nucleus and the *N*-oxide function occur simultaneously; the reduction of quinazoline-3-oxide [282], 3-methoxy-2,5-dimethylpyrazine-1-oxide [385], and cinnoline mono- and dioxide [382] is reported to take this route.

At pH < 1.5, however, *N*-oxides like cinnoline-*N*-oxide [79], 3,6-diphenylpyridazine-*N*-oxide [126], or pyrazine-*N*-oxide [385] give a two-electron polarographic wave followed by the reduction waves of the parent compounds.

Derivatives of 1-hydroxy-2,2,6,6-tetramethylpiperidine have been widely investigated electrochemically, mainly due to the interest in the nitroxyl radicals derived from them [387]. The application of *N*-hydroxyphthalimide [388] as an agent in indirect oxidations is discussed in Chap. 29.

S-*Oxides*

Both phenothiazine-*S*-oxide and *S*-dioxide are polarographically [389] reducible at rather negative values in a medium containing tetramethylammonium chloride. The electrode reaction is not known. In acid solution only the *S*-oxide is reducible; the reduction is probably preceded by a loss of water to the phenothiazonium salt.

C. Aldehydes, Ketones, and Their Derivatives

π-Electron-deficient heterocyclic rings facilitate the addition of nucleophilic reagents to a carbonyl group in conjugation with the ring; protonation of a ring nitrogen enhances this effect. Protonated or quaternized formylpyridines [390–395] are to a high degree hydrated in aqueous solution. This influences their polarographic behavior since the hydrated form, a *gem*-diol, is not reducible at the dropping mercury electrode. The wave height of such hydrated aldehydes is thus determined at a given pH by the equilibrium constant for the hydration and the rate of the dehydration reaction. Since only the protonated form is appreciably hydrated, the pKa of the compound is also of importance for interpreting the polarographic data. The dehydration reaction is both acid and base catalyzed, the latter being of importance for the quaternized aldehydes. The acid-catalyzed dehydration reaction has been shown to follow this path:

$$\begin{array}{ccccc} R\overset{+}{N}C_5H_4CH(OH)_2 & \underset{}{\overset{H^+}{\rightleftharpoons}} & \left[R\overset{+}{N}C_5H_4CH(OH)\overset{+}{O}H_2\right] & \underset{+H_2O}{\overset{-H_2O}{\rightleftharpoons}} & \left[R\overset{+}{N}C_5H_4\overset{+}{C}H{-}OH\right] \\ & & & & -H^+ \Big\downarrow\!\Big\uparrow +H^+ \\ & & & & R\overset{+}{N}C_5H_4CHO \end{array}$$

The hydration is most pronounced when the formyl group is in a position strongly activated toward nucleophilic attack; thus 2- and 4-formylpyridine are hydrated more than the 3-isomer, and 2-formylimidazole [396] and 2-formylthiazole [397] more than the 4- and 5-isomers. The quaternary derivatives are hydrated at least as strongly as the protonated parent compounds and in a broader pH region.

In strongly acid solution the diprotonated pyridinecarboxaldehydes are reduced in a two-electron reaction, whereas between pH 2 and 12, 2- and 4-formylpyridine are reduced in two steps; reduction at the potential of the first wave gives pinacols [395]. In alkaline solution pinacols and some polymeric material are formed. The stereochemistry of the reduction of acetopyridines is discussed in Chap. 27.

The electrochemistry of the azomethine derivatives, such as oximes, Schiff bases, and hydrazones, is similar to that of analogous carbocyclic compounds, Chap. 11, and is not discussed here.

D. Carboxylic Acids and Derivatives

Carboxylic acids and most of their derivatives are difficult to reduce, and a polarographic reduction of a carboxyl group requires activation by a strongly electron-attracting group. The addition of electrons may take place either to the carboxyl group or to the activating group, often depending on pH.

π-Electron deficient heterocyclic nuclei may activate a carboxyl group sufficiently to render it polarographically reducible [396,398–404]. An aldehyde is always more easily reducible than the corresponding carboxylic acid, so it is generally not possible to isolate aldehydes from a reduction of carboxylic acids. The electron-attracting groups that activate a carboxyl group toward reduction also activate the aldehyde toward nucleophilic addition reactions, and these aldehydes are in a certain pH region hydrated to a certain degree. It is thus possible in some cases to trap the easily reducible aldehyde formed during the reduction of a carboxylic acid as the nonreducible aldehyde hydrate. As some of the free, reducible aldehyde is always present in equilibrium with the hydrate, a reduction of an acid to an aldehyde must be performed at a suitable pH and at a low temperature, at which the rate of the dehydration reaction is low. The reduction must not be allowed to proceed to completion but must be stopped after the consumption of about 2 F/mol. Some dimerization and polymerization are usually observed, and in some cases these may be the dominant reactions.

The 4-pyridinecarboxylic acids [400,403], 2-carboxythiazoles [401] and 2-carboxyimidazoles [396,402], and their esters and amides [355,405] are reduced in this way. The corresponding hydrazides [406] and hydroxamic acids [43] are reduced to the amides in the first step and then further reduced. When the activating group in pyridine is in the 3-position the reduction may take place in the nucleus.

Whereas pyridine carboxamides are reduced to the carbinol at a lead cathode, in aqueous sulfuric acid the addition of titanium salts results in the aminomethylpyridine as the main product [407].

Anodic Reactions

Kolbe reactions of heterocyclic compounds have been studied in only a few cases. Anodic oxidation of 1-azabicyclo[2.2.2]octane-2-carboxylic acid under "Kolbe conditions" produced 2-methoxy-1-azabicyclo[2.2.2]octane [408]. The primary radical, formed by loss of an electron from the carboxylate ion, decarboxylates and is oxidized further to a carbonium ion, which is attacked by a methoxide ion. A similar pseudo-Kolbe reaction is found in the anodic decarboxylation of 1,2,3,4-tetrahydroisoquinoline-1-carboxylic acid derivatives to 3,4-dihydroisoquinolines [409].

If two vicinal carboxyl groups are present, the Kolbe electrolysis may result in the introduction of a double bond; this has been used in the transformation of 2-aza-3-oxo-bicyclo[2.2.2]oct-7-en-5,5-dicarboxylic acid to 2-aza-bicyclo[2.2.2]octa-5,7-dien-3-one [410].

Nitriles

The reduction of cyanopyridines exhibits special features. In acid solution the protonated species is reduced in a four-electron reaction [411–415] to the pyridylmethylamine, but in alkaline medium the carbon-carbon bond is cleaved with formation of pyridine and cyanide ion [405,416] (Chap. 24). 3-Cyanoquinoline is reduced to the 1,4-dihydro derivative [417] in slightly alkaline solution.

For the reduction of cyanopyridines to the aminomethylpyridines a lead cathode in dilute sulfuric acid containing $FeSO_4$ has been recommended [414].

E. Nitrogen-Containing Substituents

The reduction of heterocyclic C- and N-nitro and nitroso compounds exhibits generally no special features and is not discussed here; compare Chap. 9.

N-Aminopyridinium iodide in acid solution is reduced to pyridine; similar N-amino heteroaromatics are probably reduced in the same way [126, 418].

Activated amino groups as found in 4-pyridylmethylamines [419] or 2-pyridyltrimethylammonium iodide [126] are reductively removed in acid solution. The reduction of the latter requires protonation of the ring to obtain sufficient activation.

Elimination of an amino group is sometimes found as a chemical step between the transfer of electrons. The reduction of adenine (XCIX) illustrates this, Eq. (45) [420,421]:

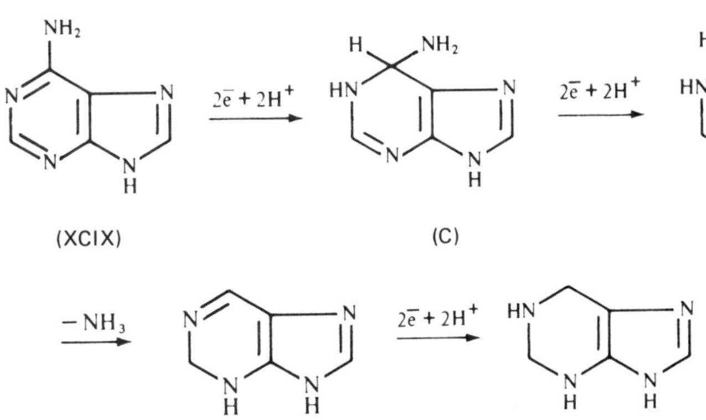

(45)

An elimination of ammonia from C to purine followed by reduction of the latter may also be possible [421]; such a reduction route is analogous to that of 4-aminoquinazoline in acid solution [369], in which quinazoline can be detected as an intermediate during the reduction.

Adenine (XCIX) gives a single well-defined voltammetric anodic wave at the pyrolytic graphite microelectrode [422]; coulometric data showed a six-electron oxidation. Exhaustive electrolysis at a fixed anode potential in aqueous 1 M acetic acid solution produced a mixture of products.

Electrochemical oxidation of 2-aminobenzothiazole in aqueous solutions at a pH between 3 and 11 consumes 2 F/mol, and 2,2'-azobisbenzothiazole is reported as the product [423].

F. Sulfur-Containing Substituents

Reduction of heterocyclic disulfides and anodic reactions of sulfhydryl compounds show no special features. Elimination of a sulfhydryl group may be the slow step in some reactions.

4-Mercaptocinnoline [79] gives a two-electron polarographic wave, but a preparative reduction consumed four electrons per molecule and produced 1,4-dihydrocinnoline. The reaction must thus involve a slow step after the first two-electron reduction, which is assumed to be loss of H_2S with formation of cinnoline; this is further reduced to the product.

A similar reaction takes place in 4-mercaptoquinazoline [369] and 2,6-dimercaptopurine [424].

2-Mercapto-4-hydroxythiazole (CI, rhodanone) is reducible in two, two-electron steps [425], the second step being found at rather negative potentials. The reaction takes place at the exocyclic sulfur and has been formulated as follows, Eq. (46):

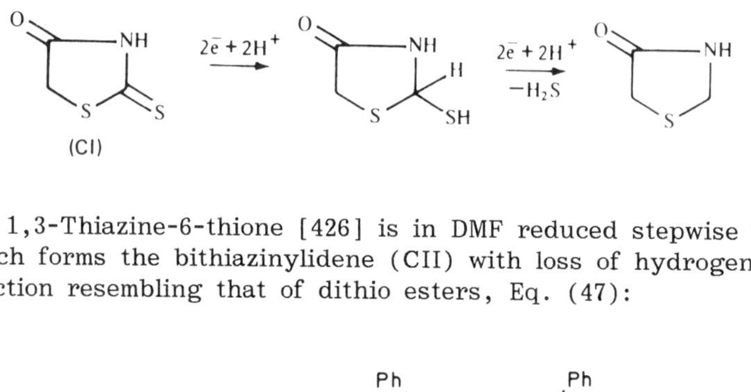

(CI)

(46)

1,3-Thiazine-6-thione [426] is in DMF reduced stepwise to the dianion, which forms the bithiazinylidene (CII) with loss of hydrogen sulfide in a reaction resembling that of dithio esters, Eq. (47):

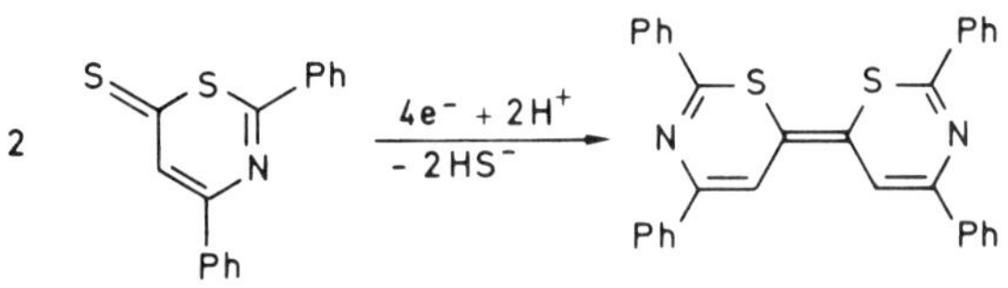

(CII)

(47)

In thiouracil and other 2-pyrimidinethiones the reduction takes place in the hetero ring rather than at the thione [427].

Cyclic voltammetric data at a pyrolytic graphite electrode of 2,6-dimercaptopurine show three peaks. The first and second correspond to a disulfide formation from the 6- and 2-mercapto groups, respectively, whereas the third peak is due to an oxidation to purine-2,6-disulfonic acid. Some disulfonic acid was obtained on preparative oxidation [428]. Oxidation of 6-thioguanine gives at low potential the disulfide, whereas oxidation at more positive potentials involves both the sulfur group and the purine ring [429].

G. Halogen Derivatives

The electrode reactions of heterocyclic halogen compounds exhibit no special features, but some may be of preparative value. Anodic halogenation may be advantageous in fluorination and iodination reactions and in halogenations in which a low, easily measured concentration of halogen is desirable.

Large-scale reductive removal of the halogens is sometimes of preparative value; thus it is convenient to remove chlorine from 4-chloroquinazoline [282], halogen from 2-halothiazoles [430,431] and iodine from 4-iodophthalazines [271] by reduction at a mercury cathode, and chlorine from 2-amino-4-chloropyrimidine at a spongy cadmium electrode [432].

Reduction of pentafluoropyridine in DMF at a mercury cathode gives perfluoro-4,4'-bipyridyl; in the presence of hydroquinone 2,3,5,6-tetrafluoropyridine is obtained. Pentachloropyridine gives very little bipyridyl derivative, but tetrachloropyridine and bis(tetrachloropyridyl)mercury in the absence of a proton donor [433].

In a mixture of THF and H_2O (1:1) pentachloropyridine is reduced at a silver cathode to 2,3,5,6-tetrachloropyridine and further to 2,3,5-trichloropyridine [434]. 3,4,5,6-Tetrachloro-2-picolinic acid may similarly be stepwise reduced [435] and chlorine reductively removed from 3-chlorobenzothiophene [436].

In DMF containing a few percent water, 4-chloro-6-(trichloromethyl)-2-phenylpyrimidine can be reduced nearly quantitatively to 4-chloro-6-(dichloromethyl)-2-phenylpyrimidine [134]. Dihalo-β-lactams may be reductively dehalogenated in DMF containing a proton donor [437].

REFERENCES

1. H. Lund, in Advances in Heterocyclic Chemistry (A. R. Katritzky and A. J. Boulton, eds.), Vol. 12, Academic, New York, 1970, p. 213; (b) H. Lund and I. Tabakovic, in Advances in Heterocyclic Chemistry (A. R. Katritzky and A. J. Boulton, eds.), Vol. 36, Academic, New York, 1984, p. 235; (c) J. E. Toomey, in Advances in Heterocyclic Chemistry (A. R. Katritzky and A. J. Boulton, eds.), Vol. 37, Academic, New York, 1984, p. 167; (d) J. Armand and J. Pinson, in Physical Methods in Heterocyclic Chemistry (R. R. Gupta, ed.), Vol. 7, Wiley, New York, 1984, p. 427; (e) H. Baumgärtel and K.-J. Retslav, in Encyclopedia of Electrochemistry of the Elements (A. J. Bard and H. Lund, eds.), Vol. 15, Dekker, New York, 1984, p. 168; H. P. Richter and P. Pfegel, Pharmazie, 40, 81 (1985).
2. A. J. Klein and D. H. Evans, J. Org. Chem., 42, 2560 (1977).

3. H. Lund and A. D. Thomsen, Acta Chem. Scand., 23, 3567 (1969).
4. F. Barba, M. D. Velasco, A. Guirado, and N. Moreno, Synth. Commun., 15, 939 (1985).
5. F. Barba, M. D. Velasco, and A. Guirado, Synthesis, 593 (1984).
6. F. Barba, M. D. Velasco, and A. Guirado, Synthesis, 625 (1981); Electrochim. Acta., 28, 259 (1983).
7. J. D. Anderson, and M. M. Baizer, Tetrahedron Lett., 511 (1966); J. D. Anderson, M. M. Baizer, and J. P. Petrovich, J. Org. Chem., 31, 3890 (1966); J. P. Petrovich, J. D. Anderson, and M. M. Baizer, J. Org. Chem., 31, 3897 (1966); M. M. Baizer, J. D. Anderson, J. H. Wagenknecht, M. R. Ort, and J. P. Petrovich, Electrochim. Acta, 12, 1377 (1967).
8. B. Vieth and W. Jugelt, Z. Chem., 21, 286 (1981); Z. Phys. Chem. (Leipzig), 265, 947 (1984).
9. B. Vieth and W. Jugelt, Z. Chem., 23, 63 (1963).
10. I. Carelli, A. Inesi, V. Carelli, M. A. Casadai, F. Liberatore, and F. M. Moracci, Synthesis, 591 (1986).
11. W. J. Begley, J. Grimshaw, and J. Trocha-Grimshaw, J. Chem. Soc., Perkin Trans. 1, 2633 (1974); J. Grimshaw and J. Trocha-Grimshaw, Tetrahedron Lett., 2601 (1975).
12. J. Grimshaw, R. J. Haslett, and J. Trocha-Grimshaw, J. Chem. Soc., Perkin Trans. 1, 2448 (1977); J. Chem. Soc., Perkin Trans. 1, 229 (1982).
13. J. Grimshaw and D. Mannus, J. Chem. Soc., Perkin Trans. 1, 2456 (1977).
14. R. Gottlieb and J. L. Neumeyer, J. Amer. Chem. Soc., 98, 7108 (1976); R. Gottlieb, J. L. Neumeyer, and E. Elmaleh, Amer. Chem. Soc. Mett., Apr. 1977, ORGN 111.
15. (a) T. Shono, K. Yoshida, K. Ando, Y. Usui, and H. Hamaguchi, Tetrahedron Lett., 4819 (1978); (b) T. Shono, Y. Usui, and H. Hamaguchi, Tetrahedron Lett., 21, 1351 (1980); (c) T. Shono, Y. Usui, T. Mizutani, and H. Hamaguchi, Tetrahedron Lett., 21, 3073 (1980); (d) T. Shono, T. Miyamoto, M. Mizukami, and H. Hamaguchi, Tetrahedron Lett., 22, 2385 (1981).
16. L. L. Miller, F. R. Stermitz, and J. R. Falck, J. Amer. Chem. Soc., 93, 5941 (1971); L. L. Miller, F. R. Stermitz, and J. R. Falck, J. Amer. Chem. Soc., 95, 2651 (1973); L. L. Miller, F. R. Stermitz, J. Y. Becker, and V. Ramachandran, J. Amer. Chem. Soc., 97, 2922 (1975); J. R. Falck, L. L. Miller, and F. R. Stermitz, Tetrahedron, 30, 931 (1974); L. L. Miller, R. F. Stewart, J. P. Gillespie, V. Ramachandran, Y. H. So, and F. R. Stermitz, J. Org. Chem., 43, 1580 (1978).
17. J. B. Kerr, T. C. Jempty, and L. L. Miller, J. Amer. Chem. Soc., 101, 7338 (1979); L. Christensen and L. L. Miller, J. Org. Chem., 46, 4876 (1981).
18. E. Kotani and S. Tobinaga, Tetrahedron Lett., 4759 (1973).
19. T. Kametani, K. Shishido, and S. Takano, J. Heterocyclic Chem., 12, 305 (1975).
20. U. Palmquist, A. Nilsson, T. Pettersson, A. Ronlan, and V. D. Parker, J. Org. Chem., 44, 196 (1979).
21. M. Sainsbury and R. F. Schinazi, J. Chem. Soc., Chem. Commun., 718 (1972).

22. I. W. Elliott, J. Org. Chem., 42, 1090 (1977).
23. H. Klünenberg, C. Schäffer, and H.-J. Schäfer, Tetrahedron Lett., 23, 4581 (1982).
24. E. Kotani and S. Tobinaga, J. Chem. Soc., Chem. Commun., 550 (1973); E. Kotani and S. Tobinaga, Tetrahedron, 30, 3027 (1974).
25. M. Sainsbury and J. Wyatt, J. Chem. Soc., Perkin Trans. 1, 661.
26. D. Krikorian, R. Vlahov, S. Parushev, M. Chinova, I. Vlahov, H. J. Schäfer, H. Duddeck, and G. Snatzke, Tetrahedron Lett., 25, 2969 (1984).
27. P. Bird, G. G. Black, H. Lloyd, M. Sainsbury, and D. I. C. Scopes, J. Chem. Soc., Perkin Trans. 1, 1197 (1986).
28. W. Eilenberg and H. J. Schäfer, Tetrahedron Lett., 25, 5023 (1984).
29. D. Koch and H. J. Schäfer, Angew. Chem., 85, 264 (1973).
29a. D. Jeya Kumar, K. M. Snow, and K. M. Smith, J. Amer. Chem. Soc., 110, 8562 (1988).
30. H. Lund, Elektrodereaktioner i Organisk Polarografi og Voltammetri, Aarhus, Stiftsbogtrykkerie, Aarhus, 1961.
31. N. L. Weinberg and E. A. Brown, J. Org. Chem., 31, 4058 (1966).
32. S. H. Pines, J. Org. Chem., 38, 3854 (1973).
33. H. Ohmori, C. Ueda, Y. Nobusue, N. Saitou, T. Yokota, and M. Masui, J. Chem. Soc., Perkin Trans. 2, 1599 (1981).
34. H. G. Thomas and H. W. Schwager, Tetrahedron Lett., 25, 4471 (1984).
35. O. Hammerich and V. D. Parker, J. Chem. Soc., Perkin Trans. 1, 1718 (1972).
36. (a) M. Lacan, K. Jakopcić, V. Rogic, S. Damoni, O. Rogic, and I. Tabakovic, Synth. Commun., 4, 219 (1974); (b) I. Tabakovic, M. Trkovnik, M. Batusic, and K. Tabakovic, Synthesis, 590 (1979).
37. Z. Saničanin and I. Tabakovič, Tetrahedron Lett., 27, 407 (1986); Electrochim. Acta, 33, 1595 (1988).
38. Z. Grujic, I. Tabakovic, and M. Trkovnik, Tetrahedron Lett., 4823 (1976).
39. C. Bratu and A. T. Balaban, Rev. Roum. Chim., 10, 1001 (1965); Chem. Abstr., 64, 12489 (1966).
40. M. Le Guyader, C. R. Acad. Sci., Ser. C, 262, 1383 (1966).
41. C. Baezner and A. Gardiol, Ber. Dtsch. Chem. Ges., 39, 2512 (1906).
42. C. Bellec, D. Bertin, R. Colau, S. Deswarte, P. Maitte, and C. Viel, J. Hetrocyclic Chem., 16, 1611, 1657 (1979).
43. H. Lund, Symp. Synthetic and Mechanistic Aspects of Electroorganic Chemistry, Durham, N.C., Oct. 1968, Preprints of Papers, p. 197; H. Lund and J. Hakl, unpublished observation, 1968.
44. H. Lund and L. G. Feoktistov, Acta Chem. Scand., 23, 3482 (1969).
45. A. Chibani, R. Hazard, M. Jubault, and A. Tallec, Bull. Soc. Chim. Fr. 795 (1987).
46. M. Jubault and E. Raoult, C. R. Acad. Sci., Ser. C, 270, 1743 (1970); M. Jubault and D. Peltier, Bull. Soc. Chim. Fr., 1561 (1972).
47. (a) R. Hazard and A. Tallec, Bull. Soc. Chim. Fr., 3040 (1973); (b) J. Hlavaty, J. Volke, O. Manousek, and V. Bakos, Electrochim. Acta, 24, 541 (1979); (c) C.-P. Maschmeier, H. Tanneberg, and H. Matschiner, Z. Chem., 21, 219 (1981). H. Matschiner, H. Biering, H. Schilling, H. Tanneberg, K. Trautner, and C. P. Mashmeyer, Ger. (East) DD 220345 (1985); Chem. Abstr., 104, 88572 (1986).

R. Hazard, M. Jubault, C. Mouats, and A. Tallec, Electrochim. Acta, 33, 1335 (1988).
48. R. Hazard, M. Jubault, C. Mouats, and A. Tallec, Electrochim. Acta, 31, 489 (1986).
49. M. Cariou, R. Hazard, M. Jubault, and A. Tallec, Can. J. Chem., 61, 2359 (1983); M. Cariou, R. Hazard, M. Jubault, and A. Tallec, J. Electroanal. Chem., 182, 345 (1985); M. Cariou, R. Hazard, M. Jubault, and A. Tallec, J. Chem. Res. (S), 184 (1986).
50. M. A. Casadei, A. Inesi, F. M. Moracci, and D. Occhialini, Tetrahedron, 44, 2601 (1980).
51. T. Shono, Y. Matsumura, S. Katch, and J. Ohshita, Chem. Lett., 1065 (1988).
52. J. Hlavaty, J. Volke, and O. Manousek, J. Electroanal. Chem., 61, 219 (1975).
53. J. Hlavaty, Coll. Czech. Chem. Commun., 50, 33 (1985).
54. (a) H. Lund, Acta Chem. Scand., 14, 359 (1960); (b) H. Lund, P. Lunde, and F. Kaufmann, Acta Chem. Scand., 20, 1631 (1966).
55. K. Brand and E. Stohr, Ber. Dtsch. Chem. Ges., 39, 4058 (1906).
56. M. Le Guyader, Bull. Soc. Chim. Fr., 1967 (1966).
57. E. Yu. Khmel'nitskaya, E. A. Zalogina, G. I. Migachev, and A. M. Andrievskii, Zh. Obshch. Khim., 47, 2603 (1977); Chem. Abstr., 88, 81028 (1978).
58. J. Hlavaty, V. Barkos, and J. Volke, J. Appl. Electrochem., 17, 1228 (1987) and references therein.
59. S. Kwee and H. Lund, Acta Chem. Scand., 23, 2711 (1969).
60. (a) I. Tabakovic, M. Lacan, and S. Damoni, Electrochim. Acta, 21, 621 (1976); (b) I. Tabakovic, M. Trkovnik, and D. Galijas, J. Electroanal. Chem., 86, 241 (1978); (c) M. Lacan, V. Rogic, I. Tabakovic, D. Galijas, and T. Solomon, Electrochim. Acta, 28, 199 (1983).
61. G. Popp, J. Org. Chem., 37, 3058 (1972).
62. U. Hess, K. Hiller, R. Schroeder, and E. Gründemann, J. Prakt. Chem., 319, 568 (1977).
63. A. Najafi and M. Sainsbury, Heterocycles, 6, 459 (1977).
64. P. Berkenkotter and R. F. Nelson, J. Electrochem. Soc., 120, 346 (1973).
65. M. D. Hawley, S. V. Tatawawadi, S. Piekarski, and R. N. Adams, J. Amer. Chem. Soc., 89, 447 (1967).
66. T. E. Young and W. T. Beidler, J. Org. Chem., 49, 4833 (1984).
67. G. Barbey, D. Delahaye, M. Lament, and C. Caullet, Electrochim. Acta, 25, 1273 (1980).
68. R. M. Elofson, F. F. Gadallah, and J. K. Laidler, Can. J. Chem., 63, 1170 (1985).
69. M. Le Guyader and D. Peltier, Bull. Soc. Chim. Fr., 2695 (1966); R. Hazard and A. Tallec, Bull. Soc. Chim. Fr., 121 (1974).
70. S. Wawzonek and S. M. Heilman, J. Org. Chem., 39, 511 (1974).
71. P. R. Moses and J. R. Chambers, J. Amer. Chem. Soc., 96, 945 (1974).
72. G. Bontempelli, F. Magno, G.-A. Mazzocchin, and R. Seeber, J. Electroanal. Chem., 63, 23 (1975); J. C. Lodmell, W. C. Anderson, M. F. Hurley, and J. Q. Chambers, Anal. Chim. Acta, 129, 49 (1981).
73. P. Jeroschewski, Z. Chem., 18, 27 (1978); P. Jeroschewski, B.

Struebing, and H. Berge, Z. Chem., 20, 102 (1980); P. Jeroschewski, Z. Chem., 21, 412 (1981); P. Jeroschewski and F. Pragst, J. Electroanal. Chem., 149, 131 (1983).
74. E. K. Moltzen, A. Senning, R. G. Hazell, and H. Lund, Acta Chem. Scand., B40, 593 (1986).
75. G. Le Guillanton, Q. T. Do, and J. Simonet, Tetrahedron Lett., 27, 2261 (1986); Stud. Org. Chem. (Amsterdam) 30, 257 (1987).
76. H. Berge, B. Strübing, S. Strasen, M. Aquila, and K. Peseke, Wiss. Z. Wilhelm-Pieck-Univ. Rostock, 27, 639 (1978); Chem. Abstr., 93, 94899 (1980); L. Contreras, S. Rivas, and R. Rozas, J. Electroanal. Chem., 177, 299 (1984).
77. A. Bewick, D. E. Coe, J. M. Mellor, and W. M. Owton, J. Chem. Soc., Perkin 1, 1033 (1985).
78. S. Torii, H. Tanaka, M. Sasaoka, and T. Shiroi, Ger. Offen. DE 3511 149 (1984); Chem. Abstr., 104, 138261 (1986).
79. H. Lund, Acta Chem. Scand., 21, 2525 (1967).
80. T. Wolfahrt, J. Prakt. Chem. [2], 65, 295 (1902).
81. E. Laviron and T. Lewandowska, Bull. Soc. Chim. Fr., 3177 (1970).
82. J. W. Rogers and W. M. H. Watson, Anal. Chim. Acta, 54, 41 (1971).
83. E. Laviron, D. Bernard, and G. Tainturier, Tetrahedron Lett., 3643 (1972).
84. D. Bernard, G. Tainturier, and E. Laviron, Bull. Soc. Chim. Fr., 1645 (1973).
85. Y. Mugnier and E. Laviron, J. Heterocyclic Chem., 14, 351 (1977).
86. J. Hlavaty, J. Volke, and O. Manousek, Electrochim. Acta, 23, 589 (1978); J. Hlavaty and J. Volke, Electrochim Acta, 29, 1399 (1984).
87. Y. Mugnier and E. Laviron, Electrochim. Acta, 25, 1329 (1980).
88. G. M. LoPresti, S. Huang, and A. A. Reidlinger, J. Electrochem. Soc., 115, 1135 (1968).
89. J. Møller, Elektrochem. Z., 10, 201 (1904).
90. Bad. Anilin- und Sodafabrik, DRP 79, 406 (1894).
91. H. Lund, First Int. Cong. Heterocyclic Chem., Albuquerque, N.M., 1967; H. Lund and N. H. Nilsson, Acta Chem. Scand., B30, 5 (1976).
92. R. Hazard and A. Tallec, Bull Soc. Chim. Fr., 679 (1975); 433 (1976).
93. K. Elbs and W. Keiper, J. Prakt. Chem. [2], 67, 580 (1903).
94. H. Lund and S. Kwee, Acta Chem. Scand., 22, 2879 (1968).
95. R. Hazard and A. Tallec, Bull. Soc. Chim. Fr., 2917 (1971).
96. M. Studnicova, V. N. Flerov, O. Fischer, and M. Potaček, J. Electroanal. Chem., 187, 297, 307 (1985).
97. R. Hazard and A. Tallec, Bull. Soc. Chim. Fr., 433 (1976).
98. H. Lund, Elektroderaktioner i Organisk Polarografi og Voltammetri, Aarhus Stiftsbogtrykkerie, Aarhus, 1961, p.136.
99. (a) M. Lacan, I. Tabakovic, and Z. Cekovic, Tetrahedron, 30, 2911 (1974); (b) I. Tabakovic, M. Trkovnik, and Z. Grujic, J. Chem. Soc., Perkin Trans. 2, 166 (1979).
100. W. Jugelt, M. Tismer, and M. Rauh, Z. Chem., 23, 29 (1983).
101. N. Henning, T. Dassler, and W. Jugelt, Z. Chem., 22, 25 (1982).
102. M. E. Niyazymbetov, V. A. Petrosyan, and E. V. Kozlovskaya, Izv. Akad. Nauk SSSR, Ser. Khim., 2151 (1986).
103. R. Rastogi, G. Dixit, and K. Zutshi, Electrochim. Acta, 29, 1345 (1984).

104. T. Troll and W. Elbe, Electrochim. Acta, 22, 615 (1977).
105. (a) C. Degrand, C. Grosdemouge, and P.-L. Compagnon, Tetrahedron Lett., 3023 (1978); (b) C. Degrand and D. Jacquin, Tetrahedron Lett., 4955 (1978); (c) C. Degrand, P.-L. Compagnon, G. Belot, and D. Jacquin, J. Org. Chem., 45, 1189 (1980); (d) C. Degrand, G. Belot, F. Gasques, and P.-L. Compagnon, Electrochim. Acta, 27, 1529 (1982).
106. J. H. Wagenknecht, J. Org. Chem., 42, 1836 (1977); J. H. Wagenknecht and G. V. Johnson, J. Electrochem. Soc., 134, 2754 (1987).
107. I. Tabakovic and S. Crljenak, Heterocycles, 16, 699 (1981).
108. W. Jugelt and L. Grubert, Z. Chem., 25, 408, 442 (1985); W. Jugelt, L. Grubert, and V. Paulss, Z. Chem., 25, 443 (1985).
109. G. Cauquis and M. Genies, Tetrahedron Lett., 3404 (1970); 3959 (1971).
110. G. Cauquis, B. Chabaud, and M. Genies, Bull. Soc. Chim. Fr., 3482, 3487 (1973); Tetrahedron Lett., 2389 (1974).
111. G. Cauquis and G. Reverdy, Tetrahedron Lett., 3267 (1977).
112. W. Jugelt and S. Schwertuer, Z. Chem., 23, 28, 409 (1983); 24, 149 (1984).
113. J. Yoshida, K. Sakaguchi, and S. Isoe, Tetrahedron Lett., 27, 6075 (1986); J. Yoshida, M. Yamamoto, and N. Kawabata, Tetrahedron Lett., 26, 6217 (1985).
114. T. Shono, N. Kise, and T. Tanabe, J. Org. Chem., 53, 1364 (1988).
115. G. Fiori, Journée d'Electrochemie, Dijon 1987, abstracts of papers 4–3; G. Palmisano, B. Danieli, G. Lesma, and G. Fiori, Tetrahedron, 44, 1545 (1988).
116. H. Baltes, L. Stork, and H. J. Schäfer, Angew. Chem., 89, 425 (1977); Chem. Ber., 112, 807 (1979).
117. G. Cauquis and J.-L. Cros, Bull. Soc. Chim. Fr., 3760, 3765 (1971).
118. E.-L. Dreher, J. Bracht, M. El-Mobayed, P. Hütter, W. Winter, and A. Rieker, Chem. Ber., 115, 288 (1982).
119. L. A. Simonson and C. K. Mann, Tetrahedron Lett., 3303 (1970).
120. R. B. A. Pardy, UK Pat. Appl. GB 2153 849 (1985); Chem. Abstr., 103, 202767 (1985).
121. J. Voss, P. Mischke, and G. Adiwidjaja, Phosphorus Sulfur, 27, 261 (1986).
122. F. Maran, E. Vianello, G. Cavicchioni, and F. D'Angeli, J. Chem. Soc., Chem. Commun., 660 (1985).
123. H. Lund, Acta Chem. Scand., 13, 249 (1959).
124. H. Lund, Acta Chem. Scand., 18, 563 (1964).
125. H. Lund and P. Lunde, Acta Chem. Scand., 21, 1067 (1967).
126. H. Lund, Discuss. Faraday Soc., 45, 193 (1968).
127. H. Lund and E. T. Jensen, Acta Chem. Scand., 24, 1867 (1970).
128. J. Pinson, J.-P. M'Packo, N. Vinot, J. Armand, and P. Bassinet, Can. J. Chem., 50, 1581 (1972).
129. H. H. Holst and H. Lund, Acta Chem. Scand., B33, 233 (1979).
130. R. Fuhlendorff and H. Lund, Acta Chem. Scand., 42B, 52 (1988).
131. C. Bellec, N. Vinot, and P. Maitte, J. Heterocyclic Chem., 23, 491 (1986).
132. H. Lund, Coll. Czech. Chem. Commun., 30, 4237 (1965).
133. H. Oelschläger and H. Hoffmann, Arch. Pharm. (Weinheim) 300, 817 (1967).
134. P. Martigny and H. Lund, Acta Chem. Scand., B33, 575 (1979).

135. B. Bujoli, M. Chena, M. Jubault, and A. Tallec, J. Electroanal. Chem., 199, 461 (1986).
136. H. Lund, Tetrahedron Lett., 3973 (1965).
137. W. R. Vaugham, D. I. McCane, and G. J. Sloan, J. Amer. Chem. Soc., 73, 2298 (1951).
138. S. Swann and N. Leonard, J. Amer. Chem. Soc., 74, 4620, 6251 (1952).
139. P. Zuman and V. Horak, Coll. Czech. Chem. Commun., 26, 176 (1961).
140. M. Libert and C. Caullet, Bull. Soc. Chim. Fr., 1947 (1971).
141. N. Clauson-Kaas, F. Limborg, and K. Glens, Acta Chem. Scand., 6, 531 (1952); N. Clauson-Kaas and Z. Tyle, Acta Chem. Scand., 6, 962 (1952).
142. N. Clauson-Kaas, N. Elming, and Z. Tyle, Acta Chem. Scand., 9, 1 (1955).
143. N. Clauson-Kaas and F. Limborg, Acta Chem. Scand., 1, 619 (1947).
144. N. Clauson-Kaas, S. Li, and N. Elming, Acta Chem. Scand., 4, 1233 (1950).
145. N. Elming, Acta Chem. Scand., 6, 572 (1952).
146. N. Elming and N. Clauson-Kaas, Acta Chem. Scand., 9, 23 (1955).
147. K. Yoshida, J. Chem. Soc., Chem. Commun., 20 (1978)
148. A. J. Baggaley and R. Brettle, J. Chem. Soc., Chem. Commun., 108 (1966)
149. K. E. Kolb and C. L. Wilson, J. Chem. Soc., Chem. Commun., 271 (1966).
150. S. Arita, Y. Nagahiro and K. Takeshita, Kogyo Kagaku Zasshi, 72, 1896 (1969); Chem. Abstr., 73, 10122 (1970).
151. K. Yoshida and T. Fueno, J. Org. Chem., 36, 1523 (1971).
152. K. Yoshida and T. Fueno, Bull. Chem. Soc. Jpn., 42, 2411 (1969); 60, 229 (1987).
153. K. Yoshida, J. Amer. Chem. Soc., 98, 254 (1976).
154. S. Torii, H. Tanaka, H. Ogo, and S. Yamasita, Bull. Chem. Soc. Jpn., 44, 1079 (1971).
155. S. Torii, H. Tanaka, and T. Okamoto, Bull. Chem. Soc. Jpn., 45, 2783 (1972).
156. E. Gird and A. T. Balaban, J. Electroanal. Chem., 4, 48 (1962).
157. (a) F. Pragst, Electrochim. Acta, 21, 497 (1976); (b) F. Pragst and R. Ziebig, Electrochim. Acta, 23, 735 (1978); (c) F. Pragst, R. Ziebig, U. Seydewitz, and G. Driesel, Electrochim. Acta, 25, 341 (1980); (d) F. Pragst, M. Janda, and I. Stibor, Electrochim. Acta, 25, 779 (1980); F. Pragst and U. Seydewitz, J. Prakt. Chem., 319, 952 (1977); F. D. Sawve and G. R. Olin, J. Amer. Chem. Soc., 102, 299 (1980).
158. M. Vajda, Advan. Polarogr., Proc. 2nd Int. Cong. Polarogr., 2, 786 (1959).
159. V. Wintgens, J. Pouliguen, and J. Kossanyi, Nouv. J. Chim., 10, 345 (1986).
160. M. Yamauchi, S. Katayama, Y. Nakashita, and T. Watanabe, Synthesis, 33 (1981).
161. (a) A. J. Harle and L. E. Lyons, J. Chem. Soc., 1575 (1950); (b) R. Patzak and L. Neugebauer, Monatsh. Chem., 82, 662 (1951); 83, 776 (1952); (c) V. S. Griffith and J. B. Westmore, J. Chem. Soc., 1704 (1962).
162. R. N. Gourley, J. Grimshaw, and P. G. Miller, J. Chem. Soc. C, 2318 (1970).

163. A. Zweig, G. Metzler, A. H. Maurer, and B. G. Roberts, J. Amer. Chem. Soc., 88, 2864 (1966).
164. (a) M. Janda, Coll. Czech. Chem. Commun., 28, 2524 (1963); (b) M. Janda and J. Radouch, Coll. Czech. Chem. Commun., 32, 2672 (1967); (c) M. Janda and L. Paviensky, Coll. Czech. Chem. Commun., 32, 2675 (1967); (d) J. Sprogl, M. Janda, and M. Valentova, Coll. Czech. Chem. Commun., 35, 148 (1970).
165. K. Yoshida, T. Sacki, and T. Fueno, J. Org. Chem., 36, 3673 (1971).
166. (a) S. G. Mairanovskii, L. I. Kosychenkov, and S. Z. Taits, Izv. Akad. Nauk. SSSR, Ser. Khim., 1382 (1980); Chem. Abstr., 93, 83354 (1980).
167. V. P. Gul'tyai, I. V. Proskurovskaya, T. Ya. Rubinskaya, A. V. Lozanova, A. M. Moiseenkov, and A. V. Semenovskii, Izv. Akad. Nauk. SSSR, Ser. Khim., 1576 (1979); Chem. Abstr., 91, 157538 (1979).
168. J. Sprogl, M. Janda, I. Stibor, J. Kos, and V. Vyskocil, Coll. Czech. Chem. Commun., 43, 2015 (1978).
169. H. V. Drushel and J. F. Miller, Anal. Chem., 30, 1271 (1958).
170. H. Lund, unpublished observation, 1968.
171. T. Lund and H. Lund, Acta Chem. Scand., B42, 269 (1988).
172. D. A. Tyssee and M. M. Baizer, J. Electrochem. Soc., 118, 1420 (1971).
173. M. B. Kass, A. P. Borsetti, and D. R. Crist, J. Amer. Chem. Soc., 95, 959 (1973).
174. R. Kossai, J. Simonet, and G. Dauphin, Tetrahedron Lett., 21, 3575 (1980).
175. M. Dennstedt, DRP 127,086 (1901); B. Sakurai, Bull. Chem. Soc. Jpn., 11, 347 (1936).
176. L. C. Craig, J. Amer. Chem. Soc., 55, 2543 (1933).
177. O. Carrasco, Gazz. Chim. Ital., 38, 301 (1908).
178. J. von Braun and W. Sobecki, Ber. Dtsch. Chem. Ges., 44, 2158 (1911).
179. J. T. Wrobel and K. M. Pazdro, Rocz. Chem., 41, 637 (1967).
180. F. Ender, E. Moisar, K. Schäfer, and H.-J. Teuber, Z. Elektrochem., 63, 349 (1959).
181. E. Kariv-Miller, D. F. Dedolph, C. M. Ryan, and T. J. Mahachi, J. Heterocylic Chem., 22, 1389 (1985).
182. G. Cauquis and M. Genies, Bull. Soc. Chim. Fr., 3220 (1967).
183. P. J. Grossi, L. Marchetti, R. Ramasseul, A. Rassat, and D. Serve, J. Electroanal. Chem., 87, 353 (1978).
184. N. L. Weinberg and E A. Brown, J. Org. Chem., 31, 4054 (1966).
185. M. Libert, C. Caullet, and G. Barbey, Bull. Soc. Chim. Fr., 536 (1973); M. Libert and C. Caullet, Bull. Soc. Chim. Fr., 800 (1974).
186. K. Yoshida, J. Amer. Chem. Soc., 99, 6111 (1977); 101, 2116 (1979).
187. L. Eberson, Acta Chem. Scand., B34, 747 (1980).
188. T. Shiba and Y. Takata, J. Org. Chem., 42, 2973 (1977).
189. G. T. Cheek and R. F. Nelson, J. Org. Chem., 43, 1230 (1978); J. M. Bobbitt, C. L. Kulkarni, and J. P. Willis, Heterocycles, 15, 495 (1981).
190. G. T. Cheek, R. F. Nelson, J. M. Bobbitt, and P. M. Scola, J. Org. Chem., 52, 5277 (1987).

191. G. T. Cheek and R. F. Nelson, J. Org. Chem., 43, 1230 (1978).
192. S. Longchamp and C. Caullet, Electrochim. Acta, 29, 1075 (1984).
193. J. M. Bobbitt, P. M. Scola, C. L. Kulkarni, A. J. DeNicola, and T. T. Chou, Heterocycles, 24, 669 (1986).
194. (a) J. F. Ambrose and R. F. Nelson, J. Electrochem. Soc., 115, 1159 (1968); (b) J. F. Ambrose, L. L. Carpenter, and R. F. Nelson, J. Electrochem. Soc., 122, 876 (1975).
195. K. Yasokouchi, I. Taniguchi, H. Yamaguchi, and M. Shiraishi, J. Electroanal. Chem., 105, 403 (1979).
196. B. Emmert, Ber. Dtsch. Chem. Ges., 46, 1716 (1913); E. P. Hart, J. Chem. Soc., 3872 (1953).
197. O. R. Brown, R. J. Butterfield, and J. P. Millington, Electrochim. Acta, 27, 1655 (1982).
198. J. Nadra, H. Givadinovitch and M. Devaud, J. Chem. Res. (S), 192 (1983).
199. W. R. Turner and P. J. Elving, Anal. Chem., 37, 467 (1965).
200. J. E. Toomey, U.S. Pat. 4,482,439 (1984); Chem. Abstr., 102, 102453 (1985).
201. M. A. TeSchinnen, Neth. Appl. NL 8502 485 (1985); Chem. Abstr., 107, 86156 (1987); F. T. B. J. Van den Brink and R. Van Hardeveld, Neth. Appl. N. L. 85 02486 (1985); Chem. Abstr., 107, 86157 (1987).
202. J. Ludvik, J. Volke, and J. Klima, Electrochim. Acta., 32, 1063 (1987); M. Largeron and M. B. Fleury, J. Electroanal. Chem., 246, 373 (1988).
203. P. C. Tompkins and C. L. A. Schmidt, Univ. Calif. Publ. Physiol., 8, 237, 247 (1944).
204. S. G. Mairanovskii, Dokl. Akad. Nauk. SSSR, 110, 593 (1956).
205. I. Carelli, M. E. Cardinali, and B. Grazia, Ann. Chim. (Rome), 67, 89 (1977); M. G. Bonicelli, M. E. Cardinali, and I. Carelli, J. Electroanal. Chem., 131, 345 (1982).
206. R. Raghavan and R. T. Iwamoto, J. Electroanal. Chem., 92, 101 (1978).
207. J. G. Gaudiello, D. Larkin, J. D. Rawn, J. J. Sosnowski, E. E. Bancroft, H. N. Blount, J. Electroanal. Chem., 131, 203 (1982).
208. F. T. McNamara, J. W. Nieft, J. F. Ambrose, and E. S. Huyser, J. Org. Chem., 42, 988 (1977).
209. T. N. Khomchenko and L. N. Nekrasov, Electrokhimiya, 23, 1134 (1987).
210. T. Misumi, S. Furahashi, and M. Shiga, Ger. Offen., 2 812 508: Chem. Abstr., 90, 22828 (1979); J. E. Toomey, U.S. Pat. 4670111 (1987); Chem. Abstr., 108, 112235 (1988).
211. (a) O. R. Brown, R. McIntyre, and P. L. H. Miller, J. Electroanal. Chem., 110, 247 (1980); (b) S. Kashti-Kaplan, J. Hermolin, and E. Kirowa-Eisner, J. Electrochem. Soc., 128, 802 (1981).
212. H. Lund and L. H. Kristensen, Acta Chem. Scand., B33, 495 (1979).
213. O. Fischer and K. Neundlinger, Ber. Dtsch. Chem. Ges., 46, 2544 (1913).
214. S. J. Leach, J. H. Baxendale, and M. G. Evans, Aust. J. Chem., 6, 395 (1953).
215. J. N. Burnett and A. L. Underwood, J. Org. Chem., 30, 1154 (1965).
216. A. J. Cunningham and A. L. Underwood, Biochemistry, 6, 266 (1967).
217. B. Janik and P. J. Elving, J. Amer. Chem. Soc., 92, 235 (1970).

218. C. O. Schmakel, K. S. V. Santhanam, and P. J. Elving, J. Electrochem. Soc., 121, 345 (1974).
219. S. Kato, J. Nakaya, and E. Imoto, Rev. Polarogr. (Jpn.), 18, 29 (1972).
220. (a) F. M. Moracci, F. Liberatore, V. Carelli, A. Arnone, I. Carelli, and M. E. Cardinali, J. Org. Chem., 43, 3420 (1978); (b) I. Carelli, M. E. Cardinali, and F. M. Moracci, J. Electroanal. Chem., 107, 391 (1980).
221. (a) Y. Ohnishi and M. Kitami, Bull. Chem. Soc. Jpn., 52, 2674 (1979); (b) Y. Ohnishi, Y. Kikuchi, and M. Kitami, Tetrahedron Lett., 3005 (1979).
222. V. Carelli, F. Liberatore, A. Casini, R. Mondelli, A. Arnone, I. Carelli, G. Rotilio, and I. Mavelli, Bioorg. Chem., 9, 342 (1980); Chem. Abstr., 94, 98618 (1981).
223. (a) B. Ke, J. Amer. Chem. Soc., 78, 3649 (1956); (b) Arch. Biochem. Biophys., 60, 505 (1956).
224. R. F. Powning and C. C. Kratzing, Arch. Biochem. Biophys., 66, 249 (1957).
225. H. Jaegfeldt, J. Electroanal. Chem., 128, 355 (1981).
226. (a) J. N. Burnett and A. L. Underwood, Biochemistry, 4, 2060 (1965); (b) A. J. Cunningham and A. L. Underwood, Arch. Biochem. Biophys., 117, 88 (1966); (c) R. W. Burnett and A. L. Underwood, Biochemistry, 7, 3328 (1968).
227. R. J. Day, S. J. Kinsey, E. T. Seo, N. Welike, and H. P. Silverman, Trnas. N. Y. Acad. Sci., 34, 588 (1972).
228. W. J. Blaedel and R. G. Haas, Anal. Chem., 42, 918 (1970); W. J. Blaedel and R. A. Jenkins, Anal. Chem., 47, 1337 (1975); I. Carelli, R. Rosati, and A. Casini, Electrochim. Acta, 26, 1695 (1981); J. Morioux and P. J. Elving, J. Amer. Chem. Soc., 102, 6533 (1980); N. Cenas, J. Kanapieniene, and J. Kulys, J. Electroanal. Chem., 189, 163 (1985); J. M. Laval, C. Bourdillon, and J. Moiroux, Biotechnol. Bioenerg., 30, 157 (1987).
229. N. E. Khomutov and V. V. Tsodikov, Elektrokhimiya, 1, 482 (1965); 2, 722 (1966).
230. J. Claret, C. Müller, J. M. Felia, and J. Alcalde, Electrochim. Acta, 32, 1431 (1987).
231. P. Fuchs, U. Hess, H. H. Holst, and H. Lund, Acta Chem. Scand., B35, 185 (1981).
232. W. H. Smith and A. J. Bard, J. Amer. Chem., Soc., 97, 6491 (1975); O. R. Brown and R. J. Butterfield, Electrochim. Acta, 27, 1663 (1982).
233. C. Degrand and H. Lund, Acta Chem. Scand., B31, 593 (1977); U. Hess, D. Huhn, and H. Lund, Acta Chem. Scand., B34, 413 (1980).
234. M. Maturova, T. Shun, J. Ctvrtnik, and F. Santavy, Coll. Czech. Chem. Commun., 25, 3321 (1960).
235. M. Kulka, J. Amer. Chem. Soc., 68, 2472 (1946).
236. J. C. Cochran and W. F. Little, J. Org. Chem., 26, 808 (1961).
237. R. C. Kaye and H. I. Stonehill, J. Chem. Soc., 27 (1951).
238. E. Sturn, H. Kiesele, and F. Daltrozzo, Chem. Ber., 111, 227 (1978).
239. J. G. Frost and J. H. Saylor, Rec. Trav. Chem. Pays-Bas, 82, 828 (1963); R. Mitzner, J. Bendig, R. Ziebig, F. Graichen, D. Kreysig, and F. Pragst, J. Prakt. Chem., 327, 242 (1985).

240. H. Lund, Acta Chem. Scand., 23, 563 (1969).
241. V. D. Bezuglyi, N. P. Shimanskaya, and E. M. Peresleni, Zh. Obshch. Khim., 34, 3540 (1964); N. P. Shimanskaya, L. A. Kotok, T. F. Alekhina, and V. D. Bezuglyi, Zh. Obshch. Khim., 43, 1445 (1973); 46, 2107 (1976).
242. W. N. Greig and J. W. Rogers, J. Electrochem. Soc., 117, 1141 (1970); 118, 1089 (1971); S. L. Smith, L. D. Cook, and J. W. Rogers, J. Electrochem. Soc., 119, 1332 (1972); V. S. Tsveniashvili, M. V. Malashkhiya, O. P. Shvaika, and N. S. Khavtasi, Zh. Obshch. Khim., 54, 624 (1984).
243. I. Surov and H. Lund, Acta Chem. Scand. B40, 831 (1986).
244. T. Shono, S. Kashimura, Y. Yamaguchi, O. Ishige, H. Uyama, and F. Kuwata, Chem. Lett., 1511 (1987); T. Shono, S. Kashimura, Y. Yamaguchi, and F. Kuwata, Tetrahedron Lett., 28, 4411 (1987).
245. E. Barrado, R. Pardo, P. S. Batanero, A. Alberola, M. A. Laguna, and F. J. Pulido, Electrochim. Acta, 33, 171 (1988).
246. K. Nyberg and R. Servin, Acta Chem. Scand., B30, 640 (1976); K. Nyberg, Synthesis, 545 (1976).
247. T. Shono, H. Hamaguchi, and Y. Matsumura, J. Amer. Chem. Soc., 97, 4264 (1975); T. Shono, Y. Matsumura, and K. Tsubata, J. Amer. Chem. Soc., 103, 1172 (1981); K. Mitzloff, K. Warning, and H. Jensen, Liebigs Ann. Chem., 1713 (1978).
248. P. E. Iversen and H. Lund, Acta Chem. Scand., 21, 389 (1967).
249. Z. N. Timofeeva, M. V. Petrova, M. Z. Girshorich, and A. V. El'tsov, Zh. Obshch. Khim., 39, 54 (1969).
250. S. Roffia and G. Feroci, J. Electroanal. Chem., 88, 169 (1978).
251. A. Ciszewski, M. Studničkova, and O. Fischer, Bioelektrochem. Bioenerg., 13, 25 (1984).
252. M. Matsui, T. Sawamura, and T. Adachi, Mem. Coll. Sci. Kyoto Imp. Univ., 15A, 151 (1932); Chem. Abstr., 26, 5264 (1932).
253. M. Jubault, A. Tallec, B. Bujoli, J.-C. Rozé, and J.-P. Pradère, Tetrahedron Lett., 26, 745 (1985).
254. B. Bujoli, M. Chehna, M. Jubault, and A. Tallec, J. Electroanal. Chem., 199, 461 (1986); B. Bujoli, M. Jubault, J.-C. Rozé, and A. Tallec, Tetrahedron, 43, 2709 (1987).
255. P. Kabasakalian and J. McGlotten, Anal. Chem., 31, 431 (1959).
256. J.-P. Billon, Bull. Soc. Chim. Fr., 1784 (1960); 1923 (1961); J.-P. Billon, Ann. Chim., 7, 183 (1962).
257. G. Cauquis, A. Deronzier, J.-L. Lepage, and D. Serve, Bull. Soc. Chim. Fr., 295, 303 (1977).
258. H. H. Rüttinger, R. Spitzner, W. Schroth, H. Matschiner, and R. Ziebig, J. Prakt. Chem., 323, 33 (1981).
259. F. J. Goodman and J. Q. Chambers, J. Org. Chem., 41, 626 (1976).
260. M. Falsig and H. Lund, Acta Chem. Scand., B34, 585, 591 (1980).
261. W. Schroth, R. Borsdorf, R. Herzschuh, and J. Seidler, Z. Chem., 10, 147 (1970).
262. G. Cauquis and M. Maurey, C. R. Acad. Sci., Ser. C, 266, 1021 (1968); Bull. Soc. Chim. Fr., 3588 (1972).
263. H. Lund, Coll. Czech. Chem. Commun., 31, 4175 (1966).
264. J. P. Kitaev and G. K. Budnikov, Coll. Czech. Chem. Commun., 30, 4178 (1965).
265. A. J. Fry, W. E. Britton, and R. Wilson, J. Org. Chem., 38, 2620 (1973).

266. N. P. Shimanskaya, N. A. Lodygin, and V. D. Bezuglyi, Zh. Obshch. Khim., 38, 929 (1968); N. P. Shimanskaya, V. D. Bezuglyi, and V. E. Bondarenko, Zh. Obshch. Khim., 39, 2171 (1969).
267. J. Grimshaw and J. Trocha-Grimshaw, J. Chem. Soc., Perkin Trans. 1, 1275 (1973).
268. F. Pragst, Z. Chem., 14, 236 (1974); J. Prakt. Chem., 315, 549 (1979).
269. F. Pragst, H. Köppel, W. Jugelt, and F. G. Weber, J. Electroanal. Chem., 60, 323 (1975).
270. F. Pragst and C. Böck, J. Electroanal. Chem., 61, 47 (1975).
271. H. Lund, Oesterreich, Chem. Z., 68, 43 (1967).
272. S. Millefiori, Ann. Chim., (Rome), 59, 15 (1969).
273. H. Lund and J. Simonet, C. R. Acad. Sci., Ser. C, 277, 1387 (1973).
274. C. Furlani, S. Bertola, and G. Murpurgo, Ann. Chim. (Rome), 50, 858 (1960).
275. H. Lund and E. T. Jensen, Acta Chem. Scand., 25, 2727 (1971).
276. H. Lund, Lecture, 12, Nordiske Kemikermøde, Trondheim, June 1965, abstract of papers, p. 33.
277. D. L. Smith and P. J. Elving, J. Amer. Chem. Soc., 84, 2741 (1962).
278. B. Janik and P. J. Elving, Chem. Rev., 68, 295 (1968).
279. G. Dryhurst, Electrochemistry of Biological Molecules, Academic Press, New York, 1977.
280. P. J. Elving, S. J. Pace, and J. E. O'Reilly, J. Amer. Chem. Soc., 95, 647 (1973).
281. D. Thevenot, J. Electroanal. Chem., 46, 89 (1973).
282. H. Lund, Acta Chem. Scand., 18, 1984 (1964).
283. T. P. Devi, C. Kalidas, and C. S. Venkatachalam, Monatsh. Chem., 115, 1279 (1984).
284. P. Pflegel and G. Wagner, Pharmazie, 22, 60, 643 (1967).
285. P. Pflegel and G. Wagner, Pharmazie, 24, 308, 384 (1969).
286. P. Pflegel and G. Wagner, Pharmazie, 27, 24 (1972).
287. L. F. Wiggins and W. S. Wise, J. Chem. Soc., 4780 (1956).
288. J. Volke, D. Dumanovic, and V. Volkova, Coll. Czech. Chem. Commun., 30, 246 (1965).
289. J. Armand, K. Chekir, and J. Pinson, Can. J. Chem., 52, 3971 (1974).
290. J. Pinson and J. Armand, C. R. Acad. Sci., Ser. C, 266, 1081 (1968).
291. G. Sartori and C. Furlani, Ann. Chim. (Rome), 45, 251 (1955).
292. M. P. Strier and J. C. Cavagnol. J. Amer. Chem. Soc., 80, 1565 (1958).
293. J. Pinson and J. Armand, C. R. Acad. Sci., Ser. C, 268, 629 (1969); Bull. Soc. Chim. Fr., 1764 (1971).
294. J. Pinson, J.-P. Launay, and J. Armand, C. R. Acad. Sci., Ser. C, 270, 1881 (1970).
295. M. Fedoronko and I. Jezo, Coll. Czech. Chem. Commun., 37, 1781 (1972).
296. O. Fischer and T. H. Thuy, Coll. Czech. Chem. Commun., 41, 1853 (1976).
297. W. Kaim, Heterocycles, 23, 1363 (1985).
298. P. Martigny, H. Lund, and J. Simonet, Electrochim. Acta, 21, 345 (1976).
299. D. K. Root, R. D. Pendarvis, and W. H. Smith, J. Org. Chem., 43,

778 (1978); R. Gottlieb and W. Pfleiderer, Liebigs Ann. Chem., 1451 (1981).
300. R. F. Nelson, D. W. Leedy, E. T. Seo, and R. N. Adams, Z. Anal. Chem., 224, 184 (1967).
301. L. A. Summers and J. E. Dickeson, J. Chem. Soc., Chem. Commun. 1183 (1967).
302. B. J. Tabner and J. R. Yandle, J. Chem. Soc. A, 381 (1968).
303. E. Laviron and L. Roullier, C. R. Acad. Sci., Ser. C, 274, 1489 (1972); L. Roullier and E. Laviron, Electrochim. Acta, 21, 421 (1976); 22, 669 (1977); 23, 773 (1978).
304. D. Lloyd, C. A. Vincent, D. J. Walton, J. P. Declercq, G. Germain, and M. van Meerssche, J. Chem. Soc., Chem. Commun., 499 (1978); D. Lloyd, C. A. Vincent, and D. J. Walton, J. Chem. Soc., Perkin Trans. 2, 668, 1441 (1980).
305. H. Oelschläger, Arch. Pharm., 296, 396 (1963); H. Oelschläger, J. Volke, and H. Hoffmann, Coll. Czech. Chem. Commun., 31, 1264 (1966); H. Oelschläger, J. Volke, H. Hoffmann, and E. Kurck, Arch. Pharm., 300, 250 (1967).
306. B. Z. Senkowski, M. S. Levin, J. R. Urbigkit, and E. G. Wollish, Anal. Chem., 36, 1991 (1964).
307. B. Maupas and M. B. Fleury, Electrochim. Acta, 26, 399 (1981); 27, 141 (1982).
308. K. Butkiewicz, J. Electroanal. Chem., 90, 271 (1978).
309. J. M. Mellor, B. S. Pons, and J. H. A. Stibbard, J. Chem. Soc., Chem. Commun., 761 (1979).
310. H. Oelschläger, J. Volke, and C. Bresch, Arch. Pharm. (Weinheim), 318, 271 (1985).
311. E. Szebenyi-Györy, V. Kovacs-Mindler, I. Hermecz, B. Podanyi, and J. Brlik, J. Appl. Electrochem., 15, 145 (1985).
312. V. Sh. Tsveniashvili, M. V. Malashkhiya, O. P. Shvaika, and V. F. Liphitskii, Zh. Obshch. Khim., 54, 915 (1984).
313. R. Schindler, H. Will, and L. Holleck, Z. Elektrochem., 63, 596 (1959).
314. M. Breant and J.-C. Merlin, Bull. Soc. Chim. Fr., 54 (1964).
315. P. Zuman, Coll. Czech. Chem. Commun., 25, 3245 (1960).
316. F. F. Medovshchikova and I. Y. Postovskii, Zh. Obshch. Khim., 24, 1989 (1954).
317. L. S. Efros and Z. V. Todres, Zh. Obshch. Khim., 27, 983 (1957).
318. V. S. Tsveniashvili, S. I. Zdanov, and Z. V. Todres, Z. Anal. Chem., 224, 389 (1967).
319. M. Falsig and P. E. Iversen, Acta Chem. Scand., B31, 15 (1977).
320. S. U. Pedersen and H. Lund, Acta Chem. Scand., B42, 319 (1988).
321. A. El Jammal, E. Graf, and M. Gross, Electrochim. Acta, 31, 1457 (1986).
322. H.-S. Chien and M. M. Labes, J. Electrochem. Soc., 133, 2509 (1986).
323. J. Armand, K. Chekir, J. Pinson, and N. Vinot, C. R. Acad. Sci., Ser. C, 281, 547 (1975); J. Armand, K. Chekir, and J. Pinson, C. R. Acad. Sci., Ser. C., 284, 391 (1977); Can. J. Chem., 56, 1804 (1978).
324. (a) B. Jambor, Acta Chim. Acad. Sci. Hung., 4, 55 (1954); (b) B. Jambor, J. Chem. Soc., 1604 (1958); (c) H. Campbell and P. O. Kane, J. Chem. Soc., 3130 (1956); (d) P. Kivalo and K. K. Musta-

kallio, Suom. Kemistil., B29, 154 (1956), B30, 214 (1957); (e) B. A. Abdel-Nabey and A. M. Kiwan, J. Electroanal. Chem., 105, 365 (1979); K. Umemoto, Bull. Chem. Soc. Jpn., 58, 2051 (1985).
325. D. L. Smith and P. J. Elving, J. Amer. Chem. Soc., 84, 1412 (1962).
326. G. Dryhurst, Bioelectrochem. Bioenerg., 14, 251 (1985).
327. P. Supramanian, M. J. Heeg, and G. Dryhurst, J. Electroanal. Chem. 197, 279 (1986).
328. D. L. McAllister and G. Dryhurst, J. Electroanal. Chem., 59, 75 (1975); D. Ege-Serpkenci and G. Dryhurst, J. Electroanal. Chem., 141, 175 (1982).
329. K. Kretzschmar and W. Jaenicke, Z. Naturforsch., 297, 285, 999 (1971).
330. S. Kwee and H. Lund, Biochim. Biophys. Acta, 297, 285 (1973); H. Lund, in Chemistry and Biology of Pteridines (W. Pfleiderer, ed.), W. de Gruyter, Berlin, 1975, p. 645.
331. S. Kwee and H. Lund, Bioelectrochem. Bioenerg., 1, 87 (1974).
332. S. Kwee and H. Lund, J. Electroanal. Chem., 104, 441 (1979).
333. J. Armand, L. Boulares, K. Chekir, and C. Bellec, Can. J. Chem., 59, 3237 (1981); J. Armand, L. Boulares, C. Bellec, and J. Pinson, Can. J. Chem., 60, 2797 (1982).
334. J. Armand, L. Boulares, C. Bellec, C. Bois, M. Philoche-Levisalles, and J. Pinson, Can. J. Chem., 62, 1028 (1984).
335. H. H. Inhoffen and P. Jäger, Tetrahedron Lett., 1317 (1964); Tetrahedron Lett., 3387 (1965); H. H. Inhoffen and R. Mahlhop, Tetrahedron Lett., 4283 (1966); H. H. Inhoffen, P. Jäger, R. Mahlhop, and C.-D. Mengler, Liebigs Ann. Chem., 704, 188 (1967).
336. D. W. Clack and N. S. Hush, J. Amer. Chem. Soc., 87, 4238 (1965).
337. A. Stanienda and G. Biebl, Z. Phys. Chem. (Frankfurt), 52, 254 (1967).
338. H. J. Callot, A. Louati, and M. Gross, Tetrahedron Lett., 21, 3281 (1980).
339. V. G. Khomyakov, S. S. Kruglikov, and V. M. Berezovskii, Zh. Obshch. Khim., 28, 2898 (1958).
340. J. E. Toomey, U.S. Pat. 4,482,439 (1984); Chem. Abstr., 102, 102453 (1985).
341. M. A. TeSchinnen, Neth. Appl. NL 85 02485 (1987); Chem. Abstr., 107, 86156 (1987); F. T. B. J. Van den Brink and R. Van Hardeveld, Neth. Appl. NL 85 02486 (1987); Chem. Abstr., 107, 86157 (1987).
342. H. Sagae, M. Fujihira, H. Lund, and T. Osa, Heterocycles, 13, 321 (1979).
343. J. D. Anderson, M. M. Baizer, and E. J. Prill, J. Org. Chem., 30, 1645 (1965).
344. V. D. Bezuglyi and T. A. Alekseeva, Zh. Anal. Khim., 20, 244 (1965).
345. V. D. Bezuglyi, V. D. Dmitrieva, and T. A. Alekseeva, Zh. Anal. Khim., 16, 477 (1961).
346. C. Prevost, P. Souchay, and J. Chauvelier, Bull. Soc. Chim. Fr., 714 (1951).
347. A. R. Katritzky and J. M. Lagowski, in Advances in Heterocyclic Chemistry (A. R. Katritzky, ed.), Vol. 1, Academic, New York, 1963, Chaps. 6 and 7; Vol. 2, Chaps. 1 and 2.
348. J. M. Bobbitt, H. Yagi, S. Shibuya, and J. T. Stock, J. Org. Chem. 36, 3006 (1971).

349. P. Kabasakalian and J. McGlotten, Anal. Chem., 31, 1091 (1959).
350. H. Lund, Acta Chem. Scand., 14, 1927 (1960).
351. O. Manousek and P. Zuman, Coll. Czech. Chem. Commun., 29, 1432 (1964).
352. T. Lund and H. Lund, Acta Chem. Scand., B38, 387 (1984); B39, 429 (1985).
353. M. Ochiai, D. Aki, A. Morimoto, T. Okada, K. Shinozaki, and Y. Asahi, Tetrahedron Lett., 2341 (1972); J. Chem. Soc., Perkin Trans. 1, 258 (1974).
354. D. A. Hall, D. M. Berri, and C. J. Schneider, J. Electroanal. Chem., 80, 155 (1977).
355. T. Nonaka, T. Kato, T. Fuchigami, and T. Sekine, Electrochim. Acta, 26, 887 (1981).
356. J. Tafel and M. Stern, Ber. Dtsch. Chem. Ges., 33, 2224 (1900); A. B. Ershler, T. S. Orekhova, and I. M. Levinson, Electrokhimiya, 15, 520 (1979).
357. R. G. Barradas, S. Fletcher, and J. D. Porter, J. Electroanal. Chem., 75, 533 (1977); S. Fletcher, J. Electroanal. Chem., 213, 219 (1986).
358. H. Lund, J. Electroanal. Chem., 202, 299 (1986).
359. T. Shono, O. Ishige, H. Uyama, and S. Kashimura, J. Org. Chem., 51, 546 (1986).
360. J. Tirouflet, R. Robin, and M. Guyard, Bull. Soc. Chim. Fr., 571 (1956).
361. A. Ryvolova, Coll. Czech. Chem. Commun., 25, 420 (1960).
362. A. Dunet and A. Willemart, Bull. Soc. Chim. Fr., 887 (1948).
363. E. Späth and F. Brench, Monatsh. Chem., 50, 349 (1928).
364. B. Sakurai, Bull. Chem. Soc. Jpn., 5, 184 (1930).
365. F. Fichter and I. Stein, Helv. Chim. Acta, 12, 821 (1929).
366. B. Czochralska and D. Shugar, Biochim. Biophys. Acta, 281, 1 (1972).
367. T. Wasa and P. J. Elving, J. Electroanal. Chem., 91, 249 (1978).
368. H. E. Baumgarten, P. L. Creger, and R. L. Zey, J. Amer. Chem. Soc., 82, 3977 (1960).
369. S. Kwee and H. Lund, Acta Chem. Scand., 25, 1813 (1971); Experientia (Suppl.) 18, 387 (1971).
370. J. Armand, Y. Armand, and L. Boulares, C. R. Acad. Sci., Ser. C, 286, 17 (1978).
371. C. Furlani, Gazz. Chim. Ital., 85, 1646 (1955).
372. J. Pinson, J.-P. M'Packo, N. Vinot, J. Armand, and P. Bassinet, Can. J. Chem., 50, 1581 (1972).
373. E. Ochiai, J. Org. Chem., 18, 534 (1953).
374. A. Foffani and E. Fornasari, Gazz. Chim. Ital., 83, 1051, 1059 (1953).
375. L. V. Varyukhina and Z. V. Pushkareva, Zh. Obshch. Khim., 26, 1740 (1956); Chem. Abstr., 51, 1960 (1957).
376. O. N. Nechaeva and Z. V. Pushkareva, Zh. Obshch. Khim., 28, 2693 (1958); Chem. Abstr., 53, 9229 (1959).
377. T. R. Emerson and C. W. Rees, J. Chem. Soc., 1923 (1962).
378. N. A. Kudryavtseva, Z. V. Pushkareva, and V. F. Gryazev, Zh. Obshch. Khim., 35, 14 (1965).
379. C. R. Warner and P. J. Elving, Coll. Czech. Chem. Commun., 30, 4210 (1965).
380. T. Kubota and H. Miyazaki, Bull. Chem. Soc. Jpn., 35, 1549 (1962); T. Kubota and H. Miyazaki, Bull. Chem. Soc. Jpn., 39, 2059 (1966).

381. H. Matschiner and H. Tanneberg, Z. Chem., 20, 263 (1980).
382. I. Suzuki, M. Nakadate, T. Nakashima, and N. Nagasawa, Tetrahedron Lett., 2899 (1966).
383. M. Tachibana, S. Sawaki, and Y. Kawazoe, Chem. Pharm. Bull., 15, 1112 (1967).
384. J. Hranilovic, D. Koruncev, and E. Gustak, Electrochem. Tech., 6, 62 (1968).
385. T. Okano and K. Ohira, Yakugaku Zasshi, 88, 1170 (1968); Chem. Abstr., 70, 16609 (1969).
386. L. Horner and H. Röder, Chem. Ber., 101, 4179 (1968).
387. N. P. Bogdanova, A. D. Cherednichenko, I. A. Avrutskaya, and M. Ya. Fioshin, Elektrokhimiya, 21, 1070 (1985); N. P. Bogdanova, E. A. Popora, I. A. Avrutskaya, and M. Ya. Fioshin, Elektrokhimiya, 21, 1095, 1214 (1985); M. Ya. Fioshin, I. A. Avrutskaya, N. P. Bogdanova, and I. I. Surov, Coll. Czech Chem. Commun., 51, 1986 (1986); M. Ya. Fioshin, I. A. Avrutskaya, I. I. Surov, and V. T. Novikov, Coll. Czech. Chem. Commun., 52, 182 (1987).
388. M. Masui, T. Ueshima, and S. Ozaki, J. Chem. Soc., Chem. Commun., 479 (1983); M. Masui, Stud. Org. Chem. (Amsterdam), 30, 137 (1987).
389. N. A. Kudryavtseva, Z. V. Pushkareva, and V. F. Gryazev, Zh. Obshch. Khim., 35, 14 (1965).
390. J. Volke, Coll. Czech. Chem. COmmun., 23, 1486 (1958).
391. J. Volke, Experientia, 13, 274 (1957).
392. J. Tirouflet and E. Laviron, C. R. Acad. Sci., Ser. C, 247, 217 (1958); E. Laviron and J. Tirouflet, C. R. Acad. Sci., Ser. C, 248, 826 (1959).
393. E. Laviron, Bull. Soc. Chim. Fr., 2325 (1961).
394. J. F. Rusling, J. P. Segretario, and P. Zuman, J. Electroanal. Chem., 143, 291 (1983).
395. J. F. Rusling and P. Zuman, J. Electroanal. Chem., 213, 245, 255 277 (1986).
396. P. E. Iversen and H. Lund, Acta Chem. Scand., 21, 279 (1967).
397. J. Tirouflet, E. Laviron, J. Metzger, and J. Boichard, Coll. Czech. Chem. Commun., 25, 3277 (1960).
398. J. Volke and V. Volkova, Coll. Czech. Chem. COmmun., 20, 1332 (1955).
399. H. H. G. Jellinek and J. R. Urwin, J. Phys. Chem., 58, 168 (1954).
400. H. Lund, Acta Chem. Scand., 17, 972 (1963).
401. P. E. Iversen and H. Lund, Acta Chem. Scand., 21, 389 (1967).
402. P. E. Iversen, Acta Chem. Scand., 24, 2459 (1970).
403. O. R. Brown, J. A. Harrison, and K. S. Sastry, J. Electroanal. Chem., 58, 387 (1975).
404. W. Zheng-Hao and H. Zhi-Bin, Electrochim. Acta, 30, 779 (1985).
405. H. Lund, Acta Chem. Scand., 17, 2325 (1963); R. Marin Galvin and J. M. Rodriguez Mellado, J. Electroanal. Chem., 250, 399 (1988); 260, 101 (1989); 265, 195 (1989).
406. H. Lund, Acta Chem. Scand., 17, 1077 (1963).
407. J. E. Toomey, Eur. Pat. Appl. EP 189678 (1986); Chem. Abstr., 105, 215760 (1986).
408. P. G. Gassman and B. L. Fox, J. Org. Chem., 32, 480 (1967).
409. J. M. Bobbitt and T. Y. Cheng, J. Org. Chem., 41, 443 (1976).
410. R. Gompper and A. Schmidt, Angew. Chem., 92, 468 (1980).
411. J. Volke, R. Kubicek, and F. Santavy, Coll. Czech. Chem. Commun., 25, 1510 (1960).

412. E. Laviron, C. R. Acad. Sci., Ser. C, 250, 3671 (1960).
413. V. Krishnan, K. Raghupathy, and H. V. K. Udupa, J. Electroanal. Chem., 88, 433 (1978).
414. J. E. Toomey, U.S. Pat. 4,482,437 (1984); Chem. Abstr., 102, 53067 (1985); J. E. Toomey, C. A. Chaney, and M. Wilcox, Stud. Org. Chem. (Amsterdam), 30, 245 (1987).
415. T. L. Antonova, L. N. Ivanovskaya, I. A. Avrutskaya, M. Ya. Fioshin, and L. I. Gorbacheva, Elektrokhimiya, 22, 546 (1986).
416. J. Volke and J. Holubek, Coll. Czech. Chem. Commun., 28, 1597 (1963).
417. D. N. Schluter, T. Biegler, E. V. Brown, and H. H. Bauer, Electrochim. Acta, 21, 753 (1976).
418. H. Lund and S. K. Sharma, Acta Chem. Scand., 26, 2329 (1972).
419. O. Manousek and P. Zuman, Coll Czech. Chem. Commun., 29, 1432 (1964).
420. D. L. Smith and P. J. Elving, J. Amer. Chem. Soc., 84, 1412 (1962).
421. S. Kwee and H. Lund, Acta Chem. Scand., 26, 1195 (1972).
422. G. Dryhurst and P. J. Elving, J. Electrochem. Soc., 115, 1014 (1968).
423. R. N. Goyal, A. Minocha, and A. R. Nautiyal, J. Electroanal. Chem., 200, 119 (1986); R. N. Goyal, Indian J. Chem., 26A, 656 (1987).
424. G. Dryhurst, J. Electrochem. Soc., 117, 1118 (1970).
425. C. Dreux, M.-L. Girard, and P. Souchay, C. R. Acad. Sci., Ser. C, 262, 1565 (1966).
426. H. H. Ruttinger, H. Matschiner, and W. Schroth, J. Prakt. Chem., 321, 274 (1979).
427. M. Wrona and B. Czochralska, J. Electroanal. Chem., 48, 433 (1973); M. Wrona, B. Czochralska, and D. Shugar, J. Electroanal. Chem., 68, 355 (1976); M. Wrona, J. Electroanal. Chem., 104, 243 (1979); M. Wrona and M. Geller, J. Electroanal. Chem., 104, 263 (1979).
428. G. Dryhurst, J. Electrochem. Soc., 117, 1113 (1970).
429. P. J. Kraske and A. Brajter-Toth, J. Electroanal. Chem., 207, 101 (1986).
430. P. E. Iversen, Synthesis, 484 (1972).
431. R. N. Gedye, Y. N. Sadana, and R. Leger, Can. J. Chem., 63, 2669 (1985).
432. K. Sugino, K. Shirai, T. Sekine, and K. Odo, J. Electrochem. Soc., 104, 667 (1957).
433. R. D. Chambers, W. K. R. Musgrave, C. R. Sargent, and F. G. Drakesmith, Tetrahedron, 37, 591 (1981).
434. D. Kyriacou, Eur. Pat. Appl. 18,069 (1980); Chem. Abstr., 94, 54933 (1981).
435. F. Y. Edamura, D. Kyriacou, and J. Love, U.S. Pat. 4,217,185 (1980); Chem. Abstr., 94, 22193 (1981); DOW Chemical Co., Jpn. Kokai Tokkyo Koho J.P. 86 03893 (1986); Chem. Abstr., 105, 14204 (1986); C. K. Bon, A. J. Kamp, and T. J. Sobieralski, U.S. Pat. 4,592,810 (1986); Chem. Abstr., 105, 69063 (1986).
436. R. M. Justice and D. A. Hall, U.S. Pat. 4,588,484; Chem. Abstr., 105, 69060 (1986).
437. M. A. Cadadei, F. M. Moracci, and A. Inesi, J. Chem. Soc., Perkin Trans. 2, 419 (1986).

19
Natural Products and Pharmaceuticals

YOSHIO BAN / Hokkaido University, Sapporo, Japan

TAMEO IWASAKI AND HIROSHI OHMIZU / Tanabe Seiyaku Company, Ltd., Osaka, Japan

I. Introduction 765
II. Amino Acids 765
III. Nucleic Acid Bases and Nucleosides 768
IV. Alkaloids 770
V. Carbon Aromatics 775
VI. Terpenoids and Steroids 777
VII. Antibiotics 779
VIII. Miscellaneous 782
References 783

I. INTRODUCTION

This chapter discusses the electrosyntheses that have appeared during the last decade of biologically active natural products and pharmaceuticals. Particular emphasis is placed on the syntheses of these compounds using an electrode reaction as a key step, which are meaningful from viewpoint of organic synthesis. The discussion is divided in seven sections as follows:

1. Amino acids
2. Nucleic acid bases and nucleosides
3. Alkaloids
4. Carbon aromatics
5. Terpenoids and steroids
6. Antibiotics
7. Miscellaneous

II. AMINO ACIDS

Some analogs of the common amino acids exhibit remarkable biological activities due to their antagonistic action. During the last decade, novel synthetic methods of these biologically active amino acids have been developed utilizing electrode reactions.

Armentomycin, which was isolated from natural sources, has demonstrated significant in vitro growth inhibitory activities against several microorganisms. This can be electrosynthesized in an optically active form by selective reduction of the one C—Cl bond of γ,γ,γ-trichloro-L-butyrine [1], which is readily derived from L-methionine, Eq. (1):

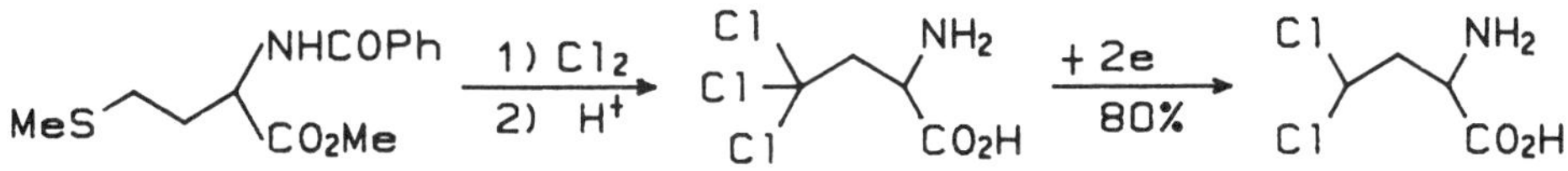

(1)

The method is applicable to a synthesis of β,γ-unsaturated γ,γ-dichlorobutyrines and threo-γ,γ-dichloro-β-hydroxybutyrines, some of which show a stronger growth inhibitory activity than armentomycin against Pseudomonas aeruginosa. The same authors take advantage of L-methionine as a starting material for a synthesis of "ethanebutol," which has been clinically used as a tuberculostatic, Eq. (2) [2]. The key step in a series of the reactions involves the cathodic reductive cleavage of the sulfonium ion.

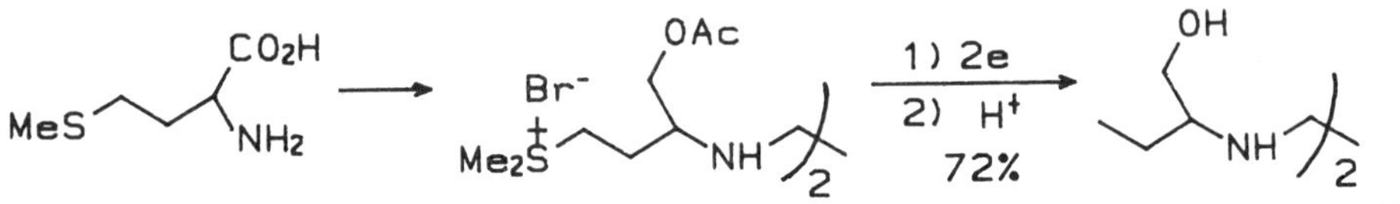

(2)

α-Functionalized α-amino acids, typified by α-hydroxy, α-methoxy, and α-acetoxy α-amino acids, are considered "cationic amino acid synthons," which are potent intermediates in the synthesis of biologically active α-substituted α-amino acids. The synthesis of these amino acids having a labile N,O-acetal skeleton can be realized by anodic functionalization based on the Kolbe reaction. The anodic oxidation of N-acylaminomalonic acid monoesters in methanol-sodium methoxide gives α-methoxy-α-amino acids in almost quantitative yields [3]. The use of acetic acid-sodium acetate yields synthetically more valuable α-acetoxy-α-amino acids [4], which react with a variety of nucleophiles under base or Lewis acid-catalyzed conditions to give the α-substituted α-amino acids having antitumoral, antibacterial, and insulin-releasing activities, Eq. (3) [5–7]:

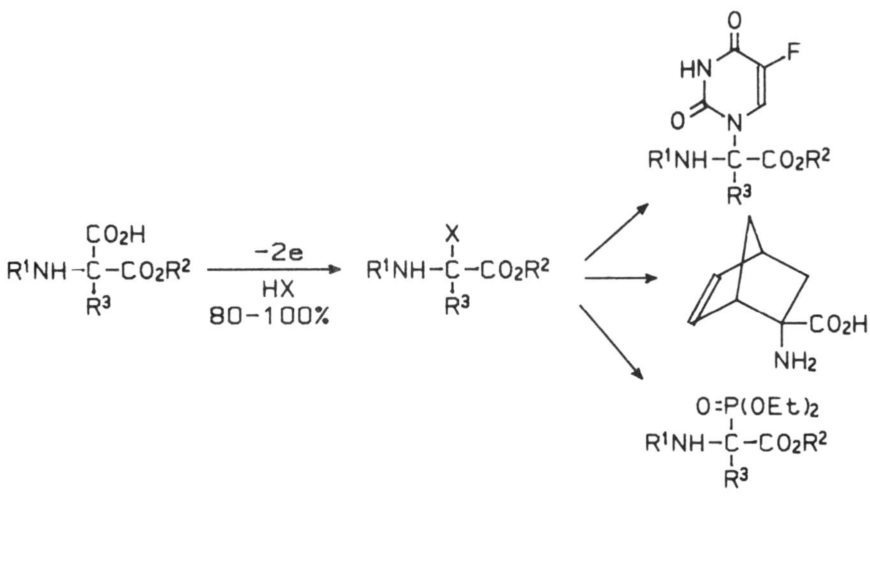

(3)

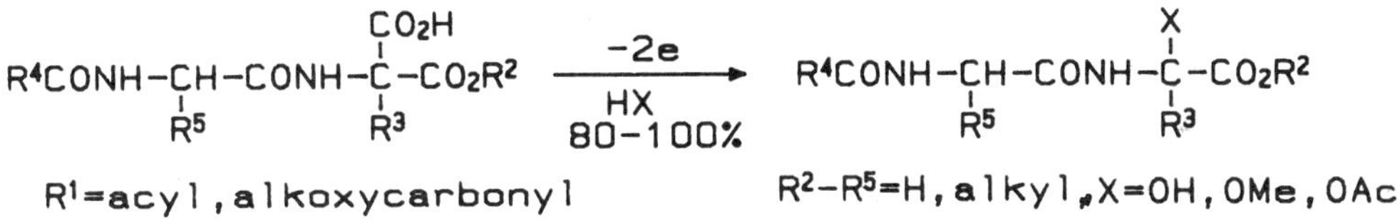

(4)

The method is applicable to the functionalization of peptides [8] in which the cleavage of the peptide bond does not occur, Eq. (4). Thus, in principle, any type of functional group may be introduced into amino acids and peptides in two steps.

The functionalization leading to α-methoxy-α-amino acids can also be achieved by anodic oxidation of N-methoxycarbonylamino acid esters in a mediatory system consisting of methanol and sodium chloride [9]. On the basis of this methodology, the basic amino acids, such as L-lysine and L-ornithine, are convertible into the unsaturated cyclic imino acids, Eq. (5), which are useful intermediates for the synthesis of optically active pyrrolidine and piperidine alkaloids [10].

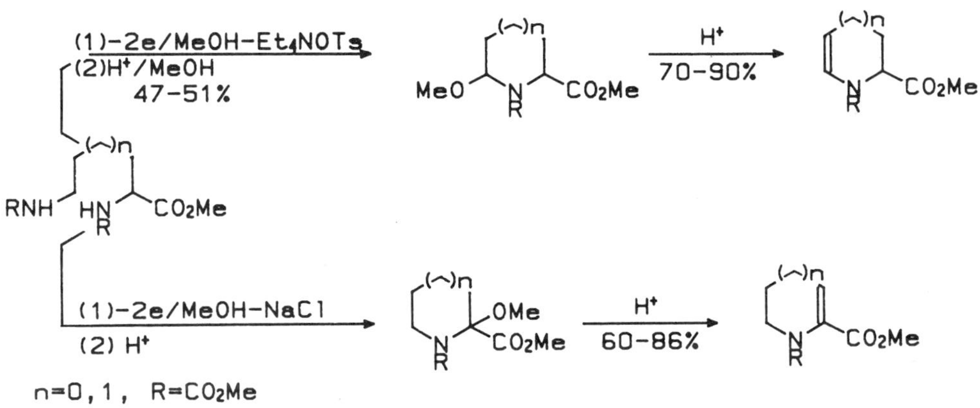

(5)

N,O-Acetals may be regarded as synthetic equivalents for aminoaldehydes, which are indispensable intermediates in the synthesis of α-amino acids. Thus, the cyclic N,O-acetals prepared by anodic oxidation of the cyclic amines are transformed into tryptophan, Eq. (6), [11,12] and lysine, Eq. (7) [13]:

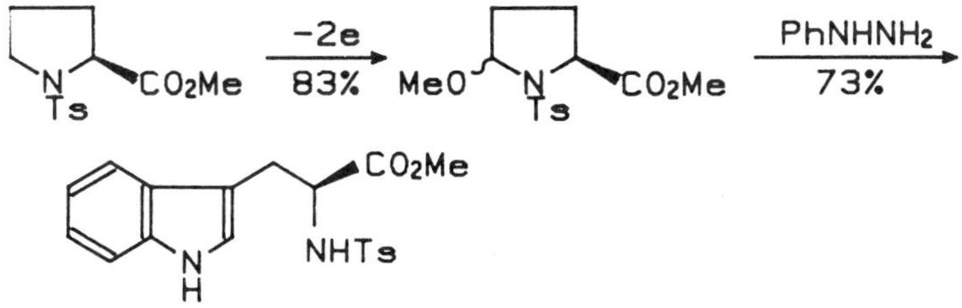

(6)

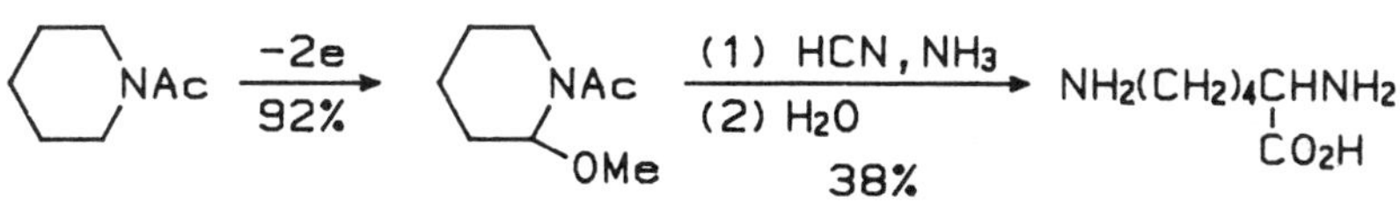

(7)

The cathodic addition of the anionic species generated from alkyl halides to imines seems to be effective in the synthesis of α-substituted α-amino acids [14]. The reduction of 3,4-methylenedixoybenzyl bromide in the presence of the Schiff base derived from benzylamine and methyl pyruvate leads to the formation of α-methyl dopa, an antihypertensive agent [15].

III. NUCLEIC ACID BASES AND NUCLEOSIDES

During the last two decades, the analogs of the nucleic acid bases and nucleosides of natural origin have been synthesized to obtain clinically useful chemotherapeutic agents.

5-Substituted uracils are well known to display antitumoral and/or antiviral activities. One synthetic approach to these compounds involves the introduction of a double bond accompanying aromatization based on Kolbe decarboxylation. The orotic acid derivatives that are readily obtained from aspartic acid are oxidized to afford the corresponding 5-substituted uracils in satisfactory yields, Eq. (8) [16].

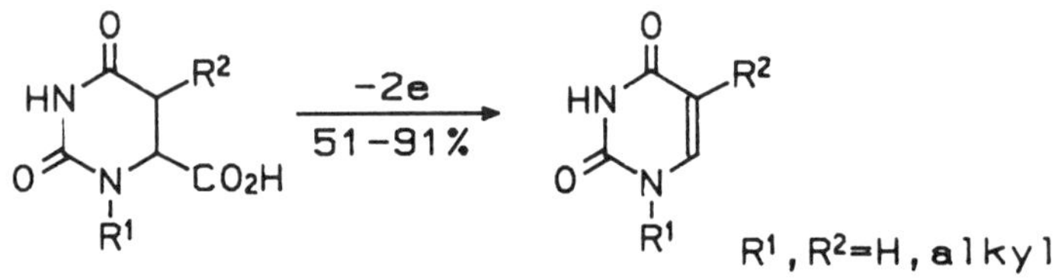

(8)

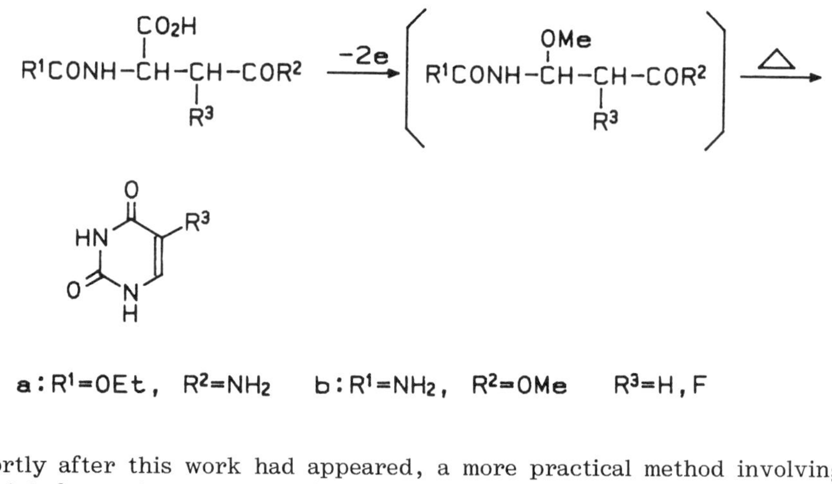

(9)

Shortly after this work had appeared, a more practical method involving the straightforward conversion of aspartic acid derivatives into uracil and its derivatives was developed, Eq. (9) [17,18]. 5-Fluorouracil, an antitumoral agent, can be prepared by this method [18]. The entire process of this method may be conducted in a single container.

1-(Tetrahydrofuran-2-yl)-5-fluorouracil (florafur), a pyrimidine antimetabolite, is an effective antitumoral agent. This can be synthesized in an excellent overall yield from furoic acid using electrode reactions as key steps, which involves both the anodic and cathodic processes, Eq. (10) [19]:

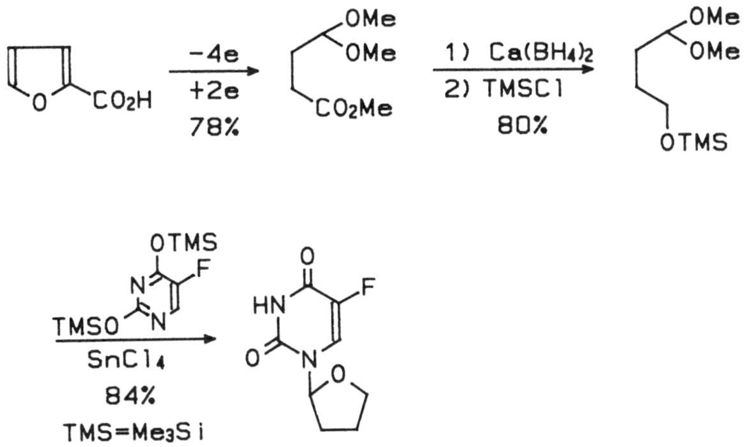

(10)

This method is highly promising in the synthesis of the 5-fluorouracil derivatives having substituted tetrahydrofurans [20]. The aza analogs in which the tetrahydrofuran ring of these compounds is replaced by azacycloalkanes and lactams can also be synthesized [21].

Modification of the sugar moiety of nucleosides is of particular interest in connection with the search for antineoplastic agents. 2',3'-Unsaturated nucleosides, which have been prepared with difficulty by conventional chemical methods, can readily be synthesized by cathodic reduction of the vicinal halohydorin acylates without appreciable cleavage of the glycosidic bonds, Eq. (11) [22]:

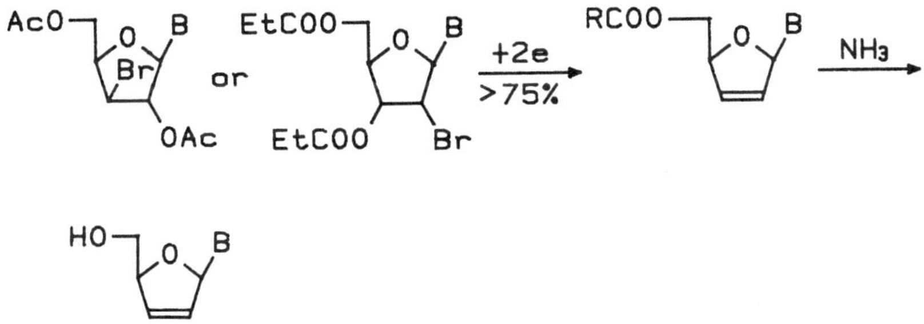

(11)

B:adenine,N⁴-acetylcytosine,thymine

Even in the case of the cathodic reduction of the nucleosides having cis-configuration of the 2',3'-substituents, no glycocidic bond cleavage takes place but the desired 2',3'-olefins are obtained. The electrochemical method should be contrasted with the chromous acetate-induced elimination reaction, in which extensive glycosidic bond cleavage leading to uracil and 2'-olefins takes place. It has been shown that this methodology is quite effective in the synthesis of some natural products, such as formycin and tubercidine [23].

Cathodic 1,3-elimination reaction of the 3',5'-diiodonucleosides yields novel nucleosides having a fused cyclopropane ring on the sugar moiety, Eq. (12) [24].

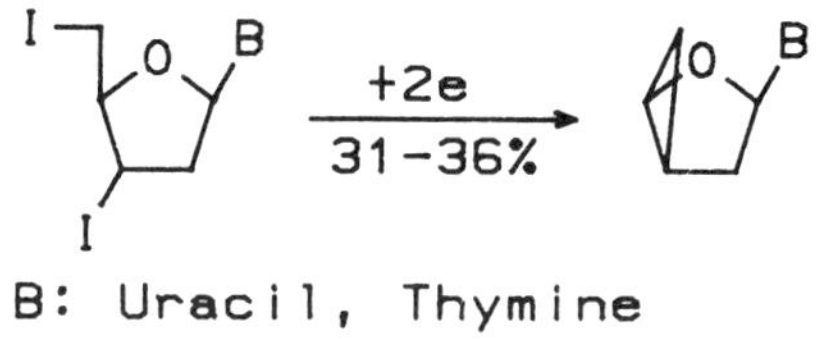

(12)

The synthesis of 4'-substituted nucleosides was also examined using an electrode reaction. The cationic species generated by Kolbe anodic oxidation of uridine 4'-carboxylic acid is trapped either by methanol or C_2-carbonyl of the pyrimidine ring to form the cyclonucleosides, Eq. (13) [25]:

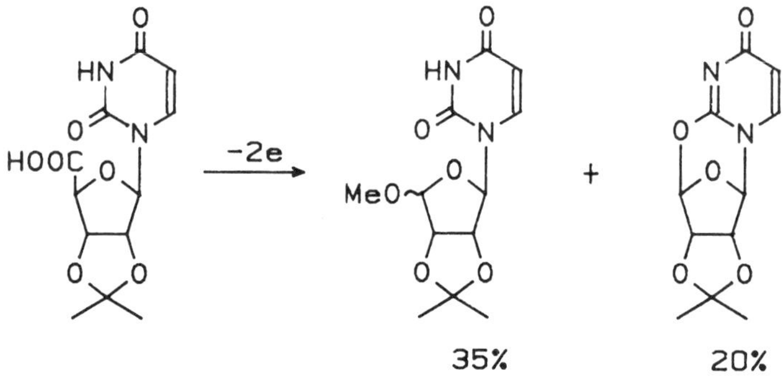

(13)

IV. ALKALOIDS

Many recent advances have been made in the electrosynthesis of alkaloids, and indeed many elegant syntheses have been reported. In this section, the methods are classified according to the types of reaction employed rather than according to the types of compound prepared.

Electrochemical aryl-aryl coupling reaction has been widely used to synthesize a certain type of alkaloid. The first approach involving intramolecular phenolic oxidative coupling has been impressively illustrated by the synthesis of O-methylflavinatine from laudanosine, Eq. (14) [26]:

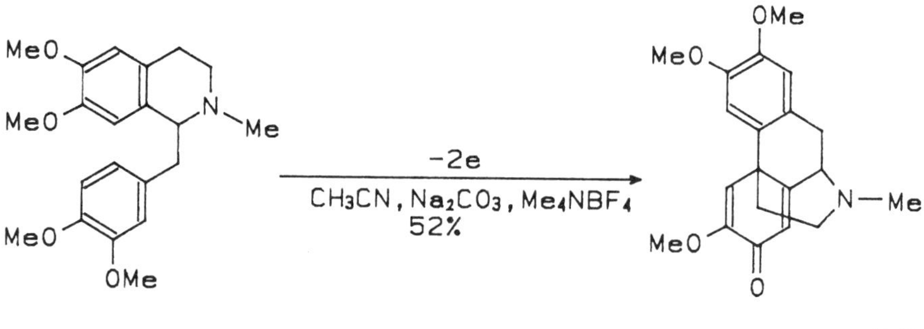

(14)

This type of anodic coupling reaction is useful in the synthesis of oxocrine, oxomaritidine, amurine, pallidine, and cryptopleuring [27,28]. How to control the regiochemistry of the coupling sites in this type of reaction has remained unsolved, however. The mechanism of this type of reaction is not always simple; under acidic conditions, the initial electron transfer takes place probably from the aromatic ring system but not the nitrogen atom [29].

Several alkaloids have been synthesized based on the recently developed technique of the intramolecular cathodic reductive coupling of aryl halides. It was shown that the addition of the radical species generated by cathodic reduction of one aryl halide to another present in the same molecule took place effectively to lead eventually to aporphines, Eq. (15) [30]:

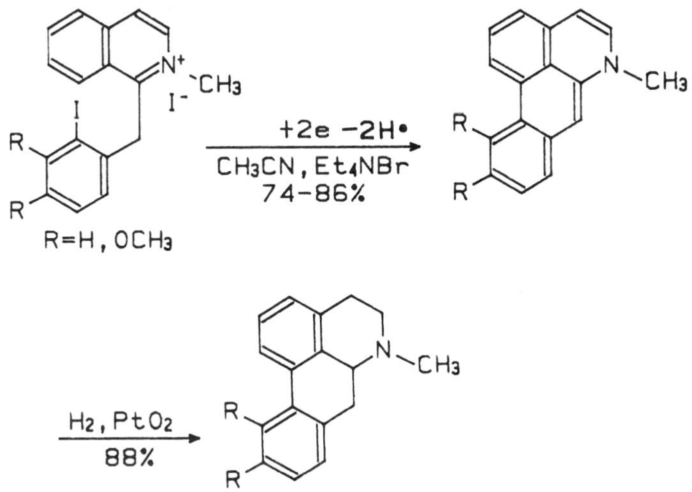

(15)

The aryl radicals formed by cathodic reduction of aryl halides are likewise intramolecularly captured by the neighboring N-aryl groups to afford the isoquinoline skeletons, Eq. (16) [31]:

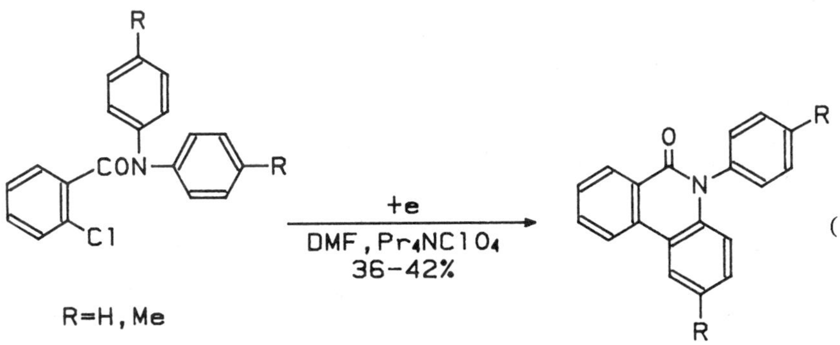

(16)

The intramolecular coupling between aryl group and functional groups other than aromatic rings can also be realized. Equation (17) illustrates the intramolecular carbon-carbon bond-forming reaction between the aryl group and the enaminoketone moiety to form a novel class of the heterocyclics [32].

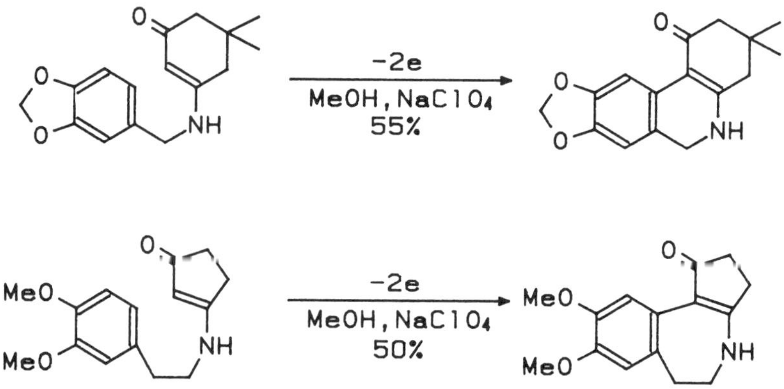

(17)

The intramolecular nucleophilic capture of the reactive intermediate generated by anodic oxidation of epinephrine by the N-methylamino group is quite effective in forming adrenochrome, which is convertible into sodium carbazochrome sulfonate, an antihemorrhagic, Eq. (18) [33]:

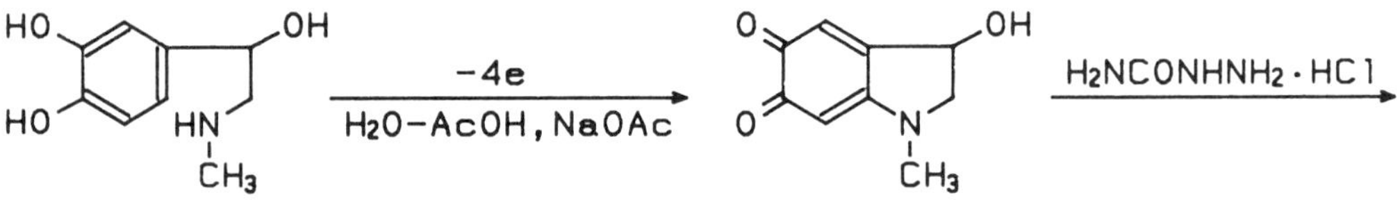

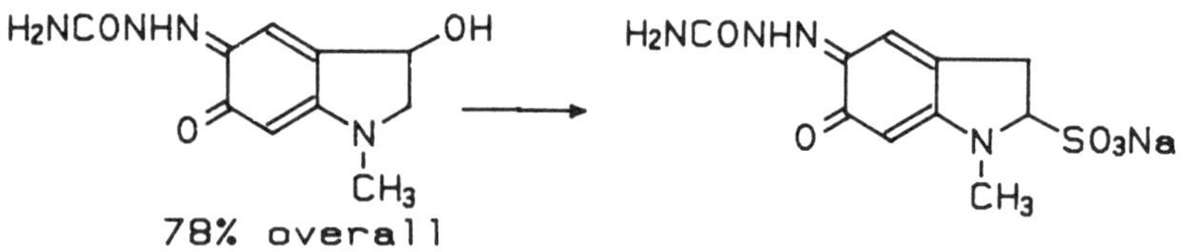

(18)

This reaction can be conducted on a large scale, which is an advantage over those using such chemical oxidants as potassium ferricyanide, manganese dioxide, and lead dioxide.

The synthetic routes involving electrochemical carbon-carbon bond forming reactions in the position α to nitrogen atoms are in many ways unsurpassed as versatile routes to a whole range of alkaloids.

The anionic species generated by cathodic reduction of imminium salts are capable of reacting with alkyl halides, furnishing isoquinoline and indol alkaloids. The following examples, Eqs. (19), are illustrative:

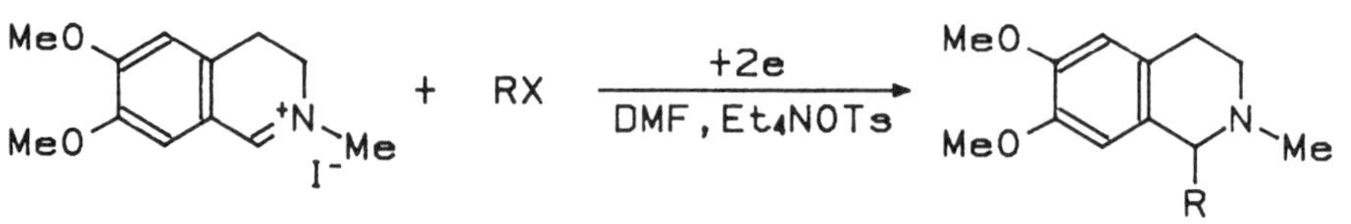

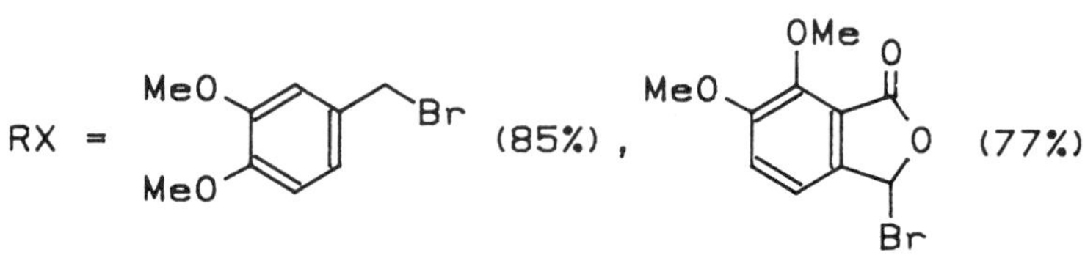

(19)

Laudanosine [34] and condrastine [35] were synthesized by cathodic reduction of 2-methyl-3,4-dihydro-6,7-dimethoxyisoquinolinium iodide in the presence of 3,4-dimethoxybenzyl bromide and 3-bromomeconine, respectively. The intramolecular variant of this reaction can likewise be realized to give berbin and yohimban alkaloids [34]. Since some of the iminium salts are readily available, this type of reaction seems to be one of the practical methods for the synthesis of this type of alkaloids.

A most remarkable methodology for syntheses of a wide range of alkaloids involves a sequence of anodic α-functionalizations of amides and nucleophilic substitution to form carbon-carbon bonds at positions α to nitrogen atom (for a review, see Ref. 36). Although this methodology has only recently appeared in print, it is already of proven value in alkaloid syntheses. Equation (20) shows the prominent use of this methodology in a formal total synthesis of vincamine [37]:

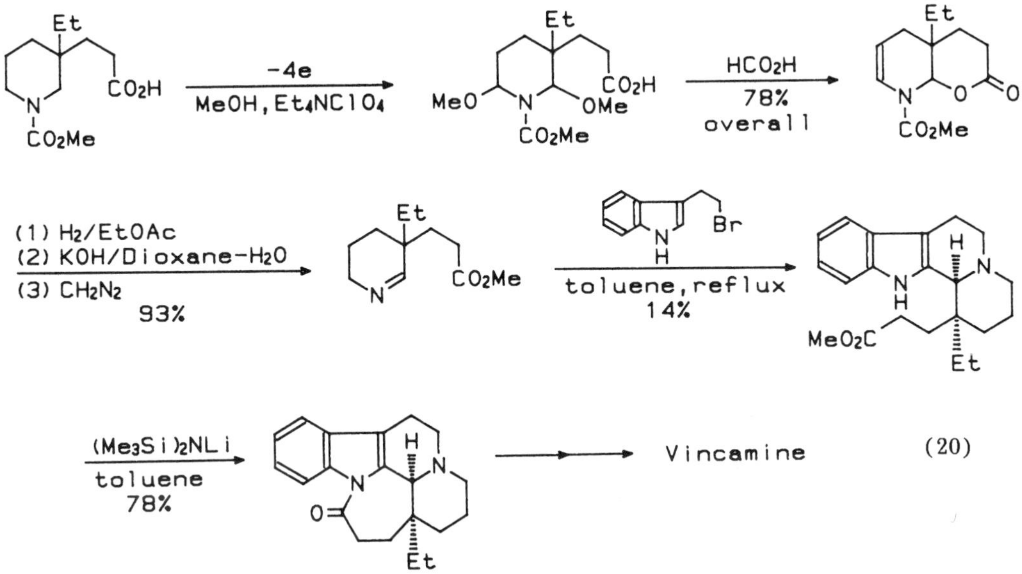

(20)

The key step in a series of the reactions involves the anodic conversion of 1-carbomethoxy-3-ethyl-3-(β-carboxyethyl)piperidine into the 2,6-dimethoxypiperidine derivative, which is susceptible to intramolecular nucleophilic substitution.

The anodic oxidation of the amides, followed by intramolecular substitution by the phenyl, Eq. (21) [38], and active methyl group, Eq. (22) [38a], yields the isoquinoline derivative and the tropanone skeleton, respectively.

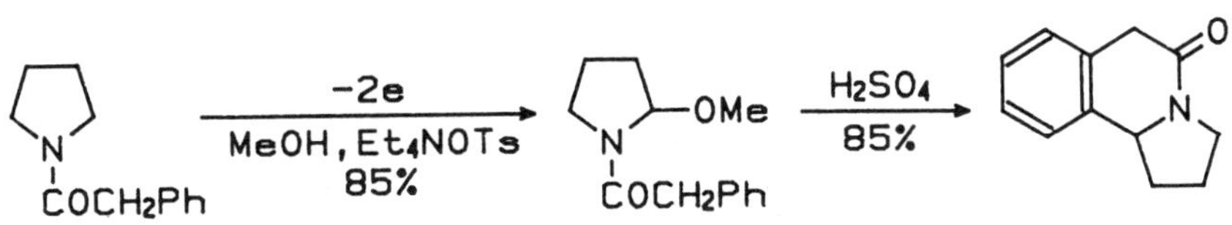

(21)

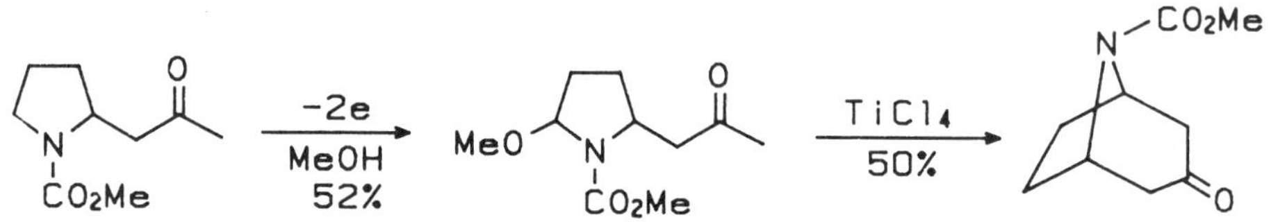

(22)

The application of this methodology has been successfully extended to the syntheses of pyrrolizidine, indolizidine, and quinolizidine alkaloids [39].

The synthesis of 1-azabicyclo[4.n.0]systems can likewise be realized by using a similar substitution reaction between furan and α-methoxycarbamate as a key reaction, in which furan acts as both the nucleophile and the precursor of the enedione system, Eq. (23) [40]:

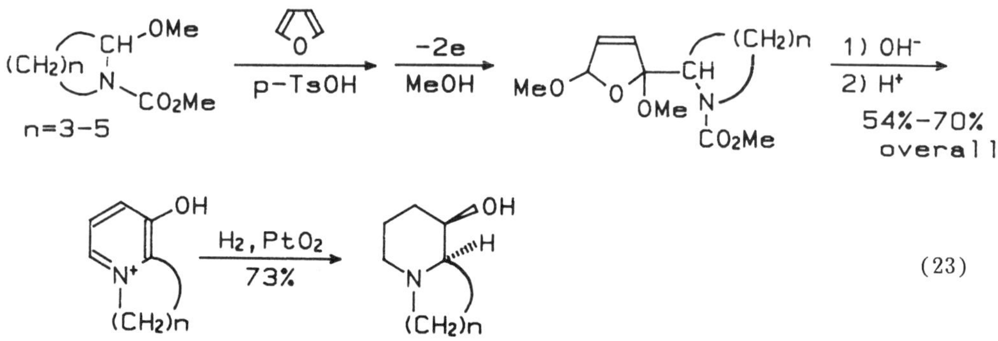

(23)

The α-hydroxycarbamate is also an effective reagent in generating the radical species on the carbon α to the nitrogen atom by cathodic reduction in the presence of chlorotrimethylsilane. For example, the cathodic reduction of N-benzyloxycarbonyl-2-hydroxypyrrolidine in the presence of chlorotrimethylsilane and dimethyl maleate, followed by hydrogenolysis yields the pyrrolizidine alkaloid in a good overall yield, Eq. (24) [41]:

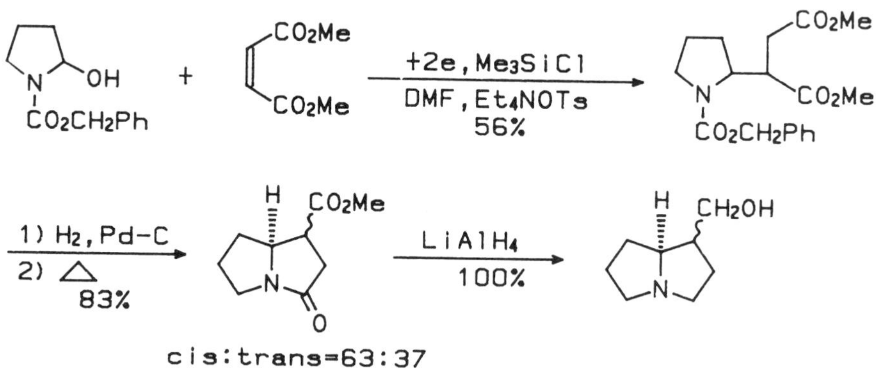

(24)

1-Benzazocinone derivative, which is envisioned as an important precursor for a synthetic strategy toward mitomycins, can be synthesized by control of crisscross annulation based on the newly developed cathodic removal of the N-tosyl protecting group with the combined assistance of anthracene and ascorbic acid, Eq. (25) [42]:

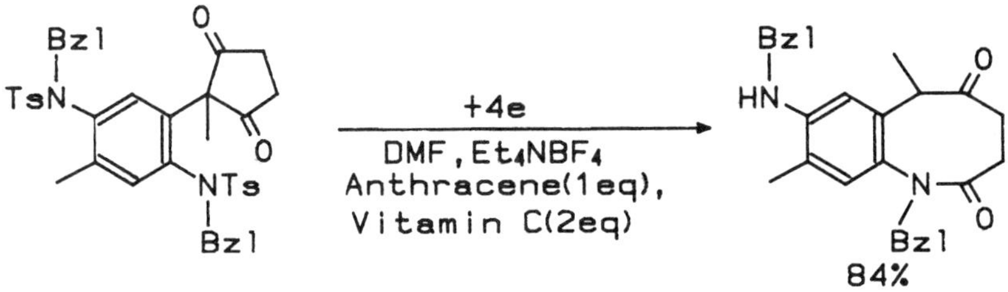

(25)

The cathodic reduction may be explained by an ECEC mechanisms, as shown in Eq. (26):

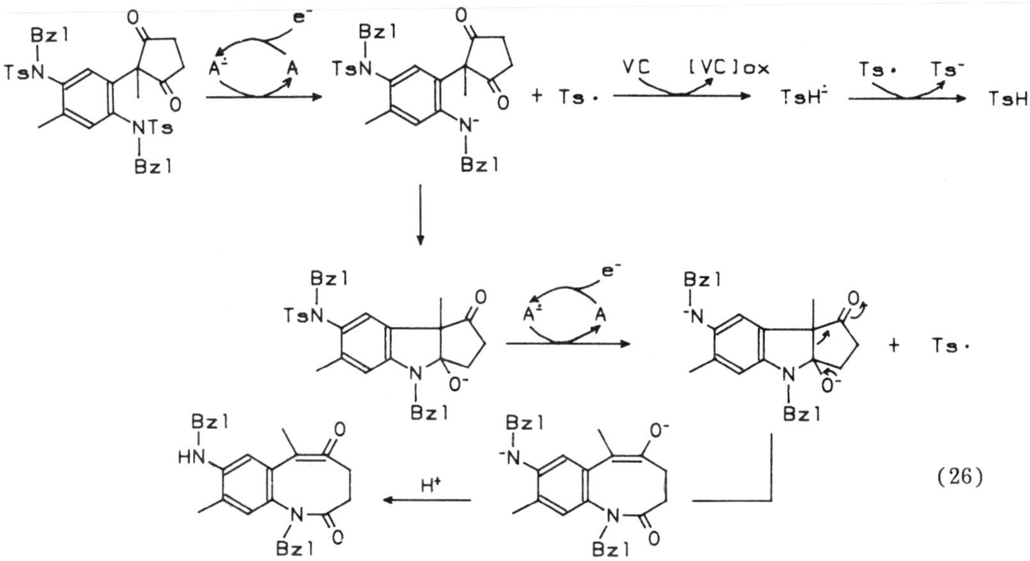

(26)

V. CARBON AROMATICS

Carbon aromatics include lignans and flavonoids. In recent years dramatic developments have been made in the electrosynthesis of lignans and neolignans. These compounds have attracted much interest over the years on account of their widespread occurrences in nature and their broad range of biological activities.

The oxidative coupling reactions of cinnamyl alcohols, cinnamic acids, and propenyl alcohols constitute biomimetic syntheses of lignans and neolignans. Anodic oxidation of 4-allyl-2-methoxyphenol in N–NaOH–MeOH yields the radical coupling product in an almost quantitative yield. On the other hand, the replacement of N–NaOH by $LiClO_4$ gives a high yield of the cation-derived product, the dienone, which cyclodimerizes spontaneously at room temperature to form the asatone compound quantitatively, Eq. (27) [43]:

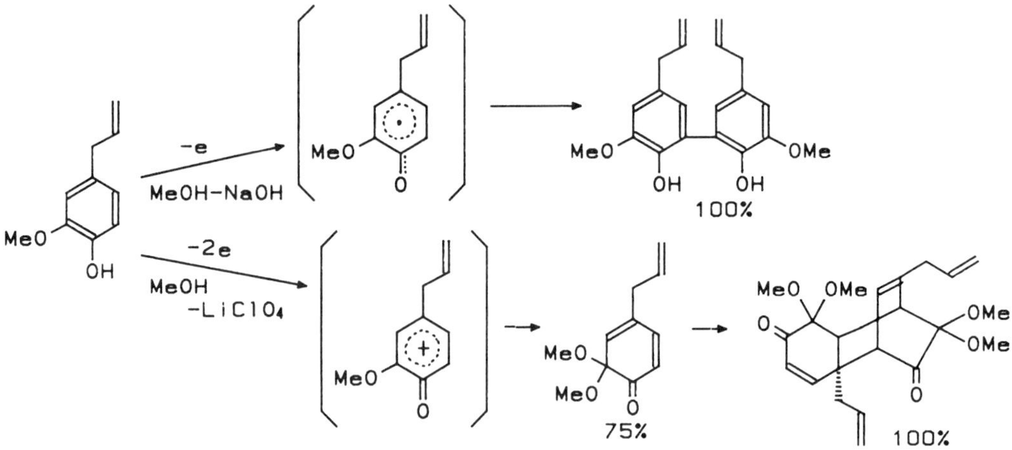

(27)

In the anodic oxidation of the cinnamic acid derivatives, the dilactone forms in a 68% yield under conditions of higher substrate concentration; in the lower substrate concentration, only the isoasatone compound is obtained, Eq. (28) [44]:

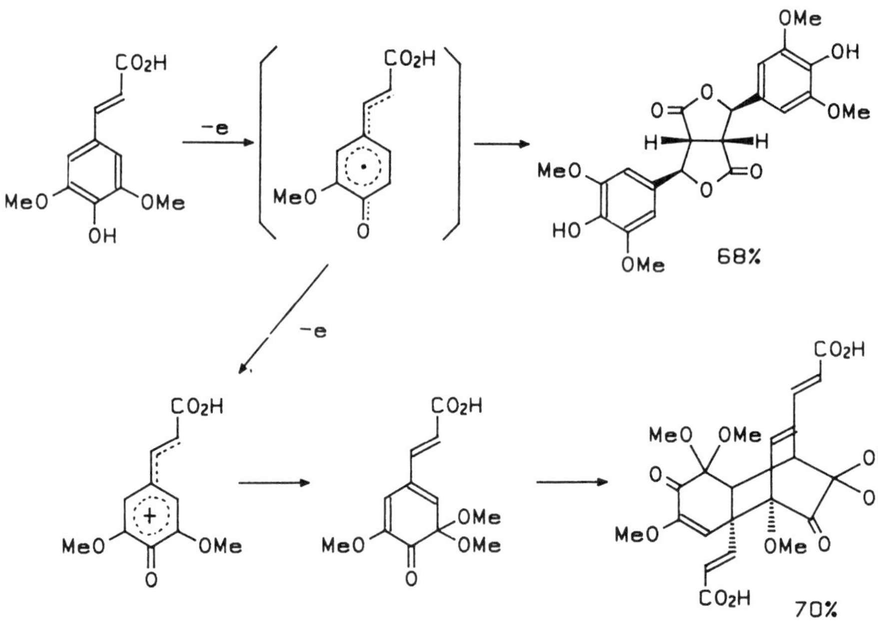

(28)

The dilactones as described here have unambiguous structures and serve as the key intermediate for a wide range of structural modifications. The dilactones, for example, are convertible under acid-catalyzed conditions to podophyllotoxin arylnaphthalene lignans with antitumoral activities.

Analogous transformations have been reported in the anodic oxidations of 2-methoxy-6-allylphenol [45], 2-methoxy-4-(2-propenyl)phenol [46], and 3,5-dimethoxy-4-hydroxycinnamic acid amide [47].

The mixed anodic coupling reactions of the different types of phenols likewise proceed in a satisfactory yield to afford aniba and magnolia neolignans, Eq. (29) [48]:

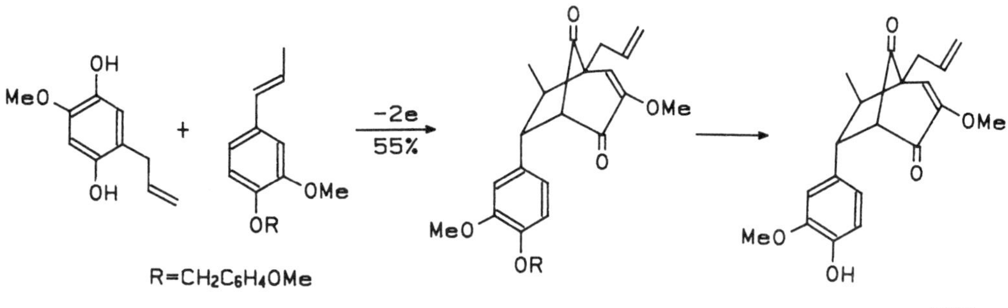

(29)

In these reactions, acidic medium is required for the reaction to proceed; under these conditions, the cationic species generated in the initial stage of the reaction is stabilized enough for an intermolecular Diels-Adler reaction to take place.

VI. TERPENOIDS AND STEROIDS

In conncetion with the production of perfumes, some electrochemical reactions have been applied to the conversion of the readily available compounds into valuable terpenoids. l-α-Pinene can be transformed into l-carvone via four steps, Eq. (30) [49]:

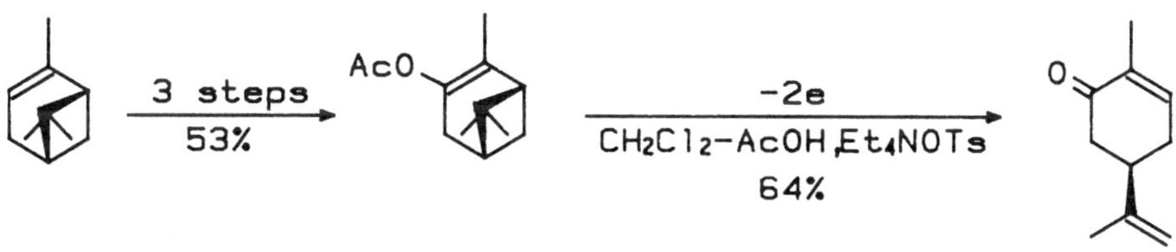

(30)

The key step involves a new type of electrode reaction, in which the enol acetate is anodically converted to the α,β-unsaturated enone. Another example is provided by Shono's group in their approach to a stereospecific transformation of l-menthone into l-citronellol [50].

The conversion of α-dihydroionol into d,l-theaspirane is also intriguing; the electrode reaction involves allylic oxidation in the selenium-mediated system, Eq. (31) [51]:

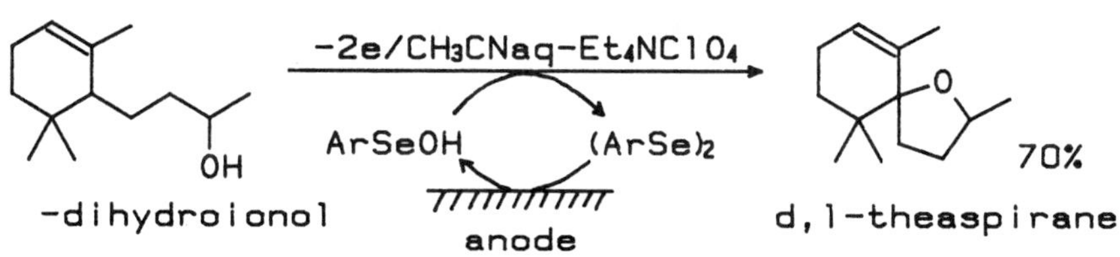

(31)

The method is suitable for the syntheses of d,l-rose oxide, d,l-marmelactone [52], and d,l-dihydroactinidiolide [53].

Electrochemical reactions seem to be quite effective for biogenetic syntheses of terpenoids. Torii et al. reported the interesting synthetic method of some isoprenoids, such as ionene, α-damascone, limonene, and terpinoline [54], by electrogenerated acid-mediated cyclization, Eq. (32):

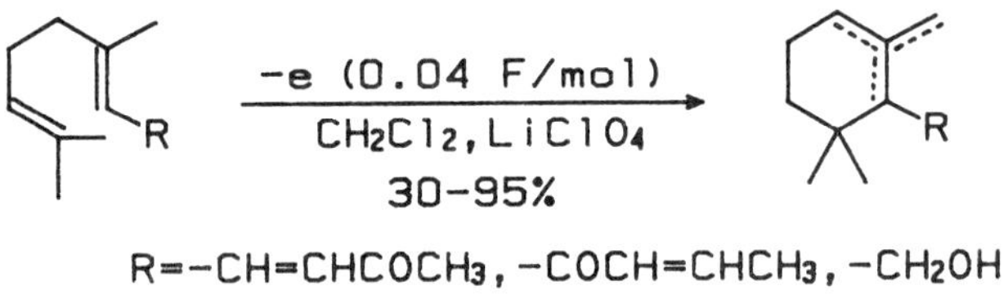

(32)

Application of electrochemical reaction to the syntheses of carotenoids has been studied extensively by Utley et al. For example, astaxanthin, an important carotenoid, was synthesized by the cathodic reduction of astacene in the presence of acetic anhydride followed by hydrogenolysis, Eq. (33) [55]:

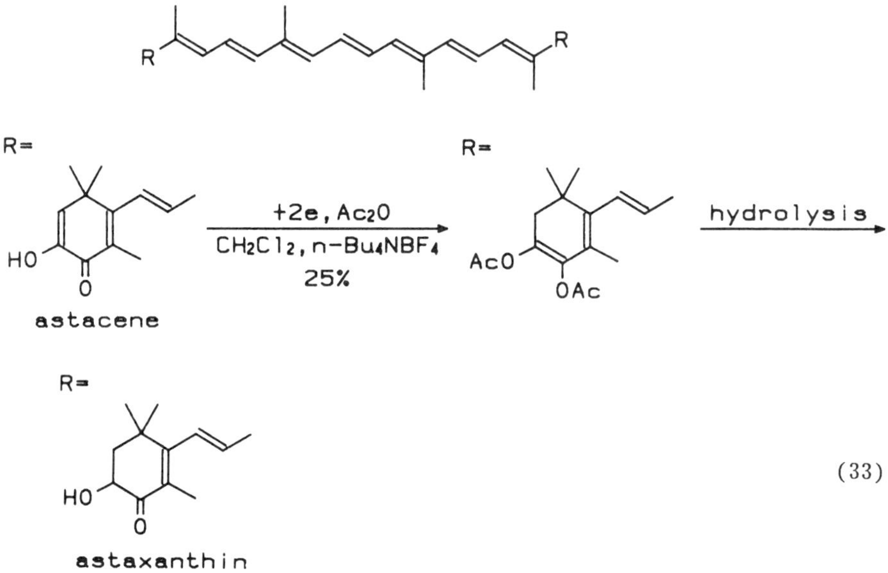

(33)

As for the syntheses of steroids that are important as clinical drugs, electrochemical functionalization of the side chains of steroids has been studied. Direct regioselective introduction of functional groups to the CH and CH_2 groups of steroids enable valuable products to be obtained in a few steps from inexpensive starting materials.

Androstan can be oxidized with remarkable regioselectivity by the newly developed reaction conditions comprising trifluoroacetic acid-trifluoroacetic anhydride in dichloromethane to yield androstan-6-one in an excellent yield, Eq. (34) [56]. The oxidation of 3-β-acetoxyandrostan also takes place regioselectively at the 17-position, Eq. (35) (see Ref. 57 and references cited therein):

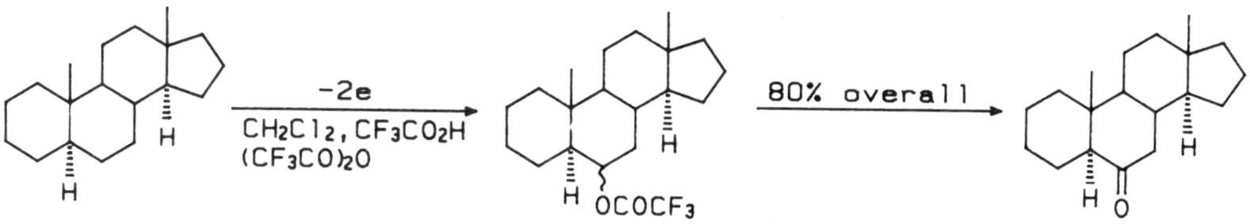

(34)

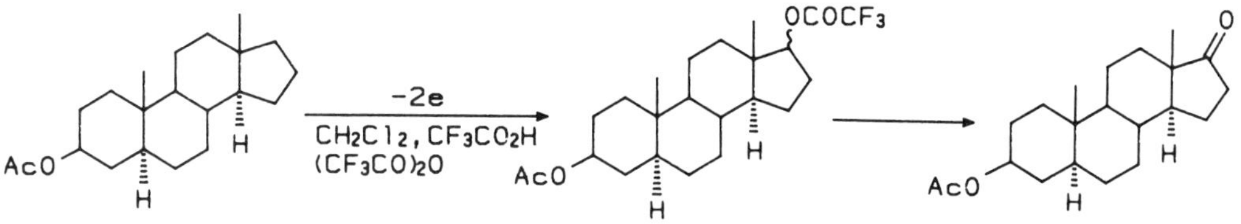

(35)

The 3-methoxyestrone is susceptible to nucleophilic attack regioselectively on the 9-position under the specified reaction conditions to afford 9-substituted 3-methoxyestrones, Eq. (36) [58]:

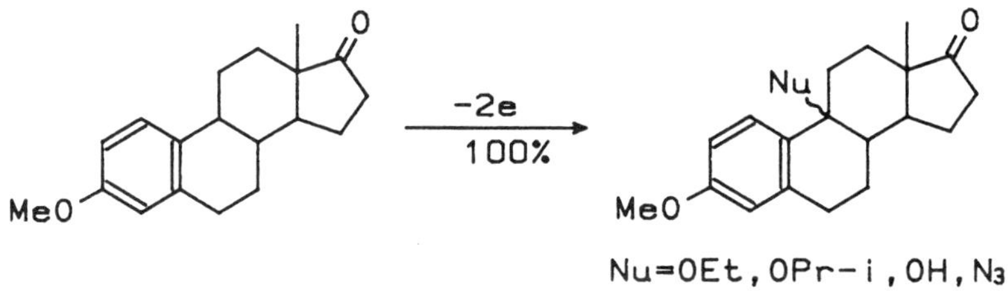

(36)

In these reactions, however, no stereoselectivity is observed. In the case of anodic cyanation of 3-methoxyestrone, the cyano anion attacks exclusively on the A ring [59].

Biomimetic functionalization of steroids was reported by Breslow et al., which was based on template-directed electrochemical chlorination. The oxidation of 3-cholestanyl m-iodobenzoate at 2.7 V versus Ag results in a remarkably regioselective chlorination at the C-9 position [60].

As to the stereochemistry of anodic oxidation of steroids, anodic oxidation bears a marked resemblance to microsomal oxidation, since both types of oxidation take place at the interface [61].

VII. ANTIBIOTICS

Among the existing antibiotics, cephalosporins and anthracyclines are of particular interest because of their significant therapeutic properties. The clinical demand for these antibiotics has provided a powerful stimulus to basic research of synthetic organic chemistry on these compounds with complex chemical structure. During the last decade, elegant and, in some cases, practical electrosyntheses have been reported.

Torii et al. have shown the straightforward transformation of penicillin into 3'-chlorocephalosporin, which is a key intermediate in the synthesis of clinically useful 3'-substituted cephalosporins. The conversion of 4-arenesulfonylthioazetidin-2-one derived from penicillin G and V into 3-chloro-Δ^3-cephems comprises electrolytic ene chlorination and base-catalyzed cyclization of the chloride, Eq. (37) [62]:

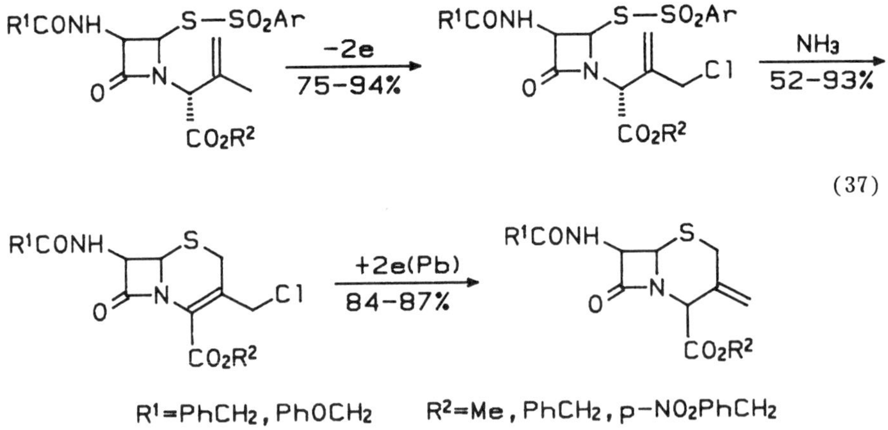

(37)

The anodic chlorination may be conducted in a two-phase system containing $CHCl_3$, aqueous NaCl, and H_2SO_4. The 3-chlorocephems can be readily converted into 3-methylenecephams by cathodic reduction in aqueous tetrahydrofuran-lithium perchlorate using a Pb cathode.

Shortly after this work appeared, Torii's group succeeded in a direct two-step conversion of thiazoline-azetidinones derived from penicillin G and V to 3'-thio-substituted cephalosporins, Eq. (38) [63]:

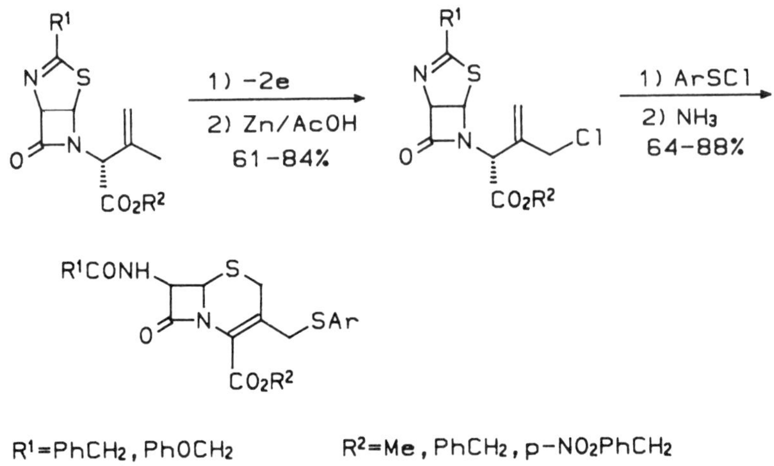

(38)

It is notable that cefazoline, a third-generation cephalosporin, was synthesized.

In ene anodic chlorination, the reaction is conducted in a two-phase system as described that brings about no appreciable cleavage of either the thiazoline or the β-lactam ring.

Swenton et al. showed that anodic oxidation is potentially an excellent method for the conversion of oxygenated aromatics with a high degree of functionality into the respective quinone monoketals, Eq. (39) [64]:

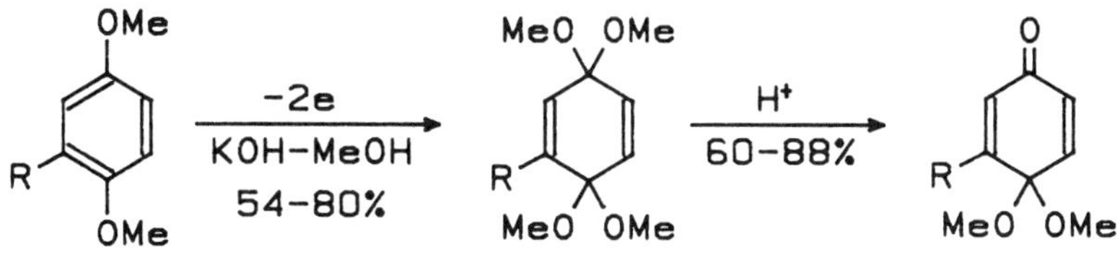

(39)

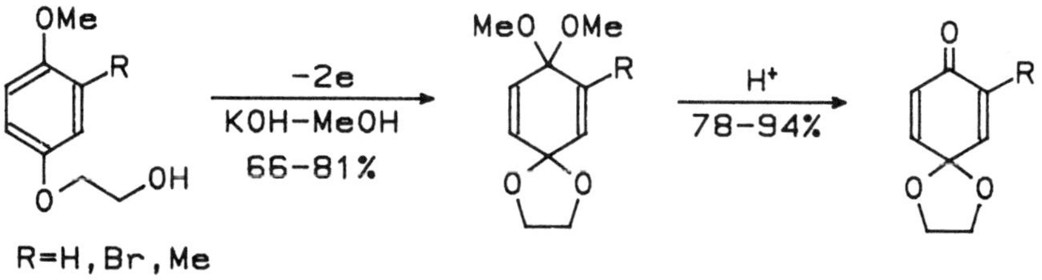

(40)

To control the regioselectivity of the quinone bisketal monohydrolysis, the complementary ketals, such as ethylene glycol ketal, Eq. (40), and mixed methyl silyl ketal, may be employed [65]. The method is applicable to a total synthesis of the aglycons of the anthracyclinones, Eq. (41) [65c]:

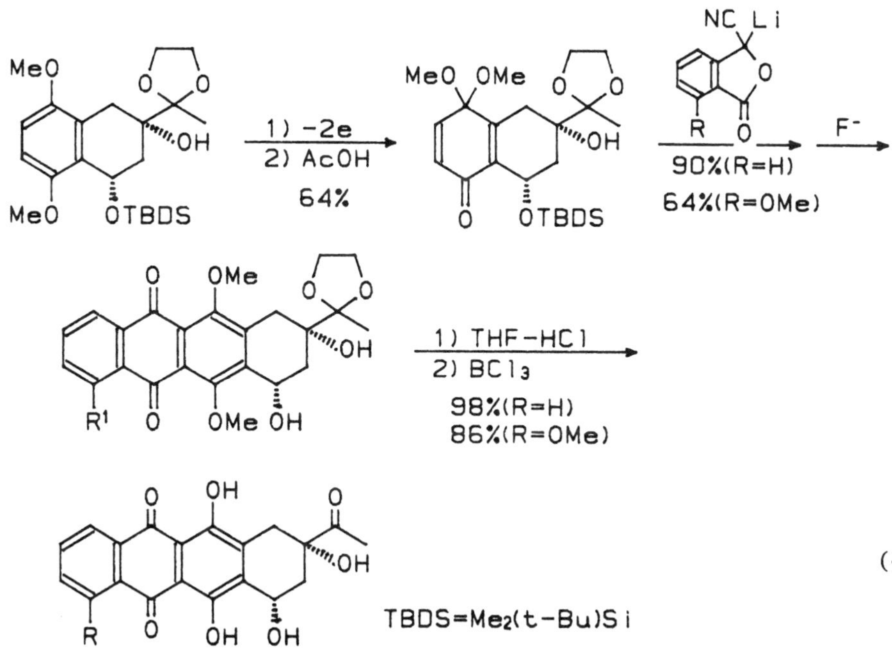

(41)

The quinone monoketal with the A-ring substitution pattern, which serves as the AB ring component of the anthracyclinones, can be prepared from the respective dimethyl ketal via anodic oxidation and subsequent regio-specific quinone bisketal hydrolysis. Annulation of the quinone monoketal with the anion of 3-cyano-1-3(H)-isobenzofuranone, followed by deblocking of the protecting group, gives 4-demethoxydaunomycinone in an excellent overall yield. This synthetic strategy, based on anodic oxidation leading to quinone monoketals, allows a convergent approach to a variety of the D ring analogs.

VIII. MISCELLANEOUS

Some papers cite attractive electrosyntheses of natural products other than those already described.

d,l-Muscone was synthesized in several steps from cyclododecanone using anodic oxidation as a key step, Eq. (42) [66]:

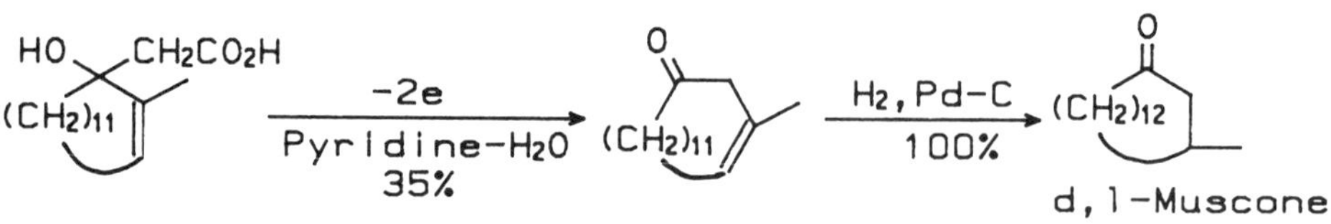

(42)

The anodic oxidation involves generation of the cationic species from the carboxylate, bringing about a ring expansion reaction.

Shono's [67] and Torii's [68] group have independently developed an impressive synthetic method of maltol starting from the furfuryl alcohol derivatives using anodic oxidation as a key step, Eq. (43):

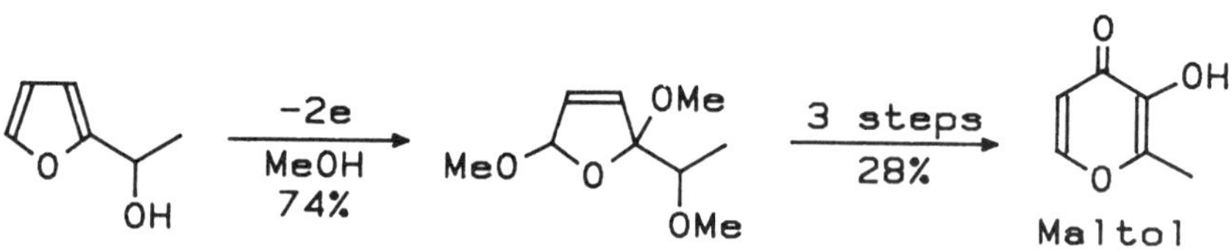

(43)

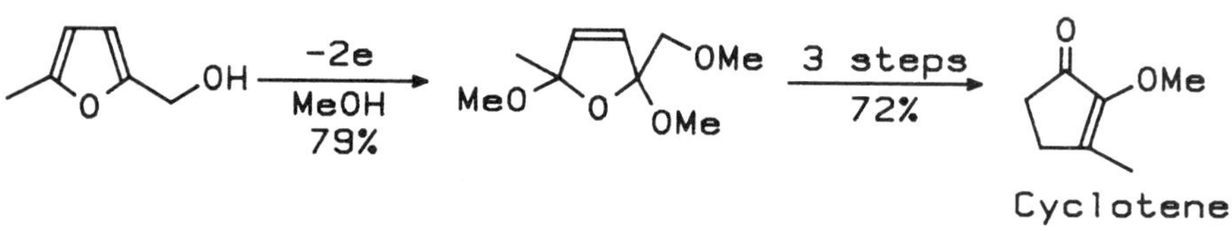

(44)

This process has now been industrialized by Ohtsuka's group.

The synthetic strategy based on anodic oxidation of furans serves for the construction of cyclopentenoids that are involved in a number of natural products, including monoterpenoids and prostanoids. 5-Methylfurfuryl alcohol can be converted to cycloten in a quite good overall yield, Eq. (44) [69]. Anodic oxidation of t-butyl furfurylacetate followed by acid-catalyzed intramolecular aldol cyclization of enedione yields the cyclopentenoid, Eq. (45) [70]:

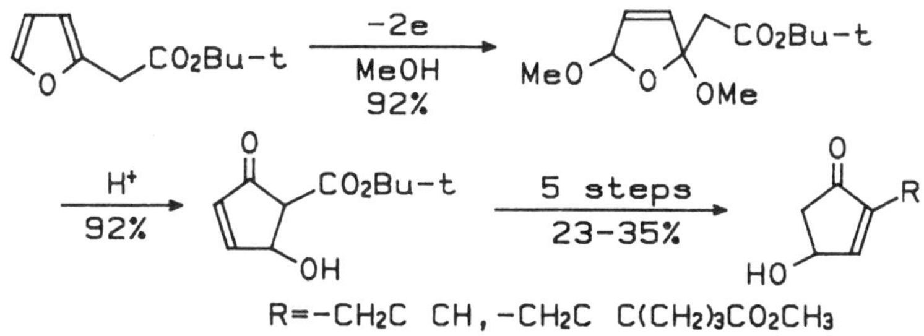

(45)

The stereoselective synthesis of 1,2,3-trisubstituted cyclopentane can be realized, Eq. (46) [71]:

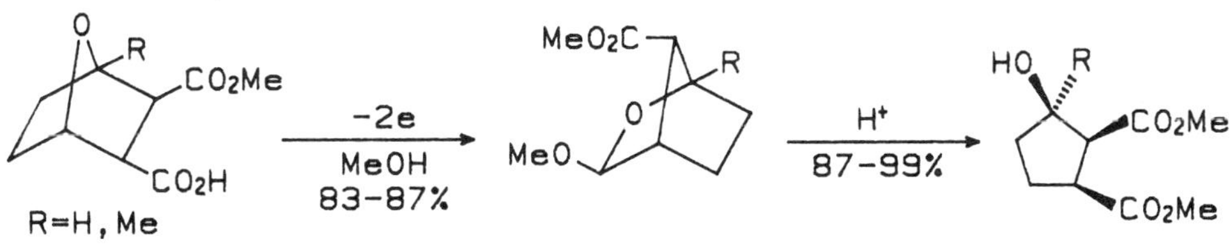

(46)

The reaction involves the Wagner-Meerwein rearrangement initiated by the carbonium ion generated by anodic decarboxylation.
Mixed Kolbe electrolysis has been nicely applied to the synthesis of sex attractant pheromones of house fly [72], the southern corn rootworm, [73], the German cockroach, [74], flour beetles [75], and the cabbage looper [76].

ACKNOWLEDGMENTS

This review chapter is dedicated to Professor Emeritus Shigehiko Sugasawa, University of Tokyo, on the occasion of his 90th birthday with our sincere gratitude for his continuous encouragement. Acknowledgment is also made to Dr. Ichiro Chibata, Vice-President, Dr. Sei-ichi Saito, Managing Director, Research and Development Executive, and Dr. Tetsuya Tosa, General Manager, and Dr. Kazuo Matsumoto, Manager, Tanabe Seiyaku Company, Ltd., for their generous permission to publish the present review.

REFERENCES

1. (a) Y. Urabe, T. Iwasaki, K. Matsumoto, and M. Miyoshi, Tetrahedron Lett., 997 (1975); (b) T. Iwasaki, Y. Urabe, Y. Ozaki, M. Miyoshi, and K. Matsumoto, J. Chem. Soc., Perkin Trans., I, 1019 (1976).
2. T. Iwasaki et al., unpublished data,
3. (a) H. Horikawa, T. Iwasaki, K. Matsumoto, and M. Miyoshi, Tetrahedron Lett., 191 (1976); (b) T. Iwasaki, H. Horikawa, K. Matsumoto, and M. Miyoshi, Bull. Chem. Soc. Jpn., 52, 826 (1979).
4. T. Iwasaki, H. Horikawa, K. Matsumoto, and M. Miyoshi, J. Org. Chem., 42, 2419 (1977).
5. (a) T. Nishitani, T. Iwasaki, Y. Mushika, and M. Miyoshi, J. Org.

Chem., 44, 2019 (1979); (b) T. Nishitani, T. Iwasaki, Y. Mushika, I. Inoue, and M. Miyoshi, Chem. Pharm. Bull., 28, 1137 (1980).
6. (a) H. Horikawa, T. Nishitani, T. Iwasaki, Y. Mushika, I. Inoue, and M. Miyoshi, Tetrahedron Lett., 4101 (1980); (b) H. Horikawa, T. Nishitani, T. Iwasaki, and I. Inoue, Tetrahedron Lett., 2193 (1983); (c) H. Horikawa, T. Iwasaki, T. Nishitani, and I. Inoue, Heterocycles., 21, 458 (1984).
7. (a) Y. Ozaki, T. Iwasaki, H. Horikawa, M. Miyoshi, and K. Matsumoto, J. Org. Chem., 44, 391 (1979); (b) Y. Ozaki, T. Iwasaki, M. Miyoshi, and K. Matsumoto, J. Org. Chem., 44, 2019 (1979).
8. T. Iwasaki et al., unpublished data,
9. T. Shono, Y. Matsumura, and K. Inoue, J. Org. Chem., 48, 1388 (1983).
10. T. Shono, Y. Matsumura, and K. Inoue, J. Chem. Soc., Chem. Commun., 1169 (1983).
11. T. Shono, Y. Matsumura, and K. Tsubata, Tetrahedron Lett., 1259 (1983).
12. K. Irie, A. Ishida, T. Nakamura, and T. Oh-ishi, Chem. Pharm. Bull., 32, 2126 (1984).
13. K. Warning, M. Mitzlaff, and H. Jensen, Liebigs Ann. Chem., 1707 (1978).
14. T. Iwasaki and K. Harada, J. Chem. Soc., Chem. Commun., 338 (1974).
15. T. Iwasaki and K. Harada, J. Chem. Soc., Perkin Trans. I, 1730 (1977).
16. K. Matsumoto, H. Horikawa, T. Iwasaki, and M. Miyoshi, Chem. Ind., 920 (1978).
17. T. Iwasaki, H. Horikawa, K. Matsumoto, and M. Miyoshi, Tetrahedron Lett., 4799 (1978).
18. T. Iwasaki, H. Horikawa, Y. Ozaki, and K. Matsumoto, Nippon Kagaku Kaishi, 1826 (1984); Chem. Abstr., 102, 1317986 (1985).
19. T. Iwasaki, T. Nishitani, H. Horikawa, and I. Inoue, J. Org. Chem., 47, 3799 (1982).
20. T. Iwasaki, T. Nishitani, H. Horikawa, and I. Inoue, Tetrahedron Lett., 1029 (1981).
21. T. Nishitani, H. Horikawa, T. Iwasaki, K. Matsumoto, I. Inoue, and M. Miyoshi, J. Org. Chem., 47, 1706 (1982).
22. T. Adachi, T. Iwasaki, I. Inoue, and M. Miyoshi, J. Org. Chem., 44, 1404 (1979).
23. R. Mengel and J. M. Seifert, Tetrahedron Lett., 4203 (1977).
24. T. Adachi, T. Iwasaki, M. Miyoshi, and I. Inoue, J. Chem. Soc., Chem. Commun., 248 (1977).
25. K. Kondo and I. Inoue, J. Org. Chem., 44, 4713 (1979).
26. (a) L. L. Miller, F. R. Stermitz, and J. R. Falck, J. Amer. Chem. Soc., 93, 5941 (1971); (b) L. L. Miller, F. R. Stermitz, and J. R. Falck, J. Amer. Chem. Soc., 93, 2651 (1973).
27. (a) E. Kotani, N. Takeuchi, and S. Tobinaga, J. Chem. Soc., Chem. Commun., 550 (1973); (b) E. Kotani and S. Tobinaga, Tetrahedron Lett., 4759 (1973); (c) T. Kametani, K. Shishido, and S. Takano, J. Heterocycl. Chem., 12, 305 (1975).
28. E. Kotani, M. Kitazawa, and S. Tobinaga, Tetrahedron, 30, 3027 (1974).
29. J. B. Kerr, T. C. Jempty, and L. L. Miller, J. Amer. Chem. Soc., 101, 7338 (1971).

30. R. Gottlieb and J. L. Neumeyer, J. Amer. Chem. Soc., 98, 7108 (1976).
31. (a) J. Grimshaw and D. Mannus, J. Chem. Soc., Perkin Trans. I, 2456 (1977); (b) W. J. Begler and J. Grimshaw, J. Chem. Soc., Perkin Trans I, 2324 (1977).
32. W. Eilenberg and H. J. Schäfer, Tetrahedron Lett., 25, 5323 (1984).
33. (a) M. Miyoshi, T. Iwasaki, S. Shigematsu, and Y. Kameyama, Japan Kokai, 83194858 (1983); (b) M. Miyoshi, T. Iwasaki, S. Shigematsu, and Y. Kameyama, Japan Kokai, 83225036 (1983).
34. T. Shono, K. Yoshida, K. Ando, Y. Usui, and H. Hamaguchi, Tetrahedron Lett., 4819 (1978).
35. T. Shono, Y. Usui, and H. Hamaguchi, Tetrahedron Lett., 135 (1980).
36. T. Shono, Tetrahedron, 40, 811 (1984).
37. (a) K. Irie, M. Okita, T. Wakamatsu, and Y. Ban, Nouveau J. Chim., 4, 275 (1980); (b) K. Irie and Y. Ban, Heterocycles, 15, 201 (1981); (c) K. Irie and Y. Ban, Heterocycles, 18, 255 (1982).
38. (a) T. Shono, Y. Matsumura, and K. Tsubata, J. Amer. Chem. Soc., 103, 1172 (1981); (b) P. D. Palasz, J. H. P. Utley, and J. D. Hardstone, Acta Chem. Scand., B38, 281 (1984).
39. (a) T. Shono, Y. Matsumura, K. Uchida, K. Tsubata, and A. Makino, J. Org. Chem., 49, 300 (1984); (b) Z. Blum, M. Ekstrom, and L. G. Wistrand, Acta Chem. Scand., B38, 297 (1984); (c) M. Okita, T. Wakamatsu, and Y. Ban, Heterocycles, 20, 401 (1983).
40. T. Shono, Y. Matsumura, K. Tsubata, K. Inoue, and R. Nishida, Chem. Lett., 21 (1983).
41. S. Torii, Recent Advances in Electroorganic Synthesis, Proceedings of the 1st International Symposium on Electroorganic Synthesis, Kurashiki, Japan, 1986, pp. 241–244.
42. K. Oda, T. Ohnuma, and Y. Ban, J. Org. Chem., 49, 953 (1984).
43. (a) M. Iguchi, A. Nishiyama, Y. Terada, and S. Yamamura, Anal. Lett., 10A, 1079 (1979); (b) M. Iguchi, A. Nishiyama, Y. Terada, and S. Yamamura, Chem. Lett., 451 (1978); (c) A. Nishiyama, H. Eto, Y. Terada, M. Iguchi, and S. Yamamura, Chem. Pharm. Bull., 31, 2820 (1983).
44. (a) R. Cooper, E. C. Levy, and D. Lavie, J. Chem. Soc., Chem. Commun., 794 (1977); (b) M. Iguchi, A. Nishiyama, H. Eto, Y. Terada, and S. Yamamura, Chem. Lett., 1397 (1979).
45. G. Buchi and C. P. Mark, J. Amer. Chem. Soc., 99, 8073 (1977).
46. (a) M. Iguchi, A. Nishiyama, M. Hara, Y. Terada, and S. Yamamura, Chem. Lett., 1015 (1978); (b) A. Nishiyama, H. Eto, Y. Terada, M. Iguchi, and S. Yamamura, Chem. Pharm. Bull., 31, 2834 (1983).
47. A. Nishiyama, H. Eto, Y. Terada, M. Iguchi, and S. Yamamura, Chem. Pharm. Bull., 31, 2845 (1983).
48. Y. Shizuri and S. Yamamura, Tetrahedron Lett., 5011 (1983).
49. T. Shono, Nishiguchi, T. Yokoyama, and M. Nitta, Chem. Lett., 433 (1975).
50. T. Shono, Y. Matsumura, K. Hibino, and S. Miyawaki, Tetrahedron Lett., 1295 (1974).
51. K. Uneyama, S. Fujibayashi, and S. Torii, Tetrahedron Lett., 26, 4637 (1985).
52. S. Torii, K. Uneyama, and M. Ono, Tetrahedron Lett., 21, 2653 (1980).
53. S. Torii, K. Uneyama, and T. Bannou, J. Amer. Chem. Soc., 103, 4606 (1980).

54. K. Uneyama, A. Ishimura, and S. Torii, Bull. Chem. Soc. Jpn., 58, 1859 (1985).
55. E. A. H. Hall, G. P. Moss, J. H. P. Utley, and B. C. L. Weedon, J. Chem. Soc., Chem. Commun., 387 (1987).
56. A. Hembrock, H. J. Schäfer, and G. Zimmermann, Angew. Chem., 97, 1408 (1985).
57. K. Ponsold and H. Kash, Z. Chem., 22, 157 (1982).
58. K. Ponsold and H. Kash, Tetrahedron Lett., 4463 (1979).
59. K. Ponsold and H. Kash, Tetrahedron Lett., 4465 (1979).
60. R. Breslow and R. Goodin, Tetrahedron Lett., 2675 (1976).
61. (a) T. Shono, T. Toda, and N. Oshino, Tetrahedron Lett., 25, 91 (1984); (b) T. Shono and S. Kashimura, J. Org. Chem., 48, 1937 (1983).
62. (a) S. Torii, H. Tanaka, N. Saitoh, T. Siroi, M. Sasaoka, and J. Nokami, Tetrahedron Lett., 2187 (1982); (b) S. Torii, H. Tanaka, T. Ohshima, and M. Sasaoka, Bull. Chem. Soc. Jpn., 59, 3975 (1986).
63. S. Torii, H. Tanaka, N. Saitoh, T. Siroi, M. Sasaoka, and J. Nokami, Tetrahedron Lett., 2495 (1982).
64. (a) J. S. Swenton, Acc. Chem. Res., 16, 74 (1983); (b) M. G. Dolson and J. S. Swenton, J. Amer. Chem. Soc., 103, 2361 (1981); (c) M. G. Dolson and J. S. Swenton, J. Org. Chem., 46, 177 (1981); (d) A. Stern and J. S. Swenton, J. Org. Chem., 52, 2763 (1987).
65. (a) J. S. Swenton, J. N. Freskos, G. W. Morrow, and A. D. Sercel, Tetrahedron, 40, 4625 (1984); (b) D. K. Anderson, C. E. Coburn, A. P. Haag, and J. S. Swenton, Tetrahedron Lett., 1329 (1983); (c) J. S. Swenton, D. K. Anderson, C. E. Coburn, and A. P. Hagg, Tetrahedron, 40, 4633 (1984).
66. T. Shono, J. Hayashi, H. Omoto, and Y. Matsumura, Tetrahedron Lett., 2667 (1977).
67. T. Shono and Y. Matsumura, Tetrahedron Lett., 1363 (1976).
68. S. Torii, H. Tanaka, T. Ando, and Y. Shimizu, Chem. Lett., 495 (1977).
69. T. Shono, Y. Matsumura, and H. Hamaguchi, J. Chem. Soc., Chem. Commun., 712 (1977).
70. (a) T. Shono, H. Hamaguchi, and K. Aoki, Chem. Lett., 1053 (1977); (b) T. Shono, Y. Matsumura, H. Hamaguchi, and K. Nakamura, Chem. Lett., 1249 (1976); (c) T. Shono, Y. Matsumutra, S. Yamane, and M. Suzuki, Chem. Lett., 1619 (1980); (d) M. B. Floyd, J. Org. Chem., 43, 1649 (1978).
71. T. Akiyama, T. Fujii, H. Ishiwari, T. Imagawa, and M. Kawanishi, Tetrahedron Lett., 2165 (1978).
72. A. K. Yadav and P. Tissot, Helv. Chim. Acta, 67, 1698 (1984).
73. R. Rossi, A. Carpita, and M. Chini, Tetrahedron, 41, 627 (1985).
74. (a) U. Jensen-korte and H. J. Schäfer, Liebigs Ann. Chem., 1532 (1982); (b) W. Seidel and H. J. Schäfer, Chem. Ber., 113, 451 (1980).
75. K. Mori, S. Kuwahara, and H. Ueda, Tetrahedron, 39, 2439 (1983).
76. W. Seidel, J. Knolle, H. J. Schäfer, Chem. Ber., 110, 3544 (1977).

20
Biomass

HELENA L. CHUM / Solar Energy Research Institute, Golden, Colorado

I. Introduction 787

II. Whole-Wood or Wood Pulps: Production of Bleached Cellulosics 791
   A. Indirect electrochemical bleaching 791
   B. Alternative bleaching agents 791
   C. Regeneration of reagents 793

III. Wood-Polymer Components 793
   A. Polymeric carbohydrates 793
   B. Lignins 794

IV. Wood-Derived Monomers 799
   A. Monosaccharides 799
   B. Lignin model compounds 800
   C. Pulping mechanisms 802

V. Conclusions 802

References 803

I. INTRODUCTION

Materials produced by plants grown on land or in water as a direct result of photosynthesis are called biomass. They are composed of a number of classes of compounds from monomeric to polymeric depending on the source. The polymeric components, the largest weight fraction of biomass, perform various functions, such as food storage (starch), support of plant structure (cellulose and hemicelluloses), plant water conduction, and elevation of crown for solar energy conversion (lignins) [1]. This chapter concentrates on the electrochemistry of lignocellulosic biomass. The types of biomass under consideration are trees, classified into gymnosperms (examples are conifers or softwoods) and angiosperms (examples are hardwoods) [2,3], and agricultural residues. Biomass, especially wood, is essentially composed of compounds containing only carbon, hydrogen, and oxygen. Woody substances are composed predominantly of carbohydrates and an aromatic polymeric component—lignin. The carbohydrates are polymers and are the most highly oxygenated constituents of wood. Lignin is also macromolecular but much less oxygenated. However, lignin is easily oxidized, a property that is often exploited in the delignification of wood. For this reason, a large

---

Dedicated to Manuel Baizer—a guide and a friend.

Table 1. Cell Wall Constituents of Biomass (%)

Biomass type	Softwood	Hardwoods	Grasses
Cellulose	41–43	43–45	35–39[a]
Lignin			
average	28	22[b]	16[b]
range	24–33	16–24	15–18
Xylan	8–14	20–30	27–34
Glucomannan	16–22[c]	3–5	None

[a]May be higher.
[b]Some soluble lignin.
[c]Contains variable galactose amounts.
Source: From Ref. 4.

fraction of this chapter is devoted to lignin chemistry and electrochemistry. Typical composition data for various cell wall components of wood biomass are given in Table 1 [1–6].

Cellulose is the most plentiful constituent of biomass. On the average, cellulose in situ contains 10,000 anhydroglucose units, but this value is greatly reduced upon isolation. Hemicelluloses consist of a mannose-containing and a xylose-containing polysaccharide. Lignin is the noncarbohydrate component of woody substances. The basic carbon skeletal structure of the monomeric unit is that of phenylpropane. Lignin is a three-dimensional macromolecule formed by endwise free radical polymerization of cinnamyl alcohols with or without methoxyl substituents in the *ortho* position to the phenolic hydroxyl group (coniferyl and sinapyl alcohols, respectively, for one- and two-methoxyl substituents; see Fig. 1). There is no consistent repeating linkage between the monomeric units. The main linkage between phenylpropane units is the β-O-4-alkyl aryl ether (ring numbering starts at the carbon linked to the propyl side chain; the side chain carbons are named α, β, and γ, respectively, from the aromatic carbon). α-O-4-ether bonds are less frequent and are used to attach lignins to the hemicelluloses. Other types of bonds link β–β (pinoresinol), β-5 (phenylcoumaran), 5-5, and β-1 carbon-carbon bonds. Table 2 compares the proportions of these types of bonds for hardwoods and softwoods [6,7].

Renewable polymeric materials are produced worldwide at about 200 trillion lb/year [8] and consumed at a rate of about 7-8 trillion lb/year. The major process converting wood into chemical grade pulp is the kraft pulping process [2,3,5,6,9], in which the chemicals sodium hydroxide and sulfide (white liquor) are employed to dissolve the lignin and hemicellulosic fractions of the wood, leaving a cellulosic pulp of high strength that still contains some residual lignins (which color paper bags brown) and hemicelluloses. After pulping, the black liquor contains the kraft lignins, hydroxycarboxylic acids, and the pulping chemicals, which are recovered by air oxidation, liquor concentration and burning in the recovery furnace, and causticization of the liquor. This complex recovery scheme is very efficient in recovering sodium and sulfur species. A small fraction of the kraft lignins (70 million lb from a total of 40 billion lb produced in the United States annually) are withdrawn from the black liquors, isolated,

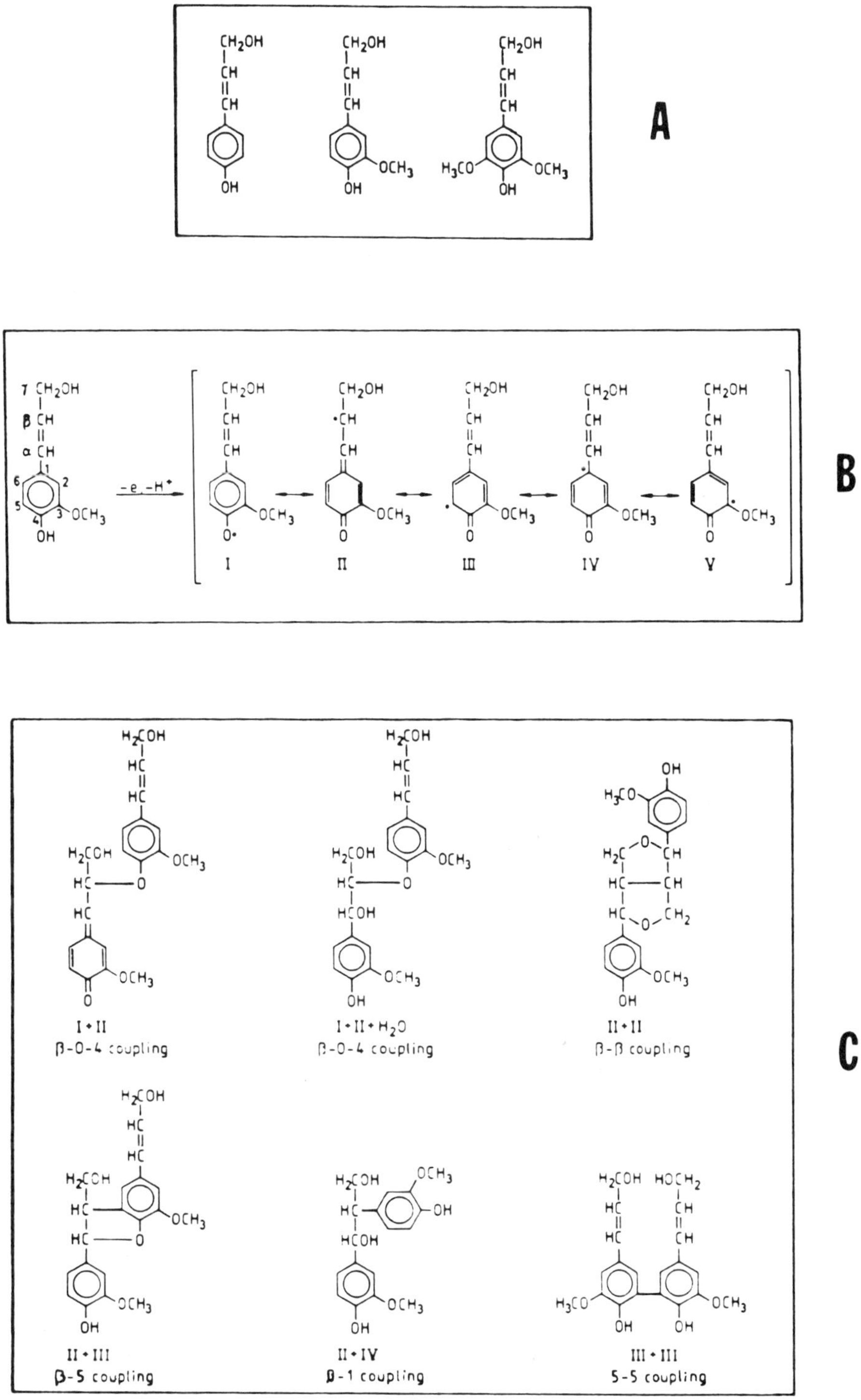

Figure 1. (A) Lignin precursor alcohols; (B) Examples of intermediate free radicals for coniferyl alcohol; (C) Examples of dimeric structures listed as β-O-4, β-β, β-5, β-1, and 5-5 coupling products in the text. (Adapted from Ref. 2.)

Table 2. Proportions of Main Different Types of Bonds Connecting Phenylpropane Units in Milled Wood Lignins From Various Species (see Fig. 1)

Type of structure	Proportions (%)		
	Guaiacyl	Syringyl	Total
Spruce (Picea abies)			
Arylglycerol-β-aryl ether (β-O-4)	48		
Biphenyl (β-β)	9.5–11		
Phenylcoumaran (β-5)	9–12		
Noncyclic benzyl (α) aryl ethers	6–8		
Pinoresinol (β-β)	2		
Diphenyl ether	3.5–4		
Condensed structures two- or six-positions	2.5–3		
Birch (Betula verrucosa)			
Arylglycerol-β-aryl ether (β-O-4)	22–28	34–39	60
Biphenyl (β-β)	4.5		4.5
Phenylcoumaran (β-5)			6
Noncyclic benzyl (α) aryl ethers			6–8
Syringaresinol (β-β)			3
Diphenyl ether	1	5.5	6.5
Condensed structures two- or six-position	1–1.5	0.5–1	1.5–2.5

Source: from Ref. 7.

purified, and chemically modified. Examples of polymeric applications are dispersants, resin coreactant, emulsifier and stabilizer, metal ion sequestering agent, controlled release carriers, component of lead cathodes for batteries, rubber stabilizers, and reinforcing agents [10].

The second main commercial pulping process is the sulfite process, which produces a cellulosic fraction by dissolution of the lignins and hemicelluloses [2,3,5,6,9]. These components are transformed into lignosulfonates and simple monosaccharides (which can be fermented to ethanol), respectively. The pulping solutions contain the active (bi)sulfite anions and various types of cations. The lignosulfonates (>1.4 billion lb/yr) produced in these processes are usually sold as very inexpensive industrial surfactants [10]. High-purity cellulosic feedstocks from this process can be converted into cellulose derivatives. The technical lignins (kraft and lignosulfonates) are abundant sources of raw materials, the single most widely available source of renewable aromatic materials [10].

Some new, emerging technologies, such as organosolv pulping [11,12], are more energy and feedstock efficient and could become important sources of products [13,14]. The acid hydrolysis of wood is often practiced in the

Soviet Union [15], primarily as high-temperature dilute acid hydrolysis processes. The products are sugar substrates for fermentation to alcohol, protein, and furfural from the acid-catalyzed degradation of xylose. The acid hydrolysis lignins produced in these processes are heavily condensed and are usually burned. The Russian literature has many examples of approaches to upgrade these lignins to increase their utilization. Other fractionation schemes based on mild acid hydrolysis are the steam explosion process (batch [16] or continuous [17]).

This chapter describes the electrochemistry of biomass and a few selected derived materials. Additional literature on this topic can be found in the monograph by Chum and Baizer [18] and the reviews by Utley [19], Chum et al. [20], and Nonaka and Baizer [21]. Some of the references have been cited from the extensive abstracts of the Institute of Paper Chemistry (IPC-AN) from the PaperChem Database (Appleton, Wisconsin).

II. WHOLE-WOOD OR WOOD PULPS: PRODUCTION OF BLEACHED CELLULOSICS

A. Indirect Electrochemical Bleaching

Kraft pulps and higher yield pulps, which contain more of the hemicelluloses and lignin, need to be bleached to be suitable for various uses, such as bleached papers and newsprint. A number of bleaching agents are employed; chlorine (C), chlorine dioxide (D), hypochlorite (H), peroxide (P), and oxygen (O) are examples of oxidizing agents employed, usually in combination with hydrolysis stages (e.g., alkaline extraction (E)] [2,3,9]. Industrial bleaching sequences can have three to seven stages, and examples for kraft pulp bleaching sequences are C-E-H, C-E-H-D (or C-E-H-P, C-E-H-H, and so on), C-E-H-P-D, C-H-E-D-E-D, and C-H-H-D-E-P. The chlorine (and in some cases the peroxide) is produced electrochemically. Several attempts have been made to replace a few bleaching steps with the in situ generation of the oxidant either from the electrolysis of chloride ions or the oxidation of ferrocyanide in the presence of oxygen. Slurries of pulp are suspended in the appropriate medium, and electrolysis is carried out using dimensionally stable (or graphite) anodes with stainless steel cathodes. Lower temperatures are used for bleaching in the presence of alkaline ferrocyanide solutions; various additives can be used to minimize cellulose degradation. Without additives and at higher temnperatures, substantial degradation of cellulose macromolecules occurs. In many patented examples, pulp yield increases have been observed, as well as a decreased amount of effluent pollutants. Pertinent literature examples are given in Table 3.

B. Alternative Bleaching Agents

On-site generation of hydrogen peroxide within the pulp mill is possible. Cells using trickle bed designs have been operated. They produce a fixed $NaOH$-H_2O_2 weight ratio of 2:1, which is higher in base than the conventional bleaching solutions (1:1). The solutions are, however, useful for some bleaching sequences [32,33].

Alternative oxidizing agents for bleaching are the peracids, principally peracetic acid. A new method has been proposed for converting the low

Table 3. Electrochemical Bleaching and Delignification of Wood Pulps

Pulp or feedstock	Electrolysis solution	Comments	References
Kraft	NaCl	Two-stage electrolytic bleaching; one at pH 0.2 and the other at pH up to 8	22
Kraft	$K_4[Fe(CN)_6]$, O_2 saturated, alkaline	Additives to reduce degradation: soluble amines, alcohols, or ethers; mediator concentration 0.004–0.4% wood; up to 65°C	23
Unbleached kraft	NaCl	Energy consumption was 0.1–0.75 kWh/kg pulp; brightness comparable to chemically bleached pulp; graphite anode; stainless steel cathode	24
Kraft	NaCl, 3–5%	30–40°C, 25 A/kg pulp at 3–4 V for 1.5–3h, at 0.01–0.05 A/cm^2; slurry consistency 2–5%	25, 26
Larchwood, high yield	NaCl, 6%	0.2 A/cm^2, 9–14% higher delignification than for equivalent bleaching	27, 28
Kraft	NaCl	5 min at 0.05 A/cm^2 best; replaced C&E stages in bleaching, had some pulp degradation; pilot-scale tests conducted; suited for larch	29
Kraft	NaOH, 4%	Best results with addition of potassium ferrocyanide carrier, 60°C; pulp yield increase of 1.5–1.8%; reduced COD effluents by 70%; EC prior to C-E-H-D-E-D sequence reduced 75% chlorine and 50% NaOCl consumption	30
Kraft	NaCl	pH 6.3–7; horizontal anode and vertical cathode; anode-cathode surface areas 1:1.5 at 0.2 A/cm^2	31

energy content black liquor hydroxycarboxylic acids into peracids for bleaching in an integrated pulping and bleaching mill. For instance, lactic acid can be oxidized to volatile acetaldehyde that can be reacted with hydrogen peroxide for the production of peracetic acid [34].

C. Regeneration of Reagents

Recovery of NaOH and Na_2SO_4 from black liquors has been investigated. The black liquors are acidified with HCl, which precipitates lignins and forms hydrogen sulfide and brine solutions. Electrolysis of these solutions to NaOH and Cl_2, followed by absorption of the H_2S into the alkaline solution, recovers the caustic-sulfide pulping chemical and generates chlorine for bleaching [35].

Partial electrolytic oxidation of pulping (white) liquors was shown to give higher pulp yields for larch [36], a species difficult to pulp and bleach. These polysulfide-rich liquors significantly increase the rate of delignification of larch [37]. Electrochemical oxidation of the green liquors can lead to semichemical pulps with higher pulp yields and reduced levels of noxious odors during processing [38]. Electrolytic recovery of solutions used in the production of cellulose derivatives, such as rayon, has been investigated. Recovery of both NaOH and H_2SO_4 uses energy, but these costs are offset by a 50% reduction in steam needed to recover Na_2SO_4 [39,40].

The two-stage organosolv process described in Ref. 13 can involve an interesting electrochemical step. The liquor from the second stage is distilled, and the resulting aqueous alkaline solution is electrolyzed in an ion-exchange separated cell. The liquor flows to the anode compartment, and water is circulated in the cathodic compartment. As the electrolysis proceeds, the anodic compartment becomes more acidic, leading to precipitation of the acid-insoluble lignins; in the cathodic compartment, the aqueous solution is regenerated for use in pulping [41].

III. WOOD-POLYMER COMPONENTS

A. Polymeric Carbohydrates

The oxidative reactions and degradations of carbohydrates have been extensively covered in the review by Green [42] in 1980, including electrochemical reactions. Only the more recent references are reviewed in this chapter.

The acid generated at the platinum anode of an electrolytic cell separated by a ceramic material from the plasma-sprayed Raney nickel cathode was used to hydrolyze cellulose slurries (12–30% by weight) in an asymmetric electrochemical batch recycle reactor operating at 160°C. Conversion of cellulose into glucose of the order of 30% was observed with a very high yield of glucose (90–98%) at 0.5 A [43]. These yields of glucose are higher than those obtained by dilute acid hydrolysis at 180–210°C or concentrated acid hydrolysis at lower temperatures, although the observed conversion of cellulose into glucose is lower [15]. Improved mass transfer in the reactors may lead to higher conversion efficiencies. This method could generate glucose from inexpensive cellulosic sources for conversion into such chemicals as sorbitol at the cathode in a paired synthesis (see Chap. 35).

Electrochemical oxidation of hemicelluloses from wood prehydrolyzates was investigated at 25–90°C, pH 5.5–10, using Pt, Pt/Ti, and other electrodes, in the 1.6–2.5 V range, with 60–100 A-min, with various potassium salts as electrolytes (Cl^-, Br^-, and SO_4^{2-}). Oxidative and hydrolytic degradation occurred (89%), with hydroxyacids (41%) the main product obtained in the chloride case [44].

B. Lignins

Kraft Process

The recent electrochemical studies of kraft lignins are assembled in Table 4. They span synthetic studies for the preparation of lignin derivatives (chlorinated and oxidized polymers), depolymerization studies to monomers, dimers, and oligomers (solvated electron reductions), and mechanistic studies and the examination of properties of lignins. For the literature prior to 1985, the reader is referred to the citations in Ref. 18.

Sulfite Process

The recent electrochemical studies of lignosulfonates are assembled in Table 5. Oxidative electrolysis of sodium lignosulfonates produced a light brown colored material with significant change in the carbonyl, carboxyl, and sulfonic acid group contents. The molecular weight distribution of the polymer is changed such that it can be used in a variety of applications in which stable soluble surfactants and dispersants are needed [57,58]. On the other hand, it is possible electrochemically to depolymerize the calcium lignosulfonates (e.g., Lignosite from Georgia Pacific) into primarily three low-molecular-weight compounds with total yields of over 60% [55]. High yields of methyl ethyl ketone from a variety of lignin sources have been obtained [52,61] after exhaustive electrolysis, thus confirming the early work of Bailey and Brooks [76]. Yoshiyama et al. [59] presented a good review of the literature of lignin electrolysis.

Table 4. Electrochemical Studies of Kraft Lignins

Electrolysis conditions and methods	Comments	Reference
Pt rotating disk anode, 0.68–0.8 V, 0.1 KCl	Up to 0.625 V electrolysis lignin oxidation and polymerization; above 0.625 V lignin degradation observed	45
Dropping mercury electrode 0.05 M each LiOH, LiCl	Butanol-, dioxane-, and diethyl ether-soluble kraft lignins of M_w = 6800, 7200, and 8300, respectively; latter two of high polydispersity; the higher the M_w the lower the diffusion coefficient	46
Potentiometry during alkaline hydrolysis	Pine lignin heated at 160°C for 5–160 min; on heating a high concentration of chromophoric groups of acceptor properties occurs; followed as a function of time	47

Table 4. (Continued)

Electrolysis conditions and methods	Comments	Reference
Electrochemical oxidation	Alkali lignin, 1 M NaOH, 50°C, 0.01 A/cm^2; HPLC analysis of 21 oxidation products	48
Electrochemical oxidation	Spruce, pine, and birch (kraft, soda, and dioxane) lignins; basic aqueous medium in presence and absence of O_2 and also of electron carriers (potassium ferrocyanide); lignins after oxidation had cleaved benzene rings, decreased phenolic groups, and increased carboxylic acid content	49
Electrochemical reduction and oxidation	Hardwood and softwood kraft lignins; no major change in molecular weight distributions on reduction at moderate potentials; at 0.5 versus SCE depolymerization is observed for hardwoods; for softwoods only in the presence of oxygen	50
Electrochemical reduction	Alkali ligning, 1 atm liquid ammonia at −33°C, at 0.007 A/cm^2; cleavage to soluble monomers and oligomers occurs; higher yields at higher reaction times; after 3 j 32% ether-soluble and 18% water-soluble lignin of oligomers obtained; partial ring hydrogenation occured	51
Anodic degradation	Lead dioxide anode, 1 M NaOH at 50°C, 5–50 mA/cm^2; Japanese red pine kraft lignin; 35 oxidation products identified and quantitated; at 30 mA/cm^2 obtained highest monomeric product yields; methyl ethyl ketone (3.2%) and catechol (0.5%), etc., with 0.2 F/g lignin	52
Electrochemical chlorination	35% HCl, graphite electrodes, undivided cell; best conditions 20°C, 1.5–2 h, acid 18–3 at 0.1–0.2 A/cm^2; up to 30% by weight of Cl introduced into the lignins; substantial demethoxylation and oxidation occur	53
Properties	Ionic and electronic conduction observed; lignin reduced from 1 M NaOH onto Hg or lignin pellets with silvered surfaces	54

Table 5. Electrochemical Studies of Lignosulfonates

Electrolysis conditions and methods	Comments	References
Electrochemical oxidation	Calcium lignosite electrolyzed in a divided cell with DMF or aqueous LiCl or LiOH; major products: vanillic acid (15%), 3-methoxy-4-hydroxyphenylacetone (16%), and a dimer (29%)	55
Electrochemical oxidation	Acid- and alkali-treated lignosulfonates and high- and low-M_W fractions from the acid type extracted with 60% ethanol; electrolytic oxidation decreases sulfo content and increases carboxyl group content; for high-M_W fractions the increase in carboxylic acid was small; decreases in methoxyl group content higher for low-M_W fractions and the sodium form	56
Electrochemical reduction	Sodium lignosulfonate in acidic media; best cathode tin, 54% starting lignin depolymerized; reduction improved by addition of 3–5% tin chloride; higher carrier concentrations led to higher M_W products; used as binders, stabilizers, fillers in resins and plastics	57, 58
Anodic degradation	Lead dioxide anode, 1 M NaOH, at 50°C, 5–50 mA/cm^2; lignosulfonates acid and sodium salts; 35 oxidation products identified and quantitated; at 50 mA/cm^2 obtained highest monomeric product yields: 50% weight conversion for the sodium salts 0.2 F/g lignin; main products catechol, vanillyl alcohol, and methyl ethyl ketone (33%) from softwoods; from hardwoods very ow yields of degradation products obtained	52, 59
Corrosion inhibitors	Good protection observed for 100–200 mg/liter lignosulfonates, 10–20 mg/liter organic phosphonates, and 5–10 mg/liter zinc	60

Table 6. Electrochemical Studies of Lignins from Acid-Catalyzed Processes

Electrolysis conditions and methods	Comments	References
Anodic degradation	Lead dioxide anode, 1 M NaOH, at 50°C, 5–50 mA/cm^2; bagasse acid hydrolysis lignins; 35 oxidation products identified and quantitated; at 5 mA/cm^2 obtained highest monomeric product yield: 58 wt% methyl ethyl ketone obtained from acid-hydrolyzed bagasse; base treated bagasse lignins at six times the current density gave 12% catechol; all conversions with <1 F/g; vanillyl alcohol could also be obtained in 7% yield	61
Electrochemical chlorination	DMF or AN with Et_4NCl; pine hydrolysis and dioxane lignins; factorial design applied, with charge, lignin concentration, anodic current density, and Cl^- concentration as variables; no noticeable oxidative degradation observed (as is in aqueous); chlorine found in the side chains (in aqueous solutions ring chlorination observed); best conditions: 3600 C/g; 140 mA/cm^2; 0.2 M Et_4NCl, and 1.5% lignin	62
Electrochemical bromination	Investigated under acid, neutral, and alkaline conditions in water or DMSO; in aqueous alkaline conditions >90% lignin degradation occurred with formation of low-molecular-weight product gels; in DMSO bromolignins with 30% Br content formed, which were not substantially degraded	63
Electrochemical bromination	DMF or AN with Et_4NBr electrolyte, acid hydrolysis, Klason, and dioxane lignins; bromination up to 7% observed at 140 mA/cm^2 for 2650 C/g; higher charges led to products with decreased Br content; oxidative degradation did not occur	64
Electrochemical iodination	Alkaline conditions best (pH 10) can lead to up to 21% incorporation of iodine into acid hydrolysis lignins	65
Electrochemical oxidation	Mechanistic studies in aqueous and DMSO solutions on graphite electrodes; formation of quinonoid structures observed	66
Anodic sulfonation	Alkaline and acid solutions of sodium sulfite employed; introduction of <0.5 mmol sulfo groups into acid hydrolysis lignins	67

Table 6. (Continued)

Electrolysis conditions and methods	Comments	References
Electrochemical oxidation	Phenol pulping spruce lignin oxidized in the presence of nitrobenzene and its derivatives; high-temperature electrolysis on Ni electrodes coupled with a high-temperature alkaline hydrolysis produced higher yields of benzaldehyde, vanillin, and vanillic acid than the nitrobenzene oxidation alone, commonly employed as an analytical method; phenol pulping spruce and straw lignins chemically characterized and found to incorporate a substantial amount of phenol in the α-position	68, 69
Electrochemical reduction	Nitrolignins reduced in alkaline solutions on Hg cathodes, electrolyses followed by functional group changes, molecular weight distributions (MWD), GC, and UV spectroscopy; nitrolignins are drilling fluid additives but have low solubility; reduction produces materials of greatly improved solubility (1.5% NaOH solubility to 72% after reduction on copper electrodes); reduced carbonyl and increase in carboxyl group content observed; electrolyses caused 6% degradation of nitrolignin; polarographic studies conducted; main electroactive groups nitro, carbonyl, quinone, and quinone methide groups; Hg and Sn best cathodes	70–72
Electrochemical reduction	Ethanol-extracted steam-exploded aspen lignins reduced on Hg cathodes; functional group changes, MWD, NMR, and IR determined as a function of reduction conditions and compared with hydrolysis effects alone; reductions performed at Ph 12 and 7 in methanol/TEAP solutions at −2.6 V versus Ag/AgCl; electrolysis leads to increased amounts of acid-soluble (AS) components of low polydispersity increased phenolic hydroxy and total hydroxy contents; carbonyl group content decreased upon reduction and activated β-O-4-alkyl aryl ether bonds reductively cleaved; acid-insoluble fractions have lower phenolic hydroxy content and higher M_w but low polydispersity;	73, 74

Table 6. (Continued)

Electrolysis conditions and methods	Comments	References
Electrochemical reduction (cont.)	at pH 12 reductive couplings occur; at pH 7 reductive couplings can be avoided; such couplings lead to lignins with increased carbon-carbon condensed structures; at pH 7 electrolysis leads to the most extensive depolymerization and lowest carbon-carbon condensation structures; over 80% lignin recovery in the two fractions obtained	
Electrochemical oxidation	Ethanol-extracted steam-exploded aspen lignins oxidized on Cu anodes in 1 M aqueous solutions as a function of applied potential (0.65 and 0.80 V versus Ag/AgCl in divided cells); functional group changes, MWD, NMR, and IR determined as a function of oxidation conditions and compared with the hydrolysis effects alone; at the lower potential C–O and C–C anodic couplings occur and the resulting lignin has higher M_W and polydispersity; at the higher potential, anodic degradations take place producing and acid-insoluble lignin of lower M_W in lower yield; high yields of recovered highly insoluble polymeric lignins obtained at lower potential	75, 76

Acid-Catalyzed Processes

The work summarized in Table 6 can be divided into (1) studies of polymeric modifications of acid hydrolysis lignins and their derivatives (such as nitrolignins) to produce useful polymeric products for a variety of applications, from prepolymers for materials to increased solubility drilling fluid additives; and (2) studies of anodic degradation for the production of selective oxidation products. An interesting use of nitrobenzene-mediated electrochemical oxidation is described [68] for the syntheses of aldehydes and acids, such as benzaldehyde, vanillin, and vanillic acid.

IV. WOOD-DERIVED MONOMERS

A. Monosaccharides

The electrochemical conversion of biomass-derived monosaccharides to a number of specialty chemicals has been reviewed by Baizer [77]. Glucose conversion to sorbitol, mannitol and gluconic acid were reviewed, as well as the main reactions of starch to dialdehyde starch (see also Ref. 18). These reactions and the recent success by Baizer and co-workers in pairing the glucose electrochemical oxidation to gluconic acid and reduction to sorbitol are described in Chap. 35.

The anodic oxidation of levoglucosan and levoglucosenone was investigated [78]. Anodic oxidation of levoglucosan in MeOH-Bu_4NClO_4 at 2.3 V at a vitreous carbon electrode gave 47% of methyl-β-D-arabinopyranoside, 9% D-arabinose, and traces of sodium formate. A similar electrolysis at pH 7 controlled by the gradual addition of sodium methoxide gave 81% D-arabinose and sodium formate. Electrolysis of levoglucosenone initiated its polymerization, and the mixture of polymers had an apparent molecular weight distribution of 600–2000.

B. Lignin Model Compounds

A number of lignin model compounds studied were reviewed by Chum and Baizer [18]. Some of the recent work involving anodic oxidations is assembled in Table 7.

Electrochemical studies of 20 dimeric lignin model compounds with α-O-4 and β-O-4 alkyl aryl ethers, β-5 (phenylcoumaran), β-1 and 5-5 carbon-carbon bonded types of structures were investigated in aqueous and nonaqueous media by polarography and voltammetry (see Fig. 1 for explanations of abbreviations) [88]. Polarographic studies were carried out in 40:60 vol% ethanol-water containing 0.2 M KCl (neutral) and aqueous 1 M NaOH (basic). The main classes of compounds investigated were as follows.

α-O-4-ethers were not electroactive even with the β-carbonyl group in neutral or basic medium under the conditions employed.

A total of 14 β-O-4-ethers were investigated with α-carbonyl groups; some ethers had derivatized phenolic hydroxyl groups either as methyl or benzyl ethers or as acetates; both guaiacyl and syringyl derivatives were investigated. Measurements were made at 50°C because of low solubility under neutral conditions. Two reductive waves were observed: the first, around −1.2 V, corresponds to the cleavage of the activated ether bond; the second corresponds to the reduction of the carbonyl group to the hydroxyl group. Preparative electrolysis at the first wave confirms the two-electron ether cleavage for selected compounds. A few dimers also had free aldehyde groups in the guaiacyl (or syringyl) group in the p-position. Those groups were also shown to be electroactive with formation of the alcoholic derivatives. Dimers with free phenolic structures and an aldehyde group gave poor amperometric curves and significant electrode blockage. One β-O-4-ether structure investigated had an α-OH group and a benzyl group protecting the phenolic group. This compound was not electroactive under these conditions. Only α-carbonyl-activated β-O-4-ether structures were easily reducible.

There were three β-1 structures containing α-carbonyl groups that were soluble and stable in neutral media. The carbonyl group is reducible under these conditions.

For the 5-5 structure, dehydrodivanillin showed a single reduction wave in both media, which corresponds to the reduction of the aldehyde functionality.

Table 7. Anodic Oxidation of Lignin Model Compounds

Electrolysis conditions and methods, comments	References
Guaiacol and 4-propylguaiacol investigated in $LiClO_4$-AN or Bu_4NClO_4-AN by cyclic voltammetry; after the first electron transfer rapid following reactions lead to polymer formation: preparative electrolyses under anhydrous conditions did not show demethylation reactions but the formation of methoxybenzoquinones and dimers and trimers of these compounds; ESR detected the short-lived radicals from the electron-transfer step	79
Catechol oxidation investigated by cyclic voltammetry in MeOH/PrOH and Bu_4NOH; electrolyses at the half-peak potential (0.82 V) followed by visible and UV spectra	80
α-Arylethers investigated in undivided cells on Pt anodes in methanolic solutions; reaction products investigated as a function of sodium methoxide concentration	81
Vanillyl alcohol oxidation and adsorption onto Pt, graphite, and Ru/Ti oxides investigated under potentiodynamic conditions; reaction order found close to 1 in the alcohol; side-chain cleavage occurred	82
Guaiacol and pyrocatechol investigated on a number of solid electrodes (Pt, Pt/Ti, Ru oxides, glassy carbon, Ti, and Co/Ti) by cyclic voltammetry, chronopotentiometry, rotating disk voltammetry, and controlled potential electrolysis	83
Model dimers β-guaiacyl ethers of veratrylpropanone-1 and veratrylpropanol-1 oxidized; ether cleavage and the formation of 10 phenolic compounds detected upon anodic oxidation in AN or aqueous AN in 1 M $LiClO_4$; oxidation of the guaiacyl fragment leads to degradation products; gradient elution reversed-phase chromatography employed in the analyses as well as adsorption chromatography	84, 85
Isoeugenol direct electrochemical oxidation on Pt mesh anode in CH_2Cl_2 (saturated Bu_4NBF_4 and lutidine) led to the formation of the β-5-dehydrodiisoeugenol (phenylcoumaran structure) in 61%; if, however, Ag_2O was used as the oxidizing agent in benzene, the dimerization yielded mostly β-O-4-alkyl aryl ether dimeric isomers (21% erythro and 31% threo) and 18% of the β-5 product	69
Vanillyl sulfonic acid oxidized in aqueous neutral and alkaline solutions; contents of methoxyl, sulfonic acid, and phenolic hydroxyl groups analyzed as a function of electrolysis conditions; in alkaline solutions both methoxyl and phenolic hydroxyl contents decreased rapidly upon electrolysis and the carboxyl content of the products increased concomitantly; the higher the pH, the faster the oxidative reactions	86
Veratryl sulfonic acid oxidized in aqueous acid and alkaline solutions; remarkable decreases in methoxyl, phenolic hydroxyl, and sulfo group content observed in alkaline solutions	87

For the β-5, dehydrodiisoeugenol and its hydrogenated derivatives were not electroactive in the neutral or basic media investigated.

The main oxidation products in acetonitrile 0.1 M Et_4NClO_4 were polymers. Indirect oxidations were also carried out by coelectrolyzing p-tribromophenylamine in the presence of the dimers. Only those with free phenolic or hydroxyl groups underwent oxidation. The isolated oxidation products were two to three times higher in molecular weight than the starting compounds. Half-wave potentials for the oxidations of these compounds were obtained.

C. Pulping Mechanisms

A key step in the delignification of wood is the cleavage of the β-O-4-alkyl aryl ether bonds of the lignin macromolecules. The key intermediates in depolymerization reactions in alkaline solutions are the lignin quinone methides [89]. In alkaline solutions these intermediates can undergo cleavage at the β-O-4-aryl ether bond, which leads to fragmentation and thus delignification, or they can be converted into stable vinyl ethers without undergoing fragmentation, thus leading to the undesirable formation of residual lignins. Anthraquinone is a well-known catalyst of the depolymerization reaction. The mechanism of catalysis of such a cleavage is a topic of considerable interest. Rival hypotheses include the postulation of quinone methide-anthraquinone adducts as key intermediates as opposed to a catalytic mechanism involving single-electron transfer between anthraquinone radical anions and quinone methides (see Ref. 89 and references therein).

In nonaqueous and mixed solvents, evidence was added to the electron transfer mechanism by such electrochemical techniques as cyclic voltammetry [90] in reactions carried out at low concentrations of quinone methides. To understand the competing reactions further, isolate reaction products in more detail, and measure intermediate reaction products, it was necessary to work at higher concentrations of quinone methides. Under these conditions and on a different time scale, it has been shown that novel lignin-model polymers can be prepared from quinone methides by base-catalyzed condensations that may be initiated chemically and electrochemically. These polymers are α- and β-bis-O-4-aryl ether bonded and have regular repetitive structures. The key bonds and functionalities are similar to those exhibited by lignins. The polydispersities of the polymers are small (1.2–1.5), and they can be prepared in the 700–7000 apparent weight-average molecular weight range. The stereochemistry of the α and β side-chain carbons of the polymer was shown to be crucially dependent on the countercation present during polymerization. In the presence of tetrabutylammonium cations, the thermodynamically more stable erythro isomer is preferentially formed. The threo isomer is preferentially formed in the presence of alkali metal cations. The results were explained in terms of the Cram empirical rule of nucleophilic addition reactions and the well-established template roles of alkali metal cations [91].

V. CONCLUSIONS

Electrochemistry applied to biomass can lead to (1) syntheses of novel polymeric materials for use as prepolymers or as materials in a variety of applications. Of particular interest are the modifications of lignin polymers because of their ease of oxidation and reduction and their low cost and large resource availability. (2) Electrolyses to monomeric compounds can lead to significant yields of a few compounds. Of particular interest is the hydrolysis of cellulose at an anode coupled to electrochemical reductions producing sorbitol. In the lignin area, the ability to generate high yields of monomeric aldehydes, such as benzaldehyde, vanillyl aldehyde, and other simple aromatic organic compounds, should continue to be explored. The utility of nitrobenzene-mediated electrochemical oxidations of lignins should be examined as a potentially valuable method for the determination of syringyl-guaiacyl ratios for lignocellulosic biomass. (3) Electrochemical mechanistic studies with lignin model compounds, namely quinone methides, can shed some light on pulping mechanisms and also lead to the preparation of novel lignin model polymers that can advance the field of lignin chemistry. (4) Electrolytic bleaching and recovery of lignins with acid generated at anodes is being carried out, primarily in industry. There are opportunities in these areas to carry out bleaching with reduced cellulose degradation and also to produce lignins, as carried out in the two-stage organosolv pulping process. Studies with electrogenerated oxygen in the presence of mediators should be continued since environmental pressure to decrease the amount of chlorinated aromatic compounds discharged in effluent streams from these industries will continue to increase.

ACKNOWLEDGMENTS

The preparation of this review was sponsored by the Biobased Materials Project of the Department of Energy's Energy Conversion and Utilization Technologies (ECUT), FWP D071. The comments of Professors Herbert Schroeder, Kyosti Sarkanen, and Manuel Baizer and Drs. J. Christie and D. Johnson are gratefully acknowledged.

REFERENCES

1. I. S. Goldstein (ed.), Organic Chemicals from Biomass, CRC Press, Boca Raton, Florida, 1981.
2. D. Fengel and G. Wegener, Wood—Chemistry, Ultrastructure, Reactions, Walter de Gruyter, New York, 1984.
3. E. Sjöström, Wood Chemistry—Fundamentals and Applications, Academic Press, New York, 1981.
4. H. L. Chum, L. D. Douglas, D. A. Feinberg, and H. A. Schroeder, Evaluation of Pretreatments of Biomass for Enzymatic Hydrolysis of Cellulose, SERI/TR-231-2183, Solar Energy Research Institute, Golden, Colorado, 1985.
5. B. L. Browning (ed.), The Chemistry of Wood, Robert Krieger Publishing Co., New York, 1975.
6. K. V. Sarkanen and C. Ludwig (eds.), Lignins: Occurance, Formation, Structure and Reactions, Wiley, New York, 1971.
7. E. Adler, Wood Sci. Technol., 11, 169–218 (1977).
8. R. F. Decker, Metall. Trans. A, 17, 5–30 (1986).
9. S. A. Rydholm, Pulping Processes, Wiley-Interscience, New York,

1965. H. F. J. Wenzl, The Chemistry and Technology of Wood, Academic Press, New York, 1970.
10. S. Y. Lin, in Progress in Biomass Conversion (D. A. Tillman and E. C. Jahn, eds.), Vol. 4, pp. 32–77 (1983) and references therein; H. L. Chum, S. K. Parker, D. A. Feinberg, J. D. Wright, P. A. Rice, S. A. Sinclair, and W. G. Glasser, The Economic Contribution of Lignins to Ethanol Production from Biomass, SERI/TR-231-2488, Solar Energy Research Institute, Golden, Colorado, 1985.
11. K. V. Sarkanen, in Progress in Biomass Conversion (K. V. Sarkanen and D. A. Tilman, eds.), Academic Press, New York, 1982, pp. 127–144.
12. Proceedings of the Solvent Pulping Conference, Appleton, Wisconsin, Doshi and Associates, Inc., Appleton, Wisconsin, 1987.
13. R. C. Myerly, M. D. Nicholson, R. Katzen, and J. M. Taylor, Chemtech, 11, 186 (1981); E. K. Pye, W. R. Klein, J. H. Lora, and M. Cronlund, The ALCELL Process, Proceedings of the Solvent Pulping Conference, Doshi and Associates, Inc., Appleton, Wisconsin, 1987, pp. 55–67.
14. J. Feckl and E. Edel, Organosolv Pulping with Addition of Alkali: Procedure and Products, Proceedings 4th International Symposium on Wood and Pulping Chemistry, April 27–30, Paris, France, Vol. 1, 1987, pp. 368–372.
15. H. R. Bungay, Energy, the Biomass Options, Wiley & Sons, New York, 1980, pp. 192–198.
16. W. H. Mason, U.S. Patent 1,578,609 (1926); T. P. Schultz, D. P. McGinnis, and C. J. Biermann, Similarities and differences in pretreating woody biomass by steam explosion, wet oxidation, autohydrolysis, and rapid steam hydrolysis/continuous extraction, in Energy from Biomass and Wastes VIII (D. L. Klass, ed.), Institute of Gas Technology, Chicago, 1984, pp. 1171–1198; K. L. Mackie, H. H. Brownell, K. L. West, and J. N. Saddler, J. Wood Chem. Technol., 5, 403 (1985).
17. R. Bender, U.S. Patent 4,136,207 (1979): J. D. Taylor, Continuous autohydrolysis, a key step in the economic conversion of forest and crop residues into ethanol, in Energy from Biomass (W. Palz, P. Chartier, D. O. Hall, eds.), Applied Science Publ. Ltd., London, 1981, pp. 330–336.
18. H. L. Chum and M. M. Baizer, Electrochemistry of Biomass and Derived Materials, American Chemical Society, Washington, D.C., 1985.
19. J. H. P. Utley, The electrochemical conversion of biomass-derived compounds, in Fundamentals of Thermochemical Biomass Conversion (R. Overend, T. Milne, and L. Mudge, eds.), Elsevier Applied Science, London, 1985, pp. 1087–1102.
20. H. L. Chum, D. W. Sopher, and H. A. Schroeder, Electrochemistry of lignin material and derived compounds, in Fundamentals of Thermochemical Biomass Conversion (R. Overend, T. Milne, and L. Mudge, eds.), Elsevier Applied Science, London, 1985, pp. 1103–1113.
21. T. Nonaka and M. M. Baizer, Denki Kagaku, 55, 13–17 (1987).
22. G. Schwab, M.-T. Lee, and J. W. Bentley, Electrochemical Bleaching of Wood Pulps, U.S. Patent 4,617,099 (December 23, 1985).
23. M. P. Godsay, M. N. Hull, and V. M. Yasnovsky, Process for the Delignification of Lignocellulosic Material with Oxygen, Ferricyanide and a Protector, U.S. Patent 4,622,100 (1986).

24. M. M. Nassar and O. A. Fadaly, J. Appl. Electrochem., 13, 663–667 (1983).
25. V. A. Demin and V. D. Davydov, Method of Bleaching Pulp, USSR Patent 912,803 (1982); IPC-AN-53-05608.
26. V. A. Demin, B. D. Bogomolov, A. P. Karmanov, and V. D. Davydov, Khim. Drev. (Riga) No. 4, 82–84, (1980); IPC-AN-51-08436.
27. L. N. Spiridonova, V. A. Babkin, M. I. Anisimova, G. S. Mikhailov, and T. P. Belova, Khim. Drev. (Riga) No. 4, 16-19 (1982); IPC-AN-53-10666.
28. V. G. Gorokhova, V. A. Babkin, L. N. Spiridonova, G. P. Vasyanovich, and A. I. Gerasimov, Khim. Drev. (Riga), No. 1, 101–103 (1986); IPC-AN-57-02209.
29. V. A. Babkin, E. N. Serdobol'skii, A. I. Gerasimov, and Y. P. Rulev, Tsellyuloza Bumaga Karton, No. 13, 10(1983); IPC-AN-55-06661.
30. E. I. Chupka and S. V. Stromskii, Bumazh. Prom., No. 1, 15–16 (1982); IPC-AN-53-03973.
31. L. N. Spiridonova, Y. K. Svyatkin, G. S. Mikhailov, V. Y. Dubinin, V. A. Babkin, and M. I. Anisimova, Method of Delignifying Cellulose-Containing Pulp, USSR Patent 889,765 (1981); IPC-AN-53-03560.
32. G. Brown, D. F. Dong, J. A. McIntyre, and R. F. Phillips, Alkaline Peroxide Solutions for the Pulp and Paper Industry, TAPPI Pulping Conf. Proc., Houston, Texas (Book 1), 1983, pp. 341–344.
33. C. Oloman, Can. Pulp Paper Ind., 33(7), 63–64 (1980).
34. H. L. Chum, M. A. Ratcliff, and P. D. Palasz, Process for Producing Peracids from Aliphatic Hydroxy Carboxylic Acids, U.S. Patent 4, 617,090 (1986).
35. H. J. Kleine, Chloride Process for Recovering Alkali from Sulfate Liquors, Braz. Patent 4,233/84 (1986); IPC-AN-57-08247.
36. V. V. Domnitskii, Y. N. Nepenin, and Y. A. Malkov, Bumazh. Prom., No. 12, 8–9 (1980); IPC-AN-52-07293.
37. Y. A. Malkov and V. V. Domnitskii, Mezhvuz. Sb. Nauch. Tr., Khim. Tekhnol. Drev. Tsellyul. (Leningrad), 45–50 (1983); IPC-AN-57-14127.
38. V. G. Kenin, V. A. Zhalina, Y. N. Nepenin, and Y. A. Malkov, Method of Producing Semichemical Pulp, USSR Patent 910,896 (1982); IPC-AN-53-04550.
39. G. D. Kovalev, A. T. Serkov, and I. Z. Eifer, Khim. Volokna, No. 3, 38–40 (1985); IPC-AN-56-09484.
40. H. Kau, DECHEMA Monographien 98 (Tech. Elektrolysen), 291–298 (1985).
41. E. Edel, J. Feckl, C. Grambow, A. Huber, and D. Wabner, Process for Obtaining Lignin from Alkaline Solutions Thereof, U.S. Patent 4,584,076 (1986).
42. J. W. Green, Carbohydrate Chemistry and Biochemistry, (W. Pigman, D. Horton, and J. D. Wander, eds.), 2nd ed., Academic Press, New York, Vol. IB, Chap. 24, 1980, pp. 1101–1166.
43. J. C. Yu, M. M. Baizer, and K. Nobe, J. Electrochem. Soc., 135, 83–87 (1988).
44. I. A. Antipova, Khim. Drev. (Riga), No. 1, 82–88 (1985); IPC-AN-56-02895.
45. M. I. Anisimova, V. A. Babkin, L. G. Kozhevnikova, L. A. Zaitseva, G. P. Vasyanovich, and N. V. Isakova, Khim. Drev. (Riga), No. 2, 72–77 (1980); IPC-AN-51-05495.

46. O. F. Gorbunova, B. D. Bogomolov, and G. M. Bogolitsyna, Isv. VUZ Lesnoi Zh., No. 2, 78–82 (1986); IPC-AN-57-05835.
47. A. D. Sergeev, A. N. Zakazov, and E. I. Chupka, Mezhvuz. Sb. Nauch. Tr., Khimiya Tekhnol. Tsellyulozy (Leningrad), No. 8, 25–30 (1981); IPC-AN-57-03274.
48. A. Yoshiyama, T. Fuchigami, T. Nonaka, T. C. Chou, and H. J. Tien, Chem. Express., 1, 212–215 (1986); CA 105(6):45021k.
49. P. D. Limosin, G. Pierre, and G. Cauquis, Holzforschung, 40, 31–36 (1986).
50. G. Cauquis, D. Limosin, and G. Pierre, First Results of the Electrochemical Study of Lignin in Aqueous Solution and in Solvent Medium, Colloque Sci. Ind. Bois 3 (Transform. Biol. Chim.), Grenoble, France, 1982, pp. 161–183.
51. G. I. Stromskaya and E. I. Chupka, Khim. Drev. (Riga), No. 2, 60–63 (1983); IPC-AN-54-07094.
52. A. Yoshiyama, T. Nonaka, H. Kawakami, T. C. Chou, and H. J. Tein, Anodic Degradation of Lignins. II. Softwood Lignins, J. Jpn. Wood Res. Soc., 34(3), 275–280 (1988).
53. V. D. Davydov and G. Y. Osipova, Khim. Drev. (Riga), No. 3, 66–71 (1982); IPC-AN-53-07021.
54. V. V. Sinitskii, I. M. Kashirskii, Y. S. Bulyshev, N. V. Afanas'ev, E. I. Chupka, and V. M. Burlakov, Khim. Drev. (Riga), No. 3, 66–71 (1980); IPC-AN-51-08458.
55. V. R. Koch, Electrolysis of Lignosite at a Reticulated Vitreous Carbon Electrode, in Fundamentals of Thermochemical Biomass Conversion (R. Overend, T. Milne, and L. Mudge, eds.), Elsevier Applied Science, London, 1985, pp. 1115–1125.
56. H. Ogawa, J. Natl. Chem. Lab. Ind. (Japan), 79, 527–532 (1984); IPC-AN-55-09894.
57. G. I. Stromskaya and G. P. Martynova, Tsellyuloza Bumaga Karton, No. 8, 15 (1982); IPC-AN-54-05084.
58. G. I. Stromskaya, G. P. Martynova, and E. I. Chupka, Zh. Prikl. Khim., 56, 2539–2543 (1983); IPC-AN-55-06399.
59. A. Yoshiyama, T. Nonaka, T. C. Chou, and H. J. Tien, Anodic degradation of lignins. I. Hardwood lignins, J. Jpn. Wood Res. Soc., 34(2), 155–161 (1988).
60. M. Khullar, S. Chakraborti, and B. Allard, Ann. Univ. Ferrara, Sez. 5, Suppl. 7/Proc. Eur. Symp. Corrosion Inhibitors 5th, 3; 815-825 (1980); IPC-AN-53-00260.
61. A. Yoshiyama, T. C. Chou, T. Nonaka, and H. J. Tien, Anodic degradation of lignins. III. Herbal lignins, J. Jpn. Wood Res. Soc., 34(3), 702–706 (1988).
62. E. I. Kovalenko, N. P. Kotenko, V. A. Smirnov, and O. V. Lyashko, Khim. Drev. (Riga), No. 5, 66-72 (1986); IPC-AN-57-12626.
63. E. I. Kovalenko, V. N. Shalimov, and N. A. Kovalenko, Zh. Prikl. Khim., 53, 850-854 (1980).
64. E. I. Kovalenko, N. P. Kotenko, V. A. Smirnov, and L. A. Saratova, Khim. Drev. (Riga), No. 6, 98-101 (1986); IPC-AN-58-00263.
65. V. N. Shalimov, E. I. Kovalenko, N. A. Kovalenko, and V. A. Smirnov, Zh. Prikl. Khim., 53, 1284-1288 (1980); IPC-AN-52-07004.
66. V. F. Korytseva, Y. V. Vodzinskii, V. Z. Latypova, and Y. M. Kargin, Izv. VUZ, Lesnoi Zh., No. 2, 82-85 (1980); IPC-AN-52-01337.
67. E. I. Kovalenko, V. A. Smirnov, I. K. Pimachkova, and O. V. Korovina, Khim. Drev. (Riga) No. 4, 67-72 (1986); IPC-AN-57-12627.

68. J. H. P. Utley, C. Z. Smith, and N. Islam, The Electrochemical Conversion of Lignins and Furan Derivatives, The Electrochemical Society Extended Abstract, Vol. 87(2), p. 1145 (1987).
69. G. E. Hawkes, C. Z. Smith, J. H. P. Utley, and H. L. Chum, Holzforschung, 40(Suppl.), 115-123 (1987).
70. A. F. Gogotov, E. I. Chupka, and M. N. Raskin, Gidroliz. Lesokhim. Prom., No. 5, 5-7 (1980); IPC-AN-52-04639.
71. A. F. Gogotov and E. I. Chupka, Khim. Drev. (Riga), No. 2, 64-67 (1983); IPC-AN-54-07081.
72. A. F. Gogotov and E. I. Chupka, Mezhvuz. Sb. Nauch. Tr. Khim. Tekhnol. Drev. Tsellyul. (Leningrad), 112-115 (1983); IPC-AN-57-09803.
73. H. L. Chum, M. Ratcliff, H. A. Schroeder, and D. W. Sopher, J. Wood Chem. Technol., 4, 505-532 (1984).
74. D. K. Johnson, S. Black, M. Ratcliff, H. L. Chum, C. A. Barnett, and W. G. Glasser, Reductive Electrolysis of Explosively Depressurized Aspen Lignin, The Electrochemical Society Extended Abstract, Vol. 84(1), pp. 475-476 (1984).
75. H. L. Chum, D. K. Johnson, M. Ratcliff, and S. Black, Electrochemical Modifications of Lignins, Proc. 1984 International Chemical Congress of Pacific Basin Societies, Honolulu, Hawaii, December 16-21, Abstract 10T19 (1984); D. K. Johnson, S. Black, and H. L. Chum, Lignin Electrolyses in Flow Systems, Proceedings 35th Meeting of the International Society of Electrochemistry, Berkeley, pp. 796-798 (1984); D. K. Johnson, S. Black, M. Ratcliff, D. Robert, and H. L. Chum, manuscript in preparation.
76. A. Bailey and H. M. Brooks, J. Am. Chem. Soc., 68, 445-449 (1946).
77. M. M. Baizer, Electrochemical Conversion of Biomass-Derived Materials to Specialty Chemicals, Proc. BIO EXPO 86, pp. 341-349 (1986).
78. C. Z. Smith, J. H. P. Utley, and H. L. Chum, J. Chem. Res., Synop., No. 3, p. 88-89 (1987).
79. F. Sundholm and G. Sundholm, Holzforschung, 36, 71-75 (1982).
80. F. Sundholm and G. Sundholm, Finn. Chem. Lett., No. 7-8, p. 85-86 (1981).
81. P. R. Janissek, V. L. Pardini, and H. Viertler, Anodic Cleavage of Lignins: Aryl Ethers as Model Compounds, The Electrochemical Society Extended Abstracts, Vol. 87(2), p. 1149 (1987).
82. V. A. Medvedev, V. A. Babkin, L. T. Gorokhova, M. I. Anishimova, S. A. Medvedeva, and L. N. Petrushenko, Khim. Drev. (Riga), No. 1, 82-86 (1982); IPC-AN-53-04836.
83. V. A. Medvedev, V. V. Yanilkin, V. A. Babkin, S. Z. Ivanova, and G. P. Vasyanovich, Khim. Drev. (Riga), No. 4, 82–86 (1985), IPC-AN-56-09679.
84. V. G. Gorokhova, V. A. Babkin, N. A. Tyukavkina, V. V. Yanilkin, and L. Y. Dubenko, Khim. Drev. (Riga), No. 4, 92-95 (1986); IPC-AN-57-12622.
85. L. Y. Dubenko, Khim. Drev. (Riga), No. 5, 70-75 (1985); IPC-AN-56-10775.
86. H. Ogawa, J. Natl. Chem. Lab. Ind. (Japan), 79, 563-569 (1984); IPC-AN-56-00270.
87. H. Ogawa, J. Natl. Chem. Lab. Ind. (Japan), 79, 571-575 (1984); IPC-AN-56-00271.

88. D. Limosin, G. Pierre, G. Cauquis, J. Chopin, M. Chadenson, and M. Hauteville, Holzforschung, 39, 91-98 (1985).
89. D. R. Dimmel, J. Wood Chem. Technol., 5, 1-15 (1984).
90. D. R. Dimmel, L. F. Perry, P. D. Palasz, and H. L. Chum, J. Wood Chem. Technol., 5, 15-36 (1985).
91. H. L. Chum, D. K. Johnson, P. D. Palasz, C. Z. Smith, and J. H. P. Utley, Macromolecules, 20, 2698-2702 (1987).

21

Organoelemental and Coordination Compounds

LORENZ WALDER / Institute of Organic Chemistry, University of Bern, Bern, Switzerland

I. Introduction 809

II. Coordination Compounds and Their Organometallic Derivatives 812
- A. N-containing macrocyclic and bidentate metal complexes 812
- B. Chelating and simple phosphine ligands, formation and cleavage of M—C bonds, and electrocatalysis 834

III. Electrosynthesis of Coordination Compounds 836

IV. Selected Transition Metal Organometallics 837
- A. Cyclopentadienyl and arene complexes 837
- B. Activation of σ- or π-bound carbon ligands in carbonyl complexes 845

V. Electrosynthesis of Transition Metal Organometallics 847

VI. Group IVB Organometallic Compounds 848
- A. Electrosynthesis and electrocatalysis 848
- B. Activation of M—C bonds 852

VII. Group III Organometallic Compounds 857
- A. Formation of M—C bonds 857
- B. Activation of M—C bonds 858

VIII. Electrochemistry of Grignard Reagents 860

IX. Group IIB Organometallic Compounds 862
- A. Formation of M—C bonds 862
- B. Activation of M—C bonds 863

References 864

I. INTRODUCTION

This chapter is concerned with the electrochemistry of organometallics and coordination compounds. Preparative and mechanistic aspects are considered. The emphasis is on compounds relevant for organic synthesis as reagents or catalysts. Bearing in mind the huge activity in the field of organometallic and coordination chemistry directed toward organic synthesis in the last decade [1,2], the coverage must be selective. An insight into the topic is provided by earlier general reviews on organometallic electrochemistry (containing extensive lists of reduction potentials) [3–11] and by monographs and reviews on the principal aspects of electron transfer reactions involving organometallic species [3,12–14] and application to organometallic and organic synthesis as well as on inner sphere electroanalysis (see also Chap. 29) [15–20]. Reviews concerning specific topics are mentioned in the corresponding sections of this chapter.

From the viewpoint of an organic chemist, an organometallic compound is a reagent or intermediate considered to undergo or induce a reaction with, it is hoped, high selectivity (i, scheme 1). Metal-carbon σ- or π-bond formation (d), possible reactions of the organic residue R during coordination (e), and the cleavage reaction (f) are the central steps. Some of these reactions need activation; others require an external driving force to proceed (a–c). As organometallic and coordination compounds are usually electroactive, electrochemistry is the technique of choice to provide this energy.

Formation of M–C bonds: Electrochemical generation of a reactive metal or ligand oxidation state can lead to M–C bond formation, for example by reductively induced oxidative addition, cathodic or anodic generation of alkyl radicals that attack the sacrificial electrode, or electrochemically induced ligand exchange.

Transformation of the coordinated carbon ligand: Trans-metallation, insertion of other ligands into M–C bonds, electrophilic or nucleophilic attack on a coordinated carbon ligand, and polymerization may be controlled by electrochemistry. Except for the first example, these reactions are important for transition metals.

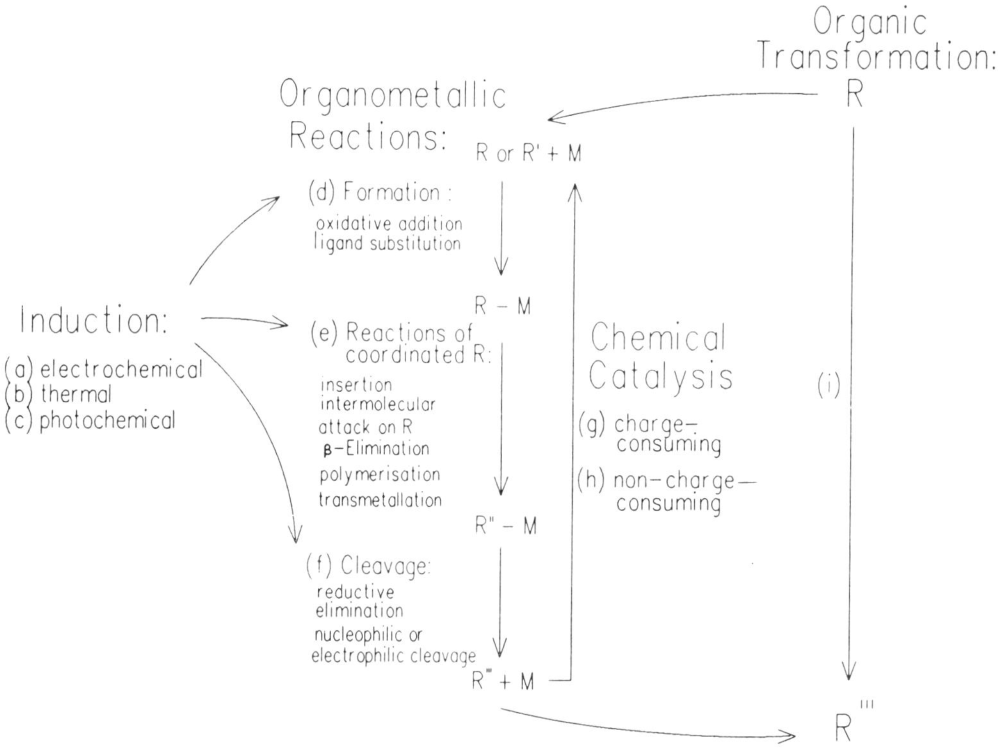

Scheme 1. Role of electrochemistry in organic synthesis involving organometallic intermediates.

Cleavage of M—C bonds: If a redox orbital with a bonding or antibonding M—C character is accessible, the M—C bond may cleave by electrochemical oxidation or reduction, respectively. The reactivity of the organometallic reagent is easily tuned by an electron transfer, that is, oxidatively induced reductive elimination possibly assisted by a nucleophile, reductively induced cleavage assisted by an electrophile, or generation of radicals by oxidation of carbanionlike organometallics.

A transformation of R to R^{III} by organometallic intermediates is catalytic in M if the metal can be recycled (scheme 1). If additionally any of the steps d through f involves electrochemistry, the process is called electrocatalytic. In some electrocatalytic reactions only short-lived organometallic intermediates exist and it may be difficult to distinguish chemical from outer sphere electrocatalysis involving no organometallic intermediates. Another detection problem may arise from electrochemically initiated but non-charge-consuming electrocatalyses (h), that is, if an oxidative and a reductive step are coupled and proceed in solution by homogeneous redox chemistry or isomerization occurs thermodynamically downhill when trace amounts of the active metal oxidation state have been formed electrochemically.

Similarly as described for organometallic compounds, the formation, transformation, and cleavage of coordination compounds can be induced by electrochemistry (for a very short review, see Secs. II.A, Porphyrins, and III). As the main concern in organic electrochemistry is carbon-carbon bond formation, coordination compounds are discussed in this chapter mostly with respect to their ability to stabilize or activate metal-carbon bonds, that is, as the corresponding organometallic derivatives.

Section II deals with organometallic derivatives of coordination compounds with a relatively rigid coordination sphere. Generally, the multidentate and nonreactive ligand limits the scope of electrochemically induced organometallic reactions to M—C bond formation (mostly reductively induced oxidative additions), simple transformations during coordination, and cleavage. Section II includes main group and transition metals and follows the trend macrocyclic-bidentate-monodentate. In Sec. IV the electrochemistry of transition metal organometallics with a multiple π-bonded carbon ligand, such as an olefin or an aromatic compound (Sec. IV.A), or with several carbon ligands, such as CO (Sec. IV.B), is discussed. The scope of ligand-centered reactions induced by electron transfer during coordination is quite large for this class of compounds and involves activation of the π-system toward electrophilic or nucleophilic attack (Sec. IV.A, Activation of cyclopentadienyl, arene, or tropylium ligands), π- and σ-transformations (Sec. IV.A, Metal-olefin, metal-alkyl transformation), and insertion reactions (Sec. IV.A, Carboxylation-decarboxylation). As the free ligands are usually less basic, formation of the M—C bond is often reversible and may occur by electrochemically induced ligand substitution. This type of reaction is separately discussed in Sec. V. The second part of the chapter deals with typical main group organometallics with emphasis on compounds with a covalent rather than ionic M—C bond: group IVB (Sec. VI), group III (Sec. VII), and group IIB (Sec. IX); the electrochemistry of Grignard reagents is separately discussed in Sec. VIII. The different sections are organized according to formation (Secs. VI through IX) and activation of the metal-carbon bond (e.g., transmetallation and cleavage, Secs. VI through IX).

II. COORDINATION COMPOUNDS AND THEIR ORGANOMETALLIC DERIVATIVES

For a recent and comprehensive review on coordination chemistry containing several chapters on electrochemistry (cited below), see Ref. 1a.

A. N-Containing Macrocyclic and Bidentate Metal Complexes

A good reason so much interest in modern electrochemistry is directed toward the study of nitrogen-containing macrocyclic coordination compounds is simply that many of these complexes play an essential role in life as biocatalysts. The most prominent class of compounds are the metal complexes of porphine-type compounds, the so-called pigments of life. All these coenzymes are derived biochemically from one common source, the uroporphinogen III [21], a structure of potential prebiotic origin [22]. Recent and concise reviews concern the genesis of these compounds (scheme 2) [23] and their action as biocatalysts [24]. The most unsaturated (and evolutionarily latest) members of this group of complexes containing the maximum of eleven

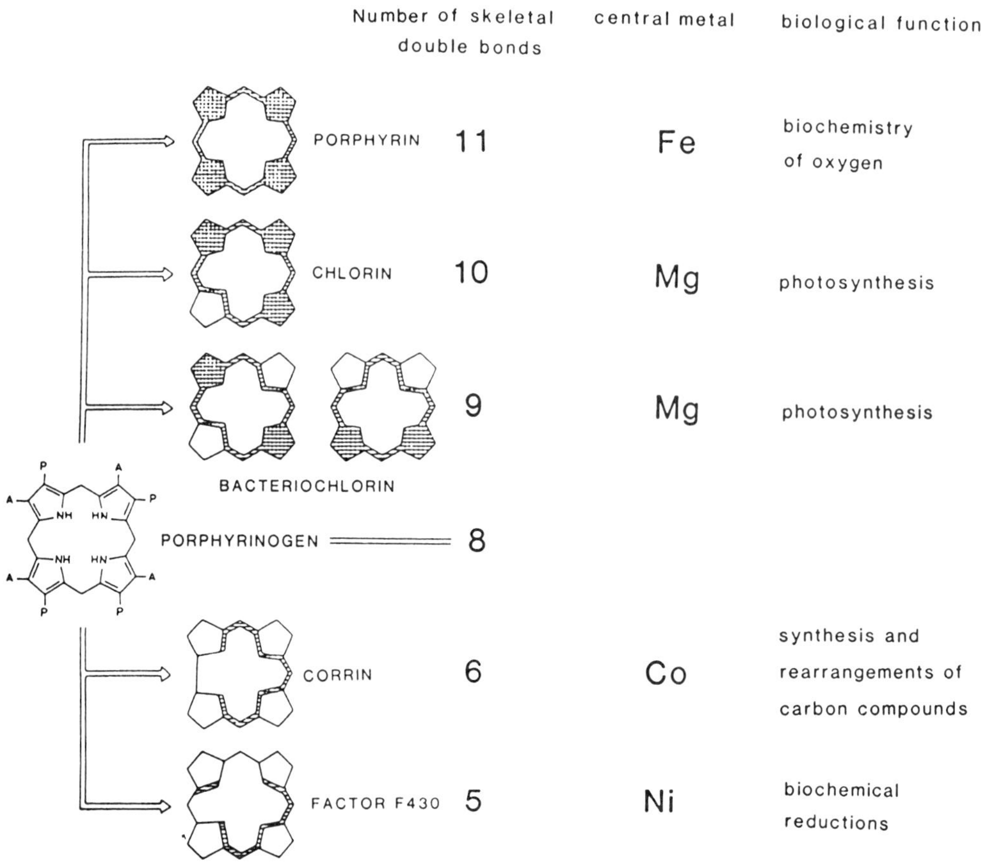

Scheme 2. Genesis of pigments of life. Dark regions symbolize π-delocalization.

skeletal double bonds are the porphyrins, responsible for the biochemistry of oxygen [25,26]. The least unsaturated (and evolutionary earliest) porphine-type compounds containing only five and six skeletal double bonds are factor F_{430} [27] and corrin, respectively [28,29]. The coenzyme F_{430} is involved in the formation of CH_4 from CO and CO_2 by methanogenic bacteria [30,31], the fundamental reactions being based on redox reactions [32–34]. The co-enzyme vitamin B_{12} acts as a biocatalyst in the formation of C—C bonds and 1,2-rearrangements [28,29,35]. Recent reviews by Lexa and Savéant [36] concern the electrochemistry of B_{12} [36] and by Scheffold et al. the application of vitamin B_{12} and related compounds as catalysts in organic electrosynthesis [37–39]. Parallel to the research on the chemistry of the pigments of life, an increasing interest in the synthesis and electrochemistry of "synthetic" macrocyclic coordination compounds developed (see the reviews on synthesis [40] and electrochemistry in aqueous [41] and nonaqueous solvents [42,43]).

Artificial macrocyclic metal complexes from phthalocyanines to cyclams span an even broader range of π-delocalization. Interestingly, the same trend in typical electrocatalytic reactivity from oxygen reduction to C—C bond formation is observed as with the natural prototypes. Accordingly, metal complexes of macrocyclic ligand systems are presented in the following section with decreasing π-delocalization.

Porphyrins

The electrochemistry of metalloporphyrins in nonaqueous media has recently been summarized by Kadish [44]. This and earlier reviews demonstrate the enormous interest in the macrocyclic ligand system known in a dozen basic variations and incorporating 55 different central atoms [45–46,47]:

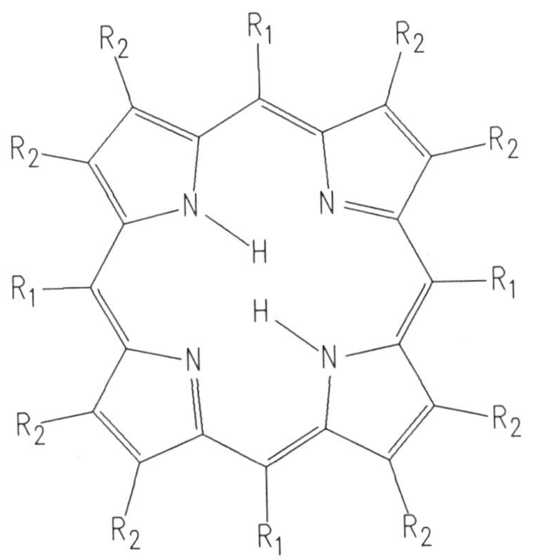

H_2P (I): R_1 = H, R_2 = H; H_2OEP (II): R_1 = H, R_2 = C_2H_5;
H_2TPP (III): R_1 = C_6H_5, R_2 = H

Metal free octaethylporphyrin (II) exhibits four redox processes in aprotic solvents [49]. The potential difference $E_{p+/p} - E_{p/p} = 2.25 \pm 0.15$ V is in

Table 1. Representative Reduction Potentials[a] of (Metallo)porphyrins

	H_2OEP [49]	H_2TPP [48]	ZnOEP [49]	CoTPP [50]
$E_{p-/p2-}$	-1.86	-1.43		
$E_{p/p-}$	-1.46	-1.05	-1.61	-1.87
$E_{M++/+}$				-0.81
$E_{M+++/++}$				0.13
$E_{p+/p}$	0.81	1.06	0.63	1.16
$E_{p++/p+}$	1.30		1.02	1.35

[a]Versus SCE measured in different organic solvents.

agreement with the LUMO/HOMO gap based on extended Hückel calculations [51]. The tetraphenylporphyrin H_2TPP (III) is more easily reduced than H_2OEP (II), and the ligand redox potentials can be tuned about 1 V by different para substituents in the benzene rings [48]. Complexation by a nonelectroactive metal(II) ion shifts the entire pattern of redox potentials to more positive or negative values compared with H_2TPP or H_2OEP. Inductive effects are responsible for these shifts [49].

Porphyrins with complexed electroactive (transition) metals show overlaid porphyrin and metal electrochemistry (Table 1). Mutual inductive effects by the metal ion and oxidation state stabilization by the square planar ligand field define the four-wave pattern and the metal potentials, respectively. Additionally, the metal potentials are strongly influenced by axial ligation. This may cause difficulty in assigning the site of redox process (metal or porphyrin) as in case of one-electron reduced iron porphyrins. Resonance structures involving the ligand radical and Fe(I) have been used to describe its ground-state electron configuration (continuous lowering of the π^*-orbital by para-substituents in $FeTPPX_4$ (X = H, Br, CN) reverses the sequence of redox orbitals (π^* above d_{z^2} for X = H) [52]). Electrochemical methods and correlations between electrochemical and spectroscopic data [52], Hammett substitution parameters [48], pK_a values of ligands [53], Gutmann donor numbers of solvents [54], and extended Hückel calculations have been used to rationalize the electrochemistry of metalloporphyrins.

$E_{1/2}$ (Red) versus $E_{1/2}$ (Ox): The linear dependence and slope 1 of the first porphyrin oxidation versus the first porphyrin reduction (each varying approximately 1 V) for a series of metals has been attributed to variations in the electrostatic forces induced by the different metal ions [49].

$E_{1/2}$ versus $4\sigma_p^-$: In the case of para-substituted MTPP (M = metal) ρ-values of 0.073 and 0.051 V are typical for ring-localized redox processes [48]. Smaller values for ρ have been observed for metal-centered redox processes in the case of para-substituted CoTTP [50].

$E_{1/2}$ versus log free axial ligand concentration: Slopes of 0 mV have been interpreted as no change in the axial coordination during a redox process, of 60 mV as a loss or gain of one ligand, and of 120 mV as a loss or gain of two axial ligands. Any slope different from 0 is indicative of a metal-centered process [55]. If the standard potential in pure solvent is available, the complex formation constant of the metalloporphyrin with axial bases can be calculated.

$E_{1/2}$ versus pK_a of axial base: A linear correlation with slopes different from 0 are typical for metal-centered redox processes. They have been observed in $FeTPP(L)_2^+$ and $FeTPP(L)_2$ reductions [53].

Formation and Cleavage of M—C Bonds Metal σ-bonded alkyl-, acyl-, or arylporphyrins have recently been reviewed by Guilard et al. [56]. In the case of the transition metals the σ-bond is stabilized by interaction of the carbon-centered sp^n (n = 2, 3) and the metal d_z2 orbitals. Synthesis is possible of the organometalloporphyrin from a d^8-metalloporphyrin (e.g., Co(I) [57]) and an organic electrophile by an oxidative addition or from a d^6-metal (e.g., Rh(III) [58]) and an organic nucleophile by an electrophilic addition (scheme 3). Both Fe(I) and Fe(0) have been shown to react with alkyl halides [59], probably from a singly and doubly occupied d_z2 orbital, respectively [52]. Two- and one-electron mechanisms as well as atom abstraction have been invoked for the formation of the metal-carbon bond, depending on the metal and organic substrates; for example, two Rh(II)OEP react with benzyl bromide by halide abstraction to $RH(III)OEP(C_7H_7)$ and Rh(III)OEPBr [60].

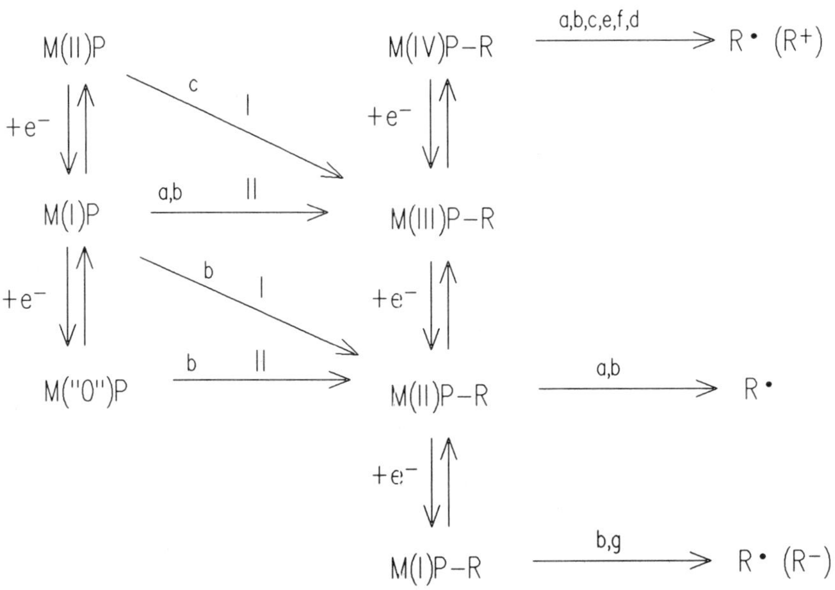

M = a) Co, b) Fe, c) Rh, d) Ge, e) Ga, f) In, g) Tl
A: + R˙ or + 1/2 RX; B: + RX

Scheme 3: Electrochemically induced formation and cleavage of the metal-carbon bond in metalloporphyrins.

Electrosynthesis of the organometalloporphyrin according to an EC (or ECE/DISP) mechanism is feasible if the reactive metal oxidation state can be formed electrochemically within the π-π^*-gap of the porphyrin in the presence of the organic substrate (the organic substrate and product should be stable at this potential). CoPR [61], FePR [59], and RhPR [62 have been prepared analytically or on a preparative scale by reduction of the metal ion to a low valence state. Double electrochemical induction of a Fe—C bond has been reported for Fe(I)P and vinyl or aryl radicals, both formed at the electrode [63].

The rate constant of the chemical step has been determined for FeP [64] and CoP [61] by cyclic voltammetry. The metal-carbon bond in σ-alkyl porphyrins can be cleaved or activated electrochemically toward electrophilic or nucleophilic attack if a redox orbital with an antibonding or bonding character is accessible within the potential range of porphyrin ligand reduction or oxidation, respectively.

Reductive cleavage of the metal-carbon bond has been reported for formal Co(II)TPPC_4H_9 generated electrochemically at −1.3 V [61], Fe(II)TPPC_4H_9 at −0.718 V [59], and Fe(I)TPPC_4H_9 at −1.986 V [59]. Tl(II)OEPCH_3 generated at −1.52 V (E_p) undergoes fast dealkylation [65]. However, the addition of electrons to the organoporphyrins of In and Ga generates relatively stable singly and doubly reduced species, and a stable anionic complex is also generated upon reduction of FeTPPC_6H_5.

Oxidative cleavage by electrochemical methods has been reported for GaTPPR (R = CH_3, C_6H_5, $C_2C_6H_5$) at 1.2−1.5 V [66], InTPPR (R = $C(CH_3)_3$, CH_3, C_6H_5) at 0.53−0.79 V]67], CoTPPC_2H_5 at 0.98 V (versus Ag/AgCl) [68], RhTPPCOCH_3 at 1.03 and 1.33 V [69], and FeTPPC_6H_5 at 0.61 V [70].

The nature of the organic moiety (radical or cation) and the influence of an organic nucleophile have not yet been studied in detail. It is interesting to note that oxidation of the organoporphyrin leads to an intramolecular migration of R from metal to a pyrrol nitrogen in the case of Co and Fe [68,70], but not for Ga and In. This biochemically important reaction (inactivation of cytochrome P_{450} [71]) can be reversed on reduction of the N-alkylmetalloporphyrin, Eq. (1):

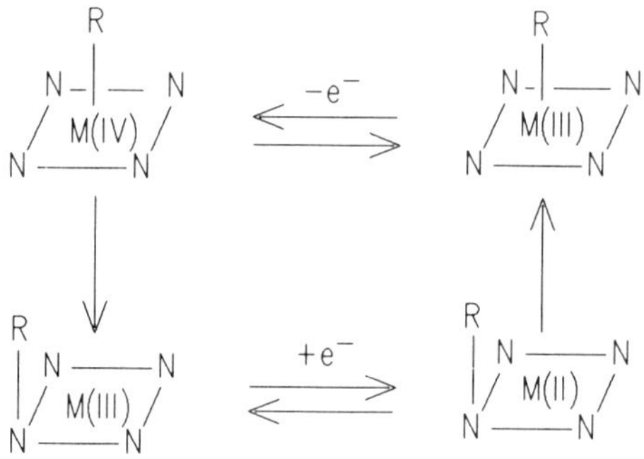

(1)

Iron (vinyl)carbenes and thioformyl and a μ-carbido dimeric FeP are all reducible at approximately −1.3 ($E_1°$) and --1.7 V ($E_2°$). Conversion of the coordinated vinyl carbene into a σ-bound vinyl moiety has been observed, Eq. (2) [72]:

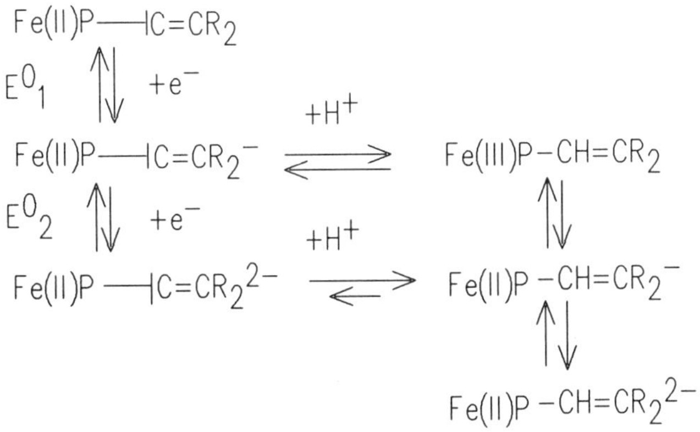

(2)

Photochemical activation of the axial alkyl ligand is also possible. Under visible light irradiation, $AlTPPC_2H_5$ [73] and $InPCH_3$ [74] exhibit Grignard-like reactivity and insert CO_2 into the metal-carbon bond.

Electrochemically induced formation and cleavage of the metal-carbon bond in organometalloporphyrins has been used for the electrocatalytic reduction of alkyl halides (M = Co[61], Fe [59]), Eq. (3):

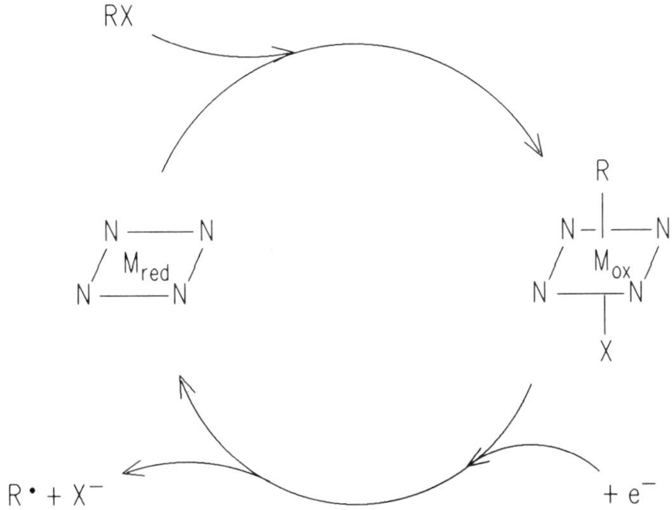

(3)

The intermediacy of organometallic derivatives during catalysis of the trans-1,2-dibromocyclohexane reduction with Ni-, Fe-, and CoOEP is supported by their catalytic efficiency. The activity was compared with that of H_2OEP^-, $CuOEP^-$, perylene$^{\bar{\cdot}}$, terephthalonitrile$^{\bar{\cdot}}$, and other aromatic anion radicals. The latter monoanions show a Brønsted-Marcus activation driving force free energy relationship typical of outer sphere redox catalysis, whereas in the case of NiOEP, FeOEP, and $CoOEP^-$, the rate-limiting step is enhanced by a factor of 1.2×10^2, 2.5×10^3, and 2.5×10^5, respectively [75]. Alkyl-NiOEP may be transiently present during catalysis; however, the compound has never been isolated. Other types of catalytic cycles, including formation and cleavage induced by electrochemical oxidation or reduction or combined with photochemical techniques, may be envisaged.

Activation of Noncarbon Axial Ligands The axial coordination chemistry of anionic and neutral ligands is well documented in the field of metalloporphyrins [44]. However, most ligands are reversibly bound and only their formation constant is changed upon a redox process. Activation of an axial ligand has been reported for oxometalloporphyrins. OxoMn(V)P epoxidizes methylstyrene [76]. The electrochemical porphyrin-centered oxidation of Mn(III)P leads after treatment with OH^- to Mn(IV)P=O. This compound and Mn(V)P=O are the active epoxidizing agents. Direct access to Mn(V)P=O in a electrocatalytic version with molecular oxygen, carboxylic acid anhydride, and the olefinic compound as reagents has been described, Eq. (4) [77]:

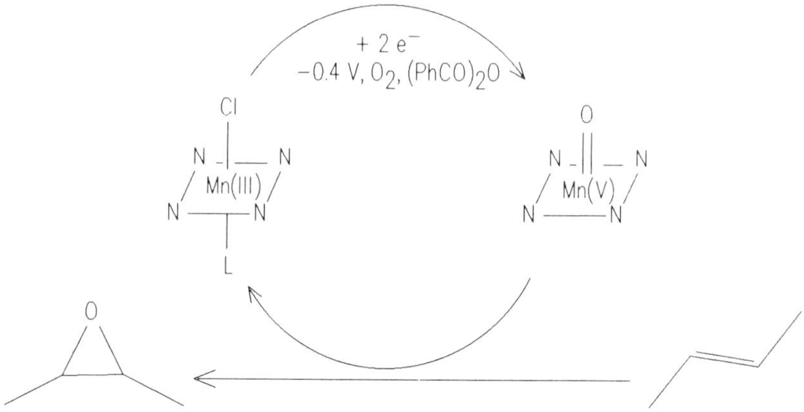

(4)

Binuclear, Polymeric, and Superstructured Porphyrins Binuclear and polymeric porphyrins are currently extensively investigated. Metalloporphyrin dimers with a multiple metal-metal bond (IV) allow a systematic study of the bond order-oxidation state relation without any change in the square planar coordination. $(RuOEP)_2$ and $(OsOEP)_2$ both exhibit two reversible metal-centered oxidations with a concomitant increase in the bond order from two to three [78]. O- [44], N- [44], and C- [72] bridged FeTPP (V) exhibit reversible one-electron oxidation and reduction. GeOEP undergoes axially double arylation and acts as a bridging ligand in the biferrocene Fc-GeOEP-Fc (VIII).

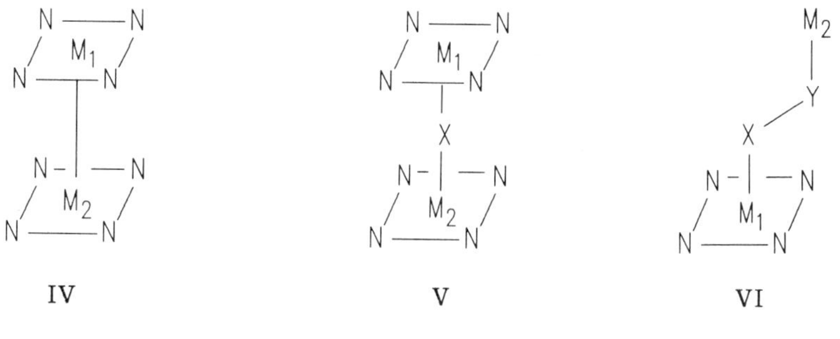

IV V VI

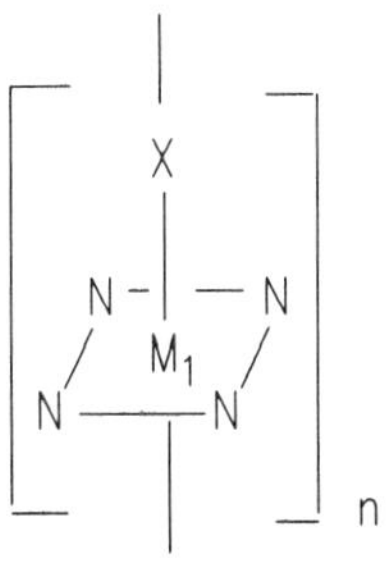

VII

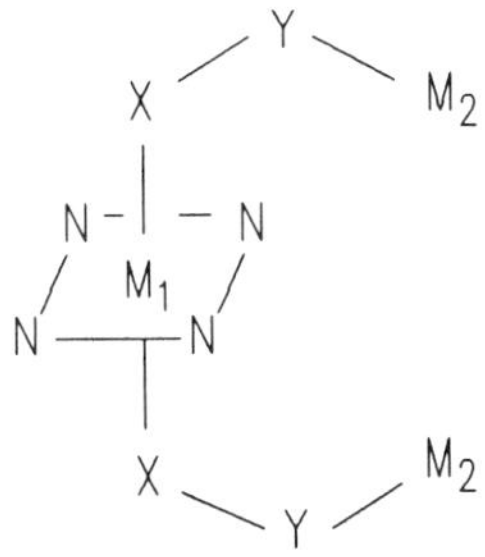

VIII

Four oxidations have been observed. The first, belonging to ferrocene-ferricenium, is shifted 350 mV negatively compared with the free redox couple and yields the mixed valence compound [79]. A tripeptide was used for the synthesis of the heterodinuclear complex Fe(III)TPPCu(II)(aib_3) (aib = tripeptide of α-aminoisobutyric acid) in which a carboxylate coordinates both the Fe(III) and Cu(II) (VI) as a model of the active site of cytochrome c oxidase [80]. Upon electrochemical reduction of Fe(III), the axially ligated Cu(II)(aib) dissociates. Coordination polymers of metalloporphyrins (VII) exhibit electronic conductivity [81]. Electrochemical methods have been used for their doping and characterization [82].

In recent years there has been considerable synthetic effort in extending the structure of the rigid porphyrins at their periphery (at R_1 or R_2) to build up so-called superstructures, that is, macromolecules with a well-defined three-dimensional arrangement. Porphyrin superstructures known today contain

1. Electroactive subunits (organic or metal ions) to assist intramolecular (light induced) charge separation [83]
2. Dimeric porphyrins acting as coordination centers and/or an electron pool for multielectron substrate conversion [84]
3. Additional ligands (e.g., crown ether) for complexation of substrate and/or other metals directed to catalysis or biomimetic active site simulation [85,86]
4. Bulky groups to discriminate small molecules [87]

Recently, asymmetry has been introduced into synthetic porphyrins by appropriate substitution of their periphery, and the first examples of chiral induction by optically active porphyrins used as homogeneous epoxidation catalysts have been reported [88,89].

Only a few of these superstructured porphyrins have been studied electrochemically, for example, the dimeric "face-to-face" porphyrins used as electrocatalysts for oxygen-reduction [84]. More recently, "strapped" and "capped" porphyrin monomers and covalently linked dimers have been investigated electrochemically [90]. In some cases the latter mutually influence the redox behavior. For "basket-handle" bridged porphyrins, local solvation of the metal by secondary amide groups and steric hindrance have been shown to determine the thermodynamics and kinetics of the redox processes [54]. Until now porphyrin superstructures have had little impact on organic electrosynthesis; however, their possible use as optical active electrocatalysts may become important in the future.

Table 2. Representative Reduction Potentials of Phthalocyanines

	H_2Pc	ZnPc	NiPc	CuPc	CoPc
$E_{Pc-/Pc2-}$ [a]	-1.06	-1.33	-1.23	-1.18	-1.80
$E_{Pc/Pc-}$ [a]	-0.66	-0.89	-0.85	-0.84	-1.40
$E_{M++/+}$ [a]					-0.37
$E_{M+++/++}$ [a]					0.77
$E_{Pc+/Pc}$ [b]	1.10	0.68	1.05	0.98	

[a] $DMF/(Pr_4N)ClO_4$ (versus SCE) [93].
[b] 1-Chloronaphthalene/TBAP (versus SCE) [94].

Phthalocyanines

The phthalocyanines (Pc) (IX) are tetraaza- and tetrabenzo derivatives of the parent porphyrins (I). They have found widespread applications as catalysts in organic chemistry, as photocatalysts and electrocatalysts, and as electrochromic material (Table 2) [91,92]. Their electrochemistry has been reviewed by Lever [91].

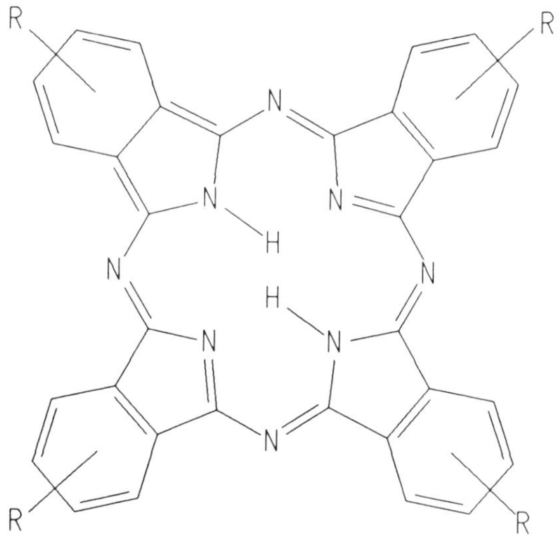

IX

The ligand system of Pc is by about 400 mV easier to reduce than the porphyrin macrocycle and the $\pi-\pi^*$ gap is smaller in Pc compared with porphyrins. The resulting potential difference $E_{Pc+/Pc} - E_{Pc/Pc-}$ is approximately 1.6–1.9 V [95]. Appropriate substitution of the benzene units leads to redox tuning of the oxidation potentials [96]. Overlaid electroactivity of metal and macrocycle is observed with such transition metals as Fe and Co. Depending on the extent of stabilization of the Co(III) oxidation state by axial ligands, reduction proceeds along path a or b in Eq. (5) [91]:

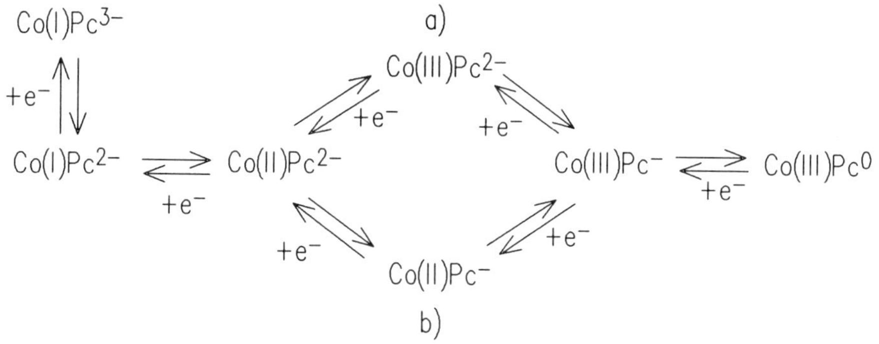

(5)

The most important derivation at the periphery of IX is its partial or complete sulfonation. The resulting tetrasulfonated Pc (R in IX = SO_3^-) is water soluble. More sophisticated structures include di- and tetrameric, organic solvent soluble derivatives of IX. They exhibit improved (multi-electron) electrocatalytic oxygen reduction [91]. Formation of mixed valence species has been observed upon reduction at −0.59 and −0.83 V for single-atom bridged dimers $(Fe(III)Pc)_2O$ and $(Fe(III)Pc)_2N$, respectively [97].

Alkyl-, allyl-, and acyl-metalloPc [Co(III)Pc-R and Fe(II or III)Pc-R] have been prepared by chemical reduction of metallated IX followed by treatment with an alkyl halide [98]. The metal-carbon bond of Co(III)Pc-R cleaves upon visible light irradiation or chemical reduction.

Metal phthalocyanines (MPc, M = Co, Fe, Pd, or Mn) reduced to the corresponding monoanions have been used by Ugi and Eckert as catalysts in organic synthesis for functional group reductions and deprotection. The catalyst was regenerated nonelectrochemically by $NaBH_4$ reduction [99]. A partially sulfonated CoPc was used as a catalyst for the specific reduction of carvone to dihydrocarvone in EtOH using Zn and NH_4Cl as electron and proton sources [39].

The low solubility renders homogeneous electrochemistry of IX difficult and favors their use as adsorbed electrocatalysts [91,92]. Typical applications of IX adsorbed on graphite electrodes have been reported, for example:

CoPc adsorbed on graphite electrodes catalyzing the reduction of CO_2 to CO at −0.9 V (versus SCE) [100]. No CO_2-CoPc adduct has been observed.

Electrocatalysis of cystein oxidation at FePc adsorbed on graphite electrodes [101].

Corrins and B_{12} Model Systems

The macrocyclic ligans systems (L) IX through XIV are synthetic models of the parent corrin ligand Xa. Most of the recent studies relevant to organic electrochemistry have been reported for the corresponding cobalt complexes. Representative redox potentials of Co-L (L = X to XIV) are listed in Table 3.

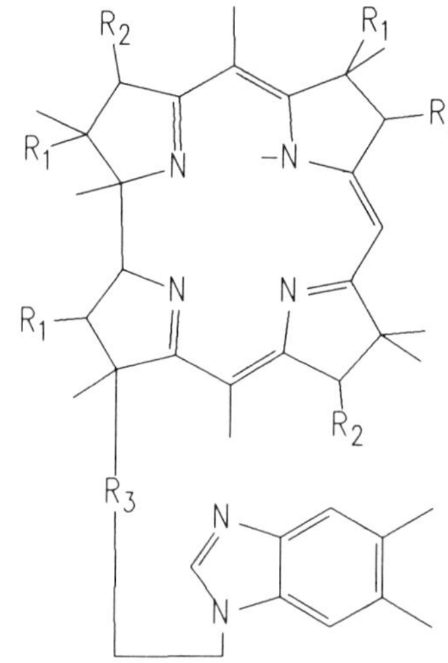

X

Xa: $R_1 = CH_2CONH_2$
$R_2 = C_2H_4CONH_2$
$R_3 = C_{11}H_{19}NO_8P$

Xb: $R_1 = CH_2COOCH_3$
$R_2 = R_3 = C_2H_4COOCH_3$

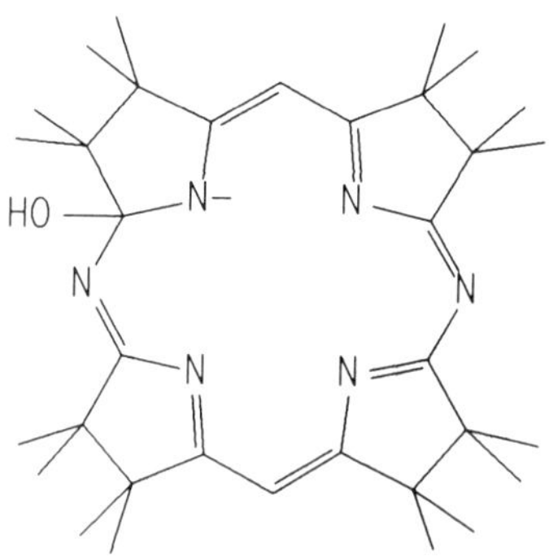

XI

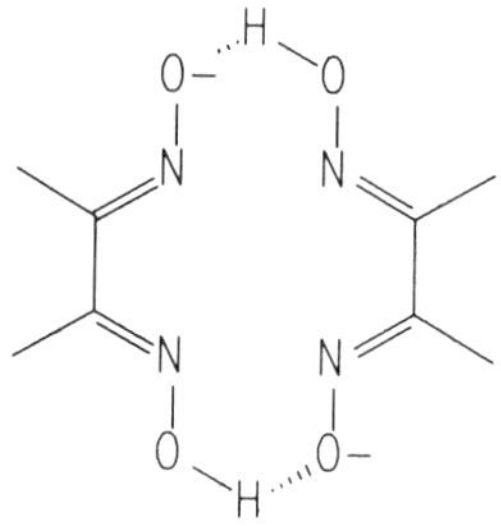

XII

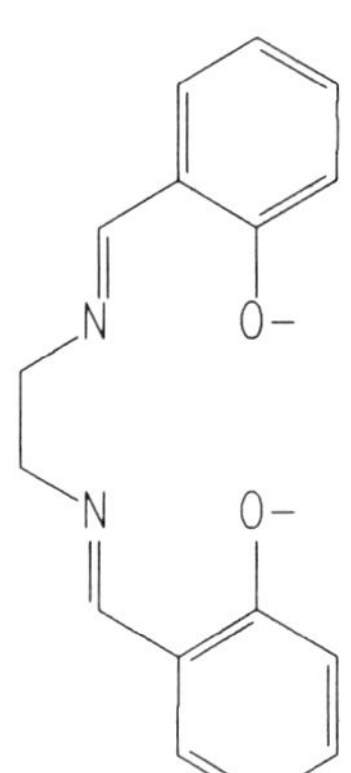

XIII

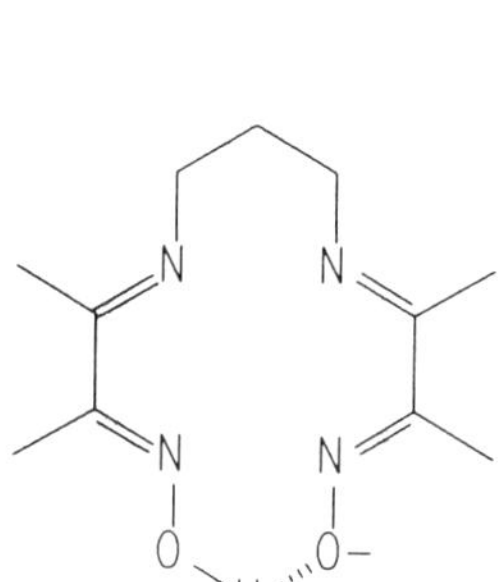

XIV

Table 3. Representative Redox Potentials of Macrocyclic Cobalt Complexes

L	Axial ligand, Co(III)		Formal redox couple			Solvent/reference electrode	Reference
	X	Y	IV/III	III/II	II/I		
Xa	OH_2	Bzm		-0.04	-0.85	H_2O /SCE	36
Xa	CH_3	Bzm		-1.60		DMF-propanol/SEC	102
Xa	CH_3	Bzm	0.69			H_2O/SCE	107
Xb	DMF	DMF(?)		0.4	-0.61	DMF/SCE	109
XI	py	DMA		0.14	-0.72	DMA/SCE	51
XI	CH_3	DMF		-1.72		DMF/Ag-Ag^+	103
XI	CH_3		0.9			MeOH/Ag-Ag^+	106
XII	Cl^-	Bzm		-0.4	-1.12	CH_3CN/SCE	110
XII	CH_3	DMF		-1.36		DMF/SCE	108
XIII	CH_3	DMF	0.62	-1.57		DMF/SCE	105
XIV	DMF	DMF		0.16	-0.79	DMF/SCE	104
XIV	CH_3	DMF		-0.85		DMF/SCE	104

The electrochemistry of vitamin B_{12} (Co-Xa) and its organometallic derivatives has been thoroughly investigated and reviewed [36]. Vitamin B_{12} models have been judged based on their redox pattern, which should mimic that of B_{12} (Table 3) [110]. More sophisticated models have been synthesized, but no electrochemistry has been reported [111–113].

Compared with porphyrins and phthalocyanines, the Co complexes of the more saturated macrocyclic ligands X through XIV are less susceptible to ligand-centered redox reactions. The formal metal oxidation states III, II, and I are electrochemically accessible. LCo(III) is a strong Lewis acid, LCo(II) a moderate one, and LCo(I) does not coordinate basic ligands axially. As a rule of thumb, LCo(III) is ligated by two, LCo(II) by one, and LCo(I) by no axial base X and Y. The redox processes are therefore connected with the loss or gain of axial ligands X and Y, Eq. (6):

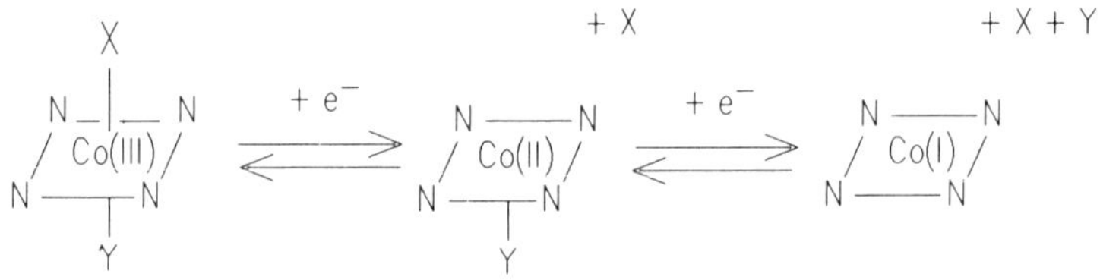

(6)

LCo complexes containing ligands X and Y, which are either stable in the solvent or undergo fast but reversible reactions, may lead to nernstian electrochemical behavior. However, ligand exchange kinetics often slow the heterogeneous electron transfer or free X undergoes irreversible reactions (e.g., X = CH_3^-). Some of the redox potentials reported in Table 3 therefore reflect both thermodynamics and kinetics.

In an exemplary series of papers, Savéant reported the electrochemistry of the naturally superstructured Co complex vitamin B_{12} (Co-Xa) with its intramolecularly bound axial base Y. He determined the rate of Co–N bond cleavage and formation as well as the connected complex formation constants and proton equilibria using CV, spectroelectrochemical, and rotating disk techniques [36]. A qualitative study on the vitamin B_{12}-heptamethyl ester (Co-Xb) demonstrates a strong dependence of E^0 on solvent polarity for the Co(III)/(II) couple and only a slight effect for the Co(II)/(I) couple [109]. Through investigations by Costa on different B_{12} models, for example, cobalt complexes of L = XII–XIV, show linear relationships between the pK_a

Table 4. Selected Second-Order Rate Constants for the Reaction of Co(I) with Organic Electrophils (liter/mol s)

	CH_3I	n-C_3H_7Br	$(CH_3CO)_2O$
Xa	3.4×10^4 [a]	1.4×10^1 [a]	2×10^{-2} [b]
XII'	2.3×10^3 [a]	1.5 [a]	
XIII		1.0 [c]	

[a]Ref. 118.
[b]Ref. 119.
[c]Ref. 120. XII' is a derivative of XII.

of axial N-containing bases or solvent donicity and the E° of the reversible Co(III)/(II) couple [114–116].

Formation and Cleavage of M–C Bonds The synthesis of organocobalt complexes starting from Co(III), (II), or (I) has been reviewed [117]. Electrosynthesis following an EC mechanism is possible at a potential between the CoL(II)/(I) and the R-CoL(III)/(II) redox couples (Table 3), Eq. (7):

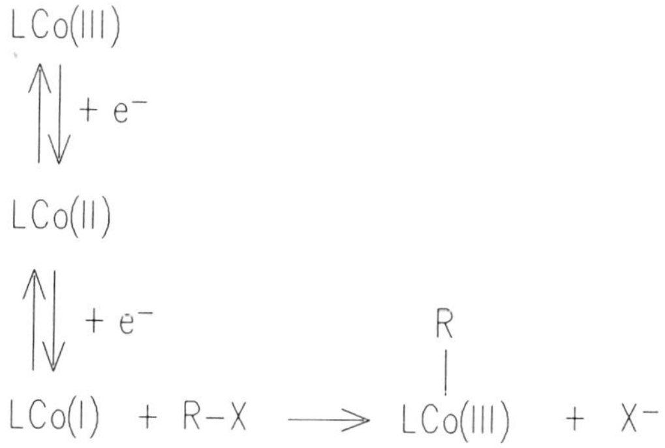

(7)

where R–X = primary or secondary alkyl halide, acid anhydride. The second-order rate constants cover the range of 1×10^{-2} to 3×10^{4} liter/mol s (Table 4).

Different mechanisms from classic S_N2 with inversion of configuration to electron transfer have been proposed. The arguments are based on the stereochemical course of the reaction [121], the RX structure-reactivity dependence [118], and the trapping of intermediate radicals [122].

Organocobalt complexes LCo(III)–R (L = X through XIV, R = alkyl) undergo reduction at −0.85 to −1.6 V (Table 3). The lifetime of the one-electron reduced species LCo(II)–R depends strongly on the macrocyclic ligand system, further axial coordination, and the solvent system [102,108]. The acidity of RH (pK_a from 35 to 41) correlates with the reduction potential of LCo(III)–R for L = XII [108], XIII, and XIV [114].

The analogous dependence for alkylcobalamins and alkylated cobesters (L = Xa and b) has only recently been reported and is valid for a pK_a range of 35 units and a potential range exceeding 500 mV [37]. Other acidity reduction potential correlations are known for Hg–R [123] and $cp(CO)_2Fe$–R [124]. Generally, homolytic cleavage of the metal-carbon bond in LCo(II)–R is observed, which leads in a first step to LCo(I) and R˙ [path f in Eq. (8)] followed by typical radical reactions. Reversible intramolecular

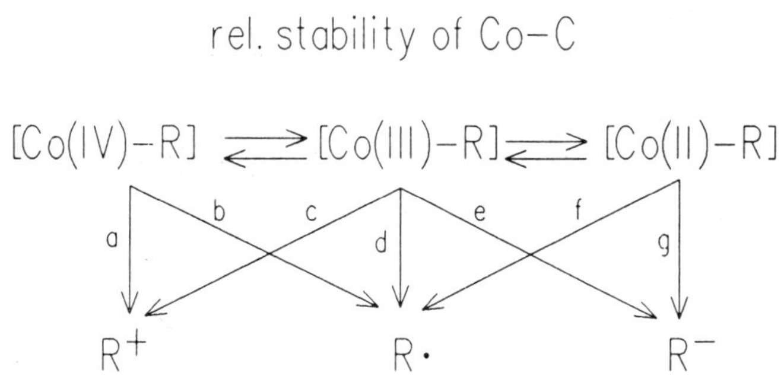

(8)

trapping of the radical has been assumed in the case of alkylcobaloximes (L = XII) [125]. Heterolytic cleavage [path g in Eq. (8)] has been proposed for the electroreduction of XI-Co(II)−R in a protic solvent leading to RH and XI-Co(II) [126].

Electrochemical oxidation of LCo(III)−R complexes is also well documented (Table 3). The one-electron intermediate LCo(IV)−R has been observed for L = XII [127−131] and XIII [129, 131], for example, as well as LRh(IV)−R (L = XIII) [131]. The oxidation potential has been correlated with pK_a [130], Taft or Hammett parameters [128,130], and metal-carbon bond length [128]. Lifetimes of LCo(IV)−R depend strongly on the solvent and the presence of nucleophilic species. Upon CPE in the presence of nucleophiles, the organic product formed is R−Nu, with inversion of configuration at the originally metal-bonded carbon. These results may be interpreted as an oxidatively induced nucleophilic assisted reductive elimination [path a in Eq. (8)].

Electrocatalysis Early studies have demonstrated that vitamin B_{12} is a catalyst for the electroreduction of alkyl halides R−X [132]. Radicals $R^{\cdot}$ are the intermediates in this reaction.

Of practical interest for organic synthesis is the vitamin B_{12}-catalyzed C−C bond formation introduced by Scheffold. Here the carbon-centered radicals derived from R−X are trapped by addition to suitable olefins leading (after reduction and protonation of the intermediate product radicals) to the hydroalkylated product, Eq. (9):

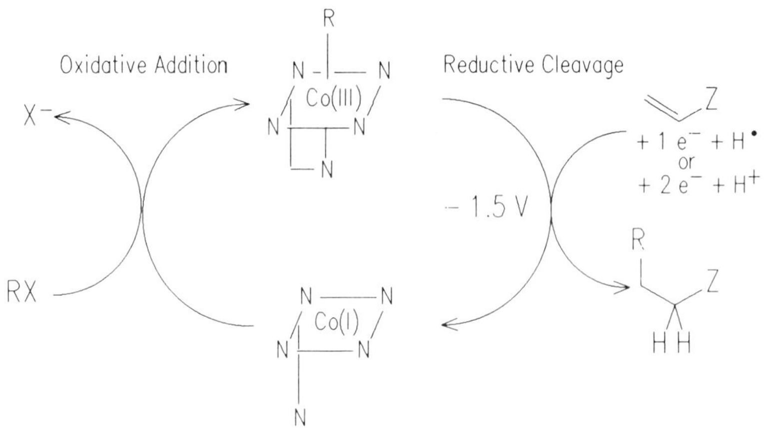

(9)

The electrode potential necessary to drive the catalysis corresponds to the reduction potential of Co−R and is in the region of −1.5 V (versus SCE). In the presence of visible light the Co−C bond of the intermediate organocobalamin is photochemically cleaved and the voltage necessary for catalysis is only −0.9 to −1 V (versus SCE), Eq. (10). This very mild C−C bond-forming reaction has been called the "OPEC" reaction (organic photoelectrocatalysis) [37]. A completely sun-powered version of the OPEC reaction has recently been reported [133]. A mechanistic study has shown that a nucleophilic acyl radical attacks the olefin for RX = acetic anhydride, as in Eq. (13), yielding an adduct radical stabilized by Co(II) [119]. The latter may be responsible for the high selectivity and suppression of polymerization.

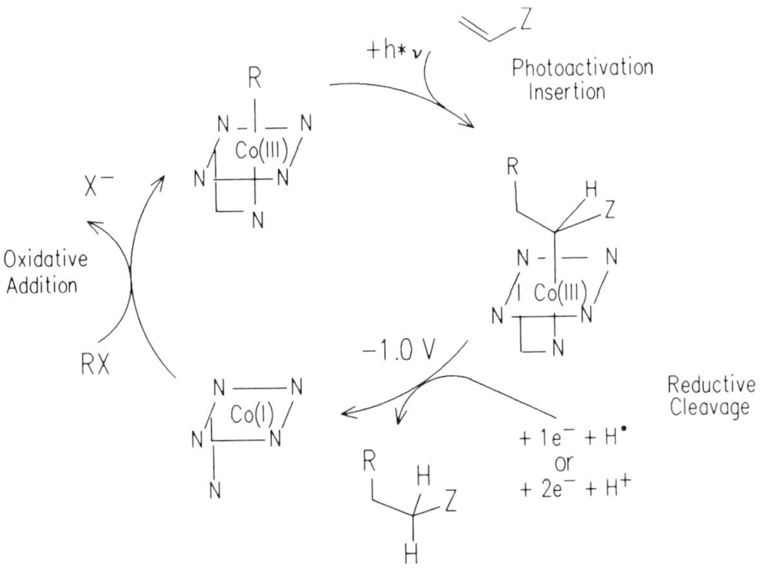

(10)

Examples of the electrochemical C−C bond formation catalyzed by cobalt complexes LCo (L = X through XIII) are shown in Eqs. (11) through (21), [38,134−141]. In some cases CPR is replaced by chemical reduction or non-catalytic conditions are used. Other examples of B_{12} catalysis include deprotection of functional groups [142], synthesis of olefins [37], or reductions coupled with 1,2-rearrangements of the carbon skeleton [140].

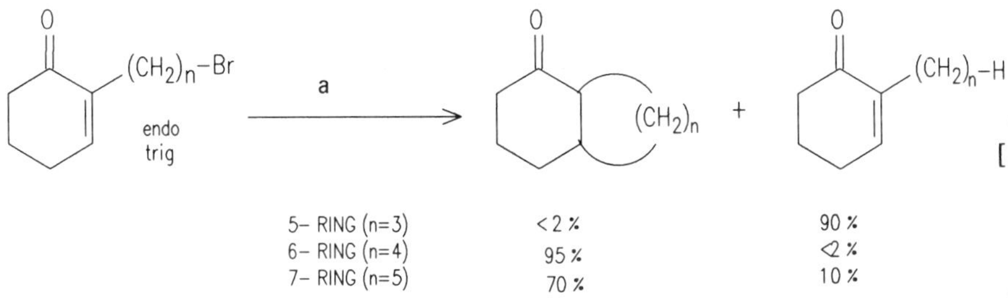

(11)
[134]

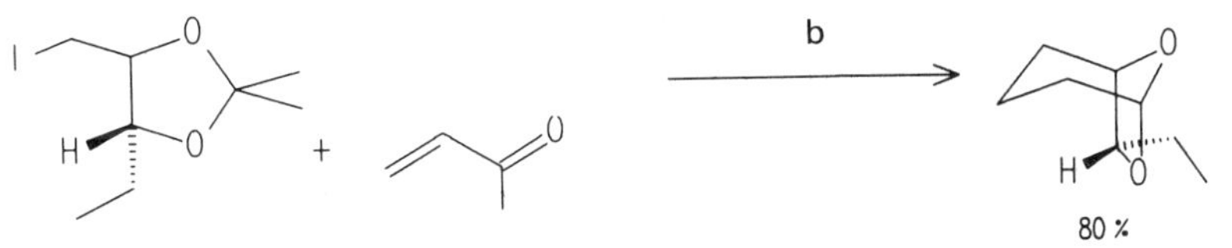

(12)
[38]

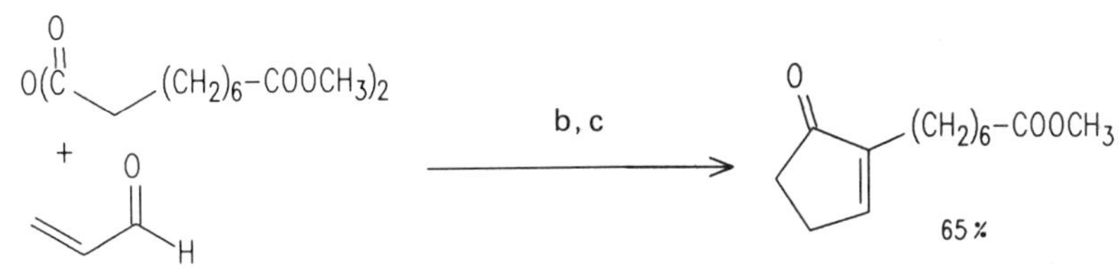

(13)
[135]

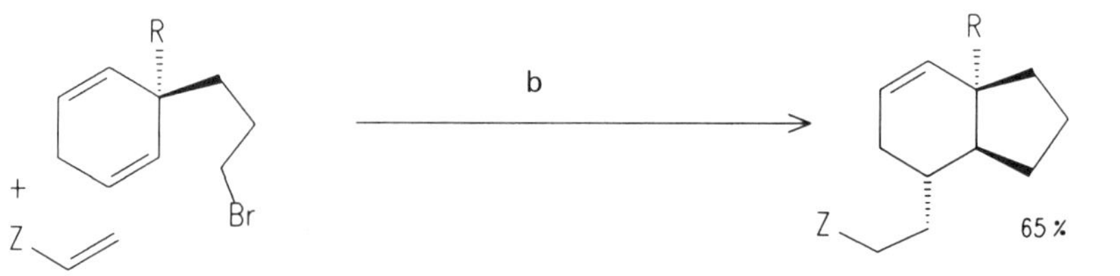

(14)
[136]

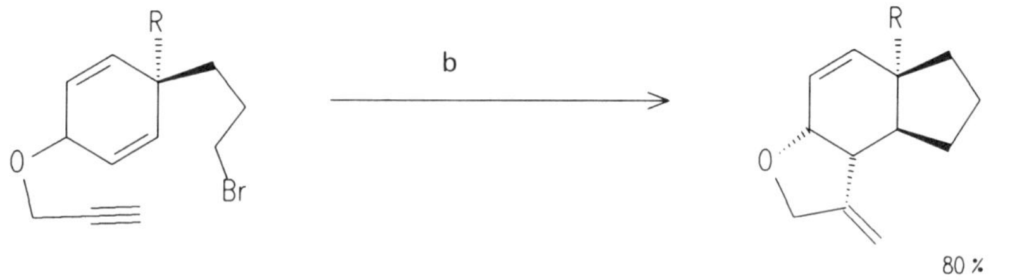

(15)
[136]

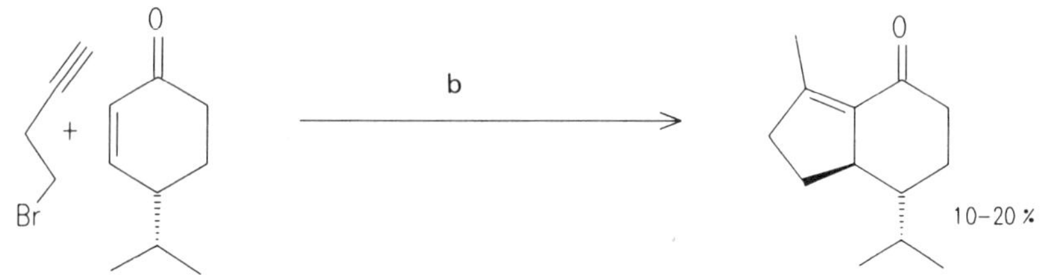

(16)
[136]

Conditions, Eqs. (11) through (16):

a. Use 5 mol per 100 mol Xa-Co or 20 mol per 100 mol XI-$CoBr_2$, DMF/$LiClO_4$/NH_4Br, −1.9 V (versus Ag/Ag^+), Hg pool, in the dark.
b. General procedure (OPEC conditions): 1–5 mol per 100 mol Xa-Co,DMF/$LiClO_4$, −0.95 V (versus SCE), carbon felt, visible light irradiation.
c. CH_3OH/$NaOCH_3$ [38].

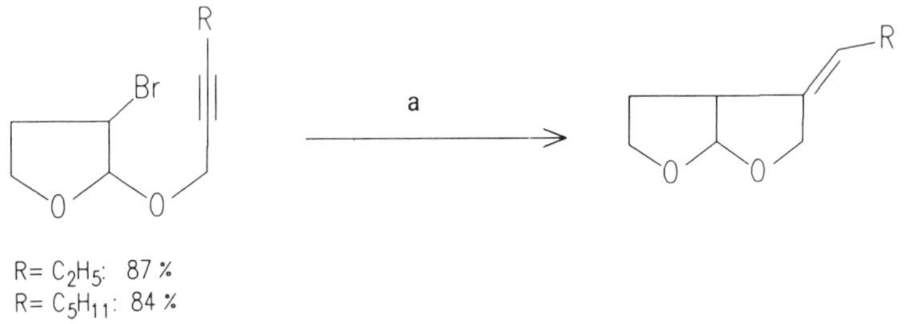

(17)
[137]

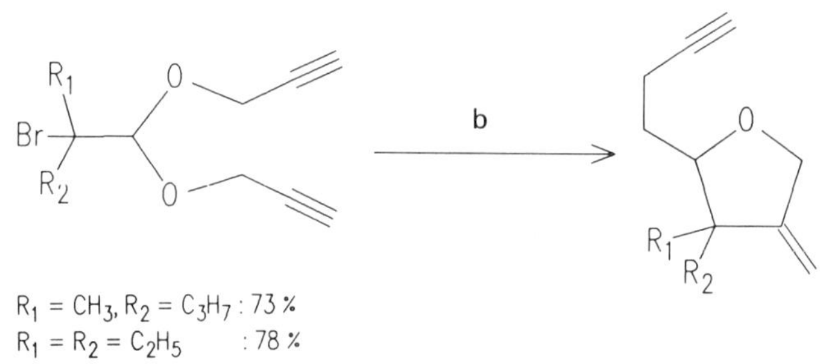

(18)
[138]

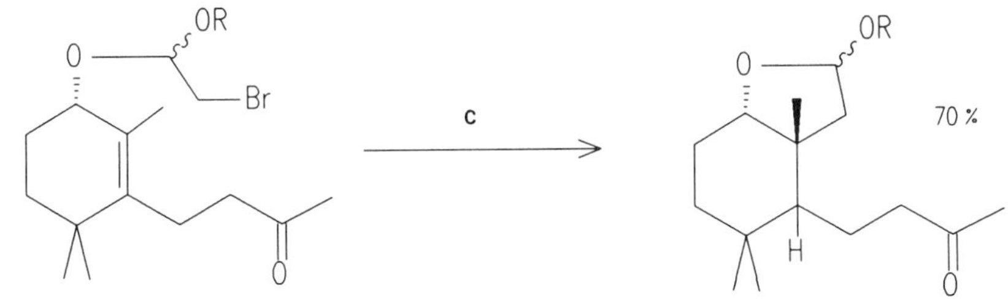

(19)
[139]

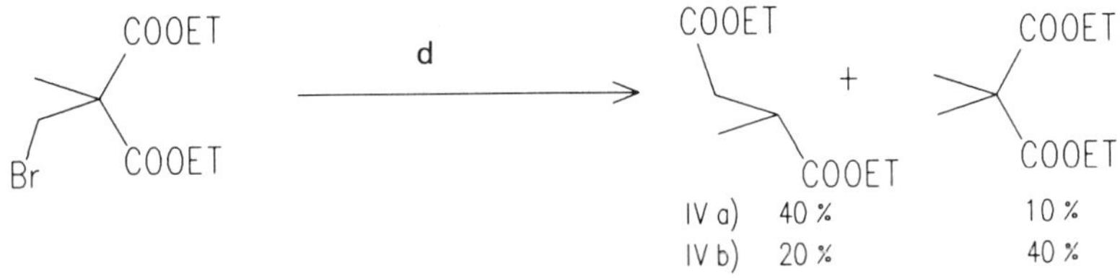

(20)
[140]

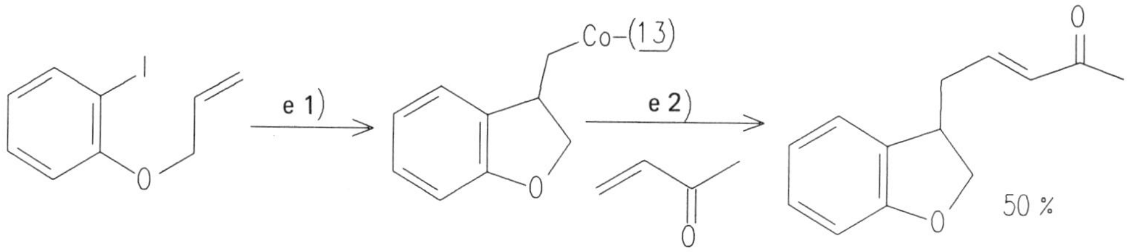

(21)
[141]

Conditions, Eqs. (17) through (21):

a. Use 30–50 mol per 100 mol XII-Co(py)Cl^- in MeOH/NaOH/Et_4NOTs galvanostatic (20 mA), separated cell, Pt electrodes, 50°C.
b. Use 10 mol per 100 mol XII-Co$(Cl^-)_2$ in EtOH/NaOH 1 Eq. $NaBH_4$.
c. Use a catalytic amount of vitamin B_{12}(X-Co) in MeOH/$LiClO_4$ at −1.8 V.
d. Use 1 mol per 100 mol Cob(II)7C_1ester (L = Xb), DMF/CH_3COOH/TBAF −1.5 V (versus SCE) two-compartment cell, Pt electrode: (1) visible light irradiation; (2) in the dark.
e1. Use 100 mol per 100 mol Co-salophen (L = derivative of XIII) THF/NaHg, 25°C;
e2. reflux in CH_2Cl_2 under visible light irradiation in the presence of 5 Eq methyl vinyl ketone.

Electrocatalysis as depicted in Eqs. (9) and (10) is clearly dominated by cobalt complexes LCo (L = X through XIV). However, Pletcher has shown that such nickel complexes as XIII-Ni catalyze R–X reductions and the reductive hydroaddition of RX to activated C,C double bonds, Eqs. (22) through (24) [143–146]:

$$n\text{-}C_8H_{17} \xrightarrow{a} n\text{-}C_{16}H_{34} + C_8H_{18} + C_8H_{16}$$ (22) [146]

	$n\text{-}C_{16}H_{34}$	C_8H_{18}	C_8H_{16}
(1)	72%	5%	2%
(2)	61%	29%	5%

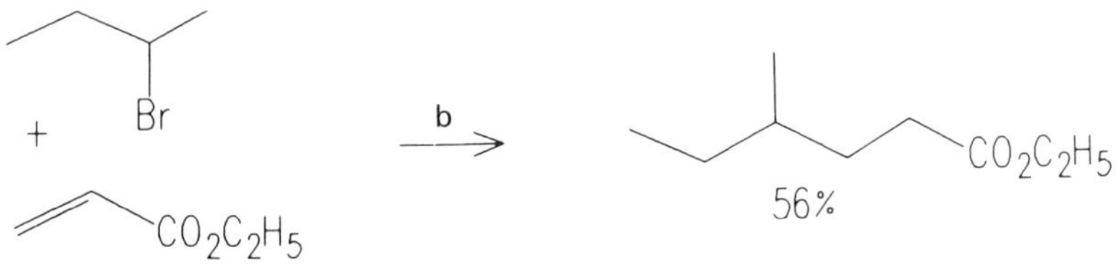

(23)
[145]

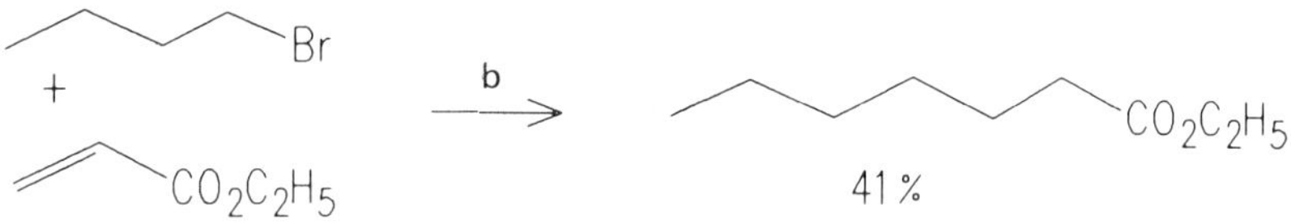

(24)
[145]

Conditions:

a. Use 10 mol per 100 mol NiL, $CH_3CN/TBABF_4$, Pt gauze, −1.75 V (versus SCE): (1) L = XIV; (2) L = XVI.
b. Use 10 mol per 100 mol Ni-XIV, 5 Eq activated olefin, $CH_3CN/TBABF_4$, Pt gauze, −1.75 V (versus SCE).

To exhibit electrocatalytic properties, the nickel ion needs generally a more saturated macrocyclic ligand system than the Co complexes. Typical ligand systems for nickel are cyclam XV and its derivatives (e.g., XVI), as well as the partially unsaturated macrocycles XVIII and XIX [147]:

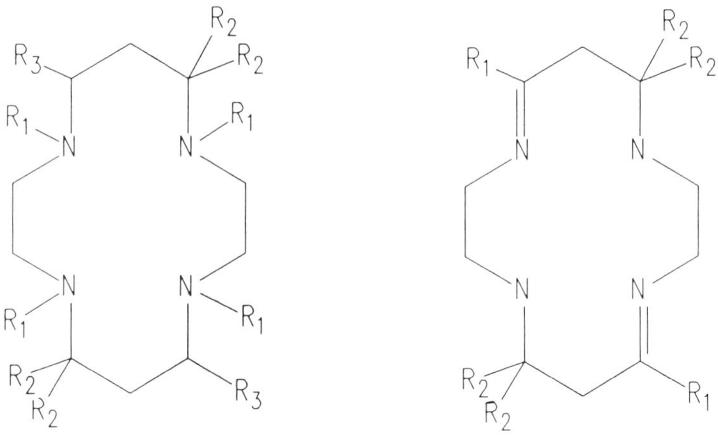

XV: $R_1 = R_2 = R_3 = H$

XVI: $R_2\ R_3 = CH_3$, $R_1 = H$

XVII: $R_1 = CH_3$, $R_2 = R_3 = H$

XVIII: $R_1 = CH_3$, $R_2 = H$

XIX: $R_1 = R_2 = CH_3$

The mechanism of catalysis has been discussed, and a short-lived organonickel complex has been invoked [148]. As originally formulated by Pletcher, the organometallic intermediate is a Ni(III)alkyl complex that may be reduced at the potential of CPR or reacts with an excess of R−X or activated olefin. Recently, Bakac and Espenson with Ram proposed formation of the strongly reducing complex LNi(II)−R (L = XVII) upon reaction of two LNi(I) species with RX, Eq. (25) [149–151]:

$$\mathrm{LNi(I)} + \mathrm{R{-}X} \longrightarrow \mathrm{LNi(II)} + \mathrm{R}^{\cdot} + \mathrm{X}^{-} \qquad (25)$$

$$\mathrm{LNi(I)} + \mathrm{R}^{\bullet} \longrightarrow \mathrm{LNi(II){-}R}$$

Cobalt and nickel complexes of XIX as well as the nickel complex XV-Ni have been reported to exhibit CO_2 reduction catalysis with possible organometallic intermediates [152,153].

Bipyridines and Phenanthrolines, Formation and Cleavage of M−C Bonds, and Electrocatalysis

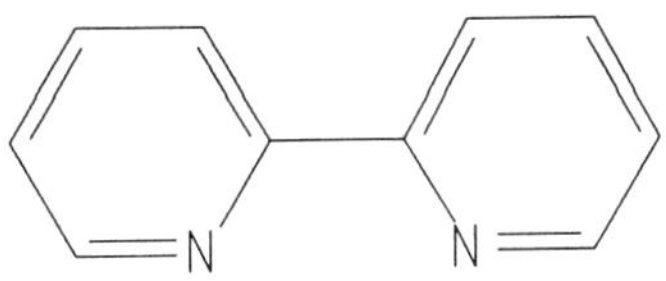

XX

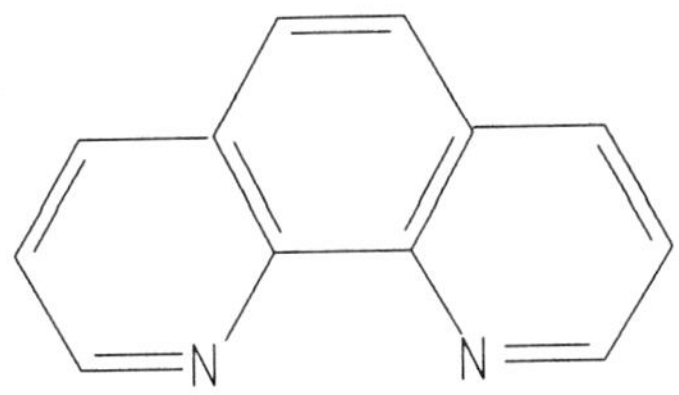

XXI

The unusual properties of $Ru(bipy)_3$ as a sensitizer for the photochemical cleavage of water has led to a vast number of investigations of its redox behavior. Ligand- and metal-localized redox orbitals have been identified. There are several reviews, for example by DeArmond's group [154]. Many synthetic variations of the basic ligand system are known:

Superstructures, such as the catenands, built from two phen ligands coordinated preferentially by a tetrahedral arrangement and stabilizing, for example, Ni(I) and Cu(I) [155].

Dimeric and polymeric metal complexes containing chelating and bridging bipy and phen derivatives, such as 2,2'-bipyrimidine, 4',7'-phenanthrolino-5',6':5,6-pyrazine, which are used as multielectron donors [156].

Organometallic derivatives containing one of the basic ligand systems and additionally a σ- or π-metal-carbon bond have been electrosynthesized and/or activated by electrochemical techniques in the case of the corresponding Fe, Co, Ni, and Pt complexes. The bipy ligand system stabilizes formal Ni(O) from which oxidative addition of alkyl halides yields the doubly alkylated Ni complex, Eq. (26). For L = bipy (XX) or phen (XXI) the same reaction pattern [Eq. (26)] is observed [157–159]. The rates of oxidative addition with alkyl bromides are 1 to 5 liters/mol s if L = bipy [158] but slower for the phenanthroline complex [159]. Upon electrochemical oxidation of the dialkyl nickel complex or reaction with an excess of RX, the metal-carbon bonds cleave and lead to the dimer in high yields. Reduction has been reported as chemically reversible on the time scale of CV [Eq. (27)] [158].

$$NiL_2^{2+} \underset{E^0_1}{\overset{+2e^-}{\rightleftarrows}} NiL_2 \underset{E^0_2}{\overset{+1e^-}{\rightleftarrows}} NiL^- \quad (26)$$

$$NiL_2 \xrightarrow{+RX} \tfrac{1}{2}\,NiLR_2 + \tfrac{1}{2}\,NiL_2^{2+}$$

$$R{-}R + NiL_2 \underset{E_{p3}}{\overset{-2e^- + L}{\longleftarrow}} NiLR_2 \underset{E^0_4}{\overset{+1e^-}{\rightleftarrows}} NiR_2 \quad (27)$$

	L = **XXa**	L = **XX1b**
E^0_1	−1.1	−1.7
E^0_2	−1.9	−2.33
E_{p3}	−0.2	−0.9[c)]
E^0_4	−1.55[c)]	−2.08[c)] (−1.9 R=Phenyl)

Conditions:

a. Versus SCE [158].
b. Versus Ag/Ag^+ [159].
c. R = alkyl.

Low-valence dialkyl Ni-complexes stabilized by bipy of phosphine ligands are known in organic synthesis as catalysts for the selective cross-coupling of Grignard reagents and alkyl halides [160]. An attractive electrochemical version has recently been published by Perichon and others [157,158]. Catalysis of the reductive coupling of alkyl halides has been performed on the plateau of the redox couples with E_2^o and E_4^o [Eqs. (26) and (27)]. Precipitation of metallic Ni may hamper an efficient catalysis.

Electrochemically generated $Ni(phen)_2$ coordinates one activated olefin. This complex exhibits an irreversible oxidation at $E_p = -0.7$ V and a chemically reversible reduction at $E° = -2.03$ V (versus Ag/Ag^+). However, upon CPE at the reduction potential, catalytic oligomerization is observed in the presence of an excess of the olefinic compound. The electron consumption is relatively high considering the non-charge-consuming propagation step [161]. The previously claimed analogous reaction using electrochemically generated $Co(bipy)_2CH_2CHCN$ as catalyst has been shown to be erroneous. The complex does not catalyze the polymerization of acrylonitrile [162]. On the other hand, Anson found that $Co(I)(bipy)_2^+$ is a catalyst for the reduction of allyl chloride at −1.4 V (versus Ag/Ag^+) to hexadiene. A π-complex observable on the time scale of CV, with a peak potential near −1.5 V undergoing and intramolecular redox process, has been formulated as an intermediate [163]. This catalysis has been adapted to an aqueous medium using $Co(bipy)_3^{2+}$ in micelles [164].

Bipy-stabilized Pt(II) metallacyclobutanes can be prepared by CPE as shwon by Kochi's group, Eq. (28) [165]:

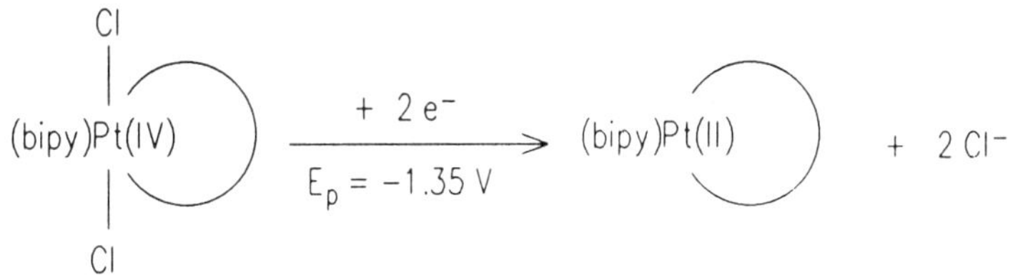

(28)

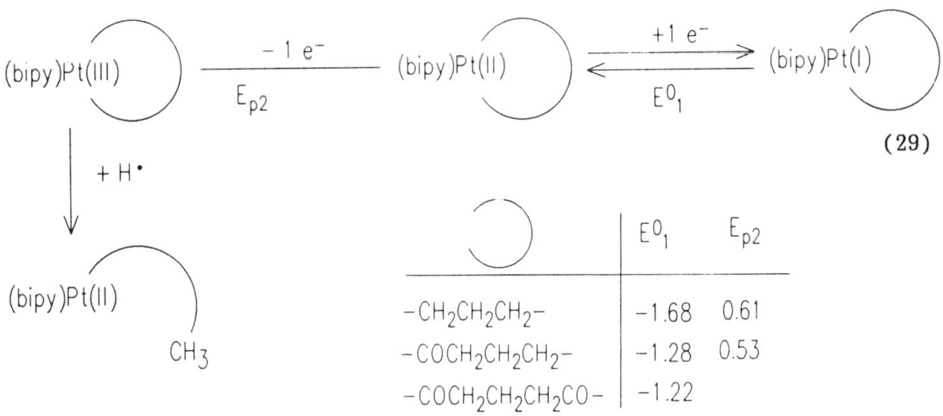

(29)

They are reversibly reduced at E_1^0 on the time scale of CV [Eq. (29)]. E_1^0 reflects anion stabilization by the carbonyl group. Electrochemical oxidation at E_{p2} leads to homolytic cleavage of a metal-carbon compound followed by hydrogen abstraction from the solvent. In an intersting mechanistic study of the first-order decomposition of dialkylbi(bipy) derivatives of Fe(II), Fe(III), and Fe(IV), Kochi showed that the organic product distribution depends crucially on the metal oxidation state, Eq. (30) [166]:

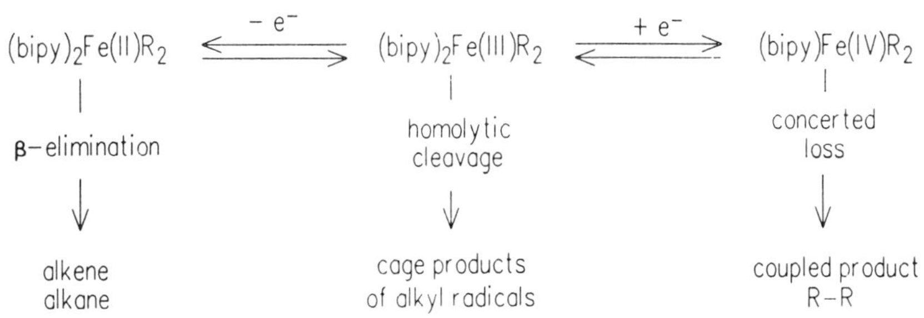

(30)

Oxo-bridged dimeric $Ru(bipy)_2$ as well as monomeric oxo derivatives of Os and Ru bipy complexes exhibit interesting two-electron oxidation electrocatalysis toward organic and inorganic substrates, as shown by Meyer and others [167,168]. Stoichiometric transformation of toluene to benzaldehyde (31%), of tetrahydrofuran to butyrolactone (41%), of cyclohexane to cyclohexanone (57%), and of norbornene to exo-2,3-epoxynorbornane has recently been reported using cis-dioxoruthenium(VI) with a modified bipy ligand [169].

There are several reports on the oxidation of $Ru(bipy)_2$-coordinated amines to imines or nitriles. For example, $Ru(bipy)_2en$ (en-ethylendiamine) undergoes a one-electron reversible oxidation on the time scale of CV at 0.96 V. In a CPE experiment the complex consumes 3.8 F/mol and $Ru(bipy)_2dim$ (dim-diimine) is obtained [170].

B. Chelating and Simple Phosphine Ligands, Formation and Cleavage of M–C Bonds, and Electrocatalysis

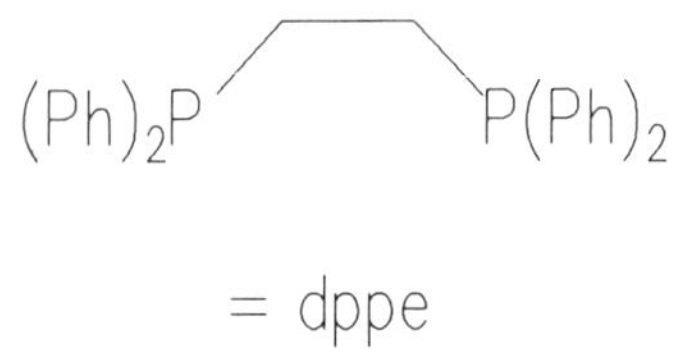

XXII

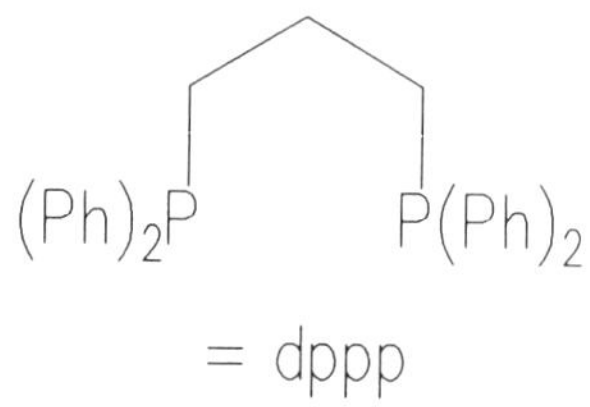

XXIII

Phosphine-stabilized low-valence metal olefin complexes have been electrosynthesized, for example by reduction of tris(pentane-2,4-dionato)iron or from cobalt chloride in the presence of phosphine and diolefin [171].

Aryl halides undergo oxidative addition with electrochemically generated, phosphine-stabilized Ni(0) to yield σ-aryl complexes, which are reducible at −2.3 V (versus Ag/Ag^+) [172]. π-Allyl complexes have been observed upon reaction of electrochemically generated Ni(0) and allyl halides in CH_3CN/PPh_3. The π-allyl–Ni(I)PPh_3X are reducible at −1.32, −1.24, and −1.2 for X = Cl, Br, and I, respectively. Oxidation occurs at 0.25 (X = Cl) and −0.3 V (X = Br) and regenerates the corresponding allyl halide [173].

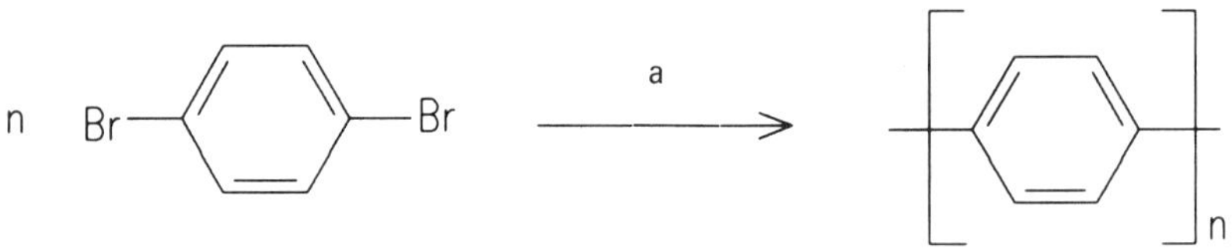

(31)

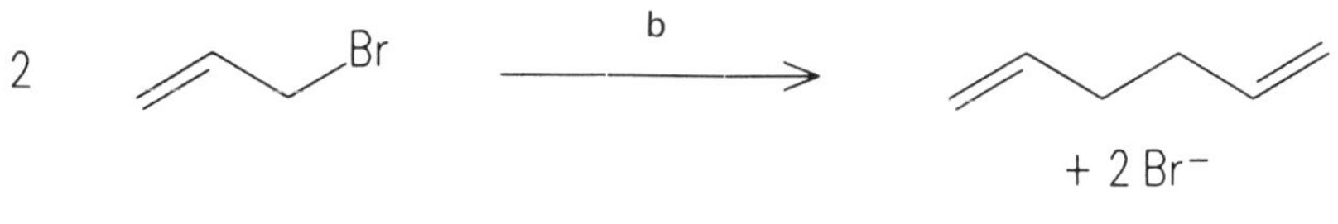

(32)

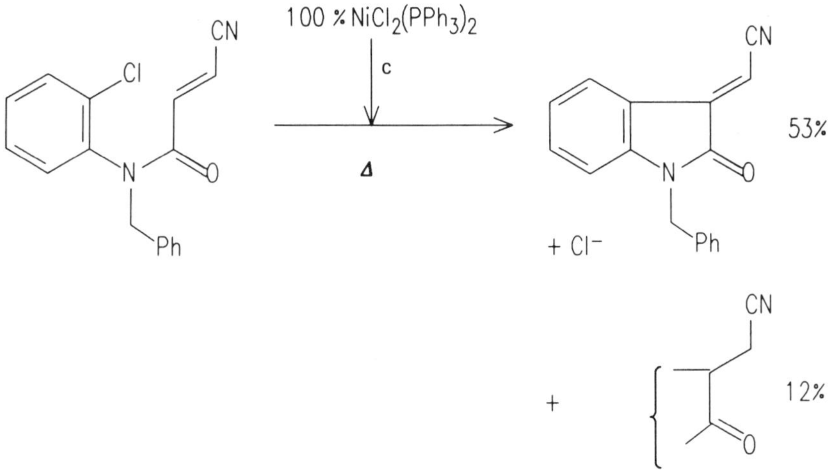

(33)

Conditions:

a. Use 10 mol per 100 mol $NiCl_2(dppe)_2$, THF/HMPT/TBAP at −2.5 V (versus Ag/Ag^+).
b. Use 5 mol per 100 mol $Ni(Cl)_4)_2$, PPh_3/CH_3CN/TBAP at −1.3 V (versus SCE).
c. DMF/TBABr at −1.9 V (versus SCE).

Interesting applications of Ni phosphine complexes as electrocatalysts under reducing conditions have been published by Fauvarque et al. [174,175] and Perichon's group [172,176]. These include oligomerization of dihalobenzenes [Eq. (31)] [174], dimerization of monohalobenzenes [172,177], or allyl halides [Eq. (32)] [173], and intramolecular [Eq. (33)] [178] or intermolecular addition [176,179] of aromatic halides to C—C double bonds, sometimes followed by $H^{\cdot}$ loss and addition of benzyl halides to CO_2 [175]. A similar electrocatalysis has recently been described by Torii. He reported higher turnover numbers in the case of a reductive dimerization of aryl halides catalyzed by a phosphine Ph(0) complex [180]. Bontempelli has recently summarized the mechanisms of the Ni phosphine-promoted dimerization of allyl halides, aryl halides, and alkyl halides as well as the isomerization process occurring upon coupling of two allyl halides [181].

$Pt(PPh_3)_2$ has been prepared electrochemically followed by reaction with CH_3I or C_6H_5COCl to yield the corresponding monomethyl or monobenzoyl derivatives [182].

Two-electron electrooxidation of the Vaska-type organometallic complex trans-$Ir(R)(CO)(PPh_3)_2$ leads to a phosphine-assisted reductive elimination, Eq. (34) [183]:

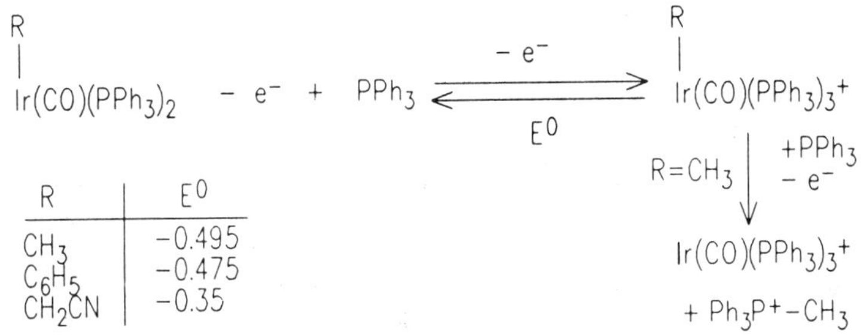

(34)

Some non-charge-consuming electrocatalytic processes have been reviewed by Mortreux and Petit [184]. The concentration of the active catalyst, a zero-valence phosphine-stabilized Ni complex, is electrochemically adjusted. It catalyzes the cyclodimerization of butadiene to cyclobutadiene and vinylcyclohexene. Other in situ electrosyntheses of catalysts reviewed in Ref. 184 include "$Fe(NO)_2$" for the cyclodimerization of butadiene, "$Cl_nW{=}CH_2$" for the metathesis of 2-pentene, and "$Rh(CO)(PPh_3)_2$" for the hydroformulation of alkenes.

A hydridophosphonitecobalt(I) complex [$CoH(PPh(OEt)_2)_4$] becomes a catalyst upon one-electron oxidation. Double-bond migration in phenylpropene was observed upon treatment of the substrate with catalytic amounts of electrogenerated $CoH(PPh(OEt)_2)_4^+$ [185].

Dinitrogen as a ligand of low-valence tertiary phosphine complexes, such as trans-$MoBr(N_2)(dppe)_2^+Br^-$, undergoes alkylation with 1,5-dibromopentane. Upon further reduction and protonation, either piperidine or N-aminopiperidine is formed according to Eq. (35) [186,187]:

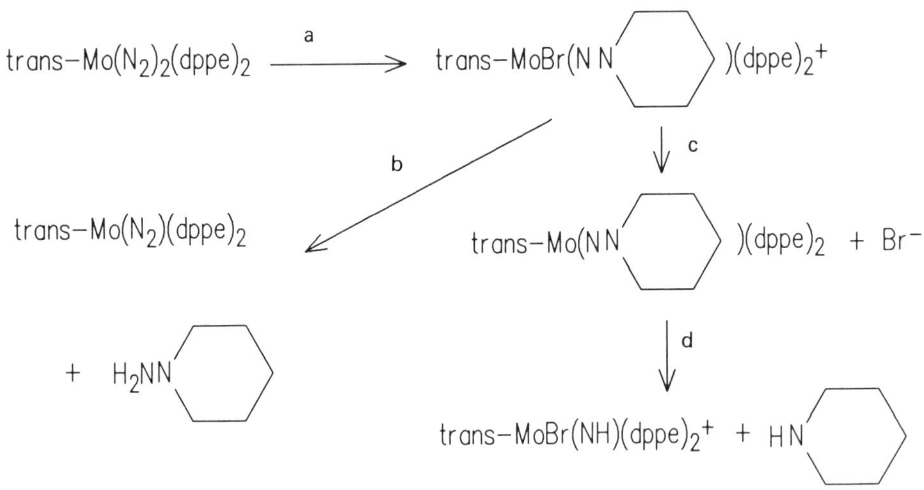

(35)

Conditions:

a. $Br(CH_2)_5Br$
b. +4 e^-, THF, −1.8 V under N_2
c. +2 e^-, THF, −1.8 V under Ar
d. HBr

Based on this chemistry an electrocatalytic nitrogen-fixing cycle has been constructed by Pickett and Leigh [188].

III. ELECTROSYNTHESIS OF COORDINATION COMPOUNDS

The electrosynthesis of a variety of coordination compounds by concomitant generation of the ligand base and the metal ion according to Eq. (36) has been reported. It involves anodic dissolution of a sacrificial metal electrode and cathodic discharge of an acidic parent ligand RH to R^- and $1/2H_2$ in an undivided cell. The base R^- should be stable in the solvent system (acetonitrile, acetone, and alcohols) and undergoes complex formation with the metal ion. The anodic dissolution of the metal may be assisted by halide ions (X^-).

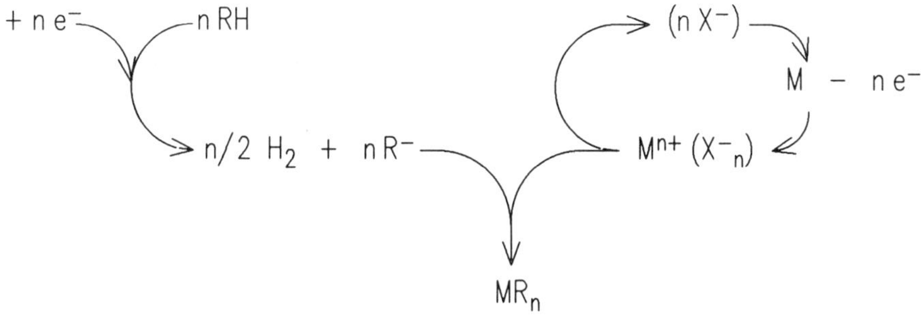

(36)

Lehmkuhl describes the electrosyntheses of Fe(II), Co(II), and Ni(II) alkoxides, phenoxides, and pentanedionates and developed cell designs for up-scaling the reaction [189]. The use of Ti, Zr, Hf, V, Cr, Mn, Fe, Co, Ni, Cu, and In anodes in the presence of acetylacetone, 2-acetyl pyrrole, dimethylglyoxime, and other bidentate proton-donating ligands leads to the corresponding coordination compounds [190]. Further electrosynthesis of β-diketonato complexes include $Cd(acac)_2phen$ [191], $Th(acac)_4$ [192], $U(acac)_4$, and $UO_2(acac)_2$ [193]. $Sn(II)(SC_6H_5)_2$ and the analogous Pb(II) compounds have been synthesized from the metal anodes in the presence of thiophenol [194]. Similarly, the thiolates of Zn, Cd, and Hg are accessible [195]. Electrolysis in the presence of further nonelectroactive ligands as bipy or phen may result in coordination compounds with mixed ligand sphere, such as $Sn(IV)(SC_6H_5)_4(bipy)$ [194] and $Ni(PPh_3)_3Cl$ [196]. The electrosyntheses of coordination compounds described earlier is closely related to that of organometallic compounds like phenylethynylcopper(I) ($C_6H_5\equiv CCu$) or cp_2Fe [198,199] obtained form metallic anodes and phenylacetylene or cyclopentadiene, respectively. In both cases the ligand anion is produced by hydrogen discharge, it is stable in the solvent system, and it is trapped by the cation.

IV. SELECTED TRANSITION METAL ORGANOMETALLICS

A. Cyclopentadienyl and Arene Complexes

The electrochemistry of cyclopentadienyl (cp) and arene π-transition metal complexes has been thoroughly studied and recently summarized (Table 5) [10]. The closed-shell complexes are oxidized to give stable 17-electron products without cleavage of metal-carbon bonds. Oxidative decomplexation can be induced by multielectron anodic reactions at very positive potentials [200]. The one-electron oxidations are rather metal than ligand centered. Because of its reversibility, the ferricenium-ferrocene couple has been recommended as an internal redox potential standard [201]. Several electrochemical measurements of ferrocene with cp derivatives have been performed to test linear free energy relationships [10]. More recent studies include correlations of $E_{1/2}$ with spectroscopic measurements, such as ^{13}C-NMR [202]. Reductions are ligand based and may lead to decomplexation (partial or complete) and/or activation of the ligand toward electrophilic attack.

Table 5. Representative Reduction Potentials of Cyclopentadienyl and Arene Metal Complexes

M	L_1	L_2	E°	Solvent/reference electrode
Cr	C_6H_6	C_6H_6	-0.81	DMF/Ag-Ag^+
Fe	$C_5H_5^-$	$C_5H_5^-$	0.31	CH_3CN/SCE
Fe	$^-C_5H_4COCH_3$	$C_5H_5^-$	0.57	CH_3CN/SCE
Fe	$C_5H_5^-$	C_6H_6	-1.62	H_2O/SCE
Co	$C_5H_5^-$	$C_5H_5^-$	-1.16	EtOH/SCE

Source: Values from Ref. 10.

Complete reductive decomplexation of a cyclic π-ligand is rare but occurs with $(cp)_2Ti(OAr)_2$ upon one-electron reduction [203]. Permethylated ferrocene, $C_5(CH_3)_5FeC_5(CH_3)_5$, has been called an "electron reservoir." The complex is another canditate for electron transfer catalysis. It is strongly reducing but does not undergo dimerization [204].

The electrosynthesis of cp_2Fe by anodic dissolution of Fe in the presence of the cathodically generated cyclopentadiene anion has been described by Tomilov and Lehmkuhl and colleagues [198,199]. Lehmkuhl also investigated a larger scale two-step electrosynthesis of cp_2Fe.

There is considerable interest in the synthesis and electrochemistry of bridged and polymeric metallocene derivatives for the study of mixed-valence compounds and conductivity, respectively [205,206]. Other developments in superstructured derivatives include ferrocene functionalized crown ethers used for electrochemically driven cation transport [207].

Organoiron and other metal complexes of aromatic compounds have found widespread applications in organic synthesis [208]. From this point of view their electrosynthesis and electrochemically induced decomplexation, as well as the tuning of reactivity toward other reagents by electrochemical reduction or oxidation, are of interest. The latter is treated in the following sections.

Activation of Cyclopentadienyl, Arene, or Tropylium Ligands

The regiospecific reduction of acyl cyclopentadienyl complexes may lead to the corresponding alcohols, Eq. (37), or pinacols depending on pH as reported for $(C_5H_4COCH_3)Fe(\eta^6\text{-arene})$ [209]. Pinacolization was also found in the case of electrochemically generated $(\eta^6\text{-}C_6H_5CH{=}CHCOC_6H_5)_2Cr^{\bar{\cdot}}$ as well as in the case of the free ligand. Complexation shifts the reduction potential negatively by only ~50 mV, and the rate of dimerization is not affected strongly [210].

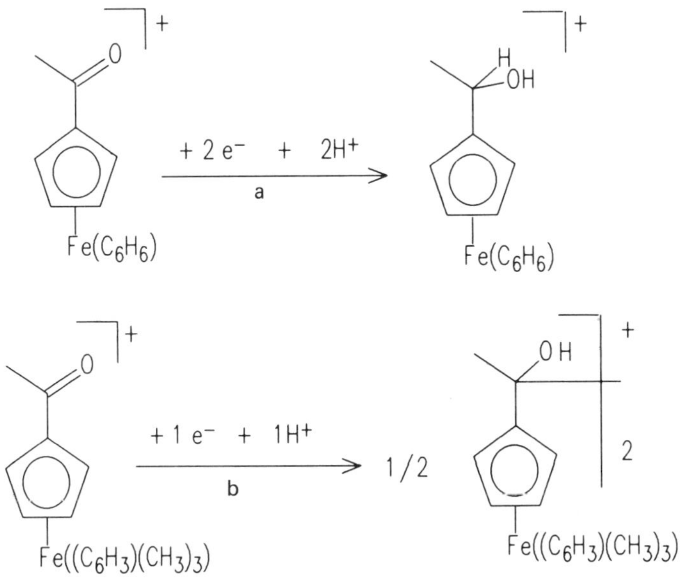

(37)

Conditions:

a. −1.0 V (versus SCE), Hg pool, H_2O/H_2SO_4, pH 0, yield 89%.
b. −1.0 V, Hg pool, H_2O/OH^-, pH 13, yield 75%

Further functional group reductions are known; for example,

Oximes of acetophenone or benzophenone complexed to $(cp)Fe^+$ undergo a four-electron reduction in H_2SO_4/H_2O/acetone at potentials approximately 150 mV more positive than the parent oximes and yield the corresponding amines [211].

CPE of haloferrocenes at −3.4 V (versus Ag/Ag^+) in mono-glyme/TBAP affords ferrocene in quantitative yield [212].

The cathodic dimerization of $(CO)_3M$(tropylium) cations (M = Cr, Mo, W) has been studied [213]. The yield of dimeric products is higher for M = Mo than for M = Cr or W and decreases if the tropylium ring bears bulky substituents. An anion-cation dimerization mechanism is proposed for complexed $C_7H_7^+$, in contrast to free $C_7H_7^+$, which dimerizes by radical-radical coupling on reduction. In the case of several arene $Cr(CO)_3$ complexes, a positive shift of the reduction potential (~0.6 V) compared with the free arenes is observed. Upon CPR the hydrodimerized products are obtained [214].

Degrand reported the electrochemical reductive O-acylation of benzophenone complexed to $Cr(CO)_3$ in the presence of acetic acid anhydride [215]. She observed the reduction of free benzophenone occurring at −1.78 and −2.37 V, whereas the corresponding reduction waves for the $Cr(CO)_3$ monocomplexed substrates are observed at −1.45 and −1.72 V, respectively. The positive shift in potential upon complexation assures an unambiguous heterogeneous electron transfer to benzophenone in the presence of the anhydride. In CPR experiments, mainly O-acylated products are obtained from complexed substrate. Electrochemical oxydative decoordination [216, 217] may follow this reductive acylation. C-alkylation of (benzophenone)-$Cr(CO)_3$ and (fluorenone)$Cr(CO)_3$ occurs upon CPR in the presence of alkyl halides, Eq. (38) [218]:

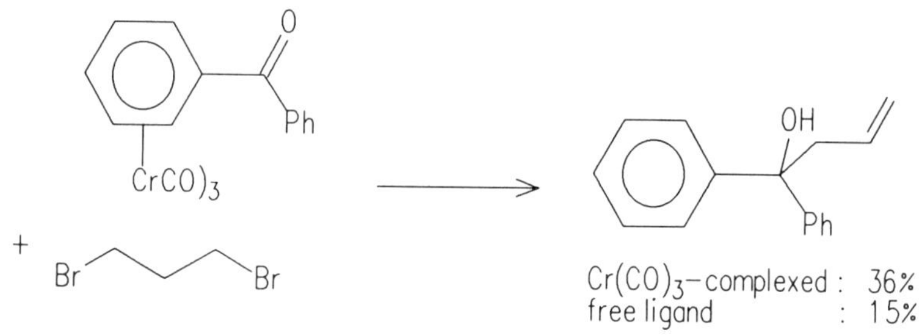

(38)

Conditions: Hg pool, −1.4 V (versus SCE), DMF/$TBAPF_6$ + 1 Eq $Br(CH_2)_3Br$.

The complexed benzophenone is approximately 300 mV easier to reduce than the free ligand. Less selectivity in the product distribution is observed for the reductive alkylation of the free ligand, for example, spiro-ether formation has additionally been reported in the latter case.

Electrochemically induced C,C bond formation is also observed upon reduction of the Co-coordinated cp or arene moiety in the presence of an organic electrophile.

Unsubstituted cobalticinium salts are reduced in two reversible steps [E_1 −0.77 V; E_2 −1.73 V (versus SCE)] [219]. CPR of cobalticinium salts in DMF at −0.77 to −1.2 V in the presence of aroyl or alkyl halides (RX) affords monosubstituted cyclopentadienyl cobalt cyclopentadienes in practically quantitative yield [219]. The reaction proceeds according to Eq. (39) via cobaltocene, which is trapped by RX in a 2:1 molar ratio.

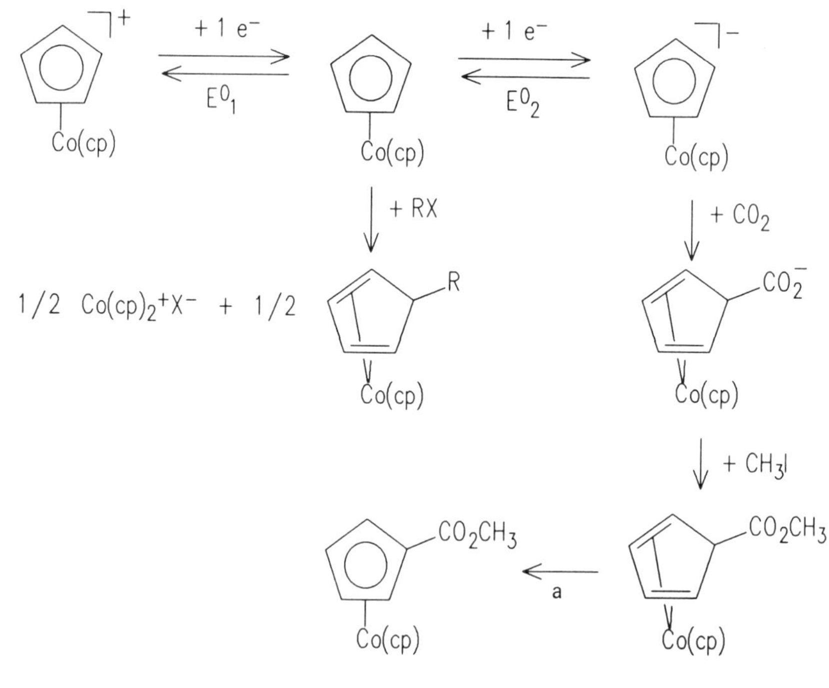

(39)

Condition: a. −2 e^-, $-H^+$, 0.05 V EtOH/H_2O, yield 10% or $Ph_3C^+BF_4^-$, yields 100%.

Electrolysis in DMF at −1.9 V in the presence of CO_2 affords cyclopentadienyl cobalt carboxycyclopentadiene, which on esterification gives the corresponding ester. Its electrochemical or chemical oxidation leads to a methoxycarbonyl cobalticinium salt.

The addition of a nucleophile to a π-complexed arene ligand to give a cyclohexadienyl transition metal complex is a well-established reaction. Electrochemical techniques have been used to promote the oxidation-deprotonation step in this type of nucleophilic substitution at a complexed arene ligand, Eq. (40) [220]:

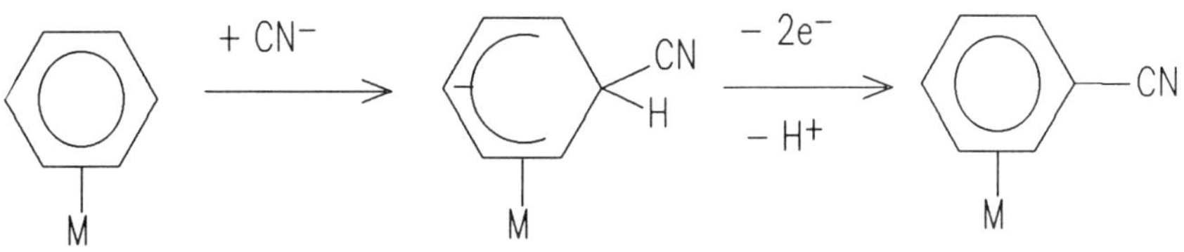

(40)

where M = Mo (μ-SMe)$_4$Mo(η^6-C_6H_6)$^{2+}$.

Conditions: PC/TEAPF$_6$, 1.9 V (versus SCE).

Metal-Olefin, Metal-Alkyl Transformations

Gubin found that cp(CO)$_2$Fe-R complexes undergo irreversible two-electron reduction in acetonitrile at a dropping mercury electrode with concomitant heterolytic cleavage of the metal-carbon bond, Eq. (41) [124]:

$$\mathrm{cp(CO)_2Fe{-}R} \xrightarrow{+\,2\,e^- \,+\, H^+} \mathrm{cp(CO)_2Fe^-} + \mathrm{RH} \tag{41}$$

It is not clear whether protonation by the solvent occurs during or after formation of the alkyl anion. As for the irreversible reduction of Hg−C compounds [123], alkylcobalamins [37] and its models [108,114] with a σ−Co−C bond, the $E_{1/2}$ of the iron alkyl complexes correlate well with the pK_a values of the respective hydrocarbons.

Gladysz and Parker have recently reported the oxidative disproportionation of the Re−C bond in cpRe(PPh_3)(NO)(CH_3) by chemical oxidants and electrochemical techniques involving a metal carbene intermediate [221].

Geiger has shown in several papers that π-complexed diolefins may undergo reversibly π to σ conversions upon electrochemical reduction [222–224]. An interesting electrochemically induced reversible insertion of two Ru atoms into the eight-carbon chain of cyclooctatetraene has been observed in the case of the triple-decker complex (cp)$_2$Ru$_2$(cot) [222]. Oxidation occurs at 0.04 V, reduction at −0.25 V (versus SCE) as determined by CV. This redox-initiated C,C bond activation may have relevance to the Ni-catalyzed tetramerization of acetylene, Eq. (42):

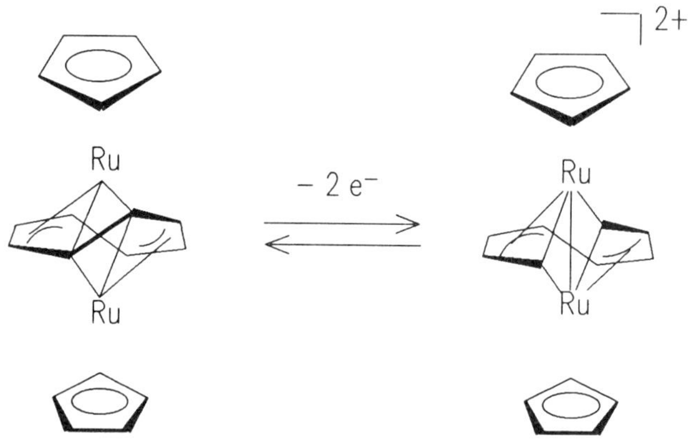

(42)

Conditions: acetone/$TBAPF_6$, 0.04 V (versus SCE), 298 K.

The Pd(I) π-radicals of cpPd(diolefin) complexes (diolefin - dibenzocyclooctatetraene) generated by cathodic reduction at −0.22 V (versus SCE) react with benzoyl peroxide as a radical source with formation of a compound containing a σ-metal-carbon bond. The same product is observed upon attack of benzoate on the same Pd(II) complex. On oxidation at 1.0 V (SCE), the benzoate group is lost and the Pd(II) complex is reformed in what is analogous to an oxidatively induced reductive elimination, Eq. (43):

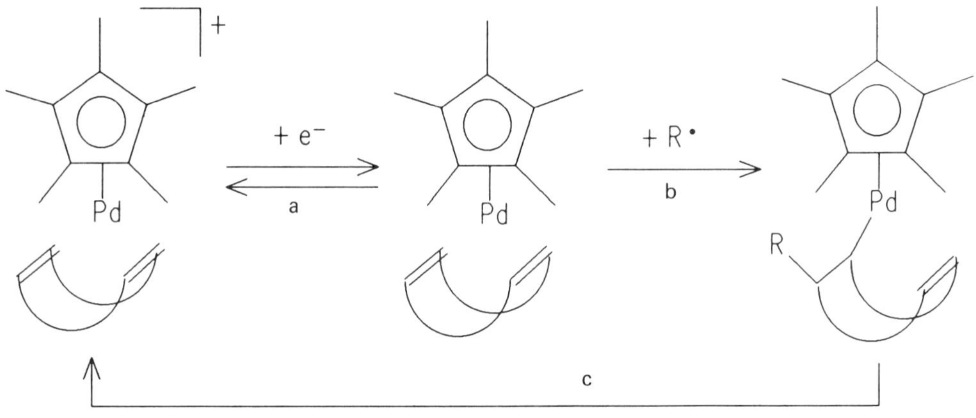

(43)

Conditions:

a. For cpPd($dbcot^+$), CH_2Cl_2/$TBAPF_6$, Pt basket −0.6 V (versus SCE), n - 1.0 e^-.
b. Add 1 Eq benzoylperoxide in CH_2Cl_2 at 243 K.
c. E_{pa} approximately 1.0 V, CH_2Cl_2/$TBAPF_6$, 258 K, n = 1.0 e^-.

The isomerization of (1,5-cyclooctadiene)cpCo to (1,3-cyclooctadiene)cpCo induced by one-electron reduction of (1,5-cod)$cpFe^+$ at −2.6 V (versus SCE) has been reported. A π-allyl hydridocobalt species has been proposed as intermediate in the isomerization step. After reoxidation the 1,3-diene isomer rearranges to the starting material [225].

The Co(III)metallacycle cp(PPh_3)Co(C_4Ph_4) undergoes reversible one-electron oxidation at 0.52 V. An irreversible one-electron reduction with a concomitant loss of PPh_3 is observed at −1.7 V (E_{pc}) in THF on the time scale of CV. Upon CPE at −1.9 V, conversion of the metallacycle to the corresponding diolefin complex has been reported, Eq. (44) [224]:

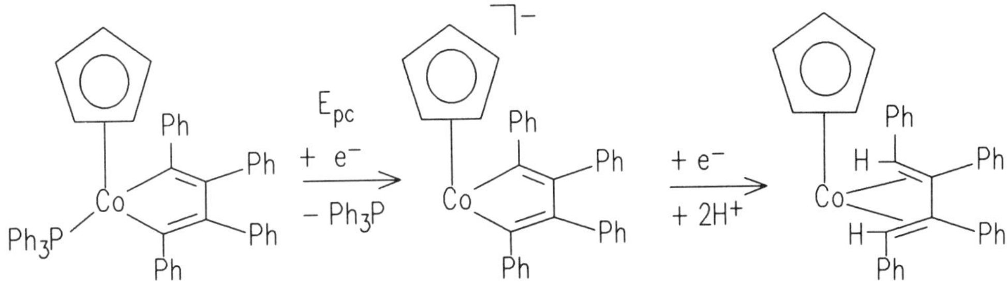

(44)

Conditions: CPE THF/$TBAPF_6$ −1.95 V (versus SCE), under N_2, n = 1.96 F/mol.

Oxidative cleavage of the metal-carbon σ-bond occurs upon chemical or electrochemical oxidation of cpRu(PPh_3)$_2$−R (R = methyl or benzyl) under CPE conditions at 0.6 V (versus Ag/AgCl) [226]. In the case of the benzyl derivatives the organic products formed in the presence of TBACl are benzyl chloride (under Ar) and bibenzyl and benzaldehyde (under air). The former product may be interpreted as an oxidatively induced nucleophilic attack at the metal-bound carbon.

Carbonylation-Decarbonylation

Transition metal acyl complexes are the prinicpal intermediates in a number of catalytic processes, such as hydroformylation of olefins and carbonylation of methanol or organic halides. However, until recently nothing was known about the electrochemical induction of carbonylation or decarbonylation processes. CPE of alkyl and acyl iron complexes has shown that both migratory insertion of CO into a metal-alkyl bond as well as decarbonylation can be induced electrochemically, Eq. (45):

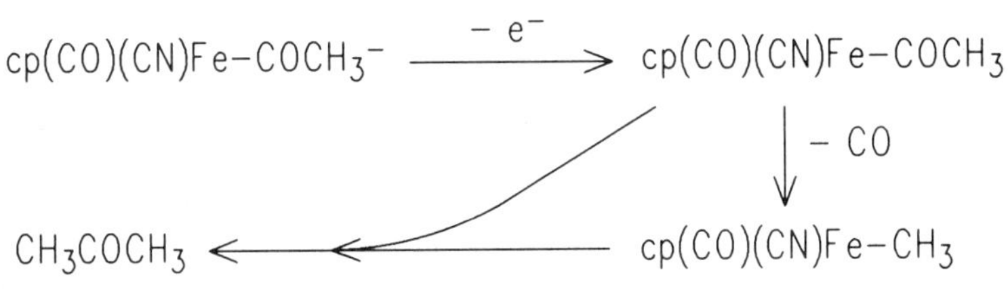

(45)

Conditions: +0.4 V (versus SCE), CH_3CN/TEAP, 20°C, 1 atm CO, high yield of acetone for low conversion.

The oxidatively induced nucleophilic attack of ethanol at the acyl carbon is evident from the products in Eq. (46) [227]:

$$cp(CO)_2Fe{-}COCH_3 \xrightarrow[a]{-3\,e^-} EtOCOCH_3 \quad 70\%$$
$$cp(CO)_2Fe{-}CH_3 \xrightarrow[b]{-2\,e^-} EtOCOCH_3 \quad 60\% \qquad (46)$$

Conditions:

a. +1.3 V (versus SCE), $CH_3CN/EtOH/TEAP$, 20°C
b. +1.0 V (versus SCE), $CH_3CN/EtOH/TEAP$, 20°C

Extensive work of Magnuson and Giering concerns the oxidatively induced alkyl to acyl migration (or CO insertion) in $cp(CO)(L)Fe{-}CH_3$ complexes [L_1 = PPh_3, $(i\text{-}pro)_3P$ and CO)] [228–230]. K^+/K^o can be calculated according to Eqs. (47) and (48):

$$\begin{array}{ccc} cp(L_1)(CO)Fe{-}CH_3 + L_2 & \underset{E^0_1}{\overset{-e^-}{\rightleftharpoons}} & [cp(L_1)(CO)Fe{-}CH_3]^+ + L_2 \\ \updownarrow K^0 & & \updownarrow K^+ \\ cp(L_1)(L_2)Fe{-}COCH_3 & \underset{E^0_2}{\overset{-e^-}{\rightleftharpoons}} & [cp(L_1)(L_2)Fe{-}COCH_3]^+ \end{array} \qquad (47)$$

$$\ln \frac{K^+}{K^o} = \frac{nF}{RT}(E^o_1 - E^o_2) \qquad (48)$$

and has been estimated to be 10^{12} for L_2 = acetonitrile [229]. Under a CO atmosphere (L_1 = PPh_3 and L_2 = CO), the authors found that E^o_2 is positive enough compared to E^o_1 and K^+ sufficiently large that an electrocatalytic situation with only a small charge consumption is established [228]. Divergent results concerning the mechanism of the insertion step have been reported. Trogler found evidence for solvent coordination (L_2 = pyridine) before migration [231], but Giering proposed migration stabilized by η^2-acyl complexation followed by solvent attack [230]. Both interpretations are based on CV experiments, and both avoid a 15-electron transition state. Miholova and Vlček reported the analogous catalysis induced by cathodic reduction of $cp(CO)_2Fe{-}CH_3$ in the presence of PPh_3. The authors used 0.15–0.2 F/mol for total conversion of the alkyl complex to the corresponding acetyl iron derivative [232].

Synthesis of β-lactams via transition metal complexes has been accomplished by oxidative degradation of the β-aminoalkyl complex $cp(CO)_2Fe{-}CH_2CHRNHR$. The mechanism includes base-induced cyclization and efficient decomplexation from formal Fe(IV) upon two, one-electron oxidations with typical acyl cation reactivity, Eq. (49) [233]:

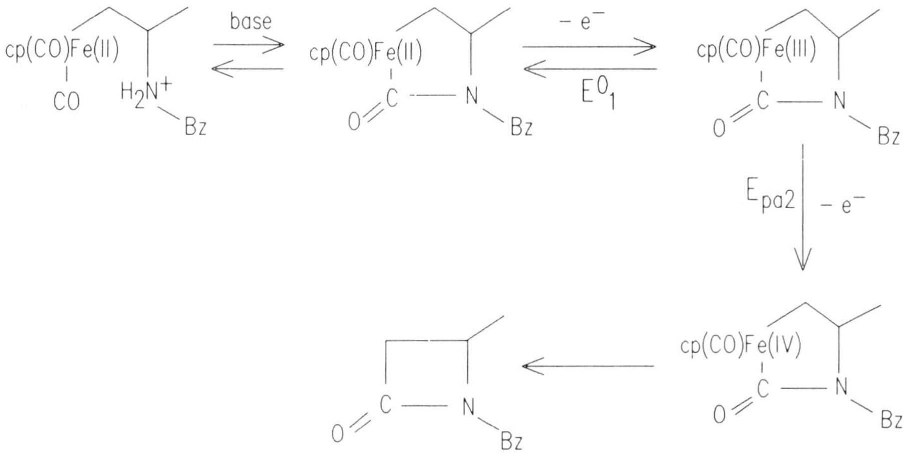

(49)

Conditions:

Analytical: CH_3CN/TBAP, 0°C, E°_1 −0.59 V, $E_{pa}{}^2$ +0.95 V (versus SCE)
Preparative: 2 Eq $(cp)_2Fe \cdot BF_4$

Reductive dimerization of two coordinated acyl ligands has been reported for $cp(COC(CH_3)_3)_2 \cdot BF_2$ [234]. Pivaloin, the organic coupling product, is formed either upon Na-Hg reduction or electrochemical reduction with 44% yield using an excess of Na or with 26% current yield based on a one-electron process, Eq. (50):

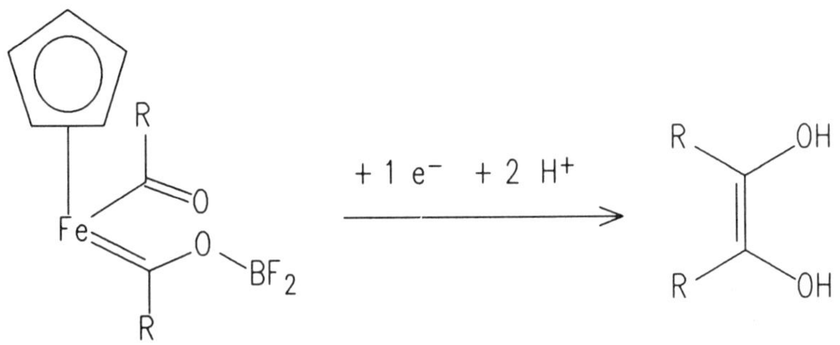

(50)

Conditions:

THF/TBAP, N_2, Hg pool, potential not specified
H_2O/NaOH reflux
H_2O/HCl

B. Activation of σ- or π-Bound Carbon Ligands in Carbonyl Complexes

The electrochemically induced formation of formyl metal carbonyls upon reduction of metal carbonyls (M = Cr, Fe) in the presence of Bu_3SnH has been shown to occur in competition with CO loss followed by dimerization, Eq. (51) [235]:

$$(CO)_nM\text{-}CO \xrightarrow{+\ e^-} (CO)_nM\text{-}CO \xrightarrow{+\ Bu_3SnH} (CO)_nM\text{-}CHO \quad (51)$$

$$(CO)_nM\text{-}CO \xrightarrow{-\ CO} 1/2\ (CO)_{2n}M_2$$

The σ-alkyl or σ-aryl transition metal pentacarbonyls of Mn and Re exhibit an irreversible two-electron reduction and yield upon protonation the corresponding alkane or arene, Eq. (52) [236]:

$$(CO)_5Mn\text{-}CH_3 \xrightarrow{+\ 2\ e^-\ +\ H^+} (CO)_5Mn^- + CH_4 \quad (52)$$

Kochi has shown that group 6 acylmetallic complexes, $(CO)_5M\text{-}COR^-$ (M = Cr, W), are converted to the hydrocarbene derivatives upon one-electron oxidation in the presence of the hydrogen source TEAP [237]. The intermediate paramagnetic acyl derivative is stable on the time scale of CV at −78°C and exhibits an ESR spectrum typical of a benzoyl-localized radical.

Reduction of $Mo(CO)_6$ in the presence of alkyl halides leads to RH and traces of R−R. The reaction is not catalytic in $Mo(CO)_6$, but an intermediate $(CO)_5Mo\text{-}R^-$ has been proposed based on coulometric and CV results [238].

Partial reduction of the olefinic ligand system of $(CO)_3Fe(\eta^4\text{-}C_4H_6)$ in the presence of a proton source has been reported [239]. Products of the type $(CO)_3Fe(\eta^3\text{-allyl})^-$ have earlier been electrosynthesized from $(CO)_3Fe(\eta^3\text{-allyl})X$ (X = Br, Cl, I), Eq. (53) [240]:

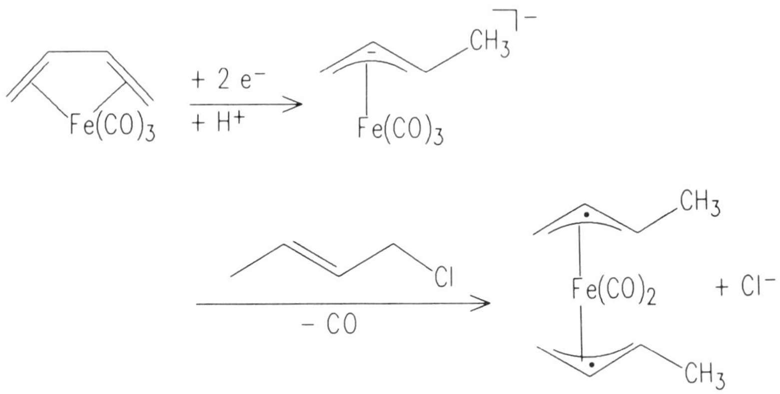

(53)

Conditions: −2.1 V (versus SCE), THF/$TBAPF_6$, phenol

Oxidative coupling of two cot ligands occurs upon one-electron oxidation of $(CO)_3M(\eta^4\text{-cot})$ for M = Fe or Ru. The further reaction path is influenced by the metal. Electrochemical techniques include CV studies, but chemical oxidants have been used for the preparative-scale dimerization, Eq. (54) [241,242]:

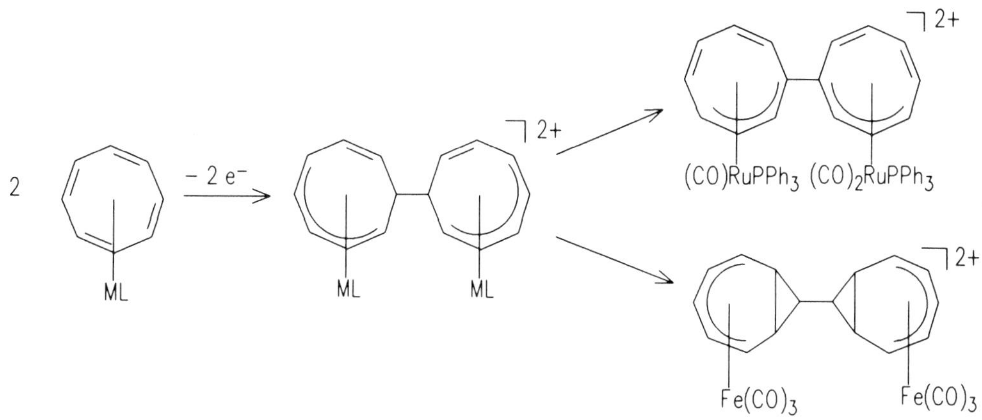

(54)

Conditions:

M = Fe, L = $(CO)_3$: 1.4 Eq $N(C_6H_4Br\text{-}p)_3/NO^{\cdot}PF_6$ in CH_2Cl_2.
M = Ru, L = $(CO)_2PPh_3$: 1 Eq $Fe(cp)_2^+$ in CH_2Cl_2.

V. ELECTROSYNTHESIS OF TRANSITION METAL ORGANOMETALLICS

The electrochemically induced formation and cleavage of σ- or π-metal-carbon bonds has been discussed in Secs. II and IV with emphasis on an overall conversion of the organic ligand. There is also great interest in the electrosynthesis of the intermediate organometallic compounds, particularly π-complexed olefins. Great efforts have been made to develop commercially attractive electrosyntheses, that is, to use both the anodic and cathodic reaction and undivided cells. The procedure consists of the following:

Generation of a reactive metal oxidation state ($M^{(n-m)+}$) in the presence of the free ligand (L) by cathodic reduction of a metal ion [M^{n+}, Eq. (55a)], possibly with concomitant dissolution of a sacrificial anode [M, Eq. (55)]

Formation of a reactive ligand oxidation state [$L^{m/q-}$, Eq. (56a)] in the presence of a metal ion M^{n+}, which may be produced at a sacrificial anode [Eq. (56)]

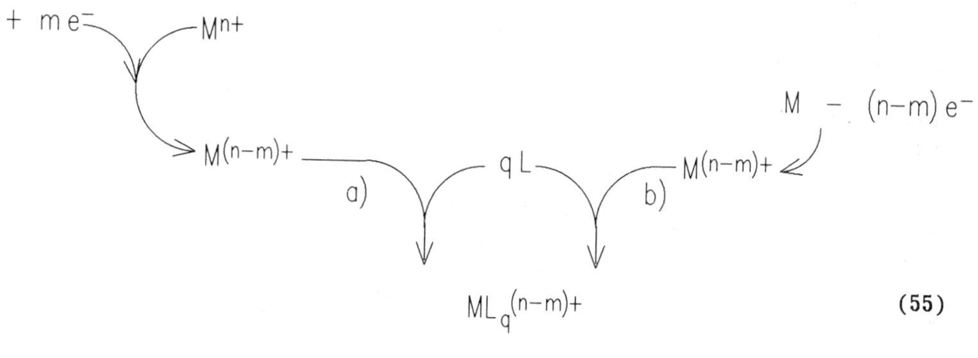

(55)

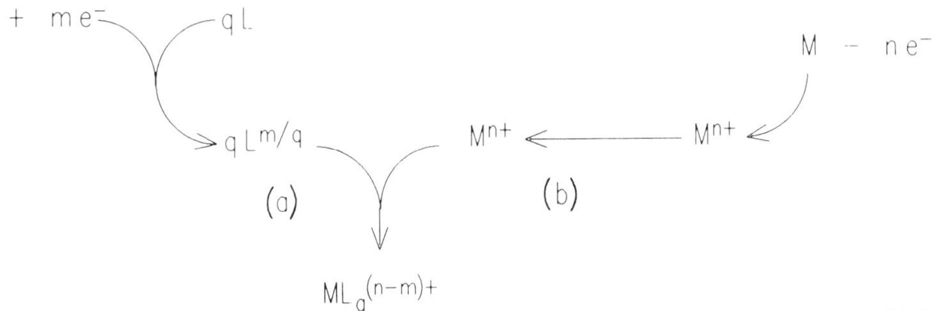

(56)

Reaction (55) was originally reported by Manahan for the formation of $Cu(COD)_2(ClO_4)_2$ from the anodic dissolution of a copper anode in the presence of 1,5-cyclooctadiene and copper(II)perchlorate in an undivided cell [243]. The scope of the reaction [Eqs. (55) and (56)] has been explored, and conditions for scaling it up have been found by Lehmkuhl and colleagues [199,244–246]. They reviewed the electrosyntheses of Ni(COT), $Ni(COD)_2$, Co(COD), $Fe(COT)_2$, $Ti(COT)_2$, $Zr(COT)_2$ and cp_2Fe, as well as possible cell designs [246,199]. The electrosyntheses depicted in Eqs. (55) and (56) are not restricted to olefin complexes; for example, phenylethynylcopper(I) [197] and mixed cyclopentadienyl carbonyl complexes [247] have been prepared according to Eq. (56).

Electrosyntheses of σ-alkyl transition metal complexes following a formal oxidative addition of an electrochemically generated low-valence metal ion and an organic electrophile [Eq. (7), scheme 3] are also feasible on a preparative scale if the resulting organometallic derivative is stable.

VI. GROUP IVB ORGANOMETALLIC COMPOUNDS

A. Electrosynthesis and Electrocatalysis

Electrosyntheses at Sacrificial Metal Electrodes

Stimulated by the commercial success of the NALCO tetraalkyllead process [248], electrosyntheses of group IVB organometallic derivatives using sacrificial electrodes—especially lead and tin—were intensively studied in the 1960s and early 1970s. Excellent reviews have been published [5,11,249], and the following presentation is a rather short summary. The electrosynthetic procedure involves the generation of alkyl radicals either by reduction of organic electrophiles (mostly electrophilic alkylating agents as alkyl halides [Eq. (57a)], and to a smaller extent activated olefins [Eq. (57b)], or by oxidation of carbanionic species as Grignard reagents at the corresponding sacrificial metal electrode. Cathodic electrosyntheses in divided cells according to Eq. (57) have been reported for PbR_4 (RX = C_2H_5I [250]) and SnR_4 (RX = CH_3I [251,252]). Reduction of an activated olefin as acrylonitrile at a tin electrode yields $(NCCH_2CH_2)_4Sn$ [253]. The outstanding example of an anodically generated organometallic compound is documented by the NALCO tetraethyllead process, Eq. (58) (see also Chap. 31) [248].

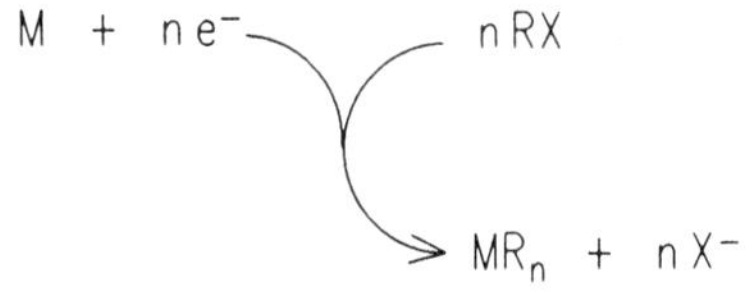

(57a)

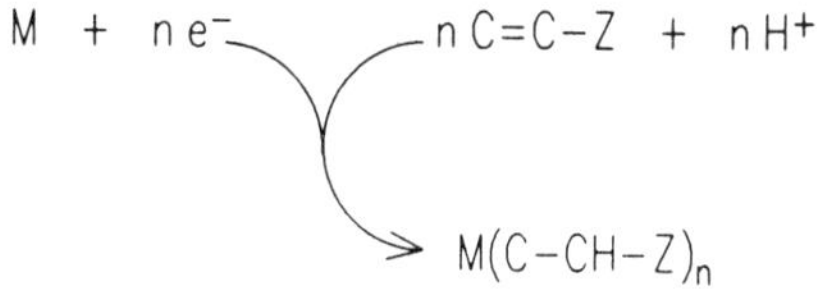

(57b)

Z: electron-withdrawing group

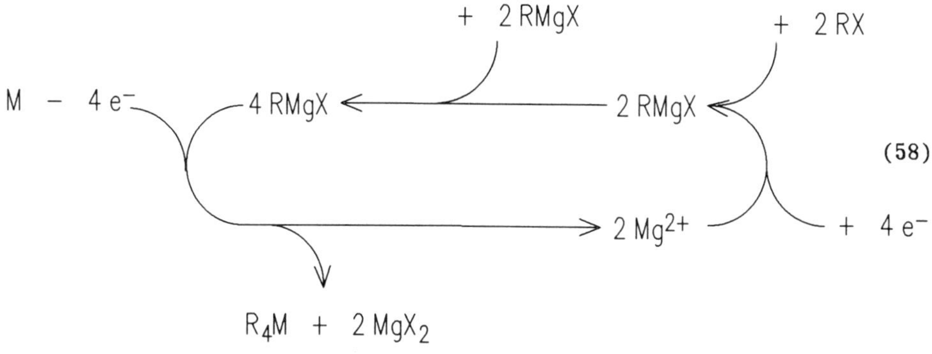

(58)

Here a mixture of the Grignard reagent and the corresponding alkyl halide is electrolyzed at the anode and cathode, respectively. Other in situ generated carbanions have been used as electroactive substrates, including sodium tetramethylaluminate [254] and diethylcadmium [255] at lead anodes for the synthesis of R_4Pb and zinc alkyls at a tin anode for the electrosynthesis of R_4Sn [256].

Electrosyntheses Using IVB Halides

Electrosynthesis starting from group IVB halides is a more general method. An elegant synthesis of tetramethylsilane and other alkyl metals (or nonmetals) from the corresponding halide derivatives on a kilogram scale has been reported [Eq. (59)] [257]. It includes Al-assisted alkylation of $SiCl_4$ [Eq. (60)] followed by electrochemical regeneration of metallic aluminum and Cl_2 [Eq. (61)].

$$3\,SiCl_4 + 12\,CH_3Cl \xrightarrow{NaCl/AlCl_3/Al} 3\,(CH_3)_4Si + 12\,Cl_2 \qquad (59)$$

$$3\,SiCl_4 + 12\,CH_3Cl + 8\,Al \longrightarrow 3\,(CH_3)_4Si + 8\,AlCl_3 \qquad (60)$$

$$8\,AlCl_3 \xrightarrow{24\,F} 8\,Al + 12\,Cl_2 \qquad (61)$$

Table 6. Electrochemical Silylation of Organic Halides

Organic halide	F/mol	Product	Yield (% isolated)
$Ph-CH=CH-CH_2Cl$	2.4	$Ph-CH=CH-CH_2SiMe$	70
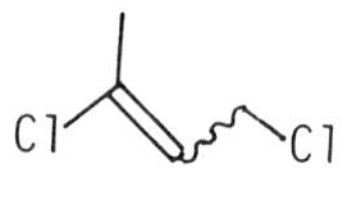	3.63		74
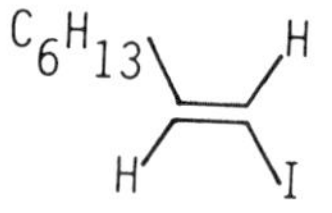	3.32	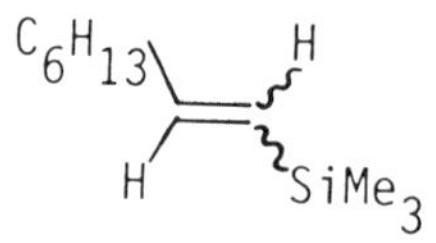	47
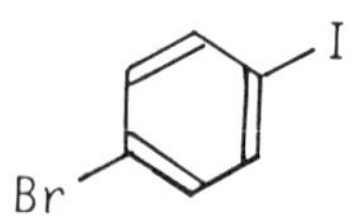	2.62	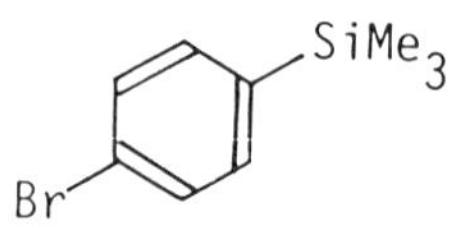	60

Source: From Ref. 259.

Electrosyntheses of organosilanes R_4Si on a millimolar scale have recently been reported by Yoshida and Shono and colleagues [258–260]. Cathodic reduction of allylic, benzylic, vinylic, or aromatic halides (R–Y) in the presence of silylating agents (R'_3Si-X) leads to the corresponding organosilanes ($R-SiR'_3$), Eq. (62) (Table 76):

$$R-Y + X-SiR'_3 \xrightarrow{+\ 2e^-} R-SiR'_3 + X^- + Y^- \qquad (62)$$

Allylic acetates are reductively converted to the corresponding allyl silanes in the presence of silylating agents and a catalytic amount of $Pd(PPh_3)_4$, Eq. (63) [261]:

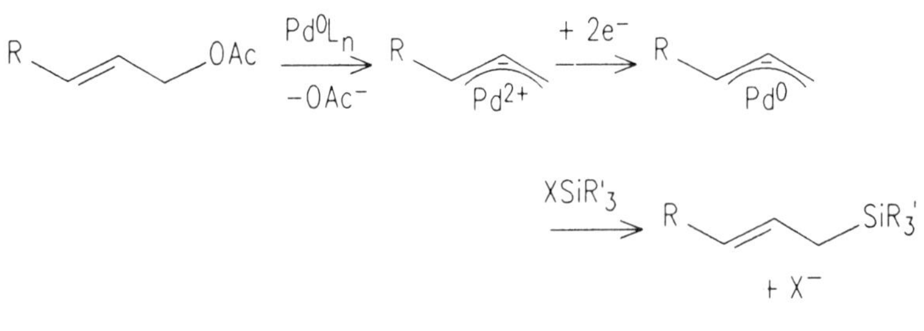

(63)

The reaction involves Pd(0)-catalyzed displacement of the acetate group, electrochemical reduction of the intermediate π-allyl Pd complex, and trapping of the allyl anion by the silylating agent.

The reductive electrochemical silylation of heteroatoms has also been described, for example O-silylation upon cathodic reduction of phthalic acid anhydride in the presence of $(CH_3)_3SiCl$ [262].

On the other hand, tin halides [263,264] or lead(II) compounds [265] are more easily reduced than are allyl bromides. The low-valence metal ions react subsequently with the organic electrophiles to yield the corresponding organometallic derivatives. These exhibit Grignard-like reactivity and may transfer the formal allyl anion to the carbonyl group of aldehydes or ketones. An electrocatalytic situation according to Eq. (64) may result [263–265].

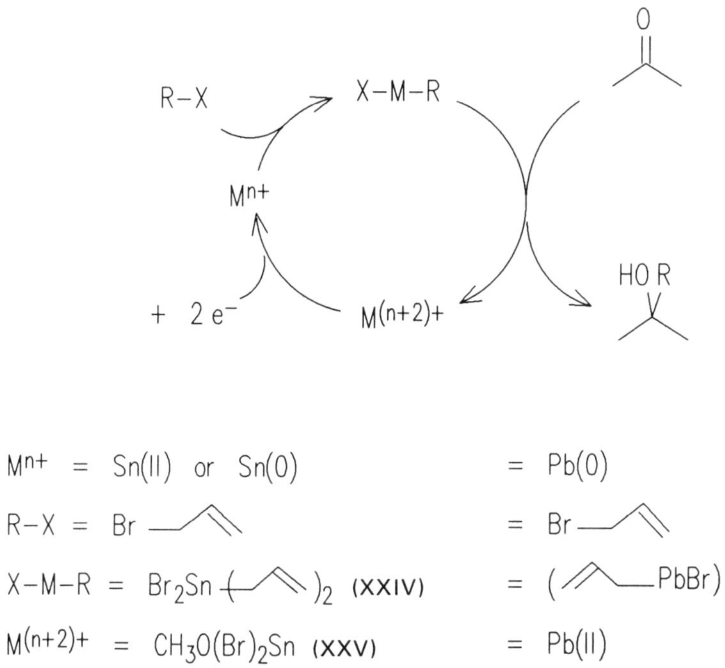

(64)

The reduction potential of the two proposed organotin intermediates (−1.0V for XXIV and −0.8 V for XXV versus Ag/Ag^+) has been determined by CV. The electrochemical version of the Grignard-type allylation of carbonyl compounds via allyltin intermediates leads to homoallylic alcohols in good yields. The reaction is carried out as a constant-current electrolysis in an undivided cell at 50°C in MeOH/AcOH. Some results are reported in Table 7.

The analogous lead-promoted carbon-carbon bond-forming reaction is also known [265–269]. In the Barbier-type reaction of aldehydes or ketones with organolead compounds leading to the corresponding alcohols, the organometallic intermediates are generated in situ from Pb^0 and such organic halides as propargyl [266] or allyl bromides [267] and 1,1,1-trichloro-2,2,2-trifluoroethane [268]. Competitive dimerization of allyl and benzyl halides occurs [269]. Lead can be used in catalytic amounts if it is recycled by chemical reducing agents, such as aluminum [267,268], or electrochemically [265].

Table 7. Tin-Catalyzed Electrosynthesis of Homoallyl Alcohols

Carbonyl compound	F/mol	Product	Yield (%)	Remarks
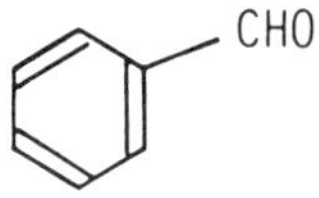	3.9	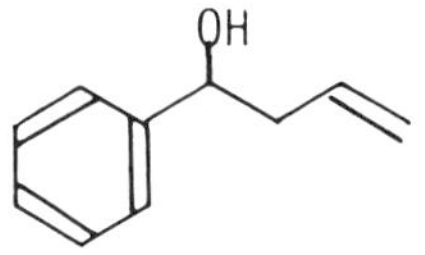	91	
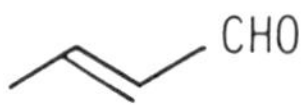	4.5	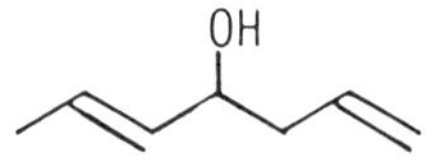	85[a]	No Michel-type addition
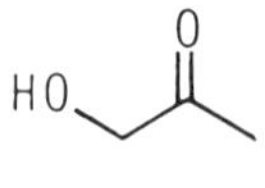	5.0	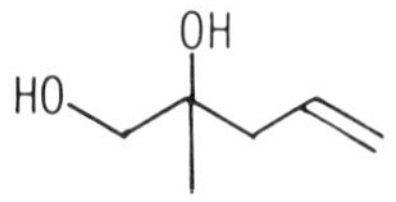	73	No protection of the OH group
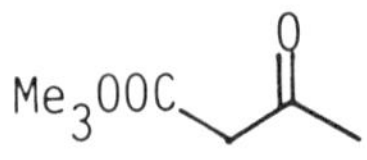	2.8	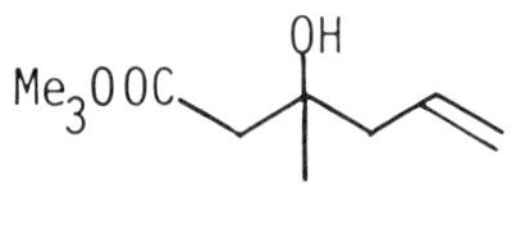	72	No reaction with ester group

[a]Electrolysis in divided cell.
Source: From Ref. 264.

B. Activation of M–C Bonds

Anodic Reactions

Oxidation of tetraalkyl derivatives of group IVB elements leads to cleavage of the metal-carbon bond. CV experiments have been interpreted in terms of an ECE mechanism whereby the anodic peak potential E_p is determined solely by electron transfer kinetics and diffusion. Kochi also reported a striking linear correlation between the E_p of the IVB-tetraalkyl compounds and their ionization potentials I_D (Table 8 and scheme 4) [270].

Table 8. Anodic Peak Potentials E_p and Ionization Potentials

Tetraethyl metals			Tetraalkyl tin		
Compound	E_p (V)	I_D (eV)	Compound	E_p (V)	I_D (eV)
Et_4Si	2.56	9.78	Me_4Sn	2.48	9.69
Et_4Ge	2.24	9.41	Et_4Sn	1.76	9.69
Et_4Sn	1.76	8.93	i-Pr_4Sn	1.51	8.46
Et_4Pb	1.26	8.13	t-Bu_2Me_2Sn	1.25	8.22

Source: From Ref. 270.

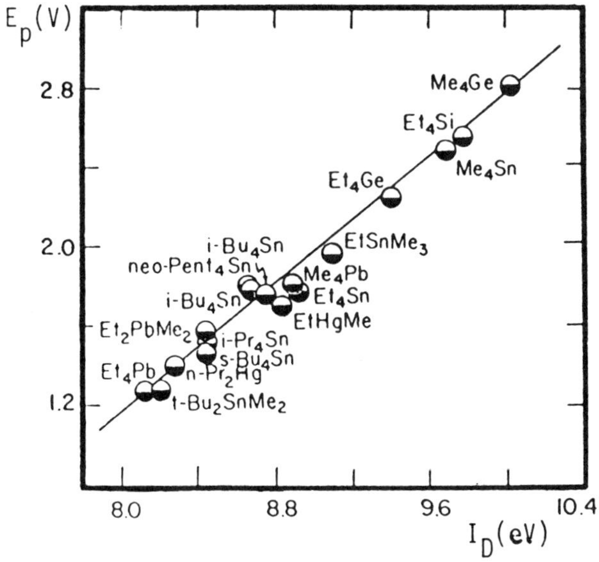

Scheme 4. Correlation of anodic peak potentials E_p and gas phase ionization potentials I_D. (From Ref. 270.)

The ionization and electrochemical oxidation occurs from a carbon-metal bonding orbital (HOMO); E_p and I_D decrease with increasing energy of the HOMO in the sequence Si > Ge > Sn > Pb and Me > Et > i-Pr > t-Bu. However, for cyclic polysilanes [$(R_1-Si-R_2)_n$; n = 3–7], electrochemical oxidation leads to Si–Si bond cleavage and E_p decreases with decreasing ring size [271].

Electrochemical oxidation of tetraphenylsilane takes place at 2.1 V versus Ag/Ag^+ in $CH_3CN/LiClO_4$; its exhaustive electrolysis in CH_2Cl_2/TBAP containing $AlCl_3$ affords black deposits at the Pt anode [272]. However, electrooxidation of allyl- and benzylsilanes in the presence of water, alcohols, or carboxylic acids leads to the corresponding alcohols, ethers, or esters in moderate to good yield, as shown by Yoshida et al. [Eq. (65)] [273]. The reaction in Eq. (65) proceeds by a one-electron oxidation-induced Si–C bond cleavage with formation of an allyl radical that is further oxidized to the cation and subsequently trapped by R–OH. The extension of this work to 1-trimethylsilyl-1,3-dienes is of considerably more impact in organic synthesis. The four-electron oxidation in methanol yields directly 1,1,4-trimethoxy-2-alkenes [Eq. (66)] [274]. Some examples are presented in Table 9.

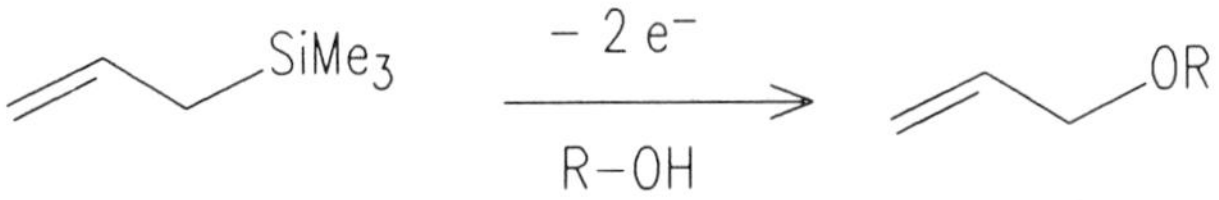

(65)

Table 9. Electrooxidation of 1-Trimethylsilyl-1,3-dienes

Organosilane	Product	Yield (% isolated)
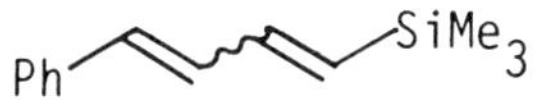	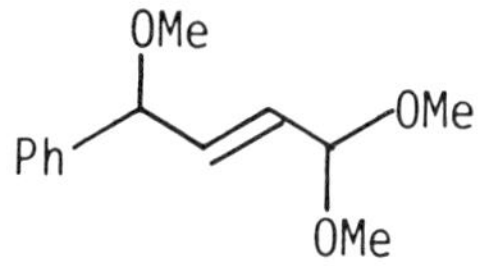	72
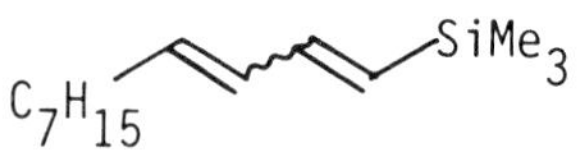	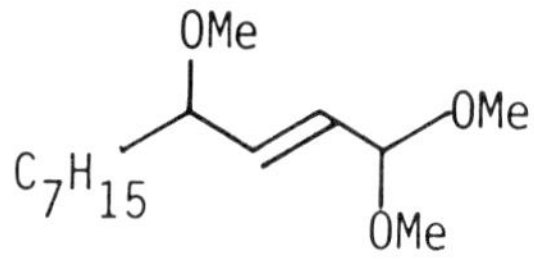	49
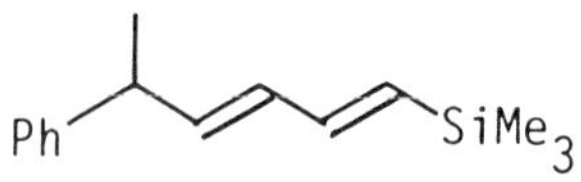	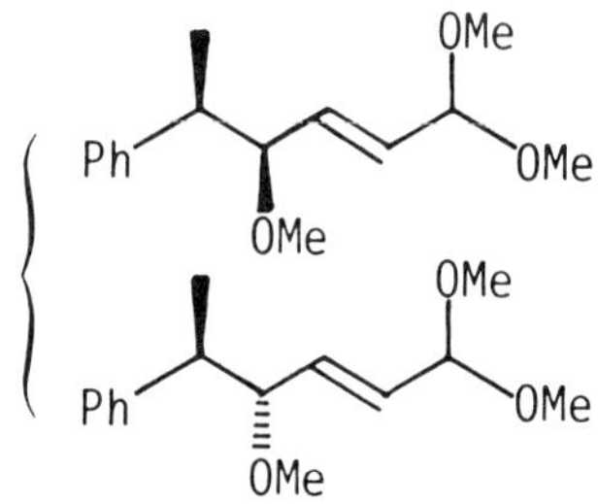	(erythro) 53 (erythro-threo = 4:1) (threo)

Source: From Ref. 274.

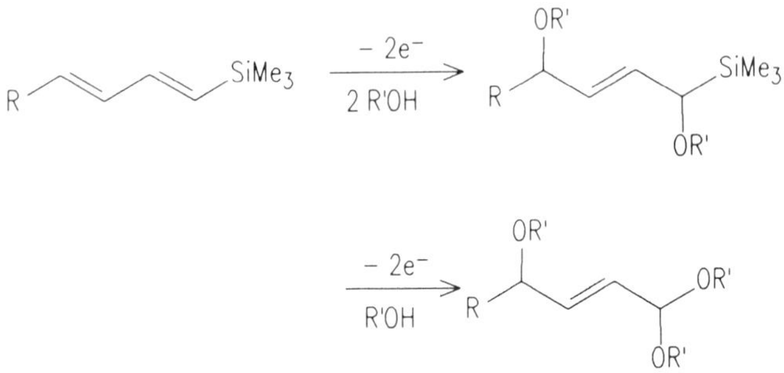

(66)

Conditions: CH_3CN/Et_4NOTs + $R^{(')}$–OH (R = H, alkyl or acyl, R' = CH_3), galvanostatic conditions (10 mA/cm^2) at carbon electrodes, undivided cells: (I) ~2 F/mol silane [273]; (II) ~4 F/mol silane [274].

Since the starting silyl-1,3-dienes may readily be prepared by Peterson-type olefination (step a, Eq. (67)] from aldehydes or ketones, the overall process (a + b + hydrolysis) represents a carbon chain extension of carbonyl compounds by a formyl vinyl anion equivalent.

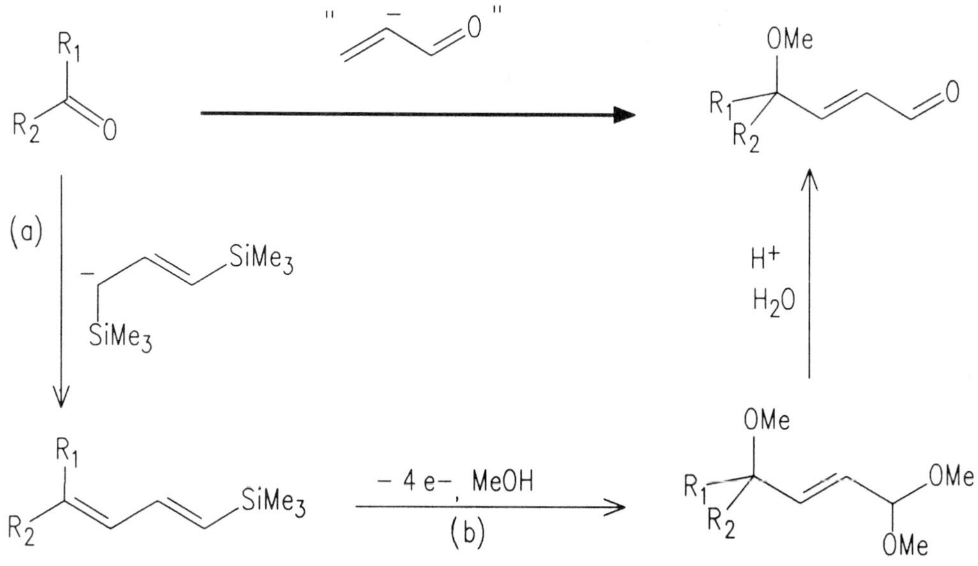

(67)

The anodic oxidation of 1-carboxylate-2-trimethylsilyl-substituted carbon skeletons leads ultimately to Si−C bond cleavage, but the initial electron transfer involves a carboxylate-based redox orbital [275a]. Similarly, oxidation of α-silylethers in alcohol leads to Si−C bond cleavage at relatively low oxidation potentials and to the corresponding acetals if a coplanar arrangement of an oxygen atom lone pair and the Si−C σ-bond is accessible [275b]. The analogous reaction with α-silyl carbamates undergoing oxidatively induced Si−C bond cleavage in methanol has also been described [275c].

Cathodic Reactions

Much information concerning reduction potentials and the reduction mechanism of group IVB organometallic derivatives is available from the early studies of Dessy et al. [276]. For compounds of the type R_nMX_{4-n} (R = alkyl, aryl; X = electronegative group like halogen or carboxylate), the reduction potential becomes more positive in the sequences Si < Ge < Sn < Pb and $R_4M < R_3MX < R_2MX_2 < R_1MX_3$ [276]. In the organohalosilanes (−1.9 to 2.2 V versus Ag/AgI) and germanes (−1.6 to 1.85 V), cleavage of the C−X bond generally takes place [277,278].

The reduction mechanism of organohalosilanes has recently been reviewed by Genies and El Omar [279]. It involves dimerization by radical-radical coupling [Eq. (68)] or by reaction of an organosilylanion with the organosilylhalide [Eq. (71); compare Eq. (62)] [280], formation of organosilane hydride [277] by H atom abstraction [Eq. (69)] or by protonation (not shown), and reaction with the electrode surface [Eq. (70)] [281].

Arylmethylsilanes and germanes yield typical Birch-type products, that is, silyl- and germyl-substituted cyclohexa-1,4-dienes, upon electroreduction, indicating that the LUMO is benzene localized. However, trimethyl(phenyl)tin undergoes concomitant cleavage of the metal-carbon bond [Eq. (72)] [282].

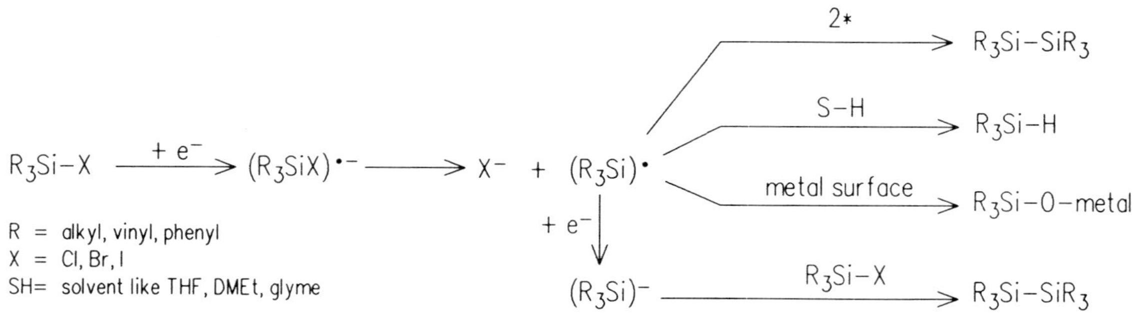

(68)

(69)

(70)

(71)

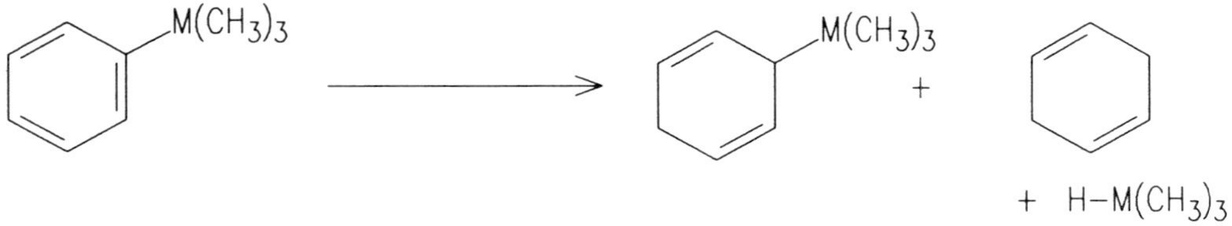

(72)

M = Si main

M = Ge main

M = Sn main

Conditions: methylamine/LiCl, −5 to −40°C, undivided cell, ~5 F/mol.

Tin-carbon bond cleavage has also been reported in the case of R_3SnCH_2I upon electrochemical reduction in the presence of cyanide ions [283]. A similar reduction mechanism as for organosilyl halides [Eqs. (68) through (71)] has been proposed for dissociated trialkyl tin hydroxides, Eq. (73) [284]:

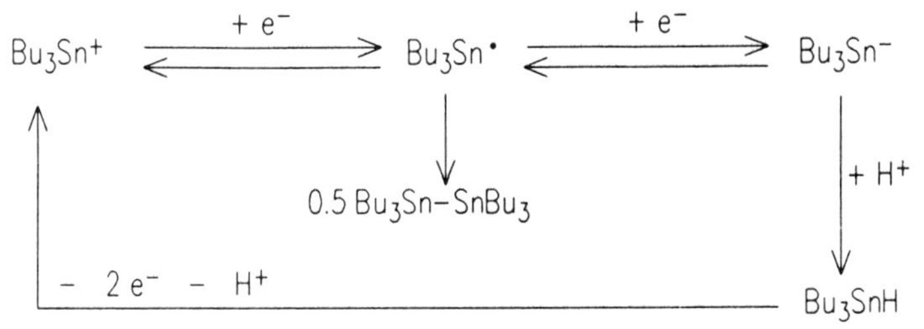

(73)

Following Eq. (73), the electrosyntheses of different organoditin compounds have been described [285,286]. Synthesis of organic compounds by way of radical-type C−C bond formation is currently one of the fastest developing fields in organic chemistry [287,288]. The reagent most used for the inter- and intramolecular hydroaddition of organic halides to olefins according to Eq. (74) is tributyl tin hydride [289].

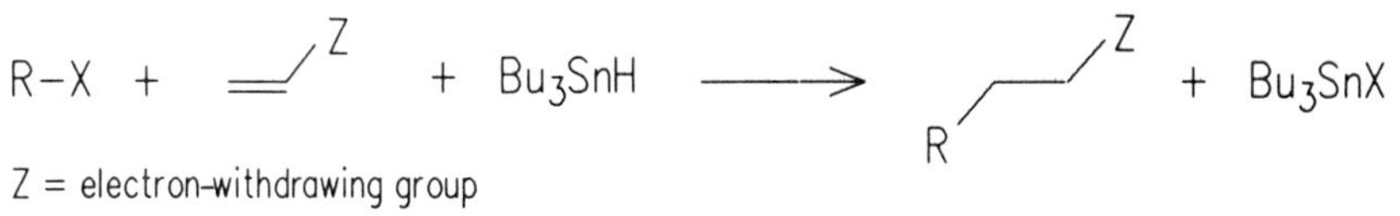

(74)

Catalytic versions using $NaBH_4$ to regenerate Bu_3SnH have been developed [290]. Electrochemical regeneration of Bu_3SnH according to Eq. (73) could be an interesting alternative way if side reactions such as its dimerization and alkylation could be circumvented.

VII. GROUP III ORGANOMETALLIC COMPOUNDS

A. Formation of M−C Bonds

Anodic processes include electrochemical oxidation of a Grignard reagent at a boron anode, yielding BR_3 [291], oxidation of $3NaAl(C_2H_5)_4$ at an Al anode, leading to $4Et_3Al$, and oxidation of dichloromethane at an aluminum anode, yielding methylene bis(aluminum) dichloride [Eq. (75)] [292,293]. A series of patents and some publications deal with the electrochemical production of Ziegler-Natta catalysts from aluminum and different transition metals [295]. More recently, an electrochemical version of the Friedel-Crafts type of alkylation of aromatics by anodic dissolution of an Al anode in the presence of an alkyl halide and an aromatic compound has been described [296,297]. The formation of the catalyst may involve organoaluminum species. The direct electrochemical synthesis of neutral and anionic organoindium halides at an indium anode has been reported [Eq. (75a)], cathodic reduction of RX at a sacrificial indium cathode may contribute to the product formation [Eq. (75b)] [294].

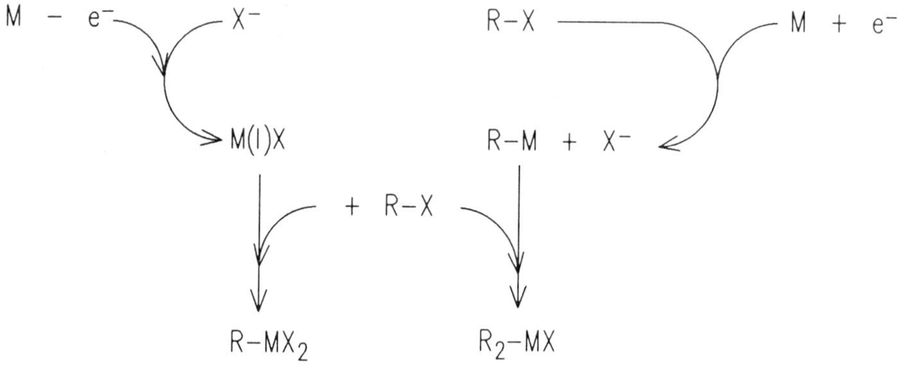

(75a and b)

$$MX_3 + R_2MX \longrightarrow 2\ RMX_2 \qquad (75c)$$

Sacrificial Al anodes in undivided cells have been found to stabilize the product carboxylates of cathodically induced carboxylations by coordination of Al(III) to $RCOO^-$ [298,299]. Similar effects have been reported with sacrificial Zn anodes [Eq. (76)].

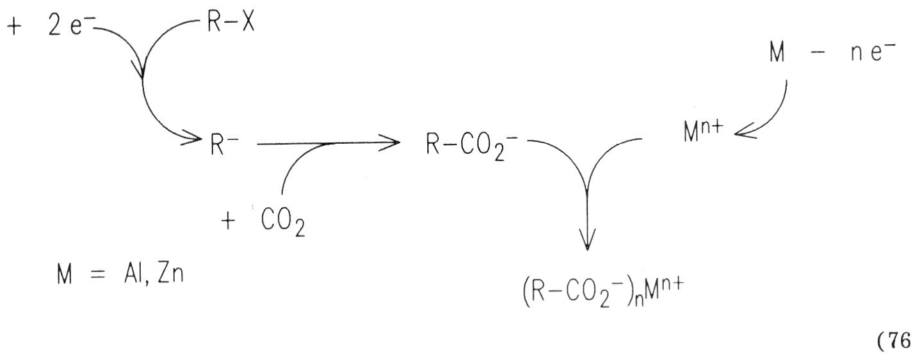

(76)

Formation of the metal complex $(R_4N)_3Ga(NCS)_6$ at a Ga anode in presence of NCS^- occurs under similar conditions, another example of the electrosynthesis of a coordination compound (Sec. III) [300].

B. Activation of M–C Bonds

The electrochemistry of BH_4^-, boranes, carboranes, and metallocarboranes has been reviewed [301]. Electrochemical oxidation of triorganoborates and boranes has been reported by Schäfer and colleagues [302,303] and Suzuki's group [304]. Dimerization of the alkyl residues is achieved in good yields, Eq. (77) [303]:

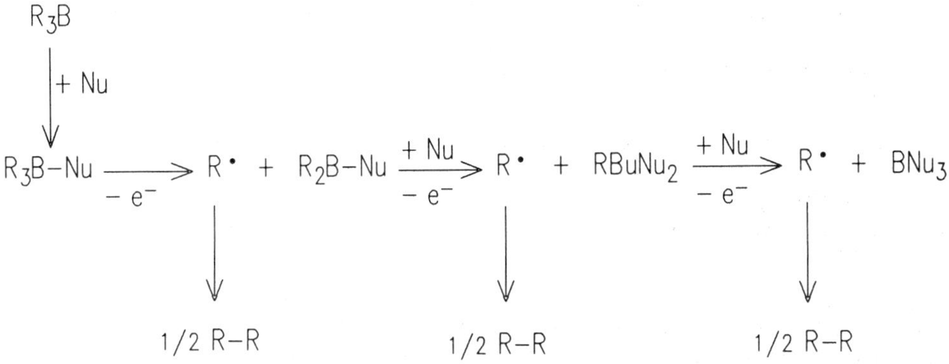

(77)

In the presence of butadiene the intermediate radicals may be trapped by the olefin, leading to different products derived from further radical intermediates [302]. The oxidation potential ot the triorganoboranes in the presence of nucleophiles (R_3BNu) depends on the type of Nu (Table 10).

Another reaction related to organic synthesis has been described by Suzuki's group. C–H acidic molecules undergoing cathodic reduction [Eq. (78b)] can be coupled with the alkyl groups of triorganoboron derivatives, as in Eq. (78f). The mechanism includes oxidative cleavage of a B–C bond, probably by iodine-induced C–B → C–C migration [Eq. (78a) and (78c-e)] [305].

$$\begin{array}{lll}
\text{(a)} & I^- - e^- \longrightarrow & 1/2\ I_2 \\
\text{(b)} & C_6H_5C{\equiv}CH + e^- \longrightarrow & C_6H_5C{\equiv}C^- + 1/2\ H_2 \\
\text{(c)} & R_3B + C_6H_5C{\equiv}C^- \longrightarrow & R_3B{-}C{\equiv}CC_6H_5 \\
\text{(d)} & R_3B{-}C{\equiv}CC_6H_5 + I_2 \longrightarrow & R_2B{-}C(R){=}C(I)C_6H_5 \\
\text{(e)} & R_2B{-}C(R){=}C(I)C_6H_5 \longrightarrow & RC{\equiv}CC_6H_5 + R_2BI \\
\text{(f)} & R_3B + C_6H_5C{\equiv}CH \longrightarrow & R{-}C{\equiv}CC_6H_5
\end{array} \qquad (78)$$

Table 10. Oxidation Potentials of Tri-n-butylboranes in the Presence of Nucleophiles (Bu_3B-Nu)

Nu	E_p (Bu_3BNu)	pK_a (Nu-H)	E_p (Nu)
THF	+1.65	-7	+2.0
Pyridine	+1.65	5.2	+2.0
NH_3	+1.2	9.24	+1.4
SCN^-	+0.9	4.0	+1.15/1.4
CN^-	+0.61	9.4	+1.2
OH^-	+0.37	15.74	+1.15

Further examples include the coupling of organoboranes with propenoic acid esters [306], nitroalkanes [307], and nitriles [308]. In two other oxidations of organoboranes a carbonium ion intermediate seems to be involved:

Formation of ethers and esters has been reported from organoborane oxidation at carbon electrodes in the presence of alcohols or carboxylates, respectively [309].

Cyclopropane was formed by oxidation of propylboronate at a Pt electrode [310].

Spin trapping experiments indicate radical intermediates in the electrochemical oxidation of n-Bu_4B^- but not Ph_4B^- [311]. Formation of biphenyl has been reported upon electrolysis of Ph_4B^- [312], but this result has been questioned [303].

Cathodic investigations of organoborane compounds with relevance to organic electrochemistry are quite scarce. Utleys' group investigated the reductive cleavage of cyclic boronates, such as XXVI, to promote olefin formation, but probably dimerization to a diborate took place [313].

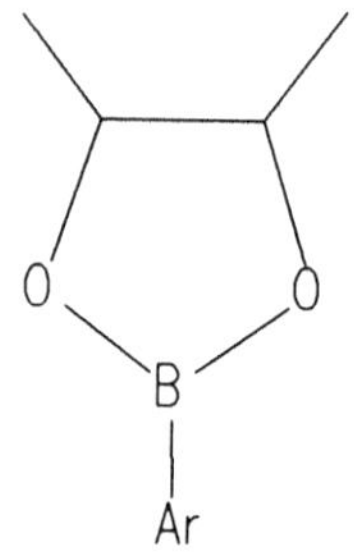

XXVI

The reduction of organothallium compounds at mercury electrodes leading to organomercury compounds has been reviewed [11].

VIII. ELECTROCHEMISTRY OF GRIGNARD REAGENTS

Ethereal solutions of Grignard reagents exhibit electrical conductivity due to the formation of the anionic and cationic species RMg^+ and $RMgX_2^-$ according to Eq. (79).

RMg^+ and $RMgX_2^-$ are also the electroactive species at the cathode [Eq. (80)] and anode [Eq. (81)], leading to metallic magnesium and alkyl radicals, respectively. The other products, R_2Mg and MgX_2, are in equilibrium with the Grignard reagent [Eq. (82)].

$$4\ RMgX \longrightarrow 2\ RMg^+ + 2\ RMgX_2^- \quad (79)$$

$$2\ RMg^+ + 2\ e^- \longrightarrow Mg + R_2Mg \quad (80)$$

$$2\ RMgX_2^- \longrightarrow 2\ R^\bullet + 2\ MgX_2 + 2\ e^- \quad (81)$$

$$R_2Mg + MgX_2 \longrightarrow 2\ RMgX \quad (82)$$

The decomposition potential of ethereal Grignard solutions [Eqs. (80) and (81)] has been discussed [314].

Grignard reagent have been widely used as a source of radicals produced at a sacrificial anode for the formation of other organometallic compounds (R_4Pb [148] and R_3B [291]).

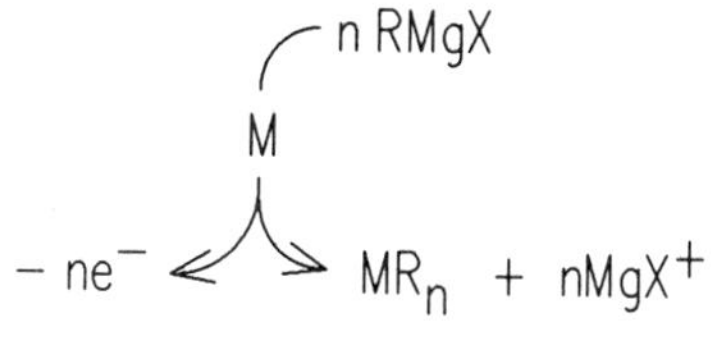

(83)

At inert anodes, the radicals can dimerize or abstract hydrogen from the solvent [315,316]. It is possible to trap the original radical with olefinic compounds, such as styrene and butadiene, Eq. (84) [317]:

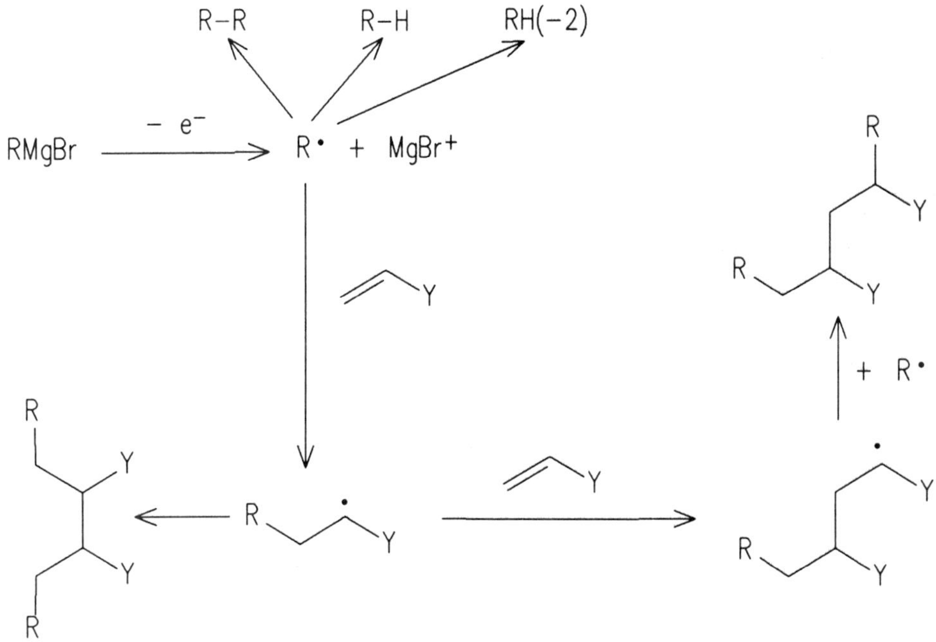

(84)

The reaction is of limited synthetic use because polymerization can easily take place and because many functionalized olefins are not stable in the presence of the Grignard reagent.

IX. GROUP IIB ORGANOMETALLIC COMPOUNDS

A. Formation of M–C Bonds

The direct electrosynthesis of group IIB coordination compounds is well documented by recent work of Tuck and colleagues. β-Diketonato complexes of Cd, thiolates of Zn, Cd, and Hg, and catecholates of Zn and Cd have been prepared according to Eq. (36), that is, by concomitant generation of the ligand anion at an inert cathode and the metal cation at a sacrificial anode in an undivided cell [191,195,318,319]. Oxidation of metallic mercury in the presence of chelating agents has been extensively studied and applied to analytical investigations [320,321]. More recent work includes studies on multielectron oxidation processes [322].

Cathodic electrosynthesis of R_2Hg occurring at sacrificial mercury electrodes in the presence of alkylating agents according to Eqs. (85) and (86) is also a well-documented process [11]. The reduction mechanism is complicated by adsorption phenomena. The yield of R_2Hg increases along the series RX = tertiary < secondary < primary [323] and X = Cl < Br < I. Additionally, a relatively positive electrode potential and neighboring groups on R, which are able to stabilize a radical, favor the formation of R_2Hg. These observations are explained by the competing reduction of $RHg^{\bullet}$, Eq. (87):

$$RX + e^- \longrightarrow RHg^{\bullet} \quad X^- \tag{85}$$

$$2\,RHg^{\bullet} \longrightarrow R_2Hg + Hg \tag{86}$$

$$RHg^{\bullet} + e^- \longrightarrow R^- + Hg \tag{87}$$

Several organometallic compounds undergoing reduction at mercury cathodes yield dialkylmercury compounds, for example alkyl lead halides [276].

Reports on the synthesis of organocadmium and aryl organozinc halides at sacrificial anodes have been published [324] and reviewed [19]. The mechanism is depicted in Eq. (88) and involves oxidation of a cathodically produced halide ion, followed by a non-charge-consuming, radical-type chain reaction at the anode.

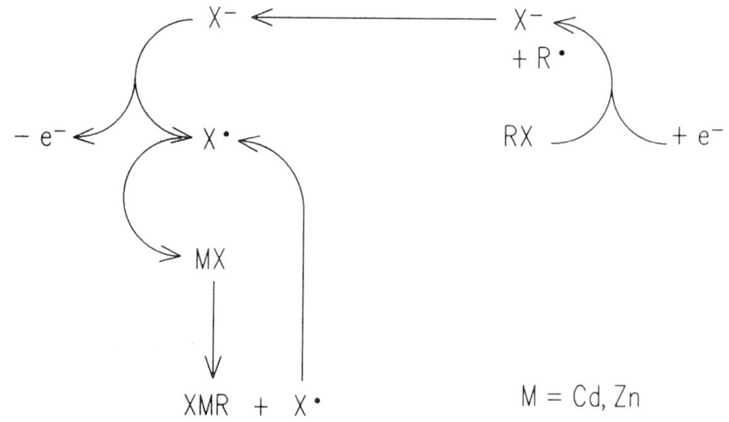

(88)

B. Activation of M–C Bonds

Cathodic Processes

The reduction of RHgX and R_2Hg at mercury electrodes is one of the best studied subject in organometallic electrochemistry. Extensive reviews dealing with electroanalytic and electropreparative aspects are available [11]. The ultimate reduction products of RHgX are R_2Hg or RH, depending on the reduction potential, Eq. (89) [325,326], and RH for R_2Hg reduction, Eq. (90):

$$\text{RHgX} \xrightarrow[E_{1/2\,(a)}]{+e^-,\ -X^-} \text{RHg}^\bullet$$

$$\tfrac{1}{2}\,R_2Hg + \tfrac{1}{2}\,Hg \longleftarrow \text{RHg}^\bullet \xrightarrow[E_{1/2\,(b)}]{+e^- + H^+} \text{RH} + \text{Hg} \qquad (89)$$

$$R_2Hg + 2\,e^- + 2\,H^+ \xrightarrow{E_{1/2(c)}} 2\,RH + Hg \qquad (90)$$

Attempts to trap an intermediate anion by intramolecular cyclization have been reported [327]. The stereochemical course of the reaction [Eqs. (85) through (90)] has been studied. Retention or racemization of configuration at the halide-bearing carbon has been observed [325].

Linear correlations of the reduction potentials $E_{1/2b,c}$ [Eqs. (89) and (90)] of organomercury compounds (RHgX and R_2Hg) with the pK_a of RH have been reported and interpreted as carbanion stability affecting both observables [123,328,329]. The dependence of $E_{1/2a}$ and $E_{1/2b}$ on the pK_a of RH has been observed by Dessy et al. for the pK_a range 35–40 [328]. Butin et al. found a linear correlation for a larger pK_a range (pK_a = 10–40) for the charge transfer-corrected reduction potential $E_{1/2}$ [329]. Similar LFERs are known for other organometallic species, that is, alkylcobalamins [37], Co-alkylated vitamin B_{12} model systems [108,114], and $cp(CO)_2Fe{-}R$ [124].

Anodic Processes

The potentials of the irreversible oxidation of R_2Hg compounds exhibit a linear correlation of E_p and the gas-phase ionization potentials (I_D) of the organometallics similar to that of several representatives of group IVB metals (scheme 4) [270]. Oxidation of organozinc and organocadmium compounds at inert anodes yields the dimerized organic products. The electrosynthesis of long-chain alkanes, diaryls, and dimeric terpenes from Zn and Cd organometallics has been compared with the products obtained from the corresponding Mg and Li compounds [330]. Hg(II) in the presence of an olefin leads to the corresponding σ-bonded organomercury compound. The organic products observed upon anodic oxidation of a mixture of olefin and Hg(II)salt (catalytic amount) in aqueous perchloric acid can be explained by Markovnikow addition of Hg(II) and water across the C–C double bond followed by electrochemically induced cation formation at the Hg-bonded carbon [331].

REFERENCES

1. (a) G. Wilkinson (ed.), Comprehensive Coordination Chemistry, Vol. 1–7, Pergamon, Oxford, 1987; (b) G. Wilkinson (ed.), Comprehensive Organometallic Chemistry, Vol. 1–9, Pergamon, Oxford, 1982.
2. J. P. Collman, L. S. Hegedus, J. R. Norton, and R. G. Finke, Principles and Application of Organotransition Metal Chemistry, University Science Books, Mill Valley, California, 1987.
3. W. E. Geiger, Prog. Inorg. Chem., 33, 275–352 (1985).
4. C. J. Pickett, in The Chemistry of Functional Groups, Vol. 2, The Chemistry of the Metal-Carbon Bond, F. R. Hartley and S. Patai, eds.), Wiley, Chichester, 1985, pp. 1–24.
5. D. A. White, in Organic Electrochemistry, 2nd ed. (M. M. Baizer and H. Lund, eds.), Dekker, New York, 1983, pp. 591–636.
6. H. L. M. Van Gaal and J. G. M. Van der Linden, Coord. Chem. Rev., 47, 41–54 (1982).
7. J. Chatt, Coord. Chem. Rev., 43, 337–347 (1982).
8. B. G. Ershow, Russ. Chem. Rev. (English Translation), 50, 1119–1133 (1983).
9. G. K. Buduikow and T. V. Troepolskaya, Russ. Chem. Rev. (English Translation), 48, 449–462 (1979).
10. M. D. Morris and G. L. Kok, in Encyclopedia of Electrochemistry of the Elements, (A. J. Bard and H. Lund, eds.), Org. Sec., Vol. 13, Dekker, New York, 1979, pp. 1–76.
11. W. J. Settineri and L. D. McKeever, in Technique of Electroorganic Synthesis, (N. L. Weinberg, ed.,), Vol. 5, Part 2, Wiley, New York, 1975, pp. 397–558.
12. L. Eberson, Electron Transfer Reactions in Organic Synthesis, Springer, Berlin, 1987.
13. J. K. Kochi, J. Organometal Chem., 300, 139–166 (1986).
14. T. J. Meyer and H. Taube, in Comprehensive Coordination Chemistry, Vol. 1, (G. Wilkinson, ed.), Pergamon, Oxford, 1987, pp. 331–384.
15. E. Steckhan, Topics Current Chem., 142, 1–69 (1987).
16. M. Tokuda, Yuki Gosei Kagaku Kyokaishi, 43, 522 (1985); CA, 103, 122508n.
17. S. Torii, Synthesis, 873 (1986).
18. E. Steckhan, Angew. Chem., 98, 681–699 (1986); Int. Ed. Engl., 25, 693 (1986).
19. D. G. Tuck, Pure Appl. Chem., 51, 2005 (1979).
20. J. A. R. van Veen and J. F. Van Baar, Rev. Inorg. Chem., 4, 293–327 (1982).
21. A. R. Battersby and E. McDonald, in Porphyrins and Metalloporphyrins (K. M. Smith, ed.), Elsevier, Amsterdam, 1975, pp. 61–122.
22. G. Ksander, G. Bold, R. Lattman, C. Lehmann, T. Früh, Y.-B. Xiang, K. Inomata, H.-P. Buser, J. Schreiber, E. Zass, and A. Eschenmoser, Helv. Chim. Acta, 70, 1115 (1987), and references cited therein.
23. A. Eschenmoser, Nova Acta Leopoldina, 55(247) (1982); CA, 97, 215813f.
24. B. Kräutler, Chimia, 41, 277 (1987).
25. A. B. P. Lever and H. B. Gray (eds.), Iron Porphyrins, Part 1 and 2, Addison-Weseley, Reading, Massachusetts, 1983.

26. M. Wikström, K. Krab, and M. Saraste, Cytochrome Oxidase, A Synthesis, Academic Press, London, 1981.
27. A. Pfaltz, B. Jaun, A. Fässler, A. Eschenmoser, R. Jaenchen, H. H. Giles, G. Diekert, and R. K. Thauer, Helv. Chim. Acta, 65, 828 (1982).
28. Z. Schneider and A. Stroinski (eds.), Comprehensive B_{12}, Walter de Gruyter, Berlin, 1982.
29. D. Dolphin, B_{12}, Vols. 1 and 2, Wiley, New York, 1982.
30. R. K. Thauer and G. Fuchs, Naturwissenschaften, 66, 89 (1979).
31. W. L. Ellefson, W. B. Whitman, and R. S. Wolfe, Proc. Natl. Acad. Sci. USA, 79, 3707 (1982).
32. C. Kratky, A. Fässler, A. Pfaltz, B. Kräutler, B. Jaun, and A. Eschenmoser, J. Chem. Soc., Chem. Commun., 1368 (1984).
33. A. Eschenmoser, Ann. N.Y. Acad. Sci., 471, 108 (1986).
34. B. Jaun and A. Pfalz, J. Chem. Soc., Chem. Commun., 1327 (1986).
35. J. Rétey and J. A. Robinson, Stereospecificity in Organic Chemistry and Enzymology, Verlag Chemie, Weinheim, 1982, pp. 185-207.
36. D. Lexa and J.-M. Savéant, Acc. Chem. Res., 16, 235 (1983).
37. R. Scheffold, S. Abrecht, R. Orlinski, H.-R. Ruf, P. Stamouli, O. Tinembart, L. Walder, and C. Weymuth, Pure Appl. Chem., 59, 363 (1987).
38. R. Scheffold, Chimia, 39, 203 (1985).
39. R. Scheffold, G. Rytz, and L. Walder, in Modern Synthetic Methods (R. Scheffold, ed.), Vol. 3, Salle, Frankfurt, Wiley, London, 1983, pp. 355-440.
40. D. St. C. Black, in Comprehensive Coordination Chemistry, Vol. 6, (G. Wilkinson, ed.), Pergamon, Oxford, 1987, pp. 155–227.
41. D. H. Vaughan, in Comprehensive Coordination Chemistry, Vol. 1, (G. Wilkinson, ed.), Pergamon, Oxford, 1987, pp. 479–491.
42. C. J. Pickett, in Comprehensive Coordination Chemistry, Vol. 1, (G. Wilkinson, ed.), Pergamon, Oxford, 1987, pp. 493–501.
43. D. H. Busch, Acc. Chem. Res., 11, 392 (1978).
44. K. M. Kadish, in Progress in Inorganic Chemistry, Vol. 34, (S. J. Lippard, ed.), Wiley, New York, 1986, pp. 435–605.
45. R. H. Felton, in The Porphyrins, Vol. 5, part C, Chap. 3, (D. Dolphin, ed.), Academic Press, New York, 1978.
46. D. G. Davis, in The Prophyrins, Vol. 5, part C, Chap. 4 (D. Dolphin, ed.), Academic Press, New York, 1978.
47. J.-H. Fuhrhop, Structure and Bonding, Vol. 18, Springer Verlag, Berlin, 1974, pp. 1–67.
48. K. M. Kadish and M. M. Morrison, J. Amer. Chem. Soc., 98, 3326 (1976).
49. J.-H. Fuhrhop, K. M. Kadish, and D. G. Davis, J. Amer. Chem. Soc., 95, 5140 (1973).
50. F. A. Walker, D. Beroiz, and K. M. Kadish, J. Amer. Chem. Soc., 98, 3484 (1976).
51. M. Zerner and M. Gouterman, Theor. Chim. Acta, 4, 44 (1966).
52. R. J. Donohoe, M. Atamian, and D. F. Bocian, J. Amer. Chem. Soc., 109, 5593 (1987).
53. K. M. Kadish and L. A. Bottomley, Inorg. Chem., 19, 832 (1980).
54. C. Gueutin, D. Lexa, M. Momenteau, J.-M. Savéant, and F. Xu, Inorg. Chem., 25, 4294 (1986).

55. L. Walder, G. Rytz, U. Vögeli, R. Scheffold, and P. Engel, Helv. Chim. Acta, 67, 1801 (1984).
56. R. Guilard, C. Lecomte, and K. M. Kadish, in Structure and Bonding, Vol. 64, Springer Verlag, Berlin, 1987, pp. 205–268.
57. H. Ogoshi, E. Watanabe, N. Koketsu, and Z. Yoshida, Bull. Chem. Soc. Japan, 49, 2529 (1976).
58. Y. Aoyama, T. Yoshida, K. Sakurai, and H. Ogoshi, Organometallics, 5, 168 (1986).
59. D. Lexa, J.-M. Savéant, and D. L. Wang, Organometallics, 5, 1428 (1986).
60. J. Setsune, Z. Yoshida, and H. Ogoshi, J. Chem. Soc. Perkin I, 983 (1982).
61. D. Lexa, J. M. Savéant, and J. P. Soufflet, J. Electroanal. Chem. 100, 159 (1979).
62. J. E. Anderson, C.-L. Yao, and K. M. Kadish, Inorg. Chem., 25, 718 (1986).
63. D. Lexa and J.-M. Savéant, J. Amer. Chem. Soc., 104, 3503 (1982).
64. D. Lexa, J. Mispelter, and J.-M. Savéant, J. Amer. Chem. Soc., 103, 6806 (1981).
65. K. M. Kadish, A. Tabard, A. Zrineh, M. Ferhat, and R. Guilard, Inorg. Chem., 26, 2459 (1987).
66. K. M. Kadish, B. Boisselier-Cocolios, A. Coutsolelos, P. Mitaine, and R. Guilard, Inorg. Chem., 24, 4521 (1985).
67. K. M. Kadish, B. Boisselier-Cocolios, P. Cocolios, and R. Guilard, Inorg. Chem., 24, 2139 (1985).
68. D. Dolphin, D. J. Halko, and E. Johnson, Inorg. Chem., 20, 4348 (1981).
69. K. M. Kadish, J. E. Anderson, C.-L. Yao, and R. Guilard, Inorg. Chem., 25, 1277 (1986).
70. D. Lancon, P. Cocolios, R. Guilard, and K. M. Kadish, J. Amer. Chem. Soc., 106, 4472 (1984).
71. I. Artaud, L. Devaucelle, J.-P. Battioni, J.-P. Girault, and D. Mansuy, J. Amer. Chem. Soc., 109, 3782 (1987).
72. J.-P. Battioni, D. Lexa, D. Mansuy, and J.-M. Savéant, J. Amer. Chem. Soc., 105, 207 (1983).
73. S. Inoue and N. Takeda, Bull. Chem. Soc. Jpn., 50, 984 (1977).
74. P. Cocolios, R. Guilard, D. Bayeul, and C. Lecomte, Inorg. Chem. 24, 2058 (1985).
75. D. Lexa, J.-M. Savéant, K. B. Su, and D. L. Wang, J. Amer. Chem. Soc., 109, 6464 (1987).
76. J. T. Groves and M. K. Stern, J. Amer. Chem. Soc., 109, 3812 (1987).
77. S. E. Creager, S. A. Raybuck, and R. W. Murray, J. Amer. Chem. Soc., 108, 4225 (1986).
78. J. P. Collman, J. W. Prodolliet and C. R. Leidner, J. Amer. Chem. Soc., 108, 2916 (1986), and references cited.
79. K. M. Kadish, Q. Y. Xu, and J. M. Barbe, Inorg. Chem., 26, 2565 (1987).
80. C. M. Elliott, N. C. Jain, B. K. Cranmer, and A. W. Hamburg, Inorg. Chem., 26, 3655 (1987).
81. B. M. Hoffman and J. A. Ibers, Acc. Chem. Res., 16, 15 (1983).
82. J. P. Collman, J. T. McDevitt, C. R. Leidner, G. T. Yee, J. B. Torrance, and W. A. Little, J. Amer. Chem. Soc., 109, 4606 (1987).

83. J. A. Cowan, J. K. M. Sanders, G. S. Beddard, and R. J. Harrison, J. Chem. Soc., Chem. Commun., 55 (1987).
84. J. P. Collman, P. Denisevich, Y. Konai, M. Marrocco, C. Koval, and F. C. Anson, J. Amer. Chem. Soc., 102, 6027 (1980).
85. S. J. Rodgers, C. A. Koch, J. R. Tate, C. A. Reed, C. W. Eigenbrot, and W. R. Scheidt, Inorg. Chem., 26, 3647 (1987).
86. A. Hamilton, J.-M. Lehn, and J. L. Sessler, J. Amer. Chem. Soc., 108, 5158 (1986).
87. J. P. Collman, J. I. Brauman, T. J. Collins, B. Iverson, and J. L. Sessler, J. Amer. Chem. Soc., 103, 2450 (1981).
88. W. R. Scheidt, C. W. Eigenbrot, M. Ogiso, and K. Hatano, Bull. Chem. Soc. Jpn., 60, 3529 (1987).
89. J. T. Groves and R. S. Myers, J. Amer. Chem. Soc., 105, 5791 (1983).
90. J. Y. Becker, D. Dolphin, J. B. Paine, and T. Wijesekera, J. Electroanal. Chem., 164, 335 (1984).
91. A. B. P. Lever, M. R. Hempstead, C. C. Leznoff, W. Liu, M. Melnik, W. A. Nevin, and P. Seymour, Pure Appl. Chem., 58, 1467 (1986).
92. F. H. Moser and A. L. Thomas, The Phthalocyanines, Vol. I, CRC Press, Boca Raton, Florida, 1983.
93. D. W. Clack, N. S. Hush, and I. S. Woolsey, Inorg. Chim. Acta, 19, 129 (1976).
94. A. Wolberg and J. Manassen, J. Amer. Chem. Soc., 92, 2982 (1970).
95. R. H. Campbell, G. A. Heath, G. T. Hefter, and R. C. S. McQueen, J. Chem. Soc., Chem. Commun., 1123 (1983).
96. V. I. Gavrilov, L. G. Tomilova, V. M. Derkacheva, E. V. Chernykh, N. T. Iotte, I. V. Shelepin, and E. A. Lukyanets, Zh. Obshch. Khim., 53, 1347 (1983).
97. L. A. Bottomley, C. Erocolani, J.-N. Gorce, G. Pennesi, and G. Rossi, Inorg. Chem., 25, 2338 (1986).
98. R. Taube, H. Drevs, and D. Steinborn, Z. Chem., 12, 425 (1978).
99. H. Eckert and Y. Kiesel, Angew. Chem., 93, 477 (1981) and references cited.
100. C. M. Lieber and N. S. Lewis, J. Amer. Chem. Soc., 106, 5033 (1984).
101. J. Zagal, C. Fierro, and R. Rozas, Contrib. Cient. Tecnol., 48, 47 (1981).
102. D. Lexa and J. M. Savéant, J. Amer. Chem. Soc., 100, 3220 (1978).
103. G. Rytz, L. Walder, and R. Scheffold, in Vitamin B_{12} (B. Zagalak and W. Friedrich, eds.), Proc. 3rd Eur. Symp. on Vitamin B_{12} and Intrinsic Factor, Walter de Gruyter, Berlin, 1979, p. 175.
104. G. Costa, A. Puxeddu and E. Reisenhofer, J. Chem. Soc., Chem. Commun., 993 (1971).
105. G. Costa, A. Puxeddu, and E. Reisenhofer, J. Chem. Soc., Dalton Trans., 1519 (1972).
106. L. Walder, Ph.D. thesis, University of Bern, 1979.
107. Harry B. Mark, personal communication, 1978.
108. G. Costa, A. Puxeddu, and C. Tavagnacco, J. Organometal. Chem., 296, 161 (1985).
109. Y. Murakami, Y. Hisaeda, A. Kajihara, and T. Ohno, Bull. Chem. Soc. Japan, 57, 405 (1984).
110. C. M. Elliott, E. Hershenhart, R. G. Finke, and B. L. Smith, J. Amer. Chem. Soc., 103, 5558 (1981).

111. P. Schulthess, D. Ammann, W. Simon, C. Caderas, R. Stepanek, and B. Kräutler, Helv. Chim. Acta, 67, 1026 (1984).
112. Y. Murakami, Y. Hisaeda, T. Ohno, and Y. Matsuda, Chem. Lett., 731 (1986).
113. H. Flohr, W. Pannhorst, and J. Rétey, Helv. Chim. Acta, 61, 1565 (1978).
114. G. Costa, Pure Appl. Chem., 30, 335 (1972).
115. G. Costa, G. Mestroni, A. Puxeddu, and E. Reisenhofer, J. Chem. Soc. (A), 2870 (1970).
116. E. Reisenhofer and G. Costa, Inorg. Chim. Acta, 49, 121 (1981).
117. K. L. Brown, in B_{12}, Vol. 1 (D. Dolphin, ed.), Wiley, New York, 1982, p. 245.
118. G. N. Schrauzer and E. Deutsch, J. Amer. Chem. Soc., 91, 3341 (1969).
119. L. Walder and R. Orlinski, Organometallics, 6, 1606 (1987).
120. A. Puxeddu and G. Costa, J. Chem. Soc., Dalton, 1285 (1982).
121. M. Okabe and M. Tada, Chem. Lett., 831 (1980).
122. M. Tada and M. Okabe, Chem. Lett., 201 (1980).
123. L. I. Denisovich and S. P. Gubin, J. Organometal Chem., 57, 109 (1973).
124. L. I. Denisovich, S. P. Gubin, Y. A. Chapovskii, and N. A. Ustrynyuk, Izv. Akad. Nauk. SSSR, Ser. Khim., 924 (1968).
125. M. D. Le Hoang, Y. Robin, J. Devynck, C. Bied-Charreton, and A. Gaudemer, J. Organometal Chem., 222, 311 (1981).
126. L. Walder, G. Rytz, K. Meier, and R. Scheffold, Helv. Chim. Acta, 61, 3013 (1978).
127. J. Halpern, M. S. Chan, J. Hanson, T. S. Roche, and J. A. Topich, J. Amer. Chem. Soc., 97, 1606 (1975).
128. G. Costa, A. Puxeddu, C. Tavagnacco, and R. Dreos-Garlatti, Inorg. Chim. Acta, 89, 65 (1984).
129. A. V. Benedetti, E. R. Dockal, H. L. Chum, and T. Rabockai, J. Electroanal. Chem., 133, 45 (1982).
130. H. L. Chum, E. R. Dockal, and T. Rabockai, J. Electroanal. Chem., 63, 197 (1975).
131. I. Levitin, A. L. Sigan, and M. E. Vol'pin, J. Chem. Soc., Chem. Commun., 469 (1975).
132. H. A. O. Hill, J. M. Pratt, M. P. O'Riordan, F. R. Williams, and R. J. P. Williams, J. Chem. Soc. (A), 1859 (1971).
133. B. Steiger, L. Walder, and R. Scheffold, Chimia, 40, 93 (1986).
134. R. Scheffold, M. Dike, S. Dike, T. Herold, and L. Walder, J. Amer. Chem. Soc., 102, 3642 (1980).
135. R. Scheffold and R. Orlinski, J. Amer. Chem. Soc., 105, 7200 (1983).
136. R. Scheffold, Nachrichten Chem. Technol., 36, 261 (1988).
137. S. Torii, T. Inokuchi, and T. Yukawa, J. Org. Chem., 50, 5875 (1985).
138. M. Okabe and M. Tada, J. Org. Chem., 47, 5382 (1982).
139. M. J. Begley, H. Bhandal, J. H. Hutchinson, and G. Pattenden, Tetrahedron Lett., 28, 1317 (1987).
140. Y. Murakami, Y. Hisaeda, T. Ozaki, T. Tashiro, T. Ohno, Y. Tani, and Y. Matsuda, Bull. Chem. Soc. Jpn., 60, 311 (1987).
141. V. F. Patel and G. Pattenden, J. Chem. Soc., Chem. Commun., 871 (1987).

142. R. Scheffold and E. Amble, Angew. Chem., 92, 643 (1980); Int. Ed. Engl., 19, 629 (1980).
143. C. Gosden and D. Pletcher, J. Organometal. Chem., 186, 401 (1980).
144. C. Gosden, J. B. Kerr, D. Pletcher, and R. Rosas, J. Electroanal. Chem., 117, 101 (1981).
145. C. Gosden and D. Pletcher, J. Organometal. Chem., 186, 401 (1980).
146. C. Gosden, K. P. Healy, and D. Pletcher, J. Chem. Soc., Dalton Trans., 972 (1978).
147. J. Y. Becker, J. B. Kerr, D. Pletcher, and R. Rosas, J. Electroanal. Chem., 117, 87 (1981).
148. D. Pletcher, Rev. Port. Quim., 23, 59 (1981).
149. M. S. Ram, A. Bakac, and J. H. Espenson, Inorg. Chem., 25, 3267 (1986).
150. A. Bakac and J. H. Espenson, J. Amer. Chem. Soc., 108, 713 (1986).
151. A. Bakac and J. H. Espenson, J. Amer. Chem. Soc., 108, 719 (1986).
152. B. Fisher and R. Eisenberg, J. Amer. Chem. Soc., 102, 7361 (1980).
153. M. Beley, J.-P. Collin, R. Ruppert, and J.-P. Sauvage, J. Amer. Chem. Soc., 108, 7461 (1986).
154. Y. Ohsawa, K. W. Hanck, and M. K. DeArmond, J. Electroanal. Chem., 175, 229 (1984) and Refs. 1–4 cited there.
155. C. O. Dietrich-Buchecker, J.-M. Kern, J.-P. Sauvage, J. Chem. Soc., Chem. Commun., 760 (1985).
156. Y. Fuchs, S. Lofters, T. Dieter, W. Shi, R. Morgan, T. C. Strekas, H. D. Gafney, and A. D. Baker, J. Amer. Chem. Soc., 109, 2691 (1987).
157. Y. Rollin, M. Troupel, D. G. Tuck, and J. Perichon, J. Organometal. Chem., 303, 131 (1986).
158. S. Mabrouk, S. Pellegrini, J.-C. Folest, Y. Rollin, and J. Perichon, J. Organometal. Chem., 301, 391 (1986).
159. W. H. Smith and Y.-M. Kuo, J. Electroanal. Chem., 188, 189 (1985).
160. K. Tamao, K. Sumitani, Y. Kiso, M. Zembayashi, A. Fujioka, S. Kodama, I. Nakajima, A. Minato, and M. Kumada, Bull. Chem. Soc. Jpn., 49, 1958 (1976).
161. W. H. Smith and Y.-M. Kuo, J. Electroanal. Chem., 188, 203 (1985).
162. S. Margel, W. Smith, and F. C. Anson, J. Electrochem. Soc., 125, 241 (1978).
163. S. Margel and F. C. Anson, J. Electrochem. Soc., 125, 1232 (1978).
164. J. F. Rusling and G. N. Kamau, J. Electroanal. Chem., 187, 355 (1985).
165. R. J. Klinger, J. C. Huffman, and J. K. Kochi, J. Amer. Chem. Soc., 104, 2147 (1982).
166. W. Lau, J. C. Huffman, and J. K. Kochi, Organometallics, 1, 155 (1982).
167. T. J. Meyer, J. Electrochem. Soc., 131, 221C (1984).
168. W. Kutner, J. A. Gilbert, A. Tomaszewski, T. J. Meyer, and R. W. Murray, J. Electroanal. Chem., 205, 185 (1986).
169. C.-M. Che and W.-H. Leung, J. Chem. Soc., Chem. Commun., 1376 (1987).
170. G. M. Brown, T. R. Weaver, F. R. Keene, and T. J. Meyer, Inorg. Chem., 15, 190 (1976).
171. W. Schäfer, H.-J. Kerrinnes, and U. Langbein, Z. Anorg. Chem., 406, 101 (1974).
172. M. Troupel, Y. Rollin, S. Sibille, and J. Perichon, J. Organometal. Chem., 202, 435 (1980).

173. G. Bontempelli, S. Daniele, and M. Fiorani, J. Electroanal. Chem., 160, 249 (1984).
174. J. F. Fauvarque, A. Digua, M.-A. Petit, and J. Savard, Makromol. Chem., 186, 2415 (!985).
175. J. F. Fauvarque, A. Jutand, and M. Francois, Nouv. J. Chim., 10, 119 (1986).
176. Y. Rollin, M. Troupel, G. Meyer, and J. Perichon, J. Electroanal. Chem., 183, 247 (1985).
177. P. W. Jennings, D. G. Pillsbury, J. L. Hall, and V. T. Brice, J. Org. Chem., 41, 719 (1976).
178. M. Mori, Y. Hashimoto, and Y. Ban, Tetrahedron Lett., 21, 631 (1980).
179. Y. Rollin, G. Meyer, M. Troupel, J. F. Fauvarque, and J. Perichon, J. Chem. Soc., Chem. Commun., 793 (1983).
180. S. Torii, H. Tanaka, and K. Morisaki, Tetrahedron Lett., 26, 1655 (1985).
181. G. Bontempelli and M. Fiorani, Ann. Chim. (Rome), 75, 303 (1985).
182. J. A. Davies, C. T. Eagle, D. E. Otis, and U. Venkataraman, Organometallics, 5, 1264 (1986).
183. S. Zecchin, G. Zotti, and G. Pilloni, J. Organometal. Chem., 294, 379 (1985).
184. A. Mortreux and F. Petit, Appl. Catal., 24, 1 (1986).
185. M. Onishi, K. Hiraki, Y. Ishida, and K. Dakeshita, Chem. Lett., 333 (1986).
186. J. Chatt, G. J. Leigh, R. A. Head, and C. J. Pickett, J. Chem. Soc., Dalton Trans., 1638 (1978).
187. W. Hussain, G. J. Leigh, and C. J. Pickett, J. Chem. Soc., Chem. Commun., 747 (1982).
188. C. J. Pickett and G. J. Leigh, J. Chem. Soc., Chem. Commun., 1033 (1981).
189. H. Lehmkuhl and W. Eisenbach, Justus Liebigs Ann. Chem., 672 (1975).
190. J. J. Habeeb, D. G. Tuck, and F. H. Walters, J. Coord. Chem., 8, 27 (1978).
191. L. Bustos, J. H. Green, J. L. Hencher, M. A. Khan, and D. G. Tuck, Can. J. Chem., 61, 2141 (1983).
192. N. Kumar and D. G. Tuck, Can. J. Chem., 60, 2579 (1982).
193. L. Matassa, N. Kumar, and D. G. Tuck, Inorg. Chim. Acta, 109, 19 (1985).
194. J. L. Hencher, M. Khan, F. F. Said, R. Sieler, and D. G. Tuck, Inorg. Chem., 21, 2787 (1982).
195. F. F. Said and D. G. Tuck, Inorg. Chim. Acta, 59, 1 (1982).
196. H. Lehmkuhl, K. Mehler, and G. Hauschild, Chem. Ber., 116, 438 (1983).
197. R. Kumar and D. G. Tuck, J. Organometal. Chem., 281, C47 (1985).
198. E. A. Chernyshev, A. V. Bukhtiarov, B. K. Kabanov, A. P. Tomilov, I. A. Rodnikov, N. A. Maier, V. L. Shirokii, and Y. A. Ol'dekop, Elektrokhimiya, 18, 239 (1982).
199. W. Eisenbach and H. Lehmkuhl, Dechema-Monogr., 98, 269 (1985).
200. S. Valcher, Ric. Sci., 8, 1007 (1965).
201. (a) G. Gritzner and J. Kuta, Pure Appl. Chem., 56, 461 (1984); (b) Electrochim. Acta, 29, 869 (1984).

202. A. G. Nagy and S. Toma, J. Organometal. Chem., 266, 257 (1984) and references cited therein.
203. A. Chaloyard, A. Dormond, J. Tirouflet, and N. El Murr, J. Chem. Soc., Chem. Commun., 214 (1980).
204. J.-R. Hamon, D. Astruc, and P. Michaud, J. Amer. Chem. Soc., 103, 758 (1981).
205. D. R. Talham and D. O. Cowan, Organometallics, 6, 932 (1987).
206. M.-T. Lee, B. M. Foxman, and M. Rosenblum, Organometallics, 4, 539 (1985).
207. T. Saji and I. Kinoshita, J. Chem. Soc., Chem. Commun., 716 (1986).
208. D. Astruc, Tetrahedron, 39, 4027 (1983).
209. E. Roman, D. Astruc, and A. Darchen, J. Organometal. Chem., 219, 221 (1981).
210. L. N. Nekrasov, L. P. Yur'eva, and S. M. Peregudova, J. Organometal. Chem., 238, 185 (1982).
211. C. Moinet and E. Raoult, J. Organometal. Chem., 229, C13 (1982).
212. T. Saji and N. Ito, Bull. Chem. Soc. Japan, 58, 3375 (1985).
213. (a) A. M. Romanin, A. Ceccon, and A. Venzo, J. Electroanal. Chem., 130, 245 (1981); (b) A. M. Romanin, A. Venzo, and A. Ceccon, J. Electroanal. Chem., 112, 147 (1980).
214. Y. Lin Chiu, A. E. G. Sant'ana, and J. H. P. Utley, Tetrahedron Lett., 28, 1349 (1987).
215. C. Degrand, M. Bikrani, and B. Gautheron, J. Organometal. Chem., 299, 111 (1986).
216. T. Ikeshoji and V. D. Parker, Acta Chem. Scand., A38, 175 (1984).
217. K. M. Doxsee, R. H. Grubbs, and F. C. Anson, J. Amer. Chem. Soc., 106, 7819 (1984).
218. C. Degrand, F. Gasquez, and P.-L. Compagnon, J. Organometal. Chem., 280, 87 (1985).
219. N. El Murr and E. Laviron, Tetrahedron Lett., 11, 875 (1975).
220. W. E. Silverthorn, J. Amer. Chem. Soc., 102, 842 (1980).
221. M. Tilset, G. S. Bodner, D. R. Senn, J. A. Gladysz, and V. D. Parker, J. Amer. Chem. Soc., 109, 7551 (1987).
222. J. Edwin, W. E. Geiger, A. Salzer, U. Ruppli, and A. L. Rheingold, J. Amer. Chem. Soc., 109, 7893 (1987).
223. G. A. Lane, W. E. Geiger, and N. G. Connelly, J. Amer. Chem. Soc., 109, 402 (1987).
224. R. S. Kelly and W. E. Geiger, Organometallics, 6, 1432 (1987).
225. J. Moraczewski and W. E. Geiger, J. Amer. Chem. Soc., 103, 4779 (1981).
226. M. F. Joseph, J. A. Page, and M. C. Baird, Organometallics, 3, 1749 (1984).
227. R. J. Klingler and J. K. Kochi, J. Organometal. Chem., 202, 49 (1980).
228. R. H. Magnuson, R. Meirowitz, S. J. Zulu, and W. P. Giering, Organometallics, 2, 460 (1983).
229. R. H. Magnuson, R. Meirowitz, S. Zulu, and W. P. Giering, J. Amer. Chem. Soc., 104, 5790 (1982).
230. M. N. Golovin, R. Meirowitz, M. M. Rahman, H. Y. Liu, A. Prock, and W. P. Giering, Organometallics, 6, 2285 (1987).
231. M. J. Therien and W. C. Trogler, J. Amer. Chem. Soc., 109, 5127 (1987).

232. D. Mihholova and A. A. Vlcek, J. Organometal. Chem., 240, 413 (1982).
233. S. Javaheri and W. P. Giering, Organometallics, 3, 1927 (1984).
234. L. C. Hall, C. M. Lukehart, and R. Srinivasan, Organometallics, 4, 2071 (1985).
235. B. A. Narayanan, C. Amatore, and J. K. Kochi, Organometallics, 5, 926 (1986).
236. L. I. Denisovich, S. P. Gubin, A. A. Ioganson, N. E. Kologova, and K. N. Anismov, Izv. Akad. Nauk. SSSR, Ser. Khim., 258 (1969).
237. R. J. Klingler, J. C. Huffman, and J. K. Kochi, Inorg. Chem., 20, 34 (1981).
238. C. J. Pickett and D. Pletcher, J. Chem. Soc., Dalton Trans., 749 (1976).
239. N. El Murr and J. D. Payne, J. Chem. Soc., Chem. Commun., 162 (1985).
240. S. P. Gubin and L. I. Denisovich, J. Organometal. Chem., 15, 471 (1968).
241. N. G. Connelly, R. L. Kelly, M. D. Kitchen, R. M. Mills, R. F. D. Stansfield, M. W. Whiteley, S. M. Whiting, and P. Woodward, J. Chem. Soc., Dalton Trans., 1317 (1981).
242. N. G. Connelly, P. G. Graham, and J. B. Sheridan, J. Chem. Soc., Dalton Trans., 1619 (1986).
243. S. E. Manahan, Inorg. Chem., 5, 2063 (1966).
244. H. Lehmkuhl, E. Janssen, S. Kintopf, W. Leuchte, and K. Mehler, Chem.-Ing.-Technol., 44, 170 (1972).
245. H. Lehmkuhl, W. Leuchte, and W. Eisenbach, Justus Liebigs Ann. Chem., 692 (1973).
246. H. Lehmkuhl, Synthesis, 377 (1973).
247. J. Grobe, B. H. Schneider, and H. Zimmermann, Z. Anorg. Allg. Chem., 481, 107 (1981).
248. G. E. Blackmar (NALCO Co), U.S. Patent 3,573178 (March 30, 1971).
249. G. Mengoli, Rev. Silicon, Germanium, Tin, Lead Compounds, 4, 59 (1979).
250. M. Fleischmann, D. Pletcher, and C. J. Vance, J. Electroanal. Chem., 29, 325 (1971).
251. O. R. Brown, E. R. Gonzalez, and A. R. Wright, Electrochim. Acta, 18, 369 (1973).
252. M. Fleischmann, G. Mengoli, and D. Pletcher, Electrochim. Acta, 18, 231 (1973).
253. O. R. Brown, E. R. Gonzalez, and A. R. Wright, Electrochim. Acta, 18, 555 (1973).
254. H. Lehmkuhl, Chem.-Ing.-Technol., 36, 612 (1964).
255. G. Mengoli and S. Daolio, J. Organometal. Chem., 131, 409 (1977).
256. G. Mengoli and S. Daolio, J. Chem. Soc., Chem. Commun., 96 (1976).
257. W. Sundermeyer and W. Verbeek, Angew. Chem., Int. Ed., 5, 1 (1966).
258. J.-I. Yoshida, K. Muraki, H. Funahashi, and N. Kawabata, J. Organometal. Chem., 284, C33 (1985).
259. J. I. Yoshida, K. Muraki, H. Funahashi, and N. Kawabata, J. Org. Chem., 51, 3996 (1986).
260. T. Shono, Y. Matsumura, S. Katoh, and N. Kise, Chem. Lett., 463 (1985).

261. S. Torii, H. Tanaka, T. Katoh, and K. Morisaki, Tetrahedron Lett., 25, 3207 (1984).
262. T. Troll and G. W. Ollmann, Tetrahedron Lett., 22, 3497 (1981).
263. S. Torii and K. Uneyama, Jpn. Kokai Tokkyo Koho JP [60,184,689]; CA, 104, 118602s.
264. K. Uneyama, H. Matsuda, and S. Torii, Tetrahedron Lett., 25, 6017 (1984).
265. H. Tanaka, S. Yamashita, T. Hamatani, T. Nakahara, and S. Torii, Stud. Org. Chem. (Amsterdam), 30 (Rec. Adv. Electroorg. Synth.), 307 (1987).
266. H. Tanaka, T. Hamatani, S. Yamashita, and S. Torii, Chem. Lett., 1461 (1986).
267. H. Tanaka, S. Yamashita, T. Hamatani, Y. Ikemoto, and S. Torii, Chem. Lett., 1611 (1986).
268. H. Tanaka, S. Yamashita, Y. Katayama, and S. Torii, Chem. Lett., 2043 (1986).
269. H. Tanaka, S. Yamashita, and S. Torii, Bull. Chem. Soc. Japan, 60, 1951 (1987).
270. R. J. Klinger and J. K. Kochi, J. Amer. Chem. Soc., 102, 4790 (1980).
271. H. Watanabe, K. Yoshizumi, T. Muraoka, M. Kato, Y. Nagai, and S. Sato, Chem. Lett., 1683 (1985).
272. E. M. Genies and F. El Omar, Electrochim. Acta, 28, 547 (1983).
273. J.-I. Yoshida, T. Murata, and S. Isoe, Tetrahedron Lett., 27, 3373 (1986).
274. J.-I. Yoshida, T. Murata, and S. Isoe, Tetrahedron Lett., 28, 211 (1987).
275. (a) D. Hermeling and H. J. Schäfer, Angew. Chem., Int. Ed. Engl., 23, 233 (1984); (b) J.-I. Yoshida, T. Murata, and S. Isoe, J. Organometal. Chem., 345, C23 (1988); (c) J.-I. Yoshida and S. Isoe, Tetrahedron Lett., 28, 6621 (1987).
276. R. E. Dessy, W. Kitchin, and T. Chivers, J. Amer. Chem. Soc., 88, 453 (1966).
277. R. J. P. Corriu, G. Dabosi, and M. Martineau, J. Organometal. Chem., 188, 63 (1980).
278. R. J. P. Corriu, G. Dabosi, and M. Martineau, J. Organometal. Chem., 186, 19 (1980).
279. E. M. Genies and F. El Omar, Electrochim. Acta, 28, 541 (1983).
280. E. Hengge and H. Firgo, J. Organometal. Chem., 212, 155 (1981).
281. A. L. Allred, C. Bradley, and T. H. Newman, J. Amer. Chem. Soc., 100, 5081 (1978).
282. C. Eaborn, R. A. Jackson, and R. Pearce, J. Chem. Soc, Perkin I, 2055 (1974).
283. M. Devaud and J.-L. Lecat, Bull. Soc. Chim. Fr., 1187 (1985).
284. B. Fleet and N. B. Fouzder, J. Electroanal. Chem., 63, 59 (1975).
285. M. Devaud, M. Engel, C. Feasson, and J.-L. Lecat, J. Organometal. Chem., 281, 181 (1985).
286. M. Engel and M. Devaud, J. Organometal. Chem., 307, 15 (1986).
287. B. Giese, Radicals in Organic Synthesis: Formation of Carbon-Carbon Bonds, Pergamon Press, London, 1986.
288. M. Ramaiah, Tetrahedron, 43, 3541 (1987).
289. W. P. Neumann, Synthesis, 665 (1987).
290. B. Giese and R. Rupauer, Justus Liebigs Ann. Chem., 231 (1987).

291. J. W. Ryznar and J. G. Premo, NALCO, U.S. Patent 3,100,181 (August 6, 1963).
292. K. Ziegler, Brennstoff-Chem., 40, 209 (1959).
293. E. H. Mottus and M. R. Ort, J. Electrochem. Soc., 117, 885 (1970).
294. J. J. Habeeb, F. F. Said, and D. G. Tuck, J. Organometal. Chem., 190, 325 (1980).
295. G. Silvestri, S. Gambino, and G. Filardo, Adv. Polym. Sci., 38, 27 (1981), and references cited therein.
296. G. Silvestri, S. Gambino, and G. Filardo, Electrochim. Acta, 32, 965 (1987).
297. G. Silvestri, S. Gambino, and G. Filardo, Stud. Org. Chem. (Amsterdam) 30, (Rec. Adv. Electroorg. Synth.), 165 (1987).
298. G. Silvestri, S. Gambino, G. Filardo, G. Greco, and A. Gulotta, Tetrahedron Lett., 25, 4307 (1984).
299. G. Silvestri, S. Gambino, and G. Filardo, Stud. Org. Chem. (Amsterdam), 30, (Rec. Adv. Electroorg. Synth.), 287 (1987).
300. C. Oldham, M. J. Taylor, and D. G. Tuck, Inorg. Chim. Acta, 100, L9 (1985).
301. J. H. Morris, H. J. Gysling, and D. Reed, Chem. Rev., 85, 51 (1985).
302. H. Schäfer and D. Koch, Angew. Chem., 84, 32 (1972).
303. G. Schlegel and H. J. Schäfer, Chem. Ber., 117, 1400 (1984).
304. T. Taguchi, M. Itoh, and A. Suzuki, Chem. Lett., 719 (1973).
305. Y. Takahashi, M. Tokuda, M. Itoh, and A. Suzuki, Chem. Lett., 461 (1980).
306. Y. Takahashi, K. Yuasa, M. Tokuda, M. Itoh, and A. Suzuki, Bull. Chem. Soc. Japan, 51, 339 (1978).
307. Y. Takahashi, M. Tokuda, M. Itoh, and A. Suzuki, Synthesis, 616 (1976).
308. Y. Takahashi, M. Tokuda, M. Itoh, and A. Suzuki, Chem. Lett., 523 (1975).
309. T. Taguchi, Y. Takahashi, M. Itoh, and A. Suzuki, Chem. Lett., 1021 (1974).
310. J. T. Keating and P. S. Skell, J. Org. Chem., 34, 1479 (1969).
311. E. C. Bancroft and H. N. Blount, J. Amer. Chem. Soc., 101, 3692 (1979).
312. W. R. Turner and P. J. Elving, Anal. Chem., 37, 207 (1965).
313. J. H. Riley, D. W. Sopher, J. H. P. Utley, and D. J. Walton, J. Chem. Res. (S), 326 (1982).
314. A. P. Tomilov, S. G. Mairanovskii, M. Y. Fioshin, and V. A. Smirnov, in, Electrochemistry of Organic Compounds, Halsted Press, London, 1972.
315. W. V. Evans and F. H. Lee, J. Amer. Chem. Soc., 56, 654 (1934).
316. W. V. Evans, R. Pearson, and D. Braithwaite, J. Amer. Chem. Soc., 63, 2574 (1941).
317. H. Schäfer and H. Küntzel, Tetrahedron Lett., 38, 3333 (1970).
318. J. L. Hencher, M. A. Khan, F. F. Said, and D. G. Tuck, Polyhedron, 4, 1263 (1985).
319. H. E. Mabrouk, D. G. Tuck, and M. A. Khan, Inorg. Chim. Acta, 129, 75 (1987).
320. C. Boudon, F. Peter, and M. Gross, J. Electroanal. Chem., 135, 93 (1982).
321. L. L. Jackson, J. Osteryoung, and R. A. Osteryoung, Anal. Chem., 52, 66 (1980).

322. C. Boudon, J. P. Gisselbrecht, M. Gross, F. Kotzyba-Hibert, and J. M. Lehn, J. Electroanal. Chem., 202, 191 (1986).
323. K. L. Vieira and D. G. Peters, J. Org. Chem., 51, 1231 (1986).
324. F. F. Said and D. G. Tuck, J. Organometal. Chem., 224, 121 (1982).
325. J. L. Webb, C. K. Mann, and H. M. Walborsky, J. Amer. Chem. Soc., 92, 2042 (1970).
326. B. C. Willett and D. G. Peters, J. Electroanal. Chem., 123, 291 (1981).
327. R. E. Dessy and S. A. Kandil, J. Org. Chem., 30, 3857 (1965).
328. R. E. Dessy, W. Kitching, T. Psarras, R. Salinger, A. Chen, and T. Chivers, J. Amer. Chem. Soc., 88, 460 (1966).
329. K. P. Butin, I. P. Beletskaya, A. N. Kashin, and O. A. Reutov, J. Organometal. Chem., 10, 197 (1967).
330. J.-L. Morgat and R. Pallaud, C. R. Acad. Sci., Ser. C, 260, 5579 (1965).
331. M. Fleischmann, D. Pletcher, and G. M. Race, J. Chem. Soc. (B), 1746 (1970).

VI

ELECTRODE REACTIONS CLASSIFIED BY REACTION TYPE

22

Electrolytic Reductive Coupling

MANUEL M. BAIZER† / University of California, Santa Barbara, and University of California, Los Angeles, California

I. Introduction 879

II. Types of Coupling Reactions 880

III. Mechanisms of Reductive Coupling 884

IV. Reductive Coupling Reactions 885
- A. Hydrocarbons 885
- B. Halogen compounds 886
- C. Nitro and derived compounds 892
- D. Nonolefinic aldehydes and ketones 893
- E. Reductive couplings of olefinic aldehydes and ketones 904
- F. α,β-Unsaturated acids 905
- G. Activated olefins 908
- H. Onium compounds 925
- I. Miscellaneous classes 928

References 932

I. INTRODUCTION

Although a variety of cathodic reactions of organic compounds were discussed in Chaps. 7 through 12, the basis of classification there was the nature of the electrophoric group. A separate examination of reductive coupling [1] as a reaction type is warranted because (1) it is synthetically important in usually advantageously providing di- and polyfunctional molecules, often in high yield, from simpler starting materials; (2) it has provided great stimulus to mechanistic investigations designed to differentiate between EC and ECE sequences (Chap. 3); (3) it has been an active area in the last two decades; and (4) it continues to present many challenging research opportunities.

All the classes of compounds discussed in Chaps. 7 through 12 and several others may under appropriate conditions form reductively coupled products. As a matter of consistency, therefore, reductive coupling reactions are discussed following the order of presentation of the several classes of electrophores in Chaps. 7 through 12. However, there are equally instructive ways of organizing the subject. For example, in a recent review of anodic and cathodic means for forming C–C bonds (including intramolecular reactions), the classification is based on the nature of the electrogenerated intermediate; anion radical, radical, carbanion [2].

†Deceased, July 1988.

Electrolytic reductive coupling (ERC) is interpreted very broadly here as a methodology for forming C–C, C–N, N–N, and N–O bonds, among others. The formation solely of C–H bonds, as in RX + proton source → $RH + X^-$, is excluded because it is more appropriately discussed under Cathodic Substitution (e.g., Chap. 8).

II. TYPES OF COUPLING REACTIONS [1]

Hydrodimerizations

Two identical molecular species (X) are joined to yield a dimer + 2H (HXXH). There are no organic cleavage products.

1. Protonation follows the coupling; for example,

$$2CH_2{=}CHCN + 2H_2O + 2e^- \longrightarrow NC(CH_2)_4CN + 2OH^- \qquad (1)$$

$$2PhCH{=}CH_2 + 2H_2O + 2e^- \longrightarrow Ph(CH_2)_4Ph + 2OH^- \qquad (2)$$

$$2PhCOCH_3 + 2H_2O + 2e^- \longrightarrow (Ph\overset{\displaystyle OH}{\overset{|}{\underset{|}{C}}}{-}CH_3)_2 + 2OH^- \qquad (3)$$

2. Protonation precedes the coupling; for example,

$$2(CH_3)_2CO + 2H^+ \longrightarrow 2(CH_3)_2\overset{+}{C}{-}OH \qquad (4)$$

$$2(CH_3)_2\overset{+}{C}{-}OH + 2e^- \longrightarrow [(CH_3)_2\underset{|}{C}{-}OH]_2 \qquad (5)$$

Mixed (or Crossed) Coupling

Two dissimilar molecules (X and X') are joined with acquisition of 2H (HXX'H).

1. Postprotonation:

$$CH_2{=}CHCN + CH_2{=}CHCOOR \xrightarrow{2e^-} NC(CH_2)_4COOR \quad 2OH^- + \text{other products} \qquad (6)$$

2. Preprotonation:

$$(CH_3)_2CO + H^+ \longrightarrow (CH_3)_2\overset{+}{C}{-}OH \qquad (7)$$

---

*This term was introduced by I. L. Knunyants, Isv. Akad. Nauk, SSSR, Otd. Khim. Nauk (2), 238 (1951).

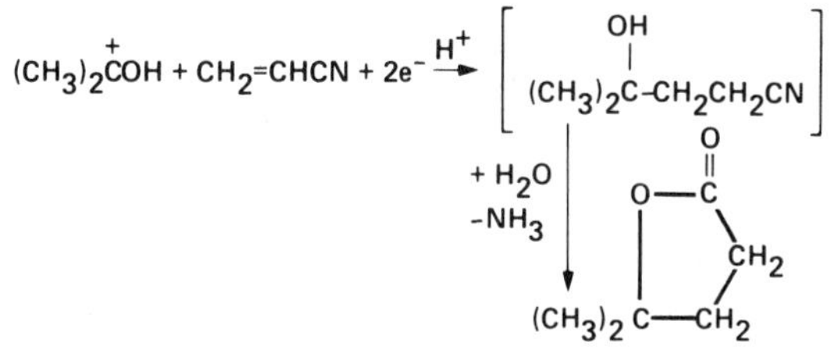

(8)

The coupling partner may be an unactivated olefin.

Compounds Condense in Solution

Electrochemically produced intermediate compounds condense in solution.

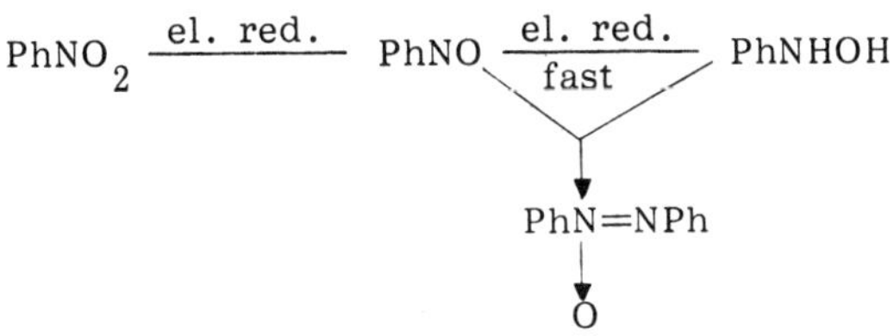

(9)

Coupling with Cleavage

1. The starting compound is reductively cleaved, and the organic fragment undergoes coupling reactions:

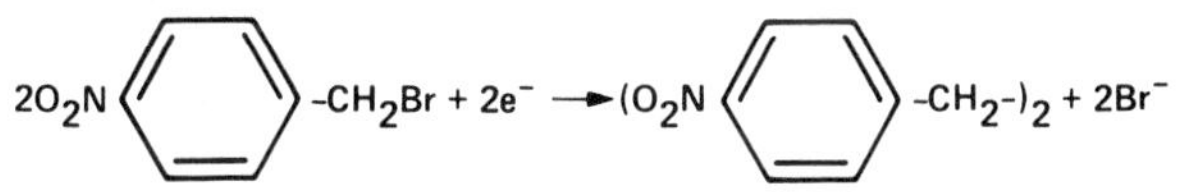

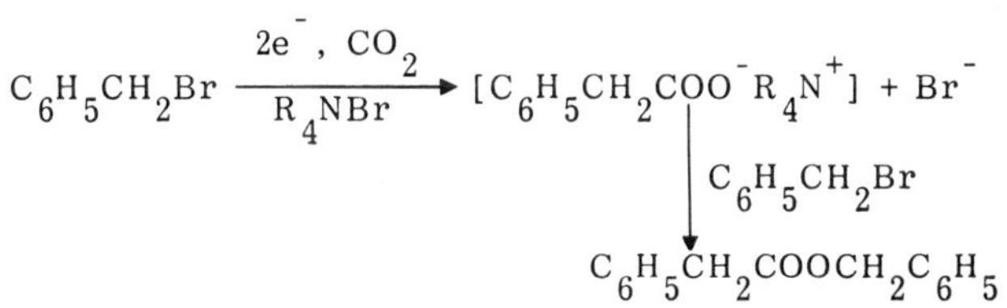

(10)

$$2PhCH_2\overset{+}{N}Me_3NO_3^- \xrightarrow[Al]{2e^-} (PhCH_2^-)_2$$

$$CH_3CN + 2e^- \longrightarrow \frac{1}{2} H_2 + (CH_2CN)^-$$

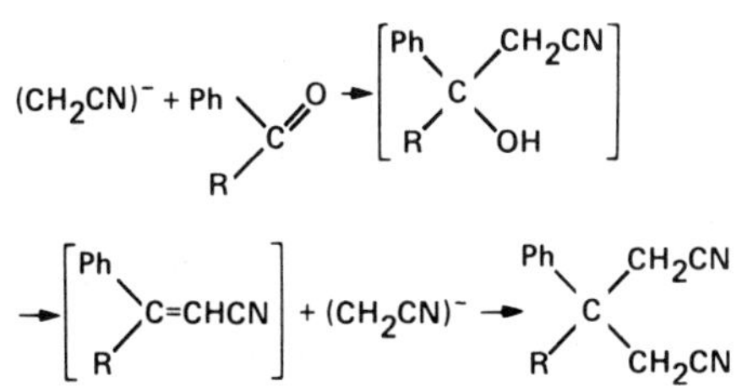

(11)

Reductive acetylation of nitro compounds to amides may fall into category 3 or 4a, depending on the mechanism.

2. The starting compound is reduced intact and then couples with a second species, which cleaves (see nucleophilic displacements).

$$\mathrm{PhCOPh + EtI \xrightarrow{2e^-} Ph_2C(OH)Et + I^-} \qquad (12)$$

$$\mathrm{PhN{=}NPh + 2MeI \xrightarrow[\text{HMPA}]{2e^-} PhN(CH_3){-}N(CH_3)Ph + 2I^-} \qquad (13)$$

3. The starting compound is reductively cleaved, and the organic fragment reacts with the coupling partner by means of a nucleophilic displacement on the latter [3].

$$\mathrm{PhCH(R')\text{-}Cl + R^2COCl \xrightarrow{2e^+} PhCH(R')\text{-}C({=}O)R^2 + 2\ Cl^-} \qquad (14)$$

Indirect ("Mediated") Reductive Couplings (see Chap. 29)

1. A compound A is reduced (usually to the anion radical $A^{\overline{\cdot}}$); the latter transfers an electron to a second component B, which participates in coupling reactions [Eq. (15)]. Under special circumstances [Eq. (16)], the initial electron transfer initiates a chain reaction whose overall result is a catalytic nucleophilic aromatic substitution.

$$\mathrm{Anthracene\ (A) \xrightarrow{1e^-} A^{\overline{\cdot}}}$$

$$\mathrm{A^{\overline{\cdot}} + CH_2{=}CHCN \longrightarrow A + (CH_2CHCN)^{\overline{\cdot}}} \qquad (15)$$

$$\mathrm{2(CH_2CHCN)^{\overline{\cdot}} \longrightarrow NC\bar{C}H(CH_2)_2\bar{C}HCN \xrightarrow{2H^+} \underset{(50\%)}{NC(CH_2)_4CN}}$$

$$\mathrm{A^{\overline{\cdot}} + (CH_2CHCN)^{\overline{\cdot}} \xrightarrow[H^+]{} \underset{(25\%)}{ACH_2CH_2CN}}$$

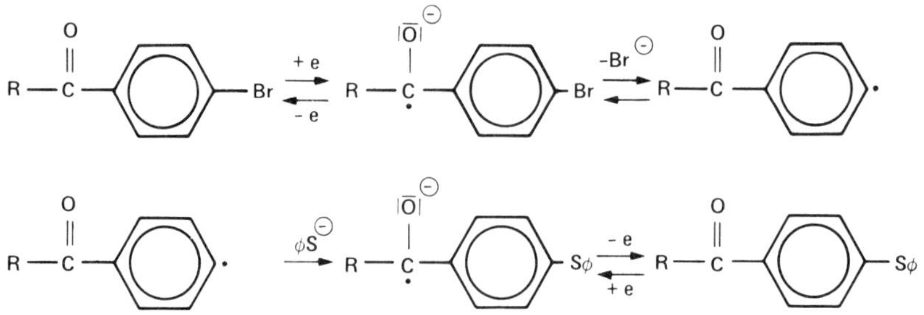

(16)

2. The reduced species is an electrogenerated base EGB^- (anion radical, anion, dianion) that abstracts a proton from a carbon (e.g., nitrogen or sulfur) acid; the resultant anion undergoes coupling reactions (e.g., Eq. (17)]. Here, too, the initial deprotonation may initiate a base-catalyzed condensation sequence [Eq. (18)].

$$(ROOC)_2C{=}C(COOR)_2 + 2e^- \longrightarrow [(ROOC)_2C{-}C(COOR)_2]^{2-}$$
$$[(ROOC)_2C{-}C(COOR)_2]^{2-} \xrightarrow{2\ PhCH_2COOR} 2(PhCHCOOR)^- + (ROOC)_2CH{-}CH(COOR)_2$$
$$2(PhCHCOOR)^- \xrightarrow{2CO_2} 2PhCH(COO^-)COOR \qquad (17)$$

$$2CH_2{=}CHCN + 2e^- \longrightarrow [NCCH(CH_2)_2CHCN]^{2-}$$
$$[NCCH(CH_2)_2CHCN]^{2-} \xrightarrow{2CH_2(COOR)_2} 2[CH(COOR)_2]^- + NC(CH_2)_4CN$$
$$2[CH(COOR)_2]^- \xrightarrow{2CH_2{=}CHCN} 2NC\bar{C}HCH_2CH(COOR)_2$$
$$2NC\bar{C}HCH_2CH(COOR)_2 \xrightarrow{2CH_2(COOR)_2} 2NCCH_2CH_2CH(COOR)_2 + 2[CH(COOR)_2]^- \qquad (18)$$

Intramolecular Reductive Coupling

This is an excellent method for preparing three-, five-, and six-membered carbocyclic and heterocyclic compounds with functionality in side chains. Cleavages may or may not be involved.

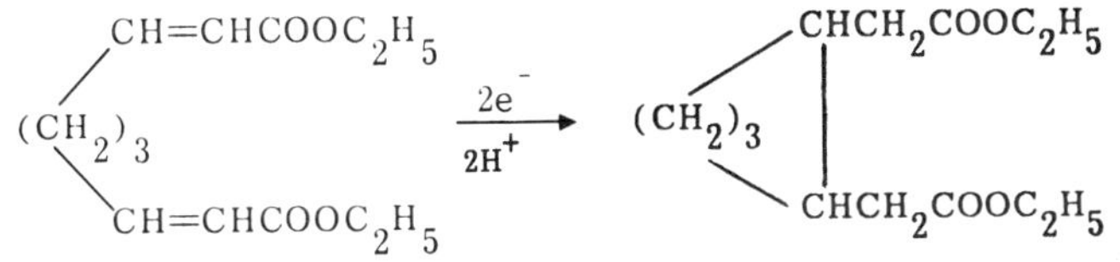

(19)

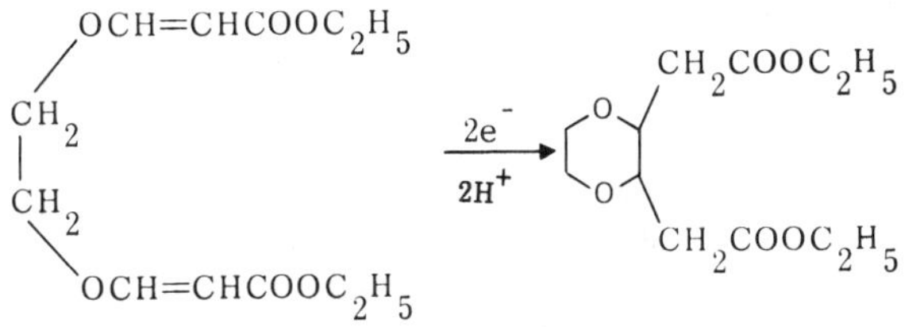

(20)

I II

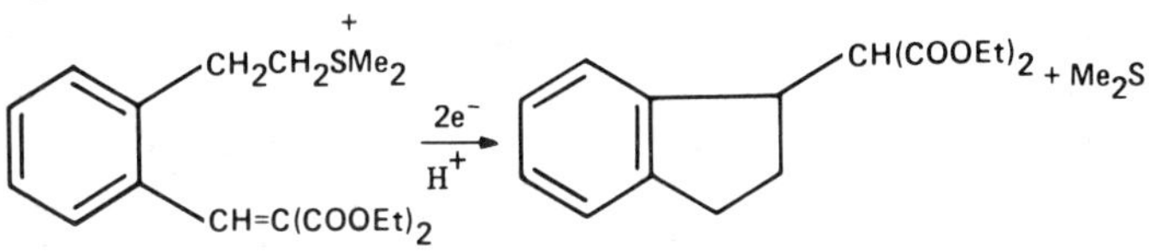

(21)

Cathodic Electropolymerization (Chap. 32)

Whenever an anion radical, anion, or dianion is generated cathodically in aprotic media in the presence of coupling parameters (e.g., activated olefins) that can be nucleophilically attacked without producing a cleavage product (i.e., category 1a or 2a but not 4b), the possibility exists for initiating oligomerization or polymerization via "living anionic" chains. Termination can occur via protonation by traces of impurity or proton transfer (chain transfer) from the monomer.

III. MECHANISMS OF REDUCTIVE COUPLING

The wide variety of reductive coupling types illustrated in Sec. II obviously involves an almost equal variety of reaction mechanisms; these are examined at appropriate places in the text. Except in the cases illustrated by Eq. (9), in which coupling almost certainly occurs as a homogeneous "bulk" reaction, the mechanisms of reductive coupling are complicated because of the heterogeneous nature of electrode reactions. Coupling at the electrode surface or in its vicinity may occur by one route, in the "bulk" by another. Coupling at low pH and with a given electrode material may proceed by a different pathway at high pH and at another electrode. Reaction of an anion radical when it is formed in the presence of a given counterion and in a given degree of association with that counterion may be by one mechanism and when the counterion and/or solvent are changed by a different mechanism. It is almost always appropriate to keep this pluralism in mind. Allied to this complication is the question of establishing the sequence of E and C steps that leads to the ultimate products. The major sequences are (C)EC, ECE(C), and perhaps EEC, but the rigorous establishment of mechanisms for reductive couplings of a given class of compounds has been achieved only occasionally.

To a great extent, incisive mechanistic investigations have become possible only with the development of sophisticated instrumental tools (Chap. 3). Polarography, cyclic voltammetry, chronopotentiometry, ESR spectroscopy, and determination of the structure of products (including their stereochemistry) all contribute to establishment of the final mechanistic conclusions. As Zuman has forcefully pointed out, however, the utilization of only one or two of these techniques in the study of a given problem may lead to completely erroneous rationalizations. Happily, model extensive investigations have now been carried out at least for 1,2-diactivated olefins [4]; it is questionable whether the conclusions reached also apply to mono-activated olefins.

As implied, the reduced intermediate directly involved in the coupling may be a radical, a radical anion, an anion, or a dianion; the coupling may be (1) radical dimerization, (2) radical anion dimerization, (3) anionic or radical attack of a radical anion upon substrate, (4) radical-radical anion addition, or (5) anionic attack upon substrate. It is particularly difficult to distinguish between those EC and ECEC mechanisms that both conceptually lead to the same symmetrical product (e.g., hydrodimer of diethyl maleate).

Success in achieving satisfactory yields in reductive coupling reactions obviously depends on minimizing the probability that the intermediate formed by electroreduction will find more suitable pathways to stable products. Radicals may react with the electrode material to yield organometallics (Chap. 21). Radicals formed from olefinic structures are reduced to anions at the same potential as that at which they were formed. Radical anions may disproportionate to dianions and starting materials. Radical anions on protonation are "free" radicals [5]. However, under proper experimental conditions, good to excellent yields of reductively coupled products have been obtained.

IV. REDUCTIVE COUPLING REACTIONS

A. Hydrocarbons

Stilbene formed 30% of its reduced dimer ("hydrodimer") upon reduction in anhydrous DMF at a mercury cathode with tetrabutylammonium electrolyte [6]. The intermediate anions or dianions are, however, protonated when acetonitrile is the solvent, so that only saturated monomer is obtained. More sterically hindered vinyl compounds (triphenylethylene and tetraphenylethylene) do not form hydrodimers.

Aromatically substituted ethylenes, which are less hindered than stilbene (e.g., styrene and α-methylstyrene), are subject to anionic polymerization (Chap. 32), and therefore reduction of these monomers to hydrodimers and other low oligomers requires careful control of proton donor concentration. Whether the dianion $C_6H_5\bar{C}HCH_2CH_2\bar{C}HC_6H_5$ responsible for initiation of "living" anionic polymerization arises by dimerization of anionic radicals (EC) or by anion radical attack upon monomer followed by a second reduction (ECE) is not yet clear [7]. α-Methylstyrene tetramers, whose structure has been the subject of some controversy [8], have been prepared by the reaction of sodium with the monomer. Electrochemical preparation, particularly since it allows a wider choice of counterions, would be interesting.

The conversion of styrene to 1,4-diphenylbutane (among other products) by electron transfer from $O_2^{\cdot-}$ to the monomer has been reported [239].

Reduction of hydrocarbons in the presence of carbon dioxide (Chap. 7) is sometimes [6] an excellent preparative method for dicarboxylic acids (e.g., stilbene → 92% *meso*-diphenylsuccinic acid) but gives poor results when the voltage required (e.g., for butadiene) is close to that for the discharge of the supporting cation [9].* These reactions proceed presumably via carbanionic intermediates. Successful reductive coupling of CO_2 and butadiene [11], if it led largely to hexenedicarboxylic acid catalytically reducible to adipic acid, would be industrially important because the latter is now made by (1) catalytic reduction of benzene to cyclohexane, (2) catalytic air oxidation of cyclohexane to a mixture of cyclohexanol and cyclohexanone, and (3) HNO_3 oxidation of this mixture.

The intriguing formation of *succinic* as well as oxalic acid by reduction of ethylene and aqueous CO_2 presents new synthetic possibilities [10].

There are now numerous examples of reductive couplings between polycyclic aromatics and yet other electrophiles. Older work has shown that stilbene and ethyl iodide gave a poor yield of 1,2-diphenylbutane [6]. Electron transfer from an alkali metal, not from an electrode, was used to prepare naphthalene anion radical and dianion, which underwent interesting coupling reactions with α-mono- and α,ω-dihalides. Evidence has been ad-

*Since CO_2 is electroreducible at approximately -2 V (SCE) at Hg, the mechanism of these trapping reactions is uncertain.

duced that the coupling is not due to a nucleophilic attack of the anion radical on RX but rather an electron transfer followed by radical coupling [12]. Anthracene has been reductively acylated [13] and also coupled with 1,3-dihalides [14]. Most interesting is the cathodic t-butylation of pyrene, which yields 52% 1-t-butylpyrene [15] due to facile autooxidation of the reduced intermediate.

Highly conjugated hydrocarbons like 9-benzalfluorene (I), 2-phenyl-1,3-butadiene, and 8,8-diphenylbenzofulvene (II), which are types known to be Michael acceptors [16], are rather easily electroreducible and form coupled products with activated olefins [17]. Thus, I and II with acrylonitrile formed III and IV, respectively. Hydrodimerization was also achieved with I but not with II.

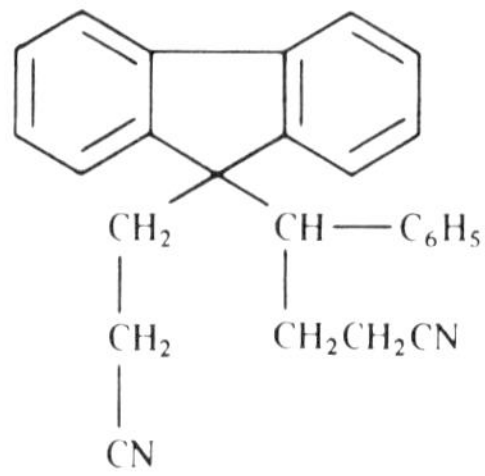
(III)

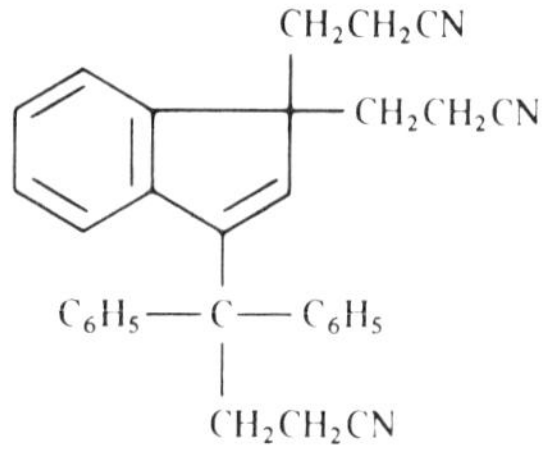
(IV)

Phenylacetylene in the presence of Bu_4N^+ has been electroreduced to phenylacetylide anion, which can be alkylated with RX or caused to react with acetone to yield $PhC{\equiv}C(OH)(CH_3)_2$ [18]. Likewise, reduction of cyclopentadiene (presumably with liberation of hydrogen) leads to the anion, which has been alkylated [19].

B. Halogen Compounds

In the fullest sequence—and very few experimental examples showing all the features have been provided—an organic halide (V) undergoes a one-electron reduction via the electrode or homogeneous electron transfer [20] to produce an intact radical ion (VI); the radical ion then fragments and the radical formed (VII) may take up another electron and form an anion (VIII) (Chap. 8).

$$\underset{V}{RX} + e^- \longrightarrow \underset{VI}{[RX]^{\overset{\cdot}{-}}} \longrightarrow \underset{VII}{R^\cdot} + X^- \xrightarrow{e^-} \underset{VIII}{R^-} + H^-$$

The possible routes to dimers are then (1) radical recombination of VII; (2) anionic attack of VIII upon V; or (3) interaction of two molecules of VI to yield R—R and $2X^-$. The bulk of the experimental evidence has been interpreted on the basis of 1 and/or 2. The common use of a mercury cathode and the frequent intervention of mercurial intermediates has complicated the elucidation of the experimental results. Many examples are cited in Chap. 8.

The halonitrobenzenes [21], because their structure and behavior are favorable for the purpose, have been studied extensively from the mechanistic point of view. They are discussed in Chap. 8. Diphenyl 4-nitrobenzyl phosphate behaves similarly [22].

Certain polyhalo compounds undergo reductive dimerization, probably via carbene intermediates. Thus, dichlorodifluoromethane was reduced with high current efficiency at 5% conversion in eutectic LiCl-KCl melts using porous C or Ni cathodes. The product was 1,1,2,2-tetrafluoroethylene [23]. Likewise, 9,9-difluorene formed difluorylidene [24,25].

1,2-Dibromobenzocyclobutene [26] and α,α,α',α'-tetrabromo-σ-xylene [27] were reduced to benzocyclobutadiene dimer:

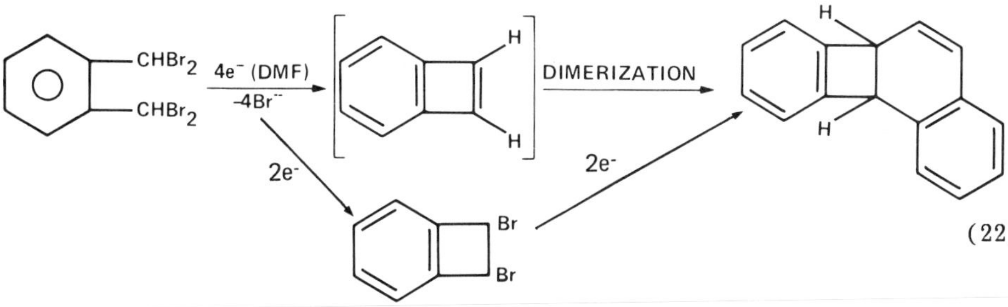

(22)

Couplings with RX in which RR is not formed are classified here as mixed reductive couplings involving intermediates VII or VIII. They may be intra- or intermolecular. In the former case, cyclic products are formed and only RX has been involved; in the latter, another moiety becomes bonded to VII or VIII.

Intramolecular

The proposal [28] that α,ω-dihaloalkenes form cycloalkenes by a $2e^-$ cleavage of one C—X bond to form a carbanion that in a concerted process displaces the second X has been largely invalidated by further incisive experimental work [29]; a two-step process with discrete carbanionic intermediates has been established.

Reductive cyclization of dihalo compounds has been used to prepare cyclopropenes [30], a propellane [31], and a tricylene [32]:

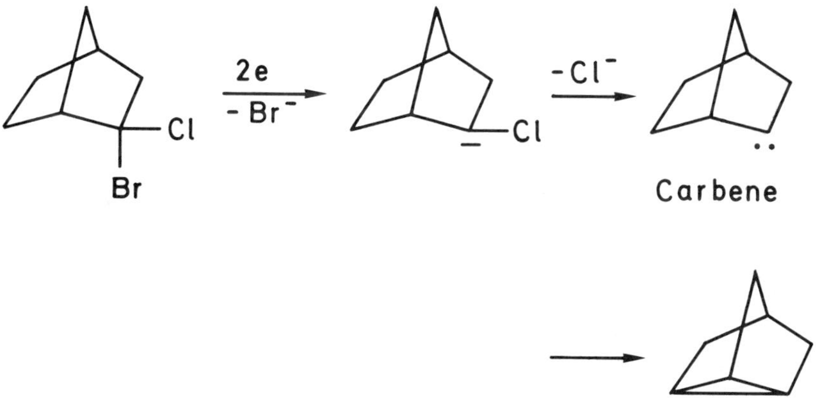

(23)

Besides being applied with elegant results to the synthesis of small-ring compounds, the method has been applied with beautiful synthetic results to the construction of complex heterocyclic systems, including some that are found in natural products, for example,

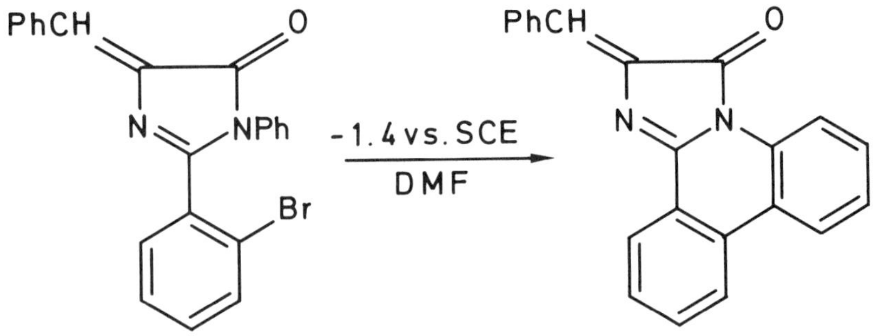

(24)

(25)

A number of 2-halo-N-methylbenzanilides (IX) in which X = Cl, Br, or I; R^1, R^2, and R^3 are H or Me or MeO; and the N-Me may be of syn or anti conformation undergo electroreduction to yield an intermediate radical (X) [33,34], which may (1) be further reduced and protonated to yield the product of IX in which X has been replaced or by H or (2) attack the adjacent ring at the C atom bearing the R_3, leading eventually to a phenanthridone, or (3) attack the C atom bearing the N-methyl, leading eventually to a diphenylcarboxamide. Highest yields of cyclized products result when X = Cl and when the aniline component bears no o-substituent.

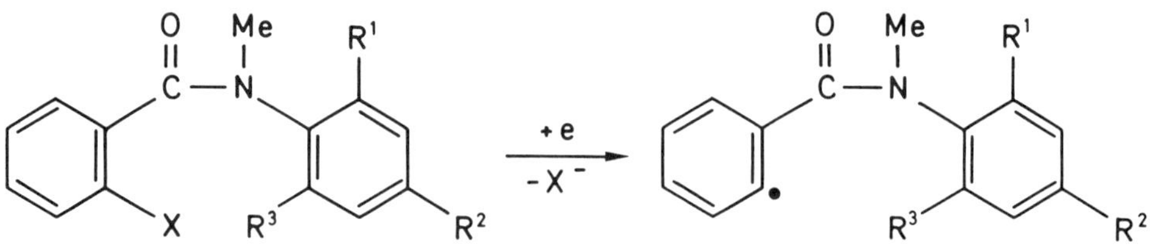

(IX) (X)

(26)

Concerning the tendency of intermediate radicals like X to undergo intramolecular attack, the Grimshaws [35] have given the following generalization: Stable radical anions that fragment in solution do not incorporate D from D_2O; unstable radical anions that fragment at the electrode surface (and the resultant radical is further reduced to the anion) do incorporate D.

The reductive cyclization of iodobenzylisoquinolinium salts to aporphine and apomorphine has been nicely illustrated in the conversion of iodopapaverine methiodide to (±)-glaucine [36].

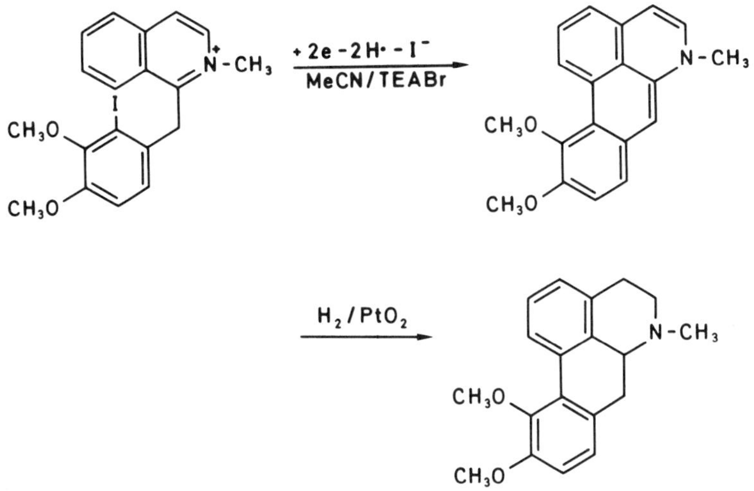

(27)

The synthesis is not only elegant but intriguing because the same product has been made from (±)-laudanosine by an electrooxidative coupling followed by chemical reactions [37].

Although all these cyclizations proceed by initial reductive cleavage of C—X bonds, other routes to cyclized products are taken by compounds that have an additional electrophore in the molecule, one that is more easily reduced than the C—X moiety; for example, 6-bromo-1-phenylhexyne and 7-bromo-1-phenyl-1-heptyne yield benzalcycloalkanes [38]. The behavior of 6-chloro-1-phenyl-1-hexyne is more complex [39]; at low concentrations (2.5×10^{-4} M) the same cycloalkane is formed, accompanied by C-Cl cleavage, presumably via intramolecular electron transfer; at higher concentrations (>0.002 M) induced isomerization to 6-chloro-1-phenyl-1,2-hexadiene and reduction led to four-, five-, and six-membered carbocycles.

Intermolecular

Many workers have used CO_2 to trap carbanions formed by $2e^-$ cleavage of RX to $R^- \rightarrow RCOO^-$. When R is benzyl or allyl, reduction of the halide in the presence of CO_2 and R_4N^+ at a cathodic voltage less negative than CO_2 reduction leads cleanly and directly to esters [40].

$$RX \xrightarrow[R_4'N^+]{2e^-} [R^- R_4'N^+] \xrightarrow{CO_2} [RCOO^- R_4'N^+] \xrightarrow{RX} RCOOR + R_4'NX \qquad (28)$$

When the starting material is a bis-halomethylaromatic, a similar procedure is a one-step synthesis of polyesters, for example,

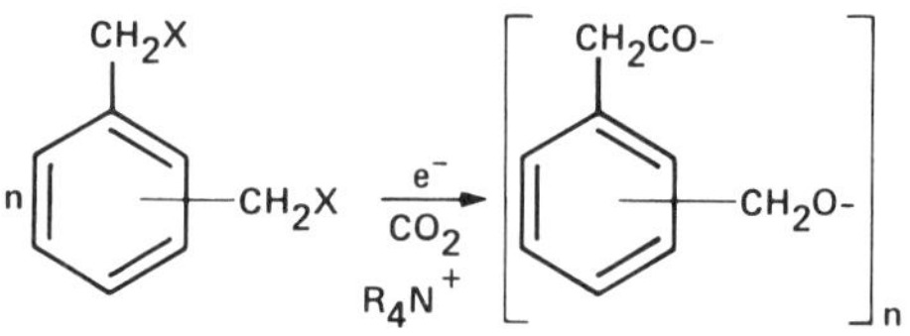

(29)

When, however, RX is an alkyl derivative, such as n-butyl bromide, and the derived anion (Bu^-) is less stable than those above, the expected ester RCOOR is formed accompanied by other products, among them dibutyl butylmalonate [41].

Reductive alkylations of hydrocarbons with RX were mentioned in Sec. IV.A.

The reductive cleavage of RX, in which R bears other functional groups, in the presence of activated olefins has yielded interesting synthetic results. α-Monohaloketones (bromoacetone and α-monochlorocyclohexanone) in the presence of acrylonitrile yielded α-mono (β-cyanoethyl)ketones in a divided cell with Pb cathodes, Pt/Ti anodes, and R_4N^+/MeOH or H_2O [42]. In a more extensive study [43], halo compounds were chosen from the group CCl_4, CCl_3COOEt, $BrCH_2COOEt$, $CH_2{=}CHCH_2Br/Cl$, $PhCH_2Br/Cl$, $Br(CH_2)_3COOEt$, $Br(CH_2)_3CN$, and $ClCH_2CN$ and Michael acceptor olefins from the group $CH_2{=}CHCN$, $CH_2{=}CHCOOEt$, and diethyl fumarate/maleate. The partners were chosen on the basis that the RX underwent a $2e^-$ reduction less negative than the voltage at which the olefin was reduced. A divided cell with a Hg cathode and DMF/R_4N^- SSE was usually used. Representative results are given below:

$$CCl_4 + CH_2{=}CHCN \xrightarrow[\text{proton donor}]{2e^-} Cl_3CCH_2CH_2CN + Cl^-$$

This became electrocatalytic (i.e., a chain reaction) when the proton source was $CHCl_3$ since $\bar{C}Cl_3$ was thereby generated. It might be interesting to use CH_2Cl_2 as the solvent.

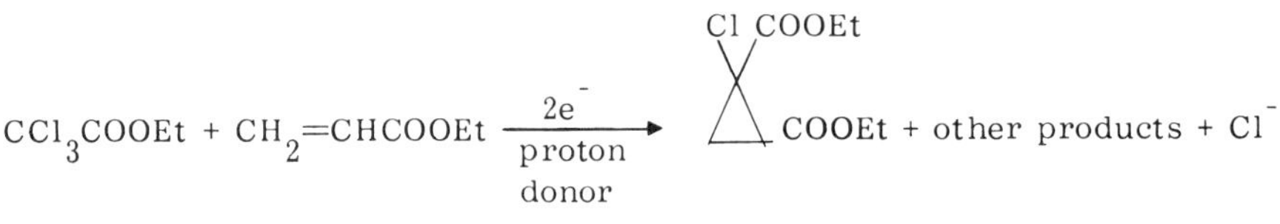

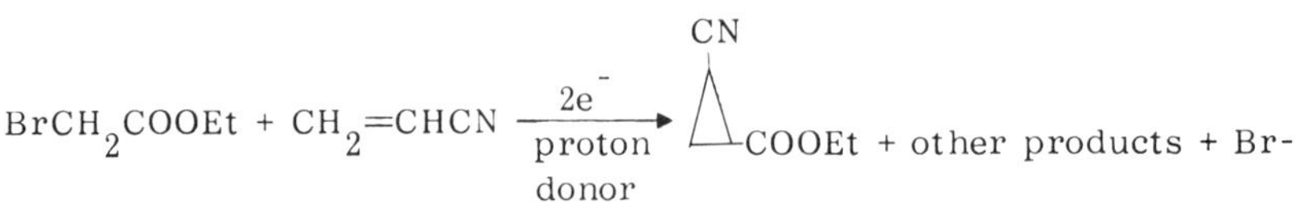

The older literature has several examples of reductive coupling between carbonyls and alkyl halides; electrolysis of benzaldehyde in the presence of ethyl iodide yielded a small quantity of the ethyl ether of phenyl ethyl carbinol [195], and electrolysis of benzophenone in the presence of ethyl iodide yielded diphenyl carbinol [196]. More recently, reductive ethylation of aliphatic ketones in Et_4NBF_4-NH_3 has been reported [197]. Cyclic and linear ketones have been alkylated with primary, secondary, and tertiary alkyl bromides [218]. α,β-Unsaturated ketones, such as EtC(COPh)=CHMe, have been formed by reduction of $PhCCl_3$ in the presence of 3-pentanone [219].

Reduction of polyhalo compounds (CCl_4 and CCl_3COOEt) in the presence of ketones yielded the expected *tert*-alcohols [43]. Condensation with aldehydes can, as above, be made into chain reactions (current efficiency approximately 10^4) when $CHCl_3$ is used as a proton donor in the reduction of CCl_4 in the presence of aldehydes [44]. Reduction of CCl_4 and $Cl_2CHOOEt$ in the presence of R(R')CHO yields the expected adducts stereoselectively [45]. A related reaction leading to olefins has been reported [46].

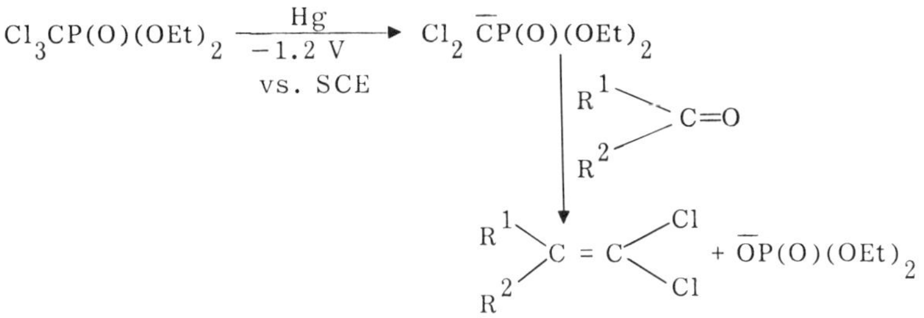

(30)

As has been mentioned [12–15], [20], electrogenerated aromatic anion radicals may homogeneously transfer an electron to RX, which cleaves; among the products are alkylated aromatics.

Reduction of benzylic or allylic halides led, in the presence of aldehydes and ketones, to good yields of the expected alcohols, such as from $PhCH_2Cl$ and Me_2CO in DMF using a Mg anode; in the presence of anhydrides moderate to good yields of ketones were obtained [48]. Alkyl benzyl ketones were formed by the reductive coupling of acyl chlorides and benzyl chlorides [49].

Reduction of α-bromopropiophenone [50] and of methyl or ethyl 2-bromo-2-phenylpropionate [51] leads, with elimination of bromide, to the dimers.

When SO_2 is reduced in the presence of halides, sulfones are formed [52]; in the presence of dihalides polysulfones are obtained [52b]. Sulfonyl halides with other electrophiles also yield sulfones [53].

The electron-induced displacement of X in ArX by nucleophiles [see Ref. 54 and Eq. (16)], similar to Bunnett's $S_{RN}1$ reaction [54d], has been demonstrated for halobenzophenones, benzonitriles, and 1-naphthalenes using thiolate and cyanide anions [54a]. The process becomes "electrocatalytic" if E_2°, the standard potential for oxidation of the product anion radical [Eq. (16)], for example, is smaller than E_1°, the standard potential for the reduction of starting material; otherwise, the penultimate anion radical must be oxidized to product separately.

A related reaction has been reported by Russian workers [55]:

$$\underline{p}\text{-}NO_2C_6H_4Cl + PhS\text{-}SPh \xrightarrow{e^-} \underline{p}\text{-}NO_2C_6H_4SPh$$

It has also been used in the preparation of $PhMC_6H_4CN$ (M = Se, Te) from $(PhM)_2$ and *p*-bromobenzonitrile [56]. Both involve double cleavages.

Allyl and benzyl silanes have been prepared by the reduction of the corresponding halides in the presence of Me_3SiCl [57].

Mention should be made of the reductive couplings that have been reported using certain N-halo compounds. N-haloimides having a suitably placed double bond in the molecule on reduction form cyclic compounds via the intermediate radical [58]. N,N-dichloro-p-toluenesulfonamide yielded products presumably through an intermediate nitrene [59].

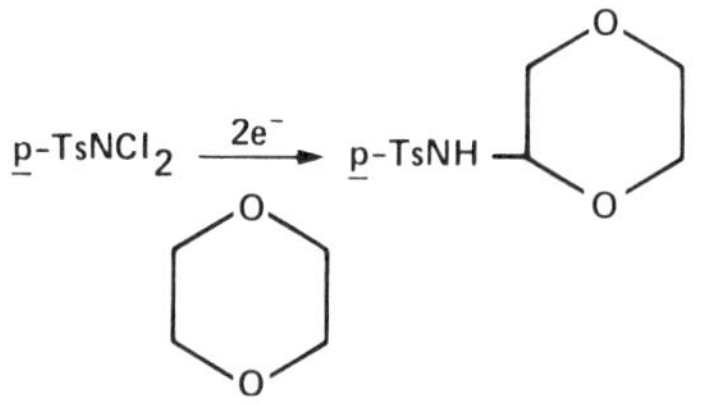

(31)

Likewise, Cl_2NCOOR reductively condensed with dioxane via the carbalkoxynitrene [60].

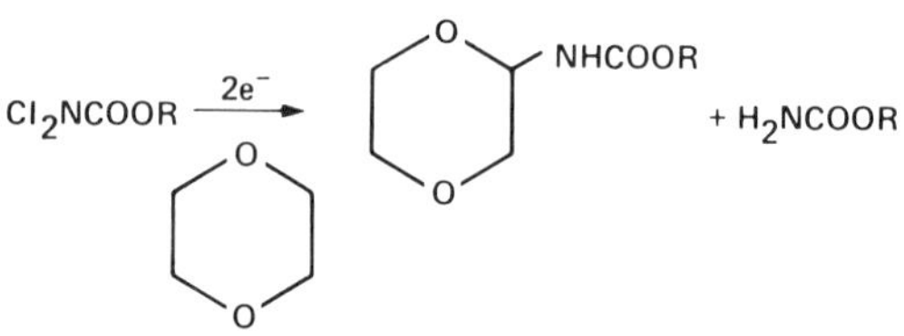

(32)

C. Nitro and Derived Compounds

The intermolecular reductive coupling of two molecules of identical nitro compounds ("homo" coupling) to azoxy, azo, and hydrazo derivatives is discussed in Chap. 9. The intramolecular reductive coupling of nitro compounds, which leads to heterocycles (e.g., 2,2'-dinitrobiphenyl to benzocinnoline), is discussed in Chap. 18. These types, as well as mixed coupling of nitro compounds with other species, furnish syntheses of compounds accessible only with difficulty by other means. Reduction with or without coupling of nitroaryl compounds also bearing other functional groups, such as halides, quaternary ammonium, and quaternary phosphonium, offers information of value for both synthesis and mechanism.

Among recent homo couplings may be mentioned the two-cell procedure [61] for preparing 4,4'-azobenzene dicarboxylic acid: p-nitrobenzoic acid is reduced in aqueous alkaline solution in the presence of Pb^{2+} to the hydrazo compound; in the second cell the hydrazo compound is oxidized at Pt, Ni, or other corrosion-resistant material. Both cells are undivided.

It has been reported [62] that two molecules of 9-nitroanthracene reductively couple through the 10-positions. Mixed coupling between $PhNO_2^{\overline{\cdot}}$ and RX is a novel route N,O-dialkylphenylhydroxylamines [63]; if RX is 1.5-dibromopentane an oxazepine is formed [63]. Reductive acylation of nitro compounds [64] analogously leads to the N,O-diacylated derivative.

Of the many intramolecular reductive couplings of nitro compounds (Chap. 18) the following may serve as examples.

A 2-nitro-2'-isothiocyanotobiphenyl has been converted to a diazepin oxide [65]:

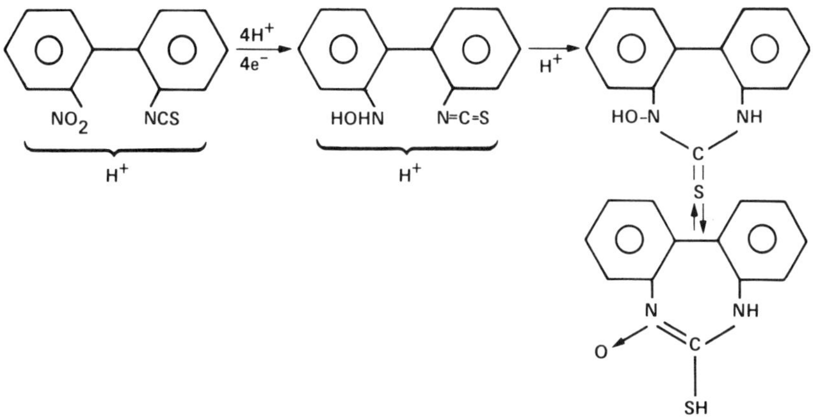

(33)

Reduction of *o*-nitrobenzylamines to the hydroxylamine stage and anodic oxidation of the latter (via the intermediate nitroso compound) is a route to indazoles [66]. An indazole monoxide is formed when 1,8-dinitronaphthalene is reduced and the intermediate is oxidized by air or an anode [67].

Reduction of γ-nitroketones leads to pyrroline *N*-oxides, which, under other conditions, can form pyrrolines and pyrrolidones [68a]. δ-Nitro ketones yield pyrrolines and pyrrolidones [68b].

Nitrophenylonium compounds, such as *m*- and *p*-triphenylphosphonium or trimethylammonium salts, give stable one-electron reduction products; the radicals formed have been observed by ESR [69]. *p*-Nitrobenzyltrimethylammonium salts, however, give a quantitative yield of *p*,*p*'-dinitrobibenzyl in HMPT at any cathode [70].

Reductive couplings with azo compounds are discussed in Sec. I.

D. Nonolefinic Aldehydes and Ketones

Hydrodimerizations

Direct and indirect (amalgam) reductive couplings of carbonyls to *gem*-diols (pinacols) have been the subject of extensive studies (see Refs. 71 and 72 and Chap. 9).

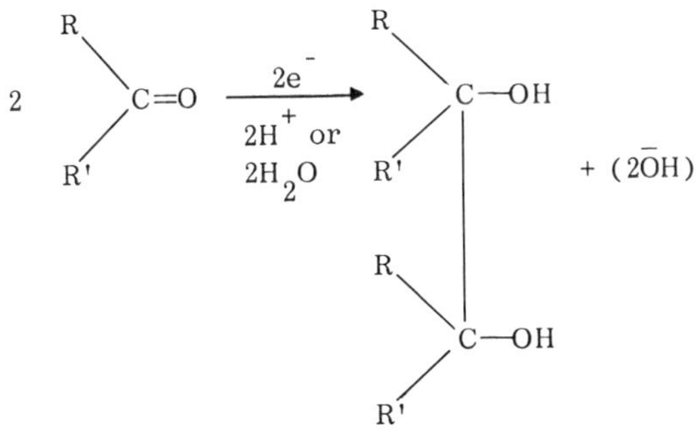

(34)

where:

R = alkyl or aryl
R' = alkyl or aryl or H

The analogous intramolecular reaction from suitably constructed diketones is also well known [73–76]. Because of the great variety of conditions that have been used by a host of workers, no simple statement can be made concerning the optimum method for achieving the dimeric products except that aromatic compounds have given more satisfactory results than aliphatic compounds. The principal side reactions are reduction to alcohol, reduction to hydrocarbon, formation of organometallics, and rearrangement of the pinacol to its pinacolone. When "active methylene" is present or is generated in the presence of alkali, additional chemical side reactions are expected.

Representative hydrodimerizations of aldehydes and ketones are given in Tables 1 and 2, respectively. Despite the long history of this particular reaction, new investigations are continuing. Interest focuses mainly on (1) the products per se or as intermediates; (2) the mechanisms of the coupling, particularly the question of dimer formation in bulk solution or at the electrode surface; (3) the influence of counterions on the pinacol/alcohol ratio [77]; (4) the factors affecting the d,l-meso ratio [77]; and (5) improvements of yields and current efficiencies.

A 1979 discovery [78] disclosed that the presence of Cr(III) strongly favors pinacolization over alcohol formation from saturated or polyene carbonyl compounds. Presumably, complex formation near the electrode between Cr(III) and the carbonyl compound is involved; the reduction potential is 0.2–0.3 V less cathodic than required in the absence of Cr(III) (see also Chap. 29). The mechanism has now been further investigated [79]. There is also enhancement of pinacolization at the expense of alcohol formation when other transition metal cations, such as Cr(III), Mn(II), Fe(II), Co(II), and Zn(II) [but not Ni(II)] are present [80]. The results with Mn(II) are described in Ref. 81 and with Fe(II) in Ref. 82.

Biacetyl, an example of 1,2-diketones, hydrodimerizes through one of the carbonyl groups [83]. DMF and a glassy C cathode were used.

Products of Special Practical Interest

Formaldehyde and acetaldehyde have been electroreduced at pH 2–5 at a graphite cathode in a KH_2PO_4 solution containing Hg^{2+} [84]. The respective hydrodimers were obtained in 46.2% yield, 24.9% current efficiency and 70.7% yield and 69.6% current efficiency. Substantial improvements have been made in the electropinacolization of acetone in both divided [85] and undivided cells [86]. At one time pinacol itself, the hydrodimer of acetone, was considered a favorable intermediate to 2,3-dimethyl-1,3-butadiene for synthetic rubber. A great effort is now being made to hydrodimerize formaldehyde to ethylene glycol, a large-tonnage, relatively inexpensive product [87]. Oxidized graphite surfaces provide the best cathodes [87a]. The current work, unlike the original [84], is facing formidable competition from attempts to prepare ethylene glycol catalytically at a competitive cost from CO + H_2.

Results Relevant to Elucidations of Mechanism

Experimentally, it has been found [88] that almost quantitative yields of pinacols were formed when acetophenones are reduced in very dry CH_3CN in the presence of R_4N^+; the authors suggest that intimate ion pairing between the anion radical and the cation is involved. Reduction of $PhCOCH_3$ at the first wave in DMF containing PhCOOH yielded the pinacol [89]; yields of the pinacol produced in aqueous alkaline solution did not benefit from the presence of R_4N^+ but were improved when EtOH or DMF was added [90]. Other solvent, salt, ion pairing, and adsorption effects have also been studied [91]. The entropy effects in the formation of anion radicals related to benzophenone have been discussed [92].

The hydrodimerization of benzaldehyde in buffered solution was investigated by linear sweep voltammetry [93]; it was concluded that the anion radical is protonated and the neutral radicals then dimerize. On the basis of the results obtained with the rotating disk, it was shown that numerous pathways are available for the anion radicals to form by-products from PhCHO, PhCOPh, and 17 derivatives [94]. Rate constants for the dimerization of aromatic radical anions and association constants for ion pair intermediates from aromatic aldehydes were determined by cyclic voltammetry in DMF and MeCN [95].

The role of medium components (DMF, DMF + H_2O, MeCN, MeCN + H_2O; Et_4N, Bu_4N, Li perchlorates, $LiNO_3$, and HOAc) in the EHD of α-thiophenaldehyde has been studied [96]. An 80% yield of hydrodimer was obtained in DMF containing equimolar Bu_4NClO_4 and $LiClO_4$. Solvation and ion pairing of the intermediate anion radicals are of critical importance in determining the course of the reaction.

There is increased agreement with the view that alcohol formation takes place at the electrode surface and pinacol formation in solution [97].

Data on the d,l-meso ratio of the pinacols obtained have often been used to support a given concept relating to the mechanism of coupling or to asymmetric induction [97,98]. Optical induction has been mediated by chiral electrodes (cited in Ref. 98), by including alkaloids in the electrolyte (cited in Ref. 97), by chiral electrolytes [98], and by a chiral solvent [98]. Many of the conclusions may have to be reassessed in view of a study [99] that analyzed the pinacol from p-hydroxybenzaldehyde by high-pressure liquid chromatography—without prior purification of the reaction mixture—and found that the d,l-meso ratio remained in the range 0.77–0.94 in spite of considerable changes in the electrolysis conditions. Previous reports had indicated that almost pure meso isomer was obtained in aqueous base and pure dl in acid [174].

The d,l-meso ratio of the pinacols obtained in EHD of 4-pyridine carboxaldehyde depended on the pH of the electrolyte and changed from 0.59 to 0.34 to 1.97 at values of 6.1, 10.6, and 13.3, respectively. This is explained by changes in surface orientation of the radical cations, radicals, and radical anions present at the respective pH levels [100].

Stereochemical data relating to the pinacol reduction are only briefly mentioned here, because a separate chapter (27) is dedicated to stereochemistry. Usually, electroreduction of aldehydes and ketones leads to a mixture of diastereoisomeric pinacols, and the stereoselectivity is rather low. There is no firm evidence that benzaldehyde reduction in alkaline medium [106,107,160] resulted in only one isomer of hydrobenzoin. However, in some cases the predominant formation of a single (probably meso) isomer was observed; for example, in the reduction of vanillin and 2-hydroxy-3-methoxybenzaldehyde [161]. It was demonstrated recently that, contrary to earlier publications [111,162], pinacolization of p-hydroxybenzaldehyde leads to a product with a d,l-meso ratio of 0.77–0.94 [99]; maybe this work will stimulate a thorough inspection of existing data.

Of the total pinacol yield of 97% in the case of p-dimethylaminobenzaldehyde, the fractions of high- and low-melting forms are 39 and 58%, respec-

Table 1. Typical Electroreductions of Aldehydes to Glycols: $-CHO + e^- + H^+ \longrightarrow -CHOH-CHOH-$

Aldehyde	Cathode; E, V (SCE)	Medium; conditions	Yield (current efficiency), %	References
Formaldehyde	Mercury, graphite	KH_2PO_4	46.2 (24.9)	84, 87
Acetaldehyde	Mercury, graphite	KH_2PO_4	70.7 (69.6)	84
Propanal	Sn	pH 9; 5–10°C, 6 A/dm^2	90 (70)	101
Valeraldehyde	Sn	pH 9; 5–10°C, 6 A/dm^2	Good	101
Heptanal	Sn	pH 9; 5–10°C, 6 A/dm^2	Good	101
Glyceraldehyde	Zn	pH 4–6; 0.04 A/dm^2	75–80 (55–70)	102
Glyoxylic acid (esters)	C, Ni, Cu, Fe, Ag, Au, Pt		Good	103, 104
Glyoxylic acid	Ag, Hg	Various pH values	Tartaric acid major	105
Benzaldehyde[a]	Pt	15% $KHSO_4$; 0.012 A/dm^2	50–65	106, 107
Benzaldehyde[b]	Pt	Alkaline		106
Benzaldehyde[c]	Pb	Acid		107, 108

Benzaldehydes[d]	Ni, Pt, Fe, Cu	Alkaline	70–90	107
m- and p-Chloro-				109
o- and p-Methyl-				109, 110
2,4-Dimethyl-				101
m-Hydroxy-		10% NaOH	20	112
p-Hydroxy-			95.2	111
4-Hydroxy-3-methoxy	Pb	NaOH	78	111
3,4-Dimethoxy-				111
p-Dimethylamino-	Hg	EtOH + 10% KOH	35–58	113
p-Acetamino-				114
Furfural	Pb	Phosphate buffer	42–63 (~48)	118
Furfural	Cd, Zn; −0.9 (NHE)	Phosphate buffer, pH 9; 5–6°C, 2 A/dm^2	ca. 85 (~50)	115, 116
2-Formylselenophene[e]	Hg	Acid or alkaline		117
Thiophene-2,5-dialdehyde[f]	Hg	EtOH, pH > 3		118

[a]Hydrobenzoin and isohydrobenzoin obtained.
[b]One isomer of the product obtained.
[c]Benzyl alcohol and hydrocarbons as by-products.
[d]Similar results were obtained with anisaldehyde, piperonal, salicylaldehyde, and vanillin.
[e]With partial degradation to selenium.
[f]The EHD of α-thiophenealdehyde in organic solvents has been reported [96].

Table 2. Typical Electroreductions of Ketones to Pinacols: $>C{=}O + e^- + H^+ \longrightarrow >COH{-}COH<$

Ketone	Cathode; E, V (SCE)	Medium; conditions	Yield (current efficiency), %	References
Acetone	Zn sponge	Alkaline; 10–16°C, 0.005 A/dm^2	50–60	374, 375
Acetone	Hg; −0.925 (RHE)	Alkaline	84 (91)	85c
Acetone	Graphite			87
Acetone	Zn	0.2 M KOH, 1000 A/m^2		120
2-Butanone	Pb	Acid		121
2-Butanone	Zn	Alkaline	25–28 (8)	122
Cyclohexanone[a,j]	Zn	Alkaline	15 (6)	123
Tetralone	Hg	MeOH, Me_4NCl	50	124
Dimedone[b]	Hg		30–70	125
Androstenedione[c]			High	126
Progesterone[c]			High	126
Acetophenone			High	127, 128
Acetophenone	Pb, Pt	Alkaline		106
Acetophenone	Pt	Acetate		129, 130
Acetophenone	Pb	Acid		121
Acetophenone	Hg	80% EtOH, KOAc	15.7 meso-, 44.8 d, l	128
Acetophenone	Sn	Aq. EtOH, KOAc	54.2 (~50)	131
Acetophenone	Mercury, graphite, Cu, Zn; −1.6	50% EtOH or 50% DMF, NaOH; 60–70°C	80 (~75)	132

Acetophenone	Hg	MeOH, H_2SO_4	~100[d]	133, 134
Acetophenone	Hg; −2.05 + −2.45	MeCN, Et_4NClO_4 (with hydrogen anode)	96	135
Acetophenone	Hg, −1.65 V	1:1 H_2O-EtOH + 0.1 M K_2SO_4	75% pinacol + 15% starting material	136
α-Trifluoroacetophenone[i]	Hg; −1.45	MeCN, $LiClO_4$ + Et_4NClO_4	35	139
Acetophenones				
p-Hydroxy-	Hg	2 N NaOH	77	113
p-Carboxy-			96	140
p-Amino-	Hg, Sn	1.67 N HCl	~60	141, 142
o-Amino-	Hg	20% EtOH + buffers	54[e], 30, 30	143, 246
p-Dimethylamino-	Hg		82	144
3-Amino-4-methoxy-		Alkaline	33	145
p-Chloro-[e]	Hg			110
p-Methoxy-	Hg			110
p-Trifluoromethyl-[e,g]	Hg			110
Propiophenones[c,g]	Hg, Cu			110
p-Methoxy-	Hg		22	144
p-Amino	Hg		57	144
p-Hydroxy-	Sn	Alkaline; 1.5–5.0 A/dm^2	90 (20)	147, 148
p-Hydroxy-	Hg	10% NaOH; 5–18 A/dm^2	79.5[f] (48)	148
p-Dimethylaminobutyrophenone	Hg		6–11	114
Benzyl phenyl ketone	Hg[c]	Alkaline		110
Cyclohexyl phenyl ketone	Hg	Alkaline		110
Mannich bases		Alkaline or acid		140, 144
Benzophenone	Pb, Hg	Acid		121, 150

Table 2. (Continued).

Ketone	Cathode; E, V (SCE)	Medium; conditions	Yield (current efficiency), %	References
Benzophenone	Cu, Fe, Ni	HOAc + H_2SO_4		151
Phenyl-p-tolyl ketone				121
Phenyl-2,4-xylyl ketone				121
p,p'-Bis-dimethylamino-phenyl ketone	Cu			128
Benzoin	Pb	Alkaline	10	106
Anisoin	Ni, Pt, Fe, Cu	Alkaline		107
Phenyl-α-naphthyl ketone				121
Cuminoin				107
Phthaloyl-α-aminobenzo-phenone				121
Phenylglyoxylic acid	Hg	DMF, Et_4NClO_4		152

Acetylpyridines	Hg	Alkaline	68–98	113, 146, 153
3-Pyridyl ethyl ketone				145
2-Acetylthiophene	Sn	Aq. DMF, KOAc		154
2-Acetylthiophene[h]				155, 156
2-Benzoylthiophene	Hg	MeCN, HOAc	68	157

[a]Cyclopentanone yielded a complex unsaturated ketone as well as the expected products [158].
[b]Dehydration product isolated; yield dependent on concentration and cathode potential.
[c]At controlled potential.
[d]Isolated as the dimethoxy derivative.
[e]54% at pH 1.5 (0.5 N HCl), 30% with phosphate buffer or 0.1 N KOH; mixture of isomers.
[f]Isolated as the pinacolone.
[g] Mixture of isomeric products obtained.
[h]The pinacolone may ultimately be isolated, especially from acid media.
[i]$RR'C_6H_3COMe$ (R = 3-F, 4-F; R' = H, OME. Cl); using Et_4NBr gave the corresponding pinacols [137].
[j]Certain tetraalkylammonium amalgams in DMF also gave the pinacol [162].
[k]Reduction of rotenone in MeCN gave the monomeric alcohol and the pinacol [159].

tively, [113]. The pinacols obtained from veratraldehyde and p-methylbenzaldehyde are approximately equimolecular mixtures of meso and d,l forms [111]. A mixture of isomers (d,l-meso ratio 3:7) was obtained from thiophene-2,5-dialdehyde [118]. A mixture of isomers is also formed from 2-benzoylthiophene [157]. It was found that on mercury, tin, and copper in alkaline medium, the d,l-meso ratio for 2,3-diphenylbutanediols-2,3 obtained by reduction of acetophenone (7-^{14}C) is equal to 2.5–3.2; in acidic medium the ratio becomes 0.9–1.4 [128]. The influence of various factors has been investigated [166]. The highest ratio (19) was obtained at Hg in dry DMF with 0.1 M Bu_4NClO_4 + 0.01 M $LiClO_4$ [167]. The nature of electrolyte and solvent has a very strong effect on the d,l-meso ratio, partly because of a change in the conditions of the formation of ion pairs, ion radicals, and carbanions [124,175].

Asymmetric induction takes place if the reduction occurs in a chiral medium (a mixture of optically active 1,4-bis(dimethylamino)-2,3-dimethoxybutane and MeOH [98d].

Dicarbonyl Compounds

The hydrodimerization of biacetyl [83] has been mentioned as an example of the behavior of 1,2-diketones. Cyclopropanediols (as diacetates) have been obtained in high yield by the reduction of several 1,3-diketones. $PhCOCRR'COR^2$ ($R = R^1$ = Me. R^2 = Ph; R = H,R' = Me, R^2 = Ph, R = R' = R^2 = Me; R' = CH_2Ph, R^2 = Me) are concerted in 80–95% yields to substituted cyclopropane-1,2-diols [168]. The trione XI behaved similarly.

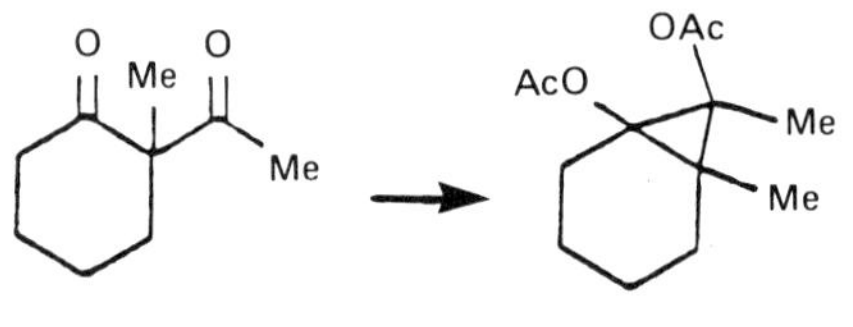

(35)

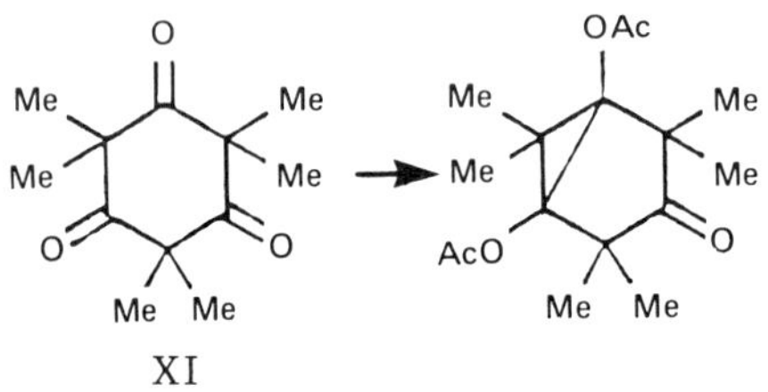

(36)

In other cases, acetylacetone additional products are formed due to pinacolization. (2,6-Diisopropyl-5,5-dimethyl-1,3-dioxane-4-ol) was formed from isobutyraldehyde and 2-methylpentanediol-1,3 from propanal in phosphate buffer (pH 11.0–11.5), probably via the aldol [169]. Plainly diols may be formed not only from diketones.

The 2,6- and 3,7-diketobicyclo [3.3.1]nonanes are reduced to dihydroxybenzene or dihydroxynoradamantane [291].

Earlier [355] cyclic glycols had been obtained from 2,2'-diaroylbiphenyls and 1,8-diaroylnaphthalenes, respectively, by reduction in alcoholic or aqueous sodium of potassium hydroxide using a mercury cathode and a lead anode. $ArCO(CH_2)_nCOAr$ yielded cyclic glycols or their derived pinacolones when n = 3; when n was 4 or 5 polymeric materials, presumably glycols were formed [362].

Reduction of 2-phenylindan-1,3-dione in acid medium which might have given an interesting bridged glycol yielded instead a stable diradical.

Reduction of acetylacetone in dilute hydrochloric yielded 3-hydroxy-3-methylbutanone-2 and 3-methylbutanone-2, (XII) and (XIII). The two methylbutanones are formed through an intramolecular pinacolization via a cyclopropanediol derivative, whereas (XII) and (XIII) are formed through inter- and intramolecular pinacolization. Traces of durene, obtained through intermolecular pinacolization, are also found [246]. Under other conditions other products may be formed

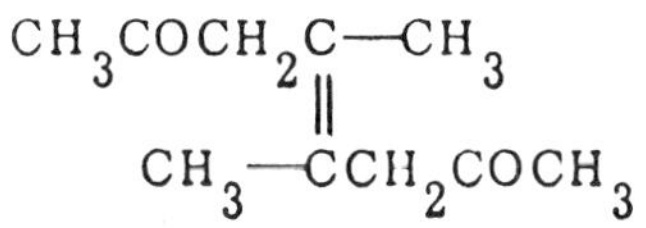

XII

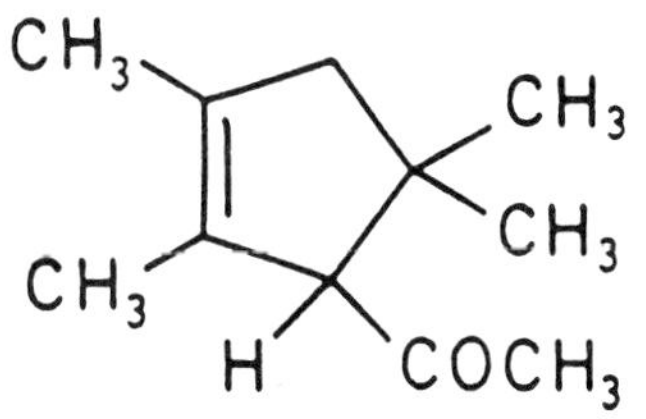

XIII

Macroelectrolysis of 1,3-diphenyl-1,3-propanedione at −1.15 V (pH 4.2) yielded the glycol XIV resulting from the involvement of only one of the two carbonyl groups: at −1.50 V mixture of isomers of XV were probably obtained in which both carbonyl groups had been reduced [171].

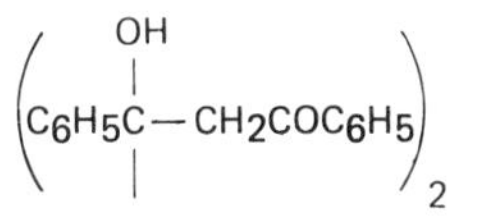

XIV

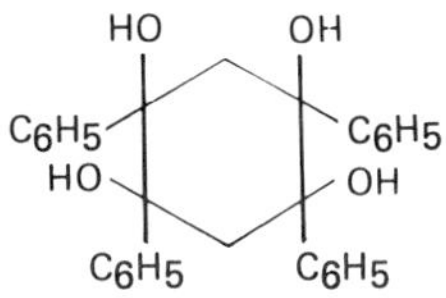

XV

A study of the electrochemical reduction of enolized 1,3-diketones (including 1,3-diphneyl-1,3-propanedione) in nonaqueous media [172] showed the following apparent 0.5 electron process:

$$4R\overset{OH}{\overset{|}{C}}{=}CH{-}COR + 2e^- \longrightarrow \left(R\overset{OH}{\overset{|}{C}}CH_2COR\right)_2 + 2R{-}\overset{O^-}{\overset{|}{C}}{=}CHCOR \qquad (40)$$

The enolized diketones themselves served as proton donors toward the intermediate anion radicals. A related study involving cyclohexadiones has also been reported (see also Chap. 30) [125].

Reductive coupling of monoenol ethers of β-diketones has been mentioned before. Unsymmetrical pinacols have been formed from aromatic ketones and enol acetates [199]. Reductive alkylation of 9,10-anthraquinone [200] and reductive acetylation of 1,4-naphthoquinones [201] have been successful.

Interesting deviations from the normal products expected from hydrodimerizations or mixed reductive couplings occur when preceding, accompanying, or following chemical reactions are associated with the electrochemical steps.

1. Dibenzoylmethane, whose methylenic H can protonate the reduction product, gives not only fractional electron coulometry but a pinacol converted by double cyclization and dehydration to a tricyclotrioxanonane XVI [202]:

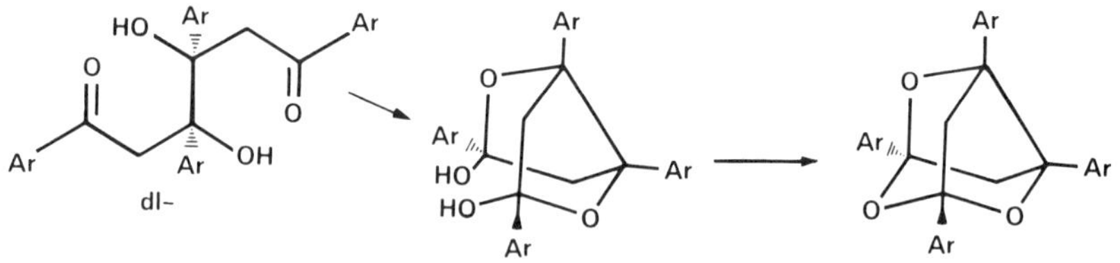

XVI (37)

2. Reduction of n-propionaldehyde at pH 11–12.5 leads to a diol that is not a pinacol but formed by a base-catalyzed aldol condensation followed by reduction of the aldehyde group [84]:

$$2CH_3CH_2CHO \rightarrow [CH_3CH_2CH(OH)CH(CHO)CH_3]$$

$$\rightarrow CH_3CH_2CH(OH)CH(CH_3)CH_2OH$$

3. Reduction of acetophenone in methanolic HCl yields the pinacol bismethyl ether directly [203].

There has been a surge of recent activity in converting carbonyl compounds to hydroxyacids by reducing the former in the presence of CO_2. Using diaphragmless cells and sacrificial Al anodes, acetaldehyde gave only 9% of the hydroxyacid; PhCHO yielded 31–40% and 6-methoxy-2-acetonaphthene 85% [204a]. Other carbonyl compounds were also used [204b]. p-Isobutylacetophenone gave a good yield of the hydroxyacid* that is an intermediate to the analgesic drug ibuprofen [205]. Benzilic acid was obtained from benzophenones (except when OH, NO_2, or Br substituents were present) at Hg]206].

In a redox system consisting of Pb(O)/Pb(II) and Br^-/Br^+, carbonyl compounds were alkylated with allyl bromide. An undivided cell was used in which there were cyclohexenes, $PbBr_2$, HOAc or DMF; the electrodes were Pt [208].

E. Reductive Couplings of Olefinic Aldehydes and Ketones

This category includes compounds in which a double bond is not conjugated with the carbonyl group and those in which it is. In the latter case the substances are activated olefins and are included in Sec. G. Likewise deferred to Sec. G is a discussion of the electrochemistry of carbonyl compounds, which elsewhere in the molecule have an activated olefinic group.

*Using Hg, Bu_4NI, and DMF, an 85% yield was obtained [207].

Although the chemistry that follows from the reduction intermediate (anion radical or anion) of an activated olefin (including α,β-unsaturated carbonyl compounds) can best be rationalized by proposing that the intermediate is a nucleophile, this explanation cannot apply when the olefinic bond of the substrate is not activated, that is, electrophilic. The apparent radical behavior of anion radicals derived from ketones is illustrated in the cyclization (intramolecular reductive coupling) in high yield of olefinic ketones to cyclopentanols

XIV
$RCH{=}CH(CH_2)_2COR'$
R,R' = alkyl

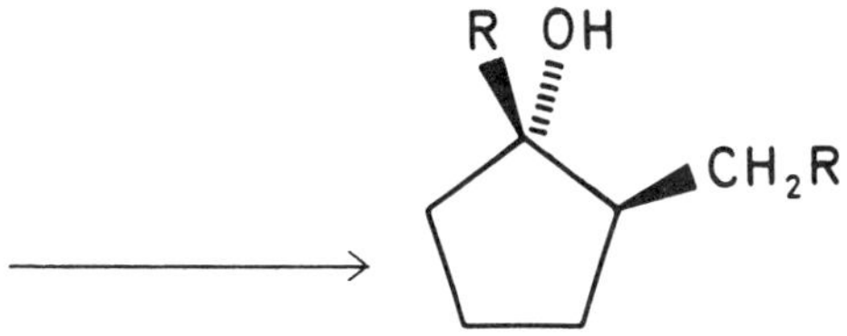

(38)

and only negligible quantites of saturated ketone. Similarly, but in lower yields, bicyclic alcohols are formed from cyclic ketones having an olefinic side chain [209]. More startling, perhaps, is the reduction of $Ph(CH_2)_3COMe$ in 70% yield to the hexahydronaphthalene (XVII) [210]:

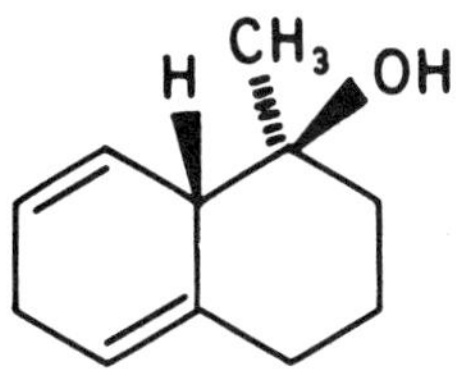

XVII

Earlier, Shono and co-workers [212] reductively cyclized nonconjugated acetylenic ketones to 2-methylenecyclopentanols. In related work, cyclopentanones and cyclohexanones with an allenic group α to the carbonyl formed bicyclic compounds with OH at the bridgehead position and a double bond in the new ring [213]:

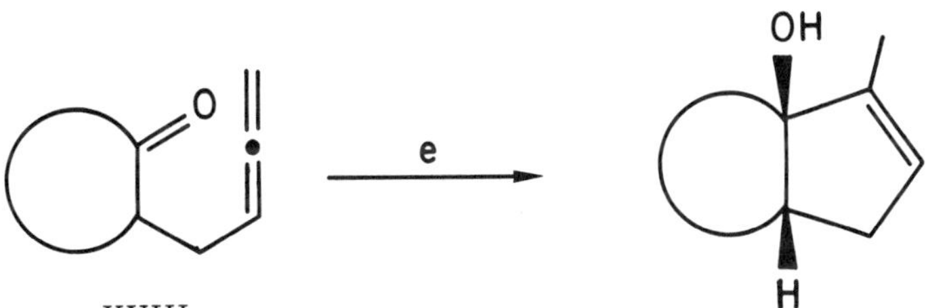

XVIII XIX (39)

F. α,β-Unsaturated Acids

C. L. Wilson [214] reductively dimerized sorbic acid to a mixture of products. In a more comprehensive study, Wilson and co-workers [215] reduced cinnamic acid in aqueous alcoholic sulfuric acid at Hg and obtained a modestly good yield of 3,4-diphenyladipic acid. The method worked well for p-methoxy-o-chloro, and p-hydroxycinnamic acids and poorly for the o-cyano-acid. It failed with o,m,p-nitrocinnamic acids and the p-amino derivative.

In a more recent application of this procedure, Wawzonek et al. [216] obtained 2,3,4,5-tetraphenyl hexanedioic acid from phenylcinnamic acid.

Derivatives of α,β-unsaturated acids (esters, amides, and others) are treated in Sec. G. It has also been reported that cathodic debrominative dimerization of ω-bromoalkanoic acids, as well as their methyl esters, gave the diacids and diamides, respectively [217].

Mixed Reductive Coupling: Intermolecular

Sugino and co-workers [177], later followed by others [178,179], have reported reductive couplings between ketones and acrylonitrile [Eq. (38)], maleic acid [Eq. (39)], pyridine [Eq. (40)], cyanamide [Eq. (41)] [307], dicyandiamide [Eq. (420)] [307], cyanourea [180], α,β-unsaturated esters [181], and activated olefins [182], and with nonactivated olefins [Eq. (43)], e.g., 1-octene [183a] cyclohexene, 1,3-butadiene, styrene, α-methylstyrene, and 1-heptene [183b]. In the latter case, tetraethylammonium p-toluenesulfonate led to higher yields of cross-coupled products than did acid [183c].

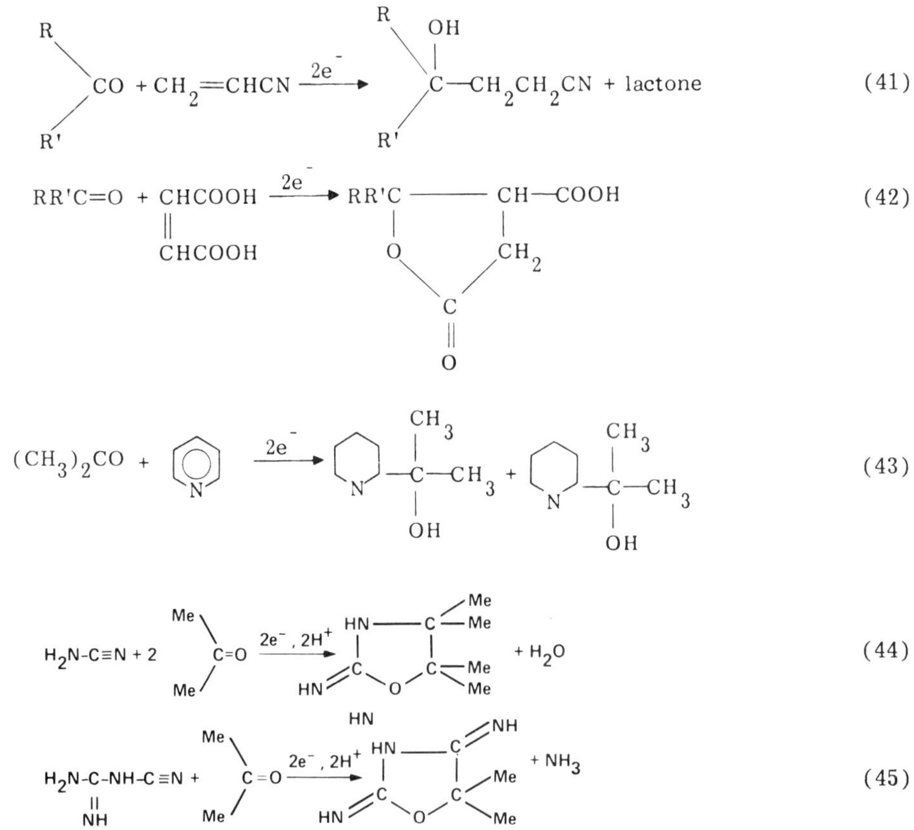

$$\begin{matrix} R \\ \quad \backslash \\ \quad CO \\ \quad / \\ R' \end{matrix} + CH_2{=}CHZ \xrightarrow{2e^-} RR'\overset{\overset{OH}{|}}{C}{-}CH_2CH_2Z \qquad (46)$$

Z = EWG is an electron withdrawing group.

In 20% sulfuric acid, acetone (R = R' = Me), methyl ethyl ketone (R = Me, R' = Et), and diethyl ketone (R = R' = Et) gave the indicated products with 70, 60, and 30% current efficiencies, respectively, at mercury [Eq. (41)]. [Benzophenone (R = R' = Ph) did not participate in cross-coupling under these conditions.] The cathode voltage required [approximately −1.2 V (SCE)] indicated reduction of the protonated ketone. When a mixture of acetone and acrylonitrile [Eq. 41] was electrolyzed in aqueous tetraethylammonium p-toluenesulfonate, it was the acrylonitrile that was reduced [approximately −1.9 V (SCE)]; the same cross-coupled products were formed with approximately 10% current efficiency together with adiponitrile and propionitrile. Moreover, if benzophenone is reduced while acrylonitrile is added slowly, the expected reductively coupled products are formed [182c].

Reduction of a mixture of acetone and allyl alcohol in aqueous alkaline solution at mercury-activated graphite gave 95–98% of 2-methyl-2,5-pentanediol [184]. An analogous condensation of acetone with vinyl chloride at −3°C using 25% H_2SO_4 in the anolyte and MeCN + Et_4NBr in the catholyte has been claimed to yield 52% of $Me_2C(OH)CH_2CH_2Cl$ [185a]. A similar adduct is formed with $F_2C{=}CFCl$ [185b].

Aldehydes have been used less frequently than ketones in the processes of Eqs. (41) and (42) and their analogs, but they do participate successfully [182a,186–188].

The coupling of aliphatic aldehydes with acrylonitrile has been particularly intensively studied [188–191]. The expected hydroxynitriles are formed.

Aliphatic aldehydes of ketones coupled with methyl acrylate in the presence of Me_3SiCl gave good yields of lactones; coupling with α,β-unsaturated nitriles gave γ-hydroxynitriles [192].

Both (C)EEC [177] and (C)ECE [193] mechanisms have been proposed for the process of Eq. (41) (R = R' = Me): the first postulates that $2e^-$ reduction of acetone yields the anion $(CH_3)_2\underline{C}OH$,* which nucleophilically attacks acrylonitrile; the second considers that the one-electron reduction product $(CH_3)_2\dot{C}OH$ adds to the 2-position of acrylonitrile and that the resulting radical is then rapidly reduced (and protonated) to the final product. The evidence for the anionic mechanism is mainly negative: (1) failure of benzophenone to participate in the reductive coupling under acid conditions, and (2) absence of polymerization, which might be expected to result from intermediacy of free radicals. The evidence for the (C)ECE mechanisms is based on data that demonstrate that a chemical reaction intervenes between the two one-electron reductions. A recent [194] reexamination of the reductive coupling of acetone with unsaturated compounds in acid solution agrees with the (C)ECE proposal.

---

*This intermediate has been proposed to be the precursor to pinacol [198] cited in Ref. 177 and to organometallics.

G. Activated Olefins*

Introduction and Background

Included in this section are compounds of types XX through XXIV:

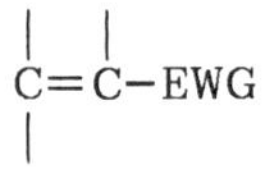

XX

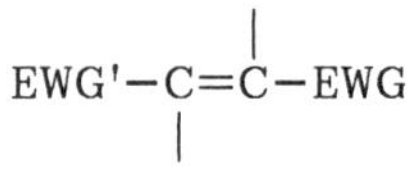

XXI

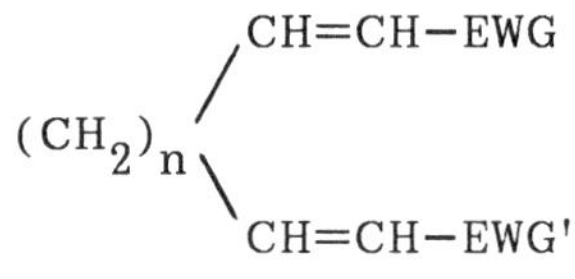

XXII

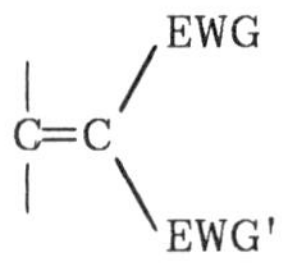

XXIII

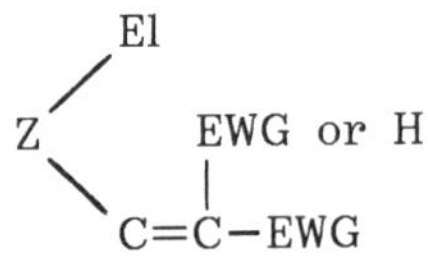

XXIV

in which EWG and EWG' are electron-withdrawing groups (e.g., carbonyl, CN, COOR, >C=C<, $CONR_2$, SO_2R, and $PO(OR)_2$. El is an electrophilic group, such as carbonyl, and Z is a connecting group. With occasional limitations due to steric effects, all these olefins are Michael acceptors [16], and this led to the original working hypothesis [22] that Michael acceptors are capable of hydrodimerization, mixed reductive coupling, and intramolecular cyclization.† The immediate stimulus for initiating this work was the report by Knunyants and co-workers [163] that showed a dramatic yield improvement, from 5 to 60%, in converting acrylonitrile (AN) to the important nylon-6,6 intermediate adiponitrile (ADN) by alkali metal amalgams. It was found, however [224], that yields of ADN from AN could be raised to about 100% by direct electrolytic reduction under certain conditions described here. In about 1961 this process went into a pilot plant, where many economic improvements were made,‡ and into full-scale commercial production in August 1965.

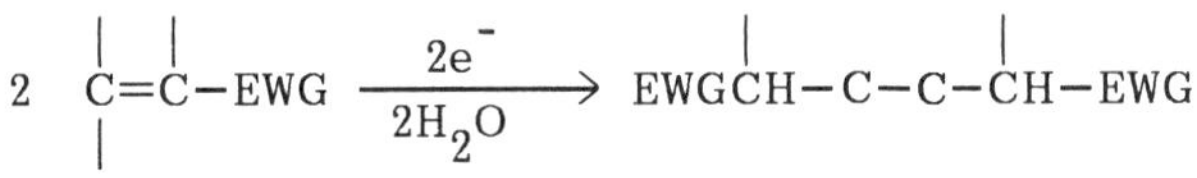

XX XXV (47)

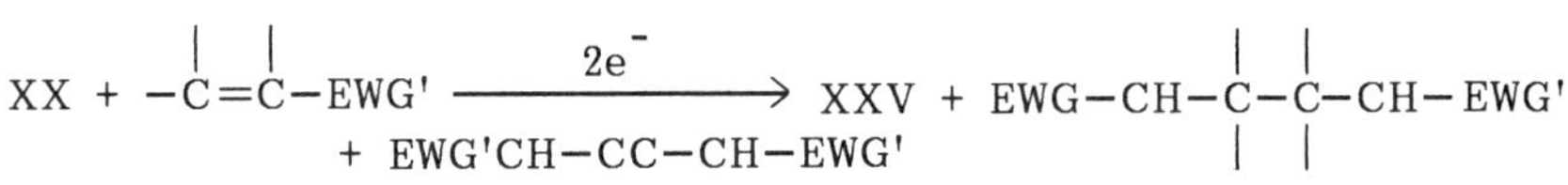

XXVI XXVII XXVIII (48)

---

*The very complex behavior of activated alkynes has only recently been reported [220]. Reductive alkylation of phenylacetylene has been known for some time. It is now reported that reduction of BuC≡CCOOEt in HMPT or DMF/Bu_4NI at a platinum cathode yielded the nonconjugated products PrC≡$CCMe_2COOEt$ in 72% yield [212].

†As illustrated in Tables 3, 4, and 5.

‡Simultaneously there have been advances elsewhere in reductive coupling via amalgams (Ref. 222 and Chap. 26) and via catalytic hydrogenation [223].

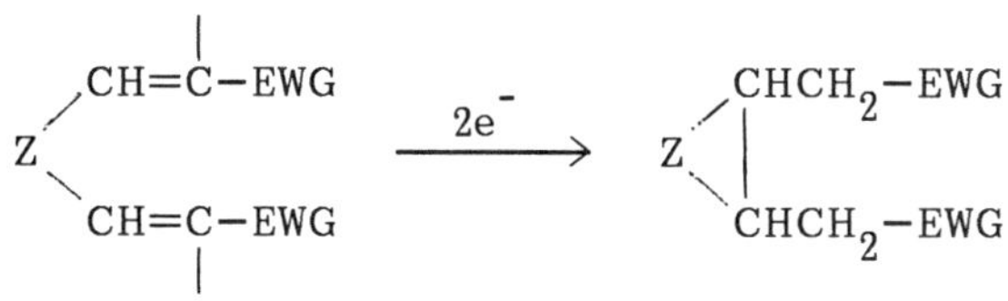

(49)

XXIX XXX

$$EWG'C{=}C{-}EWG \longrightarrow \text{mixture of coupled products}$$

XXXI

when EWG ≠ EWG'.

The scope and limitations of electrohydrodimerization (EHD) [Eq. (47)] of monoactivated olefins, mixed (or crossed) reductive couplings among them [Eq. (48)], and for electrohydrocyclization (EHC) [Eq. (49)] of Michael acceptor olefins in which Z is a bridging group, such as

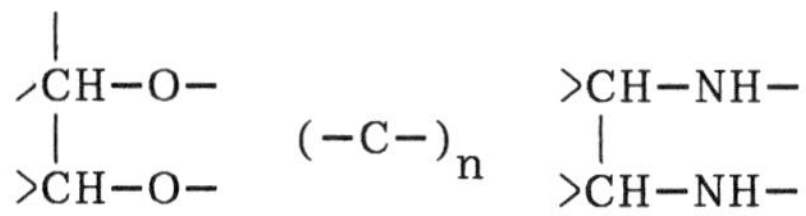

or (un)substituted ethylenes have been discussed in a series of papers and patents [224]. Representative examples of Eq. (47) are given in Table 3, of Eq. (48) in Table 4, and of Eq. (49) in Table 5. Table 3 gives a sampling of the electron-withdrawing groups that have been used. Coupling occurs almost exclusively through the β-positions; yields are generally very good [225].

Interest in the industrially most important example of Eq. (47), the manufacture of adiponitrile (ADN) from acrylonitrile (AN), continues. The invention, development, and commercialization of the Monsanto process have been chronicled [230]. The Asahi Co. variant uses emulsions rather than solution [231]. The process is used in Japan as well as elsewhere in a joint venture with Rhône-Poulenc. The effect of experimental conditions on selectivity for ADN has been reviewed [232]. About 10% savings in energy are possible using nickel cobalt oxide anodes [233]; glassy C cathodes have also been discussed [234]. Energy requirements may also be lowered, as suggested long ago, by using depolarizer reducing gases at the anode [234].

Kinetic and mechanistic studies abound [236,237]; often di- and even triactivated olefins have served as "models" of the EHD of monoactivated olefins [237] despite that the anion radicals from XXXI are much more stable than those from XX and XXVI.

The rates of the follow-up reactions of the anion radicals from XX and XXVI must be much higher than those from XXXI. The hydrodimerization of benzalmalononitrile does not seem analogous to the EHD of acrylonitrile, although it is interesting per se. The EHD of benzalmalononitriles, leading after a Ziegler-Thorpe reaction to cis- and trans-2-amino-4,5-diaryl-2-cyclopentene-1,1,3-tricarbonitriles, has been studied very carefully [164]. The major influences on the cis-trans ratios were the concentration of the supporting electrolyte and of the substrate. It is suggested that at the foot

Table 3. Representative Hydrodimerizations of Michael Acceptor Olefins[a,e]

Olefins	Operating cathode voltage[b] −E versus SCE	Current efficiency (yield), %	References
CH_2=CHCN	1.81-1.91	>90(>90)	224b
CH_2=C(CH_3)CN	2.01-2.05	75.4(-)	224, 228
CH_2=C(CH_2CH_2CN)CN	1.8-1.9	93.5(-)	224m
NC(CH_2)CH=CHCN	–	1.2-66.8[c]	226
RR'C=C$(CN)_2$			229, 308
CH_2=CHCOOEt	1.85	74-87(-)	224
$(CH_3)_2$C=CHCOOEt	2.10-2.20	66.2(-)	224
CH_3CH=CHCONEt_2	2.03-2.12	61.4(-)	224
CH_2=CH-4-pyridyl	1.6	82(-)	224n
CH_2=CHPO$(OEt)_2$	2.0-2.1		224d
C_6H_5CH=9-fluorylidene	1.5-1.6	66(-)	224n
C_6H_5CH=CHC_6H_5			227
CH_2=CHCOCH_3	1.35-1.42	-(40-70)[d]	224d
$(CH_3)_2$C=CHCOCH_3	1.35	-(0-56)[d]	228
CH_2=CHSO_2CH_3	1.72-1.88	–	224d

[a]See Eq. (47).
[b]Usually a few tenths of a volt less negative than $E_{1/2}$.
[c]Depending on cathode and electrolyte used.
[d]Strongly dependent on solvent.
[e]In the hydrodimerization of the dimethyl benzene-1,2-diacrylate it was at first thought the coupling [224h] had occurred through the α-position. However, later examination of the products by x-ray crystallography showed that the following three products had been formed [224m].

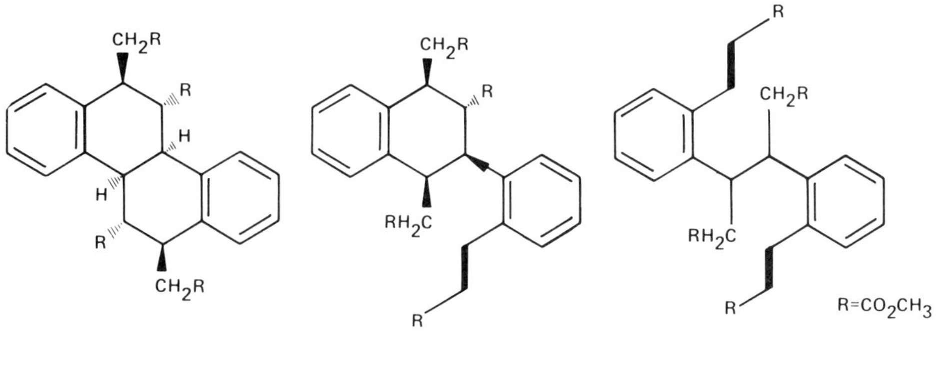

Table 4. Representative Mixed Reductive Couplings Among Activated Olefins[a]

Olefin	$-E_{1/2}$ SCE	Olefin	$-E_{1/2}$ SCE	Operating voltage −E SCE, V	Product	Reference
$CH_2=CHCOOR$[b]	1.85	$CH_2=CHCN$	1.9	1.8-1.9	$ROCO(CH_2)_4COOR + ROCO(CH_2)_4CN + NC(CH_2)_4CN$	269
$CH_2=CH-CH=CHCN$	1.5	$CH_2=CHCN$	1.9	1.5-1.7	$NC(CH_2)_6CN$ (after hydrogeneration) + some isomers	224c
CHCOOEt ‖ CHCOOEt	1.3-1.4	$CH_2=CHCN$	1.9	1.3-1.4	$(EtOCOCH_2\overset{\vert}{C}HCOOEt)_2$ + $EtOCOCH_2CH(COOEt)CH_2CH_2CN$	224
$(CH_3)_2C=CHCOCH_3$	1.50-1.79	$CH_2=CHCN$	1.9	1.65-1.70	Hydrodimers of mesityl oxide + $(CH_3)_2C(CH_2CH_2CN)CH_2COCH_3$	224
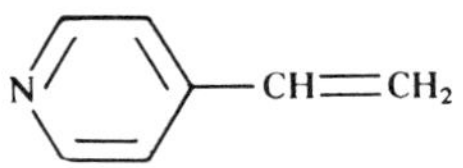	1.5	$CH_2=CHCOCH_3$	1.4	1.4-1.5	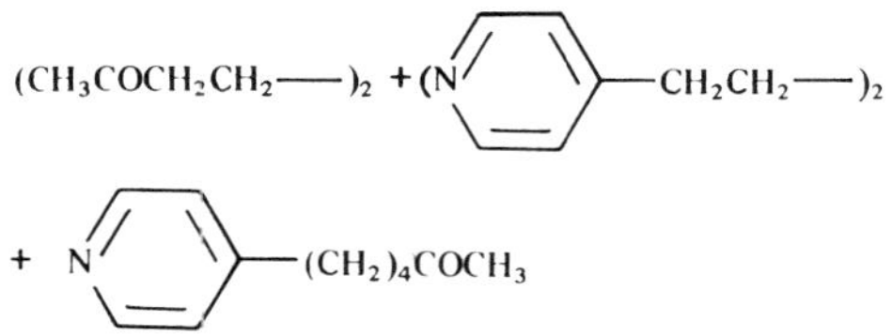	224k 270

[a]See Eq. (48).
[b]It is interesting that this one reaction gives ADN, an intermediate for nylon 6,6, alkyl cyanovalerate, which on catalytic reduction of the −CN group and concomitant or subsequent cyclization yields caprolactam (for Nylon 6) and diethyl adipates for nylon 6,6 plasticizers.

Table 5. Representative Intramolecular Reductive Coupling (Electrohydrocyclization)[a,b]

Olefin	Product	Percentage cyclization	Reference
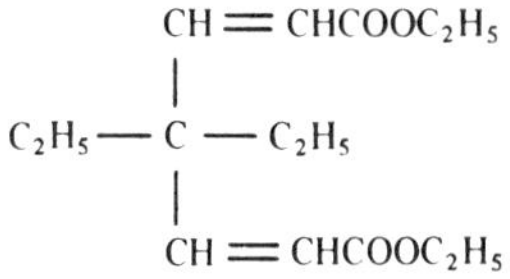	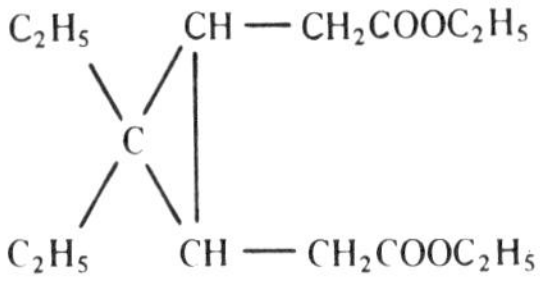	98	224e
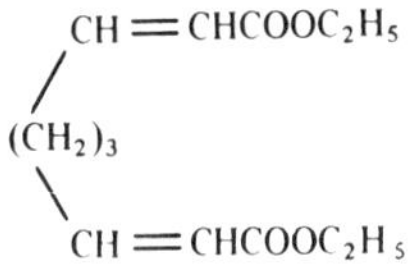	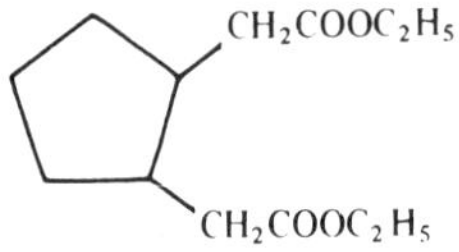	100	224e
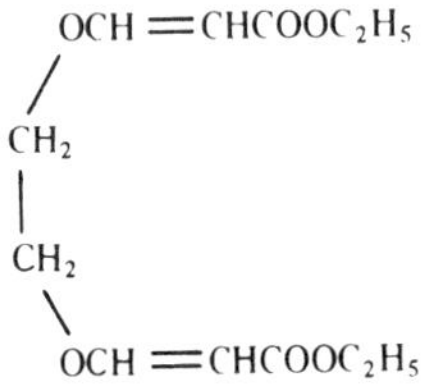	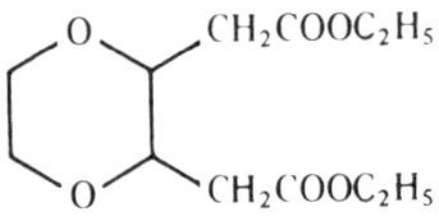	89	224e
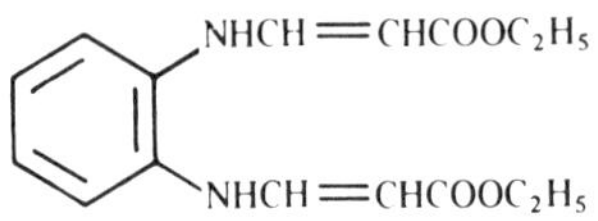	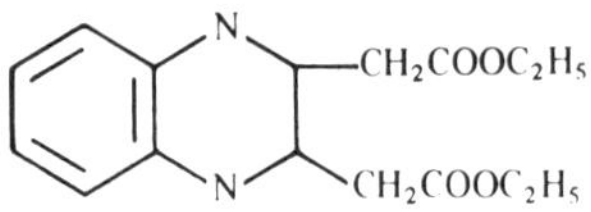	86	224e
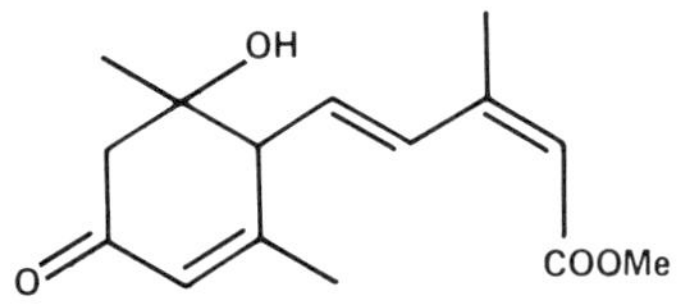	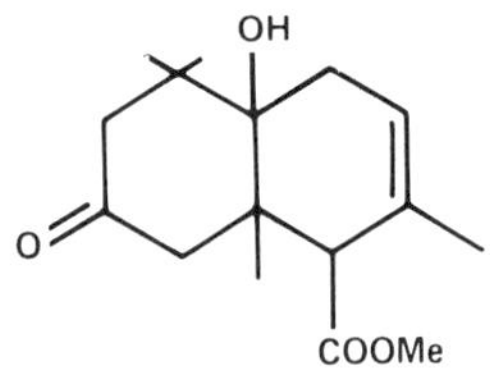	67	224n

[a]See Eq. (49).
[b]Interesting deviations from the EHC norm [Eq. (49)] have been found in the reactions with dibenzalsuccinic acid (fulgenic acid) and esters and with benzene-1,2-diacrylate esters, since the formation of a four-membered ring would have been involved in the "normal" course of the reaction. See note e in Table 3.

of the reduction wave [164] the mechanism may be ECE (reduction + coupling + a second reduction); at the crest of the wave (usually examined in electroanalytic methods), the mechanism may be EC, that is, dimerization of anion radicals. It is therefore not quite certain which mechanism is operating in the EHD of acrylonitrile on a macroscale.

α, β-Unsaturated Carbonyl Compounds

As has been mentioned, reductive coupling may occur via the carbonyl or the β-carbon atoms or both, so that the initial products may be glycols (XXXII), diketones (XXXIII), or γ-hydrocarbonyls (XXXIV) [173]:

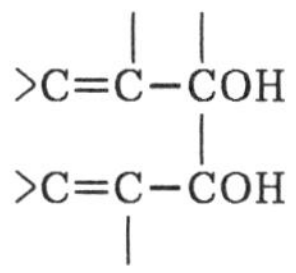

XXXII

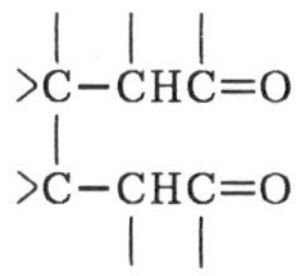

XXXIII

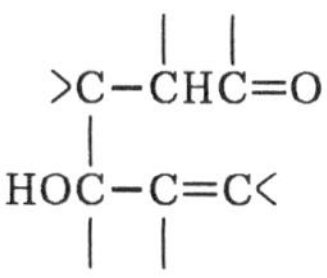

XXXIV

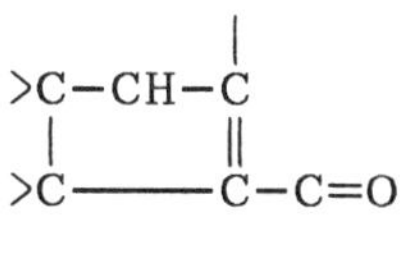

XXXV

Subsequent well-known chemical reactions may lead to ultimate products that are different from those formed upon electrolysis. For example, the dicarbonyls (XXXII) readily cyclize and the cyclic tert-alcohols undergo dehydration to α,β-unsaturated carbonyls (XXXV). As an example, acrolein on electroreduction yielded cyclopentenaldehyde [241]. The product is of the same type as the starting material, although more substituted at the β-position, so that, without adequate cathode voltage control, it too is reduced.

Representative reductive dimerizations of α,β-unsaturated carbonyl compounds are presented in Table 6 and of cross-couplings in Table 7. The double coupling of 4-pyranone is unusual [286].

*α, β-Unsaturated Esters**

These compounds usually form the normal expected hydrodimers. When, however, the dimeric anion is formed in suitably aprotic media, it can undergo intramolecular reaction† more readily than it can react with available water; a cyclic compound rather than or in addition to the hydrodimer is then obtained. The Ziegler-Thorpe reaction, Eq. (50), was reported in 1970 [216] and has been studied and verified by others [3k,164]. Esters behave analogously to nitriles.

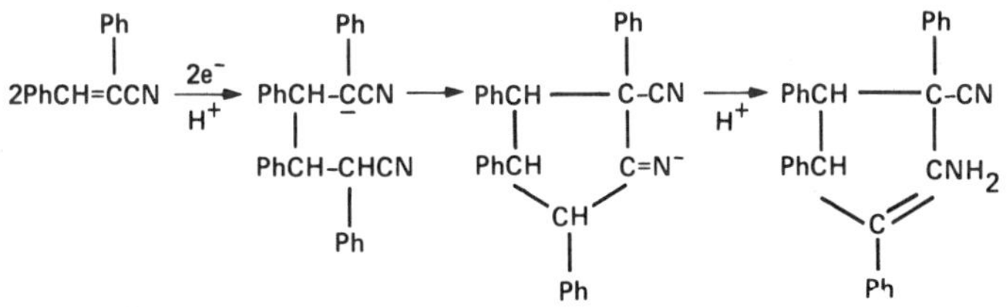

(50)

*These types of reaction continue to be studied because of the range of synthetic possibilities, mechanistic puzzles, and so on.

†Intramolecular reductive couplings (cyclizations) are considered in a separate category later in this chapter.

Table 6. Representative Reductive Dimerizations of α,β-Unsaturated Carbonyls

Carbonyl compound	Cathode	Medium	Principal products	Reference
Acrolein	Hg	Aq. Et_4N tosylate	Cyclopentenaldehyde	238
Acrolein	Hg	HOAc, KOAc, H_2O, MeCN	Adipaldehyde	176
Acrolein	Hg	pH 3.6	Tri- and tetramers	240
Methacrolein	Hg	Aq. Et_4N tosylate	2-Isopropenyl-4-methyl-2,3-dihydrofuran	241
3-Methylcrotonaldehyde	Hg	EtOH + acetate buffer	4,4-Dimethyl-2-hydroxy-5-(2-methyl-1-propenyl)tetrahydrofuran (46%) 2,7-dimethyl-2,6-octadien-4,5-diol (8%)	242
Retinal	Hg	MeCN, Bu_4NOAc + HOAc	Pinacol (18%)	243
Retinal	Hg	MeCN, Bu_4NClO_4 + malonate ester	Pinacol (85%)	244
Methyl vinyl ketone	Hg	HOAc-NaOAc-EtOH	2,7-Octanedione	245
Δ^4-3-Ketosteroids	Hg	MeOH-LiCl + aq. buffer	Pinacol	126
$\Delta^{1,4}$-3-Ketosteroids	Hg	MeOH-LiCl + aq. buffer	Pinacol	126
Mesityl oxide	Hg	CH_3CN + H_2O + KOAc HOAc + THF	4,4,5,5-Tetramethyl-2,7-octanedione	245
Styryl t-butyl ketone	Hg or Pt	DMF + Et_4NBr	2,2,9,9-Tetramethyl-5,6-diphenyl-3,8-decanedione	248
2-Cyclohexenone	Hg	CH_3CN; strongly or weakly acidic; basic	d,1- and meso-3,3'-dioxo-1,1'-dicyclohexyl (major) + 3'-oxo-1-cyclohexyl-2-cyclohexene-1-ol	249
2-Cyclohexenone[a]	Hg	MeCN, Bu_4NBF_4	Same as above, 25 and 15%, respectively	250

Benzalacetophenone	Hg	KI in DMF	Hydrotrimer	251
Benzalacetophenone	Hg	Bu_4NI in DMF	meso-1,3,4,6-Tetraphenyl-1,5-hexanedione	251
Coumarin[b]	Na-Hg	MeOH, pH 7-8	4,4-Bihydrocoumarin (two isomers)	250
Benzalacetone	Pt		Methyl phenethyl ketone	252
Chalcone[c,d]	Hg		Dimer	253
α-Ionone	Hg	MeCN + Bu_4NOAc + HOAc	Pinacol (46%)	243
β-Ionone	Hg	MeCN + Bu_4NOAc + HOAc	Pinacol (71%)	243
4-(2-Furyl)-3-buten-2-one	Hg	Aq. Me_2CO + H_2SO_4	4,5-Di(2-furyl)-2,7-octadieonone 52%	254
Dibenzylidenacetone	Hg	Acid	Di(phenylethyl) ketone	???
2,3-Substituted indenones	Hg	Aq. EtOH	Mixture of cis- and trans-indanones 80–100%	255
1,2-Bis(6-oxo-2-methylcyclohexenyl-1) ethane	Hg	80% MeCN, Et_4NCl	trans-4a,5a-Dimethyl-1,8-dioxoperhydrophenanthrene 81%	242
2,2,5,5-Tetramethyl-3-methylencyclohexanone	Hg	Aq. i-PrOH, Et_4NTs	2,2,5,5-Tetramethyl-3-methylencyclohexanol 61%	256
α-Chloro-2,6-di-tert-butylmethylenquinone	Hg	DMF, Bu_4NClO_4	α-(4-Hydroxy-3,5-di-tert-butylphenyl)-4-hydroxy-3,5-di-tert-butylacetophenone	257
2-(5-Phenyl-2-furfuryliden)-1,3-indandione	Hg	50% EtOH, NaOH	2-(5-Phenyl-2-furyl)-3-hydroxy-1-indan	258
Estratetraene	Pt	$MeNH_2$, Alk_4NTs; 0.05–0.1 A/cm^2	8,9-Dihydroestratetraene 96%	259
Benzoylacetylene	Hg	Aq. EtOH, buffer solutions	Chalcone 60%, acetylenic alcohol 40%	260
Benzoylmethylacetylene	Hg	50% EtOH, HCl	Hydrodimer 84%	261

Table 6. (Continued).

Carbonyl compound	Cathode	Medium	Principal products	Reference
Benzoylmethylacetylene	Hg	30% EtOH, pH 8.2	Hydrodimer 48% Saturated alcohol 17% Ethylenic alcohol 19% Acetylenic alcohol 8%	261
3-Benzylidenindol-2(3H)-one	Hg	40% EtOH, Et_4NBr	3-Benzylindol-2(3H)-one to 95% (isolated 50%)	259
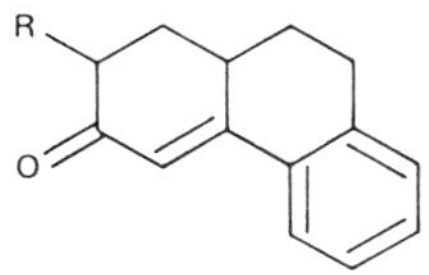			One of the possible pinacols	262
p-OHC_6H_4CHO	Hg	Undivided cell		263

[a]Studies that favor an EC mechanism for the EHD have been done in aqueous [264] and EtOH-H_2O media [265].
[b]The hydrodimer is formed mainly when the supporting electrolyte is a lithium salt [266].
[c]The rate-determining step occurs in the adsorbed state in the absence of a surfactant [267] and in the bulk in the presence of one [268].
[d]The EHD of the related compound PhCH=$CHCOCMe_3$ is assisted by metal salts in aprotic media because of the acid-base reaction between the ketone and the metallic cation acting as a Lewis acid. Some cyclization also occurs [284].

Table 7. Representative Cross-Couplings with α,β-Unsaturated Ketones

Ketone	Partner	Cathode	Medium	Principal products	Reference
Mesityl oxide	Acrylonitrile	Hg	$DMF + H_2O + Et_4\overset{+}{N}$ p-T$\bar{S}$	4,4-Dimethyl-6-cyano-2-hexanone	282
Mesityl oxide	Ethyl acrylate	Hg	HOAc-NaOAc	4,4-Dimethyl-6-carbethoxy-2-hexanone	282
Methyl vinyl	Diethyl fumarate	Hg	$DMF + H_2O + Et_4\overset{+}{N}$ p-T$\bar{S}$	Diethyl-α-(2-acetylmethyl) succinate	283

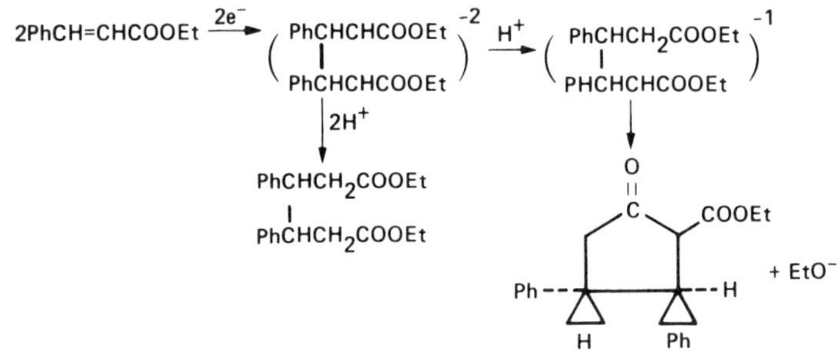

(51)

The Dieckmann product was a minor one when the EHD was carried out DMF-H_2O [271a] but major when the solvent was MeCN [272].

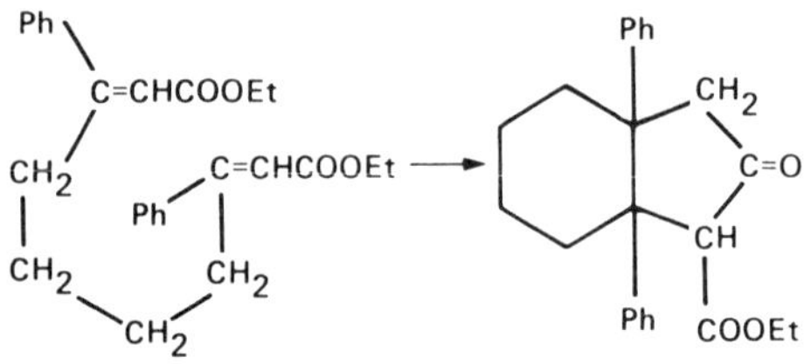

The yields of cyclized product versus linear dihydro or tetrahydro product are low because of steric hindrance and the need for very dry MeCN solvent [273].

Intramolecular reductive couplings, cathodic cyclizations, and EHC
The formation of cyclic from acyclic compounds [see, e.g., Eq. (49)] is an important chemical operation. When the newly formed ring also bears substituents that allow further elaboration by "chemical' or electrochemical means of a synthetic sequence leading to desired target molecules, the annulation step is even more important. If the final molecule is a natural product with one or more chiral centers, the stereochemical result of the ring closure is of the utmost importance.

Cathodic cyclizations of ω-dihalides and other types of halides have been cited in several places in this book (this chapter and Chap. 8), of bis-activated olefins (this chapter), of ketones with remote unsaturation (Chap. 10), the Thorpe and Dieckmann condensations accompanying EHD of certain α,β-unsaturated nitriles and esters, and of 5-phenyl-2-pentanone to a hexahydronaphthalene derivative [210] in which the keto anion radical has attacked the benzene ring.

Concerning the mechanisms of EHC, the following information is available. Two reduction waves are exhibited by XXII when, for example EWG = EWG' = COOR and Z is such that a suitable ring size can form. The relative heights of the two waves correlated well with the degree of cyclization versus acyclic reduction [224e]. The anodic shift (first wave) was ascribed to concerted one-electron reduction-ring closure; a similar phenomenon was encountered in the reduction of those ω-dihalo compounds that formed cyclized products. A mechanistic study [273] showed that at low water concentration cyclization is probably in accord with that; at higher water concentrations radical-radical coupling competes.

The general field of EHC has been reviewed [367–369].

R. D. Little and co-workers [285] are studying the benefits of having at least one electrochemical step in the synthesis of natural products from XXIV (Z = substituted alkylene groups, El = carbonyl [289]). The synthesis of sterpurene has been completed [285b].

Cathodic cyclization of a compound of type XXIV leads to a mixture of cis and trans isomers. (Overall yields are 70–80%.) When the newly formed OH and the adjacent alkoxycarbonylmethyl groups are in a cis relationship, lactonization may occur in situ or under mild conditions during workup, as illustrated in Table 8.

However, for further synthetic purposes, the trans isomer is most desired and a number of factors that may influence the trans-cis ratio have been investigated: nature of the cathode material, temperature, and proton donor. Although the results may not be of general applicability, a copper cathode and diethyl malonate have given the highest trans-cis ratio (11.6:1) in the sterpurene synthesis (scheme 1).

It has also been established that under controlled-potential conditions the enoate group is reduced before the carbonyl.

Synthetic aspects Although EHD, EHC, and their variants, all involving olefins, have provided a method of wide scope for preparing di- and polyfunctional products, often by the most convenient route, even more types of products have been made accessible by reductive coupling of activated olefins with other classes of coupling partners, usually electrophiles. Representative examples are given in Table 9. In some cases, the activated olefin is reduced and the intermediate couples with the partner; in others the unreduced activated olefin is the acceptor.

The reductive coupling of CO_2 with mono- and diactivated olefins, including ethyl and methyl cinnamates, cinnamonitrile, benzalacetone, benzalacetophenone, and fumaronitrile, has been studied mechanistically [274]. Related studies were reported earlier [275].

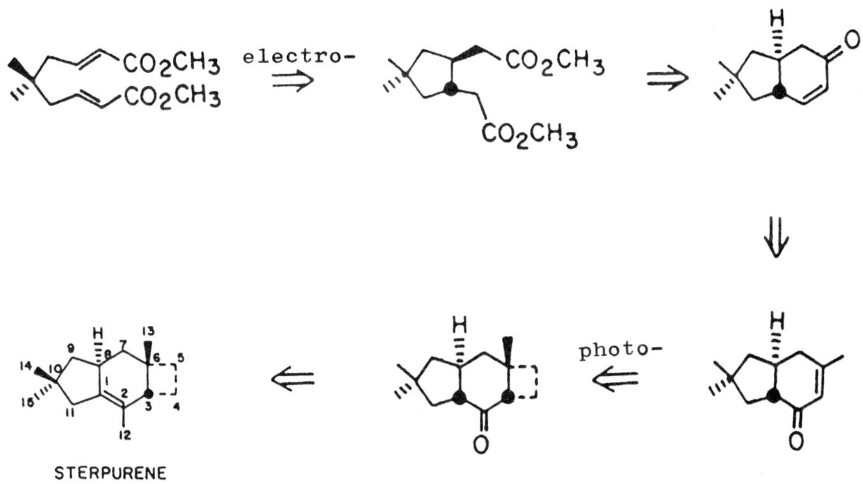

Scheme 1. Sequence of reactions resulting in sterpurene.

Table 8. Some Reductive Cyclizations of Enoate Carbonyl Compounds

Substrate	Cyclized products	Trans/cis	Total isolated yield (%)[a]
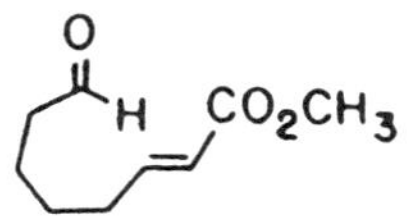	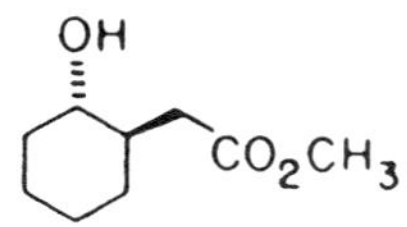 + 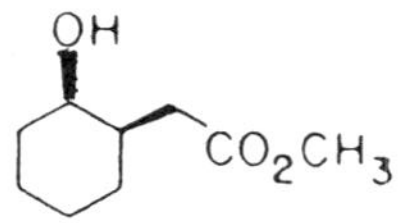 + lactone	1.8:1	72
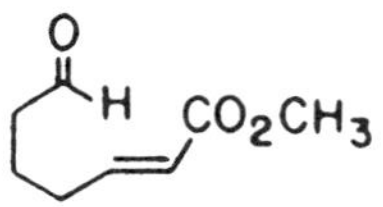	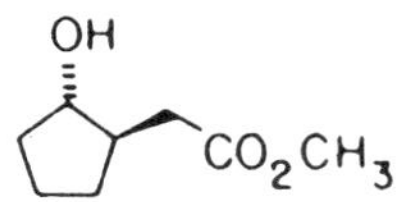 + 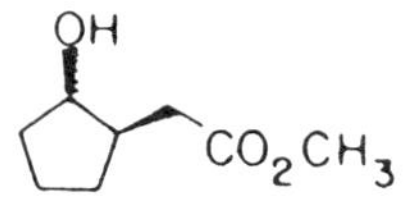 + lactone	1.4:1	70
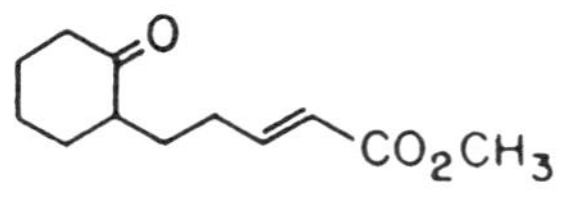	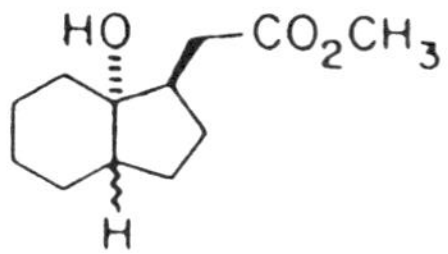 + 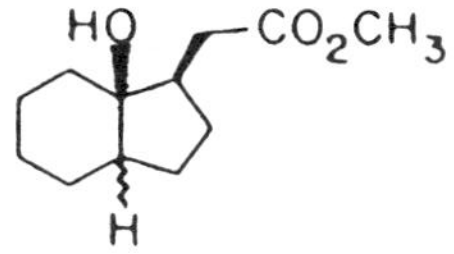 + lactone	11.4:1	79

[a]The yield of lactone is credited to that of the cis-product.

Considering hydrodimerizations first [Eq. (47)], the effect of the nature of the EWG and of the electronic and steric character of α- and β-substituents upon the reduction potential and the ease of coupling have been summarized [224a]. Electron-withdrawing groups shift the required voltage anodically, electron-donating groups cathodically. Bulky substituents may block coupling completely and favor simple saturation of the double bond. Extension of the conjugation, as in 1-cyano-1,3-butadiene [224c], causes an anodic shift; interestingly, coupling occurs mainly through the ω-position.

The achievement of a large body of synthetic results occurred in the first decade of this work and defined the scope, limitations, and vagaries of the EHD and EHC reactions; the second decade roughly saw the development or invention of analytic techniques that are finally elucidating the mechanisms of these reactions.

The widespread study of the ADN process in particular has had beneficial effects on the development of an understanding of the several procedural factors affecting the success of electrolytic reductive coupling in this series. For practical purposes a totally aqueous system, when possible, is preferred. Side reactions that must be controlled are simple saturation of the double bond, formation of organometallics, oligomerization and polymerization, nonelectrochemical cyanoethylations, saponifications, cyclizations, and electrochemical cleavages. Studies have been concerned with the effect on these alternative reactions on the (1) nature of the cathode material; (2) pH of the catholyte; (3) supporting electrolyte, particularly its cation; (4) concentration of olefin in catholyte; and (5) type of cell.

Cathode material High hydrogen overvoltage cathodes give the best results, but even within this group there are individual variations. Mercury and lead are excellent [224b], graphite is good [290], cadmium favors the formation of saturated monomers,* and tin forms substantial quantities of tetra-β-cyanoethyl tin with AN [292] and lesser quantities of organometallics with other olefins [293]. Current efficiencies for the EHD of methyl acrylate [224h] were determined at a series of cathodes: Sn > Pb > C > Ti > Zn. From the mechanistic point of view it is interesting that, whereas lead and mercury do not form organometallics in the reduction of AN, they and thallium do yield organometallics in the reduction of β-iodopropionitrile [294].

pH of catholyte The qualification must be borne in mind that in aqueous electrolyses the pH in the vicinity of the cathode may be several units higher than in the bulk of the solution. Too low a pH favors oligomerization-polymerization [295], formation of saturated monomer [296], chemical side reactions, such as hydrolysis; too high pH favors cyanoethylation of water (in the case of AN), saponifications, cyclizations [272,296], and so on. Good to excellent results have been reported in ranges of apparent pH of 3.5–9.

*These results were reached with a Cd plate in the laboratory. The pilot plant showed that certain forms of cadmium deposited on a base metal were satisfactory, less toxic, and easier to handle than Hg. This was therefore adopted for full-scale production.

Table 9. Mixed Reductive Couplings of Activated Olefins and Other Types of Partner

Activated olefin	Coupling partner	Solvent/ electrolyte system	−E versus SCE	Products formed	Reference
$CH_2{=}CHCN$	RCHO	Aq. K_3PO_4, pH 4–5	–	$RCH(OH)CH_2CH_2CN$	170
$CH_2{=}CHCN$	Me_2CO	20% H_2SO_4	1.2	$Me_2C(OH)CH_2CH_2CN$ + lactone	276
cis-$CHCOOMe{=}CHCOOMe$	CO_2	$MeCN/R_4N^+$	1.53	$[(MeOCO)_2CHCHCOOMe]_2$	277
cis-$CHCOOMe{=}CHCOOMe$	CO_2	$MeCN/R_4N^+$	1.84	$(MeOCO)_2CHCH(COOMe)_2$	277
$R^1R^2C{=}CR^3Y^a$	Ac_2O	$MeCN/R_4N^+$	2.3-2.5	$R^4COC(R^1)(R^2){-}CH(R^3)(Y)^a$	278

PhCH=CHCOOEt[b]	t-Bu halide	DMF/Bu_4N^+	2.3	Ethyl 2- and 3-t-butyl-3-phenyl-propionate	279
$C_6H_5CH{=}CH_2$	$Ph_3\overset{+}{P}CH_2CH_2CN$	DMF/none	1.58-1.60	$C_6H_5(CH_2)_4CN$	280
$CH_2{=}CHCN$	CCl_4[c]	$CHCl_3$ + H_2O/Et_4N^+	0.98	$Cl_3CCH_2CH_2CN$	281
$CH_2{=}CHCOOMe$	RCHO[d]	Pb alloy cathode		4-Butanolides	287
$CH_2{=}CHCOOMe$	RCOR'	Pb or alloy cathode H_2SO_4		Lactones[e]	288

[a] Y = COOMe, COOEt, CN.
[b] Examples with PhCH=CHCN also described. Butylation at α, β, and ring positions obtained.
[c] Many other examples of halo compounds and activated olefins.
[d] When RCHO is butyraldehyde, product is 5-propyl-2(3H)-furanone.
[e] From 2-octanone γ-methyl-γ-n-hexylbutyrolactone was obtained; from Me_2CO γ,γ-dimethylbutyrolactone cyclized by polyphosphoric acid to 3-methyl-2-cyclopentenone.

Supporting electrolyte Optimum practical results were obtained initially with strong aqueous solutions of quaternary ammonium hydrotropic salts (e.g., tetraethylammonium p-toluenesulfonate), which permitted one to dispense with or minimize the need for organic cosolvents and to eliminate the possibility that discharge of the cation occurs simultaneously with the reduction of the olefin. There is a further advantage to the quaternary ammonium cations: they are in aqueous systems preferentially adsorbed at mercury, for example, and, unlike the alkali metal cations, are not hydrated, so that the intermediate produced by reduction of the olefin has a favorable opportunity to react first with another molecule of parent rather than with water [119]. This effect is manifested even at low concentrations of supporting elctrolyte [146], even when quaternary ammonium is added to electrolytes containing alkali metal cations [165], and has been applied to a radical improvement in the results obtained from amalgam reductions (Chap. 28), it was also operative in the EHD of diethyl maleate [339].

Concentration of olefin in catholyte Optimum yields and conversions have been obtained using homogeneous solutions containing more than approximately 5% olefin. The electrolysis is run to partial conversion, with replenishment of the olefin to maintain this level. Lower concentrations and emulsions of the olefin in nonhydrotropic electrolytes are claimed to be operable [301], although there are numerous reports [299,302] that very low concentrations favor the formation of saturated monomers. Too high a concentration of olefin with a concomitant decrease in the concentration of proton donor leads to polymerization [303].

Type of cell Excellent results are obtained in divided cells [224h,224f] with cation-exchange membranes [224h]. Undivided cells are operable* and have the advantage of requiring smaller cell voltages but are subject to the restrictions that (1) the olefin and product not be substantially oxidized at the anode, and (2) the oxygen evolved at the anode in aqueous systems not promote undesirable side reactions. For economic reasons, spacing between the electrodes should be minimized. Thin polyolefin spacers [304] as well as various ingenious devices that are intermediate between totally divided and totally undivided cells, have been proposed [302].

Many variations of the EHD process for adiponitrile have been patented and are cited in Chap. 31.

The synthetic considerations discussed also apply to mixed reductive coupling among activated olefins [Eq. (48)]. In addition, the following special features obtain in this category:

1. When the two participating olefins each reduce individually at about the same cathode voltage, three coupling products are obtained.
2. When the individual reduction potentials are sufficiently far apart ($\geq$0.5 V) that the cathode voltage during the electrolysis of the mixture can be controlled to avoid reduction of the more cathodically reduced olefin (say, XX), only two products are obtained, XXV and XXVIII. Within this voltage range, more cross-coupled product XXVIII is obtained toward the negative limit of the range than at more anodic voltages (224g]†; if

*This was indicated in an early patent issued in this series.

†It has been suggested that the first-formed anion radical homogeneously transfers an electron to the second monomer and that radical coupling is the mechanism of the hydrodimerization.

the gap between XX and XXV is > about 0.5 V, no cross-coupling occurs. This result finds analogy in certain anionic homo- versus copolymerizations [247,305].

The reactions of Eq. (49) is an excellent method for producing three- to six-membered cyclic compounds (carbo and hetero) with functional groups in the side chains. In contrast to the acyloin condensation [306], it fails in the construction of larger ring compounds.

The couplings represented in Eqs. (47) and (48) represent the most complex synthetic results obtained in this series. When EWG = EWG', yields in one step are obtained of products that could otherwise be prepared only by multiple-step syntheses; when EWG ≠ EWG', reactions (47) and (48) are less useful for synthesis than for the information they provide concerning the mechanism of coupling.*

H. Onium Compounds

Although the reductive cleavage of acyclic onium compounds, particularly quaternary ammonium compounds, has been studied very extensively (Chap. 24), there have been only a few cases reported in which dimeric products were obtained from an R group that was cleaved. Electrolyses in dipolar aprotic solvents yielded 6–35% R-R when R in $RR^1R^2R^3N^+$ was cinnamyl, p-methoxybenzyl, fluorenyl, α-phenylethyl, or benzyl [309a]. Allyl quaternary ammonium and phosphonium compounds likewise reductively cleave to coupled products [309g]. It is obvious that these are the R groups that form relatively stable radicals.

Bibenzyl was formed from benzyltrimethylammonium nitrate in DMF, at Al but not at Pt [309b]; yields when HMPT was the solvent were very good [309c]. An extensive study of the perchlorate salt [309c] shows the effect of four solvents and 17 cathodes on the distribution of the reduction products toluene, bibenzyl, and N,N-dimethyl-N-(2-methylbenzyl)amine.

The establishment of the mechanism(s) for the formation of dimeric products is beset by some of the same problems involved in the reductive dimerization of organic halides (Chap. 8): dimerization of $R^\cdot$ versus anionic attack of R^- upon $RR^1R^2R^3N^+$. Since electrolysis of d-α-phenylethyltrimethylammonium nitrate gave only d,l- and meso-2,3-diphenylbutane, the former mechanism is favored in this case. It is certainly possible that coupling occurs by a radical mechanism in aprotic solvents and by an anionic mechanism in water; this view is supported by work on reduction of aziridinium compounds [309f].

Phosphonium cations follow a $1e^-$ pathway to dimers and a $2e^-$ pathway to ylids, Hofmann degradation, and phosphine oxides [310].

A Dow patent [311] claims a yield of 95% p,p'-dinitrobenzyl in the electrolysis of dimethyl p-nitrobenzylsulfonium chloride at a Cu or Pt cathode and C anode using KCl as supporting electrolyte.

Nitro-substituted quaternary ammonium and phosphonium salts are discussed in Sec. IV.C.

*It had been hoped that this procedure could be used to determine the electron-withdrawing ability of EWG versus EWG' since the coupling should occur β to the more electron-withdrawing group [47].

When the heteroatom is part of a heterocyclic ring, reductive dimerization through the partly reduced rings occurs readily via intermediate radicals; for example,

1. N-benzylnicotinamide or acid couples through the 6-positions [312]:

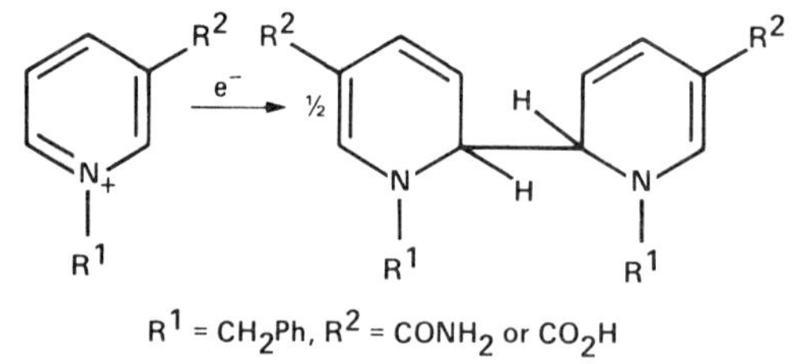

(52)

Similar behavior is reported for 1-substituted nicotinamide in MeCN or Me_2SO solvent [313].
2. N-methylpyridinium salts dimerize through the 4-position [314]. However, products that result from the reduction of pyridinium cations in DMF (0.05 M $KClO_4$) depend on the substituents in the ions. Simple N-alkyl compounds yield hydrodimers. The 2,4,6-trimethyl-N-alkyl compounds reversibly produce free radicals, whereas the 2-methyl-4,6-diphenyl-N-alkyl pyridinium ions (where the alkyl group is H, Me, Ph, 1-naphthyl, or substituted phenyl) are reduced to cyclic dienes in two steps [315].
3. 1-Methyl cyanoquinolinium perchlorates yield hydrodimers (via radical intermediates) if the −CN group is in the 2- or 3-position; when it is in the 4-position, a stable radical is formed [316].
4. 2-Methyl isoquinolinium perchlorate at −1.40 V and pH 10 gives 66% of the hydrodimer [312].
5. The behavior of 4,4'-bipyridyl derivatives depends on the degree of quaternization: the parent compound and its mono-N-methyl derivative yield hydrodimer; the N,N'-dimethyl compound yields the stable methylviologen cation radical [317].

In an analogous way, protonated N-heterocyclic compounds undergo EHD:

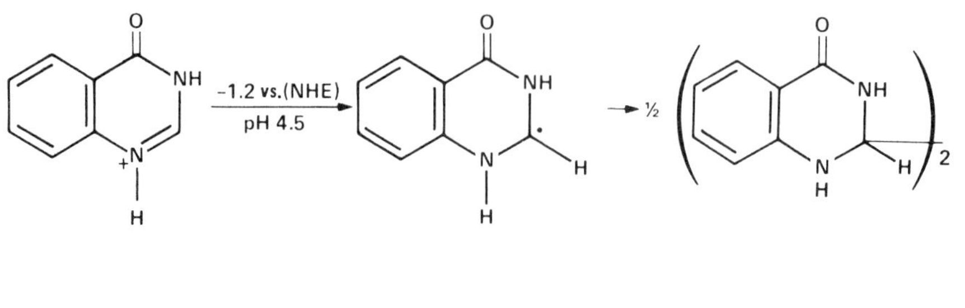

Similar reductive dimerization has been shown to occur with an acrizidinium salt and its photodimer [138]:

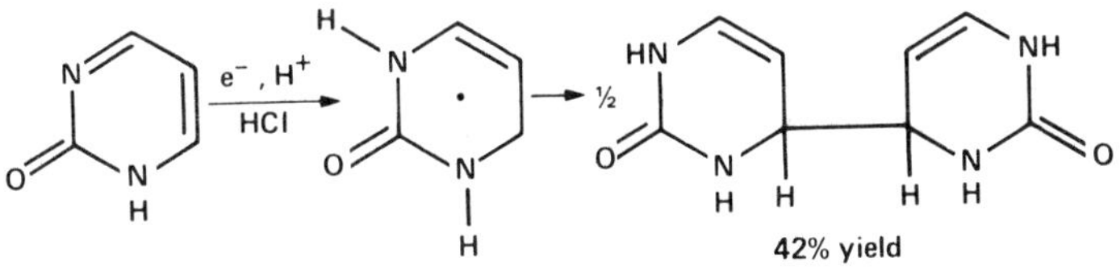

[319]

Substitution by Ph groups in the 2- or 3-position blocked dimerization [318].

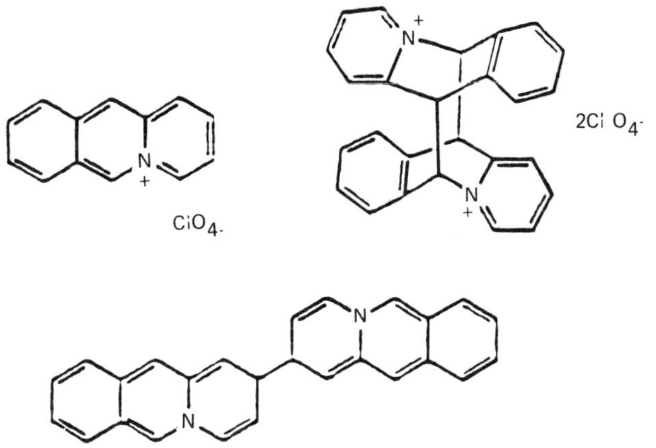

An interesting reductive dimerization has been reported in the case of a cyclopropyl <u>carbonium</u> ion (carbenium ion, carbocation):

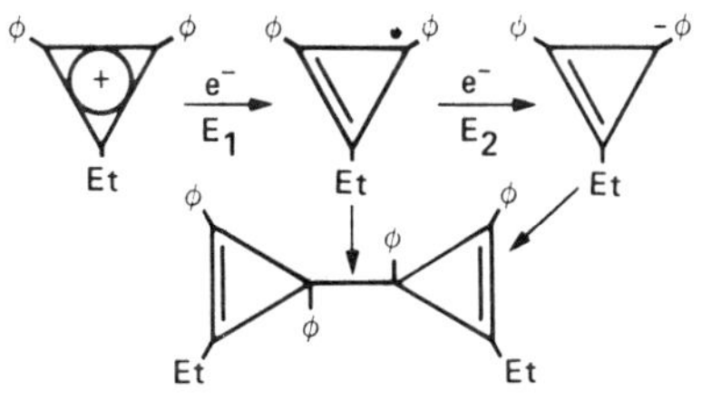

(53)

That the same product was formed at E_1 and E_2, a phenomenon that has often been observed in the reduction of organic halogen compounds, and that the juncture is not at the C atom bearing the Et group, is interpreted by the authors to mean that the carbanion reacts with the initial carbonium ion by an electron-transfer process [320].

The reductive coupling of ketones with N-heterocycles [321] is discussed in Sec. D.

The intramolecular reductive coupling of haloaromatics, which proceeds by initial cleavage of the C–X bond, is discussed in Sec. B [322].

I. Miscellaneous Classes

Immonium Salts

The system here was chosen for study because reductive coupling, when it occurred, had to proceed unequivocally by free-radical dimerization [323]:

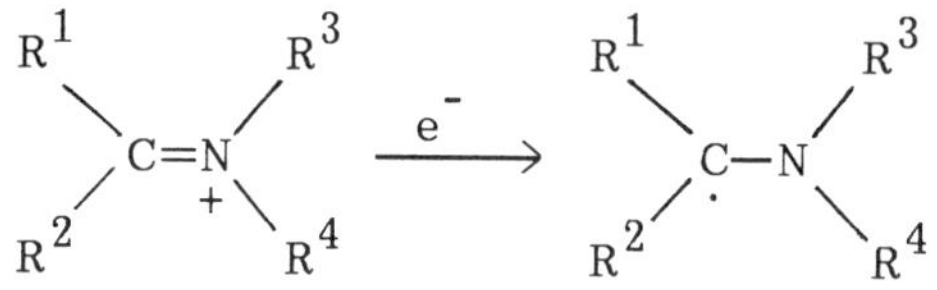

XXXVI

One-electron reduction was verified by coulometry and in certain cases by observing the ESR spectra of XXXVI. Cyclic voltammetry was used to estimate the reversibility of the system; in irreversible systems dimer was, indeed, obtained in macroelectrolysis. The dimer could be electrooxidized to starting material.

An elegant intramolecular reductive coupling leading to alkaloids makes use of immonium salts as starting material. Thus 3,4-dihydro-6,7-dimethoxy isoquinoline is quaternized with methyl 2-bromomethyl-5,6-dimethoxybenzoate in situ; electroreduction in DMF gave a quantitative yield of 8-oxotetrahydropalmatine [324].

Pyrylium Salts

The reduction of 2,4,6-triphenylpyrylium perchlorate in MeCN or DMF at −0.4 V versus SCE yields a red triphenylpyranyl radical in equilibrium with its dimer; the latter could be electrooxidized to the initial cation [325]. 2,6-Diphenylpyrylium perchlorate yielded an analogous dimer, which was converted to 2,2',6,6'-tetraphenyl-4,4'-bipyrylene [326].

Azo Compounds

Azobenzene, reduced presumably to the dianion stage, has been coupled with a number of partners, as shown in Table 10.

Schiff Bases and Imines

The hydrodimerization of Schiff bases has been reinvestigated [327]. Among the groups R and R^1 employed were

$$2RCH{=}NR^1 + 2e^- + 2H^+ \rightarrow \begin{array}{l} RCH{-}NHR^1 \\ \quad| \\ RCH{-}NHR^1 \end{array}$$

phenyl, *p*-tolyl, benzyl, cyclohexyl, and anisyl. The cathode was mercury and the catholyte a solution of tetra-*n*-butylammonium bromide in a mixture of ethanol, methyl acetate, and water (5:3:1). Except where the *d,l* product was difficult to isolate from the by-product *sec*-amine, the yields were 36–67% of an approximate 1:1 mixture of *d,l* and *meso* hydrodimers. Reduction in the presence of chiral electrolytes [327b] yielded optically inactive

Table 10. Coupling Reactions with Reduced Azobenzene

Coupling partner	Products	Reference
$CH_2{=}CHCOOEt$	1,2-Diphenyl-3-oxopyrazolidine	328
$CH_2{=}CHCN$	N-(β-cyanoethyl)-diphenylhydrazine	328
CH_3I	N-methyl and N,N'-dimethyldiphenylhydrazine	329
CO_2	N,N'-dicarboxydiphenylhydrazine	330
$Br(CH_2)_{2-5}Br$	PhN——NPh \ / $(CH_2)_{2-5}$	331

hydrodimers but promoted up to 11% optical induction in the by-product secondary amine.

When the Schiff base is reduced to a moderately stable radical anion, as from N-(p-cyanobenzal)aniline, the initial dimeric dianion dissociates into two radical anions that are reduced to monomeric dianions [332].

Several types of reductive coupling of Schiff bases are now known and illustrated in the following.

Benzalaniline and excess acrylonitrile yielded, among other products, 1,5-diphenyl-2-pyrrolidine [328,333].

Amino acids have been formed by

1. Reduction of Schiff bases in molten tetraethylammonium p-toluenesulfonate in the presence of CO_2 [334].
2. Reductive alkylation of a pyruvic ester imine

$$PhCH_2N{=}\underset{\displaystyle Me}{\underset{|}{C}}{-}COOR' \xrightarrow[R''X]{} PhCH_2NH{-}\overset{\displaystyle Me}{\overset{|}{\underset{\displaystyle R''}{\underset{|}{C}}}}COOR'$$

   followed by N-debenzylation [332].
3. Reductive coupling of PhCN with $CO_2/HCl/AlCl_3$ at Pb yielded α-phenylglycine [335].

Reductive coupling of Schiff bases with α,ω-dibromoalkanes leads to pyrrolidine and piperidine derivatives [336].

Further examples of reductive alkylation of imines are given elsewhere.

Heterocycles in which coupling is through the hetero ring are discussed elsewhere (Chap. 18). Reductive couplings of carbonyl compounds bearing heterocyclic rings, such as acetylpyridines [337], 4-(2-furyl-3-butene-2-one) [338], and 2-benzoylthiophene [157], are unexceptional.

Organometalloids

Reduction at room temperature in dimethoxyethane or similar solvent of chlorotrimethylsilane using Bu_4NClO_4 or LiCl as electrolyte, Pb or Hg as cathode, and Hg or graphite as anode gave hexamethyldisilane in 95% current efficiency [340]. This is reminiscent of the reductive coupling of triphenyltin chloride to hexaphenyltin reported some years ago [341]. Similarly, diphenylchlorophosphine has been converted to tetraphenyldiphosphine [342].

A more recent study deals with reductive dimerizations of halosilanes and halogermanes [343].

Aromatic Esters

Reduction at Hg with R_4N^+ and HOAc, without which ring-hydrogenated products are formed, yielded about 30% of the hydrodimer, for example 1,2-diphenylethylene glycol [344].

Reductive Couplings Involving Inorganic Reagents

Some of these individual topics are discussed elsewhere in a different context but are cited here for completeness.

Carbon monoxide This gas has been reduced to the cyclic tetramer squaric acid (quadratic acid) [345], whereas chemical reduction leads to dimer and hexamer (cited in Ref. 346). Reduction in liquid ammonia gave the dimer $C_2O_2^{2-}$ [347]. C_1-G_4 hydrocarbons were reported to be formed from CO at Hg in the presence of a Mn(III) pyrocathecol complex [348].

Carbon dioxide Reductive coupling to oxalate has been extensively investigated [349a,d]; formation of malic acid has been claimed, not substantiated, and recently claimed again. The use of CO_2 as an electrophile to trap anionic electrogenerated species is a useful route to carboxylates and is mentioned frequently throughout this book (e.g., Refs. 349a and 350). Interestingly recent examples are the reaction of CO_2 with the anionic species formed by electroreduction RCOCH=CClSR' to yield 30–53% RCOCH=C(SR')COOH [351] and the reductive carboxylation of ketene S,S-acetals [352]. Electrogenerated superoxide anionically attacks CO_2 to yield peroxydicarbonate C_2O_6 [353]. Reductive coupling of CO_2 and butadiene as a route to dicarboxylic acids, particularly hexenedicarboxylic acid [354], readily converted to adipic acid, is still being investigated [11]. The reductive coupling of Schiff bases with CO_2 as a route to amino acid has been extended [357].

The originally proposed ECE mechanism, based on spectroelectrochemical evidence for the reductive coupling of CO_2 itself, has been challenged [358]. Furthermore, whereas oxalate is obviously the result of C—C coupling, C—O coupling leads to CO.

CO_2 has been reduced to methane and ethylene at Cu [361].

Sulfur compounds

Carbon disulfide Reduction of CS_2 in DMF/Bu_4NBr followed by reaction with CH_3I yielded [359]

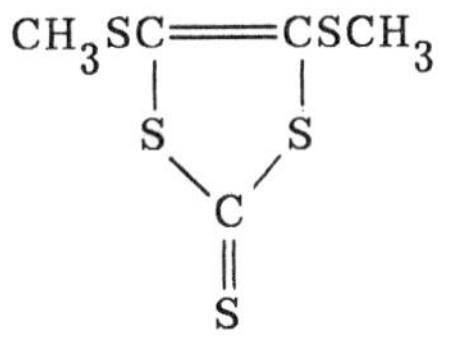

The originally proposed mechanism has been challenged and amended [360, 362]. Interest in the mechanism of this coupling [363,364b], the nature of the intermediates, and the preparation (in 75% current yield) of the tetrathioxalate product continues [365,366].

Sulfur dioxide Reduction of SO_2 to the anion radical leads to a very reactive species that may dimerize or, in reactions with organic substrates, form polymers that insulate the electrode. However, the reactivity, is moderated in the complex $SO_2^{\cdot-} \cdot xSO_2$. The latter, produced in MeCN or DMF/Et_4NBr, reacts with allylic, benzylic, and analogous activated halo compounds to give good yields of sulfones, presumably by thc following sequence [52a]:

$$SO_2^{\cdot-} + RX \longrightarrow RSO_2^{\cdot} + X^-$$

$$RSO_2^{\cdot} + SO_2^{\cdot-} \longrightarrow RSO_2^- + SO_2$$

$$RSO_2^- + RX \longrightarrow RSO_2R + X^-$$

o-Xylylene dibromide gave the (cyclic) sultone, which was converted by silica gel in ethanol to the cyclic sulfone. By control of the concentration of RX and R'X and other parameters, it was possible to adapt this synthesis to the preparation of unsymmetrical sulfones RSO_2R' almost exclusively. Selected dihalides yield polysulfones.

Sulfur An interesting and potentially useful way of incorporating S into organic compounds is the use of sulfur-carbon cathodes. Thus using Michael acceptors, such as acetylenes or thioorganic compounds, mainly thiophenes are formed [370]; methyldithiooxalamides have been made from piperidine or morpholine [371]; benzonitrile has been converted about 40% yield to thiobenzamide [356].

Miscellaneous compounds Methyl anthracene-9-carboxylate reductively couples through the 9-position [367].

In a one-electron reduction in aqueous medium, maleimide has been converted to its hydrodimer [297].

Many reductive alkylations have been reported in the past. Recent N-alkylations of amides, carbamates, and N-containing heterocycles by reduction of the substrates in the presence of alkyl halides are of synthetic interest [298].

Electrogenerated superoxide has been used to deprotonate a variety of carbon acids; the resulting carbanions couple with many electrophiles. This is discussed fully in Chap. 30.

ACKNOWLEDGMENT

The author is grateful to Henning Lund for an especially careful editing of this chapter and for supplying the camera-ready artwork.

REFERENCES

1. F. Beck, Angew. Chem., Int. Ed. Engl., 11, 760 (1972); M. M. Baizer and J. P. Petrovich, in Progress in Physical Organic Chemistry, Vol. 7 (A. Streitwieser and R. W. Taft, eds.), Wiley Interscience, New York, 1970, pp. 189–227; P. Zuman, J. Polarogr. Soc., 13, 53 (1967).
2. H. J. Schäfer, Angew. Chem., Int. Ed. Engl., 20, 911 (1981).
3. (a) C. P. Andrieux, L. Nadjo, and J.-M. Savéant, J. Electroanal. Chem., 32, 199 (1970); (b) W. V. Childs, J. T. Maloy, C. P. Keszthelyi, and A. J. Bard, J. Electrochem. Soc., 118, 874 (1971); (c) V. J. Puglisi and A. J. Bard, J. Electrochem. Soc., 119, 829 (1971); (d) L. Nadjo and J.-M. Savéant, J. Electroanal. Chem., 44, 127 (1973); (e) V. J. Puglisi and A. J. Bard, J. Electrochem. Soc., 120, 748 (1973); (f) E. Lamy, L. Nadjo, and J.-M. Saveant, J. Electroanal. Chem., 42, 189, (1973); (g) C. P. Andrieux, L. Nadjo, and J.-M. Savéant, J. Electroanal. Chem., 42, 223 (1973); (h) J. W. Hayes, I. Rusic, D. E. Smith, A. L. Booman, and J. R. Delmastro, J. Electroanal. Chem., 51, 269 (1974); (i) S. C. Rifkin and D. H. Evans, J. Electrochem. Soc., 121, 769 (1974); (j) I. B. Goldberg, D. Boyd, R. Hirasawa, and A. J. Bard, J. Phys. Chem., 78, 295 (1974); (k) L. A. Avaca and J. H. P. Utley, J. Chem. Soc., Perkin Trans. 2, 161 (1975); (l) I. Vartires, W. H. Smith, and A. J. Bard, J. Electrochem. Soc., 122, 894 (1975); (m) R. D. Grypa and J. T. Maloy, J. Electrochem. Soc., 122, 509 (1975); (n) M. J. Hazelrigg, Jr., and A. J. Bard, J. Electrochem. Soc., 122, 211 (1975); (o) L. Nadjo and J.-M. Savéant, J. Electroanal. Chem., 73, 163 (1976); (p) J.-M. Nigrette and A. J. Bard, J. Electrochem. Soc., 123, 1303 (1976); (q) L.-S. R. Yeh and A. J. Bard, J. Electrochem. Soc., 124, 189 (1977); (r) C. P. Andrieux, D. J. Brown, and J.-M. Savéant, Nouv. J. Chim., 1, 157 (1977); (s) M. Svaan and V. D. Parker, Acta Chem. Scand. (B), B39(6), 445 (1985); (t) V. D. Parker, Acta Chem. Scand. (B), B37(5) (1983); V. D. Parker, Adv. Phys. Org. Chem., 19, 131–322 (1983).
4. L. Nadjo, J.-M. Savéant, and D. Tessier, J. Electroanal. Chem., 64, 143 (1975). and the Savéant refs. in 3 above.
5. P. J. Elving, Can. J. Chem., 55, 3392 (1977).
6. S. Wawzonek, E. W. Blaha, R. Berkey, and M. E. Runner, J. Electrochem. Soc., 102, 235 (1955); S. Wawzonek and J. W. Fan, J. Amer. Soc., 68, 2541 (1946).
7. See, for example, M. Levy and M. Szwarc, J. Amer. Chem. Soc., 82, 521 (1960).
8. A. Vrancken, J. Smid, and M. Szwarc, Trans. Faraday Soc., 58, 2036 (1962); N. M. Berger, J. J. K. Boulton, B. W. Brooks, and M. J. Evans, J. Chem. Soc., Chem. Commun., 8, (1967); R. L. Williams and D. H. Richards, J. Chem. Soc., Chem. Commun., 414 (1967).
9. J. W. Loveland, U.S. Patent 3,032,489 (1962); Chem. Abstr., 57, 4470 (1962).
10. S. Gambino and G. Silvestri, Tetrahedron Lett., 3025 (1973).
11. D. Pletcher and J. T. Girault, J. Appl. Electrochem., 16(6), 791 (1986); H. T. Chen, H. J. Tien, T. T. Lai, and T. C. Chou, J. Chin. Inst. Chem. Eng., 16(1) 25 (1985); Chem. Abstr., 103, 29152g (1985).

12. J. F. Garst, Acc. Chem. Res., 4, 400 (1971); J. F. Garst, J. F. Pacifici, V. D. Singleton, M. F. Ezzel, and J. I. Morris, J. Amer. Chem. Soc., 97, 5242 (1975); S. Bank and S. P. Thomas, J. Org. Chem., 42, 2858 (1977); K. Schreiner, H. Oehling, H. E. Zieger, and I. Ingres, J. Amer. Chem. Soc., 99, 2638 (1977); S. Bank and D. A. Juckett, J. Amer. Chem. Soc., 98, 7742 (1976).
13. H. Lund, Denki Kagaku, 45, 2 (1977); Acta Chem. Scand., B31, 424 (1977).
14. E. Hobolth and H. Lund, Acta Chem. Scand., B31, 395 (1977).
15. P. E. Hansen, A. Berg, and H. Lund, Acta Chem. Scand., B30, 267 (1976).
16. E. D. Bergmann, D. Ginsburg, and R. Pappo, Organic Reactions, Vol. 10, Wiley, New York, 1959, p. 242.
17. M. M. Baizer and J. D. Anderson, J. Org. Chem., 30, 1348 (1965).
18. M. Tokuda, T. Taguchi, C. Nishio, and M. Itoh, J. Chem. Soc., Chem. Commun., 606 (1976).
19. A. V. Bukhtiarov, V. N. Golyshin, I. A. Rodnikov, O. V. Kuz'min, A. P. Tomilov, L. N. Nekrasov, and T. N. Khomchenko, Elektrokhimiya, 22(5), 663 (1986); Chem. Abstr., 105, 68883g (1986).
20. (a) J. Simonet, M. A. Michel, and H. Lund, Acta Chem. Scand., B29, 489 (1975); (b) H. Lund, M. A. Michael, and J. Simonet, Acta Chem. Scand., B28, 901 (1974); (c) C. Degrand and H. Lund, Acta Chem. Scand., B31, 593 (1977); (d) H. Lund and J. Simonet, J. Electroanal. Chem., 65, 205 (1975); (e) H. Lund, Denki Kagaku, 45, 2 (1977); (f) electron transfer from nonelectrochemically generated tetraphenylethylene dianion and RX: J. F. Garst, R. D. Roberts, and S. A. Pacifici, J. Amer. Chem. Soc., 99, 5528 (1977); electron transfer from chemically generated $Ph_2\bar{C}-\bar{N}Ph$ to a hindered RX yielded C- and p-alkyl derivatives: J. G. Smith and D. J. Mitchell, J. Amer. Chem. Soc., 99, 5045 (1977).
21. See, for example, K. J. Houser, D. E. Bartak, and M. D. Hawley, J. Amer. Chem. Soc., 95, 6033 (1973); R. F. Nelson, A. K. Carpenter, and E. T. Seo, J. Electrochem. Soc., 120, 206 (1973).
22. C. M. Pak and W. M. Gulick, Jr., Electrochim. Acta, 18, 1025 (1973).
23. R. Dietz and M. E. Peover, J. Appl. Electrochem., 3, 143 (1973).
24. R. Dietz, G. Biolchini, and W. M. Matthews, Anal. Chem., 46, 167 (1974).
25. F. M. Triebe, K. J. Borhani, and M. D. Hawley, J. Electroanal. Chem., 107, 375 (1980).
26. R. D. Rieke and P. M. Hudnall, J. Amer. Chem. Soc., 95, 2646 (1973).
27. L. Rampazzo, A. Inesi, and R. M. Bettolo, J. Electroanal. Chem., 83, 341 (1977).
28. M. R. Rifi, Tetrahedron Lett., 1043 (1969); Coll. Czech. Chem. Commun., 36, 932 (1971).
29. A. J. Fry and W. E. Britton, J. Org. Chem., 38, 4016 (1973); K. B. Wiberg and G. A. Epling, Tetrahedron Lett., 1119 (1974).
30. H. Doupeux and J. Simonet, Bull. Soc. Chim. Fr., 1219 (1972).
31. K. B. Wiberg, G. A. Epling, and M. Jason, J. Amer. Chem. Soc., 96, 912 (1974).
32. A. J. Fry and R. G. Reed, J. Amer. Chem. Soc., 94, 8475 (1972).
33. J. Grimshaw and J. Trocha-Grimshaw, Tetrahedron Lett., 2601 (1975).

34. J. Grimshaw, R. J. Haslett, and J. Trocha-Grimshaw, J. Chem. Soc., Perkin Trans. 1, 2448 (1977); J. Grimshaw and D. Mannus, J. Chem. Soc., Perkin Trans., 1, 2456 (1977).
35. J. Grimshaw and J. Trocha-Grimshaw, J. Chem. Soc., Perkin Trans. 2, 215 (1975); Tetrahedron Lett., 995 (1974).
36. R. Gottlieb and J. L. Neumeyer, 175th Meet., Am. Chem. Soc., Anaheim, Calif., Mar. 13-17, 1978, Abstr. 111; J. Amer. Chem. Soc., 98, 7108 (1976).
37. L. L. Miller, F. R. Stermitz, and J. R. Falck, J. Amer. Chem. Soc., 95, 2651 (1973); 97, 2922 (1975); S. M. Kupchan, V. Kameswaran, J. T. Lynn, D. K. Williams, and A. J. Liepa, J. Amer. Chem. Soc., 97, 5622 (1975).
38. W. M. Moore and D. G. Peters, Tetrahedron Lett., 453 (1972); G. W. McNamee, Diss. Abstr. Int. B, 37, 3917 (1976).
39. W. M. Moore, A. Salajegeh, and D. G. Peters, J. Amer. Chem. Soc., 97, 4954 (1975).
40. M. M. Baizer and J. L. Chruma, J. Org. Chem., 37, 1951 (1972).
41. M. M. Baizer and J. H. Wagenknecht, U.S. Patent 3,764,492 (Oct. 9, 1973); Chem. Abstr., 79, 142364b (1973).
42. N. Yamashina and H. Miura, Jpn. Kokai 75-151,122 (June 11, 1976); Chem. Abstr., 88, 169669c (1978).
43. F. Karrenbrock and H. J. Schäfer, Tetrahedron Lett., 1521 (1978).
44. T. Shono, N. Kise, M. Masuda, and T. Suzumoto, J. Org. Chem., 50(14), 2527 (1985).
45. T. Shono, N. Kise, and T. Suzumoto, J. Amer. Chem. Soc., 106(1), 259 (1984).
46. F. Karrenbrock, H. J. Schäfer, and I. Langer, Tetrahedron Lett., 2915 (1979).
47. J. P. Petrovich, M. M. Baizer, and M. R. Ort, J. Electrochem. Soc., 116, 749 (1969); (b) T. Hirai, I. Tari, and H. Yamagishima, Jpn. Kokai Tokkyo Koho, 78, 101, 311 (Sept. 4, 1978); Chem. Abstr., 90, 5921u (1979).
48. E. d'Incan, S. Sibille, J. Perichon, M. O. Moingeon, and J. Chaussard, Tetrahedron Lett., (35), 4175 (1985); S. Sibille, E. D'Incan, L. Leport, and J. Perichon, Tetrahedron Lett., 27(27), 3129 (1980).
49. T. Shono, I. Nishiguchi, and H. Ohmizu, Chem. Lett. (Tokyo) 1021 (1977).
50. F. Barba, M. D. Velasco, A. Guirado, and I. Barba, Electrochim. Acta, 30(9), 1119 (1985).
51. C. De Luca, A. Inesi, and L. Rampazzo, J. Chem. Soc., Perkin Trans., 2(11), 1403 (1982).
52. (a) B. Kastening and D. Knittel, J. Electroanal. Chem., 214(1-2), 221 (1986); (b) Ger. Offen. 2,500,727 of 7/15/76; Chem. Abstr., 85, P (4060z) (1976); (c) Ger. Offen. 2,629,320 of 1/15/78; Chem. Abstr., 88, P112558n (1979); (d) D. Knittel and B. Kastening, Ber. Bunsenges. Phys. Chem., 77, 833 (1973); (e) J. Appl. Electrochem., 3, 291 (1973).
53. J. G. Gourcy, G. Jimenet, and J. Simonet, C. R. Acad. Sci., Ser. C, 277, 1079 (1973).
54. (a) J. Pinson and J.-M. Savéant, J. Chem. Soc., Chem. Commun., 833 (1974); (b) W. J. M. Van Tilborg, C. J. Smit, and J. J. Scheele,

Tetrahedron Lett., 2113 (1977); (c) J. Pinson and J. M. Savéant, J. Amer. Chem. Soc., 100, 1506 (1978); (d) J. F. Bunnett, Acc. Chem. Res., 5, 139 (1975).
55. I. M. Sosonkin, T. K. Polynnikova, A. Y. Kaminskii, S. S. Gitis, and A. N. Domarev, Chem. Abstr., 82, 49222p (1975).
56. C. Degrand, J. Chem. Soc., Chem. Commun., (14), 1113 (1986).
57. T. Shono, Y. Matsumura, S. Kitoh, and N. Kise, Chem. Lett. (4), 463 (1985).
58. D. Bérubé, J. Caze, F. M. Kimmerle, and J. Lessard, Can. J. Chem., 53, 3060 (1975).
59. T. Fuchigama, T. Nonaka, and K. Iwata, J. Chem. Soc., Chem. Commun., 951 (1976).
60. T. Fuchigama and T. Nonaka, Chem. Lett. (Tokyo), 1087 (1977).
61. B. G. Soldatov et al., USSR Patent 533,590; Monomers Research Inst., USSR Patent 077757 (Nov. 25, 1974); Chem. Abstr., 83, P96747m (1975).
62. L. V. Mikhailchenko, A. S. Mendkovich, and V. P. Gul'tyai, Izv. Akad. Nauk SSSR, Ser. Khim., (9) 2158 (1985); Chem. Abstr., 105, 97136w (1986).
63. J. H. Wagenknecht, J. Org. Chem., 42, 1836 (1977); J. H. Wagenknecht and G. V. Johnson, J. Electrochem. Soc., 134, 2754 (1987).
64. L. H. Klemm, P. E. Iversen, and H. Lund, Acta Chem. Scand., B28, 593 (1974); R. D. Rieke, C. K. White, and C. M. Millren, J. Amer. Chem. Soc., 38, 6872 (1976).
65. J. Hlavaty, J. Volke, and O. Manousek, J. Electroanal. Chem., 61, 219 (1975). This type of reductive cyclization was first reported by H. Lund and L. G. Feotistov, Acta Chem. Scan., 23, 3482 (1969).
66. R. Hazard and A. Tallec, Bull. Soc. Chim. Fr., 679 (1975).
67. D. Bernard, Y. Mugnier, G. Tainturier, and E. Laviron, Bull. Soc. Chem. Fr., 2364 (1976).
68. (a) M. Cariou, R. Hazard, M. Jubault, and A. Tallec, Tetrahedron Lett., 22(40), 3961 (1981); (b) Can. J. Chem., 61(10), 2359 (1983).
69. R. D. Rieke, C. K. White, and C. M. Millren, J. Amer. Chem. Soc., 98, 6872 (1976).
70. A. D. Monvernay, J.-E. Dubois, and P. Clause, J. Electroanal. Chem., 81, 377 (1977).
71. M. M. Baizer and J. P. Petrovich, in Progress in Physical Organic Chemistry (A. Streitwieser and R. W. Taft, eds.), Vol. 7, Wiley-Interscience, New York, 1970, pp. 189–227.
72. S. Swann, Jr., in Techniques of Organic Chemistry (A. Weissberger, eds.) Vol. 2, Interscience, New York, 1956, pp. 495–486; M. J. Allen, Organic Electrode Processes, Reinhold, New York, 1958.
73. (a) J. Armand and L. Boulares, Can. J. Chem., 54, 1197 (1976); (b) T. Nonaka, A. Udagawa, and K. Odo, Chem. Lett. (Tokyo), 1261 (1975); (c) F. Ammar, C. P. Andrieux, and J. M. Savéant, J. Electroanal. Chem., 53, 407 (1974).
74. B. M. Davis, P. H. Gore, K. A. K. Lott, and E. L. Short, J. Chem. Soc., Perkin Trans. 2, 58 (1981).
75. E. Yu. Kmel'nitskaya, G. N. Vorozhtsov, K. A. Romanova, and N. S. Dokunikhin, Zh. Obshch. Khim., 49, 1859 (1979); Chem. Abstr., 91, 183827w (1979).
76. A. J. Klein, Disst. Abstr. Int. B, 40, 1674 (1979).
77. (a) N. Egashira, T. Minami, T. Kondo, and F. Hori, Electrochim.

Acta, 31(4), 463 (1986); (b) S. Thangavelu, Bull. Electrochem., 1(6), 553 (1985); (c) V. P. Gul'tyai and N. V. Surikova, Izv. Akad. Nauk SSSR, Ser. Khim., (7), 1690 (1983); Chem. Abstr., 99, 194556y (1983); (d) M. Gonet, J. Swiderski, and A. Korczynaki, Przem. Chem., 62(2), 104 (1983); Chem. Abstr., 99, 21898h (1983); (e) S. G. Maironovskii, Elektrokhimya 19(6), 838 (1983); (f) M. Honnorat and P. Martinet, Electrochim. Acta, 28(12), 1703 (1983); (g) M. Blasquez, J. M. Rodriguez-Mellardo, and J. J. Ruiz, Electrochim. Acta, 30(11), 1527 (1985).

78. D. W. Sopher and J. H. P. Utley, J. Chem. Soc., Chem. Commun. 1087 (1979).

79. D. W. Sopher and J. H. P. Utley, J. Chem. Soc., Perkin Trans. 2 (8), 1361 (1984).

80. F. Fournier and M. Fournier, Can. J. Chem., 64(5), 881 (1986); F. Fournier, J. Berthelot, and Y. L. Pascal, Can. J. Chem., 61(9), 339 (1984).

81. F. Fournier, J. Berthelot, and Y. L. Pascal. Can. J. Chem., 61(9), 2121 (1983).

82. M. T. Escot, P. Pouillen, and P. Martinet, Electrochim. Acta, 28(12), 1697 (1983).

83. S. Thangavelu, M. Noel, and K. S. Udupa, Electrochim. Acta, 31(12), 1563 (1986).

84. (a) A. P. Tomilov, B. L. Klyuev, and V. D. Nechepurnoi, J. Gen. Chem. USSR, 43, 2769 (1973); (b) J. Org. Chem. USSR, 11, 2010 (1975).

85. (a) T. T. Sugano, B. P. Schenker, J. A. Walburg, and N. Shuster, U. S. Patent 3,992,269 (Nov. 16, 1976); Chem. Abstr., 86, 23591h (1977); Belg. Patent 847,782 (Apr. 29, 1977); (b) H. Nohe, F. Beck, W. Dietmar, and E. J. Schier, Ger. Offen. 2,345,461 (Mar. 27, 1975); Chem. Abstr., 83, 9181f (1975); (c) H. Kita, S. Ishikura, and A. Katayama, Electrochim. Acta, 20, 441 (1975).

86. H. Nohe and F. Beck, Ger. Offen. 2,343,054 of 2/6/75; Chem. Abstr., 84, 1049s (1976); A. F. Mazanko, R. P. Sokolov, Z. H. Malysheva, A. P. Tomilov, and V. A. Glazov, USSR SU 947,226 of 7/30/82; Chem. Abstr., 97, 225706n (1982).

87. (a) N. L. Weinberg, U.S. Patent 4,478,694 of 10/29/84; Chem. Abstr., 102, 35368c (1985); (b) J. J. Barber, U.S. Patent 4,517,062 of 5/14/85; Chem. Abstr., 103, 13562e (1985).

88. W. J. M. Van Tilborg and C. J. Smith, Tetrahedron Lett., 3651 (1977).

89. E. Hori, Y. Takiguchi, and N. Urabe, Denki Kagaku, 40, 455 (1972); Chem. Abstr., 77, 139215g (1972).

90. E. V. Kryukova and A. P. Tomilov, J. Appl. Chem. USSR, 45, 873 (1972).

91. (a) E. J. Rudd and B. E. Conway, Trans. Faraday Soc., 67, 440 (1971); (b) A. Bewick and D. J. Brown, J. Chem. Soc., Perkin 2, 1977 (1977).

92. M. Svaan and V. D. Parker, Acta Chem. Scand., Ser. B, B36(6), 365 (1977).

93. F. Ammar, L. Nadjo, and J.-M. Savéant, J. Electroanal. Chem., 48, 146 (1973).

94. L. N. Nekrasov, L. N. Vykhodtseva, A. P. Korotkov, and L. P. Yur'yeva, Sov. Electrochem., 13, 628 (1977).

95. W. R. Fawcett and A. Lasia, Can. J. Chem., 59(23), 3256 (1981).
96. V. P. Gul'tyai, L. M. Korotova, A. S. Mendkovich, and I. V. Proskurovskaya, Izv. Akad. Nauk SSSR, Ser. Khim., 834 (1981); Chem. Abstr., 94, 216563x (1981).
97. W. J. M. Van Tilborg and C. J. Smit, Recl. Trav. Chim. Pays-Bas, 97, 89 (1978) and references cited therein.
98. (a) D. J. Brown and L. Horner, Justus Liebigs Ann. Chem., 77 (1977); (b) I. A. Titova, I. M. Levinson, V. G. Mairanovskii, and A. B. Ershler, Sov. Electrochem., 9, 408 (1973); (c) J. Kopilov, E. Kariv, and L. L. Miller, J. Amer. Chem. Soc., 99, 3450 (1977); (d) D. Seebach, H.-A. Oei, and H. Daum, Chem. Ber., 110, 2316 (1977).
99. D. F. Tomkins and J. H. Wagenknecht, J. Electrochem. Soc., 125, 372 (1978).
100. J. F. Rusling and P. Zuman, J. Org. Chem., 46, 1906 (1981).
101. V. G. Khomyakov, A. P. Tomilov, and B. G. Soldatov, Elektrokhimiya, 5, 850, 853 (1969).
102. A. P. Tomilov, A. A. Sergo, and S. L. Varshavskii, Elektrokhimiya, 1, 1126 (1965).
103. A. P. Tomilov, V. G. Khomyakov, and B. G. Soldatov, Zh. Vses. Khim. Obshchest., 14, 230 (1969); Chem. Abstr., 71, 18188 (1969).
104. E. Bauer, Z. Elektrochem., 37, 255 (1931).
105. S. Thangavelu, Bull. Electrochem., 1(6), 553 (1985).
106. H. Kaufmann, Z. Elektrochem., 2, 365 (1896); 4, 461 (1898).
107. H. D. Law, J. Chem. Soc., 89, 1512 (1906); 91, 748 (1907); 99, 1113 (1911).
108. W. Scheps, Ber. Deutsch. Chem. Ges., 46, 2564 (1913).
109. H. D. Law, J. Chem. Soc., 91, 748 (1907); 99, 1113 (1911).
110. J. H. Stocker, R. M. Jenevein, and D. J. Kern, J. Org. Chem., 34, 2810 (1969).
111. T. Arai and T. Oguri, Bull. Chem. Soc. Jpn., 33, 1018 (196); J. Grimshaw and J. S. Ramsey, J. Chem. Soc., C, 653 (1966).
112. W. S. Rapson and R. Robinson, J. Chem. Soc., 1537 (1935).
113. M. J. Allen, J. Org. Chem., 15, 435 (1950).
114. M. J. Allen, J. E. Fearn, and M. Levine, J. Chem. Soc., 2220 (1952).
115. W. O. Reichsfeld and S. K. Kudrna, Tr. Leningr. Technol. Inst. L. (Russ.) 110 (1956).
116. M. E. Manzhelei, V. I. Spektor, V. Ya. Botsan and M. N. Timotin, Zh. Obshch. Khim., 46, 1328 (1976).
117. D. Guerout and C. Caullet, C. R. Acad. Sci., Ser. C, 281, 643 (1975).
118. J. P. Salaun, M. Salaun-Bouix, and C. Caullet, C. R. Acad. Sci., Ser. C, 280, 165 (1975).
119. (a) I. E. Gillet, ibid., 40, 573 (1968); (b) L. G. Feoktistov, A. P. Tomilov, E. V. Kryukova, V. A. Klimov, and I. N. Brago, Sov. Electrochem., 3, 1352 (1967).
120. M. Gonet, J. Swiderski, and A. Korczynski, Przem. Chem., 62(2), 104 (1983); Chem. Abstr., 99, 21898h (1983).
121. K. Elbs and K. Brand, Z. Elektrochem., 45, 448 (1912).
122. A. P. Tomilov and M. I. Kalitina, Zh. Prikl. Khim., 38, 1574 (1965); Chem. Abstr., 63, 11343e (1965).
123. A. P. Tomilov and L. A. Ignat'eva, J. Appl. Chem. USSR, 38, 2651 (1965).

124. L. Horner and D. Degner, Electrochim. Acta, 19, 611 (1974).
125. E. Kariv, J. Hermolin, and E. Gileadi, J. Electrochem. Soc., 117, 342 (1970).
126. H. Lund, Acta Chem. Scand., 11, 283 (1957).
127. S. Swann, Jr., and G. H. Nelson, Trans. Electrochem. Soc., 67, 201 (1935).
128. J. H. Stocker and R. M. Jenevein, J. Org. Chem., 33, 294 (1968).
129. F. Kaulfer, Z. Elektrochem., 13, 635 (1907).
130. E. Müller, Z. Elektrochem., 16, 236 (1910).
131. S. Swann, Jr., P. Autbrose, R. Dale, R. Rowe, H. Ward, H. Kerman, and S. Axelrod., Trans. Electrochem. Soc., 85, 231 (1944).
132. E. V. Kryukova and A. P. Tomilov, Zh. Prikl. Khim., 45, 861 (1972); Chem. Abstr., 77, 42374v (1972).
133. M. D. Bhatti and O. R. Brown, J. Chem. Soc., Faraday Trans., 71, 1, 106 (1975).
134. E. J. Rudd and B. E. Conway, Trans. Faraday Soc., 67, 440 (1971).
135. C. P. Andrieux, J. M. Dumas-Bouchiai, and J. M. Saveant, J. Electroanal. Chem., 83, 255 (1979).
136. V. P. Gul'tyai and N. V. Surikova, Izv. Akad. Nauk SSSR, Ser. Khim.(7), 1690 (1983).
137. R. A. Misra and B. L. Sharma, Indian J. Chem., B, 22B(2), 160 (1983).
138. R. Mitzner, J. Bendig, R. Ziebig, F. Graichen, D. Kreysig, and F. Pragst, J. Prakt. Chem., 327(2), 241 (1985).
139. C. P. Andrieux and J.-M. Savéant, Bull. Soc. Chim. Fr., 2090 (1973).
140. M. J. Allen, J. Amer. Chem. Soc., 73, 3503 (1951).
141. M. J. Allen and A. H. Corwin, J. Amer. Chem. Soc., 72, 114 (1950).
142. N. J. Leonard, S. Swann, Jr., and G. Fuller, J. Amer. Chem. Soc., 75, 5127 (1953).
143. H. Lund and A. D. Thomsen, Acta Chem. Scand., 23, 3567, 3582 (1969).
144. M. J. Allen, J. Chem. Soc., 1598 (1951).
145. M. J. Allen, M. Cohen, W. G. Pierson, and J. A. Siragusa, J. Chem. Soc., 757 (1961).
146. A. P. Tomilov and V. A. Klimov, ibid., 3, 323 (1967); F. Beck, Ber. Bunsenges. Phys. Chem., 72, 379 (1968).
147. R. E. Juday and W. Sullivan, J. Org. Chem., 20, 617 (1955).
148. I. A. Avrutskaya, M. Ya. Fioshin, L. E. Gerasimova, and Nguen Van Tchan, Sov. Electrochem., 6, 81 (1970).
149. M. J. Allen, J. A. Siragusa, and W. G. Pierson, J. Chem. Soc., 1045 (1960).
150. S. Swann, Jr., S. W. Briggs, V. C. Neklutin, and A. J. Jerome, Trans. Electrochem. Soc., 80, 163 (1941).
151. S. Swann, Jr., Trans. Electrochem. Soc., 63, 239 (1933); 64, 313 (1933).
152. M. Ohmori and M. Takagi, Bull. Chem. Soc. Jpn., 50, 773 (1977).
153. M. J. Allen and M. Cohen, J. Electrochem. Soc., 106, 451 (1959).
154. E. V. Kryukova and A. P. Tomilov, Sov. Electrochem., 5, 508 (1969).
155. C. Caullet, J. Bessin, and J. Bodard, C. R. Acad. Sci., Ser. C, 261, 1848 (1965).
156. C. Caullet, M. Salaun, and M. Herbert, C. R. Acad. Sci., Ser. C, 261, 1848 (1965).
157. P. Foulatier, J.-P. Salaun, and C. Caullet, C. R. Acad. Sci., Ser. C, 279, 679, 779 (1974).

158. S. Goldschmidt, Angew. Chem., 69, 132 (1957).
159. G. Capobianci, G. Farnia, A. Sambaro, and M. Severin, J. Chem. Soc., Perkin Trans. 2(5), 545 (1982).
160. J. Dewar and J. Read, J. Soc. Chem. Ind. Trans., 55, 347 (1936).
161. I. A. Pearl, J. Amer. Chem. Soc., 74, 4260 (1952).
162. M. J. Allen, J. Amer. Chem. Soc., 72, 3787 (1950).
163. I. L. Knunyants and N. S. Vyazankin, Izv. Akad. Nauk USSR, Otdel. Khim. Nauk, (2) 233–240 (1957).
164. S. U. Pedersen, R. G. Hazell, and H. Lund, Acta Chem. Scand., B41, 336 (1987).
165. Y. Arad, M. Levy, I. R. Miller, and D. Vofsi, J. Electrochem. Soc., 114, 899 (1967).
166. L. Horner and R. Schneider, Tetrahedron Lett., 3133 (1973).
167. A. Bewick and D. J. Brown, J. Chem. Soc., Perkin Trans. 2, 99 (1977).
168. (a) T. J. Curphey, C. W. Amelotti, T. P. Layloff, R. L. McCartney, and J. H. Williams, J. Amer. Chem. Soc., 91, 2817 (1969); (b) T. Nonaka, A. Udagawa, and K. Odo, Chem. Lett., 1261 (1975); (c) J. Armand and L. Boulares, Can. J. Chem., 54, 1197 (1976).
169. B. F. Becker and H. P. Fritz, Chem. Ber., 108, 3292 (1975).
170. A. P. Tomilov, B. L. Klyuev, and V. D. Nechepurnoi, Zh. Obshch. Khim., 42, 2576 (1972); 43, 2792 (1973).
171. D. H. Evans and E. C. Woodbury, J. Org. Chem., 32, 2158 (1967).
172. R. C. Buchta and D. H. Evans, J. Electrochem. Soc., 117, 1494 (1970).
173. J. Wiemann, Bull. Soc. Chim. Fr., 2545 (1964).
174. J. Dewar and J. Read, J. Soc. Chem. Ind. (Trans.), 55, 347 (1936).
175. A. Bewick and H. P. Cleghorn, J. Chem. Soc., Perkin Trans. 2, 1410 (1973).
176. J. Wiemann, M. Dedieu, and D. Lelandais, Fr. Pat. 1,591,372, June 5, 1970; Chem. Abstr., 74, 70990z (1971).
177. K. Sugino and T. Nonaka, Electrochim. Acta, 13, 613 (1968); J. Electrochem. Soc., 116, 615 (1969); T. Nonaka and K. Sugino, Denki Kagaku, 38, 105 (1970); T. Nonaka and T. Sekine, Denki Kagaku, 39, 29 (1971); T. Nonaka, T. Sekine, K. Odo, and K. Sugino, Electrochim. Acta, 22, 271 (1977); M. Ferles, M. Vanka, and A. Silhankova, Coll. Czech. Chem. Commun., 34, 2108 (1969); A. P. Tomilov, B. L. Klyuev, V. D. Nechepurnoi, and N. Sh. Fuks, Sov. Electrochem., 9, 505 (1973).
178. T. Nonaka and K. Odo, Denki Kagaku, 40, 66 (1972).
179. T. Nonaka and K. Odo, Denki Kagaku, 40, 665 (1972); T. Nonaka, A. Omura, T. Fuchigami, and K. Odo, Denki Kagaku, 45, 111 (1977); Chem. Abstr., 87, 30999w (1977).
180. T. Nonaka, N. Yui, and K. Odo, Denki Kagaku, 42, 160 (1974); Chem. Abstr., 81, 25592e (1974).
181. M. Itoh, T. Taguchi, V. V. Chung, M. Takuda, and A. Suzuki, J. Org. Chem., 37, 2357 (1972).
182. (a) S. M. Makarochkina and A. P. Tomilov, J. Gen. Chem. USSR, 44, 2523 (1974); (b) A. Froling, Rec. Trav. Chim. Pays-Bas, 44, 2523 (1974); A. Froling, Rec. Trav. Chim. Pays-Bas, 93, 47 (1974); (c) M. M. Baizer and J. L. Chruma, J. Electrochem. Soc., 118, 450 (1971).

183. (a) T. Shono and M. Mitani, Nippon Kagaku Kaishi, 975 (1973); Chem. Abstr., 79, 3795g (1975); (b) M. Nicolas and R. Pallaud, C. R. Acad. Sci., Ser. C, 265, 1044 (1967); (c) M. Nicolas and R. Pallaud, C. R. Acad. Sci., Ser. C, 276, 1834 (1968).
184. A. P. Tomilov and B. L. Klyuev, Zh. Obshch. Khim., 39, 470 (1969); Chem. Abstr., 70, 129479c (1969).
185. (a) M. Rakoutz, Fr. Patent 2,138,463 (Feb. 9, 1973); (b) F. Liska, M. Nemec, and V. Dedek, Coll. Czech. Chem. Commun., 39, 580 (1974).
186. M. Seko, A. Yontanza, Y. Takahashi, F. Yomamoto, K. Harada, and S. Matsumoto, Jpn. Kokai 74-20,298 (May 23, 1974); Chem. Abstr., 82, 139427f (1975); A. P. Tomilov, B. L. Klyuev, and V. D. Nechepurnoi, J. Org. Chem. USSR, 11, 1329 (1975); A. P. Tomilov, B. L. Klyuev, V. D. Nechepurnoi, and N. Sh. Fuks, USSR Patent 427,927 (May 15, 1974); Chem. Abstr., 81, 49494h (1974).
187. T. Nonaka and K. Oda, Denki Kagaku, 41, 662 (1973); Chem. Abstr., 80, 81752v (1974).
188. A. P. Tomilov, B. L. Klyuev, and V. D. Nechepurnoi, J. Org. Chem. USSR, 11, 2010 (1975).
189. N. L. Askerov, S. I. Mekhtiev, V. M. Mamedova, and A. P. Tomilov, Zh. Prikl. Khim. (Leningrad), 51, 1173 (1978); Chem. Abstr., 89, 97047u (1978); Azerb. Khim. Zh., 121 (1978), Chem. Abstr., 89, 214869b (1978).
190. N. L. Askerov, S. I. Mekhtiev, V. M. Mamedova, A. P. Tomilov, and G. S. Ovanesova, Azerb. Khim. Zh., 10 (1978); Chem. Abstr., 90, 22280b (1979).
191. N. L. Askerov, S. I. Mekhtiev, V. M. Mamedova, and A. P. Tomilov, Zh. Obshch. Khim., 48, 863 (1978); Chem. Abstr., 89, 23742v (1978).
192. T. Shono, H. Ohmizu, S. Kawakami, and H. Sugiyama, Tetrahedron Lett., 5029 (1981).
193. O. R. Brown and K. Lister, Disc. Faraday Soc., 45, 106 (1968).
194. T. Koizumi, T. Fuchigami, Z. Kandeel, N. Sato, and T. Nonaka, Bull. Chem. Soc. Jpn., 59(3), 757 (1986).
195. J. M. Nelson and A. M. Collins, J. Amer. Chem. Soc., 48, 2256 (1924).
196. S. Wawzonek and A. Gundersen, J. Electrochem. Soc., 107, 537 (1960).
197. E. M. Abbot and A. J. Bellamy, J. Chem. Soc., Perkin Trans. 2, 254 (1978).
198. A. Yamura, T. Sekine, and K. Sugino, J. Electrochem. Soc. Jpn., 34, 110 (1966).
199. T. Shono, H. Ohmizu, and S. Kawakami, Tetrahedron Lett., 4091 (1981).
200. R. A. Misra and A. K. Yadav, Electrochim. Acta, 25, 1221 (1980).
201. V. T. Glezer, J. Stradins, and J. Dregeris, Zh. Org. Khim., 15, 1776 (1979); Chem. Abstr., 91, 183934d (1979); R. A. Misra, A. Jain, and A. K. Yadav, J. Indian Chem. Soc., 59, 289 (1982).
202. A. J. Klein and D. H. Evans, J. Org. Chem., 42, 2560 (1977).
203. M. V. D. Bhatti and O. R. Brown, J. Chem. Soc., Faraday Trans., 1, 71, 106 (1975).
204. (a) G. Silvestri, S. Gambino, and G. Filardo, Tetrahedron Lett., 27(29), 3429 (1986); (b) G. Filardo, G. Silvestri, and S. Gambino, Eu. Patent Appl. EP 189,120 of January 21, 1985; Chem. Abstr., 105, 180522q (1986).

205. (a) J. H. Wagenknecht, U.S. Patent 4,582,577 of April 15, 1986; Chem. Abstr., 105, 50861p (1986); (b) J. H. Wagenknecht, U.S. Patent 4,601,797 of December 19, 1984; Chem. Abstr., 106, 24982e (1987).
206. Y. Ikeda and E. Manda, Bull. Chem. Soc. Jpn., 58(6), 1723 (1985).
207. Y. Ikeda and E. Manda, Chem. Lett., (3), 453 (1984).
208. H. Tanaka, S. Yamashita, T. Hanatani, T. Nakahara, and S. Torii, Stud. Org. Chem. (Amsterdam), 30(Recent Adv. Electroorg. Synth.), 307 (1987); Chem. Abstr., 107, 143590f (1987).
209. T. Shono, I. Nishiguchi, H. Ohmizu, and M. Mitani, J. Amer. Chem. Soc., 100, 545 (1978).
210. T. Shono, N. Kise, T. Suzumoto, and T. Morimoto, J. Amer. Chem. Soc., 108, 4676 (1986).
211. E. Kariv-Miller and J. Mihachi, J. Org. Chem., 51(7), 1041 (1986).
212. T. Shono, I. Nishiguchi, and H. Ohmizu, Chem. Lett., 1233 (1976).
213. G. Pattenden and G. M. Robertson, Tetrahedron Lett., 24(42), 4617 (1983).
214. C. L. Wilson, Trans. Electrochem. Soc., 75, 353 (1939).
215. C. L. Wilson and K. B. Wilson, Trans. Electrochem. Soc., 84, 153 (1943); E. P. Goodings and C. L. Wilson, Trans. Electrochem. Soc., 88, 77 (1945).
216. S. Wawzonek, A. R. Zigman, and G. R. Hansen, J. Electrochem. Soc., 117, 1351 (1970).
217. M. T. Ismail, K. M. Kassen, and A. A. Abdel-Wahab, Bull. Fac. Aci. Assiut Univ., 11(1), 121 (1982); Chem. Abstr., 98, 16263x (1983).
218. A. Curulli, A. Inesi, and E. Zeuli, Electrochim. Acta, 32(7), 1117 (1987).
219. M. Steininger and H. J. Schäfer, Angew. Chem., 94(1), 75 (1982).
220. J. M. Kern and H. J. Schäfer, Electrochim. Acta, 30(1), 81 (1985).
221. E. Paspaleer and K. Batsalova, Chem. Abstr., 77, 74674, 74575 (1972).
222. See, e.g., Y. Aard, M. Levi, and D. Vofsi, Fr. Patent 1,553,195 (Jan 10, 1969); Chem. Abstr., 72, 42853x (1970) and many others. For a review see Ref. 300.
223. W. J. Linn et al. (duPont), Fr. Patent 1,572,893 (June 27, 1969); Chem. Abstr., 72, 66427n (1970); G. Agnes et al., J. Chem. Soc., Chem. Commun., 1515A (1968); ICI, U.S. Patent 3,641,106 (Feb. 8, 1972); duPont, U.S. Patent 3,655,724 (April 11, 1972); T. Sata (Mitsubishi) Jpn. Kokai 73-96518 (Dec. 10, 1973); Chem. Abstr., 80, 95302p (1974); P. F. Todd (ICI), Ger. Offen. 2,446641 (April 3, 1975); Chem. Abstr., 83, 276390r (1975); W. Strohmeier and A. Kaiser, J. Organometal. Chem., 114, 273 (1976); SOHIO, Belg. Patent 861,222 (Nov. 26, 1976); D. Wright (ICI) Ger. Offen. 2,721,808 (Dec. 8, 1977); Chem. Abstr., 88, 50326v (1978); T. Yoshino, S. Sato, H. Uchiyumi, and J. Yaki, Jpn. Tokkyo Koho, 79, 12,4501; Chem. Abstr., 91, 174,847a (1979); J. R. Jennings and F. M. L. Kelly, Ger. Offen. 2,620,592 of 11/18/76; Chem. Abstr., 86, 55002h (1977).
224. See, for example, (a) M. M. Baizer, Tetrahedron Lett., 973 (1963); (b) M. M. Baizer, J. Electrochem. Soc., 111, 215 (1964); (c) M. M. Baizer and J. D. Anderson, J. Electrochem. Soc., 111, 223 (1964); (d) M. M. Baizer and J. D. Anderson, J. Org. Chem., 30, 3138 (1965); (e) J. P. Petrovich, J. D. Anderson, and M. M. Baizer, J. Org. Chem., 31, 3897 (1966); (f) M. M. Baizer, J. D. Anderson, J. H. Wagenknecht, M. R. Ort, and J. P. Petrovich, Electrochim.

Acta, 12, 1377 (1967); (g) M. M. Baizer, J. P. Petrovich, and D. A. Tyssee, J. Electrochem. Soc., 117, 173 (1970); (f) U. S. Patent 3,249,521 of 3/3/65. Ten patents issued July 6, 1965, including ones on use of cation-exchange membranes and of cells without diaphragms. There were subsequent patents of the use of a variety of R_4N^+ supporting electrolytes including bis-quaternary ammonium salts and other activated olefins. (g) J. Andersson and L. Eberson, Nouv. J. Chim., 1, 413 (1977); (h) J. Andersson and L. Eberson, J. Chem. Soc., Chem. Commun., 565 (1976); (i) I. N. Brago, L. V. Kaabak, and A. P. Tomilov, J. Appl. Chem. USSR, 42, 1135 (1969); (j) M. M. Baizer and J. D. Andersson, J. Org. Chem., 30, 1357 (1965); (k) J. D. Anderson, M. M. Baizer, and E. J. Prill, J. Org. Chem., 30, 1645 (1965); (l) Ref. 283; (m) J. Andersson, L. Eberson, and C. Svensson, Acta Chem. Scand., B32, 234 (1978); (n) B. Terem and J. H. P. Utley, Electrochim. Acta, 24, 1181 (1979).

225. J. P. Petrovich, M. M. Baizer, and M. R. Ort, J. Electrochem. Soc., 116, 749 (1969); T. Hirai, I. Tari, and H. Yamagishima, Jpn. Kokai Tokkyo Koho, 78-101,311 (Sept. 4, 1978); Chem. Abstr., 90, 5921u (1979).
226. Y. D. Smirnov, S. K. Smirnov, and A. P. Tomilov, J. Org. Chem. USSR, 4, 208 (1968).
227. S. Wawzonek, E. W. Blaha, R. Berkey, and M. E. Runner, J. Electrochem. Soc., 102, 235 (1955); S. Wawzonek and J. W. Fan, J. Amer. Chem. Soc., 68, 2541 (1946).
228. Asai Chemical Ind. Co., Ltd., Jpn. Kokai Tokkyo Koho JP 60 52,587 of 3/25/85; Chem. Abstr., 103, 131244a (1985); S. Yiang and B. Zhang, Chem. Abstr., 103, 38027x (1985).
229. M. Sertel, A. Yildiz, and H. Baumgärtel, Electrochim. Acta, 31(12) 1625 (1986).
230. D. E. Danly, J. Electrochem. Soc., 131(10), 435C (1984), the discovery, development, and commercialization of ADN process were presented by M. M. Baizer (Part I) and D. E. Danly (Part II) as the Armstrong Memorial Lecture at the University of Southampton, England.
231. T. Isoya, K. Nakagawa, and A. Aoshima, Stud. Org. Chem. (Amsterdam), 30, 415 (1987); Chem. Abstr., 107, 134699w (1987).
232. L. Onisciu, F. Ciomos, and I. A. Silberg, Rev. Chim. (Bucharest), 36(6), 503 (1985); Chem. Abstr., 103, 131072t (1985).
233. J. C. Trocciola and D. A. Landsmann, U.S. Patent 4,596,638 of 6/24/86; Chem. Abstr., 105, 104745z (1986).
234. J. C. Trocciola, U.S. Patent 4,566,957 of 1/28/86; Chem. Abstr., 104, 158155e (1986).
235. S. Thangavelu and P. Subbish, Bull. Electrochem., 1(2), 137 (1985).
236. G. Piccardi, L. Nucci, F. Pergola, and R. Guidelli, J. Electroanal. Chem., 164(1), 145 (1984).
237. P. Margaretha and V. D. Parker, Acta Chem. Scand., Ser. B, B36(4), 260 (1982); C. Amatore, R. Guidelli, R. M. Moncelli, and J.-M. Savéant, J. Electroanal. Chem., 148(1), 25 (1983).
238. A. Misono, T. Osa, and H. Y. Takamichi, Kogyo Kagaku Zasshi, 69, 945 (1966); Chem. Abstr., 66, 75482e, (1967).
239. G. Feroci and S. Roffia, J. Electroanal. Chem., 81, 387 (1977).
240. R. Vadaszy and R. E. Cover, J. Electroanal. Chem., 49 433 (1974).
241. A. Misono, T. Osa, and T. Ueno, Nippon Kagaku Zasshi, 88, 1184 (1969); Chem. Abstr., 69, 58794c (1968).

242. L. Mandell, R. F. Daley, and R. A. Day, Jr., J. Org. Chem., 41, 4087 (1976).
243. R. E. Sioda, B. Terem, J. H. P. Utley, and B. C. L. Weedon, J. Chem. Soc., Perkin Trans. 1, 561 (1976).
244. L. A. Powell and R. M. Wightman, J. Amer. Chem. Soc., 101, 4412 (1979).
245. J. Wiemann and M. L. Bouguerra, Ann. Chim. (Paris), 2, 35 (1967).
246. A. D. Thomsen and H. Lund, Acta Chem. Scand., 25, 1576 (1971).
247. K. Hofelmann, R. Sattelmeyer, and K. Hamann, Makromol. Chem., 130, 221 (1969).
248. J. Simonet, C. R. Acad. Sci., Ser. C, 263, 1546 (1966).
249. E. Touboul, F. Weisbuch, and J. Wiemann, C. R. Acad. Sci., Ser. C, 268, 1170 (1969).
250. L. N. Lavrishcheva, Chem. Abstr., 70, 106313q (1969); P. Tissot and P. Margaretha, Helv. Chim. Acta, 60, 1472 (1977).
251. S. Wawzonek and A. Gundersen, J. Electrochem. Soc., 111, 324 (1964); J. C. Johnston, J. D. Faulkner, L. Mandell, and R. A. Day, Jr., J. Org. Chem., 41, 2611 (1976).
252. L. A. Taran, S. I. Berezina, L. G. Smolenceva, and Yu. P. Kitaev, Elektrokhimiya, 11, 601 (1975).
253. L. A. Taran, S. I. Berezina, L. G. Smolenceva, and Yu. P. Kitaev, Izv. Akad. Nauk SSSR, Ser. Khim., 2611 (1974).
254. H. Satonaka, Z. Saito, and T. Shimura, Bull. Chem. Soc. Jpn., 46, 2892 (1973).
255. J. Sarrazin and A. Tallec, Tetrahedron Lett., 1579 (1977).
256. E. Kariv and J. Hermolin, Electrochim. Acta, 18, 417 (1973).
257. L. I. Kudinova, A. A. Volod'kin, V. V. Ershov, and T. I. Prokof'eva, Izv. Akad. Nauk SSSR, Ser. Khim., 1503 (1978).
258. D. Zacharova-Kalavska, A. Perjessy, and P. Hrnciar, Coll. Czech. Chem. Commun., 38, 2461 (1973).
259. L. G. Chatten, R. W. Daisley, and C. J. Oiliff, J. Chem. Soc., Perkin Trans. 2, 469 (1973); K. Junghans, Chem. Ber., 106, 3465 (1973).
260. C. Degrand, B. Grosjean, and P.-L. Compagnon, Bull. Soc. Chim. Fr. (2), 281 (1973).
261. C. Degrand and M. Lacour, Bull. Soc. Chim. Fr. (2), 817 (1976).
262. E. Touboul and G. Dana, Tetrahedron, 31, 1925 (1975).
263. C. J. H. King, U.S. Patent 4,133,729 of Jan. 9, 1979; Chem. Abstr., 90, 112128r (1979).
264. E. Brillas and J. J. Ruiz, J. Electroanal. Chem., 215 (91-2), 293 (1986).
265. E. Brillas and A. Ortiz, J. Chem. Soc., Faraday Trans. 1, 82(2), 495 (1986); Electrochim. Acta, 30(9), 1185 (1985).
266. L. Horner and C. Franz, Z. Naturforsch, B, Anorg. Chem. Org. Chem., 40B(6), 822 (1985).
267. M. Moncelli, L. Nucci, P. Mariani, and R. Guidelli, J. Electroanal. Chem., 183(1-2), 285 (1985).
268. M. Moncelli, L. Nucci, P. Mariani, and R. Guidelli, J. Electroanal. Chem., 172(1-2), 83 (1984).
269. M. M. Baizer, J. Org. Chem., 29, 1670 (1964); S. M. Makarochkina and A. P. Tomilov, Zh. Obshch. Khim., 40, 676 (1970); Chem. Abstr., 73, 41210u (1970); K. Sugino, K. Shirai, and T. Nonaka, Bull. Chem. Soc. Jpn., 37, 1895 (1964).

270. By Li: D. H. Richards, N. F. Scilly, F. J. Williams, Chem. Ind., 1298 (1970).
271. (a) M. R. Ort and M. M. Baizer, J. Org. Chem., 31, 1646 (1966); (b) M. M. Baizer, U.S. Patent 3,390,066 (June 25, 1968); (c) J.-Y. Pape and J. Simonet, Electrochim. Acta, 23, 445 (1978).
272. L. H. Klemm and D. R. Olson, J. Org. Chem., 38, 3390 (1973).
273. (a) L. Nadjo and J.-M. Savéant, J. Electroanal. Chem., 73, 163 (1976); (b) C. P. Andrieux, D. J. Brown, and J.-M. Savéant, Nouv. J. Chim., 1, 157 (1976).
274. E. Lamy, L. Nadjo, and J.-M. Savéant, Nouv. J. Chim., 4, 21 (1979).
275. D. A. Tyssee, J. H. Wagenknecht, M. M. Baizer, and J. L. Chruma, Tetrahedron Lett., 4809 (1972); D. A. Tyssee and M. M. Baizer, J. Org. Chem., 39, 2819, 2923 (1974); D. A. Tyssee, U.S. Patent 3,945,896 (1976).
276. K. Sugino and T. Nonaka, Electrochim. Acta, 13, 613 (1968); J. Electrochem. Soc., 116, 615 (1969); T. Nonaka and K. Sugino, Denki Kagaku, 38, 105 (1970); M. M. Baizer and J. L. Chruma, J. Electrochem. Soc., 118, 450 (1971); T. Nonaka and T. Sekine, Denki Kagaku, 39, 29 (1971).
277. D. A. Tyssee and M. M. Baizer, J. Org. Chem., 39, 2819 (1974).
278. T. Shono, I. Nishiguchi, and H. Ohmizu, J. Amer. Chem. Soc., 99, 7396 (1977); H. Lund and C. Degrand, Tetrahedron Lett., 3593 (1977).
279. C. Degrand and H. Lund, Nouv. J. Chim., 1, 35 (1977).
280. J. H. Wagenknecht and M. M. Baizer, J. Org. Chem., 31, 3885 (1966).
281. M. M. Baizer and J. L. Chruma, J. Org. Chem., 37, 1951 (1972). A. P. Tomilov and L. V. Kaabak, Zh. Obshch. Khim., 33, 731 (1963); Ref. 98b and citation; also I. N. Brago, L. V. Kaabak, and A. P. Tomilov, Chem. Abstr., 67, 104513a (1967).
282. J. Wiemann and M. L. Bouguerra, C. R. Acad. Sci. Ser. C, 265, 751 (1967).
283. M. M. Baizer and J. D. Anderson, J. Org. Chem., 30, 1348 (1965).
284. J. Berthelot, Electrochim. Acta, 32(1), 179 (1987).
285. (a) D. P. Fox, R. D. Little, and M. M. Baizer, J. Org. Chem., 50(12), 2202 (1985); (b) L. Moens, M. M. Baizer, and R. D. Little, J. Org. Chem., 51, 4497 (1986).
286. G. Mabon, G. Le Guillanton, and J. Simonet, Nouv. J. Chim., 7(5), 305 (1983).
287. Asahi Chem. Ind. Co., Ltd., Jpn Kokai Tokkyo Koho JP 82,108,274 of July 6, 1982; Chem. Abstr., 97, 171368h (1982); JP 58 58,289 of April 6, 1983; Chem. Abstr., 99, 95855p (1983).
288. Asahi Chem. Ind. Co., Ltd., Jpn Kokai Tokkyo Koho, JP 82 02,883 of January 8, 1982; Chem. Abstr., 96, 207435y (198?); JP 57 95,978 of June 15, 1982; Chem. Abstr., 98, 16567z (1983); JP 82 98,683 of June 18 1982; Chem. Abstr., 97, 135782 (1982).
289. Analogous intermolecular couplings have been reported by T. Shono, H. Ohmizu, S. Kawakami, and H. Sugiyami, Tetrahedron Lett., 5029 (1980).
290. A. P. Tomilov and L. V. Kaabak, Zh. Obshch. Khim., 33, 731 (1963).
291. A. S. Mendkovich, V. N. Leibzon, S. G. Mairanovskii; M. M. Krauyshkin, T. A. Klimova, S. S. Novikov, and V. V. Sevost'yanova, Izv. Akad. Nauk SSSR, Ser. Khim., 1866 (1978).
292. L. N. Brago, L. V. Kaabak, and A. P. Tomilov, Chem. Abstr., 67, 104513a (1967).
293. D. W. Grisley, Jr., and M. M. Baizer, unpublished work.

294. A. P. Tomilov, Yu. D. Smirnov, and S. L. Varshavskii, J. Gen. Chem. USSR, 35, 390 (1965).
295. W. Kern and H. Quast, Makromol. Chem., 10, 202 (1953).
296. F. Beck, Chem.-Ing. Tech., 37, 607 (1965); T. Asahara, M. Seno, and T. Arai, Bull. Chem. Soc. Jpn., 42, 1316 (1969).
297. R. G. Barradas, S. L. Gust, and J. D. Porter, Tetrahedron Lett., 22(46), 4579 (1981).
298. T. Shono, S. Kahimura, and H. Nogussa, Chem. Lett., (3), 425 (1986).
299. J. P. Petrovich, M. M. Baizer, and M. R. Ort, J. Electrochem. Soc., 116, 749 (1969); T. Hirai, I. Tari, and H. Yamagishima, Jpn. Kokai Tokkyo Koho 78 101,311 (Sept. 4, 1978); Chem. Abstr., 90, 5921u (1979).
300. For a review, see C. Van Eygen, A. Hendrickx, J. Ramiouille, J. Walravens, and A. Verheyden, Chim. Ind. (Genie Chim.), 104, 71 (1971).
301. See, for example, Asahi Co., Fr. Patent 1,533,472 (July 19, 1968); Chem. Abstr., 71, 21705q (1969); Fr. Patent 1,531,738 (July 5, 1968); Chem. Abstr., 71, 21709u (1969).
302. L. G. Feoktistov, A. P. Tomilov, and I. G. Sevast'yanova, Sov. Electrochem., 1, 1165 (1965).
303. M. M. Baizer and J. D. Anderson, J. Org. Chem., 30, 1351 (1965), and literature cited therein.
304. C. Takagi and K. Ubukata, Chem. Abstr., 69, 66798x (1968), a review of 43 patents.
305. K. Hofelmann, R. Sattelmeyer, and K. Hamann, Makromol. Chem., 112, 300 (1968).
306. S. M. McElvain, Org. React., 4, 256 (1948).
307. T. Nonaka and K. Odo, Denki Kagaku, 40, 665 (1972); Chem. Abstr., 77, 171870 (1972).
308. Yu. D. Smirnov, S. K. Smirnov, and A. P. Tomilov, J. Org. Chem. USSR, 10, 1613 (1974).
309. (a) S. D. Ross, M. Finkelstein, and R. C. Petersen, J. Amer. Chem. Soc., 82, 1582 (1960); (b) S. D. Ross et al., J. Amer. Chem. Soc., 92, 6003 (1970); (c) A. Monvernay and P. C. Lacaze, Bull. Soc. Chim. Fr., 4206 (1970); (d) J.-E. Dubois, A. Monvernay, and P. C. Lacaze, Electrochim. Acta, 15, 315 (1970); (e) A. Monvernay-Desbene, Y. Robillard, P. C. Lacaze, and J.-E. Dubois, Bull. Soc. Chim. Fr., 548 (1975); (f) D. A. Tyssee and M. M. Baizer, J. Electrochem. Soc., 118, 1420 (1971); (g) R. N. Gedye, Y. N. Sadana, and R. Eng, J. Org. Chem., 45, 3721 (1980).
310. J.-M. Savéant and S. K. Binh, J. Org. Chem., 42, 1242 (1977).
311. R. A. Wessling and W. S. Settineri, U.S. Patent 3,480,537 (Nov. 25, 1969); Chem. Abstr., 72, 66593p (1970).
312. S. Sato, J. Nakaya, and E. Imoto, Denki Kagaku, 40, 708 (1972); Rev. Polarogr. (Jpn.), 18, 29 (1972).
313. K. S. V. Santhanam and P. J. Elving, J. Amer. Chem. Soc., 95, 5482 (1973).
314. M. Naarova and J. Volke, Coll. Czech. Chem. Commun., 38, 2670 (1973); F. R. Yu, Y. Y. Wang, and C. C. Wan, Electrochim. Acta, 30(12), 1693 (1985).
315. M. K. Polievktov, A. K. Sheinkman, and L. N. Mozorova, Khim. Geterotsikl. Soedin., 1067 (1973).

316. S. Kato, J. Nakaya, and E. Imoto, Bull. Chem. Soc. Jpn., 44, 1928 (1971).
317. L. Rouiller and E. Laviron, Electrochim. Acta, 22, 669 (1977).
318. P. Pflegel and G. Wagner, Pharmazie, 27, 24 (1972).
319. B. Czochralska and D. Shugar, Biochim. Biophys. Acta, 281, 1 (1972).
320. R. W. Johnson, T. Widlanski, adn R. Breslow, Tetrahedron Lett., 4685 (1976); R. Breslow and R. F. Drury, J. Amer. Chem. Soc., 96, 4702 (1974).
321. See, for example, M. Ferles, M. Lebl, P. Stern, and P. Trska, Coll. Czech. Chem. Commun., 40, 2183 (1975); T. Nonaka and K. Sugino, Denki Kagaku, 38, 105 (1970).
322. See, for example, R. Gottlieb and J. L. Neumeyer, J. Amer. Chem. Soc., 98, 7108 (1976).
323. C. P. Andrieux and J.-M. Savéant, Bull. Soc. Chim. Fr., 4671 (1968).
324. T. Shono, Y. Usui, T. Mizutani, and H. Hamaguchi, Tetrahedron Lett., 21, 3073 (1980).
325. F. Pragst, Electrochim. Acta, 21, 497 (1976).
326. F. Pragst and U. Seydewitz, J. Prakt. Chem., 319, 952 (1977).
327. (a) L. Horner and D. H. Skaletz, Tetrahedron Lett., 1103 (1970); (b) Justus Liebigs Ann. Chem., 1365 (1977); (c) 1210 (1975).
328. M. M. Baizer, J. D. Anderson, J. H. Wagenknecht, M. R. Ort, and J. P. Petrovich, Electrochim. Acta, 12, 1377 (1967); M. M. Baizer, U.S. Patent 3,438,877 (Apr. 15, 1969); Chem. Abstr., 71, 49534g (1969).
329. T. Troll and M. M. Baizer, Electrochim. Acta, 20, 33 (1975).
330. R. C. Hallcher and M. M. Baizer, Justus Liebigs Ann. Chem., 737 (1977).
331. T. Troll and W. Elbe, Electrochim. Acta, 22, 615 (1977)
332. T. Isakawa and K. Harada, J. Chem. Soc., Chem. Commun., 338, (1974); J. Chem. Soc. Perkin Trans. 1, 1730 (1977).
333. M. M. Baizer, U.S. Patent 3,410,769 (Nov. 12, 1968); Chem. Abstr., 70, 63589q (1969).
334. N. L. Weinberg, A. K. Hoffmann, and T. B. Reddy, Tetrahedron Lett., 2271 (1971).
335. Ube Industries Ltd., Jpn. Kokai Tokkyo Koho 80-119,185 (Sept. 12, 1980); Chem. Abstr., 93, 212419p; S. Kato, Chem. Abstr., 93, 176055j (1980).
336. C. Degrand, C. Grosdemouge, and P.-L. Compagnon, Tetrahedron Lett., 3023 (1978).
337. J. Kopilov, E. Kariv, and L. L. Miller, J. Amer. Chem. Soc., 99, 3450 (1977).
338. H. Satonaka, Z. Saito, and T. Shimura, Bull. Chem. Soc. Jpn., 46, 2892 (1973).
339. M. M. Baizer and J. P. Petrovich, J. Electrochem. Soc., 114, 1023 (1967).
340. E. Hengge and G. Litscher, Angew. Chem. Int., 15, 370 (1976).
341. R. E. Dessy, W. Kitching, and T. Chivers, J. Amer. Chem. Soc., 88, 453 (1966).
342. H. Matschiner, A. Tzschach, and R. Matuschke, E. Ger. Patent 86,394 (Dec. 12, 1973); Chem. Abstr., 77, 140281g (1972).
343. R. J. P. Corriu, F. Dabosi, and M. Matineau, J. Chem. Soc., Chem. Commun., 457 (1979).
344. L. Horner and H. Hoenel, Justus Liebigs Ann. Chem., 2036 (1977).

345. (a) G. Silvestri, S. Gambino, G. Filardo, G. Spadaro, and L. Palmisano, Electrochim. Acta, 23, 413 (1978); (b) J. J. Barber, U.S. Patent 4,461,681 of May 31, 1983; Chem. Abstr., 101, 160121x (1984).
346. G. Bockmair and H. P. Fritz, Z. Naturforsch., 30b, 330 (1975).
347. F. A. Uribe, P. R. Sharp, and A. J. Bard, J. Electroanal. Chem., 152(1-2) 173 (1983).
348. G. N. Petrova, O. N. Efimov, and V. V. Strelets, Izv. Akad. Nauk SSSR(11), 2608 (1982); Chem. Abstr., 98, 61943e (1983).
349. (a) D. A. Tyssee, J. H. Wagenknecht, M. M. Baizer, and J. L. Chruma, Tetrahedron Lett., 4809 (1972); D. A. Tyssee and M. M. Baizer, J. Org. Chem., 39, 2819, 2823 (1974); (c) D. A. Tyssee, U.S. Patent 3,945,896 (1976); (d) J. Y. Becker, B. Vainar, R. Eger, and L. Kaufman, J. Chem. Soc., Chem. Commun., 1471 (1985).
350. C. Degrand, R. Mora, and H. Lund, Acta Chem. Scand., B37, 429 (1983).
351. H. H. Rüttinger, H. Matschiner, and W. D. Gollnow, J. Prakt. Chem., 328(4), 539 (1986).
352. H. H. Rüttinger, W. D. Rudorf, and H. Matschiner, Electrochim. Acta, 30(2), 155 (1985).
353. J. L. Roberts, Jr., T. S. Calderwood, and D. T. Sawyer, J. Amer. Chem. Soc., 106(17), 4667 (1984).
354. J. W. Loveland, U.S. Patent 3,032,489 (1962); Chem. Abstr., 57, 4470 (1962); W. C. Neikam, U.S. Patent 3,344,045 (1967); Chem. Abstr., 68, 8809 (1968).
355. R. N. Gourley and J. Grimshaw, J. Chem. Soc. C, 2388 (1968).
356. H. Berge, B. Struebing, S. Strassen, and M. Aguila (E. Ger.), DD152,856 of May 11, 1978; Chem. Abstr., 225400n (1982).
357. (a) U. Hess and M. Ziebig, Pharmazie, 37(2), 107 (1982); (b) U. Hess and R. Thiele, J. Prakt. Chem., 324(3), 385 (1982).
358. A. W. B. Aylmer-Kelly, A. Bewick, P. R. Cantrell, and A. S. M. Tuxford, Faraday Disc. Chem. Soc., 56, 96 (1973); C. Amatore and J. M. Savéant, J. Amer. Chem. Soc., 103, 5021 (1981).
359. S. Wawzonek and S. M. Heilman, J. Org. Chem., 39, 511 (1974).
360. P. R. Moses and J. R. Chambers, J. Amer. Chem. Soc., 96, 945 (1974); P. Jeroschewski, Z. Chem., 18, 27 (1978); G. Bontempelli, F. Magno, G.-A. Mazzocchin, and R. Seeber, J. Electroanal. Chem., 63, 231 (1975).
361. R. L. Cook, R. C. MacDuff, and A. F. Sammells, J. Electrochem. Soc., 134(9), 2375 (1987).
362. M. F. Hurley and J. Q. Chambers, J. Org. Chem., 46, 775 (1981).
363. P. Jeroschewski, Z. Chem., 21, 412 (1981); H. Lund, E. Hoyer, and R. G. Hazell, Acta Chem. Scand., B36, 207 (1982).
364. P. Jeroschewski and F. Pragst, J. Electroanal. Chem., 149(1–2), 131 (1983).
365. (a) S. Wawzonek, H.-F. Chang, W. Everett, and M. Ryan, J. Electrochem. Soc., 130(4), 803 (1983); (b) H. Lund, E. Hoyer, and R. G. Hazell, Acta Chem. Scand., B36(3), 207 (1982), reported that reduction of CS_2 in CH_3CN saturated with KI gave tetrathiooxalate but that the latter was unstable in solution. For further characterization it could be isolated as the Ph_4P^+ salt.
366. P. Jeroschewski and H. Patzner, Ger (East), DD 200,805 of June 15, 1983; Chem. Abstr., 99, 112782n (1983). The role of CS in the electrochemical preparation of 4,5-bis(methylthio)-1,3-dithiole-2-thione is discussed by E. K. Moltzen, A. Senning, R. Hazell, and H. Lund, Acta Chem. Scand., B40, 593 (1986).

367. C. Z. Smith and J. H. P. Utley, J. Chem. Res. Synop (1), 18 (1982). See analogous coupling of 9-nitro-derivatives, Ref. 76.
368. J. D. Anderson, J. P. Petrovich, and M. M. Baizer, in Advances in Organic Chemistry (E. C. Taylor and H. Wynberg, eds.), Vol. 6, 1969, pp. 257–283. Classification mainly on the basis of electrophores.
369. Same citation as Ref. 2. This is part of a general review of anodic and cathodic C-C bond formation. Classification according to electrogenerated intermediates (radical ions, radicals, and carbanions.
370. J. Simonet and G. Le Guillanton, Bull. Soc. Chim. Fr. (2) 221 (1986). Classification mainly on the basis of functional groups.
371. G. Le Guillanton, T. Q. Do, and J. Simonet, Tetrahedon Lett., 27 (20), 2261 (1986).
372. M. L. Contreras, S. Rivas, and R. Rozas, J. Electroanal. Chem., 177(1-2), 299 (1984).
373. H. Berge, B. Struebing, S. Strassen, and M. Aguila, (East Ger.) DD 152.586 of May 11, 1978; Chem. Abstr., 96, 225400n (1982).
374. A. P. Tomilov and B. L. Klyuev, in Itogi Nauki. Elektrokhimiya, 1967, VINITI, Moscow, 1969, p. 200; Chem. Abstr., 72, 38095 (1970).
375. V. G. Khomyakov and A. P. Tomilov, Zh. Prikl. Khim., 36, 378 (1963).

23
Electrolytic Oxidative Coupling

HANS J. SCHÄFER / Organisch-Chemisches Institut, Wilhelms Universität, Münster, Federal Republic of Germany

I. Introduction 949

II. General Survey 950

III. Coupling via Radical Cations 951
- A. Aromatic compounds 951
- B. Olefins 979
- C. Intramolecular Coupling 985

IV. Coupling via Radicals 1001
- A. Kolbe dimerization 1001
- B. Coupling of anions 1005
- C. Oxidative addition of anions to double bonds 1007

V. Anodically Induced Cycloadditions 1013

References 1018

I. INTRODUCTION

For the construction of organic molecules, for example natural products and compounds for commercial use or molecules of theoretical interest, two types of reactions are used: carbon-carbon bond-forming reactions and functional group interconversions. The first assemble a large target molecule from small units; the latter introduce the functionality into the target and activate the building units for the carbon-carbon bond formation.

The large number of carbon-carbon bond-forming reactions can be summarized in four groups:

1. Polar reactions between an activated acceptor and a donor or vice versa
2. Radical couplings and additions
3. Pericyclic reactions, especially cycloadditions
4. Transition metal-catalyzed reactions

Electrochemistry can contribute to those reactions, especially to types 1 and 2. The reactive intermediates involved, namely radical ions, cations, anions, and radicals, can be generated by electron transfer at the electrode more simply, in larger variety, and more cheaply than with chemical reagents. This chapter deals with carbon-carbon bond-forming reactions at the anode.

II. GENERAL SURVEY

Anodic coupling reactions may be arranged according to the reactive intermediates involved. These are radical cations that couple either by radical dimerization [Eq. (1), path a] or by electrophilic addition to the nucleophilic starting compound [Eq. (1), path b]:

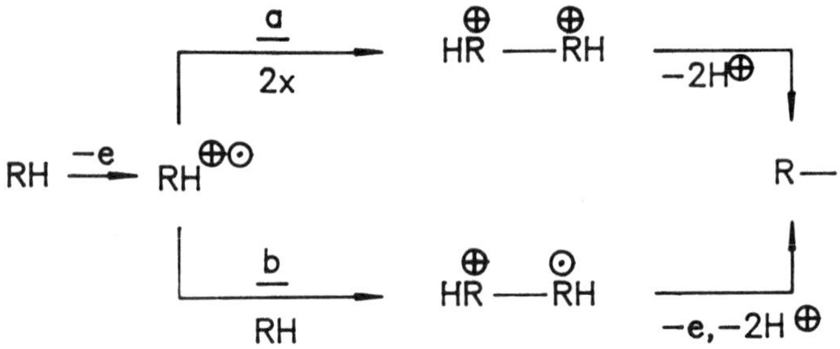

(1)

Radicals generated by oxidation of anions can either dimerize [Eq. (2), path a] or add to double bonds to yield additive monomers or dimers [Eq. (2), path b]:

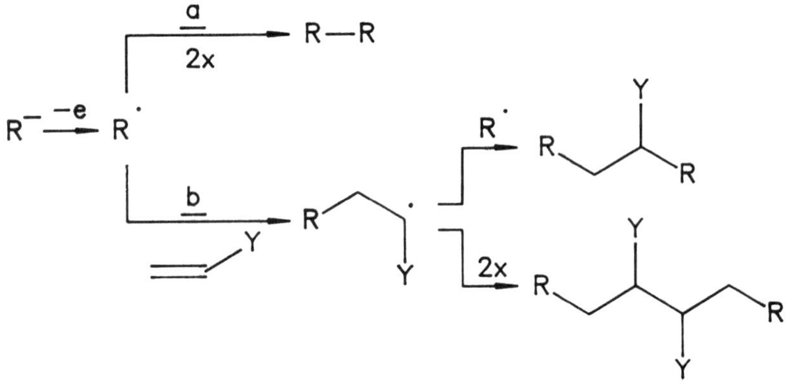

(2)

These reactions can lead to carbon-carbon or heteroatom-heteroatom bonds, and their course can be intermolecular or intramolecular. Furthermore, cycloadditions can be initiated by anodic generation of the dienophile or by inducing a chain reaction with a radical cation as dienophile.

The arrangement according to reactive intermediates makes the broad scope of these reactions more obvious and comprehensible than an arrangement according to educt and product. However, one must keep in mind that this organization is not free from arbitrary action, since the mechanisms of many anodic couplings are not accurately known.

Recent reviews on oxidative coupling at the anode are summarized in Ref. 1.

III. COUPLING VIA RADICAL CATIONS

A. Aromatic Compounds

Hydrocarbons

Representative couplings of aromatic hydrocarbons are summarized in Table 1. Alkyl-substituted aromatic hydrocarbons can be coupled to diphenyls and/or diphenylmethanes depending on their substitution pattern (Table 1, numbers 1–6). The initially formed radical cation I [Eq. (3)] reacts with the starting compound to the diphenyl (II) [Eq. (3), path a] or loses a proton to form a benzyl radical [Eq. (3), path b], which after oxidation to the cation undergoes an electrophilic aromatic substitution at the starting compound to form the diphenylmethane (III). A low charge density on an unsubstituted carbon atom of I favors path a, whereas a low charge density on a substituted carbon atom favors path b [4].

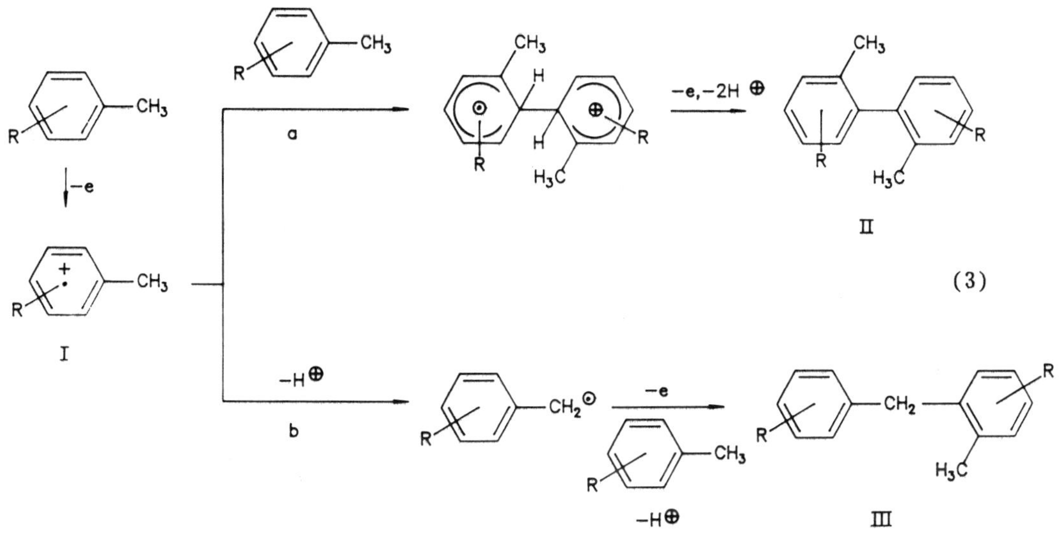

(3)

For mesitylene and durene the kinetics have been followed by specular reflectance spectroscopy [17]. The results indicated that mesitylene produces a fairly stable radical cation that dimerizes. That of durene, however, is more unstable and loses a proton to form a benzyl radical, which subsequently leads to a diphenylmethane. The stability of the radical cation increases with increasing charge delocalization, blocking of reactive sites, and stabilization by specific functional groups (phenyl, alkoxy, and amino) [18]. The complex reaction mechanisms of radical cations and methods of their investigation have recently been reviewed in detail [19].

To favor the coupling reaction, the competing side reaction of the radical cation with nucleophiles must be suppressed by the use of a medium of low nucleophilicity. The solvent of choice is dichloromethane. Especially in electroanalytic studies, neutral alumina is frequently added to suppress hydroxylation of the radical cation [20]. The reversible cyclic voltammetric behavior of radical cations is also enhanced in mixtures of methylene dichloride, trifluoroacetic acid, and trifluoroacetic anhydride (TFAn) with $TBABF_4$ as supporting electrolyte. With acetonitrile as solvent, acetamides

Table 1. Anodic Coupling of Aromatic Hydrocarbons

Number	Substrate	Conditions	Product	Yield (%)	Reference
1	p-Xylene	C, CH_2Cl_2, TFA, $TBABF_4$	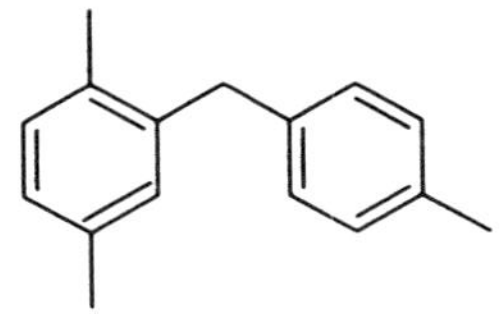	22	2
2	1,2,4-Trimethylbenzene	C, CH_3NO_2, Bu_4NBF_4	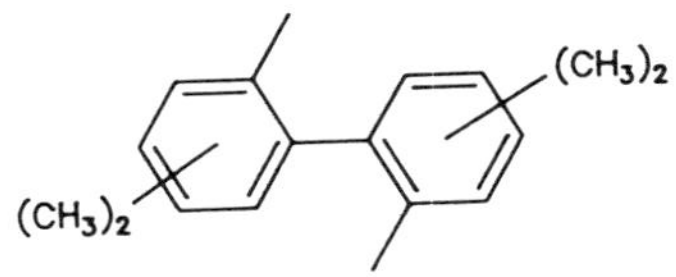	17	3
			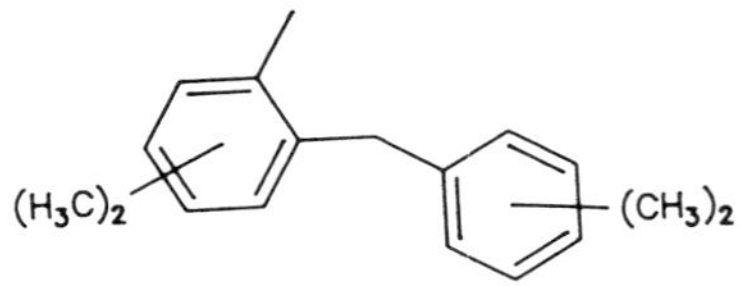	21	
3	1,3,5-Trimethylbenzene	Pt, Bu_4NBF_4, CH_3CN	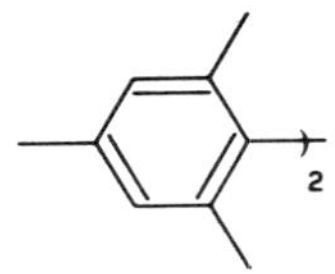	71	4
4	5-Alkyl-substituted m-xylenes	Pt, Ch_3NO_2, or MeCN, Bu_4NBF_4	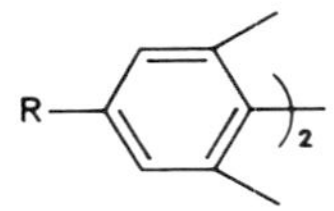 R = Me, Et, i-Pr, t-Bu	4–82	5

5	1,2,4,5-Tetramethyl-benzene	C, CH_2Cl_2, $TBABF_4$	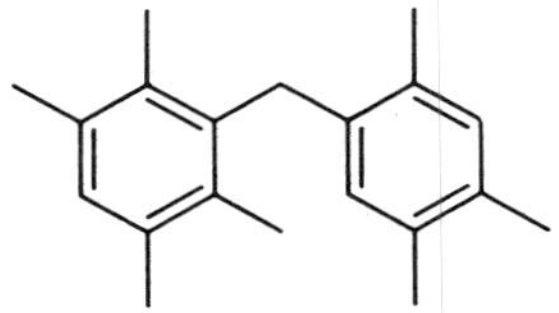	85	2
6	2-Carbomethoxy-1,3,5-trimethylbenzene	Conditions not given	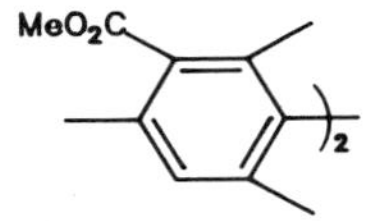	50	6
7	1-Methylnaphthalene	PbO_2 anode, acetone, H_2SO_4, water	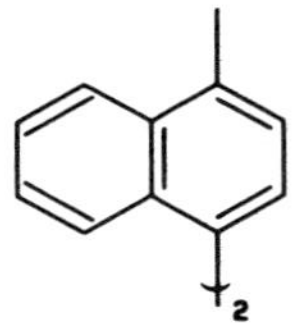	10	7
8	Anthracene	MeCN, EtOH, $LiClO_4$	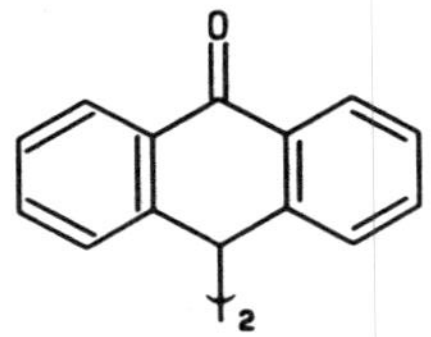	91	8
9	9,10-Dimethylanthracene	$LiClO_4$, MeCN, 2,6-dimethylpyridine	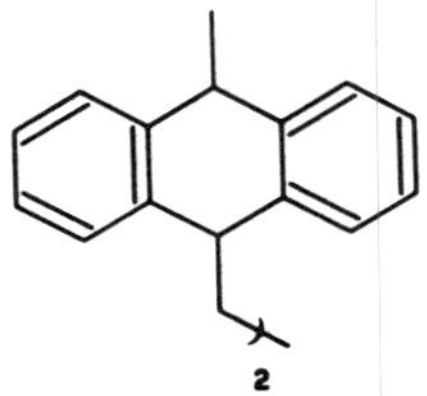	Yield not given[a]	9

Table 1. (Continued).

Number	Substrate	Conditions	Product	Yield (%)	Reference
10	9-Phenylanthracene	$TEAClO_4$, MeCN	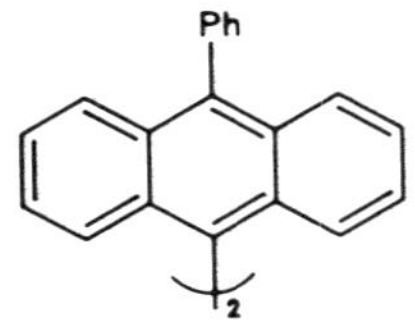	37	10
11	1,5-Dichloroanthracene	MeCN, EtOH	1,1',5,5'-Tetrachlorobianthrone	75[b]	12
Mixed coupling					
12	Naphthalene, m-Xylene	Pt, Bu_4NBF_4, MeCN, AcOH	1,1'-Binaphthyl	28	13
			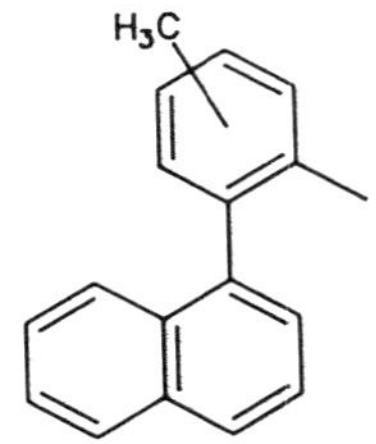	0.4	
13	Naphthalene, mesitylene	Pt, Bu_4NBF_4, MeCN, AcOH	1,1'-Binaphthyl	22	14
			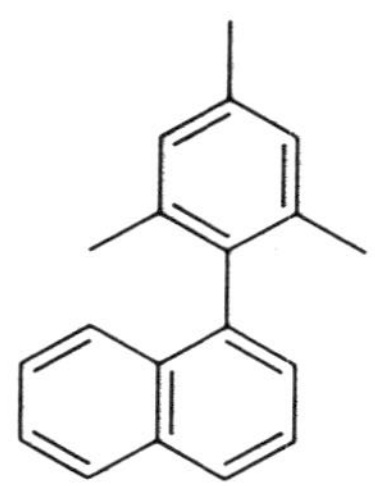	19	

14	Naphthalene, penta-methylbenzene	Pt, Bu_4NBF_4, MeCN, AcOH	1,1'-Binaphthyl 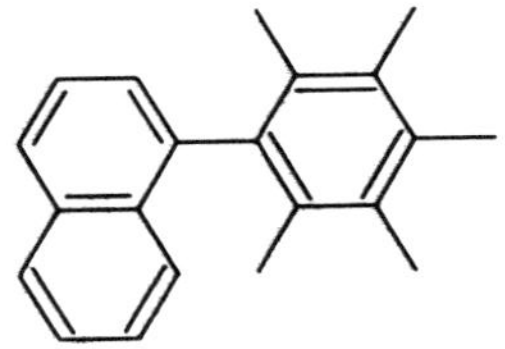	1.6 64	13
15	Hexamethylbenzene and aromatic hydrocarbons (benzene, toluene, p-xylene, mesitylene)	Pt, CH_2Cl_2: TFA (9:1), $TBABF_4$	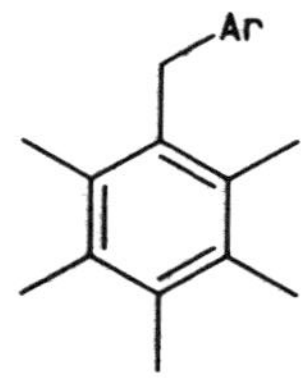	8–66	15
16	Anthracene, anisole	Pt, Bu_4NBF_4, CH_2Cl_2-TFA-TFAn (47:2:1)	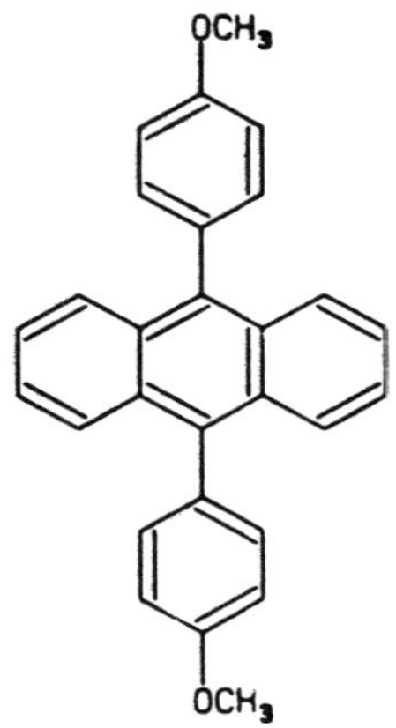	70	16

[a]In the presence of acetate ion a diarylmethane type of dimer is formed [11].
[b]The presence of EtOH is essential for the selectivity.

Table 2. Anodic Coupling of Phenols

Number	Phenol	Conditions	Product	Yield (%)	Reference
1	o-Cresol	PbO_2, H_2SO_4	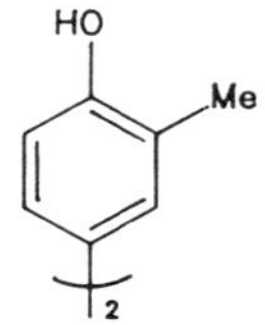	5	30
2	2,6-Dimethylphenol	Pt, MeCN, $NaClO_4$, 2,6-lutidine	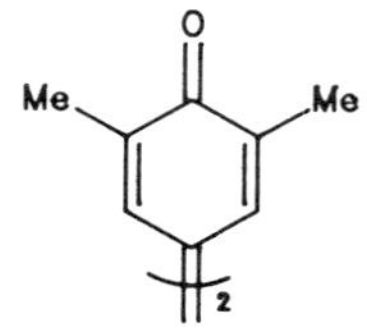	90	31
3	2,6-Di-t-butylphenol	Nickel hydroxide electrode, 70°C	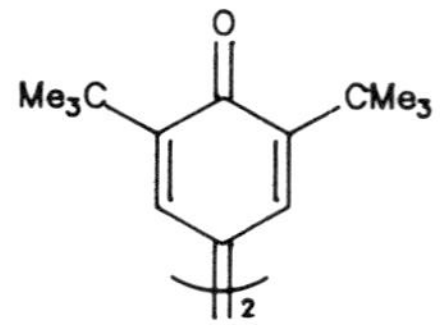	92	32
4	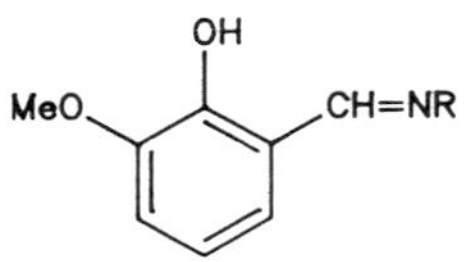 R = ^{i}Pr, ^{t}But, $CH_2CH(CH_3)CO_2H$, Me, Et, benzyl	Glassy carbon, MeCN, $NaClO_4$	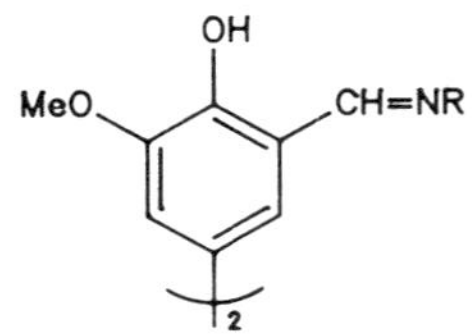	41–100	33

5	2-Acetyl-7-methoxy-8-hydroxy-1,2,3,4-tetrahydroisoquinoline	Graphite, MeCN	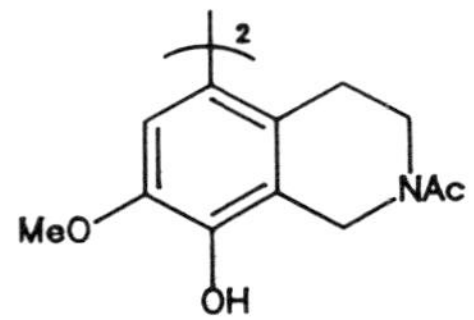	31	34
6	1,2-Dimethyl-7-hydroxy-6-methoxy-1,2,3,4-tetrahydroisoquinoline	Graphite felt, MeONa, MeCN, $TEAClO_4$	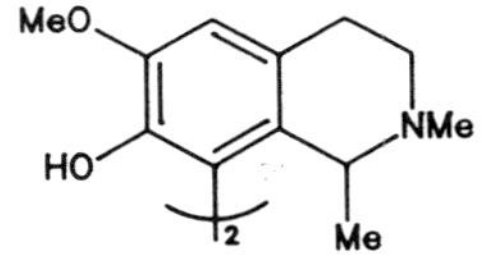	69[a]	35
7	p-Cresol	PbO_2, H_2SO_4	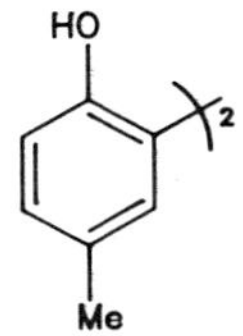	11	30
8	2-Acetyl-4-methylphenol	0.1 M NaOH, H_2O, MeOH, Pt	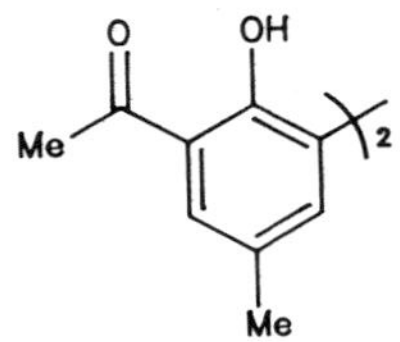	26	36
9	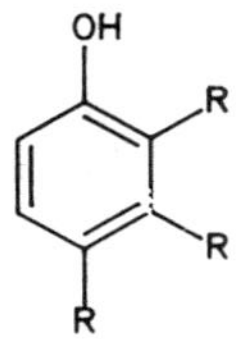 R = H, CH_3, C_2H_5	CH_3CN, H_2O, $TEABF_4$	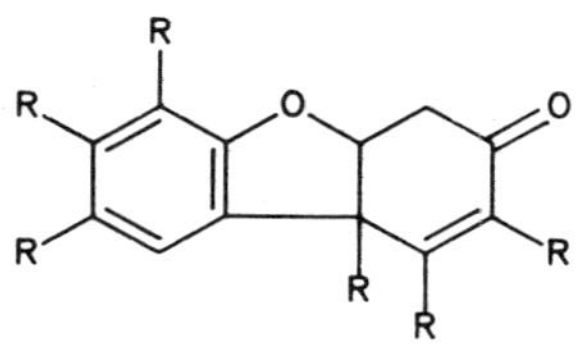	20–37	37

Table 2. (Continued).

Number	Phenol	Conditions	Product	Yield (%)	Reference
10	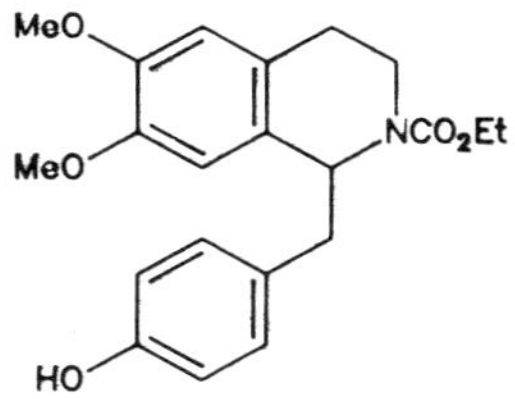	Graphite felt, MeCN, $TEAClO_4$	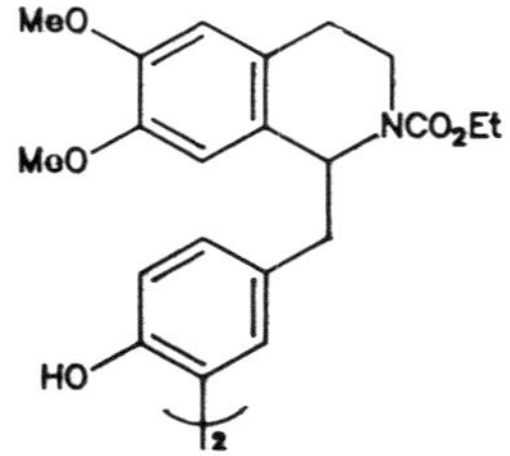	45[b]	38
11	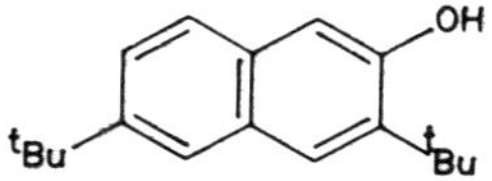	CH_3OH-ether, 1% KOH	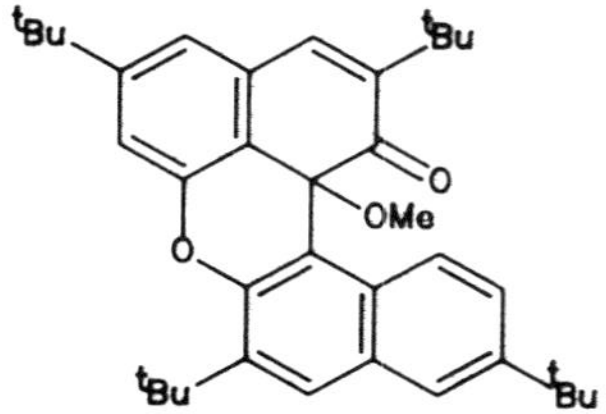	65	39
12	2-Methoxy-4-allyl-phenol	Pt, MeOH-$LiClO_4$, NaOH	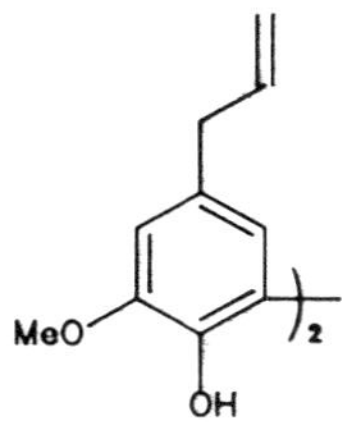	100	40
13	Vanillin	Pt, MeCN, $NaClO_4$, Et_4NOH, 0.25 V (SCE)	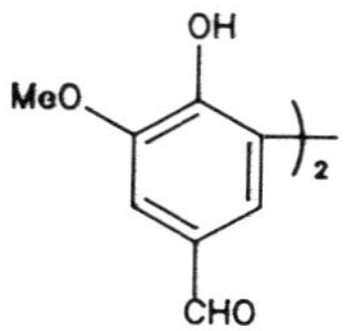	65	41

14	2,6-Di-t-butylcresol, anisole	Pt, MeCN, Bu_4NBF_4	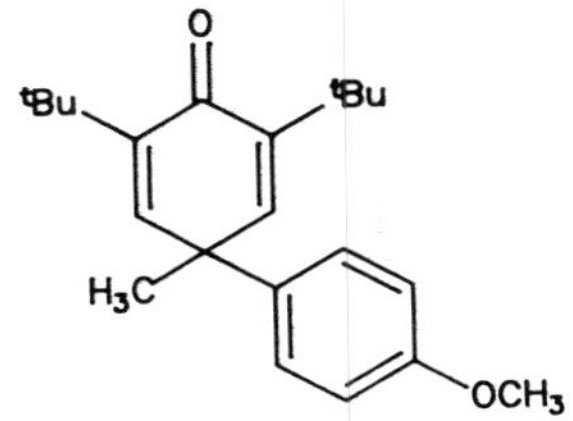	81	42
15	Phenol	Graphite, CH_3OH, benzene, $LiClO_4$	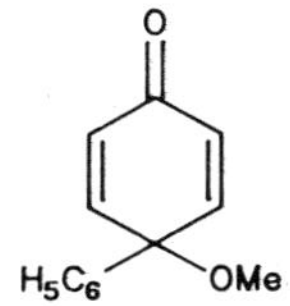	23	43
16	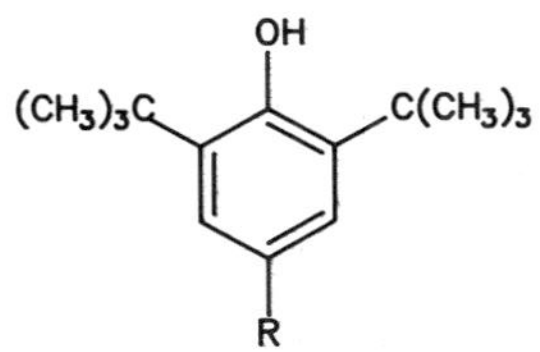 R = ^{t}Butyl, Ph, Ac, CN, CHO, CO_2H	Pt, MeCN, $NaClO_4$	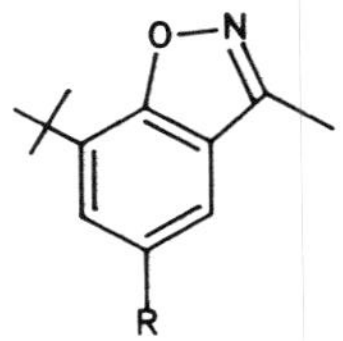	7–70	48

[a]Only one of three possible diastereomers is obtained; additionally, 7% O–C coupling product is found.
[b]Additional O–C coupling product is found.

formed in a Ritter reaction are often the major products. The selective dimerization of mesitylene in acetonitrile is exceptional (Table 1, number 3). Dichloromethane, however, is reducible at the cathode. The chloride formed thereby may react as nucleophile or may be oxidized at the anode to chlorine, which leads to unwanted chlorinated side products. This reductive cleavage can be avoided by addition of strong acids, such as trifluoroacetic acid, methane sulfonic acid, or trifluoro methane sulfonic acid [2]. These acids effectively suppress the chlorination, because H^+ is reduced at the cathode rather than dichloromethane. They also decrease the yield of biaryl products, however, and increase the yield of diphenylmethanes.

The anodic oxidation of benzene produces a mixture of polyphenylene compounds. This oligomerization can be performed in acetonitrile [21] or in liquid sulfur dioxide [22].

Mixed coupling between naphthalene and alkyl benzenes has also been demonstrated (Table 1, numbers 12–16). The relative yield of mixed coupling products increases with the basicity of the alkylbenzene; with mesitylene 19%, with tetramethyl benzene 42%, and with pentamethyl benzene 64% of mixed coupling products are obtained. This suggests an electrophilic reaction between naphthalene cation radicals and alkylbenzenes. The mixed coupling reaction of phenanthrene with anisole has been studied kinetically. The results indicate that initially a complex is formed between the phenanthrene radical cation ($PA^{+\cdot}$) and anisole, followed by an electron transfer from the complex to $PA^{+\cdot}$. The resulting PA^{2+}-anisole complex then decomposes to the product [23].

Bibenzyls can be obtained in a remarkable reaction from toluene, p-xylene, or ethylbenzene in 60–100% yield [24]. For this reaction catalytic amounts of quinones are irradiated in the presence of the alkylbenzenes and the anode potential is held at that of the quinone-hydroquinone couple. The photoexcited quinone abstracts a benzyl hydrogen to form a radical, which dimerizes. The resulting hydroquinone is reoxidized to the quinone.

Cyanation of aromatic hydrocarbons, also a carbon-carbon coupling reaction, is achieved in the case of anthracene in MeCN-Et_4NCN to yield 54% 9,10-dicyanoanthracene [25]. The cyanation is simplified when it is carried out in an emulsion system (aqueous sodium cyanide, dichloromethane, and $TBAHSO_4$). Its synthetic utility in this mode has been demonstrated for the preparation of 4-alkoxy-4-cyanobiphenyls, a class of liquid crystals [26].

Anions of CH-acidic compounds (dimethyl malonate and nitromethane) can be coupled to aromatics (benzene, toluene, naphthalene, and 1,4-dimethoxybenzene) when they are coelectrolyzed in methanol-sodium methoxide [27].

Phenols

The earlier literature on oxidative coupling of phenols is reviewed in Ref. 28 and that on anodic coupling in Ref. 29; representative examples of the coupling reactions are summarized in Table 2.

Phenols are coupled mostly in alkaline media by oxidation of the phenolates to phenoxy radicals. However, coupling in neutral or acidic medium by way of phenoxy radical cations and phenoxonium cations as intermediates is also known. To generate the phenolate ion, bases as alkali hydroxides, Et_4NOH or 2,6-lutidine is used. The anodically generated phenoxy radicals react by carbon-carbon and carbon-oxygen coupling to dimers, which additionally can be further oxidized. This frequently leads to product mixtures that are synthetically not very useful. Blocking of the 2-, 4-, or 6-position makes the coupling reaction more selective. Phenols with unsubstituted para-position usually form para,para coupling products as the major dimer (Table 2, numbers 1–5). If the 4-position is blocked ortho,ortho coupling is the main reaction (Table 2, numbers 6–13).

A series of 1,2,3,4-tetrahydroisoquinolines with phenolic OH groups has been oxidized to carbon-carbon and carbon-oxygen dimers. Optimum conditions for the coupling are the oxidation of the sodium salts in MeCN using $TEAClO_4$ as supporting electrolyte and a graphite felt anode. The products depend strongly upon the location of the phenolic OH group in the aromatic ring (Table 2, number 5) [34]. The ratio of CC to CO coupling depends also on the size of the 1-substituent in the tetrahydroisoquinoline; with the 1–H derivative predominantly CC, and with the $1-CH_3$ and $1-C_2H_5$ compound only CO coupling is found [44]. The dimerization of the tetrahydroisoquinoline, shown in Table 2, number 6, yields only one of three possible diastereomers. This outcome is best explained as a result of a surface reaction on the graphite electrode, at which only molecules of identical configuration couple. The anodic oxidation of phenols (IV) [Eq. (4)] at −300 to 250 mV (CH_3OH, 1% KOH) leads, depending on the substituents (Table 3), to bisphenols (V), to oxepinobenzofuran (VI), or to a Pummerer ketone (VII) [39].

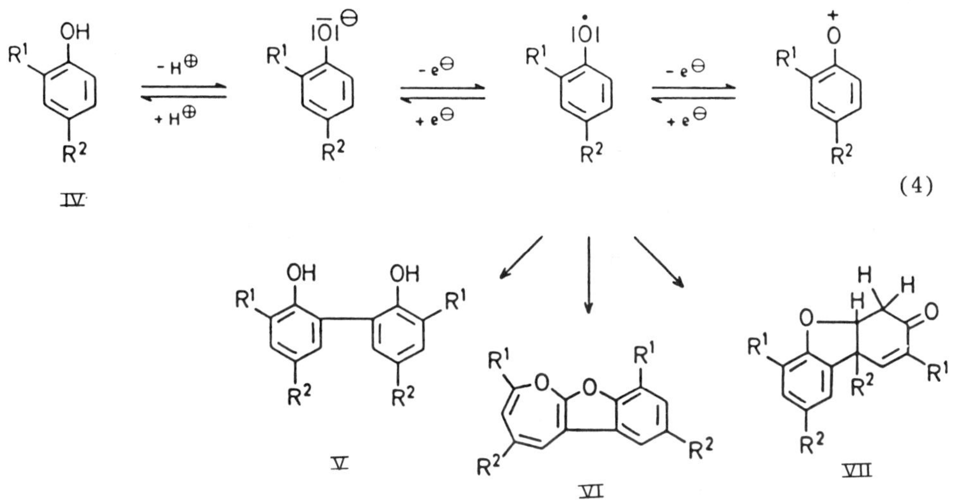

(4)

Table 3. Yields of the Oxidation IV

IV		Yield (%)		
R^1	R^2	V	VI	VII
^tBu	OEt	81	–	–
^tBu	^tBu	–	85	–
^tBu	Me	–	–	41

Ortho,ortho' coupling of the phenoxy radical produces V, its further oxidation forms via an arene oxide VI, and the ortho,para' coupling leads after a subsequent Michael addition to the Pummerer ketone (VII).

In the electrolysis of VIII and IX a mixture of CO and CC coupling products (symmetrical, unsymmetrical, aryl-olefin, aryl-aryl, and olefin-olefin coupling) is obtained; the products are related to lignans and neolignans [45]. 2,4,6-Tri-t-butylphenolate is reversibly oxidized at −0.2 V (versus SCE) to the phenoxy radical; at a higher potential (+1.0 V) the phenoxonium ion is formed irreversibly. Kinetics of the follow-up reactions are evaluated by CV [40].

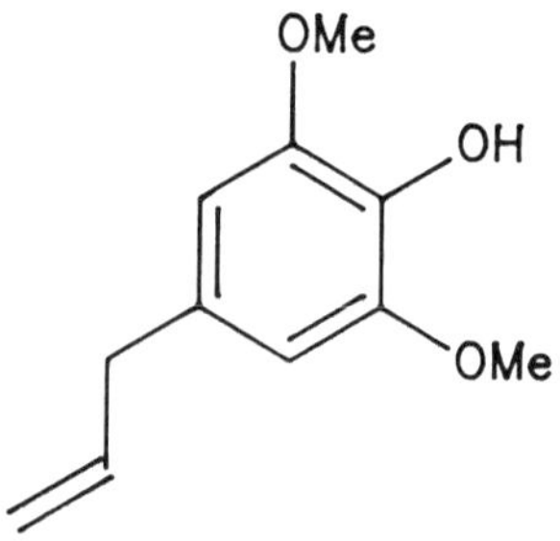

VIII

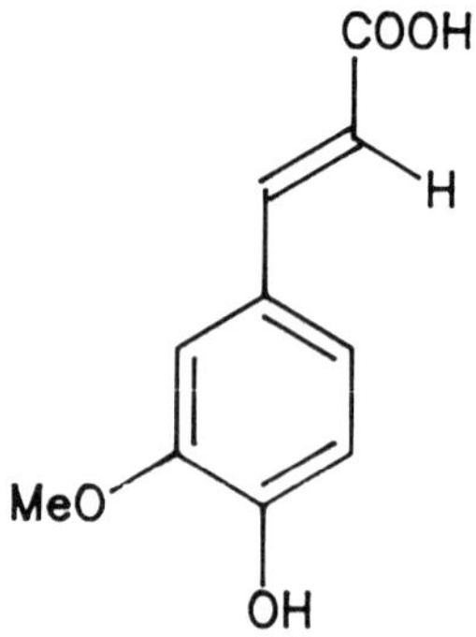

IX

In neutral or acidic medium phenols are oxidized to phenoxonium ions, which can undergo an electrophilic aromatic substitution, for example with anisole to para-methoxyphenylcyclohexadienones (Table 2, number 14). In the oxidation of X [Eq. (5)], the intermediate phenoxonium ion disproportionates to phenoxy radicals, which couple in 95% yield to the quinhydrone (XI) [47]:

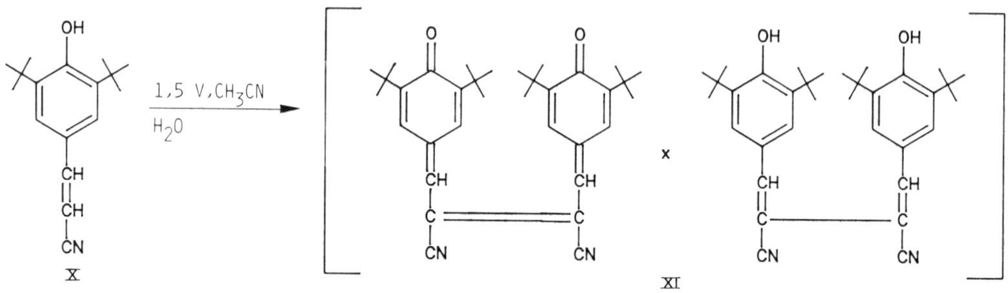

(5)

Oxidation of phenol in methanol-benzene (1:1) leads to a phenylcyclohexadienone in addition to the quinone monoacetal. The phenylation is thought to occur through electrophilic attack of the phenoxonium ion on benzene, followed by anodic methoxylation (Table 2, number 15). Anodic oxidation of 15 differently substituted 2,6-di-t-butyl-4-alkylphenols in MeCN leads to the corresponding 7-t-butyl-2-methylbenzoxazoles in 7–70% yield. The reaction is suggested to involve the oxidation to the phenoxonium ion and its electrophilic attack on MeCN. The resulting nitrilium ion undergoes an electrophilic aromatic substitution with elimination of one t-butyl group (Table 2, number 16).

Aryl Ethers

The anodic coupling of aryl ethers is reviewed in Ref. 49. Aryl ethers are more selectively coupled than phenols, for the following reasons: the carbon-oxygen coupling is made impossible and the ortho coupling and the oxidation to quinones become more difficult. A mixture of trifluoroacetic acid (TFA) and dichloromethane proved to be the most suitable electrolyte [50]. TFA enhances the radical cation stability and suppresses the nucleophilicity of water. Of further advantage is the addition of alumina or trifluoroacetic anhydride [51]. Table 4 summarizes representative examples of the aryl ether coupling.

The coelectrolysis of veratrole with anisole derivatives afforded in TFA-CH_2Cl_2 good yields of aryl-substituted triphenylene derivatives (Table 4, number 9). Product formation probably occurs by initial coupling of the veratrole cation radical with anisole to an unsymmetrical dimer. This is followed by coupling of the dimer and intramolecular cyclization to the product.

The trimethylsilyloxy (TMSO) group is stable under the coupling conditions in acetonitrile (Table 4, number 11). After oxidative dimerization the TMS-ether can be mildly hydrolyzed (H^+ and H_2O) to the phenol or converted to a dibenzofuran. 1,2-Dialkoxybenzenes have been trimerized to triphenylenes (Table 4, numbers 9, 12, and 13). The reaction product is the triphenylene radical cation, which is reduced to the final product either by zinc powder or in a flow cell consisting of a porous anode and cathode [60]. Dibenzo-crown ethers are converted by anodic oxidation to electroactive polymers. Films of these polytriphenylenes exhibit unusual doping properties [62–64].

A variety of dimethoxybiphenyls has been investigated by voltammetry. The biphenyl linkage was shown to have a remarkable stabilizing effect on both cation radicals and dications. In the series of para-substituted compounds [$MeO(C_6H_4)_nOMe$], maximum stability of ions was observed with $n = 2$ and 3 [65]. The pathways for the coupling of 1,4-dimethoxybenzene and 2,5-dimethoxytoluene to biphenyls and a dimethoxyphenyl-p-benzoquinone have been elucidated by CPE and CV [66].

Coupling at the benzylic position occurs with 9-methoxy-10-methylanthracene. The dimer of a quinone-methide is obtained in about 50% yield. The product is thought to be formed by deprotonation of the initially formed cation radical, dimerization of the benzyl radicals, and subsequent oxidative demethylation at the methoxy group (Table 4, number 14).

Cyanation of aryl ethers, also a carbon-carbon coupling reaction, proceeds in $MeCN-Et_4NCN$ with fair to excellent yields to methoxybenzonitriles. Here a methoxy group is replaced by cyanide [67].

Arylamines

The oxidation of arylamines leads to three principal products: benzidines, aminodiphenylamines, and azo compounds. In these oxidations the first intermediate is in general the radical cation of the amine. This can lose a proton, react with a nucleophile, or couple in three different ways [Eq. (6)]: (1) in a carbon-carbon coupling leading to benzidines, (2) in a carbon-nitrogen coupling with aminodiphenyl amines as products, and (3) in a

Table 4. Anodic Coupling of Arylethers

Number	Aryl ether	Conditions	Product	Yield (%)	Reference
1	Diphenyl ether	Pt, CF_3CO_2H-CH_2Cl_2 (1:2), Bu_4NBF_4	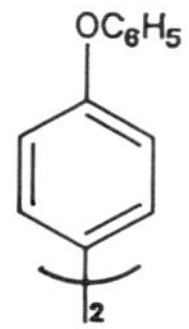	56	50
2	Anisole	Pt, CF_3CO_2H-CH_2Cl_2 (1:2), Bu_4NBF_4	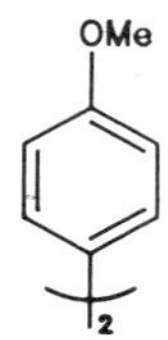	63	50
3	2-Bromo-anisole	Pt, CF_3CO_2H-CH_2Cl_2 (1:2), Bu_4NBF_4	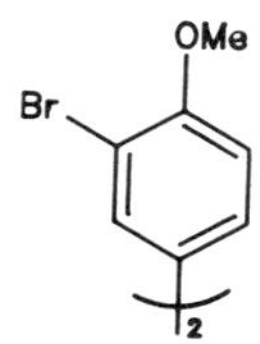	43	50
4	1,2-Dimethoxy-benzene	CF_3CO_2H-CH_2Cl_2 (1:2), Bu_4NBF_4	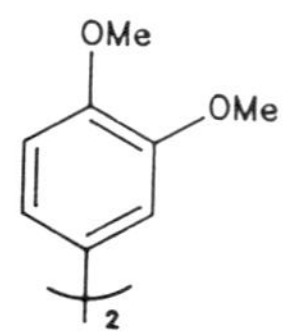	86	50

5	1,2,4-Tri-methoxy-benzene		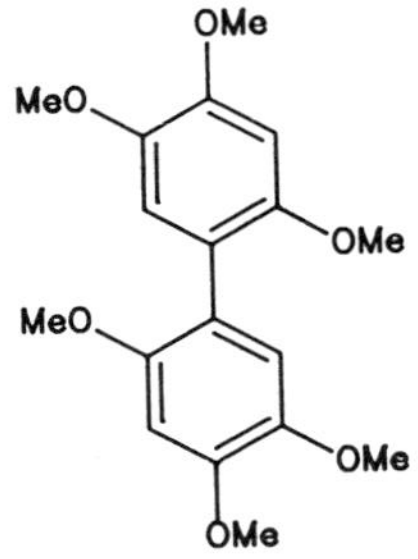	85	52
6	2,5-Dimethoxy-toluene	MeCN, water, $TEAClO_4$	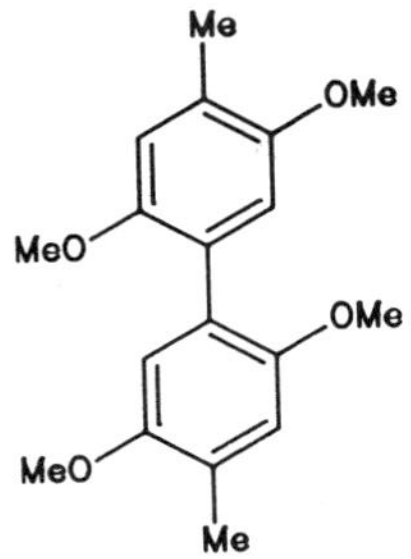	No yield given	53
7	4-Methoxyphenyl-3-methoxybenzoate	MeCN, $NaClO_4$	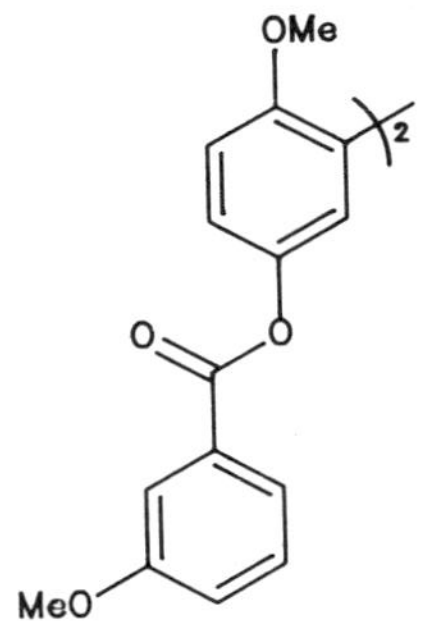	65	54

Table 4. (Continued).

Number	Aryl ether	Conditions	Product	Yield (%)	Reference
8	9-Methoxy-anthracene	MeCN, TFA, Bu_4NBF_4	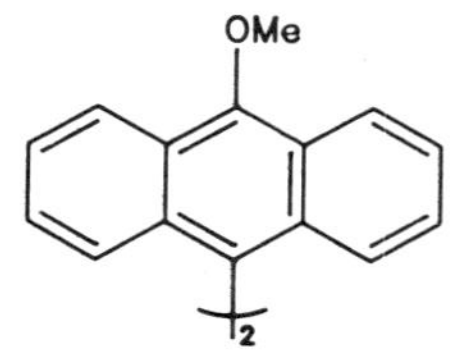	95	55
9	Anisole, 1,2-dimethoxy-benzene	MeCN, TFA, Bu_4NBF_4	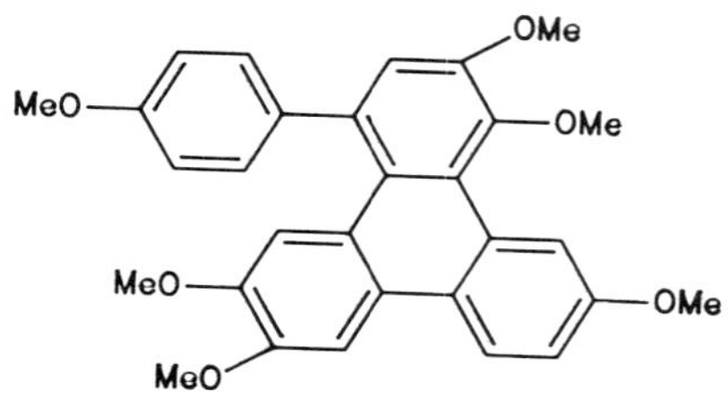	60	56
10	9-Anisyl-anthracene, anisole	CH_2Cl_2, MeCN, TFA, Bu_4NBF_4	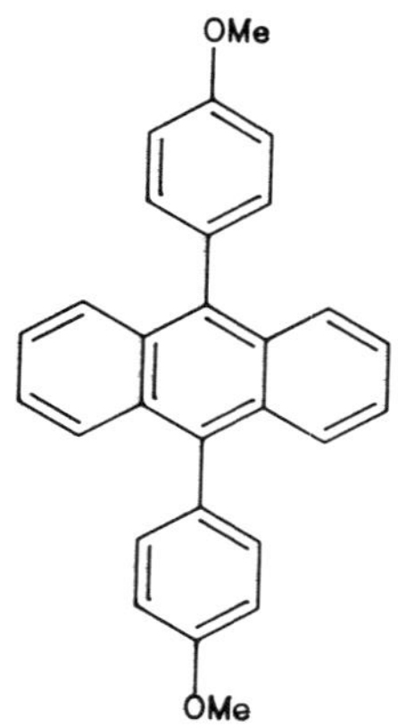	90	57

11	2-Methoxy-1,4-bis(trimethyl-siloxy)benzene	CH_3CN, $LiClO_4$	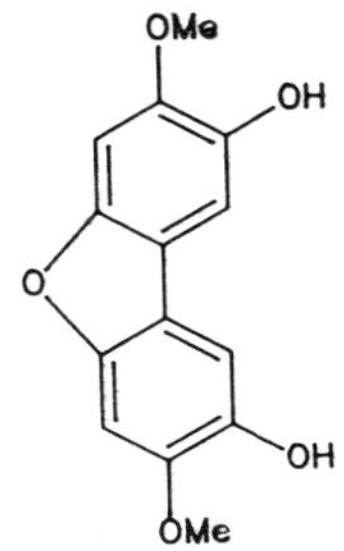	65	58
12	1,2-Dimethoxy-benzene	CH_2Cl_2, TFA, Bu_4NBF_4	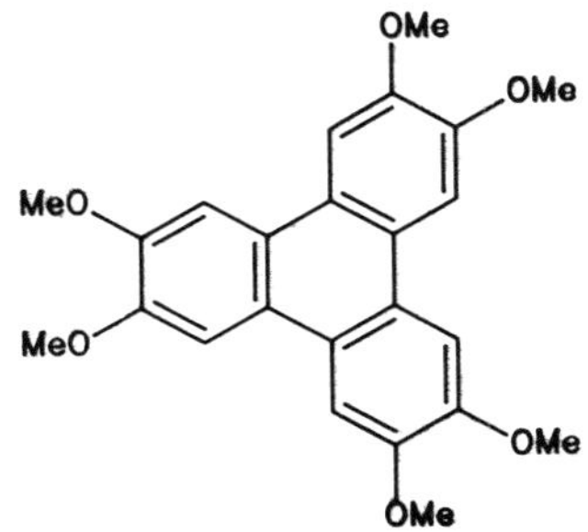	50	59
13	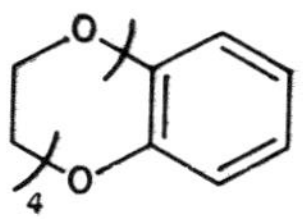	MeCN or CH_2Cl_2, TFA	Cyclotrimer	30	60
14	9-Methoxy-10-methyl-anthracene	MeCN, $LiClO_4$	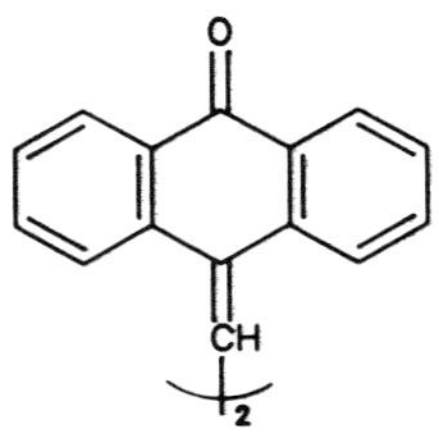	50	61

nitrogen-nitrogen coupling forming hydrazobenzenes. These products are often easily further oxidized. The earlier work is covered by a review [18], and representative examples are summarized in Table 5.

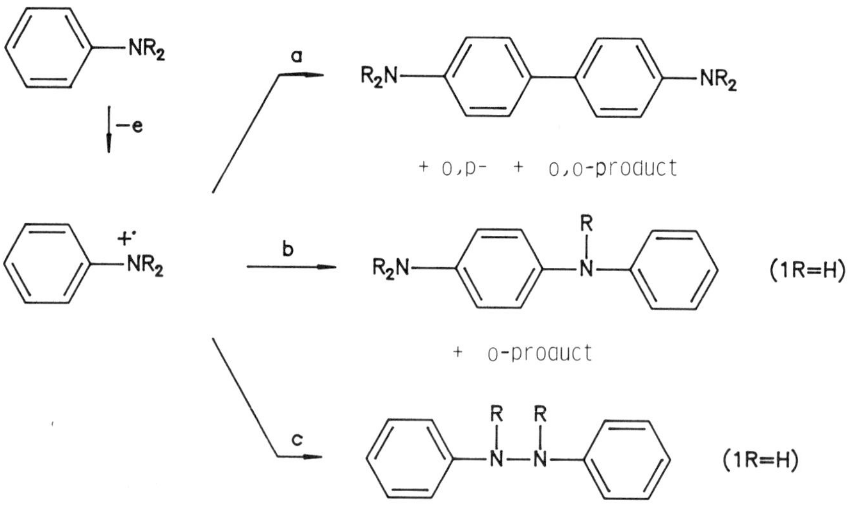

(6)

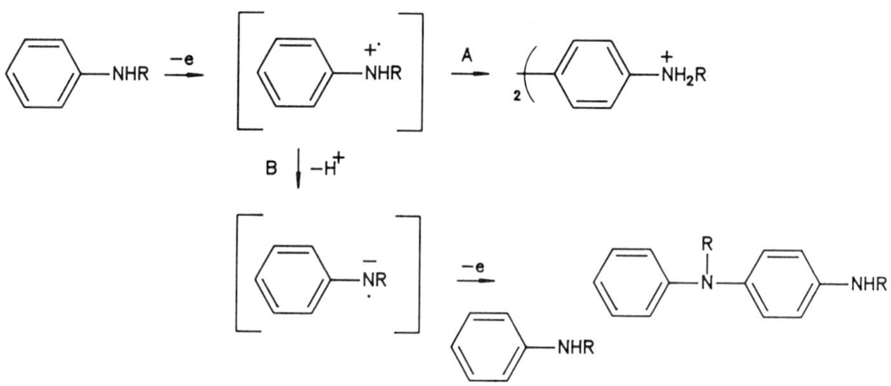

R	% A	% B
Me	50	50
Et	60	40
^{i}Prop	65	35
^{t}Bu	100	0

(7)

Triphenylamines are coupled to the tetraphenylbenzidines, which may be reversibly further oxidized to their dications (Table 5, number 1). The rates of the coupling reaction have been measured, and a fairly complete picture of the oxidation pathways has emerged [68]. Para-substituted triphenylamines cyclize to some extent to carbazoles [69].

With anilines and N-alkylanilines the manner of dimerization is influenced by the reaction conditions and the substitution of the aryl ring [70]. The benzidines predominate in strongly acidic aqueous media. The formation of benzidine is further favored by a low substrate concentration and a high current density. The benzidine derivatives may be reversibly oxidized to their dications. Diphenylamines are the major products under neutral conditions and higher pH values, a low current density, a high substrate concentration, a small alkyl group in monoalkylanilines, and a nonaqueous solvent [Eq. (7)]. Under weakly acidic conditions in aqueous medium, an additional electron discharge followed by hydrolysis produces benzoquinones in almost quantitative yield.

The electrooxidation of N,N-dimethylaniline (DMA, Table 5, number 2) has been studied by cyclovoltammetry and rotating ring-disk voltammetry in strongly acidic aqueous media [71]. The results are qualitatively consistent with a mechanism involving deprotonation of the DMA cation radical and subsequent coupling between neutral radicals. N,N-dimethylaniline forms different products [Eq. (8)] in MeCN-TEAP, depending on the substrate concentration and the electrolysis period. At low concentration (about 10^{-3} M) benzidine is the major product, at intermediate concentrations (10^{-2} M) coupling at the N-methyl group to XII occurs, and at a still higher concentration and a longer electrolysis time crystal violet (XIII) is being formed [72].

8,8'-Dianilino-5,5'-binaphthalene-1,1-disulfonates can be efficiently synthesized (25–50% yield) from the corresponding naphthylamines [74], analogously to the coupling of diphenylamines (Table 5, number 3a).

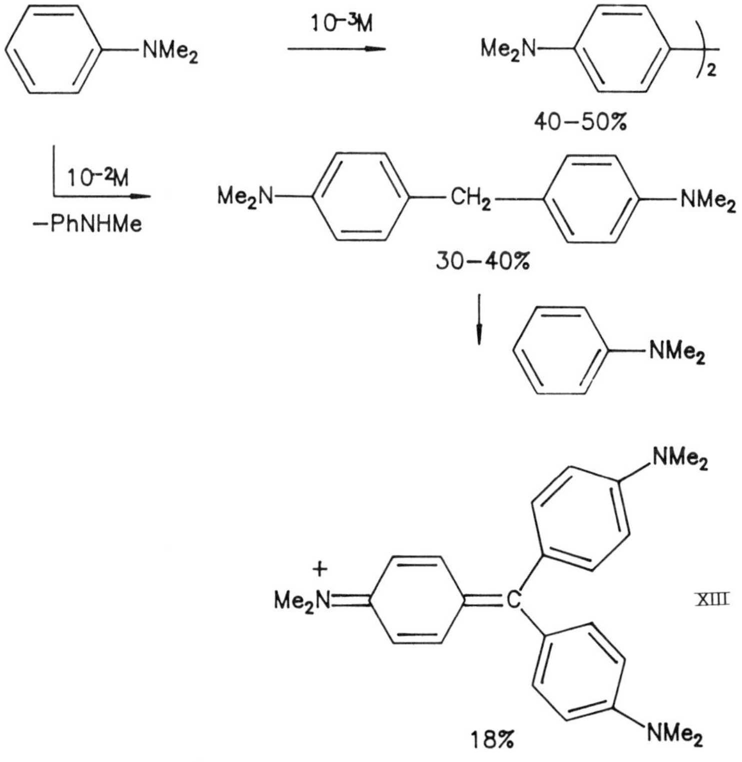

(8)

XII

XIII

Table 5. Anodic Coupling of Arylamines

Number	Arylamine	Conditions	Product	Yield (%)	Reference
1	Triphenylamine	MeCN, Et_4NClO_4, Pt	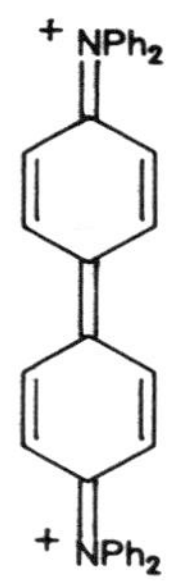	–	68
2	Dimethylaniline	MeCN, Et_4NClO_4	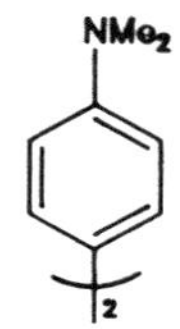	40–50	72
3a	Diphenylamine	MeCN, Et_4NClO_4	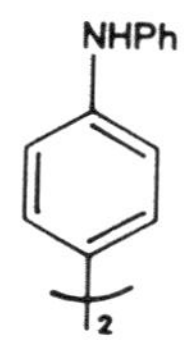	80	73

3b	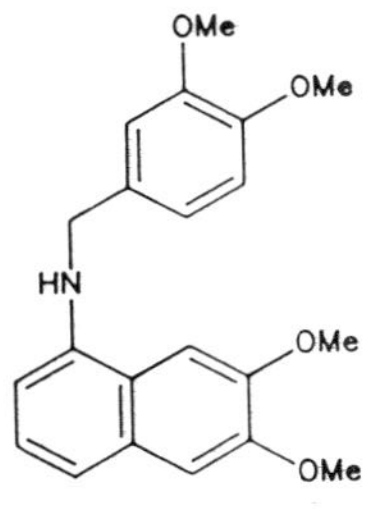	MeCN, $LiClO_4$	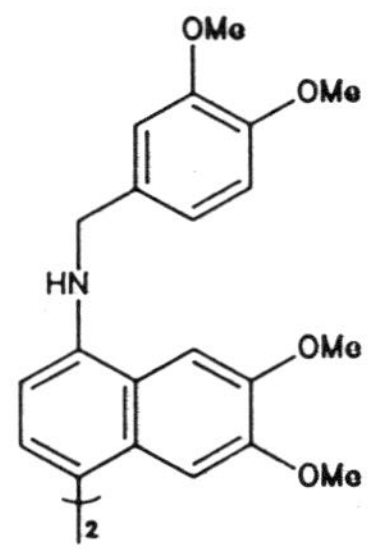	42	92
4a	p-Substituted anilines	MeCN, Pyr, $NaClO_4$, Pt		X / % NO_2 39 Cl 24 2,4−Cl 30 OMe 5	75
4b	Substituted anilines	H_2O-KOH or H_2O, MeCN, NaOAC	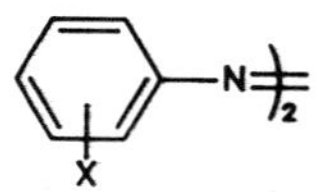	X / % H 30 2,4−NO_2 38 2,4−Cl 48 4−CH_3 37 2−SO_3H 3 4−SO_3H 33 3−Me 3	75, 77

Table 5. (Continued).

Number	Arylamine	Conditions	Product	Yield X	(%)	Reference
4b (cont.)			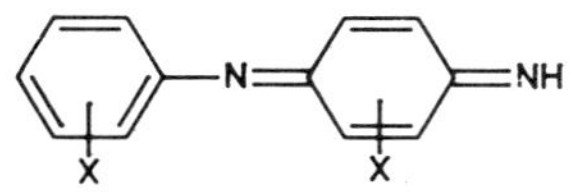	X	%	
				2−SO_3H	70	
				4−SO_3H	40	
				3−Me	37	
4c	Aniline	6 M H_2SO_4	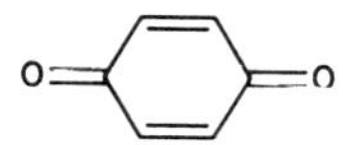		80−90	75
			Benzidine		10−20	
4d	3-Aminopyridine	Buffer at pH 9−10 in water, Pt	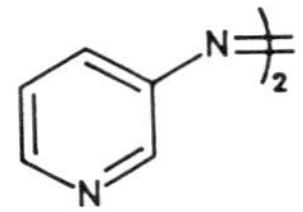		28	78
5a	N-Ethylaniline	MeCN, Et_4NClO_4	N,N'-Diethylbenzidine		70−80	75
5b	N-t-butylaniline	MeCN, Et_4NClO_4	N,N'-di-t-butylbenzidine		100	75
6	4-Chloro-dimethyl-aniline	MeCN, Bu_4NBF_4, Pt	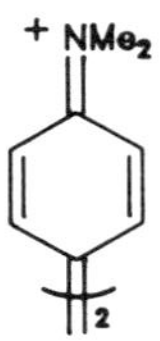		−	76

7	2-Aminofluorene	Pt, MeCN, $NaClO_4$, pyridine	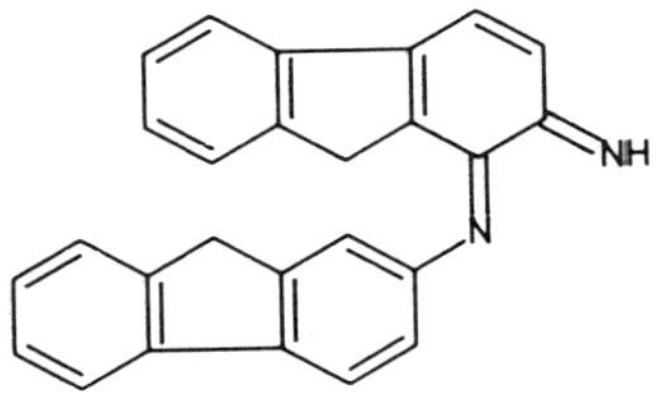	70	79
8	N,N-dimethyl-p-toluidine		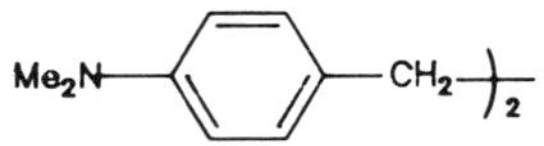	—	80
9	Carbazole	MeCN, Et_4BClO_4, Pt	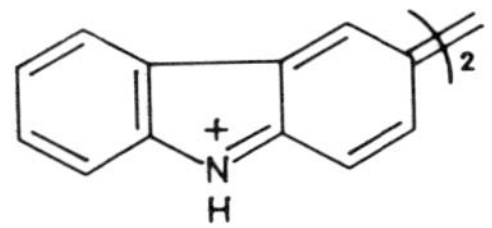	—	87
10	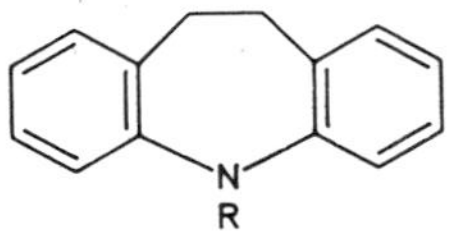 R = H, CH_3	MeCN, Bu_4NClO_4, Pt	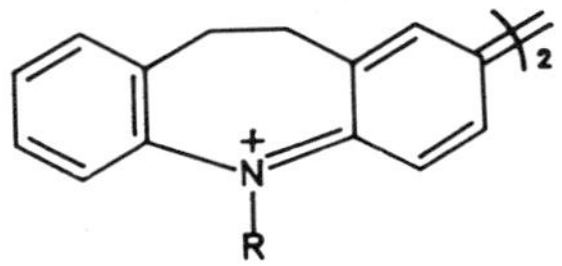	—	90
11	1,2,3,4-Tetrahydrocarbazole	MeCN, H_2O, $LiClO_4$, C	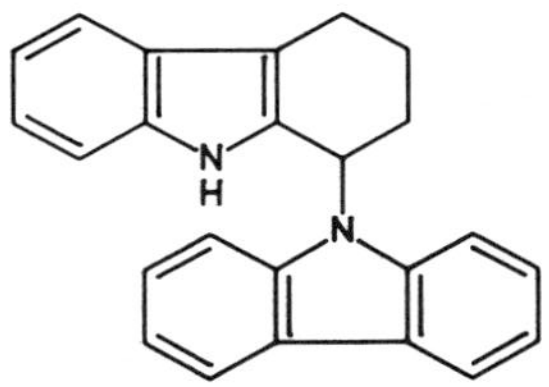	nearly quantitative	91

Table 5. (Continued).

Number	Arylamine	Conditions	Product	Yield (%)	Reference
12	1-Phenyl-pyrazol-idin-3-one	Pt, MeCN, $TEAClO_4$	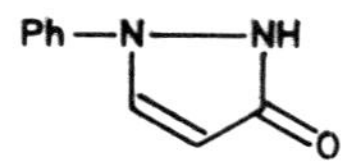 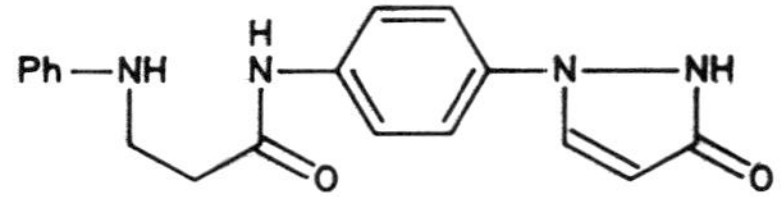	75	93
13	3-Amino-4,5-dihydro-1-phenyl pyrazole	Pt, MeCN, $TEABF_4$	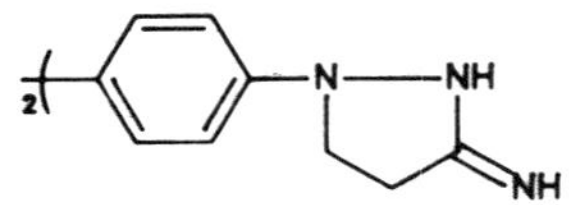	90	94a
14a	Ethyl-5-amino-3-anilino pyrazolo-4-carboxylate	MeCN, $TEAClO_4$, pyridine	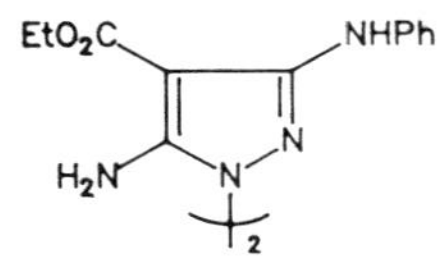	42	95
14b	1,2,3,5-Tetra-phenyl pyrrole	MeCN, $LiClO_4$	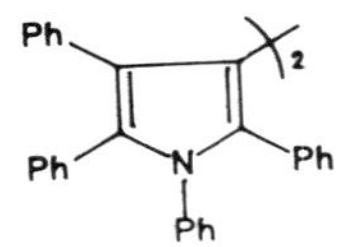	–	98
15	1,3,4,5-Tetra-phenyl-Δ^2-pyra-zoline	MeCN, $LiClO_4$, Pyr.	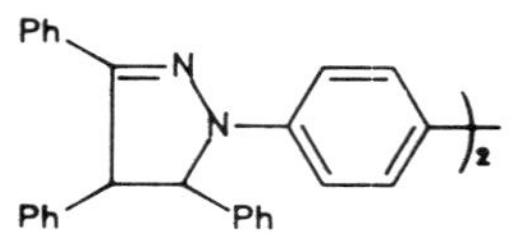	70	94b

16	2-Amino-5-ethoxycarbonyl-4-methyl-thiazole	MeCN, $NaClO_4$	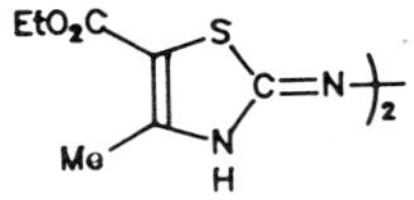	50	97
17	t-Butylamine	Li-cyclohexyl-amide, THF, $LiClO_4$	$^tBu-N=N-^tBU$	25	101
18	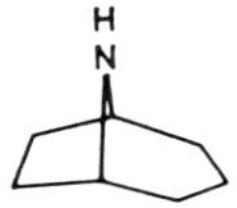	PhCN, $LiClO_4$ NaOH	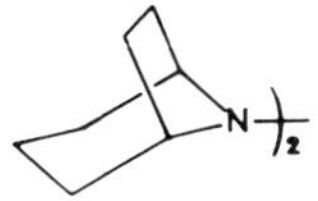	58	102
19	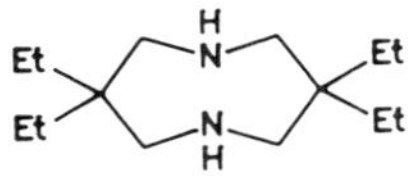	MeCN, $LiClO_4$, Et_4NOTos	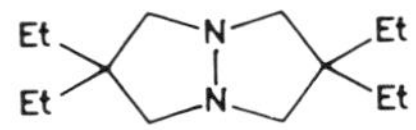	10	103
20	1,1-Dimethyl-hydrazine	MeCN, TEAP, Na_2CO_3	$(Me_2N-N\not=)_2$	85	104
21	1,1-Diphenyl-hydrazine	MeCN, $LiClO_4$, pyridine	$(Ph_2N-N\not=)_2$	100	105
22	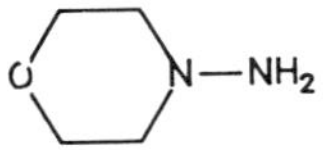	NiOOH, H_2O, K_2CO_3	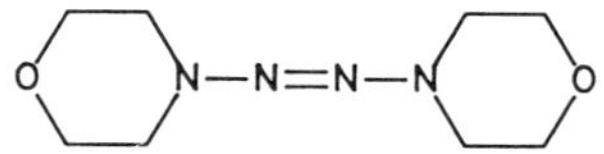	80	106
23	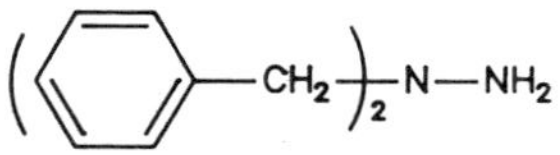	NiOOH, H_2O, K_2CO_3	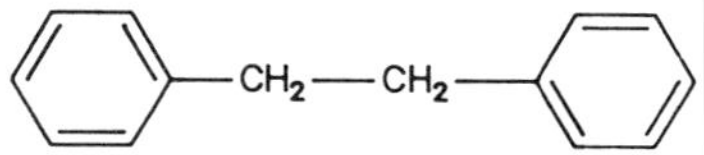	91	106

Table 5. (Continued).

Number	Arylamine	Conditions	Product	Yield (%)	Reference
24	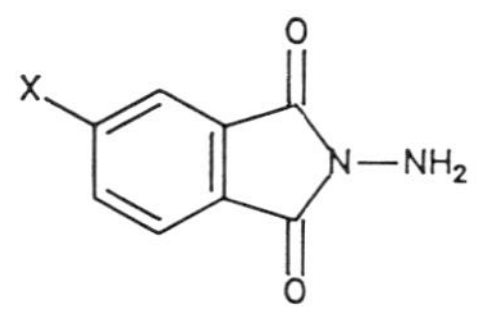 X = H, NO_2	Pt, MeCN, Bu_4NBF_4	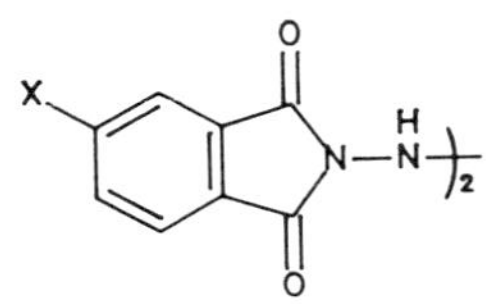	0–80	107
			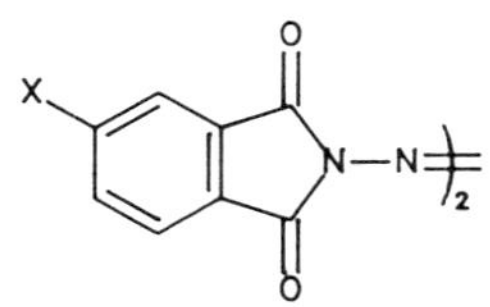	51–86	

Para-substituted anilines are oxidized in low to moderate yields to azobenzenes in basic solution (Table 5, number 4a). By elimination of one para-substituent (Cl or OMe) during the reaction, they can also couple to 4-substituted 4-aminodiphenylamines [76]. Carbon-carbon coupling can occur with 4-substituted arylamines as substrates. The para-substituent is eliminated and a para-benzidine is obtained (Table 5, number 6), ortho carbon-carbon coupling occurs (number 7), or a benzyl radical is formed, which subsequently couples (number 8). Ortho- and meta-substituted anilines yield at graphite cloth electrodes in 6 N H_2SO_4 substituted benzidines and 4-aminodiphenylamines. The latter products are converted by further oxidation and hydrolysis to 2-substituted para-benzoquinones, sometimes in substantial yields (e.g., 35%) [81].

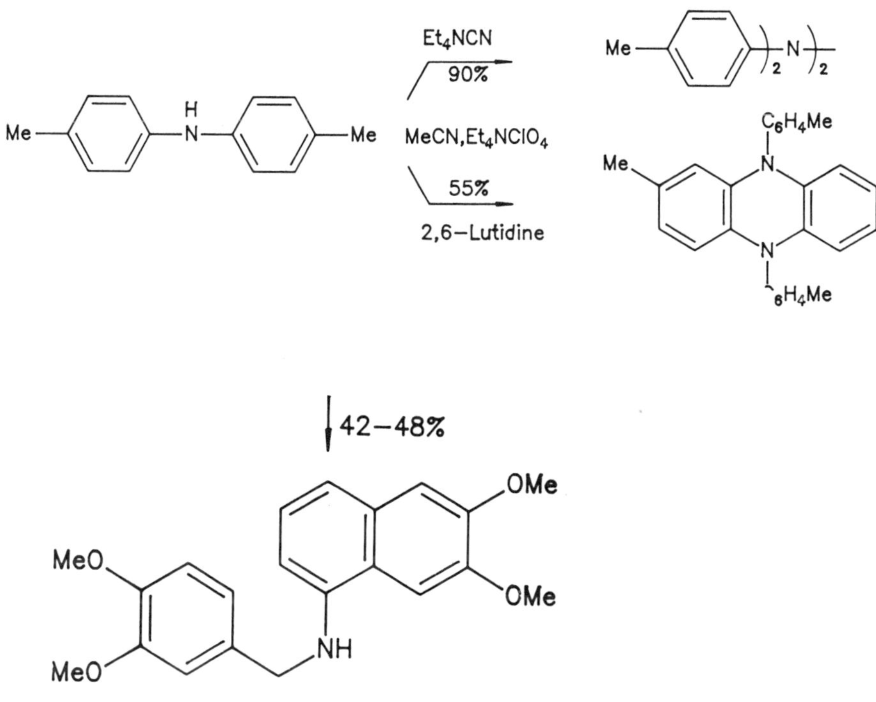

XIV

XV

(9)

XVI

Electrolysis of di-para-tolylamine in MeCN-Et_4NCN affords 90% of the hydrazine (XIV) [Eq. (9)]; with lutidine as base, however, the dihydrophenazine (XV) is formed as major product [82]. Electrolysis of N-phenyl-para-bromoaniline in basic medium (MeCN, TEAP, and Et_4NCN) produces the N,N coupling product in 70% yield, whereas in acidic medium (H_2O and $HClO_4$) the benzidine type of product is formed in 87% yield [83]. The attempted intramolecular cyclization of the substituted naphthylamines (XVI) failed and led only to an intermolecular coupling [92]. 4-Substituted diphenylamines form in a remarkable cyclodimerization 5,10-dihydro-5,10-diphenylphenazines [Eq. (10)] [84]. One requirement is that all para-positions must be substituted by groups that are not eliminated and not strongly electron withdrawing. Yields are increased by electrolyzing at low current densities. Di-(2-naphthyl)amine undergoes the same reaction anodically, but the analogous oxidation with permanganate fails [85]. Benzenesulfene anilides can be coupled to 2,7-disubstituted phenazines [Eq. (11)] [86].

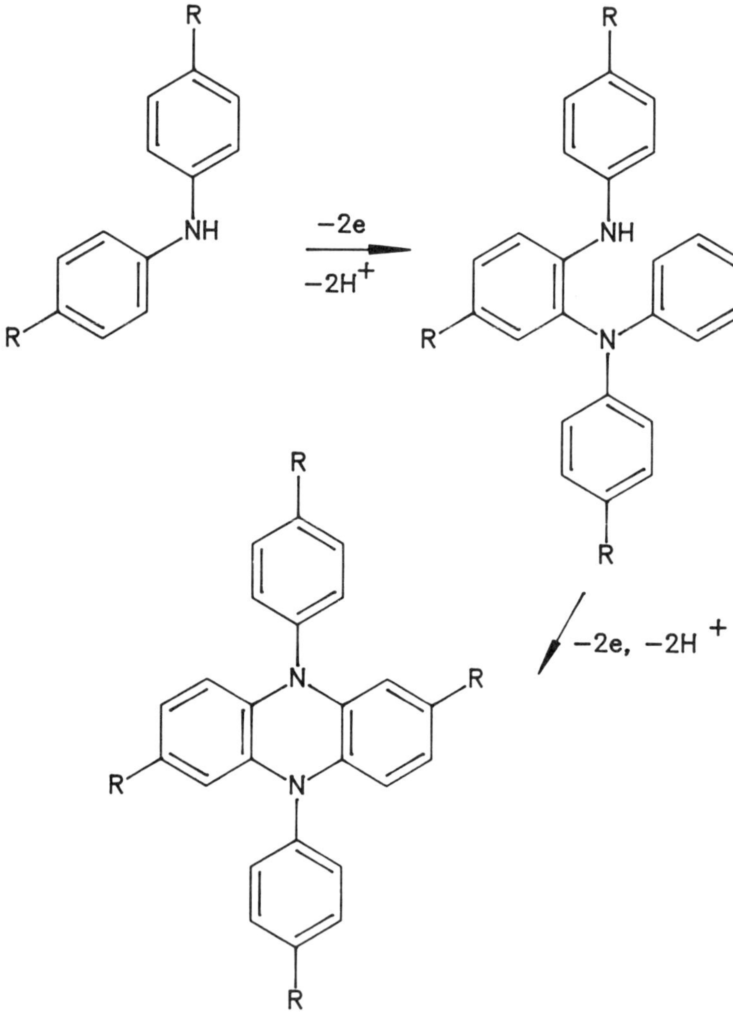

(10)

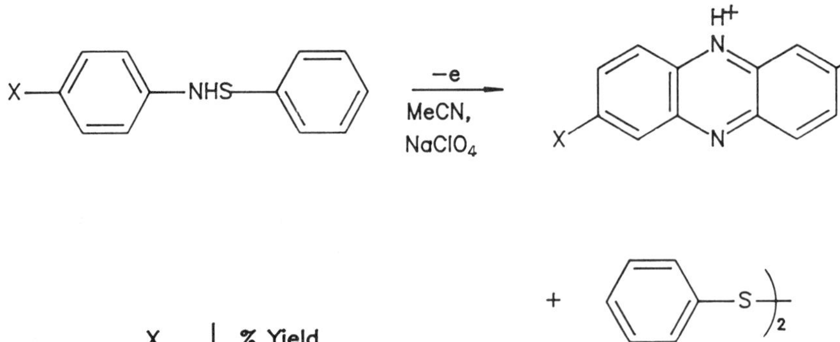

(11)

X	% Yield
OMe	39
Me	42

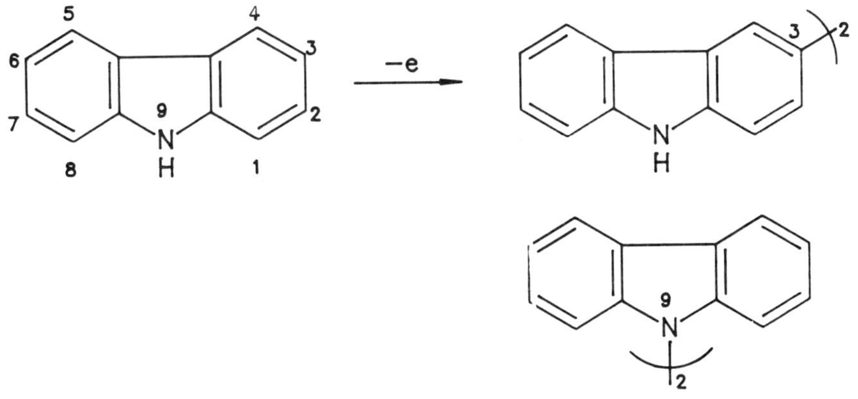

(12)

Anodic oxidation of π electron-rich N-heterocyclic compounds follows a pattern of predominant para,para' and N,N' coupling. Carbazole leads preponderantly to the oxidized form of the 3,3'- and 6,6'-dimer, together with some 9,9'-dimerization [Eq. (12) and Table 5, number 9] [87,88]. These observations are confirmed by results from the investigation of 76 ring-substituted carbazoles [89]. In basic solutions (2,4,6-collidine as base) the N,N'-coupling is often the exclusive reaction. 1-Phenylpyrazolidin-3-one forms besides the pyrazoline as major product a para carbon-nitrogen coupling product (Table 5, number 12); the corresponding imine leads to the expected para,para' dimer (Table 5, number 13). Other couplings of nitrogen heterocyclic compounds can be found in Table 5, numbers 14–16, and in a recent review [96].

Aliphatic amines are mainly converted to α-substituted products [99,100], whereby especially the α-methoxylation leads to valuable reagents for synthesis. The intermediate iminium salts can be directly trapped by silyl enol ethers to form Mannich bases [108]. If the α-position is blocked or steric conditions favor it, N,N-coupling to hydrazo or azo compounds occurs (Table 5, numbers 17–19). 1,1-Disubstitutes hydrazines are dimerized to tetrazenes in fair to excellent yields (Table 5, numbers 20–24). The intermediate diazenium ions can attack enolizable carbonyl compounds to form aza-Mannich bases [109]. Arylazonaphthols undergo anodic oxidation, producing radical cations. These couple to biphenylbisazo compounds (up to 34%) or can be trapped by anisidine to form azodiphenylamines (up to 74%) [110].

The CC coupling reactions of sulfur compounds are less numerous. One example is mentioned here. When 1,3-dithioles are oxidized in MeCN containing pyridine, tetrathiofulvalenes are obtained in moderate to fair yields (13–40%) [111].

B. Olefins

Olefins with electron-donating substituents as the alkoxy, acylamino, phenyl, or vinyl groups can be coupled in methanol to give 1,4-dimethoxy dimers and/or dienes [Eq. (13)]. The first intermediate in this coupling reaction is a radical cation, which either by electrophilic addition to the olefin and subsequent $1e^-$ oxidation (path A) or by radical dimerization (path B) leads

Table 6. Anodic Oxidation of Ethylenic Compounds

Number	Substrate	Conditions	Product	Yield (%)	Reference
1	Styrene	MeOH, NaOAc, $NaClO_4$, C	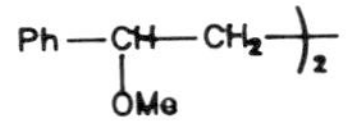	64	112
2	α-Methyl-styrene	MeOH, NaOMe, $NaClO_4$, C	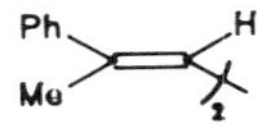	67	112
3	4,4'-Dimethyloxy-stilbene	+0.9 V, MeCN, HOAc	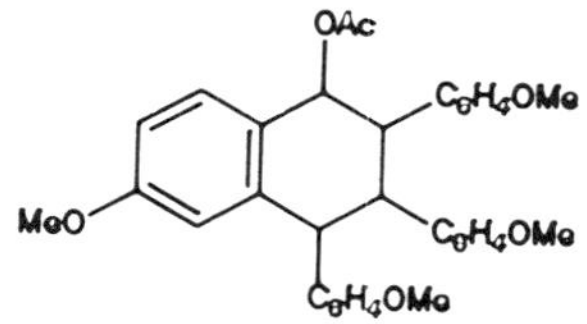	–	118
4	Styrene	CH_2Cl_2-H_2O, $TBAHSO_4$	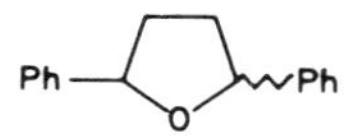	52	119
5	Vinyl ethyl ether	MeOH, NaI, C	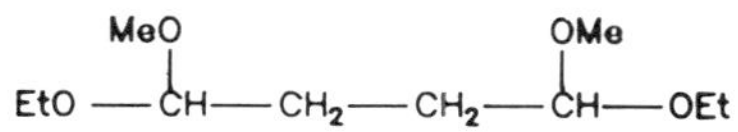	51	121
6	1-Ethoxycyclohexan	MeOH, $NaClO_4$, 2,6-lutidine, C	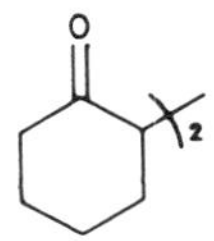	48	121
7	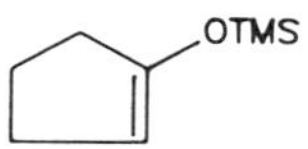	CH_2Cl_2, MeOH, Bu_4NBF_4	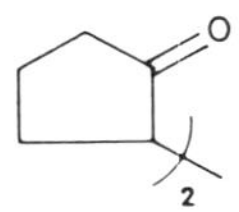	50	123

8	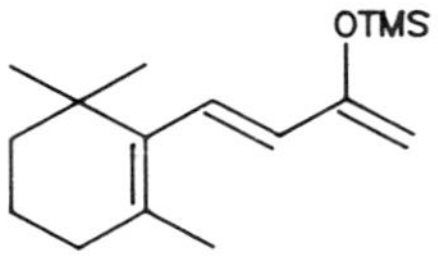	MeCN, MeOH, $LiClO_4$	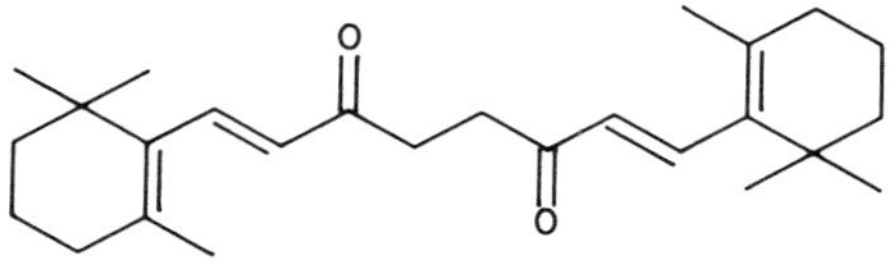	78	123
9	1,1-Bis(dimethyl-amino)-ethylene	DMF, Et_4NClO_4	$((Me_2N)_2CH{=}CH)_2$	–	125
10	Butadiene	C, MeOH, $NaClO_4$	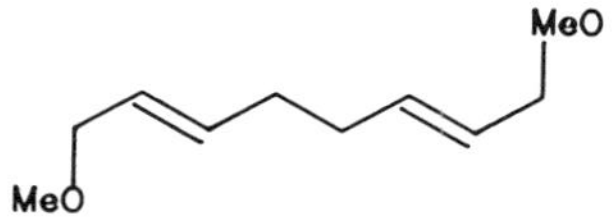	17	129a
			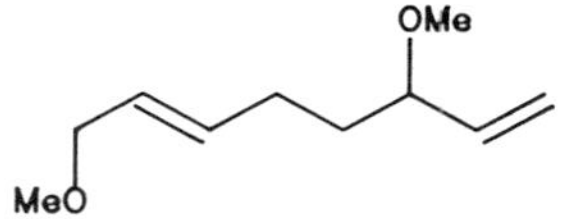	16	
			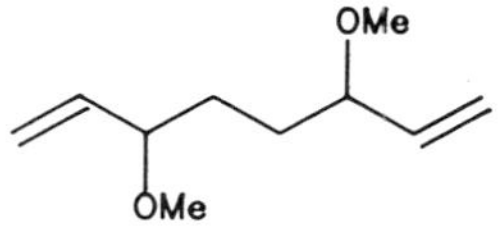	16	
11	Methylene cyclo-hexan	C, MeOH, $NaClO_4$	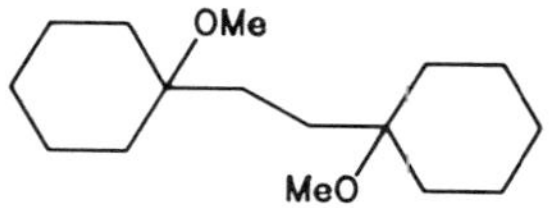	38	133
12	α-Methylstyrene, vinyl ethyl ether	C, MeOH, $NaClO_4$	Ph–C(Me)(OMe)–CH_2–CH_2–CH(OEt)(OMe)	32	134

to a dimer dication that undergoes methanolysis or deprotonation. Representative examples of this coupling reaction are summarized in Table 6.

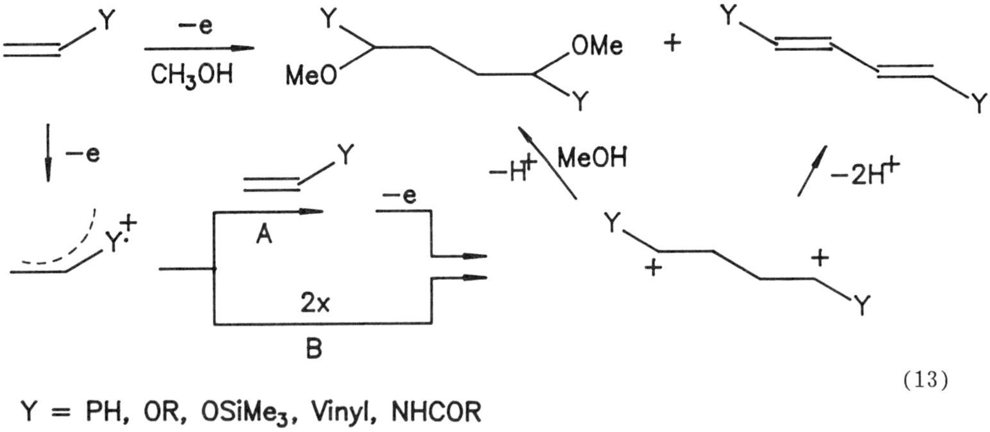

(13)

Styrene and indene derivatives [Eq. (13), Y = Ph] are dimerized to 1,4-dimethoxy-1,4-diphenylbutanes or 1,4-diphenylbutadienes (Table 6, numbers 1 and 2) [112,113]. The product distribution is in some cases strongly dependent on the anode potential and the supporting electrolyte. By substitution of sodium perchlorate for sodium camphor sulfonate or by increasing the anode potential, olefins are formed at the expense of methyl ethers. This may be explained by an increased displacement of methanol from the electrode surface through the stronger adsorbed camphor sulfonate or at higher potentials through the increased adsorption of anions compared to the adsorption of the polar solvent [99]. The surface concentrations of the reactants (solvent and styrene) govern the relative rates of solvolysis to the monomer and the addition reaction to the dimer and determine the product distribution. The styrene surface concentration is effectively influenced by the adsorption equilibria on different anode materials (C, Fe_3O_4, PbO_2, and Pt) [114]. β-Alkyl substituents drastically decrease the yield of dimers and favor the formation of dimethoxylated monomers.

For the dimerization of 4,4'-dimethoxystilbene, it has been possible to demonstrate spectroelectrochemically [115] and at the rotating ring-disk electrode [116] that the product is formed mainly by radical dimerization of the intermediate radical cations [path B, Eq. (13)]. Fast derivative CV, however, supports for the same olefin a complex ECE pathway [path A, Eq. (13)] [117]. Depending on the oxidation potential and the kind of the nucleophiles (acetate, water, or methanol), a tetrahydronaphthalene derivative (Table 6, number 3) [118], a monomer diacetate [118], a tetrahydrofuran [115], or a dimer dimethoxy compound is found. When methanol is replaced by aqueous dichloromethane or by aqueous acetonitrile emulsions as solvent, styrene (Table 6, number 4) [119a] and α-methylstyrene [119b] yield 2,5-diphenyltetrahydrofurans. In some cases cyclization occurs by electrophilic aromatic substitution (Table 6, number 3).

Enol ethers [Eq. (13), Y = OR] with half-wave potentials between $E_{1/2}$ = + 1.2 and + 1.8 V (Ag/AgCl) couple in methanol with 2,6-lutidine at a graphite anode to give the acetals of 1,4-dicarbonyl compounds in 30–60% yield (Table 6, numbers 5, 6) [120,122]. The tail-to-tail coupling of the monomers and the electrochemical reaction orders ($\nu_{\text{enol ether}}1$, $\nu_{\text{methanol}}1,0$)

support the radical cation as intermediate in the dimerization. Silyl enol ethers [Eq. (13), Y = $OSi(CH_3)_3$] can be dimerized to 1,4-dicarbonyl compounds in good yields. To suppress the methanolysis of the silyl enol ethers, MeCN-5% MeOH is used as solvent and the electrolysis is terminated within 1 h by the use of a capillary gap cell, which allows high currents (Table 6, numbers 7 and 8) [123]. In this way, unsymmetrical ketones can be coupled selectively in the α- or α'-position, since the corresponding silyl enol ethers can be prepared regioselectively [124].

Examples for the coupling of enamines are rare (Table 6, number 9). In most cases the enamines undergo methanolysis to electroinactive aminoacetals prior to the coupling reaction. Enamino ketones or enamino esters yield via dimerization of the radical cations and subsequent ring closure pyrrole derivatives [Eq. (14)] [126]:

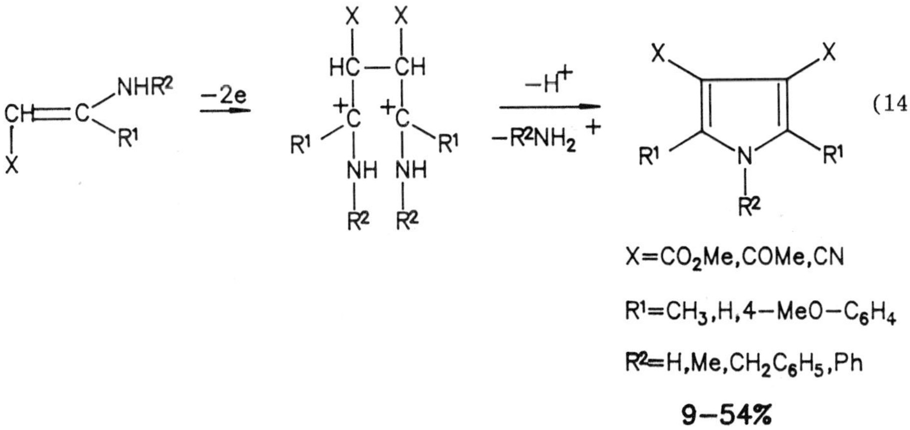

(14)

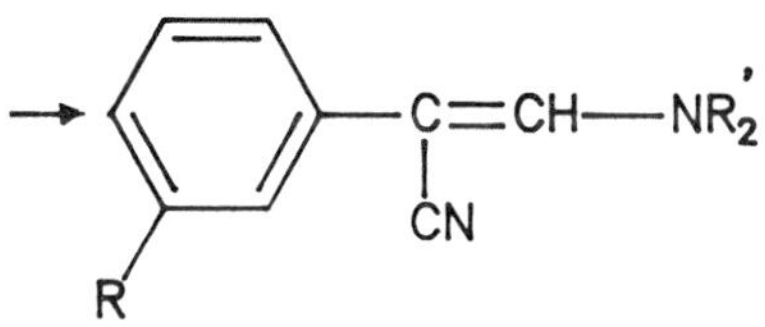

XVII

In the case of the enaminonitrile XVII, coupling through the phenyl ring is observed [127]. 1,4-Dithiafulvenes are dimerized to the dimer dication, which can be reduced to the dimer [128].

Dienes [Eq. (13), Y = vinyl] are dimerized to dimethoxyoctadienes or are dimethoxylated to dimethoxybutenes at graphite, platinum, or glassy carbon anodes in an electrolyte consisting of methanol-sodium perchlorate (Table 6, number 10) [120,129]. Dimerization is favored by graphite or carbon cloth anodes, by high olefin concentrations, and by terminally unsubstituted dienes [129,130]. The ratio of monomer to dimer is determined mainly by the anode material. Anodes with a smooth surface, such as platinum, gold, and glassy carbon, promote the formation of monomers; porous materials, such as soft graphite or carbon cloth, favor the dimers. Qualitatively this can be explained by lower real current densities at the porous surfaces. This leads to a less depleted diene concentration at the electrode surface, which favors the electrophilic addition of the radical cation [Eq. (13), path A] to the dimer, compared to the methanolysis yielding the dimethoxylated monomer. Butadienyl acetate and the β-ionone derivative XVIII could not be dimerized but yielded dimethoxylated monomers, whereby XVIII regioselectively forms 73% of the 9,10-dimethoxy product [129].

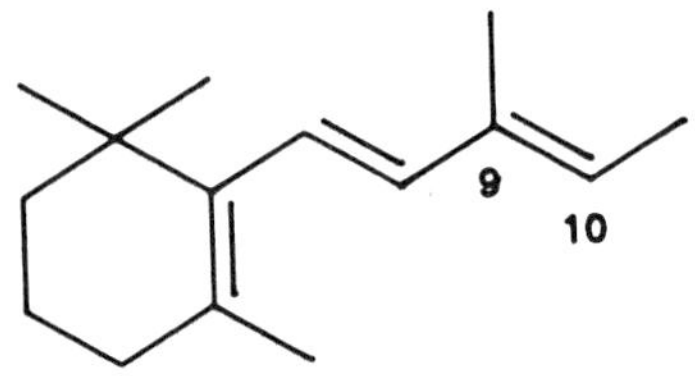

XVIII

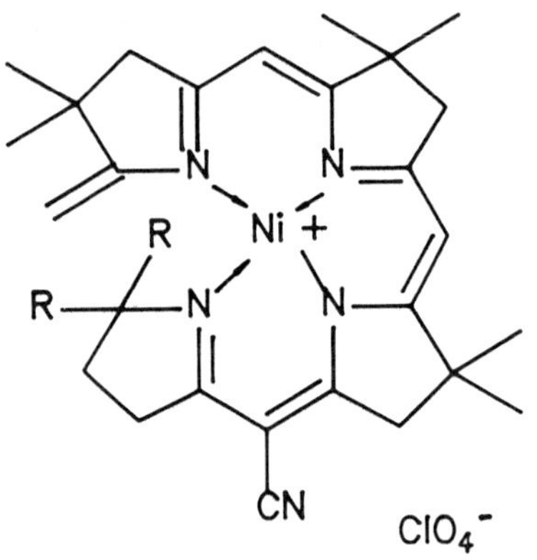

XIX

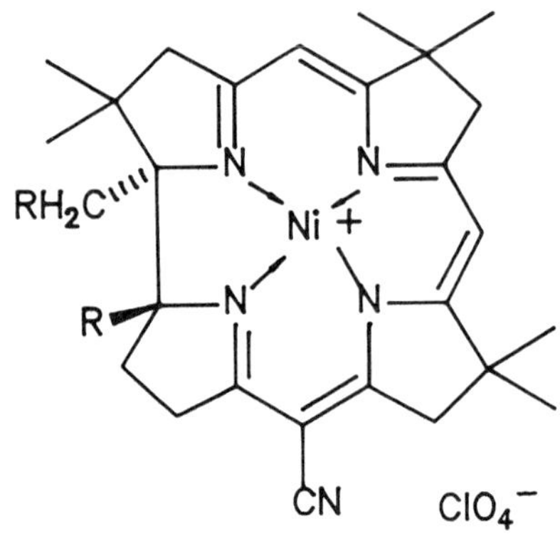

XX

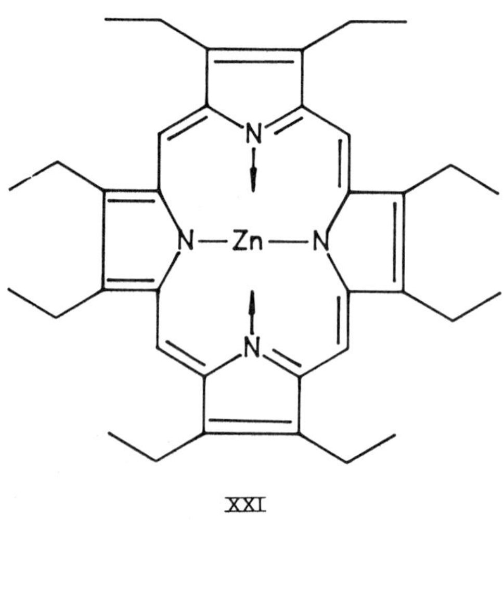

XXI

The polyene derivative XIX can be cyclized to form 18% of the nickel corrin complex XX (R = H), if it is first oxidized at +1.22 V to the radical cation, which undergoes a 1,16-hydrogen shift and then is coupled intramolecularly through cathodic reduction (−0.3 V) [131]. Zinc octaethylporphin (XXI) may be cyanated anodically at the meso-positions via the cation radical; by variation of the oxidation potential and the extent of electrolysis, it is possible to obtain optimum yields of mono-, di-, tri-, and tetracyanooctaethylporphyrins [132].

Alkenes without +M substituents can be coupled if they are dialkylated at one carbon atom of the double bond and are unsubstituted at the other (Table 6, number 11). This is a further indication that the dimerization of radical cations responds sensitively to steric hindrance.

Unsymmetrical dimers are accessible through the coelectrolysis of different olefins (Table 6, number 12).

C. Intramolecular Coupling

A large number of 1,4-bis(methoxyphenyl)alkenes (XXII) have been coupled intramolecularly to afford biphenyls with annulated rings of different size. The reaction is a method to prepare in few steps meta-cyclophanes, polycyclic hydroaromatics, and intermediates for the synthesis of steroids, heterocyclic compounds, and alkaloids. The section is organized according to an increasing number n of methylene groups in XXII, with Y being carbon or a heteroatom, mainly nitrogen.

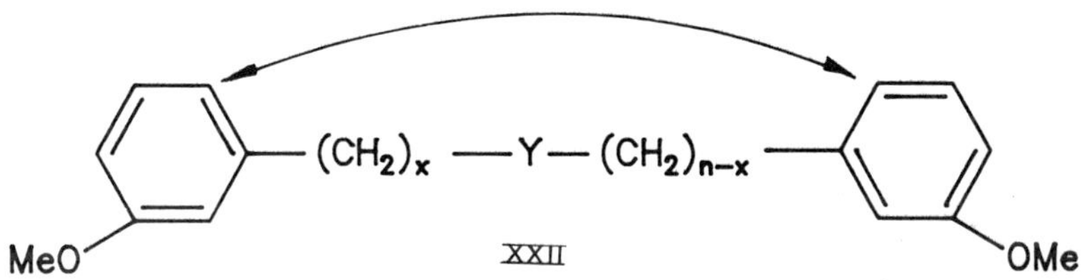

XXII

The coupling sites and yields in the anodic coupling of the diarylalkanes (XXII) are strongly dependent on the value of n, the position of substituents in the aryl ring, the choice of the electrolyte-electrode system, the pH of the medium, the current density, and the oxidation potential.

In acetonitrile, methoxybiphenyls are coupled in low yield, but almost quantitatively in dichloromethane-trifluoroacetic acid [135,136]. Yields are low in the former solvent because the cyclized products are more easily oxidized than their precursors and the oxidation products are not stable. In the presence of TFA, the cation radicals derived from the oxidation of cyclyzed XXII are stable and after reduction with zinc dust, for example, cyclized XXII may be isolated in high yield.

1,n-Bis-(3,4-dimethoxyphenyl)alkanes (XXII, Y = −) cyclize almost quantitatively, when n = 1–4 (Table 7, numbers 1–6), but when n = 6–10 only in moderate yields and then intermolecularly [136,137]. The cyclic voltammograms of unsymmetrically methoxylated diphenylalkanes exhibit two oxidation potentials, at which either intramolecularly or intermolecularly formed products arise, permitting some conclusions to be drawn about the dimerization mechanism. These observations provide strong evidence that two radical cations are necessary for aryl-aryl bond formation [140]. How-

Table 7. Intramolecular Coupling

Number	Substrate	Conditions	Product	Yield (%)	Reference
1	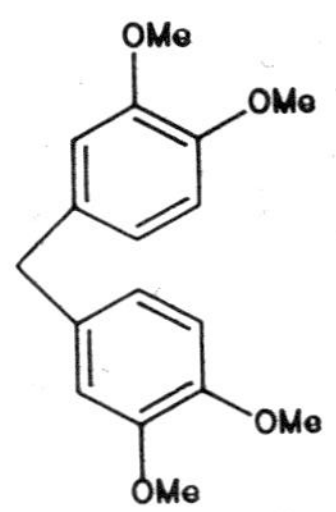	CH_2Cl_2, TFA, $TBABF_4$, Pt	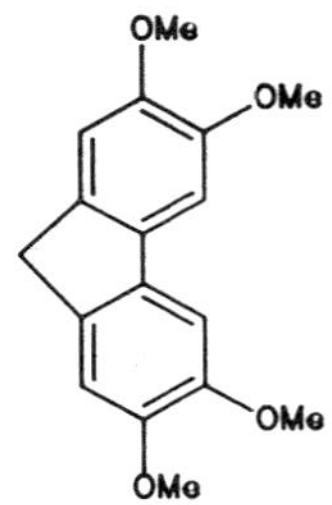	95	137
2	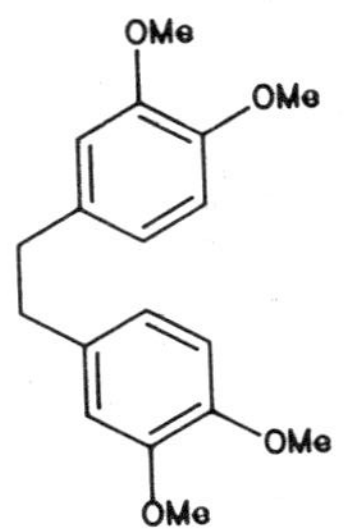	CH_2Cl_2, TFA, $TBABF_4$, Pt	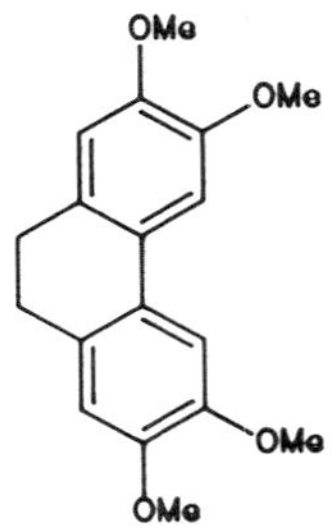	95	137
3	Tetraphenylethylene	Pt, CH_2Cl_2, TFA, Bu_4NBF_4	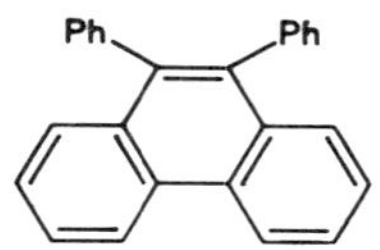	92	138

4	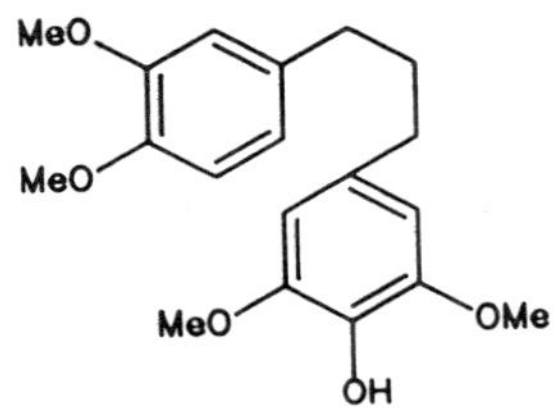	MeCN, $TMABF_4$, Pt, 0.74 V (SCE)	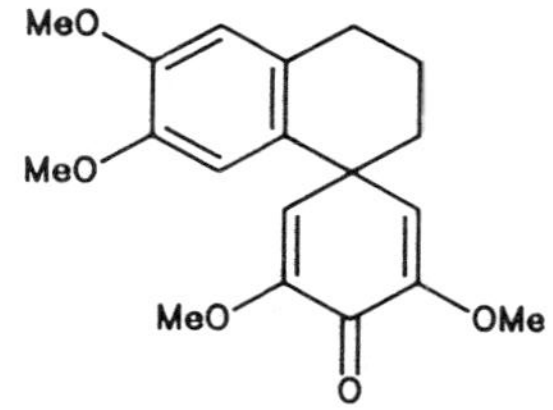	85	139
5	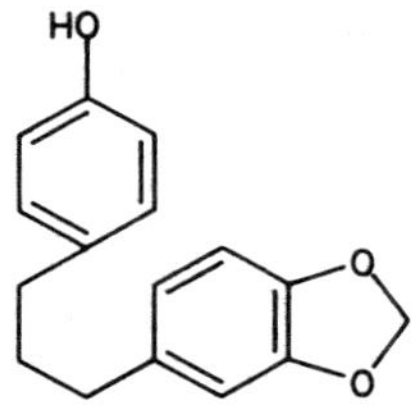	Pt, MeCN, $NaClO_4$	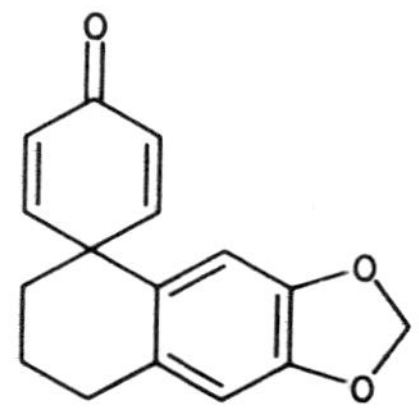	75	139
6	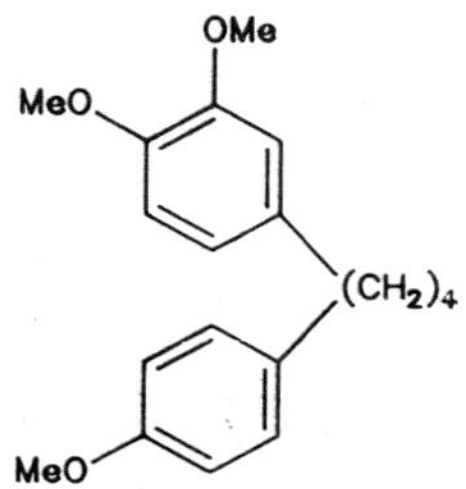	Pt, CH_2Cl_2, TFA (3:1), $TBABF_4$	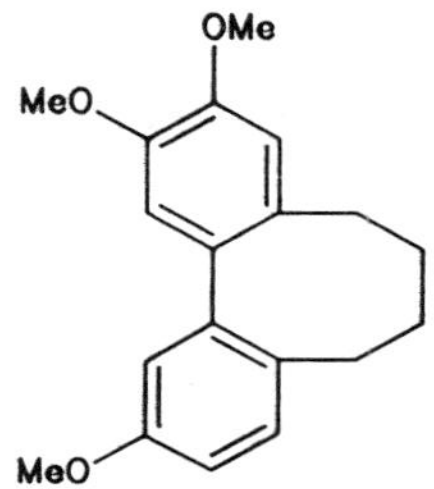	33	140
7	Norbornadiene	HOAc or MeOH, Et_4NOTos	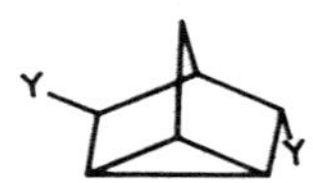	exo, endo 14–16 exo, exo 15–18	142

Table 7. (Continued).

Number	Substrate	Conditions	Product	Yield (%)	Reference
8	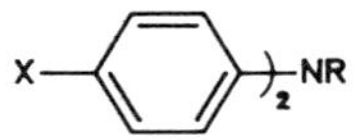 $X = CH_3$, t-Bu, Cl R = Et	MeCN, Et_4NClO_4	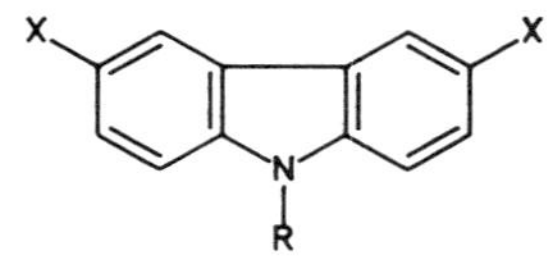	-	69
9a	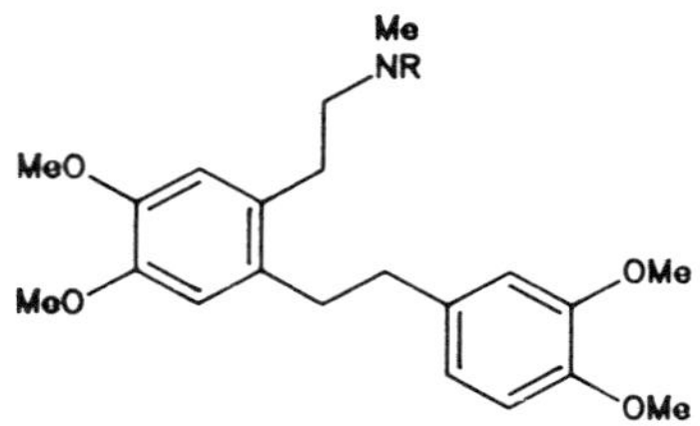	MeCN, HBF_4, Pt, R = CO_2Et, 1.0 V (SCE)	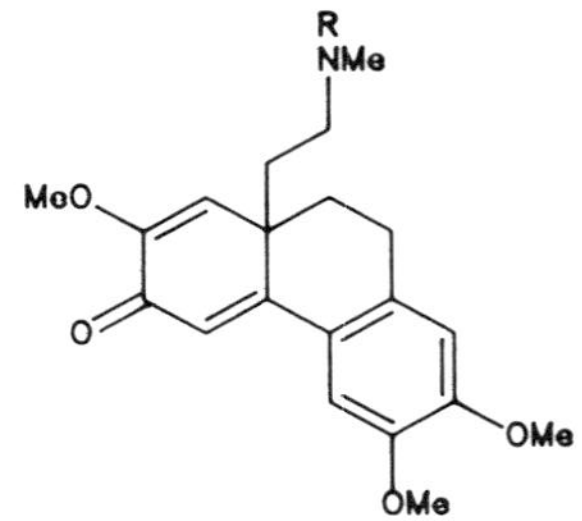	87	145
9b	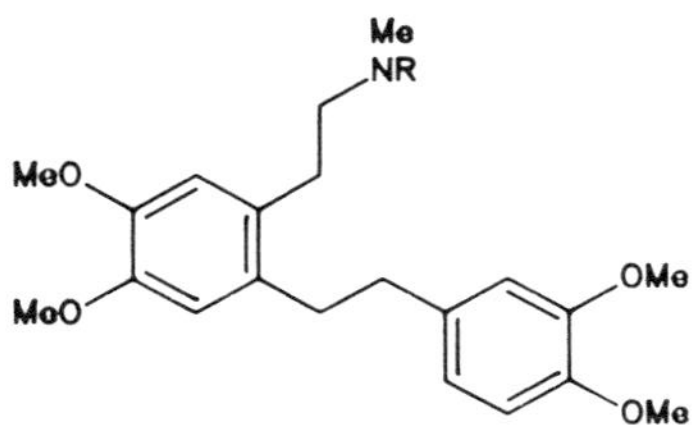	MeCN, HBF_4, Pt, R = $COCF_3$, 1.3 V (SCE)	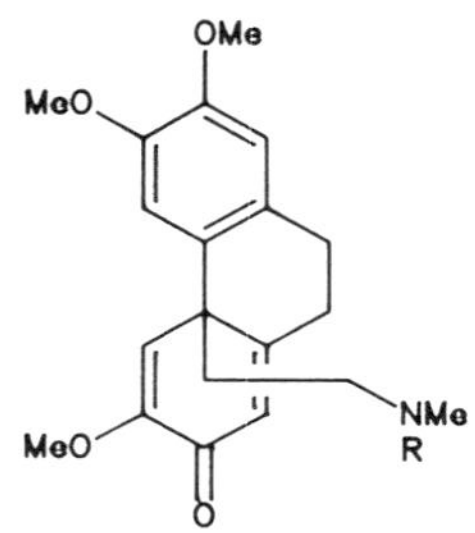	75	145

10a	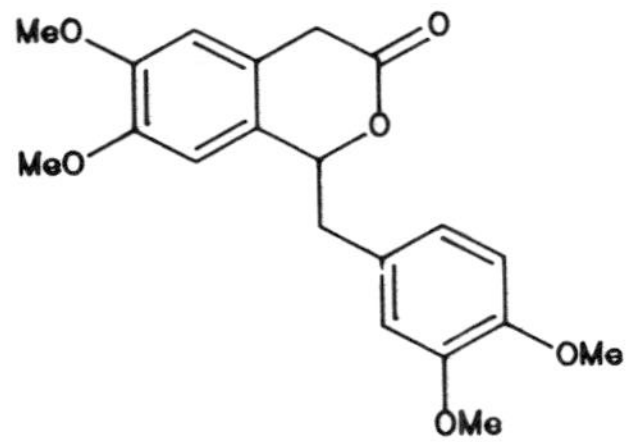	CH_2Cl_2, TFA, $TBABF_4$	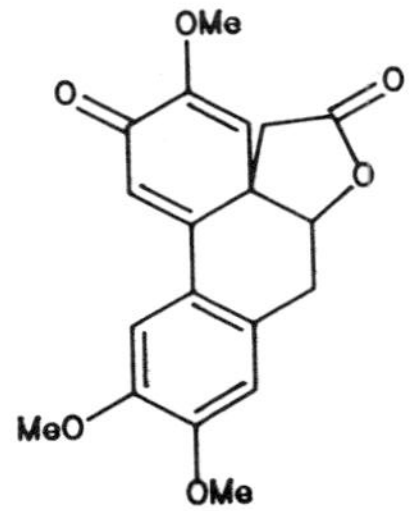	33	162
10b	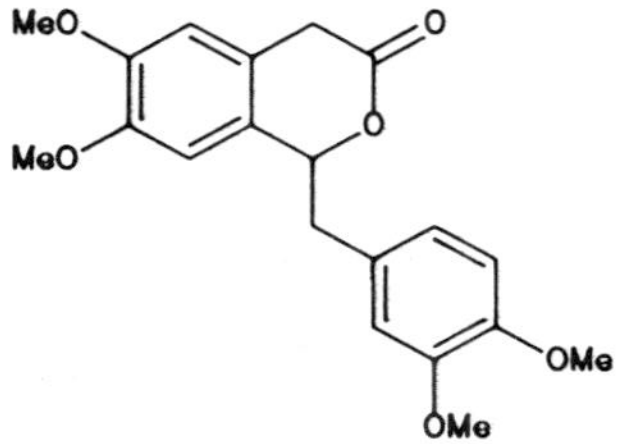	MeCN, $TBABF_4$	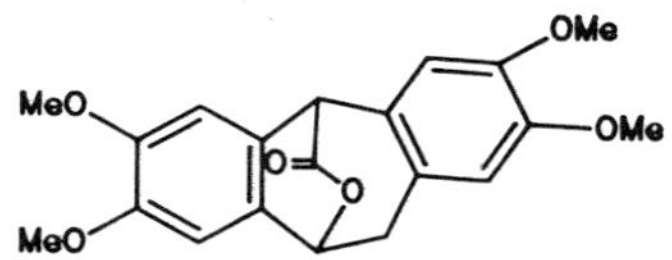	20	162
11	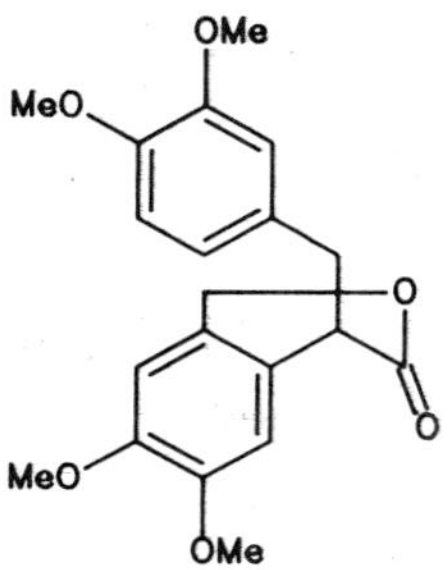	MeCN, $NaClO_4$, Pt	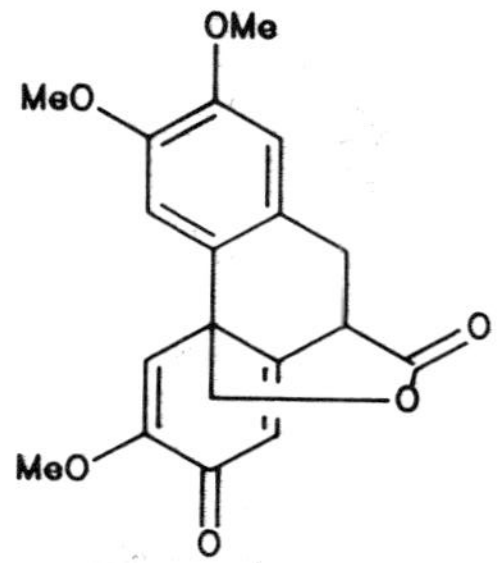	67	163

Table 7. (Continued).

Number	Substrate	Conditions	Product	Yield (%)	Reference
12	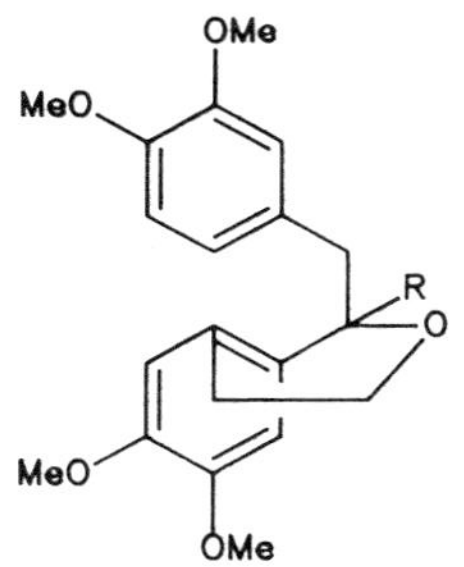	Pt, MeCN, HBF_4	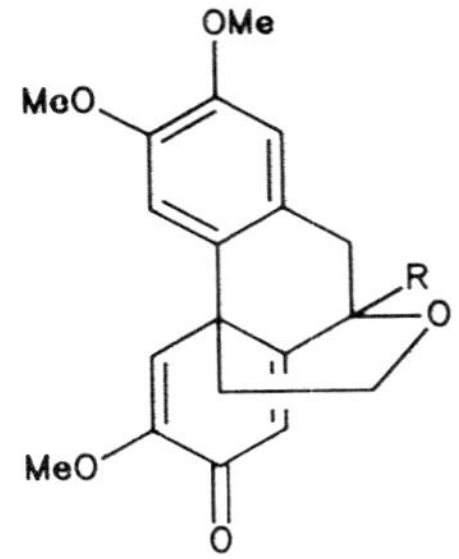	R = H 26 R = CH_3 56	161
13	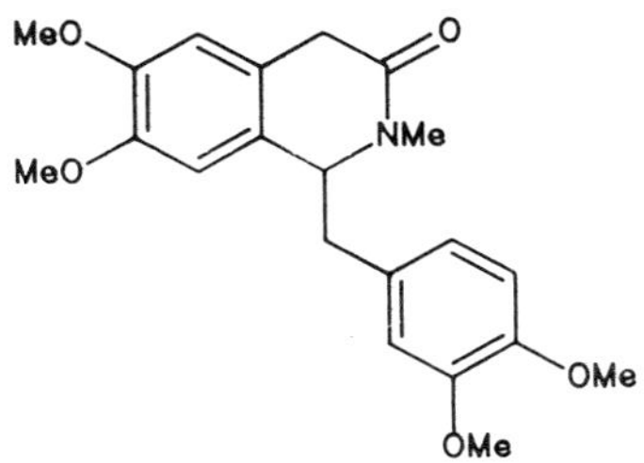	CH_2Cl_2, TFA, $TBABF_4$, Pt	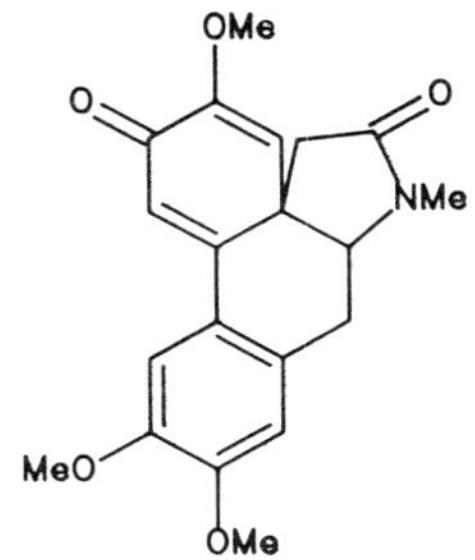	40	162
14	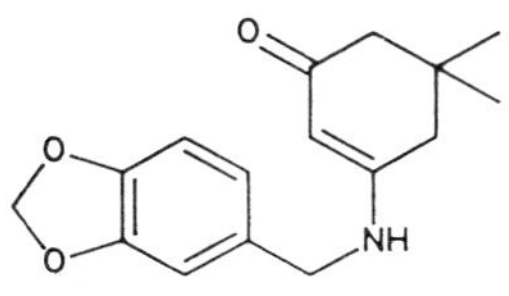	MeOH, $NaClO_4$, C	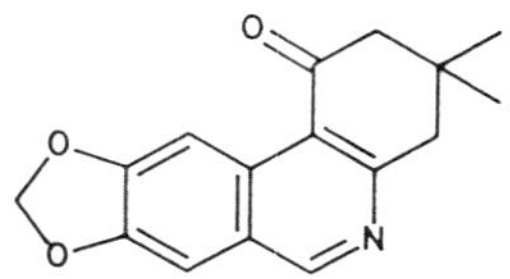	45	164

15	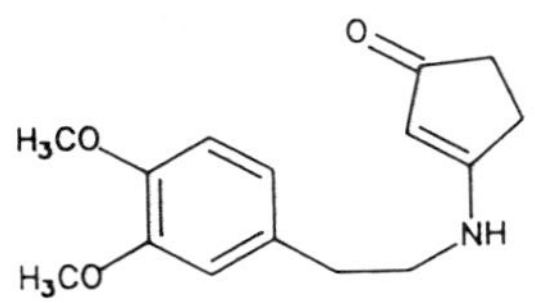	MeOH, $NaClO_4$, C	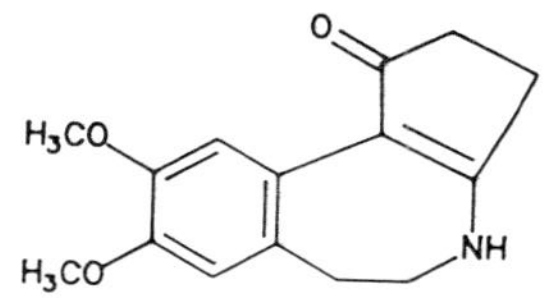	43	164
16	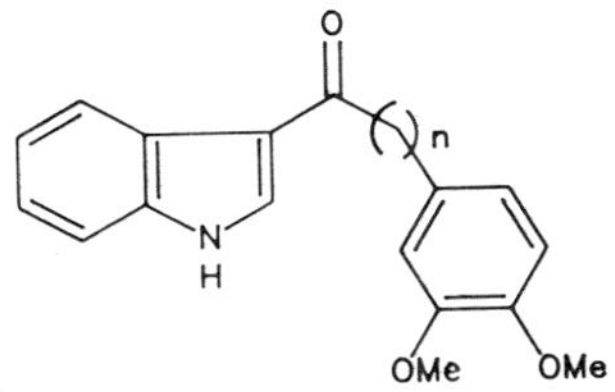	C, MeCN, $NaClO_4$	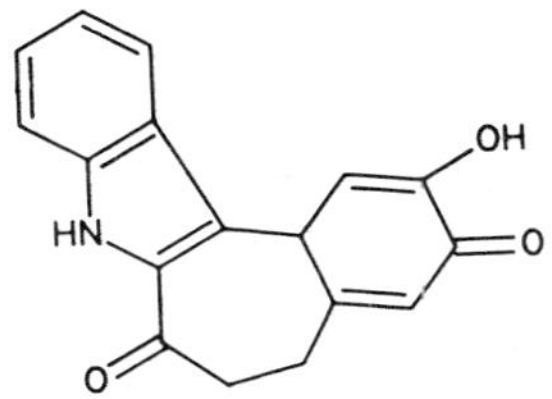	62 (for n = 2)	166
17	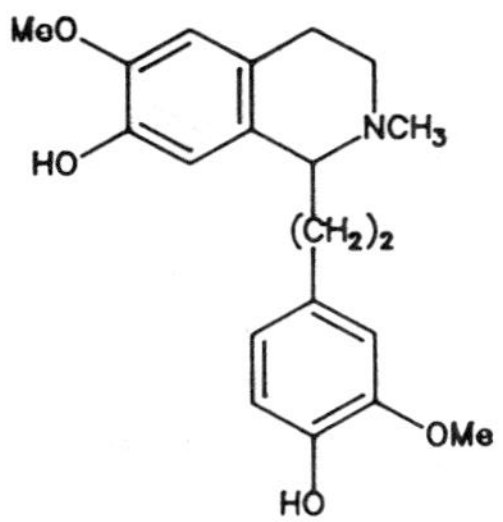	H_2O, Et_4NClO_4, C	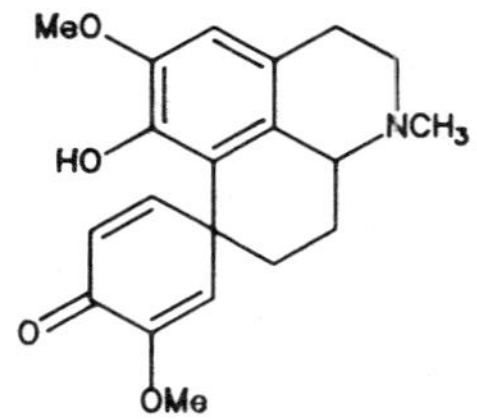	36	169
18	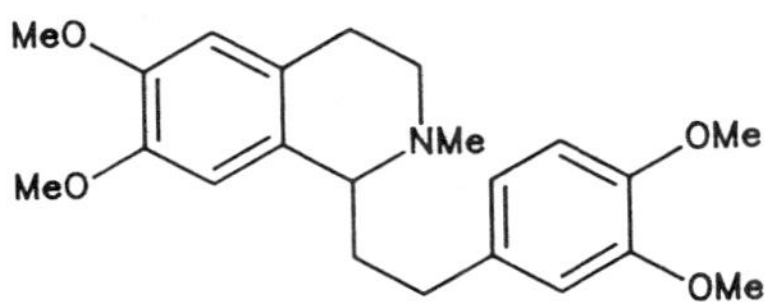	MeCN, $NaClO_4$, 1.2 V (SCE)	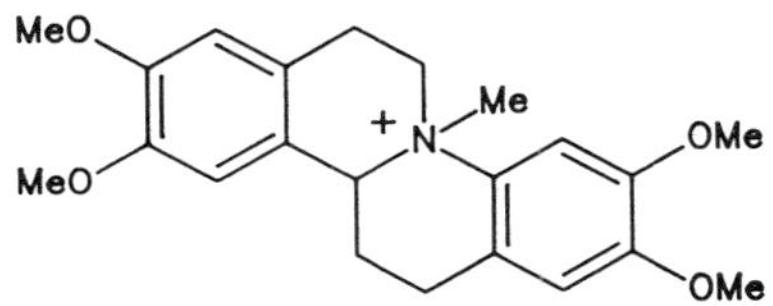	–	170

Table 7. (Continued).

Number	Substrate	Conditions	Product	Yield (%)	Reference
19	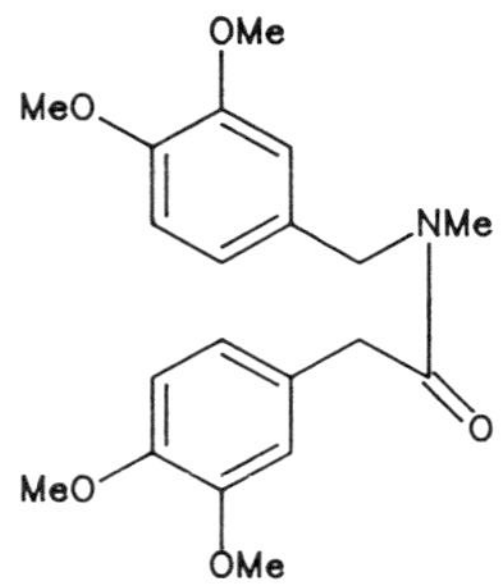	MeCN, $NaClO_4$, Pt, 1.1 V	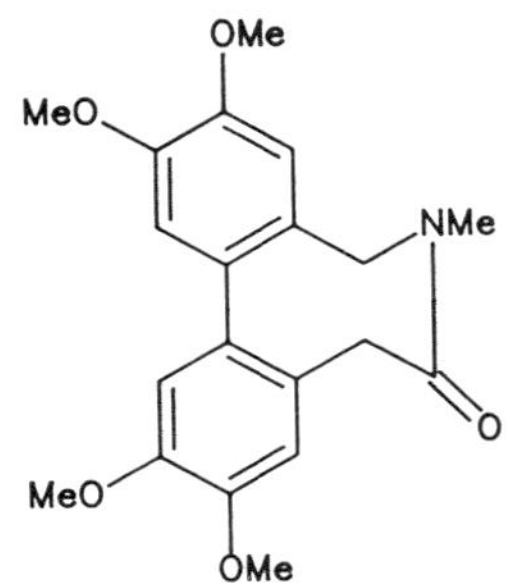	45	174

ever, kinetic studies indicate a cyclization via path A, Eq. (13), followed by a rate-determining electron transfer [141]. Diarylalkanes substituted with a phenol ether group at one end and a phenolic group at the other end afford high yields of spirodienones (Table 7, numbers 4 and 5). This anodic cyclization can follow three mechanistic pathways.

Intramolecular couplings between nonconjugated double bonds being suitably disposed for transannular reactions have been carried out with norbornadiene (XXII, n = 1; Table 7, number 7) and bicyclo[2.2.2]octadiene-2.5 [142].

Substituted di- and triphenylamines lead to carbazoles (XXII, n = 0, Y = NR; Table 7, number 8). The yields are poor, however, and the cyclization is restricted to tertiary amines with the para-position occupied by substituents not susceptible to substitution or elimination.

The 2-methyltetramethoxybibenzyl derivative XXIII cyclizes in an anodic 2,6'-coupling reaction, with a subsequent dienone-phenol rearrangement, to give an 88–98% yield of the dihydrophenanthrene XXIV [Eq. (15)], which contains the essential structural elements of the B, C, and D rings of steroids [143].

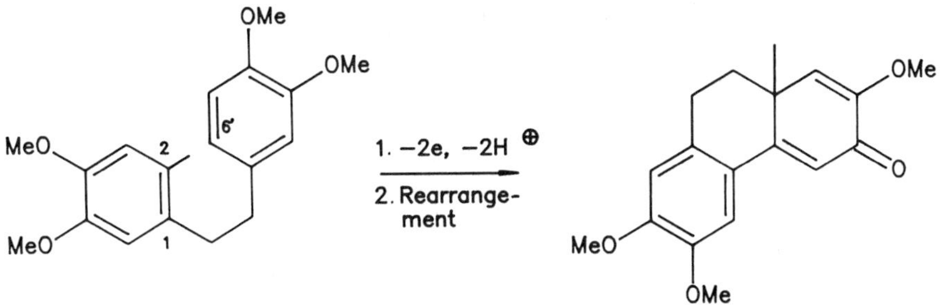

XXIII XXIV (15)

With the corresponding dimethoxybibenzyl derivative, ring closure takes place without rearrangement to form a structure analogous to the A, B, and C rings of steroids [144], but the yield is only moderate (22%) because of the lower activation of the arene system. A remote substituent can strongly influence the occurrence of the subsequent rearrangement (Table 7, numbers 9a and b).

Tetraphenylethylene is coupled to 9,10-diphenylphenanthrene (Table 7, number 3). The corresponding coupling with tri- and diphenylethylene was unsuccessful. It was proposed that the initially formed cation radical disproportionates to a dication that couples to the observed product [138].

In the case of the [2.2]-meta-cyclophane (XXV), two anisyl radical cations couple in 90% yield to give the tetrahydropyrene derivative XXVI [Eq. (16)] [146]. XXVI can be further oxidized to a pyrene [147]. [2.2]-Meta-

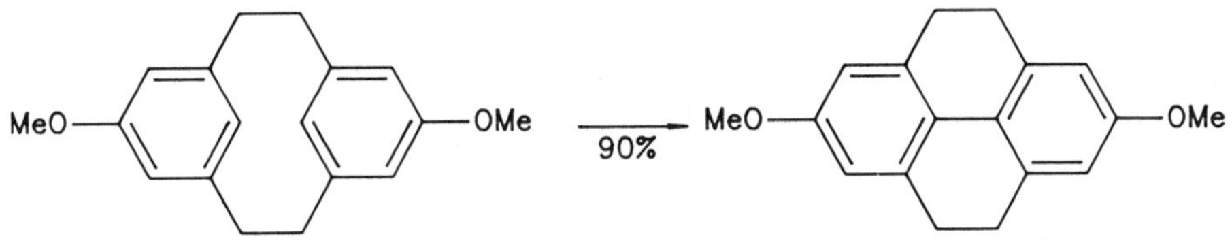

XXV XXVI (16)

cyclophanes with other substituents than methoxy, for example with CH_3, Br, CN, or NO_2, can also be cyclized [148]. The anodic cyclization of benzyltetrahydroisoquinolines XXVII [XXII, n = 2, Y = –; Eq. (17a)] to the morphinane skeleton is particularly interesting from the preparative point of view since the cyclization cannot be achieved so satisfactorily with chemical oxidants. XXVIIIa and XXVIIIb can be obtained in good yield in acetonitrile-sodium bicarbonate using a divided cell [149–153]; XXVIIIc–f can even be prepared in an undivided cell and in acidic electrolytes [154a,b], although under less mild conditions. The reaction has also been performed to prepare optically active 9-(R)-O-methylflavinantine [154]. Further oxidation of flavinantine leads to a specific hydroxylation at C10 together with a subsequent ring cleavage and reclosure to an oxohomomorphinane [155c].

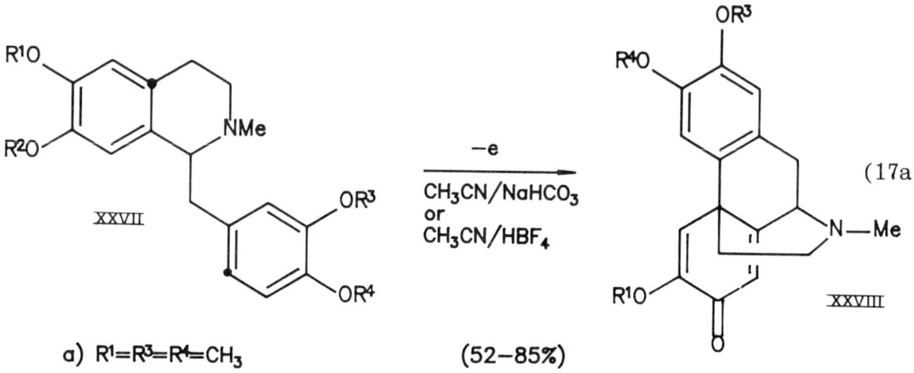

(17a)

a) $R^1=R^3=R^4=CH_3$ (52–85%)

b) $R^1=R^3=CH_3, R^4=PhCH_2$ (43–63%)

c) $R^1=R^3=CH_3, R^4=H$ (63%)

d) $R^1=R^4=CH_3, R^3=H$ (50%)

e) $R^1=CH_3$, $-R^3-R^4- = -OCH_2O-$ (80%)

f) $R^1=R^2=R^3=R^4=OCH_3$ (70%)

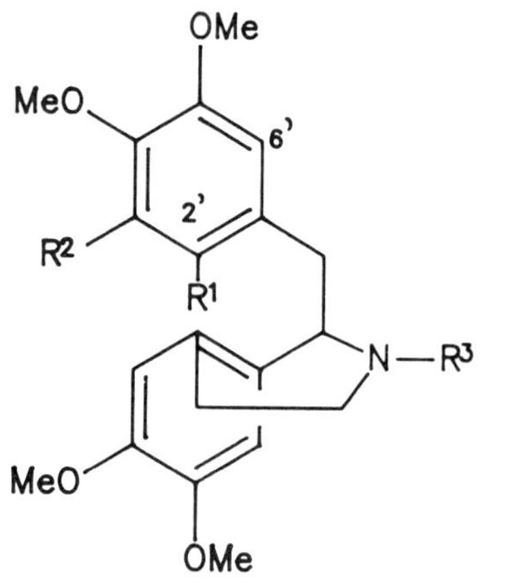

XXIX

	R^1	R^2	R^3
a	Br	H	Me
b	Cl	H	Me
c	NO_2	H	Me
d	$NHCOCH_3$	H	Me
e	H	I	Me
f	H	Br	Me
g	H	MeO	Me
h	H	H	CHO
i	H	H	Me
j	H	H	CF_3CO

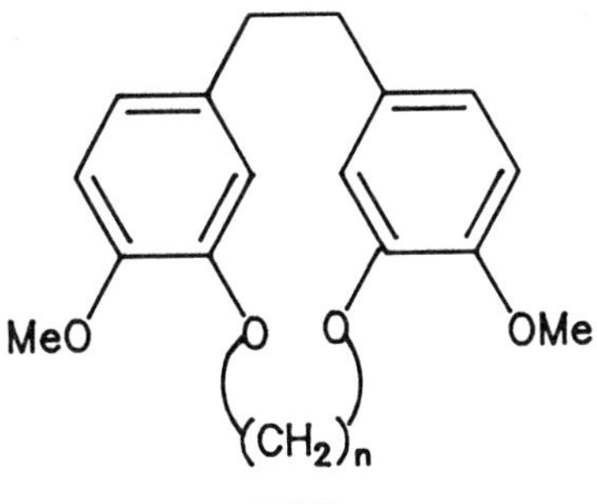

XXX

	n
a	6
b	4
c	3

It is, however, a disadvantage that in XXIXi the −I effect and the steric hindrance of the 5'-methoxy group induces a C2'−C5a coupling to the flavinantine type of compound and prevents the C6'−C5a coupling to the desired morphine type of compound. Even when the two dialkoxy-substituted phenyl rings are forced into a ortho,ortho-position as in XXX, no ortho-ortho coupling but an intermolecular dimerization occurs. When the methylene chain in XXXb,c becomes longer, intramolecular ortho,para coupling occurs [156].

Attempts to enforce the "morphine type" of coupling have so far been unsuccessful. XXIXa cyclizes at C2' with debromination [153]. XXIXb [149,153] and XXIXc,d [157] lead to cleavage at C1; in XXIXe,f the C2' ring closure is hardly hindered [149,153]; XXIXg, in which the C2' and C6' coupling are identical, forms 35−55% of the cyclization product [151,153]. N-Acyl substituents considerably affect the formation of products; N-formyl (XXIXh) is cleaved at C1, and N-trifluoroacetyl (XXIXj in 3':OCH_2Ph instead of OMe) gives the dienone (XXXIV) in good yield [157].

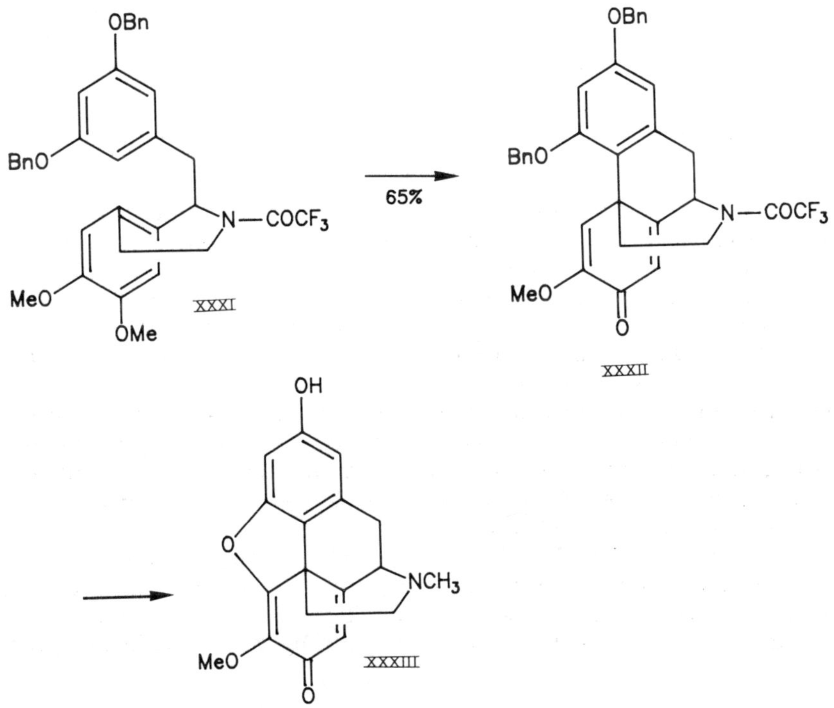

(17b)

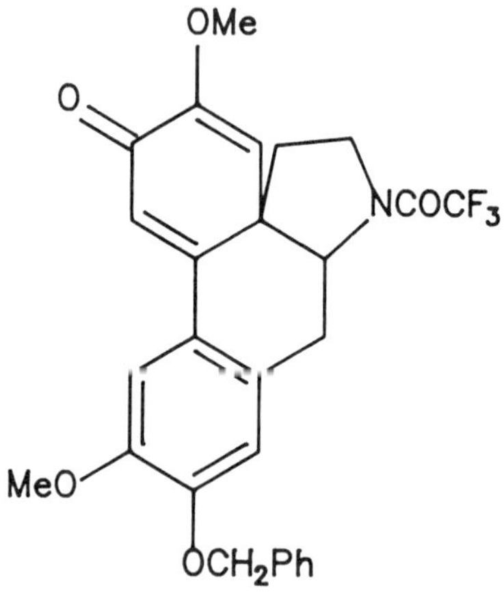

XXXIV

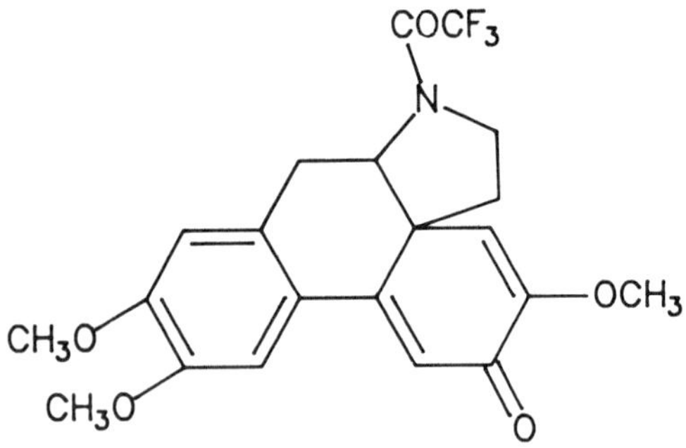

XXXV

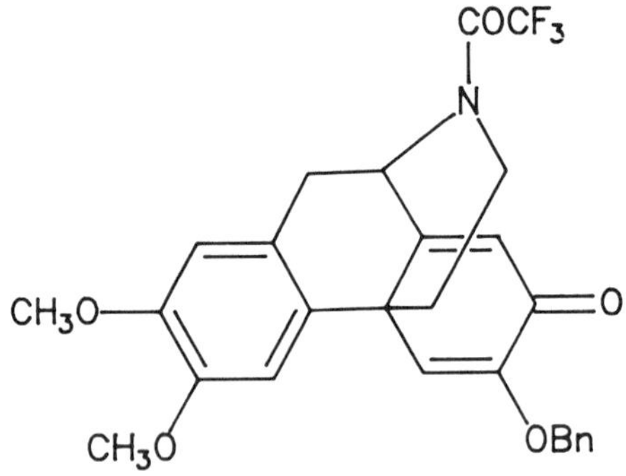

XXXVI

XXXI [Eq. (17b)], which also has two identical ortho coupling sites, can be cyclized in 65% yield to XXXII, which subsequently has been converted into 3-desoxythebain (XXXIII) [158].

The solvent exerts a strong influence on the product distribution. Starting from the same tetrahydrobenzylisoquinoline in MeCN, 82% XXXV are formed. In MeCN-MeOH the intermediate can be effectively trapped by methanolysis and 58% XXXVI are obtained [159].

A major mechanistic study seeking common denominators in the oxidations leading to XXIV, XXVI, and XXVIII has been published [152b]. A stereochemical study indicated that the yield in the cyclization is independent on the pseudoequatorial or pseudoaxial position of the benzyl substituent in the benzyltetrahydroisoquinoline [160].

Carbon analogs (XXXVII, Y = CH_2, Z = CH_2) of benzyltetrahydroisoquinolines [161] couple at C2' and C5a in good yield to give XXXVIII [Eq. (18)] or at C8a to give IXL; in acidic electrolytes ($CF_3CO_2H-CH_2Cl_2$) IXL rearranges to give XL. A 4-benzyl-substituted tetrahydroisoquinoline afforded the tetracyclic structure XLI, the higher homolog, the 4-β-phenethyl derivative cyclized to a homoisoaporphine XLII. The results have been analyzed in terms of the Baldwin rules [165].

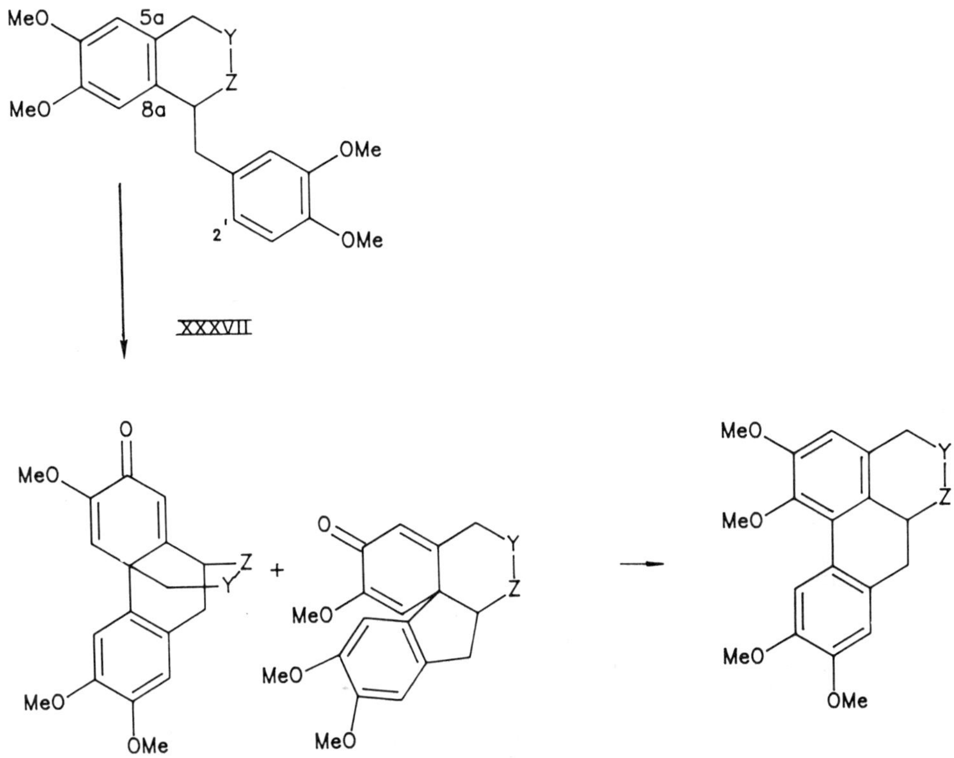

XXXVII

XXXVIII XXXIX XL

(18)

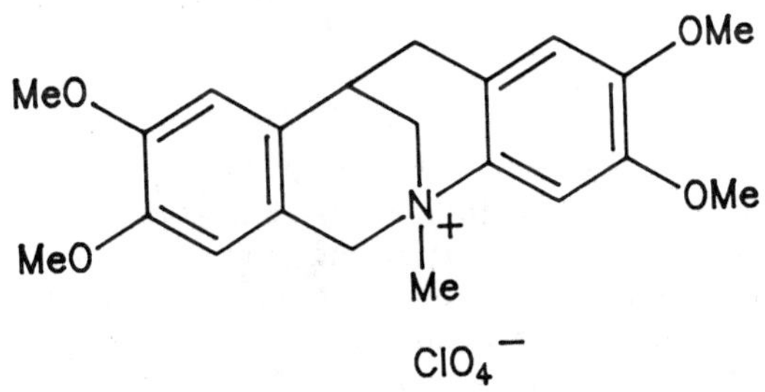

XLI

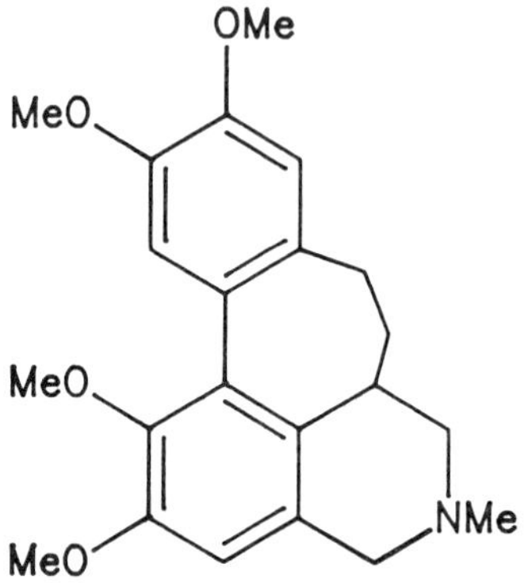

XLII

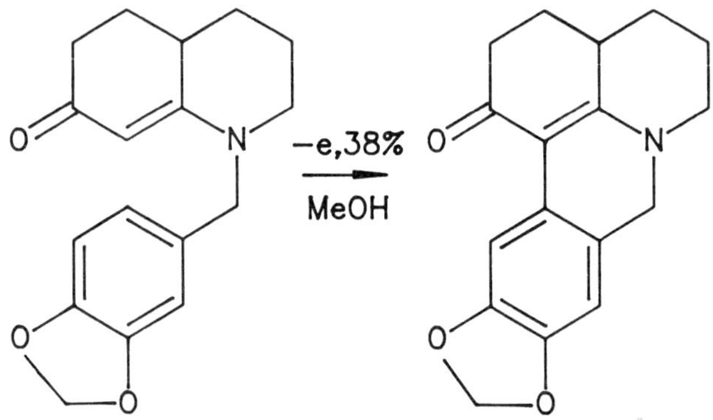

(19)

Oxygen analogs of benzyltetrahydroisoquinolines (XXXVII: Y = CO, Z = O) form in dichloromethane-TFA a spirodienone by C−2', C−8a coupling; in acetonitrile a bridged lactone was obtained (Table 7, numbers 10a and b). The isomeric isochroman-3-one (Y = O, Z = CO) undergoes an intramolecular coupling between C−6' and C−8a and migration of the 6'−8a bond to afford a γ-lactone (Table 7, number 11). Also benzylisochromanes (XXXVII: Z = O, Y = CH_2) afford in fair to good yields C−2' to C−5a coupled diphenyl products (Table 7, number 12). For XXXVII: Y = CO, Z = NCH_3, a C−2' to C−8a coupled product with subsequent rearrangement is found (Table 7, Number 13). A survey of anodic intramolecular coupling with multiple variations of XXXVII was undertaken [161]. The report offers mechanisms and a fairly complete overview of the area.

Benzyl- and 2-phenethylenaminoketones (XXII: Y = NR, n = 1,2) can be cyclized in good yields to quinolines and benzazepines (Table 7, numbers 14 and 15). The cyclization has also been used to prepare a precursor of the alkaloid lycorane (Eq. (19)] [167]. In indoles the oxidation of the heterocycle must be suppressed by electron-withdrawing substituents. This is the case in 3-keto-substituted indoles. Although the α-phenylacetyl indole (Table 7, number 16, n = 1) undergoes no cyclization, which has been analyzed in terms of a geometrically unfavored transition state [165], the β-phenylpropionyl indole (n = 2) cyclizes. The radical cation first couples to a spiro intermediate, which undergoes oxidation, bond breaking, rearrangement, and tautomerization to the indolenine.

Unsymmetrical 1,3-diaryl propanes (XXII, n = 3, Y = –), which bear phenolic hydroxyl groups, react to spirodienones, presumably via a phenoxonium ion (Table 7, numbers 4 and 5). Trimethoxy-substituted diphenylpropanes form different products depending on the reaction conditions. In MeCN-HBF_4 a spiro dienone is obtained [Eq. (20)] [140]; in dichloromethane-TFA (3:1) a seven-membered ring is formed. The latter result is explained by the low nucleophilicity of the medium and TFA catalysis of the dienone phenollike rearrangement.

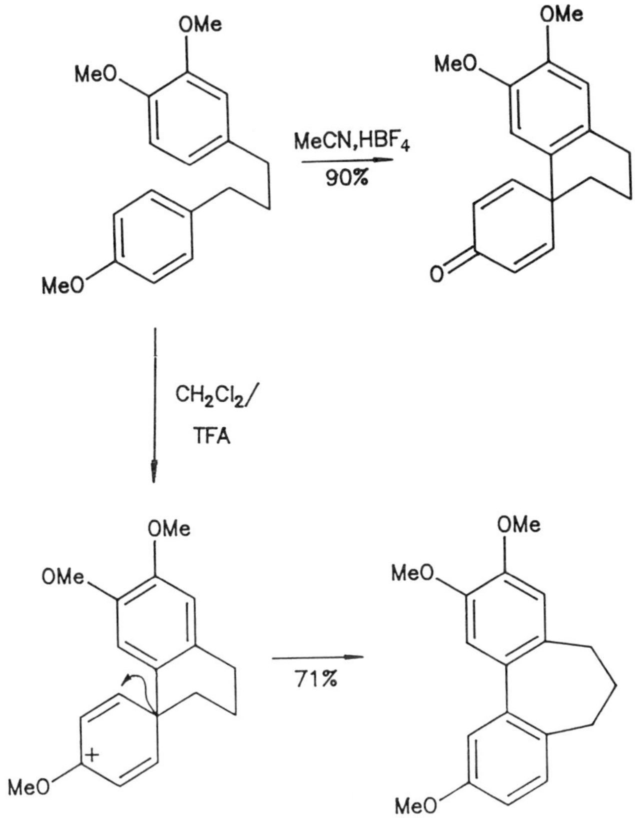

(20)

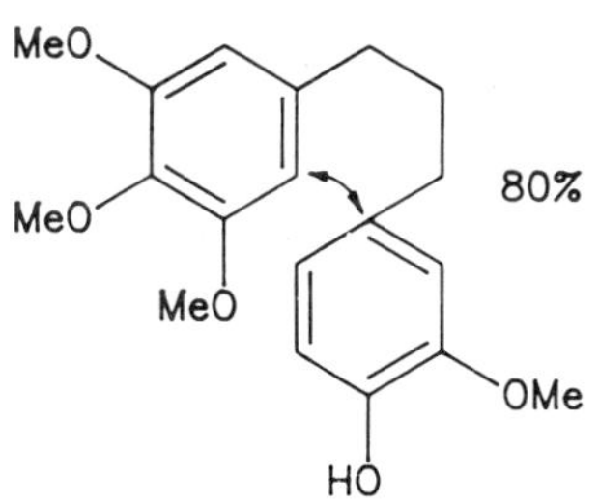

XLIII

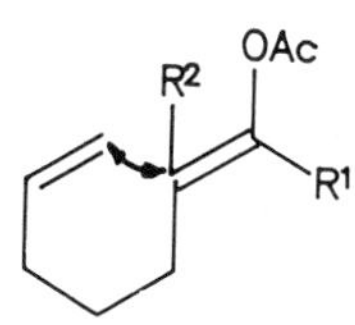

XLIV

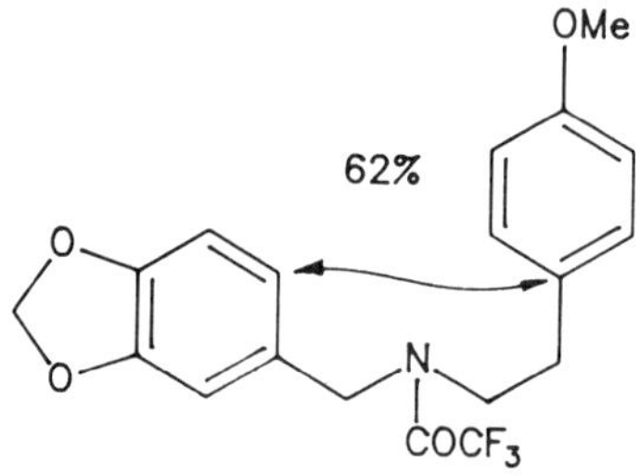

XLV

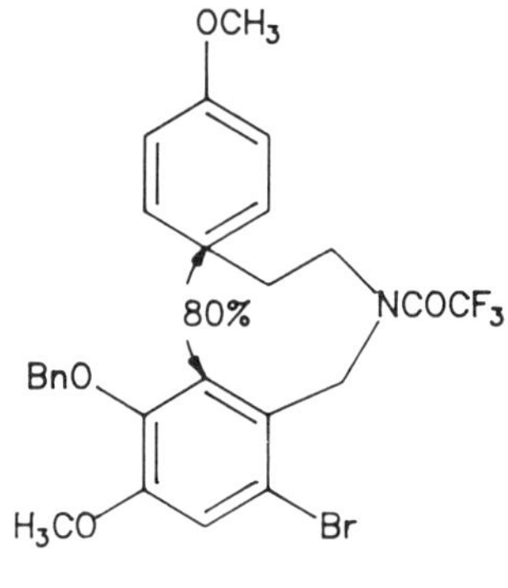

XLVI

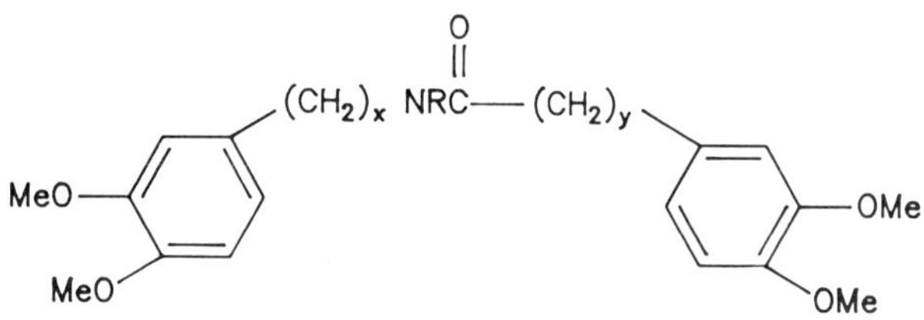

XLVII

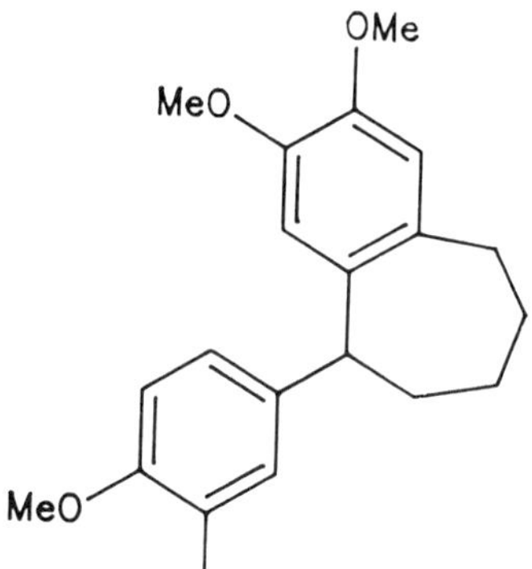

XLVIII

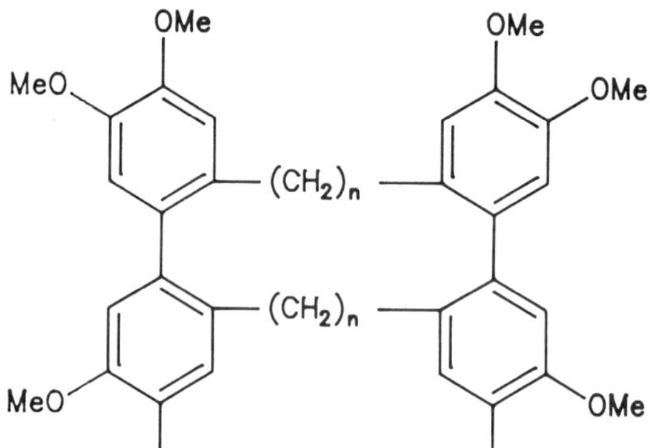

XLIX

A 2-phenethyltetrahydroisoquinoline with hydroxyl groups as substituents leads to a ortho,para-coupled spiro compound (Table 7, number 17); with alkoxy groups a tetracyclus by nitrogen-aryl coupling is obtained (Table 7, number 18). The intramolecular aryl coupling of XLIII has been used in a synthesis of colchicine [171]. Unsaturated enol acetates (XLIV) (XXII, n = 3, Y = –) have been cyclized to cyclohexenylketones [172].

1,4-Di(methoxyphenyl)butane (XXII, n = 4, Y = –) forms in dichloromethane-TFA initially a seven-membered ring followed by rearrangement to the eight-membered ring and its dimerization (Table 7, number 6). The amide XLV (XXII, n = 3, Y = $NCOCF_3$) has been cyclized to a precursor of (+)-oxocrinine [168] and XLVI to a precursor of galanthamine [173].

Diarylamides (Table 7, number 19) and XLVII can be cyclized when the amide is tertiary and the carbonyl and amino groups are not conjugated with the aromatic system (XLVII, x, y, ≠ 0) [174]. From the corresponding secondary amides only intermolecularly coupled products were obtained, which can be rationalized in terms of the preferred stereochemistry of the amides.

1,5-Bis(dimethoxyphenyl)butane (XXII: n = 5) forms a seven-membered ring XLVIII by a para-benzyl coupling [137]. With XXII (n > 6), coupling occurs intermolecularly and with partial subsequent cyclization to form up to 40-membered rings XLIX [175].

IV. COUPLING VIA RADICALS

A. Kolbe Dimerization

Radicals can be generated in preparative amounts and in an experimentally simple procedure by the use of Kolbe electrolysis. This reaction is discussed in detail in Chap. 14; for reviews see Refs. 176–182. Here, carboxylates are decarboxylated to radicals via $1e^-$ oxidation or to carbenium ions via $2e^-$ oxidation. The radicals couple or react with olefins to give "additive" dimers and/or monomers (see Sec. C). Alcohols, amides, esters, alkenes, and rearranged products are formed from the carbenium ions. The radical route is favored by platinum anodes and sometimes also by those of vitreous or baked carbon [183] and also by a high current density, by an acidic electrolyte, in which only 5–10% of the acid is neutralized, by methanol as solvent, and finally by hydrogen- or electron-attracting substituents on the α-carbon atom of the carboxylic acid [184]. Some representative Kolbe dimerizations are listed in Table 8.

If the retrosynthesis of a target compound leads to two symmetrical fragments, the Kolbe electrolysis is a favorable method for its synthesis. For example, the dimerization of L or LI is the key step in synthesis of pentacyclosqualene [192] and of α-onocerin [193], respectively. Similarly, the precursor of a perhydrophenanthrene has been synthesized [194].

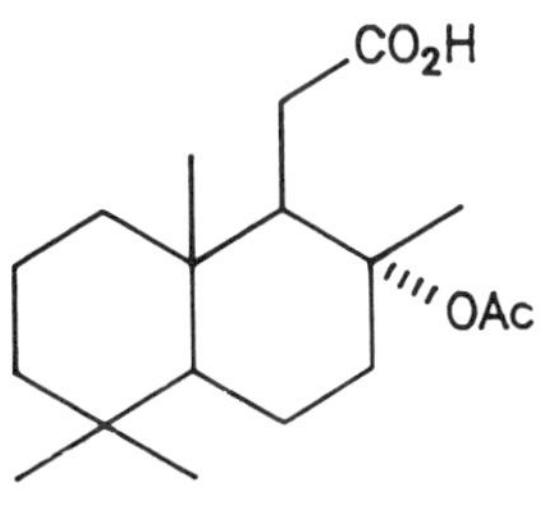

L

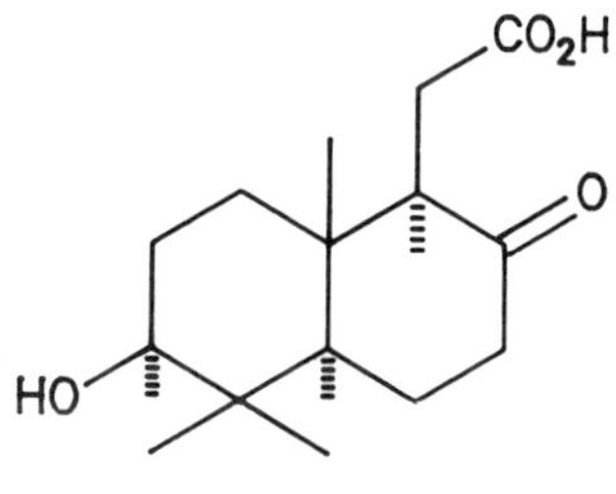

LI

Unsymmetrical compounds can also be prepared via mixed Kolbe electrolyses of different carboxylates. The disadvantageous but unavoidable formation of symmetrical dimers can be suppressed to the formation of essentially one by-product if the cheaper carboxylic acid is used in a 5- to 10-fold excess. Hydrogenated carotenoids, saturated and unsaturated fatty acids, optically active ω-hydroxycarboxylic acids, and intermediates for the preparation of muscone and humulene have been obtained in this way [177b].

Table 8. Some Examples of the Kolbe Electrolysis of Carboxylic Acids

Carboxylic acid	Yield of dimers (%)	Reference
$H_3C(CH_2)_nCO_2H$, n = 5–15	60–90	177b
$RO_2C(CH_2)_nCO_2H$, n = 4–16	45–95	177b
$(CH_3)_2CH{-}CH(CO_2Et)CO_2H$	67	185
$AcO(CH_2)_nCO_2H$, n = 3–5	73–83	186
$F(CH_2)_nCO_2H$, n = 4–10	45–70	187
$EtCO(CH_2)_4CO_2H$	75	188
Oleic acid	75	189
$R^1CO{-}CR^2R^3CO_2H$	35	190
$F_2CCl{-}CO_2H$	45	190
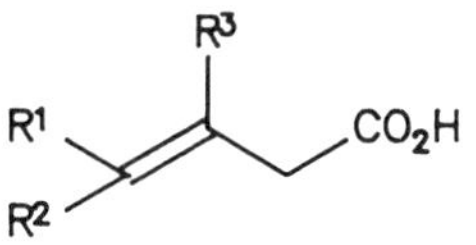 R^1: octyl, ^{i}butyl, ^{i}propyl R^2: H, Me R^3: H	15–79	191

The method has recently been applied to the synthesis of pheromones [1a, 182]. For example, muscalure (LII) has been synthesized in 80% yield [195]. Furthermore, the antagonist of muscalure, (Z)-11-heneicosene [195b], looplure (LIII) [196], brevicomin (LIV) [197], disparlure (LV) [198], the attractant of the German cockroach (LVI) [199], optically active LVII, a Trogoderma pheromone [200], and pheromone dienes and trienes, such as LVIII [201] (the arrows indicate the positions where coupling occurred) have been prepared. In addition, alkyne carboxylic esters with different chain lengths can be obtained [202], which can be hydrogenated at will to (E)- or (Z)-pheromones. Furthermore, useful intermediates for the synthesis of dicarba analogs of cystine peptides have been prepared from L-glutamate [203] and chiral building blocks from enantiomerically pure β-hydroxybutyric acid derivatives [204]. A further selection of mixed Kolbe electrolyses is summarized in Table 9.

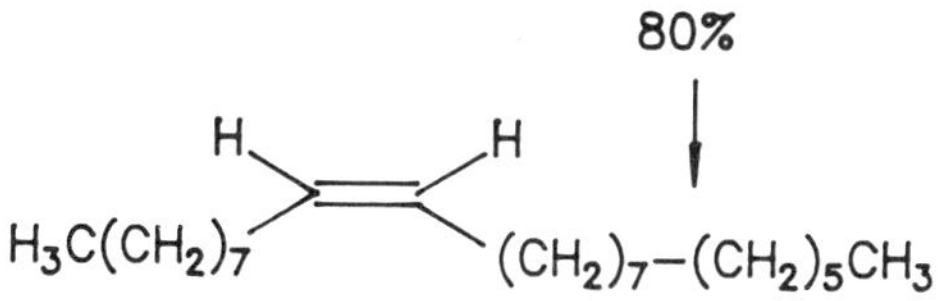

LII

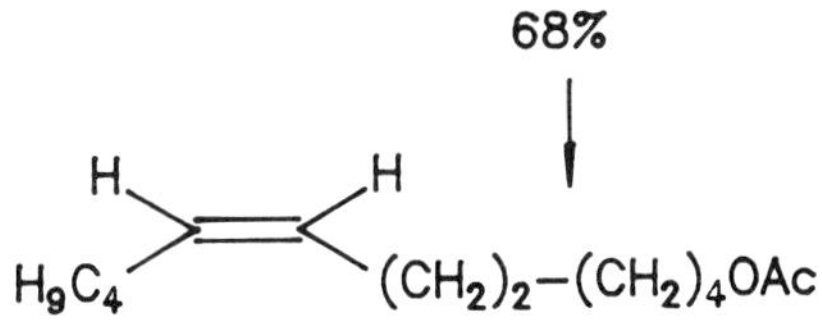

LIII

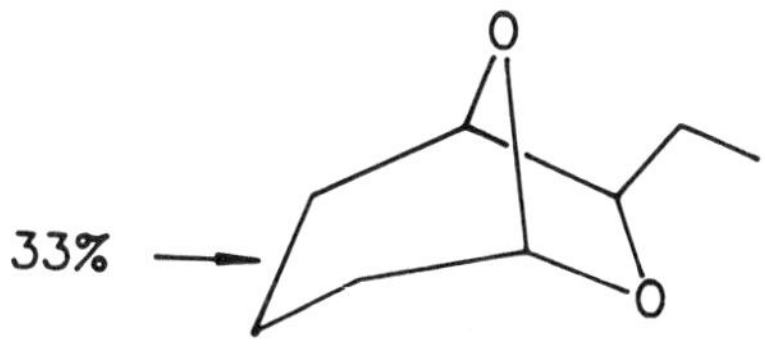

LIV

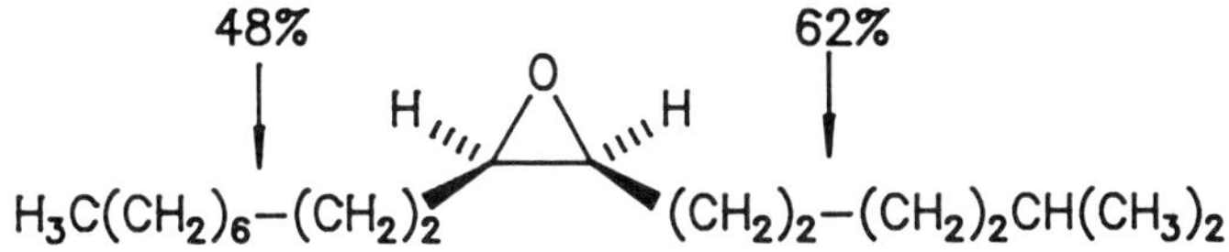

LV

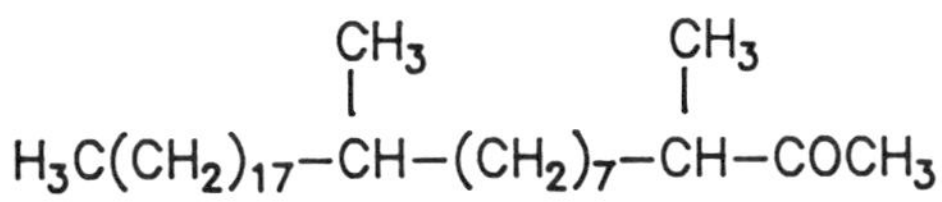

LVI

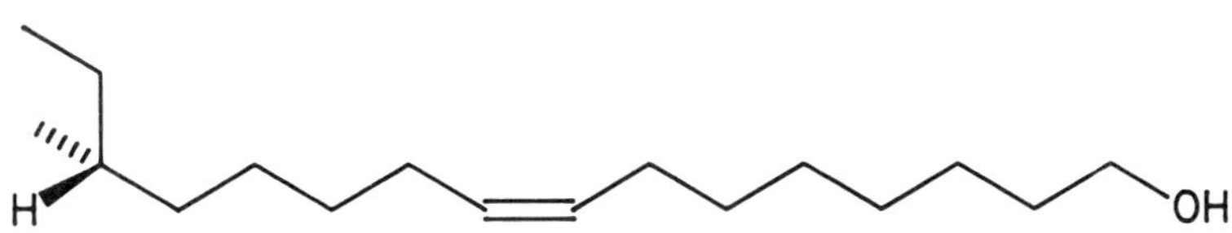

LVII

LVIII

Table 9. Selected Examples of Mixed Kolbe Electrolysis of Carboxylic Acids

Carboxylic acid		Product	
A	B	A–B (%)	Reference
$MeO_2C(CH_2)_4CO_2H$	$H_3C(CH_2)_nCO_2H$	38	205
$MeO_2C(CH_2)_2CO_2H$	$H_3C(CH_2)_7CH(CH_3)CH_2CO_2H$	35	206
$CH_3(CH_2)_2CH{=}CH(CH_2)_7CO_2H$	$HO_2C(CH_2)_6CO_2Me$	35	207
$EtO_2C(CH_2)_8CO_2H$	$Br(CH_2)_{10}CO_2H$	54	208
$(CH_3)_2CHCH{=}CHCH_2CO_2H$	$MeO_2C(CH_2)_nCO_2H$ (n = 1, 2, 4)		209
CF_3CO_2H	$EtO_2CCH_2CO_2H$	46	210
CF_3CO_2H	CD_3CO_2H	68	211

B. Coupling of Anions

In addition to the carboxylates, other anions can be dimerized at the anode, presumably via radicals. Anionized or enolized 1,3-dicarbonyl compounds and heteroanalogs of CH acids couple in satisfactory yields (Table 10, numbers 1–3) [212].

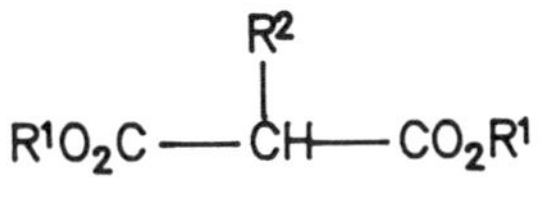

LIX

	R^1	R^2
a	Et	Me
b	Me	H
c	Et	H

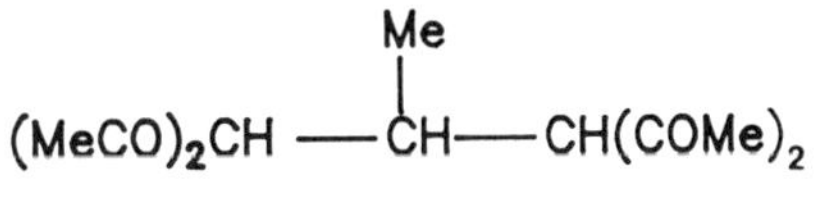

LX

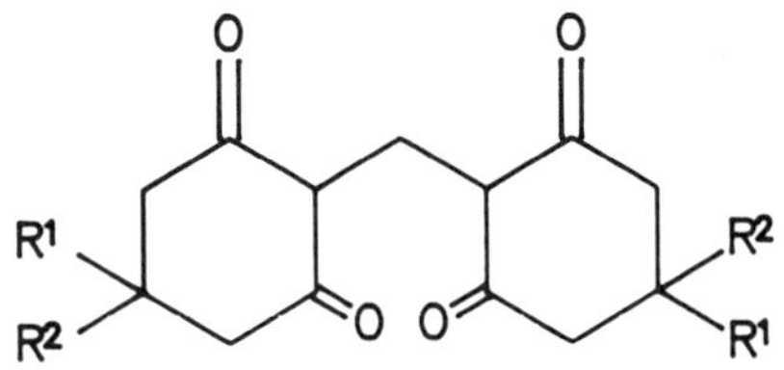

LXI

$R^1 = R^2 = Me$

$R^1 = H$, $R^2 = Ph$

$R^1 = H$, $R^2 = Me$

In the case of 1,3-dicarbonyl compounds, the solvent frequently interferes with the coupling reaction. So with diethyl sodiomalonate in ethanol [212a], methanol [212a], dimethylacetamide [212a], or HMPTA [216], besides the dimer and the trimer, the compounds LIXa–c are obtained. They are presumably formed by oxidation of the solvent to the aldehyde and its condensation with the active methylene compound. No dimer was detected in the oxidation of sodioacetoacetate in ethanol, the major product being LX [217]. Anodic oxidation of cyclic 1,3-diketones in aqueous methanolic sodium hydroxide does not yield the dimer but a product LXI formed by condensation of the starting compound with formaldehyde [218].

With some substrates, the yield is substantially improved with iodide as supporting electrolyte; an indirect electrolysis probably takes place here, with iodine as an electrocatalyst (Chap. 29) [219]. In one of the rare examples in which a useful product is obtained simultaneously at the working electrode and the counterelectrode (coupled or paired electrosynthesis, Chap. 35), diethyl malonate dimerizes at the anode to give tetraethyl ethanetetracarboxylate and ethyl acrylate couples at the cathode to give diethyl adipate in yields of over 90% in each case [220]. Furthermore, anions of nitroaliphatic compounds are coupled to give vicinal dinitroalkanes (Table 10, number 4). Grignard compounds (Table 10, number 5) and borates (Table 10,

Table 10. Selected Examples of Coupling of Anodically Generated Radicals from Carbanions

Number	Carbanion or precursor	Conditions	Product	Yield (%)	Reference
1	$EtCH(CO_2Et)_2$	EtONa, MeCN	$(EtO_2C)_2CEt_2$	55	213
2	Triethyl methanetri-carboxylate	NaOH, H_2O, acetone, EtOH	Hexaethyl ethanehexacarboxylate	50	212c, 214
3	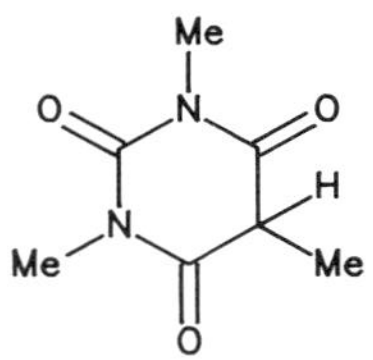	pH 1, H_2O	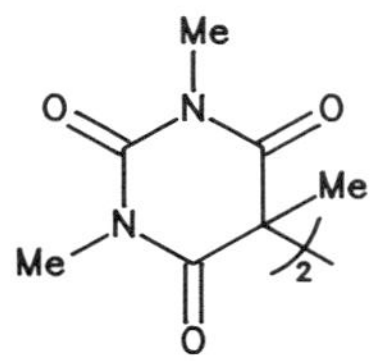	69	215
4	2-Nitrobutane	25% NaOH	3,4-Dimethyl-3,4-dinitrohexane	70	221
5	RMgBr ($R = C_5H_{11}$, $C_{18}H_{37}$, Ph)	Ether	R−R	54−60	222
6	R_3B $R = C_5H_{11}$-C_8H_{17}	MeOH, KOH	R−R	24−82	223
7	Ph−C≡C−H	THF, $LiClO_4$	$(Ph-C{\equiv}C)_2$	35	224b

number 6) couple to give alkanes, alkyne anions to give dialkynes (Table 10, number 7), and amides to give azoalkanes [224].

C. Oxidative Addition of Anions to Double Bonds

If anions R^- are oxidized in the presence of olefins, additive dimers (LXII) and substituted monomers (LXIV) are obtained [Eq. (21), Table 11, and Ref. 1a]. The products can be rationalized by the following pathway: the radical $R^\cdot$ obtained by a $1e^-$ oxidation from the anion R^- adds to the alkene to give the primary adduct LXIII, which dimerizes to give the additive dimer LXII with regiospecific head-to-head linkage of the two olefins, or couples with $R^\cdot$ to give the additive monomer LXIV. If the substituent Y in the olefin can stabilize a carbenium ion, LXIII is oxidized to the cation LXV, which reacts intra- or intermolecularly with nucleophiles to give LXVI or LXVII.

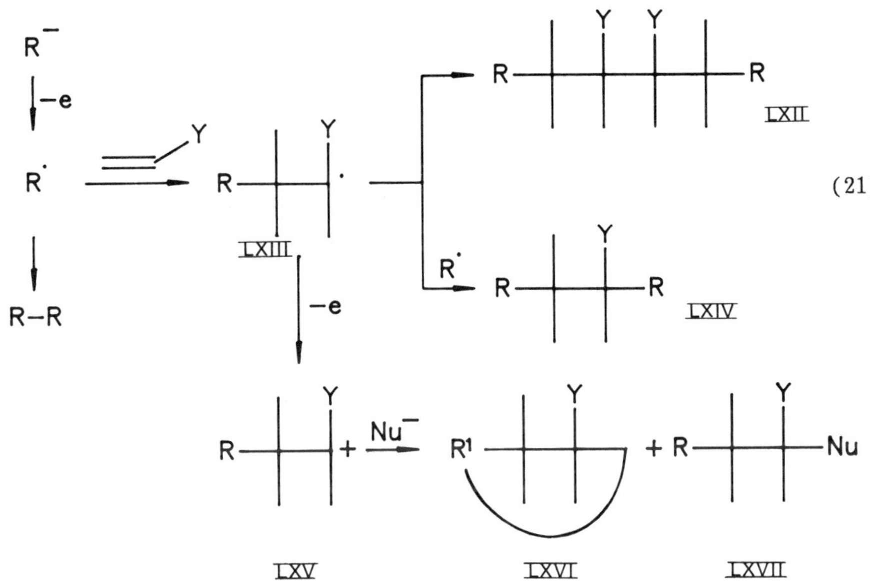

(21)

Satisfactory to good yields of adducts have been found for styrenes [Eq. (21), Y = phenyl], conjugated dienes (Y = vinyl) enamines (Y = NR_2), and enol ethers (Y = alkoxy), particularly if they are unsubstituted at the β-carbon atom. Nonactivated alkenes react less satisfactorily. In the oxidation of anionized 1,3-dicarbonyl compounds (Table 11, numbers 1–8) at potentials between 0.6 and 1.4 V (SCE) and in the presence of butadiene, only the additive dimer LXII is obtained; in the presence of ethyl vinyl ether only the disubstituted monomers LXVI or LXVII arise, but with styrene both types of products LXII and LXVI are formed. This result indicates that the primary adduct LXIII is oxidized to the carbenium ion in the case of a ethoxymethyl radical (Y = OEt) below 0.6 V, in the case of an allyl radical (Y = vinyl) above 1.4 V, and in the case of a benzyl radical (Y = phenyl) within this potential range [225].

Table 11. Addition of Anodically Generated Radicals to Alkenes

Number	Radical precursor	Olefin R−CH=CYR'	Conditions	Product	Yield (%)	Reference
Anions of 1,3-dicarbonyl compounds						
1	$MeO_2CCH(COMe)^-$	$H_2C{=}CHOEt$	MeOH, MeONa, Pt	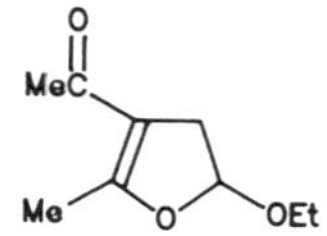	36	225
2	$(MeO_2C)_2CH^-$	$CH_2{=}CHOEt$	MeOH, MeONa, Pt	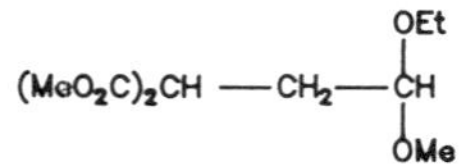	37	225
3	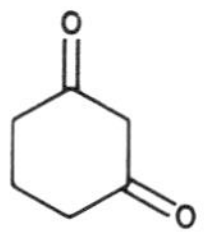	$EtOCH{=}CH_2$	EtOH, EtONa	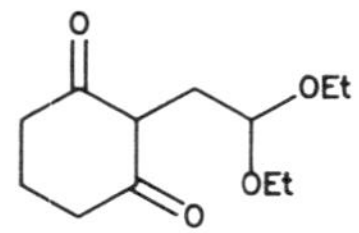 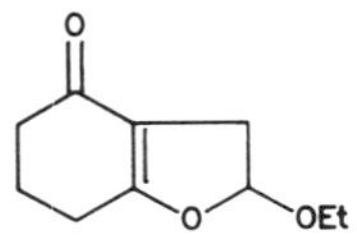	79	226
4	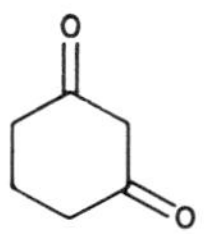	$PhCH{=}CH_2$,	MeCN, Et_4NOTos	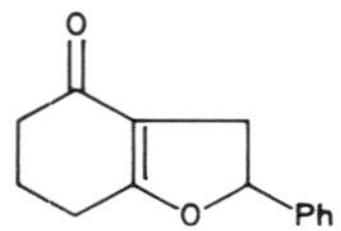	85	227

5	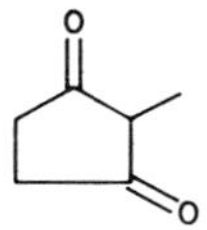	$PhCH{=}CH_2$, O_2	MeCN, Et_4NOTos	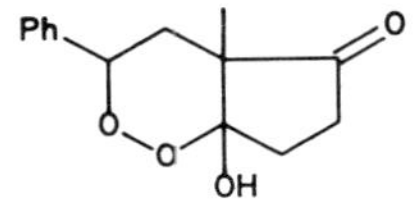	79	227
6	$CH_3COCH_2CO_2CH_3$	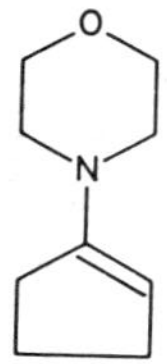	CH_3ONa, MeOH	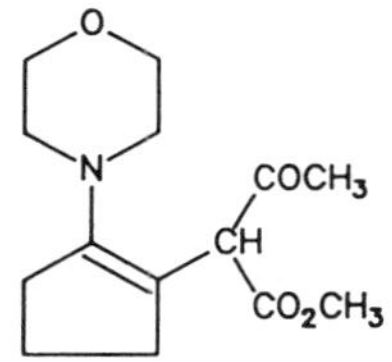	67	228
7	$(MeO_2C)_2CH^-$	Styrene	MeOH, MeONa, Pt	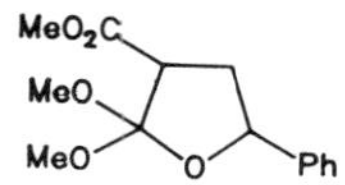	40	225
8	$(MeCO)_2CH^-$	Butadiene	MeOH, MeONa, Pt	$[(MeCO)_2CH-CH_2-CH{=}CH-CH_2-]_2$ + isomers	46	225
Other anions						
9	$(CH_3)_2C(NO_2)^-$	Styrene	MeOH, MeONa, Pt	$(CH_3)_2C(NO_2)-CH_2-CH(Ph)-OMe$	43	229
10	nBuMgBr	Styrene	$(C_2H_5)_2O$, $LiClO_4$, Pt	$(nBu-CH_2-CHPh-)_2$	29	230

Table 11. (Continued).

Number	Radical precursor	Olefin R−CH=CYR'	Conditions	Product	Yield (%)	Reference
11	$(C_6H_{13})_3B(OMe)^-$	Butadiene	MeOH, MeONa, C	$(C_6H_{13}-CH_2-CH=CH-CH_2)_2$ $C_6H_{13}-CH_2-CH=CH-CH_2-C_6H_{13}$ $C_6H_{13}-CH_2-CH=CH-CH_2OMe$ + isomers	5 14	231
Carboxylates						
12	$EtO_2CCO_2^-$	Butadiene	MeOH, Pt	$(EtO_2C-CH_2-CH=CH-CH_2)_2$ + isomers $EtO_2C-CH_2-CH=CH-CH_2-CO_2Et$	66 4	232
13	$MeO_2C(CH_2)_4-CO_2-$	Ethylene	MeOH, Pt	$[MeO_2C(CH_2)_6-]_2$ $MeO_2C(CH_2)_{10}CO_2Me$	92–95	233
14	$MeO_2CCH_2CO_2^-$	Vinyl ethyl, ether	MeOH, Pt	$[MeO_2C-(CH_2)_2-CH(OEt)-]_2$	35	234
15	$MeO_2CCH_2CO_2^-$	Styrene	MeOH, Pt	$(MeO_2C-(CH_2)_2-CHPh-)_2$	38	232
16	$MeO_2C(CH_2)_4-CO_2-$	Butadiene	MeOH, Pt	$[MeO_2C-(CH_2)_4-CH_2-CH==CH-CH_2-]_2$ $MeO_2C-(CH_2)_4-CH_2-CH=CH--CH_2-(CH_2)_4-CO_2Me$	47 47	235
17	$F_3CCO_2^-$	$CH_2=CH(CH_2)_2CH_3$	MeCN, NaOH, Pt	$CF_3-CH_2-CH(CH_2)_2CH_3$ (dimer, $\vert_2$)	40	236

18	$CH_3CO_2^-$	Methacroleine	MeCN, H_2O, Pt	$CH_3-CH_2C(CH_3)CHO$ $\vert_2$	80	237
19	$CF_3CO_2^-$	Vinyl acetate	MeCN, H_2O, NaOH	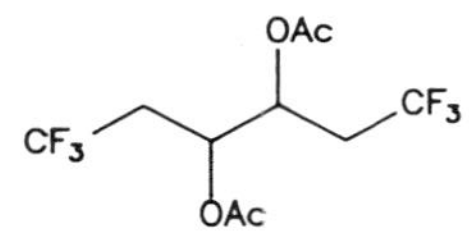	24	238
20	$CH_3CO_2^-$	Dimethyl fumarate	MeCN, H_2O	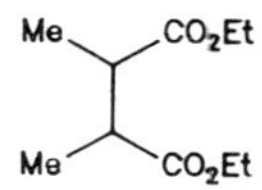	80	239
21	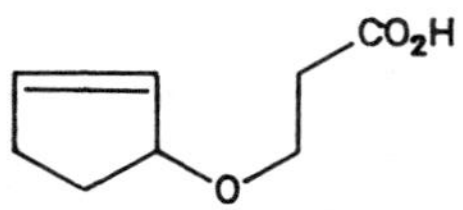	$MeO_2C(CH_2)_2CO_2H$	5% MeONa, MeOH	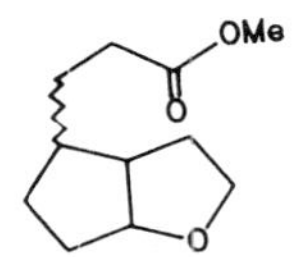	41	242
22	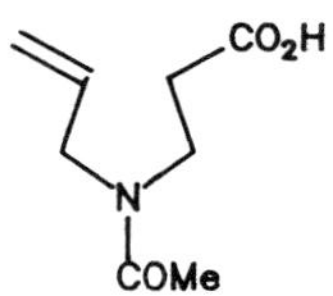	$MeO_2C(CH_2)_4CO_2H$	5% MeONa, MeOH	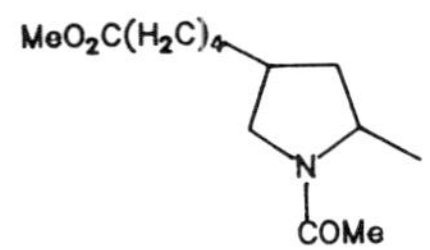	53	243

In an electroinitiated radical addition, an olefin and oxygen can be added simultaneously to the 1,3-dicarbonyl compound [Eq. (22) and Table 11, number 5] [227].

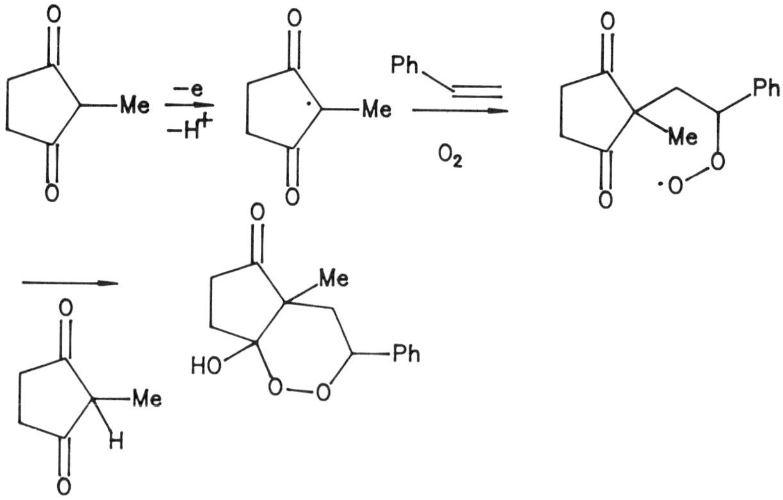

(22)

Other anions, such as the 2-nitropropanate anion, Grignard reagents, and borates, can be added to olefins (Table 11, numbers 9–11).

If carboxylates are subjected to Kolbe electrolysis in the presence of olefins, the generated Kolbe radicals add to the double bonds to afford mainly additive dimers (Table 11, numbers 12–20). In vicinal disubstituted styrenes, upon addition of the Kolbe radical $MeO_2CCH_2^{\cdot}$, the yields of adducts decrease with increasing size of the β-substituent: H = 42%, Me = 27%, Et = 11%, ^{i}Pr = 5%, ^{t}Bu = 2% [240]. The additive dimer (LXII) to monomer (LXIV) ratio can be affected to some extent by the current density i. With butadiene and ethyl oxalate (Table 10, number 12), LXII to LXIV = 15.6 for i = 0.025 mA cm^{-2}, but 0.67 for i = 0.66 mA cm^{-2}. Obviously, at low current densities the radicals $R^{\cdot}$ are trapped mainly by the olefin to give LXIII, a process that no longer occurs at higher current densities, that means a higher radical concentration, so that now the mixed coupling to give LXIV is favored. Electrolysis of a methanol solution of methyl oxalate with ethylene under pressure yielded 70–90% of the dimethyl esters of succinic, adipic, suberic, and sebacic acids. Decrease in the ethylene pressure or increase of the current density lead to a decrease in the higher esters in the product mixture [241]. The influence of mechanism and kinetic data on yields and selectivities in addition reactions of anodically generated radicals to olefins has been calculated and predictions tested in preparative electrolyses [244].

β-Allyloxy- and β-allylaminopropionates cyclize in a mixed Kolbe electrolysis (5-exo-trig cyclization) by intramolecular addition to form 3-substituted tetrahydrofurans and pyrrolidines (Table 11, numbers 21 and 22). This intermolecular addition has been used to synthesize a precursor of prostaglandin $PGF_{2\alpha}$[Eq. (23)] [245].

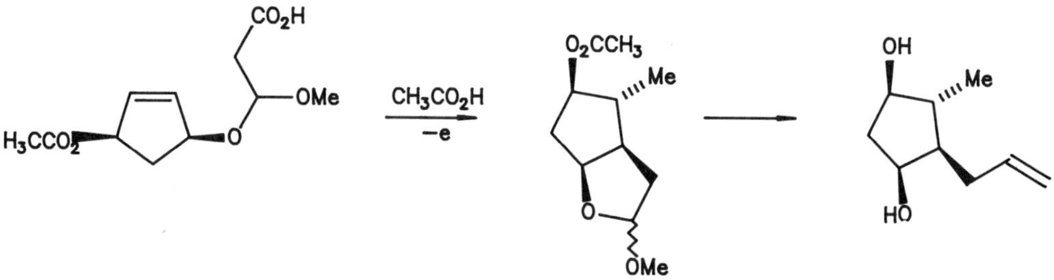

(23)

V. ANODICALLY INDUCED CYCLOADDITONS

The [4 + 2] and [2 + 2] cycloadditions can be induced by electron transfer. In a radical chain reaction the olefin is oxidized to its radical cation, which first undergoes an addition. The dimeric radical cation then accepts an electron from the olefin to undergo ring closure and to initiate a new cycle [Eq. (24)] [246]. This way vinyl carbazole (Table 12, number 1) and 5,6-dimethoxyindene [247] have been cyclized. Indenone also reacts in an electrocatalyzed [4 + 2] and [2 + 2] cycloaddition with styrene derivatives [248]. Furthermore, a series of dienes has been subjected to [4 + 2] and [2 + 2] cycloaddition with $Ar_3N^{+\cdot}$ as mediator (Table 11, numbers 2 and 3).

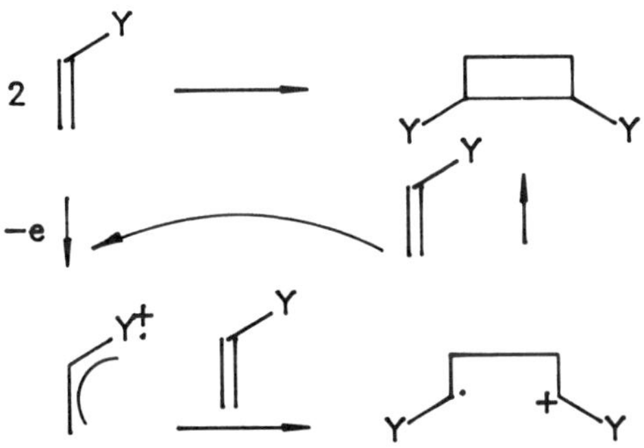

(24)

In other cases the cycloadditions are not electrocatalytic. The reactive intermediates are generated electrochemically and then react with the olefin to an adduct that stabilizes itself by reaction with a nucleophile or by deprotonation, not by electron transfer from the starting compound. This way <u>ortho</u>-phenols react with enols presumably via an <u>ortho</u>-quinone as intermediate (Table 11, number 4). A cyclohexadienyl cation is obtained from the oxidation of dimethoxyphenol (LXVIII), which undergoes a cycloaddition with a styrene derivative to afford the bicyclic compound LXIX in 80% yield [Eq. (25)] [251].

Table 12. Cycloadditions of Anodically Generated Intermediates

Number	Substrate	Olefin	Conditions	Product	Yield (%)	Reference
1	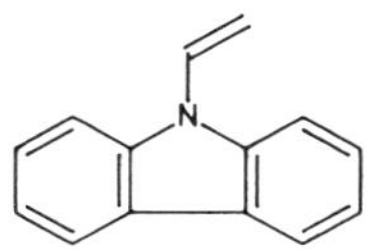		$Fe(NO_3)_3$, MeOH	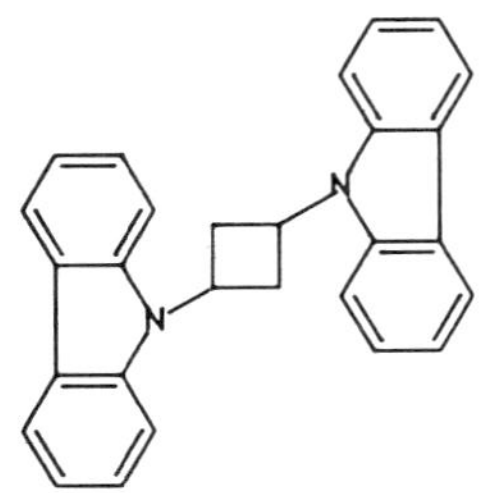	80	246
2	1,3-Cyclohexadiene		MeCN, CH_2Cl_2, lutidine, C	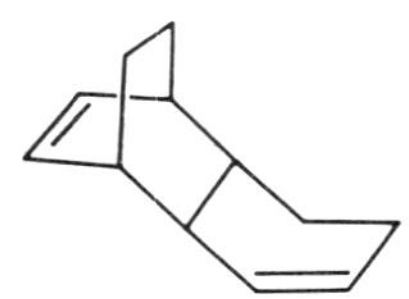	24	249

3	1,3-Cyclohexadiene	Styrene	MeCN, CH_2Cl_2, lutidine, C	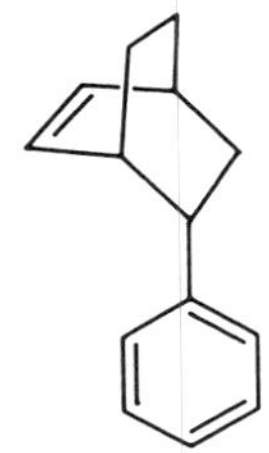	33	249
4	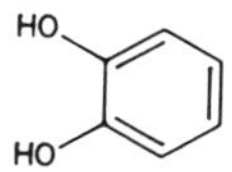	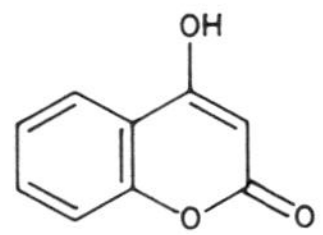	H_2O, NaOAc	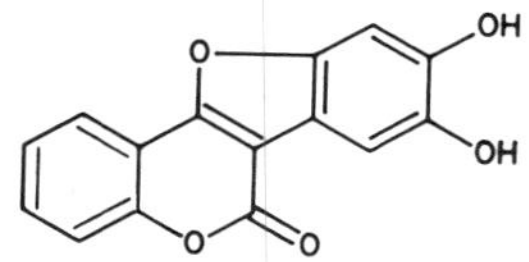	95	250
5	1,1,2-Triphenyl-hydrazine	Vinyl ethyl ether	MeCN, $LiClO_4$, Py, Pt	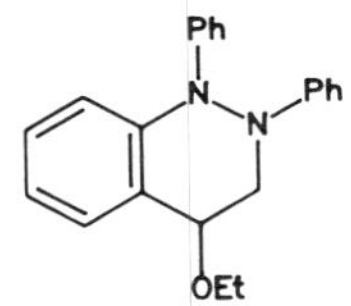	95	253

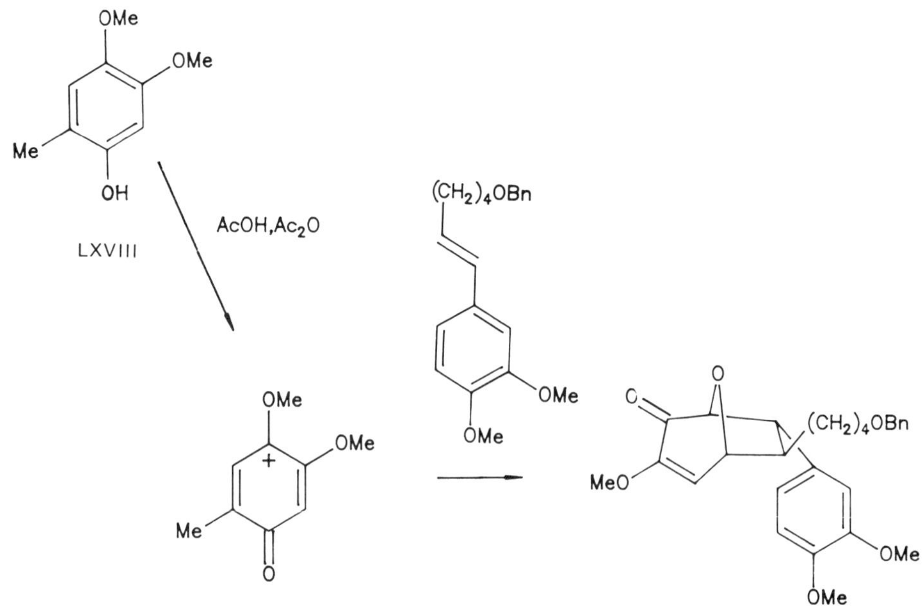

LXIX

(25)

The electrolysis of N-methylamides (LXX) results in the formation of iminium intermediates that undergo a [4 + 2] cycloaddition with olefins [Eq. (26)] [252] to afford heterocycles (LXXI) in 45–58% yield.

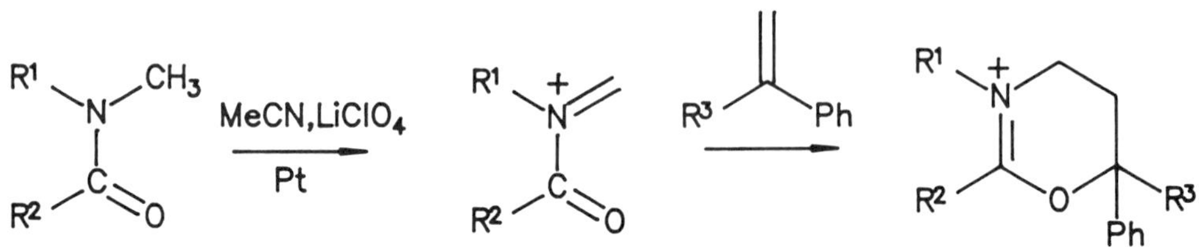

LXX LXXI

R^1	R^2	R^3
Me	H	Ph
Me	Me	Ph
H	Me	H

(26)

1,1-Diphenylhydrazine can be dehydrogenated to the diazenium ion (LXXII) [Eq. (27)], which undergoes cycloadditions with olefins (Chap. 18) [253].

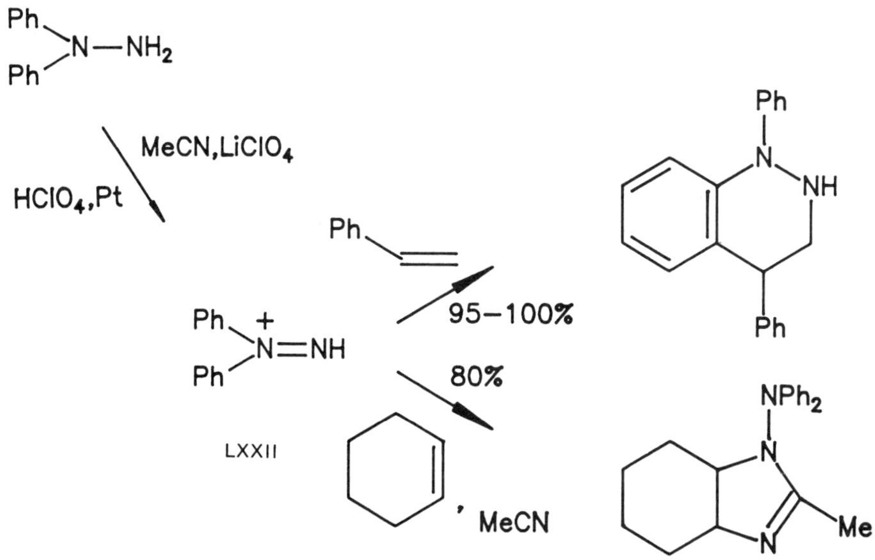

(27)

1,1,2-Triphenylhydrazine reacts in a similar way with enol ethers to cinnoline derivatives (Table 12, number 5). The electrooxidation of N-alkylhydrazines leads to an iminium ion, which can react with olefins to five-membered rings [Eq. (28)] [254]. The anodic oxidation of phenylhydrazones of benzaldehyde affords diphenylnitrilimines, which add to dipolarophilic compounds [255].

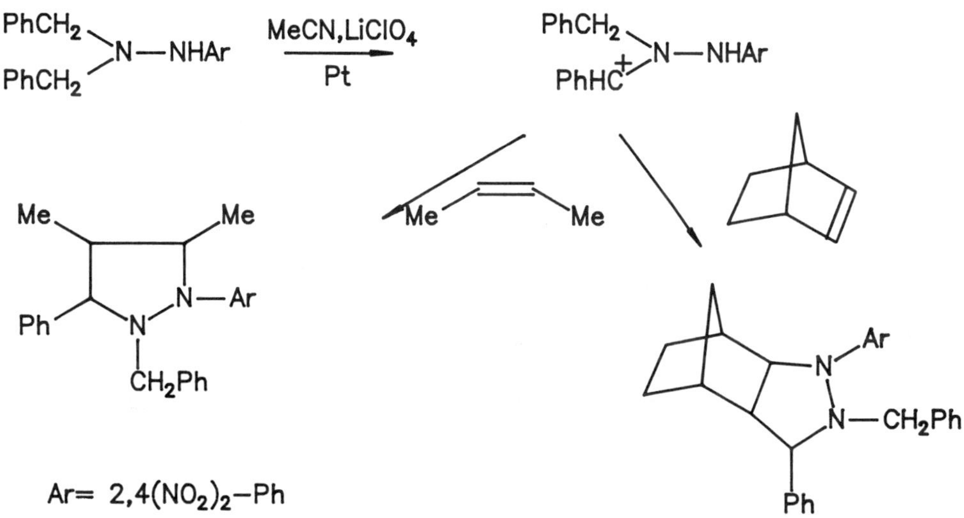

(28)

REFERENCES

1. (a) H. J. Schäfer, Angew. Chem., 93, 978 (1981); Angew. Chem. Int. Ed. Engl., 20, 911 (1981); (b) K. Yoshida, Electrooxidation in Organic Chemistry, J. Wiley, New York, 1984; (c) T. Shono, Electroorganic Chemistry as a New Tool in Organic Synthesis, Springer, Berlin, 1984; (d) M. M. Baizer, Tetrahedron, 40, 935 (1984); (e) S. Torii, Electroorganic Syntheses, Part I, Oxidations; Monographs in Modern Chemistry, Vol. 15, Kodansha, Tokyo, VCH, Weinheim, 1985; (f) H. J. Schäfer, Kontakte (Merck)17(2), (1987).
2. L. Eberson, K. Nyberg, and H. Sternerup, Acta Chem. Scand., 27, 1679 (1973).
3. K. Nyberg, Acta Chem. Scand., 25, 2499 (1971).
4. K. Nyberg, Acta Chem. Scand., 25, 534 (1971).
5. K. Nyberg, Acta Chem. Scand., 25, 2983 (1971).
6. J. A. Donohue, Electrochemical Society, Spring Meeting, May 6–11, 1984, Ext. Abstr. Vol. 84-1, p. 442.
7. F. Fichter and S. Herszbein, Helv. Chim. Acta, 11, 1264 (1928).
8. V. D. Parker, Acta Chem. Scand., 24, 3163 (1970).
9. V. D. Parker and L. Eberson, Tetrahedron Lett., 2839 (1969).
10. V. D. Parker, Acta Chem. Scand., 24, 3171 (1970).
11. V. D. Parker, J. Chem. Soc., Chem. Commun., 848 (1969).
12. A. E. Coleman, H. H. Richtol, and D. A. Aikens, J. Electroanal. Chem., 18, 165 (1968).
13. K. Nyberg, Acta Chem. Scand., 27, 503 (1973).
14. K. Nyberg, Acta Chem. Scand., 25, 3770 (1971).
15. K. Nyberg and A. Trojánek, Coll. Czech. Chem. Commun., 40, 526 (1975).
16. U. Svanholm and V. D. Parker, J. Amer. Chem. Soc., 98, 2942 (1976).
17. A. Bewick, G. J. Edwards, J. M. Mellor, and S. Pons, J. Chem. Soc., Perkin Trans. II, 1952 (1977).
18. R. N. Adams, Acc. Chem. Res., 2, 175 (1969).
19. O. Hammerich and V. D. Parker, Adv. Phys. Org. Chem., 20, 55 (1984); V. D. Parker, Acc. Chem. Res., 17, 243 (1984).
20. O. Hammerich and V. D. Parker, J. Amer. Chem. Soc., 96, 4289 (1974).
21. T. Osa, A. Yildiz, and T. Kuwana, J. Amer. Chem. Soc., 91, 3994 (1969).
22. M. Delamar, P. C. Lacaze, J. Y. Dumousseau, and J.-E. Dubois, Electrochim. Acta, 27, 61 (1982).
23. U. Svanholm and V. D. Parker, J. Amer. Chem. Soc., 98, 2942 (1976).
24. J. M. Bobbitt and J. P. Willis, J. Org. Chem., 42, 2347 (1977).
25. S. Andreades and E. W. Zahnow, J. Amer. Chem. Soc., 91, 4181 (1969).
26. L. Eberson and B. Helgee, Acta Chem. Scand., B31, 813 (1977).
27. M. T. Ismail, A. H. El-Khawaga, A. A. Abdul-Wahab, A. M. Aboel, and A. A. Khalaf, Ann. Chim. (Rome), 72, 269 (1982); Chem. Abstr., 97, 100546 (1982); M. T. Ismail, M. F. El-Zohry, and A. A. Abdel Wahab, J. Appl. Electrochem., 15, 469 (1985); Chem. Abstr., 103, 44765 (1985).
28. (a) A. I. Scott, Quart. Rev., 19, 1 (1965); (b) H. Musso, in Oxidative Coupling of Phenols (W. I. Taylor and A. R. Battersby, eds.), Dekker, New York, 1967, Chap. 1; (c) T. Kametani and K. Fukomoto, Synthesis, 657 (1972).

29. (a) H. Lund, in The Chemistry of the Hydroxyl Group (S. Patai, ed.), Wiley, New York, 1971, Vol. 1, Chap. 5; (b) A. Ronlan, in Encyclopedia of Electrochemistry of the Elements (A. J. Bard and H. Lund, eds.), Dekker, New York, 1978, Vol. 11, p. 242.
30. F. Fichter and F. Ackerman, Helv. Chim. Acta, 2, 583 (1919).
31. K. Sasaki, S. Nanao, and A. Kunai, Denki Kagaku, 45, 130 (1977); Chem. Abstr., 87, 134212 (1977).
32. H. J. Schäfer, Top. Curr. Chem., 142, 100 (1987).
33. H. Ohmori, A. Matsumoto, M. Masui, and H. Sayo, J. Electrochem. Soc., 124, 1849 (1977).
34. J. M. Bobitt, H. Yagi, S. Shibuya, and J. T. Stock, J. Org. Chem., 36, 3006 (1971).
35. J. M. Bobbitt, I. Noguchi, H. Yagi, and K. H. Weisgraber, J. Amer. Chem. Soc., 93, 3551 (1971); J. Org. Chem., 41, 845 (1976).
36. K. M. Johnston, Tetrahedron Lett., 837 (1967).
37. L. L. Miller, R. F. Stewart, J. P. Gillespie, V. Ramachandran, Y. H. So, and F. R. Stermitz, J. Org. Chem., 43, 1580 (1978).
38. J. M. Bobbitt and R. C. Hallcher, J. Chem. Soc., Chem. Commun., 543 (1971).
39. A. Rieker, H. P. Schneider, and E. Streich, in Recent Advances in Electroorganic Synthesis (S. Torii, ed.), Kadansha, Elsevier, Amsterdam, 1987, p. 57.
40. M. Iguchi, A. Nishiyama, Y. Terada, and S. Yamamura, Chem. Lett., 451 (1978).
41. F. J. Vermillion, Jr. and I. A. Pearl, J. Electrochem. Soc., 111, 1392 (1964).
42. A. Ronlán, J. Chem. Soc., Chem. Commun., 1643 (1971).
43. A. Nilsson, U. Palmquist, T. Petterson, and A. Ronlán, J. Chem. Soc., Perkin Trans., I, 696 (1978).
44. J. M. Bobbitt, K. H. Weisgraber, A. S. Steinfeld, and S. G. Weiss, J. Org. Chem., 35, 2884 (1970).
45. A. Nishiyama, H. Eto, Y. Terada, M. Iguchi, and S. Yamamura, Chem. Pharm. Bull., 31, 2820 (1983); M. Iguchi, A. Nishiyama, H. Eto, Y. Terada, and S. Yamamura, Chem. Lett., 1397 (1979).
46. J. A. Richards, P. E. Whitson, and D. H. Evans, J. Electroanal. Chem., 63, 311 (1975); D. H. Evans, P. J. Jiminez, and M. J. Kelly, J. Electroanal. Chem., 163, 145 (1984).
47. A. Rieker, DFG-(Deutsche Forschungsgemeinschaft)-Bericht "Angewandte Elektrochemie," 1979, p. 154.
48. E. L. Dreher, J. Bracht, M. El-Mobayed, P. Hütter, W. Winter, and A. Rieker, Chem. Ber., 115, 288 (1982).
49. V. D. Parker, in Encyclopedia of Electrochemistry of the Elements (A. J. Bard and H. Lund, eds.), Dekker, New York, 1978, Vol. 11, p. 276.
50. A. Ronlán, K. Bechgaard, and V. D. Parker, Acta Chem. Scand., 27, 2375 (1973).
51. O. Hammerich and V. D. Parker, Electrochim. Acta, 18, 537 (1973).
52. H. G. H. Erdtman, Proc. Roy. Soc. (London), 143(A), 191 (1934).
53. V. D. Parker and R. N. Adams, Tetrahedron Lett., 1721 (1969).
54. M. Sainsbury and J. Wyatt, J. Chem. Soc., Perkin Trans., I, 1750 (1977).
55. O. Hammerich and V. D. Parker, Acta Chem. Scand., B36, 519 (1982).
56. A. Ronlán, B. Aalstad, and V. D. Parker, Acta Chem. Scand., B36, 317 (1982).

57. U. Svanholm and V. D. Parker, Acta Chem. Scand., B34, 5 (1980).
58. R. F. Stewart and L. L. Miller, J. Amer. Chem. Soc., 102, 4999 (1980).
59. K. Beechgard and V. D. Parker, J. Amer. Chem. Soc., 94, 4749 (1972).
60. L. Le Berre, L. Angély, N. Simonet-Gueguen, and J. Simonet, J. Chem. Soc., Chem. Commun., 984 (1987).
61. V. D. Parker and L. Eberson, Tetrahedron Lett., 1449 (1972).
62. V. Le Berre, L. Angély, N. Simonet-Gueguen, and J. Simonet, Nov. J. Chim., 9, 419 (1985).
63. V. Le Berre, L. Angély, N. Simonet-Gueguen, and J. Simonet, J. Electroanal. Chem., 206, 115 (1986).
64. V. Le Berre, L. Angély, N. Simonet-Gueguen, and J. Simonet, J. Electroanal. Chem., 240, 117 (1988).
65. A. Ronlán, J. Coleman, O. Hammerich, and V. D. Parker, J. Amer. Chem. Soc., 96, 845 (1974).
66. R. P. Buck and D. E. Wagoner, J. Electroanal. Chem., 115, 89 (1980).
67. S. Andreades and E. W. Zahnow, J. Amer. Chem. Soc., 91, 4181 (1969).
68. E. T. Seo, R. F. Nelson, J. M. Fritsch, L. S. Marcoux, D. W. Leedy, and R. N. Adams, J. Amer. Chem. Soc., 88, 3498 (1966); R. F. Nelson and R. N. Adams, J. Amer. Chem. Soc., 90, 3925 (1968); L. S. Marcoux, R. N. Adams, and S. W. Feldberg, J. Phys. Chem., 73, 2611 (1969).
69. R. Reynolds, L. L. Line, and R. F. Nelson, J. Amer. Chem. Soc., 96, 1087 (1974).
70. R. L. Hand and R. F. Nelson, J. Amer. Chem. Soc., 96, 850 (1974).
71. G. Neubert and K. B. Prater, J. Electrochem. Soc., 121, 745 (1974).
72. R. Hand and R. F. Nelson, J. Electrochem. Soc., 117, 1353 (1970).
73. G. Cauquis, J. Cognard, and D. Serve, Tetrahedron Lett., 4645 (1971).
74. J. M. Hornback and H. E. Gossage, J. Org. Chem., 50, 541 (1985).
75. S. Wawzonek and T. W. McIntyre, J. Electrochem. Soc., 114, 1025 (1967); 119, 1350 (1972).
76. E. Ahlberg, B. Helgee, and V. D. Parker, Acta Chem. Scand., B34, 187 (1980).
77. (a) Y. Matsuda, A. Shono, C. Iwakura, Y. Ohshiro, T. Agawa, and H. Tamura, Bull. Chem. Soc. Jpn., 44, 2960 (1971); (b) P. G. Desideri, L. Lepri, and D. Heimler, J. Electroanal. Chem., 43, 387 (1973); 52, 93, 105 (1974).
78. P. G. Desideri, D. Heimler, and L. Lepri, J. Electroanal. Chem., 88, 407 (1978).
79. K. Yasukouchi, I. Taniguchi, H. Yamaguchi, K. Miyaguchi, and K. Horie, Bull. Chem. Soc. Jpn., 52, 3208 (1979).
80. M. Melicharek and R. F. Nelson, J. Electroanal. Chem., 26, 201 (1970).
81. R. L. Hand and R. F. Nelson, J. Electrochem. Soc., 125, 1059 (1978).
82. G. Cauquis, H. Delhomme, and D. Serve, Tetrahedron Lett., 4113 (1971).
83. D. Serve, Electrochim. Acta, 21, 1171 (1976).
84. P. Berkenkotter and R. F. Nelson, J. Electrochem. Soc., 120, 346 (1973).
85. K. Yasukouchi, I. Taniguchi, H. Yamaguchi, M. Yokoyama, and M. Murasaki, Chem. Lett., 1979, 1167.

86. H. Sayo, K. Mori, and T. Michida, Chem. Pharm. Bull., 27, 2093 (1979).
87. J. F. Ambrose and R. F. Nelson, J. Electrochem. Soc., 115, 1159 (1968); W. Lamm, F. Pragst, and W. Jugelt, J. Prakt. Chem., 317, 995 (1975).
88. J. M. Bobbitt, C. L. Kulkarni, and J. P. Willis, Heterocycles, 15, 495 (1981).
89. J. F. Ambrose, L. L. Carpenter, and R. F. Nelson, J. Electrochem. Soc., 122, 876 (1975).
90. S. N. Frank, A. J. Bard, and A. Ledwith, J. Electrochem. Soc., 122, 898 (1975).
91. J. M. Bobbitt and J. P. Willis, J. Org. Chem., 45, 1978 (1980); Mechanistic investigation, C. L. Kulkarni, B. J. Scheer, and J. F. Rusling, J. Electroanal. Chem., 140, 57 (1982).
92. J. M. Quante, F. R. Stermitz, and L. L. Miller, J. Org. Chem., 44, 293 (1979).
93. H. H. Adam, B. D. Baigrie, and T. A. Joslin, J. Chem. Soc., Perkin Trans., II, 1287 (1977).
94. (a) B. D. Baigrie, T. A. Joslin, and D. W. Sopher, J. Chem. Soc., Perkin Trans., II, 77 (1979); (b) P. Corbon, G. Barbey, A. Dupré, and C. Caullet, Bull. Soc. Chim. Fr., 1974, 768; for similar couplings, F. Pragst and C. Böck, J. Electroanal. Chem., 61, 47 (1975); I. Tabakovic, M. Lacan, and S. Damoni, Electrochim. Acta, 21, 621 (1976).
95. F. Pragst, L. Mögel, and M. Schulz, J. Prakt. Chem., 325, 148 (1983).
96. J. Armand and J. Pinson, in Physical Methods in Heterocyclic Chemistry (R. R. Gupta, ed.), Wiley, New York, 1984, pp. 427–625.
97. G. Cauquis, H. M. Fahmy, G. Pierre, and M. H. Elnagdi, J. Heterocyclic Chim., 16, 413 (1979).
98. S. Longchamp, C. Caullet, and M. Libert, Bull. Soc. Chim. Fr., 1974, 353.
99. T. Shono, Electroorganic Chemistry as a New Tool in Organic Synthesis, Springer, Berlin, 1984, p. 63.
100. S. Torii, Electroorganic Synthesis, Part I, Oxidations, Monographs in Modern Chemistry, Vol. 15, Kodansha, Tokyo, VCH, Weinheim, 1985, p. 171.
101. R. Bauer and H. Wendt, Angew. Chem., 90, 214 (1978); Angew. Chem. Int. Ed. Engl., 17, 202 (1978).
102. B. L. Laube, M. R. Asirvatham, and C. K. Mann, J. Org. Chem., 42, 670 (1977).
103. T. Fuchigami, T. Iwaoaka, T. Nonaka, and T. Sekine, Bull. Chem. Soc. Jpn., 53, 2040 (1980).
104. G. Cauquis, B. Chabaud, and M. Genies, J. Electroanal. Chem., 40, 6 (1972).
105. G. Cauquis and M. Genies, Tetrahedron Lett., 2903 (1970).
106. H. J. Schäfer, Top. Curr. Chem., 142, 102, 122 (1987).
107. T. Fuchigami, T. Sato, and T. Nonaka, Electrochim. Acta, 31, 365 (1986).
108. R. N. Renaud, D. Bérubé, and C. J. Stephens, Can. J. Chem., 61, 1379 (1983); R. N. Renaud, C. J. Stephens, and G. Brochu, Can. J. Chem., 62, 565 (1984).
109. W. Jugelt and S. Schwertner, Z. Chem., 23, 409 (1983).

110. T. W. Bentley, T. W. Richards, and M. G. Hutchings, Tetrahedron Lett., 1986, 5261.
111. F. D. Saeva, B. P. Morgan, M. W. Fichtner, and N. F. Haley, J. Org. Chem., 49, 390 (1984).
112. R. Engels, H. J. Schäfer, and E. Steckhan, Liebigs Ann. Chem., 204 (1977).
113. M. Kojima, H. Sakuragi, and K. Tokumaru, Chem. Lett., 1707 (1981).
114. V. Plzak, H. Schneider, and H. Wendt, Ber. Bunsenges. Ges., 78, 1373 (1974).
115. E. Steckhan, J. Amer. Chem. Soc., 100, 3526 (1978).
116. G. Burgbacher and H. J. Schäfer, J. Amer. Chem. Soc., 101, 7590 (1979).
117. B. Aalstad, A. Ronlan, and V. D. Parker, Acta Chem. Scand., B35, 247 (1981).
118. V. D. Parker and L. Eberson, Chem. Commun., 1969, 340; L. Eberson and V. D. Parker, Acta Chem. Scand., 24, 3553 (1970).
119. (a) E. Steckhan and H. Schäfer, Angew. Chem., 86, 480 (1974); Angew. Chem. Int. Ed. Engl., 13, 472 (1974); (b) H. Sternerup, Acta Chem. Scand., B28, 579 (1974).
120. H. Schäfer and E. Steckhan, Angew. Chem., 81, 532 (1969); Angew. Chem. Int. Ed. Engl., 8, 518 (1969); B. Belleau and Y. K. Au-Young, Can. J. Chem., 47, 2117 (1969).
121. D. Koch, H. Schäfer, and E. Steckhan, Chem. Ber., 107, 3640 (1974).
122. M. A. Le Moing, G. Le Guillanton, and J. Simonet, Electrochim. Acta, 26, 139 (1981).
123. R. Engels, Dissertation, Universität Münster, 1978.
124. H. O. House, L. J. Czuba, M. Gall, and H. D. Olmstead, J. Org. Chem., 34, 2324 (1969); R. D. Miller and D. R. McKean, Synthesis, 730 (1979); G. Olah, B. G. B. Gupta, S. C. Narang, and R. Malhotra, J. Org. Chem., 44, 4272 (1979).
125. J. M. Fritsch and H. Weingarten, J. Amer. Chem. Soc., 90, 793 (1968); J. M. Fritsch, H. Weingarten, and J. D. Wilson, J. Amer. Chem. Soc., 92, 4038 (1970).
126. D. Koch and H. J. Schäfer, Angew. Chem., 85, 264 (1973); Angew. Chem. Int. Ed. Engl., 12, 254 (1973); H. W. Handiak, Dissertation, Universität Göttingen, 1976; N. Dahrenmöller, Diplomarbeit, Universität Münster, 1976.
127. M. Masui, C. Ueda, T. Moriguchi, T. Michida, M. Katoaka, and H. Ohmori, Chem. Pharm. Bull., 32, 1382 (1984).
128. R. Mayer and H. Kröber, J. Prakt. Chem., 316, 907 (1974); C. Madec, F. Quentel, and J. Courtot-Coupez, Anal. Lett. A, 13, 33 (1980); Chem. Abstr., 92, 188216 (1980).
129. H. Baltes, E. Steckhan, and H. J. Schäfer, Chem. Ber., 111, 1294 (1978).
130. M. Katz, O. Saygin, and H. Wendt, Electrochim. Acta, 19, 193 (1974).
131. B. Kräutler, A. Pfaltz, R. Nordmann, K. O. Hodgson, J. D. Dunitz, and A. Eschenmoser, Helv. Chim. Acta, 59, 924 (1976).
132. H. J. Callot, A. Louati, and M. Gross, Tetrahedron Lett., 21, 3281 (1980).
133. K. Nehring and H. J. Schäfer, unpublished results, 1974.
134. H. Schäfer and E. Steckhan, Tetrahedron Lett., 1970, 3835.
135. A. Ronlán and V. D. Parker, J. Chem. Soc. Chem. Commun., 1567 (1970).

136. A. Ronlán, O. Hammerich, and V. D. Parker, J. Amer. Chem. Soc., 95, 7132 (1973).
137. A. Ronlán and V. D. Parker, J. Org. Chem., 39, 1014 (1974).
138. (a) U. Svanholm, A. Ronlan, and V. D. Parker, J. Amer. Chem. Soc., 96, 5108 (1974); (b) J. D. Stuart and W. E. Ohnesorge, J. Amer. Chem. Soc., 93, 4531 (1971).
139. U. Palmquist, A. Ronlán, and V. D. Parker, Acta Chem. Scand., B28, 267 (1974); U. Palmquist, A. Nilsson, V. D. Parker, and A. Ronlán, J. Amer. Chem. Soc., 98, 2571 (1976).
140. V. D. Parker and A. Ronlán, J. Amer. Chem. Soc., 97, 4714 (1975).
141. B. Aalstad, A. Ronlan, and V. D. Parker, Acta Chem. Scand., B36, 171 (1982).
142. T. Shono, A. Ikeda, J. Hayashi, and S. Hakozaki, J. Amer. Chem. Soc., 97, 4261 (1975).
143. J. R. Falck, L. L. Miller, and F. R. Stermitz, J. Amer. Chem. Soc., 96, 2981 (1974).
144. L. L. Miller, R. F. Stewart, J. P. Gillespie, V. Ramachandran, Y. H. So, and F. R. Stermitz, J. Org. Chem., 43, 1580 (1978).
145. T. Kametani, K. Shishido, and S. Takano, J. Heterocyclic Chem., 12, 305 (1975).
146. J. Y. Becker, L. L. Miller, V. Boekelheide, and T. Morgan, Tetrahedron Lett., 1976, 2939.
147. T. Sato and M. Kamada, J. Chem. Soc., Perkin Trans., II, 384 (1977).
148. T. Sato, K. Torizuka, R. Komaki, and H. Atobe, J. Chem. Soc., Perkin Trans., II, 561 (1980).
149. L. L. Miller, R. F. Stewart, J. P. Gillespie, V. Ramachandran, Y. H. So, and F. R. Stermitz, J. Org. Chem., 43, 1580 (1978).
150. L. L. Miller, F. R. Stermitz, and J. R. Falck, J. Amer. Chem. Soc., 93, 5941 (1971); 95, 2651 (1973).
151. J. Y. Becker, L. L. Miller, and F. R. Stermitz, J. Electroanal. Chem., 68, 181 (1976).
152. (a) L. L. Miller, F. R. Stermitz, J. Y. Becker, and V. Ramachandran, J. Amer. Chem. Soc., 97, 2922 (1975); (b) J. B. Kerr, T. C. Jempty, and L. L. Miller, J. Amer. Chem. Soc., 101, 7338 (1979).
153. J. R. Falck, L. L. Miller, and F. R. Stermitz, Tetrahedron, 30, 931 (1974).
154. R. E. Gawley and G. A. Smith, Tetrahedron Lett., 29, 301 (1988).
155. (a) E. Kotani and S. Tobinaga, Tetrahedron Lett., 4759 (1973); (b) T. W. Bentley and S. J. Morris, J. Org. Chem., 51, 5005 (1986); (c) L. Christensen and L. L. Miller, J. Org. Chem., 46, 4876 (1981).
156. H. J. Schäfer and W. Eilenberg, unpublished results, 1985; M. Murase and S. Tobinaga, Heterocycles, 15, 1219 (1981).
157. H. Klünenberg, Dissertation, Universität Münster, 1981.
158. C. Schlegel and H. J. Schäfer, unpublished results, 1984.
159. H. Klünenberg, C. Schäffer, and H. J. Schäfer, Tetrahedron Lett., 23, 4581 (1982).
160. D. Burnett and D. J. Hart, J. Org. Chem., 52, 5662 (1987).
161. U. Palmquist, A. Nilsson, T. Petterson, A. Ronlán, and V. D. Parker, J. Org. Chem., 44, 196 (1979).
162. I. W. Elliott, Jr., J. Org. Chem., 42, 1090 (1977).
163. A. J. Majeed, M. Sainsbury, and S. A. Hall, J. Chem. Soc., Perkin I, 833 (1984); A. J. Majeed, P. J. Patel, and M. Sainsbury, J. Chem. Soc., Perkin Trans. I, 1195 (1985); M. Sainsbury and R. F. Shinazi, J. Chem. Soc., Chem. Commun., 1972, 718.

164. H. J. Schäfer and W. Eilenberg, Tetrahedron Lett., 25, 5023 (1984).
165. M. Sainsbury, Heterocycles, 9, 1349 (1978); M. Powell and M. Sainsbury, Tetrahedron Lett., 22, 4751 (1981).
166. M. Sainsbury and J. Wyatt, J. Chem. Soc., Perkin Trans., I, 108 (1979).
167. W. Eilenberg, Dissertation, Universität Münster, 1985.
168. E. Kotani, N. Takeuchi, and S. Tobinaga, J. Chem. Soc., Chem. Commun., 550 (1973).
169. J. M. Bobitt, I. Noguchi, R. S. Ware, K. N. Chiong, and S. J. Huang, J. Org. Chem., 40, 2924 (1975).
170. A. Najafi and M. Sainsbury, Heterocycles, 6, 459 (1977).
171. E. Kotani, F. Miyazaki, and S. Tobinaga, J. Chem. Soc., Chem. Commun., 300 (1974); 1750 (1977).
172. T. Shono, I. Nishigushi, and S. Kashimura, and M. Okawa, Bull. Chem. Soc. Jpn., 51, 2181 (1978).
173. D. Krikorian, R. Vlahow, S. Parashuv, M. Khinova, I. Vlahov, H. J. Schäfer, H. Duddeck, and G. Snatzke, Tetrahedron Lett., 25, 2969 (1984).
174. M. Sainsbury and J. Wyatt, J. Chem. Soc., Perkin Trans., I, 661 (1976).
175. T. Petterson, V. D. Parker, and A. Ronlán, Acta Chem. Scand., B37, 675 (1983).
176. J. H. P. Utley, Technique of Electroorganic Synthesis, Part I, in Techniques of Chemistry, Vol. 5 (N. L. Weinberg, ed.), Wiley-Interscience, New York, 1974, p. 793.
177. (a) G. E. Svadkovaskaya and S. A. Voitkevich, Russ. Chem. Rev., 29, 161 (1960); (b) B. C. L. Weedon, Adv. Org. Chem., 1, 1 (1963).
178. L. Eberson, in The Chemistry of Carboxylic Acids and Esters (S. Patai, ed.), Wiley-Interscience, New York, 1974, Chap. 2.
179. F. Beck, Elektroorganische Chemie, Verlag Chemie, Weinheim, 1974, p. 201.
180. E. H. Pryde, in Fatty Acids (E. H. Pryde, ed.), The American Oil Chemist's Society, Champaign, Illinois, 1979, p. 478.
181. S. D. Ross, M. Finkelstein, and E. J. Rudd, Anodic Oxidation, Academic Press, New York, 1975, p. 134.
182. H. J. Schäfer, Chem. Phys. Lipids, 24, 321 (1979).
183. M. P. J. Brennan and R. Brettle, J. Chem. Soc., Perkin Trans. I, 257 (1973).
184. J. P. Coleman, R. Lines, J. H. P. Utley, and B. C. L. Weedon, J. Chem. Soc., Perkin Trans. II, 1064 (1974).
185. L. Eberson, Acta Chem. Scand., 13, 40 (1959).
186. J. Haufe and F. Beck, Chem. Ing. Tech., 42, 170 (1970).
187. F. L. M. Pattison and J. B. Stothers, R. G. Woolford, J. Amer. Chem. Soc., 78, 2255 (1956).
188. F. Fichter and S. Lurie, Helv. Chim. Acta, 16, 885 (1933).
189. J. Peterson, Z. Elektrochem., 18, 710 (1912).
190. (a) D. Lelandais and M. Chkir, Tetrahedron Lett., 3113 (1974); (b) J. P. Waefler and P. Tissot, Electrochim. Acta, 23, 899 (1978).
191. (a) R. F. Garwood, N. ud Din, C. J. Scott, and B. C. L. Weedon, J. Chem. Soc., Perkin Trans. I, 2714 (1973); (b) J. Knolle, Dissertation, Universität Münster, 1976.
192. E. J. Corey and R. R. Sauers, J. Amer. Chem. Soc., 81, 1739 (1959).
193. G. Stork, A. Meisels, and J. E. Davies, J. Amer. Chem. Soc., 85, 3419 (1963).

194. L. Mandell, R. F. Daley, and R. A. Day, Jr., J. Org. Chem., 41, 4087 (1976).
195. (a) G. W. Gribble, J. K. Sanstead, and J. W. Sullivan, J. Chem. Soc., Chem. Commun., 735 (1973); (b) W. Seidel, Dissertation, Universität Münster, 1980.
196. W. Seidel, J. Knolle, and H. J. Schäfer, Chem. Ber., 110, 3544 (1977).
197. J. Knolle and H. J. Schäfer, Angew. Chem., 87, 777 (1975); Angew. Chem. Int. Ed. Engl., 14, 758 (1975).
198. H. Klünenberg and H. J. Schäfer, Angew. Chem., 90, 777 (1978); Angew. Chem. Int. Ed. Engl., 17, 47 (1978).
199. W. Seidel and H. J. Schäfer, Chem. Ber., 113, 451 (1980).
200. U. Jensen and H. J. Schäfer, Chem. Ber., 114, 292 (1981).
201. H. J. Bestmann, K. Roth, K. Michaelis, O. Vostrowsky, H. J. Schäfer, and R. Michaelis, Liebigs Ann. Chem., 417 (1987).
202. W. Seidel and H. J. Schäfer, Chem. Ber., 113, 3898 (1980).
203. R. F. Nutt, R. G. Strachan, D. F. Veber, and F. W. Holly, J. Org. Chem., 45, 3078 (1980).
204. D. Seebach and Ph. Renaud, Helv. Chim. Acta., 68, 2342 (1985).
205. W. S. Greaves, R. P. Linstead, B. R. Shephard, S. L. S. Thomas, and B. C. L. Weedon, J. Chem. Soc., 3326 (1950).
206. R. P. Linstead, J. C. Lunt, and B. C. L. Weedon, J. Chem. Soc., 3331 (1950).
207. D. G. Bounds, R. P. Linstead, and B. C. L. Weedon, J. Chem. Soc., 3326 (1954).
208. Y. Suhura and S. Miyazaki, Bull. Chem. Soc. Jpn., 42, 3022 (1969).
209. M. Takahashi, K. Osawa, J. Ueda, K. Okada, Yakugaki Zasshi, 96, 137 (1976); Chem. Abstr., 84, 135228 (1976); J. Knolle, Dissertation, Universität Münster, 1976.
210. R. N. Renaud and P. J. Champagne, Can. J. Chem., 53, 529 (1975).
211. R. N. Renaud and D. E. Sullivan, Can. J. Chem., 50, 3084 (1972).
212. (a) R. Brettle and J. C. Parkin, J. Chem. Soc. C., 1352 (1967); (b) T. Okubo and S. Tsutsumi, Bull. Chem. Soc. Jpn., 37, 1794 (1964); (c) R. Hertwig, G. Schlegel, and H. J. Schäfer, unpublished results, 1983; (d) M. Lacan, I. Tabakovic, and M. Vukicevic, Croat. Chem. Acta, 45, 465 (1973); Chem. Abstr., 80, 33211 (1974).
213. H. G. Thomas, M. Streukens, and R. Peek, Tetrahedron Lett., 45 (1978).
214. F. Adickes, W. Brunnert, and O. Lücker, J. Prakt. Chem., 130, 163 (1931).
215. S. Kato and G. Dryhurst, J. Electroanal. Chem., 79, 391 (1977).
216. R. Brettle and D. Seddon, J. Chem. Soc. C, 1153 (1970).
217. T. D. Binns and R. Brettle, J. Chem. Soc. C, 336 (1966).
218. R. G. Cooks, D. H. Williams, K. M. Johnston, and J. D. Stride, J. Chem. Soc. C, 2199 (1968).
219. T. Okubo and S. Tsutsumi, Bull. Chem. Soc. Jpn., 37, 1794 (1964).
220. (a) H. G. Thomas and E. Lux, Tetrahedron Lett., 965 (1972); (b) M. M. Baizer and R. C. Hallcher, J. Electrochem. Soc., 123, 809 (1976).
221. (a) C. T. Bahner, Ind. Eng. Chem., 44, 317 (1952); U.S. Patent 2,485,803 (1949); Chem. Abstr., 44, 2876b (1950); (b) H. J. Schäfer, Chem. Ing. Tech., 41, 179 (1969).
222. J. L. Morgat and R. Pallaud, C. R. Acad. Sci., 260, 574, 5579 (1965).

223. G. Schlegel and H. J. Schäfer, Chem. Ber., 117, 1400 (1984).
224. (a) R. Bauer and H. Wendt, Angew. Chem., 90, 390 (1978); Angew. Chem. Int. Ed. Engl., 17, 370 (1978); (b) R. Bauer and H. Wendt, J. Electroanal. Chem., 80, 395 (1977); (c) R. Bauer and H. Wendt, Angew. Chem., 90, 214 (1978); Angew. Chem. Int. Ed. Engl., 17, 202 (1978).
225. H. Schäfer and A. Al Azrak, Chem. Ber., 105, 2398 (1972).
226. S. Torii, K. Uneyama, T. Onishi, Y. Fujita, M. Ishiguro, and T. Nishida, Europ. Patent; Chem. Abstr., 96, 181142 (1982).
227. J. Yoshida, K. Sakuguchi, S. Nakatani, S. Isoe, in Recent Advances in Electroorganic Synthesis (S. Torii, ed.), Elsevier, Amsterdam, 1985, p. 85; J. Yoshida, K. Sakaguchi, and S. Isoe, Tetrahedron Lett., 27, 6075 (1986); J. Yoshida, K. Sakaguchi, S. Isoe, and K. Hirotsu, Tetrahedron Lett., 28, 667 (1987).
228. T. Chiba, M. Okimoto, H. Nagai, and Y. Takata, J. Org. Chem., 44, 3519 (1979).
229. (a) H. Schäfer, unpublished results; Habilitation-Thesis, University of Göttingen, 1970; (b) H. Schäfer and H. W. Handiak, unpublished results, 1973, H. W. Handiak, Diplom-Thesis, University of Göttingen, 1971.
230. H. Schäfer and H. Küntzel, Tetrahedron Lett., 333 (1970).
231. H. Schäfer and D. Koch, Angew. Chem., 84, 32 (1972); Angew. Chem. Int. Ed. Engl., 11, 48 (1972).
232. H. Schäfer and R. Pistorius, Angew. Chem., 84, 893 (1972); Angew. Chem. Int. Ed. Engl., 11, 841 (1972).
233. Y. B. Vasilev, L. S. Kanevskii, K. G. Karapetyan, E. P. Kovsman, A. M. Skundin, G. A. Tarkhanov, and G. N. Freidlin, Elektrokhimiya, 14, 770 (1978); Chem. Abstr., 89, 119649 (1978).
234. H. Schäfer and L. Stork, unpublished results, 1977.
235. M. Y. Fioshin, L. A. Salmin, L. A. Mirkind, and A. G. Kornienko, Zh. Obshch. Khim., 10, 594 (1965); Chem. Abstr., 64, 1949d (1966).
236. C. J. Brookes, P. L. Coe, D. M. Owen, A. E. Pedler, and J. C. Tatlow, J. Chem. Soc., Chem. Commun., 323 (1974).
237. M. Chkir and D. Lelandais, J. Chem. Soc., Chem. Commun., 1369 (1971); Additive monomers, R. N. Renaud and P. J. Champagne, Can. J. Chem., 57, 990 (1979); 58, 1101 (1980); K. G. Karapetyan, A. A. Bezzubov, L. S. Kanevskii, A. M. Skundin, and Y. B. Vasilev, Elektrokhimiya, 12, 1623 (1976); Chem. Abstr., 86, 62705 (1977).
238. R. N. Renaud, P. J. Champagne, and M. Savard, Can. J. Chem., 57, 2617 (1979).
239. P. J. Champagne and R. N. Renaud, Can. J. Chem., 58, 1101 (1980).
240. H. Schäfer and R. Pistorius, unpublished results; R. Pistorius, Dissertation-Thesis, University of Göttingen, 1974.
241. Y. B. Vasilev, V. S. Bagotsky, E. P. Kovsman, V. A. Grinberg, L. S. Kanevskii, and V. R. Polischyuk, Elektrochim. Acta, 27, 919 (1982).
242. M. Huhtasaari, H. J. Schäfer, and L. Becking, Angew. Chem., 96, 995 (1984); Angew. Chem. Int. Ed. Engl., 23, 980 (1984).
243. L. Feldhues and H. J. Schäfer, Tetrahedron Lett., 29, 2797 (1988).
244. H. Wendt and V. Plzak, J. Electroanal. Chem., 154, 13, 29 (1983).
245. L. Feldhues and H. J. Schäfer, Tetrahedron Lett., 29, 2801 (1988).
246. A. Ledwith, Acc. Chem. Res., 5, 133 (1972).

247. L. Cedheim and L. Eberson, Acta Chem. Scand. B, 30, 527 (1976).
248. J. Delaunay, A. O.-LeMoing, J. Simonet, and L. Toupet, Tetrahedron Lett., 27, 6205 (1986).
249. J. Mlcoch and E. Steckhan, Tetrahedron Lett., 28, 1081 (1987).
250. I. Tabakovic, Z. Grujic, and Z. Bejtovic, J. Heterocyclic Chem., 20, 635 (1983).
251. Y. Shizuri, K. Suyama, and S. Yamamura, J. Chem. Soc., Chem. Commun., 63 (1986); see also Y. Shizuri, H. Shigemori, Y. Okuno, and S. Yamamura, Chem. Lett., 2097 (1986); A. Nishiyama, H. Eto, Y. Terada, M. Iguchi, and S. Yamamura, Chem. Pharm. Bull., 31, 2820 (1983).
252. M. Genies, Bull. Soc. Chim. Fr., 389 (1975).
253. G. Cauquis and M. Genies, Tetrahedron Lett., 3403 (1970); 3959, 4677 (1971); see also W. Jugelt and S. Schwertner, Z. Chem., 23, 28 (1983).
254. G. Cauquis and B. Chabaud, C. R. Acad. Sci. Paris, Ser. C, 279, 1061 (1974).
255. W. Jugelt and L. Grubert, Z. Chem., 25, 442, 443 (1985).

24

Cleavages and Deprotection

HENNING LUND / Aarhus University, Aarhus, Denmark

I. Introduction 1030

II. Carbon-Carbon Bond 1032
   A. By reduction 1032
   B. By oxidation 1033

III. Carbon-Oxygen Bond 1038
   A. By reduction 1038
   B. By oxidation 1042

IV. Carbon-Sulfur Bond 1044
   A. Sulfonium compounds 1044
   B. Thioethers and sulfoxides 1045
   C. Thiol esters and thiocyanates 1047
   D. Sulfones 1048

V. Carbon-Nitrogen Bond 1049
   A. By reduction 1049
   B. By oxidation 1051

VI. Carbon-Phosphorus Bond 1051

VII. Carbon-Silicon Bond 1053

VIII. Carbon-Halogen Bond 1053

IX. Oxygen-Oxygen Bond 1053

X. Sulfur-Sulfur Bond 1054

XI. Nitrogen-Nitrogen Bond 1055
   A. Hydrazones 1055
   B. Hydrazides and azides 1056
   C. Triazines 1056
   D. N-nitrosamines 1056

XII. Oxygen-Sulfur Bond 1056

XIII. Oxygen-Nitrogen Bond 1057

XIV. Sulfur-Nitrogen Bond 1058

XV. Sulfur-Halogen Bond 1059

XVI. Sulfur-Phosphorus Bond 1060

References 1060

I. INTRODUCTION

Electrochemical reactions resulting in cleavage of bonds are synthetically useful in many respects, for example deprotection of functional groups and generation of reactive intermediates. The bond breakage may occur in concert with electron uptake or in a follow-up reaction. The main types of electrochemical cleavage reactions are described briefly in generalized terms in this section; in Secs. II through XVI the cleavage of the different types of bonds is discussed in more detail. Several of these reactions may be performed by direct as well as by indirect electrolysis.

Reductive cleavage of a bond may occur by addition of an electron to the antibonding orbital of a single bond; in most cases the electron transfer and the breakage of the bond are synchronous, a so-called dissociative electron transfer; only for very strong single bonds (e.g., C–F) may a certain lifetime be found for the initially formed anion radical. Owing to the stretching of the bond and the solvation of the developing ion in the transition state of the electron transfer, the free energy of activation of the reaction is high (the bond and solvent reorganization energy $\lambda \sim 100$ kcal mol^{-1}) and the rate relatively low. Electrode reactions of this type usually give drawn-out, irreversible voltammetric curves. Typical examples of such reactions are the cleavage of RX [R aliphatic, X a good leaving group such as halogen, R_2S^+, R_3N^+, or R_3P^+; Eq. (1)], and RY–YR' [Y = O, S, or Se; Eq. (2)]. The radical formed in the cleavage is usually further reduced.

$$R\text{-}X + e \longrightarrow R^{\cdot} + X^- \tag{1}$$

$$R\text{-}Y\text{-}Y\text{-}R + e \longrightarrow R\text{-}Y^{\cdot} + RY^- \tag{2}$$

On accepting an electron, aromatic and heteroaromatic compounds ArX (X being a good leaving group) usually form an anion radical of a certain lifetime. The reorganization energy λ connected with the electron transfer is much lower than for reactions (1) and (2) and the voltammetric curves are more easily interpretable. This suggests that the electron is transferred primarily to the π-electron system. In some cases, especially with symmetrical compounds, the anion radical is stable, whereas the dianion cleaves, Eq. (4):

$$ArX + e \rightleftharpoons [Ar\text{-}X]^{\overline{\cdot}} \longrightarrow Ar^{\cdot} + X^- \tag{3}$$

$$Ar\text{-}Ar + e \rightleftharpoons Ar\text{-}Ar^{\overline{\cdot}} \xrightarrow{e} Ar\text{-}Ar^{2-} \longrightarrow 2\ Ar^{\overline{\cdot}} \tag{4}$$

Vinylic compounds, Eq. (5), and certain carbonyl derivatives, Eq. (6), behave similarly, also with respect to having a low λ value. In Eq. (5), X is a good leaving group; in Eq. (6) Y is OR or NRR':

$$>C=CRX + e \rightleftharpoons [>C=CRX]^{\overline{\cdot}} \longrightarrow >C=\dot{C}R + X^- \tag{5}$$

$$>C=\overset{+}{N}H-Y + e \rightleftharpoons [>C=NH-Y]^{\cdot} \longrightarrow >C=N^{\cdot} + HY \tag{7}$$

Compounds with suitably substituted carbon atoms α to an unsaturated center, such as a carbon-carbon double bond, an aryl group, Eq. (7), or a carbonyl group, Eq. (8), may lose the substituent at a less negative potential than required for the reduction of the unsaturated center. In these cases it is not clear whether the electron is added to the σ-bond or the π-electron system; for benzyl chlorides the free energy of activation seems more like that of alkyl halides than that of aryl or vinyl halides. This suggests an addition of the electron to the σ-bond.

$$ArCH_2X + e \longrightarrow ArCH_2^{\cdot} + X^- \qquad (7)$$

$$-\underset{\underset{O}{\|}}{C}-CH_2X + e \longrightarrow -\underset{\underset{O}{\|}}{C}-CH_2^{\cdot} + X^- \qquad (8)$$

Some cleavages occur via saturation of a double bond followed by elimination of HX; this is especially found in some substituted heterocycles, Eq. (9), vinyl compounds, Eq. (10), and acid derivatives, Eq. (11). The product of the elimination reaction may or may not be reduced further at the potential used to reduce the substrate.

$$\overline{\underline{|C(X)=Y|}} + 2e + 2H^+ \longrightarrow \overline{\underline{|CH(X)-YH|}} \longrightarrow HX + \overline{\underline{|CH=Y|}} \qquad (9)$$

$$RR'C=CXY + 2e + H^+ \longrightarrow RR'\bar{C}-CHXY \longrightarrow RR'C=CHY + X^- \qquad (10)$$

$$R-\underset{\underset{Y}{\|}}{C}-X + 2e + 2H^+ \longrightarrow R-\underset{\underset{YH}{|}}{CH}-X \longrightarrow R-CH=Y + HX \qquad (11)$$

The reductive cleavage of two bonds and formation of a multiple bond in a concerted or nearly concerted reaction may occur when two good leaving groups are at vicinal atoms, Eq. (12):

$$R-CHX-CHY-R' + 2e \longrightarrow RCH=CHR' + X^- + Y^- \qquad (12)$$

Reactions (1) through (12) can be performed both by direct and by indirect electrolysis; the following reaction, Eq. (13), is indirect by nature. In reaction (13) hydrogen formed at a catalytically active cathode, such as platinum or Raney nickel, cleaves the bond; in Eq. (13) Y may be OH.

$$2H^+ + 2e \longrightarrow 2H^{\cdot}$$

$$2H^{\cdot} + RY \longrightarrow RH + HY \qquad (13)$$

A bond cleavage would also occur in an elimination reaction induced by an electrogenerated base (EGB, Chap. 30).

A C–H or Y–H cleavage, which involves a simple acid-base reaction or dissociation of an acidic compound and reduction of the protons thus formed to hydrogen at an electrode with a low hydrogen overvoltage and creation of a base or nucleophile, is not discussed here, (see Chap. 30).

Oxidative cleavage may begin with a loss of an electron from a heteroatom or an anion. Removal of a nonbonded electron from a heteroatom creates a cation radical; in the absence of a suitable nucleophile or the possibility of losing a proton, the system may be stabilized by bond cleavage and formation of a double bond in a way similar to that found in mass spectrometry. In Eq. (14), Y is O, S, or NR"; $R-Y^+$ may, after further steps, be hydrolyzed to R=O or attacked by some nucleophile.

$$R-\ddot{Y}-R' - e \longrightarrow [R-\dot{Y}-R']^{+} \longrightarrow R-Y^{+} + {}^{\cdot}R' \quad (14)$$

If the radical cation can lose a proton rapidly, the radical thus formed easily loses an electron to a cation, which may cleave or accept a nucleophile.

In Kolbe electrosynthesis the carboxylate ion loses an electron and then carbon dioxide; this reaction is discussed in Chap. 14. A number of electrochemical reactions involve a cleavage of the bond between the central metal atom and one or more ligands. Such reactions are not discussed here but in Chap. 21.

II. CARBON-CARBON BOND

Electrolytic cleavage of a carbon-carbon bond may occur in the course of both cathodic and anodic reactions. During an uncontrolled anodic oxidation, many organic compounds end up as carbon dioxide and water, which involves splitting of C—C bonds, but only oxidations in which the reaction does not have the character of an extensive degradation are discussed here. The Kolbe reaction that involves a cleavage of a carbon-carbon bond has been discussed in Chap. 14.

A. By Reduction

Carbon-carbon single bonds may be cleaved when the two carbon atoms bear substituents that sufficiently stabilize the anions formed on the uptake of electrons to make them relatively weak bases. The reduction of ethanetetracarboxylate esters illustrates that these esters are reducible at rather negative potentials in DMF-10% water at a mercury cathode at 60–65°C with splitting of the central carbon-carbon bond, Eq. (15) [1]:

$$(ROOC)_2CR'-CR'(COOR)_2 + 2e \longrightarrow 2R'\bar{C}(COOR)_2 \quad (15)$$

A similar cleavage occurs on reduction of 9,9'-bianthryl in DMF; in this case the anion radical is stable but the dianion undergoes bond rupture [2]. This reaction is an example of Eq. (4), whereas Eq. (15) is probably an example of Eq. (8).

The reduction of some aryltritylketones in aprotic, deoxygenated DMF, Eq. (16), is of the same kind. The arylcarbon radical reacts with DMF and eventually forms diaryloxymethane. The primarily formed radical anion is stable on the time scale of cyclic voltammetry but cleaves slowly under preparative conditions. The cleavage is suggested [3] not to be the result of a slow disproportionation of the radical anion and cleavage of the dianion, but a slow splitting of the radical anion, Eq. (16):

$$ArCOCPh_3 + e \longrightarrow [ArCOCPh_3]^{\dot{-}} \longrightarrow ArCO^{\cdot} + Ph_3C^{-} \quad (16)$$

Pinacols with at least one electroactive group α to the alcoholic function may be reduced. In protic media there occur cleavages of the C—O and the central C—C bonds with formation of a hydrocarbon; in strongly basic media the pinacol is cleaved by the base to the anion radical of the parent ketone [4,5]. The base may be generated electrochemically.

Cathodic reduction of some aryl-substituted α-diimines in DMF at the potential of the second wave in the presence of dioxygen results in splitting of the central carbon-carbon bond [6]. Possibly the dianion of the diimine, like the dianion of some pinacols, is unstable; the anion radicals thus formed could react with O_2 or $O_2^{\cdot-}$ to the isolated amides.

Electrolytic cleavage of a carbon-carbon bond is found in certain aromatic nitriles in which the aromatic ring is π-electron deficient, due to the presence of either a heteroatom in the ring or an electron-attracting substituent on the ring. The driving force for the cleavage is the gain in resonance energy on reestablishing the aromaticity of the ring combined with the stability of the cyanide ion. Whereas 2- and 4-cyanopyridine in acid solution are reduced to aminomethylpyridines, the reduction in alkaline solution yields pyridine and cyanide ions [7–9]. Benzonitriles bearing electron-attracting groups like cyano or carbethoxy groups meta or para to the cyano group are also cleaved [10] with the loss of a cyanide ion.

Another type of cleavage during electrolysis occurs as a slow step in the reduction of certain β-diketones in DMSO [11]; the primarily formed radical dimerizes to a pinacol that, catalyzed by the base generated during the electrolysis, is cleaved in a reverse aldol condensation to an α-diketone. 1,3-Diphenyl-1,3-propanedione is thus reduced in DMSO according to Eq. (17):

$$2C_6H_5COCH_2COC_6H_5 \xrightarrow{2e\ +\ 2HA} [C_6H_5C(OH)CH_2COC_6H_5]_2 + 2A^-$$
$$\xrightarrow{[A^-]} C_6H_5COCOC_6H_5 + 2CH_3COC_6H_5 \quad (17)$$

The benzil is then further reduced to a radical ion.

A ring opening involving a carbon-carbon bond is found in the electrochemically induced sigmatropic rearrangement of 2,2,4,4-tetraalkylcyclobutane-1,3-dithione (I) to the β-dithiolactone (II); the rearrangement might involve base catalysis by $I^{\cdot-}$ rather than isomerization of $I^{\cdot-}$ to $II^{\cdot-}$ [12].

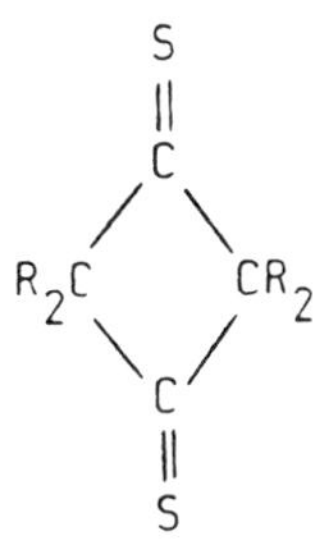

I

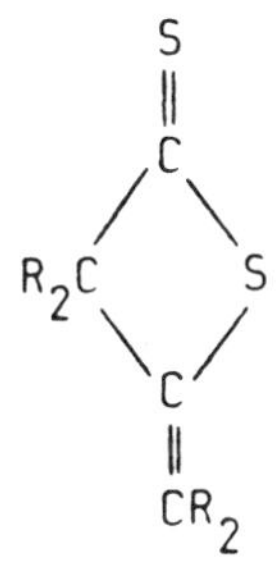

II

B. By Oxidation

Cyclopropane rings may be opened on anodic oxidation; which bond is cleaved depends on the substituents and the experimental conditions. Thus 1-ethoxy-7,7-dichlorobicyclo[4.1.0]heptane (IIIa, X = Et) [13] gives on anodic oxidation in methanol in the presence of lutidine at a carbon anode methyl

7,7-dichlorohept-6-enoate (IV), whereas 7,7-dichloro-1-trimethylsilyloxybicyclo[4.1.0]heptane (IIIb, X = Me_3Si) on oxidation in methanol at $-13°$ to $-10°$ in the presence of iron(III)nitrate [14] afforded 2-methoxycarbonylcyclohexanone (V), Eq. (18):

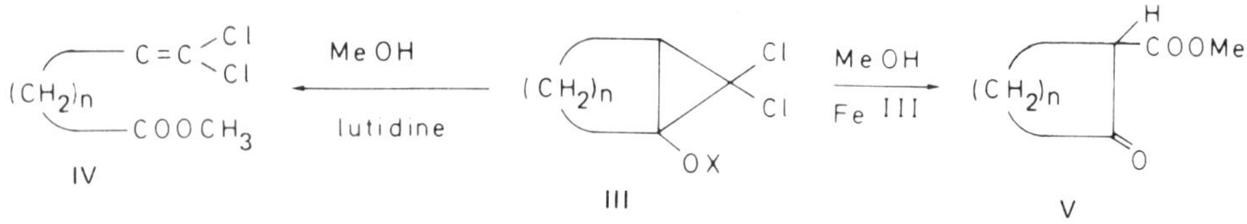

(18)

Anodic oxidation of 1-methoxy-2-phenylthiocyclopropanes (VIII) and phenylthiocycloalkanols [16] in methanol in the presence of potassium carbonate proceeds with ring cleavage between C−1 and C−2; the product is a β-[methoxy(phenylthio)methyl]-alkanone (IX), Eq. (20):

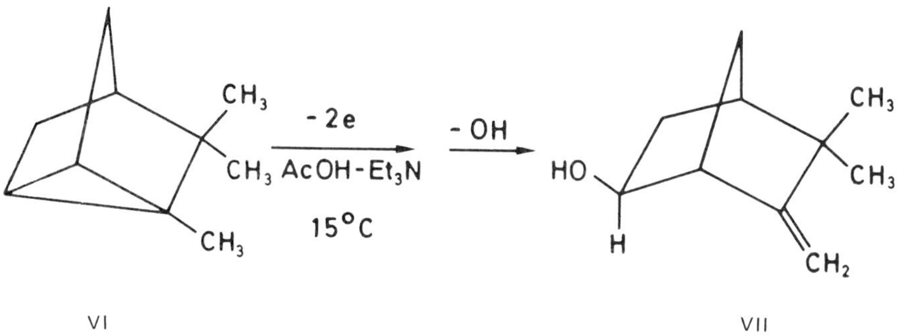

(19)

Anodic oxidation of 1-methoxy-2-phenylthiocyclopropanes (VIII) [16] in methanol in the presence of potassium carbonate proceeds with ring cleavage between C−1 and C−2; the product is a 2-[methoxy(phenylthio)methyl]-alkanone (IX), Eq. (20):

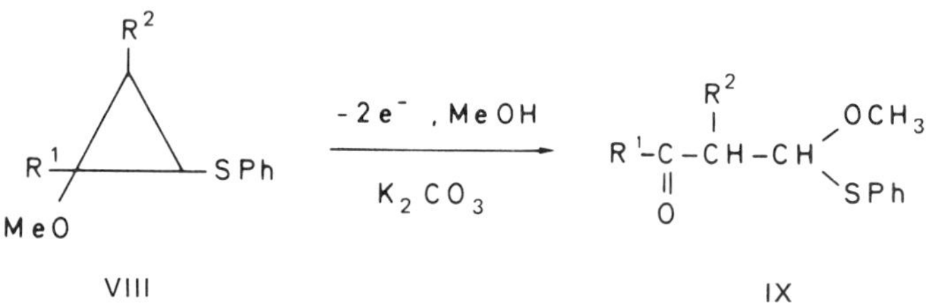

(20)

A ring enlargement involving rupture of a cyclopropane ring on anodic oxidation is exemplified in the transformation of bicyclo[4.1.0]heptane-7-carboxylic acid (X) in methanol to 3-methoxycycloheptene (XI), Eq. (21) [17]:

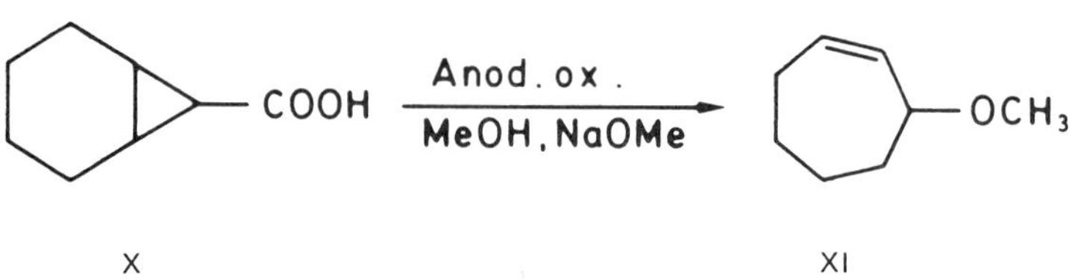

(21)

An anodic oxidation resulting in a ring enlargement for the preparation of macrorings has been developed [18]; the tosylhydrazone of bicyclo-[10.3.0]pentadec-1(12)-ene-2-one (XII) gives an anodic oxidation of some of the parent ketone (XIV), whereas the main product is the cyclopentadec-4-ynone (XIII), Eq. (22):

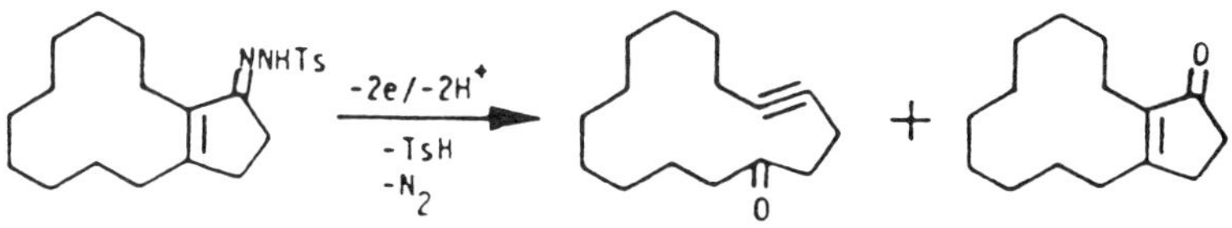

XII XIII XIV (22)

Vicinal glycols and their mono or dialkyl ethers can be cleaved at a carbon anode in methanol with TEA-OTs as supporting electrolyte to acetals or carbonyl compounds, Eq. (23). The anodic oxidation does not show any of the stereochemical limitations usually observed in the cleavage reaction by chemical oxidizing agents [19]. Vic-glycols may also be cleaved by indirect oxidation using a periodate-iodate system as mediator and lead dioxide as anode [20] or by direct oxidation at a PbO_2 electrode at pH 1.2 [21]. A similar oxidative cleavage of aziridines and 2-aminocycloalkanones may be achieved by electrogenerated "Cl^+" [22].

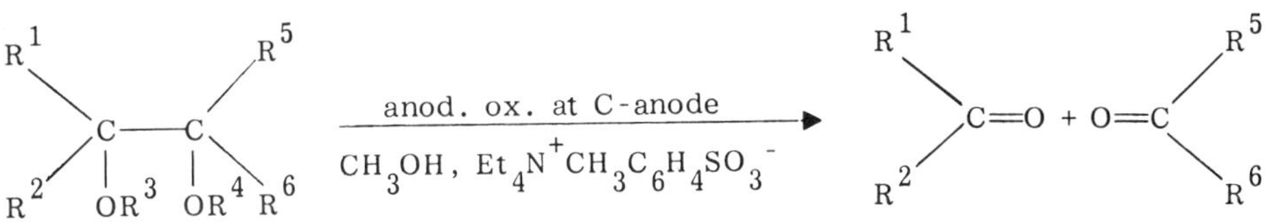

(23)

Symmetrical ketones may thus be prepared by alkylation of methyl methoxyacetate with a metal-organic compound to a dialkyl methoxymethylcarbinal followed by anodic oxidation [23].

Unsymmetrical ketones can be obtained from symmetrical ketones by anodic oxidation of their enol ethers to the α-methoxy ketone (XV) [23], reaction with R'MgX, and anodic oxidation of the glycol monomethyl ether. Alternatively, the ketone may be oxidatively aminated, alkylated at C=O, and oxidized (y = OCH_3 or NR_2), Eq. (24):

$$\underset{\text{XV}}{RCH_2-\overset{\overset{O}{\|}}{C}-\underset{\underset{y}{|}}{C}HR} \xrightarrow{R'MgX} RCH_2\underset{\underset{R'}{|}}{\overset{\overset{OH}{|}}{C}}-\underset{\underset{y}{|}}{C}H-R \xrightarrow[\text{MeOH}]{\text{anod. ox.}} RCH_2\underset{\underset{R'}{|}}{\overset{\overset{O}{\|}}{C}} + RCOOH \qquad (24)$$

Strangely enough, it has been observed that in the oxidation of the racemic and meso forms of 4,4'-dihydroxyhydrobenzoin at an anode of mild steel at pH 13.5, only the racemic form was oxidized at a measurable rate to p-hydroxybenzaldehyde, whereas the meso form was recovered unchanged [24].

Oxiranes may be ring opened in acidic methanol to α-methoxy alcohols, which can be cleaved anodically. This has been used to transform a propylene side chain to an aldehyde [25]. β-Hydroxyhydroxylamines may be cleaved to aldehydes and oximes on anodic oxidation at pH 8, Eq. (25) [26]:

$$\text{R-CHOH-CR'R''-NHOH} \xrightarrow[\text{pH 8}]{-2e\ -\ 2H^+} \text{RCHO} + \text{R'R''C=NOH} \tag{25}$$

One way of avoiding an extensive anodic degradation of organic compounds is to use acetone as a modifying medium, which is also well known from chemical oxidations, as with $KMnO_4$. Benzoin [27] is thus in sulfuric acid-acetone containing some water cleaved mainly to benzaldehyde (35%) and benzoic acid (45%); a small amount of benzil (5%) and tar (10%) is also formed. Benzil is apparently not an intermediate, as only benzoic acid (67%) was obtained from oxidation of benzil under similar conditions, Eq. (26), [27]:

$$C_6H_5CHOHCOC_6H_5 \xrightarrow{-2\bar{e}\ -\ H^+} C_6H_5CHO + [C_6H_5CO^+] \xrightarrow{H_2O} C_6H_5CHO + C_6H_5COOH \tag{26}$$

Vicinally tri- or tetramethoxy compounds may also be cleaved anodically. Thus 1,1,2-trimethoxycyclohexane obtained on electrooxidation of cyclohexanone enol ether may be oxidized to the acetal ester, $(CH_3O)_2CH(CH_2)_4COOCH_3$ [28].

Oxidation of resorcinol dimethyl ether [29] in methanolic KOH produces among other products 5,5,6,6-tetramethoxy-1,3-cyclohexadiene (XVI), and hexamethyl cis,cis-orthomuconate (XVII); XVII could be obtained from (XVI) in 77% yield on anodic oxidation, conceivably according to Eq. (27):

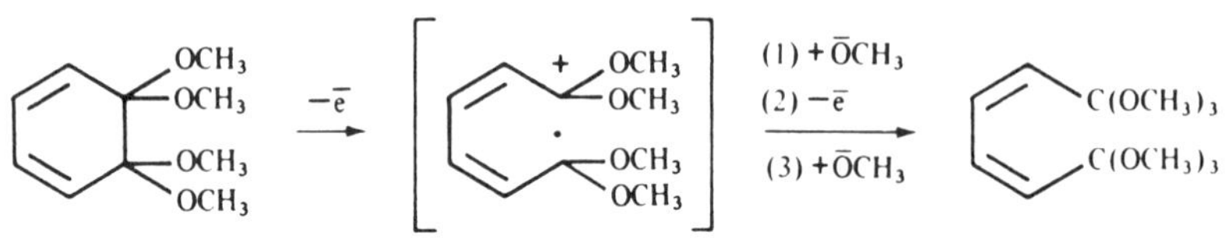

(27)

XVI XVII

It is noteworthy that only the least stable cis,cis isomer is formed; this indicates that the intermediates are short-lived and/or adsorbed on the electrode during the fixation of the geometry of the product.

Electrolysis under Kolbe conditions (Chap 14) of the γ-hydroxy acid (XVIIIa) leads to the unsaturated ketone (XIX) together with some of the Kolbe coupling products, Eq. (28) [30,31]:

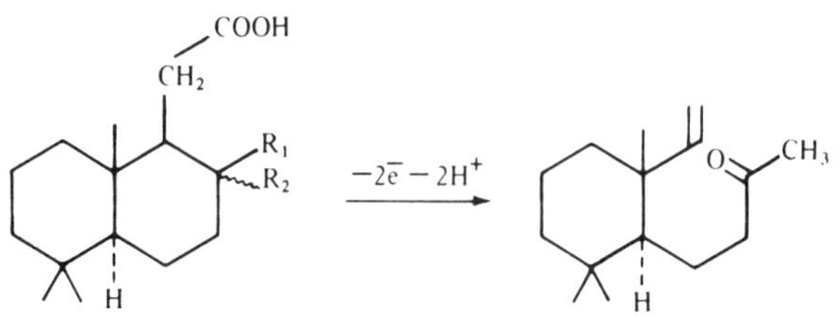

(28)

XVIII XIX

a: $R_1 = CH_3, R_2 = OH$
b: $R_1 = OH, R_2 = CH_3$
c: $R_1 = CH_3, R_2 = OAc$

A Kolbe electrolysis of a vicinal decarboxylic acid, also referred to as anodic bis-decarboxylation, may be used for introducing a double bond; an example is the transformation of 2-aza-3-oxobicyclo[2.2.2]oct-7-en-5,6-dicarboxylic acid (XVIII) to 2-azabicyclo[2.2.2]octa-5,7-dien-3-one (XIX) [31]. The C–8 epimer (XVIIIb) also yielded the cleavage product (XIX), whereas the acetoxy acid (XVIIIc) on electrolysis gave the corresponding Kolbe product. No cleavage product, neither XIX nor the enol acetate, was detected. The reaction mechanism has not been proved. The radical formed on decarboxylation of the primarily formed radical may either lose a hydrogen atom with carbon-carbon cleavage or be oxidized further to a carbonium ion, which then is cleaved with loss of a proton. The former possibility has been considered [31] the more likely as it more easily explains why the acetoxy acid is not cleaved in the same manner.

In a similar way, 1-nonyl-3-oxocyclohexanecarboxylic acid ethylene ketal (XX) has been ring opened [32] to methyl 5-nonyl-5-hexenoate (XXI) by anodic oxidation at a C anode in methanol containing K_2CO_3, Eq. (29):

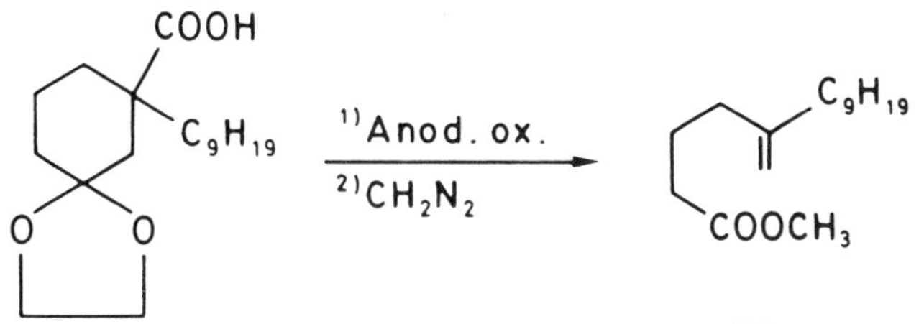

XX XXI (29)

A ring opening, which has been suggested to occur via the primarily formed radicals, is found in the anodic oxidation of the sodium bisulfite addition product of cyclohexanones (XXII) from which the lactones XXIII and XXIV were isolated in a ratio of 2:3. Electrolysis of XXII in methanol rather than in water gives methyl 5-methoxyhexanoate as the major product [33].

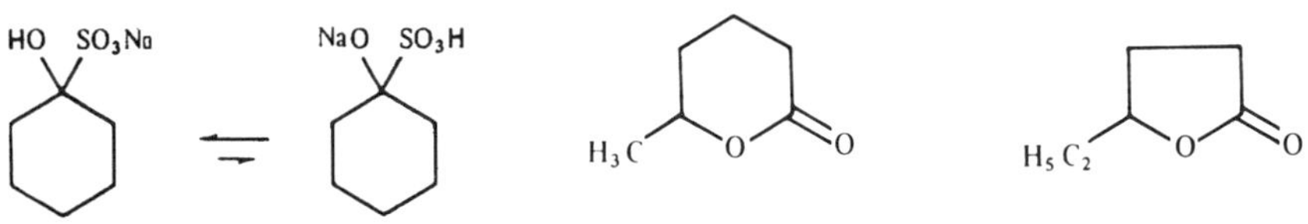

XXII XXIII XXIV

The cyclopentanone derivatives are electrolyzed similarly.

Ketones with branching in the α position may be anodically cleaved at a platinum electrode in CH_3CN-$LiClO_4$ [34]. A mechanism was proposed that involved generation of carbenium ion on α cleavage of the ketone cation radical, analogous to reactions in photochemistry and mass spectroscopy, Eq. (30):

$$R_3CCOCH_3 \xrightarrow{2e^-} R_3C^+ + CH_3CO^+ \xrightarrow[H_2O]{CH_3CN} R_3CNHCOCH_3 + CH_3COOH \tag{30}$$

Some reductively dimerized nitrogen heterocycles are oxidized anodically to their parent compounds with rupture of the C—C bond; thus the dimerized product from reduction of acridizinium (XXV) may be oxidized to acridizinium (XXVI) [35]:

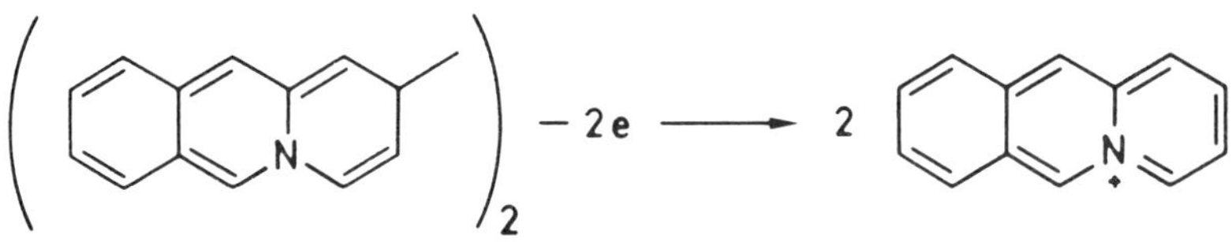

(31)

XXV XXVI

III. CARBON-OXYGEN BOND

A. By Reduction

Reductive cleavage of a C—O bond requires activation of the bond; such activation can be obtained in different ways. Conjugated carbinols, like allylic and benzylic alcohols, π-electron-deficient heterocyclic carbinols, and α-hydroxyketones and the ethers of these compounds may be reductively cleaved. Reductive elimination of two vicinal hydroxyl groups or derivatives thereof may also be possible. These reactions have been exploited in the electrochemical removal of protecting groups [36,37].

Unactivated aliphatic and aromatic hydroxyl groups may be reductively removed after esterification with a suitable inorganic acid whereby the hydroxyl group is transformed into a better leaving group. Thus the reduction of an aliphatic alcohol into an alkane may be done in a high yield by preparing the methanesulfonate of the alcohol and reducing it in DMF-Bu_4N^+ salt at a lead cathode at a rather negative potential [38]. Alternatively, the alcohol may be transformed into a reducible alkyl halide.

Aromatic hydroxyl groups may be similarly removed after transformation into an aryl diethyl phosphate (XXVII), which can be electrochemically cleaved, Eq. (32) [39]:

$$\text{ArOH} \longrightarrow \underset{\text{XXVII}}{\text{ArOP(=O)(OEt)}_2} \xrightarrow[\text{DMF}]{2e^- + H^+} \text{ArH} + {}^-\text{OP(=O)(OEt)}_2 \qquad (32)$$

The anion radical of diphenyl ether [40] and of some substituted diphenyl and naphthyl ethers [41] have been found to cleave slowly in DMF to aryl radical and phenoxide ion.

Conjugated Carbinols

α,β-Unsaturated alcohols and their derivatives can be cleaved by direct electrolysis [42,43], electrocatalytically at a platinized platinum electrode in acidic medium [44,45], and by indirect electrolysis [46]. In protic medium, direct electrocatalytic reduction usually results in formation of the saturated alcohol or hydrocarbon, whereas indirect reduction using iodide as mediator [46] gives the alkene; in electrocatalytic reduction the rupture of the C—O bond precedes the saturation of the double bond. In aprotic medium (e.g.,

DMF), the primarily formed radical anion is cleaved with loss of the oxygen function.

Indirect electrolysis of allyl alcohols can be performed at a mercury electrode at a rather positive potential [−0.2 V (SCE)], and high current densities may be used (10–20 A dm^{-2}). The first step is a transformation of the alcohol to an iodide; the double bond is shifted to the terminal position, even if it means breaking a conjugation with a phenyl ring [46]. A 20:1 mixture of 1-butene and 2-butene was obtained from crotyl alcohol, Eq. (33):

$$CH_3CH{=}CH{-}CH_2OH \xrightarrow[40-60^\circ C,\ Hg]{6\ N\ HCl,\ KI} CH_3CH_2CH{=}CH_2 \qquad (33)$$

Allyl carboxylates can be cleaved to carboxylic acids using a reduction catalyzed by Pd^0; the reaction may be used for deprotection of carboxylic acids [47].

Carbinols conjugated with aromatic systems, such as benzylic alcohols [48–51], trityl (XXVIII), tritylone (XXIX), and benzhydryl alcohols [43,51, 52], and 9-hydroxy-9-phenylanthrone, heteroaromatic systems [53], α-hydroxy ketones [54–56], α-hydroxy acids, and their derivatives [57–59] can also be cleaved at potentials less negative than that of the activating group. The N-benzyloxycarbonyl group, which cannot be removed electrochemically in methanol, may be cleaved in DMF [36]. ψ-Esters, such as o-benzoylbenzoic acid ψ-ethyl ester, which is an acylated ketal of a benzophenone, may be reduced in an acetate buffer to a phthalide [60].

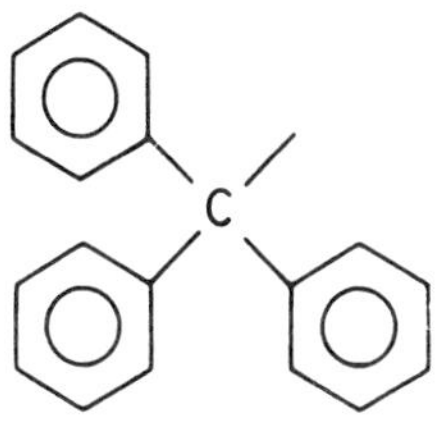

XXVIII

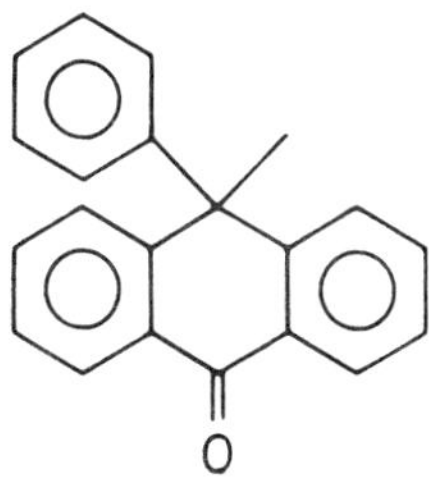

XXIX

All these systems may be used as protecting groups for alcohol functions, and the choice between them will depend, among other things, on the presence of electrophores in the protected molecule. It is usually possible to find a protecting system that is cleaved at a less negative potential than the electrophores in the protected system; otherwise, introduction of an electron-withdrawing substituent in the protecting group may be necessary. Resistance toward acid hydrolysis and hydrogenation are considerations to be taken into account when choosing a suitable protecting group.

The relative abilities of the leaving groups in the cathodic cleavage of p-methoxycarbonylbenzyl carboxylates have been measured in a DMF acetate buffer. There seemed to be no recognizable correlation between the relative rate of cleavage and either the pK_a of the conjugate acid of the leaving anion or the peak potential in cyclic voltammetry [50].

In all these cases the electron is probably initially added to the π-electron system; the molecule is then stabilized by a follow-up reaction involving the expulsion of a negatively charged oxygen-containing species. In some cases the rate constants of the reactions involved can be measured. The reductive deglycosidation of anthracyclines as adriamycin (XXX) or daunorubicin [61–65] is illustrative; the 7-deoxyatriamycinone (XXXI) is further reducible to the hydroquinone, Eq. (34):

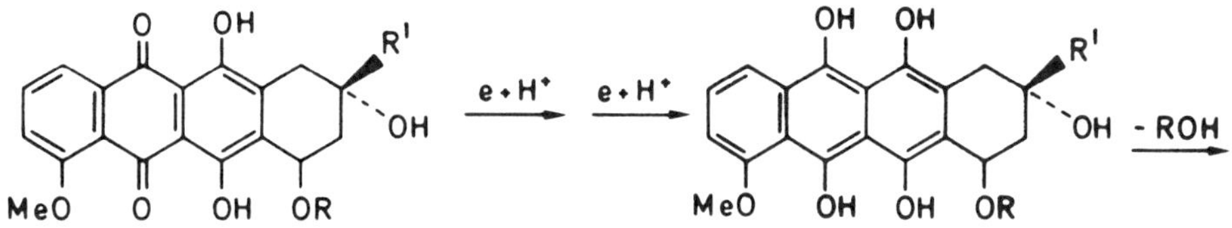

XXX

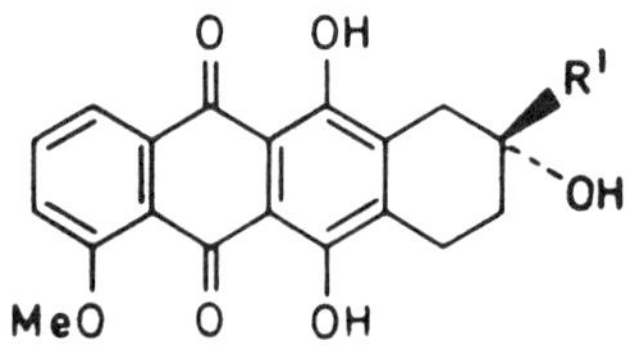

XXXI

(34)

α-Substituted Alcohols

Reductive elimination of X and Y, as X^- and Y^-, from $RR^1CX{-}CYR^2R^3$ or $RR^1CXCH_2CYR^2R^3$, where X^- and Y^- are reasonably good leaving groups, may lead to the formation of alkenes or cyclopropanes. Here some reductive eliminations are discussed in which X is an oxygen function and Y is oxygen or another heteroatom, regardless of whether C—Y is more easily cleaved than C—X.

Vicinal dioxalates are cleaved electrochemically to alkenes in DMF at a less negative potential (~ −1.2 V versus Ag/AgI) than vicinal diacetates or diols [66]. Oxalic esters of monohydric alcohols give quasi-reversible peaks in cyclic voltammetry, whereas oxalates of vicinal diols give irreversible peaks. Oxalates of the diols may be formed through a rapid, base-catalyzed transesterification using diethyl oxalate, for example [67].

Cyclic sulfates of unactivated diols, such as 2,3-butanediol, are reduced in DMF-0.4 M Et_4NClO_4 at a mercury electrode at potentials less negative than the background reduction; 2-butene (cis-trans, 5:95) was formed in moderate yields, n = 1; minor amounts of several unidentified compounds were seen in GLC [68].

Highly aryl-substituted oxiranes can be reduced in DMF, partly to the alkene; the reduction potential is more negative than that of the resulting alkene, so the yield of alkene is moderate [69]. Oxiranes activated by an α-carbonyl group may be reduced to the β-hydroxyketone. Epoxypregnone (XXXII) is thus reduced to 16-hydroxypregnone (XXXIII) in 75% aqueous ethanol containing a catalytic amount of hydrogen bromide. Eq. (35) [70]:

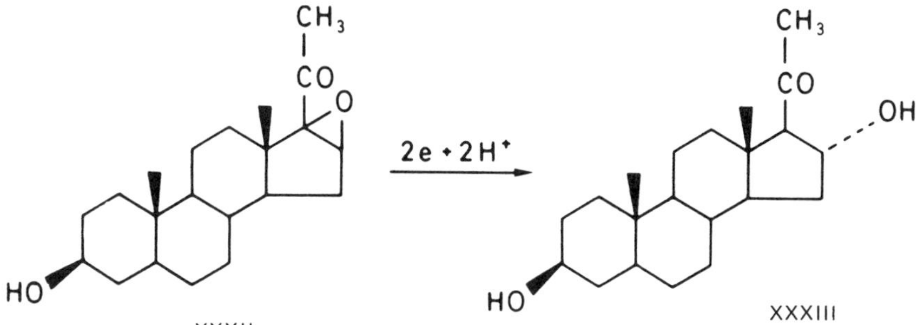

XXXII XXXIII (35)

Reductive elimination may be used for the removal of such protecting groups. By suitable changes of Y one can usually find a protecting group reducible in a desired potential range. Thus 2-haloethyl derivatives are reducible at different potentials depending on the number and kind of halogens substituted at C—2. The reduction can be made in methanol with $LiClO_4$ as supporting electrolyte according to Eq. (36) [71]:

$$PhCOOCH_2CCl_3 \xrightarrow{2e^-} PhCOOCH_2CCl_2^- \xrightarrow[(H^+)]{}$$

$$\underset{87\%}{PhCOO^-} + CH_2{=}CCl_2 + \underset{6\%}{PhCO_2CH_2CHCl_2} \qquad (36)$$

This kind of protecting group has been employed during a nucleotide synthesis using the triester method, in which the problem was the selective cleavage of one of the ester groups. The 5-trityl-protected nucleoside was first treated with 2,2,2-trichloroethylphosphoric acid dichloride and then with 2,2,2-tribromoethanol. On electrochemical reduction in acetonitrile the latter group could be removed selectively, whereas the detritylation could be done with 1% trifluoroacetic acid in methylene chloride [58].

The preparation of olefins by reductive elimination has been exploited in several cases. Thus the elimination of the hydroxyl and phenylthio groups from β-hydroxysulfides is a convenient route to olefins, Eq. (37) [72]:

$$\begin{matrix}R \\ R^1\end{matrix}\!\!>C{=}O \xrightarrow[(2)\ H_3O^+]{(1)\ PhSCH_2Li} \begin{matrix}R \\ R^1\end{matrix}\!\!>C\!<\!\begin{matrix}OH \\ CH_2SPh\end{matrix} \xrightarrow[DMF]{2e^-}$$

$$\begin{matrix}R \\ R^1\end{matrix}\!\!>C{=}CH_2 + \bar{O}H + PhS^- \qquad (37)$$

A Michael addition of thiophenol to an α,β-unsaturated carbonyl compound, followed by sodium borohydride reduction or addition of a metal-organic reagent, gives a γ-phenylthioalcohol; electrochemical reduction in dry DMF of the methanesulfonate of this alcohol yields a cyclopropane and a small amount of alkene, Eq. (38) [72]:

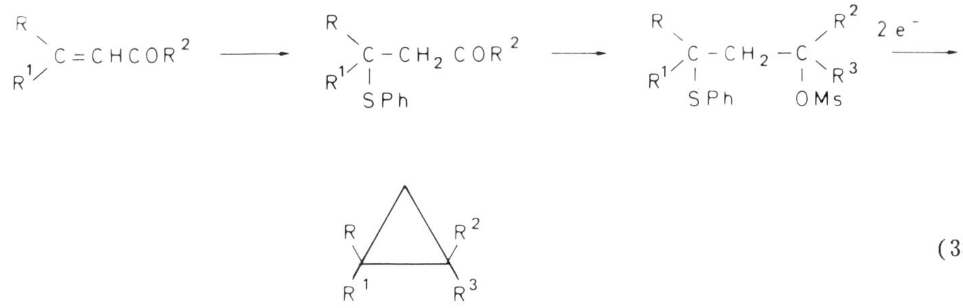

(38)

Aldehydes may be prepared from the lower homolog or a ketone or ester by reaction with the anion of formaldehyde mono- or dithioacetal. The β-hydroxythioacetal may be reduced with elimination of a hydroxyl group and phenylthiolate. The resulting enol ether or thioenol ether can be transformed into an aldehyde on acid hydrolysis or reaction with mercuric chloride [73].

During the reduction of 2-phenylthio-1,2-diphenylethanol, the stilbene obtained by the reductive elimination is more easily reducible than the substrate; it is consequently reduced to its anion radical, which in a catalytic reduction with solution electron transfer reduces the substrate [74].

A transformation of an ester group into a vinylic group may be accomplished by reacting the ester with p-tolylsulfonylmethylmagnesium iodide followed by $NaBH_4$ reduction to the β-hydroxysulfones; cathodic reduction in DMF at a lead cathode produces the vinylic compound in good yield [75].

Aldehydes may be condensed with a nitroalkane and the resulting vicinal nitroalcohol reduced to an alkene after acylation of the alcohol [76].

The reduction of some β-hydroxysulfones depends on proton activity [77]; when the proton concentration is too high, the carbanion produced after uptake of two electrons is protonated and no elimination occurs. In aprotic media the electrogenerated base may lead to a retrocondensation reaction of the β-hydroxysulfones to a sulfone and a ketone; for β-acetoxysulfones a base-catalyzed elimination may occur. The presence of a weak proton donor like phenol may thus be advisable in such cases.

1,2-Disubstituted alkenes may undergo reductive elimination to acetylenes; 1,2-diacetoxy-1,2-diphenylethylene may thus be reduced to diphenylacetylene in DMF at a mercury cathode [78]. In more protic media the diphenylacetylene is further reduced. A similar cleavage occurs during the reduction of aromatic dithio and thiol esters to diarylacetylenes [79].

Nitroalkoxybenzenes

As described in Chap. 8, certain nitroalkoxybenzenes lose the alkyl group during reduction [80] owing to a nucleophilic attack by water. The mechanism of the loss during this reduction and during the oxidation of alkoxybenzenes is similar; in both reactions the essential intermediate has a quinonoid structure, which is attacked by water.

B. By Oxidation

It was an early observation [87] that both hydroquinone mono- and dimethyl ethers on anodic oxidation in dilute sulfuric acid yield quinone, which probably results from a nucleophilic attack by water on the primary oxidation product. Similarly, the oxidation of dimethoxydurene [82] in acetonitrile produced duroquinone rather than a side-chain oxidation product similar to that obtained from p-methoxytoluene.

A special reaction of this type is the oxidation of 2,2,5,7,8-pentamethyl-6-hydroxychromane (XXXIV) for which the following reaction was suggested [83] on the basis of polarographic data, Eq. (39):

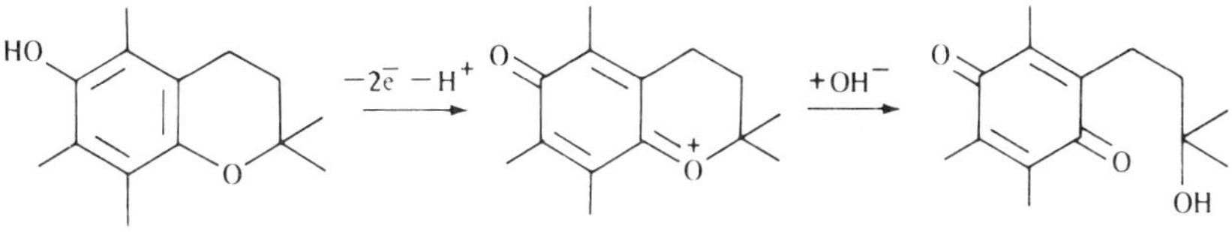

XXXIV

(39)

Later the reaction was proved by controlled potential oxidation to be reversible in acetonitrile-acetic acid containing sodium acetate [84].

Benzyl alcohols [85] and ethers [86] may be oxidized anodically. This has been employed in an anodic removal of benzylic protecting groups. Anisyl ethers of the protected alcohols were preferred because of the relatively low oxidation potential of these ethers [1.65 V (SCE)], Eq. (40) [87]:

$$CH_3OC_6H_4CH_2OR \xrightarrow[-2e^- \; - \; 2H^+ \; + \; H_2O]{aq.\ CH_3CN} CH_3OC_6H_4CHO + ROH \qquad (40)$$

The cleavage of p-methoxybenzyl ethers may be done indirectly using the stable cation radical of tris(p-bromophenyl)amine (XXXVa) as mediator in acetonitrile containing some base (2,6-dimethylpyridine or $NaHCO_3$) [88, 89]; the reaction occurs under mild conditions. Benzyl ethers cannot be cleaved by XXXVa, but cation radicals of more highly brominated triphenylamines (XXXVb, c, and d) can cleave these ethers [89]. It is thus possible to deprotect selectively groups that are protected by differently substituted benzyl groups by using different brominated derivatives of XXXV; the more bromine in the triphenylamine, the higher is its oxidation potential. Alternatively, metal-organic compounds such as di-oxo-bridged binuclear manganese compounds, may be used as mediator in MeCN [62].

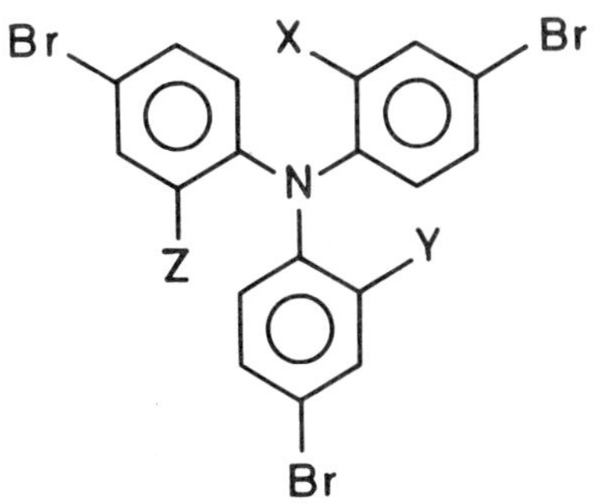

a X = Y = Z = H

b X = Y = H Z = Br

c X = H Y = Z = Br

d X = Y = Z = Br

XXXV

Protected carbonyl groups such as a 4-phenyl-1,3-dioxolane, may be deprotected by indirect oxidation in MeCN containing pyridine with N-hydroxyphthalimide as mediator [91].

IV. CARBON-SULFUR BOND

A. Sulfonium Compounds

Sulfonium salts are electrochemically reducible in both aqueous and aprotic media with cleavage of a carbon-sulfur bond; the electrochemistry of sulfonium salts has been reviewed [92,93]. The mechanism of the reduction has been discussed in several papers.

Colichman and Love [94] assumed a one-step, two-electron cleavage to the sulfide and a carbanion, but later studies pointed to a cleavage after the uptake of one electron [95,96]. In trialkylsulfonium salts a dissociative electron uptake probably takes place, whereas in aryl dialkylsulfonium salts the reduction may be a sulfur-carbon bond breaking concomitantly with the electron uptake [96] or the sulfaranyl radical, formed by addition of an electron to the aryl-π-system, may have a certain lifetime [97]. In favor of the former possibility, it is argued that the E_p values in CV are very sensitive to the electronegativity of the fragmenting radical.

The leaving group ability of different substituents in the reduction of sulfonium salts has been studied in water and MeCN [98] at a mercury electrode and in THF using potassium in graphite, KC_8 [97]. In water and MeCN [98], methyl is preferentially cleaved, which is in contrast to the findings for phosphonium or arsonium salts; in THF with electron transfer from KC_8, the following leaving group propensity was found [97]: benzyl > secondary > primary > methyl > phenyl.

Reduction of simple sulfonium salts in an aprotic solvent such as DMF at a Hg electrode at −20°C, gives a compound that has been described as an amalgam; at higher temperatures it decomposes [99]. The nature of these "amalgams" and the corresponding compounds obtained from quaternary ammonium compounds is discussed in Chap. 28.

Sulfonium salts undergo generally a coulometric two-electron reduction to sulfide and an anion; attempts to use the anion in a Michael condensation with acrylonitrile gave only moderate yields [100]. The formation of a basic anion may also be responsible for the formation of ylids [101,102] during the reduction of certain sulfonium salts. The reactions are probably, taking benzyldimethylsulfonium salt as an example, Eqs. (41) and (42):

$$PhCH_2\overset{+}{S}Me_2 + e^- \longrightarrow Me_2S + PhCH_2^{\cdot} \xrightarrow{e^-} Me_2S + PhCH_2^- \tag{41}$$

$$PhCH_2\overset{+}{S}Me_2 + PhCH_2^- \longrightarrow Ph\overset{-}{C}H-\overset{+}{S}Me_2 + PhCH_3 \tag{42}$$

In this connection it is also of interest that acrylonitrile and dimethylsulfide are formed during the electroreduction of β-cyanoethyldimethylsulfonium salts in DMSO; the acrylonitrile is formed in a base-promoted β-elimination [103]. The formation of the benzyl radical, Eq. (41), is made likely by the finding of a small yield of 1,2-diphenylethane [101,102] and that 4-nitrobenzyldimethylsulfonium salts give good yields of 4,4-dinitrobibenzyl on reduction at a mercury cathode [104].

The reduction products from triphenylsulfonium ions depend on the electrode material and the potential [105–108]. Electrolysis of triphenylsulfonium bromide at a mercury electrode at the potential of the first wave quantitatively yields diphenylsulfide and diphenylmercury. No benzene or diphenyl is formed. At potentials more negative than the second reduction wave, diphenylsulfide and benzene are formed quantitatively, provided low substrate concentrations are used. With increasing substrate concentration the yield of diphenylmercury increases at the expense of benzene. Reduction in aqueous solution at an aluminum cathode produces diphenylsulfide and some benzene; addition of DMF increases the yield of benzene, as DMF is a better hydrogen atom donor than water [107].

B. Thioethers and Sulfoxides

Unconjugated thioethers are not reducible in aqueous solution; in aprotic solvents alkyl aryl sulfides, ArSR, are reducible at relatively negative potentials (~ −2.5 V (SCE)) [109] with C−S cleavage to ArS^- and R·; the latter may be reduced further. Reduction of diphenylsulfide gives equal amounts of benzene and thiophenoxide ion, Eq. (43):

$$C_6H_5SC_6H_5 + 2e^- + H^+ \longrightarrow C_6H_5S^- + C_6H_6 \tag{43}$$

Phenacyl phenyl sulfide can be reduced in aqueous-alcoholic medium to acetophenone and thiophenol [110]; this is analgous to several other phenacyl derivatives; see Eq. (8). This reaction has been used to cleave some phenacylthio-substituted thiophenes by reduction at a mercury cathode in DMF containing 2 Eq. phenol [111].

A number of other groups can be used for the protection of thiols; benzyl thioethers of amino acids and peptides have been cleaved electrochemically at a platinum cathode in liquid ammonia [112], at mercury in MeOH-TMACl [113], and in DMF [114,115], and the trityl group has been used to protect cysteine; cysteine was recovered in 90% yield after reduction in DMF [115]. The 4-pyridylmethyl [116,117] and diphenyl-4-pyridylmethyl groups can be cleaved in acid solution [115] at a mercury cathode.

Reductive cleavage of β-ketosulfoxides is a practical way of preparing methylketones [118]; thus phenacylmethylsulfoxide, Eq. (44), gives acetophenone (74% yield) on reduction at a mercury cathode:

$$RC_6H_4COCH_2SOCH_3 + 2e + H^+ \longrightarrow RC_6H_4COCH_3 + CH_3SO^- \tag{44}$$

Reduction of (E)-1-methylsulfinyl-1-methylthio-2-phenylethene (XXXVI) and 1-methylthio-p-tolylsulfonyl-2-arylethenes in acetonitrile with excess of phenol involves a selective cleavage of one C−S bond, resulting in formation of (E)-1-methylthio-2-phenylethene (XXXVII). The dependence of the CV curves on the phenol-substrate ratio and stereochemical considerations suggested [119,120] an ECEC reaction, Eq. (45):

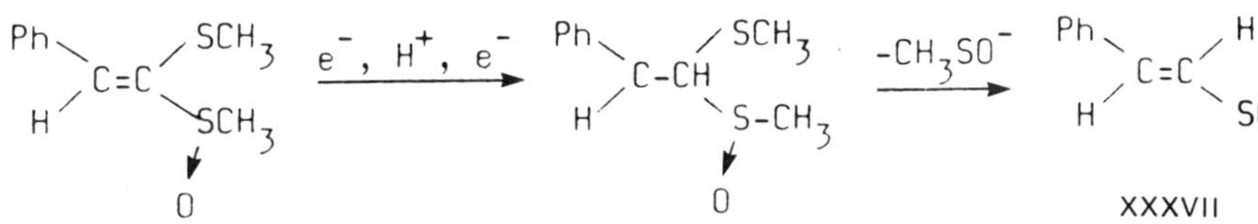

XXXVI XXXVII (45)

Reduction of the vinylic gem-disulfide (XXXVIII) in the presence of CO_2 yields the carboxylated compound (XXXIX) [121] described as an addition-elimination ECEC reaction.

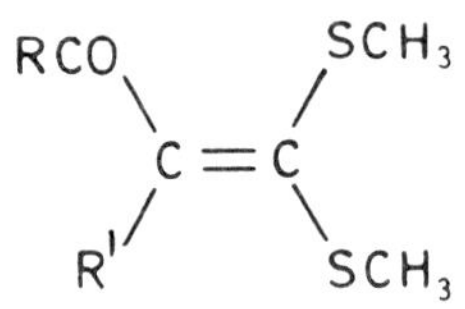

XXXVIII

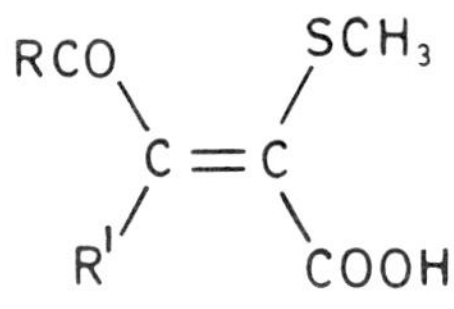

XXXIX

Anodic oxidation of thioethers usually attacks the sulfur with formation of sulfoxides, sulfones, or sulfonium compounds (Chap. 17). If R in ArSR is a good leaving group, oxidation in acetonitrile may lead to rupture of the C–S bond; $PhSCPh_3$ is thus oxidized [122] with formation of 48% PhS–SPh.

The methylthiomethyl group is used as a protecting group of hydroxy functions. A deprotection can be carried out by electrolysis in an undivided cell with Pt electrodes in AcOH containing NaOAc. The resulting acetoxymethyl ether can be hydrolized in weakly alkaline medium Eq. (46), to the hydroxy function [123]:

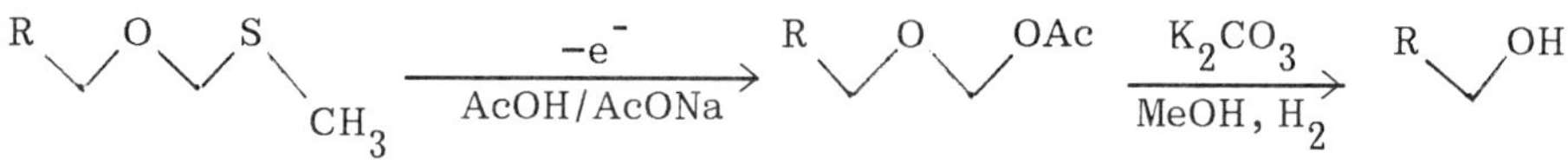

(46)

Electrooxidative desulfenylation of Michael-type thiol adducts of activated olefins may be performed in ethanol with TBABr as supporting electrolyte in an undivided cell at platinum electrodes. The activated olefin is formed in excellent yield; most of the sulfur is found as sulfinate [124].

gem-Disulfides can be oxidized anodically with cleavage of the carbon-sulfur bond [125–129]; in the presence of water a carbonyl group is formed. This has been utilized for the removal of the 1,3-dithion protecting group [127]; the 1,3-dithion derivative is oxidized anodically at a platinum electrode in aqueous acetonitrile according to Eq. (47):

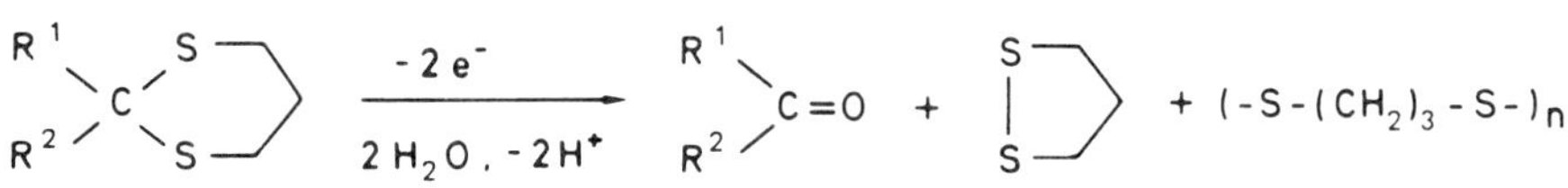

(47)

Indirect anodic cleavage of 1,3-dithianes can be done using tris(p-tolyl)-amine as electron transfer agent in acetonitrile-$LiClO_4$ containing $NaHCO_3$; the yields of recovered carbonyl compound were almost quantitative [130, 131].

C. Thiol Esters and Thiocyanates

Thiol esters are reduced at less negative potentials than carboxylic esters and can be cleaved in protic (e.g., methanol) [132] or aprotic [114,133] media. Aroyl and benzyloxycarbonyl groups may thus be used for protection of sulfhydryl groups.

Cleavage of a C–S bond in the initially formed anion radical has been shown to be the first chemical step in the electrochemically induced rearrangement of S,S-diarylbenzene-1,2-dicarbothioates (XL) to 3,3-bis(arylthio)phthalides (XLI) in DMF, Eq. (48). The reaction could be effected with 0.1 F mol^{-1} and was considered a kind of internal $S_{RN}1$-reaction [134].

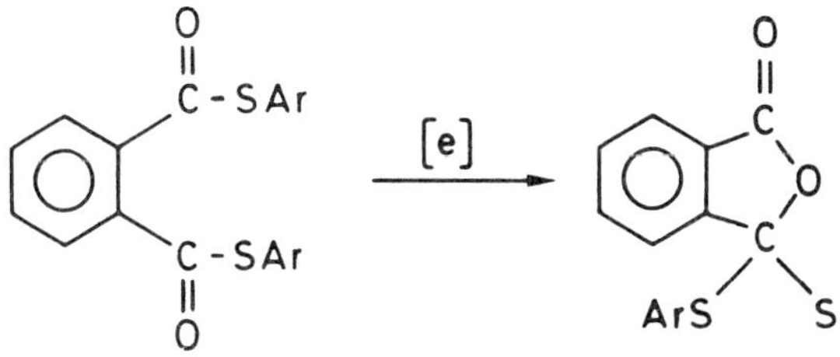

(48)

XL XLI

A tert-butylthiol group may be used as a protecting group for carboxylic acids; thus S-tert-butylthiolates can be cleaved anodically in aqueous acetonitrile under neutral conditions using a bromide as electrolyte and the thus electrogenerated bromine as oxidizing agent [135].

Triarylamine cation radicals may also be used in an indirect electrochemical oxidative cleavage of thiol esters, Eq. (49) [133]:

$$R{-}CO{-}SR' \xrightarrow[MeCN/H_2O/Na_2CO_3]{Ar_3N^{+}} RCOOH + R'S{-}SR' \qquad (49)$$

Dialkyl, alkyl aryl, and diaryl trithiocarbonates are reduced in DMF to the anion radical, which loses a thiolate ion [136–138]. The rate of the cleavage was measured with CV; for substituted diphenyltrithiocarbonates, the rate constant of the cleavage was found to vary linearly with Hammett's substituent constant σ. The anion radicals of mono- and dithiocarbonates having a carbonyl group were cleaved faster than those possessing a thiocarbonyl group.

Two types of electrode reactions were found in the reduction of thiocyanates. 2-Thiocyanatoacetophenones [110] are reduced in a two-electron reduction to the acetophenone and thiocyanate ion, Eq. (50):

$$C_6H_5COC(CH_3)_2SCN \xrightarrow{2\bar{e}} SCN^- + [C_6H_5COC(CH_3)_2]^- \xrightarrow{H^+} C_6H_5COCH(CH_3)_2 \qquad (50)$$

In most other thiocyanates the reduction results in the loss of a cyanide ion; the half-wave potential is nearly independent of pH. Macroscale reductions at a mercury cathode yielded [139] the mercaptan according to Eq. (51):

$$R{-}SCN \xrightarrow{2\bar{e}} RS^- + CN^- \tag{51}$$

D. Sulfones

Simple aliphatic sulfones, RSO_2R', are polarographically inactive both in protic and aprotic media; methylene disulfones are, however, reducible in DMF at potentials close to the decomposition potential of the medium. β-Disulfones and benzyl sulfones are rather acidic and may function as proton donors during the reduction, and their anions are generally not reducible [140]. Alkyl aryl sulfones and diaryl sulfones are reducible under protic and aprotic conditions [141–144]; the electrochemistry of sulfones has been reviewed by Chambers [93] and Simonet [147].

On reduction of alkyl aryl sulfones in methanol containing TMACl at a mercury electrode, the cleavage usually but not always [149] takes place between sulfur and the alkyl group [144–150] with formation of an arylsulfinic acid, Eq. (52):

$$ArSO_2R + 2e + H^+ \longrightarrow ArSO_2^- + RH \tag{52}$$

The propensity to act as a leaving group in diaryl sulfones or alkyl aryl sulfones is benzyl ~ allyl > alkyl > phenyl [144–147]. In unsymmetrically substituted diaryl sulfones, both steric and electronic effects play a role; substituents in the *ortho*-position promote cleavage of the neighboring C–S bond, probably by hindrance of the conjugation of the aromatic ring with the sulfone group. Thus, phenyl-2,4,6-trimethylphenylsulfone is reduced to 80% benzenesulfinic acid and 88% mesitylene [144] in methanol, and methyl-*o*-*tert*-butylphenyl sulfone is reduced predominatly (> 90%) with arylsulfur cleavage [149].

The general rule for the electronic effects on the reductive cleavage seems to be that the aryl group with the strongest conjugation to the electron-attracting sulfone group is found in the sulfinic acid. Electron-donating groups would then promote a retention of the Ar–S bond; in the reduction of phenyl *p*-methoxyphenylsulfone in methanol [144], 16% anisole and 54% benzene are formed. Electron-attracting substituents would then be expected to promote splitting of the Ar–S bond; in this connection it is of interest that methyl 4-cyanophenyl sulfone, for example, is cleaved in aqueous ethanol at pH 9.3 at a mercury cathode to benzonitrile and methylsulfinic acid, Eq. (53) [151]:

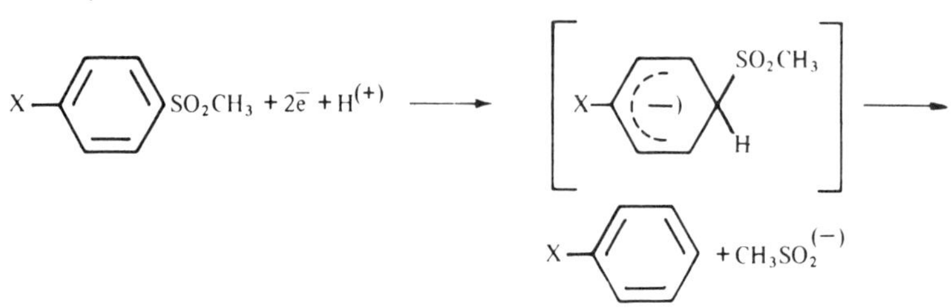

(53)

X: 4–COOH, 4—COOMe, 4—COOEt, 4—$CONH_2$, 4—CN, 4—SO_2CH_3

In the selective removal of an allylic benzenesulfonyl group in the presence of an unactivated benzenesulfonyl group, it was found that indirect electrochemical reduction employing suitable electrogenerated anion radicals as reducing agents increased the selectivity [152].

Reductive cleavage of β-ketosulfones [153], $RCOCHR'SO_2R''$, in DMF at mercury is a practical way of preparing alkyl ketones; the primarily formed anion radical cleaved to a radical $RCO\dot{C}HR'$ and a sulfinate ion $R''SO_2^-$. Since the anodic reaction was the oxidation of $R''SO_2^-$ to $R''SO_3^-$, an undivided cell could be used.

V. CARBON-NITROGEN BOND

A. By Reduction

Quaternary ammonium compounds are widely used as supporting electrolyte, and their reduction is thus of considerable interest. It was demonstrated long ago [154] that anilinium salts with three N-alkyl groups are discharged at a lead cathode. Regardless of the nature of the saturated substituents on nitrogen, benzene and the corresponding tertiary amine were always obtained. These results show that in the discharge of the ammonium ion the bond between N and phenyl is particularly weakened; N-allyl and N-benzyl bonds are weakened even more. Dimethylallyl (or benzyl) anilinium ions yield in aqueous solution at a lead cathode propylene (or toluene) and dimethylaniline [155]:

$$[(CH_3)_2RNC_6H_5]^+ \xrightarrow{2\bar{e}\ +\ H^+} C_6H_5N(CH_3)_2 + RH \qquad (54)$$

The propensity of cleavage at a mercury cathode has been found to be benzyl > allyl > phenyl > alkyl.

In many cases it has been observed that reductive cleavage of quaternary ammonium compounds at lead is minor and that a Hofmann elimination is dominant; this is brought about by an electrogenerated base.

A useful preparation is the "Emmert cleavage." This is a reductive ring opening of anilinium salts in which one of the N-alkyl groups is simultaneously attached to the 2-position of the benzene ring (XLIIa-c). The products are of the type XLIIIa-c [156,157].

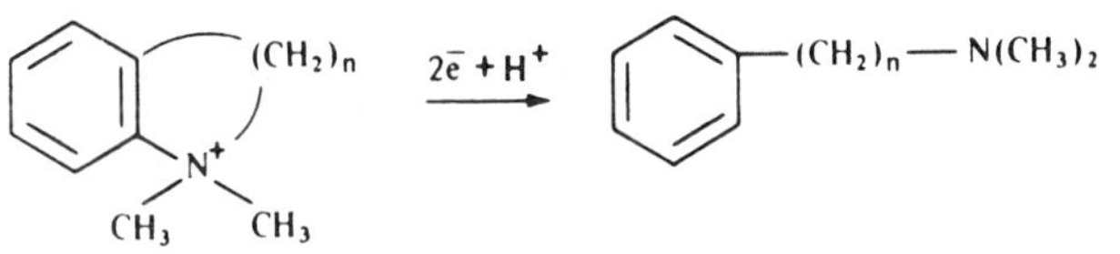

XLIIa–c XLIIIa–c (55)

The yield of XLIIIa (n = 2) is 90%; of XLIIIb (n = 3), 88%; of XLIIIc (n = 4), 90%. Tetrahydroisoquinolinium salts are not cleaved at the mercury cathode under ordinary conditions but are cleaved when NaI in liquid NH_3 is used [158].

The reduction of quaternary ammonium compounds having a phenyl, allyl, or benzyl substituent probably proceeds in a way similar to that of other compounds of the form ArX, $RCH{=}CHCH_2X$, or $ArCH_2X$, where X is a good leaving group, Eqs. (3) and (7). Radicals seem to be intermediates in the cleavage; this is in accordance with the fact that a mixture of meso- and d,l-2,2-diphenylbutane is formed in the electrolysis of chiral D-α-phenylethyl-1-trimethylammonium nitrate in DMF [159] and that ESR spectroscopy reveals the presence of radical fragments in the reduction of dimethylbenzylanilinium salts [160].

Reduction in aprotic media of ammonium compounds having four aliphatic substituents at mercury or some other metals yields "tetraalkylammonium amalgam" or "tetraalkylammonium metals." This phenomenon is discussed in Chap. 28.

Protection of amino groups as carboxamides is generally not useful; however, benzyloxycarbonyl derivatives of primary and secondary amines may be reduced at vitreous carbon cathodes in DMF at very negative potentials, with cleavage to toluene and free amine in good yield (>80%) [161,162]. However, 4-nitrobenzyloxycarbonyl derivatives may be deprotected at 1.2 V (SCE); the cleavage of the initially formed anion radical is slow on the time scale of CV but sufficiently fast for a preparative-scale reduction.

The reduction of amides [163] to alcohols proceeds through a gem-amino alcohol, which loses either ammonia (or amine), forming an aldehyde, or water, forming an imine; the aldehyde is then reduced to the alcohol, whereas the imine forms an amine on reduction. The intermediate gem-amino alcohol is also formed in the reduction of oxaziridines (Chap. 18) [164], and a similar gem-diamine during the reduction of amidines to amines [57].

3,5-Di-tert-butyl-4-oxo-1-phenyl-2,5-cyclohexadiene may be used as a protecting group in peptide synthesis; it can be introduced by electrochemical oxidation, Eq. 56 [165,166]:

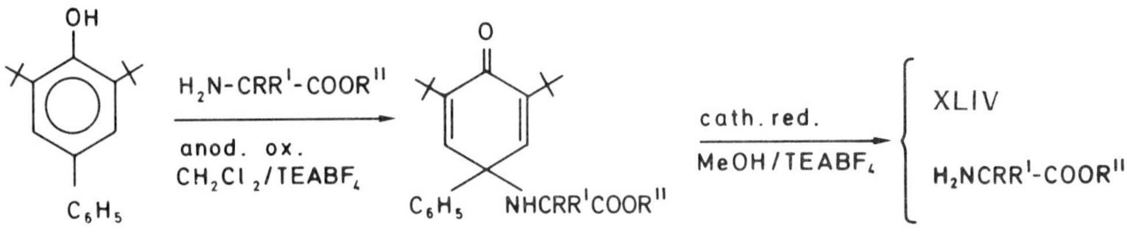

XLIV (56)

Several groups are useful for protecting amino groups, where the deprotection is a reductive elimination, Eq. (12), in which Y is an urethane and X a halogen [71,167]. Thus, for example, iodoethoxycarbonyl, trichloroethoxycarbonyl, and tribromoethoxycarbonyl are of practical importance in peptide synthesis.

Cleavage of C—N bonds has been described elsewhere, for example the reduction of gem-dinitroalkanes to nitrite ion and nitroalkane anion [168, 169] (Chap. 9) and the reduction of pyridylamines to picoline and ammonia (Chap. 18) [7,8]. The reduction of α-aminocarbonyl compounds [170–173] or α-aminonitriles [174] is similar to the reductive cleavage of the α-hydroxy compounds; the loss of amine occurs at a potential less negative than the reduction of the activating group.

If the amino group is β to the carbonyl group, a reduction takes place at the carbonyl group rather than resulting in a cleavage of the carbon-nitrogen bond; the polarographic reduction of the β-aminoketone and the α,β-unsaturated ketone obtained by elimination of the amine may be used for following the kinetics of the elimination reaction [175,176].

Electrolysis of 4-aminobenzonitrile in DMF gave a yellow solution that, from the electron paramagnetic resonance (EPR) spectrum, was identified [177] as the radical anion of 4,4'-dicyanobiphenyl; a loss of amide ion is followed by dimerization of the neutral radical.

A carbon-nitrogen bond is broken during a chemical follow-up reaction in many ring closures or ring contractions [178] of heterocyclic compounds in which a nucleophilic attack on an imine followed by an elimination of the amine or ammonia takes place. This is discussed in Chap. 18, in which is also mentioned the loss of ammonia from the primarily formed dihydro compound in the reduction of certain aminoheterocyclic compounds.

B. By Oxidation

Anodic oxidation of primary and tertiary [179] aliphatic amines in acetonitrile containing a small amount of residual water produced several compounds, including aldehydes. Tripropylamine is oxidized according to Eq. (57):

$$(CH_3CH_2CH_2)_3N \xrightarrow{-e^-} (CH_3CH_2CH)_3\dot{N}^+ \xrightarrow{-H^+} (CH_3CH_2CH_2)_2N\dot{C}HCH_2CH_3 \xrightarrow{-e^-}$$

$$(CH_3CH_2CH_2)_2\overset{+}{N}CHCH_2CH_3 \rightarrow (CH_3CH_2CH_2)_2\overset{+}{N}{=}CHCH_2CH_3 \xrightarrow{H_2O} (CH_3CH_2CH_2)_2NH + CH_3CH_2CHO + H^+ \quad (57)$$

N-Methoxyphenyl-2-azetidinones may be deprotected under mild conditions by anodic oxidation in 1:1 MeCN-H_2O containing 1% $LiClO_4$. The dication is probably the species cleaved. The benzoquinone by-product may be extracted with 10% Na_2SO_3 [180].

VI. CARBON-PHOSPHORUS BOND

Phosphonium compounds are reduced more easily than the corresponding ammonium compounds; the reduction results in a breakage of a carbon-phosphorus bond. The ease with which a group is cleaved from a phosphonium compound depends on several variables, primarily the electrode material, the potential, and the temperature [181–185].

At a mercury cathode the propensity for cleavage from a phosphonium salt in methanol increases in the sequence methyl < ethyl < n-butyl ~ phenyl < tert-butyl < allyl < benzyl; however, at a lead cathode the tendency for loss of an alkyl group from an alkyltriphenylphosphonium group is much higher than at a mercury cathode; ethyltriphenylphosphonium chloride is thus reduced in methanol at a mercury cathode to a 5:3 mixture of ethyldiphenylphosphine and triphenylphosphine, whereas 91% triphenylphosphine were isolated at a lead cathode [181].

A convenient synthetic procedure for the preparation [186] of mixed tertiary phosphines and quaternary phosphonium salts has been based on the preferential cleavage of the benzyl group from P (and As). This method also provides a simple means for preparing phosphonium salts with four different ligands. The salts, after resolution into their optical antipodes, can be cathodically cleaved and thus provide optically active tertiary phosphines. The sequence is summarized in Eq. (58):

$$
\begin{aligned}
& R_3'P \xrightarrow{R^1X} [R_3'R^1P]X \xrightarrow[R'H]{+\,2\bar{e}\,+\,H^+} R_2'R^1P \\
& \xrightarrow{R^2X} [R_2'R^1R^2P]X \xrightarrow[-\,R'H]{+\,2\bar{e}\,+\,H^+} R'R^1R^2P \xrightarrow{R^3X} \\
& [R'R^1R^2R^3P]X \xrightarrow[-\,R'H]{+\,2\bar{e}\,+\,H^+} R^1R^2R^3P \xrightarrow{R_4X} [R^1R^2R^3R^4P]X \qquad (58) \\
& [R^1R^2R^3R^4P]X \xrightarrow[\text{(2) Cathodically cleave}]{\text{(1) Resolve}} \text{Optically active } \underline{\text{tert}}\text{-phosphines}
\end{aligned}
$$

The benzyl group is especially suitable as R'. This method can also be applied to the preparation of optically active tertiary arsines [187].

Reduction of alkyltriphenylphosphonium salts in aprotic media leads to the formation of ylids [188,189]. The formation of ylids as a result of the electrogeneration of bases (EGB) [190] is discussed in Chap. 30.

Triphenylphosphine is reduced in aprotic media; the results from polarographic, voltammetric, and preparative experiments point to the following radical mechanism, Eq. (59) [191]:

$$
(C_6H_5)_3P \xrightarrow{e} [(C_6H_5)_3P]^{\overline{\cdot}} \longrightarrow (C_6H_5)_2P^- + C_6H_5^{\cdot} \qquad (59)
$$

Oxidation of the cleavage products leads to diphenylphosphonic acid and phosphine.

However, in DMF-$TBABF_4$, electrolysis of triphenylphosphine yielded butyldiphenylphosphine, tributylamine, and benzene and no biphenyl; a catalytic reduction involving $TBABF_4$ was proposed to take place during the reduction of triphenylphosphine [192]. Similar results were obtained from triphenylphosphine oxide.

Attempts to reduce triphenylphosphine oxide electrochemically to triphenylphosphine have not yet been met with success (see, however, sec. XVI); in aprotic media a carbon-phosphorus bond is cleaved in preference to the phosphorus-oxygen bond [191,192].

Quaternary arsonium compounds are in general reduced at potentials 0.3 V less negative than those of the corresponding phosphonium salts [193]. The ease of cleavage of the different groups is mostly similar to that in the phosphonium compounds, but some differences are found.

VII. CARBON-SILICON BOND

Allyl and benzyl silanes may be anodically oxidized in the presence of an alcohol, a carboxylic acid, or water to the corresponding ether, ester, or alcohol [194]. The initially formed cation radical is believed to lose the silicon moiety by attack by the nucleophile on silicon, Eq. (60):

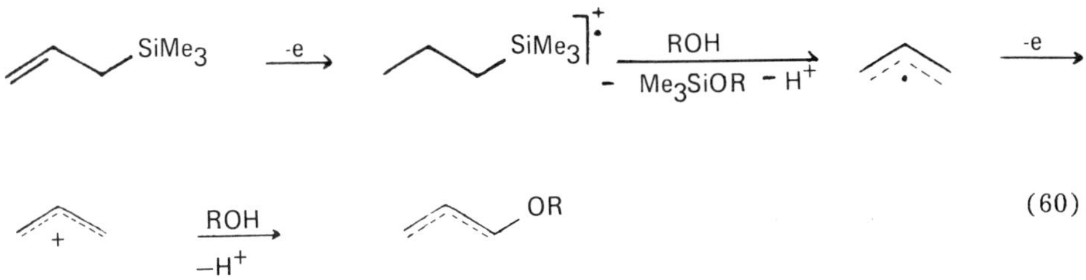

(60)

The oxidation of allyl and benzyl silanes provides a method for "umpolung" of such organosilicon compounds.

VIII. CARBON-HALOGEN BOND

Reductions resulting in cleavage of a carbon-halogen bond have been discussed in Chap. 8. Here are mentioned only some cases in which halogens have been used as protecting groups.

Vicinal dibromides can be cathodically converted into olefins (Chap. 8). Their reduction potential is shifted to less negative values with increasing alkyl substitution. This may be used as a selective monoprotection of the less alkylated double bond in dienes. The diene is converted into the tetrabromide with bromine and subsequently deprotected by controlled potential electrolysis [195].

β-Halogenethyl groups have been used as protecting groups for several types of functional groups. The deprotection can be done electrochemically by direct [71,167] and indirect [196] electrolysis. Using a vitamin B_{12}-derivative, penicillin V β-bromoethyl ester can be deprotected to penicillin V under such mild conditions that the β-lactam ring is intact [196].

IX. OXYGEN-OXYGEN BOND

Compounds containing an oxygen-oxygen single bond are generally [197] polarographically reducible. Thus, hydroperoxides [198–200], peracids [201,202], acylperacids [203], ozonoides, and aldehyde and ketone peroxides [204] can be determined polarographically. The electrode reaction, which is a cleavage of the O—O bond, has not been widely used preparatively, although the half-wave potentials of the various types of peroxides are sufficiently different to permit a selective reduction of one type in the presence of others. The presence of many alkyl groups shifts the half-wave potential to more negative values because of the electron-donating power of these groups and their shielding of the reaction center. In methanol-benzene containing LiCl, the half-wave potential [203] of di-tert-butylperoxide is higher than 2 V; dialkylperoxides, higher than 1 V, tert-butylperoxy esters, 0.8–

1.0V; hydroperoxides, 0.6–0.9 V, diacylperoxides, ~0.1 V, and peracids, ~0.05 V. The half-wave potential of an organic peroxide is thus valuable in the elucidation of its structure.

X. SULFUR-SULFUR BOND

Disulfides

A sulfur-sulfur single bond may be cleaved either anodically or cathodically. A sulfur-sulfur bond is generally reducible and such compounds as di-, tri-, and tetrasulfides [205] and thiolsulfonates [206] are polarographically active.

Disulfides are easily reducible to mercaptans; a high yield of thioglycolic acid [207] from dithiodiglycolic aicd was obtained by reduction in 2 N sulfuric acid at a lead cathode at −0.55 V (SCE). This reduction is one of the few examples of a controlled potential reduction prior to the invention of potentiostats.

The rate constant of the heterogeneous electron transfer from a platinum electrode to aromatic disulfides in DMF is low, whereas the homogeneous electron transfer from suitable anion radicals is much faster; these can be generated at potentials more negative than E^0 of the disulfides but less negative than their peak potentials [208].

Reactions of disulfides in the presence of alkylating or acylating agents give S-alkylated or S-acylated products in good yield [209]. If the reduction is done in the presence of oxygen, sulfinic acids may be formed [210].

Oxidation of diphenyldisulfide in a mixture of glacial acetic acid and concentrated hydrochloric acid at a platinum anode yielded benzenesulfonic acid; ethyl phenyl sulfide yields benzenesulfonic acid [211].

Oxidative cleavage of disulfides can be made by chemical and electrochemical means; thus anodic oxidation at platinum or carbon electrodes of different kinds of disulfides in the presence of alkenes in acetonitrile affords products of acetamidosulfenylation; with cyclic alkenes a high selectivity for trans addition is observed. The reaction may be formulated, Eq. (61):

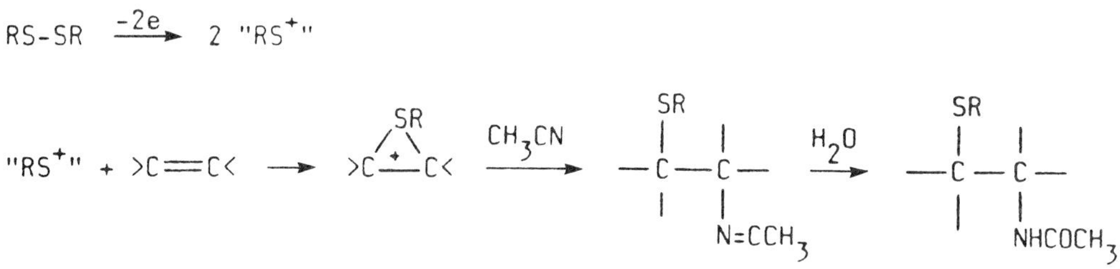

(61)

With terminal alkenes the terminal sulfides are formed with high regioselectivity [212].

This reaction is employed in the ring closure, Eq. (62) [213], where XLV is anodically oxidized in methanol. It might be surprising that a five-membered ring is formed in preference to a six-membered ring; an episulphonium ion is probably formed as an intermediate.

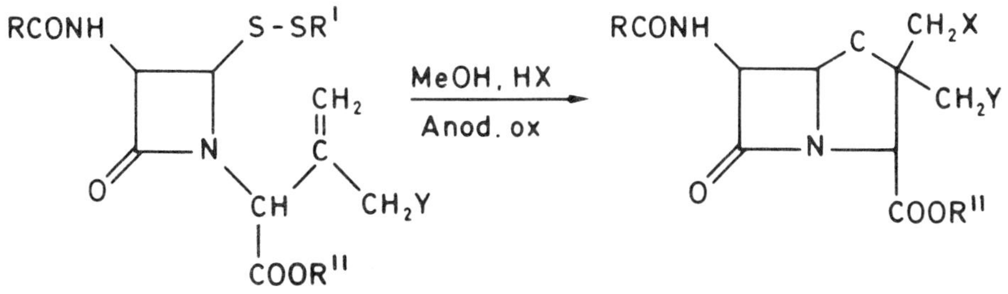

XLV (62)

XI. NITROGEN-NITROGEN BOND

Reductive splitting of a bond between a nitrogen atom and heteroatom occurs in many compounds of the type X=N−Y, where X may be carbon, oxygen, or nitrogen and Y may be oxygen or nitrogen [214]. This rule thus applies to cyclic and acyclic hydrazines, oximes, p-amino and p-hydroxyazobenzenes, triazines in acid solution, and N-nitrosamines in alkaline solution.

A. Hydrazones

Phenylhydrazones may be reduced to aniline and an amine in a four-electron reduction or to a hydrazine in a two-electron reaction, depending on pH. These results have been explained in Chap. 11.

A reduction route similar to that of phenylhydrazones [215] seems to be rather general for azomethine derivatives of hydrazine [216] as it is followed by cyclic and acyclic phenylhydrazones, semicarbazones, azines, cyclic hydrazones, and acylated cyclic and acyclic hydrazones [217] under protic conditions; in DMF, acylated hydrazones of aromatic aldehydes are reduced with saturation of the C=N bond and cleavage of the N−N bond at a more negative potential [218]. The suggestion that the cleavage of the N−N bond precedes the saturation of the azomethine bond is also an essential part of the interpretation of many of the ring contractions of heterocyclic compounds (Chap. 18).

The preprotonation step is indicated from the polarographic data and from the fact that a quaternized hydrazone like benzaldehyde trimethylhydrazonium iodide, $C_6H_5CH{=}\overset{+}{N}N(CH_3)_3I^-$, is reducible in a four-electron reduction both in alkaline and acid solutions, whereas hydrazones are generally either nonreducible or reducible in a two-electron reduction in alkaline solution [214–216].

The cleavage of an N−N bond in p-hydroxyazobenzenes [219–222] may be regarded as a variant of the general scheme if the p-hydroxyazobenzene is considered reduced in its tautomeric form, the quinone monophenylhydrazone.

Electrochemical oxidation of substituted hydrazones of benzaldehyde may result in cleavage of the nitrogen-nitrogen bond and formation of a benzonitrile [223] and oxidation products from the amine.

B. Hydrazides and Azides

Certain activated hydrazides [224], such as isonicotinic hydrazide, are reduced to the amides, Eq. (63):

$$\overset{+}{\text{H}}\text{NC}_5\text{H}_4\text{CONHNH}_2 \xrightarrow{2e^- + 3H^+} \overset{+}{\text{H}}\text{NC}_5\text{H}_4\text{CONH}_2 + \text{NH}_4^+ \quad (63)$$

The reduction of isonicotinoyl azide also produces isonicotinic amide, and most other azides are similarly reduced according to Eq. (64) [225–227]:

$$\text{R}-\overline{\text{N}}-\overset{+}{\text{N}}\equiv\text{N} \xrightarrow{2e^- + 2H^+} \text{RNH}_2 + \text{N}_2 \quad (64)$$

This reaction has been employed for the synthesis of saturated and unsaturated aminoacids by cathodic reduction of azidocinnamic esters [226]. Reduction of azides in DMF in the presence of acetic anhydride yields diacetylated amines [227]. Azide groups on activated methylene groups may, however, be reduced to azide ion and methyl compound [225].

C. Triazines

Triazines are reduced to amines and hydrazines. It has not been proved for acyclic trizines whether the cleavage of the N—N bond precedes the saturation of the N=N bond, but in benzotriazoles [228] and benzotriazines (Chap. 18) [229], the first detectable intermediate contains the N=N bond even after cleavage of the N—N bond.

D. N-Nitrosamines

As discussed in Chap. 9, N-nitrosamines, O=N—NRR', are reduced in alkaline solution with cleavage of the N—N bond according to Eq. (65) [230]:

$$\text{O}=\text{N}-\text{NRR'} \xrightarrow{2e^- + 2H^+} \tfrac{1}{2}\text{N}_2\text{O} + \text{HNRR'} \quad (65)$$

XII. OXYGEN-SULFUR BOND

Aliphatic sulfoxides are not cathodically reducible, but some give a catalytic hydrogen wave in acid solution [231,232]; at a lead cathode in alcoholic sulfuric acid diphenylsulfoxide [233] and dibenzylsulfoxide [234] have been reduced in fair yield to the sulfide. Methylphenylsulfoxide has been reduced in DMF or MeCN with phenol as proton donor at -2.3 V (SCE) to methylphenylsulfide [235].

Methyl benzenesulfenate has been reported [236] to be cleaved to thiophenol and methanol at −0.5 V (SCE) in aqueous buffer at pH 7; methyl 2-nitrobenzenesulfenate in DMF is reductively cleaved similarly [90].

Esters of arylsulfonic acids are generally reducible, and this has been employed for protection [36] of hydroxyl groups; the cleavage of tosylates [140,237–242] and nosylates (esters of 4-nitrobenzenesulfonic acid) [243] has especially been studied.

The arylsulfonate esters, Ar–SO_2–OR, are reducible at rather negative potentials; the primarily formed anion radical can in principle cleave between Ar and SO_2, between SO_2 and OR, and between O and R. An O–R cleavage is unlikely in view of the observation that tosyl esters with optically active alcohols, in which the tosyl ester is at the chiral center, are electrochemically cleaved with retention of the configuration [238].

According to a simple MO theory it has been argued [244] that the Ar–SO_2 cleavage should predominate in protic media, whereas in aprotic media the anion radical would be reduced further to the dianion, which then decomposed. In general, the cleavage SO_2–O is predominant, Eq. (66), but some ArH is also observed [241,242]. It seems, however, that the cleavage of the C–S bond to give ArH and the cleavage of the S–O bond to give alcohol are not simple competitive reactions.

$$\text{Ar–SO}_2\text{–OR} + 2e \longrightarrow \text{ArSO}_2^- + \text{RO}^- \xrightarrow{H^+} \text{ROH} \tag{66}$$

It was found that the cleavage of the primarily formed anion radical was the rate-determining step in an indirect reduction of tosylates in DMF when R was aliphatic, whereas with R aromatic the cleavage was so fast that the homogeneous electron transfer from the reduced catalyst became rate determining [242].

In the absence of protons the deprotection is complicated by nucleophilic attack of RO^- on the tosylate ester with formation of the ether, ROR, Eq. (67). This side reaction can be suppressed on addition of a proton donor, such as acetic acid. In that case aromatic anion radicals cannot be used as catalyst and Ni(acacen) was used. The difference in yield of deprotected alcohol from direct and indirect reduction was insignificant [242].

$$\text{Ar–SO}_2\text{–OR} + \text{RO}^- \longrightarrow \text{Ar–SO}_3^- + \text{ROR} \tag{67}$$

Nosylates are reduced in two steps; the first reduction gives a rather stable anion radical at a potential characteristic of substituted nitrobenzenes [243]. In the second step (−1.2 − −1.7 V versus SCE) a dianion is formed that cleaves rapidly at the S–O bond to give ROH and the arylsulfinate ion.

XIII. OXYGEN-NITROGEN BOND

The reduction of derivatives of hydroxylamine involves splitting of the N–O bond. It has been shown [245] that the cleavage of the N–O bond in oximes generally occurs prior to the saturation of the double bond. In alkaline solution at low temperature, however, some reduction of an aldoxime to a hydroxylamine takes place. A similar primary cleavage of an N–O bond is found in the reduction of amide oximes [245] and hydroxamic acids [246]. The reduction of oximes in acid solution has been formulated as Eq. (68):

$$\text{RR'C=NOH} \overset{H^+}{\rightleftarrows} \text{RR'C=}\overset{+}{\text{N}}\text{H–OH} \xrightarrow{2e\ +\ 2H^+} \text{RR'C=}\overset{+}{\text{N}}\text{H}_2 + \text{H}_2\text{O} \xrightarrow{2e^-\ +\ 2H^+} \text{RR'CH}\overset{+}{\text{N}}\text{H}_3 \tag{68}$$

Whereas oximes are generally reduced in a single polarographic four-electron wave, a few are reduced in two steps [247], and in such reactions the intermediate imine has been isolated (see Chap. 18).

The hypothesis that oximes with a positive charge on nitrogen are reductively cleaved at N–O is supported by the observation that after quaternization, isoxazoles or 4,5-dihydroisoxazoles can be cleaved in neutral solution without reduction of the C=N bond [248]. As isoxazolines can be made conveniently from alkenes and oximes (through nitrile oxides), this is a convenient way of preparing β-hydroxyketones, Eq. (69); the use of CPE makes it possible to tolerate substituents that are attacked by other reducing agents.

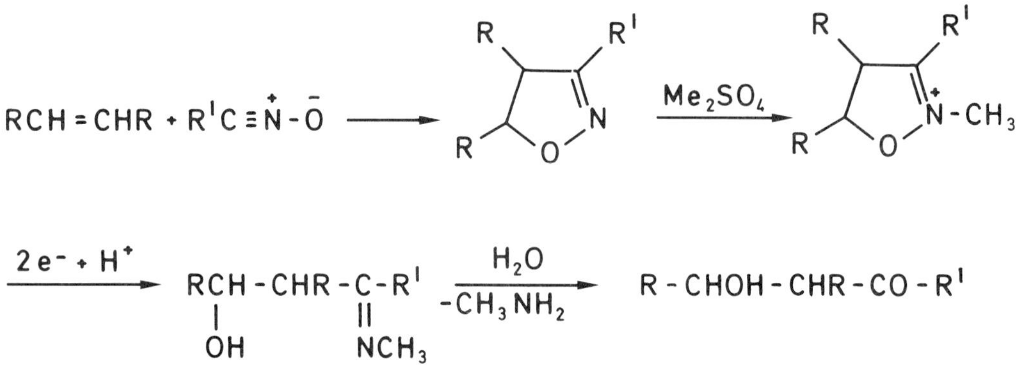

(69)

Nitrile-N-oxides are reducible at rather positive potentials to the nitriles; thus 2,4,6-trimethylbenzonitrile-N-oxide is reduced at −0.7 V at pH 6 in aqueous ethanol to the corresponding nitrile [249].

XIV. SULFUR-NITROGEN BOND

The reduction of sulfonamides has been extensively studied in connection with the protection of amino groups [237–248,250–255]; the cleavage of N-tosyl groups probably follows Eq. (70):

$$RR'NTs + e \longrightarrow RR'NTs^{\cdot -} \longrightarrow RR'N^{\cdot} + Ts^-$$

$$RR'N^{\cdot} + e \longrightarrow RR'N^- \xrightarrow{H^+} RR'NH \qquad (70)$$

In the presence of a suitable proton donor, n = 2 and Eq. (70) is followed; the yield of deprotected amine is usually good to excellent. The reduction potential at mercury or glassy carbon is rather negative (in DMF typically −2.3 to −2.4 V versus SCE), and this must be taken into consideration if other reducible groups are present in the protected molecule.

In aprotic media complications of a different kind may arise. If the protected amine is primary, the resulting sulfonamide is a fairly good proton donor that can protonate the resulting substituent amide ion, and n = 1 is found; on workup 50% of the sulfonamide is recovered, Eq. (71) [252]:

$$2RNH{-}Ts + 2e \longrightarrow RNH_2 + Ts^- + (RNTs)^- \qquad (71)$$

When no hydrogen atoms are bonded to nitrogen the electrogenerated base (EGB), the amide ion, may still cause a side reaction, if one of the substituents on the nitrogen has an α-hydrogen. The EGB may induce a β-elimination with formation of an aldimine or ketimine, Eq. (72); the imine may be reducible at the potential used for cleavage of the sulfonamide [253].

$$RR'CH{-}NR''{-}Ts \xrightarrow[-Ts^-]{2e^-} (RR'CHNR'')^- \xrightarrow{RR'CH{-}NR''{-}Ts} RR'CHNHR'' + RR'C{=}NR'' \qquad (72)$$

The reduction of sulfonamides has been investigated in methanol [237–239], DMF [240,242], and MeCN [241]; the reductive cleavage of N-tosyl amino acids and peptides has been studied in methanol at pH 11 at different electrodes; lead was the preferred material for deprotection [250,251].

Besides being used for the deprotection of amino acids and peptides, electrochemical cleavage has been employed in the synthesis of polyaza and polyaza-polyoxa ligands [254]; the deprotection may be performed by direct as well as by indirect electrolysis.

Indirect electrolysis using anion radicals of aromatic hydrocarbons or Ni(acacen) as electron carrier has made it possible to determine the E^0 of some tosylamides in DMF and also the rate of cleavage of the sulfonamide anion radicals [242]. The results were similar to those for the tosylates; when the amine was aliphatic, the cleavage of the anion radical was the rate-determining step, whereas the homogeneous electron transfer from the electron donor became rate determining when the amine was aromatic.

A useful system for indirect electrolysis of sulfonamides was reported to be anthracene as mediator in the presence of ascorbic acid as proton donor [255].

XV. SULFUR-HALOGEN BOND

Both sulfenyl [256,257] and sulfonyl chlorides [258–265] are electrochemically reducible with cleavage of the sulfur-halogen bond.

Sulfenyl chlorides are reported [256] to give a two-electron polarograhic wave; preparative reduction in acetonitrile at a platinum cathode shows an overall one-electron reduction to disulfide, Eq. (73) [256]:

$$2PhSCl + 2e^- \longrightarrow PhSSPh + 2Cl^- \qquad (73)$$

Chloride ion, as well as DMF, catalyzes the reaction of phenylsulfenyl chloride with water to a phenylthiosulfinate, which spontaneously undergoes decomposition to the corresponding disulfide and thiosulfonate, Eq. (74):

$$2PhSCl + H_2O \xrightarrow{Cl^-} PhSOSPh \longrightarrow \tfrac{1}{2}\,(PhSSPh + PhSO_2SPh) \qquad (74)$$

The reduction of sulfonyl chlorides may lead to sulfinates [261–265], thiosulfonates [258,259,265], disulfides, and mercaptans depending on the conditions; in the presence of alkylating agents [263,264], sulfones or sulfides are among the products.

The sulfinate is the immediate product in the absence of a strong proton donor; under sufficiently acidic conditions reduction to mercaptans may take place. The thiosulfonate may be formed on thioacylation of the mercaptan or dismutation of the sulfinic acid; the thiosulfonate is reducible under the conditions employed, so the final product is the thiol.

XVI. SULFUR-PHOSPHORUS BOND

Electrochemical regeneration of triphenylphosphine from triphenylphosphine oxides may be performed by transforming the oxide to the sulfide with P_2S_5, methylating it into the phosphonium compound, and cleaving it electrochemically, Eq. (75) [266]:

$$Ph_3PO \longrightarrow Ph_3PS \longrightarrow Ph_3\overset{+}{P}SCH_3 \xrightarrow{2e^-} Ph_3P + CH_3S^- \qquad (75)$$

REFERENCES

1. D. A. White and J. H. Wagenknecht, J. Electrochem. Soc., 128, 1470 (1981).
2. O. Hammerich and J. M. Savéant, J. Chem. Soc., Chem. Commun., 938 (1979).
3. J. Dalaunay, A. Orliac-Le Moing, and J. Simonet, Electrochim. Acta, 30, 1109 (1985).
4. M.-A. Michel, G. Mousset, J. Simonet, and H. Lund, Electrochim. Acta, 20, 143 (1975).
5. H. Lund, J. Mol. Catal., 38, 203 (1986).
6. K. Boujlel and J. Simonet, Tetrahedron Lett., 26, 3005 (1985).
7. H. Lund, Acta Chem. Scand., 17, 2325 (1963).
8. (a) J. Volke and J. Holubek, Coll. Czech. Chem. Commun., 28, 1597 (1963); (b) J. Volke, R. Kubicek, and F. Santavý, Coll. Czech. Chem. Commun., 25, 1510 (1960); (c) J. Volke and A. M. Kardos, Coll. Czech. Chem. Commun., 33, 2560 (1968).
9. A. M. Kardos, P. Valenta, and J. Volke, J. Electroanal. Chem., 12, 84 (1966).
10. (a) O. Manousek, P. Zuman, and O. Exner, Coll. Czech. Chem. Commun., 33, 3979 (1968); P. Zuman and O. Manousek, Coll. Czech. Chem. Commun., 34, 1580 (1969); (b) A. Gennaro, F. Maran, and E. Vianello, J. Electroanal. Chem., 185, 353 (1985).
11. R. C. Buchta and D. H. Evans, Anal. Chem., 40, 2181 (1968).
12. L. Kistenbrügger, C. P. Klages, and J. Voss, J. Chem. Res. S, 320 (1979).
13. M. Klehr and H. J. Schäfer, Angew. Chem., 87, 173 (1975).
14. S. Torii, T. Okamoto, and N. Ueno, J. Chem. Soc., Chem. Commun., 293 (1978).
15. Y. Matsubara, T. Uchida, T. Ohnishi, K. Kanehira, Y. Fujita, T. Hirashima, and I. Nishiguchi, Tetrahedron Lett., 26, 4513 (1985).
16. S. Torii, T. Inokuchi, and N. Takahasi, J. Org. Chem., 43, 5020 (1978); S. Torii, T. Inokuchi, and Y. Araki, Synth. Commun., 17, 1797 (1987).

17. A. Mizuno, K. Takagi, N. Shirai, Y. Kuniwaki, and K. Togami, Kenkyo Kiyo-Wakayama Kogyo Koto Senmon Gakko, 19, 48 (1984); Chem. Abstr., 102, 166355 (1985).
18. F. Delay, U.S. Patent 4,448,650 (1984); Chem. Abstr., 101, 90463 (1984); F. Delay, L. Limacher, N. Bédert, and P. Tissot, Journées Electrochimie, Dijon 1987, Abstracts 4–19.
19. T. Shono, Y. Matsumura, T. Hasimoto, K. Hibino, H. Hamaguchi, and T. Aoki, J. Amer. Chem. Soc., 97, 2546 (1975).
20. A. Yoshiyama, T. Nonaka, M. M. Baizer, and T.-C. Chou, Bull. Chem. Soc. Jpn., 58, 201 (1985).
21. C. W. Nam, J. H. Sohn, and S. C. Park, Hwahak Konghak, 22, 245 (1984); Chem. Abstr., 102, 193763 (1985).
22. S. Torii, I. Inokuchi, S. Takagishi, E. Sato, and H. Tsujiyama, Chem. Lett., 1469 (1987).
23. T. Shono, H. Hamaguchi, and Y. Matsumura, and K. Yoshida, Tetrahedron Lett., 3625 (1977); T. Shono, I. Nishiguchi, and M. Nitta, Chem. Lett., 1319 (1976).
24. J. H. Wagenknecht, J. Electrochem. Soc., 126, 983 (1979).
25. K. Uneyama, Y. Masatsugu, and S. Torii, Bull. Chem. Soc. Jpn., 58, 2361 (1985).
26. S. Ozaki, S. Nishiguchi, and M. Masui, Chem. Pharm. Bull., 32, 2609 (1984).
27. H. D. Law, J. Chem. Soc., 1437 (1906).
28. T. Shono, Y. Matsumura, H. Hamaguchi, T. Imanishi, and K. Yoshida, Bull. Chem. Soc. Jpn., 51, 2179 (1978).
29. B. Belleau and N. L. Weinberg, J. Amer. Chem. Soc., 85, 2525 (1963).
30. E. J. Corey, R. R. Sauers, and S. Swann, J. Amer. Chem. Soc., 79, 5826 (1957); E. J. Corey and R. R. Sauers, J. Amer. Chem. Soc., 89, 1732 (1959).
31. R. Gompper and A. Schmidt, Angew. Chem., 92, 468 (1980).
32. P. G. M. Wuts and M.-C. Cheny, J. Org. Chem., 51, 2844 (1986).
33. M. Oyama and M. Ohno, Tetrahedron Lett., 5201 (1966).
34. J. Y. Becker, L. L. Miller, and T. M. Siegel, J. Amer. Chem. Soc., 97, 849 (1975).
35. R. Mitzner, J. Bendig, R. Ziebig, F. Graichen, D. Kreysig, and F. Pragst, J. Prakt. Chem., 327, 241 (1985).
36. V. G. Mairanovskii, Angew. Chem., 88, 283 (1976).
37. M. I. Montenegro, Electrochim. Acta, 31, 607 (1986).
38. T. Shono, Y. Matsumura, K. Tsubata, and Y. Sugihara, Tetrahedron Lett., 2157 (1979).
39. T. Shono, Y. Matsumura, K. Tsubata, and Y. Sugihara, J. Org. Chem., 44, 4508 (1979).
40. T. A. Thornton, N. F. Woolsey, and D. E. Bartak, J. Amer. Chem. Soc., 108, 6497 (1986).
41. M. D. Koppang, N. F. Woolsey, and D. E. Bartak, J. Amer. Chem. Soc., 107, 4692 (1985).
42. (a) V. G. Mairanovskii and G. I. Samokhvalov, Elektrokhimiya, 2, 717 (1966); (b) V. G. Mairanovskii, L. A. Vakulova, and G. I. Samokhvalov, Electrokhimiya, 3, 23, 611 (1967).
43. H. Lund, M.-A. Michel, G. Mousset, and J. Simonet, Electrochim. Acta, 19, 629 (1974).
44. G. Horanyi, Electrochim. Acta, 31, 1095 (1986) and references cited therein.

45. G. Horanyi, G. Inzelt, and K. Torkos, J. Electroanal. Chem., 101, 101 (1979).
46. T. Lund and H. Lund, Acta Chem. Scand., B38, 387 (1984); B39, 429 (1985).
47. S. Torii, H. Tanaka, T. Katoh, and K. Morisaka, Tetrahedron Lett., 25, 3207 (1984).
48. V. G. Mairanovskii, L. N. Fokina, L. A. Vakulova, and G. I. Samokhvalov, Zh. Obshch. Khim., 36, 1345 (1966).
49. (a) J. Coleman, H. G. Gilde, J. H. P. Utley, and B. C. L. Weedon, J. Chem. Soc., Chem. Commun., 738 (1970); (b) J. Coleman, Naser-ud-din, H. G. Gilde, J. H. P. Utley, B. C. L. Weedon, and L. Eberson, J. Chem. Soc., Perkin Trans. II, 1903 (1973); (c) R. F. Garwood, Naser-ud-din, and B. C. L. Weedon, J. Chem. Soc., Perkin Trans. I, 2471 (1975).
50. N. Berenjian and J. H. P. Utley, J. Chem. Soc., Chem. Commun., 550 (1979).
51. C. van der Stouwe and H. J. Schäfer, Tetrahedron Lett., 2643 (1979); Chem. Ber., 114, 946 (1981).
52. (a) V. G. Mairanovskii, A. Ya. Veinberg, and G. I. Samokhvalov, Zh. Obshch. Khim., 38, 666 (1968); (b) V. G. Mairanovskii, N. F. Loginova, and S. J. Melnik, Elektrokhimiya, 9, 1174 (1973).
53. O. Manousek and P. Zuman, Coll. Czech. Chem. Commun., 29, 1432 (1964); R. Camble, R. Garner, and G. T. Young, J. Chem. Soc. C, 1911 (1969).
54. M. Fedoronko, Chem. Anal., 3, 573 (1958).
55. P. Kabasakalian and J. McGlotten, Anal. Chem., 31, 1091 (1959).
56. H. Lund, Acta Chem. Scand., 14, 1927 (1960).
57. P. O. Kane, Z. Anal. Chem., 173, 50 (1960).
58. J. Engels, Angew. Chem., 91, 155 (1979); Liebigs Ann. Chem., 557 (1980).
59. R. F. Erickson and C. M. Fischer, J. Org. Chem., 35, 1604 (1970).
60. H. Lund, Acta Chem. Scand., 14, 359 (1960).
61. G. M. Rao, J. W. Lown, and J. A. Plambeck, J. Electrochem. Soc., 125, 534 (1978).
62. A. Gref, G. Balavoine, H. Riviere, and C. P. Andrieux, Nouv. J. Chim., 8, 615 (1984).
63. K. Kano, T. Konse, K. Hasegawa, B. Uno, and T. Kubota, J. Electroanal. Chem., 225, 187 (1987).
64. V. Malatesta, S. Penco, N. Sacchi, L. Valentini, A. Vigevani, and F. Arcamone, Can. J. Chem., 62, 2845 (1984).
65. D. Tresselt, W. Ihn, G. Horn, and H. Berg, Pharmazie, 39, 417 (1984).
66. D. W. Sopher and J. H. P. Utley, J. Chem. Soc., Chem. Commun., 134 (1981); K. G. Ellis, Nazar-ul-Islam, D. W. Sopher, J. H. P. Utley, H. L. Chum, and M. Ratcliff, J. Electrochem. Soc., 134, 3058 (1987).
67. Nazar-ul-Islam, D. W. Sopher, and J. H. P. Utley, Tetrahedron, 43, 959, 2741 (1987).
68. T. Nonaka and M. M. Baizer, Electrochim. Acta, 28, 661 (1983).
69. K. Boujlel and J. Simonet, Electrochim. Acta, 24, 481 (1979).
70. A. V. Kamernitskii, I. G. Reshetora, E. I. Chernoburova, S. G. Mairanovskii, and N. K. Lisitsina, Izv. Akad. Nauk SSSR, Ser. Khim., 1901 (1984); Chem. Abstr., 102, 62512 (1985).

71. M. F. Semmelhack and G. E. Heinsohn, J. Amer. Chem. Soc., 94, 5139 (1972).
72. T. Shono, Y. Matsumura, S. Kashimura, and H. Kyntoku, Tetrahedron Lett., 1205, 2807 (1978).
73. T. Shono, Y. Matsumura, and S. Kashimura, Tetrahedron Lett., 1545 (1980).
74. P. Martigny and J. Simonet, J. Electroanal. Chem., 81, 407 (1977).
75. T. Shono, Y. Matsumura, and S. Kashimura, Chem. Lett., 69 (1978).
76. A. Petsom and H. Lund, Acta Chem. Scand., B34, 614 (1980).
77. S. Gambino, P. Martigny, G. Mousset, and J. Simonet, J. Electroanal. Chem., 90, 105 (1978).
78. (a) P. Martigny, M.-A. Michel, and J. Simonet, J. Electroanal. Chem., 73, 373 (1976); (b) C. Adams, N. M. Kamkar, and J. H. P. Utley, J. Chem. Soc., Perkin Trans. II, 1767 (1979).
79. M. Falsig and H. Lund, Acta Chem. Scand., B34, 585 (1980).
80. M. Le Guyader, Bull. Soc. Chim. Fr., 1858 (1966); A. Tallec, Ann. Chim., 3, 347 (1968).
81. F. Fichter and W. Dietrich, Helv. Chim. Acta, 7, 131 (1924).
82. V. D. Parker, J. Chem. Soc., Chem. Commun., 610 (1969).
83. L. I. Smith, I. M. Kolthoff, S. Wawzonek, and P. M. Rouff, J. Amer. Chem. Soc., 63, 1018 (1941).
84. V. D. Parker, J. Amer. Chem. Soc., 91, 5380 (1969).
85. H. Lund, Acta Chem. Scand., 11, 990 (1957).
86. E. Mageda, L. L. Miller, and J. Wolf, J. Amer. Chem. Soc., 94, 6812 (1972).
87. S. M. Weinreb, G. A. Epling, R. Comi, and M. Reitano, J. Org. Chem., 40, 1356 (1975).
88. W. Schmidt and E. Steckhan, Angew. Chem., 90, 717 (1978).
89. W. Schmidt and E. Steckhan, Angew. Chem., 91, 850, 851 (1979).
90. Z. V. Todres, Electrokhimiya, 24, 563 (1988).
91. M. Masui, T. Kawaguchi and S. Ozaki, J. Chem. Soc., Chem. Commun., 1484 (1985).
92. J. Grimshaw, in The Chemistry of the Sulfonium Group (C. J. M. Stirling, ed.), Wiley, New York, 1981, p. 141.
93. J. Q. Chambers, in Encyclopedia of Electrochemistry of the Elements (A. J. Bard and H. Lund, eds.), Dekker, New York, 1978, p. 329.
94. E. L. Colichman and D. L. Love, J. Org. Chem., 18, 40 (1953).
95. A. Lüttringhaus and H. Machatzke, Liebigs Ann. Chem., 671, 165 (1964).
96. F. D. Saeva and B. P. Morgan, J. Amer. Chem. Soc., 106, 4121 (1984).
97. P. Beak and T. A. Sullivan, J. Amer. Chem. Soc., 104, 4450 (1982).
98. O. Spohn, Thesis, University of Mainz, 1967.
99. ICI Ltd., Fr. Patent 1565 481 (1969); Chem. Abstr., 72, 74129 (1970).
100. M. M. Baizer, J. Org. Chem., 31, 3847 (1966).
101. T. Shono and M. Mitani, Tetrahedron Lett., 687 (1969).
102. T. Shono, T. Akazawa, and M. Mitani, Tetrahedron, 29, 817 (1973).
103. J. H. Wagenknecht and M. M. Baizer, J. Electrochem. Soc., 114, 1095 (1967).
104. R. A. Wessling and W. J. Settineri, U.S. Patent 3480527 (1969); Chem. Abstr., 72, 66593 (1970).
105. W. A. Bonner, J. Amer. Chem. Soc., 74, 5078 (1952).
106. P. S. McKinney and S. Rosenthal, J. Electroanal. Chem., 16, 261 (1968).

107. M. Finkelstein, R. C. Petersen, and S. D. Ross, J. Electrochem. Soc., 110, 422 (1963).
108. H. J. Bär, Z. Phys. Chem. (Leipzig), 243, 398 (1970).
109. R. Gerdil, J. Chem. Soc. B, 1971 (1966).
110. H. Lund, Acta Chem. Scand., 14, 1927 (1960).
111. T. Fuchigami, Z. E.-S. Kandeel, and T. Nonaka, Bull. Chem. Soc. Jpn., 59, 338 (1986).
112. D. A. Ives, Can. J. Chem., 47, 3697 (1969).
113. T. Iwasaki, K. Matsumoto, M. Matsuoka, and M. Mioshi, Bull. Inst. Chem. Res., Kyoto Univ., 50, 220 (1972); Chem. Abstr., 77, 165065 (1972).
114. N. F. Loginova and V. G. Mairanovskii, Zh. Obshch. Khim., 44, 1838 (1974).
115. V. G. Mairanovskii and N. F. Loginova, Zh. Obshch. Khim., 41, 2581 (1972); 43, 207 (1973).
116. A. Gosden, D. Stevenson, and G. T. Young, J. Chem. Soc., Chem. Commun., 1123 (1972).
117. A. Gosden, R. Macrae, and G. T. Young, J. Chem. Res., 5, 22 (1977).
118. B. Lamm and B. Samuelsson, Acta Chem. Scand., 23, 691 (1969).
119. A. Kunugi, T. Hagi, and T. Hirai, Electrochim. Acta, 30, 1049 (1985).
120. A. Kunugi, K. Abe, T. Hagi, and T. Hirai, Bull. Chem. Soc. Jpn., 59, 2009 (1986); A. Kunugi, T. Ikeda, T. Hirai, and K. Abe, Electrochim. Acta, 33, 905 (1988); A. Kunugi, T. Norimasa, K. Abe, and T. Hirai, Bull. Chem. Soc. Jpn., 62, 2055 (1989).
121. H. H. Rüttinger, W. D. Rudorf, and H. Matschiner, Electrochim. Acta, 30, 155 (1985).
122. K. Uneyama and S. Torii, Tetrahedron Lett., 329 (1971).
123. T. Mandai, H. Yasunaga, M. Kawada, and J. Otera, Chem. Lett., 715 (1984).
124. M. Kimura, S. Matsubara, Y. Sawaki and H. Iwamura, Tetrahedron Lett., 27, 4177 (1986).
125. J.-G. Gourcy, G. Jeminet, and J. Simonet, J. Chem. Soc., Chem. Commun., 634 (1974).
126. J.-G. Gourcy, P. Martinet, J. Simonet, and G. Jeminet, Tetrahedron, 37, 1495 (1986).
127. Q. N. Porter and J. H. P. Utley, J. Chem. Soc., Chem. Commun., 255 (1978); J. Chem. Soc., Perkin Trans. I, 973 (1984).
128. R. S. Glass, A. Petsom, G. S. Wilson, R. Martinez, and E. Juaristi, J. Org. Chem., 51, 4337 (1986).
129. A. Lebouc, J. Simonet, J. Gelas, and A. Dehbi, Synthesis, 320 (1987).
130. M. Platen and E. Steckhan, Tetrahedron Lett., 21, 511 (1980).
131. M. Platen and E. Steckhan, Chem. Ber., 117, 1679 (1984).
132. L. Horner and R.-J. Singer, Chem. Ber., 101, 3329 (1968).
133. E. Steckhan, Angew. Chem., 98, 681 (1986).
134. K. Praefcke, C. Weichsel, M. Falsig, and H. Lund, Acta Chem. Scand., B34, 403 (1980).
135. M. Kimura, S. Matsubara, and Y. Sawaki, J. Chem. Soc., Chem. Commun., 1619 (1984).
136. M. Falsig, H. Lund, L. Nadjo, and J.-M. Savéant, Acta Chem. Scand., B34, 685 (1980).
137. M. Falsig, H. Lund, L. Nadjo, and J.-M. Savéant, Nouv. J. Chim., 4, 445 (1980).
138. M. Falsig and H. Lund, Acta Chem. Scand., B34, 545 (1980).

139. K. Schwabe and J. Voigt, Z. Elektrochem., 56, 44 (1952).
140. J.-G. Gourcy, G. Jeminet and J. Simonet, Bull. Soc. Chim. Fr., 2982 (1972); J.-Y. Pape and J. Simonet, Electrochim. Acta, 23, 445 (1978).
141. S. G. Mairanovskii and M. B. Neiman, Dokl. Akad. Nauk SSSR, 87, 805 (1952); Chem. Abstr., 47, 6271f (1953).
142. E. S. Levin and A. P. Shestov, Dokl. Akad. Nauk SSSR, 96, 999 (1954); Chem. Abstr., 48, 13471e (1954).
143. L. Horner and H. Nickel, Chem. Ber., 89, 1681 (1956).
144. L. Horner and H. Neumann, Chem. Ber., 98, 1715 (1965).
145. L. Horner and E. Meyer, Liebigs Ann. Chem., 2053 (1975).
146. L. Horner and E. Meyer, Ber. Bunsenges. Phys. Chem., 79, 136, 143 (1975).
147. J. Simonet, in The Chemistry of Sulphones and Sulphoxides (S. Patai, Z. Rappoport, and C. Stirling, eds.), Wiley, New York, 1988, p. 1001.
148. J. Simonet and G. Jeminet, Bull. Soc. Chim. Fr., 2754 (1971).
149. N. Djeghidjegh and J. Simonet, Electrochim. Acta, 34, 1615 (1989).
150. B. Lamm and K. Ankner, Acta Chem. Scand., B31, 375 (1977); K. Ankner, B. Lamm, and J. Simonet, Acta Chem. Scand., B31, 742 (1977).
151. (a) O. Manousek, O. Exner, and P. Zuman, Coll. Czech. Chem. Commun., 33, 3988 (1968); (b) O. Manousek, O. Exner, and P. Zuman, Coll. Czech. Chem. Commun., 4000 (1968); (c) G. Jeminet and J. Simonet, C. R. Acad. Sci., Ser. C, 272, 661 (1971).
152. J. Simonet and H. Lund, Acta Chem. Scand., B31, 909 (1977).
153. B. Lamm and B. Samuelsson, Acta Chem. Scand., 24, 561 (1970); J. Chem. Soc. D, 1010 (1970).
154. B. Emmert, Ber. Dtsch. Chem. Ges., 42, 1507 (1909); 45, 430 (1912).
155. L. Horner and A. Mentrup, Liebigs Ann. Chem., 646, 49 (1961).
156. L. Horner and H. Röder, Chem. Ber., 101, 4179 (1968).
157. H. Krauch and W. Kunz, Reaktionen der Organischen Chemie, 3rd ed., Dr. A. Hüthig-Verlag, Heidelberg, 1966, p. 203, for analogous Emde Na-Hg method.
158. (a) J. T. Wrobel, K. M. Pazdro, and Z. Bien, Chem. Ind., 1760 (1966); (b) J. T. Wrobel and A. R. Krawczyk, Chem. Ind., 656 (1969).
159. (a) M. Finkelstein, R. C. Petersen, and S. D. Ross, J. Amer. Chem. Soc., 81, 2361 (1959); (b) S. D. Ross, M. Finkelstein, and R. C. Petersen, J. Amer. Chem. Soc., 82, 1582 (1960); (c) S. D. Ross, M. Finkelstein, and R. C. Petersen, J. Amer. Chem. Soc., 92, 6003 (1970); (d) A. Monvernay and P. C. Lacaze, Bull. Soc. Chim. Fr., 4206 (1970).
160. J. S. Mayell and A. J. Bard, J. Amer. Chem. Soc., 85, 421 (1963).
161. V. G. Mairanovskii, S. Ya Melnik, A. Ya. Veinberg, and G. I. Samokhvalov, USSR 226628 (1966); Chem. Abstr., 71, 2957 (1969).
162. H. L. S. Maya, M. J. Medeiros, M. I. Montenegro, and D. Pletcher, J. Electroanal. Chem., 200, 363 (1986); J. Chem. Soc., Perkin Trans., 409 (1988); M. A. Casadei and D. Pletcher, Synthesis, 1118 (1987); L. Grehn, K. Gunnarsson, H. L. S. Maia, M. I. Montenegro, L. Pedro, U. Ragnarsson, J. Chem. Res. (S), 399 (1988).
163. H. Lund, Acta Chem. Scand., 17, 2325 (1963).
164. H. Lund, Acta Chem. Scand., 23, 563 (1969).
165. M. H. Khalifa, G. Jung, and A. Rieker, Liebigs Ann. Chem., 1068 (1982).
166. M. H. Khalifa and A. Rieker, Tetrahedron Lett., 25, 1027 (1984).

167. E. Kasafirek, Tetrahedron Lett., 2021 (1972).
168. M. Masui and H. Sayo, J. Chem. Soc., 4773, 5323 (1961); 1733 (1962).
169. J. Armand, Bull. Soc. Chim. Fr., 543 (1966).
170. P. Zuman and V. Horak, Coll. Czech. Chem. Commun., 26, 176 (1961).
171. J. Stradins, I. Tutane, A. Arens, and G. Vanags, Zh. Obshch. Khim., 35, 1327 (1965).
172. E. Kariv, J. Hermolin, and I. Rubinstein, Tetrahedron, 27, 3707 (1971).
173. J. Armand, L. Boulares, and P. Souchay, C. R. Acad. Sci., Ser. C, 276, 97 (1973).
174. P. Zuman, O. Manousek, and V. Horak, Coll. Czech. Chem. Commun., 29, 2906 (1964).
175. P. Zuman and V. Horak, Coll. Czech. Chem. Commun., 27, 187 (1962).
176. P. Carsky, P. Zuman, and V. Horak, Coll. Czech. Chem. Commun., 29, 3044 (1964).
177. P. H. Rieger, I. Bernal, W. H. Reinmuth, and G. K. Fraenkel, J. Amer. Chem. Soc., 85, 683 (1963).
178. H. Lund, Adv. Heterocycl. Chem., 12, 213 (1970); H. Lund and I. Tabakovic, Adv. Heterocycl. Chem., 36, 235 (1984).
179. K. K. Barnes and C. K. Mann, J. Org. Chem., 32, 1474 (1967); P. J. Smith and C. K. Mann, J. Org. Chem., 34, 1821 (1969).
180. E. G. Corley, S. Karady, L. N. Abramson, D. Ellison, L. M. Weinstock, Tetrahedron Lett., 29, 1497 (1988).
181. L. Horner and A. Mentrup, Liebigs Ann. Chem., 646, 65 (1961).
182. L. Horner and U.-M. Duda, Phosphorus, 119 (1975).
183. L. Horner and J. Röder, Phosphorus, 147 (1976).
184. L. Horner and J. Röder, Phosphorus, 2067 (1977).
185. L. Horner and J. P. Bercz (with cooperation of C. V. Bercz), Tetrahedron Lett., 5783 (1966), report the preparation of optically active biphosphines.
186. L. Horner, H. Winkler, A. Rapp, A. Mentrup, H. Hoffman, and P. Beck, Tetrahedron Lett., 161 (1961); for cleavage of optically active phosphonium salts containing benzyl groups by means of $LiAlH_4$ see L. Horner and M. Ernst, Chem. Ber., 103, 318 (1970).
187. L. Horner and H. Fuchs, Tetrahedron Lett., 203 (1962).
188. J. H. Wagenknecht and M. M. Baizer, J. Org. Chem., 31, 3885 (1966).
189. T. Shono and M. Mitani, J. Amer. Chem. Soc., 91, 2728 (1968).
190. P. E. Iversen and H. Lund, Tetrahedron Lett., 3523 (1969).
191. K. S. V. Santhanam and A. J. Bard, J. Amer. Chem. Soc., 90, 1118 (1968).
192. J.-M. Savéant and S. K. Binh, J. Electroanal. Chem., 88, 27 (1978).
193. L. Horner and J. Haufe, Chem. Ber., 101, 2903 (1968).
194. J.-I. Yoshida, T. Murata, and S. Isoe, Tetrahedron Lett., 27, 3373 (1986).
195. U. Husstedt and H. J. Schäfer, Synthesis, 964, 966 (1979).
196. R. Scheffold and E. Amble, Angew. Chem., 92, 643 (1980).
197. O. Hammerich, in Encyclopedia of Electrochemistry of the Elements (A. J. Bard and H. Lund, eds.), Vol. 11, Dekker, New York, 1978, pp. 316–328.
198. D. A. Skoog and A. B. H. Lauwzecha, Anal. Chem., 28, 825 (1956).
199. M. L. Whisman and B. H. Eccleston, Anal. Chem., 30, 1638 (1958).
200. S. Hayano and N. Shinozuka, Bull. Chem. Soc. Jpn., 43, 2039 (1970).
201. W. E. Parker, C. Ricciuti, C. L. Ogg, and D. Swern, J. Amer. Chem. Soc., 77, 4037 (1955).

202. W. E. Parker, L. P. Witnauer, and D. Swern, J. Amer. Chem. Soc., 79, 1929 (1957).
203. D. Swern and L. S. Silbert, Anal. Chem., 35, 880 (1963).
204. B. N. Moryganov, A. I. Kalinin, and L. N. Mikhotova, Zh. Obshch. Khim., 32, 3476 (1962).
205. J. H. Karchmer and M. T. Walker, Anal. Chem., 26, 271 (1954).
206. D. Barnard, M. B. Evans, G. McHiggins, and J. F. Smith, Chem. Ind., 20, (1961).
207. E. Larson, Ber. Dtsch. Chem. Ges., 61, 1439 (1928).
208. J. Simonet, M. Carriou, and H. Lund, Liebigs Ann. Chem. 1665 (1981).
209. P. E. Iversen and H. Lund, Acta Chem. Scand., B28, 827 (1974).
210. C. Degrand and H. Lund, Acta Chem. Scand., B33, 512 (1979).
211. F. Fichter and W. Wenk, Ber. Dtsch. Chem. Ges., 45, 1373 (1912).
212. A. Bewick, D. E. Coe, J. C. Mellor, and W. M. Owton, J. Chem. Soc., Perkin Trans. I, 1033 (1985).
213. S. Torii, H. Tanaka, M. Sasaoka, T. Shiroi, and S. Oto, Ger. Offen DE 3511 149 (1985); Chem. Abstr., 104, 138261 (1986).
214. H. Lund, Acta Chem. Scand., 13, 249 (1959).
215. H. Lund and E. T. Jensen, Acta Chem. Scand., 25, 2727 (1971).
216. H. Lund, in Chemistry of the Carbon-Nitrogen Double Bond (S. Patai, ed.), Wiley, New York, 1970, p. 505 ff.
217. H. Lund, Electrochim. Acta, 28, 395 (1983).
218. Yu. P. Kitaev, T. V. Troepol'skaya, L. V. Ermolaeva, and E. N. Munin, Izv. Akad. Nauk. SSSR, Ser. Khim., 1736 (1985).
219. H. A. Laitinen and T. J. Kneip, J. Amer. Chem. Soc., 78, 736 (1956).
220. L. Holleck, D. Jannakondakis, and A. Wildman, Electrochim. Acta, 12, 1523 (1967).
221. T. M. Florence, Aust. J. Chem., 18, 609 (1965); J. Electroanal. Chem., 52, 115 (1974).
222. H. Lund, PCT Int. Appl. WO 86 03194 (1986); Chem. Abstr., 105, 190654 (1986).
223. V. Z. Latypova, M. Yu. Kitaeva, Yu. M. Kargin, and A. V. Il'yasov, Izv. Akad. Nauk SSSR, Ser. Khim., 958 (1986).
224. H. Lund, Acta Chem. Scand., 17, 1077 (1963).
225. H. Lund, Oesterreich. Chem.-Z., 68, 43 (1967).
226. D. Knittel, Monatsh. Chem., 115, 1335 (1984).
227. D. Knittel, Monatsh. Chem., 115, 523 (1984); 116, 1133 (1985).
228. S. Kwee and H. Lund, Acta Chem. Scand., 22, 2879 (1968).
229. S. Kwee and H. Lund, Acta Chem. Scand., 23, 2711 (1969).
230. H. Lund, Acta Chem. Scand., 11, 990 (1957).
231. L. Suchomelova, J. Zýka, and P. Zuman, J. Electroanal. Chem., 12, 194 (1966).
232. G. Gavioli, G. Davolio, and E. S. Guidetti, J. Electroanal. Chem., 33, 1 (1971).
233. F. Fichter and F. Braun, Ber. Dtsch. Chem. Ges., 47, 1526 (1914).
234. F. Fichter and P. Sjöstedt, Ber. Dtsch. Chem. Ges., 43, 3422 (1910).
235. P. Chiorboli, G. Davolio, G. Gavioli, and M. Salvaterra, Electrochim. Acta, 12, 767 (1967).
236. B. J. Lindberg, Acta Chem. Scand., 24, 1110 (1970).
237. L. Horner and H. Neumann, Chem. Ber., 98, 1715, 3462 (1965).
238. L. Horner and R.-J. Singer, Chem. Ber., 101, 3329 (1968).
239. L. Horner and R.-J. Singer, Liebigs Ann. Chem., 723, 1 (1969).

240. V. G. Mairanovskii and N. F. Loginova, Zh. Obshch. Khim., 41, 2581 (1971).
241. P. Yousefzadeh and C. K. Mann, J. Org. Chem., 33, 2716 (1968).
242. H. L. S. Maya, M. J. Medeiros, M. I. Montenegro, D. Court, and D. Pletcher, J. Electroanal. Chem., 164, 347 (1984).
243. D. Pletcher and N. R. Stradiotto, J. Electroanal. Chem., 186, 211 (1985).
244. R. Gerdil, Helv. Chim. Acta, 56, 196 (1973).
245. H. Lund, Acta Chem. Scand., 13, 249 (1959).
246. H. Lund, Osterreich, Chem. Z., 68, 43 (1967).
247. H. Lund, Acta Chem. Scand., 18, 563 (1964).
248. I. Surov and H. Lund, Acta Chem. Scand., B40, 831 (1986).
249. T. Kubota, S. Hiramatsu, K. Kano, B. Uno, and H. Miyazaki, Chem. Pharm. Bull., 32, 3830 (1984).
250. K. Okumura, T. Iwasaki, M. Matsuoka, and K. Matsumoto, Chem. Ind. (London), 929 (1971); T. Iwasaki, K. Matsumoto, M. Matsuoka, and M. Miyoshi, Bull. Chem. Soc. Jpn., 46, 852 (1973).
251. R. C. Roemmele and H. Rapoport, J. Org. Chem., 53, 2367 (1988).
252. R. Kossai, G. Jeminet, and J. Simonet, Electrochim. Acta, 22, 1395 (1977).
253. A. Lebouc, P. Martigny, R. Carlier, and J. Simonet, Tetrahedron, 41, 1251 (1985).
254. R. Kossai, J. Simonet, and G. Jeminet, Tetrahedron Lett., 1059 (1979); R. Kossai, B. Emir, J. Simonet, and G. Mousset, J. Electroanal. Chem., 270, 253 (1989).
255. K. Oda, T. Ohnuma, and Y. Ban, J. Org. Chem., 49, 953 (1984).
256. M. I. Kalinkin, N. T. Ioffe, and K. P. Butin, Izv. Akad. Nauk. SSSR, Ser. Khim., 2072 (1969).
257. G. Bontempelli, F. Magno, R. Seeber, and G. A. Mazzocchim, J. Electroanal. Chem., 87, 73 (1978).
258. N. Urabe and K. Yasukochi, Denki Kagaku, 24, 360 (1956); 25, 17 (1957); Chem. Abstr., 51, 14447 (1957).
259. G. M. Nichols, Electrochem. Soc. Meet., New York, May 1969, Extended Abstracts, p. 364.
260. F. Fichter and W. Bernoulli, Z. Elektrochem., 13, 310 (1907).
261. S. G. Mairanovskii and M. B. Neiman, Dokl. Akad. Nauk. SSSR, 79, 85 (1951); Chem. Abstr., 46, 28 (1952).
262. L. Horner and H. Nickel, Chem. Ber., 89, 1681 (1956).
263. G. Jeminet and J. Simonet, C. R. Acad. Sci., Ser. C, 272, 1661 (1971).
264. J.-G. Gourcy, G. Jeminet, and J. Simonet, C. R. Acad. Sci., Ser. C, 277, 1079 (1973).
265. G. Jeminet, J. Simonet, and J.-G. Gourcy, Bull. Soc. Chim. Fr., 1102 (1974).
266. J. L. Lecat and M. Devand, Tetrahedron Lett., 28, 5821 (1987).

25
Anodic Substitution

LENNART EBERSON / Division of Organic Chemistry, University of Lund, Lund, Sweden

JAMES H. P. UTLEY / Queen Mary and Westfield College, University of London, London, England

OLE HAMMERICH / H. C. Ørsted Institute, University of Copenhagen, Copenhagen, Denmark

I. Introduction 1069

II. Formation of Carbon-Oxygen Bonds 1071
- A. Hydroxylation 1071
- B. Alkoxylation 1073
- C. Acyloxylation 1080
- D. Substitution by nitrate ion 1083

III. Formation of Carbon-Sulfur Bonds: Thiocyanation 1084
- A. Acetamidation 1084
- B. N-cyanomethylation and amination 1087
- C. Nitration 1088
- D. Substitution by cyanate ion 1088
- E. Pyridination 1089

V. Formation of Carbon-Carbon Bonds 1089
- A. Cyanation 1089
- B. Carbomethoxylation 1092
- C. Inter- and intramolecular coupling with C–C bond formation 1092

VI. Formation of Carbon-Halogen Bonds 1094
- A. Fluorination 1094
- B. Chlorination, bromination, and iodination 1094

References 1095

I. INTRODUCTION

Anodic substitution reactions are common in organic electrochemistry. The overall reaction may be represented as in Eq. (1):

$$R\text{–}E + Nu^- \longrightarrow R\text{–}Nu + E^+ + 2e^- \quad (1)$$

The nucleophile [Nu^- (or Nu)] may include H_2O, ROH, OH^-, RO^-, $RCOO^-$, NO_3^-, SCN^-, $SeCN^-$, CH_3CN, NO_2^-, N_3^-, OCN^-, pyridine, halide ion, and CN^-; the electrophile [E^+ (or E)] is most commonly H^+, but could be a carbocation or an alkoxonium ion. The case in which E is CO_2 (the Kolbe reaction) is treated in Chap. 14 and is not dealt with here.

Anodic substitution reactions therefore allow a net substitution by a nucleophile, a reaction not easily achieved by ordinary chemical means. Specifically, they make possible the direct formation of a C–O or C–N bond by substitution of a C–H bond. This is of considerable synthetic interest.

The mechanisms of anodic substitution reactions can be divided into indirect and direct processes. In an indirect process, the anode serves merely as a convenient source of an oxidant, for example Cl_2 by oxidation of Cl^- or $NO_3^{\bullet}$ by oxidation of NO_3^-, or an organic radical, for example methoxy radical from methoxide ion, which then attacks the organic substrate in a reaction similar to that taking place when the reagent is generated in any other way. The indirect processes also encompass reactions in which an organic compound is oxidized by an electrochemically generated oxidant, the reduced form of which is continuously reoxidized and therefore, in principle, needs only to be present in catalytic amounts. In the direct process, oxidation of the organic substrate takes place at the anode with formation of a positively charged species (cation radical or dication), which undergoes follow-up reactions with nucleophiles and/or bases present in the electrolyte, possibly with further electrochemical or homogeneous electron transfer steps on the reaction path (see Chap. 13). Essentially the same mechanistic pattern is found in oxidative substitution driven by some high-valence inorganic ions, such as Ce(IV), Co(III), Ag(II), and W(VI) [1-3], and in photosubstitution of aromatic compounds carrying electron-donating substituents [4,5]. In some cases the evidence does not allow a clear-cut assignment of mechanism. In this chapter we emphasize reactions proceeding according to the direct mechanism, since in many cases the indirect processes do not differ in any important respects from analogous homogeneous processes.

In the context of the direct mechanisms, several perplexing features concerning nuclear and side-chain substitution of aromatic compounds have been highlighted [6-8], and in the search for explanations attempts have been made to estimate, by thermochemically based calculations, free energies of activation for reactions between cation radicals and nucleophiles. Longstanding puzzles in this area include the dependence of the ratio of side chain to nuclear substituted products on the nucleophile; it is seen that for alkylbenzenes, side-chain acetoxylation predominates in acetic acid (AcOH) unless AcO^- is present, when nuclear substitution becomes important. Conversely, anodic cyanation occurs at the nucleus only.

Recently it was pointed out by Pross [9] that the direct bond formation between a nucleophile and a cation radical is "forbidden" from the viewpoint of the CM model [10,11]. The configuration describing the product is that of a doubly excited species and hence its formation should be a high-energy forbidden process. Since some direct bond formation steps of this type seem to be very fast [12], other factors must intervene as well in determining the reactivity of cation radicals (see Ref. 13 for other references on this subject).

In this chapter the results are presented in a largely pragmatic form; mechanistic aspects are discussed in more detail in Chap. 13. Apart from the leading references given here, the subject has been reviewed extensively [14-19]. The reader is also referred to the following chapters in this book: Chap. 15 (Nitrogen-Containing Compounds), Chap. 16 (Oxygen-Containing Compounds), Chap. 17 (Sulfur-Containing Compounds), Chap. 26 (Anodic Fluorination), and Chap. 29 (Electrogenerated Reagents).

II. FORMATION OF CARBON-OXYGEN BONDS

The possibility of the direct introduction of a carbon-oxygen bond in a substitution process according to Eq. (1) is of considerable preparative interest, since in principle cheap starting materials could be used. Consequently, it would be possible to prepare phenols using only hydrocarbon, water, and electricity in the process, Eq. (2). However, apart from the practical problems associated with this type of process, there are theoretical grounds why this reaction should not be feasible in general.

$$Ar{-}H + H_2O \longrightarrow Ar{-}OH + 2H^+ + 2e^- \quad (2)$$

In this case the starting material is considerably more difficult to oxidize than the product; under comparable conditions benzene is oxidized at a potential ~1.3 V more anodic than phenol. Consequently, it is observed that the reaction cannot be stopped at the phenol stage but proceeds to benzoquinone [20,21]. The same type of reasoning pertains to a greater or lesser degree to nuclear substitution processes involving C–O bond formation in aromatic compounds, since the additional oxygen function in the product usually decreases its oxidation potential compared with that of the starting material. However, the problem may to some extent be overcome if the degree of substrate conversion is kept low (e.g., the anodic oxidation of phenol to hydroquinone [22]) or if the oxidation product is continuously removed (e.g., the indirect oxidation of methylarenes to the corresponding carbonyl compounds [23]).

An important exception is anodic trifluoroacetoxylation; the CF_3COO group thus introduced effectively deactivates the aromatic nucleus toward further oxidation (see later).

A. Hydroxylation

The anodic oxidation of aromatic and alkylaromatic hydrocarbons at Pt or PbO_2 anodes in aqueous media generally leads to complex mixtures of products [24,25]. In most cases acidic electrolytes are employed, and acetone, an alcohol, or AcOH is added as a cosolvent to obtain a homogeneous medium. Mechanistically, these reactions are not very well understood. It has been suggested that at Pt anodes a layer of platinum oxide is the oxidizing agent; however, a mechanism involving oxidation of the substrate with initial formation of cationic species would also account for the products. According to this latter scheme the primary product would be a phenol from a nuclear hydroxylation and a benzyl alcohol from a side-chain hydroxylation of an aromatic compound. Both are easily oxidized in secondary reactions that, qualitatively at least, may account for the observed product distributions.

The anodic hydroxylation of phenols at PbO_2 in aqueous H_2SO_4 has been studied in detail [26]. PbO_2 was found to be superior as an anode to C, Ni, or Pt, and in this case at least it seems likely that chemical oxidation to phenoxonium ions is by a continuously regenerated surface species. Substitution is invariably at the 4-position to give benzoquinones; if the 4-position is already occupied, then 4-substituted 4-hydroxycyclohexadienones are formed, Eq. (3) (see also Chap. 16):

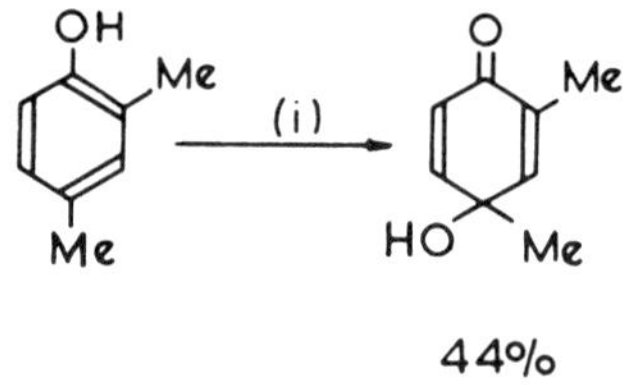

(3)

This particular reaction may sometimes be preparatively useful; for example, 4-allylphenols give good yields of 4-allyl-4-hydroxycyclohexa-2,5-dienones. Similarly, mesitylene is converted to 2,4,6-trimethyl-4-hydroxycyclohexa-2,5-dienone upon anodic oxidation in a mixture of H_2O, MeCN, and H_2SO_4 [27].

Anodic hydroxylation of quinones, substituted at the 2-position with an electron-withdrawing group, may be achieved at low potentials in aqueous solution [28]. In this case it is likely that nucleophilic attack is on the intermediate quinone, Eq. (4):

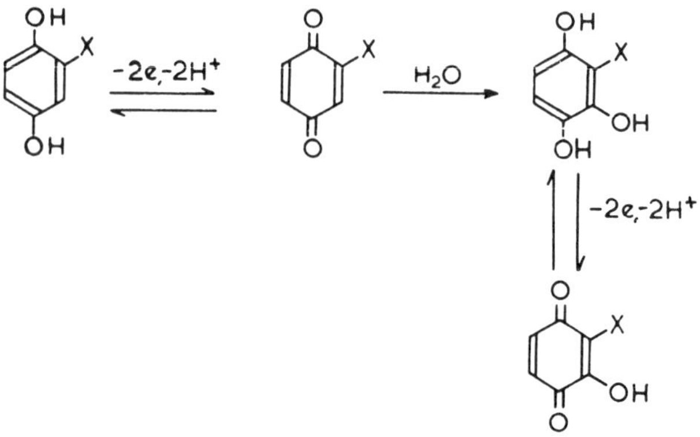

(4)

Anodic hydroxylation of furans provides a high-yield route to substituted maleic anhydrides, Eq. (5) [29]:

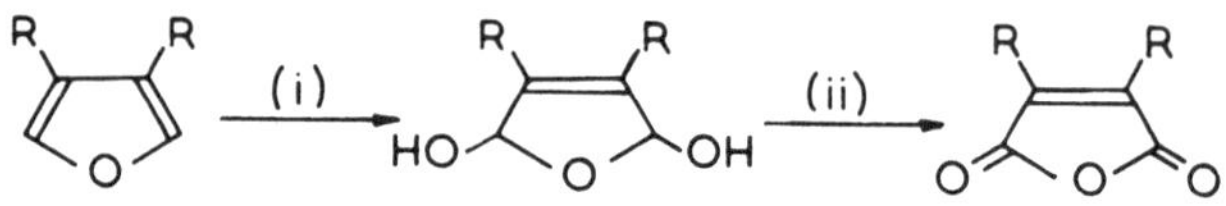

(5)

The introduction of a hydroxy group in an aliphatic compound already containing a simple oxygen function, such as an ether, does not significantly affect the oxidation potential. Consequently, the products of anodic hydroxylation of aliphatic ethers may survive the conditions necessary for the oxidation of substrate. For example, tetrahydrofuran has successfully been converted to 2-hydroxytetrahydrofuran by anodic oxidation in 1 M H_2SO_4 at Pt, Eq. (6) [30]:

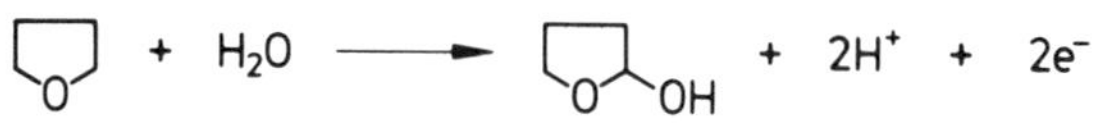

(6)

In nonaqueous solvents, water is an impurity that on a preparative scale even rigorous drying fails to remove completely [31]. Hydroxylation products are therefore common by-products. An example is the oxidation of alkylaromatic compounds in a nominally anhydrous solution of $NaClO_4$ in MeCN, Eqs. (7) and (8), which gives mainly the product of acetamidation [32,33]:

$$\text{Ar-CH}_3 \xrightarrow[-\text{H}^+]{-2e^-} \text{Ar-CH}_2^+ \xrightarrow{\text{H}_2\text{O}} \text{Ar-CH}_2\text{OH} \quad (\text{Ar-CHO}) \tag{7}$$

$$\text{Ar-CH}_2^+ \xrightarrow{\text{CH}_3\text{CN}} \text{Ar-CH}_2\text{N}{=}\overset{+}{\text{C}}\text{CH}_3 \xrightarrow{\text{H}_2\text{O}} \text{Ar-CH}_2\text{NHCOCH}_3 \tag{8}$$

One of the side reactions gives the corresponding benzyl alcohol (and its oxidation product, the aldehyde); the alcohol is formed at a considerably lower potential than the anodic limit of the solvent-supporting electrolyte, so at least in this case a direct mechanism for hydroxylation appears to operate. The anion of the supporting electrolyte has been found to have a profound influence on the product distribution [34]. For instance, exchanging ClO_4^- for BF_4^- changes the ratio between $ArCH_2NHCOCH_3$ and $ArCH_2OH$ by a factor of 400!

Effective nuclear hydroxylation has been achieved indirectly by trifluoroacetoxylation with subsequent hydrolysis (see later).

In summary, direct or indirect anodic hydroxylation of aromatic compounds has not yet been developed as a preparatively useful reaction, although it is an important (and sometimes embarrassing) side reaction in most anodic oxidations owing to the presence of trace amounts of water in the electrolytes. The reaction is difficult to study quantitatively by electrochemical methods. Some preparative advantage may be gained for cases in which further oxidation of the hydroxyl derivative gives a single product, such as a quinone, in good yield [20,21,26,35]. In addition to the quinone syntheses, methyl groups on aromatic nuclei may be oxidized to carboxyl groups via side-chain hydroxylation [14,25].

B. Alkoxylation

Alkoxylation may be achieved by anodic oxidation in an alcohol, often methanol (MeOH) containing a suitable electrolyte, such as KOH, NaOMe, NaCN, $NaBF_4$, or NH_4NO_3 [16-19,36]. Substrates that have been alkoxylated in this manner include aromatic compounds (both nuclear and side-chain alkoxylation is known), alkenes (allylic alkoxylation), _N_-alkylsubstituted amides, _N_,_N_-dialkylamines, and ethers. In many cases alkoxylation by substitution is a side reaction only to concurrent alkoxylation by addition [14,16,18,37].

Methanolic electrolytes have been shown to undergo oxidation at lower potentials than many aromatic substrates, which strongly suggests that an indirect mechanism operates in the α-methoxylation of toluene, ethylbenzene, and cumene [38]. The nature of the anion present greatly affects the yield of α-methoxylation product, the effectiveness decreasing in the series:

$$F^- \approx ClO_4^- > CN^- \approx OH^- > Cl^- \approx CH_3O^- > Br^-$$

The mechanistic possibilities are given in Eqs. (9) through (12):

$$CH_3OH \longrightarrow HOCH_2^{\bullet} + H^+ + e^- \tag{9}$$

or

$$CH_3O^- \longrightarrow CH_3O^{\cdot} + e^-$$

$$RH + X^{\cdot} \longrightarrow HX + R^{\cdot} \longrightarrow R^+ + e^- \quad (10)$$

($X = HOCH_2^{\cdot}$ or $CH_3O^{\cdot}$).

$$R^{\cdot} + CH_3O^{\cdot} \longrightarrow ROCH_3 \quad (11)$$

or

$$R^+ + CH_3OH \longrightarrow ROCH_3 + H^+ \quad (12)$$

One would not expect F^- or ClO_4^- to be oxidized at all under the conditions used, which suggests that MeOH is the electroactive species, forming a radical intermediate that then functions as $X^{\cdot}$ in Eq. (10). The radical $HOCH_2^{\cdot}$ is the most likely candidate since the O–H bond in MeOH is much stronger than the C–H bond. In alkaline medium, MeO^- is probably the electroactive species, and $MeO^{\cdot}$ would then be $X^{\cdot}$. Further support for the indirect mechanism is indicated by the fact that α-methoxylation of tetralin in MeOH-NaCN takes place at potentials as low as 0.5 V versus SCE, whereas the half-wave potential $E_{1/2}$ of tetralin must be around 1.9 V [39].

The anodic methoxylation of phenols [40,41], alkylanisoles [42], and methoxynaphthalenes [37] has been particularly carefully studied. The many products of methoxylation of phenols are further oxidized under the reaction conditions used. Experimental conditions can be found, however, that in some cases allow selective formation of para-methoxylated dienones, ortho-methoxylated dienones, or dimeric products. It is believed that methoxylation involves attack of MeOH on anodically generated phenoxonium ions, whereas dimeric products arise by combination of phenoxyl radicals, Eq. (13):

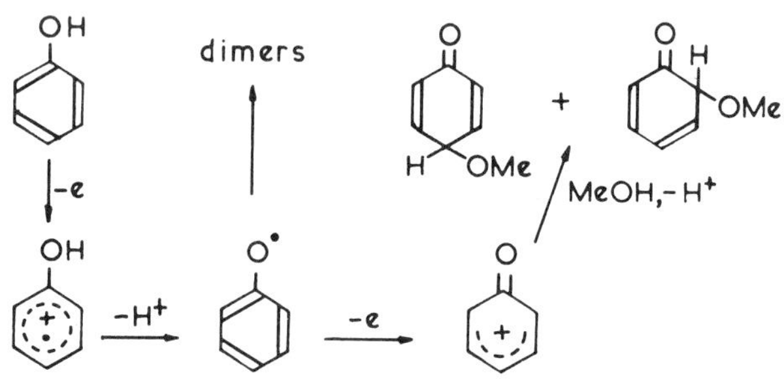

(13)

For the oxidation of alkylanisoles in MeOH containing various electrolytes [42], the current-potential curves can be used to support the scheme given for p-methylanisole in Eq. (14):

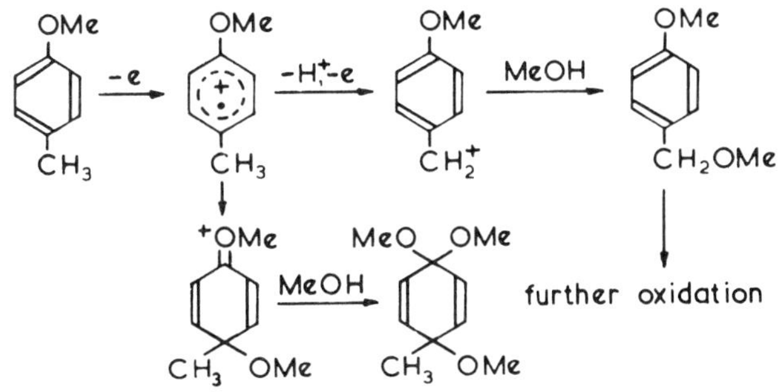

(14)

A general conclusion from this work is that proton loss from the side chain is rapid and nuclear attack competes only if it is fast by virtue of being a radical-radical attack (as proposed for methoxylation) or a cation radical-strong nucleophile reaction (as proposed for cyanation and acetoxylation).

Anodic alkoxylation of alkoxyarenes (ArOR) that do not carry substituents having an α-hydrogen atom generally results in mixtures of nuclear substitution and addition products. Nothing definitive can be said about the mechanistic details, the discussions of which have included all three possible ways to form the C–O bond, that is, (1) $ArOR^{+\cdot} + RO^-$, (2) $ArOR^{+\cdot} + RO^\cdot$, and (3) $ArOR + RO^\cdot$ (see Chap. 16 and references cited therein for further details).

Evidence for a mechanism involving electron transfer from the substrate to the anode has been obtained in the methoxylation of N,N-dimethylaniline, which takes place in MeOH–KOH with formation of I, Eq. (15), as the major product [43,44]:

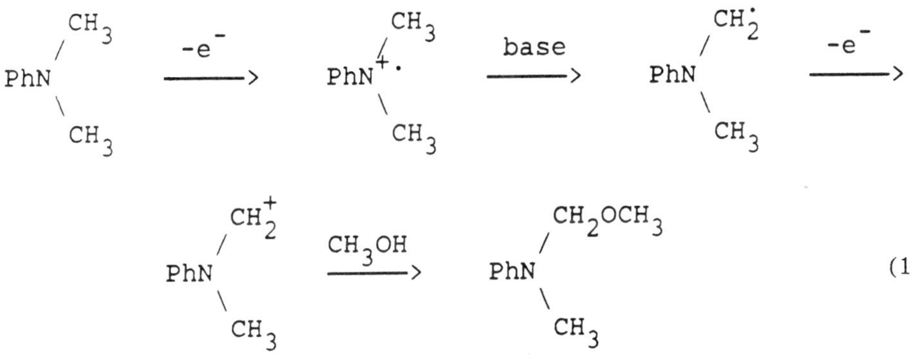

(15)

I

From the current-voltage curves it can be shown that the amine is oxidized at lower potentials than the solvent-supporting electrolyte. Similar behavior was observed in $MeOH–NH_4NO_3$, from which a dimer (II) is isolated as the predominant product. The mechanism in both cases must be an initial one-electron transfer from N,N-dimethylaniline to form the cation radical, which then undergoes different reactions depending on the basicity of the medium. With strong base present, proton abstraction gives a neutral radical, which is oxidized to a cation, Eq. (15); with no strong base present, dimerization of the cation may take place, Eq. (16), or the cation radical may attack a molecule of the starting material.

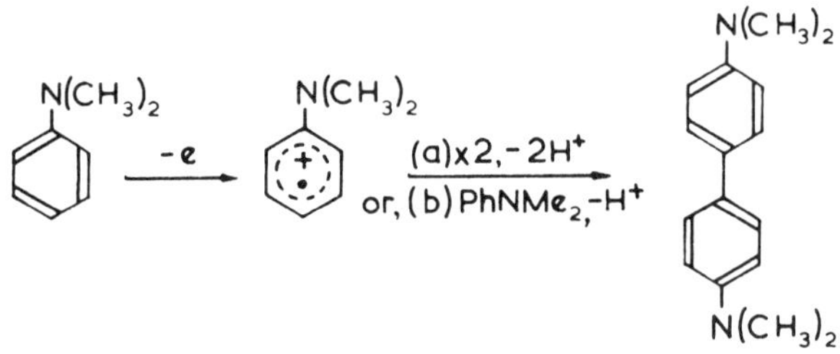

(16)

(II)

N,N-Dimethylbenzylamine exhibits similar behavior on electrolysis in MeOH-KOH [45,46]. The unusual preference for substitution in the methyl group over that in the methylene group may be due to adsorption [43], but these experiments were at potentials at which amine and MeO^- were being discharged [44]. Controlled potential electrolysis (CPE) in the absence of MeO^- resulted only in the production of protonated starting material with some dealkylation [46]. The implication is that in the original experiments $MeO^{\bullet}$ is somehow involved.

A reaction that has attracted much attention is the alkoxylation of N,N-dialkylamides [47]. The prototype of this process is the oxidation of N,N-dimethylamides, Eq. (17), in ROH-NH_4NO_3 [48]:

$$R^1CON(CH_3)_2 + R^2OH \xrightarrow{52\text{-}82\%} R^1CON\begin{matrix} / CH_3 \\ \backslash CH_2OR^2 \end{matrix} + H^+ + 2e^- \qquad (17)$$

R^1 = H, Me, Ph

R^2 = Me, Et, Bu

The use of a nitrate salt results in better yields and a cleaner reaction than is observed with NaOR. In MeOH–NH_4NO_3, current-potential curves at a Pt anode are not affected by the addition of the substrate, indicating that the substrate is more difficult to oxidize. A mechanism involving the initial discharge of NO_3^- to form $NO_3^{\bullet}$ is considered the most likely in this case, Eq. (18):

$$NO_3^- \xrightarrow{-e^-} NO_3^{\bullet} \xrightarrow[-HNO_3]{RCON(CH_3)_2} RCON\begin{matrix} / CH_2^{\bullet} \\ \backslash CH_3 \end{matrix} \xrightarrow{-e^-}$$

$$\underset{\text{III}}{RCON\begin{matrix} / CH_2^+ \\ \backslash CH_3 \end{matrix}} \xrightarrow{CH_3OH} RCON\begin{matrix} / CH_2OCH_3 \\ \backslash CH_3 \end{matrix} \qquad (18)$$

When NaOMe is used as supporting electrolyte, for example, the reaction most likely involves the substrate cation radical as the primary intermediate, which by deprotonation and oxidation is converted to III. However, some doubt has been cast about the nature of III. For instance, the intramolecular alkoxylation expected for a series of hydroxyamides, Eq. (19), did not take place in MeCN; only polymeric material was formed [49]. When the reaction was instead carried out in MeOH only the "normal" methoxylation products were formed, Eq. (20). These, however, could easily be converted to the cyclized products by addition of acid, a reaction known to involve carbocations like III. The conclusions drawn from these studies are that either III is not formed during the anodic methoxylation of the hydroxyamides, for not so obvious reasons, does not undergo the intramolecular alkoxylation if it is formed.

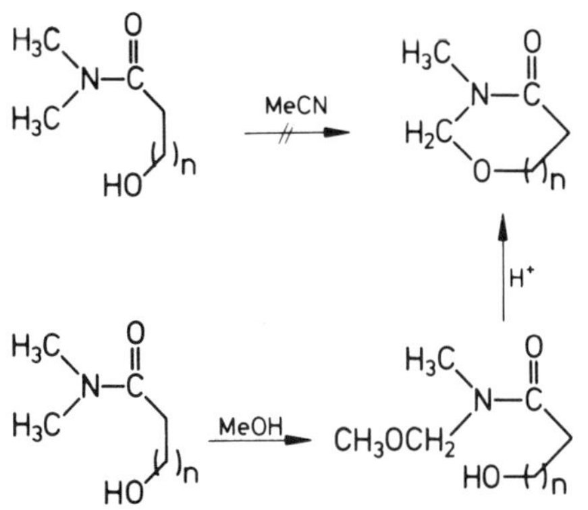

(19)

(20)

Numerous examples of the application of this preparatively useful reaction have been reported. Illustrative examples include the synthesis of eneamides and enecarbamates [50,51] and a large series of pyrrolidine, piperidine, and perhydroazepine derivatives [50,52-55]. Attempts to methoxylate smaller ring systems, like N-acetylaziridine and N-formylazetidine, failed and resulted in ring-opened products [56]. The reaction lends itself well to operation on a relatively large scale [57-61]. For instance, several N-formyl derivatives have been anodically methoxylated on a 0.5 kg scale using a capillary gap cell with a graphite anode [58,60]. An example is given in Eq. (21). Polyalkoxylation of N,N-dimethylformamide (DMF) has also been achieved [59].

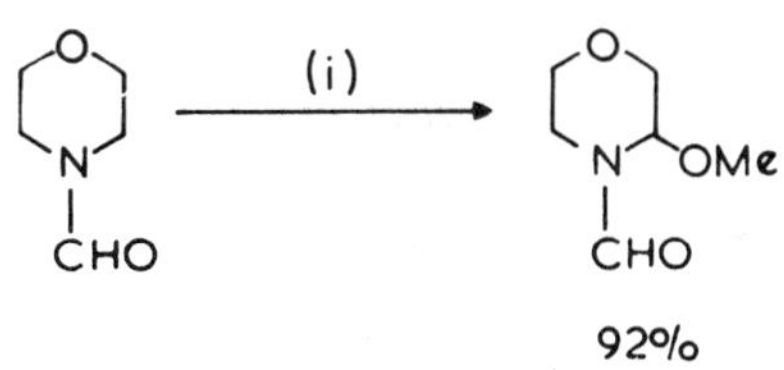

(21)

(i) C anode, MeOH, 8.4 $F mol^{-1}$

The selectivity and stereochemistry of the anodic methoxylation of the cyclic amides is noteworthy. First, N-formyl-2-methylpiperidine gives the 6-methoxysubstituted product derived from the secondary carbocation (IV), despite that loss of the proton in the 2-position would give the tertiary carbocation (V), which intuitively would be expected to be the more stable intermediate. Second, 4-substituted N-formylpiperidines give the corresponding 2-methoxy derivative with the methoxy group occupying an axial position [62]. Both these experimental facts were attributed to the steric constraints imposed by the N-formyl group.

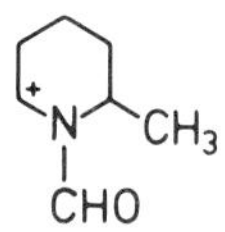

(IV)

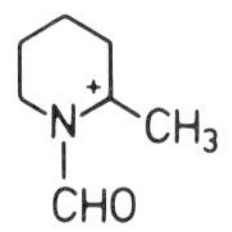

(V)

The anodic methoxylation of amides is not restricted to simple formic and AcOH derivatives but has also been performed with sulfonamides and amidophosphates as starting materials [63].

In all the examples given so far, the substrate carries at least one N-α-hydrogen atom. The anodic oxidation of fully substituted amides, like N,N-di-*tert*-butylformamide and N-formyl-2,2,6,6-tetramethylpiperidine in MeOH, would be expected to follow a different pathway. The products isolated after 12-14 F mol^{-1}, methyl N-*tert*-butyl carbamate and N-methoxycarbonyl-2,2,6,6-tetramethylpiperidine, respectively, Eq. (22) [64], indicated that the primarily formed substrate cation radical loses the formyl proton. Further oxidation of the neutral radical (VI) leads to the cation (VII), which may either undergo cleavage, Eq. (22a), or nucleophilic attack by the solvent, Eq. (22b).

$$R_2NCHO \xrightarrow{-e^-} R_2NCHO^{+\cdot} \xrightarrow{-H^+} \underset{\text{VI}}{R_2NCO^{\cdot}} \xrightarrow{-e^-} \underset{\text{VII}}{R_2\overset{+}{N}{=}C{=}O}$$

$$\text{(a) VII} \longrightarrow R{-}N{=}C{=}O + R^+ \xrightarrow[-H^+]{2CH_3OH} R{-}NHCOOCH_3 + ROCH_3 \qquad (22)$$

$$\text{(b) VII} \xrightarrow[-H^+]{CH_3OH} R_2NCOOCH_3$$

Saturated ethers can be α-methoxylated by electrolysis in MeOH-NaOMe or MeOH-NH_4NO_3 in 10-30% yield, Eq. (23) [65]. Yields are considerably higher, 45-70%, for the heterocyclic 1,3-benzodioxoles [66].

$$RO{-}\overset{|}{\underset{|}{C}}H + CH_3O^- \longrightarrow RO{-}\overset{|}{\underset{|}{C}}{-}OCH_3 + H^+ + 2e^- \qquad (23)$$

Related to these reactions is the anodic methoxylation of sulfides carrying an electron-withdrawing substituent, such as cyano [67] or trifluoromethyl [68] in the α-position, Eq. (24). Yields are generally between 50 and 90%.

$$\mathrm{X{-}CH_2{-}S{-}R + CH_3O^- \longrightarrow X{-}\underset{\displaystyle OCH_3}{\underset{|}{C}H}{-}S{-}R + H^+ + 2e^-} \qquad (24)$$

$$\mathrm{X = CN,\ CF_3}$$

Allylic methoxylation is often the result of anodic oxidation of alkenes and dienes in methanolic media [69,70]. Cyclopropanes behave similarly [71]. The reaction is usually accompanied by products of addition and combination, and the products may all be accommodated in the mechanistic scheme given in Eq. (25).

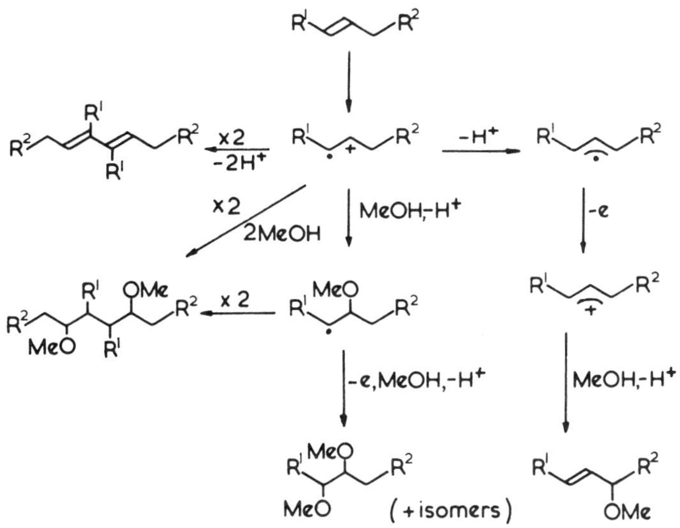

(25)

The yields for allylic methoxylation are only moderate (~25%) and compare unfavorably with the analogous allylic acetoxylation (see later) [72]. Anodic reactions of alkenes and dienes, which give both combination and methoxylation, can be remarkably efficient [70], and some key examples are given in Eqs. (26) and (27):

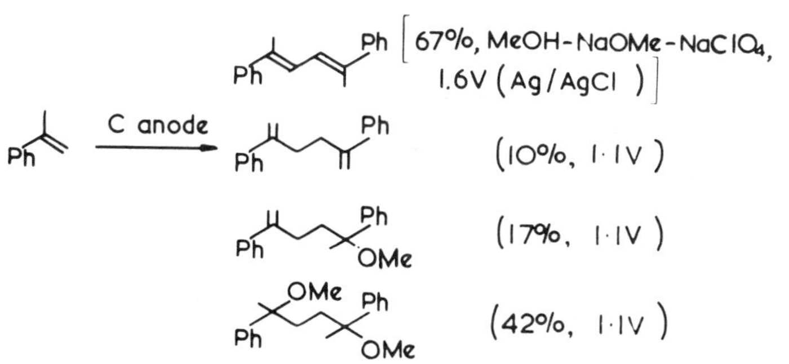

(26)

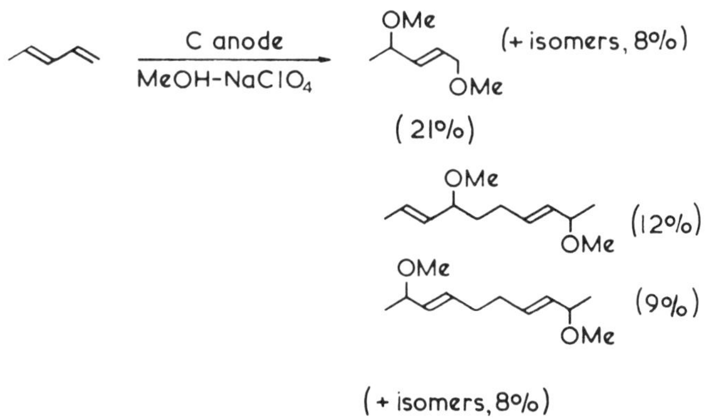

(27)

In summary, for many substrates, alkoxylation appears to follow an indirect mechanism involving the initial formation of radical species by discharge of solvent and/or supporting electrolyte. For easily oxidized substrates the direct mechanism may operate and in these cases products of coupling are often found.

C. Acyloxylation

Substitution by an acyloxy group can be accomplished by electrolysis of aromatic and alkylaromatic compounds [2,16-19,73], alkenes [72,74], benzyl ethers [75], alkyl aryl sulfides [68,76], or N-alkyl-substituted amides [48, 53,77] in the presence of a carboxylate in a suitable solvent. For trifluoroacetoxylation (see later), a variety of relatively deactivated aromatic substrates or aliphatic compounds may be employed. The most commonly encountered process is acetoxylation, which takes place in AcOH or AcOH–Ac_2O at C, Pt, or PbO_2 anodes. The supporting electrolyte in this case is normally NaOAc, although Me_4NNO_3, Bu_4NBF_4, $NaClO_4$, and others can be employed in certain cases. Other acyloxylations using carboxylates whose corresponding acids cannot be used as solvent have been performed in DMF [77] or MeCN [78,79].

Acyloxylation does not involve attack by the acyloxy radical as was originally believed [80-82]; AcOH–NaOAc solutions are not easily oxidized (see Chap. 14), and the lifetime of the acyloxy radical is very short, of the order of 10^{-9} s. It can be shown [83] that compounds that can be acetoxylated successfully also show a polarographic wave at the rotating Pt anode in this medium. CPE at a potential around $E_{1/2}$ of the substrate unambiguously shows that this wave is associated with the acetoxylation reaction. Compounds with estimated $E_{1/2}$ far above the anodic limit of AcOH–NaOAc do not undergo acetoxylation [84]. Yields of acetoxylation products vary greatly [2,16-19,73,85], depending on the oxidation potential of the product(s) formed. As a general rule, they tend to be low in nuclear acetoxylation of aromatic compounds and fair to good in acetoxylations at activated C atoms.

A characteristic feature of the reaction is that nuclear acetoxylation of aromatic compounds requires the presence of AcO^- [84], whereas acetoxylation of the other types of substrates takes place with either AcO^- or other anions present [48,72,86]. Both nuclear and side-chain acetoxylation may be rationalized according to the ECE-type mechanism given in Eq. (28):

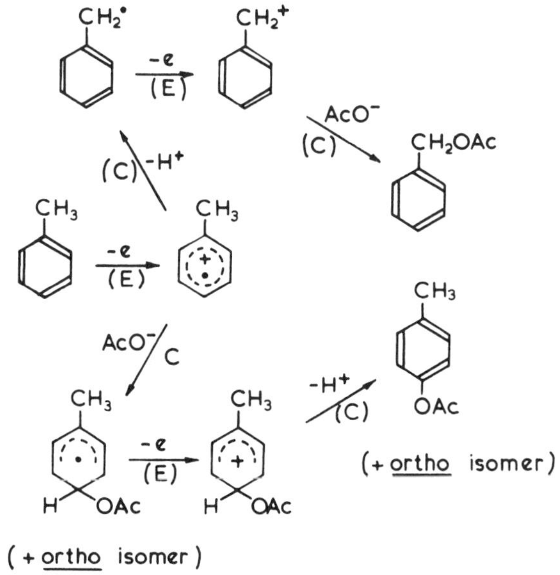

(28)

It may well be that for nuclear attack on the cation radical a good nucleophile (i.e., AcO^-) is required, whereas several species are sufficiently basic to accept the proton from the side chain of the cation radicals, which are known to be acids of considerable strength (see Chap. 13). It is noteworthy that, when comparisons can be made, there is a close similarity of product distributions for oxidation by an anode and $S_2O_8^{2-}$, Ce(IV), Ag(II), and W(VI) complexes [2,3]. These latter reagents are assumed to produce cation radicals by one-electron oxidation.

The isomeric distribution for anodic acetoxylation of a number of monosubstituted benzenes has been determined [84]. The reaction closely resembles ordinary electrophilic aromatic substitution processes, perhaps on the side of low-selectivity reactions. The isotope effect k_H/k_D for nuclear acetoxylation in anisole was found to be 1.0, whereas for α substitution in ethylbenzene a value of 2.6 was observed. The interpretation of these values is not straightforward [8].

From a preparative point of view it is of interest to note that selective nuclear substitution may be obtained when the oxidation is carried out in an undivided cell using a cathode made of Pd/C [73]. With this experimental setup, the side-chain acetoxylated product is reduced back to starting material at the cathode. The nuclear substituted product is unaffected and accumulates during the electrolysis.

The use of emulsion systems (e.g., AcOH–NaOAc–CH_2Cl_2–ArH) with Bu_4NHSO_4 as a phase transfer system does not apparently offer any preparative advantage for acetoxylation [87], although the method can greatly facilitate other anodic substitutions (e.g., cyanation; see later).

Aromatic compounds, in particular those carrying electron-withdrawing substituents, may be oxidized in the presence of CF_3COO^- with consequent trifluoroacetoxylation [88-93]. For compounds PhX, X may be $(CH_3)_3C$, COOMe, NO_2, CF_3, COMe, COPh, and CN; the trifluoroacetoxy derivatives produced by anodic oxidation in CH_2Cl_2–CF_3COOH–Et_4NBF_4 [86], CF_3COOH–$NaOOCCF_3$ [89], or CH_3NO_2–CF_3COOH [91] are hydrolyzed during workup to phenols. Yields of the phenols so produced are in the range of 20–80%.

and substantial amounts of the meta-substituted phenols are usually formed. For chlorobenzene, prolonged electrolysis in $CF_3COOH-NaOOCCF_3$, followed by hydrolysis, leads to good yields of the corresponding resorcinol and pyrocatechol derivatives [92].

A number of intramolecular acyloxylations have been reported [83,94-101], two examples of which are given in Eqs. (29) and (30):

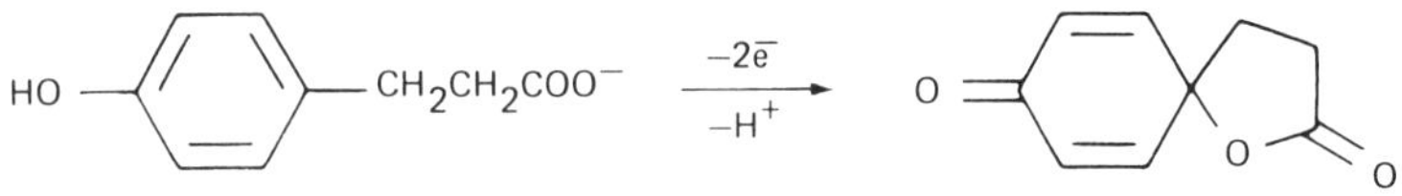

(29)

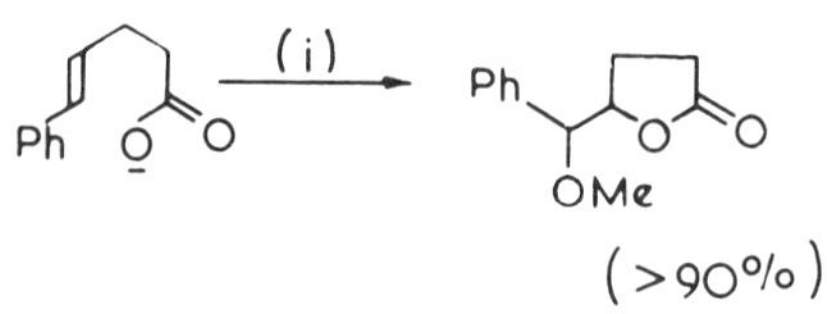

(30)

(i) Pt anode, MeOH

The products are best rationalized in terms of intramolecular attack on the cation radical formed [101].

Acyloxylation of a number of benzyl ethers has been reported, Eq. (31). A hemiacetal acetate is presumably formed and subsequently hydrolyzed during the workup procedure to give the aldehyde [75]:

$$\text{Ar-CH}_2\text{OCH}_3 \xrightarrow{-2e^-} \text{Ar-}\overset{+}{\text{C}}\text{HOCH}_3 \xrightarrow{\text{CH}_3\text{COO}^-}$$

$$\text{Ar-CH}\begin{matrix}/\text{OCOCH}_3 \\ \backslash\text{OCH}_3\end{matrix} \xrightarrow{\text{H}^+,\ \text{H}_2\text{O}} \underset{30\text{-}70\%\ \text{yield}}{\text{Ar-CHO}} \qquad (31)$$

Thioethers, such as alkyl aryl sulfides, containing an α-trifluoromethyl group are α-acetoxylated in good to excellent yields [68]. In the absence of the electron-withdrawing substituent the yields are very poor [68] and give a mixture of the S-oxide and the α-acetoxylated product [76]. However, if the oxidation is carried out in boiling $AcOH-Ac_2O-NaOAc$ the corresponding α-acetoxy derivatives may be obtained in almost quantitative yields (see Chap. 17, Table 5) [76]. Under these conditions the S-oxide, if formed at all, is converted to the desired α-acetoxy derivative in a Pummerer reaction parallel to the direct electrochemical acetoxylation. This illustrates that reactions that at room temperature give rise to mixtures of products may be synthetically useful when carried out at higher temperatures at which the thermodynamically more stable products are formed.

Acetoxylation of alkenes takes place in allylic position [72,74] concurrently with the addition of acetoxy groups across the double bond or double-bond system. The mechanism is almost certainly analogous to that discussed previously for methoxylation (25). Because AcO^- is difficult to oxidize, the anodic discharge of simple, unactivated alkenes may be achieved in its presence. Subsequent reactions of the cationic intermediates are not very selective, but several experimental parameters may be controlled and the method can compare well to the few competing chemical methods. For instance, anodic acetoxylation of methyl oleate may be optimized [74] to give mono-acetoxylation (60%) or diacetoxylation (85%). Other examples of possible synthetic interest are given in Eqs. (32) and (33):

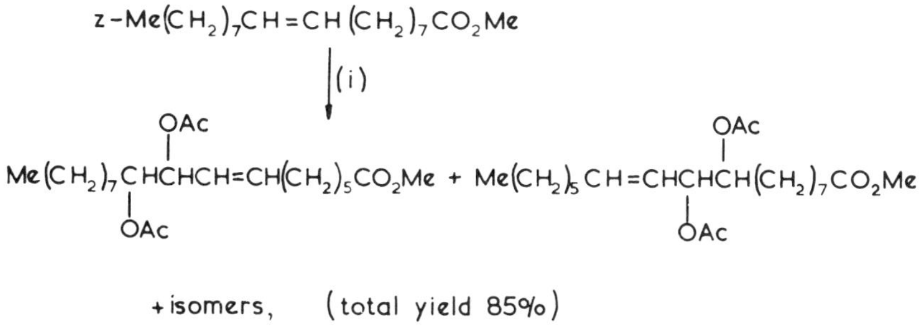

(32)

(i) Pt anode, AcOH-$LiClO_4$, constant current

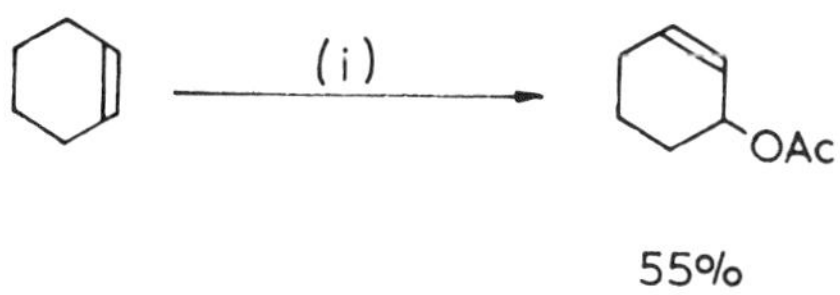

(33)

(i) C anode, AcOH-Et_4NOTs, constant current

In CF_3COOH containing R_4NBF_4, the oxidation of n-alkanes is possible, with the carbocations so formed being trapped to yield secondary trifluoroacetates [102]. Anodic trifluoroacetoxylation of long-chain aliphatic ketones [103] or carboxylic acids [104] gives rise to mixtures of positional isomers. The substitution pattern for the ketones indicates that the reaction proceeds via intramolecular abstraction of a hydrogen atom from the 5-position by the carbonyl group. This reaction has been observed to take place during the oxidation of aliphatic ketones in MeCN (see later). The trifluoroacetoxylation of carboxylic acids is less selective and the product distribution gives little indication of the mechanism.

The acyloxylation of N-alkyl-substituted amides has received considerable attention [48,53,77] from both the mechanistic and the preparative points of view. In electrolytes containing alkali metal carboxylates the direct mechanisms probably operates, whereas when a nitrate salt is the supporting electrolyte an indirect mechanism involving hydrogen abstraction from the N-alkyl group by anodically generated $NO_3^{\bullet}$ [see methoxylation, Eqs. (18) through (20)] is indicated. The same mechanistic problem is encountered in side-chain acetoxylation of alkylaromatic compounds in the presence of NO_3^- [86,105,106]. The method for large-scale substitution into N-formyl derivatives, which works so well for methoxylation, fails when applied to acetoxylation, probably because of the acid sensitivity of the products [58].

D. Substitution by Nitrate Ion

Concurrently with side-chain acetoxylation of alkylaromatic compounds in AcOH-NH_4NO_3, formation of benzyl nitrates, $ArCH_2ONO_2$, is observed in about the same yield as the $ArCH_2OAc$ [86,105,106]. It seems likely that

NO_3^- and AcOH compete as nucleophiles for an intermediate benzyl cation, although an indirect mechanism cannot be ruled out completely because of the complexity of the supporting electrolyte behavior [107].

III. FORMATION OF CARBON-SULFUR BONDS: THIOCYANATION

Phenols and aromatic amines are thiocyanated on electrolysis in an aqueous solution of SCN^- [25]. The reaction is indirect since there is compelling evidence that the thiocyanating reagent must be $(SCN)_2$ formed by anodic oxidation of SCN^- [108]. In MeCN solution, oxidation of SCN^- and $SeCN^-$ in the presence of phenol of various aromatic amines leads to thiocyanation and selenocyanation in 55-80% yields [109].

The direct reaction between a cation radical derived from an aromatic hydrocarbon (naphthalene) and SCN^- has been studied by mixing the salt $(C_{10}H_8)_2PF_6$ with a solution of Bu_4NSCN in CH_2Cl_2 at $-78°C$ [110]. Two major products were isolated, the kinetically controlled 1-naphthyl thiocyanate (28%) and the thermodynamically controlled 2-naphthyl isothiocyanate (8%). The mechanistic details leading to this product distribution are not clear.

IV. FORMATION OF CARBON-NITROGEN BONDS

A. Acetamidation

The acetamidation reaction has been partly discussed [Sec. II.A, Eq. (7)]. Methyl-substituted benzenes are particularly good substrates for side-chain acetamidation, and they have featured in numerous product and mechanistic studies [16-19,111]; ethylbenzene and isopropylbenzene give other products predominantly [112]. Hexamethylbenzene has been a favored substrate for mechanistic investigations of acetamidation [111,113], and in this case there is evidence [114] that on the time scale of conventional cyclic voltammetry (CV) proton loss is rapid from hexamethylbenzene cation radical but relatively slow from that of hexaethylbenzene.

The use of MeCN containing ClO_4^- or BF_4^- electrolytes allows the attainment of very anodic potentials (~3.6 V versus Ag/Ag^+) and consequently very energetic carbocations may be produced in such solutions. In addition, the presence of $(CF_3CO)_2O$ removes traces of water [115] and suppresses the hydroxylation reaction that so often obscures the results of mechanistic studies (see Sec. II.A). In rigorously dried MeCN, acetamidation of anthracene has been achieved (85% yield) and CV evidence was adduced [115] in favor of the scheme in Eq. (34):

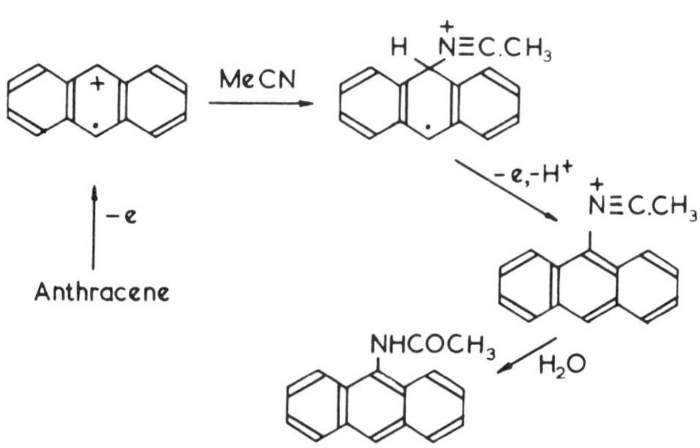

(34)

The rapid reaction of the anthracene cation radical with MeCN is nicely illustrated by the observation of an irreversible, two-electron peak in the voltammogram at +1.2 V versus SCE, whereas in $CH_2Cl_2-(CF_3CO)_2O$ the corresponding peak is reversible and corresponds to one-electron oxidation.

At a Pt anode, aromatic carbonyl compounds undergo fairly efficient nuclear acetamidation in moist MeCN with Et_4NBF_4 as supporting electrolyte [90].

n-Alkanes can be oxidized in $MeCN-R_4NBF_4$ solutions (see Chap. 13). For alkane substrates, the observation of well-defined oxidation waves, the consumption of 2 F mol^{-1}, and the formation of acetamides at secondary carbon atoms support the mechanism outlined in Eq. (35) for n-octane [102]. In some cases prior cleavage of carbon-carbon bonds is observed, again in a manner that suggests a carbocation intermediate, Eq. (36) [116].

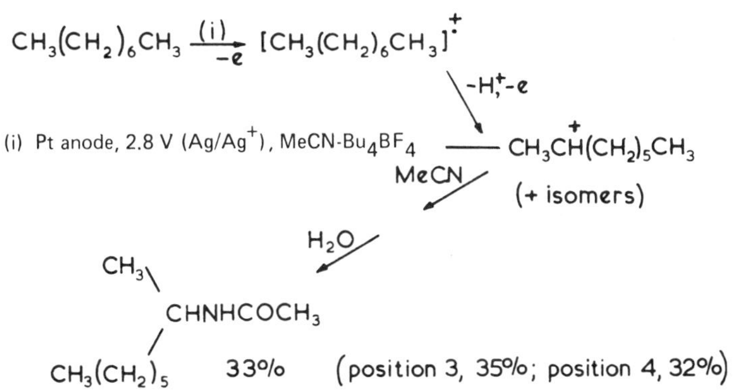

(35)

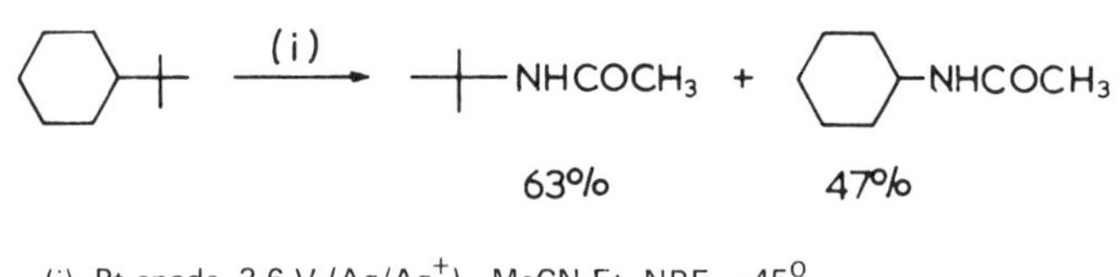

(36)

In the wake of this early work, several investigations of anodic acetamidation of polycyclic hydrocarbons were initiated; substituted adamantanes were much used in this work because of their availability and their relative ease of oxidation [117-121]. The results are generally explicable in terms of expected stabilities of intermediate carbocations and competition between proton loss and C−C bond cleavage. Difunctionalization of adamantanes is also possible using anodic acetamidation, Eq. (37) [121]:

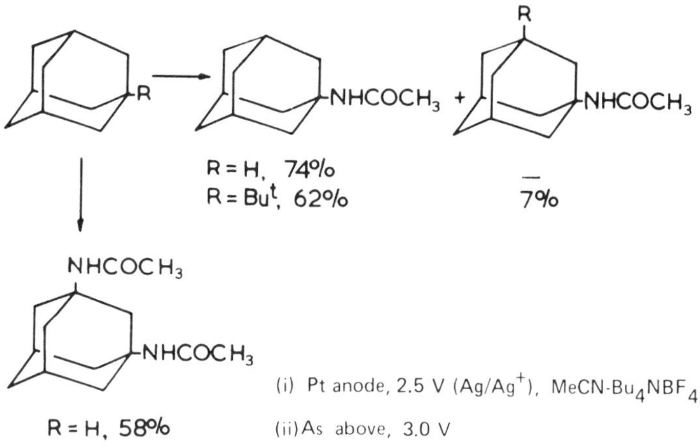

(37)

(i) Pt anode, 2.5 V (Ag/Ag^+), $MeCN\text{-}Bu_4NBF_4$

(ii) As above, 3.0 V

Esters, ketones, and alcohols may also undergo substitution in MeCN [122-127]. Some of these substrates are oxidized in a region where oxidation of the electrolyte is concurrent, and certainly for esters, it may well be that an indirect mechanism operates with necessary initial abstraction of hydrogen by a radical derived from the electrolyte [123]. Examples are given in Eq. (38). In one case, the anodic oxidation of 2-octanone in MeCN containing $LiClO_4$, the reaction mechanism has unambiguously been demonstrated to involve specific abstraction of one hydrogen atom at C-5 by the carbonyl oxygen in a six-membered ring cyclic transition state, reminiscent of the first chemical step in the gas-phase McLafferty rearrangement [127]. Oxidation of 5,5-dideuterio-2-octanone gave exclusively monodeuteriated acetamido-2-octanones, whereas oxidation of 4,4- (VIII) and 6,6-dideuterio-2-octanone (IX) gave products that all retained both deuterium atoms. In acetamidated VIII, these remained at C-4. In the 5-acetamido product resulting from IX, the two deuterium atoms remained at C-6, whereas the 6- and 7-acetamido products both appeared to have one D at C-5 and one at C-6 [see Chap. 16, Eqs. (75) through (77)].

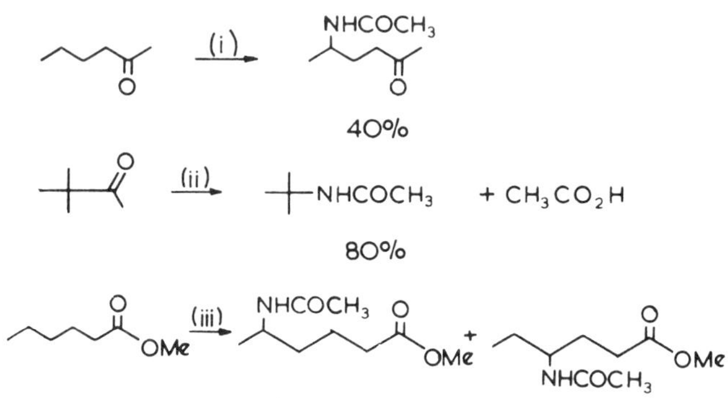

(38)

(i) Pt anode, 2.3 V (Ag/Ag^+), $MeCN\text{-}LiClO_4$.

(ii) As above, 2.15 V.

(iii) Pt anode, MeCN, 2.7 V (Bu_4NCl or Et_4NClO_4) or 3.1 V (Et_4NBF_4)

Acetamidation is also a result of the anodic oxidation in MeCN solution of carboxylates (see Chap. 14) and alkyl halides [117,128-131]. Much discussion and experimentation have centered on the likely mechanism of cleavage of the alkyl halides, particularly the iodides. From the extent of rearrangement of cations supposedly produced by alkyl iodide oxidation compared with that for the same cations produced by solvolysis of tosylates, it has been suggested that in electrolysis MeCN assists cation formation [129]. The S_N2 character of this process receives support from the 20% inversion observed for oxidation of optically active 2-iodooctane, Eq. (39):

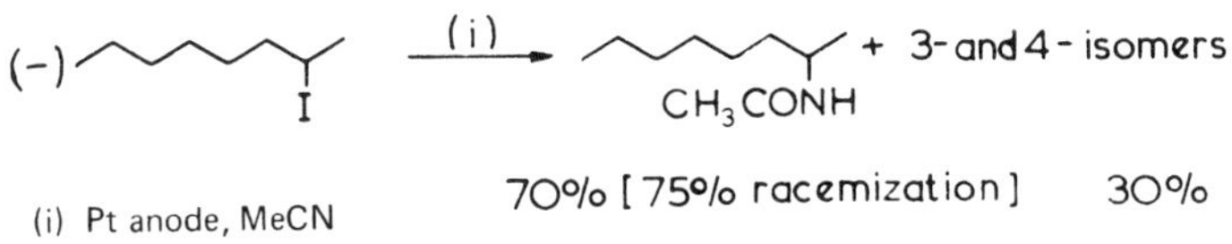

(39)

An alternative mechanism must be considered, however, which is based on the known anodic oxidation at the potentials used for alkyl iodide cleavage of iodine to an iodine(I) species [130]. The process given in Eq. (40), which is catalytic in iodine, is therefore plausible, given the initial generation of some iodine by direct cleavage.

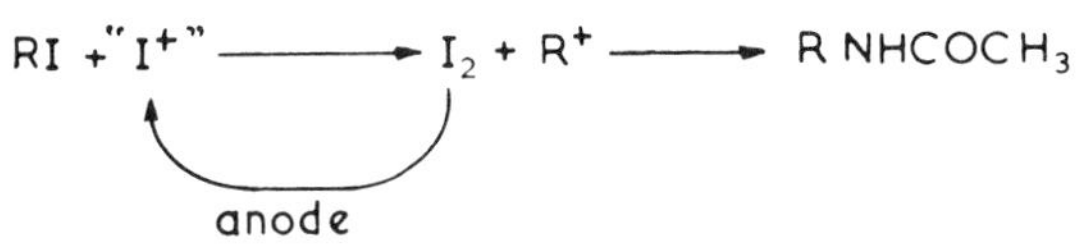

(40)

B. N-Cyanomethylation and Amination

The solvent, MeCN, may be incorporated in the product in an altogether different type of process, which leads to formation of substituted aminoacetonitriles. For instance, the products (XI) could be isolated after anodic oxidation of 2,2,6,6-tetramethylpiperidine and the corresponding morpholine in MeCN containing sodium perchlorate under oxygen-free conditions [132]. The reaction was suggested to follow the mechanism given in Eq. (41):

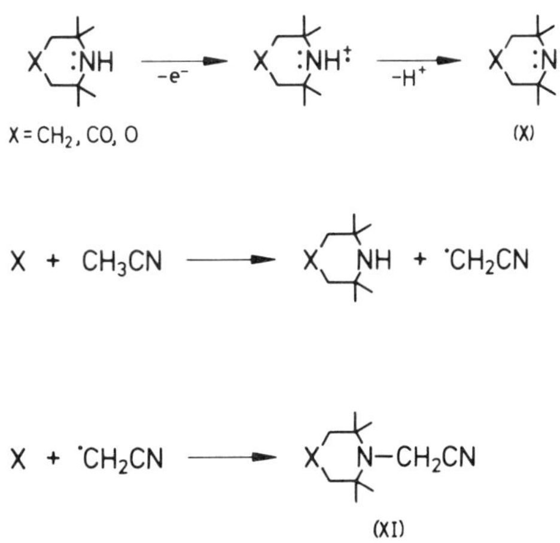

(41)

Essentially the same mechanism has been proposed for anodic α-amination of tetrahydrofuran, which may be accomplished by oxidation of R_2NH or R_2NLi, for example, in this solvent [133]. The best yields (70-80%) were observed when R_2NH was piperidine.

C. Nitration

The formation of nitro compounds upon electrolysis of a suspension of an aromatic hydrocarbon in dilute aqueous HNO_3 has been known for a long time [24]. It has been interpreted in terms of formation of a high concentration of HNO_3 near the anode surface, in which region an ordinary electrophilic nitration process would take place. Studies of the anodic oxidation of NO_3^- in nitromethane [134] seem to confirm this assumption; N_2O_5 is the oxidation product, probably formed via reactions (42) and (43):

$$2NO_3^- \longrightarrow 2NO_2^+ + O_2 + 4e^- \tag{42}$$

$$NO_2^+ + NO_3^- \longrightarrow N_2O_5 \tag{43}$$

In aqueous solution this would lead to the formation of HNO_3.

Another mechanism of anodic nitration involves the generation of a cation radical in the presence of NO_2, Eq. (44):

$$ArH \xrightarrow{-e^-} ArH^{+\cdot} \xrightarrow{NO_2} Ar^+{\left<{}^{H}_{NO_2}\right.} \xrightarrow{-H^+} ArNO_2 \tag{44}$$

Using naphthalene and methyl-substituted naphthalene cation radicals as substrates in CH_2Cl_2–Bu_4NPF_6, it was shown [135] that this coupling reaction is more selective than nitration by acetyl nitrate in $(CH_3CO)_2O$ or N_2O_4 in CH_2Cl_2.

The anodic oxidation of solutions of aromatic hydrocarbons and $AgNO_2$ in MeCN gives mononitration with a substitution pattern held to support initial formation of a cation radical [136]. Alkenes may similarly be oxidized in the presence of $AgNO_2$, and depending on the alkene, nitro products are formed by substitution or addition. In probably the best example, coelectrolysis of anthracene and $AgNO_2$ at 1.3 V versus SCE gave 9-nitroanthracene in 47% yield (current efficiency 28%), and from 1,1-diphenylethylene, electrolyzed at 1.5 V, 1,1-diphenyl-2-nitroethylene was obtained in 40% yield. In each case 1 F mol^{-1} was consumed. However, since NO_2^- is more easily oxidized than both anthracene and 1,1-diphenylethylene, it cannot be excluded that NO_2^- is oxidized at the anode to give NO_2/N_2O_4, in itself a good nitrating reagent in homogeneous medium [135].

Aliphatic gem-dinitro compounds can be prepared by anodic oxidation of a nitroparaffin in aqueous alkaline solution in the presence of an excess of NO_2^- [137]. Yields are considerably better when Ag^+ is the oxidizing agent [138], so that an indirect oxidation process involving the generation of Ag^+ at an Ag anode was found to give excellent results [139].

D. Substitution by Cyanate Ion

Both nuclear and side-chain substitutions of alkylaromatics by NCO^- have been observed [140]. Thus, electrolysis of anisole in MeOH-KOCN results in substitution of hydrogen by a methylcarbamyl group, via the isocyanate. The reaction has been pictured as a direct oxidation process. Yields are fairly low (<20%).

E. Pyridination

Nitrogen bases (e.g., amines) are not generally good nucleophiles for anodic substitution reactions, as they are themselves easily oxidized (see Chap. 15). Pyridine bases, which are difficult to oxidize (E_{pa} = 1.4 V versus Ag/Ag^+ [141]), provide exceptions. The vinylic proton of the very easily oxidized tris(p-methoxyphenyl)ethylene may be substituted anodically by pyridine to give the corresponding pyridinium salt [142]. Similar anodic pyridination of Schiff's bases and N-methyl-4-methoxybenzanilide [143] has been reported and investigated in detail.

The reaction has been put to good preparative use for the formation in fair yields (~50%) of some heterocyclic zwitter-ions, Eq. (45) [144]. This reaction also provides an example in which the leaving group is a carbocation rather than a proton.

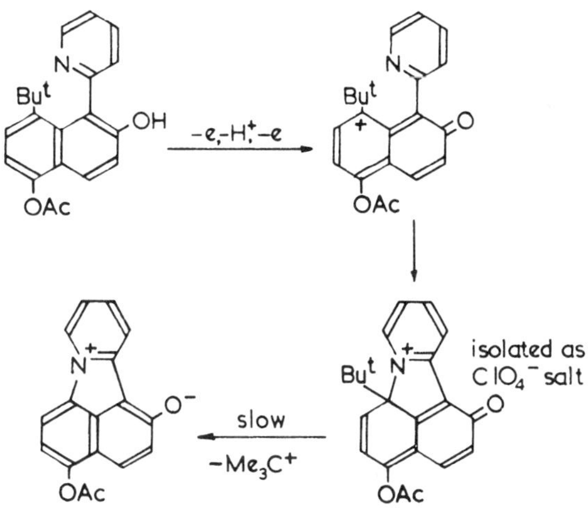

(45)

V. FORMATION OF CARBON-CARBON BONDS

A. Cyanation

Anodic cyanation is a reaction of considerable preparative interest since it allows the direct introduction of a cyano group into an aromatic ring [145-157]. Contrary to the situation in acyloxylation and alkoxylation, the product is more resistant toward oxidation than the starting material owing to the strong electron-withdrawing effect of the cyano group, and in principle good yields should be obtainable. The reaction is often run in MeOH–NaCN [145,150], which gives concurrent methoxylation [158]. This is avoided if $MeCN–Et_4NCN$ is used [151]. An important advance is the use of emulsions with phase-transfer agents [153-156].

The reaction was initially believed to be a homolytic substitution reaction by anodically generated cyano radicals [145]. However, although significant oxidation of CN^- takes place at a potential as low as 0.5 V versus SCE, the cyanation process occurs only in the region around or above $E_{1/2}$ of the substrate. This strongly implies a direct mechanism [146,150,151] analogous to that for acetoxylation.

Isomer distribution for cyanation of aromatic compounds has been determined in a number of cases [148-151]. The selectivity of the cyanation reaction is higher than that of the anodic acetoxylation process. This high regioselectivity has been put to good use for the preparation of 2-cyanopyrroles and 2-cyanoindoles, Eq. (46) [157]. If the 2-position is blocked, cyanation occurs at the 3-position. The efficient substitution for many pyrroles, imidazoles [159], and indoles studied contrasts markedly with the addition reactions observed for furans (see Chap. 16). However, this difference may only be apparent. At least for two of the cases cited earlier [157], it has been demonstrated [160] that the cyano-substituted pyrroles arise as a result of elimination during workup of the 2,5-addition product originally formed.

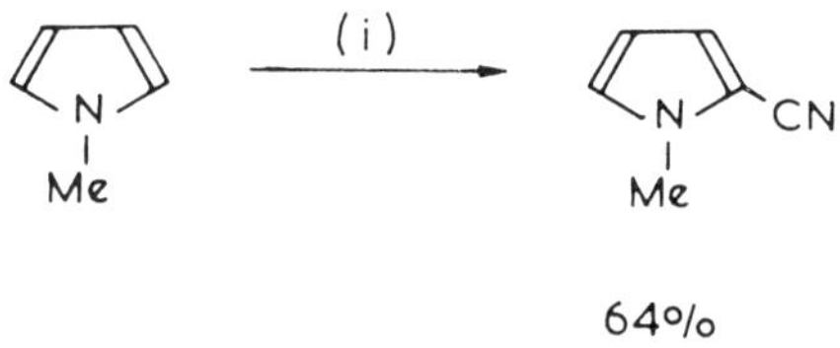

(46)

(i) Pt anode, 1.0 V (sce), 2 $F mol^{-1}$, MeOH-NaCN

Controlled potential (0.45 V versus SCE) oxidation of zinc octaethylporphyrin in DMF–Et_4NClO_4 in the presence of Et_4NCN results in the almost quantitative formation (>95%) of the 5-cyano derivative, whereas attempts to introduce more than one cyano group by oxidation at higher potentials resulted in mixtures of di-, tri-, and tetra-substituted products [161].

The direct nature of attack of CN^- on cation radicals of aromatic compounds has been demonstrated by CV [152]. The reversible one-electron oxidation of anthracene becomes irreversible in the presence of CN^-, and 2 F mol^{-1} electrolysis gives a mixture of cyano and isocyano addition across the 9,10-position. The best conditions for nuclear cyanation involve the use of emulsions, typically CH_2Cl_2–H_2O–NaCN, with a phase-transfer agent. This has been described in detail for the monocyanation of naphthalene and anisole with isolated yields in the region 50-70% [153,154]. For dimethoxybenzenes the yields are somewhat lower [154].

For trans-stilbene and trans-4,4-dimethoxystilbene, cyanation in an emulsion system gives for the former a mixture of addition and substitution products (including substitution of the vinylic protons), whereas for the latter substitution of the vinylic protons is the predominant reaction [156].

Alkylaromatic compounds do not give any side-chain substitution product under conditions that produce nuclear cyanation products [148-150]; this continuing mystery was referred to in the introduction to this chapter, together with recent attempts at an explanation.

In most cases cyanation involves loss of a proton, but in some cases an alkoxyl group may be displaced with very good yields of substitution products [151,155]. Thus, 1,2- and 1,4-dimethoxybenzene undergo cyanation by displacement of one of the methoxy groups, presumably via an ECE-type mechanism, Eq. (28). On the other hand, anisole and 1,3-dimethoxybenzene do not undergo methoxyl displacement but, instead, give nuclear substitution in the usual manner. The presence of an activating group in the ortho and para positions thus seems essential for methoxyl group displacement. However, the selectivity is less pronounced when the reactions are carried out in aqueous micellar solution [162]. The fate of the displaced alkoxyl group is not always clear; the observation of aldehyde and alcohol among the products of anodic cyanation of 4,4'-dialkoxybiphenyls [155] and the proven loss of benzyloxy radical from the cation radical of dibenzyl ether [163] support route a in Eq. (47). The cyanation of the 4,4'-dialkoxybiphenyls, Eq. (47), shows the use of emulsion systems to good advantage; the products include an important class of liquid crystals.

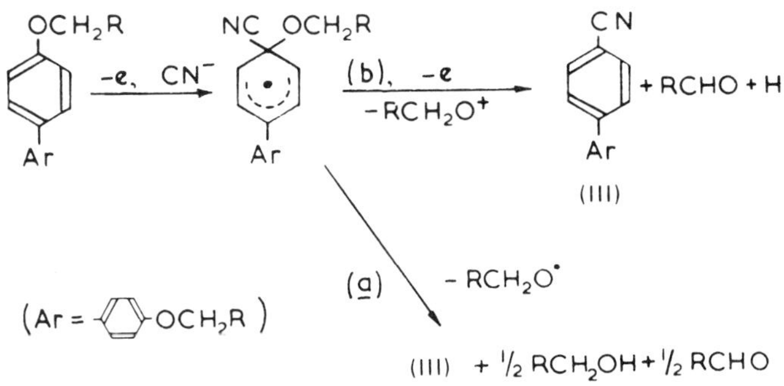

(47)

Tertiary amines are also substrates for the cyanation reaction [151,154, 164], substitution taking place at the carbon atom α to the nitrogen atom. In this case substitution occurs preferentially at a primary carbon atom in mixed amines, such as N-ethyl-N-methylaniline, Eq. (48). This is not so for cyclic tertiary amines, Eq. (49) [164].

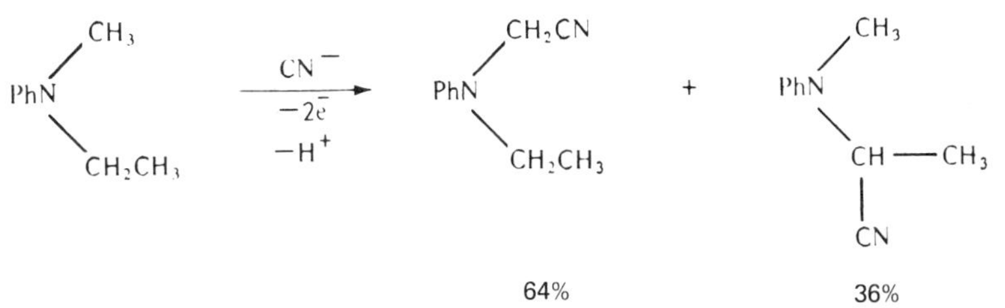

(48)

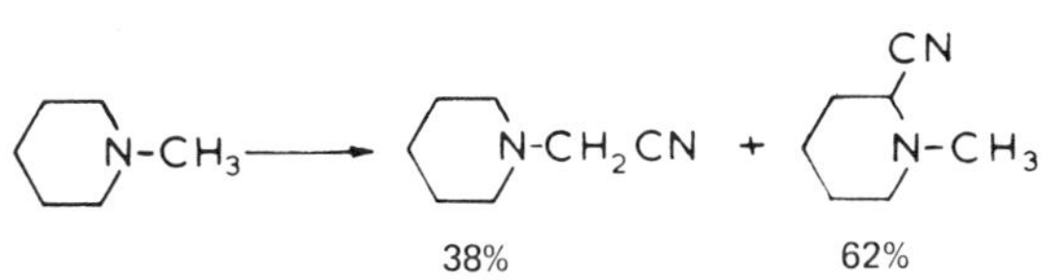

(49)

In this study, current-potential curves have been used to show that discharge of triethylamine, for example, occurs prior to the oxidation of the MeOH–NaCN electrolyte. Again, anodic cyanation in emulsion systems offers preparative advantage [154].

The fate of the anodically generated cyano radical in the absence of any substrate has been investigated [151] in several solvents. ESR spectroscopy shows that a relatively stable tricyanomethylenimine anion radical is ultimately formed, presumably via the pathway shown in Eq. (50):

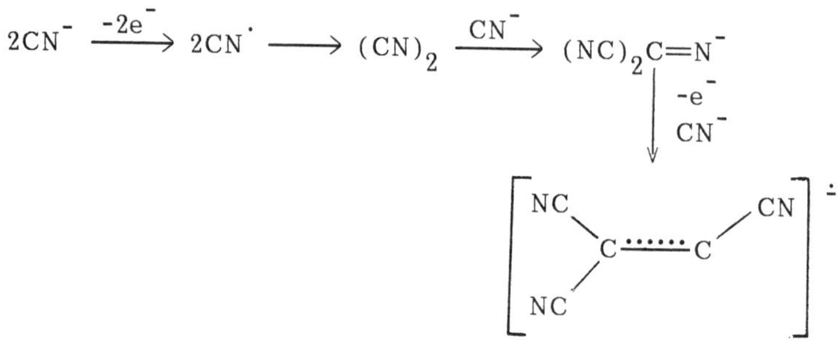

(50)

B. Carbomethoxylation

An unusual and interesting carbomethoxylation reaction occurs during electrolysis of an arylolefin in MeOH–NaOMe under CO pressure between two Pt electrodes, Eq. (51) [165]:

$$C_6H_5\overset{\overset{R}{|}}{C}{=}CH_2 + CO + CH_3O^- \xrightarrow{Pt_x(CO)_y^-} C_6H_5\overset{\overset{R}{|}}{C}{=}CHCOOCH_3 + 2e^- \qquad (51)$$

It has been shown that the Pt cathode dissolves to give a Pt(CO) complex. This is transformed at the anode to a carbomethoxy-Pt (CO) complex, which attacks the arylolefin to give the carbomethoxylation product. Hence, this reaction is of the indirect type.

C. Inter- and Intramolecular Coupling with C–C Bond Formation

Reaction conditions for anodic substitution may often be arranged so that the starting material is the most nucleophilic species present. An example of this has already been given, Eq. (45). Consequently, many preparatively useful inter- and intramolecular reactions may be so accomplished (see Chap. 23 for a more detailed discussion of coupling reactions).

An obvious case is the oxidation of electron-rich hydrocarbons in non-nucleophilic media, typically CH_2Cl_2–R_4NBF_4 or CH_2Cl_2 admixed with strong acids, such as CF_3COOH, CH_3SO_3H, or CF_3SO_3H [166,167]. The cation radicals formed by one-electron oxidation either couple or attack the starting material; an example is given in Eq. (52). Such couplings have also been achieved using a large-scale capillary gap cell with a graphite anode [167].

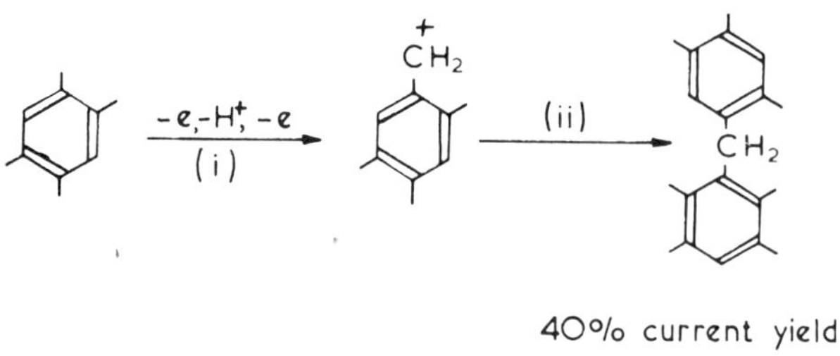

(52)

40% current yield

(i) Pt anode, CH_2Cl_2-CF_3CO_2H, Bu_4NBF_4, constant current

(ii) Durene.

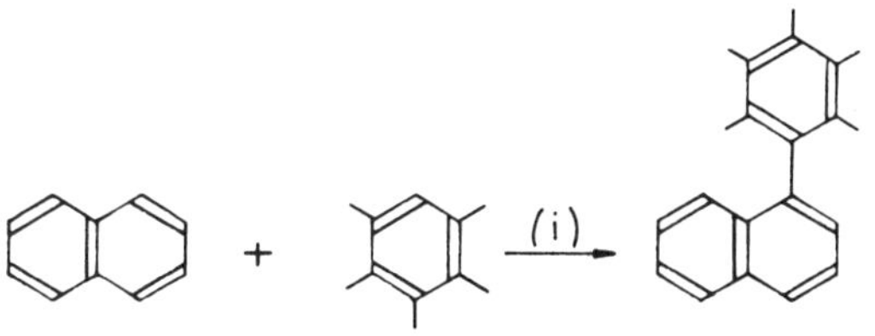

(52)

54% current yield

(i) Graphite anode, MeCN-HOAc, Bu_4NBF_4, constant current

Intramolecular coupling of aromatic systems based on this principle has been particularly fruitful. Phenols or aromatic ethers are usually involved: if the activation of the aromatic system is sufficient, the reactions do not require especially nonnucleophilic solvents. In the simplest case anisole may be anodically coupled in $CH_2Cl_2-CF_3COOH$ to give, after workup in the presence of Zn dust, 4,4'-dimethoxybiphenyl in 60% yield [168]. Much mechanistic information comes from detailed investigations of the intramolecular coupling reactions, such as Eqs. (53) and (54), of diarylalkanes [169, 170] and metacyclophanes [171,172]. For phenols the burden of evidence is in favor of intramolecular electrophilic attack by a phenoxonium ion, Eq. (53).

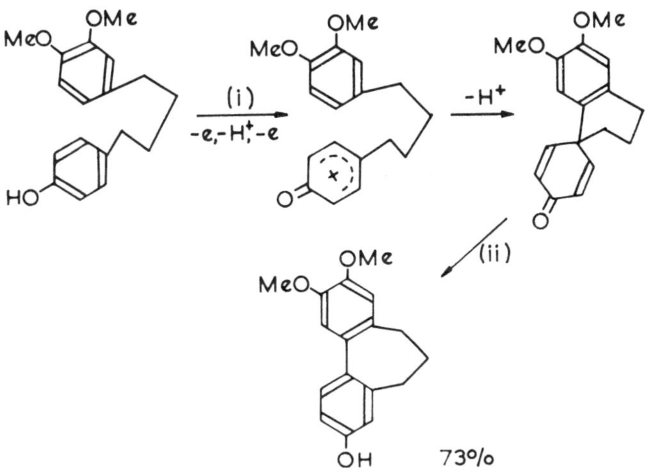

(53)

(i) Pt anode, 1.25 V (sce), MeCN-$NaClO_4$, 2Fmol^{-1}

(ii) CF_3CO_2H

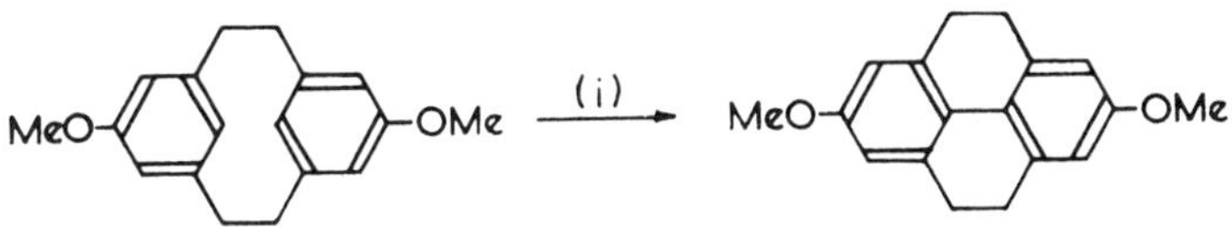

(54)

>90%

(i) Pt anode, 0.65 V (Ag/Ag^+), MeCN-$LiClO_4$

The anodic method has proved to be remarkably successful in accomplishing the sorts of oxidative couplings associated with alkaloid synthesis [173-176]. An illustrative example is given in Eq. (55):

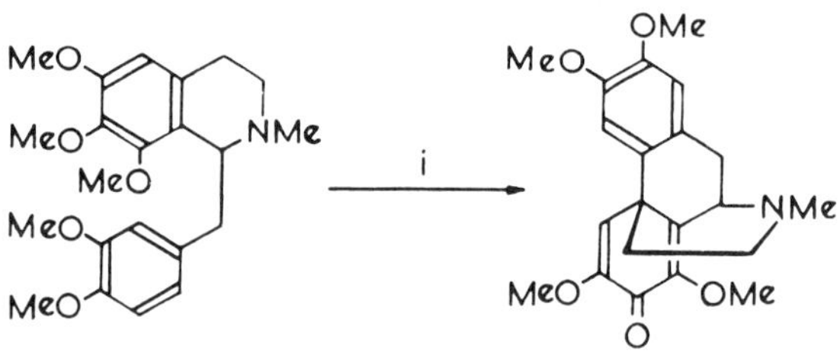

(55)

35%

(i) Pt anode, 1.1 V (Ag/Ag^+), MeCN-$LiClO_4$, ≤ 2 $Fmol^{-1}$

VI. FORMATION OF CARBON-HALOGEN BONDS

A. Fluorination

The virtually unique suitability of electrochemical methodology for preparing fluorinated compounds [177-185] warrants the inclusion of a separate treatment of this subject (Chap. 26).

B. Chlorination, Bromination, and Iodination

The oxidation of halide ions to molecular halogen is relatively easy; consequently, the mechanism of anodic halogenation, brought about by oxidation of organic substrates in the presence of halides, is not always clear [186-188]. In many cases it may involve homogenous halogenation by anodically generated halogen. In other cases the halide nucleophilically attacks a cation radical.

The oxidation of I_2 in MeCN produces a very reactive albeit ill-defined iodine(I) species [130,189]. The production of this species, in either MeCN or CH_2Cl_2 in the presence of a variety of substituted benzenes, results [189] in efficient iodination, as in Eq. (56). Plots of log k versus σ^+ are linear, which is consistent with conventional homogeneous electrophilic aromatic substitution.

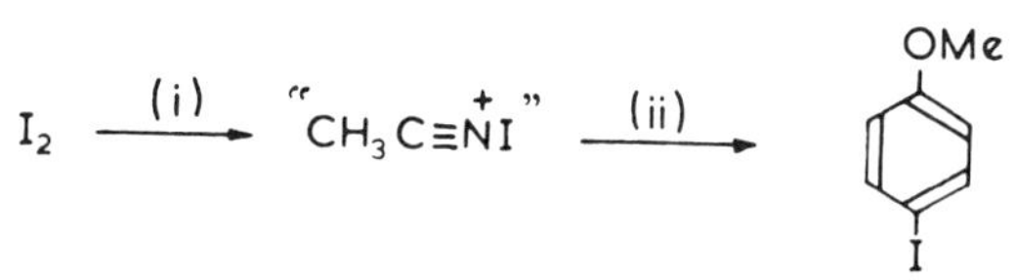

(56)

(i) Pt anode, 1.9 V (Ag/Ag^+), MeCN-$LiClO_4$, 2.4 $Fmol^{-1}$

90%

(ii) Anisole

Formation of a C–I bond is also the result of anodic oxidation of iodobenzene in the presence of benzene [190,191]. Two reactions are in competition, Eqs. (57) and (58); a careful investigation of the electrode kinetics supports the mechanism given and also allows optimization of conditions to the point at which a 95% yield of the diphenyliodonium cation is obtained [191]. Hydrolysis of this species completes a route from benzene to phenol.

$$\mathrm{PhI} \xrightleftharpoons{-e^-} \mathrm{PhI}^{+\cdot} \xrightarrow{\mathrm{PhH},\,-e^-,\,-H^+} \mathrm{Ph_2I^+} \longrightarrow \mathrm{PhOH} \qquad (57)$$

$$\mathrm{PhI}^{+\cdot} \xrightarrow{\mathrm{PhI}^{+\cdot}\ (\text{or PhI})} \text{dimeric products} \qquad (58)$$

REFERENCES

1. L. Eberson, Electron Transfer Reactions in Organic Chemistry, Springer-Verlag, Berlin, 1987, and references cited therein; Adv. Phys. Org. Chem., 18, 79 (1982); J. Amer. Chem. Soc., 105, 3192 (1983); L. Eberson, L. Jönsson, and O. Sänneskog, Acta Chem. Scand., B39, 113 (1985); C. Walling, C. Zhao, and G. M. El-Taliawi, J. Org. Chem., 48, 4910 (1983).
2. E. Baciocchi, L. Eberson and, C. Rol, J. Org. Chem., 47, 5106 (1982); E. Baciocchi, D. Bartoli, C. Rol, R. Ruzziconi and, G. V. Sebastiani, J. Org. Chem., 51, 3587 (1986).
3. L. Jönsson and L. G. Wistrand, J. Chem. Soc., Perkin Trans. I, 669 (1979); K. Nyberg and L.-G. Wistrand, J. Org. Chem., 43, 2613 (1978).
4. J. Cornelisse and E. Havinga, Chem. Rev., 75, 353 (1975); E. Havinga and J. Cornelisse, Pure Appl. Chem., 47, 1 (1976).
5. C. Párkányi, Pure Appl. Chem., 55, 331 (1983).
6. L. Eberson, Z. Blum, B. Helgée, and K. Nyberg, Tetrahedron, 34, 731 (1978).
7. L. Eberson and K. Nyberg, Acta Chem. Scand., B32, 235 (1978).
8. L. Eberson, L. Jönsson, and L.-G. Wistrand, Acta Chem. Scand., B32, 520 (1978).
9. A. Pross, J. Amer. Chem. Soc., 108, 3537 (1986).
10. A. Pross, Adv. Phys. Org. Chem., 21, 99 (1985).
11. S. S. Shaik, Prog. Phys. Org. Chem., 15, 197 (1985).
12. V. D. Parker and M. Tilset, J. Amer. Chem. Soc., 109, 2521 (1987).
13. O. Hammerich and V. D. Parker, Adv. Phys. Org. Chem., 20, 55 (1984).
14. N. L. Weinberg, (ed.), Techniques of Chemistry, Vol. 5, Parts 1 and 2; Techniques of Electroorganic Synthesis, Wiley-Interscience, New York, 1974.
15. N. L. Weinberg and B. V. Tilak (eds.), Techniques of Chemistry, Vol. 5, Part 3; Techniques of Electroorganic Synthesis, Wiley-Interscience, New York, 1982.
16. T. Shono, Tetrahedron, 40, 811 (1984); Electroorganic Chemistry as a New Tool in Organic Synthesis, Springer, Berlin, 1984.
17. K. Yoshida, Electrooxidation in Organic Chemistry, Wiley-Interscience, New York, 1984.
18. S. Torii, Electroorganic Syntheses, Part I; Oxidations, VCH, Weinheim, 1985.
19. L. Eberson and K. Nyberg, Adv. Phys. Org. Chem., 12, 1 (1976); Tetrahedron, 32, 2185 (1976).

20. J. S. Clarke, R. E. Ehigamusoe, and A. T. Kuhn, J. Electroanal. Chem., 70, 333 (1976).
21. A. Aramata, J. Electroanal. Chem., 182, 197 (1985).
22. F. H. Covitz, U.S. Patent 3,509,031 (1970); Chem. Abstr., 73, 11582w (1970).
23. H. Wendt and H. Schneider, J. Appl. Electrochem., 16, 134 (1986).
24. F. Fichter, Organische Elektrochemie, Steinkopf, Dresden, 1942.
25. N. L. Weinberg and H. R. Weinberg, Chem. Rev., 68, 449 (1968).
26. A. Nilsson, A. Ronlán, and V. D. Parker, J. Chem. Soc., Perkin Trans. I, 2337 (1975).
27. E. Oberrauch and L. Eberson, J. Appl. Electrochem., 16, 575 (1986).
28. L. Papouchado, G. Petrie, and R. N. Adams, J. Electroanal. Chem., 38, 389 (1972).
29. J. Froborg, G. Magnusson, and S. Thoren, J. Org. Chem., 40, 122 (1975); 40, 1595 (1975).
30. B. Wermeckes, F. Beck, and H. Schulz, Tetrahedron, 43, 577 (1987); B. Wermeckes and F. Beck, Chem. Ber., 120, 1679 (1987).
31. R. N. Adams, Acc. Chem. Res., 2, 175 (1969); V. D. Parker and L. Eberson, J. Chem. Soc., Chem. Commun., 441 (1972); O. Hammerich and V. D. Parker, Electrochim. Acta, 18, 537 (1973).
32. L. Eberson and B. Olofsson, Acta Chem. Scand., 23, 2355 (1969).
33. V. D. Parker and B. E. Burgert, Tetrahedron Lett., 2411 (1968).
34. K. Nyberg, J. Chem. Soc., Chem. Commun., 774 (1969).
35. I. M. Dalrymple and J. P. Millington, J. Appl. Electrochem., 16, 885 (1986).
36. H. J. Kim, K. Kusakabe, S. Hokazono, S. Morooka, and Y. Kato, J. Appl. Electrochem., 17, 1213 (1987).
37. M. G. Dolson and J. S. Swenton, J. Amer. Chem. Soc., 103, 2361 (1981).
38. K. Sasaki, H. Urata, K. Uneyama, and S. Nagura, Electrochim. Acta, 12, 137 (1967).
39. V. D. Parker and B. E. Burgert, Tetrahedron Lett., 2415 (1968).
40. A. Nilsson, U. Palmquist, T. Pettersson, and A. Ronlán, J. Chem. Soc., Perkin Trans. I, 696 (1978).
41. A. Nilsson, A. Ronlán, and V. D. Parker, Tetrahedron Lett., 1107 (1975).
42. A. Nilsson, U. Palmquist, T. Pettersson, and A. Ronlán, J. Chem. Soc., Perkin Trans. I, 708 (1978).
43. N. L. Weinberg and E. A. Brown, J. Org. Chem., 31, 4058 (1966).
44. N. L. Weinberg and T. B. Reddy, J. Amer. Chem. Soc., 90, 91 (1968).
45. N. L. Weinberg, J. Org. Chem., 33, 4326 (1968).
46. P. J. Smith and C. K. Mann, J. Org. Chem., 33, 316 (1968).
47. T. Shono, Tetrahedron, 40, 811 (1984) and references cited therein.
48. S. D. Ross, M. Finkelstein, and R. C. Petersen, J. Amer. Chem. Soc., 88, 4657 (1966); E. J. Rudd, M. Finkelstein, and S. D. Ross, J. Org. Chem., 37, 1763 (1972).
49. Z. Blum and K. Nyberg, Acta Chem. Scand., B36, 165 (1982).
50. T. Shono, Y. Matsumura, K. Tsubata, Y. Sugihara, S. Yamane, T. Kanazawa, and T. Aoki, J. Amer. Chem. Soc., 104, 6697 (1982).
51. S. Torii, T. Inokuchi, and M. Kubota, J. Org. Chem., 50, 4157 (1985).

52. T. Shono, Y. Matsumura, T. Kanazawa, M. Habuka, K. Uchida, and K. Toyoda, J. Chem. Res. (M) 2876 (1984); T. Shono, Y. Matsumura, K. Tsubata, and K. Uchida, J. Org. Chem., 51, 2590 (1986).
53. M. Malmberg and K. Nyberg, Acta Chem. Scand., B38, 85 (1984).
54. D. Duquet, R. E. A. Callens, F. A. M. Borremans, and M.J. O. Anteunis, Bull. Soc. Chim. Belg., 94, 51 (1985).
55. M. Okita, M. Mori, T. Wakamatsu, and Y. Ban, Heterocycles, 23, 247 (1985).
56. Z. Blum, M. Malmberg, and K. Nyberg, Acta Chem. Scand., B35, 739 (1981).
57. L. Cedheim, L. Eberson, B. Helgée, K. Nyberg, R. Servin, and H. Sternerup, Acta Chem. Scand., B29, 617 (1975).
58. K. Nyberg and R. Servin, Acta Chem. Scand., B30, 640 (1976).
59. M. Finkelstein, K. Nyberg, S. D. Ross, and R. Servin, Acta Chem. Scand., B32, 182 (1978).
60. L. Eberson, J. Hlavaty, L. Jönsson, K. Nyberg, R. Servin, H. Sternerup, and L.-G. Wistrand, Acta Chem. Scand., B33, 113 (1979).
61. P. D. Palasz, J. H. P. Utley, and J. D. Hardstone, Acta Chem. Scand., B38, 281 (1984).
62. P. D. Palasz, J. H. P. Utley, and J. D. Hardstone, J. Chem. Soc., Perkin Trans. II, 807 (1984).
63. T. Shono, Y. Matsumura, K. Tsubata, K. Uchida, T. Kanazawa, and K. Tsuda, J. Org. Chem., 49, 3711 (1984).
64. Z. Blum and K. Nyberg, Acta Chem. Scand., B35, 743 (1981).
65. T. Shono and Y. Matsumura, J. Amer. Chem. Soc., 91, 2803 (1969).
66. H. G. Thomas and A. Schmitz, Synthesis, 31 (1985).
67. M. Kimura, K. Koie, S. Matsubara, Y. Sawaki, and H. Iwamura, J. Chem. Soc., Chem. Commun., 122 (1987).
68. T. Fuchigami, Y. Nakagawa, and T. Nonaka, Tetrahedron Lett., 27, 3869 (1986).
69. T. Shono and A. Ikeda, J. Amer. Chem. Soc., 94, 7892 (1972); Chem. Lett., 311 (1976); T. Shono, A. Ikeda, J. Hayashi, and S. Hakozaki, J. Amer. Chem. Soc., 97, 4261 (1975).
70. R. Engels, H. J. Schäfer, and E. Steckhan, Liebigs Ann. Chem., 204 (1977); J. Baltes, E. Steckhan, and H. J. Schäfer, Chem. Ber., 11, 1294 (1978).
71. T. Shono and Y. Matsumura, Bull. Chem. Soc. Jpn., 48, 2861 (1975); J. Org. Chem., 35, 4157 (1970); T. Shono, Y. Matsumura, and Y. Makagawa, J. Org. Chem., 36, 1771 (1971).
72. T. Shono and T. Kosaka, Tetrahedron Lett., 6207 (1968).
73. L. Eberson and E. Oberrauch, Acta Chem. Scand., B35, 193 (1981).
74. C. Adams, E. N. Frankel, and J. H. P. Utley, J. Chem. Soc., Perkin Trans. I, 353 (1979).
75. R. F. Garwood, Naser-ud-din, and B. C. L. Weedon, J. Chem. Soc., Chem. Commun., 923 (1968).
76. K. Almdal and O. Hammerich, Sulfur Lett., 2, 1 (1984).
77. L. Rand and F. Mohar, J. Org. Chem., 30, 3156, 3885 (1965).
78. K. Koyama, K. Yoshida, and S. Tsutsumi, Bull. Chem. Soc. Jpn., 39, 516 (1966); K. Koyama, T. Ebara, and S. Tsutsumi, Bull. Chem. Soc. Jpn., 41, 2668 (1968).
79. J. F. K. Wilshire, Aust. J. Chem., 16, 432 (1963).
80. R. P. Linstead, J. C. Blunt, B. C. L. Weedon, and B. R. Shephard, J. Chem. Soc., 3624 (1952).

81. C. L. Wilson and W. T. Lippincott, J. Amer. Chem. Soc., 78, 4291 (1956); J. Electrochem. Soc., 103, 672 (1956).
82. D. R. Harvey and R. O. C. Norman, J. Chem. Soc., 4860 (1964).
83. L. Eberson and K. Nyberg, J. Amer. Chem. Soc., 88, 1686 (1966); Acta Chem. Scand., 18, 1568 (1964).
84. L. Eberson, J. Amer. Chem. Soc., 89, 4669 (1967).
85. L. Eberson and A. Webber, Acta Chem. Scand., B36, 53 (1982).
86. S. D. Ross, M. Finkelstein, and R. C. Petersen, J. Amer. Chem. Soc., 89, 4088 (1967).
87. L. Eberson and B. Helgée, Acta Chem. Scand., B32 157 (1978).
88. U. Svanholm and V. D. Parker, Tetrahedron Lett., 471 (1972).
89. Z. Blum, L. Cedheim, K. Nyberg, and L. Eberson, Acta Chem. Scand., B29, 715 (1975).
90. Y.-H. So, J. Y. Becker, and L. L. Miller, J. Chem. Soc., Chem. Commun., 262 (1975).
91. Y.-H. So and L. L. Miller, Synthesis, 468 (1976).
92. G. Bockmair, H. P. Fritz, and H. Gebauer, Electrochim. Acta, 23, 21 (1978).
93. R. P. Kreh, M. E. Tadros, H. M. Hand, M. P. Cockerham, and E. K. Smith, J. Appl. Electrochem., 16, 440 (1986).
94. H. Iwasaki, L. A. Cohen, and B. Witkop, J. Amer. Chem. Soc., 85, 3701 (1963).
95. A. I. Scott, P. A. Dodson, F. McCapra, and M. B. Meyers, J. Amer. Chem. Soc., 85, 3702 (1963).
96. W. A. Bonner and F. D. Mango, J. Org. Chem., 29, 430 (1964).
97. W. J. Koehl, J. Org. Chem., 32, 614 (1967).
98. F. M. Banda and R. Brettle, J. Chem. Soc., Perkin Trans. I, 1907 (1974).
99. A. Laurent, E. Laurent, and M. Thomalla, C. R. Acad. Sci. Ser. C, 274, 1537 (1972).
100. C. J. Myall, D. Pletcher, and C. Z. Smith, J. Chem. Soc., Perkin Trans. I, 2035 (1976).
101. C. Adams, N. Jacobsen, and J. H. P. Utley, J. Chem. Soc., Perkin Trans. II, 1071 (1978).
102. D. B. Clark, M. Fleischmann, and D. Pletcher, J. Chem. Soc., Perkin Trans. II, 1578 (1973).
103. C. B. Campbell and D. Pletcher, Electrochim. Acta, 23, 923 (1978).
104. A. Hembrock, H. J. Schäfer, and G. Zimmermann, Angew. Chem. Int. Ed. Engl., 24, 1055 (1985).
105. S. D. Ross, M. Finkelstein, and R. C. Petersen, J. Org. Chem., 35, 781 (1970).
106. K. Nyberg, Acta Chem. Scand., B24, 473 (1970).
107. R. R. Rao, S. B. Milliken, S. L. Robinson, and C. K. Mann, Anal. Chem., 42, 1076 (1970).
108. G. Cauquis and G. Pierre, C. R. Acad. Sci., Ser. C, 266, 883 (1968).
109. G. Cauquis and G. Pierre, C. R. Acad. Sci., Ser. C, 272, 609 (1971).
110. H. P. Fritz and P. Ecker, Chem. Ber., 114, 3643 (1981).
111. A. Bewick, G. J. Edwards, J. M. Mellor, and B. S. Pons, J. Chem. Soc., Perkin Trans. II, 1952 (1977).
112. K. Koyama, T. Susuki, and S. Tsutsumi, Tetrahedron, 23, 2665 (1967).
113. A. E. Coleman, H. H. Richtol, and D. A. Aikens, J. Electroanal. Chem., 18, 165 (1968).

114. V. D. Parker, J. Electroanal. Chem., 22, 277 (1969).
115. O. Hammerich and V. D. Parker, J. Chem. Soc., Chem. Commun., 245 (1974).
116. J. Y. Becker, L. L. Miller, and T. M. Siegel, J. Chem. Soc., Chem. Commun., 341 (1974).
117. V. R. Koch and L. L. Miller, Tetrahedron Lett., 693 (1973).
118. L. L. Miller, V. R. Koch, T. Koenig, and M. Tuttle, J. Amer. Chem. Soc., 95, 5075 (1973).
119. F. Vincent, R. Tardivel, P. Mison, and P. von R. Schleyer, Tetrahedron, 33, 325 (1977).
120. G. J. Edwards, S. R. Jones, and J. M. Mellor, J. Chem. Soc., Perkin Trans. II, 505 (1977).
121. A. Bewick, G. J. Edwards, S. R. Jones, and J. M. Mellor, J. Chem. Soc., Perkin Trans. I, 1831 (1977).
122. L. L. Miller and V. Ramachandran, J. Org. Chem., 39, 369 (1974).
123. L. L. Miller and M. Katz, J. Electroanal. Chem., 72, 329 (1976).
124. J. Y. Becker, L. R. Byrd, L. L. Miller, and Y.-H. So, J. Amer. Chem. Soc., 97, 853 (1975).
125. J. Y. Becker, L. L. Miller, and T. M. Siegel, J. Amer. Chem. Soc., 97, 849 (1975).
126. E. A. Mayeda, J. Amer. Chem. Soc., 97, 4012 (1975).
127. S. Hammerum and O. Hammerich, Tetrahedron Lett., 5027 (1979).
128. A. Laurent and R. Tardivel, C. R. Acad. Sci., Ser. C, 271, 324 (1970); 272, 8 (1971); A. Laurent, E. Laurent, and R. Tardivel, Tetrahedron Lett., 4861 (1973).
129. A. Laurent, E. Laurent, and R. Tardivel, Tetrahedron, 30, 3423, 3431 (1974).
130. L. L. Miller and B. F. Watkins, Tetrahedron Lett., 4495 (1974).
131. F. Vincent, R. Tardivel, and P. Mison, Tetrahedron Lett., 603 (1975); Tetrahedron, 32, 1681 (1976).
132. H. Ohmori, C. Ueda, K. Yamagata, M. Masui, and H. Sayo, J. Chem. Soc., Perkin Trans. II, 1065 (1987).
133. T. Fuchigami, T. Sato, and T. Nonaka, J. Org. Chem., 51, 366 (1986).
134. G. Cauquis and D. Serve, C. R. Acad. Sci., Ser. C, 262, 1516 (1966).
135. L. Eberson and F. Radner, Acc. Chem. Res., 20, 53 (1987) and references cited therein.
136. A. Laurent, E. Laurent, and P. Locher, Electrochim. Acta, 20, 857 (1975).
137. C. T. Bahner, Ind. Eng. Chem., 44, 317 (1952).
138. R. B. Kaplan and H. Schechter, J. Amer. Chem. Soc., 83, 3535 (1961).
139. C. M. Wright and D. R. Levering, Tetrahedron, 19(Suppl. 1), 3 (1963).
140. V. D. Parker and B. E. Burgert, Tetrahedron Lett., 3341 (1968).
141. W. R. Turner and P. J. Elving, Anal. Chem., 37, 467 (1965).
142. V. D. Parker and L. Eberson, J. Chem. Soc., Chem. Commun., 451 (1969).
143. M. Masui and H. Ohmori, J. Chem. Soc., Perkin Trans. II, 1112 (1973); M. Masui, H. Ohmori, H. Sayo, A. Ueda, and C. Ueda, J. Chem. Soc., Perkin Trans. II, 1180 (1976).
144. G. Popp, J. Org. Chem., 37, 3058 (1972).

145. K. Koyama, T. Susuki, and S. Tsutsumi, Tetrahedron Lett., 627 (1965).
146. V. D. Parker and B. E. Burgert, Tetrahedron Lett., 4065 (1965).
147. K. Koyama, T. Susuki, and S. Tsutsumi, Tetrahedron, 23, 2675 (1967); T. Susuki, K. Koyama, A. Omori, and S. Tsutsumi, Bull. Chem. Soc. Jpn., 41, 2663 (1968).
148. S. Tsutsumi and K. Koyama, Disc. Faraday Soc., 45, 247 (1968).
149. K. Yoshida and T. Fueno, J. Chem. Soc., Chem. Commun., 711 (1970).
150. L. Eberson and S. Nilsson, Disc. Faraday Soc., 45, 242 (1968).
151. S. Andreades and E. W. Zahnow, J. Amer. Chem. Soc., 91, 4181 (1969).
152. V. D. Parker and L. Eberson, J. Chem. Soc., Chem. Commun., 441 (1972).
153. L. Eberson and B. Helgée, Chem. Scripta, 5, 47 (1974).
154. L. Eberson and B. Helgée, Acta Chem. Scand., B29, 451 (1975).
155. L. Eberson and B. Helgée, Acta Chem. Scand., B31, 813 (1977).
156. L. Eberson and B. Helgée, Acta Chem. Scand., B31, 313 (1978).
157. K. Yoshida, J. Amer. Chem. Soc., 99, 6111 (1977).
158. K. Yoshida and T. Fueno, Bull. Chem. Soc. Jpn., 60, 229 (1987) and references cited therein.
159. K. Yoshida and H. Kitabayashi, Bull. Chem. Soc. Jpn., 60, 3693 (1987).
160. L. Eberson, Acta Chem. Scand., B34, 747 (1980).
161. H. J. Callot, A. Louati, and M. Gross, Bull. Soc. Chim. Fr., II-317 (1983).
162. E. Laurent, G. Rauniyar, and M. Thomalla, Nouv. J. Chim., 6, 515 (1982).
163. R. Lines and J. H. P. Utley, J. Chem. Soc., Perkin Trans. II, 803 (1977).
164. T. Chiba and Y. Takata, J. Org. Chem., 42, 2973 (1977).
165. T. Inoue and S. Tsutsumi, J. Amer. Chem. Soc., 87, 3525 (1965); Bull. Chem. Soc. Jpn., 38, 2122 (1965).
166. K. Nyberg, Acta Chem. Scand., B24, 1609 (1970); L. Eberson, K. Nyberg, and H. Sternerup, Acta Chem. Scand., B27, 1679 (1973).
167. L. Eberson, K. Nyberg, and H. Sternerup, Chem. Scripta, 3, 12 (1973).
168. A. Ronlán, K. Bechgaard, and V. D. Parker, Acta Chem. Scand., B27, 2375 (1973).
169. U. Palmquist, A. Ronlán, and V. D. Parker, Acta Chem. Scand., B28, 267 (1974); U. Palmquist, A. Nilsson, V. D. Parker, and A. Ronlán, J. Amer. Chem. Soc., 98, 2571 (1976); A. Ronlán, and V. D. Parker, J. Org. Chem., 39, 1014 (1974).
170. J. R. Falck, L. L. Miller, and F. R. Stermitz, J. Amer. Chem. Soc., 96, 2981 (1974).
171. J. Y. Becker, L. L. Miller, V. Boekelheide, and T. Morgan, Tetrahedron Lett., 2939 (1976).
172. T. Sato and M. Kamada, J. Chem. Soc., Perkin Trans. II, 384 (1977).
173. E. Kotani, M. Kitazawa, and S. Tobinaga, Tetrahedron, 30, 3027 (1974); E. Kotani, F. Miyazaki, and S. Tobinaga, J. Chem. Soc., Chem. Commun., 300 (1974).
174. J. R. Falck, L. L. Miller, and F. R. Stermitz, Tetrahedron, 30, 931

(1974); L. L. Miller, R. F. Stewart, J. P. Gillespie, V. Ramachandran, Y. H. So, and F. R. Stermitz, J. Org. Chem., 43, 1580 (1978).
175. J. M. Bobbitt, I. Noguchi, and R. S. Ware, J. Org. Chem., 40, 2924 (1975); J. M. Bobbitt, I. Noguchi, H. Yagi, and K. A. Weisgraber, J. Org. Chem., 41, 845 (1976).
176. M. Sainsbury and J. Wyatt, J. Chem. Soc., Perkin Trans. I, 661 (1976); J. Chem. Soc., Perkin Trans. II, 1750 (1977).
177. J. Burdon and J. C. Tatlow, Adv. Fluorine Chem., 1, 129 (1960).
178. S. Nagase, Fluorine Chem. Rev., 1, 77 (1967).
179. W. V. Childs in Ref. 15, p. 341.
180. V. S. Plashkin and Y. P. Dolnakov, J. Appl. Chem. USSR, 44, 1181 (1971).
181. H. M. Fox, F. N. Ruehlen, and W. V. Childs, J. Electrochem. Soc., 118, 1246 (1971).
182. B. Chang, H. Yanase, K. Nakanishi, and N. Watanabe, Electrochim. Acta, 16, 1179 (1971).
183. C. J. Ludman, E. McCarron, and R. F. O'Malley, J. Electrochem. Soc., 119, 874 (1972); J. F. Carpenter, L. H. Ekes, P. F. King, H. A. Mariani, M. M. Zadeh, R. F. O'Malley, and V. J. Roman, J. Electrochem. Soc., 130, 2170 (1983); R. F. O'Malley, H. A. Mariani, D. R. Buhler, and D. M. Jerina, J. Org. Chem., 46, 2816 (1981); P. F. King and R. F. O'Malley, J. Org. Chem., 49, 2803 (1984).
184. A. Bensadat, G. Bodennec, E. Laurent, and R. Tardivel, Tetrahedron Lett., 3799 (1977); E. Laurent, R. Tardivel, and H. Thiebault, Tetrahedron Lett., 24, 903 (1983).
185. I. N. Rozhkov, Russ. Chem. Rev., 45, 615 (1976).
186. S. R. Forsyth, D. Pletcher, and K. P. Healy, J. Appl. Electrochem., 17, 905 (1987) and references cited therein.
187. J. P. Millington, J. Chem. Soc. B, 982 (1969).
188. G. Faita, M. Fleischmann, and D. Pletcher, J. Electroanal. Chem., 25, 455 (1970).
189. L. L. Miller and B. F. Watkins, J. Amer. Chem. Soc., 98, 1515 (1976).
190. L. L. Miller and A. K. Hoffmann, J. Amer. Chem. Soc., 89, 593 (1967).
191. H. Hoffelner, H. W. Lorch, and H. Wendt, J. Electroanal. Chem., 66, 183 (1975).

26
Anodic Fluorination

W. V. CHILDS, LEIF CHRISTENSEN, F. W. KLINK, and C. F. KOLPIN / 3M Company, St. Paul, Minnesota

I. Introduction 1103

II. Electrochemical Fluorination in Anhydrous HF: The Simons Process 1104
- A. Experimental overview 1104
- B. Synthctic aspects 1106
- C. Mechanistic considerations 1108

III. Electrochemical Fluorination in Molten KF·2HF: The Phillips Process 1113
- A. Operational summary 1113
- B. Process rationalization 1115

IV. Partial Electrochemical Fluorination 1117

V. Summary 1123

References 1124

I. INTRODUCTION

The unique characteristics and commercial significance of anodic fluorination* warrant a separate chapter on this example of anodic substitution and/or addition. Over the last 50 years three general methodologies for electrochemical fluorination (ECF) have developed.

The Simons ECF process is the oldest technique. It remains a significant route to many perfluorinated organics. A solution or dispersion of organic substrate is electrolyzed at a Ni anode in anhydrous hydrogen fluoride (AHF). The cell products are mostly perfluorinated, and materials of commercial significance [1] include perfluoroalkanes, perfluorocarbonylfluorides, perfluorosulfonylfluorides, perfluorotrialkylamines, and perfluoroalkyl ethers. Fluorinated by-products associated with cleavage, degradaticn, rearrangement, and dimerization of the reactant are typically observed. The mechanism of Simons ECF appears to involve the generation of carbocationic intermediates and high-valence nickel fluorides.

---

*CAUTION: Many of the reagents used and products formed during anodic fluorination are extremely hazardous. It is essential that the hazards are reviewed by competent professionals, that proper materials of construction are employed, that adequate safety precautions are established, and that suitable medical resources are available.

The Phillips ECF process utilizes a porous carbon anode in a molten $KF \cdot 2HF$ electrolyte. The substrate, typically a gas or volatile liquid not particularly soluble in the electrolyte, is introduced through the anode and undergoes statistical replacement of hydrogen with fluorine. The products range from monofluoro to perfluoro compounds and include alkane, chloroalkane, carbonyl fluoride, and ester derivatives. The mechanism is believed to involve the generation of elemental fluorine followed by "in situ" reaction with the organic substrate via a free radical scheme. Thus, the process can be classified as an indirect anodic substitution reaction.

Partial ECF is typically carried out in an inert solvent in the presence of an organic substrate and a source of fluoride ions. The products are mono- and/or difluorinated and result from the generation of a carbocationic species followed by reaction with the fluoride nucleophile. Hence, the mechanism has much in common with other anodic substitution processes (see Chap. 25).

The general destructive nature of the Simons process has prompted the term "hard fluorination," describing electrolysis in AHF, compared with the "mild fluorination" conditions employed during the partial ECF [2]. Such terminology is not continued in this chapter. Several excellent reviews have appeared on anodic fluoridation [3–7].

II. ELECTROCHEMICAL FLUORINATION IN ANHYDROUS HF: THE SIMONS PROCESS

Electrochemical fluorination using nickel anodes to electrolyze a solution of a dissolved organic compound in an AHF electrolyte was discovered by Joseph Simons in the late 1930s and first published in 1949 [8]. Since that time the Simons process has been extensively investigated and commercially practiced; the literature contains many excellent reviews of Simons ECF [2–6,9–13]. No attempt is made here to duplicate these previous summaries. Instead, following an overview of the Simons process, the discussion focuses on information that has become available since I. N. Rozhkov's review [2].

A. Experimental Overview

Simons ECF may be characterized as an electrolysis leading to perfluorination. All hydrogens on the substrate are replaced with fluorine, and all unsaturated carbon-carbon bonds are saturated with fluorine. Some retention of functional groups is observed, but considerable carbon skeletal rearrangement and substrate degradation occur. This electrochemical process is characterized by the formation of a conductive passivating layer of metal fluorides covered by a relatively thick layer of polymerized organic and partially fluorinated products. These two layers form the anode film, which is largely insoluble in the HF electrolyte. The formation of this anode film is crucial to the fluorination process. It appears that this film, not the bulk electrolyte, serves as the solvent medium for the fluorination reaction.

The net anodic reaction for the ECF of a hydrocarbon requires the transfer of two electrons per fluorine atom for substitution and the transfer of one electron per fluorine atom for addition (e.g., double-bond saturation).

For the perfluorination of a saturated, aliphatic hydrocarbon, the overall reaction may be written as

$$C_mH_n + nHF \longrightarrow C_mF_n + nH_2$$

The consumption of fluoride (with accompanying generation of H^+) in the anodic reaction causes this region to become more acidic than the bulk electrolyte—increasing the likelihood of acid-catalyzed degradation of the organic substrate. The net cathodic reaction is generation of hydrogen. Overall, one HF molecule is consumed for each fluorine incorporated into the reactant molecule.

A diagram of a typical Simons ECF system is shown in Figure 1. The majority of reported investigations have been conducted employing a pack of alternating anode and cathode sheets immersed in a common HF electrolyte [3,8,9]. However, studies have also been conducted using membrane divided cells [14], rotating disk cells [14–16], rotating ring-disk cells [17–19], and cells using porous anodes [3,20], hollow anodes [3,8] or rotating cylindrical anodes [3].

Nickel or a nickel alloy is the material of choice for the anode. Nickel and to a lesser extent platinum [2] are the only materials that form a reasonably conductive, passivating film when anodically polarized in AHF. The cathodes may be fabricated of nickel, a nickel alloy, or iron. If desired, a reference electrode, such as palladium/hydrogen [14–16,21], Cu/CuF_2 [16–17,22,23], or Hg/Hg_2F_2[16,22,24], may be incorporated in the system.

The electrolysis is conducted in a constant-current, constant cell voltage, or constant anode potential mode. Owing to the large voltage drop over the anode film [18,25], the constant cell voltage mode approximates the constant anode potential mode. Cell voltages of 4.5–7.0 V are typical for this process. Linear rather than Tafel behavior is dominant during steady-state electrolysis. Reported current densities are in the range of 1–25 mA/cm^2 based on the planar projected electrode area.

Anhydrous HF boils at 19.5°C at 1 atm pressure, thus limiting ECF from higher operating temperatures at ambient pressure. Experiments are typically performed at electrolyte temperatures of −10°C up to slightly above the boiling temperature of AHF.

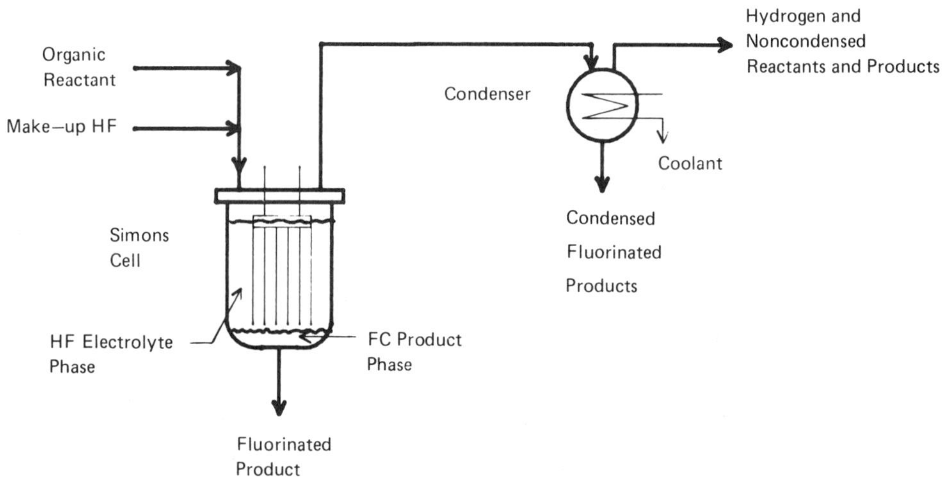

Figure 1. Typical Simons ECF process.

Anhydrous HF is usually shipped in steel cylinders; to reduce trace quantities of water and corrosion products, it is often desirable to distill the material before use. Preelectrolysis of the AHF may be used to further purify the electrolyte, in which case it is desirable to employ a separate set of preelectrolysis electrodes. Care should always be taken to maintain the electrolyte free of water, as oxygen difluoride, ozone, and, depending on the organic substrate, other powerful oxidants are produced during the electrolysis of H_2O in HF.

The acidity and high dielectric constant of AHF lead to the protonation of many organic compounds and the dissociation of the resulting ion pairs to form highly conductive solutions [4,14]. These compounds include amines, ethers, sulfonyl and acyl halides, alcohols, and esters. A range of reactant concentrations may be employed, although typically the reactant concentration is less than 16% by weight [9].

Some of the least soluble organic materials in AHF are saturated or halogenated hydrocarbons. In such cases, additives have been used to enhance conductivity. Alkali metal and alkaline earth fluorides may be used as inert additives. A variety of other substances have been used, which themselves are fluorinated simultaneously with the substrate [9]. The first inert additives can lead to higher anodic corrosion rates [8,9] and cause the conductivity to remain high even after the organic reactant is consumed; the latter additives produce a by-product and reduce the current efficiency. Agitation or external forced convection improves the uniformity of the ECF environment [26] and increases the effectiveness of the additives.

Product recovery is quite simple; in contrast to the substrates, most fluorochemicals are relatively insoluble in the electrolyte and the product may be removed from the bottom of the cell or recovered from traps in the exiting vapor stream. During cryogenic recovery of product from vapor streams, care should be taken to consume or absorb potential oxidants from the vapor prior to the product traps.

B. Synthetic Aspects

A variety of classes of organic compounds may be fluorinated by Simons ECF with various degrees of success. These classes include organic acid halides, ethers, esters, amines, amino ethers, aliphatic hydrocarbons, halohydrocarbons, and divalent and hexavalent sulfur compounds [9,27]. Cleavage and dimerization do occur, and in general the yield of desired product decreases with increasing carbon chain length.

Several recent studies have cast new light on some of the synthetic possibilities.

A series of thiols were fluorinated [28] to prepare the corresponding perfluoroalkylsulfur pentafluoride analogs. Extensive cleavage occurred, especially at the carbon-sulfur bond producing the corresponding perfluoroalkane and sulfur hexafluoride. The highest yield reported for the series was 12% for perfluorobutylsulfur pentafluoride. Sizable quantities of sulfur and polymeric tarry materials were found in the cells following the experiments. This is in accordance with Simons' earlier observation [29] that thiols are unstable in anhydrous HF.

Formation of cyclic ethers during electrolyses of acid fluorides or esters during Simons ECF is well documented [9]. A recent series of papers further explores this phenomenon [30–35]. Alcohols, such as 2-methyl-1-pentanol, were electrolyzed to yield a product mix consisting primarily of perfluoroalkanes, together with small amounts of perfluorinated cyclic ether and the corresponding perfluorinated acid fluoride [30].

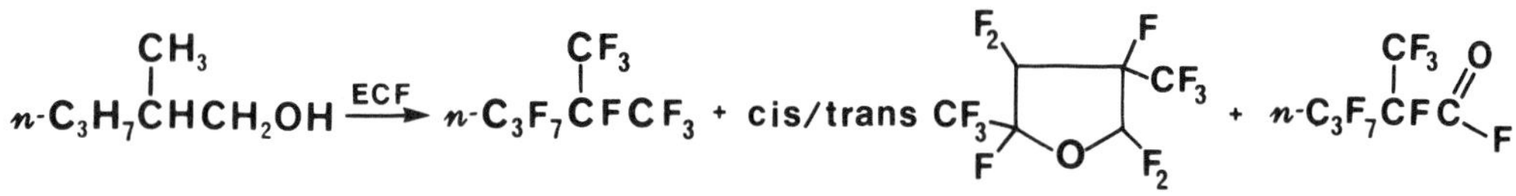

(1)

α Alkyl-substituted acid chlorides or methyl esters were fluorinated to yield the analogous perfluoroacid fluorides and 5- and 6-membered ring perfluorocyclic ethers in good yields [31]. Tar formation was reported to be reduced with the methyl ester feeds compared with that found with the analogous acid chlorides.

Methyl cyclohexen-2-ylbutyrate was fluorinated to produce perfluorobicyclic ethers [33]. The presence of the double bond in the ring is thought to promote attack at the ring rather than at the pendent alkane.

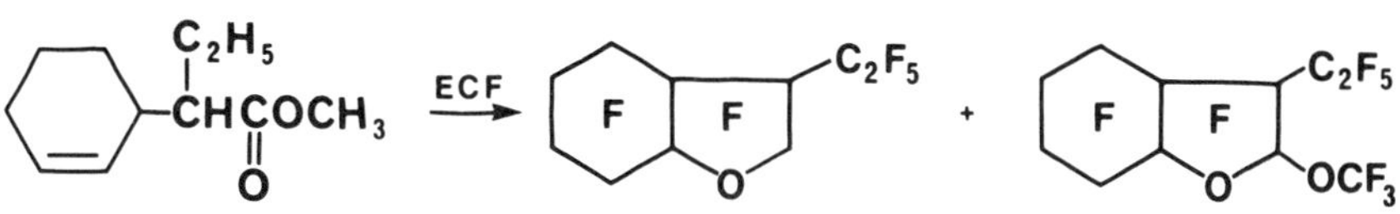

(2)

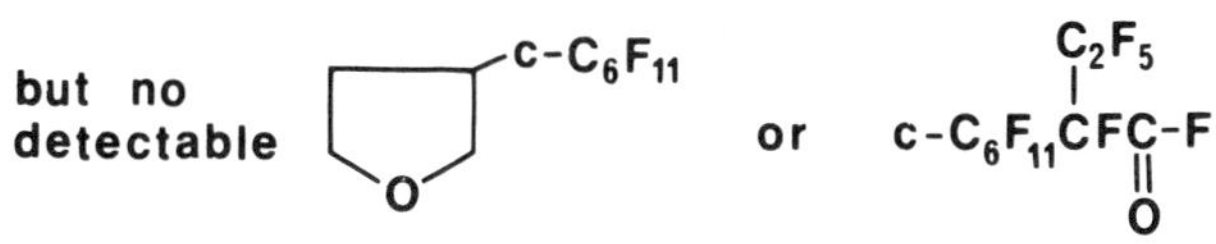

The fluorination of cycloalkyl-substituted acid derivatives produced both perfluorobicyclic and perfluoromonospiro cyclic ethers [34]. The primary by-products were 7- and 8-membered ring-substituted perfluorocycloalkanes. The nature of the products is controlled by the number of carbon atoms between the ring and the carbonyl group and by the length of the carbon side chain. For example, cyclization, illustrated by Eq. (3), does not occur unless formation of that 5- or 6-membered cyclic ether ring is possible.

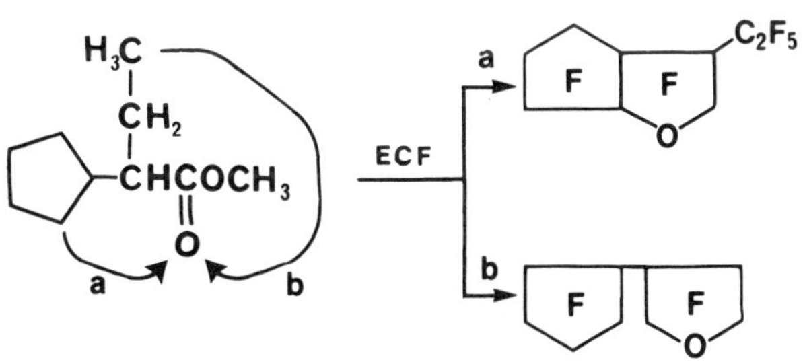

(3)

In these experiments, fluorination products were collected at three points: from the bottom of the cell; in a −78°C trap; and in a −196°C trap. The relative amounts of product recovered from each collection point, really a measure of the extent of cleavage, was very dependent on the structure of the organic feed. The total quantity of fluorochemical product was similar in all cases, but the percentage recovered from the cell bottom varied from 13% for methylcyclohexylacetate to 86% for methyl-3-cyclohexyl-2-methylpropionate. It was also reported that the corresponding methyl esters undergo ECF at lower and more stable voltages than their corresponding acid chlorides. Even for the preparation of the small molecule perfluoromalonylfluoride by ECF, our laboratory has found that molar yields from the diester are higher than those obtained from the diacid chloride.

Recently, α-cyclohexenyl-substituted carboxylic acid esters have been evaluated as a source of perfluorobicyclic ethers [35]. It was proposed that the presence of the double bond in the ring would stabilize an intermediate radical,

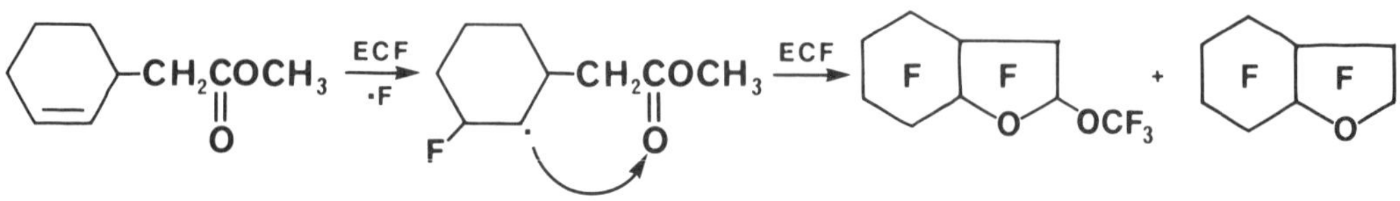

(4)

facilitating the ring closure necessary to form the desired cyclic ethers. Good yields are reported.

A new class of compounds, perfluoroamino ethers, was prepared by Simons ECF [27]. Amino ethers were utilized to investigate the relationship of structural features on yield. It was found that the amino groups improved the conductivity and yields over that obtained with only the ether groups present. n-Propylamino, piperidino, morpholino, and pyrrolidino groups were associated with low cleavage rates; methylamino and diisopropylamino groups promote extensive cleavage of the parent structure.

C. Mechanistic Considerations

Several recent studies have focused on the mechanism of Simons ECF and further illuminated the key role of the anode film. Mechanistic pathways involving radical intermediates, radical cations and carbocations, high-valence nickel fluorides, and bound fluorine have been proposed [4,9,12]. It is probable that no single mechanism totally describes anodic fluorination in anhydrous HF.

A typical Simons ECF experiment consists of at least four overlapping phases:

1. The rapid electrochemical oxidation of the nickel anode surface to form nickel fluorides
2. The growth on this surface of a relatively insoluble, polymeric, organic containing layer of macroscopic dimension
3. The pseudo-steady-state fluorination of the organic substrate dissolved in the electrolyte, during which time the reactants diffuse into the film, adsorb, and are electrochemically oxidized, and the products diffuse back out into the bulk
4. The exhaustion of the organic substrate in the film or the fouling of the anode film, which dramatically decreases the apparent conductivity and usually results in termination of the experiment

Failure to segregate the results of the second and third phases, as is often the case with batch experiments, has no doubt contributed to the notorious irreproducibility reported in Simons ECF investigations [36,37]. For example, the current efficiency measured during the film formation periods can be substantially lower than that observed later during the "steady-state" portion of the experiment. Additionally, comparisons between studies are complicated by the difficulty of conducting the first and second phases in a reproducible manner. Reproducibility of Simons ECF experiments and eventual attainment of pseudo-steady-state fluorination are critically dependent on proper growth and conditioning of the anode film [20]. Further, this film growth and conditioning dictate the particular current voltage curve, which serves as the operating line for the steady-state fluorination experienced during phase 3.

Fluorine is evolved from and incorporated within [12] the film formed during the anodic polarization of nickel in AHF in the absence of organic substrate. In rotating ring-disk studies in AHF, fluorine generation from the anodic nickel disk electrode has been detected in the form of a fluorine reduction peak on the ring [17–19]. Even if the disk electrode is switched to open circuit, fluorine continues to appear at the ring until the anode potential decays [17]. The charge associated with the reduction of this film was investigated as a function of anodization potential and anodization time [16]. The results indicated that at anode potentials above approximately 4.5 V versus H_2, the charge increased linearly with time, reaching a potential- and time-independent limiting plateau after 75–100 s. During cyclic voltammetry in anhydrous HF, a reduction peak below 2.8 V versus Hg/Hg_2F_2 is observed during the cathodic scan [16]. This peak displays typical EC behavior as a function of scan rate. The height decreases more rapidly with decreasing sweep rate than diffusion alone would predict; the peak disappears completely at low sweep rates. This is a result of the competing electrochemical and chemical discharge processes of the film.

The anode film formed in the presence of organic is different and typically supports a greater current density than a film formed in AHF or AHF/KF alone. For this reason, we believe studies of the anode film formed in anhydrous HF in the absence of organic substrate are limited in their applicability to Simons ECF. For example, the addition of benzoyl chloride was found to increase the conductivity of the anode film [38]. At constant voltage, 15–20 mA cm^{-2} was measured in the presence of the organic substrate. Only 1–3 mA cm^{-2} was observed without the substrate in a control run using enough KF to produce an electrolyte of the same conductivity. The presence of organic material within the anode film increases the rate of electron transfer.

An induction period has been reported in several investigations [4,5]. During the induction period, current flows, some fluorine is generated, and the anode film is formed, but virtually no fluorinated products are produced. Even in the absence of organic material, the corrosion of the nickel anodes is insufficient to account for the quantity of charge passed [18,25]. The corrosion current is further diminished by the presence of organic material [39]. Formation of the polymeric layer on the anode requires the electrochemical oxidation of organic substrate and is responsible for a portion of the charge passed during an induction period. This polymeric layer formed on the anode was analyzed and found to exhibit a low degree of fluorination [14]. The delayed appearance of perfluorinated product as a separate phase may be explained both by this induction period and by product solubility in the electrolyte.

Once the anode film is formed, the current in Simons ECF displays a nearly linear [16,20] dependence on the voltage. However, the linear behavior of the process does not appear to be due to the ohmic resistance of the electrolyte [25] and is more likely due to the resistance of the anode film [18]. The anode film acts as a barrier to electron transfer [12]; the anode potential at the outer edge of the nickel salt layer is very different from the potential of the base metal. For example, this layer on nickel prevents the electrochemical oxidation of quinols at potentials less than the discharge potential observed for fluorine evolution [40].

It has been postulated that fluorine bound loosely in the anode film is the fluorinating agent in Simons ECF [12,17]. In a rotating ring-disk study [17], no reducible species, that is, fluorine, is detected at the ring electrode following the addition of organic substrate. Nevertheless, it is doubtful that fluorine is the primary fluorinating agent in Simons ECF, as typically the sequence of fluorine substitution is ordered, not distributed statistically as observed for direct fluorination. Fluorine is consumed more rapidly, as measured by potential decay, during open-circuit discharge of the cell in the presence of organic material than in AHF alone [17]. We believe the rate for this chemical fluorination step is too slow for this pathway to dominate. However, high-valence nickel or a loose elemental fluorine-nickel fluoride complex could contribute to the composite fluorination process [16].

The cation radical mechanism (EC_bEC_n) initially proposed by Burdon et al. [9,41], and supported by work described by Rozhkov [5] and Gambaretto et al. [38], appears to most closely predict the product distributions observed during the Simons ECF of many species. In the proposed EC_bEC_n mechanism, the substitution of each fluorine requires the four-step sequence

$$R{-}CH_2{-}R' \xrightarrow[(E)]{-e} R{-}\overset{+\cdot}{C}H_2{-}R' \xrightarrow[(C_b)]{-H^+} R{-}\dot{C}H{-}R' \xrightarrow[(E)]{-e}$$

$$R{-}\overset{+}{C}H{-}R' \xrightarrow[(C_N)]{+F^-} R{-}CHF{-}R'$$

In the first step, the reactant molecule adsorbed on the anode is electrochemically oxidized to form a radical cation, which subsequently loses a proton in an acid-base reaction. The resulting radical is electrochemically oxidized during the third step to yield the corresponding cation, which then undergoes nucleophilic addition to complete the sequence. With this scheme, formation of the observed by-products may be explained by rearrangements of the carbocation intermediates or reactions of the radical intermediates.

Results obtained for the oxidation of trifluoroacetic acid in HF using nickel and platinum electrodes argue the importance of adsorption in Simons ECF [5]. With platinum, the dimer (C_2F_6) formed from two CF_3 radicals comprised 98% of the product; with nickel, CF_4 was the dominant product. The authors suggest this results from strong adsorption of the organic substrate on nickel favoring further oxidation over dimerization. In a voltammetric and tensammetric investigation, further evidence of the adsorption of alkane sulfonylfluorides and fluoride ions on the nickel anode was reported [42].

The carbocation formed as an intermediate during the EC_bEC_N mechanism favors ring contraction over ring expansion during fluorination. A typical example of this is reported for the ECF of several chloroalkylmethylpiperidines and chloroalkylhexamethyleneimines [43].

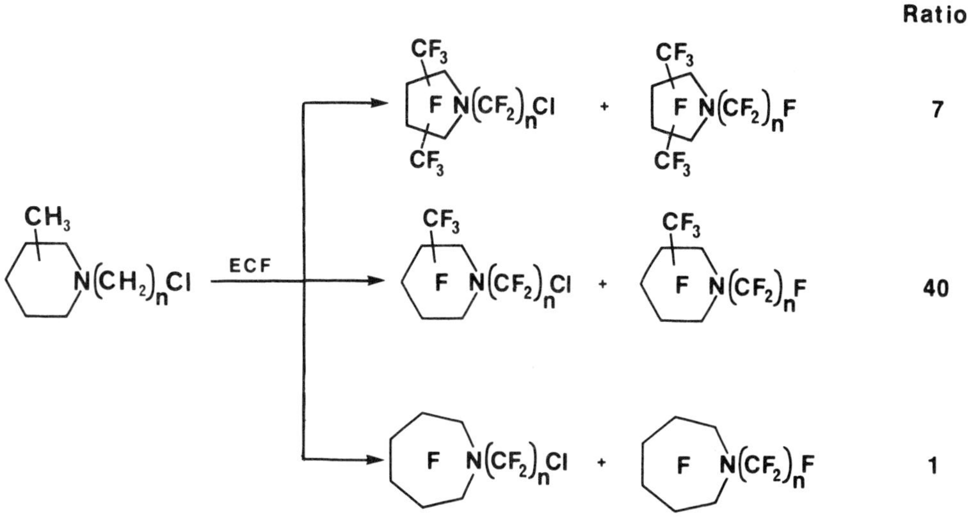

(5)

The isomerization of the product from ECF experiments using feeds of n- and iso-butyrylchloride also supports a dominant EC_bEC_N mechanism [38]. Only 4.5% isomerization was found with the n-reactant, but 36% rearrangement to n-perfluorobutyrylfluoride was observed using the iso reactant. This isomerization appears to result from formation of a carbocation intermediate. A radical mechanism could not account for the observed product distribution [38].

In a study of the ECF of some cyclic tertiary amines, residual hydrogen was reported to be only on the ring carbon α to the nitrogen [44].

(6)

The EC_bEC_N cation radical mechanism would predict this to be the last site to fluorinate; nitrogen is positively charged in HF and cation radical formation would likely occur as far from the nitrogen as possible.

In a mechanistic study using N-methylmorpholine, the ECF reaction was reported to be highly selective [45]. A large excess of the organic reactant was employed to more effectively isolate partially fluorinated intermediates. Mono- and difluorinated products revealed that initial fluorination adjacent to the oxygen was predominant. Only a small amount of one partially fluorinated compound contained fluorine on the ring carbon α to the nitrogen [38]. After incorporation of an average of 2.4 fluorine atoms per mole of n-methylmorpholine, the partially fluorinated products were isolated and were characterized to be as follows:

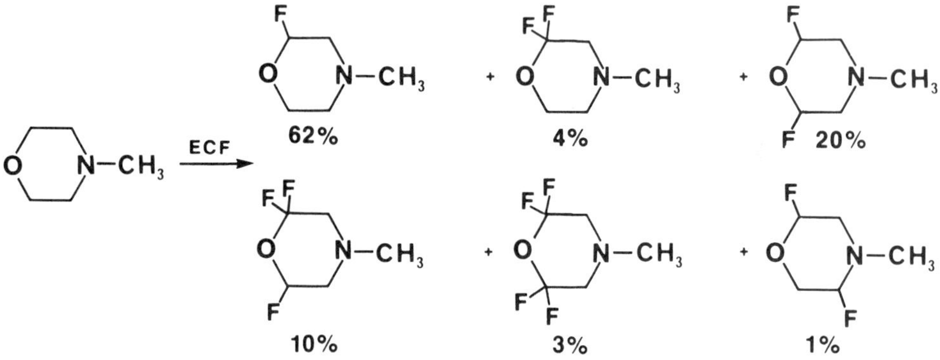

(7)

Initial fluorination near the oxygen weakens the basicity of the nitrogen and allows fluorination of the methyl group of N-methylmorpholine without the high degree of cleavage observed in the ECF of trimethylamine. With the exception of the perfluorinated N-methylmorpholine product, no fluorines were found on the methyl group.

A free radical fluorination mechanism cannot be completely excluded and must be considered a secondary side reaction that becomes more important with increases in the potential of the anodic electron transfer reaction [38]. Product distribution is frequently found to be a function of cell voltage, which is evidence of competing mechanisms. Results reported for the ECF of benzoyl fluoride displayed decreasing yields of the desired $C_6F_{11}COF$ product with increasing cell voltage [38]. Similar behavior was reported during the ECF of propene, in which fragmentation was found to be dependent on anode potential [20].

The ECF of some compounds appears to occur in a single pass, but the fluorination of others more closely resembles a sequential process in which only one or two fluorine atoms are substituted per pass through the anodic reaction zone. With N-methylmorpholine, it was found that after only enough current had passed to introduce an average of only 1.5 atoms of fluorine per molecule, 6.5% of the organic substrate was perfluorinated but 61% remained unreacted [45]. Evidence of single-pass fluorination was also observed in attempts to prepare C_3F_7H from propene as a precursor to C_3F_6 [20]. The propensity of the propene reactant to perfluorinate prevented the isolation of useful yields of the C_3F_7H monohydride. Conversely, with tripropylamine, no perfluorinated product was detected after enough current had passed to introduce an average of 2.5 atoms of fluorine per molecule [46]. The majority of the fluorinated product molecules contained only one or two fluorine atoms. After passage of 50% of the theoretical quantity of current necessary to perfluorinate 1,3-propanedisulfonylfluoride, no perfluorinated product could be identified [47].

Mass transport of organic reactant from the bulk electrolyte to the outside of the anode film does not limit the rate of the electrochemical reaction for conventional concentrations [16,17]. However, if film conductivity is maintained and if the organic concentration in the electrolyte is sufficiently depleted, predictable mass transport limitations from the bulk are observed [25]. This does not exclude the possibility of separate mass transport limitations across the anode film. However, if proper film formation is not accomplished or an unsuitable organic feed is employed, steady-state fluorination is not realized and cell conductivity deteriorates.

III. ELECTROCHEMICAL FLUORINATION IN MOLTEN KF·2HF: THE PHILLIPS PROCESS

The development of this process at Phillips Petroleum Company is described elsewhere [7]. Substrates of choice should be at least moderately volatile and not particularly soluble in the KF·2HF electrolyte. Suitable substrates include alkanes, cycloalkanes, chloroalkanes, acyl fluorides, and esters. The anode is porous carbon, and the process is thought to involve the electrolytic generation of elemental fluorine and reaction of that fluorine with substrate within the porous carbon anode. The products appear to come from the statistical replacement of substrate hydride with fluoride [48] and range from monofluoro to perfluoro.

The process is efficient for many fluorinations. Ethane is frequently run as a model substrate. It runs well, and the product fluoroethanes can be analyzed completely. The products of the fluorination of ethane are as follows:

Product	Mol per 100 mol
Fluoroethanes	96
Fluorobutanes	4
Fluoromethanes	Trace
Olefins	Trace

All the fluoroethanes are produced. Butane itself is not found. Identified fluorobutanes range from butyl fluoride to perfluorobutane. The level of fluoromethanes is barely above the level of methane in the ethane feed. Ethylene and vinyl fluoride are produced.

Other easily run substrates include 1,2-dichloroethane (80% retention of 1,2-dichloro structure); acetyl fluoride (85% yield of fluorinated acetylfluorides); tetrafluorocyclobutane (90% retention of structure); and isopropyltrifluoroacetate (to the perfluoro ester and subsequently hexafluoroacetone). Current efficiencies typically range from 80 to 100%.

This process is a useful complement to the Simons process for many volatile substrates and products.

A. Operational Summary

A typical laboratory cell is shown in Fig. 2. The mild steel cell case is the cathode. Hydrogen from the cathode lifts the electrolyte upward in the anode/cathode gap; it escapes into the vapor space; and the electrolyte returns through the downcomer. The cell case is jacketed (not shown) for temperature control. The KF·2HF electrolyte melts at 55–60°C, and the electrolysis is run at 90–100°C. The temperature control fluid must be tempered to prevent the electrolytes freezing on the case.

It is thought that elemental fluorine is generated as the anode reaction. The anode is porous carbon (not graphite). It is not wetted by the electrolyte. Substrate is introduced continuously into the network of pores through the gas cap cavity at the bottom of the anode; from there it passes up through the network of pores and reacts with fluorine as it moves up.

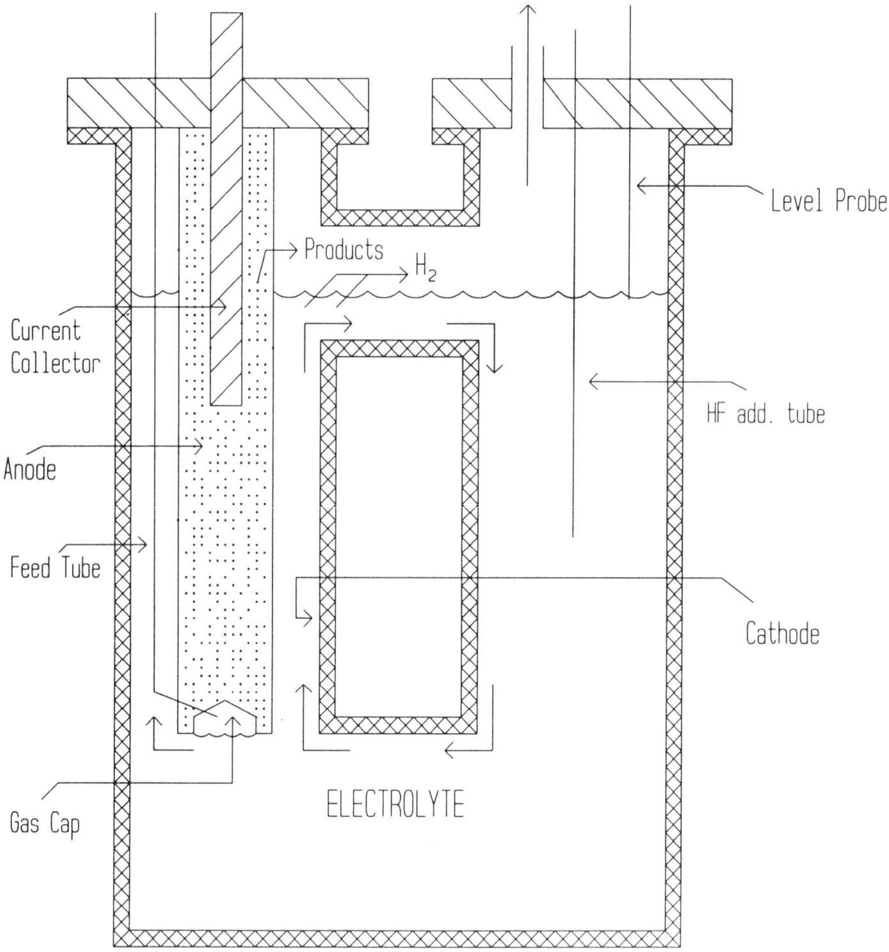

Figure 2. Laboratory cell, 60 A.

An anode 3.50 cm in diameter and immersed 30 cm runs easily at 60 A. The anode/cathode gap is about 1.2 cm.

Two power supplies are used. The power supply for routine operation can deliver 100 A at 12 V with a peak-to-peak ripple of 1%. The depolarization power supply can deliver 100 A at 100 V.

The electrolyte is nominally KF·2HF (41.7 ± 0.2 wt% HF). It is prepared by the addition of HF vapor to KF·HF. (Addition of HF vapor to KF is fraught with hazard, as is the addition of liquid HF to KF·HF.) Sometimes 0.1 wt% lithium fluoride is added because of a belief that it helps reduce erratic polarization.

Substrate addition rate and current level determine the per pass replacement of hydride. Typically, 20–50% hydride is replaced per pass. Partially fluorinated materials can be recycled to the cell. Complete replacement of hydride in a single pass is possible with excess current [7].

Polarization has a meaning that dates back to the early years of fluorine generation with carbon anodes [49]. (Polarization has been identified with the anode effect noted in molten salt electrolysis, but it differs in a number of ways. In particular the onset of polarization is not predictable.) The cell is operated at constant current, and the terminal voltage varies from 7 to 9 V, with a hardware limit of 12–15 V. On occasion the voltage increases rapidly to the limit and the current falls to half or less of the set value. The cell is said to be polarized. The cell can be depolarized by permitting the voltage to rise at constant current. Normally it rises to 60–80 V. If the cell is permitted to operate at this voltage for about 90 s and then turned off and back on, it operates normally and is said to be depolarized.

The frequency of polarization is somewhat erratic and depends on the choice of carbon, the current density, and the feed. The initial polarization occurs soon after start-up; the elapsed time between subsequent polarizations may be measured in days or even weeks.

The cause of polarization is not clear. It apparently is due to the formation of a gas (perhaps elemental fluorine) film over much of the anode surface. This film blocks the passage of current through the anode/electrolyte interface. The film may form because the carbon becomes highly nonwetting, or it may form because paths for fluorine to pass from the surface into the anode interior become blocked.

The processes involved in depolarization are not clear. At 60–80 V and 200 mA cm^{-2}, 12–16 W cm^{-2} must be dissipated. After a depolarization treatment the anode surface is shiny. This treatment could remove the highly nonwetting surface, or it could open plugged passages.

B. Process Rationalization

The porous carbon anode is the heart of the system. A good combination of physical properties, polarization behavior, and commercial availability is found in Union Carbide's PC25, PC45, and PC60 filter carbons. PC25 shows the best polarization behavior, but it is friable and difficult to work with. PC45 and PC60 are mechanically better.

The $KF \cdot 2HF$ electrolyte does not wet a carbon anode, but it can be forced into the pores by hydrostatic pressure. It is thought that once inside the pores the electrolyte moves about freely.

The process works best with reasonably volatile feeds that are relatively insoluble in the electrolyte and produce volatile products.

Given this background, the process can be understood as follows.

Elemental fluorine is generated at the anode surface and enters the network of pores in the anode. Substrate is introduced into the anode through the gas cap at the bottom of the anode and moves through the anode. The substrate does not bubble out into the electrolyte. As the substrate moves through the anode, it reacts with the fluorine. The heat of reaction, perhaps 400 kJ per gram-equivalent of hydrogen replaced with fluorine, is rejected to the carbon, to the electrolyte within the anode, and thence to the bulk electrolyte and the temperature control system. Temperature measurements have shown the anode to be as much as 30°C hotter than the bulk electrolyte. Fluorinated and unreacted substrate escape from the anode above the electrolyte, mix with hydrogen from the cathode, and leave the cell.

Figure 3 shows a likely electrolyte-substrate-anode relationship. Note the electrolyte bulging into the anode, the fluorine moving into the anode,

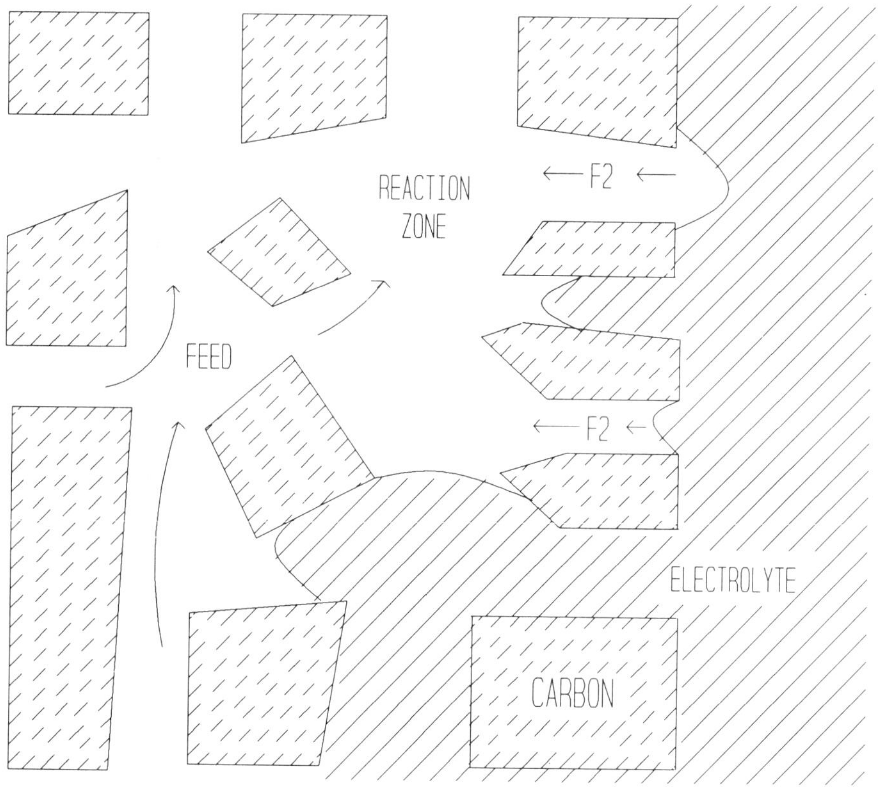

Figure 3. Electrolyte vapor-carbon normal relationship.

the substrate moving through the network of pores and reacting with fluorine, and the electrolyte moving in the network of pores.

Free radical processes can explain the observed products. For example,

Main radical processes:

$$CH_3{-}CH_3 + F_2 \longrightarrow CH_3{-}CH_2\cdot + F^\cdot + HF$$

$$CH_3{-}CH_2\cdot + F^\cdot \longrightarrow CH_3{-}CH_2F$$

Radical disproportionation:

$$2CH_3{-}CH_2\cdot \longrightarrow CH_2{=}CH_2 + CH_3{-}CH_3$$

Production of dimers:

$$CH_3{-}CH_2\cdot + CH_2{=}CH_2 \longrightarrow CH_3{-}CH_2{-}CH_2{-}CH_2\cdot$$

$$CH_3{-}CH_2{-}CH_2{-}CH_2\cdot + F^\cdot \longrightarrow CH_3{-}CH_2{-}CH_2{-}CH_2F$$

(If dimers were produced by simple radical-radical dimerization, butane should be formed from ethane. Butane is not found.)

The process can be summarized as follows:

1. The KF·2HF electrolyte does not wet the porous carbon anode.
2. Elemental fluorine is generated at the anode surface.
3. The fluorine enters the anode.
4. The substrate reacts with fluorine within the anode.
5. The observed products can be explained from free radical substitution, disproportionation, and dimerization.

IV. PARTIAL ELECTROCHEMICAL FLUORINATION

In addition to electrolysis in anhydrous hydrogen fluoride, anodic fluorination of organic substrates has also been accomplished in aprotic solvents (acetonitrile, sulfolane, nitromethane, and others) in the presence of fluoride ion. Such conditions mostly lead to mono- and/or difluorinated materials [5]. The reactions are most often carried out in an undivided cell via controlled potential at a platinum or graphite anode. The source of fluoride ion, which usually also serves as supporting electrolyte, includes HF [50a, 51], AgF [52], $R_3N{\cdot}HF$ [50], $R_4NF{\cdot}xHF$ [50b,53–55], and even R_4NBF_4 [56]. Since the concentration of fluoride is kept relatively low (0.1–1.0 M), the reactions can be carried out in ordinary glass cell equipment without special safety precautions.

The anode potential is normally kept slightly above the first half-wave potential of the substrate, usually well below the discharge potential of fluoride ion (>2.9 V versus SCE at Pt in acetonitrile [57]). Therefore, a pathway involving initial formation of a cation radical followed by attack from fluoride ion (ECEC) is often suggested:

$$RH \xrightarrow{-e^-} RH^{\dot{+}} \xrightarrow[-H^+]{-e^-,\ F^-} RF \tag{8}$$

The product distribution during anodic substitution generally reflects the stability and selectivity of the intermediate cationic species toward all nucleophiles present. In addition to fluoride ion, these may include the substrate itself, solvent, and adventitious impurities. In water the fully hydrated fluoride ion is a weak nucleophile. In aprotic solvents F^- becomes a strong nucleophile and HF a weak acid. The reactivity of F^- is, however, especially sensitive to the water content of the electrolysis system [58]. Rigorous purification and drying of the solvent and electrolyte are therefore necessary to optimize the formation of fluorinated products.

Partial fluorination of aromatics has received the most attention and shows all the features of an ECE reaction between an electrogenerated aromatic cation radical and a fluoride nucleophile (i.e., substitution and addition), in addition to other well-established reactions associated with such cation radicals (for a detailed discussion, see Chap. 13). The situation is well illustrated in the case of anthracenes (Table 1) [55,59–62].

Cyclic voltammetry under aprotic conditions reveals that, upon addition of fluoride ion, the electrooxidation of 9,10-diphenylanthracene changes from a reversible one-electron formation of a cation radical to an irreversible two-electron process. Preparative electrolysis results in the isolation of 9,10-difluoro-9,10-dihydro-9,10-diphenylanthracene as the sole product in

Table 1. Partial Electrochemical Fluorination of Some Anthracenes

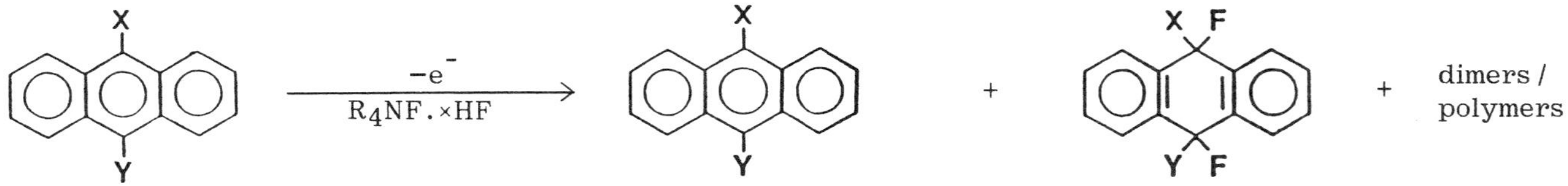

Substrate	E_a V versus Ag/Ag^+	Yield %		References
X = Y = Ph	0.95	—	X = Y = Ph 75	[55,59]
X = Ph Y = H	1.0	X = Ph Y = F minor	X = Ph Y = F major	[60]
X = Y = H	1.0	X = Y Y = F 15 X = Y = F trace	—	[60]
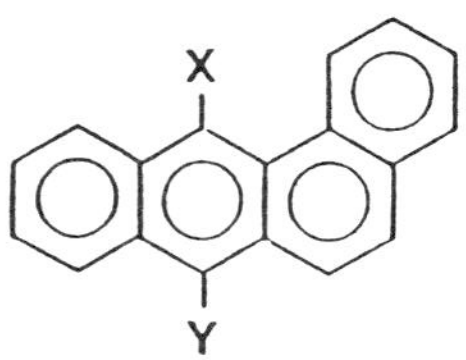 X = Y = H	1.1	X = H Y = F } 40 X = F Y = H } X = Y = F[a]	—	[61]

[a] Isolated as benz[a]anthracene-7,12-dione.

high yield [59]. As the stability of the aromatic cation radical decreases, its selectivity toward fluoride diminishes, and the yield of fluorinated products is lowered and side reactions (dimerization, polymerization) become more prominent.

Much controversy was created by a suggestion that the reaction of cation radicals from polycyclic aromatic hydrocarbons (PAHs) with fluoride ion is disallowed according to the laws of orbital interaction [55,59,62–64]. Since then, the independent fluorination of PAH has been demonstrated, albeit in low yield, by reaction of isolated PAH cation radicals with fluoride ions [65,66]. This apparent discrepancy has been explained by evoking the formation of an intermediate complex between nucleophile and cation radical from which a "nucleophile-assisted" second electron transfer step takes place [64] or by suggesting that fluorination of certain PAHs is charge controlled rather than frontier orbital controlled [62].

The electrofluorination of benzene and naphthalene to yield primarily monofluoro derivatives has been carried out in acetonitrile-$R_4NF \cdot 3HF$ [53, 54,57].

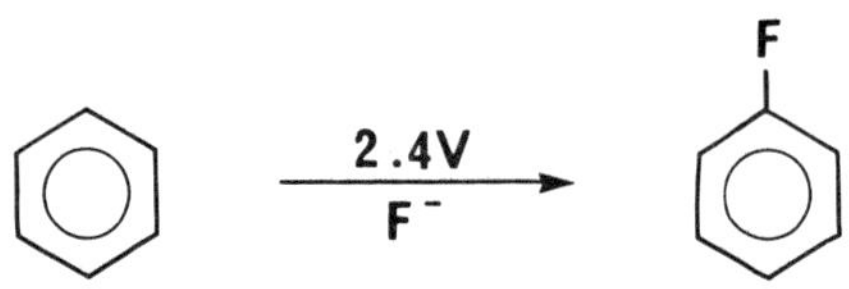

(9)

(10)

A platinum anode in an undivided cell was utilized. Frequent pulsing of the anode potential between working potential and zero is critical to avoid interference from buildup of a nonconducting polymer film at the anode. Some workers have reported difficulties in repeating these results [55,64].

The oxidation of naphthalene in a capillary gap cell using a graphite anode at constant current in an acetonitrile-$Et_3N \cdot HF$ electrolyte [50a] has been described as well. Polarity reversal was also necessary in this system to avoid anode film formation. The product was a mixture of 1-fluoronaphthalene (21% current efficiency) and 2-fluoronaphthalene (1% current efficiency).

It has recently been shown in pyridine-HF or in trialkylamine-HF that partial electrofluorination of aromatics occurs without anode passivation, thus eliminating the need for polarity reversal or pulsing [50b]. It appears that in these solvents, F^- is a stronger nucleophile, and HF a weaker acid, than in acetonitrile.

Partial electrofluorination of substituted benzenes reportedly leads to products via H substitution, ipso substitution, or addition [50b,67]:

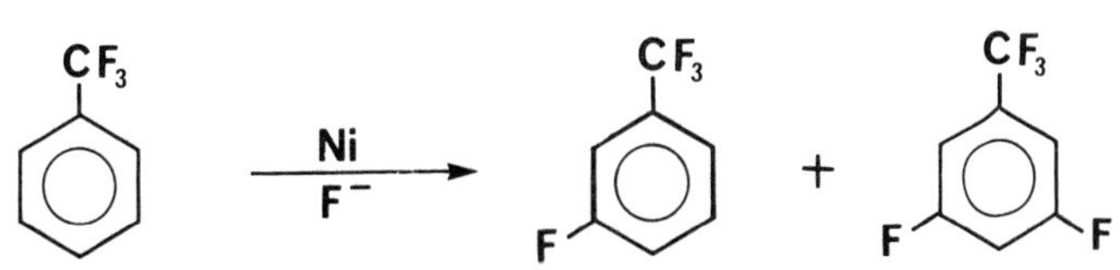

(11)

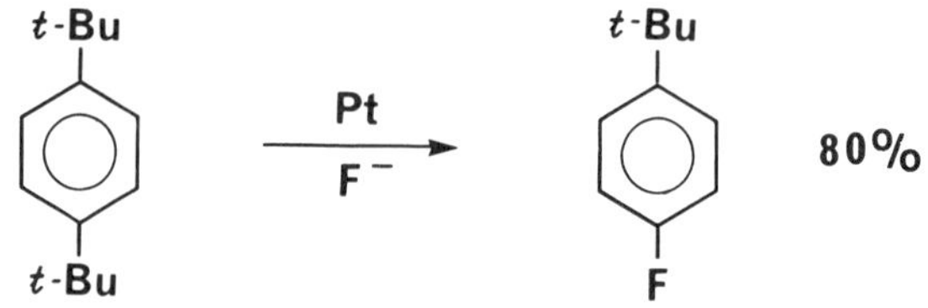

(12)

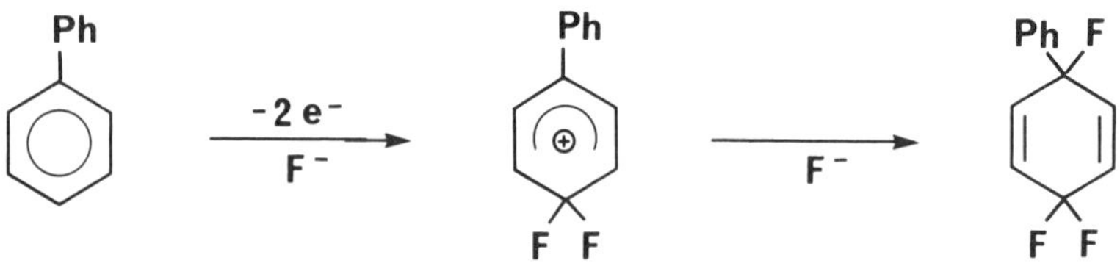

(13)

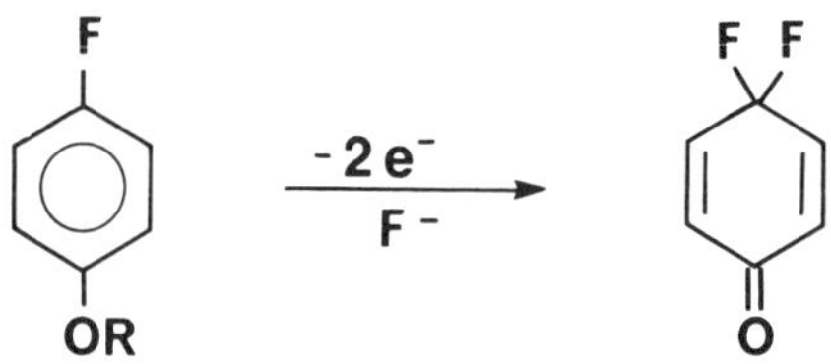

(14)

Electrofluorination of pyridine at 2.5 V versus Ag/Ag+ produces 2-fluoropyridine in 22% yield [68]:

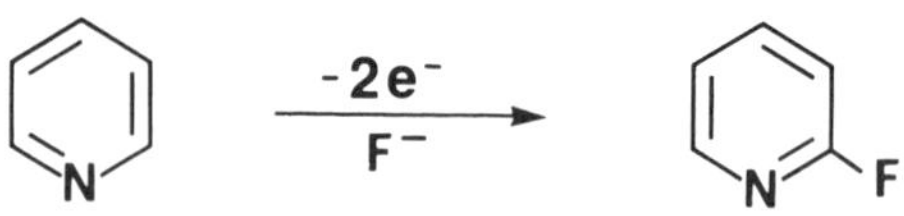

(15)

Oxidation of alkylaromatics in acetonitrile usually results in the formation of benzyl cations via rapid proton loss.

In the presence of fluoride ions, a follow-up competitive reaction with fluoride or solvent (acetamidation) leads to substitution:

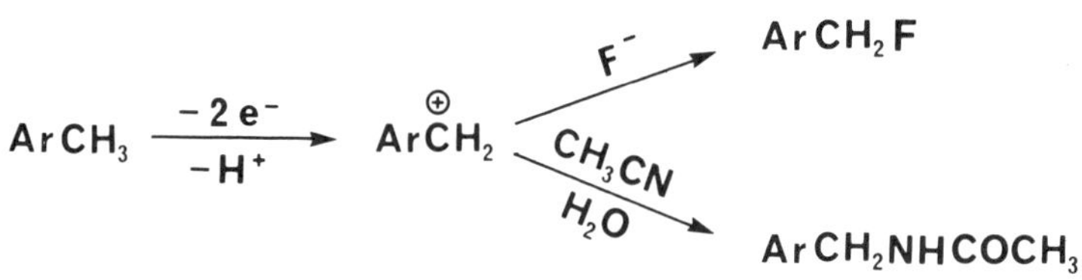

(16)

The reactivity order between fluoride and acetonitrile appears classic; the more stable carbocations are more selective toward fluoride ion [69–71]:

$ArCH_3$	fluorination/acetamidation	combined yield %	
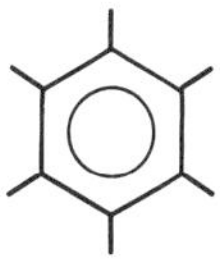	88:12	58	
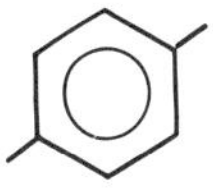	25:75	39	(17)
	0:100	8.5	

Interestingly, in contrast to these results, nuclear substitution of toluene has also been reported [50a,72]. o-Fluorotoluene (7% current efficiency) resulted from electrolysis in an acetonitrile-HF mixture [50a].

Anodic benzylic substitution by fluoride has proven synthetically useful in the following transformation, selectively introducing one or two fluorines [73]:

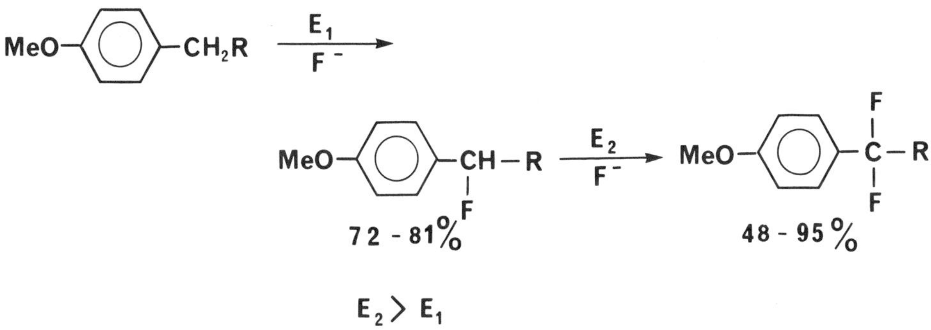

R = $COCH_3$, COAr, COOEt, CN (18)

The anodic oxidation of olefins on platinum in the presence of fluoride ion leads to addition via an ECEC mechanism:

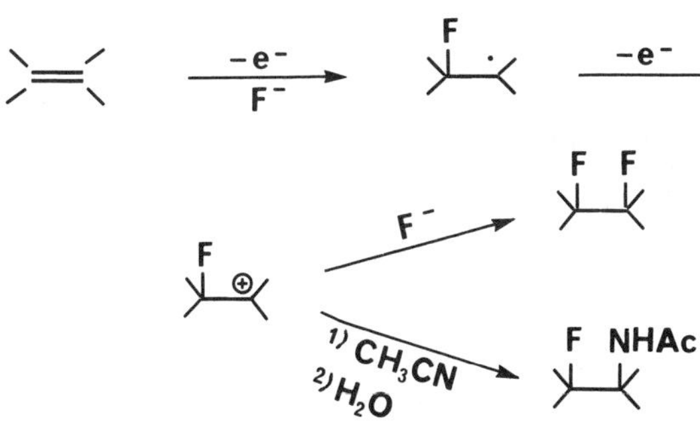

(19)

Competitive reaction of the intermediate cation with the solvent results in overall fluoroacetamidation, in addition to formation of difluorides. The product ratio is directly related to the stability of the intermediate cation. Both reactions are stereoselective. Additionally, the fluoroacetamidation is regiospecific [52,69,71,72,74]:

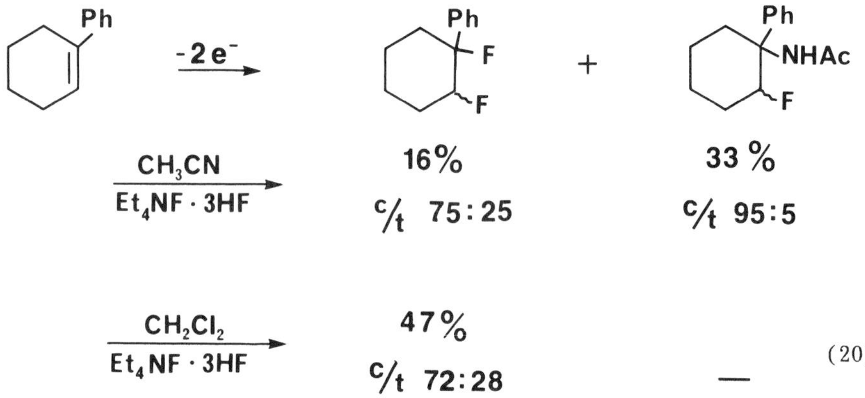

(20)

In the less nucleophilic solvent dichloromethane, the yield of the difluoride is significantly increased [71]. The stereoselectivity has been ascribed to an adsorbed cation intermediate available to nucleophilic attack from only one side. In comparison, the solution fluorination with XeF_2 is totally nonstereoselective [71].

Anodic oxidation of α-acetoxystyrenes in fluoride containing acetonitrile provides a useful route to α-fluoroketones in isolated yields of 44–63% [75]:

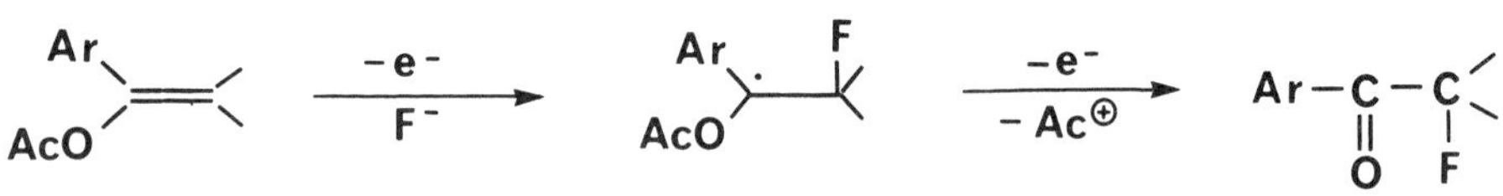

(21)

The anodic oxidation of trialkyl- or triarylphosphines, stibines or arsines generates the corresponding difluoro derivatives in good to excellent yield in the presence of fluoride-producing anions [76–78]:

$$R_3P \xrightarrow[F^-]{-2e^-} R_3PF_2 \qquad 69-82\% \tag{22}$$

$$R_3Sb \longrightarrow R_3SbF_2 \qquad 90\% \tag{23}$$

$$R_3As \longrightarrow R_3AsF_2 \qquad 30\% \tag{24}$$

R = alkyl, benzyl, aryl

Oxidation of secondary phosphines follows the same pattern and phosphinic fluorides are isolated following hydrolysis [79]:

$$R_2PH \xrightarrow[F^-]{-2e^-,\,-H^+} R_2PF \xrightarrow[F^-]{-2e^-}$$

$$R_2PF_3 \xrightarrow{H_2O} R_2P(O)F \quad 83-90\% \tag{25}$$

An alkyl group is replaced by fluorine in the anodic oxidation of tetraalkyl- or trialkylarylsilanes in the presence of fluoride ion [80]:

$$R_3Si-R \xrightarrow[F^-]{-2e^-} R_3Si-F \tag{26}$$

$$R = \text{alkyl, benzyl, aryl}$$

The replacement is selective and determined by the stability of the leaving alkyl group.

An ECEC mechanism for partial electrofluorination in aprotic organic media seems well established. The intermediacy of cationic species in the Simons process is also well documented. The most significant difference between fluorination on Ni in HF and on Pt in CH_3CN is the absence of perfluorination in the latter case. It has been suggested [45,67e] that this difference is due to varying degrees of adsorption of the substrate on the anode. Changing conditions from a Ni anode to a Pt anode, and from AHF to an aprotic electrolyte containing fluoride ion, result in a change from a fully fluorinated to a monofluorinated substrate. Indeed, as an illustration of such a concept of gradual change, partial fluorination of alkyl bromides at a Pt anode in AHF has been reported [81], as well as both partial fluorination and perfluorination at a Ni anode in organic dilute HF [45,67e,82].

V. SUMMARY

Three electrochemical fluorination processes have been discussed. Partial fluorination in the presence of fluoride ion and organic solvent occurs exclusively via initial electrochemical oxidation of the substrate. Electrochemical fluorination in molten KF·2HF is actually the reaction of an organic substrate with electrochemically generated fluorine within the confines of a porous carbon anode. The electrochemical processes in Simons ECF are believed to include both direct substrate oxidation and generation of elemental fluorine.

In partial electrochemical fluorination, potentials sufficient for fluorine discharge are inaccessible because of solvent oxidation (the anodic limit). In the molten KF·2HF cell, the organic substrate is confined to the interior of the anode and there is no evidence for its direct electrochemical oxidation. The Simons process operates at potentials known to be adequate for the discharge of fluorine. Under normal Simons operating conditions, direct oxidation of the organic substrate is the dominant anode reaction.

REFERENCES

1. W. H. Pearlson, J. Fluorine Chem., 32, 29 (1986).
2. I. N. Rozhkov, in Organic Electrochemistry (M. M. Baizer and H. Lund, eds.), 2nd ed., Dekker, New York, 1983, p. 805.
3. S. Nagase, in Fluorine Chemistry Reviews (P. Tarrent, ed.), Vol. I, Dekker, New York, 1967, p. 77.
4. N. L. Weinberg, in Technique of Electroorganic Synthesis (N. L. Weinberg, ed.), part II, John Wiley & Sons, New York, 1975, p. 1.
5. I. N. Rozhkov, Uspekhi Khim., 45, 1222 (1976); Chem. Abstr., 85, 142131x (1976).
6. E. Hollitzer and P. Sartori, Chem. Ing. Tech., 58, 31 (1986).
7. W. V. Childs, in Technique of Electroorganic Synthesis (N. L. Weinberg, ed.), Vol. V, Part III, John Wiley & Sons, New York, 1982, p. 341.
8. J. H. Simons, Trans. Electrochem. Soc., 95, 47 (1949); J. H. Simons, H. T. Francis, and J. A. Hogg, Trans. Electrochem. Soc., 95, 53 (1949); J. H. Simons, W. H. Pearlson, T. J. Brice, W. A. Wilson, and E. D. Dresdner, Trans. Electrochem. Soc., 95, 59 (1949); J. H. Simons and R. D. Dresdner, Trans. Electrochem. Soc., 95, 64 (1949).
9. J. Burson and J. C. Tatlow, in Advances in Fluorine Chemistry (M. Stacey, J. C. Tatlow, and A. G. Sharpe, eds.), Vol. 1, Butterworth's, London, 1960, p. 129.
10. A. J. Rudge, in Industrial Electrochemical Processes (A. T. Kuhn, ed.), Dekker, New York, 1967, p. 71.
11. A. P. Tomilov, S. G. Mairanovskii, M. Ya. Fioshin, and V. A. Smirnov, Electrochemistry of Organic Compounds (J. Schmorak, trans.), Wiley, New York, 1972, p. 417.
12. N. Watanabe, J. Fluorine Chem., 22, 205 (1983).
13. E. Hollitzer and P. Sartori, Chem. Ing. Tech., 58, 31 (1986).
14. F. Huba, E. B. Yeager, and G. A. Olah, Electrochim. Acta, 24, 489 (1979).
15. F. Huba, Ph.D Thesis, Case Western Reserve University, Cleveland, Ohio, 1977.
16. J. S. Clarke and A. T. Kuhn, J. Electroanal. Chem., 85, 299 (1977).
17. M. Novak and J. Boa, J. Electroanal. Chem., 109, 179 (1980).
18. F. W. Klink, Ph.D Thesis, Case Western Reserve University, Cleveland, Ohio, 1985.
19. M. Novak, J. Jeko, and J. Boa, Stud. Org. Chem., 30, 67 (1987).
20. F. G. Drakesmith and D. A. Hughes, J. Fluorine Chem., 32, 103 (1986).
21. A. Doughty, M. Fleischmann, and D. Pletcher, J. Electroanal. Chem., 51, 329 (1974).
22. G. G. Koerber and T. De Vries, J. Amer. Chem. Soc., 74, 5008 (1952).
23. B. Burrows and R. Jasinski, J. Electrochem. Soc., 115, 348 (1968).
24. N. Hackerman, E. N. Snavely, Jr., and L. D. Fiel, Electrochim. Acta, 12, 535 (1967).
25. F. W. Klink, D. J. Wasser, and C. C. Liu, J. Fluorine Chem., 32, 89 (1986).
26. D. Lines and H. Sutcliffe, J. Fluorine Chem., 17, 423 (1981).
27. G. G. I. Moore, J. C. Hansen, T. M. Barrett, J. E. Waddell, K. M. Jewell, T. A. Kestner, and R. M. Payfer, J. Fluorine Chem., 32, 41 (1986).

28. H. Baba, K. Kodaira, S. Nagase, and T. Abe, Bull. Chem. Soc. Jpn., 51, 1891 (1978).
29. J. H. Simons, in Fluorine Chemistry (J. H. Simons, ed.), Vol. 1, Academic Press, New York, 1950, p. 239.
30. T. Abe, S. Nagase, and H. Baba, Bull. Chem. Soc. Jpn., 49, 1888 (1976).
31. T. Abe, K. Kodaira, H. Baba, and S. Nagase, J. Fluorine Chem., 12, 1 (1978).
32. T. Abe and S. Nagase, J. Fluorine Chem., 13, 519 (1979).
33. T. Abe, E. Hayashi, H. Baba, and S. Nagase, Chem. Lett., 121 (1980).
34. T. Abe, H. Baba, E. Hayashi, and S. Nagase, J. Fluorine Chem., 23, 123 (1983).
35. T. Abe, E. Hayashi, H. Baba, and S. Nagase, J. Fluorine Chem., 25, 419 (1984).
36. V. Y. Kazakov, L. A. Savelev, R. A. Dzerzhinskaya, E. A. Shiskkin, and N. L. Gudimov, Zhur. Prikl. Khim., 41, 2212 (1968); Chem. Abstr., 70, 25173s (1969).
37. R. D. Dresdner and J. A. Young, J. Amer. Chem. Soc., 81, 574 (1959).
38. G. P. Gambaretto, M. Napoli, L. Conte, A. Scipioni, and R. Armelli, J. Fluorine Chem., 27, 149 (1985).
39. D. J. Wasser, P. S. Johnson, F. W. Klink, F. Kucera, and C. C. Liu, J. Fluorine Chem., 35, 557 (1987).
40. A. Doughty, M. Fleischmann, and D. Pletcher, J. Electroanal. Chem., 51, 456 (1974).
41. J. Burdon, I. W. Parsons, and J. C. Tatlow, Tetrahedron, 28, 43 (1972).
42. G. Cauquis, B. Keita, and G. Pierre, J. Electroanal. Chem., 100, 205 (1979).
43. E. Hayashi, T. Abe, H. Baba, and S. Nagase, J. Fluorine Chem., 26, 417 (1984).
44. L. Conte, C. Fraccaro, M. Napoli, and M. Mistrorigo, J. Fluorine Chem., 34, 183 (1986).
45. G. P. Gambaretto, M. Napoli, C. Fraccaro, and L. Conte, J. Fluorine Chem., 19, 427 (1982).
46. L. Conte, M. Napoli, and G. P. Gambaretto, J. Fluorine Chem., 30, 89 (1985).
47. E. Hollitzer and P. Sartori, J. Fluorine Chem., 35, 329 (1987).
48. W. V. Childs and D. J. Fall, paper no. 1345 at the 172nd meeting of the Electrochemical Society, fall 1987.
49. A. J. Rudge, The Manufacture and Use of Fluorine and Its Compounds, Oxford University Press, Oxford, 1962.
50. (a) F. Beck, Ger. Offen. 2516355 (1976); Chem. Abstr., 86, 23587m (1977); (b) D. W. Sopher and J. H. H. Meurs, G.B. 2191480A (1987); Chem. Abstr., 108, 212559n (1988).
51. H. Schmidt adn H. D. Schmidt, J. Prakt. Chem., 2, 250 (1955).
52. H. Schmidt and H. Meinert, Angew. Chem., 72, 109 (1960).
53. I. L. Knunyants, I. N. Rozhkov, A. V. Bukhtiarov, M. M. Gol'din, and R. V. Kudryavtsev, Izv. Akad. Nauk SSSR, Ser. Khim., 1207 (1970); Chem. Abstr., 73, 65752y (1970).
54. I. N. Rozhkov, A. V. Bukhtiarov, N. D. Kuleshova, and I. L. Knunyants, Dokl. Akad. Nauk SSSR, 193, 1322 (1970); Chem. Abstr., 74, 70878u (1971).

55. C. J. Ludman, E. M. McCarron, and R. F. O'Malley, J. Electrochem. Soc., 119, 874 (1972).
56. V. R. Koch, L. L. Miller, D. B. Clark, M. Fleischmann, T. Joslin, and D. Pletcher, J. Electroanal. Chem., 43, 318 (1973).
57. I. N. Rozhkov, A. V. Bukhtiarov, and I. L. Knunyant, Izv. Akad. Nauk USSR, Ser. Khim., 1130 (1972); Chem. Abstr., 77, 82878u (1972).
58. (a) I. N. Rozhkov and I. L. Knunyants, Dokl. Akad. Nauk SSSR, 199, 614 (1971); Chem. Abstr., 76, 7291y (1972); (b) I. N. Rozhkov and N. D. Kuleshova, Izv. Akad. Nauk, SSSR, 9, 2048 (1976); Chem. Abstr., 86, 120912w (1977).
59. I. N. Rozhkov, N. P. Gambaryan, and E. G. Galpern, Tetrahedron Lett., 4819 (1976).
60. J. F. Carpenter, L. H. Ekes, P. F. King, H. A. Mariani, M. Mirza Zadeh, R. F. O'Malley, and V. J. Roman, J. Electrochem. Soc., 130, 2170 (1983).
61. R. F. O'Malley, H. A. Mariani, D. R. Buhler, and D. M. Jerina, J. Org. Chem., 46, 2816 (1981).
62. P. F. King, and R. F. O'Malley, J. Org. Chem., 49, 2803 (1984).
63. (a) L. Eberson, J. Chem. Soc., Chem. Commun., 826 (1975); (b) L. Eberson, and K. Nyberg, Tetrahedron, 32, 2185 (1976).
64. L. Eberson, Z. Blum, B. Helgee, and K. Nyberg, Tetrahedron, 34, 731 (1978).
65. H. J. Shine, B. K. Bandlish, and M. T. Stephenson, Tetrahedron Lett., 733 (1978).
66. M. T. Stephenson and H. J. Shine, J. Org. Chem., 46, 3139 (1981).
67. (a) I. N. Rozhkov, A. V. Bukhtiarov, E. G. Galpern, and I. L. Knunyants, Dokl. Akad. Nauk SSSR, 199, 369 (1971); Chem. Abstr., 75, 147087r (1971); (b) I. Y. Aliev, I. N. Rozhkov, and I. L. Knunyants, Izv. Akad. Nauk SSSR, Ser. Khim., 1430 (1973); Chem. Abstr., 79, 91647m (1973); (c) I. Y. Aliev, I. N. Rozhkov, and I. L. Knunyants, Izv. Akad. Nauk SSSR, Ser. Khim., 2390 (1974); Chem. Abstr., 82, 66043r (1973); (d) I. N. Rozhkov and I. Y. Aliev, Tetrahedron, 31, 997 (1975); (e) V. A. Shreider and I. N. Rozhkov, Izv. Akad. Nauk SSSR, Ser. Khim., 676 (1979); Chem. Abstr., 91, 65188r (1979).
68. J. R. Ballinger, F. W. Teare, B. M. Bowen, and E. S. Garnett, Electrochim. Acta, 30, 1075 (1985).
69. A. Bensadat, G. Bodennec, E. Laurent, and R. Tardivel, Tetrahedron Lett., 3799 (1977).
70. A. Bensadat, G. Bodennec, E. Laurent, and R. Tardivel, Nouv. J. Chim., 4, 453 (1980).
71. A. Bansadat, G. Bodennec, E. Laurent, and R. Tardivel, J. Fluorine Chem., 20, 333 (1982).
72. I. N. Rozhkov, I. Y. Aliev, and I. L. Knunyants, Izv. Akad. Nauk SSSR, 6, 1418 (1976); Chem. Abstr., 85, 101051c (1976).
73. G. Alvernhe, A. Laurent, E. Laurent, and R. Tardivel, Ann. Chim. Fr. (Paris), 9, 659 (1984).
74. (a) A. Bensadat, G. Bodennec, E. Laurent, and R. Tardivel, Nouv. J. Chim., 5, 127 (1981); (b) A. Bensadat, E. Laurent, and R. Tardivel, Nouv. J. Chem., 5, 397 (1981); (c) E. Laurent, H. Lefranc, and R. Tardivel, Nouv. J. Chem., 8, 345 (1984).
75. E. Laurent, R. Tardivel, and H. Thiebault, Tetrahedron Lett., 24, 903 (1983).

76. (a) E. V. Nikitin, et al., Izv. Akad. Nauk SSSR, Ser. Khim., 1686 (1986); Chem. Abstr., 106, 196505t (1987); (b) E. V. Nikitin, et al., SU 1100897 (1986); Chem. Abstr., 105, 209198b (1986).
77. E. V. Nikitin, A. A. Kazakova, O. V. Parakin, and Y. M. Kargin, Zh. Obshch. Khim., 52, 2027 (1982); Chem. Abstr., 97, 204879f (1982).
78. (a) E. V. Nikitin, et al., Zh. Obshch. Khim., 53, 230 (1983); Chem. Abstr., 98, 115708c (1983); (b) E. V. Nikitin, et al., SU 1146304 (1985); Chem. Abstr., 103, 71514 (1985).
79. E. V. Nikitin, et al., Zh. Obshch. Khim., 52, 2721 (1982); Chem. Abstr., 98, 107424s (1983).
80. I. Y. Aliev, I. N. Rozhkov, and I. L. Knunyants, Tetrahedron Lett., 2469 (1976).
81. J. Badoz-Lambling, A. Thiebault, and P. Oliva, Electrochim. Acta, 24, 1029 (1979).
82. V. A. Shreider and I. N. Rozhkov, Izv. Akad. Nauk SSSR, Ser. Khim., 673 (1979); Chem. Abstr., 90, 212219m (1979).

VII

STEREOCHEMISTRY

27

Stereochemistry of Organic Electrode Processes

TSUTOMU NONAKA / Tokyo Institute of Technology, Yokohama, Japan

I. Introduction 1132

II. Stereocontrol in Electrochemical Reaction 1132
- A. Steric and polar factors 1132
- B. Conformational and configurational stabilities 1133
- C. Thermodynamic and kinetic control 1134
- D. Stereoselective synthesis 1134

III. Stereoselective and Stereospecific Electroreduction 1135
- A. Olefinic compounds 1135
- B. Acetylenic compounds 1141
- C. Aromatic compounds 1142
- D. Carbonyl compounds 1142
- E. α,β-Oxygen compounds 1147
- F. Nitrogen compounds 1147
- G. Sulfur compounds 1150
- H. Halogen compounds 1151
- I. Other heteroatom compounds 1155
- J. Organometallic compounds 1156

IV. Stereoselective and Stereospecific Electrooxidation 1156
- A. Carboxylic acids 1156
- B. Acetoxylation 1160
- C. Methoxylation 1161
- D. Acetamidation 1163
- E. Other addition reactions to olefins 1164
- F. Intermolecular reactions of olefins and/or aromatics 1166
- G. Active hydrogen compounds (carbon acids) 1166
- H. Heteroatom compounds 1167
- I. Organometallic compounds 1168

V. Asymmetric Reduction and Oxidation 1168
- A. Reactions of chiral compounds 1168
- B. Asymmetric synthesis 1171
- C. Optical resolution 1174

VI. Electrochemical Stereoisomerization 1174

VII. Stereoisomeric Effects on Thermodynamics and Kinetics of Electrode Processes 1175
- A. Cathodic processes 1175
- B. Anodic processes 1178

References 1179

I. INTRODUCTION

It can be stated without doubt that stereochemistry is one of the most interesting and important fields in organic chemistry and has brilliantly advanced to date. Although organic electrochemistry has also rapidly developed since the 1960s, its stereochemical aspect has advanced very slowly. One of the earliest works concerning the stereochemistry of an electrochemical reaction may have been made by Clemo and Smith [1] in 1928: They reduced p-dimethylaminobenzaldehyde at a cathode in aqueous sulfuric acid and separated one of the two diastereomeric glycols as a product. Although the stereochemistry of this kind of reductive dimerization of carbonyl compounds was afterward studied extensively and made considerably clearer, many of later workers who investigated other reactions did not pay much attention to the stereochemistry of their products.

However, particularly in the last 10 years, the situation seems to be changing, since many convenient and useful techniques for stereoisomeric analysis have been developed and have become familiar not only to organic chemists but also to electrochemists.

An electrochemical reaction is a unique oxidation-reduction that takes place in a unique heterogeneous interface between an electrode and an electrolytic solution, as is described here in detail. Therefore, the stereochemical feature of electrochemical reaction must be also unique. However, there is a problem. Generally it is not easy to rationalize simply and clearly a stereochemical course involved in a given electrochemical reaction on the basis of experimental results and theoretical considerations, since the manner of stereocontrol is very complicated in general. Thus, despite a number of rules, it is also not easy to predict the stereochemical results in advance. Although it is likely that this problem will be solved in the near future, fortunately it is not so difficult to improve the stereochemical results empirically (optimization) from the aspect of stereoselective synthesis. Several hundred original reports published in the past and some general [2-8] and topical [9-17] reviews should be very helpful for this type of investigation.

II. STEREOCONTROL IN ELECTROCHEMICAL REACTION

A. Steric and Polar Factors

Substrate and/or intermediate species absorb on an electrode surface and orient themselves so that their least hindered sides face the electrode unless there is another effect, such as a polar one. This may be the simplest steric factor governing the stereochemistry of reactions. There may also be more complicated steric effects that result in conformational or configurational change of the species, since the electrode on which the species adsorb strongly may behave as if it were a very bulky substituent. For example, visualize a substituted cyclohexane molecule. The substituent must be preferentially in the equatorial position in a free state (I), but it may be in the axial position in an adsorbing state (II), if the electrode is bulkier in the practical effectiveness of the steric hindrance than the substituent.

(I) (II)

In addition to the steric interaction of adsorbed species with the electrode, the species can sterically interact with the same and different species adsorbed on the electrode surface.

An electrode interface has a layered structure in which an nonuniform electric field (potential slope) is generated by polarization of the electrode. An extremely strong electric field of around 10^8 V cm^{-1} in the most inner layer, the so-called electron transfer layer (see Chap. 2), which is very thin, 10 Å or less, should cause a variety of polar effects. Since not only the electron transfer step but also adsorption and some of chemical steps involved in an electrolytic reaction take place in the electron transfer layer, the electrochemical reaction should be strongly influenced by polar factors. The orientation of polar adsorbed species, such as ions and dipoles, are electrostatically influenced, and consequently the stereochemistry of their reactions is also controlled by this kind of electrostatic factor.

A conformational or configurational change may also occur at the electrode interface. For example, when a planar radical (III) receives one electron from a cathode to give the corresponding anion, the original configuration of the radical may be favorably inverted by such a strong electric field to give the inverted anion (IV), as shown in Eq. (1):

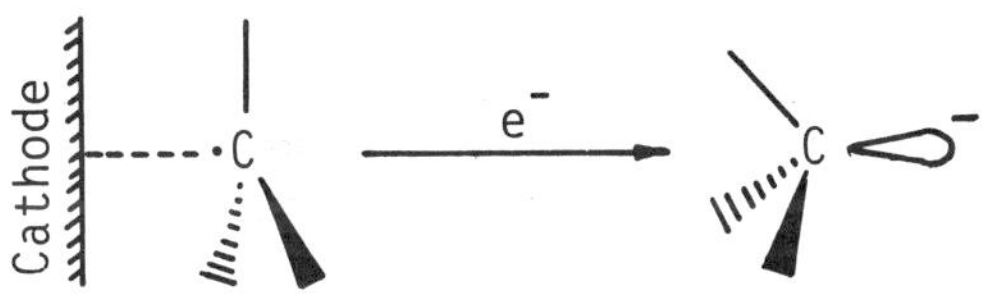

(1)

(III) (IV)
Planar radical Inverted anion

B. Conformational and Configurational Stabilities

The stereochemistry of reactions is controlled by conformational and configurational stabilities of intermediate species in the transition states. Therefore, the stereochemical course of the reactions must be discussed on the basis of the intermediates and reaction mechanism that have been presumably clarified by electrochemically and organic chemically reasonable ways. Of course, to the contrary, it is also likely that a stereochemical result makes the mechanism more clear. In this sense, a stereochemical study can be useful as one of methods of analyzing the mechanism.

Anyway, it should be noted that the conformational and configurational stabilities of the species in an electrode interface may be quite different from those in free states, owing to the peculiar steric and polar factors. In other words, in a given situation, the conformation or configuration at the electrode interface may be the reverse of that in the free state.

The relation between the rate of each reaction step and the conformational or configurational stability of the intermediate species must also be considered to rationalize the stereochemical results. Suppose the electrochemical reduction [Eq. (2)] of an alkyl halide (R–X) to the corresponding alkane (R–H) via an anion radical [$(R-X)^{-}$], radical (R), and anion (R^{-}) mechanism (ECEC). The order of configurational stability may be R–H > $(R-X)^{-}$ > R^{-} > R in general, but the stability of R is not always lower than that of R^{-} when the adsorption of R^{-} is weakened by electrostatic repulsion by the cathode. The situation is more complicated when the stereoisomerization of the intermediates must be considered. For instance, if the reaction rate of an unstable intermediate is large enough, its isomerization can actually be neglected. In contrast, the isomerization of the other intermediates with very small rates cannot be neglected even though they were considerably stable in general.

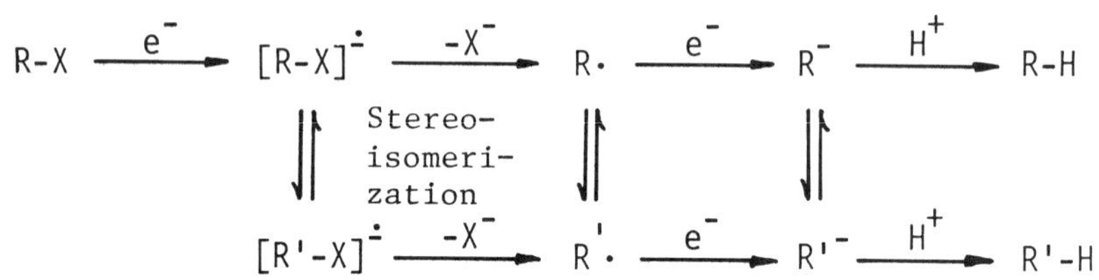

(2)

C. Thermodynamic and Kinetic Control

It is important in a mechanistic discussion of stereochemical results to know whether the stereochemistry of products is kinetically or thermodynamically controlled. However, in many reports this point is usually ignored; moreover, only a few reports have mentioned the thermodynamic equilibrium ratio of stereoisomeric products. The former point can be experimentally examined without great difficulty, but it may not always be easy to determine the ratio. For the latter, much more attention to the stereochemical literature must be paid by electrochemists engaging in stereochemical studies.

D. Stereoselective Synthesis

How to increase the stereoselectivity of a desired product is the most important problem from a synthetic aspect. For this purpose, empirical procedures seem to be more useful at present rather than theoretical or mechanistic ones, because the manner of stereocontrol in electrochemical reactions is too complicated, as described earlier. In other words, the optimization of reaction conditions can be practically achieved by doing many experiments under different conditions. It has been well known that stereochemical results are affected by a variety of the conditions, such as electrode material, electrode potential, current density, supporting electrolyte, solvent, additive, and temperature. Modification of electrodes is also useful for changing stereochemical results.

It is strongly desired that the theoretical considerations are much more helpful in designing electrochemical stereoselective synthesis in near future. Therefore, in this chapter the theoretical and mechanistic problems of the stereochemistry of electrochemical reactions are discussed as well as stereoselective synthesis.

III. STEREOSELECTIVE AND STEREOSPECIFIC ELECTROREDUCTION

A. Olefinic Compounds

Hydrogenation to Alkanes

Since olefin molecules can have stereoisomers (cis and trans; Z and E), the interesting stereochemistry, which is not only stereoselectivity but also stereospecificity, of their hydrogenation should be observed when the products also have stereoisomers (d,l and meso; threo and erythro). However, in only a few works could the stereospecificity be observed. Rosenthal and coworkers [18] found that dimethylmaleic and dimethylfumaric acids were reduced at a mercury cathode to form d,l- and meso-α,α'-dimethylsuccinic acids, respectively. This is direct evidence that the hydrogenation proceeds through trans addition of hydrogen and may be the first experimental evidence of this point.

α,β-Disubstituted cyclic olefins give the corresponding stereoisomeric (cis and trans) products in the hydrogenation, but studies from the aspect of stereoselective synthesis have been very rarely reported [19]. On the other hand, Sarrazin and Tallec [20,21] reported remarkable stereochemical results in the hydrogenation of some 2,3-disubstituted indenones to the corresponding indanones at various pH values and indicated a tentative interpretation of the results. The most remarkable result was obtained using 2,3-dipheylindenone, as shown in Eq. (3). This suggests that the hydrogenation proceeds through the cis addition in a strongly acidic medium but with trans addition in approximately neutral media.

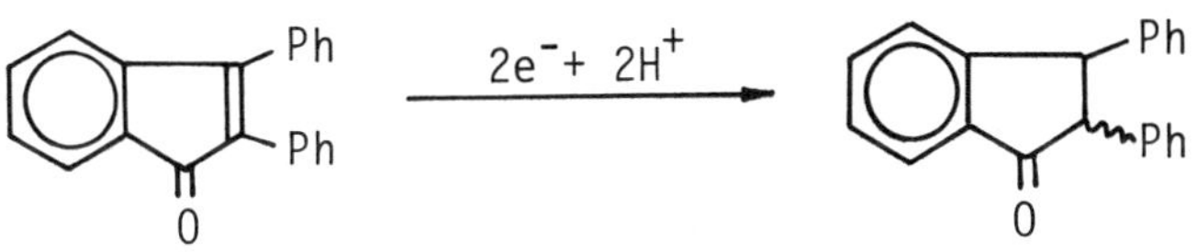

(3)

100%(*cis/trans*: ∞) in 0.25 M H_2SO_4

100%(*cis/trans*: 0) at pH 4.7 - 9.3

The hydrogenation of conjugated dienes can also have stereochemical features; however, either synthetically [22,23] or mechanistically [24-26] valuable studies have rarely been reported.

A characteristic nature of the cathodically initiated copolymerization of isoprene-maleate was found [27]. The cathodic polymerization afforded a copolymer with an alternating structure with 1,4-isoprene units only and the ratio (cis 1,4) / (trans 1,4) = 7; chemical polymerization led a mixture of alternating block and cyclic structures. Stereoregularity has been found in a few studies of the electropolymerization of vinyl compounds.

Peters and coworkers [28] reported that some trans-monoolefins were found as by-products in the reduction of a long-chain accumulated diene with a chloro group at the ω-position.

Generally speaking, the stereochemistry of electrochemical olefin hydrogenation has been poorly studied from both the synthetic and mechanistic aspects.

Hydrodimerization

Intermolecular dimerization The intermolecular hydrodimerization of olefins gives stereoisomeric (d,l and meso) products. Moreover, the stereospecificity can be discussed by using stereoisomerically (cis and trans; Z and E) stated olefins.

It is well known that Baizer and coworkers established the most brilliant industrial electroorganic synthesis of the hydrodimerization of acrylonitrile to adiponitrile. They extended this hydrodimerization to a variety of activated olefins and in some cases [29-31] paid a little attention to the stereochemistry of products. Therefore their stereochemical data were not enough to discuss the stereochemical course of reaction. Afterward, Klemm and Olson [32] tried to make a working hypothesis in the hydrodimerization of cinnamates by considering an orientated adsorption of anion radical intermediates on a cathode surface, but this was not persuasive becuase of a lack of experimental data on the stereochemistry of both the starting olefins and products.

Simonet and coworkers [33] reported the stereochemistry of hydrodimers derived from 2-substituted 3-(2-furyl)-acrylonitriles in detail. The meso-hydrodimers were formed preferentially, and the d,l-dimers were predominant when the phenyl instead of the furyl derivatives were used. However, this remarkable substituent effect was not explained.

More fundamental studies were made earlier. House and coworkers [34] found that the hydrodimerization of an α,β-unsaturated ketone (enone) was at least partially stereospecific, with the d,l-dimer formed from the trans-enone and mainly the meso-dimer formed from the cis-enone. They postulated hydrogen bonds bridging the two radical anions coupling between their β-carbon atoms to explain the stereospecificity observed. Kanetsuna and Nonaka [35] also examined both the stereoselectivity and stereospecificity of the hydrodimerization in detail. It was indicated that either the reaction mechanism or the stereochemical course is changed by the nature (protolytic or aprotic) of the reaction media. Particularly, the latter is greatly influenced by the manner of adsorption of an anion radical intermediate on the cathode. It was also postulated to explain the experimental results that the anion radical adsorbs at the β-position, where the radical p-electron is delocalized [Eq. (4)] and the adsorption is affected by the nature of the media. Recently, Lund and coworkers [36] have reported interesting stereochemical results in the reductive dimerization-cyclization of benzylidenemalononitrile, but the reaction pathway has not been stereochemically clarified.

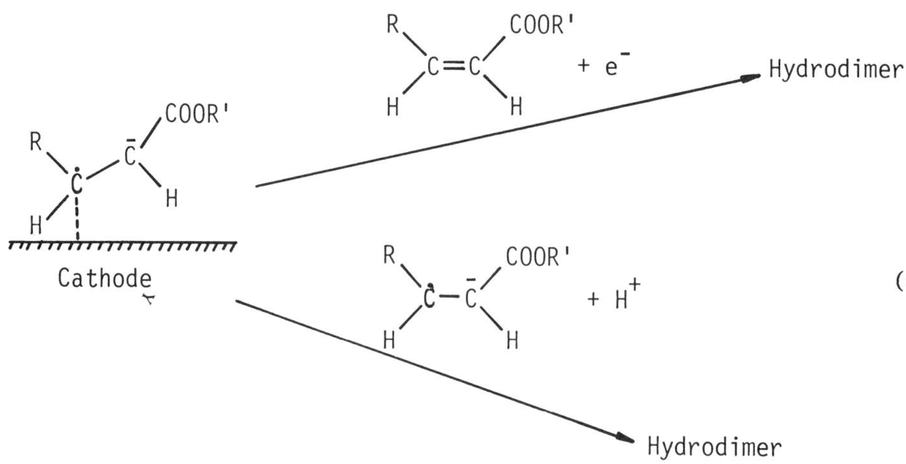

(4)

As described, the stereochemical course of the hydrodimerization of activated olefins has remained somewhat ambiguous.

Although the cross-reductive coupling between different activated olefins should also afford the formation of stereoisomeric (threo and erythro) products, no study has been reported so far. However, it was reported that acrylates gave the corresponding trans-unsaturated carboxylic acids (V) almost predominantly by reduction in the presence of butadiene [Eq. (5)] [37]. This reaction may be valuable from the aspect of stereoselective synthesis.

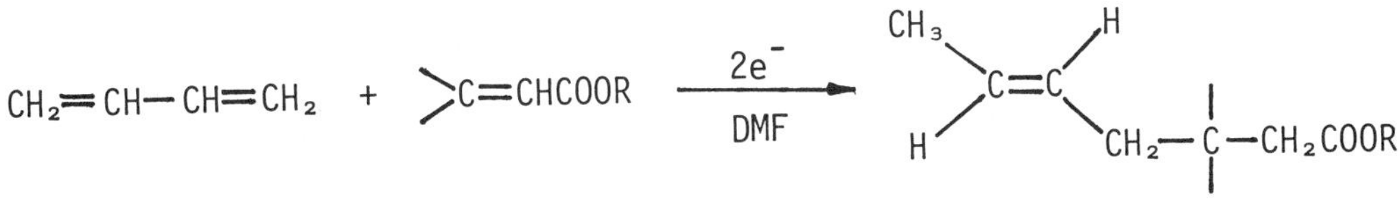

(V) (5)

Intramolecular cyclization Since the development of the hydrodimerization of activated olefins in aqueous solutions containing quaternary ammonium salt electrolytes, some workers have paid attention to the stereochemistry of reductive intramolecular cyclization of diolefins as well as the intermolecular dimerization of monoolefins. The reaction is presented in Eq. (6), where the trans-cyclized products are always predominant [38-46]. These results seem interesting from the aspect of the stereoselective synthesis of cyclic compounds, although the stereochemical course of each reaction has not been made clear in most cases because of uncertainty about the stereochemistry of the starting diolefins themselves.

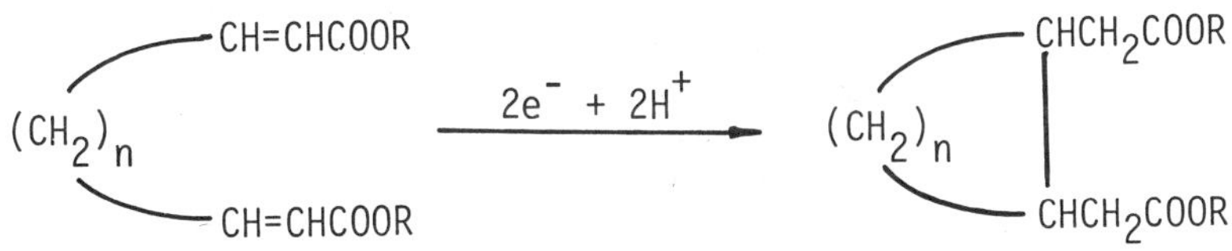

(6)

n = 1 - 4 37 - 87 % (*cis*/*trans*: 0 - 0.93)

It is noticeable that the cis-trans ratio of the reaction products is influenced strongly by proton donor and cathode material [39]. The ratio as well as the yield increases when a stronger proton donor is used. This may suggest that the protonation of anionic intermediate species is an important key step in either cyclization or stereocontrol. The fast protonation of the configurationally unstable anionic species favorably results in the formation of the less stable cis products.

Benzene-1,2-diacrylate gave quite a different type of stereoisomeric cyclized product, which originated from two molecules of the starting diacrylate, along with a normal intermolecular hydrodimer with a meso form [42,43]. A dissymmetrical diolefin was also found to give a mixture of stereoisomeric cyclization products, but their formation ratios could not be determined [44].

Recently, a unique cathodically initiated intermolecular cyclization of α,β-unsaturated ketones has been reported by Fournier and coworkers [45,46]. An unsaturated ketone (VI) gave nonstereoselectively two stereoisomeric products (VIIa and VIIb) in 92% yield, consuming ~1 F mol^{-1} of electricity when electrolyzed at a mercury cathode in a dry DMF solution of $LiClO_4$ [Eq. (7)]; in a Bu_4NBr solution a cyclized trimer (VIII) was obtained as the sole product by passing a catalytic amount of electricity. The configuration of VIII is as shown in Eq. (8). On the other hand, an analog of VI gave stereoselectively a different type of dimer in 100% yield under similar conditions. The dimer and trimer (VIII) are formed through a sort of cathodically initiated radical oligomerization, but the detailed mechanism has not been clarified.

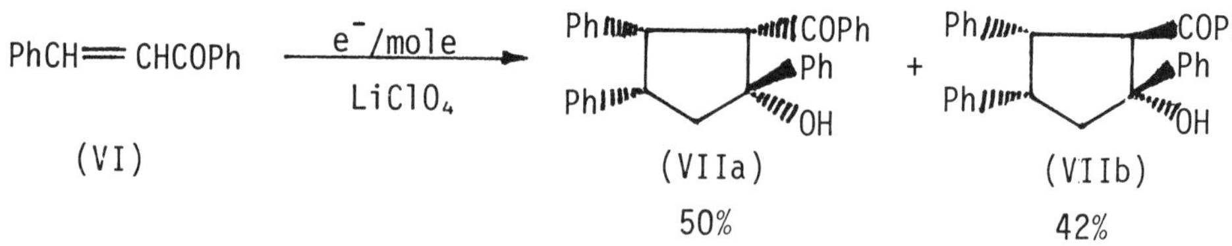

(7)

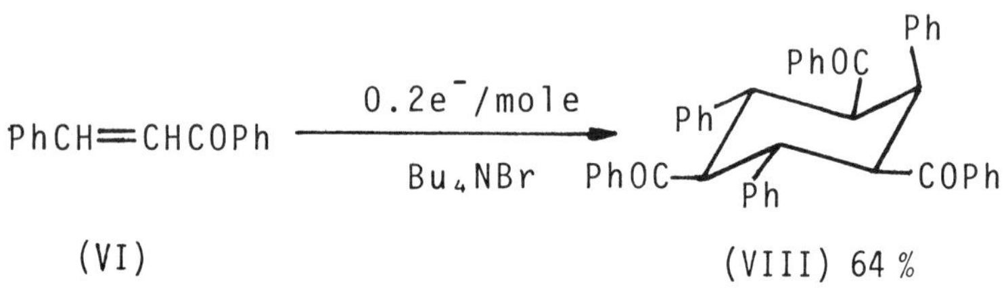

(8)

Cross-coupling and cyclization Reductive cross-coupling of olefins with anionic and radical species formed by electroreduction of other molecules is also important from both the synthetic and the mechanistic aspects.

Acylation with acetic anhydride. Shono and coworkers [47] reported that the reduction of ethyl 1-cyclohexenecarboxylate (IX) in an anhydrous acetonitrile solution containing tetraethylammonium tosylate in the presence of acetic anhydride afforded the cis and trans products (X) in good yield and moderate stereoselectivity [Eq. (9)]:

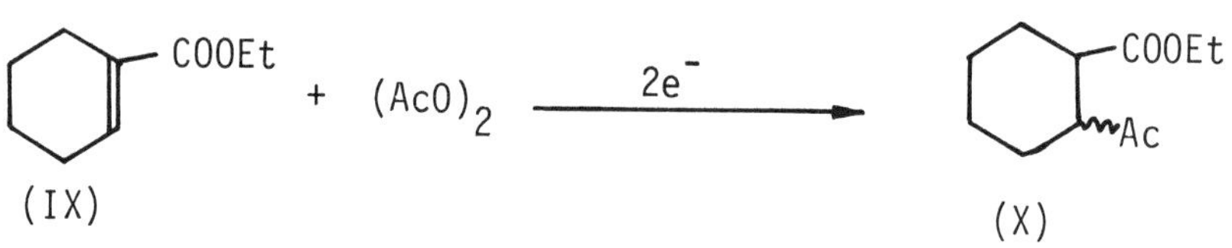

(9)

74%(*cis/trans*: 0.32)

Lund and coworkers [48,49] also extended this reaction to a variety of activated olefins, but in most cases the acylation occurred with either low stereoselectivity or low product selectivity.

Carboxylation with carbon dioxide. It is well known that olefins, when reduced in the presence of carbon dioxide, are dicarboxylated in high yields. The stereochemistry of the reaction was first mentioned by Wawzonek and coworkers [50], and then poor stereospecificity was found using cis- and trans-stilbenes as starting olefins by Dietz and Peover [51].

Cross-coupling with carbonyl compounds. The electroreductive cross-coupling of olefins with carbonyl compounds, which are initially reduced to radical or anion intermediates to attack the olefins, is synthetically useful. The stereochemistry of this type of intermolecular reaction has lately been more rarely examined [52-54] than that of intramolecular cyclization [55-61].

Kanetsuna and Nonaka [52] examined in detail the stereochemistry of final cyclic products (XII) derived by hydrolysis from initially formed cross-coupling products (XI) of crotonic acid derivatives with dissymmetrical carbonyl compounds [Eq. (10)]. The cis-trans ratio of XII varied in a wide range of 2.2-0.22 depending on the molecular structure of the starting compounds and electrolytic conditions. Stereochemistry around a C–C bond formed by the coupling reaction was discussed in this case, but in the coupling of 1-cyclohexene carboxylic acid derivatives with a symmetrical ketone (acetone) the stereochemistry of a C–C single bond originated to the corresponding olefinic double bond was examined [53]. In the latter [Eq. (11)], the cis-trans ratio of the products also varied, but the less stable cis products were always predominant. This result can give a helpful suggestion for discussing stereocontrol in this reaction. In fact, Kanetsuna and Nonaka [53] clarified the stereochemical course on the basis of this result.

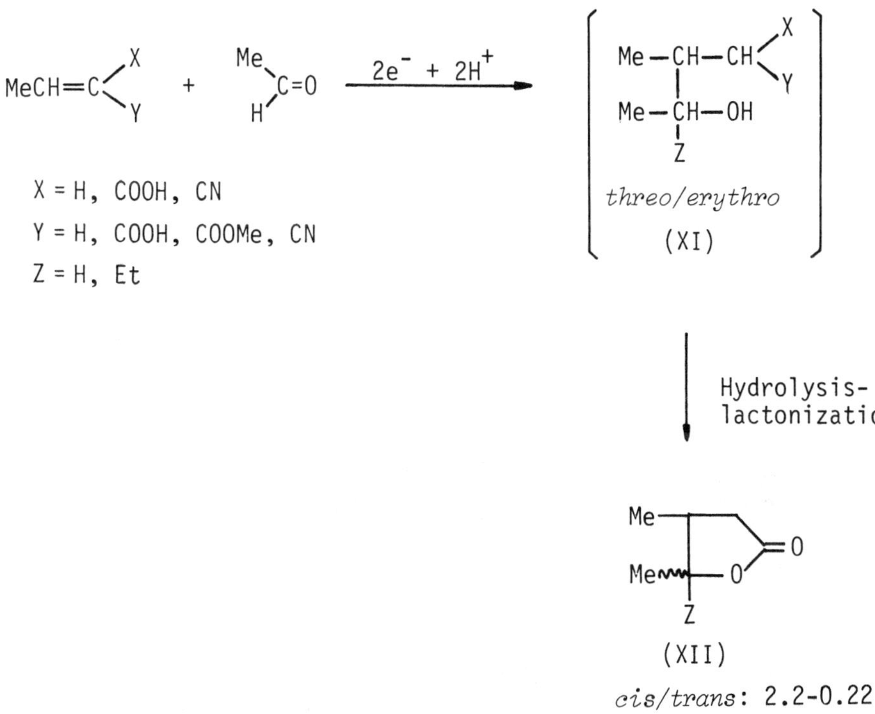

(10)

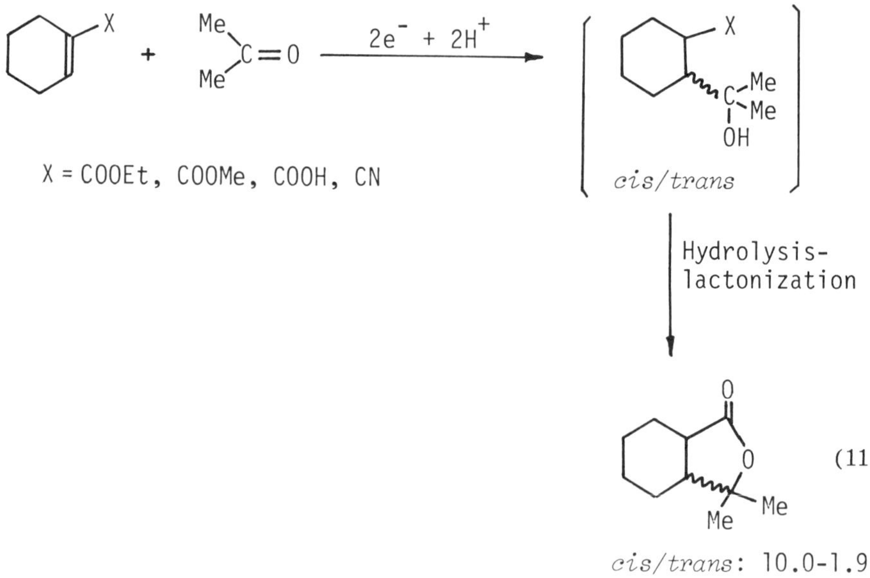

(11)

The stereochemistry of intramolecular cyclization of olefinic carbonyl compounds has been extensively studied [55-60]. Particularly in the case of unactivated olefins with carbonyl functions, which are reduced more easily than the C=C double bonds, the stereoselectivity is so high that only single isomers form, Eq. (12) [55,56]:

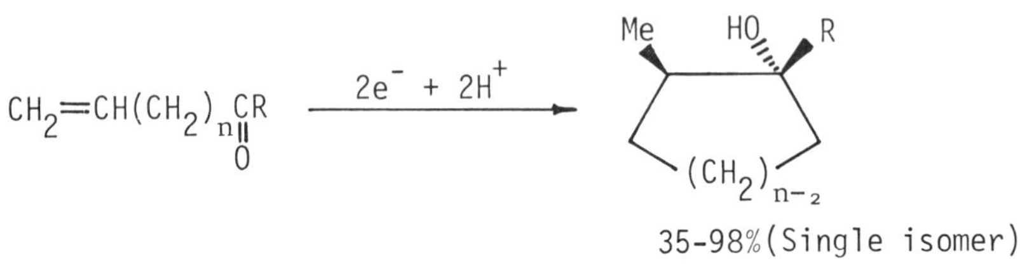

(12)

Compared with unactivated olefins, activated olefins with carbonyl groups are reductively cyclized with somewhat little lower stereoselectivity [58,59].

The reductive cyclization of unsaturated carbonyl compounds can also be achieved by using chemical reducing reagents, such as Al-amalgam, Mg-$TiCl_4$, and Na-NH_3, but in general not only chemical yields of cyclized products but also stereoselectivity are lower than those of the electroreductive cyclization.

Pattenden and Robertson [60] have recently reported a very interesting cathodic reduction of a variety of terminal allenic ketones that were highly stereoselectively cyclized through the exo mode, producing five-membered rings. It is particularly notable that the products formed in the electrochemical and chemical (sodium naphthalenide) reductions are two structural isomers with the same configuration, Eq. (13):

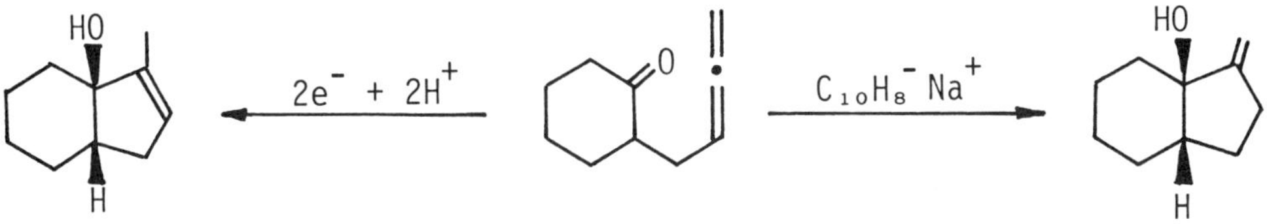

(13)

A quite different type of reductive cyclization of olefins with 1,4-dihaloalkanes was also stereochemically studied [61]. In this case the reaction is initiated by the reduction of the C=C double bonds.

B. Acetylenic Compounds

The stereochemical feature of electrolytic hydrogenation of acetylenes to the corresponding olefins changes drastically with electrolytic conditions, such as supporting electrolyte, solvent, and cathode material.

In acidic aqueous solutions, disubstituted acetylenes are smoothly hydrogenated on so-called hydrogen-active cathode materials, such as spongy nickel [62] and silver-palladium alloy [63], to give almost exclusively the cis-olefins. Kita and Nakajima [64] found that the cis-trans ratio of 6.3 of 2-butene formed from 2-butyne at a palladium black cathode was not influenced by either cathode potential or pressure. On the other hand, Nonaka and coworkers [65] examined the cis-trans ratios of dimethyl-2-butenedioate formed from acetylene dicarboxylic acid dimethyl ester on a palladium black cathode and catalyst at various pH levels and found that an increase in pH resulted in a decrease in the ratio in the both electrolytic and catalytic hydrogenations, but the former was less strongly influenced by pH. They also performed similar comparative studies with olefins and ketones from a stereochemical aspect. Kato and coworkers [66] reported the predominant formation of cis-olefins in alkaline solutions.

The stereochemistry of olefins formed from acetylenes at a mercury cathode in quaternary ammonium salt solution is sometimes greatly and in a complicated manner affected by conditions [28,67]. Horner and Röder [67] reported that the yield of trans-2-butene-1,4-diol formed from 2-butyne-1, 4-diol vaired in a wide range of 23-82% with changes in the concentration of the supporting electrolyte (Me_4NCl), cosolvent, current density, and temperature.

The hydrogenation of acetylenes in amine solvents containing alkali metal salts gives almost exclusively the trans-olefins [68,69]. However, extremely complicated results may be obtained due to skeletal isomerization of the reaction intermediates [69].

Shono and coworkers [70] reported an interesting reductive cyclization of acetylenic ketones [XIII, Eq. (14)]. The cyclized products (XIV) were formed somewhat stereoselectively in good yields.

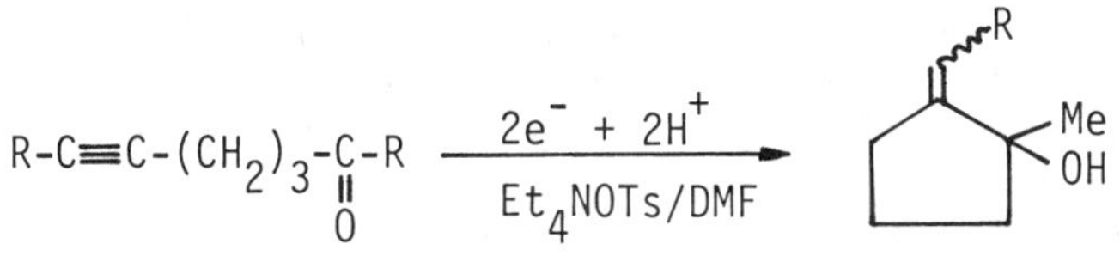

(14)

(XIII) (XIV)

53-60%(*E*/*Z*: 0.5)

C. Aromatic Compounds

Aromatic hydrocarbon nuclei can be hydrogenated partially or completely under various electrolytic conditions.

The most practical hydrogenation is that of o-phthalic acid to dihydrophthalic acid. o-Phthalic acid is efficiently hydrogenated at high hydrogen overvoltage cathodes, such as mercury and lead in acidic solutions, to give almost exclusively trans-dihydrophthalic acid [71-75].

In contrast to phthalic acid, benzene nuclei without electron-withdrawing substituents, such as a carboxyl group, are hydrogenated by using hydrogen-active cathodes in acidic solutions. For example, o-methylphenol is stereoselectively hydrogenated on a Rh/C cathode to give cis-2-methylcyclohexanol [42-83% (cis-trans ratio, 2.3-6.7)] as the major isomer [76].

Naphthols are also hydrogenated to the corresponding decanols. β-Naphthol gives trans-2-decanol (70% yield) in an acidic solution [77], but α-naphthol is hydrogenated in HMPA to cis-1-decanol (47% yield) [78]. The latter is regarded as a hydrogenation induced by cathodically generated solvated electrons.

Phenanthrene is hydrogenated at a mercury cathode in DMF containing Bu_4NI in the presence of carbon dioxide to give trans-9,10-dicarboxy-9,10-dihydrophenanthrene (43% yield) [79].

D. Carbonyl Compounds

The stereochemistry of the cathodic reaction of carbonyl compounds has been extensively studied from synthetic and mechanistic aspects. In this section the reaction is discussed in two parts: hydrogenation to alcohols and hydrodimerization to pinacols.

Hydrogenation to Alcohols

Acyclic carbonyl compounds A set of stereochemical studies of the reduction of 1,2-diphenyl-1-propanone to the corresponding diastereomeric (threo and erythro) alcohols has been reported. Earlier work was done by Mandel and coworkers [80], who reduced the propanone at a mercury cathode in a buffered solution (pH 8) and separated the erythro-alcohol only. Afterward, Diaz and coworkers [81] determined that both isomers were formed at a threo-erythro ratio of 0.45. They also suggested that such a low stereoselectivity would be due to proton exchange at the α-position of anion radical intermediates formed by the first one-electron transfer. In contrast, Fedoronko [82] has described that any chemical change does not occur at the α-position of D-(-)-1-phenyl-1-hydroxy-2-propanone during the reduction, since the product is an isomeric mixture of the optically active D-threo- and D-erythro-1-phenyl-1,2-propanediols. This refutation is acceptable if any intramolecular asymmetric induction does not occur in the reduction.

Nonaka and coworkers [83] avoided the problem of intramolecular asymmetric induction by using racemic mixtures of ketones. They reduced such ketones at a mercury cathode and found the stereochemistry of the products was affected greatly by the molecular structure of the starting ketones and the pH of the electrolytic solutions used [Eq. (15)]. It is noticeable that when R^1 = cyclo-Hex and R^2 = Me, the anti-Cram isomer was formed as a major product. This suggests that stereochemical interaction between the cathode and intermediates contributes to stereocontrol in the reaction. In addition, considering the equilibrium threo-erythro ratios of the alcohols, it is concluded that the stereochemistry of reaction is kinetically controlled. The stereochemical course of the reaction was discussed in detail, and it was suggested that key points for the stereocontrol are the adsorption step of the starting ketone molecules on the cathode surface and the configurational stability of one-electron transferred intermediates (anion radicals) at the cathode interface.

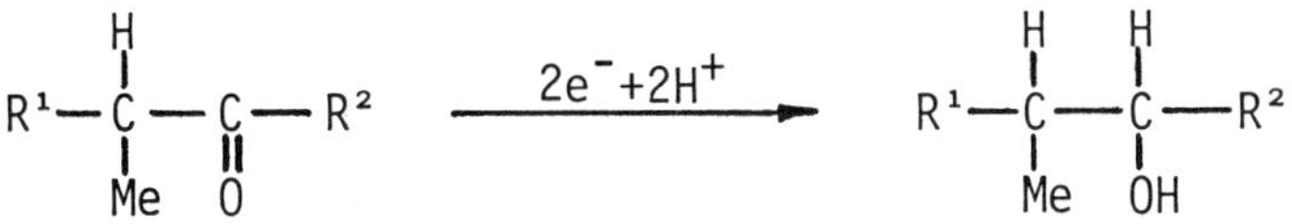

(15)

R^1 = Ph, R^2 = Ph ; 8-91%(*threo/erythro*: 0.2-0.7)

R^1 = Ph, R^2 = Me ; 3-82%(*threo/erythro*: 1.3-1.4)

R^1 = c-Hex, R^2 = Me; 3- 7%(*threo/erythro*: 0.5-0.9)

This reduction is also interesting for stereoselective synthesis [84].

Similar studies using another type of starting ketone were also reported by Nonaka and Kusayanagi [85], Eq. (16), and Lund and Hansen [86]. In this case, the stereochemical course can be more clearly explained.

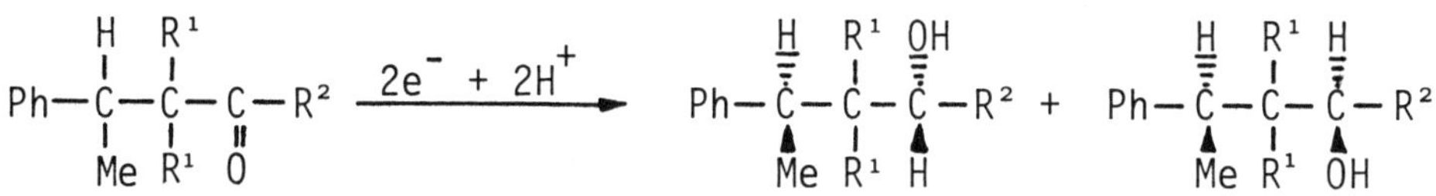

A *B*

R^1 = Me, R^2 = Ph; 75-100%(*A/B*: 0.64-0.76)

R^1 = Me, R^2 = Me; 1- 7%(*A/B*: 0.80-0.87)

R^1 = H, R^2 = Ph; 12-100%(*A/B*: 1.13-1.44) (16)

Quite different stereochemical results were obtained when hydrogen-active cathodes, such as platinum and palladium blacks, were used instead of the mercury cathode [65] in the reduction given in Eq. (15).

Cyclic carbonyl compounds Most of the stereochemical studies on the hydrogenation of cyclic carbonyl compounds to the cyclic alcohols have been performed using substituted cyclohexanones [65,87-95]: namely, the stereochemistry of the cyclohexanols formed are discussed by examining the cis-trans geometry and/or ax-eq conformation of the hydroxy groups. The stereoselectivity of the reaction varies drastically depending on such electrolytic conditions as pH, solvent system, proton donor, cathode material, cathodic potential, supporting electrolyte cation, and temperature [65,87-90,92-94]. In one case, 100% stereoselectivity for a stereoisomeric cyclohexanol is achieved; in other cases another isomer is formed in 100% stereoselectivity. However, it is not easy to clarify the stereochemical course of the hydrogenation of monocycloalkanone systems, except for t-butylcyclohexanones, because the conformation of the intermediates is not fixed. t-Butyl groups substituted on a cyclohexane ring are always at the equatorial position. Therefore, for instance, it can be reasonably explained that the less stable cis-4-t-butylcyclohexanol with an ax-OH is formed predominantly in electrolytic solutions containing bulky cations, such as $MgCl^+$ and $ZnClO_4^+$, which make tight ion pairs with carbanions, Eq. (17) [90-92]:

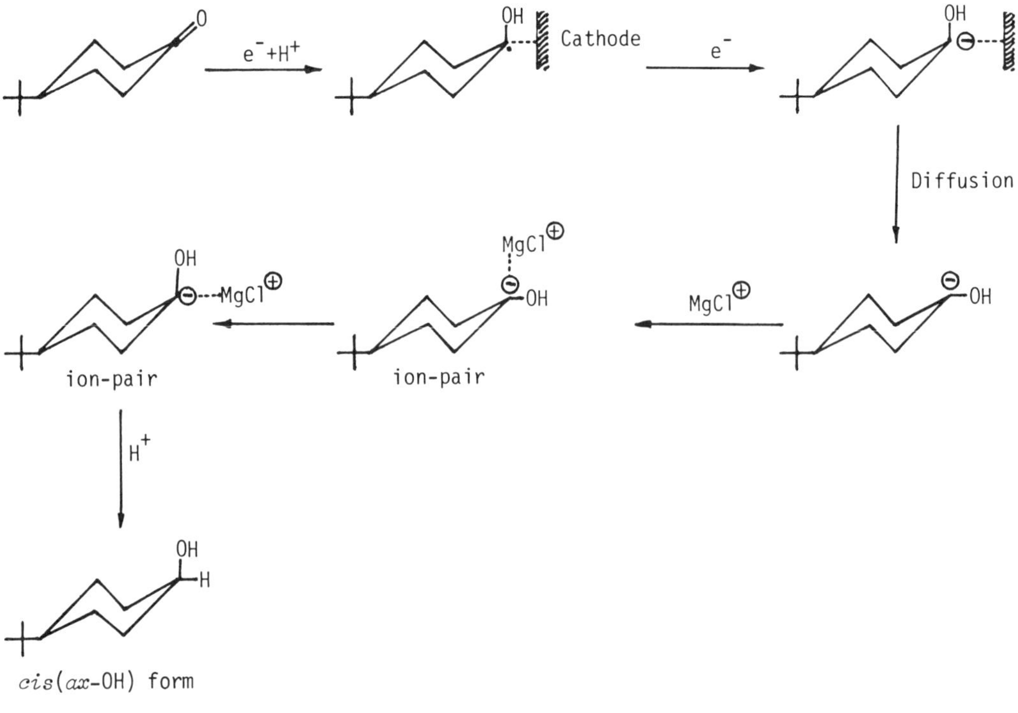

Since polycyclic systems are also conformationally fixed, their stereochemical course may be easily rationalized [89,90,92,96,97]

Stereochemical studies of the hydrogenation of cyclopentanones have been less frequently reported [94,98].

Hydrodimerization to Pinacols

Intermolecular dimerization As mentioned in Sec. I, the earliest stereochemical study in electroorganic chemistry may be that dealing with the intermolecular hydrodimerization of an aromatic carbonyl compound [1]. A number of reports of the hydrodimerization of aromatic carbonyl compounds were published, and a basic concept of stereocontrol was established [99-133]. The stereochemistry (d,l and meso) of pinacols is generally controlled by steric and nonsteric interactions between two radicals (illustrated below) coupling with each other, but not by interaction with the cathode [119], because electrochemical hydrogenation gives almost the same stereochemical results as nonelectrochemical hydrogenation in homogeneous media. The radicals XV, XVI, and XVII are favorably formed by one-electron transfer in neutral, acidic, and aprotic media, respectively. The steric interaction is brought about by a steric hindrance effect (repulsion) between substituents of the radicals and nonsteric interactions are attributed to hydrogen bonding (attraction) and electrostatic force (repulsion). These stereocontrol factors are influenced by various reaction conditions, and consequently the stereochemistry of reaction must be influenced in very delicate and complicated ways.

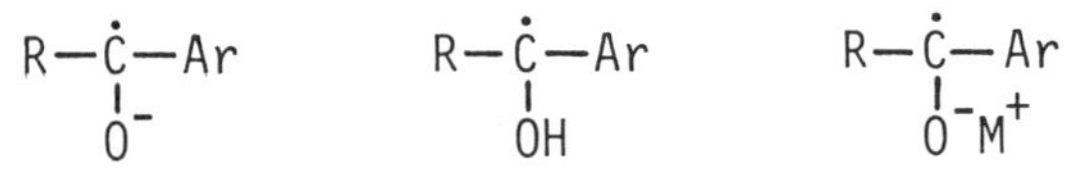

(XV) (XVI) (XVII)

In recent years, some interesting results have been reported on this subject. Although most stereochemical investigations of hydrodimerization have been performed using aromatic carbonyl compounds with unsubstituted and substituted phenyl groups, some examples dealing with other aromatic groups have also appeared [114,121-124]. The hydrodimerization of simple aliphatic carbonyl compounds generally proceeds with very low yields, and no stereochemical study has been reported so far.

However, α,β-unsaturated carbonyl compounds are stereoselectively hydrodimerized to give the corresponding unsaturated glycols in a variety of yields [126-131].

Since the olefinic part of the α,β-unsaturated carbonyl compounds is conjugated with the carbonyl group, the original geometry of the olefinic part is not always retained in the anion radical intermediates (XV, XVI, and XVII). Consequently, the stereochemistry of the final products is expected to be complicated. Additionally, the stereochemistry may become more complicated for other reasons, particularly in the case of cyclic unsaturated carbonyl compounds. Touboul and Dana [125,128] analyzed six stereoisomeric glycols formed from a variety of substituted tetrahydro-2H-phenanthrones and discussed the stereochemical course of reaction. For example, racemic and optically active XVIII gave the six and three isomers, respectively, of which XIX was major over a wide range of pH, Eq. (18):

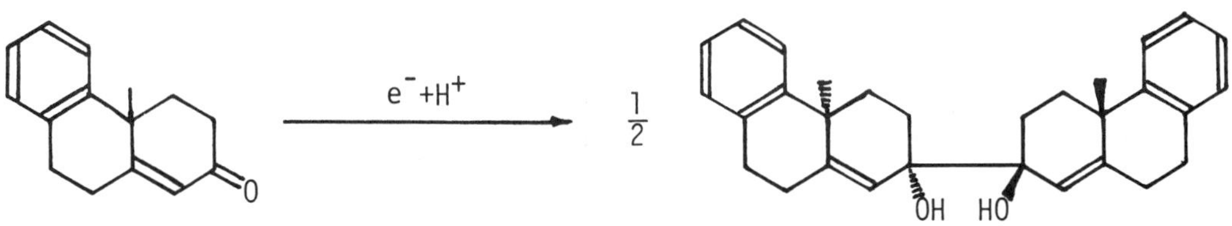

(XVIII) *cis-threo-cis* (Major)

(XIX) (18)

In the hydrodimerization of 2,3-dimethylindanone, 11 hydrodimers, including 8 stereoisomers, were analyzed [130]. The stereochemistry of hydrodimers formed from 2-methylindenone in an oxygen-saturated DMF solution should also be complicated, but it is known only that the stereochemistry around the central C−C bond is a threo form [131]. Another complication in the stereochemistry of a hydrodimer was observed in the dimerization of benzoin at a mercury-plated copper cathode in a basic aqueous ethanol solution: a single stereoisomeric hydrodimer was formed in 100% stereoselectivity, Eq. (19) [132]:

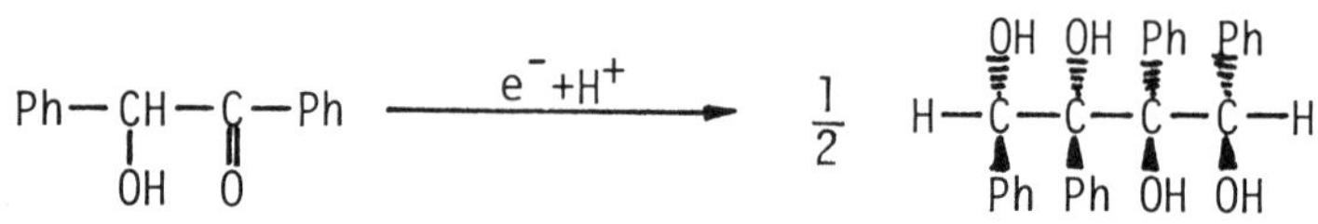 (19)

erythro-threo-erythro

Although the stereochemistry of hydrodimers is greatly affected by the structure of the starting carbonyl compounds and electrolytic conditions, the less stable d,l isomers are very rarely formed in d,l-meso ratios higher than 1.5 in the hydrodimerization of simple aromatic carbonyl compounds. Rusling and Zuman [122] reported that 4-pyridylcarboxaldehyde gave the hydrodimer in the d,l-meso ratio of 1.92 in a strongly alkaline solution. Furthermore, Smith and Utley [133] found benzaldehyde was hydrodimerized at a mercury cathode in a dry DMF solution of Bu_4NBF_4 to give the d,l-meso ratio of 14; the addition of β-cyclodextrin drastically decreased the ratio to 0.33.

Indirect electroreduction using redox mediatory systems sometimes gives quite different stereochemical results from usual direct reduction [134,135].

The stereochemistry of hydrodimerization of aliphatic carbonyl compounds has been very rarely studied. Kariv and coworkers [136] reported that an aliphatic ketone gave an almost equal quantity of the isomeric hydrodimers at a mercury cathode in a mixture of 2-propanol and water containing Et_4NOTs.

Intramolecular cyclization Some aromatic diketones were stereoselectively hydrocyclized under various electrolytic conditions, which influenced strongly the stereochemistry of the cylic diols formed, Eq. (20) [137-141]:

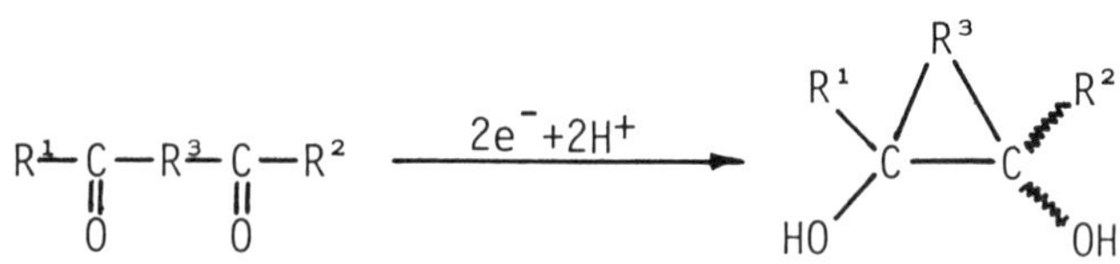 (20)

$R^1 = R^2 = C_6H_5$, R^3 = CHMe ; 80% (*cis/trans*: ?)[140]
$R^1 = R^2 = C_6H_5$, $R^3 = CMe_2$; 80% (*cis/trans*: ∞)[140]
$R^1 = C_6H_5$, R^2 = Me, $R^3 = CMe_2$; 80% (*cis/trans*: 1)[140]
$R^1 = R^2 = C_6H_5$, $R^3 = (CH_2)_3$; 80% (*cis/trans*: ∞)[138]
$R^1 = R^2 = 4\text{-}HOC_6H_4$, $R^3 = (CH_2)_3$; 82-89%(*cis/trans*:0.9-24.3)[137]
$R^1 = R^2 = 4\text{-}HOC_6H_4$, $R^3 = (CH_2)_4$; 64% (*cis/trans*: ∞)[137]

In one case [141], the stereochemical course of reaction could be somewhat clearly explained.

The hydrocyclization of Eq. (20) [$R^1 = R^2 = C_6H_5$, $R^3 = (CH_2)_3$] could also similarly occur with anodically generated unipositive magnesium ion (Mg^+) to afford the cis-cyclic diol in 87% yield [142].

Only one example of a stereochemical study of the hydrocyclization of an aliphatic carbonyl compound has been reported by Curphey and coworkers [143], but the stereoselectivity is extremely low.

E. α,β-Oxygen Compounds

Some classes of doubly oxygen-functionalized compounds at both the α- and β-positions are reduced to give stereoisomeric products [144-150].

Cathodic reduction of benzil in acidic and neutral solutions led to the formation of cis- and trans-stilbenediols, which were not separated from the catholyte but could be detected polarographically [144]. Variation in the cis-trans ratio was discussed in terms of acceleration of the trans → cis stereoisomerization of primarily formed trans-stilbenediol species in the high electric field of the cathode interface (double layer). The cis-trans ratio could be determined by trapping the chemically unstable stilbene diols as the O,O'-dimethyl derivatives during electrolysis in the presence of dimethyl sulfate [145]. The cis isomer as expected was always predominant, Eq. (21):

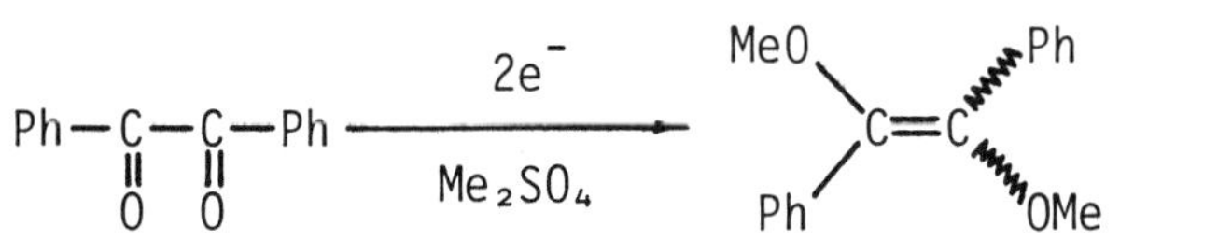

(21)

cis/trans: 1.0-5.5

Shono and coworkers [146] found that 1,3-dimethanesulfonates underwent the electroreductive cyclization in DMF containing quaternary ammonium salts to give stereoselectively the corresponding cyclopropanes in a good yield, Eq. (22):

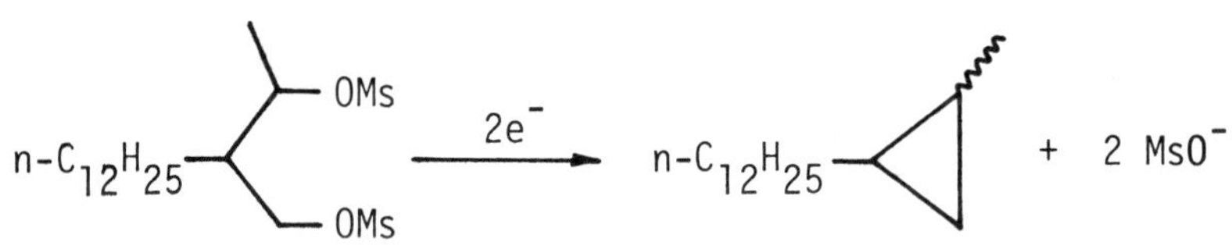

(22)

76% (*cis/trans*: 0.33)

Nonaka and coworkers [147,148] found a stereoselective electroreductive elimination reaction of cyclic sulfates of 1,2-diols to the trans-olefins and proposed a reaction mechanism based on the experimental results obtained. This reaction may be not only stereoselective but also stereospecific. An epoxide was also stereoselectively and regioselectively reduced to the mono-alcohol [149,150].

F. Nitrogen Compounds

Imines and Immoniums

Cyclic imines are highly stereoselectively hydrogenated. As a typical example, the hydrogenation of the Schiff base of aniline with camphor was reported by Fry and Reed, Eq. (23) [151]. An unsubstituted analog with norcamphor gave a little less stereoselectivity. The electrohydrogenation gives the same stereochemical result as sodium reduction in ethanol, but platinum-catalyzed hydrogenation does the reverse.

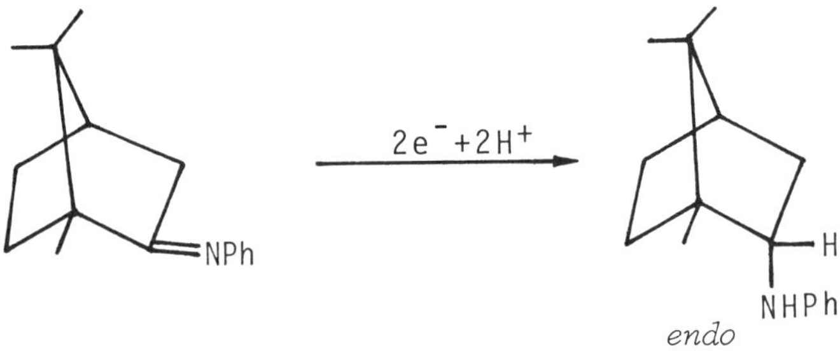

(23)

An acyclic diimine was hydrogenated at a mercury cathode to give a mixture of the d,l- and meso-diamines in 70% yield [152].

Imines can be stereoselectively hydrodimerized to 1,2-diamines in a manner similar to the hydrodimerization of carbonyl compounds to pinacols [153, 154]. The stereochemistry of the diamines was examined in detail by Horner and Skaletz [153]. Bicyclic oximes also give highly stereoselectively, but in a manner different from the bicyclic imine of Eq. (23), the corresponding amines [Eq. (24)]: the stereochemical feature of the electrochemical reduction is the same as that of the $LiAlH_4$ reduction, but opposite to the sodium reduction [155].

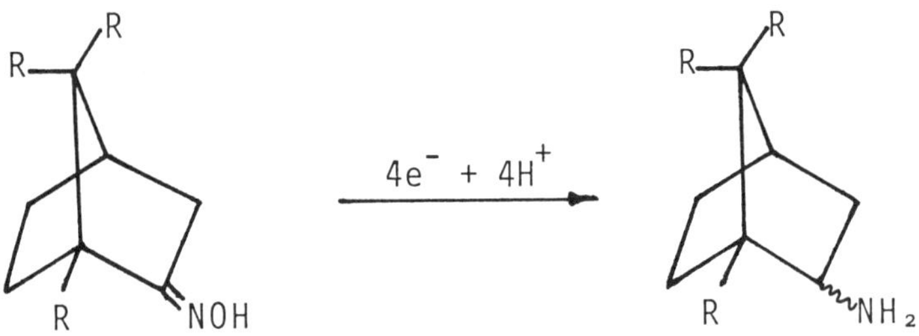

(24)

R = H; *exo*/*endo*: 0.01

R = Me; *exo*/*endo*; ∞

Substituted cyclic immonium salts are also electrochemically hydrogenated in aprotic solvents to give stereoselectively the isomeric cyclic amines [156]: The stereoselectivity is influenced by solvent and supporting electrolyte. Immonium cations can be somewhat stereoselectively hydrodimerized with a certain class of bromo compounds [157].

Azacyclopropane and Azo Compounds

The azacyclopropane ring of mitomycin C was opened in methanol under electrochemical reductive conditions, but the stereoselectivity of methoxyamine formed was low [158]. The electroreduction of azobenzene in the presence of benzyltrimethylphosphonium bromide and benzaladehyde led in a Wittig-type reaction to afford the stereoselective formation of stilbens (cis/trans ratio 1.5; yield 98%) [159].

Quaternary Ammoniums

Quaternary ammonium cations are cathodically decomposed at negative potentials in aprotic solvents to generate tertiary radicals that couple with each other to give the isomeric dimers. Such a radical derived from a trimethylbenzylammonium cation seemed to be free from steric interaction with the cathode because the stereoisomeric ratio of the dimers was almost equal to that of the same product obtained by a well-known free radical process in a homogeneous medium, Eq. (25) [160]:

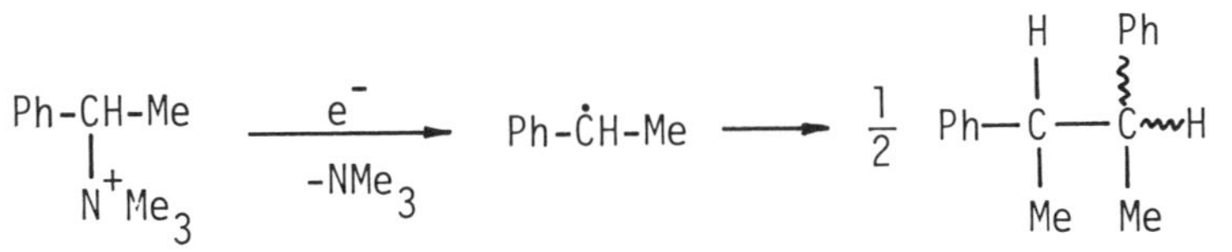

(25)

dl/meso: 1.17

Iversen and Madsen [161] reported a synthetically interesting electroreductive ring-opening reaction of cyclic quaternary ammonium salts. The stereochemistry of products is greatly affected by pH, Eq. (26):

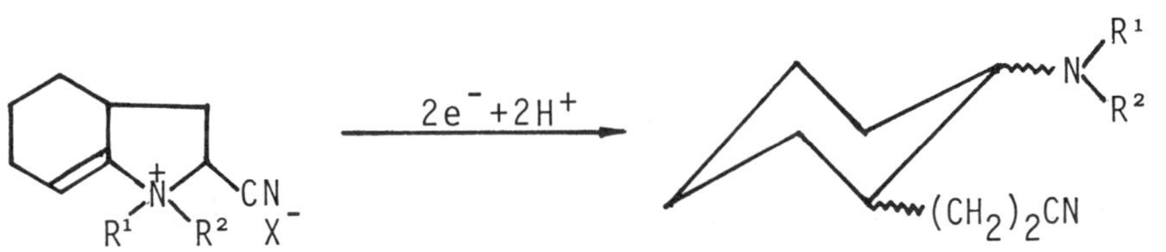

(26)

$R^1 = R^2 = Me$;
$R^1, R^2 = (CH_2)_4$, $(CH_2)_2O(CH_2)_2$
$X = Cl$, ClO_4

0-82%(*cis*/*trans*: 0.2-10.1)

Others

Trans-azobenzene was Raman spectroscopically detected as an adsorbed species on a silver cathode in the reduction of nitrobenzene [162]. Hydroxylamine intermediates derived from the electroreduction of γ-nitroketones undergo intramolecular cyclization, and the cyclized products are further reduced to stereoisomerically substituted pyrrolidines [163].

Knittel [164] found that electroreductive acetylation of benzoyl azide led to the isolation of both rotational isomers of N-acetylbenzamide formed.

A stereospecific nature of the cathodically initiated anionic polymerization of isocyanates in DMF containing tetrabutylammonium salts was mentioned [13].

G. Sulfur Compounds

Olefin Formation Reaction

Sulfur compounds (XX) with oxygen functions at the β-positions are reductively desulfurized to give stereoisomeric olefins, Eq. (27) [165-167]. The reaction is highly stereoselective but not stereospecific. Methyl stylyl sulfoxide and sulfone are also desulfurized in the presence of tertiary butyl bromide to give E-phCH=CHCMe$_3$ [168].

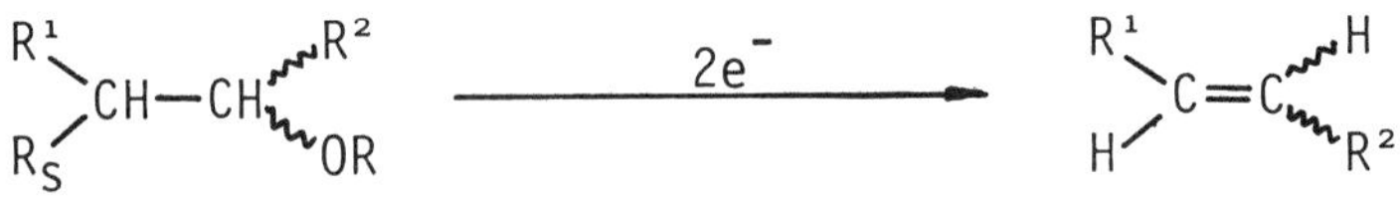

(27)

(XX)

erythro-XX	$R^1 = R^2 = Ph$, R = Ac, R_S = SPh	92%(*cis*/*trans*: 0) [165]
erythro-XX	$R^1 = R^2 = Ph$, R = H, $R_S = O_2SPh$	40%(*cis*/*trans*: 0) [166]
threo-XX	$R^1 = R^2 = Ph$, R = H, $R_S = O_2SPh$	10%(*cis*/*trans*: 0) [166]
erythro-XX	$R^1 = R^2 = Ph$, R = Ac, $R_S = O_2SPh$	83%(*cis*/*trans*: 0.19) [166]
threo-XX	$R^1 = R^2 = Ph$, R = Ac, $R_S = O_2SPh$	95%(*cis*/*trans*: 0.19) [166]
unknown-XX	$R^1 = C_9H_{18}$, $PhCH_2CH_2$, R^2 = Me, R = H, $R_S = O_2SPh$	50-52%(*cis*/*trans*: 0.3-0.4) [169]

Some ring-opening reactions of cyclic sulfur compounds to olefins proceed stereoselectively via desulfurization. 1,2-Diphenylthiirene dioxide is reduced to eliminate sulfur dioxide, giving trans-stilbene in 30% yield [169]. Functionalized 1,2-dithiole-3-thiones are also ring opened in the presence of alkylating reagents [170].

The reductive coupling of dithiobenzoate esters results in the predominant formation of Z-1,2-dialkylthio-1,2-diphenylethylenes in good yields when alkyl halides are added after completion of the electrolysis [Eq. (28)] [171]. The reaction mechanism via a dianion intermediate species is postulated, but the stereochemical course has remained ambiguous.

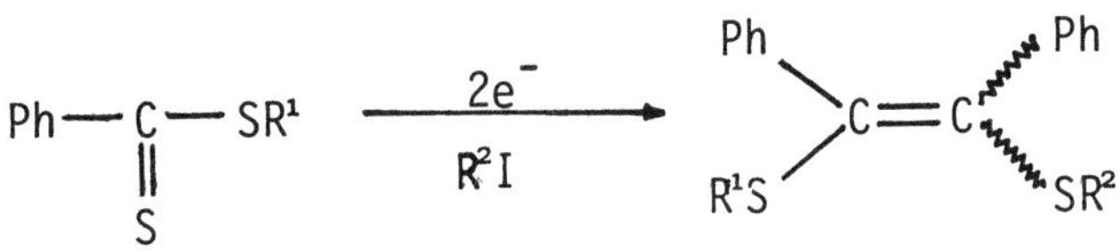

(28)

$R^1 = R^2$ = Me	24-70%(*E*/*Z*: 0-0.06)
$R^1 = R^2$ = Et	56%(*E*/*Z*: 0.22)
R^1 = Me, R^2 = Et	46%(*E*/*Z*: 0.21)
R^1 = Et, R^2 = Me	10%(*E*/*Z*: 0.00)

Cyclization Reaction

Shono and coworkers [172] reported a synthetically useful reductive cyclization reaction of γ-hydroxyphenylsulfides to lead to the formation of stereoisomeric cyclopropanes, but the isomeric ratio was not determined. Ethylene bis-dithiobenzoate is also reductively cyclized in dry acetonitrile containing methyl iodide to give nonstereoselectively d,l- and meso-dithiane derivatives in a good yield [173].

H. Halogen Compounds

A variety of stereoselective and stereospecific reductions of halogen compounds have been reported as described here.

Cyclic Halides

Cyclic monohalides The stereochemistry of reduction of monohalides, except chiral halides (see Sec. V), has been scarcely reported so far. Tallec and coworkers [174] have recently found that the reduction of cis- and trans-cyclopropyl bromides proceed stereospecifically in a tetrabutylammonium salt solution. On the other hand, in an ammonium (NH_4^+) salt solution the reversed stereochemistry of the products is observed.

Monocyclic gem-dihalides 1-Substituted-2,2-dibromocyclopropanes (XXI) are stereoselectively reduced to the corresponding diastereomeric monobromides (XXIIa and XXIIb) in good yields [Eq. (29)] [175]. Although the stereochemistry of the monobromides formed is strongly influenced by the substituents (R^1 and R^2) of XXI and electrolytic conditions, such as pH, supporting electrolyte in aqueous ethanol, and cathode potential, XXIIa (erythro-type isomer) is always predominant.

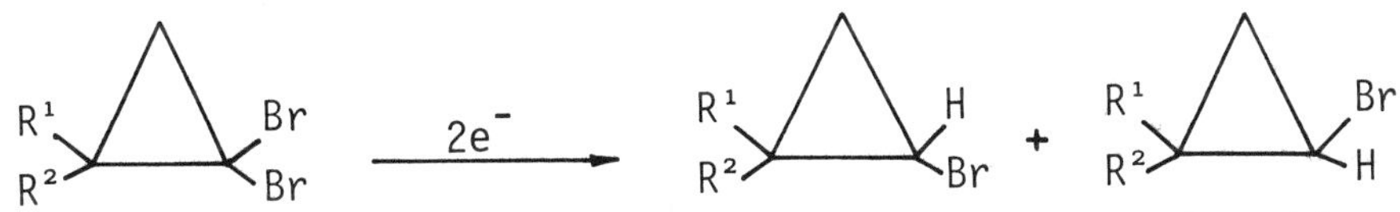

(XXI) (XXIIa) (XXIIb)

R^1 = H, R^2 = Ph ; XXIIa/XXIIb: 1.6-5.3

R^1 = Me, R^2 = Ph ; XXIIa/XXIIb: 1.1-4.6

R^1 = Me, R^2 = COOH ; XXIIa/XXIIb: 1.9-6.1

R^1 = Me, R^2 = COOMe; XXIIa/XXIIb: 1.8-15.7

R^1 = Me, R^2 = COOEt; XXIIa/XXIIb: 1.6-99.0

R^1 = Ph, R^2 = COOH ; XXIIa/XXIIb: 1.0-10.1

R^1 = Ph, R^2 = COOMe; XXIIa/XXIIb: 1.1-2.2

R^1 = Ph, R^2 = COOEt; XXIIa/XXIIb: 0.9-1.8

(29)

A similar diastereoselective reaction has been found in the reduction of 3-substituted 4,4-dichlorosuccinimides [176]. The stereochemistry of the products in an acidic medium is opposite to that in an alkaline medium.

Bicyclic gem-dihalides Bicyclic *gem*-dihalocyclopropanes (XXIII) always give predominantly the corresponding *endo*-monohalocyclopropanes (XXIV), which seem to be formed by the reduction of the less-hindered *exo*-halogen against the cathode in an adsorption state [Eq. (30)], although the isomeric ratio varies with the molecular structure of XXIII and electrolytic conditions, such as solvent and supporting electrolyte [177,178]. It is noticeable that an *exo*-bromo-*endo*-chlorocyclopropane (XXVa) normally gives the *endo*-chloride (XXVIa) as a major isomeric product, but the *endo*-bromo-*exo*-chlorocyclopropane (XXVb) gives the *exo*-chloride (XXVIb) and a mixture of XXVIa and XXVIb gives both isomers in an equal amount under the same conditions, Eq. (31). Considering that alkyl bromides are more easily reduced than chlorides in general, this suggests that the reduction of *gem*-dihalocyclopropanes may occur at either halogen atom rather than only on the less-hindered *exo*-halogen. Stereoisomerization between *exo*- and *endo*-monochloroanionic intermediates may also be considered for explaining the experimental results, Eq. (32) [178].

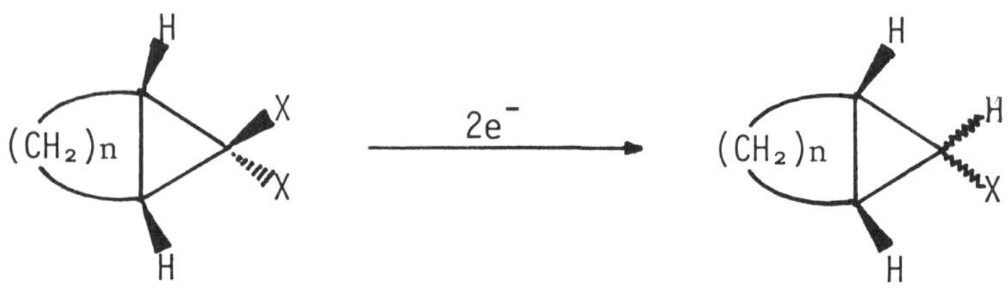

(XXIII) (XXIV) (30)

n = 4, X = Br	*endo/exo*: 1.1-5.0[177,178]
n = 4, X = Cl	*endo/exo*: 1.9-∞ [178]
n = 6, X = Br	*endo/exo*: ∞ [177]

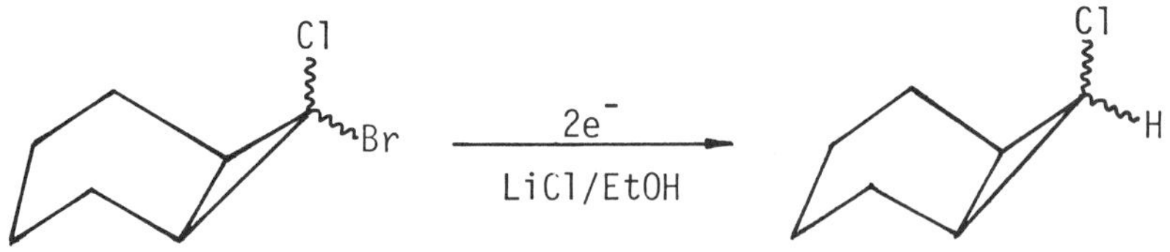

(31)

exo-Br, *endo*-Cl (XXVa)	XXVIa (*endo/exo*: 2.6)
endo-Br, *exo*-Cl (XXVb)	XXVIb (*endo/exo*: 0.38)
XXVa + XXVb	XXVIa + XXVIb (*endo/exo*: 0.97)

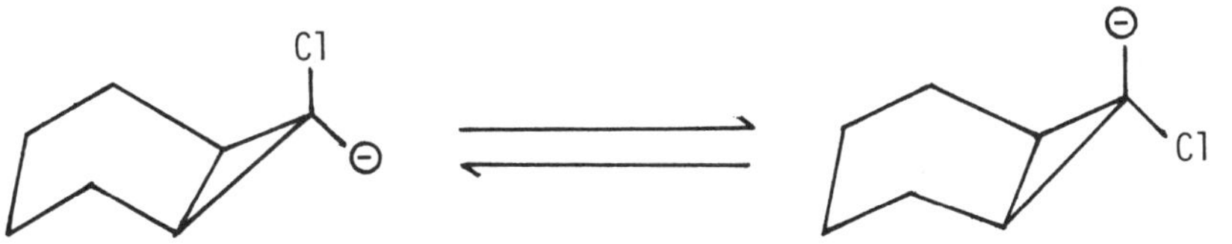

(32)

Similar stereochemical studies also have been made in detail using gem-dihalonorboranes and also bicyclic monohalides [179-181].

Acyclic Polyhaloalkanes

It was found that vic-dihaloalkanes were highly stereospecifically reduced at a mercury cathode to the corresponding olefins in high yields, Eq. (33) [182,183]. The resulting electrochemistry suggests that the elimination of halogen atoms from the trans-periplanar conformation is strongly preferred. The detailed stereochemical course of the reaction has recently been discussed by Brown and coworkers [184]. The stereospecific reduction of α, α'-dibromosuccinates to fumarates and maleates was also reported earlier [185].

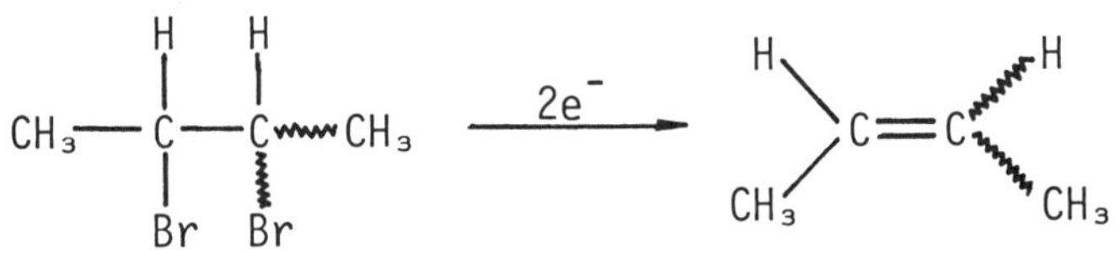

(33)

dl 90-95%(*cis*/*trans*: 29-∞)

meso 95%(*cis*/*trans*: 0)

It is known that pyridinium tribromide converts olefins to specific anti-dibromination. Therefore, the vic-dibromides thus obtained can be cathodically reduced to the original olefins without geometrical changes of molecular structure [186]. This stereospecific bromination-debromination may be useful as an olefin protection-deprotection method.

Monomeric and polymeric N-alkylpyridine-4-carboxylates serve as redox mediators for the indirect electroreduction of vic-dihalides to trans-olefins [187,188].

1,1,2,2-Tetrachloroalkanes also give stereoselectively the corresponding 1,2-dichloro olefins in which the E-(trans)-isomers are major [189,190]. Trichloromethylmethanols are reduced with electrogenerated chromium (II) ion to give Z-monochlorovinyl compounds in good yields [191]. When a carboxy function is at the α-position, and E-double bond is predominantly formed.

Acyclic Haloolefins

Monohalomaleic and fumaric acids and their esters are reductively dehalogenated with complete retention of the original molecular geometry in a wide range of pH, with the exception that bromomaleic acid gives mixtures of maleic and fumaric acids at pH 0.6-5.8 [192,193]. Cis- and trans-3-iodo-3-hexenes are also similarly deiodinated in DMF, but the stereospecificity is lower [194].

α,β-Dihalomaleic acid is also dehalogenated to give maleic acid in high stereoselectivity in strongly acidic media; mixtures with fumaric acid are formed in weakly acidic media [18]. Dihalofumaric acid always gives fumaric acid.

Intermolecular Dimerization

Monohalides Kanetsuna and Nonaka [195] reported that ethyl α-chlorophenylacetate was reductively dimerized and the d,l-meso ratio (0.4-1.4) of diethyl 2,3-diphenylsuccinate was influenced by the cathode material. A similar study was also reported by de Luca and coworkers [196]. 3-Halo-6-nitro-5-cholestenes [197] and aroyl chlorides [198] are also stereoselectively dimerized.

Polyhalides Stereoselective dimerization of a gem-dihalide was achieved in indirect electroreduction. α,α-Dibromotoluene was reductively dimerized by a Cr(II)/Cr(III) redox mediatory system to stilbene (cis/trans 0.18) [199]. α,α,α-Trichlorotoluene also gives 1,2-dichlorostilbene (cis/trans: 1) [190].

A vic-dihaloadamantane undergoes the intermolecular reductive dimerization via unstable adamantene species to give stereoisomeric products in 80-90% yields, Eq. (34) [200]. An α,ω-dibromo derivative of o-xylene gives stereoisomeric dimerization products via unstable o-quinodimethane species [201].

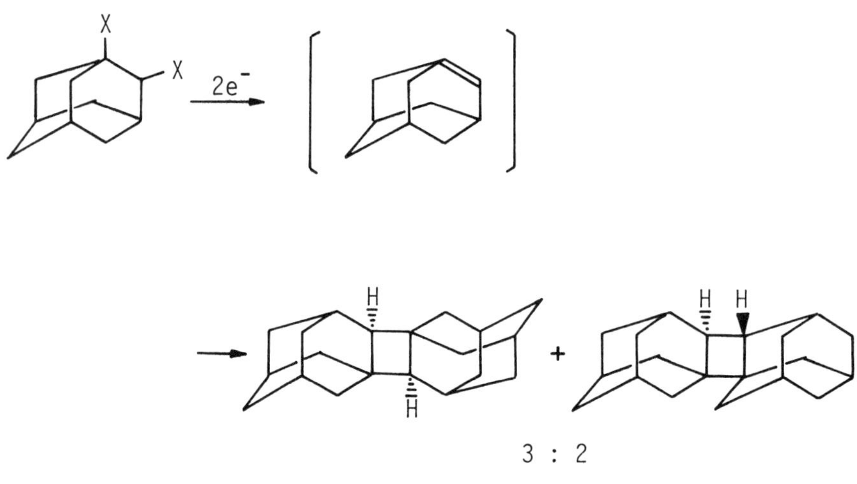

(34)

Intramolecular Cyclization

The d,l- and meso-2,4-dibromopentanes are reductively cyclized at a mercury cathode in DMSO in high yields, but neither the stereoselectivity nor stereospecificity is high, Eq. (35) [202,204]. A cyclic dibromide also gives a similar result [203]. On the other hand, 1,3-dibromo-1,3-dimethoxycarbonylpropane gives the corresponding trans-cyclopropane in high stereoselectivity, Eq. (35) [205].

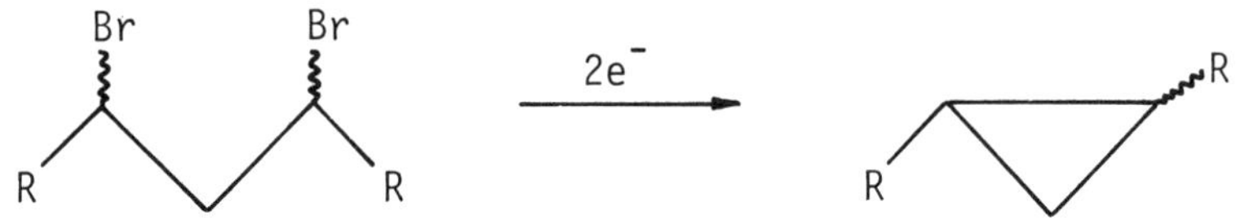

(35)

dl	; R = Me	; Hg/DMSO:	83.5 %	(*cis/trans*: 1.1)	[204]
meso	: R = Me	; Hg/DMSO:	86-91 %	(*cis/trans*: 0.8-0.9)	[204]
Unknown;	R = COOMe;	Pt/THF ;	60-74 %	(*cis/trans*: 0.25)	[205]
Unknown;	R = COOMe;	Pt/DMF :	?	(*cis/trans*: 0.07)	[205]
Unknown;	R = COOMe;	Pt/THF :	?	(*cis/trans*: 0.11)	[205]
Unknown;	R = COOMe;	Pt/DMF :	?	(*cis/trans*: 0.04)	[205]

The reduction of α,α'-dihaloketones may lead to the formation of stereoisomeric cyclopropanones, but the stereoselectivity of not only themselves but also their decomposition products could not be examined because of their instability under electrolytic conditions [206,207]. On the other hand, α,α'-dihalophosphinates were found to give stereoisomeric olefins via unstable oxyphosphirane species, but the stereoselectivity was low [208]. A similar result was also obtained in the reduction of α,α'-dibromodibenzylsulfone [209].

Others

1,2-Dihalonorbornanes gave stereoisomeric mixtures of reductive adducts with furan via reactive norbornene species [210]. Shono and coworkers [211] reported the diastereoselective addition of cathodically generated dichloromethyl anion to α-branching aldehydes. The reaction proceeds catalytically, Eq. (36):

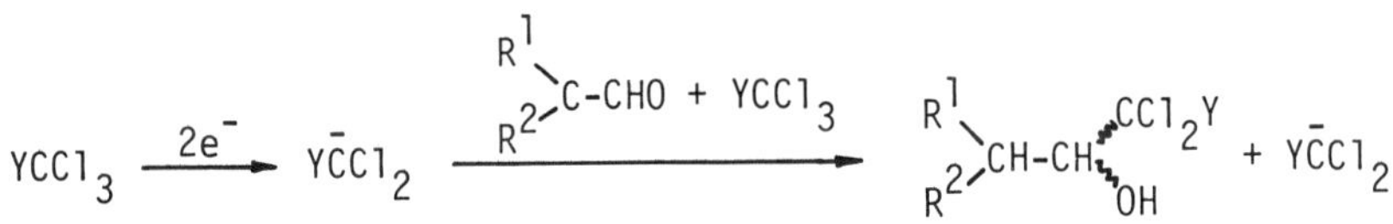

(36)

Y = Cl, COOMe; R^1, R^2 = Alkyl, Aryl: 25-89 % (*syn/anti*: 5-∞)

Y = Cl, COOMe; R^1, R^2 = OAc, 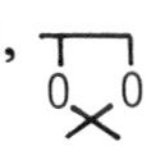 : 52-76 % (*syn/anti*: 0.05-0.50)

I. Other Heteroatom Compounds

Only a few stereochemical studies of electroreduction of compounds containing heteroatoms other than oxygen, nitrogen, sulfur, and halogens have been reported so far.

It has been reported that the electroreduction of mixtures of triphenylalkylphosphonium salts and aliphatic aldehydes at an aluminum cathode in acetonitrile gives exclusively the cis-olefins in good yields, Eq. (37) [212]. This reaction may be regarded as an electrochemical Wittig-like reaction.

$$Ph_3\overset{+}{P}CH_2R^1 + R^2CHO \xrightarrow{2e^-} \underset{R^1}{\overset{H}{>}}C{=}C\underset{R^2}{\overset{H}{<}} \qquad (37)$$

$R^1 = C\ H_7,\ R^2 = Et_2CH$: 82 % (*cis*/*trans*: ∞)

$R^1 = (CH_2)COOMe,\ R^2 = C_9H_{19}$: 69 % (*cis*/*trans*: ∞)

J. Organometallic Compounds

Stereoselective electroreduction of some organometal complexes has been reported.

The reduction of α-oxotetramethylene-cyclopentadienyl-iron$^+$-η^6-benzenes in an acidic solution occurs specifically from the exo-side to give only the endo-alcohol, Eq. (38), but in alkaline solution the corresponding stereoisomeric (d,l and meso) pinacols are formed but the stereochemistry has not been determined [213].

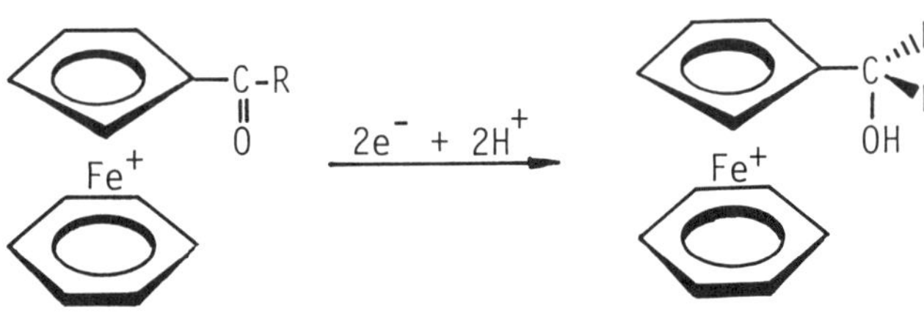

(38)

R = Me, Ph *endo*-OH

Arewgoda and coworkers [214] reported the cathodic preparation of stereoisomers of $(CF_3)_6C_6Co_2(CO)_2\ [P(OMe)_3]_2$ complex.

IV. STEREOSELECTIVE AND STEREOSPECIFIC ELECTROOXIDATION

A. Carboxylic Acids

Kolbe-type Reaction

Dimerization Some studies have dealt with the stereochemistry of Kolbe dimerization products [215-220]. Some of them confirmed the formation of stereoisomeric (d,l and meso) dimers but did not determine the stereoisomeric ratio [215,216].

Eberson and coworkers [217,218] examined the d,l-meso ratio of isomeric dimers from a variety of carboxylic acids in various solvents (H_2O, MeOH, AN, and DMF), Eq. (39). The ratio was found to depend on substituents (R^1 and R^2) of the acids and solvents. The persulfate oxidation of the acids to the dimers was also performed and the ratio was determined, but this comparative study does not seem to be helpful for discussing the stereochemical course of reaction because of the significant dependence of the ratio on the amounts of the persulfate reagent used. It is noted that the low ratios are obtained in aprotic solvents, such as DMF and AN.

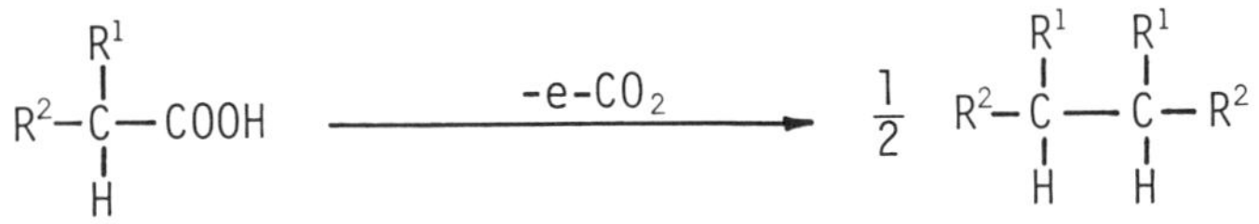

(39)

R^1 = *t*-Bu, R^2 = CN — 9-61 % (*dl*/*meso*: 0.6-1.4)

R^1 = *t*-Bu, R^2 = COOEt — 41-74 % (*dl*/*meso*: 1.0)

R^1 = *t*-Bu, R^2 = $CONH_2$ (MeOH) — 55 % (*dl*/*meso*: 0.5)

R^1 = *i*-Pr, R^2 = CN(MeOH) — 25 % (*dl*/*meso*: 0.9)

R^1 = *c*-Hex, R^2 = CN(MeOH) — 22 % (*dl*/*meso*: 0.6)

Hawkes and coworkers [219] mentioned the stereochemistry of three stereoisomeric dimers from 4-t-butylcyclohexane carboxylic acid; Kanetsuna and Nonaka [220] found that the ratio of the isomeric dimers from ethyl phenylmalonate was not influenced by the kind of anode material.

Kolbe cross-dimerization has also been stereochemically studied mainly from a somewhat practical aspect, in which it is aimed to prepare stereoisomeric long-chain olefins from saturated and unsaturated carboxylic acids [221,222]. The stereoisomeric ratio (Z/E: 2.4-10.1) is influenced by current density (100-250 mA cm^{-2}) and temperature (−16 to 60°C). Although the pure Z acid was used as the starting material, the product was not the pure Z olefin but a mixture with the E olefin. This indicates that stereoisomerization around the double bond at the γ-position can occur in the Kolbe-type decarboxylation. Hawkes and coworkers [219] also reported that cross-coupling between acetic and 4-t-butylcyclohexane carboxylic acids proceeded nonstereoselectively.

Radical addition to olefins The following four types of addition reactions of anodically generated radicals X· (RCOO· and R·) from RCOOH have been stereochemically studied:

1. Addition of two X· to one olefin molecule, Eq. (40) [223-227].
2. Addition of two X· to two olefin molecules, Eq. (41) [225,228]. This addition reaction prefers activated olefins.
3. Intramolecular addition in unsaturated carboxylic acids, Eq. (42) [229].
4. Addition of two X· to one diene molecule, Eq. (43) [230].

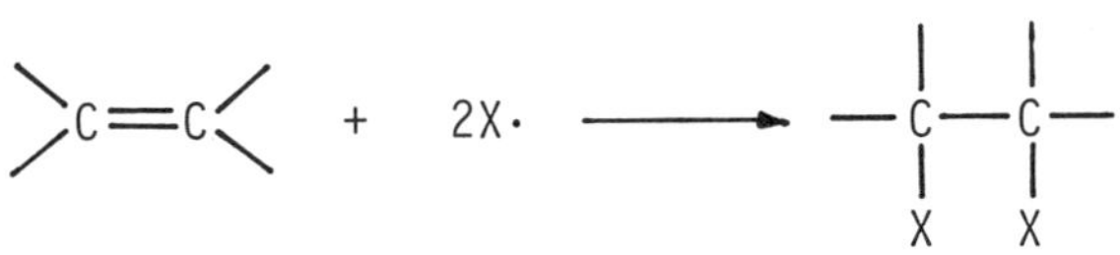

(40)

Z/*E* — *dl*/*meso*, *erythro*/*threo* (Acyclic), *cis*/*trans* (Cyclic)

Z/E *dl/meso, erythro/threo* (41)

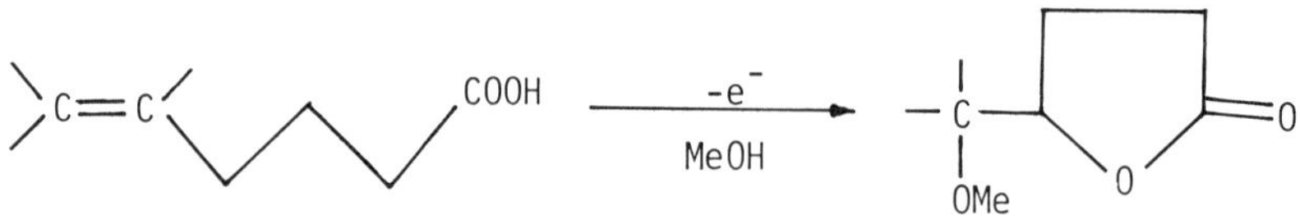

Z/E *dl/meso* (42)

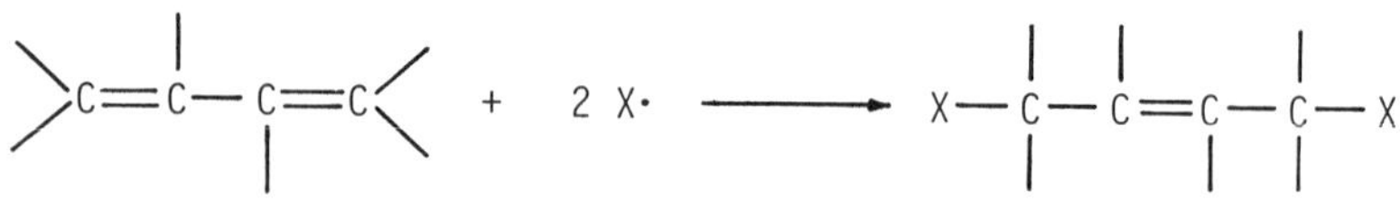

Z/E,E/Z *E/Z* (43)

Non-Kolbe-type Reaction

Monocarboxylic acids Carboxyl groups of cyclic monocarboxylic acids are anodically exchanged by nucleophiles, such as methoxide ion, to give stereoisomeric products [219,231-233]. 4-Substituted cyclohexane carboxylic acids give the stereoisomeric methoxylated products, Eq. (44) [219]. The reaction is nonstereospecific. Pyrrolidine-2-carboxylic acids are also methoxylated in high yields, but the stereoselectivity is generally low [231-233].

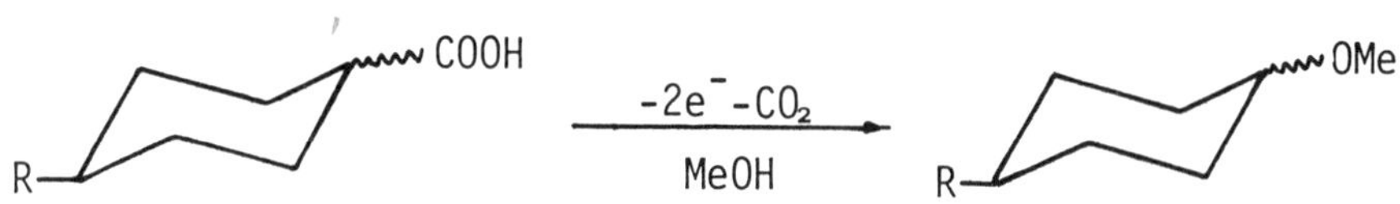

(44)

R = *t*-Bu, *trans*: 10.5-27.1 % (*cis/trans*: 0.3-0.4)
R = *t*-Bu, *cis* : 30.8 % (*cis/trans*: 0.3)
R = Ph, *trans* : 13.2 % (*cis/trans*: 0.5)

Utley and Yates [234] reported an interesting stereochemistry and regiochemistry of methoxylated products derived from 4-phenylcyclohexe-2-enecarboxylic acid, Eq. (45):

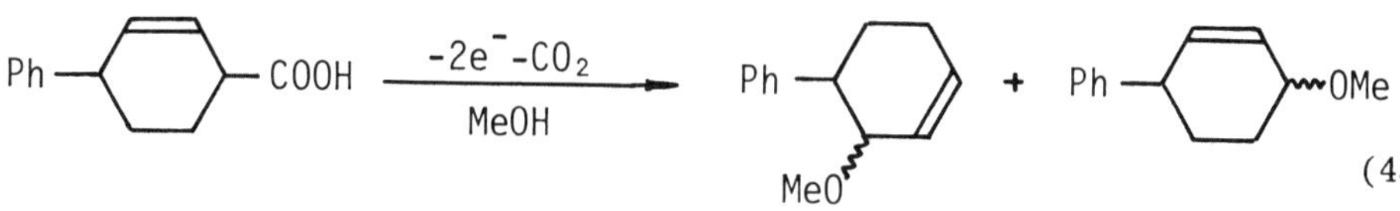

(45)

26 % (*cis/trans*: 0.4) 49 % (*cis/trans*: 1.5)

Stereoisomeric (exo and endo) bicycloalkane carboxylic acids are stereospecifically methoxylated, hydroxylated, and amidated [235-238]. Cyclic monocarboxylic acids are also anodically decarboxylated to give stereoselectively cycloalkenes in high yields, Eq. (46) [239,240]:

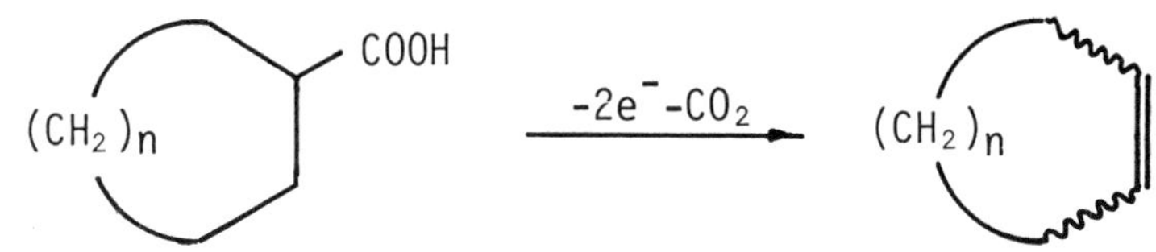

(46)

n = 5: 87 % (*cis*/*trans*: ∞)
6: 75 % (*cis*/*trans*: ∞)
7: 88 % (*cis*/*trans*: 2.1)
17: 89 % (*cis*/*trans*: 0.6)

Substituted cyclopropane carboxylic acids undergo a ring-opening reaction under conditions for anodic methoxylation and give stereoisomeric products [241,242]. The stereospecificity of the reaction depends in a complicated way on the structure of the acids.

Dicarboxylic acids 1,2-Dicarboxylic acids give stereoselectively but nonstereospecifically olefins [243,244]: d,l- and meso-α,β-diphenylsuccinic acids gave trans-stilbene as the sole isomeric product in 63 and 36% yield, respectively [243].

On the other hand, anodic oxidation of perhydrodiphenic acids results in the stereospecific formation of lactones, Eq. (47) [245]. A similar lactonization was achieved using α- and β-truxilic acids (cis- and trans-cyclobutane-1,2-dicarboxylic acids) [246]; 3,4-disubstituted cis- and trans-cyclobutane-1,3-dicarboxylic acids gave stereoisomeric (cis and trans) mixtures of 2,4-disubstituted bicyclobutanes but the stereoselectivity was not determined [247].

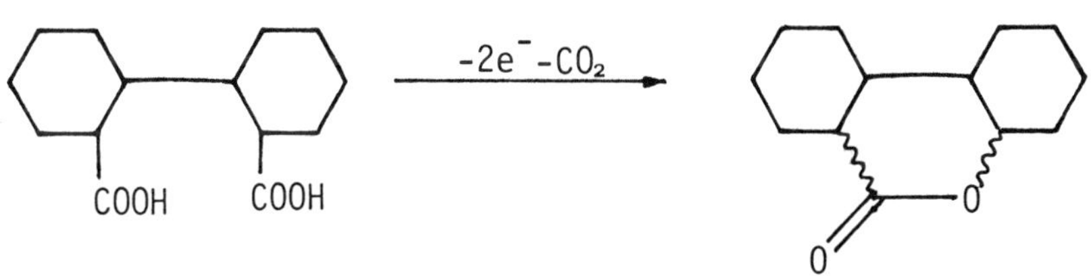

(47)

trans-anti-trans and *trans-syn-trans*	45 % (*cis*/*trans*: 0)
cis-syn-cis	30 % (*cis*/*trans*: ∞)

B. Acetoxylation

Acetoxylation occurs in either of the following ways: (1) nucleophilic attack of methoxide ion on cationic intermediate species formed by electron elimination of substrates, and (2) radical attack of the acetoxy radical generated by electron elimination of acetate ion to substrates.

Olefins

Monoolefins Diacetoxylation of monoolefins sometimes proceeds stereoselectively and/or stereospecifically [248-254]. The stereoselectivity is influenced by the structure of the starting olefins and electrolytic conditions. Under different conditions, acetohydroxylation also occurs [255].

Diolefins 1,2- and/or 1,4-diacetoxylation of conjugated dienes occurs stereoselectively, but the stereoselectivity of the latter is generally low [256-258]: Both diacetoxylations of cyclohexadiene occur simultaneously [256].

Benzyl and Allyl Compounds

The stereochemistry of acetoxylation at benzyl positions of substituted indanes and acenaphthenes was investigated in detail, Eq. (48) [259-261]. The cis-trans ratio of diacetoxylated products is significantly larger in the anodic processes than in chemical in homogeneous reaction media, indicating that adsorption at least partially controls the stereochemistry of the anodic acetoxylation.

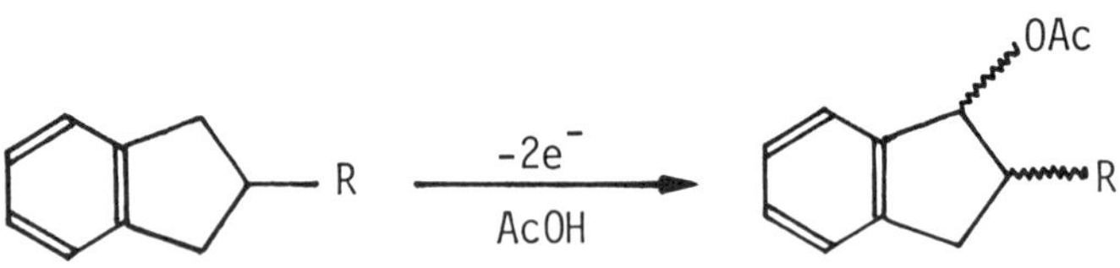

(48)

R = *t*-Bu: 21-87 % (*cis*/*trans*: 0.02-0.19)[259]
R = Me : 51-80 % (*cis*/*trans*: 0.59-1.08)[261]

It has been reported that the anodic acetoxylation at allylic positions proceeds with high stereoselectivity (88-93%), as does microsomal acetoxylation (88-93% stereoselectivity). A chemical method with perbenzoic acid gives selectivity as low as 74-80% [262]. The stereochemistry of two sets of stereoisomeric acetoxylated products of α-pinene was made clear by Shono and Ikeda [263].

Cyclopolyenes

Cyclooctatetraene gave stereoisomeric (cis and trans) bicyclo [4,2,0] octane-7,8-diols and their diacetates in water and acetic acid, respectively, Eq. (49) [264,265]:

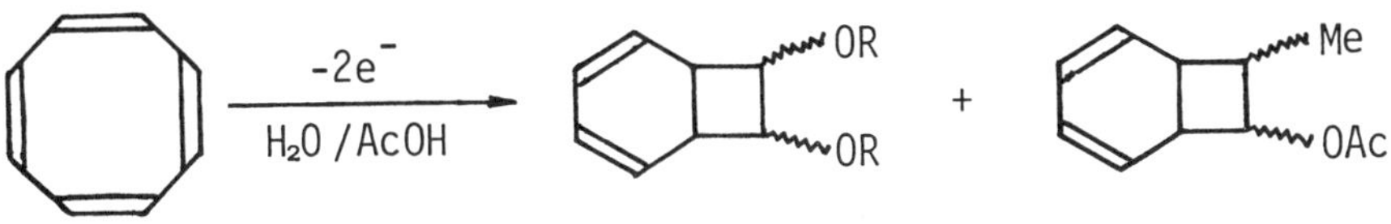

(49)

R = H : 49 % (*cis*/*trans*: 0.7) -

R = Ac : 71 % (*cis*/*trans*: 1.6) (*cis*/*trans*: 0.4)

The anodic acetoxylation of anthracene resulted in the predominant formation of trans-9,10-dihydroanthracene [266].

C. Methoxylation

Olefins

Acyclic olefins are stereoselectively and sometimes stereospecifically dimethoxylated [249,267-272]. A typical example is in Eq. (50) [267,269]:

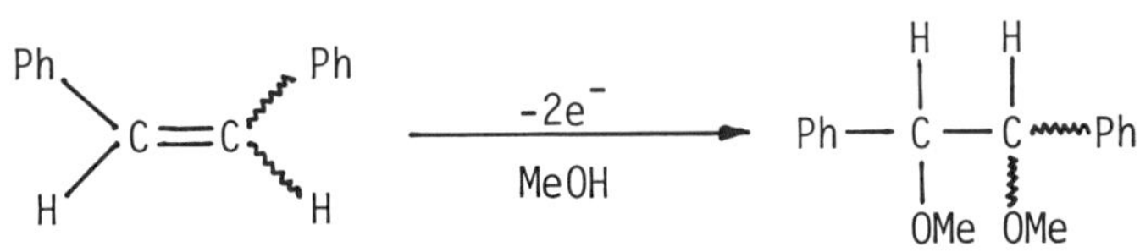

(50)

cis	64 % (*dl*/*meso*: 2.2)
trans	45 % (*dl*/*meso*: 0.7)

The anodic dimethoxylation of simple cyclic olefins, such as cyclohexenes, results in the predominant formation of trans-dimethoxycyclohexanes [250, 272,273]. However, Barba and coworkers [274] recently reported that the stereochemistry of 1,2-dimethoxy-1,2-dihydroacenaphthene derived from acenaphthene is drastically influenced by the anode material. Palasz and Utley [275] have reported that N-acetylpiperidines are methoxylated via the enamide intermediate species formed by the oxidation. 1,3-Dienes are also 1,4-dimethoxylated in low stereoselectivity [258]. The anodic methoxylation of α-pinene proceeds similarly to the acetoxylation to result in the formation of two sets of stereoisomeric products [263].

Furans

Since furans are anodically 2,5-dimethoxylated in high yields to give the corresponding cis- and trans-2,5-dimethoxy-2,5-dihydrofurans, which are regarded as equivalents to the E- and Z-unsaturated 1,2-dicarbonyl compounds, this anodic methoxylation is important and useful in synthetic chemistry.

The stereochemistry (cis and trans) of the methoxylated furans have been examined [276-279]. In most cases the trans isomers are major, Eq. (51):

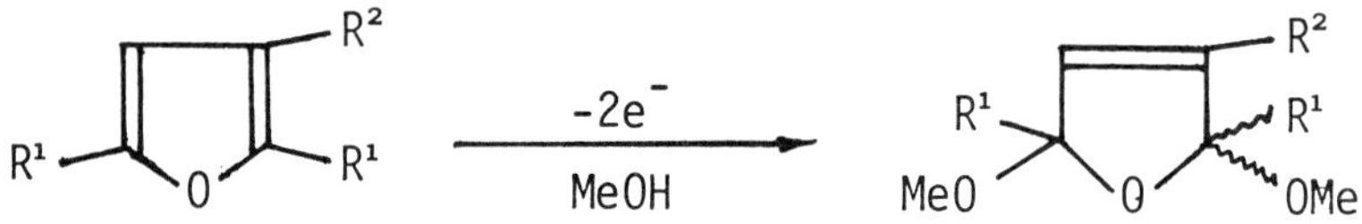

(51)

$R^1 = R^2 = H$	73 % (*cis/trans*: ?) [276]
$R^1 = R^2 = Me$	41 % (*cis/trans*: 1.0) [277]
$R^1 = H$; $R^2 = COOMe$	100 % (*cis/trans*: 0.37) [278]
$R^1 = (CH_2)_3OH$; $R^2 = H$	70 % (*cis/trans*: ∞) [279]

2,5-Dimethylfuran was also anodically cyanomethoxylated in the presence of cyanide ion to give 2,5-dimethyl-2-methoxy-5-cyano-2,5-dihydrofuran at the cis-trans ratios of 1.46-2.89, which varied with the type of aromatic additive in anodic solutions [280].

2-Furylacetic acid derivatives are 2,5-dimethoxylated and then followed by the elimination of methanol to lead stereoisomeric olefinic products (XXVII) [281]:

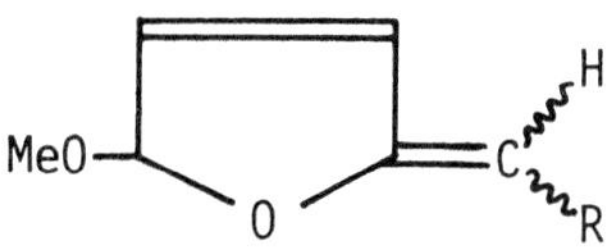

R = CN, COOMe, COOEt, COOH

(XXVII)

The further oxidative methoxylation of 2,5-dimethoxydihydrofurans in situ leads to the stereoselective formation of ring-opened olefins [282-286]. A typical example [282] is in Eq. (52):

(52)

R = COOH, CHO, CH_2OH, CH_2OAc

84 % (*Z/E*: 0.04) in NH_4Br

24-87 % (*Z/E*: ∞) in Et_4NClO

Aromatics

Aromatic hydrocarbon rings can also be anodically dimethoxylated to give the trans-dimethoxy dihydro derivatives in high or moderate stereoselectivity [287-290].

Others

Benzyl and allyl positions are stereoselectively monomethoxylated or sometimes hydroxylated [291,292].

Bicyclic cyclopropyl compounds are dimethoxylated to give stereoisomeric (cis and trans) cyclopropane ring-opened products, Eq. (53) [293,294]. Shono and coworkers [295] found a different type of ring-opening reaction of a cyclobutyl compound through anodic methoxylation.

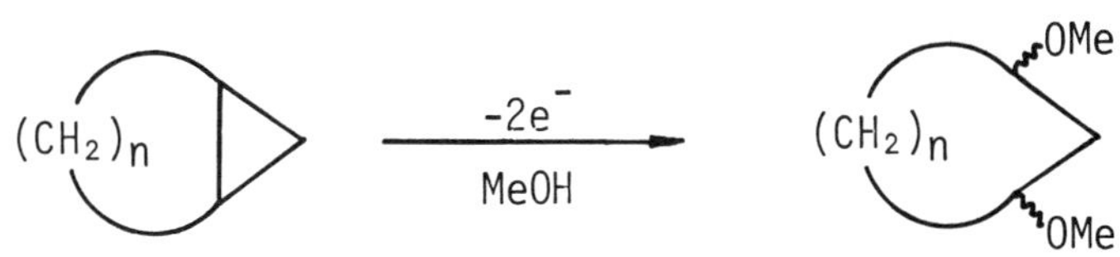

(53)

n = 3: 35 % (*cis/trans*: 1.3)

n = 4: 21-23 % (*cis/trans*: 1.1-1.3)

Skeleton-changing reactions of complicated polycyclic compounds are also induced by anodic methoxylation [296-299]: The product yields are not high, but high stereoselectivities are obtained in most cases.

The α-positions of cyclic amides are stereoselectively methoxylated, Eq. (54) [300]:

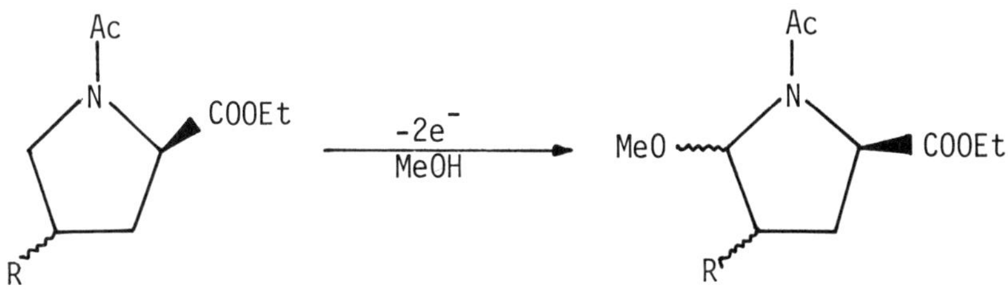

(54)

R = *trans*-OH:COOEt : 60 % (*cis/trans*-MeO:COOEt : 2.2)

R = *cis*-OAc:COOEt : 50 % (*cis/trans*-MeO:COOEt : 1.9)

Recently, Noyori and Kurimoto [301] have found an interesting anodic ether exchange reaction of aryl glycosides. They described that this is a rare example of S_{ON} 1-type transformation reaction, but the reaction may be EGA (electrogenerated acid)-catalyzed. The reaction is stereoselective but not always stereospecific.

D. Acetamidation

The anodic acetamidation of olefins proceeds smoothly in wet acetonitrile containing suitable nucleophiles to lead the formation of stereoisomeric products.

Haloacetamidation

The stereospecificity, stereoselectivity, and regioselectivity of the fluoroacetamidation of a variety of olefins have been studied in detail from synthetic and mechanistic aspects by Tardivel and coworkers, Eq. (55) [302-304]. The significant stereochemical results observed may be rationalized by considering coplanar adsorption of carbonium cation intermediates derived from the olefins on the anode surface.

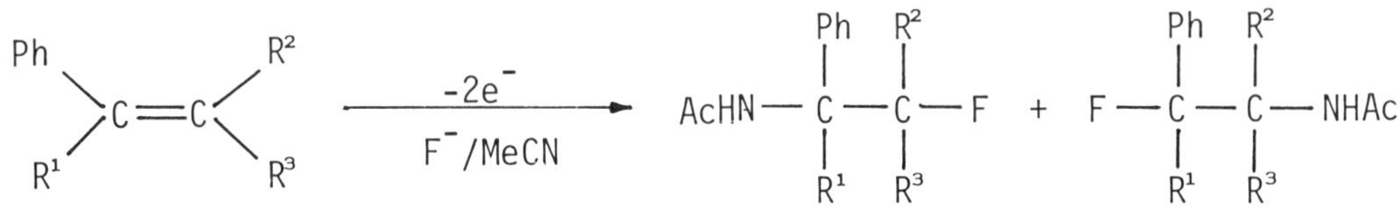

$R^1 = R^2 = H$, $R^3 = Ph$: 29 % (*erythro*/*threo*: 0.28) 0 % [302]
$R^1 = R^3 = H$, $R^2 = Ph$: 45 % (*erythro*/*threo*: 0.25) 0 % [302]
$R^2 = H$, $R^1 = R^3 = Me$: 29 % (*erythro*/*threo*: 3.00) 0 % [302]

(55)

Stereoselective chloroacetamidation [305,306] and iodoacetamidation [307] have also been reported.

Nitroacetamidation

Bloom and coworkers [308] reported in detail the stereochemistry and regiochemistry of the nitroacetamidation of conjugated dienes that were electrolyzed in acetonitrile containing N_2O_2 as a precursor of electrogenerated nitronium ion at an anode. The reaction proceeded efficiently in most cases.

Thioacetamidation

Bewick and coworkers [309,310] found that the anodic thioacetamidation of cycloalkenes occurred highly stereoselectively in acetonitrile containing disulfides, Eq. (56):

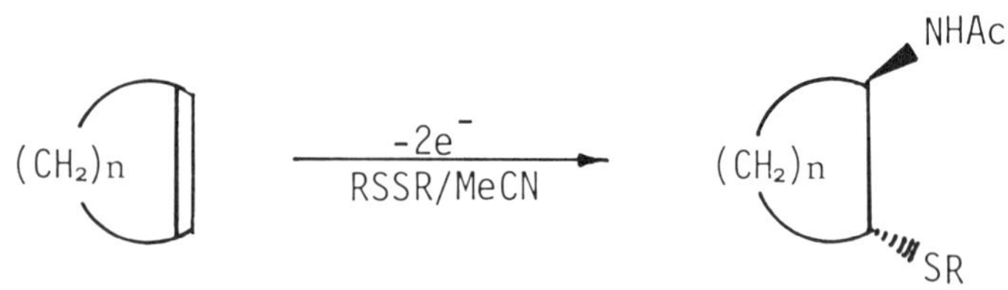

(56)

n = 3; R = Ph: 56 %
n = 4; R = Ph: 50 %
n = 4; R = Me: 58 %

E. Other Addition Reactions to Olefins

Halogenation

Olefins are anodically dihalogenated, sometimes in very high stereoselectivity.

Laurent and coworkers [304] reported the difluorination of 3-t-butylindene, but the stereoselectivity was not high. On the other hand, the dichlorination of olefins, dienes, and acetylenes are highly stereoselective and gives almost exclusively single stereoisomers of products, Eqs. (57) through (59) [306, 311]. The bromination also gives single stereosiomers of vic-dibromo compounds [312-314]. No study of the stereochemistry of anodic iodination has been reported.

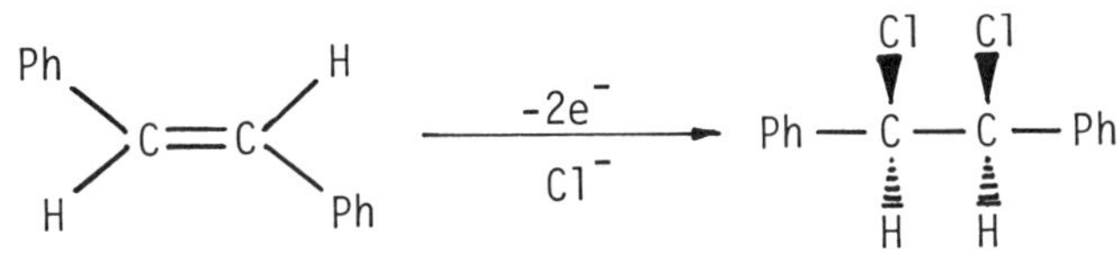

meso

65 % [305]

(57)

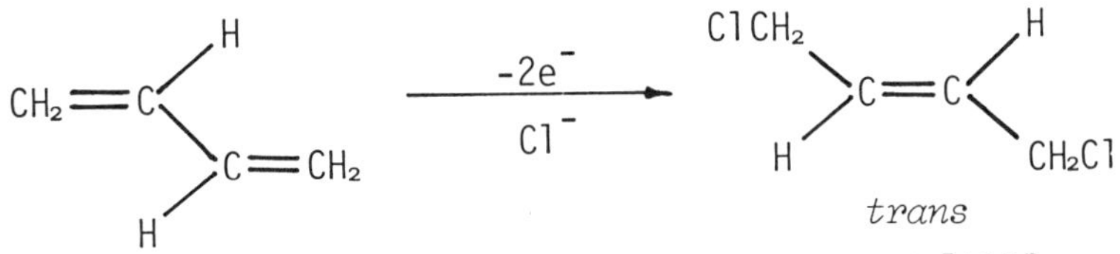

trans

51 % [309]

(58)

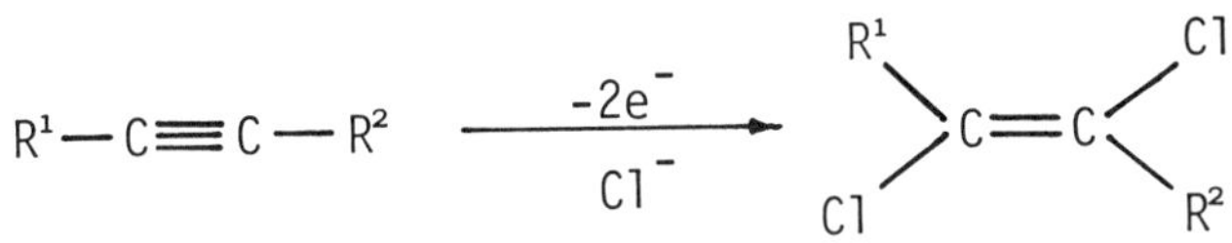

trans

R^1 = HOCH , Ph(HO)CH; R^2 = H: 60-70 % [305]

R^1 = R^2 = Ph : 40 % [305]

(59)

Cyanation, Thiocyanation, and Carboxylation

The anodic dicyanation and dithiocyanation of olefins occurs in electrolytic solutions containing cyanide ion and thiocyanogen [$(NCS)_2$], respectively, but the stereochemistry of any products has not been examined [315,316]. The selenoalkoxylation of olefins, such as cyclohexene, Eq. (60), cyclopentadiene, and indene, proceeds highly stereoselectively and regioselectively to give trans products in high yields [317].

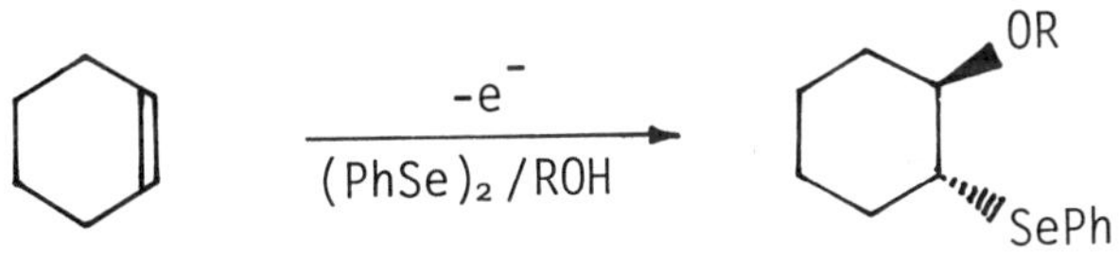

(60)

R = H, Me, Ac 96 - 98 %

Inoue and Tsutsumi [318] reported that styrenes were anodically carbonylated with carbon monoxide in methanol in the presence of a platinum-carbonyl complex ($[Pt_x(CO)_y]_n$) to give trans-methyl cinnamates.

F. Intermolecular Reactions of Olefins and/or Aromatics

Dimerization

Olefins [319] and aromatics [320] are oxidatively dimerized to give stereoselectively the dienes and biaryls (rotamers), respectively.

Some types of EGA-catalyzed dimerizations, which proceed stereoselectively, have recently been reported [251,321-324].

Maleic anhydride is stereoregularly polymerized by anodic initiation to give mainly cis-poly(vinyleneketoanhydride) [325].

Other Reactions

Some interesting stereoselective anodic reactions of unsaturated hydrocarbons have been known, although the stereoselectivity is not always high [326-328]. The reaction seems to proceed via cation radicals of the diene. Ando and coworkers [327] investigated the stereochemical consequence of anode (probably EGA) catalyzed oxygenation of stereoisomeric (syn and anti) di-t-butyl (bicyclo[3,3,1]non-9-ylidene)s in detail compared with the photosensitized one.

G. Active Hydrogen Compounds (Carbon Acids)

Few studies of the stereochemistry of anodic oxidative dimerization of active hydrogen compounds (carbon acids) and their salts (carbanions) have been reported.

Schäfer and Alazrak [329] reported that the radical formed by the one-electron oxidation of sodium acetylacetonate added to butadiene gave trans, trans-tetradeca-5,9-diene-2,13-dione in 40% yield. Kanetsuna and Nonaka [220,330] studied comparatively the stereochemistry of diethyl 2,3-diphenylsuccinate formed in the oxidation of (1) ethyl phenylacetate (in the presence of iodide ion), (2) its sodium salt, and (3) ethyl hydrogen phenylmalonate [Eq. (61)]. A phenylethoxycarbonylmethyl radical may be a common intermediate in these three reactions that undergoes the coupling to form the succinate.

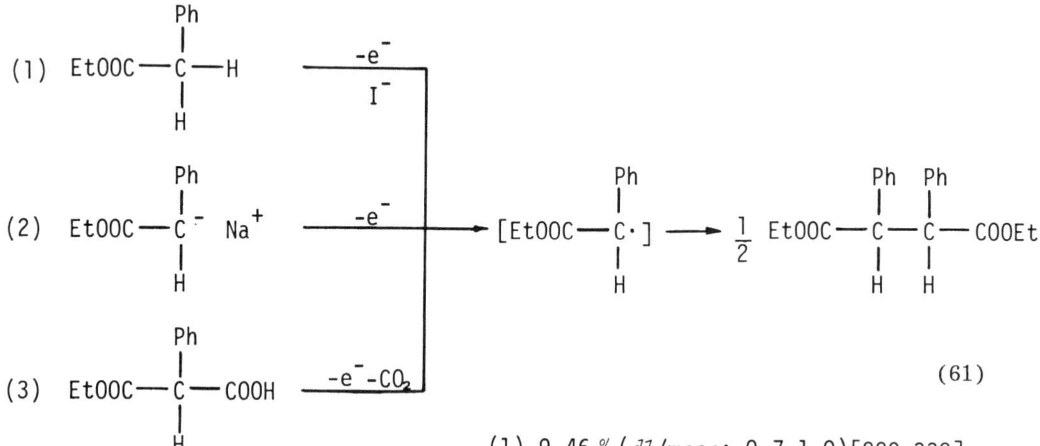

(61)

(1) 9-46 % (*dl/meso*: 0.7-1.0)[220,330]

(2) 1 - 4 % (*dl/meso*: 1.4-1.8)[220]

(3) 4-17 % (*dl/meso*: 0.9-1.0)[220]

H. Heteroatom Compounds

Nitrogen Compounds

A few interesting reactions of nitrogen compounds from a stereochemical aspect have been reported. Cauquis and Genies [331] investigated the reaction of an anodically formed diazenium ion ($Ph_2N^+{=}NH$) from N,N-diphenylhydrazine with cis- and trans-1,2-disubstituted olefins, but the stereochemistry of N-diphenylaminomonoazacyclopropanes could not be completely determined because of their instability. On the other hand, Tokuda and coworkers [332] have reported a stereochemically clear result that lithium alkenylamides undergo a highly stereoselective cyclization by anodic oxidation to give cis-1-methyl-2,5-disubstituted pyrrolidines in reasonable yields, Eq. (62). The cyclization seems to proceed via neutral amino radicals.

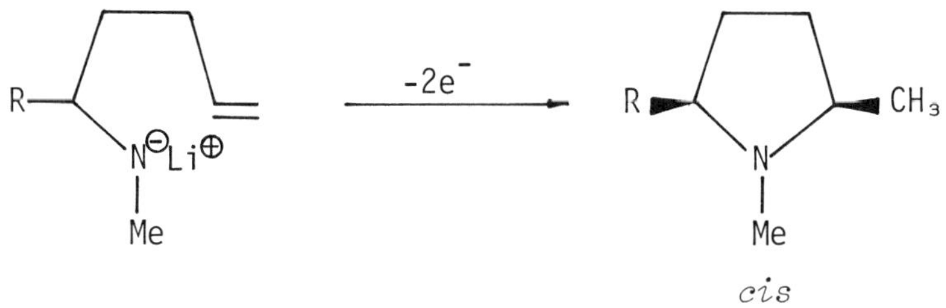

(62)

R = Me, Et, Ph, 4-MeC_6H_4, 4-$MeOC_6H_4$: 31-52 %

It has been also found that ditosylates of α,ω-diamines are anodically cyclized in methanol containing halide ions to give stereoisomeric N-tosyl-2-methoxy-2-tosylaminopyrrolidines (or piperidines) [333].

Sulfur Compounds

Stereoselective anodic methoxylation [334] and ring-opening reaction of thiophenes leading to olefin formation [335,336] have been reported. Anodic oxidation of saturated cyclic sulfur compounds has also interesting stereochemical features. Kimura and coworkers [337] reported that a 4-substituted thiane was stereoselectively oxidized to the trans-thianeoxide, which was always the major isomeric product, although the isomeric ratio was affected by electrolyte and solvent used, Eq. (63). The oxidation of 2,5-substituted-1,3-dithianes to the cis- and trans-4-substituted-1,2-dithiolanes is stereoselective but not stereospecific [338].

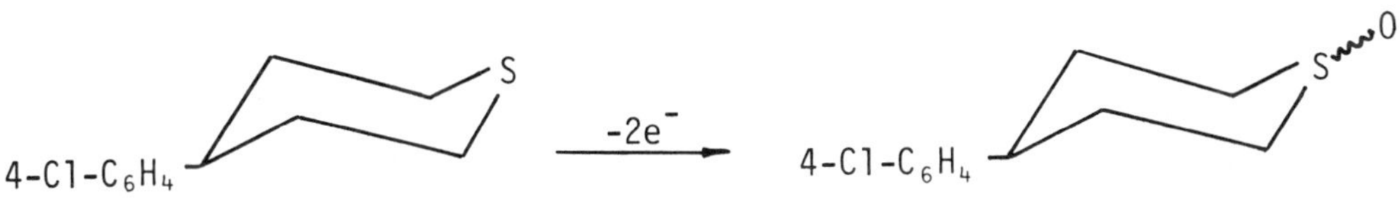

ca. 50 % (*cis*/*trans*: 0.08-0.52)

There is an interesting example of stereoselective anodic oxidation of an cyclic sulfur compound. Kimura and coworkers [339] have also reported that the electrooxidative desulfenylation of Michael-type thiol adducts of α, β-unsaturated esters and ketones and also nitriles results in the formation of E- and Z-olefins as major products, respectively, Eq. (64):

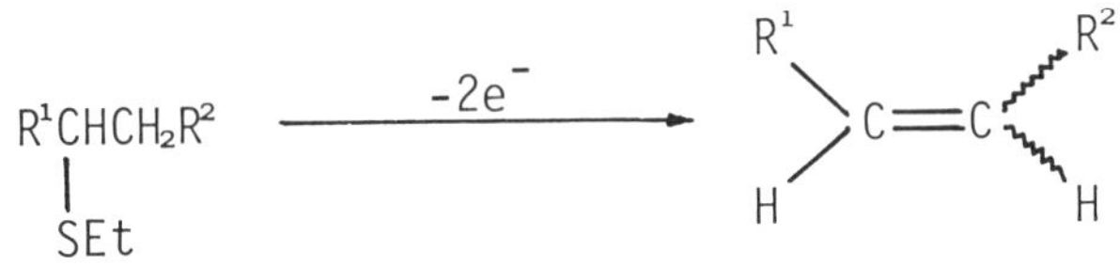

(64)

R^1 = Me, Ph; R^2 = COOEt, COOMe, COMe: 24-98 % (*Z*/*E*: 0.0)

R^1 = Me, Pr; R^2 = CN : 28-67 % (*Z*/*E*: 1.1-2.0)

Silicon Compounds

Stereochemistry of electrooxidation of silicon compounds has been very rarely reported. Yoshida and coworkers [340] have found that stereoisomeric mixtures of 1-trimethylsilyl-1,3-dienes are oxidized in methanol to give trimethoxylated trans-olefins as single stereoisomers.

The stereoselective EGA-catalyzed reaction of trimethylsilyl enolates and dimethylacetals has also been reported [341].

I. Organometallic Compounds

Stereoselective synthesis of organometallic complexes has been achieved in the oxidative addition of aryl halides to triethylphosphine nickel(O) complexes, leading to the exclusive formation of trans-aryl nickel(II)halide complexes [342]. Electron-transfer reactions on the Fe of cis- and trans-$[\eta\text{-}C_5H_5Fe(CO)SR]_2$ occur stereospecifically, in other words, with no stereoisomerization on changing the oxidation state of the Fe [343,344]. In the electrochlorination of a ligand (R) of the $\eta\text{-}C_5H_5Fe(CO)R$ complex, the stereochemistry is retained [345].

V. ASYMMETRIC REDUCTION AND OXIDATION

A. Reactions of Chiral Compounds

Electrochemical reactions that occur at chiral centers and also give chiral products are mentioned here.

Reduction of Halides

Cathodic reduction of chiral (optically active) haloalkanes to the chiral alkanes has been studied mainly from a mechanistic aspect. The configurational change (retention and inversion) during the reduction is discussed.

The reduction of chiral 1-substituted 1-halo-2,2-diphenylcyclopropanes at a mercury cathode to the chiral cyclopropanes has been reported by some workers [346-350]. The configurational change is so complicated and influenced by subsitutent at the 1-position, solvent, pH, supporting electrolyte, and cathode potential [Eq. (65)] that it is difficult to obtain a universal rationalization of the stereochemical mechanism of the reaction.

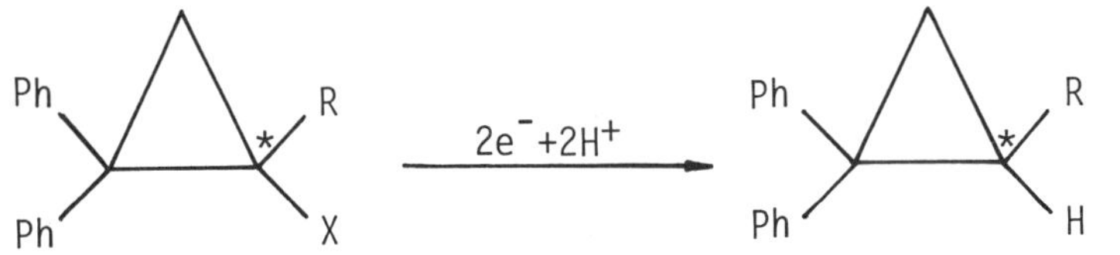

(65)

R	X	Retention(%)	Inversion(%)
Me	Br	0-63 [346-348]	-
Me	I	44-53 [347,348]	-
COOH	Br	0-53 [346,350]	4-37 [346,350]
COOMe	Br	4-74 [350]	5-59 [346,350]

The reduction of chiral open-chain halides has also been reported. Czochralska [351] reduced chiral 2-phenyl-2-chloropropionic acid to obtain inverted phenylpropionic acid in high optical purities. However, this result could not be reproduced by another investigation [3]. Nonaka and coworkers [352] have also investigated the reduction of a chiral chloroalkane and attempted to rationalize the stereochemical course of the reaction by considering the configurational stability of the radical intermediate species in the cathode interface. Fry and Britton [203] reported that optically pure (+)-(1R,2R)-1,2-dimethylcyclopropane was obtained in the reductive intramolecular cyclization of (+)-(2S,4S)-2,4-dibromopentane.

Chiral alkyl iodides give the corresponding dialkyl mercuries in good yields under certain electroreductive conditions, but the dialkyl mercuries are completely racemized [348]. Optically active isobutyl mercury bromide is racemized by repeating a electroreduction-oxidation cycle via diisobutyl mercury [353,354].

Reduction of Alcohols and Amines

Electroreduction of alcohols and amines hardly occur in general. Only peculiar alcohols and amines can be reduced to the corresponding alkanes.

Erickson and Fischer [355] reported that optically active O-benzoylatrolatic acid and its methyl ester were reduced to 2-phenylpropionic acids with almost complete loss of optical activity. On the other hand, Nonaka and coworkers [356] found that 1-(2-pyridyl)alkanols were reduced to 2-alkylpyridines with configurational retention in an acidic medium; 1-(4-pyridyl) alkanols gave racemic 4-alkylpyridines. The stereochemical course was discussed considering perpendicular adsorption of pyridylalkanol molecules and the configurational stability of carbanion intermediates, Eq. (66) [356]:

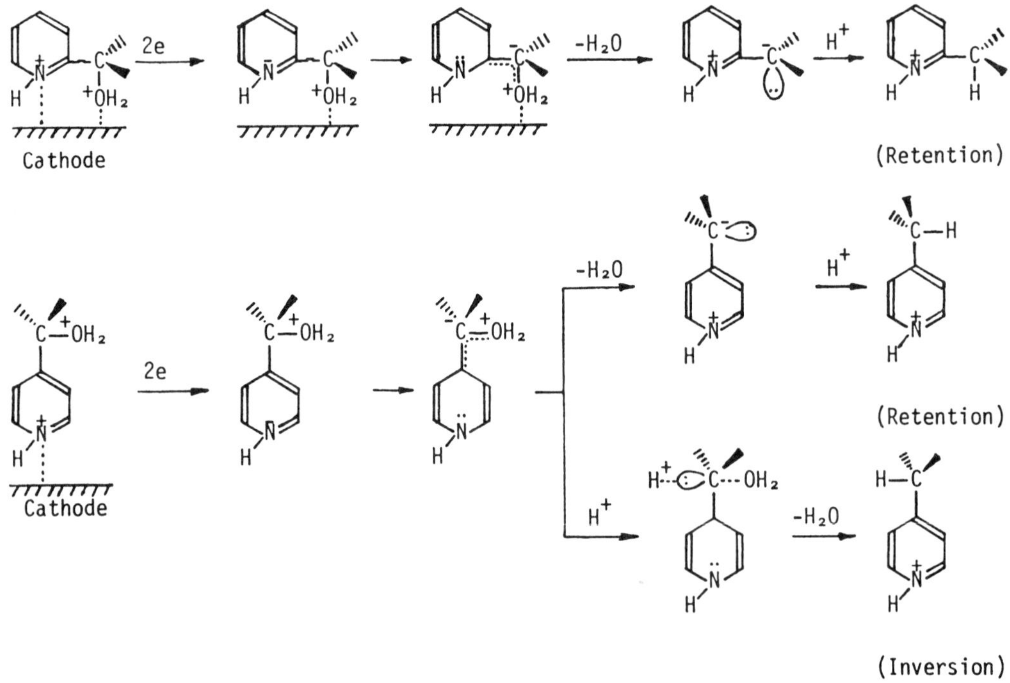

(66)

Electroreductive deprotection of tosylates [357] and sulfonamides [358] of chiral alcohols and amines, respectively, has been achieved without racemization.

Reduction of Oniums

Optically active quaternary arsonium [359,360] and phosphonium [361] salts are electroreductively decomposed to tertiary arsines and phosphines, respectively, with retention of configuration. Since the tertiary products can be converted again into different quaternary salts with alkylating reagents, this decomposition reaction may be useful for mutual transformation between a variety of optically active quaternary salts [360].

Oxidation of Carboxylic Acids

Optically active α-chiral carboxylic acids give racemic Kolbe dimers [362-364]. Similar results are obtained in mixed Kolbe dimerization [365,366]. Addition products of radicals derived from such carboxylic acids to olefins are also racemic [367,368]. Racemization is also observed in anodic methoxylation of carboxylic acids [235]. These facts may suggest that Kolbe intermediate radicals interact so weakly with the anode surface that their configuration cannot be stabilized during the reaction. However, it can also be stated that the radicals interact sterically with the anode so that the geometry of the dimers is influenced by the anode material (see Sec. IV).

β-Chiral carboxylic acids give optically active dimers in high optical purity [369,370].

It is interesting that anodic methoxylation of optically active N-protected α-amino acids with one more chiral center and subsequent chemical alkylation affords the final products in high optical purity, although the diastereomeric purity of the intermediate methoxylated products do not seem so high [371, 372]. Intramolecular asymmetric induction may occur in the chemical alkylation step.

Oxidation of Heteroatom Compounds

Shono and coworkers [373] reported than an α-chiral alcohol was converted into the corresponding chloride by diphenylsulfide-mediated anodic reaction. Anodic cyclization of α-amino acids in methanol has also been reported [374].

B. Asymmetric Synthesis

Asymmetric synthesis is a very important area in organic chemistry. However, it must be noted that electrochemical asymmetric synthesis has been immature compared with chemical, catalytic, and emzymatic methods, although much interest has arisen and many efforts have been made since the first report in this area by Grimshaw and coworkers [375] in 1967. Although the history is short, a number of papers have been published and some methodologies have already been established. Particularly in recent years, new methodological developments have been successful. This point was reviewed by Nonaka [16].

Use of Chiral Electrode Adsorbents

In most studies, optically active alkaloids have been used as asymmetry inducers in this method.

The asymmetric reduction of activated olefins was performed using prochiral cumarin derivatives [375-377]. A maximum asymmetric yield reported is 17%, Eq. (67) [376], and an asymmetry induction mechanism has also been discussed [377].

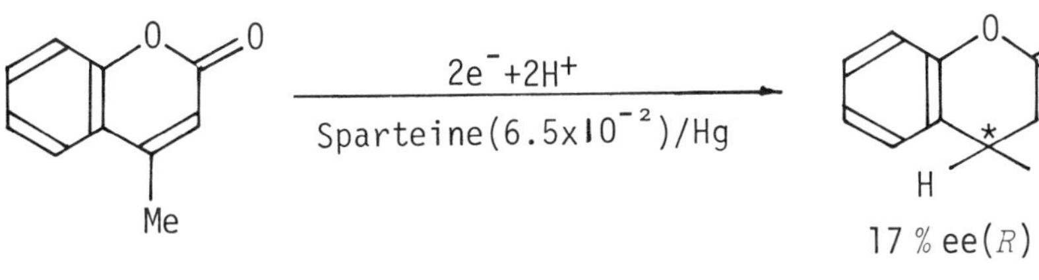

(67)

The asymmetric reduction of prochiral ketones to chiral alcohols has been extensively investigated [378-389], and 48.4% maximum asymmetric yield has been reported using 2-acetylpyridine [382]. The induction mechanism has also been discussed [380,381].

The asymmetric reduction of C=N double bonds has been investigated using prochiral oximes [388-392]. The maximum asymmetric yield is 18% [389]. A mechanistic study was also reported [388].

gem-Dihalides are also asymmetrically reduced to chiral monohalides [393-395], and a maximum asymmetric yield of 44.3% has been reported, Eq. (68) [394]:

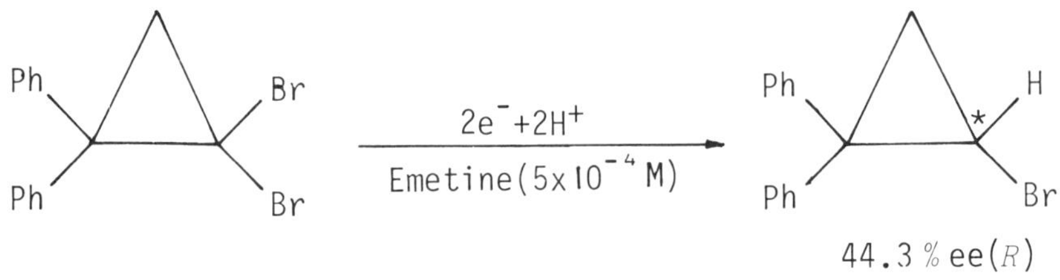 (68)

Not only asymmetric yields of products but also their absolute configurations vary drastically with cathode potential, supporting electrolyte, solvent, pH, cathode material, concentration, and other factors that influence adsorption. General and theoretical discussions of an asymmetry induction mechanism have been made on the basis of adsorption phenomena of inducers and substrates [396-399]. It may sometimes be a disadvantage in asymmetric synthesis that a reaction is too sensitive to a variety of conditions. Therefore, it has been attempted to replace alkaloids with other inducers, such as optically active amino acids [400].

Use of Chiral Supporting Electrolytes

This method was first developed in 1968 by Horner and Degner [401] and has been applied to the asymmetric reduction of ketones [401-409] and imines [400,410,411] only. Maximum asymmetric yields reported for the former and the latter are 20% [406] and 8.95% [411], respectively. A higher yield (20.6%) was obtained in the hydrodimerization of a ketone [409].

Mechanistic studies [405,406,411] have been insufficiently made. The sensitivity of this type of asymmetric reduction to reaction conditions seems less high than that of the reaction using chiral electrode adsorbents. However, it is a problem that lower asymmetric yields are obtained using much larger amounts of asymmetry inducers, which must play the role of supporting electrolyte.

Use of Chiral Media

Seebach and Oei [412,413] reported the asymmetric hydrodimerization of acetophenone (a maximum asymmetric yield of 6.4%) in a chiral cosolvent. The use of small amounts of chiral crown ethers [408] or cyclodextrins [414] may have more potential, although any significant asymmetry induction has not been observed to date.

Use of Chiral Modified Electrodes

The use of chiral modified electrodes (see Chap. 33) may be the most promising method for electrochemical asymmetric synthesis because of (1) use of extremely small amounts of inducers, (2) low oversensitivity to reaction conditions, (3) variety of kind, and (4) possible application to both cathodic and anodic reactions, although there are some problems to be solved.

Chemically modified electrodes (CME) Miller and coworkers [415,416] found that prochiral carbonyl compounds were reduced asymmetrically to chiral alcohols on a cathode chemically modified with (S)-phenylglycine. This may be the first example of a chemically modified electrode. The maximum asymmetric yield obtained was 14.5% in the reduction of 4-acetylpyridine.

A similar electrode modified with (+)-camphoric acid oxidized asymmetrically p-tolyl methylsulfide to the sulfoxide in 2.5% asymmetric yield [417]. However, no asymmetry induction could be observed in the reduction of carbonyl compounds on other modified cathodes with α-chiral amines [418].

Irreversibly adsorbed electrodes Osa and coworkers [419,420] developed a chiral Raney nickel electrode on which (S,S)-(-)-tartaric acid was irreversibly adsorbed and applied it to the asymmetric reduction of carbonyl compounds. A maximum asymmetric yield of about 20% was obtained in the reduction of methyl acetacetate.

Polymer-caoted electrodes Nonaka and coworkers [421-426] prepared a variety of electrodes coated with optically active poly(amino acid)s and applied them to both asymmetric reduction and oxidation. Maximum asymmetric yields in the reduction and oxidation were 43% [421] and 93% [426], respectively. The mechanism of asymmetry induction was also discussed [422]. The reproducibility of asymmetric yield in the sulfide oxidation was not high when poly(amino acid)s prepared from different lots were used [16,427].

Clay-metal chelate-coated electrodes Yamagishi and Aramata [428] prepared a clay-coated electrode that incorporated optically active tris-(1,10-phenanthroline)ruthenium(II) into the montmorinite thin film. In the oxidation of sulfides, a maximum asymmetric yield of 20% was obtained.

Intercalated graphite electrodes Simonet and coworkers [429] prepared a new type of chiral electrode by forming a lamellar compound ($C_nNR_4^*$) of graphite and optically active ammonium by electrochemical intercalation. This electrode resulted in a maximum asymmetric yield of 9.3% in the reduction of phenyl t-butyl ketone to the alcohol.

Intramolecular Asymmetry Induction

Intramolecular asymmetry induction is not a unique method characterized by electrochemistry. The reduction of optically active alcohol esters of α-keto [430] and unsaturated [431] acids and oximes [432] and the oxidation of olefins [433] have been reported. A maximum asymmetric yield of 21.8% was obtained in the reduction of (-)-menthyl phenylglyoxylate [430]. Nonaka and coworkers [434,435] found that amino acid N-carboxy anhydrides were polymerized with various electrogenerated base catalysts to give the poly (amino acid)s with highly chiral secondary structures in high yields.

In simulation of the intramolecular asymmetric reduction using chiral but optically inactive (racemic) substrates, such as ketones [83,84] and gem-dihalopropanes [436], asymmetric yields higher than 80% could be obtained.

It was also found that the intramolecular asymmetric reduction could be improved in asymmetric yield by using an optically active supporting electrolyte [411,437,438].

Electroreductive asymmetric alkylation of chiral substrates has also been reported [439,440].

New Assymetry Induction Methods

Recently, asymmetric electroreduction methods based on new principles have been developed.

Electroenzymatic reduction Electroenzymatic asymmetric reduction of activated olefins has been achieved by the methylviologen-mediated indirect electroreduction in the presence of enoate reductase [441]. (E)-2-Phenylsuccinic acid gave (R)-2-methyl-3-phenylpropionic acid in 96% asymmetric yield. Electromicrobial asymmetry induction has also been proposed.

Electromagnetic reduction Takahashi and coworkers [442] have reported an epoch-making finding, that is, the occurrence of asymmetric electroreduction of α-keto acids in a magnetic field. This is absolute asymmetry induction in the absence of any chemical inducers. This phenomenon is very interesting in relation to asymmetry induction and concentration in the ancient earth [443]. However, these results have not been reporduced by any other investigators to date.

C. Optical Resolution

Electrochemical optical resolution of racemic mixtures is achieved by using enantiomer-differentiating oxidation or reduction.

Komori and Nonaka [444] reported the first example of an electrochemical enantiomer-differentiating reaction: when racemic 2,2-dimethyl-1-phenyl-1-propanol was oxidized at a poly-(L-valine)-coated anode, 43% optically pure (S)-(-)-propanol was recovered as an unreacted part. Yamagishi and Aramata [445] also found electrooxidative optical resolution of racemic Co(1,10-phenanthroline)$_3^{2+}$ complex by the chiral clay-coated anode, and Yoshinaga and coworkers [446] optically electroreductively resolved racemic Co(acetylacetonato)$_3$ by using optically active supporting electrolytes.

VI. ELECTROCHEMICAL STEREOISOMERIZATION

As shown in Eq. (69), a stable stereoisomer (A) can be converted into another stable isomer (B) through electrochemical oxidation-reduction or reduction-oxidation, if an unstable stereoisomeric intermediate ($A^{\pm n}$) is rapidly stereoisomerized to another isomeric intermediate ($B^{\pm n}$). Such an electrochemical stereoisomerization sometimes can be easily observed under cyclic voltammetric conditions; for preparative isomerization reactions, AC electrolysis at a suitable frequency or undivided cell electrolysis may be available.

$$A \xrightarrow{\pm ne^-} [\, A^{\pm n} \underset{\text{Stereoisomerization}}{\longrightarrow} B^{\pm n} \,] \xrightarrow{\mp ne^-} B \qquad (69)$$

cis to trans Isomerization in Maleates

Yeh and Bard [447] observed the isomerization of dialkyl maleates to the fumarates through the reduction-oxidation by cyclic voltammetry and rotating ring-disk electrode techniques, Eq. (70):

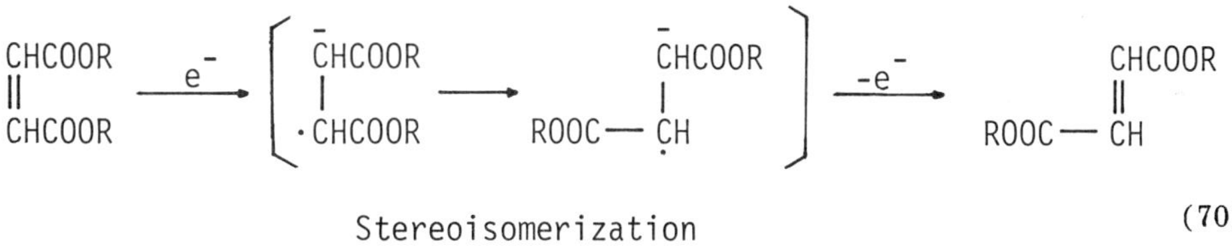

(70)

cis to trans Isomerization in Organometallic Complexes

Recently, electrochemically induced cis to trans positional stereoisomerization of ligands has been found on oxidation-reduction of several types of octahedral complexes of groups VIB and VIIB metals [448-458]. The key step is isomerization of cationic cis intermediates formed by one-electron elimination.

Tub to Chair Isomerization in Cyclooctatetraene (COT) Ligands

COT ligands in COT-cobalt complexes undergo the tub to chair conformational isomerization induced by electrochemical reduction-oxidation via unstable planar COT^- or COT^{2-} [459-465].

A to B Form Isomerization in Bianthrones

A form bianthrones are isomerized to the B form by both electrochemical reduction-oxidation and oxidation-reduction via unstable anionic and cationic species with the bifolded structures, respectively [9,466-476].

VII. STEREOISOMERIC EFFECTS ON THERMODYNAMICS AND KINETICS OF ELECTRODE PROCESSES

It is sometimes observed that stereoisomers have different oxidation and reduction potentials and electron transfer rates. Such phenomena are effectively utilized for clarification of reaction mechanisms and analysis of stereoisomers. The stereoisomeric effect on adsorption has been very rarely studied [477].

A. Cathodic Processes

Hydrocarbons

Polarographic studies of stereoisomeric α,β-diactivated olefins have been made since an early time [478-482]. Generally, half-wave potentials $E_{1/2}$ of the first wave of trans isomers may be somewhat less negative than those of the corresponding cis isomers, although the behaviors of the second waves and limiting current i_l are not simple. Kita and coworkers [483] have reported in a kinetic study that the electron transfer rate of a cis isomer is several times larger than that of the trans isomer.

Z-isomers of monoactivated olefins may be more easily reduced than the E isomers in aqueous solutions, but no significant differences are observed in nonaqueous media [484-487].

Differences in $E_{1/2}$ and i_l between stereosiomeric nonactivated olefins have also been observed [488].

The cis- and trans-bicyclo [6.1.0]nona-2,4,6-trienes (XXVIII) have almost equal $E_{1/2}$ values, but the lifetime of anion radicals formed from trans-XXVIII in aprotic media is much shorter than that from cis-XXVIII because of the contribution of nonclassic structure [489].

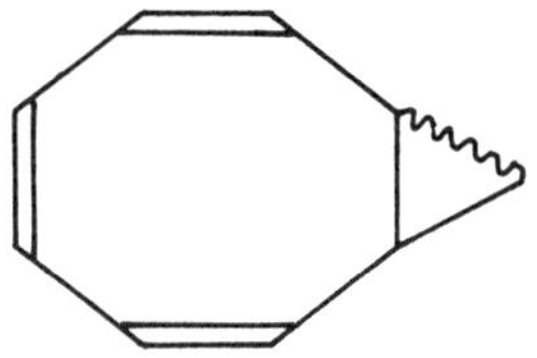

(XXVIII)

Halogen Compounds

Less stable ax- and exo-monohalo derivatives of monocyclo- and bicycloalkanes, respectively, have remarkably less negative $E_{1/2}$ values than their eq- and endo-halo isomers [490-492]. However, this is not always applicable to polycyclic systems, such as steroids, because the positional effect is sometimes dominant [493,494].

In the reduction of acyclic vic-dihalo compounds, it is not easy to find a clear tendency in stereoisomeric effect [495,496]; in cyclic vic-dihalides a regularity can be found [496-498]. Particularly, Závada and coworkers [496] found a very interesting law; that is, there is a pronounced dependence of $E_{1/2}$ on the torsion dihedral angle between the two C–Br bonds of vic-dibromides. Such a significant difference in $E_{1/2}$ values between stereoisomers of vic-dihalides is applicable to stereoisomeric analysis [499,500].

No difference between stereoisomers of cyclic 1,3-dihalide was observed [501].

A significant difference in $E_{1/2}$ was found between cis- and trans-1,2-dihalo olefins [189,193,502,503]: trans isomers are more easily reduced. Moreover, sometimes the reaction course is different for isomers that give the same product [189].

Recently, Lund and coworkers [504] reported an interesting kinetic phenomenon in the single-electron transfer from vinyl anions generated by indirect electroreduction of stereoisomeric (cis and trans) vinyl bromides.

Other Heteroatom Compounds

Carbonyl compounds Kugatova-Shemyakina and coworkers [505] reported an interesting stereoisomeric effect in the reduction of a formyl group. Compound XXIX is reduced with more difficulty than XXX owing to an interaction between the formyl group and the double bond or to a difference in the reaction rate of the ax and eq substituents; the semicarbazones of XXIX and XXX, however, do not afford any difference because the bulky azomethine groups of the both compounds are always equatorial.

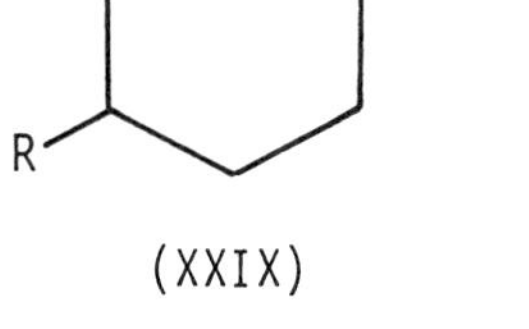

(XXIX)

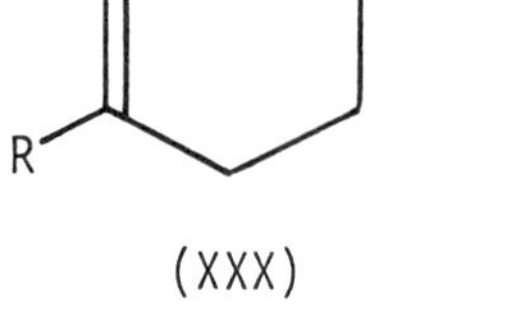

(XXX)

Nitrogen compounds Stereoisomeric (erythro and threo) o-nitro-α-(nitromethyl)benzylalcohols (XXXI) undergo the two-step electroreduction and subsequent air oxidation to give cinnolines (XXXIII) and/or their 1-N-oxides (XXXIV) as final products, Eq. (71) [506]. The first wave $E^1_{1/2}$ values of erythro-XXXI are less negative by 10-50 mV than those of the threo-isomers; the second wave $E^2_{1/2}$ values of the former ones are more negative than those of the latter. Since the configuration of intermediate XXXII should be the same as XXXI, this is stereochemically interesting.

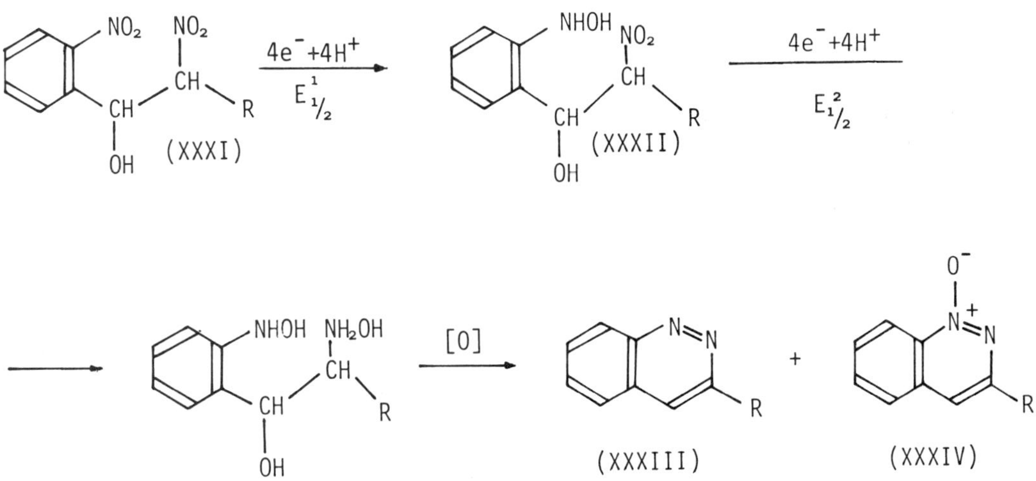

R = Me,	*erythro*:	$E^1_{1/2}$ = -0.54 V ,	$E^2_{1/2}$ = -0.92 V	0%	56-63%
Me,	*threo*:	-0.55	-0.91	0	67
Ph,	*erythro*:	-0.47	-0.83	20	49
Ph,	*threo*:	-0.52	-0.82	16	47

(71)

The $E_{1/2}$ of cis-azocyclohexane is less negative by about 200 mV at 2 < pH < 9 than that of the trans isomer, but they are almost equal in aprotic media [507]. On the other hand, aromatic diazo compounds show a quite different behavior. Isomeric (cis and trans) azobenzenes are reduced at the almost same potential at pH < 11 because of a fast cis-trans isomerization during the first stage of the reduction, but at pH > 11 the cis isomer is reduced at a less negative potential than the trans isomer [508].

Organometallic compounds 2-Substituted cyclohexyl mercury halides are reductively led to the carbanions. A significant difference between the $E^2_{1/2}$ values of cis and trans forms may be rationalized as due to differences in the conformational stability of carbanions formed or in some charge-dipole interaction in the transition state leading to the carbanions [509].

The trans isomers of (η^6-stilbene)$Cr(CO)_3$ and (η^6,η^6-stilbene)$[Cr(CO)_3]_2$ have less negative $E_{1/2}$ values than the cis isomers [510].

B. Anodic Processes

Hydrocarbons

Arnold and Wong [511] have reported that the oxidation potentials of cis- and trans-1,2-diphenylcyclopropanes have simple linear correlations with both their ionization energies and charge transfer absorption energies with tetracyanoethylene as an acceptor. Similar correlations have also been observed in cis- and trans-2,3-diphenyloxiranes.

Alcohols

Nonaka and coworkers [512] compared the anodic oxidation rates of ax- and eq-hydroxyl groups of a variety of cyclic alcohols. ax-Alcohols with no substituent at the 2-positions are more easily oxidized than the eq-alcohols; for 2-substituted alcohols the situation is reversed. This is rationalized as due to a different contribution of adsorption between the isomers on an anode.

It is well known that cis-cyclic diols are much more easily oxidized with chemical oxidizing reagents, such as chromic and periodic acids, than the trans-diols. However, such a significant difference is not observed in the anodic oxidative cleavage [513-515]. It is interesting to note that cis- and trans-1,2-cyclohexanediols give adipaldehyde and adipoin, respectively, at a platinum anode, but at a lead dioxide anode, adipaldehyde only is formed from both diols [514]. 2-Aminoalcohols are also anodically cleaved to give the corresponding carbonyl compounds as main products, along with some amounts of by-products [515,516]. Masui and coworkers [516] reported that threo- and erythro-amino alcohols were oxidized at equal potentials to give a somewhat different product selectivity.

Other Heteroatom Compounds

Nelsen and coworkers [517] detected conformational equilibria in eq,eq- and ax,eq-N,N'-disubstituted cyclic hydrazines from their oxidation potentials. The anodic oxidation reactions of trans- and cis-1,3-diisopropyl-2,4-bis(diisopropylamino)cyclodiphospha(III)azanes (XXXV) are quite different [518]: The trans isomer is reversibly oxidized at 0.53 V versus SCE to form a stable cation radical; the cis isomer undergoes a completely irreversible oxidation at a more positive potential because an unstable cation radical is formed.

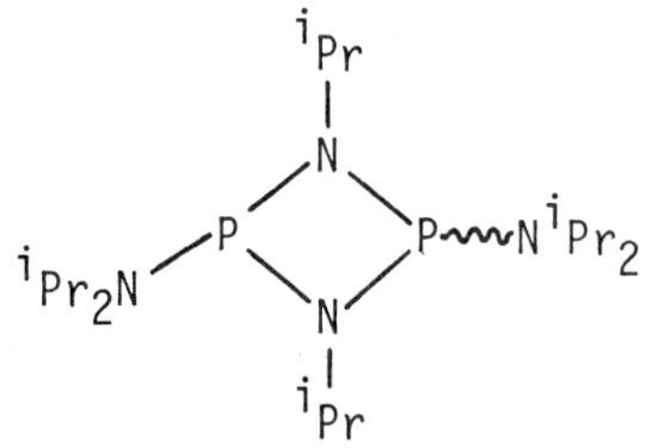

(XXXV)

The oxidation potential of 2-endo-methylthio-6-endo-substituted norbornanes is strongly influenced by the substituents because of 2,6-space interaction, but there is no such influence in the 6-exo-substituted isomer [519-521].

REFERENCES

1. G. R. Clemo and J. M. Smith, J. Chem. Soc., 2423 (1928).
2. L. G. Feoktistov, in Progress in Electrochemistry of Organic Compounds, Vol. 1 (A. A. Frumkin and A. A. Ershler, eds.), Plenum Press, New York, 1971, p. 135.
3. A. J. Fry, in Organic Electrochemistry (A. Davison, M. J. S. Dewar, K. Hafner, E. Heilbronner, U. Hofmann, K. Niedenzu, H. Schäfer, and G. Wittig, eds.), Springer-Verlag, Heidelberg, 1972, p. 1.
4. L. Eberson and L. Horner, in Organic Electrochemistry, 1st ed. (M. M. Baizer, ed.), Dekker, New York, 1973, p. 869.
5. T. Nonaka, Kagaku to Kogyo, 32, 399 (1979).
6. T. Nonaka and T. Fuchigami, Yukigosei Kagaku Kyokaishi, 37, 372 (1973).
7. T. Nonaka, Kagakukogyo, 82, 401 (1981).
8. J. T. Andersson, J. H. Stocker, and L. Horner, in Organic Electrochemistry, 2nd ed. (M. M. Baizer and H. Lund, eds.), Dekker, New York, 1983, p. 903.
9. Z. R. Grabowski, B. Czochralska, A. Vincenz-Chodkowska, and M. S. Balasiewicz, Disc. Faraday Soc., 45, 145 (1968).
10. K. P. Butin, Russ. Chem. Rev., 40, 525 (1971).
11. B. E. Conway and E. J. Rudd, in Techniques of Electroorganic Synthesis, Part 1 (N. L. Weinberg, ed.), Wiley-Interscience, New York, 1974, p. 210.
12. N. L. Weinberg, in Techniques of Electroorganic Synthesis, Part 1 (N. L. Weinberg, ed.), Wiley-Interscience, New York, 1974, p. 254.
13. B. L. Funt and J. Tanner, in Techniques of Electroorganic Synthesis, Part 2, (N. L. Weinberg, ed.), Wiley-Interscience, New York, 1975, p. 598.
14. W. E. Britton, in Topics in Organic Electrochemistry (A. J. Fry and W. E. Britton, eds.), Plenum Press, New York, 1986, p. 227.
15. A. Yamagishi and A. Aramata, Hyomen, 24, 442 (1986); Chem. Abstr., 105, 190186x (1986).
16. T. Nonaka, Shokubai, 29, 579 (1987).
17. A. Tallec, Bull. Soc. Chim. Fr., 743 (1985).
18. I. Rosenthal, J. R. Hayes, A. J. Martin, and P. J. Elving, J. Amer. Chem. Soc., 80, 3050 (1958).
19. K. Junghans, Chem. Ber., 109, 395 (1976).
20. J. Sarrazin and A. Tallec, Tetrahedron Lett., 1579 (1977).
21. J. Sarrazin and A. Tallec, Nouv. J. Chim., 3, 571 (1979).
22. L. H. Klemm, D. R. Olson, and D. V. White, J. Org. Chem., 36, 3740 (1971).
23. A. Froeling, F. Hornung, and R. O. de Jongh, Rec. Trav. Chim. Pay-Bas, 105, 123 (1986).
24. H. Kata, N. Kubota, and K. Shimazu, Electrochim. Acta, 26, 1185 (1981).
25. N. Kubota and H. Kita, Electrochim. Acta, 27, 861 (1982).
26. G. Farnia, F. Marcuzzi, G. Melloni, and G. Sandoná, J. Amer. Chem. Soc., 106, 6503 (1984).
27. C. Andrei and V. Bărboiu, Rev. Roum. Chim., 24, 11 (1979); Chem. Abstr., 90, 204630y (1979).
28. W. M. Moore, A. Salajegheh, and D. G. Peters, J. Amer. Chem. Soc., 97, 4954 (1975).
29. M. M. Baizer and J. D. Anderson, J. Electrochem. Soc., 111, 223 (1964).

30. J. H. Wagenknecht and M. M. Baizer, J. Org. Chem., 31, 3885 (1966).
31. J. P. Petrovich, M. M. Baizer, and M. R. Ort, J. Electrochem. Soc., 116, 749 (1969).
32. L. H. Klemm and D. R. Olson, J. Org. Chem., 38, 3390 (1973).
33. J. Delaunay, A. Lebouc, G. Le Guillanton, L. M. Gomes, and J. Simonet, Electrochim. Acta, 27, 287 (1982).
34. K. W. Bowers, R. W. Giese, J. Grimshaw, H. O. House, N. H. Kolodny, K. Kronberger, and D. K. Roe, J. Amer. Chem. Soc., 92, 2783 (1970).
35. H. Kanetsuna and T. Nonaka, Denkikagaku, 49, 526 (1981); Chem. Abstr., 95, 219861b (1981).
36. S. U. Pedersen, R. G. Hazell, and H. Lund, Acta Chem. Scand., B41, 336 (1987).
37. H. G. Thomas and F. Thönnessen, Chem. Ber., 112, 2786 (1979).
38. J. D. Anderson, M. M. Baizer, and J. P. Petrovich, J. Org. Chem., 31, 3890 (1966).
39. L. Moëns, M. M. Baizer, and R. D. Little, J. Org. Chem., 51, 4497 (1986).
40. J. D. Anderson and M. M. Baizer, Tetrahedron Lett., 511 (1966).
41. L. Mandell, R. F. Daley, and R. A. Day, Jr., J. Org. Chem., 41, 4087 (1976).
42. J. Andersson and L. Eberson, J. Chem. Soc., Chem. Commun., 565 (1976).
43. J. Andersson, L. Eberson, and C. Svensson, Acta Chem. Scand., B32, 234 (1978).
44. B. Terem and J. H. P. Utley, Electrochim. Acta, 24, 1081 (1979).
45. F. Fournier, J. Berthelot, and J.-J. Basselier, Tetrahedron, 41, 5667 (1985).
46. F. Fournier, D. Davoust, and J.-J. Basselier, Tetrahedron, 41, 5677 (1985).
47. T. Shono, I. Nishiguchi, and H. Ohmizu, J. Amer. Chem. Soc., 99, 7396 (1977).
48. H. Lund and C. Degrand, Acta Chem. Scand., B33, 57 (1979).
49. C. Degrand, R. Mora, and H. Lund, Acta Chem. Scand., B37, 429 (1983).
50. S. Wawzonek, E. W. Blaha, R. Berkey, and M. E. Runner, J. Electrochem. Soc., 102, 235 (1955).
51. R. Dietz and M. E. Peover, Disc. Faraday Soc., 45, 154 (1968).
52. H. Kanetsuna and T. Nonaka, Denkikagaku, 47, 422 (1979); Chem. Abstr., 92, 41169e (1980).
53. H. Kanetsuna and T. Nonaka, Denkikagaku, 47, 480 (1979); Chem. Abstr., 92, 163453p (1980).
54. T. Shono, H. Ohmizu, S. Kawakami, and H. Sugiyama, Tetrahedron Lett., 21, 5092 (1980).
55. T. Shono and M. Mitani, J. Amer. Chem. Soc., 93, 5284 (1971).
56. T. Shono, I. Nishiguchi, H. Ohmizu, and M. Mitani, J. Amer. Chem. Soc., 100, 545 (1978).
57. T. Shono, N. Kise, T. Suzumoto, and T. Morimoto, J. Amer. Chem. Soc., 108, 4676 (1986).
58. D. P. Fox, R. D. Little, and M. M. Baizer, J. Org. Chem., 50, 2202 (1985).
59. R. D. Little, D. P. Fox, L. Van Hifte, G. Sowell, R. Dannecker, R. L. Wolin, L. Moën, and M. M. Baizer, 172 ECS Meeting, Honolulu, 1987, Abstract 1330.

60. G. Pattenden and G. M. Robertson, Tetrahedron, 41, 4001 (1985).
61. C. Degrand, P.-L. Compagnon, G. Belot, and D. Jacquin, J. Org. Chem., 45, 1189 (1980).
62. K. N. Campbell and E. E. Young, J. Amer. Chem. Soc., 65, 965 (1943).
63. J. B. Lee and P. Cashmore, Chem. Ind., 1758 (1966).
64. H. Kita and H. Nakajima, J. Chem. Soc., Faraday Trans. I, 77, 2105 (1981).
65. T. Nonaka, M. Takahashi, and T. Fuchigami, Bull. Chem. Soc. Jpn., 56, 2584 (1983).
66. J. Kato, M. Sakuma, and T. Yamada, Denkikagaku, 25, 331 (1957); Chem. Abstr., 52, 4469a (1958).
67. L. Horner and H. Röder, Liebigs Ann. Chem., 723, 11 (1969).
68. L. J. Throop, U.S. Patent 3,444,057; Chem. Abstr., 71, 50359k (1969).
69. R. A. Benkeser and C. A. Tincher, J. Org. Chem., 33, 2727 (1968).
70. T. Shono, I. Nishiguchi, and H. Ohmizu, Chem. Lett., 1233 (1976).
71. S. Ono, Nippon Kagaku Zasshi, 75, 1195 (1954).
72. P. C. Condit, Ind. Eng. Chem., 48, 1252 (1956).
73. H. Suter, H. Nohe, F. Beck, W. Bruegel, and H. Aschenbrenner, U.S. Patent 3,542,656 (1970).
74. F. Beck, Electrochim. Acta, 20, 361 (1975).
75. F. Mizutani, N. Sato, and T. Sekine, Denkikagaku, 45, 418 (1977); Chem. Abstr., 86, 23488e (1977).
76. L. L. Miller and L. Christensen, J. Org. Chem., 43, 2059 (1978).
77. R. A. Misra and B. L. Sharma, Electrochim. Acta, 24, 727 (1979).
78. R. A. Misra and A. Jain, Trans. SAEST, 19, 189 (1984); Chem. Abstr., 102, 131674v (1985).
79. S. Wawzonek and D. Wearring, J. Amer. Chem. Soc., 81, 2067 (1959).
80. L. Mandell, R. M. Powers, and R. A. Day, Jr., J. Amer. Chem. Soc., 80, 5284 (1958).
81. A. F. Diaz, Y. Y. Cheng, and M. Ochoa, J. Amer. Chem. Soc., 99, 6319 (1977).
82. M. Fedoronko, Coll. Czech. Chem. Commun., 37, 3897 (1972).
83. T. Nonaka, Y. Kusayanagi, and T. Fichigami, Electrochim. Acta, 25, 1679 (1980).
84. T. Nonaka and Y. Kusayanagi, Rep. Asahi Glass Found. Ind. Technol., 36, 185 (1980); Chem. Abstr., 94, 199672q (1981).
85. T. Nonaka and Y. Kusayanagi, Electrochim. Acta, 26, 893 (1981).
86. H. Lund and J. N. Hansen, in Proc. 1st International Symp. Electroorg. Synth. (S. Torii, ed.), 1987, p. 249.
87. P. Anziani, A. Aubry, and R. Cornubert, C. R. Acad. Sci., 225, 878 (1947).
88. J. P. Coleman, R. J. Kobylecki, and J. H. P. Utley, J. Chem. Soc., Chem. Commun., 104 (1971).
89. T. Shono and M. Mitani, Tetrahedron, 28, 4747 (1972).
90. J. P. Coleman, R. J. Holman, and J. H. P. Utley, J. Chem. Soc., Perkin Trans. II, 879 (1976).
91. R. J. Holman and J. H. P. Utley, Tetrahedron Lett., 1553 (1974).
92. R. J. Holman and J. H. P. Utley, J. Chem. Soc., Perkin Trans. II, 884 (1976).
93. T. Nonaka, S. Wachi, and T. Fuchigamin, Chem. Lett., 47 (1977).
94. G. Le Guillanton, Tetrahedron Lett., 2567 (1978).

95. T. Nonaka, M. Takahashi, and T. Fuchigami, Denkikagaku, 51, 129 (1983); Chem. Abstr., 98, 187917t (1983).
96. P. Kabasakalian, J. McGlotten, A. Basch, and M. D. Yudis, J. Org. Chem., 26, 1738 (1961).
97. C. Agami, A. Kazakos, and J. Levisalles, Tetrahedron Lett., 4073 (1977).
98. A. Perjéssy, D. Zacharová-Kalavská, and M. Lácová, Coll. Czech. Chem. Commun., 37, 1160 (1972).
99. A. J. Harle and L. E. Lyons, J. Chem. Soc., 1575 (1950).
100. M. J. Allen, J. Org. Chem., 15, 435 (1950).
101. H. Lund, Acta Chem. Scand., 11, 283 (1957).
102. P. Bladon, J. W. Cornforth, and R. H. Jaeger, J. Chem. Soc., 863 (1958).
103. S. Wawzonek and A. Gundersen, J. Electrochem. Soc., 111, 324 (1964).
104. J. Grimshaw and J. S. Ramsey, J. Chem. Soc. (C), 653 (1966).
105. D. H. Evans and E. C. Woodbury, J. Org. Chem., 32, 2158 (1967).
106. J. Armand, G. Dana, and F. Valentini, Bull. Soc. Chim. Fr., 4581 (1968).
107. J. H. Stocker and R. M. Jenevein, J. Org. Chem., 33, 294 (1968).
108. J. H. Stocker and R. M. Jenevein, J. Org. Chem., 33, 2145 (1968).
109. M. J. Simonet and M. G. Champetier, C. R. Acad. Sci., Ser. C, 267, 1548 (1968).
110. H. Lund and A. D. Thomsen, Acta Chem. Scand., 23, 3567 (1969).
111. A. D. Thomsen and H. Lund, Acta Chem. Scand., 23, 3582 (1969).
112. V. J. Puglisi, G. L. Clapper, and D. H. Evans, Anal. Chem., 41, 279 (1969).
113. J. H. Stocker, R. M. Jenevein, and D. H. Kern, J. Org. Chem., 34, 2810 (1969).
114. J. H. Stocker and R. J. Jenevein, J. Org. Chem., 34, 2807 (1969).
115. A. Bewick and H. P. Cleghorn, J. Chem. Soc., Perkin Trans. II, 1410 (1973).
116. W. J. M. van Tilborg and C. J. Smit, Tetrahedron Lett., 3651 (1977).
117. A. Bewick and D. J. Brown, J. Chem. Soc., Perkin Trans. II, 99 (1977).
118. D. F. Tomkins and J. H. Wagenknecht, J. Electrochem. Soc., 125, 372 (1978).
119. M. D. Birkett and A. T. Kuhn, Electrochim. Acta, 25, 273 (1980).
120. W. J. M. van Tilborg and C. J. Smit, Rec. J. Roy. Netherl. Chem. Soc., 98, 532 (1979).
121. E. Roman, D. Astruc, and A. Darchen, J. Chem. Soc., Chem. Commun., 512 (1976).
122. J. F. Rusling and P. Zuman, J. Org. Chem., 46, 1906 (1981).
123. V. P. Gul'tyai and L. M. Korotaeva, Izv. Akad. Nauk SSSR, Ser. Khim., 165 (1982); Chem. Abstr., 96, 141941b (1982).
124. T. Lund and H. Lund, Acta Chem. Scand., B39, 429 (1985).
125. E. Touboul and G. Dana, Tetrahedron, 31, 1925 (1975).
126. R. E. Sioda, B. Terem, J. H. P. Utley, and B. C. L. Weedon, J. Chem. Soc., Perkin Trans. I, 561 (1976).
127. A. J. Klein and D. H. Evans, J. Org. Chem., 42, 2560 (1977).
128. E. Touboul and G. Dana, J. Org. Chem., 44, 1397 (1979).
129. L. Mandell, H. Hamilton, and R. A. Day, Jr., J. Org. Chem., 45, 1710 (1980).

130. M.-A. O.-L. Moing, J. Delaunay, and J. Simonet, Nouv. J. Chim., 8, 217 (1984).
131. J. Delaunay, A. O.-L. Moing, A. Lebouc, J. Simonet, and L. Toupet, J. Chem. Res. (S), 418 (1986).
132. R. E. Juday and W. L. Waters, Tetrahedron Lett., 4321 (1975).
133. C. Z. Smith and J. H. P. Utley, J. Chem. Soc., Chem. Commun., 492 (1981).
134. D. W. Sopher and J. H. P. Utley, J. Chem. Soc., Perkin Trans. II, 1361 (1984).
135. F. Fournier, J. Berthelot, and Y. L. Pascal, Tetrahedron, 40, 339 (1984).
136. E. Kariv, B. J. Cohen, and E. Gileadi, Tetrahedron, 27, 805 (1971).
137. R. N. Gourley and J. Grimshaw, J. Chem. Soc. (C), 2388 (1968).
138. F. Ammar, C. P. Andrieux, and J. M. Savéant, J. Electroanal. Chem., 53, 407 (1974).
139. T. Nonaka, A. Udagawa, and K. Odo, Chem. Lett., 1261 (1975).
140. J. Armand and L. Boulares, Can. J. Chem., 54, 1197 (1976).
141. T. Nonaka and M. Asai, Bull. Chem. Soc. Jpn., 51, 2976 (1978).
142. W. D. Hoffman, W. E. McEwen, and J. Kleinberg, Tetrahedron, 5, 293 (1959).
143. T. J. Curphey, C. W. Amelotti, T. P. Layloff, R. L. McCartney, and J. H. Williams, J. Amer. Chem. Soc., 91, 2817 (1969).
144. A. Vincenz-Chodkowska and Z. R. Grabowski, Electrochim. Acta, 9, 789 (1964).
145. T. Troll, H. Leffler, and W. Elbe, Electrochim. Acta, 24, 969 (1979).
146. T. Shono, Y. Matsumura, K. Tsubata, and Y. Sugihara, J. Org. Chem., 47, 3090 (1982).
147. T. Nonaka and M. M. Baizer, Electrochim. Acta, 28, 661 (1983).
148. T. Nonaka, S. Kihara, T. Fuchigami, and M. M. Baizer, Bull. Chem. Soc. Jpn., 57, 3160 (1984).
149. E. L. Shapiro and M. J. Gentles, J. Org. Chem., 46, 5017 (1981).
150. A. V. Kamernitskii, I. G. Reshetova, E. I. Chernoburnova, S. G. Mairanovskii, and N. K. Lisitsina, Izv. Akad. Nauk SSSR, Ser. Khim. (Engl. Ed.) 1736 (1984); Chem. Abstr., 102, 62512b (1985).
151. A. J. Fry and R. G. Reed, J. Amer. Chem. Soc., 91, 6448 (1969).
152. R. W. Koch and R. E. Dessy, J. Org. Chem., 47, 4452 (1982).
153. L. Horner and D. H. Skaletz, Liebigs Ann. Chem., 1210 (1975).
154. B. Czochralska, D. Shugar, S. K. Arora, R. B. Bates, and R. S. Cutler, J. Amer. Chem. Soc., 99, 2583 (1977).
155. A. J. Fry and J. H. Newberg, J. Amer. Chem. Soc., 89, 6374 (1967).
156. J. B. Kerr and P. E. Iversen, Acta Chem. Scand., B32, 405 (1978).
157. T. Shono, Y. Usui, and H. Hamaguchi, Tetrahedron Lett., 21, 1351 (1980).
158. H. Kohn, N. Zein, X. Q. Lin, J.-Q. Ding, and K. M. Kadish, J. Amer. Chem. Soc., 109, 1833 (1987).
159. P. E. Iversen and H. Lund, Tetrahedron Lett., 3523 (1969).
160. S. D. Ross, M. Finkelstein, and R. C. Petersen, J. Amer. Chem. Soc., 82, 1582 (1960).
161. P. E. Iversen and J. O. Madsen, Tetrahedron, 30, 3477 (1974).
162. H. Shindo, J. Chem. Soc., Faraday Trans. I, 82, 45 (1986).
163. M. Cariou, R. Hazard, M. Jubault, and A. Tallec, Can. J. Chem., 61, 2359 (1983).
164. D. Knittel, Monatsh. Chem., 117, 679 (1986).

165. P. Martigny and J. Simonet, J. Electroanal. Chem., 81, 407 (1977).
166. S. Gambino, P. Martigny, G. Mousset, and J. Simonet, J. Electroanal. Chem., 90, 105 (1978).
167. T. Shono, Y. Matsumura, and S. Kashimura, Chem. Lett., 69 (1978).
168. H. Lund and D. Degrand, C. R. Hebd. Seances Acad. Sci., Ser. C, 287, 535 (1978); Chem. Abstr., 90, 186514g (1979).
169. A. J. Fry, K. Ankner, and V. K. Handa, J. Chem. Soc., Chem. Commun., 120 (1981).
170. A. Darchen, P. Berthelot, C. Vaccher, M. N. Viana, M. Debaert, and J. L. Burgot, J. Heterocyclic Chem., 23, 1603 (1986).
171. G. Adiwidjaja, L. Kistenbrügger, and J. Voss, J. Chem. Res. (S), 88 (1981).
172. T. Shono, Y. Matsumura, S. Kashimura, and H. Kyutoku, Tetrahedron Lett., 1205 (1978).
173. J. Voss, C. Von Bülow, T. Drews, and P. Mischke, Acta Chem. Scand., B37, 519 (1983).
174. R. Hazard, S. Jaouannet, and A. Tallec, Bull. Soc. Chim. Belg., 94, 199 (1985); Chem. Abstr., 103, 104316u (1985).
175. R. Hazard, S. Jaouannet, and A. Tallec, Electrochim. Acta, 28, 1095 (1983).
176. M. F. Chasle-Pommeret, S. Jaouannet, A. Lebouc, and A. Tallec, Electrochim. Acta, 29, 1287 (1984).
177. A. J. Fry and R. H. Moore, J. Org. Chem., 33, 1283 (1968).
178. R. E. Erickson, R. Annino, M. D. Scanlon, and G. Zon, J. Amer. Chem. Soc., 91, 1767 (1969).
179. A. J. Fry and R. G. Reed, J. Amer. Chem. Soc., 93, 553 (1971).
180. A. J. Fry and R. G. Reed, J. Amer. Chem. Soc., 94, 8475 (1972).
181. D. P. G. Hamon and K. R. Richards, Aust. J. Chem., 36, 2243 (1983).
182. J. Casanova and H. R. Rogers, J. Org. Chem., 39, 2408 (1974).
183. H. Lund and E. Hobolth, Acta Chem. Scand., B30, 895 (1976).
184. O. R. Brown, P. H. Middleton, and T. L. Threlfall, J. Chem. Soc., Perkin Trans. II, 955 (1984).
185. P. J. Elving, I. Rosenthal, and A. J. Martin, J. Amer. Chem. Soc., 77, 5218 (1955).
186. U. Husstedt and H. J. Schäfer, Tetrahedron Lett., 22, 623 (1981).
187. T. Takiguchi and T. Nonaka, Nippon Kagaku Kaishi, 1147 (1985); Chem. Abstr., 103, 149410m (1985).
188. T. Lund, S. U. Pedersen, H. Lund, K. M. Cheung, and J. H. P. Utley, Acta Chem. Scand., B41, 285 (1987).
189. W. H. Jura and R. J. Gaul, J. Amer. Chem. Soc., 80, 5402 (1958).
190. J.-P. Gisselbrecht and H. Lund, Acta Chem. Scand., B39, 773 (1985).
191. R. Wolf and E. Steckhan, J. Chem. Soc., Perkin Trans. I, 733 (1986).
192. P. J. Elving, I. Rosenthal, J. R. Hayes, and A. J. Martin, Anal. Chem., 33, 330 (1961).
193. R. Annino, R. J. Boczkowski, D. J. Bolton, W. E. Geiger, Jr., D. T. Jackson, Jr., and J. Mahler, J. Electroanal. Chem., 38, 403 (1972).
194. A. J. Fry and M. A. Mitnick, J. Amer. Chem. Soc., 91, 6207 (1969).
195. H. Kanetsuna and T. Nonaka, Denki Kagaku, 49, 129 (1981); Chem. Abstr., 95, 15080m (1981).
196. C. de Luca, A. Inesi, and L. Rampazzo, J. Chem. Soc., Perkin Trans. II, 1403 (1982).

197. T. Sato, T. Wada, Y. Komeichi, and M. Kainosho, Bull. Chem. Soc. Jpn., 58, 1452 (1985).
198. A. Guirado, F. Barba, C. Manzanera, and M. D. Velasco, J. Org. Chem., 47, 142 (1982).
199. J. Wellmann and E. Steckhan, Synthesis, 901 (1978).
200. D. Lenoir, W. Kornrumpf, and H. P. Fritz, Chem. Ber., 116, 2390 (1983).
201. C. de Luca, A. Inesi, and L. Rampazzo, J. Chem. Soc., Perkin Trans. II, 1821 (1983).
202. A. J. Fry and W. E. Britton, Tetrahedron Lett., 4363 (1971).
203. A. J. Fry and W. E. Britton, J. Org. Chem., 38, 4016 (1973).
204. M. R. Rifi, J. Amer. Chem. Soc., 89, 4442 (1967).
205. S. Satoh, M. Itoh, and M. Tokuda, J. Chem. Soc., Chem. Commun., 481 (1978).
206. A. J. Fry and J. J. O'Dea, J. Org. Chem., 40, 3625 (1975).
207. W. J. M. van Tilborg, R. Plomp, R. de Ruiter, and C. J. Smit, Rec. J. Roy. Netherl. Chem. Soc., 99, 206 (1980).
208. A. J. Fry and L.-L. Chung, Tetrahedron Lett., 645 (1976).
209. A. J. Fry, K. Ankner, and V. Handa, Tetrahedron Lett., 22, 1791 (1981).
210. E. Stamm, L. Walder, and R. Keese, Helv. Chim. Acta, 61, 1545 (1978).
211. T. Shono, N. Kise, and T. Suzumoto, J. Amer. Chem. SOc., 106, 259 (1984).
212. T. Shono, S. Kashimura, M. Mizukami, and S. Matsumura, 47th Annual Meeting Chem. Soc. Jpn., Abstract II, 1983, p. 931.
213. E. Roman, D. Astruc, and A. Darchen, J. Organometal. Chem., 219, 221 (1981).
214. C. M. Arewgoda, B. H. Robinson, and J. Simpson, J. Chem. Soc., Chem. Commun., 284 (1982).
215. T. D. Binns, R. Brettle, and G. B. Cox, J. Chem. Soc. (C), 584 (1968).
216. L. Eberson, J. Org. Chem., 27, 2329 (1962).
217. L. Eberson and S. Nilsson, Acta Chem. Scand., 22, 2453 (1968).
218. L. Eberson, S. Gränse, and B. Olofsson, Acta Chem. Scand., 22, 2462 (1968).
219. G. E. Hawkes, J. H. P. Utley, and G. B. Yates, J. Chem. Soc., Perkin Trans. II, 1709 (1976).
220. H. Kanetsuna and T. Nonaka, Denki Kagaku, 48, 444 (1980); Chem. Abstr., 94, 22101b (1981).
221. M. Huhtasaari, H. J. Schäfer, and H. Luftmann, Acta Chem. Scand., B37, 537 (1983).
222. A. K. Yadav and P. Tissot, Helv. Chim. Acta, 67, 1698 (1984).
223. K. Koyama, T. Ebara, T. Tani, and S. Tsutsumi, Can. J. Chem., 47, 2484 (1969).
224. H. Schäfer and R. Pistorius, Angew. Chem., Int. Ed., 11, 841 (1972).
225. C. J. Brookes, P. L. Coe, A. E. Pedler, and J. C. Tatlow, J. Chem. Soc., Perkin Trans. I, 202 (1978).
226. R. N. Renaud and P. J. Champagne, Can. J. Chem., 57, 990 (1979).
227. P. J. Champagne and R. N. Renaud, Can. J. Chem., 58, 1101 (1980).
228. M. Chkir and D. Lelandais, J. Chem. Soc., Chem. Commun., 1369 (1971).

229. F. M. Banda and R. Brettle, J. Chem Soc., Perkin Trans. I, 1907 (1974).
230. R. V. Lindsey, Jr., and M. L. Peterson, J. Amer. Chem. Soc., 81, 2073 (1959).
231. H. Horikawa, T. Iwasaki, K. Matsumoto, and M. Miyoshi, J. Org. Chem., 43, 335 (1978).
232. P. Renaud and D. Seebach, Synthesis, 424 (1986).
233. P. Renaud and D. Seebach, Helv. Chim. Acta, 69, 1704 (1986).
234. J. H. P. Utley and G. B. Yates, J. Chem. Soc., Perkin Trans. II, 395 (1978).
235. E. J. Corey, N. L. Bauld, R. T. La Londe, J. Casanova, Jr., and E. T. Kaiser, J. Amer. Chem. Soc., 82, 2645 (1960).
236. P. G. Gassmann and F. V. Zalar, J. Amer. Chem. Soc., 88, 2252 (1966).
237. E. Laurent and M. Thomalla, Bull. Soc. Chim. Fr., Part 1, 834 (1977).
238. T. Imagawa, S. SUgita, T. Akiyama, and M. Kawanishi, Tetrahedron Lett., 22, 2569 (1981).
239. P. S. Wharton, G. A. Hiegel, and R. V. Coombs, J. Org. Chem., 28, 3217 (1963).
240. J. G. Traynham and J. S. Dehn, J. Amer. Chem. Soc., 89, 2139 (1967).
241. T. Shono, I. Nishiguchi, S. Yamane, and R. Oda, Tetrahedron Lett., 1965 (1969).
242. L. B. Rodewald and M. C. Lewis, Tetrahedron, 27, 5273 (1971).
243. E. J. Corey and J. Casanova, Jr., J. Amer. Chem. Soc., 85, 165 (1963).
244. H. H. Westberg and H. J. Dauben, Jr., Tetrahedron Lett., 5123 (1968).
245. C. G. Overberger and P. Kabasakalian, J. Amer. Chem. Soc., 79, 3182 (1957).
246. W. W. Paudler, R. E. Herbener, and A. G. Zeiler, Chem. Ind. (London), 1909 (1965).
247. A. F. Vellturo and G. W. Griffin, J. Org. Chem., 31, 2241 (1966).
248. F. D. Mango and W. A. Bonner, J. Org. Chem., 29, 1367 (1964).
249. S. Tsutsumi, K. Koyama, T. Inoue, T. Ebara, T. Tani, and T. Inoue, 17th CITCE Meeting, Extended Abstract, 109 (1966).
250. T. Shono and T. Kosaka, Tetrahedron Lett., 6207 (1968).
251. L. Cedheim and L. Eberson, Acta Chem. Scand., B30, 527 (1976).
252. G. Le Guillanton and M.-A. Le Moing, Electrochim. Acta, 24, 1203 (1979).
253. J. Delaunay, J. Simonet, and L. Toupet, J. Chem. Soc., Chem. Commun., 1093 (1986).
254. T. Shono, Y. Matsumura, O. Onomura, T. Kanazawa, and M. Habuka, Chem. Lett., 1104 (1984).
255. V. D. Parker and L. Eberson, J. Chem. Soc., Chem. Commun., 340 (1969).
256. A. J. Baggaley and R. Brettle, J. Chem. Soc. (C), 2055 (1968).
257. T. Shono, I. Nishiguchi, and M. Okawa, Chem. Lett., 573 (1976).
258. T. Shono and A. Ikeda, Chem. Lett., 311 (1976).
259. L. Eberson and H. Sternerup, Acta Chem. Scand., 26, 1431 (1972).
260. J. P. Dirlan and L. Eberson, Acta Chem. Scand., 26, 1454 (1972).
261. H. Sternerup, Acta Chem. Scand., B28, 969 (1974).
262. T. Shono, T. Toda, and N. Oshino, Tetrahedron Lett., 25, 91 (1984).

263. T. Shono and A. Ikeda, J. Amer. Chem. Soc., 94, 7892 (1972).
264. M. Finkelstein, R. C. Petersen, and S. D. Ross, Tetrahedron, 23, 3875 (1967).
265. L. Eberson, K. Nyberg, M. Finkelstein, R. C. Petersen, S. D. Ross, and J. J. Uebel, J. Org. Chem., 32, 16 (1967).
266. V. D. Parker, Acta Chem. Scand., 24, 3162 (1970).
267. T. Inoue, K. Koyama, T. Matsuoka, K. Matsuoka, and S. Tsutsumi, Tetrahedron Lett., 1409 (1963).
268. T. Inoue and S. Tsutsumi, Bull. Chem. Soc. Jpn., 38, 661 (1965).
269. T. Inoue, K. Koyama, T. Matsuoka, and S. Tsutsumi, Bull. Chem. Soc. Jpn., 40, 162 (1967).
270. E. Steckhan, J. Amer. Chem. Soc., 100, 3526 (1978).
271. M. Iguchi, A. Nishiyama, M. Hara, Y. Terada, and S. Yamamura, Chem. Lett., 1015 (1978).
272. R. Brettle and J. R. Sutton, J. Chem. Soc., Perkin Trans. I, 1947 (1975).
273. K. Yoshida, T. Kanbe, and T. Fueno, J. Org. Chem., 42, 2313 (1977).
274. A. Guirado, F. Barba, and J. A. Franco, Electrochim. Acta, 27, 1621 (1982).
275. P. D. Palasz and J. H. P. Utley, J. Chem. Soc., Perkin Trans. II, 807 (1984).
276. N. Clauson-Kaas, F. Limborg, and K. Glens, Acta. Chem. Scand., 6, 531 (1952).
277. S. D. Ross, M. Finkelstein, and J. J. Uebel, J. Org. Chem., 34, 1018 (1969).
278. I. Stibor, J. Šrogl, M. Janda, N. Piricǒcá, and K. Vlažný, Coll. Czech. Chem. Commun., 47, 3261 (1982).
279. A. A. Ponomarev and I. A. Markushina, J. Gen. Chem. USSR (Engl. Ed.), 33, 3892 (1963).
280. K. Yoshida, J. Chem. Soc., Chem. Commun., 20 (1978).
281. M. Janda, J. Šrogl, E. Körblová, and I. Stibor, Coll. Czech. Chem. Commun., 45, 1361 (1980).
282. H. Tanaka, Y. Kobayashi, and S. Torii, J. Org. Chem., 41, 3482 (1976).
283. S. Torii and H. Tanaka, Japan Kokai 7823925 (1978); Chem. Abstr., 89, 108169p (1978).
284. T. Iwasaki, T. Nishitani, H. Horikawa, and I. Inoue, J. Org. Chem., 47, 3799 (1982).
285. O. Kawana, Y. Nakamura, and Y. Yoshihiro, Nippon Kagaku Kaishi, 1747 (1983); Chem. Abstr., 100, 156445t (1984).
286. S. Torii, H. Tanaka, Y. Tsutsui, T. Ohshima, and S. Yamashita, Chem. Lett., 1535 (1986).
287. R. E. Sioda, J. Phys. Chem., 72, 2322 (1968).
288. N. L. Weinberg and B. Belleau, Tetrahedron, 29, 279 (1973).
289. F. Barba, A. Guirado, and I. Barba, J. Org. Chem., 49, 3022 (1984).
290. F. Barba, A. Guirado, and I. Barba, J. Chem. Res. (S), 228 (1986).
291. T. Shono, A. Ikeda, J. Hayashi, and S. Hakozaki, J. Amer. Chem. Soc., 97, 4261 (1975).
292. L. Christensen and L. L. Miller, J. Org. Chem., 46, 4876 (1981).
293. T. Shono, Y. Matsumura, and Y. Nakagawa, J. Org. Chem., 36, 1771 (1971).
294. P. G. Gassman and R. Yamaguchi, Tetrahedron, 38, 1113 (1982).

295. T. Shono, Y. Matsumura, T. Hashimoto, K. Hibino, and H. Hamaguchi, T. Aoki, J. Amer. Chem. Soc., 97, 2546 (1975).
296. M. Iguchi, A. Nishiyama, Y. Terada, and S. Yamamura, Tetrahedron Lett., 4511 (1977).
297. M. Iguchi, A. Nishiyama, Y. Terada, and S. Yamamura, Chem. Lett., 451 (1978).
298. Y. Matsubara, T. Uchida, T. Ohnishi, K. Kanehira, Y. Fujita, T. Hirashima, and I. Nishiguchi, Tetrahedron Lett., 26, 4513 (1985).
299. A. Hembrock, H. J. Schäfer, and G. Zimmermann, Angew. Chem., 97, 1048 (1985).
300. M. Thaning and L.-G. Wistrand, Helv. Chim. Acta, 69, 1711 (1986).
301. R. Noyori and I. Kurimoto, J. Org. Chem., 51, 4320 (1986).
302. A. Bensadat, G. Bodennec, É. Laurent, and R. Tardivel, Nouv. J. Chim., 5, 127 (1981).
303. A. Bensadat, É. Laurent, and R. Tardivel, Nouv. J. Chim., 5, 397 (1981).
304. É. Laurent, H. Lefranc, and R. Tardivel, Nouv. J. Chim., 8, 345 (1984).
305. A. Klemer and G. Nicolaus, Chem. Ber., 112, 3063 (1979).
306. M. Verniette, C. Daremon, and J. Simonet, Electrochim. Acta, 23, 929 (1978).
307. N. L. Weinberg and A. K. Hoffman, Can. J. Chem., 49, 740 (1971).
308. A. J. Bloom, M. Fleischmann, and J. M. Mellor, J. Chem. Soc., Perkin Trans. I, 79 (1986).
309. A. Bewick, D. E. Coe, J. M. Mellor, and D. J. Walton, J. Chem. Soc., Chem. Commun., 51 (1980).
310. A. Bewick, D. E. Coe, J. M. Mellor, and W. M. Owton, J. Chem. Soc., Perkin Trans. I, 1033 (1985).
311. Y. Takasu, T. Matsuda, A. Shimizu, M. Harada, M. Saito, and H. Watanabe, Denki Kagaku, 50, 975 (1982); Chem. Abstr., 99, 60830z (1983).
312. P. Pouillen, R. Minko, M. Verniette, P. Martinet, and M. A. M. Martre, Electrochim. Acta, 24, 1189 (1979).
313. S. Torii, K. Uneyama, H. Tanaka, T. Yamanaka, T. Yasuda, M. Ono, and Y. Kohmoto, J. Org. Chem., 46, 3312 (1981).
314. K. G. Ellis and R. E. W. Jansson, J. Appl. Electrochem. 13, 657 (1983).
315. L. Eberson and B. Helgée, Acta Chem. Scand., B32, 313 (1978).
316. G. Cauquis and G. Pierre, Tetrahedron, 34, 1475 (1978).
317. S. Torii, K. Uneyama, and M. Ono, Tetrahedron Lett., 21, 2741 (1980).
318. T. Inoue and S. Tsutsumi, J. Amer. Chem. Soc., 87, 3525 (1965).
319. R. Engels, H. J. Schäfer, and E. Steckhan, Liebigs Ann. Chem., 204 (1977).
320. J. M. Bobbitt, I. Noguchi, H. Yagi, and K. H. Weisgraber, J. Org. Chem., 41, 845 (1976).
321. Y. Shizuri, K. Nakamura and S. Yamamura, J. Chem. Soc., Chem. Commun., 530 (1985).
322. K. Uneyama, A. Isimura, and S. Torii, Bull. Chem. Soc. Jpn., 58, 1859 (1985).
323. J. Delaunay, A. Orliac-Le Moing, J. Simonet, and L. Toupet, Tetrahedron Lett., 27, 6205 (1986).
324. J. Mlcoch and E. Steckhan, Tetrahedron Lett., 28, 1081 (1987).

325. N. K. Pattanayak and B. Nayak, Makromol. Chem., 187, 1337 (1986).
326. H. Baltes, L. Stork, and H. J. Schäfer, Liebigs Ann. Chem., 818 (1979).
327. W. Ando, Y. Kabe, and T. Takata, J. Amer. Chem. Soc., 104, 7314 (1982).
328. P. Bird, G. G. Black, H. Lloyd, M. Sainsbury, and D. I. C. Scopes, J. Chem. Soc., Perkin Trans. I, 1197 (1986).
329. H. Schäfer and A. Alazrak, Angew. Chem., Int. Ed., 7, 474 (1968).
330. H. Kanetsuna and T. Nonaka, Denki Kagaku, 49, 100 (1981); Chem. Abstr., 95, 61280z (1981).
331. G. Cauquis and M. Genies, Tetrahedron Lett., 3959 (1971).
332. M. Tokuda, Y. Yamada, T. Takagi, and H. Suginome, Tetrahedron Lett., 26, 6085 (1985).
333. T. Shono, Y. Matsumura, S. Katoh, K. Inoue, and Y. Matsumoto, Tetrahedron Lett., 27, 6083 (1986).
334. J. Šrogl, M. Janda, I. Stibor, J. Kos, and V. Vyskočil, Coll. Czech. Chem. Commun., 43, 2015 (1978).
335. M. Némec, M. Janda, and J. Šrogl, Coll. Czech. Chem. Commun., 38, 3857 (1973).
336. I. Izumi and I. Yamamoto, Kenkyu Kiyo-Nara Kogyo Koto Senmon Gakko, 18, 75 (1982); Chem. Abstr., 99, 52862w (1983).
337. M. Kimura, N. Kuriki, M. Inaishi, and Y. Sawaki, Tetrahedron Lett., 45, 4665 (1984).
338. R. S. Glass, A. Petsom, and G. S. Wilson, J. Org. Chem., 51, 4337 (1986).
339. M. Kimura, S. Matsubara, Y. Sawaki, and H. Iwamura, Tetrahedron Lett., 27, 4177 (1986).
340. J. Yoshida, T. Murata, and S. Isoe, Tetrahedron Lett., 28, 211 (1987).
341. S. Torii, T. Inokuchi, S. Takagishi, H. Horike, and F. Akahoshi, 127th Electrochem. Soc. Meeting, Abstract 1346 (1987).
342. T. T. Tsou and J. K. Kochi, J. Amer. Chem. Soc., 101, 6319 (1979).
343. P. D. Frisch, M. K. Lloyd, J. A. McCleverty, and D. Seddon, J. Chem. Soc., Dalton Trans., 2268 (1973).
344. J. A. de Beer and R. J. Haines, J. Chem. Soc., Dalton Trans., 2341 (1973).
345. W. N. Rogers, J. A. Page, and M. C. Baird, Inorg. Chem., 20, 3521 (1981).
346. R. Annino, R. E. Erickson, J. Michalovic, and B. McKay, J. Amer. Chem. Soc., 88, 4424 (1966).
347. C. K. Mann, J. L. Webb, and H. M. Walborsky, Tetrahedron Lett., 2249 (1966).
348. J. L. Webb, C. K. Mann, and H. M. Walborsky, J. Amer. Chem. Soc., 92, 2042 (1970).
349. S. Jaouannet, R. Hazard, and A. Tallec, J. Electroanal. Chem., 111, 397 (1980).
350. R. Hazard, S. Jaouannet, E. Raoult, and A. Tallec, Nouv. J. Chim., 6, 325 (1982).
351. B. Czochralska, Chem. Phys. Lett., 1, 239 (1967).
352. T. Nonaka, T. Ota, and K. Odo, Bull. Chem. Soc. Jpn., 50, 419 (1977).
353. B. H. M. Billinge and B. G. Gowenlock, J. Chem. Soc. (C) 1201 (1962).

354. R. E. Dessey, W. Kitching, T. Psarras, R. Salinger, A. Chen, and T. Chivers, J. Amer. Chem. Soc., 88, 460 (1966).
355. R. E. Erickson and C. M. Fischer, J. Org. Chem., 35, 1604 (1970).
356. T. Nonaka, T. Ota, and T. Fuchigami, Bull. Chem. Soc. Jpn., 50, 2965 (1977).
357. L. Horner and R.-J. Singer, Chem. Ber., 101, 3329 (1968).
358. L. Horner and H. Neumann, Chem. Ber., 98, 3462 (1965).
359. L. Horner and H. Fuchs, Tetrahedron Lett., 203 (1962).
360. L. Horner and H. Fuchs, Tetrahedron Lett., 1573 (1963).
361. L. Horner, H. Fuchs, H. Winkler, and A. Rapp, Tetrahedron Lett., 965 (1963).
362. E. S. Wallis and F. H. Adams, J. Amer. Chem. Soc., 55, 3838 (1933).
363. L. Eberson, G. Ryde-Petterson, Acta Chem. Scand., B27, 1159 (1973).
364. L. Eberson, Electrochim. Acta, 12, 1473 (1967).
365. L. Eberson, K. Nyberg, R. Servin, and I. Wennerbeck, Acta Chem. Scand., B30, 186 (1976).
366. L. Eberson, K. Nyberg, and R. Servin, Acta Chem. Scand., B30, 906 (1976).
367. W. B. Smith and H.-G. Gilde, J. Amer. Chem. Soc., 81, 5325 (1959).
368. W. B. Smith and H.-G. Gilde, J. Amer. Chem. Soc., 83, 1355 (1961).
369. J. Takahashi, K. Mori, and M. Matsui, Agric. Biol. Chem., 43, 1605 (1979); Chem. Abstr., 91, 174752r (1979).
370. D. Seebach and P. Renaud, Helv. Chim. Acta, 68, 2342 (1985).
371. P. Renaud and D. Seebach, Angew. Chem., 98, 836 (1986).
372. P. Renaud and D. Seebach, Helv. Chim. Acta, 69, 1704 (1986).
373. T. Shono, Y. Matsumura, J. Hayashi, and M. Usui, Denki Kagaku, 51, 131 (1983); Chem. Abstr., 98, 187943y (1983).
374. T. Shono, Y. Matsumura, K. Tsubata, and K. Uchida, J. Org. Chem., 51, 2590 (1986).
375. R. N. Gourley, J. Grimshaw, and P. G. Millar, J. Chem. Soc., Chem. Commun., 1278 (1967).
376. R. N. Gourley, J. Grimshaw, and P. G. Millar, J. Chem. Soc. (C), 2318 (1970).
377. E. Raoult, J. Sarrazin, and A. Tallec, Bull. Soc. Chim. Fr., II, 420 (1981).
378. E. Kariv, H. A. Terni, and E. Gileadi, J. Electrochem. Soc., 120, 639 (1973).
379. E. Kariv, H. A. Terni, and E. Gileadi, Electrochim. Acta, 18, 433 (1973).
380. M. Jubault, E. Raoult, and D. Peltier, Electrochim. Acta, 19, 865 (1974).
381. M. Jabault, E. Raoult, and D. Peltier, Electrochim. Acta, 21, 407 (1976).
382. J. Hermolin, J. Kopilov, and E. Gileadi, J. Electroanal. Chem., 71, 245 (1976).
383. J. Kopilov, E. Kariv, and L. L. Miller, J. Amer. Chem. Soc., 99, 3450 (1977).
384. K. Köster, P. Piemenschneider, and H. Wendt, Israel J. Chem., 18, 141 (1979).
385. H. Wendt, Angew. Chem., Int. Ed. Engl., 21, 256 (1982).
386. K. Köster and H. Wendt, J. Electroanal. Chem., 138, 209 (1982).
387. K. Köster and H. Wendt, J. Electroanal. Chem., 157, 113 (1983).
388. M. Jubault, E. Raoult, and D. Peltier, Electrochim. Acta, 25, 1359 (1980).

389. M. Jubault, J. Chem. Soc., Chem. Commun., 953 (1980).
390. M. Jubault and E. Raoult, J. Chem. Soc., Chem. Commun., 250 (1977).
391. M. Jubault, E. Raoult, and D. Peltier, J. Chem. Soc., Chem. Commun., 232 (1979).
392. M. Jubault, A. Lebouc, and A. Tallec, Electrochim. Acta, 27, 1339 (1982).
393. R. Hazard, S. Jaouannet, and A. Tallec, Tetrahedron Lett., 1105 (1979).
394. R. Hazard, S. Jaouannet, and A. Tallec, Tetrahedron, 38, 93 (1982).
395. A. Tallec, R. Hazard, A. Le Bouc, and J. Grimshaw, J. Chem. Res. (S), 342 (1986).
396. R. Bennes, M. Jubault, M. Privat, and P. Vanel, Electrochim. Acta, 26, 1079 (1981).
397. E. Tronel-Peyroz and D. Schuhmann, J. Electroanal. Chem., 129, 265 (1981).
398. J. Sarrazin, J. Simonet, and A. Tallec, Electrochim. Acta, 27, 1763 (1982).
399. K. Köster, H. Wendt, and A. Lebouc, J. Electroanal. Chem., 157, 89 (1983).
400. T. Nonaka, Y. Kumai, M. Ito, and T. Fuchigami, Asahi Glass Found. Res. Rep., 32, 155 (1978); Chem. Abstr., 91, 98972v (1979).
401. L. Horner and D. Degner, Tetrahedron Lett., 5889 (1968).
402. L. Horner and D. Degner, Tetrahedron Lett., 1245 (1971).
403. L. Horner, D. Degner, and D. Skaletz, Chem.-Ing.-Tech., 44, 209 (1972).
404. L. Horner and R. Schnedier, Tetrahedron Lett., 3133 (1973).
405. L. Horner and D. Degner, Electrochim. Acta, 19, 611 (1974).
406. D. Brown and L. Horner, Liebigs Ann. Chem., 77 (1977).
407. L. Horner and W. Brich, Liebigs Ann. Chem., 710 (1978).
408. L. Horner and W. Brich, Chem. Ber., 111, 574 (1978).
409. W. J. M. van Tilborg and C. J. Smit, Rec. J. Roy. Netherl. Chem. Soc., 97, 89 (1978).
410. L. Horner and D. H. Skaletz, Tetrahedron Lett., 3679 (1970).
411. L. Horner and D. H. Skaletz, Liebigs Ann. Chem., 1365 (1977).
412. D. Seebach and H. A. Oei, Angew. Chem. Int. Ed. Engl., 14, 634 (1975).
413. D. Seebach, H.-A. Oei, and H. Daum, Chem. Ber., 110, 2316 (1977).
414. T. Nonaka, unpublished observation.
415. B. F. Watkins, J. R. Behling, E. Kariv, and L. L. Miller, J. Amer. Chem. Soc., 97, 3549 (1975).
416. B. E. Firth, L. L. Miller, M. Mitani, T. Rogers, J. Lennox, and R. W. Murray, J. Amer. Chem. Soc., 98, 8271 (1976).
417. B. E. Firth and L. L. Miller, J. Amer. Chem. Soc., 98, 8272 (1976).
418. L. Horner and W. Brich, Liebigs Ann. Chem., 1354 (1977).
419. M. Fujihira, A. Yokozawa, H. Kinoshita, and T. Osa, Chem. Lett., 1089 (1982).
420. A. Yokozawa, M. Fujihira, and T. Osa, 4th Electroorg. Chem. Symp., Abstract 26 (1983).
421. A. Abe, T. Nonaka, and T. Fuchigami, J. Amer. Chem. Soc., 105, 3630 (1983).
422. T. Nonaka, S. Abe, and T. Fuchigami, Bull. Chem. Soc. Jpn., 56, 2778 (1983).

423. S. Abe, T. Fuchigami, and T. Nonaka, Chem. Lett., 1033 (1983).
424. S. Abe and T. Nonaka, Chem. Lett., 1541 (1983).
425. T. Komori and T. Nonaka, J. Amer. Chem. Soc., 105, 5690 (1983).
426. T. Komori and T. Nonaka, J. Amer. Chem. Soc., 106, 2656 (1984).
427. T. Nonaka, Hyomen, 25, 283 (1987).
428. A. Yamagishi and A. Aramata, J. Electroanal. Chem., 191, 449 (1985).
429. J. Berthelot, M. Jubault, and J. Simonet, Electrochim. Acta, 28, 1719 (1983).
430. M. Jubault, E. Raoult, and D. Peltier, Electrochim. Acta., 22, 67 (1977).
431. T. Nonaka, M. Ito, and T. Fuchigami, Asahi Glass Found. Res. Rep., 34, 227 (1979); Chem. Abstr., 95, 24103y (1981).
432. H. Tang, W. Chai, L. Zhang, H. Jing, Y. Geng, L. Fan, O. Zhang, and H. Peng, Yaoxue Xuebao, 16, 440 (1981); Chem. Abstr., 97, 91844v (1982).
433. V. Montiel, M. López-Segura, and A. Aldaz, Electrochim. Acta, 29, 1123 (1984).
434. T. Komori, T. Nonaka, and T. Fuchigami, Chem. Lett., 11 (1986).
435. T. Komori, T. Nonaka, T. Fuchigami, and K. Zhang, Bull. Chem. Soc. Jpn., 60, 3315 (1987).
436. R. Hazard, S. Jaouannet, and A. Tallec, Bull. Soc. Chem. Fr., II, 263 (1983).
437. S. Kwee and H. Lund, J. Electroanal. Chem., 116, 693 (1980).
438. M. Jubault, E. Raoult, and D. Peltier, Electrochim. Acta, 26, 287 (1981).
439. T. Fuchigami, A. Sato, and T. Nonaka, Chem. Express, 1, 363 (1986).
440. T. Shono, H. Ohmizu, and N. Kise, Tetrahedron Lett., 23, 4801 (1982).
441. H. Simon, H. Günther, J. Bader, and W. Tischer, Angew. Chem., 93, 897 (1981).
442. F. Takahashi, K. Tomii, and H. Takahashi, Electrochim. Acta, 31, 127 (1986).
443. H. Sugiyama, Kagaku-to-Kogyo, 40, 155 (1987).
444. T. Komori and T. Nonaka, Chem. Lett., 509 (1984).
445. A. Yamagishi and A. Aramata, J. Chem. Soc., Chem. Commun., 452 (1984).
446. K. Yoshinaga, T. Kito, and K. Ohkubo, Nippon Kagaku Kaishi, 165 (1986); Chem. Abstr., 104, 195273n (1986).
447. L.-S. R. Yeh and A. J. Bard, J. Electrochem. Soc., 124, 189 (1977).
448. F. L. Wimmer, M. R. Snow, and A. M. Bond, Inorg. Chem., 13, 1617 (1974).
449. A. M. Bond, R. Colton, and J. J. Jackowski, Inorg. Chem., 14, 274 (1975).
450. R. D. Rieke, H. Kojima, and K. Öfele, J. Amer. Chem. Soc., 98, 6735 (1976).
451. A. M. Bond, R. Colton, and M. J. McCormick, Inorg. Chem., 16, 155 (1977).
452. A. M. Bond, B. S. Grabaric, and Z. Grabaric, Inorg. Chem., 17, 1013 (1978).
453. A. M. Bond, B. S. Grabaric, and J. J. Jackowski, Inorg. Chem., 17, 2153 (1978).
454. A. M. Bond, R. Colton, and M. E. McDonald, Inorg. Chem., 17, 2842 (1978).

455. R. D. Rieke, H. Kojima, and K. Öfele, Angew. Chem., 92, 550 (1980); Int. Ed. Engl., 19, 538 (1980).
456. A. M. Bond, D. J. Darensbourg, E. Mocellin, and B. J. Stewart, J. Amer. Chem. Soc., 103, 6827 (1981).
457. K. A. Conner and R. A. Walton, Organometallics, 2, 169 (1983).
458. J. Courtot-Coupez, M. Guéguen, J. E. Guerchais, F. Y. Pétillon, J. Talarmin, and R. Mercier, J. Organometal. Chem., 312, 81 (1986).
459. J. Moraczewski and W. E. Geiger, Jr., J. Amer. Chem. Soc., 101, 3407 (1979).
460. B. Tulyathan and W. E. Geiger, Jr., J. Electroanal. Chem., 109, 325 (1980).
461. J. Moraczewski and W. E. Geiger, Jr., J. Amer. Chem. Soc., 103, 4779 (1981).
462. G. Lane and W. E. Geiger, Organometallics, 1, 401 (1982).
463. J. Moraczewski and W. E. Geiger, Jr., J. Amer. Chem. Soc., 100, 7429 (1978).
464. T. A. Albright, W. E. Geiger, Jr., J. Moraczewski, and B. Tulyathan, J. Amer. Chem. Soc., 103, 4787 (1981).
465. M. Grzeszczuk, D. E. Smith, and W. E. Geiger, Jr., J. Amer. Chem. Soc., 105, 1772 (1983).
466. Z. R. Grabowski and M. S. Balasiewicz, Trans. Faraday Soc., 64, 3346 (1968).
467. B. A. Olsen and D. H. Evans, J. Amer. Chem. Soc., 103, 839 (1981).
468. E. Ahlberg, O. Hammerich, and V. D. Parker, J. Amer. Chem. Soc., 103, 844 (1981).
469. P. Neta and D. H. Evans, J. Amer. Chem. Soc., 103, 7041 (1981).
470. O. Hammerich and V. D. Parker, Acta Chem. Scand., B35, 395 (1981).
471. B. A. Olsen, D. H. Evans, and I. Agranat, J. Electroanal. Chem., 136, 139 (1982).
472. D. H. Evans and X. Naixian, J. Electroanal. Chem., 133, 367 (1982).
473. D. H. Evans and R. W. Busch, J. Amer. Chem. Soc., 104, 5057 (1982).
474. D. H. Evans and N. Xie, J. Amer. Chem. Soc., 105, 315 (1983).
475. T. Matsue and D. H. Evans, J. Electroanal. Chem., 163, 137 (1984).
476. D. H. Evans and A. Fitch, J. Amer. Chem. Soc., 106, 3039 (1984).
477. B. B. Damaskin, O. A. Petrii, and V. V. Batrakov, Adsorption of Organic Compounds on Electrodes, Plenum Press, New York, 1971.
478. S. Wawzonek, R. C. Reck, W. W. Vaught, Jr., and J. W. Fan, J. Amer. Chem. Soc., 67, 1300 (1945).
479. R. Pasternak, Helv. Chim. Acta, 31, 753 (1948).
480. P. J. Elving and C. Teitelbaum, J. Amer. Chem. Soc., 71, 3916 (1949).
481. P. J. Elving and I. Rosenthal, Anal. Chem., 26, 1454 (1954).
482. R. Fujita and T. Sekine, Denki Kagaku, 50, 491 (1982); Chem. Abstr., 97, 100523u (1982).
483. H. Kita, T. Nakamura, H. Itoh, and H. Kano, Electrochim. Acta, 23, 405 (1978).
484. J. P. Petrovich, J. D. Anderson, and M. M. Baizer, J. Org. Chem., 31, 3897 (1966).
485. K. W. Bowers, R. W. Giese, J. Grimshaw, H. O. House, N. H. Kolodny, K. Kronberger, and D. K. Roe, J. Amer. Chem. Soc., 92, 2783 (1970).

486. G. Le Guillanton and M. Cariou, Electrochim. Acta, 22, 619 (1977).
487. M. Cariou, G. Mabon, and G. Le Guillanton, Tetrahedron, 39, 1551 (1983).
488. S. Wawzonek, E. W. Blaha, R. Berkey, and M. E. Runner, J. Electrochem. Soc., 102, 235 (1955).
489. L. B. Anderson, M. J. Broadhurst, and L. A. Paquette, J. Amer. Chem. Soc., 95, 2198 (1973).
490. A. M. Wilson and N. L. Allinger, J. Amer. Chem. Soc., 83, 1999 (1961).
491. F. L. Lambert, A. H. Albert, and J. P. Hardy, J. Amer. Chem. Soc., 86, 3155 (1964).
492. N. Limosin and E. Laviron, Bull. Soc. Chim. Fr., 4189 (1969).
493. V. Delaroff, M. Bolla, and M. Legrand, Bull. Soc. Chim. Fr., 1912 (1961).
494. P. Kabasakalian and J. McGlotten, Anal. Chem., 34, 1440 (1962).
495. M. G. McKeon, J. Electroanal. Chem., 3, 402 (1962).
496. J. Závada, J. Krupicka, and J. Sicher, Coll. Czech. Chem. Commun., 28, 1664 (1963).
497. M. R. Rifi, Coll. Czech. Chem. Commun., 36, 932 (1971).
498. S. F. Nelsen, E. F. Travecedo, and E. D. Seppanen, J. Amer. Chem. Soc., 93, 2913 (1971).
499. A. J. Klein and D. H. Evans, J. Amer. Chem. SOc., 101, 757 (1979).
500. K. M. O'Connell and D. H. Evans, J. Amer. Chem. Soc., 105, 1473 (1983).
501. M. R. Rifi, Tetrahedron Lett., 1043 (1969).
502. I. Rosenthal, J. R. Hayes, A. J. Martin, and P. J. Elving, J. Amer. Chem. Soc., 80, 3050 (1958).
503. S. G. Mairanovskii and L. D. Bergel'son, Russ. J. Phys. Chem., 34, 112 (1960); Chem. Abstr., 55, 21913g (1961).
504. N. Gatti, S. U. Pedersen, and H. Lund, Acta Chem. Scand., B42, 11 (1988).
505. G. P. Kugatova-Shemyakina, G. M. Nikolaev, S. G. Mairanovskii, and N. V. Surikova, Tetrahedron Lett., 1725 (1971).
506. H. Lund and N. H. Nilsson, Acta Chem. Scand., B30, 5 (1976).
507. C. Degrand and G. Belot, Electrochim. Acta, 23, 71 (1978).
508. E. Laviron and Y. Mugnier, J. ELectroanal. Chem., 111, 337 (1980).
509. R. E. Dessy, W. Kitching, T. Psarras, R. Salinger, A. Chen, and T. Chivers, J. Amer. Chem. Soc., 88, 460 (1966).
510. R. D. Rieke, S. N. Milligan, and L. D. Schulte, Organometallics, 6, 699 (1987).
511. D. R. Arnold and P. C. Wong, Can. J. Chem., 57, 2098 (1979).
512. T. Nonaka, K. Abe, and T. Sekine, Denki Kagaku, 47, 184 (1979); Chem. Abstr., 91, 65223y (1979).
513. T. Shono, Y. Matsumura, T. Hashimoto, K. Hibino, H. Hamaguchi, and T. Aoki, J. Amer. Chem. Soc., 97, 2546 (1975).
514. A. Yoshiyama, T. Goto, and T. Sekine, Denki Kagaku, 49, 532 (1981); Chem. Abstr., 96, 13015r (1982).
515. T. Fuchigami, T. Nonaka, C. Watanabe, A. Yoshiyama, and T. Sekine, Denki Kagaku, 51, 812 (1983); Chem. Abstr., 100, 164026x (1984).
516. M. Masui, Y. Kamada, E. Sasaki, and S. Ozaki, Chem. Pharm. Bull., 30, 1234 (1982).
517. S. F. Nelsen, L. Echegoyen, and D. H. Evans, J. Amer. Chem. Soc., 97, 3530 (1975).

518. A. F. Diaz, O. J. Scherer, and K. Andres, J. Chem. Soc., Chem. Commun., 982 (1980).
519. R. S. Glass, J. R. Duchek, J. T. Klug, and G. S. Wilson, J. Amer. Chem. Soc., 99, 7349 (1977).
520. G. S. Wilson, D. D. Swanson, J. T. Klug, R. S. Glass, M. D. Ryan, and W. K. Musker, J. Amer. Chem. SOc., 101, 1040 (1979).
521. R. S. Glass, B. R. Coleman, U. D. G. Prabhu, W. N. Setzer, and G. S. Wilson, J. Org. Chem., 47, 2761 (1982).

VIII

INDIRECT REACTIONS

28
Amalgam and Related Reductions

HENNING LUND / Aarhus University, Aarhus, Denmark

I. Introduction 1199
II. Olefinic and Aromatic Compounds 1200
III. Halogen Compounds 1204
IV. Organomercurials 1205
V. Nitro Compounds and Derivatives 1206
VI. Carbonyl Compounds and Derivatives 1206
VII. Acids and Derivatives 1208
VIII. Cleavage Reactions (Other Than Carbon-Halogen) 1209
IX. Polymerization Reactions 1211
References 1211

I. INTRODUCTION

Amalgams are alloys between mercury and a metal; however, compounds that are not proper amalgams, but function in a similar way as amalgams are also treated in this chapter. Thus, "alloys" between tetraalkylammonium and mercury or some other metals and also some interstitial compounds, for example with graphite, are included in this review, even if these compounds have only a limited lifetime and are formed only as intermediates during direct electrolysis.

Amalgams function as electron donors in the same manner as a mercury electrode, and in many cases a similar product distribution is obtained by direct electrolysis at a mercury electrode and by an amalgam reduction. However, simultaneously with the electron transfer through the double layer a transfer of cations takes place, and this may influence the reaction, for example by ion-pair formation.

Compared with direct electrolysis, reduction via an electrolytically produced amalgam may have some advantages. The preparation of an amalgam is more easily coupled with a useful anodic process, such as production of chlorine, than is direct electrolysis. Often a higher current density can be employed. If the production of and reduction by amalgam are performed separately, a higher degree of freedom is possible with respect to choice of solvent, temperature, and pressure in the second step. Amalgam reductions may be run either as batch operations or continuous processes.

It is, however, more difficult to change the reducing power of an amalgam than to change the potential of an electrode, but by working with different metals and different concentrations of metal and by varying the rate of stirring (which influences the size of the amalgam droplets and thus the effective surface of the amalgam) it is possible to control within certain limits the degree of reduction by an amalgam.

The oxidation potential of an amalgam is related to its reducing power [1–4]; the oxidation potentials in N-methylpyrrolidone (NMP) of the metal amalgams becomes more negative in the sequence Ca > Sr > Br > K ~ Na > Li [3], and in agreement with this, Ca(Hg) is the weakest and Li(Hg) the strongest reducing agent among the amalgams mentioned. Lithium amalgam is thus the only one able to reduce in NMP 8-hydroxyquinoline to its dianion [3].

Sodium amalgam is the most widely employed amalgam, but sometimes other amalgams may be used to advantage. Aluminum amalgam is thus useful for reduction of alkali-sensitive compounds, but it cannot be prepared in situ by electrolysis of aqueous solutions of aluminum salts. Calcium amalgam has been employed when a weakly reducing agent is desired.

Tetraalkylammonium amalgams [TAA(Hg)] [5–9], especially N,N-dimethylpyrrolidinium amalgam [DMP(Hg)] [7–9], have been shown to be useful intermediates in the electrochemical reduction of certain substrates that are reducible only with difficulty. The nature of these compounds, whose empirical formula has been given as $R_4N^+Hg_4^-$ [10] or DMP(Hg_5) [11], is not clear; the original term "amalgam" is not appropriate for a stochiometric compound [11], and they have been suggested [10] to be Zintl ion salts [12,13].

DMP(Hg) is formed in a reversible one-electron process on reduction of $DMP^+BF_4^-$ in DMF or diglyme at mercury electrode; the cyclic voltammogram of DMP is analogous to that of a metal forming an amalgam. DMP(Hg) is insoluble in DMF but reacts with DMF with a half-life of about 6 h at 5°C. DMP(Hg) is especially useful in promoting one-electron reductions; its reducing ability can be shown by the enhancement of the peak current in CV and limiting current in polarography on addition of certain substrates that are reducible only with difficulty.

Similar compounds with quaternary ammonium cations may be formed with other metals, such as lead, tin, antimony, and bismuth [14]; furthermore, TAA ions may make interstitial compounds with graphite [15,16] that have reducing properties similar to those of the tetraakylammonium metals.

II. OLEFINIC AND AROMATIC COMPOUNDS

Reduction of C=C double bonds may result either in saturation of the double bond or in hydrodimerization; the latter reaction is most common in compounds that can act as Michael acceptors.

Reduction of an unpolarized C=C double bond is exemplified in the conversion of biphenylenediphenylethylene to the corresponding ethane [17]. The reduction was performed in boiling amyl alcohol that is poorly suited for a direct electrolysis but adequate for dissolving the substrate.

The reduction of cinnamylidene fluorene by aluminum amalgam in moist ether to 1-fluorenyl-3-phenyl-1-propene (I), a 1,4-addition, exemplifies the use of Al(Hg) rather than Na(Hg) for the reduction of alkali-sensitive compounds; in the presence of Na(Hg) I might be tautomerized to 3-phenyl-1-propylidene fluorene [18]. Another example is the reduction of a cumulene, 1,1,6,6-tetraphenyl-1,2,3,4,5-hexapentaene to 1,1,6,6-tetraphenyl-1,2,4,5-hexatetraene by Al(Hg) [19].

Of more interest than the saturation of double bonds is the hydrodimerization reaction (Chap. 22); sodium amalgam hydrodimerization of α,β-unsaturated ketones were reported early [20], and the product distribution resembled that obtained from direct electrolytic reduction (Chap. 10).

The first sodium amalgam-induced dimerization of acrylonitrile to adiponitrile [21] was performed in aqueous solution, and the yield was discouragingly low (5%). Mainly, a base-induced addition of water resulting in β,β'-bis(cyanoethyl) ether occurred together with a reduction to propionitrile.

Higher yields (30–60%) of a hydrodimerization with alkali metal amalgams were later reported [22,23] in the reduction of α,β-unsaturated esters and nitriles. The different alkali metal amalgams often gave different yields of hydrodimerized products; the hydrodimerization of acrylonitrile using potassium amalgam thus yielded 62% adiponitrile, whereas sodium amalgam under similar conditions gave only 37%.

After the development of the electrolytic reductive coupling of acrylonitrile to adiponitrile (Chap. 22), many investigations have been directed toward developing an alternative, indirect electrolytic process. The electrolytic hydrodimerization reaction in neutral solution [24,25] is critically dependent on the proton activity at the electrode. If the proton activity is too high, the product is predominantly propionitrile, whereas oligo- or polymerization occurs at too low a proton activity. The establishment of a reaction layer with a suitable proton activity in which conditions are favorable for a hydrodimerization of acrylonitrile to adiponitrile is, thus, of paramount importance. This can be accomplished in different ways.

One possibility is to add quaternary ammonium salts to an aqueous reaction mixture; the positive ions accumulate in the double layer [26–29] around the amalgam and establish here a layer of low water content. The protonation may, thus, be slowed down sufficiently so the dimerization can compete successfully with it.

The unavoidable presence of sodium ions in the double layer during a Na(Hg) reduction introduces differences compared to the electrolytic hydrodimerization with respect to hydration and ion pairing.

The effect of the quaternary ammonium salt depends also on the substrate and on the electrode material [30]. Addition of tetraethylammonium salts thus sharply increases the yield of dimer product from acrylonitrile, has little effect upon the hydrodimerization of methyl acrylate, and does not affect the hydrodimerization of mesityl oxide.

It has been suggested [31] that quaternary ammonium amalgams are intermediates in the hydrodimerization of AN in an acetonitrile-water-CO_2 medium; other explanations [28,29,32], however, have also been considered. A layer of quaternary ammonium (Q^+) salts could first be produced by anodic dissolution of a sufficient amount Q^+ from the amalgam.

Another way of producing a low proton activity at the amalagam surface is to add a suitable polar, aprotic compound to the medium. Suitable compounds include urea and its derivatives, amides, lactams, sulfoxides, and sulfones, such as 1,3-dimethylurea, ethyleneurea, biuret, formamide, _N_-methyl formamide, acetamide, δ-butyrolactam, dimethyl sulfoxide, di-_n_-butyl sulfoxide, and dimethyl sulfone [33,34]. DMF or acetonitrile, besides some of the compounds just mentioned, may be used as solvents [35,36]. One function of the added compound is to establish an aprotic reaction layer at the amalgam surface by selective adsorption of the compound on the amalgam;

another is to solvate the ion pair (consisting of the radical ion and the cation) in a suitable way.

Hydrodimerization of AN by Na(Hg) may also take place in liquid ammonia [37] in the presence of a suitable proton donor; the dimerization is faster than the amination. If water is used as proton donor, ADN is formed in 80% yield, but when NH_4Cl is used, the yield is raised to nearly 100%. The ratio of PN to ADN depends on the ratio between NH_4Cl and AN; when the ratio is smaller than unity, some polymer is formed because there is not enough proton donor to terminate the anionic polymerization.

In the hydrodimerization of methacrylonitrile (MeAN) by Na(Hg) in liquid NH_3, NH_4Cl is too strong a proton donor; a mixture of isobutyronitrile (IBM) and dimethyladiponitrile (diMeADN) is formed. Phenol is a suitable weak proton donor in this case, and when the ratio of phenol to monomer is unity, only hydrodimerization is observed.

The ratio of the apparent rate constants of the formation of ADN and diMeADN from the monomers is about 2 in NH_3, whereas it is about 100 in Diglyme [38] or Diglyme-DMSO; this means that on cross-coupling of AN and MeAN in NH_3, all the possible combinations are obtained, whereas in Diglyme-DMSO no diMeADN is found. An explanation for this difference may be that the monomer in phenol-NH_3 exists in a protonated form.

Many patents have been issued to groups working at ICI, UCB, and the Weizmann Institute on the amalgam hydrodimerization of AN and related compounds [39].

Reaction between Na(Hg) and acrylic acid (AA) in water, dioxane, or Diglyme produces propionic acid (PA); however, when DMSO (containing about 10% water) is used as solvent, adipic acid (ADA) is the main product [40].

To achieve a high conversion it is necessary to add hydrochloric acid during the reaction. Under such conditions it is possible to obtain 70% ADA and 20–25% PA. Addition of HCl increases the rate of the reaction but not the ratio of PA to ADA; this is because it is the free acid, not the acrylate ion, that is reduced. This is in accordance with the less negative reduction potential of the free acid compared with that of the acrylate ion.

The hydrodimerization of acrylic acid to adipic acid is a tail-to-tail addition; hydrodimerization of methacrylic acid and crotonic acid produces mainly the head-to-head and head-to-tail dimers [40].

Another example of this type of reaction is the hydrodimerization of dialkylfulvenes to dicyclopentadienyl alkanes by alkali metal amalgams in an aqueous-alcoholic medium [41,42].

Polycyclic aromatic hydrocarbons, such as anthracene (II) and phenanthrene, can be reduced at elevated temperatures (80°C) at constant current to different hydroderivatives in an aqueous medium containing $(TBA^+)OH^-$ (2.1 M). The following products were obtained in yields depending on the number of F mol^{-1} and the cell construction: 9,10-dihydroanthracene (III), 1,4,5,8,9,10-hexahydroanthracene (IV), 1,2,3,4,5,6-hexahydroanthracene (V), 1,2,3,4,5,6,7,8-octahydroanthracene (VI), and 1,2,3,4,5,6,7,8,9,10-decahydroanthracene (VII). Using more than 10 F mol^{-1} IV was the main product. Further reduction of IV to VI and VII could be obtained after isomerization of IV to V by reflux in the alkaline solution for 48 h, followed by further electrolysis [43].

Phenanthrene can be reduced under similar conditions; no isomerization step is necessary to obtain octahydro- and decahydrophenanthrene [43].

Benzene [44], anisole, 1,2,3,4-tetrahydro-6-methoxynaphthalene [45], and estradiol-3-methyl ether [46] have been reduced in a similar cathodic Birch reduction in water [44], aqueous diglyme, and aqueous tetrahydrofuran with TBA^+ as electrolyte at room temperature; TBA(Hg) has been proposed [45,46] to be a mediator in the reduction, Eq. (1):

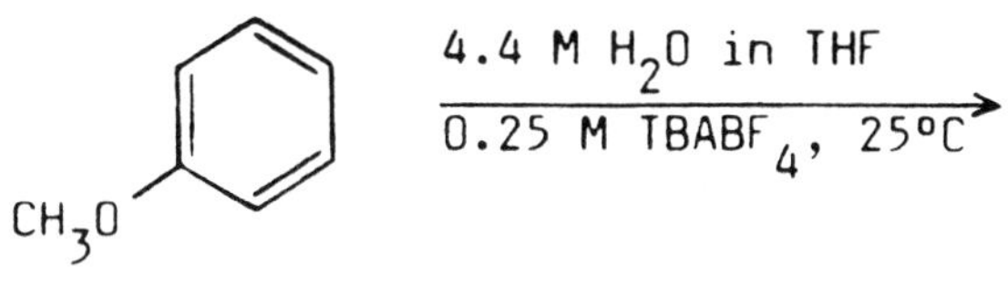

(1)

Alkynes may be reduced to alkenes using aqueous THF-$TBABF_4$; thus 1-hexyne gave 93% 1-hexene using 2.8 F mol^{-1}; however, attempts to reduce 17-α-ethynylestradiol-3-methyl ether (VIII) to the alkene without reducing the aromatic ring failed, but in contrast to the classic Birch reduction the ethynyl group was not lost, Eq. (2) [46]:

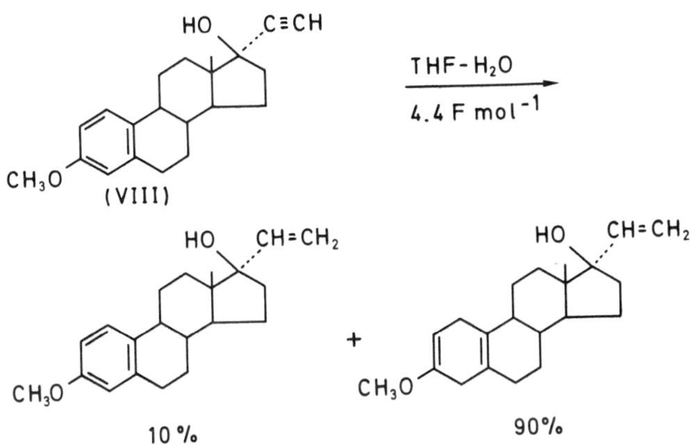

(2)

Reductions in aqueous THF-$TBABF_4$ of benzo[b]thiophen [47], benzofurans, and 1-methylindole [48] have been shown to yield the 2,3-dihydro compounds; this is probably a direct electrochemical reduction, whereas the further reduction involving ring opening of the hetero ring seems to be mediated by TBA(Hg).

The nucleus of certain phenolic compounds can be reduced by Na(Hg) in slightly acid solution containing boric acid. The latter probably acts as a complexing agent besides maintaining a suitable pH.

Resorcinol can be reduced in the nucleus by Na(Hg) [49] in slightly acid solution at elevated temperature (90°C); the acidity can be maintained by bubbling CO_2 through the solution, Eq. (3);

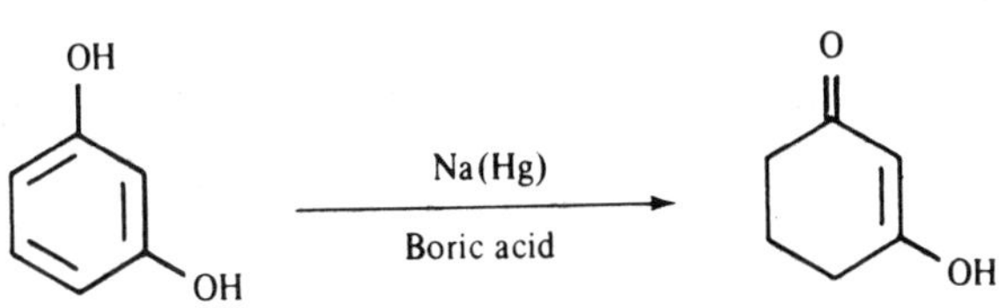

(3)

The process has not been reported by direct electrolysis, but it may be possible to do it by applying a suitable potential somewhat more negative than the decomposition potential of the medium at that pH in the absence of the substrate.

III. HALOGEN COMPOUNDS

An amalgam reduction of a halogen compound may result in cleavage of the carbon-halogen bond, a coupling reaction, a reductive elimination, or formation of an organic mercury compound.

Many halogen compounds are reduced to hydrocarbons by Na(Hg). This may be used to obtain reduction of isolated double bonds that are unreactive toward Na(Hg) by addition of hydrogen bromide to the double bond followed by amalgam reduction of the halogen compound to a hydrocarbon; the transformation of 4-pentenoic acid via bromovaleric acid to valeric acid is an example [50].

A reductive coupling of a halogen compound may be exemplified by the reaction of β-chloropropionitrile with Na(Hg) in the presence of some acrylonitrile and a small amount of water to adiponitrile; yields up to 90% have been obtained [51–53].

Treatment of vicinal halogen compounds with sodium amalgam results in many cases in a reduction-elimination reaction; 4,5-dibromovaleric acid thus forms 4-pentenoic acid on treatment with Na(Hg) [50]. 1,1,2-Trichloro-1,2,2-trifluoroethane and 1,2-dichloro-1,1,2,2-tetrafluoroethane react with Na(Hg) similarly with formation of 1-chloro-1,2,2-trifluoroethylene and tetrafluoroethylene, respectively [54,55].

A special reaction of this type is the formation of benzyne. o-Fluorobromobenzene reacts with lithium amalgam in ether or furan with an intermediate formation of benzyne. In the former medium diphenylene and small amounts of triphenylene and 9,10-dimercura-9,10-dihydroanthracene were formed, whereas the major product (76%) in furan was 1,4-dihydronaphthalene-1,4-endoxide [56].

1,3-Difluorobenzene (IX) can be reduced to fluorobenzene (X), which in turn can be reduced further to benzene using constant current in diglyme with TAA salts as electrolyte [57]. The best conditions for a selective reduction of IX to X was a relative low current density (1.25 mA cm^{-2}), low concentration of $(DMP)BF_4$ (0.01 M + 0.1 M $TBABF_4$), and low water concentration (0.5%).

An attack on a C–F bond may operate in the corrosion of poly(tetrafluoroethylene) (XI); the reaction of XI with amalgams of Li, Na, and K has been suggested [58] to proceed by an electrochemical mechanism in the solid phase, with the alkali metal acting as an anode and the polymer as a cathodic-active material typical of a galvanic corrosion.

Polymerization of p-dichlorobenzene in xylene and of α,α'-dichloro-p-xylene in dioxane takes place on reduction with sodium or potassium amalgam; yields of 60–70% polymerized products are obtained [59].

Treatment of methyl-chloroacetate in acetone containing a small amount of phenyltrimethylammonium chloride with sodium amalgam is reported [60] to give a fair (35–40%) yield of methyl β,β-dimethylglycidate. The reaction is not a reduction and is probably catalyzed by a reductively generated base, Eq. (4):

$$CH_3COCH_3 + ClCH_2CO_2CH_3 \xrightarrow{Na(Hg)} (CH_3)_2C\overset{O}{\frown}CHCOOCH_3 \quad (4)$$

Reduction of benzyl halides by Na(Hg) in propylene carbonate in the presence of CO_2 gave the Na-salt of the corresponding phenylacetic acid [61].

IV. ORGANOMERCURIALS

A good yield of dialkylmercury compounds [62] may be obtained from alkyl halides and Na(Hg); too high a Na content in Na(Hg) lowers the yield of dialkylmercury due to increased formation of the Wurtz coupling product. The reaction may be formulated as follows, Eq. (5):

$$RX + Hg \longrightarrow RHgX \xrightarrow{Na} \tfrac{1}{2}R_2Hg + \tfrac{1}{2}Hg \quad (5)$$

The ratio of the dialkylmercury to the Wurtz coupling product is dependent on the halogen; thus n-butyl iodide yields 80-81% dibutylmercury and 10% octane, whereas n-butyl chloride gave 11–17% Bu_2Hg and 65–70% octane.

Treatment of 1,2-dibromobenzene or its derivatives with Na(Hg) gave trimeric phenylenemercury derivatives [63].

Sodium amalgam reduction of organomercurials in alkaline deuterium oxide gives products with stereospecific retention during the replacement of mercury by deuterium [64]. This was shown in the reduction of cis-8-acetoxymercuri- (XIIa) and cis-8-chloromercuridibenzobicyclo[2.2.2]octadien-7-ol acetates (XIIb), trans-8-chloromercuridibenzobicyclo[2.2.2]octadien-7-ol, and some norbornyl derivatives. Reduction of these compounds with sodium borodeuteride gave products without stereospecific deuterium incorporation [64].

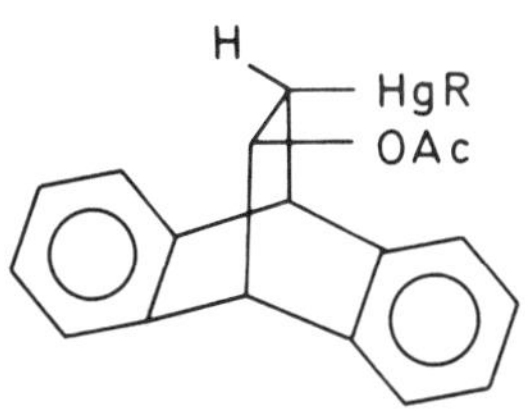

XIIa R = OAc

XIIb R = Cl

Hex-5-enyl-1-mercuric bromide (XIII) is reduced to 1-hexene (XIV) by sodium amalgam, whereas sodium borohydride, lithium aluminum hydride, and tri-n-butyltin hydride gave a mixture of XIV and methylcyclopentane. This is taken as evidence that sodium amalgam reduction of alkylmercury salts does not proceed via a free radical pathway [65].

Reduction of XIII with sodium amalgam in a mixture of tetrahydrofuran and deuterium oxide gave >96% incorporation of deuterium [65]. Sodium amalgam is thus preferable as reducing agent for organomercurials for stereospecific deuterium labeling and for synthesis in systems that are prone to free radical rearrangements.

Substituted 2-aminoethylmercuric chlorides, prepared by addition of $HgCl_2$ to an alkene in the presence of an amine, have been reported [66] to give the substituted 2-aminoethane in moderate yield, whereas direct reduction neither in DMF nor in 50% ethanol produced detectable amounts of the amine [67].

V. NITRO COMPOUNDS AND DERIVATIVES

Nitro compounds are reduced to amalgams in a manner similar to that found at a mercury electrode. Aliphatic nitro compounds [68] are thus reduced to alkylhydroxylamines by $NH_4(Hg)$, whereas aromatic nitro compounds may form amines, hydrazo, azo, or azoxy compounds and even nitroso compounds [68,69]. Aliphatic, tertiary vicinal dinitro compounds may be reduced to alkenes by elimination of the nitro groups as nitrite [70].

Reduction of nitrobenzene with Al(Hg) in ether at controlled temperature yields phenylhydroxylamine [71].

Azobenzene has been manufactured by reduction of a suspension of nitrobenzene by Na(Hg). Further reduction to hydrazobenzene is possible if higher temperatures and concentrations of alkali are employed [72]. The corresponding arsenic compounds may be prepared from a phenylarsenoxide by amalgam reduction; thus p-aminophenylarsenoxide yields 80% $H_2NC_6H_4As{=}AsC_6H_4NH_2$ on treatment with 4% Na(Hg) in methanol [73].

It is interesting that treatment of some aromatic nitro compounds with ammonium amalgam yields among other products nitrosobenzenes [69]. 2-Nitrobenzoic acid thus gives about 45% 2-nitrosobenzoic acid together with anthranilic acid (10%), and 2-nitrotoluene (5 g) yields 0.6 azoxytoluene, 2 g 2-nitrosotoluene, and 1.5 g o-toluidine on treatment with $NH_4(Hg)$.

N-nitroso and N-nitro compounds may both be reduced to hydrazines. In this case care must be taken to avoid a reductive cleavage of the nitrogen-nitrogen bond [72,74].

VI. CARBONYL COMPOUNDS AND DERIVATIVES

On reaction with Na(Hg), saturated aldehydes or ketones may form either alcohols or pinacols [75–77]; the reaction is similar to the electrolytic reduction (Chap. 10).

The use of Al(Hg) in ether for the reduction of alkali-sensitive compounds may give higher yield than Na(Hg); thus, diethyl oxaloacetate is reduced to diethyl malate by Al(Hg) in 70–80% yield, whereas Na(Hg) gives about 50% yield [71].

α,β-Unsaturated ketones of the structure $RR'C{=}CHCOCH_3$ yield on treatment with sodium amalgam a mixture of products, including tetrahydrofurans (XV) and dihydrofuran derivatives (XVI), glycols (XVII), ε-diketones (XVIII), and ketols (XIX) [78–80]. The ratio between the products depends on R and R'; bulky groups as found in steroids favor formation of glycols rather than of ε-diketones [81]. The reactions leading to these products are similar to those discussed in Chap. 10.

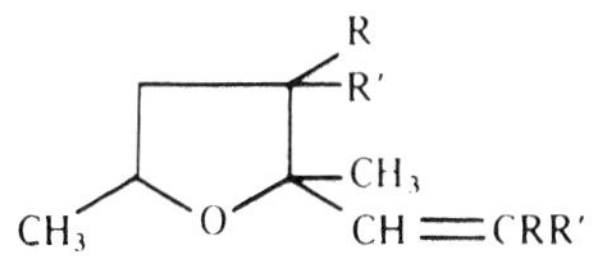

(XV)

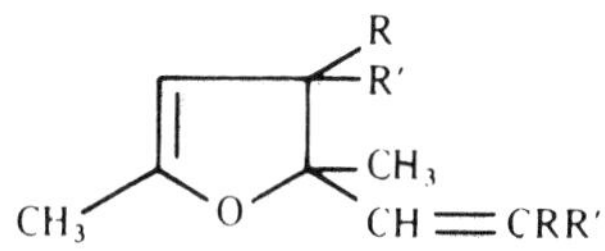

(XVI)

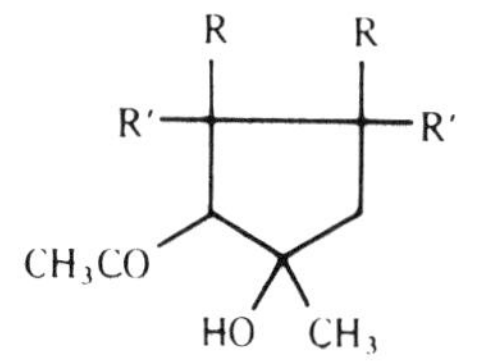

(XIX)

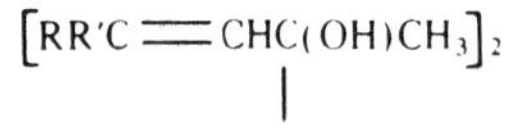

(XVI)

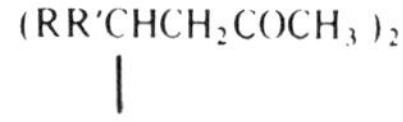

(XVII)

Reduction with Na(Hg) in toluene has been used to prepare p-hydroxy di- and triphenylmethyl radicals from the corresponding galvinols [82].

Whereas the reduction of cyclohexanone (XX) in DMF-0.1 M $TBABF_4$ gives cyclohexanol in good yield, addition of $(DMP)BF_4$ (0.01 M) changes the product to the pinacol (95%). The cyclic voltammetric peak of $(DMP)BF_4$ in DMF-$TBABF_4$ is enhanced on addition of XX, which suggests a reduction of XX by DMP(Hg) and regeneration of DMP^+ [83]. That TBA^+ does not function in the same manner as DMP^+ could suggest that the reduction potential of TBA^+ is so negative that a direct reduction of XX occurs.

Similarly, reduction of 6-heptene-2-one (XXI) at Hg in DMF-0.1 M TBA^+ yields the alcohol, whereas reduction in the presence of catalytic amounts of DMP^+ yields cis-1,2-dimethylcyclopentanol (95%) [83]. The cyclization is suggested to occur at the stage of the anion radical of XXI since in diglyme-0.5% H_2O DMP^+ promoted the formation of the pinacol of XXI. The cyclization is similar to that described by Shono et al. [84], who reduced XXI in MeOH-dioxane at a carbon cathode with 2 M TEA^+ as electrolyte; see Chap. 10.

The reduction of carbonyl derivatives as oximes and hydrazones is similar to the electrolytic reaction. Oximes are reduced to amines [51], whereas hydrazones in alkaline solution may form hydrazines; benzalhydrazine can thus be reduced to benzylhydrazine [85], benzalazine to benzaldehyde benzylhydrazone or further to dibenzylhydrazine [86,87], and benzaldehyde semicarbazone to benzylsemicarbazide [88].

Isoxazolines may be regarded as cyclic oximes; the reduction of 3,5-diphenylisoxazoline (XXIIa) with $LiAlH_4$ gave 95:5 erythro-threo 1,3-diphenyl-3-aminopropanol (XXIIIa) [89], whereas reduction with Na(Hg) [89,90] in methanol yielded the ratio 2:3 of erythro-threo XXIIIa, Eq. (6). Similar results were obtained with 3-phenyl-5-methylisoxazoline (XXIIb), whereas 5-phenyl-3-methylisoxazoline (XXIIc) was not reducible.

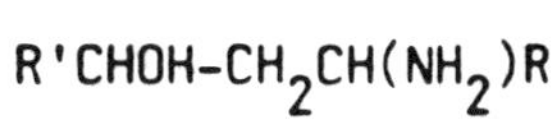

(XXII) (XXIII)

a R = R' = C_6H_5 erythro:threo 2:3

b R = C_6H_5 R' = CH_3

c R = CH_3 R' = C_6H_5

5-(Bromomethyl)isoxazolines, prepared from a nitrile oxide and an allylbromide, may be reductively ring opened to β,γ-enoximes (XXIV); XXIV may be reduced in aqueous THF with Al(Hg) to the homoallylamine (XXV), Eq. (7) [91]:

$$CH_2{=}CR{-}CH_2{-}C({=}NOH)R' \longrightarrow CH_2{=}CR{-}CH_2{-}CH(NH_2)R' \qquad (7)$$

XXIV XXV

VII. ACIDS AND DERIVATIVES

Esters of fatty acids are only to some extent reduced to the alcohols by sodium amalgam. Aldonolactones and their esters, however, are reduced to aldoses by Na(Hg) [92]. Optimal conditions are found at pH 3–3.5 and temperatures below 15°C; 2.5–3 atoms of Na [as 2.5% Na(Hg)] are used per mole of lactone. Yields of different aldoses range from 50 to 84% [93,94].

The reduction of aromatic carboxylic acids by Na(Hg) may take place in the nucleus or in the carboxyl group. Phthalic acid is reduced to trans-3,5-cyclohexadiene-1,2-dicarboxylic acid [95] or 2,5-cyclohexadiene-1,2-dicarboxylic acid [96], whereas certain phenolic carboxylic acids may be reduced to aldehydes.

Salicylic acid and some other acids were reduced by Na(Hg) to the aldehyde [97–100] (40–60%) in a slightly acid medium containing p-toluidine and boric acid; the reaction mixture was kept slightly acid by addition of boric acid. A similar electrolytic reduction was not successful unless the mercury cathode was allowed to take up some sodium, thus making the reaction an indirect reaction [101].

The reduction by Na(Hg) of several o-hydroxynaphthalenecarboxylic acids in slightly acid solution containing boric acid has been investigated. 1-Hydroxynaphthalene-2-carboxylic acid [102] is, like salicylic acid, reduced to the aldehyde (62% yield), isolated as the Schiff base, whereas 2-hydroxynaphthalene-3-carboxylic acid [103] under similar conditions yields mainly 2-hydroxy-1,2,3,4-tetrahydronaphthalene-3-carboxylic acid (75–80%) and 5% of the aldehyde. When hydrochloric acid rather than CO_2 was used to neutralize the base formed during the reduction, the 2-oxo-1,2,3,4-tetrahydronaphthalene-3-carboxylic acid [104] was isolated (60%). On heating, the compound, a β-keto acid, decarboxylated.

A loss of phenolic hydroxyl group occurs during the reduction of 1-hydroxynaphthalene-4-carboxylic acid with Na(Hg) in the boric acid medium from which 1,4-dihydronaphthoic acid is isolated [105]. The primary reduction probably yields 1,4-dihydro-1-hydroxy-4-naphthoic acid, which loses water to 1-naphthoic acid; this compound is under these conditions reduced mainly to the 1,4-dihydro derivative, whereas 2-naphthoic acid predominantly gives the 1,2-dihydro compound.

Reduction of an aromatic acid chloride by Li(Hg) or another alkali metal amalgam in diethyl ether or THF yields the enediol diester (XXVI) [106]; further reduction by Li(Hg) yields diarylacetylene (XXVII), Eq. (8) [107]:

$$4\ ArCOCl + 4e\ (Li(Hg)) \longrightarrow ArC(OCOAr){=}C(OCOAr)Ar\ (Z + E)$$

XXVI

$$XXVI + 2e\ (Li(Hg)) \longrightarrow ArC{\equiv}CAr + 2\ ArCOO^- \qquad (8)$$

XXVII

Carbon dioxide may be reduced by alkali metal amalgams in alcohol to alkali metal formate [108] in a yield of 85–90%; the rate increases with the CO_2 pressure and is dependent on the alkali metal, Na(Hg) < Li(Hg) < K(Hg).

An attempt has been made [109] to hydrodimerize CO_2 to oxalic acid, but only in hexane and with Na(Hg) was sodium oxalate obtained in a reasonable yield (40%).

In other cases a wide variety of products, ranging from simple organic acids such as formic and oxalic acid to totally reduced CO_2 in the form of carbon black, have been obtained by reduction of CO_2 by an appropriate alkali metal amalgam [110].

VIII. CLEAVAGE REACTIONS (OTHER THAN CARBON-HALOGEN)

A reaction that involves both a carbon-carbon and a nitrogen-nitrogen cleavage is found in the reduction of 2-phenyl-4-carboxytriazole by Na(Hg) in slightly alkaline solution [111]. The reaction probably starts with a reduction in the ring, which then opens up and loses cyanide ion according to Eq. (9):

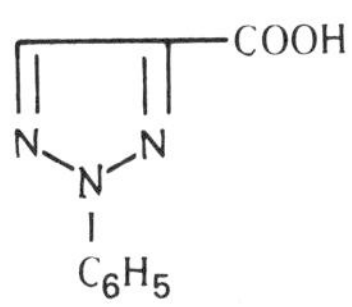

$$\xrightarrow{\text{Na(Hg)}} C_6H_5NHNHCH_2COO^- + CN^- \qquad (9)$$

Treatment of benzopinacol with Na(Hg) in a mixture of anhydrous ether and benzene gives a dark blue solution that fades. Benzhydrol can be isolated in a quantitative yield. It has been suggested that the disodium salt of benzopinacol disproportionates with C—C cleavage into benzhydrol and benzophenone; the latter is reduced by Na(Hg) to benzopinacol, which then decomposes [112].

A cleavage of carbon-oxygen bonds caused by amalgam reduction is found, for example, in the reduction of dihydroxyacetoxime to isopropylamine [113] and 2-hydroxy-4-phenyl-3-butenoic acid to 4-phenyl-3-butenoic acid [114]. These reactions are analogous to those discussed in Chap. 24.

Whereas sodium amalgam is unreactive toward phenyltrimethylammonium chloride, benzyltrimethylammonium chloride is cleaved to toluene and trimethylamine. A similar reaction is responsible for the reductive ring opening of N,N-dimethyltetrahydroquinolinium chloride to 2-propyl-N,N-dimethylaniline [115].

The reductive cleavage of achiral and optically active quaternary phosphonium and arsonium salts with alkali metal amalgams to form tertiary phosphines and arsines succeeds in high yield with retention of configuration [116]. The reduction with the amalgams was found to give better yields than the conventional cathodic cleavage.

Sodium amalgam has been used to cleave sulfur-sulfur bonds. Solutions or suspensions of organic di- or polysulfides can be reduced with Na(Hg) in the presence of an alkali sulfide to the mercaptide [117].

Certain sulfonic acids may be cleaved by Na(Hg); thus 1-aminonaphthalene-3,8-disulfonic acid can be reductively cleaved to 1-aminonaphthalene-3-sulfonic acid. The sulfonic group in the 8-position is more easily removed reductively than sulfonic groups in the 2-, 4-, or 6-positions [72].

β-Ketosulfoxides react rapidly and cleanly with aluminum amalgam in a mixture of 90% tetrahydrofuran and 10% water to yield the ketones, Eq. (10) [118].

$$RCOCH_2SOCH_3 \xrightarrow[(H_2O)]{Al(Hg)} RCOCH_3 + CH_3SH \qquad (10)$$

For preparation on a larger scale the direct electrolytic cleavage of β-ketosulfoxides or β-ketosulfones [119] has been found preferable.

Alkyl arylsulfonates and diaryl sulfones are cleaved by alkali metal amalgams to the corresponding sulfinate salts. Some sterically hindered arylsulfonates, however, undergo cleavage at the ring-S bond [120].

Selective demesylation of 2-O-(methylsulfonyl)-D-mannopyranoside derivatives has been made with Na(Hg) in 98.6% yield in diethyl ether containing 2-propanol [121].

The N—O bond of Diels-Alder adducts of hydroxamic acids with 1,3-dienes may be cleaved by Al(Hg) or Na(Hg). Thus XXVIII was reduced to XXIX by Al (Hg) in quantitative yield, Eq. (11) [122]:

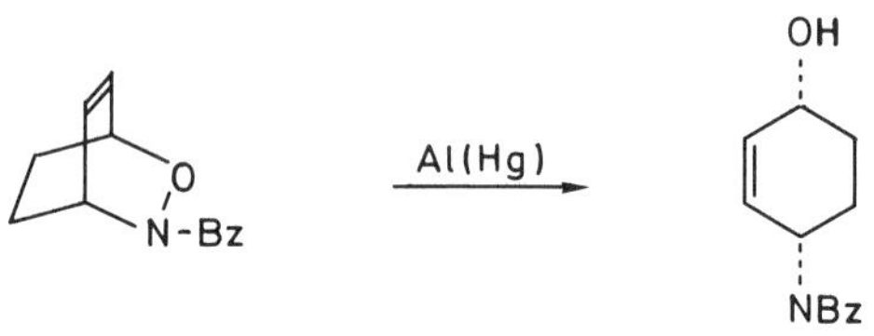

(11)

(XXVIII) (XXIX)

Symmetrical alkynes and symmetrical and unsymmetrical olefins may be prepared by reductive elimination of suitable groups by amalgams; the preparation of diarylalkynes from aromatic acid chlorides with Li(Hg) was described earlier [106,107].

One method of preparing alkenes is a reductive elimination using sodium amalgam of a substituted 2-phenylsulfonylethanol (XXX), obtained by addition of phenylalkylsulfone to a ketone using a basic organometallic catalyst [123].

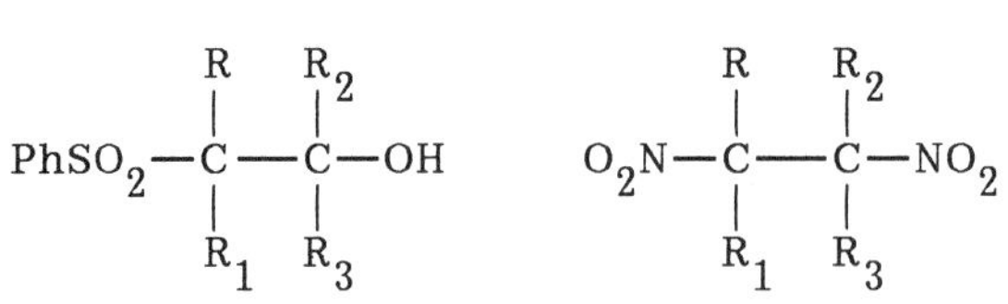

XXX XXXI

Another one is to prepare vicinal, tertiary nitro compounds (XXXI) by oxidative dimerization of the mononitro anion, followed by treatment of XXXI with calcium amalgam in hexamethylphosphoramide [70] at room temperature.

IX. POLYMERIZATION REACTIONS

Finally, it might be mentioned that Na(Hg) can act as a catalyst in a polymerization of acetaldehyde to a polyvinyl alcohol-type polymer [124,125].

The polymerization may be performed in a cooled vessel at high pressure under nitrogen with vigorous stirring. The product was suggested to be a linear polymer of the type $CH_3(CHOHCH_2)_nCHO$, but later investigations suggested a branched structure [126].

An oligomerization of ethylene in the presence of cyclopentadienyltitanium dichloride, $C_5H_5TiCl_2$, is possible using Na(Hg). When the ratio of ethylene to $C_5H_5TiCl_2$ was 12, a mixture of 80% n-butane and 20% n-hexane was obtained [127].

Isobutylene polymers and copolymers can be prepared by polymerization in the presence of vanadium tetrachloride or titanium tetrachloride and sodium amalgam [128].

REFERENCES

1. T. Mussini, P. Longhi, and S. Rondinini, Pure Appl. Chem. 57, 169 (1985).
2. Y. T. Tur'yan, Zh. Obshch. Khim. 57, 1381 (1986).
3. M. Brenant and J. L. Sue, J. Electroanal. Chem. 40, 346 (1972).
4. L. Horner and R. E. Schmitt, Z. Naturforsch., B: Anorg. Chem., Org. Chem. 37B, 1163 (1982).
5. H. N. McCoy and W. C. Moore, J. Amer. Chem. Soc. 33, 273 (1911).
6. J. D. Littlehailes and B. J. Woodhall, J. Chem. Soc. Chem. Commun. 665 (1967); Disc. Faraday Soc. 45, 187 (1968).
7. E. Kariv-Miller, J. Nanjundiah, and K. E. Swenson, J. Electroanal. Chem. 167, 141 (1984).
8. E. Kariv-Miller and R. Andruzzi, J. Electroanal. Chem. 187, 175 (1985).
9. V. Svetlicić and E. Kariv-Miller, J. Electroanal. Chem. 209, 91 (1986).
10. E. Garcia, A. H. Cowley, and A. J. Bard, J. Amer. Chem. Soc. 108, 6082 (1986).
11. E. Kariv-Miller and V. Svetlicić, J. Electroanal. Chem. 205, 319 (1986); C. M. Ryan, V. Svetlicić, and E. Kariv-Miller, J. Electroanal. Chem. 219, 247 (1987). E. Kariv-Miller, V. Svetlicić, and P. B. Lawin, J. Chem. Soc., Faraday Trans. I 83, 1169 (1987).
12. E. Zintl, J. Goubeau, and W. Dullenkopf, Z. Phys. Chem. A, 154, 1 (1931); H. Schäfer, B. Eisenmann, and W. Müller, Angew. Chem. 85, 742 (1973).
13. R. G. Teller, L. J. Krause, and R. C. Haushalter, Inorg. Chem. 22, 1809 (1983); J. D. Corbett, Chem. Rev. 85, 383 (1985).
14. E. Kariv-Miller, P. B. Lawin, and Z. Vajtner, J. Electroanal. Chem. 195, 435 (1985); E. Kariv-Miller and P. B. Lawin, J. Electroanal. Chem. 247, 345 (1988).
15. J. Simonet and H. Lund, J. Electroanal. Chem. 75, 719 (1975); E. H. Hall, J. Simonet, and H. Lund, J. Electroanal. Chem. 100, 197 (1979).
16. G. Bernard and J. Simonet, J. Electroanal. Chem. 112, 117 (1980); J. Berthelot, M. Jubault, and J. Simonet, J. Chem. Soc., Chem. Commun. 759 (1982).

17. V. Kaufman, Ber. Dtsch. Chem. Ges. 29, 73 (1896).
18. J. Thiele and Henk, Liebigs Ann. Chem. 347, 290 (1906).
19. R. Kuhn and H. Fischer, Chem. Ber. 94, 3060 (1961).
20. C. Harries and F. Kaiser, Ber. Dtsch. Chem. Ges. 31, 1806 (1898).
21. O. Bayer, Angew. Chem. 61, 229 (1949).
22. I. L. Knunyants and N. S. Vyazankin, Dokl. Akad. Nauk SSSR 113, 112 (1957); Chem. Abstr. 51, 14637 (1957).
23. I. L. Knunyants and N. S. Vyanzankin, Izv. Akad. Nauk SSSR, Otd. Khim. Nauk 238 (1957); Chem. Abstr. 51, 11235 (1957).
24. M. M. Baizer, Tetrahedron Lett. 973 (1963).
25. F. Beck, Chem.-Ing.-Tech. 37, 607 (1965).
26. L. G. Feoktistov, A. P. Tomilov, and I. G. Sevast'yanova, Elektrokhimiva 1, 1300 (1965).
27. M. M. Baizer and J. P. Petrovich, J. Electrochem. Soc. 114, 1023 (1967).
28. F. Beck, Z. Elektrochem. 72, 379 (1968).
29. I. E. Gillet, Bull. Soc. Chim. Fr. 2919 (1968).
30. A. P. Tomilov, E. V. Kryukova, V. A. Klimov, and I. N. Brago, Elektrokhimiya 3, 1501 (1967).
31. J. D. Littlehailes and B. J. Woodhall, Disc. Faraday Soc. 45, 187 (1968).
32. F. Beck and I. E. Gillet, Disc. Faraday Soc. 45, 234 (1968).
33. F. Matsuda, Tetrahedron Lett. 6193 (1966).
34. A. Katchalsky, D. Vofsi, and J. I. Padova, Isr. Patent 13,842 (1961); Chem. Abstr. 56, 9977 (1962); A. Katchalsky, D. Vofsi, and J. I. Padova, Isr. Patent 13,844 (1961); Chem. Abstr. 56, 11450 (1961); ICI, Neth. Appl. 6,504,863 (1965); Chem. Abstr. 64, 12557 (1966); ICI, Neth. Appl. 6,609,103 (1967); Chem. Abstr. 66, 115334 (1967); UCB, Neth. Appl. 6,609,241 (1967); Chem. Abstr. 67, 53724 (1967); A. Misonou and T. Naga, Jpn. Patent 6,907,925 (1969); Chem. Abstr. 71, 35536 (1969).
35. Y. Arad, M. Levy, H. Rosen, and D. Vofsi, J. Polym. Sci., A-1, 7, 2159 (1969).
36. M. Kawamata, Kokyo Kagaku Zasshi 74, 2063, 2479 (1971); Chem. Abstr. 76, 3379, 100103 (1972).
37. Y. Arad, M. Levy, H. Rosen, and D. Vofsi, J. Polym. Sci., B, Polym. Lett. 7, 197 (1969).
38. H. Rosen, Y. Arad, M. Levy, and D. Vofsi, J. Amer. Chem. Soc. 91, 1425 (1969).
39. For example, UCB Fr. Patent 1,505,954 (1967); Chem. Abstr. 69, 95999 (1968); UCB, S. Afr. Patent 6,704,827 (1968); Chem. Abstr. 70, 37239 (1969); Toyo Koatsu Ind., Fr. Patent 1,535,374 (1968); Chem. Abstr. 71, 38378 (1969); ICI of Australia and New Zealand, Fr. Patent 1,543,322 (1968); Chem. Abstr. 72, 12163 (1970); Y. Arad, M. Levy, and D. Vofsi, Fr. Patent 1,553,195 (1969); Chem. Abstr. 72, 42853 (1970); ICI, U.S. Patent 3,458,559 (1969); Chem. Abstr. 71, 123625 (1969); ICI of Australia and New Zealand, U.S. Patent 3,523,967 (1971); Chem. Abstr. 74, 12646 (1971).
40. Y. Arad, M. Levy, and D. Vofsi, J. Org. Chem. 34, 3709 (1969).
41. V. A. Smirnov, V. G. Kul'nevich, I. V. Chenikov, G. N. Soltovets, and S. V. Babicheva, USSR Patent 271,500 (1970); Chem. Abstr. 73, 120180 (1970).

42. G. N. Soltovets, I. V. Chenikov, S. V. Babicheva, V. A. Smirnov, V. G. Kul'nevich, L. A. Dulan, and A. V. Kovaleva, Tr. Krasnodar. Polytekh. Inst. 38, (1971); Chem. Abstr. 78, 71519 (1973).
43. E. Kariv-Miller and R. I. Pacut, Tetrahedron 42, 2185 (1986).
44. J. P. Coleman and J. H. Wagenknecht, J. Electrochem. Soc. 128, 322 (1981).
45. K. E. Swenson, D. Zemach, C. Nunjundiah, and E. Kariv-Miller, J. Org. Chem. 48, 1777 (1983).
46. E. Kariv-Miller, K. E. Swenson, G. K. Lehman, and R. Andruzzi, J. Org. Chem. 50, 556 (1985).
47. R. I. Pacut and E. Kariv-Miller, J. Org. Chem. 51, 3468 (1986).
48. E. Kariv-Miller, D. F. Dedolph, C. M. Ryan, and T. J. Mahachi, J. Heterocyclic Chem. 22, 1389 (1985).
49. G. Merling, Liebigs Ann. Chem. 278, 20 (1894).
50. A. Messerschmidt, Liebigs Ann. Chem. 208, 92 (1881).
51. Knapsack-Grieshiem A.-G., Ger. Patent 1,151,791 (1963); Chem. Abstr. 60, 1604 (1964).
52. A. Misono, T. Osa, T. Asano, and T. Suzuki, Bull. Chem. Soc. Jpn. 41, 2434 (1968).
53. J. Hasegawa, I. Kojima, and M. Micuo, Jpn. Kokai 73-07, 616 (1973); Chem. Abstr. 79, 41997 (1973).
54. Farbenfabriken Bayer, Ger. Patent 847,442 (1950); 857,956 (1950); 875,808 (1950).
55. E. Montecatini, It. Patent 866,611 (1970); Chem. Abstr. 78, 135631 (1973).
56. G. Wittig and L. Pohmer, Chem. Ber. 89, 1334 (1956).
57. E. Kariv-Miller And Z. Vajtner, J. Org. Chem. 50, 1394 (1985).
58. J. Jansta, F. P. Dousek, and J. Riha, J. Appl. Polym. Sci. 19, 3201 (1975).
59. T. A. Kazintseva, L. M. Lagushin, and A. A. Yakubovich, Tr. Novocherk. Polyteckh. Inst. 132 (1972); Chem. Abstr. 79, 148291 (1973).
60. M. Kawamata and S. Fujikake, Jpn. Kokai 72 38,928; Chem. Abstr. 78, 71487 (1973).
61. L. Morechini, A. Del Vesco, and G. Carrara, It. Patent 894,274 (1971); Chem. Abstr. 83, 27924 (1975).
62. N. N. Mel'nikov, Zh. Obshch. Khim. 16, 2065 (1946).
63. N. A. A. Al-Jabar and A. G. Massey, J. Organometal. Chem. 275, 9 (1984).
64. F. R. Jensen, J. J. Miller, S. J. Christol, and R. S. Beckley, J. Org. Chem. 37, 4341 (1972).
65. R. P. Quirk and R. E. Lea, J. Amer. Chem. Soc. 98, 5973 (1976).
66. R. K. Freidlina and N. S. Kochet Kova, Bull. Acad. Sci. USSR, Classe Sci. Chim. 128 (1945); Chem. Abstr. 40, 3450 (1946).
67. J. H. Wagenknecht, J. Electrochem. Soc. 119, 1494 (1972).
68. S. Takaki and T. Ueda, J. Pharm. Soc. Jpn. 58, 276, 533 (1938).
69. S. Takaki, T. Ueda, and T. Sakaguti, J. Pharm. Soc. Jpn. 58, 938 (1938); Chem. Abstr. 33, 2113 (1939).
70. N. Kornblum and L. Cheng, J. Org. Chem. 42, 2944 (1977).
71. H. Wislicenus and L. Kaufmann, Ber. Dtsch. Chem. Ges. 28, 1323 (1885).
72. W. Funke, Chem.-Ing.-Tech. 35, 336 (1963).

73. P. Ehrlich and A. Bertheim, Ber. Dtsch. Chem. Ges. 44, 1260 (1911).
74. Ger. Patent 1,130,821 (1960); Chem. Abstr. 57, 12320 (1962); Belg. Patent 612,838 (1961); Chem Abstr. 58, 4586 (1963).
75. S. Takaki and T. Ueda, J. Pharm. Soc. Jpn. 58, 44, 141, 152, 156 353, 422, 648 (1938); Chem. Abstr. 32, 4149, 5777, 6635, 8384 (1938).
76. S. Takaki and T. Tamai, J. Pharm. Soc. Jpn. 58, 701 (1938); Chem. Abstr. 33, 536 (1939).
77. E. Otto, Prax. Naturwiss., Teil 3 22, 12 (1973).
78. N. V. Thoai and J. Wiemann, C. R. Acad. Sci., Ser. C 247, 212 (1958).
79. J. Wiemann and R. Nahum, C. R. Acad. Sci., Ser. C 238, 2091 (1954).
80. R. Nahum, C. R. Acad. Sci., Ser. C 243, 849 (1956).
81. R. Dulou, J. Chopin, and Y. Raoul, Bull. Soc. Chim. Fr. 616 (1951).
82. B. Kirste, W. Harrer, and H. Kurreck, J. Amer. Chem. Soc. 107, 20 (1985).
83. E. Kariv-Miller and T. J. Mahachi, J. Org. Chem. 51, 1041 (1986); J. E. Swartz, T. J. Mahachi, and E. Kariv-Miller, J. Amer. Chem. Soc. 110, 3622 (1988).
84. T. Shono, I. Nishiguchi, H. Ohmizu, and M. Mitani, J. Amer. Chem. Soc. 100, 545 (1978).
85. T. Curtius, Ber. Dtsch. Chem. Ges. 33, 2459 (1900).
86. T. Curtius, J. Prakt. Chem. 62, 83 (1900).
87. T. Curtius, J. Prakt. Chem. 85, 393 (1912).
88. S. Kessler and H. Rupe, Ber. Dtsch. Chem. Ges. 45, 26 (1912).
89. V. Jäger and V. Buss, Liebigs Ann. Chem. 101 (1980).
90. W. Stühmer and W. Heinrich, Chem. Ber. 84, 224 (1951).
91. V. Jäger, H. Grund, and W. Schwab, Angew. Chem. 91, 91 (1979).
92. E. Fischer and O. Piloty, Angew. Chem. 24, 4214 (1891).
93. N. Sperber, H. E. Zaugg, and W. M. Sandstrom, J. Amer. Chem. Soc. 69, 915 (1947).
94. A. Korczynsky, L. Piszczek, and J. Swiderski, Zesz. Nauk Politech. Slask., Chem. 199 (1973); Chem. Abstr. 81, 85208 (1974).
95. A. Baeyer, Liebigs Ann. Chem. 269, 145 (1892).
96. H. Von Metnitz, O. Bleh, and A. Lenz, Ger Offen. 2,053,289 (1972); Chem. Abstr. 77, 62486 (1972).
97. H. Weil, Ber. Dtsch. Chem. Ges. 41, 4147 (1908).
98. H. Weil, M. Traun, and S. Marcel, Ber. Dtsch. Chem. Ges. 55, 2664 (1922).
99. R. R. Davies and H. H. Hodgson, J. Soc. Chem. Ind. 62, 128 (1943).
100. R. Dobrzeniecka, J. Kaszubska, and W. Jacobson, Chem. Stosow. 20, 265 (1976); Chem. Abstr. 86, 72119 (1977).
101. C. Mettler, Ber. Dtsch. Chem. Ges. 41, 4148 (1908).
102. L. N. Lavrishcheva, N. M. Przhiyalgovskaya, S. A. Voitkevich, and V. N. Belov, Zh. Obshch. Khim. 27, 1264 (1957); Chem. Abstr. 52, 2819h (1958).
103. N. M. Przhiyalgovskaya, L. N. Lavrishcheva, G. T. Mondodoev, and V. N. Belov, Zh. Obshch. Khim. 31, 2321 (1961); Chem. Abstr. 56, 3427e (1962).
104. N. M. Przhiyalgovskaya, L. N. Lavrishcheva, and V. N. Belov, Zh. Obshch. Khim. 27, 1266 (1957); Chem. Abstr. 52, 3750h (1958).
105. N. M. Przhiyalgovskaya, G. T. Mondodoev, and V. N. Belov, Zh. Obshch. Khim. 31, 3375 (1961); Chem. Abstr. 57, 3373e (1962).

106. L. Horner and K. Dickerhof, Chem. Ber. 116, 1603 (1983).
107. L. Horner and K. Dickerhof, Chem. Ber. 116, 1615 (1983).
108. V. A. Smirnov, V. D. Semchenko, D. P. Semchenko, and L. I. Antropov, Zh. Prikl. Khim. 40, 2345 (1967); Chem. Abstr. 68, 63071 (1968).
109. V. A. Tushnalobova, V. A. Smirnov, and D. P. Semchenko, Zh. Prikl. Khim. 41, 275 (1968); Chem. Abstr. 68, 110961 (1968).
110. R. W. Treharre and C. M. Cox, U.S. Clearinghouse Fed. Sci. Tech. Inf., A.D. 1968, A.D. 679597; Chem. Abstr. 70, 99413 (1969).
111. A. Jonas and H. von Pechmann, Liebigs Ann. Chem. 262, 277 (1891).
112. W. E. Bachmann, J. Amer. Chem. Soc. 55, 355, 770 (1933).
113. O. Piloty and O. Ruff, Ber. Dtsch. Chem. Ges. 30, 1656 (1897).
114. G. H. Lahib and P. Cordier, C. R. Acad. Sci., Ser. C 241, 1051 (1955).
115. H. Emde, Liebigs Ann. Chem. 391, 88 (1912).
116. L. Horner and K. Dickerhof, Phosphorus Sulfur 15, 213 (1983).
117. I. G. Farbenind. A.-G., Ger. Patent 737,334 (1943); Chem. Abstr. 38, 3666 (1944).
118. E. J. Corey and M. Chaykobsky, J. Amer. Chem. Soc. 86, 1639 (1964).
119. B. Lamm and B. Samulesson, Acta Chem. Scand. 23, 691 (1969); 24, 561 (1970).
120. L. Horner and R. E. Schmitt, Phosphorus Sulfur 13, 301 (1982).
121. K. T. Wabster, R. Eby, and C. Schuerch, Carbohydrate Res. 123, 335 (1983).
122. G. E. Keck, S. Fleming, D. Nickell, and P. Weider, Synth. Commun. 9, 281 (1979).
123. M. Julia, Belg. Patent 818,336 (1974); Chem. Abstr. 83, 27519 (1975).
124. T. Imoto and T. Matsubara, J. Polym. Sci. 56, S-5 (1962).
125. T. Imoto and T. Matsubara, J. Polym. Sci., A 2, 4573 (1964).
126. M. Modena, S. Carraro, and G. Corri, J. Polym. Sci., B. Polym. Lett. 4, 613 (1966).
127. K. Shikata, Y. Miura, S. Nakao, and K. Azuma, Kogyo Kagaku Zasshi 68, 2266 (1965); Chem. Abstr. 64, 15721 (1966).
128. M. Marek and L. Toman, Ger. Offen. 2,140,448 (1972); Chem. Abstr. 76, 154480 (1972).

29

Electrogenerated Reagents

JACQUES SIMONET / Laboratory of Electrochemistry, University of Rennes, Rennes, France

I. Introduction 1217

II. Electrogenerated Species Used as Homogeneous Redox Reagents 1218
- A. General aspects of indirect electron exchanges 1218
- B. Pure redox catalysis: Use of organic mediators 1221
- C. Redox catalysis: An efficient tool for deprotecting groups in organic chemistry 1225
- D. Autocatalysis phenomena 1227
- E. Metallic cations as redox carriers 1228
- F. Reaction activation and catalytic role of the electron ($\overrightarrow{E}C\overleftarrow{E}$-type processes) 1237
- G. Indirect oxidations performed in the presence of halide ions 1242
- H. Electrogenerated superoxide 1245
- I. Perturbed redox catalysis: Electrogenerated nucleophiles 1247

III. Indirect Reductions by Means of Solvated Electrons 1253
- A. Use of solvated electrons in the presence of a proton donor 1254
- B. Reduction in amines 1255

References 1256

I. INTRODUCTION

Most electrochemical reactions essentially take place at the electrode interface by a heterogeneous, primary electron transfer. Other electron transfer reactions in solution, such as disproportionation, are not excluded. However, the heterogeneous electron exchange process is often complicated and consequently slowed by the intrinsic nature of the interface. Therefore, the scope of electrochemistry can be widened by the use of organic or inorganic mediators (electron carriers between the conducting interface and the solution) that thereby act as reducing or oxidizing reagents. Ideal mediators may then be regarded as an ionic or molecular electrode of a well-defined potential where inhibiting processes are effectively excluded. However, other kinds of electrogenerated species possessing either redox and/or nucleophilic or electrophilic properties may react with organic substrates by mechanisms in which electron transfer does not seem to play the major role.

Electrogenerated intermediates in general react in liquid electrolytes. More sophisticated indirect electrochemical and catalytic processes have been recently developed principally by using electrodes coated with conducting organic polymers. The electron itself (possibly transferred by indirect means) may contribute to initiate organic reactions according to so-called electrocatalytic processes. Last, the reactions of electrogenerated nucleophiles and electrophiles in solution toward solvent or electrolyte are not discussed in this chapter as they are dealt with in detail in other parts of this book.

II. ELECTROGENERATED SPECIES USED AS HOMOGENEOUS REDOX REAGENTS

A. General Aspects of Indirect Electron Exchanges

The electrochemical method allows the reduction or the oxidation of a mediator (or electron carrier) able to exchange electron(s) with a less active organic species at a given electrode (scheme 1). (The process of regeneration of the mediator is assumed to occur in good yield and then the indirect electron exchange, here a redox catalytic process, can happen either in solution or at the interface in the broadest sense.)

Polarographic "catalytic currents" are well known, especially in inorganic electrochemistry with classic examples, such as the reoxidation of ferrous ions—formed cathodically from ferric ions—by hydrogen peroxide or hydroxylamine. However, it appears obvious that the term "catalysis" is used too often quite ambiguously and its use in organic electrochemistry needs to be clarified; a distinction should be made between the two main kinds of catalysis. This differentiation has been emphasized by Andrieux et al. [1,2].

The term "redox catalysis" is used when the redox couple P/Q merely plays the role of an electron carrier to substrate X principally in homogeneous phase (there is no formation of an adduct of any kind between the organic species and the catalyst). The electron exchange between the two soluble entities Q and X occurs only by means of a so-called outer sphere

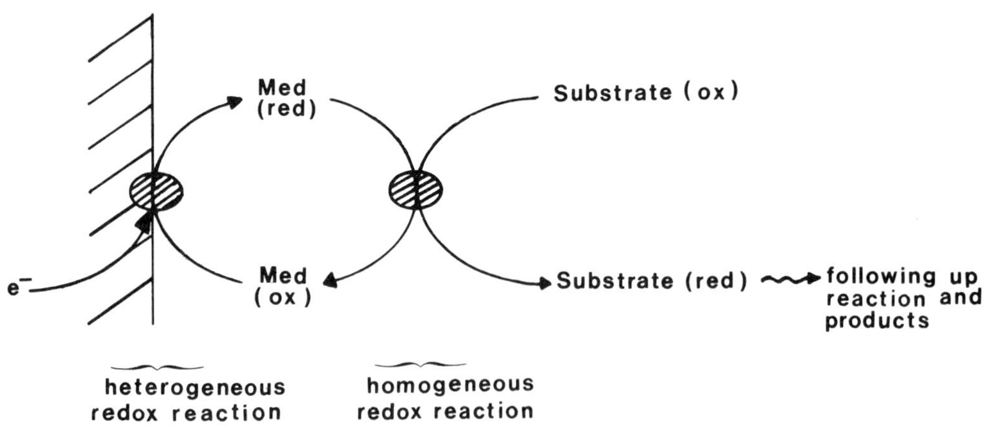

Scheme 1.

process (see Chap. 2). It is, for example, the case of the indirect reduction of aromatic halides by anion radicals electrogenerated from a properly chosen mediator (often an aromatic hydrocarbon or a ketone).

$$P \overset{+ne^-}{\rightleftharpoons} Q \qquad (1)$$

$$Q + X \underset{k_{(-2)}}{\overset{k_{(2)}}{\rightleftharpoons}} P + Y \qquad (2)$$

$$Y \xrightarrow{k} Z \text{ (products)} \qquad (3)$$

The term "chemical catalysis" is applied when the regeneration process is electrochemical and proceeds through the formation of an adduct that decomposes before or after an additional electron transfer. The electron may be transferred either from the electrode or from a reducing (or oxidizing) species in solution to regenerate one member of the catalyst couple P/Q. The indirect reduction of acidic protons in the presence of nitrogen aromatic heterocycles may provide a good example.

$$P \overset{+ ne^-}{\rightleftharpoons} Q \qquad (1)$$

$$Q + X \rightleftharpoons QX \qquad (4)$$

$$QX \longrightarrow P + \text{(Products)} \qquad (5)$$

and/or

$$QX + n'e^- \longrightarrow Q + \text{(Products)'} \qquad (6)$$

Catalysis of the reaction brings in both cases the possibility of a decrease in the overvoltage for a given current density. Furthermore, for the redox-catalyzed process, the change in the product distribution can be expected to be small, whereas in the case of chemical catalysis, the different possibilities for the formation and decomposition of the adduct (depending on the nature of the catalyst) may lead to a change in selectivity.

In this chapter the reader's attention is focused mainly on pure redox catalysis (Q is regenerated in high yields during the electrolysis) of organic compounds named X according to the process given in reactions (2) and (3). What are the conditions that make the observation of such catalytic currents possible? The only thermodynamic requirement is that the standard potential E^o_{PQ} is lower (in case of a reduction) than the standard potential corresponding to E^o_{XZ} taking into account the overall reaction X $\longrightarrow$ Z. Andrieux et al. [1,2] have emphasized that the catalytic efficiency does not always depend on the difference $E^o_{PQ} - E^{1/2}_X$, since the half-wave potential of X is influenced by different factors, such as the rate of the heterogeneous electron transfer to X ($X \longrightarrow Y$ at a given electrode) and/or the rate of the subsequent chemical reactions ($Y \longrightarrow Z$). Consequently, it has been proposed that a catalysis occurs (for a reduction) when one of the following sequences is fulfilled:

$$E^o_{XZ} > E^o_{XY} > E^o_{PQ} > E^{1/2}_X \tag{7}$$

$$E^o_{XZ} > E^o_{PQ} > E^{1/2}_X > E^o_{XY} \tag{8}$$

With regard to redox catalysis, it appears that the situation is more favorable, from a kinetic point of view, in case (7) than in case (8). Case (7) favors a displacement of equilibrium (2) to the right-hand side; the sequence corresponds to an indirect reduction of small molecules or ions in which the electron transfer involves a large change in solvation or structure. It is also the case of very bulky biological molecules in which the electroactive group for steric reasons is hindered from undergoing electron transfer at the electrode. On the other hand, large conjugated molecules, in which the change of solvation with the electron transfer is negligible, are more likely to be reduced according to case (8). Here equilibrium (2) remains displaced to the left-hand side when the solution electron transfer (SET) is not associated with a chemical reaction, which in certain cases is the rate-determining step. In other words, it may be expected, for a given difference $E^o_{PQ} - E^{1/2}_X$, that redox catalysis is more efficient in case (7). These considerations may explain why most of the observed cases of redox catalysis involve small inorganic molecules. Moreover, the use of redox catalysis theory allows one to obtain thermodynamic values, such as the standard potential of very unstable species (in the case of aromatic bromides, chlorides, and fluorides [3]), and also to determine the lifetimes of ion radicals down to the nanosecond time range (Chap. 2) [4].

In practice, many preparative reactions do not follow the previous scheme at a reasonable current density because of side reactions. At the working electrode a modification of the surface may be observed that may change the nature and the overvoltage of the basic electrochemical reaction. This is, for example, the case when the mediator is an inorganic ion, leading during the electrolysis to either deposition or etching of the surface. Furthermore, the common use of mercury cathodes sometimes causes the unintentional formation of amalgams, which also may play the role of reducing reagents at the interface. Other kinds of electrogenerated reagents (not necessarily electron carriers) may be formed and the careful interpretation of the product distribution after workup may be difficult.

In the following, examples are presented of those indirect electrochemical reactions that occur according to either a one-stage process (mainly, a case of redox catalysis with an organic mediator) or a two-stage process (a case in which the chemical reaction is not performed in the cell). In the latter case it is worth noting that the process does not necessarily obey the previous catalysis scheme [Eqs. (1) through (6)], the X compound being possibly more electroactive than the "mediator," which is then a chemical reagent conveniently regenerated by electrochemical means. These schemes (1) permit the use of what may be expensive reagents in catalytic rather than in stoichiometric quantities, and (2) eliminate the environmentally important problem of disposing of spent reagents.

B. Pure Redox Catalysis : Use of Organic Mediators

Enhancement of polarographic currents, due to regeneration of the organic depolarizer by means of an electron exchange with another species in solution has been investigated in detail only in the last few years. One of the first examples of this type of organic catalysis was reported by Fry [5] and was concerned with the indirect reduction of n-butyl chlorides by the electrogenerated radical anion of naphthalene in a nonaqueous solution. Since that time, a large number [6–23] of papers have given further information on the homogeneous electron transfer, with, in certain cases, a useful theoretical analysis of the problem [1,2] for EC and ECE catalytic processes.

The most classic case of redox catalysis is the indirect cleavage of a carbon-halogen bond according to the process described in Eqs. (9) through (17) (A is the organic catalyst, RX is the organic species able to cleave after a single electron transfer, and SH is the solvent):

$$A + e^- \rightleftharpoons A^{\overline{\cdot}} \quad (E^o) \tag{9}$$

$$A^{\overline{\cdot}} + RX \underset{k_2}{\overset{k_1}{\rightleftharpoons}} A + RX^{\overline{\cdot}} \tag{10}$$

$$RX^{\overline{\cdot}} \xrightarrow{k_c} R^{\cdot} + X^- \tag{11}$$

$$R^{\cdot} + A^- \longrightarrow R^- + A \quad \text{(solution)} \tag{12}$$

or

$$R^{\cdot} + e^- \longrightarrow R^- \quad \text{(electrode)} \tag{12'}$$

$$R^{\cdot} + RX^{\overline{\cdot}} \longrightarrow R^- + RX \quad \text{(disproportionation)} \tag{13}$$

$$R^{\cdot} + R^{\cdot} \longrightarrow R{-}R \quad \text{(dimerization)} \tag{14}$$

$$R^{\cdot} + HS \longrightarrow RH + S^{\cdot} \tag{15}$$

$$S^{\cdot} + A^{\overline{\cdot}} \longrightarrow S^- + A \tag{16}$$

$$S^{\cdot} + RX^{\overline{\cdot}} \longrightarrow S^- + RX \tag{17}$$

Reactions (9) through (12) of this mechanism, which present the main reaction pathway concerning RX, correspond to the two-electron conversion of RX into R^-. The mediator A (or homogeneous catalyst or electron carrier) is assumed here not to be consumed (therefore no fast reaction of $A^{\overline{\cdot}}$ with the intermediates is includes in the scheme). Consequently it is totally regenerated (high turnover) during the duration of the electrolysis. Such a case seems to be [12] the indirect reduction, of benzyl chloride by anthracene anion radical, for example (Fig. 1). In this case, a continuous increase in the cathodic current is observed, at least for moderate ratios of [RX] to [A].

When the rate of the chemical reaction (11) following the solution electron transfer is high, it can be considered the driving force of the overall process. The rate of the catalysis is therefore determined by the electron ex-

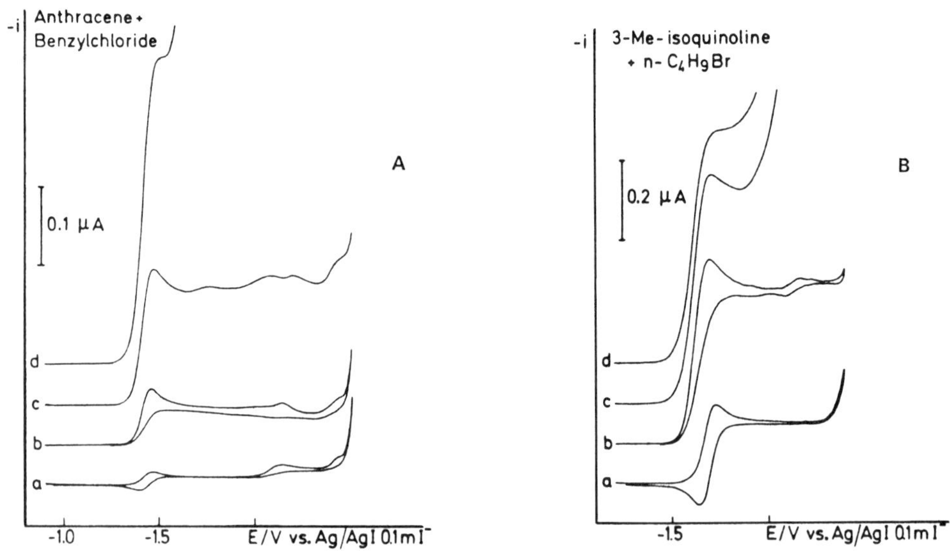

Figure 1. (A) Cyclic voltammograms of anthracene (3.4×10^{-4} M) in DMF-0.1 M TBAI in the presence of different concentrations of benzyl chloride. (a) 0 M; (b) 7.3×10^{-4} M; (c) 3.7×10^{-3} M; (d) 1.85×10^{-2} M benzyl chloride. Sweep rate 10 mV s^{-1}. (B) Cyclic voltammograms of 3-methylisoquinoline (5.8×10^{-3} M) in DMF/0.1 M TBAI in the presence of different concentrations of butyl bromide. (a) 0 M; (b) 5×10^{-3} M; (c) 3.2×10^{-2} M; (d) 6.4×10^{-2} M butyl bromide. Sweep rate 10 mV s^{-1}. (From Ref. 12.)

change reaction; k_1 depends on the potential difference ΔE^o between the standard potentials of two redox couples $A/A^{\cdot-}$ and $RX/RX^{\cdot-}$ (see Chap. 2). The latter is very difficult to ascertain by conventional electrochemical methods, and it is easier to analyze the catalytic efficiency as a function of the half-wave potential $\Delta E_{1/2}$, although $E_{1/2(RX)}$ (determined for an overall two-electron cleavage) is in general somewhat less cathodic than the standard potential corresponding to the first charge transfer. Figure 2 shows the variation of log k_1 with $\Delta E_{1/2}$ and ΔE^o to illustrate the indirect reduction of chlorobenzene, considered an RX type of species in which $R^{\cdot}$ is of low reactivity toward A and/or its reduced forms. The reducing species are anion radicals of several aromatic hydrocarbons and heterocycles. In Fig. 2a, k_1 is calculated [6] using polarographic limiting currents by the Koutecky method [24], and in Fig. 2b, k_1 is determined by a more careful and widely usable approach in voltammetry, taking into account the results of the Hush-Marcus theory of outer sphere homogeneous and heterogeneous electron transfers [(Refs. 1, 2, and 19) and references given therein].

From Fig. 2b it appears that for small values of $\Delta E_{1/2}$, the solution electron transfer is no longer diffusion controlled. A nonlinear increase in log k_1 with decreasing ΔE is characteristic of activation control of the electron transfer. Moreover, an analysis of the results allowed, according to Savéant's theory [1,2,19], indirect determination of the standard potential of RX. For example, for chlorobenzene, the standard potential E^o is shifted +66 mV compared with the polarographic half-wave potential. The E^o values for other two-electron cleavages have been determined by this method [25,

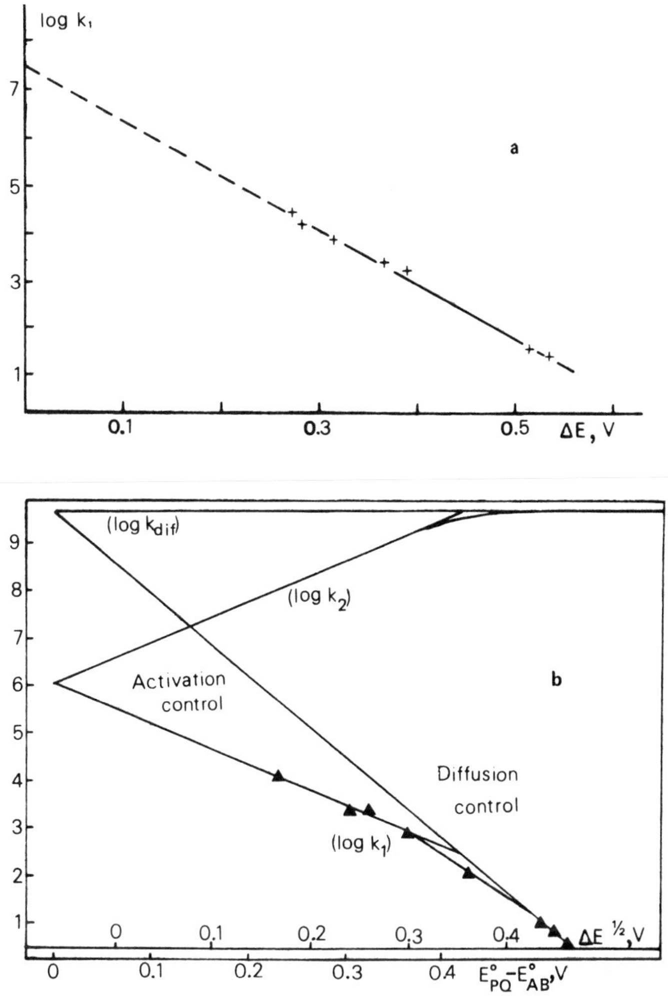

Figure 2. (a) Dependence on ΔE of the logarithm of the rate constants of the chlorobenzene reoxidation of the following anion radicals (with increasing ΔE): biphenyl, 1-methylnaphthalene, naphthalene, 2-methylphenanthrene, phenanthrene, p-terphenyl, and benzonitrile. (From Ref. [6]). (b) Rate constants of the homogeneous electron exchange between chlorobenzene and redox catalysts as functions of the half-wave potential difference and of the standard potential difference. With increasing ΔE: biphenyl, naphthalene, dibenzothiophene, phenanthrene, p-toluonitrile, m-toluonitrile, p-terphenyl, and benzonitrile. (From Refs. 1 and 2.)

26]. When the chemical step [reaction (11)] is slower, the overall process is no longer kinetically governed by the indirect electron transfer but by the rate of the follow-up reaction. In such cases, k_c may be calculated [19]. The redox catalysis is therefore of use in the determination of rates of very fast irreversible chemical reactions (e.g., those occurring in cleavage reactions); determination of such rate constants can easily be made either by linear sweep voltammetry [27] or by all kinds of stationary electrochemical techniques [28], even in the cases of small excess factors (Chaps. 2 and 3).

Fast cleavage reactions associated with solution electron transfer processes usually provide the driving force of the redox catalysis. Beside organic halides (X = I, Br, or Cl), aromatic sulfones and sulfides, activated acetates, ethers, and alcohols may be cleaved by indirect electrolysis. In general, organic species reduced or oxidized according to an ECE, Disp, or EC process (with a reasonably fast chemical reaction) are candidates for exhibiting redox catalysis. One should discriminate carefully between two-electron and one-electron schemes (rate of reduction of $R^{\cdot}$ by the mediator), however, before calculating the k values belonging to the indirect process. Compounds possessing no potential leaving group but that react quickly after SET have been reported as catalytic processes. Examples include the following:

Indirect reduction of activated olefins [12] (e.g., anthracene in the presence of acrylonitrile). The SET is then followed by a fast dimerization or polymerization.

Indirect opening of some rigid small rings, such as epoxides and methylene cyclopropanes [29].

Indirect hydrogenation, in which the electron transfer is followed by a protonation reaction with residual water (A^- is assumed not to be affected by the presence of the proton donor), for example the indirect reduction of thiophenes [17] and of conjugated dienes and trienes [30] using aromatic hydrocarbons as mediators. Here the catalytic enhancement was strongly dependent on the availability of protons given by the solvent and the driving force was assumed to be the fast protonation of the corresponding basic intermediates formed from the polyenes on electron transfer.

Indirect cathodic elimination as in the conversion of 1,2-dichloro-1,2-diphenylethane into stilbene [31].

Indirect decomposition of tetraalkylammonium [8,32,33] and trialkyl sulfonium [32] cations.

Indirect oxidative decarboxylation of carboxylates [18] in the presence of tris(p-bromophenyl)amine. Since the radical cation of this mediator is not deprotonated rapidly, other kinds of catalytic oxidations have also been proposed by Schmidt and Steckhan [34]; the fast chemical reaction is then, after the SET, a deprotonation reaction substantially accelerated by the presence of a base, such as 2,6-lutidine.

Advantageous change in the product distribution in the indirect cleavage of vinyl halides (e.g., the removal of bromine from the(carcinogenic)5-acetyl-10-bromo-5H-dibenz[b,f]azepine]35].

C. Redox Catalysis: An Efficient Tool for Deprotecting Groups in Organic Chemistry

Several deprotection reactions have been conducted by indirect means. In most cases the purpose was to use the indirect electrolyses to obtaining a better selectivity with higher current yield. For example, the oxidative debenzylation of some benzyl ethers permits [36] the cleavage into alcohols to be achieved in good to very good yields. It is worth noting that this method of deprotection does not attack double bonds, in contrast to more classic reductive debenzylation procedures [37]. The anodic cleavage of C—O bonds is here promoted by the action of nonnucleophilic bases, such as lutidine or collidine. In such oxidations a mediator must be used that forms a long-lived cation radical that must not (or only slowly) react with the solvent or the base. In practice few types of organic compounds are useful as mediators for oxidations. Thus, Schmidt and Steckhan [38] demonstrated the usefulness of a wide range of differently substituted triphenylamines with different E° (from about +1 to +2 V versus NHE) as mediators in oxidations; thus a mediator with an E° suitable for a given oxidation may be selected (scheme 2).

The advantage in terms of selectivity of the indirect deprotection of benzyl ethers has been exemplified [36] in the case of the anisyl ether of 4-phenyl-3-butenol in which the substrate possesses two electrophores whose potentials are slightly different. In contrast to inefficient direct oxidation, indirect electrochemical cleavage occurs with a yield of 80%. Benzyl esters have also been cleaved successfully with this method [39].

The scissions of C—S bonds may be achieved anodically under mild conditions with the help of a mediator. Such a possibility was used in the indirect anodic deprotection of carbonyl groups by oxidizing the corresponding

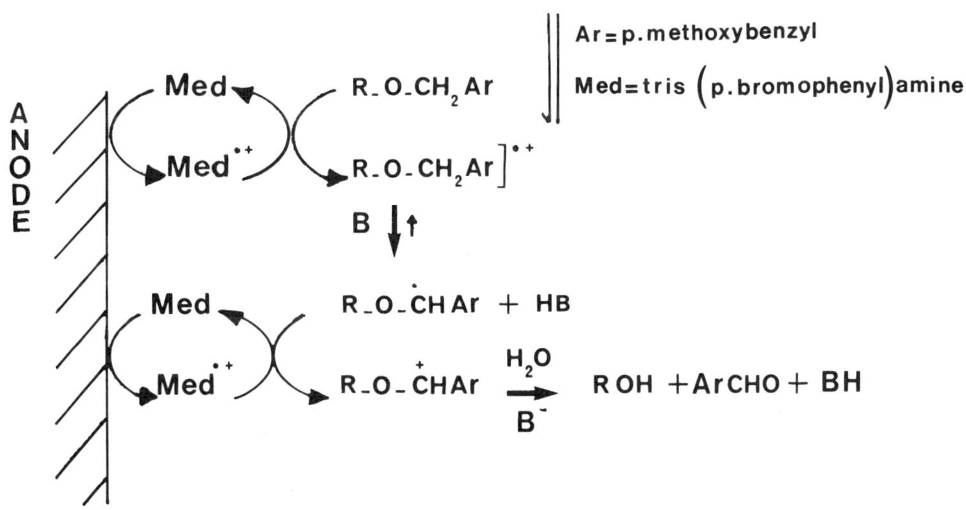

Scheme 2.

1,3-dithianes and 1,3-dithiolanes. Similar results were also reported for gem-dithioacetals [40].

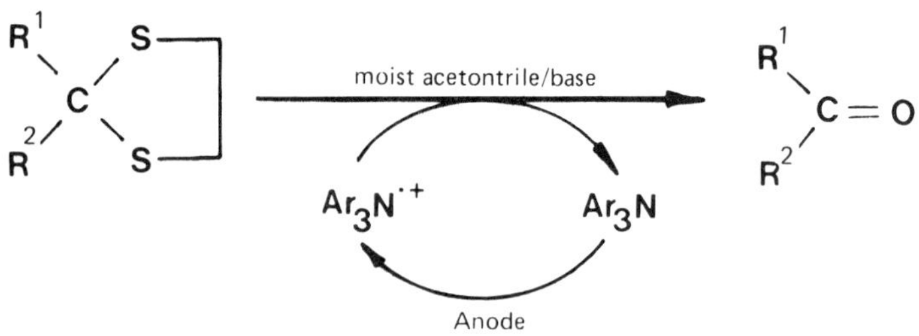

(18)

Other indirect reactions concerning the anodic cleavage of C−S bonds have been exploited, such as the formation of disulfide bridges in polycystinyl [41] peptides and the oxidation of thioesters [42].

Cathodic deprotections are also feasible when using organic mediators. An important use of indirect electrolyses is that it allows reaction beyond the cathodic limit. It is therefore possible to reduce substances whose reduction potential is more negative than that of the electrolyte. Thus, if a mercury cathode is used, the aprotic solvent-alkali metal salt system appears to be inconvenient since many compounds are cathodically cleaved and/or deprotected at potentials beyond that of alkali-metal amalgam formation. Nevertheless, in certain cases the use of lithium salt as electrolyte is necessary (e.g., to avoid the possibility of a Hofmann degradation of the tetraalkylammonium ion by electrogenerated bases). Under such experimental conditions, the cathodic synthesis of some aza and aza-oxa ligands [43] has been successfully achieved from the corresponding and readily available polysulfonamides. Pyrene, stilbene, and anthracene were shown to be convenient mediators. Kinetic parameters concerning such indirect deprotections of amines are also available [44].

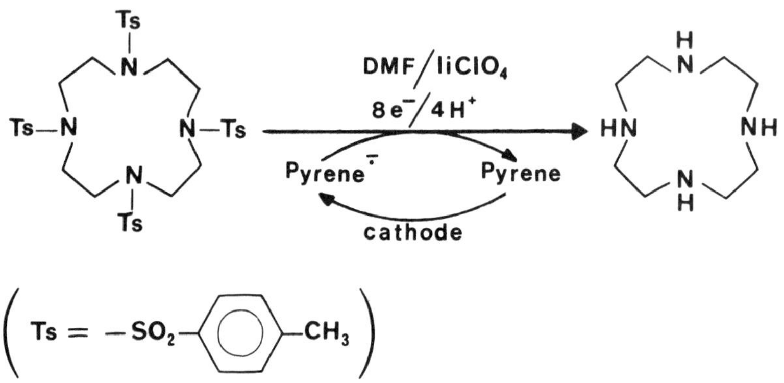

(19)

Arylsulfonyl groups, required to render the α-carbon atom more nucleophilic, can be removed by means of indirect reductions [45]. The increase in the selectivity compared with that obtained with a working electrode of large area can be observed with well-chosen mediators. In scheme 3 [20], the reduction of disulfones denoted XRY (X and Y are the two sulfonyl

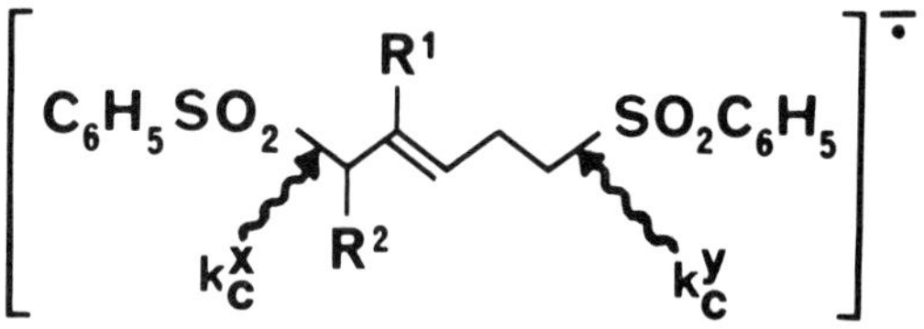

Scheme 3.

moieties differently activated) was investigated: for such structures, the difference in peak potentials corresponding to the electrochemical cleavage of X^- and Y^- anions is rather small (of the order of 150 mV). Therefore, under conventional electrochemical conditions, a very selective cleavage of the more activated group when using an ordinary laboratory cell is difficult to achieve. The main reactions in the indirect reduction are assumed to be

$$A + e^- \rightleftharpoons A^{\overline{\cdot}} \tag{20}$$

$$A^{\overline{\cdot}} + YRX \underset{}{\overset{K_x(K_y)}{\rightleftharpoons}} A + YRX^{\overline{\cdot}}(XRY^{\overline{\cdot}}) \tag{21}$$

$$YRX^{\overline{\cdot}}(XRY^{\overline{\cdot}}) \xrightarrow{k_x(k_y)} YR^{\cdot} + X^-(XR^{\cdot} + Y^-) \tag{22}$$

$$A^{\overline{\cdot}} + YR^{\cdot}(XR^{\cdot}) \xrightarrow{\text{fast}} A + YR^-(XR^-) \tag{23}$$

To obtain the best selectivity, the mediator A must be chosen so that K_Xk_X has a reasonable magnitude, whereas K_Yk_Y should be nearly negligible. Thus, when the disulfone anion radicals are formed from anthracene anion radical, the allylic sulfonyl group is cleaved first and then the fairly stable blue color of $A^{\overline{\cdot}}$ appears at $2e^-$ per mole. This demonstrates that electron exchange with the second sulfonyl moiety is much slower. This experimental fact is in agreement with other experimental data taking into account a saturated sulfone alone. However, it is worth noting that sulfones in strongly aprotic media may be sufficiently acidic to lead to some protonation of the anion radical of the mediator [46].

D. Autocatalysis Phenomena

Examples of autocatalyzed cathodic reactions have also been provided in which the first two-electron reaction (mainly a cleavage) is followed by a rapid elimination process leading to an unsaturated compound more easily reduced than the substrate itself (scheme 4). This has been observed when some α,α'-disubstituted 1,2-diphenylethanes [21] or analogous epoxides are reduced in nonaqueous media [47].

Such autocatalyzed redox reactions can, in the case of rather fast SET, be followed by means of voltammetric techniques (abnormally large variation of $d \log E_p/d \log v$, where v is the sweep rate) and chronopotentiometry ("unexpected" maximum in E–t curves). It thus appears that a careful stereochemical study of an elimination process (possibly influenced by the

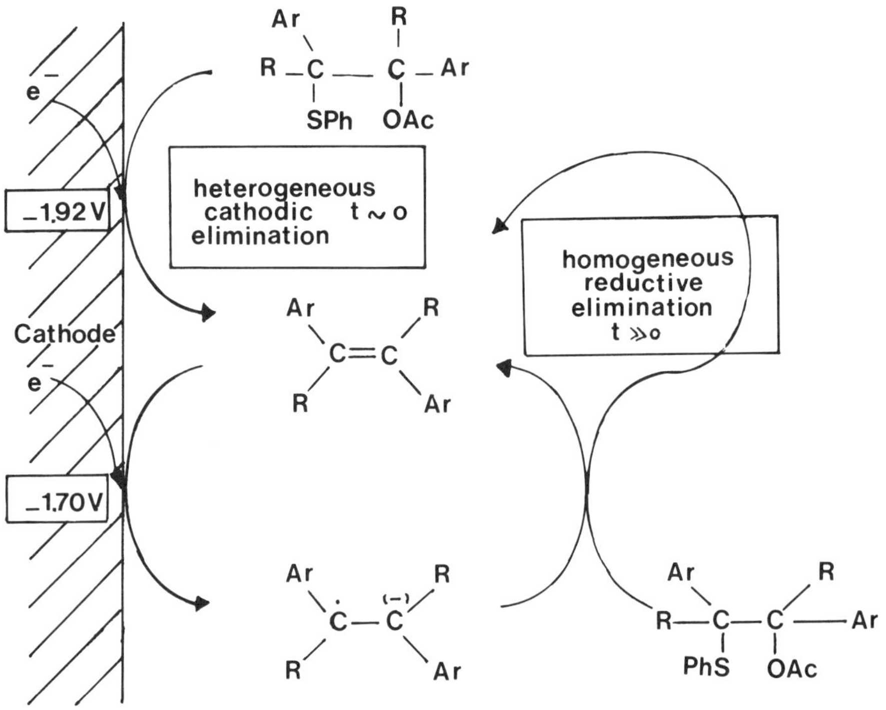

Scheme 4.

physical nature of the interface) must take into account a possible concomitant occurence of a homogeneous reaction. Knowledge of the redox potential of the olefin may furnish valuable information on the product of elimination.

Sometimes anions (either carbanions from a cleavage reaction and/or the associated leaving group) may themselves reduce the starting material. Therefore, similar autocatalyzed processes are observed in which the rate of the indirect reduction increases owing to the continuous accumulation [48,49] of a reducing species; the process is then compared to an "avalanche" or a "rolling snowball" to underscore the growing importance with time of the homogeneous electron transfer.

E. Metallic Cations as Redox Carriers

During the three last decades, techniques for indirect electrolyses using two-stage processes (also called "ex-cell" processes) have been developed. The inorganic reagent, acting in a separate chemical reactor either as oxidizing or reducing species, is generally regenerated in an electrochemical cell after its separation from organic products (Fig. 3). The ideal system works when a continuous and constant stream of the electrogenerated redox reagent from the cell is fed to the reactor at the same rate as the solution of the organic species [51].

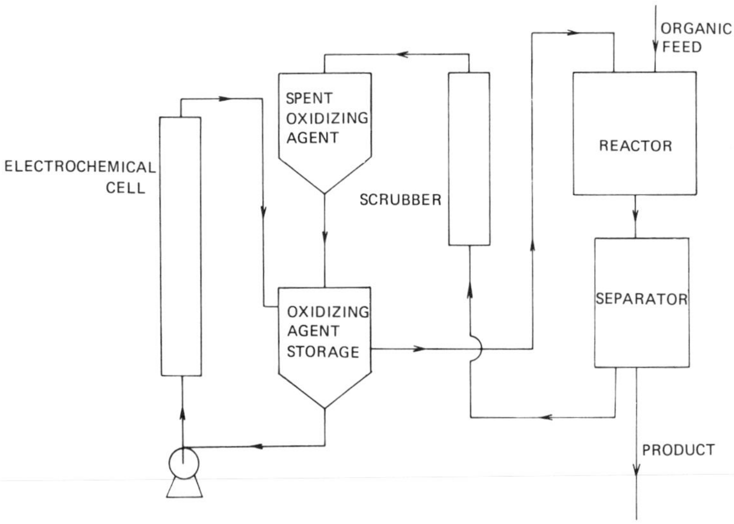

Figure 3. Flowsheet for an indirect electrochemical oxidation process with a separate chemical reactor. (From Ref. 50.)

A number of one-stage (in-cell) or two-stage (ex-cell) electrochemical conversions of aromatic systems using inorganic oxidizing mediators are listed in Table 1. The list is not exhaustive.

Among electron carriers used for indirect oxidation reactions, cerium salts (Ce(IV) + e^- $\rightleftharpoons$ Ce(III), E^o = + 1.44 V versus NHE) appear to be of interest when a mild oxidation must be conducted. Substituted toluenes are converted into the corresponding aldehydes in high yield [68–71]. Acidic solutions are required (such as aqueous AcOH, aqueous methane sulfonic acid, or aqueous trifluorosulfonic acid). Additional oxidations giving good yields in methylene chloride using platinized titanium anodes have also been reported. The conversion of aromatic compounds into quinones may also be conducted by means of electrogenerated ceric ions. For example, 1,4-naphthoquinone is obtained from naphthalene on oxidation using a two-phase system [72,73].

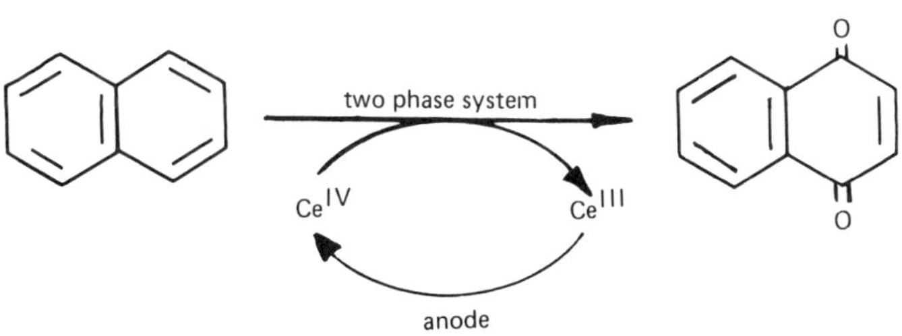

(24)

In the latter example [72], the oxidizing mixture appears to be rather complex since the presence of Ag^{2+} is assumed to be essential. The formation of 3,3-dimethylglutaric acid (yield 90%) from cyclohexane-1,3 dione [74]

Table 1. Some Examples of Indirect Oxidative Conversions by Means of Electrogenerated Inorganic Mediators[a]

Type of oxidative conversion	Basic catalyst/electrolyte	Electrode (in-cell) or (ex-cell)	Reaction	References
ArH → quinone	$Cr_2O_7^{2-}/H_2SO_4$	Pt/in-cell	Anthracene → anthraquinone	42,52,53
	Cr^{6+}-Ag^{2+}/H_2SO_4			
	Ce^{+}/H_2SO_4			
	$(NH_4)_2Ce(NO_3)_6/HNO_3$	Pt/ex-cell	Anthracene → anthraquinone	54
	Ce^{4+}/H_2SO_4	PbO_2/ex-cell	Naphthalene →naphthaquinone	55
	Mn^{3+}/H_2SO_4	PbO_2/in-cell	Aniline → benzoquinone	56
$ArCH_3$ → ArCHO	Mn^{3+}/H_2SO_4	PbO_2/in-cell	Toluene → benzaldehyde	57–59
	$Ce^{4+}/HClO_4$	TiPt/in-cell	Toluene → benzaldehyde	60

	Ce^{4+}/CF_3COOH	Pt/in-cell	Substituted toluene → substituted benzaldehyde	58,61
	Mn^{3+}/H_2SO_4	PbO_2/in-cell		
$ArCH_3$ → ArCOOH or Ar-Alkyl → ArCOOH	Chromic acid	PbO_2	Oxidation of paranitrotoluene, toluene, xylenes	62
	Cr^{3+}-Ag^{+}/H_2SO_4	PbO_2/ex-cell	o-Toluenesulfonamide → saccharin	53,63
	Cr^{6+}	Pt/in-cell		
$Ar-CH_2-R$ → ArCOR	MnO_2/H_2SO_4	MnO_2	Tetralin → α-tetralone	64
>N−CH_3	RuO_2/aq. NaCl + CCl_4 (double mediator system)	Pt or Ni/in-cell (two phase)	Amine → amide	65,66
−C−OH (with single bonds above and below C)	Same	Same	Primary alcohols → acid Secondary alcohols → ketones	65,66

[a]Additional references and details may be found in Ref. 67.

and that of D-arabinose (yield 54%) from calcium D-gluconate [75] using ceric mediators have also been reported.

The use of the Mn^{3+}/Mn^{2+} couple (E = + 1.51 V versus NHE) for indirect oxidations enables conversion under mild conditions of methyl or methylene groups directly attached to a phenyl ring. The corresponding carbonyl function is formed [57,58,61,64]. However, more drastic oxidations leading to carboxylic acids are generally performed by chromic acid whose electrogeneration is catalyzed by chromium salts (at the anode Ag^{2+} is formed). The interest of that type of oxidation is exemplified principally by the preparation of saccharin from o-toluene sulfonamide (see Table 1). Chromic acid formed by electrochemical means may also be used in the conversion of anthracene into anthraquinone.

The indirect oxidation of olefins is usually done in the presence of the couples Tl^{3+}/Tl^{+} (E^{o} = + 1.25 V versus NHE), Pd^{2+}/Pd^{0} (E^{o} = + 0.83 V), or Os^{8+}/Os^{6+} (E = + 0.43 V). Oxidations by Tl^{3+} [76,77] and Pd^{2+} [76–78] are known to lead preferentially to carbonyl compounds. Other examples are the use of Tl^{3+} for obtaining diols [79] or Os^{8+} if cleavage of the ethylenic bond is desired [80]. Both carbonyl compounds and corresponding carboxylic acids are obtained when using Hg^{2+} ions as redox mediators (see Ref. 81 and 82 and references cited therein).

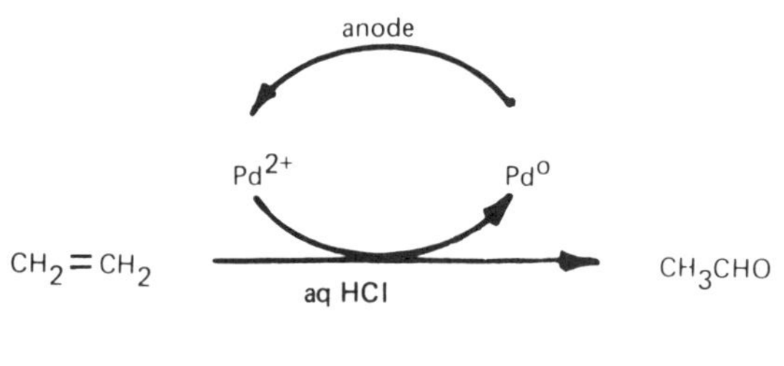

(25)

Moreover, one can say that any soluble high-valence inorganic species may potentially be a candidate as a valuable mediator for indirect electrochemical oxidation. The catalytic power can in some cases be greatly improved by using a double-mediator system (scheme 5) [65].

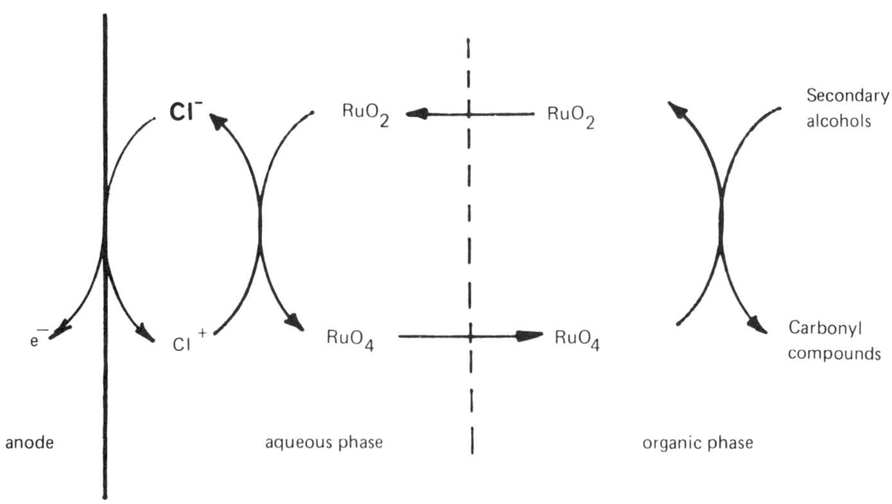

Scheme 5.

The correct choice of the electrode material (allowing a better regeneration of the "oxygen" carrier), and a selective extraction of reagents by an organic phase may improve the usefulness of the process.

On the other hand, inorganic redox mediators also appear to be useful for achieving indirect reductions of organic substrates (Table 2). Thus electrogenerated low-valence metal cations can be employed in both aqueous and aprotic dipolar solvents. In this latter category, great changes occur in the solvation of nascent low-valence (or even zero-valence) metal ions, which might give them a peculiar reactivity not found in organic chemistry. The use of environmentally undesirable transition metal ions in catalytic amounts with a continuous regeneration in situ is obviously of interest.

Thus, data concerning the use of salts formed directly at the cathode as reducing species are available. Many are well established; for example, the indirect reduction of nitrobenzene [87] in the presence of stannous ions appears to slow the reduction of the intermediate hydroxylamine and thus favor its rearrangement into p-aminophenol. There are also some advantages in conducting the Fenton reaction (hydroxylation of aromatic rings with reduced forms of oxygen) by indirect means in the presence of either copper salts [85,86] or iron salts [94]. The use of a double-mediator system [Cu(I)/Fe(III)] makes the reaction more effective. Similarly, the mixed reduction of hydroxylamine by Ti(III) can lead to a valuable source of aminyl radical: the transient radical can be trapped [88,95,96] in situ by the alkenes and aromatic derivatives. For example, the mixed electrolyses of 1,3-butadiene and hydroxylamine in the presence of titanium salts leads to diaminodimers, such as diaminooctadienes.

Electrogenerated Cr(II) has special reducing properties. In situ electrogenerated Cr(II) ions [Cr(III)/Cr(II) couple: $E^\circ = -0.41$ V versus NHE] have been widely used for various kinds of reactions, acetylene (totally electroinactive under usual conditions) may be reduced to ethylene in acidic media [90]. Similar reactions are found for nonactivated acetylenic derivatives using an ex-cell system (see Ref. 41). Steckhan [42,82,97–99] has developed several indirect cleavage reactions, essentially of C—Cl and C—Br bonds, with simultaneous dimerization, for example the reductive coupling of allylic bromides into the corresponding dienes and also the interesting reductive rearrangements of trichloromethyl-substituted secondary alcohols or ethers leading in one step to vinyl chlorides of Z configuration.

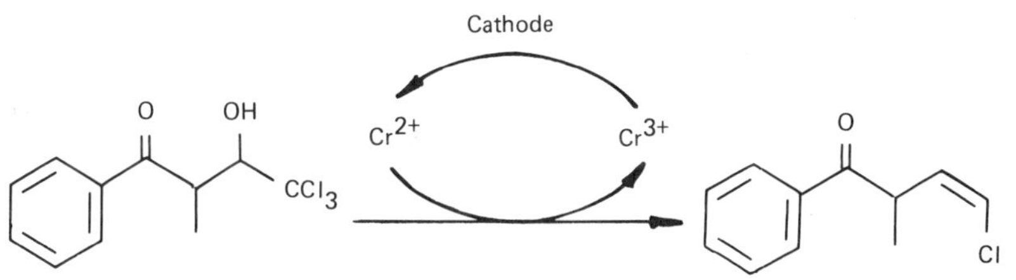

(26)

In the case of geminal dibromocyclopropanes, the indirect reduction by cathodically generated Cr^{2+} in anhydrous DMF affords the corresponding allenes. Such a process is assumed to proceed via an inner sphere electron transfer mechanism [99].

Further more, the addition of chromium salts to electrolysis solutions may give unexpected results. Thus, as was first demonstrated by Martinet et al. [91], pinacolization reactions may be carried out successfully by re-

Table 2. Some Indirect Reduction or Oxidation Reaction of Organic Substrates via Electrogenerated Inorganic Mediators

Heterogeneous electrochemical reaction	Redox potential[a] (V)	Examples of indirect electrochemical reactions → products (working electrode material)	References
$Ce^{4+} + e^- = Ce^{3+}$	1.44	Oxidation of substituted methylbenzenes → corresponding monoaldehyde and carboxylic acid[b]	50,58,61
		Electrochemical synthesis of saccharin from o-toluene sulfonamide	62
$Mn^{3+} + e^- = Mn^{2+}$	1.51	Oxidation of tetralin → α-tetralone	
$Tl^{2+} + 2e^- = Tl^+$	1.25	Mild oxidation of propylene → acetone	76,77
$Pd^{2+} + 2e^- = Pd$	0.83	Oxidation of olefins[b] → carbonyl compounds and degradation	76,78
$OsO_4 + 8H^+ + 8e^- = Os + 4H_2O$	0.85	Oxidation of acetylene → oxalic acid	76,78
		Oxidation of olefins → 1,2-glycols	76.83

$2AgO + H_2O + 2e^- = Ag_2O + 2OH^-$		0.60^c	Oxidation of aliphatic amines → nitriles	76,84
$2Hg^{2+} + 2e^- = Hg_2^{2+}$		0.90	Oxidation of propylene → acrolein	81
$Cu^{2+} + e^- = Cu^+$		0.16	Hydroxylation of aromatic rings in the presence of oxygen	85,86
$Sn^{4+} + 2e^- = Sn^{2+}$		0.15	Reduction of nitrobenzene → aminophenol	87
$Ti^{3+} + e^- = Ti^{2+}$		-2.0	Electrochemical formation of Ti^{2+} allowing the reduction of hydroxylamines, sulfoxides	88,89
$Ti^{2+} + 2e^- = Ti$		-1.63		
$Cr^{3+} + e^- = Cr^{2+}$		-0.41	Reduction of acetylene → ethylene	90
			Reduction of aromatic and polyethylenic ketones → pinacols	91,92
			Cleavage of carbon-halogen bonds	93

[a]Referred to NHE.
[b]Preferably according to a two-stage process.
[c]In this case, the electron transfer takes place at the anode interface, probably through an electrocatalytic mechanism.
[d]Also using cobaltic, manganic ions, or insoluble oxides, such as lead or manganese dioxides.

ducing activated carbonyl compounds in DMF in the presence of Cr^{3+}. Since this first intriguing and promising work, several papers [92,100,101] have considerably enlarged the potentiality of that coupling method and clarified the specific action of the redox couple Cr(III)/Cr(II); the components may be solvated by a donor aprotic dipolar solvent, its residual water, and by the carbonyl compound itself. A quite convincing explanation of the role of those metallic cations and their efficiency in the reduction-dimerization overall process has more recently been proposed by Sopher and Utley [102]. Thus, chromium(II) does not act as a redox agent but catalyzes the formation of a chromium(III) complex in which the carbonyl compound is involved: Cr(III) may be considered an acidic species, making the reduction of the carbonyl group easier, which moreover favors the pinacol formation.

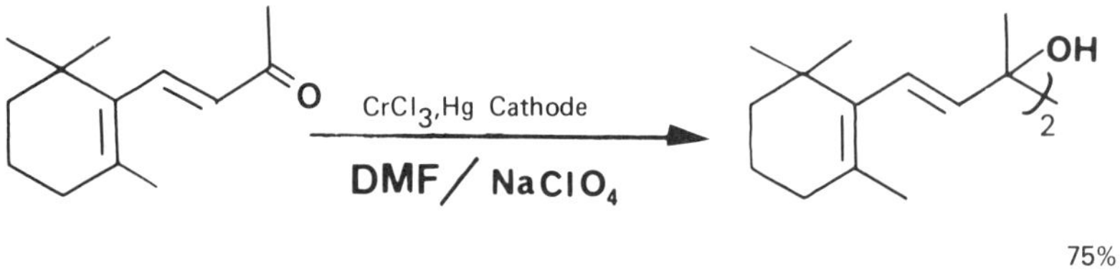

(27)

$$[Cr^{III}L_6] \rightleftharpoons [Cr^{II}L_6] \qquad \text{[L : donating species]} \tag{28}$$

$$[Cr^{II}L_6] \xrightleftharpoons{x\,R_2CO} [Cr^{II}(R_2CO)_xL_{6-x}] \xrightarrow{Cr^{III}L_6} [Cr^{III}(R_2CO)_xL_{6-x}] \tag{29}$$

(30)

The electrochemical method may also produce "zero" metals from transition metal cations: these metals can be used either as reducing species or as catalysts (scheme 6). Thus cathodically generated Zn(0), Cu(0), Fe(0), and Sn(0) are of interest [103,104] in indirect reductive dehalogenations or reduction of nitro groups in two-phase systems at high current density. The use of cathodically generated Ni(0) and Pd(0) in polar solvents is of great interest in new coupling reactions involving aromatic rings (symmetrical dimerization of aryl moieties from the corresponding halide [105,106] or cross-couplings leading to bisaryls). Proposed reaction mechanisms are rather complex since they must in principle take into account the cathodic formation of tetracoordinate Ni(0), the reductive insertion of the metal within the aromatic carbon-halide bond, and the increase by electron transfer of the nucleophilic behavior of the organometallic transient toward the starting aryl halide.

On the other hand, an excess of another electrophile in solution may modify, via the electrochemical activation of $ArNiXL_2$-complex, the chemical route. Thus addition of carbon dioxide affords [107,108] the carboxylation of the aromatic nucleus. The similar preparation of pharmaceutically important 2-arylpropionic acids has been reported [109] to proceed in good yields from 1-arylethylhalides.

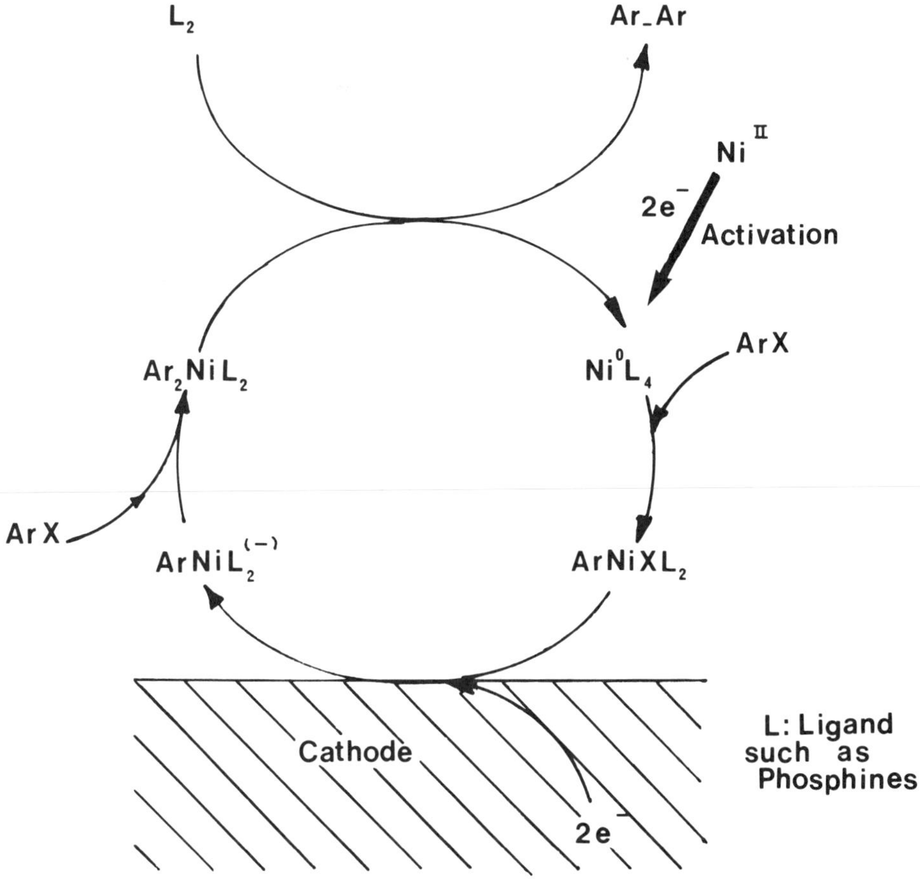

Scheme 6.

F. Reaction Activation and Catalytic Role of the Electron ($\overrightarrow{E}C\overleftarrow{E}$-type Processes)

Kornblum [110] and Bunnett [111] first demonstrated clearly that nucleophilic substitution reactions may occur in certain cases by a chain process initiated by a single electron transfer (scheme 7). The initial step is a reaction between the "nycleophile" (here acting as a reducing reagent) and the "electrophile." Such reactions with organic anions as electron-donating species may be general. In addition to photochemical induction processes, dissolved metals and solvated electrons are useful for initiating such reactions. Reactions of this are conveniently initiated and studied by means of electrochemistry [112]. The first example showing the role of the cathode in the activation of aromatic nucleophilic substitutions ($S_{RN}1$) was provided by Pinson and Savéant [113]; ArX compounds (e.g., p-bromobenzophenone) were reduced in the presence of a nonelectroactive nucleophile (here benzenethiolate). The aromatic nucleophilic substitution then obeys the following schemes:

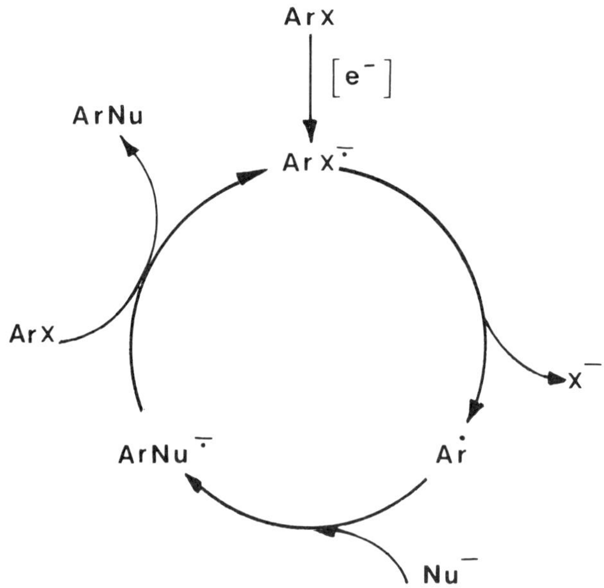

Scheme 7.

Inducing reaction:

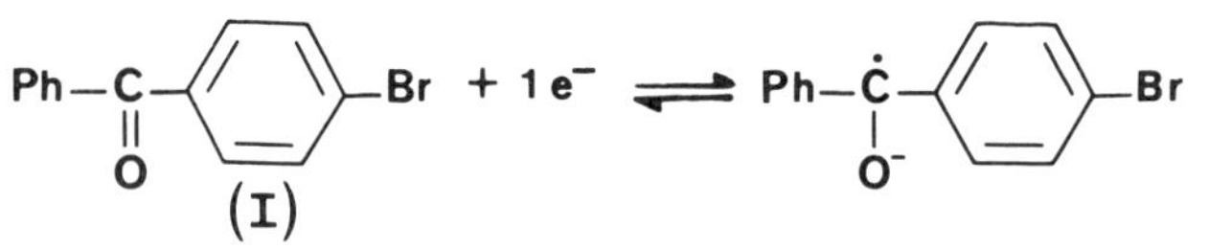

Standard potential E_1^0 (31)

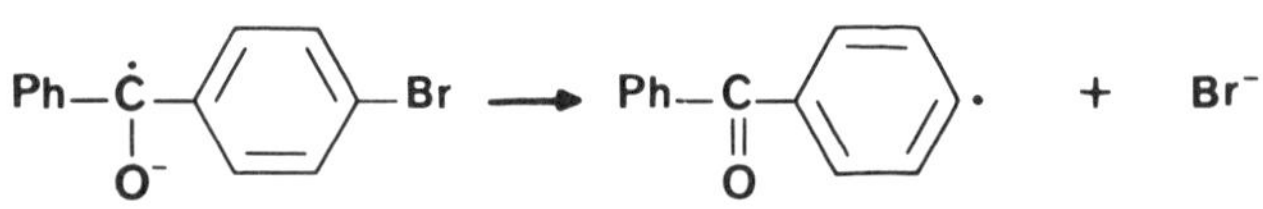

(32)

(33)

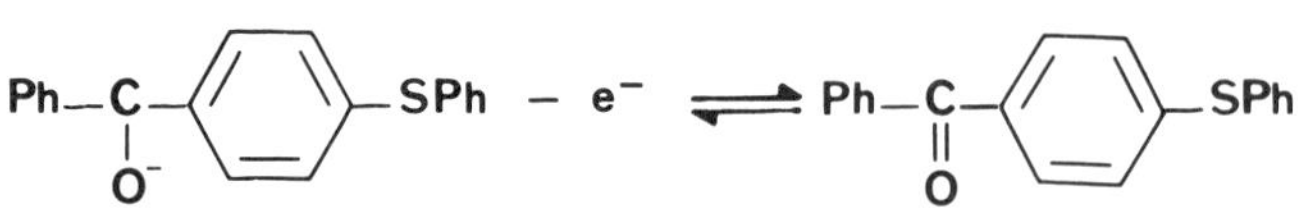

Standard potential E_2^0 (34)

Propagation reaction:

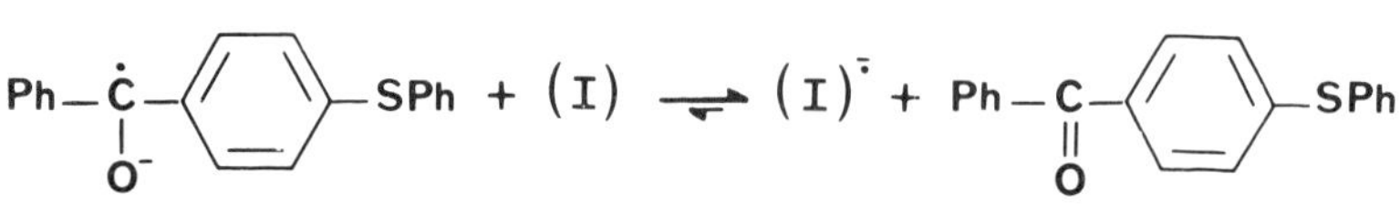

(35)

Generally speaking, the reduction or activation of ArX in the presence of a soft nucleophile Nu^- leads to the formation of ArNu. Given the thermodynamic requirement of $E^o_{ArX} > E^o_{ArNu}$, the yield of substitution is highly dependent on the ability of the medium to donate hydrogen atoms. Therefore most analytic studies concerning $S_{RN}1$ reactions were carried out in liquid ammonia [114], which is a poor hydrogen atom donor. The determination of absolute and relative rate constants for sufficiently activated aryl radicals toward nucleophiles [such as PhS^- and $(EtO)_2PO^-$, and $CH_3COCH_2^-$] leads to the conclusion that the rates are close to that of diffusion control.

The direct reduction of an aryl radical Ar$^{\cdot}$ formed in the electrode vicinity may be advantageously avoided by using an electron carrier [115]. The advantage of such an activation is apparent [116] in the synthesis of diphenyls substituted by both electron-donating and electron-withdrawing groups.

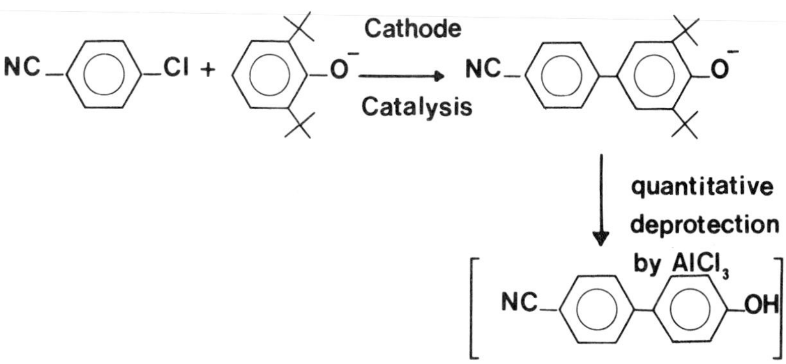

(36)

Direct method: 22% yield
Indirect method: 78% yield

The mediated electrochemical induction has also been used successfully with high yield in the synthesis of phenylselenobenzonitriles and phenyltellurobenzonitriles (scheme 8) [117].

Similarly, another type of nucleophilic substitution may be initiated by oxidative electron transfer (the $S_{ON}2$ type of reaction). The substitution has been found to follow such a scheme [118].

In addition to initiating substitution reactions, a cathode or anode may be used for catalyzing chemical reactions proceeding via electron transfer chains [119]. Efficient cathodic $\overrightarrow{E}C\overleftarrow{E}$ chain reactions are involved in cis-trans isomerism only if two conditions are fulfilled: (1) the standard potentials of the two isomers are sufficiently different ($\Delta E^o > 0.1$ V), and (2) the less electroactive isomer is the more stable. Therefore, in the absence of fast cleavage or protonation reaction, kinetic factors of chemical conversion coupled with the cathodic electron transfer (as activation step) may start a catalytic process. An example of such a process is the isomerization of enol ethers [30],

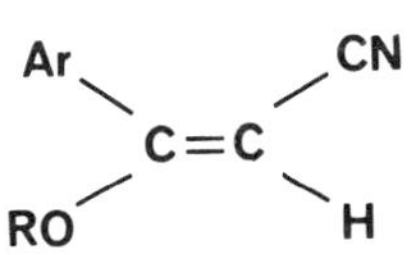

(Z and E)

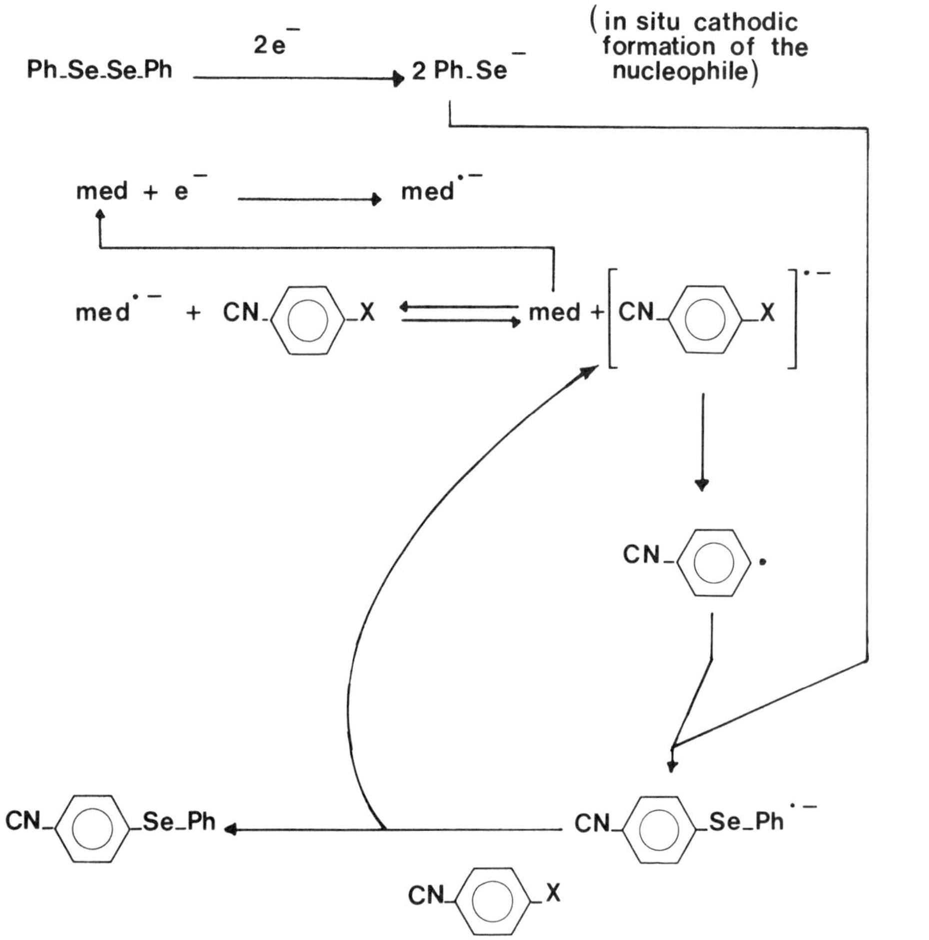

Scheme 8.

in which the less stable Z form is more easily reduced than the E form ($\Delta E^o \simeq 0.2$ V).

initiation $$Z \underset{}{\overset{e^-}{\rightleftharpoons}} Z^{\cdot-} \qquad E^0_Z \tag{37}$$

$$Z^{\cdot-} \xrightarrow{\text{fast}} E^{\cdot-} \tag{38}$$

chain propagation in solution $$E^{\cdot-} + Z \rightleftharpoons E + Z^{\cdot-} \qquad E^0_Z > E^0_E \tag{39}$$

Other examples have been provided by the anodically initiated isomerization of several dicarbene metal carbonyl complexes [120]. Electrochemical induction of chemical reactions can also be successfully used for conversion in NH_3 of alcoholates into ketones with simultaneous reduction of aromatic halides [121], the tetramerization of aziridines [122], or the ligand substitution [123] in organometallic compounds.

Cyclodimerization of an indene, rationalized as a [2 + 2] addition, probably induced by oxidative electron transfer, was described by Cedheim and Eberson [124]. Other kinds of [2 + 2] and [4 + 2] additions, the latter understood as an electron transfer-catalyzed Diels-Alder reaction, were recently proposed (scheme 9) [125–127]. For example, the mixed oxidation of substituted indenones and substituted styrenes may lead to the corresponding four- and six-membered ring adducts and their oxidation products. Because the activation needs the formation of a radical cation (then playing the role of an acceptor), these types of processes should instead be named [1 + 4] or [3 + 2] additions; in such a way the assumed activating species would be stated precisely.

The electron transfer-catalyzed addition may be pertubed by fast follow-up reactions of the transient radical cation. Such a situation is illustrated by a second process of cyclization with formation of an adjacent butenolide ring (in which the donor possesses an ester group in the β position). The overall process then occurs in two steps: the first corresponds to the expected activation, but in this case the radical cation of the adduct does not react in a chain (here a homogeneous oxidation of the ene) but undergoes a further two-electron cyclization step according to scheme 10.

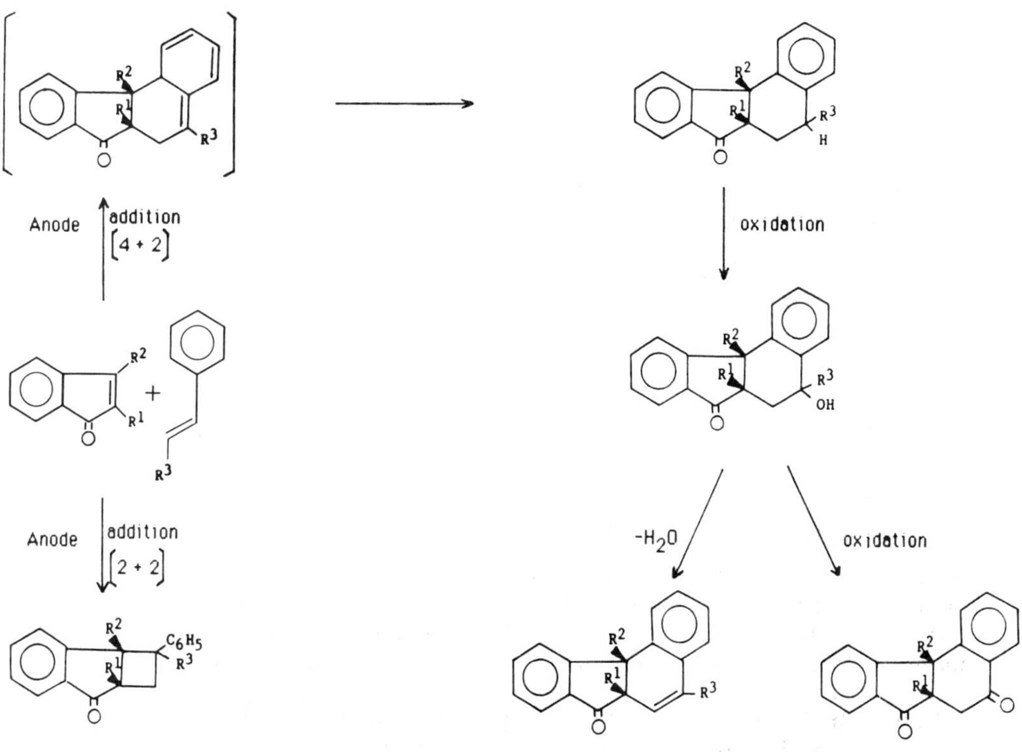

Scheme 9.

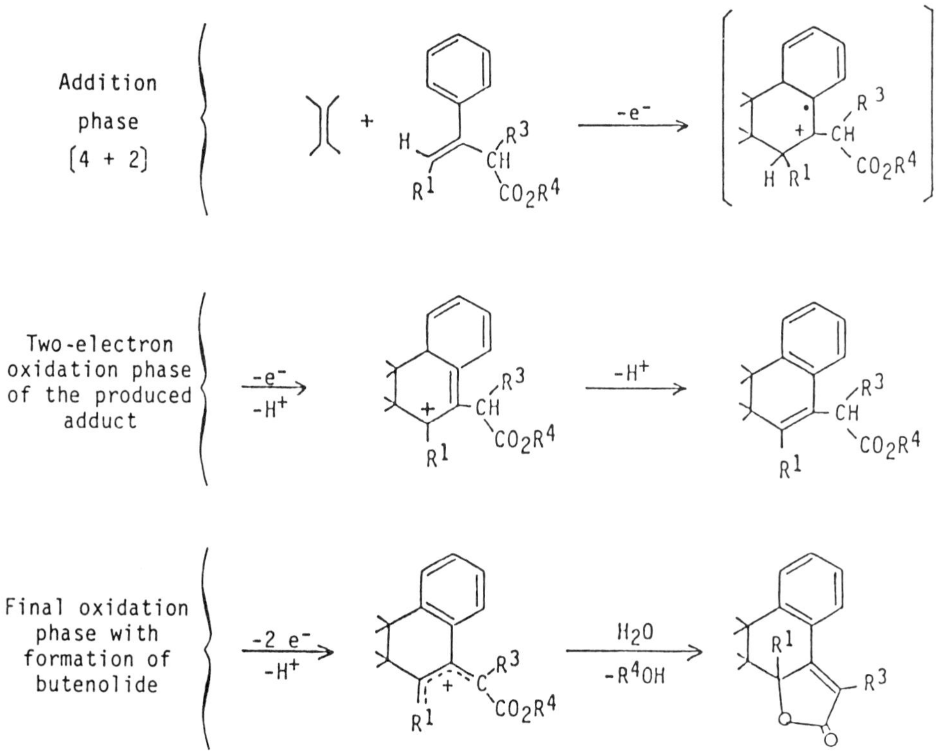

Scheme 10.

It should be noted [128] that attempts to produce [2 + 2] cyclodimerizations of indenones failed; the objective was an electrochemical route to fruxones. The main product of this reaction came from addition of residual water to the oxidized dimeric form.

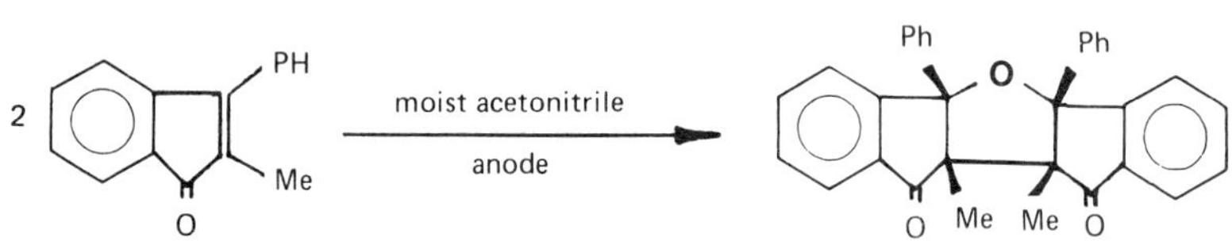

(40)

G. Indirect Oxidations Performed in the Presence of Halide Ions

Oxidations carried out in the presence of iodide and bromide ions have been extensively developed during the last decade. They may often be used with advantage within a potential range in which other kinds of catalyst are not numerous (e.g., organic mediators) or need to be used ex-cell. Moreover, active oxidized species obtained from I^- and Br^- (assumed generally to be "I^+" and "Br^+") are of a very broad applicability and are very active, even in water, toward substrates that are more difficult to oxidize than the halides. In most cases it has been demonstrated that such electrooxidized species often play a specific role at or near the anode, whereas the same halogens in bulk solution are unreactive. Only reactions offering that advantage are reviewed here.

Oxidants obtained directly from easily oxidized halides often react chemically with organic substrates; the resulting adduct may then undergo a reaction that regenerates the halide ion. Indirect oxidations of this kind are essentially inner sphere processes, as previously defined.

Many examples of reactions that may involve electrogenerated chlorine and bromine are in the literature. Thus, alkenes may be converted into halohydrins [129–132]: ethylene chlorohydrin can be formed [129] in 90% yield at a graphite anode, but subsequent elimination to give ethylene oxide depends critically on pH [135]. A similar electrochemical conversion of propene into propylene oxide [134] has been reported; a porous titanium anode coated with Pt/Ir is required. More complex olefins have also been transformed into epoxides; in these cases aqueous organic solvents were used with dissolved sodium bromide [135,136].

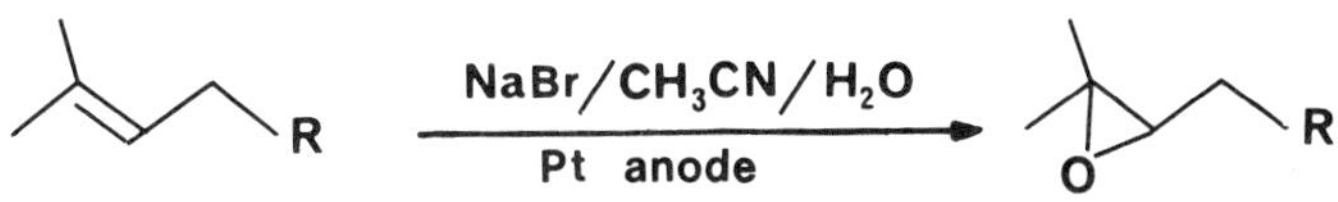

(41)

The presence of halide ions in the anolyte may efficiently help the oxidative addition of the solvent to an unsaturated compound; the addition may be followed by a bond cleavage [137]. The conversion of furan into either 2,5-dimethoxy-2,5-dihydrofuran [138,139] or 2,5-diacetoxy-2,5-dihydrofuran [140] has been shown to be possible in methanol or an acetic acid-acetonitrile mixture, respectively.

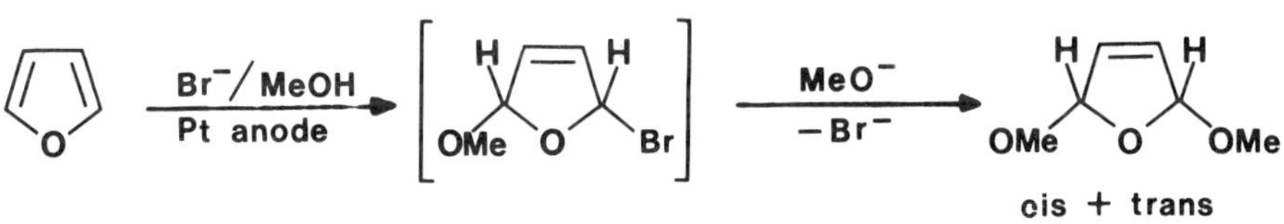

(42)

The oxidative coupling of activated C—H bonds has been shown to be catalyzed by the addition of iodide [141]. The indirect oxidation of alcohols in the presence of Br^- or I^- may, depending on experimental conditions, lead to carbonyl or carboxylic compounds [142,143]. Therefore, glucose can be oxidized in industrial scale to gluconic acid by oxidation using aqueous $NaBr/CaCO_3$ solutions [144]. Similarly, an efficient anodic conversion of isosorbide into the corresponding monoketone has been reported [145] using an aqueous NaBr solution.

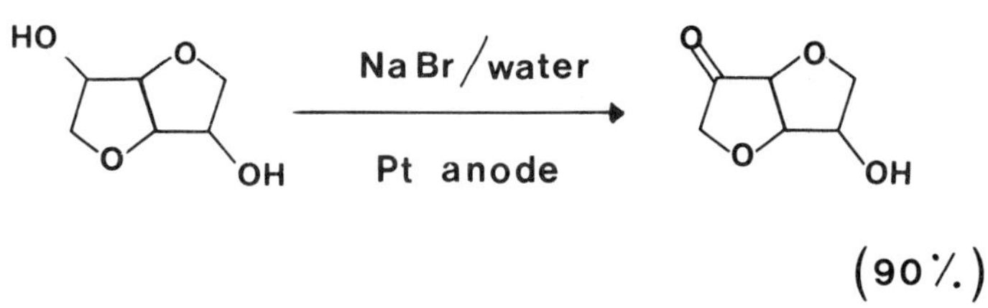

(43)

More complex oxidation processes can be carried out. Selenation and sulfuration reactions have been demonstrated using PhSeSePh/Br^- or RSSR/Br^- mixtures. In such reactions, anodically oxidized bromine species polarize Se—Se or S—S bonds positively leading to additions or substitutions in "adequate" yields [137]. In most cases, simultaneous addition of nucleophilic solvents is observed. Adducts so formed may themselves be anodically oxidized to give valuable allyl ethers, Eqs. (44) and (45).

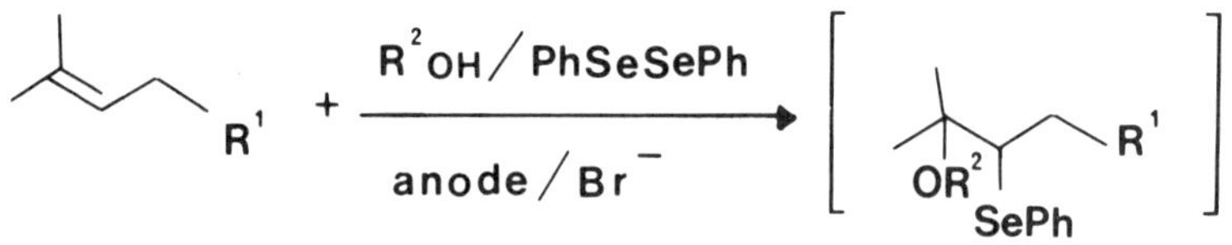

(44)

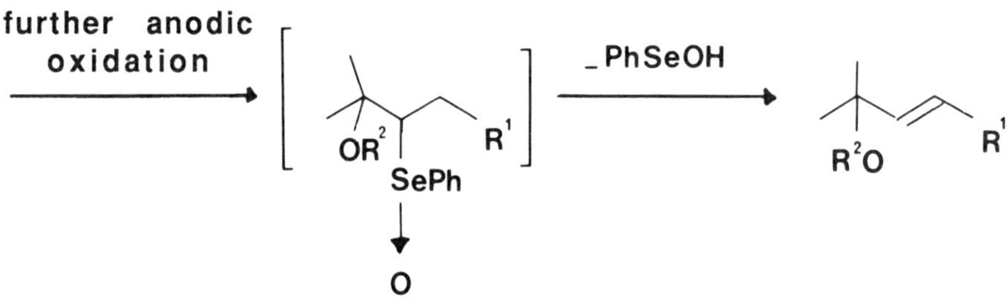

(45)

Such allyloxylation reactions allow, for example, the formation of (±)-rose oxide from citronellol [135]. The synthesis of butenolides can be performed efficiently by starting from β-unsaturated acids [146] as described in Eq. (46):

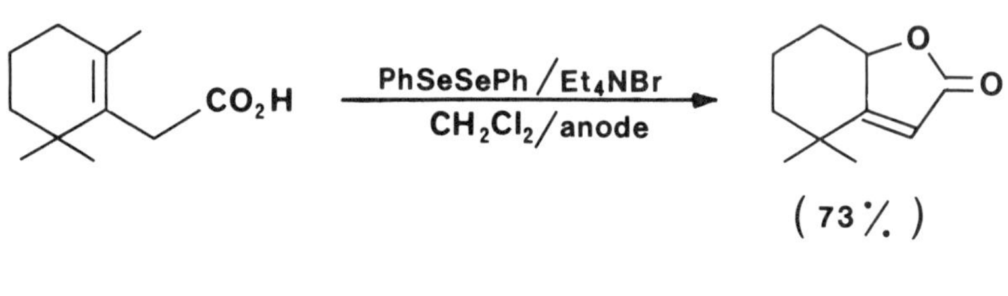

(46)

The formation of nitrogen-sulfur bonds, as in the synthesis of sulfenimides and sulfenimines [147], and phosphorous-sulfur bonds [148], is achieved by electrolyses of activated N—H and P—H compounds in the presence of an R—SS—R/Br^- mixture, Eqs. (47) and (48):

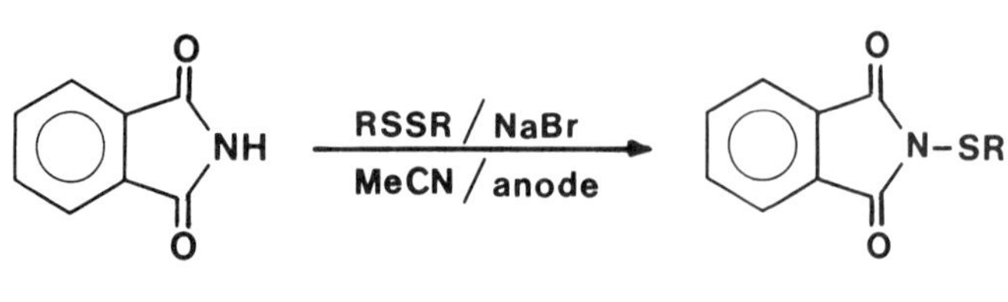

(47)

(90—93%)

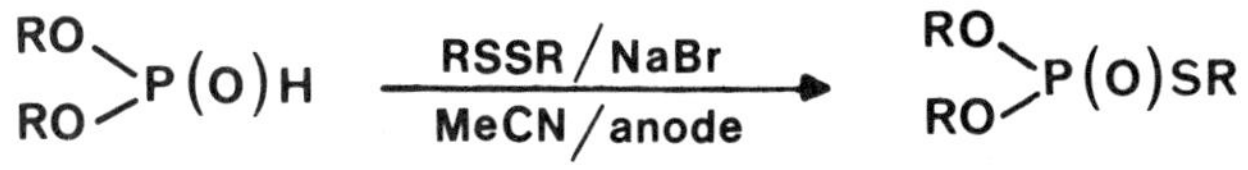

(48)

(45–97%)

Indirect oxidations in the presence of bromide may offer even more advantages, especially when experimental conditions have not yet been precisely defined. For example, the use of polymeric electrodes modified with poly-4-vinylpyridine hydrobromide, for example, may be used to effect the conversion of methylene into carbonyl [149] and sulfides into sulfoxides [150].

H. Electrogenerated Superoxide

The advantage of using the electrochemical method is often in producing intermediates that are not readily accessible by conventional means. The electrogeneration of transient nucleophiles and electrophiles is an example. The formation of superoxide ions by the cathodic reduction of oxygen in aprotic solvents (DMF, DMSO, and MeCN) is another case. Thus, in the absence of an efficient protonating reagent, oxygen gives a quasi-reversible system at −0.85 V versus SCE at commonly used cathodes (Hg, Au, Pt, graphite, or glassy carbon) in the presence of nonacidic cations, such as tetraalkylammonium salts [151–153]. The use of alkali metal salts as electrolyte cannot be recommended since the corresponding superoxides are often insoluble, especially LiO_2. Its precipitation at the electrode interface is inhibiting [152,154]. The relatively long half-life of superoxide (about 40 min at room temperature [155,156] allows reactions (even moderately slow ones) to occur in situ in the catholyte solution with continuous generation of the reagent; the rate of formation of $O_2^{\cdot-}$ is limited by the fairly low solubility of oxygen in the usual solvents.

Superoxide may act either as a nucleophile or as an electron donor. The chemistry of superoxide is also complicated by the fact that its protonated form $^{\cdot}O_2H$ is a moderately powerful oxidant. Consequently, the behavior of superoxide appears to be highly dependent on the proton activity of the solution. Alternatively, the acidity of the electroactive organic substrate may be important, depending on whether it reacts rapidly with the superoxide ion at or near to the cathode. Some reviews [157–159] attempt to shed some light on the organic chemistry of $O_2^{\cdot-}$ and its alkali metal salts, for which both solubility and nucleophilicity may be increased by the use of well-chosen ligands, such as crown ethers [157,158,160]. Note that continuous cathodic reduction of oxygen may lead directly to tetraalkylammonium superoxide, which is more soluble and more reactive than the alkali metal salts of $O_2^{\cdot-}$, which in any case must be used in situ in a polar solvent.

Electrogenerated superoxide reacts as a moderately strong nucleophile [51,155,161]; thus, the reaction with alkyl halides leading to dialkyl peroxides has been reported. Reactions toward Michael acceptors have also been described in the literature [155,160], such as the epoxidation of cyclohexen-3-one (yield 30%). However, such reactions are possible only for double bonds possessing a rather high activation; with less activated double bonds, such as chalcones [160] and benzalacetones, oxidative cleavage of the double bond can occur, which leads to the corresponding acids. Similarly, aromatic ketones [162] may be converted into carboxylic acids and imines [163] into amides.

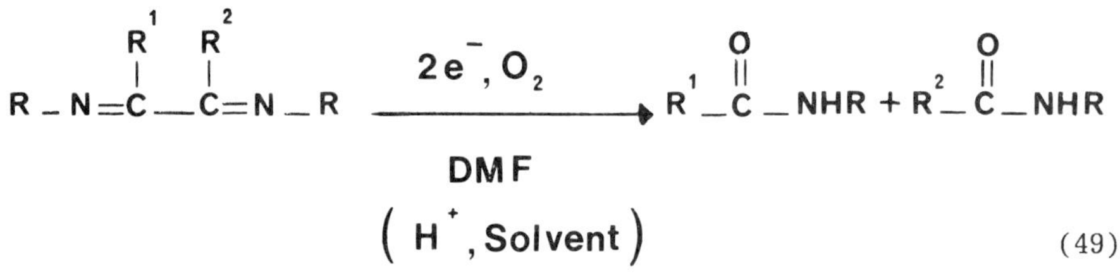

(49)

Moreover, superoxide may react with some activated aromatic halides to give through substitution the corresponding phenols [164,165]. A consequence of the results and reactions described earlier is that for electrolyses conducted in aprotic solutions, adventitious oxygen may quite often cause large discrepancies, not only in coulometric measurements but also in product distribution. Such interference has been shown [166] to be commonplace, especially for the reactions of activated ketones, carbon acids, and aromatic hydrocarbons.

Reactions of superoxide with activated C—H functions afford the formation of ketones or acids. Superoxide consequently acts as a strong oxidant toward organic substrates of a certain acidity [167–169]. However, within the malonate series, the reaction seems to be rather mild [170]: diethyl methylmalonate is thus oxidized by $O_2^{\bar{}}$ to diethylmethyltartronate in high yield [170].

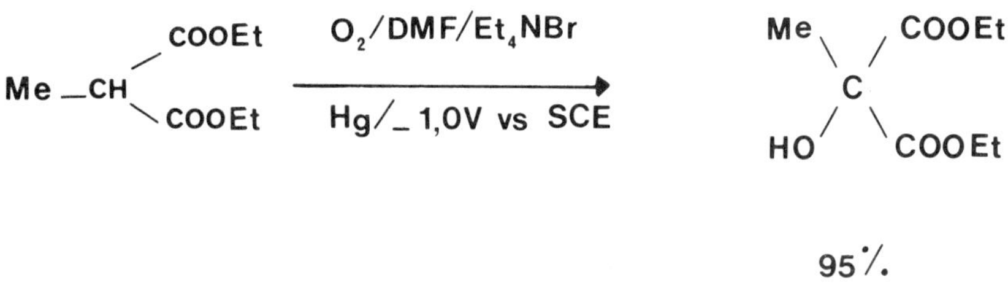

(50)

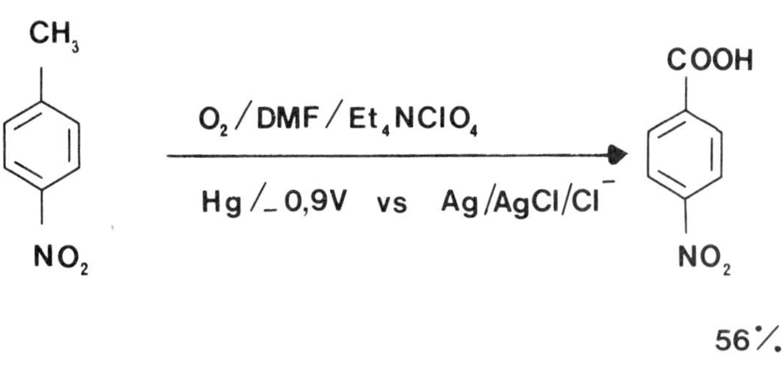

(51)

In such conversions, the primary chemical reaction appears to be the protonation of the electrogenerated base (here the superoxide) by means of the organic substrate.

Finally, the selectivity of CH hydroxylation by means of superoxide can be well illustrated by the reduction of some acidic indanones in the presence of oxygen. In one step, the CH hydroxylation and the reductive coupling of the carbonyl moiety can be achieved in high yield and surprising selectivity to a tetrol [67].

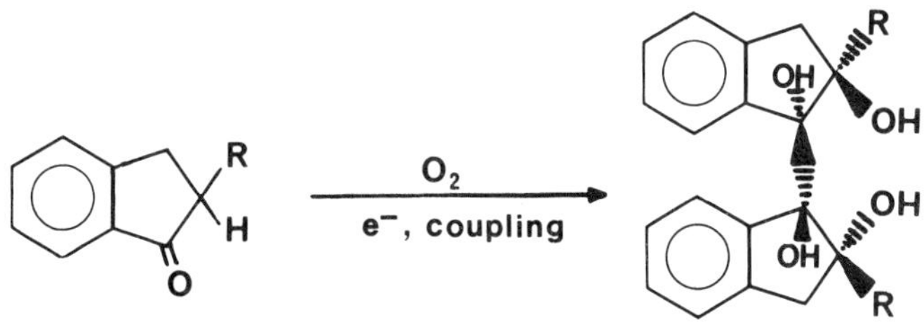

(52)

Finally, superoxide is able to catalyze [171] the decomposition of hydrogen peroxide according to the reaction of Haber and Weiss [172]. This reaction (which may be regarded as an electrochemical Fenton's reaction) allows in the presence of hydrogen peroxide the simultaneous in situ reduction of oxygen and hydroxylation of aromatic rings by means of strongly oxidizing hydroxyl radicals [173].

H_2O_2 may be present in solution or formed from superoxide when the organic substrate possesses a sufficient acidity. Reactions of this kind may be also initiated by using an appropriate inorganic [Cu(I)/Cu(II)] redox pair [174]. Addition of ferrous ions may help to achieve such a reaction [175].

$$O_2 + e^- \rightleftharpoons O_2^{\cdot -} \qquad (53)$$

$$H_2O_2 + O_2^{\cdot -} \longrightarrow OH^{\cdot} + OH^- + O_2 \qquad (54)$$

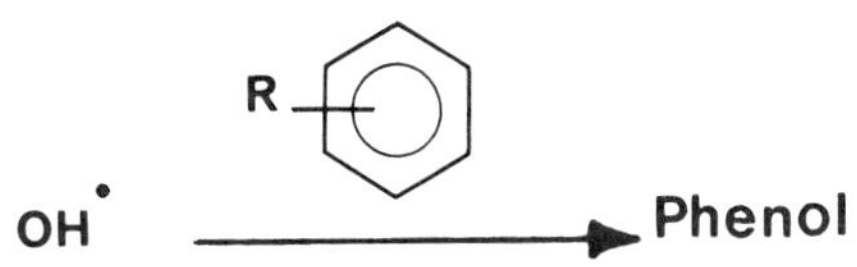

(55)

I. Perturbed Redox Catalysis: Electrogenerated Nucleophiles

So far, only redox catalyses in which the mediator is regenerated in high yield after the SET have been described. However, quite frequently the redox catalysis process competes with side reactions that lead to a decay of the concentration of the mediator. In such cases, the SET rate is not proportional to the concentration of the organic substrate and a limiting value of the catalytic current is observed. A typical example [11,12] is the indirect reduction of aliphatic organic halides RX with organic mediators (Fig. 2b and Table 3). For a given concentration of RX, it appears from mixtures of naphthalene and butyl chloride that the limiting current depends on the nature of the R group; the catalytic efficiency decreased in the order n-alkyl > sec-alkyl > tert-alkyl, and the largest effect was found experimentally for t-BuX compounds.

Electrolysis of an "A compound" in a mixture of two "RX compounds" may be interrupted after a certain quantity of electricity had been passed (e.g., 2 F mol^{-1} calculated on the basis of the concentration of A, here naphthalene) and product analysis demonstrated that for the reductive alkylation of the mediator, the best yields were obtained with t-BuX (70–80%), n-BuX compounds being less efficient (10–20% of alkylation).

Table 3. Examples of Conversion by Means of Organic Mediators

Type of Substrate A	Mediator used	Nature of the driving force of the catalysis (interest in synthesis)	References
Reductive redox catalysis			
Aromatic halides ArX	Benzonitrile or anthracene (depending on X)	Fast cleavage of the C-X bond (the indirect formation of Ar· is useful in some $S_{NR}1$ reactions)	3,4,12,25, 57
Activated olefins 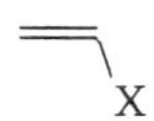	Anthracene	Fast coupling of radical anion or enone or unsaturated nitrile (indirect dimerizations)	12
Aromatic sulfones $-\overset{\mid}{\underset{\mid}{C}}-SO_2Ar$	Anthracene, pyrene	Cleavage of C—S bond (hydrocarbon deprotection)	20
Aromatic sulfonamides $R^1R^2NSO_2Ar$ $R^1N(SO_2Ar)_2$	Pyrene	Cleavage of N—S bond (deprotection of amines: synthesis of polyamines as macro ligands)	43
Bianthrone	Quinone	Fast rotation of the radical anion into the thermodynamically more stable form	176,177
Small rings (epoxides, cyclopropanes)	Biphenyl	Fast opening of the anion radical	29,47
Onium cations of moderate acidity	Naphthalene/$\overset{+}{N}R_4$	Fast cleavage of C—N and C—S bond	32
	Acenaphthylene/$\overset{+}{S}R_3$	Synthesis of R· radical in solution	32

α,α'-Dihalide derivatives	Aromatic hydrocarbons chosen as a function of more easily cleaved C—X bond	Cathodic elimination in solution	15,21
Polyenes (dienes and trienes)	Naphthalene, anthracene	Fast protonation of basic anion radicals by residual acidic impurities	31
Oxidative redox catalysis			
Benzylic ethers	Tris (p-bromophenyl)amine	Fast deprotonation of radical cation in the presence of a soluble base	173
Activated thioethers	Triarylamines	Oxidative cleavage of the C—S bond	41
Gem-dithioacetals 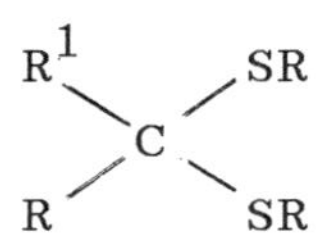	9,10-Diphenylanthracene Triarylamines	Rapid decomposition of radical cation by cleavage (R = aromatic) or nucleophilic attack (R = aliphatic); (oxidative deprotection of carbonyl compounds)	41,179 41
Carboxylate ions	Triarylamines	Kolbe reaction in homogeneous phase	18

Table 4. Some Examples of Redox Catalytic Systems Followed by Alkylation Reaction in the Presence of Alkyl Halides and Onium Salts

Organic mediator	Reactive species (reduction in solution)	Nature of the alkylation	Compounds (yields, %)	Remarks	References
Fluorenone	MeCl (saturated)	Dialkylation	9-Methyl-9-methoxyfluorene	General reaction of activated ketones	180
Anthracene	$ClCH_2CH_2CH_2Br$	Cyclization	1,2-Cyclopenteno-1,2-dihydroanthracene (41)	In DMF-n-Bu_4NI (0.1 M)	14
Pyrene	t-BuCl t-$BuSO_2Ar$	Specific monoalkylation	1-Tertiobutylpyrene (52)	Reoxidation of the dihydro form during the work-up; similar reaction with quinoline (35)	181
Anthraquinone α-diketone	MeCl (saturated)	O-dimethylation	Dimethoxyanthracene (80); cis- and trans-dimethoxyethylene (50)	For diketones, competition between C and O alkylation	166
Activated olefins	t-BuCl or t-BuBr	Monoalkylation	Different sites of alkylation along the chain	Alkylation in competition with the dimerization reaction	181-183
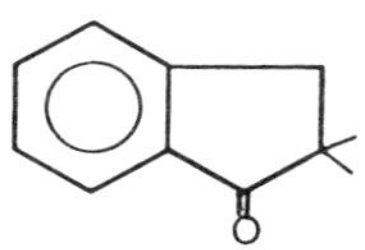	t-BuBr	t-Butylation of aromatic ring	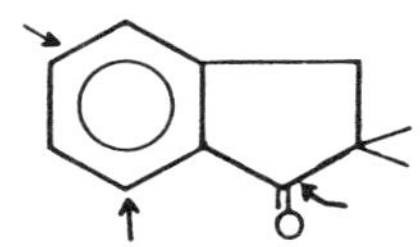 (60)	Possible poly-t-butylation	184
Naphthalene	t-$Bu\overset{+}{N}Me_3$ as electrolyte	Only monotertiobutylation	Mixture of tertiobutyltetrahydronaphthalenes (80)	Similar reaction of butylation with Bu_4N^+ (low yields, <1%)	8,32
Aromatic hydrocarbons	$\overset{+}{S}R_3$	Different kinds of alkylation	Mono-t-butylation of acephtylene and benzophenone by t-$Bu\overset{+}{S}Me_2$	Quite general reaction with $\overset{+}{S}R_3$; not apllicable to $^+PR_4$ because of its acidity	32

A review by Lund [14] has listed numerous alkylation reactions by means of electrolyses of different compounds in aprotic organic solvent, in the presence of organic halides. Other recent papers [8–12] describe further possibilities for such mixed electrolyses and highlight applications for organic synthesis (see some examples of cathodic alkylation by organic halides or onium salts given in Table 4.)

Most of the reactions may be understood as a radical coupling following the solution electron transfer.

$$RX^{\cdot -} \longrightarrow R^{\cdot} + X^{-} \quad (56)$$

$$A^{\cdot -} + R^{\cdot} \longrightarrow AR^{-} \quad (57)$$

$$AR^{-} \xrightarrow[\text{Proton Source}]{H^{+}} ARH$$

$$AR^{-} \xrightarrow[\text{electrophile}]{RX} AR_2 \quad (58)$$

The relative extent of dialkylation depends on the electrophilicity of RX (and the nucleophilicity of AR^-) when a relatively fast SET ($\Delta E < 0.5$ V) is the primary reaction; other mechanisms may also satisfactorily explain the distribution of products. For instance, adduct formation between the alkyl radical and the mediator is possible and must be considered; further reduction of $AR^{\cdot}$ may take place either by electron transfer or by abstraction of a hydrogen atom from the solvent. In other cases, the mediator anion radical or dianion may act as a nucleophile since a partial inversion of configuration of some optically active secondary RX compounds has been found [185] after workup under experimental conditions similar or identical to those of the electrolyses. Table 4 presents some alkylation reactions following a SET. The reaction scheme may be complicated by the fact that reduced form(s) of the mediator may act as a "reducing nucleophile" toward RX. The SET may then be rate determining in aliphatic nucleophilic substitutions [186], and/or R^- generated in solution may be added to an electrophilic mediator, such as an activated ketone [187]. Last, radicals $R^{\cdot}$ formed after SET may be trapped rapidly and completely by spin markers and therefore analyzed [188]. This method is of especially high interest when redox catalysis allows cleavage of an RX species, when RX is normally reduced more negatively than the spin marker (here an aromatic nitrone).

Additionally, it is also worth mentioning that electrolytically generated anion radicals react with carbon dioxide [12], leading to an N-functionalization of azines and diazines. A similar type of reaction with CO_2 has also been proposed [14,180,189], and a reaction in the presence of acetic anhydride was described [190].

$$A^{\cdot -} + CO_2 \longrightarrow A + CO_2^{\cdot -} \quad (59)$$

$$2\,CO_2^{\cdot -} \longrightarrow \text{oxalate ion} + \text{other products} \quad (60)$$

$$A^{\cdot -} + CO_2^{\cdot -} \longrightarrow ACO_2^{2-} \xrightarrow{CO_2} A(CO_2)_2^{2-} \quad (61)$$

or

$$A^{\cdot -} + CO_2 \longrightarrow ACO_2^{\cdot -} \xrightarrow[e^-]{CO_2} A(CO_2)_2^{2-} \quad (62)$$

Here, special mention must be made of the carboxylation of organic halides to give the corresponding carboxylic acids, often in good yield. The proposed scheme involves the formation of an electrogenerated transient nucleophile by means of the two-electron scission of the carbon-halide bond. Best yields are obtained when using an undivided cell with a sacrificial anode (Mg, Zn, and Al) in DMF [191,192].

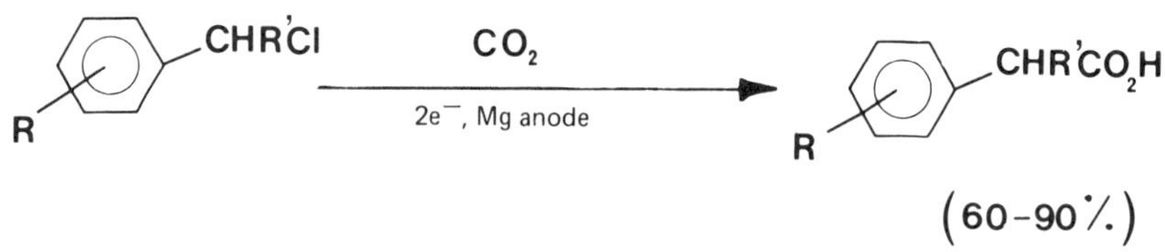

(63)

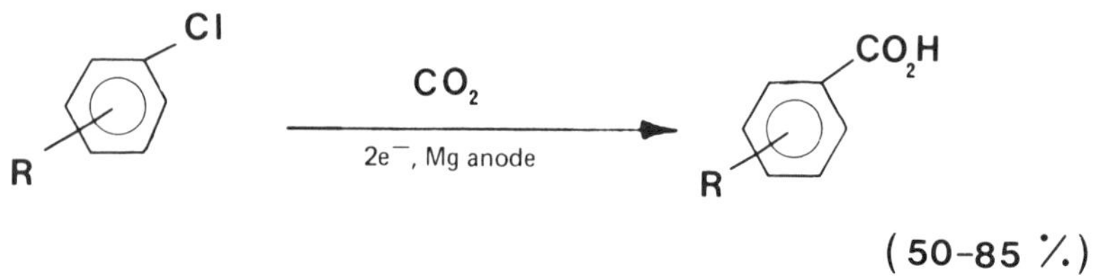

(64)

Generally speaking, the reduction of the mediator (overall a two-electron process) may be represented as an electrochemical generation of nucleophiles. A few examples of reductive alkylation and acylation (here, the primary electron exchange in solution is not demonstrated or looks improbable) are listed in Eqs. (65) and (66) [166,193–195]:

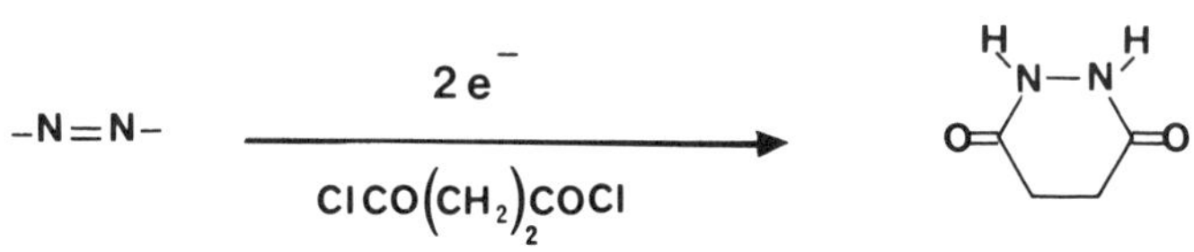

(65)

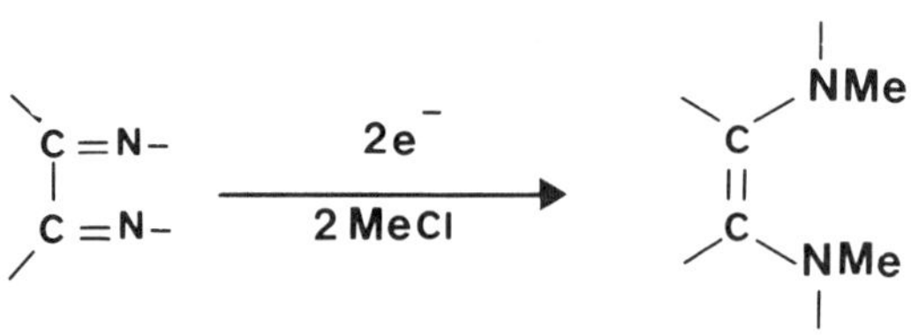

(66)

Activated olefins were shown to be monoacylated when reduced in the presence of acetic anhydride [196–199].

III. INDIRECT REDUCTIONS BY MEANS OF SOLVATED ELECTRON

At rather negative potentials, electrons may be released from the cathode in contact with the electrolyte although no organic acceptor is present. The electron may then end up in a potential energy "trap" possibly induced by its own polarization field. Such electrons possess strong reducing properties and are called solvated electrons [200,201].

Reductions by means of electrochemically produced solvated electron may be carried out under well-defined conditions. The reduction then occurs by homogeneous electron transfer but without an organic or inorganic electron carrier present as defined previously. In this section only indirect reductions by solvated electrons in which their presence as intermediates has been demonstrated by means of physical methods, such as UV and EPR spectroscopy, are included.

The stability of solvated electrons in some polar solvents (liquid ammonia, HMPA, and amines) can be explained either by a solvent cavity in which the electron is trapped or by the presence of an "expanded orbital" occupied by the electron; this orbital could involve the hydrogen atoms of the nearest solvent sphere around the metal cation (here used as electrolyte).

The electrochemical formation of solvated electrons (e_S) is usually very convenient since solutions of known concentration of (e_S) can be prepared under coulometric control and consequently used in a strictly stoichiometric ratio. The generation of solvated electrons (usually understood as an electrochemical dissolution in a given polar solvent HS) can be represented generally as Eq. (67):

$$e^- \text{ cathode} + \text{HS} \rightleftharpoons e_{HS} \tag{67}$$

Such a reaction can be considered reversible. However, a slow reduction of the solvent may take place with hydrogen evolution. Therefore, indirect reductions by means of solvated electrons would be expected to occur only in solvents of quite low proton activity.

The cathode potential necessary for the production of solvated electrons is rather negative; the standard potential of the hydrated electron has been calculated to be −2.68 V versus NHE. Also, in other solvents compatible with its formation, very negative potentials must be used; for example, in liquid ammonia the generation of e_{HS} is achieved at −2.74 V versus Ag/$AgNO_3$ (0.1 M) [202], but the dissolution standard potential measured in HMPA was found to be −3.44 V versus Ag/$AgClO_4$ (0.1 M) [203]. Similarly in methylamine (t^o -50^oC), a potential of −2.90 V versus Ag/$AgNO_3$ was reported [204].

To render electron exchanges between e_{HS} and organic substrates irreversible, it appears necessary to add to the solvent some proton donor known to react very slowly with the solvated electron. Mixtures of HMPA-ethanol and HMPA-acetic acid have been used successfully [205] as proton-rich solutions.

The most commonly used electrolyte in electrolysis mediated by solvated electrons is lithium chloride. The reasons are both its high solubility and the advantage of having in solution lithium ions known to form tight ion pairs. Tetrabutylammonium salts have also been employed successfully, however.

The electrode material may in certain cases affect the product distribution and the current yield, especially when reductions are run in amines. However, in other cases the material was shown not to be critical when such different materials as carbon, copper, platinum, lead, and aluminum were used as cathode.

A. Use of Solvated Electrons in the Presence of a Proton Donor

The use of hexamethylphosphorictriamide (HMPA) in the electrolytic reduction of substrates, such as benzene derivatives, aliphatic ketones, amides, and olefines, involves the intermediate formation of solvated electrons since very negative potentials must be reached. The substrates are normally difficult to reduce. However, it is important not to add so much proton donor to the solvent that the rate of its reaction with the solvated electrons becomes too high. Generally, ethanol can be used as cosolvent [206] and the best stability of the intermediate solvated electron combined with a maximum protonating efficiency was found to be the mixture of HMPA-ethanol (33–66 mol mol^{-1}); the relatively high concentration of EtOH does not protonate e_{HS} because HMPA appears to be selectively adsorbed on the cathode interface and the reaction of e_{HS} with the substrate is fast. With such a ratio of ethanol, a maximum current efficiency was found [207].

The reduction of benzene itself may then be achieved in ethanol-HMPA (66.6 mol% ethanol) at −2.4 V (versus Ag-wire electrode) at an aluminum cathode [206]; this gives a mixture of 22% cyclohexadiene, 10% cyclohexene, and 67% cyclohexane with an overall current efficiency of 95% [206]. The electrolysis must be carried out with a large excess of benzene. However, product distribution may depend on both the concentration of ethanol and the current density. As previously studied under chemical conditions (lithium metal added to ethanol in ammonia [208]), the product distribution would depend on the rates of reduction of benzene (K_b) and of cyclohexene (K_c) by the reducing species (here the solvated electron), the nature of the solvent, and the efficiency of the proton donor in the considered solvent. In ethylenediamine, the ratio K_b/K_c was found to be equal to 200, and this value explains why reduction stops when an isolated double bond has been formed. In contrast, in HMPA-ethanol mixtures, the ratio K_b/K_c is only 1.4; this may explain the lack of selectivity and high yield of cyclohexane when a large excess of proton donor is used. However, the product distribution may be spectacularly changed to give selective formation of cyclohexadiene when lower current densities or a limited concentration of proton donor is used [206].

Under similar conditions, reduction of other aromatic compounds may be achieved. Phenol may be converted into cyclohexanol [207], and tetralin is reduced to decalin as the major product [209]. Other aromatic systems known to be quite easily reduced under conventional conditions in electrochemistry have also been tested using HMPA-ethanol-lithium salts as medium. The effective reduction of naphthalene and anthracene requires very negative potentials for the working cathode, suggesting that the potential-determining step is the release of electrons from the electrode to the solvent. Under such conditions, highly hydrogenated forms are isolated [205]. For example, anthracene is mainly reduced to 1,4,5,8,9,10-hexahydroanthracene and 1,4,9,10-tetrahydroanthracene.

Carbonyl derivatives and olefins are reduced by electrochemically produced solvated electrons. Camphor and 4-t-butylcyclohexanone thus form the corresponding alcohols in HMPA-ethanol; the proportion of epimeric alcohols would suggest that the reduction is thermodynamically controlled [210]. Surprisingly enough, sterically hindered alkenes are reduced at a rate only slightly lower than that of cyclohexene [206].

Acetamide may be converted into ethylamine and ethanol. In such reductions, the ratio of ethylamine to ethanol was found to be strongly dependent on the nature of the protonating reagent; better yields of amine were obtained when HCl was used as proton source [211].

B. Reduction in Amines

The Birch and Benkeser reactions of some unsaturated organic compounds (see Ref. 212 and references therein), which consists of a reduction by sodium or lithium in amines, can be mimicked electrochemically in the presence of an alkali salt electrolyte. The cathodic reaction is not the deposition of alkali metal on the solid electrode but the formation of solvated electrons. Most of the reactions described were performed in ethylenediamine [213] or methylamine [204]. A feature of these studies is variety introduced by the use of a divided or undivided cell. In a divided cell, the product distribution appears to be the same as that in the classic reduction by metal under similar conditions. In contrast, in an undivided cell the corresponding ammonium salt is formed at the anode; it plays the role of an in situ generated proton donor; under such conditions, the proton concentration during the electrolysis remains nearly unchanged, and this may affect the results of the electrolyses.

The reduction of aromatic hydrocarbons under these conditions has also been much studied. As expected, the reduction of benzene in methylamine containing lithium chloride is dramatically dependent on the design of the cell. Thus, reduction in an undivided cell affords 1,4-dihydrobenzene as a major product, but in a divided cell (the two compartments are separated by asbestos) cyclohexene is formed. Such fundamental differences have been ascribed to the fact that in the divided cell the isomerization of unconjugated dienes (and therefore a further reduction) should be favored by the increase in the basicity. Thus, the reduction of toluene under similar experimental conditions gives 2,5-dihydrotoluene in 95% yield in an undivided cell and a mixture of methylcyclohexenes in 83% yield [214] in a divided cell. The same monoolefins were obtained chemically in a sodium-water-ammonia reduction of toluene at 0°C under pressure [215]. Other results concerning the electrolytic reduction of substituted benzenes dissolved in amines confirm that the influence of the separator is crucial [212].

It is of interest to compare the relative reactivity of solvated electrons toward an isolated double or triple bond and an aromatic ring. Thus, the electrolytic reduction of acetylenes in undivided cells (2-octyne, 3-octyne and 5-octyne) gave mainly the corresponding trans-olefins accompanied by only small amounts of the cis isomers [216]. Reductions of unconjugated arylalkanes (5-phenyl-2-pentyne) by solvated electrons gave the corresponding trans-olefin in good yield, although small amounts of products of further reduction were also obtained.

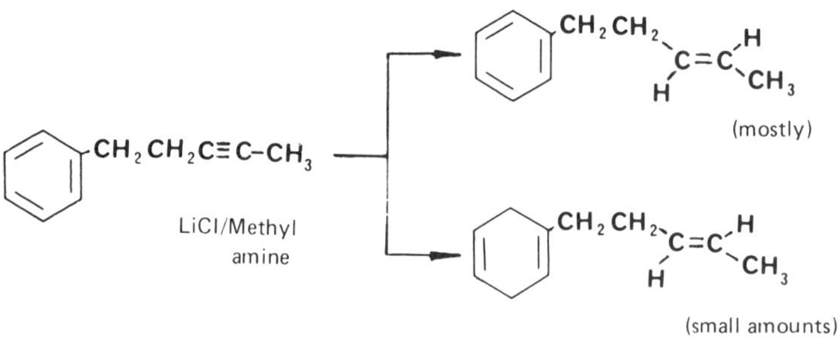

(68)

Although allylbenzene is readily reduced using 2 F mol^{-1} to propylbenzene in methylamine in an undivided cell, the addition of a proton donor (such as ethanol) changes the product distribution; 2,5-dihydropropylbenzene is then the main product [217].

Amides of aliphatic acids are converted into alcohols in methylamine [218]. It appears that this is the only general reduction of all kinds of amides to alcohols.

REFERENCES

1. C. P. Andrieux, J. M. Dumas-Bouchiat, and J. M. Savéant, J. Electroanal. Chem., 87, 39 (1978).
2. C. P. Andrieux, J. M. Dumas-Bouchiat, and J. M. Savéant, J. Electroanal. Chem., 87, 55 (1978).
3. C. P. Andrieux, C. Blocman, J. M. Dumas-Bouchiat, and J. M. Savéant, J. Amer. Chem. Soc., 101, 3432 (1979); C. P. Andrieux, C. Blocman, and J. M. Savéant, J. Electroanal. Chem., 105, 413 (1979).
4. C. P. Andrieux, C. Blocman, J. M. Dumas-Bouchiat, F. M'Halla, and J. M. Savéant, J. Amer. Chem. Soc., 102, 3806 (1980).
5. A. J. Fry, Synthetic Organic Electrochemistry, Harper & Row, New York, 1972, pp. 67–68, and reference by J. Sease cited therein.
6. H. Lund, M. A. Michel, and J. Simonet, Acta Chem. Scand., B28, 900 (1974).
7. J. Sease and C. Reed, Tetrahedron Lett., 393 (1975).
8. H. Lund, M. A. Michel, and J. Simonet, Acta Chem. Scand., B29, 217 (1975).
9. V. G. Mairanovskii and N. F. Loginova, Zh. Obshch. Khim., 45, 2112 (1975).
10. S. Margel and M. Levy, J. Electroanal. Chem., 56, 529 (1974).
11. J. Simonet, M. A. Michel, and H. Lund, Acta Chem. Scand., B29, 489 (1975).
12. H. Lund and J. Simonet, J. Electroanal. Chem., 65, 205 (1975).
13. S. Margel and F. C. Anson, J. Electrochem. Soc., 125, 1232 (1978).
14. H. Lund, Denki Kagaku, 45, 2 (1977).
15. E. Hobolth and H. Lund, Acta Chem. Scand., B30, 895 (1976).
16. V. G. Mairanovskii, N. F. Loginova, and I. A. Titova, Dokl. Akad. Nauk. SSSR, 223, 643 (1975).
17. S. G. Mairanovskii, L. I. Kosychenko, and S. Z. Taits, Elektrokhimiya, 8, 2350 (1977).
18. E. Steckhan and W. Schmidt, J. Electroanal. Chem., 89, 215 (1978).

19. C. P. Andrieux, J. M. Dumas-Bouchiat, and J. M. Savéant, J. Electroanal. Chem., 88, 43 (1978).
20. J. Simonet and H. Lund, Acta Chem. Scand., B31, 909 (1977).
21. P. Martigny and J. Simonet, J. Electroanal. Chem., 81, 407 (1977).
22. V. G. Mairanovskii, Angew. Chem. Int. Ed. Engl., 15, 281 (1976).
23. Yu. V. Bendersky and V. G. Mairanovskii, J. Electroanal. Chem., 79, 401 (1977).
24. J. Koutecky, Coll. Czech. Chem. Commun., 18, 311 (1953).
25. D. Zan, Thesis, Université de Paris, 1984.
26. K. Boujlel and J. Simonet, J. Electroanal. Chem., 144, 437 (1983).
27. C. Andrieux, C. Blocman, J. M. Dumas-Bouchiat, F. M'Halla, and J. M. Savéant, J. Electroanal. Chem., 113, 19 (1980).
28. C. P. Andrieux, J. M. Dumas-Bouchiat, and J. M. Savéant, J. Electroanal. Chem., 113, 1 (1980).
29. K. Bouhlel, J. Simonet, J. P. Barnier, C. Girard, and J. M. Conia, J. Electroanal. Chem., 117, 161 (1980).
30. R. Carlier, K. Boujlel, and J. Simonet, J. Electroanal. Chem., 221, 275 (1987).
31. T. Lund, S. U. Petersen, H. Lund, K. M. Cheung, and J. H. P. Utley, Acta Chem. Scand., B41, 285 (1987).
32. P. Martigny and J. Simonet, J. Electroanal. Chem., 101, 275 (1979).
33. J. M. Savéant and Su Khac Binh, J. Electroanal. Chem., 88, 27 (1978).
34. W. Schmidt and E. Steckhan, Sandbjerg meeting on organic electrochemistry, Abstract, Sept. 1978, p. 67.
35. N. Gatti, W. Jugelt, and H. Lund. Acta Chem. Scand., B41, 646 (1987).
36. W. Schmidt and E. Steckhan, Angew. Chem., 90, 717 (1978); 91, 850 (1979); 91, 851 (1979).
37. J. F. W. McOmie, Protective Groups in Organic Chemistry, Plenum, London, 1973, p. 98.
38. W. Schmidt and E. Steckhan, Chem. Ber., 113, 577 (1980).
39. S. Dapperheld and E. Steckhan, Angew. Chem., 90, 717 (1978).
40. P. Martigny and J. Simonet, J. Electroanal. Chem., 111, 113 (1980).
41. M. Platen and E. Steckhan, Liebigs Ann. Chem., 1563 (1984).
42. E. Steckhan, Tpr. Curr. Chem., 142 (1987).
43. R. Kossai, G. Jeminet, and J. Simonet, Tetrahedron Lett., 1059 (1979); R. Kossai, Thesis, Clermont-Fd, France, 1978.
44. R. Kossai, Electrochim. Acta, 31, 1643 (1986).
45. J. Simonet, in The Chemistry of Sulphones and Sulphoxides, (S. Patai, Z. Rappoport, and C. J. M. Stirling, eds.), John Wiley, New York, 1988, Chap. 22, p. 1001.
46. M. Chaquiq El Badre and J. Simonet, to be published.
47. K. Boujlel and J. Simonet, Electrochim. Acta, 24, 481 (1979).
48. C. P. Andrieux, A. Merz, J. M. Savéant, and R. Tomahogh, J. Amer. Chem. Soc., 196, 1957 (1984).
49. N. Schultz and E. Steckhan, to be published (from Ref. 173).
50. I. M. Dalrymple, A. R. Jones, and J. P. Millington, Electrochem. Soc. Meeting, Extended Abstract, Washington, D.C., 1976, p. 690.
51. F. Magno and G. Bontempelli, J. Electroanal. Chem., 68, 337 (1976).

52. K. Shirai and K. Sugino, Denki Kagaku, 25, (1967); Chem. Abstr., 52, 8794f (1958).
53. L. C. Jiang and D. Pletcher, J. Electroanal. Chem., 152, 157 (1983).
54. R. Clarke, A. Kuhn, and E. Okoh, Chem. Br., II, 59 (1975).
55. Y. Takahashi, S. Ikuta, and Y. Nakanishi, (Nippon Shokubai Kagaku Kogyo Co. Ltd.), Jpn. Kokai Tokkyo Koho JP 6 152 382 (8652382); Chem. Abstr., 104, 232913u (1986).
56. G. A. Kokarev, M. Ya. Fioshin, and E. V. Gramova, Electrokhimiya, 17, 1380 (1981).
57. R. Ramaswamy, M. S. Venkatachalapathy, and H. V. K. Udupa, J. Electrochem. Soc., 110, 202 (1963).
58. S. Chidabaram, M. S. V. Pathy, and H. V. K. Udupa, J. Electrochem. Soc. India, 17, 95 (1968); Chem. Abstr., 70, 2561a (1969); Bull. Chem. Soc. Jpn., 35, 1751 (1962).
59. J. H. Wagenknecht, J. P. Ideman, R. C. Hallcher, D. E. McMakins, T. F. Rogers, and W. G. Wagner, J. Appl. Electrochem., 13, 535 (1983).
60. K. Kramer, P. M. Robertson, and N. Ibl, J. Appl. Electrochem., 10, 29 (1980).
61. M. Marroco and G. Brilmyer, J. Org. Chem., 48, 1487 (1983).
62. F. Lapicque and A. Stork, Electrochim. Acta, 30, 1247 (1985).
63. H. Lowe, Br. Patent 174,913 (1922); Chem. Abstr., 16, 1710 (1922); W. Klages, Swiss. Patent 78,277 (1918); Chem. Abstr., 13, 2374 (1919). K. Durkes, Ger. Patent 920,186 (1954); Chem. Abstr., 52, 19624 (1958).
64. P. L. Matlock and M. R. Sander, (Union Carbide Corporation), USP 4277318 (1980); Chem. Abstr., 95, 88237C (1981).
65. S. Torii, T. Inokuchi, and T. Sugiura, J. Org. Chem., 51, 155 (1986).
66. Osaka Yuki Kagaku Kogyo Co. Ltd., Jpn. Kokai Tokkyo Koho JP 59,166,687, JP 59,157,291, 84,157,291, JP 59,129,790, JP 84,129,790.
67. A. Orliac-Le Moing, J. Delaunay, and J. Simonet, Electrochim. Acta, 32, 1769 (1987).
68. S. Torii, H. Tanaka, I. Inokuchi, S. Nakane, M. Akada, N. Saito, and T. Sirakawa, J. Org. Chem., 47, 1647 (1982).
69. R. Kreh and R. Spotnitz (W. R. Grace and Co.) U.S. Patent 4,639,298 (1987); Chem. Abstr., 106, 127956d (1987).
70. R. Kreh and R. Spotnitz (W. R. Grace and Co.) U.S. Patent 4,647,349, (1987); Chem. Abstr., 106, 164600d (1987).
71. G. Kreysa and H. Medin, J. Appl. Electrochem., 16, 757 (1986).
72. J. Skarzewski, Tetrahedron, 40, 4997 (1984).
73. Y. Takahashi, S. Ikuta, and Y. Nakanishi, (Nippon Shokubai Kagaku Kogyo Co. Ltd.), Jpn. Kokai Tokkyo Koho JP 61,52,382, 86,52,382; Chem. Abstr., 104, 232913u (1986).
74. J. Diegiec, Chem. Stosow, 29, 263 (1985).
75. J. Kubala, S. Kucar, and K. Linek, Czech CS225, 576 (1985); Chem. Abstr., 105, 172968z (1986).
76. V. A. Shepelin, Electrokhimiya, 11, 1767 (1975).
77. (a) F. Goodridge and C. J. H. King, Trans. Faraday Soc., 66, 2889 (1970); (b) J. Appl. Chem. Biotechnol., 21, 208 (1971).
78. D. L. Klass, U.S. Patent 3,147,203 (1964); Chem. Abstr., 61, 13195 (1964).

79. J. S. Mayell, Ind. Engl. Chem., Prod. Res. Dev., 7, 129 (1968); Chem. Abstr., 69, 18517 (1968).
80. R. Kaldova and E. Tvirzicka, J. Electroanal. Chem., 34, 515 (1972); S. Z. Ali, A. Trojanek, and R. Kaldova, J. Electroanal. Chem., 52, 85 (1974).
81. M. Fleischmann and D. Pletcher, R. Inst. Chem. REv., 2, 87 (1969).
82. E. Steckhan, Angew. Chem., Int. Ed. Engl., 25, 683 (1986).
83. J. S. Mayell, U.S. Patent 3,555,084 (Jan. 12, 1971); Chem. Abstr., 74, 6025a (1971).
84. N. A. Hampson, J. B. Lee, and K. I. MacDonald, Electrochim. Acta, 17, 921 (1972).
85. H. Staudinger and V. Ulrich, Z. Naturforsch., 19b, 409 (1964).
86. C. M. Wright, R. F. Heck, and W. Yokoyama, Electrochem. Soc. Meeting, Extented Abstract, New York, May 1969, p. 361.
87. H. C. Rance and J. M. Coulson, Elctrochim. Acta, 14, 283 (1969).
88. H. Lund and G. Ferroci, Acta Chem. Scand., B30, 651 (1976).
89. O. Christofis, J. J. Habeeb, R. S. Steevensz, and D. G. Tuck, Can. J. Chem., 56, 2269 (1978).
90. W. Barker and D. Kennedy, Trans. Electrochem. Soc., 92, 391 (1947); G. Noyori, Jpn. Patent 4536 (1950); Chem. Abstr., 47, 3327c (1953).
91. P. Pouillen, M. Perrin, and P. Martinet, C. R. Acad. Sci. Ser. C, 284, 955 (1977).
92. D. W. Sopher and J. H. P. Utley, J. Chem. Soc., Chem. Commun., 23, 1087 (1979).
93. J. Wellmann and E. Steckhan, Synthesis, 901 (1978).
94. E. Steckhan and J. Wellmann, Angew. Chem., 88, 306 (1976); J. Wellmann and E. Steckhan, Chem. Ber., 110, 3561 (1977).
95. G. Farnia, G. Sondana, and G. Vianello, J. Electroanal. Chem., 88, 147 (1978).
96. T. Chibata and Y. Takata, Bull. Chem. Soc. Jpn., 51, 1418 (1978).
97. J. Wellmann and E. Steckhan, Synthesis, 901 (1978).
98. R. Wolf and E. Steckhan, J. Chem. Soc., Perkin Trnas., 733 (1986).
99. R. Wolf and E. Steckhan, J. Electroanal. Chem., 130, 367 (1981).
100. M. Perrin, P. Pouillen, G. Mousset, and P. Martinet, Electrochim. Acta, 24, 73 (1979); Tetrahedron, 36, 221 (1979).
101. F. Fournier, J. Berthelot, and Y. L. Pascal, Tetrahedron, 40, 2121 (1983); F. Fournier and M. Fournier, Can. J. Chem., 64, 881 (1986).
102. D. W. Sopher and J. H. P. Utley, J. Chem. Soc., Perkin Trans., 1361 (1984).
103. D. Pletcher, M. Razaq, and G. D. Smilgin, J. Appl. Electrochem., 11, 601 (1981).
104. N. E. Gunawardena and D. Pletcher, Acta Chem. Scand., b37, 549 (1983).
105. J. Perichon and J. F. Fauvarque, J. Chem. Res. Synop., 24, (1980); J. Organometal. Chem., 202, 435 (1980).
106. S. Torii, H. Tanaka, and K. Morisaki, Tetrahedron Lett., 26, 1655 (1985).
107. J. F. Fauvarque, M. A. Petit, F. Pfluger, A. Jutand, C. Chevrot, and M. Troupel, Makromol. Chem., Rapid Commun., 4, 455 (1983).
108. M. Troupel, Y. Rollin, J. Perichon, and J. F. Fauvarque, Nouv. J. Chim., 5, 621 (1981).
109. J. F. Fauvarque, A. Jutand, C. Chevrot, F. Pfluger, and M. Troupel,

(Compagnie Générale d'électricité, S.A.), Fr. Demande, 2530266 (1984); J. F. Fauvarque, A. Jutand, and M. Francois, J. Appl. Electrochem., 18, 109 (1988).
110. N. Kornblum, Angew. Chem. Int. Ed. Engl., 14, 734 (1975).
111. J. F. Bunnett, Acc. Chem. Res., 11, 413 (1978) and references cited therein.
112. J. M. Savéant, Acc. Chem. Res., 13, 323 (1980) and references cited therein.
113. J. Pinson and J. M. Savéant, J. Chem. Soc., Chem. Commun., 933 (1974); J. Amer. Chem. Soc., 100, 1507 (1978).
114. C. Amatore, M. A. Oturan, J. Pinson, J. M. Savéant, and A. Thiebault, J. Amer. Chem. Soc., 184, 25 (1985); A. Amatore, C. Conbellas, J. Pinson, M. A. Oturan, S. Robveille, J. M. Savéant, and A. Thiebault, J. Amer. Chem. Soc., 107, 4846 (1985).
115. J. E. Swartz and T. T. Stenzal, J. Amer. Chem. Soc., 106, 2520 (1984).
116. N. Alam, C. Amatore, C. Combellas, A. Thiebault, and J. N. Verpeaux, Tetrahedron Lett., 49, 6171 (1987).
117. C. Degrand, J. Electroanal. Chem., 238, 239 (1987); C. Degrand, R. Prest, and P. L. Compagnon, J. Org. Chem., 52, 52 (1987).
118. L. Eberson and L. Jonsson, J. Chem. Soc., Chem. Commun., 1187 (1980).
119. L. R. Yeh and A. J. Bard, J. Electroanl. Chem., 70, 157 (1975); 81, 233 (1977); L. R. Yeh and A. J. Bard, J. Electrochem. Soc., 124, 189 (1977).
120. R. D. Ricke, H. Kojima, and K. Ofele, J. Electroanal. Chem., 98, 6735 (1976).
121. C. Amatore, J. Badoz-Lambling, C. Bonnel-Huyges, J. Pinson, J. M. Savéant, and A. Thiebault, J. Amer. Chem. Soc., 104, 1979 (1982).
122. R. Kossai, J. Simonet, and G. Dauphin, Tetrahedron Lett., 3575 (1980); J. Electroanal. Chem., 139, 211 (1982).
123. J. W. Hershberger, R. J. Klinger, and J. K. Kochi, J. Amer. Chem. Soc., 104, 3034 (1982); A. Darchen, C. Mahe, and H. Patin, Nouv. J. Chim., 6, 539 (1982); B. A. Narayanan, C. Amatore, and J. K. Kochi, J. Chem. Soc., Chem. Commun., 397 (1983).
124. L. Cedheim and L. Eberson, Acta Chem. Scand., b30, 527 (1976).
125. J. Delaunay, A. Orliac, J. Simont, and L. Toupet, Tetrahedron Lett., 27, 6205 (1986).
126. J. McCoch and E. Steckhan, Tetrahedron Lett., 28, 1081 (1987).
127. J. Delaunay, A. Orliac, and J. Simonet, Tetrahedron (1988) in press.
128. J. Delaunay and J. Simonet, to be published.
129. R. Haddon, Br. Patent 140, 831 (1917); P. McElroy, U.S. Patent 1,253,615 (1918).
130. N. Ibl and A. Selvig, Chem.-Ing.-Tech., 42, 180 (1970).
131. J. W. Cornforth, F. Jones, K. H. Clifford, and D. T. Green, Electrochem. Soc. Meeting, Chester, England, April 1969.
132. S. K. Battacharyya, M. S. Muthanna, and A. D. Patankar, J. Sci. Ind. Res. India, 11B, 365 (1952).
133. J. A. M. LeDuc, U.S. Patent 3,288,693 (1966); Chem. Abstr., 66, 43288 (1967).

134. W. Kroening and P. Konrad, Ger. Patent 1,290,926 (1969); Chem. Abstr., 70, 10636 (1969).
135. S. Torii, K. Uneyama, M. Ono, H. Tazawa, and S. Matsunami, Tetrahedron Lett., 4661 (1979); S. Torii, K. Uneyama, and M. Ono, Tetrahedron Lett., 2741 (1980); 2653 (1980); S. Torii, K. Uneyama, H. Tanaka, T. Yamanaka, T. Yasuda, M. Ono, and Y. Kohmoto, J. Org. Chem., 46, 3312 (1981).
136. J. Yoshida, J. Mashimoto, and N. Kawabata, J. Org. Chem., 47, 3575 (1982).
137. T. Iwasaki, T. Nishitani, H. Horikawa, and I. Inoue, J. Org. Chem., 47, 3799 (1982).
138. P. Nedenskov, N. Elming, J. T. Nielsen, and N. Clauson-Kaas, Acta Chem. Scand., 9, 17 (1955) and references cited therein.
139. D. Degner, in Technique of Chemistry, Vol. 5, Part 3 (N. L. Weinberger and B. V. Tilak, eds.), Chap. V, Wiley, New Yor, 1982, p. 273.
140. T. Atsuno, Jpn. Kokai Tokkyo Koho 5803988 (1983); Chem. Abstr., 98, 224280c (1983).
141. H. G. Thomas and E. Lux, Tetrahedron Lett., 965 (1972).
142. G. Dixit, R. Rastogi, and K. Zutshi, Electrochim. Acta, 27, 561 (1982).
143. J. Yoshida, R. Nakai, and N. Kawabata, J. Org. Chem., 45, 5269 (1980).
144. C. Jackson and A. T. Kuhn, in Industrial Electrochemical Processes, (A. T. Kuhn, ed.), Elsevier, Amsterdam, 1971, pp. 513–519.
145. F. Jacquet, A. Gaset, J. Simonet, and G. Lacoste, Electrochim. Acta, 30, 477 (1985).
146. S. Torii, K. Uneyama, and M. Ono, J. Amer. Chem. Soc., 103, 4606 (1981).
147. S. Torii, H. Tanaka, and M. Ikida, J. Org. Chem., 44, 1554 (1979).
148. S. Torii, H. Tanaka, S. Hamano, N. Tada, J. Nakami, and M. Sasaoka, Chem. Lett., 1823 (1984).
149. J. Yoshida, K. Ogura, and N. Kawabata, J. Org. Chem., 49, 3419 (1983).
150. J. Yoshida, H. H. Solfuku, and N. Kawabata, Bull. Chem. Soc. Jpn., 56, 1243 (1983).
151. D. L. Maricle and W. G. Hodgson, Anal. Chem., 31, 1562 (1965); M. E. Peover and B. S. White, Electrochim Acta, 11, 1061 (1966); D. T. Sawyer and J. L. Roberts, J. Electroanal. Chem., 12, 90 (1966).
152. E. L. Johnson, K. L. Pool, and R. E. Hamm, Anal. Chem. 38, 183 (1966); 39, 888 (1967).
153. J. San Filippo, C. I. Chern, and J. S. Valentine, J. Org. Chem., 41, 1077 (1976).
154. J. Simonet, unpublished observations.
155. R. Dietz, A. E. J. Forno, B. E. Larcombe, and M. E. Peover, J. Chem. Soc.B, 816 (1970).
156. W. Slongl, Chem. Commun., 420 (1965).
157. A. A. Frimer, I. Rosenthal, and S. Hoz, Tetrahedron Lett., 4631 (1977).
158. E. Lee-Ruff, Chem. Soc. Rev., 6, 195 (1977).
159. S. Torii, Synthesis, 873 (1986).
160. I. Rosenthal and A. Frimer, Tetrahedron Lett., 3731 (1975); A. Frimer and I. Rosenthal, Tetrahedron Lett., 2809 (1976).

161. M. V. Merritt and D. T. Sawyer, J. Org. Chem., 35, 2157 (1970).
162. K. Boujlel and J. Simont, Tetrahedron Lett., 1063 (1979).
163. K. Boujlel and J. Simonet, Tetrahedron Lett., 26, 4814 (1985).
164. M. Gareil, J. Pinson, and J. M. Savéant, Nouv. J. Chim., 5, 311 (1981).
165. F. Magno, G. Bontempelli, M. M. Andrenzzi Sedea, J. Electroanal. Chem., 97, 85 (1979).
166. J. Simonet and H. Lund, Bull. Soc. Chim. Fr., 2547 (1975).
167. M. Tezuka, Y. Ohkatsu, and T. Osa, Bull. Soc. Chim. Jpn., 48, 1471 (1975).
168. H. Sagae, M. Fujihara, T. Osa, and H. Lund, Chem. Lett., 793 (1977).
169. H. Sagae, H. Fujihara, H. Lund, and T. Osa, Bull. Chem. Soc. Jpn., 53, 1537 (1980).
170. P. N. Allen, U. Hess, C. S. Forte, and M. M. Baizer, Synth. Commun., 12, 123 (1982).
171. I. M. Kolthoff and J. Jordan, J. Amer. Chem. Soc., 74, 4801 (1952).
172. F. Haber and J. Weiss, Proc. R. Soc., A147, 332 (1934).
173. J. M. Maissant, C. Bouchoule, P. Canesson, and M. Blanchard, J. Mol. Catal., 18, 189 (1983).
174. C. M. Wright, R. F. Heck, and W. Yokoyama, Electrochem. Soc. Meeting, New York, May 1969.
175. T. Matsue, M. Fujihira, and T. Osa, J. Electrochem. Soc., 128, 2565 (1981).
176. D. H. Evans and N. Xie, J. Electroanal. Chem., 133, 367 (1982); J. Amer. Chem. Soc., 105, 315 (1983).
177. J. E. Evans, T. Kuwana, M. T. Henne, and G. P. Royer, J. Electroanal. Chem., 80, 409 (1977).
178. W. Schmidt and E. Steckhan, Angew. Chem., 90, 717 (1978).
179. M. Platen and E. Steckhan, Tetrahedron Lett., 21, 511 (1980); Chem. Ber., 117, 1679 (1984).
180. D. Michelet, Fr. Demand. 244030; Chem. Abstr., 94. 164. 821K (1981).
181. P. E. Hansen, A. Berg, and H. Lund, Acta Chem. Scand., B30, 267, (1976).
182. C. Degrand and H. Lund, Nouv. J. Chim., 1, 35 (1977).
183. C. Degrand and H. Lund, Acta Chem. Scand., B31, 593 (1977).
184. A. M. Orliac-Le Moing, J. Delaunay, A. Lebouc, and J. Simonet, Tetrahedron, 41, 4483 (1985).
185. M. Malissard, J. P. Mazaleyrat, and Z. Welvart, J. Amer. Chem. Soc., 99, 6933 (1977).
186. T. Lund and H. Lund, Acta Chem. Scand., B40, 470 (1986); B42, 269 (1988).
187. T. Lund and H. Lund, Acta Chem. Scnad., B41, 93 (1987).
188. N. Djeghidjegh, M. Chaquiq El Badre, and J. Simonet, J. Electroanal. Cham., 246, 457 (1988).
189. P. Fuchs, U. Hess, H. Holzt, and H. Lund, Acta Chem. Scand., B35, 185 (1981).
190. H. Lund and J. Simonet, C. R. Acad. Sci., Ser. C, 227, 1387 (1973); P. Martigny, H. Lund, and J. Simonet, Electrochim. Acta, 21, 345 (1976).
191. G. Silvestri, S. Gambino, G. Filardo, G. Greco, and A. Gulotta, Tetrahedron Lett., 25, 4307 (1984).

192. O. Sock, M. Troupel, and J. Perichon, Tetrahedron Lett., 26, 1509 (1985); J. Poulichen, M. Heintz, O. Sock, and M. Troupel, J. Chem. Educ., 63, 1013 (1986).
193. C. Degrand and D. Jacquin, Tetrahedron Lett., 4955 (1978).
194. H. Lund and J. Simonet, Bull. Soc. Chim. Fr., 1843 (1973).
195. K. Boujlel and J. Simonet, Tetrahedron Lett., in press.
196. T. Shono, I. Nishiguchi, and H. Ohmizu, J. Amer. Chem. Soc., 99, 7396 (1977).
197. T. Shono, I. Nishiguchi, and Hl Ohmizu, Chem. Lett., 1021 (1977).
198. H. Lund and C. Degrand, Tetrahedron Lett., 3593 (1977).
199. H. Lund and C. Degrand, Acta Chem. Scand., B33, 57 (1979).
200. E. J. Hart and M. Anbar, The Hydrated Electron, Wiley-Interscience, New York, 1970; G. A. Kenney and D. C. Walker, in Electroanalytical Chemistry (A. J. Bard, ed.), Vol. 5, Dekker, New York, 1971, p.1.
201. E. J. Hart, S. Gordon, and E. M. Frielden, J. Phys. Chem., 70, 150 (1966).
202. T. Teherani, K. Itaya, and A. J. Bard, Nouv. J. Chim., 2, 481 (1978); A. J. Bard, K. Itaya, R. E. Malpas, and T. Teherani, J. Phys. Chem., 84, 1262 (1980).
203. Y. Kanzaki and S. Aoyagui, J. Electroanal. Chem., 36, 297 (1972); 51, 19 (1974).
204. Y. Harima, H. Kurihara, and S. Aoyagui, J. Electroanal. Chem., 124, 103 (1981).
205. L. A. Avaca and A. Bewick, J. Chem. Soc., Perkin Trans., 2, 1709 (1972); N. M. Alpatova and S. E. Zabusova, Electrokhimiya, 16, 1262 (1980).
206. H. W. Sternberg, R. E. Markby, I. Wnder, and D. M. Mohilner, J. Chem. Soc., 89, 186 (1967); 91, 4191 (1969).
207. R. A. Misra and A. K. Yadav, Bull. Chem. Soc. Jpn., 55, 347 (1982).
208. A. P. Krapcho and A. A. Bothner-By, J. Amer. Chem. Soc., 81, 3658 (1959).
209. H. W. Sternberg, R. E. Markby, and I. Wender, J. Electrochem. Soc., 110, 425 (1963).
210. R. J. Holman and J. H. P. Utley, J. Chem. Soc., Perkin Trans., 2, 884 (1976); S. E. Zabusova, A. P. Tomilov, T. F. Filimonova, and N. M. Alpatova, Electrokhimiya, 16, 970 (1980).
211. L. A. Avaca and A. Bewick, J. Chem. Soc., Perkin Trans., 1712 (1972).
212. E. M. Kaiser, Synthesis, 391 (1972).
213. H. W. Sternberg, R. E. Markby, I. Wnder, and D. M. Mohilner, J. Electrochem. Soc., 110, 425 (1963).
214. R. A. Benkeser, E. M. Kaiser, and R. F. Lambert, J. Amer. Chem. Soc., 86, 5272 (1964).
215. Br. Patent 1025 426 (1964), Chiyoda Chemical Engineering and Construction Co. Ltd.; Chem. Abstr., 64, 19446 (1966).
216. R. A. Benkeser and C. Tincher, J. Org. Chem., 33, 3970 (1968).
217. R. A. Benkeser and S. Mels, J. Org. Chem., 34, 3970 (1969).
218. R. A. Benkeser, H. Watanabe, S. J. Mels, and M. A. Sabol, J. Org. Chem., 35, 1210 (1970).

30

Electrogenerated Bases and Acids

MANUEL M. BAIZER† / University of California, Los Angeles, and University of California, Santa Barbara, California

I. Electrogenerated Bases 1265
  A. Synthetic objectives 1266
  B. Development of EGB chemistry 1267
  C. Azobenzenes as probases 1267
  D. Other probases 1268
  E. Some syntheses mediated by EGBs 1269
II. Electrocarboxylation of Carbon Acids 1272
  A. Oxygenation of carbon acids with electrogenerated superoxide 1273
  B. Miscellaneous synthetic applications of EGBs 1275
III. Electrogenerated Acids 1277
  A. Types of EGA-promoted reaction 1280
  B. Effect of changes in the solvent-electrolyte system 1280
References 1280

I. ELECTROGENERATED BASES

Cathodic reduction of many types of organic compounds (e.g., activated olefins, aromatic azo compounds, imines, allylic and benzylic halides, and carbon tetrachloride) results in the formation at an intermediate stage of anion radicals, anions, and monomeric and dimeric dianions. These negatively charged species may function as (1) nucleophiles, and many examples of this behavior are found in this book; (2) electron transfer reagents, an increasingly recognized type of reaction [1]; and (3) as bases, that is, electrogenerated bases (EGBs). The progenitors of the EGBs are not themselves bases; they have been named probases [2a].

It has long been implicitly recognized that in coupling reactions in which these species are used as nucleophiles their simultaneous behavior as bases, for example toward residual water or any other proton donors in the system, leads to undesired by-products. They may be avoided largely by working in rigorously aprotic systems.* The converse is true when one wishes to use these negatively charged species as EGBs: their potency as nucleo-

†Deceased, July 1988.
*This measure is not always appropriate; e.g., reductive couplings with activated olefins (Chapter 22) in strictly aprotic media would lead overwhelmingly to polymerization.

philes must be suppressed, for example, when appropriate, by steric blocking of positions near the negative charge. A review on EGB has recently appeared [2b].

A. Synthetic Objectives

The use of EGBs permits one to bring electrochemically into reaction substrates that are not or are only with great difficulty electroreducible but that have "acidic" protons on C, N, S and perhaps other elements. A very favorable aspect of this methodology is that in many cases (examples follow) the EGBH can be reconverted simply to its precursor PB so that the latter, even if expensive, becomes economical since it is essentially used in catalytic amounts. This is especially important when the EGB can deprotonate an organic acid of high pK_a since the alternative, using a strong chemical base, involves expensive reagents that can be used only once. At present, EGBs can deprotonate substrates with pK_a levels of maximum value of approximately 25; the prospects for going beyond this number are good. There remains the puzzling question of how strong a base may be usefully generated without deprotonating some constituent of the SSE (solvent-supporting electrolyte) system.

$$PB \longrightarrow EGB^-$$

$$EGB^- + RH \longrightarrow EGBH + R^-$$

$$R^- + El \longrightarrow REl^-$$

$$EGBH \longrightarrow H+ \text{ (scavenged) } + PB \qquad \text{for recycle}$$

where El = electrophile. Ideally this is accomplished in an undivided cell.

It is often desirable, unless R^- is exceptionally stable, that the electrophile with which it is to react be present in situ. Since the electrophiles themselves, such as CO_2, are often electroreducible, it is necessary in order to minimize by-product formation that the PB be more easily reduced than is the El (within potentiostatic control). Further, it is R^-, not EGB^-, that is to react with the electrophile. Means for minimizing the latter reaction have already been mentioned.

The reconversion of EGBH to PB may be done by air (hydrazobenzene → azobenzene) or by direct or indirect anodic oxidation.

In many cases the pK_a of EGBH and therefore the base strength of some EGBs, especially anion radicals, which have been used successfully in synthesis, is not known. A crude approximate estimation may be made empirically by running the cyclic voltammogram (CV) of the PB under strictly aprotic conditions. If at least one of the two waves sometimes detected is reversible, a small amount of a carbon acid of known high pK_a is added and the CV reexamined. If reversibility is still found, a carbon acid of lower pK_a is added, and so on, until a given compound produces irreversibility. If the gap between the two pK_a of the reference substances added has not been too large, the effective pK_a of EGBH can at least be bracketed between two values.

Although in a literal sense hydroxy and alkoxy, when made by cathodic reduction of water and alcohol, respectively, are electrogenerated bases, they are not included in the term "EGBs" now used by many people. This is in contrast to the term "electrogenerated acids" (EGAs, see later), which encompasses not cation radicals and cations, for example, but protons.

B. Development of EGB Chemistry

The fact that an acid-base reaction and therefore an EGB was involved was first noted in a study of the reductive cleavage of $Ph_3P^+(CH_2)_nCN$ [3]. When n = 1, an apparent one-electron transfer was involved. Examination of the products, however (including the first electrochemical preparation of an ylid), led to the hypothesis that the electrochemical step involved <u>two</u> electrons but that a second molecule of starting material became involved in an acid-base reaction (self- or autoprotonation)*:

$$\underset{\text{I}}{Ph_3\overset{+}{P}CH_2EWG} + 2e^- \longrightarrow Ph_3 + \underset{\text{II}}{(CH_2EWG)^-} \tag{1}$$

$$\text{I} + \text{II} \longrightarrow Ph_3P{=}CHEWG + CH_3EWG \tag{2}$$

where EWG = electron withdrawing group, in this case −CN. This was followed by publication of a similar study [4] in which, in addition, the intermediate ylids were used effectively in Wittig reactions in situ (see Table 1). A variety of electrochemical Wittig reactions based on EGB chemistry have been published by the Utley group (Table 1) [5].

The detailed mechanism of reactions of type (1) has been reported [6]. Other PBs analogous to I (when EWG = CN and II is CH_2CN^-) have also been studied: Ph_3As^+ salts, substituted triarylphosphonium salts, and $PhSO_2CH_2CN$. Their relative acidities were estimated voltammetrically, and it was found that I (EWG = CN) is a stronger acid than HOAc [8].

The paper describing the first electrochemical preparation of a S-ylid followed [9] soon after the publication of Ref. 3. Thus prepared S-ylids in the presence of aldehydes yielded epoxides in low to moderate yields [10].

In these cases formation of the EGBs requires the cleavage of the PBs; they can be reconstituted only in separate multistep operations. However, the use of PBs that must be cleaved to provide their EGBs is still feasible if (1) the protonated EGB is more valuable than the PB, and (2) the PB can be simply reconstituted. An example of type 1 is the use of $CCl_4 \rightarrow {}^-CCl_3 \rightarrow CHCl_3$ [12]; type 2 is under investigation [13].

C. Azobenzenes as Probases

A seminal paper that avoided the problems of cleavage described the use of azobenzenes as a PB [11]. With $Ph_3\overset{+}{P}CH_2Ph$ as the carbon acid and in the presence of PhCHO, reduction of azobenzene initiated a Wittig reaction resulting in a 98% yield of stilbene (<u>cis</u> and <u>trans</u>).

Azobenzene has been used as a PB by a number of subsequent investigators, including Refs. 12, 14, and 15. Since the anion radical (one-electron reduction), the dianion (two-electron reduction), and the half-protonated dianion may serve as nucleophiles as well as bases (see Ref. 16), it is sometimes essential for synthetic success to introduce bulky groups

*Autoprotonation was responsible for the amusing result that the reduction of dibenzoylmethane required only 0.5 electron per mole [9].

ortho to the azo linkage [2,17–19]. However, these substituted azo components are more prone than is azobenzene itself to reductive cleavage of the N–N bond.

The suggestion was made some time ago that in the presence of proton donors (or Li^+), the azobenzene anion radical undergoes disproportionation:

$$PhN{=}NPh \xrightarrow{e^-} Ph\dot{N}{-}\bar{N}Ph \xrightarrow{e^-} Ph\bar{N}{-}\bar{N}Ph \tag{3}$$

$$2Ph\dot{N}{-}\bar{N}Ph \longrightarrow PhN{=}NPh + Ph\bar{N}{-}\bar{N}Ph \tag{4}$$

This phenomenon has the synthetic advantage, mentioned in Sec. A, that it permits the carbon acid in the presence of the electrophile to be deprotonated at a less cathodic potential than is necessary to generate the dianion.

The question of the relative basicities of the various species generated by the reduction of azobenzene has recently been examined in detail [20]. The authors report increasing basicity in the order $(PhNNPh)^{2-}$ > $(PhNHNPh)^-$ > $(PhN{=}NPh)^{\dot{-}}$.

D. Other Probases

A number of other PBs from which EGBs can be prepared without cleavage have been examined: tetralkyl ethenetetracarboxylates [17], ArCOR [21], PhCOOR and $PhCOOCH_2Ar$ [22], $(CH_2O)^-$ [23], 2-pyrrolidinone [24], oxygen [25], hydrogen (Pt cathode) [26], aliphatic aldehydes [27], and activated olefins [14]. Those that have been used successfully in synthesis are mentioned later.

The potential use of H_2 as a PB had intriguing possibilities (a pK_a of 25 is cited in Ref. 28). Initial exploratory work was done in 1973 at the University of Southampton [29]. Fleischmann suggested the direct reduction of H_2 to H^- at a palladium cathode; Baizer proposed that the supporting electrolyte be a R_4NX so that the expected R_4NH would, unlike metal hydrides, presumably be soluble in the medium and, further, that Ph_3CH ($pK_a \simeq 32$) be used as an internal indicator (or possibly as probase) since the hydrocarbon is colorless and its derived anion blood red. These details are mentioned because this background has been incorrectly cited in Ref. 26.

If R_4NH could indeed be made, the ensuing reactions would be catalytic, so that a little H_2 would be necessary only to initiate the sequence

$$H_2 \xrightarrow[2R_4NX]{2e^-} 2R_4NH + 2X^- \tag{5}$$

$$R_4NH + RH \longrightarrow R^-R_4N^+ + H_2 \quad \text{for recycle} \tag{6}$$

$$R^- + \text{electrophiles} \longrightarrow \text{products} \tag{7}$$

The reduction occurred at a potential very close to that for the discharge of R_4N^+. A red coloration was indeed seen in the vicinity of the cathode. However, interpretation of the results was equivocal; strong bases could have been formed by the discharge of R_4N^+ or by OH^- due to residual traces of water. Attempts to resolve this ambiguity by using very anhydrous media and working in a glove box failed [30].

Dioxygen is an interesting probase. In aprotic media it is easily reduced (~ −1.0 V versus SCE at Hg). Since in a laboratory H cell where the oxygen is continuously bubbled through the catholyte at atmospheric pressure only a portion of it is reduced, it is inevitable that the most favored consequence is the reaction of the carbanion formed by deprotonation of a carbon acid (by the EGB $O_2^{\cdot -}$) with unreduced oxygen.

Although $O_2^{\cdot -}$, the conjugate base of $\cdot HO_2$ (pK_2 4.88), is not a strong base per se, disproportionation in the presence of proton donors leads to the formation of strong bases [31] capable of deprotonating CH_3CN and n-BuOH (pK_a 33.3) [32]:

$$O_2 + e^- \longrightarrow O_2^{\cdot -}$$

$$2O_2^{\cdot -} + BH \longrightarrow O_2 + HO_2^- + B^-$$

The products formed from the carbanions and dioxygen, via intermediate electron transfer steps [33], are oxygenated species formed <u>under reducing conditions</u> and often difficult to prepare by alternative means.

Some carbon acids, such as CH_3CN [34], PhC≡CH [35], and cyclopentadiene [36], have been reduced directly at Pt cathodes presumably with liberation of hydrogen. The so-formed carbanions, however, have thus far been used only as nucleophiles. Their use as EGBs would have the intriguing feature of being an indirect deprotonation without by-products. Alternatively, the H_2/Pt could be used for electrocatalytic reduction of nonelectroactive compounds.

E. Some Syntheses Mediated by EGBs

Wittig Reactions

These syntheses were the first to be exploited using EGB chemistry and were referred to earlier. Table 1 provides representative examples.

Early in the development of EGB chemistry, Iversen [37] reported the equivalent of the Sommelet-Hauser rearrangement:

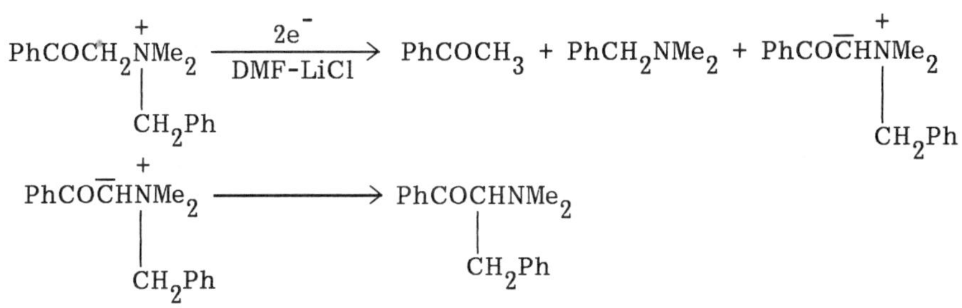

Michael Reactions

Almost all of these are base catalyzed [38]. A compound containing "active" hydrogen, the donor DonH, is deprotonated by the base; the resulting anion adds to an activated olefin, the acceptor; the adduct anion deprotonates a new molecule of DonH, and the cycle continues:

Table 1. Electrochemical Wittig Reactions

Carbon-acid (also Pro-ylid)	Probase	Carbonyl compound	Product	Yield (%)	Reference
$Ph_3\overset{+}{P}CH_2PhBr^-$	Self[a]	PhCHO	PhCH=CHPh	72	4
		C_3H_7CHO	$PhCH=CHC_3H_7$	95	4
$Ph_3\overset{+}{P}CH_2COOMeCl^-$	Self[a]	PhCHO	PhCH=CHCOOMe	75	4
$Ph_3\overset{+}{P}CH_2Ph$	PhN=NPh	PhCHO	PhCH=CHPh	98	11
$Ph_3\overset{+}{P}CH_2Ph$	Fl=C(CN)COOEt[b]	PhCHO	PhCH=CHPh	94	5[c]
$Ph_3\overset{+}{P}CH_2Ph\ \ NO_3^-$		PhCH=CHCHO	1,4-Diphenylbutadiene	26	5[c]

[a]Reductive cleavage yields the base.
[b]Fl = 9-fluorenylidene.
[c]Other examples given.

Table 2. Electrochemically Initiated Michael Reactions

Probase	Donor	Acceptor	Yield of 1:1 adduct (%)	Reference
Acceptor	$CH_2(COOEt)_2$	$CH_2{=}CHCN$	18.5	14
Acceptor	$CH_2(COOEt)_2$	$(EtOCO)_2C{=}C(COOEt)_2$	93	14
Acceptor	$CH_2(COOEt)_2$	$CH_2{=}CHCOOEt$	77	14
PhN=NPh	$CH_2(COOEt)_2$	$CH_2{=}CHCOOEt$	75	14
PhN=NPh	$CH_3CH_2NO_2$	$CH_2{=}C(CH_3)COOMe$	>60[a]	40
PhN=NPh	$CH_3CH_2NO_2$	2-Cyclopentenone	>57[a]	40

[a] The yield of the nitro adduct, which was an intermediate, was not determined; it was treated with electrogenerated superoxide and converted to a ketone.

$$\mathrm{DonH} + \mathrm{B}^- \longrightarrow \underset{\mathrm{III}}{\mathrm{Don}^-} + \mathrm{BH} \tag{8}$$

$$\mathrm{III} + \overset{|}{\underset{|}{\mathrm{C}}}{=}\overset{|}{\mathrm{C}}{-}\mathrm{EWG} \longrightarrow \underset{\mathrm{IV}}{(\mathrm{Don}{-}\overset{|}{\underset{|}{\mathrm{C}}}{-}\overset{|}{\underset{|}{\mathrm{C}}}{-}\mathrm{EWG})^-} \tag{9}$$

$$\mathrm{IV} + \mathrm{DonH} \longrightarrow \mathrm{III} \cdots$$

Many bases have been used as catalysts. In addition, some very weak bases that are, however, good nucleophilic reagents, such as tert-phosphines, have also been found to be effective [39]. The required base, needed in only very small amounts, can be generated electrochemically by using as PBs either the acceptor itself or an added substance, such as azobenzene. Only a very brief period of electrolysis is necessary; the quantities of by-product EGBH produced are negligible. Illustrative examples are given in Table 2.

When the acceptors likewise have "active" hydrogen on the carbon atom adjacent to an EWG, they are simultaneously donors*; the EGB-initiated Michael reaction leads, among other products, to dimers. This was illustrated some years ago [14] using cyclohex-2-en-1-one and diethylglutarate as substrates. In a recent very interesting example [41], reduction using 0.20 F mol^{-1} of 1-phenyl-1-penten-3-one initiates two successive Michael

*Although we do not wish to proliferate the use in the literature of "cutesy" terms like "wolf-and-lamb" and "father-son" reactions, we mention that these compounds are hermaphrodites.

reactions leading to cyclized products. When $LiClO_4$ was the supporting electrolyte, 100% of V was formed; when Bu_4NBr was used, equal parts (34%) of V and VI resulted together with some polymers. VI is a bicyclic compound and presumably arose from an intramolcular aldol condensation of VII, an isomer of V.

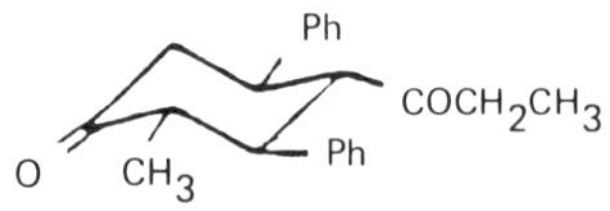

V

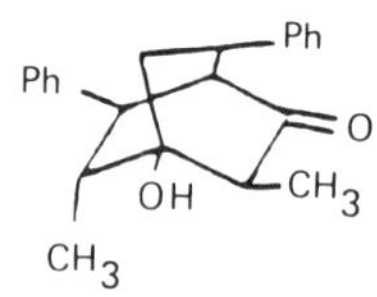

VI

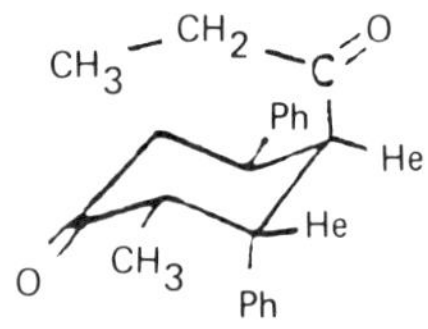

VII

II. ELECTROCARBOXYLATION OF CARBON ACIDS

This section does not deal with electrophoric compounds that are reduced to anion radicals and anions, intermediates that then react with CO_2 either nucleophilically or by electron transfer with CO_2. It concerns carboxylation of substrates with no electrophores or those that are reduced only with great difficulty (very cathodic potential). The necessity for using PBs that are reduced more easily, preferably much more easily, than CO_2 itself (~ −2.2 V versus SCE) has been mentioned in Sec. I.A. Hindered azobenzenes and $(ROCO)_2C{=}C(COOR)_2$ have been shown to be satisfactory. In addition, the PBs can be regenerated from the EGBHs by oxidation with air or anodically generated bromine, respectively. Results are summarized in Table 3.

A. Oxygenation of Carbon Acids with Electrogenerated Superoxide

The general principles have been given in Sec. I.C. Depending on the functional groups in the carbon acid, it may, while being oxygenated by $O_2^{\overline{\cdot}} + O_2$, (1) remain intact or (2) be simultaneously cleaved [18]. It is well known that carboxylic esters are saponified by superoxide [31a].

Examples of both types of behavior are given in Table 4.

It should be noted that these oxygenations are often autocatalytic, that is, require only a fraction of a faraday per mole to reduce enough O_2 to start the sequence that is then continued, presumably by the reactions

$$RH + O_2^{\overline{\cdot}} \longrightarrow R^- + HO_2^{\cdot} \tag{10}$$

$$R^- + O_2 \longrightarrow\!\!\!\rightarrow\!\!\!\rightarrow RO_2^- \tag{11}$$

$$RO_2^- + R^- \longrightarrow 2RO^- \tag{12}$$

$$RO^- + RH \longrightarrow ROH + R^- \cdots \tag{13}$$

Table 3. Some Electrocarboxylations of Carbon Acids

Carbon acid	Probase	Product	Yield	Reference
Ethyl phenylacetate	$(MeOCO)_2C{=}C(COOMe)_2$	Diethyl phenylmalonate[a]	—	2a
Ethyl phenylacetate	$(n\text{-}BuOCO)_2C{=}C(COOn\text{-}B\mu)_2$	Diethyl phenylmalonate	78	2a
9-Phenylfluorene	$(EtOCO)_2C{=}C(COOEt)_2$	Sodium 9-phenylfluorene-9-carboxylate	74-81[b]	2a

[a]Major products were the Michael adducts. The esters were formed by addition of EtI to the catholyte directly after the electrolysis.

[b]In three successive runs. In run 2 what had been the original anolyte now regenerated PB and served as the catholyte, and so on. NaBr plus 18-Crown-6 as supporting electrolyte.

Table 4. Oxygenation of Carbon Acids by the EGB $O_2^{\cdot-}$ Followed by O_2

Carbon acid	Products	Yield (%)	Reference
$CH_2(COOEt)_2$	Complex mixture	—	25[a]
$MeCH(COOEt)_2$	$MeC(OH)(COOEt_2)$	95	25[a]
$PhCH(COOEt)_2$	$PhC(OH)(COOEt)_2$ +	7	25[a]
	PhCOCOOEt[a]	74	
$Me_2CHCOOMe$	$Me_2C(OH)COOH$[b]	40	25[b]
$Ph_2CHCOOMe$	$Ph_2C(OH)COOMe$ +	45	25[b]
	Ph_2CO[a]		
Ph_2CHCN	Ph_2CO	95	25[b]
$PhCH(CH_3)CN$	$PhCOCH_3$ +	72	25[b]
	Starting material	20	
$PhCH_2CN$	PhCOOH	89	25[b]
$PhCH_2CN$	$PhCH(OAc)_2$[c]	54	25[b]
	Ph_2CO[a]	54	25[b]
$Ph(CH_3)CHCONEt_2$	$Ph(CH_3)C(OH)CONEt_2$ +	53	25[b]
	$PhCOCH_3$ +	Trace	
	Starting material	30	
Ph_2CHSO_2Ph	PhCOPh	90	25[b]
Nitrocyclohexane[d]	Cyclohexanone		25[c]
2-Nitro-5-oxo-hexane[d]	2,5-Dioxohexane	82–86	25[c]
Methyl 2-methyl-4-nitropentanoate[d]	Methyl 2-methyl-4-oxo-pentanoate	76–98	25[c]

[a]Oxygenation and cleavage.
[b]Oxygenation and saponification.
[c]Ac_2O present.
[d]These are simple clean alternatives to the classic Nef reaction. The use of this oxygenation step following the addition of nitronate anions to Michael acceptors is discussed in Sec. IE and shown in Table 2.

Electrogenerated $O_2^{\overset{\cdot}{-}}$ accompanied by O_2 can also be used to epoxidize activated olefins [25d]. The reaction presumably occurs by nucleophilic attack of a hydroperoxide anion on the olefin followed by elimination of alkoxides [44].

$$ROO^- + \mathrm{>C{=}C{-}EWG} \longrightarrow \mathrm{-\underset{OOR}{\overset{|}{C}}{-}\overset{(-)}{C}{-}EWG} \longrightarrow \mathrm{-\underset{\;\;O}{C{-}C}{-}EWG} + OR^- \qquad (14)$$

When the substrate is hermaphrodite, e.g. cyclohex-2-en-1-one, it itself can furnish, through deprotonation and oxygenation, the ROO^- of Eq. (14). Even in these cases, and certainly when the substrate has no active hydrogen, better results are obtained if ROO^- is generated from a separately supplied carbon acid. Coproducts arise from the simultaneously oxygenated carbon acids. The process is more efficient if the coproduct is a noncleaved product, such as $CH_3C(OH)(COOEt)_2$ from $CH_3CH(COOEt)_2$, than a cleaved product, such as Ph_2CO from Ph_2CHCN.

So far [25d] the epoxidation of only α,β-unsaturated ketones has been studied. Some of the results are summarized in Table 5.

B. Miscellaneous Synthetic Applications of EGBs

The acetoacetic ester (Claisen) synthesis using ethyl acetate as the carbon acid and 2,2',4,4',6,6'-hexaisopropylazobenzene as PB has been reported [45]. In related work in which aliphatic aldehydes were themselves the PBs, self-condensation (aldol) yielded α,β-unsaturated carbonyl compounds in good (72–76%) yields [27]. The reactions are autocatalytic: very little electricity is required. The use of other carbonyl compounds as PBs has already been mentioned [21–23].

The paired reactions acrylate to adipate cathodically plus malonate to bimalonate anodically also involve EGB chemistry [46]. The anionic reductively coupled species formed from acrylate abstract protons from malonate, thus becoming adipates; the resulting malonate anions are (indirectly) oxidatively coupled at the anode (see Chap. 35).

The one useful N-alkylation that has been reported [19] uses, for example, 2,2'-di-*t*-butylazobenzene as the probase, *N*-(2,6-dimethyl-1-cyclohexen-1-yl)-2-chloracetamide as the acid, and $ClCH_2OC_2H_5$ as the alkylating agent. The product was obtained in 41% yield and with 82% current efficiency; "chemical" methods using a variety of basic condensation reagents had given extremely poor results. Numerous C-alkylations have been accomplished, such as the formation, ad libitum, of 9-methyl- or 9,9-dimethylfluorene using azobenzene as PB [16]; the conversion of n-BuCl to 1-cyclopentane using 2,4,6,2',4',6'-hexaisopropylazobenzene as the PB and CH_3CN as the carbon acid [47]; the similar conversion of 1,3-dichloropropane to 1,5-dicyanopentane; the alkylation at the α-methylene group of methyl arylacetates using 2-pyrrolidone as the PB [24b]; the deprotonation of CH_3CN presumably by H^- generated cathodically at Pt and the addition of the $(CH_2CN)^-$ to isobutyraldehyde to give the expected product [26]; and the deconjugative addition of alkyl iodides to acetylenic esters presumably caused by alkyl iodides also serving as PBs [48].

Table 5. EGB-Promoted Epoxidation of α,β-Unsaturated Ketones

Enone	Carbon acid	Yield (%) epoxide	Yield (%) recovered enone	Coproduct
2-Cyclohexen-1-one	0	40		
	Ph_2CHCN	89	Trace	Ph_2CO
4,4-Dimethyl-2-cyclohexen-1-one	0	9.5	81	Ph_2CO
	Ph_2CHCN	31	59	Ph_2CO
PhCH=CHCOPh	$CH_3CH(COOEt)_2$	90	Trace	$CH_3C(OH)(COOEt)_2$
	Ph_2CHCN	84	Trace	Ph_2CO

In addition to the cyanomethylations, these should be mentioned: (1) the use of PhCOR as a probase, the EGB of which deprotonates CH_3CN. The $(CH_2CN)^-$ then adds to the ketone, yielding a β-hydroxynitrile after loss of water, becomes a Michael acceptor to which $(CH_2CN)^-$ adds, yielding finally $PhC(R)(CH_2CN)_2$ [21]; (2) cyanomethylation as a competitive reaction in the reduction of α,β-unsaturated nitriles in CH_3CN [49]; and the addition of electrogenerated $(CH_2CN)^-$ to fluorenone and its Schiff bases [50], to fluorenone and its anil in the presence of acylating agents [51], and to fluorene-9-ylideneacetonitrile [15].

Using 2-pyrrolidinone as a PB, the Cieckmann [52]* and Stobbe condensations have also been effected [52]. When PhN=NPh was used as a PB, dimethyl ω-bromoalkylomalonate underwent reductive cyclization to give, for example, cyclopentanes and cyclohexanes in >80% yield [12a].

A most unusual EGB, formaldehyde anion radical $(CH_2O)^-$, has recently been reported [23,54]. Nitroalkylbenzenes are converted to side chain mono- and/or bishydroxymethylated derivatives. Two alternative procedures are described: (1) a DMF solution of the nitroalkylbenzene and paraformaldehyde is reduced at constant current [23]; and (2) the paraformaldehyde solution alone is reduced first and then the nitroalkylbenzene is added [54]. In (2) the $(CH_2O)_n$ is unequivocally reduced first; in (1) it may be the nitrocompound that is reduced first and transfers an electron to CH_2O to form $(CH_2O)^-$.

These successful syntheses raise the question about whether in the conversion of nitroalkylbenzenes (and methylpyridines and methylpyridine N-oxides) to their corresponding acids by electrogenerated O_2^- the latter is acting, as reported [55], as a radical or as an EGB.

III. ELECTROGENERATED ACIDS

It is well known that during electrolyses in divided cells of water-containing solutions the catholyte becomes alkaline owing to the depletion of protons, and the anolyte, conversely, accumulates an excess of hydronium ions. The excess base and acid, respectively, are in high concentration only in the vicinity of the electrodes; they tend to be dissipated by convection and diffusion, among other processes. In undivided cells, particularly if stirring is used, the local differences in pH tend to be minimized. The effective acidity of the acid generated at the anode (EGA) in aqueous-organic media or in organic media containing even only minor amounts of water depends along with other factors upon buffering effects in undivided cells and on solvation and counterion effects in both types of cells.

Note that EGAs are protic acids, not the counterparts of EGAs; that is, they are not cation radicals or cations, for example.

The literature in this field has been recently reviewed [56]. Although the earlier work formulated the principles mentiond earlier and examples are scattered through Part IV of this book, it was not part of a deliberate sustained effort to use EGAs advantageously in electroorganic syntheses. The latter objective has been pursued with notable success mainly by Torii and his co-workers.

---

*The Dieckmann condensation was also observed [53] during the hydrodimerization of ethyl cinnamate; the product was 2-carbethoxy-3,4-diphenylcyclopentanone.

Table 6. EGA-Catalyzed Reactions of Epoxides VIII at a Platinum Anode

Solvents	Electrolyte	Cell	F mol^{-1}	Products	Yield (%)	Reference
$7.5CH_2Cl_2 + 0.5THF$	$LiClO_4$	A[a]	0.03	IX[b]	91[c]	56, 57
$7.5CH_2Cl_2 + 0.5THF$	$LiClO_4$	B[d]	0.03	IX[e]		57c
THF	$NaClO_4$	A	0.04	IX	90	57c
THF	$Mg(ClO_4)_2$	A	0.04	IX	79	57c
THF	Et_4NClO_4	A	0.5	IX	89	57c
CH_2Cl_2	Et_4NOTs	A	3.1	XI	57	57c
CH_2Cl_2	Et_4NBr	A	3.7	No IX		57c
MeCN, dry		A		No IX, all VIII[f]		57c
MeOH, dry		A		No IX		57c
CH_2Cl_2	CF_3COOLi	A		No IX cleavage		57c
Me_2CO	$LiClO_4$[g]	A	0.1	X[h]	87	57c
Me_2CO	$Mg(ClO_4)_2$	A	0.1	X	90	57c

Me_2CO	$NaClO_4$	A	0.16	X	84	57c
Me_2CO	$LiBF_4$	A	0.1	X	81	57c
Me_2CO	Et_4NClO_4		1.7	X	37	57c
Me_2CO	Et_4NBr		2.7	Recovered VIII		57c
Me_2CO	Et_4NOTs		2.7	Recovered VIII		57c
$ClCH_2CH_2Cl$	$TsONa + TsONEt_4$	A	0.3 + 0.15[h]	XI, IX, VIII[i]	74, 7, 13	57d
$ClCH_2CH_2Cl$	$LiClO_4$	A		XI, IX	2, 91	57d
$ClCH_2CH_2Cl$	$LiBF_4$	A		IX, IX	28, 50	57d

[a]Undivided.
[b]Plus 8% XI.
[c]Same yield if solution is preelectrolyzed for 1 min and VIII added afterward and allowed to react for 3 min.
[d]Divided.
[e]Only product in anolyte.
[f]R in VIII = $-(CH_2)_2CH(Me)(CH_2)_2OCH_2Ph$. Other R in [57c].
[g]Epoxide, 1 mol per 100 mol.
[h]R = $-(CH_2)_2C(CH_3)(OAc)C{\equiv}CH$.
[i]Preelectrolysis. Then add VIII and let stand 1 h. Electrolyze again 1 h, and let stand 4 h.

A. Types of EGA-Promoted Reactions

Among the reactions mediated by EGAs are (1) isomerization of epoxides [57], (2) aromatization of a precursor to vitamin K_3 (2-methyl-1,4-naphthohydroquinone) [58]; and cyclization of d,l-nerolidol to d,l-bisabol [59]. A recent publication [60], which, like [58e], is not from the Torii group, discloses the monoalkylation of hydroquinone with 86% conversion, 99% selectivity, and 85% yield. Pt electrodes are used; the reaction is probably EGA catalyzed.

B. Effect of Changes in the Solvent-Electrolyte System

The results obtained in a study of category (1) provide an impressive illustration of the differences in products and their yields that result from changes in the solvent-electrolyte system and the operating conditions. These occurred at an invariable anode Pt.

The epoxides (VIII) can be converted to the ketone (IX), the acetonide (X), or the allylic alcohol (XI) under the conditions shown in Table 6.

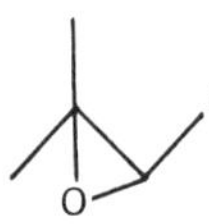

VIII

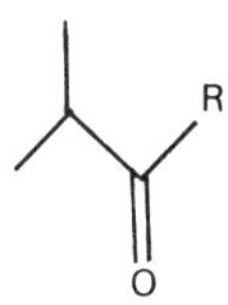

IX

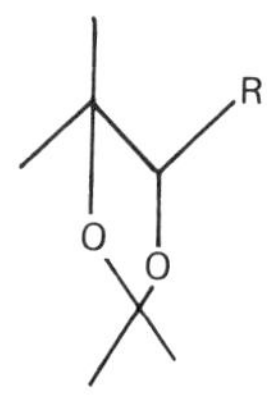

X

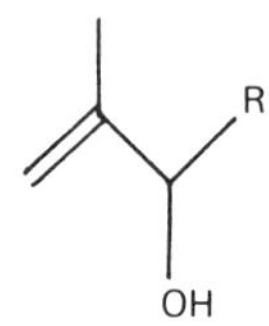

XI

REFERENCES

1. H. Lund, M.-A. Michel, and J. Simonet, Acta Chem. Scand. B28, 900 (1974); H. Lund and J. Simonet, J. Electroanal. Chem., 65, 205 (1975).
2. (a) R. C. Hallcher and M. M. Baizer, Liebigs Ann. Chem., 737 (1977); (b) J. H. P. Utley, Top. Curr. Chem., 142, 131 (1987).
3. J. H. Wagenknecht and M. M. Baizer, J. Org. Chem., 31, 3885 (1966).
4. T. Shono and M. Mitani, J. Amer. Chem. Soc., 90, 2728 (1968).
5. R. R. Mehta, V. L. Pardini, and J. H. P. Utley, J. Chem. Soc., Perkin Trans. I 2921 (1982) and references therein.
6. J.-M. Savéant and K. B. Su, Bull. Soc. Chim. Fr., 3549 (1972).
7. R. C. Buchta and D. H. Evans, J. Electrochem. Soc., 117, 1494 (1970).
8. A. J. Bellamy and (in part) I. S. MacKurdy, J. Chem. Soc., Perkin Trans II, 1093 (1981).
9. J. H. Wagenknecht and M. M. Baizer, J. Electrochem. Soc., 114, 1095 (1967).
10. (a) T. Shono and M. Mitani, Tetrahedron Lett., 687 (1969); (b) T. Shono, T. Akagawa, and M. MItani, Tetrahedron, 29, 817 (1973).
11. P. E. Iversen and H. Lund, Tetrahedron Lett., 3523 (1969).
12. (a) S. T. Nugent, M. M. Baizer, and R. D. Little, Tetrahedron Lett., 23, 1339 (1982); (b) M. M. Baizer, R. D. Goodin, and R. C. Hallcher, U.S. Patent 4,132,611 of 1/2/79; Chem. Abstr., 90, 112137 (1979).
13. M. M. Baizer and M. Sugawara, unpublished work; T. Fuchigami, work in progress.
14. M. M. Baizer, J. L. Chruma, and D. A. White, Tetrahedron Lett., 5209 (1973).

15. C. Degrand, P. L. Compagnon, and F. Gazques, J. Chem. Soc., Chem. Commun., 383 (1983).
16. T. Troll and M. M. Baizer, Electrochim. Acta, 20, 35 (1975).
17. R. C. Hallcher, D. A. White, and M. M. Baizer, J. Electrochem. Soc., 126, 404 (1979).
18. M. M. Baizer, Tetrahedron, 40, 953 (1984).
19. R. D. Goodin, R. C. Hallcher, and M. M. Baizer, U. S. Patent 4,248,678 of February 3, 1981; Chem. Abstr., 94, 147556 (1981).
20. S. Chang and M. D. Hawley, J. Org. Chem., 50, 3388 (1985).
21. E. M. Abbott, A. J. Bellamy, and J. Kerr, Chem. Ind., 828 (1974).
22. L. Kistenbrugger, P. Mischke, J. Voss, and G. Wiegand, Liebigs Ann. Chem., 461 (1980).
23. S. Torii, Y. Murakami, H. Tanaka, and K. Okamoto, J. Org. Chem., 51, 3143 (1986).
24. (a) T. Shono, S. Kashimura, K. Ishizaki, and O. Ishige, Chem. Lett., 1311 (1983); (b) T. Shono, S. Kashimura, and H. Nogusa, J. Org. Chem., 49, 2043 (1984); (c) T. Shono, Mie Jpn. Kokai Tokkyo Joho JP 59,219,484 of Dec. 1984; Chem. Abstr. 103, 149601z (1985).
25. (a) P. M. Allen, U. Hess, C. S. Foote, and M. M. Baizer, Synth. Commun., 12, 123 (1982); (b) M. Sugawara, M. M. Baizer, W. T. Monte, R. D. Little, and U. Hess, Acta Chem. Scand., B37, 509 (1983); (c) W. T. Monte, M. M. Baizer, and R. D. Little, J. Org. Chem., 48, 103 (1983); (d) M. Sugawara and M. M. Baizer, J. Org. Chem., 48, 4931 (1983).
26. S. Pons and S. B. Khoo, Electrochim. Acta, 27, 1161 (1982).
27. T. Shono, S. Kashimura, and K. Ishizaki, Electrochim. Acta, 29, 603 (1984).
28. F. R. Hartley, CHEMTECH, 10, 688 (1980).
29. M. Fleischmann, M. M. Baizer, and I. Tabakovic (later followed up by D. Masheder), unpublished work (1973).
30. J. P. Coleman and M. M. Baizer, unpublished work at the Monsanto Co., 1974.
31. (a) D. T. Sawyer and J. S. Valentine, Acc. Chem. Res., 14, 400 (1981); (b) J. Wilshire and D. T. Sawyer, Acc. Chem. Res., 12, 105 (1979).
32. D.-H. Chin, G. Chericato, E. J. Nanni, Jr., and D. T. Sawyer, J. Amer. Chem. Soc., 104, 1296 (1982); D. T. Sawyer and M. J. Gibian, Tetrahedron, 35, 1471 (1979).
33. G. A. Russell, E. G. Janzen, A. G. Bemis, E. J. Geels, A. J. Moye, S. Mak, and E. T. Strom, Adv. Chem., 51, 112 (1965).
34. E. M. Abbot, A. J. Bellamy, J. B. Kerr, and I. S. MacKurdy, J. Chem. Soc., Perkin Trans II, 425 (1982) and earlier papers by the Bellamy group.
35. M. Tokuda, T. Taguchi, O. Nishio, and M. Itoh, J. Chem. Soc., Chem. Commun., 606 (1976).
36. A. V. Bukhtiarov, V. N. Golyshin, O. V. Kuz'min, A. P. Tomilov, and E. A. Chernyshev, Zh. Obshch. Khim., 57, 151 (1987); Chem. Abstr., 107, 58319e (1987); A. V. Bukhtiarov, E. A. Chernyshev, O. V. Kuz'min, B. K. Kabanov, S. S. Ninov'eva, and A. P. Tomilov, Zh. Obshch. Khim., 55, 395 (1985); A. V. Bukhtiarov, E. A. Chernyshev, O. V. Kuz'min, B. K. Kabanov, V. N. Golyshin, and A. P. Tomilov, Dokl. Akad. Nauk SSSR, 278, 1127 (1984); Chem. Abstr., 102, 102325b (1985).
37. P. E. Iverssen, Tetrahedron Lett., 55 (1971).
38. E. D. Bergmann, D. Ginsberg, and R. Pappo, Org. React., 10, 179 (1959).

39. D. A. White and M. M. Baizer, Tetrahedron Lett., 3597 (1973) and references therein.
40. W. T. Monte, M. M. Baizer, and R. D. Little, J. Org. Chem., 48, 803 (1983).
41. F. Fournier, D. Davoust, and J.-J. Basselier, Tetrahedron, 41, 5677 (1985).
42. R. C. Hallcher, D. A. White, and M. M. Baizer, J. Electrochem. Soc., 126, 404 (1979).
43. M. M. Baizer, CHEMTECH, 15, 178 (1985).
44. H. O. House, Modern Synthetic Reactions, 2nd ed., Benjamin Cummings Publishing Co., Menlo Park, CA, 1972, pp. 307 ff.
45. R. C. Hallcher, R. D. Goodin, and M. M. Baizer, U. S. Patent 4,293,393 of Oct. 6, 1981.
46. (a) H. G. Thomas and E. Lux, Tetrahedron Lett., 965 (1972); (b) M. M. Baizer and R. C. Hallcher, J. Electrochem. Soc., 123, 809 (1976).
47. R. C. Hallcher, R. D. Goodin, and M. M. Baizer, U.S. Patent 4,293,392 of Oct. 6, 1981; Chem. Abstr., 95, 212007 (1981).
48. M. Tokuda, and O. Nishio, J. Org. Chem., 50, 1592 (1985).
49. A. J. Ballamy, J. B. Kerr, C. J. McGregor, and I. S. MacKirdy, J. Chem. Soc., Perkin Trans. II, 161 (1982).
50. C. Degrand, P. Compagnon, and F. Gasquez, J. Org. Chem., 47, 4586 (1983).
51. C. Degrand, G. Belot, P. L. Compagnon, and F. Gasquez, Can. J. Chem., 61, 2581 (1983).
52. T. Shono, Kagaku (Kyoto), 40, 830 (1985); Chem. Abstr., 104, 195247g (1986).
53. J. P. Petrovich, M. M. Baizer, and M. R. Ort, J. Electrochem. Soc., 116 (1969).
54. S. Torii, H. Tanaka, and Y. Murakami, Jpn. Kokai Tokkyo Koho JP 61 91,382 of 9 May 1986; Chem. Abstr., 105, 226017j (1986).
55. H. Sagae, M. Fujihira, T. Osa, and H. Lund, Chem. Lett. (Tokyo), 793 (1977); H. Sagae, M. Fujihira, H. Lund, and T. Osa, Heterocycles, 13, 321 (1979); H. Sagae, Bull. Chem. Soc. Jpn., 53, 1537 (1980).
56. S. Torii, Electroorganic Syntheses, Methods and Applications, Part I: Oxidations, Kodansha, VCH, 1985, Chap. 13, pp. 315 ff; K. Uneyama, Top. Curr. Chem., 142, 167 (1987).
57. (a) T. Shono, Y. Matsumura, T. Hashimoto, K. Hibino, H. Hamaguchi, and T. Aiki, J. Amer. Chem. Soc., 97, 2546 (1975); (b) S. Torii, K. Uneyama, H. Tanaka, and T. Kohmoto, J. Org. Chem., 46, 3312 (1981); (c) K. Uneyama, A. Isimura, K. Fujii, and S. Torii, Tetrahedron Lett., 24, 2857 (1983); (d) K. Uneyama, N. Nisiyama, and S. Torii, Tetrahedron Lett., 25, 4137 (1984); (e) J. Delaunay, A. Lebouc, A. Tallec, and J. Simonet, J. Chem. Soc., Chem. Commun., 387 (1982).
58. S. Torii, H. Tanaka, and S. Nakane, Bull. Chem. Soc. Jpn., 55, 1673 (1982).
59. K. Uneyama, Y. Takakuwa, T. Ueda, and S. Torii, Chem. Lett., 529 (1984).
60. H. Takamatsu, Y. Takakuwa, and S. Tsuchiya, Jpn. Kokai Tokkyo Koho JP 6,164,891 of 3 April 1986; Chem. Abstr. 105, 104738z (1986).

IX

PRESENT AND FUTURE APPLICATIONS

31

Industrial Electroorganic Chemistry

D. E. DANLY and C. J. H. KING / Monsanto Chemical Company, Pensacola, Florida

I. Introduction 1285

II. Cell Design 1287
  A. Materials 1287
  B. Transport processes 1292
  C. Reaction engineering 1299
  D. Cell geometry 1303
  E. Economic considerations 1314

III. Industrial Electroorganic Processes 1316
  A. General 1316
  B. Commercial processes for adiponitrile 1317
  C. Nalco process for tetraalkyllead 1327
  D. Electrochemical route to sebacic acid 1329

IV. Conclusions 1331

References 1332

I. INTRODUCTION

Although the history of the electrochemical industry can be traced to the early 1800s, few processes for the electrolytic production of organic chemicals have been commercialized (large or small scale). This cannot be attributed to the lack of organic molecules that can be synthesized via an electrode, as attested by the size of this volume. Various authors have listed a number of factors to explain the lack of applied electroorganic synthesis, and it would appear that no single item is dominant. Briefly, factors that have been cited are the following.

Education: Few organic chemists have been introduced to the idea of using an electrode in their synthesis methods or given some understanding of electroorganic reactions.

Availability of equipment: Until the mid-1960s, chemists were expected to build their own potentiostats and fabricate laboratory cells based on intuition and equipment available.

Materials: Prior to the 1950s, cell fabrication for commercial use lacked the availability of easily formed, lightweight, insulator, chemically resistant plastics. Coupled to this was the slow development of anolyte and catholyte dividers.

Engineering: The electrochemical engineering part of chemical engineering has been slow to develop but is now becoming well established in the literature.

With many of these constraints out of the way, it would be expected that in the immediate past and in the future that electroorganic synthesis is set for wide commercial utilization. If this is occurring it is in a limited way only, however, although it is difficult to assess because of the secrecy applied to commercial activities. One fundamental problem not mentioned here is that the electron is not one of the cheapest reagents. Electroorganic synthesis often competes with oxidations and reductions using hydrogen and oxygen. Then there is always that second reaction at the counterelectrode, which at the least adds to the energy costs, if not an unwanted by-product unless it is part of a paired synthesis.

So where are the commercial opportunities for electroorganic synthesis? An interesting discussion in this area was organized by the Electric Power Research Institute (EPRI) in 1981 [1]. No simple conclusion was reached, but it is seen that the electrode process is not the answer to most organic synthesis routes in spite of the wide range of reaction types that can be carried out. One is left to conclude that for an organic electrode reaction to become commercial then one, or preferably more, of the following criteria should be met when compared to the nonelectrochemical method: (1) has a high product selectivity coupled with a simple recovery process from the electrolyte; (2) uses cheaper feedstocks; (3) eliminates one or more process steps, probably by carrying out a difficult chemical reaction; (4) avoids waste disposal and/or affords effective recycling of a reagent; and (5) obtains desired products from both anode and cathode (difficult to achieve). Much of this is common to all chemical processing!

Although the research electroorganic chemist is accustomed to considering such variables as electrode material, current density, electrolyte composition, pH, and temperature, certain other information is important in the development of a viable cell process. This is where the discipline of electrochemical engineering comes into play. Electrochemical engineering as a subdiscipline of chemical engineering began to take form in the mid-1950s. Electrochemical engineering brings together electrode kinetics and cell thermodynamics from electrochemistry, potential theory from physics, and transport theory from chemical engineering. This results in two significant parts of electrochemical engineering, namely transport processes and reaction engineering, and these are discussed in this chapter. Consideration of these phenomena is not only important for producing a working commercial cell, but can be a key in demonstrating a good laboratory-scale synthesis. Poor current distribution, mass transfer, or mixing in a laboratory cell can result in a high yield of by-products.

In this chapter space allows only a brief introduction to the topic of cell design together with examples of some practiced electroorganic synthesis processes, or processes developed to the point of commercialization. The references that follow immediately give an in-depth report of the subject matter, and more specific items are referenced in each section.

In the field of electrochemical engineering, comprehensive texts have been produced by Pickett [2], Fahidy [3], and Coeuret and Storck [4]. A two-volume work has been produced by Rousar et al. [5]. Two landmark works of a mathematical nature are by Levich [6] and Newman [7]. Based on a DECHEMA experimental course, Heitz and Kreysa have produced a practical monograph [8] on the subject. An introductory text emphasizing organic synthesis is given by Danly in Ref. 9. Works slanted more toward the inorganic side of electrochemical engineering are by Hine [10] and Mantell [11] and the series edited by Yeager and Salkind [12]. Several series that carry electrochemical engineering topics are given in Refs. 13–16.

Texts dealing with the technology of electrochemical cells and processes have been prepared by Kuhn [17], Gallone [18], Pletcher [19], and Landau et al. [20]. A limited series of reviews of the patent literature related to electrochemical technology has been assembled by Le Duc [21]. Books dealing with electroorganic processes have been published by Weinberg and Tilak [22], Tomilov et al. [23], Frumkin and Ershler [24], and Beck [25]. The review article by Fioshin and Tomilov [26] is of interest. Finally, a useful bibliography covering the entire field of electroorganic synthesis from 1801 to 1975 has been prepared by Alkire and Swann [27].

II. CELL DESIGN

A. Materials

Cell Structure

Electroorganic syntheses in the main are carried out at close to room temperature, and therefore a wide range of plastics is available for cell construction. Cell frames or vessels are made from the full range of readily available polymers (e.g., polypropylene, polyethylene, PVDF, PTFE, and other thermoplastic materials) and the cell parts sealed with elastomers (e.g., viton, neoprene, EPDM, and PTFE encased). Since the cell contents are likely to have a solvent action, these materials must be tested for long-term stability in the electrolyte. This should not only consider whether the materials dissolve but also their dimensional stability. It should also be noted that some polymers and elastomers have fillers that have conductivity, and hence materials should be also screened for conductivity. Fabrication can of course be by machining, but injection molding is preferred for multiple manufacture. Where strength is required, some form of external steel reinforcing can be used.

Electrodes

The scaleup of an electroorganic reaction from the laboratory cell to the commercial system poses two types of problems with regard to electrodes. The first is the selection of the lowest cost form of the material that gives the desired electrode reactions and structural integrity. The second is maintenance of the active life of the electrodes. An in-depth discussion of electrode materials for organic synthesis can be found in Chap. 6 of this volume and in Chap. IV of Ref. 22. A general account of electrodes for industrial processes is given in Chap. 14 of Ref. 17.

Electrode Selection Although the choice of electrode materials found in the research studies may have been limited to one or two materials, a greater range of materials may need to be investigated for use in large-scale cells. Also it must be remembered that two electrodes must be selected and that the counterelectrode should have the lowest possible overvoltage. In the undivided cell the latter is critical to avoid side reactions at the counterelectrode as well as to keep the cell energy consumption down.

A list of widely used cathode materials includes Hg, Pb, Al, Zn, Ni, Fe, Cu, Sn, Cd, and C. Mercury is not an attractive electrode material for large cells because of its mechanical inconvenience. Therefore Pb or an amalgam is chosen. However, even Pb has mechanical problems because of its softness and tendency to deform. One means of overcoming this is by hot dip or electroplating on copper, steel, or another rigid base metal. Alternatively, it may be possible to introduce a few percent of an alloying element, such as copper, antimony, or calcium, to harden the lead and improve the dimensional stability. This technique requires some caution since alloying elements can markedly affect by-product formation. In electrolytic processes in which the cathode material is not critical (hydrogen evolution), steel is often employed because of its low cost and high mechanical strength. An alternative to steel, the more expensive Raney Ni-coated cathode (as developed for chloralkali cells), may be useful if a lower hydrogen overvoltage is needed.

The list of anode materials is shorter than for cathodes because corrosion problems are usually encountered. Pt, C, and PbO_2 are often-used anodes, however. For organics more easily oxidized, Ag, Au, Cu, Fe, and Ni anodes may find application. At high pH, the aqueous electrolytes Fe and Ni are relatively stable and have been widely used. For an anodic process such as the Kolbe synthesis, which gives the best results on smooth platinum, the choice is narrowed to platinum-clad or platinum-plated titanium or hard carbon [28]. Although carbon electrodes are brittle, cells can be designed to accomodate them.

The PbO_2 electrode can be formed by one of three methods [29]. A lead surface when anodized forms a coating of dioxide. However, for this to have stability it has been found necessary to alloy the lead. Various binary and ternary alloys have been formulated to stabilize the coating. The most common is 1% Ag, which may have Ca added. Another common alloying metal is Sb, which has been used up to 15%. Other metals that have been used are Co, Te, As, Se, Cu, Tl, and Sn [17,29,30]. The stability of the PbO_2 surface from these alloys in an electroorganic synthesis must be checked for each new electrolyte. Organic compounds are likely to change the morphology of the PbO_2 [31]. A second approach to the PbO_2 anode may be that used by the storage battery industry. Here compressed lead dioxide is supported on a lead alloy grid. A third, and probably preferred method is the anodic plating of lead dioxide from lead nitrate-containing electrolytes [32]. This has met with some success on substrates of graphite and titanium. The plating method is probably the more attractive solution for a technical-scale lead dioxide anode. Another oxide that may be used technically as an oxygen evolver, but also has fabrication problems, is magnetite. Methods of preparing this anode have been reviewed by Hayes and Kuhn [33]. Given magnetite's ceramic nature, the anodes are probably best fabricated as a coating on titanium.

Anodes for oxygen evolution in alkaline electrolytes are generally made of steel or nickel. For anodic halogenation of organics, graphite or nickel is frequently employed as anode, the graphite being readily available from suppliers to the caustic chlorine industry. With the invention of the so-called DSA (dimensionally stable anode) by Beer for chlorine evolution, the approach to anode construction has taken a shift. The Beer electrode for chlorine is a coating of RuO_2 and TiO_2 on titanium. This fulfills the needs of the industry beautifully, with a corrosion-resistant mechanically tough chlorine evolver. However, this electrode is not as long-lived as an oxygen generator. A better formulation for this is IrO_2 and TiO_2 on titanium (or tantalum). It should be noted that the IrO_2-based anode, although working well as an oxygen evolver in inorganic systems, has shown a short life in the presence of organics, even at a low concentration. So again trial-and-error testing is needed before a coated anode can be decided upon. A number of DSA formulations have been described in the patent literature, some of which can be found in the survey by Le Duc [34]. Companies that offer coated anodes (and may include noble metal on titanium) are Permelec in Europe, Conradty in West Germany, ICI in the United Kingdom, and Eltech and Engelhard in the United States.

Electrode Stability In the commercialization of an electroorganic process, specification of the electrode materials and geometry is necessary, but this may still leave technical obstacles to a viable process. The nature of the active electrode surface is likely to change chemically or physically with time, altering the selectivity and/or current efficiency of the reaction. In general, the electrode is best viewed as a "catalyst" in the process, and should be costed and assessed accordingly (e.g., fractions of an electrode per metric ton of product).

The composition of the surface of an electrode may change owing to the leaching out of a component from an alloy. Alternatively, the surface may become contaminated by insoluble organic by-products or electrodeposition of extraneous metals. Although electrode surface contamination is not always recognized as a problem in the laboratory experiments with short runs and reagent-grade chemicals, it may prove to be a major consideration in developing a commercial process. Means of circumventing electrode surface contamination include (1) routine mechanical or chemical cleaning, (2) periodic current reversal, (3) stringent raw material specifications with regard to offending impurities, (4) purge and processing of a recycle electrolyte stream, or (5) replacement of electrodes.

Mechanical deterioration of the electrode surface may occur owing to either corrosion or erosion. Selection of electrode material is of course the primary means of minimizing such problems. Clues to corrosion problems can be obtained by studying potential-pH plots (Pourbaix diagrams). These point to the conditions that may lead to corrosion [35]. For best results the Pourbaix diagram should be generated to include all components of the electrolyte [36]. The diagrams for metal and water alone can give only a very approximate idea of corrosion conditions. Included in the possible corrosion conditions should be those of an open circuit. These conditions arise when power is removed from a cell, for example for maintenance; then it must be known how long the electrolyte can be left inside the cells. Corrosion may also be caused by intermediates of the organic electrode process. Both anode and cathode may be attacked by their respective radical ions, to form organometallics.

Erosion may be due to high rates of circulation of electrolyte containing fine solid particles of reaction products or electrode corrosion material. In this case removal of the solids external to the cell may contain this problem. Porous surfaces can contribute their own problems caused by nonuniformity of the electrode/electrolyte interface. When electrode coatings are used, possible dislocation of these can occur. A new bonding strategy may be the solution. In commercial cells it is obviously undesirable to have to remove electrodes for refurbishing, but this could well be the solution and still fall within the bounds of a viable process.

Diaphragms

In many electroorganic systems the use of a diaphragm to separate anolyte and catholyte is usually necessary to cut down on side reactions caused by the incompatability of a cell's two half-cell reactions. In cells evolving hydrogen and oxygen (or chlorine), a diaphragm may be needed to avoid handling explosive gas mixtures. Selection of a diaphragm for a large-scale cell often presents a serious obstacle to the development of a commercially feasible process. The preferred electrolysis diaphragm possesses the following properties: (1) low electrical resistance; (2) limited interdiffusion of anolyte and catholyte constituents other than current carrying ions; (3) susceptibility to plugging and fouling (with consequent resistance increase); (4) good dimensional stability; (5) long operating life; and (6) low cost. As expected, the chances of finding all these qualities in a single material are small.

Two types of diaphragm materials are available: permeable and permselective. The permeable material is a porous matrix that offers a limited barrier to all species in an electrolyte, with little or no change in electrolyte transport numbers. An attractive feature of these materials is that their cost may be up to an order of magnitude lower than that of the permselective membranes. Permselective materials, or ion-exchange membranes, have the ability to allow the passage of ions of one charge with a high degree of exclusion of ions of the opposite charge. Thus cation-exchange membranes allow greater mobility to cations than anions, so that the bulk of the current transfer is the movement of cations from the anolyte to the catholyte. Similarly, anion-exchange membranes are used if movement of cations is to be suppressed.

Permeable Diaphragms In some cases a small amount of flow between the anolyte and catholyte can be tolerated or may even be desired, in which case a permeable separator is entirely satisfactory. In simple laboratory cells alundum cups and glass frits have been used. Alundum and other forms of unglazed porcelain are not, however, very practical for use in production cells because they are fragile. As alternatives, a variety of porous media are available in a range of porosities. These have been manufactured for filters or separators in electrical batteries. The following is an outline of available materials that might be used: (1) filter clothes of glass, polyethylene, polypropylene, PVC, or PTFE; (2) nonwoven fabrics of polyethylene, polypropylene, or copolymers of vinyl chloride and acrylonitrile; and (3) porous thermoplastics of polyethylene, polypropylene, PVC, PTFE, polycarbonate, or isoprene rubber.

Both the electrical and diffusional resistance of porous materials are related to their permeability, porosity, pore size, and thickness. Methods of evaluating and correlating the electrical conductivity of different porous media have been described by MacMullin and Muccini [37]. The hydrophobic character of some membranes can make them unsuitable for many applications, and it should be remembered that too high a current density on a membrane can lead to local boiling of the electrolyte.

Permselective Diaphragms There has been considerable activity in recent years on development of permselective diaphragms (ion-exchange membranes) for use in electrodialysis, especially chloralkali processes. Several general reviews [38–41] have been published on commercially available membranes. One approach to the preparation of thin ion-exchange membranes involves the casting of a finely divided powder of ion-exchange resin in a plastic binder. These membranes have several disadvantages resulting from their heterogeneous character; namely, relatively large channels usually exist through the matrix between resin particles leading to high rates of electroosmotic water or solvent transfer and molecular diffusion.

Homogeneous ion-exchange membranes, on the other hand, are produced that are essentially a continuous polymeric structure. One matrix used by several membrane manufacturers is a copolymer of styrene and divinylbenzene to produce a cross-linked structure. This is then sulfonated for cation exchange or aminated for anion exchange. The styrene and divinylbenzene can also be grafted to a polymer film and then sulfonated or aminated to produce another series of membranes. The polymer is frequently reinforced with a screen of Saran, Teflon, glass, or PVC to improve mechanical properties.

For the more rigorous environments of the chloralkali industry, perfluorinated matrixes have been converted to ion-exchange membranes. The first of these was trade named Nafion, a sulfonated membrane for cation exchange, and marketed by DuPont. Because these membranes have excellent chemical and physical stability, as well as success in the chloralkali industry, they have found application in electroorganic synthesis cells. DuPont now has a range of perfluorosulfonic acid membranes for various cell applications (Ref. 20, p. 73). Other companies have followed DuPont in this field. Asahi Glass (with Flemion) and Asahi Chemical of Japan produce perfluorocarboxylic acid-type cation-exchange membranes. Tokuyama Soda of Japan also produce a fluorocarbon-based cation-exchange resin. Many of these perfluoro membranes are surface treated to perform well in highly alkaline conditions.

The RAI Research Corporation (United States) produces a range of cation- and anion-exchange membranes on polyfluoroethylene or polyethylene matrixes. RAI uses radiation grafting techniques to build on the ion-exchange properties. Typical features of commercially available membranes are given in Table 1. Properties of ion-exchange membranes can be varied by altering (1) polymer composition, (2) polymer porosity, (3) exchange capacity, (4) backing material, and (5) thickness.

Table 1. Typical Properties of Commercial Ion-Exchange Membranes

Thickness, mm	0.1–1.0
Width, cm	90–160
Length, cm	90–any
Resistance,[a] Ω-cm	1.5–20
Transport number[a]	0.8–0.99
Water content, %	10–60
Mullen burst strength, kg cm^{-2}	2–20
Cost, \$ m^{-1}	50–500

[a]Measured in 0.5 N NaCl.

Thus, it is conceivable to tailor a membrane to fit the specific requirements of an electroorganic synthesis. This probably requires trial-and-error testing together with the cooperation of the membrane vendor. Examples of membrane evaluations are given in Refs. 22 (Chap. VI) and 31. Here it is seen that mechanical properties are important, closely followed by problems of unwanted transport of solvent or reactant. Not only does this latter constitute a loss of material, it could result in the degradation of the counterelectrode.

B. Transport Processes

Electrochemical reactions, being heterogeneous, are governed by various transport phenomena, and these are important features in the design of a commercial electroorganic cell system. As with other heterogeneous reactions, the electrochemical reaction is impacted by mass and heat transport. However, the electrochemical reaction is unique in that it also has charge transport characteristics. These topics are briefly covered here, and the reader is referred to more comprehensive works in this area [2–8,42].

Mass Transport

Mass transfer in the form of a limiting current is probably the most investigated physical phenomena in an electrode process. A good extensive review of this is given by Selman and Tobias [43]. The basic concepts of mass transfer at an electrode have been introduced in Chap. 1 of this book. To simplify the analysis of the limiting current characteristics of various operating cell designs, an empirical approach is taken. The concept of similarity is used with the correlation of dimensionless groups. Starting with the limiting current relationship

$$i_{lim} = \frac{nFD_j(C_j - C_j[O, t])}{\delta_n} \tag{1}$$

where D_j is the diffusion coefficient of species j, n is the number of electrons in the reaction, F is the Faraday, and C_j and C_j [O, t] are the concentrations of reactant in the bulk and at the electrode surface, and δ_n is the Nearnst diffusion layer thickness. At a sufficiently high overpotential, Eq. (1) becomes

$$i_{lim} = \frac{nFD_jC_j}{\delta_n} \tag{2}$$

or

$$i_{lim} = nFk_mC_j \tag{3}$$

where k_m is the local mass transfer coefficient. The mass transfer coefficient is described by the dimensionless Sherwood number Sh,

$$Sh = \frac{k_m L}{D_j} \tag{4}$$

where L is a characteristic length. Correlations of various dimensionless groups have been established for various cell configurations [43], which enables cell designs to be compared and scaled up. By way of example, the parallel plate electrode is considered.

For mass transfer under free convection, the Sh number is related to the dimensionless numbers Schmidt (Sc) and Grasshof (Gr) as

$$Sh = a(ScGr)^b \tag{5}$$

where a < 1, and b falls between 0.25 and 0.33. The Sc and Gr numbers are given by

$$Sc = \frac{\nu}{D_j} \tag{6}$$

and

$$Gr = \frac{gL^3 \, \Delta\rho}{\nu^2 \rho} \tag{7}$$

where ν is the kinematic viscosity, ρ is the density, g is the gravitational acceleration, and $\Delta\rho$ is the density difference between the phases. For free convection at a vertical planar electrode and for short lengths, the exponent in Eq. (5) is $\frac{1}{4}$ and the constant 0.67. Then for scaleup of the limiting current in the direction of fluid flow, the $\frac{1}{4}$ exponent leads to the conclusion that the limiting current is proportional to 1 over the $\frac{1}{4}$ power of the distance up the electrode.

For the horizontal configuration the exponent of Eq. (5) becomes 1/3, which results in the limiting current density being independent of electrode length. For an electrode under free convection but with gas evolution, the Sherwood number correlation is similar to that seen under forced convection.

It is probably fair to say that for most electroorganic synthesis applications forced convection environments provide the best cell conditions. Flow can be under laminar conditions (Reynolds number Re below 1000) or under turbulent flow (Re > 2500). Considering the parallel plate electrode cell (a duct), the Reynolds number is

$$\mathrm{Re} = \frac{UL}{\nu} = \frac{Ud}{\nu} \tag{8}$$

where U is the electrolyte velocity and L is the characteristic length of 4 × cross-sectional area/perimeter of duct, which equals d, known as the hydraulic diameter. For laminar flow the following correlation has been developed:

$$\mathrm{Sh}_{AV} = 1.85X\left(\mathrm{ReSc}\,\frac{d}{L}\right)^{1/3} \tag{9}$$

and L is the electrode length. An aspect ratio term X (width w/electrode spacing s) is used to accommodate cells in which width becomes significant An understanding of the significance of the X term can be found in Table 2, where some values are tabulated.

Equation (9) leads to the conclusion that the limiting current under these conditions is inversely proportional to the 1/3 power of the distance along the electrode. This does not accommodate any entrance effects, and a further correlation is required for this region.

Turbulent flow at a planar electrode in a duct is dominated by eddy diffusion as a transport mechanism, but close to the electrode transport is by diffusion across a laminar sublayer. This sublayer is much thinner than the layer under laminar flow, which results in higher mass transfer rates under turbulent conditions. Under turbulent flow correlations take the form

$$\mathrm{Sh} = a\mathrm{Re}^{b}\mathrm{Sc}^{1/3} \tag{10}$$

where b ranges 0.8 to 0.9 and a falls between 0.1 and 0.16. Examining Eq. (10) shows that the limiting current is expected to be independent of distance in the direction of electrolyte flow, assuming an essentially constant reactant concentration.

Table 2. Values of X in Eq. (9) for Some Aspect Ratios w/s

w/s	X
20	0.9829
10	0.9671
5	0.9388
3	0.9071
2	0.8760
1	0.8214

Methods that have been proposed for improving mass transfer rates in large-scale cells are the following:

1. Rotation of a cylindrical or disk electrode, including wiping of the surface
2. External pumping of electrolyte:
   a. Open channels
   b. Channels with turbulence promoters
3. Turbine or propellor agitators
4. Fluidized bed of electrode particles
5. Fluidized bed of nonconducting particles
6. Vibration of electrode
7. Gas sparging

Although rotating-type electrodes are attractive in laboratory work, they run the risk of being an expensive solution to the mass transfer problem in commercial cells. Although the various technologies are effective, each method tends to have a penalty. Costs come in the form of operating, investment, maintenance, and crevice corrosion.

In many electrode reactions it is as important to transfer the products away from the vicinity of the electrode as it is to move the reactants to the electrode. For example, cathodic reactions in aqueous solution generate base, which through this local shift in pH can lead to unwanted side reactions. An example of this is given by Danly [44,45] in the electrohydrodimerization of acrylonitrile. Here the formation of bis-cyanoethyl ether, whose formation is pH controlled, was minimized by increasing the mass transfer rate at the cathode.

In the preceding example, Monsanto achieved the increased mass transfer by increased electrolyte velocity. With the same electrode reaction, Asahi employed turbulence promoters in the catholyte compartment [46], similar to those used in dialysis cells, to achieve the required mass transfer rates.

Rotating cylinder electrodes have attracted some attention [47] as an approach to the mass transfer problem in cell design for electroorganic synthesis. Vibration of an electrode pair at 10–100 Hz and 0.5–1.5 mm amplitude has been shown by Beck and Guthke [48] to reduce the problem of concentration polarization in the synthesis if dimethylsebacate from monomethyladipate. No commercial-scale operation of this cell design was reported. A general discussion of mass transfer problems with a range of cell designs is given by Houghton and Kuhn [49].

Charge Transport

If the current distribution along an electrode is not uniform, side reactions can occur. These can be unwanted by-product formation or localized corrosion of the electrode. The analysis of charge transport in cell design addresses the problem of current distribution. The path of current flow in a cell depends on cell geometry, conductivity of the electrolyte and electrodes, activation overpotential, and concentration overpotential. When these factors are analyzed (which is not trivial even for simple geometries [7]), three types of current distribution can be described, and these are outlined here. In these analyses only a simple parallel plate cell is considered, and Fig. 1 is a diagram of this cell.

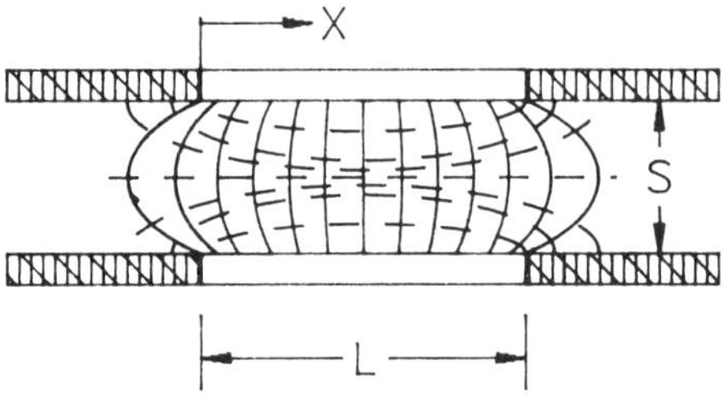

Figure 1. Current distribution at planar electrodes in an insulated channel.

Primary Current Distribution Current flow in this category is influenced by geometric factors only and is normal to the lines of equipotential. This is shown in Fig. 1, and under these conditions the potential field satisfies Laplace's equation:

$$\nabla^2\phi = \frac{\delta^2\Phi}{\delta x^2} + \frac{\delta^2\Phi}{\delta y^2} + \frac{\delta^2\Phi}{\delta z^2} = 0 \tag{11}$$

where ϕ is the electric potential. Applying the appropriate boundary conditions, current distribution for this simple electrode system is given by

$$\frac{i}{i_{AV}} = \frac{\varepsilon \cosh \varepsilon/k \ (\tanh^2 \varepsilon)}{[\sinh^2 \varepsilon - \sinh^2 (2x - L)\varepsilon/L]^{0.5}} \tag{12}$$

where $\varepsilon = \pi L/2s$ and ε is the complete elliptical integral of the first kind, L is the electrode length, x is distance along electrode, i_{AV} is the average current density, and s the electrode spacing. Some results of this analysis are shown in Fig. 2 for a few L/s values. In a commercial cell L/s is expected to be more than 100, and poor current distribution is expected only at the very edges of an electrode. Even with very high values of L/s, the very edges of the electrode show high current density. At the edges of the electrode the primary current distribution is theoretically infinite. For an electrode not mounted flush in a cell [7], then, depending on the angle the insulator makes with the electrode, the current density at the edge is lower, and for a crevice approach zero. This latter situation of the crevice explains the tendency for corrosion to occur where plastic spacers meet an electrode.

Secondary Current Distribution The effect of secondary current distribution occurs when activation overvoltage alone is superimposed on the primary current distribution. For the primary current distribution to be achieved at the edge of the electrode, high overpotentials are required. Since the electrode is essentially unipotential, this requires a redistribution

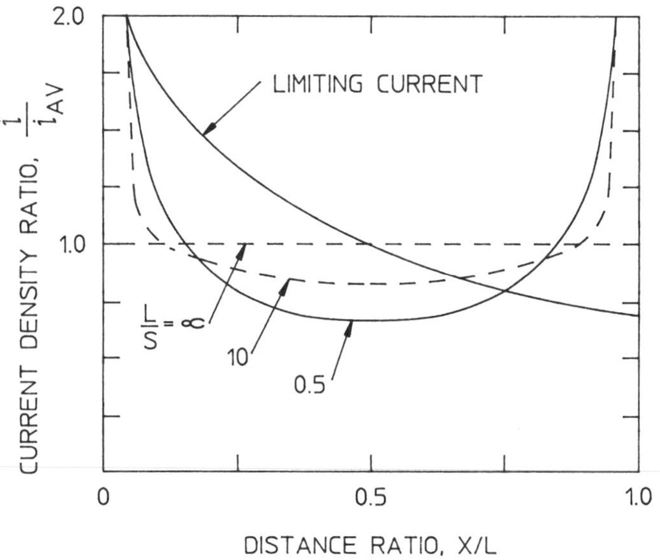

Figure 2. Primary current distribution for cell in Fig. 1 with and without limiting current.

of electrolyte potential, which in turn redistributes the current. So the resultant of the influence of the activation overvoltage is that the primary current distribution tends to be evened out. An example of this is shown in Fig. 3.

Quite often a linear polarization relationship is applicable; a solution of the Laplace equation can then be made using a dimensionless polarization parameter P:

$$P = \frac{K}{s}\frac{d\eta}{di} \tag{13}$$

where K is the electrical conductivity, η is the overpotential, and s the distance between the electrodes. When the polarization resistance is high relative to the resistance through the electrolyte (s/K), a nearly uniform current density results. The effect of changing P values where the interelectrode distance is high compared with the electrode width is shown in Fig. 3. Similar results can be obtained when the length parameter varied is the interelectrode distance. For scaleup of cells the polarization parameter should be kept constant. This is not practical, and provided interelectrode gaps are small, then increasing the electrode length has little effect on current distribution (s dominates the relationship).

Tertiary Current Distribution When the overpotential influence is that of concentration, the current distribution is again impacted; this occurs as the limiting current density takes effect. This has been analyzed [7], and the case for full limiting current is shown in Fig. 2. In the case of

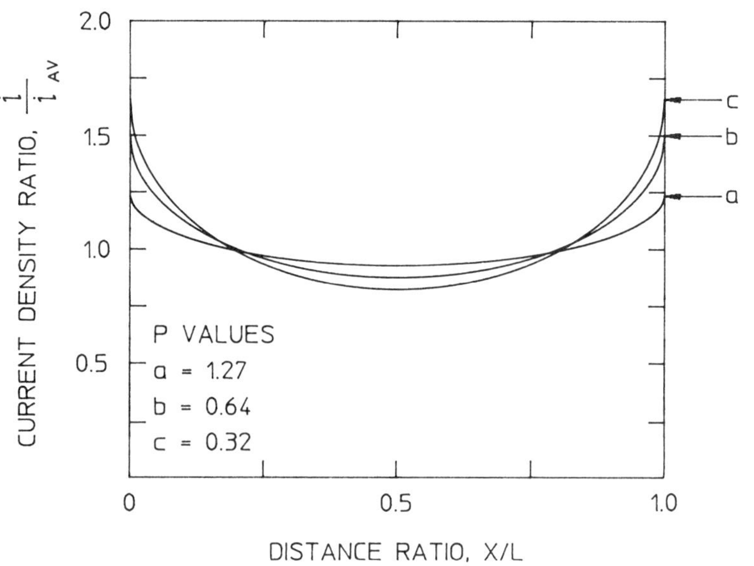

Figure 3. Secondary current distribution for cell in Fig. 1 using polarization parameter P.

a partial limiting current, a current distribution between the limiting current case in Fig. 2 and that of the secondary current distribution shown in Fig. 3 is obtained.

Secondary and tertiary current distribution are further complicated when there is gas evolution along the cell track, because of the nonuniform electrolyte resistance. An example of investigations in this area can be found in Refs. 50–52. The gas void fraction impact on electrolyte resistance, and hence cell voltage, has been investigated by a number of workers (for example Ref. 52).

Heat Transfer

Although cells are designed to minimize electrical power consumption, heat removal is probably required because of the electrical resistive heating of the electrolyte. This becomes a convenient, although not the most economical, method of cell temperature control. For the commercial electrolysis system, heat removal is generally carried out by one of the following means:

1. Circulation of electrolyte through external heat exchangers
2. Internal cooling with coils, jackets, or tubes, which may act as electrodes
3. External or internal evaporative cooling

Each of these methods has found some application in a commercial context.

When electrolyte is circulated outside the cell, external coolers are clearly attractive. This is the case, for example, with the Monsanto hydrodimerization of acrylonitrile process. Since coolers are usually made of conductors (steels or graphite), they must be placed a distance away from the cell to avoid acting as an electrode and the risk of corroding. Also, they should probably be placed upstream of the cell so as not to pressurize the cell or have poor performance because of any gas fraction in the electrolyte.

Internal cooling has been applied in the Nalco process for tetraalkyllead (see Sec. III.C and Fig. 15). The Nalco cell is essentially a shell and tube heat exchanger in which the tubes form the outer electrode of a packed bed cell. Coolant is applied to the shell side. The cell designed for isobutyric acid synthesis [53] uses cooling coils that also act as cathodes (see Fig. 9).

Evaporative cooling has been applied in the URBK process for hydroquinone from benzene [54]. Heat is removed by vacuum evaporation from aqueous sulfuric catholyte. This is also used as a means of water balance control. Electrofluorination cells also use the evaporative technique by refluxing hydrogen fluoride in the cell tank (see Chap. 26).

Other aspects of cooling system design follow standard chemical engineering practice [42,55].

C. Reaction Engineering

Electrochemical reaction engineering, as does its counterpart in chemical engineering, chemical reaction engineering [56], considers the performance of the overall cell design in carrying out a reaction. Here electrode kinetics are joined with the physical environment of the reaction to give a description of the reaction system. The electrode configuration and reactant flow patterns are taken into account. Space here allows only a brief qualitative introduction to this subject area; in-depth treatments can be found in Refs. 2–4 and 8.

Electrochemical Cells as Reactors

The electrochemical reactor readily parallels the chemical reactor, and two types are used as ideal models. One is the plug flow reactor (PFR), which presumes that concentration changes occur only in the direction of (electrolyte) flow, and thus the reactants and products move through the reactor in a pluglike manner. There is no mixing in this model. The other assumes "perfect" mixing with uniform composition in all zones of the reactor. This is referred to as a back-mixed or stirred tank reactor (STR). For the electrochemical version of the abbreviations, an E is inserted to give PFER and STER. These are illustrated in Fig. 4. As expected, these idealized models are not achieved in practice. Mixing occurs in the PFER and imperfect mixing occurs in the STER. Thus reactors are often described as having either PFER or STER features and models are approximate or become overly complicated.

These two reactor types are influenced by their mode of operation. The PFER can be used as shown in Fig. 4a, or a recirculation loop can be added so that the inlet composition contains product. This latter configuration makes it possible to shorten the length of the reactor and have higher mass transfer rates. The change in composition in these reactors is shown in Fig. 6a and is seen to depend on electrode length (or flow rate). The plug flow model can be useful in estimating electrode lengths. A recent example of this reactor model analysis can be found in Ref. 57.

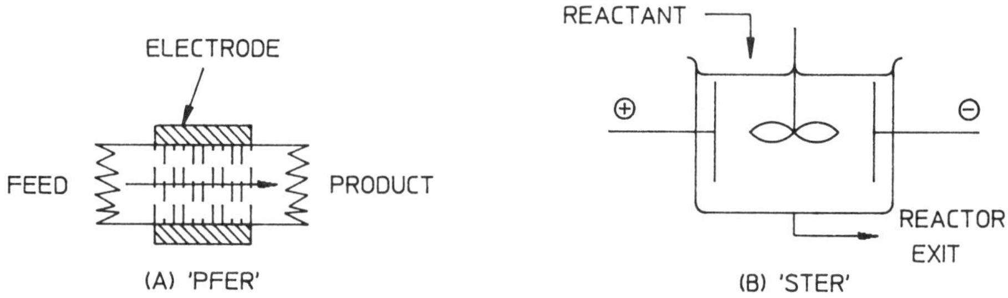

Figure 4. Schematic models of (A) plug flow electrochemical reactor, and (B) stirred tank electrochemical reactor.

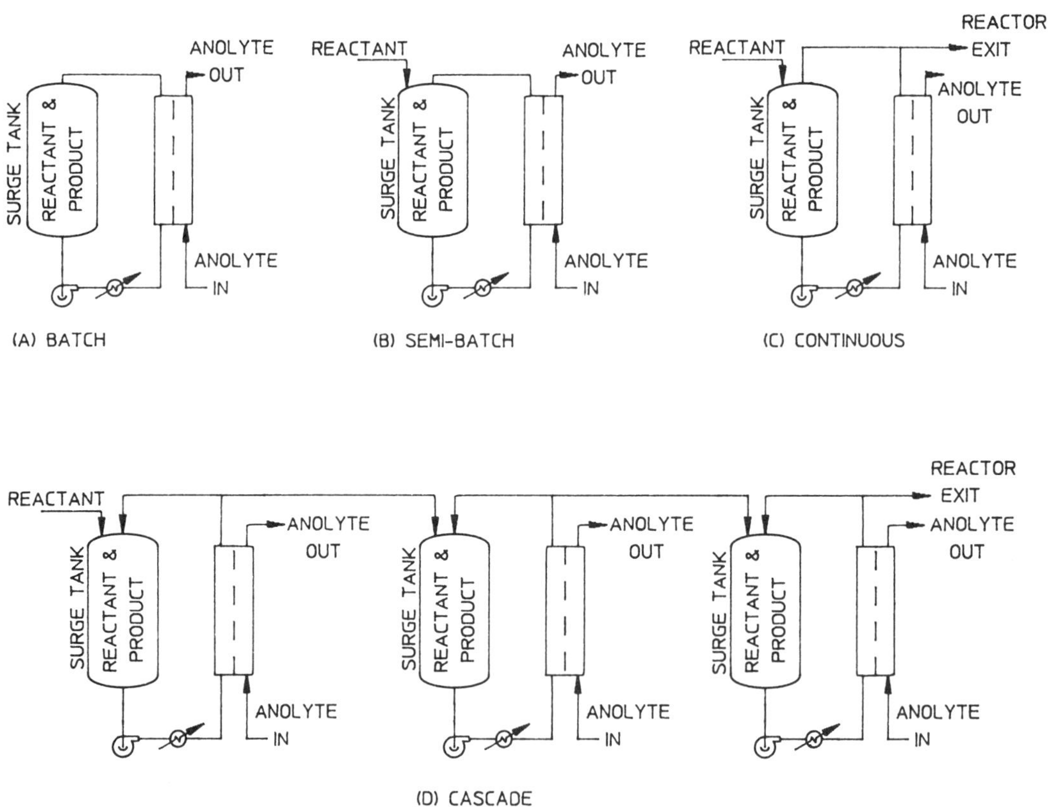

Figure 5. Configurations of the stirred tank electrochemical reactor.

The STER can be configured for four modes of operation, and these are illustrated in Fig. 5. In these schematic drawings a narrow gap divided cell is depicted. This is shown in a circulation loop containing a surge tank and heat exchanger. This is probably the most likely general cell arrangement for the STER because of the dominant need for low cell resistance and good mass transfer through forced convection. In general, parallel plates give acceptable current distribution.

For the batch reactor a single charge of reactant is made and after due reaction time the charge, which is now product, is removed. The profile of this reactor is shown in Fig. 6b, where it is seen that conversion depends on time. The semibatch reactor configuration in Fig. 5b has the same feature except the reactant composition in the batch is kept constant with time. The continuous mode of operation of the STER is shown in Fig. 5c and its reaction profile shown in Fig. 6c. It is seen that the compositions around this reactor are constant with time. With this arrangement the first step in product refining is recovery of reactant for recycling to the reactor. The final scheme for the STER is the linking of continuous reactors in a cascade, and this is shown in Fig. 5d. This arrangement leads to a combined overall performance similar to that of the PFER, with each reactor at a point on the curves of Fig. 6a. An example of some current work on modeling the STER is given by White et al. [58].

The choice of reactor configuration and mode of operation depends on economics and the key characteristics of the electrode process. In reaction schemes that involve electrochemical regeneration of a reactant, reaction engineering design can be critical when reaction rates must be matched. An example is the regeneration of Mn^{3+} in the reaction sequence of butadiene to sorbic acid [59]. For commercial cell operation continuous operation is preferred. Downtime and labor needs for batch operation make this mode economically unattractive except on the smallest scale. The continuous operation is usually stable and results in the best yields. When a feature of the reaction significantly determines yield, this determines the reactor choice. An example is the electrohydrodimerization of acrylonitrile to adiponitrile. Here there is an optimum concentration of acrylonitrile in the electrolyte for minimum by-product formation [60]. Therefore the continuous STER-type reactor is used and the product refining first goes through a reactant recovery step [61]. Electrochemical cell design in the context of reaction engineering and electroorganic synthesis has been reviewed by Jansson [62] and Lee [63].

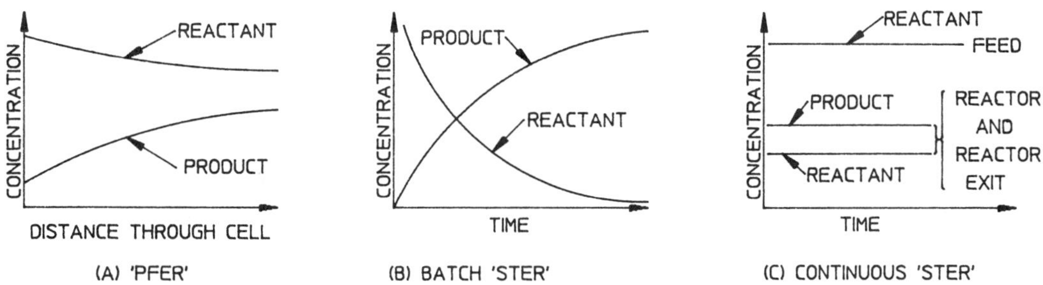

Figure 6. Schematic representations of reactor compositions for modes of operation.

Scaleup of Electrochemical Reactors

The scaleup of a reactor or process to full scale frequently uses the intermediate scale of the pilot plant, and this applies to the electrochemical counterpart. Scaleup in chemical engineering has made use of dimensional analysis to simplify and generalize a multivariant system. This may be applied to electrochemical systems but has shown limitations and is best used in conjunction with mathematical models. Scaleup often involves pursuing a few critical parameters. In general for electrochemical cells these are current distribution and cell resistance. The scaleup of these was referred to in Sec. II.B. The characteristics of electrolytic process scaleup have been given by MacMullin [64,65], and more recently the topic has been reviewed by Selman [66].

In standard chemical engineering, scaleup is sometimes achieved by small multiples of reactors. In the case of electrochemical cells this is normal and the multiples large. This is an attractive form of scaleup because it gives reliable performance but it loses the cost advantage of the large reactor. With multiple cells come options in hydraulic and electrical connection.

In a given stack of cells the electrolyte may be run in parallel from one manifold to another, as shown in Fig. 7a. Alternatively, electrolyte can flow in series, as shown in Fig. 7b. Parallel flow has the general advantages of low pressure drop, small temperature rise, low conversion per pass, and lower gas fraction problems when these are applicable. The disadvantages of parallel flow are possibilities of nonuniform flow distribution, large volumetric flows, and low conversion per pass.

Electrical connection to the cell bank can be monopolar as in Fig. 8a or bipolar as in Fig. 8b. The advantages of bipolar cell connection are fewer bus-bar connections, more uniform current in each cell, no problem of current distribution down a plate because of plate resistance, better matching to rectifier equipment, and lower voltage losses because of fewer connections. Disadvantages of the bipolar arrangement are current losses due to bypass currents in the manifolds, cell malfunction disrupting a whole bank of cells, and high d.c. voltages that present a hazard.

Over the years, bypass currents (also known as shunt, leakage, or parasitic currents) in bipolar cell stacks have been modeled by a number of workers. Models have been based on an analysis of an analog resistance network of the cell bank. A comprehensive and recent modeling is

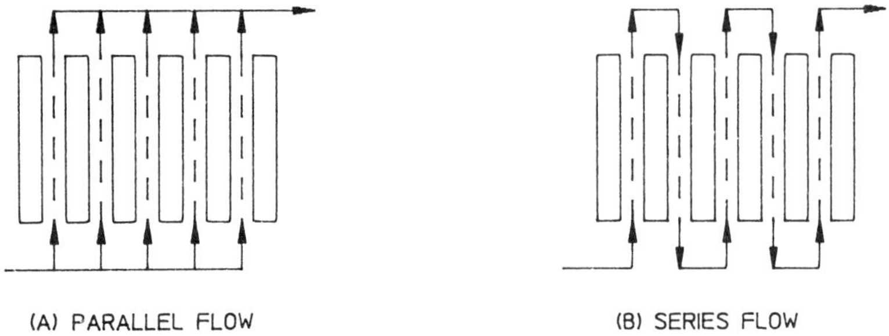

Figure 7. Electrolyte flow through parallel plate cells.

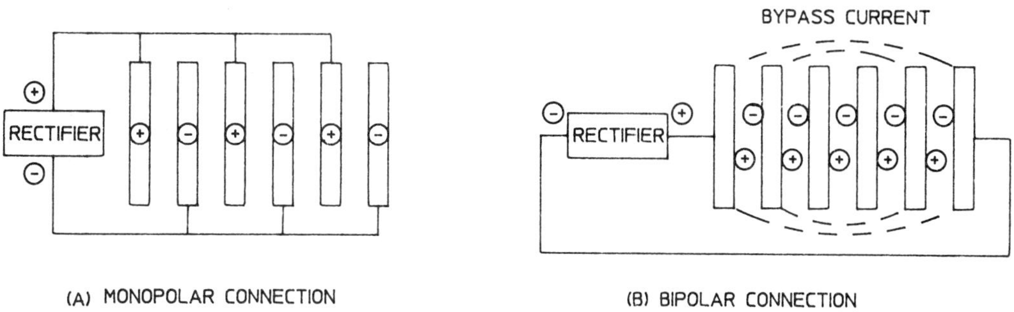

Figure 8. Electrical connections to parallel plate cells.

given by White et al. [67], although a simpler but more approximate model may be used [68]. This latter has been supported experimentally [69] and shows some general characteristics of bypass currents. The fraction current leakage goes up with approximately the square or the number of cells but goes down with increased current density (because of the nonlinearity of over-potential) and increased electrode length. To achieve low current bypass, cell inlet and exit manifold design is a compromise between constricted high-resistance channeling and tolerable pressure drop. It is in this manifold region that corrosion of exposed metal may occur owing to considerable voltage gradients.

D. Cell Geometry

In carrying out the mechanical design of a cell for commercial use, cost is clearly a factor. Thus a goal in cell design should be to keep it as simple as possible, which should result in a cell that is cheap to construct and run and less subject to malfunction. The critical features discovered in the laboratory experiments about the electrode process obviously dominate the cell design. Many electrochemical reactions benefit from uniform electrode potential, short interelectrode gaps, and good mixing and mass transport. To achieve this it is difficult to depart from the narrowly spaced rectangular plates with turbulent flow electrolyte. Recent reviews of electrolytic cell designs can be found in Refs. 70–72. Some cell designs are reviewed here.

Cell Configurations

The various cell designs found in the literature often make use of design features in standard chemical engineering hardware. This is especially true if that equipment has a desired feature, for example mass transfer, heat transfer, or gas absorption. There are the familiar filter press cells, which are made more sophisticated with internal manifolds like plate heat exchangers. The fluidized and packed bed catalytic reactors, and others, are known.

<u>Tank Cells</u> The use of plate electrodes immersed in a lined, rectangular tank represents a direct extention of laboratory beaker cells. The tank

may be fitted with a cover for gas collection or vapor control. The tank cell is used in batch operations. It has the attraction of being simple to design and is usually cheap. They are not the most suitable cells for large-scale operation or when forced convection is needed. Tank cells have mainly found application in the plating and metal winning industry but not very much in electroorganic synthesis. Some cells used for the latter are described here.

A 1000 A tank cell has been described for semicommercial production of isobutyric acid from isopropanol [53]. The body of the 70 × 40 × 40 cm cell is constructed from sheet vinyl using welded joints (Fig. 9). Several rows of anodes of lead dioxide electroplated on a nickel grid are suspended from the tank cover. A vinyl frame covers the edge of each anode to protect the nickel substrate from attack by the dilute sulfuric acid electrolyte. These frames could be eliminated if passivating base metals, such as titanium or tantalum, are used. Planar coils of copper are positioned between the rows of anodes to serve the dual function of cathode and cooler.

To overcome mass transfer problems in tank cells, rotating cylinders or rotating disks have been used. An example of this work for electroor-

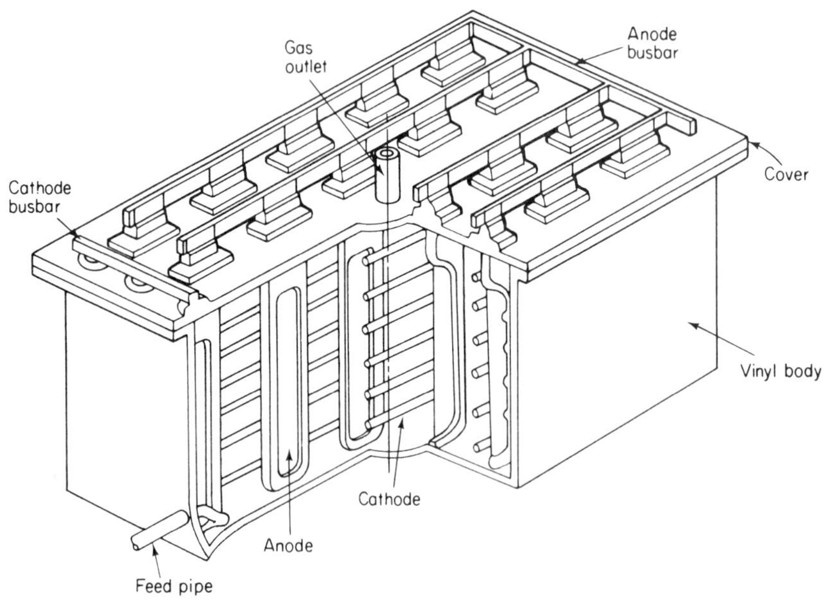

Figure 9. Kryschenko cell for synthesis of isobutyric acid. (From Ref. 53.)

ganic synthesis is given by Udupa and coworkers [47,73]. The reduction of salicylic acid to salicylaldehyde is carried out at three cylindrical copper cathodes, 10 cm in diameter, rotated at 1800 rpm in an 80 × 35 × 28 cm tank. The anodes are lead alloy and are caged in a microporous rubber diaphragm. The unit operates at 300 A. The same worker has constructed a 3000 A rotating disk cell for converting nitrobenzene to hydrazobenzene, which consisted of disks mounted on a drive shaft. The counterelectrodes were placed in compartments around two vertical disk shafts. The current density distribution on the disks is probably very nonuniform.

Two-Dimensional Flow-Through Cells In this category falls the simplest and least expensive cell design, the undivided parallel plate cell with electrolyte flow via some form of manifold. Electrical connection may be bipolar to the cell stack. An example of such a cell design is that of Foreman and Veatch [74], and an exploded view is given in Fig. 10. Note that in this simple structure electrolyte flow is in series and it is not easily adapted for divided cell operation.

The Foreman and Veatch cell has the possible problem of sealing the electrolyte path. This is avoided by placing the cells in a vessel; the best known example of this is due to Beck and Guthke [75]. A version of this cell is shown in Fig. 11. It consists of a stack of annular bipolar electrodes in which the electrolyte is fed through the center and flows radially out. Synthesis experience with this cell at BASF is described in Ref. 76. This cell design has problems of current bypass at the inner and outer edge of the annular cells; when this has become a serious problem insulator edges have been fitted. This cell stack has parallel electrolyte flow but, again, is not readily adaptable to divided cell operation.

Cell design work at Monsanto has also moved in the direction of vessel-enclosed electrode stacks. In one design [77], square rather than circular electrodes are set in a cylindrical vessel, as shown in Fig. 12. The four segment cavities form manifolds. The plastic electrode spacers insulate the

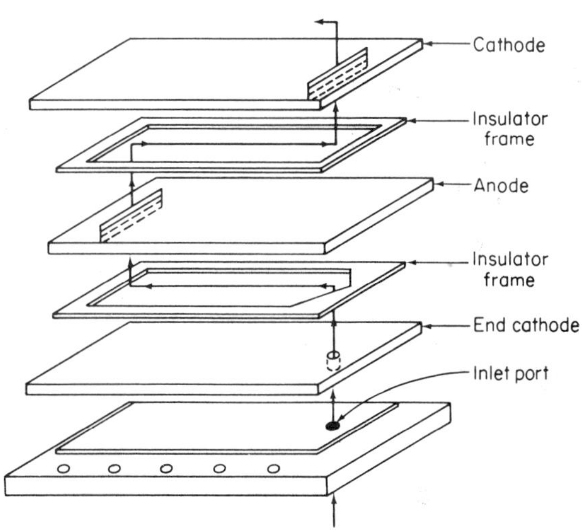

Figure 10. Foreman and Veatch cell. (From U.S. Patent 3,119,770, 1964.)

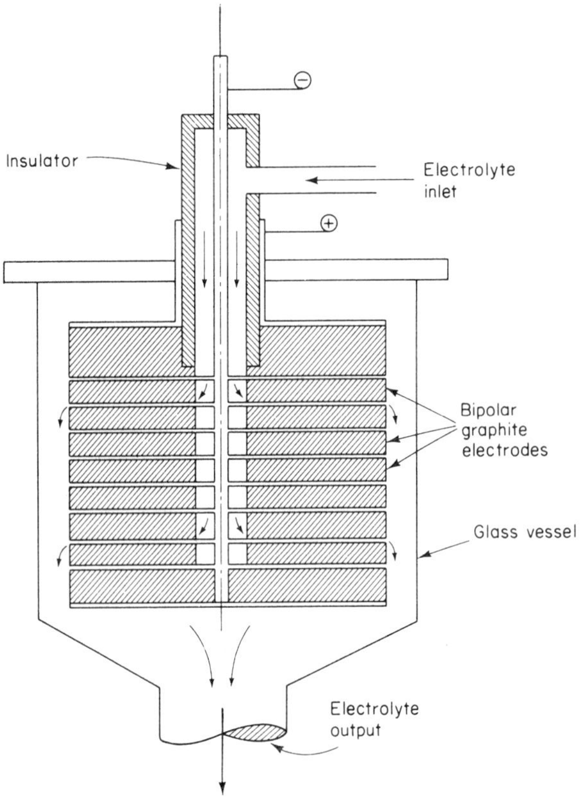

Figure 11. BASF capillary gap cell. (From W. German Patent 1,804,809, 1968.)

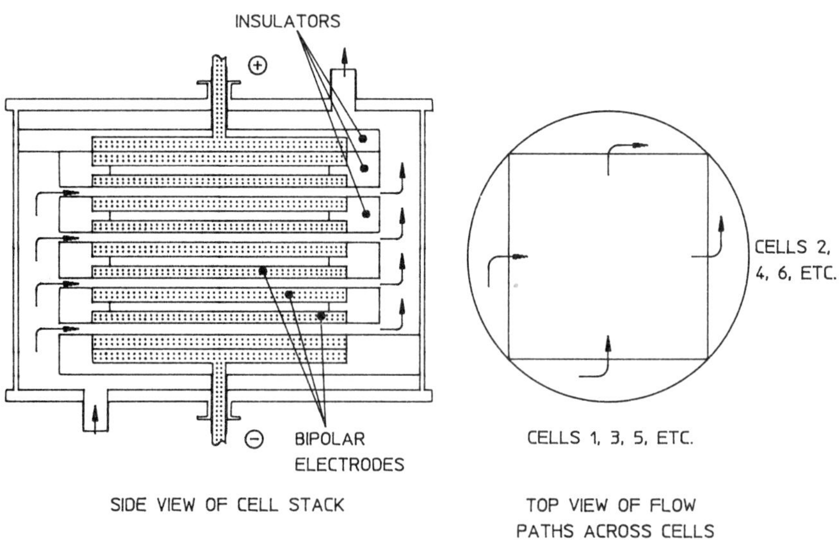

Figure 12. Monsanto stack cell. (From U.S. Patent 4,406,768, 1983.)

cell edges to minimize current bypass in the bipolar operation. This cell has the advantage over the disk stack cell of uniform electrolyte flow and square electrodes that can be economically cut from sheets. An interesting variation of this cell design is that it can be adapted for divided cell operation [78]. In this mode the four manifolds are paired to give anolyte and catholyte flow paths and alternate electrodes replaced by diaphragm material. A further geometric variation of the stacked electrode system is achieved by placing the electrodes vertically in the cylindrical vessel, as shown in Fig. 13. This cell is used in EHD technology as described in Sec. III.B.

These last three cell designs also have an advantage in maintenance. The cell stacks can be assembled prior to an electrode change-out. Then when the electrodes need to be changed the new and old stacks are swapped with little downtime.

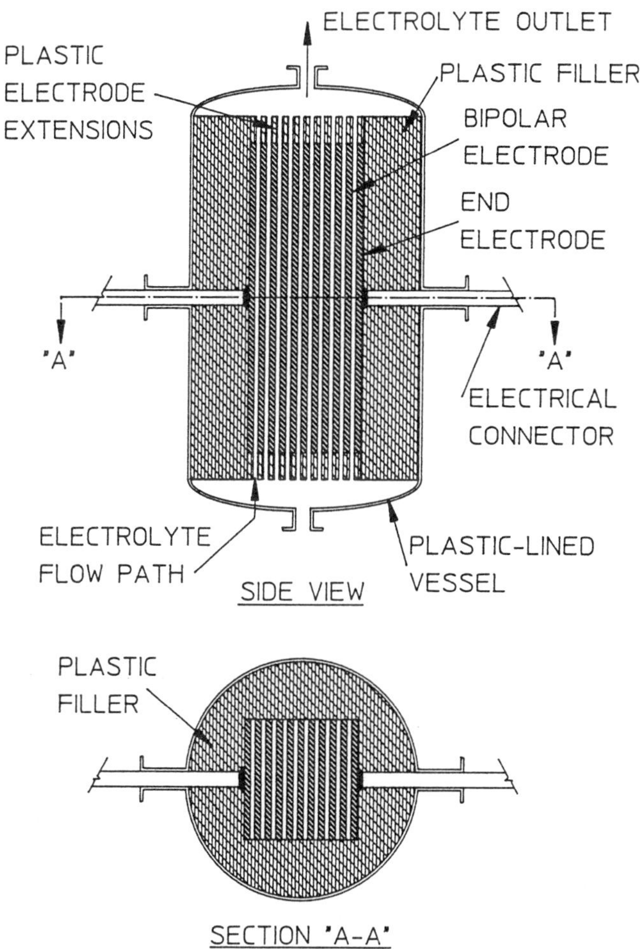

Figure 13. Monsanto EHD process stack cell. (From Ref. 9.)

The divided flow cells often follow the plate and frame construction. An example of the cell unit used at Monsanto is shown in Fig. 14. In this design [44,79] the anode and cathode are attached to opposite sides of a polypropylene plate and are connected by lugs for bipolar cell operation. The diaphragm, an ion-exchange membrane, is mounted in its own polypropylene frame and mounted between the electrode frames. Plastic spacers are used to keep the membranes from touching the electrodes. An open-mesh polypropylene screen, fused to the anode side of the membrane frame, serves the dual function of providing reinforcement against outward distortion from hydraulic forces and preventing the membrane from touching the anode. Thin plastic strips are fastened to the cathode on about 2.5 cm centers to support the membrane at the desired distance from the electrodes, about 1.5–3 mm. The membrane is held against the cathode strips by anolyte hydraulic pressure to give the desired cathode compart-

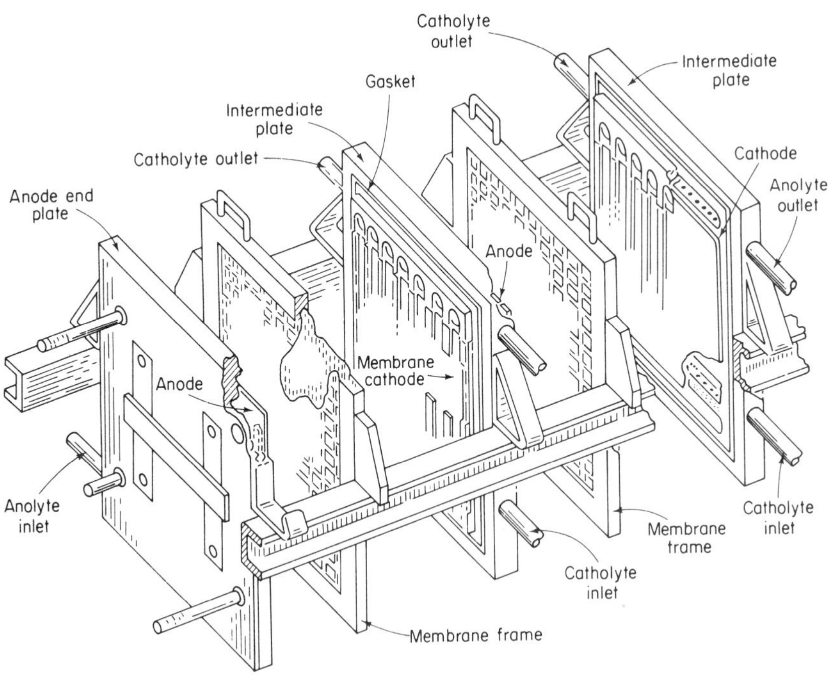

Figure 14. Monsanto plate-and-frame cell. (From U.S. Patent 3,660,259, 1972.)

ment width. Electrolyte is distributed through the electrode frames. Electrolyte flow is in parallel and under turbulent flow. The operation of this cell is described in Sec. III.B.

Other workers have designed and used plate and frame cells. BASF has made routine use of such cells [80]. Also, these cells can now be readily purchased and these are described in Sec. II.D, Commercially Available Cells.

A novel plate cell has been designed known as the Swiss-roll cell and has been pilot-planted for the manufacture of an intermediate in the manufacture of vitamin C and has been considered for other reactions [81]. (The process is outlined in Table 4 in Sec. III.A.) The cell is fabricated from a multilayer stack of the following sheets: (1) a polypropylene cathode spacer cloth, (2) a cathode foil or net, (3) an ion-exchange membrane if desired, (4) a polypropylene anode spacer cloth, and (5) an anode foil or net. These sheets are assembled in the sequence 12134543 to give a 5–6 mm thick sandwich rolled up around a rod of anode material and inserted in a cylindrical PVC container. The central rod acts as an anode connector, and connection to the cathode is by a bolt through the side of the container. This electrode is clearly very compact, with a high electrode area to electrode volume ratio, and has good mass transfer characteristics because of the turbulence promoter spacer material. A disadvantage of the unit is the inevitable voltage loss along the thin electrodes, and thus it is better for low current density applications. Electrical connection is monopolar, and several units are connected together to give series or parallel electrolyte flow.

Three-dimensional Electrode Cells For a number of years these cells have attracted attention as electrode systems that give a high electrode area to cell volume ratio. They also can have good mass transfer characteristics. These cells parallel the heterogeneous catalytic reactors, being packed or fluidized beds of electrode material. Each of these may also function as monopolar or bipolar electrodes. For both these electrical configurations, the current distribution on the electrode surface is nonuniform, which may be undesirable. Electrolyte pressure drops through an electrode bed can be large.

An attraction of a particulate electrode is the possibility of handling electrode material as a process stream. This has been done on a large scale in the Nalco Chemical process for the manufacture of tetraalkylead [82,83]. This process is discussed in Sec. III.B. The original cell for this process took the general form of a tube and shell heat exchanger and is shown in Figure 15. The steel tube cells are 7.5 cm in diameter and 15 m long and filled with lead pellets. The lead shot is made anodic and held away from the steel cathode by a woven polypropylene diaphragm. The lead shot is topped up while it is consumed in the process. The shell side is used for coolant.

Commercially Available Cells

It was not long ago that a major obstacle to pilot plant testing was the design and construction of a commercially viable cell. This has now changed, and several companies market cells for general use in the electroorganic synthesis field. These are all parallel plate cells fitted in a press with or without diaphragm and with forced flow convection. They are offered with

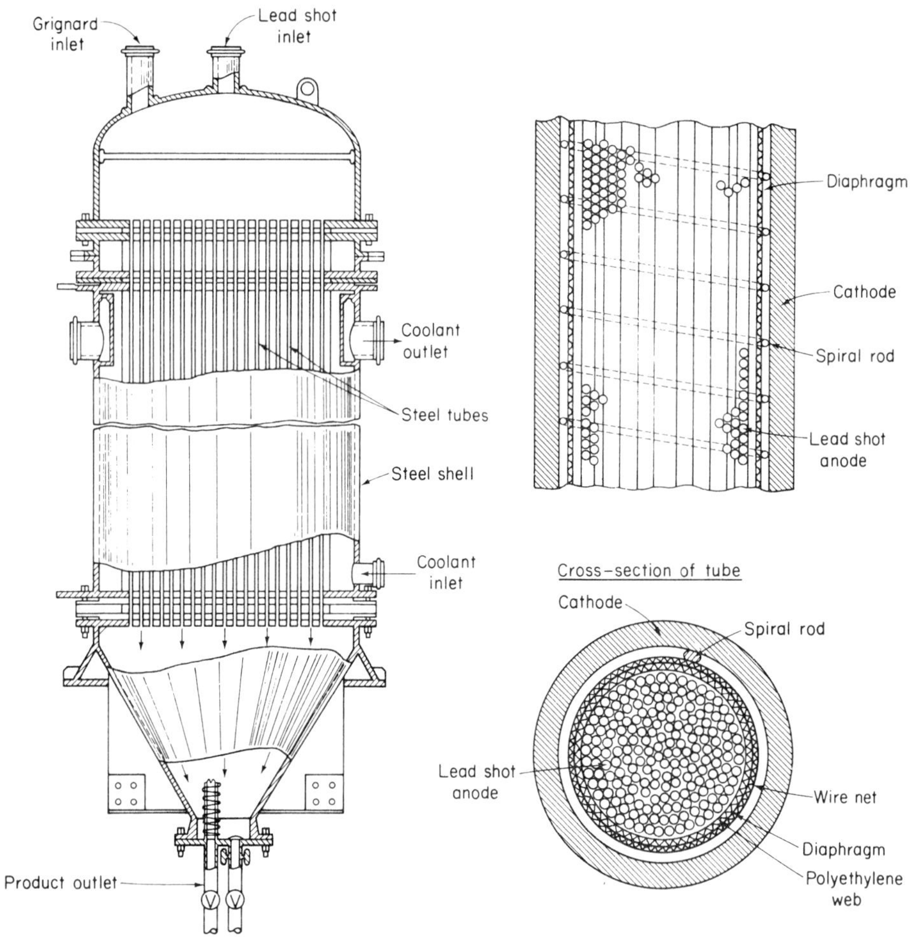

Figure 15. Nalco packed bed cell. (From Ref. 83.)

a variety of electrode materials (pure metals, PbO_2, DSA, graphite, and alloys) and materials of construction (polypropylene, PVC, PVDF, PTFE, and various elastomers). Given here is a description of cells for which brochures were obtained.

ElectroCell AB. The development of these cells came about through the cooperation of the Swedish National Development Co. (SU) and the foundation for Industrial Organic Chemistry at the University of Lund, Sweden (ElectroCell AB, P.O. Box 34 S-18400, Akersberga, Sweden). The cells are now marketed by ElectroCell AB, which is jointly owned by SU and Investment AB Eken. The cells are of the plate and frame design using internal manifolding. Multiple cell modules can be fit together, as shown in Fig. 16, which is of the ElectroSynCell model. Three sizes of modular cells are available, and the statistics of these are compared in Table 3. Electrolyte flow is in parallel, and turbulence promoters are used to en-

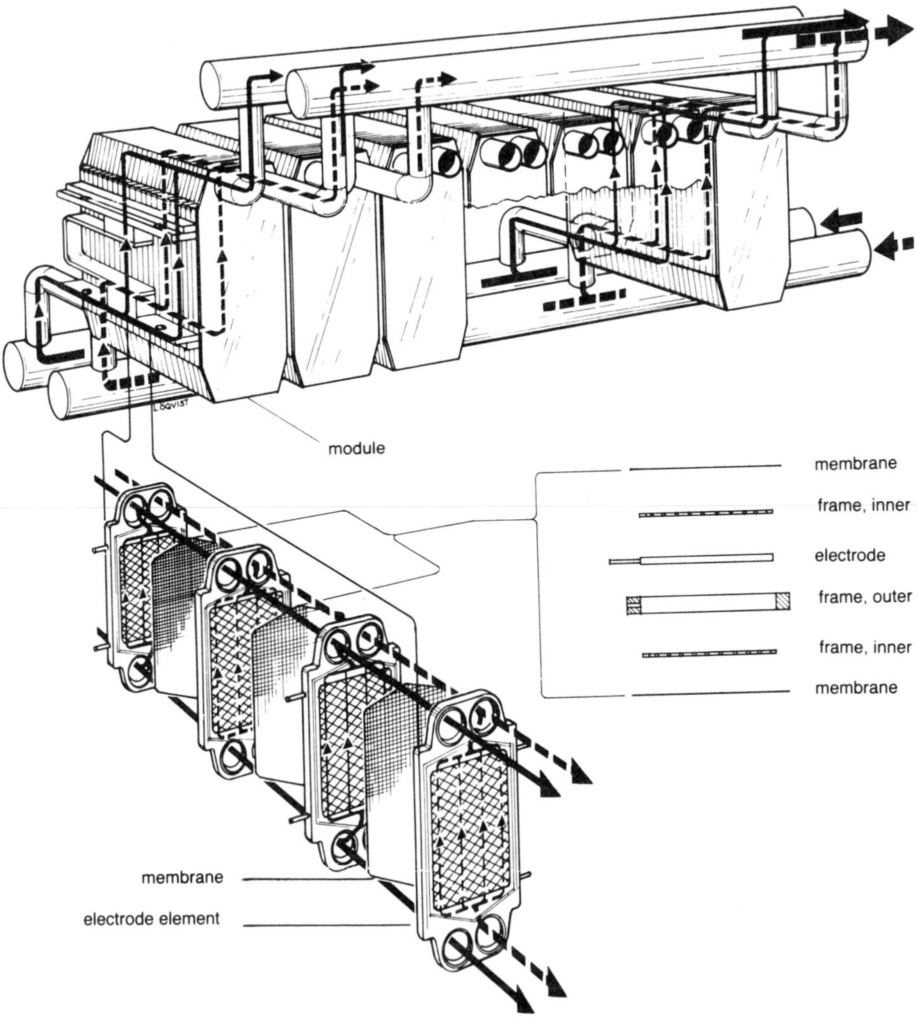

Figure 16. The ElectroSynCell of ElectroCell AB.

hance mass transfer. Electrical connection is monopolar, and pilot plant ancillary equipment is available.

The ElectroSynCell has been demonstrated in the production of glyoxylic acid from oxalic acid with current efficiencies of 84%. This, coupled with the low cell voltage, gave a power usage of 4.3 kWh kg^{-1} glyoxylic acid formed. For an indirect electrochemical process the oxidation of the cerous to ceric ion has also been performed with 90% current efficiency [84,85].

ICI. Some years back ICI introduced the FM21 cell to the chloralkali industry. This successful cell has since been improved and has found some application in electroorganic synthesis (ICI C&P Ltd., Electrochemical Technology, P.O. Box 14, The Heath, Runcorn, Cheshire, England) and is known as the FM21-SP (superior performance) cell. The design takes advantage of modern concepts and components. An overall view of the FM21 cell is shown in Fig. 17. The cell can be used divided or undivided and uses internal manifolding of electrolyte, which is fed to the cell pack in parallel. Electrodes can be flat plate, bladed, or other enhanced surface area forms. The electrodes are spaced as low as 2 mm and have a nominal projected area of 0.21 m^2. Electrical connection is monopolar via a copper bus attached along the long edge of the electrode. The cathode connection is on top and the anode on the underside. A laboratory version of the FM21 cell is available, the FM01-LC cell. This cell is aimed at electrode process development prior to scale up to the FM21-SP unit. Support process equipment is available. Several organic processes have been described as run in this cell [86–88].

Electrocatalytic. The cell manufactured by Electrocatalytic Inc. (2 Milltown Court, Union, NJ 07470) [89] was patented [90] and developed by the Electricity Council, Capenhurst, England. This cell addresses the problem of placing parallel plate electrodes close together but remaining able to have external manifolding. This is done by "dishing" the electrodes as in Fig. 18, which shows the divided and undivided configurations. The electrode and diaphragm spacing is 4 mm, with three single-electrode areas of 0.05, 0.175, and 1.0 m^2 available. The external manifolding by hoses makes possible a ready change between series or parallel electrolyte flow. The large

Table 3. Characteristics of Cells by ElectroCell AB

Cell	Electro MP Cell	Electro-SynCell	Electro-ProdCell
Electrode area (min) m^2	0.01	0.04	0.4
Module area (max) m^2	0.2	1.04	16
Current density (max) KA m^{-2}	4	4	4
Interelectrode gap, mm	6–12	5	0.5–4
Electrolyte flow per cell, L min^{-1}	1–5	5–15	10–30
Flow rate per cell, m s^{-1}	0.03–0.3	0.2–0.6	0.15–0.45

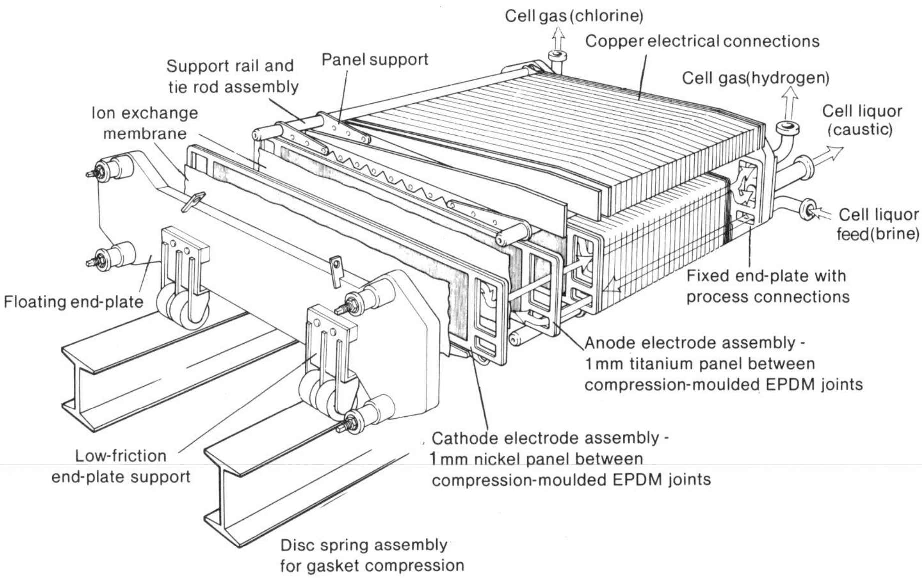

Figure 17. The FM21 cell of ICI.

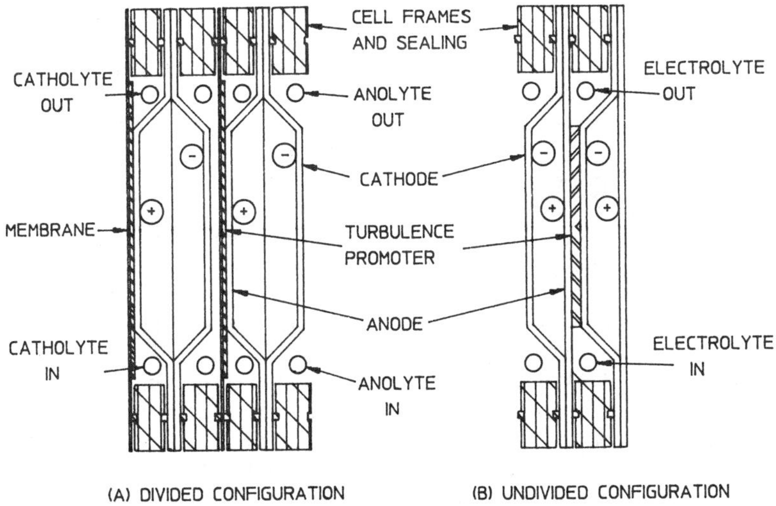

Figure 18. The DEM cell of Steetley Engineering.

manifolds make bipolar as well as monopolar electrical connections. High volumetric flow rates can be used for good mass transfer, but turbulence promoters are also used. The cells are available as a pilot plant package with pumps, tanks, piping, and so on.

E. Economic Considerations

There is relatively little published literature dealing with the economics of electroorganic processing. A comprehensive analysis of the cost of two types of plate and frame cells has been presented by Keating and Sutlic of duPont [91]. Published cost data for four electroorganic syntheses have been reviewed by Jansson, and a rule of thumb for predicting the total annual processing cost is given [92]. Fitzjohn [93] presents a comparison between electrochemical and conventional chemical processing with principal emphasis on the capital requirements. Economic optimization of electrochemical production facilities has been described by Despić [94] and Keller [95], both of whom present analytic approaches for determination of the conditions minimizing total product cost. The special report by Danly compares electroorganic processing cost for plate and frame cells and undivided stack cells [96].

In the commercialization of an electrolytic process for the production of organic chemicals, the major economic considerations are (1) reaction selectivity, (2) electrolysis system investment, and (3) power usage. Compared with an inorganic electrochemical process, such as caustic-chlorine manufacture, the reaction selectivity of the electroorganic process is likely to be relatively more important and the cost of power a less important factor. In general, the capital cost of the electrolysis system is greater for organics than for inorganics per unit of production capacity because of the greater complexity of the cells needed to achieve high selectivities and the smaller scale of production.

The major items of equipment in an electrolysis system for large-scale electroorganic processing are (1) electrolytic cells, (2) rectifiers and buses, (3) heat exchangers, and (4) electrolyte circulating pumps. If the selectivity and current efficiency are relatively constant over a range of current densities, selection of the most economical cell current involves a balance between the cost of power per unit of product and the cost of capital for the electrolysis system. Since cell voltage, and hence power usage per unit mass of product, increases with current density, savings result-from a reduction in the number of cells at high current densities are offset to a varying degree by increased cost for electricity and high investment in the electrical system and in heat exchangers. Selection of the economic optimum depends on such factors as (1) unit cost of power, (2) required rate of return or interest rate on capital, (3) slope of the voltage-current density curve, and (4) relative costs of the major equipment types.

For any given electroorganic reaction, there is an optimal current density that minimizes the total cost of the product. As an illustration, the components of the production costs for a hypothetical organic material produced in a 10,000 metric ton year^{-1} facility are shown in Fig. 19A. The electrolysis reaction was assumed to involve two electrons and require the use of a plate and frame divided cell. The product molecular weight was taken as 100. The electrical power cost and the electrolyte cooling cost increase directly with current density (as a result of the linear voltage-current density relationship). The cost of the capital (taken here as 30%

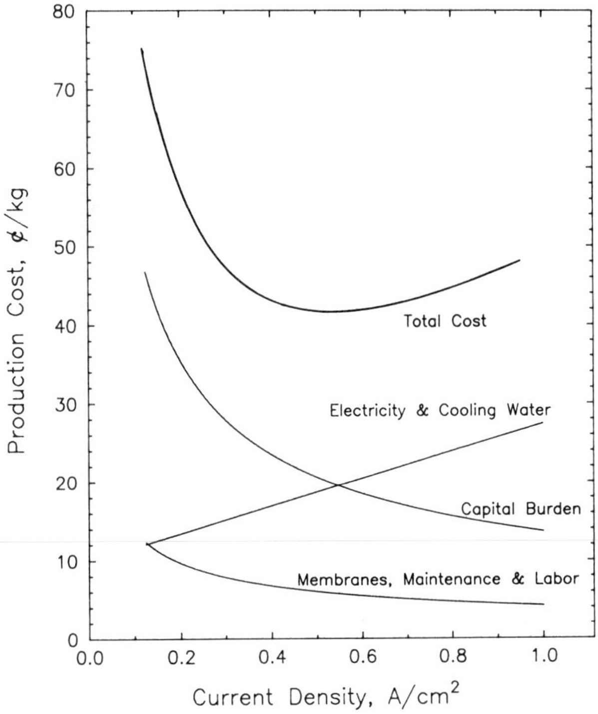

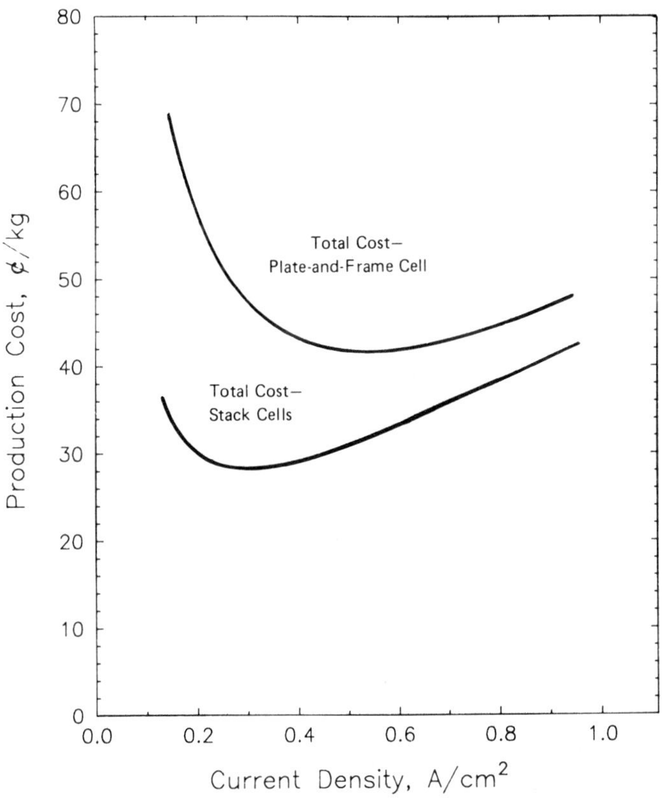

Figure 19. Economic optimization: (A) conversion costs using plate and frame cell; (B) comparison of plate and frame versus stack cells.

of the electrolysis investment) falls sharply with increasing current density at the low end of the range but tends to level out at very high current densities. The cost of membranes, maintenance, and labor falls slightly with increasing current density but is not a major cost factor in this scale plant. The total conversion cost, excluding raw materials, is the sum of these three component costs and has a minimum value at a current density of about 0.5–0.6 Ac m^{-2}.

For electroorganic reactions that can be conducted in undivided cells, the optimum current density is generally lower than in the preceding example. This results from the significantly smaller contribution of capital to the total cost, since the cost of undivided cells may be as much as an order of magnitude lower than that of plate and frame cells. The cell voltage, electrical power cost, and cooling capital are also lower, further improving process economics. The production costs for the same electroorganic product made in a stacked-electrode undivided cell are shown in Fig. 19B. The optimum current density is at 0.25 A cm^{-2}, and the production cost is about two-thirds that for the divided cell system. It is therefore important for the electroorganic researcher to define a synthesis system that does not require a diaphragm, if at all possible.

III. INDUSTRIAL ELECTROORGANIC PROCESSES

A. General

The electrochemical preparation of ingredients or precursors of certain pharmaceuticals and perfumes has been of commercial interest for a number of years. However, production of organic chemicals at multithousand ton per year rates is only relatively recent. One of the first large plants was that of the Atlas Powder Company, which was built in 1937 for the manufacture of 1400 metric ton $year^{-1}$ of sorbitol and mannitol by cathodic reduction of glucose [97]. This early success was rather short-lived and the process was replaced within a few years by a high-pressure catalytic hydrogenation route. It was not until 1964 that a truly large-scale electroorganic manufacturing facility was put on-stream with the startup at Freeport, Texas, of Nalco's tetraalkyllead plant. The capacity of the original installation was 14,000 metric ton $year^{-1}$ tetramethyllead or 18,000 metric ton $year^{-1}$ tetraethyllead. The following year Monsanto began production of adiponitrile electrolytically at Decatur, Alabama, at rates of the same order of magnitude as Nalco's [82]. Further details of the Monsanto adiponitrile process and the Nalco tetraalkyllead facility are given in Secs. III.B and C.

Electroorganic processes that have been investigated on a pilot-plant or semicommercial scale cover a wide range of reaction types. In addition to the reductive coupling of acrylonitrile and the preparation of sorbitol or mannitol, examples of other cathodic processes considered for commercialization include the reduction of (1) phthalic acids to the corresponding dihydrophthalic acids [98], benzene to 1,4-cyclohexadiene [117–119], (2) naphthalene to dihydronaphthalene [102], and (3) the reductive coupling of acetone to yield 2,3-dimethyl-2-3-butanediol (or pinacol) [103]. Anodic processes examined at least through the pilot-plant stage include (1) the oxidation of propylene to propylene oxide [104,105], (2) the Kolbe synthesis of dimethylsebacate from monomethyladipate [106–108], and (3) oxidation of 1,4-butynediol to acetylene dicarboxylic acid [116]. Electrochemical fluorination of organics has been practiced by the 3M Company at Hastings, Minnesota, since 1951 [109]. Two electrofluorination products of industrial importance are perfluorooctanoic acid and perfluorooctanesulfonic acid.

More than 40 electrochemical processes have been developed by the Central Electrochemical Research Institute (CECRI) in India, and many of these have been commercialized on a relatively small scale. Some of the industrial products include p-aminobenzoic acid, p-aminophenol, p-nitrobenzoic acid, succinic acid, benzyl alcohol, benzaldehyde, salicylaldehyde, benzidine, and saccharin [110]. Many of these materials have been produced on a scale of 100–500 kg day^{-1} using rotating electrode cells, operated in the range of 1000–3000 A.

Otsuka Chemical Company in Japan has operated several different electroorganic processes on a semicommercial scale, using plate and frame and tube-type cells at operating currents of 4500–6000 A [111]. Their first venture in the field involved a process for synthesizing maltol using furfural as the starting material. Maltol is a food additive and flavor enhancer, produced in nature from sugar. The key electrochemical step involves oxidation of α-methylfurfuryl alcohol at the C-2 and C-5 positions to give the cyclic acetal. This step is followed by an acid-catalyzed ring expansion, epoxidation with hydrogen peroxide, and an acid-catalyzed rearrangement to yield maltol.

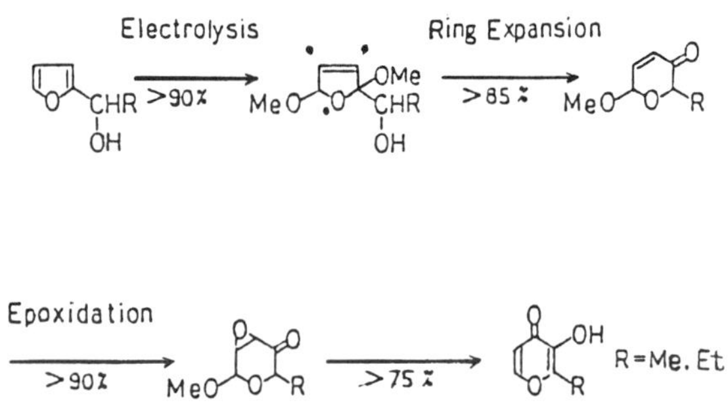

Otsuka has also developed highly selective processes for electrochemical oxidation of a variety of methylbenzene derivatives to the corresponding benzyl alcohol or aldehyde. For example, they describe the oxidation of 3,4,5-trimethoxytoluene to 3,4,5-trimethoxytolyl alcohol at a carbon anode achieving a reaction selectivity of 87%.

A summary of cell operating conditions and performance for a number of commercial or near-commercial electroorganic processes is given in Table 4. In general, these are characterized by high product selectivity (>90%), relatively high current efficiency (>70%), and low electrical power usage (1.5–5 kWh kg^{-2}). Current densities are typically in the range 10–100 A dm^{-2}.

B. Commercial Processes for Adiponitrile

Unquestionably the biggest success story to date in industrial electroorganic processing is the production of adiponitrile by the electrohydrodimerization (EHD) of acrylonitrile. The commercial importance of adiponitrile

Table 4. Cell Operating Conditions and Performance of Some Commercial and

Product	Reactants	Company	Cell type	Anode material	Cathode material	Diaphragm material	Temperature (°C)
Adiponitrile	Acrylonitrile	Monsanto	P&F[a]	Pb alloy	Pb	Ion exchange	55
		Monsanto	Tank	Steel	Cd	None	55
		Asahi Chemical	P&F	Pb-Sb	Pb	Ion exchange	50
		Asahi Chemical	P&F	Ni-steel	Pb	None	50
		BASF	P&F	PbO_2	C	None	30
		UCB-MCI		Fe oxides	C	None	20
		Phillips Petroleum	P&F	Pb	Pb	None	50
Dimethylsebacate	Monomethyladipate	Asahi Chemical	P&F	Pt/Ti	Steel	None	55
		BASF	Tank	Pt	Ni	None	42
		USSR	Tank	Pt/Ti	Stainless steel	None	62
Tetraalkllead	RCl, Pb, RMgCl	Nalco	Packed bed	Pb	Steel	Polyethylene fabric	45
	Pb, $KAlR_4$	Ziegler	Tank	Pb	Hg	None	110
Propylene oxide	Propylene	Kellogg	Tank	C	Steel	Asbestos	
		Bayer	Tank	C	Steel	Acrylic fabric	52
Acetylene dicarboxylic acid	1,4-Butynediol	BASF	P&F	PbO_2	Pb	Cation exchange	10
1,2-Dihydrophthalic acid	o-Phthalic acid	BASF	P&F	Pb	Pb	Cation exchange	30–40
1,4-Cyclohexadiene	Benzene	Asahi Chemical	Tank	Pt	Hg	Cation exchange	65
		Mitsubishi Chemical	Tank	Pt	Hg	Cation exchange	

Near-Commercial Electroorganic Processes

Current density (A dm^{-2})	Cell voltage	Product selectivity (%)	Current efficiency (%)	Power usage (kWh kg^{-1})	Status	Reference
20–100	6–12	90–93	90–92	6.6	Commercial	112
20	4	87–89	80–85	2.3	Commercial	100
10–20	5–6	85–95	80–95	4	Commercial	101
20	4	92	90	2.4	Commercial	113
7–10	4–5	82–92	80–83	3	Pilot plant	48
4–15	5–6	72–92	84–93	3.7	Pilot plant	114
20	4	91	90	2.2	Laboratory	60
20	13–16	90–92	86–90	4	Pilot plant	107
25	12–15	80–85	60–70	5–6	Pilot plant	106
6	12	82	58	6	Semicommercial	108
1.5–3	15–30	96–99	175	4–8	Commercial	82
46	2.5	100	100	1	Pilot plant	115
80–110	3.5–4.0	82–86	70–80	5	Pilot plant	104
2–5	3–5	90	88	4–5	Pilot plant	105
5		70	50	(19)	Semicommercial	116
10	10	94	75	4.3	Semicommercial	98
8		93	67	(5)	Pilot plant	117
		96	52	(6)	Pilot plant	118

Table 4. (Continued)

Product	Reactants	Company	Cell type	Anode material	Cathode material	Diaphragm material	Temperature (°C)
		Monsanto	Tank	Steel	Hg	None	
1,4-Dihydronaphthalene	Naphthalene	Standard Oil	Tank	C	Pb	None	30–50
Dihydro-2-methoxynaphthalene	2-Methoxy-naphthalene	Hoechst	P&F	C	Hg/Cu	Cation exchange	40–50
Pinacol	Acetone	BASF	Tank	PbO_2/C	C	None	20
		Diamond Shamrock	Tank	DSA	Pb	Cation exchange	
Diacetone-2-ketogulonic acid (vitamin C precursor)	Diaceto-L-sorbose	Hoffman-LaRoche	Swiss roll	Ni	Ni	None	60
Ethylene glycol	Formaldehyde	Electrosynthesis Co.	P&F	Pt/Ti	C	Cation exchange	55
Maltol precursor	α-Methylfurfural	Otsuka Chemical	P&F	C	Stainless steel	None	5–10
3,4,5-trimethoxytoluene	3,4,5-trimethoxytolyl alcohol	Otsuka Chemical	Tube cell	C	Stainless steel	None	5–10

[a]P&F = plate and frame.

[b]Power usages in parentheses based on 5 V per cell.

Current density (A dm^{-2})	Cell voltage	Product selectivity (%)	Current efficiency (%)	Power usage (kWh kg^{-1})	Status	Reference
2	5	90	100	3.4	Pilot plant	119
20–40	6–8	92	65	4–5	Pilot plant	102
20	8	95	70	4	Pilot plant	120
10	7–10	90	20–30	15	Semicommercial	121
16	5–6		70–80	4–5	Pilot plant	103
40	2.2	93	70	1.2	Pilot plant	122
10	5–6	>95	98–100	4–5	Pilot plant	123
10	7–15	92	97.5	3–5	Pilot plant	111
3–5	5–10	87	74	2–4	Pilot plant	111

stems from its use as a precursor for hexamethylenediamine, one of the monomers required for the manufacture of nylon-66 polymer. In 1987 the worldwide production of nylon-66 was over 1.8 million metric ton year^{-1}, corresponding to more than 800,000 metric ton year^{-1} adiponitrile. The importance of the production cost of adiponitrile to the economics of making nylon is well recognized and has been responsible for considerable research activity on the preparation of this intermediate. The possibility of electroreductive coupling of acrylonitrile to adiponitrile was reported in the 1940s, but it was not until the discovery in 1960 by Monsanto's Baizer of the importance of the supporting electrolyte that commercially attractive selectivities and current efficiencies were realized. Baizer found that through the use of concentrated solutions of certain quaternary ammonium salts (QAS), such as tetraethylammonium-p-toluenesulfonate, yields of adiponitrile of about 90% could be achieved at lead or mercury cathodes [124].

The principal cathode reaction is

$$2CH_2{=}CHCN + 2H_2O + 2e^- \longrightarrow NC(CH_2)_4CN + 2OH^-$$

Water is decomposed at the anode:

$$H_2O \longrightarrow 2H^+ + \tfrac{1}{2}O_2 + 2e^-$$

Major by-products are formed according to the following reactions:

$$\underset{\text{AN}}{CH_2{=}CHCN} + 2H_2O + 2e^- \longrightarrow \underset{\text{propionitrile}}{CH_3CH_2CN} + 2OH^-$$

$$\underset{\text{AN}}{3CH_2{=}CHCN} + 2H_2O + 2e^- \longrightarrow \underset{\text{trimer}}{NCCH(CH_2)_3CN} + 2OH^-$$

(with the CH bearing a $(CH_2)_2CN$ branch: $NCCH[(CH_2)_2CN](CH_2)_3CN$)

Following Baizer's discovery of the feasibility of synthesizing adiponitrile via the electrohydrodimerization of acrylonitrile, an intensive process development effort was undertaken at Monsanto's Technical Center in Pensacola, Florida [112]. This work led to the construction and startup in 1965 of a 14,500 metric ton year^{-1} facility in Decator, Alabama, for the production of adiponitrile via this technology.

Monsanto Divided-Cell EHD Process

The Monsanto divided cell EHD system employs a bank of bipolar, plate and frame diaphragm cells of the type shown in Fig. 14 and described in Sec. II.D, Cell Configurations: Two-dimensional flow-through cells. The catholyte consists of an aqueous solution of acrylonitrile and adiponitrile in 40% wt tetraethylammonium ethyl sulfate. The anolyte is a dilute solution of sulfuric acid. Lead is used as a cathode material, and lead with 1% silver is used for the anode. Electrolyte is pumped through each cell in parallel, and both anolyte and catholyte are recirculated through external exchangers for removal of the heat generated in the cells.

The electrolysis system of the original Monsanto EHD facility consisted of 16 cell presses, each containing 24 bipolar cells (Fig. 20). The cell banks were operated at a current of about 3000 A (45 A dm^{-2}) and 300 V. Four separate catholyte systems were employed to minimize the effect of an upset in catholyte composition in the event of a severe diaphragm rupture.

Monsanto Undivided Cell EHD Process

The divided cell process commercialized by Monsanto in 1965 had several shortcomings:

1. The concentrated QAS solution had a relatively high electrical resistance and hence high IR drop.
2. The ion-exchange membrane imposed an additional IR drop and also represented a significant replacement cost.
3. The plate and frame cell was expensive to construct and maintain.

Figure 20. Original Monsanto adiponitrile cell room. (From Ref. 112.)

Clearly, an undivided cell process using a more conductive electrolyte offered an opportunity for significant improvements in the process economics. Monsanto developed an undivided cell EHD system that differed in many respects from its divided cell process [99,100]. It involves electrolysis of a dilute solution of acrylonitrile in a mixed sodium phosphate-borate electrolyte using a cadmium cathode and a carbon steel anode (Table 5). As with the divided cell process, the presence of a quaternary ammonium salt is essential for high adiponitrile selectivities, but only a very low concentration is required. Sodium EDTA is added to the electrolyte to chelate the iron and cadmium entering the electrolyte via electrode corrosion and thus prevent cathode fouling.

The Monsanto undivided EHD cells consist of a bipolar stack of carbon steel sheets, electroplated on one face with a thin layer of cadmium. Plastic spacers set the electrode gap at 2 mm. From 100 to 200 electrodes are placed in a single stack, and the assembly is installed in a pressure vessel,

Table 5. Comparison of Monsanto EHD Processes

	Divided cell	Undivided cell
Cathode	Pb	Cd
Anode	Pb (1% Ag)	Steel
Membrane	Ionics CR 61	None
Temperature, °C	50	55
Electrode gap, inch	0.28	0.07
Electrolyte velocity, ft s^{-1}	6	3–5
Electrolyte composition		
QAS		
Cation	Et_4N	bisQuat[a]
Anion	Ethyl sulfate	Phosphate
QAS Concentration, %w	40	0.4
Current density, A dm^{-2}	45	20
Voltage drops		
Reaction EMF	2.50	2.50
Overpotential	1.22	0.87
Catholyte IR drop	5.47	0.47
Anolyte IR drop	0.77	—
Membrane IR drop	1.69	—
	11.65	3.83
Power usage, kWh lb^{-1}	3.0	1.1

[a]Hexamethylenebis(ethyldibutylammonium) cation.

designed to provide uniform flow of electrolyte through the stack (Fig. 13). The cell stack is operated at a current density of 20 A dm^{-2} and 3.8 V per cell; power usage at 2.4 kWh kg^{-1} is less than half that for the divided cell process.

For recovery of the adiponitrile product, a side stream of catholyte is withdrawn from each circulating system, cooled, and countercurrently contacted with excess acrylonitrile in a multistage extraction column (Fig. 21). The acrylonitrile-adiponitrile extract stream is then back-extracted with water in a similar column for removal of dissolved QAS. The organic stream leaving the second extraction column is sent to a stripping column, where the acrylonitrile is distilled overhead and recycled to the process, following removal of by-product propionitrile. The bottoms from the acrylonitrile stripper are further refined by distillation to produce a 99+% pure product, suitable for hydrogenation into hexamethylenediamine.

A portion of the aqueous electrolyte is purged from the electrolysis circulating system to remove the organic and metal ion impurities that can adversely affect the reaction selectivity and current efficiency. This stream is extracted with acrylonitrile for recovery of adiponitrile, concentrated by evaporation, and crystallized for recovery of the phosphate and borate salts. The cadmium in the mother liquor from crystallization is effectively removed by precipitation.

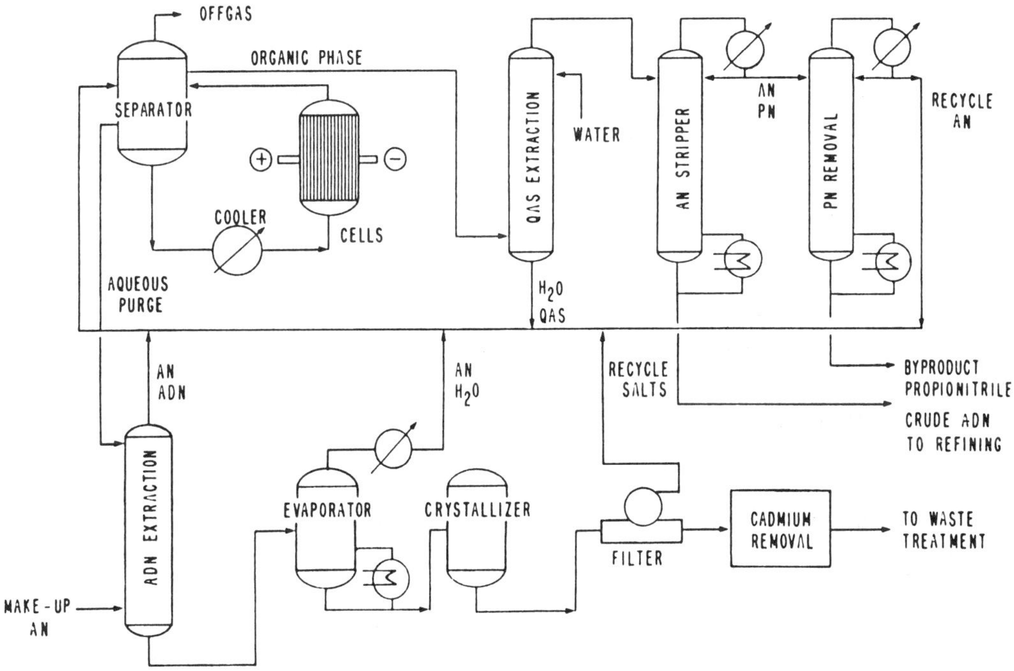

Figure 21. Monsanto undivided cell EHD process flow sheet. (From Ref. 100.)

Asahi Chemical EHD Processes

An alternative electrolyte system for the electroreductive coupling of acrylonitrile was developed by Asahi Chemical Industries in Japan in the late 1960s. In the Asahi divided cell process, the catholyte consists of an emulsion of acrylonitrile and electrolysis products in a 10% aqueous solution of tetraethylammonium sulfate. The concentration of the acrylonitrile in the aqueous phase is only about 2% wt, compared with levels of 15–20% wt employed in the original Monsanto process. Claimed advantages of the Asahi process are simplicity in separation of the adiponitrile from the catholyte and ease in recovery of the quaternary ammonium salt and purification of the adiponitrile. As in the original Monsanto process, a cation-exchange membrane is employed with dilute sulfuric acid in the anode compartment. Lead containing 6% antimony serves as the cathode, and the same alloy with 0.7% silver added is used for the anode [101]. Selectivities to adiponitrile of 91% are reported at current efficiencies of 88–89%. A 30,000 metric ton year^{-1} adiponitrile plant was put into operation by Asahi in 1971 at Nobeoka City, Japan.

Recent publications by Asahi describe an undivided cell process employing a lead alloy cathode, a nickel-steel anode, and an electrolyte composed of an emulsion of 20% wt of an oil phase and 80% wt of an aqueous phase [113]. The aqueous phase is 10% wt K_2HPO_4, 3% wt $K_2B_4O_7$, and 2% wt $(EtBu_3N)_2HPO_4$. The oil phase is about 28% wt AN and 50% wt ADN, the balance consisting of by-products and water. The cell is operated at a current density of 20 A dm^{-2} at 50°C. The electrolyte is circulated across the cathode surface at a superficial velocity of 1.5 m s^{-1}.

A yield of ADN of 92% is claimed at a current efficiency of 91%. Anode corrosion is about 0.2 mg (A-h)$^{-1}$. Asahi is reported to have commercialized their improved EHD process in 1987. The Asahi divided cell EHD process was licensed to Rhone Poulenc for use in their Brazilian subsidiary, Rhodia. Worldwide EHD plant capacities are given in Table 6.

Table 6. Worldwide EHD Plant Capacities

Company	Startup year	Estimated capacity (10^3 metric tons per year)
Monsanto Company, Decatur, Alabama	1965	90
BASF AG, Seal Sands, Teeside, UK (former Monsanto-Montedison joint venture)	1978	90
Asahi Chemical, Nobeoka, Japan	1971	30
Rhodia, Camaciri, Brazil	1984	30

C. Nalco Process for Tetraalkyllead

The second greatest success story for electroorganic processing in terms of total tons of product is the Nalco process for tetraalkyllead. Although environmental concerns about lead antiknock additives in gasoline forced closure of Nalco's commercial plant in the early 1980s, this facility operated for nearly 20 years to produce something of the order of 300,000 metric tons of tetraalkyllead. The process is significant in that it represents a very large scale demonstration of the manufacture of organometallic materials via electrolysis.

Prior to the introduction by Nalco Chemical in 1964 of its electrolytic process for tetraalkyllead, this important gasoline additive was produced via a 40-year-old process involving the reaction of methyl or ethyl halide with sodium-lead alloy. Nalco combined the large-scale production of Grignard reagent with a novel electrolysis system to pioneer this new process and installed facilities to produce tetramethyllead (TML) or tetraethyllead (TEL) at its plant in Freeport, Texas [82]. A simplified flow sheet of the Nalco TML process is shown in Fig. 22.

The first step is the production of the Grignard reagent from magnesium metal turnings and alkyl halide in a mixture of anhydrous ethers, such as tetrahydrofuran and the diethyl ether of tetraethylene glycol. The reaction is carried out in several 30 m^3 propane-cooled reactors, operating at about 2 kg cm^{-2} pressure and 35–40°C. Yields above 98% are realized.

The reactor product, containing excess alkyl chloride, is fed to ten 30 m^3 electrolysis vessels containing a bank of long steel tubes filled with lead shot. The lead constitutes a sacrificial anode and is replaced through a feed arrangement in the top head of the vessel [83]. The Nalco cell is described in Sec. II.D, Cell Configurations: Three-dimensional electrode

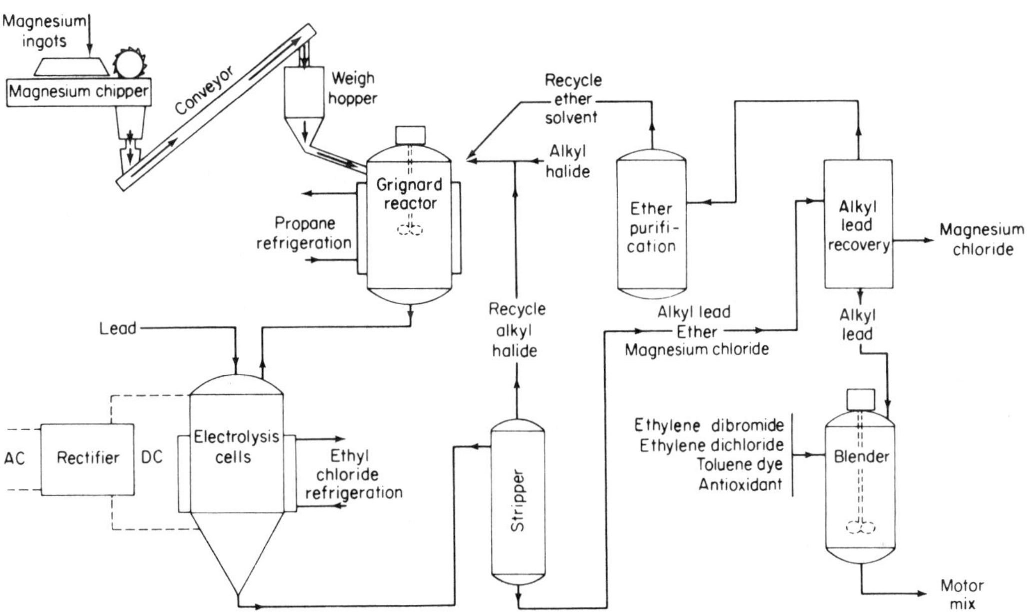

Figure 22. Nalco process flow sheet. (From Ref. 82.)

cells (Fig. 15). The electrolysis is conducted at 40–50°C, cooling being provided by an ethyl chloride refrigeration system. Current densities are typically 1.5–3.0 A dm^{-2} at 15–30 V.

The electrode reactions may be written as follows, assuming dissociation of the Grignard reagent into two ions, R^- and $MgCl^+$. Anode reaction:

$$4R^- + Pb \longrightarrow R_4Pb + 4e^-$$

Cathode reaction:

$$4MgCl^+ + 4e^- \longrightarrow 2Mg + 2MgCl_2$$

Although the theoretical cathode reaction shows formation of a mole of free magnesium for each mole of magnesium chloride, the excess alkyl halide added to the feed combines with the magnesium to re-form Grignard reagent equivalent to half of that consumed. Thus, the overall reaction is

$$2RMgCl + 2RCl + Pb \longrightarrow R_4Pb + 2MgCl_2$$

If the concentration of alkyl halide in the electrolyte is too high, there is a loss of reactants through Wurtz-Fittig type side reactions, such as

$$RCl + R'MgCl \longrightarrow RR' + MgCl_2$$

On the other hand, insufficient levels of alkyl halide allow magnesium to be plated out on the cathode, ultimately bridging the cathode-anode and causing short circuits. By proper control of concentrations, temperature, and flow rate, yields up to 96% on the Grignard reagent can be achieved at conversion levels as high as 99%, with yield based on the magnesium consumed of 95% [125].

The effluent from the cells is stripped with natural gas to remove unreacted methyl chloride, which is subsequently reabsorbed in ether for recycle to the Grignard reactors. The liquid phase from the stripper is contacted with water in a series of extraction stages for separation of the magnesium chloride, TML, and mixed ethers. The ethers are purified by azeotropic distillation and gas drying for recycling to the process. By using different alkyl groups on the Grignard reagent and the alkyl chloride fed to the cells, the process may be employed to prepare mixtures of tetraalkylleads, such as Me_4Pb, Me_3EtPb, Me_2Et_2Pb, and Et_4Pb.

D. Electrochemical Route to Sebacic Acid

Sebacic acid, an important intermediate in the manufacture of polyamide resins, has an estimated worldwide demand of about 20,000 metric ton year^{-1}. The conventional method of preparation is via alkaline hydrolysis of castor oil, which has historically shown some wide price fluctuations. For this reason, there has been considerable interest in recent years in an electrochemical route to sebacic acid based on adipic acid as the starting material. The electrochemical step involves the anodic coupling of the monomethyl ester of adipic acid to form dimethylsebacate (Kolbe or Brown-Walker reaction).

Thus the reaction sequence from adipic acid to sebacic acid is in three steps:

1. Formation of half-ester of adipic acid:

$$HOOC(CH_2)_4COOH + CH_3OH \longrightarrow CH_3OOC(CH_2)_4COOH + H_2O$$

2. Kolbe coupling of half-ester:

$$CH_3OOC(CH_2)_4COO^- + H^+ \longrightarrow 1/2CH_3OOC(CH_2)_8COOCH_3 + CO_2 + 1/2H_2$$

3. Hydrolysis of diester:

$$CH_3OOC(CH_2)_8COOCH_3 + 2H_2O \longrightarrow HOOC(CH_3)_8COOH + 2CH_3OH$$

Pilot-plant studies of the electrochemical route to sebacic acid have been carried out by BASF [106], Asahi Chemical Industry [107], and a group in the USSR [108]. A comparison of the electrolysis conditions employed by each is given in Table 4.

The yields of dimethylsebacate reported for the BASF and USSR processes are in the range of 80–85%; Asahi claims product yields as high as 92% at current efficiencies in the range of 85–90%. The key to achieving these superior results is evidently the use of very high electrolyte velocities through the cell (3–4 m s^{-1}). The electrolyte is a 20% wt aqueous solution of monomethyl adipate, with 30 mol per 100 mol of the organic acidity neutralized with sodium hydroxide. The electrolysis temperature is 55°C. The Asahi process employs a bipolar plate-and-frame cell with a platinum-plated titanium anode and a steel cathode. The cells are typically operated at a current density of 20 A dm^{-2} and 14 V, with an electrode spacing of 2 mm. A flow diagram for the manufacture of sebacic acid from

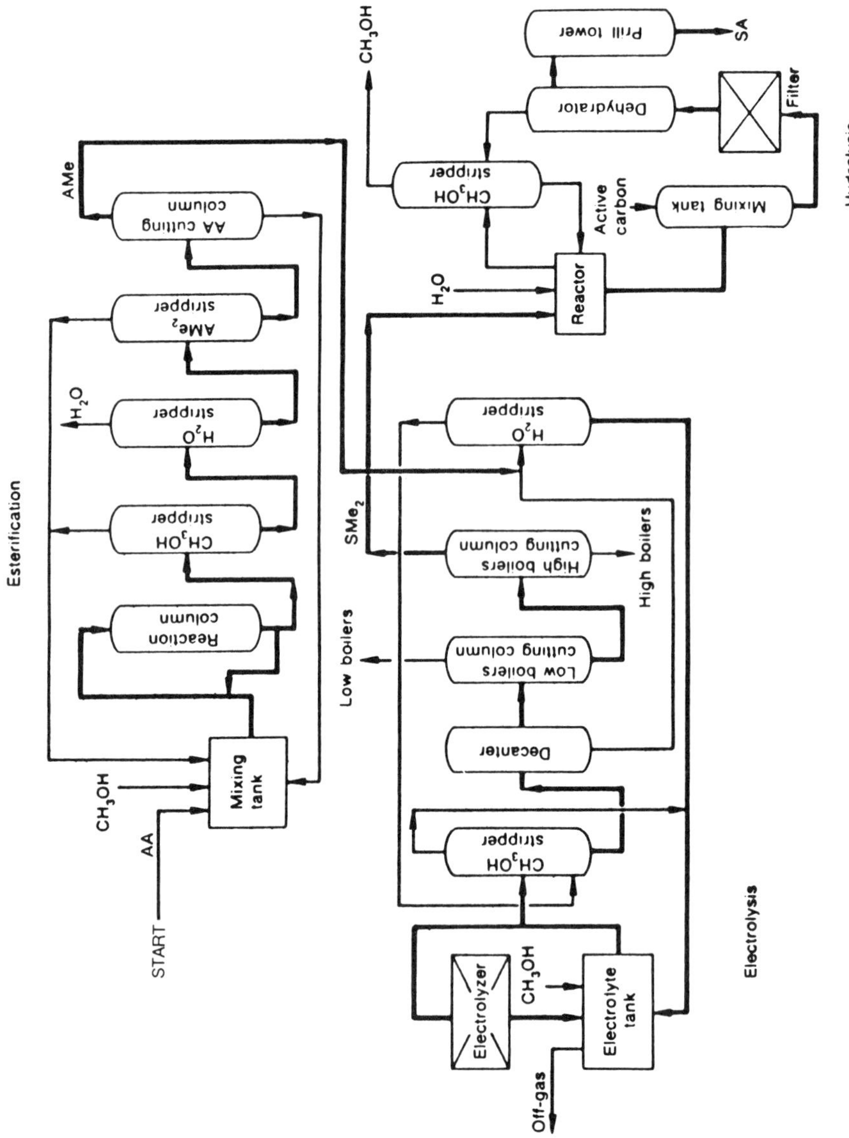

Figure 23. Asahi sebacic acid process flow sheet. (From Ref. 126.)

adipic acid is shown in Fig. 23. Raw material and energy requirements for the three-step process are as follows:

Material, energy	Requirements per ton sebacic acid
Adipic acid	1.8 ton
Methanol	0.1 ton
Electrolysis power	2600 kWh
Steam	13 ton

An economic study prepared by Chem Systems in 1976 showed the electrochemical route to have slightly better than a 10% cost advantage over the conventional process from castor oil in a 11,000 metric ton $year^{-1}$ plant. Asahi has demonstrated this technology in a 100 metric ton $year^{-1}$ pilot plant, but to date has not commercialized the process. It is reported that sebacic acid has been produced on a commercial scale in the USSR by the electrochemical approach.

IV. CONCLUSIONS

It was not long ago that the possibility of cheap electrical power via nuclear energy generation was forecast as just around the corner. With this some workers considered that the application of electroorganic synthesis would expand significantly. It is now known that nuclear power generation has probably had the opposite effect on electrical power costs. The argument might be that this has led to curtailment of the wider application of electroorganic synthesis. In general terms the authors do not agree. In working with the material that went into this chapter, it is evident that electroorganic synthesis has wide but special applications. The chemist and technologist who do not evaluate the possible electrochemical route (if only on paper) may well miss a commercial opportunity. The application of electroorganic synthesis and the fundamental work behind it continue to grow but at a pace less than inspirational. However, the growth and interest appear steady, and the future for electroorganic synthesis is not without promise.

REFERENCES

1. SRI International, Workshop on the Status of Industrial Organic Electrochemistry, EPRI EM-2173, Project 1086-9, Proceedings 1981, Research Reports Center, Box 50490, Palo Alto, CA.
2. D. J. Pickett, Electrochemical Reactor Designs, Elsevier, New York, 1977.
3. T. Z. Fahidy, Principles of Electrochemical Reactor Analysis, Elsevier, New York, 1985.
4. F. Coeuret and A. Storck, Elements de Génie Electrochimique, Lavoisier, Paris, 1984.
5. I. Rousar, K. Micka, and A. Kimla, Electrochemical Engineering, Vols I and II, Elsevier, New York, 1986.
6. V. Levich, Physicochemical Hydrodynamics, Chapters II, VI, IX, and X, Prentice Hall, New York, 1962.
7. J. S. Newman, Electrochemical Systems, Prentice-Hall, New York, 1973.
8. E. Heitz and G. Kreysa, Principles of Electrochemical Engineering, VCH Publishers, Inc., Deerfield Beach, FL, 1986.
9. D. E. Danly, Emerging Opportunities for Electroorganic Processes, Dekker, New York, 1984.
10. F. Hine, Electrode Processes and Electrochemical Engineering, Plenum Press, New York, 1985.
11. C. L. Mantell, Electrochemical Engineering, McGraw-Hill, New York, 1960.
12. E. Yeager and A. J. Salkind (eds.), Techniques of Electrochemistry, Vols I, II, and III, Wiley Interscience, New York, 1972, 1973, and 1978.
13. R. Alkire (ed.), AIChE Symp. Series, 75 (185) (1979). R. Alkire and T. Beck (eds.), AIChE Symp. Series, 77 (204) (1981). R. Alkire and D.-T. Chin (eds.), AIChE Symp. Series, 79 (229) (1983). R. E. White, R. F. Savinell, and A. Schneider (eds.), AIChE Symp. Series, 83 (254) (1987).
14. J. O'M. Bockris, B. E. Conway, and E. Yeager (eds.), Comprehensive Treatise of Electrochemistry, Vols 1–5; J. O'M. Bockris, B. E. Conway, E. Yeager, and S. Savangdpani (eds.), Vol. 6; J. O'M. Bockris, B. E. Conway, E. Yeager, U. M. Khan, and R. E. White (eds.), Vol. 7, Plenum Press, New York, 1980–1983.
15. P. Delahay and C. W. Tobias (eds.), Advances in Electrochemistry and Electrochemical Engineering, Vols. 1–9; H. Gerischer and C. W. Tobais (eds.), Vols. 10–13, Wiley Interscience, New York, 1961–1984.
16. A. Weissberger (gen. ed.), Technique of Electroorganic Synthesis, Vol. 5; N. L. Weinberg (ed.), Parts I and II; N. L. Weinberg and B. V. Tilak (eds.), Part III, Wiley Interscience, New York, 1974, 1974 and 1982.
17. A. Kuhn (ed.), Industrial Electrochemical Processes, Elsevier, New York, 1971.
18. P. Gallone, Trattato di Ingegneria Electtrochimica, Tamburini Editore, Milano, 1973.
19. D. Pletcher, Industrial Electrochemistry, Chapman and Hall, New York, 1982.
20. U. Landau, E. Yeager, and D. Kortan (eds.), Electrochemistry in Industry, Plenum Press, New York, 1982.

21. J. A. M. Le Duc, Electrochemical Industries and Technology, International Electrochemical Institute, Milburn, N.J., 1976–1978.
22. N. L. Weinberg and B. V. Tilak (eds.), Technique of Electroorganic Synthesis, Part III, John Wiley & Sons, New York, 1982.
23. A. P. Tomilov, S. G. Mairanovskii, M. Ya. Fioshin, and V. A. Smirnov, Electrochemistry of Organic Compounds (trans. J. Schmorak), Halsted, New York, 1972.
24. A. N. Frumkin and A. B. Ershler (eds.), Progress in Electrochemistry of Organic Compounds (trans. D. E. Hayler; trans. ed., P. Zuman), Plenum Press, London, 1971.
25. F. Beck, Elektroorganische Chemie, Grundlagen und Anwendungen, Verlag Chemie, Weinheim, 1974.
26. M. Ya. Fioshin and A. P. Tomilov, Int. Chem. Eng., 9, 375 (1969); Br. Chem. Eng., 16, 154 (1971).
27. R. C. Alkire and S. Swann, Jr., Bibliography of Electroorganic Synthesis, 1801–1975, The Electrochemical Society, Princeton, N.J., 1980.
28. C. J. H. King, K. Lister, and R. E. Plimley, Trans. Int. Chem. Eng., 53 (1), 20 (1975).
29. J. P. Carr and N. A. Hampson, Chem. Rev., 72 (6), 679 (1972).
30. Ionics, Inc., W. Ger. Patent 2,036,881 (July 24, 1970); Chem. Abstr., 74, 82500c (1971).
31. H. Nohe and H. Hannebaum, AIChE Symp. Series, 75 (185), 109 (1979).
32. D. Gilroy, J. Appl. Electrochem., 12, 171 (1982).
33. M. Hayes and A. T. Kuhn, J. Appl. Electrochem., 8, 327 (1978).
34. J. A. M. Le Duc, Industrial Electrodes 1976, Electrochemical Industries and Technology, Vol. 3, International Electrochemical Institute, Milburn, N.J., 1976.
35. M. Pourbaix, Atlas of Electrochemical Equilibrium in Aqueous Solutions, NACE (CEBELLOR), Houston, 1974.
36. D. C. Silverman, Corrosion-NACE, 38, 541 (1982).
37. R. B. MacMullin and G. A. Muccini, Am. Inst. Chem. Eng. J., 2, 393 (1956).
38. A. T. Kuhn (ed.), Industrial Electrochemical Processes, Elsevier, New York, 1971, p. 590.
39. G. Richter, Int. Chem., 17, 57 (1977); Chem.-Ing.-Tech., 47, 909 (1975).
40. W. Pusch, Chem.-Ing.-Tech., 47, 914 (1975).
41. S. C. Stitson, C & E News, March 15, 22 (1982).
42. R. B. Bird, W. E. Stewart, and E. N. Lightfoot, Transport Phenomena, John Wiley & Sons, New York, 1960.
43. J. R. Selman and C. W. Tobias, Advances in Chemical Engineering, (T. B. Drew et al., ed.), Vol. 10, Academic Press, New York, 1978, p. 211.
44. D. E. Danly and R. W. McWhorter (to Monsanto), Br. Patent 1,196, 631; Chem. Abstr., 71, 871883 (1969).
45. D. E. Danly, Chem. Ind. (13), 439 (1979).
46. M. Seko, Chem. Econ. Eng. Rev., 7 (10), 20 (October 1975); Chem. Abstr., 84, 60033 (1976).
47. H. V. K. Udupa, AIChE Symp. Series, 75 (185), 26 (1979).
48. F. Beck and H. Guthke, Chem.-Ing.-Tech., 41, 943 (1969).

49. R. W. Houghton and A. T. Kuhn, J. Appl. Electrochem., 4, 173 (1974).
50. C. W. Tobias, J. Electrochem. Soc., 106, 833 (1959).
51. J. E. Funk and J. F. Thorpe, J. Electrochem. Soc., 116, 48 (1969).
52. Y. Nisliki, K. Aoki, K. Tokuda, and H. Matsuda, J. Appl. Electrochem., 16, 615 (1986).
53. K. I. Kryshchenko, M. Ya. Fioshin, N. G. Bakhchisaraytsʹyan, and G. A. Kokarev, Sov. Chem. Ind. (7), 12 (1969); Chem. Abstr., 71, 101253 (1969).
54. M. Fremery, H. Hoever, and G. Schwarzlose, Chem.-Ing.-Tech., 46, 635 (1974).
55. G. Rogers and Y. Mayhew, Engineering Thermodynamics, Work and Heat Transfer, 3rd ed., Longman, 1980.
56. O. Levenspiel, Chemical Reaction Engineering, John Wiley & Sons, New York, 1972.
57. F. Lapicque and A. Storck, J. Appl. Electrochem., 15, 925 (1985).
58. T. V. Nyuyen, C. W. Walton, and R. E. White, J. Electrochem. Soc., 133, 1130 (1986).
59. S. V. Ho, J. Appl. Electrochem., 17, 398 (1987).
60. W. V. Childs and H. C. Walters, AIChE Symposium Series, 75 (185), 19 (1979).
61. C. J. H. King, AIChE Symp. Series, 79 (229) (1983).
62. R. E. W. Jansson, Chem. Eng. Sci., 35, 1979 (1980).
63. J. Lee, AIChE Symp. Series, 79 (229), 111 (1983).
64. R. B. MacMullin, Electrochem. Technol., 1, 5 (1963).
65. R. B. MacMullin, Electrochem. Technol., 2, 106 (1964).
66. J. R. Selman, AIChE Symp. Series, 79 (229), 101 (1983).
67. R. E. White, C. W. Walton, H. S. Burney, and R. N. Beaver, J. Electrochem. Soc., 133, 485 (1986).
68. J. C. Burnett and D. E. Danly, AIChE Symp. Series, 75 (185), 8 (1979).
69. C. J. H. King and D. E. Danly, Extended Abstracts Electrochemical Society, Vol. 82-1, Abstract p. 648, Montreal, May 9–14, 1982.
70. R. J. Marshall and F. C. Walsh, Surface Technol., 24, 45 (1985).
71. C. Oloman, AIChE Symp. Series, 79 (229), 68 (1983).
72. F. Goodridge, Phil. Trans. R. Soc., Lond., A302, 275 (1981).
73. S. Ganesan, K. S. Udupa, G. S. Subramanian, and H. V. K. Udupa, Chem. Age India, 13, 346 (1964).
74. R. W. Foreman and F. Veatch (to Standard Oil of Ohio), U.S. Patent 3,119,760 (Jan. 28, 1964); Chem. Abstr., 60, 10221d (1964).
75. F. Beck and H. Guthke, Chem.-Ing.-Tech., 41, 943 (1969).
76. F. Wenisch, H. Nohe, H. Hannebaum, R. K. Horn, M. Stroezel, and D. Degner, AIChE Symp. Series, 75 (185), 14 (1979).
77. C. J. H. King (to Monsanto Co.), U.S. Patent 4,406,768 (1983).
78. C. J. H. King (to Monsanto Co.), U.S. Patent 4,500,403 (1985).
79. D. E. Danly and C. R. Campbell, Technique of Electroorganic Synthesis, Part III (N. L. Weinberg and B. V. Tilak, eds.), John Wiley & Sons, New York, 1982, p. 302.
80. H. Nohe and H. Hannebaum, AIChE Symp. Series, 75 (185), 109 (1979).
81. P. Robertson, P. Cettou, D. Matic, F. Scwager, A. Storck, and N. Ibl, AIChE Symp. Series, 75 (185), 115 (1979).
82. L. L. Bott, Hydrocarbon Proc., 44 (1), 115 (1965).

83. G. E. Braithwaite (to Nalco Chemical), U.S. Patent 3,391,067 (1968).
84. L. Carlsson, B. Sandegren, and D. Simonsson, J. Electrochem. Soc., 130, 342 (1983).
85. L. Carlsson, H. Holmberg, B. Johanson, and A. Nilsson, Technique of Electroorganic Synthesis, Part III (N. L. Weinberg and B. V. Tilak, eds.), John Wiley & Sons, New York, 1982, p. 179.
86. A. Thompson, Recent Advances in Electroorganic Synthesis (S. Tori, ed.), Elsevier, Amsterdam, 1987, p. 449.
87. I. J. M. Girvan and W. N. Brooks, Recent Advances in Electroorganic Synthesis (S. Tori, ed.), Elsevier, Amsterdam, 1987, p. 445.
88. W. N. Brooks, I.Ch.E. Symp. Series (98), 1 (1986).
89. D. Hughes, Chem. Eng. (440), 17 (1987).
90. J. P. Millington and I. M. Dalrymple, Electricity Council, Br. Patent 2,098,238 (1984).
91. K. B. Keating and V. D. Sutlic, AIChE Symp. Series, 75 (185), 76 (1974).
92. R. E. W. Jansson, J. Appl. Electrochem., 12, 163 (1982).
93. J. L. Fitzjohn, Chem. Eng. Prog., 71 (2), 85 (1975).
94. A. R. Despić, Bull. Soc. Chim. (Belgrade), 42, 693 (1977).
95. R. Keller, Electrochimica Acta, 25, 303 (1980).
96. D. E. Danly, Emerging Opportunities for Electroorganic Processes, Dekker, New York, 1984, pp. 188–208.
97. R. L. Taylor, Chem. Met. Eng., 44, 588 (1937).
98. H. Nohe, Chem.-Ing.-Tech., 46, 594 (1974).
99. D. E. Danly, AIChE Symp. Series, 77 (204), 39 (1981).
100. D. E. Danly, Hydrocarbon Proc., 161 (April 1981).
101. M. Seko, A. Yomiyama, S. Ogawa, T. Miyake, R. Komori, and N. Ogawa (to Asahi Kasei); U.S. Patent 3,682,792 (Aug. 8, 1972).
102. J. F. Connolly (to Standard Oil Co.), U.S. Patent 4,022,673 (May 10, 1977).
103. T. T. Sugano, B. A. Schenker, J. A. Walburg, and N. Shuster, (to Diamond Shamrock Corp.), U.S. Patent 3,992,269 (Nov. 16, 1976).
104. J. A. M. Le Duc (to Pullman, Inc.), U.S. Patent 3,342,717 (Sept. 19, 1967).
105. W. Kroenig and J. Grolig (to Farbenfabriken Bayer), U.S. Patent 3,451,905 (1969).
106. F. Beck, W. Himmele, J. Haufe, and A. Brunold, (to BASF AG), U.S. Patent 3,652,430 (Mar. 28, 1972).
107. T. Isoya, R. Kakuta, and C. Kawamura (to Asahi Kasei), U.S. Patent 3,896,011 (July 22, 1975).
108. E. P. Kovsman, Sov. Chem. Ind. (1), 13 (Jan. 1973).
109. J. H. Simons and T. J. Brice, in Fluorine Chemistry (J. H. Simons, ed.), Vol. II, Academic Press, New York, 1967, p. 77.
110. H. V. K. Udupa and K. S. Udupa, Technique of Electroorganic Synthesis, Part III (N. L. Weinberg and B. V. Tilak, eds.), John Wiley & Sons, New York, 1982, pp. 385–422.
111. M. Taniguchi, Recent Advances in Electroorganic Synthesis, (S. Tori, ed.), Elsevier, Amsterdam, 1987, p. 433.
112. M. M. Baizer and D. E. Danly, Chemtech., 10 (3), 161; (5), 302 (1980).
113. K. Nakagawa and Y. Nagamori (to Asahi Chemical Industry Co., Ltd.); Japanese Patent 62(1987)-27583 (Feb. 5, 1987); Recent Advances in Electroorganic Synthesis (S. Tori, ed.), Elsevier, Amsterdam, 1987, p. 415.

114. C. Van Eygen, A. Hendricks, J. Ramioulle, J. Walravens, and A. Verheyden, Chem. Ind., 104 (1), 71 (1971).
115. K. ZIegler and H. Lehmkuhl, U.S. Patent 3,372,097 (Mar. 5, 1968).
116. D. Degner, Technique of Electroorganic Synthesis, Part III, (N. L. Weinberg and B. V. Tilak, eds.), John WIley & Sons, New York, 1982, p. 251.
117. T. Hatayama, Y. Hamano, and T. Yamamoto (to Asahi Kasei), U.S. Patent 3,700,572 (Oct. 24, 1972).
118. A. Minono and T. Osa (to Mitsubishi Chemical Industries Ltd.), U.S. Patent 3,485,726 (Dec. 23, 1969).
119. J. P. Coleman and J. H. Wagenknecht, J. Electrochem. Soc., 128, 322 (1981).
120. D. H. Skaletz (to Hoechst AG.), U.S. Patent 4,115,216 (Sept. 19, 1978).
121. H. Nohe and F. Beck (to BASF AG.), U.S. Patent 3,899,401 (Aug. 12, 1975).
122. P. Seiler and P. M. Robertson, Chimia, 36, 305 (1982).
123. N. L. Weinberg (to SKA Associates), U.S. Patent 4,478,694 (Oct. 23, 1984).
124. M. M. Baizer, J. Electrochem. Soc., 111, 215 (1964).
125. D. G. Braithwaite and L. L. Bott (to Nalco Chemical Co.), U.S. Patent 3,372,097 (Mar. 5, 1968).
126. M. Seko, A. Yomiyama, and T. Isoya, Hydrocarbon Proc., 117 (December 1979).

32
Electrochemical Polymerization

B. LIONEL FUNT / Simon Fraser University, Burnaby, Vancouver, British Columbia, Canada

I. Introduction 1337

II. Comparison of Electrochemistry of Large and Small Molecules 1338
- A. Factors influencing reaction rates 1338
- B. Solution properties 1339

III. Experimental Techniques 1340
- A. Control of electrical parameters 1340
- B. Reaction medium 1344
- C. Field strength 1346

IV. Initiation of Chain Reactions 1346
- A. Free radical initiation 1346
- B. Ionic polymerization 1349
- C. Anionic polymerization 1350
- D. Cationic polymerization 1351
- E. Coordination polymerization 1352
- F. Copolymerization 1353

V. Stepwise Reactions 1355
- A. Kolbe electrolysis 1355
- B. Oxidative coupling reactions 1356

VI. Polymeric Coatings 1358

VII. Conclusion 1358

Symbols 1359

References 1359

I. INTRODUCTION

The goal of most electroorganic synthesis is the efficient production of a pure, stoichiometric compound of unique structure and properties. Such a definition is not applicable to macromolecular syntheses. The polymeric molecules that are formed span a considerable region of molecular weights. Their conformation in solution is variable. Their chain structure may be linear or branched.

The physical properties of polymeric materials reflect this variability. Polymers differing only in molecular weight range from viscous liquids to brittle solids to strong structural materials.

The reaction profile with time also differs in polymerization processes from that in ordinary reactions. Instead of a mixture of reactant, product, and solvent, the polymerizing solution may contain reactive and reacting centers covering a range of chain lengths. If the polymerization proceeds by successive coupling reactions, larger entities predominate as the reaction progresses and the increases in molecular weight become more pronounced.

II. COMPARISON OF ELECTROCHEMISTRY OF LARGE AND SMALL MOLECULES

A. Factors Influencing Reaction Rates

The primary event of transferring an electron at the electrode is the same whether a simple coupling or disproportionation ensues or if a chain is initiated. However, the latter is reflected in a chemical magnification of the initial electrochemical event. Literally thousands of monomer units may react and grow on a chain initiated by a single electron transfer. The electrochemical efficiency expressed as molecules reacted per electron transferred is correspondingly large.

Many important differences from simple electroorganic synthesis can be inferred from this observation. The current density required to form a quantity of polymer is much less than for other electroorganic preparations conducted for the same period. Therefore, the demands imposed on the designs of cells and ancillary equipment are much less stringent. Working conditions of mA cm^{-2} frequently produce quantities of product requiring A cm^{-2} in conventional electrolysis. On the industrial scale of production this is reflected in a much lower energy cost per unit weight of product. Indeed, the reduction of energy costs by several orders of magnitude in comparison with other types of electrochemical processes greatly reduces their significance relative to other factors considered in feasibility assessment.

The molecular weights of soluble polymers can be determined by independent methods. The number of chains initiated per electron transferred is deduced from the weight of product formed and the number of repeat units in the chain. This provides a valuable insight into the true electrical efficiency of the process and the "cleanness" of the reaction. However, such an analysis cannot generally provide unequivocal answers: it requires a knowledge of the termination mechanism. If chain coupling occurs at the termination step, as is common in free radical reactions, then two electrons are required for each resulting chain. However, if termination by disproportionation occurs only one electron is required for each chain produced. In many instances both processes occur simultaneously and the ratio of disproportionation to coupling varies with temperature and other conditions [1].

Another phenomenon, chain transfer, complicates the interpretation of molecular weight data.

$$M + X \longrightarrow X\cdot + P \tag{1}$$

where X is a chain transfer agent.

A chain growing by addition polymerization through a free radical or ionic mechanism can react with a chain transfer agent to produce a "dead" polymer and a new growing chain. Consequently, a number of polymer molecules may be produced by an initial single electron transfer at the electrode. The process of chain transfer can be evaluated kinetically, and tabulated rate constants for chain transfer are available for many monomers. Transfer reactions to solvent, monomer, and polymer can occur, but chain transfer to solvent or to supporting electrolyte is the most probable in electropolymerization. The choice of solvent is particularly important for cationic polymerization in which chain transfer is the dominant method of chain termination.

The polymers formed exhibit a statistical distribution of molecular weights. The distribution reflects the mechanism of formation and specific conditions employed. One can infer kinetically the distributions expected from free radical or ionic or condensation reactions. In general, free radical polymerization produces a molecular weight that is relatively constant over the course of the reaction. Each chain is initiated, grows, and is terminated with a lifetime of approximately 1 s per chain. Although depletion of monomer during the course of the reaction gradually leads to changes in rate and in molecular weight, the first approximation of relatively constant molecular weight during the polymerization has been confirmed experimentally [2]. At late stages in the reaction other complications arise due to the gel effect and an abrupt increase in viscosity at high conversion of monomer to polymer. The decreased rate of termination that results from these factors produces an increase in rate of reaction as the monomer concentration nears exhaustion.

The molecular weight pattern is dramatically different for stepwise addition reactions. In condensation polymerization the molecular weight is relatively low until the reaction is within a few precent of completion and rises precipitously at the end of the reaction. As polymerization proceeds, higher oligomeric products dominate the reaction medium, and their coupling produces the high molecular weights obtained at the final reaction steps.

In general, a quantitative prediction of the molecular weight distribution is not feasible. The determination of absolute electrical efficiencies based on number of chains produced versus the electrons transferred is subject to a corresponding uncertainty. Nevertheless, such comparisons of electrochemical yield provide considerable insight into the mechanisms of the processes.

B. Solution Properties

The distinctive solution properties of polymers affect the conditions of their electrochemical preparation. The most obvious difference from simple solutions is the much higher viscosity produced by a given weight of polymer compared with that obtained with the same quantity of monomer. Solutions of relatively low concentration are viscous, and heat transfer and stirring are affected by the several orders of magnitude increase in viscosity that occurs during polymerization. This behavior is intrinsic to the structure of the polymers and results from chain entanglement during the process of flowing.

A second important factor that influences electrochemical response is the very low rate of diffusion of macromolecules. When addition to the polymer chain must occur at the electrode, it is necessary for the reactant to diffuse to the electrode and for the product to diffuse away. The rate of diffusion varies as $M^{1/2}$ for long chains [3], and the response in electroanalytic methods also reflects this behavior.

From such a model one can expect that for an electroactive group anchored in a polymer chain, but free from interactions with neighboring groups, the potential at which a response is obtained is unchanged but the current is greatly reduced. The ratio of diffusion coefficients of the polymer molecule to that of the parent electroactive compound indicate the predicted decrease in current.

In general, the electrochemical response obtained from a macromolecule represents the summation of single electron transfers from the electroactive groups along the chain. Thus ferrocene and polyvinylferrocene [4], benzoquinone and polyvinylbenoquinone [5], and anthraquinone and polymers with anthraquinone moieties [6] produce voltammetric waves almost identical in pattern that occur at essentially the same potentials. Only rarely does the response represents a contribution from the backbone chain. However, this occurs with polypyrrole and other conducting polymers in which additional charges become stored in the resonance structure of the polymer backbone [7].

The dependence of the current on the diffusional characteristics of the macromolecule has been demonstrated directly. The molecular weights of a series of monodisperse polymers were calculated from their limiting diffusion currents and were in excellent agreement with values obtained by experimental measurement [8].

III. EXPERIMENTAL TECHNIQUES

A. Control of Electrical Parameters

A choice between constant current and constant potential conditions is required. Each method has particular merit. Constant current conditions ensure a constant current density and a fixed, controlled rate of reaction and presumably of polymer production. Only a simple, two-electrode configuration need be employed. The Faradaic efficiency is calculated directly. The product of $i \times t$ is obtained without resorting to separate coulometric measurement.

However, the procedural simplicity may be accompanied by kinetic complexity. As the reaction proceeds the potential on the working electrode changes. New electrochemical processes may occur at these higher potentials, and a change in the composition or character of the product may result. For simple systems constant current density may be the method of choice, but its limitations must be considered.

Constant potential electrolysis is conducted with a three-electrode system consisting of a working electrode, reference electrode, and counterelectrode. The potential between working and reference electrodes is maintained by a potentiostat. An electronic feedback circuit continually compares and adjusts the potential with respect to a reference electrode.

Several factors are considered in choosing the appropriate potential: the threshold reaction potential determined by voltammetry; the potential at which interfering reactions may occur; and the desired rate of polymer formation. A cyclic voltammogram of the solution provides the salient facts on the appropriate range of potentials. A comparison with the solution free of monomer shows the upper limits of potential imposed by the solvent or electrolyte. Such a comparison may also indicate the presence of impurities, such as water, if residual current is obtained at low potentials.

The dependence of currrent on potential is described by the Butler-Volmer equation:

$$i = i_0\left(e^{(1-\beta)\eta}\frac{F}{RT} - e^{-\beta\eta F/RT}\right) \qquad (2)$$

In the working region where potentials exceed the equilibrium potential by at least several hundred millivolts, the Tafel relation

$$\ell n\ i = \ell n\ i_0 - \frac{\alpha F \eta}{RT} \tag{3}$$

is a simplified statement of this behavior.

Hence an exponential dependence of current on β is predicted for the region of high η but below the mass transport limiting values. For electroanalytic purposes, the diffusion limited region at the peak of the voltammetric curve is chosen as a basis for comparison.

In electrosynthesis, lower current densities may be desirable and positions one-quarter or half-way up the rising curve are sometimes chosen. When coating of the electrodes is to be avoided, the use of lower current densities may prevent the formation of cross-linked insoluble polymer. Cross-linking is probably due to the generation of a high concentration of active species, such as free radicals near the electrode. If attack on the backbone of the polymer chain occurs, radical centers are formed on the polymer molecule. These couple to form interchain cross-links and result in an insoluble three-dimensional matrix.

Interrupted Current Methods

If the rate of mass transport to the electrodes is insufficient to maintain a steady concentration of reactant, changes in concentration may produce undesirable effects. The morphology of a solid deposit may be changed, competitive reactions may be favored, and electrical conditions may exceed prescribed limits. Various methods of interrupting or modifying the pattern of imposed potential have been employed to allay this behavior. A repetitive linear cycling of potential between fixed limits has been employed in forming uniform layers of a polymer film for a modified electrode [10]. Presumably the cycling of the system provides an opportunity for diffusion to replenish the monomer consumed and provides a lower current density for the process. A simple count of the number of sweeps provides a measure of the thickness of the film. A fast pulse of high-intensity current was employed by Mengoli et al. [11] to produce propylene sulfide of high molecular weight by a cationic mechanism.

Superimposed alternating current on a direct current is sometimes employed in plating processes to provide improved deposits, and a.c. electrolysis to produce a polymer has also been reported [12]. Presumably the peak current region during a cycle is effective in electrolysis and the intervening chemical reaction is kinetically or thermodynamically irreversible on the opposite cycle.

Current Reversal Methods

A square-wave cycle can be employed to produce an active intermediate that is allowed to polymerize or react for a desired time and is then destroyed or terminated by a pulse of reverse polarity. Such methods are particularly appropriate for polymerizations in which chain termination does not occur readily. Current reversal is also used to form polymer at both electrodes and to minimize the effects of depletion of reactant at the electrodes. In recent work on Kolbe coupling reactions, successive current reversal was employed [13].

Mass Transport

The depletion of monomer at the electrode requires a movement of ions to maintain charge balance and of monomer to provide an adequate concentration of reactant. Mass transfer occurs by diffusion, which results from a concentration gradient, by migration or movement of charged particles in an electric field, and by natural or forced convection due to a gross movement of solution.

The control of solution hydrodynamics is essential for time-dependent, kinetic analysis but is not critical for preparative methods. The quantity of material deposited is predicted exactly by Faraday's law, irrespective of variations in current. However, the rate of reaction or the instantaneous current is very sensitive to the methods of mass transport. It is impossible to maintain a linear diffusion profile between the electrodes in the bulk of the solution by static methods. The effects of thermal convection and slight mechanical vibrations frustrate the establishment of such a condition.

It is therefore necessary to maintain a reproducible hydrodynamic pattern or to employ very low current densities. If the current density is such that the diffusion can compensate for the rate of reactant removal, static electrolysis still provides reproducible data. For higher currents and for kinetic analysis some form of stirring is required.

The best defined conditions are obtained with rotating electrodes. The rotating disk electrode provides a hydrodynamic pattern that brings material up from the bulk of the solution and spreads it parallel to the disk surface [14]. The current is a function of the concentration and of the rotation rate as described by the Levich equation

$$i_{lim} = 0.62nFAC_0D_0^{2/3}\nu^{1/6}\omega^{1/2} \quad (4)$$

The same methodology is applicable to larger scale laboratory electrolyses. The rotating wire and rotating cylinder electrodes also provide defined mass transport conditions.

Movement of the solution to a stationary electrode is the more common and simpler technique and is adequate for polymer preparation. In general, magnetic stirrers and common laboratory mixing systems do not provide a reliable rate of mass transport. However, controlled stirring, in conditions in which the mass transport rate greatly exceeds the rate of monomer consumption, may provide empirically reproducible conditions that are adequate for a comparison of rates of polymer formation.

Voltammetry

The choice of operating potential and of general parameters for electrolysis are best determined by obtaining a voltammogram of the solution. This indicates the equilibrium potential for the reaction, the potential at which a diffusion limited current is obtained, the reversibility of the reaction, and the possible occurrence of multiple electron transfer steps.

A cyclic voltammogram is obtained by applying a triangular potential wave to the working electrode with respect to a reference electrode. For ideal Nernstian behavior the peak separation is 57 mV for a single electron transfer process, and the current response is

$$i_p = 0.466nFA\left(\frac{nF\nu}{RT}\right)^{1/2} D_0^{1/2}C_0 \quad (5)$$

Considerable additional information can be obtained from a qualitative inspection of a voltammogram. If successive peaks are obtained, they indicate that distinct additional electron transfers occur. The absence of a reverse peak may indicate the intervention of a chemical step after the initial electron transfer. Adsorbed species are evidenced by lack of peak separation of cathodic and anodic peaks [15].

A more sophisticated voltammetric technique, which is sensitive to and measures reaction intermediates, employs a rotating ring-disk electrode system. A voltammetric determination with a rotating ring-disk system is based on a steady-state flow of reactant brought to the electrode. The diffusion limited region is evidenced as a plateau, and the current does not diminish as the potential is increased further.

When the potential of the ring electrode is controlled at a value at which the initial product can be converted reversibly to reactant, a constant ring current is obtained. However, the intervention of a chemical reaction to form a new product results in lowering of the ring current. The variation in ring current serves to detect and quantify the intervening chemical reactions.

Solvents

The choice of solvent impinges on many aspects of electrochemical polymerization. A conducting medium must be provided for the passage of current. A mutual solubility of monomer, polymer, and supporting electrolyte is generally required. However, the solvent must be chemically compatible with the polymerization reaction. Proton-donating solvents quench anionic polymerization and therefore must be avoided in such reactions. Solvents prone to chain transfer, such as CCl_4, must not be used if high molecular weights are required.

The solvent also influences the morphology of the polymer chains. In "good" solvents the chains are extended; in "poor" solvents chain-to-chain attraction dominates and a tightly coiled structure results. The secondary effects of this behavior are the precipitation of polymer from the solution and more commonly a coating of the electrode with insoluble polymer. The choice of solvents and solvent pairs can be made in a semiempirical manner from cohesive energy densities and solubility parameters [3]. This approach provides a systematic method for choosing solvents for polymers that have been formed and for selecting solvents to minimize coating of the electrodes during reaction.

The cohesive energy density of a solvent is obtained from its latent heat of vaporization E:

$$\frac{\Delta E}{\Delta V} = \delta^2 \tag{6}$$

The heat of mixing of a solution is

$$\Delta H = V(\delta_1 - \delta_2)^2 \phi_1 \phi_2 \tag{7}$$

where δ_1 and δ_2 are the solubility parameters and ϕ_1 and ϕ_2 the mole fractions.

For the best solubility $\Delta H \longrightarrow 0$ and $\delta_1 \approx \delta_2$. The solubility parameters are additive, and mixtures of solvents can be used to obtain a better match of parameters and to maximize the solubility of a polymer.

Some selected values of solubility parameters of solvents, monomers, and polymers are listed in Table 1.

B. Reaction Medium

Reactions in Emulsions

The requirements of high electrical conductivity of the medium and solubility of reactants and products are frequently in conflict. A number of innovative attempts have been made to overcome this dilemma. Polymerization in emulsion provides a two-phase system in which the initiator is generated in the aqueous phase and induces polymerization in the dispersed oil phase [16-18]. In free radical reactions, growth occurs within a micelle formed by a surfactant. The entry of a radical initiates polymerization of the monomer. The polymerization continues within the particle until another radical enters and terminates the growing chain. Under typical reaction

Table 1. Solubility Parameters

	δ ($J^{1/2}cm^{-3/2}$)
Solvents	
Acetic acid	20.7
Acetonitrile	24.3
Dimethylformamide	24.5
Formamide	39.3
Hexamethylphosphoramide	21.5
Methylene chloride	19.8
Nitrobenzene	20.5
Propylene carbonate	27.2
Propylene glycol	25.8
Pyridene	21.9
Water	47.9
Monomers	
Acrylonitrile	21.5
Isoprene	15.1
Methyl methacrylate	18.0
Styrene	19.0
Vinyl chloride	16.0
Polymers	
Cellulose	32.1
Polyacrylonitrile	25.6
Polybutadrene	17.4
Polyisoprene	16.6
Poly(methyl methacrylate)	18.6
Polystyrene	18.6
Poly(vinyl chloride)	19.8

Source: Adapted from J. Brandrup and E. H. Immergut, Polymer Handbook, Interscience, New York, 1967, p. iv-345.

conditions a radical enters the micelle once every 10 s. Thus 10 s of polymer growth is followed by an equal period of inactivity before a new chain grows for 10 s [19].

Reactions in Bulk Monomer

The addition of a supporting electrolyte to a liquid monomer is sufficient in certain instances to provide a reaction medium for electrosynthesis. Thus methyl methacrylate was polymerized at graphite electrodes by the passage of current through the bulk monomer containing HNO_3 or $(C_4H_9)_4NO_3$ as supporting electrolyte [20].

Electrode Materials

Electrodes that are chemically inert in the solution may nevertheless influence the rate of reaction and the type of product formed. This behavior is explicable in terms of the Butler equation. The equilibrium potential values at zero current flow differ considerably from those obtained when kinetic factors influence the attainment of a particular current density. The overpotential is a manifestation of these factors, and it in turn is influenced by the exchange current density i_0, Eq. (2). The rate of product formation and the ratio of products formed can both depend dramatically on the type of electrode material. For example, the evolution of H_2 in aqueous systems is 10^9 faster on Pt than on Hg [21].

The polymerization of methyl methacrylate on various metals was investigated, and lead cathodes were found to be particularly efficient [22]. In this instance H atom participation was not a factor in the initiation process, which depended on polymeric peroxides associated with the monomer.

Supporting Electrolyte

The supporting electrolyte serves to increase the conductivity of the solution by providing a relatively high ionic concentration. Ideally, the transport numbers of the anion and cation furnished should be equal so that unequal migration rates do not produce concentration changes. In electroanalytic applications the supporting electrolyte compensates for concentration changes due to migration of the substrate. The relatively high concentration of the supporting electrolyte swamps the effects of migration of the substrate. Consequently, in a quiescent solution, diffusion of the substrate becomes the only significant means of mass transport and the diffusion limited currents are solely a function of the bulk concentration of the reactants.

In electroorganic applications the solubility of the supporting electrolyte and the range of potentials over which it is stable emerge as the major criteria for its selection. Relatively large anions and cations provide solubility in relatively nonpolar systems. Tetraalkylammonium cations and perchlorate and tetrafluoborate anions are representative of the substrates employed.

Cell Design

Cell designs are dictated by the analytic and preparative requirements. For laboratory preparations, glass cells with sintered glass disks separating anode and cathode compartments in an H-type geometry are commonly employed. A reference electrode compatible with the organic solvents employed is incorporated in the working electrode compartment. The design of the cell should accomodate the needs for minimum RI drop, through compact geometry, isolation of anodic from cathodic products by use of a glass frit or ion-exchange membrane, constant temperature, and adequate control of mass transport hydrodynamics by controlled movement of electrode or electrolyte. A more detailed discussion of principles and methods is presented in Chap. VI and is generally applicable to these systems.

C. Field Strength

The effects of variation in potential should be distinguished from those due to high electric fields. As the potential impressed on the electrodes increases, so does the electric field. At low electric fields (less than 100 V cm^{-1}), the effects are described by the Butler equation (2); that is, the rates of the electrochemical reactions are increased exponentially. However, at high imposed field strengths > 1 kV, secondary processes occur due to the distortion of the electrical environment of the individual ions. Both acceleration [23] and retardation [8] of polymerization have been reported. These effects are most clearly evidenced in ionic polymerizations. The distortion of the solvent sheath of the propagating macroion can produce considerable changes in rate because the rate of polymerization of a free ion is very much higher than that of an associated ion pair [24].

IV. INITIATION OF CHAIN REACTIONS

A. Free Radical Initiation

The first complete account of an electropolymerization reaction appeared in 1949 [25]. It described the free radical polymerization of methyl methacrylate initiated by cathodically generated hydrogen produced by the electrolysis of aqueous sulfuric acid. Free radicals can also be generated at the anode by oxidations, such as those associated with the Kolbe synthesis [13]. The oxidation of acetate ions produces methyl radicals by the reaction

$$CH_3COO^- - e \longrightarrow CH_3^{\cdot} + CO_2 \tag{8}$$

Such radicals react with vinyl monomers to initiate the polymerization process. However, the likelihood of recombination within the solvent cage is high and consequently the fraction of radicals generated that participates as initiators of polymerization is low. The basic steps in the polymerization process are as follows:

Chain initiation:

$$R - e \longrightarrow R^{\cdot} \tag{9}$$

Propagation:

$$R\cdot + M \longrightarrow M^{\cdot} \tag{10}$$

$$M + M\cdot \longrightarrow M^{\cdot} \tag{11}$$

Chain transfer:

$$M^{\cdot} + X \longrightarrow X + \text{polymer} \tag{12}$$

Termination:

$$M\cdot + M\cdot \longrightarrow M_{m+n} \tag{13}$$

or

$$M_m^{\bullet} + M_n^{\bullet} \longrightarrow M_m + M_n \tag{14}$$

The progress of the reaction is not controlled by the electrochemical process except in the act of generating the initial radical. At a constant current the rate of radical generation is constant and monomer reacts at a rate proportional to its concentration. An increase in current results in a higher rate of polymer production, but this is accompanied by a decrease in molecular weight.

The kinetics predicted by reactions (9) to (14) in the absence of chain transfer can be simply stated as

$$R_p = k_p[M\cdot][M] \tag{16}$$

$$R_i = \theta \frac{i}{FV} \tag{17}$$

$$DP = \frac{R_p}{R_i} \tag{18}$$

$$R_t = k_t[M\cdot]^2 = R_i \tag{19}$$

$$R_p = \frac{k_p}{k_t^{1/2}} R_i^{1/2}[M] \tag{20}$$

where DP is the degree of polymerization, that is, the number of mer units in the plymer molecule, θ is the initiating efficiency, and v is the volume.

When the reaction is initiated electrochemically, the kinetics can be analyzed in an analogous fashion.

The rate of polymer formation R_p is governed by the propagation step. The consumption of monomer in the initiation step is negligible for long chains:

$$R_p = k_p[M^{\bullet}][M] \tag{21}$$

where [M] is the monomer concentration, $[M^{\bullet}]$ is the concentration of radicals, and k_p is the rate constant for propagation. The rate of termination by either termination or disproportionation in the steady state equals the rate of initiation, and this equals the current in an efficient process.

$$R_t = k_t[M^{\bullet}]^2 = R_i \tag{22}$$

On the basis of this mechanism one infers the following:

1. The average molecular weight changes only gradually as the monomer is consumed.

2. The rate of polymer formation is proportional to monomer concentration and to the square root of the current.
3. The molecular weight is inversely proportional to the rate of the reaction and therefore varies as $i^{-1/2}$.
4. Polymerization should cease approximately 1 s after the electrolysis is discontinued.

Although the general validity of this mechanism has been verified for electrochemical polymerization [26,27], various complications have become apparent in reactions initiated by the Kolbe synthesis. Olaj et al. [28] consider that combination of primary radicals occurs immediately at the electrode. Hence, the efficiency factor is strongly dependent on the current and decreases as $i^{1/2}$. In such circumstances the rate of polymerization, which varies as $R_i^{1/2}$ [Eq. (20)], shows a current dependence of $i^{1/4}$. Such behavior is apparent in the detailed studies reported of the polymerization of methyl methacrylate in dimethylformamide and acetic acid.

A profound effect of O_2 has been noted. Inhibition of methyl methacrylate polymerization by O_2 is commonly encountered in chemically initiated polymerizations and is due to interaction with the radical intermediates. However, Federova found that with stringent exclusion of O_2 the reaction is also inhibited. He suggested the formation of HO_2 and H_2O_2 intermediates with OH radicals as the ultimate initiators of reaction in acid solution [29].

Postpolymerization after the cessation of current flow has been reported [30]. This has been attributed to occluded radicals [30] and to peroxide formation [29]. Low current efficiencies are believed due to the radical recombination reactions associated with the radical generating step. An insight into the complex series of competitive reactions is obtained from the studies of Kolbe reactions in the absence of any added monomer [13]. The high concentration of radicals at the electrode and their high reactivity results in a series of radical-radical and radical-molecule interactions that compete with the polymerization process and reduce the supply of radicals available for initiation of chain polymerization.

The reaction of a cathodically reduced metal ion and an organic peroxide has been utilized to generate a controlled flux of radicals for the initiation of free radical polymerizations [31].

$$Fe^{3+} + e^- \longrightarrow Fe^{2+} \tag{23}$$

$$Fe^{2+} + ROOR \longrightarrow Fe^{3+} + RO^{\bullet} + RO^- \tag{24}$$

A similar pattern has been employed with an inorganic system. The reaction of cuprous ions with persulfate was utilized to compare the reactivity of various monomers to the same radical intermediate. The mechanism

$$Cu^{2+} + e^- \longrightarrow Cu^+ \tag{25}$$

$$Cu^+ + S_2O_8{}^{2-} \xrightarrow{k_1} Cu^{2+} + SO_4{}^{\bullet-} + SO_4{}^{2-} \tag{26}$$

$$Cu^{2+} + SO_4{}^{\bullet-} \xrightarrow{k_2} SO_4{}^{2-} + Cu^{2+} \tag{27}$$

$$M + SO_4{}^{\bullet-} \xrightarrow{k_3} {}^{\bullet}MSO_4{}^{2-} \tag{28}$$

was assumed, and the reaction was examined with the rotating ring-disk electrode. When monomer is added, reaction (28) competes with reaction (27) and there is an increase in ring current and a decrease in disk current. At high monomer concentrations reaction (27) is swamped by reaction (28), and because reaction (26) is independent of monomer concentration both i_r and i_d are reduced to half their original values. At intermediate concentrations of monomer the variation of i_r and i_d with concentration enables the evaluation of k_2/k_3 = [M]/[Cu(I)]. The relative reactivities of several acrylic monomers were obtained in this way, and an order of magnitude variation in reactivity toward the $SO_4^{\cdot}$ radical was shown [32,33]. Redox initiators provide a more efficient route to free radical polymerization than that provided by Kolbe radicals. Primary radical combination is diminished and the wide spectrum of coupled organic products is avoided.

B. Ionic Polymerization

The kinetics of ionic polymerization have been summaried recently by Olaj [34]. Some ionic polymerizations occur with the virtual absence of termination. These "living polymerizations" are of particular interest as they facilitate structural modification of the resultant polymer. Block copolymers and polymers with narrow molecular weight distribution have been prepared with nonterminating systems.

We consider the kinetics for reactions with and without termination and for preparations with continuous and with pulsed currents.

For continuous initiation (constant current) in systems with termination we have

$$R_p = -\frac{d[M]}{dt} = \frac{k_p}{k_t}[M]R_i \tag{29}$$

Hence the reaction proceeds at a constant rate at constant current when the monomer concentration is maintained at its initial value. The integrated form of the first-order reaction at constant current is

$$-\ln\frac{[M]}{[M_o]} = \frac{k_p}{kt}R_it = Kt \tag{30}$$

where K is a constant.

For continuous initiation (constant current) in systems without termination,

$$R_p = k_pM\int_0^t R_i\,d\tau = k_p[M]R_it \tag{31}$$

and

$$\ln\frac{[M]}{[M_o]} = 0.5k_pR_it^2 \tag{32}$$

It is seen that the rate of reaction increases with time as more and more active centers are formed. However, at later stages the depletion of monomer forces the conversion time curve to flatten.

For ionic polymerization with pulsed initiation, the kinetics are expressed as

$$R_p = -\frac{d[M]}{dt} = k_p R_i [M] t_p \, e^{-k_t t} \tag{33}$$

$$-\ln \frac{[M]}{[M_0]} = \frac{k_p}{kt} R_i t_p \left(1 - e^{-k_t t}\right) \tag{34}$$

where t_p is the pulse duration.

When k_t is small the reaction proceeds as a living polymerization. The product $R_i t_p$ represents the number of "living ends" generated by the pulse, and all of these add monomer until monomer is depleted.

If termination is not negligible the reaction shows features of the "dead end" polymerizations described by Tobolsky for free radical systems [35]. The initiator is depleted before the monomer is reacted completely.

Experimental data for various ionic polymerizations are now considered in terms of this kinetic model.

C. Anionic Polymerization

The electrochemical reduction of a monomer either directly or through the formation of an intermediate can lead to a nonterminating anionic reaction. In this type of reaction each electron transferred at the electrode produces a single chain. Each chain grows until the monomer is exhausted but a living chain end remains. This reacts further with additional monomer or with a new monomer added to the reaction medium. In the latter case a block copolymer is formed. The length of the two blocks is proportional to the molar ratios of the parent monomers employed sequentially.

The rate of the polymerization is given simply by

$$R_p = k_p [M][LE] \tag{35}$$

which is the limit of Eq. (34) when

$$R_i t_p = [LE] = \frac{i}{FV} \theta \tag{36}$$

The simplicity of the mechanism is counterbalanced by a need for stringent purity of materials and scrupulous exclusion of proton-donating impurities during the course of reaction (24). However, under such conditions, a remarkable degree of control can be exercised over the rate of the reaction and the product composition. By imposing a pattern of positive and negative pulses separated by a chosen time interval, the chains are initiated, grow to a chosen length, and are terminated. The technique produces polymers with a controlled molecular weight distribution [36,37] and of controlled molecular weight and predetermined yield.

The data in Table 2 are indicative of the control that can be attained. The molecular weights were calculated from the initial concentrations of living ends generated and the time period for propagation, in a reaction intended to produce polymer with two distinct ranges of molecular weight. A fraction of the active centers was terminated electrochemically and the remainder polymerized further to form the high-molecular-weight portion.

Table 2. Comparison of Experimental and Calculated Molecular Weights of Polybutadiene Formed by Programmed Pulses in Anionic Polymerization

High MW component				Low MW component			
Calculated		Found		Calculated		Found	
M_n	Yield (wt%)	M_n	Yield (wt%)	M_n	Yield (wt%)	M_n	Yield (wt%)
50,000	50	50,000	55	10,000	50	7500	45
50,000	67	52,000	70	10,000	33	8000	30
50,000	33	55,000	41	10,000	67	8000	59

Anionic polymerizations have been conducted with continuous current flow in systems in which the anions are less stable and have a short life [38,39]. Shapoval identified the active centers in methyl methacrylate polymerization [38]. He demonstrated that the carbanions were associated as ion pairs whose interaction was dependent on the radius of the cation. Simionescu polymerized several phenylacetylenes [40]. Mengoli investigated the anionic polymerization of propylene oxide [41]. Mano and Calafate polymerized carbazole and compared the polymer structures to those obtained chemically [42]. Funt and Yu obtained kinetic results in accord with Eq. (32) for the polymerization of styrene [43].

D. Cationic Polymerization

Oxidation at the cathode produces active cationic centers that initiate polymerization. The oxidation of a supporting electrolyte containing anions, such as BF_4^- or ClO_4^-, can form compounds, such as BF_3 or $HClO_4$, that are common chemical initiators of cationic polymerization. Alternatively, direct electron transfer to monomer may occur at potentials below that required for oxidation of the electrolyte. The kinetic distinction between these possibilities is frequently difficult to determine experimentally. Relatively stable, long-lived cations of aromatic hydrocarbons can be generated by electrolysis and alter used to initiate cationic polymerization.

The kinetics of cationic polymerization are strongly affected by the ionic environment. Traces of water act as cocatalysts. Rates of polymerization differ considerably in different solvents. Chain transfer is a dominating process that limits chain growth and, therefore, the molecular weight of the product at ambient temperatures. At low temperatures, chain transfer is diminished and both rate and molecular weight increase with decrease in temperature. These are manifestations of the influence of such factors as the dielectric constant of the medium on ion pairing and of the much higher reactivity of free ions in comparison with ion pairs.

These characteristics are exemplified in the extensive studies reported of electroinitiated cationic polymerizations. Mengoli and Valcher [44] polymerized trioxane in ethylene chloride solution with perchlorate electrolytes. A short pulse of current initiated a polymerization that progressed at a rate proportional to the monomer concentration and the charge. However, in methylene chloride continuous passage of current was required for reaction.

Related studies [45-48] indicate that $HClO_4$ is the initiating agent and that a similar reaction pattern occurs in bulk polymerization in the absence of solvent [46], in acetonitrile, in nitrobenzene, and in benzonitrile [47]. These investigators have also provided detailed reaction mechanisms for other monomers, including styrene [49] and ethyl and isobutyl vinyl ethers [50,51].

The polymerization of trioxane was also investigated by Bhadani and Ghose [52], who followed the rate of polymerization by dilatometry and showed that the rate was higher in CH_2Cl_2 than in $PhNO_2$. The cationic polymerization of indene has been investigated in detail [53-56], and the activation energy of 8.1 kcal was determined. The effects of various metallic anodes and of several solvents on the rate of reaction were described.

Cerrai and coworkers have investigated the initiating step in cationic polymerizations [57-61]. Direct electron transfer to anethole was demonstrated [57], whereas the occurrence of parallel initiation processes was proposed initially [59]. Akbulut has also reported direct electron transfer to monomer in the cationic polymerization of styrene in acetonitrile [62] and of epoxides in ethylene chloride [63,64].

Cationic polymerization can be induced by various radical cations formed by electrolysis. These may be prepared in a separate step or formed in situ. Mengoli and Vidotto generated radical cations of 9,10-diphenylanthracene and of perylene [65]. Funt and Glasel and colleagues employed a rotating ring-disk electrode in a study of radical cation interactions with monomers and determined rate constants for the initiation process [66,67].

E. Coordination Polymerization

Synthesis of Ziegler-Natta Catalysts

Organometallic constituents of polymerization catalysts have been prepared electrochemically. Anodic dissolution of metallic electrodes of Al or of transition metals produces the organoaluminum or corresponding transition metal components.

The reaction employed to form $(Cl_2Al)_2CH_2$, which is a cocatalyst in Ziegler-Natta polymerizations, is exemplary of the method.

Cathode:

$$CH_2Cl_2 + e^- \longrightarrow {}^{\bullet}CH_2Cl + Cl^- \tag{37}$$

$$CH_2Cl_2 + 2e^- \longrightarrow :CH_2 + 2Cl^- \tag{38}$$

Anode:

$$Cl^- \longrightarrow e^- + Cl^{\bullet} \tag{39}$$

$$Cl^{\bullet} + Al \longrightarrow AlCl \tag{40}$$

$$2CH_2: \longrightarrow CH_2 = CH_2 \tag{41}$$

$$CH_2: + CH_2Cl_2 \longrightarrow ClCH_2{-}CH_2Cl \tag{42}$$

$$AlCl + CH_2Cl_2 \longrightarrow Cl_2AlCH_2Cl \tag{43}$$

$$AlCl + Cl_2AlCH_2Cl \longrightarrow Cl_2AlCH_2AlCl_2 \quad (44)$$

A similar approach was employed to produce the transition metal components of the catalyst. The aluminum and transition metal products can be produced in the same cell by the simultaneous use of electrodes of each material or by employing an alloy electrode containing both metals. A large number of transition compounds has been investigated, including those of Ti, Cr, Fe, Mn, Ni, Co, and Zn in the form of acetylacetonates.

The catalysts were prepared in situ by Silvestri and coworkers [68,69] or separately and subsequently introduced into the reaction system [70]. Polymerization of olefins, such as ethylene, have been conducted with these systems.

Coordination Compounds

Lewis acids, such as zinc halides, coordinate with various monomers to form donor and acceptor complexes. These possess enhanced reactivity and produce polymers of high molecular weights. Phillips et al. investigated the polymerization of n-vinylcarbazole complexed with $ZnBr_2$ [71] and the copolymerization of styrene and diethylfumarate [72].

The combination of a complexed donor and an acceptor monomer generally produces alternating copolymers. Styrene has been copolymerized with acrylonitrile [75] and with methyl methacrylate [76] to form such copolymers.

An interesting use of the products of electrolysis was employed by Lemkuhl et al. [77], who synthesized metal chelates that are proposed as polymerization catalysts.

Cathode:

$$ROH + e^- \longrightarrow RO^- + 1/2H_2 \quad (44)$$

Anode:

$$L \longrightarrow L^{n+} + ne^- \quad (45)$$

The products of the anodic and cathodic reactions were mixed and the catalysts formed by the reaction

$$L^{n+} + RO^- \longrightarrow L(O - R)_n \quad (46)$$

where L = Co, Fe, Ni, Mn, Sb, Cu, and Au and the complexing agents are acetyl acetonate and alcoholate amines [68].

F. Copolymerization

When a solution containing two monomers is polymerized, the resultant product may contain sequences of units of each monomer in the same chain. The units may be arranged randomly or in a strictly alternating order. They may occur as blocks of each monomer or as grafts onto a parent chain. Important variations in physical properties are produced by such variations in the composition and structure of the final product.

A general mechanism for copolymerization is

$$-M_1^{\bullet} + M_1 \xrightarrow{k_{11}} -M_1^{\bullet} \tag{47}$$

$$-M_1^{0} + M_2 \xrightarrow{k_{12}} -M_2^{\bullet} \tag{48}$$

$$-M_2^{\bullet} + M_1 \xrightarrow{k_{21}} -M_1^{\bullet} \tag{49}$$

$$-M_2^{\bullet} + M_2 \xrightarrow{k_{22}} -M_2^{\bullet} \tag{50}$$

The homopolymerization steps have rate constant k_{11} and k_{22} and the cross-polymerization steps k_{12} and k_{21}. Under steady-state conditions,

$$d[M_1^{\bullet}]\ dt = \frac{d[M_2^{\bullet}]}{dt} \tag{51}$$

and

$$\frac{d[M_1]}{d[M_2]} = \frac{(1 + r_1)[M_1]/[M_2]}{(1 + r_2)[M_2]/[M_1]} \tag{52}$$

where $r_1 = k_{11}/k_{12}$ and $r_2 = k_{22}/k_{21}$.

It should be noted that the composition is unaffected by the rate of reaction but only by concentration ratios and the values of the reactivity ratios (rate constants). Furthermore, because the cross-reactions (48) and (49) occur repetitively in radical or ionic reactions, initiation of either monomer should result in both being incorporated in the polymer. Formation of a radical of either monomer ensures a concentration of reactive radicals of both types through the cross-reaction steps (48) and (49). For the concentration of $M_1^{\dot{-}}$ and $M_2^{\dot{-}}$ to remain constant, the rates of interconversion must be equal and $k_{21}[M_2^{\dot{-}}][M_1] = k_{12}[M_1^{\dot{-}}][M_2]$. Thus it appears that variation in the electrochemical parameters should change the rate of copolymer formation but not the composition of product obtained. For direct electron transfer, a change in the electrode potential of the working electrode is expected to increase the rate of each monomer's initiation in the same exponential ratio. If only one monomer is initiated directly, the rate of incorporation of both monomers is still increased in accordance with the cross-polymerization steps. Only the secondary factor β in Eq. (2), which is the symmetry factor of the activated complex, is susceptible to variation in this model. However, several investigators have reported the effects of applied potential on the composition of copolymeric product. Akbulut et al. investigated indene-polymerized with methoxystyrene, with 2-bromostyrene [78], with α-methylstyrene [79], and 4-methoxystyrene and styrene [80]. In all these, changes in reactivity ratios with potential were reported. Benzuglyi and Shapvalov [81] also reported a dependence of composition on applied potential in the copolymerization of methyl methacrylate with styrene. An elaboration of the theoretical base is required to accommodate these findings.

Block Copolymers

Blcok copolymers containing continuous sequences of units of one monomer followed by sequences of another have been formed electrochemically. For example, a solution of 1,3-dioxolane was electrolyzed in bulk and the mixture of dioxolane and electrolyte was added to molten trioxane. The reaction continued in the latter with the formation of a block copolymer. Living anionic polymerization also produces block polymers. A second monomer is introduced to a solution of living ends, and the polymerization then continues to form a block in which the chain length of the second component is determined by its relative concentration.

Graft Copolymers

A graft of a second polymer onto an existing polymer backbone produces a significant change in physical properties of the product. The grafting of methyl methacrylate, acrylonitrile, and methyl acrylate onto rayon has been investigated. The degree of grafting increased with an increase in current density, but the molecular weight of the grafted chain decreased to a limiting as the current was increased [82]. An electrochemical grafting of methyl methacrylate onto cellulose has also been reported [83]. A graft of methyl methacrylate to rubber latex produced a rubber with hard and soft forms [84]. Funt and Hsu employed benzoylated styrenes as starting materials. The benzophenone sites were reduced, and the macroradical ions thus formed were reacted with either acrylonitrile or methyl methacrylate with the resultant formation of grafts of these monomers on the backbone [85].

V. STEPWISE REACTIONS

A. Kolbe Electrolysis

The Kolbe coupling of dicarboxylic acids has only recently been studied in detail [13,86-89]. This coupling reaction is a prototype of electrochemical stepwise growth to form long chains:

$$nHOOC(CH_2)_mCOOH \longrightarrow n(CH_2)_m + 2nCO_2 + nH_2$$

In a series of definitive papers, Gozlan and Zilkha reported the electrochemical polymerization of adipic acid (m = 4), suberic acid (m = 6), azelaic acid (m = 7), and sebacic acid (m = 8). The balance of products formed was determined in each case by detailed analysis by gas chromatography, mass spectrometry, and microanalysis. A general comparison of the yields of products is shown in Fig. 1. The yield of polymer was approximately 30-35% of total reaction product in each case and was influenced by choice of solvent. The polymer yield increased in passing from methanol to methanol-pyridine (1:4) and also had a greater hydrocarbon structure in the latter solvent.

The effect of increase in chain length between the carboxyl groups was noted in this work. For example, C_{24} hydrocarbons were formed from the oligomerization of six molecules of adipic acid, four of suberic, or three of sebacic. It was concluded that the main factor governing the yield of hydrocarbons of the same chain length is the number of molecules that must condense to form them. Thus the fraction of C_{14} hydrocarbons formed from the

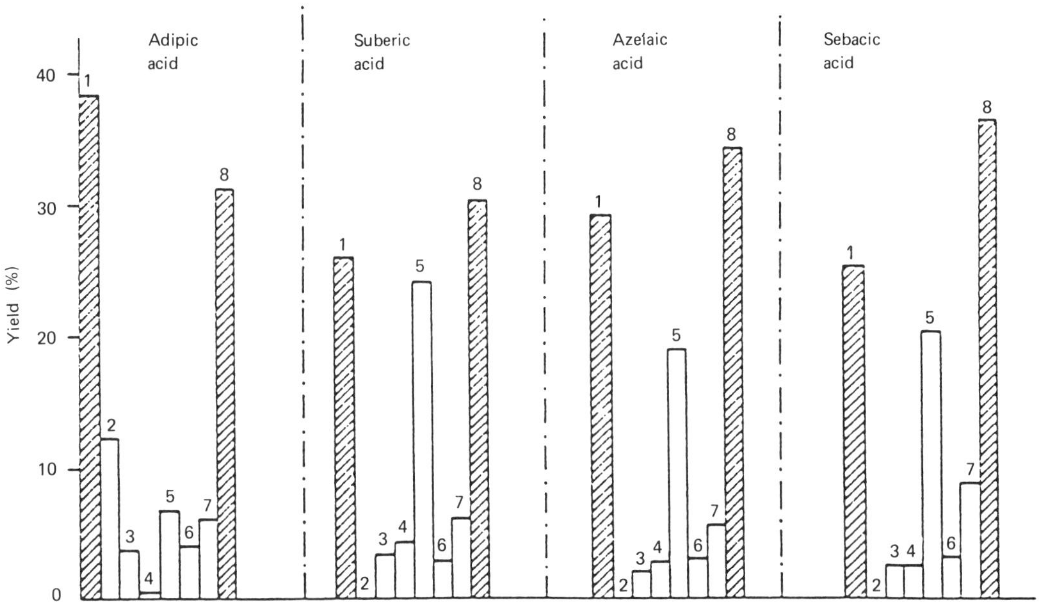

Figure 1. Comparison of the yield of all products obtained in the electrolysis of various dicarboxylic acids in methanol-pyridine (1:1): (1) hydrocarbons, (2) 7- and δ-valerolactone, (3) methyl esters, (4) methyl ethers, (5) carboxylic acids, (6) benzene fraction, (7) methanol fraction, and (8) polymer. (Courtesy of Pergamon Publishers.)

condensation of two molecules of azelaic acid corresponds to the number of C_{16} hydrocarbons formed from sebacic acid. The polymer yields (Fig. 1) were about the same for this series of acids, and all polymers were insoluble and infusible up to 300°C. However, differences in structure were apparent and differing degrees of hydrolysis with KOH were found.

B. Oxidative Coupling Reactions

Judged by the number of published reports and patents, the production of polymers by oxidative coupling is the most widely employed method of electrochemical polymerization in the 1980s. Its popularity stems from its use for the formation of electrically conducting polymers from a variety of common heterocyclics (Chap. 33) [90]. The mechanism described by Diaz et al. [91] is essentially as shown in scheme 1.

The reaction pattern consists of a repetitive generation of radical cations, their coupling, and a subsequent rearrangement with proton expulsion. Each unit added to the chain requires the passage of two electrons for formation, with a further fractional charge required for oxidation of the polymer.

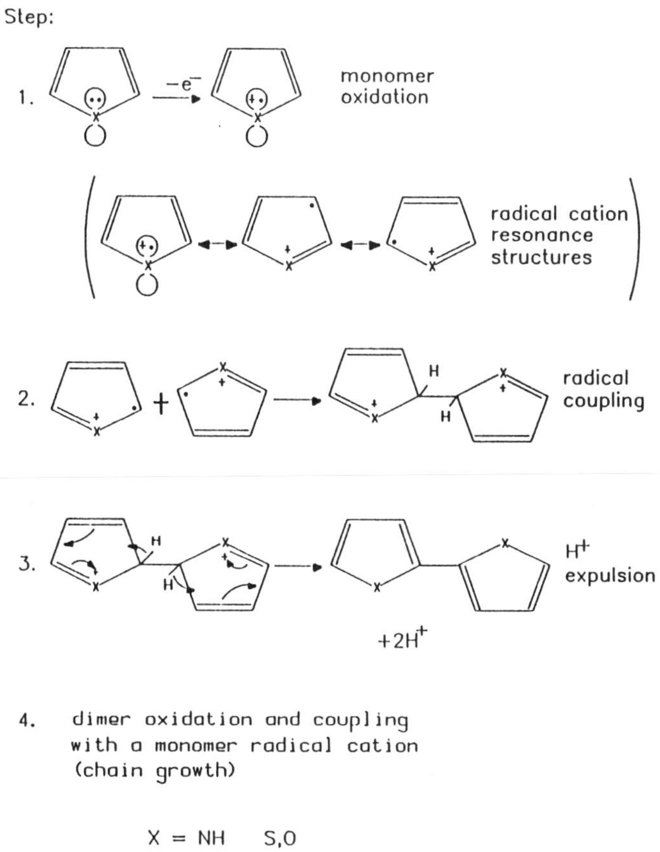

Scheme 1.

A considerable number of heterocyclic polymers have been prepared. These include single-ring heterocycles, such as pyrrole, thiophene, and furan [92], and heterocyclics, such as aniline [93].

The oxidative coupling of benzene to form poly-(p-phenylene) was achieved by Satoh et al. [94].

The copolymerization of these polymers differs from that of vinyl compounds. The initiation of each monomer occurs independently, and each has a threshold potential. In contrast to chain polymerization, the initiation of one monomer does not result in the reaction of another in the same solution. Consequently, the oxidation potentials of the reactants must be similar to obtain significant incorporation of each of the components in the final product. The copolymerization of pyrrole with bithiophene has been described, and the reaction course was characterized in terms of reactivity ratios of the reacting monomers [95].

Soluble conducting polymers of methoxythiophene have been reported recently [96]. These may provide a basis for determining the relationships between structure and molecular weight and the resultant physical and electrical properties of these interesting materials.

VI. POLYMERIC COATINGS

Metallic substrates have been coated with films polymerized in situ. The coating process is self-healing and provides films with structural integrity. As in other electroplating processes, pinholes or exposed portions of the substrate attract a higher current density and an irregular surface can be covered completely. Acrylic monomers [97], phenols, and phenolic derivatives [98,99] have been employed. Polymerizations of benzonitrile [100] and aminophthalic acids [101,102] and the formation of polyimides [103] are representative of electrolyses of nitrogen-containing polymers employed for surface coatings.

The reduction of corrosion of steel by the application of electropolymerized coatings has been investigated intensively [104]. Prior adsorption of the monomer and a choice of solvent in which the polymer is much less soluble than the monomer are two characteristics sought for such systems.

The films can be insolubilized by a cross-linking agent. With $ZnCl_2$, acrylamide and N,N'-methylenebisacrylamide have been electropolymerized to form a cross-linked polymer.

Methods of controlling film thickness and adhesion have been explored in the detailed studies described by Subramainan [105]. Acrylic acid, N-methylolacrylamide, maleic anhydride, methyl methacrylate, phenyl glycidyl ether, and styrene were polymerized on Al, Cu, and steel. Polymer formation did not occur in the absence of electrolyte, even at very high field strengths. However, free radical or ionic electropolymerization produced coherent films whose morphology was studied and categorized by electron micrographic analysis [105].

The coating of graphite fibers to improve their bonding characteristics has been reported. Flexible coatings were obtained with methyl methacrylate; stiff coatings were prepared with styrene or acrylonitrile [106]. The choice of coatings can be tailored to the mechanical requirements of the product.

In another unusual application, fluidized and packed bed electrodes of Al shot were coated with poly(diacetoneacrylamide) by electrochemical polymerization. The current density increased with increase in flow velocity during the polymerization [107].

Polymeric films in which functionally desired groups are incorporated have been prepared and have some important uses. A striking example is the formation of polymer-modified electrodes (see Chap. 33). The electrode is altered to provide reaction sites with optical, chemical, and electrochemical characteristics governed by the molecules immobilized on the surface. The resulting electrode may favor diverse such processes as photosensitization, catalysis of fast reaction kinetics, inhibition of corrosion, or stereoregulation of chirality. Recent reviews reflect the interest in this topic [108-112].

VII. CONCLUSION

Although the overall pattern of electrochemical polymerization is clear, a number of intriguing questions remain. The new and varied uses of polymers formed by electrochemical methods may soon focus interest on the chemistry of their formation. The applications of conducting polymers are particularly exciting and include macromolecular circuit elements, batteries, and electrochromic devices. However, the details of the reaction mechanism are not yet defined. The molecular weight profile with time, the variation of chain length with current density, and the change in properties with structure remain open to conjecture.

Further elucidation is also required of electrochemical effects on the composition of copolymers. The reported effects of potential on the composition of copolymers formed by chain reactions do not fit the present theoretical framework satisfactorily and present a challenge for future studies.

SYMBOLS

L.	metal
M	monomer
$M^{\bullet}$	polymer radical
$R^{\bullet}$	radical
V	volume
LE	living ends
i_0	exchange current density
δ	solubility parameter
X	chain transfer agent
ω	rotation rate
ν	sweep rate, V s^{-1}

REFERENCES

1. J. C. Bevington, H. W. Melville, and R. P. Taylor, J. Polym. Sci., 14, 463 (1954).
2. K. F. O'Driscoll, Pure Appl. Chem., 53, 617 (1981).
3. C. Tanford, Physical Chemistry of Macromolecules, Wiley, New York, 1961, Chap. 6.
4. J. B. Flanagan, S. Margel, A. J. Bard, and F. C. Anson, J. Amer. Chem. Soc., 100, 4248 (1978).
5. B. L. Funt, L. C. Hsu, and P. M. Hoang, J. Polym. Sci., Polym. Chem. Ed., 19, 203 (1981).
6. B. L. Funt and P. M. Hoang, J. Electrochem. Soc., 131, 2295 (1984).
7. G. B. Street, Handbook of Conducting Polymers, Vol. 1, Dekker, New York, 1986, p. 265.
8. B. L. Funt, L. C. Hsu, P. M. Hoang, and J. P. Martenot, J. Polym. Sci., Polym. Chem. Ed., 20, 109 (1982).
9. J. O'M. Bockris and A. Reddy, Modern Electrochemistry, Plenum Press, New York, 1970, Chap. 8.
10. H. D. Abruna, P. Denisevich, M. Umanu, and T. J. Mayer, J. Amer. Chem. Soc., 103, 1 (1981).
11. G. Mengoli, F. Furlanello, and G. Vidotti, Makromol. Chem., 177, 819 (1976).
12. N. Kuman and A. Q. Contractor, Trans. SAEST, 21, 93 (1986).
13. A. Gozlan, G. Again, S. Vardi, and A. Zilkha, Eur. Polym. J., 20, 759 (1984).
14. A. Bard and L. R. Faulkner, Electrochemical Methods, Wiley, New York, 1980, p. 280.
15. E. Laviron, L. Roullier, and C. Degrand, J. Electroanal. Chem., 154, 229 (1983).

16. B. M. Tidswell and A. W. Train, Br. Polym. J., 7, 409 (1975).
17. B. M. Tidswell and A. W. Train, Br. Polym. J., 7, 417 (1975).
18. B. M. Tidswell and A. W. Train, Br. Polym. J., 7, 429 (1975).
19. D. C. Blakley, Emulsion Polymerization, Applied Science Publishers, London, 1975.
20. G. Pistoia and G. M. Inge, J. Polym. Sci., Polym. Chem. Ed., 16, 1659 (1978).
21. J. O'M. Bockris and A. Reddy, Modern Electrochemistry, Plenum Press, New York, 1970, p. 1661.
22. G. Pistoia, A. Ricci, and M. A. Voso, J. Appl. Polym. Sci., 20, 2441 (1976).
23. P. Cerrai, G. Guerra, L. Nucci, and M. Tricoli, Eur. Polym. J., 21, 505 (1985).
24. M. Szwarc, Carbanions, Living Polymers and Electron Transfer Processes, Interscience, New York, 1968, p. 212.
25. C. L. Wilson, Rec. Chem. Prog., 10, 25 (1949).
26. G. Parravano, J. Amer. Chem. Soc., 73, 628 (1951).
27. T. W. Breitenbach and C. Srna, Pure Appl. Chem., 14, 245 (1962).
28. O. F. Olaj, J. W. Breitenbach and F. Sommer, Makromol. Chem., 148, 135 (1971).
29. A. J. Fedorova, Likero-tung, J. V. Skelepin, and J. Russ, J. Phys. Chem., 38, 920 (1964).
30. G. Pistoia and R. Benedetti, Electrochim. Acta, 23, 861 (1978).
31. I. M. Kolthoff and L. L. Ferstandig, J. Polym. Sci., 5, 563 (1951).
32. L.-C. Hsu, A. Glasel, J. F. Skinner, and B. L. Funt, J. Electrochem. Soc., 126, 941 (1979).
33. B. L. Funt and L.-C. Hsu, Polymer, 20, 81 (1979).
34. O. F. Olaj, Makromol. Chem., Macromol. Symp., 8, 235 (1987).
35. R. H. Gobran, B. M. Berenbaum, and A. V. Tobolsky, J. Polym. Sci., 46, 431 (1960).
36. B. L. Funt and D. H. Richardson, J. Polym. Sci., A1, 1055 (1970).
37. B. L. Funt and V. Hornof, J. Polym. Sci., 9, 2171 (1971).
38. G. S. Shapoval, J. Macromol. Sci., Chem., A17, 453 (1982).
39. V. Farafonov, M. Grovu, and C. Simionescu, J. Polym. Sci., Polym. Chem. Ed., 15, 2041 (1977).
40. C. Simionescu and M. Grovu, ANgew. Makromol. Chem., 111, 149 (1983).
41. G. Mengoli, G. Vidotto, and F. Furlanetto, Makromol. Chem., 37, 203 (1970).
42. E. B. Mano and B. A. L. Calafate, J. Polym. Sci., Polym. Chem. Ed., 21(3), 829 (1983).
43. B. L. Funt and K. C. Yu, J. Polym. Sci., 62, 359 (1962).
44. G. Mengoli and S. Valcher, Eur. Polym. J., 10, 959 (1974).
45. G. Mengoli and G. Vidotto, J. Appl. Polym. Sci., 18, 3095 (1974).
46. G. Mengoli and G. Vidotto, Makromol. Chem., 175, 893 (1974).
47. G. Mengoli and G. Vidotto, Makromol. Chem., 165, 137 (1973).
48. G. Mengoli and G. Vidotto, Makromol. Chem., 165, 149 (1973).
49. G. Mengoli and G. Vidotto, Eur. Polymer J., 8, 661 (1972).
50. G. Mengoli and G. Vidotto, Makromol. Chem., 153, 57 (1972).
51. G. Mengoli and G. Vidotto, Makromol. Chem., 139, 293 (1970).
52. S. Bhadeni and A. Ghose, Trans. Soc. Adv. Electrochem. Sci. Technol., 9, 13 (1973).

53. S. N. Bhadani, P. P. Baranwal, and Y. K. Prasad, Makromol. Chem., 179, 1623 (1978).
54. S. N. Bhadani, P. P. Baranwal, and H. K. Prasad, Makromol. Chem., 178, 2637 (1977).
55. S. N. Bhadani, and P. P. Baranwal, Makromol. Chem., 178, 1049 (1977).
56. Y. Kikuchi and H. Furuda, J. Polym. Sci., Polym. Chem. Ed., 11, 2709 (1973).
57. P. Cerrai, G. Guerra, M. Tricoli, and L. Nucci, Eur. Polym. J., 16, 867 (1980).
58. P. Cerrai, G. Guerra, and M. Tricoli, Eur. Polym. J., 16, 153 (1979).
59. P. Cerrai, G. Guerra, and M. Tricoli, Eur. Polym. J., 12, 247 (1976).
60. P. Cerrai, P. Giusti, G. Guerra, and M. Tricoli, Eur. Polym. J., 11, 101 (1975).
61. P. Cerrai, P. Giusti, and M. Tricoli, Eur. Polym. J., 10, 1195 (1979).
62. A. Akbulut, J. E. Fernandez, and R. L. Birke, J. Polym. Sci., Polym. Chem. Ed., 13, 133 (1975).
63. U. Akbulut, L. Toppare, A. Usanmaz, and A. Onal, Makromol. Chem. Rapid Commun., 4, 259 (1983).
64. U. Akbulut, L. Toppare, and L. Turker, Makromol. Chem., 184, 1861 (1983).
65. G. Mengoli and G. Vidotto, J. Electroanal. Chem., 75, 595 (1977).
66. B. L. Funt, W. Severs, and A. Glasel, J. Polym. Sci., Polym. Chem. Ed., 14, 2763 (1976).
67. A. Glasel, K. Murray, B. L. Funt, Makromol. Chem., 177, 3345 (1976).
68. G. S. Silvestri, S. Gambino, and G. Filarda, Adv. Polym. Sci., 38, 27 (1981).
69. R. Ercoli, M. Guainazzi, and G. Silvestri, J. Chem. Soc., Chem. Commun., 927 (1967).
70. E. H. Mottus, and M. O. Ort, J. Electrochem. Soc., 117, 885 (1970).
71. D. C. Phillips, D. H. Davies, and J. D. Smith, Makromol. Chem., 169, 177 (1973).
72. D. C. Phillips, D. H. Davies, and J. D. Smith, Makromol. Chem., 164, 317 (1972).
73. D. C. Phillips, D. H. Davies, and J. D. Smith, Macromolucules, 5, 674 (1972).
74. D. C. Phillips, D. H. Davies, and J. D. Smith, J. Polym. Sci., A-1, 3267 (1972).
75. B. L. Funt, J. McGregor, and J. Tanner, J. Polym. Sci. B, 8, 699 (1970).
76. B. L. Funt, J. McGregor, and J. Tanner, J. Polym. Sci. B, 8, 695 (1970).
77. H. Lehmkuhl and W. Eisenbach, Liebigs Ann. Chem., 672 (1975).
78. L. Toppare, S. Eren, and U. Akbulut, Polymer, 25, 1655 (1984).
79. L. Toppare, S. Eren, and U. Akbulut, J. Polym. Sci., Polym. Chem. Ed., 22, 2941 (1984).
80. L. Toppare, S. Eren, and U. Akbulut, Br. Polym. J., 16, 71 (1984).
81. V. O. Benzuglyi and V. A. Shapvalov, Elektrokhimiya, 13, 984 (1977).
82. T. Skwarski and T. Mikolajczyk, Monatsch. Chem., 107, 1103 (1976).
83. M. L. Dupraz and J. Rouger, Cellul. Chem. Technol., 7, 41 (1973).
84. L. P. Mendis and K. P. W. Pemawansa, and S. Hettiarachchi, Polymer, 18, 100 (1977).

85. B. L. Funt and L. C. Hsu, J. Polym. Sci., Polym. Chem. Ed., 18, 1957 (1980).
86. A. Gozlan and A. Zilkha, Eur. Polym. J., 20, 1199 (1984).
87. A. Gozlan and A. Zilkha, Eur. Polym. J., 23, 133 (1987).
88. A. Gozlan and A. Zilkha, Eur. Polym. J., 23, 143 (1987).
89. A. Gozlan and A. Zilkha, Eur. Polym. J., 23, 393 (1987).
90. A. F. Diza, K. K. Kanazawa, and G. P. Gardini, J. Chem. Soc., Chem. COmmun., 635, (1979).
91. A. M. Genies, G. Bidan, and A. F. Diaz, J. Electroanal. Chem., 149, 101 (1983).
92. G. Tourillon and G. Garnier, J. Electroanal. Chem., 135, 173 (1982).
93. A. Diaz and J. A. Logan, J. Electroanal. Chem., 111, 111 (1983).
94. M. Satoh, K. Kaneto, and K. Yoshino, J. Chem. Soc., Chem. · Commun., 22, 1629 (1985).
95. B. L. Funt, E. M. Peters, and J. D. Van Dyke, J. Polym. Sci. A, 24, 1529 (1986).
96. R. L. Elsenbaumer, K. Y. Jen, and K. Oboodi, Synthetic Metals, 15, 169 (1986).
97. G. Mengoli and B. W. Tidswell, Polymer, 16, 881 (1975).
98. D. E. Danly, Hydrocarbon Proc., 48, 159 (1969).
99. W. F. H. Borman, Can. Patent 800592 (1968).
100. R. V. Subramainan and G. K. Barg, Polym. Bull., 1, 421 (1979).
101. D. C. Phillips, S. Spewock, and W. M. Alvinc, J. Polym. Sci., Polym. Chem. Ed., 14, 1137 (1976).
102. D. C. Phillips and W. M. Alvino, J. Polym. Sci., Polym. Chem. Ed., 14, 1157 (1976).
103. D. C. Phillips, J. Electrochem. Soc., 119, 1645 (1972).
104. G. Mengoli and B. M. Tidswell, Polymer, 16, 881 (1975).
105. R. V. Subramanian, Adv. Polym. Sci., 33, 33 (1979).
106. B. K. Garg, R. A. V. Raff, and R. V. Subramanian, J. Appl. Polym. Sci., 22, 65 (1978).
107. R. Mahaling, F. A. Teng, and R. V. Subramanian, J. Appl. Polym. Sci., 22, 3587 (1978).
108. J. Zak and T. Kuwana, J. Electroanal. Chem., 149, 645 (1983).
109. R. W. Murray, Acc. Chem. Res., 13, 135 (1980).
110. R. W. Murray, in Electroanalytical Chemistry, Vol. 13, Dekker, New York, 1984.
111. K. D. Snell and A. G. Keenan, J. Chem. Soc., Chem. Soc. Rev., 259 (1979).
112. W. J. Albery and A. R. Hillman, Chem. Soc. Ann. Rep. Prog. Chem., Sec. C, 78, 377 (1978).

33

Chemically Modified Electrodes and Conducting Polymers

A. F. DIAZ / IBM Research Division, San Jose, California

I. Introduction 1363

II. Covalently Attached Monolayers 1364
- A. Silylated oxide surfaces 1365
- B. Characterization of silylated surfaces 1368
- C. Electrochemical characteristics 1368
- D. Derivatized carbon surfaces 1370

III. Chemisorbed Monolayers 1371

IV. Chemisorbed Electroactive Polymer Films 1372

V. Electropolymerized Polymer Films 1375
- A. Electrochemical stoichiometry 1376
- B. Monomer structure-reactivity considerations 1380
- C. Cell conditions 1382

VI. Polymer-Ceramic Coatings 1390

VII. Ad-atoms 1390

VIII. Concluding Remarks 1390

References 1392

I. INTRODUCTION

The activity in chemically modified electrodes has been growing over the last 15 years and is motivated by the need to develop well-defined surfaces for studying electron exchange processes and for controlling electrochemical reactions. This area has grown very rapidly, and a large number of materials can now be used to modify an electrode. The preparative methods used to produce the electrode coatings are limited, and most procedures can be classified in three general categories. One category involves placement of a covalently bonded layer of small molecules on the electrode. The most commonly used reagents for this modification are organosilanes in which the silyl functionality bonds to the surface oxide of the electrode. The surface densities of the resulting coatings are up to a monolayer, but submonolayer coverages often result. The layer has order in which the molecules are oriented by the bonding interaction between the silyl group and the surface oxide. The organosilane is bifunctional, and the group at the other end of the molecule is subsequently used to attach electroactive structures. The other category involves coating the electrode with a relatively thick layer of a polymer. In one approach, electroactive-conductive polymer film coatings can be generated in situ by the electrooxidation of aromatic compounds. These films are prepared from benzenoid, nonbenzenoid, and heterocyclic aromatics, in particular from the derivatives of pyrrole, thiophene, carbazole, azulene, pyrene, triphenylene, and aniline. Alternatively, a fully formed polymer with the desired chemical structure can be applied to the electrode by dip or spin coating using a dilute solution of the polymer. Finally, modifications can be accomplished by the irreversible adsorption of large aromatic compounds and polymers.

The interest in chemically modified electrodes stems from the fact that these electrodes may impact on many different technology areas, such as electrocatalysis, electrosensors, electrosynthesis, electrophotosensitization, energy storage, electrorelease of reagents, membrane barriers, display devices, and optical shutters. The latter applications are associated with polymeric film coatings. Not surprisingly, each area has its own set of criteria for the materials, and the properties of the modified electrode must be optimized for each application. However, all the applications require a material with high stability and good chemical selectivity or efficiency. The details of the technical applications that have been considered are not reviewed here since they are covered in several recent review articles [1-6]. Two review articles by Murray et al. summarize the design of electrode surface coatings and their use for studying the electron transfer processes [1] and their application to electroanalysis [2]. Two other articles by the same authors [3,4] summarize the design, fabrication, and electrochemical characterization of polymeric electrode microstructures using electroactive polymer films. Fujihara has also summarized the electrochemical processes and electrode applications associated with electrodes modified with monolayers and polymer films [5]. Finally, the electrode applications of electroactive-conductive polymers are summarized in a recent article [6]. This chapter reviews the preparation routes to modified electrode surfaces. It is not an exhaustive literature report but instead stresses the studies on covalently attached monolayers and electropolymerized conductive coatings, which are two areas that have received a considerable amount of attention. This chapter summarizes a selection of studies that help provide a view of the characteristics and applicability of the reactions used to modify electrode surfaces.

II. COVALENTLY ATTACHED MONOLAYERS

Much of the activity on electrodes with covalently attached monolayers occurred between 1975 and 1982, and this area has been the subject of several excellent review articles [1-5]. Electrode modifications have been made with metal, semiconductor, and carbon electrodes. In the modification of the metal [7,8] and semiconductor [9-12], the electrodes make use of the reaction between the surface oxide layer and a reactive organosilane. Although electrodes that are bulk oxides, such as SnO_2 and TiO_2, can be used directly, the surface of the metal and semiconductor electrodes, such as Pt, Au, Si, and Ge, must first be oxidized to produce a surface oxide layer. Organosilanes are used because they provide a convenient chemical bridge between the electrode and an electroactive molecule of interest, even though there is little electrochemical interest in the structures themselves. The total surface modification process normally involves several steps in which the organosilane is first attached to the electrode by treatment of the electrode surface with a reactive organosilane ($RSiX_3$ or R_2SiX_2; X = Cl or OR) to form a metal-oxygen-silane bond. An example of the reaction is shown in Eq. (1) using an alkyltriethoxysilane derivative:

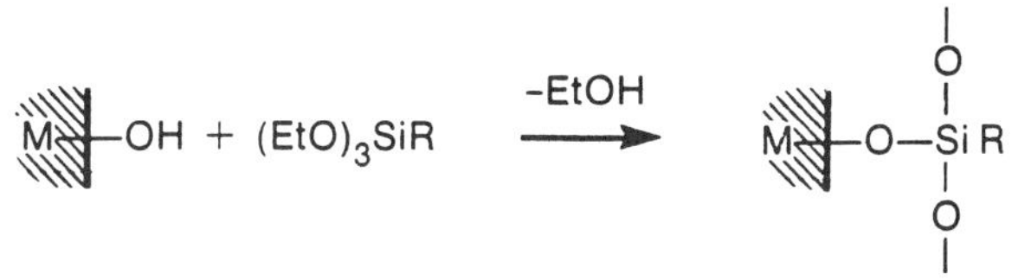

(1)

The second step in the preparation involves the attachment of the electroactive center, as by acylation of a terminal amino group, as shown in Eq. (2):

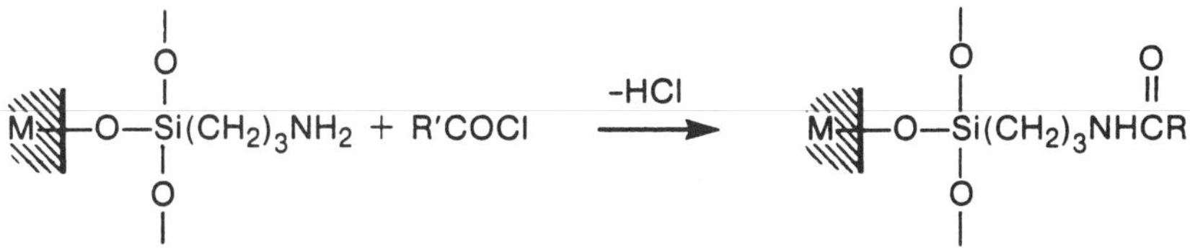

(2)

Carbon electrodes lend themselves to a greater variety of modifications [13-21]. Glassy carbon can be derivatized without a prior surface activation treatment [14], or more commonly, as with the metal electrodes, the surface can be pretreated to generate an oxidized layer [13,15-17,19]. Oxidation of the carbon surface produces carboxylic acid groups that can be derivatized. Examples of these processes are shown in Eqs. (3) and (4), where the oxidized surface is activated with thionyl chloride before derivatization with the amine:

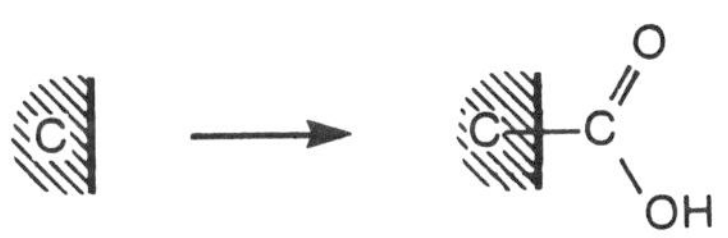

(3)

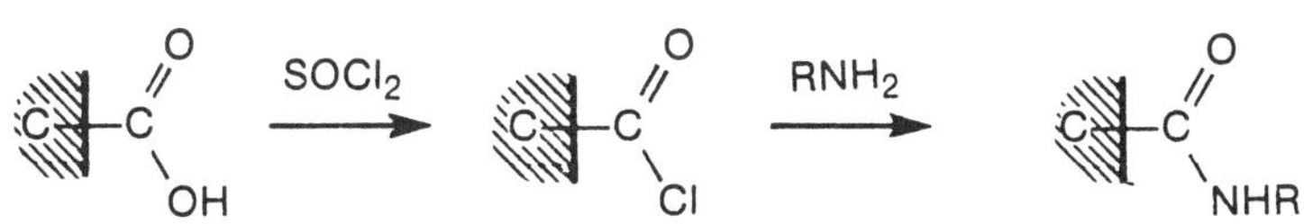

(4)

A. Silylated Oxide Surfaces

The number of procedures developed for preparing the organosilane monolayers are few; however, a consideration common to all the modification procedures is the preparation of the electrode surface. This includes a proper cleaning and drying procedure to ensure bonding of the organosilane to the surface OHs. The cleaning procedures usually involve several extractions and rinses with kerosenes, mineral acids, ethanol, and water to remove grease and other contaminants from the surface. The surfaces are then dried at elevated temperatures. Electrodes that do not have a good oxide layer on the surface must be pretreated. Platinum electrodes are anodized in aqueous acid solutions following the procedures reported by Murray and colleagues [7,9]. The surface of n-type Si is activated by etching the electrode in concentrated HF and subsequently immersing the electrode in 10 M NaOH [22]. RuO_2 surfaces have been prepared by spraying a dilute $RuCl_3$ solution on the support electrode and fully oxidizing the RuO_2 layer by heating in air [8].

Derivatization of the electrode surfaces is often accomplished by exposing the electrode to a dilute solution of the organosilane. Any organic aprotic solvent that can be properly dried can be used for this procedure. Some of the more common solvents are octanes, benzene, toluene, methylene chloride, THF, and acetonitrile. Because of the sensitivity of the reaction to moisture, it is important that the reaction be performed under anhydrous conditions through the final rinses, which remove unreacted organosilane, since any remaining reagent hydrolyzes with moisture and produces polymeric siloxane deposits on the electrode surface. The conditions for treatment of the electrode surface range from exposure to a 1% alkylsilane solution for 10-30 min at room temperature [10] to refluxing overnight in a 10% alkylsilane solution [12].

Although a wide variety of alkylchlorosilanes and alkylethoxysilanes can be used to modify electrodes, much of the work has been performed with 3-aminopropyltrimethoxysilane, 3-(2-aminoethyl)aminopropyltrimethoxysilane, and β-trichlorosilyl-2-ethylpyridine because the terminal N-functionality permits subsequent attachment of electroactive centers or affiliation of metal ion centers. Reaction on the amino group is normally via an acylation reaction carried out using a carboxylic acid derivative of the redox compound and a coupling agent, such as dicyclohexylcarbodiimide (DCC) [23], cyanuric chloride [17], or heating in the presence of 2,6-lutidine [7,15]. In principle, derivatization of the amino group can be accomplished using the acyl halide [15] or anhydride of the carboxylic acid, but this approach is rarely used because of the limited availability of these derivatives. With this combination of reactions, electrode assemblies have been prepared with various interesting electroactive structures, such as the derivatives of pyrazoline (1) [23], dihydroxybenzene (II) [7], ferrocene (III) [22,24,25,26], dinitrobenzoate (IV) [7,15], viologen (V) [17], and porphyrin (VI) [15,26,27]. In addition, several metal ion complexes have been immobilized on the amine and the ethylenediamine fuctionalities [12,28]. Using an analogous procedure, 3-thiolpropyltrimethoxysilane was used to attach an iron cluster $[Fe_4S_4(SR)_4]^{2-}$ to a SnO_2 surface [29].

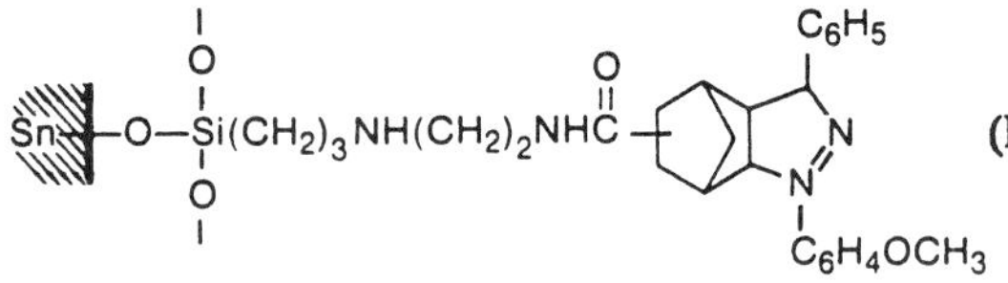

(I)

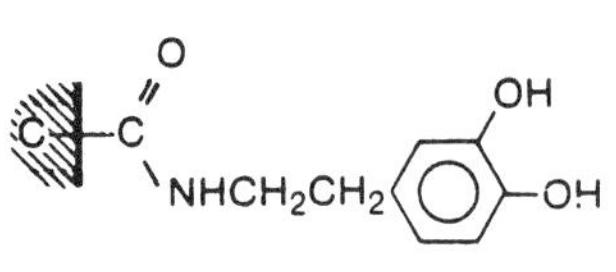

(II)

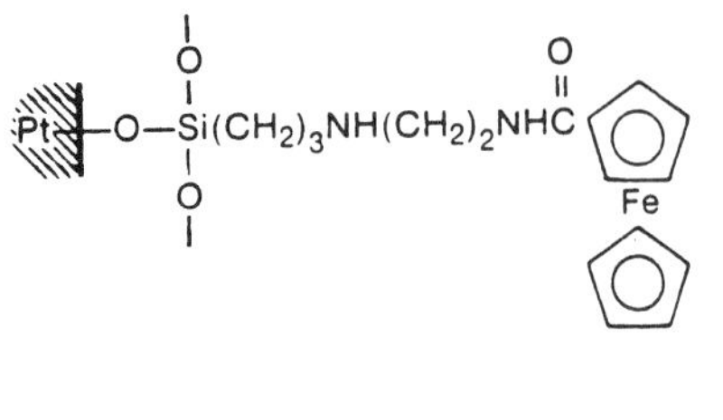

(III)

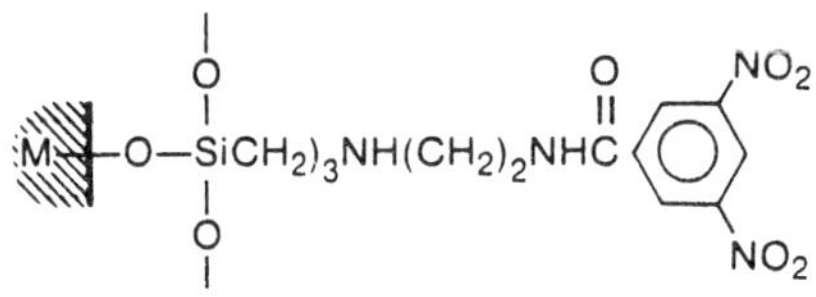

(IV)

M = Pt, Ru

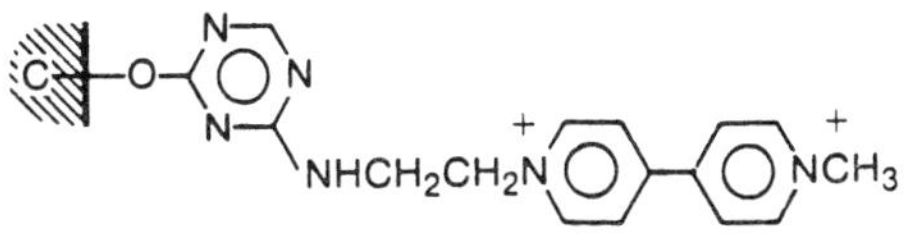

(V)

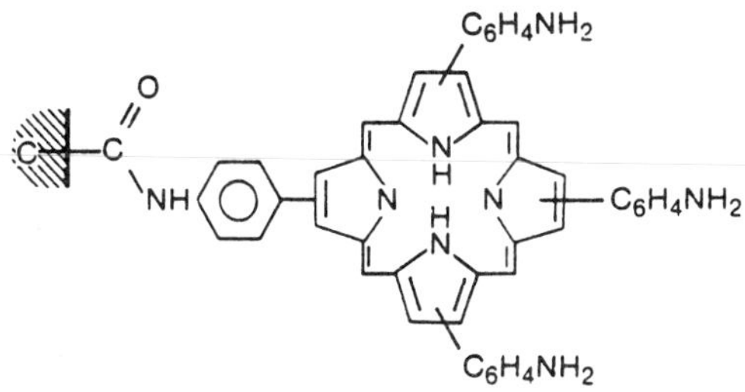

(VI)

Not all modifications involving surface silylation reactions require the use of aminoalkylsilane as bridging compounds. With the availability of a good synthetic capability, the silyl function can be substituted directly on the redox structure of interest. For example, the chlorosilane derivatives of ferrocene (VII and VIII) were prepared and attached directly on platinum [22,24,25]. The use of the 3-aminopropylsilyl structure to immobilize structures on the electrode surface has been extended to involve a preformed polymer [30]. Poly(methacryl chloride) was reacted with a 3-aminopropylsilyl-derivatized SnO_2 surface and subsequently reacted with hydroxymethylferrocene (IX). The resulting electrode coating had four to five times greater density of electroactive centers than a monolayer. The coating was stable in acetonitrile for several hours.

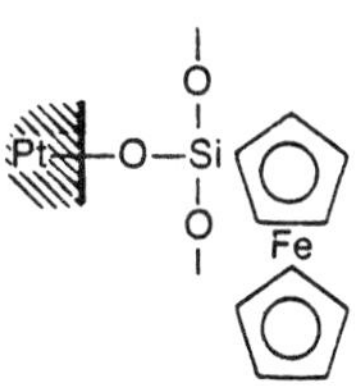

(VII)

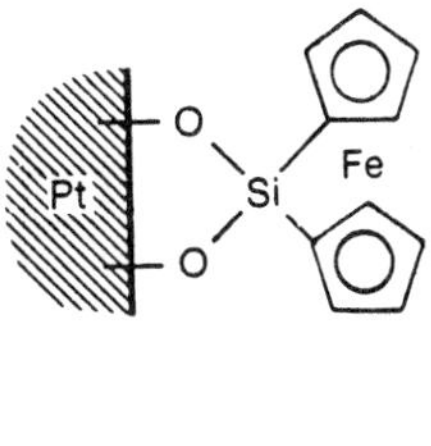

(VIII)

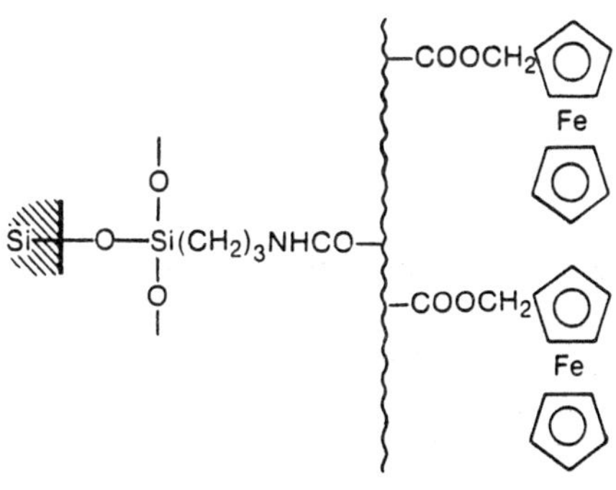

(IX)

In addition, the method can be used to build up thick polymeric layers by polymerizing the reactive alkylsilane with the addition of water to produce polyalkylsiloxane surface coatings [31,32]. For example, treatment of platinum electrode with 3-(2-aminoethyl)-aminopropylsilyl with moisture present produced a polymeric film that could be derivatized with ferrocenecarboxylic acid [31]. The resulting electroactive film had ~50 times greater density of electroactive centers and was more stable than the corresponding monolayer. As is apparent, this method is quite general and has little dependence on the nature of the oxide surface and on the structure of the electroactive derivative. Redox compounds with a large variation in structure and redox potential can be attached to electrodes.

B. Characterization of Silylated Surfaces

ESCA has been a useful tool for determining the presence of molecular monolayers. This is especially true for the surface modifications involving the alkylsilanes, which are not electroactive and cannot be analyzed electrochemically. The analyses have been used on transparent SnO_2 electrodes treated with a series of trichloroalkylsilanes and triethoxyalkylsilanes in which the alkyl group is aminopropyl, chloropropyl, pyridylethyl, and 3-(2-amino)aminopropyl. In these analyses, the signals for the silicon and nitrogen atoms are the most informative since the carbon signal is usually obscured by contamination. The method helps monitor the incorporation of metal ions immobilized on the amine, ethylenediamine, and porphyrin ligands on the surface monolayer [12]. Inelastic electron tunneling spectroscopy (IETS) [33] has also been used to analyze an aluminum surface modified with triethoxyvinylsilane. The spectrum, which resembles an IR spectrum, shows the absorption of the CH and C=C groups. In addition, it reveals the extensive removal of surface OH and shows that some ethoxysilane groups survive the silylation reaction. These groups are available for reaction with neighboring molecules to form dimers and oligomers. Regarding the surface coverage, estimates are available from electrochemical measurements with electroactive monolayers, and the coverage for a covalently attached monolayer is ~30% that for an adsorbed monolayer [10]. This estimate results from a comparison of a SiO_2 surface modified with [^{14}C]octadecyltrichlorosilane ($\Gamma = 1.5 \times 10^{14}$ molecules per cm^2) with a surface coated with a cadmium [^{14}C]arachidite Langmuir monolayer ($\Gamma = 5 \times 10^{14}$ molecules per cm^2). This result suggests that the attached molecules are loosely packed, and the difference between the two molecular layers probably reflects the steric requirements of the nonlinear covalent bonds.

C. Electrochemical Characteristics

Almost all electrochemical studies on modified electrodes have been performed using cyclic voltammetry, but differential pulse voltammetry [15,19,34,35] and second harmonic a.c. voltammetry [11] have also been used. The surface-localized redox structures can be oxidized and reduced as shown with the ferrocene derivative in Eq. (5), and the electrochemical response of surface species is predictable from the measurements of these structures in solution [26].

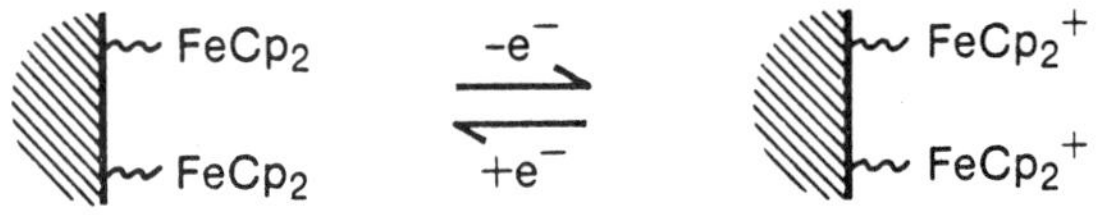

(5)

$Cp = C_5H_5$

The formal potentials have been compiled for a large number of surface-localized redox centers, and there is a good correspondence with the solution analog [26]. The series includes compounds with widely different chemical structures and with redox potentials that encompass a 3.0 V range. This is shown schematically in the summary figure by Murray [1], which is shown in part in Fig. 1. Figure 1 shows the position of the redox waves for redox structures covalently attached to platinum and carbon. The cyclic voltammograms have the expected characteristics for a reaction in the absence of diffusion. The i_p values scale with the sweep rate and the peaks

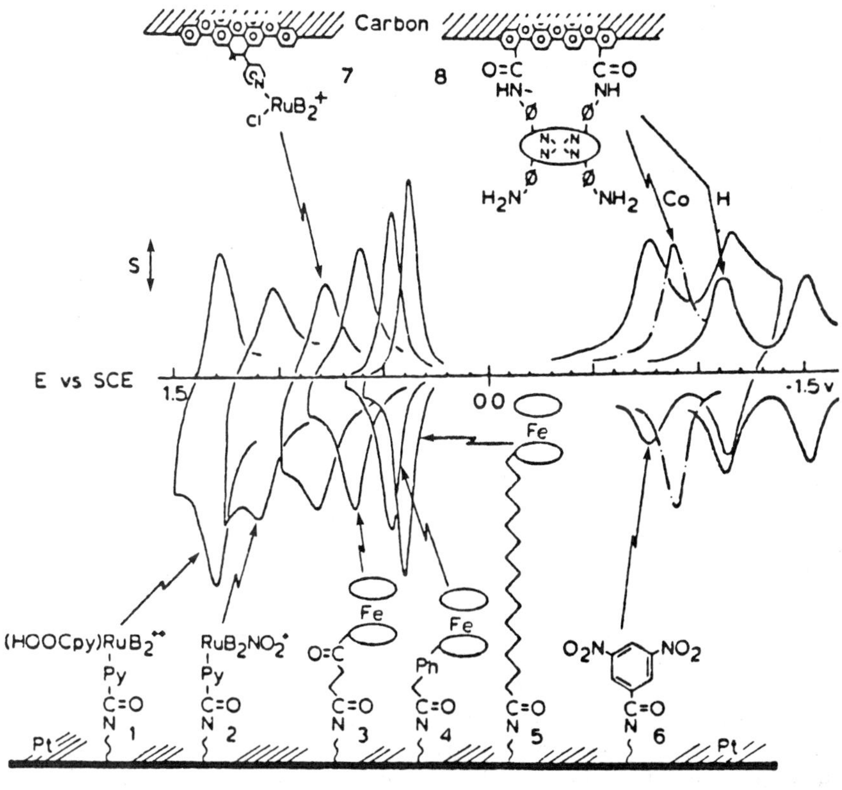

Figure 1. Cyclic voltammograms of selected modified carbon and platinum electrode surface structures. Structure 7 is 2,2'-bipyridyl and structure 8 is the α,β surface isomer for tetrakis-(p-aminophenyl)porphyrin. Voltammograms at positive E are in acetonitrile, except 2, which is −80°C butyronitrile; those at negative E are in DMSO. The current densities are 3-70 $\mu A\ cm^{-2}$, Γ equals 0.14-9 × 10^{14} molecules per cm^2, and scan rates are 0.1-0.2 $V\ s^{-1}$. (From Ref. 1.)

are reasonably symmetrical, but the peak width at half-height is usually greater than the expected 90.6 mV owing to the effect of surface activities [34]. The surface densities of the monolayer, estimated with coatings containing electroactive structures, are in the range $0.2 - 2 \times 10^{14}$ molecules per cm^2. These estimates are based on the geometric area; the actual coverage is lower and depends on the surface roughness [7,10,17]. The oxidation-reduction reactions are simplified by the absence of diffusion but can still be complicated because redox centers attached to flexible chemical chains containing three to seven carbon atoms are subject to thermal molecular motions and must fold to place the redox center close to the electrode surface.

D. Derivatized Carbon Surfaces

In contrast to the metal oxide surfaces, the surface of carbon electrodes can be modified by direct derivatization of functional groups and reactive centers present on the surface. The oxidized surfaces contain carbonyl and carboxylic acid functions that can be derivatized, but the oxygen free surfaces contain "reactive bonds." Without pretreatment, glassy carbon and graphite surfaces can be modified by derivatizing the surface carbonyls with dinitrophenylhydrazine in ethanolic acid solutions to produce the hydrazone derivative [14]. Oxidization of the surface of graphite electrodes by heating in air [13,16] or vacuum [15] or by exposure to an RF oxygen plasma produces surface carboxylic acid groups [16]. These acid groups can be activated by treatment with thionyl chloride or diisopropylcarbodiimide for reaction with amine reagents [13,15,16,19], as shown in Eqs. (3) and (4). Alternatively, the oxidized surface can be reduced with $LiAlH_4$ to produce carbinols, which can then be reacted with cyanuric chloride and amine reagents, Eqs. (6) and (7) [17]:

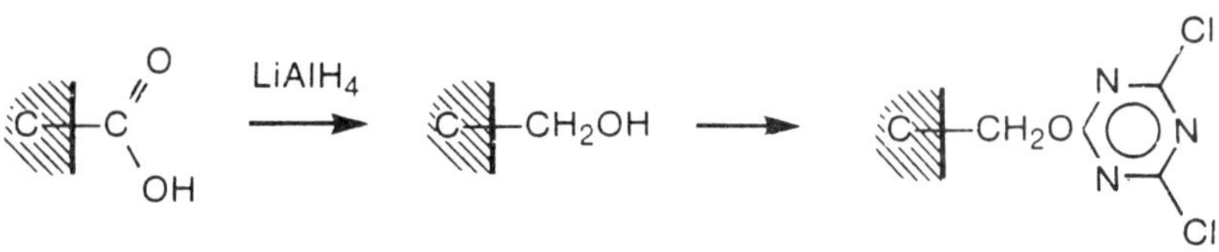

(6)

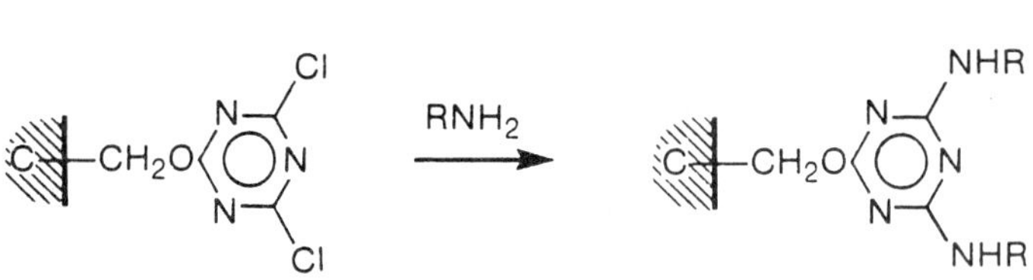

(7)

Finally, carbon surfaces can be made reactive by the removal of the surface oxide. This can be accomplished by the mechanical abrasion of the surface in a solution of a reactive olefinic reagent, which can undergo cycloaddition reactions with the reactive centers formed on the surface [19]. This procedure was used to attach vinylferrocene and vinylpyridine of the surface of glassy carbon. Alternatively, the edge plane of pyrolytic graphite and glassy carbon can be activated by etching with an RF argon plasma to remove surface oxide [18,19,21] and subsequently exposed to an amine reagent, such as ammonia, ethylenediamine, or diethylenetriamine

[18,21]. The surface-attached amino groups can then be derivatized with a carboxylic derivative and dicyclohexylcarbodiimide or used to immobilize metal cations. Pyrolysis of carbon fibers at 1020°C in vacuo also removes surface oxide and generates acitve sites that undergo cycloaddition reactions with a variety of olefins, including ethylene, vinyl bromide, and acrylates [20].

As with the metal oxide electrodes, the presence of the surface monolayers can be determined by ESCA, and the electrochemical response of the attached electroactive centers or of metal centers immobilized on the amine ligands are predictable from the measurements of the same structures in solution as were observed with the metal oxide electrodes [26]. The surface density of the attached redox centers as measured by cyclic voltammetry is approximately 0.1-9 × 10^{14} molecules per cm^2 [17,19,21].

III. CHEMISORBED MONOLAYERS

The first systematic studies of modified electrodes were reported by Lane and Hubbard [35]. It was shown that monolayer coverages of a variety of allyl derivatives could be adsorbed on platinum and carbon electrodes. These layers were prepared by dip coating or vapor deposition, and the molecular layer could be addressed electrochemically either by using an electroactive molecule, such as vinyldihydroquinone, or by a subsequent chemical transformation. For example, the allylamine layer was used to immobilize a platinum bromide complex. Extension of this approach to substituted pyridine and quinone structures has provided one of the cleanest systems for studying the details of structure and orientation of molecules on electrodes [36].

More recently, this approach has been extended to the irreversible adsorption of large aromatic compounds on graphite electrodes for electrochemical studies [37,38]. For example, 9,10-phenanthrene quinone (X) and iron(III)tetraphenylporphyrin chloride (XI) are adsorbed onto a graphite electrode by immersing the electrode in an aqueous acid solution [37]. The adsorption process is slow and may take an hour for the surface to saturate even with a stirred solution. With reversible redox compounds the adsorption process can often be conveniently followed to completion by observing the development of the redox peaks. This approach is shown in Fig. 2 for the adsorption of 9,10-phenanthrenequinone onto a carbon electrode from an aqueous trifluoroacetic acid solution (E° = 175 mV versus SCE) [37]. With this technique, ligands for metal complexes can be affiliated to the graphite surface by adsorbing large aromatic molecules that have the ligand as a substituent (XII) can be used to immobilize $Ru(NH_3)_5X^{2-}$ complexes [38].

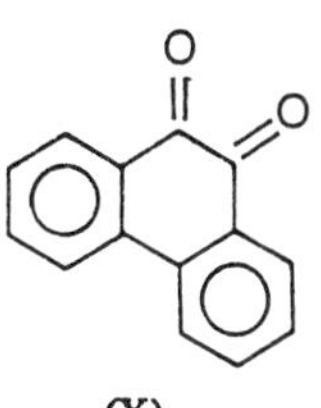

(X)

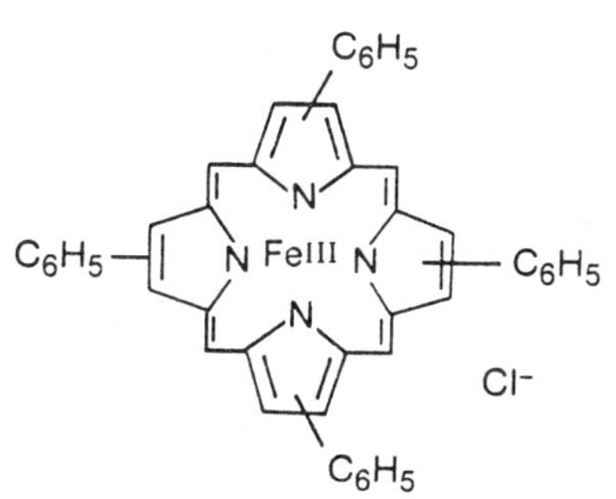

(XI)

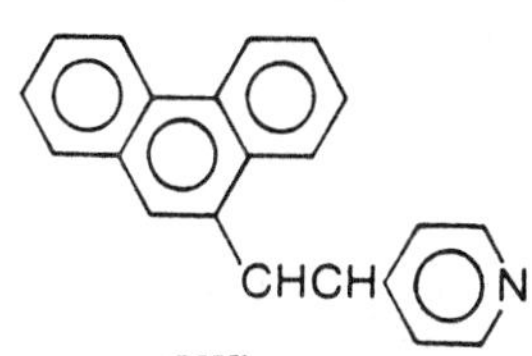

(XII)

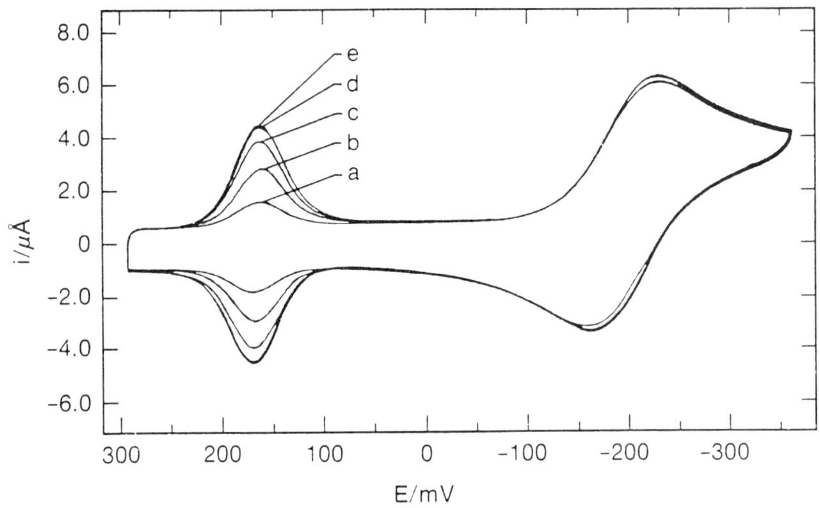

Figure 2. Cyclic voltammograms obtained with pyrolytic graphite electrode (0.2 cm^2) in a solution containing 0.3 mM $Ru(NH_3)_6^{3+}$ and 0.001 mM 9,10-phenanthrenequinone. Scan rate: 0.02 V s^{-1}; supporting electrolyte: 1 M CF_3COOH. Minutes of exposure of the freshly cleaved electrode to the solution with stirring by bubbling nitrogen between scans: (a) 2, (b) 8, (c) 20, (d) 47, (e) 84. (From Ref. 37.)

With this method, the immobilization of very large molecules, for example dicobalt cofacial porphyrin, has been achieved [39].

IV. CHEMISORBED ELECTROACTIVE POLYMER FILMS

Extension of this approach has led to modification of electrode surfaces by chemisorbing thick films of electroactive polymers. With this material change, the loading of electroactive centers on the surface of an electrode can be impressively increased from a monolayer, ~10 Å thick, up to the equivalent of 10^5 monolayers with μm thick films. The polymer films are applied and retained on the electrode surface by adsorption in a solvent, where it has low solubility. Polymer film coatings have been applied by dip coating [40], spin coating [41,42], adsorption from solution [43], and electrochemical precipitation [44]. Many of these polymers are electroactive and are classified as redox polymers in which the electroactive centers are pendent groups on the saturated backbone of the polymer and have localized electronic states. Interaction between vicinal redox centers in the film occurs by electron self-exchange in which electrons hop between the reduced and the oxidized sites [45]. For example, electrodes have been coated with polymers containing electroactive structures like ferrocene (XIII) [46], tetrathiafulvalene (TTF) (XIV) [42], and quinone (XV) [47]. All the electroactive centers can be oxidized and reduced; however, the peaks are very distorted because of the complicated kinetics. This can be seen with the reaction of the TTF polymer (XIV) shown in Fig. 3 [42]. Although TTF in solution exhibits two reversible waves at 0.33 and 0.7 V versus SCE, the cyclic voltammogram

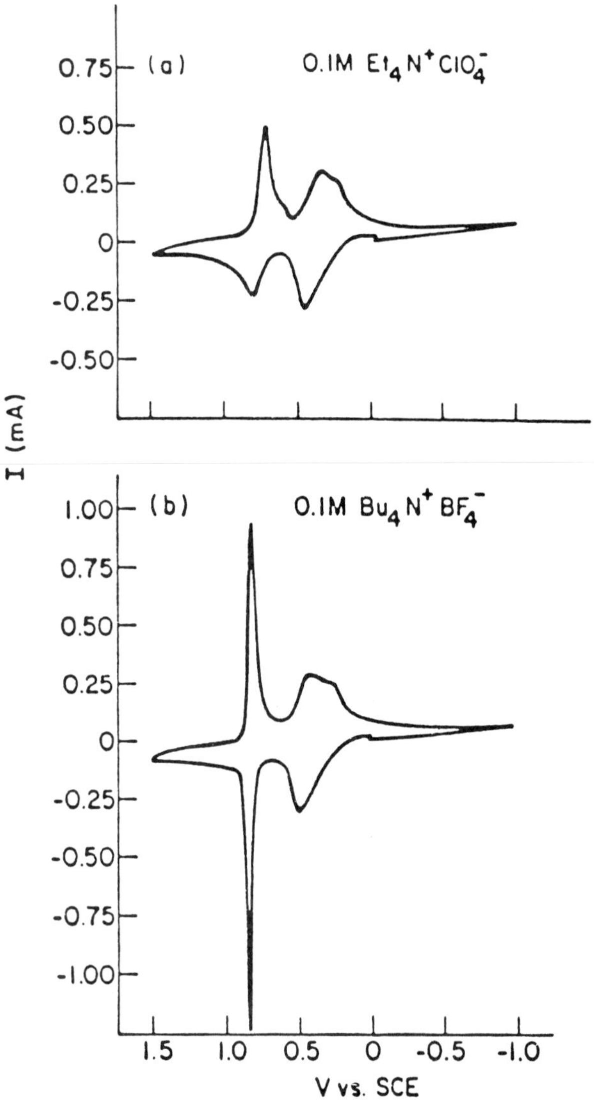

Figure 3. Cyclic voltammograms of a 400 Å TTF polymer film at 0.02 V s^{-1} in acetonitrile with two different electrolytes. (From Ref. 42.)

for the polymer-bonded TTF centers is complicated because of phase changes and the kinetics of the reaction, which are strongly influenced by the anion of the electrolyte. With thicker films, 4000 Å, there is an added complication in which the insoluble polymer film completely insulates the electrode until it becomes wetted with the electrolyte. It normally takes 5-10 voltage cycles to induce this change. Ion-exchange ploymers have also been used as electrode film coatings. Although these polymers are not electroactive, electroactive ions can be introduced into the polymer by ion exchange of the mobile counterion that accompanies the polymer for charge neutrality. Polyvinylpyridinium perchlorate is an example of a material that has been used successfully to modify electrodes and can be converted to the ferricyanide polymer (XVI) [43]. Polyacrylic acid and sulfonated polymers, such as polyvinylsulfonate, Nafion (XVII) and Gore-tex, have been used to produce films which will incorporate positively charged electroactive structures [48-50].

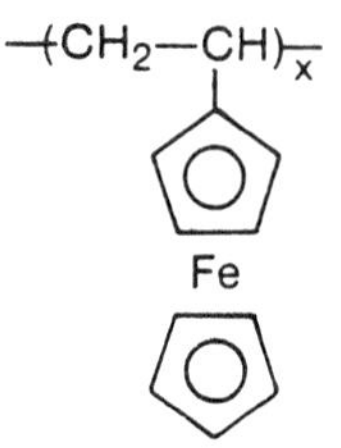

(XIII)

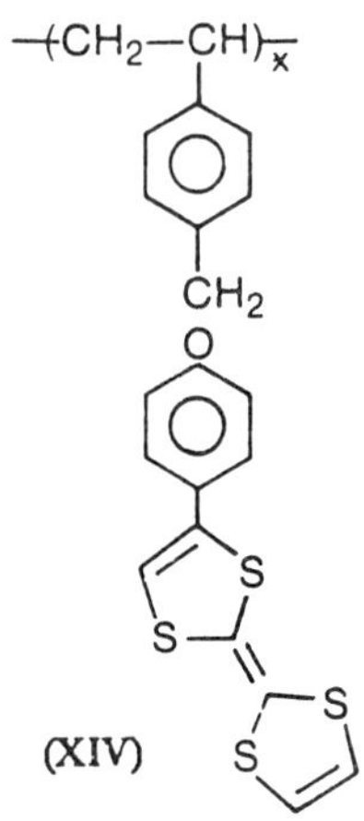

(XIV)

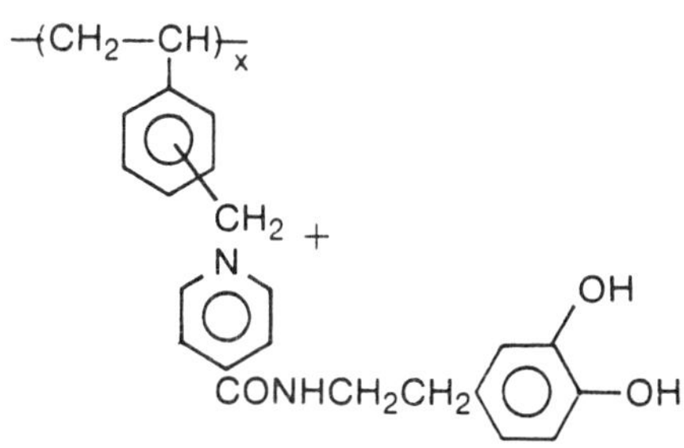

(XV)

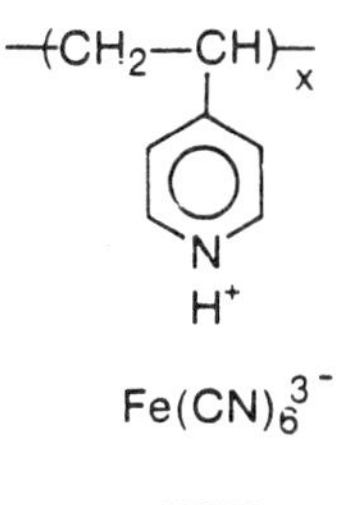

(XVI)

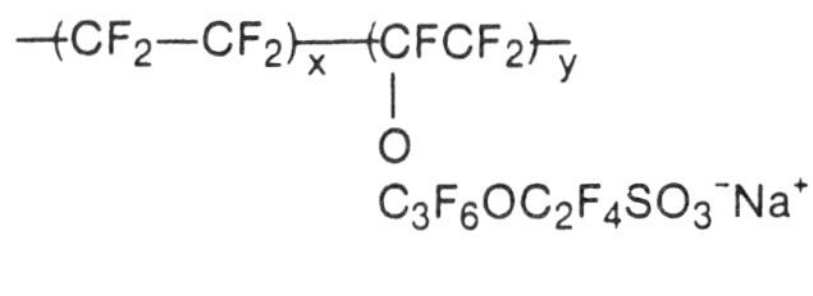

(XVII)

In general, there is good electrical interaction between the electrode and the electroactive centers in the polymer film whether present as a pendent group on the polymer or as a mobile counterion. The limitation in the use of these films is that they can be solublized; therefore the electrodes must be used in solvents in which the polymer has little or no solubility.

Finally, polymer film coatings have been prepared by plasma discharge polymerization of vinylferrocene, polynitrostyrene, and polyvinylpyridine [1,21].

V. ELECTROPOLYMERIZED POLYMER FILMS

The electropolymerization reactions of aromatic compounds, such as pyrrole, thiophene, azulene, and aniline, which produce polymer film coatings, provide a convenient route to modified electrodes. These reactions produce polymers that are electroactive and electrically conductive [51-55]. In early studies, the coatings produced from pyrrole and aniline were insoluble, powdery materials that are difficult to manage and characterize. These materials are known as "pyrrole black" and "aniline black," respectively [6,56]. More recently, procedures for the preparation of these materials as films have been reported. The film coatings produced by these reactions consist of the polymer in the oxidized state (~65% by weight) and the anion of the electrolyte (~35% by weight). The resulting linear polyaromatic polymer chains consist of linked aromatic units, which in the case of pyrrole and thiophene are primarily 2,5-coupled as shown in Eq. (8); in the case of azulene, the cooupling is via the 1- and 3-positions. With aniline, the polymer is primarily para-coupled, with both head-to-tail and head-to-head coupling as shown in Eq. (9), but ortho-coupling is also present [57]. It is structurally similar to the aniline octamers emeraldine and nigraniline. The electropolymerization reaction of aniline produces benzidine and 4-aminodiphenylamine as soluble by-products. In every case, the coupling occurs at the carbon atoms, which are known to be the most reactive toward addition and substitution reactions. There are two characteristics common to the aromatic compounds that successfully produce electroactive-conductive films on the electrode. They oxidize at relatively low anodic potentials (E_{pa} values below 2.1 V) and are susceptible to electrophilic substitution reactions. These characteristics are a fair guide in determining whether an aromatic compound will electropolymerize to produce electroactive films, even though the reaction proceeds via radical coupling.

These film coatings are electroactive and can be switched between the neutral, nonconducting state and the oxidized, conducting state, as shown in Eqs. (8) and (9):

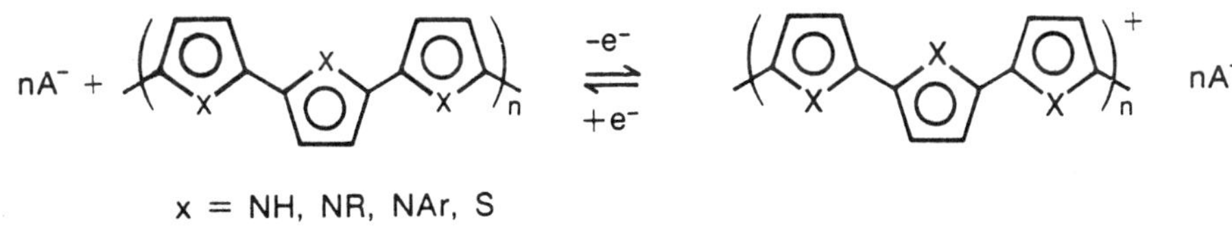

(8)

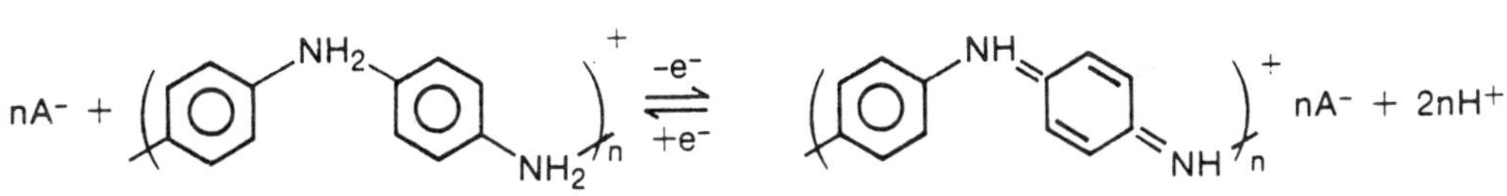

(9)

The materials are also electrochromic, and the redox reaction is accompanied by a color change from yellow to green-black to a pale copper, respectively. The reaction is often chemically reversible and can be driven repeatedly without loss of electroactivity, especially in well-contained cells in the absence of oxygen and impurities and with the proper choice of solvent and electrolyte salt.

A. Electrochemical Stoichiometry

The reactions that produce electroactive films have electrochemical stoichiometry is in the range 2.06-2.5 F mol^{-1} of monomer as can be seen in Table 1. This stoichiometry is formalized in Eq. (10) using the reaction of pyrrole and thiophene as an example. The equation combines the stoichiometry for the formation of the polymer, Eq. (11), and the charge associated with the oxidation of the polymer, Eq. (12).

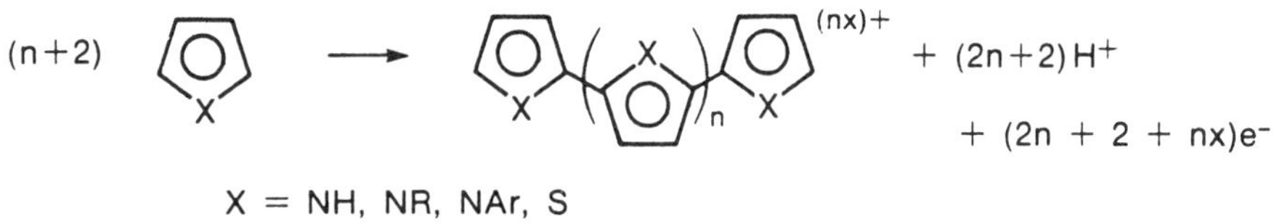

(10)

(11)

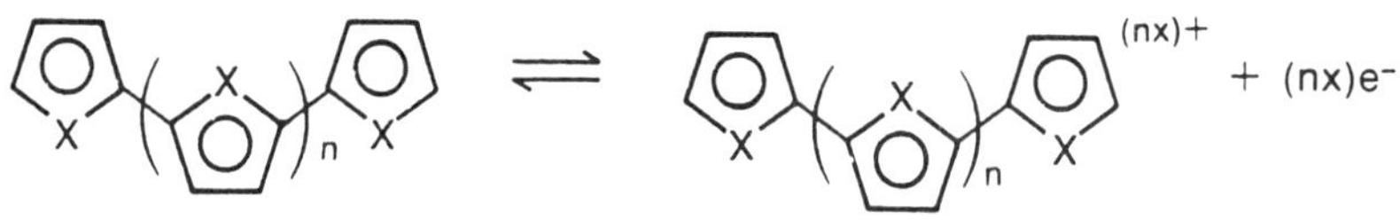

(12)

Since the molecular weights of the polymers are sufficiently high [58,59], the stoichiometry for Eq. (11) is close to 2. The electropolymerization reaction of aniline in aqueous sulfuric acid solutions using a constant potential was described as a bimolecular process involving a radical cation intermediate with an αn value equal to 1.2-1.3 and n = 2 [57-60]. The variation in the apparent stoichiometry results from oxidation of the polymer, in which each polymer has a different level of oxidation. The analog reactions can be written for azulene and aniline.

The polymerization reaction is very complicated [61] and is a sequence of oxidation, coupling, and deprotonation reactions. It is an extension of the ECE mechanism, and the $E(CE)_n$ notation was suggested [62]. The steps in the reaction have been described for pyrrole and azulene [62] and are shown in Fig. 4. The initial step involves the oxidation of the monomer to produce the radical cation (a), and the monomers must be continually oxidized during the reaction (b). At the anodic potentials employed, the concentration of radical cations of the monomer and oligomers at the anode

Table 1. Electrochemical Data for Several Monomer-Polymer Combinations

	Monomer		Polymer	
Structure	E_{pa} (V)[a]	n value	E_{pa} (V)[a]	Oxidation level
Aniline	1.4	2.2-2.3	0.2	0.25-0.33[b]
Pyrrole	1.2	2.2-2.4	−0.2	0.25-0.33
Bipyrrole	0.55	—	—	—
Terpyrrole	0.26	—	—	—
Thiophene	2.07	2.06	0.96	0.06-0.3
Bithiophene	1.31	2.22	1.06	0.22
Terthiophene	1.05	—	—	—
Thienothiophene[c]	—	—	1.0-1.5	0.06
Dithienothiophene[d]	—	—	0.5-0.8	0.44-0.5
Azulene	0.91	2.2	0.65-0.7	0.25
Benzene[e]	1.1	—	0.55	0.11
Pyrene	1.23	2.31	1.1	0.31
Carbazole	1.30	2.45	0.95	0.45
Indole	—	—	—	0.2-0.3
Furan	—	—	—	0.20

Source: Unless otherwise specified, values are summarized from Tables 1, 4, and 12 of Ref. 62.
[a]Values versus SCE.
[b]Ref. 124.
[c](TT), Ref. 81.
[d](DTT), Ref. 82.
[e]Reference 78.

surface is high and the concentration of the neutral species is low. These conditions favor the radical cation dimerization process. The combination of two radical cations produces the dication of the dihydrodimer, which subsequently deprotonates to produce the aromatic dimer (c). The dimer is next oxidized, coupled, and deprotonated to produce the next higher homolog (d and e). Oxidation of the polymer and incorporation of counter-anions results in the polymer-anion composite film (f). The final step in the reaction is also not clear; however, there is some reaction with moisture to produce oxygenated material (g).

This stoichiometry can be used to control the amount of film produced under conditions in which side reactions do not interfere. This control has been used with all the polymers, and the reactions can be easily followed by monitoring the redox peaks for the developing film. This is shown in Fig. 5 for the preparation of a polyaniline film on a platinum surface in aqueous sulfuric acid.

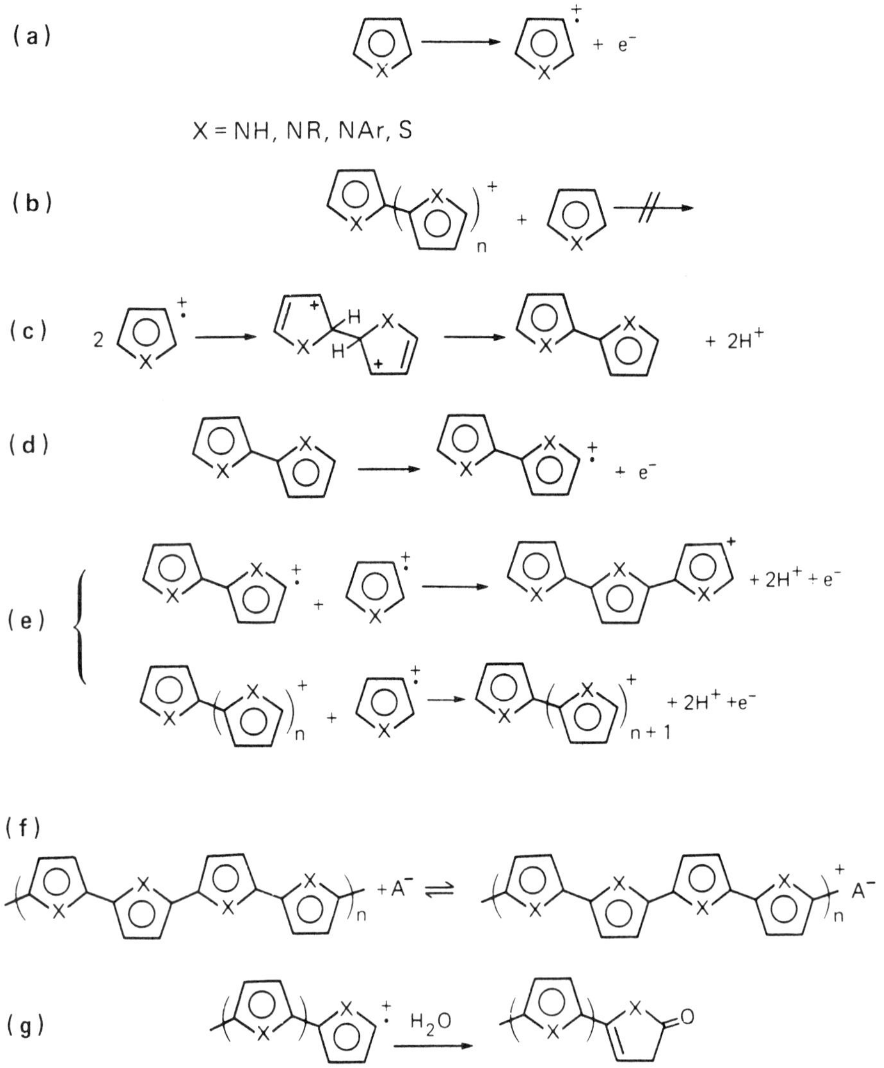

Figure 4. Reaction steps in the polymerization of heteroaromatics to produce electroactive or conductive film coatings on the electrode surface. (From Ref. 62.)

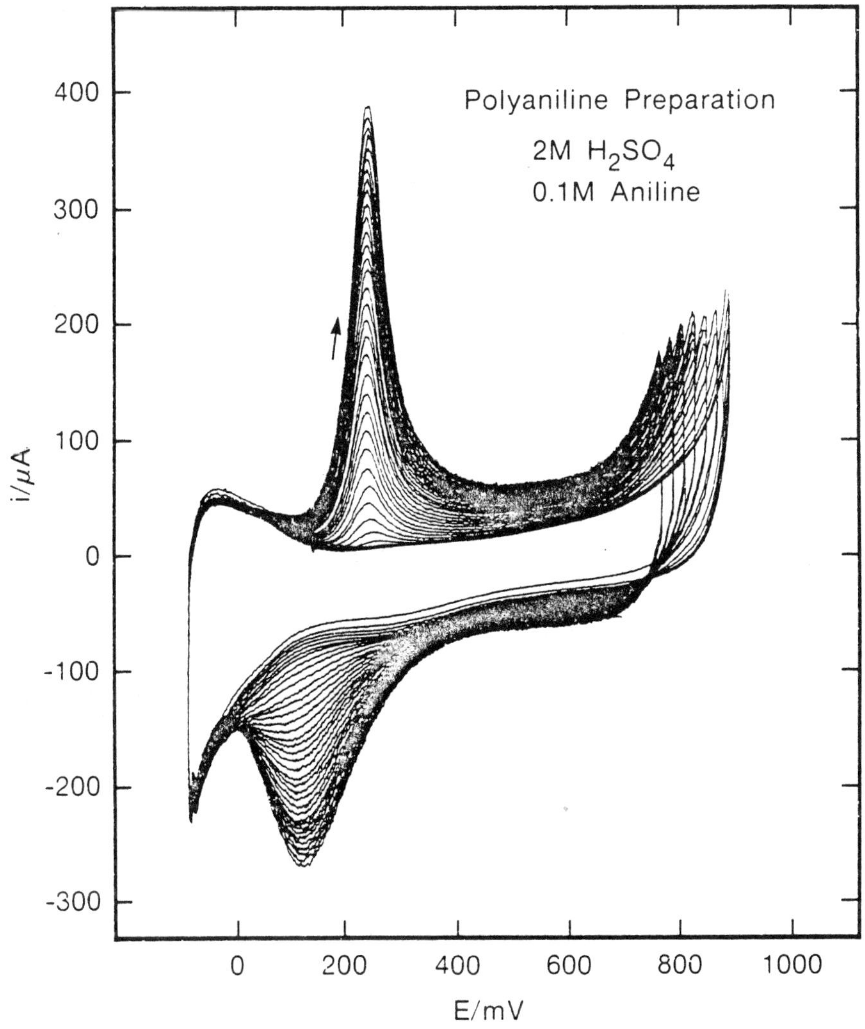

Figure 5. Cyclic voltammogram of the aniline polymerization reaction on a platinum electrode using a 2 M sulfuric acid aqueous solution containing 0.2 M aniline, and with continuous voltage cycling at 0.05 V s^{-1}. (From Ref. 136.)

B. Monomer Structure-Reactivity Considerations

Electroactive films can be prepared from a number of substituted derivatives of pyrrole, thiophene, and azulene. At present, only the derivatives of thiophene offer a commercially available selection. With the derivatives of pyrrole and thiophene, the 2- and 5-positions must remain unsubstituted so as not to block the coupling reaction. Pyrrole and thiophene derivatives with substituents in the 3- or 4-position polymerize to give good-quality films. Accordingly, both the 1- and 3-positions of azulene must remain unblocked to allow polymerization [55,63]. Azulene with methyl groups on the seven-membered ring also polymerizes, although the resulting films have lower conductivities. Three characteristics of the substituted monomers control the quality of the film: (1) the stability of the intermediate, (2) steric effects, and (3) site of coupling. These can be used as an approximate guide to predict the success of the reaction.

The stability of the incipient cation is probably the most important consideration for the formation of polymer film. Radical cations with intermediate stability seem to possess the chemical selectivity that favors the radical coupling reaction to produce polymer film. The more stable cations diffuse away from the electrode surface and produce soluble products. The less stable cations react indiscriminately with solvent and other nucleophiles in the region of the electrode surface. Thus, the yield of polymer is reduced and the polymer molecular weight is lowered. This effect is seen with the thiophene series in which the monomer produces a polymeric film of good quality and with high conductivity but the dimer and the trimer, which produce a more stable radical cation intermediate, produce powders with lower conductivities [64]. Bromothiophene and iodothiophene, which oxidize at the same potentials as thiophene, produce film (with lower conductivities). 3-Thiophenecarboxylic acid, 3-cyanothiophene, and 3-nitrothiophene, which oxidize at higher potentials, do not yield polymers [62].

The methyl substituent in the 3- and 4-positions influences the conductivity of the film [55,59,65]. The films from 3-methylpyrrole [55] and 3,4-dimethylpyrrole [59] have mechanical properties similar to those of the polypyrrole films but are less conducting (by a factor of approximately 10) and have a small shift in the E_{pa} values (< 100 mV). Methyl groups in the 3- or 4-position of thiophene increase the conductivity and decrease the oxidation potential of the polymer [55,66]. In a recent report by Tanaka et al. [67], polythiophene, poly-(3-methylthiophene), and poly-(3-ethylthiophene) films had conductivities of 190, 510, and 270 Ω^{-1} cm^{-1}, respectively. The variations in the conductivity of the substituted polymers is attributed to the balance of increasing amounts of α,α-coupling, and decreasing ring coplanarity due to steric effects [62,67-69]. With pyrrole, N-alkyl substituents influence both the coupling step in the polymerization reaction and the conformation of the polymer in which the loss of planarity of the aromatic rings reduces its active properties [54,70]. The yield and quality of thin films from the N-alkylpyrrole derivatives are progressively worse as the size of the group increases from hydrogen to methyl to unadecyl [70,71]. There is a progressive reduction in the oxidation level of the polymer (from 0.3 to 0.1 charges per aromatic unit) and the conductivity (from 30-100 to 10^{-5} Ω^{-1} cm^{-1}) [65]. The best films resulted with the hexyl substituent or with the 1,3-propylene chain containing two terminal pyrrole groups [70]. The linear alkyl substituents provide a means of introducing a hydroxy or tosyl group [70]. The t-butyl or cylclohexyl substituent produces only a brown-black powder. Finally, the poly-(N-arylpyrrole) films have been prepared in acetonitrile although with some difficulty because of the high voltages required to oxidize the monomer. These monomers have E_p values in the range 1.4-1.8 V versus SCE [72]. The poly-(N-phenylpyrrole) tetrafluoroborate films are less oxidized (~0.15 charges per pyrrole unit) and less conducting than polypyrrole. Phenylpyrrole derivatives with p-nitro, p-methyl, and p-methoxy substituents yield thin films in a fairly clean process with minimum coloration of the solution due to soluble side products [72].

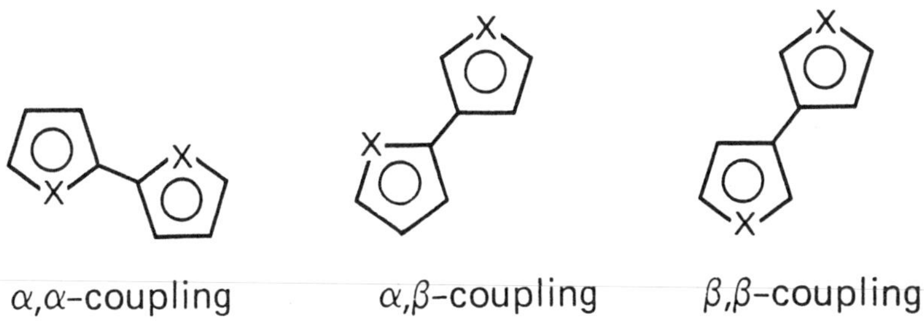

Additional electroactive centers can be introduced into the film by polymerizing monomers containing electroactive substituents. For example, the polymerization of N-p-nitrophenylpyrrole produces a film with two separate redox potentials that correspond to the oxidation of the polypyrrole backbone and reduction of the nitrophenyl groups [72]. This is shown in Fig. 6. Recently, Bidan et al. reported the preparation of polypyrrole containing a nitroxide [73] and a viologen group [74]. Several attempts to polymerize transition metal complexes containing pyrrole [75,76] or aniline [77]

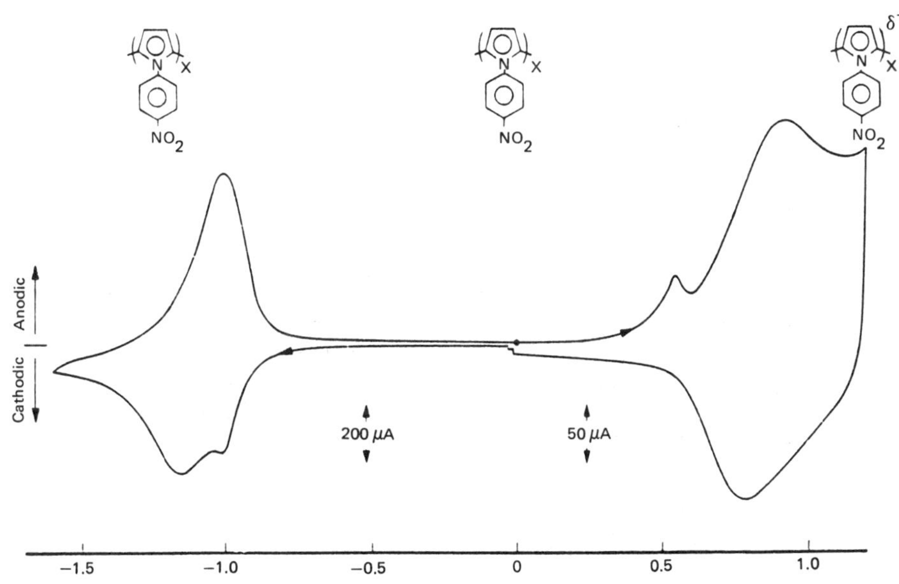

Figure 6. Cyclic voltammogram of poly-(N,p-nitrophenylpyrrole) on platinum in CH_3CN containing 0.1 M Et_4NBF_4. (From Ref. 72.)

ligands have produced thin films that did not have an electroactive polymer backbone and were poorly conducting. Although some polymerization occurs, the reaction is again limited by the large size of the substituents. As with polypyrrole, the polyphenylene can be further modified by nitration and sulfonation [78]. Attempts in our laboratory to produce electroactive polymers with substituted aniline derivatives have been unsuccessful, but such films can be prepared from aniline with o-hydroxy and amino substituents [79]. Other polyaromatic films have been reported, but the studies have not included substituent effects. These include the electroactive-electrochromic films from the electrooxidation of pyrene [62], indole [62], poly(thieno[3,2-b]thiophene) (PTT) [80], and poly(dithieno-[3,2-b,2',3'-d]-thiophene) (PDTT) [81] and the electroactive films from pyrene and triphenylene, which have poor mechanical properties and are fairly resistive [62,82]. Electroactive films have been generated from the electrooxidation of selenophene [83] and carbazole [55,84] derivatives.

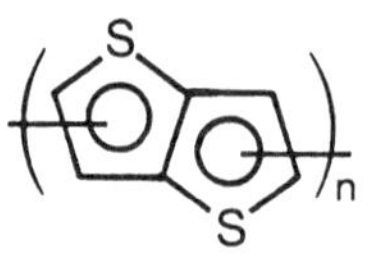

(PTT)

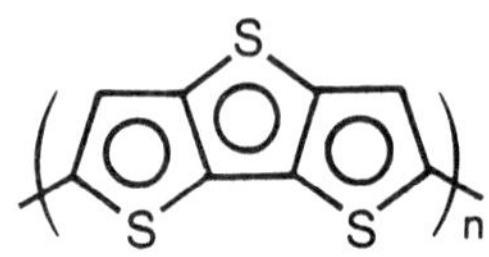

(PDTT)

C. Cell Conditions

The preparation of the electroactive-conductive films from pyrrole, thiophene, azulene, and aniline is quite straightforward. As described later in more detail, the films prepared from the pyrrole, thiophene, and azulene derivatives are best prepared in aprotic solvents and the preparation conditions are almost identical. The best films have been prepared in a divided cell using three electrodes in which the counterelectrode is separated from the working electrode. In the majority of these cases, the films are produced so readily that the only serious limitations are the nucleophilic nature of the electrolyte solution and the nature of the working electrode. The films can often be produced using a simple two-electrode, undivided cell powered by a d.c. battery [62]. However, films prepared in this manner are of poorer quality because of the complications that this cell configuration presents. The choice of the working electrode is a critical consideration because it must not oxidize under the anodic conditions employed to produce the films. Thus, most of the available films are prepared using a platinum or a gold electrode. However, films have also been prepared using silicon, gallium arsenide [85-87], and graphite electrodes [88]. As described later, the electrolyte and the applied potential have a strong influence on the electropolymerization reaction and the quality of the film.

Solvent

In selecting a solvent for the preparation of films from the derivatives of pyrrole, thiophene, and azulene, the sensitivity of the reaction toward nucleophiles must be recognized. Aprotic solvents are the best candidates for these preparations because the nucleophilicity of the electrolyte can be kept low. Among these, acetonitrile has been the most commonly used solvent. The best polypyrrole films were generated in acetonitrile, methylene chloride, butanone, and propylene carbonate [62]. In the propylene carbonate electrolyte, the films prepared at −20°C are more conducting (300 Ω^{-1} cm^{-1}) than the films prepared at 20°C (97 Ω^{-1} cm^{-1}), possibly due to a more regular polymeric structure [89]. The best polythiophene films are obtained in nitrobenzene [64,90] and propylene carbonate [67], and the more conducting films result from the preparations at −15°C [67]. Nucleophiles, such as pyridine, can completely eliminate the formation of films [62], and nucleophilic solvents like dimethylformamide, dimethylsulfoxide, and hexamethylphosphoramide do not permit film formation unless the solvent is buffered by the addition of a protic acid [62]. In aqueous media, high salt concentrations are necessary to produce polypyrrole films with good mechanical properties and conductivities [62]. Films prepared in ethanol and aqueous-aprotic solvent mixtures have intermediate conductivities and good mechanical properties when the solutions contain mineral acids that are nonnucleophilic and dissociated [62]. For example, films are not produced in hydrochloric acid and phosphoric acid electrolytes.

In acetonitrile, small amounts of water have a great influence on the rate of the reaction [61] and the mechanical and electrical properties of the polypyrrole film [62]. As the water content is increased from 0.01 to 0.1 M, the rate of the reaction is enhanced by a factor of 10-100 [61]. The conductivity of thicker films of polypyrrole decreased from 100 to 0.5 Ω^{-1} cm^{-1} [91] as the water content increased from 1 to 33% [91], then increased slowly up to 50 Ω^{-1} cm^{-1} with additional amounts of water. All the films are continuous and have good mechanical properties. With thiophene, added water inhibits film formation owing to the formation of a passivating layer on the electrode [61].

Continuous films of polyaniline are best prepared in aqueous sulfuric solution when the applied voltage is swept between −100 and 900 mV versus SCE (91,92). As the film forms, peaks for the redox reaction of polyaniline appear at 250 mV and the anodic voltage limit is gradually reduced to minimize the hydrolytic degradation of the film [85,93-98]. The polymerization reaction is actually faster on the polymer surface than on platinum, a characteristic of the reaction that permits the anodic limit to be reduced. The hydrolytic decomposition of the film at anodic voltages produces benzoquinones detected by the oxidation peak at 500 mV. The amount of switching charge developed per cycle is 0.1 mC cm^{-2}. A wide variety of acid electrolytes can be used in the preparation of polyaniline [99]. Good films are prepared in all the strong acid electrolytes. Films prepared in aqueous sulfuric, fluorosulfonic, trifluoromethylsulfonic, hydrochloric, and nitric acid produce cyclic voltammograms that are effectively superimposable with E_{pa} and E_{pc} values at 250 and 130 mV versus SCE, respectively [99]. The films prepared in aqueous hydrofluoric and trifluoroacetic acid contain decomposition product because the polymerization is slow on the polyaniline surface and the anodic limit cannot be reduced during the preparation. Electroactive polyaniline films can also be prepared and switched in aprotic electrolyte [99,100]. However, films with good switching behavior result only when both organic salt and protic acid are present in the electrolyte solution [99]. As seen in Table 2, the position and the width of the peaks are quite sensitive to the solvent-electrolyte combination of the solution, and the E_p values respond to both the polarity and the basicity of the solvents.

Electroactive films with improved electroactivity have been prepared in strong acid media using a platinum, gold, or glassy carbon working electrode

Table 2. Cyclic Voltammetry Data for Polyaniline in Several Electrolyte Solutions[a]

Solvent	Electrolyte	Anodic E_p[a]	Anodic $i_p/A\nu$[b]	Cathodic E_p[a]	Cathodic $i_p/A\nu$[b]
Water	2 M H_2SO_4	250	13.5	120	8.5
Ethanol	2 M H_2SO_4	380	13	250	12.5
	1 M CF_3COOH-0.1 M Bu_4NBF_4	170	4.5	80	6
2,2,2-Trifluoroethanol	2 M H_2SO_4	370	12	230	7.5
	1 M CF_3COOH-0.1 M Bu_4NBF_4	500	8	260	7
Acetic acid	2 M H_2SO_4	420	15	320	12.5
	1 M CF_3COOH-0.1 M Bu_4NBF_4	380	4.8	200	5.5
Acetonitrile	2 M H_2SO_4	450	13.5	360	12
	1 M CF_3COOH-0.1 M Bu_4NBF_4	540	5.5	260	6.5

Source: From Ref. 99.
[a]Values in mV.
[b]Values in mA-s cm^{-2} V.

and a polymer-coated Ag/AgCl or Pd/H reference electrode [78,101]. Polypyrrole films have been prepared in $AlCl_3/NH_4Cl$ molten salt mixtures at room temperature [102,103]; polyaniline has been prepared in eutectic mixtures of ammonium fluoride and hydrogen fluoride [100,104]. In the case of the polyphenylene polymer, electroactive films are produced from the electrooxidation of benzene only in strong acid media, such as HF [78,105] liquid SO_2, $HCl/AlCl_3$ [106], and HF/SbF_5 [101].

Effect of the Anion

The main considerations in the choice of the electrolyte salt are the solubility, degree of dissociation, and nucleophilicity. Since most of the preparations are performed in aprotic solvents, tetraalkylammonium salts and protic acids are commonly used. Lithium, sodium, and potassium salts are poor choices because they are highly aggregated and poorly soluble in aprotic solvents. The highly nucleophilic anions, such as halides, hydroxide, alkoxide, cyanide, acetate, and benzoate, interfere with the reaction and increase the production of soluble products that color the reaction solution [62]. The concentration of the electrolyte can also be important. Although the effect is poorly understood, more conductive films are generally produced with higher concentrations of electrolyte. The conductivity of both polypyrrole films prepared in aqueous sodium toluenesulfonate and polythiophene films prepared in polycarbonate solutions containing Et_4NPF_6 show an initial increase in the conductivity with increasing electrolyte concentration [67,107].

The composite nature of these films, ~70% polymer and 30% anion, permits the incorporation of electroactive and catalytic species as anions. For example, polypyrrole films containing ferricyanide anion were prepared by polymerizing pyrrole in an acetonitrile solution containing the tetraalkylammonium salt of the anion. In the same manner, polypyrrole films containing iron [88] and cobalt [108] phthalocyanine-tetrasulfonate anion were prepared in aqueous solution. The electrochemical or catalytic response of the electroactive anion is best when it is incorporated into the film during the preparation. Attempts to incorporate the anion by ion exchange with the preformed polymeric film does not work well. Finally, the films do not have long-term stability because the electroactive anion eventually exists from the film via ion exchange during the voltage cycling process and does not reincorporate very easily.

Effect of the Applied Voltage

The properties of the electroactive-conductive films can be changed by varying the electrochemical parameters, even while maintaining the same electrolyte conditions. The conductivity of polypyrrole films, prepared potentiostatically in aqueous solutions, varies with the potential applied during the polymerization, and there is an optimum potential for producing highly conductive films [107]. Polypyrrole films prepared in aqueous sodium toluenesulfonate solutions had conductivies of 150, 500, and 100 Ω^{-1} cm^{-1} when the applied potential was 0.6, 0.75, and 1.0 V versus SCE, respectively [107]. The mechanical properties of the films begin to degrade at the higher voltages. Similar results are observed in the preparation of poly(methylthiophene) when varying the current densities. Polythiophene films prepared in 0.03 M Et_4NPF_6 in acetonitrile had conductivities of 450, 500, and 450 Ω^{-1} cm^{-1} when the current densities were 5, 10, and 15 mA cm^{-2}, respectively [67]. As described earlier, polyaniline films that are continuous and have good electroactive properties can be prepared by continuously cycling the voltage up to 0.8-0.9 V versus SCE [92,94], or potentiostatically [93]. These films, however, show the additional peak at 500 mV versus SCE due to the hydrolytic product unless the anodic voltage limit and the effective current density are controlled [99].

Electroactive Characteristics

The polyaromatic films listed in Table 1 are electroactive and have good electrical contact and adhesion to the support electrode. The films are strongly adherent and insoluble in almost every solvent; therefore, the electrode can be removed from the preparation solution and mounted in another cell without concern about detachment of the film. The films can be switched between the nonconducting and the conducting state with chemical reversibility, and the electroactive response is optimum with films having thicknesses less than ~1 μm. The anion does not influence the redox potential but does influence the kinetics of the redox reaction. The reaction is limited by the resistance of the film when in the neutral form and by the mobility of the anion in the film [109]. Very complicated cyclic voltammograms result that represent the combined faradaic and capacitive currents [62]. For example, as seen in Fig. 7, the switching reaction of polypyrrole in acetonitrile shows gross differences with the different anions in the electrolyte.

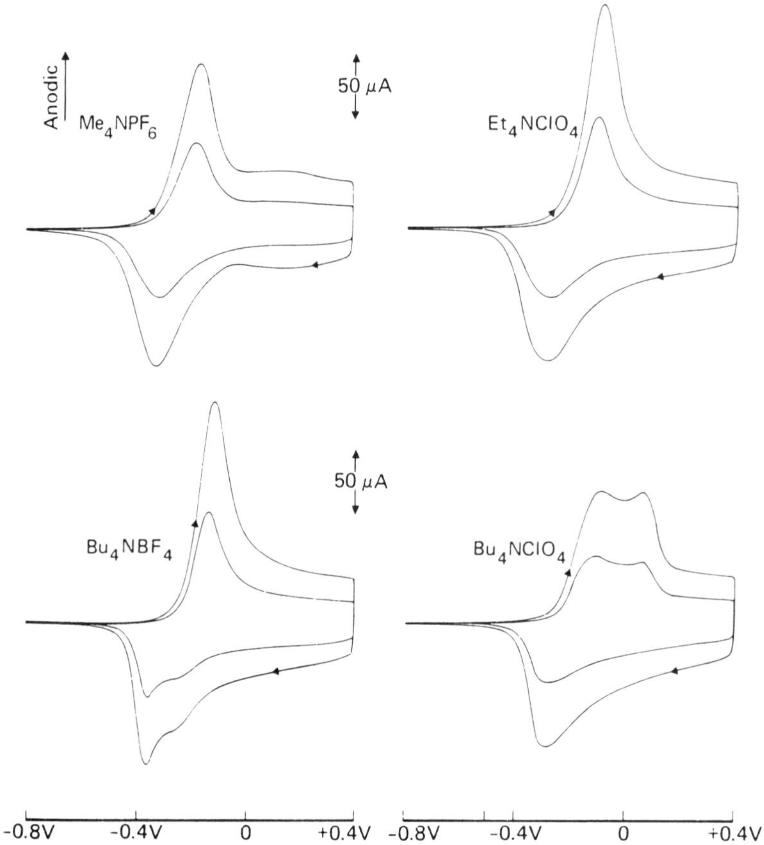

Figure 7. Cyclic voltammograms of polypyrrole (20 nm thick) on a platinum electrode in CH_3CN containing various electrolyte salts. Scan rates: 0.05 and 0.1 V s^{-1}. (From Ref. 137.)

The redox potential of the films is influenced by substituents on the aromatic structure, and a linear correlation exists between the E_{pa} values for the substituted monomers and the corresponding polymers [62]. This is probably the most important factor when considering the use of these films for real applications, since the redox potential establishes the stability of the film in air. Polypyrrole, which oxidizes at −0.2 V versus SCE, is air sensitive and irreversibly oxidizes in air to lose all electroactivity. The stability of the film to air is improved with substituents on the nitrogen because the oxidation potential is shifted to 0.45-0.6 V with N-alkyl groups [53,54] and to 0.7-0.9 V with N-aryl groups [54]. This added stability greatly simplifies the handling and storage procedures for these films. Polythiophene and its derivatives have oxidation potentials at 0.72-1.33 V [62], and are fairly stable in air. Polymers from the other aromatic compounds oxidize in the range 0.7-1.1 V (see Table 1) and are also stable in air.

The cyclic voltammogram for polyaniline in 2 M sulfuric acid containing no aniline shows redox peaks at 120-250 and 800-900 mV [92,97-99].

Oxidation-reduction of polyaniline introconverts the benzenoid- and quinoid-like structures, as shown in Eq. (9) [60,98,100]. Unlike the reactions of polypyrrole and polythiophene [62], the polyaniline reaction has very little sensitivity to the anion [99]. The chronocoulometric response of polyaniline is quite fast and is strongly dependent on the magnitude of the voltage step and the acid concentration. The switching times are in tens of milliseconds and are significantly less than for polypyrrole (hundreds of milliseconds) [54].

Freestanding Polymer Electrodes

The thick amorphous films prepared from the electrochemical oxidation of pyrrole, thiophene, and azulene in acetonitrile electrolyte can be peeled from the electrode and handled as freestanding films [54,106]. Although these thick films are poorly electroactive, they can be used as freestanding electrodes. These films are produced in the conducting form and are composites containing the polymer in the oxidized state (0.1-0.3 charges per ring) plus the anion at 10-35% by weight. The level of oxidation of the polymer controls the amount of anion in the film. As seen in Table 3, various anions can be used to adjust the conducting properties of the film [65]. Tetraalkylammonium salts were used in the preparation of these films. With polypyrrole, the conductivity of the films can be varied from 0.01 to 200 Ω^{-1} cm^{-1} by the anion; the best conductivity is observed with the perfluoride, perchlorate, and toluenesulfonate anions. The conductivities are temperature dependent and increase with temperature. With the polythiophene films, the range in conductivity is from 0.0001 to 10-20 Ω^{-1} cm^{-1} [62].

The polypyrrole films are generally flexible and reasonably durable [54]. The anion also affects the tensile strength (1200-8600 psi) and Young's modulus (70,000-350,000 psi) of the films. Films containing toluenesulfonate, perchlorate, and tetrafluoroborate anions are hard and strong and stretch very little (4-8%). Polypyrrole films prepared in propylene carbonate at $-20^{\circ}C$ are more elastic (67% elongation) and more conducting, especially when stretched (100 Ω^{-1} cm^{-1}) [89]. Regarding storage under ambient conditions, polypyrrole films are stable for 12-18 months, but polythiophene films are less stable and slowly lose conductivity over several days [99].

Polymer Blends

The most recent direction in the study of modified electrodes involves electropolymerization of pyrrole, aniline, and thiophene on an electrode coated with a film of an inert, nonelectroactive "host" polymer. The interest in these polymer blends stems from the desire to modify the mechanical properties and the permeability of the films. Two different methods are used to prepare these films. There is a two-step method in which the electrode is first coated with an inert host polymer by the usual techniques (spin coating, dip coating, and others) and the electrode is subsequently used to grow electrochemical polymers [48,99,110-116]. The polymerization reaction occurs at the electrode surface, since the inert films are sufficiently porous and permit the penetration of electrolyte. This approach to new materials was first demonstrated with the incorporation of a conducting charge transfer salt, TTF, inside the polymer Nafion [117-120] and is proving to be quite general for the preparation of polymer blends. The other method is

Table 3. Influence of Anion on Film Conductivity

Anion (Et_4X)	Oxidation level	Conductivity (ohm^{-1} cm^{-1})
	Polyaniline	
Hydrogen sulfate[a]	0.25-0.33	0.1, 5
	Polypyrrole	
Tetrafluoroborate Hexafluoroarsenate Hexafluorophosphate	0.25-0.32	30-100
Perchlorate	0.30	60-200
Hydrogen sulfate	0.30	0.3
Fluorosulfonate	–	0.01
Trifluoromethylsulfonate	0.31	0.3-1
p-Bromobenzenesulfonate	0.33	50
p-Toluenesulfonate	0.32	20-100
Trifluoroacetate	0.25	12
	Polythiophene	
Tetrafluoroborate	0.05-0.3	0.02-20
Hexafluorophosphate	0.06	0.02
Perchlorate	0.13-0.3	10-20
Hydrogen sulfate[a]	0.1	0.0001
TCNQ anion	0.02	0.0001
Trifluoromethylsulfonate	0.3	10-20
	Polybithiophene	
Hydrogen sulfate[a]	0.22	0.1
	Poly-(3-methylthiophene)	
Hexafluorophosphate	0.12	1
Perchlorate	0.25	10-30
Trifluoromethylsulfonate	0.3	30-100
	Poly-(3,4-dimethylthiophene)	
Trifluoromethylsulfonate	0.3	10-50
	Polyazulene	
Tetrafluoroborate Hexafluorophosphate Perchlorate Toluenesulfonate	0.25	0.02-1
	Polypyrene	
Perchlorate	0.31	0.1-1[b]
	Polycarbazole	
Perchlorate	0.45	0.001[b]

Source: Values summarized from Tables 5, 9, 10, and 12 of Ref. 62.
[a] H_2SO_4.
[b] Pressed pellets.

one step and uses an electrolyte solution that contains both the aromatic heterocyclic compound and the inert host polymer. During the electropolymerization process, there is an incorporation of the inert polymer within the electroactive polymer.

Electroactive polymer blends prepared using the two-step method include combinations of polypyrrole with poly(vinyl chloride) [110-113], poly(vinyl alcohol) [114], and Nafion [115,116,121] and of polyaniline with poly(methyl methacrylate) [99], poly(butyl methacrylate) [99], and polycarbonate [99]. In practice, host polymer films that are 1-2 μm thick have been used, and the preparations are performed in solutions in which the inert polymer is not solubilized. In these reactions, the monomer plus electrolyte diffuses across the film and reacts at the metal electrode surface. The polymerization rate is depressed when the host polymer is neutral and depends on the film thickness [99,110-113,121] but does not appear to be affected by the presence of the ionic polymers [115,116,121]. The electroactive-conductive polymer grows from the electrode across the host polymer film [99,110] and often produces domains of pure electroactive-conductive polymer coexisting with domains of host polymer [99,110]. The cyclic voltammogram for these blends is similar to the response for films on clean electrodes, with small differences appearing in the position and the shape of the redox peaks [99,115]. However, the rate of the switching reaction is generally slower. For example, the Pt-polyaniline-poly(methyl methacrylate) blends switch six times slower than polyaniline on bare platinum under the same conditions [99]. An important characteristic of these materials is that the conductive polymer can be grown to any desired thickness within the blend [99,110,112, 121]. Thus, films can be prepared with different conducting geometries, such as films with an evenly distributed bulk conductivity, with only one conducting surface, or with both surfaces conducting but no bulk conductivity. This can be accomplished because the thickness of the electroactive film can be adjusted relative to the thickness of the inert polymer film. With the polypyrrole poly(vinyl chloride) blends [110-113], DePaoli et al. [111] have shown that the mechanical properties of these are improved and resemble those of pure poly(vinyl chloride). The properties were further improved by the addition of poly(chloroprene) as a plasticizer.

Film coatings consisting of polymer blends can also be produced in one step by electropolymerizing pyrrole or thiophene in an electrolyte solution containing an inert host polymer. Film coatings have been prepared with such combinations as polyaniline with poly(vinyl sulfonic acid) [99], polypyrrole with poly(methyl methacrylic acid) [120], various sulfonic acid polymers [120] and latex [122], and poly(methylthiophene) with poly(methyl methacrylate) [123]. A possible advantage provided by these blends is that they are soluble. Thus, they can be dissolved and cast on any desired surface. The cyclic voltammetry and chronocoulometry characteristics of the blends with polypyrrole and polyaniline can be very similar to those for the unblended electroactive polymer when the cell resistance is properly compensated for IR loss. The preparation of a poly-(3-methylthiophene)-poly-(methyl methacrylate) film in a nitrobenzene-dichloromethane solution is interesting because the neutral polymer is incorporated into the film during its formation even though it is nonpolar and does not have anionic centers. The cyclic voltammogram of the thin film measured in aqueous lithium perchlorate solution closely resembles that for poly(methylthiophene). The conductivity of the thicker films was reported to be different for the two faces of the film. The film side originally against the electrode is 10 times more conducting than the side that faced the electrolyte. The mechanical properties of this material remain to be determined.

VI. POLYMER-CERAMIC COATINGS

The studies reviewed in the preceding sections are based on the use of organic structures to modify electrode surfaces. The materials used are often selected by the availability of the redox centers on molecules or polymers with a well-defined bonding structure. Surface modifications have also been accomplished using thick layers of clay and zeolite materials mixed with electroactive molecules. Although these electrode-modifying layers have a less well defined chemical structure, they produce a similar electrochemical response and have improved stability. For example, porous aluminum oxide films on electrodes have been coated with poly(vinyl pyridine) and used to immobilize electroactive reagents [124]. In other studies, poly(vinyl alcohol)-clay [125,126] and polymer-zeolite [127] coatings have been applied to electrode surfaces and characterized electrochemically. Similarly, the electrochemical behavior of molecular redox structures, such as viologen and metal chelates, have been studied in zeolites [128] and clays [129-132].

VII. AD-ATOMS

For completeness, we mention briefly the studies with electrode surfaces modified with ad-atoms, which have been used for electrocatalysis. Much of this work has been generated from the laboratories of Shibata and Motoo [133-135], where electrooxidation reactions have been studied on platinum electrons with ad-atoms, such as Bi, Fe, Se, S, Tl, Ag, Hg, Pb, Ge, Sn, Sb, As, and Ru. These ad-atoms adsorb oxygen, and their presence in submonolayer concentrations catalyzes the electrooxidation reactions of acetaldehyde, formic acid, dioxolane, carbon monoxide, and so on.

VIII. CONCLUDING REMARKS

Many studies have been reported in which electrodes are modified by molecular monolayers and by relatively thick polymer films. Many of the electrode modifications have been successfully used to address mechanistic issues and, less successfully, to impact on a technology. The limitation with the latter usually results from the poor stability of the electrode assembly. Instability due to the loss of adhesion and desorption from the electrode is as much a limitation as the chemical or thermal stability of the electroactive structure. The loss of monolayers covalently attached to the electrode occurs because of the hydrolytic cleavage of the Si—O—M bond. The preparation of electroactive monolayers on metal oxide surfaces using organosilanes normally involves two steps and has wide general applicability. The process can be used successfully with a wide variety of materials and is effectively independent of the nature of the metal oxide surface and the structure of the electroactive center. Molecular monolayers have been particularly useful for studying the mechanism of the electron exchange processes. This is especially true with the adsorbed species in which there is no diffusion and little or no thermal motion. The adsorbed polymers are considerably more durable but can only be used with solvents when they are poorly soluble to avoid dissolution from the electrode.

The generality of the procedure for preparing linear polyaromatic polymers as films that are electroactive and conductive is fairly well developed. These films are generated in situ, and the thickness can be controlled electrochemically. At present, most studies address the derivatives of pyrrole, thiophene, azulene, and aniline. Films with the best mechanical and active properties are produced in solutions containing nonnucleophilic solvent and electrolyte anions. A wide variety of organic aprotic solvents, electrolyte salts, and acidified aqueous solutions are available for selection. Because the anion of the electrolyte is incorporated in the film, it can be used to modify the film properties. The solvent and temperature also have a substantial influence on the conductivity and quality of the films. Many factors in the preparation influence the properties of the films; therefore, care must be exercised when comparing results from different laboratories. There are some clear advantages in using conducting polymer films to modify electrodes. These films are usually quite stable in both the oxidized and neutral state and are well adhered to the support electrode. For additional information, the reader is referred to the recent review articles in Refs. 138 and 139.

Besides using these modified electrodes for mechanistic studies and electrochemical analysis, the preliminary results from several laboratories demonstrated the utility of these electrodes in preparative electrochemistry. The reader is referred to Chap. 27 for illustrative examples of the use of chemically modified electrodes in organic synthesis. The results from these studies are very encouraging, for example, the generation of electrochemical products with a stereospecificity, but the performance of the electrodes needs development. The amount of the reaction specificity and the stability of the electrode assembly must be improved. With regard to the influence of the modifying layer on the specificity of the reaction, the minimal thickness of the monolayer assembly may inherently limit its effectiveness. This is certainly the case for those reactions in which the reaction step that controls stereospecificity in the product occurs in the region of the electrode removed from the electrode interface. In addition, the hydrolytic instability of the covalently attached monolayers is always a concern. In contrast to the monolayer assemblies, electrodes modified with a thick layer of polymer could be more easily developed for successful use in electrosynthesis. The electrodes with adsorbed polymers are considerably more durable, and the thickness of the layer can be designed to include the complete reaction volume. Electrodes with polymer coatings are currently limited to use in solvents in which the polymers are poorly soluble and do not dissolve. The solubility limitation is not a serious concern and should be easily resolved. For example, the polymer could be cross-linked after application to the electrode. Finally, the use of conducting polymers in electrosynthesis seems particularly attractive. This application will gain interest with the development of stable conductive polymers with a highly open structure and a high surface area. In principle, a completely open polymer structure in which the individual polymer chains are exposed would be a very attractive electrode. The conductive polymers are fairly stable and are subject to chemical modification with pendent substituents.

The polymer blends consisting of an electroactive-conductive polymer plus an inert host or anionic polymer appear to be the correct approach for improving the mechanical properties of these films. The work in this area is preliminary, and a full characterization of the electroactive properties and processibility of these materials is needed. In considering future work, several areas of improvement are evident: (1) developing electroactive-conductive films that incorporate other active centers that can be made to interact electronically with the conductive polymer backbone; (2) improving the scope of the preparative procedure by understanding the relationship between the monomer structure and the characteristics of the resulting film to predict the desired film properties; and (3) improving the physical properties to permit processing of these materials.

REFERENCES

1. R. W. Murray, Acc. Chem. Res., 13, 135 (1980).
2. R. W. Murray, A. G. Ewing, and R. A. Durst, Anal. Chem., 59, 379 (1987).
3. R. W. Murray, Ann. Rev. Mater. Sci., 14, 145 (1984).
4. C. E. D. Chedsey and R. W. Murray, Science, 231, 25 (1986).
5. M. Fijihara, Topics in Organic Electrochemistry (A. J. Fry and W. E. Britton, eds.), Plenum Press, New York, 1986, pp. 255–294.
6. A. Diaz, J. F. Rubinstein, and H. B. Mark, Jr., Adv. Polym. Sci., 84, 113 (1988).
7. J. R. Lenhard and R. W. Murray, J. Electroanal. Chem., 78, 195 (1977).
8. P. R. Moses and R. W. Murray, J. Electroanal. Chem., 77, 393 (1977).
9. P. R. Moses and R. W. Murray, J. Amer. Chem. Soc., 98, 7435 (1976).
10. A. F. Diaz and K. K. Kanazawa, IBM J. Res. Dev., 23, 239 (1979).
11. A. F. Diaz, Proceedings from Symposium on Silylated Surfaces (D. E. Leyden and W. T. Collins, eds.), Gordon and Breach, New York, 1987, p. 137.
12. P. R. Moses, L. Wier, and R. W. Murray, Anal. Chem., 47, 1882 (1975).
13. B. F. Watkins, J. R. Behling, E. Kariv, and L. L. Miller, J. Amer. Chem. Soc., 97, 3549 (1975).
14. M. Elliott and R. W. Murray, Anal. Chem., 48, 1247 (1976).
15. J. C. Lennox and R. W. Murray, J. Electroanal. Chem., 78, 395 (1977).
16. D. C.-S. Tse and T. Kuwana, Anal. Chem., 50, 1315 (1978).
17. D. C.-S. Tse, T. Kuwana, and G. P. Royer, J. Electroanal. Chem., 98, 345 (1979).
18. N. Oyama, A. P. Brown, and F. C. Anson, J. Electroanal. Chem., 87, 435 (1978).
19. N. Oyama and F. C. Anson, J. Electronanal. Chem., 88, 289 (1978).
20. S. Mazur, T. Matusinovic, and K. Cammann, J. Amer. Chem. Soc., 99, 3888 (1977).
21. R. Nowak, F. A. Schultz, M. Umana, H. Abruna, and R. W. Murray, J. Electroanal. Chem., 94, 219 (1978).
22. M. S. Wrighton, R. G. Austin, A. B. Bocarsly, J. M. Bolts, O. Hass, K. D. Legg, L. Nadjo, and M. C. Palazzotto, J. Amer. Chem. Soc., 100, 1602 (1978).

23. A. Diaz, J. Amer. Chem. Soc., 99, 5838 (1977).
24. J. M. Bolts and M. S. Wrighton, J. Amer. Chem. Soc., 100, 5257 (1978).
25. M. S. Wrighton, R. G. Austin, A. B. Bocarsly, J. M. Bolts, O. Hass, K. D. Legg, L. Nadjo, and M. C. Palazzotto, J. Electroanal. Chem., 87, 429 (1978).
26. J. R. Lenhard, R. Rocklin, H. Abruna, K. Willman, K. Kuo, R. Kowak, and R. W. Murray, J. Amer. Chem. Soc., 100, 5213 (1978).
27. M. Fujihara, T. Kubota, and T. Osa, J. Electroanal. Chem., 119, 379 (1981).
28. K. N. Kuo and R. W. Murray, J. Electroanal. Chem., 131, 37 (1982).
29. R. J. Burt, G. J. Leigh, and C. J. Pickett, J. Chem. Soc., Chem. Commun., 940 (1976).
30. K. Itaya and A. J. Bard, Anal. Chem., 50, 1487 (1978).
31. J. R. Lenhard and R. W. Murray, J. Amer. Chem. Soc., 100, 7870 (1978).
32. J. A. Bruce and M. S. Wrighton, J. Amer. Chem. Soc., 104, 74 (1982).
33. A. F. Diaz, U. Hetzler, and E. Kay, J. Amer. Chem. Soc., 991, 6780 (1977).
34. A. P. Brown and F. C. Anson, Anal. Chem., 49, 1589 (1977).
35. R. F. Lane and A. T. Hubbard, J. Phys. Chem., 77, 1401, 1411 (1973).
36. A. T. Hubbard, J. L. Stickney, M. P. Soriaga, V. K. F. Chia, S. D. Rosasco, B. C. Schardt, T. Solomun, D. Song, J. H. White, and A. Wieckowski, J. Electroanal. Chem., 168, 43 (1984).
37. A. P. Brown, C. Koval, and F. C. Anson, J. Electroanal. Chem., 72, 379 (1976).
38. A. P. Brown and F. C. Anson, J. Electroanal. Chem., 83, 203 (1977).
39. J. P. Collman, P. Denisevich, Y. Konai, M. Morrocco, C. Koval, and F. C. Anson, J. Amer. Chem. Soc., 102, 6027 (1980).
40. M. R. Van DeMark and L. L. Miller, J. Amer. Chem. Soc., 100, 437, 3223 (1978).
41. F. B. Kaufman and E. M. Engler, J. Amer. Chem. Soc., 101, 547 (1979).
42. F. B. Kaufman, A. H. Schroder, E. M. Engler, S. R. Kramer, and J. Q. Chambers, J. Amer. Chem. Soc., 102, 483 (1980).
43. N. Oyama and F. C. Anson, J. Amer. Chem. Soc., 101, 739 (1979).
44. A. Merz and A. J. Bard, J. Amer. Chem. Soc., 100, 3222 (1978).
45. P. G. Pickup and R. W. Murray, J. Amer. Chem. Soc., 105, 4510 (1983).
46. S. Chao, J. L. Robbins, and M. S. Wrighton, J. Amer. Chem. Soc., 105, 181 (1983).
47. A. N. K. Lau and L. L. Miller, J. Amer. Chem. Soc., 105, 5271 (1983).
48. R. M. Penner and C. R. Martin, J. Electrochem. Soc., 132, 514 (1985).
49. F.-R. F. Fan and A. J. Bard, J. Electrochem. Soc., 133, 301 (1986).
50. N. Oyama and F. C. Anson, J. Electrochem. Soc., 127, 247 (1980).
51. A. Dall'Olio, Y. Dascola, V. Varacca, and V. Bocchi, Comp. Rend., C267, 433 (1968).

52. A. F. Diaz, K. K. Kanazawa, and G. P. Gardini, J. Chem. Soc., Chem. Commun., 635 (1979).
53. A. F. Diaz, J. I. Castillo, J. A. Logan, and W. Y. Lee, J. Electroanal. Chem., 129, 115 (1981).
54. A. F. Diaz and K. K. Kanazawa, in Extended Linear Chain Compounds (J. S. Miller, ed.), Plenum, New York, 1982, p. 417.
55. J. Bargon, S. Mohmand, and R. J. Waltman, IBM J. Res. Dev., 27, 330 (1983).
56. G. P. Gardini, Adv. Heterocyclic Chem., 15, 95 (1973) and references therein.
57. J. Bacon and R. N. Adams, J. Amer. Chem. Soc., 90, 6596 (1968).
58. A. F. Diaz, J. I. Crowley, J. Bargon, G. P. Gardini, and J. B. Torrance, J. Electroanal. Chem., 121, 355 (1981).
59. A. Nazzal and G. B. Street, J. Chem. Soc., Chem. Commun., 83 (1984).
60. D. M. Mohilner, R. N. Adams, and W. J. Argersinger, Jr., J. Amer. Chem. Soc., 84, 3618 (1962).
61. A. J. Downard and D. Pletcher, J. Electroanal. Chem., 206, 139, 147 (1986).
62. A. Diaz and J. Bargon, in Handbook of Conducting Polymers (T. A. Skotheim, ed.), Vol. 1, Marcel Dekker, Inc., 1986, p. 82.
63. J. Bargon, S. Mohmand, and R. J. Waltman, Mol. Cryst. Liq. Cryst., 93, 279 (1983).
64. J. Roncali, F. Garnier, M. Lemaire, and R. Garreau, Synthesis Metals, 15, 323 (1986).
65. M. Salmon, A. F. Diaz, A. J. Logan, M. Krounbi, and J. Bargon, Mol. Cryst. Liq. Cryst., 83, 1297 (1983).
66. F. Garnier, G. Tourillon, M. Gazard, and J. C. Dubois, J. Electroanal. Chem., 148, 299 (1983).
67. M. Aki Sato, S. Tanaka, and K. Kaeriyama, Synthesis Metals, 14, 279 (1986).
68. K. Sanechika, T. Yamamoto, and A. Yamamoto, J. Polym. Sci., Polym. Lett. Ed., 20, 365 (1982).
69. K. Tamaka, T. Shichiri and T. Yamabe, Synthesis Metals, 14, 271 (1986).
70. G. Bidan and M. Guglielmi, Synthesis Metals, 15, 49 (1986).
71. A. Diaz, J. Castillo, K. K. Kanazawa, J. A. Logan, M. Salmon, and O. Fajardo, J. Electroanal. Chem., 133, 237 (1981).
72. M. Salmon, A. Diaz, and J. Goitia, Chemically Modified Surfaces, in Catalysis and Electrocatalysis (J. S. Miller, ed.), ACS Symposium Series, 192, 65 (1982).
73. G. Bidan and D. Limosin, Ann. Phys. Coll., 1, Suppl. 1, Vol. 11, 5 (1986).
74. G. Bidan, A. Deronzier, and J. C. Moutet, J. Chem. Soc., Chem. Commun., 1185 (1984).
75. J. G. Eaves, H. S. Munro, and D. Parker, J. Chem. Soc., Chem. Commun., 684 (1985).
76. S. Cosnier, A. Deronzier, and J. C. Moutet, J. Electroanal. Chem., 193, 193 (1985).
77. C. D. Ellis, L. D. Margerum, R. W. Murray, and T. J. Meyer, Inorg. Chem., 22, 1283 (1983).
78. I. Rubinstein, J. Electrochem. Soc., 130, 1506 (1983).

79. N. Oyama, private communication, 1988.
80. R. Danieli, C. Taliani, R. Zamboni, G. Giro, M. Bisierni, M. Mastragostino, and A. Testoni, Synthesis Metals, 13, 325 (1986).
81. T. R. Jow, K. Y. Jen, R. L. Elsenbaumer, L. W. Shacklette, M. Angelopoulos, and M. P. Cava, Synthesis Metals, 14, 53 (1986).
82. M. E. Peover and B. S. White, J. Electroanal. Chem., 13, 93–99 (1967).
83. R. H. Shacklette, Chem. Rev., 82, 209 (1982).
84. J. F. Ambrose, L. L. Carpenter, and R. F. Nelson, J. Electrochem. Soc., 122, 876 (1975) and references therein.
85. R. Noufi, A. J. Frank, and A. J. Nozik, J. Amer. Chem. Soc., 103, 1849 (1981).
86. R. Noufi, D. Tench, and L. F. Warren, J. Electrochem. Soc., 127, 2310 (1980).
87. G. Horowitz, G. Tourillon, and F. Garnier, J. Electrochem. Soc., 131, 151 (1984).
88. R. A. Bull, F.-R. F. Fan, and A. J. Bard, J. Electrochem. Soc., 130, 1636 (1983).
89. M. Ogasawara, K. Funahashi, T. Demura, T. Hagiwara, and K. Iwata, Synthesis Metals, 14, 61 (1986).
90. J. Roncali and F. Garnier, Macro. J. Chem., 10, 4 (1986).
91. A. F. Diaz and B. Hall, IBM J. Res. Dev., 27, 342 (1983).
92. A. Diaz and J. A. Logan, J. Electroanal. Chem., 111, 111 (1980).
93. E. W. Paul, A. J. Ricco, and M. S. Wrighton, J. Phys. Chem., 89, 1441 (1985).
94. A. Kitani, J. Izumi, J. Yano, Y. Hiromoto, and K. Sasaki, Bull. Chem. Soc. Jpn., 57, 2254 (1984).
95. R. Noufi, A. J. Nozik, J. White, and L. F. Warren, J. Electrochem. Soc., 129, 2261 (1982).
96. T. Kobayashi, H. Yoneyama, and H. Tamura, J. Electroanal. Chem., 177, 293 (1984).
97. T. Kobayashi, H. Yoneyama, and H. Tamura, J. Electroanal. Chem., 161, 419 (1984).
98. T. Kobayashi, T. Yoneyama, and H. Tamura, J. Electroanal. Chem., 177, 281 (1984).
99. A. F. Diaz and J. C. Lacroix, Proceedings from Symposium on Electrochemistry of Polymer Layers, Makromolekulare Chemie/Macromolecular Symposia, 8, 17 (1987).
100. E. M. Genies and C. Tsintavis, J. Electroanal. Chem., 195, 109 (1985).
101. G. Brilmyer and R. Jasinski, J. Electrochem. Soc., 129, 1950 (1982).
102. P. G. Pickup and R. A. Osteryoung, J. Amer. Chem. Soc., 106, 2294 (1984).
103. P. G. Pickup and R. A. Osteryoung, J. Electroanal. Chem., 195, 271 (1985).
104. E. M. Genies and C. Tsintavis, J. Electroanal. Chem., 200, 127 (1986).
105. A. F. Shepard and B. F. Dannels, J. Polym. Sci. A-1, 4, 511 (1966).
106. D. G. Walker and N. E. Wisdom, Jr., U.S. Patent 3,437,569 (1969); Chem. Abstr., 70(26) 115722z; N. E. Wisdon, Jr., U.S. Patent 3,437,570 (1969); Chem. Abstr., 70(26) 115723a.
107. M. Satoh, K. Kaneto, and K. Yoshino, Synthesis Metals, 14, 289 (1986).

108. T. Skotheim, M. Velasquez-Rosenthal, and C. A. Linkous, J. Chem. Soc., Chem. Commun., 612 (1985); Synthesis Metals, 15, 219 (1986).
109. E. M. Genies, G. Bidan, and A. Diaz, J. Electrochem. Soc., 149, 101 (1983).
110. M.-A. DePaoli, R. J. Waltman, A. F. Diaz, and J. Bargon, J. Chem. Soc., Chem. Commun., 1015 (1984).
111. M.-A. DePaoli, R. J. Waltman, A. F. Diaz, and J. Bargon, J. Polym. Sci., Poly. Chem. Ed., 23, 1687 (1985).
112. O. Niwa and T. Tamamura, J. Chem. Soc., Chem. Commun., 817 (1984).
113. T. T. Wang, S. Tasaka, R. S. Hutton, and P. V. Lu, J. Chem. Soc., Chem. Commun., 1343 (1985).
114. S. E. Lindey and G. B. Street, Synthesis Metals, 10, 67 (1984/1985).
115. F.-R. F. Fan and A. J. Bard, J. Electrochem. Soc., 133, 301 (1986).
116. R. M. Penner and C. R. Martin, J. Electrochem. Soc., 132, 514 (1985).
117. T. P. Henning, H. S. White, and A. J. Bard, J. Amer. Chem. Soc., 103, 3937 (1981).
118. T. P. Henning, H. S. White, and A. J. Bard, J. Amer. Chem. Soc., 104, 5862 (1982).
119. T. P. Henning and A. J. Bard, J. Electrochem. Soc., 1930, 613 (1983).
120. N. Bates, M. Cross, R. Lines, and D. Walton, J. Chem. Soc., Chem. Commun., 871 (1985).
121. R. M. Penner and C. R. Martin, J. Electrochem. Soc., 133, 310 (1986).
122. S. J. Jasne and C. K. Chiklis, Synthesis Metals, 151, 175 (1986).
123. J. Roncali and F. Garnier, J. Chem. Soc., Chem. Commun., 784 (1986).
124. C. J. Miller and M. Marcin, J. Electroanal. Chem., 207, 49 (1986).
125. P. K. Ghosh and A. J. Bard, J. Amer. Chem. Soc., 105, 5691 (1983).
126. D. Ege, D. K. Ghosh, J. R. White, J.-F. Equey, and A. J. Bard, J. Amer. Chem. Soc. 107, 5644 (1985).
127. R. W. Murray, R. J. Nowak, and D. R. Rolison, J. Electroanal. Chem., 164, 205 (1984).
128. Z. Li and T. E. Mallouk, J. Phys. Chem., 91, 643 (1987).
129. A. Yamagishi and A. Aramota, J. Electroanal. Chem., 191, 449 (1985).
130. J. R. White and A. J. Bard, J. Electroanal. Chem., 197, 233 (1986).
131. N. Oyama and F. C. Anson, J. Electroanal. Chem., 199, 467 (1986).
132. K. Itaya and A. J. Bard, J. Phys. Chem., 89, 5565 (1985).
133. M. Shibata and S. Motoo, (a) J. Phys. Chem., 187, 151 (1985); (b) 188, 111 (1985); (c) 193, 217 (1985).
134. M. Watanabe, M. Shibata, and S. Motoo, J. Phys. Chem., 187, 161 (1985).
135. M. Shibata and S. Motoo, J. Phys. Chem., 194, 275 (1985).
136. J. C. LaCroix and A. F. Diaz, J. Electrochem. Soc., 135, 1457 (1989).
137. A. F. Diaz, J. I. Castillo, J. A. Logan, and W. Y. Lee, J. Eelctroanal. Chem., 129, 115 (1981).
138. F. Beck, Electrochim. Acta., 33, 839 (1988).
139. J. Heinze, K. Hinkelmann and M. Land, DECHEMA monograph, 112, 75 (1989).

34
Photoelectrochemistry

MARYE ANNE FOX / University of Texas, Austin, Texas

I. Introduction 1397
II. Principles of Photoelectrochemistry 1398
A. Comparing an electrochemical cell with a photoelectrochemical cell 1398
B. Dispersal of the semiconductor 1403
C. Excited state properties 1404
D. Thermodynamics 1405
E. Solvent effects 1406
F. Mechanism 1407
G. Function group reactivity 1408
H. Stability of the semiconductor 1409
I. Sensitization 1409
III. Selective Reactivity via Photoelectrochemical Methods 1410
A. Adsorption 1410
B. Potential control 1410
C. Current control 1411
D. Competitive pathways from photogenerated radical ions 1412
IV. Future Directions for Photoelectrochemistry 1413
V. Photoreactions of Electrogenerated Radical Ions 1413
VI. Acknowledgments 1415
References 1415

I. INTRODUCTION

Photoelectrochemistry refers to a subset of redox reactions initiated by adsorption of a photon at the electrode/electrolyte interface. We discuss in this chapter the similarities and differences encountered in photoelectrochemical conversions and those that take place at more conventional dark electrodes. As with all electrochemical transformations, photoelectrochemical reactions offer the experimental chemist control of reactivity from three venues: adsorption, potential control, and current control. We discuss in this chapter how each of these electrochemical features provides unique capability in photoelectrochemical cells. Since photoelectrochemical events are driven by light, the kinetics for interfacial electron transfer must always compete with excited-state degradation products. Furthermore, back electron transfer between the oxidized and reduced intermediates formed on the irradiated surface can compete with diffusion of the redox intermediates from the surface into the bulk or with trapping by specifically added reagents. The surface also influences the relative stability of the redox intermediates generated, and this altered reactivity can then be reflected in differential reactivity from that observed with the same intermediates in solution.

An important aspect of photoelectrochemistry is that the electrochemical cell can be "miniaturized." Thus both anodic and cathodic sites can be introduced on a single particle, whose dimensions can vary from subcolloidal to a single crystal in size. This variation then allows conducting photoelectrochemical experiments without the usual devices for maintaining potential or analyzing current flow.

This brief review cannot do justice to the large number of photoelectrochemical transformations that have been investigated in the last 15 years. Since this text deals with organic transformations, the principles discussed in this chapter are illustrated with examples chosen from organic chemistry. After discussing the principles involved in photoelectrochemical interfacial electron transfer, we consider some representative oxidative, reductive, and pericyclic reactions, exemplifying the principles of this reaction by considering 1,1-diphenylethylene and its oxidative reactions before giving a brief overview of functional reactivity. These principles apply equally to inorganic transformations, and extensive work involving the solar energy applications of simple inorganic molecule (e.g., water splitting, carbon dioxide reduction, and the water-gas shift reaction) are abundant. Even by restricting our attention to organic transformations, the review would exceed space limitations if a comprehensive discussion were involved. We therefore refer the interested reader to a number of reviews that address on a more detailed level specific reactions that can be accomplished by photoelectrosynthetic transformations at irradiated semiconductor surfaces [1–5].

Finally, this chapter briefly considers some areas for further development. Among the most promising of such directions are the use of semiconductor particles as light-sensitive constituents of multicomponent devices. We also consider recent developments in pairing of semiconductors as a method for improved charge separation and, in chemically modified form, as rationally designed sites for enhanced chemical selectivity. Several commercially important applications of photoelectrochemistry are briefly discussed.

II. PRINCIPLES OF PHOTOELECTROCHEMISTRY

A. Comparing an Electrochemical Cell with a Photoelectrochemical Cell [6–16]

An electrochemical cell consists of two electrodes, one of which is rendered electron rich by an externally applied potential, to which, as a consequence, electron-deficient species migrate and accept an electron. A cathodic current then flows across the electrode/electrolyte interface as this reduction proceeds. The second electrode is rendered electron deficient by application of the same external potential difference, so that electron-rich species migrate toward its surface. Electrons flow from the adsorbed molecule to the electrode, inducing an anodic current. The completed electrochemical cell thus requires a conductive pathway for electrons to flow from the anode to the cathode as well as an electrolyte to allow charge equilibration within the solution phase.

If the two electrodes are immersed in a common electrolyte, the potential difference between the two electrodes must be supplied by an external potentiostat, which can be controlled at the will of the experimenter. The potential difference between these two electrodes is usually compared to a reference electrode, and the cell can be further described by connection with an ammeter or a voltmeter. If separate half-cells are used for the anodic and cathodic compartments, the potential difference between the two electrodes can be controlled by adjusting the composition of the separate electrolyte solutions in the half-cells, and hence the Fermi level of each electrode. The Fermi level can be defined as that energy at which the probability of a level being occupied by an electron is one-half; that is it is equally probable that the level is occupied or vacant. The energy of this level depends on the orbital energies of the metal or semiconductor and the number of electrons that occupy the bands. The potential gradient between the two electrodes is thus chemical, and as long as a physical mixture of the two half-cells is inhibited, the induced electrochemical gradient can allow flow of electrons through an external circuit.

In contrast, the photoelectrochemical cell need not have a bias induced either by an external potentiostat or by chemical gradients in the respective half-cells. A potential difference can be created specifically by adsorption of light.

Let us consider an electrochemical cell constructed by dipping a light-responsive single crystalline semiconductor electrode and a metal counter-electrode into a common electrolyte containing a reversible oxidation-reduction couple. We connect the semiconductor and the metal electrode by an external wire and observe current and potential changes induced by irradiation and chemical reactions that occur at the two electrodes.

Such a cell is shown schematically in Fig. 1. If the metal electrode is irradiated by an external light source, electrons are promoted from a filled level in the metal to a vacant level. Since a metal, by definition, is composed of a continuum of states, the excited electron lies at some potential above the initial Fermi level. Its relaxation to its original potential is virtually instantaneous, however, since there is no barrier to charge recombination that allows kinetic separation of the photoexcited electron from its original source. Thus the possibility that the electron excited at the surface of a metal could be transferred to an adsorbed solution phase molecule is extremely low. The energy adsorbed in the photon is thus dissipated as heat, no potential difference is generated, and no photoeffects are observed. This situation can best be illustrated as in Fig. 2.

If, however, the semiconductor electrode is irradiated, we observe first that only selected wavelengths are effective at promoting an electron from the filled orbitals to the vacant ones. By definition, a semiconductor

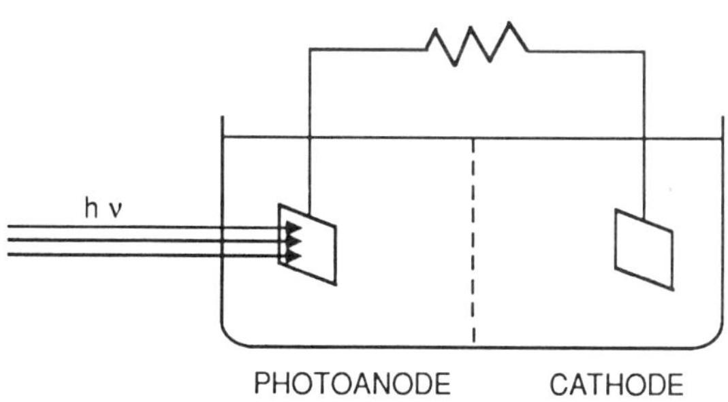

Figure 1. A photoelectrochemical cell.

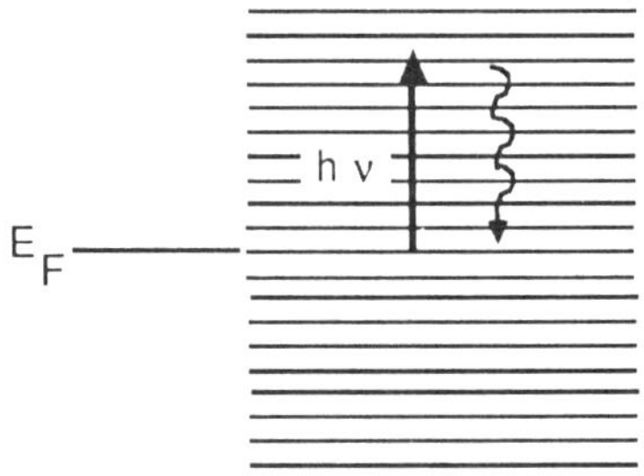

Figure 2. Excitation of a metal electrode.

does not have a continuum of states but rather has band structure. The filled level, referred to as the valence band, consists of an energetically closely spaced array of orbitals deriving from the valence electrons of the material in point. A gap exists between the top edge of this band E_{VB} and the lower edge of a similar closely spaced array of orbitals that are unoccupied in the ground state, E_{CB}. Since population of these latter orbitals makes the material conductive, this band is referred to as the conduction band. The gap separating the valence band from the conduction band (Fig. 3), is referred to as the band gap, E_g. Only if a photon is spectrally matched with the accessible band levels of the semiconductor does it possess sufficient energy to cause promotion of an electron from the valence band to the conduction band (Fig. 3). Every semiconductor should therefore have a wavelength onset for adsorption characteristic of that particular material.

If a photon of sufficient energy is used to cause promotion of an electron from the valence band to the conduction band, we generate an electron-hole pair. That is, a vacancy is created in the valence band from its normal occupancy, and an electron is placed in the conduction band. Recombination of the electron-hole pair is, of course, energetically highly favorable. Since the semiconductor lacks the state continuity found in a metal, however, quantum restrictions slow the rate of recombination so that the hole can, at least in principle, act as a photoanode and the electron can, in principle, act as a photocathode, if site-specific localization of these respective redox reagents can be attained.

An intrinsic, that is, undoped, semiconductor is one in which the Fermi level bisects the edges of the valence and conduction bands (Fig. 4).

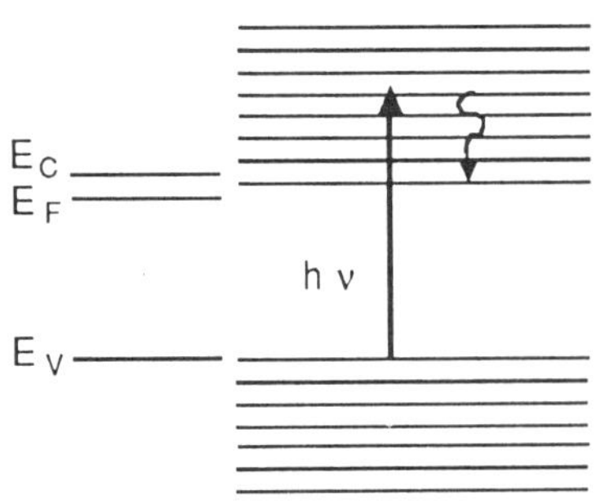

Figure 3. Excitation of a semiconductor electrode.

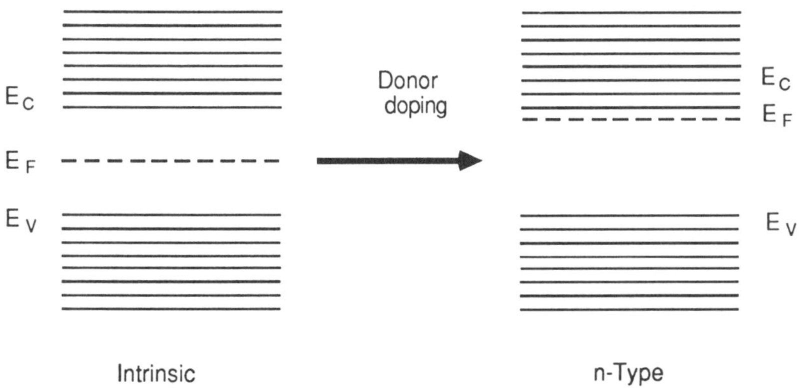

Figure 4. Response of the Fermi level to n-type doping.

Most semiconductors used in photoelectrochemical cells are doped, either naturally or by specific addition of dopants. A dopant is a material that changes the electron density of a material. An electron-rich dopant confers a small partial negative charge on a semiconductor: such a small n-type semiconductor thus has partial occupancy of its conduction band, and the Fermi level moves from the position halfway between the valence and conduction bands to a position quite near the conduction band edge. A positive dopant, generating a p-type semiconductor, has the opposite effect, with the Fermi level moving to a position just above the valence band edge.

If an n-type semiconductor is immersed into an electrolyte containing a redox couple, interfacial electron flow occurs to equilibrate the Fermi levels of the semiconductor bulk with the solution phase E^0. This is illustrated in Fig. 5 [7]. As this equilibration occurs, bending of the bands from the bulk to the original positions at the surface occurs. This band bending assists in charge separation in the photoelectrochemical experiment. Adsorption of a photon near the interface between the solid semiconductor and the liquid electrolyte promotes an electron from the valence band to the conduc-

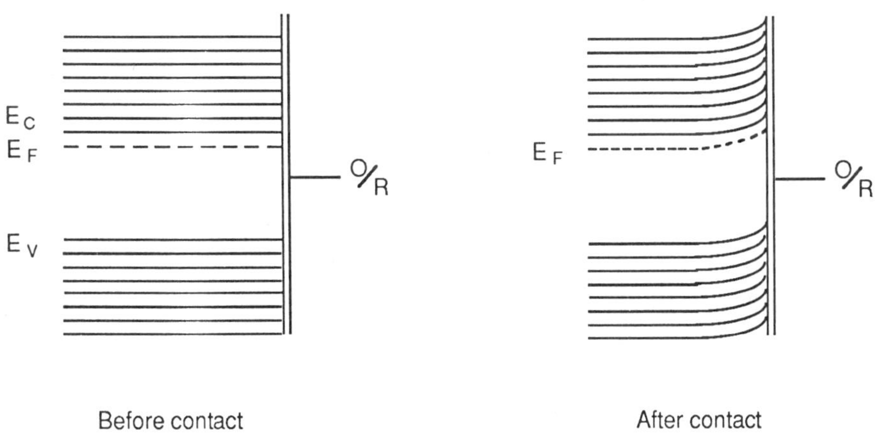

Figure 5. Equilibration of an n-type semiconductor electrode with a dissolved redox couple.

tion band. By virtue of band bending, this electron moves away from the interface into the bulk of the solution, falling to the lower energy level of the equilibrated bulk. Thus, the surface of an irradiated n-type semiconductor becomes electron deficient and acts as a photoanode. If the electrons from the bulk of the semiconductor are bled off at a counterelectrode, net chemistry is observed in which an adsorbed donor transfers an electron to fill the photogenerated hole and the photogenerated electron moves to the counterelectrode, where it is able to effect a reduction. As long as the electrolyte contains a reversible couple, no net chemistry is observed in the solution. In the absence of observable chemical change, the photoelectrochemical events can be detected by the development of a photovoltage or a photopotential. These effects thus bear analogy to solid-state photovoltaic devices, and an electrochemical cell based on these principles can be referred to as a liquid-phase photovoltaic (Fig. 6) [6].

If one separates the photoanodic reaction from the dark cathodic reaction by conducting the reaction in a divided half-cell, for example, then separate oxidation and reduction reactions can be observed. Oxidation of an adsorbed donor at the irradiated semiconductor interface generates a single electron-oxidized intermediate, that is, a radical cation from a neutral donor. Single-electron reduction of a coupled acceptor at the dark cathode generates a singly reduced product, that is, a radical anion from a neutral acceptor. The net observable chemistry is thus transfer of an electron from the donor to the acceptor induced indirectly by light adsorbed by the semiconductor. If the oxidation potential of the donor lies below that of the reduction potential of the acceptor, as is commonly the case for most accessible donor-acceptor pairs, the net electron transfer shown in Fig. 7 [6] represents a storage of electrical energy as the cation radical-anion radical pair. This cell thus bears analogy to photosynthesis in terms of energy storage, and such a cell may be referred to as a liquid-phase photoelectrosynthetic cell.

If we simply invert the oxidation potential of the donor and the acceptor, we can construct a cell as shown in Fig. 8. In this cell, again under band gap irradiation, the photogenerated electron-hole pair induces oxidation of an adsorbed donor and reduction of an acceptor adsorbed at the counterelectrode surface. If the oxidation potential of the donor is negative compared with the reduction potential of the acceptor, this photoelectrochemical cell simply provides a new catalytic route for what would be an exothermic

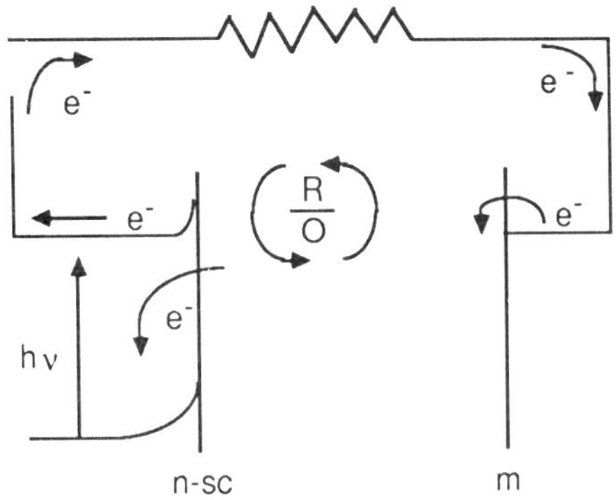

Figure 6. A photovoltaic photoelectrochemical cell.

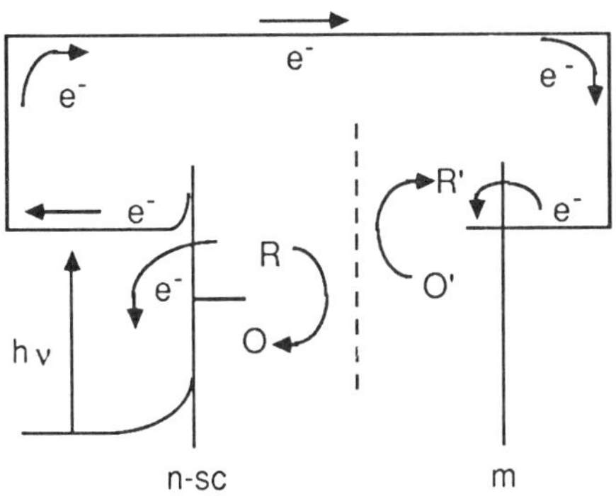

Figure 7. A photoelectrosynthetic electrochemical cell.

electron transfer; that is, the electron transfer within the donor-acceptor pair is favored thermodynamically but perhaps is inhibited in the ground state by an energy barrier. The overcoming of such a barrier represents a kinetic acceleration, in which the semiconductor electrode or electrolyte is regenerated, ready for another cycle. It is thus analogous to a catalytic conversion and can therefore be properly termed a photocatalytic cell [6].

In terms of organic reactivity, the latter two types of cells are of greatest interest. Since the photoelectrosynthetic cells provide access to a new reactivity that is not available without excitation of the semiconductor, they have provided the major focus for such studies. (This is not to denigrate the importance of photocatalytic applications, which are treated in detail elsewhere.)

B. Dispersal of the Semiconductor

The photoelectrochemical cell can be "short-circuited" by depositing an inert metal on the powdered semiconductor. Such an aggregate is represented in Fig. 9 [2,17]. This powder includes a site for both photoactivated electron transfer at the irradiated semiconductor electrolyte interference and collection of the photogenerated electrons. Although these metallized

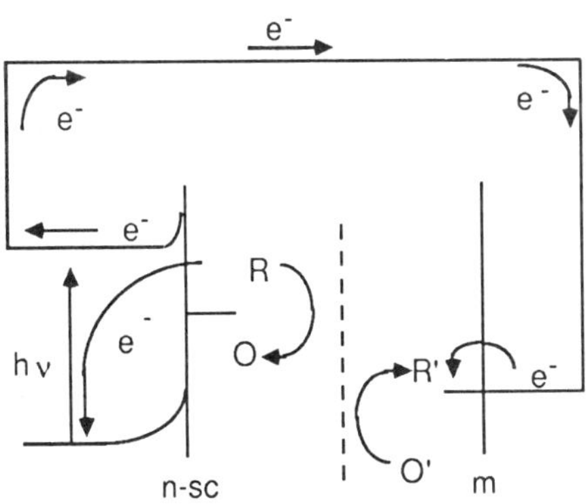

Figure 8. A photoelectrocatalytic electrochemical cell.

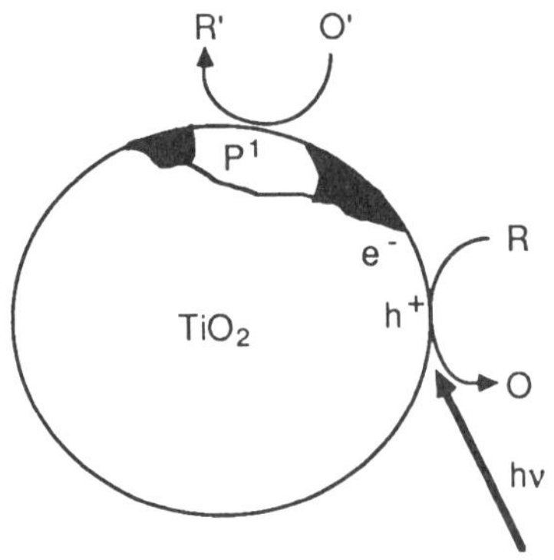

Figure 9. Redox exchange on an irradiated metallized semiconductor powder.

powders involve electron-hole pair separation in the same way as a photoelectrochemical cell, no external current flow need accompany these chemical transformations.

In fact, even unmodified powders can themselves sometimes function as active photocatalysts. For photoelectrochemical conversions to occur on untreated surfaces, either the oxidation or reduction must occur readily on the dark material, thus scavenging one of the photogenerated charge carriers. That is, there must be a kinetic preference for either the oxidation or the reduction. An alternative explanation for the effectiveness of native powders can be suggested as involving defect sites, which can act as charge accumulation centers in parallel to purposely introduced metal sites.

Obviously these powders are simpler to work with than a full electrochemical cell in that they are easily moved and inexpensive and do not require extensive electrochemical analytical instrumentation. Furthermore, irradiated powders allow only small separation of the oxidation and reduction sites on a common surface so that the primary redox products are generated in close proximity. Subsequent dark chemistry may then occur, as the oxidized and reduced species migrate toward each other, leading either to interesting reactions or to undesired complications. Furthermore, depending on particle size, the complete development of a space charge region at the interface of such a powder may be absent.

If one conceptually considers making the particle smaller and smaller, one eventually reaches a colloidal state. Colloidal forms of metal oxide semiconductors can often be prepared by hydrolysis of an appropriate organometallic precursor, and the resulting suspension may represent the ultimate miniature photoelectrochemical cell [18–21].

C. Excited State Properties

The formation of an electron-hole pair on an irradiated semiconductor is analogous to the formation of an excited state from a typical organic compound [1]. It has been recognized for some time that the promotion of an electron in a typical organic molecule from its highest occupied molecular orbital (HOMO) to its lowest-lying unoccupied molecular orbital (LUMO) generates a species with different redox properties from those encountered in the ground state. That is, it is much easier to remove or add an electron from/to the excited state than from/to the ground state. This enhanced redox reactivity follows directly from the orbital population levels. An enhanced electron affinity is attained by creation of a vacancy in the HOMO, and reduced ionization potential is achieved by virtue of the lower energy required to promote an electron from a LUMO to the vacuum than from its ground state highest occupied MO.

The formation of an electron-hole pair, then, is quite analogous to formation of an excited state. Thermal reversal of the electronic promotion regenerates the ground state, and back electron transfer in the electron-hole pair also quenches the enhanced redox activity. As we shall see, the kinetic competition for interfacial electron transfer must be won if efficient chemistry is to evolve compared with the rate of electron-hole recombination either at the interface or in the bulk of the semiconductor. Since the electron-hole pair typically lives only of the order of nanoseconds [22], this means that any effective donor or acceptor must be in contact with the semiconductor at the moment the electron-hole pair is generated.

D. Thermodynamics

The accessible potentials that can be obtained for the electron-hole pair generated by irradiation of a given semiconductor are governed by the respective valence band and conduction band edges for that material. Although the band positions in common semiconductors are subject to shifts with pH, solvent, temperature, and so on, we can use the relative band positions under similar circumstances to predict the thermicity of the requisite electron transfer. The band positions for some common semiconductors are summarized in Table 1 [12], together with onset wavelengths. Judicious

Table 1. Band Positions[a] in Some Common Semiconductors

Semiconductor	E_{VB} (V versus SCE)	E_{CB} (V versus SCE)	Eg (V)	λ(Eg) (nm)
TiO_2	3.0	0.0	3.0	410
SnO_2	3.8	+0.3	3.5	350
ZnO	2.9	-0.1	3.0	410
$SrTiO_3$	3.0	-0.2	3.2	390
CdS	2.2	-0.2	2.4	520
CdSe	1.4	-0.3	1.7	730
CdTe	1.4	0.0	1.4	890
GaAs	0.8	-0.6	1.4	890
GaP	1.4	-0.9	2.3	540
InP	1.2	-0.1	1.3	950

[a]Approximate positions in water at pH 1.
Source: From Ref. 3.

choice of the semiconductor allows wavelengths varying from the long-wavelength ultraviolet (UV) through the visible region to be used and the potentials attainable to be adjusted appropriately. Single electron transfer from an adsorbed donor competes with electron-hole recombination for any adsorbate whose oxidation potential lies less positive than the valence band edge, and an adsorbed acceptor whose reduction potential lies less negative than the conduction band edge can be similarly reduced. Many oxidations can be thus conducted on irradiated TiO_2, since the band edge (3.0 V vs. SCE) allows the oxidation of virtually any organic substrate that contains either nonbonded electrons or a conjugated π system.

E. Solvent Effects

The band edge positions can also be adjusted by controlling the particle size. Quantization effects have recently been reported that allow control of the onset wavelength and of the band positions of several common semiconductors [23–28].

In aqueous solution, however, direct oxidation of the substrate must compete with solvent oxidation. The formation of hydroxy radicals by single-electron oxidation of water is usually the dominant process in aqueous solutions. Since the hydroxy radical is highly reactive, high yields of oxidation and oxygenation products are often obtained. Its high reactivity has a drawback, however, since the hydroxy radical is very unselective and attacks virtually any substrate [29]. If it is incorporated into an organic substrate, it effectively increases the adsorptivity of that product so that the partially oxidized substrate does not diffuse from the irradiated surface. Such an intermediate is then further oxidized, and overoxidation, sometimes to the point of complete mineralization, is often observed in water [30]. Of course, this complete destruction of organic materials is useful if the goal is to purify a water stream, but if one is seeking to do selective organic redox chemistry, it can be disastrous.

The shift from a reactive solvent (e.g., water) to an inert solvent (e.g., acetonitrile) can take care of this problem. Since acetonitrile is oxidized outside the potential window of the band gap of common semiconductors, the adsorbed substrate preferentially participates in the primary electron exchange.

In nonaqueous solvent, the potentials shift compared with those seen in water at pH 1. In fact, the conduction band becomes sufficiently negative to cause reduction of oxygen, so that interfacial electron transfer to capture the photogenerated electron-hole pair generates an organic cation radical and a superoxide, respectively. That these are formed at sites in reasonably close proximity opens the possibility for new reactions in inert solvent. In fact, many reactions have been reported in which a variety of organic substrates become oxygenated upon sensitized irradiation of semiconductor suspensions in the presence of oxygen. An example of such electron transfer chemistry can be seen in Fig. 10, where the respective band positions for TiO_2 immersed in acetonitrile are compared with the oxidation potential of 1,1-diphenylethylene and the reduction potential of oxygen in the same medium [31]. In fact, irradiation of TiO_2 suspended in oxygenated acetonitrile containing 1,1-diphenylethylene leads to virtually quantitive formation of benzophenone, an oxygenation product in which the C=C double bond has been cleaved [32].

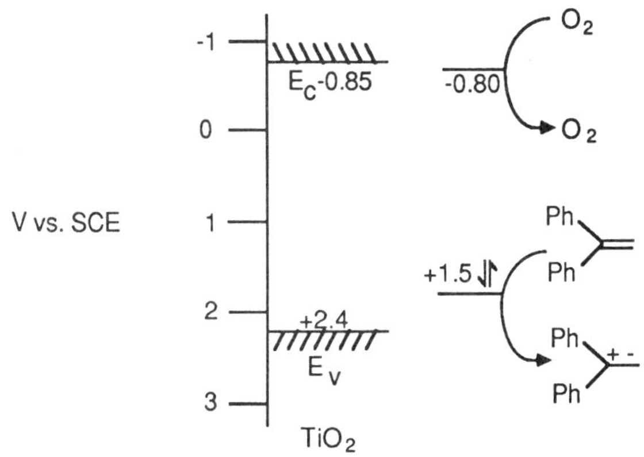

Figure 10. Energy levels for diphenylthylene oxidation on irradiated TiO_2.

F. Mechanism

As these arguments suggest, organic oxidations should be particularly facile on irradiated semiconductor surfaces. We can look in some detail at the oxidation of 1,1-diphenylethylene as a typical example of a photoinduced oxygenation. The thermodynamic considerations discussed earlier suggest that irradiation of a TiO_2 suspension gives rise to an electron-hole pair. Interfacial electron transfer from an adsorbed olefin and adsorbed oxygen could generate the olefin cation radical and superoxide.

Although superoxide is highly reactive, control experiments establish that superoxide solubilized in acetonitrile is ineffective at attacking neutral 1,1-diphenylethylene in the dark [32]. Furthermore, the superoxide scavengers, present in solution are ineffective at inhibiting the reactivity [33]. Thus, superoxide, if it is formed, never escapes from the irradiated surface, but rather must interact as it is generated on the irradiated semiconductor.

Back electron transfer between the olefin cation radical and superoxide could potentially generate singlet oxygen, since spin restrictions are relaxed if free radical ions are indeed formed. That is, a doublet, on reduction, can generate either a singlet or a triplet. Singlet oxygen, were it formed in this way, would indeed oxidize 1,1-diphenylethylene to benzophenenone. There is, however, no evidence for the involvement of singlet oxygen as a general route, since the chemistry of singlet oxygen differs significantly from that observed on irradiated semiconductors.

Tetramethylethylene, for example, reacts with singlet oxygen by an ene reaction, generating an allylic hydroperoxide as a major product [34]. In contrast, irradiation of titanium dioxide in the presence of tetramethylethylene gives a complex mixture in which oxidative cleavage, parallel to that observed with 1,1-diphenylethylene, represents the major route. Similarly, Spada and Foote have shown that singlet oxygen is unreactive toward trans-stilbene, unless electron transfer pathways are involved [35]. To our knowledge there is no unambiguous demonstration that singlet oxygen is ever formed by reversible electron transfer on irradiated semiconductor surfaces.

A third possibility is that the organic cation radical formed by oxidative capture of the photogenerated hole at the irradiated surface is trapped either by co-generated superoxide or by co-adsorbed triplet oxygen. In the former route, one may expect products to result from chemical bond formation between the cation radical-anion radical pair, which are both paramagnetic and of opposite charge. In the latter route, precedent exists for the formation of dioxetane intermediates of stable olefin cation radicals [36], as in the characterization by Nelsen and coworkers of a dioxetane cation radical from adamantylidene cation radical [37]. If a dioxetane is formed, either in neutral form or as a cation radical, the TiO_2 surface can function in an additional role, that is, as a Lewis acid catalyst, to induce decomposition of the dioxetane. Since no chemiluminescence could be observed in these reactions, apparently Lewis acid catalysis provides a nonradiative route for cleavage of this high energy intermediate. That TiO_2 can indeed function in this way can be demonstrated by independent synthesis of the dioxetane derived from 1,1-diphenylethylene, which does indeed decompose to benzophenone upon stirring in the dark on titanium dioxide.

Additional evidence that an olefin cation radical is the critical intermediate in these reactions can be found by competition experiments involving para-substituted diphenylethylenes, in which a linear free energy relationship is observed, consistent with formation of an intermediate involving the development of positive charge [38]. Furthermore, the transient formation of cation radicals can be observed when a colloidal titanium dioxide suspension is flashed in the presence of an olefin [39]. Although this experiment does not necessarily prove that the cation radical is the critical intermediate in the transformation, since it requires the extrapolation of similar photochemical reactivity to colloids and powders, it nonetheless does demonstrate that single electron transfer can be induced by photoexcitation of the semiconductor.

G. Functional Group Reactivity

To the extent that single-electron oxidation is mechanistically important in these reactions, we can use the thermodynamic criteria discussed earlier, together with the known reactions of cation radicals, to predict new organic oxidation products that may be formed from a variety of substrates. For example, a wide variety of oxidations have been described using photoelectrochemical methods. Alcohols, for example, can be oxidized to the corresponding carbonyl compounds in high yield [40–61]. In fact, primary alcohols can be selectively oxidized to the corresponding aldehydes, without appreciable overoxidation, whereas poorer yields were obtained with secondary aliphatic aromatic alcohols [56]. Similarly, phenols can be converted to quinones, and further hydroxylated species and amines are oxidized to imines and coupling products [61]. N-alkylation of amines has also been reported in the presence of alcohols on irradiated TiO_2 [62–68]. Amines bearing α-hydrogens react by carbon-carbon coupling at the α-positions, as do ethers and some substituted alcohols [53]. With toluidines, azo coupling products are formed [69].

Amides are oxidized to either imides or α, β-unsaturated amides [70], and thioethers convert cleanly to the corresponding sulfoxides [71]. Thiols can be converted to disulfides [72], and Stevens rearrangement has been observed with dimethylsulfide [73]. Amino acids have been shown to form oligopeptides [74], and photoelectrochemical synthesis of amino acids from simple molecules (ammonia, methane, and water) has been reported [75–77]. Oxidative decarboxylation of carboxylic acid is reasonably efficient [78–86], as is carbon-carbon coupling reactions from allyl ethers [87–89]. Aromatic hydrocarbons can be oxidatively cleaved either on a side chain or in the ring [90–95]. Even saturated hydrocarbons can be induced to become oxygenated or to participate in isotope exchange on irradiated platinum/titanium dioxide catalysts [96–99].

Although reductions of organic substrates are less frequent, some examples are available. Some of these involve photogeneration of hydrogen on platinized semiconductor particles as in the hydrogenation of olefins, vinyl ethers, and α, β-unsaturated enones and alkynes [100–102]. Nonetheless, the N=N double bond of diaryl azo compounds can be reduced [103,104], and scattered reports of carbonyl double-bond reduction have appeared [105].

Apart from net redox chemistry, the reversible formation of radical ions allows formation of formal pericyclic reaction products. For example, ring openings are known of strained saturated hydrocarbons, and formal [2 + 2] cycloadditions have been described for phenyl vinyl ether [106] and N-vinyl carbazole [107]. The cyclization of nonconjugated dienes, such as norbornadiene, have also been reported [108]. A recent example involves a 1,3-sigmatropic shift [109].

Transient formation of cation radicals is presumably also involved in the geometric isomerization of substituted styrenes [110], although a break in the Hammett plot suggested a change in the rate-determining step with a change in the electron-donating ability of the substituent [111].

H. Stability of the Semiconductor

One of the major problems encountered in using semiconductors as initiators for redox chemistry is the instability of some irradiated surfaces in contact with electrolytes. In addition to inducing interfacial electron transfer, the semiconductor itself may participate in redox activity, either corroding or generating an insulating layer at the surface. This lack of stability is particularly troublesome with small band gap semiconductors that adsorb strongly in the visible region.

Recent advances in the modification of electrodes, powders, and films by chemical attachment (see Chap. 33) [112,113] or adsorptive deposition of photoactive or electroactive catalysts [114,115] could address these difficulties. By attaching a chemical catalyst, one can competitively improve the operational lifetime of a given semiconductor by suppressing photocorrosion and improving the rate of a desired redox reaction.

I. Sensitization

The attachment of adsorptive molecules. that is, dyes, allows another mode of photoactivation. The dye adsorbs a photon generating an excited state in which a sufficiently higher energy orbital is populated to allow direct injection of an electron into the conduction band edge (Fig. 11) [1]. The dye thus becomes oxidized and can act chemically with nucleophiles either by bond formation or by electron transfer. In the latter case, the donor is oxidized, regenerating the ground state of the sensitizer ready to participate in another sensitization cycle. After these events have transpired, charge separation has been accomplished to the same end as would have been attained by direct excitationof the semiconductor, followed by interfacial electron transfer from an adsorbed donor. (Compare Figs. 7 and 11).

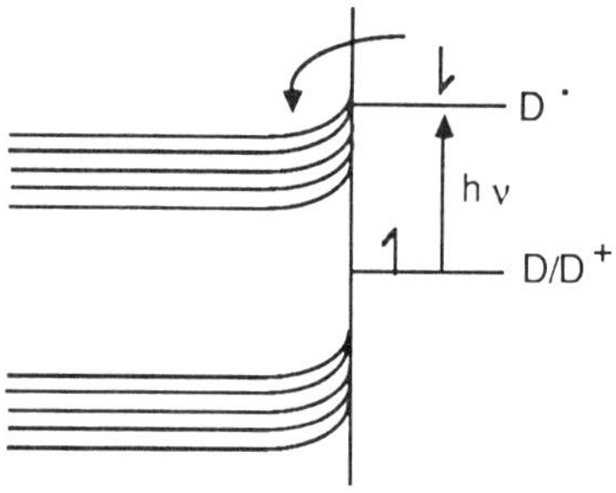

Figure 11. Oxidative electron exchange at a dye-sensitized n-type semiconductor electrode.

III. SELECTIVE REACTIVITY VIA PHOTOELECTROCHEMICAL METHODS

A. Adsorption

From kinetic considerations it is known that pre-adsorption of an active substrate is necessary for efficient photoelectrochemical activation. Adsorption forces operative on the surface of metal oxide semiconductors can be quite complex. In addition to acid and base sites that exist on TiO_2 [116,117], at which strong hydrogen bonding interaction is accomplished [118–120], both charge transfer and π bonding interactions [116,121] have been suggested as responsible for controlling adsorption properties. Even saturated hydrocarbons bind to oxide surfaces through nonspecific dispersive forces [116].

Preferential adsorption allows specific activation of the better adsorbate from a mixture. Such effects have been considered important in controlling the relative rates of photooxidation on semiconductor surfaces since adding very small amounts of co-solvent additives allows higher reactivity [122] and higher selectivity in the oxidation of a more strongly adsorbed site over a less strongly adsorbed one, as in the 7:1 preference for oxidation of the primary site over the secondary site in 1,4-pentanediol, Eq. (1) [123]:

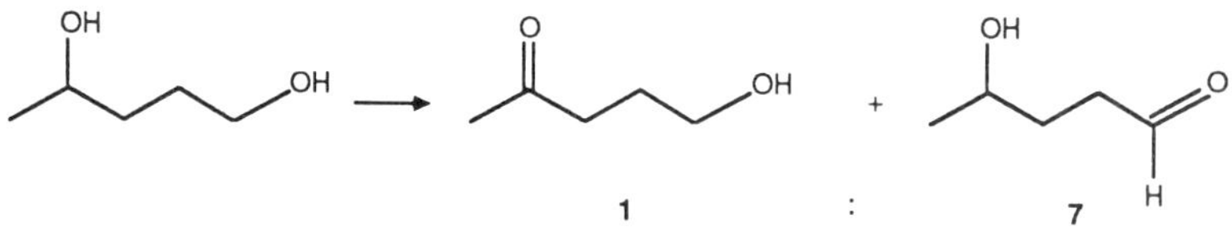

(1)

B. Potential Control

Selectivity may also be achieved in organic redox transformations by adjustment of the band positions of the irradiated semiconductor. Switching from TiO_2 to CdS as the irradiated sensitizer appreciably alters the oxidizability of the photogenerated hole. Since TiO_2 has a more positive valence band edge, it should be the more active photocatalyst. Consistent with this idea, photocatalytic oxidation of lactic acid from platinized TiO_2 produces decarboxylation, but that on platinized CdS leads to pyruvic acid by oxidation of the alcohol group, Eq. (2) [124]:

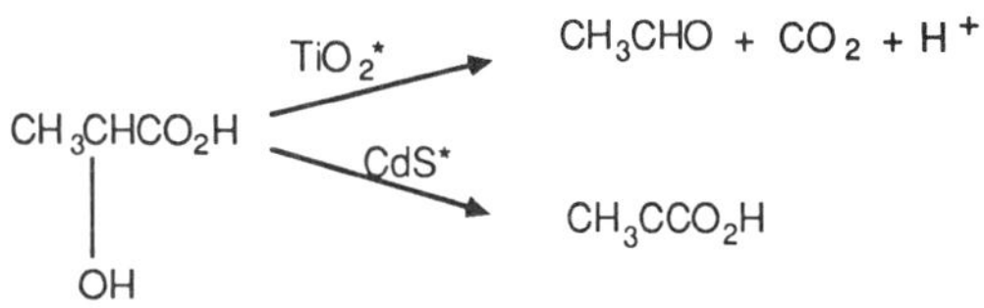

(2)

The relative importance of shifts in valence band position and of adsorption differences on the two semiconductors, however, is still ambiguous. Shifts in product distribution in oligopeptide formation were shown to be partially controllable by choice of sensitizer: diglycine is formed twice as efficiently on platinized TiO_2 as on platinized CdS, and pentaglycine is formed four times faster on CdS/Pt than on TiO_2/Pt [74].

C. Current Control

In principle, instantaneous switching of oxidizability is obtainable on an irradiated semiconductor. For example, in a conventional electrochemical cell held at a highly oxidizing potential, oxidation of an adsorbed donor occurs readily. If the product of that oxidation is itself oxidized at the applied potential, a second electron also flows. This is a relatively common occurrence for substrates that undergo geometry change after the initial oxidation, either by cleaving a bond or by twisting about a multiple bond.

In contrast, photoelectrochemical activation forms a highly oxidizing surface-confined hole, but interfacial electron transfer from an adsorbed donor effectively fills the hole and blocks further oxidation. Thus, if the ratio of the number of adsorbed molecules to the number of photogenerated holes on an individual surface is high, photoelectrochemical reactivity is restricted to single-electron oxidation, even in systems that react by multiple-electron oxidation in conventional routes.

An example of this specific chemistry can be seen in the electrochemical decarboxylation of vicinal diacids. Such substrates have been long used for synthetic purposes as protecting groups for double bonds, since upon electrochemical deprotection, a two-electron oxidation occurs, effecting bis-decarboxylation and uncovering a C=C double bond. In contrast, when platinized titanium dioxide is irradiated in the presence of such a vicinal diacid, a major product is monodecarboxylation, Eq. (3):

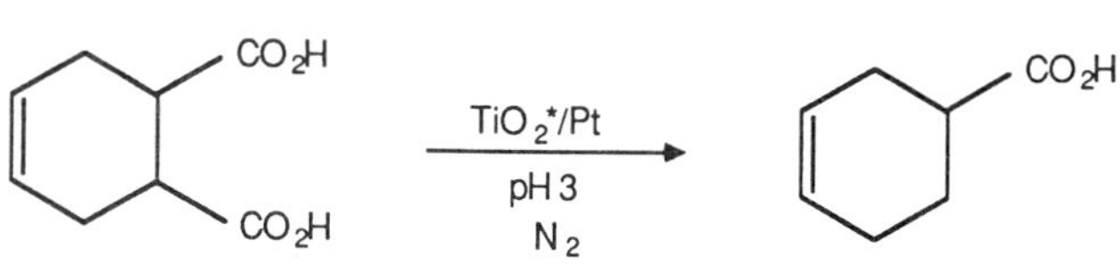

(3)

Here monodecarboxylation occurs, generating a surface-adsorbed radical whose further oxidation is inhibited by the instantaneous change in surface potential that accompanies interfacial electron transfer. This radical apparently is ultimately reduced and protonated to give the observed product.

This control, of course, is dependent on incident light flux. At high incident light flux, charge can actually accumulate in the semiconductor powder, and electron-loaded particles have been described in the literature [125,126].

D. Competitive Pathways from Photogenerated Radical Ions

Photoelectrochemical routes also provide a method for controlling competitive reactions. Radical ions are known to undergo a variety of possible reactions [127]. Formation of a cation radical from an organic donor on an irradiated semiconductor surface gives rise to a chemistry different from that in the highly oxidizing environment encountered at the surface of a poised metal electrode or of the same cation radical formed in solvated form in homogeneous solution, Eq. (4) [128]:

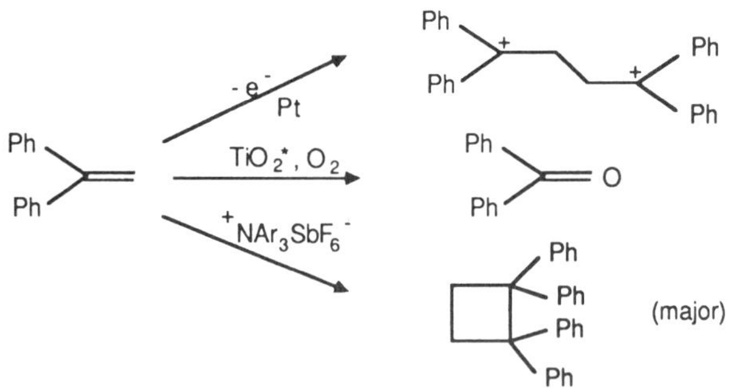

(4)

Thus the photoelectrochemical catalyst acts not only as a vehicle to generate the radical ion but also provides an environment to control its subsequent chemistry.

Surface control of competitive secondary reactions from cation radicals generated by irradiation of semiconductor suspensions can also be seen in the divergent chemistry observed from 1-methylnaphthalene. Selective ring cleavage is observed on irradiated TiO_2 suspensions; the corresponding radical cation formed in homogeneous solution gives exclusive side-chain oxidation, Eq. (5) [95]:

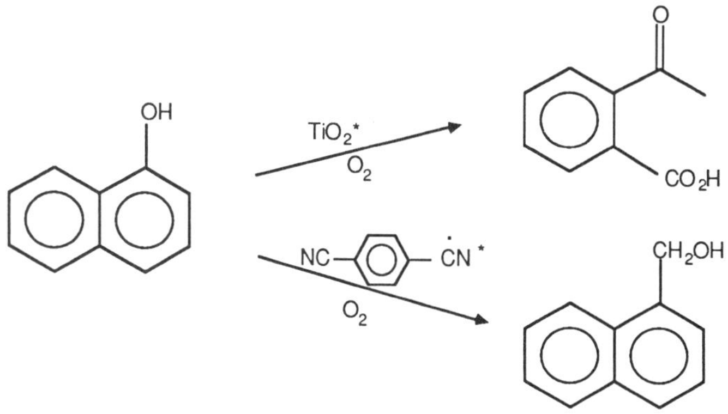

(5)

Finally, the judicious choice of solvent can influence quite dramatically the course of a given photoelectrochemical reaction [122].

IV. FUTURE DIRECTIONS FOR PHOTOELECTROCHEMISTRY

These photoelectrochemical principles provide a new vehicle for the control of organic redox chemistry. Elaborating the methods of chemoselectivity discussed earlier represents a major direction for continuing investigations in this field.

Environmental issues can also be addressed by photoelectrochemistry. The conversion of acetic acid to methane [78], the oxidative decomposition of biomass, the photodegradation of organochlorine contaminants [129,130], and the photochemical recovery of toxic or precious metals [131] represent applications of solar energy conversion directed to important societal uses. That such reactions may be involved in the dissolution of minerals [132], in the generation of biologically active oxidants [133] or metabolites [134], or in the reduction of chemisorbed nitrogen on titania-rich desert sands [135] all represent areas of interest to the community of biologists.

Industrial applications of organic photoelectrochemical transformations involve film sensitization [136], construction of supported semiconductor systems [137], paired semiconductor junctions [138], the design of multi-panel systems for unassisted water splitting [139], the construction of analytical sensing devices [140], the purification of water streams [141], and other applications in imaging technology.

V. PHOTOREACTIONS OF ELECTROGENERATED RADICAL IONS

The reactions discussed so far have been initiated by absorption of a photon by a light-responsive electrode, producing electrons and holes that initiate redox chemistry in adsorbed donors or acceptors. An alternative approach to the combination of electrochemical and photochemical techniques involves studies of either the excited states of the radical ions formed by electro-oxidation or electroreduction or of the redox properties of excited molecules. Because of the short lifetimes of the excited states of typical organic molecules, the first approach has been explored more extensively, but such experiments remain relatively rare.

Redox reactions of polycyclic aromatics often produce radical ions sufficiently stable that their photochemistry can be explored. For example, the radical anions of perylene and pyrene and the radical cation of thianthrene can be excited photochemically as they are generated at a poised electrode [142]. If a current-potential curve is generated during the irradiation, an enhanced current flow is observed. Since flash photolysis studies showed that the photoexcited pyrene radical anion ejects an electron [143], a mechanistic pathway proposed in Eq. (6):

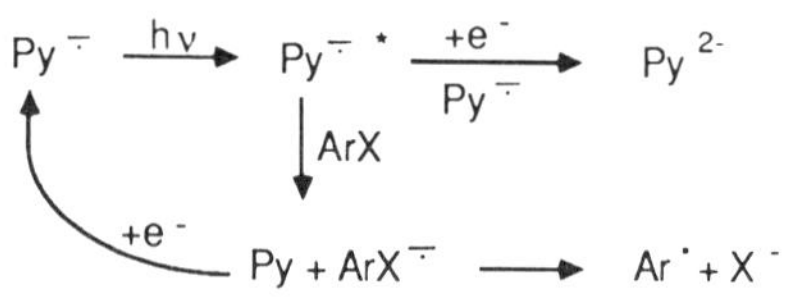

(6)

was suggested to explain the higher current. If the ejected electron is taken up by another radical anion, the dianion formed disproportionates, generating a new higher concentration of redox active reagent at the electrode surface. The photoinduced ejection of electrons from anions seems to be a general phenomenon [144], so analogous pathways may be broadly applicable.

The magnitude of the current enhancement depends on the rate of the follow-up chemical reaction with an acceptor, pronounced effects being observed if the electron transfer from the excited radical ion is fast enough to compete with back donation to the electrode [145]. The rates of such electron transfers, as measured by fluorescence quenching, are reasonably predicted by Marcus theory, or the Rehm-Weller equation, if large solvent or bond reorganization energies are introduced [146]. Implicit in this observation is the reasonable surmise that the solvent reorganization of the radical ion is greater than that required by the quencher. For example, a 10 kcal mol^{-1} reorganization energy for the anthraquinone radical anion and a value of 15.3 kcal mol^{-1} for the thianthrene cation radical are found by an energy-weighted maximum overlap calculation [147].

An additional advantage of this enhanced electron transfer from an excited radical anion is in the possible avoidance of overreduction. With aryl halides, for example, electroreduction generates an anion radical, which rapidly dissociates, forming an aryl radical whose reduction potential is less negative than its precursor. At the poised potential of the electrode, it is therefore converted to an aryl anion. By the indirect photolysis of a anion radical of a more easily reduced acceptor, such as anthraquinone, the overreduction can be avoided. When the anthraquinone radical anion is excited in the presence of bromobenzene, for example, the phenyl radical can be efficiently trapped as an adduct to anthracene [148].

Photoexcitation of cation radicals also initiates electron transfer, since the excited state is also more easily reduced than its ground state. Excitation of the phenothiazine cation radical, for example, allows oxidative activation of diphenylethylene, which initiates cyclization and dimerization sequences [149,150]. No reaction is observed in the dark between the cation radical and the olefin. Photorearrangements can also be observed from excited cation radicals, as in the benzidine rearrangement of the tetraphenylhydrazine cation radical [151].

The general area of photochemistry of radical ions is in its infancy, and many novel interesting transformations are now beginning to appear. We can anticipate a much more thorough discussion of such reactions in the next edition of this volume.

Finally, we consider the use of electrochemical methods to assist in photoconversions. In their excited states, quinones are reduced via either hydrogen atom transfer or sequential electron-proton transfer. By either route, hydroquinones are produced as a donor substrate is oxidized. For example, the oxidative coupling of toluene to bibenzyl and the oxidation of 2-propanol to acetone can be accomplished by photoexcitation of a quinone. If the hydroquinone formed in this reaction is electrochemically reconverted to the quinone, one may observe many turnovers of the donor for each sensitizer molecule [152]. In this technique, then, the photolysis is conducted in an electrochemical cell poised at a potential for dark reoxidation of the photoproduct. This route, although limited to reversible redox sensitizers, thus allows high turnover numbers in sensitized photoredox reactions.

VI. ACKNOWLEDGMENTS

Our program in organic photoelectrochemistry and chemically modified electrodes has been supported by the U.S. Department of Energy, Office of Basic Energy Sciences. The investigation of surface adsorbed radical ions has been supported by the U.S. National Science Foundation, and the use of irradiated semiconductor suspensions for decontamination purposes is supported by the Army Research Office. Our investigation of practical applications of photoelectrochemistry has been funded by the Texas Advanced Technology Research Program.

REFERENCES

1. M. A. Fox, Top. Org. Electrochem., 1, 177 (1986).
2. M. A. Fox, Top. Curr. Chem., 142, 71 (1987).
3. M. A. Fox, Acc. Chem. Res., 16, 314 (1983).
4. M. A. Fox and P. Pichat, in Photoinduced Electron Transfer (M. Chanon and M. A. Fox, eds.), Elsevier Scientific, AMsterdam, in press, 1988.
5. M. A. Fox, in Photocatalysis (N. Serpone and E. Pelizzetti, eds.), John Wiley and Sons, New York, in press, 1988.
6. A. J. Bard, J. Photochem., 10, 59 (1979).
7. A. J. Bard, Science, 207, 139 (1980).
8. A. J. Bard, J. Phys. Chem., 86, 172 (1982).
9. R. Memming, in Electroanalytical Chemistry (A. J. Bard, ed.), Dekker, New York, 1979.
10. M. S. Wrighton, Acc. Chem. Res., 12, 303 (1979).
11. H. O. Finklea, J. Chem. Ed., 60, 325 (1983).
12. A. J. Nozik, Faraday Disc., 70, 7 (1980).

13. T. Freund and W. P. Gomes, Catal. Rev., 3, 1 (1969).
14. J. A. Turner, J. Chem. Ed., 60, 327 (1983).
15. B. Parkinson, J. Chem. Ed., 60, 327 (1983).
16. A. J. Nozik, Ann. Rev. Phys. Chem., 29, 189 (1978).
17. M. D. Ward, J. R. White, and A. J. Bard, J. Amer. Chem. Soc., 105, 27 (1983).
18. C. Leygraf, M. Hendewerk, and G. A. Somorjai, Proc. Natl. Acad. Sci. USA, 79, 5739 (1982) and references therein.
19. M. Graetzel, Acc. Chem. Res., 14, 376 (1981).
20. J. Kuczynski and J. K. Thomas, Chem. Phys. Lett., 88, 445 (1982).
21. J. Moser and M. Graetzel, Helv. Chim. Acta, 65, 1435 (1982).
22. Y. Nosaka and M. A. Fox, J. Phys. Chem., 90, 6521 (1986).
23. R. Rossetti and L. Brus, J. Phys. Chem., 90, 558 (1986).
24. A. Henglein, Ber. Bungenges. Phys. Chem., 86, 301 (1982).
25. J. J. Ramsden and M. Graetzel, J. Chem. Soc., Faraday Trans. I, 80, 919 (1984).
26. A. J. Nozik, F. Williams, M. Nenadovic, T. Rajh, and O. I. Micic, J. Phys. Chem., 89, 397 (1985).
27. R. Rossetti, R. Hull, J. M. Gibson, and L. E. Brus, J. Chem. Phys., 82, 552 (1985).
28. J. M. Nedeljkovic, M. T. Nenadovic, O. I. Micic, and A. J. Nozik, J. Phys. Chem., 90, 12 (1986).
29. I. Izumi, F. R. F. Fan, and A. J. Bard, J. Phys. Chem., 85, 218 (1981).
30. I. Izumi, W. W. Dunn, K. O. Wilbourn, F. R. F. Fan, and A. J. Bard, J. Phys. Chem., 84, 3207 (1980).
31. T. Kanno, T. Oguchi, H. Sakuragi, and K. Tokumaru, Tetrahedron Lett., 21, 467 (1980).
32. M. A. Fox and C. C. Chen, J. Amer. Chem. Soc., 103, 6757 (1981).
33. J. San Fillippo, C. I. Chern, and J. S. Valentine, J. Org. Chem., 41, 1077 (1976).
34. G. O. Schenck and K. Schulte-Elte, Liebigs Ann. Chem., 618, 185 (1958).
35. L. T. Spada and C. S. Foote, J. Amer. Chem. Soc., 102, 393 (1980).
36. M. A. Fox and C. C. Chen, J. Comp. Chem., 4, 488 (1983).
37. S. F. Nelsen, D. L. Kapp, R. Akaba, and D. H. Evans, J. Amer. Chem. Soc., 108, 6863 (1986).
38. M. A. Fox and C. C. Chen, Tetrahedron Lett., 24, 547 (1983).
39. M. A. Fox, B. A. Lindig, and C. C. Chen, J. Amer. Chem. Soc., 104, 5828 (1982).
40. R. B. Cundall, R. Rudham, and M. Salim, J. Chem. Soc., Faraday Trans. I, 72, 1642 (1976); R. H. Hussein and R. Rudham, J. Chem. Soc., Faraday Trans. I, 80, 2817 (1984).
41. I. A. Ichou, M. Formenti, B. Pommier, and S. J. Teichner, J. Catal., 9, 293 (1985); M. Kawai, T. Kawai, S. Naito, and K. Tamaru, Chem. Phys. Lett., 110, 58 (1984).
42. T. Kawai and T. Sakata, J. Chem. Soc., Chem. Commun., 694 (1980).
43. T. Kawai and T. Sakata, Nature, 286, 474 (1980).
44. P. Pichat, J. M. Herrmann, J. Disdier, H. Bourbon, and M. N. Mozzanega, Nouv. J. Chim., 5, 627 (1981).
45. P. Pichat, M. N. Mozzanega, J. Disdier, and J. M. Herrmann, Nouv. J. Chim., 6, 559 (1982); P. Pichat, J. M. Herrmann, H. Courbon, J. Disdier, and M. N. Mozzenega, Can. J. Chem. Eng., 60, 27 (1982).

46. I. A. Ichou, M. Formenti, and S. J. Teichner, Stud. Surf. Sci. Catal., 19, 297 (1984).
47. S. Teratani, J. Nakamichi, K. Taya, and K. Tanaki, Bull. Chem. Soc. Jpn., 55, 1688 (1982).
48. T. Carlson and G. L. Griffin, J. Phys. Chem., 90, 5896 (1986).
49. S. Naito, J. Chem. Soc., Chem. Commun., 1211 (1985).
50. K. Domen, S. Naito, T. Ohnishi, and K. Tamaru, Chem. Lett., 555 (1982); L. T. Prahov, J. Disdier, J. M. Herrmann, and P. Pichat, Int. J. Hydrogen Energy, 9, 397 (1984).
51. I. M. Fraser and J. R. MacCallum, J. Chem. Soc., Faraday Trans. I, 82, 607 (1986).
52. I. M. Fraser and J. R. MacCallum, J. Chem. Soc., Faraday Trans. I, 82, 2747 (1986).
53. S. Yanagida, T. A. Zuma, H. Kawakami, H. Kizumoto, and H. Sakurai, J. Chem. Soc., Chem. Commun., 21 (1984).
54. S. Yanagida, H. Kawakami, K. Hashimoto, T. Sakata, C. Pac, and H. Sakurai, Chem. Lett., 1449 (1984).
55. S. Yanagida, T. Azuma, and H. Sukurai, Chem. Lett., 1069 (1982).
56. F. H. Hussein, G. Pattenden, R. Rudham, and J. J. Russell, Tetrahedron Lett., 3363 (1984).
57. T. Sakata and T. Kawai, Chem. Phys. Lett., 80, 341 (1981).
58. O. Enea, Electrochim. Acta. 31, 405 (1986); O. Enea and A. J. Bard, Nouv. J. Chim., 9, 691 (1985).
59. S. Nishimoto, B. Ohtani, and T. Kagiya, J. Chem. Soc., Faraday Trans. I, 81, 2467 (1985).
60. S. Nishimoto, B. Ohtani, H. Shirai, and T. Kagiya, J. Chem. Soc. Perkin Trans. II, 661 (1986).
61. K. Okamoto, Y. Yamamoto, H. Tanaka, M. Tanaka, and A. Itaya, Bull. Chem. Soc. Jpn., 58, 2015 (1985).
62. M. A. Fox and M. J. Chen, J. Amer. Chem. Soc., 105, 4497 (1983); M. A. Fox and J. N. Younathan, Tetrahedron, 42, 6285 (1986).
63. B. Ohtani, H. Osaki, S. Nishimoto, and T. Kagiya, Chem. Lett., 1075 (1986).
64. S. Nishimoto, B. Ohtani, T. Yoshikawa, and T. Kagiya, J. Amer. Chem. Soc., 105, 7180 (1983).
65. S. Yanagida, H. Kizumoto, Y. Ishimaru, C. Pac, and H. Sakurai, Chem. Lett., 141 (1985).
66. B. Ohtani, H. Osaki, S. Nishimoto, and T. Kagiya, J. Amer. Chem. Soc., 108, 308 (1986).
67. B. Ohtani, H. Osaki, S. Nishimoto, and T. Kagiya, Tetrahedron Lett., 2019 (1986).
68. H. Miyama, Y. Nosaka, T. Fukushima, and H. Toi, J. Photochem., 1553 (1986).
69. M. A. Hema, V. Ramakrishnan, and J. C. Kuriacose, Ind. J. Chem., 15B, 947 (1977); M. A. Hema, V. Ramakrishnan, and J. C. Kuriacose, Ind. J. Chem., 16B, 619 (1978).
70. J. W. Pavlik an S. Tantayanon, J. Amer. Chem. Soc., 103, 6755 (1981).
71. R. S. Davidson and J. E. Pratt, Tetrahedron Lett., 24, 547 (1983).
72. J. D. Spikes, Photochem. Photobiol., 34, 549 (1981).
73. W. Hoyer and K. D. Asmus, Sixth Int. Conf. Photochem. Conver., Storage Solar Energy, Abstract C-55 (1986).

74. J. Onoe, T. Kawai, and S. Kawai, Chem. Lett., 1667 (1985).
75. H. Reiche and A. J. Bard, J. Amer. Chem. Soc., 101, 3127 (1979).
76. W. W. Dunn, Y. Aikawa, and A. J. Bard, J. Amer. Chem. Soc., 103, 6893 (1981).
77. T. Kawai, M. Fujii, S. Kambe, and S. Kawai, Electrochem. Soc., Jpn. Abstract E214, 211 (1983).
78. B. Kraeutler and A. J. Bard, J. Amer. Chem. Soc., 100, 2339 (1978).
79. B. Kraeutler and A. J. Bard, J. Amer. Chem. Soc., 100, 5985 (1978).
80. H. Reiche, W. W. Dunn, K. Wilbourn, F. R. F. Fan, and A. J. Bard, J. Phys. Chem., 84, 3207 (1980).
81. D. N. Furlong, D. Wells, and W. H. F. Sasse, Aust. J. Chem., 39, 757 (1986).
82. H. Yoneyama, Y. Takao, H. Tamura, and A. J. Bard, J. Phys. Chem., 87, 1417 (1983).
83. C. D. Jaeger and A. J. Bard, J. Phys. Chem., 83, 3146 (1979).
84. J. M. Herrmann, M. N. Mozzanega, and P. Pichat, J. Photochem., 22, 333 (1983).
85. K. Miyahita, T. Sakata, K. Nakamura, T. Kawai, and T. Sakata, Photochem. Photobiol., 39, 151 (1984).
86. I. Willner and Z. Goren, J. Chem. Soc., Chem. Commun., 172 (1986).
87. J. Buecheler, N. Zeug, and H. Kisch, Angew. Chem., 94, 792 (1982).
88. S. Yanagida, T. Azuma, Y. Midori, C. Pac, and H. Sakurai, J. Chem. Soc., Perkin Trans. II, 1487 (1985).
89. N. Zueg, J. Buecheler, and H. Kisch, J. Amer. Chem. Soc., 107, 1459 (1985).
90. M. Fujihira, Y. Satoh, and T. Osa, Nature, 293, 206 (1981).
91. M. Fijihara, Y. Satoh, and T. Osa, Bull. Chem. Soc. Jpn., 55, 666 (1982).
92. Y. Simamura, H. Misawa, T. Oguchi, T. Kanno, H. Sakuragi, and K. Tokumaru, Chem. Lett., 1691 (1983).
93. K. Takagi, T. Fujioka, Y. Sawaki, and H. Iwamura, Chem. Lett., 913 (1985).
94. M. Fujihara, Y. Satoh, and T. Osa, J. Electroanal. Chem., 126, 277 (1981).
95. M. A. Fox, C. C. Chen, and J. N. Younathan, J. Org. Chem., 49, 1969 (1984).
96. K. Hashimoto, T. Kawai, and T. Sakata, J. Phys. Chem., 88, 4083 (1984).
97. H. Courbon, J. M. Herrmann, and P. Pichat, J. Catal., 72, 129 (1981).
98. P. Pichat, Amer. Chem. Sympos. Ser., 278, 21 (1985).
99. M. N. Mozzanega, J. M. Herrmann, and P. Pichat, Tetrahedron Lett., 2965 (1977).
100. M. A. Fox, C. C. Chen, K. H. Park, and J. N. Younathan, Amer. Chem. Soc. Sympos. Ser., 278, 69 (1985).
101. A. J. Frank, Z. Goren, and I. Willner, J. Chem. Soc., Chem. Commun., 1029 (1985).
102. H. Yamataka, N. Seto, J. Ichihara, T. Hanafusa, and S. Teratani, J. Chem. Soc., Chem. Commun., 788 (1985).
103. G. T. Brown and J. R. Darwent, J. Phys. Chem., 88, 4955 (1984).
104. G. T. Brown and J. R. Darwent, J. Chem. Soc., Faraday Trans. I, 80, 1631 (1984).

105. S. Yanagida, Y. Ishimaru, and C. Pac, J. Amer. Chem. Soc., in press (1986).
106. R. A. Barber, P. deMayo, and K. Okada, J. Chem. Soc., Chem. Commun., 1073 (1982).
107. H. Al-Ekabi and P. deMayo, Tetrahedron, in press (1987).
108. S. Lahiry and C. Haldar, Solar Energy, 37, 71 (1986).
109. P. deMayo and G. Wenska, Tetrahedron, in press (1987).
110. H. Al-Ekabi and P. deMayo, J. Chem. Soc., Chem. Commun., 1231 (1984).
111. T. Hasegawa and P. deMayo, J. Chem. Soc., Chem. Commun., 1534 (1985).
112. R. M. Murray, Acc. Chem. Res., 13, 135 (1980).
113. (a) A. J. Bard, J. Chem. Ed., 60, 302 (1983); (b) M. S. Wrighton, J. Chem. Ed., 60, 335 (1983).
114. T. Kuwana and W. R. Heineman, Acc. Chem. Res., 9, 241 (1976).
115. J. W. Schultze and M. A. Habib, J. Appl. Electrochem., 9, 255 (1979).
116. (a) A. V. Kiselev and A. V. Uvarov, Surf. Sci., 6, 399 (1967); (b) G. D. Parfitt and I. J. Wiltshire, J. Phys. Chem., 68, 3545 (1964).
117. H. P. Boehm, Disc. Faraday Soc., 52, 264 (1971).
118. I. Carrizosa and G. Munuera, J. Catal., 49, 174 (1977).
119. Y. Suda, T. Morimoto, and M. Nagao, Langmuir, 3, 99 (1987).
120. G. Ramis, G. Busca, and V. Lorenzelli, J. Chem. Soc., Faraday Trans. I, 83, 1591 (1987).
121. (a) G. Busca, H. Saussey, O. Saur, J. C. Lavalley, and V. Lorenzelli, Appl. Catal., 14, 245 (1985); (b) N. E. Tret'yakov and V. N. Filimonov, Kinet. Katal., 13, 735 (1972).
122. D. D. Sackett and M. A. Fox, J. Phys. Org. Chem., 1, 103 (1988).
123. M. A. Fox, H. Ogawa, and J. Muzyka, Proc. Sympos. Photoelectrochem. Electrosyn. Semiconduct. Mater., 88-14, 9 (1988).
124. H. Harada, T. Sakata, and T. Ueda, J. Amer. Chem. Soc., 107, 1773 (1985).
125. D. Bahnemann, A. Henglein, J. Lilie, and L. Spankel, J. Phys. Chem., 88, 711 (1984).
126. D. Duonghang, J. Ramsden, and M. Graetzel, J. Amer. Chem. Soc., 104, 2977 (1982).
127. A. J. Bard, A. Ledwith, and H. J. Shine, Adv. Phys. Org., 13, 156 (1976).
128. M. A. Fox, C. C. Chen, K. H. Park, and J. N. Younathan, Amer. Chem. Soc. Sympos. Ser., 278, 69 (1985).
129. D. Cesareo, A. deDomenico, S. Marchini, L. Passerini, and M. L. Tosato, in Homogeneous and Heterogeneous Photocatalysis (E. Pelizzetti and N. Serpone, eds.), D. Reidel, Dordrecht, 1986, p. 593.
130. D. F. Ollis, in Homogeneous and Heterogeneous Photocatalysis (E. Pelizzetti and N. Serpone, eds.), D. Reidel, Dordrecht, 1986, p. 651.
131. E. Borgarello, N. Serpone, M. Barbeni, E. Pelizetti, P. Pichat, J. M. Herrmann, and M. A. Fox, J. Photochem., 107, 373 (1987).
132. T. D. Waite and F. M. M. Morel, J. Colloid Interface Sci., 102, 121 (1984).
133. D. W. Bahnemann, C. Kormann, and M. R. Hoffman, Prep. Amer. Chem. Soc., Div. Petrol. Chem., 31, 519 (1986).

134. F. Baykut, G. Benlioglu, and G. Kaykut, in Homogeneous and Heterogeneous Photocatalysis (E. Pelizzetti and N. Serpone, eds.), D. Reidel, Dordrecht, 1986, p. 161.
135. G. N. Schrauzer, T. Guth, J. Salchi, N. Strampach, L. Hui, and M. R. Palmer, in Homogeneous and Heterogeneous Photocatalysis (E. Pelizzetti and N. Serpone, eds.), D. Reidel, Dordrecht, 1986, p. 161.
136. H. Gerischer and F. Willig, Top. Curr. Chem., 61, 33 (1976).
137. A. W. H. Mau, C. B. Huang, N. Kakuta, A. J. Bard, M. A. Fox, S. E. Webber, and J. M. White, J. Amer. Chem. Soc., 106, 6537 (1984).
138. N. Serpone, E. Borgarello, and M. Graetzel, J. Chem. Soc., Chem. Commun., 34 (1984).
139. E. Smothkin, A. J. Bar, A. Campion, M. A. Fox, T. Mallouk, S. E. Webber, and J. M. White, J. Phys. Chem., 90, 4604 (1986).
140. M. A. Fox and T. P. Tien, Anal. Chem., 60, 2278 (1988).
141. B. G. Oliver and J. H. Carey, in Homogeneous and Heterogeneous Photocatalysis (E. Pelizzetti and N. Serpone, eds.), D. Reidel, Dordrecht, 1986, p. 629.
142. H. Lund and H. S. Carlsson, Acta Chem. Scand., B32, 505 (1978).
143. K. A. Engdahl and G. Ramme, Chem. Phys. Lett., 41, 100 (1976).
144. M. A. Fox, Chem. Rev., 79, 253 (1979).
145. H. S. Carlsson and H. Lund, Acta Chem. Scand., B34, 409 (1980).
146. J. Eriksen, H. Lund, and A. I. Nyvad, Acta Chem. Scand., B37, 459 (1983).
147. J. Eriksen, K. A. Jorgensen, J. Linderberg, and H. Lund, J. Amer. Chem. Soc., 106, 5083 (1984).
148. P. Nelleborg, H. Lund, and J. Eriksen, Tetrahedron Lett., 26, 1773 (1985).
149. J. C. Moutet and G. Reverdy, Tetrahedron Lett., 2389 (1979).
150. J. C. Moutet and G. Reverdy, Nouv. J. Chim., 7, 105 (1983).
151. U. Svanholm and V. D. Parker, J. Amer. Chem. Soc., 94, 5507 (1972).
152. J. M. Bobbitt and J. P. Willis, J. Org. Chem., 42, 2348 (1977).

35
Paired Electrosynthesis

MANUEL M. BAIZER† / University of California, Santa Barbara, and University of California, Los Angeles, California

I. Introduction 1421
II. Examples of Paired Synthesis: Case Histories 1422
A. Synthesis of an intermediate (I) easily converted to II 1422
B. Paired oxidation of glucose to gluconate and reduction of glucose to sorbitol 1428
III. Concluding Remarks 1429
References 1429

I. INTRODUCTION

At first glance the title of this chapter may appear to contain a redundancy since by the very nature of the methodology all electrochemical reactions must occur in pairs: the number of electrons added at the cathode (for reduction) must simultaneously be removed at the anode (for oxidation). In the bulk of the literature, however, attention is focused on one of the pairs; the accompanying reaction at the counter electrode is chosen to be innocuous, not to interfere with starting materials, intermediates, or products. The "sacrificial" reactions at the counterelectrode are often the liberation of oxygen or hydrogen. "Paired reactions" in the sense used here are those in which both cathodic and anodic reactions contribute usefully to the formation of the final product(s) [1]. It is desirable also for the product(s) to be useful, although it is impossible for an organic chemical pair to emulate the effectiveness of caustic chlorine. The pairs proposed by Fioshin [2] are formally correct but yield no desirable products.

The proposal to devise paired syntheses is an old one but was not deliberately followed until about the last decade. The growing literature on this subject is being more frequently cited [3].

An outstanding example of a paired synthesis that leads to only one major product is the conversion of propylene to propylene oxide. (This has been a much studied reaction. Only recent papers, which refer to earlier work, are cited in Ref. 4). Propylene is passed into a cell in which aqueous sodium halide, preferably bromide, is electrolyzed. The olefin is hypohalogenated at or near the anode; the hypohalite reacts with the simultaneously generated sodium hydroxide at the cathode to yield the epoxide and regenerate the electrolyte. This procedure has at least the environmental advantage that it avoids the disposal of calcium halides when, in the nonelectrochemical process, lime is used for the ring-closure step.

†Deceased.

Paired syntheses leading to two products are more plentiful, as in the reductive dimerization of acrylate esters to adipate diesters and the simultaneous oxidation of malonic ester anions to bimalonates [1a].

Inventing paired syntheses, particularly if they are to be carried out in undivided cells, presents many intriguing challenges. The raw materials before electrolysis must be compatible. The intermediates (or products) formed at one electrode must not react irreversibly with the intermediates or products formed at the other electrode. Above all, an intermediate must not, before reacting, return to the counterelectrode and undo the electron transfer(s) that was so laboriously wrought. If the operation is to be a viable synthesis, it cannot be a large-scale cyclic voltammogram; otherwise we would have an exercise "full of sound and fury, signifying nothing."

When a single organic raw material is charged, its anodic product may be the one to react at the cathode (in propylene oxide synthesis, 2,3-butanediol→acetoin→2-butanone) [1c], or it may be itself oxidized and reduced (glucose→gluconate, glucose→sorbitol) [1d].

It has been pointed out that paired syntheses in undivided cells leading to two products engender separation problems. However, since developments in separation science are proceeding rapidly, we must have faith that separation scientists will be as ingenious in devising economical schemes as was the inventor of the paired synthesis.

The following two examples, drawn from the author's own experience, illustrate the problems that were encountered in devising these particular paired syntheses and how the difficulties were overcome. The first process was carried out in a divided H cell with a sintered glass divider; the second was achieved in a continuous flow cell without diaphragm. Other examples are listed in Table 1.

II. EXAMPLES OF PAIRED SYNTHESIS: CASE HISTORIES

A. Synthesis of an Intermediate (I) Easily Converted to II [19]

The final product desired was II, a candidate nonphosphate-containing biodegradable detergent builder. It had to be as cost effective as the long-used workhorse sodium tripolyphosphate (STP). (At this stage many potential processes are guillotined. The French may say c'est la vie even when it is applied to death. Almost always rejected chemical processes do not enjoy the acclaim that "still-life" art has, however.)

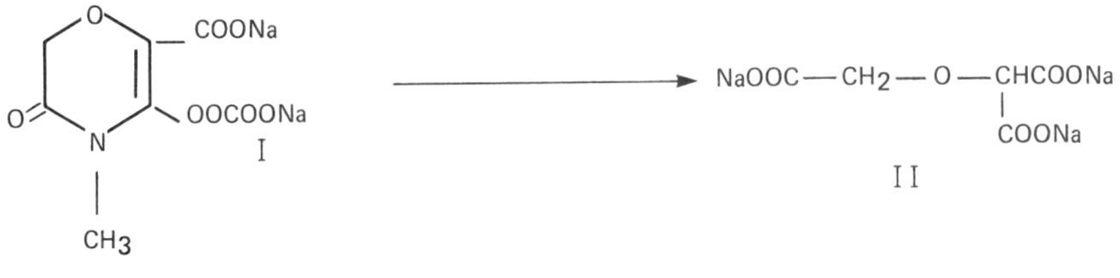

From the economic point of view it was obvious the II could not be prepared from a malonic acid precursor. Therefore the starting material had to be the relatively inexpensive diglycolic acid (III) or derivatives, such as the diester (IV) or the cyclic imide (V) obtained from IV by $MeNH_2$.

$HOCOCH_2OCH_2COOH$ III

$ROCOCH_2OCH_2COOR$ IV

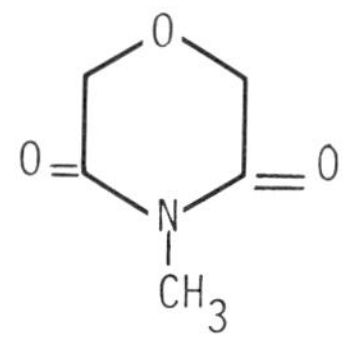

V

The problem was to carboxylate the $-CH_2-$ group adjacent to the carbonyl group.

Colleagues had already established that carboxylation of V took place more readily than that of IV and had developed a process using V, a large excess of phenoxide as base, and CO_2. The phenol had to be recovered and reconverted to the sodium salt, and the latter thoroughly dried. An electrochemical alternative would have to be at least competitive.

It was an irresistible attraction to bring to bear on this problem our previous experience with EGBs (Chapter 30) and electrocarboxylation [9,10]. A well-designed paired synthesis would avoid the messy recycling and produce only minor amounts of innocuous by-product(s).

The use of EGBs involves the following stages:

$$\text{PB} \xrightarrow{ne^-} \text{EGB}^-$$

$$\text{EGB}^- + \text{RH} \longrightarrow \text{EGBH} + \text{R}^-$$

$$\text{R}^- + \text{El} \longrightarrow \text{product}$$

$$\text{EGBH} + \text{proton scavenger} + 2e^- \text{ (or indirect)} \longrightarrow \text{PB} + \text{HX}$$

where PB = probase, RH = proton donor, and El = electrophile. In this case the PB had to be chosen, RH was V, and the proton scavenger was Na_2CO_3 ($\rightarrow NaHCO_3$).

For an electrocarboxylation to be successful, the following criteria must be met: the PB must be reduced anodically at CO_2 (that is, at a less negative potential), and EGB^- must not react or react reversibly with El.

Of the several PB candidates tried, VI was the most convenient. It is easily reduced in a one- or two-electron step to its EGB^-, and it does not react with CO_2.

$$\underset{\text{VI}}{(\text{EtOCO})_2\text{C}{=}\text{C}(\text{COOEt})_2} \xrightarrow[2H^+]{2e^-} \underset{\text{VII}}{(\text{EtOCO})_2\text{CH}{-}\text{CH}(\text{COOEt})_2}$$

The EGBH is easily oxidized to VI by anodically generated bromine.

The following system was therefore chosen: CH_3CN(S) solvent, Et_4NBr + NaBr + 18-Crown-6 as electrolyte (SE) in the cathode compartment through which CO_2 was continuously bubbled. The anolyte contained Na_2CO_3 and, for the first run only, to prevent the anode from quixotically destroying any of the SSE system, an equivalent of VII. The cathode was Hg, the anode a Pt foil, the temperature 45–50°C, and the cathode voltage −1.2 V versus SCE. The electrolysis was run to completion. The anolyte was stirred with activated Al_2O_3 and it and Na_2CO_3 and $NaHCO_3$ removed by filtration. The

Table 1 Some Paired Syntheses

Example	Product(s) obtained	Anode	Cathode	Type of cell[a]	Anode reaction	Cathode reaction	References
1	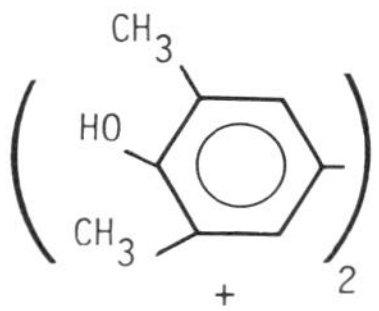 + $(EtOCOCH_2CH_2-)_2$	Graphite	Hg	D	2,6-Dimethylphenol Oxidatively coupled at 4-position	EHD of acrylate ester	1b
2	$MeOCO(CH_2)_8COOMe$ + $(EtOCOCH_2CH_2-)_2$	Graphite	Pb/graphite	U	Crum Brown-Walker[b]	EHD of acrylate ester	5
3	$HOOCC{\equiv}CCOOH$ + dihydrophthalic acid	Ti/PbO_2 with Ti carbide interlayer		U	$HOCH_2C{\equiv}CCH_2OH$	Phthalic acid $\longrightarrow$ dihydrophthalic acid	6
4	Gluconate + sorbitol	Graphite	Raney Ni	U	Glucose $\longrightarrow$ gluconate via Br^+	Glucose $\longrightarrow$ sorbitol via electrocatalytic reduction	1e
5	2-Butanone	Graphite	Pb/Hg or Zn/Hg	U	2,3-Butanediol $\longrightarrow$ acetoin; indirect	Acetoin $\longrightarrow$ 2-butanone	1c

6	Dialkyl malonate + tetraalkylbimalonate	—	—	U	$CH_2(COOR)_2 \longrightarrow [(ROCO)_2CH-]_2$ Indirect via I^+	EHD of acrylate ester	1a, 8
7	t-Butylhydroquinones	PbO_2	Pb	D	t-Butylphenols $\longrightarrow$ Butylquinones	t-Butylquinones $\longrightarrow$ butylhydroquinones	11
8	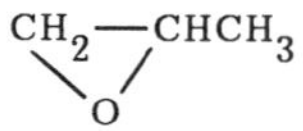	—	—	U	X^- (preferably Br $\longrightarrow$ NaXO)	$2Na^+ \longrightarrow 2Na$ $2Na + 2H_2O \longrightarrow 2NaOH + H_2$	
9	Ferrocene	Fe	—	U	Fe $\longrightarrow$ Fe ions	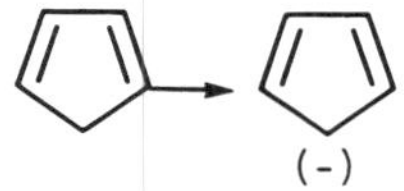	12
10	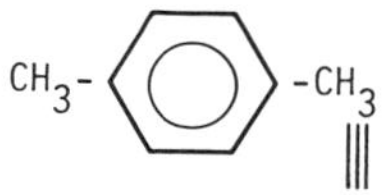 p-Xylene	Graphite	Graphite	U	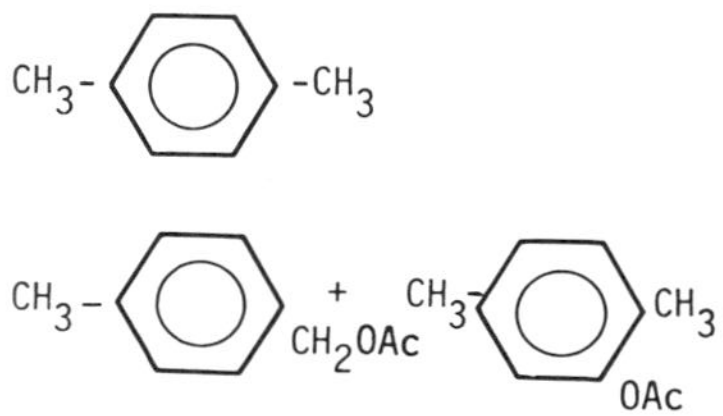	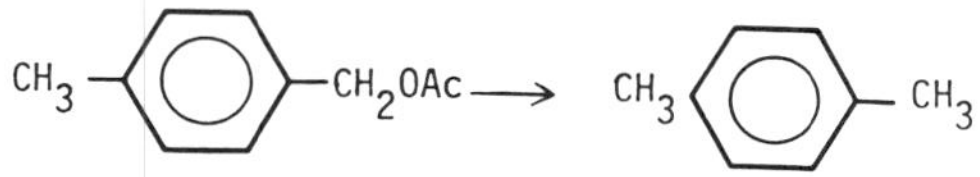	7
11	RNHCOOMe	Pt	Pt	U	$Br^- \longrightarrow Br^+$ $Br^+ + RCONH_2 + MeO^- \longrightarrow$ RNHCOOMe via RNCO	$K^+ \longrightarrow K^o$ $K^o + MeOH \longrightarrow MeO^- + K^+ + \frac{1}{2}H_2$	13

Table 1 (Continued)

Example	Product(s) obtained	Anode	Cathode	Type of cell[a]	Anode reaction	Cathode reaction	References
12	Pr_2SO_2	—	—	D	$Br^- + PrOH \longrightarrow PrBr$ + other products	$2PrBr + SO_2 \longrightarrow 2Pr_2SO_2 + 2Br^-$	14
13	$(MeO)_2CHCH_2-CH_2COOMe$	Graphite	Graphite	U	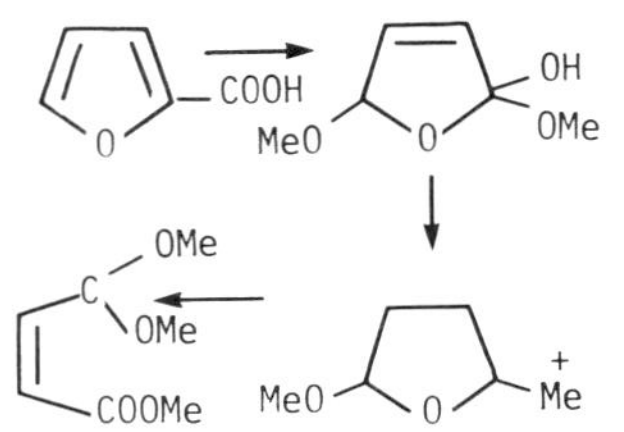	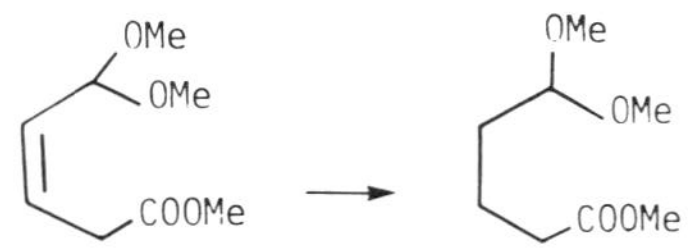	15
14	$PhC\equiv CR$	—	—	U	$I^- \longrightarrow I \quad I + R_3B \longrightarrow RI$	$PhC\equiv CH \longrightarrow PhC\equiv C^-$	16
15	$RCH_2CH(R)-COOEt$	Pt	Pt	U	$I^- + Br^-$ in presence of R_3B, $CH_2=CHCOOEt$	$CH_2=CHCOOEt \longrightarrow (CH_2-CHCOOEt)^{\bar{\cdot}} \xrightarrow{RX} RCH_2\bar{C}HCOOEt \xrightarrow{RX} RCH_2CH(R)COOEt$	17

16	Indazoles	—	—	U	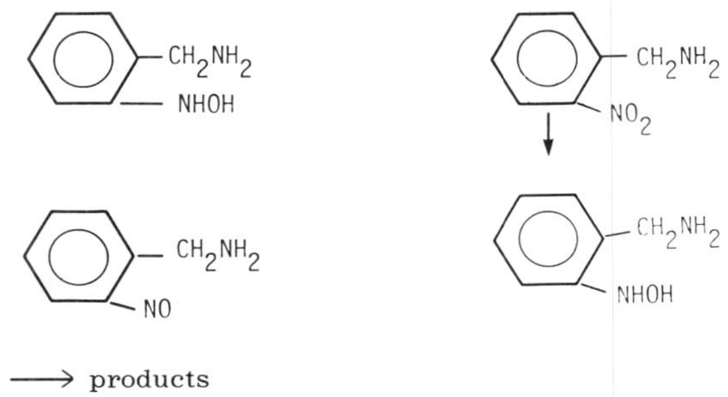	18, 23[d]

[a]U = undivided; D = divided.
[b]A variant of the Kolbe reaction in which half-esters are oxidatively decarboxylated; the resulting radicals couple to yield diesters.
[c]Thus the unwanted isomer is reduced to starting material for recycling.
[d]When the number of electrons supplied at the cathode is not exactly the same as the number removed at the anode, and extra electrode should be used.

I that had precipitated in the catholyte was also removed by filtration. Treatment of I with NaOH yielded II and the $MeNH_2$ for recycling.

For the second run the former anolyte was charged with fresh V and used as the catholyte. In three successive runs using this procedure, the following yields (current efficiencies), both as percentages, were obtained: 87 (70), 85 (58), 88 (62).

Engineering cost estimators who tend to look at "new" technology with a jaundiced eye agreed that this process was at least competitive with the nonelectrochemical process. (Perhaps this is engendered by the biliousness indigenous to the type of activity. After all, it does take a certain amount of gall to estimate the cost of operating on a large scale a type of process with which one is completely unfamiliar.) The cost effectiveness of II compared to STP? Other companies found at least temporarily other cheaper solutions.

B. Paired Oxidation of Glucose to Gluconate and Reduction of Glucose to Sorbitol

The indirect oxidation by anodically generated bromine of glucose to gluconate has been and is used commercially. The electrochemical reduction to sorbitol (and mannitol) was for some years an industrial process until it was replaced by a catalytic one. If the two could be paired in a single undivided flow cell, the economics for each would be improved. (Despite that the process had been invented by careful, honorable laboratory chemists, the estimators included provisions for one after another of the steps going wrong.)

Preliminary experiments were done in H cells, and it was established that (1) Br^+ oxidation of glucose at graphite was indeed very efficient, (2) gluconate was not reduced at the cathode when glucose was present, and (3) fructose is more easily reduced to sorbitol (and mannitol) than glucose and therefore a pH high enough to affect this isomerization had to be avoided.

However, cathodic yields and current efficiencies were unsatisfactory using the "best" of the cathodes (amalgamated Zn) previously employed; the deficiency did <u>not</u> arise from the cycling of Br^- from anode to cathode. (Admittedly an additional fillip was provided to this project by the prospect of finally replacing a catalytic process by an electrochemcial one; the contrary has been the nemesis of many an electrochemical process.) A summary of the cathodes tried and the results reported is given in Ref. 20.

It was rediscovered that not gluconate but gluconolactone is the first oxidation product of glucose [21], that the lactone is easily reduced to glucose, and that it is not instantly saponified to gluconate. Interposing a holding tank in the circulating system to allow time for the saponification to occur helped a little.

There was still the problem that although calcium gluconate was the desired anodic end product, Ca^{2+} could not be used in the supporting electrolyte. To circumvent these and other difficulties, it was decided to carry out the cathodic reduction electrocatalytically. Chiba et al. [22] reported the successful electrocatalytic hydrogenation at Raney Ni of a number of aromatic carbonyl compounds but failed in the reduction of aliphatic carbonyls because at the relatively high pH they were using aldol condensations and other problems occurred. Since we expected to operate at lower pH, it was anticipated that the latter side reactions would not occur. Also, now calcium salts could be used as supporting electrolytes.

Thus it was that almost 100% yields and current efficiencies were obtained for both products using a Raney Ni cathode, calcium bromide electrolyte, pH 5–7, 60°C, 250–500 mA per 10 g of cathode, graphite chip anodes, 1.6 M glucose solution, and a volumetric flow rate of 100 ml m^{-1} [12].

III. CONCLUDING REMARKS

The question has frequently arisen concerning the scale of production of paired cathodic and anodic products. Both may not be useful to the same extent. However, it is possible to choose one partner whose product is required in very large amount (e.g., adiponitrile or adipate esters) and to produce the second one on the scale desired. The first product, at a good price, should find its way into the hands of manufacturers by other means.

Table 1 summarizes some of the paired syntheses that have been reported recently.

REFERENCES

1. (a) M. M. Baizer and R. C. Hallcher, J. Electrochem Soc., 123(6), 809 (1976); (b) M. M. Baizer, R. C. Hallcher, and O. Deex, Research disclosure 15903, July 1977, p. 4; (c) M. M. Baizer, T. Nonaka, K. Park, Y. Saito, and K. Nobe, J. Appl. Electrochem., 14, 197 (1984); (d) P. N. Pintauro, D. K. Johnson, K. Park, M. M. Baizer, and K. Nobe, J. Appl. Electrochem., 14, 209 (1984); (e) K. Park, P. N. Pintauro, M. M. Baizer, and K. Nobe, J. Electrochem. Soc., 132(8), 1850 (1985).
2. M. Ya. Fioshin, Khim. Prom., 44, 882 (1968); Chem. Abstr., 70, 63434k (1969).
3. (a) A. T. Kuhn, in Industrial Electrochemical Processes (A. T. Kuhn, ed.), Elsevier, Amsterdam, 1971, pp. xi–xii; (b) M. M. Baizer, in Proceedings of the Symposium on Electrochemical Processes and Plant Design (R. C. Alkire, T. R. Beck, and R. D. Varjian, eds.), The Electrochemical Society, Pennington, N. J., 1983, pp. 31–41; (c) M. M. Baizer, Tetrahedron, 40, 935 (1984).
4. (a) A. Manji and C. W. Oloman, J. Appl. Electrochem., 17(3), 532 (1987); (b) T. C. Chou, J. J. Jow, and S. T. Yu, Chem. Eng. Commun., 42(4–6), 191 (1986); Chem. Abstr., 104, 207713w (1986); (c) F. Goodridge, S. Harrison, and R. E. Plimley, J. Electroanal. Chem., 214(1–2), 283 (1986); (d) R. C. Alkire and J. D. Lisius, J. Electrochem. Soc., 132(8), 1879 (1985); (e) K. G. Ellis and R. E. W. Jansson, J. Appl. Electrochem., 11(4), 531 (1981).
5. T. E. Rogers, R. E. Jansson, and M. M. Baizer, U.S. Patent 4,457,813 of July 3, 1984; Chem. Abstr., 101, 100219 (1984).
6. D. Degner, in Technique of Electro-Organic Synthesis (N. Weinberg, ed.), Part II, Wiley-Interscience, New York, 1975, pp. 559–665.
7. L. Eberson and E. Oberrauch, Acta Chem. Scand., B35, 193 (1981); E. Oberrauch and L. Eberson, Fr. Demande 2,486,968 of 15 July 1980; Chem. Abstr., 96, 18673t (1982).

8. H. G. Thomas and E. Lux, Tetrahedron Lett., 965 (1972); D. K. Johnson and R. E. W. Jansson, J. Chem. Technol. Biotechnol., 30(4), 200 (1981); Chem. Abstr., 93, 113509m (1980); J. Electrochem. Soc., 128(9), 1885 (1981).
9. M. M. Baizer and J. L. Chruma, J. Org. Chem., 37, 1951 (1972).
10. M. M. Baizer and R. C. Hallcher, Liebigs Ann. Chem., 737 (1977).
11. S. Takahashi and H. Uchida, Jpn. Kokai Tokkyo Koho, JP 61,106,784 of 24 May 1986; Chem. Abstr., 105, 215765a (1986).
12. E. A. Chernyshev, A. V. Bukhtiarov, B. K. Kabanov, A. P. Tomilov, N. A. Maier, V. L. Shirokii, and Yu. A. Ol'dekop, Elektrokhimiya, 10(2), 239 (1982).
13. T. Shono, Y. Matsumura, S.-I. Yamane, and S. Kashimura, Chem. Lett., 565 (1982).
14. H. J. Wille, D. Knittel, B. Kastening, and J. Mergel, J. Appl. Electrochem., 10, 489 (1980).
15. T. Iwasaki, T. Nishitani, H. Horikawa, and I. Inoue, J. Org. Chem., 47(19), 3799 (1982).
16. Y. Takahashi, M. Tokuda, M. Itoh, and A. Suzuki, Chem. Lett., 461 (1980).
17. Y. Takahashi, K. Yuaza, M. Tokuda, M. Itoh, and A. Suzuki, Bull. Chem. Soc., Jpn., 51(1), 339 (1978).
18. R. Hazard and A. Tallec, Bull. Soc. Chim. Fr., (3–4), 679 (1975).
19. R. C. Hallcher, D. A. White, and M. M. Baizer, J. Electrochem. Soc., 126, 404 (1979); M. M. Baizer, CHEMTECH, 178 (1985).
20. H. L. Chum and M. M. Baizer, The Electrochemistry of Biomass and Derived Materials, American Chemical Society, Monograph 183, American Chemical Society, Washington, D.C., 1985, pp. 230–233.
21. N. Federonko, Adv. Carbohydr. Chem. Biochem., 29, 107 (1974).
22. T. Chiba, M. Okimoto, H. Nagai, and Y. Takata, Bull. Chem. Soc. Jpn., 56, 719 (1983).
23. C. Moinet and A. Tallec, Electrochim. Acta, 31, 1 (1986); J. Appl. Electrochem., 16, 819 (1986).

36
Challenges and Opportunities

MANUEL M. BAIZER† / University of California, Los Angeles, and University of California, Santa Barbara, California

I. Introduction 1431

II. Specific Challenges and Opportunities 1432
   A. Use of undivided rather then divided cells 1432
   B. Orientation of organic molecules at electrodes: Implications for synthesis 1433
   C. Alternative, non-petroleum-based raw materials 1433
   D. Catalytic electrodes 1435
   E. Complexed substrates 1435
   F. Miscellaneous areas in need of further R&D 1436

References 1436

I. INTRODUCTION

A prominent organic chemist once said that the ultimate objective of his field was the synthesis of life. As organic electrochemists we have a more modest but nonetheless idealistic first objective: to replace as many of the present "chemical" processes that use reagents with cleaner, milder, more selective electrochemical alternatives. It is inevitable that in the course of this work many unexpected reactions and phenomena are discovered.

We need better understanding of the phenomena at the basic fundamental level (adsorption and orientation of organic molecules at electrode surfaces; heterogeneous and homogeneous electron transfers; nature of first-formed intermediates; and double-layer structure at least insofar as it affects "branching" of follow-up reactions). At the applied and technological level we need even better cells and systems for economical processing; means for efficiently handling dilute solutions that may be required both to take advantage of a particular orientation of organic molecules at electrode surfaces and to exploit future electroenzymological reactions.

Increasingly it will become necessary for the (synthetic) organic electrochemist to work as a member of a team comprising physical chemists, biochemists, separations specialists, catalyst chemists, solid-state physicists, metallurgists, and perhaps others. At all but the very earliest exploratory stages, electrochemical engineers should become involved. The greatest advances in the field are yet to come, and they will come only by multidisciplinary efforts.

A challenge that must be met before any of these objectives can be accomplished is this: How can cooperation between several people with different areas of expertise be accomplished in an academic environment? Is it

†Deceased.

not a contradiction in terms to expect a harmonius chorus of prima donnas? The team style of research and development groups has now become fairly common in industrial laboratories. Perhaps academe can learn something from industry? Industry has certainly been happy for generations to adapt to its own purposes the fundamental work done at universities and institutes. Pride and parochialism, when they are obstacles, must be overcome. The American automobile industry is now adopting the Japanese management style.

It is unfortunate that in the United States at least funding on a national level for research and development for the conversion of non-petroleum-based materials to fuels, energy, and chemicals decreases as the price of a barrel of crude oil decreases. Coupled with the alleged desire to gain independence from and the need to import this vital feedstock is the purpose of maintaining "stability" and the political status quo in the supplier countries. Thus the immense need for long-term R&D to use effectively and economically as basic raw materials shale oil (fossil), coal (fossil), and biomass (renewal) is met only with sporadic effort. Relevant work in electrochemistry, which will be a component of the total technology needed, is likewise receiving erratic financial support.

These challenges confront us not only because we are organic electrochemists but because we are scientists, citizens of our respective countries and of the one world of science, and members of groups working in several milieus (academic, industrial, research, institutional, and others). We must each, in whatever way seems appropriate, create opportunities for better interdisciplinary cooperation and for far-sighted advancement of this field into the complex areas that surely lie ahead.

It is extremely gratifying to discover an organic electrochemical synthesis that is unique; that is, the product cannot be made by any other means. Less euphoric but still exhilarating is a synthesis that is easier, cleaner, and more economical than an alternative nonelectrochemical process. It seems to be an inescapable conclusion that the greatest advantage is taken of organic electrochemistry when one employs the feature of this methodology that is truly unique, not possessed by any other, namely, the inevitable, unavoidable genetic property of electrochemical reactions in cells that they must occur in pairs (Chap. 35).

II. SPECIFIC CHALLENGES AND OPPORTUNITIES

What are some of the specific challenges for organic electrochemists and engineers that offer opportunities for important advances in this science and technology? A sampling follows, not presented in any order of priority.

A. Use of Undivided Rather Than Divided Cells [1]

Although advances in the composition and fabrication of separators, when they are unavoidably essential, is very desirable, it is the author's bias that many more reactions are carried out in divided cells than are necessary. The savings in energy and complexity of operations involved in undivided cells is obvious. There have already been a number of dramatic demonstrations of substitution of undivided for divided cells: in the hydrodimerization of acrylonitrile to adiponitrile (Chap. 31), in the reduction of benzene to 1,4-

cyclohexadiene [2], and in the paired oxidation of glucose to gluconic acid and reduction of glucose to sorbitol [3]. This stratagem places additional burdens on schemes for separating product(s) from by-product(s), from solvent and electrolyte; however, advances in separations science are proceeding apace and should be adapted (or new ones invented) to meet the needs of a particular case.

B. Orientation of Organic Molecules at Electrodes: Implications for Synthesis

Hubbard and co-workers have published many fundamental studies [4] of the orientation of aromatic hydrocarbons and of aromatics bearing hetero atoms (e.g., O, N, and S) in the rings or on side chains at clean platinum anodes. The orientation of a particular molecule (e.g., flat or perpendicular to the electrode surface) depends on the applied voltage, on the concentration of substrate, and on the presence or absence of particularly strongly adsorbed species, such as iodide. One must infer that the oxidation products of a molecule adsorbed in a particular orientation are different from those when it is adsorbed in a different orientation. The demonstration of these differences is one of the longer range objectives of the Hubbard group, but many difficult problems must be solved before this can be achieved. Very dilute solutions have been involved so far. The analytic problems have been formidable. How can the results be translated to a synthetically useful scale? Cell designs and suitable systems must be devised. Successful solutions to these problems would add a new dimension to selectivity in organic electrochemical synthesis.

C. Alternative, Non-petroleum-based Raw Materials

Two types of end products are desired: (1) fuels and (2) organic chemicals and their transformation products. In category (1), CH_3OH and hydrocarbons are target molecules. The raw materials have the attractive feature that they need not be derived from petroleum, but it remains desirable that less energy be consumed in converting them to the fuels named than the latter can furnish on combustion. For category 2 it is most desirable that products of more than one carbon atom be obtained.

Simple Molecules: CO and CO_2

CO has been electrochemically reduced to CH_3OH [5],* to CH_4 [6], to C_{1-2} hydrocarbons and ethylene [7], and to C_{1-4} hydrocarbons [8]. Clearly a a great deal more work, applying a multiplicity of techniques in concert, is necessary before these reductions are more than interesting laboratory observations.

Chemical products or intermediates made by electrochemical reduction of CO <u>alone</u> are squaric acid, the cyclic tetramer [9] and the recently described dimer [10] that arises by reaction of CO with solvated electrons in liquid NH_3.

---

*It is significant that only Ref. 5a reports simple reduction at Pt in CH_3CN containing residual water. The Ogura papers [5b–g] used photocells, chemically modified electrodes, or CMEs plus Fe(II), Fe(III), or Cr(III) complexes.

Electroorganic syntheses in which CO is a coreagent have been recently reviewed (Ref. 1, p. 946ff.).

Although the reductions of CO_2 in simple systems to formate in aqueous and to oxalate in nonaqueous media have been known for some years now [11], they remain the subject of investigations [12]. However, a problem of potential great industrial importance remains to be solved: the reduction of CO_2 to oxalate (or its further reduction products glyoxylic acid, glycolic acid, and d,l-tartaric acid) in aqueous media at moderate overvoltages. Again it is plain that multiple techniques will have to be employed simultaneously.

Electrolytic reductive couplings with CO_2 as a coreagent continue to provide a useful route to carboxylic acids [13]. Many of these studies also use in situ catalysts and devote more attention than has previously been the case to the effects of several different cathode materials.

Because of the temper of the times, an unprecedented degree of attention is being paid to the technical possibility of reducing CO_2 to fuel or fuel components, CO [14],* CH_3OH, and CH_4 [15], preferably in aqueous media. Could the CO be reductively dimerized in situ to the more valuable ethylene glycol? It has been done catalytically.

Complex Structures: Coal and Biomass

Many technological advances, nonelectrochemical in nature, have been made in rendering coal a more attractive fuel source: means for providing water or oil slurries that can be transported by pipeline, alkylation and partial hydrogenation for increasing the solubility of some coal constituents, and others. Despite an early promising start [16], research on the electrochemistry of coal remained sluggish until about 1980. A 1978 symposium volume [17] does not have a single entry relating to electrochemistry. This situation was soon rectified [18] and continues to improve [19]. Here, too, the major emphasis has been on the use of coal as a fuel, including the Fe(II)/Fe(III)-mediated oxidation [20] of coal at anodes to provide potentially economic hydrogen at cathodes [21].

There have been many developments this last decade in the exploration of so-called C_1 chemistry. The raw materials are CO and H_2 obtained by drastic decomposition of coal. Elaborate syntheses, usually catalytic, have been devised for converting this mixture to a variety of organic products. This is "syngas chemistry"; it seems, however, a sin to degrade coal all the way down to these simple materials when the coal structure has so many units that are or can easily be converted to industrial organic chemicals.† What is needed is a means for selectively cleaving bonds, a task electrochemistry is especially qualified to achieve [22].

Biomass (Chap. 20 and Ref. 11) is likewise composed of complex modules that can be degraded either severely to yield simple building blocks or more gently to provide useful products. Electrochemistry has already demonstrated its role (Chap. 20). Much remains to be done before biomass is "a potential resource for meeting all our chemical needs in place of petrochemicals" [23].

---

*The reduction of CO_2 in DMSO at Au or Hg electrodes produced Co and carbonate [19a]. How these products as well as oxalate may arise from the intermediate CO_2 anion radical was explained by Amatore and Savéant [14b].

†This must give the readers who remember the 1920s a sense of déjà vu since up until that time coal was a major feedstock for the industry.

D. Catalytic Electrodes

Unfortunately the term "electrocatalysis" is used by organic electrochemists in several senses, which include means for lowering the overvoltage for a given electrode reaction and for using, preferably in situ, electroregenerable reagents ("mediators" or "electron transfer agents"). Some of these are discussed in Chaps. 29, 30, and 33.

The focus of attention here is only on the synthetic advantages of using as cathode metals, which in a proper state of subdivision are known to be catalysts for nonelectrochemical hydrogenation. Previous examples of the use of Raney nickel and Pt and Rh black are cited in Ref. 24. In Ref. 24 glucose was successfully reduced to sorbitol even in the presence of Ca^{2+} at Raney nickel; other methods failed to give satisfactory results. One method of reactivating the catalytic cathode has been described [24b]. "Dimensionally stable" electrodes with Raney nickel-like surfaces plus Devarda copper alloy have also been described [25], for example, in the reduction of nitrobenzene to aniline. Kuwana and Kao discuss the incorporation of catalytic metal particles within a porous polymer layer of glassy carbon [26]. Renewable Ni surfaces deposited on graphite or vitreous carbon have been used in the electrocatalytic reduction of carbonyl compounds [27]. There has been a recent review of the electrocatalytic reductive cleavage of certain carbonyl compounds and alcohols to hydrocarbons [28]. This is relevant to biomass-derived materials.

The use of catalytic metal cathodes allows the use of lower cathode potentials than required for other cathodes and allows the reaction to proceed at reasonable rates at ambient temperatures and pressure, thus avoiding the use of costly equipment and the hazards associated with stored hydrogen.

E. Complexed Substrates

This refers to many types of complexation, some of which have recently been cited (Ref. 1, p. 958). Here we refer to another type that has led to interesting synthetic results but requires further elucidation of mechanism and definition of scope: the participation of transition metal cations, not themselves reduced under the experimental conditions used and therefore not "mediators," in the reduction of carbonyl compounds, α,β-unsaturated esters, and undoubtedly many other electrophores.

In the presence of Cr(III) and a minimum amount of Cr(II), a variety of aromatic ketones and aldehydes (including β-ionone and retinal) are reduced to their respective pinacols. It has been proposed that a complex between the carbonyl compound and Cr(III) is the species being reduced [29].

The bis-(α,β-unsaturated) ester I has been reductively cyclized to *trans*- and *cis*-II in 58–87% yield. The *trans*-*cis* ratio varied with cathode material (Hg, glassy C, or Cu) and with proton donor present (HOAc/H_2O and diethyl malonate) but the highest *trans*-*cis* ratio thus obtained was 11.6:1. Inclusion of $CeCl_3$ in the medium (CH_3CN) increased the *trans*-*cis* ratio to 14.8:1 and the rate of reduction nearly fivefold [30].

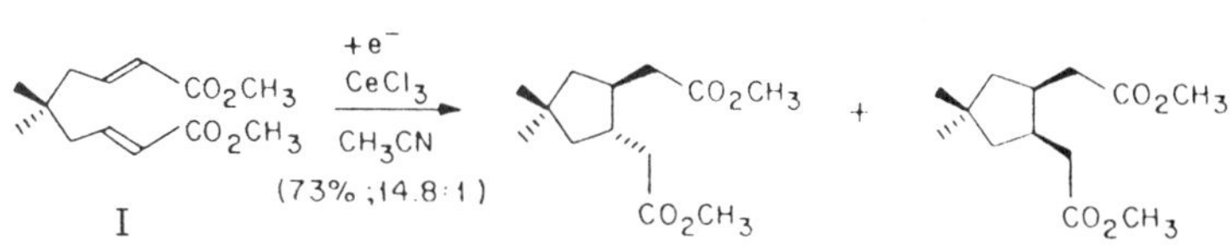

I (1)

F. Miscellaneous Areas in Need of Further R&D

Limitations of space permit us to mention only briefly two other areas that offer scientific and technological opportunities and to cite relevant leading references.

Bioelectrochemistry and particularly the use of electrochemical reactions instead of cofactors in conjunction with enzymatic systems (which we call electroenzymology) may provide means for specific syntheses not attainable by other means [31]. For example, p-hydroxybenzaldehyde has been obtained in about 80% yield from p-creosol [31d].

Examples of electrogenerative processes in which both chemicals (including organic chemicals) and electricity are simultaneously produced have been reviewed [32]. The authors mention some deterrents to the use of this technology but express optimism concerning the outlook for the next 10 years.

REFERENCES

1. M. M. Baizer, Tetrahedron, 40(6), 935 (1984).
2. J. P. Coleman and J. H. Wagenknecht, J. Electrochem. Soc., 128, 322 (1981).
3. K. Park, P. N. Pintauro, M. M. Baizer, and K. Nobe, J. Electrochem. Soc., 132(8), 1850 (1985); J. Appl. Electrochem., 16, 941 (1986).
4. A. T. Hubbard, Report 1985, AFOSR-TR-85-0627m, Order No. AD-A158621; Chem. Abstr., 104, 193875m (1986).
5. (a) G. Pierre and A. Ziade, Nouv. J. Chim., 10(4–5), 233 (1986); (b) K. Ogura and M. Takagi, Sol. Energy, 37(1), 41 (1986); Chem. Abstr., 105, 87397t (1986); (c) K. Ogura and S. Yamasaki, J. Chem. Soc., Faraday Trans. I, 81, 267 (1985); (d) J. Mol. Catal., 31, 49 (1985); (e) K. Ogura and S. Yamasaki, J. Chem. Soc., Dalton Trans. (12), 2499 (1985); (f) K. Ogura and M. Takagi, J. Electroanal. Chem., 195(2), 2571 (1985); (g) K. Ogura and H. Watanabe, J. Chem. Soc., Faraday Trans. I, 81(7), 1569 (1985).
6. (a) V. V. Strelets and V. V. Tsarev, Kinet. Katal., 25(4), 821 (1984); Chem. Abstr., 101, 198825t (1984); (b) T. M. Gür and R. A. Huggins, Science, 219, 967 (1983).
7. G. A. Kolyagin, V. G. Danilov, V. L. Korienko, I. A. Kedrunskii, and S. P. Gubin, Elektrokhimiya, 19(7), 1004 (1983); Chem. Abstr., 99, 95730u (1983).
8. G. N. Petrova, O. N. Efimov, and V. V. Strelets, Izv. Akad. Nauk SSSR, Ser. Khim., (11), 2608 (1982); Chem. Abstr., 98, 61943e (1983).
9. (a) G. Silvestri, S. Gambino, G. Filardo, and L. Palmisano, Electrochim. Acta, 23, 413 (1978); (b) J. J. Barber, U.S. Patent 4,461,681 of 31 May 1983; Chem. Abstr., 101, 160121x (1984).
10. F. A. Uribe, P. Sharp, and A. J. Bard, J. Electroanal. Chem., 152, (1–2), 173 (1983).
11. H. L. Chum and M. M. Baizer, The Electrochemistry of Biomass and Derived Mateials, ACS Monograph 183, American Chemical Society, Washington, D.C., 1985, Chap. 3.
12. (a) N. M. Mahmood and D. Masheder, UK Patent Appl. GB 2,171,115 of 20 August 1986; Chem. Abstr., 105, 21563y (1986); (b) J. Y.

Becker, B. Vainas, R. Eger, and L. Kaufman, J. Chem. Soc., Chem. Commun. (21), 1471 (1985); (c) Y. Hori, K. Kikuchi, and S. Suzuki, Chem. Lett.(11), 1695 (1985); (d) K. Ito, S. Ikeda, N. Yamauchi, T. Iida, and T. Takagi, Bull. Chem. Soc. Jpn., 58(10), 3027 (1985); (e) C. M. Bolinger, B. P. Sullivan, D. Conrad, N. Story, and T. J. Meyer, J. Chem. Soc., Chem. Commun.(12), 796 (1985); (f) T. Fukaya, M. Kodaka, and M. Sugiura, Chem. Abstr., 105, 68905r (1986); (g) P. G. P. Ang, A. F. Sammels, and A. Morduchowitz, U.S. Patent 4,595,465 of 17 June 1986; Chem. Abstr., 105, 87537p (1986); (h) X. Cao, Y. Mu, and L. Luan, Chem. Abstr., 104, 195348r (1986); (i) D. J. Pearce and D. Pletcher, J. Electroanal. Chem., 197(1–2), 317 (1986); (j) M. H. Miles, A. N. Fletcher, G. E. McManis, and L. V. Spreer, J. Electroanal. Chem., 190(1–2), 157 (1985).

13. (a) G. Silvestri, S. Gambino, and G. Filardo, Tetrahedron Lett., 27(29), 3429 (1986); (b) S. Torii, H. Tanaka, T. Hamatani, K. Morisaki, A. Jutand, F. Pfluger, and J. F. Fauvarque, Chem. Lett. (2), 169 (1986); (c) J. H. Wagenknecht, U.S. Patent 4,582,577 of 15 April 1986; Chem. Abstr., 105, 50861p (1986); (d) G. Filardo, G. Silvestri, and S. Gambino, Eur. Patent Appl. E. P. 189,120 of 30 July 1986; Chem. Abstr., 105, 180522g (1986); (e) M. A. Casadei, F. M. Moracci, M. Franco, and A. Inesi, J. Chem. Soc., Perkin Trans. II,(3), 419 (1986); (f) Y. Ikeda and E. Manda, Bull. Chem. Soc. Jpn., 58(6), 1723 (1985); J. C. Folest, J. M. Duprilot, J. Perichon, Y. Robin, and J. Devynck, Tetrahedron Lett., 26(22), 2633 (1985); (g) H. Matschiner and H. H. Rüttinger, J. Prakt. Chem., 327, 45 (1985); (h) H. H. Rüttinger, W. D. Reidorf, and H. Matschiner, Electrochim. Acta, 30(2), 155 (1985); J. H. Wagenknecht, U.S. Patent 4,601,797 of 22 July 1986, Chem. Abstr., 106, 24982e (1986).

14. (a) L. V. Haynes and D. T. Sawyer, Anal. Chem., 39, 332 (1967); (b) C. Amatore and J.-M. Savéant, J. Amer. Chem. Soc., 103, 5021 (1981).

15. (a) D. L. DuBois and A. Miedaner, J. Amer. Chem. Soc., 109(1), 113 (1987); (b) K. W. Frese, Jr., S. C. Leach, and D. P. Summers, U.S. Patent 4,609,441 of 2 Sept. 1986; Chem. Abstr., 105, 109209q (1986); (c) K. Ogura, Proc. Electrochem. Soc., 86–87, 591 (1986); (d) M. Beley, J. P. Collin, R. Ruppert, and J. P. Sauvage, J. Amer. Chem. Soc., 108(24), 7461 (1986); (e) D. P. Summers, S. Leach, and K. W. Frese, Jr., J. Electroanal. Chem., 205(1–2), 219 (1986); (f) K. Ogura and I. Yoshida, J. Mol. Catal., 34(1), 67 (1986); (g) D. J. Pearce and D. Pletcher, J. Electroanal. Chem., 197(1–2), 317 (1986); (h) K. Ogura and M. Takagi, J. Electroanal. Chem., 206(1-2) 209 (1986); (i) J. Electroanal. Chem., 201(2), 359 (1986); (j) Y. Hori, K. Kikuchi, A. Murata, and S. Suzuki, Chem. Lett. (6), 897 (1986); (k) K. Ogura, J. Appl. Electrochem., 16(5), 732 (1986); (l) M. Beley, J. P. Collin, J. P. Sauvage, J. P. Petit, and P. Chartier, J. Electroanal. Chem., 206(1–2), 333 (1986); (n) B. P. Sullivan, C. M. Bolinger, D. Conrad, W. J. Vining, and T. J. Meyer, J. Chem. Soc., Chem. Commun. (20), 1414 (1985); (m) K. W. Frese, Jr., S. C. Leach, and D. P. Summers, U.S. Patent 4,609,440 of 2 Sept. 1986; Chem. Abstr., 105, 199208p (1986).

16. H. W. Sternberg and I. Wender, Report 5943, U.S. Dept. of the Interior, 1962.
17. J. W. Larsen (ed.), Organic Chemistry of Coal, ACS Symposium Series, No. 71, American Chemical Society, Washington, D.C., 1978.
18. J. B. Pierce, Research on the Electrochemistry of Coal-Derived Liquids, ACS Symposium Series 169, 331–345 (1981).
19. (a) G. Kreysa, Electrochemistry of Coal, Nachr. Chem. Tech. Lab., 34(10), 967–968, 970–972 (1986); Chem. Abstr., 105, 229394r (1986); (b) M. Santoa and R. P. Baldwin, J. Appl. Electrochem., 16(2), 203 (1986); (c) G. Kreysa and W. Kochanek, DECHEMA Monogr., 98, 479 (1985); Chem. Abstr., 104, 91836g (1986); (d) L. Franke and A. Bosch, Ger. (East) DD 223,173 of 5 June 1985; Chem. Abstr., 103, 198470q (1985).
20. (a) G. Okada, V. Gurnsway, and J. O'M. Bockris, J. Electrochem. Soc., 128(10), 2097 (1981); (b) S. B. Lalvani, Chem. Eng. Commun., 48(1–3), 117 (1986); Chem. Abstr., 106, 10470z (1986); (c) G. Kreysa and W. Kochanek, J. Electrochem. Soc., 132(9), 2084 (1985).
21. R. W. Coughlin and M. Farooque, J. Appl. Electrochem., 10, 729 (1980) and references therein.
22. (a) S.-M. Park, J. Electrochem. Soc., 131(9), 363 (1984); (b) L. Schanne and M. W. Haenel, Tetrahedron Lett., 4245 (1979); (c) C. J. Collins, H.-P. Hombach, B. Maxwell, M. C. Woody, and B. M. Benjamin, J. Amer. Chem. Soc., 102(2), 851 (1980); (d) R. W. Coughlin and G. Kreysa, Chem.-Ing.-Tech., 55(1), 8 (1983).
23. I. S. Goldstein (ed.), Organic Chemicals from Biomass, CRC Press, Baton Rouge, Florida, 1981.
24. (a) K. Park, P. N. Pintauro, M. M. Baizer, and K. Nobe, J. Electrochem. Soc., 132(8), 1850 (1985); (b) J. Appl. Electrochem., 16, 941 (1986).
25. (a) J. Lessard and G. Belot, U.S. Patent 4,584,069 of 22 Feb. 1985; Chem. Abstr., 105, 14281r (1986); (b) G. Belot, S. Desjardins, and J. Lessard, Tetrahedron Lett., 25(4), 5347 (1984).
26. T. Kuwana and W. H. Kao, U.S. Patent 4,541,905 of 13 December 1983; Chem. Abstr., 103, 202755f (1985).
27. M. J. Lain and D. Pletcher, Electrochim. Acta, 30(1), 99, 109 (1987).
28. G. Horányi, Electrochim. Acta, 31(9), 1095 (1986).
29. D. W. Sopher and J. H. P. Utley, J. Chem. Soc., Chem. Commun., 1087 (1979).
30. L. Moëns, M. M. Baizer, and R. D. Little, J. Org. Chem., 51, 4497 (1986).
31. (a) H. Assefa and E. F. Bowden, Biochem. Biophys. Res. Commun., 139(3), 1003 (1986); Chem. Abstr., 105, 221522k (1986); (b) F. A. Armstrong, H. A. O. Hill, and N. J. Walton, Quart. Rev. Biophys., 18(3), 261 (1985); Chem. Abstr., 105, 221086w (1986); C. Bourdillon, J. M. Laval, and D. Thomas, J. Electrochem. Soc., 133(4), 706 (1986); (d) H. A. O. Hill, B. N. Oliver, D. J. Page, and D. J. Hopper, J. Chem. Soc., Chem. Commun. (21), 1469 (1985); (e) P. M. Allen and R. W. Bowen, Trends Biotechnol., 3(6), 145 (1985); Chem. Abstr., 103, 213252r (1985).
32. S. H. Langer, J. C. Card, and M. J. Foral, Pure Appl. Chem., 58 (6), 895 (1986).

Author Index

Numbers in parentheses are reference numbers and indicate that an author's work is referred to although his name is not cited in the text. Italic numbers give the page on which the complete reference is listed.

Aalbersberg, W., 332(14), 334(14), *353*

Aalstad, B., 123(37–39), 125(37–39), 133(37), 135(38), 136(37–39), 140(37), *196*, 139(86), *198*, 345(134, 135), *357*, 507(30), 519(30), 524(30), *529*, 966(56), *1019*, 982(117), *1022*, 993(141), *1023*

Abakumov, G. A., 617(10), *650*

Abbot, E. M., 296(329), *324*, 903(197), *940*, 1268(21), 1269(34), 1275(21), 1277(21), *1281*

Abdel, A. A., 960(27), *1018*

Abdel-Nabeg, B. A., 736(324c), *759*

Abdel-Wahab, A. A., 905(217), *941*, 960(27), *1018*

Abe, A., 1173(421), *1191*

Abe, K., 1045(120), *1064*, 1178(512), *1194*

Abe, S., 476(92), *481*, 1173(422), *1191*, 1173(423, 424), *1192*

Abe, T., 1106(28, 30–35), 1107(30, 31, 33, 34), 1108(35), 1111(43), *1125*

Aboel, A. M., 960(27), *1018*

Abramson, L. N., 1051(180), *1066*

Abrecht, S., 813(37), 825(37), 826(37), 827(37), 841(37), 863(37), *865*

Abruna, H. D., 1341(10), *1359*, 1365(21), *1392*, 1366(26), 1368(26), 1369(26), *1393*, 1370(21), 1371(21, 26), 1374(21), *1392*

Ackerman, F., 956(30), *1019*

Adachi, T., 728(252), *757*, 769(22), 770(24), *784*

Adams, H. H., 974(93), *1021*

Adams, C., 1042(78b), *1063*, 1080(74), 1082(74), *1097*, 1082(101), *1098*

Adams, F. H., 1170(362), *1190*

Adams, R. N., 154(136), *199*, 165(169), *200*, 168(174), 169(177), 170(178), *201*, 284(206), *320*, 286(370), 301(366, 370), *325*, 350(109), *356*, 411(82), *429*, 453(221), 455(221), *463*, 489(64), *498*, 510(48), 511(56), *530*, 520(112), 521(112), 524(131), *532*, 589(31, 35–37), *608*, 592(54, 57–59), 593(60), *609*, 594(79, 84), *610*, 620(28, 29), 621(29), 625(28), *650*, 620(46, 47), 621(47, 53), *651*, 621(74), 625(74), *652*, 711(65), *750*, 733(300), *759*, 951(18), *1018*, 965(53), *1019*, 968(68), 970(68), *1020*, 1072(28, 31), *1096*, 1375(57), 1376(57, 60), 1387(60), *1394*

Adami, F., 179(233), *203*, 640(163), *655*

Adanuvor, P. K., 165(167), *200*

Adhami, G., 284(217), 290(217), *320*

Adickes, F., 1066(214), *1025*

Adiwidjaja, G., 717(121), *752*, 1150(171), *1184*

Adkins, H., 453(214, 220), *463*

Adler, E., 788(7), *803*

Adzhemyan, Ts. A., 282(199), 285(199), 290(199), *319*

Afanas'ev, N. V., 794(54), 795(54), *806*

Again, G., 1341(13), 1346(13), 1348(13), 1355(13), *1359*

Agami, C., 1144(97), *1182*

Agawa, T., 971(77a), *1020*

Agmon, N., 47(61b), *117*

Agnes, G., 908(223), *941*

Agranat, I., 1175(471), *1193*

Aguila, M., 275(133), *317*, 712(76), *751*, 902(373), *948*, 931(356), *947*

Ahlberg, E., 68(89), *118*, 123(31–33, 36, 39), 126(31–33, 36, 39), 127(33,36), 133(36), 136(37–39), 140(36, 37), 142(36), *196*, 124(45), 127(61), 128(66), 133(75), 140(75), *197*, 150(119), *199*, 184(252), 186(252), *203*, 188(269), *204*, 345(136), *357*, 506(15), 511(15), *529*, 511(73), 513(73), *531*, 593(71), *610*, 972(76), 977(76), *1020*, 1175(468), *1193*

Aihara, S., 519(110), *532*

Aikawa, Y., 1408(76), *1418*

Aikens, D. A., 511(69), *531*, 954(12), *1018*, 1084(113), *1098*

Aiki, T., 1278(57a), 1280(57a), *1282*

Akaba, R., 516(88, 89), 519(89), *531*, 519(110), *532*, 1408(37), *1416*

Akada, M., 1229(68), *1258*

Akagawa, T., 1267(10), *1280*

Akahoshi, F., 597(108), *611*, 602(145), *613*, 1168(341), *1189*

Akashi, T., 367(29), *395*

Akazawa, T., 1044(102), *1063*

Akbulut, U., 1352(62), *1361*, 1352(63, 64), 1354(78–80), *1361*

Aki, O., 489(56), *498* 740(353), *761*,

Akimoto, K., 153(132), *199*

Akiyama, T., 570(160–163), *579*, 782(71), *786*, 1159(238), *1186*

Alam, N., 78(100), *119*, 226(24b, c), *232*, 1239(116), 1260

Al Azrak, A., 235(19), *246*, 1007(225), 1008(225), 1009(225), *1026*, 1166(329), *1189*

Alberola, A., 726(245), *757*

Albert, A. H., 376(111), *397*, 1176(491), *1194*

Alberg, W. J., 144(100), *198*, 165(164), *200*, 172(184), 174(193– 200), *201*, 195(274), *204*, 193(285), *205*, 1358(112), *1362*
Albisson, A., 434(19), *457*
Albrecht, S., 239, 240(44), *247*
Albright, C. H., 388(163), *399*
Albright, T. A., 1175(464), *1193*
Alcalde, J., 725(230), *756*
Alcaraz, M. L., 153(130), *199*
Alciaturi, C. E., 518(107), *532*
Aldaz, A., 388(161), *399*, 645(209, 210), *656*, 1173(433), *1192*
Alder, R. W., 233(1), *245*
Al-Ekabi, H., 1409(107, 110), *1419*
Alekhina, T. F., 726(241), *757*
Alekseeva, T. A., 739(344, 345), *760*
Ali, S. Z., 1232(80), *1259*
Ali, Z., 410(75c), *428*
Aliev, I. Y., 1119(67b–d), 1121(72), 1122(72), *1126*, 1123(80), *1127*
Al-Jabar, N. A. A., 1205(63), *1213*
Alkaitas, S. A., 686(145), *696*
Alkire, R., 273(123a, b, e), *317*, 1287(13), *1332*, 1287(27), *1333*, 1421(4d), 1429
Allard, B., 796(60), *806*
Allen, M. J., 259(31), 263(31), 309(31), *313*, 505(3), *528*, 897(113, 114), 899(113, 114), 900(113), 902(113), *937*, 899(140, 141, 144, 145), 901(153), *938*, 895(162), 901(162), *939*, 1144(100), *1182*, 1246(170), *1262*
Allen, P. M., 1268(25a), 1274(25a), *1281*, 1436(31e), *1438*
Allendorfer, R. D., 190(278), *204*, 192(283), *205*, 303(393), *326*, 334(36), *354*, 335(51a, b), 342(51), 345(51), 346(51), *354*, 337(69), *355*
Allinger, N. L., 1176(490), *1194*
Allison, A. C., 332(16), *353*
Allred, A. L., 855(281), *873*
Almdal, K., 123(30), 133(30), *196*, 677(100), *695*, 1080(76), 1082(76), *1097*
Almgren, C. W., 519(108), *532*
Almir6n, M. A., 163(155), *200*
Alpatova, N. M., 495(127), *500*, 1253(205), 1254(205), 1255(210), *1267*
Alvernhe, G., 1121(73), *1126*
Alvino, W. M., 1358(101, 102), *1362*
Alwair, K., 389(172), *399*
Amatore, C., 24(29), *115*, 30(36a–c, 376), 31(36b–c), 34(47), *116*, 47(36), 50(36), 51(36), 52(36), *117*, 58(73a), 68(87), *118*, 73(93c, d), 77(96), 78(100), 81(102, 103), 82(106), 108(102, 103), *119*, 140(93, 95, 96), 142(93, 95), 143(93), 146(105, 106), 148(93, 95), 150(97, 115), *198*, 208(4a), 211 (7a), 214(8a, 9), 220(7a), 220(14– 20), 224(17), *231*, 226(24), 228(15), *232*, 344(117), 345(117), 345(117), *356*, 345(133, 139), 346(152), 350(152), *357*, 349(179, 180, 182), *358*, 506(18, 19), 508(18, 19), 509(18, 19), 511(18, 19), 513(19), 517(19), *529*, 511(74), 513(74), *531*, 618(23), *650*, 845(235), *872*, 909(237), *942*, 1239(114, 116), 1241(121, 123), *1260*, 1434(14b), *1437*
Amble, E., 393(206), *400*, 827(142), *869*, 1053(196), *1066*
Ambrose, J. F., 723(194a, b), 724(208), *755*, 973(87), 979(87, 89), *1021*, 1382(84), *1395*
Amjad, M., 277(151), *318*
Ammann, D., 824(111), *868*
Ammar, F., 123(25), 133(25), 143(25), *196*, 435(43), *457*, 894(73c), *935*, 1146(138), *1183*
Ammelotti, C. W., 490(82), *499*, 902(168a), *939*, 1146(143), *1183*
Anantharaman, P. N., 265(75), *315*, 417(126), 418(130b), *430*
Anbar, M., 1253(200), *1263*
Anbry, A., 443(118), *460*
Anderson, C. W., 179(237), *203*
Anderson, D. K., 781(65b, c), *786*
Anderson, F. L., 644(195), *656*
Anderson, J. D., 5(17), *8*, 236(21), *246*, 348(174), *358*, 468(27), *479*, 703(7), 739(343), *760*, *748*, 739(343), *760*, 886(17), *933*, 908(224c–e), 910(224c–e), 911(224c–e), 912(224e), 918(224e), *941*, 908(224g, h, j, k, m), 910(224g, h, j, k, m), 911(224g, h, j, k, m), 924(g, h), *942*, 917(283), *944*, 924(303), *945*, 929(328), *946*, 1136(29), *1179*, 1137(38), 1137(38, 40), *1180*, 1175(484), *1193*
Anderson, J. E., 816(62, 69), *866*
Anderson, J. L., 180(241), *203*
Anderson, J. N., 918(368), *948*
Andersen, J. R., 295(318), *323*
Anderson, L., 337(68), *355*
Anderson, L. B., 125(48), *196*, 1176(489), *1194*
Anderson, M. R., 366(23), *394*
Anderson, R. N., 620(34, 36, 37), *651*
Anderson, W. C., 712(72), *750*
Andersson, J. T., 387(158), *398*, 1132(8). *1179*, 1137(42, 43), *1180*
Ando, K., 704(15a), 726(15a), *748*, 773(34), *785*
Ando, T., 782(68), *786*
Ando, W., 686(146), 687(146), *696*, 1166(327), *1189*
Andreades, S., 528(147), *533*, 589(138), *608*, 960(25), *1018*, 963(67), *1020*, 1089(151), 1090(151), 1091(151), *1100*
Andrei, C., 1135(27), *1179*
Andrenzzi Sedea, M. M., 1246(165), *1262*
Andres, K., 1178(518), *1195*
Andricacos, P. C., 284(219), 290(219), *320*, 290(272), *322*
Andrieux, C. P., 17(14, 16), *115*, 64(81), 68(81), 72(81), *118*, 73(93b), 74(94), 76(95a), 97(94), 108(94), *119*, 122(15), 123(15), *195*, 123(23–25), 133(23–25), 143(25), *196*, 128(68), 133(15), *197*, 223(21), *232*, 259(29), 310(29), *313*,

[Andrieux, C. P.]
344(113–116), *356*, 363(6), 366(6, 22), 379(6), 384(6), *394*, 379(127), 384(127), *398*, 435(44), *457*, 443(122), *460*, 467(24), *479*, 469(39–41), *480*, 597(101), *611*, 882(3a, g, r), 885(3a, g, r), *932*, 894(73c), *935*, 899(135, 139), *938*, 918(273b), *944*, 928(323), *946*, 1040(62), *1062*, 1043(90), *1063*, 1146(138), *1183*, 1218(1, 2), 1219(1, 2), 1220(3, 4), *1256*, 1221(19), *1257*, 1221(1, 2), 1222(1, 2), *1256*, 1222(19), *1257*, 1223(1, 2), *1256*, 1224(19, 27, 28), 1228(48), *1257*, 1248(3, 4), *1256*, 1265(1), *1280*
Andrievskii, A. M., 709(57), *750*
Andruzzi, R., 239(41), *247*, 1200(8), *1211*, 1203(46), *1213*
Anet, F. A. L., 342(105), *356*
Ang, P. G. P., 1434(12g), *1437*
Angelopoulos, M., 1377(81), 1382(81), *1395*
Angély, L., 963(60, 62–64), 967(60), *1020*
Angerstein-Kozlowska, H., 179(228), *202*
Anisimov, K. N., 846(236), *872*
Anisimova, M. I., 792(27, 31), 794(45), *805*, 801(82), *807*
Anklam, E., 668(61), *693*
Ankner, K., 337(66a), *355*, 1048(149, 150), *1065*, 1150(169), *1184*, 1155(209), *1185*
Annino, R., 375(101), 377(118), *397*, 1152(178), 1153(193), *1184*, 1168(346), 1169(346), *1189*, 1176(193), *1184*
Anson, F. C., 150(120, 122), 151(120), *199*, 284(214), *320*, 308(417), *327*, 655(126). 636(126), *654*, 819(84), *897*, 832(162), 833(163), *869*, 839(217), *871*, 1221(13), *1256*, 1340(4), *1359*, 1365(18, 19), 1368(19), 1370(18, 19), *1392*, 1370(34), *1393*, 1371(18, 19), *1392*, 1371(37, 38), 1372(37, 39, 43), 1374(43, 50), *1393*, 1390(131), *1396*
Anteunis, M. J. O., 1077(54), *1097*
Antipova, I. A., 794(44), *805*
Antonova, T. L., 496(136), *501*, 745(415), *763*
Antropov, L. I., 445(161, 166), *461*, 1209(108), *1215*
Anziani, P., 443(118), *460*, 1143(87), *1181*
Aoe, K., 602(163), *613*
Aoki, K., 153(131, 132), *199*, 782(70a), *786*, 1298(52), *1334*
Aoki, T., 1035(19), *1061*, 1077(50), *1096*, 1163(295), *1188*, 1178(513), *1194*
Aoshima, A., 909(231), *942*
Aoyagi, R., 542(49), 548(76), *576*
Aoyagui, S., 1253(203, 204), 1255(204), *1263*
Aoyama, Y., 815(58), *866*
Appenrodt, J., 349(185), *359*
Appleby, A. J., 644(202), *656*
Arad, Y., 299(352), *324*, 1201(35), 1202(37–40), *1212*
Arai, T., 260(43), *314*, 436(66), *458*, 443(120), *460*, 895(111), 897(111), 902(111), *937*, 921(296), *945*
Aramata, A., 640(152), *654*, 1071(21), 1073(21), *1096*, 1132(15), *1179*, 1173(428), 1174(445), *1192*, 1390(129), *1396*
Arapakos, P. G., 495(126), *500*
Arbic, W., 540(35), *575*
Arbuzor, A. E., 471(54), *480*, 474(77), *481*
Arcamone, F., 1040(64), *1062*
Archer, M. D., 174(193, 194), *201*
Ard, Y., 908(222), *941*
Arena, J. V., 384(143), *398*
Arens, A., 1050(171), *1066*
Arewgoda, C. M., 1156(214), *1185*
Argersinger, W. J., Jr., 589(31), *608*, 1376(60), 1387(60), *1394*
Argyropoulos, J. N., 14(4), *114*
Arita, S., 721(150), *753*
Arlt, D., 602(144), *612*
Armand, J., 405(31–34, 37–39), 406(38, 40), 407(38, 44), 408(31), 409(38), *427*, 407(52, 60, 61), *428*, 437(133), 443(120), *460*, 476(88, 90), *481*, 701(1d), *747*, 719(128), *752*, 733(289, 290, 293, 294), *758*, 736(323), 738(323), *759*, 738(333, 334), *760*, 741(370), 742(372), *761*, 894(73a), *935*, 902(168c), *939*, 979(96), *1021*, 1050(169, 173), *1066*, 1144(106), *1182*, 1146(140), *1183*
Armand, Y., 741(370), *761*
Armelli, R., 1109(38), 1110(38), 1111(38), 1112(38), *1125*
Armstrong, F. A., 1436(31b), *1438*
Armstrong, N. R., 435(22), *457*, 620(33, 39), *651*
Arnold, D. R., 517(103), 526(103), *532*, 1178(511), *1194*
Arnone, A., 725(220a, 222), *756*
Arora, S. K., 1148(154), *1183*
Artaud, I., 816(71), *866*
Artes, R., 511(66), *530*
Arvia, A. J., 284(211), *320*, 645(205), *656*
Asahara, T., 260(43), *314*, 271(100), *316*, 487(38), *497*, 927(296), *945*
Asahi, Y., 474(78), 475(78), *481*, 489(56), *498*, 740(353), *761*
Asai, M., 1146(141), *1183*
Asano, T., 1204(52), *1213*
Aschenbrenner, H., 486(23), *497*, 1142(73), *1181*
Ashby, E. L., 14(4, 6), *114*
Asirvatham, M. R., 584(21), *608*, 975(102), *1021*
Askerov, N. L., 907(189–191), *940*
Asmus, K.-D., 660(8), *692*, 663(41), 668(61, 62), *693*, 668(63), 669(68–70), *694*, 1408(73), *1417*
Asseta, H., 1436(31a), *1438*
Asthana, M., 294(314), *323*, 229(350), *324*
Astle, M. J., 309(424), *327*
Aston, W. J., 645(207), *656*
Astrue, D., 838(204, 208, 209), *871*, 1144(121), 1145(121), *1182*, 1156(213), *1185*
Atamian, M., 814(52), 815(52), *865*

Aten, A. C., 332(20, 22), 345(20), *353*
Atkins, D. L., 344(127), *357*
Atkins, P. W., 36(49), 38(50), 63(50), *116*
Atkinson, E. R., 424(66), *431*
Atkinson, J. G., 475(85), *481*
Atobe, H., 994(148), *1023*
Atsuno, T., 1243(140), *1261*
Attwood, P., 639(139), *654*
Aubry, A., 1143(87), *1181*
Austen, D. E., 434(9), *456*
Austin, R. G., 1366(22), *1392*, 1366(25), *1393*, 1367(22), *1392*, 1367(25), *1393*
Autbrose, P., 898(131), *938*
Au-Young, Y. K., 525(138), *533*, 982(120), 983(120), *1022*
Avaca, L. A., 162(149), 163(155, 160, 161), *200*, 335(50), *354*, 369(59), *395*, 882(3k), 885(3k), *932*, 1253(205), 1254(205), 1255(211), *1263*
Avilov, V. B., 567(146), *579*
Avrutskaya, I. A., 273(120), *317*, 404(29a–d), 408(29a–d), *427*, 408(58, 59), 409(67), *428*, 424(170), *432*, 466(6), *479*, 470(49), 476(49), *480*, 490(84, 85), *499*, 743(387), *762*, 745(415), *763*, 899(148), *938*
Awaji, H., 682(124), *695*
Awata, T., 484(9), *496*
Awerjanow, S. F., 290(270), *322*
Axelrod, S., 898(131), *938*
Ayabe, Y., 132(73), *197*
Aylmer-Kelly, A. W. B., 184(253), *203*, 930(358), *947*
Ayres, J. T., 296(324), *323*
Azuma, K., 1211(127), *1215*
Azuma, T., 1408(53, 55), *1417*, 1408(88), *1418*

Baastad, F., 438(77), 446(77), 450(77), *458*
Baba, H., 1106(28, 30, 31, 33–35), 1107 (30, 31, 33, 34), 1108(35), 1111(43), *1125*
Babicheva, S. V., 1202(41), *1212*
Babievskii, K. K., 404(29a–c), 408(29a–c), *427*, 408(58,59), *428*
Babkin, V. A., 792(27, 28, 29, 31), 794(45), *805*, 801(82–84), *807*
Bachman, G. B., 309(424), *327*
Bachmann, W. E., 1209(112), *1215*
Bachiocchi, E., 244(76), *248* 1070(2), 1080(2), 1081(2), *1095*
Backer, H. J., 424(173), *432*
Backhurst, J. R., 268(89a), *315*
Bacon, J., 589(35), *608*, 620(28), 625(28), *650*, 1375(57), 1376(57), *1394*
Bacquet, C., 542(47) *576*, 569(153, 154) *579*
Bader, J., 1174(441), *1192*
Badoz-Lambling, J., 21(21c), *115*, 284(222), *320*, 620(41), *651*, 1123(81), *1127*, 1241(121), *1260*
Bae, D. H., 687(177), *697*
Baeyer, A., 1208(95), *1214*
Baezner, C., 706(41), *749*
Bagchi, R. N., 190(281), 193(281), *204*
Baggaley, A., J., 721(148), *753*, 1160(256), *1186*
Bagotzky, V. S., 639(136, 137), *654*, 1012(241), *1026*
Bahary, W., 340(79), *355*
Bahnemann, D. W., 1412(125), 1413(133), *1419*
Bahner, C. T. 1006(221a), *1025*, 1088(137), *1099*
Bähr, K., 349(186), *359*
Baigrie, B. D., 974(93, 94a), *1021*
Bailey, A., 794(76), 799(76), *807*
Bailey, S. I., 282(201), *319*, 302(383), *325*, 616(4), 619(4), 620(4), *650*
Bailey, W. F., 376(115), 378(115,122), 397
Baird, M C., 843(226), *871*, 1168(345), *1189*
Baizer, M. M., 5(18), *8*, 236(21), *246*, 239(45), *247*, 240(55), *248*, 257(15), *313*, 275(131), *317*, 305(403), 306(408), *326*, 308(420), *327*, 342(97), 350(97), *356*, 347(171), 348(171,174), *358*, 349(183), *359*, 367(32), 370(32), 373(32), 388(32), *395*, 370(67), *396*, 410(73), *428*, 433(3), 452(3), *456*, 439(92), 440(92), 441(92), *459*, 437(135a), *460*, 451(206), 456(206), *462*, 468(27), *479*, 484(9), *496*, 490(72), *498*, 542(38), *575*, 641(168), 643(168), *655*, 703(7), *748*, 722(172), *754*, 739(343), *760*, 791(18,21), 794(18), *804*, 793(43), *805*, 799(77), *807*, 879(1), 885(1), *932*, 886(17), *933*, 889(40), 890(41), 891(47), *934*, 893(71), *935*, 906(182c), 907(182c), *939*, 908(224a–f), 910(224a–f), 912(224e), 918(224e), 911(224a–f), 921(224a–c), 924(224f), *941*, 908(224g, j, k), 909(245), 910(224g, j, k), 911(224g, j, k), *942*, 911(269), *943*, 917(283), 918(271a, b), 919(275), 919(285a, b), 921(923), 922(276, 277), 923(280, 281), *944*, 924(299), 924(303), 925(309f), *945* 929(328–330, 333), *946*, 930(349a, b), *947*, 931(368), *948*, 950(1d), *1018*, 1005(220b), *1025*, 1035(20), *1061*, 1040(68), *1062*, 1044(100, 103), *1063*, 1052(188), *1066*, 1136(29), *1179*, 1136(30, 31), 1137(38–40), 1139(58, 59), 1140(58, 59), *1180*, 1147(147, 148), *1183*, 1175(484), *1193*, 1201(24, 27), *1212*, 1246(170), *1262*, 1265(2), 1267(3, 9, 12, 13), *1280*, 1267(14, 16), 1268(17–19, 25a, c, d, 29, 30), *1281*, 1268(2), *1280*, 1268(14), *1281*, 1271(14), *1281*, 1271(39, 40), *1282*, 1272(18) *1281*, 1273(2), *1280*, 1274(25a, c, d), 1275(16, 19, 25d), *1281*, 1275(45, 46b, 47), 1277(53), *1282*, 1319(112), *1335*, 1321(124), *1336*, 1322(112, 124), 1323(112), *1335*, 1422(1a, c, d), *1429*, 1422(19), *1430*, 1424(1b, 1c, 1e, 5), 1425(1a), 1429(20), 1428(20), *1430*, 1432(1), 1433(3), *1436*, 1421(1a–e, 3b, c), *1429*, 1434(11), *1436*, 1435(24a, 30), *1438*
Bakac, A., 831(149–151), *869*
Baker, A. D., 831(156), *869*
Baker, R., 233(1), 245

Bakhchisaraytsyan, N. G., 1299(53), 1304(53), *1334*
Bakos, V., 707(47b), *749*, 709(58), *750*
Balaban, A. T., 706(39), *749*, 721(156), *753*
Balachandran, U. S., 239(39), *247*
Balakrishnan, T. D., 265(71), 271(71), *315*, 271(102), *316*, 417(121), *430*
Balasiewicz, M. S., 163(156), 164(156), *200*, 1132(9), 1175(9), *1179*, 1175(466), *1193*
Balavoine, G., 1040(62), *1062*, 1043(90), *1063*
Baldwin, R. P., 1434(19b), *1438*
Ballinger, J. R. 1120(68), *1126*
Ball, M. C., 276(149), *318*
Baloy, G., 290(279), *322*
Baltes, H., 240(53), *248*, 516(87), 517(87), 525(87), *531*, 716(116), *752*, 981(129), 984(129), *1022*, 1166(326), *1189*
Baltes, J., 1079(70), *1097*
Baltruschat, H., 179(232), *203*
Bampfield, H. A., 288(256), 292(256), *321*
Ban, Y., 603(149, 152, 153), *613*, 773(37a–c), 774(39c), 775(42), *785*, 835(178), *870*, 1058(255), 1059(255), *1068*, 1077(55), *1097*
Bancroft, E. E., 193(289), *205*, 511(55), *530*, 724(207), *755* 860(311), *874*
Bancroft, W. D., 272(116) *316*
Banda, F. M., 1082(98), *1098*
Bandlish, B. K., 686(160, 162, 165), *697*, 1119(65), *1126*
Bank, S., 391(186, 187), *399*, 886(12), 891(12), *933*
Banks, C. H., 309(436), *327*
Bannou, T., 664(54), 666(54), *693*, 778(53), *785*
Bansal, K. M., 434(20), *437*
Bar, A. J., 1413(139) *1420*
Bär, H. J., 1045(108), *1064*
Baranowska, I., 184(250), 186(250), *203*
Baranwal, P. P., 1352(53–55), *1361*
Barba, F., 384(144–146), 385(144), 386(144, 148–152) 387(151) *398*, 388(161), *399*, 492(98), *499*, 523(125), *532*, 563(132), *578*, 645(204), *656*, 703(4–6), *748*, 891(50), *934*, 1154(198), *1185*, 1161(274), 1162(289, 290), *1187*
Barba, I., 386(152), 387(152), *398*, 523(125), *532*, 891(50), *934*, 1162(289,290), 1187
Barbe, J. M. 819(79), *866*
Barbeni, M., 1413(131), *1419*
Barber, J. J., 894(87b), 898(87b), *936*, 930(345b), *947*, 1433(9b), 1436
Barber, R. A., 1409 (106), *1419*
Barbey, G., 712(67), *750*, 723(185), *754*, 974(94b), *1021*
Barboiu, V., 1135(27), *1179*
Bard, A. J., 13(1), 46(1a), *114*, 20(20), 21(21a), 22(22a), 22(23), *115*, 26(33), *116*, 55(69), 57(72), *117*, 59(76), 64(80), 65(83b), 68(86), *118*, 77(97), 85(107), *119*, 122(4), 123(4), 146(4), 150–152(4), 157(4), 165(4), 179(4), 180(4), *195*, 125(47), *196*, 148(109), *198*, 153(128), *199*, 171(181), 172(181), 173(181), 172(182), 173(182), *201*, 190(275–277), 191(275), 192(275), 194(277), 195(276), *204*, 193(288), *205*, 209(5b), 210(6a), 212(6), *231*, 263(55), 293(55), *314*, 270(97), 303(97), 304(97), *316*, 282(204), 285(204), *319*, 288(258), *321*, 293(305), *323*, 302(385, 389), *326*, 309(434), *327*, 332(19), 333(19), *353*, 336(59), 342(59), *354*, 340(78, 86b), *355*, 341(93), *356*, 344(120, 123, 125, 128), *357*, 350(195), *359*, 370(68), *396*, 401(3), *426*, 411(77), *428*, 413(93), *429*, 423(161), *431*, 506(11, 25), 507(32), 510(11), 511(11), 519(32), *529*, 511(57), *530*, 629(92), *652*, 726(232), *756*, 882(3b, c, e, j, l, n, p, q), 885(3b, c, e, j, l, n, p, q), *932*, 930(347), *947*, 973(90), *1021*, 1050(160), *1065*, 1052(191), *1066*, 1174(447), *1192*, 1200(10), *1211*, 1239(119), *1260*, 1340(4), 1342(14), *1359*, 1367(30), 1372(44), 1374(49), *1393*, 1382(88), 1385(88), *1395*, 1387(115,117–119), 1389(115), *1389*, 1390(125, 126, 130, 132), *1396*, 1398(6–8), 1401(7), 1402(6), 1403(6), *1415*, 1403(17), 1406(29,30), *1416*, 1408(58), *1417*, 1408(75, 76, 78–80, 82, 83) *1418*, 1409(113a), 1412(127), *1419*, 1413(137), *1420*, 1433(10), *1436*
Barek, A., 511(73), 513(73), 531
Barendrecht, E., 417(118), 422(118), *430*
Barg, G. K., 1358(100, 106), *1362*
Bargon, J., 341(90), *356*, 1375(55), 1376(58, 62), 1377(62), 1378(62), 1380(55, 62, 63, 65), 1382(55, 62), 1383(62), 1384(62), 1385(62), 1386(62), 1387(62, 65), *1394*, 1388(62), *1394*, 1389(110, 111), *1396*
Barker, D. J., 363(5), *394*
Barker, W., 1233(90), 1235(90), *1259*
Barnard, D., 1054(206), *1067*
Barnard, G. W., 541(36), *575*
Barnes, D., 435(34), *457*, 436(64), *458*, 452(208a), *462*
Barnes, J. H., 467(26), *479*, 474(72), *481*
Barnes, K. A., 582(1, 2), 583(1), *607*, 586(29), *608*
Barnes, K. K., 289(263), *321*, 375(104), *397*, 454(227), *463*, 1051(179), *1066*
Barnett, C. A., 798(74), *807*
Barnier, J. P., 1224(29), 1248(29), *1257*
Barnum, C., 660(9), 662(9), *692*
Barradas, R. G., 494(119), *500*, 924(297), 931(297), *945*
Barrado, E., 726(245), *757*
Barrett, T. M., 1106(27), 1108(27), *1124*
Barry, C., 265(69), *315*, 684(131), *696*
Barry, J. E., 584(17), *608*, 648(232), *657*
Bartak, D. E., 193(286, 287), *205*, 373(83, 84, 85)

[Bartak, D. E.]
374(83, 86), *396*, 418(133), 419(133), *430*, 886(21), *933*, 1038(40, 41), *1061*
Bartlett, P. N., 174(196), *201*
Bartoli, D. 1070(2), 1080(2), 1081(2), *1095*
Barzan, J. C., 284(211), *320*
Basch, A., 1144(96), *1182*
Basolo, F., 51(68), *117*
Basselier, J.-J., 1137(45, 46), *1180*, 1271(41), *1282*
Bassinet, P., 406(40), *427*, 476(88), *481*, 719(128), *752*, 742(372), *761*
Batanero, P. J., 726(245), *757*
Bates, N., 1387(120), 1389(120), *1396*
Bates, R. B., 233(5d), 234(5d), 240(5d), *246*, 1148(154), *1183*
Bates, R. G., 284(220), *320*, 301(375), *325*
Batley, G. E., 423(157), *431*
Batrakov, V. V., 1175(477), *1193*
Battersby, A. R., 812(21), *864*
Battioni, J.-P., 816(71, 72), 818(72), *866*
Batusic, M., 706(36b), 710(36b), 712(36b), 714(36b), *749*
Batzalova, K., 453(215b), *463*
Baucke, F. G. K., 301(377), *325*
Bauer, E., 896(104), *937*
Bauer, H. H., 745(417), *763*
Bauer, R., 410(75a–c), *428*, 441(143), *460*, 583(9), 584(9), *607*, 975(101), *1021*, 1006(224a–c), 1007(224a–c), *1026*
Bauld, N. L., 350(202), *359*, 569(150), *579*, 1159(235), 1170(235), *1186*
Baumane, L. C., 434(14), *457*
Baumann, H., 336(62), 339(62), *354*
Baumgärtel, H., 340(82a), *355*, 348(177), *358*, 496(130), *500*, 701(1e), *747*, 910(229), *942*
Baumgarten, H. E., 740(368), *761*
Bausch, M. J., 686(144), *696*
Baxendale, J. H., 725(214), *755*
Bayer, F., 1204(54), *1213*
Bayer, O., 1201(21), *1212*
Bayer, R. E., 293(313), *323*
Bayeul, D., 817(74), *866*
Baykut, F., 1413(134), *1420*
Bazoin, M., 297(336), 286(336), *324*
Beak, P., 1044(97), *1063*
Beau, D., 418(129), *430*
Beaver, R. N., 1303(67), *1334*
Bechgaard, K., 511(64), *530*, 621(78), 625(78), 629(91), *652*, 963(50), *1019*, 1093(168), *1100*
Beck, F., 234(15), *246*, 257(17), 265(17), 268(17), 271(17), 307(17), *313*, 259(34), 263(34), *314*, 268(86), *315*, 275(125), *317*, 276(148), 277(157, 158), *318*, 307(416), *327*, 346(157a), *358*, 349(183), *359*, 441(151), *461*, 486(23), *497*, 542(39), *575*, 544(55), *576*, 640(153), *654*, 641(171), *655*, 647(225), *657*, 879(1), 885(1), *932*, 894(85b, 86), *936*, 921(296), *945*, 1001(179), 1002(186), *1024*, 1072(30, 31), *1096*, 1117(50a), 1121(50a), *1125*, 1142(73, 74), *1181*, 1201(25, 28, 32), *1212*, 1287(25), 1295(48), *1333*, 1305(75), *1334*, 1316(106), 1319(48), *1333*, 1319(106), *1335*, 1321(121), *1336*, 1329(106), *1335*, 1391(138), 1396
Beck, G., 686(145), *696*
Beck, P., 1052(186), *1066*
Beck, T., 1287(13), *1332*
Beck, T. R., 486(25), *497*
Becker, J. Y., 307(415), *327*, 362(4), *394*, 392(201), 393(201), *400*, 514(82), 515(82), *531*, 523(127), *532*, 584(18), *608*, 643(189–191), *656*, 690(192), *698*, 704(16), *748*, 819(90), *867*, 930(349d), *947*, 993(146), *1023*, 1037(34), *1061*, 1085(116), 1086(124, 125), *1099*, 1093(171), *1100*, 1434(12b), *1436*
Becker, Y., 1081(90), 1085(90), *1098*
Becking, L., 550(93, 94), 551(95), *577*, 1011(242,243), 1012(245), *1026*
Beckley, R. S., 1205(64), *1213*
Beckwith, A. J. L., 347(172), 348(172), *358*
Beddard, G. S., 819(83), *867*
Beden, B., 639(140), 640(154), *654*, 640(161), *655*
Bédert, N., 1035(18), *1061*
Bednarski, M. D., 621(83), 627(83), *652*
Beechgard, K., 967(59), *1020*
Beer, H. B., 277(155), *318*
de Beer, J. A., 1168(344), *1189*
Begley, W. J., 703(11), *748*, 771(31b), *785*, 827(139), 829(139), *868*
Behling, J. R., 441(150), *461*, 1172(415), *1191*, 1365(13), 1370(13), *1392*
Behr, B., 435(52), *458*
Behrin, E., 309(429), *327*
Beidler, W. T., 711(66), *750*
Bejtovic, Z., 1015(250), *1027*
Beland, F. A., 380(133), 384(140, 141), *398*
Belanger, G., 639(145), *654*, 641(181), *655*
Belen'kaya, N., 438(84), 446(84), *459*
Beletskaya, I. P., 366(25), *394*, 376(112), *397*, 863(329), 875
Beley, M., 831(153) *869*, 1434(15d, 1), *1437*
Belikov, V. M., 403(17), *426*, 404(27a–c), 408(27a–c), *427*, 408(58, 59), *428*
Bellama, J. M., 282(204), 285(204), *319*
Bellamy, A. J., 131(72), *197*, 150(118), *198*, 296(329), *324*, 903(197), *940*, 1267(8), *1280*, 1268(21), 1269(34), 1275(21), 1277(21), *1281*, 1277(49), 1282
Bellavance, M. I., 174(203), *202*
Belleau, B., 525(138), *533*, 629(94, 95), *653*, 648(227), *657*, 982(120), 983(120), *1022*, 1036(29), *1061*, 1162(288), *1187*
Bellec, C., 706(42), *749*, 719(131), *752*
Belokoneva, N. A., 366(25), *394*, 376(112), *397*
Belot, G., 275(130), *317*, 410(72), *428*, 441(152b), *461*, 715(105c, d), *752*, 1139(61), 1141(61), *1181*, 1177(507), *1194*, 1277(51), *1282*, 1435(25a, b), *1438*
Belov, V. N., 490(74, 75), *499*, 1208(102–105), *1214*
Belova, T. P., 792(27), *805*

Belozerskii, I., 438(84), 446(84), *459*
Beltowska-Brzezinska, M., 641(165), *655*
Bemis, A. G., 1269(33), *1281*
Bender, R., 791(17), *804*
Bender, Y. A., 434(14), *457*
Benders, J., 441(139), *460*
Bendersky, Yu. V., 1221(23), *1257*
Bendig, J., 726(239), *756*, 926(138), *938*, 1038(35), *1061*
Benedetti, A. V., 826(129), *868*
Benedetti, R., 1348(30), *1360*
Ben Hadid, A., 616(3), 620(3), *650*
Benjamin, B. M., 1434(22c), *1438*
Benkeser, R. A., 239(41), *247*, 271(101), *316*, 293(309,310), *323*, 493(111), *500*, 1141(69), *1181*, 1255(214, 216), 1255(217, 218), *1263*
Benlioglu, G., 1413(134), *1420*
Bennes, R., 1172(396), *1191*
Bennett, C. E., 387(155), *398*
Bennion, D. N., 279(179), *319*, 293(307), *323*, 420(140), *431*
Benoit, R. L., 300(357), 286(357), *325*
Benoliel, R. W., 447(179), *462*
Bensadat, A., 1094(184), *1101*, 1120(69–71), 1122(69, 71, 74a, b), *1126*, 1163(302, 303), 1164(302), *1188*
Bentley, J. W., 792(22), *804*
Bentley, T. W., 979(110), *1022*, 994(155b), *1023*
Benzuglyi, V. O., 1354(81), *1361*
Beran, P., 165(166), 169(166), *200*
Bercz, J. P., 1051(185), *1066*
Berdinskii, I. S., 474(70), *481*
Berenbaum, B. M., 1350(35), *1360*
Berenjian, N., 1039(50), *1062*
Berezina, S. I., 915(252, 253), *943*
Berezovskii, V. M., 739(339), *760*
Berg, A., 3(12), *8*, 346(150), *357*, 886(15), 891(15), *933*, 1250(181), *1262*
Berg, H., 1040(65), *1062*
Berg, P., 277(154), *318*, 645(203), *656*
Berge, H., 275(132, 133, 136, 137), *317*, 712(73, 76), *751*, 931(356), *947*, 902(373), *948*
Bergel'son, L. D., 1176(503), *1194*
Berger, N. M., 885(8), *932*
Bergman, I., 284(216), 286(216), *320*, 291(293), *322*, 334(37), 345(37), *354*, 345(143,146), *357*
Bergmann, E. D., 351(209), *359*, 886(15, 16), 891(15), *933*, 1269(38), *1282*
Berkenkotter, P., 592(51), *609*, 711(64), *750*, 977(84), *1020*
Berkey, R., 291(296), *322*, 332(15), 348(15), *353*, 885(6), *932*, 910(227), *942*, 1138(50), *1180*, 1175(488), *1194*
Bernal, I., 495(128), *500*, 1051(177), *1066*
Bernard, D., 713(83, 84), *751*, 893(67), *935*
Bernard, G., 1200(16), *1211*
Berndt, A., 411(81), *429*
Berneth, H., 194(293), *205*, 526(139), *533*
Bernoulli, W., 1059(260), *1068*
Bernstein, P., 689(190), 698
Berntsson, P., 307(411), *326*
Beroiz, D., 814(50), *865*
Berri, D. M., 740(354), *761*
Bertheim, A., 1206(73), *1214*
Berthelot, J., 894(80, 81), *936*, 916(284), *944*, 1137(45), *1180*, 1146(135), *1183*, 1150(170), *1184*, 1173(429) *1192*, 1200(16), *1211*, 1236(101), *1259*
Bertin, D., 706(42), *749*
Bertocci, U., 284(209), *320*
Bertola, S., 731(274), *758*
Bertozzi, R. J., 300(358), 307(358), *325*
Bertram, J., 285(218), 290(218), *320*, 290(265, 266), *321*, 514(79), 515(79, 98), 516(79), 518(98), *531*
Bérubé, D., 891(58), *935*, 979(108), *1021*
Berzins, T., 154(133), *199*, 307(415), *327*
Bessière, J., 291(284), *322*
Bessin, J., 901(155), *938*
Bestmann, H. J., 1002(201), *1025*
Bethell, D., 140(91, 92), 150(117), *198*, 478(103–105, 107, 108, 110), *482*
Bettolo, R. M., 887(27), *933*
Bevington, J. C., 1338(1), *1359*
Bewick, A., 31(386), *116*, 179(223), *202*, 184(253), *203*, 309(428), *327*, 335(50), *354*, 506(20–24), 511(20–24), 512(20–24), 514(21), *529*, 511(72), 512(72), 515(85), *531*, 639(140), *654*, 663(46), 664(46, 48–50), 665(48–50), *693*, 713(77), *751*, 895(91b), *936*, 902(167), 902(175), *939*, 930(358), *947*, 951(17), *1018*, 1054(212), *1067*, 1084(111), *1098*, 1085(121), *1099*, 1144(115, 117), *1182*, 1164(309, 310), 1165(309), *1188*, 1253(205), 1254(205), 1255(211), *1263*
Bey, B. B., 486(31), *497*
Bezilla, B. M., Jr., 151(127), *199*
Bezrukova, Z. P., 276(141), *317*
Bezuglyi, V. D., 434(4), *456*, 726(241), *757*, 729(266), *758*, 739(344, 345), *760*
Bezzubov, A. A., 1011(237), *1026*
Bhadani, S. N., 306(405), *326*
Bhadeni, S., 1352(52), *1360*, 1352(53–55), *1361*
Bhandal, H., 827(139), 829(139), *868*
Bhat, G. A., 525(136), *533*
Bhatti, M. V. D., 525(136), *533*, 899(133), *938*, 904(203), *940*
Bhattacharyya, D. N., 335(46), *354*
Bhattacharyya, S. K., 1243(132), *1260*
Bialecka, E., 307(412), *327*
Bianco, E., 660(10), *692*
Bibson, H. W., 467(25), *479*
Bidan, G., 584(13), 590(13), *608*, 1356(91), *1362*, 1380(70), 1381(73, 74), *1394*, 1385(109), *1396*
Bieber, R., 451(207), 453(207), *462*
Biebl, G., 738(337), *760*
Bied-Charreton, C., 826(125), *868*

Biegler, T., 745(417), *763*
Bielstich, W., 639(142), *654*
Bien, Z., 1049(158a), *1065*
Biering, H., 707(47c), *749*
Biermann, C. J., 791(16), *804*
Bighi, C., 424(181), *432*
Bikrani, M., 240(51a), *248*, 839(215), *871*
Bil'dinov, K. N., 290(277), *322*
Bilewicz, R., 363(10), 366(10), *394*
Bilger, R., 341(92), *356*
Billinge, B. H. M., 1169(353), *1189*
Billon, J.–P., 296(327), 286(327), *324*, 686(143), *696*, 728(256), *757*
Billups, W. E., 242(60), *248*
Binh, S. K., 925(310), *945*, 1052(192), *1066*, 1224(33), *1257*
Binkley, W., 439(86), *459*
Binns, T. D., 1005(217), *1025*, 1156(215), *1185*
Biolchini, G., 887(24), *933*
Birch, A. J., 239(40), *247*, 345(129), *357*
Bird, P., 704(27), *749*, 1166(328), *1189*
Bird, R. B., 1292(42), 1299(42), *1333*
Birke, R. L., 344(127), *357*, 1352(62), *1361*
Birkett, M. D., 1144(119), *1182*
Birr, E. J., 280(182), 295(182), *319*
Bisierni, M., 1382(80), *1395*
Bixon, M., 46(60), *117*
Bjelica, L., 276(142), *317*
Bjerrum, J., 292(297), *322*
Blanberg, D., 346(150), *357*
Black, D. St. C., 813(40), *865*
Black, G. G., 704(27), *749*, 1166(328), *1189*
Black, S., 798(74), 799(75), *807*
Blackham, A. V., 410(74), *428*, 583(10), *607*
Blackmar, G. E., 848(248), *872*
Blackwood, D. J., 179(232), *203*, 278(175), *319*, 332(15), 348(15), *353*
Bladon, P., 1144(102), *1182*
Blaedel, W. J., 725(228), *756*
Blaha, E. W., 332(15), 348(15), *353*, 885(6), *932*, 910(227), *942*, 1138(50), *1180*, 1175(488), *1194*
Blakley, D. C., 1345(19), *1360*
Blanchard, M., 1247(173), 1249(173), *1262*
Blankespoor, R. L., 488(53), *498*, 689(187, 191), *698*
Blázquez, M., 161(147), 163(147, 159), *200*, 894(77g), *936*
Bleh, O., 1208(96), *1214*
Blocman, C., 64(81), 68(81), 72(81), *118*, 344(113, 116), *356*, 379(127), 384(127), *398*, 1220(3, 4), 1248(3, 4), *1256*, 1224(27), *1257*
Bloodworth, A. J., 649(233), *657*
Bloom, A. J., 1164(308), *1188*
Bloomfield, J. J., 537(15), 564(15), 565(15), *575*
Blout, E. R., 453(215a), *463*
Blount, H. N., 179(220), *202*, 187(260), 188(261), *204*, 188(267, 268), *204*, 193(289), *205*, 270(96), 277(96), *316*, 411(87), 412(87), *429*, 511(55), *530*, 520(113–115), 521(113–115, 117), 522(114), *532*, 686(156, 161), *697*, 724(207), *755*, 860(311), *874*
Blubaugh, E. A., 282(204), 285(204), *319*
Blum, Z., 345(145), *357*, 509(41), *530*, 604(158), *613*, 774(396), *785*, 1070(6), *1095*, 1077(49), *1096*, 1077(56), 1078(64), *1097*, 1081(81), *1098*, 1119(64), *1126*
Blunt, J. C., 1080(80), *1097*
Boa, J., 1105(17, 19), 1109(17, 19), 1110(17), 1112(17), *1124*
Bobbitt, J. M., 567(148, 149), *579*, 723(189, 190), *754*, 723(193), *755*, 739(348), *760*, 745(409), *762*, 960(24), *1018*, 957(34, 35), 961(34, 38, 44), *1019*, 973(91), 979(88, 91), *1021*, 991(169), *1024*, 1094(175), *1101*, 1166(320), *1188*, 1415(152), *1420*
Bobilliart, F., 290(267), *321*, 290(268), *322*, 515(94), 518(94), *531*
Bocarsly, A. B., 1366(22), *1392*, 1366(25), *1393*, 1367(27), *1392*, 1367(25), *1393*
Bocchi, V., 1375(51), *1393*
Bocian, D. F., 814(52), 815(52), *865*
Böck, C., 729(270), *758*, 974(94b), *1021*
Bock, H., 346(148), *357*, 663(42, 47), *693*, 673(83), *694*
Bockmair, G., 930(346), *947*, 1081(92), 1082(92), *1098*
Bockris, J. O'M., 46(58), *7*, 13(1), *114*, 20(20), 22(22b), 22(24), 23(26), 24(27a), 42(27a), *115*, 26(34a), *116*, 2(1), *117*, 1287(14), *1332*, 1345(21), *1360*, 1434(20a), *1438*
Boczkowski, R. J., 1153(193), 1176(193), *1184*
Bodard, J., 901(155), *938*
Boden, D. P., 301(376), *325*
Bodennec, G., 1094(184), *1101*, 1120(69–71), 1122(69, 71, 74a), *1126*, 1163(302), 1164(302), *1188*
Bodner, G. S., 841(221), *871*
Boehm, H. P., 1410(117), *1419*
Boehme, W., 442(110), *459*
Boekelheide, V., 334(40), *354*, 993(146), *1023*, 1093(171), *1100*
Boer, H., 487(39), *497*
de Boer, E., 333(27), *353*
de Boer, T. J., 620(42), *651*
Bogan, S., 311(438), *327*
Bogar, F. D., 284(240), *321*
Bogdanova, N. P., 743(387), *762*
Bogolitsyna, G. M., 794(46), *806*
Bogomolov, B. D., 792(26), *805*, 794(46), *806*
Boichard, J., 744(397), *762*
Bois, C., 738(334), *760*
Boisselier-Cocolios, B., 816(66, 67), *866*
Bold, G., 812(22), *864*
Bolhuis, P. A., 521(118), *532*
Bolikal, D., 455(239), *464*
Bolinger, C. M., 1434(12e, n), *1437*
Bolla, M., 1176(493), *1194*
Bolton, D. J., 1153(193), 1176(193), *1184*
Bolts, J. M., 1366(22), *1392*, 1366(23, 25), *1393*, 1367(22), *1392*, 1367(23, 25), *1393*
Bon, C. K., 389(170), *399*, 747(435), *763*

Bond, A. M., 190(281), 193(281), *204*, 662(33, 35, 36), *693*, 1175(448, 449, 451–454), *1192*, 1175(456), *1193*
Bondarenko, V. E., 729(266), *758*
Bonicelli, M. G., 724(205), *755*
Bonifacic, M., 663(41), *693*, 669(68, 69, 70), *694*
Bonnel-Huyges, C., 1241(121), *1260*
Bonner, W. A., 1045(105), *1063*, 1082(96), *1098*, 1160(248), *1186*
Bontempelli, G., 663(44), *693*, 675(95, 96), 676(98), *695*, 712(72), *750*, 834(173), 835(173), 835(181), 931(360), *947*, 1059(257), *1068*, 1228(51), *1257*, 1245(51), 1246(165), *1262*
Booman, G. L., 209(5c), *231*, 309(433), *327*, 434(21), *457*, 882(31), 885(31), *932*
Boone, J. R., 14(6), *114*
Bordet, J., 268(90), *315*
Bordwell, F. G., 686(144), *696*
Borgarello, E., 1413(131), *1419*, 1413(138), *1420*
Borghesani, G., 424(181), *432*
Borhani, K. J., 140(90), *198*, 346(153b), 350(153b), *358*, 375(91), *396*, 478(100–102), *482*, 887(25), *933*
Borsdorf, R., 684(130), *696*, 728(261), *757*
Borman, W. F. H., 1358(99), *1362*
Borremans, A. M., 1077(54), *1097*
Borsetti, A. P., 722(173), *754*
Bosch, A., 1434(19d), *1438*
Bothner-By, A. A., 1254(208), *1263*
Botsan, V. Y., 440(99), *459*
Botsanand, V. Y., 897(116), *937*
Bott, L. L., 1309(82), 1316(82), 1319(82), *1334*, 1321(125), *1336*, 1327(82), *1334*, 1328(125), *1336*
Bottomley, L. A., 814(53), 815(53), *865*, 821(97), *867*
Bouchoule, C., 1247(173), 1249(173), *1262*
Boudon, C., 862(320), *874*, 862(322), *875*
Bougverra, M. L., 914(245), *943*, 917(282), *944*
Boujlel, K., 1033(6), *1060*, 1040(69), *1062*, 1224(26, 29, 30), 1227(47), *1257*, 1239(30), *1260*, 1245(162, 163), *1262*, 1248(29, 47), *1257*, 1252(195), *1263*
Boulares, L., 437(133), 443(120), *460*, 476(88, 90), *481*, 738(333, 334), *760*, 741(370), *761*, 894(73a), *935*, 902(168c), *939*, 1050(173), *1066*, 1146(140), *1183*
Boulton, J. J. K., 885(8), *932*
Bounds, D. G., 1004(207), *1025*
Bourdillon, C., 725(228), *756*, 1436(31b), *1438*
Bouveau, J., 603(149), *613*
Bowden, E. F., 1436(31a), *1438*
Bowen, B. M., 1120(68), *1126*
Bowen, R. W., 1436(31e), *1438*
Bowen, W. R., 174(195–197), *201*
Bower, W. J., 635(123), *653*
Bowers, K. W., 1136(34), *1180*, 1175(485), *1193*
Boyd, D., 190(277), 194(277), *204*, 882(3j), 885(3j), *932*
Boyd, J. W., 638(132), *654*
Boyer, R. F., 351(207, 209), *359*
Bracht, J., 717(118), *752*, 959(48), *1019*
Bradfield, A. E., 291(294), *322*
Bradley, C., 855(281), *873*
Brago, I. N., 436(68), *458*, 908(244i), 910(224i), 911(224i), *942*, 923(281), *944*, 1201(30), *1212*
Brähler, G., 673(83), *694*
Braithwaite, G. E., 861(316), *874*, 1309(83), 1310(83), *1335*, 1321(125), *1336*, 1327(83), *1335*, 1328(125), *1336*
Brajter-Toth, A., 747(429), *763*
Brand, K., 413(96), *429*, 442(103), 443(103), 445(103), *459*, 448(186), *462*, 467(15), *479*, 709(55), *750*, 898(121), 899(121), 900(121), *937*
Brändström, A., 307(411), *326*
Bratu, C., 706(39), *749*
Brauman, J. I., 819(87), *867*
Braun, F., 670(71), *694*, 1056(233), *1067*
Brauer, E., 173(191), *201*
Brdicka, R., 163(158), *200*
Bréant, M., 297(336, 337), 286(336), *324*, 417(123), *430*, 734(314), *759*
Breckenridge, R. J., 620(33), *651*
Bredt, J., 442(115, 116), *460*
Breitenbach, J. W., 555(105), *577*, 589(32), *608*, 1348(27, 28), *1360*
Brenant, M., 1200(3), *1211*
Brench, F., 740(363), *761*
Brenet, J., 278(172), *319*
Brennan, M. P. J., 276(143), *318*, 1001(183), *1024*
Bresch, C., 734(310), *759*
Breslow, R., 242(61), *248*, 309(422), *327*, 340(79, 80, 81a, b), *355*, 779(60), *786*, 927(320), *946*
Brettle, R., 276(143), *318*, 563(131), *578*, 721(148), *753*, 1001(183), *1024*, 1005(212a, 216, 217), 1006(212a), *1025*, 1082(98), *1098*, 1156(215), *1185*, 1157(229), 1160(256), *1186*, 1161(272), *1187*
Brewis, D. M., 363(5), *394*
Brewster, J. D., 180(241), *203*, 293(313), *323*
Brezina, M., 474(75), *481*, 646(212), *656*
Brice, T. J., 1104(8), 1105(8), 1106(8), *1124*, 1317(109), *1335*
Brice, V. T., 835(177), *870*
Brich, W., 1172(407, 408), 1173(418), *1191*
Briggs, S., 899(150), *938*
Brillas, E., 370(69), *396*, 455(236), *464*, 916(264, 265), *943*
Brilmyer, G., 344(128), *357*, 1231(61), 1232(61), 1234(61), *1258*, 1384(101), *1395*
Briner, E., 425(191), *432*
Brintzinger, H., 407(47), *427*
Britt, W. J., 685(137), *696*
Britton, W. E., 369(53, 54), *395*, 391(185), *399*, 621(55), *651*, 621(76), 625(76), *652*, 729(265), *757*, 887(29), *933*, 1132(14), *1179*, 1154(202, 203), 1169(203), *1185*
Britz, D., 126(53), *196*
Brlik, J., 734(311), *759*

Broadhurst, M. J., 337(68), *355*, 1176(489), *1194*
Brochu, G., 979(108), *1021*
Brockmann, C. J., 287(245a), *321*
Brodsky, A. E., 455(230), *463*
Broman, R. F., 188(266), *204*
Brookes, C. J., 553(88), *577*, 1010(236), *1026*, 1157(225), *1185*
Brooks, B. W., 885(8), *932*
Brooks, H. M., 794(76), 799(76), *807*
Brooks, M. Y., 151(125), *199*, 288(253), 292(253), *321*
Brooks, W. W., 1312(87, 88), *1335*
Brown, A. C., 536(3), *574*
Brown, A. P., 270(94), *316*, 1365(18), 1370(18), *1392*, 1370(34), *1393*, 1371(18), *1392*, 1371(37, 38), 1372(37), *1393*
Brown, D. J., 441(149a), *461*, 882(3r), 885(3r), *932*, 895(91b), *936*, 895(98a), *937*, 902(167), *939*, 918(273b), *944*, 1144(117), *1182*, 1172(406), *1191*
Brown, E. A., 584(14), 589(39), *608*, 601(127), *612*, 630(104), 632(104), 633(104), *653*, 705(31),*749*, 723(184), 724(184), *754*, 1075(43), 1076(43), *1096*
Brown, E. V., 745(417), *763*
Brown, G., 793(32), *805*
Brown, G. M., 38(51a), *117*, 273(123d), *317*, 834(170), *869*
Brown, G. T., 1409(103, 104), *1418*
Brown, J. M., 233(1), *245*
Brown, K. L., 825(117), *868*
Brown, O. R., 309(428), *327*, 363(7), *394*, 368(46), 369(58), *395*, 373(80), *396*, 485(16), 489(16), 492(16), *496*, 635(121), *653*, 723(197), 724(211a), *755*, 726(232), *756*, 744(403), *762*, 848(251, 253), *872*, 899(133), *938*, 904(203), *940*, 907(193), *940*, 1153(184), *1184*
Brown, R., 284(226), 303(226), *320*
Brown, T. G., 669(66), *694*
Brownell, H. H., 791(16), *804*
Browning, B. L., 788(5), 790(5), *803*
Bruce, J. A., 1368(32), *1393*
Bruckenstein, S., 58(73c), *118*, 165(170), *200*, 168(173), *201*, 174(202–206), *202*, 190(278), *204*, 294(314), *323*, 621(74), 625(74), *652*
Bruegel, W., 486(23), *497*, 1142(73), *1181*
Brummer, S. B., 297(333), *324*
Brum, G., 680(120), 681(120), *695*
Brunel, D., 691(193), *698*
Brunnert, W., 1006(214), *1025*
Brunold, A., 1316(106), 1319(106), 1329(106), *1335*
Brunschwig, B. S., 50(66), *117*
Brus, L., 1406(23, 27), *1416*
Bruzco de Milano, D. L., 664(51), 665(51), *693*
Bry, B., 300(362), *325*
Bub, F. P., 616(1). *650*
Bubnov, N. N., 455(231), *463*
Buchi, G., 776(45), *785*
Buchta, R. C., 906(172), *939*, 1033(11), *1060*
Buck, R. P., 963(65), *1020*
Buckland, S. J., 184(257), *204*
Budniok, A., 641(176), 642(176), *655*
Budnikov, G. K., 436(70, 71), 445(71), *458*, 473(69), *481*, 474(77), *481*, 729(264), *757*, 809(9), 864
Buecheler, J., 1408(87, 89), *1418*
Buhler, D. R., 1094(183), *1101*, 1117(61), 1118(61), *1126*
Buhmann, H. R., 602(144), *612*
Buisson, C., 297(336, 337), 286(336), *324*
Bujoli, B., 719(135), *753*, 728(253, 254), *757*
Bukhtiarov, A. V., 474(73), *481*, 837(198), 838(198), *870*, 886(19), *933*, 1117(53, 54), 1119(53, 54, 57, 67a), *1125*, 1117(57), *1126*, 1269(36), *1281*, 1425(12), *1430*
Bull, R. A., 1382(88), 1385(88), *1395*
Bulyshev, Y. S., 794(54), 795(54), *806*
Bungay, H. R., 791(15), 793(15), *804*
Bunnett, J. F., 78(99a), *119*, 175(209), *202*, 194(291), *205*, 8911(54d), *935*, 1237(111), *1260*
Burdon, J., 1094(177), *1101*, 1110(41), *1125*
Burfield, D. R., 285(242), *321*, 295(320a–d), 296(320a–d), *323*
Burgbacher, B., 139(88), *198*, 982(116), *1022*
Burgert, B. E., 528(145), *533*, 1072(33), 1074(39), *1096*, 1088(140), *1099*, 1089(146), *1100*
Burgot, J. L., 1150(170), *1184*
Burke, L. D., 640(151), *654*
Burlakov, V. M., 794(54), 795(54), *806*
Burnett, D., 996(160), *1023*
Burnett, J. C., 1303(68), *1334*
Burnett, J. N., 725(215), *755*, 725(226a, c), *756*
Burney, H. S., 1303(67), *1334*
Burrows, B., 281(195), 285(195), 290(195), *319*, 1105(23), *1124*
Burson, J., 1104(9), 1105(9), 1106(9), 1108(9), 1110(9), *1124*
Burt, R. J., 1366(29), *1393*
Burzyk, J., 641(176), 642(176), *655*
Busca, G., 1410(120, 121a), *1419*
Busch, D. H., 813(43), *865*
Busch, R., 337(70), 350(70), *355*, 342(100), *356*
Busch, R. W., 1175(473), *1193*
Buschmann, H. W., 639(142, 144), *654*
Buschow, K. H. J., 335(45), *354*
Buser, H.-P., 812(22), *864*
Buss, V., 1207(89), *1214*
Bussey, R. J., 686(155), *697*
Bustos, L., 837(191), 862(191), *870*
Buthker, C., 332(20), 345(20), *353*

Butin, K. P., 366(25), *394*, 376(112), *397*, 863(329), *875*, 1059(256), *1068*, 1132(10), *1179*
Butkiewicz, K., 734(307), *759*
Butler, A. F., 424(166), *431*
Butler, J. A. V., 64(82a), 65(82a), *118*
Butler, J. N., 278(164), *318*, 280(189), 281(193), 299(189), *319*, 299(351), *324*, 301(369), *325*
Butterfield, R. J., 723(197), *755*, 726(232), *756*
Buvet, R., 284(223), 285(231), 286(223), 302(223), *320*
Buzykin, B. I., 466(7), *479*, 495(122), *500*
Bykova, L. N., 288(248), 295(248), 296(248), *321*
Byrd, L. R., 511(65), *530*, 643(190, 191), *656*, 1086(124), *1099*
Bystrikova, E. F., 424(170), *432*

Cadadei, M. A., 747(437), *763*
Caderas, C., 824(111), *868*
Caillet, A., 302(384), *326*
Caiola, A., 301(374), *325*
Cairns, E. J., 290(272), *322*, 505(1), *528*
Calafate, B. A. L., 1351(42), *1360*
Calderwood, T. S., 930(353), *947*
Callens, R. E. A., 1077(54), *1097*
Callingham, M., 300(359), *325*
Callot, H. J., 738(338), *760*, 985(132), *1022*, 1090(161), *1100*
Camacho, L., 161(147), 162(149), 163(147, 155, 160, 161), *200*
Cambell, K. A., 526(139), *533*
Camble, R., 1039(53), *1062*
Cameron, C., 620(33, 39), *651*
Cammann, K., 1365(20), 1371(20), *1392*
Campbell, C. B., 1083(103), *1098*, 1141(62), *1181*
Campbell, C. R., 1308(79), *1334*
Campbell, H., 736(324c), *759*
Campbell, R. H., 820(95), *867*
Campion, A., 1413(139), *1420*
Campion, J. J., 24(27d), 42(27d), *115*, 278(171), *319*
Canesson, P., 1247(173), 1249(173), *1262*
Canfield, N. D., 659(5), 680(5), *692*, 670(72), 671(72), 673(80), *694*, 680(115), 681(115), *695*
Cannon, R. D., 39(55b, c), *117*
Cantrell, P. R., 930(358), *947*
Cantrill, P. R., 184(253), *203*
Cao, X., 1434(12h), *1437*
Capobianco, G., 224(22), *232*, 346(152), 350(152), *357*, 346(156), *358*, 494(115), *500*, 901(157), *939*
Capparelli, M. P., 630(112), 633(112), *653*
Card, J. C., 1435(32), *1438*
Cardinali, M. E., 724(205), *755*, 725(220a, b, 222), *756*
Carelli, I., 703(10), *748*, 724(205), *755*, 725(220a, b, 222, 228), *756*
Carelli, V., 703(10), *748*, 725(220a, 222), *756*
Carey, J. H., 1413(141), *1420*
Cariou, M., 597(104,105), 598(104,105), 599(104), *611*, 648(230), *657*,893(68a), *935*, 1054(208), *1067*, 1149(163), 1175(486,487), *1183*, *1194*
Carlier, R., 597(104), 598(104), 599(104), *611*, 1058(253), 1059(253), *1068*, 1224(30), *1257*, 1239(30), *1260*
Carlin, R. T., 303(393), *326*
Carlsen, L., 295(318), *323*
Carlson, T., 1408(48), *1417*
Carlsson, L., 1312(84, 85), *1335*, 1414(142, 145), *1420*
Carlsson, S., 307(411), *326*
Caroll, G. L., 410(73), *428*
Carpenter, A. K., 418(134), *430*, 590(42), *609*, 886(21), *933*
Carpenter, J. F., 1094(183), *1101*, 1117(60), 1118(60), *1126*
Carpenter, L. L., 723(194b), *755*, 979(89), *1021*, 1382(84), *1395*
Carpita, A., 547(74), *576*, 783(73), *786*
Carr, J. P., 1288(29), *1333*
Carrahar, P., 487(36), *497*
Carraro, S., 1211(126), *1215*
Carrasco, O., 722(177), *754*
Carre, B., 290(271), *322*, 515(100), *532*
Carrizosa, I., 1410(118), *1419*
Carroll, S., 284(239), *320*
Carroll, W. F. Jr., 363(8, 9), 364(9), 366(9), 371(9), *394*, 376(114), 378(125), *397*, 378(126), *398*
Carruthers, W. C., 243(71, 74), *248*
Carsky, P., 1051(176), *1066*
Carter, M. T., 485(14), *496*
Casadei, M. A., 390(181), *399*, 489(55), *498*, 703(10), *748*, 708(50), *750*, 1050(162), *1065*, 1434(13e), *1437*
Casanova, J., 368(37), 365(55), *395*, 374(88), *396*, 378(121), *397*, 387(156), *398*, 1153(182), *1184*
Casnova, J., Jr., 569(150), *579*, 1159(235, 243), 1170(235), *1186*
Casey, A. T., 662(33, 35, 36), *693*
Cashmore, P., 1141(63), *1181*
Casini, A., 725(222), 725(228), *756*
Cass, A. E. G., 645(207), *656*
Casselberry, R. L., 411(87), 412(87), *429*
Cassidy, J., 514(80), 515(80), *531*
Casson, D., 350(200), *359*
Castellonese, P., 302(388), *326*
Castillo, J. I., 1375(53), 1380(71), 1394(53), *1394*, 1386(137), *1396*
Caullet, C., 424(171), *431*, 444(129), *460*, 712(67), *750*, 720(140), 723(140), *753*, 723(185), *754*, 723(192), *755*, 897(117, 118), *937*, 901(155–157), *938*, 902(165), *939*, 929(339), *946*, 974(94b, 98), *1021*
Cauquis, G., 193(284), *205*, 265(69), *315*, 280(184), 281(184), 284(184), 286(184), 300(184), 304(184), *319*, 283(236), 295(236), 304(236), *320*, 305(400), *326*, 522(122), *532*, 591(46, 48, 50), 592(48, 50), *609*, 599(116), *611*, 661(19, 23), *692*, 682(125), 684(131), 685(141), 686(151), *696*, 716(109–111), 717(117), *752*, 723(182), *754*, 728(257, 262), *757*, 795(49, 50), *806*,

[Cauquis, G.]
800(88), *808*, 970(73), 977(82), *1020*, 975(97, 104, 105), *1021*, 1015(253), 1016(253), 1017(254), *1027*, 1084(108, 109), *1098*, 1088(134), *1099*, 1110(42), *1125*, 1165(316), *1188*, 1167(331), *1189*
Cava, M. P., 682(125), *696*, 1377(81), 1382(81), *1395*
Cavagnol, J. C., 733(292), *758*
Cavicchioni, G., 717(122), *752*
Caze, J., 891(58), *935*
Ceccon, A., 839(213a, b), *871*
Cedheim, L., 267(76b), *315*, 509(41), *530*, 527(141), *533*, 601(139), *612*, 1013(247), *1027*, 1077(57), *1097*, 1081(89), *1098*, 1160(251), 1166(251), *1186*, 1241(124), *1260*
Cekovix, Z., 441(147), *461*, 714(99a), *751*
Cenas, N., 646(217), *657*, 725(228), *756*
Cerefice, S. A., 486(24), *497*
Cerles, M., 515(96), *531*
Cerrai, P., 1346(23), *1360*, 1352(57–61), *1361*
Cessac, J., 350(202), *359*
Cesareo, D., 1413(129), *1419*
Cettou, P., 1309(81), *1334*
Chabaud, B., 599(116), *611*, 716(110), *752*, 975(104), *1021*, 1017(254), *1027*
Chadenson, M., 800(88), *808*
Chadwick, A. T., 190(274), *204*
Chakraborti, S., 796(60), *806*
Chakravarti, S. K., 662(34), *693*
Chai, W., 1173(432), *1192*
Chaloyard, A., 838(203), *871*
Chambers, C. A., 621(80, 81), 627(80, 81), *652*
Chambers, J. Q., 621(80, 81), 627(80,81), *652*, 635(119), *653*, 659(1, 5), 660(1), 662(1), 680(5), *692*, 670(72), 671(72), 673(80– 82), *694*, 676(99), 680(115), 681(115), *695*, 683(127, 128), *696*, 689(188, 189), 690(188, 189), *698*, 712(72), *750*, 728(259), *757*, 903(362), 931(362), *947*, 1044(93), 1048(93), *1063*, 1372(42), 1373(42), *1393*
Chambers, J. R., 712(71), *750*, 931(360), *947*
Chambers, R. D., 388(169), *399*, 747(433), *763*
Chami, Z., 380(130), *398*
Champagne, P. J., 548(77), 553(77, 87), *576*, 1004(210), *1025*, 1011(237–239), *1026*, 1157(226, 227), *1185*
Champetier, M. G., 1144(109), *1182*
Chan, M. S., 826(127), *868*
Chance, R. R., 340(84), 341(84), *355*
Chandra, S., 635(121), *653*
Chandross, E. A., 344(122), *357*, 510(47), *530*
Chaney, C. A., 745(414), *763*
Chang, B., 1094(182), *1101*
Chang, H.–F., 931(362a), *947*
Chang, P., 376(110), *397*
Chang, R., 334(36), 337(69) *354*, *355*
Chang, S., 1268(20), *1281*
Chao, J. C., 257(12), *313*
Chao, S., 1372(46), *1393*
Chapman, D. L., 59(75b), *118*
Chapovskii, Y. A., 825(124), 841(124), *868*
Chaquiq El Badre, M., 1227(46), *1257*, 1251(188), *1262*
Charlot, G., 21(21c), *115*
Chartier, P., 1434(151), *1437*
Chasle-Pommeret, M. F., 1152(176), *1184*
Chatt, J., 809(7), *864*, 836(186), *870*
Chatten, L. G., 915(259), *916*(259), *943*
Chaturvedi, U. S., 662(27), *692*
Chaussard, J., 421(145), *431*, 891(48), *934*
Chauvelier, J., 739(346), *760*
Chaykobsky, M., 1210(118), *1215*
Che, C.–M., 834(169), *869*
Chedsey, C. E. D., 1314(4), *1392*
Cheek, G. T., 386(147), *398*, 491(96), *499*, 723(189, 190), *754*, 723(191), *755*
Chehna, M., 719(135), *753*, 728(254), *757*
Chekanova, V. D., 276(141), *317*
Chekir, K., 733(289), *758*, 736(323), 738(323), *739*, 738(333), *760*
Chemla, M., 73(93c), *119*
Chen, A., 863(328), *875*, 1169(354), *1190*, 1177(509), *1194*
Chen, C. C., 1406(32), 1407(32), 1408(36, 38, 39), *1416*, 1409(95, 100), 1412(95), *1418*, 1412(128), *1419*
Chen, C. H., 685(139), *696*
Chen, H. T., 885(11), *932*
Chen, M. J., 1408(62), *1417*
Chen, W.–Y., 621(52), *651*
Cheng, H.–Y., 621(53), *651*, 646(218), *657*
Cheng, L., 1206(70), *1213*
Cheng, L. K., 541(36), *575*, 547(75), *576*
Cheng, M.–C., 569(156), *579*, 1037(32), *1061*
Cheng, S., 125(49), *125*
Cheng, T. Y., 567(149), *579*, 745(409), *762*
Cheng, Y. Y., 1142(81), *1181*
Chenikov, I. V., 1202(41), *1212*, 1202(42), *1213*
Cherednichenko, A. D., 743(387), *762*
Chericato, G., 1269(32), *1281*
Chern, C. I., 1245(153), *1261*, 1407(33), *1416*
Chernenko, A. A., 342(103), *356*
Chernoburova, E. I., 1040(70), *1062*, 1147(150), *1183*
Cherykh, E. V., 820(96), *867*
Chernyshev, E. A., 837(198), 838(198), *870*, 1269(36), *1281*, 1425(12), *1430*
Cheung, K. M., 392(193), *400*, 1153(188), 1154(188), *1184*, 1224(31), 1249(31), *1257*
Chevrot, C., 1236(107, 109), *1259*
Chia, V. K. F., 1371(36), *1393*
Chiappardi, D. M., 411(87), 412(87), *429*
Chiba, T., 275(129), *317*, 441(152a), *461*, 583(12), *607*, 596(90), *610*, 597(107), *611*, 1009(228), *1026*, 1091(164), *1100*, 1428(22), *1430*
Chibani, A., 707(45), *749*
Chibata, T., 1233(96), *1259*
Chibrikin, V. M., 411(80), *429*
Chidabaram, S., 1230(58), 1231(58), 1232(58), 1234(58), *1258*

Chien, H.–S., 736(322), *759*
Chiklis, C. K., 1389(122), *1396*
Child, W. C., 291(295), *322*
Childs, W. V., 148(109), *198*, 882(3b), 885(3b), *932*, 1094(179, 181), *1101*, 1104(7), 1113(7), 1114(7), *1124*, 1113(48), *1125*, 1301(60), 1319(60), *1334*
Chilesotti, A., 273(121), 274(121), *317*
Chin, D.–H., 1269(32), *1281*
Chin, D. T., 257(12), *313*, 1287(13), *1332*
Chini, M., 547(74), *576*, 783(73), *786*
Chinova, M., 704(26), *749*
Chiong, K. N., 991(169), *1024*
Chiorboli, P., 1056(235), *1067*
Chiu, Y. L., 240(516), *248*
Chivers, T., 284(225), 301(225), 302(225), *320*, 855(276), 862(276), *873*, 863(328), *875*, 930(341), *946*, 1169(354), *1190*, 1177(509), *1194*
Chkir, M., 552(82), *577*, 565(141), *578*, 1002(190a), *1024*, 1011(237), *1026*, 1157(228), *1185*
Chodkowski, J., 163(154), *200*
Chopin, J., 800(88), *808*, 1206(81), *1214*
Chou, T. C., 437(135b), 441(135b), *460*, 540(32), *575*, 641(168), 643(168), *655*, 794(52, 59, 61), 795(48, 52), 796(52, 59), 797(61), *806*, 885(11), *932*, 1035(20), *1061*, 1421(4b), *1429*
Chou, T. T.–T., 151(125), *199*, 723(193), *755*
Choukroun, H., 691(193), *698*
Christensen, L., 704(17), 708(17), *748*, 994(155c), *1025*, 1142(76), *1181*, 1162(292), *1187*
Christensen, P. A., 179(234, 235), *203*
Christensen, T. B., 2(7), *8*
Christie, J. H., 150(121), *199*
Christofis, O., 1235(89), *1259*
Christol, S. J., 1205(64), *1213*
Chruma, J. L., 308(420), *327*, 347(171), 348(171), *358*, 367(32), 370(32), 373(32), 388(32), *395*, 889(40), *934*, 906(185c), 907(182c), *939*, 919(275), 922(276), 923(281), *944*, 930(349a), *947*, 1267(14), 1268(14), 1271(14), *1281*
Chu, W., 340(81a), *355*
Chum, H. L., 433(3), 452(3), *456*, 489(61), *498*, 788(4), *803*, 790(10), 791(18, 20), 794(18), *804*, 793(34), *805*, 798(69, 73, 74), 799(75), 800(78), 801(69), *807*, 802)90, 91), *808*, 826(129, 130), *868*, 1428(20), *1430*, 1434(11), *1436*
Chung, L. L., 334(40, 41), 335(41), 342(41), *354*, 1155(208), *1185*
Chung, V. V., 906(181), *939*
Chupka, E. I., 792(30), *805*, 794(47, 54, 58), 795(51, 54), 796(58), *806*, 798(70–72), *807*
Chupp, J. P., 494(113), *500*
Churilina, A. P., 488(50), *498*
Cihalik, J., 284(212), 291(212), *320*, 291(292), 286(292), *322*
Ciomas, F., 909(232), *942*
Cisak, A., 263(59, 60), *314*, 284(210), 298(210), *320*
Ciszewski, A., 727(251), *757*
Citterio, A., 237(34), *247*
Clack, D. W., 738(336), *760*, 820(93), *867*
Claghorn, H. R., 902(175), *939*
Clapper, G. L., 436(57), *458*, 1144(112), *1182*
Clardy, J., 662(37), *693*
Clarenbach, S., 173(191, 192), *201*
Claret, J., 725(230), *756*
Clark, D. B., 515(85), *531*, 515(101), *532*, 1083(102), 1085(102), *1098*, 1117(56), *1126*
Clarke, J. S., 290(283), 291(283), *322*, 1071(20), 1073(20), *1096*, 1105(16), 1109(16), 1110(16), 1112(16), *1124*
Clarke, R., 1230(54), *1258*
Claus, H., 239(46), *247*
Clause, P., 893(70), *935*
Clauson-Kass, N., 720(141), 721(142–144, 146), *753*, 1161(276), *1187*, 1243(138), *1261*
Cleary, J. A., 366(23), *394*, 370(74), 372(74), *396*
Cleghorn, H. P., 1144(115), *1182*
Clement, A. A., 415(109), *430*
Clemo, G. R., 447(177), *462*, 1132(1), 1144(1), *1179*
Clennan, E. L., 519(108), *532*
Clifford, A. F., 260(44), *314*, 290(282), *322*, 292(299), 293(299), *323*
Clifford, K. H., 1243(131), *1260*
Coburn, C. E., 781(65b, c), *786*
Coche, L., 445(154), *461*
Cochran, J. C., 726(236), *756*
Cockerham, M. P., 1081(93), *1098*
Cockrell, J. R., 175(209), *202*
Cocolios, P., 816(67, 70), 817(74), *866*
Coe, D. E., 663(46), 664(46, 48–50), 665(48–50), *693*, 713(77), *751*, 1054(212), *1067*, 1164(309, 310), 1165(309), *1188*
Coe, P. L., 553(88), 554(91), *577*, 1010(236), *1026*, 1157(225), *1185*
Coetzee, J. F., 24(27d), 42(27d), *115*, 278(171), 279(178), 281(191), 282(178, 203), 285(203), 295(203), *319*, 289(261), *321*, 295(317, 321), *323*, 296(328), *324*, 300(358, 361), 307(358), *325*, 484(7), *496*, 506(16), *529*
Coevret, F., 268(90), *315*, 1286(4), 1292(4), 1299(4), *1332*
Coffen, D. L., 673(80), *694*, 680(115), 681(115), *695*, 689(188), 690(188), *698*
Cogley, D. R., 280(189), 299(189), *319*, 301(369), *325*
Cognard, J., 591(46), *609*, 970(73), *1020*
Cohen, B. J., 1146(136), *1183*
Cohen, H., 901(153), *938*
Cohen, L. A., 1082(94), *1098*
Cohen, M., 899(145), 901(145), *938*
Colau, R., 706(42), *749*
Cole, R. C., 291(285), *322*
Coleman, A. E., 511(69), *531*, 954(12), *1018*, 1084(113), *1098*
Coleman, B. R., 1178(521), *1195*
Coleman, G. C., 486(34), *497*

Coleman, J. P., 285(218), 290(218), *320*, *321*, 290(269), *322*, 375(97), *396*, 375(98), *987*, 483(4), 488(4), 493(4), 495(4), *496*, 489(58), *498*, 515(97, 98), 518(97, 98), *531*, 540(30, 31), 563(30), *575*, 628(90), *652*, 963(65), 1001(184), *1020*, *1024*, 1039(49a,b), 1062, 1143(88, 90), 1144(90), *1181*, 1203(44), *1213*, 1268(30), *1281*, 1321(119), *1336*, 1372(39), *1393*, 1433(2), *1436*
Coles, B. A., 190(274, 279), 192(279), 192(279), 193(279), *204*, 194(294, 295), *205*
Colichman, E. L., 1044(94), *1063*
Collin, J.–P., 831(153), *869*, 1434(15d, l), *1437*
Collins, A. M., 903(195), *940*
Collins, C. J., 1434(22c), *1438*
Collins, T. J., 635(126), 636(126), *654*, 819(87), *867*
Collman, J. P., 809(2), *864*, 818(78), 819(78), *866*, 819(84, 87), *867*
Colton, C. K., 644(193, 195–197), 645(193, 196), *656*
Colton, R., 1175(449, 451, 454), *1192*
Combellas, C., 78(100), *119*, 226(24), *232*, 1239(114, 116), *1260*
Comi, R., 6(20), *8*, 1043(87), *1063*
Commeyrus, A., 290(271), *322*
Compagnon, P.–L., 456(243, 245), *464*, 715(105a, c, d), *752*, 839(218), *871*, 915(260), *943*, 929(336), *946*, 1139(61), 1141(61), *1181*, 1239(117), *1260*, 1267(15), 1277(15), *1281*, 1277(50, 51), *1282*
Compton, R. G., 193(285), *193*, 165(168), 173(168), *200*, 174(200), *201*, 190(274, 279, 280), 192(279), 193(279), *204*, 194(294, 295), *205*
Conant, J. B., 282(202), 285(202), 291(202), *319*
Condit, P. C., 265(68), *314*, 458(76), *458*, 486(20), *497*, 1142(72), *1181*
Conia, J. M., 1224(29), 1248(29), *1257*
Conley, R. F., 292(300), *323*
Conner, K. A., 1175(457), *1193*
Connolly, J. F., 1316(102), 1321(102), *1335*
Connelly, N. G., 841(223), *871*, 846(241, 242), *872*
Connous, T. F., 122(12), *195*, 258(22), *313*
Conrad, D., 1434(12e, n), *1437*
Conte, L., 1109(38), 1110(38), 1111(38, 44, 45), 1112(38, 45, 46), 1123(45), *1125*
Contractor, A. Q., 1341(12), *1359*
Contrerns, M. L., 275(134), *317*, 931(372), *948*
Convert, O., 405(31, 34), 408(31), *427*, 408(60), *428*
Conway, B. E., 122(7), 123(7), 157(7), 179(7), *195*, 179(228), *202*, 285(221), *320*, 895(91a), *936*, 1132(11), *1179*, 1287(14), *1332*
Cook, H. J., 411(78a), *429*
Cook, L. D., 726(242), *757*
Cook, R. L., 930(361), *947*
Cooks, R. G., 1005(218), *1025*
Coombs, D. M., 478(11), *482*
Coombs, R. V., 571(164), *579*, 1159(239), *1186*
Cooney, R. P., 179(225), *202*
Cooper, R., 776(44a), *785*
Copenhafer, R. A., 335(53), 336(53), 339(53), *354*
Corbett, J. D., 1200(13), *1211*
Corbon, P., 974(94b), *1021*
Cordier, P., 1209(114), *1215*
Cordoba, M. H., 153(130), *199*
Corey, E. J., 544(53), *576*, 564(134), 566(134), *578*, 569(150), *579*, 1001(192), *1024*, 1036(30), *1061*, 1159(235, 243), 1170(235), *1186*, 1210(118), *1215*
Corley, E. G., 1051(180), *1066*
Cornelisse, J., 1070(4), *1095*
Cornforth, J. W., 1144(102), *1182*, 1243(131), *1260*
Cornubert, R., 443(118), *460*, 1143(87), *1181*
Corri, G., 1211(126), *1215*
Corrigan, D. A., 144(102), *198*, 404(25), *427*
Corriu, R. J. P., 855(277, 278), *873*, 930(343), *946*
Corwin, A. H., 899(141), *938*
Cosnier, S., 1381(76), *1394*
Costa, G., 826(104, 105, 108), 863(108), *867*, 824(120), 825(114– 116), 826(128), 841(114), 863(114), *868*
Costa, J. M., 370(69), *396*
Cottrell, P. T., 283(234), *320*, 388(168), *399*, 667(55), *693*
Coughlin, R. W., 1434(21, 22d), *1438*
Coulson, J. M., 268(89a), 271(89a), *315*, 1233(87), 1235(87), *1259*
Couper, A. M., 190(274), *204*
Courbis, P., 193(290), *205*
Courbon, H., 1409(97), 1408(44,45), *1416*, *1418*
Couret, C., 411(76), *428*
Court, D., 1056(242), 1057(242), 1058(242), 1059(242), *1068*
Courtneidge, J. L., 506(27), *529*
Courtot-Coupez, J., 301(368), *325*, 685(135), *696*, 983(128), *1022*, 1175(458), *1193*
Coutsolelos, A., 816(66), *866*
Couture, R., 648(227), *657*
Cover, R. E., 914(240), *942*
Covitz, F. H., 375(92), *396*, 621(66), 625(66), *652*, 1071(22), *1096*
Cowan, D. O., 334(35), *354*, 680(118), 681(118), 682(118), *695*, 838(205), *871*
Cowan, J. A., 819(83), *867*
Cowley, A. H., 411(77), *428*, 1200(10), *1211*
Cox, C. M., 1209)110), *1215*
Cox, F. W., 453(214), *463*
Cox, G. B., 1156(215), *1185*
Craig, L. C., 722(176), *754*
Cramer, J., 602(146), 603(146), *613*
Cranmer, B. K., 819(80), *866*
Creager, S. E., 818(77), *866*
Creason, S. L., 174(201), *202*, 309(430), *327*
Creger, P. L., 740(368), *761*
Creighton, H. J., 439(85, 88), *459*
Creutz, C., 50(66), *117*
Criegee, R., 15(10a), *114*
Crist, D. R., 722(173), *754*
Crljenak, S., 596(92, 97), *610*, 715(107), *752*
Cronlund, M., 790(13), *804*

Crooks, R. M., 270(97), 303(97), 304(97), *316*
Cros, J.-L., 717(117), *752*
Cross, M., 1387(120), 1389(120), *1396*
Crossman, F., 351(209), *359*
Crowley, J. I., 341(90), *356*, 1376(58), *1394*
Crumbliss, A. L., 645(208), *656*
Cruz, J. L., 162(149), *200*
Csizi, G., 277(157), *318*
Ctvrtnik, J., 726(234), *756*
Cuccia, C., 292(301), 303(301), *323*
Culcasi, M., 411(76), *428*
Cundall, R. B., 1408(40), *1416*
Cunkle, G. T., 691(195), *698*
Cunningham, A. J., 725(226b), *756*
Cunningham, G. P., 295(321), *323*, 725(216), *755*
Curphey, T. J., 447(181), 450(180), *462*, 490(82), *499*, 902(168a), *939*, 1146(143), *1183*
Curtius, T., 1207(85–87), *1214*
Curulli, A., 903(218), *941*
Cushman, M. R., 179(237), *203*
Cutler, R. S., 1148(154), *1183*
Czapla, S., 468(31), *479*
Czochralska, B., 740(366), *761*, 746(427), *763*, 927(319), *946*, 1132(9), 1175(9), *1179*, 1148(154), *1183*, 1169(351), *1189*, 1175(9), *1170*

Dabosi, F., 930(343), *946*
Dabosi, G., 855(277, 278), *873*
Dahm, R. H., 275(138), *317*, 363(5), *394*
Dahrenmöller, N., 983(126), *1022*
Daisley, R. W., 915(259), 916(259), *943*
Dakeshita, K., 836(185), *870*
Dalavnay, J., 1032(3), *1060*
Dale, R., 898(131), *938*
Daley, R. F., 914(242), *943*, 1001(194), *1025*, 1137(41), *1180*
Dalgaard, G. A. P., 468(30), *479*
Dalgleish, D. T., 620(34, 35, 37, 38), *651*
Dall'Olio, A., 1375(51), *1393*
Dalrymple, I. M., 1073(35), *1096*, 1229(50), 1234(50), *1257*, 1312(90), *1335*
Daltrozzo, F., 726(238), *756*
Damange-Guerin, G., 620(41), *651*
Damani, S., 596(94, 96), *610*
Damaskin, B. B., 435(53), *458*, 1175(477), *1193*
Damoni, S., 706(36a), 710(36a), 712(36a), 714(36a), *749*, 710(60a), 715(60a), *750*, 974(94b), *1021*
Dana, G., 916(262), *943*, 1144(106, 125, 128), 1145(125, 128), *1182*
Danen, W. C., 193(286), *205*
D'Angeli, F., 717(122), *752*
Daniele, S., 834(173), 835(173), *870*
Danieli, B., 716(115), *752*
Danieli, R., 1382(80), *1395*
Danilov, V. G., 1433(7), *1436*
Danly, D. E., 909(230), *942*, 1287(9), *1332*, 1295(44, 45), *1333*, 1303(68, 69), *1334*, 1307(9), *1332*, 1308(79), *1334*, 1308(44), *1333*, 1314(96), 1316(99, 100), 1319(100, 112), 1322(112), 1324(99, 100), 1325(100), *1335*, 1358(98), *1362*
Dannecker, R., 1139(59), 1140(59), *1180*
Dannels, B. F., 1384(105), *1395*
Dannhauser, W., 291(285), *322*
Daolio, S., 849(255, 256), *872*
Dapperheld, S., 245(81), *249*, 593(64, 66, 67), *609*, 593(74), 594(76), *610*, 599(119), *611*, 1225(39), *1257*
Darchen, A., 267(78), 273(78), 288(78), *315*, 273(123f), *317*, 412(89), 413(90, 91), 417(91), 421(90), *429*, 415(110, 111), 418(131), *430*, 838(209), *871*, 1144(121), 1145(121), *1182*, 1150(170), *1184*, 1156(213), *1185*, 1241(123), *1260*
Daremon, C., 1164(306), *1188*
Darensbourg, D. J., 1175(456), *1193*
Darwent, J. R., 1409(103, 104), *1418*
Daschbach, J., 278(175), *319*
Dascola, Y., 1375(51), *1393*
Dassler, T., 715(101), *751*
Date, Y., 375(90), *396*
Daub, G. H., 344(125), *357*
Dauben, H. J., 565(144), *578*, 1159(244), *1186*
Dauer, A., 369(63), *369*
Daum, H., 895(98c), *937*, 1172(413), *1191*
Dauphin, G., 722(174), *754*, 1241(122), *1260*
Dauplaise, D., 662(37), *693*
Dautartas, M. F., 647(221), *657*
Davidson, R. S., 1408(71), *1417*
Davies, A. G., 506(27), *529*
Davies, D. H., 1353(71, 72), *1361*
Davies, J. A., 835(182), *870*
Davies, J. D., 162(150), *200*, 335(43), *354*
Davies, J. E., 544(54), *576*, 1001(193), *1024*
Davies, R. R., 1208(99), *1214*
Davis, B. M., 894(74), *935*
Davis, D. G., 660(10), *692*, 813(46, 49), 814(49), *865*
Davis, G., 645(207), *656*
Davis, J. E., 188(270), *204*
Davis, S. M., 411(78a), *429*
Davolio, G., 1056(232, 235), *1067*
Davoust, D., 1137(46), *1180*, 1271(41), *1282*
Davydor, V. D., 792(25, 26), *805*, 794(53), 795(53), *806*
Day, M. J., 194(294, 295), *205*
Day, R. A., Jr., 434(17), 435(17, 35), *457*, 452(210), 453(210), 455(210), *463*, 914(242), 915(251), *943*, 1001(194), *1025*, 1137(41), *1180*, 1142(80), *1181*, 1144(129), 1145(129), *1182*
Day, R. J., 725(227), *756*
Deakin, M. R., 68(88), *118*, 82(106), *119*, 208(4a), *231*
DeArmond, M. K., 831(154), *869*
Debaert, M., 1150(170), *1184*
DeBoer, E., 332(14), 334(14), *353*
Debye, P., 32(45a), 33(45a), *116*
Decker, R. F., 788(8), *803*
Declercq, J. P., 733(304), *759*
Dedek, V., 907(185b), *940*
Dedolph, D. F., 722(181), *754*, 1203(48), *1213*
Deex, D., 1421)1a), 1424(1b), *1429*
de Ficquelmont, A. M., 298(343), 286(343), *324*

DeFranco, R. J., 377(119), *397*
De Giovani, W. F., 641(169), 643(169), *655*
Degner, D., 243(73), *248*, 436(61), *458*, 593(68), *609*, 593(70), *610*, 604(157), *613*, 898(124), *938*, 1172(401–403, 405), *1191*, 1243(139), *1261*, 1305(76), *1334*, 1319(116), *1336*, 1424(6), *1429*
Degrand, C., 5(15), *8*, 239(42a), 240(42a), *247*, 240(51a), *248*, 258(21), *313*, 410(72), 428, 456(243–245), *464*, 599(113), *611*, 715(105a–d), *752*, 726(233), *756*, 839(215, 218), *871*, 886(20c), 891(20c), *933*, 891(56), *935*, 915(260, 261), 916(261), *943*, 922(278), 923(279), *944*, 929(336), *946*, 930(350), *947*, 1054(210), *1067*, 1138(48, 49), 1150(168), *1180*, 1139(61), 1141(61), *1181*, *1184*, 1177(507), *1194*, 1239(117), *1260*, 1250(182, 183), *1262*, 1252(193, 198, 199), *1263*, 1267(15), 1277(15), *1281*, 1277(50, 51), *1282*, 1343(15), *1359*
Degtiarev, L. S., 455(230), *463*
Dehayes, L. J., 661(20), *692*
Dehn, J. S., 1159(240), *1186*
DeJuan, M., 163(161), *200*
DeJongh, R. O., 1135(23), *1179*
de la Fuente, J., 544(56), *576*
Delahay, P., 154(133), *199*, 1287(15), *1332*
Delahaye, D., 712(67), *750*
Delamar, M., 960(22), *1018*
Delaroff, V., 1176(493), *1194*
Delaunay, J., 1013(248), *1027*, 1136(33), *1180*, 1144(130, 131), 1145(130, 131), *1183*, 1160(253), *1186*, 1166(323), *1188*, 1160(253), *1186*, 1166(323), *1188*, 1231(67), *1258*, 1241(125, 127), 1242(128), *1260*, 1246(67), *1258*, 1250(184), *1262*, 1278(57e), 1280(57e), *1282*
Delay, F., 1035(18), *1061*
Delhomme, H., 591(48, 50), 592(48, 50), *609*, 977(82), *1020*
Delmar, M., 302(390), *325*
Delmastro, J. R., 434(21), *457*, 882(3h), 885(3h), *932*
De Luca, C., 891(51), *934*
Demand, F., 1250(180), 1251(180), *1262*
Demange-Guerin, G., 302(384), *326*
Demin, V. A., 792(25, 26), *805*
Demortier, A., 293(305), *323*
Demura, T., 1383(89), 1387(89), *1395*
DeNicola, A. J., 723(193), *755*
Denisevich, P., 819(84), *867*, 1341(10), *1359*, 1372(39), *1393*
Denisovich, L. I., 825(123, 124), 841(123, 124), 863(123), *868*, 846(236, 240), *872*
Dennstedt, M., 722(175), *754*
Deno, N. C., 242(60), *248*
DeNora, O., 277(156), *318*
DePaoli, M.–A., 1387(110, 111), 1389(110, 111), *1396*
Derjaquin, B. V., 257(18a), *313*
Derkacheva, V. M., 820(96), *867*
Deronzier, A., 635(125), 636(125), *653*, 661(23), *692*, 728(257), *757*, 1381(74, 76), *1394*
Desbarres, J., 300(357), 286(357), *325*
DeSchepper, R. E., 630(111, 112), 633(111, 112), *653*
Desideri, P. G., 971(776), 972(78), *1020*
Desjardins, S., 275(130), *317*, 441(152b), *461*, 1435(25b), *1438*
Desmond, M. M., 508(34), 509(34), 514(34), 516(34), *529*
Despic, A. R., 1314(94), *1335*
Dessy, R. E., 284(225), 301(225), 302(225), *320*, 855(276), 862(276), *873*, 863(327, 328), *875*, 930(341), *946*, 1148(152), *1183*, 1169(354), *1190*, 1177(509), *1194*
Deswarte, S., 405(37), *427*, 408(56, 61–63), *428*, 706(42), *749*
Detty, M. R., 672(79), *694*
Deutsch, E., 824(118), *868*
Devaucelle, L., 816(71), *866*
Devaud, M., 723(198), *755*, 857(283, 285, 286), *873*, 1060(266), *1068*
Devi, T. P., 732(283), *758*
DeVicente, M. S., 150(124), *199*
DeVries, T., 280(188), 293(188), *319*, 285(207), *320*, 311(439), *327*, 1105(22), *1124*
Devynck, J., 290(271), *322*, 515(99, 100), 518(99), *532*, 616(3), 620(3), *650*, 826(125), *868*, 1434(13f), *1437*
Dewar, J., 895(160, 174), 898(160), *939*
Dewar, M. J. S., 24(28), *115*, 334(39), *354*, 508(44), *530*
Dey, A. N., 301(371), *325*
Dey, B. B., 414(101), 414(102b), *429*, 418(128b), *430*
Dhar, D. N., 453(222), *463*
Diaz, A. F., 341(90), *356*, 564(137), 567(137), *578*, 1142(81), *1181*, 1178(518), *1195*, 1356(6, 10, 11), 1366(10), *1392*, 1366(24), *1393*, 1368(10, 11), *1392*, 1368(33), *1393*, 1370(10), 1375(6), *1392*, 1375(52–54), 1376(58–62), 1377(62), 1378(62), *1394*, 1379(136), *1396*, 1380(54, 62, 65, 71), 1381(72), 1382(62), 1383(62), *1394*, 1383(91, 92, 99), *1395*, 1384(62), 1385(62), *1394*, 1385(92, 99), *1395*, 1385(109), *1396*, 1386(53, 54, 62), *1394*, 1386(92, 99), *1395*, 1386(137), *1396*, 1387(54, 62, 65), *1394*, 1387(99), *1395*, 1387(110, 111), *1396*, 1388(62), *1394*, 1389(99), *1395*, 1389(110, 111), *1396*
Dickens, J. E., 51(68), *117*
Dickerhof, K., 492(97), *499*, 1208(106), 1210(106), *1214*, 1208(107), 1209(116), 1210(107), *1215*
Dickeson, J. E., 733(301), *759*
Dickinson, T., 260(38), *314*, 538(18), 542(18), *575*

Diefenderfer, A. J., 369(56), *395*, 373(79), *396*, 388(167), *399*
Diegiec, J., 1229(74), *1258*
Diekert, G., 813(27), *865*
Dieleman, J., 335(45), *354*
Dieter, T., 831(156), *869*
Dietmar, W., 894(85b), *936*
Dietrich, M., 336(56), 343(56), 351(56), *354*, 506(12), 511(12), *529*
Dietrich, W., 1042(81), *1063*
Dietrich-Buchecker, C. O., 831(155), *869*
Dietz, H. J., 437(137), *460*
Dietz, R., 332(21), 333(25), *353*, 341(94), 350(94), *356*, 347(169), 348(169), *358*, 887(23, 24), *933*, 1138(51), *1180*, 1245(155), *1261*
Digua, A., 835(174), *870*
Dike, M., 827(134), *868*
Dike, S., 827(134), *868*
Dimeler, G. R., 508(33, 34), 509(34), 514(34), 516(34), *529*
Dimitrieva, W. D., 434(4), *456*
Dimmel, D. R., 802(89, 90), *808*
Dimroth, K., 616(5), 618(5), 621(5), *650*
Ding, J.–Q., 1148(158), *1183*
Dingle, T. W., 336(60), 339(60), *354*
Dirlam, J. P., 387(156), *398*, 1160(260), *1186*
Disdier, J., 1408(44, 45), *1416*, 1408(50), *1417*
Divisek, J., 195(296), *205*
Dixit, G., 637(127), *654*, 715(103), *751*, 1243(142), *1261*
Dixon, W. T., 617(16, 17), *650*
Djeghidjegh, N., 1251(188), *1262*
Djurhuus, A., 307(411), *326*
Dmitrieva, V. D., 739(345), *760*
Do, Q. T., 275(135), *317*, 712(75), *751*, 931(371), *948*
Doane, L. M., 647(224), *657*
Dobrzeniecka, R., 1208(100), *1214*
Dockal, E. R., 826(129, 130), *868*
Dodin, G., 297(341), 298(341), *324*
Dodson, P. A., 621(69), 625(69), *652*, 1082(95), *1098*
Dogonadze, R. R., 32(44b–f), 63(44), *116*, 39(52), *117*, 342(103), *356*
Dohrmann, J. K., 195(297), *205*
Doi, J. T., 668(60), *693*
Doi, K., 572(165), *579*
Dojlido, J., 435(51), *458*
Dokunikhin, N. S., 894(75), *935*
Dolnakov, Y. P., 1094(180), *1101*
Dolphin, D., 305(399), *326*, 813(29), *865*, 816(68), *866*, 819(90), *867*
Dolson, M. G., 630(99, 107), 632(107), *653*, 781(64b, c), *786*, 1073(37), 1074(37), *1096*
Domarev, A. N., 891(55), *935*
Dombro, R. A., Jr., 270(98), 304(98), *316*
Domen, K., 1408(50), *1417*
de Domenico, A., 1413(129), *1419*
Dominguez, M., 161(147), 163(147), *200*, 645(209, 210), *656*
Domnitskii, V. V., 793(36, 37), *805*
Doney, J. J., 685(139), *696*
Dong, D. F., 793(32), *805*
Dong, S., 646(215), 647(223), *657*
Donohoe, R. J., 814(52), 815(52), *865*
Donohue, J. A., 953(6), *1018*
Dormond, A., 838(203), *871*
Dougherty, J. A., 369(56), *395*
Doughty, A. G., 282(200), 285(200), 290(200), *319*, 1105(21), *1124*, 1110(40), *1125*
Douglas, A. F., 268(91), *316*
Douglas, L. D., 788(4), *803*
Doupeux, H., 370(70, 71), *396*, 445(159), 449(159), *461*, 887(30), *933*
Dousek, F. P., 1204(58), *1213*
Dowalo, F., 403(15), *426*
Downard, A. J., 1376(61), 1383(61), *1394*
Doxsee, K. M., 839(217), *871*
Doyle, A. M., 377(120), *397*
Doyle, M. P., 689(187, 191), *698*
Drake, R. F., 644(201), *656*
Drakesmith, F. G., 382(136), *398*, 388(169), *399*, 487(36), *497*, 747(433), *763*, 1105(20), 1109(20), 1110(20), 1112(20), *1124*
Drash, A. L., 645(206), *656*
Dregeris, J., 903(201), *940*
Dreher, E.–L., 621(59), 624(59), *651*, 717(118), *752*, 959(48), *1019*
Dreos-Garlatti, R., 826(128), *868*
Dresdner, R. D., 1104(8), 1105(8), 1106(8), *1124*, 1109(37), *1125*
Dreux, C., 746(425), *763*
Drevs, H., 821(98), *867*
Drews, T., 491(90), *499*, 1151(173), *1184*
Driesel, G., 721(157c), *753*
Drogowska, M., 435(52), *458*
Drury, J. S., 172(184), 174(199), *201*
Drury, R. F., 309(422), *327*, 927(320), *946*
Drushel, H. V., 722(169), *754*
Druy, M. A., 331(9), 341(9), *353*
Dryden, H. L., 442(111), 444(111), *459*, 437(131), 441(131), *460*
Dryhurst, G., 260(36b), 263(36b), *314*, 660(13), *692*, 732(279), 736(279), 737(279), *758*, 736(326, 327), 737(328), *760*, 746(422, 424), 747(428), *763*, 1006(215), *1025*
Dubenko, L. Y., 801(84, 85), *807*
Dubinin, V. Y., 792(31), *805*
DuBois, D. L., 1434(15a), *1437*
Dubois, J. C., 1380(66), *1394*
Dubois, J.–E., 179(233), *203*, 262(56), 263(56), *314*, 286(343), 297(341), 298(343), *324*, 302(390), *326*, 640(163), *655*, 893(70), *935*, 925(309d, e), *945*, 960(22), *1018*
Duchek, J. R., 667(59), *693*, 1178(519), *1195*
Dücker, G., 179(160), *185*
Duda, U.–M., 1051(182), *1066*

Duddeck, H., 704(26), *749*, 1000(173), *1024*
Dudek, V., 449(197), *462*
Dukhin, S. S., 257(18a), *313*
Dulan, L. A., 1202(42), *1213*
Dullenkopf, W., 1200(12), *1211*
Dulou, R., 1206(81), *1214*
Dumanovic, D., 732(288), *758*
Dumas-Bouchiat, J. M., 64(81), 68(81), 72(81), *118*, 259(29), 310(29), *313*, 344(113–116), *356*, 379(127), 384(127), *398*, 899(135), *938*, 1218(1, 2), 1219(1, 2), 1220(3, 4), *1256*, 1221(19), *1257*, 1221(1, 2), 1222(1, 2), *1256*, 1222(19), *1257*, 1223(1, 2), *1256*, 1224(19, 27, 28), *1257*, 1248(3, 4), *1256*
Dumousseau, J. Y., 960(22), *1018*
Dunet, A., 740(362), *761*
Dunitz, J. D., 985(131), *1022*
Dunn, W. W., 1406(30), *1416*, 1408(76, 80), *1418*
Dunlap, R. B., 280(187), 284(187), 285(187), 297(187), *319*
Dunning, J. S., 420(140), *431*
Duonghang, D., 1412(126), *1419*
Dupin, M., 297(336), 286(336), *324*
Dupraz, M. L., 1355(83), *1361*
Dupré, A., 974(94b), *1021*
Duprilot, J. M., 1434(13f), *1437*
Durham, J. E., 441(148), 449(148), 461, 490(81), *499*
Durst, R. A., 282(204), 285(204), *319*, 1314(2), *1392*
Duquet, D., 1077(54), *1097*
Durkes, K., 1231(63), *1258*
Dushchenko, V. P., 257(18b), *313*
Dusserre, A., 190(271), *204*, 686(173), *697*
Duty, R. C., 367(31), 375(31), *395*, 373(77), *396*
Dvali, V. G., 32(44c), 63(44), *116*
Dvorák, V., 591(45), *609*
Dyukkiev, E. F., 539(22), *575*
Dzerzhinskaya, R. A., 1109(36), *1125*

Eaborn, C., 855(282), *873*
Eachus, R. S., 672(79), *694*
Eagle, C. T., 835(182), *870*
Eaves, J. G., 1381(75), *1394*
Ebara, T., 1080(78), *1097*, 1157(223), *1185*, 1160(249), 1161(249), *1186*
Eberson, L., 3(11), *8*, 31(40), *116*, 49(63), *117*, 122(3), *195*, 233(2), 234(2a), *245*, 234(16b), *246*, 236(25, 26), 237(31, 33, 35), *247*, 242(62), *248*, 260(42), *314*, 266(76a), 267(76), *315*, 288(249, 250), 292(249, 250), 295(249), *321*, 291(291), 286(291), *322*, 342(104), *356*, 345(145), *357*,375(98), *397*, 387(156), *398*, 483(5), 493(5), 495(5), *496*, 507(31), 508(38), 519(31), *529*, 508(41), 510(54), 511(67), *530*, 511(70, 76), 512(70), *531*, 521(120), 522(121, 123), 523(130), *532*, 524(133), 527(141), 528(146), *533*, 539(19), 542(44), 543(44), *575*, 593(72), *610*, 723(187), *754*, 601(139, 141), *612*, 809(12), *864*, 908(224g, h, m), 910(224g, h, m), 911(224g, h, m), 924(224g, h), *942*, 952(2), 953(2, 9), 960(2, 26), *1018*, 967(61), *1020*, 980(118), 982(118), *1022*, 1001(178), 1002(185), *1024*, 1013(247), *1027*, 1039(49b), *1062*, 1070(1, 2, 6–8, 19), *1095*, 1072(27, 31, 32), *1096*, 1073(19), *1095*, 1077(57, 60), 1080(2, 19), *1095*, 1080(73), *1097*, 1080(83–85), *1098*, 1081(2), 1081(8), *1095*, 1081(73), *1097*, 1081(84, 89), 1082(83), *1098*, 1084(19), *1095*, 1088(135), 1089(142), *1099*, 1089(150, 152–156), 1090(150, 152–156, 160), 1091(154, 155), 1092(166, 167), *1100*, 1119)63a, b, 64), *1126*, 1132(4), *1179*, 1137(42, 43), *1180*, 1156(216–218), *1185*, 1160(251, 255, 259, 260), *1186*, 1160(265), *1187*, 1165(315), *1188*, 1166(251), *1186*, 1170(363–366), *1190*, 1239(118), 1241(124), *1260*, 1425(7), *1429*
Eby, R., 1210(121), *1215*
Eccleston, B. H., 1053(199), *1066*
Echegoyen, L., 1178(517), *1194*
Ecker, P., 511(66), *530*, 1084(110), *1098*
Eckert, H., 821(99), *867*
Edamoto, T., 258(27), *313*
Edamura, F. Y., 389(174), *399*, 747(435), *763*
Edel, E., 790(14), *804*, 793(41), *805*
Edgell, R. G., 174(194, 197), *201*
Edwards, G. J., 506(20–22), 511(20–22), 512(20–22), 514(21), *529*, 511(72), 512(72), 514(81), 515(85), 517(81), *531*, 951(17), *1018*, 1084(111), *1098*, 1085(120, 121), *1099*
Edwin, J., 841(222), *871*
Efimov, O. N., 930(348), *947*, 1433(8), *1436*
Efrima, S., 46(60), *117*
Efros, L. S., 735(317), *759*
Egashira, N., 381(135), *398*, 894(77a), *936*
Ege, D., 1390(126), *1396*
Eger, R., 930(349d), *947*, 1434(12b), *1437*
Ege-Serpkenci, D., 737(328), *760*
Eggelte, H. J., 649(233), *657*
Eggert, G., 301(373), *325*
Eggins, B. R., 455(228), *463*, 621(81, 82), 627(81, 82), *652*
Egsgaard, H., 295(318), *323*
Ehigamusoe, R. E., 1071(20), 1073(20), *1096*
Ehrlich, P., 1206(73), *1214*
Eichinger, G., 301(372), *325*
Eifer, I. Z., 793(39), *805*
Eigenbrot, C. W., 819(85), 819(88), *867*
Eilenberg, W., 599(111), *611*, 704(28), *749*, 771(32), *785*, 995(156), *1023*, 990(164), 998(167), *1024*
Einhorn, J., 542(47), *576*, 569(153, 154), *579*
Eisenbach, W., 837(189, 199), 838(199), 848(199), *870*, 848(245), *872*, 1353(77), *1361*
Eisenberg, M., 256(9), *313*
Eisenberg, R., 392(197), *400*, 831(152), *869*
Eisenmann, B., 1200(12), *1211*

Ekeda, T., 1045(120), *1064*
Ekes, L. H., 1094(183), *1101*, 1117(60), 1118(60), *1126*
Ekstrom, M., 774(39b), *785*
El-Baradic, H. Y. F., 474(71), *481*
Elbe, W., 929(331), *946*
Elbe, W., 715(104), *752*, 1147(145), *1183*
Elbs, K., 414(97), *429*, 417(116b), *430*, 442(103), 443(103), 445(103), *459*, 448(186), *462*, 714(93), *751*, 898(121), 899(121), 900(121), *937*
Eldikov, V. N., 376(112), *397*
Elbe, W. 929(331), *946*
Elgy, G. M., 630(102), *653*
Eliason, R., 123(34, 35), 126(34, 35), 127(34, 35), *196*, 150(116), *198*
El Jammal, A., 736(321), *759*
El-Khawaga, A. H., 960(27), *1018*
Ellefson, W. L., 813(31), *865*
Ellias, L., 301(365), *325*
Elliot, C. M., 819(80), *866*, 823(110), 824(110), *867*
Elliot, I. W., Jr., 704(22), *749*, 989(162), 990(162), *1023*
Elliott, M., 1365(14), 1370(14), *1392*
Ellis, C. D., 1381(77), *1394*
Ellis, K. G., 489(61), *498*, 1164(314), *1188*, 1421(4e), *1429*
Ellis, S. R., 276(149), *318*, 288(251), 292(251), *321*
Ellison, D., 1051(180), *1066*
Elmaleh, E., 703(14), *748*
Elming, N., 630(108), 633(108), *653*, 721(142, 144–146), *753*, 1243(138), *1261*
El-Mobayed, M., 717(118), *752*, 959(48), *1019*
El Murr, N., 838(203), 840(219), *871*, 846(239), *872*
Elnagdi, M. H., 975(97), *1021*
Elofson, R. M., 424(167), *431*, 453(220), *463*, 475(85), *481*, 712(68), *750*
El Omar, F., 853(272), 855(279), *873*
Elsenbaumer, R. L., 1357(96), *1362*, 1377(81), 1382(81), *1395*
El-Shamy, H. K., 260(44), *314*
El-Sheikh, F., 659(3), *692*
El-Taliawi, G. M., 1070(1), *1095*
El'tsov, A. V., 727(249), *757*
Elving, P. J., 260(36b), 263(36b), *314*, 263(59), *314*, 276(143), *318*, 284(210), 298(210), *320*, 299(349), 286(349), *324*, 369(62, 64), *396*, 387(155), *398*, 388(162–164), *399*, 435(25), *457*, 724(199), 725(217), *755*, 725(218, 228), *756*, 732(277, 278, 280), 736(278), *758*, 736(325), *760*, 740(367), 742(379), *761*, 745(420), 746(421), *763*, 860(312), *874*, 885(5), *932*, 926(313), *945*, 1089(141), *1099*, 1135(18), *1179*, 1153(185, 192), *1184*, 1154(18), *1179*, 1175(480, 481), *1193*, 1176(502), *1194*
El-Zohry, M. F., 960(27), *1018*
Emde, H., 1209(115), *1215*
Emeléus, H. J., 260(44), *314*
Emerson, T. R., 742(377), *761*
Emirdzhanova, G. N., 484(10), *496*
Emmert, B., 723(196), *755*, 1049(154), *1065*
Ender, F., 722(180), *754*
Enea, O., 1408(58), *1417*
Eng, R., 925(309g), *945*
Engdahl, K. A., 1414(143), *1420*
Engel, M., 857(285, 286), *873*
Engel, P., 815(55), *866*
Engel, P. S., 687(177), *697*
Engelmann, S., 437(137), *460*
Engels, H.-W., 340(74), *355*
Engels, J., 1039(58), 1041(58), *1062*
Engels, R., 240(53), *248*, 306(404), 311(404), *326*, 347(166), 348(166), *358*, 516(87), 517(87), 525(87), *531*, 980(112, 123), 981(123), 982(112), 983(123), *1022*, 1079(70), *1097*, 1166(319), *1188*
Engler, E. M., 682(126), 683(128), *696*, 1372(41, 42), 1373(42), *1393*
English, F. A., 486(33), *497*
Enkelmann, V., 511(68), *530*
Ensman, R. E., 55(70), *117*
Enyo, M., 640(152), *654*, 641(170), *655*
Epimakhov, W. N., 434(15), *457*
Epling, G. A., 6(20), *8*, 369(52), *395*, 378(123), *397*, 887(29, 31), *933*, 1043(87), *1063*
Equey, J.-F., 1390(126), *1396*
Ercoli, R., 438(79), 442(79), 443(79), *458*, 1353(69), *1361*
Erdey-Gruz, T., 64(82b), 65(82b), *118*, 272(108), *316*
Erdtman, H. G. H., 965(52), *1019*
Eren, S., 1354(78–80), *1361*
Erickson, R. E., 375(101), 377(118), *397*, 489(57), *498*, 1039(59), *1062*, 1152(178), *1184*, 1168(346), 1169(346), 1169(355), *1190*
Eriksen, J., 1414(146–148), *1420*
Erlikh, Y. I., 435(54), *458*
Ermolaeva, L. V., 473(62), 474(62), *480*, 1055(218), *1067*
Ernst, S., 644(194, 200), 645(194), *656*
Erocolani, C., 821(97), *867*
Ershler, A. B., 436(59), *458*, 441(149b), *461*, 495(120), *500*, 740(356), *761*, 895(98b), *937*, 1287(24), *1333*
Ershov, V. V., 455(231, 232), *463*, 915(257), *943*
Ershow, B. G., 809(8), *864*
Erzhanova, M. S., 439(97), 440(97), *459*
Esch, T., 593(64, 65), *593*, 594(83), *610*
Eschenmoser, A., 812(22, 23), *864*, *865*, 813(27, 32, 33), 985(131), *1022*
Escherich, F., 443(124), *460*
Escot, M.-T., 425(192), *432*, 894(82), *936*
Escudie, J., 411(76), *428*
Espenson, J. H., 831(149–151), *869*
Etemad, S., 682(126), *696*
Eto, H., 621(87), 627(87), *652*, 755(43c), 776(44b, 46a, 47), *785*, 962(45), *1019*, 1013(251), *1027*
Evans, D. H., 126(58), 129(70), *197*, 144(102), *198*, 307(414), *327*, 331(7), *353*, 342(100), 343(107), *356*, 346(154), 350(154), *358*, 368(47, 48), *395*, 404(25), *427*, 433(1), *456*, 435(47),

[Evans, D. H.]
457, 436(57), 458, 516(89), 519(89), 531, 616(2, 6, 7, 9), 617(6, 7, 9), 619(7), 621(6), 624(6), 650, 702(2), 747, 882(3i), 885(3i), 932, 905(171), 906(172), 913(171), 939, 903(202), 940, 962(46), 1019, 1033(11), 1060, 1144(105, 112, 127), 1145(127), 1182, 1175(467, 469, 471– 476), 1193, 1176(499, 500), 1178(517), 1194, 1248(176), 1262, 1408(37), 1416
Evans, J. F., 188(267), 204, 520(114, 115), 521(114, 115, 117), 522(114), 532, 647(221), 657, 686(156, 161), 697, 1248(177), 1262
Evans, M. B., 1054(206), 1067
Evans, M. G., 725(214), 755
Evans, M. J., 885(8), 932
Evans, S., 290(280), 322
Evans, W. V., 861(315, 316), 874
Everett, W., 931(365a), 947
Evilia, R. F., 388(167), 399
Evins, C. V., 126(59), 197
Ewing, A. G., 70(91a), 118, 647(222), 657, 1314(2), 1392
Ewing, G. D., 335(54), 336(61), 354, 340(88), 355
Exner, O., 476(94), 477(94, 97), 482, 1033(10a), 1060, 1048(151a, b), 1065
Ezzel, M. F., 368(45), 395, 886(12), 891(12), 933

Fabre, J.–M., 680(120), 681(120), 682(123), 695
Fabre, P. L., 290(271), 322, 515(99, 100), 518(99), 532
Fadaly, O. A., 792(24), 805
Fadeeva, T. P., 438(83), 459
Fagan, R., 180(239), 189(239), 203
Fahidy, T. Z., 258(26), 313, 1286(3), 1292(3), 1299(3), 1332
Fahmy, H. M., 466(8), 479, 975(97), 1021
Fainzil'berg, A. A., 406(42), 427, 407(49, 51), 428
Faita, G., 1094(188), 1101
Fajardo, O., 1380(71), 1394
Falat, L., 646(218), 657
Falck, J. R., 704(16), 748, 770(26a, b), 784, 889(37), 934, 993(143), 994(150, 153), 995(153), 1023, 1093(170), 1094(174), 1100
Fall, D. J., 1113(48), 1125
Falsig, M., 39(55e), 117, 489(68), 498, 491(89, 91), 492(91), 499, 728(260), 757, 735(319), 759, 1042(79), 1063, 1047(134, 136–138), 1064
Fan, F.–R. F., 1374(49), 1393, 1382(88), 1385(88), 1395, 1387(115), 1389(115), 1396, 1406(29, 30), 1416, 1408(80), 1418
Fan, F. W., 332(11b), 345(11), 353
Fan, J. W., 885(6), 932, 910(227), 942, 1175(478), 1193
Fan, L., 1173(432), 1192
Fanghänel, E., 679(112), 680(112), 681(116), 682(112, 116), 683(116), 695
Farafonov, V., 1351(39), 1360
Farbwerke Hoechst, A.–G., 491(88), 499
Farina, G., 2(4), 7, 224(22), 232, 342(98), 346(98), 356, 347(152), 350(152), 357, 346(155, 156), 358, 494(115), 500, 901(159), 939, 1135(26), 1179, 1233(95), 1259
Farooque, M., 1434(21), 1438
Farr, F. R., 350(202), 359
Farrington, G. C., 305(399), 326
Farwell, S. O., 380(133), 384(140, 141), 398
Fässler, A., 813(27, 32), 865
Fatiadi, A. J., 647(224), 657
Fatouros, N., 73(93c), 119
Faulkner, J. D., 452(210), 453(210), 455(210), 463
Faulkner, J. L., 915(251), 943
Faulkner, L. R., 122(4), 123(4), 146(4), 150(4), 151(4), 152(4), 157(4), 165(4), 179(4), 180(4), 195, 344(120), 357, 1342(14), 1359
Faulkner, R. L., 13(1), 114, 20(20), 21(21a), 22(22a, 23), 115, 26(33a), 116, 46(1a), 55(69), 57(72), 117, 59(76), 65(83b), 68(86), 85(107), 118, 209(5b), 231
Fauvarque, J. F., 240(50), 248, 298(348), 324, 393(204, 205), 400, 835(174, 175, 179), 870, 1236(105, 107–109), 1259, 1236(109), 1260, 1434(13b), 1437
Fauvelot, G., 193(284), 205
Fauero, P., 434(10), 456
Fawcett, W. R., 895(95), 937
Fearn, J. E., 897(114), 899(114), 937
Feasson, C., 857(285), 873
Feckl, J., 790(14), 804, 793(41), 805
Federlin, P., 387(153, 154), 398
Federonko, M., 437(134), 460, 490(83), 499, 733(295), 758, 1039(54), 1062, 1142(82), 1181, 1428(21), 1430
Fedkiw, P. S., 257(12), 313
Fedorova, A. J., 1348(29), 1360
Fedorovich, N. W., 435(53), 458
Feess, H., 288(257), 292(257), 321
Feinberg, B., 621(74), 625(74), 652
Feinberg, D. A., 788(4), 803, 790(10), 804
Feldberg, S. W., 123(29), 133(29), 170(29), 196, 170(178), 201, 350(109), 356, 593(60), 609, 968(68), 970(68), 1020
Feldhues, M., 237(28), 247, 545(62–64), 576
Feldhues, U., 277(153), 318, 583(8), 607
Felia, J. M., 725(230), 756
Felton, R. H., 813(45), 865
Feng, E., 307(413), 327
Fengel, D., 787(2), 788(2), 789(2), 790(2), 791(2), 803
Feoktistov, L. G., 272(109), 316, 362(2), 366(2), 394, 368(39–43), 395, 707(44), 708(44), 749, 924(302), 945, 1132(2), 1179, 1201(26), 1212
Ferhat, M., 816(65), 866
Ferles, M., 906(177), 907(177), 927(321), 939, 946
Fernandez, J. E., 1352(62), 1361

Feroci, G., 409(68), *428*, 727(250), *757*, 914(239), *942*, 1233(88), 1235(88), *1259*
Ferstandig, L. L., 1348(31), *1360*
Fichigami, T., 1142(83), 1173(83), *1181*
Fichter, F., 3(10), *8*, 272(115), *316*, 401(1), 411(1), *426*, 407(45, 46), *427*, 414(100), *429*, 417(119), *430*, 422(152), *431*, 444(128), *460*, 449(194, 196), *462*, 483(1), 493(1), 495(1), 496(1), *496*, 490(79, 80), *499*, 505(2), *528*, 542(45), 551(96), *576*, 571, 621(88), *652*, 660(6), *692*, 663(43, 45), 667(43, 45), 675(45), *693*, 670(71), *694*, 740(365), *761*, 953(7), *1018*, 956(30), *1019*, 1002(188), *1024*, 1042(81), *1063*, 1054(211), 1056(233, 234), *1067*, 1059(260), *1068*, 1071(24), 1088(24), *1096*
Fichtner, M. W., 671(78), *694*, 979(111), *1022*
Fiel, L. D., 284(232), 290(232), *320*, 1105(24), *1124*
Field, E. W., 438(78), 446(78), 447(78), *458*
Field, M., 333(28b), *353*
Field, N. J., 174(193), *201*
Fields, E. K., 486(24), *497*
Fields, M., 260(47), 263(47), *314*, 453(215a), *463*
Fierro, C., 821(101), *867*
Figeys, H. P., 349(184), *359*
Figeys, M., 349(184), *359*
Filardo, G., 240(49), *247*, 277(159), 311(159), *318*, 292(301), 303(301), *323*, 438(79), 442(79), 443(79), *459*, 857(295–297), 858(298, 299), *874*, 904(204a, b), *940*, 930(345a), *947*, 1252(191), *1262*, 1353(68), *1361*, 1433(9a), *1436*, 1434(13a,d), *1437*
Filatova, I. M., 404(29b, c). *427*
Filimonov, V. N., 1410(121b), *1419*
Filimonova, T. F., 1255(210), *1263*
Filonova, A. D., 390(177), *399*
Findlay, A., 439(91), *459*
Finke, R. G., 809(2), *864*, 823(110), 824(110), *867*
Finkelstein, M., 240(52), 242(52), *248*, 260(40), *314*, 273(118), *316*, 528(149), *533*, 539(24), 542(41), *575*, 584(17), *608*, 601(135–138), 602(135–136), *612*, 603(154), *613*, 638(134), *654*, 648(232), *657*, 925(309a), *945*, 1001(181), *1024*, 1045(107), *1064*, 1050(159a–c). *1065*, 1076(48), 1080(48), *1096*, 1077(59), *1097*, 1080(86), 1081(86), 1083(48, 86, 105), *1098*, 1149(160), *1183*, 1160(264, 265), 1161(277), *1187*
Finklea, H. O., 1398(11), *1415*
Fiorani, M., 834(173), 835(173, 181), *870*
Fiori, G., 716(115), *752*
Fioshin, M. Ya., 259(32), 263(32), 309(32), *313*, 273(120), *317*, 404(29a–d), 408(29a–d), *427*, 408(58, 59), 409(67), *428*, 424(170), *432*, 445(158), *461*, 466(6), *479*, 470(49), 476(49), *480*, 490(84, 85), *499*, 496(136), *501*, 552(81), *577*, 743(387), *762*, 745(415), *763*, 861(314), *874*, 899(148), *938*, 1010(235), *1026*, 1104(11), *1124*, 1230(56), *1258*, 1287(23, 26), *1333*, 1299(53), 1304(53), *1334*, 1421(2), *1429*
Firgo, H., 855(280), *873*
Firth, B. E., 675(91), *694*, 1172(416), 1173(417), *1191*
Fischer, C. M., 489(57), *498*, 1039(59), *1062*, 1169(355), *1190*
Fischer, E., 1208(92), *1214*
Fischer, H., 234(13), *246*, 1200(19), *1212*
Fischer, O., 423(164), *431*, 594(82), *610*, 714(96), *751*, 725(213), *755*, 727(251), *757*, 733(296), *758*
Fishbein, R., 242(60), *248*
Fisher, B., 831(152), *869*
Fisher, F. S., 174(195–197), *201*
Fitch, A., 129(70), *197*, 1175(476), *1193*
Fitzjohn, J. L., 1314(93), *1335*
Flanagan, J. B., 1340(4), *1359*
Flannery, B. P., 122(8), *195*
Flarsheim, W. M., 270(97), 303(97), 304(97), *316*
Fleet, B., 474(80), *481*, 857(284), *873*
Fleischhaver, J., 557(112), *577*
Fleischmann, M., 104(109), *119*, 128(63), *197*, 179(226), *202*, 179(230), *203*, 268(89a, b), 271(89a, b), *315*, 270(94), *316*, 277(151), *318*, 282(200), 285(200), 290(200), *319*, 285(218), 290(218), *320*, 288(256), 290(265, 266), 292(256), *321*, 303(397), *326*, 496(139), *501*, 514(79, 80), 515(79, 80, 83, 98), 516(79), 518(98), *531*, 515(101), *532*, 583(7), *607*, 641(177–179), 642(177), *655*, 848(250, 252), *872*, 863(331), *875*, 1083(102), *1098*, 1094(188), *1101*, 1105(21), *1124*, 1110(40), *1125*, 1117(56), *1126*, 1164(308), *1188*, 1232(81), 1235(81), *1235*, 1268(29), *1281*
Fleming, S., 1210(122), *1215*
Flerov, V. N., 423(164), *431*, 714(96), *751*
Fletcher, A. N., 1434(12j), *1437*
Fletcher, F. A., 447(184), *462*
Fletcher, S., 494(119), *500*, 740(357), *761*
Fleury, D., 474(82), *481*
Fleury, M. B., 474(82), *481*, 724(202), *755*, 734(307), *759*
Flohr, H., 824(113), *868*
Florence, T. M., 423(157, 158), *431*, 1055(221), *1067*
Floss, J. G., 307(416), *327*
Floyd, M. B., 782(70d), *786*
Foffani, A., 742(374), *761*
Fokina, L. N., 1039(48), *1062*
Folest, J. C., 277(160), 311(160), *318*, 831(158), 832(158), *869*, 1434(13f), *1437*
Foley, J. F., 182(245), *203*
Foley, J. K., 179(221), *202*, 184(257), *204*, 296(330), *324*
Foley, R. T., 284(240), *321*

Foote, C. S., 1268(25a), 1274(25a), *1281*, 1407(35), *1416*
Foral, M. J., 1435(32), *1438*
Forcier, G. A., 295(322), *323*
Foreman, R. W., 1305(74), *1334*
Formenti, M., 1408(41), *1416*, 1408(46), *1417*
Fornasari, E., 742(374), *761*
Forno, A. E. J., 1245(155), *1261*
Forrest, N. P., 620(38), *651*
Forsyth, S. R., 1094(186), *1101*
Forte, C. S., 1246(170), *1262*
Foulatier, P., 444(129), *460*, 929(339), *946*
Foulds, A. W., 174(196, 197), *201*
Fournari, P., 424(174–176), 425(176), *432*
Fournier, F., 894(80, 81), *936*, 1137(45, 46), *1180*, 1146(135), *1183*, 1236(101), *1259*, 1271(41), *1282*
Fournier, M., 894(80), *936*, 1236(101), *1259*
Fouzder, N. B., 857(284), *873*
Fox, B. L., 744(408), *762*
Fox, D. P., 919(285a), *944*
Fox, H. M., 1094(181), *1101*, 1139(58, 59), 1140(58, 59), *1180*
Fox, M. A., 516(88), *531*, 526(139), *533*, 1398(1–5), 1403(2), 1404(1), 1405(3), *1415*, 1405(22), 1406(32), 1407(32), 1408(36, 38, 39), *1416*, 1408(62), *1417*, 1409(95, 100), 1412(95), *1418*, 1410(122, 123), 1412(128), 1413(122, 131), *1419*, 1413(137, 139, 140), 1414(144), *1420*
Foxman, B. M., 838(206), *871*
Fraccaro, C., 1111(44, 45), 1112(45), 1123(45), *1125*
Fraenkel, G. K., 411(83), *429*, 434(11), *456*, 495(128), *500*, 1051(177), *1066*
Franaszczuk, K., 639(147), *654*
Francais-Habert, T., 418(129), *430*
Francis, G. D., 645(207), *656*
Francis, H. T., 1104(8), 1105(8), 1106(8), *1124*
Franco, J. A., 1161(274), *1187*
Franco, M., 1434(13e), *1437*
Francois, M., 240(50), *248*, 298(348), *324*, 835(175), *870*, 1236(109), *1260*
Frank, A. J., 1382(85), 1383(85), *1395*, 1409(101), *1418*
Frank, S. N., 973(90), *1021*
Franke, J., 594(77), *610*
Franke, L., 1434(19d), *1438*
Frankel, E. N., 1080(74), 1082(74), *1097*
Franz, C., 275(128), *317*, 916(266), *943*
Fraser, I. M., 1408(51, 52), *1417*
Freeman, F., 15(9), *114*
Freidlin, G. N., 540(33), 541(33), *575*, 1010(233), *1026*
Freidlina, R. K., 1206(66), *1213*
Fremery, M., 1299(54), *1334*
Frese, Jr., K. W., 1434(15b, e, m), *1437*
Freskos, J. N., 781(65a), *786*
Freund, T., 1398(13), *1416*
Friedlander, H. Z., 450(203), *462*
Friedrich, P., 511(66), *530*
Friedrichs, G., 488(42), 489(42), *497*
Frielden, E. M., 1253(201), *1263*
Frimer, A. A., 1245(157, 160), *1261*
Frisch, P. D., 1168(343), *1189*
Fritsch, J. M., 301(366), *325*, 511(56), *530*, 592(58), *609*, 597(99, 100), *611*, 968(68), 970(68), *1020*, 981(125), *1022*
Fritz, H. P., 276(146, 147), *318*, 291(284), *322*, 511(58, 66), *530*, 515(92, 93), 518(92), *531*, 930(346), *947*, 1081(92), 1082(92), 1084(110), *1098*, 1154(200), *1185*
Froborg, J., 1072(29), *1096*
Fröling, A., 906(182b), *939*, 1135(23), *1179*
Frommer, J. E., 340(84), 341(84), *355*
Frost, J. G., 726(23a), *756*
Früh, T., 812(22), *864*
Frumkin, A. N., 171(180), *201*, 1287(24), *1333*
Fry, A., 303(391), *326*, 334(32), *353*, 334(40, 41), 335(41), 336(60), 339(60), 342(41), *354*, 337(73), 340(89), *355*, 345(144), *357*, 364(11), 365(17), *394*, 368(44), 369(53, 54, 61), *395*, 374(89), *396*, 376(116), 377(117), *397*, 380(132), 382(132), 387(157, 158), *398*, 387(159), 391(185), *399*, 467(23), *479*, 476(90), *481*, 729(265), *757*, 887(29, 32), *933*, 1132(3), *1179*, 1147(151), 1148(155), *1183*, 1150(169), 1152(177), 1153(179, 180, 194), *1184*, 1154(202, 203), 1155(206, 208, 209), *1185*, 1169(3), *1179*, 1169(203), *1185*, 1221(5), *1256*
Fuchigami, T., 436(63), *458*, 476(92), *481*, 484(9), *496*, 584(22), *608*, 678(101), *695*, 740(355), *761*, 795(48), *806*, 891(59), 892(60), *935*, 907(194), *940*, 975(103), 976(107), *1021*, 1045(111), *1064*, 1079(68), 1080(68), 1082(68), *1097*, 1087(133), *1099*, 1132(6), *1179*, 1141(65), 1143(93), *1181*, 1143(95), *1182*, 1143(65), *1181*, 1147(148), *1183*, 1169(356), 1170(359–361), *1190*, 1172(400), 1173(421, 422), *1191*, 1173(423, 431, 434, 435, 439), *1192*, 1178(515), *1194*, 1267(13), *1280*
Fuchs, G., 813(30), *865*
Fuchs, H., 1052(187), *1066*
Fuchs, P., 389(173), *399*, 726(231), 730(231), *756*, 1251(189), *1262*
Fuchs, Y., 831(156), *869*
Fueno, T., 592(52, 53), *609*, 633(114), *653*, 721(151, 152), *753*, 722(165), *754*, 1089(149), 1090(149), *1100*, 1161(273), *1187*
Fuente, T., 153(129), *199*
Fuhlendorff, R., 470(48), *480*, 477(96), *482*, 719(130), 733(130), *752*
Fuhrhop, J.–H., 240(47), *247*, 813(47, 49), 814(49), *865*

Fujibayashi, S., 777(51), *785*
Fujihira, M., 272(114), 275(114), *316*, 393(211), *400*, 416(114), *430*, 443(123), 445(123), 455(123), *460*, 739(342), *760*, 1173(419, 420), *1191*, 1246(168, 169), 1247(175), *1262*, 1277(55), 1282, 1314(5), *1392*, 1366(27), *1393*, 1409(90, 91, 94), *1418*
Fujii, K., 1278(57c), 1279(57c), 1280(57c), *1282*
Fujii, M., 1408(77), *1418*
Fujii, T., 570(161), *579*, 782(71), *786*
Fujikake, S., 1204(60), *1213*
Fujinaga, T., 162(153), *200*, 298(344, 346), *324*, 301(378), *325*, 306(407), *326*, 455(229), *463*
Fujioka, A., 832(160), *869*
Fujioka, T., 1409(93), *1418*
Fujishima, A., 301(380), *325*
Fujita, R., 1175(482), *1193*
Fujita, Y., 1008(226), *1026*, 1034(15), *1060*, 1163(298), *1188*
Fukaya, T., 1434(12f), *1437*
Fukomoto, K., 960(28c), *1018*
Fuks, N. S., 906(177), 907(177), *939*, 907(186), *940*
Fukunaga, T., 307(415), *327*
Fukuoka, A., 641(170), *655*
Fukuoka, T., 574(169), *579*
Fukushima, T., 1408(68), *1417*
Fuller, G., 899(142), *938*
Fuller, G. B., 664(49), 665(49), *693*
Fulmer, E. I., 453(218), *463*
Funahashi, H., 850(258, 259), *872*
Funahashi, K., 1383(89), 1387(89), *1395*
Funk, J. E., 1298(51), *1334*
Funke, W., 1206(72), 1209(72), *1213*
Funt, B. L., 7(23), *8*, 306(405), *326*, 350(195), *359*, 555(107), *577*, 1132(13), *1179*, 1340(5, 6, 8), 1346(8), *1359*, 1349(32, 33), 1350(36, 37), 1351(43), *1360*, 1352(66, 67), 1353(75, 76), *1361*, 1355(85), 1357(95), *1362*
Furahashi, S., 724(210), *755*
Furlanetto, F., 1341(11), 1351(41), *1359*, *1360*
Furlani, C., 154(135), *199*, 731(274), 733(291), *758*, 741(371), *761*
Furlong, D. N., 1408(81), *1418*
Furth, B., 407(52), *428*
Furuda, H., 1352(56), *1361*

Gabriel, J., 557(112), *577*
Gabriel, W., 275(126), *317*, 276(148), *318*
Gadallah, F. F., 712(68), *750*
Gafney, H. D., 831(156), *869*
Gak, E. Z., 258(25), *313*
Gal, J.-Y., 297(340), 298(340), *324*
Gale, R. J., 303(392), *326*
Galijas, D., 596(95), *610*, 710(60 b, c), 715(60 b, c), *750*
Gall, M., 1022(124), *1022*
Gallardo, I., 366(22), *394*
Gallone, P., 268(88), 269(88), *315*, 1287(18), *1322*
Gallusser, F., 195(297), *205*
Galluzzo, M., 438(79), 442(79), 443(79), *458*
Galpern, E. G., 1117(59), 1118(59), 1119(59, 67a), *1126*
Galsworthy, P. J., 140(91), *198*, 478(103), *482*
Galus, Z., 157(146), 159(146), 161(146), 162(151), *200*, 168(174), *201*, 422(150), *431*, 435(51), *458*
Galvez, J., 153(129, 130), *199*
Gambaretto, G. P., 1109(38), 1110(38), 1111(45), 1111(38), 1112(38, 45, 46), 1123(45), *1125*
Gambaryan, N. P., 1117(59), 1118(59), 1119(59), *1126*
Gambert, R., 496(130), *500*
Gambino, S., 240(49), *247*, 277(159), 311(159), *318*, 292(301), 303(301), *323*, 857(295–297), 858(298, 299), *874*, 885(10), *932*, 904(204a, b), *940*, 930(345a), *947*, 1042(77), *1063*, 1150(166), *1184*, 1252(191), *1262*, 1353(68), *1361*, 1433(9a), *1436*, 1434(13a, d), *1437*
Gan, G. H., 295(320c), 296(320c), *323*
Ganesan, S., 1305(73), *1334*
Gannett, P. M., 691(195), *698*
Gantheron, B., 240(51a), *248*
Gantz, E. C., 488(47), *498*
Ganushchak, N. I., 424(172), *432*
Gaon, I., 596(92), *610*
Garber, C., 280(186), *319*
Garcia, E., 1200(10), *1211*
Garcia, P. A., 645(204), *656*
Garcia-Blanco, F., 163(160), *200*
Gardini, G. P., 341(90), *356*, 1356(90), *1362*, 1375(52,56), 1376(58), *1394*
Gardiol, A., 706(41), *749*
Gardlik, J. M., 335(54), *354*
Gardner, C. W., 279(178), 282(178, 203), 285(203), 295(203), *319*
Gardner, H. J., 476(86), *481*
Gareil, M., 140(95, 96), 142(95), 148(95), 150(95), *198*, 344(117), 345(117), *356*, 345(139), *357*, 380(130), *398*, 1246(164), *1262*
Garito, A. F., 682(125), *696*
Garland, C. E., 424(166), *431*
Garner, R., 1039(53), *1062*
Garnett, E. S., 1120(68), *1126*
Garnier, F., 688(178), *697*, 1382(87), *1395*, 1383(64), *1394*, 1389(123), *1396*
Garnier, G., 1357(92), *1362*, 1380(64, 66), *1394*
Garreau, D., 65(84e), *118*, 73(93d), *119*, 126(56), 128(68), *197*, 150(115), *198*, 349(182), *358*
Garreau, R., 688(178), *697*, 1380(64), 1383(64), *1394*
Garrett, P. E., 680(115), 681(115), *695*
Garst, J. F., 350(196), *359*, 368(45), *395*, 886(12, 20f), 891(12, 20f), *933*
Garwood, R. F., 543(52), *576*, 1002(191a), *1024*, 1039(49c), *1062*, 1080(75), 1082(75), *1097*
Gaset, A., 1243(145), *1261*
Gasques, F., 715(105d), *752*, 839(218), *871*
Gassman, P. G., 514(78), 517(78), *531*, 585(23), *608*, 744(408), *762*, 1159(236), *1186*, 1163,(294), *1187*
Gastisa-Michelcic, B., 195(296), *205*
Gatterman, L., 414(99), 416(99), *429*, 416(112), 417(122), *430*

Gatti, N., 392(196), *400*, 466(2, 3), *479*, 1176(504), *1194*, 1224(35), *1257*
Gaudemer, A., 826(125), *868*
Gaudiello, J. G., 724(207), *755*
Gaul, R. J., 369(65), *396*, 1153(189), 1176(189), *1184*
Gautheron, B., 258(21), *313*, 839(215), *871*
Gavar, R. A., 434(14), *457*
Gavioli, G., 1056(232, 235), *1067*
Gavrilov, V. I., 820(96), *867*
Gawley, R. E., 994(154), *1023*
Gazard, M., 1380(66), *1394*
Gazetdinova, N. G., 495(122), *500*
Gazques, F., 1267(15), 1277(15), *1281*, 1277(50, 51), *1282*
Gebauer, H., 511(58, 66), *530*, 1081(92), 1082(92), *1098*
Gedye, R. N., 747(431), *763*, 925(309g), *945*
Geels, E. J., 1269(33), *1281*
Geer, R. D., 380(133), 384(140, 141), *398*
Geiger, W. E., Jr., 809(3), *864*, 841(222–224), 842(225), 843(224), *871*, 1153(193), 1176(193), *1184*, 1175(459–465), *1193*
Geisel, H., 621(59), 624(59), *651*
Geist, B., 426(196), *432*
Geller, M., 746(427), *763*
Geng, L., 70(91a), *118*, 1173(432), *1192*
Genies, E. M., 190(271), *204*, 525(135), *533*, 584(13), 590(13), *608*, 686(173), *697*, 716(109, 110), *752*, 723(182), *754*, 853(272), 855(279), *873*, 975(104, 105), *1021*, 1015(253), 1016(252, 253), *1027*, 1167(331), *1189*, 1383(100), 1384(100, 104), 1387(100), *1395*, 1385(109), *1396*, 1356(91), *1362*
Gennaro, A., 240(49), *247*, 496(129), *500*, 1033(10b), *1060*
Gentles, M. J., 1147(149), *1183*
Geoffroy, M., 13(2b), 17(2b), *114*
Georgans, W. P., 476(86), *481*
George, A. B., 272(116), *316*
George, M. V., 239(39), *247*
George, R., 17(15b), *115*
Gerasimov, A, I., 792(28, 29), *805*
Gerasimova, L. E., 273(120), *317*, 899(148), *938*
Gerastopolov, B. G., 287(245b), *321*
Gerdil, R., 369(57), *395*, 1045(109), *1064*, 1057(244), 1058(244), *1068*
Gerischer, H., 346(157a), *358*, 1287(15), *1332*, 1413(136), *1420*
Germain, A., 290(271), *322*, 691(193), *698*
Germain, G., 733(304), *759*
Germain, S., 268(89c), 271(89c), *315*
German, E. D., 32(44c), 63(44), *116*, 39(55f), *117*
German, L. S., 368(39, 43), *395*, 551(100), *577*
Gerson, F., 340(87), *355*, 519(109), *532*
Geske, D. H., 155(138), *199*, 281(194), 284(194), *319*, 334(34), *554*, 514(77), *531*, 597(98), *611*, 680(114), *695*
Ghe, A. M., 345(132), *357*
Ghoneim, M. M., 474(71), *481*
Ghose, A., 1352(52), *1360*
Ghosh, P. K., 1390(125, 126), *1396*
Giacometti, G., 434(10), *456*
Giannici, B., 438(79), 442(79), 443(79), *458*
Gibian, M. J., 1269(32), *1281*
Gibson, J. M., 1406(27), *1416*
Giering, W. P., 844(228–230), *871*, 844(233), *872*
Giese, B., 233(4b), 234(4b), 241(4b), *246*, 240(48), *247*, 857(287, 290), *873*
Giese, R. W., 1136(34), *1180*, 1175(485), *1193*
Gilberg, J. C., 193(288), *205*
Gilbert, J. A., 834(168), *869*
Gilde, H.-G., 375(97), *396*, 375(98), *397*, 489(58), *498*, 555(104), *577*, 1039(49a, b), *1062*, 1170(367, 368), *1190*
Gilendi, E., 267(77), *315*, 436(60), 443(60), *458*, 441(126), 444(126), *460*, 898(125), 906(125), *938*, 1146(136), *1183*, 1171(378, 379, 382), *1190*
Giles, H. H., 813(27), *865*
Gilles, J.-M., 336(62), 339(62), *354*
Gillespie, J. P., 704(16), *748*, 957(37), *1019*, 993(144), 994(149), 995(149), *1023*, 1094(174), *1101*
Gillet, I. E., 1201(29,32), *1212*
Gilroy, D., 285(221), *320*, 1288(32), *1333*
Giménez-Huguet, M., 292(299), 293(299), *323*
Ginsburg, D., 886(16), 908(16), *933*, 1269(38), *1282*
Giner, J., 644(193, 195–197), 645(193, 196), *656*
Ginley, D. S., 179(234), *203*
Ginzburg, O. G., 424(170), *432*
Ginzel, K. D., 593(70), *610*, 604(162), 605(162), 607(162), *613*
Giordan, J., 663(47), 664(47), *693*
Giordano, C., 237(34), *247*
Giordano, M. C., 284(211), *320*, 594(81), *610*
Gipson, S. L., 635(126), 636(126), *654*
Giral, L., 680(120), 681(120), 682(123), *695*
Girard, C., 1224(29), 1248(29), *1257*
Girard, M.-L., 746(425), *763*
Girault, J.-P., 816(71), *866*, 885(11), *932*
Gird, E., 721(156), *753*
Giro, G., 1382(80), *1395*
Girshorich, M. Z., 727(249), *757*
Girvan, I. J. M., 1312(87), *1335*
Gisselbrecht, J. P., 375(94, 95), *396*, 862(322), *875*, 1153(190), 1154(190), *1184*
Gitis, S. S., 891(55), *935*
Giusti, P., 1352(60, 61), *1361*
Givadinovitch, H., 723(198), *755*
Given, P. H., 334(38), *354*, 434(9), *456*
Gladys, J. A., 841(221), *871*
Glasel, A., 1349(32), *1360*, 1352(66, 67), *1361*
Glass, R. S., 667(59), *693*, 668(64), 669(65), 670(77), *694*, 685(137), *696*, 686(157), 687(157), *697*, 1046(128), *1064*, 1167(338), *1189*, 1178(519, 521), *1195*
Glasser, W. G., 790(10), *804*, 798(74), *807*
Glazov, V. A., 894(86), *936*

Glens, K., 720(141), *753*, 1161(276), *1187*
Glezer, V. T., 903(201), *940*
Glukhoedov, N. P., 491(93), *499*
Gnéguen, M., 1175(458), *1193*
Göbl, M., 668(62), *693*, 668(63), 669(69), *694*
Gobran, R. H., 1350(35), *1360*
Godsay, M. P., 792(23), *804*
Goel, A. B., 14(4), *114*
Gogolin, H. P., 600(122), *611*, 641(167), *655*
Gogotov, A. F., 798(70–72), *807*
Gohee, Y., 599(116), *611*
Goitia, J., 1381(72), *1394*
Gol-Av, E., 537(14), 564(14), 565(14), *575*
Goldberg, I. B., 190(273, 275–277), 191(275), 192(275), 194(277), 195(276), *204*, 882(3j), 885(3j), *932*
Goldberger, J., 442(107), *459*
Gol'din, M. M., 272(109), *316*, 368(39–43), *395*, 1117(53), 1119(53), *1125*
Goldschmidt, S., 237(27), *247*, 555(103), *577*, 901(158), *939*
Goldstein, I. S., 787(1), 788(1), *803*, 1435(23), *1438*
Gollnow, W. D., 930(351), *947*
Golovin, M. N., 844(230), *871*
Golyshin, V. N., 474(73), *481*, 886(19), *933*, 1269(36), *1281*
Gomathi, H., 646(216), *657*
Gomes, L. M., 1136(33), *1180*
Gomes, W. P., 1398(13), *1416*
Gomez Nieto, M. A., 470(47), *480*
Gompper, R., 661(22), *692*, 745(410), *762*, 1036(31), 1037(31), *1061*
Goncalves, J. M., 25(29), 72(29), *115*, 506(18), 508(18), 509(18), 511(18), *529*
Gonet, M., 894(77d), *936*, 898(120), *937*
González, E. R., 369(58, 59), *395*, 848(251, 253), *872*
González-Arjona, D., 163(162), *200*
Goodin, R. D., 122(11), *195*, 148(11), *198*, 193(288), *205*, 488(51), *498*, 494(113), *500*, 779(60), *786*, 1267(12b), *1280*, 1268(19), 1275(19), *1281*, 1275(45, 47), *1282*
Goodings, E. P., 905(215), *941*
Goodridge, F., 2(5), *8*, 259(33), 263(33), *313*, 268(89a, c), 271(89a, c), *315*, 414(103b), *429*, 485(17), *496*, 728(259), *757*, 1232(77a), 1234(77a), *1258*, 1303(72), *1334*, 1421(4c), *1429*
Goodrow, M. H., 668(60), *693*
Gorbacheva, L. I., 745(415), *763*
Gorbachuk, I. T., 257(186), *313*
Gorbunova, O. F., 794(46), *806*
Gorce, J.-N., 821(97), *867*
Gordienko, L. L., 455(230), *463*
Gordon, J. E., 303(395), *326*
Gordon, K. M., 234(12), *246*
Gordon, S., 1253(201), *1263*
Gore, P. H., 894(74), *935*
Goren, Z., 1408(86), 1409(101), *1418*
Gores, G. J., 193(287), *205*
Gorocholinskii, J., 448(190), *462*
Gorokhova, L. T., 801(82), *807*
Gorokhova, V. G., 792(28), *805*, 801(84), *807*
Gosden, A., 1045(116, 117), *1064*
Gosden, C., 392(198, 200, 202), 393(198), *400*, 829(143–146), 830(145), *869*
Gossage, H. E., 969(74), *1020*
Gosser, D. K., 288(252), 292(252), *321*
Goto, M., 135(80), *197*, 1178(514), *1194*
Gottesfeld, S., 179(228), *202*
Gottlieb, R., 703(14), *748*, 733(299), *759*, 771(30), *784*, 889(36), *934*, 927(322), *946*
Goubeau, J., 1200(12), *1211*
Gough, D. A., 644(195), *656*
Gough, P., 288(251), 292(251), *321*
Gough, T. A., 332(16), 333(24), *353*
Gould, R. M., 273(123e), *317*
Gourcy, J.-G., 662(30), *692*, 670(75), 671(75), *694*, 891(53), *934*, 1046(125, 126), *1064*, 1048(140), 1056(140), *1065*, 1059(264, 265), 1059(264, 265), 1060(264), *1068*
Gourley, R. N., 236(22), *246*, 441(127), 444(127), *460*, 721(162), *753*, 1146(137), *1183*, 1171(375, 376), *1190*
Gourney, J. G., 489(60), *498*
Gouterman, M., 814(51), 823(51), *865*
Gouy, G., 59(75a), *118*
Govindachari, T. R., 414(101), *429*
Gowenlock, B. G., 1169(353), *1189*
Gowland, F. W., 14(7), *114*
Goyal, R. N., 746(423), *763*
Gozlan, A., 544(58–61), *576*, 1341(13), 1346(13), 1348(13), 1355(13), *1359*, 1355(86–89), *1362*
Grabaric, B. S., 1175(452, 453), *1192*
Grabaric, Z., 1175(452), *1192*
Grabner, E. W., 173(191, 192), *201*
Grabowski, Z. R., 163(156), 164(156), *200*, 434(8), 435(8), *456*, 441(142), *460*, 638(130), *654*, 1132(9), 1175(9), *1179*, 1147(144), *1183*, 1175(466), *1193*
Gracon, B., 273(123a), *317*
Graf, E., 736(321), *759*
Graham, P. G., 846(242), *872*
Graichen, F., 726(239), *756*, 926(138), *938*, 1038(35), *1061*
Graizon, M., 474(76), *481*
Grambow, C., 793(41), *805*
Gramova, E. V., 1230(56), *1258*
Gramstad, T., 260(45), *314*
Gränse, S., 237(33), *247*, 1156(218), *1185*
Grätzel, M., 434(20), *457*, 686(145), *696*, 1404(19, 21), 1406(25), *1416*, 1412(126), *1419*, 1413(138), *1420*
Graves, P. R., 179(230), *203*
Gray, D., 350(195), *359*
Gray, H. B., 813(25), *864*
Grazia, B., 724(205), *755*
Gream, G. E., 570(158), *579*
Greaves, W. S., 1004(205), *1025*
Greco, G., 858(298), *874*, 1252(191), *1262*
Green, D. C., 680(119), 681(119), *695*, 683(128, 129), *696*
Green, D. T., 1243(131), *1260*

Green, J. H., 837(191), 862(191), *870*
Green, J. W., 793(42), *805*
Greene, T. W., 244(79), *249*
Greeness, M., 135(79), *197*
Gref, A., 1040(62), *1062*, 1043(90), *1063*
Greig, W. N., 421(148), 422(148), *431*, 726(242), *757*
Greusand, M., 424(175), *432*
Gribble, G. W., 1002(195a), *1025*
Griffin, G. L., 1408(48), *1417*
Griffin, G. W., 1159(247), *1186*
Griffith, V. S., 721(161c), *753*
Grimme, W., 337(70), 350(70), *355*
Grimshaw, J. T., 226(23), *232*, 236(22), *246*, 373(81), *396*, 383(138, 139), *398*, 389(172), *399*, 436(62), *458*, 441(127), 444(127), *460*, 448(189), *462*, 659(4), *692*, 703(11–13), *748*, 721(162), *753*, 729(267), *758*, 771(31a, b), *785*, 888(33), 888(34), *933*, 888(34, 35), *934*, 895(111), 897(111), 902(111), *937*, 1044(92), *1063*, 1136(34), *1180*, 1144(104), *1182*, 1146(137), *1183*, 1171(375, 376), *1190*, 1171(395), *1191*, 1175(485), *1193*
Grinberg, V. A., 267(85), 268(85), 271(85), 307(85), *315*, 539(22), *575*, 1012(241), *1026*
Grisley, D. W., Jr., 921(293), *944*
Gritzner, G., 278(174), 282(174), *319*, 832(201), *870*
Grobe, J., 848(247), *872*
Groebel, B.-T., 630(98), *653*
Grolig, J., 1316(105), 1319(105), *1335*
Gromova, E. V., 490(84, 85), *499*
Gronchi, G., 193(290), *205*, 411(76), *428*
Groot, W. G. F., 271(106), *316*
Grosdemouge, C., 715(105a), *752*, 926(336), *946*
Grosjean, B., 456(243), *464*, 915(260), *943*
Gross, M., 285(227), 293(227), *320*, 736(321), *759*, 738(338), *760*, 862(320), *874*, 862(322), *875*, 985(132), *1022*, 1090(161), *1100*
Gross, W., 333(29), *353*
Grosse, A., 239(43), *247*
Grosse Brinkhaus, K. H., 593(64, 68), *609*, 593(69, 73), *610*, 635(124), 636(124), *653*
Grossi, P. J., 723(183), *754*
Groth, J. L., 376(110), *397*
Grover, E. R., 621(83), 627(83), *652*
Groves, J. T., 818(76), *866*, 819(89), *867*
Grovu, M., 1351(39, 40), *1360*
Grubbs, R. H., 839(217), *871*
Grubert, L., 715(107), *752*, 1017(255), *1027*
Grubina, L. M., 282(199), 285(199), 290(199), *319*
Grujic, Z., 621(51), *651*, 706(38), *749*, 1015(250), *1027*
Grund, H., 1208(91), *1214*
Gründemann, E., 711(62), *750*
Grunwald, E., 301(369), *325*
Grunwald, E. J., 31(41e), *116*
Gruver, G. A., 183(246), *203*
Gryazev, V. F., 742(378), *761*, 743(389), *762*
Grypa, R. D., 151(126), *199*, 882(3m), 885(3m), *932*
Grzeszczuk, M., 342(96), 343(96), *356*, 1175(465), *1193*
Guainazzi, M., 1353(69), *1361*
Guarino, E., 292(301), 303(301), *323*
Gubin, S. P., 825(123, 124), 841(123, 124), 863(123), *868*, 846(236, 240), *872*, 1433(7), *1436*
Gudimov, N. L., 1109(36), *1125*
Guerchais, J. E., 1175(458), *1193*
Guerout, D., 897(117), *937*
Guerra, G., 1346(23), *1360*, 1352(57–60), *1361*
Gueutin, C., 814(54), 819(54), *865*
Guglielmi, M., 1380(70), *1394*
Guidelli, R., 150(124), *199*, 157(145), 161(145), *200*, 909(237), *942*, 916(267, 268), *943*
Guidetti, E. S., 1056(232), *1067*
Guilard, R., 815(56), 816(65, 67, 69, 70), 817(74), *866*
Guirado, A. 384(144–146), 385(144–146), 386(144), 386(144, 148– 152), 387(151), *398*, 492(98), *499*, 523(125), *532*, 703(4–6), *748*, 891(50), *934*, 1154(198), *1185*, 1161(274), 1162(289, 290), *1187*
Gulick, W. M., 886(22), 908(22), *933*
Gulotta, A., 858(298), *874*, 1252(191), *1262*
Gultyai, V. P., 413(94), *429*, 435(50), *458*, 439(94), *459*, 456(242), *464*, 488(49, 50), 492(49), *498*, 722(167), *754*, 892(62), *935*, 894(77c), *936*, 895(96), 897(119), *937*, 899(136), *938*, 1144(123), 1145(123), *1182*
Gunawardena, N. E., 1236(104), *1259*
Gundersen, A., 452(210), 453(210), 455(210), *463*, 895(196), 903(196), *940*, 915(251), *943*, 1144(103), *1182*
Günther, H., 1174(441), *1192*
Gupta, A. K., 285(229), *320*, 425(184), *432*
Gupta, B. G. B., 983(124), *1022*
Gupta, S. L., 418(132), *430*
Gür, T. M., 1433(66), *1436*
Gurnsway, V., 1434(20a), *1438*
Gusakova, O. Y., 435(53), *458*
Gust, S. L., 924(297), 931(297), *945*
Gustak, E., 271(103), *316*, 742(384), *762*
Gustavii, K., 307(411), *326*
Guth, T., 1413(135), *1420*
Guthke, H., 257(17), 265(17), 268(17), 271(17), 307(17), *313*, 307(416), *327*, 1295(48), 1319(48), *1333*, 1305(75), *1334*
Gutmann, V., 285(227), 293(227), *320*, 288(247), *321*
Guy, G., 301(374), *325*
Guyard, M., 494(117), *500*, 740(360), *761*
Gwrado, A., 563(132), *578*
Gysling, H. J., 858(301), *874*

Haag, A. P., 781(65b, c), *786*
Haas, H., 237(27), *247*

Haas, R. G., 725(228), *756*
Habeeb, J. J., 837(190), *870*, 857(294), *874*, 1235(89), *1259*
Haber, F., 401(2), *426*, 413(95), *429*, 1247(172), *1262*
Habermann, W. 307(41b), *327*
Habib, M. A., 1409(115), *1419*
Habuka, M., 1077(52), *1097*, 1160(254), *1186*
Hackerman, N., 284(232), 290(232), *320*, 1105(24), *1124*
Haddon, R. C., 680(117), 681(117), 682(117), 683(117), *695*, 1243(129), *1260*
Haenel, M. W., 1434(22b), *1438*
Häfliger, O., 474(74), *481*
Hafner, K., 337(71), *355*
Haggerty, J., 445(163), *461*
Hagi, T. 1045(119, 120), *1064*
Hagiwara, T., 1383(89), 1387(89), *1395*
Hahl, H., 449(195), *462*, 490(76), *499*
Hahn, F., 640(154), *654*
Haines, R. J., 1168(344), *1189*
Hakl, J., 707(43), 719(43), 727(43), 734(43), *749*
Hakozaki, S., 516(86), 517(86), *531*, 987(142), 993(142), *1023*, 1079(69), *1097*, 1162(291), *1187*
Haldar, C., 1409(108), *1419*
Hale, J. M., 63(77), *118*
Hales, R., 439(88), *459*
Haley, N. F., 671(78), 672(79), *694*, 979(111), *1022*
Halko, D. J., 816(68), *866*
Hall, B., 1383(91), *1395*
Hall, D. A., 390(180), *399*, 489(54), *498*, 740(354), *761*, 747(436), *763*
Hall, E. A. H., 778(55), *786*, 1200(15), *1211*
Hall, J. L., 835(177), *870*
Hall, K. J., 174(197), *201*
Hall, L. C., 845(234), *872*
Hall, S. A., 989(163), *1023*
Hallcher, R. C., 308(420), *327*, 490(72), *498*, 929(330), *946*, 961(38), *1019*, 1005(220b), *1025*, 1230(59), *1258*, 1265(2), 1267(12b), *1280*, 1268(17, 19), *1281*, 1268(2), 1273(2), *1280*, 1275(19), *1281*, 1275(45–47), *1282*, 1421(1a, b), *1429*, 1422(1a), *1429*, 1422(19), *1430*, 1424(1b), 1425(1a), *1429*
Halpern, J., 826(127), *868*
Halton, B., 184(257), *204*
Halvorsen, J., 188(269), *204*
Hamaguchi, H., 404(27), *427*, 597(106), *611*, 600(125), 602(144), *612*, 603(150), *613*, 704(15a–d), 726(15a–d), *748*, 727(247), *757*, 773(34, 35), *785*, 782(69, 70a, b), *786*, 928(324), *946*, 1035(19, 23), 1036(28), *1061*, 1148(157), *1183*, 1163(295), *1188*, 1178(513), *1194*, 1278(57a), 1280(57a), *1282*
Hamann, C. H., 234(9), *246*, 644(194), 645(194), *656*
Haman, K., 925(247), *943*, 925(305), *945*
Hamano, S.-I., 600(120), *611*, 1244(148), *1261*
Hamano, Y., 1319(117), *1336*
Hamatani, T., 851(265–267), *873*
Hamburg, A. W., 819(80), *866*
Hamet-Elfakir, C., 424(171), *431*
Hamilton, A., 819(86), *867*
Hamilton, H., 1144(129), 1145(129), *1182*
Hamilton, M. A., 414(103b), *429*
Hamilton, R., 226(23), *232*
Hamilton, R. G., 123(21), 133(21), *196*
Hamm, R. E., 1245(152), *1261*
Hammer, R. N., 280(190), 284(190), *319*
Hammerich, O., 123(30), 125(51), 133(30), *196*, 125(46, 51), 127(62), 138(85), 140(46), *197*, 140(97, 98), 141(97, 98), 142(97, 98), 143(99), 148(97, 98), 150(97, 116), *198*, 150(119), *199*, 163(157), *200*, 174(207), 175(207), 179(214), 190(207), *202*, 234(7b), *246*, 284(237), 294(237), *320*, 291(287, 288), 302(288), *322*, 331(8a), 333(8a), 349(8a), *353*, 345(140–142), *357*, 349(178), *358*, 350(199), *359*, 506(6, 8), 509(6), 510(6), 511(6), *528*, 506(26), *529*, 511(59), *530*, 518(105), 521(119), *532*, 596(93), *610*, 617(21, 22), 620(22), *650*, 628(90), *652*, 643(192), *656*, 659(2), *692*, 677(100), *695*, 684(132–134), 685(133, 136) 686(150, 153), 687(150, 153), *696*, 686(171), 689(185, 186), *697*, 706(35), *749*, 951(19, 20), *1018*, 963(51), 966(55), *1019*, 963(65), *1020*, 1032(2), *1060*, 1053(197), *1066*, 1070(13), *1095*, 1072(31), *1096*, 1080(76), 1082(76), *1097*, 1084(115), 1086(127), *1099*, 1175(468, 470), *1193*
Hammerum, S., 518(105), *532*, 643(192), *656*, 1086(127), *1099*
Hammi, M., 73(93d), *119*, 150(115), *198*
Hamnett, A., 179(234, 235), *203*
Hamon, D. P. G., 1153(181), *1184*
Hamon, J.-R., 838(204), *871*
Hampson, N. A., 190(274), *204*, 276(149), *318*, 639(138), *654*, 1235(84), *1259*, 1288(29), *1333*
Hanada, V., 1155(209), *1185*
Hanafey, M. K., 148(110), 188(110), *198*
Hanafusa, T., 1409(102), *1418*
Hanatani, T., 904(208), *941*, 1434(13b), *1437*
Hanck, K. W., 831(154), *869*
Hand, R. L., 309(432), *327*, 587(30), 588(30), 589(30), *608*, 590(42), *609*, 969(70, 72), 977(81), *1020*
Hand, H. M., 1081(93), *1098*
Handa, V. K., 1150(169), *1184*
Handiak, H. W., 983(126), *1022*, 1009(229b), *1026*
Handoo, K. L., 140(91), *198*, 478(103), *482*
Hannebaum, H., 604(157), *613*, 1288(31), 1292(31), *1333*, 1305(76), 1309(80), *1334*
Hansen, G. R., 905(216), *941*
Hansen, J. C., 1106(27), 1108(27), *1124*
Hansen, J. N., 1143(86), *1181*
Hansen, P. E., 3(12), *8*, 346(150), *357*, 1250(181), *1262*
Hanson, D. C., 288(259), *321*

Hanson, J., 826(127), *868*
Hanus, V., 163(158), *200*
Hapiot, P., 128(68), *197*
Hara, M., 621(86), 627(86), *652*, 1161(271), *1187*
Harada, H., 1411(124), *1419*
Harada, K., 375(99, 100), *397*, 468(32, 33), *479*, 907(186), *940*, 929(332), *946*
Harada, M., 1164(311), *1188*
Hardstone, J. D., 604(159), *613*, 773(38b), *785*, 1077(61), 1078(62), *1097*
Hardy, J. P., 376(111), *397*, 1176(491), *1194*
Harima, Y., 1253(204), 1255(204), *1263*
Harland, R. G., 165(168), 173(168), *200*
Harle, A. J., 721(161a), *753*, 1144(99), *1182*
Harms, R. H., 682(123), *695*
Harnden, R. M., 659(5), 680(5), *692*, 673(81, 82), *694*
Harrar, J. E., 257(13), *313*, 309(429, 435), *327*
Harrer, W., 1207(82), *1214*
Harries, C., 1201(20), *1212*
Harris, F. E., 291(286), *322*
Harrison, J. A., 485(16), 489(16), 492(16), *496*, 635(121), *653*, 744(403), *762*
Harrison, R. J., 819(83), *867*
Harrison, S., 1421(4c), *1429*
Hart, D. J., 996(160), *1023*
Hart, E. J., 1253(200, 201), *1263*
Hart, E. P., 723(196), *755*
Hartley, F. R., 1268(28), *1281*
Hartwig, W., 243(68), *248*
Hartzell, C. R., 182(244), 183(244), *203*
Harvey, D. R., 1080(82), *1098*
Harwood, R. 621(82), 627(82), *652*
Hasegawa, K., 1040(63), *1062*, 1204(53), *1213*, 1409(111), *1419*
Hashimoto, K., 1408(54), *1417*, 1409(96), *1418*
Hashimoto, T., 1163(295), *1188*, 1178(513), *1194*, 1278(57a), 1280(57a), *1282*
Hashimoto, Y., 835(178), *870*
Hashmall, J. A., 24(28), *115*, 508(44), *530*
Hasimoto, T., 1035(19), *1061*
Haslett, R. J., 703(12), *748*, 888(34), *934*
Hass, O., 1366(22), *1392*, 1366(25), *1393*, 1367(22), *1392*, 1367(25), *1393*
Hassett, J. W., 672(79), *694*
Hatano, K., 819(88), *867*
Hatayama, T., 1319(117), *1336*
Hatsumura, H., 698(196), *698*
Haufe, J., 1002(186), *1024*, 1052(193), *1066*, 1316(106), 1319(106), 1329(106), *1335*
Hauffe, W., 641(166), *655*
Hauschild, G., 837(196), *870*
Hauser, J. J., 680(117), 681(117), 682(117), 683(117), *695*
Haushalter, R. C., 1200(13), *1211*
Häussermann, C., 414(104), *429*
Hauteville, M., 800(88), *808*
Havinga, E., 1070(4), *1095*
Hawkes, G. E., 563(127), *578*, 798(69), 801(69), *807*, 1156(219), 1157(219), 1158(219), *1185*
Hawkridge, F. M., 179(220), *202*, 270(96), 277(96), *316*
Hawley, M. D., 125(49), *196*, 140(90), 150(114), *198*, 154(136), *199*, 193(286), *205*, 346(153b), 350(153b), *358*, 362(3), 366(3), 382(3), *394*, 373(83, 84, 85), 374(83), 379(91), *396*, 418(133), 419(133), *430*, 467(26), *479*, 471(53), 473(53), *480*, 474(72), *481*, 478(100–102, 106, 109), *482*, 590(43), *609*, 594(84), *610*, 620(49), *651*, 621(79), 625(79), *652*, 711(65), *750*, 886(21), 887(25), *933*, 1268(20), *1281*
Hayano, S., 1053(200), *1066*
Hayashi, E., 1106(33–35), 1107(33, 34), 1108(35), 1111(43), *1125*
Hayashi, J., 6(22), *8*, 243(69), *248*, 516(86), 517(86), *531*, 598(109), *611*, 643(185), *655*, 643(188), *656*, 781(66), *786*, 987(142), 993(142), *1023*, 1079(69), *1097*, 1162(291), *1187*, 1171(373), *1190*
Hayashi, S., 439(89), *459*
Hayashi, T., 487(37), *497*, 570(157), *579*
Hayes, J. R., 369(62, 64), *396*, 1135(18), *1179*, 1153(192), *1184*, 1154(18), *1179*, 1176(502), *1194*
Hayes, M., 1288(33), *1333*
Haynes, L. V., 1434(14a), *1437*
Hays, J. W., 434(21), *437*, 882(3h), 885(3h), *932*
Hazard, R., 375(105), 376(106–108), *397*, 418(136), 422(155), 423(163), *431*, 707(45, 47a, c), *749*, 707(48), 708(49), 712(69), *750*, 714(92, 95, 97), *751*, 893(66, 68a), *935*, 1149(163), *1183*, 1151(174, 175), *1184*, 1168(349, 350), 1169(350), *1189*, 1171(393–395), *1191*, 1173(436), *1192*, 1427(18), *1430*
Hazelrigg, M. J., Jr., 882(3n), 885(3n), *932*
Hazeldine, R. N., 260(44, 45), *314*
Hazell, R. G., 468(30), *479*, 712(74), *751*, 909(164), *939*, 930(363), 931(363, 365b, 366), *947*, 1136(36), *1180*
Head, R. A., 836(186), *870*
Headridge, J. B., 300(359), *325*
Healy, K. P., 288(253), 292(253), *321*, 392(198, 199), 393(198), *400*, 829(146), *869*, 1094(186), *1101*
Heath, G. A., 411(77), *428*, 820(95), *867*
Hebd, C. R., 1150(168), *1184*
Hebert, M., 901(156), *938*
Heck, R. F., 1233(86), 1235(86), *1259*, 1247(174), *1262*
Heckner, K. H., 589(32), *608*
Hedges, R. M., 26(31c), *115*
Hedstrand, D. M., 689(187), *698*
Heeg, M. J., 736(327), *760*
Heeger, A. J., 331(9), 341(9), *353*
Hefter, G. T., 820(95), *867*
Hegedus, L. S., 809(2), *864*
Heilbronner, E., 26(31e), *115*, 346(148), *357*
Heimann, W., 616(1), *650*
Heilman, S. M., 712(70), *750*, 930(359), *947*

Heimler, D., 971(77b), 972(78), *1020*
Heineman, W. R., 13(1), *114*, 179(215, 216, 219, 220, 229), 180(215), *202*, 182(244), 183(244), *203*, 188(266), *204*, 270(96), 277(96), *316*, 297(338), *324*, 1409(114), *1419*
Heinrich, W., 1207(90), *1214*
Heinsohn, G. E., 367(30), *395*, 489(63), *498*
Heintz, M., 4(13), *8*, 1252(192), *1263*
Heinze, J., 128(65), *197*, 234(7c), *246*, 296(331), *324*, 332(10), 333(26, 31), 334(33), 336(33), 338(10, 26, 33), 341(10, 31, 33), 349(26), *353*, 334(42), 335(42), 336(56), 337(42), 339(42), 343(56), 351(56), *354*, 337(70, 72), 338(72), 340(85, 86a), 341(72), 350(70), *355*, 341(92), *356*, 351(205), *359*, 506(12), 511(12), *529*, 511(68), *530*, 589(33), *608*, 682(123), *695*, 1391(139), *1396*
Heitbaum, J., 179(232), *203*, 301(373), *325*, 640(156), 641(166), *655*, 644(194, 200), 645(194), *656*
Heitz, E., 268(92), *316*, 1286(8), 1292(8), 1299(8), *1332*
Heinsohn, G. E., 1041(71), 1050(71), 1053(71), *1062*
Hejrovsky, J., 272(110), *316*
Helgée, B., 3(11), *8*, 267(76b), *315*, 288(249, 250), 292(249, 250), 295(249), *321*, 345(145), *357*, 593(71), *610*, 601(139), *612*, 960(26), *1018*, 972(76), 977(76), *1020*, 1070(6), *1095*, 1077(57), *1097*, 1081(87), *1098*, 1089(153–156), 1090(153–156), 1091(154, 155), *1100*, 1119(64), *1126*, 1165(315), *1188*
von Helmholtz, H. L. F., 58(74), *118*
Hema, M. A., 1408(69), *1417*
Hembrock, A., 242(57), *248*, 778(56), *786*, 1083(104), *1098*, 1163(299), *1188*
Hempstead, M. R., 820(91), 821(91), *867*
de Hemptinne, X., 272(113), *316*, 439(93), *459*, 445(170), *461*
Hencher, J. L., 837(191, 194), 862(191), *870*, 862(318), *874*
Hendewerk, M., 1404(18), *1416*
Hendricks, A., 924(300), *945*, 1319(114), *1336*
Hengge, E., 855(280), *873*, 930(340), *946*
Henglein, A., 434(20), *457*, 1406(27), *1416*, 1412(125), *1419*
Henk, F., 1200(18), *1212*
Henkel, U., 673(83), *694*
Henne, B. J., 374(86), *396*
Henne, M. T., 1248(177), *1262*
Hennig, G., 445(162), *461*
Henning, N., 715(101), *751*
Henton, D. R., 630(103), 631(103), 633(103), *653*
Hevas, A. M., 162(149), 163(160, 161), *200*
Herbener, R. E., 1159(246), *1186*
Herlem, M., 284(217), 290(217), *320*, 290(267), *321*, 290(268), *322*, 293(303, 304), *323*, 515(94, 95, 96), 518(94, 95), *531*
Hermann, C. H., 644(200), *656*
Hermecz, I., 734(311), *759*
Hermeling, D., 564(133), 566(133), *578*, 855(275a), *873*
Hermolin, J., 441(126), 444(126), *460*, 724(211b), *755*, 898(125), 906(125), *938*, 915(256), *943*, 1050(172), *1066*, 1171(382), *1190*
Herold, T., 827(134), *868*
Herrmann, J. M., 1408(44, 45), *1416*, 1408(50), *1417*, 1408(84), 1409(97, 99), *1418*, 1413(131), *1419*
Hershberger, J. W., 1241(123), *1260*
Hershenhart, E., 823(110), 824(110). *867*
Herszbein, S., 953(7), *1018*
Hertwig, R., 1005(212c), 1006(212c), *1025*
Herz, J., 291(290), *322*
Herzberg, H. H., 437(137), *460*
Herzschuh, R., 684(130), *696*, 728(261), *757*
Hess, U., 389(173), *399*, 468(31), *479*, 468(35, 36, 38), *480*, 595(88), *610*, 711(62), *750*, 726(231, 233), 730(231), *756*, 930(357a, b), *947*, 1246(170), 1251(189), *1262*, 1268(25a, b), 1274(25a, b), *1281*
Hetzler, U., 1368(33), *1393*
Hewgill, F. R., 616(4), 620(4), *650*
Heyer, J., 593(67), *609*, 599(119), *611*, 604(162), 605(162), 607(162), *613*
Heyrovsky, J., 157(140), 161(140), *199*, 256(2), *312*, 256(10), *313*, 452(212), *463*
Heyrovsky, M., 409(66), *428*, 411(85, 86), 421(86), *429*, 422(151), 424(165), *431*
Hibber, H., 445(165), *461*
Hibino, K., 777(50), *785*, 1035(19), *1061*, 1163(295), *1188*, 1178(513), *1194*, 1278(57a), 1280(57a), *1282*
Hickling, A., 260(39), *314*, 309(425), *327*
Hiddleston, J. N., 268(91), *316*
Hiegel, G. A., 571(164), *579*, 1159(239), *1186*
Higgins, I. J., 645(207), *656*
Hill, H. A. O., 645(207, 208), *656*, 826(132), *868*, 1436(31b, c), *1438*
Hill, I. R., 179(226), *202*
Hiller, K.-O., 668(62), *693*
Hiller, K., 711(62), *750*
Hillman, A. R., 174(197, 200), *201*, 179(234), *203*, 1358(112), *1362*
Hills, G. J., 285(228), *320*
Himmele, W., 1316(106), 1319(106), 1329(106), *1335*
Hindelang, F., 276(146, 147), *318*
Hine, F., 1287(10), *1332*
Hinklemann, K., 336(56), 343(56), 351(56), *354*
Hinman, A. S., 180(238), *203*
Hino, H., 563(129), *578*
Hirabayashi, Y., 661(16), *692*
Hiraki, K., 836(185), *870*
Hirai, S., 574(169), *579*
Hirai, T., 891(47b), *934*, 909(225), *942*, 924(299), *945*, 1045(119, 120), *1064*
Hiramatsu, M., 541(37), *575*
Hiramatsu, S., 1058(249), *1068*

Hirasawa, R., 190(277), 194(277), *204*, 882(3j), 885(3j), *932*
Hirashima, T., 1034(15), *1060*, 1163(298), *1188*
Hirata, Y., 600(121), *611*
Hirayama, M., 488(45), *497*, 492(101), *499*
Hiromoto, Y., 1383(94), 1385(94), *1395*
Hirschon, A. S., 669(66), *694*
Hirokawa, T., 620(32), *651*
Hiromatsu, H., 258(28), *313*
Hirotsu, K., 1008(227), 1012(227), *1026*
Hisaeda, Y., 823(109), 824(109), *867*, 824(112), 827(140), 829(140), *868*
Hitchman, M. L., 165(164), *200*, 174(199), *201*
Hlavaty, J., 267(76d), *315*, 601(141), *612*, 707(47b), *749*, 708(52), 709(58), *750*, 713(86), *751*, 892(65), *935*, 1077(60), *1097*
Ho, F. L., 179(228), *202*
Ho, S. V., 1301(59), *1334*
Hoang, M., 1340(5, 6, 8), 1346(8), *1359*
Hobolth, E., 344(111), 347(111), *356*, 368(37), *395* 392(191), *399*, 886(14), 891(14), 1153(183), *1184*, 1221(15), 1249(15), *1256*
Hodgson, H. H., 1208(99), *1214*
Hodgson, K. O., 985(131), *1022*
Hodgson, W. G., 284(224), 286(224), *320*, 365(16), *394*, 404(22), *427*, 1245(151), *1261*
Hodler, M., 489(60), *498*
Hoenl, H., 930(344), *946*
Hoever, H., 1299(54), *1334*
Hofelmann, K., 925(247), *943*, 925(305), *945*
Hofer, H., 536(5), *574*
Hoffelner, H., 675(97), *695*, 1094(191), *1101*
Hoffman, B. M., 819(81), *866*
Hoffmann, A. K., 284(224), 286(224), *320*, 365(16), *394*, 404(22), *427*, 468(34), *480*, 662(38), *693*, 929(334), *946*, 1094(190), *1101*, 1164(307), *1188*
Hoffmann, H., 719(133), 733(133), *752*, 734(305), *759*, 1052(186), *1066*
Hoffmann, J. C., 486(29), *497*
Hoffmann, M. R., 1413(133), *1419*
Hoffmann, W. D., 1146(142), *1183*
Hogg, J. A., 1104(8), 1105(8), 1106(8), *1124*
Hoier, S. N., 179(234), *203*
Hoijtink, G. J., 26(31d), *115*, 331(1, 3), 332(1, 13, 14, 20, 22), 333(27), 334(1, 3, 14), 345(13, 14, 20), 346(13), *353*, 335(45), *354*, 344(121), 345(130, 131), 346(149), *357*, 508(42), *530*
Hokazono, S., 1073(36), *1096*
Holbrook, W. B., 209(5c), *231*, 309(433), *327*
Holland, F. S., 273(122), *317*
Holleck, L., 409(64), *428*, 411(86), 421(86), *429*, 419(137), 420(137, 141), 421(141), 422(151, 156), 424(165), *431*, 424(180), *432*, 435(30), *457*, 436(67), *458*, 441(146), *461*, 467(20), *479*, 734(313), *759*, 1055(220), *1067*
Hollitzer, E., 1104(6, 13), 112(47), *1125*
Holloway, R., 350(202), *359*
Holly, F. W., 546(67), *576*, 1002(203), *1025*
Holman, R. J., 554(90), *577*, 1143(90–92), 1144(90, 92), *1181*, 1255(210), *1263*
Holmberg, H., 1312(85), *1335*
Holmes, T. J., 680(121), *695*
Holst, H. H., 389(173), *399*, 719(129), *752*, 726(231), 730(231), *756*, 1251(189), *1262*
Holt, P. T., 155(139), 156(139), *199*, 685(142), *696*
Holton, D. M., 617(18), *650*
Holub, F. F., 271(104), *316*
Holubek, J., 388(166), *399*, 745(416), *763*, 1033(8a), 1050(8a), *1060*
Holy, N. L., 331(6), 335(6), *353*
Hombach, H.-P., 1434(22c), *1438*
Honda, K., 153(131), *199*, 301(380), *325*
Honl, H., 488(41), *497*
Honnorat, M., 894(77f), *936*
Hoog, H., 442(108), 447(108), *459*, 450(201), *462*
Hopper, D. J., 1436(31d), *1438*
Hoque, A. K. M. M., 687(176, 177), *697*
Horak, V., 720(139), *753*, 1050(170, 174), 1051(175, 176), *1066*
Horányi, G., 277(150), *318*, 446(173), 450(202), *462*, 640(159, 160), 641(172), *655*, 1038(44), *1061*, 1038(45), *1062*, 1435(28), *1438*
Hori, F., 381(135), *398*, 894(77a), 894(89), *936*,
Hori, Y., 1434(12c, 15j), *1437*
Horie, K., 973(79), *1020*
Horie, S., 543(50), *576*
Horikawa, H., 558(117, 118), 559(119, 121, 122), 562(119, 125, 126), 564(136), 567(118, 136), *578*, 567(145, 147), 573(167), *579*, 601(128), *612*, 766(3a, b, 4), *783*, 766(6a, c, 7a), 768(16–19), 769(20, 21), *784*, 1158(231), *1186*, 1162(284), *1187*, 1243(137), 1244(137), *1261*, 1426(15), *1430*
Horike, H., 1168(341), *1189*
Horine, P. A., 386(147), *398*, 491(96), *499*
Horiuti, J., 272(111), *316*
Horn, G., 1040(65), *1062*
Horn, R. K., 1305(76), *1334*
Hornback, J. M., 969(74), *1020*
Horner, L., 236(22), *246*, 263(62), *314*, 275(128), *317*, 436(61), *458*, 441(149a), *461*, 467(21, 22), *479*, 488(41), *497*, 492(97), *499*, 493(108, 109), *500*, 742(386), *762*, 895(98a), *937*, 898(124), *938*, 902(166), *939*, 916(266), *943*, 928(327a), 930(344), *946*, 1047(132), *1064*, 1048(143–147), 1049(155, 156), *1065*, 1051(181– 185), 1052(186, 187, 193), *1066*, 1056(237–239), 1057(238), 1058(237–239), 1059(237–239), *1067*, 1059(262), *1068*, 1132(4, 8), *1179*, 1141(67), *1181*, 1148(153), *1183*, 1170(357–361), *1190*, 1172(401–408, 410, 411), 1173(411, 418), *1191*, 1200(4), *1211*, 1208(106),

[Horner, L.] *1214*, 1208(107), 1209(116), 1210(107, 120), *1215*, 1210(106), *1214*
Horner, M., 351(203), *359*
Hornof, V., 1350(37), *1360*
Hornung, F., 1135(23), *1179*
Horowitz, G., 1382(87), *1395*
Houghton, D. S., 675(89, 90), *694*
Houghton, R. W., 1295(49), *1334*
House, H. O., 307(413), *327*, 983(124), *1022*, 1136(34), *1180*, 1175(485)), *1193*, 1275(44), *1282*
Houser, K. J., 886(21), *933*
Houts, J. J., 661(15), *692*
Howell, J. O., 25(29), 72(29), *115*, 55(70), *117*, 70(91b), *118*, 128(67), 129(67), *197*, 506(17, 18), 508(18), 509(18), 510(17), 511(17, 18), *529*
Howie, J. K., 661(15), *692*
Hoyer, E., 930(363), 931(363, 365b), *947*
Hoyer, W., 1408(73), *1417*
Hoz, S., 1245(157), *1261*
Hranilovic, J., 271(103), *316*, 742(384), *762*
Hrnciar, P., 915(258), *943*
Hrubesch, A., 486(23), *497*
Hsu, L. C., 1340(5, 8), 1346(8), *1359*, 1349(32, 33), *1360*, 1355(85), *1362*
Huang, C. B., 1413(137), *1420*
Huang, S., 713(88), *751*
Huang, S. J., 991(169), *1024*
Huang, S.-R., 421(143), *431*
Huba, F., 1105(14, 15), 1106(14), 1109(14), *1124*
Hubbard, A. T., 1368(35), 1371(36), *1393*, 1433(4), *1436*
Huber, A., 793(41), *805*
Huber, W., 337(65, 66b, 70), *355*, 351(204), *359*
Hubnall, P. M., 887(26), *933*
Huebert, B. J., 335(52), 342(52), *354*
Huffman, J. W., 13(2a), *114*
Huffman, J. C., 833(165, 166), *869*, 846(237), *872*
Huggins, R. A., 1433(6b), *1436*
Hughes, D. A., 1105(20), 1109(20), 1110(20), 1112(20), *1124*, 1312(89), *1335*
Huhn, D., 726(233), *756*
Huhtasaari, M., 545(79), 550(93), *576*, 1011(242), *1026*, 1157(221), *1185*
Hui, L., 1413(135), *1420*
Hull, M. N., 689(190), *698*, 792(23), *804*
Hull, R., 1406(27), *1416*
Hull, V. J., 686(170), *697*
Hulsmeyer, K., 493(110), *500*
Humffray, A. A., 675(89, 90), *694*, 686(147), *696*
Hummel, H.-U., 661(21), *692*
Hummelstedt, L. E., 373(78), *396*
Hünig, S., 194(293), *205*, 351(203), *359*, 526(139), *533*, 682(122), *695*
Hurley, M. F., 712(72), *750*, 903(362), 931(362), *947*
Hush, N. S., 41(57c), *117*, 63(78), *118*, 342(101), *356*, 373(76), *396*, 738(336), *760*, 820(93), *867*
Huss, R., 276(146, 147), *318*
Hussain, W., 836(187), *870*
Hussein, R. H., 1408(40), *1416*, 1408(56), *1417*
Husstedt, U., 243(75), *248*, 1053(195), *1066*, 1153(186), *1184*
Hutchings, M. G., 979(110), *1022*
Hutchins, C. S., 334(41), 335(41), 342(41), *354*
Hutchinson, J. H., 827(139), 829(139), *868*
Hütter, P., 717(118), *752*, 959(48), *1019*
Hutton, R. S., 1387(113), 1389(113), *1396*
Huyser, E. S., 724(208), *755*

Ibers, J. A., 819(81), *866*
Ibl, N., 258(20), *313*, 268(87), *315*, 486(29), *497*, 1243(130), *1260*, 1309(81), *1334*
Ibrisagic, Z., 288(253, 254), 292(253, 254), *321*
Ichidawa, M., 641(170), *655*
Ichihara, J., 1409(102), *1418*
Ichou, I. A., 1408(41), *1416*, 1408(46), *1417*
Ideman, J. P., 1230(59), *1258*
Idriss, K. A., 683(127), *696*
Ignat'eva, L. A., 442(105), 443(105), *459*, 898(123), *937*
Iguchi, M., 621(84–87), 627(84–87), *652*, 775(43a–c), 776(44b, 46a, b, 47), *785*, 958(40), 962(45), *1019*, 1013(251), *1027*, 1161(271), *1187*, 1163(296, 297), *1188*
Ihn, W., 1040(65), *1062*
Iizuka, T., 601(129), *612*
Ikada, T., 489(56), *498*
Ikeda, A., 516(86), 517(86), *531*, 523(126), *532*, 987(142), 993(142), *1023*, 1079(69), *1097*, 1160(258), *1186*, 1160(263), *1187*, 1161(258), *1186*, 1161(263), 1162(291), *1187*
Ikeda, S., 1434(12d), *1437*
Ikeda, Y., 556(108), *557*, 560(123), *578*, 904(206, 207), *941*, 1434(13f), *1437*
Ikemoto, Y., 851(267), *873*
Ikeshoji, T., 839(216), *871*
Ikida, M., 1244(147), *1261*
Ikuta, S., 1229(73), 1230(55), *1258*
Il'yasov, A. V., 473(63), *480*, 473(67), *481*, 488(45), *497*, 1055(223), *1067*
Imagawa, T., 570(160–163), *579*, 782(71), *786*, 1159(238), *1186*
Imanishi, T., 597(106), *611*, 648(229), *657*, 1036(28), *1061*
Imbeaux, J. C., 126(56), 135(81), *197*
Imberger, H. E., 686(147), *696*
Imoto, E., 725(219), *756*, 926(312), *945*, 926(316), *946*
Imoto, T., 1211(124, 125), *1215*
Inaba, H., 686(146), 687(146), *696*
Inaba, M., 267(84), 268(84), 271(84), 288(84), 307(84), *315*
Inaishi, M., 667(58), *693*, 1167(337), *1189*
Inayama, S., 565(139, 140), *578*
d'Incan, E., 277(160), 277(161), 311(160), *318*, 891(48), *934*
Inesi, A., 390(181), *399*, 703(10), *748*, 708(50), *750*, 747(437), *763*, 887(27), *933*, 891(51), *934*, 903(218), *941*, 1154(196), *1184*, 1154(201), *1185*, 1434(13e), *1437*
Ingall, G. B., 365(20), *394*

Inge, G. M., 1345(20), *1360*
Ingram, D. S., 434(9), *456*
Ingres, I., 886(12), 891(12), *933*
Inhoffen, H. H., 738(335), *760*
Inokuchi, T., 243(70), *248*, 393(207), *400*, 537(11, 12), 563(11, 12), 572(11), *574*, 561(124), *578*, 569(152), *579*, 586(25), 601(25), *608*, 597(108), *611*, 602(145), *612*, 827(137), 828(137), *868*, 1034(16), 1035(22), *1060*, *1061*, 1077(51), *1096*, 1168(341), *1189*, 1229(68), 1231(65), 1232(65), *1258*
Inomata, K., 812(22), *864*
Inoue, I., 559(119), 562(119), *578*, 573(167), *579*, 766(5b, 6), 767(9–11), 768(19), 769(20–22), 770(24,25), *784*, 1162(284), *1187*, 1243(137), 1244(137), *1261*, 1426(15), *1430*
Inoue, K., 589(40), 590(40), *608*, 598(109), *611*, 603(156), 604(160, 161), 605(160), *613*, 643(188), *656*, 774(40), *785*, 1167(333), *1189*
Inoue, S., 817(73), *866*
Inoue, T., 528(144), *533*, 1092(165), *1100*, 1160(249), 1161(249), *1186*, 1161(267–269), *1187*, 1165(318), *1188*
Inzelt, G., 403(18), *426*, 1038(45), *1062*
Ioffe, B. V., 471(57), *480*
Ioffe, N. T., 820(96), *867*, 1059(256), *1068*
Ioganson, A. A., 846(236), *872*
Ippoliti, J. T., 582(5), *607*
Iqakura, C., 971(77a), *1020*
Irmen, W., 340(89), *355*
Irei, H., 679(106), *695*
Irei, K., 603(149), 607(163), *613*, 767(12), *784*, 773(37a–c), *785*
Isakawa, T., 929(332), *946*
Isakova, N. V., 794(45), *805*
Ishida, A., 767(12), *784*
Ishida, Y., 836(185), *870*
Ishifuku, K., 496(131), *500*
Ishige, O., 726(244), 728(244), *757*, 740(359), *761*, 1268(24a), *1281*
Ishiguro, M., 1008(226), *1026*
Ishikawa, A., 640(150), *654*
Ishikura, S., 442(100), 445(100), *459*, 445(167), *461*, 894(85c), *936*
Ishimaru, Y., 1408(65), *1417*, 1409(105), *1419*
Ishiwari, H., 570(161, 162), *579*, 782(71), *786*
Ishiwata, S., 455(234), *464*
Ishizaki, K., 1268(24a, 27), 1275(27), *1281*
Isimura, A., 1166(322), *1188*, 1278(57c), 1279(57c), 1280(57c), *1282*
Islam, N., 489(61), *498*, 798(68), 799(68), *807*, 1040(67), *1062*
Ismail, M. T., 403(21), *427*, 905(217), *941*, 960(27), *1018*
Isoe, S., 679(108), *695*, 716(113), *752*, 853(273, 274), 854(274), 855(275b, c), *873*, 1008(227), 1012(227), *1026*, 1053(194), *1066*, 1168(340), *1189*
Isoya, T., 909(231), *942*, 1316(107), 1319(107), *1335*, 1321(126), *1336*, 1329(107), *1335*, 1330(126), *1336*
Issa, R. M., 474(71), *481*
Itaya, A., 1408(61), *1417*
Itaya, K., 340(86b), *355*, 1253(202), *1263*, 1367(30), *1393*, 1390(131), *1396*
Ito, K., 435(31), *457*, 1434(12d), *1437*
Ito, M., 280(183), 297(183), 307(183), *319*
Ito, N., 839(212), *871*
Itoh, H., 1175(483), *1193*
Itoh, M., 392(192), *399*, 858(304), 859(305), 860(306–309), *874*, 886(18), *933*, 906(181), *939*, 1154(205), 1155(205), *1185*, 1172(400), *1191*, 1173(431), *1192*, 1269(35), *1281*, 1426(16, 17), *1430*
Itoh, T., 643(188), *656*
Ivanitskii, B. G., 257(18b), *313*
Ivanova, S. Z., 801(83), *807*
Ivanova, V. K., 466(4), 473(4), *479*, 471(52), *480*
Ivanovskaya, L. N., 496(136), *501*, 745(415), *763*
Iversen, P. E., 2 (7, 8, 9), 3(7, 8, 9), *8*, 261(50), 263(50, 61), 311(61), *314*, 308(418, 419), *327*, 389(175), *399*, 402(6, 8), 403(16), 404(8), 410(6), 423(16), *426*, 409(69, 71), *428*, 417(124), 418(124), *430*, 425(182, 184), *432*, 469(42), *480*, 485(15), 492(15), *496*, 599(114), *611*, 727(248), *757*, 735(319), *759*, 743(396), 744(396, 401, 402), *762*, 747(430), *763*, 892(64), *935*, 1052(190), *1066*, 1054(209), *1067*, 1148(156, 159), 1149(161), *1183*, 1267(11), *1280*, 1269(37), *1282*, 1270(11), *1280*
Iverson, B., 819(87), *867*
Ives, D. A., 1045(112), *1064*
Ives, D. J. G., 21(21b), *115*, 278(163), *318*
Iwai, K., 686(166), *697*
Iwakura, C., 258(27), *313*, 621(55), *651*
Iwamoto, R. T., 278(168), *318*, 284(206), *320*, 724(206), *755*
Iwamura, H., 678(102), 679(111), *695*, 1079(67), *1097*, 1167(339), *1189*, 1409(93), *1418*
Iwaoka, T., 584(22), *608*, 975(103), *1021*
Iwasaki, F., 643(188), *656*
Iwasaki, H., 1082(94), *1098*
Iwasaki, M., 50(67b), *117*
Iwasaki, T., 375(99, 100), *397*, 468(32, 33), *479*, 558(117, 118), 559(119, 121, 122), 562(119, 125, 126), 564(136), 567(118, 136), *578*, 567(145, 147), 573(167), *579*, 601(128), *612*, 766(1a, b, 2– 5), *783*, 766(6, 7), 767(8), 768(14–19), 769(20–22), 770(24), 772(33a, b), *784*, 1045(113), *1064*, 1058(250), *1068*, 1158(231), *1186*, 1162(284), *1187*, 1243(137), 1244(137), *1261*, 1426(15), *1430*
Iwasita, T., 594(81), *610*, 639(143, 144), *654*
Iwata, K., 891(59), *935*, 1383(89), 1387(89), *1395*
Iyoda, M., 337(65), *355*
Izgaryshev, N. A., 265(67), *314*
Izumi, I., 1167(336), *1189*

Izumi, J., 1383(94), 1385(94), *1395*, 1406(29, 30), *1416*
Izumi, M., 537(6), 573(6), *574*
Izumi, U., 445(153), *461*
Izutsu, K., 162(153), *200*, 280(183), 297(183), 298(346), 301(378), 306(407), *324*, *325*, *326*, 307(183), *319*, 455(229), *463*

Jackowski, J. J., 1175(449, 453), *1192*
Jackson, C., 1243(144), *1261*
Jackson, Jr., D. T., 1153(193), 1176(193), *1184*
Jackson, L. L., 862(321), *874*
Jackson, R. A., 855(282), *873*
Jacob, C., 267(79), 273(79), 288(79), *315*
Jacob, G., 425(185, 186), *432*
Jacobsen, N., 1082(101), *1098*
Jacobson, W., 1208(100), *1214*
Jacquet, F., 1243(145), *1261*
Jacquin, D., 715(105b, c), *752*, 1139(61), 1141(61), *1181*, 1252(193), *1263*
Jaeger, C. D., 1408(83), *1418*
Jaeger, D. A., 455(239), *464*
Jaeger, R. H., 1144(102), *1182*
Jaegfeldt, H., 725(225), *756*
Jaenchen, R., 813(27), *865*
Jaenicke, W., 737(329), *760*
Jäger, P., 738(335), *760*
Jäger, V., 1207(89), 1208(91), *1214*
Jagur-Grodzinski, J., 331(5), 343(15), *352*, 333(28b), *353*
Jain, A., 903(201), *940*, 1142(78), *1181*
Jain, N. C., 819(80), *866*
Jakopcic', K., 596(94), *610*, 706(36a), 710(36a), 712(36a), 714(36a), *749*
Jallabert, C., 214(9), *231*
Jambor, B., 736(324a, b), *759*
James, J. C., 284(216), 286(216), *320*, 291(293), *322*
Jan, C.-C., 180(240, 242), *203*
Janda, M., 689(179–183), *697*, 721(157d), *753*, 722(164a–d, 168), *754*, 1161(278), 1167(281), *1187*, 1167(334, 335), *1189*
Janik, B., 725(217), *755*, 732(278), 736(278), *758*
Janissek, P. R., 801(81), *807*
Janiszewska, L., 303(393), *326*
Jannakovdakis, D., 422(156), *431*, 640(157), *655*, 646(213), *656*, 1055(220), *1067*
Janssen, E., 336(58), *354*, 848(244), *872*
Janssen, L. J., 417(118), 422(118), *430*
Jansson, R. E. W., 208(1), *231*, 259(35), *314*, 1164(314), *1188*, 1301(62), *1134*, 1314(92), *1335*, 1421(4e), 1424(5), *1429*
Jansta, J., 1204(58), *1213*
January, J. R., 140(90), *198*, 478(100, 102), *482*
Janz, G. J., 21(21b), *115*, 278(163), *318*
Janzen, E. G., 193(289), *205*
Jaovannet, S., 375(105), 376(106–108), *397*, 1151(174, 175), 1152(176), *1184*, 1168(349, 350), 1169(350), *1189*, 1171(393, 394), *1191*, 1173(436), *1192*
Jasinski, R. J., 281(195), 285(195), 290(195), *319*, 284(238, 239), *320*, 1105(23), *1124*, 1384(101), *1395*
Jasne, S. J., 1389(122), *1396*
Jason, M. E., 887(31), *933*, 376(115), 378(115, 122, 123), *397*
Jaun, B., 813(27, 32, 34), *865*
Javaheri, S., 844(233), *872*
Jayaraman, K., 414(102c), *429*
Jeffrey, E. A., 467(13, 14), *479*
Jeftić, Lj., 435(51), *458*, 524(131), *532*
Jeko, J., 1105(19), 1109(19), *1124*
Jellinek, H. H. G., 744(399), *762*
Jeminet, G., 662(30), *692*, 670(75), 671(75), *694*, 891(53), 934, 1046(125, 126), *1064*, 1048(140, 148, 151c), 1056(140), *1065*, 1058(252, 254), 1059(263–265), 1060(263, 264), *1068*, 1226(43), 1248(43), *1257*
Jempty, T. C., 704(17), 708(17), *748*, 771(29), *784*
Jen, K. Y., 1357(96), *1362*, 1377(81), 1382(81), *1395*
Jenevein, R. M., 897(110), *937*, 898(128), 899(146), 900(128), 901(146), 902(128), *938*, 1144(107, 108, 113, 114), 1145(114), *1182*
Jenkins, R. A., 725(228), *756*
Jennings, J. R., 908(223), *941*
Jennings, P. W., 835(177), *870*
Jennings, W. B., 630(102), *653*
Jensen, B. S., 125(50, 52), *196*, 260(41), 294(41), *314*, 284(237), 294(237), *320*, 331(8b), 335(8b), 338(8b), 349(8b), *353*, 336(55), 342(55), *354*, 341(95), 350(95), *356*, 413(92), *429*, 453(223), *463*, 506(9), *528*
Jensen, E. T., 472(58), *480*, 719(127), *752*, 731(275), *758*, 1055(215), *1067*
Jensen, F. R., 1205(64), *1213*
Jensen, H., 603(148), *613*, 727(247), *757*, 767(13), *784*
Jensen, N. J., 375(93), *396*
Jensen, U., 546(69), *576*, 1002(200), *1025*
Jensen-Korte, U., 546(68), 547(68), *576*, 783(74a), *786*
Jentzen, E. G., 1269(33), *1281*
Jeong, J., 599(118), *611*
Jeremic, D., 596(92), *610*
Jerina, D. M., 1094(183), *1101*, 1117(61), 1118(61), *1126*
Jernigan, J. C., 70(91a), *118*
Jerome, A., 899(150), *938*
Jeroschewski, P., 173(185), *201*, 275(136, 137), *317*, 930(363), 931(360, 363, 364, 366), *947*, 712(73), *750*
Jewell, K. M., 1106(27), 1108(27), *1124*
Jeyakumar, D., 705(29a), *749*
Jezo, I., 733(295), *758*
Jezorek, J. R., 294(315), *323*, 346(153a), 350(153a), *358*
Jiang, L. C., 1230(53), 1231(53), *1258*
Jimenez, P. J., 616(9), 618(9), *650*, 962(46), *1019*
Jiménez, C., 163(161), *200*
Jiménez, M., 161(147), 163(147), *200*
Jing, H., 1173(432), *1192*
Johansen, O., 467(14), *479*
Johanson, B., 1312(85), *1335*
Johnson, D. A., 423(157), *431*
Johnson, D. C., 107(110), *119*
Johnson, D. K., 439(92), 440(92), 441(92), *459*, 798(74), 799(75), *807*,

[Johnson, D. K.]
802(91), *808*, 1421(1d), 1422(1c), *1429*
Johnson, E., 816(68), *866*
Johnson, E. L., 1245(152), *1261*
Johnson, G. V., 417(125), *430*, 715(106), *752*, 892(63), *935*
Johnson, H. L., 486(34), *497*
Johnson, J. H., 290(280), *322*
Johnson, P. S., 1109(39), *1125*
Johnson, R. E., 300(364), *325*
Johnson, R. W., 340(76, 80), *355*, 621(83), 627(83), *652*, 927(320), *946*
Johnston, J. C., 452(210), 453(210), 455(210), *463*
Johnston, K. M., 957(36), *1019*, 1005(218), *1025*
Johnston, K. P., 270(97), 303(97), 304(97), *316*
Jonas, A., 1209(111), *1215*
Jones, A. R., 1229(50), 1234(50), *1257*
Jones, F., 1243(131), *1260*
Jones, G. C., 621(67), 625(67), *652*
Jones, J. G., 290(276), *322*
Jones, P. R., 486(18), *497*
Jones, S. B., 175(209), *202*
Jones, S. L., 309(426), *327*
Jones, S. R., 514(81), 515(85), 517(81), *531*, 1085(120, 121), *1099*
Jönsson, L., 31(40), *116*, 234(16b), *246*, 236(26), *247*, 267(76d), *315*, 511(76), *531*, 601(141), *612*, 1070(1, 3, 8), 1081(3, 8), *1095*, 1077(60), *1097*, 1260(118), *1260*
Jordan, J., 515(95), 518(95), *531*, 1247(171), *1262*
Jørgensen, K. A., 1414(147), *1420*
Joseph, M. F., 843(226), *871*
Joshi, G. C., 453(222), *463*, 542(48), *576*
Joslin, T. A., 974(93, 94a), *1021*, 1117(56), *1126*
Jow, J.-J., 437(135b), 441(136b), *460*
Jow, T. R., 1377(81), 1382(81), *1395*, 1421(4b), *1429*
Juaristi, E., 670(77), *694*, 1064(128), *1064*
Jubault, M., 421(144), *431*, 476(91), *481*, 707(45, 46, 47c), 709(46), *749*, 707(48), 708(49), *750*, 719(135), *753*, 728(253, 254), *757*, 893(68a), *935*, 1149(163), *1183*, 1171(380, 381, 388), *1190*, 1171(389–392), 1172(396), *1191*, 1173(429, 430, 438), *1192*, 1200(16), *1211*
Juckett, D. A., 391(186, 187), *399*, 886(12), 891(12), *933*
Juday, R. E., 899(147), *938*, 1144(132), 1145(132), *1183*
Jugelt, W., 392(196), *400*, 466(2), *479*, 703(8, 9), *748*, 715(100, 101), *751*, 715(108), 716(112), *752*, 729(269), *758*, 973(87), 979(87, 109), *1021*, 1015(253), 1016(253), 1017(255), *1027*, 1224(35), *1257*
Juillard, J., 297(335), *324*
Julia, M., 1210(123), *1215*
Jung, G., 621(61), *652*, 1050(165), *1065*
Jungers, J. C., 272(113), *316*, 439(93), *459*
Junggren, U., 307(411), *326*
Junghans, K., 915(259), 916(259), *943*, 1135(19), *1179*
Jura, W. H., 284(224), 286(224), *320*, 365(16), *394*, 369(65), *396*, 404(22), *427*, 1153(189), 1176(189), *1184*
Jürgens, W., 449(192), *462*, 490(77), *499*
Justice, J. C., 26(34b), *116*
Justice, M. C., 26(34b), *116*
Justice, R. M., Jr., 390(180), *399*, 747(436), *763*
Justand, A., 81(102), 108(102), *119*, (8a, 9), *231*, 240(50), *248*, 298(348), *324*, 835(175), *870*, 1236(107, 109), *1259*, 1236(109), *1260*, 1434(13b), *1437*
Juul, M., 403(18), *426*

Kaabak, L. V., 263(63), *314*, 366(21), *394*, 908(224i), 910(224i), 911(224i), *942*, 921(290), 923(281), *944*
Kabachnik, M. I., 263(63), *314*, 366(21), *394*
Kabanov, B. K., 837(198), 838(198), *870*, 1269(36), *1281*, 1425(12), *1430*
Kabanova, S. D., 273(120), *317*
Kabasakalian, P., 728(255), *757*, 739(349), *761*, 1039(55), *1062*, 1144(96), *1182*, 1159(245), *1186*, 1176(494), *1194*
Kabe, Y., 686(146), 687(146), *696*, 1166(327), *1189*
Kadirgan, F., 640(161), *655*
Kadish, K. M., 179(236), *203*, 813(44, 49), 814(45, 50, 53), 815(53), *865*, 815(56), 816(62, 65–67, 69, 70), 818(44), 819(79), *866*, 1148(158), *1183*
Kadysh, V., 441(139), *460*
Kagan, M. Y., 438(73), 446(73), *458*
Kagechika, K., 559(120), 562(120), *578*
Kagiya, T., 467(12), *479*, 1408(59, 60, 63, 64, 66, 67), *1417*
Kahimura, S., 924(298), *945*
Kaim, W., 733(297), *758*
Kainosho, M., 1154(197), *1185*
Kaiser, A., 908(223), *941*
Kaiser, E. M., 239(41), *247*, 271(101), *316*, 293(309), *323*, 1255(212, 214), *1263*
Kaiser, E. T., 569(150), *579*, 1159(235), 1170(235), *1186*
Kaiser, F., 1201(20), *1212*
Kajihaam, A., 823(109), 824(109), *867*
Kakác, B., 477(97), *482*
Kakuta, N., 1413(137), *1420*
Kakuta, R., 1316(107), 1319(107), 1329(107), *1335*
Kaldova, R., 1232(80), *1259*
Kalidas, C., 732(283), *758*
Kalinin, A. I., 1053(204), *1067*
Kalinin, Y. K., 539(22), *575*
Kalinkin, M. I., 1059(256), *1068*
Kalinowski, M., 434(8), 435(8), *456*, 683(130), *654*
Kalitina, M. I., 442(104), *459*, 898(122), *937*
Kalvoda, R., 436(70, 71), 445(71), *458*
Kamada, M., 993(147), *1023*, 1093(172), *1100*
Kamada, Y., 600(124), *611*, 600(126), 601(129), *612*, 1178(516), *1194*
Kamardinova, B. B., 439(97), 440(97), *459*
Kamau, G. N., 485(14), *496*, 833(164), *869*
Kambara, S., 306(406), *326*, 349(188), *359*
Kambe, S., 1408(77), *1418*

Kamdee, Z., 907(194), *940*
Kamernitskii, A. V., 1040(70), *1062*, 1147(150), *1183*
Kametani, T., 704(19), *748*, 771(27c), *784*, 960(28c), *1018*, 988(145), *1023*
Kameyama, Y., 772(33a, b), *785*
Kaminskii, A. Y., 891(55), *935*
Kamkar, N. M., 489(64), *498*, 1042(78b), *1063*
Kamigauchi, T., 556(108, 109), *577*, 560(123), *578*
Kamp, A. J., 389(170), *399*, 747(435), *763*
Kampmeier, J. A., 392(197), *400*
Kanakam, R., 265(70), 271(70), *315*
Kanapieniene, J., 725(228), *756*
Kanazawa, K. K., 1356(90), *1362*, 1314(10), 1366(10), *1392*, 1368(10), 1370(10), *1392*, 1375(52, 54), 1380(54, 71), 1386(54), 1387(54), *1394*
Kanazawa, T., 603(155), *613*, 1077(50), *1096*, 1077(52), 1078(63), *1097*, 1160(254), *1186*
Kanbe, T., 1161(273), *1187*
Kandeel, Z. E.-S., 1045(111), *1064*
Kandil, S. A., 863(327), *875*
Kane, P. O. 469(44), *480*, 736(324c), *759*, 1039(57), 1050(57), *1062*
Kanehira, K., 1034(15), *1060*, 1163(298), *1188*
Kaneko, H., 271(100), *316*
Kaneto, K., 1357(94), *1362*, 1384(107), 1385(107), *1395*
Kanetsuna, H., 1136(35), 1139(52, 53), *1180*, 1154(195), *1184*, 1156(220), 1157(220), *1185*, 1166(330), *1189*, 1166(220), *1185*
Kanevskii, L. S., 247(141), *317*, 1010(233), 1011(237), 1012(241), *1026*
Kanno, T., 1406(31), *1416*, 1409(92), *1418*
Kannuck, R. M., 282(204), 285(204), *319*
Kano, H., 1175(483), *1193*
Kano, K., 1040(63), *1062*, 1058(249), *1068*
Kanzaki, Y., 174(206), *202*, 1253(203), *1263*
Kao, W. H., 1435(26), *1438*
Kaplan, M. L., 680(117), 681(117), 682(117), 683(117), *695*
Kaplan, R. B., 1088(138), *1099*
Kapovits, I., 676(99), *695*
Kapp, D. L., 516(89), 519(89), *531*, 519(109), *532*, 649(234), *657*, 1408(37), *1416*
Karabinas, P., 646(213), *656*
Karady, S., 1051(180), *1066*
Karakatsanis, C. G., 299(356), *325*
Karapetyan, K. G., 1010(233), 1011(237), *1026*
Karchmer, J. H., 1054(205), *1067*
Kardash, I. E., 491(93), *499*
Kardos, A. M., 469(46), *480*, 495(124), *500*, 1033(8c, 9), 1050(8c), *1060*
Kargin, Yu. M., 287(244), 305(244), *321*, 435(40–42), 436(40), 441(40), *457*, 441(141), *460*, 454(208d–f, g), 453(208e, f), *462*, 473(63), *480*, 488(45), *497*, 797(66), *806*, 1055(223), *1067*, 1122(77), *1127*
Kargina, N. M., 452(208d–f, g), 453(208e, f), *462*
Kariv, E., 236(22), *246*, 436(60), 443(60), *458*, 441(126), 444(126), *460*, 441(149c, 150), *461*, 895(98c), *937*, 898(125), 906(125), *938*, 915(256), *943*, 929(337), *946*, 1050(172), *1066*, 1146(136), *1183*, 1171(378, 379, 383), *1190*, 1172(415), *1191*, 1365(13), 1370(13), *1392*
Kariv-Miller, E., 239(41), *247*, 380(129), 381(129), *398*, 722(181), *754*, 1200(7–9, 11, 14), *1211*, 1202(43), 1203(45–48), 1204(57), *1213*, 1207(83), *1214*
Karmanov, A. P., 792(26), *805*
Kariyama, K., 1380(67), 1383(67), 1384(67), 1385(67), *1394*
Karrenbrock, F., 367(33), *395*, 890(43), 891(46), *934*
Karube, I., 234(10), *246*
Kasafirek, E., 1050(167), 1053(167), *1065*
Kash, H., 778(57), 779(58, 59), *786*
Kashimura, S., 404(27), *427*, 493(112), 494(112), *500*, 589(40), 590(40), *608*, 726(244), 728(244), *757*, 740(359), *761*, 779(61b), *786*, 1000(172), *1024*, 1041(72), 1042(73, 75), *1063*, 1150(167), 1151(172) *1184*, 1155(212), *1185*, 1268(24a, b, 27), 1275(24b, 27), *1281*, 1425(13), *1430*
Kashin, A. N., 863(329), *875*
Kashirskii, I. M., 794(54), 795(54), *806*
Kashti-Kaplan, S., 724(211b), *755*
Kasparek, F., 477(99), *482*
Kass, M. B., 722(173), *754*
Kassen, K. M., 905(217), *941*
Kastening, B., 190(272), *204*, 195(296), *205*, 422(151), *431*, 435(30), *457*, 467(20), *479*, 891(52a, d), 931(52a), *934*, 1426(14), *1430*
Kaszubska, J., 1208(100), *1214*
Kata, H., 1135(24, 25), *1179*
Kataoka, M., 50(67a), *117*, 597(103), *611*
Katayama, A., 442(100), 445(100), *459*, 445(167), *461*, 894(85c), *936*
Katayama, S., 721(160), *753*
Katayama, Y., 851(268), *873*
Katch, S., 708(51), *750*
Katchalsky, A., 1201(34), *1212*
Katiyar, S. S., 453(221), 455(221), *463*
Kato, M., 853(271), *873*
Kato, S., 468(37), 470(50), *480*, 725(219), *756*, 926(316), *946*, 1006(215), *1025*
Kato, T., 740(355), *761*
Kato, Y., 1073(36), *1096*, 1141(66), *1181*
Katoaka, M., 983(127), *1022*
Katoh, S., 603(156), *613*, 850(260), *872*, 1167(333), *1189*
Katoh, T., 404(28), *427*, 850(261), *873*, 1039(47), *1062*
Katritzky, A. R., 739(347), *760*
Katz, M., 518(104), 523(128), *532*, 648(231), *657*, 983(130), *1022*, 1086(123), *1099*
Katz, T. J., 333(23), *353*
Katzen, R., 790(13), *804*
Katzer, K., 557(113), *577*

Kau, H., 793(40), *805*
Kaufman, F. B., 411(78a), *429*, 467(19), *479*, 683(128), *696*, 709(546), *750*, 1372(41, 42), 1373(42), *1393*
Kaufman, L., 930(349d), *947*, 1206(71), *1213*, 1434(12b), *1437*
Kaufman, V., 1200(17), *1212*
Kaufmann, H., 446(175), 447(175), *462*, 898(106), 900(106), *937*
Kaulen, H., 277(153), *318*
Kaulen, J., 641(174), 642(174), *655*
Kaulfer, F., 898(129), *938*
Kaurova, G. I., 282(199), 285(199), 290(199), *319*
Kawabata, N., 643(187), *655*, 667(57), *693*, 716(113), *752*, 850(258, 259), *872*, 1243(136, 143), 1245(149, 150), *1261*
Kawada, M., 537(6, 8, 9, 10), 572(8), 573(6, 10), *574*, 679(106, 107, 109, 110), *695*, 1046(123), *1064*
Kawaguchi, T., 1044(91), *1063*
Kawai, M., 1408(41), *1416*
Kawai, S., 1408(74, 77), 1411(74), *1418*
Kawai, T., 1408(41–43), *1416*, 1408(57), *1417*, 1408(74, 77, 85), 1409(96), 1411(74), *1418*
Kawakami, H., 794(52), 795(52), *806*, 1408(53, 54), *1417*
Kawakami, S., 903(199), 907(192), *940*, 919(289), *944*, 1139(54), *1180*
Kawamata, M., 1201(36), *1212*, 1204(60), *1213*
Kawamata, T., 565(139, 140), *578*
Kawamura, C., 1316(107), 1319(107), 1329(107), *1335*
Kawamura, G., 640(150), *654*
Kawana, O., 1162(285), *1187*
Kawanisi, M., 570(160–163), *579*, 782(71), *786*, 1159(238), *1186*
Kawasaki, K., 675(93), *694*
Kawase, T., 682(124), *695*
Kawasumi, M., 542(49), 548(76), *576*
Kawazoe, Y., 742(383), *763*
Kay, E., 1368(33), *1393*
Kaye, R. C., 726(237), *756*
Kaykut, G., 1413(134), *1420*
Kazakos, A., 1144(97), *1182*
Kazakov, V. Ya., 284(233), *320*, 1109(36), *1125*
Kazakova, A. A., 1122(77), *1127*
Kazarinov, V. E., 267(85), 268(85), 271(85), 307(85), *315*, 539(22), *575*
Kazintseva, T. A., 1204(59), *1213*
Ke, B., 725(223), *756*
Keating, J. T., 860(310), *874*
Keating, K. B., 1314(91), *1335*
Keats, G. H., 449(199), *462*
Keck, G. E., 1210(122), *1215*
Kedrunskii, I. A., 1433(7), *1436*
Keenan, A. G., 1358(111), *1362*
Keene, F. R., 834(170), *869*
Keese, R., 1155(210), *1185*
Keinan, E., 242(58), *248*
Keiper, W., 714(93), *751*
Keita, B., 1110(42), *1125*
Keith, H., 17(15b), *115*
Keller, R., 1314(95), *1335*
Kellett, P. M., 292(302), 293(302), *323*
Kelly, F. M. L., 908(223), *941*
Kelly, M. J., 297(338), *324*, 616(9), 618(9), *650*, 962(46), *1019*
Kelly, R. C., 537(13), 562(13), 564(13), 565(13), *575*
Kelly, R. L., 846(241), *872*
Kelly, R. S., 841(224), 843(224), *871*
Kemula, W., 422(149), *431*, 434(8), 435(8), *456*, 638(130), *654*
Kenin, V. G., 793(38), *805*
Kennedy, D., 1233(90), 1235(90), *1259*
Kenney, G. A., 1253(200), *1263*
Kerman, H., 898(131), *938*
Kern, D. H., 897(110), 899(146), 901(146), *937*, *938*
Kern, J.-M., 831(155), *869*, 908(220), *941*
Kern, W., 921(295), *945*
Kerns, C., 417(117), *430*
Kerr, I. S., 193(285), *205*
Kerr, J. B., 296(329), *324*, 392(201), 393(201), *400*, 469(42), *480*, 704(17), 708(17), 771(29), *748*, 829(144), 830(147), *869*, 1148(156), *1183*, 1268(21), 1275(21), *1281*, 1269(34), *1281*, 1277(49), *1282*
Kerrinnes, H.-J., 834(171), *869*
Kessel, S., 557(111), *577*
Kessler, S., 1207(88), *1214*
Kestner, T. A., 1106(27), 1108(27), *1124*
Keszthelyi, C. P., 148(109), *198*, 882(3b), 885(3b), *932*
Keumi, T., 527(142), *533*
Keys, D. E., 687(177), *697*
Khalaf, A. A., 906(27), *1018*
Khalifa, M. H., 621(59), 624(59), *651*, 621(61), *652*, 638(129), *654*, 1050(165, 166), *1065*
Khan, M. A., 563(131), *578*, 837(191, 194), 862(191), *870*, 862(318, 319), *874*
Khan, U. M., 1287(14), *1332*
Kharashvil, R. L., 284(241), *321*
Khavtasi, N. S., 284(241), *321*, 726(242), *757*
Khazova, O. A., 639(136, 137), *654*, 644(198), 645(198), *656*
Khinova, M., 1000(173), *1024*
Khmel'nitskaya, E. Yu., 709(57), *750*
Khodair, A. I. A., 659(3), *692*
Khomchenko, T. N., 724(209), *755*, 886(19), *933*
Khomutov, N. E., 438(83), *459*, 725(229), *756*
Khomyakov, V. G., 273(119), *317*, 436(74), 438(74), *458*, 445(169), *461*, 739(339), *760*, 898(101, 103), 897(101), *937*
Khoo, S. B., 182(245), 184(255), *203*, 346(157b), *358*, 514(80), 515(80), *531*, 1268(26), 1275(26), *1281*
Khrishnan, V., 496(137), *501*
Khullar, M., 796(60), *806*
Kiesel, Y., 821(99), *867*
Kiesele, H., 295(319), 296(325), *323*, 333(30), 349(30), *353*, 335(49), 350(49), *354*, 344(118), 351(118), *356*, 346(154), 350(154), *358*, 352(210), *359*, 726(238), *756*
Kiesslich, G., 682(122), *695*
Kihara, S., 1147(148), *1183*
Kikuchi, K., 1434(12c, 15j), *1437*

Kikucki, Y., 490(87), *499*, 725(221b), *756*, 1352(56), *1361*
Kilduff, J. E., 411(77), *428*
Killer, R., 288(259), *321*
Kilpatrick, M., 290(276), 290(279), *322*
Kim, H. J., 1073(36), *1096*
Kim, J.-Y., 293(313), *323*
Kim, K., 686(159, 167, 168, 170), *697*
Kim, M. H., 122(13), 163(13), *195*
Kimball, G., 445(162), *461*
Kimla, A., 1286(5), 1292(5), *1332*
Kimmerle, F. M., 891(58), *935*
Kimura, K., 543(50), *576*
Kimura, M., 667(58), *693*, 678(102), 679(111), *695*, 1046(124), 1047(135), *1064*, 1079(67), *1097*, 1167(337, 339), *1189*
King, C. J. H., 259(33), 263(33), *313*, 916(263), *943*, 1232(77a), 1234(77a), 1288(28), *1333*, 1301(61), 1303(69), 1305(77), 1307(78), *1334*
King, P. F., 1094(183), *1101*, 1117(60, 62), 1118(60), 1119(62), *1126*
Kinlen, P. J., 122(11), *195*, 148(11), *198*, 488(51), *498*
Kinoshita, H., 488(44), *497*, 1173(419), *1191*
Kinoshita, I., 838(207), *871*
Kinsey, S. J., 725(227), *756*
Kintopf, S., 336(58), *354*, 848(244), *872*
Kipnes, S. M., 309(423), 286(423), *327*
Kirchhof, G., 445(164), *461*
Kirk, D. W., 621(68), 625(68), *652*
Kirkland, S., 284(238), *320*
Kirillova, A. S., 435(27), *457*
Kirilyus, I. V., 445(156), *461*
Kirowa-Eisner, E., 724(211b), *755*
Kirrman, A., 387(153), *398*
Kirst, W., 273(117), *316*, 442(113), 449(113), *460*
Kirste, B., 1207(82), *1214*
Kisch, H., 1408(87, 89), *1418*
Kise, N., 367(35, 36), *395*, 433(2), *456*, 716(114), *752*, 850(260), *872*, 891(44, 45), *934*, 891(44, 45), *934*, 891(57), *935*, 904(210), 918(210), *941*, 1139(57), 1140(57), *1180*, 1155(211), *1185*, 1173(440), *1192*
Kiselev, A. V., 1410(116a), *1419*
Kishi, K., 403(14), *426*
Kishore, N., 418(132), *430*
Kiso, Y., 832(160), *869*
Kissinger, P. T., 13(1), *114*, 155(139), 156(139), *199*, 179(229), *202*, 685(142), *696*
Kistenbrügger, L., 488(52), 490(70), *498*, 1033(12), *1060*, 1150(171), *1184*, 1268(22), 1275(22), *1281*
Kisza, A., 303(396), *326*
Kita, H., 442(100), 445(100), *459*, 445(167), *461*, 894(85c), *936*, 1141(64), *1181*, 1175(483), *1193*
Kitabayashi, H., 1090(159), *1100*
Kitaev, Yu. P., 466(4, 7), 473(4), *479*, 471(52, 54, 55), 473(61, 62), 474(62), *480*, 473(64–67, 69), 474(77), *481*, 476(93), *482*, 495(122), *500*, 729(264), *757*, 915(252, 253), *943*, 1055(218), *1067*
Kitaeva, M. Yu., 473(63), *480*, 1055(223), *1067*
Kitagawa, I., 556(108, 109), *577*, 560(123), *578*
Kitagawa, T., 411(82), *429*
Kitahara, S., 379(128), *398*
Kitami, M., 725(221a, b), *756*
Kitani, A., 646(219), *657*, 1383(94), 1385(94), *1395*
Kitazawa, M., 601(143), 602(143), *612*, 771(28), *784*, 1094(173), *1100*
Kitchen, M. D., 846(241), *872*
Kitching, W., 284(225), 301(225), 302(225), *370*, 855(276), 862(276), *873*, 863(328), *875*, 930(341), *946*, 1169(354), *1190*, 1177(509), *1194*
Kito, T., 1174(446), *1192*
Kitoh, S., 891(57), *935*
Kitsuya, Y., 563(129), *578*
Kivalo, P., 736(324d), *759*
Kiwan, A. M., 736(324e), *759*
Kizumoto, H., 1408(53, 65), *1417*
Klages, C. P., 1033(12), *1060*
Klar, G., 685(138), *696*
Klasinc, L., 24(29), 72(29), *115*, 506(18), 508(18), 509(18), 511(18), *529*
Klass, D. L., 1232(78), 1234(78), *1258*
Klehr, M., 1033(13), *1060*
Klein, A. J., 368(47), *395*, 702(2), *747*, 894(76), *936*, 903(202), *940*, 1144(127), 1145(127), *1182*, 1176(499), *1194*
Klein, W. R., 790(13), *804*
Kleine, H. J., 793(35), *805*
Kleinberg, J., 1146(142), *1183*
Klem, R., 565(142), *578*
Klemer, A., 1164(305), 1165(305), *1188*
Klemm, L. H., 417(124), 418(124), *430*, 892(64), *935*, 918(272), 921(272), *944*, 1135(22), *1179*, 1136(32), *1180*
Klima, J., 724(202), *755*
Klima, K., 485(13), *496*
Klimov, V. A., 263(65), *314*, 1201(30), *1212*
Klimova, T. A., 378(124), *397*, 403(12), *426*, 443(119), *460*
Klinger, J., 179(228), *202*
Klingler, R. J., 67(85b), 72(85b), *118*, 833(165), *869*, 843(227), *871*, 846(237), *872*, 852(270), 853(270), *873*, 1241(123), *1260*
Klink, F. W., 1105(18, 25), 1109(18, 25), 1110(18, 25), 1112(25), *1124*, 1109(39), *1125*
Klopman, G., 373(82), *396*
Klug, J. T., 600(123), *611*, 668(59), *693*, 668(64), *694*, 1178(519, 520), *1195*
Klünenberg, H., 546(71), *576*, 601(143), 602(143), *612*, 704(23), *749*, 995(157), 996(159), *1023*, 1002(198), *1025*
Klyuev, B. L., 263(64), *314*, 435(49), 436(9), 894(84a), 896(84a), 903(84a), *936*, 902(170, 176), 906(177), 907(177), 922(170), *939*, 907(184, 186, 188), *940*
Kmel'nitskaya, E. Y., 894(75), *935*
Knapsack-Grieshiem, A.-G., 1204(51), 1207(51), *1213*

Kneip, T. J., 422(154), 423(154), *431*, 1055(219), *1067*
Knittel, D., 426(195, 196), *432*, 685(138), *696*, 891(52a, d), 931(52a), *934*, 1056(226, 227), *1067*, (164), *1183*, 1426(14), *1430*
Knolle, J., 547(72, 73), *576*, 783(76), *786*, 1002(191b), *1024*, 1002(196, 197), 1004(209), *1025*
Knopp, J. A., 291(295), *322*
Knunyants, I. L., 551(100), *577*, 908(163), *939*, 1117(53, 54), *1125*, 1117(57, 58a), *1126*, 1119(53, 54), *1125*, 1119(57, 67a–c), 1121(72), 1122(72), *1126*, 1123(80), *1127*, 1201(22, 23), *1212*
Kobayashi, H., 240(52), 242(52), *248*
Kobayashi, K., 365(15), 376(15), *394*
Kobayashi, S., 686(146), 687(146), *696*
Kobayashi, T., 1383(96–98), 1386(97, 98), 1387(98), *1395*
Kobayashi, Y., 537(10), 573(10), *574*, 573(166), *579*, 1162(282), *1187*
Kobe, K. A., 486(30), *497*
Kobylecki, R. J., 1143(88), *1181*
Koch, C. A., 819(85), *867*
Koch, D., 240(53), *248*, 374(86), *396*, 597(102), *611*, 648(228), *657*, 704(29), *749*, 858(302), *874*, 980(121), 983(126), *1022*, 1010(231), *1026*
Koch, R. W., 1148(152), *1183*
Koch, V. R., 515(84), *531*, 517(102), *532*, 794(55), 796(55), *806*, 1085(117, 118), 1086(117), *1099*, 1117(56), *1126*
Kochanek, W., 1434(19c, 20c), *1438*
Kochetkova, N. S., 1206(66), *1213*
Kochi, J. K., 15(10c), 16(10c), 18(10c), *114*, 25(29), *115*, 30(36a–c, 37a, b), 31(36b–c), 34(47), *116*, 47(36), 50(36), 51(36), 52(36), *117*, 67(85b), 72(85b), *118*, 233(4a), 234(4a), 241(4a), *245*, 243(65), *248*, 424(168), *431*, 506(18, 19), 508(18, 19), 509(18, 19), 511(18, 19), 513(19), 517(19), *529*, 511(74), 513(74), *531*, 809(13), *864*, 833(165, 166), *869*, 843(227), *871*, 845(235), 846(237), *872*, 852(270), 853(270), *873*, 1168(342), *1189*, 1241(123), *1260*
Kodaira, K., 1106(28, 31), 1107(31), *1125*
Kodaka, M., 1434(12f), *1437*
Kodama, S., 832(160), *869*
Kodintseva, E. Yu., 470(49), 476(49), *480*
Kodoma, T., 302(382), *325*
Koehl, W. J., Jr., 260(36a), 263(36a), *314*, 569(151), *579*, 1082(97), *1098*
Koenig, T., 517(102), *532*, 1085(118), *1099*
Koepp, H.-M., 278(166), *318*
Koeppe, C. E., 193(287), *205*
Koerber, G. G., 1105(22), *1124*
Kohmoto, T., 1278(57b), 1280(57b), *1282*
Kohmoto, Y., 1164(313), *1188*, 1243(135), 1244(135), *1261*
Kohn, H., 1148(158), *1183*
Koie, K., 678(102), *695*, 1079(67), *1097*
Koizumi, T., 907(194), *940*
Kojima, H., 64(80), *118*, 1175(450), *1192*, 1175(455), *1193*, 1240(120), *1260*
Kojima, I., 1204(53), *1213*
Kojima, M., 982(113), *1022*
Kok, G. B., 680(118), 681(118), 682(118), *695*
Kok, G. L., 809(10), 837(10), 838(10), *864*
Kokarev, G. A., 1230(56), *1258*, 1299(53), 1304(53), *1334*
Koketsu, N., 815(57), *866*
Kokkinidis, G., 278(162), *318*, 640(157), *655*
Kolb, D. M., 179(228), *202*
Kolb, K. E., 721(149), *753*
Kolbe, H., 535(1), *574*
Kolfeubach, J. J., 453(218), *463*
Kolodny, N. H., 1136(34), *1180*, 1175(485), *1193*
Kologova, N. E., 846(236), *872*
Kolthoff, I. M., 157(143), 161(143), *199*, 256(1, 3a), 309(1), 311(3), *312*, 276(139), *317*, 278(169), *318*, 285(230), *320*, 299(354), *324*, 454(224, 225), *463*, 484(7, 8), *496*, 660(9), 662(9, 25, 26), *692*, 1043(83), *1063*, 1247(171), *1262*, 1348(31), *1360*
Kolyagin, G. A., 1433(7), *1436*
Komaki, R., 994(148), *1023*
Komarek, V., 594(82), *610*
Komeichi, Y., 1154(197), *1185*
Komori, R., 1316(101), 1319(101), 1326(101), *1335*
Komori, T., 675(92), *694*, 1173(425, 426, 434, 435), 1174(444), *1192*
Konai, Y., 819(84), *867*, 1372(39), *1393*
Kondo, K., 770(25), *784*
Kondo, T., 894(77a), *936*
Kondranina, V. Z., 435(40–42), 436(40), 441(40), *457*, 441(141), *460*
König, P., 640(159, 160), *655*
Königshofen, K., 339(75), 340(75), *355*
Königstein, J., 437(134), *460*
Konoshita, H., 272(114), 275(114), *316*
Konrad, P., 1243(134), *1261*
Konse, T., 1040(63), *1062*
Konz, E., 630(97, 98), *653*
Kopetelova, M. M., 257(18a), *313*
Kopilov, J., 236(22), *246*, 441(126), 444(126), *460*, 441(149c), *461*, 895(98c), *937*, 929(337), *946*, 1171(382, 383), *1190*
Koppang, M. D., 1038(41), *1061*
Köppel, H., 729(269), *758*
Koppert, K., 416(112), *430*
Koranyi, G., 486(35), *497*
Körblová, E., 1162(281), *1187*
Korczynski, A., 894(77d), *936*, 898(120), *937*, 1208(94), *1214*
Korienko, V. L., 1433(7), *1436*
Korinek, K., 227(151), *318*, 583(7), *607*, 641(177–179), 642(177), *655*
Kormann, C., 1413(133), *1419*
Korn, S. R., 288(251), 292(251), *321*
Kornblum, N., 1206(70), 1210(70), *1213*, 1237(110), *1260*

Kornienko, A. G., 552(81), *577*, 1010(235), *1026*
Kornrumpf, W., 1154(200), *1185*
Korotaeva, L. M., 1144(123), 1145(123), *1182*
Korotkov, A. P., 895(94), *937*
Korotoava, L. M., 488(49, 50), 492(49), *498*, 895(96), 897(119), *937*
Korovina, O. V., 797(67), *806*
Korshunov, I. A., 435(27), *457*, 484(6), *496*, 617(10), *650*
Korsun, A. D., 434(13), *456*
Kortar, D., 256(7), *312*, 1287(20), 1291(20), *1332*
Koruncev, D., 271(103), *316*, 742(384), *762*
Koryta, J., 646(212), *656*, 660(11, 12), *692*
Korytseva, V. F., 797(66), *806*
Koryttsev, N. K., 425(194), *432*
Korzeniewski, C., 182(245), *203*, 296(330), *324*
Kos, J., 689(183), *697*, 722(168), *754*, 1167(334), *1189*
Kosaka, T., 1079(72), 1080(72), 1082(72), *1097*, 1160(250), 1161(250), *1186*
Koski, W. S., 334(35), *354*, 488(48), 491(48), 494(48), *498*, 510(50), *530*
Kossai, R., 585(24), *608*, 722(174), *754*, 1058(252, 254), *1068*, 1226(43, 44), 1248(43), *1257*, 1241(122), *1260*
Kossanyi, J., 407(52), *428*, 721(159), 722(159), *753*
Köster, K., 1171(384, 386, 387), *1190*, 1172(399), *1191*
Kostyanovskii, R. G., 411(80), *429*
Kosychenkov, L. I., 722(166), *754*, 1221(17), 1224(17), *1256*
Kotani, E., 601(143), 602(143), *612*, 704(18), *748*, 704(24), *749*, 771(27a, b, 28), *784*, 994(155a), 1000(168, 171), *1024*, 1094(173), *1100*
Kotenko, N. P., 797(62, 64), *806*
Kothe, G., 340(82a), *355*
Kotok, L. A., 726(241), *757*
Kotzyba-Hilbert, F., 862(322), *875*
Koutecky, J., 163(158), *200*, 1222(24), *1257*
Kovac, N., 438(72), *458*
Kovacs-Mindler, V., 734(311), *759*
Koval, C., 819(84), *867* 1371(37), 1372(37, 39), *1393*
Kovalenko, E. I., 797(62–64, 67), *806*
Kovalenko, N. A., 797(63), *806*
Kovalev, G. D., 793(39), *805*
Kovaleva, A. V., 1202(42), *1213*
Kovel'chuk, E. P., 424(172), *432*
Kovelesky, A. C., 687(176), *697*
Kovsman, E. P., 247(141), *317*, 540(33), 541(33), *575*, 1010(233), 1012(241), *1026*, 1316(108), 1319(108), 1329(108), *1335*
Kowak, R., 1366(26), 1368(26), 1369(26), 1371(26), *1393*
Koyama, K., 528(143), *533*, 1080(78), *1097*, 1084(112), *1098*, 1089(145, 147, 148), 1090(148), *1100*, 1157(223), *1185*, 1160(249), 1161(249), *1186*, 1161(267, 269), *1187*
Kozhevnikova, L. G., 794(45), *805*
Kozlovskaya, E. V., 715(102), *751*
Krab, K., 813(26), *865*
Kraft, L., 15(10a), *114*
Kraft, T. E., 635(126), 636(126), *654*
Kramer, K., 1230(60), *1258*
Kramer, M. K., 388(162), *399*
Kramer, S. R., 1372(42), 1373(42), *1393*
Krapcho, A. P., 1254(208), *1263*
Krashnikova, G. S., 551(100), *577*
Kraske, P. J., 747(429), *763*
Kratky, C., 813(32), *865*
Kratochvil, B., 280(186), *319*, 305(401), *326*
Kratzing, C. C., 725(224), *756*
Krauch, H., 1049(157), *1065*
Krause, L. J., 1200(13), *1211*
Kräutler, B., 812(24), *864*, 813(32), *865*, 824(111), *868*, 985(131), *1022*, 1408(78, 79), 1413(78), *1418*
Krawczyk, A. R., 1049(158b), *1065*
Krayushkin, M. M., 378(124), *397*, 403(12), *426*, 443(119), *460*
Kreh, R. P., 1081(93), *1098*, 1229(69, 70), *1258*
Kreider, L. W., 280(187), 284(187), 285(187), 297(187), *319*
Kreshkov, A. P., 484(10), *496*
Kretlow, W. J., 645(211), 646(211), *656*
Kretzschmar, K., 737(329), *760*
Kreysa, G., 268(92), *316*, 1229(71), *1258*, 1286(8), 1292(8), 1299(8), *1332*, 1434(19a, c), 1434(19a, c, 20c, 22d), *1438*
Kreysig, D., 726(239), *756*, 926(138), *938*, 1038(35), *1061*
Krieger, R. L., 365(17), 388(17), 376(17), *394*
Krikorian, D., 704(26), *749*, 1000(173), *1024*
Krishnamurthy, G. S., 414(102b), *429*
Krishnan, V., 745(413), *763*
Kristensen, L. H., 724(212), *755*
Kristian, P., 469(46), *480*
Kröber, H., 983(128), *1022*
Kroening, W., 1243(134), *1261*, 1316(105), 1319(105), *1335*
Krohn, K., 240(54), *248*
Kröhnke, C., 511(68), *530*
Kronberger, K., 1136(34), *1180*, 1175(485), *1193*
Kroon, J. L., 311(439), *327*
Kropotova, N. V., 435(55), 436(55), 442(101), 447(101), *458*, *459*
Krounbi, M., 1394(65), 1387(65), *1394*
Kruglikov, S. S., 739(339), *760*
Krupicka, J., 376(109), *397*, 1176(496), *1194*
Krygowski, T. M., 162(151), *200*, 422(150), *431*
Kryshchenko, K. I., 1299(53), 1304(53), *1334*
Kryukov, V. G., 438(73), 446(73), *458*

Kryukova, E. V., 895(90), *936*, 898(132), 901(154), *938*, 1201(30), *1212*
Ksander, G., 812(22), *864*
Kubala, J., 1232(75), *1258*
Kubicek, R., 745(411), *762*, 1033(8b), 1050(8b), *1060*
Kubota, M., 1077(51), *1096*, 1135(24, 25), *1179*
Kubota, T., 476(95), 477(95), *482*, 542(49), 548(76), *576*, 597(108), *611*, 602(145), *612*, 742(380), *761*, 1040(63), *1062*, 1058(249), *1068*, 1366(27), *1393*
Kucar, S., 1232(75), *1258*
Kucera, F., 1109(39), *1125*
Kuczynski, J., 1404(20), *1416*
Kuder, J. E., 467(25), *479*
Kudinova, L. I., 455(232), *463*, 915(257), *943*
Kudo, T., 640(150), *654*
Kudrna, S. K., 440(98), 447(98), *459*, 897(115), *937*
Kudryavtsev, R. V., 1117(53), *1125*, 1119(53), *1125*
Kudryavtseva, N. A., 742(378), *761*, 743(389), *762*
Kuempel, J. R., 280(187), 284(187), 285(187), 297(187), *319*
Kugatova-Shemyakina, M. M., 1176(505), *1194*
Kuhn, A. T., 290(283), 291(283), *322*, 1071(20), 1073(20), *1096*, 1105(16), 1109(16), 1110(16), 1112(16), *1124*, 1144(119), *1182*, 1230(54), *1258*, 1243(144), *1261*, 1287(17), 1288(17), *1332*, 1288(33), 1290(38), *1333*, 1295(49), *1334*, 1421(3a), *1429*
Kuhn, R., 1200(19), *1212*
Kuhr, W. G., 55(70), *117*
Kuleshova, N. D., 1117(54), *1125*, 1117(58b), *1126*, 1119(54), *1125*
Kulka, M., 726(235), *756*
Kulkarni, C. L., 723(189), *754*, 723(193), *755*, 973(91), 979(88), *1021*
Kul'nevich, V. G., 1202(41), *1212*, 1202(42), *1213*
Kulys, J., 646(217), *657*, 725(228), *756*
Kumada, M., 832(160), *869*
Kumai, Y., 1172(400), *1191*
Kuman, N., 1341(12), *1359*
Kumar, N., 837(192, 193), *870*
Kumar, R., 837(197), *870*
Kunai, A., 961(31), *1019*
Kunimatsu, K., 639(140, 141), 640(152), *654*
Kuniwaki, Y., 1034(17), *1061*
Küntzel, H., 302(381b), *325*, 861(317), *874*, 1009(230), *1026*
Kunugi, A., 1045(119, 120), *1064*
Kunz, W., 1049(157), *1065*
Kuo, K., 683(129), *696*, 1366(26, 28), 1368(26), 1369(26), 1371(26), *1393*
Kuo, Y.-M., 831(159), 832(159, 161), *869*
Kupchan, S. M., 889(37), *934*
Kurek, E., 734(305), *759*
Kuriacose, J. C., 1408(69), *1417*
Kurihara, H., 1253(204), 1255(204), *1263*
Kuriki, N., 667(58), *693*, 1167(337), *1189*
Kurimoto, I., 1163(301), *1188*
Kuroiwa, S., 487(38), *497*
Kurreck, H., 1207(82), *1214*
Kurz, J. G., 31(41d), *116*
Kusakabe, K., 1073(36), *1096*
Kusayanagi, Y., 1142(83), 1143(84, 85), 1173(83, 84), *1181*
Kusumoto, M., 561(124), *578*
Kusumoto, Y., 537(8), 572(8), *574*, 679(109), *695*
Kuta, J., 155(140, 141), 158(140, 141), 161(140, 141), *199*, 256(2), *312*, 278(174), 282(174), *319*, 837(201), *870*
Kutner, W., 834(168), *869*
Kuwahara, S., 783(75), *786*
Kuwana, T., 179(216, 217), 181(217), *202*, 182(243, 244), 183(244, 246), *203*, 188(261, 266, 268), *204*, 270(96), 277(96), *316*, 646(215), *657*, 960(21), *1018*, 1248(177), *1262*, 1358(108), *1362*, 1365(16, 17), 1366(17), 1370(16, 17), 1371(17), *1392*, 1409(114), *1419*, 1435(26), *1438*
Kuwata, F., 726(244), 728(244), *757*
Kuwata, K., 597(98), *611*
Kuz'min, O. V., 474(73), *481*, 886(19), *933*, 1269(36), *1281*
Kuznetsov, A. M., 32(44b–e), 63(44), *116*, 39(55f), *117*, 342(103), *356*
Kuznetsova, Z. B., 435(27), *457*, 484(6), *496*
Kozovenko, S. V., 287(244), 305(244), *321*
Kwak, S., 410(74), *428*, 583(10), *607*
Kwee, S., 408(53), *428*, 423(162), *431*, 709(59), 710(59), 736(59), *750*, 714(94), 735(94), *751*, 737(330–332), *760*, 741(369), *761*, 745(421), *763*, 1056(228, 229), *1067*, 1173(437), *1192*
Kyntoku, H., 1041(72), *1063*
Kyriacou, D., 389(171, 174), *399*, 747(434, 435), *763*
Kyrides, L. P., 437(109), 442(109), *459*
Kyutoku, H., 1151(172), *1184*

Labes, M. M., 736(322), *759*
Lacan, M., 441(147), *461*, 596(94–96), *610*, 706(36a), 710(36a), 712(36a), 714(36a), *749*, 710(60a, c), 715(60a, c), *750*, 714(99a), *751*, 974(94b), *1021*, 1005(212d), 1006(212d), *1025*
Lacaze, P. C., 262(56), 263(56), *314*, 298(343), 286(343), *324*, 302(388, 390), *326*, 640(163), *655*, 925(309c–e), *945*, 960(22), *1018*, 1050(159d), *1065*
Lachenal, D., 661(19), *692*
Lacoste, G., 1243(145), *1261*
Lacoste, R. J., 367(28), *395*
Lacour, M., 456(244, 245), *464*, 915(261), 916(261), *943*
Lácová, M., 1144(98), *1182*
LaCroix, J. C., 1379(136), *1396*, 1383(99), 1385(99), 1386(99), 1387(99), 1389(99), *1395*
Ladygin, A., 471(52), *480*
Lae, H. D., 897(109), *937*
Lagendijk, A., 350(198), *359*
Lagowski, J. J., 280(190), 284(190), *319*, 289(264), *321*
Lagowski, J. M., 739(347), *760*
Lagu, A., 346(153a), 350(153a), *358*

Laguna, M. A., 726(245), *757*
Lahib, G. H., 1209(114), *1215*
Lahiry, S., 1409(108), *1419*
Lai, T. T., 885(11), *932*
Laidler, J. K., 712(68), *750*
Lain, M. J., 441(152c), *461*, 1435(27), *1438*
Laitinen, H. A., 284(205), 285(205), 286(205), 293(205), 295(205), *319*, 292(298), 293(298), 296(298), *323*, 332(11a), 345(11), *353*, 422(154), 423(154), *431*, 1055(219), *1067*
Lalithambika, M., 453(222), *463*
LaLonde, R. T., 569(150), *579*, 1159(235), 1170(235), *1186*
Lalvani, S. B., 1434(20b), *1438*
Lambert, F. L., 365(15, 20), 376(15), *394*, 376(111), *397*, 1176(491), *1194*
Lambert, R. F., 239(41), *247*, 271(101), *316*, 1255(214), *1263*
Lament, M., 712(67), *750*
Lamm, B., 259(30), *313*, 307(411), *326*, 337(66a), *355*, 1045(118), *1064*, 1048(149, 150), 1049(153), *1065*, 1210(119), *1215*
Lamm, W., 973(87), 979(87), *1021*
Lamont, M., 437(132), *460*
Lamoureux, C., 267(80, 81), 271(81), 273(80, 81), 288(80, 81), *315*, 418(136), 421(147), *431*
Lamy, C., 639(140), 640(154), *654*, 640(161), *655*
Lamy, E., 882(3f), 885(3f), *932*, 919(274), *944*
Lancon, D., 816(70), *866*
Land, E. J., 617(14, 15), 618(24), *650*
Landaez-Machado, H. J., 267(78), 273(78), 288(78), *315*, 273(123f), *317*
Landau, U., 256(7), *312*, 267(77), *315*, 1287(20), 1291(20), *1332*
Landsmann, D. A., 909(233), *942*
Lane, G. A., 841(223), *871*, 1175(462), *1193*
Lane, R. F., 1368(35), *1393*
Langbein, U., 834(171), *869*
Langendijk, S. L., 442(108), 447(108), *459*, 450(201), *462*
Langer, I., 890(43), 891(46), *934*
Langer, S. H., 1435(32), *1438*
Langhus, D. L., 184(251), 185(251), *203*
Langushin, L. M., 1204(59), *1213*
LaPerriere, D. M., 297(334), *324*, 363(8, 9), 364(9), 366(9), 371(9), *394*, 370(74), 372(74), *396*
Lapicque, F., 418(135), *431*, 1231(62), 1234(62), *1258*, 1299(57), *1334*
Lapkowski, M., 184(249, 250), 186(250), *203*
Larcombe, B. E., 332(21), *353*, 1245(155), *1261*
Largeron, M., 724(202), *755*
Larkin, D., 724(207), *755*
Larscheid, M. E., 334(36), *354*, 337(69), *355*
Larsen, J. W., 1434(17), *1438*
Larson, E., 1054(207), *1067*
Larson, R. C., 284(206), 295(206), *320*
Larsson, B., 522(123), *532*, 593(72), *610*
Lasia, A., 162(152), *200*, 435(46), *457*, 895(95), *937*
Lasserre, F., 104(109), *119*
Latham, R. J., 275(138), *317*
Latimer, W. M., 24(27b), 42(27b), *115*
Lattman, R., 812(22), *864*
Latypova, V. Z., 473(63), *480*, 797(66), *806*, 1055(223), *1067*
Latypova-Kondranina, V. Z., 452(208g), *462*
Lau, A. N. K., 488(53), *498*, 1372(47), *1393*
Lau, W., 833(166), *869*
Laube, B. L., 584(21), *608*, 975(102), *1021*
Lauer, G., 150(121), *199*
Launay, J.-P., 733(294), *758*
Laurent, A., 1082(99), *1098*, 1086(128, 129), 1087(129), 1088(136), *1099*, 1121(73), *1126*
Laurent, E., 288(255), 292(255), *321*, 290(281), 291(281), *322*, 1082(99), *1098*, 1086(128, 129), 1087(129), 1088(136), *1099*, 1090(162), *1100*, 1094(84), *1101*, 1120(69–71), 1121(73), 1122(69, 71, 74a–c, 75), *1126*, 1159(237), *1186*, 1163(302–304), 1164(302, 304), *1188*
Laurinovics, E., 441(139), *460*
Laursen, S. A., 127(62), *197*
Lauwzecha, A. B. H., 1053(198), *1066*
Lavachielli, M., 126(57), 135(57), *197*
Laval, J. M., 725(228), *756*, 1436(31b), *1438*
Lavalley, J. C., 1410(121a), *1419*
Lavie, D., 776(44a), *785*
Lavine, B. K., 180(242), *203*
Láviron, E., 78(101), *119*, 285(243), *321*, 350(194), *359*, 376(113), *397*, 388(165), *399*, 406(43), *427*, 424(174–176), 425(176), *432*, 713(81, 83–85, 87), *751*, 733(303), *759*, 743(392, 393), 744(397), *762*, 745(412), *763*, 840(219), *871*, 893(67), 926(317), *935*, *946*, 1176(492), *1194*, 1177(508), *1194*, 1343(15), *1359*
Lavrishcheva, L. N., 914(250), 915(250), *943*, 1208(102–104), *1214*
Law, H. D., 439(95), 445(95), *459*, 467(16), *479*, 895(107), 896(107), 897(107), 900(107), *937*, 1036(27), *1061*
Lawin, P. B., 1200(11, 14), *1211*
Lawless, J. G., 373(83), 374(83), *396*, 418(133), 419(133), *430*
Layloff, T., 411(82), *429*, 447(181), 450(180), *462*, 453(221), *463*, 490(82), *499*, 902(168a), *939*, 1146(143), *1183*
Lea, R. E., 1205(65), *1213*
Leach, S. C., 1434(15b, e, m), *1437*
Leach, S. J., 725(214), *755*
Lead, J. M., 340(78), *355*
Lebedev, A. V., 474(73), *481*
Le Berre, L., 963(60, 62–64), 967(60), *1020*

Lebl, M., 927(321), *946*
Le Bot, 537(16), 564(16), 565(16), *575*
Lebouc, A., 1046(129), *1064*, 1058(253), 1059(253), *1068*, 1136(33), *1180*, 1144(131), 1145(131), *1183*, 1152(176), *1184*, 1171(392, 395), 1172(399), *1191*, 1250(184), *1262*, 1278(57e), 1280(57e), *1282*
Lecat, J.–L., 857(283, 285), *873*, 1060(266), *1068*
Lecomte, C., 815(56), 817(74), *866*
Le Duc, J. A. M., 1287(21), 1289(34), *1333*, 1316(104), 1319(104), *1335*
Ledwith, A., 506(25), *529*, 973(90), *1021*, 1013(246), 1014(246), *1026*, 1412(127), *1419*
Lee, C. L., 335(46), *354*
Lee, F. H., 861(315), *874*
Lee, J., 1301(63), *1334*
Lee, J. B., 1141(63), *1181*, 1235(84), *1259*
Lee, K.–H., 295(320a), 296(320a), *323*
Lee, M.–T., 792(22), *804*, 838(206), *871*
Lee, P. I., 257(19), *313*
Lee, W.–K., 687(176, 177), *697*
Lee, W. Y., 1375(53), 1386(53), *1394*, 1386(137), *1396*
Leeds, M. W., 402(5), 403(10), *426*
Leedy, D. W., 494(116), *500*, 592(54), *609*, 621(75), 625(75), *652*, 733(300), *759*, 968(68), 970(68), *1020*
Lee-Ruff, E., 1245(158), *1261*
Leffler, H., 470(51), *480*, 491(95), *499*, 1147(145), *1183*
Lefranc, H., 1122(74), *1126*, 1163(304), 1164(304), *1188*
Leftin, J. H., 537(14), 564(14), 565(14), *575*
Leger, J. M., 640(161), *655*
Leger, R., 747(431), *763*
Legg, K. D., 1366(22), *1392*, 1366(25), *1393*, 1367(22), *1392*, 1367(25), *1393*
LeGoff, P., 268(90), *315*
Legrand, M., 1176(493), *1194*
LeGuillanto, G., 275(135), *317*, 369(63), *396*, 434(4), *456*, 437(132), *460*, 488(40), *497*, 678(104), *695*, 712(75), *751*, 913(286), *944*, 931(370, 371), *948*, 982(122), *1022*, 1136(33), *1180*, 1143(94), 1144(94), *1181*, 1160(252), *1186*, 1175(486, 487), *1194*
Le Guyader, M., 263(52), *314*, 415(106), *429*, 415(107, 110), *430*, 706(40), *749*, 709(56), 712(69), *750*, 1042(80), *1063*
Lehmann, C., 812(22), *864*
Lehmann, G. K., 239(41), *247*, 1203(46), *1213*
Lehmann, O., 441(146), *461*
Lehmkuhl, H., 336(58), *354*, 837(189, 196, 199), 838(199), 848(199), *870*, 848(244–246), 849(254), *872*, 1319(115), *1336*, 1353(77), *1361*
Lehn, J.–M., 819(86), *867*, 862(322), *875*
Lehnert, W., 565(143), *578*
Lehnsen, J. E., 674(84), *694*
Le Hoang, M. D., 826(125), *868*
Leibzon, V. N., 378(124), *397*, 403(12, 17), *426*, 443(119), *460*
Leicher, W., 237(27), *247*
Leidner, C. R., 818(78), 819(82), *866*
Leigh, G. J., 836(186–188), *870*, 1366(29), *1393*
Leland, J. K., 411(77), *428*
Lelandais, D., 542(47), *576*, 552(82), *577*, 565(141), *578*, 569(153, 154), *579*, 1002(190a), *1024*, 1011(237), *1026*, 1157(228), *1185*
Lemaire, M., 688(178), *697*, 1380(64), 1383(64), *1394*
Lemmon, W. R., 290(282), *322*
Lemoine, P., 285(227), 293(227), *320*
Le Moing, M. A., 982(122), *1022*, 1160(252), *1186*
Lenhard, J. R., 683(129), *696*, 1314(7), 1365(7), 1366(7), *1392*, 1366(26), 1368(26, 31), 1369(26), *1393*, 1370(7), *1392*, 1371(26), *1393*
Lenkait, J. A., 190(274), *204*
Lennox, J. C., 1172(416), *1191*, 1365(15), 1366(15), 1368(15), 1370(15), *1392*
Lenoir, D., 1154(200), *1185*
Lentsch, S. E., 641(182), 642(182), 643(186), *655*
Lenz, A., 1208(96), *1214*
Leonard, J. E., 643(186), *655*
Leonard, N. J., 442(111), 444(111), *459*, 437(131), 441(131), *460*, 720(138), *753*, 899(142), *938*
Leone, J. T., 435(25), *457*
Lepage, J.–L., 728(257), *757*
Leport, L., 277(160), 277(161), 311(160), *318*, 891(48), *934*
Lepri, L., 971(77b), 972(78), *1020*
Lerstrup, K. A., 680(118), 681(118), 682(118), *695*
Lesma, G., 716(115), *752*
Lessard, J., 275(130), *317*, 441(152b), *461*, 891(58), *935*, 1435(25a, b), *1438*
Leuchte, W., 848(244, 245), *872*
Leung, M., 281(192), 284(192), 291(192), *319*, 291(290), *322*
Leung, W.–H., 834(169), *869*
Leung, W.–P., 282(201), *319*, 302(383), *325*
Levanon, H., 350(194), *359*
Levenspiel, O., 1299(56), *1334*
Lever, A. B. P., 813(25), *864*, 820(91), 821(91), *867*
Levering, D. R., 1088(139), *1099*
Leveson, L. L., 478(111), *482*
Levi, M., 908(222), *941*
Levich, V. G., 32(44a, e), 63(44), *116*, 39(52), *117*, 81(104b, 105), *119*, 168(175), *201*, 1286(6), 1292(6), *1332*
Levin, E. S., 1048(142), *1065*
Levin, G., 349(176), *358*
Levin, M. S., 734(306), *759*
Levin, Y. L., 488(45), *497*
Levine, M., 897(114), 899(114), *937*
Levine, R. D., 47(61b,c)
Levinson, I. M., 436(59), *458*, 441(149b), *461*, 495(120), *500*, 740(356), *761*, 895(98b), *937*
Levisalles, J., 1144(97), *1182*
Levitin, I., 826(131), *868*
Levy, E. C., 776(44a), *785*
Levy, M., 299(352), *324*, 391(183), *399*, 452(208b), *462*, 885(7), *932*, 1201(35),

[Levy, M.]
1202(37–40), *1212*, 1221(10), 1251(10), *1256*
Levy, P. F., 541(36), *575*, 547(75), *576*
Levy, S. V., 336(57), *354*
Lew, B., 439(86), *459*
Lewandowska, T., 713(81), *751*
Lewis, M. C., 543(51), *576*, 1159(242), *1186*
Lewis, N. S., 821(100), *867*
Lexa, D., 234(7d), *246*, 813(36), 814(54), 819(54), 823(36), 824(36), *865*, 815(59), 816(59, 61, 63, 64, 72), 817(75), 818(72), *866*, 823(102), *867*
Leygraf, C., 1404(18), *1416*
Leznoff, C. C., 820(91), 821(91), *867*
L'Her, M., 301(368), *325*
Li, C., 186(259), *204*
Li, F., 647(223), *657*
Li, S., 721(144), *753*
Li, Z., 1390(128), *1396*
Liao, C.-S., 621(81), 627(81), *652*, 676(99), *695*
Libby, W. F., 35(48), *116*
Liberatore, F., 703(10), *748*, 725(220a, 222), *756*
Libert, M., 663(46), 664(46), *693*, 720(140), 723(140), *753*, 723(185), *754*, 974(98), *1021*
Lieber, C. M., 821(100), *867*
Liepa, A. J., 889(37), *934*
Lightfoot, E. N., 1292(42), 1299(42), *1333*
Lilie, J., 1412(125), *1419*
Lillie, E. D., 349(176), *358*
Limacher, L., 1035(18), *1061*
Limaye, P. A., 542(48), *576*
Limborg, F., 720(141), 721(143), *753*, 1161(276), *1187*
Limosin, D., 635(125), 636(125), *653*, 800(88), *808*
Limosin, N., 376(113), *397*, 406(43), *427*, 1176(492), *1194*, 1381(73), *1394*
Limosin, P. D., 795(49, 50), *806*
Lin, D. K., 15(9), *114*
Lin, E.-C., 596(89), *610*
Lin, S. Y., 790(10), *804*
Lin, X. Q., 1148(158), *1183*
Lin Chiu, Y., 839(214), *871*
Lindberg, B. J., 1056(236), *1067*
Linderberg, J., 1414(147), *1420*
Lindey, S. E., 1387(114), 1389(114), *1396*
Lindig, B. A., 1408(39), *1416*
Lindsey, Jr., R. V., 1157(230), *1186*
Line, L. L., 592(56), *609*, 968(69), 988(69), *1020*
Linek, K., 437(134), *460*, 1232(75), *1258*
Lines, D., 1106(26), *1124*
Lines, R., 70(91b), *118*, 125(50), *196*, 128(64), *197*, 260(41), 294(41), *314*, 301(379), *325*, 341(95), 350(95), *356*, 453(223), *463*, 506(9), *528*, 506(10), 509(10), 511(10), *529*, 540(31), *575*, 638(131), *654*, 1001(184), *1024*, 1091(163), *1100*, 1387(120), 1389(120), *1396*
Lingane, J. J., 22(22c), *115*, 157(143), 161(143), *199*, 177(211), *202*, 210(6b), 212(6b), *231*, 256(1, 4), 309(1, 4), *312*, 260(47), 263(47), *314*, 309(426), *327*, 454(224, 225), *463*
Linkous, C. A., 1385(108), *1396*
Linn, W. J., 908(223), *941*
Linnell, W. S., 291(295), *322*
Linstead, R. P., 260(37), *314*, 1004(205–207), *1025*, 1080(80), *1097*
Liphitskii, V. F., 734(312), *759*
Lipkin, D., 332(12a), *353*
Lippincott, W. T., 1080(81), *1098*
Lipsztajn, M., 162(151), *200*, 275(125), *316*, 303(393), *326*, 422(150), *431*
Lisitsina, N. K., 1040(70), *1062*, 1147(150), *1183*
Lisius, J. D., 1421(4d), *1429*
Liska, F., 907(185b), *940*
Lissi, E. A., 544(56), *576*
Lister, K., 485(17), *496*, 907(193), *940*, 1288(28), *1333*
Litscher, G., 930(340), *946*
Little, R. D., 410(73), *428*, 919(285a, b), *944*, 1137(39), 1139(58, 59), 1140(58, 59), *1180*, 1267(12a), *1280*, 1268(25b, c), *1281*, 1271(40), *1282*, 1274(25b, c), *1281*, 1435(30), *1438*
Little, W. A., 819(82), *866*
Little, W. F., 726(236), *756*
Littlehailes, J. D., 1200(6), *1211*, 1201(31), *1212*
Liu, C. C., 645(206), *656*, 1105(25), 1109(25), *1124*, 1109(39), 1112(25), *1124*
Liu, H. Y., 844(230), *871*
Liu, W., 820(91), 821(91), *867*
de Llano, C., 334(39), *354*
Lloyd, D., 733(304), *759*
Lloyd, H., 704(27), *749*, 1166(328), *1189*
Lloyd, M. K., 1168(343), *1189*
Löb, W., 414(97), *429*, 417(116a), *430*
Locher, P., 1088(136), *1099*
Lodmell, J. C., 712(72), *750*
Lodygin, N. A., 729(266), *758*
Loev, B., 403(15), *426*
Lofters, S., 831(156), *869*
Logan, J. A., 1357(93), *1362*, 1380(65, 71), 1386(53), 1387(65), *1394*, 1383(92), 1385(92), 1386(92), *1395*, 1386(137), *1396*
Logemann, H., 235(18), *246*
Loginova, N. F., 489(59), *498*, 1039(52b), *1062*, 1045(114, 115), 1047(114), *1064*, 1056(240), 1058(240), 1059(240), *1068*, 1221(9, 16), 1251(9), *1256*
Lögler, P., 160, *182*
Lomölder, R., 237(29), *247*
Longchamp, S., 723(192), *755*, 974(98), *1021*
Longhi, P., 1200(1), *1211*
Longworth, J. W., 344(122), *357*
Lopez, J., 340(87), *355*, 519(109), *532*
López-Segura, M., 1173(433), *1192*
LoPresti, G. M., 421(143), *431*, 713(88), *751*
Lora, J. H., 790(13), *804*
Lorah, E., 280(186), *319*
Lorch, H. W., 1094(191), *1101*
Lorenz, W. J., 616(1), *650*
Lorenzelli, V., 1410(120, 121a), *1419*
Lorimer, S. E., 165(167), *200*
Lotsch, W., 340(82b), *355*
Lott, K. A. K., 894(74), *935*
Louati, A., 738(338), *760*, 985(132), *1022*, 1090(161), *1100*
Loucka, T., 646(212), *656*
Loutfy, R. O., 434(5), *456*
Love, D. L., 1044(94), *1063*

Love, J., 389(174), *399*, 747(435), *763*
Loveland, J. W., 347(168, 173), 348(168, 173), *358*, 508(33), *529*, 885(9), *932*, 930(354), *947*
Lowe, H., 1231(63), *1258*
Lown, J. W., 1040(61), *1062*
Lozanova, A. V., 722(167), *754*
Lu, P. V., 1387(113), 1389(113), *1396*
Luan, L., 1434(12h), *1437*
Lubyanitskii, I. Ya., 408(55), *428*
de Luca, C., 1154(196), *1184*, 1154(201), *1185*
Lücker, O., 1006(214), *1025*
Ludman, C. J., 1094(183), *1101*, 1117(55), 1118(55), 1119(55), *1126*
Ludvik, J., 724(202), *755*
Ludwig, C., 788(6), 790(6), *803*
Luftmann, H., 545(79), *576*, 1157(221), *1185*
Lukehart, C. M., 845(234), *872*
Lukes, R., 448(190), 449(197), *462*
Lukyanets, E. A., 820(96), *867*
Lund, H., 3(12), 4(15), *8*, 21(21d), *115*, 39(55e), 56(71), 58(71), 70(71), *117*, 216(11, 13), *231*, 239(42a), 240(42a), *247*, 263(49, 61), 310(49), 311(61), *314*, 279(181), 295(181), 298(181), *319*, 292(297), *322*, 305(398), *326*, 308(418), *327*, 344(110–112), 347(110, 111), 348(112), *356*, 347(165, 170), 348(165, 170), *358*, 368(37), *395*, 375(93–95), *396*, 329(191), 389(173, 175), 390(182), 391(188, 189), *399*, 392(193–196), *400*, 401(3), 402(6, 8), 403(18), 410(6), *426*, 408(53, 54, 57), 409(54, 68), *428*, 411(79), *429*, 416(113, 114), 417(124), 418(124), *430*, 423(159, 162), *431*, 424(177, 179), 425(179, 193), 426(193), *432*, 437(136), 441(140), 450(136), *460*, 445(159), 449(159), *461*, 449(191), *462*, 456(241), *464*, 465(1), 466(1, 2, 3, 5), 467(1), 468(28, 29), 469(5), 470(1), 471(1), 472(1), 474(1), 475(1), 476(1), 477(1–3), *479*, 469(45), 470(48), 472(58–60), 475(59), 476(80), *480*, 473(68), 474(81), 475(81, 83), 476(81, 87), *481*, 477(96, 98), *482*, 484(11), 485(15), *496*, 489(68), *498*, 491(89, 91), 492(91), *499*, 492(107), 495(121), *500*, 505(5), 508(5), 509(5), 510(5), *528*, 620(48), *651*, 635(120), *653*, 701(1a, b), 719(1a), *747*, 702(3), *748*, 705(30), 707(43, 44), 708(44), 719(43), 726(30), 727(43), 734(43), *749*, 709(54a, b, 59), 710(59), 719(59), 736(59), *750*, 712(74), 713(79, 91), 714(94, 98), 718(79), 730(79), 735(94), 740(79), 741(79), 743(79), 746(79), *751*, 718(123, 124), 719(125–127, 129, 130, 132, 134), 729(126), 730(123, 125, 126), 731(127), 732(127), 734(130), 735(126), 736(126), 741(132), 742(126), 743(126), 745(126), 747(134), *752*, 719(136), *753*, 722(170, 171), *754*, 724(212), *755*, 726(231, 233), *756*, 726(240, 243), 727(248), 728(260), 729(263, 271), *757*, 729(271), 730(271, 273), 731(275), 732(271, 273, 275, 276, 282), 733(273, 298), 743(282), 747(271, 282), *758*, 735(320), *759*, 737(330–332), 739(342), *760*, 739(350), 740(352, 358), 741(369), 746(369), *761*, 743(396), 744(396, 400, 401, 405, 406), 745(405), *762*, 745(418, 421), 746(421), *763*, 886(13, 14, 20a–e), 891(13, 14), *933*, 892(64), *935*, 898(126), 899(143), *938*, 909(164), 913(173), *939*, 914(246), *943*, 922(278), 923(279), *944*, 930(350, 363), 931(363, 366), *947*, 960(29a), *1019*, 1032(4, 5), 1033(7), 1050(7), *1060*, 1038(43), 1039(43), *1061*, 1038(46), 1039(46, 56, 60), *1062*, 1042(79), 1043(85), *1063*, 1045(110), 1047(110, 134, 136–138), *1064*, 1049(152), 1050(163, 164), *1065*, 1051(178), 1052(190), *1066*, 1054(208–210), 1055(214–217, 222), 1056(224, 225, 228–230), *1067*, 1057(245, 246), 1058(245–248), *1068*, 1136(36), 1138(48, 49), *1180*, 1143(86), *1181*, 1144(101, 110, 111, 124), 1145(124), *1182*, 1148(159), *1183*, 1150(168), 1153(183, 188, 190), 1154(188, 190), *1184*, 1173(437), *1192*, 1176(504), 1177(506), *1194*, 1200(15), *1211*, 1221(6, 8, 11, 12, 14, 15), 1222(6), 1223(6), 1224(8, 12), 1246(166, 168, 169), 1247(11, 12), 1248(12), 1249(15), 1250(8, 14), 1251(8, 11, 12, 14), *1256*, 1221(20), 1224(31), 1226(20), 1248(20), 1249(31), *1257*, 1233(88), 1235(88), *1259*, 1250(181–183), 1251(186, 187, 189, 190), 1252(166), *1262*, 1252(194, 198, 199), *1263*, 1267(11), 1270(11), *1280*, 1414(142, 145–148), *1420*
Lund, T., 344(112), 347(112), 348(112), *356*, 392(193–195), *400*, 437(136), 450(136), *460*, 722(171), *754*, 740(352), *761*, 1038(46), 1039(46), *1062*
Lunde, P., 467(19), *479*, 709(54b), *750*, 719(125), 730(125), *752*
Lunt, J. C., 260(37), *314*, 1004(206), *1025*
Luque de Castro, M. D., 470(47), *480*
Lurie, S., 542(45), *576*, 1002(188), *1024*
Lushnikov, V. I., 276(141), *317*
Luss, H. R., 672(79), *694*
Lüttke, W., 409(64), *428*
Lüttringhaus, A., 1044(95), *1063*

Lux, E., 1005(220a), *1025*, 1275(46a), *1282*
Lyashko, O. V., 797(62), *806*
Lynn, T. T., 889(37), *934*
Lyons, L. E., 721(161a), *753*, 1144(99), *1182*
Lyons, L. R., 447(179), *462*

Mabon, G., 913(286), *944*, 1175(487), *1194*
Mabrouk, H. E., 862(319), *874*
Mabrouk, S., 831(158), 832(158), *869*
Macachi, T. J., 1203(48), *1213*, 1207(83), *1214*
Macartney, D. H., 50(66), *117*
MacCallum, J. R., 1408(51, 52), *1417*
MacDiarmid, A. G., 331(9), 341(9), *353*
MacDonald, D. D., 122(5), 123(5), 125(5), 126(5), 146(5), 150(5), 151(5), 152(5), 154(5), 157(5), *195*
MacDonald, K. I., 1235(84), *1259*
MacDougall, B., 179(228), *202*
MacDuff, R. C., 930(361), *947*
Machatzke, H., 1044(95), *1063*
Machida, K., 641(170), *655*
Machion, P. D., 662(31), 670(31), 671(31), *692*
Machtinger, M., 300(363), *325*
MacInnes, D., Jr., 331(9), 341(9), *353*
Mackie, K. L., 791(16), *804*
MacKirdy, I. S., 296(329), *324*, 1267(8), *1280*, 1269(34), *1281*, 1277(49), *1282*
MacMullin, R. B., 270(99), *316*, 1290(37), *1333*, 1302(64, 65), *1334*
Macrae, R., 1045(117), *1064*
Madec, C., 685(135), *696*, 983(128), *1022*
Madkour, L. M., 474(71), *481*
Madsen, J. O., 1149(161), *1183*
Maeda, H., 621(56), *651*
Mageda, E., 1043(86), *1063*
Magno, F., 660(14), *692*, 663(44), *693*, 675(95, 96), 676(98), *695*, 712(72), *750*, 931(360), *947*, 1059(257), *1068*, 1228(51), *1257*, 1245(51), 1246(165), *1262*
Magnoli, D. E., 31(41a, b), *116*
Magnuson, R. H., 844(228, 229), *871*
Magnusson, G., 1072(29), *1096*
Mahachi, T. J., 722(181), *754*, 1207(83), *1214*
Mahaling, R., 1358(107), *1362*
Mahe, C., 1241(123), *1260*
Mahedevan, R., 336(60), 339(60), *354*
Mahler, J., 1153(193), 1176(193), *1184*
Mahlhop, R., 738(335), *760*
Mahling, S., 668(62), *693*
Mahmood, N. M., 1434(12a), *1436*
Mahoney, L. R., 619(25), *650*
Mahoney, M. R., 179(225), *202*
Maia, H. L. S., 489(55), *498*
Maier, G., 526(139), *533*
Maier, N. A., 837(198), 838(198), *870*, 1425(12), *1430*
Mairanovskii, S. G., 244(78), *249*, 259(32), 263(32), 309(32), *313*, 378(124), *397*, 390(177), *399*, 403(12, 17), *426*, 406(41, 42), *427*, 407(49), *428*, 435(26, 28, 32, 33, 36), *457*, 436(58, 59), *458*, 439(94), *459*, 441(149b), *461*, 443(119), *460*, 445(158), *461*, 456(242), *464*, 489(59), *498*, 722(166), *754*, 724(204), *755*, 861(314), *874*, 894(77e), *936*, 895(98b), *937*, 1038(36, 42a, b), 1039(36), *1061*, 1039(48, 52a, b), *1062*, 1040(70), *1062*, 1045(114, 115), 1047(114), *1064*, 1048(141), *1065*, 1050(161), *1065*, 1056(240), 1058(240), *1068*, 1059(240, 261), 1104(11), *1124*, 1147(150), *1183*, 1176(503, 505), *1194*, 1221(9, 16, 17, 22, 23), *1257*, 1224(17), *1256*, 1251(9), *1256*, 1287(23), *1333*
Maissant, J. M., 1247(173), 1249(173), *1262*
Maitte, P., 706(42), *749*, 719(131), *752*
Majda, M., 647(220), *657*
Majeed, A. J., 989(163), *1023*
Majeski, E. J., 524(132), *533*
Mak, S., 1269(33), *1281*
Makagawa, Y., 1079(71), *1097*
Makarochkina, S. M., 906(182a), *939*, 911(269), *943*
Maki, A., 334(34), *354*
Makino, A., 774(39a), *785*
Makosza, M., 307(412), *327*
Malachesky, P. A., 169(177), *201*, 176(210), *202*, 510(48), *530*
Malashkhiya, M. V., 284(241), *321*, 726(242), *757*, 734(312), *759*
Malatesta, V., 1040(64), *1062*
Malhotra, R., 983(124), *1022*
Malissard, M., 1251(185), *1262*
Malkov, Y. A., 793(36, 37, 38), *805*
Mallen, E. F., 179(234), *203*
Maller, R. K., 418(128b), *430*
Mallouk, T. E., 1390(128), *1396*, 1413(139), *1420*
Malmberg, M., 1077(53, 56), 1080(53), 1083(53), *1097*
Maloy, J. T., 122(10), *195* 148(109), *198*, 151(126, 127), *199*, 882(3b, m), 885(3b,m), *932*
Malpas, R. E., 1253(202), *1263*
Malsch, K.-D., 526(139), *533*
Malysheva, Z. H., 894(86), *936*
Malyugina, N. I., 425(194), *432*
Mamedova, V. M., 907(189–191), *940*
Manahan, S. E., 848(243), *872*
Manapova, L. Z., 287(244), 305(244), *321*
Manassen, J., 820(94), *867*
Manda, E., 904(206, 207), *941*, 1434(13f), *1437*
Mandai, T., 570(159), *579*, 679(106, 107), *695*, 1046(123), *1064*
Mandell, L., 452(210), 453(210), 455(210), *463*, 914(242), 915(251), *943*, 1001(194), *1025*, 1137(41), *1180*, 1142(80), *1181*, 1144(129), 1145(129), *1182*
Mango, F. D., 1082(96), *1098*, 1160(248), *1186*
Mani, S. R., 686(160, 165, 168), 687(174), *697*
Manji, A., 1421(4a), *1429*
Mann, C. K., 70(90), *118*, 283(234, 235), *320*, 286(262), 289(262, 263),

[Mann, C. K.]
321, 296(324), *323*, 375(102, 103, 104), *397*, 454(227), *463*, 582(1–4), 583(1, 4), *607*, 584(16, 21), 586(29), *608*, 600(123), *611*, 601(134), *612*, 667(55), *693*, 717(119), *752*, 863(325), *875*, 975(102), *1021*, 1051(179), *1066*, 1056(241), 1057(241), 1058(241), 1059(241), *1068*, 1168(347, 348), 1169(347, 348), *1189*, 1076(46), *1096*, 1084(107), *1098*
Manning, D. T., 555(106), *577*
Manning, G., 520(112), 521(112), *532*
Manning, M. J., 630(100), *653*
Mannus, D., 703(13), *748*, 771(31a), *785*
Mano, E. B., 1351(42), *1360*
Manousek, O., 495(123), *500*, 707(47b), *749*, 708(52), *750*, 713(86), *751*, 739(351), *761*, 745(419), *763*, 892(65), *935*, 1033(10a), *1060*, 1039(53), *1062*, 1048(151a), *1065*, 1050(174), *1066*
Mansfield, W. C., 26(35), *116*
Mansuy, D., 816(71, 72), 818(72), *866*
Mantell, C. L., 1287(11), *1332*
Manzanera, C., 384(144), 385(144), 386(144), *398*, 1154(198), *1185*
Manzhelei, M. E., 440(99), *459*, 897(116), *937*
Maran, F., 342(98), 346(98), *356*, 717(122), *752*, 1033(10b), *1060*
Marathe, W., 168(172), *201*
Marcel, S., 1208(98), *1214*
March, J., 14(8), 18(8), *114*, 243(66), *248*
Marchetti, L., 723(183), *754*
Marchini, S., 1413(129), *1419*
Marcin, M., 1377(124), 1390(124), *1396*
Marcott, C., 182(245), *203*
Marcoux, L. S., 145(103, 104), *198*, 169(177), 170(178), *201*, 301(366), *325*, 350(109), *356*, 510(48, 53), 511(53, 56), *530*, 592(58), *609*, 968(68), 970(68), *1020*
Marcus, M. F., 590(43), *609*, 620(49), *651*, 621(79), 625(79), *652*
Marcus, R. A., 31(41), 32(41, 43a–e), 63(43), *116*, 39(55a), 41(57a), 47(61a), *117*, 342(102), 343(108), 344(119), *356*
Marcuzzi, F., 346(155, 156), *358*, 1135(26), *1179*
Margaretha, P., 4(14), *8*, 630(96, 105), 632(105), *653*, 909(237), *942*, 914(250), 915(250), *943*
Marek, M., 1211(128), *1215*
Margel, S., 391(183), *399*, 832(162), 833(163), *869*, 1221(10, 13), 1251(10), *1256*, 1340(4), *1359*
Margerum, L. D., 1381(77), *1394*
Mariani, H. A., 1094(183), *1101*, 1117(60, 61), 1118(60, 61), *1176*
Mariani, P., 916(267, 268), *943*
Maricle, D. L., 284(224), 286(224), *320*, 365(16), *394*, 404(22), *427*, 1245(151), *1261*
Marin Galvin, R., 744(405), *762*
Marincic, L., 644(193, 196, 197), 645(193, 196), *656*
Marincic, N., 285(221), *320*
Mark, C. P., 776(45), *785*
Mark, H. B., Jr., 294(315), *323*, 344(126), *357*, 346(153a), *358*, 823(107), *867*, 1314(6), 1375(6), *1392*
Markby, R. E., 285(208), *320*, 293(311, 312), *323*, 297(339), *324*, 1254(206, 209), 1255(206, 213), *1263*
Markina, Z. V., 453(217), *463*
Markova, I. G., 445(171), *461*
Markowitz, J. M., 388(164), *399*
Markushina, I. A., 1161(279), *1187*
Marple, L. W., 282(198), 297(198), *319*, 373(78), *396*
Marquarding, D., 436(67), *458*
Marquet, B., 290(281), 291(281), *322*
Marquez, J., 414(103a), *429*, 518(107), *532*
de Márquez, O. P., 518(107), *532*
Marr, D. H., 633(113), *653*
Marrocco, M., 819(84), *867*, 1231(61), 1232(61), 1234(61), *1258*, 1372(39), *1393*
Marschoff, C. M., 518(107), *532*
Marshall, R. J., 1303(70), *1334*
Marsikova, D., 646(212), *656*
Martenot, J. P., 1340(8), 1346(8), *1359*
Martigny, P., 670(75, 76), 671(75), *694*, 719(134), *752*, 733(298), *758*, 1042(74, 77, 78a), 1058(253), 1059(253), *1063*, *1068*, 1150(165, 166), *1184*, 1221(21), 1224(32), 1226(40), 1227(21), 1248(32), 1249(21), 1250(32), *1257*, 1251(190), *1262*Martin, A. J., 369(62, 64), *396*, 1135(18), 1154(18), *1179*, 1153(185, 192), *1184*, 1176(502), *1194*
Martin, C. R., 1374(48), 1387(48), *1393*, 1387(116), 1389(121), *1396*
Martin, J., 384(145, 146), 385(145, 146), *398*, 492(98), *499*
Martin, P., 1164(312), *1188*
Martinet, P., 425(192), *432*, 662(29), 663(29), *692*, 894(77f, 82), *936*, 1046(126), *1064*, 1233(91), 1235(91), 1236(100), *1259*
Martinchek, G. A., 190(278), *204*
Martineau, M., 855(277, 278), *873*
Martinez, R., 670(77), *694*, 1046(128), *1064*
Martinman, J., 300(360), *325*
Martre, M. A. M., 425(192), *432*, 1164(312), *1188*
Martynova, G. P., 794(57, 58), *806*
Maruyamor, M., 384(142), *398*
Marvel, C. S., 450(203), *462*
Mas, A., 680(120), 681(120), *695*
Mas, F., 150(123), *199*
Masatsuga, Y., 1036(25), *1061*
Maschmeier, C.–P., 707(47c), *749*
Masheder, D., 1268(29), *1281*, 1434(12a), *1436*
Mashimoto, J., 1243(136), *1261*
Maslova, L. I., 473(66), *481*
Mason, M. G., 672(79), *694*

Mason, W. H., 791(16), *804*
Massebien, M.–C., 277(161), *318*
Massey, A. G., 1205(63), *1213*
Mastragostino, M., 123(26), 133(26), *196*, 1382(80), *1395*
Masuda, M., 891(44), *934*
Masui, M., 403(13, 14), *426*, 404(23, 24, 26), 405(35, 36), *427*, 425(187), *432*, 474(79), 475(78), *481*, 583(11), *607*, 594(85), 595(86, 87), *610*, 597(103), 599(112, 115–118), 600(124), *611*, 600(126), 601(126), *612*, 621(56), *651*, 621(63), 624(63), *652*, 705(33), *749*, 743(388), *762*, 961(33), *1019*, 983(127), *1022*, 1044(91), *1063*, 1050(168), *1066*, 1087(132), 1089(143), *1099*, 1178(516), *1194*
Masuno, M., 487(38), *497*
Matassa, L., 837(193), *870*
Mather, W. B., 284(214), *320*, 308(417), *327*
Matic, D., 258(20), *313*, 1309(81), *1334*
Matlock, P. L., 1258(64), 1232(64), *1258*
Matschiner, H., 309(431), *327*, 403(11), *426*, 437(137), *460*, 707(47c), *749*, 728(258), *757*, 742(381), *762*, 746(426), *763*, 930(342), *946*, 930(351, 352), *947*, 1046(121), *1064*, 1434(13g, h), *1437*
Matsen, F. A., 26(31c), *115*
Matsubara, S., 678(102), 679(111), *695*, 1046(124), 1047(135), *1064*, 1079(67), *1097*, 1167(339), *1189*
Matsubara, T., 1211(124, 125), *1215*
Matsubara, Y., 1034(15), *1060*, 1163(298), *1188*
Matsuda, F., 1201(33), *1212*
Matsuda, H., 132(73), *197*, 153(131, 132), *199*, 851(264), 852(264), *873*, 1298(52), *1334*
Matsuda, T., 1164(311), *1188*
Matsuda, Y., 824(112), 827(140), 829(140), *868*, 971(77a), *1020*
Matsue, T., 275(127), *317*, 379(128), *398*, 443(123), 445(123), 455(123), *460*, 455(240), *464*, 1175(475), *1193*, 1247(175), *1262*
Matsui, M., 442(112), *460*, 728(252), *757*, 1170(369), *1190*
Matsui, Y., 375(90), *396*, 555(102), *577*
Matsumato, Y., 603(156), *613*
Matsumoto, A., 595(87), *610*, 621(63), 624(63), *652*, 678(103), *695*, 956(33), *1019*
Matsumoto, K., 558(117, 118), 559(119, 121, 122), 562(119, 125, 126), 564(136), 567(118, 136), *578*, 567(145, 147), *579*, 601(128), *612*, 766(1a, b, 3, 4), *783*, 766(7a, b), 768(16–18), 769(21), *784*, 1045(113), *1064*, 1058(250), *1068*, 1158(231), *1186*
Matsumoto, S., 448(185), *462*, 490(86), *499*, 907(186), *940*
Matsumoto, Y., 1167(333), *1189*
Matsumura, M., 570(157), *579*
Matsumura, N., 679(109), *695*
Matsumura, Y., 6(22), *8*, 240(52), 242(52), 243(69), *248*, 517(90, 91), *531*, 589(40), 590(40), *608*, 597(106), 598(109), *611*, 600(125), 602(144, 145), *612*, 603(150, 155, 156), 604(160, 161), 605(160), 607(164), *613*, 643(185), *655*, 643(188), *656*, 647(226), 648(229), *657*, 708(51), *750*, 727(247), *757*, 767(9–11), *784*, 773(38a), 774(39a, 40), 777(50), *785*, 781(66), 782(67, 69, 70b, c), *786*, 850(260), *872*, 891(57), *935*, 1035(19, 23), 1036(28), 1038(38, 39), *1061*, 1041(72), 1042(73, 75), *1063*, 1077(50), *1096*, 1077(52), 1078(63, 65), 1079(71), *1097*, 1147(146), *1183*, 1150(167), 1151(172), *1184*, 1155(212), *1185*, 1160(254), *1186*, 1163(293), *1187*, 1163(295), *1188*, 1167(333), *1189*, 1171(373, 374), *1190*, 1178(513), *1194*, 1278(57a), 1280(57a), *1282*, 1425(13), *1430*
Matsunami, S., 1243(135), 1244(135), *1261*
Matsuoka, M., 1045(113), *1064*, 1058(250), *1068*
Matsuoka, T., 528(144), *533*, 1161(267, 269), *1187*
Matsuura, M., 537(8), 572(8), *574*
Matsuura, N., 278(173), *319*
Matsuyama, Y., 675(93), *694*
Matthews, W. M., 887(24), *933*
Mattineau, M., 930(343), *946*
Maturova, M., 726(234), *756*
Matuschke, R., 930(342), *946*
Matusinovic, T., 1365(20), 1371(20), *1392*
Mau, A. W. H., 1413(137), *1420*
Maugh, T., 284(226), 303(226), *320*
Maupas, B., 734(307), *759*
Maurer, A. H., 674(85), *694*, 721(163), 723(163), *754*
Maurey, M., 684(131), *696*, 728(262), *757*
Maurey-Mey, M., 685(141), 686(151), *696*
Mavelli, I., 725(222), *756*
Maxwell, B., 1434(22c), *1438*
May, J. A., 486(30), *497*
Maya, H. L. S., 1050(162), *1065*, 1056(242), 1057(242), 1058(242), 1059(242), *1068*
Mayausky, J. S., 187(264), 188(264), *204*, 686(154), *696*
Mayeda, E. A., 302(387), *326*, 351(206), *359*, 508(45), 511(62), *530*, 584(17), *608*, 635(122), 648(122), *653*, 637(128), 638(128), *654*, 1086(126), *1099*
Mayell, J. S., 1050(160), *1065*, 1232(79), 1234(83), *1259*
Mayer, F., 486(33), *497*
Mayer, R., 983(128), *1022*
Mayer, T. J., 1341(10), *1359*, 1381(78), *1394*
Mayhew, Y., 1299(55), *1334*
de Mayo, P., 1409(106, 107, 109–111), *1419*
Mazaleyrat, J. P., 1251(185), *1262*
Mazanko, A. F., 894(86), *936*
Mazume, T., 661(16), *692*
Mazur, D. J., 275(125), *316*
Mazur, S., 1365(20), 1371(20), *1392*
Mazur, Y., 242(58), *248*

Mazzocchin, G.–A., 663(44), *693*, 675(95), 676(98), *695*, 712(72), *750*, 931(360), *947*, 1059(257), *1068*
McAllister, D. L., 737(328), *760*
McCane, D. I., 720(137), *753*
McCarpa, F., 621(69), 625(69), *652*, 1082(95), *1098*
McCarron, E., 1094(183), *1101*, 1117(55), 1119(55), *1126*
McCartney, R. L., 490(82), *499*, 902(168a), *939*
McCleverty, J. A., 1168(343), *1189*
McCoch, J., 1241(126), *1260*
McComsey, H. J., Jr., 301(367), *325*
McCormick, M. J., 1175(451), *1192*
McCoy, H. N., 1200(5), *1211*
McCreery, R. L., 179(227), *202*, 180(239, 240), 189(239), *203*, 188(263), 187(264), 188(263, 264), 188(263, 264, 265), *204*, 630(103), 631(103), 633(103), *653*, 686(154), *696*
McCullough, L. L., 351(208), *359*
McCullough, R. D., 680(118), 681(118), 682(118), *695*
McCurdy, C. W., 180(242), *203*, 188(263), *204*
McDevitt, J. T., 819(82), *866*
McDonald, E., 812(21), *864*
McDonald, M. E., 1175(454), *1192*
McDonald, R. N., 140(90), *198*, 478(100–102, 106), *482*
McDowall, L. J., 478(107), *482*
McElroy, P., 1243(129), *1260*
McElvain, S. M., 925(306), *945*
McEwen, W. E., 1146(142), *1183*
McGinnis, D. P., 791(16), *804*
McGlotten, J., 728(255), *757*, 739(349), *761*, 1039(55), *1062*, 1144(96), *1182*, 1176(494), *1194*
McGregor, C. J., 296(329), *324*, 1277(49), *1282*
McGregor, J., 1353(75, 76), *1361*
McGuire, D. K., 295(321), *323*
McHiggins, G., 1054(206), *1067*
McHugh, M. A., 270(98), 304(98), *316*
McIntire, G. L., 411(87), 412(87), *429*, 686(156), *697*
McIntyre, J. A., 793(32), *805*
McIntyre, J. D. E., 179(228), *202*
McIntyre, R., 724(211a), *755*
McIntyre, T. W., 971(75), 972(75), *429*, 422(153), *431*, 589(34), *608*
McKay, B., 375(101), *397*, 1168(346), 1169(346), *1189*
McKean, D. R., 983(124), *1022*
McKee, R. H., 287(245a, b), *321*, 305(402), *326*
McKeever, L. D., 349(175), *358*, 809(11), 860(11), 862(11), 863(11), *864*
McKelvay, J. M., 672(79), *694*
McKeon, M. G., 1176(495), *1194*
McKinney, P. S., 1045(106), *1063*
McKinney, T. M., 190(273), *204*
McLarnon, F. R., 258(24), *313*, 290(272), *322*
McMakins, D. E., 1230(59), *1258*
McManis, G. E., 1434(12j), *1437*
McMasters, D. L., 280(187), 284(187), 285(187), 297(187), *319*
McMillan, G. W., 265(66), *314*, 403(9), *426*
McNamara, F. T., 724(208), *755*
McNamee, G. M., 297(334), *324*, 363(9), 364(9), 366(9), 371(9), *394*, 889(38), *934*
McNicol, B. D., 639(138, 139), *654*
McOmie, J. F. W., 1225(37), *1257*
McQueen, R. C. S., 820(95), *867*
McQuillan, A. J., 179(225), *202*
McWhorter, R. W., 1295(44), 1308(44), *1333*
Medeiros, M. J., 489(55), *498*, 1050(162), *1065*, 1056(242), 1057(242), 1058(242), 1059(242), *1068*
Medin, H., 1229(71), *1258*
Medovshchikova, F. F., 735(316), *759*
Medvedev, V. A., 801(82, 83), *807*
Medvedeva, S. A., 801(82), *807*
Meerholz, K., 336(56), 343(56), 351(56), *354*, 337(72), 338(72), 341(72), *355*, 341(92), *356*
van Meerssche, M., 733(304), *759*
Mehler, K., 837(196), *870*, 848(244), *872*
Mehta, R. R., 1267(5), 1270(5), *1280*
Mei, Q., 184(257), *204*
Meier, E. P., 621(81), 627(81), *652*, 635(119), *653*
Meier, K., 826(126), *868*
Meinert, H., 293(308), *323*, 1117(52), 1122(52), *1125*
Meinwald, J., 662(37), *693*
Meirowitz, R., 844(228–230), *871*
Meisels, A., 544(54), *576*, 1001(193), *1024*
Meisinger, R. H., 336(57), *354*
Meisters, A., 467(13, 14), *479*
Meites, L., 25(30), *115*, 157(142), 158(142), 159(142), 161(142), *199*, 311(438), *327*
van Meji, P. H., 333(27), *353*
Mekhtiev, S. I., 907(189–191), *940*
de Mele, M. F. L., 645(205), *656*
Melicharek, M., 590(42, 44), *609*, 973(80), *1020*
Melkonian, G. A., 467(20), *479*
Mellado, J. M. R., 148(112), 150(112), *198*
Melloni, G., 346(155, 156), *358*, 1135(26), *1179*
Mellor, J. M., 31(38b), *116*, 506(20–24), 511(20–24), 512(20–24), 514(21), *529*, 511(72), 512(72), 514(81), 515(85), 517(81), *531*, 663(46), 664(46, 48–51), 665(48–51), *693*, 713(77), *751*, 734(309), *759*, 951(17), *1018*, 1054(212), *1067*, 1084(111), *1098*, 1085(120, 121), *1099*, 1164(308–310), 1165(309), *1188*
Melnik, M., 820(91), 821(91), *867*,
Mel'nik, S. Ya, 1050(161), *1065*

Mel'nik, S. Y., 489(59), *498*, 1039(52), *1062*
Mel'nikov, N. N., 1205(62), *1213*
Mels, S. J., 293(310), *323*, 493(111), *500*, 1255(217, 218), *1263*
Melville, H. W., 1338(1), *1359*
Mémec, M., 1167(335), *1189*
Memming, R., 184(254), *203*, 1398(9), *1415*
Mendelson, W. L., 489(65), *498*
Mendis, L. P., 1355(84), *1361*
Mendkovich, A. S., 378(124), *397*, 403(12), *426*, 413(94), *429*, 435(50), *458*, 443(119), *460*, 488(50), *498*, 892(62), *935*, 895(96), 897(119), *937*
Mengel, R., 769(23), *784*
Mengler, C.-D., 738(335), *760*
Mengoli, G., 2(4), *7*, 450(204), *462*, 496(139), *501*, 848(249, 252), 849(255, 256), *872*, 1341(11), *1359*, 1351(41), 1351(44), 1352(45- 51), *1360*, 1352(65), *1361*, 1358(97, 104), *1362*
Mentrup, A., 263(62), *314*, 1049(155), *1065*, 1051(181), 1052(186), *1066*
Mercier, R., 1175(458), *1193*
Mergel, J., 1426(14), *1430*
Merkel, G., 275(132, 137), *317*
Merlin, J.-C., 299(355), *325*, 417(123), *430*, 734(314), *759*
Merling, G., 1203(49), *1213*
Merritt, M. V., 680(114), *695*, 1245(161), *1262*
Merwin, W. J., 296(326), *323*
Merz, A., 367(26), *394*, 370(68), *396*, 1228(48), *1257*, 1265(1), *1280*, 1372(44), *1393*
Meschke, P., 1268(22), 1275(22), *1281*
Meshkova, O. W., 434(4), *456*
Messerschmidt, A., 1204(50), *1213*
Mestroni, G., 825(115), *868*
Metcalf, H. F., 263(54), *314*, 424(178), *432*
Mettler, C., 486(21, 26, 32), 488(21, 32, 43), *497*, 489(21, 32, 43), *498*, 1208(101), *1214*
Metzger, A., 340(87), *355*
Metzger, J., 744(397), *762*
Metzler, G., 721(163), 723(163), *754*
Meul, T., 337(65), *355*
Meurs, J. H. H., 1117(50b), 1119(50b), 1121(50b), *1125*
Meyer, E., 1048(145–147), *1065*
Meyer, G., 393(205), *400*, 835(176, 179), *870*
Meyer, T. J., 50(65), *117*, 641(169), 643(169), *655*, 809(14), *864*, 834(167, 168, 170), *869*, 1434(12e, n), *1437*
Meyers, M. B., 621(69), 625(69), *652*, 1082(95), *1098*
M'Halla, F., 70(92a–c), *118*, 220(15), *231*, 226(25), 227(25), *232*, 299(353), *324*, 344(113, 115), *356*, 382(137), *398*, 1220(4), 1248(4), *1256*, 1224(27), *1257*
Michael, A., 349(185), *359*
Michaelis, K., 1002(201), *1025*
Michaelis, R., 243(64), *248*, 569(155), *579*, 1002(201), *1025*
Michalchenko, L. V., 413(94), *429*
Michalovic, J., 375(101), *397*, 1168(346), 1169(346), *1189*
Michaud, P., 838(204), *871*
Michel, M.-A., 344(110), 347(110), *356*, 390(182), 391(188), *399*, 445(159), 449(159), *461*, 449(191), *462*, 886(20a, b), 891(20a, b), *933*, 1032(4), *1060*, 1038(43), 1039(43), *1061*, 1042(78a), *1063*, 1221(6, 8, 11), 1222(6), 1223(6), 1224(8), 1250(8), 1251(8, 11), *1256*, 1247(11), *1262*
Michelet, D., 1250(180), 1251(180), *1262*
Michida, T., 597(103), *611*, 691(196), *698*, 977(86), *1021*, 983(127), *1022*
Micic, O. I., 1406(26, 28), *1416*
Micka, K., 1286(5), 1292(5), *1332*
Micuo, M., 1204(53), *1213*
Middleton, P. H., 368(46), *395*, 1153(184), *1184*
Midori, Y., 1408(88), *1418*
Miedaner, A., 1434(15a), *1437*
Migachev, G. I., 709(57), *750*
Miholova, D., 844(232), *872*
Mikami, H., 404(27), *427*
Mikawa, H., 663(39), *693*
Mikhailchenko, L. V., 892(62), *935*
Mikhailov, G. S., 792(27, 31), *805*
Mikheev, V. V., 474(73), *481*
Mikhelashvii, I. L., 456(242), *464*
Mikhotova, L. N., 1053(204), *1067*
Mikolajczyk, T., 1355(82), *1361*
Miles, M. H., 292(302), 293(302), *323*, 1434(12j), *1437*
Millar, P. G., 1171(375, 376), *1190*
Millaver, H., 630(98), *653*
Millefiori, S., 729(272), *758*
Miller, B., 165(170), *200*, 168(173), 173(189, 190), *201*, 174(202- 204), *202*, 663(40), *693*
Miller, C. J., 1377(124), 1390(124), *1396*
Miller, G. T., 258(23), 275(23), *313*
Miller, J. F., 722(169), *754*
Miller, J. J., 1205(64), *1213*
Miller, L. L., 236(22), *246*, 242(56), *248*, 291(284), *322*, 302(387), *326*, 334(36), *354*, 337(69), *355*, 351(206, 207, 209), *359*, 369(60), *395*, 441(149c, 150), *461*, 488(53), *498*, 508(45), 511(62, 65), *530*, 514(82), 515(82, 84), *531*, 517(102), 518(104), *532*, 584(18), *608*, 634(115), 635(122), 648(122), *653*, 638(132), *654*, 643(189–191), *656*, 646(219), *657*, 675(91), *694*, 704(16, 17), *748*, 770(26a, b), 771(29), *784*, 889(37), *934*, 895(98c), *937*, 929(337), *946*, 957(37), *1019*, 967(58), *1020*, 971(92), 977(92), *1021*, 993(143, 144, 146), 994(149–153, 155c), 995(149, 151, 153), 999(149), *1023*, 1037(34), *1061*, 1043(86), *1063*, 1081(90, 91), 1085(90), *1098*, 1085(116, 117),

[Miller, L. L.]
1086(117, 122–125, 130), 1087(130), 1094(130), *1099*, 1093(170, 171), 1094(174), *1100*, 1094(189, 190), *1101*, 1117(56), *1126*, 1142(76), *1181*, 1162(292), *1187*, 1171(383), *1190*, 1172(415, 416), 1173(417), *1191*, 1365(13), 1370(13), *1392*, 1372(40, 47), *1393*
Miller, P. G., 441(127), 444(127), *460*, 721(162), *753*
Miller, P. L. H., 724(211a), *755*
Miller, R. D., 983(124), *1022*
Miller, T. G., 236(22), *246*
Miller, W. N., 685(137), *696*
Millet, C., 599(113), *611*
Milligan, S. N., 1177(510), *1194*
Milliken, S. B., 283(235), *320*, 1084(107), *1098*
Millington, J. P., 723(197), *755*, 1073(35), *1096*, 1094(187), *1101*, 1229(50), 1234(50), *1257*, 1312(90), *1335*
Millren, C. M., 893(69), *893*, 892(64), *935*
Mills, J. L., 125(48), *196*
Mills, R. M., 846(241), *872*
Minami, T., 894(77a), *936*
Minato, A., 832(160), *869*
Minato, I., 543(50), *576*
Minisci, F., 237(34), *247*
Minko, R., 1164(312), *1188*
Minocha, A., 746(423), *763*
Minono, A., 1319(118), *1336*
Mioshi, M., 1045(113), *1064*
Mirkind, L. A., 552(81), *577*, 1010(235), *1026*
Mironova, O. Y., 495(122), *500*
Misawa, H., 1409(92), *1418*
Mischke, P., 488(52), *498*, 491(90), *499*, 717(121), *752*, 1151(173), *1184*
Mishra, S. C., 425(184), *432*
Mison, P., 1085(119, 131), *1099*
Misono, A., 302(382), *325*, 346(151), *357*, 914(238, 241), *942*, 1204(52), *1213*
Misonou, A., 1201(34), *1212*
Misovic, V., 288(254), 292(254), *321*
Mispelter, J., 816(64), *866*
Misra, R. A., 425(184), *432*, 456(241), *464*, 563(128), *578*, 901(137), *938*, 903(200, 201), *940*, 1142(77, 78), *1181*, 1254(207), *1263*
Missol, D., 276(146, 147), *318*
Mistrorigo, M., 1111(44), *1125*
Misumi, T., 724(210), *755*
Mitaine, P., 816(66), *866*
Mitani, M., 451(205), 453(205), 456(205), *462*, 906(183a), *940*, 904(209), *941*, 1044(101, 102), *1063*, 1052(189), *1066*, 1139(55, 56), 1140(55, 56), *1180*, 1143(89), 1144(89), *1181*, 1172(416), *1191*, 1207(84), *1214*, 1267(4, 10), 1270(4), *1280*
Mitchell, D. J., 886(20f), 891(20f), *933*
Mitchell, R. H., 336(60), 339(60), *354*
Mitnick, M. A., 369(61), *395*, 380(132), 382(132), *398*, 1153(194), *1184*
Mitzlaff, M., 602(146), 603(146, 148), 603(146, 148, 151), *613*, 767(13), *784*
Mitzloff, K., 727(247), *757*
Mitzner, R., 726(239), *756*, 926(138), *938*, 1038(35), *1061*
Miura, H., 890(42), *934*
Miura, Y., 1211(127), *1215*
Miyaguchi, K., 973(79), *1020*
Miyahita, K., 1408(85), *1418*
Miyake, T., 1316(101), 1319(101), 1326(101), *1335*
Miyama, H., 1408(68), *1417*
Miyamoto, T., 704(15d), 726(15d), *748*
Miyashi, T., 526(140), *533*
Miyawaki, S., 777(50), *785*
Miyazaki, F., 1000(171), *1024*, 1094(173), *1100*
Miyazaki, H., 476(95), 477(95), *482*, 742(380), *761*, 1004(208), *1025*, 1058(249), *1068*
Miyzaki, S., 1004(208), *1025*
Miyoshi, M., 558(117, 118), 559(119, 121, 122), 562(119, 125, 126), 564(136), 567(118, 136), *578*, 567(145, 147), *579*, 601(128), *612*, 766(1a, b, 4, 5), *783*, 766(6a, 7a, b), 768(16, 17), 769(21, 22), 770(24), *784*, 772(33a, b), *785*, 1058(250), *1068*, 1158(231), *1186*
Mizoguchi, M., 6(22), *8*, 243(69), *248*, 643(185), *655*
Mizuguchi, J., 448(185), *462*
Mizuguchi, K., 537(11), 563(11), 572(11), *574*
Mizukami, M., 1155(212), *1185*
Mizuno, A., 1034(17), *1061*
Mizuno, M., 682(125), *696*
Mizuno, S., 600(121), *611*, 601(133), *612*
Mizutani, F., 1142(75), *1181*
Mizutani, T., 704(15c), 726(15c), *748*, 928(324), *946*
Mlcoch, J., 1014(249), 1015(249), *1027*, 1166(324), *1188*
Mocellin, E., 1175(456), *1193*
Modena, M., 1211(126), *1215*
Moe, N. S., 124(44), *196*, 279(177), *319*, 291(287), *322*, 295(323), 295(323), 297(323), *323*, 684(132), *696*
Moëns, L., 919(285b), *944*, 1137(39), 1139(59), 1140(59), *1180*, 1435(30), *1438*
Moest, M., 443(124), *460*, 536(5), *574*
Mögel, L., 974(95), *1021*
Mohan, H., 668(61), *693*
Mohar, F., 1080(77), 1083(77), *1097*
Mohar, A. F., 539(25), *575*
Mohilner, D. M., 239(41), *247*, 293(312), *323*, 297(339), *324*, 589(31), *608*, 1254(206), 1255(206, 213), *1263*, 1376(60), 1387(60), *1394*
Mohmand, S., 1375(55), 1380(55, 63), 1382(55), *1394*
Moinet, C., 257(14, 16), 263(16), 264(16), 271(16), *313*, 267(78– 81), 273(78–81), 288(78–81), *315*, 273(123f, g), *317*, 412(89), 413(90, 91), 417(91), 421(90), *429*, 419(136), 421(147), *431*, 425(185, 186), *432*, 839(211), *871*, 1427(23), *1430*
Moingeon, M. O., 891(48), *934*

Moisar, E., 722(180), *754*
Moiseenkov, A. M., 722(167), *754*
Moldenhaver, W., 471(56), *480*
Molina, A., 153(129, 130), *199*
Møller, J., 713(89), *751*
Möllers, F., 184(254), *203*
Mollin, J., 477(99), *482*
Moltzen, E. K., 712(74), *751*, 931(366), *947*
Momenteau, M., 814(54), 819(54), *865*
Moncelli, M. R., 150(124), *199*, 909(237), *942*, 916(267, 268), *943*
Mondelli, R., 725(222), *756*
Mondodoev, G. T., 490(74, 75), *499*, 1208(103, 105), *1214*
Mönig, J., 668(62), *693*, 668(63), *694*
Monoszon, A. M., 282(197), 285(197), *319*
Monte, W. T., 1268(25b), 1274(25b), *1281*, 1271(40), *1282*
Montecatini, E., 1204(55), *1213*
Montenegro, M. I., 129(69), *197*, 208(4b), *231*, 489(55), *498*, 1038(37), *1061*, 1050(162), *1065*, 1056(242), 1057(242), 1058(242), 1059(242), *1068*
Montiel, V., 1173(433), *1192*
Montmillon, H., 444(128), *460*
Monvernay, A., 262(56), 263(56), *314*, 893(70), *893*, 925(309c–e), *945*, 1050(159d), *1065*
Moore, G. G. I., 1106(27), 1108(27), *1124*
Moore, G. R., 15(9), *114*
Moore, R. H., 377(117), *377*, 1152(177), *1184*
Moore, W. C., 1200(5), *1211*
Moore, W. M., 370(73), 371(73), *396*, 889(38, 39), *934*, 1135(28), 1141(28), *1179*
Mora, R., 390(178, 179), *399*, 930(350), *947*, 1138(49), *1180*
Moracci, F. M., 390(181), *399*, 703(10), *748*, 708(50), *750*, 725(220a, b), *756*, 747(437), *763*, 1434(13e), *1437*
Moraczewski, J., 842(225), *871*, 1175(459, 461, 463, 464), *1193*
Morduchowitz, A., 1434(12g), *1437*
Morel, F. M. M., 1413(132), *1419*
Moreno, N., 386(151), 387(151), *398*, 388(161), *399*, 703(4), *748*
Morgan, B. P., 671(78), *694*, 979(111), *1022*, 1044(96), *1063*
Morgan, R., 831(156), *869*
Morgan, T., 993(146), *1023*, 1093(171), *1100*
Morgan, W. L., 370(72), *396*
Morgat, J.-L., 863(330), *875*, 1006(222), *1025*
Mori, K., 783(75), *786*, 977(86), *1021*, 1170(369), *1190*
Mori, M., 559(120), 562(120), *578*, 603(152, 153), *613*, 835(178), *870*, 1077(55), *1097*
Mori, Y., 476(95), 477(95), *482*
Moriguchi, T., 597(103), *611*, 983(127), *1022*
Morimoto, A., 489(56), *498*, 740(353), *761*
Morimoto, T., 433(2), *456*, 904(210), 918(210), *941*, 1139(57), 1140(57), *1180*, 1410(119), *1419*
Morin, M.-C., 640(154), *654*
Morioux, J., 725(228), *756*
Morisaki, K., 835(180), *870*, 850(261), *873*, 1039(47), *1062*, 1236(106), *1259*, 1434(13b), *1437*
Morita, M., 455(237, 238), *464*
Morizur, J.-P., 407(52), *428*
Moro, A. D., 346(155), *358*
Morooka, S., 1073(36), *1096*
Morozova, I. D., 473(67), *481*, 488(45), *497*
Morpurgo, G., 154(135), *199*
Morra, B. S., 511(68), *530*
Morris, J. H., 858(301), *874*
Morris, J. I., 368(45), *395*, 886(12), 891(12), *933*
Morris, M. D., 809(10), 837(10), 838(10), *864*
Morris, S., 644(193), 645(193), *656*, 994(155b), *1023*
Morrow, G. W., 781(65a), *786*
Mortensen, J., 332(10), 333(31), 334(33), 336(33), 338(10, 33), 341(10, 31, 33), *353*, 334(42), 335(42), 337(42), 339(42), *354*, 340(86a), *355*, 506(12), 511(12), *529*, 682(123), *695*
Mortreux, A., 836(184), *870*
Moryganov, B. N., 1053(204), *1067*
Moser, F. H., 820(92), 821(92), *867*
Moser, J., 1404(21), *1416*
Moses, P. R., 673(81, 82), *694*, 683(129), *696*, 689(188, 189), 690(188, 189), *698*, 712(71), *750*, 931(360), *947*, 1314(8, 9, 12), 1365(9), 1366(8, 12), 1368(12), *1392*
Moshier, R. W., 453(213), *463*
Mosley, S. E., 275(138), *317*
Moss, G. P., 778(55), *786*
Motoo, S., 639(146, 148), 640(149), *654*, 1390(133–135), *1396*
Mottus, E. H., 857(293), *874*, 1353(70), *1361*
Mouats, C., 707(47c), *749*, 707(48), 708(49), *750*
Mousset, G., 193(290), *205*, 434(19), *457*, 445(159), 449(159), *461*, 449(191), *462*, 1032(4), *1060*, 1038(43), 1039(43), *1061*, 1042(77), *1063*, 1150(166), *1184*, 1236(100), *1259*
Moutet, J.-C., 445(154), *461*, 525(135), *533*, 635(125), 636(125), *653*, 1381(74, 76), *1394*, 1414(149, 150), *1420*
Moyer, B. A., 641(169), 643(169), *655*
Moys, A. J., 1269(33), *1281*
Mozorova, L. N., 926(315), *945*
Mozzanega, M. N., 1408(44, 45), *1416*, 1408(84), 1409(99), *1418*
M'Packo, J.-P., 719(128), *752*, 742(372), *761*
Mpandasoni Banda, F., 1157(229), *1186*
Mu, Y., 1434(12b), *1437*
Mubarak, M. S., 306(410), *326*, 364(14), 366(14, 23), *394*, 367(27), *395*
Muccini, G. A., 1290(37), *1333*
Muck, P. L., 494(116), *500*, 539(26), *575*
Mueller, B., 437(137), *460*
Mueller, T. R., 126(59), *197*
Mugnier, Y., 285(243), *321*, 350(194), *359*, 713(85, 87), *753*, 893(67), *935*, 1177(508), *1194*

Mukai, T., 526(140), *533*
Mukherjee, L. M., 284(220), *320*, 294(314), *323*, 299(350), *324*, 301(375, 376), *325*
Mukherji, S. K., 425(184), *432*
Mukhtarov, A. Sh., 471(52), *480*
Müllen, K., 333(31), 341(31), *353*, 337(65, 66b, 70), 339(75), 340(74, 75, 89), *355*, 351(204, 205), *359*
Müller, C., 725(230), *756*
Müller, E., 436(69), 445(69), 447(69), *458*, 448(187), *462*, 898(130), *938*
Müller, H. P., 602(144), *612*
Muller, N., 550(83, 84, 86), 552(83–86), *577*
Muller, R. H., 179(218), *202*
Müller, U., 243(64), *248*, 569(155), *579*
Müller, W., 1200(12), *1211*
Munin, E. N., 473(61, 62), 474(62), *480*, 1055(218), *1067*
Munoz, E., 162(149), 163(155), *200*
Munro, H. S., 1381(75), *1394*
Munuera, G., 1410(118), *1419*
Murai, N., 570(160), *579*
Murakami, K., 384(142), *398*
Murakami, Y., 417(115), *430*, 573(168), *579*, 823(109), 824(109), *867*, 824(112), 827(140), 829(140), *868*, 1268(23), 1275(23), 1277(23), *1281*, 1277(54), *1282*
Muraki, K., 850(258, 259), *872*
Muraoka, T., 853(271), *873*
Murasaki, M., 977(85), *1020*
Murase, M., 995(156), *1023*
Murata, A., 1434(15j), *1437*
Murata, T., 853(273, 274), 854(274), 855(275b), *873*, 1053(194), *1066*, 1168(340), *1189*
Murata, Y., 686(148, 149), *696*
Murdoch, J. R., 31(41a–c), *116*
Murphy, D., 617(16–18), *650*
Murphy, O. J., 640(151), *654*
Murpurgo, G., 731(274), *758*
Murray, K., 1352(67), *1361*
Murray, R. W., 70(91a), *118*, 175(209), *202*, 234(16a), *246*, 393(210), *400*, 683(129), *696*, 818(77), *866*, 834(168), *869*, 1172(416), *1191*, 1314(1–4, 7–9, 12), 1358(109, 110), 1365(7, 9, 14, 15, 21), 1366(7, 8, 12, 15), *1362*, 1368(12, 15), 1369(1), 1370(7, 14, 15, 21), 1371(21), 1374(1, 21), *1392*, 1366(26, 28), 1368(26, 31), 1369(26), 1371(26), 1372(26), *1393*, 1381(77), *1394*, 1390(127), *1396*, 1409(112), *1419*
Musgrave, O. C., 635(118), *653*
Musgrave, W. K. R., 388(169), *399*, 747(433), *763*
Musha, S., 263(51), *314*
Mushika, Y., 766(5a, b), *783*, 766(6a), *784*
Musker, W. K., 668(60), *693*, 668(64), 669(66, 67), *694*
Mussini, T., 1200(1), *1211*
Musso, H., 960(28b), *1018*
Mustakallio, K. K., 736(324d), *759*
Muthanna, M. S., 1243(132), *1260*
Muthukumarau, A., 496(137), *501*
Muzychenko, L. A., 490(85), *499*
Muzyka, J., 1410(123), *1419*
Myall, C. J., 1082(100), *1098*
Myerly, R. C., 790(13), *804*
Myers, H., 17(12), *114*
Myers, R. S., 819(89), *867*

Naarman, H., 341(91), *356*
Naarova, M., 926(314), *945*
Nabe, K., 437(135a), *460*
Nadjo, L., 39(55e), *117*, 67(85a), 72(85a), *118*, 73(93b), *119*, 122(15), 123(15), *195*, 123(22, 23), 132(22), 133(22, 23), 134(22), 135(22), *196*, 126(56), 133(15), 135(82), *197*, 435(43), *457*, 882(3a, d, f, g, o), 884(4), 885(3a, d, f, g, o, 4), *932*, 895(93), *936*, 918(273a), 919(274), *944*, 1047(136, 137), *1064*, 1366(22), 1367(22), *1392*, 1366(25), 1367(25), *1393*
Nadra, J., 723(198), *755*
Naga, T., 1201(34), *1212*
Nagahiro, Y., 721(150), *753*
Nagai, H., 275(129), *317*, 441(152a), *461*, 596(90), *610*, 597(107), *611*, 1009(228), *1026*, 1428(22), *1430*
Nagai, Y., 853(271), *873*
Nagakura, S., 258(28), *313*, 541(37), *575*
Nagamori, Y., 1319(113), 1326(113), *1335*
Nagao, M., 367(29), *395*, 1410(119), *1419*
Nagasawa, N., 742(382), 743(382), *762*
Nagase, S., 290(275), *322*, 1094(178), *1101*, 1104(3), 1105(3), *1124*, 1106(28, 30–35), 1107(30, 31, 33, 34), 1108(35), 1111(43), *1125*
Nagaura, S., 528(143), *533*, 1073(38), *1096*
Nagy, A. G., 837(202), *871*
Nagy, F., 641(172), *655*
Nagy, G., 277(150), *318*
Nahum, R., 1206(79, 80), *1214*
Nairns, D. P., 331(9), 341(9), *353*
Naito, S., 1408(41), *1416*, 1408(49, 50), *1417*
Naito, T., 263(51), *314*
Naixian, X., 1175(472), *1193*
Najafi, A., 711(63), *750*, 991(170), *1024*
Nakabayashi, S., 301(380), *325*
Nakadate, M., 742(382), 743(382), *762*
Nakagaki, R., 258(28), *313*, 541(37), *575*
Nakagawa, K., 467(12), *479*, 909(231), *942*, 1319(113), 1326(113), *1335*
Nakagawa, M., 337(65), *355*
Nakagawa, Y., 678(101), *695*, 1079(68), 1080(68), 1082(68), *1097*, 1163(293), *1187*
Nakahama, S., 306(406), *326*, 349(188), *359*
Nakahara, T., 851(265), *873*, 904(208), *941*
Nakai, R., 643(187), *655*, 1243(143), *1261*
Nakajima, H., 1141(64), *1181*
Nakajima, I., 832(160), *869*
Nakajima, T., 50(67a), *117*
Nakami, J., 1244(148), *1261*
Nakamichi, J., 1408(47), *1417*
Nakamura, K., 621(65), 624(65), *652*, 782(70b), *786*, 1166(321), *1188*, 1408(85), *1418*
Nakamura, T., 767(12), *784*, 1175(483), *1193*

Nakamura, Y., 1162(285), *1187*
Nakane, S., 621(50), *651*, 1229(68), 1280(58), *1282*
Nakanishi, K., 1094(182), *1101*
Nakanishi, Y., 1229(73), 1230(55), *1258*
Nakano, J., 487(38), *497*
Nakao, S., 1211(127), *1215*
Nakashima, T., 742(382), 743(382), *762*
Nakashita, Y., 721(160), *753*
Nakatani, S., 1008(227), 1012(227), *1026*
Nakaya, J., 488(44, 46), *497*, 725(219), *756*, 926(312), *945*, 926(316), *946*
Nam, W. C., 1035(21), *1061*
Nanao, S., 956(31), *1019*
Nanbu, H., 553(89), *577*
Nanis, L., 258(24), *313*
Nanjundiah, J., 1200(7), *1211*
Nanni, Jr., E. J., 1269(32), *1281*
Nanya, S., 601(133), *612*
Napoli, M., 1109(38), 1110(38), 1111(38, 44, 45), 1112(38, 45, 46), 1123(45), *1125*
Narang, S. C., 527(142), *533*, 983(124), *1022*
Narayanan, B. A., 845(235), *872*, 1241(123), *1260*
Naser-ud-din, 375(98), *397*, 1039(49b, c), *1062*
Naser-ud-Din, N., 543(52), *576*
Nassar, M. M., 792(24), *805*
Nath, B., 455(239), *464*
Nautiyal, A. R., 746(423), *763*
Nayak, B., 1166(325), *1189*
Nazzal, A., 1376(59), 1380(59), *1394*
Nechaeva, O. N., 742(376), *761*
Nechepurnoi, V. D., 894(84a), 896(84a), 903(84a), *936*, 902(170, 176), 906(177), 907(177), 922(170), *939*, 907(186, 188), *940*
Nedelec, J. Y., 277(160), 311(160), *318*
Nedeljkovic, J. M., 1406(28), *1416*
Nedenskov, P., 1243(138), *1261*
Nedungadi, P. A. K., 425(184), *432*
Negoda, E. A., 617(13), *650*
Nehring, K., 981(133), *1022*
Neikam, W. C., 347(168), 348(168), *358*, 508(34), 509(34), 514(34), 516(34), *529*, 930(354), *947*
Neimann, M. B., 453(217), *463*, 1048(141), *1065*, 1059(261), *1068*
Neklutin, V., 899(150), *938*
Nekrasov, L. N., 171(180), *201*, 434(13), *456*, 435(45), *457*, 435(48), *458*, 724(209), *755*, 838(210), *871*, 886(19), *933*, 895(94), *937*
Nelleborg, P., 1414(148), *1420*
Nelsen, S. F., 492(102), *499*, 516(89), 519(89), *531*, 519(109), *532*, 582(5), *607*, 649(233, 234), *657*, 691(195), *698*, 1176(498), 1178(517), *1194*, 1408(37), *1416*
Nelson, A. H., 898(127), *938*
Nelson, J. M., 903(195), *940*
Nelson, L. V., 278(168), *318*
Nelson, R. F., 125(48), *196*, 174(201), *202*, 286(370), 301(370), *325*, 309(430, 432), *327*, 418(134), *430*, 586(27), 587(30), 588(30), 589(30), 590(27, 41), *608*, 590(42, 44), 592(51, 56, 58, 59), *609*, 711(64), *750*, 723(189, 190), *754*, 723(191, 194a, b), *755*, 733(300), *759*, 886(21), *933*, 968(68, 69), 969(70, 72), 970(68), 973(80), 977(81, 84), 988(69), *1020*, 973(87), 979(87, 89), *1021*, 1382(84), *1395*
Nemec, I., 591(45), *609*
Nemec, M., 907(185b), *940*
Nenadovic, M., 1406(26, 28), *1416*
Nepenin, Y. N., 793(36, 38), *805*
Netta, P., 350(194), *359*, 1175(469), *1193*
Neubert, G., 969(71), *1020*
Neugebauer, L., 721(161b), *753*
Neumann, H. M., 51(68), *117*, 493(109), *500*, 1048(144), *1065*, 1170(358), *1190*
Neumann, W. P., 243(67), *248*, 857(289), *873*
Neumeyer, J. L., 703(14), *748*, 771(30), *785*, 889(36), *934*, 927(322), *946*
Neundlinger, K., 725(213), *755*
Nevin, W. A., 820(91), 821(91), *867*
Newberg, J. H., 476(90), *481*, 1148(155), *1183*
Newman, J., 168(171, 172), *201*, 273(123c, d), *317*
Newman, J. S., 81(104a), *119*, 208(2), 209(5a), *231*, 1286(7), 1292(7), 1295(7), 1296(7), 1297(7), *1332*
Newman, T. H., 855(281), *873*
Ng, P. K., 273(123b), *317*
Nguyen-Van-Thanh, N., 899(148), *938*
Nichols, G. M., 1059(259), *1068*
Nicholson, M. D., 790(13), *804*
Nicholson, M. M., 667(56), *693*
Nicholson, R. S., 123(17–21), 129(18), 132(18), 133(18), *196*, 126(54), 131(71), *197*
Nickel, H., 1048(143), *1066*, 1059(262), *1068*
Nickell, D., 1210(122), *1215*
Nicolas, M., 906(183b, c), *940*
Nicolaus, G., 1164(305), 1165(305), *1188*
Nieft, J. W., 724(208), *755*
Nielsen, J. T., 1243(138), *1261*
Nielsen, M. F., 123(30), 133(30), *196*, 125(46), 127(62), 140(46), *197*, 140(97, 98), 141(97, 98), 142(97, 98), 143(99), 148(97, 98), 150(97), *198*, 345(140–142), *357*
Nigenda, S. E., 527(142), *533*
Nigrette, J.–M., 882(3p), 885(3p), *932*
Nigrey, P. J., 331(9), 341(9), *353*
Nikitin, E. V., 1122(76a, b, 77, 78a, b), 1123(79), *1127*
Nikolaev, G. M., 1176(505), *1194*
Nikolaeva, N. N., 644(198), 645(198), *656*
Nikulin, V. N., 452(208d–f), 453(208e, f), *462*
Nilsson, A., 617(20), 620(30), 621(30), 624(30), *650*, 620(31), 621(31, 60), 624(31, 60), *651*, 630(106), 632(106), 633(106), *653*, 704(20), *748*, 959(43), *1019*, 987(139), 990(161), 996(161), 998(161), *1023*,

[Nilsson, A.]
1071(26), 1073(26), 1074(40–42), *1096*, 1089(150), 1090(150), 1093(169), *1100*, 1156(217), *1185*, 1312(85), *1335*
Nilsson, N. H., 1177(506), *1194*
Nilsson, S., 528(146), *533*
Nishida, R., 774(40), *785*
Nishida, T., 1008(226), *1026*
Nishiguchi, I., 5(16), *8*, 239(426), 240(426), *247*, 374(87), *396*, 404(30), *427*, 451(205), 453(205), 456(205), *462*, 585(23), *608*, 882(3), 885(3), *932*, 891(49), *934*, 904(209), 905(212), 908(212), *941*, 922(278), *944*, 1000(172), *1024*, 1034(15), *1060*, 1035(23), *1061*, 1138(47), 1139(56), 1140(56), *1180*, 1141(70), *1181*, 1159(241), 1160(257), *1186*, 1163(298), *1188*, 1207(84), *1214*, 1252(196, 197), *1263*
Nishiguchi, S., 1036(26), *1061*
Nishiguchi, T., 777(49), *785*
Nishimoto, S., 467(12), *479*, 1408(59, 60, 63, 64, 66, 67), *1417*
Nishio, K., 267(82), 271(82), 288(82), 301(82), 307(82), *315*
Nishio, O., 392(192), *399*, 886(18), *933*, 1269(35), *1281*, 1275(48), *1282*
Nishitani, T., 559(119), 562(119), *578*, 573(167), *579*, 766(5a, b), *783*, 766(6a–c), 768(19), 769(20, 21), *784*, 1162(284), *1187*, 1243(137), 1244(137), *1261*, 1426(15), *1430*
Nisiyama, A., 621(84–87), 627(84–87), *652*, 775(43a–c), 776(44b, 46a, b, 47), *785*, 958(40), 962(45), *1019*, 1013(251), *1027*, 1161(271), *1187*, 1163(296, 297), *1188*
Nishiyama, N., 1278(57d), 1279(57d), 1280(57d), *1282*
Nisliki, Y., 1298(52), *1334*
Nitta, M., 777(49), *785*, 1035(23), *1061*
Niwa, O., 1387(112), 1389(112), *1396*
Niyazymbetov, M. E., 715(102), *751*
Nobe, K., 275(131), *317*, 439(92), 440(92), 441(92), *459*, 793(43), *805*, 1421(1c–e), 1422(1c, d), 1424(1c, e), *1429*, 1433(3), *1436*, 1435(24a), *1438*
Nobusue, Y., 705(33), *749*
Noel, M., 417(126b), *430*, 437(138), *460*, 894(83), *936*
Nogami, T., 663(39), *693*
Noguchi, I., 957(35), *1019*, 991(169), *1024*, 1094(175), *1101*, 1166(320), *1188*
Nogusa, H., 404(27), *427*, 493(112), 494(112), *500*, 924(298), *945*, 1268(24), 1275(24b), *1281*
Nohe, H., 486(23), *497*, 894(85b, 86), *936*, 1142(73), *1181*, 1288(31), 1292(31), *1333*, 1305(76), 1309(80), *1334*, 1316(98), 1319(98), *1335*, 1321(121), *1336*
Noike, Y., 447(183), 448(183), *462*
Nokami, J., 537(6, 8, 9, 10), 572(8), 573(6, 10), *574*, 600(120), *611*, 679(109, 110), *695*, 780(62a, 63), *786*
Nomura, T., 162(153), *200*, 306(407), *326*, 455(229), *463*
Nomura, Y., 404(26b), *427*
Nonaka, T., 436(63), *458*, 437(135a), *460*, 476(92), *481*, 484(9), *496*, 496(134), *500*, 540(32), *575*, 584(22), *608*, 641(168), 643(168), *655*, 675(92), *694*, 678(101), *695*, 740(355), *761*, 791(21), *804*, 794(52, 59, 61), 795(48, 52), 796(52, 59), 797(61), *806*, 891(59), 892(60), 894(73b), *935*, 902(168b), 906(177, 180), 907(177), *939*, 907(187, 194), *940*, 911(269), *943*, 922(276), *944*, 906(307), *945*, 927(321), *946*, 975(103), 976(103), *1021*, 1035(20), *1061*, 1040(68), *1062*, 1045(111), *1064*, 1079(68), 1080(68), 1082(68), *1097*, 1087(133), *1099*, 1132(5–7, 16), 1171(9), 1173(16), *1179*, 1136(35), 1139(52, 53), *1180*, 1141(65), 1142(83), 1143(65, 84, 85, 93), 1173(83, 84), *1181*, 1143(95), *1182*, 1146(139, 141), 1147(147, 148), *1183*, 1153(187), 1154(195), *1184*, 1156(220), 1157(220), 1166(220, 330), *1185*, 1169(352), *1189*, 1169(356), *1190*, 1172(400, 414), 1173(421, 422), *1191*, 1173(423–427, 431, 434, 435, 439), 1174(444), *1192*, 1178(512, 515), *1194*, 1421(1c), 1422(1c), 1424(1c), *1429*
Nonhebel, D. C., 620(33–39), *651*
Nordblom, G. D., 508(45), *530*
Nordio, P. I., 434(10), *456*
Nordmann, R., 985(131), *1022*
Norman, N. C., 411(77), *428*
Norman, R. O. C., 347(172), 348(172), *358*, 1080(82), *1098*
Normant, H., 298(342), *324*
North, A. M., 38(51b), *117*
Norton, J. R., 809(2), *864*
Nosaka, Y., 1405(22), *1416*, 1408(68), *1417*
Nose, K., 425(187), *432*, 599(118), *611*
Noufi, R., 1382(85, 86), 1383(85, 95), *1395*
Novák, M., 446(173), *462*, 1105(17, 20), 1109(17, 19), 1110(17), 1112(17), *1124*
Novikov, S. S., 378(124), *397*, 403(12), *426*, 443(119), *460*
Novikov, V. T., 404(29c, d), 408(27c, d), *427*, 408(58, 59), *428*, 466(6), *479*, 470(49), 476(49), *480*, 490(84, 85), *499*, 743(387), *762*
Novoa, J. J., 437(130), *460*, 455(236), *464*
Nowak, R., 1365(21), 1370(21), 1371(21), 1374(21), *1392*, 1390(127), *1396*
Noyes, R. M., 24(27c), 42(27c), *115*
Noyori, R., 1163(301), *1188*, 1233(90), 1235(90), *1259*
Nozaki, T., 455(234), *464*

Nozik, A. J., 1382(85), 1383(85, 95), *1395*, 1398(12, 16), 1405(12), *1415*, 1406(26, 28), *1416*
Nucci, L., 916(267, 268), *943*, 1346(23), *1360*, 1352(57), *1361*
Nugent, S. T., 1267(12a), *1280*
Nunjundiah, C., 1203(45), *1213*
Nunome, K., 50(67b), *117*
Nutt, R. F., 546(67), *576*, 1002(203), *1025*
Nyberg, K., 237(31, 35), *247*, 240(52), 242(52, 62), *248*, 260(42), *314*, 266(76a), 267(76), *315*, 286(291), 291(291), *322*, 302(386), 311(386), *325*, 345(145), *357*, 483(5), 493(5), 495(5), *496*, 507(31), 508(36), 509(38), 519(31), *529*, 523(129, 130), *532*, 528(150), *533*, 542(44), 543(44), *575*, 601(138–142), 602(142), 603(138), *612*, 604(158), *613*, 727(246), *757*, 951(4), 952(2–5), 953(2), 954(13, 14), 955(13), 960(2), *1018*, 1070(3, 6, 7, 19), 1073(19), 1080(19), 1081(3), 1084(19), *1095*, 1073(34), 1077(49), *1096*, 1077(53, 56–60), 1078(64), 1080(53, 83), 1081(89), 1082(83), 1083(53, 58), *1097*, 1083(106), *1098*, 1092(166, 167), *1100*, 1119(63b, 64), *1126*, 1160(265), *1187*, 1170(365, 366), *1190*
Nygard, B., 423(160), *431*, 662(28), *692*
Nyman, C. J., 284(205), 285(205), 293(205), 295(205), 286(205), *319*
Nyuyen, T. V., 1301(58), *1334*
Nyvad, A. I., 1414(146), *1420*

Oae, S., 537(7), *574*
Oberhoff, J., 442(108), 447(108), *459*, 450(201), *462*
Oberlin, R., 258(20), *313*
Oberrauch, E., 1072(27), *1096*, 1080(73), 1081(73), *1097*, 1425(7), *1429*
Oboodi, K., 1357(96), *1362*
O'brien, T. J. P., 145(103), *198*
Obushak, N. D., 424(172), *432*
Occhialini, D., 708(50), *750*
Ochi, N., 537(6), 573(6), *574*
Ochiai, E., 742(373), *761*
Ochiai, M., 489(56), *498*, 740(353), *761*
Ochoa, M., 1142(81), *1181*
O'Connell, K. M., 343(107), *356*, 368(48), *395*, 1176(500), *1194*
Oda, R., 1159(241), *1186*
Odaira, Y., 543(50), *576*
O'Dea, J. J., 387(157), *398*, 1155(206), *1185*
Odo, K., 747(432), *763*, 755(42), *785*, 894(73b), 902(168b), 906(177, 180), 907(177), *939*, 907(187), *940*, 775(42), 906(307), *945*, 1058(255), 1059(255), *1068*, 1146(139), *1183*, 1169(352), *1189*
O'Donnell, J. F., 296(324), *323*, 601(134), *612*
O'Driscoll, K. F., 1339(2), *1359*
Oehling, H., 886(12), 891(12), *933*
Oei, H. A., 288(260), *321*, 441(149d), *461*, 895(98d), *937*, 1172(412, 413), *1191*
Oelschläger, H., 719(133), 733(133), *752*, 734(305, 310), *759*
Öfele, K., 1175(450), *1192*, 1175(455), *1193*, 1240(120), *1260*
Oganesyan, A. S., 445(161), *461*
Ogasawara, M., 1383(89), 1387(89), *1395*
Ogawa, H., 794(56), 796(56), *806*, 801(86, 87), *807*, 1410(123), *1419*
Ogawa, N., 1316(101), 1319(101), 1326(101), *1335*
Ogawa, S., 271(105), *316*, 1316(101), 1319(101), 1326(101), *1335*
Ogg, C. L., 1053(201), *1066*
Ogiso, M., 819(88), *867*
Ogle, C. A., 233(5d), 234(5d), 240(5d), *246*
Ogo, H., 721(154), *753*
Ogoshi, H., 815(57, 58, 60), *866*
Oguchi, T., 1406(31), *1416*, 1409(92), *1418*
Ogumi, Z., 267(82, 84), 268(84), 271(82, 84), 288(82, 84), 301(82), 307(82, 84), *315*
Ogura, K., 1245(149), *1261*, 1433(5b, c, e–g), *1436*, 1434(15c, f, h, k), *1437*
Ogura, T., 436(66), *458*, 895(111), 897(111), 902(111), *937*
Ohashi, S.–I., 267(84), 268(84), 271(84), 288(84), 307(84), *315*
Ohira, K., 742(385), 743(385), *762*
Oh-Ishi, T., 767(12), *784*
Ohkatsu, Y., 1246(167), *1262*
Ohki, S., 447(183), 448(183), *462*
Ohkubo, K., 1174(446), *1192*
Ohman, V., 551(97), *577*
Ohmizu, H., 5(16), *8*, 239(426), 240(426), *247*, 367(35), *395*, 374(87), *396*, 451(205), 453(205), 456(205), *462*, 589(40), 590(40), *608*, 882(3), 885(3), *932*, 891(49), *934*, 903(199), 907(192), *940*, 904(209), 905(212), 908(212), *941*, 919(289), 922(278), *944*, 1138(47), 1139(54, 56), 1140(56), *1180*, 1141(70), *1181*, 1173(440), *1192*, 1207(84), *1214*, 1252(196, 197), *1263*
Ohmori, H., 425(187), *432*, 474(79), 475(79), *481*, 556(109), *577*, 594(85), 595(86, 87), *610*, 597(103), 599(118), *611*, 621(56), *651*, 705(33), *749*, 956(33), *1019*, 983(127), *1022*, 1087(132), 1089(143), *1099*
Ohmori, M., 443(117), *460*, 621(63), 624(63), *652*, 900(152), *938*
Ohnesorge, W. E., 524(132), 525(134), *533*, 986(138b), 993(138b), *1023*
Ohnishi, S., 663(39), *693*
Ohnishi, T., 1034(15), *1060*, 1163(298), *1188*, 1408(50), *1417*
Ohnishi, Y., 725(221a, b), *756*
Ohno, A., 284(226), 303(226), *320*
Ohno, M., 1037(33), *1061*
Ohno, T., 823(109), 824(109), *867*, 824(112), 827(140), 829(140), *868*

Ohnuma, T., 775(42), *785*, 1058(255), 1059(255), *1068*
Ohsaka, T., 620(32), *651*
Ohsawa, Y., 831(154), *869*
Ohshima, T., 780(62b), *786*, 1162(286), *1187*
Ohshiro, Y., 971(77a), *1020*
Ohshita, J., 708(51), *750*
Ohtani, B., 467(12), *479*, 1408(59, 60, 63, 64, 66, 67), *1417*
Oida, T., 564(135), 567(135), *578*
Oiliff, C. J., 915(259), 916(259), *943*
Oka, H., 555(101), *577*
Okabe, M., 825(121, 122), 827(138), 828(138), *868*
Okada, G., 1434(20a), *1438*
Okada, K., 1004(209), *1025*, 1409(106), *1419*
Okada, T., 740(353), *761*
Okamoto, G., 272(111), *316*
Okamoto, H., 496(131), *500*, 640(150), *654*
Okamoto, K., 1268(23), 1275(23), 1277(23), *1281*, 1408(61), *1417*
Okamoto, T., 539(21), *575*, 563(129, 130), 564(135), *578*, 721(155), *753*, 1034(14), *1060*
Okano, T., 742(385), 743(385), *762*
Okawa, M., 1000(172), *1024*, 1160(257), *1186*
Okawara, R., 537(6, 9), 573(6), *574*, 679(110), *695*
Okhlobystina, L. V., 407(49), *458*
Okimoto, M., 275(129), *317*, 441(152a), *461*, 596(90), *610*, 597(107), *611*, 1009(228), *1026*, 1428(22), *1430*
Okinaka, Y., 256(3a), *312*
Okita, M., 603(149), *613*, 773(37a), 774(39b), *785*, 1077(55), *1097*
Okoh, E., 1230(54), *1258*
O'Konski, C. T., 291(286), *322*
Oku, M., 336(57), *354*
Okubo, T., 1005(212b, 219), 1006(212b), *1025*
Okumoto, H., 677(105), 679(105), *695*
Okumura, K., 1058(250), *1068*
Okuno, Y., 1013(251), *1027*
Okuyama, T., 686(155), *697*
Olah, G. A., 233(3), 234(3), *245*, 983(124), *1022*, 1105(14), 1106(14), 1109(14), *1124*
Olaj, O. F., 1348(28), 1349(34), *1360*
Ol'dekop, Y. A., 837(198), 838(198), *870*, 1425(12), *1430*
Oldfield, J. W., 268(89b), 271(89b), *315*
Oldham, C., 858(300), *874*
Oldham, K. B., 135(78, 79, 80), *197*, 148(111), 150(111), *198*, 373(76), *396*
O'Leary, B. F., 621(83), 627(83), *652*
Oleinik, A. V., 425(194), *432*, 445(171), *461*
Olin, G. R., 721(157d), *753*
Oliva, P., 1123(81), *1127*
Oliver, B. G., 1413(141), *1420*
Oliver, B. N., 1435(31d), 1436(31d), *1438*
Ollis, D. F., 1413(130), *1419*
Ollmann, G. W., 470(51), *480*, 491(95), *499*, 494(118), *500*, 851(262), *873*
Olmstead, H. D., 983(124), *1022*
Olmstead, M. L., 123(20, 21), 133(20, 21), *196*
Olofsson, B., 237(33), *247*, 511(70), 512(70), *531*, 1072(32), *1096*, 1156(218), *1185*
Oloman, C., 793(33), *805*, 1303(71), *1334*, 1421(4a), *1429*
Olsen, B. A., 1175(467, 471), *1193*
Olson, D. C., 295(316), *323*
Olson, D. R., 918(272), 921(272), *944*, 1135(22), *1179*, 1136(32), *1180*
Olton, W. M., 1164(310), *1188*
Olver, J. W., 295(322), *323*
O'M. Bockris, J., 122(7), 123(7), 157(7), 179(7), *195*, 285(213), *320*, 335(44), *354*
O'Malley, R. F., 1094(183), *1101*, 1117(55, 60–62), 1118(60, 61), 1119(55, 62), *1126*
Omori, A., 1089(147), *1100*
Omoto, H., 570(157), *579*, 781(66), *786*
Omoto, M., 278(173), *319*
Onal, A., 1352(63), *1361*
Onisciu, L., 909(232), *942*
Onishi, M., 836(185), *870*
Onishi, T., 1008(226), *1026*
Ono, M., 664(52–54), 666(52–54), *693*, 778(52), *785*, 1164(313), 1165(317), *1188*, 1243(135), 1244(135, 146), *1261*
Ono, S., 486(19, 27), 487(37), 489(19), *497*, 488(44, 46), *497*, 1142(71), *1181*
Ono, Y., 434(6), *456*
Onoe, J., 1408(74), 1411(74), *1418*
Onomura, O., 602(145), *612*, 1160(254), *1186*
Onstott, E., 438(77), 446(77), 450(77), *458*
Opekar, F., 165(166), 169(166), *200*
Orange, O., 424(171), *432*
Orchard, A. F., 174(197), *201*
O'Reilly, J. E., 732(280), *758*
Orekhova, T. S., 495(120), *500*, 740(356), *761*
O'Riordan, M. P., 826(132), *868*
Orliac-Le Moing, A., 1013(248), *1027*, 1032(3), 1144(130, 131), 1145(130, 131), *1060*, 1166(323), *1183*, *1188*, 1231(67), 1246(67), *1258*, 1241(125, 127), *1260*, 1250(184), *1262*
Orlinski, R., 239(44), 240(44), *247*, 492(99), *499*, 813(37), 825(37), 826(37), 827(37), 841(37), 863(37), *865*, 824(119), 826(119), 827(135), *868*
Orlova, L. N., 466(4), 473(4), *479*, 471(52), *480*
Oroshnik, W., 489(66), *498*
Ort, M. R., 239(45), *247*, 451(206), 456(206), *462*, 468(27), *479*, 857(293), *874*, 891(47), *934*, 909(225), *942*, 918(271a), *944*, 924(299), *945*, 929(328), *946*, 1277(53), *1282*, 1353(70), *1361*
Ortiz, A., 916(265), *943*,
Orton, K. J. P., 291(294), *322*
Osa, T., 272(114), 275(114), *316*, 275(127), *317*, 302(382), *325*, 346(151), *357*, 379(128), *398*, 416(114), *430*, 443(123), 445(123), 455(123), *460*,

[Osa, T.]
445(155), *461*, 455(240), *464*, 739(342), *760*, 914(238, 241), *942*, 960(21), *1018*, 1173(419, 420), *1191*, 1204(52), *1213*, 1246(167–169), 1247(175), *1262*, 1277(55), *1282*, 1319(118), *1336*, 1366(27), *1393*, 1409(90, 91), *1418*
Osaki, H., 1408(63, 66, 67), *1417*
Osawa, K., 1004(209), *1025*
Oshino, N., 584(19), *608*, 779(61a), *786*, 1160(262), *1186*
Osipova, G. Y., 794(53), 795(53), *806*
Ossendorfová, N., 660(12), *692*
Osteryoung, J., 153(132), *199*, 363(10), *394*, 862(321), *874*
Osteryoung, R. A., 150(120, 121), 151(120), *199*, 303(392, 393, 394), *326*, 862(321), *874*, 1384(102, 103), *1395*
Ota, T., 1169(352), *1189*, 1169(356), *1190*
Otega, P., 290(271), *322*
Otera, J., 679(106, 107), *695*, 1046(123), *1064*
Oth, J. F. M., 336(62), 339(62), *354*, 336(64), 338(67), 339(75), 340(74, 75), *355*
Otis, D. E., 835(182), *870*
Oto, S., 1054(213), *1067*
Otte, K. B., 309(431), *327*
Otto, D., 272(112), *316*
Otto, E., 1206(77), *1214*
Oturan, M. A., 81(103), 109(103), *119*, 1239(114), *1260*
Oudard, J. F., 303(393), *326*
Oudrhiri-Hassani, M., 691(193), *698*
Ovanesova, G. S., 907(190), *940*
Overberger, C. G., 1159(245), *1186*
Owen, D. M., 553(88), *577*, 1010(236), *1026*
Owlia, A., 122(12), *195*
Owton, W. M., 664(50), 665(50), *693*, 713(77), *751*, 1054(212), *1067*
Oyama, M., 1037(33), *1061*
Oyama, N., 620(32), *651*, 1365(18, 19), 1368(19), 1370(18, 19), 1371(18, 19), *1392*, 1372(43), 1374(42), *1393*, 1382(79), *1395*, 1390(131), *1396*
Ozaki, S., 599(112, 115–117), 600(124), *611*, 600(126), 601(129), *612*, 743(388), *762*, 1036(26), *1061*, 1044(91), *1063*, 1178(516), *1194*
Ozaki, T., 827(140), 829(140), *868*
Ozaki, Y., 766(16), *783*, 766(7a, b), *784*

Pac, C., 1408(54, 65), *1417*, 1408(88), *1418*, 1409(105), *1419*
Pace, S. J., 732(280), *758*
Pachori, R., 425(184), *432*
Pacifici, J. A., 368(45), *395*, 886(12), 891(12), *933*
Pacifici, J. G., 350(196), *359*
Pacifici, S. A., 886(20f), 891(20f), *933*
Pack, J. G., 647(222), *657*
Packard, R. T., 179(227), *202*
Pacut, R. I., 1202(43), *1213*
Padilla, A. G., 686(165), *697*
Padmanabhan, G. R., 281(191), *319*, 295(321), *323*
Padova, J. I., 1201(34), *1212*
Paffett, M. T., 344(125), *357*
Page, D. J., 190(280), 195(280), *204*, 645(208), *656*, 1436(31d), *1438*
Page, J. A., 843(226), *871*, 1168(345), *1189*
Pagni, R. M., 506(28), *529*
Pagsberg, P., 341(95), 350(95), *356*
Pai, B. R., 418(128b), *430*
Paine, J. B., 819(90), *867*
Pak, C. M., 886(22), 908(22), *933*
Pake, G. E., 332(12b), *353*
Pakulski, M., 411(77), *428*
Palasz, P. D., 604(159), *613*, 773(38b), *785*, 793(34), *805*, 802(90, 91), *808*, 1077(61), 1078(62), *1097*, 1161(275), *1187*
Palazzotto, M. C., 1366(22), 1367(22), *1392*, 1366(25), 1367(25), *1393*
Palffy, G., 486(35), *497*
Pallaud, R., 863(330), *875*, 906(183b, c), *940*, 1006(222), *1025*
Pal'm, U. V., 435(54), *458*
Palmer, J. L., 410(74), *428*, 583(10), *607*
Palmer, M. R., 1413(135), *1420*
Palmisano, G., 716(115), *752*
Palmisano, L., 930(345a), *947*, 1433(9a), *1436*
Palmquist, U., 617(20), *650*, 620(31), 621(31), 624(31), *651*, 621(71, 72), 625(71, 72), *652*, 630(106), 632(106), 633(106), *653*, 704(20), *748*, 959(43), *1019*, 987(139), 990(161), 996(161), 998(161), *1023*, 1074(40, 42), *1096*, 1093(169), *1100*
Pannhorst, W., 824(113), *868*
Panzer, R. E., 276(143), *318*
Papadopoulos, A., 604(162), 605(162), 607(162), *613*
Pape, J.–Y., 918(271c), *944*, 1048(140), 1056(140), *1065*
Papouchado, L., 593(60), *609*, 620(28, 29), 621(29), 625(28), *650*, 620(46, 47), 621(47), *651*, 1072(28), *1096*
Pappo, R., 886(16), 908(16), *933*, 1269(38), *1282*
Paquette, L. A., 335(54), 336(57, 61), *354*, 337(68), 340(88), *355*, 1176(489), *1194*
Parakin, O. V., 1122(77), *1127*
Paranov, V. V., 407(50), *428*
Parashuv, S., 1000(173), *1024*
Pardini, V. L., 662(31), 670(31), 671(31), *692*, 801(81), *807*, 1267(5), 1270(5), *1280*
Pardo, R., 726(245), *757*
Pardy, R. B. A., 717(120), *752*
Parfitt, G. D., 1410(116b), *1419*
Park, K., 275(131), *317*, 439(92), 440(92), 441(92), *459*, 437(135a), *460*, 1412(128), *1419*, 1421(1c–e), 1422(1c, d), 1424(1c, e), *1429*, 1433(3), *1436*, 1435(24a), *1438*
Park, K. H., 1409(100), *1418*
Park, S. C., 1035(21), *1061*
Park, S.–M., 153(129), *199*, 179(234), *203*, 344(125), *357*, 1434(22a), *1438*
Párkányi, C., 508(43), *530*, 1070(5), *1095*
Parker, A. J., 287(246), 288(246), *321*

Parker, D. P., 184(252), 186(252), *203*, 1381(75), *1394*
Parker, E., 439(90), *459*
Parker, S. K., 790(10), *804*
Parker, V. D., 26(31f), *115*, 31(38a), *116*, 68(89), 70(91b), *118*, 122(1, 2), 123(1, 2), *195*, 122(16), 123(30–42), 126(30–42), 127(33, 36), 133(36, 40, 42), 135(38), 136(37–39), 139(40), 140(36, 37, 40, 46), 142(36), *196*, 124(45, 46, 50, 51, 52), 127(61), 128(64, 66), 133(74, 75), 134(76), 138(76, 85), 140(75), *197*, 139(86), 140(89, 91, 92, 94, 97, 98), 148(97, 113), 150(97, 116, 117), *198*, 150(119), *199*, 163(157), *200*, 170(179), *201*, 174(207, 208), 175(207), 177(212, 213), 190(207), *202*, 184(252), 186(252), *203*, 188(269), *204*, 234(7a, b), *246*, 260(41), 294(41), *314*, 278(176), *319*, 284(237), 294(237), *320*, 291(287, 288), 302(288), *322*, 301(379), *325*, 331(8a, b), 333(8a), 335(8b), 338(8b), 349(8a, b), *353*, 336(55), 342(55), *354*, 341(95), 350(95), *356*, 345(134–138, 140–142), *357*, 349(178), *358*, 413(92), *429*, 453(223), *463*, 478(103–105, 107, 108, 110), *482*, 506(6–9), 509(6), 510(6), 511(6–8), *528*, 506(10, 13–15, 26), 507(30, 31), 508(13, 37), 509(10), 510(14), 511(10, 15), 519(30, 31), 524(30), *529*, 509(40), 510(40, 52, 54), 511(40, 59–61, 64), *530*, 511(71, 73, 75), 512(71), 513(73, 75), 514(71), *531*, 519(111), 520(112, 116), 521(112, 116, 119, 120), 524(111), *532*, 524(133), 528(145), *533*, 593(71), 596(93), *610*, 617(20–22), 620(22, 30), 621(30), 624(30), *650*, 620(44), 621(57, 58, 60), 624(58, 60), *651*, 621(71, 72, 77, 78), 625(71, 72), 628(90), 629(91, 93), *652*, 634(116, 117), 635(116), *653*, 659(2), *692*, 684(132–134), 685(133, 136), 686(150, 153), 687(150, 153, 175), *696*, 686(171, 172), 689(184–186), *697*, 704(20), *748*, 706(35), *749*, 839(216), 841(221), *871*, 882(3s, t), 885(3s, t), *932*, 895(92), *936*, 909(237), *942*, 951(19, 20), 953(8, 9), 954(10), 955(11, 16), 960(23), *1018*, 963(49–51), 965(53), 966(55, 56), *1019*, 963(65), 966(57), 967(59, 61), 972(76), 977(76), *1020*, 980(118), 982(117, 118), 985(135), *1022*, 985(136, 137, 140), 986(138a), 987(139, 140), 990(161), 993(141), 996(161), 998(161), 1001(137), *1023*, 1001(175), *1024*, 1043(82, 84), *1063*, 1070(12, 13), *1095*, 1071(26), 1072(31, 33), 1073(26), 1074(39, 41), *1096*, 1081(88), *1098*, 1084(114, 115), 1088(140), 1089(142), *1099*, 1089(146, 152), 1090(152), 1093(168, 169), *1110*, 1160(255), *1186*, 1161(266), *1187*, 1175(468, 470), *1193*, 1415(151), *1420*
Parker, W. E., 1053(201), *1066*, 1053(202), *1067*
Parkin, J. C., 1005(212a), 1006(212a), *1025*
Parkinson, B., 1398(15), *1416*
Parravano, G., 1348(26), *1360*
Parsons, I. W., 1110(41), *1125*
Parsons, R., 276(142), *317*
Parushev, S., 704(26), *749*
Pascal, Y. L., 894(80, 81), *936*, 1146(135), *1183*, 1236(101), *1259*
Paspaleev, E., 453(215b), *463*, 476(89b, c), *481*
Passerini, L., 1413(129), *1419*
Pasternak, R., 260(48), 263(48), *314*, 435(24), 452(24), *457*, 1175(479), *1193*
Patankar, A. D., 1243(132), *1260*
Patel, P. J., 989(163), *1023*
Patel, V. F., 827(141), 829(141), *868*
Patel, V. V., 682(126), *696*
Pathy, M. S. V., 265(70), 271(70), *315*, 425(190), *432*, 1230(58), 1231(58), 1232(58), 1234(58), *1258*
Patin, H., 1241(123), *1260*
Pattanayak, N. K., 1166(325), *1189*
Pattenden, G., 827(139, 141), 829(139, 141), *868*, 905(213), *941*, 1139(60), 1140(60), *1181*, 1408(56), *1417*
Pattison, F. L. M., 1002(187), *1024*
Patzak, R., 721(161b), *753*
Patzner, H., 931(366), *947*
Paudler, W. W., 1159(246), *1186*
Paul, D. E., 332(12a, b), *353*
Paul, E. W., 1383(93), 1385(93), *1395*
Paulss, V., 715(108), *752*
Pauson, P. L., 620(33–38), *651*
Pavelich, B. J., 180(238), *203*
Paviensky, L., 689(181), *697*, 722(164c), *754*
Pavlik, J. W., 1408(70), *1417*
Pavlov, V. N., 435(26, 33, 36), 436(36), *457*
Payfer, R. M., 1106(27), 1108(27), *1124*
Payne, D. A., 621(67), 625(67), *652*
Payne, J. D., 846(239), *872*
Pazdro, K. M., 722(179), *754*, 1049(158a), *1065*
Pearce, D. J., 1434(12i, 15g), *1437*
Pearce, R., 855(282), *873*
Pearl, I. A., 619(26), 621(26), 624(26), *650*, 901(161), *939*, 958(41), *1019*
Pearlson, W. H., 1103(1), 1104(8), 1105(8), 1106(8), *1124*
Pearson, R. G., 522(124), *532*, 861(316), *874*
von Pechmann, H., 1209(111), *1215*
Peck, W. F., Jr., 179(228), *202*
Pedersen, C. T., 177(213), *202*, 689(184–186), *697*
Pedersen, S. U., 392(193), *400*, 466(3), *479*, 735(320), *759*, 909(164), *939*, 1136(36), *1180*, 1153(188), 1154(188), *1184*, 1176(504), *1194*, 1224(31), 1249(31), *1257*

Pedler, A. E., 377(120), *397*, 553(88), *577*, 630(102), *653*, 1010(236), *1026*, 1157(225), *1185*
Pedriali, R., 424(181), *432*
Peek, R., 1006(213), *1025*
Peerce, P. J., 282(204), 285(204), *319*
Peet, N. P., 307(413), *327*
Pelizetti, E., 1413(131), *1419*
Pellegrini, S., 831(158), 832(158), *869*
Peltier, D., 257(14, 16), 263(16), 271(16), *313*, 263(52), *314*, 418(131), *430*, 707(46), 709(46), *749*, 712(69), *750*, 1171(380, 381, 388), *1190*, 1171(391), *1191*, 1173(430, 438), *1192*
Pemawansa, K. P. W., 1355(84), *1361*
Pemberton, J. E., 511(55), *530*, 686(156), *697*
Penco, S., 1040(64), *1062*
Pendarvis, R. D., 733(299), *758*
Peng, H., 1173(432), *1192*
Penner, R. M., 1374(48), 1387(48), *1393*, 1387(116), 1389(116), 1389(121), *1396*
Pennesi, G., 821(97), *867*
Penney, T., 682(126), *696*
Penzlin, G., 240(47), *247*
Peover, M. E., 23(25), *115*, 39(55d), *117*, 162(150), *200*, 331(2, 4), 332(2), 334(2, 4), 342(2), *352*, 332(16), 333(24, 25), *353*, 335(43), *354*, 341(94), 350(94), *356*, 347(169), 348(169), *358*, 411(84), *429*, 434(9), *456*, 491(92), *499*, 508(46), 510(48), *530*, 887(23), *933*, 1138(51), *1180*, 1245(151, 155), *1261*, 1377(82), 1382(82), *1395*
Peregudova, S. M., 838(210), *871*
Peresleni, E. M., 726(241), *757*
Pérez-Hiyareda, G., 57, *90*
Periasamy, M., 525(136), *533*
Perichon, J. L., 4(13), *8*, 277(160, 161), 311(160), *318*, 284(223), 285(231), 286(223), 302(223), *320*, 366(24), *394*, 393(204, 205, 209), *400*, 831(157, 158), 832(157, 158), 834(172), 835(172), *869*, 835(176, 179), *870*, 891(48), *934*, 1236(105, 108), *1259*, 1252(192), *1263*, 1434(13f), *1437*
Perjessy, A., 915(258), *943*, 1144(98), *1182*
Perkin, W. H., 442(115), *460*
Perone, S. P., 123(43), *196*, 126(59, 60), *197*, 441(144), *461*, 645(211), 646(211), *656*
Perrin, L. L., 157(144), 161(144), 162(144), *199*
Perrin, M., 1233(91), 1235(91), 1236(100), *1259*
Perry, L. F., 802(90), *808*
Person, M., 390(178, 179), *399*, 418(129), *430*
Persson, B., 662(28), *692*, 691(194), *698*
Peseke, K., 275(133), *317*, 712(76), *751*
Petek, M., 621(74), 625(74), *652*
Peter, F., 862(320), *874*
Peter, J. P., 1434(15l), *1437*
Peters, D. G., 297(334), *324*, 306(410), *326*, 363(8, 9), 364(9, 14, 18, 19), 366(9, 14, 23), 371(9), *394*, 367(23), *395*, 370(73–75), 371(73), 372(74, 75), *396*, 376(114), 378(125), *397*, 378(126), *398*, 862(323), 863(326), *875*, 889(38, 39), *934*, 1135(28), 1141(28), *1179*
Peters, E., 311(438), *327*, 1357(95), *1362*
Petersen, R. L., 240(52), 242(52), *248*, 260(40), *314*, 273(118), *316*, 528(149), *533*, 539(24), *575*, 601(135), 602(135), *612*, 925(309a), *945*, 1045(107), *1064*, 1050(159a–c), *1065*, 1076(48), 1080(48), *1096*, 1080(86), 1081(86), 1083(48, 86, 105), *1098*, 1149(160), *1183*, 1160(264, 265), *1187*
Peterson, J., 542(46), *576*, 1002(189), *1024*
Peterson, M. L., 1157(230), *1186*
Peterson, P., 418(134), *430*
Peterson, R. A., 307(414), *327*
Pétillon, F. Y., 1175(458), *1193*
Petit, F., 836(184), *870*
Petit, G., 291(284), *322*
Petit, M.–A., 835(174), *870*, 1236(107), *1259*
Petrie, G., 620(29), 621(29), *650*, 620(46, 47), 621(47), *651*, 1072(28), *1096*
Petrii, O. A., 1175(477), *1193*
Petrosyan, V. A., 406(41, 42), *427*, 407(49, 50, 51), *428*, 715(102), *751*
Petrov, S. I., 288(248), 295(248), 296(248), *321*
Petrova, A. A., 265(67), *314*
Petrova, G. N., 930(348), *947*, 1433(8), *1436*
Petrova, M. V., 727(249), *757*
Petrovich, J. P., 5(18), *8*, 239(45), *247*, 257(15), *313*, 306(408), *326*, 349(183), *359*, 370(67), *396*, 468(27), *479*, 703(7), *748*, 879(1), 885(1), *932*, 891(47), *934*, 893(71), *935*, 908(224e), 910(224e), 911(224e), 912(224e), 918(224e), *941*, 908(224g), 909(225), 910(224g), 911(224g), 924(224g), *942*, 924(299), *945*, 929(328), *946*, 931(368), *948*, 1136(31), 1137(38), *1180*, 1175(484), *1193*, 1201(27), *1212*, 1277(53), *1282*
Petrushenko, L. N., 801(82), *807*
Petsom, A., 408(57), *428*, 416(113), *430*, 670(77), *694*, 1042(26), *1063*, 1046(128), *1064*, 1167(338), *1189*
Pettersson, T., 620(31), 621(31), 624(31), *651*, 630(106), 632(106), 633(106), *653*, 704(20), *748*, 959(43), *1019*, 990(161), 996(161), 998(161), *1023*, 1001(175), *1024*, 1074(40, 42), *1096*
Pfaltz, A., 813(27, 32, 34), *865*, 985(131), *1022*
Pflegel, P., 732(284–286), *758*, 927(318), *946*, 930(355), *947*
Pfleiderer, W., 733(299), *759*
Pflügler, F., 81(102), 108(102), *119*, 214(8a, 9), *231*, 1236(107, 109), *1259*, 1434(13b), *1437*
Pham, M.–C., 179(233), *203*, 640(163), *655*

Phelps, J., 302(385), *326*, 350(195), *359*, 507(32), 519(32), *529*, 511(57), *530*, 629(92), *652*
Philip, R. H., Jr., 453(221), 455(221), *463*
Philippe, R., 299(355), *325*
Phillips, D. C., 1353(71, 72), *1361*, 1358(101–103), *1362*
Phillips, R. F., 793(32), *805*
Philoche-Levisalles, M., 738(334), *760*
Photis, J. M., 336(61), *354*
Piasecki, A., 639(147), *654*
Pichat, P., 1398(4), *1415*, 1408(44, 45), *1416*, 1408(50), *1417*, 1408(84), 1409(97–99), *1418*, 1413(131), *1419*
Picket, P., 300(357), 286(357), *325*
Pickett, C. J., 809(4), *864*, 813(42), *865*, 836(186–188), *870*, 846(238), *872*, 1366(29), *1393*
Pickett, D. J., 1286(2), 1292(2), 1299(2), *1332*
Pickup, P. G., 1372(45), *1393*, 1384(102, 103), *1395*
Piekarski, S., 711(65), *750*
Piemenschneider, P., 1171(384), *1190*
Pienemann, T., 466(10), *479*
Pierce, J. B., 1434(18), *1438*
Pierre, G., 795(49, 50), *806*, 802(88), *808*, 975(97), *1021*, 1084(108, 109), *1098*, 1110(42), *1125*, 1165(316), *1188*, 1433(5a), *1436*
Pierron, P., 402(4), *426*
Pierson, C., 242(60), *248*
Pierson, W. G., 899(145), 901(145), *938*
Pietrzyk, D. J., 454(226), *463*
Pilloni, G., 835(183), *870*
Pillsbury, D. G., 835(177), *870*
Piloty, O., 1208(92), *1214*, 1209(113), *1215*
Pimachkova, I. K., 797(67), *806*
Pincombe, C. F., 570(158), *579*
Pines, A., 537(14), 564(14), 565(14), *575*
Pines, S. H., 705(32), *749*
Pinson, J., 70(92a, b), *118*, 73(93d), 78(100), 81(103), 109(103), *119*, 150(115), *198*, 220(16), *231*, 226(24a, b, 25), 227(25), *232*, 299(353), *324*, 349(179, 180, 182), *358*, 380(130), 382(137), *398*, 407(44), *427*, 476(88), *481*, 701(1d), *747*, 719(128), *752*, 733(289, 290, 293, 294), *758*, 736(323), 738(323), *759*, 738(333, 334), *760*, 742(372), *761*, 891(54a, c), *935*, 979(96), *1021*, 1237(113, 114), 1241(121), *1260*, 1246(164), *1262*
Pintauro, P. N., 275(131), *317*, 439(92), 440(92), 441(92), *459*, 1421(1d, e), 1422(1d), 1424(1e), *1429*, 1433(3), *1436*, 1435(24a), *1438*
Piricócá, N., 1161(278), *1187*
Pistoia, G., 1345(20, 22), 1348(30), *1360*
Pistorius, R., 237(36), *247*, 552(80), *577*, 602(146), 603(146), *613*, 630(97, 98), *653*, 1009(232), 1012(240), *1026*, 1157(224), *1185*
Piszczek, L., 1208(94), *1214*
Pitti, C., 290(268), *322*
Pitti, S., 515(95, 96), 518(95), *531*
Pitzer, K. S., 24(27b), 42(27b), *115*
Pivovarova, S. F., 452(208f, g), 453(208f), *462*
Pla, M., 150(123), *199*
Plambeck, J. A., 1040(61), *1062*
Plashkin, V. S., 1094(180), *1101*
Platen, M., 593(67), *609*, 599(119), *611*, 670(73), *694*, 1046(130, 131), *1064*, 1226(41), 1233(41), 1249(41), *1257*, 1249(179), *1262*
Platonova, E. P., 617(12, 13), *650*
Plesch, P. H., 340(77b), *355*
Pleskov, V. A., 278(165), 284(165), *318*, 279(180), 282(197), 285(197), 295(180), *319*
Pletcher, D., 129(69), *197*, 208(4b), 216(12), *231*, 234(8), *246*, 256(5a), 259(5a), *312*, 270(94), *316*, 277(151), *318*, 282(200), 285(200), 290(200), *319*, 285(218), 290(218), *320*, 288(251, 253), 290(265, 266), 292(251, 253), *321*, 290(269), *322*, 300(359), *325*, 303(397), 304(397), *326*, 392(198–202), 393(198, 201), *400*, 414(103a), *429*, 441(152c), 449(172), 450(172), *461*, 489(55), *498*, 496(139), *501*, 511(65), *530*, 514(79), 515(79, 83, 97, 98), 516(79), 518(97, 98), *531*, 515(101), *532*, 583(7), *607*, 630(110), 633(110), *653*, 641(177–179), 642(177), *655*, 829(143–146), 830(145, 147, 148), *869*, 846(238), 848(250, 252), *872*, 863(331), *875*, 885(11), *932*, 1050(162), *1065*, 1056(242, 243), 1057(242), 1058(242, 243), 1059(242), *1068*, 1082(100), 1083(102, 103), 1085(102), *1098*, 1094(186, 188), *1101*, 1105(21), *1124*, 1110(40), *1125*, 1117(56), *1126*, 1230(53), 1231(53), *1258*, 1232(81), 1235(81), 1236(103, 108), *1259*, 1287(19), *1332*, 1376(61), 1383(61), *1394*, 1434(12i, 15g), *1437*, 1435(27), *1438*
Plichon, V., 284(215), *320*
Plieninger, H., 565(143), *578*
Plimley, R. E., 268(89a), 271(89a), *315*, 485(17), *496*, 1288(28), *1333*, 1421(4c), *1429*
Plomp, R., 387(160), *399*
Plotkin, E. V., 645(207), *656*
Plzak, V., 211(7c–e), *231*, 235(20), *246*, 982(114), *1022*, 1012(244), *1026*
Pocius, A., 646(217), *657*
Podanyi, B., 734(311), *759*
Pohmer, L., 1204(56), *1213*
Pointeau, R., 332(18), 345(18), *353*
Pokhodenko, V. D., 617(12, 13), *650*
Polievktov, M. K., 445(171), *461*, 926(315), *945*
Polishchuk, V. R., 368(39, 43), *395*, 1012(241), *1026*
Polynnikova, T. K., 891(55), *935*
Pommier, B., 1408(41), *1416*
Ponomarev, A. A., 1161(279), *1187*
Pons, B. S., 31(38b), *116*, 506(21, 23, 24), 511(21, 23, 24), 512(21, 23, 24), 514(21), *529*, 734(309), *759*, 1084(111), *1098*

Pons, J. W., 278(175), *319*
Pons, S., 128(63), *197*, 179(221, 222), 182(222), *202*, 179(232), 182(245), 184(255, 256), *203*, 184(257), *204*, 278(175), *319*, 296(330), *324*, 346(157b), *358*, 514(80), 515(80), *531*, 951(17), *1018*, 1268(26), 1275(26), *1281*
Ponsold, K., 778(57), 779(58, 59), *786*
Pool, K. L., 1245(152), *1261*
Popora, E. A., 743(387), *762*
Popov, A. I., 281(194), 284(193), 285(193), 297(193), 299(193), *319*
Popovych, O., 278(170), *319*
Popp, F. D., 483(3), 485(3), 493(3), 495(3), *496*
Popp, G., 621(62, 70), 624(62), 625(70), *652*, 710(61), *750*, 1089(144), *1099*
Porter, G., 617(14, 15), 618(24), *650*
Porter, J. D., 494(119), *500*, 740(357), *761*, 924(297), 931(297), *945*
Porter, Q. N., 662(31), 670(31), 671(31), *692*, 670(74), *694*, 1046(127), *1064*
Porter, W. R., Jr., 686(162), *697*
Portis, L. C., 600(123), *611*
Posey, F. A., 273(123d), *317*
Postovskii, I. Ya., 735(316), *759*
Posyagin, E. Yu., 474(70), *481*
Potacek, M., 423(164), *431*, 714(96), *751*
Pouillen, P., 425(192), *432*, 894(82), *936*, 1164(312), *1188*, 1233(91), 1235(91), 1236(100), *1259*
Poulatier, P., 901(157), *938*, 902(165), *939*
Pouliquen, J., 721(159), 722(159), *753*, 1252(192), *1263*
Pourbaix, M., 1289(35), *1333*
Powell, J. S., 39(55d), *117*, 411(84), *429*
Powell, L. A., 914(244), *943*
Powell, M., 997(165), 998(165), *1024*
Powers, R. M., 434(17), 435(17, 35), *457*, 1142(80), *1181*
Powers, T. A., 340(89), *355*, 374(89), *396*
Powning, R. F., 725(224), *756*
Pozdeeva, A. A., 435(39), *457*
Prabhu, U. D. G., 1178(521), *1195*
Pradác, J., 660(11, 12), *692*
Pradère, J.-P., 728(253), *757*
Praefcke, K., 489(68), *498*, 1047(134), *1064*
Pragst, F., 173(185), *201*, 290(270), *322*, 712(73), *751*, 721(157a–d), *753*, 726(239), *756*, 729(268–270), *758*, 926(138), *938*, 928(325, 326), *946*, 931(364), *947*, 973(87), 974(94b, 95), 979(87), *1021*, 1038(35), *1061*
Prahov, L. T., 1408(50), *1417*
Prasad, Y. K., 1352(53, 54), *1361*
Prater, K. B., 171(181), 172(181, 183), 173(181), *201*, 969(71), *1020*
Pratt, J. E., 1408(71), *1417*
Pratt, J. M., 826(132), *868*
Pratt, K. W., 169(176), *201*
Pravednikov, A. N., 491(93), *499*
Prelog, V., 474(74), *481*
Premo, J. G., 857(291), 861(291), *874*
Prentice, G. A., 270(98), 304(98), *316*
Presnova, N. T., 276(141), *317*
Press, W. H., 122(8), *195*
Prest, R., 1239(117), *1260*
Prevost, C., 739(346), *760*
Prilezhajeva, E. N., 456(242), *464*
Prill, E. J., 739(343), *760*, 908(224k), 910(224k), 911(224k), *942*
Privat, M., 1172(396), *1191*
Procher, H., 661(21), *692*
Prock, A., 844(230), *871*
Prodolliet, J. W., 818(78), *866*
Prokof'ev, A. I., 411(80), *429*, 455(231), *463*
Prokof'eva, T. I., 455(231, 232), *463*, 915(257), *943*
Proske, G., 453(216), *463*
Proskurovskaya, I. V., 439(94), *459*, 722(167), *754*, 895(96), 897(119), *937*
Pross, A., 1070(9, 10), *1095*
Proudfoot, G. M., 621(54), *651*
Pruiksma, R., 180(239, 240), 189(239), *203*
Pryadkina, R. G., 276(141), *317*
Pryde, E. H., 1001(180), *1024*
Przhiyalgovskaya, N. M., 490(74, 75), *499*
Przhiyalgovskaya, S. A., 1208(102–105), *1214*
Psarras, T., 863(328), *875*, 1169(354), 1177(509), *1194*
Puglisi, V. J., 172(182), 173(182), *201*, 436(57), *458*, 882(3c, e), 885(3c, e), *932*, 1144(112), *1182*
Pujol, L., 411(76), *428*
Pujol, M. T., 455(236), *464*
Pulido, F. J., 726(245), *757*
Pulidori, F., 424(181), *432*
Pusch, W., 1290(40), *1333*
Pushkareva, Z. V., 742(375, 376, 378), *761*, 743(389), *762*
Puxeddu, A., 823(104, 105, 108), 863(108), *867*, 824(120), 825(115), 826(128), *868*
Puy, J., 150(123), *199*
Pye, E. K., 790(13), *804*
Pykhtina, F. E., 491(93), *499*
Pysh, E. S., 508(35), *529*

Qiang-Sheng, F., 163(163), *200*
Quante, J. M., 971(92), 977(92), *1021*
Quast, H., 682(122), *695*, 921(295), *945*
Quentel, F., 983(128), *1022*
Quinn, R. K., 435(22), *457*
Quirk, R. P., 1205(65), *1213*

Raabe, G., 557(112), *577*
Raaen, H. P., 290(278), *322*
Rabai, J., 676(99), *695*
Rabockai, T., 826(129, 130), *868*
Race, G. M., 863(331), *875*
Radick, P., 565(142), *578*
Radner, F., 511(67), *530*, 1088(135), *1099*
Radouch, J., 689(180), *697*, 722(164b), *754*
Raff, R. A. V., 1358(106), *1362*
Raghavan, R., 724(206), *755*
Raghupathy, K., 496(137), *501*, 745(413), *763*
Rahman, M. M., 844(230), *871*
Rainis, A., 335(48), *354*
Rajargopalan, S. C., 414(101), *429*
Rajh, T., 1406(26), *1416*
Rakoutz, M., 907(185a), *940*
Ram, M. S., 831(149), *869*

Ramachandran, V., 242(56), *248*, 704(16), *748*, 957(37), *1019*, 993(144), 994(149, 152), 995(149), 999(149), *1023*, 1086(122), *1099*, 1094(174), *1101*
Ramaiah, M., 857(288), *873*
Ramakrishnan, V., 1408(69), *1417*
Ramasseul, R., 723(183), *754*
Ramaswamy, R., 265(74), 271(74), *315*, 1230(57), 1232(57), 1232(57), 1248(57), *1258*
Ramioville, J., 924(300), *945*, 1319(114), *1336*
Ramis, G., 1410(120), *1419*
Ramme, G., 1414(143), *1420*
Rampazzo, L., 887(27), *933*, 891(51), *934*, 1154(196), *1184*, 1154(201), *1185*
Ramsden, J. J., 1406(25), *1416*, 1412(126), *1419*
Ramsey, J. S., 373(81), *396*, 436(62), *458*, 895(111), 897(111), 902(111), *937*, 1144(104), *1182*
Rance, H. C., 1233(87), 1235(87), *1259*
Rand, L., 539(25), *575*, 1080(77), 1083(77), *1097*
Rank, B., 15(10a), *114*
Ran'ko, Ts. B., 404(29c), 408(29C), *427*
Rao, G. M., 1040(61), *1062*
Rao, G. P., 646(216), *657*
Rao, G. S., 239(40), *247*
Rao, M. L. B., 644(201), *656*
Rao, M. V., 418(130a), *430*
Rao, R. R., 283(235), *320*, 1084(107), *1098*
Rao, V. S., 426(196), *432*
Raoul, Y., 1206(81), *1214*
Raoult, E., 267(83), 268(83), 271(83), 288(83), 307(83), *315*, 421(144), *431*, 476(91), *481*, 707(46), 709(46), *749*, 839(211), *871*, 1168(350), 1169(350), *1189*, 1171(377, 380, 381, 388), *1190*, 1171(390, 391), *1191*, 1173(430, 438), *1192*
Rapoport, H., 1058(251), *1068*
Rapp, A., 1052(186), *1066*, 1170(361), *1190*
Rapson, W. S., 477(178), *462*, 897(112), *937*
Rasaq, M., 449(172), 450(172), *461*
Rashwan, F., 336(56), 343(56), 351(56), *354*
Raskin, M. N., 798(70), *807*
Rassat, A., 723(183), *754*
Rastogi, R., 637(127), *654*, 715(103), *751*, 1243(142), *1261*
Ratcliff, M., 489(61), *498*, 793(34), *805*, 798(73, 74), 799(75), *807*
Ratnikova, L. A., 408(58), *428*
Rauh, M., 715(100), *751*
Rauniyar, G., 288(255), 292(255), *321*, 1090(162), *1100*
Rautenstrauch, V., 13(2b), 17(2b), *114*
Rawn, J. D., 724(207), *755*
Ray, L.-S., 435(23), *457*
Raybuck, S. A., 818(77), *866*
Raynolds, P. W., 630(100), *653*
Razaq, M., 1236(103), *1259*
Rea, E. J. F., 448(189), *462*
Read, H. J., 447(180), 450(180), *462*
Read, J., 895(160, 174), 898(160), *939*
Read, R. R., 445(165), *461*, 447(184), *462*
Rebattu, J.-M., 297(336), 286(336), *324*
Reck, R. C., 1175(478), *1193*
Reddy, A. K. N., 13(1), *114*, 20(20), 22(22b, 24), 23(26), 24(27a), 42(27a), *115*, 26(34a), *116*, 46(59), *117*, 335(44), *354*, 1345(21), *1360*
Reddy, T. B., 285(230), *320*, 299(354), *324*, 299(356), *325*, 468(34), *480*, 484(8), *496*, 528(148), *533*, 589(39), *608*, 929(334), *946*, 1075(44), 1076(44), *1096*
Redey, L., 486(35), *497*
Redpath, D., 537(14), 564(14), 565(14), *575*
Reed, C. A., 819(85), *867*, 1221(7), *1256*
Reed, D., 858(301), *874*
Reed, R. C., 391(184), *399*, 586(26), *608*, 594(80), *610*
Reed, R. G., 377(116), *397*, 380(132), 382(132), *398*, 467(23), *479*, 887(32), *933*, 1147(151), *1183*, 1153(179, 180), *1184*
Rees, C. W. 742(377), *761*
Reeves, R. M., 276(142), *317*
Refalo, G., 424(176), 425(176), *431*
Regna, P., 439(87), *459*
Rehling, H., 603(151), *613*
Rehm, D., 50(64), *117*
Reiche, H., 1408(75, 80), *1418*
Reicher, M., 275(125), *316*
Reichsfeld, W. O., 440(98), 447(98), *459*, 897(115), *937*
Reidlinger, A. A., 421(143), *431*, 713(88), *751*
Reidorf, W. D., 1434(13h), *1437*
Reilley, C. N., 148(110), 188(110), *198*, 155(139), 156(139), *199*, 188(262), *204*, 270(95), 296(95), *316*, 511(63), *530*, 685(142), *696*
Reimann, H., 277(154), *318*, 645(203), *656*
Reinhart, J., 488(47), *498*
Reinmuth, W. H., 122(14), *195*, 154(134, 137), 155(137), *199*, 333(23), *353*, 340(79), *355*, 495(128), *500*, 1051(177), *1066*
Reisenhofer, E., 823(104, 105), *867*, 825(115, 116), *868*
Reissig, H. U., 601(132), 602(132), *612*
Reitano, M., 6(20), *8*, 1043(87), *1063*
Reitz, N. C., 621(62), 624(62), *652*
Renaud, P., 557(114), *577*, 557(115), 558(116), 562(114–116), *578*, 601(128), *612*, 1002(204), *1025*, 1170(370), 1171(371, 372), *1190*
Renaud, R. N., 539(23), *575*, 548(77, 78), 553(77, 87), *576*, 979(108), *1021*, 1004(210, 211), *1025*, 1011(237–239), *1026*, 1157(226, 227), *1185*, 1158(232, 233), *1186*
Reshetova, I. G., 1040(70), *1062*, 1147(150), *1183*
Rétey, J., 813(35), *865*, 824(113), *868*
Retslav, K.-J., 701(1e), *747*
Reutov, O. A., 863(329), *875*, 376(112), *397*
Reverdy, G., 525(135), *533*, 716(111), *752*, 1414(149, 150), *1420*
Reynolds, G. A., 685(139), *696*

Reynolds, R., 592(56), *609*, 968(69), 988(69), *1020*
Reynolds-Warnhoff, P., 14(3b), *114*
Rheingold, A. L., 841(222), *871*
Rhodes, R. K., 179(236), *203*
Ricci, A., 1345(22), *1360*
Ricciuti, C., 1053(201), *1066*
Ricco, A. J., 1383(93), 1385(93), *1395*
Rice, P. A., 790(10), *804*
Richards, D. H., 885(8), *932*, 911(270), *944*
Richards, J. A., 616(6, 7), 617(6, 7), 618(7), 619(7), 621(6), 624(6), *650*, 962(46), *1019*
Richards, K. R., 1153(181), *1184*
Richards, T. W., 979(110), *1022*
Richardson, D., 306(405), *326*, 1350(36), *1360*
Richter, A. M., 679(112), 680(112), 682(112), *695*
Richter, G., 1290(39), *1333*
Richtol, H. H., 511(69), *531*, 954(12), *1018*, 1084(113), *1098*
Riddiford, A. C., 165(165), *200*
Ridgway, T. H., 148(110), 188(110), *198*, 188(262), *204*
Rieger, P. H., 335(51a), 342(51), 345(51), 346(51), *354*, 388(168), *399*, 411(83), *429*, 495(128), *500*, 1051(177), *1066*
Rieke, R. D., 194(292), *205*, 335(53), 336(53), 339(53), *354*, 887(26), *933*, 892(64), 893(69), *935*, 1175(450), *1192*, 1175(455), *1193*, 1177(510), *1194*, 1240(120), *1260*
Riekena, R., 369(60), *395*
Rieker, A., 184(256), *203*, 616(8), 617(8, 11, 19), 621(11), *650*, 621(59), 624(59), *651*, 621(61), *652*, 638(129), *654*, 717(118), *752*, 958(39), 959(48), 961(39), 962(47), *1019*, 1050(165, 166), *1065*
Rieker, E.-L., 621(73), 625(73), *652*
Riemenschneider, P., 648(231), *657*
Rifi, M. R., 5(19), *8*, 263(53), *314*, 362(1), *394*, 368(49–51), 377(49, 51), 378(51), *395*, 483(2), 493(2), 495(2), 496(2), *496*, 887(28), *933*, 1154(204), 1155(204), *1185*, 1176(497, 501), *1194*
Rifkin, S. C., 882(3i), 885(3i), *932*
Rigatti, G., 434(10), *456*
Rigaudy, J., 193(284), *205*
Rigo, A., 518(106), *532*
Riha, J., 1204(58), *1213*
Riley, J. H., 489(62), *498*, 860(313), *874*
Rimmelin, P., 616(3), 620(3), *650*
Rinderspacher, M., 621(88), *652*
Ritchie, I. M., 282(201), *319*, 302(383), *325*, 616(4), 619(4), 620(4), *650*, 621(54), *651*
Rivas, S., 275(134), *317*, 931(372), *948*
Rivière, H., 1040(62), *1062*, 1043(90), *1063*
Robbins, J. L., 1372(46), *1393*
Robert, D., 799(75), *807*
Roberts, B. G., 674(85), *694*, 721(163), 723(163), *754*
Roberts, J. L., Jr., 930(353), *947*, 1245(151), *1261*
Roberts, R. D., 886(20f), 891(20f), *933*
Robertson, G. M., 905(213), *941*, 1139(60), 1140(60), *1181*
Robertson, P. M., 258(20), *313*, 268(87), *315*, 277(152, 154), *318*, 486(29), *497*, 641(175), *655*, 645(203), *656*, 1230(60), *1258*, 1309(81), *1334*, 1321(122), *1336*
Robillard, Y., 925(309e), *945*
Robin, R., 494(117), *500*, 740(360), *761*
Robin, Y., 826(125), *868*, 1434(13f), *1437*
Robinson, B. H., 1156(214), *1185*
Robinson, C. C., 453(220), *463*
Robinson, J., 104(109), *119*, 179(230), *203*, 303(392), *326*
Robinson, J. A., 813(35), *865*
Robinson, R., 897(112), *937*
Robinson, R. S., 188(263, 265), *204*
Robinson, S. L., 283(235), *320*,1084(107), *1098*
Robles, F., 153(129), *199*
Robveille, S., 1239(114), *1260*
Roche, T. S., 826(127), *868*
Rocklin, R., 1366(26), 1368(26), 1369(26), 1371(26), *1393*
Röder, H., 742(386), *762*, 1049(156), *1065*, 1141(67), *1181*
Röder, J., 1051(183, 184), *1066*
Rodewald, L. B., 543(51), *576*, 1159(242), *1186*
Rodgers, S. J., 819(85), *867*
Rodinov, W. M., 489(67), *498*
Rodnikov, I. A., 837(198), 838(198), *870*, 886(19), *933*
Rodriguez-Mellado, J. M., 163(159), *200*, 744(405), *762*, 894(77g), *936*
Roe, D. K., 284(226), 303(226), *320*, 1136(34), *1180*, 1175(485), *1193*
Roemmele, R. C., 1058(251), *1068*
Roffia, S., 126(57), 135(57), *197*, 727(250), *757*, 914(239), *942*
Rogatinskaya, S. L., 484(10), *496*
Rogers,G., 1299(55), *1334*
Rogers, H. H., 290(280), *322*
Rogers, H. R., 368(37), 369(55), *395*, 378(121), *397*, 1153(182), *1184*
Rogers, J. W., 421(148), 422(148), *431*, 713(82), *751*, 726(242), *757*
Rogers, L. B., 373(78, 79), *396*
Rogers, T., 1172(416), *1191*
Rogers, T. E., 1230(59), *1258*, 1424(5), *1429*
Rogers, W., 286(423), 309(423), *327*, 1168(345), *1189*
Rogic, O., 596(95), *610*, 706(36a), 710(36a), 712(36a), 714(36a), *749*
Rogic, V., 596(94, 95), *610*, 706(36a), 710(36a), 712(36a), 714(36a), *749*, 710(60c), 715(60c), *750*
Roholl, H., 277(153), *318*
Rol, C., 1070(2), 1080(2), 1081(2), *1095*
Rolison, D., 128(63), *197*, 1390(127), *1396*

Rollin, Y., 393(204, 205, 209), *400*, 831(157, 158), 832(157, 158), 834(172), 835(172), *869*, 835(176, 179), *870*, 1236(108), 1259
Romaini, A. M., 496(129), *500*
Roman, E., 838(209), *871*, 1144(121), 1145(121), *1182*, 1156(213), *1185*
Roman, V. J., 1094(183), *1101*, 1117(60), 1118(60), *1126*
Romanin, A., 494(115), *500*, 839(213a, b), *871*
Romanova, K. A., 894(75), *935*
Roncali, J., 688(178), *697*, 1380(64), 1383(64), *1394*, 1389(123), *1396*
Rondinini, S., 1200(1), *1211*
Ronlán, A., 138(85), *197*, 139(86), *198*, 170(179), *201*, 336(55), 342(55), *354*, 507(30), 519(30), 524(30), *529*, 511(61), *530*, 519(111), 524(111), *532*, 617(20, 21), 620(30), 621(30), 624(30), *650*, 620(31), 621(31, 57, 58, 60), 624(31, 58, 60), *651*, 621(64, 71, 72, 89), 624(64), 625(71, 72), 628(90), *652*, 630(106), 632(106), 633(106), 634(117), *653*, 748(20), *748*, 959(42, 43), 960(29b), 963(50), 966(56), *1019*, 963(65), *1020*, 982(117), 985(135), *1022*, 985(136, 137, 140), 986(138a), 987(139, 140), 990(161), 993(141), 996(161), 998(161), 1001(137), *1023*, 1001(175), *1024*, 1071(26), 1073(26), 1074(40–42), *1096*, 1093(168, 169), *1110*
Root, D. K., 733(299), *758*
Rosasco, S. D., 1371(36), *1393*
Rosati, R., 725(228), *756*
Rosen, H., 299(352), *324*, 1201(35), 1202(37, 38), *1212*
Rosenbach, J., 486(33), *497*
Rosenblum, M., 838(206), *871*
Rosenthal, I., 367(28), *395*, 369(62, 64), *396*, 388(162, 163), *399*, 1135(18), 1154(18) *1179*, 1153(185, 192), *1184*, 1175(481), *1193*, 1176(502),*1194*, 1245(157, 160), *1261*
Rosenthal, S., 1045(106), *1063*
Ross, J. W., 276(140), *317*
Ross, P. N., 284(219), 290(219), *320*, 290(272), *322*
Ross, S. D., 240(52), 242(52), *248*, 260(40), *314*, 273(118), *316*, 528(149), *533*, 542(41), *575*, 584(17), *608*, 601(135–138), 602(135, 136), *612*, 603(154), *613*, 638(134), *654*, 648(232), *657*, 925(309a, b), *945*, 1001(181), *1024*, 1045(107), *1064*, 1050(159a–c), *1065*, 1076(48), 1080(48), *1096*, 1077(59), *1097*, 1083(48, 86, 105), *1098*, 1080(86), 1081(86), *1183*, 1149(160), 1160(264, 265), 1161(277), *1187*
Rosser, A., 540(35), *575*
Rossetti, R., 1406(23, 27), *1416*
Rossi, G., 821(97), *867*
Rossi, M., 151(125), *199*
Rossi, P., 180(242), *203*
Rossi, R., 547(74), *576*, 783(73), *786*
Rossow, A. C., 453(220), *463*
Roth, K., 1002(201), *1025*
Rotilio, G., 725(222), *756*
Rouff, P. M., 1043(83), *1063*
Rouger, J., 1355(83), *1361*
Rouget, R., 421(145), *431*
Roullier, L., 733(303), *759*, 926(317), *946*, 1343(15), *1359*
Rousar, I., 1286(5), 1292(5), *1332*
Rousseau, K., 305(399), *326*
Rowbottom, J. D., 563(131), *578*
Rowe, R., 898(131), *938*
Royer, G. P., 1248(177), *1262*, 1366(17), 1366(17), 1370(17), 1371(17), *1392*
Rozas, R., 275(134), *317*, 392(201, 202), 393(201), *400*, 544(56), *576*, 821(101), *867*, 829(144), 830(147), *869*, 931(372), *948*
Rozé, J.-C., 728(253, 254), *757*
Rozgaite, J., 646(217), *657*
Rozhkov, I. N., 1094(185), *1101*, 1104(2, 5), 1105(2), 1109(5), 1110(5), 1117(5), *1124*, 1117(53, 54, 57, 58a, b, 59), 1118(59), 1119(53, 54), *1125*, 1119(57, 59, 67a–d), 1121(77), 1122(72), 1123(67e), *1126*, 1123(80, 82), *1127*
Rubinskaya, T. Y., 439(94), *459*, 488(49, 50), 492(49), *498*, 722(167), *754*
Rubinstein, I., 411(88), 413(88), *429*, 1050(172), *1066*, 1377(78), 1382(78), 1384(78), *1394*
Rubinstein, J. F., 1314(6), 1375(6), *1392*
Rudd, E. J., 285(221), *320*, 542(41), *575*, 601(136), 602(136), *612*, 603(154), *613*, 638(134), *654*, 648(232), *657*, 895(91a), *936*, 1001(181), *1024*, 1076(48), 1080(48), 1083(48), *1096*, 1132(11), *1179*
Rudenko, A. P., 290(270), *322*
Rudge, A. J., 1104(10), *1124*, 1114(49), *1125*
Rudhan, R., 1408(40), *1416*, 1408(56), *1417*
Rudorf, W. D., 930(352), *947*, 1046(121), *1064*
Rueda, M., 163(162), *200*
Ruehlen, F. N., 1094(181), *1101*
Rüestschi, P., 424(169), *432*
Ruf, H.-R., 239(44), 240(44), *247*, 813(37), 825(37), 826(37), 827(37), 841(37), 863(37), *865*
Ruff, O., 1209(113), *1215*
de Ruiter, R., 387(160), *399*
Ruiz, J. J., 163(159, 162), *200*, 645(210), *656*, 894(77g), *936*, 916(264), *943*
Rulev, Y. P., 792(29), *805*
Runner, M. E., 290(279), *322*, 332(15), 348(15), *353*, 885(6), *932*, 910(227), *942*, 1138(50), *1180*, 1175(488), *1194*
Ruoff, P. M., 454(224), *463*
Rupauer, R., 857(290), *873*
Rupe, H., 1207(88), *1214*
Ruppert, R., 831(153), *869*, 1434(15d), *1437*
Ruppli, U., 841(222), *871*
Rusakowa, M. S., 411(78b), *429*
Rüttinger, H. H., 1434(13g, h), *1437*
Ruzic, I., 434(21), *437*, 882(3h), 885(3h), *932*

Rusling, J. F., 122(9, 12), *195*, 151(125), *199*, 258(22), *313*, 288(252), 292(252), *321*, 384(142), *398*, 485(14), *496*, 743(394, 395), 744(395), *762*, 833(164), *869*, 895(100), *937*, 973(91), *1021*, 1144(122), 1145(122), 1146(122), *1182*
Russ, J., 1348(29), *1360*
Russell, G. A., 434(12), *456*, 455(233), *464*, 1269(33), *1281*
Russell, J. J., 1408(56), *1417*
Russell, P. G., 438(72), *458*
Russell, R. K., 336(57), *354*
Ruth, W., 275(136), *317*
Rüttinger, H. H., 728(258), *757*, 746(426), *763*, 930(351, 352), *947*, 1046(121), *1064*
Ruzziconi, R., 1070(2), 1080(2), 1081(2), *1095*
Ryan, C. M., 722(181), *754*, 1200(11), *1211*, 1203(48), *1213*
Ryan, M. D., 107(110), *119*, 435(47), *457*, 621(52), *651*, 668(64), 669(65), *694*, 931(365a), *947*
Ryde-Petterson, G., 237(31), *247*
Rydholm, S. A., 788(9), 790(9), 791(9), *803*
Rytz, G., 813(39), 821(39), *865*, 815(55), *866*, 823(103), *867*, 826(126), *868*
Ryvolova-Kejharova, A., 435(34), *457*, 452(208a), 436(64), *458*, *462*, 494(114), *500*, 740(361), *761*
Ryznar, J. W., 857(291), 861(291), *874*

Sabol, M. A., 493(111), *500*, 1256(218), *1263*
Saboureau, C., 4(13), *8*
Sacchi, N., 1040(64), *1062*
Sackett, D. D., 1410(122), 1413(122), *1419*
Sacki, T., 722(165), *754*
Sadana, Y. N., 747(431), *763*, 925(309g), *945*
Saddler, J. N., 791(16), *804*
Sadler, J. L., 125(47), *197*, 423(161), *431*
Saeva, F. D., 671(78), *694*, 685(139), *696*, 721(1572), *753*, 979(111), *1022*, 1044(96), *1063*
Sagae, H., 416(114), *430*, 739(342), *760*, 1246(168, 169), *1262*, 1277(55), *1282*
Said, F. F., 837(194, 195), 862(195), *870*, 857(294), 862(318), *874*, 862(324), *875*
Sainsbury, M., 239(37), *247*, 601(143), 602(143), *612*, 704(21), *748*, 704(25, 27), *749*, 711(63), *750*, 965(54), *1019*, 989(163), *1023*, 991(166, 170), 992(174), 997(165), 998(165), 1001(174), *1024*, 1094(176), *1101*, 1166(328), *1189*
Saito, M., 1164(311), *1188*
Saito, N., 1229(68), *1258*
Saito, S., 607(163), *613*
Saito, Y., 437(135a), *460*, 1421(1c), 1422(1c), 1424(1c), *1429*
Saito, Z., 915(254), *943*, 929(338), *946*
Saitoh, N., 780(62a, 63), *786*
Saitou, N., 705(33), *749*
Saji, T., 838(207), 839(212), *871*
Sakaguchi, K., 716(113), *752*, 1008(227), 1012(227), *1026*
Sakaguti, T., 1206(69), *1213*
Sakai, W., 271(107), *316*
Sakamoto, M., 646(214), *656*
Sakata, T., 1408(42, 43), *1416*, 1408(54, 57), *1417*, 1408(85), 1409(96), *1418*, 1411(124), *1419*
Sakuma, M., 1141(66), *1181*
Sakura, S., 298(344, 346), *324*
Sakuragi, H., 519(110), *532*, 982(113), *1022*, 1406(31), *1416*, 1409(92), *1418*
Sakurai, B., 443(120), *460*, 491(94), *499*, 722(175), *754*, 740(364), *761*
Sakurai, H., 496(131), *500*, 1408(53–55, 65), *1417*, 1408(88), *1418*
Sakurai, I., 381(135), *398*
Sakurai, K., 815(58), *866*
Salajegheh, A., 370(73), 371(73), *396*, 889(39), *934*, 1135(28), 1141(28), *1179*
Salajka, Z., 689(182), *697*
Salaün, J.-P., 444(129), *460*, 897(118), *937*, 901(157), *938*, 902(165), *939*, 929(339), *946*
Salaun-Bovix, M., 897(118), *937*
Salchi, J., 1413(135), *1420*
Salem, L., 50(67c), *117*
Salim, M., 1408(40), *1416*
Salinger, R., 863(328), *875*, 1169(354), *1190*, 1177(509), *1194*
Salkind, A. J., 1287(12), *1332*
Salmin, L. A., 552(81), *577*, 1010(235), *1026*
Salmon, M., 1380(65, 71), 1381(72), 1387(65), *1394*
Saltykova, L. I,. 496(134), *500*
Salun, M., 901(156), *938*
Salvaterra, M., 1056(235), *1067*
Salzberg, H. W., 281(192), 284(192), 291(192), *319*, 291(290), *322*
Salzer, A., 841(222), *871*
Sambaro, A., 901(159), *939*
Sammells, A. F., 930(361), *947*, 1434(12g), *1437*
Samokhvalov, G. I., 1038(42a, b), *1061*, 1039(48, 52a), *1062*, 1050(161), *1065*
Samuelsson, B., 259(30), *313*, 307(411), *326*, 1045(118), *1064*, 1049(153), *1065*, 1210(119), *1215*
Sanchez-Burgos, E., 645(209), *656*
Sandegren, B., 1312(84), *1335*
Sander, M. R., 1231(64), 1232(64), *1258*
Sanders, J. K. M., 819(83), *867*
Sanderson, J. E., 541(36), *575*, 547(75), *576*
Sandford, R. W., 620(29), 621(29), *650*
Sandman, D. J., 680(121), *695*
Sando, H., 542(49), 548(76), *576*
Sandona, G., 342(98), 346(98), *356*, 346(152), 350(152), *357*, 346(156), *358*, 1135(26), *1179*
Sandstrom, W. M., 1208(93), *1214*
Sanechika, K., 1380(68), *1394*
San Filippo, J., 1245(153), *1261*, 1407(33), *1416*
Sanicanin, Z., 706(37), *749*
Sänneskog, O., 1070(1), *1095*
Sanstead, J. K., 1002(195a), *1025*

Santa-Cruz, T. D., 344(127), *357*
Santana, A. E. G., 240(51b), *248*, 839(214), *871*
Santavy, F., 726(234), *756*, 745(411), *762*, 1033(8b), 1050(8b), *1060*
Santhanam, K. S. V., 210(6a), 212(6a), *231*, 302(385), *326*, 309(434), *327*, 332(19), 333(19), *353*, 511(57), *530*, 725(218), *756*, 926(313), *945*, 1052(191), *1066*
Santic, I., 288(254), 292(254), *321*
Santoa, M., 1434(19b), *1438*
Santos, E., 639(143), *654*
Sanz, F., 150(123), *199*
Saranceno, R. A., 647(222), *657*
Sarai, E., 280(183), 297(183), 307(183), *309*
Saraste, M., 813(26), *865*
Saratova, L. A., 797(64), *806*
Sarfarazi, F., 184(257), *204*
Sargent, C. R., 388(169), *399*, 747(433), *763*
Sargent, J. W., 290(282), *322*
Sarkanen, K. V., 788(6), 790(6), *803*, 790(11), *804*
Sarrazin, J., 267(83), 268(83), 271(83), 288(83), 307(83), *315*, 456(246), *464*, 915(255), *943*, 1135(20, 21), *1179*, 1171(377), *1190*, 1172(398), *1191*
Sartori, G., 733(291), *758*
Sartori, P., 260(46), *314*, 290(274), *322*, 1104(6, 13), *1124*, 1112(47), *1125*
Sarubin, M., Ya., 290(270), *322*
Sasaki, E., 600(126), *612*, 1178(516), *1194*
Sasaki, K., 528(143), *533*, 956(31), *1019*, 1073(38), *1096*, 1383(94), *1385*(94), *1395*
Sasaoka, M., 600(120), *611*, 713(78), *751*, 780(62a, b, 63), *786*, 1054(213), *1067*, 1244(148), *1261*
Sasse, W. H. F., 1408(81), *1418*
Sastry, K. S., 485(16), 489(16), 492(16), *496*, 744(403), *762*
Sata, K., 565(139), *578*
Sata, T., 908(223), *941*
Sato, A., 1173(439), *1192*
Sato, E., 586(25), 601(25), *608*, 1035(22), *1061*
Sato, K., 526(140), *533*
Sato, M. A., 284(222), *320*, 1380(67), 1383(67), 1384(62), 1385(62), *1394*
Sato, N., 263(57), *314*, 367(29), *395*, 539(20), *575*, 907(194), *940*, 1142(75), *1181*
Sato, S., 853(271), *873*, 908(223), *941*, 926(312), *945*
Sato, T., 976(107), *1021*, 993(147), 994(148), *1023*, 1087(133), *1099*, 1093(172), *1100*, 1154(197), *1185*
Sato, Y., 640(162), *655*
Satoh, M., 1357(94), *1362*, 1384(107), 1385(107), *1395*
Satoh, S., 1154(205), 1155(205), *1185*
Satoh, Y., 1409(90, 91, 94), *1418*
Satonaka, H., 915(254), *943*, 929(388), *946*
Sattelmeyer, R., 925(247), *943*, 925(305), *945*
Savers, R. R., 544(53), *576*, 564(134), 566(134), *578*, 1001(192), *1024*, 1036(30), *1061*
Saur, O., 1410(121a), *1419*
Saura, R., 153(129), *199*
Saussey, H., 1410(121), *1419*
Sauvage, J.-P., 831(153, 155), *869*, 1434(15d, l), *1437*
Savangdpani, S., 1287(14), *1332*
Savard, J., 835(174), *870*
Savard, M., 553(87), *577*, 1011(238), *1026*
Savéant, J. M., 17(14, 15a, 16), *115*, 39(54, 55e), *117*, 64(81), 68(81), 72(81), 65(84a), 67(85a), 68(87), 70(92a–c), 72(85a), *118*, 73(93a–c, 94), 76(15a, 95a), 77(96, 98), 78(100), 81(103), 97(94), 108(94), *119*, 122(15), 123(15), *195*, 123(22–25, 27, 28), 132(22), 133(22–25, 27, 28), 134(22), 135(22), 143(25), *196*, 126(56), 128(68), 133(15), 135(81–84), 143(83), *197*, 140(93, 95, 96), 142(93, 95), 143(93), 144(101), 146(105, 106), 148(93, 95), 150(93, 95, 115), *198*, 211(7a, f), 220(7a, 14–20), 223(21), 224(17), 228(15), *231*, 226(24a, b, 25), 227(25), *232*, 234(7d), *246*, 259(29), 310(29), *313*, 293(306), *323*, 299(353), *324*, 344(113–117), 345(117), *356*, 345(133, 139), 346(152), 350(152), *357*, 349(179–181), *358*, 350(199), *359*, 363(6), 366(6, 22), 379(6), 384(6), *394*, 379(127), 380(130), 382(137), 384(127), *398*, 435(43, 44), *457*, 443(122), *460*, 467(24), *479*, 469(39–41), *480*, 597(101), *611*, 618(23), *650*, 813(36), 814(54), 819(54), 823(36), 824(36), *865*, 815(59), 816(59, 61, 63, 64, 72), 817(75), 818(72), *866*, 823(102), *867*, 882(3a, d, f, g, o, r), 884(4), 885(3a, d, f, g, o, r, 4), *932*, 891(54a, c), 894(73c), 895(93), *936*, 899(135, 139), *938*, 909(237), *942*, 918(273a, b), 919(274), *944*, 925(310), *945*, 928(323), *946*, 1032(2), *1060*, 1047(136, 137), *1064*, 1052(192), *1066*, 1146(138), *1183*, 1218(1, 2), 1219(1, 2), 1220(3, 4), 1221(1, 2), 1222(1, 2), 1223(1, 2), 1248(3, 4), *1256*, 1221(19), 1222(19), 1224(19, 27, 28, 33), 1228(48), *1257*, 1237(112–114), 1241(121), *1260*, 1246(164), *1262*, 1265(1), 1267(6), *1280*, 1434(14b), *1437*
Savel'ev, L. A., 1109(36), *1125*
Savin, V. I., 466(4), 473(4), *479*, 471(52), *480*
Savinell, R. F., 1287(13), *1332*
Savushkina, N. N., 496(136), *501*
Sawa, S., 496(132), *500*
Sawaki, S., 742(383), *762*
Sawaki, Y., 667(58), *693*, 678(102), 679(111), *695*, 1046(124), 1047(135), *1064*, 1079(67), *1097*, 1167(337, 339), *1189*, 1409(93), *1418*
Sawamura, T., 728(252), *757*

Sawin, W. I., 452(208f), 453(208f), *462*
Sawyer, D. T., 661(15), *692*, 930(353), *947*, 1245(151), *1261*, 1245(161), *1262*, 1269(31a, b, 32), 1272(31a), *1281*, 1434(14a), *1437*,
Saxena, R. S., 662(27), *692*
Saygin, Oe., 523(128), *532*, 983(130), *1022*
Saylor, J. H., 726(239), *756*
Sayo, H., 403(14), *426*, 404(23, 24, 26), 405(35, 36), *427*, 583(11), *602*, 599(116), *611*, 691(196), *698*, 961(33), *1019*, 977(86), *1021*, 1050(168), *1066*, 1087(132), 1089(143), *1099*
Scanlon, M. D., 377(118), *397*, 1152(178), *1184*
Schaab, W. B., 292(300), 293(313), *323*
Schafer, D. E., 663(40), *693*
Schäfer, H. J., 6(21), *8*, 233(6c), 234(11, 14), 235(17a, 19), 239(14), 243(14), 244(14), *246*, 237(28, 29, 36), 239(43, 46), *247*, 240(53), 242(57), 243(64, 75), *248*, 277(153), *318*, 302(381a, b), *325*, 347(166), 348(166), *358*, 367(33), *395*, 375(96), *396*, 403(20), *427*, 466(10), *479*, 516(87), 517(87), 525(87), *531*, 525(137), *533*, 542(43), 545(43), *575*, 545(62–64, 79), 546(66, 68– 71), 547(66, 72, 73), *576*, 550(93, 94), 551(95), 552(80), *577*, 564(133), 566(133), *578*, 569(155), *579*, 583(6, 8), *607*, 597(102), 599(111), *611*, 601(143), 602(143), *612*, 641(173, 174), 642(174), *655*, 648(228), *657*, 704(23, 26, 28, 29), *749*, 716(116), *752*, 771(32), *785*, 778(56), 783(74a, b, 76), *786*, 855(275a), *873*, 858(302, 303), 861(317), *874*, 879(2), 880(2), *932*, 890(43), 891(46), *934*, 903(219), 908(220), *941*, 950(1a, f), 1002(1a), *1018*, 956(32), *1019*, 975(106), *1021*, 980(112, 119a, 121), 981(129, 133, 134), 982(112, 116, 120), 983(120, 126, 129), 984(129), *1022*, 995(156), 996(158, 159), *1023*, 990(164), 1000(173), 1001(182), 1002(182), *1024*, 1002(196–202), 1005(212c), 1006(212c, 221b), *1025*, 1006(223), 1007(225), 1008(225), 1009(225, 229a, b, 230), 1010(231, 232, 234), 1011(242, 243), 1012(240, 245), *1026*, 1033(13), *1060*, 1039(51), *1062*, 1053(195), *1066*, 1079(70), *1097*, 1083(104), *1098*, 1153(186), *1184*, 1157(221, 224,), *1185*, 1163(299), 1166(319), *1188*, 1166(326,329), *1189*, 1200(12), *1211*
Schäfer, K., 722(180), *754*
Schäfer, W., 486(33), *497*, 834(171), *869*
Schäffer, C., 704(23), *749*
Schall, C., 273(117), *316*, 442(113), 449(113), *460*, 449(198), *462*
Schanne, L., 1434(22b), *1438*
Schardt, B. C., 1371(36), *1393*
Schecter, H., 1088(138), *1099*
Scheele, J. J., 891(54b), *935*
Scheer, B. J., 151(125), *199*, 973(91), *1021*
Scheffold, R., 239(44), 240(44), *247*, 393(206), *400*, 492(99), *499*, 813(37–39), 821(39), 825(37), 826(37), 827(37, 38), 828(38), 841(37), 863(37), *865*, 815(55), *866*, 823(103), *867*, 826(126, 133), 827(134–136), 828(136), 829(141), *868*, 827(142), *869*, 1053(196), *1066*
Scheidt, W. R., 819(85, 88), *867*
Schenck, G. O., 1407(34), *1416*
Schenk, R., 333(31), 341(31), *353*
Schenker, B. A., 1316(103), 1321(103), *1335*
Schenker, B. P., 894(85a), *936*
Schepps, W., 438(75), 445(75), 446(75), 447(75), *458*, 446(176), *462*, 896(108), *937*
Scherer, O. J., 1178(518), *1195*
Scheutzow, D., 682(122), *695*
Schier, E. J., 894(85b), *936*
Schiff, H. I., 301(365), *325*
Schilling, H., 707(47c), *749*
Schiller, J. G., 645(206), *656*
Schinazi, R. F., 704(21), *748*
Schindewolf, U., 333(29), *353*
Schindler, R., 409(64), *428*, 421(146), *431*, 424(180), *432*, 734(313), *759*
Schlecker, R., 673(83), *694*
Schlegel, C., 601(143), 602(143), *612*, 996(158, 159), *1023*
Schlegel, G., 858(303), *874*, 1005(212c), 1006(212c), *1025*, 1006(223), *1026*
Schleich, D. M., 527(142), *533*
Schleich, K., 277(154), *318*, 645(203), *656*
Schlenk, W., 349(185), 351(209), *359*
Schlesener, C. J., 30(36a–c, 37b), 31(36b, c), 34(47), 47(36), 50(36), 51(36), 52(36), *116*, 506(19), 508(19), 509(19), 511(19), 513(19), 517(19), *529*, 511(74), 513(74), *531*
Schletter, I., 537(13), 562(13), 564(13), 565(13), *575*
Schleyer, P. von R., 233(3), 234(3), *245*
Schluter, D. N., 745(417), *763*
Schmakel, C. O., 725(218), *756*
Schmalzl, P. W., 638(132), *654*
Schmeisser, M., 260(46), *314*, 290(274), *322*
Schmidt, A., 745(410), *762*, 1036(31), 1037(31), *1061*
Schmidt, C. L. A., 413(95), *429*, 724(203), 725(203), *755*
Schmidt, F. C., 292(300), *323*
Schmidt, H., 293(308), *323*, 419(137), 420(137, 141), 421(141), *431*, 1117(51, 52), 1122(52), *1125*
Schmidt, H. D., 1117(51), *1125*
Schmidt, P. P., 63(79), *118*, 128(63), *197*
Schmidt, R. M., 26(31e), *115*
Schmidt, W., 593(63, 65), 594(63), *609*, 593(69), *610*, 635(124), 636(124), *653*, 1043(88, 89), *1063*, 1221(18), 1224(18), 1249(18), *1256*, 1224(34), 1225(36, 38), *1257*
Schmitt, R. E., 1200(4), *1211*, 1210(120), *1215*
Schmitz, A., 1078(66), *1097*

Schmitz, K., 442(114), 445(114), *460*
Schmüser, W., 492(106), *500*
Schneider, A., 1287(13), *1332*
Schneider, B. H., 848(247), *872*
Schneider, C. J., 740(354), *761*
Schneider, E., 407(47), *427*
Schneider, H., 235(20), *246*, 278(167), *318*, 958(39), 961(39), *1019*, 982(114), *1022*, 1071(23), *1096*
Schneider, K.-A., 526(139), *533*
Schneider, R., 902(166), *939*, 1172(404), *1191*
Schneider, Z., 813(28), *865*
Schöber, G., 285(227), 293(227), *320*
Schoen, W., 410(75c), *428*
Scholl, P. C., 641(182), 642(182), 643(186), *655*
Scholz, F., 190(281), 193(281), *204*
van Schooten, J., 332(13, 14), 334(14), 345(13, 14), 346(13), *353*
Schrage, H., 594(77, 78), *610*
Schrauzer, G. N., 825(118), *868*, 1413(135), *1420*
Schreiber, J., 812(22), *864*
Schreiner, K., 886(12), 891(12), *933*
Schröder, G., 336(62), 339(62), *354*, 338(67), *355*
Schroeder, A. H., 1372(42), 1373(42), *1393*
Schroeder, H. A., 788(4), *803*, 791(20), *804*, 798(73), *807*
Schroeder, R., 711(62), *750*
Schroth, W., 684(130), *696*, 728(258, 261), *757*, 746(426), *763*
Schubert, U., 511(66), *530*
Schuerch, C., 1210(121), *1215*
Schuhmann, D., 1172(397), *1191*
Schukat, G., 679(112), 680(112), 681(116), 682(112, 116), 683(116), *695*
Schulte, L. D., 1177(510), *1194*
Schulte-Elte, K., 1407(34), *1416*
Schulthess, P., 824(111), *868*
Schultz, F. A., 1365(21), 1370(21), 1371(21), 1374(21), *1392*
Schultz, N., 593(67), *609*, 599(119), *611*, 1228(49), *1257*
Schultz, T. P., 791(16), *804*
Schultze, J. W., 1409(115), *1419*
Schulz, H., 234(15), *246*, 277(158), *318*, 483(3), 485(3), 493(3), 495(3), *496*, 641(171), *655*, 647(225), *657*, 1072(30), *1096*
Schulz, M., 974(95), *1021*
Schulz, R. C., 263(58), *314*
Schumacher, K., 593(67), *609*, 599(119), *611*
Schumacher, P. R., 662(31), 670(31), 671(31), *692*
Schumaker, R. R., 683(128), *696*
Schuettenberg, A., 345(144), *357*
Schunk, K., 445(170), *461*
Schwab, G., 792(22), *804*
Schwab, W., 1208(91), *1214*
Schwabe, K., 1048(139), *1065*
Schwager, F., 268(87), *315*, 1309(81), *1334*
Schwager, H. W., 705(34), *749*
Schwager, I., 332(17), 334(17), *353*
Schwarz, W. M., 146(107, 108), *198*, 309(437), *327*
Schwarzlose, G., 1299(54), *1334*
Schwertner, S., 716(112), *752*, 979(109), *1021*, 1015(253), 1016(253), *1027*
Scilly, N. F., 911(270), *944*
Scipioni, A., 1109(38), 1110(38), 1111(38), 1112(38), *1125*
Scoggins, D, I., 590(42), *609*
Scoggins, R., 387(159), *399*
Scola, P. M., 723(190), *754*, 723(193), *755*
Sconavacca, A. M., 596(91), *610*
Scopes, D. I. C., 704(27), *749*
Scott, A. I., 621(69), 625(69), *652*, 960(28a), *1018*, 1082(95), *1098*
Scott, B. A., 682(126), 683(128), *696*
Scott, C. J., 543(52), *576*, 662(37), *693*, 1002(191a), *1024*
Scott, K., 485(17), *496*
Scott, L. D. L., 645(207), *656*
Scott, M. K., 495(126), *500*
Scott, R. L., 148(110), 188(110), *198*
Scott, W. Jr., 257(12), *313*
Scrimager, C., 661(20), *692*
Sealy, G. R., 190(280), 195(280), *204*
Sease, J. W., 376(110), *397*, 391(184), *399*, 1221(7), *1256*
Sebastiani, G. V., 1070(2), 1080(2), 1081(2), *1095*
Seddon, D., 1005(216), *1025*, 1168(343), *1189*
Sedlet, J., 280(188), 293(188), *319*, 285(207), *320*
Seebach, D., 288(260), *321*, 441(149d), *461*, 557(114), *577*, 557(115), 558(116), 562(114–116), *578*, 601(128), *612*, 673(83), *694*, 895(98d), *937*, 1002(204), *1025*, 1158(232, 233), *1186*, 1170(370), 1171(371, 372), *1190*, 1172(412, 413), *1191*
Seeber, R., 676(98), *695*, 712(72), *750*, 931(360), *947*, 1059(257), *1068*
Segretario, J. P., 434(4), *456*, 743(394), *762*
Seiber, J. N., 369(66), *396*
Seidel, W., 546(66, 70), 547(66, 72), *576*, 783(74b, 76), *786*, 1002(195b, 196, 202), *1025*
Seidler, J., 684(130), *696*, 728(261), *757*
Seifert, J. M., 769(23), *784*
Seiler, P., 277(154), *318*, 645(203), *656*, 1321(122), *1336*
Sekine, T., 263(57), *314*, 434(18), 436(65), 447(65), 449(65), 457, *458*, 490(87), *499*, 539(20), *575*, 584(22), *608*, 740(355), *761*, 747(432), *763*, 906(177), 907(177), 907(198), *939*, *940*, 922(276), *944*, 975(103), *1021*, 1142(75), *1181*, 1175(482), *1193*, 1178(512, 514, 515), *1194*
Seko, M., 271(105), *316*, 907(186), *940*, 1295(46) *1333*, 1316(101), 1319(101), 1326(101), *1335*, 1321(126), 1330(126), *1336*
Selman, J. R., 1292(43), 1293(43), *1333*, 1302(66), *1334*

Selvig, A., 1243(130), *1260*
Semakhina, N. I., 435(41, 42), *457*
Semchenko, D. P., 1209(108, 109), *1215*
Semchenko, V. D., 1209(108), *1215*
Semenova, A. D., 435(55), 436(55), *458*, 442(101), 447(101), *459*
Semenovskii, A. V., 722(167), *754*
Semkov, A., 663(42), *693*
Semmelhack, M. F., 367(30), *395*, 377(119), *397*, 489(63), *498*, 1041(71), 1050(71), 1053(71), *1062*
Senkowski, B. Z., 734(306), *759*
Senn, D. R., 841(221), *871*
Senning, A., 712(74), *751*, 931(366), *947*
Seno, M., 260(43), *314*, 271(100), *316*, 924(296), *945*
Seo, E. T., 592(58), *609*, 725(227), *756*, 733(300). *759*, 886(21), *933*, 968(68), 970(68), *1020*
Sep, W. J., 620(42), *651*
Seppanen, E. D., 1176(498), *1194*
Ser, S., 409(65), *428*, 475(84), *481*
Sera, A., 404(30), *427*
Sercel, A. D., 781(65a), *786*
Serdobol'skii, E. N., 792(29), *805*
Sergeev, A. D., 794(47), *806*
Sergo, A. A., 438(80), *459*, 896(102), *937*
Serkov, A. T., 793(39), *805*
Serpone, N., 1413(131), *1419*, 1413(138), *1420*
Sertel, M., 496(130), *500*, 910(229), *942*
Serve, D., 280(184), 281(184), 284(184), 300(184), 286(184), 304(184), *319*, 283(236), 295(236), 304(236), *320*, 305(400), *326*, 522(122), *532*, 591(46–50), 592(47–50, 55), *609*, 723(183), *754*, 728(257), *757*, 970(73), 977(82, 83), *1020*, 1088(134), *1099*
Servin, R., 237(31), *247*, 267(76b–d), *315*, 601(138–141), 603(138), *612*, 727(246), *757*, 1077(57–60), 1083(58), *1097*, 1170(365, 366), *1190*
Sessler, J. L., 819(86, 87), *867*
Sestili, L., 662(32), *692*
Seto, N., 1409(102), *1418*
Setsune, J., 815(60), *866*
Settineri, W. J., 809(11), 860(11), 862(11), 863(11), *864*, 925(311), *945*, 1044(104), *1063*
Setzer, W. N., 1178(521), *1195*
Sevast'yanova, I. G., 924(302), *945*, 1201(26), *1212*
Sevast'yanova, V. V., 378(124), *397*, 403(12), *426*, 443(119), *460*
Severin, M. G., 224(22), *232*, 346(152), 350(152), *357*, 901(159), *939*
Severs, W., 1352(66), *1361*
Seyama, T., 271(107), *316*
Seydewitz, U., 721(157c, d), *753*, 928(326), *946*
Seymour, P., 820(91), 821(91), *867*
Shacklette, L. W., 1377(81), 1382(81), 1382(83), *1395*
Shaik, S. S., 1070(11), *1095*
Shain, I., 123(17–20), 129(18), 132(18), 133(18), *196*, 146(107, 108), *198*, 257(13), *313*, 276(140), *317*, 309(437), *327*
Shalimar, V. N., 797(63), *806*
Sham, T. K., 50(66), *117*
Shao, R., 370(74, 75), 372(74, 75), *396*
Shapiro, E. L., 1147(149), *1183*
Shapoval, G. S., 1351(38), *1360*
Shapvalov, V. A., 1354(81), *1361*
Sharifian, H., 621(68), 625(68), *652*
Sharma, B. L., 901(137), *938*, 1142(77), *1181*
Sharma, S. K., 424(177), *432*, 745(418), *763*
Sharp, J. H., 620(46), *651*
Sharp, P. R., 930(347), *947*, 1433(10), *1436*
Sharpless, K. B., 234(12), *246*
Shatzke, G., 704(26), *749*
Shatzmiller, S., 441(126), 444(126), *460*
Shavkunov, S. P., 290(277), *322*
Shaw, J.-W., 540(32), *575*
Shaw, R. W., 687(177), *697*
Shchenikova, M. K., 484(6), *496*
Shearer, T. R., 280(187), 284(187), 255(187), 297(187), *319*
Sheinkman, A, K., 926(315), *945*
Shelepin, I. V., 820(96), *867*
Shelton, R. N., 689(188), 690(188), *698*
Shepard, A. F., 1384(105), *1395*
Shepelin, V. A., 1232(76), 1234(76), 1235(76), *1258*
Shephard, B. R., 1004(205), *1025*, 1080(80), *1097*
Sheridan, J. B., 846(242), *872*
Sherman, E. O., 295(316), *323*
Shestov, A. P., 1048(142), *1065*
Shi, C.-N., 288(252), 292(252), *321*
Shi, W., 831(156), *869*
Shiba, T., 723(188), *754*
Shibasaki, M., 559(120), 562(120), *578*
Shibata, M., 639(148), *654*, 1390(133–135), *1396*
Shibuya, S., 739(348), *760*, 957(34), 961(34), *1019*
Shichiri, T., 1380(69), *1394*
Shields, T. M., 373(84), *396*
Shiga, M., 724(210), *755*
Shigematsu, S., 772(33a, b), *785*
Shigemori, H., 1013(251), *1027*
Shima, G., 439(96), 445(96), 446(96), 448(96), *459*, 448(188), 449(188), *462*, 455(235), *464*
Shimanskaya, N. P., 726(241), *757*, 729(266), *758*
Shimazu, K., 1135(24), *1179*
Shimizu, A., 1164(311), *1188*
Shimizu, K., 487(38), *497*
Shimizu, N., 565(140), *578*
Shimizu, S., 442(112), *460*
Shimizu, Y., 782(68), *786*
Shimura, T., 915(254), *943*, 929(338), *946*
Shinazi, R. F., 989(163), *1023*
Shindo, H., 1149(162), *1183*
Shine, H. J., 506(25), *529*, 660(7), 686(7), *692*, 686(148, 149), *696*, 686(155, 158–160, 162–170), 687(174, 176, 177), *697*,

[Shine, H. J.]
1119(65, 66), *1126*, 1412(127), *1419*
Shiner, V. J., Jr., 14(3a), *114*
Shinozaki, K., 489(56), *498*, 740(353), *761*
Shinozuka, N., 1053(200), *1066*
Shirai, K., 747(432), *763*, 911(269), *943*, 1230(52), *1258*, 1408(60), *1417*
Shirai, N., 1034(17), *1061*
Shiraishi, M., 723(195), *755*
Shirakawa, K., 560(123), *578*
Shiroi, T., 713(78), *751*, 1054(213), *1067*
Shiroki, H., 271(105), *316*
Shirokii, V. L., 837(198), 838(198), *870*, 1425(12), *1430*
Shishido, K., 704(19), *748*, 771(27c), *784*, 988(145), *1023*
Shishkin, E. A., 1109(36), *1125*
Shizuri, Y., 621(65), 624(65), *652*, 777(48), *785*, 1013(251), *1027*, 1166(321), *1188*
Shoemaker, C. E., 292(298), 293(298), 296(298), *323*
Shono, T., 5(16), 6(22), *8*, 233(6b), *246*, 236(24a), 239(42b), 240(42b), *247*, 240(52), 242(52), 243(69), *248*, 244(77), *249*, 367(34, 35), *395*, 374(87), *396*, 404(27), *427*, 433(2), *456*, 451(205), 453(205), 456(205), *462*, 493(112), 494(112), *500*, 516(86), 517(86, 90, 91), *531*, 523(126), *532*, 570(157), *579*, 584(19), 589(40), 590(40), *608*, 597(106), 598(109), *611*, 600(125), 601(130, 143), 602(130, 143, 144, 145), 602(147), 603(150, 155, 156), 604(160, 161), 605(160), 607(164), *613*, 630(109), 633(109), *653*, 643(185), *655*, 643(188), *656*, 647(226), 648(229), *657*, 704(15a–d), 726(15a–d), *748*, 708(51), *750*, 716(114), *752*, 726(244), 727(247), 728(244), *757*, 740(359), *761*, 767(9–11), *784*, 773(34–36, 38a), 774(39a, 40), 777(49, 50), 779(61a, b), 781(66), *785*, 779(61a, b), 781(66), 782(67, 69, 70a–c), *786*, 850(260), *872*, 882(3), 885(3), *932*, 891(44, 45, 49), *934*, 891(57), *935*, 903(199), 906(183a), 907(192), *940*, 904(209, 210), 905(212), 908(212), *941*, 919(289), 922(278), *944*, 924(298), *945*, 928(324), *946*, 950(1c), *1018*, 971(77a), *1020*, 979(99), 982(99), *1021*, 987(142), 993(142), *1023*, 1000(172), *1024*, 1035(19, 23), 1036(28), 1038(38, 39), *1061*, 1041(72), 1042(73, 75), 1044(101, 102), *1063*, 1070(16), 1073(16), 1080(16), 1084(16), *1095*, 1076(47), 1077(50), *1096*, 1079(69, 71, 72), 1080(72), 1082(72), *1097*, 1138(47), 1139(54–57), 1140(55–57), *1180*, 1141(70), 1143(89), 1144(89), *1181*, 1147(146), 1148(157), *1183*, 1150(167), 1151(172), *1184*, 1155(211, 212), *1185*, 1159(241), 1160(250, 254, 257, 258, 262), 1161(250, 258), *1186*, 1160(263), 1161(263), 1162(291), 1163(293), *1187*, 1163(295), *1188*, 1167(333), *1189*, 1171(373, 374), *1190*, 1173(440), *1192*, 1178(513), *1194*, 1207(84), *1214*, 1252(196, 197), *1263*, 1267(4, 10a, b), 1270(4), *1280*, 1268(24a–c, 27), 1275(24b), *1281*, 1277(52), 1278(57a), *1282*, 1280(57a), *1282*, 1425(13), *1430*
Shore, S. G., 125(48), *196*
Short, E. L., 894(74), *935*
Short, R. T., 639(139), *654*
Showa Denko, K. K., 456(247), *464*
Shreider, V. A., 1119(67e), 1123(67e), *1126*, 1123(82), *1127*
Shugar, D., 740(366), *761*, 746(427), *763*, 927(319), *946*, 1148(154), *1183*
Shukla, S. S., 151(125), *199*, 288(252), 292(252), *321*
Shul'zhenko, G. I., 540(33), 541(33), *575*
Shun, T., 726(234), *756*
Shuster, N., 894(85a), *936*, 1316(103), 1321(103), *1335*
Shvaika, O. P., 726(242), *757*, 734(312), *759*
Sibille, S., 277(160), 277(161), 311(160), *318*, 834(172), 835(172), *869*, 891(48), *934*
Sicher, J., 376(109), *397*, 1176(496), *1194*
Siefker, J. R., 293(313), *323*
Siegel, T. M., 344(126), *357*, 514(82), 515(82), *531*, 643(189), *656*, 1037(34), *1061*, 1085(116), 1086(125), *1099*
Siegerman, H., 509(39), *530*
Sieler, R., 837(194), *870*
Sigan, A. L., 826(131), *868*
Silber, J. J., 686(155, 158, 169), *697*
Silberg, J. A., 909(232), *942*
Silbermann, F., 414(97), *429*
Silbert, L. S., 1053(203), *1067*
Silhanko, A., 906(177), 907(177), *939*
Silkina, R. P., 445(156), *461*
Silverman, D. C., 1289(36), 1290(36), *1333*
Silverman, H. P., 725(227), *756*
Silverthorn, W. E., 841(220), *871*
Silvestri, G. S., 240(49), *247*, 277(159), 311(159), *318*, 292(301), 303(301), *323*, 857(295–297), 858(298, 299), *874*, 885(10), *932*, 904(204a, b), *940*, 930(345a), *947*, 1252(191), *1262*, 1353(68, 69), *1361*, 1433(9a), *1436*, 1434(13a, d), *1437*
Simamura, Y., 1409(92), *1418*
Simek, J., 284(212), 291(212), *320*, 286(292), 291(292), *322*
Simionescu, C., 1351(39, 40), *1360*
Simmons, W., 519(108), *532*
Simmrock, K. H., 243(72), *248*
Simon, H., 471(56), *480*, 1174(441), *1192*
Simon, J. M., 300(358), 307(358), *325*, 336(60), 339(60), *354*
Simon, W., 824(111), *868*
Simonet, J., 193(290), *205*, 216(11, 13), *231*, 273(123g), 275(135), *317*, 344(110), 347(110), *356*, 347(170),

[Simonet, J.] 348(170), *358*, 370(70, 71), *396*, 390(182), 391(188, 189), *399*, 407(44), *427*, 437(19), *457*, 445(159), 449(159), *461*, 449(191), *462*, 452(209), 453(209), *463*, 468(28, 29), *479*, 585(24), *608*, 597(104, 105), 598(104, 105), 599(104), *611*, 648(230), *657*, 662(29, 30), 663(29), *692*, 670(75, 76), 671(75), *694*, 678(104), *695*, 712(75), *751*, 722(174), *754*, 730(273), 731(273), 733(298), *758*, 886(20a, b, d), 887(30), 891(20a, b, d), *933*, 891(53), *934*, 913(286), 918(271c), *944*, 931(370, 371), *948*, 963(60, 62–64), 967(60), *1020*, 982(122), *1022*, 1013(248), *1027*, 1032(3, 4), 1033(6), *1060*, 1038(43), 1039(43), *1061*, 1040(69), *1062*, 1042(74, 77, 78a), *1063*, 1046(125, 126, 129), *1064*, 1048(140, 148, 150, 151), 1049(152), 1056(140), *1065*, 1054(208), *1067*, 1058(252, 254), 1059(253, 254, 263–265), 1060(263, 264), *1068*, 1136(33), *1180*, 1144(109), *1182*, 1144(130, 131), 1145(130, 131), *1183*, 1150(165, 166), *1184*, 1160(253), *1186*, 1164(306), 1166(323), *1188*, 1172(398), *1191*, 1173(429), *1192*, 1200(15, 16), *1211*, 1224(8, 12), *1224*, 1221(6, 8, 11, 12), 1222(6), 1247(11, 12), 1248(12), 1250(8), 1251(8, 11, 12), *1256*, 1221(20, 21), 1224(26, 29, 30, 32), 1226(20, 40, 43, 45), 1227(21, 46, 47), 1248(20, 29, 32, 43, 47), 1249(21), 1250(32), *1257*, 1231(67), 1246(67), *1258*, 1239(30), 1241(122, 125, 127), 1242(128), *1260*, 1243(145), 1245(154), *1261*, 1245(162, 163), 1246(162, 163), 1250(166, 184), 1251(188, 190), 1252(166), *1262*, 1252(194, 195), *1263*, 1278(57e), 1280(57e), *1282*
Simonet-Gueguen, N., 963(60, 62–64), 967(60), *1020*
Simons, J. H., 290(273), *322*, 1104(8), 1105(8), 1106(8), *1124*, 1106(29), *1125*, 1317(109), *1335*
Simonson, L. A., 717(119), *752*
Simonsson, D., 1312(84), *1335*
Simpson, J., 1156(214), *1185*
Sims, J., 565(142), *578*
Sinclair, S. A., 790(10), *804*
Singer, R.-J., 493(108), *500*, 1047(132), *1064*, 1170(357), *1190*
Singleton, V. D., 368(45), *395*, 886(12), 891(12), *933*
Sinisterra, J. V., 163(155), *200*
Sinitskii, V. V., 794(54), 795(54), *806*
Sin-rhu, L., 163(163), *200*
Sioda, R. E., 334(35), *354*, 422(149), *431*, 488(48), 491(48), 494(48), *498*, 510(50, 51), *530*, 914(243), 915(243), *943*, 1144(126), 1145(126), *1182*, 1162(287), *1187*
Siragusa, J. A., 899(145), 901(145), *938*
Sirakawa, T., 1229(68), *1258*
Siroi, T., 780(62a, 63), *786*
Sjöstedt, P., 3(10), *8*, 663(45), 667(45), 675(45), *693*, 1056(234), *1067*
Sjöström, E., 787(3), 788(3), 790(3), 791(3), *803*
Skala, V., 495(125), *500*
Skaletz, D. H., 467(21, 22), *479*, 1172(403, 410, 411), 1173(411), *1191*, 1321(120), *1336*, 1148(153), *1183*
Skasietz, D. H., 928(327a), *946*
Skarzewski, J., 1229(72), *1258*
Skelepin, J. V., 1348(29), *1360*
Skell, P. S., 860(310), *874*
Skikata, K., 1211(127), *1215*
Skinner, J. F., 1349(32), *1360*
Skoog, D. A., 1053(198), *1066*
Skornyakov, V. V., 438(83), *459*
Skotheim, T. A., 340(83), 341(83), *355*, 1385(108), *1396*
Skov, E., 644(199), *656*
Skrebkova, I. M., 473(64–67), *481*
Skully, J. P., 180(239), 189(239), *203*
Skundin, A. M., 539(22), *575*, 1010(233), 1011(237), *1026*
Skwarski, T., 1355(82), *1361*
Slansky, C. M., 24(27b), 42(27b), *115*
Slates, V., 333(28a), *353*
Sleigh, J. H., 550(91), *577*
Sleszynski, N., 434(4), *456*
Sliwa, H., 537(16), 564(16), 565(16), *575*
Sloan, G. J., 720(137), *753*
Slobbe, J., 565(138), *578*
Slomkowski, S., 349(176), *358*
Slongl, W., 1245(156), *1261*
Slotterbeck, O., 442(106), 445(106), *459*
Slovetskii, V. I., 406(42), *427*, 407(49, 50, 51), *428*
Small, L. F., 282(202), 285(202), 291(202), *319*
Smart, B. E., 307(415), *327*
Smid, J., 335(46), *354*, 885(8), *932*
Smilgin, G. D., 1236(103), *1259*
Smirnov, K. M., 368(40), *395*
Smirnov, S. K., 910(226), *942*, 925(308), *945*
Smirnov, V. A., 259(32), 263(32), 309(32), *313*, 445(157, 158, 166), *461*, 797(62, 64, 67), *806*, 861(314), *874*, 1104(11), *1124*, 1202(41), *1212*, 1202(42), *1213*, 1209(108, 109), *1215*, 1287(23), *1333*
Smirnov, Yu. D., 410(70), *428*, 434(16), *457*, 496(133, 134), *500*, 496(135), *501*, 910(226), *942*, 921(294), 925(308), *945*
Smirnova, N. G., 445(166), *461*
Smit, C. J., 306(404), 311(404), *326*, 387(160), *399*, 891(54b), *935*, 894(88), *936*, 895(97), *937*, 1144(116, 120), *1182*, 1172(409), *1191*
Smith, B. L., 823(110), 824(110), *867*
Smith, C. Z., 239(45), *247*, 277(151), *318*, 798(68, 69), 799(68), 800(78), 801(69), *807*, 802(91), *808*, 918(367), *948*, 1082(100), *1098*, 1144(133), 1146(133), *1183*
Smith, D. E., 333(23), *353*, 335(52), 342(52), *354*, 342(96), 343(96), *356*, 434(21), *457*, 882(3h),

[Smith, D. E.]
885(3h), *932*, 1175(465), *1193*
Smith, D. J., 689(191), *698*
Smith, D. L., 732(277), *758*, 736(325), *760*, 745(420), *763*
Smith, E. K., 1081(93), *1098*
Smith, F. L., 293(313), *323*
Smith, G. A., 994(154), *1023*
Smith, G. B. L., 402(5), 403(10), *426*
Smith, J. D., 1353(71, 72), *1361*
Smith, J. F., 1054(206), *1067*
Smith, J. G., 886(20f), 891(20f), *933*
Smith, J. M., 447(177), *462*, 1132(1), 1144(1), *1179*
Smith, K. M., 705(29a), *749*
Smith, L. I., 454(224), *463*, 1043(83), *1063*
Smith, P. J., 582(3, 4), 583(4), *607*, 584(16), *608*, 1051(179), *1066*, 1076(46), *1096*
Smith, S. L., 726(242), *757*
Smith, W. H., 77(97), *119*
Smith, W. B., 555(104, 106), *577*, 1170(367, 368), *1190*
Smith, W. H., 263(55), 293(55), *314*, 336(59), 342(59), *354*, 413(93), *429*, 726(232), *756*, 733(299), *758*, 831(159), 832(159, 161, 162), *869*, 882(31), 885(31), *932*
Smithers, R. H., 295(320a–d), 296(320a–d), *323*
Smittenberg, J., 442(108), 447(108), *459*, 450(201), *462*
Smolenceva, L. G., 915(252, 253), *943*
Smolvchowsky, Z., 32(45b), 33(45b), *116*
Smothkin, E., 1413(139), *1420*
Smyrl, W. H., 282(196), 284(196), 299(196), *319*
Snavely, Jr., E. N., 1105(24), *1124*
Snavely, E. S., 284(232), 290(232), *320*
Snatzke, G., 1000(173), *1024*
Snell, K. D., 1358(111), *1362*
Snow, K. M., 705(29a), *749*
Snow, M. R., 1175(448), *1192*
So, Y.-H., 291(284), *322*, 643(191), *656*, 704(16), *748*, 957(37), *1019*, 993(144), 994(149), 995(149), 999(149), *1023*, 1081(90, 91), 1085(90), *1098*, 1086(124), *1099*, 1094(174), *1101*
Sobecki, W., 722(178), *754*
Sobieralski, T. J., 389(170), *399*, 747(435), *763*
Sobkowski, J., 639(147), *654*
Sock, O., 4(13), *8*, 277(160), 311(160), *318*, 285(227), 293(227), *320*, 366(24), *394*, 1252(192), *1263*
Soeldner, J. S., 644(193, 195–197), 645(193, 196), *656*
Sofranko, J. A., 392(197), *400*
Sofuku, H., 667(57), *693*, 1245(150), *1261*
Soga, T., 375(90), *396*
Sohm, J. C., 301(374), *325*
Sohn, J. H., 1035(21), *1061*
Sokolov, R. P., 894(86), *936*
Sokolova, E., 640(155), *655*
Sokol'skii, D. V., 439(97), 440(97), *459*, 445(156), *461*
Soldatov, B. G., 273(119), *317*, 436(74), 438(74), *458*, 892(61), *935*, 896(101, 103), 897(101), *937*
Solis, V., 594(81), *610*
Solodovnikov, S. P., 455(231), *463*
Solomon, T., 596(95), *610*, 710(60c), 715(60c), *750*, 1371(36), *1393*
Solon, E., 153(128), *199*
Soltovets, G. N., 1202(41), *1212*, 1202(42), *1213*
Somlo, F., 486(22), *497*
Sommer, F., 1348(28), *1360*
Sommer, J., 616(3), 620(3), *650*
Somorjai, G. A., 1404(18), *1416*
Sondana, G., 1233(95), *1259*
Sondheimer, F., 336(64), 339(64), *355*
Song, D., 1371(36), *1393*
Sonobe, T., 570(162), *579*
Sopher, D. W., 489(61, 62), 490(71), *498*, 791(20), *804*, 798(73), *807*, 860(313), *874*, 894(78, 79), 898(78), *936*, 974(94a), *1021*, 1040(66, 67), *1062*, 1117(50b), 1119(50b), 1121(50b), *1125*, 1146(134), *1183*, 1235(92), 1236(92, 102), *1259*, 1435(29), *1438*
de Sorgo, M., 685(140), *696*
Soriaga, M. P., 1371(36), *1393*
Soroka, M., 686(164), *697*
Sosnowski, J. J., 724(207), *755*
Sosonkin, I. M., 891(55), *935*
Sotnikova, N. N., 488(45), *497*
Souchay, P., 405(33, 39), 406(40), *427*, 409(65), *428*, 474(76), 475(84), 476(88), *481*, 739(346), *760*, 746(425), *763*, 1050(173), *1066*
Soufflet, J. P., 816(61), *866*
Soulier, J.-L., 542(47), *576*
Sowell, G., 1139(59), 1140(59), *1180*
Spada, L. T., 1407(35), *1416*
Spadaro, G., 930(345a), *947*
Spanier, J., 135(78), *197*
Spankel, L., 1412(125), *1419*
Späth, E., 740(363), *761*
Spektor, V. I., 440(99), *459*, 897(116), *937*
Speiser, B., 184(256), *203*, 616(8), 617(8, 11, 19), 621(11), *650*
Spencer, C., 439(86), *459*
Sperber, N., 1208(93), *1214*
Spewock, S., 1358(101), *1362*
Spikes, J. D., 1408(72), *1417*
Spitzner, R., 728(258), *757*
Spiridonova, L. A., 456(242), *464*, 792(27, 28, 31), *805*
Spoerri, P. E., 489(66), *498*
Spohn, O., 1044(98), *1063*
Spotnitz, R., 1229(69, 70), *1258*
Spreer, L. O., 1434(12j), *1437*
Spreter, V. C., 425(191), *432*
Spritzer, L., 452(208c), *462*
Spritzer, M. S., 301(367), *325*
Sprogl, J., 689(182, 183), *697*, 722(164d, 168), *754*, 1161(278), 1162(281), *1187*, 1167(334, 335), *1189*
Spurlock, S., 565(142), *578*
Srinivasan, S., 438(72), *458*, 845(234), *872*
Srna, C., 555(105), *577*, 1348(27), *1360*
von Stackelberg, M., 453(219), *463*
Stamm, E., 1155(210), *1185*
Stamouli, P., 239(44), 240(44), *247*, 813(37), 825(37), 826(37), 827(37), 841(37), 863(37), *865*
Stang, P. J., 184(257), *204*
Stanienda, A., 738(337), *760*

Stansfield, R. F. D., 846(241), *872*
Stapelfeldt, H. E., 441(144), *461*
Starkova, T. F., 284(233), *320*
Starud, N., 179(232), *203*
Stasko, A., 340(77b), *355*
Staudinger, H., 1233(85), 1235(85), *1259*
St. Buduroff, C. R., 476(89a), *481*
Steck, E. A., 442(110), *459*
Steckel, T. P., 643(186), *655*
Steckhan, E., 139(87), 190(87), *198*, 182(243), 184(247, 248), 189(248), 190(248), *203*, 239(38), 243(38), *247*, 240(53), *248*, 393(208), *400*, 516(87), 517(87), 525(87), *531*, 525(137), *533*, 593(61–68), 594(63), *609*, 593(69, 70, 74), 594(76–78), *610*, 598(110), 599(119), *611*, 604(162), 605(162), 607(162), *613*, 635(124), 636(124), *653*, 648(228), *657*, 670(73), *694*, 810(15, 18), *864*, 980(112, 119a, 121), 981(129, 134), 982(112, 115, 120), 983(120, 129), 984(129), *1022*, 1014(249), 1015(249), *1027*, 1043(88, 89), *1063*, 1046(130, 131), 1047(133), *1064*, 1079(70), *1097*, 1153(191), *1184*, 1154(199), *1185*, 1161(270), *1187*, 1166(319, 324), *1188*, 1221(18), 1224(18), 1249(18), *1256*, 1224(34), 1225(36, 38, 39, 41, 42), 1228(49), 1230(42), 1233(41, 42), 1249(41), *1257*, 1232(82), 1233(82, 94, 97– 99), 1235(93), *1259*, 1241(126), *1260*, 1249(179), *1262*
Steelink, C., 621(55), *651*, 621(76), 625(76), *652*
Steevensz, R. S., 1235(89), *1259*
Steffek, D. J., 691(195), *698*
Steiger, B., 826(133), *868*
Stein, I., 740(365), *761*
Stein, V., 663(42), *693*
Steinberg, M., 438(72), *458*
Steinberger, N., 434(11), *456*
Steinborn, D., 821(98), *867*
Steinert, A., 309(431), *327*
Steinfeld, A. S., 961(44), *1019*
Steiniger, M., 245(80), *249*, 375(96), *396*, 604(157), *613*, 903(219), *941*
Stella, L., 242(59), *248*
Stender, K.-W., 685(138), *696*
Stenzal, T. T., 1239(115), *1260*
Stenzl, H., 449(194, 196), *462*, 490(79, 80), *499*
Stepanek, R., 824(111), *868*
Stepanov, A., 445(164), *461*
Stephens, C. J., 979(108), *1021*
Stephenson, M. T., 1119(65, 66), *1126*
Stermitz, F. R., 584(18), *608*, 704(16), *748*, 770(26a, b), *784*, 889(37), *934*, 957(37), *1019*, 971(92), 977(92), *1021*, 993(143, 144), 994(149–153), 995(149, 151, 153), 999(149), *1023*, 1093(170), 1094(174), *1100*
Stern, A., 781(64d), *786*
Stern, M., 740(356), *761*, 818(76), *866*
Stern, P., 927(321), *946*
Sternberg, H. W., 239(41), *247*, 285(208), *320*, 293(311, 312), *323*, 297(339), *324*, 1254(206, 209), 1255(206, 213), *1263*, 1434(16), *1438*
Sternerup, H., 266(76a), 267(76), *315*, 523(130), *532*, 601(139, 141), *612*, 952(2), 953(2), 960(2), *1018*, 980(119b), 982(119b), *1022*, 1077(57, 60), *1067*, 1092(166, 167), *1100*, 1160(259, 261), *1186*
Steuber, F. W., 616(5), 618(5), 621(5), *650*
Stevenson, D., 1045(116), *1064*
Stewart, B. J., 1175(456), *1193*
Stewart, R. F., 634(115), *653*, 704(16), *748*, 961(37), *1019*, 967(57), *1020*, 993(144), 994(149), 995(149), 999(149), *1023*, 1094(174), *1101*
Stewart, W. E., 1292(42), 1299(42), *1333*
Stibbard, J. H. A., 734(309), *759*
Stibor, I., 689(182, 183), *697*, 721(157d), *753*, 722(168), *754*, 1161(278), 1162(281), *1187*, 1167(334), *1189*
Stickney, J. L., 1371(36), *1393*
Stitson, S. C., 1290(41), *1333*
Stocesova, D., 418(128a), *430*
Stock, J., 377(119), *397*, 957(34), 961(34), *1019*
Stocker, J. H., 897(110), *937*, 898(128), 899(146), 900(128), 901(146), 902(128), *938*, 1132(8), *1179*, 1144(107, 108, 113, 114), 1145(114), *1182*
Stocker, R., 272(115), *316*
Stöckl, E., 555(103), *577*
Stohr, E., 709(55), *750*
Stonehill, H. I., 726(237), *756*
Stopskij, V. S., 471(57), *480*
Storck, A., 418(135), *431*, 1231(62), 1234(62), *1258*, 1286(4), 1292(4), 1299(4), *1332*, 1299(57), 1309(81), *1334*
Stork, G., 544(54), *576*, 1001(193), *1024*
Stork, L., 716(116), *752*, 1010(234), *1026*, 1166(326), *1189*
Story, N., 1434(12e), *1437*
Störzbach, M., 332(10), 338(10), 341(10), *353*
Stösser, R., 190(281), 193(281), *204*
Stothers, J. B., 1002(187), *1024*
Stotz, R., 590(42), *609*
van der Stouwe, C., 6(21), *8*
Stowell, J. C., 233(5c), 234(5c), 240(5c), *246*
Strachan, E., 617(15), *650*
Strachan, R. S., 546(67), *576*, 1002(203), *1025*
Stracke, W., 453(219), *463*
Stradins, J. P., 434(14), 435(37, 38), *457*, 441(139), *460*, 441(145), *461*, 903(201), *940*, 1050(171), *1066*
Stradiotto, N. R., 369(59), *395*, 1056(243), 1058(243), *1068*
Strampach, N., 1413(135), *1420*
Strasen, S., 275(133), *317*, 712(76), *751*, 931(356), *947*, 902(373), *948*
Street, G. B., 1340(7), *1359*, 1376(59), 1380(59), *1394*, 1387(114), 1389(114), *1396*
Streich, E., 958(39), 961(39), *1019*
Streitwieser, A., 26(31a), *115*, 332(17), 334(17), *353*, 346(147), *357*
Strekas, T. C., 831(156), *869*

Strelets, V. V., 930(348), *947*, 1433 (6a, 8), *1436*
Strelow, H., 278(166, 167), *318*
Streukens, M., 1006(213), *1025*
Stricks, W., 662(25, 26), *692*, 662(34), *693*
Stride, J. D., 1005(218), *1025*
Striebel, K. A., 290(272), *322*
Strier, M. P., 733(292), *758*
Strobel, W., 263(58), *314*
Stroezel, M., 1305(76), *1334*
Strohmeier, W., 908(223), *941*
Stroinski, A., 813(28), *865*
Strojek, J. W., 183(246), 184(249), *203*, 686(152), *696*
Strom, E. T., 434(12), *456*, 1269(33), *1281*
Stromskaya, G. I., 794(57, 58), 795(51), 796(57, 58), *806*
Stromskii, S. V., 792(30), *805*
Strope, E., 621(53), *651*
Strübing, B., 275(133, 136), *317*, 712(73, 76), *751*, 931(356), *947*, 902(373), *948*
Strzelecka, H., 214(9), *231*
Stuart, J. D., 524(132), 525(134), *533*, 986(138b), 993(138b), *1023*
Stucky, G., 601(128), *612*
di Studio, C., 1154(196), *1184*
Studnickova, M., 423(164), *431*, 714(96), *751*, 727(251), *757*
Stühmer, W., 1207(90), *1214*
Sturn, E., 726(238), *756*
Sturtevant, J. M., 311(438), *327*
Stutts, K. J., 68(88), *118*
Su, K. B., 17(14), *115*, 76(95a), *119*, 363(6), 366(6, 22), 379(6), 384(6), *394*, 817(75), *866*, 1267(6), *1280*
Su, T.-Y., 403(19), *426*
Suan, Nguen, 435(45), *457*
Subba, R. G., 345(129), *357*
Subramanian, G. S., 2(6), *8*, 265(71–73), 271(71, 73), *315*, 271(102), *316*, 417(120, 121), 422(120), *430*, 1305(73), *1334*
Subramanian, R. V., 1358(100, 105–107), *1362*
Suchomelova, L., 1056(231), *1067*
Suda, K., 678(103), *695*
Suda, Y., 1410(119), *1419*
Sue, J. L., 1200(3), *1211*
Sueoka, T., 537(8), 572(8), *574*, 679(109), *695*
Sugano, T. T., 894(85a), *936*, 1316(103), 1321(103), *1335*
Sugawara, M., 1267(13), *1280*, 1268(25d), 1274(25d), 1275(25d), *1281*
Sugihara, Y., 602(145), *612*, 1038(38, 39), *1061*, 1077(50), *1096*, 1147(146), *1183*
Sugimoto, I., 682(124), *695*
Sugino, K., 263(57), *314*, 425(189), *432*, 434(18), *457*, 436(65), 447(65), 449(65), *458*, 439(89), *459*, 496(134), *500*, 539(20), *575*, 747(432), *763*, 906(177), 907(177), *939*, 907(198), *940*, 911(269), *943*, 922(276), *944*, 927(321), *946*, 1230(52), *1258*
Suginome, H., 584(20), *608*, 1167(332), *1189*
Sugita, S., 570(163), *579*, 1159(238), *1186*
Sugiura, M., 1434(12f), *1437*
Sugiura, T., 1231(65), 1232(65), *1258*
Sugiyama, H., 907(192), *940*, 919(289), *944*, 1139(54), *1180*, 1174(443), *1192*
Sugiyama, K., 686(163), *697*
Sugusawa, S., 490(86), *499*
Suhura, Y., 1004(208), *1025*
Suib, S. L., 288(252), 292(252), *321*
Sullivan, B. P., 301(371), *325*, 1434(12e, n), *1437*
Sullivan, D. E., 548(78), *576*, 1004(211), *1025*
Sullivan, J. M., 288(259), *321*
Sullivan, J. W., 1002(195a), *1025*
Sullivan, O. E., 539(23), *575*
Sullivan, T. A., 1044(97), *1063*
Sullivan, W., 899(147), *938*
Sumitani, K., 832(160), *869*
Sümmermann, W., 340(82a), *355*
Summers, D. P., 1434(15b, e, m), *1437*
Summers, L. A., 733(301), *759*
Sundermeyer, W., 849(257), *872*
Sundholm, F., 801(79, 80), *807*
Sundholm, G., 638(133), *654*, 641(180), 643(183), *655*, 801(79, 80), *807*
Supramanian, P., 736(327), *760*
Surikova, N. V., 894(77c), *936*, 899(136), *938*, 1176(505), *1194*
Surov, I. I., 470(49), 476(49), *480*, 726(243), *757*, 743(387), *762*, 1058(248), *1068*
Susuki, T., 528(144), *533*, 1084(112), *1098*, 1089(145), *1100*
Sutcliffe, H., 1106(26), *1124*
Suter, H., 486(23), *497*, 1142(73), *1181*
Sutherland, C., 557(110), *577*
Sutherland, J. O., 689(189), 690(189), *698*
Sutin, N., 32(42, 43e), 63(43), *116*, 38(51a), 39(52), 41(56), 45(58), 49(62), 50(66), *117*
Sutlic, V. D., 1314(91), *1335*
Suttie, A. B., 620(40), 621(40), *651*
Sutton, J. R., 1161(272), *1187*
Suvorov, N. N., 424(170), *432*
Suyama, K., 1013(251), *1027*
Suyatkin, Y. K., 792(31), *805*
Suzuki, A., 858(304), 859(305), 860(306–309), *874*, 906(181), *939*, 1426(16, 17), *1430*
Suzuki, I., 742(382), 743(382), *762*
Suzuki, M., 782(70c), *786*
Suzuki, S., 346(147), *357*, 1434(12c, 15j), *1437*
Suzuki, T., 620(32), *651*, 1204(52), *1213*
Suzumoto, T., 367(36), *395*, 433(2), *456*, 891(44, 45), *934*, 904(210), 918(210), *941*, 1139(57), 1140(57), *1180*, 1155(211), *1185*
Svaan, M., 506(14), 510(14), *529*
Svadkovaskaya, G. E., 1001(177a), *1024*
Svanholm, U., 170(179), *201*, 174(207, 208), 175(207), 190(207), *202*, 506(7), 511(7), *528*, 509(40), 510(40), 511(40, 60), *530*, 519(111), 520(116), 524(111), *532*, 621(78), 625(78), 629(93), *652*, 686(171, 172), 687(175), *697*, 955(16), 960(23), *1018*, 966(57), *1020*, 986(138a), 993(138a), *1023*, 1081(88), *1098*, 1415(151), *1420*

Svensmark, B., 127(61), *197*, 184(252), 186(252), *203*, 629(93), *652*,
Svensson, C., 908(224m), 910(224m), 911(224m), *942*, 1137(43), *1180*
Svetlicic, V., 1200(9, 11), *1211*
Sviderski, J., 1208(94), *1214*
Swaan, M., 895(92), *936*
Swain, C. G., 260(47), 263(47), *314*, 284(226), 303(226), *320*
Swan, D., 104(109a), *119*
Swann, S., Jr., 2(2), *7*, 438(77, 78), 446(77, 78), 450(77), *458*, 439(90), 442(111), 444(111), *459*, 437(131), 441(131), 444(125), *460*, 446(174), 447(179), 450(203), *462*, 720(138), *753*, 882(3s), 885(3s), *932*, 893(72), *935*, 898(127, 131), 899(142, 150), 900(151), *938*, 1036(30), *1061*, 1287(27), *1333*
Swanson, D. D., 668(64), 669(65), *694*, 1178(520), *1195*
Swartz, J. E., 1207(83), *1214*, 1239(115), *1260*
Swelim, A., 659(3), *692*
Swenson, K. E., 239(41), *247*, 1200(7), *1211*, 1203(45, 46), *1213*
Swenton, J. S., 630(99–101, 103, 107, 111), 631(103), 632(107), *653*, 781(64a–d, 65a–c), *786*, 1073(37), 1074(37), *1096*
Swern, D., 1053(201), *1066*, 1053(202, 203), *1067*
Swiderski, J., 894(77d), *936*, 898(120), *937*
Symons, M. C. R., 17(15c), *115*
Synnott, J. C., 281(193), 282(193), 284(193), 285(193), 297(193), 299(193), *319*, 301(369), *325*
Szebenyi-Györy, E., 734(311), *759*
Szwarc, M., 331(5), 343(5), *352*, 333(28a, b), *353*, 335(46–48), 343(46), *354*, 340(86b), *355*, 349(176), *358*, 349(187), *359*, 685(140), *696*, 885(7, 8), *932*, 1346(24), *1360*
Tabakovic, I., 288(254), 292(254), *321*, 441(147), *461*, 596(92, 94–97), *610*, 621(51), *651*, 701(1b), *747*, 706(36a, b, 37, 38), 710(36a), 712(36a), 714(36a), *749*, 710(60a–c), 715(60a–c), *750*, 714(99a, b), *751*, 715(107), *752*, 974(94b), *1021*, 1005(212d), 1009(212d), *1025*, 1015(250), *1027*, 1051(178), *1066*, 1268(29), *1281*
Tabakovic, K., 706(36b), 710(36b), 712(36b), 714(36b), *749*
Tabard, A., 816(65), *866*
Tabner, B. J., 350(200), *359*, 733(302), *759*
Taboury, F., 661(24), *692*
Tachibana, M., 742(383), *762*
Tachikawa, H., 344(123), *357*
Tacussel, J., 263(52), *314*, 309(427), *327*
Tada, M., 825(121, 122), 827(138), 828(138), *868*
Tada, N., 600(120), *611*, 1244(148), *1261*
Tadros, M. E., 1081(93), *1098*
Tafel, J., 65(83), *118*, 274(124), *317*, 442(102), *459*, 442(114), 445(114), *460*, 445(168), 447(168), *461*, 446(176), 447(182), 449(192, 193, 195), *462*, 488(42), 489(42), *497*, 490(76–78), *499*, 740(356), *761*
Taguchi, T., 392(192), *399*, 858(304), 860(309), *874*, 886(18), *933*, 904(181), *939*, 1269(35), *1281*
Tainturier, G., 258(21), *313*, 713(83, 84), *751*, 893(67), *935*
Taits, S. Z., 722(166), *754*, 1221(17), 1224(17), *1256*
Takagi, C., 925(304), *945*
Takagi, K., 1034(17), *1061*, 1409(93), *1418*
Takagi, M., 443(117), *460*, 900(152), *938*, 1443(5b, f), *1436*, 1434(15h), *1437*
Takagi, T., 584(20), *608*, 1167(332), *1189*
Takagishi, S., 586(25), 601(25), *608*, 1035(22), *1061*, 1168(341), *1189*
Takahashi, F., 1174(442), *1192*
Takahashi, H., 275(127), *317*, 1174(442), *1192*
Takahashi, M., 1004(209), *1025*, 1141(65), 1143(65), *1181*, 1143(95), *1182*, 1170(369), *1190*
Takahashi, S., 1425(11), *1430*
Takahashi, Y., 490(87), *499*, 526(140), *533*, 859(305), 860(306– 309), *874*, 907(186), *940*, 1229(73), 1230(55), *1258*, 1426(16, 17), *1430*
Takahasi, N., 1034(16), *1060*
Takaki, S., 1206(68, 69), *1213*, 1206(75, 76), *1214*
Takakuwa, Y., 1280(59, 60), *1282*
Takamatsu, H., 1280(60), *1282*
Takamura, K., 646(214), *656*
Takamura, T., 640(162), *655*
Takano, N., 455(237, 238), *464*
Takano, S., 704(19), *748*, 771(27c), *784*, 988(145), *1023*
Takao, Y., 1408(82), *1418*
Takasu, Y., 1164(311), *1188*
Takata, Y., 275(129), *317*, 441(152a), *461*, 583(12), *607*, 596(90), *610*, 597(107), *611*, 723(188), *754*, 1009(228), *1026*, 1091(164), *1100*, 1166(327), *1189*, 1233(96), *1259*, 1428(22), *1430*
Takeda, A., 555(101, 102), *577*, 573(168), *579*
Takeda, N., 817(73), *866*
Takehara, Z. I., 267(84), 271(84), 288(84), 307(84), *305*
Takei, K., 600(121), *611*
Takeno, N., 455(237, 238), *464*
Takeshita, K., 721(150), *753*
Takeuchi, N., 601(143), 602(143), *612*, 771(27a), *784*, 1000(168), *1024*
Takeuchi, Z., 278(173), *319*
Takiguchi, Y., 894(89), *936*, 1153(187), *1184*
Takuda, M., 906(181), *939*
Takyu, C., 686(146), 687(146), *696*
Talarmin, J., 1175(458), *1193*
Talaty, E. R., 434(12), *456*
Talham, D. R., 838(205), *871*
Taliani, C., 1382(80), *1395*

Tallec, A., 267(80, 81, 83), 268(83), 271(81, 83), 273(80, 81), 288(80, 81, 83), 307(83), *315*, 375(105), 376(106–108), *397*, 415(108), 420(108), *430*, 418(136), 419(138), 420(139, 142), 421(142), 422(155), 423(163), *431*, 425(185, 186), *432*, 456(246), *464*, 707(45, 47a, c), *749*, 707(48), 708(49), 712(69), *750*, 714(92, 95, 97), *751*, 719(135), *753*, 728(253, 254), *757*, 893(66, 68a), *935*, 915(255), *943*, 1132(17), 1135(20, 21), *1179*, 1149(163), *1183*, 1151(174, 175), 1152(176), *1184*, 1168(349, 350), 1169(350), *1189*, 1171(377), *1190*, 1171(392–395), 1172(398), *1191*, 1173(436), *1192*, 1278(57e), 1280(57e), *1282*, 1427(18, 23), *1430*
Tamai, T., 1206(76), *1214*
Tamaka, K., 1380(69), *1394*
Tamamura, T., 1387(112), 1389(112), *1396*
Tamao, K., 832(160), *869*
Tamaru, K., 1408(41), *1416*, 1408(50), *1417*
Tamblyn, W. H., 689(187), *698*
van Tammelen, E. E., 565(142), *578*
Tamura, H., 258(27), *313*, 621(55), *651*, 971(77a), *1020*, 1383(96–98), 1386(97, 98), 1387(98), *1395*, 1408(82), *1418*
Tan, A. S. C., 295(320d), 296(320d), *323*
Tanabe, T., 716(114), *752*
Tanaka, A., 425(187), *432*, 599(118), *611*
Tanaka, H., 404(28), *427*, 417(115), *430*, 537(9, 10), 572(9), 573(10), *574*, 539(21), *575*, 570(159), 573(166), 574(169), *579*, 600(120), *611*, 621(50), *651*, 663(39), *693*, 677(105), 679(105, 110), *695*, 713(78), *751*, 721(154, 155), *753*, 780(62a, b, 63), 782(68), *786*, 835(180), *870*, 850(261), 851(265–269), *873*, 904(208), *941*, 1039(47), *1062*, 1054(213), *1067*, 1162(282, 283, 286), *1187*, 1164(313), *1188*, 1229(68), *1258*, 1236(106), *1259*, 1243(135), 1244(135, 147, 148), *1261*, 1268(23), 1275(23) 1277(23), *1281*, 1277(54), 1278(57b), 1280(57b, 58), *1282*, 1408(61), *1417*, 1434(13b), *1437*
Tanaka, K., 1408(47), *1417*
Tanaka, M., 24(27e), 42(27e), *115*, 1408(61), *1417*
Tanaka, N., 276(139), *317*, 662(26), *692*
Tanaka, S., 1380(67), 1383(67), 1384(67), 1385 (62), *1394*
Tanaka, T., 542(49), 548(76), *576*, 607(163), *613*
Tanford, C., 1339(3), 1343(3), *1359*
Tang, H., 1173(432), *1192*
Tani, H., 404(30), *427*
Tani, T., 1157(223), *1185*, 1160(249), 1161(249), *1186*
Tani, Y., 827(140), 829(140), *868*
Tanida, G., 563(129), *578*
Taniguchi, I., 723(195), *755*, 973(79), 977(85), *1020*
Taniguchi, M., 1317(111), 1321(111), *1335*
Tanimoto, Y., 258(28), *313*, 541(37), *575*
Tanneberg, H., 403(11), *426*, 707(47c), *749*, 742(381), *762*
Tanner, D., 336(63), *355*
Tanner, J., 1132(13), *1179*, 1353(75, 76), *1361*
Tantayanon, S., 1408(70), *1417*
Taran, L. A., 495(122), *500*, 915(252, 253), *943*
Tardivel, R., 290(281), 291(281), *322*, 1085(119), 1086(128, 129, 131), 1087(129) *1099*, 1094(184), *1101*, 1120(69–71), 1121(73), 1122(69, 71, 74a–c, 75), *1126*, 1163(302, 303, 304), 1164(302, 304), *1188*
Tari, I., 891(47b), *934*, 909(225), *942*, 924(299), *945*
Tarkhanov, G. A., 276(141), *317*, 1010(233), *1026*
Tarle, M., 374(88), *396*
Tasaka, S., 1387(113), 1389(113), *1396*
Tasaka, Y., 663(39), *693*
Tashiro, T., 336(60), 339(60), *354*, 827(140), 829(140), *868*
Tassin, M., 421(145), *431*
Tatawawadi, S. V., 711(65), *750*
Tate, J. R., 819(85), *867*
Tatlow, J. C., 377(120), *397*, 553(88), 554(91), *577*, 1010(236), *1026*, 1094(177), *1101*, 1104(9), 1105(9), 1106(9), 1108(9), 1110(9), *1124*, 1100(41), *1125*, 1157(225), *1185*
Taube, H., 17(12), *114*, 809(14), *864*
Taube, R., 821(98), *867*
Tavagnacco, C., 823(108), 841(108), 863(108), *867*, 826(128), *868*
Taya, K., 1408(47), *1417*
Taylor, B. S., 282(202), 285(202), 291(202), *319*
Taylor, J. D., 791(17), *804*
Taylor, J. M., 790(13), *804*
Taylor, K., 363(7), *394*
Taylor, M. J., 858(300), *874*
Taylor, R. L., 1316(97), *1335*
Taylor, R. P., 1338(1), *1359*
Tazawa, H., 1243(135), 1244(135), *1261*
Teare, F. W., 1120(68), *1126*
Teasley, M. F., 649(233, 234), *657*
Tedoradze, G. A., 435(36), 436(36), *457*
Teherane, T., 340(78), *355*, 1253(202), *1263*
Teichmann, H., 403(11), *426*
Teichner, S. J., 1408(41), *1416*, 1408(46), *1417*
Teitelbaum, C., 1175(480), *1193*
Teles, I., 640(159), *655*
Teller, R. G., 1200(13), *1211*
Temkin, H., 662(37), *693*
Tench, D., 1382(86), *1395*
Teng, F. A., 1358(107), *1362*
Tennakoor, L. L. K., 288(256), 292(256), *321*
Terada, Y., 621(84–86), 627(84–86), *652*, 775(43a–c), 776(44b, 46a, b, 47), *785*, 958(40), 962(45), *1019*, 1013(251), *1027*, 1161(271), *1187*, 1163(296, 297), *1188*
Teratini, S., 1408(47), *1417*, 1409(102), *1418*

Terauchi, S., 599(118), *611*
Teraud, W. W., 435(37), *457*
Terem, B., 489(60), *498*, 908(224n), 910(224n), 911(224n), 912(224n), *942*, 914(243), 915(243), *943*, 1137(44), *1180*, 1144(126), 1145(126), *1182*
Terni, H. A., 436(60), 443(60), *458*, 1171(378, 379), *1190*
TeSchinner, M. A., 724(201), 739(200), *755*, 739(341), *760*
Tessier, D., 65(84), 68(87), *118*, 126(56), 135(82, 84), 143(83), *197*, 144(101), *198*, 435(44), *457*, 884(4), 885(4), *932*
Testa, A. C., 122(14), *195*, 154(137), 155(137), *199*
Testoni, A., 1382(80), *1395*
Teter, J. W., 296(326), *323*
Teuber, H.-J., 722(180), *754*
Teukolsky, S. A., 122(8), *195*
Tezuka, M., 1246(167), *1262*
Thackeray, J. R., 662(33, 35, 36), *693*
Thal, A., 349(185), *359*
Thangavelu, S., 437(138), *460*, 894(77b, 83), 896(105), *936*, *937*
Thaning, M., 1163(300), *1188*
Thaver, R. K., 813(27, 30), *865*
Theophilou, N., 341(91), *356*
Therdteppitak, A., 122(10), *195*
Therien, M. J., 844(231), *871*
Thevenot, D., 732(281), *758*
Thiebault, A., 78(100), 81(103), *119*, 220(16), *231*, 226(24), *232*, 290(267), *321*, 290(268, 281), 291(281), *322*, 293(304, 306), *323*, 380(130), *398*, 515(94, 96), 518(94), *531*, 1123(81), *1127*, 1239(114, 116), 1241(121), *1260*
Thiebault, H., 1094(184), *1101*, 1122(75), *1126*
Thiele, J., 1200(18), *1212*
Thiele, R., 468(38), *480*, 930(357b), *947*
Thielmann, F., 350(197), *359*
Thimm, K., 490(70), *498*
Thirsk, H. R., 373(80), *396*
Thoai, N. V., 1206(78), *1214*
Thomalla, M., 288(255), 292(255), *321*, 1082(99), *1098*, 1090(162), *1100*, 1159(237),*1186*
Thomas, A. L., 820(92), 821(92), *867*
Thomas, D., 1436(31b), *1438*
Thomas, F. G., 278(169), *318*
Thomas, H. G., 544(57), *576*, 557(111–113), *577*, 705(34), *749*, 1005(220a), 1006(213), *1025*, 1078(66), *1097*, 1137(37), *1180*, 1243(141), *1261*, 1275(46a), *1282*
Thomas, J. K., 1404(20), *1416*
Thomas, S. L. S., 1004(205), *1025*
Thomas, S. P., 886(12), 891(12), *933*
Thompson, A., 1312(86), *1335*
Thompson, C. D., 284(240), *321*
Thompson, M. S., 641(169), 643(169), *655*
Thomsen, A. D., 441(140), *460*, 702(3), 727(3), *748*, 899(143), *938*, 914(246), *943*, 1144(110, 111), *1182*
Thomson, D., 291(296), *322*
Thönnessen, F., 1137(37), *1180*
Thoren, S., 1072(29), *1096*
Thornton, B., 373(80), *396*
Thornton, T. A., 1038(40), *1061*
Threlfall, T. L., 368(46), *395*, 1153(184), *1184*
Throop, L., 450(200), *462*, 1141(68), *1181*
Thrope, J. F., 1298(51), *1334*
Thulin, B., 337(66a), *355*
Thumm, G., 367(26), *394*
Thuy, T. H., 733(296), *758*
Tidswell, B. M., 1344(16–18), *1360*, 1358(97, 104), *1362*
Tiedemann, W. H., 279(179), *319*, 293(307), *323*
Tien, H. J., 794(52, 59, 61), 795(48, 52), 796(52, 59), 797(61), *806*, 885(11), *932*
Tien, T. P., 1413(140), *1420*
Tilak, B. V., 256(6), *312*, 1070(15), *1095*, 1287(16), *1332*, 1287(22), 1292(22), *1333*
van Tilborg, W. J. M., 387(160), *399*, 1144(116, 120), *1182*, 1155(207), *1185*, 1172(409), *1191*
Tilset, M., 511(75), 513(75), *531*, 841(221), *871*, 1070(12), *1095*
Timmell, T., 256(11), *313*
Timofeeva, Z. N., 727(249), *757*
Timofin, M. N., 440(99), *459*, 897(116), *937*
Tincher, C. A., 1141(69), *1181*, 1255(216), *1263*
Tinembart, O., 239(44), 240(44), *247*, 813(37), 825(37), 826(37), 827(37), 841(37), 863(37), *865*
Tinker, L. A., 302(389), *326*, 506(11), 510(11), 511(11), *529*
Tirouflet, J., 494(117), *500*, 740(360), *761*, 743(392), 744(397), *762*, 838(203), *871*
Tischer, W., 1174(441), *1192*
Tismer, M., 715(100), *751*
Tissot, P., 4(14), *8*, 545(65), 546(65), *576*, 630(96, 105), 632(105), *653*, 783(72), *786*, 914(250), 915(250), *943*, 1002(190b), *1024*, 1035(18), *1061*, 1157(222), *1185*
Titova, I. A., 436(59), *458*, 441(149b), *461*, 895(98b), *937*, 1221(16), *1256*
Tobias, C. W., 282(196), 284(196), 299(196), *319*, 301(377), *325*, 1287(15), *1332*, 1292(43), 1293(43), *1333*, 1298(50), *1334*
Tobinaga, S., 601(143), 602(143), *612*, 704(18), *748*, 704(24), *749*, 771(27a, b, 28), *784*, 994(155a, 156), *1023*, 1000(168, 171), *1024*, 1094(173), *1100*
Tobolsky, A. V., 1350(35), *1360*
Toda, T., 584(19), *608*, 779(61a), *786*, 1160(262), *1186*
Todd, P. F., 908(223), *941*
Todres, Z. V., 735(317, 318), *759*
Töfl, W., 661(22), *692*
Togami, K., 1034(17), *1061*
Tohjima, K., 559(120), 562(120), *578*
Toi, H., 1408(68), *1417*
Tökes, L., 450(200), *462*
Tokuda, K., 153(131, 132), *199*, 174(205), *202*
Tokuda, M., 392(192), *399*, 584(20), *608*, 310(16), *864*, 859(305), 860(306–308),

[Tokuda, M.] *874*, 886(18), *933*, 1154(205), 1155(205), *1185*, 1167(332), *1189*, 1269(35), *1281*, 1275(48), *1282*, 1298(52), *1334*, 1426(16, 17), *1430*
Tokumaru, K., 519(110), *532*, 982(113), *1022*, 1406(31), *1416*, 1409(92), *1418*
Tokuno, Y., 621(56), *651*
Toma, S., 837(202), *871*
Tomahogh, R., 1228(48), *1257*
Tomakazu, T., 445(155), *461*
Toman, L., 1211(128), *1215*
Tomaszewski, A., 834(168), *869*
Tomat, R., 518(106), *532*
Tomii, K., 1174(442), *1192*
Tomilov, A. P., 259(32), 263(32), 309(32), *313*, 263(63–65), *314*, 273(119), *317*, 366(21), *394*, 368(40), *395*, 410(70), *428*, 434(16), *457*, 435(49), 436(49, 68, 74), *458*, 438(80), 442(104, 105), 443(105), *459*, 445(158, 169), *461*, 474(73), *481*, 495(127), 496(133, 134), *500*, 496(135), *501*, 837(198), 838(198), *870*, 861(314), *874*, 886(19), *933*, 894(84, 86), 895(90), 896(84), 903(84), *936*, 896(101–103), 897(101), 898(122, 123), *937*, 898(132), 901(154), *938*, 902(170, 176), 906(177, 182a), 907(177), 922(170), *939*, 907(184, 186, 188–191), *940*, 908(224i), 910(224i, 226), 911(224i), *942*, 911(269), *943*, 921(290), 923(281), *944*, 921(294), 924(302), 925(308), *945*, 1104(11), *1124*, 1201(26, 30), *1212*, 1255(210), *1263*, 1269(36), *1281*, 1287(23, 26), *1333*, 1425(12), *1430*
Tomilova, L. G., 820(96), *867*
Tomkins, D. F., 895(99), *937*, 1144(118), *1182*
Tommasi, 438(81), *459*
Tomov, N. R., 259(35), *314*
Tomazane, K., 537(12), 562(12), *574*
Tompkins, P. C., 724(203), 725(203), *755*
Tondler, F., 442(107), *459*
Toomey, J. E., 701(1c), 723(1c), *747*, 724(200, 210), 739(200), *755*, 739(340), *760*, 744(407), *762*, 745(414), *763*
Topich, J. A., 826(127), *868*
Toppare, L., 1352(63, 64), 1354(78–80), *1361*
Tordo, P., 193(290), *205*, 411(76), *428*
Torii, S., 233(6a), 236(23a), *246*, 243(63, 70), *248*, 393(203, 207), *400*, 404(28), *427*, 417(115), *430*, 537(7, 9–12), 563(11, 12), 572(9, 11), 573(10), *574*, 539(21, 27), 540(27), 542(27), *575*, 555(101, 102), *577*, 561(124), 563(129, 130), 564(135), *578*, 569(152), 570(159), 573(166), 574(169), *579*, 586(25), 601(25), *608*, 597(108), 600(120), *611*, 601(131), 602(131, 145), *612*, 621(50), *651*, 664(52–54), 666(52–54), *693*, 674(86, 87), 675(88, 93), *694*, 677(105), 679(105, 110), *695*, 713(78), *751*, 721(154, 155), *753*, 774(41), 777(51), 778(52, 53), *785*, 778(54), 780(62a, b, 63), 782(68), *786*, 810(17), *864*, 827(137), 828(137), *868*, 835(180), *870*, 850(261), 851(263–269), 852(264), *873*, 904(208), *941*, 950(1e), *1018*, 979(99), 982(99), *1021*, 1008(226), *1026*, 1034(14, 16), *1060*, 1035(22), 1036(25), *1061* 1039(47), *1062*, 1046(122), *1064*, 1054(213), *1067*, 1070(18), 1073(18), 1080(18), 1084(18), *1095*, 1077(51), *1096*, 1162(282, 283, 286), *1187*, 1164(313), 1165(317), 1166(322), *1188*, 1168(341), *1189*, 1229(68), 1231(65), 1232(65), *1258*, 1236(106), *1259*, 1243(135), 1244(135, 146–148), 1245(159), *1261*, 1268(23), 1275(23), 1277(23), *1281*, 1277(54, 56), 1278(56, 57b–d), 1279(57c, d), 1280(57b–d, 58, 59), *1282*, 1434(13b), *1437*
Toriyama, K., 50(67b), *117*
Torizuka, K., 994(148), *1023*
Torkos, K., 1038(45), *1062*
Torp, E. C., 363(8), *394*
Torrance, J. B., 341(90), *356*, 819(82), *866*, 1376(58), *1394*
Torreilles, E., 680(120), 681(120), *695*
Torzo, F., 494(115), *500*
Tosato, M. L., 1413(129), *1419*
Töteberg-Kaulen, S., 593(67), *609*, 599(119), *611*
Touboul, E., 914(249), 916(262), *943*, 1144(125, 128), 1145(125, 128), *1182*
Toupet, L., 648(230), *657*, 1013(248), *1027*, 1160(253), *1186*, 1166(323), *1188*, 1241(125), *1260*
Toure, V., 452(208b), *462*
Tourillon, G., 1357(92), *1362*, 1380(66), *1394*, 1382(87), *1395*
Touster, J., 303(391), *326*
Townsend, J., 332(12a, b), *353*
Toyoda, K., 1077(52), *1097*
Toyota, A., 50(67a), *117*
Trachtenberg, I., 270(97), 303(97), 304(97), *316*
Train, A. W., 1344(16–18), *1360*
Trainham, J. A., 273(123c, d), *317*
Traun, M., 1208(98), *1214*
Trautner, K., 707(47c), *749*
Travecedo, E. F., 1176(498), *1194*
Traynham, J. G., *1159*(240), *1186*
Traynor, S., 335(54), *354*, 340(99), *355*
Treharre, R. W., 1209(110), *1215*
Tremillon, B., 21(21c), *115*, 300(362, 363), *325*, 515(99), 518(99), *532*
Tresselt, D., 1040(65), *1062*
Tret'yakov, N. E., 1410(121b), *1419*
Trevellick, P. R., 179(235), *203*
Tricoli, M., 1346(23), *1360*, 1352(57–61), *1361*
Triebe, F. M., 375(91), *396*, 467(26), *479*, 471(53), 473(53), *480*, 478(102), *482*, 887(25), *933*
Trifonova, G. V., 424(172), *432*
Trifonova, L. I., 471(52), *480*

Trinajstic, N., 24(28), *115*, 508(44), *530*
Trivedi, L. D., 447(181), 450(180), *462*
Trkovnik, M., 621(51), *651*, 706(36b, 38), 710(36b), 712(36b), 714(36b), *749*, 710(60b), 715(60b), *750*, 714(99b), *751*
Trocciola, J. C., 909(233, 234), *942*
Trocha-Grimshaw, J., 226(23), *232*, 383(138, 139), *398*, 703(11, 12), *748*, 729(267), *758*
Troepol'skaya, T. V., 471(55), 473(61, 62), 474(62), *480*, 476(93), *482*, 809(9), *864*, 1055(218), *1067*
Trogler, W. C., 844(231), *871*
Trojanek, A., 1232(80), *1259*
Troll, T., 342(97), 350(97), *356*, 470(51), *480*, 491(95), *499*, 494(118), *500*, 715(104), *752*, 851(262), *873*, 929(329, 331), *946*, 1147(145), *1183*, 1267(16), 1275(16), *1281*
Tronel-Peyroz, E., 1172(397), *1191*
Troupe, M., 4(13), *8*, 277(160), 311(160), *318*, 366(24), *394*, 393(204, 205, 209), *400*, 831(157), 832(157), 834(172), 835(172), *869*, 835(176, 179), *870*, 1236(107–109), *1259*, 1252(192), *1263*
Trska, P., 927(321), *946*
Trümpler, G., 424(169), *431*, 451(207), 453(207), *462*
Tsarev, V. V., 1433(6a), *1436*
Tse, D. C.-S., 1365(16, 17), 1366(17), 1370(16, 17), 1371(17), *1392*
Tsintavis, C., 1383(100), 1384(100), 104), 1387(100), *1395*
Tsodikov, V. V., 725(229), *756*
Tsou, T. T., 1168(342), *1189*
Tsou, Y.-M., 270(97), 303(97), 304(97), *316*
Tsubata, K., 240(52), 242(52), *248*, 602(145), *612*, 603(155), 607(164), *613*, 727(247), *757*, 773(38a), 774(39a, 40), *785*, 1038(38, 39), *1061*, 1077(50), *1096*, 1077(52), 1078(63), *1097*, 1147(146), *1183*, 1171(374), *1190*
Tsuchiya, M., 519(110), *532*
Tsuchiya, S., 1280(60), *1282*
Tsuda, K., 1078(63), *1097*
Tsuji, K., 438(82), *459*
Tsujiyama, H., 561(124), *578*, 586(25), 601(25), *608*, 1035(22), *1061*
Tsukitani, Y., 404(23), *427*
Tsunaga, M., 621(55), *651*
Tsuruta, H., 572(165), *579*
Tsutsui, Y., 1162(286), *1187*
Tsutsumi, S., 528(144), *533*, 1005(212b, 219), 1006(212b), *1025*, 1080(78), *1097*, 1084(112), *1098*, 1089(145, 147, 148), 1090(148), *1100*, 1157(223), *1185*, 1160(249), 1161(249), *1186*, 1161(267, 268, 269), *1187*, 1165(318), *1188*
Tsveniashvili, V. Sh., 284(241), *321*, 726(242), *757*, 734(312), 735(318), *759*
Tuck, D. G., 393(209), *400*, 810(19), 862(19), *864*, 831(157), 832(157), *869*, 837(190–195, 197), 862(191, 195), *870*, 857(294), 858(300), 862(318, 319), *874*, 862(324), *875*
Tulyathan, B., 1175(460, 464), *1193*
Tung, R., 335(48), *354*
Turker, L., 1352(64), *1361*, 1398(14), *1416*
Turner, A. D., 174(193, 195), *201*
Turner, A. P. F., 234(10), *246*, 645(207), *656*
Turner, W. R., 724(199), *755*, 860(312), *874*, 1089(141), *1099*
Tur'yan, Ya. I., 411(78b), *429*, 1200(2), *1211*
Tushnalobova, V. A., 1209(109), *1215*
Tutane, I., 1050(171), *1066*
Tutsumi, S., 1092(165), *1100*
Tuttle, M., 517(102), *532*, 1085(118), *1099*
Tütülkoff, N., 476(89a, b), *481*
Tuxford, A. M., 184(253), *203*, 930(358), 947
Tvirzicka, E., 1232(80), *1259*
Tyle, Z., 720(141), 721(142), *753*
Tyssee, D. A., 257(15), *313*, 347(171), 348(171), *358*, 364(12), *394*, 722(172), *754*, 908(224g), 910(224g), 911(224g), 914(224g), *942*, 919(275), 922(227), *944*, 925(309f), *945*, 930(349a–c), *947*
Tyukavkina, N. A., 801(84), *807*
Tzschach, A., 930(342), *946*

Ubukata, K., 925(304), *945*
Uchida, H., 1425(11), *1430*
Uchida, K., 240(52), 242(52), *248*, 603(155), 607(164), *613*, 774(39a), *785*, 1077(52), 1078(63), *1097*, 1171(374), *1190*
Uchida, M., 267(84), 268(84), 271(84), 288(84), 307(84), *315*, 640(149), *654*
Uchida, T., 1034(15), *1060*, 1163(298), *1188*
Uchiyumi, H., 908(223), *941*
Udagawa, A., 894(73b), *935*, 902(168b), *939*, 1146(139), *1183*
ud Din, N., 1002(191a), *1024*
Udupa, H. V. K., 2(6), *8*, 265(70–75), 271(70, 71, 73), *315*, 271(102), *316*, 414(102a–c), *429*, 417(120, 121, 126), 418(127, 130), 422(120), *430*, 486(31), *497*, 496(137), *501*, 745(413), *763*, 1230(57, 58), 1231(58), 1232(57, 58), 1234(58), 1248(57), *1258*, 1295(47), 1305(47), *1333*, 1305(73), *1334*, 1317(110), *1335*
Udupa, K. S., 2(6), *8*, 265(71–73), 271(71, 73), *315*, 271(102), *316*, 414(102c), *429*, 417(120, 121), 422(120), *430*, 437(138), *460*, 894(83), *936*, 1305(73), *1334*, 1317(110), *1335*
Uebel, J. J., 1160(265), 1161(277), *1187*
Ueda, A., 1089(143), *1099*
Ueda, C., 597(103), 599(118), *611*, 621(56), *651*, 705(33), *749*, 983(127), *1022*, 1087(132), 1089(143), *1099*
Ueda, H., 783(75), *786*
Ueda, J., 1004(209), *1025*

Ueda, K., 550(92), 553(89), *577*
Ueda, T., 1206(68, 69), *1213*, 1206(75), *1214*, 1280(59), *1282*, 1411(124), *1419*
Ueno, N., 1034(14), *1060*
Ueno, T., 914(238, 241), *942*
Ueno, Y., 434(6), *456*
Ueshima, T., 743(388), *762*
Ulrich, V., 1233(85), 1235(85), *1259*
Umana, M., 1365(21), 1370(21), 1371(21), 1374(21), *1392*
Umanu, M., 1341(10), *1359*
Umemoto, K., 278(173), *319*, 736(324e), *759*
Underkofler, L. A., 453(218), *463*
Underkofler, W. L., 309(437), *327*
Underwood, A. L., 725(215, 216), *755*, 725(226a–c), *756*
Uneyama, K., 236(23b), *246*, 528(143), *533*, 537(7), *574*, 550(92), 553(89), *577*, 664(52–54), 666(52–54), *693*, 674(86, 87), 675(88, 93), *694*, 777(51), 778(52, 53), *785*, 778(54), *786*, 851(263, 264), 852(264), *873*, 1008(226), *1026*, 1036(25), *1061*, 1046(122), *1064*, 1073(38), *1096*, 1164(313), 1165(317), 1166(322), *1188*, 1243(135), 1244(135, 146), *1261*, 1278(57b–d), 1279(57c, d), 1280(57b–d, 59), *1282*
Unger,H., 309(431), *327*
Unni, A. K. R., 301(365), *325*
Uno, B., 1040(63), *1062*, 1058(249), *1068*
Unwin, P. R., 165(168), 173(168), *200*
Urabe, N., 271(107), *316*, 894(89), *936*, 1059(258), *1068*
Urabe, Y., 766(1a, b), *783*
Urata, H., 528(143), *533*, 1073(38), *1096*
Urbigkit, J. R., 734(306), *759*
Uribe, F. A., 303(394), *326*, 930(347), *947*, 1433(10), *1436*
Urwin, J. R., 744(399), *762*
Usanmaz, A., 1352(63), *1361*
Ustavshchikov, R. F., 411(78b), *429*
Ustrynyuk, N. A., 825(124), 841(124), *868*
Usui, M., 598(109), *611*
Usui, Y., 704(15a–c), 726(15a–c), *748*, 773(34, 35), *785*, 928(324), *946*, 1148(157), *1183*, 1171(373), *1190*
Utley, J. H. P., 17(13), *115*, 236(24b), 237(30, 32), 239(45), *247*, 240(51b), 243(64), *248*, 308(421), *327*, 375(97), *396*, 375(98), *397*, 392(193), *400*, 489(58, 60–62, 64), 490(71), *498*, 537(17), 540(29–31), 563(17, 30), 564(17), *575*, 554(90), *577*, 563(127), *578*, 604(159), *613*, 638(131), *654*, 662(31), 670(31), 671(31), *692*, 670(74), *694*, 773(38b), *785*, 778(55), *786*, 791(19), *804*, 798(68, 69), 799(68), 800(78), 801(69), *807*, 802(91), *808*, 839(214), *871*, 860(313), *874*, 882(3k), 885(3k), *932*, 894(78, 79), 898(78), *936*, 908(224n), 910(224n), 911(224n), 912(224n), *942*, 914(243), 915(243), *943*, 931(367), *948*, 1001(176, 184), *1024*, 1039(49a, b, 50), 1040(66, 67), *1062*, 1042(78b), *1063*, 1046(127), *1064*, 1077(61), 1078(62), 1080(74), 1082(74), *1097*, 1082(101), *1098*, 1091(163), *1100*, 1137(44), *1180*, 1143(88, 90–92), 1144(90, 91), *1181*, 1144(126), 1145(126), *1182*, 1144(133), 1146(133, 134), *1183*, 1153(188), 1154(188), *1184*, 1156(219), 1157(219), 1158(219), *1185*, 1158(234), *1186*, 1161(275), *1187*, 1224(31), 1249(31), *1257*, 1235(92), 1236(92, 102), *1259*, 1255(210), *1263*, 1267(5), 1270(5), *1280*, 1435(29), *1438*
Uvarov, A. V., 1410(116a), *1419*
Uyama, H., 726(244), 728(244), *757*, 740(359), *761*
Uziel, Z., 686(152), *696*

Vaccher, C., 1150(170), *1184*
Vadaszy, R., 914(240), *942*
Vafina, A. A., 473(63), *480*, 473(67), *481*
Vagina, G. A., 476(93), *482*
Vainas, B., 930(349d), *947*, 1434(12b), *1437*
Vajda, M., 721(158), *753*
Vajtfner, Z., 380(129), 381(129), *398*, 1204(57), *1213*
Vakulova, L. A., 1038(42b), *1061*, 1039(48), *1062*
Valcarcel, M., 470(47), *480*
Valcher, S., 345(132), *357*, 837(200), *870*, 1351(44), *1360*
Valenta, P., 434(7), *456*, 1033(9), *1060*
Valentine, J. S., 1245(153), *1261*, 1269(31), 1272(31a), *1281*, 1407(33), *1416*
Valentini, F., 405(33), *427*, 1144(106), *1182*
Valentini, L., 1040(64), *1062*
Valentova, M., 722(164d), *754*
Vanags, G., 1050(171), *1066*
Van Baar, J. F., 810(20), *864*
Vance, C. J., 848(250), *872*
van Dalen, E., 126(55), *197*
Van de Mark, M. R., 596(89, 91), *610*, 641(182), 642(182), 643(186), *655*, 1372(40), *1393*
Vanden Born, H. W., 126(58), *197*
Van den Brink, F. T. B. J., 724(201), *255*, 739(341), *760*
Vanderborgh, N. E., 435(22), *457*
Van der Linden, J. G. M., 809(6), *864*
van der Stouwe, C., 1039(51), *1062*
Van Duyne, R. P., 179(224), *202*, 188(262), *204*, 270(95), 296(95), *316*, 511(63), *530*
Van Dyke, D. A., 689(187, 191), *698*
Van Dyke, J. D., 1357(95), *1362*
Vanel, P., 1172(396), *1191*
Van Eygen, C., 924(300), *945*, 1319(114), *1336*
Van Gaal, H. L. M., 809(6), *864*
Van Galen, D. A., 150(114), *198*, 474(72), *481*, 478(106, 109), *482*

Van Hardeveld, R., 724(201), *755*, 739(341), *760*
Van Hifte, L., 1139(59), 1140(59), *1180*
Vanka, M., 906(177), 907(177), *939*
Van Koppenhagen, J. E., 647(220), *657*
van Tilborg, W. J. M., 306(404), 311(404), *326*, 891(54b), *935*, 894(88), *936*, 895(97), *937*
van Veen, J. A. R., 810(20), *864*
Varacca, V., 1375(51), *1393*
Vardi, S., 1341(13), 1346(13), 1348(13), 1355(13), *1359*
Varshavskii, S. L., 263(63), *314*, 366(21), *394*, 438(80), *459*, 496(133), *500*, 896(102), *937*, 921(294), *945*
Vartires, I., 882(31), 885(31), *932*
Varyukhinma, L. V., 742(375), *761*
Vasil'ev, Yu. B., 267(85), 268(85), 271(85), *305*, 247(141), *307*, 539(22), *575*, 639(136, 137), *654*, 644(198), 645(198), *656*, 1010(233), 1011(237), 1012(241), *1026*
Vasil'eva, A. A., 617(10), *650*
Vasyanovich, G. P., 792(28), 794(45), *805*, 801(83), *807*
Vaugham, W. R., 720(137), *753*
Vaughan, D. H., 813(41), *865*
Vaught, Jr., W. W., 1175(478), *1193*
Vavricka, S., 409(66), *428*, 411(85, 86), 421(86), *429*, 422(151), 424(165), *431*
Veatch, F., 1305(74), *1334*
Veber, D. F., 546(67), *576*, 1002(203), *1025*
Veber, M., 214(9), *231*
Vegnere, V. Ya., 441(145), *461*
Veinberg, A. Y., 1039(52a), *1062*, 1050(161), *1065*
Velasco, M. D., 384(144), 385(144), 386(144, 148–152), 387(151), *398*, 388(161), *399*, 703(4–6), *748*, 891(50), *934*, 1154(198), *1185*
Velasco, R., 645(204), *656*
Velasquez-Rosenthal, M., 1385(108), *1396*
Vellturo, A. F., 1159(247), *1186*
Veniette, M., 1164(306, 312), *1188*
Venkatachalam, C. S., 732(283), *758*
Venkatachalapathy, M. S., 265(74), 271(74), *315*, 1230(57), 1232(57), 1248(57), *1258*
Venkataraman, U., 835(182), *870*
Venzo, A., 839(213a, b), *871*
Verbeek, W., 849(257), *872*
Vergnes, F., 268(90), *315*
Verheyden, A., 924(300), *945*, 1319(114), *1336*
Verhoeven, J. W., 620(42), *651*
Vermillion, F. J., Jr., 619(26), 621(26), 624(26), *650*, 958(41), *1019*
Verpeaux, J. N., 78(100), *119*, 226(24b, c), *232*, 1239(116), *1260*
Vertes, G., 277(150), *318*, 640(160), 641(172), *655*
Vertyulina, L. N., 425(194), *432*
Vetler, K. J., 22(22d), *115*, 272(112), *316*
Vetterling, W. T., 122(8), *195*
Viana, M. Nazareth, 1150(170), *1184*
Vianello, E., 2(4), *7*, 73(93a), *119*, 123(28), 133(28), *196*, 224(22), *232*, 240(49), *247*, 346(152), 350(152), *357*, 496(129), *500*, 717(122), *752*, 1033(10b), *1060*, 1233(95), *1259*
Videiko, A. F., 410(70), *428*
Videla, H. A., 645(205), *656*
Vidotto, G., 450(204), *462*, 1341(11), *1359*, 1351(41), 1352(45–51), *1360*, 1352(65), *1361*
Vieira, K. L., 306(410), *326*, 365(18, 19), 366(23), 388(18), *394*, 367(27), *395*, 862(323), *875*
Viel, C., 706(42), *749*
Vieil, E., 686(151), *696*
Vielstich, W., 234(9), *246*, 639(143, 144), *654*, 641(165), *655*
Viertler, H., 662(31), 670(31), 671(31), *692*, 801(81), *807*
Vieth, B., 703(8, 9), *748*
Vigevani, A., 1040(64), *1062*
Vijh, A. K., 640(158), *655*
Vilambi, N. R. K., 257(12), *313*
Vincent, C. A., 733(304), *759*
Vincent, F., 1085(119, 131), *1099*
Vincenz-Chodkovska, A., 441(142), *460*, 1132(9), 1175(9), *1179*, 1147(144), *1183*
Vining, W. J., 1434(15n), *1437*
Vinnikov, Yu. Ya., 290(277), *322*
Vinot, N., 719(128, 131), *752*, 736(323), 738(323), *759*, 742(372), *761*
Virgili, J., 455(236), *464*
Visco, R. E., 173(189), *201*, 297(332), *324*, 344(122), *357*, 510(47), *530*
Vlahov, I., 704(26), *749*, 1000(173), *1024*
Vlahov, R., 704(26), *749*, 1000(173), *1024*
Vlazny, K., 1161(278), *1187*
Vlcek, A. A., 844(232), *872*
Vodzinskii, Yu. V., 617(10), *650*, 797(66), *806*
Vofsi, D., 299(352), *324*, 908(222), *941*, 1201(34, 35), 1202(37–40), *1212*
Vogel, E., 336(63), 339(75), 340(74, 75), *355*
Vogel, P., 233(3), 234(3), *245*
Vögeli, U., 815(55), *866*
Vögtle, F., 594(77, 78), *610*
Voigt, J., 1048(139), *1065*
Voitkevich, S. A., 1001(177a), *1024*, 1208(102), *1214*
Voitovich, Ya. N., 284(233), *320*
Volke, J., 388(166), *399*, 469(46), *480*, 474(75), *481*, 485(13), *496*, 495(124, 125), *500*, 707(47b), *749*, 708(52), 709(58), *750*, 713(86), *751*, 724(202), *755*, 732(288), *758*, 734(305, 310), *759*, 743(390, 391), 744(398), 745(411), *762*, 745(416), *763*, 892(65), *935*, 926(314), *945*, 1033(8a–c, 9), 1050(8a–c), *1060*
Volkova, V., 474(75), *481*, 485(13), *496*, 732(288), *758*, 744(398), *762*
Volmer, M., 64(82b), 65(82b), *118*, 272(108), *316*
Volodina, T. A., 424(170), *432*

Volod'kin, A. A., 455(231, 232), *463*, 915(257), *943*
Vol'pin, M. E., 826(131), *868*
Volz, H., 340(82b), *355*
von Braun, J., 722(178), *754*
Von Bülow, C., 491(90), *499*, 1151(173), *1184*
Von Frauenhofer, J. A., 309(436), *327*
Von Metnitz, H., 1208(96), *1214*
von Schleyer, P., 1085(119), *1099*
Vorontsova, L. M., 287(244), 305(244), *321*
Vorotynsev, M. A., 32(44b), 63(44), *116*
Vorozhtsov, G. N., 894(75), *935*
Voso, M. A., 1345(22), *1360*
Voss, H., 245(80), *249*
Voss, J., 488(52), 490(67, 70), *498*, 491(90), 492(103, 104), *499*, 429(105), 493(110), *500*, 717(121), *752*, 1033(12), *1060*, 1150(171), 1151(173), *1184*, 1268(22), 1275(22), *1281*
Vostrowsky, O., 1002(201), *1025*
Vovchenko, G. D., 442(101), 447(101), *459*
Vrancken, A., 885(8), *932*
de Vries, W. T., 126(55), *197*
Vuaille, M. J., 300(363), *325*
Vukicevic, M., 1005(212d), 1006(212d), *1025*
Vyazankin, N. S., 908(163), *939*, 1201(22, 23), *1212*
Vykhodceva, L. N., 435(48), *458*, 895(94), *937*
Vyskocil, V., 722(168), *754*, 1167(334), *1189*

Waack, R., 349(175), *358*
Wabner, D. W., 276(146, 147), *318*, 793(41), *805*
Wabster, K. T., 1210(121), *1215*
Wachi, S., 436(63), *458*, 1143(93), *1181*
Wada, S., 555(102), *577*, 573(168), *579*
Wada, T., 1154(197), *1185*
Waddell, J. E., 1106(27), 1108(27), *1124*
Waefler, J. P., 1002(190b), *1024*
Wagenknecht, J. H., 122(11), *195*, 148(11), *198*, 347(171), 348(171), *358*, 364(13), 365(13), *394*, 373(77), *396*, 380(131), 381(134), *398*, 391(190), *399*, 414(105), *429*, 417(125), *430*, 468(27), *479*, 486(28), *497*, 488(51), *498*, 490(73), *499*, 703(7), *748*, 715(106), *752*, 890(41), *934*, 892(63), *935*, 895(99), *937*, 904(205a, b), *941*, 919(275), 923(280), *944*, 929(328), *946*, 930(349a), *947*, 1032(1), *1060*, 1035(24), *1061*, 1044(103), *1063*, 1052(188), *1066*, 1136(30), *1180*, 1144(118), *1182*, 1203(44), 1206(67), *1213*, 1230(59), *1258*, 1267(3, 9), *1280*, 1321(119), *1336*, 1433(2), *1436*, 1434(13c, h), *1437*
Wagner, C., 256(8), *303*
Wagner, G., 732(284–286), *758*, 927(318), *946*, 930(355), *947*
Wagner, W. G., 1230(59), *1258*
Wagoner, D. E., 963(66), *1020*
Waisman, E., 32(43c), 63(43), *116*, 39(55a), *117*
Waite, T. D., 1413(132), *1419*
Wakamatsu, T., 603(149, 152, 153), *613*, 773(37a), 774(39c), *785*, 1077(55), *1097*
Waki, M., 278(173), *319*
Walborsky, H. M., 375(102, 103), *397*, 863(325), *875*, 1168(347, 348), 1169(347, 348), *1189*
Walburg, J. A., 894(85a), *936*, 1316(103), 1321(103), *1335*
Walden, P., 280(182), 295(182), *319*
Waldner, L., 239(44), 240(44), *247*, 813(37, 39), 821(39), 825(37), 826(37), 827(37), 841(37), 863(37), *865*, 815(55), *866*, 824(119), 826(119, 126, 133), 827(134), *868*, 823(103, 106), *867*, 1155(210), *1185*
Waldyke, M. J., 689(191), *698*
Walker, A. T. S., 268(93), *316*
Walker, D. C., 1253(200), *1263*
Walker, D. G., 1384(106), 1387(106), *1395*
Walker, F. A., 814(50), *865*
Walker, M. T., 1054(205), *1067*
Waller, A. M., 165(168), 173(168), *200*
Walling, C., 536(4), *574*, 1070(1), *1095*
Wallis, E. S., 1170(362), *1190*
Walreavens, J., 924(300), *945*, 1319(114), *1336*
Walsh, F. C., 1303(70), *1334*
Walter, R. I., 594(75), *610*
Walters, F. H., 837(190), *870*
Walters, H. C., 1301(60), 1319(60), *1334*
Waltman, R. J., 1375(55), 1380(55, 63), 1382(55), *1394*, 1387(110, 111), 1389(110, 111), *1396*
Walton, C. W., 1301(58), 1303(67), *1334*
Walton, D. J., 489(62), *498*, 664(48), 665(48), *693*, 733(304), *759*, 860(313), *874*, 1164(309), 1165(309), *1188*, 1387(120), 1389(120), *1396*, 1436(31b), *1438*
Walton, R. A., 1175(457), *1193*
Wan, C. C., 926(314), *945*
Wang, D. L., 815(59), 816(59), 817(75), *866*
Wang, H. C., 349(176), *358*
Wang, T. T., 1387(113), 1389(113), *1396*
Wang, Y. Y., 926(314), *945*
Ward, G. A., 551(98, 99), *577*
Ward, H., 898(131), *938*
Ward, M. D., 1403(17), *1416*
Ware, R. S., 991(169), *1024*, 1094(175), *1101*
Warner, C. R., 742(379), *761*
Warner, D. E., 680(121), *695*
Warnhoff, E. W., 14(36b), *114*
Warning, K., 603(148, 151), *613*, 727(247), *757*, 767(13), *784*
Warren, C. B., 537(15), 564(15), 565(15), *575*
Warren, L. F., 1382(86), 1383(95), *1395*
Warren, S., 240(47), *247*
Wasa, T., 263(51), *314*, 740(367), *761*
Wasielewski, M. R., 340(81b), *355*
Wassen, J., 339(75), 340(75), *355*

Wasser, D. J., 1105(25), 1109(25), 1110(25), 1112(25), *1124*, 1109(39), *1125*
Wasserman, B., 685(140), *696*
Watanabe, C., 1178(515), *1194*
Watanabe, E., 815(57), *866*
Watanabe, H., 493(111), *500*, 853(271), *873*, 1256(218), *1263*, 1433(5g), *1436*
Watanabe, M., 639(146), 640(149), *654*, 1390(134), *1396*
Watanabe, N., 1094(182), *1101*, 1104(12), 1108(12), 1109(12), 1110(12), *1124*, 1164(311), *1188*
Watanabe, T., 258(28), *313*, 541(37), *575*, 721(160), *753*
Waters, W. A., 15(10b), *114*
Waters, W. L., 1144(132), 1145(132), *1183*
Watkins, B. F., 441(150), *461*, 1086(130), 1087(130), 1094(130), *1099*, 1094(189), 1172(415), *1172*, 1365(13), 1370(15), *1392*
Watson, W. M. H., 713(82), *751*
Wawzonek, S., 291(296), *322*, 332(11a, b, 15), 345(11), 348(15), *353*, 347(167), 348(167), *358*, 367(31), 375(31), *395*, 373(77), *396*, 380(131), 381(134), *398*, 403(19), *426*, 422(153), *431*, 441(148), 449(148), *461*, 452(210), 453(210), 454(224), 455(210), *463*, 490(81), *499*, 589(34), *608*, 712(70), *750*, 885(6), *932*, 895(196), 903(196), *940*, 905(215), *941*, 910(227), *942*, 915(251), *943*, 930(359), 931(365a), *947*, 971(75), 972(75), *1020*, 1043(83), *1063*, 1138(50), *1180*, 1142(79), *1181*, 1144(103), *1182*, 1175(478), *1193*, 1175(488), *1194*
Wayner, D. D. M., 517(103), 526(103), *532*
Wearring, D., 347(167), 348(167), *358*, 1142(79), *1181*
Weaver, R. D., 311(440), *327*
Weaver, T. R., 834(170), *869*
Webb, J. L., 375(102, 103), *397*, 863(325), *875*, 1168(347, 348), 1169(347, 348), *1189*
Webber, A., 1080(85), *1098*
Webber, S. E., 1413(137, 139), *1420*
Weber, F. G., 729(269), *758*
Weedon, B. C. L., 260(37), *314*, 375(97), *396*, 375(98), *397*, 489(58), *498*, 540(28, 30, 31), 542(42), 545(42), 563(30), *575*, 543(52), *576*, 778(55), *786*, 914(243), 915(243), *943*, 1001(184), 1002(177b, 191a), *1024*, 1004(205–207), *1025*, 1039(49a–c), *1062*, 1080(75), 1082(75), *1097*, 1144(126), 1145(126), *1182*
Wegener, G., 511(68), *530*, 787(2), 788(2), 789(2), 790(2), 791(2), *803*
Weichsel, C., 489(68), *498*, 1047(134), *1064*
Weider, P., 1210(122), *1215*
Weijland, W. P., 333(27), *353*
Weil, H., 1208(97, 98), *1214*
Weinberg, E. P., 275(125), *317*
Weinberg, H. R., 275(125), *317*, 505(4), *528*, 586(28), 601(28), *608*, 620(27), 640(27), 645(27), *650*, 1071(25), 1073(25), 1076(43), 1084(25), *1096*
Weinberg, N. L., 256(6), *312*, 267(77), *315*, 275(125), *316*, 468(34), *480*, 505(4), *528*, 528(148), *533*, 542(40), *575*, 584(14, 15), 586(28), 589(39), 601(28), *608*, 601(127), *612*, 620(27), 640(27), 645(27), *650*, 629(94, 95), 630(104), 633(113), *653*, 705(31), *749*, 723(184), 724(184), *754*, 894(87a), 898(87a), *936*, 929(334), *946*, 1036(29), *1061*, 1070(14, 15), *1095*, 1071(25), *1096*, 1073(14), *1095*, 1075(43, 44), 1076(44, 45), 1084(25), *1096*, 1104(4), 1106(4), 1108(4), 1109(4), *1124*, 1132(12), *1179*, 1162(288), *1187*, 1164(307), *1188*, 1287(16), *1332*, 1287(22), 1292(22), *1333*, 1321(123), *1336*
Weiner, S. A., 619(25), *650*
Weiner, St. A., 434(12), *456*
Weingarten, H., 597(99, 100), *611*, 981(125), *1022*
Weiniger, J. L., 271(104), *316*
Weinreb, S. M., 6(20), *8*, 1043(87), *1063*
Weinstock, L. M., 1051(180), *1066*
Weisbuch, F., 914(249), *943*
Weisgraber, K. H., 957(35), 961(44), *1019*, 1094(175), *1101*, 1166(320), *1188*
Weiss, J., 506(29), *529*, 1247(172), *1262*
Weiss, S. G., 961(44), *1019*
Weissberger, A., 1287(16), *1332*
Weissman, S. I., 332(12a, b), *353*
Welike, N., 725(227), *756*
Weller, A., 50(64), *117*, 344(124), *357*
Wellmann, J., 393(208), *400*, 1154(199), *1185*, 1233(94, 97), 1235(93), *1259*
Wells, D., 1408(81), *1418*
Welvart, Z., 1251(185), *1262*
Wender, I., 285(208), *320*, 293(311, 312), *323*, 297(339), *324*, 1254(206, 209), 1255(206, 213), *1263*, 1434(16), *1438*
Wendt, H., 211(7b–e), *231*, 235(17b, 20), *246*, 278(166), *318*, 288(257), 292(257), *321*, 410(75a–c), *428*, 523(128), *532*, 583(9), 584(9), *607*, 600(122), *611*, 641(167), *655*, 648(231), *657*, 675(94), *694*, 675(97), *695*, 975(101), *1021*, 982(114), 983(130), *1022*, 1006(224a–c), 1007(224a–c), 1012(244), *1026*, 1071(23), *1096*, 1094(191), *1101*, 1171(384–387), *1190*, 1172(399), *1191*
Wenisch, F., 1305(76), *1334*
Wenk, W., 663(43), 667(43), *693*, 1054(211), *1067*
Wenker, H., 469(43), *480*, 496(138), *501*
Wennerbeck, I., 1170(365), *1190*
Wennerström, O., 336(63), *355*, 337(66a, b), *355*
Wenska, G., 1409(109), *1419*

Wentworth, W. E., 17(15b), *115*
Wermeckes, B., 640(153), *654*, 647(225), *657*, 1072(30, 31), *1096*
Wernberg, A. A., 672(79), *694*
Wessling, R. A., 925(311), *945*, 1044(104), *1063*
West, K. L., 791(16), *804*
Westberg, H. H., 565(143, 144), *578*, 1159(244), *1186*
Westmore, J. B., 721(161c), *753*
Wettig, K., 436(56), *458*
Weymouth, C., 239(44), 240(44), *247*, 813(37), 825(37), 826(37), 827(37), 841(37), 863(37), *865*
Whalen, R., 242(60), *248*
Wharton, P. S., 571(164), *579*, 1159(239), *1186*
Whisman, M. L., 1053(199), *1066*
White, B. S., 510(49), *530*, 1245(151), *1261*, 1377(82), 1382(82), *1382*
White, C. K., 194(292), *205*, 892(64), *935*, 893(69), *935*
White, D. A., 308(420), *327*, 490(72, 73), *499*, 643(184), *655*, 809(5), *864*, 1032(1), *1060*, 1267(14), 1268(14, 17), 1271(14, 39), *1281*, 1422(19), *1429*
White, D. V., 1135(22), *1179*
White, H. S., 1387(117, 118), *1396*
White, J. H., 1371(36), *1393*, 1383(95), *1395*
White, J. M., 1413(137, 139), *1420*
White, J. R., 1390(126, 130), *1396*, 1403(17), *1416*
White, R. E., 122(7), 123(7), 157(7), 179(7), *195*, 165(167), *200*, 1287(13, 14), *1332*, 1301(58), 1303(67), *1334*
Whiteley, M. W., 846(241), *872*
Whiting, S. M., 846(241), *872*
Whitman, W. B., 813(31), *865*
Whitnack, G. C., 263(54), *314*, 311(440), *327*, 424(178), *432*, 488(47), *498*
Whitson, P. E., 126(58), *197*, 616(6), 617(6), 621(6), 624(6), *650*, 962(46), *1019*
Whittaker, D., 14(3a), *114*
Wibaut, J. P., 442(108), 447(108), *459*, 450(201), *462*, 487(39), *497*
Wiberg, K. B., 234(11), *246*, 369(52), *395*, 376(115), 378(115), *397*, 378(122, 123), *397*, 887(29, 31), *933*
Widlanski, T., 340(80), *355*, 927(320), *946*
Wieckowski, A., 1371(36), *1393*
Wiegand, G., 488(52), *498*, 493(110), *500*, 1268(22), 1275(22), *1281*
Wiemann, J., 452(211), *463*, 914(245, 249), *943*, 917(282), *944*, 1206(78, 79), *1214*
Wier, L., 1314(12), 1366(12), 1368(12), *1392*
Wierlemann, B., 598(110), *611*
Wigfield, D. C., 14(7), *114*
Wiggins, L. F., 732(287), *758*
Wightman, R. M., 24(29), *115*, 55(70), 58(73b), *117*, 68(88), 70(91b), *118*, 82(106), *119*, 128(67), 129(67), *197*, 175(209), *202*, 208(3, 4a), *231*, 506(17, 18), 508(18), 509(18), 510(17), 511(17, 18), *529*, 586(26), *608*, 594(80), *610*, 914(244), *943*
Wijesekera, T., 819(90), *867*
Wikström, M., 813(26), *865*
Wilbourn, K. O., 1406(30), *1416*, 1408(80), *1418*
Wilcox, M., 745(414), *763*
Wildenau, A., 422(156), *431*, 1055(220), *1067*
Wilhelm, S., 639(142, 144), *654*
Wilkins, R., 260(39), *314*
Wilkinson, F., 276(149), *318*
Wilkinson, G., 809(1a, b), *864*
Will, H., 734(313), *759*
Willars, M. J., 639(138), *654*
Wille, H. J., 1426(14), *1430*
Willemart, A., 740(362), *761*
Willett, B. C., 297(334), *324*, 363(8, 9), 364(9), 366(9), 371(9), *394*, 370(73), 371(73), *396*, 863(326), *875*
Williams, D. H., 1005(218), *1025*
Williams, D. K., 889(37), *934*
Williams, D. R., 680(115), 681(115), *695*, 689(189), 690(189), *698*
Williams, E. B., Jr., 686(157), 687(157), *697*
Williams, F., 1406(26), *1416*
Williams, F. D., 555(107), *577*
Williams, F. J., 911(270), *944*
Williams, F. R., 826(132), *868*
Williams, J. H., 490(82), *499*, 902(168a), *939*
Williams, P., 288(256), 292(256), *321*
Williams, R. J. P., 826(132), *868*
Williams, R. L., 885(8), *932*
Williams, R. V., 336(60), 339(60), *354*
Williams, V. H., 439(91), *459*
Willig, F., 1413(136), *1420*
Willis, J. P., 567(148), *579*, 723(189), *754*, 960(24), *1018*, 973(91), *1021*, 979(88), *1021*, 1415(152), *1420*
Willman, K., 1366(26), 1368(26), 1369(26), *1393*
Willner, I., 1408(86), 1409(101), *1418*
Willsau, J., 640(156), *655*
Wilshire, J., 1269(31b), *1281*
Wilshire, J. F. K., 1080(79), *1097*
Wilson, A. M., 1176(490), *1194*
Wilson, C. L., 414(102a), *429*, 445(160), *461*, 721(149), *753*, 905(214, 215), *941*, 1080(81), *1098*, 1346(25), *1360*
Wilson, E. R., 539(26), *575*
Wilson, G. S., 107(110), *119*, 184(251), 185(251), *203*, 186(259), *204*, 234(10), *246*, 668(59), *693*, 668(64), 669(65), 670(77), *694*, 685(137), *696*, 686(157), 687(157), *697*, 1046(128), *1064*, 1167(338), *1189*, 1178(519, 520), *1195*, 1178(521), *1195*
Wilson, J. D., 597(99), *611*, 981(125), *1022*
Wilson, J. M., 123(20), 133(20), *196*
Wilson, K., 445(160), *461*
Wilson, K. B., 905(215), *941*
Wilson, R., 729(265), *757*
Wilson, R. M., 649(234), *657*
Wilson, W. A., 1104(8), 1105(8), 1106(8), *1124*

Wiltshire, I. J., 1410(116b), *1419*
Wimmer, F. L., 1175(448), *1192*
Wingard, L. B., Jr., 645(206), *656*
Winkel, A., 453(216), *463*
Winkler, H., 1052(186), *1066*, 1170(361), *1190*
Winograd, N., 179(217), 181(217), *202*, 188(261, 270), *204*, 270(96), 277(96), *316*
Winslow, N. M., 258(23), 275(23), *313*
Winter, W., 717(118), *752*, 959(48), *1019*
Wintersberger, K., 307(416), *327*
Wintgens, V., 721(159), 722(159), *753*
Wipf, D. O., 58(73b), *118*, 208(3), *231*
Wipplinger, C., 417(116b), *430*
Wisdom, N. E., 1384(106), 1387(106), *1395*
Wise, W. S., 732(287), *758*
Wislicensus, H., 1206(71), *1213*
Wisser, K., 616(1), *650*
Wistrand, L. G., 31(40), *116*, 267(76d), *315*, 511(76), *531*, 601(141), *612*, 774(396), *785*, 1070(3, 8), 1081(3, 8), *1095*, 1077(60), *1097*, 1163(300), *1188*
Witesides, T., 565(142), *578*
Witkop, B., 1082(94), *1098*
Wittchen, H., 195(297), *205*
Wittig, G., 1204(56), *1213*
Wolberg, A., 820(94), *867*
Wolin, R. L., 1139(59), 1140(59), *1180*
Wolf, J. F., 635(122), 648(122), *653*, 1043(86), *1063*
Wolf, R., 1153(191), *1184*, 1233(98, 99), *1259*
Wolfahrt, T., 713(80), *751*
Wolfe, R. S., 813(31), *865*
Wolfrom, M. L., 439(86), *459*
Wolfson, S. K., Jr., 644(202), 645(206), *656*
Wollish, E. G., 734(306), *759*
Wong, C. L., 30(37a), *116*
Wong, P. C., 1178(511), *1194*
Wong, M. Y. H., 14(36), *114*
Woo, E. P., 336(64), 339(64), *355*
Wood, J., 802(89), *808*
Woodard, F. E., 122(11), *195*, 148(11), *198*, 488(51), *498*
Woodbury, E. C., 905(171), 913(171), *939*, 1144(105), *1182*
Woodhall, B. J., 1200(6), *1211*, 1201(31), *1212*
Woods, R., 492(100), *499*, 661(17, 18), *692*
Woodward, P., 846(241), *872*
Woodward, R. B., 350(201), *359*
Woody, M. C., 1434(22c), *1438*
Woolford, R. G., 540(34, 35), *575*, 1002(187), *1024*
Woolsey, I. S., 820(93), *867*
Woolsey, N. F., 1038(40, 41), *1061*
Worry, G., 32(43c), *117*, 39(55a), 63(43), *116*
Wowchenko, G. D., 435(55), 436(55), *458*
Wragg, A. A., 268(93), *316*
Wright, A. R., 848(251, 253), *872*
Wright, C. M., 551(98, 99), *577*, 1088(139), *1099*, 1233(86), 1235(86), *1259*, 1247(174), *1262*
Wright, D., 908(223), *941*
Wright, J. D., 790(10), *804*
Wrighton, M. S., 1366(22), *1392*, 1366(23, 25), *1393*, 1367(22), *1392*, 1367(23, 25, 32), 1372(46), *1393*, 1383(93), 1385(93), *1385*, 1398(10), *1415*, 1409(113b), *1419*
Wrobel, J. T., 722(179), *754*, 1049(158a, b), *1065*
Wrona, M., 746(427), *763*
Wu, C. N., 633(113), *653*
Wudl, F., 663(40), *693*, 680(117), 681(117), 682(117), 683(117), *695*
Würminghausen, T., 291(284), *322*, 515(92, 93), 518(92), *531*
Wurts, A., 535(2), *574*
Wuts, P. G. M., 557(110), *577*, 569(156), *579*, 1037(32), *1061*
Wyatt, J., 601(143), 602(143), *612*, 704(25), *749*, 965(54), *1019*, 991(166), *1024*, 992(174), 1001(174), *1024*, 1094(176), *1101*
Wychick, D., 467(25), *479*
Wyckoff, J. C., 242(60), *248*
Wynne-Jones, W. F. K., 260(38), *314*, 538(18), 542(18), *575*

Xiang, Y.–B., 812(22), *864*
Xie, N., 1175(474), *1193*, 1248(176), *1262*
Xu, F., 814(54), 819(54), *865*
Xu, Q. Y., 819(79), *866*

Yadav, A. K., 456(241), *464*, 545(65), 546(65), *576*, 563(128), *578*, 783(72), *786*, 903(200), *940*, 1157(222), *1185*, 1254(207), *1263*
Yadawa, A., 903(201), *940*
Yagi, H., 739(348), *760*, 957(34, 35), 961(34), *1019*, 1094(175), *1101*, 1166(320), *1188*
Yakamichi, T., 302(382), *325*
Yaki, J., 908(223), *941*
Yakubovich, A. A., 1204(59), *1213*
Yamabe, T., 1380(69), *1394*
Yamada, M., 620(43), *651*
Yamada, T., 275(127), *317*, 445(155), *461*, 455(240), *464*, 1141(66), *1181*
Yamada, Y., 584(20), *608*, 1167(332), *1189*
Yamagata, K., 1087(132), *1099*
Yamagishi, A., 620(43), *651*, 686(146), 687(146), *696*, 1132(15), *1179*, 1173(428), 1174(445), *1192*, 1390(129), *1396*
Yamagishi, T., 346(151), *357*
Yamagishima, H., 891(47b), *934*, 909(225), *942*, 924(299), *945*
Yamaguchi, H., 723(195), *755*, 973(79), 977(85), *1020, 755*
Yamaguchi, N., 1434(12d), *1437*
Yamaguchi, R., 514(78), 517(78), *531*, 1163(294), *1187*
Yamaguchi, Y., 726(244), 728(244), *757*
Yamakawa, E., 425(187), *432*, 599(118), *611*
Yamamoto, H., 585(23), *608*
Yamamoto, I., 1167(336), *1189*
Yamamoto, K., 1380(68), *1394*
Yamamoto, M., 716(113), *752*
Yamamoto, T., 336(60), 339(60), *354*, 537(6), 573(6), *574*, 1319(117), *1336*, 1380(68), *1394*

Yamamoto, Y., 602(145), *612*, 1408(61), *1417*
Yamamura, S., 621(65, 84–87), 624(65), 627(84–87), *652*, 775(43a–c), 776(44b, 46a, b, 47), 777(48), *785*, 958(40), 962(45), *1019*, 1013(251), *1027*, 1161(271), *1187*, 1163(296, 287), 1166(321), *1188*
Yamanaka, T., 1164(313), *1188*, 1243(135), 1244(135), *1261*
Yamane, S., 782(70c), *786*, 1077(50), *1096*, 1159(241), *1186*, 1425(13), *1430*
Yamani, S.–I., 598(109), *611*
Yamasaki, R. B., 374(88), *396*
Yamasaki, S., 1433(5c, e), *1436*
Yamashina, N., 890(42), *934*
Yamashita, H., 267(82), 271(82), 288(82), 301(82), 307(82), *315*
Yamashita, M., 425(189), *432*
Yamashita, S., 721(154), *753*, 851(265–269), *873*, 904(208), *941*, 1162(286), *1187*
Yamauchi, T., 486(27), *497*
Yamauchi, M., 721(160), *753*
Yamataka, H., 1409(102), *1418*
Yamazaki, A., 367(35), *395*
Yamazaki, M., 537(11), 563(11), 572(11), *574*
Yamazaki, N., 306(406), *326*, 349(188), *359*
Yamazaki, T., 640(152), *654*
Yamura, A., 434(18), *457*, 436(65), 447(65), 449(65), *458*, 907(198), *940*
Yanagida, S., 1408(53–55, 65), *1417*, 1408(88), *1418*, 1409(105), *1419*
Yanase, H., 1094(182), *1101*
Yandle, J. R., 733(302), *759*
Yang, N. C., 508(35), *529*
Yang, S. L., 333(28b), *353*
Yanilkin, V. V., 801(83, 84), *807*
Yano, J., 1383(94), 1385(94), *1395*
Yao, C.–L., 816(62, 69), *866*
Yao, S. J., 644(202), 645(206), *656*
Yasnovsky, V. M., 792(23), *804*
Yasukouchi, K., 723(195), *755*, 973(79), 977(85), *1020*, 1059(258), *1068*
Yasuda, T., 1164(313), *1188*, 1243(135), 1244(135), *1261*
Yasunaga, H., 679(107), *695*, 1046(123), *1064*
Yates, D. A., 184(247), *203*
Yates, G. B., 237(32), *247*, 537(17), 563(17), 564(17), *575*, 563(127), *578*, 1156(219), 1157(219), 1158(219), *1185*, 1158(234), *1186*
Yatziv, S., 690(192), *698*
Yeager, E., 122(7), 122(123), 122(157), 179(7), *195*, 256(7), *312*, 1287(12, 14, 20), 1291(20), *1332*, 1105(14), 1106(14), 1109(14), *1124*
Yeager, H., 280(185), *319*, 305(401), *326*
Yee, G. T., 819(82), *866*
Yeh, L.–S., 341(93), *356*, 882(3g), 885(3g), *932*, 1174(447), *1192*, 1239(119), *1260*
Yiang, S., 910(228), *942*
Yijima, C., 403(13), *426*, 678(103), *695*
Yildiz, A., 349(177), *358*, 496(130), *500*, 910(229), *942*, 960(21), *1018*
Yokota, T., 705(33), *749*
Yokoyama, M., 977(85), *1020*
Yokoyama, T., 777(49), *785*
Yokoyama, W., 1233(86), 1235(86), *1259*, 1247(174), *1262*
Yokozawa, A., 272(114), 275(114), 275(127), *316*, *317*, 443(123), 445(123), 455(123), *460*, 455(240), *464*, 1173(419, 420), *1191*
Yomamoto, F., 907(186), *940*
Yomiyama, A., 1316(101), 1319(101), *1335*, 1321(126), *1336*, 1326(101), *1335*, 1330(126), *1336*
Yoneda, S., 682(124), *695*
Yoneyama, H., 1383(96–98), 1386(97, 98), 1387(98), *1395*, 1408(82), *1418*
Yontanza, A., 907(186), *940*
Yorgiyadi, S., 675(97), *695*
Yoshida, I., 1434(15f), *1437*
Yoshida, J.–i., 643(187), *655*, 667(57), *693*, 679(108), *695*, 716(113), *752*, 850(258, 259), *872*, 853(273, 274), 855(275b, c), *873*, 1008(227), 1012(227), *1026*, 1053(194), *1066* 1168(340), *1189*, 1243(136, 143), 1245(149, 150), *1261*
Yoshida, K., 597(106), *611*, 600(125), *612*, 633(114), *653*, 638(135), *654*, 648(229), *657*, 704(15a), 726(15a), *748*, 721(147, 151–153), *753*, 722(165), 723(186), *754*, 773(34), *785*, 950(1b), *1018*, 1035(23), 1036(28), *1061*, 1070(17), 1073(17), 1080(17), *1095*, 1080(78), *1097*, 1084(17), *1095*, 1089(149, 157, 158), 1090(149, 157, 159), *1100*, 1161(273), 1162(280), *1187*
Yoshida, M., 271(105), *316*
Yoshida, T., 367(29), *395*, 572(165), *579*
Yoshida, Y., 592(52, 53), *609*
Yoshida, Z., 682(124), *695*, 815(57, 58, 60), *866*
Yoshihiro, Y., 1162(285), *1187*
Yoshikawa, M., 556(108, 109), *577*, 560(123), *578*
Yoshikawa, T., 1408(64), *1417*
Yoshinaga, K., 1174(446), *1192*
Yoshino, K., 1357(94), *1362*, 1384(107), 1385(107), *1395*
Yoshino, T., 908(223), *941*
Yoshiyama, A., 641(168), 643(168), *655*, 794(52, 59, 61), 795(48, 52), 796(52, 59), 797(61), *806*, 1035(20), *1061*, 1178(514, 515), *1194*
Yoshizawa, S., 267(82), 271(82), 288(82), 301(82), 307(82), *315*
Yoshizumi, K., 853(271), *873*
Younathan, J. N., 1408(62), *1417*, 1409(95, 100), 1412(95), *1418*, 1412(128), *1419*
Young, E. E., 1141(62), *1181*
Young, G. T., 1039(53), *1062*, 1045(116, 117), *1064*
Young, J. A., 1109(37), *1125*
Young, M. P., 150(114), *198*, 478(106), *482*
Young, T. E., 711(66), *750*
Yousefzadeh, P., 1056(241), 1057(241), 1058(241), 1059(241), *1068*
Yu, F. R., 926(316), *945*
Yu, J. C., 793(43), *805*
Yu, K. C., 1351(43), *1360*
Yu, S. T., 1421(4b), *1429*

Yuasa, K., 860(306), *874*, 1426(17), *1430*
Yudis, M. D., 1144(96), *1182*
Yuech, A., 621(52), *651*
Yui, N., 906(180), *939*
Yukawa, T., 243(70), *248*, 393(207), *400*, 827(137), 828(137), *868*
Yur'eva, L. P., 838(210), *871*, 895(94), *937*
Yuzhakova, O. A., 411(80), *429*
Yvernault, T., 297(340), 298(340), *324*

Zabolotny, E. R., 350(196), *359*
Zabusova, S. E., 495(127), *500*, 1253(205), 1254(205), 1255(210), *1263*
Zachariasse, K., 344(124), *357*
Zacharova-Kalavska, D., 915(258), *943*, 1144(98), *1182*
Zadeh, M. M., 1094(183), *1101*, 1117(60), 1118(60), *1126*
Zagal, J., 821(101), *867*
Zagorski, Z. P., 256(3b), *312*
Zahnow, E. W., 528(147), *533*, 589(38), *608*, 960(25), *1018*, 963(67), *1020*, 1089(151), 1090(151), 1091(151), *1100*
Zahradnik, R., 508(43), *530*
Zaidenberg, A. Z., 539(22), *575*
Zaitsev, P. M., 408(55), *428*
Zaitseva, L. A., 794(45), *805*
Zaitseva, Z. V., 408(55), *428*
Zak, J., 1358(108), *1362*
Zakazov, A. N., 794(47), *806*
Zalar, F. V., 1159(236), *1186*
Zalogina, E. A., 709(57), *750*
Zamboni, R., 1382(80), *1395*
Zann, D., 17(16), *115*, 223(21), *232*, 1222(25), *1257*, 1248(25), *1257*
Zapata, J., 153(129), *199*
Zaporozhets, E. V., 404(29a, b), 408(29a, b), *427*
Zapp, J. A., 286(347), *324*
Zaripova, R. M., 473(63), *480*
Zass, E., 812(22), *864*
Zasshi, Y., 1004(209), *1025*
Zaugg, H. E., 1208(93), *1214*
Závada, J., 376(109), *397*, 1176(496), *1194*
Zawodzinski, T. A., 303(393), *326*
Zecchin, S., 675(95), 689(95), *695*, 835(183), *870*
Zefirov, N. S., 376(112), *397*
Zeiler, A. G., 1159(246), *1186*
Zein, N., 1148(158), *1183*
Zemach, D., 239(41), *247*, 1203(45), *1213*
Zembayashi, M., 832(160), *869*
Zenkin, A. A., 366(25), *394*
Zerner, M., 814(51), 823(51), *865*
Zeug, N., 1408(87, 89), *1418*
Zeuli, E., 903(218), *941*
Zey, R. L., 740(368), *761*
Zhalina, V. A., 793(38), *805*
Zhang, B., 910(228), *942*
Zhang, K., 1173(435), *1192*
Zhang, L., 1173(432), *1192*
Zhang, O., 1173(432), *1192*
Zhang, X., 288(258), *321*
Zhao, C., 1070(1), *1095*
Zhdanov, S. I., 340(77a), *355*, 435(39), *457*, 735(318), *759*
Zheng-Hao, W., 744(404), *762*
Zhi-Bin, H., 744(404), *762*
Zhuravleva, V. N., 267(85), 268(85), 271(85), 307(85), *315*
Ziade, A., 1433(5a), *1436*
Zibrova, N. A., 439(97), 440(97), *459*
Ziebig, M., 468(36), *480*, 595(88), *610*, 930(357a), *947*
Ziebig, R., 721(157b, c), *753*, 726(239), *756*, 728(258), *757*, 926(138), *938*, 1038(35), *1061*
Zieger, H. E., 886(12), 891(12), *933*
Ziegler, H. W., 407(47), *428*
Ziegler, K., 349(186), 350(197), 351(209), *359*, 857(292), *874*, 1319(115), *1336*
Zigman, A. R., 905(216), *941*
Zilkha, A., 544(58–61), *576*, 1341(13), 1346(13), 1348(13), 1355(13), *1359*, 1355(86–89), *1362*
Zimmermann, G., 778(56), *786*, 1083(104), *1098*, 1163(299), *1188*
Zimmermann, H., 340(82a), *355*, 848(247), *872*
Zinger, B., 690(192), *698*
Zinov'eva, S. S., 1269(36), *1281*
Zinh, E., 1200(12), *1211*
Zjawiony, I., 641(176), 642(176), *655*
Zolotar, R. N., 567(146), *579*
Zolotovitski, Y. M., 435(36), 436(36), *457*
Zon, G., 377(118), *397*, 1152(178), *1184*
Zotti, G., 835(183), *870*
Zrineh, A., 816(65), *866*
Zulu, S. J., 844(228, 229), *871*
Zuman, P., 2(3), *7*, 25(30), 26(31b), *115*, 26(32a), *116*, 163(154), *200*, 434(4), *456*, 435(34), *457*, 436(64), *458*, 452(208a–c), *462*, 466(9), *479*, 474(80), *481*, 476(94), 477(94), *482*, 485(12), *496*, 494(114), 495(123), *500*, 635(123), *653*, 720(139), *753*, 734(315), *759*, 739(351), *761*, 743(394, 395), 744(395), *762*, 745(419), *763*, 879(1), 885(1), *932*, 895(100), *937*, 1033(10a), *1060*, 1039(53), *1062*, 1048(151a, b), *1065*, 1050(170, 174), 1051(175, 176), *1066*, 1056(231), *1067*, 1144(122), 1145(122), 1146(122), *1182*
Zutshi, K., 425(184), *432*, 637(127), *654*, 715(103), *751*, 1243(142), *1261*
Zvorukina, V. C., 489(67), *498*
Zweig, A., 662(38), *693*, 674(84, 85), *694*, 721(163), 723(163), *754*
Zyka, J., 591(45), *609*, 1056(231), *1067*

Subject Index

N,O-Acetals, from methoxylation, 767
Acetamidation, 511, 555, 1084
2-Acetamidosulfides, from alkenes, 664
Acetic acid, 484
  as solvent, 281
Acetone, reduction of, 445
Acetonitrile
  purification, 285
  reduction of, 496
  as solvent, 285
Acetonylacetone, reduction of, 441
α-Acetoxy-α-amino acids, 766
Acetoxylation, 555, 1080
  of alkenes, 1082
  α-aminocarboxylate acids, 562
Acetylacetone, reduction of, 441
Acetylene, 1318
Acetylenic halides, reduction of, 370–372
Acridine, 725
Activation, 1238
Activation energy (see ET theories)
Acyl halides, reduction of, 384, 385, 483
Acyl peroxide, pulse electrolysis of, 545
Acyl radicals, coupling of, 384
Acylation, reductive, 239
Acylation on O, 839
  reductive, 839
Acyloxylation, 1080
  of N-alkyl-substituted amides, 1083
  intramolecular, 561
Ad-atoms, catalysis by, 268, 1390
Addition, diastereoselective, 1155
Adiponitrile, 1318
  commercial processes, 1317
Adrenochrome, 772
Adsorption, 68–69, 263
  of carbonyl compounds, 435
[Adsorption]
  of organic halides, 363, 364, 370
  in photoelectrochemistry, 1410
  stereoisomeric effect of, 1175, 1176
Adsorption waves, in polarography, 160
Affinity
  definition, 23
  standard potential, 23–24
Alcohols
  aliphatic, oxidation of, 641
  from carbonyl compounds, 436, 437
  indirect oxidation, 1232
  oxidation of, 6, 635, 638, 641
  oxidation at NiOOH–anodes, 641
Aldehydes
  heterocyclic, 743
  nonolefinic, 433
  oxidation of, 638, 643
  reduction of, 486
  reduction to alcohols, 438
Aldonolactones, reduction by amalgam, 1208
Alicyclic halides, reduction of, 375–378
Aliphatic amines, methoxylation of, 979
Aliphatic nitriles, 495
Alkaloids, 770
Alkanes, oxidation of strained polycyclic, 517
Alkene
  cation radical, 519
  reaction with O_2, 519
Alkenes
  from vicinal dihalides, 368
  from vinyl halides, 369
Alkoxyalcohols, oxidation of, 648
Alkoxybenzenes, oxidation of, 633
Alkoxylation, 1073
  of N,N-dialkylamides, 1076

Alkyl 4-imidazolecarboxylates, reduction of, 489
N-Alkylanilines, oxidation potentials, 586
Alkyl benzoates, reduction of, 489
Alkyl halides, reduction of, 362 ff.
Alkyl radicals, cathodic coupling of, 391, 392
Alkylation, reductive, 840
Alkylbenzenes
  ionization potential vs standard potential, 25
  oxidation by iron(III) phenantrolines, 30–31, 50–53
  peak potential vs standard potential, 72–73
Alkylhydroxylamines, 409
  from nitroalkanes, 402
N-Alkylphthalimides, 492
N-Alkylpyrrole derivatives, films from, 1380
Alkynes, from vicinal dihaloalkenes, 369
Allyl alcohols, indirect electrolysis, 1039
Allylic compounds, halogenated, reduction of, 370
Alumina, activated, 333
Amalgam
  aluminum, 1200
  calcium, 1200
  deuterium labeling, 1205
  N,N-dimethylpyrrolidinium, 1200
  oxidation potential of, 1200
Amalgams, quaternary ammonium, 380, 1200
Amide oximes, reduction of, 477
Amides
  α-acyloxylation, 602
  α-alkoxylation, 602
  α-oxidation, 602
  anodic oxidation, 601
  reduction of, 483
Amidines, reduction of, 469
Amidophosphates, α-methoxylation, 603
Amination, 1087
Amines
  aliphatic
    anodic oxidation, 583
[Amines]
    mechanistic scheme, 582
  primary
    α-branched, oxidation to ketones, 583
    anodic oxidation, 582
    oxidation to hydrazines, 583
    reaction mechanism, 583
  secondary, oxidation of, 583
  as solvent, 1255, 1256
  tertiary
    dealkylation of, 583, 584
    oxidation of, 583
Amino acids, 765
  from amino alcohols, 600
  anodic oxidation, 600
  carboxylation of azomethines, 468
  electrosynthesis of, 375
  α-methoxylation of protected, 604
  from Schiff bases, 468
Amino alcohols, anodic oxidation, 600
p-Aminoazobenzenes, 422
4-Aminodiphenylamine, from aniline, 588
4-Aminomethylpyridine, 493
p-Aminophenol, preparation of, 413
Aminophenols, anodic oxidation, 594
o-Aminophenyldiimide, 735
o-Aminophenylhydrazine, 735
Ammonia, as solvent, 282
Androstan-6-one, from androstan, 778
Anhydrides, 491
  reduction of, 483
Aniline derivatives, anodic oxidation, 587
Aniline polymerization, CV of, 1379
Anilines
  dimerization of, 969
  oxidation products, 588
Anion radicals
  from organic halides, 363, 373, 374, 380, 381–382, 383, 384, 389, 390, 391
  protonation of, 140, 162
Anions
  dimerization of, 1005

[Anions]
in preparation of CME, 1384
in supporting electrolyte, 294
Anisoles, oxidation of, 626
Annulenes, reduction, 336
Anode material, 265, 1288
oxygen overvoltage, 262
Anodes
composite, 267
dimensionally stable (DSA), 267, 1288
sacrificial, 267
titanium with lead oxide, 266
Anodic coupling, 1006
Anodic deprotection, of alcohols, 6
Anodic fluorination, 1103
Anodic substitution
by cyanate ion, 1088
by cyanide, 1089
by nitrate ion, 1083
mechanism of, 1070
Anthracene, 347
acylation, 348
oxidation of, 510, 524
Anthracyclines, 779
Anthracyclinones, 781
Anthranilic acid, reduction of, 486
Anthranils, 727
Anthraquinone, 1340
Antibiotics, 779
Aporphines, 771
Aprotic solvents, 284
preparation of CME in, 1382
Armentomycin, 766
Aromatic amines, anodic oxidation, 586
Aromatic esters, 488
Aromatic halides
dimerization of, 223–226
hydrogen atom transfer in, 226–229
mechanism of reduction, 17
reduction of, 379 ff.
Aromatic hydrocarbons
acetoxylation, 236
methoxy substituted, 629
oxidation of alternant, 507
voltammetry of, 505
Aromatic hydroxylamines, oxidation of, 599
Aromatic nitriles, 495
Aromatic substitution, oxidative, 236
Aromatic sulfonates, as supporting electrolyte, 295
Aryl esters, oxidation of, 628
Aryl ethers
anodic coupling of, 963
cyanation, 963
oxidation of, 628
Aryl halides
coupling, 1237
intramolecular coupling, 771
preparation of heterocycles, 703
1-Arylindene, sigmatropic 1,5-shifts, 351
Ascorbic acid, oxidation of, 645
Astaxanthin, 778
Asymmetric induction, 1142, 1171–1174
yield, 1171–1174
Asymmetric reduction, in magnetic field, 1174
Asymmetric synthesis, 1171–1172
Asymmetry inducer, 1171–1172
Autocatalysis, 1220, 1228
Ax-eq conformation, 1143, 1145, 1153, 1170
Azines
reduction of, 470
unsymmetrical, reduction of, 470
Aziridine derivatives, 722
Azo compounds, 410
from anilines, 589
cis-Azobenzene, anion radicals, 350
Azobenzene
as probase, 1267
reduction of, 422
Azoxy compounds, 410
Azoxybenzene, formation of, 417
Azoxybenzenes, reduction of, 422
Azulene, 337

Barbier reaction, 851
Batch reactor, 1301
Benzalazine, reduction of, 470
Benzamide, reduction to benzylamine, 493
Benzhydryl halides, reduction of, 375

Benzidines
from anilines, 587
from N,N-dialkylanilines, 589
Benzil, reduction of, 441
Benzoate esters, reduction of, 488
Benzodithiins, oxidation of, 683
Benzofurazan, 734
Benzoic acid, reduction of, 486
Benzonitrile, 495
Benzophenone anil, reductive alkylation, 468
Benzopyrroles, 722
Benzoquinone, 1340
p-Benzoquinone, from aniline, 588
Benzotriazole
formation of, 714
reductive acetylation, 735
Benzotriazoles, reduction of, 735
S-Benzoyl-L-cysteine, 491
Benzoylacetonitrile, 488
Benzyl alcohols
anodic oxidation of, 635
α-substituted, 637
Benzyl anion, coupling of, 374, 375
Benzyl halides, reduction of, 373 ff.
Benzylhydroxylamines, oxidation to oximes, 599
Benzylic halides, reduction of, 372–375
N-Benzylnitrilium ions, 512
Benzyloxycarbonyl, 1050
reduction of, 489
Benzyltetrahydroisoquinolines, anodic cyclization of, 994
Benzyne, from o-dibromobenzene, 381
Berbin alkaloids, 773
9,9'-Bianthryl, cleavage, 350
Bibenzyl, from toluene, 960
Bicyclo[1.1.0]butanes, 351
Biphenyls
cation radicals of, 628
dications of, 629
methoxy substituted, 628
Birch reaction, electrolytic, 1255, 1256
Birch reduction, by amalgam, 1203
Bis(biphenyl)chromium, 272
Bis-sulfides, cyclic, 668
Bisacetamidation, 512
Bisulfite addition complex, 486
Bleaching, electrochemical, 791, 792
Bond bending, 1401
Bond position, of semiconductors, 1405
Borate complex, 486
Boundary layer
diffusion, 167
hydrodynamic, 167
Bromination, electrochemical, 1094
Brown-Walker reaction, 536
Brønsted coefficient in ET, 46–48
Bulk electrolysis
charge consumption in, 212–213
current intensity, 212
equivalent rate constant, 210
homogeneous, 217–219
homogeneous equivalent, 211
optimization of, 216, 217
Butenolides, 1242
Butler-Volmer rate law, 64, 67 84
in dimensionless form, 92
N-tert-Butylhydroxylamine hydrochloride, 2
1-tert-Butylpyrene, 3
Butyric acid, reduction to butanol, 487
Butyronitrile, as solvent, 286

CA (see chronoamperometry), 145
Cadmium, as cathode material, 264
Calomel electrode, 270
Capillary gap cell, 1306
Carbamates
anodic oxidation, 602
α-oxidation, 602
Carbanions, 1006
Carbazole, oxidation of, 979
Carbenium ions, electrogenerated, 555
Carbohydrates, oxidation of, 644
Carbomethoxylation, 1092
Carbon
as anode material, 266
as cathode material, 265

Carbon anode, in fluorination, 1104
Carbon dioxide, 4
  addition to anion radicals, 348
  cathodic coupling of, 366, 370, 373, 374, 380, 389, 390
  reduction catalyzed by phthalocyanins/metal, 821
  for trapping anion radicals, 366, 370, 373, 374, 380, 389, 390
Carbon disulfide, reduction of, 712
Carbon electrodes, derivatized, 1365
Carbon tetrachloride, reduction of, 367
Carbon-carbon bond formation
  catalyzed by bipyridines/metal, 832
  catalyzed by cyclams/metal, 830
  catalyzed by corrins/metal, 826
  catalyzed by phosphines/metal, 835
  catalyzed by porphyrines/metal, 817
  catalyzed by Sn(O) or Pb(O), 851
Carbon-carbon bonds, formation of, 702
Carbon-halogen bond, cathodic cleavage of, 361–394
Carbon-nitrogen bonds, formation of, 706
Carbon-oxygen bonds, formation of, 705
Carbonyl compounds
  coupling, 1236
  olefinic, 451
  reduction mechanism, 433
  reduction products, 436
  α,β-unsaturated, 451
Carbonyl groups, reduction to methylene, 446
Carbonylation-decarbonylation, catalyzed by Fe-carbonyl compounds, 843
Carboxylation, 4, 1251, 1252
Carboxylic acids
  from alkyl monohalides, 366
  from allyl halides, 370
[Carboxylic acids]
  from benzyl halides, 373, 374
  decarboxylation, 237
  formation of, 366, 370, 373, 374, 380, 389, 390
  from halobenzenes, 386
  from halogenated lactams, 390
  reduction of, 483
  β-trimethylsilyl, 566
Carotenoids, 778
1-Carvone, from α-pinene, 777
Catechols, oxidation of, 621
Cathode materials, 263
  hydrogen overvoltage, 262
Cathodes, sacrificial, 265
Cation radicals
  deprotonation of, 512
  reaction with fluoride, 1119
Cations, in supporting electrolyte, 296
Cell
  electrochemical, for ESR spectroscopy, 190
  packed bed, 1310
  thin-layer, for spectroelectrochemistry, 180
  for voltammetry, 123
Cell configurations, 1303
Cell design, for industrial electrolysis, 1287
Cell geometry, 1303
Cells
  capacitance, 58–59
  commercially available, 1309
  divided, 250
  electrical equivalent circuit, 53–56
  filter press, 257
  with porous electrodes, 257
  potential distribution, 246
  SPE, 257
  for supercritical liquids, 260
  three dimensional, 1309
  with three-dimensional electrodes, 258
  with two-dimensional electrodes, 249
  two-dimensional flow-through, 1305
  undivided, 250
Cellulose, 788
  conversion into glucose, 793
Cephalosporin, 489, 779

Chain reactions
  initiation of, 1346
  free radical, 1346
Chain transfer
  in cationic polymerization, 1339
  in polymerization, 1346
Chargetransfer (see also Electron Transfer, ET)
  heterogeneous rate constant, 132
  irreversible, 130–132
  nernstian, 130, 131
  quasi-reversible, 131
  reactions
    chemical reactions coupled to, 133
    classification of, 130
Charge transport, 1295
Chemical potential, 82
Chemical reaction coupled to ET
  classification of, 73–78
  consecutive reaction, 75
  in homogeneous solutions, 30–31
  in redox catalysis, 75
  mediated electron transfer, 75
  microscopic reversibility, 78–79
  multi-electron reaction, 76
  preceding reaction, 74–75
  zero-electron reaction, 77–78
Chemically modified electrodes (see CME)
Chiral center, alkylation of, 1171
Chiral cosolvent, 1172
Chiral electrode adsorbents, 1171
Chlorination, electrochemical, 1094
Chloroacetamidation, stereoselective, 1164
Chronoabsorptometry, 187
  potential-step, 187
Chronoamperometry, 145
  double-potential-step (DPSCA), 145, 146
    application of, 140
  principle of, 104–107
Chronocoulometry, 150
  double-potential-step, 150, 151
Chronopotentiometry, 152
  current-reversal, 152
Cinnamoyl chloride, 492
Cinnoline, formation of, 713, 741
Cinnolines, 730
[Cinnolines]
  reduction of, 718
Cis-trans geometry, 1143, 1145, 1153, 1170
1-Citronelle, from 1-menthone, 777
Clay, in CME, 1390
Cleavage
  of alcohols, 1038
  of arylsulfonate esters, 1057
  of benzylic alcohols, 1039
  of carbon-carbon bond, 1032
  of carbon-nitrogen bond, 1049
  of carbon-oxygen bond, 1038
  of carbon-phosphorus bond, 1051
  of conjugated carbinols, 1038
  of cyclopropane rings, 1033
  of dioxalates, 1040
  of disulfides, 1054
  of ethanetetracarboxylate esters, 1032
  of gem-disulfides, 1046
  by indirect oxidation, 1043
  of isoxazoles, 1058
  of nitrogen-nitrogen bond, 1055
  of oximes, 1057
  of p-methoxybenzyl ethers, 1043
  of peroxides, 1053
  of phenols, 1038
  of pinacols, 1035
  of quaternary ammonium, 1049
  of sulfonamides, 1058
  of sulfones, 1048
  of sulfonium compounds, 1044
  of thioethers, 1045
  of thiol esters and thiocyanates, 1047
  types of, 1030
CME (chemically modified electrodes)
  applications of, 1391
  characterization of, 1368
  chemisorbed monolayers, 1371
  chemisorbed polymer films, 1372
  covalently attached monolayers, 1364
  cyclic voltammetry at, 1368
  derivatized carbon surfaces, 1370
  electrochemical characteristics, 1368
  Nafion, 1374

[CME]
polymer films, 1375
silylated oxide surfaces, 1365
sulfonate polymers, 1374
Cobaliticinium salts, in reductive alkylation, 840
Cocatalyst, water as, 1351
Collision frequency
for heterogeneous ET, 62–63
for homogeneous ET, 38
Combination, of alkyl radicals, 365
Compact layer, 59
Competition parameter, definition of, 220
Composite electrodes, as cathode material, 265
Concentration profiles, 141
simulated, 142
Conductive films, preparation, 1382
Configuration, at the electrode, 1133, 1138, 1140, 1170, 1172, 1177
Configurational stability, 1143, 1169, 1177
Conformation, at the electrode, 1133, 1143, 1153
Conformational equilibria, 1178
Conformational stability, 1132, 1134
Connection
monopolar, 1302
bipolar, 1302
Convection, 81–82
forced, 164
Coordination compounds, 812, 837
organometallics, 847
bipyridines, 831
carbonyls, 843, 845
corrins and B_{12}-models, 821
cyclams, 830
cyclopentadienyl, arene, tropylium, 838
phenanthrolines, 831
phosphines, 834
phthalocyanines, 820
porphyrines, 815
Copolymers
block, 1355
graft, 1355
Copolymerization
by donor and acceptor monomers, 1352
to form alternating copolymers, 1352
of pyrrole and bithiophene, 1357
Cottrell equation, 145
Coulombic repulsion, 76
Coulometry, 176
at constant current, 177
at constant potential, 176
Counterelectrode reaction, 300
Counterion, effects of, 334
Coupling
of anions, 1005
of aromatic amines, 587
of aromatic hydrocarbons, 951, 952
of aryl ethers, 963
of arylamines, 963
with cleavage, 881
crossed, 880
intramolecular, 985, 986
of methoxybiphenyls, 985
mixed, 960
of olefins, 979
phenolic, 770
of phenols, 960
of radicals (see radicals)
via radical cations, 951
via radicals, 1001
Coupling reactions
anodic, 523
intermolecular, 715
CPSV (see also Voltammetry, convolution potential sweep)
Current
adsorption, during polarography, 160
catalytic, during polarography, 160
diffusion, during polarography, 160
interrupted, methods, 1341
kinetic, during polarography, 160
limiting, 157
parasitic, 1302
reversal, methods, 1341
Current density, in polymerization, 1338

Current distribution
  in cells, 1295
  primary, 1296
  secondary, 1296
  tertiary, 1297
Current intensity, definition of, 61–62
  in bulk electrolysis, 212
  and heterogeneous rate constant, 62–64, 67, 84
  and mass transfer, 83–85
  limiting value in steady state methods, 96
  sign convention, 84n
Current ratio, normalized, 148
Current-potential, linear analysis, 136
Current-time curves, 163
CV (see Voltammetry, cyclic)
Cyanation, anodic, 528, 1089
9-Cyanoanthracene, electrodimerization, 349
N-Cyanomethylation, 1087
2-Cyanopyridine, 495
4-Cyanopyridine, 495
Cyanopyridine, reduction of, 745
Cyclic macromolecules, diffusion of, 1340
Cyclic voltammetry, 123, 136
  of polymers, 1340
  principle of, 107–108
Cyclization
  of acetylenic halides, 371, 372
  of α,ω-dihaloalkanes, 368
  of halobenzamides, 383
  intramolecular, 519
  stereoselective, 1167
Cycloaddition, 1241
  anodically induced, 527, 1013
Cycloalkanes, from vinyl halides, 369
Cyclodimerization, 1242
1,4-Cyclohexadiene, 1318
Cyclohexadienones
  formation of, 621, 1071
  rearrangement of, 624
Cyclonucleosides, 770
Cyclooctatetraene, reduction, 335
Cyclopentenoids, 782
Cyclopropane carboxylic acids, 543
Cyclopropenium ion, 340
D-Ribono-γ-lactone, reduction of, 490
D-Ribose, 490
DCV (see Voltammetry, derivative cyclic)
Decarboxylation
  concerted, 567
  by peroxidisulfate/Ag/I, 237
Decarboxylation-desilylation, 566
Decoordination, oxidative, 839
Dehydrohalogenation
  of acetylenic halides, 371, 372
  of alkyl halides, 363, 365
Depolarizer, 246
Deprotection, 1029ff, 1225ff
  catalyzed by Phthalocyanins/metal, 821
  without racemization, 1170
Deprotonation, of cations radicals, 512
Devarda Cu, in catalytic hydrogenation, 442
Diacetone-2-ketogulonic acid, 1320
Di-1-naphthylglyoxal, 490
Diacylperoxides, alkyl radicals from, 237
1,2-Dialkoxybenzenes, trimerization of, 963
1,4-Dialkoxybenzenes, oxidation of, 630
β-Dialkylaminocinnamonitrile, 493
N,N-Dialkylanilines
  anodic oxidation, 589
  anodic potentials, 586
  α-substitution, 589
Diaphragm, 247, 1290
  ion-exchange membranes, 261
  permeable, 1290
  permselective, 1290
  porous materials, 261
Diarylalkenes, anodic coupling of, 985
Diarylpropane, coupling of, 999
Diarylsulfides, oxidation of, 674
Diastereomeric purity, 1171
Diazepines, 733
Diazirines, 729
Diazoalkanes, reduction of, 477
9-Diazofluorene, reduction of, 140
Diazonium salts, reduction of, 424
Dibromopyromellitic anhydride, 491

α,α'-Dibromoxylenes, reduction of, 375
Dicarboxylic acid, 1318
 geminal, 573
1,2-Dicarboxylic acids, decarboxylation of, 566
Dichloroformoxime, 407
1,3-Dicyanobenzene, 496
Dienes, dimerization of, 983
Diethyl adipate, 5
Diethyl oxalate, 490
 reduction of, 489
Diethyl phthalate, 489
Diethyl terephthalate, reduction of, 489
Diffuse layer, 59
Diffusion (see also Fick's laws)
 average length of, 86–87
 at electrodes, 60, 61
 of macromolecules, 1339
 in mass transfer, 1342
Diffusion coefficient
 macroscopic, definition, 83
 microscopic, origin, 85–86
Diffusion-current constants, 158
Diffusion layer, thickness, 87
Diffusion layer, thickness of, 164
Diffusion limit, rate constant
 control by, 32, 48
 definition of, 33–34
Dihalides
 aliphatic, 367, 368–369
 aromatic, 386
Dihaloalkanes, reduction of, 369
1,3-Dihaloketones, cyclization of, 388
Dihydro-2-methoxynaphthalene, 1320
1,4-Dihydronaphthalene, 1320
1,4-Dihydrocinnolines, 730
1,2-Dihydrophthalazine, 731
1,4-Dihydroquinoxaline, acetylation of, 733
Dihydroxylamine, as intermediate, 413
Dimensionless analysis
 in electrochemistry, 91–94
 principle of, 90–91
 variables, 94
Dimerization
 to heterocycles, 703
 oxidative, 846
 radicals (see Radicals)
[Dimerization]
 reductive, 839, 845
 stereoselective, 1154
1,3-Dimethoxybenzene, 3
1,4-Dimethoxybenzene, 4
2,4-Dimethoxybenzonitrile, 3
1,2-Dimethoxyethane, as solvent, 291
2,6-Dimethoxypyridine, oxidation of, 724
4,4'-Dimethoxystilbene, dimerization of, 982
Dimethyl sebacate, 1318
 by Kolbe reaction, 1329
Dimethylsulfoxide, as solvent, 289
Dimethyl terephthalate, 488
N,N-Dimethylaniline, electrooxidation, 969
N,N-Dimethylbenzamide, 492
1,3-Dimethylbicyclobutane, 5
Dimethylformamide, as solvent, 286
2,2-Dinitroalkanes, 405
Dinitrobenzenes, 419
2,2'-Dinitrobiphenyl, 713
1,1-Dinitroethane, 405
Dinitronaphthalenes, 421
Dioxygen, as probase, 1269
Diphenylacetylene, reduction of, 348
Diphenylamide, 495
Diphenylamines
 anodic oxidation, 590
 aromatic substitution, 592
 cyclization to carbazoles, 592
9,10-Diphenylanthracene, oxidation of, 510
N,N'-Diphenylbenzidines, from diphenylamines, 591
Diphenyldiselenide, 666
1,1-Diphenylene, oxidation of, 525
1,1-Diphenylhydrazine, in cycloadditions, 1016
Diselenides, oxidation of, 664
Disproportionation
 of alkyl radicals, 365
 of anion radicals, 335
 of benzylmercury radicals, 373
Disulfides, oxidation of, 662, 664
1,3-Ditellurole, oxidation of, 672
1,4-Dithiins, oxidation of, 683
Dithioacetals, oxidation of, 670
Dithiocarbamates, oxidation of, 661

DLSV (see Voltammetry, derivative linear sweep)
DME (see Mercury electrode, dropping)
Double layer, 58–59
DPSCA (see Chronoamperometry, double-potential-step)
DSA (dimensionally stable anode), 267, 1288

ECF (electrochemical fluorination), 1103
ECE reactions, 221, 1237, 1238
Edge effects, during CV, 128
EGB (electrogenerated base)
  for electrocarboxylation, 1272
  formation of ylids, 1267
  in paired reactions, 1275
  in Wittig reactions, 1267
EGBs electrocarboxylation, in paired synthesis, 1423
Electrocatalysis, 1237, 1238
  by ad-atoms, 1390
Electrochemical catalysis, 1219
Electrochemical mechanisms
  average concentration in, 220, 221
  CE mechanism, 74–75
  Dim1, Dim2, or Dim3, 221
  DISP mechanism, 221
  EC mechanism, 221
  ECC, ECDim mechanism, 221
  ECDisp, ECE mechanism, 221
  EE mechanism, 76
  square-scheme representation of, 78–79
Electrochemical reactors, scale-up of, 1302
Electrochemiluminescence, 344
Electrode
  chirally modified, 1172
  clay-coated, 1173
  of the first kind, 269
  optically transparent (OTE), 179
  polymer, 1173
  rotating, 1342
  rotating disk, 165, 174
  rotating double-ring, 173
  rotating ring-disk (RRDE), 171
  silver/silver ion, 269
[Electrode]
  split-ring disk, 173
Electrodecarboxylation, formation of C=C bonds, 564
Electrode materials, 261, 1287
  catalytic effect, 262
Electrode modification, 58–69
Electrode stability, 1289
Electrode surfaces, derivatization, 1366
Electrodes
  chemically modified surfaces, 267
  modified by ad-atoms, 268, 1390
  optically transparent, 267
  porous, 263
  of the second kind, 270
  silylation of, 1365
Electroenzymatic reduction, 1174
Electrogenerated acids, 1277
Electrogenerated base, 493, 1265ff
Electrogenerated bases (see EGB)
Electrogenerated reagents, for reduction of organic halides, 390–393
Electrohydrodimerization (EHD), (see also Hydrodimerization)
  scope and limitations, 909
Electrolysis, removal of salts, 302
Electrolyte, KF, 2HF, in fluorination, 1104
Electrolytes
  as proton donors, 367
  effect of, on reduction of alkyl halides, 366–367, 368
Electron hole pair, 1400
Electron transfer (see also ET theories)
  activation control of, 34–35
  diffusion limit in, 32–35
  heterogeneous, 344
  homogeneous, 513, 1219ff
  outer sphere vs. inner sphere, 16–18
  rate constant (see also ET theories) 62–65
Electron transfer catalysis, chemical reactions of, 77–78
Electron transfer theories
  activation energy (see Heterogeneous ET, Homogeneous ET)
  activation energy vs. driving force, 40, 45, 46–48, 63
  activation rate constant, 36

[Electron transfer theories]
adiabaticity vs. nonadiabaticity, 36, 45
cross-relationship in, 48–49
energy diagrams in, physical meaning of, 43–45
energy stabilization in, 36, 45
Franck-Condon principle in, 36, 41–43
free energies in, 45–46
heterogeneous ET, 60
homogeneous ET, 69
inner-shell reorganization energy, 40, 50, 63
inverted rate constants, 48–49
isotopic rate constants, 64
outer-shell reorganization energy, 40, 41–43, 63–64
potential energy and nuclear coordinates, 35–37, 43–45
quadratic free energy relationships (QFER), 46–48, 50–53
rate constants, 62–65
reorganization energy, physical meaning of, 43–45
solvent fluctuations in, 41–43
transfer coefficient, α, 64–66
transmission coefficient in, 36, 45
Electrons, apparent number of, 145
Electroorganic processes, commerical, 1318
Electroosmosis, 247
ElectroSynCell, 1311
Electrosynthesis, economic considerations, 1314
Enamide intramolecular cyclization, 598
Enamines
anodic oxidation, 597
coupling of, 983
Enaminoesters, anodic oxidation, 597
Enaminoketones, anodic oxidation, 597
Enecarbamates, anodic oxidation, 597
Enol ethers, coupling of, 982
Epinephrine, 772
Epoxidation, catalyzed by porphyrins/metal, 818
ESCA, in characterization of CME, 1368
Esters
aromatic, 487
reduction of, 483
Ethanetetracarboxylate, 490
Ethers
oxidation of aliphatic, 647
oxidation of unsaturated, 648
Ethyl 2-butylacetoacetate, Tafel rearrangement of, 490
Ethyl 2-thiazolecarboxylate, 489
Ethyl 3-phenyl-4-oxopentanoate, 5
Ethyl acrylate, 5
Ethyl benzoate, 488
Ethyl cinnamate, 5
Ethyl glyoxylate, 489
Ethyl phthalate, 489
1-Ethyl-4-carboxypyridinium, 484
Ethylene glycol, 1320
from formaldehyde, 894, 1320
Ethylenediamine, as solvent, 283
Ex-Cell processes, 1229
Experimental setup for voltammetry, 124

Faradaic impedance, physical nature of, 53–55
Feedback, positive, 126
Fermi level, 1399
Ferrocene, 1340
as reference electrode, 272
in CME, 1366
FFT, noise reduction by, 127
Fick's laws of diffusion
first law, 83
second law, 87
Film conductivity, influence of anion, 1388
Flavonoids, 775
Flow cells, 255
divided, 257
undivided, 257
Fluorene, of anion radical, cleavage of C–H, 350
Fluorination, 1094
of acyl fluorides, 1113
of acyl halides, 1106
of alcohols, 1106
of alkanes, 1113
of amines, 1106
in aprotic solvents, 1117

[Fluorination]
of benzoyl fluoride, 1112
of chloroalkanes, 1113
of cycloalkanes, 1113
ECF, 1103
of esters, 1106, 1113
of ethers, 1106
free radicals in, 1112
hard, 1104
induction period in, 1109
mechanistic considerations, 1108
mild, 1104
partial, 1117ff
process rationalization, 1115
the Phillips process, 1113ff
the Simons process, 1104ff
of sulfonyl halides, 1106
of thiols, 1106
Fluorosulfonic acid
as solvent, 280
reference electrode in, 280
5-Fluorouracil, 768
Folic acid, 737
Formal potential, definition of, 26
experimental significance of, 27–29
kinetic origin of, 73
Formation and cleavage of M-C bond, 809
in Grignard reagents, 860
of group IVB, 848
of group III, 857
of group IIB, 862
Formation of dianions, 332
Fourier transform analysis (FFT), 122
Free radicals
in polymerization, 1338
reactions, 1344
Frumkin correction, 65
Ftorafur, 768
Furan, methoxylation of, 720
Furan 2,5-dicarboxylate, reduction of, 489
2-Furoic acid, 573

Glucose
oxidation of, 644
in paired reactions, 1428
reduction of, 1316
Glyoxylic acid, 485
Graphite, surface oxides in CME, 1370
Graphite electrode, intercalated, 1173
Grasshof (Gr) number, 1293

Half-peak potentials, precision of measurement, 125
Half-reaction, 53–56, 65
Half-wave potential $E_{1/2}$, 97–100, 101–104
Haloalkenes, reduction of chiral, 1168
α-Halocarbonyl compounds, reduction of, 386
Halocyclopropyl, reduction of, 375–376
Halogenated pyridine, reduction of, 388, 389
Hammett correlations, of porphyrines, 814
Hard Fluorination, 1104
Heat transfer, in cells, 1298
Helmholtz layer (see Compact layer)
Hemicelluloses, 788
oxidation of, 793
Heteroaromatics, polymerization of, 1378
N-Heterocyclic compounds, anodic oxidation, 979
Heterocyclic halides, reduction of, 388–390
Heterogeneous ET rate constant
intrinsic standard, 64
true standard, 64
Hexafluorophosphate, 295
Hexamethylbenzene, oxidation of, 512
Hexamethylphosphoramide, as solvent, 287
Hofer-Moest reaction, 536
Hofmann elimination, tetraalkylammonium cation in, 367
HOMO/LUMO
of phthalocyanines, 820
of porphyrines, 814
Homoallyl alcohols, tin catalyzed synthesis, 852
Hydraulic diameter, 1294
Hydrazides, 410
Hydrazines, reduction of, 483
Hydrazobenzene, reduction of, 423

Hydrazones
acylated, reduction of, 473
anodic oxidation, 594
reduction of, 471
tautomerization of, 471
Hydrocarbon cation radicals, 520
Hydrocarbons
anodic voltammetry, 508
benzenoid, 334
from carbonyl compounds, 449
DCV of, 508
direct oxidation, 506
LSV of, 508
redox potentials of, 338
SHAC of, 508
Hydrodimerization, 880
of aldehydes, 893
by amalgam, 1201, 1202
cathode material, 921
concentration of olefin, 924
of ketones, 893
pH of catholyte, 921
preparation of heterocycles, 703
reversible, 349
supporting electrolyte, 924
Hydrogen fluoride
reference electrode in, 280
as solvent, 280
Hydrogen overvoltage, 262
mechanism, 262
Hydrogenations, cathodic, 347
Hydroquinone monoesters, oxidation of, 627
Hydroquinone silyl ethers, 634
Hydroquinones, oxidation of, 616, 634
p-Hydroxyazobenzenes, 422
α-Hydroxycarbamates, 774
3-Hydroxycinnoline, 740
4-Hydroxycinnoline, 741
Hydroxylamino group, as nucleophile, 705
Hydroxylation, anodic, 1071

Ilkovic equation, 159
Imidazole-2-carboxylic acids, 485
Imides, reduction of, 483
Imidic esters, reduction of, 469
Imines
anodic oxidation, 594
reduction of, 466
Imino ether, 496
Imminium salts
reduction of, 469
reductive coupling, 772
Indene, dimerization of, 982
Indenes, reduction, 346
Indole, electrochromic film, 1382
Indoles, from cinnolines, 718
Indolizidine alkaloids, 774
Intramolecular coupling, of aromatic systems, 1093
Iodination, 1094
Iodonium ion, as electron carrier, 6
Ion pair, in polarography, 162
Ionization potential, definition of, 23
standard potential, 23
Irreversible ET, 66, 85
IRS (internal reflecting spectroscopy), 180
Isobutylvinyl ether, 1352
Isomerization, 842, 1239–1240
Isonicotinic acid, 744
Isonicotinic amide, 744
Isoprenoids, 778
Isoquinoline, 725
Isothiazole derivatives, 727
Isothiocyanates
oxidation of, 690
reduction of, 469
Isoxazoles, 726

Junction potential, 21

Ketocarboxylic acids, rearrangements of, 569
Ketones
heterocyclic, 743
nonolefinic, 433
oxidation of, 643
preprotonation, 434
reduction to alcohols, 442
Kinetic waves
during LSV, 133
during polarography, 160
Kinetic zone diagram, 222
Kolbe reaction, 237, 1001
alkyl radicals in, 237
anode material, 539
anode potential jumps, 538

[Kolbe reaction]
cathode material, 539
competitive adsorption, 540
constant-current density, 538
cross-coupling, 545
in electropolymerization, 1348
hydrogen abstraction, 551
mixed, 1004
preparation of pheromones, 1002
solvent system, 540
supporting electrolytes, 540
synthetic equivalents, 537
Tafel plots, 538

L-Cysteine, 491
Lactams
alkoxylation, 603
anodic oxidation, 602
reduction of, 483
reduction of halogenated, 389–390
β-Lactams, formation of, 844
Lactones, reduction of, 483
Laminar flow, 1293
Laudanosine, oxidation of, 704
LCP (linear current potential analysis), 135
Lead
as anode material, 266
as cathode material, 264
as electrode material, 1288
Lead dioxide, as anode material, 260, 1288
Lignan models, oxidation of, 962
Lignans, 775
Lignin, 788
electrolysis of, 794
Kraft, 788
Kraft process, 794
model compounds, 800
structure, 789, 790
sulfite process, 794
Limiting current, in electrode processes, 1292
Limonene, 778
Linear sweep voltammetry, 123
Liquid junction, 269
LSV (linear sweep voltammetry), 123

Magnesium anode, 4
Magnetic field, enhancement of current, 248
Magnetite, as anode material, 1288
Maltol, 782, 1320
from furfural, 1317
Mannitol, from glucose, 1316
Marcus theory, 44ff
Mass transfer, 12, 60, 85–87, 248
fundamental aspects of, 80
physical processes of, 81–83
Mass transfer rate, 67, 71
Mass transport, 1292
Mediator, 30
Mediators, reduction of halides, 391–393
Membranes
ion-exchange, 1291
permselective, 1290
4-Mercaptocinnoline, 746
Mercury
as cathode material, 264
dropping electrode (DME), 157
Mesitylene, oxidation of, 523
Metal alkyls (see also Organomercury compounds)
from halides, 366
Methanol
as solvent, 282
oxidation of, 639, 640
α-Methoxy-α-amino acids, 766
Methoxyarenes, oxidation of, 137
Methoxylation, 528, 555, 1073
of alkylanisoles, 1074
of N,N-dimethylaniline, 1075
of methoxynaphthalenes, 1074
of phenols, 1074
Methyl benzoate, 488
Methyl 1,2-dihydro-2-naphthoates, reduction of, 490
Methyl 1,4-dihydro-1-naphthoate, reduction of, 490
Methyl 1,4-dihydro-2-naphthoates, reduction of, 490
Methyl 2-naphthoate, reduction of, 490
Methyl methacrylate, electropolymerization, 1345
3-Methyl-2-oxazolidinone, as solvent, 287
Methylamine, as solvent, 283
Methylbenzenes, voltammetry of, 511

Methylene chloride, as solvent, 292
O-Methylflavinatine, from laudanosine, 770
N-Methylmorpholine, mechanism of ECF, 1111
N-Methylphenothiazine, oxidation of, 687
N-Methylphthalimide, 494
Methylpyridines, oxidation of, 739
N-Methylpyridinium-2-carboxylic acid, 485
N-Methylpyridinium-3-carboxylic acid, 485
N-Methylpyrrolidone, 495
  as solvent, 287
N-Methylsuccinimide, 495
Methylthioarenes, oxidation potentials of, 674
Microscopic reversibility, 79
Migration, 60, 82
Molecular weight
  changes during polymerization, 1347
  of polymer, 1338
Monoethyl phthalate, 491
Monomers
  structure-reactivity considerations, 1380
  vinyl, 1346
Mononitrobenzenes, 411
MSRS (see Spectroscopy, modulated specular reflectance)
d,l-Muscone, 782

Nalco process, 1299, 1327
N-Oxides, heterocyclic, 742
Naphthalene, 347
Naphtyridines, 733
Neopolarogram, 135
Nernst law, 20, 26, 66, 71, 85, 96
Nernstian ET, 66, 85
Neutral alumina, removal of impurities by addition of, 125, 333
Nickel, as anode material, 267
Nickel anodes
  in fluorination, 1104
  passivating layer, 1104
Nicotinic acid, 744
Nicotinic amide, 744
N-Nitramines, reduction of, 426
Nitrate radical, electrogenerated, 518
Nitration, anodic, 1088
Nitric acid esters, 411
Nitriles, cleavage, 1033
Nitriles
  from nitroalkenes, 404
  heterocyclic, 745
Nitro compounds
  aromatic, 411
  halogenated, 381, 382
2-Nitro-2-methylpropane, 404
Nitroalkanes
  α-halogenated, 406
  α-substituted, 405
  α,β-unsaturated, 404
  β-substituted, 408
  aci-form, 403
  primary and secondary, 402
  tertiary, 403
9-Nitroanthracene, reductive dimerization, 413
o-Nitrobenzalacetylacetone, 707
p-Nitrobenzoic acid, from p-nitrotoluene, 416
t-Nitrobutane, 2
Nitrogen fixation, electrocatalytic, 836
Nitrogen-nitrogen bond, hydrogenolysis, 718
Nitrohydrazones, 407
Nitrolic acid, 406
Nitromethane, as solvent, 290
Nitronic acid, 405
2-Nitrophenylisothiocyanate, 708
N-p-Nitrophenylpyrrole, polymerization, 1381
o-Nitrophenylthiocyanate, 708
N-Nitrosamides, reduction of, 425
Nitroso compounds, 421
  aliphatic, 409
N-Nitrosourea, reduction of, 425
β-Nitrostyrene, 404
p-Nitrotoluene, reduction of, 415, 416, 417
Non-Kolbe reaction, 243, 536
Nonbenzenoid, aromatic, 337
Norbonane halides, reduction of, 376–378
NPSV (see Voltammetry, normalized potential sweep)

Nuclear coordinates in ET, 36–40, 43–45
Nucleic acid bases, 768
Nucleophiles, electrogenerated, 1247
Nucleophilic substitution, aromatic, 841
Nucleosides, 768
  unsaturated, 769

OCR (relaxation, open circuit), 187
2,3-Octanedione, 490
2-Octanone, oxidation of, 643
Ohmic drop, 53, 55–58, 83
Olefins
  anodic coupling of, 979
  from vicinal dihalides, 368
  from vinyl halides, 369
Oligomerization
  catalyzed by bipyridines/metal, 832
  catalyzed by phosphines/metal, 835
Optical activity, loss of, 1169
Optical purity, 1169–1171
Optical resolution, electrochemical, 1174
Organic halides
  acetylenic, 370–372
  allenic, 370–372
  indirect reduction, 390–394
  mechanisms of reduction, 362, 363, 373, 374, 380
  reductions
    α,ω-dihalides, 368–369
    gem-aliphatic, 367
    vic-aliphatic, 368
    alkyl, 362 ff.
    allylic, 370
    vinylic, 369
Organoaluminium compounds, 857
Organoboron compounds, 857, 860
Organogermanes, 852, 855
Organoindium compounds, 857
Organolead compounds, 848
  from alkyl halides, 366
Organomercury compounds
  from 1,4-dihalonorbonanes, 378
  from acetylenic halides, 370
  from alkyl monohalides, 363, 364, 365
[Organomercury compounds]
  from allyl halides, 370
  from benzyl halides, 373
  from dihaloalkanes, 369
Organometallic compounds
  from halides, 366
  containing lead, 366
  containing mercury, 363, 364, 365, 369, 370, 371, 372
Organosilanes, 850, 852, 855, 1364
Organotin compounds, 848, 852, 855
  from alkyl halides, 366
Orthothioesters, oxidation of, 672
Orthothiooxalate, oxidation of, 672
OTE (optically transparent electrode), 179
Outer-Helmholtz plane (OHP), 59
Oxadiazoles, 734
Oxalic acid, 485
Oxaziridines, 726
Oxazoles, 726
Oxidation
  of arylamines, 963
  asymmetric, 1173
  catalyzed by bipyridines/metal, 833
  catalyzed by phthalocyanins/metal, 821
  mechanism of indirect, 507
Oxidation potentials
  alicyclic hydrocarbons, 514
  aliphatic hydrocarbons, 514
  correlation with UV, 508
  correlation with Hückel MO, 508
  correlation with ionization potentials, 508
Oxidative coupling
  in electropolymerization, 1356
  to form electrically conducting polymers, 1356
Oximes
  acylated, reduction of, 476
  alkylated, reduction of, 476
  anodic oxidation, 594
  from nitroalkenes, 404
  reduction of, 475
Oxygen, inhibition of polymerization, 1348
  as probase, 1269
Oxygen, removal of, 301
Oxygen overvoltage, 262
Oxygen reduction, catalyzed by porphyrines/metal, 819

Oxygen-nitrogen bond, hydrogenolysis of, 718

Paired oxidation, glucose to gluconate, 1428
Paired reduction, glucose to sorbitol, 1428
Paired syntheses, table of, 1424
Paired synthesis
  of detergent builder, 1422
  propylene oxide, 1421
Paraconic acids, electrochemical rearrangements of, 569
Pattern recognition, deviation, 122, 151
Peak potential in CV, 106
Pentachloropyridine, 747
Pentachloropyridine, organomercury compounds from, 389
Pentafluorobenzoic acid, reduction of, 486
Pentafluorobenzyl alcohol, 487
Pentafluoropyridine, 747
Perchlorate, in supporting electrolyte, 294
Perfluorinated organics, 1103
Perfluoroamino ethers, 1108
Pericyclic reactions, photoelectrochemically induced, 1409
Peroxides, oxidation of, 649
Perylene
  alkylation of, 347
  cation radicals, 510
pH-Potential diagrams, 26–29
Phathalic anhydride, 488, 491
Phathalides, 491
Phenacyl bromides,
  furans from, 386
  pyridazine from, 387
Phenacyl halides, reduction of, 384
Phenacyl radicals, combination of, 386
Phenanthrene, reduction of, 348
9,10-Phenanthrenoquinone, in CME, 1371
Phenazine, reduction of halogenated, 389
Phenazines, 732
Phenol, oxidation of, 962
Phenolate ions, oxidation of, 616
Phenols, cation radicals of, 617
Phenols
  alkenyl, oxidation of, 627
  oxidation of, 616
Phenothiazine, oxidation mechanism of, 686
Phenoxonium, formation of, 618
Phenoxonium ions
  deprotonation of, 620
  reactions with nucleophiles, 621
Phenoxy radicals
  dimerization of, 618
  disproportionation of, 619
  reversible formation of, 616
Phenyl benzoate, 489
N-Phenyl-4-pyridincarboxamide, 492
Phenylacetic acid, 487
3-Phenylbenzo-1,2,4-triazine, synthesis of, 709
Phenylenediamine, anodic oxidation, 594
Phenylhydrazones
  oxidation of, 597, 710, 712
  reduction of, 471
Phenylhydroxylamine
  condensation reactions, 417
  as nucleophile, 417
  reduction of, 417
  substitution of, 413
N-Phenylphthalimide, 494
2-Phenylpyrimidines, ring contraction, 719
Pheromones, by Kolbe electrosynthesis, 1002
Phosphines, optically active, 1052
Photocatalytic cell, 1403
Photoelectrochemistry, 1397
Photoelectrosynthetic cell, 1402
Photovoltaic effect, 1402
Phthalazines, 731
Phthalazinones, 741
Phthalic acid, 485
Phthalide, 485, 486
Phthalimide, 494, 740
Phthalonitrile, 496
Phthaloyl chloride, cyclization of, 385
Picolinic acid, reduction of halogenated, 389
Pinacol, 1320
Pinacolization, influence of Cr(III), 894

Pinacols
d,l-meso ratio, 895
from ketones, 436
by reductive coupling, 893
Piperidine alkaloids, 767
Plate-and-frame cell, 1308
Platinum
as anode material, 266
anodized, 1365
Podophyllotoxin, 776
Polarization, in fluorination, 1114
Polarization parameter, 1297
Polarography, 157
applications of, 161
derivative, 163
differential pulse, 163
Polyacetylene, 341
Polyaniline
CV data, 1384
films of, 1383
Polyanions, formation of, 332
Poly-(3-ethylthiophene), 1380
Polymer blends, 1387
polyaniline, polycarbonate, 1389
polypyrrole, Nafion, 1389
polypyrrole, poly(vinylalcohol), 1389
polypyrrole, polyvinylchloride, preparation of, 1387
Polymer electrodes, properties, 1387
Polymer film
polynitrostyrene, 1374
polyvinylferrocene, 1374
polyvinylpyridine, 1374
Polymer morphology, 1343
Polymerization
addition, 1338
anionic, 1350
cationic, 1351, 1352
cationic, effect of solvent, 1339
coordination, 1352
electrodecarboxylative, 544
free radicals, 1338
ionic, 1338
via Ziegler-Natta catalysts, 1352
Polymerization, reaction mechanism, 1376
Polymers
conducting, 340
[Polymers]
containing tetrathiafulvalene, 1372
containing quinone, 1372
Poly-(3-methylthiophene), 1380
Poly(pyrrole), 1340
conducting state, 1375
CV of, 1386
electrochemical stoichiometry, 1376
films of, 1383
films, incorporation of anions, 1385
Poly(thiophene)
conducting state, 1375
conductivity of, 1380
electrochemical stoichiometry, 1376
Polyvinylferrocene, 1340
Polyvinylbenzoquinone, 1340
Polynitroalkanes, 405
Porphines, 738
Porphyrin, in CME, 1366
Postpolymerization, in electrochemical polymerization, 1347
Potential distribution in cells, 53–56, 58, 83
Potential, half-wave, 157
Potential range for solvents, 276
Potentials peak, precision of measurement, 125
Potentiostat, role and principle of, 55
Pourbaix diagram, 1289
Probase, azobenzenes, 1267
Probases, 1265
Propagation, polymer chains, 1346
2-Propanol, oxidation of, 641
Propellanes, from bicyclic dihalides, 378
Propylene carbonate, as solvent, 291
Propylene oxide, 1318
Protecting group, 1050
2-haloethyl derivatives, 1041
4-nitrobenzyloxycarbonyl, 1050
for alcohols, 1039
nosylates, 1056
tosylat, 1054, 1056
trityl, 1039
tritylone, 1039
Proton activity, in solvents, 273

Protonation, anion radicals, 345
Pseudonitroles, 406
Pteridine, 736
Pulping
  organosolv, 790
  sulfite process, 790
Purification, of solvent, 279
Purine, 736
Pyrazines, 732
Pyrazinopyrazines, 738
Pyrazoles, 729
Pyrazoline, in CME, 1366
Pyrene, electrochromic film, 1382
Pyridazines, 729
Pyridazinones, 730
Pyridination, anodic, 1089
Pyridine, 288
  derivatives, 723
Pyridine-2-carboxylic acid, 485
  reduction to picoline, 487
Pyridine-4-carboxamide, reduction of, 492
Pyridine-4-carboxylic acid, 485
  reduction to picoline, 487
Pyridine-4-thiocarboxamide, 493
4-Pyridinecarboxylic acid hydrazide, 495
Pyridinecarboxylic acids, 744
4-Pyridinemethanol, 492, 493
Pyridininium salts, reduction of, 728
Pyridopyrimidines, 733
Pyrimidines, 732
Pyrimidones, 740
Pyrroles, 722
Pyrrolidine alkaloids, 767
2-Pyrrolidinone, as probase, 1268, 1277
Pyrrolizidine alkaloids, 774

Quater-wave potentials, 153
Quaternary ammonium compounds
  amalgams of, 380
  effect of, 366–368
  as proton donors, 367
Quinazoline, reduction of halogenated, 389
Quinazolines, 732
Quinoline, 725
  reduction of halogenated, 389
Quinone imine, from p-aminophenol, 589
Quinone monoketals, 781
Quinones
  $E_{1/2}$ values, 454
  hydroxylation of, 1072
Quinoxaline, reduction of halogenated, 389
Quinoxalines, 732

Radical addition
  electroinitiated, 1012
  intramolecular, 550
  in Kolbe reaction, 550
Radical anions, formation of, 332
  (see also Anion radicals)
Radical cations
  formation by electrolysis, 1352
  formation of polymer film, 1380
Radical clocks, 80
Radical ions
  photogenerated, 1412
  photoreactions of, 1413
Radicals
  from acetylenic halides, 371
  from acyl halides, 384–385
  addition to double bonds, 1007
  from alkyl halides, 362 ff.
  from allenic halides, 370
  from allyl halides, 370
  from benzyl halides, 373
  combination of, 365, 373, 374, 375, 384–385, 386–387
  disproportionation of, 365, 373
  from monohalobenzenes, 379, 383
  from vinyl halides, 369
Raney Ni, in catalytic hydrogenation, 442
Raney Ni electrode, chiral, 1173
Rate constant
  in dimensionless form, 91–94
  in electrochemical equations, 87–89
  of heterogeneous ET, 64
RDE, application of, 169
Reaction coordinates, 36–39, 43–45
Reaction order, apparent, 141
Reaction orders, determination of
  by cyclic voltammetry, 109
  by steady-state methods, 103

Reactions, stereochemical course, 1133, 1136, 1137, 1139, 1143, 1144, 1145, 1150, 1153, 1169
Reactor configuration, 1301
Reactor
  plug flow (PFR), 1299
  stirred tank (STR), 1299
Rearrangement
  anodic, 526
  Wagner-Meerwein, 570
Redox catalysis, 39, 75, 216, 217, 226, 1220
Redox electrodes, as reference electrode, 272
Redox polymers, CV of, 1372
Redox reagents, 1218
Reduction
  asymmetric, 1171–1174
  of carbonyl, 838
    to alcohols, 838
    to pinacols, 838
  with inversion, 1168
  at irradiated semiconductors, 1409
  with retention, 1168
Reductive coupling (see also Coupling, cathodic)
  α,β-unsaturated acids, 905
  α,β-unsaturated carbonyl compounds, 913
  α,β-unsaturated esters, 913
  activated olefins, 908
  dicarbonyl compounds, 902
  of halogen compounds, 886
  of hydrocarbons, 885
  indirect, 882
  intermolecular, 883, 887, 889, 904, 918
  mechanisms of, 884
  mixed, 906
  of olefinic aldehydes, 904
  of olefinic ketones, 902
Reference electrode
  Bis(biphenyl)chromium, 272
  calomel electrode, 270
  electron electrode, 270
  ferrocenium ion/ferrocene, 272
  lithium electrode, 270
  mercury/mercurous sulfate, 271
  position of, 247
  redox system, 272
  silver/silver chloride, 271
[Reference electrode]
  sodium amalgam electrode, 270
  zinc amalgam electrode, 270
Reference electrodes, 268, 274
  calibration of potential scales, 22
  construction of, 124
  definition of, 21–23
  "floating", 22
  standard potentials of, 22
Regression, nonlinear, 148, 150
Regression techniques, nonlinear, 122
Relaxation open-circuit (OCR), 187
Resistance, ohmic, 247
Reversible ET, 66, 70–72
Reynolds number, 165, 1293
Ring closure, to heterocycles, 702
Ring contractions, 717
Ring enlargement, cyclic ketones, 569
Ring expansions, of heterocycles, 719
Ring opening, pericyclic, 350
RRDE (see Electrode, rotating ring-disk)
RSS (rapid scan spectroscopy), 183
Rubrene, 511

Sacrificial electrode, 265, 848, 858, 861, 862
Salicylaldehyde, from salicylic acid, 486
Salicylic acid, 486
  borate complex, 486
  reduction by amalgam, 1208
Salts, as electrolytic media, 292
Sand equation, 153
Saturated carbonyls, $E_{1/2}$ of, 453
Schiff bases
  amino acids from, 375
  reduction of, 466
Schmidt (Sc) number, 1293
Sebacic acid, 1329
  Kolbe dimerization of, 544
Selenation, 1244
Selenocyanation, anodic, 1084
Selenophenol, oxidation of, 661
Self-protonation, of anion radicals, 345

Semicarbazones
  anodic oxidation, 594
  oxidative cyclization, 705
  reduction of, 474
Semiconductors, 267
Sensitization, 1409
Sign convention, current, 84
Silver/silver ion, as reference electrode, 269
Silyl enol ethers, dimerization of, 983
Simons ECF process, diagram, 1105
Simulation, digital, 123
Singlet oxygen, at irradiated semiconductors, 1407
Slow ET, 67, 70–72, 85
Solubility parameters, 1343
  table, 1344
Solution resistance, uncompensated, 126
Solvated electron, 495, 1253
Solvent, cohesive energy density of, 1343
Solvent effects on photoelectrochemistry, 1406
Solvents
  aprotic, 284
  dissolving power, 277
  for electrolysis, 273
  protic, 279
  proton activity, 273
  usable range, 275
Sorbitol, from glucose, 1316
Space-time yield, 249
Spectroelectrochemistry, 179
  cells for, 179
  kinetic measurements by, 186
Spectroscopy
  ESR, 190
  internal reflection (IRS), 180
  modulated specular reflectance (MSRS), 183
  rapid scan (RSS), 183
  specular reflectance (SRS), 180, 506
Spin-traps, application of, 193
Square schemes, 78–79
SRS (see Spectroscopy, specular reflectance)
Stack cell, 1307
Standard potential
  definition of, 18–21
  extrapolation between solvents, 25
Stereochemical interactions, 1142
Stereochemical mechanism, 1168
Stereochemistry
  in reduction of benzyl halides, 374
  in reduction of dihalonorbornanes, 376–377
  in reduction of vicinal dihaloalkanes, 368
Stereocontrol, in reactions, 1134, 1139, 1142, 1144, 1145
Stereoisomeric analysis, 1132, 1176
Stereoisomerization, 1134, 1147, 1152, 1157, 1168, 1174, 1175
Stereoisomers, 1135, 1142, 1143, 1145, 1146, 1147, 1149, 1151, 1156, 1161, 1164, 1168, 1174–1178
Stereoregularity, 1135
Stereoselective synthesis, 1134, 1135, 1137, 1143, 1168
Stereoselectivity of product, 1134–1136, 1138, 1140, 1142–1146, 1150–1156, 1169
Stereospecificity, 1135, 1136, 1138, 1153, 1154, 1159, 1163
Steric hindrance, 1132, 1144
Steric interaction, 1144, 1149
Steroids, 777
  reduction of, 450
cis-Stilbene, isomerization, 350
Stilbene, 347
  reduction of, 348
Stilbenediol dibenzoate, 491
Styrene, 1352
  dimerization of, 982
Sulfenamides, oxidation of, 691
Sulfenic acids, oxidation of, 691
α-Sulfenylcarboxylic acids, electrodecarboxylation of, 572
Sulfides
  dialkyl, 667
  oxidation of, 667, 674
    in acetic acid, 679
    in methanol, 679
Sulfinic acids, oxidation of, 691
Sulfolane, as solvent, 290

Sulfonamides, α-methoxylation, 603
Sulfonic acid, oxidation of, 691
Sulfonium salts, from sulfides, 675
Sulfuration, 1244
Sulfuric acid, as solvent, 279
Supercritical fluids, cyclic voltammetry, 293
Superoxide, in photoelectrochemistry, 1407
Superoxide ion, 1245ff, 1272
Supporting electrolyte, 294
  in electropolymerization, 1345
Supporting electrolytes, chiral, 1172
Swiss-roll cell, 1309
Sydnones, 734
$S_{ON}2$, 1239
$S_{RN}1$, 1238, 1239

Tafel plot, kinetic origin of, 65–67
Tafel rearrangement, 449
Tank cells, 1303
Terpenoids, 777
Terpinoline, 778
Tetra-p-anisylethylene
  cation radicals of, 629
  stable dication, 519
Tetraalkylammonium ions, 296
Tetraalkyllead, 1318
  from alkyl halides, 366
  Nalco process, 1327
Tetraalkyltetrathioethylene, 491
Tetracene, 511
3,4,5,6-Tetrachloro-2-picolinic acid, 747
Tetrachlorophthalic anhydride, 491
2,3,5,6-Tetrachloropyridine, 747
Tetrafluoborate, in supporting electrolyte, 294
2,3,5,6-Tetrafluorobenzoic acid, reduction of, 486
Tetrahydrofuran
  hydroxylation of, 1072
  as solvent, 291
Tetrahydroisoquinoline, dimerization of, 961
1,2,3,4-Tetrahydrophthalazine, 731
Tetralkyl ethenetetracarboxylates, as probase, 1268
3,3,6,6-Tetramethoxy-1,4-cyclohexadiene, 4
Tetramethyl tin, 496
Tetramethylurea, replacement for HMPA, 288
Tetranitromethane, 406
Tetraphenylethylene
  intramolecular coupling, 524
  oxidation of, 184, 519
  reduction, 346, 350
Tetrathiafulvalene, oxidation potentials, 681
Tetrathiafulvalene (TTF), oxidation of, 679
Tetrathioethylene, oxidation potentials, 680
Tetrathioethylenes, oxidation of, 679
Tetrazolium salts, 736
1,3,4-Thiadiazoles, 735
Thianthrene, oxidation of, 174, 683
Thiazines, 728
Thiazole derivatives, 727
Thiazole halides, reduction of, 389
Thiazole-2-carboxylic acid, 485
Thioacids, oxidation of, 661
Thiobenzanilide, oxidation of, 712
Thiocarbonyl, oxidation of, 689
Thiocyanation, anodic, 1084
Thiols, oxidation of, 660, 662
Thiophene, oxidation of, 721
Thiophene 2,5-dicarboxylate, reduction of, 489
Thiophene halides, reduction of, 389–390
Thiophenes
  conducting polymers of, 688
  oxidation of, 688
Thiosemicarbazone, reduction of, 474
Tin, as cathode material, 264
α-Tocopherol, model compound, 626
N-Tosyl group, removal of, 775
Transfer coefficient α, 64–66, 159
Transfer coefficient, potential dependence of, 144
Transition metals
  arene complexes, 837

[Transition metals]
  cyclopentadienyl complexes, 837
Transition time, 153
Transition-metal complexes, as mediators, 391
Triarylamines
  as mediators, 593
  potential range, 593
Triazanaphthalenes, 736
Triazines, 736
Tribenzylamine, α-substitution, 384
Trichloromethyl anion
  cathodic coupling of, 367
  generation of, 367
Trichloromethylbenzene, cathodic coupling of, 375
Trichloronitromethane, 407
1,3,5-Tricyanobenzene, 496
Trifluoroacetic acid, as solvent, 282
Trifluoroacetoxylation, 1081
3-Trifluoromethylbenzoic acid, 4
3,4,5-Trimethoxytoluene, 1320
Trimethylacetic acid, oxidation of, 237
Trimethylamine, α-substitution, 584
Trimethylsilyloxy (TMSO), as protecting group, 963
Trioxane, polymerization, 1351
Triphenylamines, coupling of, 968
1,1,3-Triphenylindene, rearrangement, 351
Tris-(4-Bromophenyl)amine, as redox catalyst, 585
Tris-(4-Bromophenyl)ammoniumyl, 593
Trithiocarbonates, 491
Tritylone esters, cathodic cleavage of, 6, 1039
Tryptophan, 767
TTF (see Tetrathiafulvalene)
Turbulant flow, at a planar electrode, 1294

Ultramicroelectrodes, 128
  CV at, 128
Umpolung
  of donors, 240
[Umpolung]
  reductive, 240
α,β-Unsaturated carbonyls
  $E_{1/2}$ of, 453
  reduction mechanism, 452
α,β-Unsaturated carboxylic acids, Kolbe reaction of, 543
α-β-Unsaturated ketones
  polarografic reduction, 454
  synthetic aspects, 455
Uracils, 768
Urethanes, from formamides, 604
Uronic acid, oxidation of, 560

Vicinal diols, esters of, 489
Vincamine, 773
Vinyldihydroquinone, in CME, 1371
Viologen, in CME, 1366
Vitamin A, 489
Voltage sources, 299
Voltammetry
  convolution, application of, 143
  convolution, potential sweep (CPSV), 135
  cyclic, 107, 123
    applications of, 136
  derivative cyclic (DCV), 126
  derivative linear sweep (DLSV), 127
  linear sweep, applications of, 136
  normalized potential sweep (NPSV), 135, 136
  for optimizing parameters in electrolysis, 216, 1342
  second harmonic ac (SHAC), 506

Water
  as solvent, 281
  determination of, 274
Work terms
  in diffusion limit rate constants, 33
  in heterogeneous ET, 62–65
  in homogeneous ET, 33, 34, 38, 39, 46, 47
Working curves, 142, 143, 146, 155, 170, 172, 187

Xanthates, oxidation of, 661

Ylide, from tetramethyl-
ammonium cation, 367
Yohimban alkaloids, 773
Young's modulus, 1387

Zintl ion salts, 1200